CAMBRIDGE LIBRARY COLLECTION

Books of enduring scholarly value

Physical Sciences

From ancient times, humans have tried to understand the workings of the world around them. The roots of modern physical science go back to the very earliest mechanical devices such as levers and rollers, the mixing of paints and dyes, and the importance of the heavenly bodies in early religious observance and navigation. The physical sciences as we know them today began to emerge as independent academic subjects during the early modern period, in the work of Newton and other 'natural philosophers', and numerous sub-disciplines developed during the centuries that followed. This part of the Cambridge Library Collection is devoted to landmark publications in this area which will be of interest to historians of science concerned with individual scientists, particular discoveries, and advances in scientific method, or with the establishment and development of scientific institutions around the world.

A Mathematical and Philosophical Dictionary

Born into a Newcastle coal mining family, Charles Hutton (1737–1823) displayed mathematical ability from an early age. He rose to become professor of mathematics at the Royal Military Academy and foreign secretary of the Royal Society. First published in 1795–6, this two-volume illustrated encyclopaedia aimed to supplement the great generalist reference works of the Enlightenment by focusing on philosophical and mathematical subjects; the coverage ranges across mathematics, astronomy, natural philosophy and engineering. Almost a century old, the last comparable reference work in English was John Harris' *Lexicon Technicum*. Hutton's work contains manyhistorical and biographical entries, often with bibliographies, including many for continental analytical mathematicians who would have been relatively unfamiliar to British readers. These features make Hutton's *Dictionary* a particularly valuable record of eighteenth-century science and mathematics. Volume 2 ranges from *kalendar* to *zone*. Among the other topics covered are knots, Newton, magnets, and the Moon.

A Mathematical
and
Philosophical Dictionary

*Containing an Explanation of the Terms,
and an Account of the Several Subjects,
Comprized under the Heads Mathematics, Astronomy,
and Philosophy, Both Natural and Experimental*

VOLUME 2

CHARLES HUTTON

CAMBRIDGE
UNIVERSITY PRESS

CAMBRIDGE
UNIVERSITY PRESS

University Printing House, Cambridge, CB2 8BS, United Kingdom

Cambridge University Press is part of the University of Cambridge.
It furthers the University's mission by disseminating knowledge in the pursuit of
education, learning and research at the highest international levels of excellence.

www.cambridge.org
Information on this title: www.cambridge.org/9781108077712

© in this compilation Cambridge University Press 2015

This edition first published 1795
This digitally printed version 2015

ISBN 978-1-108-07771-2 Paperback

This book reproduces the text of the original edition. The content and language reflect
the beliefs, practices and terminology of their time, and have not been updated.

Cambridge University Press wishes to make clear that the book, unless originally published
by Cambridge, is not being republished by, in association or collaboration with,
or with the endorsement or approval of, the original publisher or its successors in title.

A

MATHEMATICAL AND PHILOSOPHICAL

DICTIONARY.

VOL. II.———PART I.

A

MATHEMATICAL and PHILOSOPHICAL
DICTIONARY:

CONTAINING

AN EXPLANATION OF THE TERMS, AND AN ACCOUNT OF THE SEVERAL SUBJECTS,

COMPRIZED UNDER THE HEADS

MATHEMATICS, ASTRONOMY, and PHILOSOPHY

BOTH NATURAL AND EXPERIMENTAL:

WITH AN

HISTORICAL ACCOUNT OF THE RISE, PROGRESS, AND PRESENT STATE OF THESE SCIENCES:

ALSO

MEMOIRS OF THE LIVES AND WRITINGS OF THE MOST EMINENT AUTHORS,

BOTH ANCIENT AND MODERN,

WHO BY THEIR DISCOVERIES OR IMPROVEMENTS HAVE CONTRIBUTED TO THE ADVANCEMENT OF THEM.

IN TWO VOLUMES.

WITH MANY CUTS AND COPPER-PLATES.

═══════════

By CHARLES HUTTON, LL.D. F.R.S.

OF LONDON AND EDINBURGH, AND OF THE PHILOSOPHICAL SOCIETIES OF HAARLEM AND AMERICA;
AND PROFESSOR OF MATHEMATICS IN THE ROYAL MILITARY ACADEMY, WOOLWICH.

═══════════

VOL. II.

LONDON:

PRINTED FOR J. JOHNSON, IN ST. PAUL'S CHURCH-YARD; AND G. G. AND J. ROBINSON,
IN PATERNOSTER-ROW.

MDCCXCV.

A

PHILOSOPHICAL and MATHEMATICAL

DICTIONARY.

K.

KALENDAR. See CALENDAR.

KALENDS. See CALENDS.

KEILL (Dr. JOHN), an eminent mathematician and philofopher, was born at Edinburgh in 1671, and ftudied in the univerfity of that city. His genius leading him to the mathematics, he made a great progrefs under David Gregory the profeffor there, who was one of the firft that had embraced and publicly taught the Newtonian philofophy. In 1694 he followed his tutor to Oxford, where, being admitted of Baliol College, he obtained one of the Scotch exhibitions in that college. It is faid he was the firft who taught Newton's principles by the experiments on which they are founded: and this it feems he did by an apparatus of inftruments of his own providing; by which means he acquired a great reputation in the univerfity. The firft public fpecimen he gave of his fkill in mathematical and philofophical knowledge, was his *Examination of Dr. Burnet's Theory of the Earth; with Remarks on Mr. Whifton's New Theory;* which appeared in 1698. Thefe theories were defended by their refpective authors; which drew from him, in 1699, *An Examination of the Reflections on the Theory of the Earth,* together with *A Defence of the Remarks on Mr. Whifton's New Theory.* Dr. Burnet was a man of great humanity, moderation, and candour; and it was therefore fuppofed that Keill had treated him too roughly, confidering the great difparity of years between them. Keill however left the doctor in poffeffion of that which has fince been thought the great characteriftic and excellence of his work: and though he difclaimed him as a philofopher, yet allowed him to be a man of a fine

VOL. II.

imagination. " Perhaps, fays he, many of his readers will be forry to be undeceived about his theory ; for, as I believe never any book was fuller of miftakes and errors in philofophy, fo none ever abounded with more beautiful fcenes and furprifing images of nature. But I write only to thofe who might expect to find a true philofophy in it: they who read it as an ingenious romance, will ftill be pleafed with their entertainment."

The year following, Dr. Millington, Sedleian profeffor of natural philofophy in Oxford, who had been appointed phyfician to king William, fubftituted Keill as his deputy, to read the lectures in the public fchool. This office he difcharged with great reputation ; and, the term of enjoying the Scotch exhibition at Baliolcollege now expiring, he accepted an invitation from Dr. Aldrich, dean of Chrift-church, to refide there.

In 1701, he publifhed his celebrated treatife, intitled, *Introductio ad Veram Phyficam,* which is fuppofed to be the beft and moft ufeful of all his performances. The firft edition of this book contained only fourteen lectures ; but to the fecond, in 1705, he added two more. This work was defervedly efteemed, both at home and abroad, as the beft introduction to the Principia, or the new mechanical philofophy, and was reprinted in different places ; alfo a new edition in Englifh was printed at London in 1736, at the inftance of M. Maupertuis, who was then in England.

Being made Fellow of the Royal Society, he publifhed, in the Philof. Tranf. 1708, a paper on the Laws of Attraction, and its phyfical principles: and being offended at a paffage in the *Acta Eruditorum* of Leipfic, where Newton's claim to the firft invention of the me-

B   thod

thod of Fluxions was called in queſtion, he warmly vindicated that claim againſt Leibnitz. In 1709 he went to New-England as treaſurer of the Palatines; and ſoon after his return in 1710, he was choſen Savilian profeſſor of aſtronomy at Oxford. In 1711, being attacked by Leibnitz, he entered the liſts with that mathematician, in the diſpute concerning the invention of Fluxions. Leibnitz wrote a letter to Dr. Hans Sloane, then ſecretary to the Royal Society, requiring Keill, in effect, to make him ſatisfaction for the injury he had done him in his paper relating to the paſſage in the *Acta Eruditorum:* he proteſted, that he was far from aſſuming to himſelf Newton's method of Fluxions; and therefore deſired that Keill might be obliged to retract his falſe aſſertion. On the other hand, Keill deſired that he might be permitted to juſtify what he had aſſerted. He made his defence to the approbation of Newton, and other members of the Society. A copy of this was ſent to Leibnitz; who, in a ſecond letter, remonſtrated ſtill more loudly againſt Keill's want of candour and ſincerity; adding, that it was not fit for one of his age and experience to engage in a diſpute with an upſtart, who acted without any authority from Newton, and deſiring that the Royal Society would enjoin him ſilence. Upon this, a ſpecial committee was appointed; who, after examining the facts, concluded their report with " reckoning Mr. Newton the inventor of Fluxions; and that Mr. Keill, in aſſerting the ſame, had been no ways injurious to Mr. Leibnitz." The whole proceedings upon this matter may be ſeen in Collins's *Commercium Epiſtolicum*, with many valuable papers of Newton, Leibnitz, Gregory, and other mathematicians. In the mean time Keill behaved himſelf with great firmneſs and ſpirit; which he alſo ſhewed afterwards in a Latin epiſtle, written in 1720, to Bernoulli, mathematical profeſſor at Baſil, on account of the ſame uſage ſhewn to Newton: in the title-page of which he put the arms of Scotland, viz. a Thiſtle, with this motto, *Nemo me impune laceſſit.*

About the year 1711, ſeveral objections being urged againſt Newton's philoſophy, in ſupport of Des-Cartes's notions of a plenum, Keill publiſhed a paper in the Philoſ. Tranſ. on the Rarity of Matter, and the Tenuity of its Compoſition. But while he was engaged in this diſpute, queen Anne was pleaſed to appoint him her Decipherer; and he continued in that place under king George the Firſt till the year 1716. The univerſity of Oxford conferred on him the degree of M D. in 1713; and, two years after, he publiſhed an edition of Commandine's Euclid, with additions of his own. In 1718 he publiſhed his *Introductio ad Veram Aſtronomiam:* which was afterwards, at the requeſt of the ducheſs of Chandos, tranſlated by himſelf into Engliſh; and, with ſeveral emendations, publiſhed in 1721, under the title of *An Introduction to True Aſtronomy*, &c. This was his laſt gift to the public; being this ſummer ſeized with a violent fever, which terminated his life Sept. 1, in the 50th year of his age.

His papers in the Philoſ. Tranſ. above alluded to, are contained in volumes 26 and 29.

KEILL (Dr. *James*), an eminent phyſician and philoſopher, and younger brother of Dr. John Keill above mentioned, was alſo born in Scotland, in 1673. Having travelled abroad, on his return he read lectures on Anatomy with great applauſe in the univerſities of Oxford and Cambridge, by the latter of which he had the degree of M. D. conferred upon him. In 1703 he ſettled at Northampton as a phyſician, where he died of a cancer in the mouth in 1719. His publications are

1. An Engliſh tranſlation of Lemery's Chemiſtry.

2. On Animal Secretion, the quantity of Blood in the Human Body, and on Muſcular Motion.

3. A treatiſe on Anatomy.

4. Several pieces in the Philoſ. Tranſ. volumes 25 and 30.

KEPLER (JOHN), a very eminent aſtronomer and mathematician, was born at Wiel, in the county of Wirtemberg, in 1571. He was the diſciple of Mæſtlinus, a learned mathematician and aſtronomer, of whom he learned thoſe ſciences, and became afterwards profeſſor of them to three ſucceſſive emperors, viz. Matthias, Rudolphus, and Ferdinand the 2d.

To this ſagacious philoſopher we owe the firſt diſcovery of the great laws of the planetary motions, viz. that the planets deſcribe areas that are always proportional to the times; that they move in elliptical orbits, having the ſun in one focus; and that the ſquares of their periodic times, are proportional to the cubes of their mean diſtances; which are now generally known by the name of Kepler's Laws. But as this great man ſtands as it were at the head of the modern reformed aſtronomy, he is highly deſerving of a pretty large account, which we ſhall extract chiefly from the words of that great mathematician Mr. Maclaurin.

Kepler had a particular paſſion for finding analogies and harmonies in nature, after the manner of the Pythagoreans and Platoniſts; and to this diſpoſition we owe ſuch valuable diſcoveries, as are more than ſufficient to excuſe his conceits. Three things, he tells us, he anxiouſly ſought to find out the reaſon of, from his early youth; viz, Why the planets were 6 in number? Why the dimenſions of their orbits were ſuch as Copernicus had deſcribed from obſervations? And what was the analogy or law of their revolutions? He ſought for the reaſons of the two firſt of theſe, in the properties of numbers and plane figures, without ſucceſs. But at length reflecting, that while the plane regular figures may be infinite in number, the regular ſolids are only five, as Euclid had long ago demonſtrated: he imagined, that certain myſteries in nature might correſpond with this remarkable limitation inherent in the eſſences of things; and the rather, as he found that the Pythagoreans had made great uſe of thoſe five regular ſolids in their philoſophy. He therefore endeavoured to find ſome relation between the dimenſions of theſe ſolids and the intervals of the planetary ſpheres; thus, imagining that a cube, inſcribed in the ſphere of Saturn, would touch by its ſix planes the ſphere of Jupiter; and that the other four regular ſolids in like manner fitted the intervals that are between the ſpheres of the other planets: he became perſuaded that this was the true reaſon why the primary planets were preciſely ſix in number, and that the author of the world had determined their diſtances from the ſun, the centre of the ſyſtem, from a regard to this analogy. Being thus poſſeſſed, as he thought, of the grand ſecret of the Pythagoreans, and greatly pleaſed with his diſcovery, he publiſhed it in 1596, under the title of *Myſterum Coſmographicum;* and was for ſome time ſo charmed with it, that he ſaid

he

he would not give up the honour of having invented what was contained in that book, for the electorate of Saxony.

Kepler sent a copy of this book to Tycho Brahe, who did not approve of those abstract speculations concerning the system of the world, but wrote to Kepler, first to lay a solid foundation in observations, and then, by ascending from them, to endeavour to come at the causes of things. Tycho however, pleased with his genius, was very desirous of having Kepler with him to assist him in his labours: and having settled, under the protection of the emperor, in Bohemia, where he passed the last years of his life, after having left his native country on some ill usage, he prevailed upon Kepler to leave the university of Gratz, and remove into Bohemia, with his family and library, in the year 1600. But Tycho dying the next year, the arranging the observations devolved upon Kepler, and from that time he had the title of Mathematician to the Emperor all his life, and gained continually more and more reputation by his works. The emperor Rudolph ordered him to finish the tables of Tycho Brahe, which were to be called the *Rudolphine Tables*. Kepler applied diligently to the work: but unhappy are those learned men who depend upon the good-humour of the intendants of the finances; the treasurers were so ill-affected towards our author, that he could not publish these tables till 1627. He died at Ratisbon, in 1630, where he was soliciting the payment of the arrears of his pension.

Kepler made many important discoveries from Tycho's observations, as well as his own. He found, that astronomers had erred, from the first rise of the science, in ascribing always circular orbits and uniform motions to the planets; that, on the contrary, each of them moves in an ellipsis which has one of its foci in the sun: that the motion of each is really unequable, and varies so, that a ray supposed to be always drawn from the planet to the sun describes equal areas in equal times.

It was some years later before he discovered the analogy there is between the distances of the several planets from the sun, and the periods in which they complete their revolutions. He easily saw, that the higher planets not only moved in greater circles, but also more slowly than the nearer ones; so that, on a double account, their periodic times were greater. Saturn, for example, revolves at the distance from the sun 9½ times greater than the earth's distance from it; and the circle described by Saturn is in the same proportion: but as the earth revolves in one year, so, if their velocities were equal, Saturn ought to revolve in 9 years and a half; whereas the periodic time of Saturn is about 29 years. The periodic times of the planets increase, therefore, in a greater proportion than their distances from the sun: but yet not in so great a proportion as the squares of those distances; for if that were the law of the motions, (the square of 9½ being 90¼), the periodic time of Saturn ought to be above 90 years. A mean proportion between that of the distances of the planets, and that of the squares of those distances, is the true proportion of the periodic times; as the mean between 9½ and its square 90¼, gives the periodic time of Saturn in years. Kepler, after having committed several mistakes in determining this analogy, hit upon it at last, May the 15, 1618; for he is so particular as to mention the precise

day when he found that "The squares of the periodic times were always in the same proportion as the cubes of their mean distances from the sun."

When Kepler saw, according to better observations, that his disposition of the five regular solids among the planetary spheres, was not agreeable to the intervals between their orbits, he endeavoured to discover other schemes of harmony. For this purpose, he compared the motions of the same planet at its greatest and least distances, and of the different planets in their several orbits, as they would appear viewed from the sun; and here he fancied that he found a similitude to the divisions of the octave in music. These were the dreams of this ingenious man, which he was so fond of, that, hearing of the discovery of four new planets (the satellites of Jupiter) by Galileo, he owns that his first reflections were from a concern how he could save his favourite scheme, which was threatened by this addition to the number of the planets. The same attachment led him into a wrong judgment concerning the sphere of the fixed stars: for being obliged, by his doctrine, to allow a vast superiority to the sun in the universe, he restrains the fixed stars within very narrow limits. Nor did he consider them as suns, placed in the centres of their several systems, having planets revolving round them; as the other followers of Copernicus have concluded them to be, from their having light in themselves, from their immense distances, and from the analogy of nature. Not contented with these harmonies, which he had learned from the observations of Tycho, he gave himself the liberty to imagine several other analogies, that have no foundation in nature, and are overthrown by the best observations. Thus from the opinions of Kepler, though most justly admired, we are taught the danger of espousing principles, or hypotheses, borrowed from abstract sciences, and of applying them, with such freedom, to natural enquiries.

A more recent instance of this fondness, for discovering analogies between matters of abstract speculation, and the constitution of nature, we find in Huygens, one of the greatest geometricians and astronomers any age has produced: when he had discovered that satellite of Saturn, which from him is still called the Huygenian satellite, this, with our moon, and the four satellites of Jupiter, completed the number of six secondary planets then discovered in the system; and because the number of primary planets was also six, and this number is called by mathematicians a perfect number (being equal to the sum of its aliquot parts, 1, 2, 3,) Huygens was hence induced to believe that the number of the planets was complete, and that it was in vain to look for any more. This is not mentioned to lessen the credit of this great man, who never perhaps reasoned in such a manner on any other occasion; but only to shew, by another instance, how ill-grounded reasonings of this kind have always proved. For, not long after, the celebrated Cassini discovered four more satellites about Saturn, not to mention the two more that have lately been discovered to that planet by Dr. Herschel, with another new primary planet and its two satellites, besides many others, of both sorts, as yet unknown, which possibly may belong to our system. The same Cassini having found that the analogy, discovered by Kepler, between the periodic times and the distances from the centre, takes place in

the

the leffer fyftems of Jupiter and Saturn, as well as in the great folar fyftem; his obfervations overturned that groundlefs analogy which had been imagined between the number of the planets, both primary and fecondary, and the number fix: but eftablifhed, at the fame time, that harmony in their motions, which will afterwards appear to flow from one real principle extended over the univerfe.

But to return to Kepler; his great fagacity, and continual meditations on the planetary motions, fuggefted to him fome views of the true principles from which thefe motions flow. In his preface to the Commentaries concerning the planet Mars, he fpeaks of gravity as of a power that was mutual between bodies, and tells us, that the earth and moon tend towards each other, and would meet in a point, fo many times nearer to the earth than to the moon, as the earth is greater than the moon, if their motions did not hinder it. He adds, that the tides arife from the gravity of the waters towards the moon. But not having notions fufficiently juft of the laws of motion, it feems he was not able to make the beft ufe of thefe thoughts; nor does it appear that he adhered to them fteadily, fince in his Epitome of Aftronomy, publifhed many years after, he propofes a phyfical account of the planetary motions, derived from different principles.

He fuppofes, in that treatife, that the motion of the fun on his axis, is preferved by fome inherent vital principle; that a certain virtue, or immaterial image of the fun, is diffufed with his rays into the ambient fpaces, and, revolving with the body of the fun on his axis, takes hold of the planets, and carries them along with it in the fame direction; like as a loadftone turned round near a magnetic needle, makes it turn round at the fame time. The planet, according to him, by its inertia, endeavours to continue in its place, and the action of the fun's image and this inertia are in a perpetual ftruggle. He adds, that this action of the fun, like his light, decreafes as the diftance increafes; and therefore moves the fame planet with greater celerity when nearer the fun, than at a greater diftance. To account for the planet's approaching towards the fun as it defcends from the aphelion to the perihelion, and receding from the fun while it afcends to the aphelion again, he fuppofes that the fun attracts one part of each planet, and repels the oppofite part; and that the part attracted is turned towards the fun in the defcent, and the other towards the fun in the afcent. By fuppofitions of this kind, he endeavoured to account for all the other varieties of the celeftial motions.

But, now that the laws of motion are better known than in Kepler's time, it is eafy to fhew the fallacy of every part of this account of the planetary motions. The planet does not endeavour to ftop in confequence of its inertia, but to perfevere in its motion in a right line. An attractive force makes it defcend from the aphelion to the perihelion in a curve concave towards the fun: but the repelling force, which he fuppofed to begin at the perihelion, would caufe it to afcend in a figure convex towards the fun. There will be occafion to fhew afterwards, from Sir Ifaac Newton, how an attraction or gravitation towards the fun, alone produces the effects, which, according to Kepler, required both an attractive and repelling force; and that the virtue

which he afcribed to the fun's image, propagated into the planetary regions, is unneceffary, as it could be of no ufe for this effect, though it were admitted. For now his own prophecy, with which he concludes his book, is verified; where he tells us, that "the difcovery of fuch things was referved for the fucceeding ages, when the author of nature would be pleafed to reveal thefe myfteries."

The works of this celebrated author are many and valuable; as,

1. His *Cofmographical Myftery*, in 1596.
2. *Optical Aftronomy*, in 1604.
3. *Account of a New Star in Sagittarius*, 1605.
4. *New Aftronomy*; or, *Celeftial Phyfics*, in Commentaries on the planet Mars.
5. *Differtations*; with the *Nuncius Siderius* of Galileo, 1610.
6. *New Gauging of Wine Cafks*, 1615. Said to be written on occafion of an erroneous meafurement of the wine at his marriage by the revenue officer.
7. *New Ephemerides*, from 1617 to 1620.
8. *Copernican Syftem*, three firft books of the, 1618.
9. *Harmony of the World*; and three books of *Comets*, 1619.
10. *Cofmographical Myftery*, 2d edit. with Notes, 1621.
11. *Copernican Aftronomy*; the three laft books, 1622.
12. *Logarithms*, 1624; and the *Supplement*, in 1625.
13. His *Aftronomical Tables*, called the *Rudolphine Tables*, in honour of the emperor Rudolphus, his great and learned patron, in 1627.
14. *Epitome of the Copernican Aftronomy*, 1635.

Befide thefe, he wrote feveral pieces on various other branches, as *Chronology, Geometry of Solids, Trigonometry*, and an excellent treatife of *Dioptrics*, for that time.

KEPLER's LAWS, are thofe laws of the planetary motions difcovered by Kepler. Thefe difcoveries in the mundane fyftem, are commonly accounted two, viz. 1ft, That the planets defcribe about the fun, areas that are proportional to the times in which they are defcribed, namely, by a line connecting the fun and planet: and 2d, That the fquares of the times of revolution, are as the cubes of the mean diftances of the planets from the fun. Kepler difcovered alfo that the orbits of the planets are elliptical.

Thefe difcoveries of Kepler, however, were only found out by many trials, in fearching among a great number of aftronomical obfervations and revolutions, what rules and laws were found to obtain. On the other hand, Newton has demonftrated, *a priori*, all thefe laws, fhewing that they muft obtain in the mundane fyftem, from the laws of gravitation and centripetal force; viz, the firft of thefe laws refulting from a centripetal force urging the planets towards the fun, and the 2d, from the centripetal force being in an inverfe ratio of the fquare of the diftance. And the elliptic form of the orbits, from a projectile force regulated by a centripetal one.

KEPLER's *Problem*, is the determining the true from the mean anormly of a planet, or the determining its place, in its elliptic orbit, anfwering to any given time; and fo named from the celebrated aftronomer Kepler, who firft propofed it. See ANOMALY.

The

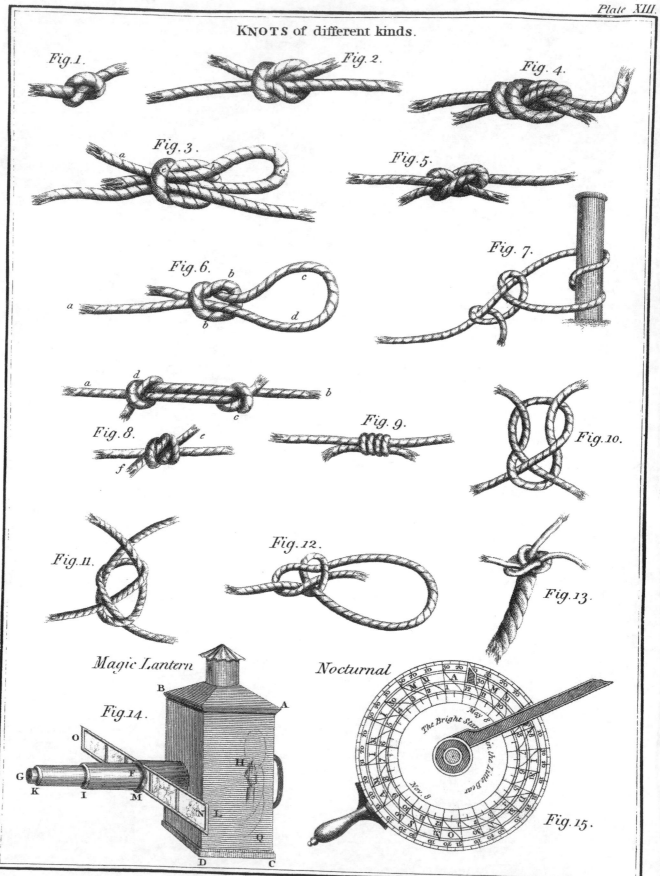

Plate XIII.

KNOTS of different kinds.

Fig.1. Fig.2. Fig. 4.

Fig. 3. Fig. 5.

Fig.6. Fig. 7.

Fig. 8. Fig. 9. Fig.10.

Fig.11. Fig.12. Fig. 13.

Magic Lantern Nocturnal

Fig.14.

The general ftate of the problem is this: To find the pofition of a right line, which, paffing through one of the foci of an ellipfis, fhall cut off an area which fhall be in any given proportion to the whole area of the ellipfis; which refults from this property, that fuch a line fweeps areas that are proportional to the times.

Many folutions have been given of this problem, fome direct and geometrical, others not: viz, by Kepler, Bulliald, Ward, Newton, Keill, Machin, &c. See Newton's Princip. lib. 1. prop. 31, Keill's Aftron. Lect. 23, Philof Tranf abr. vol. 8. pa. 73, &c.

In the laft of thefe place , Mr. Machin obferves, that many attempts have been made at different times, but with no great fuccefs, towards the folution of the problem propofed by Kepler: To divide the area of a femicircle into given parts, by a line drawn from a given point in the diameter, in order to find an univerfal rule for the motion of a body in an elliptic orbit. For among the feveral methods offered, fome are only true in fpeculation, but are really of no fervice; others are not different from his own, which he judged improper. And as to the reft, they are all fo limited and confined to particular conditions and circumftances, as ftill to leave the problem in general untouched. To be more particular; it is evident, that all conftructions by mechanical curves are feeming folutions only, but in reality unapplicable; that the roots of infinite feries are, on account of their known limitations in all refpects, fo far from being fufficient rules, that they ferve for little more than exercifes in a method of calculation. And then, as to the univerfal method, which proceeds by a continued correction of the errors of a falfe pofition, it is no method of folution at all in itfelf; becaufe, unlefs there be fome antecedent rule or hypothefis to begin the operation (as fuppofe that of an uniform motion about the upper focus, for the orbit of a planet; or that of a motion in a parabola for the perihelion part of the orbit of a comet, or fome other fuch), it would be impoffible to proceed one ftep in it. But as no general rule has ever yet been laid down, to affift this method, fo as to make it always operate, it is the fame in effect as if there were no method at all. And accordingly in experience it is found, that there is no rule now fubfifting but what is abfolutely ufelefs in the elliptic orbits of comets; for in fuch cafes there is no other way to proceed but that which was ufed by Kepler: to compute a table for fome part of the orbit, and in it examine if the time to which the place is required, will fall out any where in that part. So that, upon the whole, it appears evident, that this problem, contrary to the received opinion, has never yet been advanced one ftep towards its true folution.

Mr. Machin then proceeds to give his own folution of this problem, which is particularly neceffary in orbits of a great excentricity; and he illuftrates his method by examples for the orbits of Venus, of Mercury, of the comet of the year 1682, and of the great comet of the year 1680, fufficiently fhewing the univerfallty of the method.

KEY, in Mufic, is a certain fundamental note, or tone, to which the whole piece, be it concerto, fonata, cantata, &c, is accommodated; and with which it ufually begins, but always ends.

KEYS denote alfo, in an organ, harpfichord, &c, the pieces of wood or ivory which are ftruck by the fingers, in playing upon the inftrument.

KEYSTONE, the middle vouffoir, or the arch ftone in the top, or immediately over the centre of an arch — The length of the keyftone, or thicknefs of the archivolt at top, is allowed by the beft architects, to be about the 15th or 16th part of the fpan.

KILDERKIN, a kind of liquid meafure, containing two firkins, or 18 gallons, beer-meafure, or 16 alemeafure.

KING-piece, or KING-poft, is a piece of timber fet upright in the middle, between two principal rafters, and having ftruts or braces going from it to the middle of each rafter.

KIRCH (CHRISTIAN FREDERIC), of Berlin, a celebrated aftronomer, was born at Guben in 1694. He acquired great reputation in the obfervatories of Dantzic and Berlin. Godfrey Kirch his father, and Mary his mother, alfo acquired confiderable reputation by their aftronomical obfervations. This family corresponded with all the learned focieties of Europe, and their aftronomical works are in great repute.

KIRCHER (ATHANASIUS), a famous philofopher and mathematician, was born at Fulde in 1601. He entered into the fociety of the Jefuits in 1618, and taught philofophy, mathematics, the Hebrew and Syriac Languages, in the univerfity of Wirtfburg, with great applaufe, till the year 1631. He retired to France on account of the ravages committed by the Swedes in Franconia, and lived fome time at Avignon. He was afterwards called to Rome, where he taught mathematics in the Roman college, collected a rich cabinet of machines and antiquities, and died in 1680, in the 80th year of his age.

The quantity of his works is immenfe, amounting to 22 volumes in folio, 11 in quarto, and three in octavo; enough to employ a man for a great part of his life even to tranfcribe them. Moft of them are rather curious than ufeful; many of them vifionary and fanciful; and it is not to be wondered at, if they are not always accompanied with the greateft exactnefs and precifion. The principal of them are,

1. *Prælufiones Magneticæ.*
2. *Primitiæ Gnomonicæ Catoptricæ.*
3. *Ars magna Lucis et Umbræ.*
4. *Mufurgia Univerfalis.*
5. *Obelifcus Pamphilius.*
6. *Oedipus Ægyptiacus;* 4 volumes folio.
7. *Itinerarium Extaticum.*
8. *Obelifcus Ægyptiacus;* 4 volumes folio.
9. *Mundus Subterraneus.*
10. *China Illuftrata.*

KNOT, a tye, or complication of a rope, cord, or ftring, or of the ends of two together. There are divers forts of knots ufed for different purpofes, which may be explained by fhewing the figures of them open, or undrawn, thus. 1. Fig. 1, plate xiii. is a *Thumb knot.* This is the fimpleft of all. It is ufed to tye at the end of a rope, to prevent its opening out: it is alfo ufed by taylors &c. at the end of their thread.

Fig. 2, a *Loop knot.* Ufed to join pieces of rope &c. together.

Fig. 3, a *Draw knot,* which is the fame as the laft; only one end or both return the fame way back, as *a b c d.*

a b c d. By drawing at *a*, the part *b c d* comes through, and the knot is loosed.

Fig. 4, a *Ring knot*. This serves also to join pieces of cord &c together.

Fig. 5 is another knot for tying cords together. This is used when any cord is often to be loosed.

Fig. 6, a *Running knot*, to draw any thing close. By pulling at the end *a*, the cord is drawn through the loop *b*, and the part *c d* is drawn close about a beam, &c.

Fig. 7 is another knot, to tye any thing to a post. And here the end may be put through as often as you please.

Fig. 8, a *Very small knot*. A thumb knot is first made at the end of each piece, and then the end of the other is passed through it. Thus, the cord *a c* runs through the loop *d*, and *b d* through *c*; and then drawn close by pulling at *a* and *b*. If the ends *e* and *f* be drawn, the knot will be loosed again.

Fig. 9, a *Fisher's knot*, or *Water knot*. This is the same as the 4th, only the ends are to be put twice through the ring, which in the former was but once; and then drawn close.

Fig. 10, a *Meshing knot*, for nets; and is to be drawn close.

Fig. 11, a *Barber's knot*, or a knot for cawls of wigs; and is to be drawn close.

Fig. 12, a *Bowline knot*. When this is drawn close, it makes a loop that will not slip, as fig. 7; and serves to hitch over any thing.

Fig. 13, a *Wale knot*, which is made with the three strands of a rope, so that it cannot slip. When the rope is put through a hole, this knot keeps it from slipping through. When the three strands are wrought round once or twice more, after the same manner, it is called *crowning*. By this means the knot is made larger and stronger. A thumb knot, N°. 1, may be applied to the same use as this.

KNOTS mean also the divisions of the log line, used at sea. These are usually 7 fathom, or 42 feet asunder; but should be 8⅓ fathom, or 50 feet. And then, as many knots as the log line runs out in half a minute, so many miles does the ship sail in an hour; supposing her to keep going at an equal rate, and allowing for yaws, leeway, &c.

KOENIG (SAMUEL), a learned philosopher and mathematician, was a Swiss by birth, and came early into eminence by his mathematical abilities. He was professor of philosophy and natural law at Franeker, and afterwards at the Hague, where he became also librarian to the Stadtholder, and to the Princess of Orange; and where he died in 1757.

The Academy of Berlin enrolled him among her members; but afterwards expelled him on the following occasion. Maupertuis, the president, had inserted in the volume of the Memoirs for 1746, a discourse upon the Laws of Motion; which Koenig not only attacked, but also attributed the memoir to Leibnitz. Maupertuis, stung with the imputation of plagiarism, engaged the Academy of Berlin to call upon him for his proof; which Koenig failing to produce, he was struck out of the academy. All Europe was interested in the quarrel which this occasioned between Koenig and Maupertuis. The former appealed to the public; and his appeal, written with the animation of resentment, procured him many friends. He was author of some other works, and had the character of being one of the best mathematicians of the age.

L.

LAG

LABEL, a long thin brass ruler, with a small sight at one end, and a central hole at the other; commonly used with a tangent-line on the edge of a circumferentor, to take altitudes, and other angles.

LACERTA, *Lizard*, one of the new constellations of the northern hemisphere, added by Hevelius to the 48 old ones, near Cepheus and Cassiopeia.

This constellation contains, in Hevelius's catalogue 10 stars, and in Flamsteed's 16.

LACUNAR, an arched roof or cieling; more especially the planking or flooring above the porticos.

LADY-*Day* the 25th of March, being the Annunciation of the Holy Virgin.

LAGNY (THOMAS FANTET *de*), an eminent French mathematician, was born at Lyons. Fournier's Euclid, and Pelletier's Algebra, by chance falling in

LAG

his way, developed his genius for the mathematics. It was in vain that his father designed him for the law; he went to Paris to deliver himself wholly up to the study of his favourite science. In 1697, the Abbé Bignon, protector-general of letters, got him appointed professor-royal of Hydrography at Rochfort. Soon after, the duke of Orleans, then regent of France, fixed him at Paris, and made him sub-director of the General Bank, in which he lost the greatest part of his fortune in the failure of the Bank. He had been received into the ancient academy in 1696; upon the renewal of which he was named Associate-geometrician in 1699, and pensioner in 1723. After a life spent in close application, he died, April 12, 1734.

In the last moments of his life, and when he had lost all knowledge of the persons who surrounded his bed,

bed, one of them, through curiofity, afked him, what is the fquare of 12 ? To which he immediately replied, and without feeming to know that he gave any anfwer, 144.

De Lagny particularly excelled in arithmetic, algebra, and geometry in which he made many improvements and difcoveries. He, as well as Leibnitz, invented a binary arithmetic, in which only two figures are concerned. He rendered much eafier the refolution of algebraic equations, efpecially the irreducible cafe in cubic equations; and the numeral refolution of the higher powers, by means of fhort approximating theorems.—He delivered the meafures of angles in a new fcience, called *Goniometry*; in which he meafured angles by a pair of compaffes, without fcales, or tables, to great exactnefs; and thus gave a new appearance to trigonometry.—*Cyclometry*, or the meafure of the circle, was alfo an object of his attention; and he calculated, by means of infinite feries, the ratio of the circumference of a circle to its diameter, to 120 places of figures.—He gave a general theorem for the tangents of multiple arcs. With many other curious or ufeful improvements, which are found in the great multitude of his papers, that are printed in the different volumes of the Memoirs of the Academy of Sciences, viz, in almoft every volume, from the year 1699, to 1729.

LAKE, a collection of water, inclofed in the cavity of fome inland place, of a confiderable extent and depth. As the Lake of Geneva, &c.

LAMMAS-DAY, the 1ft of Auguft; fo called, according to fome, becaufe lambs then grow out of feafon, as being too large. Others derive it from a Saxon word, fignifying *loaf-mafs*, becaufe on that day our forefathers made an offering of bread prepared with new wheat.

It is celebrated by the Romifh church in memory of St. Peter's imprifonment.

LAMPÆDIAS, a kind of bearded comet, refembling a burning lamp, being of feveral fhapes; for fometimes its flame or blaze runs tapering upwards like a fword, and fometimes it is double or treble pointed.

LANDEN (JOHN), an eminent mathematician, was born at Peakirk, near Peterborough in Northamptonfhire, in January 1719. He became very early a proficient in the mathematics, for we find him a very refpectable contributor to the Ladies Diary in 1744; and he was foon among the foremoft of thofe who then contributed to the fupport of that fmall but valuable publication, in which almoft every Englifh mathematician who has arrived at any degree of eminence for the beft part of this century, has contended for fame at one time or other of his life. Mr. Landen continued his contributions to it at times, under various fignatures, till within a few years of his death.

It has been frequently obferved, that the hiftories of literary men confift chiefly of the hiftory of their writings; and the obfervation was never more fully verified, than in the prefent article concerning Mr. Landen.

In the 48th volume of the Philofophical Tranfactions, for the year 1754, Mr. Landen gave " An Inveftigation of fome theorems which fuggeft feveral very remarkable properties of the Circle, and are at the fame time of confiderable ufe in refolving Fractions,

the denominators of which are certain Multinomials, into more fimple ones, and by that means facilitate the computation of Fluents." This ingenious paper was delivered to the Society by that eminent mathematician Thomas Simpfon of Woolwich, a circumftance which will convey to thofe who are not themfelves judges of it, fome idea of its merit.

In the year 1755, he publifhed a volume of about 160 pages, intitled *Mathematical Lucubrations.* The title to this publication was made choice of, as a means of informing the world, that the ftudy of the mathematics was at that time rather the purfuit of his leifure hours, than his principal employment: and indeed it continued to be fo, during the greateft part of his life; for about the year 1762 he was appointed agent to Earl Fitzwilliam, an employment which he refigned only two years before his death. Thefe Lucubrations contain a variety of tracts relative to the rectification of curve lines, the fummation of feries, the finding of fluents, and many other points in the higher parts of the mathematics.

About the latter end of the year 1757, or the beginning of 1758, he publifhed propofals for printing by fubfcription, *The Refidual Analyfis*, a new Branch of the Algebraic art: and in 1758 he publifhed a fmall tract, entitled *A Difcourfe on the Refidual Analyfis*; in which he refolved a variety of problems, to which the method of fluxions had ufually been applied, by a mode of reafoning entirely new: he alfo compared thefe folutions with others derived from the fluxionary method; and fhewed that the folutions by his new method were commonly more natural and elegant than the fluxionary ones.

In the 51ft volume of the Philofophical Tranfactions, for the year 1760, he gave *A New Method of computing the Sums of a great number of Infinite Series.* This paper was alfo prefented to the Society by his ingenious friend the late Mr. Thomas Simpfon.

In 1764, he publifhed the firft book of *The Refidual Analyfis*. In this treatife, befides explaining the principles which his new analyfis was founded on, he applied it, in a variety of problems, to drawing tangents, and finding the properties of curve lines; to defcribing their involutes and evolutes, finding the radius of curvature, their greateft and leaft ordinates, and points of contrary flexure; to the determination of their cufps, and the drawing of afymptotes: and he propofed, in a fecond book, to extend the application of this new analyfis to a great variety of mechanical and phyfical fubjects. The papers which were to have formed this book lay long by him; but he never found leifure to put them in order for the prefs.

In the year 1766, Mr. Landen was elected a Fellow of the Royal Society. And in the 58th volume of the Philofophical Tranfactions, for the year 1768, he gave *A fpecimen of a New Method of comparing Curvilinear Areas*; by means of which many areas are compared, that did not appear to be comparable by any other method: a circumftance of no fmall importance in that part of natural philofophy which relates to the doctrine of motion.

In the 60th volume of the fame work, for the year 1770, he gave *Some New Theorems* for computing the Whole Areas of Curve Lines, where the Ordinates are expreffed

expreffed by Fractions of a certain form, in a more concife and elegant manner than had been done by Cotes, De Moivre, and others who had confidered the fubject before him.

In the 61ft volume, for 1771, he has invefligated feveral new and ufeful theorems for computing certain fluents, which are affignable by arcs of the conic fections. This fubject had been confidered before, both by Maclaurin and d'Alembert; but fome of the theorems that were given by thefe celebrated mathematicians, being in part expreffed by the difference between an hyperbolic arc and its tangent, and that difference being not directly attainable when the arc and its tangent both become infinite, as they will do when the whole fluent is wanted, although fuch fluent be finite; thefe theorems therefore fail in thefe cafes, and the computation becomes impracticable without farther help. This defect Mr. Landen has removed, by affigning the *limit* of the difference between the hyperbolic arc and its tangent, while the point of contact is fuppofed to be removed to an infinite diftance from the vertex of the curve. And he concludes the paper with a curious and remarkable property relating to pendulous bodies, which is deducible from thofe theorems. In the fame year he publifhed *Animadverfions on Dr. Stewart's Computation of the Sun's Diftance from the Earth.*

In the 65th volume of the Philofophical Transactions, for 1775, he gave the invefligation of a General Theorem, which he had promifed in 1771, for finding the Length of any Curve of a Conic Hyperbola by means of two Elliptic Arcs: and he obferves, that by the theorems there invefligated, both the elaftic curve and the curve of equable recefs from a given point, may be conftructed in thofe cafes where Maclaurin's elegant method fails.

In the 67th volume, for 1777, he gave " A New Theory of the Motion of bodies revolving about an axis in free fpace, when that motion is difturbed by fome extraneous force, either percuffive or accelerative." At that time he did not know that the fubject had been treated by any perfon before him, and he confidered only the motion of a fphere, fpheroid, and cylinder. After the publication of this paper however he was informed, that the doctrine of rotatory motion had been confidered by d'Alembert; and upon procuring that author's *Opufcules Mathematiques,* he there learned that d'Alembert was not the only one who had confidered the matter before him; for d'Alembert there fpeaks of fome mathematician, though he does not mention his name, who, after reading what had been written on the fubject, doubted whether there be any folid whatever, befide the fphere, in which any line, paffing through the centre of gravity, will be a permanent axis of rotation. In confequence of this, Mr. Landen took up the fubject again; and though he did not then give a folution to the general problem, viz, " to determine the motions of a body of any form whatever, revolving without reftraint about any axis paffing through its centre of gravity," he fully removed every doubt of the kind which had been ftarted by the perfon alluded to by d'Alembert, and pointed out feveral bodies which, under certain dimenfions, have that remarkable property. This paper is given, among many others equally curious, in a volume of *Memoirs,* which

he publifhed in the year 1780. That volume is alfo enriched with a very extenfive appendix, containing *Theorems for the Calculation of Fluents;* which are more complete and extenfive than thofe that are found in any author before him.

In 1781, 1782, and 1783, he publifhed three fmall Tracts on the Summation of Converging Series; in which he explained and fhewed the extent of fome theorems which had been given for that purpofe by De Moivre, Stirling, and his old friend Thomas Simpfon, in anfwer to fome things which he thought had been written to the difparagement of thofe excellent mathematicians. It was the opinion of fome, that Mr. Landen did not fhew lefs mathematical fkill in explaining and illuftrating thefe theorems, than he has done in his writings on original fubjects; and that the authors of them were as little aware of the extent of their own theorems, as the reft of the world were before Mr. Landen's ingenuity made it obvious to all.

About the beginning of the year 1782, Mr. Landen had made fuch improvements in his theory of Rotatory Motion, as enabled him, he thought, to give a folution of the general problem mentioned above; but finding the refult of it to differ very materially from the refult of the folution which had been given of it by d'Alembert, and not being able to fee clearly where that gentleman in his opinion had erred, he did not venture to make his own folution public. In the courfe of that year, having procured the Memoirs of the Berlin Academy for 1757, which contain M. Euler's folution of the problem, he found that this gentleman's folution gave the fame refult as had been deduced by d'Alembert; but the perfpicuity of Euler's manner of writing enabled him to difcover where he had differed from his own, which the obfcurity of the other did not do. The agreement, however, of two writers of fuch eftablifhed reputation as Euler and d'Alembert made him long dubious of the truth of his own folution, and induced him to revife the procefs again and again with the utmoft circumfpection; and being every time more convinced that his own folution was right, and theirs wrong, he at length gave it to the public, in the 75th volume of the Philofophical Transactions, for 1785.

The extreme difficulty of the fubject, joined to the concife manner in which Mr. Landen had been obliged to give his folution, to confine it within proper limits for the Transactions, rendered it too difficult, or at leaft too laborious a tafk for moft mathematicians to read it; and this circumftance, joined to the eftablifhed reputation of Euler and d'Alembert, induced many to think that their folution was right, and Mr. Landen's wrong; and there did not want attempts to prove it; particularly a long and ingenious paper by the learned Mr. Wildbore, a gentleman of very diftinguifhed talents and experience in fuch calculations; this paper is given in the 80th volume of the Philofophical Transactions, for the year 1790, in which he agrees with the folutions of Euler and d'Alembert, and againft that of Mr. Landen. This determined the latter to revife and extend his folution, and give it at greater length, to render it more generally underftood. About this time alfo he met by chance with the late Frifi's *Cofmographiæ Phyficæ et Mathematicæ*; in the fecond part of

which

which there is a folution of this problem, agreeing in the refult with thofe of Euler and d'Alembert. Here Mr. Landen learned that Euler had revifed the folution which he had given formerly in the Berlin Memoirs, and given it another form, and at greater length, in a volume publifhed at Roftoch and Gryphifwald in 1765, intitled, *Theoria Motûs Corporum Solidorum feu Rigidorum.* Having therefore procured this book, Mr. Landen found the fame principles employed in it, and of courfe the fame conclufion refulting from them, as in M. Euler's former folution of the problem. But notwithftanding that there were thus a coincidence of at leaft four moft refpectable mathematicians againft him, Mr. Landen was ftill perfuaded of the truth of his own folution, and prepared to defend it. And as he was convinced of the neceffity of explaining his ideas on the fubject more fully, fo he now found it neceffary to lofe no time in fetting about it. He had for feveral years been feverely afflicted with the ftone in the bladder, and towards the latter part of his life to fuch a degree as to be confined to his bed for more than a month at a time : yet even this dreadful diforder did not extinguifh his ardour for mathematical ftudies ; for the fecond volume of his *Memoirs,* lately publifhed, was written and revifed during the intervals of his diforder. This volume, befides a folution of the general problem concerning rotatory motion, contains the refolution of the problem relating to the motion of a Top ; with an inveftigation of the motion of the Equinoxes, in which Mr. Landen has firft of any one pointed out the caufe of Sir Ifaac Newton's miftake in his folution of this celebrated problem ; and fome other papers of confiderable importance. He juft lived to fee this work finifhed, and received a copy of it the day before his death, which happened on the 15th of January 1790, at Milton, near Peterborough, in the 71ft year of his age.

LARBOARD, the left hand fide of a fhip, when a perfon ftands with his face towards the head.

LARMIER, in Architecture, a flat fquare member of the cornice below the cimafium, and jets out fartheft ; being fo called from its ufe, which is to difperfe the water, and caufe it to fall at a diftance from the wall, drop by drop, or, as it were, by tears ; *larme* in French fignifying a tear.

LATERAL EQUATION, in Algebra, is the fame with fimple equation. It has but one root, and may be conftructed by right lines only.

LATION, is ufed by fome, for the tranflation or motion of a body from one place to another.

LATITUDE, in Geography, or Navigation, the diftance of a place from the equator ; or an arch of the meridian, intercepted between its zenith and the equator. Hence the Latitude is either north or fouth, according as the place is on the north or fouth fide of the equator : thus London is faid to be in 51° 31′ of north latitude.

Circles parallel to the equator are called *parallels of latitude,* becaufe they fhew the latitudes of places by their interfections with the meridian.

The Latitude of a place is equal to the elevation of the pole above the horizon of the place : and hence thefe two terms are ufed indifferently for each other.

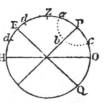

This will be evident from the figure, where the circle ZHQP is the meridian, Z the zenith of the place, HO the horizon, EQ the equator, and P the pole ; then is ZE the latitude, and PO the elevation of the pole above the horizon. And becaufe PE is = ZO, being each a quadrant, if the common part PZ be taken from both, there will remain the latitude ZE = PO the elevation of the pole.—Hence we have a method of meafuring the circumference of the earth, or of determining the quantity of a degree on its furface ; for by meafuring directly northward or fouthward, till the pole be one degree higher or lower, we fhall have the number of miles in a degree of a great circle on the furface of the earth ; and confequently multiplying that by 360, will give the number of miles round the whole circumference of the earth.

The knowledge of the Latitude of the place, is of the utmoft confequence, in geography, navigation, and aftronomy ; it may be proper therefore to lay down fome of the beft ways of determining it, both by fea and land.

1ft. One method is, to find the Latitude of the pole, to which it is equal, by means of the pole ftar, or any other circumpolar ftar, thus : Either draw a true meridian line, or find the times when the ftar is on the meridian, both above and below the pole ; then at thefe times, with a quadrant, or other fit inftrument, take the altitudes of the ftar ; or take the fame when the ftar comes upon your meridian line ; which will be the greateft and leaft altitude of the ftar : then fhall half the fum of the two be the elevation of the pole, or the latitude fought.—For, if *abc* be the path of the ftar about the pole P, Z the zenith, and HO the horizon : then is *a*O the altitude of the ftar upon the meridian when above the pole, and *c*O the fame when below the pole ; hence, becaufe *a*P = *c*P, therefore *a*O + *c*O = 2OP, hence the height of the pole OP, or latitude of Z, is equal to half the fum of *a*O and *c*O.

2d. A fecond method is by means of the declination of the fun, or a ftar, and one meridian altitude of the fame, thus : Having, with a quadrant, or other inftrument, obferved the zenith diftance Z*d* of the luminary ; or elfe its altitude H*d*, and taken its complement Z*d* ; then to this zenith diftance, add the declination *d*E when the luminary and place are on the fame fide of the equator, or fubtract it when on different fides, and the fum or difference will be the latitude EZ fought. But note, that all altitudes obferved, muft be corrected for refraction and the dip of the horizon, and for the femidiameter of the fun, when that is the luminary obferved.

Many other methods of obferving and computing the Latitude may be feen in Robertfon's Navigation ; fee book 5 and book 9. See alfo the Nautical Almanac for 1771.

Mr. Richard Graham contrived an ingenious inftrument for taking the latitude of a place at any time of the day. See Philof. Tranf. N°. 435, or Abr. vol. 8. pa. 371.

LATITUDE, in Astronomy, as of a star or planet, is its distance from the ecliptic, being an arch of a circle of latitude, reckoned from the ecliptic towards its poles, either north or south. Hence, the astronomical latitude is quite different from the geographical, the former measuring from the ecliptic, and the latter from the equator, so that this latter answers to the declination in astronomy, which measures from the equinoctial.

The sun has no latitude, being always in the ecliptic; but all the stars have their several latitudes, and the planets are continually changing their latitudes, sometimes north, and sometimes south, crossing the ecliptic from the one side to the other; the points in which they cross the ecliptic being called the *nodes* of the planet, and in these points it is that they can pass over the face of the sun, or behind his body, viz, when they come both to this point of the ecliptic at the same time.

Circle of LATITUDE, is a great circle passing through the poles of the ecliptic, and consequently perpendicular to it, like as the meridians are perpendicular to the equator, and pass through its poles.

LATITUDE, *of the Moon, North ascending*, is when she proceeds from the ascending node towards her northern limit, or greatest elongation.

LATITUDE, *North descending*, is when the moon returns from her northern limit towards the descending node.

LATITUDE, *South descending*, is when she proceeds from the descending node towards her southern limit.

LATITUDE, *South ascending*, is when she returns from her southern limit towards her ascending node.

And the same is to be understood of the other planets.

Heliocentric LATITUDE, of a planet, is its latitude, or distance from the ecliptic, such as it would appear from the sun.—This, when the planet comes to the same point of its orbit, is always the same, or unchangeable.

Geocentric LATITUDE, of a planet, is its latitude as seen from the earth.—This, though the planet be in the same point of its orbit, is not always the same, but alters according to the position of the earth, in respect to the planet.

The latitude of a star is altered only by the aberration of light, and the secular variation of latitude.

Difference of LATITUDE, is an arc of the meridian, or the nearest distance between the parallels of latitude of two places. When the two latitudes are of the same name, either both north or both south, subtract the less latitude from the greater, to give the difference of latitude; but when they are of different names, add them together for the difference of latitude.

Middle LATITUDE, is the middle point between two latitudes or places; and is found by taking half the sum of the two.

Parallax of LATITUDE. See PARALLAX.

Refraction of LATITUDE. See REFRACTION.

LATUS RECTUM, in Conic Sections, the same with parameter; which see.

LATUS *Transversum*, of the hyperbola, is the right line between the vertices of the two opposite sections; or that part of their common axis lying between the two opposite cones; as the line DE. It is the same as the transverse axis of the hyperbola, or opposite hyperbolas.

LATUS *Primarium*, a right line, DD, or EE, drawn through the vertex of the section of a cone, within the same, and parallel to the base.

LEAGUE, an extent of three miles in length. A nautical league, or three nautical miles, is the 20th part of a degree of a great circle.

LEAP-YEAR, the same as BISSEXTILE; which see. It is so called from its leaping a day more that year than in a common year; consisting of 366 days, and a common year only of 365. This happens every 4th year, except only such complete centuries as are not exactly divisible by 4; such as the 17th, 18th, 19th, 21st &c. centuries, because 17, 18, 19, 21, &c, cannot be divided by 4 without a remainder.

To find Leap Year, &c. Divide the number of the year by 4; then if 0 remain, it is leap year; but if 1, 2, or 3 remain, it is so many after leap-year.

Or the rule is sometimes thus expressed, in these two memorial verses:

Divide by 4; what's left shall be,
For leap-year 0; for past, 1, 2, or 3.

Thus if it be required to know what year 1790 is:
then 4) 1790 (447
2 remains:
so that 2 remaining, shews that 1790 is the 2d year after leap-year. And to find what year 1796 is:
then 4) 1796 (449
here 0 remaining, shews that 1796 is a leap-year.

LEAVER. See LEVER.

LEE, a term in Navigation, signifying that side, or quarter, towards which the wind blows.

LEE-WAY, of a Ship, is the angle made by the point of the compass steered upon, and the real line of the ship's way, occasioned by contrary winds and a rough sea.

All ships are apt to make some lee-way; so that something must be allowed for it, in casting up the log-board. But the lee-way made by different ships, under similar circumstances of wind and sails, is different; and even the same ship, with different lading, and having more or less sail set, will have more or less lee-way. The usual allowances for it are these, as they were given by Mr. John Buckler to the late ingenious Mr. William Jones, who first published them in 1702 in his *Compendium of Practical Navigation.* 1st, When a ship is close-hauled, has all her sails set, the sea smooth, and a moderate gale of wind, it is then supposed she makes little or no lee-way. 2d, Allow one point, when it blows so fresh that the small sails are taken in. 3d, Allow two points, when the topsail must be close reefed. 4th, Allow two points and a half, when one topsail must be handed. 5th, Allow three points and a half, when both topsails must be taken in. 6th, Allow four points, when the fore-course is handed. 7th, Allow five points, when trying under the mainsail only. 8th, Allow six points, when both main and fore-courses are taken in.

9th,

9th, Allow seven points, when the ship tries a-hull, or with all sails handed.

When the wind has blown hard in either quarter, and shifts across the meridian into the next quarter, the lee-way will be lessened. But in all these cases, respect must be had to the roughness of the sea, and the trim of the ship. And hence the mariner will be able to correct his course.

LEGS, *of a Triangle.* When one side of a triangle is taken as the base, the other two are sometimes called the legs. The term is often used too for the base and perpendicular of a right-angled triangle, or the two sides about the right angle.

Hyperbolic LEGS, are the ends of a curve line that partake of the nature of the hyperbola, or having asymptotes.

LEIBNITZ (GODFREY-WILLIAM), an eminent mathematician and philosopher, was born at Leipsic in Saxony in 1646. At the age of 15, he applied himself to mathematics at Leipsic and Jena; and in 1663, maintained a thesis *de Principiis Individuationis.* The year following he was admitted Master of Arts. He read with great attention the Greek philosophers; and endeavoured to reconcile Plato with Aristotle, as he afterwards did Aristotle with Des Cartes. But the study of the law was his principal view; in which faculty he was admitted Bachelor in 1665. The year following he would have taken the degree of Doctor; but was refused it on pretence that he was too young, though in reality because he had raised himself several enemies by rejecting the principles of Aristotle and the Schoolmen.

Upon this he repaired to Altorf, where he maintained a thesis *de Casibus Perplexis,* with such applause, that he had the degree of Doctor conferred on him.

In 1672 he went to Paris, to manage some affairs at the French Court for the baron Boinebourg. Here he became acquainted with all the Literati, and made farther and confiderable progress in the study of mathematics and philosophy, chiefly, as he says, by the works of Pascal, Gregory St. Vincent, and Huygens. In this course, having observed the imperfection of Pascal's arithmetical machine, he invented a new one, as he called it, which was approved of by the minister Colbert, and the Academy of Sciences, in which he was offered a seat as a member, but refused the offers made to him, as it would have been necessary to embrace the Catholic religion.

In 1673, he came over to England; where he became acquainted with Mr. Oldenburg, secretary of the Royal Society, and Mr. John Collins, a distinguished member of the Society; from whom it seems he received some hints of the method of fluxions, which had been invented, in 1664 or 1665, by the then Mr. Isaac Newton.

The same year he returned to France, where he resided till 1676, when he again passed through England, and Holland, in his journey to Hanover, where he proposed to settle. Upon his arrival there, he applied himself to enrich the duke's library with the best books of all kinds. The duke dying in 1679, his successor Ernest Augustus, then bishop of Osnaburgh, shewed Mr. Leibnitz the same favour as his predecessor

had done, and engaged him to write the History of the House of Brunswick. To execute this task, he travelled over Germany and Italy, to collect materials. While he was in Italy, he met with a pleasant adventure, which might have proved a more serious affair. Passing in a small bark from Venice to Mesola, a storm arose; during which the pilot, imagining he was not understood by a German, whom, being a heretic, he looked on as the cause of the tempest, proposed to strip him of his cloaths and money, and throw him overboard. Leibnitz hearing this, without discovering the least emotion, drew a set of beads from his pocket, and began turning them over with great seeming devotion. The artifice succeeded; one of the sailors observing to the pilot, that, since the man was no heretic, he ought not to be drowned.

In 1700 he was admitted a member of the Royal Academy of Sciences at Paris. The same year the elector of Brandenburg, afterwards king of Prussia, founded an academy at Berlin by his advice; and he was appointed perpetual President, though his affairs would not permit him to reside constantly at that place. He projected an academy of the same kind at Dresden; and this design would have been executed, if it had not been prevented by the confusions in Poland. He was engaged likewise in a scheme for an universal language, and other literary projects. Indeed his writings had made him long before famous over all Europe, and he had many honours and rewards conferred on him. Beside the office of Privy Counsellor of Justice, which the elector of Hanover had given him, the emperor appointed him, in 1711, Aulic Counsellor; and the czar made him Privy Counsellor of Justice, with a pension of 1000 ducats. Leibnitz undertook at the same time to establish an academy of sciences at Vienna; but the plague prevented the execution of it. However, the emperor, as a mark of his favour, settled a pension on him of 2000 florins, and promised him one of 4000 if he would come and reside at Vienna; an offer he was inclined to comply with, but was prevented by his death.

Meanwhile, the History of Brunswick being interrupted by other works which he wrote occasionally, he found, at his return to Hanover in 1714, that the elector had appointed Mr. Eccard for his colleague in writing that history. The elector was then raised to the throne of Great Britain, which place Leibnitz visited the latter end of that year, when he received particular marks of friendship from the king, and was frequently at court. He now was engaged in a dispute with Dr. Samuel Clarke, upon the subjects of free-will, the reality of space, and other philosophical subjects. This was conducted with great candour and learning; and the papers, which were published by Clarke, will ever be esteemed by men of genius and learning. The controversy ended only with the death of Leibnitz, Nov. 14, 1716, which was occasioned by the gout and stone, in the 70th year of his age.

As to his character and person: He was of a middle stature, and a thin habit of body. He had a studious air, and a sweet aspect, though near-sighted. He was indefatigably industrious to the end of his life. He eat and drank little. Hunger alone marked the time of his meals, and his diet was plain and strong.

He

He had a very good memory, and it was said could repeat the Æneid from beginning to end. What he wanted to remember, he wrote down, and never read it afterwards. He always professed the Lutheran religion, but never went to sermons; and when in his last sickness his favourite servant desired to send for a minister, he would not permit it, saying he had no occasion for one. He was never married, nor ever attempted it but once, when he was about 50 years old; and the lady desiring time to consider of it, gave him an opportunity of doing the same: he used to say, " that marriage was a good thing, but a wise man ought to consider of it all his life."

Leibnitz was author of a great multitude of writings; several of which were published separately, and many others in the memoirs of different academies. He invented a binary arithmetic, and many other ingenious matters. His claim to the invention of Fluxions, has been spoken of under that article. Hanschius collected, with great care, every thing that Leibnitz had said, in different passages of his works, upon the principles of philosophy; and formed of them a complete system, under the title of G. G. *Leibnitzii Principia Philosophiæ more geometrico demonstrata* &c, 1728, in 4to. There came out a collection of our author's letters in 1734 and 1735, intitled, *Epistolæ ad diverses theologici, juridici, medici, philosophici, mathematici, historici, & philologici argumenti e MSS. auctores: cum annotationibus suis primum divulgavit Christian Cortholtus.* But all his works were collected, distributed into classes by M. Dutens, and published at Geneva in six large volumes 4to, in 1768, intitled, *Gothofredi Guillelmi Leibnitii Opera Omnia &c.*

LEIBNITZIAN PHILOSOPHY, or the Philosophy of Leibnitz, is a system formed and published by its author in the last century, partly in emendation of the Cartesian, and partly in opposition to the Newtonian philosophy. In this philosophy, the author retained the Cartesian subtile matter, with the vortices and universal plenum; and he represented the universe as a machine that should proceed for ever, by the laws of mechanism, in the most perfect state, by an absolute inviolable necessity. After Newton's philosophy was published, in 1687, Leibnitz printed an Essay on the celestial motions in the Act. Erud. 1689, where he admits the circulation of the ether with Des Cartes, and of gravity with Newton; though he has not reconciled these principles, nor shewn how gravity arose from the impulse of this ether, nor how to account for the planetary revolutions in their respective orbits. His system is also defective, as it does not reconcile the circulation of the ether with the free motions of the comets in all directions, or with the obliquity of the planes of the planetary orbits; nor resolve other objections to which the hypothesis of the vortices and plenum is liable.

Soon after the period just mentioned, the dispute commenced concerning the invention of the method of Fluxions, which led Mr. Leibnitz to take a very decided part in opposition to the philosophy of Newton. From the goodness and wisdom of the Deity, and his principle of a *sufficient reason*, he concluded, that the universe was a perfect work, or the best that could possibly have been made; and that other things, which are evil or incommodious, were permitted as necessary consequences of what was best: that the material sys-

tem, considered as a perfect machine, can never fall into disorder, or require to be set right; and to suppose that God interposes in it, is to lessen the skill of the author, and the perfection of his work. He expresly charges an impious tendency on the philosophy of Newton, because he asserts, that the fabric of the universe and course of nature could not continue for ever in its present state, but in process of time would require to be re-established or renewed by the hand of its first framer. The perfection of the universe, in consequence of which it is capable of continuing for ever by mechanical laws in its present state, led Mr. Leibnitz to distinguish between the quantity of motion and the force of bodies; and, whilst he owns in opposition to Des Cartes that the former varies, to maintain that the quantity of force is for ever the same in the universe; and to measure the forces of bodies by the squares of their velocities.

Mr. Leibnitz proposes two principles as the foundation of all our knowledge; the first, that it is impossible for a thing to be, and not to be at the same time, which he says is the foundation of speculative truth; and secondly, that nothing is without a *sufficient reason* why it should be so, rather than otherwise; and by this principle he says we make a transition from abstracted truths to natural philosophy. Hence he concludes that the mind is naturally determined, in its volitions and elections, by the greatest apparent good, and that it is impossible to make a choice between things perfectly like, which he calls *indiscernibles*; from whence he infers, that two things perfectly like could not have been produced even by the Deity himself: and one reason why he rejects a vacuum, is because the parts of it must be supposed perfectly like to each other. For the same reason too, he rejects atoms, and all similar parts of matter, to each of which, though divisible *ad infinitum*, he ascribes a *monad* (Act. Lipsiæ 1698, pa. 435) or active kind of principle, endued with perception and appetite. The essence of substance he places in action or activity, or, as he expresses it, in something that is between acting and the faculty of acting. He affirms that absolute rest is impossible, and holds that motion, or a sort of *nisus*, is essential to all material substances. Each monad he describes as representative of the whole universe from its point of sight; and yet he tells us, in one of his letters, that matter is not a substance, but a *substantiatum*, or *phenomené bien fondé*. See also Maclaurin's View of Newton's Philosophical Discoveries, book 1, chap. 4.

LEMMA, is a term chiefly used by mathematicians, and signifies a proposition, previously laid down to prepare the way for the more easy apprehension of the demonstration of some theorem, or the construction of some problem.

LEMNISCATE, the name of a curve in the form of the figure of 8. If we call A P, *x*; P Q, *y*, and the constant line A B or A C, *a*; the equation $ay = x\sqrt{aa - xx}$, or $a^2y^2 = a^2x^2 - x^4$, expressing a line of the 4th degree, will denote a lemniscate, having a double point in the point A. There may be other lemniscates, as the ellipse of Cassini, &c; but that above defined is the simplest of them.

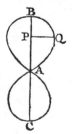

It eafily appears that this curve is quadrable. For fince $ay = x\sqrt{a^2-x^2}$, therefore the fluxion of the curve or yx is $= \dfrac{x}{a}\dot{x}\sqrt{a^2-x^2}$; the fluent of which is $\frac{1}{3}a^2 - \dfrac{1}{3a}\cdot\overline{a^2-x^2}|^{\frac{3}{2}}$ for the general area of the curve; which, when x is $= a$, becomes barely $\frac{1}{3}a^2 = AQB$.

LENS, a piece of glafs or other tranfparent fubftance, having its two furfaces fo formed that the rays of light, in paffing through it, have their direction changed, and made to converge and tend to a point beyond the lens, or to become parallel after converging or diverging, or laftly to diverge as if they had proceeded from a point before the lens. Some lenfes are convex, or thicker in the middle; others concave, or thinner in the middle; while others are plano-convex, or plano-concave; and fome again are convex on one fide and concave on the other, which are called menifcufes, the properties of which fee under that word. When the particular figure is not confidered, a lens that is thickeft in the middle is called a convex lens; and that which is thinneft in the middle is called a concave lens, without farther diftinction.

Thefe feveral forms of lenfes are reprefented in the annexed figure:

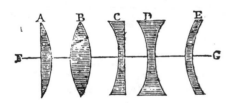

where A, B are convex lenfes, and C, D, E are concave ones; alfo A is a plano-convex, B is convexo-convex, C is plano-concave, D is concavo-concave, and E is a menifcus.

In every lens, the right line perpendicular to the two furfaces, is called the Axis of the lens, as F G; the points where the axis cuts the furface, are called the Vertices of the lens; alfo the middle point between them is called the Centre; and the diftance between them, the Diameter.

Some confine lenfes within the diameter of half an inch; and fuch as exceed that thicknefs, they call Lenticular Glaffes.

Lenfes are either blown or ground.

Blown LENSES, are fmall globules of glafs, melted in the flame of a lamp or taper. See Microfcope.

Ground LENSES, are fuch as are ground or rubbed into the defired fhape, and then polifhed. For a method of grinding them, and defcription of a machine for that purpofe, fee Philof. Tranf. vol. xli. pa. 555, or Abr. viii. 281.

Maurolycus firft delivered fomething relative to the nature of lenfes; but we are chiefly indebted to Kepler for explaining the doctrine of refraction through mediums of different forms, the chief fubftance of which may be comprehended in the cafes following.

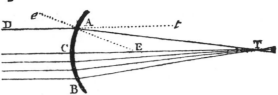

Let DA be a ray of light falling upon a convex denfe medium, having its centre at E. When the ray arrives at A, it will not proceed in the fame direction At; but it will be there bent, and thrown into a direction AT, nearer the perpendicular AE. In the fame manner, another ray falling on B, at an equal diftance on the other fide of the vertex C, and parallel to the former ray DA, will be refracted into the fame point T. And it will alfo be found, that all the intermediate parallel rays will converge to the fame point, very nearly.

On the other hand, if the rays fall parallel on the infide of this denfer medium, as in the fig below, they will tend from the perpendicular EAf; and converge to a point T in the air, or any rarer medium. Alfo the ray incident on B, at the fame diftance from the vertex C, will converge to the fame place T, together with all the intermediate parallel rays.

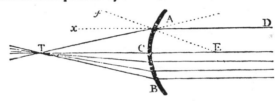

Since therefore rays are made to converge when they pafs either from a rarer or a denfer medium terminated by a convex furface, and converge again when they pafs from the fame medium convex towards the rarer, a lens which is convex on both fides muft, on both accounts, make parallel rays converge to a point beyond it. Thus, the parallel rays between A and B, falling upon

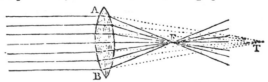

the convex furface of the glafs AB, would in that denfe medium have converged to T; but that medium being terminated by another convex furface, they will be made more converging, and be collected at fome place F, nearer to the lens.

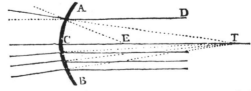

Again, to explain the effects of a concave glafs, let AB be the concave fide of a denfe medium, the centre of concavity being at E. In this cafe, DA will be refracted

fracted towards the perpendicular EA; and fo likewife will the ray incident at B; in confequence of which they will diverge from one another within the denfe medium. The intermediate rays will alfo diverge more or lefs, as they recede from the axis TC; which, being in the perpendicular, will go ftraight on.

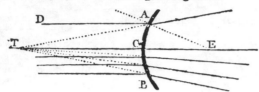

If the rays be parallel within the denfe medium, they will diverge when they pafs from thence into a rarer medium, through a concave furface. For the ray DA will be refracted from the perpendicular AE, as will alfo the ray that is incident at B, together with all the intermediate rays, in proportion to their diftance from the axis or central ray TC.

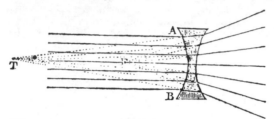

Therefore, if a denfe medium, as the glafs AB, be terminated by two concave furfaces, parallel rays paffing through it will be made to diverge by both the fides of it. Thus the firft furface AB will make them diverge as if they had come from the point T; and with the effect of the fecond furface added to this, they will diverge as from a nearer point, F.

It was Kepler, who by thefe inveftigations firft gave a clear explanation of the effects of lenfes, in making the rays of a pencil of light converge or diverge. He fhewed that a plano-convex lens makes rays, that were parallel to its axis, meet at the diftance of the diameter of the fphere of convexity; but that if both fides of the lens be equally convex, the rays will have their focus at the diftance of the radius of the circle correfponding to that degree of convexity. But he did not inveftigate any rule for the foci of lenfes unequally convex. He only fays, in general, that they will fall fomewhere in the medium, between the foci belonging to the two different degrees of convexity. It is to Cavalerius that we owe this inveftigation: he laid down this rule, As the fum of both the diameters is to one of them, fo is the other to the diftance of the focus. And it is to be noted that all thefe rules, concerning convex lenfes, are applicable to thofe that are concave, with this difference, that the focus is on the contrary fide of the glafs. See Montucla, vol. 2, pa. 176; or Prieftley's Hift. of Vifion, pa. 65, 4to.

Upon this principle it was not difficult to find the foci of pencils of rays iffuing from any point in the axis of the lens; fince thofe that are parallel will meet in the focus; and if they iffue from the focus, they will be parallel on the other fide. If they iffue from a point

between the focus and the glafs, they will continue to diverge after paffing the lens, but lefs than before; while thofe that come from beyond the focus, will converge after paffing the glafs, and will meet in a place beyond the oppofite focus. This philofopher particularly obferved, that rays which iffue from twice the diftance of the focus, will meet at the fame diftance on the other fide. The moft important of thefe obfervations have been already illuftrated by proper figures, and from them the reft may be eafily conceived. Later optical writers have affigned the diftances at which rays will meet, that iffue from any other place in the axis of a lens; but Kepler was too much intent upon his aftronomical and other purfuits, to give much attention to geometry. But, from the whole, Montucla gives the following rule concerning this fubject: As the excefs of the diftance of the object from the glafs, above the diftance of the focus, is to the diftance of the focus; fo is this diftance, to the place of convergency beyond the glafs. And the fame rule will find the point of divergency, when the rays iffue from any place between the lens and the focus: for then the excefs of the diftance of the object from the glafs, above that of the focus, is negative, which is the fame diftance taken the contrary way. Montucla, vol. 2, pa. 177.

And from the principle above-mentioned, it will not be difficult to underftand the application of lenfes, in the rationale of telefcopes and microfcopes. On thefe principles too is founded the ftructure of refracting burning glaffes, by which the fun's light and heat are exceedingly augmented in the focus of the lens, whether convex or plano-convex; fince the rays, falling parallel to the axis of the lens, are reduced into a much narrower compafs; fo that it is no wonder they burn fome bodies, melt others, and produce other extraordinary phenomena.

In the Philof. Tranf. vol. xvii. 960, or the Abr. i. 191, Dr. Halley gives an ingenious inveftigation of the foci of rays refracted through any lenfes, nearly as follows ·

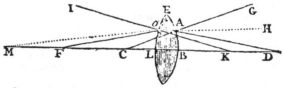

Let BEL be a double convex lens, C the centre of the fegment EB, and K the centre of the fegment EL; BL the thicknefs or diameter of the lens, and D a point in the axis; it is required to find the point F, or focus, where the rays proceeding from D fhall be collected, after being refracted through the lens at A and a, points very near to the axis BL. Put the diftance DA or DB$=d$, the radius CA or CB$=r$, the radius Ka or KL$=R$; alfo the thicknefs of the lens BL$=t$, and m to n the ratio of the fine of the angle of incidence DAG to the fine of the refracted angle HAG or CAM; or m to n will be the ratio of thofe angles themfelves nearly, fince very fmall angles are to each other in the fame ratio as their fines. Hence

m is as the angle DAG or DAC,
n is as the angle HAG or MAC,
and becaufe in this cafe the fides are as their oppo-

fits

site angles, therefore DC : DA : : ∠DAC : ∠C,

or $d+r : d :: m : \frac{dm}{d+r}$ which is as the ∠C;

from this take n or the ∠ MAC,

and there remains $\frac{dm - dn - rn}{d+r}$ as the ∠M;

hence again ∠M : ∠C :: CA : MA or MB,

that is $\frac{dm - dn - rn}{d+r} : \frac{dm}{d+r} :: r : \frac{mdr}{m-n \cdot d - nr}$ = MA

or MB; which shews in what point the rays would be collected after one refraction, viz, when nr is less than $\overline{m-n} \cdot d$. But when nr is = $\overline{m-n} \cdot d$, the point would be at an infinite distance, or the rays will be parallel to the axis; and when nr is greater than $\overline{m-n} \cdot d$, then MB is negative, or M falls on the other side of the lens beyond D, and the rays still continue to diverge after the first refraction.

The point M being now found, to or from which the rays proceed after the first refraction, and BM − BL being thus given, which call D, by a process like the former it follows that FL, or the focal distance sought, is equal to $\frac{nDR}{m-n \cdot D + mR} = f$. And here, instead of

D substituting MB − LB or $\frac{mdr}{m-n \cdot d - nr} - t$, and

putting p for $\frac{n}{m-n}$, the same theorem will become

$\frac{(mpdr - ndt + nprt) \times R}{mdr + mdR - mprR - \overline{m-n} \cdot dt + nrt} = f,$

the focal distance sought, in its most general form, including the thickness of the lens; being the universal rule for the foci of double convex glasses exposed to diverging rays.

But if t the thickness of the lens be rejected, as not sensible, the rule will be much shorter,

viz, $\frac{pdrR}{dr + dR - prR} = f.$

If therefore the lens consist of glass, whose refraction is as 3 to 2, it will be $\frac{2drR}{dr + dR - 2rR} = f.$ And if it be of water, whose refraction is as 4 to 3, it will be $\frac{3drR}{dr + dR - 3rR} = f.$ But, if the lens could be made of diamond, whose refraction is as 5 to 2, it would be $\frac{2drR}{3dr + 3R - 2rR} = f.$

If the incident rays, instead of diverging, be converging, the distance DB or d will be negative, and then the theorem for a double convex glass lens will be $\frac{-2drR}{-dr - dR - 2rR}$ or $\frac{2drR}{dr + dR + 2rR} = f$, in which case therefore the focus is always on the other side of the glass.

And if the rays be parallel, as coming from an infinite distance, or nearly so, then will d be negative, as well as the terms in the theorem in which it is found; and therefore, the other term prR will be nothing in respect of those infinite terms; and by omitting it, the theorem will be $\frac{pdrR}{dr + dR} = \frac{prR}{r + R} = f,$ or for glass $\frac{2rR}{r + R} = f.$

And here if $r = R$, or the two sides of the glass be of equal convexity, this last will become barely $\frac{2r^2}{2r}$ or barely $r = f$ the focus, which therefore is in the centre of the convexity of the lens.

If the lens be a meniscus of glass; then, making r negative, the theorem is

$\frac{-2drR}{-dr + dR + 2rR}$ or $\frac{2drR}{dr - dR - 2rR} = f$ for diverging rays,

$\frac{-2drR}{-dr + dR - 2rR}$ or $\frac{2drR}{dr - dR + 2rR} = f$ for converging rays,

and $\frac{-2rR}{-r + R}$ or $\frac{2rR}{r - R} = f$ for parallel rays.

If the lens be a double concave glass, r and R will be both negative, and then the theorem becomes

$\frac{-2drR}{dr + dR + 2rR} = f$ for diverging rays, always negative;

$\frac{-2drR}{dr + dR \times 2rR} = f$ for converging rays;

and $\frac{-2rR}{r + R} = f$ for parallel rays.

And here, if the radii of curvature r and R be equal, this last will be barely $-r = f$ for parallel rays falling on a double concave glass of equal curvature.

Lastly, when the lens is a plano convex glass; then, r being infinite, the theorem becomes

$\frac{2dR}{d - 2R} = f$ for diverging rays,

$\frac{2dR}{d + 2R} = f$ for converging rays,

and $2R = f$ for parallel rays.

The theorems for parallel rays, as coming from an infinite distance, take place in the common refracting telescopes. And those for converging rays are chiefly of use to determine the focus resulting from any sort of lens placed in a telescope, between the focus of the object-glass and the glass itself; the distance between the said focus of the object-glass and the interposed lens being made = − d; while those for diverging rays are chiefly of use in microscopes, reading glasses, and other cases in which near objects are viewed.

It is evident that the foregoing general theorem will serve to find any of the other circumstances, as well as the focus, by considering this as given. Thus, for instance, suppose it be required to find the distance at which an object being placed, it shall by a given lens be represented as large as the object itself; which is of singular use in viewing and drawing them, by transmitting the image through a glass in a dark room, as in the camera obscura, which gives not only the true figure and shades, but the colours themselves as vivid as the life. Now in this case d is = f, which makes the theorem become $pdrR = d^2r + d^2R - pdrR$, and

this

this gives $d = \frac{2prR}{r+R}$. But if the two convexities belong to equal spheres, so as that $r = R$, then it is $d = pr$, or $= 2r$ when the lens is glass. So that if the object be placed at the diameter of the sphere distant from the lens, then the focus will be as far distant on the other side, and the image as large as the object. But if the glass were a plano-convex, the same distance would be just twice as much.

Again, recurring to the first general theorem, including t, the thickness of the lens; let the lens be a whole sphere; then $t = 2r$, and $r = R$; and hence the theorem reduces to $\frac{mpdr - 2ndr - 2npr^2}{2nd + 2nr - mpr} = f$.

And here if d be infinite, the theorem contracts to $\frac{mp - 2n}{2n}r$ or $\frac{2n - m}{2m - 2n}r = f$; or for glass $\frac{1}{2}r = f$: shewing that a sphere of glass collects the sun's rays at half the radius of the sphere without it. And for a sphere of water, the focus is at the distance of a whole radius.

For another example; when a hemisphere is exposed to parallel rays; then d and R being infinite, and $t = r$, the theorem becomes $\frac{mp - n}{m}r, = \frac{nn}{m^2 - mn}r = f$. That is, in glass it is $\frac{4}{3}r$, and in water $\frac{7}{4}r$.

Several other corollaries may be deduced from the foregoing principles. As,

1st. That the thickness of the lens, being very small, the focus will remain the same, whether the one side or the other be exposed to the rays.

2d. If a luminous body be placed in a focus behind a lens, whether plano-convex, or convex on both sides; or whether equally or unequally so; the rays become parallel after refraction, as the refracted rays become what were before the incident rays. And hence, by means of a convex lens, or a little glass bubble full of water, a very intense light may be projected to a great distance. Which furnishes us with the structure of a lamp or lantern, to throw an intense light to an immense distance: for a lens, convex on both sides, being placed opposite to a concave mirror, if there be placed a lighted candle or wick in the common focus of both, the rays reflected back from the mirror to the lens will be parallel to each other; and after refraction will converge, till they concur at the distance of the radius, after which they will again diverge. But the candle being likewise in the focus of the lens, the rays it throws on the lens will be parallel; and therefore a very intense light meeting with another equally intense, at the distance of the diameter from the lens, the light will be surprising: and though it afterwards decrease, yet the parallel and diverging rays going a long way together, it will be very great at a great distance. Lanterns of this kind are of considerable service in the night time, to discover remote objects; and are used with success by fowlers and fishermen, to collect their prey together, that so it may be taken.

If it be required to have the light, at the same time, transmitted to several places, as through several streets, &c, the number of lenses and mirrors must be increased.

3d. The images of objects are shewn inverted in the focus of a convex lens: nor is the focus of the sun's rays any thing else, in effect, but the image of the sun inverted. Hence, in solar eclipses, the sun's image, eclipsed as it is, may be burnt by a large lens on a board, &c, and exhibit a very entertaining phenomenon.

4th. If a concave mirror be so placed, as that an inverted image, formed by refraction through a lens, be found between the centre and the focus, or even beyond the centre, it will again be inverted by reflection, and so appear erect; in the first case beyond the centre, and in the latter between the centre and the focus. And on these principles the camera obscura is constructed.

5th. The image of an object, delineated beyond a convex lens, is of such a magnitude, as it would be of, were the object to shine into a dark room through a small hole, upon a wall, at the same distance from the hole, as the focus is from the lens.—When an object is less distant from a lens than the focus of parallel rays, the distance of the image is greater than that of the object; otherwise, the distance of the image is less than that of the object: in the former case, therefore, the image is larger than the object; in the latter, it is less.

When the images are less than the objects, they will appear more distinct and vivid; because then more rays are accumulated into a given space. But if the images be made greater than the objects, they will not appear distinctly; because in that case there are fewer rays which meet after refraction in the same point; whence it happens, that rays proceeding from different points of an object, terminate in the same point of an image, which is the cause of confusion. Hence it appears, that the same aperture of a lens may be admitted in every case, if we would keep off the rays which produce confusion. However, though the image be then more distinct, when no rays are admitted but those near the axis, yet for want of rays the image is apt to be dim.

6th. If the eye be placed in the focus of a convex lens, an object viewed through it, appears erect, and enlarged in the ratio of the distance of the object from the eye, to that of the eye from the lens, if it be near; but infinitely if remote.

7th. An object viewed through a concave lens, appears erect, and diminished in a ratio compounded of the ratios of the space in the axis between the point of incidence, and the point to which an oblique ray would pass without refraction, to the space in the axis between the eye and the middle of the object; and the space in the same axis between the eye and the point of incidence, to the space between the middle of the object and the point to which the oblique ray would pass without refraction.

Finally, it may be observed, that the very small magnifying glasses used in microscopes, most properly come under the denomination of lens, as they most approach to the figure of the lentil, a seed of the vetch or pea kind, from whence the name is derived; but the reading glasses, and burning glasses, and all that magnify, come under the same denomination; for their surfaces are convex, although less so. A drop of water is a lens, and it will serve as one; and many have used it by way of

lens

Plate XIV.

LEVELS.

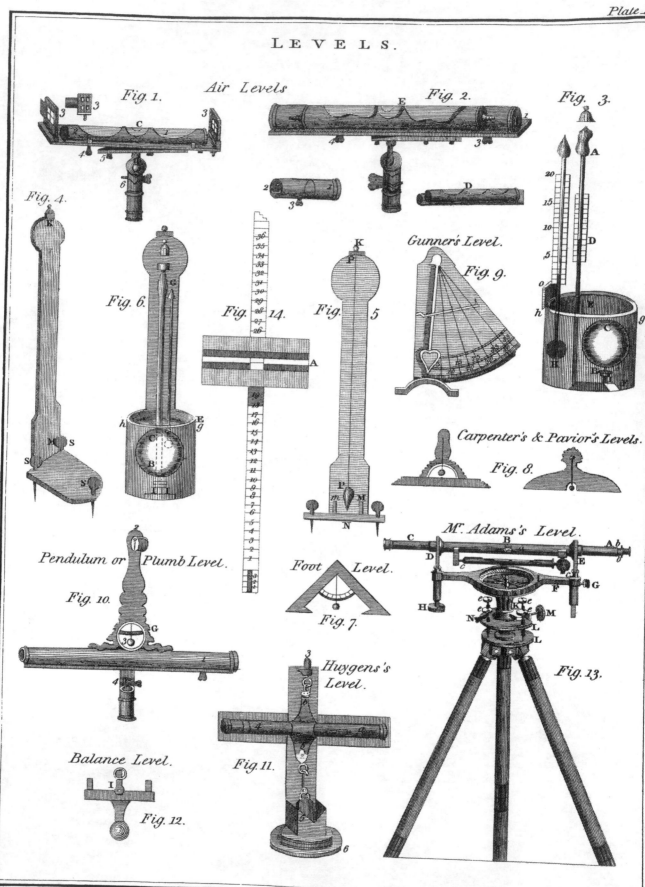

Air Levels

Fig. 1.

Fig. 2.

Fig. 3.

Fig. 4.

Fig. 6.

Fig. 14.

Fig. 5.

Gunner's Level.

Fig. 9.

Carpenter's & Pavior's Levels.

Fig. 8.

Pendulum or Plumb Level.

Fig. 10.

Foot Level.

Fig. 7.

Mr. Adams's Level.

Fig. 13.

Balance Level.

Huygens's Level.

Fig. 11.

Fig. 12.

lens in their microfcopes. A drop of any tranfparent fluid, inclofed between two concave glafies, acquires the fhape of a lens, and has all its properties. The cryftalline humour of the eye is a lens exactly of this kind; it is a fmall quantity of a tranflucent fluid, contained between two concave and tranfparent membranes, called the coats of the eye; and it acts as the lens made of water would do, in an equal degree of convexity.

LEO, *the Lion*, a confiderable conftellation of the northern hemifphere, being one of the 48 old conftellations, and the 5th fign of the zodiac. It is marked thus ♌, as a rude fketch of the animal.

The Greeks fabled that this was the Nemæan lion, which had dropped from the moon, but being flain by Hercules, was raifed to the heavens by Jupiter, in commemoration of the dreadful conflict, and in honour of that hero. But the hieroglyphical meaning of this fign, fo depicted by the Egyptians long before the invention of the fables of Hercules, was probably no more than to fignify, by the fury of the lion, the violent heats occafioned by the fun when he entered that part of the ecliptic.

The ftars in the conftellation Leo, in Ptolomy's catalogue are 27, befides 8 unformed ones, now counted in later times in the conftellation Coma Berenices, in Tycho's 30, in that of Hevelius 49, and in Flamfteed's 95; one of them, of the firft magnitude, in the breaft of the Lion, is called Regulus, and Cor Leonis, or Lion's Heart.

LEO *Minor, the Little Lion*, a conftellation of the northern hemifphere, and one of the new ones that were formed out of what were left by the ancients, under the name of Stellæ Informes, or unformed ftars, and added to the 48 old ones. It contains 53 ftars in Flamfteed's catalogue.

Cor LEONIS, *Lion's heart*, a fixed ftar, of the firft magnitude, in the fign Leo; called alfo Regulus, Bafilicus, &c.

LEPUS, *the Hare*, a conftellation of the fouthern hedemifphere, and one of the 48 old conftellations.

The Greeks fabled, that this animal was placed in the heavens, near Orion, as being one of the animals which he hunted. But it is probable their mafters, the Egyptians, had fome other meaning in this hieroglyphic.

The ftars in the conftellation Lepus, in Ptolomy's catalogue are 12, in Tycho's 13, and in Flamfteed's 19.

LEUCIPPUS, a celebrated Greek philofopher and mathematician, who flourifhed about the 428th year before Chrift. He was the firft author of the famous fyftem of atoms and vacuums, and of the hypothefis of ftorms; fince attributed to the moderns.

LEVEL, an inftrument ufed to make a line parallel to the horizon, and to continue it out at pleafure; and by this means to find the true level, or the difference of afcent or defcent between two or more places, for conveying water, draining fens, &c.

There are feveral inftruments, of different contrivance and matter, invented for the perfection of levelling, as may be feen in De la Hire's and Picard's treatifes of Levelling, in Biron's treatife on Mathematical Inftruments, alfo in the Philof. Tranf. and the Memoirs de l Acad. &c. But they may be reduced to the following kinds.

Water-LEVEL, that which fhews the horizontal line by means of a furface of water or other fluid; founded VOL. II.

on this principle, that water always places itfelf level or horizontal.

The moft fimple kind is made of a long wooden trough or canal; which being equally filled with water, its furface fhews the line of level. And this is the chorobates of the ancients, defcribed by Vitruvius, lib. viii. cap. 6.

The water-level is alfo made with two cups fitted to the two ends of a ftraight pipe, about an inch diameter, and 3 or 4 feet long, by means of which the water communicates from the one cup to the other; and this pipe being moveable on its ftand by means of a ball and focket, when the two cups fhew equally full of water, their two furfaces mark the line of level.

This inftrument, inftead of cups, may alfo be made with two fhort cylinders of glafs three or four inches long, faftened to each extremity of the pipe with wax or maftic. The pipe is filled with common or coloured water, which fhews itfelf through the cylinders, by means of which the line of Level is determined; the height of the water, with refpect to the centre of the earth, being always the fame in both cylinders. This level, though very fimple, is yet very commodious for levelling fmall diftances. See the method of preparing and ufing a water-level, and a mercurial Level, annexed to Davis's quadrant, for the fame purpofe, by Mr. Leigh, in Philof. Tranf. vol. XL. 417, or Abr. viii. 362.

Air-LEVEL, that which fhews the line of Level by means of a bubble of air inclofed with fome fluid in a glafs tube of an indeterminate length and thicknefs, and having its two ends hermetically fealed: an invention, it is faid, of M. Thevenot. When the bubble fixes itfelf at a certain mark, made exactly in the middle of the tube, the cafe or ruler in which it is fixed, is then level. When it is not level, the bubble will rife to one end.— This glafs-tube may be fet in another of brafs, having an aperture in the middle, where the bubble of air may be obferved.—The liquor with which the tube is filled, is oil of tartar, or aqua fecunda; thofe not being liable to freeze as common water, nor to rarefaction and condenfation as fpirit of wine is.

There is one of thefe inftruments with fights, being an improvement upon that laft defcribed, which, by the addition of other apparatus, becomes more exact and commodious. It confifts of an air-Level, n° 1, (*fig.* 1, *Plate XIV*) about 8 inches long, and about two thirds of an inch in diameter, fet in a brafs tube, 2, having an aperture in the middle, C. The tubes are carried in a ftrong ftraight ruler, of a foot long; at the ends of which are fixed two fights, 3, 3, exactly perpendicular to the tubes, and of an equal height, having a fquare hole, formed by two fillets of brafs croffing each other at right angles; in the middle of which is drilled a very fmall hole, through which a point on a level with the inftrument is feen. The brafs tube is faftened to the ruler by means of two fcrews; the one of which, marked 4, ferves to raife or deprefs the tube at pleafure, for bringing it towards a level. The top of the ball and focket is rivetted to a fmall ruler that fprings, one end of which is faftened with fprings to the great ruler, and at the other end is a fcrew, 5, ferving to raife and deprefs the inftrument when nearly level.

But this inftrument is ftill lefs commodious than the following one: for though the holes be ever fo fmall, yet they will ftill take in too great a fpace to determine the point of Level precifely.

D   Fig.

Fig. 2, is a *Level with Telefcopic Sights,* firft invented by Mr. Huygens. It is like the laft; with this difference, that inftead of plain fights, it carries a telefcope, to determine exactly a point of Level at a confiderable diftance. The fcrew 3, is for raifing or lowering a little fork, for carrying the hair, and making it agree with the bubble of air when the inftrument is Level ; and the fcrew 4, is for making the bubble of air, D or E, agree with the telefcope. The whole is fitted to a ball and focket, or otherwife moved by joints and fcrews.—It may be obferved that a telefcope may be added to any kind of Level, by applying it upon, or parallel to, the bafe or ruler, when there is occafion to take the level of remote objects : and it poffeffes this advantage, that it may be inverted by turning the ruler and telefcope half round ; and if then the hair cut the fame point that it did before, the operation is juft. Many varieties and improvements of this inftrument have been made by the more modern opticians.

Dr. Defaguliers propofed a machine for taking the difference of Level, which contained the principles both of a barometer and thermometer ; but it is not accurate in practice : Philof. Tranf. vol. xxxiii. pa. 165, or Abr. vol. vi. 271. *Fig.* 3, 4, 5, 6.

Mr. Hadley too has contrived a Spirit Level to be fixed to a quadrant, for taking a meridian altitude at fea, when the horizon is not vifible. See the defcription and figure of it in the Philof. Tranf. vol. xxxviii. 167, or Abr. viii. 357. Various other Spirit Levels, and Mercurial Levels, are alfo invented and ufed upon different occafions.

Reflecting LEVEL, that made by means of a pretty long furface of water, reprefenting the fame object inverted, which we fee erect by the eye; fo that the point where thefe two objects appear to meet, is on a Level with the place where the furface of the water is found. This is the invention of M. Mariotte.

There is another reflecting Level, confifting of a polifhed metal mirror, placed a little before the object glafs of a telefcope, fufpended perpendicularly. This mirror muft be fet at an angle of 45 degrees; in which cafe the perpendicular line of the telefcope becomes a horizontal line, or a line of Level. Which is the invention of M. Caffini.

*Artillery Foot-*LEVEL is in form of a fquare (fig. 7), having its two legs or branches of an equal length ; at the junction of which is a fmall hole, by which hangs a plummet playing on a perpendicular line in the middle of a quadrant, which is divided both ways from that point into 45 degrees.

This inftrument may be ufed on other occafions, by placing the ends of its two branches on a plane ; for when the plummet plays perpendicularly over the middle divifion of the quadrant, the plane is then Level.

To ufe it in Gunnery, place the two ends on the piece of artillery, which may be raifed to any propofed height, by means of the plummet, which will cut the degree above the Level. But this fuppofes the outfide of the cannon is parallel to its axis, which is not always the cafe ; and therefore they ufe another inftrument now, either to fet the piece Level, or elevate it at any angle ; namely a fmall quadrant, with one of its radii continued out pretty long, which being put into the infide of the cylindrical

bore, the plummet fhews the angle of elevation, or the line of Level. See *Gunner's* QUADRANT.

Carpenter's, Bricklayer's, or *Pavior's* LEVEL, confifts of a long ruler, in the middle of which is fitted at right angles another broader piece, at the top of which is faftened a plummet, which when it hangs over the middle line of the 2d or upright piece, fhews that the bafe or long ruler is horizontal or Level. Fig. 8.

Mafon's LEVEL, is compofed of 3 rules, fo jointed as to form an ifofceles triangle, fomewhat like a Roman A ; from the vertex of which is fufpended a plummet, which hangs directly over a mark in the middle of the bafe, when this is horizontal or Level. Fig. 8.

Plumb or *Pendulum* LEVEL, faid to be invented by M. Picard; fig. 10. This fhews the horizontal line by means of another line perpendicular to that defcribed by a plummet or pendulum. This Level confifts of two legs or branches, joined at right angles, the one of which, of about 18 inches long, carries a thread and plummet ; the thread being hung near the top of the branch, at the point 2. The middle of the branch where the thread paffes is hollow, fo that it may hang free every where : but towards the bottom, where there is a fmall blade of filver, on which a line is drawn perpendicular to the telefcope, the faid cavity is covered by two pieces of brafs, with a piece of glafs G, to fee the plummet through, forming a kind of cafe, to prevent the wind from agitating the thread. The telefcope, of a proper length, is fixed to the other leg of the inftrument, at right angles to the perpendicular, and having a hair ftretched horizontally acrofs the focus of the object-glafs, which determines the point of Level, when the ftring of the plummet hangs againft the line on the filver blade. The whole is fixed by a ball and focket to its ftand.

Fig. 12, is a *Balance* LEVEL ; which being fufpended by the ring, the two fights, when in equilibrio, will be horizontal, or in a Level.

Some other Levels are alfo reprefented in plate xiv.

LEVELLING, the art or act of finding a line parallel to the horizon at one or more ftations, to determine the height or depth of one place with refpect to another ; for laying out grounds even, regulating defcents, draining moraffes, conducting water, &c.

Two or more places are on a true level when they are equally diftant from the centre of the earth. Alfo one place is higher than another, or out of level with it, when it is farther from the centre of the earth — and a line equally diftant from that centre in all its points, is called the *line of true level.* Hence, becaufe the earth is round, that line muft be a curve, and make a part of the earth's circumference, or at leaft parallel to it, or concentrical with it ; as the line BCFG, which has all its points equally diftant from A the centre of the earth ; confidering it as a perfect globe.

But the line of fight BDE &c given by the operations of levels, is a tangent, or a right line perpendicular to the femidiameter of the earth at the point of contact B, rifing always higher above the true line of level,

the

the farther the diſtance is, is called the *apparent line of level.* Thus, CD is the height of the apparent level above the true level, at the diſtance BC or BD; alſo EF is the exceſs of height at F; and GH at G; &c. The difference, it is evident, is always equal to the exceſs of the ſecant of the arch of diſtance above the radius of the earth.

The common methods of levelling are ſufficient for laying pavements of walks, or for conveying water to ſmall diſtances, &c: but in more extenſive operations, as in levelling the bottoms of canals, which are to convey water to the diſtance of many miles, and ſuch like, the difference between the true and the apparent level muſt be taken into the account.

Now the difference CD between the true and apparent level, at any diſtance BC or BD, may be found thus: By a well known property of the circle 2AC + CD : BD : : BD : CD; or becauſe the diameter of the earth is ſo great with reſpect to the line CD at all diſtances to which an operation of levelling commonly extends, that 2AC may be ſafely taken for 2AC + CD in that proportion without any ſenſible error, it will be 2AC : BD : : BD : CD which therefore is $= \frac{BD^2}{2AC}$ or $\frac{BC^2}{2AC}$ nearly; that is, the difference between the true and apparent level, is equal to the ſquare of the diſtance between the places, divided by the diameter of the earth; and conſequently it is always proportional to the ſquare of the diſtance.

Now the diameter of the earth being nearly 7958 miles; if we firſt take BC = 1 mile, then the exceſs $\frac{BC^2}{2AC}$ becomes $\frac{1}{7958}$ of a mile, which is 7·962 inches, or almoſt 8 inches, for the height of the apparent above the true level at the diſtance of one mile. Hence, proportioning the exceſſes in altitude according to the ſquares of the diſtances, the following Table is obtained, ſhewing the height of the apparent above the true level for every 100 yards of diſtance on the one hand, and for every mile on the other.

Diſt. or BC	Dif. of Level, or CD
Yards	Inches
100	0·026
200	0·103
300	0·231
400	0·411
500	0·643
600	0·925
700	1·260
800	1·645
900	2·081
1000	2·570
1100	3·110
1200	3·701
1300	4·344
1400	5·038
1500	5·784
1600	6·580
1700	7·425

Diſt. or BC	Dif. of Level, or CD	
Miles	Feet	Inc.
¼	0	0½
½	0	2
¾	0	4½
1	0	8
2	2	8
3	6	0
4	10	7
5	16	7
6	23	11
7	32	6
8	42	6
9	53	9
10	66	4
11	80	3
12	95	7
13	112	2
14	130	1

By means of theſe Tables of reductions, we can now level to almoſt any diſtance at one operation, which the ancients could not do but by a great multitude; for, being unacquainted with the correction anſwering to any diſtance, they only levelled from one 20 yards to another, when they had occaſion to continue the work to ſome conſiderable extent.

This table will anſwer ſeveral uſeful purpoſes. Thus, firſt, to find the height of the apparent level above the true, at any diſtance. If the given diſtance be contained in the table, the correction of level is found on the ſame line with it: thus at the diſtance of 1000 yards, the correction is 2·57, or two inches and a half nearly; and at the diſtance of 10 miles, it is 66 feet 4 inches. But if the exact diſtance be not found in the table, then multiply the ſquare of the diſtance in yards by 2·57, and divide by 1000000, or cut off 6 places on the right for decimals; the reſt are inches: or multiply the ſquare of the diſtance in miles by 66 feet 4 inches, and divide by 100. 2ndly, To find the extent of the viſible horizon, or how far can be ſeen from any given height, on a horizontal plane, as at ſea, &c. Suppoſe the eye of an obſerver, on the top of a ſhip's maſt at ſea, be at the height of 130 feet above the water, he will then ſee about 14 miles all around. Or from the top of a cliff by the ſea-ſide, the height of which is 66 feet, a perſon may ſee to the diſtance of near 10 miles on the ſurface of the ſea. Alſo, when the top of a hill, or the light in a lighthouſe, or ſuch like, whoſe height is 130 feet, firſt comes into the view of an eye on board a ſhip; the table ſhews that the diſtance of the ſhip from it is 14 miles, if the eye be at the ſurface of the water; but if the height of the eye in the ſhip be 80 feet, then the diſtance will be increaſed by near 11 miles, making in all about 25 miles, diſtance.

3dly, Suppoſe a ſpring to be on one ſide of a hill, and a houſe on an oppoſite hill, with a valley between them; and that the ſpring ſeen from the houſe appears by a levelling inſtrument to be on a level with the foundation of the houſe, which ſuppoſe is at a mile diſtance from it; then is the ſpring 8 inches above the true level of the houſe; and this difference would be barely ſufficient for the water to be brought in pipes from the ſpring to the houſe, the pipes being laid all the way in the ground.

4th, If the height or diſtance exceed the limits of the table: Then, firſt, if the diſtance be given, divide it by 2, or by 3, or by 4, &c, till the quotient come within the diſtances in the table; then take out the height anſwering to the quotient, and multiply it by the ſquare of he diviſor, that is by 4, or 9, or 16, &c, for the height required: So if the top of a hill be juſt ſeen at the diſtance of 40 miles; then 40 divided by 4 gives 10, to which in the table anſwers 66⅓ feet, which being multiplied by 16, the ſquare of 4, gives 1061⅓ feet for the height of the hill. But when the height is given, divide it by one of theſe ſquare numbers 4, 9, 16, 25, &c, till the quotient come within the limits of the table, and multiply the quotient by the ſquare root of the diviſor, that is by 2, or 3, or 4, or 5, &c, for the diſtance ſought: So when the top of the pike of Teneriff, ſaid to be almoſt 3 miles or 15840 feet high, juſt comes into view at ſea; divide 15840 by 225, or the ſquare of 15, and the quotient

 is

is 70 nearly; to which in the table anfwers, by pro-portion, nearly 10$\frac{2}{7}$ miles; then multiplying 10$\frac{2}{7}$ by 15, gives 154 miles and $\frac{2}{7}$, for the diftance of the hill.

Of the Practice of Levelling.

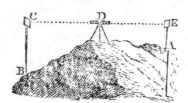

The operation of Levelling is as follows. Suppofe the height of the point A on the top of a mountain, above that of B, at the foot of it, be required. Place the level about the middle diftance at D, and fet up pickets, poles, or ftaffs, at A and B, where perfons muft attend with fignals for raifing and lowering, on the faid poles, little marks of pafteboard or other mat-ter. The level having been placed horizontally by the bubble, &c, look towards the ftaff A E, and caufe the perfon there to raife or lower the mark, till it appear through the telefcope, or fights, &c, at E: then mea-fure exactly the perpendicular height of the point E above the point A, which fuppofe 5 feet 8 inches, fet it down in your book. Then turn your view the other way, towards the pole B, and caufe the perfon there to raife or lower his mark, till it appear in the vifual line as before at C; and meafuring the height of C above B, which fuppofe 15 feet 6 inches, fet this down in your book alfo, immediately under the num-ber of the firft obfervation. Then fubtract the one from the other, and the remainder 9 feet 10 inches, will be the difference of level between A and B, or the height of the point A above the point B.

If the point D, where the inftrument is fixed, be exactly in the middle between the points A and B, there will be no neceffity for reducing the apparent level to the true one, the vifual ray on both fides be-ing raifed equally above the true level. But if not, each height muft be corrected or reduced according to its diftance, before the one corrected height is fub-tracted from the other; as in the cafe following.

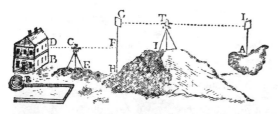

When the diftance is very confiderable, or irregular, fo that the operation cannot be effected at once placing of the level; or when it is required to know if there be a fufficient defcent for conveying water from the fpring A to the point B; it will be neceffary to per-form this at feveral operations. Having chofen a pro-per place for the firft ftation, as at I, fix a pole at the point A near the fpring, with a proper mark to flide up and down it, as L; and meafure the diftance from

A to I. Then the level being adjufted in the point let the mark L be raifed or lowered till it is feen through the telefcope or fights of the level, and mea-fure the height AL. Then having fixed another pole at H, direct the level to it, and caufe the mark G to be moved up or down till it appear through the inftru-ment: then meafure the height GH, and the diftance from I to H; noting them down in the book. This done, remove the level forwards to fome other emi-nence as E, from whence the pole H may be viewed, as alfo another pole at D; then having adjufted the level in the point E, look back to the pole H; and managing the mark as before, the vifual ray will give the point F; then meafuring the diftance HE and the height HF, note them down in the book. Then, turning the level to look at the next pole D, the vi-fual ray will give the point D; there meafure the height of D, and the diftance EB, entering them in the book as before. And thus proceed from one fta-tion to another, till the whole is completed.

But all thefe heights muft be corrected or reduced by the foregoing table, according to their refpective diftances; and the whole, both diftances and heights, with their corrections, entered in the book in the fol-lowing manner.

Back-fights.				Fore-fights.		
Difts.	Hts.	Cors.		Difts.	Hts.	Cors.
yds	ft in.	inc.		yds	ft in.	inc
IA 1650	AL 11 3	7·0		IH 1265	HG 19 5	4·0
EH 940	HF 10 7	2·2		EB 900	BD 8 1	2·1
2590	21 10	9·2		2165	27 6	6·1
	9·2			2590	6·1	
	21 0·8			Diff. 4755	26 11·9	
					21 0·8	
				Whole Dif. of level	5 11·1	

Having fummed up all the columns, add thofe of the diftances together, and the whole diftance from A to B is 4755 yards, or 2 miles and 3 quarters near-ly. Then, the fums of the corrections taken from the fums of the apparent heights, leave the two corrected heights; the one of which being taken from the other, leaves 5 feet 11·1 inc. for the true difference of level fought between the two places A and B, which is at the rate of an inch and half nearly to every 100 yards, a quantity more than fufficient to caufe the water to run from the fpring to the houfe.

Or, the operation may be otherwife performed, thus: Inftead of placing the level between every two poles, and taking both back-fights and fore-fights; plant it firft at the fpring A, and from thence obferve the level to the firft pole; then remove it to this pole, and ob-ferve the 2d pole; next move it to the 2d pole, and ob-ferve the 3d pole; and fo on, from one pole to another, always taking foreward fights or obfervations only. And then at the laft, add all the corrected heights to-gether,

ether, and the sum will be the whole difference of level ought.

Dr. Halley suggested a new method of levelling performed wholly by means of the barometer, in which the mercury is found to be suspended at so much the less height, as the place is farther remote from the centre of the earth; and hence the different heights of the mercury in two places give the difference of level. This method is, in fact, no other than the method of measuring altitudes by the barometer, which has lately been so successfully practised and perfected by M. De Luc and others; but though it serves very well for the heights of hills, and other confiderable altitudes, it is not accurate enough for determining small altitudes, to inches and parts. See the Barometrical Measurement of Altitudes.

LEVELLING *Poles*, or *Staves*, are instruments used in levelling, serving to carry the marks to be observed, and at the same time to measure the heights of those marks from the ground. They usually consist each of two long wooden rulers, made to slide over each other, and divided into feet and inches, &c.

LEVER, a straight bar of iron or wood, &c, supposed to be inflexible, supported on a fulcrum or prop by a single point, about which all the parts are moveable.

The Lever is the first of those simple machines called *mechanical powers*, as being the simplest of them all; and is chiefly used for raising great weights to small heights.

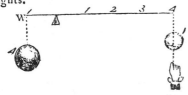

The Lever is of three kinds. First the common sort, where the weight intended to be raised is at one end of it, our strength or another weight called the power is at the other end, and the prop or fulcrum is between them both. In stirring up the fire with a poker, we make use of this Lever; the poker is the Lever, it rests upon one of the bars of the grate as a prop, the incumbent fire is the weight to be overcome, and the pressure of the hand on the other end is the force or power. In this, as in all the other machines, we have only to increase the distance between the force and the prop, or to decrease the distance between the weight and the prop, to give the operator the greater power or effect. To this kind of Lever may also be referred all scissars, pincers, snuffers, &c. The steel yard and the common balance are also Levers of this kind.

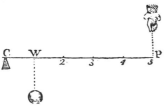

In the Lever of the 2d kind the prop is at one end, the force or power at the other, and the weight to be raised is between them. Thus, in raising a waterplug in the streets, the workman puts his iron bar or Lever through the ring or hole of the plug, till the end of it reaches the ground on the other side; then making that the prop, he lifts the plug with his force or strength at the other end of the Lever. In this Lever too, the nearer the weight is to the prop, or the farther the power from the prop, the greater is the effect. To this 2d kind of Lever may also be referred the oars and rudder of a boat, the masts of a ship, cutting knives fixed at one end, and doors, whose hinges serve as a fulcrum.

In the Lever of the third kind, the power acts between the weight and the prop; such as a ladder raised by a man somewhere between the two ends, to rear it against a wall, or a pair of tongs, &c.

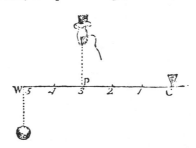

It is by this kind of Lever too that the muscular motions of animals are performed, the muscles being inserted much nearer to the centre of motion, than the point where is placed the centre of gravity of the weight to be raised; so that the power of the muscle is many times greater than the weight it is able to sustain. And in this third kind of Lever, to produce a balance between the power and weight, the power or force must *exceed* the weight, in the same proportion as it is nearer the prop than the weight is; whereas in the other two kinds, the power is less than the weight, in the same proportion as its distance is greater: that is, universally, the power and weight are each of them reciprocally as their distance from the prop; as is demonstrated below.

Some authors make a 4th sort of what is called a bended Lever, such as a hammer in drawing a nail, &c.

In all Levers, the universal property is, that the effect of either the weight or the power, to turn the Lever about the fulcrum, is directly as its intensity and its distance from the prop, that is as di, where d denotes the distance, and i the intensity, strength, or weight, &c, of the agent. For it is evident that at a double distance it will have a double effect, at a triple distance a triple effect, and so on; also that a double intensity produces a double effect, a triple a triple, and so on: therefore universally the effect is as di the product of the two. In like manner, if D be the distance of another power or agent, whose intensity is I, then is DI the effect of this also to move the Lever. And if these two agents act against each other on the Lever, and their effects be supposed equal, or the Lever kept in equilibrio by the equal and contrary effects of these two agents; then is $DI = di$, which equation resolves into this analogy, viz, $D : d :: i : I$; that is, the distances of the agents from the prop, are reciprocally or inverfely as their intenfities, or the power is to the weight, as the diftance of the latter is to the diftance of the former.

Writers on mechanics commonly demonftrate this proportion in a very abfurd manner, viz, by fuppofing the Lever put into motion about the prop, and then inferring that, becaufe the momenta of two bodies are equal, when placed upon the Lever at fuch diftances, that thefe diftances are reciprocally proportional to the weights of the bodies, that therefore this is alfo the proportion in cafe of an equilibrium ; which is an attempt abfurdly to demonftrate a thing fuppofing the contrary, that a body is at reft, by fuppofing it to be in motion. I fhall therefore give here a new and univerfal demonftration of the property, on the pure principles of reft and preffure, or force only. Thus, let PW be a lever, C the prop, and P and W any two forces acting on the lever at the points P and W, in the directions PO, WO ; then if CE and CD be the perpendicular diftances of the directions of thefe forces from the prop C, it is to be demonftrated that P : W :: CD : CE. In order to which join CO, and draw CB parallel to WO, and CF parallel to PO. Then will CO be the direction of the preffure on the prop, otherwife there could not be an equilibrium, for the directions of three forces that keep each other in equilibrium, muft neceffarily meet in the fame point. And becaufe any three forces that keep each other in equilibrium, are proportional to the three fides of a triangle formed by drawing lines parallel to the directions of thefe forces ; therefore the forces on P, C, and W, are as the three lines BO, CO, CB, which are in the fame direction, or parallel to them ; that is the force P is to the force W, as BO or its equal CF is to CB. But the two triangles CDF, CEB are fimilar, and have their like fides proportional,

viz, CF : CB :: CD : CE ;

and becaufe it was CF : CB :: P : W ;

therefore by equality P : W :: CD : CE ;

that is each force is reciprocally proportional to the diftance of its direction from the fulcrum. And it will be found that this demonftration will ferve alfo for the other kinds of Levers, by drawing the lines as directed. Hence if any given force P be applied to a Lever at A ; its effect upon the Lever, to turn it about the centre of motion C, is as the length of the arm CA, and the fine of the angle of direction CAE. For the perp. CE is as CA $\times$ fin. $\angle$A.

In any analogy, becaufe the product of the extremes is equal to that of the means ; therefore the product of the power by the diftance of its direction is equal to the product of the weight by the diftance of its direction. That is, P $\times$ CE = W $\times$ CD.

If the Lever, with the two weights fixed to it, be made to move about the centre C ; the momentum of the power will be equal to that of the weight ; and the weights will be reciprocally proportional to their velocities.

When the two forces act perpendicularly on the Lever, as two weights &c ; then, in cafe of an equilibrium, E coincides with P, and D with W ; and the diftances CP, CW, taken on the Lever, or the diftances of the power and weight, from the fulcrum, are reciprocally proportional to the power and weight.

In a ftraight Lever, kept in equilibrio by a weight and power acting perpendicularly upon it ; then, of thefe three, the power, weight, and preffure on the prop, any one is as the diftance of the other two.

And hence too P + W : P :: ED : CD ,

and P + W : W :: ED : CP ;

that is, the fum of the weights is to either of them, as the fum of their diftances is to the diftance of the other.

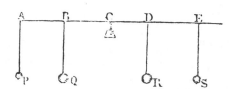

Alfo, if feveral weights P, Q, R, S, &c, act on a ftraight Lever, and keep it in equilibrio ; then the fum of the products on one fide of the prop, will be equal to the fum on the other fide, made by multiplying each weight by its diftance from the prop ; viz, P . AC + Q . BC = R . DC + S . EC + &c.

Hitherto the Lever has been confidered as a mathematical line void of weight or gravity. But when its weight is confidered, it is to be done thus : Find the weight and the centre of gravity of the Lever alone, and then confider it as a mathematical line, but having an equal weight fufpended by that centre of gravity ; and fo combine its effect with thofe of the other weights, as above.

Upon the foregoing principles depends the nature of fcales and beams for weighing all bodies. For, if the diftances be equal, then will the weights be equal alfo ; which gives the conftruction of the common fcales. And the Roman ftatera, or fteel-yard, is alfo a Lever, but of unequal arms or diftances, fo contrived that one weight only may ferve to weigh a great many, by fliding it backwards and forwards to different diftances upon the longer arm of the Lever. See BALANCE, &c.

Alfo upon the principle of the Lever depends almoft all other mechanical powers and effects. See WHEEL-AND-AXLE, PULLEY, WEDGE. SCREW, &c.

LEVITY, the privation or want of weight in any body, when compared with another that is heavier ; and in this fenfe it is oppofed to gravity. Thus cork, and moft forts of wood that float in water, have Levity with refpect to water, that is, are lefs heavy. The fchools maintained that there is fuch a thing as pofitive and abfolute Levity ; and to this they imputed the rife and buoyancy of bodies lighter in fpecie than the bodies in which they rife and float. But it is now well known that this happens only in confequence of the heavier and denfer fluid, which, by its fuperior gravity, gains the loweft place, and raifes up the lighter body by a force which is equal to the difference of their gravities. It was demonftrated by Archimedes, that a folid body will float any where in a fluid of the fame fpecific

gravity ;

gravity; and that a lighter body will always be raifed up in it.

LEUWENHOEK (ANTONY), a celebrated Dutch philofopher, was born at Delft in 1632; and acquired a great reputation throughout all Europe, by his experiments and difcoveries in Natural Hiftory, by means of the microfcope. He particularly excelled in making glaffes for microfcopes and fpectacles; and he was a member of moft of the literary focieties of Europe; to whom he fent many memoirs. Thofe in the Philofophical Tranfactions, and in the Paris Memoirs, extend through many volumes; the former were extracted, and publifhed at Leyden, in 1722. He died in 1723, at 91 years of age.

LEYDEN PHIAL, in Electricity, is a glafs phial or jar, coated both within and without with tin foil, or fome other conducting fubftance, that it may be charged, and employed in a variety of ufeful and entertaining experiments. Or even flat glafs, or any other fhape, fo coated and ufed, has alfo received the fame denomination. Alfo a vacuum produced in fuch a jar, &cs has been named the *Leyden Vacuum.*

The Leyden Phial has been fo called, becaufe it is faid that M. Cunæus, a native of Leyden, firft contrived, about the clofe of the year 1745, to accumulate the electrical power in glafs, and ufe it in this way. But Dr. Prieftley afferts that this difcovery was firft made by Von Kleift, dean of the cathedral in Camin; who, on the 4th of November 1745, fent an account of it to Dr. Liebeikuhn at Berlin: however, thofe to whom Kleift's account was communicated, could not fucceed in performing his experiments. The chief circumftances of this difcovery are ftated by Dr. Prieftley in the following manner.

Profeffor Muffchenbroek and his friends, obferving that electrified bodies, when expofed to the common atmofphere, which is always replete with conducting particles of various kinds, foon loft the moft part of their electricity, imagined that if the electrified bodies fhould be terminated on all fides by original electrics, they might be capable of receiving a ftronger power, and retaining it a longer time. Glafs being the moft convenient electric for this purpofe, and water the moft convenient non-electric, they at firft made thefe experiments with water in glafs bottles: but no confiderable difcovery was made, till M. Cunæus, happening to hold his glafs veffel in one hand, containing water, which had a communication with the prime conductor by means of a wire; and with the other hand difengaging it from the conductor, when he fuppofed the water had received as much electricity as the machine could give it, was furprifed by a fudden and unexpected fhock in his arms and breaft. This experiment was repeated, and the firft accounts of it publifhed in Holland by Meffrs. Allamand and Muffchenbroek; by the Abbé Nollet and M. Monnier, in France; and by Meffrs. Gralath and Rugger, in Germany. M. Gralath contrived to increafe the ftrength of the fhock, by altering the fhape and fize of the phial, and alfo by charging feveral phials at the fame time, fo as to form what is now called the *electrical battery.* He likewife made the fhock to pafs through a number of perfons connected in a circuit from the outfide to the infide of the phial. He alfo obferved that a cracked phial would not re-

ceive a charge: and he difcovered what is now called the *Refiduum of a charge.*

Dr. Watfon, about this time, obferved a circumftance attending the operation of charging the phial, which, if purfued, might have led him to the difcovery which was afterwards made by Dr. Franklin. He fays, that when the phial is well electrified, and you apply your hand to it, you fee the fire flafh from the outfide of the glafs, wherever you touch it, and it crackles in your hand. He alfo obferved, that when a fingle wire only was faftened about a phial, properly filled with warm water, and charged; upon the inftant of its explofion, the electrical corrufcations were feen to dart from the wire, and to illuminate the water contained in the phial. He likewife found that the ftroke, in the difcharge of the phial, was, *cæteris paribus,* as the points of contact of the non-electrics of the outfide of the glafs; which led to the method of coating glafs: in confequence of which he made experiments, from whence he concluded, that the effect of the Leyden phial was greatly increafed by, if not chiefly owing to, the number of points of non-electric in contact within the glafs, and the denfity of the matter of which thefe points confifted; provided the matter was, in its own nature, a ready conductor of electricity. He farther obferved, that the explofion was greater from hot water inclofed in glaffes, than from cold, and from his coated jars warmed, than when cold.

Mr. Wilfon, in 1746, difcovered a method of giving the fhock to any particular part of the body, without affecting the reft. He alfo increafed the ftrength of the fhock by plunging the phial in water, which gave it a coat of water on the outfide as high as it was filled within. He likewife found, that the law of accumulation of the electric matter in the Leyden phial, was always in proportion to the thinnefs of the glafs, the furface of the glafs, and that of the non-electrics in contact with its outfide and infide. He made alfo a variety of other experiments with the Leyden phial, too long here to be related.

Mr. Canton found, that when a charged phial was placed upon electrics, the wire and coating would give a fpark or two alternately, and that by a continuance of the operation the phial would be difcharged; though he did not obferve that thefe alternate fparks proceeded from the two contrary electricities difcovered by Dr. Franklin.

The Abbé Nollet made feveral experiments with this phial. He received a fhock from one, out of which the air had been exhaufted, and into which the end of his conductor had been inferted. He afcribed the force of the glafs, in giving a fhock, to that property of it, by which it retains it more ftrongly than conductors do, and is not fo eafily divefted of it as they are. It was he alfo who firft tried the effect of the electric fhock on brute animals: and he enlarged the circuit of its conveyance.

M. Monnier, it has been faid, was the firft who difcovered that the Leyden phial would retain its electricity for a confiderable time after it was charged; and that in time of froft he found it continued for 36 hours. It is remarkable too that both the French and Englifh philofophers made feveral experiments, which, with a fmall degree of attention, would have led them to the difcovery of the different qualities of the electricity on the

the contrary fides of the glafs. But this difcovery was referved for the ingenious Dr. Franklin; who, in explaining the method of charging the Leyden phial, obferves, that when one fide of the glafs is electrified plus, or pofitively, the other fide is electrified minus, or negatively: fo that whatever quantity of fire is thrown upon one fide of the glafs, the fame quantity is drawn out of the other; and in an uncharged phial, none can be thrown into the infide, when none can be taken from the outfide; and that there is really no more electric fire in the phial after it is charged than before; all that can be done by charging, being only to take from one fide, and convey to the other. Dr. Franklin alfo obferved that glafs was not impervious to electricity, and that as the equilibrium could not be reftored to the charged phial by any internal communication, it muft neceffarily be done by conductors externally joining the infide and the outfide. Thefe capital difcoveries he made by obferving, that when a phial was charged, a cork ball fufpended by filk, was attracted by the outfide coating, when it was repelled by a wire communicating with the infide, and *vice verfa*. But the truth of this principle appeared more evident, when he brought the knob of the wire, communicating with the outfide coating, within a few inches of the wire communicating with the infide coating, and fufpended a cork ball between them; for then the ball was attracted by them alternately, till the phial was difcharged.

Dr. Franklin alfo fhewed, that when the phial was charged, one fide loft exactly as much as the other gained, in reftoring the equilibrium. Hanging a fine linen thread near the coating of an electrical phial, he obferved that whenever he brought his finger near the wire, the thread was attracted by the coating; for as the fire was drawn from the infide by touching the wire, the outfide drew in an equal quantity by the thread. He likewife proved, that the coating on one fide of a phial received juft as much electricity, as was emitted from the difcharge of the other, and that in the following manner:—He infulated his rubber, and then hanging a phial to his conductor, he found it could not be charged, even when his hand was held conftantly to it; becaufe, though the electric fire might leave the outfide of the phial, there was none collected by the rubber to be conveyed to the infide. He then took away his hand from the phial, and forming a communication by a wire from the outfide coating to the infulated rubber, he found that it was charged with eafe. In this cafe it was plain, that the very fame fire which left the outfide coating, was conveyed to the infide by the way of the rubber, the globe, the conductor, and the wire of the phial. This new theory of charging the Leyden phial, led Dr. Franklin to obferve a greater variety of facts, relating both to the charging and difcharging it, than other philofophers had attended to. And this maxim, that it takes in at one furface, what it lofes at the other, led Dr. Franklin to think of charging feveral phials together with the fame trouble, by connecting the outfide of one with the infide of another; by which the fire that was driven out of the firft would be received by the fecund, &c. By this means he found, that a great number of jars might be charged with the fame labour as one only; and that they might be charged equally

5

high, were it not that every one of them receives the new fire, and lofes its old, with fome reluctance, or rather that it gives fome fmall refiftance to the charging. And on this principle he firft conftructed an electrical battery.

When Dr. Franklin firft began his experiments on the Leyden phial, he imagined that the electric fire was all crowded into the fubftance of the non electric, in contact with the glafs. But he afterwards found that its power of giving a fhock lay in the glafs itfelf, and not in the coating, by the following ingenious analyfis of the phial. To find where the ftrength of the charged bottle lay, having placed it upon a glafs, he firft took out the cork and the wire; but not finding the virtue in them, he touched the outfide coating with one hand, and put a finger of the other into the mouth of the bottle; when the fhock was felt quite as ftrong as if the cork and wire had been in it. He then charged the phial again, and pouring out the water into an empty bottle which was infulated, he expected that if the force refided in the water, it would give the fhock; but he found it gave none. He therefore concluded that the electric fire muft either have been loft in decanting, or muft remain in the bottle; and the latter he found to be true; for, upon filling the charged bottle with frefh water, he found the fhock, and was fatisfied that the power of giving it refided in the glafs itfelf. The fame experiment was made with panes of glafs, laying the coating on lightly, and charging it, as the water had been before charged in the bottle, when the refult was precifely the fame. He alfo proved in other ways that the electric fire refided in the glafs. See Franklin's Letters and Obfervations, &c. Alfo Prieftley's Hift. of Electricity, vol. i, pa. 191, &c.

From this account of Dr. Franklin's method of analyzing the Leyden phial, the manner of charging and difcharging it, with the reafon of the procefs, are eafily underftood. Thus, placing a coated phial near the prime conductor, fo that the knob of its wire may be in contact with it; then upon turning the winch of the machine, the index of the electrometer, E, fixed to the conductor, will gradually rife as far as 90° nearly, and there reft; which fhews that the phial has received its full charge: then holding the difcharger by its glafs handle, and applying one of its knobs to the outfide coating of the phial, the other being brought near the knob of the wire, or near the prime conductor which communicates with it, a report will be heard, and luminous fparks will be feen between the difcharger and the conducting fubftances communicating with the fides of the phial; and by this operation the phial will be difcharged. But, inftead of ufing the difcharger, if a perfon touch the outfide of the phial with one hand, and bring the other hand near the wire of the phial, the fame fpark and report will take place, and a fhock will be felt, affecting the wrifts and elbows, and the breaft too when the fhock is ftrong: a fhock may alfo be given to any fingle part of the body, if that part alone be brought into the circuit. If a number of perfons join hands, and the firft of them touch the outfide of the phial, while the laft touches the wire communicating with the infide, they will all feel the fhock at the fame time. If the coated phial be held by the wire, and the outfide coating be prefented to the prime conductor,

conductor, it will be charged as readily ; but only with this difference, that in this cafe the outfide will be pofitive, and the infide negative ; alfo if the prime conductor, by being connected with the rubber of the machine, be electrified negatively, the phial will be charged in the fame manner ; but the fide that touches the conductor will be electrified negatively, and the oppofite fide will be electrified pofitively. But, by infulating the phial, and repeating the fame procefs, the index of the electrometer will foon rife to 90°, yet the phial will remain uncharged; becaufe the outfide, having no communication with the earth, &c, cannot part with its own electricity, and therefore the infide cannot acquire an additional quantity : but when a chain, or any other conductor, connects the outfide of the phial with the table, the phial may be charged as before. Moreover, if a phial be infulated, and one fide of it, inftead of being connected with the earth, be connected with the infulated rubber, whilft the other fide communicates with the prime conductor, the phial will be expeditioufly charged; becaufe that whilft the rubber exhaufts one fide, the other fide is fupplied by the prime conductor ; and thus the phial is charged with its own electricity ; or the natural electric matter of one of its fides is thus thrown upon the other fide. This laft experiment may be diverfified by infulating the phial, and placing it with its wire at the diftance of about half an inch from the prime conductor, and holding the knob of another wire at the fame diftance from its outfide coating ; then, upon turning the machine, a fpark will be obferved to proceed from the prime conductor to the wire of the phial, and another fpark will pafs at the fame time from the outfide coating to the knob of the wire prefented towards it : and thus it appears that as a quantity of the electric matter is entering the infide of the phial, an equal quantity of it is leaving the outfide. If the wire prefented to the outfide of the phial be pointed, it will be feen illuminated with a ftar ; but if the pointed wire be connected with the coating of the phial, it will appear illuminated with a brufh of rays. See *Charge*, *Electrical Shock*, *Experiments*, &c.

Mr. Cavallo has defcribed the conftruction of a phial which, being charged by an electrical kite, in examining the ftate of the clouds, or in any other way, may be put into the pocket, and which will retain its charge for a confiderable time. A phial of this kind has been kept in a charged ftate for fix weeks. See his *Electricity*, pa. 340. Many other curious experiments with the Leyden phial may be feen in the books above cited, as alfo in the volumes of the Philof. Tranf. and elfewhere. In this laft-mentioned work, Mr. Cavallo defcribes a method of repairing coated phials that have cracked by any means. He firft removes the outfide coating from the fractured part, and then makes it moderately hot, by holding it to the flame of a candle ; and whilft it remains hot, he applies burning fealing-wax to the part, fo as to cover the fracture entirely ; obferving that the thicknefs of this wax coating may be greater than that of the glafs. Laftly, he covers all the fealing-wax, and alfo part of the furface of the glafs beyond it, with a compofition made with four parts of bees-wax, one of refin, one of turpentine, and a very little oil of olives ; this being fpread upon a piece of oiled filk, he applies it in the manner of a plafter. In this way feve-

ral phials have been fo effectually repaired, that after being frequently charged, they were at laft broken by a fpontaneous difcharge, but in a different part of the glafs. Philof. Tranf. vol. 68, pa. 1011.

LIBRA, *Balance*, one of the mechanical powers. See BALANCE.

LIBRA is alfo one of the 48 old conftellations, and the 7th fign of the zodiac, being oppofite to Aries, and marked like a part of a pair of fcales, thus ♎. The figure of the balance was probably given to this part of the ecliptic, becaufe when the fun arrives at this part, which is at the time of the autumnal equinox, the days and nights are equal, as if weighed in a balance.

The ftars in this conftellation are, according to Ptolomy 17, Tycho 10, Hevelius 20, and Flamfteed 51.

LIBRA alfo denotes the ancient Roman pound, which was divided into 12 unciæ, or ounces, and the ounce into 24 fcruples. It feems the mean weight of the fcruple was nearly equal to $17\frac{1}{2}$ grains Troy, and confequently the libra, or pound, 5040 grains. It was alfo the name of a gold coin, equal in value to 20 denarii. See Philof. Tranf. vol. 61, pa. 462.

The French livre is derived from the Roman libra, this being ufed in France for the proportions of their coin till about the year 1100, their fols being fo proportioned as that 20 of them were equal to the libra. By degrees it became a term of account, and every thing of the value of 20 fols was called a livre.

LIBRATION, *of the Moon*, is an apparent irregularity in her motion, by which fhe feems to librate, or waver, about her own axis, one while towards the eaft, and again another while towards the weft. See MOON, and EVECTION. Hence it is that fome parts near the moon's weftern edge at one time recede from the centre of the difc, while thofe on the other or eaftern fide approach nearer to it ; and, on. the contrary, at another time the weftern parts are feen to be nearer the centre, and the eaftern parts farther from it : by which means it happens that fome of thofe parts, which were before vifible, fet and hide themfelves in the hinder or invifible fide of the moon, and afterwards return and appear again on the nearer or vifible fide.

This Libration of the moon was firft difcovered by Hevelius, in the year 1654 ; and it is owing to her equable rotation round her own axis, once in a month, in conjunction with her unequal motion in the perimeter of her orbit round the earth. For if the moon moved in a circle, having its centre coinciding with the centre of the earth, whilft it turned on its axis in the precife time of its period round the earth, then the plane of the fame lunar meridian would always pafs through the earth, and the fame face of the moon would be conftantly and exactly turned towards us. But fince the real motion of the moon is about a point confiderably diftant from the centre of the earth, that motion is very unequal, as feen from the earth, the plane of no one meridian conftantly paffing through the earth.

The Libration of the moon is of three kinds.

1ft, Her libration in longitude, or a feeming to-and-again motion according to the order of the figns of the zodiac. This libration is nothing twice in each periodical month, viz, when the moon is in her apogeum, and when in her perigeum ; for in both thefe cafes the plane of her meridian, which is turned towards us, is directed alike towards the earth.

2d, Her libration in latitude; which arises from hence, that her axis not being perpendicular to the plane of her orbit, but inclined to it, sometimes one of her poles and sometimes the other will nod, as it were, or dip a little towards the earth, and confequently she will appear to librate a little, and to shew fometimes more of her spots, and fometimes less of them, towards each pole. Which libration, depending on the position of the moon, in respect to the nodes of her orbit, and her axis being nearly perpendicular to the plane of the ecliptic, is very properly said to be in latitude. And this also is completed in the space of the moon's periodical month, or rather while the moon is returning again to the same position, in respect of her nodes.

3d, There is also a third kind of libration; by which it happens that although another part of the moon be not really turned to the earth, as in the former libration, yet another is illuminated by the fun. For since the moon's axis is nearly perpendicular to the plane of the ecliptic, when she is most foutherly, in respect of the north pole of the ecliptic, some parts near to it will be illuminated by the fun; while, on the contrary, the fouth pole will be in darkness. In this cafe, therefore, if the fun be in the same line with the moon's fouthern limit, then, as she proceeds from conjunction with the fun towards her afcending node, she will appear to dip her northern polar parts a little into the dark hemifphere, and to raise her fouthern polar parts as much into the light one. And the contrary to this will happen two weeks after, while the new moon is defcending from her northern limit; for then her northern polar parts will appear to emerge out of darkness, and the fouthern polar parts to dip into it. And this feeming libration, or rather these effects of the former libration in latitude, depending on the light of the fun, will be completed in the moon's fynodical month. Greg. Aftron. lib. 4, fect. 10.

LIBRATION *of the Earth*, is a term applied by fome aftronomers to that motion, by which the earth is fo retained in its orbit, as that its axis continues conftantly parallel to the axis of the world.

This Copernicus calls the *motion of libration*, which may be thus illuftrated: Suppofe a globe, with its axis parallel to that of the earth, painted on the flag of a maft, moveable on its axis, and conftantly driven by an eaft wind, while it fails round an ifland, it is evident that the painted globe will be fo librated, as that its axis will be parallel to that of the world, in every fituation of the ship.

LIFE-ANNUITIES, are fuch periodical payments as depend on the continuance of fome particular life or lives. They may be diftinguifhed into Annuities that commence immediately, and fuch as commence at fome future period, called *reverfionary life-annuities*.

The value, or prefent worth, of an annuity for any propofed life or lives, it is evident, depends on two cir-

cumftances, the intereft of money, and the chance or expectation of the continuance of life. Upon the former only, it has been fhewn, under the article ANNUITIES, depends the value or prefent worth of an annuity certain, or that is not fubject to the continuance of a life, or other contingency; but the expectation of life being a thing not certain, but only poffeffing a certain chance, it is evident that the value of the certain annuity, as ftated above, muft be diminifhed in proportion as the expectancy is below certainty: thus, if the, prefent value of an annuity certain be any fum, as fuppofe 100l. and the value or expectancy of the life be $\frac{1}{2}$, then the value of the life-annuity will be only half of the former, or 50l; and if the value of the life be only $\frac{1}{3}$, the value of the life-annuity will be but $\frac{1}{3}$ of 100l, that is 33 . 6s. 8d; and fo on.

The meafure of the value or expectancy of life, depends on the proportion of the number of perfons that die, out of a given number, in the time propofed; thus, if 50 perfons die, out of 100, in any propofed time, then, half the number only remaining alive, any one perfon has an equal chance to live or die in that time, or the value of his life for that time is $\frac{1}{2}$; but if $\frac{2}{3}$ of the number die in the time propofed, or only $\frac{1}{3}$ remain alive, then the value of any one's life is $\frac{1}{3}$; and if $\frac{3}{4}$ of the number die, or only $\frac{1}{4}$ remain alive, then the value of any life is but $\frac{1}{4}$; and fo on. In these proportions then muft the value of the annuity certain be diminifhed, to give the value of the like life annuity.

It is plain therefore that, in this bufinefs, it is neceffary to know the value of life at all the different ages, from fome table of obfervations on the mortality of mankind, which may fhew the proportion of the perfons living, out of a given number, at the end of any propofed time; or from fome certain hypothefis, or affumed principle. Now various tables and hypothefes of this fort were given by the writers on this fubject, as Dr. Halley, Mr. Demoivre, Mr. Thomas Simpfon, Mr. Dodfon, Mr. Kerffeboom, Mr. Parcieux, Dr. Price, Mr. Morgan, Mr. Baron Maferes, and many others. But the fame table of probabilities of life will not fuit all places; for long experience has fhewn that all places are not equally healthy, or that the proportion of the number of perfons that die annually, is different for different places. Dr. Halley computed a table of the annual deaths as drawn from the bills of mortality of the city of Breflaw in Germany, Mr. Smart and Mr. Simpfon from thofe of London, Dr. Price from thofe of Northampton, Mr. Kerffeboom from thofe of the provinces of Holland and Weft-Friefland, and M. Parcieux from the lifts of the French tontines, or long annuities, and all thefe are found to differ from one another. It may not therefore be improper to infert here a comparative view of the principal tables that have been given of this kind, as below, where the firft column fhews the age, and the other columns the number of perfons living at that age, out of 1000 born, or of the age 0, in the firft line of each column.

TABLE

T A B L E I.

Shewing the Number of Perſons living at all Ages, out of 1000 that had been born at ſeveral Places, viz.

Ages.	Vienna.	Berlin.	London.	Norwich.	North-ampton.	Breſlaw.	Branden-burg.	Holy-Croſs.	Holland.	France.	Vaud, Switzer-land.
0	1000	1000	1000	1000	1000	1000	1000	1000	1000	1000	1000
1	542	633	680	798	738	769	775	882	804	805	811
2	471	528	548	651	678	658	718	762	768	777	765
3	430	485	492	595	585	614	687	717	736	750	735
4	400	434	452	566	562	585	664	682	709	727	715
5	377	403	426	544	544	563	642	659	689	711	701
6	357	387	412	526	530	546	622	636	676	697	688
7	344	376	397	511	518	532	607	618	664	686	677
8	337	367	388	500	510	523	595	604	652	676	667
9	331	361	380	490	504	515	585	595	646	667	659
10	326	356	373	481	498	508	577	589	639	660	653
11	322	353	367	474	493	502	570	585	633	654	648
12	318	350	361	469	488	497	564	581	627	649	643
13	314	347	356	464	484	492	559	577	621	644	639
14	310	344	351	460	480	488	554	573	616	639	635
15	306	341	347	455	475	483	549	569	611	635	631
16	302	338	343	451	470	479	544	565	606	631	626
17	299	335	338	446	465	474	539	560	601	626	622
18	295	332	334	442	459	470	535	555	596	621	618
19	291	328	329	437	453	465	531	550	590	616	614
20	287	324	325	432	447	461	527	545	584	610	610
21	284	320	321	426	440	456	522	539	577	604	606
22	280	315	316	421	433	451	517	532	571	598	602
23	276	310	310	415	426	446	512	525	566	592	597
24	273	305	305	409	419	441	507	518	559	536	592
25	269	297	299	404	412	436	502	512	551	580	587
26	265	293	294	398	405	431	498	506	543	574	582
27	261	287	288	392	398	426	495	501	535	568	577
28	256	281	283	385	391	421	492	496	526	562	572
29	251	275	278	378	384	415	489	491	517	556	567
30	247	269	272	372	378	409	486	486	508	550	563
31	243	264	266	366	372	403	482	481	499	544	558
32	239	259	260	361	366	397	477	476	490	438	553
33	235	254	254	355	360	391	422	471	482	532	548
34	231	249	248	350	354	384	467	466	474	526	544
35	226	243	242	344	348	377	462	460	467	520	539
36	221	237	236	338	342	370	456	454	460	514	533
37	216	230	230	333	336	363	450	447	453	508	527
38	211	223	224	327	330	356	444	440	446	503	520
39	205	216	218	322	324	349	438	433	439	497	513
40	199	209	214	317	317	342	432	426	432	492	506
41	194	203	207	311	310	335	427	418	425	487	500
42	189	197	201	306	303	328	422	410	419	482	494
43	185	192	194	300	296	321	417	401	413	476	488
44	181	187	187	294	289	314	412	393	407	471	482
45	176	182	180	287	282	307	407	386	400	466	476
46	171	177	174	281	275	299	400	379	393	460	469
47	165	172	167	274	268	291	394	372	386	455	461
48	159	167	159	263	261	283	388	365	378	449	451
49	153	162	153	261	254	275	381	359	370	443	441
50	147	157	147	255	247	267	374	353	362	436	431
51	142	152	141	248	239	259	367	347	354	429	422
52	137	147	135	242	232	250	359	340	345	422	414
53	133	142	130	235	225	241	351	333	336	414	406

Ages.	Vienna.	Berlin.	London.	Norwich.	North-ampton.	Breſlaw.	Branden-burg.	Holy-Croſs.	Holland.	France.	Vaud, Switzer-land.
54	128	137	125	228	218	232	343	326	327	406	397
55	123	132	120	221	211	224	334	318	318	397	388
56	117	127	116	213	204	216	324	310	309	388	377
57	111	121	111	206	197	209	314	301	300	379	364
58	106	115	106	199	190	201	304	292	291	369	348
59	101	109	101	191	183	193	293	283	282	359	331
60	96	103	96	184	176	186	282	273	273	349	314
61	91	97	92	177	169	178	271	263	264	339	299
62	87	92	87	169	162	170	260	253	255	329	286
63	82	88	83	161	155	163	248	243	245	318	274
64	77	84	78	153	148	155	236	233	235	307	262
65	72	80	74	144	141	147	224	223	225	296	250
66	67	75	70	136	134	140	213	213	215	285	236
67	62	70	65	128	127	132	202	203	205	273	220
68	57	65	61	119	120	124	190	193	195	260	202
69	52	60	56	111	113	117	178	182	185	246	184
70	48	55	52	103	106	109	166	171	175	232	168
71	44	51	47	94	99	101	153	161	165	218	153
72	40	47	43	86	92	93	138	151	155	195	140
73	36	43	39	79	85	85	122	142	145	188	129
74	33	39	35	71	78	77	107	134	135	173	119
75	30	35	32	64	71	69	93	126	125	158	109
76	27	32	28	57	64	61	80	119	114	144	98
77	24	29	25	50	58	53	68	112	103	129	85
78	21	26	22	43	52	45	59	105	92	115	71
79	18	23	19	37	46	38	51	98	82	102	58
80	16	20	17	32	40	32	44	90	72	88	46
81	14	18	14	27	34	26	38	81	62	75	36
82	12	16	12	23	28	22	32	71	53	63	29
83	10	14	10	19	23	18	25	61	45	53	24
84	8	12	8	16	19	15	21	51	38	44	20
85	7	10	7	13	16	12	15	41	31	36	17
86	6	8	6	10	13	9	11	32	25	28	14
87	5	7	5	8	11	6	8	24	19	21	11
88	4	6	4	6	8	4	6	17	14	16	9
89	3	5	3	5	6	2	4	11	10	12	7
90	2	4	2	4	4	1	3	7	7	8	5

Theſe tables ſhew that the mortality and chance of life are very various in different places; and that therefore, to obtain a ſufficient accuracy in this buſineſs, it is neceſſary to adapt a table of probabilities or chances of life, to every place for which annuities are to be calculated; or at leaſt one ſet of tables for large towns, and another for country places, as well as for the ſuppoſition of different rates of intereſt.

Several of the foregoing tables, as they commenced with numbers different from one another, are here reduced to the ſame number at the beginning, viz, 1000 perſons, by which means we are enabled by inſpection, at any age, to compare the numbers together, and immediately perceive the relative degrees of vitality at the ſeveral places. The tables are alſo arranged according to the degree of vitality amongſt them; the leaſt, or that at Vienna, firſt; and the reſt in their order, to the higheſt, which is the province of Vaud in Switzerland. The authorities upon which theſe tables depend, are as they here follow. The firſt, taken from Dr. Price's Obſervations on Reverſionary payments, is formed from the bills at Vienna, for 8 years, as given by Mr. Suſmilch, in his *Gottliche* Ordnung; the 2d, for Berlin, from the ſame, as formed from the bills there for 4 years, viz, from 1752 to 1755; the 3d, from Dr. Price, ſhewing the true probabilities of life in London, formed from the bills for ten years, viz, from 1759 to 1768; the 4th, for Norwich, formed by Dr. Price from the bills for 30 years, viz, from 1740 to 1769; the 5th, by the ſame, from the bills for Northampton; the 6th, as deduced by Dr. Halley, from the bills of mortality at Breſlaw; the 7th ſhews the probabilities of life in a country pariſh in Brandenburg, formed from the bills for 50 years, from 1710 to 1759, as given by Mr. Suſmilch; the 8th ſhews the probabilities of life in the pariſh of Holy-Croſs, near Shrewſbury, formed from a regiſter kept by the Rev. Mr. Garſuch, for 20 years, from 1750 to 1770; the 9th, for Holland,

Holland, was formed by M. Kerffeboom, from the regifters of certain annuities for lives granted by the government of Holland, which had been kept there for 125 years, in which the ages of the feveral annuitants dying during that period had been truly entered; the 10th, for France, were formed by M. Parcieux, from the lifts of the French tontines, or long annuities, and verified by a comparifon with the mortuary regifters of feveral religious houfes for both fexes; and the 11th, or laft, for the diftrict of Vaud in Switzerland, was formed by Dr. Price from the regifters of 43 parifhes given by M. Muret, in the Bern Memoirs for the year 1766.

Now from fuch lifts as the foregoing, various tables have been formed for the valuation of annuities on fingle and joint lives, at feveral rates of intereft, in which the value is fhewn by infpection. The following are thofe that are given by Mr. Simpfon, in his Select Exercifes, as deduced from the London bills of mortality.

TABLE II.

Shewing the Value of an Annuity on One Life, or Number of Years Annuity in the Value, fuppofing Money to bear Intereft at the feveral Rates of 3, 4, and 5 per cent.

Age.	Years value at 3 per cent.	Years value at 4 per cent.	Years value at 5 per cent.	Age.	Years value at 3 per cent.	Years value at 4 per cent.	Years value at 5 per cent.
6	18·8	16·2	14·1	41	13·0	11·4	10·2
7	18·9	16·3	14·2	42	12·8	11·2	10·1
8	19·0	16·4	14·3	43	12·6	11·1	10·0
9	19·0	16·4	14·3	44	12·5	11·0	9·9
10	19·0	16·4	14·3	45	12·3	10·8	9·8
11	19·0	16·4	14·3	46	12·1	10·7	9·7
12	18·9	16·3	14·2	47	11·9	10·5	9·5
13	18·7	16·2	14·1	48	11·8	10·4	9·4
14	18·5	16·0	14·0	49	11·6	10·2	9·3
15	18·3	15·8	13·9	50	11·4	10·1	9·2
16	18·1	15·6	13·7	51	11·2	9·9	9·0
17	17·9	15·4	13·5	52	11·0	9·8	8·9
18	17·6	15·2	13·4	53	10·7	9·6	8·8
19	17·4	15·0	13·2	54	10·5	9·4	8·6
20	17·2	14·8	13·0	55	10·3	9·3	8·5
21	17·0	14·7	12·9	56	10·1	9·1	8·4
22	16·8	14·5	12·7	57	9·9	8·9	8·2
23	16·5	14·3	12·6	58	9·6	8·7	8·1
24	16·3	14·1	12·4	59	9·4	8·6	8·0
25	16·1	14·0	12·3	60	9·2	8·4	7·9
26	15·9	13·8	12·1	61	8·9	8·2	7·7
27	15·6	13·6	12·0	62	8·7	8·1	7·6
28	15·4	13·4	11·8	63	8·5	7·9	7·4
29	15·2	13·2	11·7	64	8·3	7·7	7·3
30	15·0	13·1	11·6	65	8·0	7·5	7·1
31	14·8	12·9	11·4	66	7·8	7·3	6·9
32	14·6	12·7	11·3	67	7·6	7·1	6·7
33	14·4	12·6	11·2	68	7·4	6·9	6·6
34	14·2	12·4	11·0	69	7·1	6·7	6·4
35	14·1	12·3	10·9	70	6·9	6·5	6·2
36	13·9	12·1	10·8	71	6·7	6·3	6·0
37	13·7	11·9	10·6	72	6·5	6·1	5·8
38	13·5	11·8	10·5	73	6·2	5·9	5·6
39	13·3	11·6	10·4	74	5·9	5·6	5·4
40	13·2	11·5	10·3	75	5·6	5·4	5·2

TABLE III.

Shewing the Value of an Annuity for Two Joint Lives, that is, for as long as they exist together.

Age of Younger	Age of Elder	Value at 3 per cent.	Value at 4 per cent.	Value at 5 per cent.
10	10	14.7	13.0	11.6
	15	14.3	12.7	11.3
	20	13.8	12.2	10.8
	25	13.1	11.6	10.2
	30	12.3	10.9	9.7
	35	11.5	10.2	9.1
	40	10.7	9.6	8.6
	45	10.0	9.0	8.1
	50	9.3	8.4	7.6
	55	8.6	7.8	7.1
	60	7.8	7.2	6.6
	65	6.9	6.5	6.1
	70	6.1	5.8	5.5
	75	5.3	5.1	4.9
15	15	13.9	12.3	11.0
	20	13.3	11.8	10.5
	25	12.6	11.2	10.1
	30	11.9	10.6	9.5
	35	11.2	10.0	9.0
	40	10.4	9.4	8.5
	45	9.6	8.8	8.0
	50	8.9	8.2	7.5
	55	8.2	7.6	7.0
	60	7.5	7.0	6.5
	65	6.8	6.4	6.0
	70	6.0	15.7	5.4
	75	5.2	5.0	4.8
20	20	12.8	11.3	10.1
	25	12.2	10.8	9.7
	30	11.6	10.3	9.2
	35	10.9	9.8	8.8
	40	10.2	9.2	8.4
	45	9.5	8.6	7.9
	50	8.8	8.0	7.4
	55	8.1	7.5	6.9
	60	7.4	6.9	6.4
	65	6.7	6.3	5.9
	70	6.0	5.7	5.4
	75	5.2	5.0	4.8
25	25	11.8	10.5	9.4
	30	11.3	10.1	9.0
	35	10.7	9.6	8.6
	40	10.0	9.1	8.2
	45	9.4	8.5	7.8
	50	8.7	7.9	7.3
	55	8.0	7.4	6.8
	60	7.3	6.8	6.3
	65	6.6	6.2	5.8
	70	5.9	5.6	5.3
	75	5.1	4.9	4.7

Age of Younger	Age of Elder	Value at 3 per cent.	Value at 4 per cent.	Value at 5 per cent.
30	30	10.8	9.6	8.6
	35	10.3	9.2	8.3
	40	9.7	8.8	8.0
	45	9.1	8.3	7.6
	50	8.5	7.8	7.2
	55	7.9	7.3	6.7
	60	7.2	6.7	6.2
	65	6.5	6.1	5.7
	70	5.8	5.5	5.2
	75	5.1	4.9	4.7
35	35	9.9	8.8	8.0
	40	9.4	8.5	7.7
	45	8.9	8.1	7.4
	50	8.3	7.6	7.0
	55	7.7	7.1	6.6
	60	7.1	6.5	6.1
	65	6.4	6.0	5.6
	70	5.7	5.4	5.1
	75	5.0	4.8	4.6
40	40	9.1	8.1	7.3
	45	8.7	7.8	7.1
	50	8.2	7.4	6.8
	55	7.6	6.9	6.4
	60	7.0	6.4	6.0
	65	6.4	5.9	5.5
	70	5.7	5.4	5.1
	75	5.0	4.8	4.6
45	45	8.3	7.4	6.7
	50	7.9	7.1	6.5
	55	7.4	6.7	6.2
	60	6.8	6.3	5.8
	65	6.3	5.8	5.4
	70	5.6	5.3	5.0
	75	4.9	4.7	4.5
50	50	7.6	6.8	6.2
	55	7.2	6.5	6.0
	60	6.7	6.1	5.7
	65	6.2	5.7	5.3
	70	5.5	5.2	4.9
	75	4.8	4.6	4.4
55	55	6.9	6.2	5.7
	60	6.5	5.9	5.5
	65	6.0	5.6	5.2
	70	5.4	5.1	4.8
	75	4.7	4.5	4.3
60	60	6.1	5.6	5.2
	65	5.7	5.3	4.9
	70	5.2	4.9	4.6
	75	4.6	4.4	4.2
65	65	5.4	5.0	4.7
	70	4.9	4.6	4.4
	75	4.4	4.2	4.0
70	70	4.6	4.4	4.2
	75	4.2	4.0	3.9
75	75	3.8	3.7	3.6

TABLE

T A B L E IV.

For the Value of an Annuity upon the Longer of Two Given Lives.

Age of Younger	Age of Elder	value at 3 per cent.	Value at 4 per cent.	Value at 5 per cent.	Age of Younger	Age of Elder	Value at 3 per cent.	Value at 4 per cent.	Value at 5 per cent.
10	10	23.4	19.9	17.1	30	30	19.3	16.6	14.5
	15	22.9	19.5	16.8		35	18.8	16.2	14.2
	20	22.5	19.1	16.6		40	18.4	15.9	14.0
	25	22.2	18.8	16.4		45	18.1	15.6	13.8
	30	21.9	18.6	16.2		50	17.8	15.4	13.6
	35	21.6	18.4	16.1		55	17.4	15.1	13.4
	40	21.4	18.3	16.0		60	17.0	14.8	13.2
	45	21.2	18.2	15.9		65	16.6	14.5	12.9
	50	20.9	18.0	15.8		70	16.1	14.1	12.6
	55	20.7	17.8	15.7		75	15.6	13.7	12.2
	60	20.4	17.6	15.5	35	35	18.3	15.8	13.8
	65	20.1	17.4	15.3		40	17.8	15.4	13.5
	70	19.8	17.2	15.1		45	17.4	15.1	13.3
	75	19.5	16.9	14.8		50	17.1	14.8	13.1
15	15	22.8	19.3	16.7		55	16.7	14.5	12.9
	20	22.3	18.9	16.4		60	16.3	14.2	12.7
	25	21.9	18.6	16.2		65	15.8	13.8	12.4
	30	21.6	18.3	16.0		70	15.3	13.4	12.0
	35	21.3	18.1	15.9		75	14.8	13.0	11.6
	40	21.1	17.9	15.7	40	40	17.3	15.0	13.3
	45	20.9	17.8	15.6		45	16.8	14.6	13.0
	50	20.7	17.6	15.4		50	16.3	14.2	12.7
	55	20.4	17.4	15.3		55	15.9	13.9	12.4
	60	20.1	17.2	15.2		60	15.4	13.5	12.1
	65	19.8	16.9	15.0		65	14.9	13.1	11.8
	70	19.4	16.6	14.7		70	14.5	12.7	11.4
	75	18.9	16.3	14.4		75	14.0	12.3	11.0
20	20	21.6	18.3	15.8	45	45	16.2	14.2	12.8
	25	21.1	17.9	15.5		50	15.7	13.8	12.5
	30	20.7	17.6	15.3		55	15.2	13.4	12.1
	35	20.4	17.4	15.1		60	14.7	12.9	11.7
	40	20.1	17.2	15.0		65	14.1	12.5	11.4
	45	19.9	17.0	14.9		70	13.6	12.0	11.0
	50	19.6	16.8	14.7		75	13.1	11.6	10.6
	55	19.4	16.6	14.5	50	50	15.0	13.3	12.1
	60	19.1	16.3	14.3		55	14.5	12.9	11.7
	65	18.7	16.0	14.1		60	13.9	12.4	11.3
	70	18.2	15.7	13.8		65	13.3	12.0	10.9
	75	17.7	15.3	13.5		70	12.8	11.5	10.5
25	25	20.3	17.4	15.1		75	12.3	11.0	10.1
	30	19.8	17.0	14.9	55	55	13.6	12.4	11.3
	35	19.4	16.7	14.7		60	13.0	11.9	10.9
	40	19.2	16.5	14.5		65	12.4	11.3	10.5
	45	18.9	16.3	14.3		70	11.8	10.8	10.0
	50	18.7	16.1	14.2		75	11.3	10.3	9.5
	55	18.4	15.9	14.0	60	60	12.2	11.2	10.5
	60	18.0	15.6	13.8		65	11.5	10.6	10.0
	65	17.6	15.3	13.6		70	10.9	10.1	9.5
	70	17.2	15.0	13.3		75	10.3	9.5	9.0
	75	16.7	14.6	12.9	65	65	10.7	10.0	9.4
						70	10.0	9.4	8.9
						75	9.3	8.7	8.3
					70	70	9.2	8.6	8.2
						75	8.4	7.9	7.6
					75	75	7.6	7.2	6.9

The ufes of thefe tables may be exemplified in the following problems.

PROB. I. *To find the Probability or Proportion of Chance, that a perfon of a Given Age continues in being a propofed number of years.*—Thus, fuppofe the age be 40, and the number of years propofed 15; then, to calculate by the table of the probabilities for London, in tab. I. againft 40 years ftands 214, and againft 55 years, the age to which the perfon muft arrive, ftands 120, which fhews that, of 214 perfons who attain to the age of 40, only 120 of them reach the age of 55, and confequently 94 die between the ages of 40 and 55: It is evident therefore that the odds for attaining the propofed age of 55, are as 120 to·94, or as 9 to 7 nearly.

PROB. 2. *To find the Value of an Annuity for a propofed Life.*—This problem is refolved from tab. 2, by looking againft the given age, and under the propofed rate of intereft; then the corresponding quantity fhews the number of years·purchafe required. For example, if the given age be 36, the rate of intereft 4 per cent, and the propofed annuity L250. Then in the table it appears that the value is 12·1 years purchafe, or 12·1 times L250, that is L3025.

After the fame manner the anfwer will be found in any other cafe falling within the limits of the table. But as there may fometimes be occafion to know the values of lives computed at higher rates of intereft than thofe in the table, the two following practical rules are fubjoined; by which the problem is refolved independent of tables.

Rule 1. When the given age is not lefs than 45 years, nor greater than 85, fubtract it from 92; then multiply the remainder by the perpetuity, and divide the product by the faid remainder added to 2½ times the perpetuity; fo fhall the quotient be the number of years purchafe required. Where note, that by the *perpetuity* is meant the number of years purchafe of the fee-fimple; found by dividing 100 by the rate per cent at which intereft is reckoned.

Ex. Let the given age be 50 years, and the rate of intereft 10 per cent. Then fubtracting 50 from 92, there remains 42; which multiplied by 10 the perpetuity, gives 420; and this divided by 67, the remainder increafed by 2½ times 10 the perpetuity, quotes 6·3 nearly, for the number of years purchafe. Therefore, fuppofing the annuity to be L100, its value in prefent money will be L630.

Rule 2. When the age is between 10 and 45 years; take 8 tenths of what it wants of 45, which divide by the rate per cent increafed by 1·2; then if the quotient be added to the value of a life of 45 years, found by the preceding rule, there will be obtained the number of years purchafe in this cafe. For example, let the propofed age be 20 years, and the rate of intereft 5 per cent. Here taking 20 from 45, there remains 25; 8/15 of which is 20; which divided by 6·2, quotes 3·2: and this added to 9·8, the value of a life of 45, found by the former rule, gives 13 for the number of years purchafe that a life of 20 ought to be valued at.

And the conclufions derived by thefe rules, Mr. Simpfon adds, are fo near the true values, computed

4

from real obfervations, as feldom to differ from them by more than 1/10 or 1/12 of one year's purchafe.

The obfervations here alluded to, are thofe which are founded on the London bills of mortality. And a fimilar method of folution, accommodated to the Breflaw obfervations, will be as follows, viz. " Multiply the difference between the given age and 85 years by the perpetuity, and divide the product by 8 tenths of the faid difference increafed by double the perpetuity, for the anfwer." Which, from 8 to 80 years of age, will commonly come within lefs than ⅛ of a year's purchafe of the truth.

PROB. 3. *To find the Value of an Annuity for Two Joint Lives, that is, for as long as they both continue in being together.*—In table 3, find the younger age, or that neareft to it, in column 1, and the higher age in column 2; then againft this laft is the number of years purchafe in the proper column for the intereft. *Ex.* Suppofe the two ages be 20 and 35 years; then the value

is 10·9 years purchafe at 3 per cent.
or 9·8 - - at 4 per cent.
or 8·8 - - at 5 per cent.

PROB. 4. *To find the Value of the Annuity for the Longeft of Two Lives, that is, for as long, as either of them continues in being.*—In table 4, find the age of the youngeft life, or the neareft to it, in col. 1, and the age of the elder in col. 2: then againft this laft is the anfwer in the proper column of intereft.—*Ex.* So, if the two ages be 15 and 40; then the value of the annuity upon the longeft of two fuch lives,

is 21·1 years purchafe at 3 per cent.
or 17·9 - - 4 per cent.
or 15·7 - - 5 per cent.

N. B. In the laft two problems, if the younger age, or the rate of intereft, be not exactly found in the tables, the neareft to them may be taken, and then by proportion the value for the true numbers will be nearly found.

Rules and tables for the values of three lives, &c, may alfo be feen in Simpfon, and in Baron Maferes's Annuities, &c. All thefe calculations have been made from tables of the real mortuary regifters, differing unequally at the feveral ages. But rules have alfo been given upon other principles, as by De Moivre, upon the fuppofition that the decrements of life are equal at all ages; an affumption not much differing from the truth, from 7 to 70 years of age.

LIFE-ANNUITIES, payable half-yearly, &c.—Thefe are worth more than fuch as are payable yearly, as computed by the foregoing rules and tables, on the two following accounts: Firft, that parts of the payments are received fooner; and 2dly, there is a chance of receiving fome part or parts of a whole year's payment more than when the payments are only made annually. Mr. Simpfon, in his Select Exercifes, pa. 283, obferves, that the value of thefe two advantages put together, will always amount to ¼ of a year's purchafe for half-yearly payments, and to ⅜ of a year's purchafe for quarterly payments; and Mr. Maferes, at page 233 &c of his Annuities, by a very elaborate calculation, finds the former difference to be nearly ¼ alfo. But Dr. Price, in an Effay in the Philof. Tranf. vol. 66, pa. 109, pa. 109, states the same differences only

at $\frac{2}{10}$ for half-yearly payments,

and $\frac{3}{10}$ for quarterly payments :

And the Doctor then adds some algebraical theorems for such calculations.

LIFE-ANNUITIES, *secured by Land.*—These differ from other life-annuities only in this, that the annuity is to be paid up to the very day of the death of the age in question, or of the person upon whose life the annuity is granted. To obtain the more exact value therefore of such an annuity, a small quantity must be added to the same as computed by the foregoing rules and observations, which is different according as the payments are yearly, half-yearly, or quarterly, &c; and are thus stated by Dr. Price in his Essay quoted above ; viz, the addition·

is $\dfrac{y}{2n}$ for annual payments,

or $\dfrac{h}{4n}$ for half-yearly payments,

or $\dfrac{q}{8n}$ for quarterly payments :

where n is the complement of the given age, or what it wants of 86 years; and y, h, q are the respective values of an annuity *certain* for n years, payable yearly, half-yearly, or quarterly. And, by numeral examples, it is found that the first of these additional quantities is about $\frac{2}{10}$, the second $\frac{1}{10}$, and the 3d half a tenth of one year's purchase.

Complement of LIFE. See COMPLEMENT.

Expectation of LIFE. See EXPECTATION.

Insurance or Assurance on LIVES. See ASSURANCES *on Lives.*

LIGHT, that principle by which objects are made perceptible to our sense of seeing ; or the sensation occasioned in the mind by the view of luminous objects.

The nature of Light has been a subject of speculation from the first dawnings of philosophy. Some of the earliest philosophers doubted whether objects became visible by means of any thing proceeding from them, or from the eye of the spectator. But this opinion was qualified by Empedocles and Plato, who maintained, that vision was occasioned by particles continually flying off from the surfaces of bodies, which meet with others proceeding from the eye ; while the effect was ascribed by Pythagoras solely to the particles proceeding from the external objects, and entering the pupil of the eye. But Aristotle defines Light to be the act of a transparent body, considered as such : and he observes that Light is not fire, nor yet any matter radiating from the luminous body, and transmitted through the transparent one.

The Cartesians have refined considerably on this notion ; and hold that Light, as it exists in the luminous body, is only a power or faculty of exciting in us a very clear and vivid sensation ; or that it is an invisible fluid present at all times and in all places, but requiring to be set in motion, by a body ignited or otherwise properly qualified to make objects visible to us.

Father Malbranche explains the nature of Light from a supposed analogy between it and sound.—

Thus he supposes all the parts of a luminous body are in a rapid motion, which, by very quick pulses, is constantly compressing the subtle matter between the luminous body and the eye, and excites vibrations of pression. As these vibrations are greater, the body appears more luminous ; and as they are quicker or slower, the body is of this or that colour.

But the Newtonians maintain, that Light is not a fluid *per se*, but consists of a great number of very small particles, thrown off from the luminous body by a repulsive power with an immense velocity, and in all directions. And these particles, it is also held, are emitted in right lines : which rectilinear motion they preserve till they are turned out of their path by some of the following causes, viz, by the attraction of some other body near which they pass, which is called *inflection*; or by passing obliquely through a medium of different density, which is called *refraction*; or by being turned aside by the opposition of some intervening body, which is called *reflection* ; or, lastly, by being totally stopped by some substance into which they penetrate, and which is called their *extinction*. A succession of these particles following one another, in an exact straight line, is called a *ray of Light* ; and this ray, in whatever manner its direction may be changed, whether by refraction, reflection, or inflection, always preserves a rectilinear course till it be again changed ; neither is it possible to make it move in the arch of a circle, ellipsis, or other curve. For the above properties of the rays of Light, see the several words, REFRACTION, REFLECTION, &c.

The *velocity* of the particles and rays of Light is truly astonishing, amounting to near 2 hundred thousand miles in a second of time, which is near a million times greater than the velocity of a cannon-ball. And this amazing motion of Light has been manifested in various ways, and first, from the eclipses of Jupiter's satellites. It was first observed by Roemer, that the eclipses of those satellites happen sometimes sooner, and sometimes later, than the times given by the tables of them ; and that the observation was before or after the computed time, according as the earth was nearer to, or farther from Jupiter, than the mean distance. Hence Roemer and Cassini both concluded that this circumstance depended on the distance of Jupiter from the earth ; and that, to account for it, they must suppose that the Light was about 14 minutes in crossing the earth's orbit. This conclusion however was afterward abandoned and attacked by Cassini himself. But Roemer's opinion found an able advocate in Dr. Halley ; who removed Cassini's difficulty, and left Roemer's conclusion in its full force. Yet, in a memoir presented to the Academy in 1707, M. Maraldi endeavoured to strengthen Cassini's arguments ; when Roemer's doctrine found a new defender in Mr. Pound. See Philos. Trans. number 136, also Abridg. vol. 1, pa. 409 and 422, and Groves, Phys. Elem. number 2636. It has since been found, by repeated experiments, that when the earth is exactly between Jupiter and the sun, his satellites are seen eclipsed about $8\frac{1}{4}$ minutes *sooner* than they could be according to the tables ; but when the earth is nearly in the opposite point of its orbit, these eclipses happen about $8\frac{1}{4}$ minutes *later* than the tables predict them. Hence

F   then

then it is certain that the motion of Light is not inflantaneous, but that it takes up about 16½ minutes of time to pafs over a fpace equal to the diameter of the earth's orbit, which is at leaft 190 millions of miles in length, or at the rate of near 200,000 miles per fecond, as above-mentioned. Hence therefore Light takes up about 8¼ minutes in paffing from the fun to the earth; fo that, if he fhould be annihilated, we would fee him for 8¼ minutes after that event fhould happen; and if he were again created, we fhould not fee him till 8¼ minutes afterwards. Hence alfo it is eafy to know the time in which Light travels to the earth, from the moon, or any of the other planets, or even from the fixed ftars when their diftances fhall be known; thefe diftances however are fo immenfely great, that from the neareft of them, fuppofed to be Sirius, the dog-ftar, Light takes up many years to travel to the earth: and it is even fufpected that there are many ftars whofe Light have not yet arrived at us fince their creation. And this, by-the-bye, may perhaps fometimes account for the appearance of new ftars in the heavens.

It may be juft obferved that Galileo firft conceived the notion of meafuring the velocity of Light; and a defcription of his contrivance for this purpofe, is in his Treatife on Mechanics, pa. 39. He had two men with Lights covered; the one was to obferve when the other uncovered his Light, and to exhibit his own the moment he perceived it. This rude experiment was tried at the diftance of a mile, but without fuccefs, as may naturally be imagined: and the members of the Academy Del Cimento repeated the experiment, and placed their obfervers, to as little purpofe, at the diftance of 2 miles.

But our excellent aftronomer, Dr. Bradley, afterwards found nearly the fame velocity of Light as Roemer, from his accurate obfervations, and moft ingenious theory, to account for fome apparent motions in the fixed ftars; for an account of which, fee ABERRATION of Light. By a long feries of thefe obfervations, he found the difference between the true and apparent place of feveral fixed ftars, for different times of the year; which difference could no otherwife be accounted for, than from the progreffive motion of the rays of Light. From the mean quantity of this difference he ingenioufly found, that the ratio of the velocity of Light to the velocity of the earth in its orbit, was as 10313 to 1, or that Light moves 10313 times fafter than the earth moves in its orbit about the fun; and as this latter motion is at the rate of 18¹¹⁄₁₂ miles per fecond nearly, it follows that the former, or the velocity of Light, is at the rate of about 195000 miles in a fecond; a motion according to which it will require juft 8′ 7″ to move from the fun to the earth, or about 95 millions of miles.

It was alfo inferred, from the foregoing principles, that Light proceeds with the fame velocity from all the ftars. And hence it follows, if we fuppofe that all the ftars are not equally diftant from us, as many arguments prove, that the motion of Light, all the way it paffes through the immenfe fpace above our atmofphere, is equable or uniform. And fince the different methods of determining the velocity of Light thus agree in the refult, it is reafonable to conclude

that, in the fame medium, Light is propagated with the fame velocity after it has been reflected, as before.

For an account of Mr. Melville's hypothefis of the different velocities of differently coloured rays, fee COLOUR.

To the doctrine concerning the materiality of Light, and its amazing velocity, feveral objections have been made; of which the moft confiderable is, That as rays of Light are continually paffing in different directions from every vifible point, they muft neceffarily interfere with each other in fuch a manner, as entirely to confound all diftinct perception of objects, if not quite to deftroy the whole fenfe of feeing: not to mention the continual wafte of fubftance which a conftant emiffion of particles muft occafion in the luminous body, and thereby fince the creation muft have greatly diminifhed the matter in the fun and ftars, as well as increafed the bulk of the earth and planets by the vaft quantity of particles of Light abforbed by them in fo long a period of time.

But it has been replied, that if Light were not a body, but confifted in mere preffion or pulfion, it could never be propagated in right lines, but would be continually inflected ad umbram. Thus Sir I. Newton: "A preffure on a fluid medium, i. e. a motion propagated by fuch a medium, beyond any obftacle, which impedes any part of its motion, cannot be propagated in right lines, but will be always inflecting and diffufing itfelf every way, to the quiefcent medium beyond that obftacle. The power of gravity tends downwards; but the preffure of water arifing from it tends every way with an equable force, and is propagated with equal eafe and equal ftrength, in curves, as in ftrait lines. Waves, on the furface of the water, gliding by the extremes of any very large obftacle, inflect and dilate themfelves, ftill diffufing gradually into the quiefcent water beyond that obftacle. The waves, pulfes, or vibrations of the air, wherein found confifts, are manifeftly inflected, though not fo confiderably as the waves of water; and founds are propagated with equal eafe, through crooked tubes, and through ftrait lines; but Light was never known to move in any curve, nor to inflect itfelf ad umbram."

It muft be acknowledged, however, that many philofophers, both Englifh and Foreigners, have recurred to the opinion, that Light confifts of vibrations propagated from the luminous body, through a fubtle etherial medium.

The ingenious Dr. Franklin, in a letter dated April 23, 1752, expreffes his diffatisfaction with the doctrine, that Light confifts of particles of matter continually driven off from the fun's furface, with fo enormous a fwiftnefs. "Muft not, fays he, the fmalleft portion conceivable, have, with fuch a motion, a force exceeding that of a 24 pounder difcharged from a cannon? Muft not the fun diminifh exceedingly by fuch a wafte of matter; and the planets, inftead of drawing nearer to him, as fome have feared, recede to greater diftances through the leffened attraction? Yet thefe particles, with this amazing motion, will not drive before them, or remove, the leaft and flighteft duft they meet with; and the fun appears to continue of his ancient dimenfions, and his attendants move in their ancient orbits." He therefore conjectures that all the phenomena of

Light

Light may be more properly folved, by fuppofing all fpace filled with a fubtle elaftic fluid, which is not vifible when at reft, but which, by its vibrations, affects that fine fenfe in the eye, as thofe of the air affect the groffer organs of the ear; and even that different degrees of the vibration of this medium may caufe the appearances of different colours. Franklin's Exper. and Obferv. 1769, pa. 264.

The celebrated Euler has alfo maintained the fame hypothefis, in his Theoria Lucis & Colorum. In the fummary of his arguments againft the common opinion, recited in Acad. Berl. 1752, pa. 271, befides the objections above-mentioned, he doubts the poffibility, that particles of matter, moving with the amazing velocity of Light, fhould penetrate tranfparent fubftances with fo much eafe. In whatever manner they are tranfmitted, thofe bodies muft have pores, difpofed in right lines, and in all poffible directions, to ferve as canals for the paffage of the rays: but fuch a ftructure muft take away all folid matter from thofe bodies, and all coherence among their parts, if they do contain any folid matter.

Doctor Horfley, now Bp. of Rochefter, has taken confiderable pains to obviate the difficulties ftarted by Dr. Franklin. Suppofing that the diameter of each particle of Light does not exceed one millionth of one millionth of an inch, and that the denfity of each particle is even three times that of iron, that the Light of the fun reaches the earth in 7', at the diftance of 22919 of the earth's femidiameters, he calculates that the momentum or force of motion in each particle of Light coming from the fun, is lefs than that in an iron ball of a quarter of an inch diameter, moving at the rate of lefs than an inch in 12 thoufand millions of millions of years. And hence he concludes, that a particle of matter, which probably is larger than any particle of Light, moving with the velocity of Light, has a force of motion, which, inftead of exceeding the force of a 24 pounder difcharged from a cannon, is almoft infinitely lefs than that of the fmalleft fhot difcharged from a pocket piftol, or lefs than any that art can create. He alfo thinks it poffible, that Light may be produced by a continual emiffion of matter from the fun, without any fuch wafte of his fubftance as fhould fenfibly contract his dimenfions, or alter the motions of the planets, within any moderate length of time. In proof of this, he obferves that, for the production of any of the phenomena of Light, it is not neceffary that the emanation from the fun fhould be continual, in a ftrict mathematical fenfe, or without any interval; and likewife that part of the Light which iffues from the fun, is continually returned to him by reflection from the planets, as well as other Light from the funs of other fyftems. He proceeds, by calculation, to fhew that in 385,130,000 years, the fun would lofe but the 13232d part of his matter, and confequently of the gravitation towards him, at any given diftance; which is an alteration much too fmall to difcover itfelf in the motion of the earth, or of any of the planets. He farther computes that the greateft ftroke which the retina of a common eye fuftains, when turned directly to the fun in a bright day, does not exceed that which would be given by an iron fhot, a quarter of an inch diameter, and moving only

at the rate of $16\frac{1}{6}$ inches in a year; whereas the ordinary ftroke is lefs than the 2084th part of this. See Philof. Tranf. vol. 60 and 61.

In anfwer to the difficulty refpecting the non-interference of the particles of Light with each other, Mr. Melville obferves (Edinb. Eff. vol. 2), there is probably no phyfical point in the vifible horizon, that does not fend rays to every other point, unlefs where opaque bodies interpofe. Light, in its paffage from one fyftem to another, often paffes through torrents of Light iffuing from other funs and fyftems, without ever interfering, or being diverted from its courfe, either by it, or by the particles of that elaftic medium, which it has been fuppofed by fome is diffufed through all the mundane fpace. To account for this fact, he fuppofes that the particles of Light are incomparably rare, even when they are the moft denfe, or that their diameters are incomparably lefs than their diftance from one another: which obviates the objection urged by Euler and others againft the materiality of Light, from its influence in difturbing the freedom and perpetuity of the celeftial motions. Bofcovich and fome others folve the difficulty concerning the non-interference of the particles of Light, by fuppofing that each particle is endued with an infuperable impulfive force; but in this cafe, their fpheres of impulfion would be more likely to interfere, and on that account they be more liable to difturb one another.

M. Canton fhews (Philof. Tranf. vol. 58, p. 344), that the difficulty of the interference will vanifh, if a very fmall portion of time be allowed between the emiffion of every particle and the next that follows in the fame direction. Suppofe, for inftance, that a lucid point in the fun's furface emits 150 particles in a fecond of time, which, he obferves, will be more than fufficient to give continual Light to the eye, without the leaft appearance of intermiffion; yet ftill the particles of fuch a ray, on account of their great velocity, will be more than 1000 miles behind each other, a fpace fufficient to allow others to pafs in all directions without any perceptible interruption. And if we adopt the conclufions drawn from the experiments on the duration of the fenfations excited by Light, by the chevalier D'Arcy, in the Acad. Scienc. 1765, who ftates it at the 7th part of a fecond, an interval of more than 20,000 miles may be admitted between every two fucceffive particles.

The doctrine of the materiality of Light is farther confirmed by thofe experiments, which fhew, that the colour and inward texture of fome bodies are changed by being expofed to the Light.

Of the Momentum, or Force, of the Particles of Light. Some writers have attempted to prove the materiality of Light, by determining the momentum of their component particles, or by fhewing that they have a force fo as, by their impulfe, to give motion to light bodies. M. Homberg, Ac. Par. 1708, Hift. pa. 25, imagined, that he could not only difperfe pieces of amianthus, and other light fubftances, by the impulfe of the folar rays, but alfo that by throwing them upon the end of a kind of lever, connected with the fpring of a watch, he could make it move fenfibly quicker; from which, and other experiments, he inferred the weight of the particles of Light. And Hartfoecker made pretenfions

fions of the fame nature. But M. Du Fay and M. Mairan made other experiments of a more accurate kind, without the effects which the former had imagined, and which even proved that the effects mentioned by them were owing to currents of heated air produced by the burning glaffes ufed in their experiments, or fome other caufes which they had overlooked.

However, Dr. Prieftley informs us, that Mr. Michell endeavoured to afcertain the momentum of Light with ftill greater accuracy, and that his endeavours were not altogether without fuccefs. Having found that the inftrument he ufed, acquired, from the impulfe of the rays of light, a velocity of an inch in a fecond of time, he inferred that the quantity of matter contained in the rays falling upon the inftrument in that time, amounted to no more than the 12 hundred millionth part of a grain. In the experiment, the Light was collected from a furface of about 3 fquare feet; and as this furface reflected only about the half of what fell upon it, the quantity of matter contained in the folar rays, incident upon a fquare foot and a half of furface, in a fecond of time, ought to be no more than the 12 hundred millionth part of a grain, or upon one fquare foot only, the 18 hundred millionth part of a grain. But as the denfity of the rays of Light at the furface of the fun, is 45000 times greater than at the earth, there ought to iffue from a fquare foot of the fun's furface, in one fecond of time, the 40 thoufandth part of a grain of matter; that is, a little more than 2 grains a day, or about 4,752,000 grains, which is about 670 pounds avoirdupois, in 6000 years, the time fince the creation; a quantity which would have fhortened the fun's femidiameter by no more than about 10 feet, if it be fuppofed of no greater denfity than water only.

The *Expanfion* or *Extenfion* of any portion of Light, is inconceivable. Dr. Hook fhews that it is as unlimited as the univerfe; which he proves from the immenfe diftance of many of the fixed ftars, which only become vifible to the eye by the beft telefcopes. Nor, adds he, are they only the great bodies of the fun or ftars that are thus liable to difperfe their Light through the vaft expanfe of the univerfe, but the fmalleft fpark of a lucid body muft do the fame, even the fmalleft globule ftruck from a fteel by a flint.

The *Intenfity* of different Lights, or of the fame Light in different circumftances, affords a curious fubject of fpeculation. M. Bouguer, Traité de Optique, found that when one Light is from 60 to 80 times lefs than another, its prefence or abfence will not be perceived by an ordinary eye; that the moon's Light, when fhe is 19° 16′ high above the horizon, is but about ⅓ of her Light at 66° 11′ high; and when one limb juft touched the horizon, her Light was but the 2000th part of her Light at 66° 11′ high; and that hence Light is diminifhed in the proportion of 3 to 1 by traverfing 7469 toifes of denfe air. He found alfo, that the centre of the fun's difc is confiderably more luminous than the edges of it; whereas both the primary and fecondary planets are more luminous at their edges than near their centres: That, farther, the Light of the fun is about 300,000 times greater than that of the moon; and therefore it is no wonder that philofophers have had fo little fuccefs in their attempts to collect the Light of the moon with burning-glaffes;

for, fhould one of the largeft of them even increafe the Light 1000 times, it will ftill leave the Light of the moon in the focus of the glafs, 300 times lefs than the intenfity of the common Light of the fun.

Dr. Smith, in his Optics, vol. 1, pa. 29, thought he had proved that the Light of the full moon would be only the 90,900th part of the full day Light, if no rays were loft at the moon. But Mr. Robins, in his Tracts, vol. 2, pa. 225, fhews that this is too great by one half. And Mr. Michell, by a more eafy and accurate mode of computation, found that the denfity of the fun's Light on the furface of the moon is but the 45,000th part of the denfity at the fun; and that therefore, as the moon is nearly of the fame apparent magnitude as the fun, if fhe reflected to us all the Light received on her furface, it would be only the 45,000th part of our day Light, or that which we receive from the fun. Admitting therefore, with M. Bouguer, that the moon Light is only the 300,000th part of the day or fun's Light, Mr. Michell concludes that the moon reflects no more than between the 6th and 7th part of what fhe receives.

Dr. Gravefande fays, a lucid body is that which emits or gives fire a motion in right lines, and makes the difference between Light and heat to confift in this, that to produce the former, the fiery particles muft enter the eye in a rectilinear motion, which is not required in the latter: on the contrary, an irregular motion feems more proper for it, as appears from the rays coming directly from the fun to the tops of mountains, which have not near that effect with thofe in the valley, agitated with an irregular motion, by feveral reflections.

Sir I. Newton obferves, that bodies and Light act mutually on one another; bodies on Light, in emitting, reflecting, refracting, and inflecting it; and Light on bodies, by heating them, and putting their parts into a vibrating motion, in which heat principally confifts. For all fixed bodies, he obferves, when heated beyond a certain degree, do emit Light, and fhine; which fhining &c appears to be owing to the vibrating motion of their parts; and all bodies, abounding in earthy and fulphureous particles, if fufficiently agitated, emit Light, which way foever that agitation be effected. Thus, fea water fhines in a ftorm; quickfilver, when fhaken in vacuo; cats or horfes, when rubbed in the dark; and wood, fifh, and flefh, when putrefied.

Light proceeding from putrefcent animal and vegetable fubftances, as well as from glow-worms, is mentioned by Ariftotle. And Bartholin mentions four kinds of luminous infects, two of which have wings: but in hot climates it is faid they are found in much greater numbers, and of different fpecies. Columna obferves, that their Light is not extinguifhed immediately on the death of the animal. The firft diftinct account that occurs of Light proceeding from putrefcent animal flefh, is that which is given by Fabricius ab Aquapendente in 1592, de Vifione &c, pa. 45. And Bartholin gives an account of a fimilar appearance, which happened at Montpelier in 1641, in his treatife De Luce Animalium.

Mr. Boyle fpeaks of a piece of fhining rotten wood, which was extinguifhed in vacuo; but upon re-admitting the air, it revived again, and fhone as before;

though

though he could not perceive that it was increafed in condenfed air. But in Birch's Hiftory of the Royal Soc. vol. 2, pa. 254, there is an account of the Light of a fhining fifh, which was rendered more vivid by putting the fifh into a condenfing engine. The fifh called Whitings were thofe commonly ufed by Mr. Boyle in his experiments: though in a difcourfe read before the R. Soc. in 1681, it was afferted that, of all fifhy fubftances, the eggs of lobfters, after they had been boiled, fhone the brighteft. Birch's Hift. vol. 2, pa. 70. In 1672 Mr. Boyle accidentally obferved Light iffuing from flefh meat; and, among other remarks on this fubject, he obferves that extreme cold extinguifhes the Light of fhining wood; probably becaufe extreme cold checks the putrefaction, which is the caufe of the Light. The fhell fifh called Pholas, is remarkable for its luminous quality. The *luminoufnefs of the Sea* has been alfo a fubject of frequent obfervation. See *Ignis fatuus*, *Phofphorus*, and *Putrefaction*, &c.

Mr. Hawkfbee, and many writers on the fubject of electricity fince his time, have produced a great variety of inftances of the artificial production of Light, by the attrition of bodies naturally not luminous; as of amber rubbed on woollen cloth in vacuo; of glafs on woollen, of glafs on glafs, of oyfter fhells on woollen, and of woollen on woollen, all in vacuo. On the feveral experiments of this kind, he makes thefe following reflections: that different forts of bodies afford Light of various kinds, different both in colour and in force; that the effects of an attrition are various, according to the different preparations and treatment of the bodies that are to endure it; and that bodies which have yielded a particular Light, may be brought by friction to yield no more of that Light.

M. Bernoulli found by experiment, that mercury amalgamated with tin, and rubbed on glafs, produced a confiderable Light in the air; that gold rubbed on glafs, exhibited the fame in a greater degree; but that the moft exquifite Light of all was produced by the attrition of a diamond, this being equally vivid with that of a burning coal brifkly agitated with the bellows. See ELECTRICITY, &c.

Of the Attraction of Light. That the particles of Light are attracted by thofe of other bodies, is evident from numerous experiments. This phenomenon was obferved by Sir I. Newton, who found, by repeated trials, that the rays of Light, in their paffage near the edges of bodies, are diverted out of the right lines, and always inflected or bent towards thofe bodies, whether they be opaque or tranfparent, as pieces of metals, the edges of knives, broken glaffes, &c. See INFLECTION and RAYS. The curious obfervations that had been made on this fubject by Dr. Hook and Grimaldi, led Sir I. Newton to repeat and diverfify their experiments, and to purfue them much farther than they had done. For a particular account of his experiment and obfervations, fee his treatife on Optics, pa. 293 &c.

This action of bodies on Light is found to exert itfelf at a fenfible diftance, though it always increafes as the diftance is diminifhed; as appears very fenfibly in the paffage of a ray between the edges of two thin planes at different apertures; which is attended with this peculiar circumftance, that the attraction of one edge is increafed as the other is brought nearer it.

The rays of Light, in their paffage out of glafs into a vacuum, are not only inflected towards the glafs, but if they fall too obliquely, they will revert back again to the glafs, and be totally reflected. Now the caufe of this reflection cannot be attributed to any refiftance of the vacuum, but muft be entirely owing to fome force or power in the glafs, which attracts or draws back the rays as they were paffing into the vacuum. And this appears farther from hence, that if you wet the back furface of the glafs with water, oil, honey, or a folution of quickfilver, then the rays which would otherwife have been reflected, will pervade and pafs through that liquor; which fhews that the rays are not reflected till they come to that back furface of the glafs, nor even tid they begin to go out of it; for if, at their going out, they fall into any of the aforefaid mediums, they will not then be reflected, but will perfift in their former courfe, the attraction of the glafs being in this cafe counterbalanced by that of the liquor.

M. Maraldi profecuted experiments fimilar to thofe of Sir I. Newton on inflected Light. And his obfervations chiefly refpect the inflection of Light towards other bodies, by which their fhadows are partially illuminated. Acad. Paris 1723, Mem. p. 159. See alfo Prieftley's Hift. pa. 521 &c.

M. Mairan, without attempting the difcovery of new facts, endeavoured to explain the old ones, by the hypothefis of an atmofphere furrounding all bodies; and confequently two reflections and refractions of Light that impinges upon them, one at the furface of the atmofphere, and the other at the furface of the body itfelf. This atmofphere he fuppofed to be of a variable denfity and refractive power, like the air.

M. Du Tour fucceeded Mairan, and imagined that he could account for all the phenomena by the help of an atmofphere of an uniform denfity, but of a lefs refractive power than the air furrounding all bodies. Du Tour alfo varied the Newtonian experiments, and difcovered more than three fringes in the colours produced by the inflection of light. He farther concludes that the refracting atmofpheres, furrounding all kinds of bodies, are of the fame fize; for when he ufed a great variety of fubftances, and of different fizes too, he always found coloured ftreaks of the fame dimenfions. He alfo obferves, that his hypothefis contradicts an obfervation of Sir I. Newton, viz, that thofe rays are the moft inflected which pafs the neareft to any body. Mem. de Math. & de Phyf. vol. 5, pa. 650, or Prieftley's Hift. pa. 531.

M. Le Cat found that objects fometimes appear magnified by means of the inflection of Light. Looking at a diftant fteeple, when a wire, of a lefs diameter than the pupil of his eye, was held pretty near to it, and drawing it feveral times between that object and his eye, he was furprifed to find that every time the wire paffed before his eye, the fteeple feemed to change its place, and fome hills beyond the fteeple feemed to have the fame motion, juft as if a lens had been drawn between them and his eye. This difcovery led him to feveral others depending on the inflection of the rays of Light. Thus, he magnified fmall objects, as the head of a pin, by viewing them through a fmall hole in a card; fo that the rays which formed the image muft
necesfarily

necessarily pass so near the circumference of the hole, as to be attracted by it. He exhibited also other appearances of a similar nature. Traite des Sens, pa. 299. Priestley ubi supra, pa 537.

Reflection and Refraction of Light. From the mutual attraction between the particles of Light and other bodies, arise two other grand phenomena, besides the inflection of Light, which are called the reflection and refraction of Light. It is well known that the determination of bodies in motion, especially elastic ones, is changed by the interposition of other bodies in their way: thus also Light, impinging on the surfaces of bodies, should be turned out of its course, and beaten back or reflected, so as, like other striking bodies, to make the angle of its reflection equal to the angle of incidence. This, it is found by experience, Light does; and yet the cause of this effect is different from that just now assigned: for the rays of Light are not reflected by striking on the very parts of the reflecting bodies, but by some power equally diffused over the whole surface of the body, by which it acts on the Light, either attracting or repelling it, without contact: by which same power, in other circumstances, the rays are refracted; and by which also the rays are first emitted from the luminous body; as Newton abundantly proves by a great variety of arguments. See REFLECTION and REFRACTION.

That great author puts it past doubt, that all those rays which are reflected, do not really touch the body; though they approach it infinitely near; and that those which strike on the parts of solid bodies, adhere to them, and are as it were extinguished and lost. Since the reflection of the rays is ascribed to the action of the whole surface of the body without contact, if it be asked, how it happens that all the rays are not reflected from every surface; but that, while some are reflected, others pass through, and are refracted? the answer given by Newton is as follows:—Every ray of Light, in its passage through any refracting surface, is put into a certain transient constitution or state, which in the progress of the ray returns at equal intervals, and disposes the ray at every return to be easily transmitted through the next refracting surface, and between the returns to be easily reflected by it: which alteration of reflection and transmission it appears is propagated from every surface, and to all distances. What kind of action or disposition this is, and whether it consists in a circulating or vibrating motion of the ray, or the medium, or something else, he does not enquire; but allows those who are fond of hypotheses to suppose, that the rays of Light, by impinging on any reflecting or refracting surface, excite vibrations in the reflecting or refracting medium, and by that means agitate the solid parts of the body. These vibrations, thus produced in the medium, move faster than the rays, so as to overtake them; and when any ray is in that part of the vibration which conspires with its motion, its velocity is increased, and so it easily breaks through a refracting surface; but when it is in a contrary part of the vibration, which impedes its motion, it is easily reflected; and thus every ray is successively disposed to be easily reflected or transmitted by every vibration which meets it. These returns in the disposition of any ray to be reflected, he calls *fits of easy reflection;* and the returns

in the disposition to be transmitted, he calls *fits of easy transmission;* also the space between the returns, *the interval of the fits.* Hence then the reason why the surfaces of all thick transparent bodies reflect part of the Light incident upon them, and refract the rest, is that some rays at their incidence are in fits of easy reflection, and others of easy transmission. For *the properties of reflected Light,* see REFLECTION, MIRROR, &c.

Again, a ray of Light, passing out of one medium into another of different density, and in its passage making an oblique angle with the surface that separates the mediums, will be refracted, or turned out of its direction; because the rays are more strongly attracted by a denser than by a rarer medium. That these rays are not refracted by striking on the solid parts of bodies, but that this is effected without a real contact, and by the same force by which they are emitted and reflected, only exerting itself differently in different circumstances, is proved in a great measure by the same arguments by which it is demonstrated that reflection is performed without contact. See REFRACTION, LENS, COLOUR, VISION, &c.

LIGHTNING, a large bright flame, shooting swiftly through the atmosphere, of momentary or very short duration, and commonly attended with thunder.

Some philosophers accounted for this awful natural phenomenon in this manner, viz, that an inflammable substance is formed of the particles of sulphur, nitre, and other combustible matter, which are exhaled from the earth, and carried into the higher regions of the atmosphere, and that by the collision of two clouds, or otherwise, this substance takes fire, and darts out into a train of Light, larger or smaller according to the strength and quantity of the materials. And others have explained the phenomenon of Lightning by the fermentation of sulphureous substances with nitrous acids. See THUNDER.

But it is now universally allowed, that Lightning is really an electrical explosion or phenomenon. Philosophers had not proceeded far in their experiments and enquiries on this subject, before they perceived the obvious analogy between Lightning and electricity, and they produced many arguments to evince their similarity. But the method of proving this hypothesis beyond a doubt, was first proposed by Dr Franklin, who, about the close of the year 1749, conceived the practicability of drawing Lightning down from the clouds. Various circumstances of resemblance between Lightning and electricity were remarked by this ingenious philosopher, and have been abundantly confirmed by later discoveries, such as the following: Flashes of Lightning are usually seen crooked and waving in the air; so the electric spark drawn from an irregular body at some distance, and when it is drawn by an irregular body, or through a space in which the best conductors are disposed in an irregular manner, always exhibits the same appearance: Lightning strikes the highest and most pointed objects in its course, in preference to others, as hills, trees, spires, masts of ships, &c; so all pointed conductors receive and throw off the electric fluid more readily than those that are terminated by flat surfaces: Lightning is observed to take and follow the readiest and best conductor; and the same is the case with electricity in the discharge of the Leyden phial;

8

phial; from whence the doctor infers, that in a thunder-storm, it would be safer to have one's cloaths wet than dry: Lightning burns, diffolves metals, rends fome bodies, fometimes ftrikes perfons blind, deftroys animal life, deprives magnets of their virtue, or reverfes their poles; and all thefe are well-known properties of electricity.

But Lightning alfo gives polarity to the magnetic needle, as well as to all bodies that have any thing of iron in them, as bricks &c; and by obferving afterwards which way the magnetic poles of thefe bodies lie, it may thence be known in what direction the ftroke paffed. Perfons are fometimes killed by Lightning, without exhibiting any vifible marks of injury; and in this cafe Sig. Beccaria fuppofes that the Lightning does not really touch them, but only produces a fudden vacuum near them, and the air rufhing violently out of their lungs to fupply it, they cannot recover their breath again: and in proof of this opinion he alleges, that the lungs of fuch perfons are found flaccid; whereas thefe are found inflated when the perfons are really killed by the electric fhock. Though this hypothefis is controverted by Dr. Prieftley.

To demonftrate however, by actual experiment, the identity of the electric fluid with the matter of Lightning, Dr. Franklin contrived to bring Lightning from the heavens, by means of a paper kite, properly fitted up for the purpofe, with a long fine wire ftring, and called an electrical kite, which he raifed when a thunder-ftorm was perceived to be coming on: and with the electricity thus obtained, he charged phials, kindled fpirits, and performed all other fuch electrical experiments as are ufually exhibited by an excited glafs globe or cylinder. This happened in June 1752, a month after the electricians in France, in purfuance of the method which he had before propofed, had verified the fame theory, but without any knowledge of what they had done. The moft active of thefe were Meffrs. Dalibard and Delor, followed by M. Mazeas and M. Monnier.

In April and June 1753, Dr. Franklin difcovered that the air is fometimes electrified negatively, as well as fometimes pofitively; and he even found that the clouds would change from pofitive to negative electricity feveral times in the courfe of one thunder-guft. This curious and important difcovery he foon perceived was capable of being applied to practical ufe in life, and in confequence propofed a method, which he foon accomplifhed, of fecuring buildings from being damaged by Lightning, by means of CONDUCTORS. See the word.

Nor had the Englifh philofophers been inattentive to this fubject: but, for want of proper opportunities of trying the neceffary experiments, and from fome other unfavourable circumftances, they had failed of fuccefs. Mr. Canton, however, fucceeded in July 1752; and in the following month Dr. Bevis and Mr. Wilfon obferved near the fame appearances as Mr. Canton had done before. By a number of experiments Mr. Canton alfo foon after obferved that fome clouds were in a pofitive, while fome were in a negative ftate of electricity; and that the electricity of his conductor would fometimes change, from one ftate to the other, five or fix times in lefs than half an hour.

But Sig. Beccaria difcovered this variable ftate of thunder clouds, before he knew that it had been obferved by Dr. Franklin or any other perfon; and he has given a very exact and particular account of the external appearances of thefe clouds. From the obfervations of his apparatus within doors, and of the Lightning abroad, he inferred, that the quantity of electric matter in a common thunder ftorm, is inconceivably great, confidering how many pointed bodies, as fpires, trees, &c, are continually drawing it off, and what a prodigious quantity is repeatedly difcharged to or from the earth. This matter is in fuch abundance, that he thinks it impoffible for any cloud or number of clouds to contain it all, fo as either to receive or difcharge it. He obferves alfo, that during the progrefs and increafe of the ftorm, though the lightning frequently ftruck to the earth, the fame clouds were the next moment ready to make a ftill greater difcharge, and his apparatus continued to be as much affected as ever; fo that the clouds muft have received at one part, in the fame moment when a difcharge was made from them in another. And from the whole he concludes, that the clouds ferve as conductors to convey the electric fluid from thofe parts of the earth that are overloaded with it, to thofe that are exhaufted of it. The fame caufe by which a cloud is firft raifed, from vapours difperfed in the atmofphere, draws to it thofe that are already formed, and ftill continues to form new ones, till the whole collected mafs extends fo far as to reach a part of the earth where there is a deficiency of the electric fluid, and where the electric matter will difcharge itfelf on the earth. A channel of communication being thus formed, a frefh fupply of electric matter is raifed from the overloaded part, which continues to be conveyed by the medium of the clouds, till the equilibrium of the fluid is reftored between the two places of the earth. Sig. Beccaria obferves, that a wind always blows from the place from which the thunder-cloud proceeds; and it is plain that the fudden accumulation of fuch a prodigious quantity of vapours muft difplace the air, and repel it on all fides. Indeed many obfervations of the defcent of Lightning, confirm his theory of the manner of its afcent; for it often throws before it the parts of conducting bodies, and diftributes them along the refifting medium, through which it muft force its paffage; and upon this principle the longeft flafhes of Lightning feem to be made, by forcing into its way part of the vapours in the air. One of the chief reafons why thefe flafhes make fo long a rumbling, is that they are occafioned by the vaft length of a vacuum made by the paffage of the electric matter: for although the air collapfes the moment after it has paffed, and that the vibration, on which the found depends, commences at the fame moment; yet when the flafh is directed towards the perfon who hears the report, the vibrations excited at the nearer end of the track, will reach his ear much fooner than thofe from the more remote end; and the found will, without any echo or repercuffion, continue till all the vibrations have fucceffively reached him.

How it happens that particular parts of the earth, or the clouds, come into the oppofite ftates of pofitive and negative electricity, is a queftion not abfolutely determined: though it is eafy to conceive that when particular clouds, or different parts of the earth, poffefs oppofite poſite electricities, a diſcharge will take place within a certain diſtance: or the one will ſtrike into the other, and in the diſcharge a flaſh of Lightning will be ſeen. Mr. Canton queries whether the clouds do not become poſſeſſed of electricity by the gradual heating and cooling of the air; and whether air ſuddenly rarefied, may not give electric fire to clouds and vapours paſſing through it, and air ſuddenly condenſed receive electric fire from them.——Mr. Wilcke ſuppoſes, that the air contracts its electricity in the ſame manner that ſulphur and other ſubſtances do, when they are heated and cooled in contact with various bodies. Thus, the air being heated or cooled near the earth, gives electricity to the earth, or receives it from it; and the electrified air, being conveyed upwards by various means, communicates its electricity to the clouds.—Others have queried, whether, ſince thunder commonly happens in a ſultry ſtate of the air, when it ſeems charged with ſulphureous vapours, the electric matter then in the clouds may not be generated by the fermentation of ſulphureous vapours with mineral or acid vapours in the air.

With regard to places of ſafety in times of thunder and Lightning, Dr. Franklin's advice is, to ſit in the middle of a room, provided it be not under a metal luſtre ſuſpended by a chain, ſitting on one chair, and laying the feet on another. It is ſtill better, he ſays, to bring two or three mattreſſes or beds into the middle of the room, and folding them double, to place the chairs upon them; for as they are not ſo good conductors as the walls, the Lightning will not be ſo likely to paſs through them: but the ſafeſt place of all, is in a hammock hung by ſilken cords, at an equal diſtance from all the ſides of a room. Dr. Prieſtley obſerves, that the place of moſt perfect ſafety muſt be the cellar, and eſpecially the middle of it; for when a perſon is lower than the ſurface of the earth, the Lightning muſt ſtrike it before it can poſſibly reach him. In the fields, the place of ſafety is within a few yards of a tree, but not quite near it. Beccaria cautions perſons not always to truſt too much to the neighbourhood of a higher or better conductor than their own body; ſince he has repeatedly found that the Lightning by no means deſcends in one undivided track, but that bodies of various kinds conduct their ſhare of it at the ſame time, in proportion to their quantity and conducting power. See Franklin's Letters, Beccaria's Lettre dell' Ellettriceſſimo, Prieſtley's Hiſt. of Electric., and Lord Mahon's Principles of Electricity.

Lord Mahon obſerves that damage may be done by Lightning, not only by the main ſtroke and lateral exploſion, but alſo by what he calls the returning ſtroke; by which is meant the ſudden violent return of that part of the natural ſhare of electricity which had been gradually expelled from ſome body or bodies, by the ſuperinduced elaſtic electrical preſſure of the electrical atmoſphere of a thunder cloud.

Artificial LIGHTNING, an imitation of real or natural Lightning by gunpowder, aurum fulminans, phoſphorus, &c, but eſpecially the laſt, between which and Lightning there is much more reſemblance than the others.

Phoſphorus, when newly made, gives a ſort of artificial Lightning viſible in the dark, which would ſur-

priſe thoſe not uſed to ſuch a phenomenon. It is uſual to keep this preparation under water; and if it is deſired to ſee the corruſcations to the greateſt advantage, it ſhould be kept in a deep cylindrical glaſs, not more than three quarters filled with water. At times the phoſphorus will ſend up corruſcations, which will pierce through the incumbent water, and expand themſelves with great brightneſs in the upper or empty part of the glaſs, and much reſembling Lightning. The ſeaſon of the year, as well as the newneſs of the phoſphorus, muſt concur to produce theſe flaſhes; for they are as common in winter as Lightning is, though both are very frequent in warm weather. The phoſphorus, while burning, acts the part of a corroſive, and when it goes out reſolves into a menſtruum, which diſſolves gold, iron, and other metals; and Lightning, in like manner, melts the ſame ſubſtances.

LIKE Quantities, or *Similar Quantities*, in Algebra, are ſuch as are expreſſed by the ſame letters, to the ſame power, or equally repeated in each quantity; though the numeral coefficients may be different.

Thus $4a$ and $5a$ are Like quantities,

as are alſo $3a^2$ and $12a^2$,

and alſo $6bxy^2$ and $10bxy^2$.

But $4a$ and $5b$, or $3a^2b$ and $10a^2b^2$, &c, are unlike quantities; becauſe they have not every where the ſame dimenſions, nor are the letters equally repeated. —Like quantities can be united into one quantity, by addition or ſubtraction; but unlike quantities can only be added or ſubtracted by placing the ſigns of theſe operations between them.

LIKE *Signs*, in Algebra, are the ſame ſigns, either both poſitive or both negative. But when one is poſitive and the other negative, they are unlike ſigns.

So, $+ 3ab$ and $+ 5cd$ have Like ſigns,

as have alſo $- 2a^2c$ and $- 2ax^2$;

but $+ 3ab$ and $- 5cd$ have unlike ſigns,

as alſo $- 2ax$ and $3ax$.

LIKE *Figures*, or *Arches*, &c, are the ſame as *Similar* figures, arches, &c. See SIMILAR.

All Like figures have their homologous lines in the ſame ratio. Alſo Like plane figures are in the duplicate ratio, or as the ſquares of their homologous lines or ſides; and Like ſolid figures are in the triplicate ratio, or as the cubes of their homologous lines or ſides.

LILLY (WILLIAM), a noted Engliſh aſtrologer, born in Leiceſterſhire in 1602. His father was not able to give him farther education than common reading and writing; but young Lilly being of a forward temper, and endued with ſhrewd wit, he reſolved to puſh his fortune in London; where he arrived in 1620, and, for a preſent ſupport, articled himſelf as a ſervant to a mantua-maker in the pariſh of St. Clement Danes. But in 1624 he moved a ſtep higher, by entering into the ſervice of Mr. Wright in the Strand, maſter of the Salters company, who not being able to write, Lilly among other offices kept his books. On the death of his maſter, in 1627, Lilly paid his addreſſes to the widow, whom he married with a fortune of 1000l. Being now his own maſter, he followed the bent of his inclinations, which led him to follow the puritanical preachers. Afterwards, turning his mind to judicial aſtrology, in 1632 he became pupil, in that art, to one Evans, a

profligate

profligate Welfh parfon; and the next year gave the public a fpecimen of his fkill, by an intimation that the king had chofen an unlucky horofcope for the coronation in Scotland. In 1634, getting a manufcript copy of the *Ars Noticia* of Cornelius Agrippa, with alterations, he drank in the doctrine of the magic circle, and the invocation of fpirits, with great eagernefs, and practifed it for fome time; after which he treated the myftery of recovering ftolen goods, &c, with great contempt, claiming a fupernatural fight, and the gift of prophetical predictions; all which he well knew how to turn to good advantage.

Mean while, he had buried his firft wife, purchafed a moiety of 13 houfes in the Strand, and married a fecond wife, who, joining to an extravagant temper a termagant fpirit, which he could not lay, made him unhappy, and greatly reduced his circumftances. With this uncomfortable yokemate he removed, in 1636, to Herfham in Surrey, where he ftaid till 1641; when, feeing a profpect of fifhing in troubled waters, he returned to London. Here having purchafed feveral curious books in this art, which were found on pulling down the houfe of another aftrologer, he ftudied them inceffantly, finding out fecrets contained in them, which were written in an imperfect Greek character; and, in 1644, publifhed his *Merlinus Anglicus*, an almanac, which he continued annually till his death, and feveral other aftrological works; devoting his pen, and other labours, fometimes to the king's party, and fometimes to that of the parliament, but moftly to the latter, raifing his fortune by favourable predictions to both parties, fometimes by prefents, and fometimes by penfions: thus, in 1648, the council of ftate gave him in money 50l. and a penfion of 100l. per annum, which he received for two years, and then refigned it on fome difguft. By his advice and contrivance, the king attempted feveral times to make his efcape from his confinement: he procured and fent the aqua-fortis and files to cut the iron bars of his prifon windows at Carifbrook caftle; but ftill advifing and writing for the other party at the fame time. Mean while he read public lectures on aftrology, in 1648 and 1649, for the improvement of young ftudents in that art; and in fhort, plied his bufinefs fo well, that in 1651 and 1652 he laid out near 2000l. for lands and a houfe at Herfham.

During the fiege of Colchefter, he and Booker were fent for thither, to encourage the foldiers; which they did by affuring them that the town would foon be taken; which proved true in the event.—Having, in 1650, written publicly that the parliament fhould not continue, but a new government arife; agreeably to which, in his almanac for 1653, he afferted that the parliament ftood upon a ticklifh foundation, and that the commonalty and foldiery would join together againft them. Upon which he was fummoned before the committee of plundered minifters; but, receiving notice of it before the arrival of the meffenger, he applied to his friend Lenthal the fpeaker, who pointed out the offenfive paffages. He immediately altered them; attended the committee next morning, with 6 copies printed, which fix alone he acknowledged to be his; and by that means came off with only 13 days cuftody by the ferjeant at arms. This year he was engaged in a difpute with Mr. Thomas Gataker.—In 1665 he was

indicted at Hicks's-hall, for giving judgment upon ftolen goods; but was acquitted. And in 1659, he received, from the king of Sweden, a prefent of a gold chain and medal, worth about 50l. on account of his having mentioned that monarch with great refpect in his almanacs of 1657 and 1658.—After the Reftoration, in 1660, being taken into cuftody, and examined by a committee of the houfe of commons, touching the execution of Charles the 1ft, he declared, that Robert Spavin, then Secretary to Cromwell, dining with him foon after the fact, affured him it was done by cornet Joyce. The fame year he fued out his pardon under the broad feal of England; and afterwards continued in London till 1665; when, upon the raging of the plague there, he retired to his eftate at Herfham. Here he applied himfelf to the ftudy of phyfic, having, by means of his friend Elias Afhmole, procured from archbifhop Sheldon a licence to practife it, which he did, as well as aftrology, from thence till the time of his death. —In October 1666 he was examined before a committee of the houfe of commons concerning the fire of London, which happened in September that year. A little before his death, he adopted for his fon, by the name of *Merlin junior*, one Henry Coley, a taylor by trade; and at the fame time gave him the impreffion of his almanac, which had been printed for 36 years fucceffively. This Coley became afterwards a celebrated aftrologer, publifhing in his own name, almanacs, and books of aftrology, particularly one intitled *A Key to Aftrology.*

Lilly died of a palfy 1681, at 79 years of age; and his friend Mr. Afhmole placed a monument over his grave in the church of Walton upon Thames.

Lilly was author of many works. His *Obfervations on the Life and Death of Charles late King of England,* if we overlook the aftrological nonfenfe, may be read with as much fatisfaction as more celebrated hiftories; Lilly being not only very well informed, but ftrictly impartial. This work, with the Lives of Lilly and Afhmole, written by themfelves, were publifhed in one volume, 8vo, in 1774, by Mr. Burman. His other works were principally as follow:

1. Merlinus Anglicus junior.—2. Supernatural Sight. —3. The White King's Prophecy.—4. England's Prophetical Merlin: all printed in 1644.—5. The Starry Meffenger, 1645.—6. Collection of Prophecies, 1646.—7. A Comment on the White King's Prophecy, 1646.—8. The Nativities of Archbifhop Laud and Thomas earl of Strafford, 1646.—9. Chriftian Aftrology, 1647: upon this piece he read his lectures in 1648, mentioned above.—10. The third book of Nativities, 1647.—11. The World's Cataftrophe, 1647.— 12. The Prophecies of Ambrofe Merlin, with a Key, 1647.—13. Trithemius, or the Government of the World by Prefiding Angels, 1647.—14. A treatife of the Three Suns feen in the winter of 1647, printed in 1648.—15. Monarchy or no Monarchy, 1651.— 16. Obfervations on the Life and Death of Charles, late king of England, 1651; and again in 1651, with the title of Mr. William Lilly's True Hiftory of king James and king Charles the 1ft, &c.—17. Annus Tenebrofus; or, the Black Year. This drew him into the difpute with Gataker, which Lilly carried on in his Almanac in 1654.

LIMB,

LIMB, the outermoft border, or graduated edge, of a quadrant, aftrolabe, or fuch like mathematical inftrument.

The word is alfo ufed for the arch of the primitive circle, in any projection of the fphere in plano.

LIMB alfo fignifies the outermoft border or edge of the fun or moon; as the upper Limb, or edge; the lower Limb; the preceding Limb, or fide; the following Limb.—Aftronomers obferve the upper or lower Limb of the fun or moon, to find their true height, or that of the centre, which differs from the others by the femidiameter of the difc.

LIMBERS, in Artillery, a fort of advanced train, joined to the carriage of a cannon on a march. It is compofed of two fhafts, wide enough to receive a horfe between them, called the *fillet horfe*: thefe fhafts are joined by two bars of wood, and a bolt of iron at one end, and mounted on a pair of rather fmall wheels. Upon the axle-tree rifes a ftrong iron fpike, which is put into a hole in the hinder part of the train of the gun-carriage, to draw it by. But when a gun is in action, the Limbers are taken off, and run out behind it.—See the dimenfions and figure of it in Müller's Treatife of Artillery, pa. 187.

LIMIT, is a term ufed by mathematicians, for fome determinate quantity, to which a variable one continually approaches, and may come nearer to it than by any given difference, but can never go beyond it; in which fenfe a circle may be faid to be the Limit of all its infcribed and circumfcribed polygons: becaufe thefe, by increafing the number of their fides, can be made to be nearer equal to the circle than by any fpace that can be propofed, how fmall foever it may be.

In Algebra, the term *Limit* is applied to two quantities, of which the one is greater and the other lefs than fome middle quantity, as the root of an equation, &c. And in this fenfe it is ufed when fpeaking of the Limits of equations, a method by which their folution is greatly facilitated.

LIMIT *of Diftinct Vifion*, in Optics. See *Diftinct Vifion*.

LIMIT *of a Planet*, has been fometimes ufed for its greateft heliocentric latitude.

LIMITED *Problem*, denotes a problem that has but one folution, or fome determinate number of folutions: as to defcribe a circle through three given points that do not lie in a right line, which is limited to one folution only; to divide a parallelogram into two equal parts by a line parallel to one fide, which admits of two folutions, according as the line is parallel to the length or breadth of the parallelogram; or to divide a triangle in any ratio by a line parallel to one fide, which is limited to three folutions, as the line may be parallel to any of the three fides.

LINE, in Geometry, a quantity extended in length only, without either breadth or thicknefs.

A Line is fometimes confidered as generated by the flux or motion of a point; and fometimes as the limit or termination of a fuperficies, but not as any part of that furface, however fmall.

Lines are either *right* or *curved*. A *right*, or ftraight Line, is the neareft diftance between two points, which are its extremes or ends; or it is a Line which has in every part of it the fame direc-

tion or pofition. But a *curve Line* has in every part of it a different direction, and is not the fhorteft diftance between its extremes or ends.

Right LINES are all of the fame fpecies; but curves are of an infinite number of different forts. As many may be conceived as there are different compound motions, or as many as there may be different relations between their ordinates and abfciffes. See CURVES.

Again, *Curve* LINES are ufually divided into *geometrical* and *mechanical*.

Geometrical Lines, are thofe which may be found exactly in all their parts. See GEOMETRICAL LINE.

Mechanical Lines are fuch as are not determined exactly in all their parts, but only nearly, or tentatively. But

Des Cartes, and his followers, define geometrical Lines to be thofe which may be exprefled by an algebraical equation of a determinate or finite degree; called its *locus*. And mechanical Lines, fuch as cannot be exprefled by fuch an equation.

But others diftinguifh the fame Lines by the name *algebraical* and *tranfcendental*.

Lines are alfo divided into orders, by Newton, according to the number of interfections which may be made by them and a right Line, viz, the 1ft, 2d, 3d, 4th, &c, order, according as they may be cut by a right Line, in 1, or 2, or 3, or 4, &c, points. In this way of confidering them, the right Line only is of the 1ft order, being but one in number; the 2d order contains 4 curves only, being fuch as may be cut from a cone by a plane, viz, the circle, the ellipfe, the hyperbola, and the parabola; the lines of the 3d order have been enumerated by Newton, in a particular treatife, who makes their number amount to 72; but Mr. Stirling found 4 others, and Mr. Stone 2 more; though it is difputed by fome whether thefe 2 laft ought to be accounted different from fome of Newton's, or not. See Newton's Enumer. Lin. Tertii Ordin. alfo Stirling's Lineæ Tert. Ordin. Newtonianæ Oxon. 1717, 8vo. and Philof. Tranf. number 456, &c. Again,

Algebraical Lines are divided into different orders according to the power or degree of their equations. So, the fimple equation $a + by + cx = 0$ or equation of the 1ft degree, denotes the 1ft order or right line; the equation $a + by + cx + dyy + exy + fxx = 0$, of the 2d degree, denotes the Lines of the 2d order; and the equation

$$a + by + cx + dyy + exy + fxx + gy^3 + bxy^2 + ix^2y + bx^3 = 0$$

of the 3d degree, exprefles the Lines of the 3d order; and fo on. See Cramer's Introd. à l'Analyfe des Lignes Courbes.

Lines, confidered as to their pofitions, are either *parallel*, *perpendicular*, or *oblique*. And the conftruction and properties of each of thefe, fee under the refpective terms.

LINE alfo denotes a French meafure of length, being the 12th part of an inch, or the 144th part of a foot.

In *Aftronomy*,

LINE *of the Apfes*, or *Apfides*, the Line joining the two apfes, or the longer axis of the orbit of a planet.

Fiducial Line, the index line or edge of the ruler, which paffes through the middle of an aftrolabe, or other inftrument, on which the fights are fitted, and marking the divifions.

Horizontal

Horizontal Line, a Line parallel to the horizon.

LINE *of the Nodes*, that which joins the nodes of the orbit of a planet, being the common section of the plane of the orbit with the plane of the ecliptic.

In *Dialling*,

Horizontal Line, is the common section of the horizon and the dial-plate.

Horary, or *Hour Lines*, are the common intersections of the hour-circles of the sphere with the plane of the dial.

Equinoctial Line is the common intersection of the equinoctial and the plane of the dial.

In *Fortification*, *Line* is sometimes used for a ditch, bordered with its parapet: and sometimes for a row of gabions, or sacks of earth, extended lengthwife on the ground, to serve as a shelter against the enemy's fire.

When the trenches were carried on within 30 paces of the glacis, they drew two Lines, one on the right, and the other on the left, for a place of arms.

Lines are commonly made to shut up an avenue or entrance to some place; the sides of the entrance being covered by rivers, woods, mountains, morasses, or other obstructions, not easy to be passed over by an army. When they are constructed in an open country, they are carried round the place to be defended, and resemble the Lines surrounding a camp, called Lines of circumvallation. Lines are also thrown up to stop the progress of an army; but the term is most used for the Line which covers a pass that can only be attacked in front.

When lines are made to cover a camp, or a large tract of land, where a considerable body of troops is posted, the work is not made in one straight, or uniformly bending Line; but, at certain distances, the Lines project in saliant angles, called redents, redans, or flankers, towards the enemy. The distance between these angles is commonly between the limits of 200 and 260 yards; the ordinary flight of a musket ball, point blank, being commonly within those limits; though muskets a little elevated will do effectual service at the distance of 360 yards.

Fundamental Line, is the first Line drawn for the plan of a place, and which shews its area.

Central Line, is the Line drawn from the angle of the centre to the angle of the bastion.

Line of Defence, &c. See DEFENCE &c.

Line of Approach, or *Attack*, signifies the work which the besiegers carry on under cover, to gain the moat, and the body of the place.

Line of Circumvallation, is a Line or trench cut by the besiegers, within cannon-shot of the place, which ranges round the camp, and secures its quarters against any relief to be brought to the besieged.

Line of Contravallation, is a ditch bordered with a parapet, serving to cover the besiegers on the side next the place, and to stop the sallies of the garrison.

Lines of Communication are those which run from one work to another.

Line of the Base, is that which joins the points of the two nearest bastions.

To *Line* a work, signifies to face it, as with brick or stone; for example, to strengthen a rampart with a firm wall, or to encompass a parapet or moat with good turf, &c.

LINE, in Geography and Navigation, is emphatically used for the Equator or Equinoctial Line.

The seamen use to baptize their fresh men, and passengers, the first time they cross the Line: that is, to dip them in the sea, suspended by a rope from the yardarm, unless they compound for it, by giving something to drink.

In *Perspective*,

The *Geometrical Line*, is a right Line drawn in any manner on the geometrical plane.

Terrestrial or *Fundamental Line*, is the common intersection of the geometrical plane and plane of the picture.

Line of the Front, is any Line parallel to the terrestrial Line.

Vertical Line, is the section of the vertical and draft planes.

Visual Line, is the Line or ray conceived to pass from the object to the eye.

Objective Line, is any Line drawn on the geometrical plane, whose representation is sought for in the draught or picture.

Line of Measures, is used by Oughtred, and others, to denote the diameter of the primitive circle, in the projection of the sphere in plano, or that Line in which falls the diameter of any circle to be projected.

LINEAR NUMBERS, are such as have relation to length only; such, for example, as express one side of a plane figure; and when the plane figure is a square, the linear number is called a root.

LINEAR PROBLEM, is one that can be solved geometrically by the intersection of two right lines. This is called a simple problem, and is capable of only one solution.

LIQUID, a fluid which wets or smears such bodies as are immersed in it, arising from some configuration of its particles, which disposes them to adhere to the surfaces of bodies contiguous to them. Thus water, oil, milk, &c, are Liquids, as well as fluids; but quicksilver is not a Liquid, but simply a fluid.

LISLE (WILLIAM DE), a very learned French geographer, was born at Paris in 1675. His father being much occupied in the same way, young Lisle began at 9 years of age to draw maps, and soon made a great progress in this art. In 1699 he first distinguished himself to the public, by giving a map of the world, and other pieces, which procured him a place in the Academy of Sciences, 1702. He was afterwards appointed geographer to the king, with a pension, and had the honour of instructing the king himself in geography, for whose particular use he drew up several works. De Lisle's reputation was so great, that scarcely any history or travels came out without the embellishment of his maps. Nor was his name less celebrated abroad than in his own country. Many sovereigns in vain attempted to draw him out of France. The Czar Peter, when at Paris on his travels, paid him a visit, to communicate to him some remarks upon Muscovy; but more especially, says Fontenelle, to learn from him, better than he could any where else, the extent and situation of his own dominions. De Lisle died of an apoplexy in 1726, at 51 years of age. Beside the excellent maps he published, he wrote

many

many pieces in the Memoirs of the Academy of Sciences.

LIST, or LISTEL, a small square moulding, serving to crown or accompany larger mouldings; or on occasion to separate the flutings of columns.

LITERAL ALGEBRA. See ALGEBRA.

LIZARD, in Aftronomy. See LACERTA.

LOADSTONE, or MAGNET; which fee.

LOCAL *Problem*, is one that is capable of an infinite number of different folutions; becaufe the point, which is to folve the problem, may be indifferently taken within a certain extent; as fuppofe any where in fuch a line, within fuch a plane figure, &c, which is called a *geometrical Locus*.

A Local problem is *fimple*, when the point fought is in a right line; *plane*, when the point fought is in the circumference of a circle; *folid*, when it is in the circumference of a conic fection; or *furfolid*, when the point is in the perimeter of a line of a higher kind.

LOCAL MOTION, or *Loco-Motion*, the change of place: See MOTION.

LOCI, the plural of LOCUS, which fee.

LOCUS, is fome line by which a local or indeterminate problem is folved; or a line of which any point may equally folve an indeterminate problem.

Loci are expreffed by algebraic equations of different orders according to the nature of the Locus. If the equation is conftructed by a right line, it is called *Locus ad rectum*; if by a circle, *Locus ad circulum*; if by a parabola, *Locus ad parabolam*; if by an ellipfis, *Locus ad ellipfim*; and fo on.

The Loci of fuch equations as are right lines or circles, the ancients called *plane loci*; and of thofe that are conic fections, *folid loci*; but fuch as are curves of a higher order, *furfolid loci*. But the moderns diftinguifh the Loci into orders according to the dimenfions of the equations by which they are expreffed, or the number of the powers of indeterminate or unknown quantities in any one term: thus, the equation $ay = bx + c$ denotes a Locus of the 1ft order, but $y^2 = ax$, or $= ax - x^2$, &c, a Locus of the 2d order, and $y^3 = a^2x$, or $= ax^2 - x^3$, &c, a Locus of the 3d order, and fo on; where x and y are unknown or indeterminate quantities, and the others known or determinate ones; alfo x denotes the abfcifs, and y the ordinate of the curve or line which is the Locus of the equation.

For inftance, fuppofe two variable or indeterminate right lines AP, AQ, making any given angle PAQ between

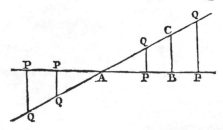

them, where they are fuppofed to commence, and to extend indefinitely both ways from the point A: then calling any AP, x, and its correfponding ordinate

PQ, y, continually changing its pofition by moving parallel to itfelf along the indefinite line AP; alfo in the line AP affume AB $= a$, and from B draw BC parallel to PQ and $= b$: then the indefinite line AQ is called in general a geometrical Locus, and in particular the Locus of the equation $y = \dfrac{bx}{a}$; for whatever point Q is, the triangles ABC, APQ are always fimilar, and therefore AB : BC :: AP : PQ, that is $a : b :: x : y$, and therefore $y = \dfrac{bx}{a}$ is the equation to the right line AQ, or AQ is the Locus of the equation $y = \dfrac{bx}{a}$.

Again, if AQ be a parabola, the nature of which is fuch, that AB : AP :: BC² : PQ², or $a : x :: b^2 : y^2$, and therefore $y^2 = \dfrac{b^2x}{a}$ is the equation which has the parabola for its Locus, or the parabola is the Locus to every equation of this form $y^2 = \dfrac{b^2x}{a}$.

Or if AQ be a circle, having its radius AB $= a$, the nature of which is this, that PQ² $=$ AP . PD, or $y^2 = x . \overline{2a - x}$ or $2ax - x^2$; therefore the Locus of the equation of this form $y^2 = 2ax - x^2$, is always a circle.

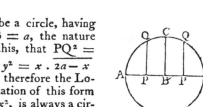

In like manner it will appear, that the ellipfe is the Locus to the equation $y^2 = \dfrac{c^2}{t^2} \times \overline{tx - x^2}$, and the hyperbola the Locus to the equation $y^2 = \dfrac{c^2}{t^2} \times \overline{tx + x^2}$; where t is the tranfverfe, and c the conjugate axis of the ellipfe or hyperbola.

All equations, whofe Loci are of the firft order, may be reduced to one of the 4 following forms:

1ft $y = \dfrac{bx}{a}$; 2d $y = \dfrac{bx}{a} + c$; 3d $y = \dfrac{bx}{a} - c$; 4th $y = c - \dfrac{bx}{a}$;

where the letter c denotes the diftance that the ordinates commence from the line AP, either on the one fide or the other of it, according as the fign of that quantity is $+$ or $-$.

All Loci of the 2d degree are conic fections, viz, either the parabola, the circle, ellipfis, or hyperbola. Therefore when an equation is given, whofe Locus is of the 2d degree, and it is required to draw that Locus, or, which is the fame thing, to conftruct the equation generally; bring over all the terms of the equation to one fide, fo that the other fide be o; then to know which of the conic fections it denotes, there will be two general cafes, viz, either when the rectangle xy is in the equation, or when it is not in it.

Cafe 1. When the term xy is not in the propofed equation. Then, 1ft, if only one of the fquares x^2, y^2

x^2, y^2 be found in it, the Locus will be a parabola. 2d, If both the fquares be in it, and if they have the fame fign, the Locus will be a circle or an ellipfe. 3d, But if the figns of the fquares x^2, y^2 be different, the Locus will be an hyperbola, or the oppofite hyperbolas.

Cafe 2. When the rectangle xy is in the propofed equation; then 1ft, If neither of the fquares x^2, y^2, or only one of them be in the equation, the Locus will be an hyperbola between the afymptotes. 2d, If both x^2 and y^2 be in it, having different figns, the Locus will be an hyperbola, having the abfciffes on its diameter. 3d, If both the fquares be in it, and with the fame fign, then if the coefficient of x^2 be greater than the fquare of half the coefficient of xy, the Locus will be an ellipfe; if equal, a parabola; and if lefs, an hyperbola.

This method of determining geometric Loci, by reducing them to the moft compound or general equations, was firft publifhed by Mr. Craig, in his Treatife on the Quadrature of Curves, in 1693. It is explained at large in the 7th and 8th books of l'Hofpital's Conic Sections. See this fubject particularly illuftrated in Maclaurin's Algebra. The method of Des Cartes, of finding the Loci of equations of the 2d order, is a good one, viz, by extracting the root of the equation. See his Geometry; as alfo Stirling's Illuftratio Linearum Tertii Ordinis. The doctrine of thefe Loci is likewife well treated by De Witt in his Elementa Curvarum. And Bartholomæus Intieri, in his Aditus ad Nova Arcana Geometrica delegenda, has fhewn how to find the Loci of equations of the higher orders. Mr. Stirling too, in his treatife above-mentioned, has given an example or two of finding the Loci of equations of 3 dimenfions. Euclid, Apollonius, Ariftæus, Fermat, Viviani, have alfo written on the fubject of Loci.

LOG, in Navigation, is a piece of thin board, of a fectoral or quadrantal form, loaded in the circular fide with lead fufficient to make it fwim upright in the water; to which is faftened a line of about 150 fathoms, or 300 yards long, called the Log-line, which is divided into certain fpaces, called Knots, and wound on a reel which turns very freely, for the line to wind eafily off.

The ufe of the Log, or Log line, is to meafure the velocity of the fhip, or rate at which fhe runs, which is done from time to time, as the foundation upon which the fhip's reckoning, or finding her place, is kept; and the practice is to heave the Log into the fea, with the line tied to it, and obferve how much of the line is run off the reel, while the fhip fails, during the fpace of half a minute, which time is meafured by a fand-glafs made to run that time very exactly. About 10 fathoms of ftray or wafte line is left next the Log before the knotting or counting commence, that fpace being ufually allowed to carry the Log out of the eddy of the fhip's wake.

The ufing of the Log for finding the velocity of the fhip, is called *Heaving the Log*, and is thus performed: One man holds the reel, and another the half-minute glafs; an officer of the watch throws the Log over the fhip's ftern, on the lee-fide, and when he obferves the ftray line, and the firft mark is going off,

5

he cries *turn!* when the glafs-holder inftantly turns the glafs crying out *done!* then watching the glafs, the moment it is run out he fays *ftop!* upon which the reel being quickly ftopt, the laft mark run off fhews the number of knots, and the diftance of that mark from the reel is eftimated in fathoms: then the knots and fathoms together fhew the diftance run in half a minute, or the diftance per hour nearly, by confidering the knots as miles, and the fathoms as decimals of a mile: thus if 7 knots and 4 fathoms be obferved, then the fhip runs at the rate of 7.4 miles an hour.

It follows, therefore, that the length of each knot, or divifion of the line, ought to be the fame part of a fea mile, as half a minute is of an hour, that is $\frac{1}{120}$th part. Now it is found that a degree of the meridian contains nearly 366,000 feet, therefore $\frac{1}{60}$ of this, or a nautical mile, will be 6100 feet; the $\frac{1}{120}$th of which, or 51 feet nearly, fhould be the length of each knot, or divifion of the Log-line. But becaufe it is fafer to have the reckoning rather before the fhip than after it, therefore it is ufual now to make each knot equal to 8 fathoms or 48 feet. But the knots are made fometimes to contain only 42 feet; and this method of dividing the Log-line was founded on the fuppofition, that 60 miles, of 5000 feet each, made a degree; for $\frac{1}{120}$th of 5000 is $41\frac{2}{3}$, or in round numbers 42 feet. And although many mariners find by experience that this length of the knot is too fhort, yet rather than quit the old way, they ufe fand-glaffes for half-minute ones that run only 24 or 25 feconds. The fand, or half-minute glafs, may be tried by a pendulum vibrating feconds, in the following manner: On a round nail or peg, hang a thread or fine ftring that has a mufket ball fixed to one end, carefully meafuring between the centre of the ball and the ftring's loop over the nail $39\frac{1}{8}$ inches, being the length of a fecond pendulum; then make it fwing or vibrate very fmall arches, and count one for every time it paffes under the nail, beginning at the fecond time it paffes; and the number of fwings made during the time the glafs is running out, fhews the feconds in the glafs.

It is not known who was the inventor of this method of meafuring the fhip's way, or her rate of failing; but no mention of it occurs till the year 1607, in an Eaft-India voyage, publifhed by Purchas; and from that time its name occurs in other voyages in his collections; after which it became famous, being noticed both by our own authors, and by foreigners; as by Gunter in 1623; Snellius, in 1624; Metius, in 1631; Oughtred, in 1633; Herigone, in 1634; Saltonftall, in 1636; Norwood, in 1637; Fournier, in 1643; and almoft all the fucceeding writers on navigation of every country. Various improvements have lately been made of this inftrument by different perfons.

LOGARITHM, from the Greek λογος *ratio*, and αριθμος *number*; q. d. *ratio of numbers*, or perhaps rather *number of ratios*; the indices of the ratios of numbers to one another; or a feries of numbers in arithmetical proportion, correfponding to as many others in geometrical proportion, in fuch fort that 0 correfponds to, or is the index of 1, in the geometricals. They have been devifed for the eafe of large arithmetical calculations.

Thus

Thus,

$$
\left\{
\begin{array}{l}
0, \ 1, \ 2, \ 3, \ 4, \ \&c.\ \text{indices or Logarithms,} \\
1, \ 2, \ 4, \ 8, \ 16, \ \&c, \\
\text{or } 2^0, \ 2^1, \ 2^2, \ 2^3, \ 2^4, \ \&c, \\
1, \ 3, \ 9, \ 27, \ 81, \ \&c, \\
\text{or } 3^0, \ 3^1, \ 3^2, \ 3^3, \ 3^4, \ \&c, \\
1, \ 10, \ 100, \ 1000, \ 10000, \ \&c, \\
\text{or } 10^0, 10^1, 10^2, 10^3, \ 10^4, \ \&c,
\end{array}
\right.
$$
the geometrical progreffions, or common numbers.

Where the fame indices, or Logarithms, ferve equally for any geometric feries; and from which it is evident, that there may be an endlefs variety of fets of Logarithms to the fame common numbers, by varying the 2d term 2, or 3, or 10, &c of the geometric feries; as this will change the original feries of terms whofe indices are the numbers 1, 2, 3, &c; and by interpolation the whole fyftem of numbers may be made to enter the geometrical feries, and receive their proportional Logarithms, whether integers or decimals.

Or the Logarithm of any given number, is the index of fuch a power of fome other number, as is equal to the given one. So if N be $= r^n$, then the Logarithm of N is n, which may be either pofitive or negative, and r any number whatever, according to the different fyftems of Logarithms. When N is 1, then n is $= 0$, whatever the value of r is; and confequently the Logarithm of 1 is always 0 in every fyftem of Logarithms. When n is $= 1$, then N is $= r$; confequently the root r is always the number whofe Logarithm is 1, in every fyftem. When r is $= 2 \cdot 718281828459$ &c, the indices are the hyperbolic Logarithms; fo that n is always the hyperbolic Logarithm of $\overline{2 \cdot 718 \ \&c})^n$. But in the common Logarithms, r is $= 10$; fo that the common Logarithm of any number, is the index of that power of 10 which is equal to the faid number; fo the common Logarithm of $N = 10^n$, is n the index of the power of 10; for example, 1000, being the 3d power of 10, has 3 for its Logarithm; and if 50 be $= 10^{1 \cdot 69897}$, then is $1 \cdot 69897$ the common Logarithm of 50. And hence it follows that this decimal feries of terms

1000, 100, 10, 1, ·1, ·01, ·001,

or 10^3, 10^2, 10^1, 10^0, 10^{-1}, 10^{-2}, 10^{-3},

have 3, 2, 1, 0, -1, -2, -3,

refpectively for the Logarithms of thofe terms.

The Logarithm of a number contained between any two terms of the firft feries, is included between the two correfponding terms of the latter; and therefore that Logarithm will confift of the fame index, whether pofitive or negative, as the fmaller of thofe two terms, together with a decimal fraction, which will always be pofitive. So the number 50 falling between 10 and 100, its Logarithm will fall between 1 and 2, being indeed equal to $1 \cdot 69897$ nearly: alfo the number ·05 falling between the terms 1 and ·01, its Logarithm will fall between -1 and -2, and is indeed $= -2 + \cdot 69897$, the index of the lefs term together with the decimal 69897. The index is alfo called the Characteriftic of the Logarithms, and is always an integer, either pofitive or negative, or elfe $= 0$; and it fhews what place is occupied by the firft fignificant figure of the given number, either above or below the place of units, being in the former cafe + or pofitive; in the latter – or negative.

7

When the characteriftic of a Logarithm is negative, the fign – is commonly fet over it, to diftinguifh it from the decimal part, which, being the Logarithm found in the tables, is always pofitive: fo $-2 + 69897$, or the Logarithm of 05, is written thus $\overline{2} \cdot 69897$. But on fome occafions it is convenient to reduce the whole expreffion to a negative form; which is done by making the characteriftic lefs by 1, and taking the *arithmetical complement* of the decimal, that is, beginning at the left hand, fubtract each figure from 9, except the laft fignificant figure, which is fubtracted from 10; fo fhall the remainders form the Logarithm wholly negative: thus the Logarithm of ·05, which is $\overline{2} \cdot 69897$ or $-2 + \cdot 69897$, is alfo expreffed by $-1 \cdot 30103$, which is all negative. It is alfo fometimes thought more convenient to exprefs fuch Logarithms entirely as pofitive, namely by only joining to the tabular decimal the complement of the index to 10: and in this way the above Logarithm is expreffed by $8 \cdot 69897$; which is only increafing the indices in the fcale by 10.

The Properties of Logarithms.—From the definition of Logarithms, either as being the indices of a feries of geometricals, or as the indices of the powers of the fame root, it follows that the multiplication of the numbers will anfwer to the addition of their Logarithms; the divifion of numbers, to the fubtraction of their Logarithms; the raifing of powers, to the multiplying the Logarithm of the root by the index of the power; and the extracting of roots, to the dividing the Logarithm of the given number by the index of the root required to be extracted.

So, 1ft,

Log. ab or of $a \times b$ is $= \log. a + \log. b$,

Log. 18 or of 3×6 is $= \log. 3 + \log. 6$,

Log. $5 \times 9 \times 73$ is $= \log. 5 + \log. 9 + \log. 73$.

Secondly,

Log. $a \div b$ is $= \log. a - \log. b$,

Log. $18 \div 6$ is $= \log. 18 - \log. 6$,

Log. $79 \times 5 \div 9$ is $= \log. 79 + \log. 5 - \log. 9$,

Log. $\frac{1}{2}$ or $1 \div 2$ is $= l. 1 - l. 2 = 0 - l. 2 = -l. 2$,

Log. $\frac{1}{n}$ or $1 \div n$ is $= l. 1 - l. n = -l. n$.

Thirdly,

Log. r^n is $= n$ l. r; Log. $r^{\frac{1}{n}}$ or of $\sqrt[n]{r}$ is $= \frac{1}{n}$ l. r;

Log. $r^{\frac{m}{n}}$ is $= \frac{m}{n}$ l. r; Log. 2^6 is $= 6 l. 2$; log. $2^{\frac{1}{3}}$ or

of $\sqrt[3]{2}$ is $= \frac{1}{3}$ l. 2; and Log. $2^{\frac{3}{5}}$ is $= \frac{3}{5}$ l. 2.

So that any number and its reciprocal have the fame Logarithm, but with contrary figns; and the fum of the Logarithms of any number and its reciprocal, or complement, is equal to 0.

Hiftory and Conftruction of Logarithms.—The properties of Logarithms hitherto mentioned, or of arithmetical indices to powers or geometricals, with their various ufes and properties, as above-mentioned, are taken notice of by Stifelius, in his Arithmetic; and indeed they were not unknown to the ancients; but they come all far fhort of the ufe of Logarithms in

Trigo.

Trigonometry, as firſt diſcovered by John Napier, baron of Merchiſton in Scotland, and publiſhed at Edinburgh in 1614, in his Mirifici Logarithmorum Canonis Deſcriptio; which contained a large canon of Logarithms, with the deſcription and uſes of them; but their conſtruction was reſerved till the ſenſe of the Learned concerning his invention ſhould be known. This work was tranſlated into Engliſh by the celebrated Mr. Edward Wright, and pub-liſhed by his ſon in 1616. In the year 1619, Robert Napier, ſon of the inventor of Logarithms, publiſhed a new edition of his late father's work, together with the promiſed Conſtruction of the Logarithms, with other miſcellaneous pieces written by his father and Mr. Briggs. And in the ſame year, 1619, Mr John Spei-dell publiſhed his New Logarithms, being an improved form of Napier's.

All theſe tables were of the kind that have ſince been called hyperbolical, becauſe the numbers expreſs the areas between the aſymptote and curve of the hyperbola. And Logarithms of this kind were alſo ſoon after publiſhed by ſeveral other perſons; as by Urſinus in 1619, Kepler in 1624, and ſome others.

On the firſt publication of Napier's Logarithms, Henry Briggs, then profeſſor of Geometry in Greſham College in London, immediately applied himſelf to the ſtudy and improvement of them, and ſoon publiſhed the Logarithms of the firſt 1000 numbers, but on a new ſcale, which he had invented, viz, in which the Logarithm of the ratio of 10 to 1 is 1, the Logarithm of the ſame ratio in Napier's ſyſtem being 2·30258 &c; and in 1624, Briggs publiſhed his Arithmetica Loga-rithmica, containing the Logarithms of 30,000 natu-ral numbers, to 14 places of figures beſides the index, in a form which Napier and he had agreed upon to-gether, which is the preſent form of Logarithms; alſo in 1633 was publiſhed, to the ſame extent of figures, his Trigonometria Britannica, containing the natural and logarithmic ſines, tangents, &c.

With various and gradual improvements, Logarithms were alſo publiſhed ſucceſſively, by Gunter in 1620, Wingate in 1624, Henrion in 1626, Miller and Nor-wood in 1631, Cavalerius in 1632 and 1643, Vlacq and Rowe in 1633, Frobenius in 1634, Newton in 1658, Caramuel in 1670, Sherwin in 1706, Gardiner in 1742, and Dodſon's Antilogarithmic Canon in the ſame year; beſides many others of leſſer note; not to mention the accurate and comprehenſive tables in the Tables Portative, and in my own Logarithms lately publiſhed, where a complete hiſtory of this ſcience may be ſeen, with the various ways of conſtructing them that have been invented by different authors.

In Napier's conſtruction of Logarithms, the natural numbers, and their Logarithms, as he ſometimes called them, or at other times the artificial numbers, are ſup-poſed to ariſe, or to be generated, by the motions of points, deſcribing two lines, of which the one is the natural number, and the other its Logarithm, or artifi-cial. Thus, he conceived the line or length of the radius to be deſcribed, or run over, by a point moving along it in ſuch a manner, that in equal portions of time it generated, or cut off, parts in a decreaſing geometrical progreſſion, leaving the ſeveral remainders, or ſines, in geometrical progreſſion alſo; whilſt another

point deſcribed equal parts of an indefinite line, in the ſame equal portions of time; ſo that the reſpective ſums of theſe, or the whole line generated, were al-ways the arithmeticals or Logarithms of the aforeſaid natural ſines. In this idea of the generation of the Logarithms and numbers, Napier aſſumed 0 as the Logarithm of the greateſt ſine or radius; and next he limited his ſyſtem, not by aſſuming a particular value to ſome aſſigned number, or part of the radius, but by ſuppoſing that the two generating points, which, by their motions along the two lines, deſcribed the natu-ral numbers and Logarithms, ſhould have their veloci-ties equal at the beginning of thoſe lines. And this is the reaſon that, in his table, the natural ſines and their Logarithms, at the complete quadrant, have equal dif-ferences or increments; and this is alſo the reaſon why his ſcale of Logarithms happens accidentally to agree with what have ſince been called the hyperbolical Loga-rithms, which have likewiſe numeral differences equal to thoſe of their natural numbers at the beginning; except only that theſe latter increaſe with the natural numbers, while his on the contrary decreaſe; the Logarithm of the ratio of 10 to 1 being the ſame in both, namely 2·30258509 &c.

Having thus limited his ſyſtem, Napier proceeds, in the poſthumous work of 1619, to explain his conſtruc-tion of the Logarithmic canon. This he effects in various ways, but chiefly by generating, in a very eaſy manner, a ſeries of proportional numbers, and their arithmeticals or Logarithms; and then finding, by pro-portion, the Logarithms to the natural ſines from thoſe of the natural numbers, among the original propor-tionals; a particular account of which may be ſeen in my book of Logarithms above mentioned.

The methods above alluded to, relate to Napier's or the hyperbolical ſyſtem of Logarithms, and indeed are in a manner peculiar to that ſort of them. But in an appendix to the poſthumous work, mention is made of other methods, by which the common Logarithms, agreed upon by him and Briggs, may be conſtructed, and which it appears were written after that agree-ment. One of theſe methods is as follows: Having aſſumed 0 for the Logarithm of 1, and 1000 &c for the Logarithm of 10; this Logarithm of 10, and the ſucceſſive quotients, are to be divided ten times by 5, by which diviſions there will be obtained theſe other ten Logarithms, namely 2000000000, 400000000, 80000000, 16000000, 3200000, 640000, 128000, 25600, 5120, 1024; then this laſt Logarithm, and its quotients, being divided ten times by 2, will give theſe other ten Logarithms,

viz; 512, 256, 128, 64, 32, 16, 8, 4, 2, 1. And the numbers anſwering to theſe twenty Lo-garithms are to be found in this manner, viz, Ex-tract the 5th root of 10 (with ciphers), then the 5th root of that root, and ſo on for ten continual extractions of the 5th root: ſo ſhall theſe ten roots be the natural numbers belonging to the firſt ten Loga-rithms above found, in dividing continually by 5. Next, out of the laſt 5th root is to be extracted the ſquare root, then the ſquare root of this laſt root, and ſo on for ten ſucceſſive extractions of the ſquare root: ſo ſhall theſe laſt ten roots be the natural numbers cor-reſponding to the Logarithms or quotients ariſing from

the

the laſt ten diviſions by the number 2. And from theſe twenty Logarithms, 1, 2, 4, 8, &c, and their natural numbers, the author obſerves that other Logarithms and their numbers may be formed, namely by adding the Logarithms, and multiplying their correſponding numbers. But, beſides the immenſe labour of this method, it is evident that this proceſs would generate rather an antilogarithmic canon, ſuch as Dodſon's, than the table of Briggs.

Napier next mentions another method of deriving a few of the primitive numbers and their Logarithms, namely, by taking continually geometrical means, firſt between 10 and 1, then between 10 and this mean, and again between 10 and the laſt mean, and ſo on; and then taking the arithmetical means between their correſponding Logarithms.

He then lays down various relations between numbers and their Logarithms, ſuch as, that the products and quotients of numbers, anſwer to the ſums and differences of their Logarithms; and that the powers and roots of numbers, anſwer to the products and quotients of the Logaithms when multiplied or divided by the index of the power or root, &c; as alſo that, of any two numbers, whoſe Logarithms are given, if each number be raiſed to the power denoted by the Logarithm of the other, the two reſults will be equal; thus, if x be the Logarithm of any number X, and y the Logarithm of Y, then is $X^y = Y^x$. Napier then adverts to another method of making the Logarithms to a few of the prime integer numbers, which is well adapted to the conſtruction of the common table of Logarithms: this method eaſily follows from what has been ſaid above, and it depends on this property, that the Logarithm of any number in this ſcale, is one leſs than the number of places or figures contained in that power of the given number whoſe exponent is 10000000000, or the Logarithm of 10, at leaſt as to integer numbers, for they really differ by a fraction, as is ſhewn by Mr. Briggs in his illuſtrations of theſe properties; printed at the end of this Appendix to the Conſtruction of Logarithms.

Kepler gave a conſtruction of Logarithms ſomewhat varied from Napier's. His work is divided into two parts: In the firſt, he raiſes a regular and purely mathematical ſyſtem of proportions, and the meaſures of them, demonſtrating both the nature and principles of the conſtruction of Logarithms, which he calls the *meaſures of ratios:* and in the ſecond part, he applies thoſe principles in the actual conſtruction of his table, which contains only 1000 numbers and their Logarithms. The fundamental principles are briefly theſe: That at the beginning of the Logarithms, their increments or differences are equal to thoſe of the natural numbers: that the natural numbers may be conſidered as the decreaſing coſines of increaſing arcs: and that the ſecants of thoſe arcs at the beginning have the ſame differences as the coſines, and therefore the ſame differences as the Logarithms. Then, ſince the ſecants are the reciprocals of the coſines of the ſame arcs, from the foregoing principles, he eſtabliſhes the following method of raiſing the firſt 100 Logarithms, to the numbers 1000, 999, 998, &c, to 900; viz, in this manner: Divide the radius 1000, increaſed with ſeven ciphers, by each of theſe numbers ſeparate-

ly, and the quotients will be the ſecants of thoſe arcs which have the diviſors for their coſines; continuing the diviſion to the 8th figure, as it is in that place only that the arithmetical and geometrical means differ. Then by adding continually the arithmetical means between every two ſucceſſive ſecants, the ſums will be the the ſeries of Logarithms. Or by adding continually every two ſecants, the ſucceſſive ſums will be the ſeries of the double Logarithms. He then derives all the other Logarithms from theſe firſt 100, by common principles.

Briggs firſt adverts to the methods mentioned above, in the Appendix to Napier's Conſtruction, which methods were common to both theſe authors, and had doubtleſs been jointly agreed upon by them. He firſt gives an example of computing a Logarithm by the property, that the Logarithm is one leſs than the number of places or figures contained in that power of the given number whoſe exponent is the Logarithm of 10 with ciphers. Briggs next treats of the other general method of finding the Logarithms of prime numbers, which he thinks is an eaſier way than the former, at leaſt when many figures are required. This method conſiſts in taking a great number of continued geometrical means between 1 and the given number whoſe Logarithm is required; that is, firſt extracting the ſquare root of the given number, then the root of the firſt root, the root of the 2d root, the root of the 3d root, and ſo on, till the laſt root ſhall exceed 1 by a very ſmall decimal, greater or leſs according to the intended number of places to be in the Logarithm ſought: then finding the Logarithm of this ſmall number, by eaſy methods deſcribed afterwards, he doubles it as often as he made extractions of the ſquare root, or, which is the ſame thing, he multiplies it by ſuch power of 2 as is denoted by the ſaid number of extractions, and the reſult is the required Logarithm of the given number; as is evident from the nature of Logarithms.

But as the extraction of ſo many roots is a very troubleſome operation, our author deviſes ſome ingenious contrivances to abridge that labour, chiefly by a proper application of the ſeveral orders of the differences of numbers, forming the firſt inſtance of what may called the *differential method;* but for a particular deſcription of theſe methods, ſee my Treatiſe of Logarithms, above quoted, pag. 65 &c.

Mr. James Gregory, in his Vera Circuli Hyperbolæ Quadratura, printed at Padua in 1667, having approximated to the hyperbolic aſymptotic ſpaces by means of a ſeries of inſcribed and circumſcribed polygons, from thence ſhews how to compute the Logarithms, which are analogous to the areas of thoſe ſpaces: and thus the quadrature of the hyperbolic ſpaces became the ſame thing as the computation of the Logarithms. He here alſo lays down various methods to abridge the computation, with the aſſiſtance of ſome properties of numbers themſelves, by which the Logarithms of all prime numbers under 1000 may be computed, each by one multiplication, two diviſions, and the extraction of the ſquare root. And the ſame ſubject is farther purſued in his Exercitationes Geometricæ. In this latter place, he firſt finds an algebraic expreſſion, in an infinite ſeries, for the Logarithm of $\dfrac{1+a}{1}$, and then the like for the

Logarithm

Logarithm of $\dfrac{1}{1-a}$; and as the one feries has all its terms pofitive, while thofe of the other are alternately pofitive and negative, by adding the two together, every 2d term is cancelled, and the double of the other terms gives the Logarithm of the product of

$\dfrac{1+a}{1}$ and $\dfrac{1}{1-a}$, or the Logarithm of the $\dfrac{1+a}{1-a}$, that is of the ratio of $1-a$ to $1+a$: thus, he finds,

firft $a - \frac{1}{2}a^2 + \frac{1}{3}a^3 - \frac{1}{4}a^4$ &c $=$ log of $\dfrac{1+a}{1}$,

and $a + \frac{1}{2}a^2 + \frac{1}{3}a^3 + \frac{1}{4}a^4$ &c $=$ log. of $\dfrac{1}{1-a}$,

theref. $2a + \frac{2}{3}a^3 + \frac{2}{5}a^5 + \frac{2}{7}a^7$ &c $=$ l. of $\dfrac{1+a}{1-a}$,

Which may be accounted Mr. James Gregory's method of making Logarithms.

In 1668, Nicholas Mercator publifhed his Logarithmotechnia, five Methodus Conftruendi Logarithmos, nova, accurata, & facilis; in which he delivers a new and ingenious method for computing the Logarithms upon principles purely arithmetical; and here, in his modes of thinking and expreffion, he clofely follows the celebrated Kepler, in his writings on the fame fubject; accounting Logarithms as the meafures of ratios, or as the number of ratiunculæ contained in the ratio which any number bears to unity. Purely from thefe principles, then, the number of the equal ratiunculæ contained in fome one ratio, as of 10 to 1, being fuppofed given, our author fhews how the Logarithm, or meafure, of any other ratio may be found. But this, however, only by-the-bye, as not being the principal method he intends to teach, as his laft and beft. Having fhewn, then, that thefe Logarithms, or numbers of fmall ratios, or meafures of ratios, may be all properly reprefented by numbers, and that of 1, or the ratio of equality, the Logarithm or meafure being always 0, the Logarithm of 10, or the meafure of the ratio of 10 to 1, is moft conveniently reprefented by 1 with any number of ciphers; he then proceeds to fhew how the meafures of all other ratios may be found from this laft fuppofition: and he explains thefe principles by fome examples in numbers.

In the latter part of the work, Mercator treats of his other method, given by an infinite feries of algebraic terms, which are collected in numbers by common addition only. He here fquares the hyperbola, and finally finds that the hyperbolic Logarithm of $1 + a$, is equal to the infinite feries $a - \frac{1}{2}a^2 + \frac{1}{3}a^3 - \frac{1}{4}a^4$ &c; which may be confidered as Mercator's quadrature of the hyperbola, or his general expreffion of an hyperbolic Logarithm, in an infinite feries.

And this method was farther improved by Dr. Wallis, in the Philof. Tranf. for the year 1668. The celebrated Newton invented alfo the fame feries for the quadrature of the hyperbola, and the conftruction of Logarithms, and that before the fame were given by Gregory and Mercator, though unknown to one another, as appears by his letter to Mr. Oldenburg, dated October 24, 1676. The explanation and conftruction of the Logarithms are alfo farther purfued in his Fluxions, publifhed in 1736 by Mr. Colfon.

Dr. Halley, in the Philof. Tranf. for the year 1695,

gave a very ingenious effay on the conftruction of Logarithms, intitled, "A moft compendious and facile method for conftructing the Logarithms, and exemplified and demonftrated from the nature of numbers, without any regard to the hyperbola, with a fpeedy method for finding the number from the given Logarithm."

Inftead of the more ordinary definition of Logarithms, viz, 'numerorum proportionalium æquidifferentes comites,' the learned author adopts this other, 'numeri rationum exponentes,' as better adapted to the principle on which Logarithms are here conftructed, confidering them as the number of ratiunculæ contained in the given ratios whofe Logarithms are in queftion. In this way he firft arrives at the Logarithmic feries before given by Newton and others, and afterwards, by various combinations and fections of the ratios, he derives others, converging ftill fafter than the former. Thus he found the Logarithms of feveral ratios, as below, viz, when multiplied by the modulus peculiar to the fcale of Logarithms,

$q - \frac{1}{2}q^2 + \frac{1}{3}q^3 - \frac{1}{4}q^4$ &c, the Log. of 1 to $1+q$,

$q + \frac{1}{2}q^2 + \frac{1}{3}q^3 + \frac{1}{4}q^4$ &c, the Log. of 1 to $1-q$,

$\dfrac{x}{a} - \dfrac{x^2}{2a^2} + \dfrac{x^3}{3a^3} - \dfrac{x^4}{4a^4}$ &c, the Log. of a to b, or

$\dfrac{x}{b} + \dfrac{x^2}{2b^2} + \dfrac{x^3}{3b^3} + \dfrac{x^4}{4b^4}$ &c, the fame Log. of a to b, or

$\dfrac{2x}{z} + \dfrac{2x^3}{3z^3} + \dfrac{2x^5}{5z^5} + \dfrac{2x^7}{7z^7}$ &c, the fame Log. of a to b,

$\dfrac{x^2}{2z^2} + \dfrac{x^4}{4z^4} + \dfrac{x^6}{6z^6} + \dfrac{x^8}{8z^2}$ &c, the Log. of $\sqrt{ab}$ to $\frac{1}{2}z$, or

$\dfrac{1}{y^2} + \dfrac{1}{3y^6} + \dfrac{1}{5y^{10}} + \dfrac{1}{7y^{14}}$ &c, the fame Log. of $\sqrt{ab}$ to $\frac{1}{2}y$;

where a, b, q, are any quantities, and the values of x, y, z, are thus, viz, $x = b - a$, $z = b + a$, $y = ab + \frac{1}{4}z^2$.

Dr. Halley alfo, firft of any, performed the reverfe of the problem, by affigning the number to a given Logarithm; viz,

$\dfrac{b}{a} = 1 + l + \frac{1}{2}l^2 + \dfrac{1}{2\cdot3}l^3 + \dfrac{1}{2\cdot3\cdot4}l^4$ &c, or

$\dfrac{a}{b} = 1 - l + \frac{1}{2}l^2 - \dfrac{1}{2\cdot3}l^3 + \dfrac{1}{2\cdot3\cdot4}l^4$ &c.

where l is the Logarithm of the ratio of a the lefs, to b the greater of any two terms.

Mr. Abraham Sharp of Yorkfhire made many calculations and improvements in Logarithms, &c. The moft remarkable of thefe were, his quadrature of the circle to 72 places of figures, and his computation of Logarithms to 61 figures, viz, for all numbers to 100, and for all prime numbers to 1100.

The celebrated Mr. Roger Cotes gave to the world a learned tract on the nature and conftruction of Logarithms: this was firft printed in the Philof. Tranf. No 338, and afterwards with his Harmonia Menfurarum in 1722, under the title Logometria. This tract has juftly been complained of, as very obfcure and intricate, and the principle is fomething between that of Kepler and the method of Fluxions. He invented the terms Modulus and Modular ratio, this being the ratio

of $1 + \dfrac{1}{1} + \dfrac{1}{2} + \dfrac{1}{2\cdot3} + \dfrac{1}{2\cdot3\cdot4} + \dfrac{1}{2\cdot3\cdot4\cdot5}$ &c to 1 or

of 1 to $1 - \dfrac{1}{1} + \dfrac{1}{2} - \dfrac{1}{2\cdot3} + \dfrac{1}{2\cdot3\cdot4} - \dfrac{1}{2\cdot3\cdot4\cdot5}$

&c;

&c; that is the ratio of 2·718281828459 &c to 1,

  or the ratio of 1 to 0·367879441171 &c; the modulus of any fyſtem being the meaſure or Logarithm of that ratio, which in the hyp. Logarithms is 1, and in Briggs's or the common Logarithms is 0·434294481903 &c.

The learned Dr. Brook Taylor gave another method of computing Logarithms in the Philof. Tranf. No. 352, which is founded on theſe three principles, viz, 1ſt, That the fum of the Logarithms of any two numbers is the Logarithm of the product of thoſe numbers; 2d, That the Logarithm of 1 is 0, and confequently that the nearer any number is to 1, the nearer will its Logarithm be to 0; 3d, That the product of two numbers or factors, of which the one is greater and the other lefs than 1, is nearer to 1, than that factor is which is on the fame ſide of 1 with itſelf; ſo of the two numbers $\frac{2}{3}$ and $\frac{4}{3}$, the product $\frac{8}{9}$ is lefs than 1, but yet nearer to it than $\frac{2}{3}$ is, which is alfo lefs than 1.— And on theſe principles he founds an ingenious, though not very obvious, approximation to the Logarithms of given numbers.

In the Philof. Tranf. a Mr. John Long gave a method of conſtructing Logarithms, by means of a ſmall table, ſomething in the manner of one of Briggs's methods for the fame purpoſe.

Alſo in the Philof. Tranf. vol. 61, a tract on the conſtruction of Logarithms is given by the ingenious Mr. William Jones. In this method, all numbers are conſidered as ſome certain powers of a conſtant determined root : thus, any number x is conſidered as the z power of any root r, or $x = r^z$ is taken as a general expreſſion for all numbers in terms of the conſtant root r and a variable exponent z. Now the index z being the Logarithm of the number x, therefore to find this Logarithm, is the fame thing as to find what power of the radix r is equal to the number x.

An elegant tract on Logarithms, as a comment on Dr. Halley's method, was alfo given by Mr. Jones in his Synopfis Palmariorum Matheſeos, publiſhed in the year 1706.

In the year 1742, Mr. James Dodſon publiſhed his Anti-logarithmic Canon, containing all Logarithms under 100,000, and their correfponding natural numbers to eleven places of figures, with all their differences and the proportional parts; the whole arranged in the order contrary to that uſed in the common tables of numbers and Logarithms, the exact Logarithms being here placed firſt, and their correfponding neareſt numbers in the columns oppoſite to them.

And in 1767, Mr. Andrew Reid publiſhed an " Eſſay on Logarithms," in which he ſhews the computation of Logarithms from principles depending on the binomial theorem, and on the nature of the exponents of powers, the Logarithms of numbers being here conſidered as the exponents of the powers of 10. In this way he brings out the uſual ſeries for Logarithms, and exemplifies Dr. Halley's conſtruction of them. But for the particulars of this, and the methods given by the other authors, we muſt refer to the hiſtorical preface to my treatiſe on Logarithms.

Befides the authors above-mentioned, many others have treated on the ſubject of Logarithms; among the principal of whom are Leibnitz, Euler, Maclaurin,

Wolfius, Keill, and profeſſor Simſon in an ingenious geometrical tract on Logarithms, contained in his poſthumous works, elegantly printed at Glaſgow in the year 1776, at the expence of the learned Earl Stanhope, and by his lordſhip diſpoſed of in prefents among gentlemen moſt eminent for mathematical learning.

For the defcription and uſes of Logarithms in numeral calculations, with the ſhorteſt method of conſtructing them, ſee the Hiſtorical Introduction to my Logarithms, pa. 124 & ſeq.

Briggs's or *Common* LOGARITHMS, are thoſe that have 1 for the Logarithm of 10, or which have 0·4342944819 &c for the modulus; as has been explained above.

Hyperbolic LOGARITHMS, are thoſe that were computed by the inventor Napier, and called alfo ſometimes *Natural Logarithms*, having 1 for their modulus, or 2·302585092994 &c for the Logarithm of 10. Theſe have ſince been called Hyperbolical Logarithms, becauſe they are analogous to the areas of a right-angled hyperbola between the aſymptotes and the curve. See LOGARITHMS, alfo HYPERBOLA and ASYMPTOTIC SPACE.

Logiſtic LOGARITHMS, are certain Logarithms of ſexagefimal numbers or fractions, uſeful in aſtronomical calculations. The Logiſtic Logarithm of any number of ſeconds, is the difference between the common Logarithm of that number and the Logarithm of 3600, the ſeconds in 1 degree.

The chief uſe of the table of Logiſtic Logarithms, is for the ready computing a proportional part in minutes and ſeconds, when two terms of the proportion are minutes and ſeconds, or hours and minutes, or other ſuch ſexagefimal numbers. See the Introd. to my Logarithms, pa. 144.

Imaginary LOGARITHM, a term uſed in the Log. of imaginary and negative quantities; ſuch as $- a$, or $\sqrt{-a^2}$ or $a\sqrt{-1}$. The fluents of certain imaginary expreſſions are alfo Imaginary Logarithms; as of

$$\frac{\dot{x}}{x\sqrt{-1}}, \text{ or of } \frac{a\dot{x}}{cx\sqrt{-1}}, \text{ &c.}$$

See Eüler Analyf. Infin. vol. i. pa. 72, 74.

It is well known that the expreſſion $\frac{\dot{x}}{x}$ reprefents the fluxion of the Logarithm of x, and therefore the fluent of $\frac{\dot{x}}{x}$ is the Logarithm of x; and hence the fluent of $\frac{\dot{x}}{x\sqrt{-1}}$ is the Imaginary Logarithm of x.

However, when theſe Imaginary Logarithms occur in the ſolutions of problems, they may be transformed into circular arcs or ſectors; that is, the Imaginary Logarithm, or imaginary hyperbolic ſector, becomes a real circular ſector. See Bernoulli Oper. tom. i, pa. 400, and pa. 512. Maclaurin's Fluxions, art. 762. Cotes's Harmon. Menf. pa. 45. Walmeſley, Anal. des Mef. pa. 63.

LOGARITHMIC, or LOGISTIC CURVE, a curve ſo called from its properties and uſes, in explaining and conſtructing the Logarithms, becauſe its ordinates are in geometrical progreſſion, while the abſciſſes are in arithmetical progreſſion; ſo that the abſciſſes are as the Logarithms of the correfponding ordinates. And hence the

the curve will be conftructed in this manner: Upon any right line, as an axis, take the equal parts AB, BC, CD, &c, or the arithmetical progreffion AB, AC, AD, &c; and at the points A, B, C, D, &c, erect the perpendicular ordinates AP, BQ, CR, DS, &c, in a geometrical progreffion; fo is the curve line drawn through all the points P, Q, R, S, &c, the Logarithmic, or Logiftic Curve; fo called, becaufe any abfcifs AB, is as the Logarithm of its ordinate BQ. So that the axis ABC &c is an afymptote to the curve.

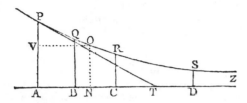

Hence, if any abfcifs $AN = x$, its ordinate $NO = y$, $AP = 1$, and $a =$ a certain conftant quantity, or the modulus of the Logarithms; then the equation of the curve is $x = a \times$ log. of $y = $ log. y^a.

And if the fluxion of this equation be taken, it will be $\dot{x} = \dfrac{a\dot{y}}{y}$; which gives this proportion,

$$\dot{y} : \dot{x} :: y : a$$

but in any curve $\dot{y} : \dot{x} :: y :$ the fubtangent AT; and therefore the fubtangent of this curve is everywhere equal to the conftant quantity a, or the modulus of the Logarithms.

To find the Area contained between two ordinates. Here the fluxion of the area A or $y\dot{x}$ is $y \times \dfrac{a\dot{y}}{y} = a\dot{y}$;

and the correct fluent is $A = a \times \overline{AP - y}$ $= a \times \overline{AP - NO} = a \times PV = AT \times PV$. That is, the area APON between any two ordinates, is equal to the rectangle of the conftant fubtangent and the difference of the ordinates. And hence, when the abfcifs is infinitely long, or the farther ordinate equal to nothing, then the infinitely long area APZ is equal AT $\times$ AP, or double the triangle APT.

For the Solid formed by the curve revolved about its axis AZ. The fluxion of the folid is $\dot{s} = py^2\dot{x} = py^2 \times \dfrac{a\dot{y}}{y} = pay\dot{y}$, where p is $= 3\cdot1416$; and the correct fluent is $s = \frac{1}{2}pa \times \overline{AP^2 - y^2} = \frac{1}{2}p \times AT \times \overline{AP^2 - NO^2}$, which is half the difference between two cylinders of the common altitude a or AT, and the radii of their bafes AP, NO. And hence fuppofing the folid infinitely long towards Z, where y or the ordinate is nothing, the infinitely long folid will be equal to $\frac{1}{2}pa \times AP^2 = \frac{1}{2}p \times AT \times AP^2$, or half the cylinder on the fame bafe and its altitude AT.

It has been faid that Gunter gave the firft idea of a curve whofe abfciffes are in arithmetical progreffion, while the correfponding ordinates are in geometrical progreffion, or whofe abfcifs are the Logarithms of their ordinates; but I do not find it noticed in any part of his writings. This curve was afterwards confidered by others, and named the Logarithmic or Logiftic

Curve by Huygens in his Differtatio de Caufa Gravitatis, where he enumerates all the principal properties of it, fhewing its analogy to Logarithms. Many other learned men have alfo treated of its properties; particularly Le Seur and Jacquier, in their Comment on Newton's Principia; Dr. John Keill, in the elegant little Tract on Logarithms fubjoined to his edition of Euclid's Elements; and Francis Mafcres Efq. Curfitor Baron of the Exchequer, in his ingenious Treatife on Trigonometry: fee alfo Bernoulli's Difcourfe in the Acta Eruditorum for the year 1696, pa. 216; Guido Grando's Demonftratio Theorematum Huygeneanorum circa Logifticam feu Logarithmicam Lineam; and Emerfon on Curve Lines, pa. 19.—It is indeed rather extraordinary that this curve was not fooner announced to the public, fince it refults immediately from Napier's manner of conceiving the generation of Logarithms, by only fuppofing the lines which reprefent the natural numbers as placed at right angles to that upon which the Logarithms are taken.

This curve greatly facilitates the conception of Logarithms to the imagination, and affords an almoft intuitive proof of the very important property of their fluxions, or very fmall increments, namely, that the fluxion of the number is to the fluxion of the Logarithm, as the number is to the fubtangent; as alfo of this property, that if three numbers be taken very nearly equal, fo that their ratios may differ but a little from a ratio of equality, as the three numbers 10000000, 10000001, 10000002, their differences will be very nearly proportional to the Logarithms of the ratios of thofe numbers to each other: all which follows from the Logarithmic arcs being very little different from their chords, when they are taken very fmall. And the conftant fubtangent of this curve is what was afterwards by Cotes called the Modulus of the Syftem of Logarithms.

LOGARITHMIC, or *Logiftic, Spiral,* a curve conftructed as follows. Divide the arch of a circle into any equal parts AB, BD, DE, &c; and upon the radii drawn to the points of divifion take Cb, Cd, Ce, &c, in a geometrical progreffion; fo is the curve Abde &c the Logarithmic Spiral; fo called, becaufe it is evident that AB, AD, AE, &c, being arithmeticals, are as the the Logarithms of CA, Cb, Cd, Ce, &c, which are geometricals; and a Spiral, becaufe it winds continually about the centre C, coming continually nearer, but without ever really falling into it.

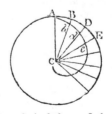

In the Philof. Tranf. Dr. Halley has happily applied this curve to the divifion of the meridian line in Mercator's chart. See alfo Cotes's Harmonia Menf., Guido Grando's Demonft. Theor. Huygen., the Acta Erudit. 1691, and Emerfon's Curves, &c.

LOGISTICS, or LOGISTICAL ARITHMETIC; a name fometimes employed for the arithmetic of fexagefimal fractions, ufed in aftronomical computations.

This name was perhaps taken from a Greek treatife of Barlæmus, a Monk, who wrote a book of Sexagefimal Multiplication, which he called Logiftic. Voffius places this author about the year 1350, but he miftakes the work for a Treatife on Algebra.

The fame term however has been ufed for the rules of

of computations in Algebra, and in other species of Arithmetic: witness the Logistics of Vieta and other writers.

Shakerly, in his Tabulæ Britannicæ, has a Table of Logarithms adapted to sexagesimal fractions, and which he calls Logistical Logarithms; and the expeditious arithmetic, obtained by means of them, he calls Logistical Arithmetic.

LOGISTICAL *Curve, Line,* or *Spiral,* the same as the Logarithmic, which see.

LONG (ROGER), D.D. master of Pembroke hall in Cambridge, Lowndes's professor of astronomy in that university, &c, was author of a well-known and much approved treatise of astronomy, and the inventor of a remarkably curious astronomical machine. This was a hollow sphere, of 18 feet diameter, in which more than 30 persons might sit conveniently. Within side the surface, which represented the heavens, was painted the stars and constellations, with the zodiac, meridians, and axis parallel to the axis of the world, upon which it was easily turned round by a winch. He died, December 16, 1770, at 91 years of age.

A few years before his death, Mr. Jones gave some anecdotes of Dr. Long, as follows: " He is now in the 88th year of his age, and for his years vegete and active. He was lately put in nomination for the office of vice-chancellor: he executed that trust once before, I think in the year 1737. He is a very ingenious person, and sometimes very facetious. At the public Commencement, in the year 1713, Dr. Greene (master of Bennet college, and afterwards bishop of Ely) being then vice-chancellor, Mr. Long was pitched upon for the tripos performance; it was witty and humorous, and has passed through divers editions. Some that remembered the delivery of it, told me, that in addressing the vice-chancellor (whom the university wags usually styled *Miss Greene*), the tripos-orator, being a native of Norfolk, and assuming the Norfolk dialect, instead of saying, *Domine Vice-Cancellarie,* archly pronounced the words thus, *Domina Vice-Cancellaria;* which occasioned a general smile in that great auditory. His friend the late Mr. Bonfoy of Rpton told me this little incident: ' That he and Dr. Long walking together in Cambridge in a dusky evening, and coming to a short *post* fixed in the pavement, which Mr. Bonfoy in the midst of chat and inattention, took to be a *boy* standing in his way, he said in a hurry, ' Get out of my way, boy!' ' *That boy, Sir,* said the Doctor very calmly and slily, *is a* post *boy, who turns out of his way for nobody.*' I could recollect several other ingenious repartees if there were occasion. One thing is remarkable, he never was a hale and hearty man, always of a tender and delicate constitution, yet took great care of it: his common drink water; he always dines with the Fellows in the Hall Of late years he has left off eating flesh-meats; in the room thereof, puddings; vegetables, &c; sometimes a glass or two of wine."

LONGIMETRY, the art of measuring lengths or distances, both accessible and inaccessible, forming a part of what is called Heights and Distances, being an application of geometry and trigonometry to such measurements.

As to accessible lengths, they are easily measured by

3

the actual application of a rod, a chain, or wheel, or some other measure of length.

But inaccessible lengths require the practice and properties of geometry and trigonometry, either in the measurement and construction, or in the computation. For example, Suppose it were required to know the length or distance between the two places A and B, to which places there is free access, but not to the intermediate parts, on account of water or some other impediment; measure therefore, from A and B, the distances to any convenient place C, which suppose to be thus, viz, AC = 735, and BC = 840 links; and let the angle at C, taken with a theodolite or other instrument, be 55° 40'. From these measures the length or distance AB may be determined, either by geometrical measurement, or by trigonometrical computation. Thus, first, lay down an angle C = 55° 40', and upon its legs set off, from any convenient scale of equal parts, CA = 735, and CB = 840; then measure the distance between the points A and B by the same scale of equal parts, which will be found to be 740 nearly.

of

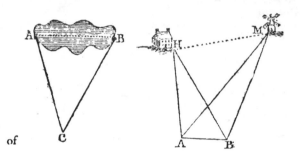

Or this by calculation,

840   180° − 55° 40' = 124° 20', its half 62° 10'.
735

Sum 1575	-	-	-	1·1972806
Dif. 105	-	-	-	0·0211893
Tang. 62° 10'	-	-	10·2773793
Tang. 7 11$\frac{11}{14}$	-	-	9·1012880
f. Sum or ∠ A = 99° 21'$\frac{11}{14}$		9·9711092
to f. ∠ C = 55 40		9·9168593
So BC = 840		0·9242793
To AB = 741·2		0·8699404

For a 2d Example—Suppose it were required to find the distance between two inaccessible objects, as between the house and mill, H and M; first measure any convenient line on the ground, as AB, 300 yards; then at the station A take the angles BAM = 58° 20'; and MAH = 37°; also at the station B take the angles ABH = 53° 30', and HBM = 45° 15'; from hence the distance or length MH may be found, either by geometrical construction, or by trigonometrical calculation, thus:

First draw a line AB of the given length of 300, by a convenient scale of equal parts; then at the point A lay down the angles BAM and MAH of the magnitudes

nitudes above given; and also at the point B the given angles ABH and HBM: then by applying the length HM to the same scale of equal parts, it is found to be nearly 480 yards.

Otherwise, by calculation. First, by adding and subtracting the angles, there is found as below:

37° 00'	58° 20'	53° 30
58 20	53 30	45 15
53 30	45 15	sum 98 45 ∠ ABM

sums	148 50	157 05	
from	180 00	180 00	

∠ AHB 31 10   22 55 ∠ AMB

Then,

as sin. AHB : sin. ABH :: AB : AH = 465·9776,
and, as sin. AMB : sin. ABM :: AB : AM = 761·4655;

their sum is 1227·4431
and their diff. 295·4879

Then as sum AM + AH : to dif. AM − AH ::
tang. ½ AHM + ½ AMH = 71° 30'
to tang. ½ AHM − ½ AMH = 35 44

the dif. of which is AMH = 35 46.

Lastly,

as s. AMH : s. MAH :: AH : HM = 479·7933, the distance sought.

LONGITUDE *of the Earth*, is sometimes used to denote its extent from west to east, according to the direction of the equator. By which it stands contra-distinguished from the Latitude of the earth, which denotes its extent from one pole to the other.

LONGITUDE *of a Place*, in Geography, is its longitudinal distance from some first meridian, or an arch of the equator intercepted between the meridian of that place and the first meridian.

LONGITUDE *in the Heavens*, as of a star, &c, is an arch of the ecliptic, counted from the beginning of Aries, to the place where it is cut by a circle perpendicular to it, and passing through the place of the star.

LONGITUDE *of the Sun or Star from the next equinoctial point*, is the degrees they are distant from the beginning of Aries or Libra, either before or after them; which can never exceed 180 degrees.

LONGITUDE, *Geocentric, Heliocentric*, &c, the Longitude of a planet as seen from the earth, or from the sun. See the respective terms.

LONGITUDE, *in Navigation*, is the distance of a ship, or place, east or west, from some other place or meridian, counted in degrees of the equator. When this distance is counted in leagues, or miles, or in degrees of the meridian, and not in those proper to the parallel of Latitude, it is usually called Departure.

An easy practicable method of finding the Longitude at sea, is the only thing wanted to render the Art of Navigation perfect, and is a problem that has greatly perplexed mathematicians for the last two centuries: accordingly most of the commercial nations of Europe have offered great rewards for the discovery of it; and in consequence very considerable advances have been made towards a perfect solution of the problem, especially by the English.

In the year 1598, the government of Spain offered a reward of 1000 crowns for the solution of this problem; and soon after the States of Holland offered 10 thousand florins for the same. Encouraged by such offers, in 1635, M. John Morin, professor of mathematics at Paris, proposed to cardinal Richlieu, a method of resolving it; and though the commissioners, who were appointed to examine this method, on account of the imperfect state of the lunar tables, judged it insufficient, cardinal Mazarin, in 1645, procured for the author a pension of 2000 livres.

In 1714 an act was passed in the British parliament, allowing 2000l. towards making experiments; and also offering a reward to the person who should discover the Longitude at sea, proportioned to the degree of accuracy that might be attained by such discovery; viz, a reward of 10,000l. if it determines the Longitude to one degree of a great circle, or 60 geographical miles; 15,000l. if it determines the same to two-thirds of that distance; and 20,000l. if it determines it to half that distance; with other regulations and encouragements. 12 Ann. cap. 15. See also stat. 14 Geo. II, cap. 39, and 26 Geo. II, cap. 25. But, by stat. Geo. III, all former acts concerning the Longitude at sea are repealed, except so much of them as relates to the appointment and authority of the commissioners, and such clauses as relate to the publishing of nautical almanacs, and other useful tables; and it enacts, that any person who shall discover a method for finding the Longitude by means of a time-keeper, the principles of which have not hitherto been made public, shall be entitled to the reward of 5000l. if it shall enable a ship to keep her Longitude, during a voyage of 6 months, within 60 geographical miles, or one degree of a great circle; to 7500l. if within 40 geographical miles, or two-thirds of a degree of a great circle; or to a reward of 10,000l. if within 30 geographical miles, or half a degree of a great circle. But if the method shall be by means of improved solar and lunar tables, the author of them shall be entitled to a reward of 5000l. if they shew the distance of the moon from the sun and stars within 15″ of a degree, answering to about 7' of Longitude, after making an allowance of half a degree for the errors of observation, and after comparison with astronomical observations for a period of 18½ years, or during the period of the irregularities of the lunar motions. Or that in case any other method shall be proposed for finding the Longitude at sea, besides those before mentioned, the author shall be entitled to 5000l. if it shall determine the Longitude within one degree of a great circle, or 60 geographical miles; to 7500l. if within two-thirds of that distance; and to 10,000l. if within half the said distance.

Accordingly, many attempts have been made for such discovery, and several ways proposed, with various degrees of success. These however have been chiefly directed to methods of determining the difference of time between any two points on the earth; for the Longitude of any place being an arch of the equator intercepted between two meridians, and this arc being proportional to the time required by the sun to move from the one meridian to the other, at the rate of 4 minutes of time to one degree of the arch, it follows that the difference of time being known, and turned

into

into degrees according to that proportion, it will give the Longitude.

This meafurement of time has been attempted by fome perfons by means of clocks, watches, and other automata: for if a clock or watch were contrived to go uniformly at all feafons, and in all places and fituations; fuch a machine being regulated, for inftance, to London or Greenwich time, would always fhew the time of the day at London or Greenwich, wherever it fhould be carried to; then the time of the day at this place being found by obfervations, the difference between thefe two times would give the difference of Longitude, according to the proportion of one degree to 4 minutes of time.

Gemma Frifius, in his tract De Principiis Aftronomiæ et Geographiæ, printed at Antwerp in 1530, it feems firft fuggefted the method of finding the Longitude at fea by means of watches, or time-keepers; which machines, he fays, were then but lately invented. And foon after, the fame was attempted by Metius, and fome others; but the ftate of watch-making was then too imperfect for that purpofe. Dr. Hooke and Mr. Huygens alfo, about the year 1664, applied the invention of the pendulum-fpring to watches; and employed it for the purpofe of difcovering the Longitude at fea. Some difputes however between Dr. Hooke and the Englifh Miniftry prevented any experiments from being made with watches conftructed by him; but many experiments were made with fome conftructed by Huygens; particularly Major Holmes, in a voyage from the coaft of Guinea in 1665, by one of thefe watches predicted the Longitude of the ifland of Fuego to a great degree of accuracy. This fuccefs encouraged Huygens to improve the ftructure of his watches, (fee Philof. Tranf. for May 1669); but experience foon convinced him, that unlefs methods could be difcovered for preferving the regular motion of fuch machines, and preventing the effects of heat and cold, and other difturbing caufes, they could never anfwer the intention of difcovering the Longitude, and on this account his attempts failed.

The firft perfon who turned his thoughts this way, after the public encouragement held out by the act of 1714, was Henry Sully, an Englifhman; who, in the fame year, printed at Vienna, a fmall tract on the fubject of watch-making; and afterwards removing to Paris, he employed himfelf there in improving time-keepers for the difcovery of the Longitude. It is faid he greatly diminifhed the friction in the machine, and rendered uniform that which remained: and to him is principally to be attributed what is yet known of watch-making in France: for the celebrated Julien le Roy was his pupil, and to him owed moft of his inventions, which he afterwards perfected and executed: and this gentleman, with his fon, and M. Berthoud, are the principal perfons in France who have turned their thoughts this way fince the time of Sully. Several watches made by thefe laft two artifts, have been tried at fea, it is faid with good fuccefs, and large accounts have been publifhed of thefe trials.

In the year 1726 our countryman, Mr. John Harrifon, produced a time-keeper of his own conftruction, which did not err above one fecond in a month, for 10 years together: and in the year 1736 he had a machine tried in a voyage to and from Lifbon; which was the means of correcting an error of almoft a degree and a half in the computation of the fhip's reckoning. In confequence of this fuccefs, Mr. Harrifon received public encouragement to proceed, and he made three other time-keepers, each more accurate than the former, which were finifhed fucceffively in the years 1739, 1758, and 1761; the laft of which proved fo much to his own fatisfaction, that he applied to the commiffioners of the Longitude to have this inftrument tried in a voyage to fome port in the Weft Indies, according to the directions of the ftatute of the 12th of Anne above cited. Accordingly, Mr. William Harrifon, fon of the inventor, embarked in November 1761, on a voyage for Jamaica, with this 4th time-keeper or watch; and on his arrival there, the Longitude, as fhewn by the time-keeper, differed but one geographical mile and a quarter from the true Longitude, deduced from aftronomical obfervations. The fame gentleman returned to England, with the time-keeper, in March 1762; when he found that it had erred, in the 4 months, no more than 1 54″$\frac{1}{2}$ in time, or 28$\frac{4}{5}$ minutes of Longitude; whereas the act requires no greater exactnefs than 30 geographical miles, or minutes of a great circle, in fuch a voyage. Mr. Harrifon now claimed the whole reward of 20,000l, offered by the faid act: but fome doubts arifing in the minds of the commiffioners, concerning the true fituation of the ifland of Jamaica, and the manner in which the time at that place had been found, as well as at Portfmouth; and it being farther fuggefted by fome, that although the time-keeper happened to be right at Jamaica, and after its return to England, it was by no means a proof that it had been always fo in the intermediate times; another trial was therefore propofed, in a voyage to the ifland of Barbadoes, in which precautions were taken to obviate as many of thefe objections as poffible. Accordingly, the commiffioners previoufly fent out proper perfons to make aftronomical obfervations at that ifland, which, when compared with other correfponding ones made in England, would determine, beyond a doubt, its true fituation: and Mr. William Harrifon again fet out with his father's time keeper, in March 1764, the watch having been compared with equal altitudes at Portfmouth, before he fet out, and he arrived at Barbadoes about the middle of May; where, on comparing it again by equal altitudes of the fun, it was found to fhew the difference of Longitude, between Portfmouth and Barbadoes, to be 3^h 55^m 3^s; the true difference of Longitude between thefe places, by aftronomical obfervations, being 3^h 54^m 20^s; fo that the error of the watch was 43^s, or 10′ 45″ of Longitude. In confequence of this, and the former trials, Mr. Harrifon received one moiety of the reward offered by the 12th of Queen Anne, after explaining the principles on which his watch was conftructed, and delivering this as well as the three former to the Commiffioners of the Longitude, for the ufe of the public: and he was promifed the other moiety of the reward, when other time-keepers fhould be made, on the fame principles, either by himfelf or others, performing equally well with that which he had laft made. In the mean time, this laft time-keeper was fent down to the Royal Obfervatory at Greenwich, to be tried there under the direction of the Rev. Dr. Mafkelyne, the Aftronomer Royal. But it did

did not appear, during this trial, that the watch went with the regularity that was expected; from which it was apprehended, that the performance even of the same watch, was not at all times equal; and consequently that little certainty could be expected in the performance of different ones. Moreover, the watch was now found to go faster than during the voyage to and from Barbadoes, by 18 or 19 seconds in 24 hours: but this circumstance was accounted for by Mr. Harrison; who informs us that he had altered the rate of its going by trying some experiments, which he had not time to finish before he was ordered to deliver up the watch to the Board. Soon after this trial, the Commissioners of Longitude agreed with Mr. Kendal, one of the watch-makers appointed by them to receive Mr. Harrison's discoveries, to make another watch on the same construction with this, to determine whether such watches could be made from the account which Mr. Harrison had given, by other persons, as well as himself. The event proved the affirmative; for the watch produced by Mr. Kendal, in consequence of this agreement, went considerably better than Mr. Harrison's did. Mr. Kendal's watch was sent out with Capt. Cook, in his 2d voyage towards the south pole and round the globe, in the year 1772, 1773, 1774, and 1775; when the only fault found in the watch was, that its rate of going was continually accelerated; though in this trial, of 3 years and a half, it never amounted to $14''\frac{1}{2}$ a day. The consequence was, that the House of Commons in 1774, to whom an appeal had been made, were pleased to order the 2d moiety of the reward to be given to Mr. Harrison, and to pass the act above mentioned. Mr. Harrison had also at different times received some other sums of money, as encouragements to him to continue his endeavours, from the Board of Longitude, and from the India Company, as well as from many individuals. Mr. Arnold and some other persons have since also made several very good watches for the same purpose.

Others have proposed various astronomical methods for finding the Longitude. These methods chiefly depend on having an ephemeris or almanac suited to the meridian of some place, as Greenwich for instance, to which the Nautical Almanac is adapted, which shall contain for every day computations of the times of all remarkable celestial motions and appearances, as adapted to that meridian. So that, if the hour and minute be known when any of the same phenomena are observed in any other place, whose Longitude is desired, the difference between this time and that to which the time of the said phenomenon was calculated and set down in the almanac, will be known, and consequently the difference of Longitude also becomes known, between that place and Greenwich, allowing at the rate of 15 degrees to an hour.

Now it is easy to find the time at any place, by means of the altitude or azimuth of the sun or stars; which time it is necessary to find by such means, both in these astronomical modes of determining the Longitude, and in the former by a time-keeper; and it is the difference between that time, so determined, and the time at Greenwich, known either by the time-keeper or by the astronomical observations of celestial phenomena, which gives the difference of Longitude, at the rate above-

mentioned. Now the difficulty in these methods lies in the fewness of proper phenomena, capable of being thus observed; for all slow motions, such as belong to the planet Saturn for instance, are quite excluded, as affording too small a difference, in a considerable space of time, to be properly observed; and it appears that there are no phenomena in the heavens proper for this purpose, except the eclipses or motions of Jupiter's satellites, and the eclipses or motions of the moon, viz, such as her distance from the sun or certain fixed stars lying near her path, or her Longitude or place in the zodiac, &c. Now of these methods,

1st, That by the eclipses of the moon is very easy, and sufficiently accurate, if they did but happen often, as every night. For at the moment when the beginning, or middle, or end of an eclipse is observed by a telescope, there is no more to be done but to determine the time by observing the altitude or azimuth of some known star; which time being compared with that in the tables, set down for the happening of the same phenomenon at Greenwich, gives the difference in time, and consequently of Longitude sought. But as the beginning or end of an eclipse of the moon cannot generally be observed nearer than one minute, and sometimes 2 or 3 minutes of time, the Longitude cannot certainly be determined by this method, from a single observation, nearer than one degree of Longitude. However, by two or more observations, as of the beginning and end &c, a much greater degree of exactness may be attained.

2d, The moon's place in the zodiac is a phenomenon more frequent than that of her eclipses; but then the observation of it is difficult, and the calculus perplexed and intricate, by reason of two parallaxes; so that it is hardly practicable, to any tolerable degree of accuracy.

3d, But the moon's distances from the sun, or certain fixed stars, are phenomena to be observed many times in almost every night, and afford a good practical method of determining the Longitude of a ship at almost any time; either by computing, from thence, the moon's true place, to compare with the same in the almanac; or by comparing her observed distance itself with the same as there set down.

It is said that the first person who recommended the finding the Longitude from this observed distance between the moon and some star, was John Werner, of Nuremberg, who printed his annotations on the first book of Ptolomy's Geography in 1514. And the same thing was recommended in 1524, by Peter Apian, professor of mathematics at Ingolstadt; also about 1530, by Oronce Finé, of Briançon; and the same year by the celebrated Kepler, and by Gemma Frisius, at Antwerp; and in 1560, by Nonius or Pedro Nunez.

Nor were the English mathematicians behind hand on this head. In 1665 Sir Jonas Moore prevailed on king Charles the 2d to erect the Royal Observatory at Greenwich, and to appoint Mr. Flamsteed his astronomical observer, with this express command, that he should apply himself with the utmost care and diligence to the rectifying the table of the motions of the heavens, and the places of the fixed stars, in order to find out the so much desired Longitude at sea, for perfecting the Art of Navigation. And to the fidelity and industry

with

with which Mr. Flamſteed executed his commiſſion, it is that we are chiefly indebted for that curious theory of the moon, which was afterwards formed by the immortal Newton. This incomparable philoſopher made the beſt poſſible uſe of the obſervations with which he was furniſhed; but as theſe were interrupted and imperfect, his theory would ſometimes differ from the heavens by 5 minutes or more.

Dr. Halley beſtowed much time on the ſame object; and a Starry Zodiac was publiſhed under his direction, containing all the ſtars to which the moon's appulſe can be obſerved; but for want of correct tables, and proper inſtruments, he could not proceed in making the neceſſary obſervations. In a paper on this ſubject, in the Philoſ. Tranſ. number 421, he expreſſes his hope, that the inſtrument juſt invented by Mr. Hadley might be applied to taking angles at ſea with the deſired accuracy. This great aſtronomer, and after him the Abbe de la Caille, and others, have reckoned the beſt aſtronomical method for finding the Longitude at ſea, to be that in which the diſtance of the moon from the ſun or from a ſtar is uſed; for the moon's daily motion being about 13 degrees, her hourly mean motion is above half a degree, or one minute of a degree in two minutes of time; ſo that an error of one minute of a degree in poſition will produce an error of 2 minutes in time, or half a degree in Longitude. Now from the great improvements made by Newton in the theory of the moon, and more lately by Euler and others on his principles, profeſſor Mayer, of Gottengen, was enabled to calculate lunar tables more correct than any former ones; having ſo far ſucceeded as to give the moon's place within one minute of the truth, as has been proved by a compariſon of the tables with the obſervations made at the Greenwich obſervatory by the late Dr. Bradley, and by Dr. Maſkeline, the preſent Aſtronomer Royal; and the ſame have been ſtill farther improved under his direction, by the late Mr. Charles Maſon, by ſeveral new equations, and the whole computed to tenths of a ſecond. Theſe new tables, when compared with the above mentioned ſeries of obſervations, a proper allowance being made for the unavoidable error of obſervation, ſeem to give always the moon's Longitude in the heavens correctly within 30 ſeconds of a degree; which greateſt error, added to a poſſible error of one minute in taking the moon's diſtance from the ſun or a ſtar at ſea, will at a medium only produce an error of 42 minutes of Longitude. To facilitate the uſe of the tables, Dr. Maſkelyne propoſed a nautical ephemeris, the ſcheme of which was adopted by the Commiſſioners of Longitude, and firſt executed in the year 1767, ſince which time it has been regularly continued, and publiſhed as far as for the year 1800. But as the rules that were given in the appendix to one of thoſe publications, for correcting the effects of refraction and parallax, were thought too difficult for general uſe, they have been reduced to tables. So that, by the help of the ephemeris, theſe tables, and others that are alſo provided by the Board of Longitude, the calculations relating to the Longitude, which could not be performed by the moſt expert mathematician in leſs than four hours, may now be completed with great eaſe and accuracy in half an hour.

As this method of determining the Longitude depends on the uſe of the tables annually publiſhed for this purpoſe, thoſe who wiſh for farther information are referred to the inſtructions that accompany them, and particularly to thoſe that are annexed to the *Tables requiſite to be uſed with the Aſtronomical and Nautical Ephemeris*, 2d edit. 1781.

4th. The phenomena of Jupiter's ſatellites have commonly been preferred to thoſe of the moon, for finding the Longitude; becauſe they are leſs liable to parallaxes than theſe are, and beſides they afford a very commodious obſervation whenever the planet is above the horizon. Their motion is very ſwift, and muſt be calculated for every hour. Theſe ſatellites of Jupiter were no ſooner announced by Galileo, in his Syderius Nuncius, firſt printed at Venice in 1610, than the frequency of their eclipſes recommended them for this purpoſe; and among thoſe who treated on this ſubject, none was more ſucceſsful than Caſſini. This great aſtronomer publiſhed, at Bologna, in 1688, tables for calculating the appearances of their eclipſes, with directions for finding the Longitudes of places by them; and being invited to France by Louis the 14th, he there, in the year 1693, publiſhed more correct tables of the ſame. But the mutual attractions of the ſatellites rendering their motions very irregular, thoſe tables ſoon became uſeleſs for this purpoſe; inſomuch that they require to be renewed from time to time; a ſervice which has been performed by ſeveral ingenious aſtronomers, as Dr. Pound, Dr. Bradley, M. Caſſini the ſon, and more eſpecially by Mr. Wargentin, whoſe tables are much eſteemed, which have been publiſhed in ſeveral places, as alſo in the Nautical Almanacs for 1771 and 1779.

Now, to find the Longitude by theſe ſatellites; with a good teleſcope obſerve ſome of their phenomena, as the conjunction of two of them, or of one of them with Jupiter, &c; and at the ſame time find the hour and minute, from the altitudes of the ſtars, or by means of a clock or watch, previouſly regulated for the place of obſervation; then, conſulting tables of the ſatellites, obſerve the time when the ſame appearance happens in the meridian of the place for which the tables are calculated; and the difference of time, as before, will give the Longitude.

The eclipſes of the firſt and ſecond of Jupiter's ſatellites are the moſt proper for this purpoſe; and as they happen almoſt daily, they afford a ready means of determining the Longitude of places at land, having indeed contributed much to the modern improvements in geography; and if it were poſſible to obſerve them with proper teleſcopes, in a ſhip under ſail, they would be of great ſervice in aſcertaining its Longitude from time to time. To obviate the inconvenience to which theſe obſervations are liable from the motions of the ſhip, a Mr. Irwin invented what he called a marine chair; this was tried by Dr. Maſkelyne, in his voyage to Barbadoes, when it was not found that any benefit could be derived from the uſe of it. And indeed, conſidering the great power requiſite in a teleſcope proper for theſe obſervations, and the violence, as well as irregularities in the motion of a ſhip, it is to be feared that the complete management of a teleſcope on ſhip-board, will always remain among the deſiderata in this part of nautical ſcience. And farther, ſince all methods that depend on the phenomena of the heavens have alſo this other defect, that they cannot be obſerved at all-times, this

renders

renders the improvement of time keepers an object of the greater importance.

Many other schemes and proposals have been made by different persons, but most of them of very little or no use ; such as by the space between the flash and report of a great gun, proposed by Messrs. Whiston and Ditton ; and another proposed by Mr. Whiston, by means of the inclinatory or dipping needle ; besides a method by the variation of the magnetic needle, &c, &c.

LONGITUDE *of Motion*, is a term used by Dr. Wallis for the measure of motion, estimated according to its line of direction ; or it is the distance or length gone through by the centre of any moving body, as it moves on in a right line.

The same author calls the measure of any motion, estimated according to the line of direction of the vis motrix, the *Altitude* of it.

LONGOMONTANUS (CHRISTIAN), a learned astronomer, born in Denmark in 1562, in the village of Longomontum, whence he took his name. Vossius, by mistake, calls him Christopher. Being the son of a poor man, a plowman, he was obliged to suffer, during his studies, all the hardships to which he could be exposed, dividing his time, like the philosopher Cleanthes, between the cultivation of the earth and the lessons he received from the minister of the place. At length, at 15 years old, he stole away from his family, and went to Wiburg, where there was a college, in which he spent 11 years ; and though he was obliged to earn his livelihood as he could, his close application to study enabled him to make a great progress in learning, particularly in the mathematical sciences.

From hence he went to Copenhagen ; where the professors of that university soon conceived a very high opinion of him, and recommended him to the celebrated Tycho Brahe ; with whom Longomontanus lived 8 years, and was of great service to him in his observations and calculations. At length, being very desirous of obtaining a professor's chair in Denmark, Tycho Brahe consented, with some difficulty, to his leaving him ; giving him a discharge filled with the highest testimonies of his esteem, and furnishing him with money for the expence of his long journey from Germany, whither Tycho had retired.

He accordingly obtained a professorship of mathematics in the university of Copenhagen in 1605 ; the duty of which he discharged very worthily till his death, which happened in 1647, at 85 years of age.

Longomontanus was author of several works, which shew great talents in mathematics and astronomy. The most distinguished of them, is his *Astronomica Danica*, first printed in 4to, 1621, and afterwards in folio in 1640, with augmentations. He amused himself with endeavouring to square the circle, and pretended that he had made the discovery of it ; but our countryman Dr. John Pell attacked him warmly on that subject, and proved that he was mistaken.——It is remarkable that, obscure as his village and father were, he contrived to dignify and eternize them both ; for he took his name from his village, and in the title page to some of his works he wrote himself *Christianus Longomontanus Severini filius*, his father's name being Severin or Severinus.

LOXODROMIC CURVE, or SPIRAL, is the same

as the Rhumb line, or path of a ship sailing always on the same course in an oblique direction, or making always the same angle with every meridian. It is a species of logarithmic spiral, described on the surface of the sphere, having the meridians for its radii.

LOXODROMICS, the art or method of oblique sailing, by the loxodromic or rhumb line.

LOZENGE, an oblique-angled parallelogram ; being otherwise called a rhombus, or a rhomboides.

LUBIENIETSKI (STANISLAUS), a Polish gentleman, born at Cracow, in 1623, and educated with great care by his father. He was learned in astronomy, and became a celebrated Socinian minister. He took great pains to obtain a toleration from the German princes for his Socinian brethren. His endeavours however were all in vain ; being himself persecuted by the Lutheran ministers, and banished from place to place ; till at length he was banished out of the world, with his two daughters, by poison, in 1675, his wife narrowly escaping.

We have, of his writing, *A History of the Reformation in Poland*; and a Treatise on Comets, intitled *Theatrum Cometicum*, printed at Amsterdam in 2 volumes folio ; which is a most elaborate work, containing a minute historical account of every single comet that had been seen or recorded.

LUCIDA CORONÆ, a fixed star of the 2d magnitude, in the northern crown. See CORONA *Borealis*.

LUCIDA HYDRÆ. See COR *Hydræ*.

LUCIDA LYRÆ, a bright star of the first magnitude in the constellation Lyra.

LUCIFER, a name given to the planet Venus, when she appears in the morning before sunrise.

LUMINARIES, a term used for the sun and moon, by way of eminence, for their extraordinary lustre, and the great quantity of light they give us.

LUNA, the Moon ; which see.

LUNAR, something relating to the moon.

LUNAR *Cycle*, or *Cycle of the Moon*. See CYCLE.

LUNAR *Method for the Longitude*, a method of keeping or finding the Longitude by means of the moon's motions, particularly by her observed distances from the sun and stars ; for which, see the article LONGITUDE.

LUNAR *Month*, is either Periodical, Synodical, or Illuminative. Which see ; also MONTH.

LUNAR *Year*, consists of 354 days, or 12 synodical months, of $29\frac{1}{2}$ days each. See YEAR.

In the early ages, the lunar year was used by all nations ; the variety of course being more frequent and conspicuous in this planet, and consequently better known to men, than those of any other. The Romans regulated their year, in part, by the moon, even till the time of Julius Cæsar. The Jews too had their lunar month and year.

LUNAR *Dial*, *Eclipse*, *Horoscope*, and *Rainbow*. See the several substantives.

LUNATION, the period or time between one new moon and another ; it is also called the synodical month, consisting of 29 days 12 hrs. 44m. 3 sec. 11 thirds ; exceeding the periodical month by 2 ds. 5 hrs. 0 m. 55 sec.

LUNE, or LUNULA, or little moon, is a geometrical

end figure, in form of a crescent, terminated by the arcs of two circles that intersect each other within.

Though the quadrature of the whole circle has never been effected, yet many of its parts have been squared. The first of these partial quadratures was that of the Lunula, given by Hippocrates of Scio, or Chios; who, from being a shipwrecked merchant, commenced geometrician. But although the quadrature of the Lune be generally ascribed to Hippocrates, yet Proclus expressly says it was found out by Oenopidas of the same place. See Heinius in Mem. de l'Acad. de Berlin, tom. ii. pa. 410, where he gives a dissertation concerning this Oenopidas. See also CIRCLE, and QUADRATURE.

The Lune of Hippocrates is this: Let ABC be a semicircle, having its centre E, and ADC a quadrant, having its centre F; then the Figure ABCDA, contained between the arcs of the semicircle and quadrant, is his Lune; and it is equal to the right-angled triangle ACF, as is thus easily proved. Since $AF^2 = 2AE^2$, that is, the square of the radius of the quadrant equal to double the square of the radius of the semicircle; therefore the quadrantal area ADCFA is = the semicircle ABCEA; from each of these take away the common space ADCEA, and there remains the triangle ACF = the Lune ABCDA.

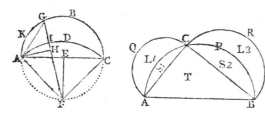

Another property of this Lune, which is the more general one of the former, is, that if FG be any line drawn from the point F, and AH perpendicular to it; then is the intercepted part of the Lune AGIA = the triangle AGH cut off by the chord line AG; or in general, that the small segment AKGA is equal to the trilineal AIHA. For, the angle AFG being at the centre of the one circle, and at the circumference of the other, the arcs cut off AG, AI are similar to the wholes ABC, ADC, therefore the small seg. AKGA is to the semisegment AIH, as the whole semicircle ABCA to the semisegment or quadrant ADCF, that is in a ratio of equality.

Again, if ABC (fig. 2) be a triangle, right angled at C, and if semicircles be described on the three sides as diameters; then the triangle T (ABC) is equal to the sum of the two Lunes L1, L2. For, the greatest semicircle is equal to the sum of both the other two; from the greatest semicircle take away the segments S1 and S2, and there remains the triangle T; also from the two less semicircles take away the same two segments S1 and S2, and there remains the two Lunes L1 and L2; therefore the triangle T = L1 + L2 the two Lunes.

LUNETTE, in Fortification, an inveloped counterguard, or mound of earth, made beyond the second ditch, opposite to the place of arms; differing from the ravelins only in their situation. Lunettes are usually made in wet ditches, and serve the same purpose as fausse-brays, to defend the passage of the ditch.

LUPUS, the *Wolf*, a southern constellation, joined to the Centaur, containing together 19 stars in Ptolomy's catalogue, but 24 in the Britannic catalogue.

LYNX, a constellation of the northern hemisphere, composed by Hevelius out of the unformed stars. In his catalogue it consists of 19 stars, but in the Britannic 44.

LYONS (ISRAEL), a good mathematician and botanist, was the son of a Polish Jew silversmith, and teacher of Hebrew at Cambridge in England, where he was come to settle, and where young Lyons was born, 1739. He was a very extraordinary young man for parts and ingenuity; and shewed very early in life a great inclination to learning, particularly mathematics, on which account he was much patronized by Dr. Smith, master of Trinity college. About 1755 he began to study botany, which he continued occasionally till his death; in which he made a considerable progress, and could remember not only the Linnæan names of almost all the English plants, but even the synonyma of the old botanists; and he had prepared large materials for a *Flora Cantabrigiensis*, describing fully every part of each plant from the specimen, without being obliged to consult, or being liable to be misled by, former authors.

In 1758, he obtained much celebrity by publishing *A Treatise on Fluxions*, dedicated to his patron, Dr. Smith; and in 1763, *Fasciculus Plantarum circa Cantabrigiam, &c.* In the same year, or the year before, he read Lectures on Botany at Oxford with great applause, to at least 60 pupils; but he could not be prevailed on to make a long absence from Cambridge.

Mr. Lyons was some time employed as one of the computers of the Nautical Almanac; and besides he received frequent other presents from the Board of Longitude for his own inventions.——He had studied the English history; and could quote whole passages from the Monkish writers verbatim. He could read Latin and French with ease, but wrote the former ill. He was appointed by the Board of Longitude to sail with Capt. Phipps, in his voyage towards the North Pole, in 1773, as astronomical observator; and he discharged that office to the satisfaction of his employers. After his return from this voyage, he married, and settled in London, where he died of the meazles in about two years.

At the time of his death he was engaged in preparing for the press, a complete edition of all the works of the late learned Dr. Halley; a work very much wanted. —His *Calculations in Spherical Trigonometry abridged*, were printed in the Philos. Transf. vol. 65, for the year 1775, pa. 470.—After his death, his name appeared in the title-page of *A Geographical Dictionary*, the astronomical parts of which were said to be " taken from the papers of the late Mr. Israel Lyons of Cambridge, author of several valuable mathematical productions, and astronomer in lord Mulgrave's voyage to the northern hemisphere."—The astronomical and other mathematical calculations, printed in the account of captain Phipps's voyage towards the north pole, mentioned above, were made by Mr. Lyons. This appeared afterwards, by the acknowledgment of captain Phipps, when Dr. Horsley detected a material error in some part of of them, in his *Remarks on the Observations made in the late Voyage, &c,* 1774.

" The Scholar's Instructor, or Hebrew Grammar, by Israel Lyons, Teacher of the Hebrew Tongue in the University of Cambridge," the 2d edit. &c, 1757, 8vo, was the production of his father; as was also another Treatise printed at the Cambridge press, under the title

of " Observations and Enquiries relating to various parts of Scripture History, 1761."

LYRA, the *Harp,* a constellation in the northern hemisphere, containing 10 stars in Ptolomy's catalogue, 11 in Tycho's, 17 in Hevelius's, and 21 in the Britannic catalogue.

M.

MAC

M, In *Astronomical Tables,* &c, is used for *Meridional* or southern; and sometimes for *Meridian,* or mid-day.—In the Roman numeration, it denotes 1000, one thousand.

MACHINE, denotes any thing that serves to augment, or to regulate moving powers: or it is any body destined to produce motion, so as to save either time or force. The word, in Greek, signifies an *Invention,* or *Art* : and hence, in strictness, a machine is something that consists more in art and invention, than in the strength and solidity of the materials; for which reason it is that the inventors of machines are called *Ingenieurs,* or *engineers.*

Machines are either simple or compound. The simple machines are the seven mechanical powers, viz, the lever, balance, pulley, wheel-and-axle, wedge, screw, and inclined plane; which are otherwise called the simple mechanic powers.

These simple machines serve for different purposes, according to the different structures of them; and it is the business of the skilful mechanist to choose them, and combine them, in the manner that may be best adapted to produce the desired effect. The lever is a very handy machine for many purposes, and its power immediately varied as the occasion may require; when weights are to be raised only a little way, such as stones out of quarries, &c. On the other hand, the wheel-and-axle serves to raise weights from the greatest depth, or to the greatest height. Pulleys, being easily carried, are therefore much employed in ships. The balance is useful for ascertaining an equality of weight. The wedge is excellent for separating the parts of bodies; and being impelled by the force of percussion, it is incomparably greater than the other powers. The screw is useful for compressing or squeezing bodies together, and also for raising very heavy weights to a small height; its great friction is even of considerable use, to preserve the effect already produced by the machine.

Compound MACHINE, is formed from these simple machines, combined together for different purposes. The number of compound machines is almost infinite; and yet it would seem that the Ancients went far beyond the Moderns in the powers and effects of them; especially their machines of war and architecture.

Accurate descriptions and drawings of machines

MAC

would be a very curious and useful work. But to make a collection of this kind as beneficial as possible, it should contain also an analysis of them; pointing out their advantages and disadvantages, with the reasons of the constructions; also the general problems implied in these constructions, with their solutions, should be noticed. Though a complete work of this kind be still wanting, yet many curious and useful particulars may be gathered from Strada, Besson, Beroaldus, Augustinus de Ramellis, Bockler, Leupold, Beyer, Limpergh, Van Zyl, Perault, and others; a short account of whose works may be found in Wolfii Commentatio de Præcipuis Scriptis Mathematicis; Elem. Mathef. Univ. tom. 5, pa. 84. To these may be added, Belidor's Architecture Hydraulique, Desaguliers's Course of Experimental Philosophy, and Emerson's Mechanics. The Royal Academy of Sciences at Paris have also given a collection of machines and inventions approved of by them. This work, published by M. Gallon, consists of 6 volumes in quarto, containing engraved draughts of the machines, with their descriptions annexed.

MACHINE, *Architectonical,* is an assemblage of pieces of wood so disposed as that, by means of ropes and pulleys, a small number of men may raise great loads, and lay them in their places: such as cranes, &c.——It is hard to conceive what sort of machines the Ancients must have used to raise those immense stones found in some of the antique buildings; as some of those still found in the walls of Balbeck in Turkey, the ancient Heliopolis, which are 63 feet long, 12 feet broad, and 12 feet thick, and which must weigh 6 or 7 hundred tons a piece.

Blowing MACHINE. See BELLOWS.

Boyleian MACHINE. Mr. Boyle's Air-Pump.

Electrical MACHINE. See ELECTRICAL *Machine.*

Wind MACHINE. See ANEMOMETER, and WIND *Machine.*

Hydraulic, or *Water* MACHINE, is used either to signify a simple Machine, serving to conduct or raise water; as a sluice, pump, and the like, or several of these acting together, to produce some extraordinary effect; as the

MACHINE *of Marli.* See MARLI. See also FIRE-*engine,* STEAM-*engine,* and WATER-*works.*

Military MACHINES, among the Ancients, were of

three

three kinds: the first serving to launch arrows, as the scorpion; or javelins, as the catapult; or stones, as the balista; or fiery darts, as the pyrabolus: the 2d sort serving to beat down walls, as the battering ram and terebra: and the 3d sort to shelter those who approach the enemy's wall, as the tortoise or testudo, the vinea, and the towers of wood. See the respective articles.

The Machines of war now in use, consist in artillery, including cannon, mortars, petards, &c.

MACLAURIN, (Colin), a most eminent mathematician and philosopher, was the son of a clergyman, and born at Kilmoddan in Scotland, in the year 1698. He was sent to the university of Glasgow in 1709; where he continued five years, and applied to his studies in a very intense manner, and particularly to the mathematics. His great genius for mathematical learning discovered itself so early as at 12 years of age; when, having accidentally met with a copy of Euclid's Elements in a friend's chamber, he became in a few days master of the first 6 books without any assistance: and it is certain, that in his 16th year he had invented many of the propositions which were afterwards published as part of his work intitled *Geometria Organica*. In his 15th year he took the degree of Master of Arts; on which occasion he composed and publicly defended a thesis on the power of gravity, with great applause. After this he quitted the university, and retired to a country seat of his uncle, who had the care of his education; his parents being dead some time. Here he spent two or three years in pursuing his favourite studies; but, in 1717, at 19 years of age only, he offered himself a candidate for the professorship of mathematics in the Marischal College of Aberdeen, and obtained it after a ten days trial, against a very able competitor.

In 1719, Mr. Maclaurin visited London, where he left his *Geometria Organica* to print, and where he became acquainted with Dr. Hoadley then bishop of Bangor, Dr. Clarke, Sir Isaac Newton, and other eminent men; at which time also he was admitted a member of the Royal Society: and in another journey, in 1721, he contracted an intimacy with Martin Folkes, Esq. the president of it, which continued during his whole life.

In 1722, lord Polwarth, plenipotentiary of the king of Great Britain at the congress of Cambray, engaged Maclaurin to go as a tutor and companion to his eldest son, who was then to set out on his travels. After a short stay at Paris, and visiting other towns in France, they fixed in Lorrain; where he wrote his piece, On the Percussion of Bodies, which gained him the prize of the Royal Academy of Sciences for the year 1724. But his pupil dying soon after at Montpelier, he returned immediately to his profession at Aberdeen. He was hardly settled here, when he received an invitation to Edinburgh; the curators of that university being desirous that he should supply the place of Mr. James Gregory, whose great age and infirmities had rendered him incapable of teaching. He had here some difficulties to encounter, arising from competitors, who had good interest with the patrons of the university, and also from the want of an additional fund for the new professor; which however at length were all surmounted, principally by the means of Sir Isaac Newton. Accordingly, in Nov. 1725, he was introduced into the university; as was at the same time his learned colleague and intimate friend, Dr. Alexander Monro, professor of anatomy. After this, the Mathematical classes soon became very numerous, there being generally upwards of 100 students attending his Lectures every year; who being of different standings and proficiency, he was obliged to divide them into four or five classes, in each of which he employed a full hour every day from the first of November to the first of June. In the first class he taught the first 6 books of Euclid's Elements, Plane Trigonometry, Practical Geometry, the Elements of Fortification, and an Introduction to Algebra. The second class studied Algebra, with the 11th and 12th books of Euclid, Spherical Trigonometry, Conic Sections, and the general Principles of Astronomy. The third went on in Astronomy and Perspective, read a part of Newton's Principia, and had performed a course of experiments for illustrating them: he afterwards read and demonstrated the Elements of Fluxions. Those in the fourth class read a System of Fluxions, the Doctrine of Chances, and the remainder of Newton's Principia.

In 1734, Dr. Berkley, bishop of Cloyne, published a piece called The Analist; in which he took occasion, from some disputes that had arisen concerning the grounds of the fluxionary method, to explode the method itself; and also to charge mathematicians in general with infidelity in religion. Maclaurin thought himself included in this charge, and began an answer to Berkley's book: but other answers coming out, and as he proceeded, so many discoveries, so many new theories and problems occurred to him, that instead of a vindicatory pamphlet, he produced a Complete System of Fluxions, with their application to the most considerable problems in Geometry and Natural Philosophy. This work was published at Edinburgh in 1742, 2 vols 4to; and as it cost him infinite pains, so it is the most considerable of all his works, and will do him immortal honour, being indeed the most complete treatise on that science that has yet appeared.

In the mean time, he was continually obliging the public with some observation or performance of his own, several of which were published in the 5th and 6th volumes of the Medical Essays at Edinburgh. Many of them were likewise published in the Philosophical Transactions; as the following: 1. On the Construction and Measure of Curves, vol. 30.—2. A New Method of describing all kinds of Curves, vol. 30.—3. On Equations with Impossible Roots, vol. 34.—4. On the Roots of Equations, &c. vol. 34.—5. On the Description of Curve Lines, vol. 39.—6. Continuation of the same, vol. 39.—7. Observations on a Solar Eclipse, vol. 40.—8. A Rule for finding the Meridional Parts of a Spheroid with the same Exactness as in a Sphere, vol. 41.—9. An Account of the Treatise of Fluxions, vol. 42.—10. On the Bases of the Cells where the Bees deposit their Honey, vol. 42.

In the midst of these studies, he was always ready to lend his assistance in contriving and promoting any scheme which might contribute to the public service. When the earl of Morton went, in 1739, to visit his estates in Orkney and Shetland, he requested Mr. Maclaurin to assist him in settling the geography of those countries, which is very erroneous in all our maps; to examine their natural history, to survey the coasts, and to take the measure of a degree of the meridian. Maclaurin's

8

laurin's family affairs would not permit him to comply with this requeſt: he drew up however a memorial of what he thought neceſſary to be obſerved, and furniſhed proper inſtruments for the work, recommending Mr. Short, the noted optician, as a fit operator for the management of them.

Mr. Maclaurin had ſtill another ſcheme for the improvement of geography and navigation, of a more extenſive nature; which was the opening a paſſage from Greenland to the South Sea by the North Pole. That ſuch a paſſage might be found, he was ſo fully perſuaded, that he uſed to ſay, if his ſituation could admit of ſuch adventures, he would undertake the voyage, even at his own charge. But when ſchemes for finding it were laid before the parliament in 1741, and he was conſulted by ſeveral perſons of high rank concerning them, and before he could finiſh the memorials he propoſed to ſend, the premium was limited to the diſcovery of a north-weſt paſſage: and he uſed to regret that the word Weſt was inſerted, becauſe he thought that paſſage, if at all to be found, muſt lie not far from the pole.

In 1745, having been very active in fortifying the city of Edinburgh againſt the rebel army, he was obliged to fly from thence into England, where he was invited by Dr. Herring, archbiſhop of York, to reſide with him during his ſtay in this country. In this expedition however, being expoſed to cold and hardſhips, and naturally of a weak and tender conſtitution, which had been much more enfeebled by cloſe application to ſtudy, he laid the foundation of an illneſs which put an end to his life, in June 1746, at 48 years of age, leaving his widow with two ſons and three daughters.

Mr. Maclaurin was a very good, as well as a very great man, and worthy of love as well as admiration. His peculiar merit as a philoſopher was, that all his ſtudies were accommodated to general utility; and we find, in many places of his works, an application even of the moſt abſtruſe theories, to the perfecting of mechanical arts. For the ſame purpoſe, he had reſolved to compoſe a courſe of Practical Mathematics, and to reſcue ſeveral uſeful branches of the ſcience from the ill treatment they often met with in leſs ſkilful hands. Theſe intentions however were prevented by his death; unleſs we may reckon, as a part of his intended work, the tranſlation of Dr. David Gregory's Practical Geometry, which he reviſed, and publiſhed with additions, in 1745.

In his lifetime, however, he had frequent opportunities of ſerving his friends and his country by his great ſkill. Whatever difficulty occurred concerning the conſtructing or perfecting of machines, the working of mines, the improving of manufactures, the conveying of water, or the execution of any public work, he was always ready to reſolve it. He was employed to terminate ſome diſputes of conſequence that had ariſen at Glaſgow concerning the gauging of veſſels; and for that purpoſe preſented to the commiſſioners of the exciſe two elaborate memorials, with their demonſtrations, containing rules by which the officers now act. He made alſo calculations relating to the proviſion, now eſtabliſhed by law, for the children and widows of the Scotch clergy, and of the profeſſors in the univerſities, entitling them to certain annuities and ſums, upon the voluntary

annual payment of a certain ſum by the incumbent. In contriving and adjuſting this wiſe and uſeful ſcheme, he beſtowed a great deal of labour, and contributed not a little towards bringing it to perfection.

Of his works, we have mentioned his *Geometria Organica*, in which he treats of the deſcription of curve lines by continued motion; as alſo of his piece which gained the prize of the Royal Academy of Sciences in 1724. In 1740, he likewiſe ſhared the prize of the ſame Academy, with the celebrated D. Bernoulli and Euler, for reſolving the problem relating to the motion of the tides from the theory of gravity: a queſtion which had been given out the former year, without receiving any ſolution. He had only ten days to draw this paper up in, and could not find leiſure to tranſcribe a fair copy; ſo that the Paris edition of it is incorrect. He afterwards reviſed the whole, and inſerted it in his Treatiſe of Fluxions; as he did alſo the ſubſtance of the former piece. Theſe, with the Treatiſe of Fluxions, and the pieces printed in the Medical Eſſays and the Philoſophical Tranſactions, a liſt of which is given above, are all the writings which our author lived to publiſh. Since his death, however, two more volumes have appeared; his *Algebra*, and his *Account of Sir Iſaac Newton's Philoſophical Diſcoveries*. The Algebra, though not finiſhed by himſelf, is yet allowed to be excellent in its kind; containing, in no large volume, a complete elementary treatiſe of that ſcience, as far as it has hitherto been carried; beſides ſome neat analytical papers on curve lines. His Account of Newton's Philoſophy was occaſioned in the following manner:—Sir Iſaac dying in the beginning of 1728, his nephew, Mr. Conduitt, propoſed to publiſh an account of his life, and deſired Mr. Maclaurin's aſſiſtance. The latter, out of gratitude to his great benefactor, cheerfully undertook, and ſoon finiſhed, the Hiſtory of the Progreſs which Philoſophy had made before Newton's time; and this was the firſt draught of the work in hand; which not going forward, on account of Mr. Conduitt's death, was returned to Mr. Maclaurin. To this he afterwards made great additions, and left it in the ſtate in which it now appears. His main deſign ſeems to have been, to explain only thoſe parts of Newton's philoſophy, which have been controverted. and this is ſuppoſed to be the reaſon why his grand diſcoveries concerning light and colours are but tranſiently and generally touched upon; for it is known, that whenever the experiments, on which his doctrine of light and colours is founded, had been repeated with due care, this doctrine had not been conteſted; while his accounting for the celeſtial motions, and the other great appearances of nature, from gravity, had been miſunderſtood, and even attempted to be ridiculed.

MACULÆ, in Aſtronomy, are dark ſpots appearing on the luminous ſurfaces of the ſun and moon, and even ſome of the planets.

The Solar Maculæ are dark ſpots of an irregular and changeable figure, obſerved in the face of the ſun. Theſe were firſt obſerved in November and December of the year 1610, by Galileo in Italy, and Harriot in England, unknown to, and independent of each other, ſoon after they had made or procured teleſcopes. They were afterwards alſo obſerved by Scheiner, Hevelius, Flamſteed, Caſſini, Kirch, and others. See Philoſ. Tranſ. vol. 1, pa. 274, and vol. 64, pa. 194.

There

There have been various observations made of the phenomena of the solar maculæ, and hypotheses invented for explaining them. Many of these maculæ appear to consist of heterogeneous parts; the darker and denser being called, by Hevelius, nuclei, which are encompassed as it were with atmospheres, somewhat rarer and less obscure; but the figure, both of the nuclei and entire maculæ, is variable. These maculæ are often subject to sudden mutations: In 1644 Hevelius observed a small thin macula, which in two days time grew to ten times its bulk, appearing also much darker, and having a larger nucleus: the nucleus began to fail sensibly before the spot disappeared; and before it quite vanished, it broke into four, which re-united again two days after. Some maculæ have lasted 2, 3, 10, 15, 20, 30, but seldom 40 days; though Kirchius observed one in 1681, that was visible from April 26th to the 17th of July. It is found that the spots move over the sun's disc with a motion somewhat slacker near the edge than in the middle parts; that they contract themselves near the limb, and in the middle appear larger; that they often run into one in the disc, though separated near the centre; that many of them first appear in the middle, and many disappear there; but that none of them deviate from their path near the horizon; whereas Hevelius, observing Mercury in the sun near the horizon, found him too low, being depressed 27″ beneath his former path.

From these phenomena are collected the following consequences. 1. That since Mercury's depression below his path arises from his parallax, the maculæ, having no parallax from the sun, are much nearer him than that planet.

2. That, since they rise and disappear again in the middle of the sun's disc, and undergo various alterations with regard both to bulk, figure, and density, they must be formed *de novo*, and again dissolved about the sun; and hence some have inferred, that they are a kind of solar clouds, formed out of his exhalations; and if so, the sun must have an atmosphere.

3. Since the spots appear to move very regularly about the sun, it is hence inferred, that it is not that they really move, but that the sun revolves round his axis, and the spots accompany him, in the space of 27 days 12 hours 20 minutes.

4. Since the sun appears with a circular disc in every situation, his figure, as to sense, must be spherical.

The magnitude of the surface of a spot may be estimated by the time of its transit over a hair in a fixed telescope. Galileo estimates some spots as larger than both Asia and Africa put together: but if he had known more exactly the sun's parallax and distance, as they are known now, he would have found some of those spots much larger than the whole surface of the earth. For, in 1612, he observed a spot so large as to be plainly visible to the naked eye; and therefore it subtended an angle of about a minute. But the earth, seen at the distance of the sun, would subtend an angle of only about 17″: therefore the diameter of the spot was to the diameter of the earth, as 60 to 17, or $3\frac{1}{2}$ to 1 nearly; and consequently the surface of the spot, if circular, to a great circle of the earth, as $12\frac{1}{4}$ to 1, and to the whole surface of the earth, as $12\frac{1}{4}$ to 4, or nearly 3 to 1. Gassendus observed a spot whose breadth was

$\frac{1}{7}$ of the sun's diameter, and which therefore subtended an angle at the eye of above a minute and a half; and consequently its surface was above seven times larger than the surface of the whole earth. He says he observed above 40 spots at once, though without sensibly diminishing the light of the sun.

Various opinions have been formed concerning the nature, origin, and situation of the solar spots; but the most probable seems to be that of Dr. Wilson, professor of practical astronomy in the university of Glasgow. By attending particularly to the different phases presented by the umbra, or shady zone, of a spot of an extraordinary size that appeared on the sun, in the month of November 1769, during its progress over the solar disc, Dr. Wilson was led to form a new and singular conjecture on the nature of these appearances; which he afterwards greatly strengthened by repeated observations. The results of these observations are, that the solar maculæ are cavities in the body of the sun; that the nucleus, as the middle or dark part has usually been called, is the bottom of the excavations; and that the umbra, or shady zone surrounding it, is the shelving sides of the cavity. Dr. Wilson, besides having satisfactorily ascertained the reality of these immense excavations in the body of the sun, has also pointed out a method of measuring the depth of them. He estimates, in particular, that the nucleus, or bottom of the large spot above mentioned, was not less than a semidiameter of the earth, or about 4000 miles below the level of the sun's surface; while its other dimensions were of a much larger extent. He observed that a spot near the middle of the sun's disc, is surrounded equally on all sides with its umbra; but that when, by its apparent motion over the sun's disc, it comes near the western limb, that part of the umbra which is next the sun's centre gradually diminishes in breadth, till near the edge of the limb it totally disappears; whilst the umbra on the other side of it is little or nothing altered. After a semirevolution of the sun on his axis, if the spot appear again, it will be on the opposite side of the disc, or on the left hand, and the part of the umbra which had before disappeared, is now plainly to be seen; while the umbra on the other side of the spot, seems to have vanished in its turn; being hid from the view by the upper edge of the excavation, from the oblique position of its sloping sides with respect to the eye. But as the spot advances on the sun's disc, this umbra, or side of the cavity, comes in sight; at first appearing narrow, but afterwards gradually increasing in breadth, as the spot moves towards the middle of the disc. Which appearances perfectly agree with the phases that are exhibited by an excavation in a spherical body, revolving on its axis; the bottom of the cavity being painted black, and the sides lightly shaded.

From these, and other observations, it is inferred, that the body of the sun, at the depth of the nucleus, emits little or no light, when seen at the same time, and compared with that resplendent, and probably, in some degree, fluid substance, that covers his surface.

This manner of considering these phenomena, naturally gives rise to many curious speculations and inquiries. It is natural, for instance, to inquire, by what great commotion this refulgent matter is thrown up on all sides, so as to expose to our view the darker part of

the

the fun's body, which was before covered by it ? what is the nature of this fhining matter ? and why, when an excavation is made in it, is the luftre of this fhining fubftance, which forms the fhelving fides of the cavity, fo far diminifhed, as to give the whole the appearance of a fhady zone, or darkifh atmofphere, furrounding the denuded part of the fun's body ? On thefe, and many other fubjects, Dr. Wilfon has advanced fome ingenious conjectures ; for which fee the Philof. Tranf. vol. 64, art. 1. See alfo fome remarks on this theory, by Mr. Woolafton, in the fame vol. pa. 337, &c.

MADRIER, in Artillery, is a thick plank, armed with plates of iron, and having a cavity fufficient to receive the mouth of a petard, with which it is applied againft a gate, or any thing elfe intended to be broken down.

This term is alfo applied to certain flat beams, fixed to the bottom of a moat, to fupport a wall.

There are alfo Madriers lined with tin, and covered with earth ; ferving as defences againft artificial fires, in lodgments, &c, where there is need of being covered overhead.

MÆSTLIN (MICHAEL), in Latin Mæftlinus, a noted aftronomer of Germany, was born in the duchy of Wittemberg ; but fpent his youth in Italy, where he made a fpeech in favour of Copernicus's fyftem, which brought Galileo over from Ariftotle and Ptolomy, to whom he was before wholly devoted. He afterwards returned to Germany, and became profeffor of mathematics at Tubingen ; where, among his other fcholars, he taught the celebrated Kepler, who has commended feveral of his ingenious inventions, in his Aftronomia Optica.

Mæftlin publifhed many mathematical and aftronomical works ; and died in 1590.—Though Tycho Brahe did not affent to Mæftlin's opinion, yet he allowed him to be an extraordinary perfon, and deeply fkilled in the fcience of aftronomy.

MAGAZINE, a place in which ftores are kept, of arms, ammunition, provifions, &c.

Artillery MAGAZINE, or the Magazine to a field battery, is made about 25 or 30 yards behind the battery, towards the parallels, and at leaft 3 feet under ground, to receive the powder, loaded fhells, port fires, &c —Its roof and fides fhould be well fecured with boards, to prevent the earth from falling in : it has a door, and a double trench or paffage funk from the magazine to the battery, the one to enter, and the other to go out at, to prevent confufion. Sometimes traverfes are made in the paffages, to prevent ricochet fhot from entering the magazine.

Powder-MAGAZINE, is the place where powder is kept in large quantities. Authors differ very much with regard to the fituation and conftruction of thefe magazines ; but all agree, that they ought to be arched and bomb proof. In fortifications, they were formerly placed in the rampart ; but of late they have been built in different parts of the town. The firft powder-magazines were made with Gothic arches : but M. Vauban finding thefe too weak, conftructed them of a femicircular form, the dimenfions being 60 feet long within, and 25 feet broad ; the foundations are 8 or 9 feet thick, and 8 feet high from the foundation to the fpring of the arch ; alfo the floor 2 feet from the ground, to keep it from dampnefs.

It is a conftant obfervation, that after the centering of femicircular arches is ftruck, they fettle at the crown, and rife up at the hances, even with a ftraight horizontal extrados ; and ftill much more fo in powder-magazines, where the outfide at top is formed, like the roof of a houfe, by inclined planes joining in an angle over the top of the arch, to give a proper defcent to the rain ; which effects are exactly what might be expected from the true theory of arches. Now, this fhrinking of the arches, as it muft be attended with very bad confequences, by breaking the texture of the cement after it has in fome degree been dried, and alfo by opening the joints of the vouffoirs at one end, fo a remedy is provided for this inconvenience, with regard to bridges, by the arch of equilibration, in my book on the Principles of Bridges : but as the ill confequences of it are much greater in powder-magazines, in queftion 96 of my Mathematical Mifcellany, I propofed to find an arch of equilibration for them alfo ; which queftion was there refolved both by Mr. Wildbore and myfelf, both upon general principles, and which I illuftrated by an application to a particular cafe, which is there conftructed, and accompanied with a table of numbers for that purpofe. Thus, if ALKMB reprefent a vertical tranfverfe fection of the arch, the roof forming an angle LKM of $112° 37'$, alfo PC an ordinate parallel to the horizon taken in any part, and IC perpendicular to the fame ; then for properly conftructing the curve fo as to be the ftrongeft, or an arch of equilibration in all its parts, the correfponding values of PC and CI will be as in the following table, where thofe numbers may denote any lengths whatever, either inches, or feet, or half-yards.

Value of PC	Value of IC
1	7·031
2	7·125
3	7·264
4	7·501
5	7·789
6	8·164
7	8·574
8	9·078
9	9·663
10	10·333

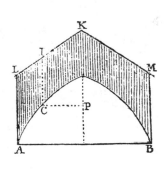

MAGAZINE, or *Powder-Room*, on fhip-board, is a clofe room or ftore-houfe, built in the fore or after part of the hold, in which to preferve the gunpowder for the ufe of the fhip. This apartment is ftrongly fecured againft fire, and no perfon is allowed to enter it with a lamp or candle. it is therefore lighted, as occafion requires, by means of the candles or lamps in the light-room contiguous to it.

MAGELLANIC-CLOUDS, whitifh appearances like clouds, feen in the heavens towards the fouth pole, and having the fame apparent motion as the ftars. They are three in number, two of them near each other.—The largeft lies far from the fouth pole ; but the other two are not many degrees more remote from it than the neareft confpicuous ftar, that is, about 11 degrees.

Mr. Boyle conjectures that if these clouds were seen throuth a good telescope, they would appear to be multitudes of small stars, like the milky way.

MAGIC LANTERN, an optical machine, by means of which small painted images are represented on the wall of a dark room, magnified to any size at pleasure. This machine was contrived by Kircher, (see his Ars Magna Lucis and Umbræ, pa. 768) ; and it was so called, because the images were made to represent strange phantasms, and terrible apparitions, which have been taken for the effect of magic, by such as were ignorant of the secret.

This machine is composed of a concave speculum, from 4 to 12 inches diameter, reflecting the light of a candle through the small hole of a tube, at the end of which is fixed a double convex lens of about 3 inches focus. Between the two are successively placed, many small plain glasses, painted with various figures, usually such as are the most formidable and terrifying to the spectators, when represented at large on the opposite wall.

Thus, (Pl. 13, fig. 14) ABCD is a common tin lantern, to which is added a tube FG to draw out. In H is fixed the metallic concave speculum, from 4 to 12 inches diameter ; or else, instead of it, near the extremity of the tube, there must be placed a convex lens, consisting of a segment of a small sphere, of but a few inches in diameter. The use of this lens is to throw a strong light upon the image ; and sometimes a concave speculum is used with the lens, to render the image still more vivid. In the focus of the concave speculum or lens, is placed the lamp L; and within the tube, where it is soldered to the side of the lantern, is placed a small lens, convex on both sides, being a portion of a small sphere, having its focus about the distance of 3 inches. The extreme part of the tube FM is square, and has an aperture quite through, so as to receive an oblong frame NO passing into it ; in which frame there are round holes, of an inch or two in diameter. Answering to the magnitude of these holes there are drawn circles on a plain thin glass ; and in these circles are painted any figures, or images, at pleasure, with transparent water colours. These images fitted into the frame, in an inverted position, at a small distance from the focus of the lens I, will be projected on an opposite white wall of a dark room, in all their colours, greatly magnified, and in an erect position. By having the instrument so contrived, as that the lens I may move on a slide, the focus may be made, and consequently the image appear distinct, at almost any distance.

Or thus : Every thing being managed as in the former case, into the sliding tube FG, insert another convex lens K, the segment of a sphere rather larger than I. Now, if the picture be brought nearer to I than the distance of the focus, diverging rays will be propagated as if they proceeded from the object ; wherefore, if the lens K be so placed, as that the object be very near its focus, the image will be exhibited on the wall, greatly magnified.

MAGIC SQUARE, is a square figure, formed of a series of numbers in arithmetical progression, so disposed in parallel and equal ranks, as that the sums of each row, taken either perpendicularly, horizontally, or diagonally, are equal to one another. As the annexed square, form-

ed of these nine numbers, 1, 2, 3, 4, 5, 6, 7, 8, 9, where the sum of the three figures in every row, in all directions, is always the same number, viz 15. But if the same numbers be placed in this natural order, the first being 1, and the last of them a square number, they will form what is called a natural square. As in the first 25 numbers, viz, 1, 2, 3, 4, 5, &c to 25.

4	9	2
3	5	7
8	1	6

Natural Square.

1	2	3	4	5
6	7	8	9	10
11	12	13	14	15
16	17	18	19	20
21	22	23	24	25

Magic Square.

16	14	8	2	25
3	22	20	11	9
15	6	4	23	17
24	18	12	10	1
7	5	21	19	13

where every row and diagonal in the magic square makes just the sum 65, being the same as the two diagonals of the natural square.

It is probable that these magic squares were so called, both because of this property in them, viz, that the ranks in every direction make the same sum, appeared extremely surprising, especially in the more ignorant ages, when mathematics passed for magic, and because also of the superstitious operations they were employed in, as the construction of talismans, &c ; for, according to the childish philosophy of those days, which ascribed virtues to numbers, what might not be expected from numbers so seemingly wonderful !

The Magic Square was held in great veneration among the Egyptians, and the Pythagoreans their disciples, who, to add more efficacy and virtue to this square, dedicated it to the then known seven planets divers ways, and engraved it upon a plate of the metal that was esteemed in sympathy with the planet. The square thus dedicated, was inclosed by a regular polygon, inscribed in a circle, which was divided into as many equal parts as there were units in the side of the square ; with the names of the angels of the planet, and the signs of the zodiac written upon the void spaces between the polygon and the circumference of the circumscribed circle. Such a talisman or metal they vainly imagined would, upon occasion, befriend the person who carried it about him.

To Saturn they attributed the square of 9 places or cells, the side being 3, and the sum of the numbers in every row 15 : to Jupiter the square of 16 places, the side being 4, and the amount of each row 34 : to Mars the square of 25 places, the side being 5, and the amount of each row 65 : to the Sun the square with 36 places, the side being 6, and the sum of each row 111 : to Venus the square of 49 places, the side being 7, and the amount of each row 175 : to Mercury the square with 64 places, the side being 8, and the sum of each

each row 260: and to the Moon the square of 81 places, the side being 9, and the amount of each row 369. Finally, they attributed to imperfect matter, the square with 4 divisions, having 2 for its side; and to God the square of only one cell, the side of which is also an unit, which multiplied by itself, undergoes no change.

However, what was at first the vain practice of conjurers and makers of talismans, has since become the subject of a serious research among mathematicians. Not that they imagine it will lead them to any thing of solid use or advantage; but rather as it is a kind of play, in which the difficulty makes the merit, and it may chance to produce some new views of numbers, which mathematicians will not lose the occasion of.

It would seem that Eman. Moschopulus, a Greek author of no high antiquity, is the first now known of, who has spoken of magic squares: he has left some rules for their construction; though, by the age in which he lived, there is reason to imagine he did not look upon them merely as a mathematician.

In the treatise of Cornelius Agrippa, so much accused of magic, are found the squares of seven numbers, viz, from 3 to 9 inclusive, disposed magically; and it is not to be supposed that those seven numbers were preferred to all others without some good reason: indeed it is because their squares, according to the system of Agrippa and his followers, are planetary. The square of 3, for instance, belongs to Saturn; that of 4 to Jupiter; that of 5 to Mars; that of 6 to the Sun; that of 7 to Venus: that of 8 to Mercury; and that of 9 to the Moon.

M. Bachet applied himself to the study of magic squares, on the hint he had taken from the planetary squares of Agrippa, as being unacquainted with Moschopulus's work, which is only in manuscript in the French king's library; and, without the assistance of any author, he found out a new method for the squares of uneven numbers; for instance, 25, or 49, &c; but he could not succeed with those that have even roots.

M. Frenicle next engaged in this subject. It was the opinion of some, that although the first 16 numbers might be disposed 20922789888000 different ways in a natural square, yet they could not be disposed more than 16 ways in a magic square; but M. Frenicle shewed, that they might be thus disposed in 878 different ways.

To this business he thought fit to add a difficulty that had not yet been considered; which was, to take away the marginal numbers quite around, or any other circumference at pleasure, or even several of such circumferences, and yet that the remainder should still be magical.

Again he inverted that condition, and required that any circumference taken at pleasure, or even several circumferences, should be inseparable from the square; that is, that it should cease to be magical when they were removed, and yet continue magical after the removal of any of the rest. M. Frenicle however gives no general demonstration of his methods, and it often seems that he has no other guide but chance. It is true, his book was not published by himself, nor did it appear till after his death, viz, in 1693.

In 1703 M. Poignard, canon of Brussels, published a treatise on sublime magic squares. Before his time there had been no magic squares made, but for serieses of natural numbers that formed a square; but M. Poignard made two very considerable improvements. 1st, Instead of taking all the numbers that fill a square, for instance, the 36 successive numbers, which would fill all the cells of a natural square whose side is 6, he only takes as many successive numbers as there are units in the side of the square, which in this case are 6; and these six numbers alone he disposes in such manner, in the 36 cells, that none of them occur twice in the same rank, whether it be horizontal, vertical, or diagonal; whence it follows, that all the ranks, taken all the ways possible, must always make the same sum; and this method M. Poignard calls repeated progressions. 2d, Instead of being confined to take these numbers according to the series and succession of the natural numbers, that is in arithmetical progression, he takes them likewise in a geometrical progression; and even in an harmonical progression, the numbers of all the ranks always following the same kind of progression: he makes squares of each of these three progressions repeated.

M. Poignard's book gave occasion to M. de la Hire to turn his thoughts to the same subject, which he did with such success, that he greatly extended the theory of magic squares, as well for even numbers as those that are uneven; as may be seen at large in the Memoirs of the Royal Academy of Sciences, for the years 1705 and 1710. See also Saunderson's Algebra, vol. 1, pa. 354, &c; as also Ozanam's Mathematical Recreations, who lays down the following easy method of filling up a magic square.

To form a magic square of an odd number of terms in the arithmetic progression 1, 2, 3, 4, &c. Place the least term 1 in the cell immediately under the middle, or central one, and the rest of the terms, in their natural order, in a descending diagonal direction, till they run off either at the bottom, or on the side: when the number runs off at the bottom, carry it to the uppermost cell, that is not occupied, of the same column that it would have fallen in below, and then proceed descending diagonalwise again as far as you can, or till the numbers either run off at bottom or side, or are interrupted by coming at a cell already filled: now when any number runs off at the right-hand side, then bring it to the farthest cell on the left-hand of the same row or line it would have fallen in towards the right-hand: and when the progress diagonalwise is interrupted by meeting with a cell already occupied by some other number, then descend diagonally to the left from this cell till an empty one is met with, where enter it; and thence proceed as before.

Thus, to make a magic square of the 49 numbers 1, 2, 3, 4, &c. First place the 1 next below the centre cell, and thence descend to the right till the 4 runs off at the bottom, which therefore carry to the top corner on the same column as it would have fallen in; but as runs off at the side, bring it to the beginning of the second line, and

22	47	16	41	10	35	4
5	23	48	17	42	11	29
30	6	24	49	18	36	12
13	31	7	25	43	19	37
38	14	32	1	26	44	20
21	39	8	33	2	27	45
46	15	40	9	34	3	28

and thence defcend to the right till they arrive at the cell occupied by 1 ; carry the 8 therefore to the next diagonal cell to the left, and fo proceed till 10 run off at the bottom, which carry therefore to the top of its column, and fo proceed till 13 runs off at the fide, which therefore bring to the beginning of the fame line, and thence proceed till 15 arrives at the cell occupied by 8 ; from this therefore defcend diagonally to the left ; but as 16 runs off at the bottom, carry it to the top of its proper column, and thence defcend till 21 run off at the fide, which is therefore brought to the beginning of its proper line ; but as 22 arrives at the cell occupied by 15, defcend diagonally to the left, which brings it into the 1ft column, but off at the bottom, and therefore it is carried to the top of that column ; thence defcending till 29 runs off both at bottom and fide, which therefore carry to the higheft unoccupied cell in the laft column ; and here, as 30 runs off at the fide, bring it to the beginning of its proper column, and thence defcend till 35 runs off at the bottom, which therefore carry to the beginning or top of its own column ; and here, as 36 meets with the cell occupied by 29, it is brought from thence diagonally to the left ; thence defcending, 38 runs off at the fide, and therefore it is brought to the beginning of its proper line ; thence defcending, 41 runs off at the bottom, which therefore is carried to the beginning or top of its column ; from whence defcending, 43 arrives at the cell occupied by 36, and therefore it is brought down from thence to the left ; thence defcending, 46 runs off at the fide, which therefore is brought to the beginning of its line ; but here, as 47 runs off at the bottom, it is carried to the beginning or top of its column, from whence defcending with 48 and 49, the fquare is completed, the fum of every row and column and diagonal making juft 175.

There are many other ways of filling up fuch fquares, but none that are eafier than the above one.

It was obferved before, that the fum of the numbers in the rows, columns and diagonals, was 15 in the fquare of 9 numbers, 34 in a fquare of 16, 65 in a fquare of 25, &c ; hence then is derived a method of finding the fums of the numbers in any other fquare, viz, by taking the fucceffive differences till they become equal, and then adding them fucceffively to produce or find out the amount of the following fums. Thus,

Side	Cells	Sums	Diffs.		
0	0	0		0	
1	1	1	1		3
2	4	5	4	3	3
3	9	15	10	6	3
4	16	34	19	9	3
5	25	65	31	12	3
6	36	111	46	15	3
7	49	175	64	18	3
8	64	260	85	21	3
9	81	369	109	24	3
10	100	505	136	27	3
				30	

having ranged the fides and cells in two columns, and a few of the firft fums in a third column, take the firft differences of thefe, which will be 1, 4, 10, 19, &c, as in the 4th column ; and of thefe take the differences 0, 3, 6, 9, 12, &c, as in the 5th column ; and again, of thefe the differences 3, 3, 3 &c, as in the 6th or laft column. Then, returning back again, add always 3, the conftant laft or 3d difference, to the laft found of the 2d differences, which will complete the remainder of the column of thefe, viz, 15, 18, 21, 24, &c : then add thefe 2d differences to the laft found of the 1ft differences, which will complete the column of thefe, viz, giving 31, 46, 64, &c : laftly, add always thefe correfponding 1ft differences to the laft found number or amount of the fums, and the column of fums will thus be completed.

Again, like as the terms of an arithmetical progreffion arranged magically, give the fame fum in every row &c, fo the terms of a geometrical feries arranged magically give the fame product in every row &c, by multiplying the numbers continually together ; fo this progreffion 1, 2, 4, 8, 16, &c, arranged as in the margin, gives, for each continual product, 4096 in every row &c, which is juft the cube of the middle term, 16.

8	256	2
4	16	64
128	1	32

Alfo, the terms of an harmonical progreffion being ranged magically, as in the margin, have the terms in each row &c in harmonical progreffion.

1260	840	630
504	420	360
315	280	252

The ingenious Dr. Franklin, it feems, carried this curious fpeculation farther than any of his predeceffors in the fame way. He conftructed both a *magic fquare of fquares*, and a *magic circle of circles*, the defcription of which is as follows. The magic fquare of fquares is formed by dividing the great fquare as in fig. 1, Pl. 15. The great fquare is divided into 256 little fquares, in which all the numbers from 1 to 256, or the fquare of 16, are placed, in 16 columns, which may be taken either horizontally or vertically. Their chief properties are as follow :

1. The fum of the 16 numbers in each column or row, vertical or horizontal, is 2056.

2. Every half column, vertical and horizontal, makes 1028, or juft one half of the fame fum 2056.

3. Half a diagonal afcending, added to half a diagonal defcending makes alfo the fame fum 2056 ; taking thefe half diagonals from the ends of any fide of the fquare to the middle of it ; and fo reckoning them either upward or downward ; or fideways from right to left, or from left to right.

4. The fame with all the parallels to the half diagonals, as many as can be drawn in the great fquare : for any two of them being directed upward and downward, from the place where they begin, to that where they end, their fums ftill make the fame 2056. Alfo the fame holds true downward and upward ; as well as if taken fideways to the middle, and back to the fame fide again. Only one fet of thefe half diagonals and their parallels, is drawn in the fame fquare upward and downward ; but another fet may be drawn from any of the other three fides.

5. The four corner numbers in the great fquare added to the four central numbers in it, make 1028, the half

Plate XV

Fig. 1.
MAGIC Square of Squares.

200	217	232	249	8	25	40	57	72	89	104	121	136	153	168	181
58	39	26	7	250	231	218	199	186	167	154	135	122	103	90	71
198	219	230	251	6	27	38	59	70	91	102	123	134	155	166	187
60	37	28	5	252	229	220	197	188	165	156	133	124	101	92	69
201	216	233	248	9	24	41	56	73	88	105	120	137	152	169	184
55	42	23	10	247	234	215	202	183	170	151	138	119	106	87	74
203	214	235	246	11	22	43	54	75	86	107	118	139	150	171	182
53	44	21	12	245	236	213	204	181	172	149	140	117	108	85	76
205	212	237	244	13	20	45	52	77	84	109	116	141	148	173	180
51	46	19	14	243	238	241	206	179	174	147	112	115	110	83	78
207	210	239	242	15	18	47	50	79	82	111	114	143	146	175	178
49	48	17	16	241	240	209	208	177	176	145	144	113	112	81	80
196	221	228	253	4	29	36	61	68	93	100	125	132	157	164	189
62	35	30	3	254	227	222	195	190	163	158	131	126	99	94	67
194	223	226	255	2	31	34	63	66	95	98	127	130	159	162	191
64	33	32	1	256	225	224	193	192	161	160	129	128	97	96	65

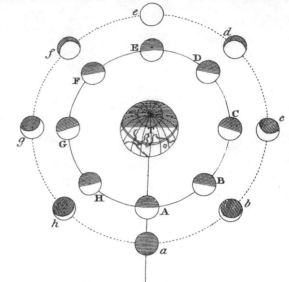

MOON'S Phases.
Fig. 3.

MAGIC Circle of Circles.

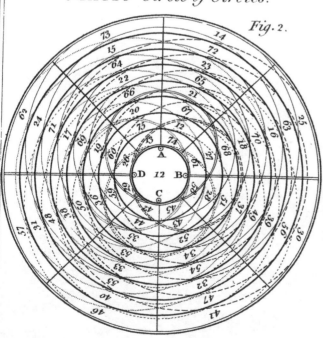

Fig. 2.

Face of the MOON.

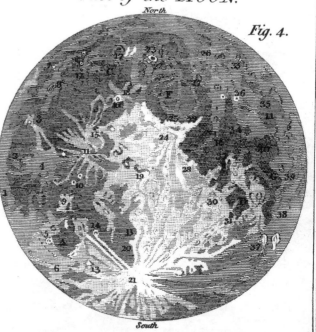

Fig. 4.

North

South

half sum of any vertical or horizontal column, which contains 16 numbers; and also equal to half a diagonal or its parallel.

6. If a square hole, equal in breadth to four of the little squares or cells, be cut in a paper, through which any of the 6 little cells in the great square may be seen, and the paper be laid upon the great square; the sum of all the 16 numbers, seen through the hole, is always equal to 2056, the sum of the 16 numbers in any horizontal or vertical column.

The *Magic Circle of Circles*, fig. 2, pl. 15, by the same author, is composed of a series of numbers, from 12 to 75 inclusive, divided into 8 concentric circular spaces, and ranged in 8 radii of numbers, with the number 12 in the centre; which number, like the centre, is common to all these circular spaces, and to all the radii.

The numbers are so placed, that 1st, the sum of all those in either of the concentric circular spaces above mentioned, together with the central number 12, amount to 360, the same as the number of degrees in a circle.

2. The numbers in each radius also, together with the central number 12, make just 360.

3. The numbers in half of any of the above circular spaces, taken either above or below the double horizontal line, with half the central number 12, make just 180, or half the degrees in a circle.

4. If any four adjoining numbers be taken, as if in a square, in the radial divisions of these circular spaces; the sum of these, with half the central number, make also the same 180.

5. There are also included four sets of other circular spaces, bounded by circles that are excentric with regard to the common centre; each of these sets containing five spaces; and the centres of them being at A, B, C, D. For distinction, these circles are drawn with different marks, some dotted, others by short unconnected lines, &c; or still better with inks of divers colours, as blue, red, green, yellow.

These sets of excentric circular spaces intersect those of the concentric, and each other; and yet, the numbers contained in each of the excentric spaces, taken all around through any of the 20, which are excentric, make the same sum as those in the concentric; namely 360, when the central number 12 is added. Their halves also, taken above or below the double horizontal line, with half the central number, make up 180.

It is observable, that there is not one of the numbers but what belongs at least to two of the circular spaces; some to three, some to four, some to five: and yet they are all so placed, as never to break the required number 360, in any of the 28 circular spaces within the primitive circle. They have also other properties. See Franklin's Exp. and Obs. pa. 350, edit. 4to, 1769; or Ferguson's Tables and Tracts, 1771, pa. 318.

MAGICAL *Picture*, in Electricity, was first contrived by Mr. Kinnersley, and is thus made: Having a large mezzotinto with a frame and glass, as of the king for instance, take out the print, and cut a pannel out of it, near two inches distant from the frame all around; then with thin paste or gum-water, fix the border that is cut off on the inside of the glass, pressing it smooth and close; then fill up the vacancy by gilding the glass well with leaf gold or brass. Gild likewise the inner edge of the back of the frame all round, except the top part, and form a communication between that gilding and the gilding behind the glass; then put in the board, and that side is finished. Next turn up the glass, and gild the foreside exactly over the back gilding and when it is dry, cover it by pasting on the pannel of the picture that has been cut out, observing to bring the corresponding parts of the border and picture together, by which means the picture will appear entire, as at first, only part behind the glass, and part before.

Hold the picture horizontally by the top, and place a small moveable gilt crown on the king's head. If now the picture be moderately electrified, and another person take hold of the frame with one hand, so that his fingers touch its inside gilding, and with the other hand endeavour to take off the crown, he will receive a violent blow, and fail in the attempt. If the picture were highly charged, the consequence might be as fatal as that of high treason. The operator, who holds the picture by the upper end, where the inside of the frame is not gilt, to prevent its failing, feels nothing of the shock, and may touch the face of the picture without danger. And if a ring of persons take the shock among them, the experiment is called the conspirators. See Franklin's Exper. and Observ. pa. 30.

MAGINI (JOHN-ANTHONY), or MAGINUS, professor of mathematics in the university of Bologna, was born at Padua in the year 1536. Magini was remarkable for his great assiduity in acquiring and improving the knowledge of the mathematical sciences, with several new inventions for these purposes, and for the extraordinary favour he obtained from most princes of his time. This doubtless arose partly from the celebrity he had in matters of astrology, to which he was greatly addicted, making horoscopes, and foretelling events, both relating to persons and things. He was invited by the emperor Rodolphus to come to Vienna, where he promised him a professor's chair, about the year 1597; but not being able to prevail on him to settle there, he nevertheless gave him a handsome pension.

It is said, he was so much addicted to astrological predictions, that he not only foretold many good and evil events relative to others with success; but even foretold his own death, which came to pass the same year: all which he represented as under the influence of the stars. Tomasini says, that Magini, being advanced to his 61st year, was struck with an apoplexy, which ended his days; and that a long while before, he had told him and others, that he was afraid of that year. And Roffeni, his pupil, says, that Magini died under an aspect of the planets, which, according to his own prediction, would prove fatal to him; and he mentions Riccioli as affirming that he said, the figure of his nativity, and his climacteric year, doomed him to die about that time; which happened in 1618, in the 62d year of his age.

His writings do honour to his memory, as they were very considerable, and upon learned subjects. The principal were the following: 1. His Ephemeris, in 3 volumes, from the year 1580 to 1630.—2. Tables of Secondary Motions.—3. Astronomical, Gnomonical, and Geographical Problems.—4. Theory of the Planets, according to Copernicus.—5. A Confutation of Scaliger's Dissertation concerning the Precession of the

K 2   Equinox.

Equinox.—6. A Primum Mobile, in 12 books.—7. A Treatife of Plane and Spherical Trigonometry.—8. A Commentary on Ptolomy's Geography.—9. A Chorographical Defcription of the Regions and Cities of Italy, illuftrated with 60 maps; with fome other papers on Aftrological fubjects.

MAGNET, MAGNES, the *Loadftone;* a kind of ferruginous ftone, refembling iron ore in weight and colour, though rather harder and heavier; and is endued with divers extraordinary properties, attractive, directive, inclinatory, &c. See MAGNETISM.

The Magnet is alfo called *Lapis Heraclæus*, from Heraclea, a city of Magnefia, a port of the ancient Lydia, where it was faid it was firft found, and from which it is ufually fuppofed that it took its name. Though fome derive the word from a fhepherd named *Magnes*, who firft difcovered it on Mount Ida with the iron of his crook. It is alfo called *Lapis Nauticus*, from its ufe in navigation; alfo *Siderites*, from its virtue in attracting iron, which the Greeks call σιδηρος.

The Magnet is ufually found in iron mines, and fometimes in very large pieces, half magnet, half iron. Its colour is different, as found in different countries. Norman obferves, that the beft are thofe brought from China and Bengal, which are of an irony or fanguine colour; thofe of Arabia are reddifh; thofe of Macedonia, blackifh; and thofe of Hungary, Germany, England, &c, the colour of unwrought iron. Neither its figure nor bulk are conftant or determined; being found of all fhapes and fizes.

The Ancients reckoned five kinds of Magnets, different in colour and virtue: the Ethiopic, Magnefian, Bœotic, Alexandrian, and Natolian. They alfo took it to be male and female: but the chief ufe they made of it was in medicine; efpecially for the cure of burns and defluxions of the eyes.—The Moderns, more happy, take it to conduct them in their voyages.

The moft diftinguifhing properties of the Magnet are, That it attracts iron, and that it points towards the poles of the world; and in other circumftances alfo dips or inclines to a point beneath the horizon, directly under the pole; it alfo communicates thefe properties, by touch, to iron. By means of which, are obtained the mariner's needles, both horizontal, and inclinatory or dipping needles.

The Attractive Power of the MAGNET, was known to the Ancients, and is mentioned even by Plato and Euripides, who call it the *Herculean ftone*, becaufe it commands iron, which fubdues every thing elfe: but the knowledge of its directive power, by which it difpofes its poles along the meridian of every place, or nearly fo, and caufes needles, pieces of iron, &c, touched with it, to point nearly north and fouth, is of a much later date; though the difcoverer himfelf, and the exact time of the difcovery, be not now known. The firft mention of it is about 1260, when it has been faid that Marco Polo, a Venetian, introduced the mariner's compafs; though not as an invention of his own, but as derived from the Chinefe, who it feems had the ufe of it long before; though fome imagine that the Chinefe rather borrowed it from the Europeans.

But Flavio de Gira, a Neapolitan, who lived in the 13th century, is the perfon ufually fuppofed to have the beft title to the difcovery; and yet Sir G. Wheeler mentions, that he had feen a book of aftronomy much older, which fuppofed the ufe of the needle; though not as applied to the purpofes of navigation, but of aftronomy. And in Guiot de Provins, an old French poet, who wrote about the year 1180, there is an exprefs mention made of the loadftone and the compafs; and their ufe in navigation obliquely hinted at.

The Variation of the MAGNET, or needle, or its deviation from the pole, was firft difcovered by Sebaftian Cabot, a Venetian, in 1500; and the variation of that variation, or change in its direction, by Mr. Henry Gellibrand, profeffor of aftronomy in Grefham college, about the year 1625.

Laftly, the Dip or inclination of the needle, when at liberty to play vertically, to a point beneath the horizon, was firft difcovered by another of our countrymen, Mr. Robert Norman, about the year 1576.

The Phenomena of the MAGNET, are as follow: 1, In every Magnet there are two poles, of which the one points northwards, the other fouthwards; and if the Magnet be divided into ever fo many pieces, the two poles will be found in each piece. The poles of a Magnet may be found by holding a very fine fhort needle over it; for where the poles are, the needle will ftand upright, but no where elfe.—2, Thefe poles, in different parts of the globe, are differently inclined towards a point under the horizon.—3, Thefe poles, though contrary to each other, do help mutually towards the Magnet's attraction, and fufpenfion of iron. —4, If two Magnets be fpherical, one will turn or conform itfelf to the other, fo as either of them would do to the earth; and after they have fo conformed or turned themfelves, they endeavour to approach or join each other; but if placed in a contrary pofition, they avoid each other.—5, If a Magnet be cut through the axis, the fegments or parts of the ftone, which before were joined, will now avoid and fly each other.—6, If the Magnet be cut perpendicular to its axis, the two points, which before were conjoined, will become contrary poles; one in the one, and one in the other fegment.—7, Iron receives virtue from the Magnet by application to it, or barely from an approach near it, though it do not touch it; and the iron receives this virtue varioufly, according to the parts of the ftone it is made to touch, or even approach to.— 8, If an oblong piece of iron be anyhow applied to the ftone, it receives virtue from it only lengthways.—9, The Magnet lofes none of its own virtue by communicating any to the iron; and this virtue it can communicate to the iron very fpeedily: though the longer the iron joins or touches the ftone, the longer will its communicated virtue hold; and a better Magnet will communicate more of it, and fooner, than one not fo good. —10, Steel receives virtue from the Magnet better than iron.—11, A needle touched by a Magnet will turn its ends the fame way towards the poles of the world, as the Magnet itfelf does.—12, Neither loadftone nor needles touched by it do conform their poles exactly to thofe of the world, but have ufually fome variation from them: and this variation is different in divers places, and at divers times in the fame places.— 13, A loadftone will take up much more iron when armed, or capped, than it can alone. (A loadftone is faid to be armed, when its poles are furrounded with

plates

plates of steel : and to determine the quantity of steel to be applied, try the Magnet with several steel bars ; and the greatest weight it takes up, with a bar on, is to be the weight of its armour.) And though an iron ring or key be suspended by the loadstone, yet this does not hinder the ring or key from turning round any way, either to the right or left.—14, The force of a loadstone may be variously increased or lessened by variously applying to it, either iron, or another loadstone.—15, A strong Magnet at the least distance from a smaller or a weaker, cannot draw to it a piece of iron adhering actually to such smaller or weaker stone ; but if it come to touch it, it can draw it from the other : but a weaker Magnet, or even a small piece of iron, can draw away or separate a piece of iron contiguous to a larger or stronger Magnet.—16, In these northern parts of the world, the south pole of a Magnet will raise up more iron than its north pole.—17, A plate of iron only, but no other body interposed, can impede the operation of the loadstone ; either as to its attractive or directive quality.—18, The power or virtue of a loadstone may be impaired by lying long in a wrong position, as also by rust, wet, &c ; and may be quite destroyed by fire, lightning, &c.—19, A piece of iron wire well touched, upon being bent round in a ring, or coiled round on a stick, &c, will always have its directive virtue diminished, and often quite destroyed. And yet if the whole length of the wire were not entirely bent, so that the ends of it, though but for the length of one-tenth of an inch, were left straight, the virtue will not be destroyed in those parts ; though it will in all the rest.—20, The sphere of activity of Magnets is greater and less at different times. Also, the variation of the needle from the meridian, is various at different times of the day.—21, By twisting a piece of wire touched with a Magnet, its virtue is greatly diminished ; and sometimes so disordered and confused, that in some parts it will attract, and in others repel ; and even, in some places, one side of the wire seems to be attracted, and the other side repelled, by one and the same pole of the stone.—22, A piece of wire that has been touched, on being split, or cleft in two, the poles are sometimes changed, as in a cleft Magnet ; the north pole becoming the south, and the south the north : and yet sometimes one half of the wire will retain its former poles, and the other half will have them changed.—23, A wire being touched from end to end with one pole of a Magnet, the end at which you begin will always turn contrary to the pole that touched it : and if it be again touched the same way with the other pole of the Magnet, it will then be turned the contrary way.—24, If a piece of wire be touched in the middle with only one pole of the Magnet, without moving it backwards or forwards ; in that place will be the pole of the wire, and the two ends will be the other pole.—25, If a Magnet be heated red hot, and again cooled either with its south pole towards the north in a horizontal position, or with its south pole downwards in a perpendicular position, its poles will be changed.—26, Mr. Boyle (to whom we are indebted for the following magnetical phenomena) found he could presently change the poles of a small fragment of a loadstone, by applying them to the opposite vigorous poles of a large one.—27, Hard iron tools well tempered,

when heated by a brisk attrition, as filing, turning, &c, will attract thin filings or chips of iron, steel, &c ; and hence we observe that files, punches, augres, &c, have a small degree of magnetic virtue.—28, The iron bars of windows, &c, which have stood a long time in an erect position, grow permanently magnetical ; the lower ends of such bars being the north pole, and the upper end the south pole.—29, A bar of iron that has not stood long in an erect posture, if it be only held perpendicularly, will become magnetical, and its lower end the north pole, as appears from its attracting the south pole of a needle : but then this virtue is transient, and by inverting the bar, the poles change their places. In order therefore to render the quality permanent in an iron bar, it must continue a long time in a proper position. But fire will produce the effect in a short time : for as it will immediately deprive a loadstone of its attractive virtue ; so it soon gives a verticity to a bar of iron, if, being heated red hot, it be cooled in an erect posture, or directly north and south. Even tongs and fireforks, by being often heated, and set to cool again in a posture nearly erect, have gained this magnetic property. Sometimes iron bars, by long standing in a perpendicular position, have acquired the magnetic virtue in a surprising degree. A bar about 10 feet long, and three inches thick, supporting the summer beam of a room, was able to turn the needle at 8 or 10 feet distance, and exceeded a loadstone of 3½ pounds weight : from the middle point upwards it was a north pole, and downwards a south pole. And Mr. Martin mentions a bar, which had been the beam of a large steel-yard that had several poles in it.—30, Mr. Boyle found, that by heating a piece of English oker red-hot, and placing it to cool in a proper posture, it manifestly acquired a magnetic virtue. And an excellent Magnet, belonging to the same ingenious gentleman, having lain near a year in an inconvenient posture, had its virtue greatly impaired, as if it had been by fire.—31, A needle well touched, it is known, will point north and south : if it have one contrary touch of the same stone, it will be deprived of its faculty ; and by another such touch, it will have its poles interchanged.—32, If an iron bar have gained a verticity by being heated red-hot and cooled again, north and south, and then hammered at the two ends ; its virtue will be destroyed by two or three smart blows on the middle.—33, By drawing the back of a knife, or a long piece of steel-wire, &c, leisurely over the pole of a loadstone, carrying the motion from the middle of the stone to the pole ; the knife or wire will attract one end of a needle ; but if the knife or wire be passed from the said pole to the middle of the stone, it will repel the same end of the needle.—34, Either a Magnet or a piece of iron being laid on a piece of cork, so as to float freely on water ; it will be found, that, whichsoever of the two is held in the hand, the other will be drawn to it : so that iron attracts the Magnet as much as it is attracted by it ; action and re-action being always equal. In this experiment, if the Magnet be set afloat, it will direct its two poles to the poles of the world nearly.—35, A knife &c touched with a Magnet, acquires a greater or less degree of virtue, according to the part it is touched on. It receives the strongest virtue, when it is drawn leisurely from the

handle

handle towards the point over one of the poles. And if the same knife thus touched, and thus possessed of a strong attractive power, be retouched in a contrary direction, viz, by drawing it from the point towards the handle over the same pole, it immediately loses all its virtue.—36, A Magnet acts with equal force in vacuo as in the open air.—37, The smallest Magnets have usually the greatest power in proportion to their bulk. A large Magnet will seldom take up above 3 or 4 times its own weight, while a small one will often take up more than ten times its weight. A Magnet worn by Sir Isaac Newton in a ring, and which weighed only 3 grains, would take up 746 grains, or almost 250 times its own weight. A magnetic bar made by Mr. Canton, weighing 10 oz. 12 dwts, took up more than 79 ounces; and a flat semicircular steel Magnet, weighing 1 oz. 13 dwts, took up an iron wedge of 90 ounces.

Armed MAGNET, denotes one that is capped, cased, or set in iron or steel, to make it take up a greater weight, and also more readily to distinguish its poles. For the methods of doing this, see Mr. Michell's book on this subject.

Artificial MAGNET, is a bar of iron or steel, impregnated with the magnetic virtue, so as to possess all the properties of the natural loadstone, and be used instead of it. How to make Magnets of this kind, by means of a natural Magnet, and even without the assistance of any Magnet, was suggested many years since by Mr. Savary, and particularly described in the Philos. Transf. number 414. See also Abridgment, vol. 6, pa. 260. But as his method was tedious and operose, though capable of communicating a very considerable virtue, it was little practised. Dr. Gowin Knight first brought this kind of Magnets to their present state of perfection, so as to be even of much greater efficacy than the natural ones. But as he refused to discover his methods upon any terms whatever (even, as he said, though he should receive in return as many guineas as he could carry), these curious and valuable secrets in a great measure died with him. The result of his method however was first published in the Philos. Transf. for 1744, art. 8, and for 1745, art. 3. See also the vol. for 1747, art. 2. And in the 69th vol. Mr. Benjamin Wilson has given a process, which at least discovers one of the leading principles of Dr. Knight's art. The method, according to Mr. Wilson, was as follows. Having provided a great quantity of clean iron filings, he put them into a large tub that was more than one-third filled with clean water; he then, with great labour, shook the tub to and fro for many hours together, that the friction between the grains of iron, by this treatment, might break or rub off such small parts as would remain suspended in the water for some time. The water being thus rendered very muddy, he poured it into a clean iron vessel, leaving the filings behind; and when the water had stood long enough to become clear, he poured it out carefully, without disturbing such of the sediment as still remained, which now appeared reduced almost to impalpable powder. This powder was afterwards removed into another vessel, to dry it. Having, by several repetitions of this process, procured a sufficient quantity

of this very fine powder, the next thing was to make a paste of it, and that with some vehicle containing a good quantity of the phlogistic principle; for this purpose, he had recourse to linseed oil, in preference to all other fluids. With these two ingredients only, he made a stiff paste, and took great care to knead it well before he moulded it into convenient shapes. Sometimes, while the paste continued in its soft state, he would put the impression of a seal; one of which is in the British Museum. This paste so moulded was then set upon wood, or a tile, to dry or bake it before a moderate fire, being placed at about one foot distance. He found that a moderate fire was most proper, because a greater degree of heat would make the composition crack in many places. The time requisite for the baking or drying of this paste, was usually about 5 or 6 hours, before it attained a sufficient degree of hardness. When that was done, and the several baked pieces were become cold, he gave them their magnetic virtue in any direction he pleased, by placing them between the extreme ends of his large magazine of artificial magnets, for a few seconds. The virtue they acquired by this method was such, that, when any of those pieces were held between two of his best ten guinea bars, with its poles purposely inverted, it immediately of itself turned about to recover its natural direction, which the force of those very powerful bars was not sufficient to counteract. Philos. Transf. vol. 65, for 1779.

Methods for artificial Magnets were also discovered and published by the Rev. Mr. John Michell, in a Treatise on Artificial Magnets, printed in 1750, and by Mr. John Canton, in the Philos. Transf. for 1751. The process for the same purpose was also found out by other persons, particularly by Du Hamel, Hist. Acad. Roy. 1745 and 1750, and by Marul Uitgeleeze Natuurkund. Verhand. tom. 2, p. 261.

Mr. Canton's method is as follows: Procure a dozen of bars; 6 of soft steel, and 6 of hard, the former to be each 3 inches long, a quarter of an inch broad, and 1-20th of an inch thick; with two pieces of iron, each half the length of one of the bars, but of the same breadth and thickness; and the 6 hard bars to be each 5½ inches long, half an inch broad, and 3-20ths of an inch thick, with two pieces of iron of half the length, but the whole breadth and thickness of one of the hard bars; and let all the bars be marked with a line quite around them at one end. Then take an iron poker and tongs (fig. 1, plate 16), or two bars of iron, the larger they are, and the longer they have been used, the better; and fixing the poker upright between the knees, hold to it, near the top, one of the soft bars, having its marked end downwards by a piece of sewing silk, which must be pulled tight by the left hand, that the bar may not slide: then grasping the tongs with the right hand, a little below the middle, and holding them nearly in a vertical position, let the bar be stroked by the lower end, from the bottom to the top, about ten times on each side, which will give it a magnetic power sufficient to lift a small key at the marked end: which end, if the bar were suspended on a point, would turn towards the north, and is therefore called the north pole; and the unmarked end is, for the same reason, called

Plate XVI.

ARTIFICIAL MAGNETS.

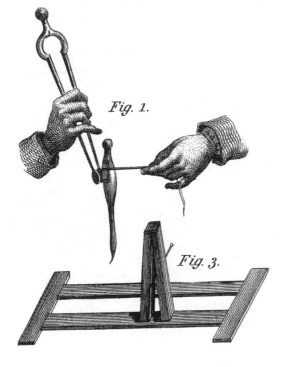

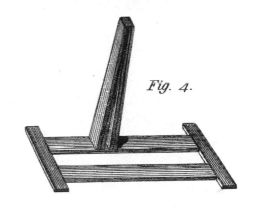

Fig. 1.

Fig. 2.

Fig. 3.

Fig. 4.

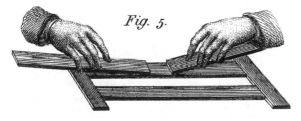

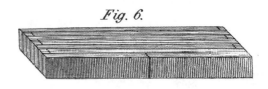

Fig. 5.

Fig. 6.

NEPER'S BONES

Fig. 7.

Fig. 8.

called the fouth pole. Four of the foft bars being impregnated after this manner, lay the two (fig. 2) parallel to each other, at a quarter of an inch diftance, between the two pieces of iron belonging to them, a north and a fouth pole againft each piece of iron : then take two of the four bars already made magnetical, and place them together fo as to make a double bar in thickne, the north pole of one even with the fouth pole of the other ; and the remaining two being put to thefe, one on each fide, fo as to have two north and two fouth poles together, feparate the north from the fouth poles at one end by a large pin, and place them perpendicularly with that end downward on the middle of one of the parallel bars, the two north poles towards its fouth end, and the two fouth poles towards its north end : flide them three or four times backward and forward the whole length of the bar ; then removing them from the middle of this bar, place them on the middle of the other bar as before directed, and go over that in the fame manner ; then turn both the bars the other fide upwards, and repeat the former operation : this being done, take the two from between the pieces of iron ; and, placing the two outermoft of the touching bars in their ftead, let the other two be the outermoft of the four to touch thefe with; and this procefs being repeated till each pair of bars have been touched three or four times over, which will give them a confiderable magnetic power. Put the half-dozen together after the manner of the four (fig. 3), and touch them with two pair of the hard bars placed between their irons, at the diftance of about half an inch from each other; then lay the foft bars afide, and with the four hard ones let the other two be impregnated (fig. 4), holding the touching bars apart at the lower end near two tenths of an inch; to which diftance let them be feparated after they are fet on the parallel bar, and brought together again before they are taken off : this being obferved, proceed according to the method defcribed above, till each pair have been touched two or three times over. But as this vertical way of touching a bar, will not give it quite fo much of the magnetic virtue as it will receive, let each pair be now touched once or twice over in their parallel pofition between the irons (fig. 5), with two of the bars held horizontally, or nearly fo, by drawing at the fame time the north end of one from the middle over the fouth end, and the fouth of the other from the middle over the north end of a parallel bar ; then bringing them to the middle again, without touching the parallel bar, give three or four of thefe horizontal ftrokes to each fide. The horizontal touch, after the vertical, will make the bars as ftrong as they poffibly can be made, as appears by their not receiving any additional ftrength, when the vertical touch is given by a great number of bars, and the horizontal by thofe of a fuperior magnetic power.

This whole procefs may be gone through in about half an hour; and each of the large bars, if well hardened, may be made to lift 28 Troy ounces, and fometimes more. And when thefe bars are thus impregnated, they will give to a hard bar of the fame fize its full virtue in lefs than two minutes ; and therefore will anfwer all the purpofes of Magnetifm in navigation and experimental philofophy, much better than the loadftone, which has not a power fufficient to impregnate

3

hard bars. The half dozen being put into a cafe (fig. 6), in fuch a manner as that no two poles of the fame name may be together, and their irons with them as one bar, they will retain the virtues they have received; but if their power fhould, by making experiments, be ever fo far impaired, it may be reftored without any foreign affiftance in a few minutes. And if, perchance, a much larger fet of bars fhould be required, thefe will communicate to them a fufficient power to proceed with ; and they may, in a fhort time, by the fame method, be brought to their full ftrength.

MAGNETISM, the quality or conftitution of a body, by which it is rendered magnetical, or a magnet, fenfibly attracting iron, and giving it a meridional direction.

This is a tranfient power, capable of being produced, deftroyed, or reftored.

The Laws of MAGNETISM.

Thefe laws are laid down by Mr. Whifton in the following propofitions.——1, The Loadftone has both an attractive and a directive power united together, while iron touched by it has only the former; i. e. the magnet not only attracts needles, or fteel filings, but alfo directs them to certain different angles, with refpect to its own furface and axis ; whereas iron, touched with it, does little or nothing more than attract them ; ftill fuffering them to lie along or ftand perpendicular to its furface and edges in all places, without any fuch fpecial direction.

2. Neither the ftrongeft nor the largeft magnets give a better directive touch to needles, than thofe of a lefs fize or virtue : to which may be added, that whereas there are two qualities in all magnets, an attractive and a directive one ; neither of them depend on, or are any argument of the ftrength of the other.

3. The attractive power of magnets, and of iron, will greatly increafe or diminifh the weight of needles on the balance ; nay, it will overcome that weight, and even fuftain fome other additional alfo: while the directive power has a much fmaller effect. Gaffendus indeed, as well as Merfennus and Gilbert, affert that it has none at all : but by miftake ; for Whifton found, from repeated trials on large needles, that after the touch they weighed lefs than before. One of 4584$\frac{1}{4}$ grains, loft 2$\frac{1}{2}$ grains by the touch ; and another of 65726 grains weight, no lefs than 14 grains.

4. It is probable that iron confifts almoft wholly of the attractive particles; and the magnet, of the attractive and directive together ; mixed, probably, with other heterogeneous matter ; as having never been purged by the fire, which iron has; and hence may arife the reafon why iron, after it has been touched, will lift up a much greater weight than the loadftone that touched it.

5. The quantity and direction of magnetic powers, communicated to needles, are not properly, after fuch communication, owing to the magnet which gave the touch ; but to the goodnefs of the fteel that receives it, and to the ftrength and pofition of the terreftrial loadftone, whofe influence alone thofe needles are afterwards fubject to, and directed by : fo that all fuch needles, if good, move with the fame ftrength, and point to the fame angle, whatever loadftone they may have been excited by, provided it be but a good one. Nor does it

feem

ſeem that the touch does much more in magnetical caſes, than attrition does in electrical ones; i. e. ſerving to rub off ſome obſtructing particles, that adhere to the ſurface of the ſteel, and opening the pores of the body touched, and ſo make way for the entrance and exit of ſuch effluvia as occaſion or aſſiſt the powers we are ſpeaking of. Hence Mr. Whiſton takes occaſion to obſerve, that the directive power of the loadſtone ſeems to be mechanical, and to be derived from magnetic effluvia, circulating continually round it.

6. The abſolute attractive power of different armed loadſtones, is, *cæteris paribus*, not according to either the diameters or ſolidities of the loadſtones, but according to the quantity of their ſurfaces, or in the duplicate proportion of their diameters.

7. The power of good magnets unarmed, ſenſibly equal in ſtrength, ſimilar in figure and poſition, but unequal in magnitude, is ſometimes a little greater, ſometimes a little leſs, than in the proportion of their ſimilar diameters.

8. The loadſtone attracts needles that have been touched, and others that have not been touched, with equal force at diſtances unequal, viz. when the diſtance of the former is to the diſtance of the latter, as 5 to 2.

9. Both poles of a magnet equally attract needles, till they are touched; then it is, and then only, that one pole begins to attract one end, and repel the other: though the repelling pole will ſtill attract upon contact, and even at very ſmall diſtances.

10. The attractive power of loadſtones, in their ſimilar poſition to, but different diſtances from, magnetic needles, is in the ſeſquiduplicate proportion of the diſtances of their ſurfaces from their needles reciprocally; or as the mean proportionals between the ſquares and the cubes of thoſe diſtances reciprocally; or as the ſquare roots of the 5th powers of thoſe diſtances reciprocally. Thus, the magnetic force of attraction, at twice the diſtance from the ſurface of the loadſtone, is between a 5th and 6th part of the force at the firſt diſtance; at thrice the diſtance, the force is between the 15th and 16th part; at four times the diſtance, the power is the 32d part of the firſt; and at ſix times the diſtance, it is the 88th part. Where it is to be noted, that the diſtances are not counted from the centre, as in the laws of gravity, but from the ſurface: all experience aſſuring us, that the magnetic power reſides chiefly, if not wholly, in the ſurfaces of the loadſtone and iron; without any particular relation to any centre at all. The proportion here laid down was determined by Mr. Whiſton, from a great number of experiments by Mr. Hawkſbee, Dr. Brook Taylor, and himſelf; meaſuring the force by the chords of thoſe arcs by which the magnet at ſeveral diſtances draws the needle out of its natural direction, to which chords, as he demonſtrates, it is always proportional. The numbers in ſome of their moſt accurate trials, he gives in the following Table, ſetting down the half chords, or the ſines of half thoſe arcs of declination, as the true meaſures of the force of magnetic attraction.

Diſtances in inches.	*Degrees of inclination.*	*Sines of $\frac{1}{2}$ arcs.*	*Seſquiduplicate ratio.*
20	2	175	466
$14\frac{8}{9}$	4	349	216
$13\frac{3}{8}$	6	523	170
$12\frac{2}{8}$	8	697	138
$11\frac{1}{8}$	10	871	105
$10\frac{1}{4}$	12	1045	87
$9\frac{1}{4}$	14	1219	70

Other perſons however have found ſome variations in the proportions of magnetic force with reſpect to diſtance: Thus, Newton ſuppoſes it to decreaſe nearly in the triplicate ratio of the diſtance: Mr. Martin obſerves, that the power of his loadſtone decreaſes in the ſeſquiduplicate ratio of the diſtances inverſely: but Dr. Helſham and Mr. Michell found it to be as the ſquare of the diſtance inverſely: while others, as Dr. Brook Taylor and M. Muſchenbroek, are of opinion, that this power follows no certain ratio at all, and that the variation is different in different ſtones.

11. An inclinatory, or dipping-needle, of 6 inches radius, and of a priſmatic or cylindric figure, when it oſcillates along the magnetic meridian, performs there every mean vibration in about 6″ or 360‴, and every ſmall oſcillation in about 5″½, or 330‴: and the ſame kind of needle, 4 feet long, makes every mean oſcillation in about 24″, and every ſmall one in about 22″.

12. The whole power of Magnetiſm in this country, as it affects needles a foot long, is to that of gravity nearly as 1 to 300; and as it affects needles 4 feet long, as 1 to 600.

13. The quantity of magnetic power accelerating the ſame dipping-needle, as it oſcillates in different vertical planes, is always as the coſines of the angles made by thoſe planes with the magnetic meridian, taken on the horizon.

Thus, in eſtimating the quantity of force in the horizontal and in the vertical ſituations of needles at London, it is found that the latter, in needles of a foot long, is to the whole force along the magnetic meridian, as 96 to 100; and in needles 4 feet long, as 9667 to 10000: whereas, in the former, the whole force in needles of a foot long, is as 28 to 100; and in thoſe of 4 feet long, as 256 to 1000. Whence it follows, that the power by which horizontal needles are governed in theſe parts of the world, is but the quarter of the power by which the dipping-needle is moved.

Hence alſo, as the horizontal needle is moved only by a part of the power that moves the dipping-needle; and as it only points to a certain place in the horizon, becauſe that place is the neareſt to its original tendency of any that its ſituation will allow it to tend to; whenever the dipping-needle ſtands exactly perpendicular to the horizon, the horizontal needle will not reſpect one point of the compaſs more than another, but will wheel about any way uncertainly.

14. The time of oſcillation and vibration, both in dipping and horizontal needles, that are equally good, is as their length directly; and the actual velocities of their points along their arcs, are always unequal. And hence, magnetical needles are, *cæteris paribus*, ſtill better, the longer they are; and that in the ſame proportion with their lengths.

Of

Of the Caufes of MAGNETISM. Though many authors have propofed hypothefes, or written concerning the caufe of Magnetifm, as Plutarch, Defcartes, Boyle, Newton, Gilbert, Hartfoeker, Halley, Whifton, Knight, Beccaria, &c; nothing however has yet appeared that can be called a fatisfactory folution of its phenomena: It is certain indeed, that both natural and artificial electricity will give polarity to needles, and even reverfe their poles; but though from this it may appear probable that the electric fluid is alfo the caufe of Magnetifm, yet in what manner the fluid acts while producing the magnetical phenomena, feems to be quite unknown.

Dr. Knight indeed deduces from feveral experiments the following propofitions, which he offers, not fo much to explain the nature of the caufe of Magnetifm, as the manner in which it acts: the magnetic matter of a loadftone, he fays, moves in a ftream from one pole to the other internally, and is then carried back in a curve line externally, till it arrive again at the pole where it firft entered, to be again admitted: the immediate caufe why two or more magnetical bodies attract each other, is the flux of one and the fame ftream of magnetical matter through them; and the immediate caufe of magnetic repulfion, is the conflux and accumulation of the magnetic matter. Philof. Tranf. vol. 44, pa. 665.

Mr. Michell rejects the motion of a fubtle fluid; but though he propofed to publifh a theory of Magnetifm eftablifhed by experiments, no fuch theory has appeared.

Signor Beccaria, from obferving that a fudden ftroke of lightning gives polarity to Magnets, conjectures, that a regular and conftant circulation of the whole mafs of the electric fluid from north to fouth may be the original caufe of Magnetifm in general. This current he would not fuppofe to arife from one fource, but from feveral, in the northern hemifphere of the earth: the aberration of the common centre of all the currents from the north point, may be the caufe of the variation of the needle; the period of this declination of the centre of the currents, may be the period of the variation; and the obliquity with which the currents ftrike into the earth, may be the caufe of the dipping of the needle, and alfo why bars of iron more eafily receive the magnetic virtue in one particular direction. Lettre dell' Elettricifmo, pa. 269; or Prieftley's Hift. Elec. vol. 1, pa. 409. See alfo Cavallo's Treatife on Magnetifm.

MAGNIFYING, is the making of objects appear larger than they ufually and naturally appear to the eye; whence convex lenfes, which have the power of doing this, are called Magnifying Glaffes.

The Magnifying power of denfe mediums of certain figures, was known to the Ancients; though they were far from underftanding the caufe of this effect. Seneca fays, that fmall and obfcure letters appear larger and brighter through a glafs globe filled with water; and he abfurdly accounts for it by faying, that the eye flides in the water, and cannot lay hold of its object. And Alexander Aphrodifenfis, about two centuries after Seneca, fays, that the reafon why apples appear large when immerfed in water, is, that the water which is contiguous to any body is affected with

the fame quality and colour; fo that the eye is deceived in imagining the body itfelf larger. But the firft diftinct account we have of the Magnifying power of glaffes, is in the 12th century, in the writings of Roger Bacon, and Alhazen; and it is not improbable that from their obfervations the conftruction of fpectacles was derived. In the Opus Majus of Bacon, it is demonftrated, that if a tranfparent body, interfperfed between the eye and an object, be convex towards the eye, the object will appear magnified.

MAGNIFYING *Glafs*, in Optics, is a fmall fpherical convex lens; which, in tranfmitting the rays of light, inflects them more towards the axis, and fo exhibits objects viewed through them larger than when viewed by the naked eye. See MICROSCOPE.

MAGNITUDE, any thing made up of parts locally extended, or continued; or that has feveral dimenfions; as a line, furface, folid, &c. Quantity is often ufed as fynonymous with Magnitude. See QUANTITY.

Geometrical MAGNITUDES, are ufually, and moft properly, confidered as generated or produced by motion; as lines by the motion of points, furfaces by the motion of lines, and folids by the motion of furfaces.

Apparent MAGNITUDE, is that which is meafured by the optic or vifual angle, intercepted between rays drawn from its extremes to the centre of the pupil of the eye. It is a fundamental maxim in optics, that whatever things are feen under the fame or equal angles, appear equal; and vice verfa.—The apparent Magnitudes of an object at different diftances, are in a ratio lefs than that of their diftances reciprocally.

The apparent Magnitudes of the two great luminaries, the fun and moon, at rifing and fetting, are a phenomenon that has greatly embarraffed the modern philofophers. According to the ordinary laws of vifion, they fhould appear the leaft when neareft the horizon, being then fartheft from the eye; and yet it is found that the contrary is true in fact. Thus, it is well known that the mean apparent diameter of the moon, at her greateft height in the meridian, is nearly 31' in round numbers, fubtending then an angle of that quantity as meafured by any inftrument. But, being viewed when fhe rifes or fets, fhe feems to the eye as two or three times as large as before; and yet when meafured by the inftrument, her diameter is not found increafed at all.

Ptolomy, in his Almageft, lib. 1, cap. 3, taking for granted, that the angle fubtended by the moon was really increafed, afcribed the increafe to a refraction of the rays by vapours, which actually enlarge the angle under which the moon appears; juft as the angle is enlarged by which an object is feen from under water: and his commentator Theon explains diftinctly how the dilatation of the angle in the object immerfed in water is caufed. But it being afterwards difcovered, that there is no alteration in the angle, another folution was ftarted by the Arab Alhazen, which was followed and improved by Bacon, Vitello, Kepler, Peckham, and others. According to Alhazen, the fight apprehends the furface of the heavens as flat, and judges of the ftars as it would of ordinary vifible objects extended upon a wide plain; the eye fees then under equal angles indeed, but withal perceives a difference in their

diftances,

diſtances, and (on account of the ſemidiameter of the earth, which is interpoſed in one caſe, and not in the other) it is hence induced to judge thoſe that appear more remote to be greater. Some farther improvement was made in this explanation by Mr. Hobbes, though he fell into ſome miſtakes in his application of geometry to this ſubject: for he obſerves, that this deception operates gradually from the zenith to the horizon; and that if the apparent arch of the ſky be divided into any number of equal parts, thoſe parts, in deſcending towards the horizon, will ſubtend an angle that is gradually leſs and leſs. And he was the firſt who expreſsly conſidered the vaulted appearance of the ſky as a real portion of a circle.

Des Cartes, and from him Dr. Wallis, and moſt other authors, account for the appearance of a different diſtance under the ſame angle, from the long ſeries of objects interpoſed between the eye and the extremity of the ſenſible horizon; which makes us imagine it more remote than when in the meridian, where the eye ſees nothing in the way between the object and itſelf. This idea of a great diſtance makes us imagine the luminary the larger; for an object being ſeen under any certain angle, and believed at the ſame time very remote, we naturally judge it muſt be very large, to appear under ſuch an angle at ſuch a diſtance. And thus a pure judgment of the mind makes us ſee the ſun, or the moon, larger in the horizon than in the meridian; notwithſtanding their diameters meaſured by any inſtrument are really leſs in the former ſituation than the latter.

James Gregory, in his Geom. Pars Univerſalis, pa. 141, ſubſcribes to this opinion: Father Mallebranche alſo, in the firſt book of his Recherche de la Verité, has explained this phenomenon almoſt in the expreſſion of Des Cartes: and Huygens, in his Treatiſe on the Parhelia, tranſlated by Dr. Smith, Optics, art. 536, has approved, and very clearly illuſtrated, the received opinion. The cauſe of this fallacy, ſays he, in ſhort, is this; that we think the ſun, or any thing elſe in the heavens, farther from us when it is near the horizon, than when it approaches towards the vertex, becauſe we imagine every thing in the air that appears near the vertex to be farther from us than the clouds that fly over our heads; whereas, on the other hand, we are uſed to obſerve a large extent of land lying between us and the objects near the horizon, at the farther end of which the convexity of the ſky begins to appear; which therefore, with the objects that appear in it, are uſually imagined to be much farther from us. Now when two objects of equal magnitude appear under the ſame angle, we always judge that object to be larger which we think is remoter. And this, according to them, is the true cauſe of the deception in queſtion. It is really aſtoniſhing that an hypotheſis ſo palpably falſe ſhould ever be held and maintained by ſuch eminent men; for it is daily ſeen that the moon or ſun, when near the horizon, very ſuddenly change their magnitude, as they aſcend or deſcend, though all the intervening objects are ſeen juſt as before: and that the luminary appears largeſt of all when feweſt objects appear on the earth, as in a thick fog or miſt. It is no wonder therefore that other reaſons have been aſſigned for this remarkable phenomenon.

Accordingly Gaſſendus was of opinion, that this effect ariſes from hence; that the pupil of the eye, being always more open as the place is more dark, as in the morning and evening, when the light is leſs, and beſides the earth being then covered with groſs vapours, through a longer column of which the rays muſt paſs to reach the horizon; the image of the luminary enters the eye at a greater angle, and is really painted there larger than when the luminary is higher. See APPARENT *Diameter* and *Magnitude*.

F. Gouge advances another hypotheſis, which is, that when the luminaries are in the horizon, the proximity of the earth, and the groſs vapours with which they then appear enveloped, have the ſame effect with regard to us, as a wall, or other denſe body, placed behind a column; which in that caſe appears larger than when inſulated, and encompaſſed on all ſides with an illuminated air.

The commonly received opinion has been diſputed, not only by F. Gouge, who obſerves, Acad. Sci. 1700, pa. 11, that the horizontal moon appears equally large acroſs the ſea, where there are no objects to produce the effect aſcribed to them; but alſo by Mr. Molyneux, who ſays, Philoſ. Tranſ. abr. vol. 1, pa. 221, that if this hypotheſis be true, we may at any time increaſe the apparent magnitude of the moon, even in the meridian; for, in order to divide the ſpace between it and the eye, we need only to look at it behind a cluſter of chimneys, the ridge of a hill, or the top of a houſe, &c. He makes alſo the ſame obſervation with F. Gouge, above mentioned, and farther obſerves, that when the height of all the intermediate objects is cut off; by looking through a tube, the imagination is not helped, and yet the moon ſeems ſtill as large as before. However, Mr. Molyneux advances no hypotheſis of his own.

Biſhop Berkley ſuppoſed, that the moon appears larger near the horizon, becauſe ſhe then appears fainter, and her beams affect the eye leſs. And Mr. Robins has recited ſome other opinions on this ſubject, Math. Tracts, vol. 2, pa. 242.

Dr. Deſaguliers has illuſtrated the doctrine of the horizontal moon, Philoſ. Tranſ. abr. vol. 8, pa. 130, upon the ſuppoſition of our imagining the viſible heavens to be only a ſmall portion of a ſpherical ſurface, and conſequently ſuppoſing the moon to be farther from us in the horizon than near the zenith; and by ſeveral ingenious contrivances he demonſtrated how liable we are to ſuch deceptions. The ſame idea is purſued ſtill farther by Dr. Smith, in his Optics, where he determines that, the centre of the apparent ſpherical ſegment of the ſky lying much below the eye, or the horizon, the apparent diſtance of its parts near the horizon was about 3 or 4 times greater than the apparent diſtance of its parts over head; from which reaſon it is, he infers, that the moon always appears the larger as ſhe is lower, and alſo that we always think the height of a celeſtial object to be more than it really is. Thus, he determined, by meaſuring the actual height of ſome of the heavenly bodies, when to his eye they ſeemed to be half way between the horizon and the zenith; that their real altitude was then only 23°: when the ſun was about 30° high, the upper always appeared leſs than the under; and he thought that it was conſtantly greater when the ſun was 18° or 20° high. Mr. Robins, in

his

his Tracts, vol. 2, pa. 245, shews how to determine the apparent concavity of the sky in a more accurate and geometrical manner; by which it appears, that if the altitude of any of the heavenly bodies be 20°, at the time when it seems to be half way between the horizon and the zenith, the horizontal distance will be hardly less than 4 times the perpendicular distance; but if that altitude be 28°, it will be little more than 2 and a half.

Dr. Smith, having determined the apparent figure of the sky, thus applies it to explain the phenomenon of the horizontal moon, and other similar appearances in the heavens. Suppose the arc ABC to re-

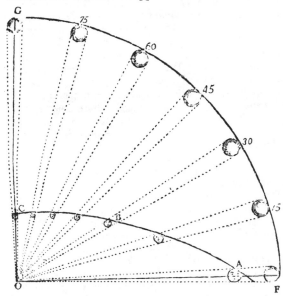

present that apparent concavity; then the diameter of the sun and moon would seem to be greater in the horizon than at any altitude, measured by the angle AOB, in the ratio of its apparent distances, AO, BO. The numbers that express these proportions he reduced into the annexed table, answering to the corresponding altitudes of the sun or moon, which are also exactly represented to the eye in the figure, in which the moon, placed in the quadrantal arc FG described about the centre O, are all equal to each other, and represent the body of the moon in the heights there noted, and the unequal moons in the concavity ABC

The alt. of the sun or moon in degrees.	Apparent diameters or distances.
00	100
15	68
30	50
45	40
60	34
75	31
90	30

are terminated by the visual rays coming from the circumference of the real moon, at those heights to the eye, at O. Dr. Smith also observes, that the apparent concave of the sky, being less than a hemisphere, is the cause that the breadths of the colours in the inward and outward rainbows, and the interval between the bows, appear least at the top, and greater at the bottom. This

theory of the horizontal moon is also confirmed by the appearances of the tails of comets, which, whatever be their real figure, magnitude, and situation in absolute space, do always appear to be an arc of the concave sky. Dr. Smith however justly acknowledges that, at different times, the moon appears of very different magnitudes, even in the same horizon, and occasionally of an extraordinary large size; which he is not able to give a satisfactory explanation of. Smith's Optics, vol. 1, pa. 63, &c, Remarks, pa. 53.

MAIGNAN (Emanuel), a religious minim, and one of the greatest philosophers of his age, was born at Thoulouse in 1601. Like the famous Pascal, he became a complete mathematician without the assistance of a teacher; and filled the professor's chair at Rome in 1636, where, at the expence of Cardinal Spada, he published his book De Perspectiva Horaria, in 1648. Upon this book, Baillet, in his Life of Des Cartes, has the following passage: " M. Carcavi acquainted Des Cartes, that there was at Rome one father Maignan, a minim, of greater learning and more depth than father Mersenne, who made him expect some objections against his principles. This father's proper name was Emanuel, and his native place Thoulouse: but he lived at that time at Rome, where he taught divinity in the convent of the Trinity upon Mount Pincio, which they otherwise call the convent of the French minims." Maignan returned to Thoulouse in 1650, and was created Provincial. His knowledge in mathematics, and physical experiments, were very early known; especially from a dispute which arose between him and father Kircher, about the invention of a catoptrical work.

The king, who in 1660 amused himself with the machines and curiosities in the father's cell, made him offers by Cardinal Mazarin, to draw him to Paris; but he humbly desired to spend the remainder of his days in a cloyster.—He published a Course of Philosophy, in 4 volumes 8vo, at Thoulouse, in 1652; to the second edition of which, in folio, 1673, he added two Treatises; the one against the vortices of Des Cartes, the other upon the speaking trumpet invented by Sir Samuel Morland.—He formed a machine, which shewed, by its movements, that Des Cartes's supposition concerning the manner in which the universe was formed, or might have been formed, and concerning the centrifugal force, was entirely without foundation.

Thus this great philosopher and divine passed a life of tranquillity, in writing books, making experiments, and reading lectures. He was frequently consulted by the most eminent philosophers; and has had a thousand answers to make, either by writing or otherwise. Never was mortal less inclined to idleness. It is said that he even studied in his sleep; for his very dreams employed him in problems, which he pursued sometimes till he came to a solution or demonstration; and he has frequently been awaked out of his sleep of a sudden, by the exquisite pleasure which he felt upon discovery of it. The excellence of his manners, and his unspotted virtues, rendered him no less worthy of esteem, than his genius and learning.—It is said that he composed with great ease, and without any alterations at all.—He died at Thoulouse in 1676, at 75 years of age.

MALLEABLE, the property of a solid ductile body, from which it may be beaten, forged, and ex-

tended

tended under the hammer, without breaking, which is a property of all metals.

MANFREDI (EUSTACHIO), a celebrated astronomer and mathematician, born at Bologna in 1674. His genius was always above his age. He was a tolerable poet, and wrote ingenious verses while he was but a child. And while very young he formed in his father's house an academy of youth of his own age, who became the Academy of Sciences, or the Institute, there. He became Professor of Mathematics at Bologna in 1698, and Superintendant of the waters there in 1704. The same year he was placed at the head of the College of Montalte, founded at Bologna for young men intended for the church. In 1711 he obtained the office of Astronomer to the Institute of Bologna. He became member of the Academy of Sciences of Paris in 1726, and of the Royal Society of London in 1729; and died the 15th of February 1739.—His works are:

1. *Ephemerides Motuum Cœlestium ab anno 1715 al annum 1750*; 4 volumes in 4to.—The first volume is an excellent introduction to astronomy; and the other three contain numerous calculations. His two sisters were greatly assisting to him in composing this work.

2. *De Transitu Mercurii per Solem, anno 1723*. Bologna 1724, in 4to.

3. *De Annuis Inerrantium Stellarum Aberrationibus*, Bologna 1729, in 4to.—Besides a number of papers in the Memoirs of the Academy of Sciences; and in other places.

MANILIUS (MARCUS), a Latin astronomical poet, who lived in the reign of Augustus Cæsar. He wrote an ingenious poem concerning the stars and the sphere, called *Astronomicon;* which, not being mentioned by any of the ancient poets, was unknown, till about two centuries since, when it was found buried in some German library, and published by Poggius. There is no account to be found of this author, but what can be drawn from his poem; which contains a system of the ancient astronomy and astrology, together with the philosophy of the Stoics. It consists of five books; though there was a sixth, which has not been recovered. In this work, Manilius hints at some opinions, which later ages have been ready to glory in as their own discoveries. Thus, he defends the fluidity of the heavens, against the hypothesis of Aristotle: he asserts that the fixed stars are not at all in the same concave superficies of the heavens, and equally distant from the centre of the world: he maintains that they are all of the same nature and substance with the sun, and that each of them has a particular vortex of its own: and lastly, he says, that the milky way is only the united lustre of a great many small imperceptible stars; which indeed the Moderns now see to be such through their telescopes.

The best editions of Manilius are, that of Joseph Scaliger, in 4to, 1600; that of Bentley, in 4to, 1738, and that of Edmund Burton, Esq. in 8vo, 1783.

MANOMETER, or MANOSCOPE, an instrument to shew or measure the alterations in the rarity or density of the air.

The Manometer differs from the barometer in this, That the latter only serves to measure the *weight* of the atmosphere, or of the column of air over it; but the former, the density of the air in which it is found; which density depends not only on the weight of the atmosphere, but also on the action of heat and cold, &c. Authors however often confound the two together; and Mr. Boyle himself has given a very good Manometer of his contrivance, under the name of a Statical Barometer, consisting of a bubble of thin glass, about the size of an orange, which being counterpoised when the air was in a mean state of density, by means of a nice pair of scales, sunk when the atmosphere became lighter, and rose as it grew heavier.

The Manometer used by captain Phipps, in his voyage towards the North Pole, consisted of a tube of a small bore, with a ball at the end. The barometer being at 29·7, a small quantity of quicksilver was put into the tube, to take off the communication between the external air, and that confined in the ball and the part of the tube below this quicksilver. A scale is placed on the side of the tube, which marks the degrees of dilatation arising from the increase of heat in this state of the weight of the air, and has the same graduation as that of Fahrenheit's thermometer, the point of freezing being marked 32. In this state therefore it will shew the degrees of heat in the same manner as a thermometer. But when the air becomes lighter, the bubble inclosed in the ball, being less compressed, will dilate itself, and occupy a space as much larger as the compressing force is less; therefore the changes arising from the increase of heat, will be proportionably larger; and the instrument will shew the differences in the density of the air, arising from the changes in its weight and heat. Mr. Ramsden found, that a heat equal to that of boiling water, increased the magnitude of the air, from what it was at the freezing point, by $\frac{414}{1000}$ of the whole. Hence it follows, that the ball and the part of the tube below the beginning of the scale, is of a magnitude equal to almost 414 degrees of the scale. If the height of both the Manometer and thermometer be given, the height of the barometer may be thence deduced, by this rule;

as the height of the Manometer increased by 414,
to the height of the thermometer increased by 414,
so is 29·7, to the height of the barometer;

or if *m* denote the height of the Manometer, and *t* the height of the thermometer; then

$$m + 414 : t + 414 :: 29\text{·}7 : \frac{t + 414}{m + 414} \times 29\text{·}7,$$

which is the height of the barometer.

Another kind of Manometer was made use of by colonel Roy, in his attempts to correct the errors of the barometer; which is described in the Philos. Transf. vol. 67, pa. 689.

MANTELETS, a kind of moveable parapet, or screen, of about 6 feet high, set upon trucks or little wheels, and guided by a long pole; so that in a siege it may be driven before the pioneers, and serve as blinds, or screens, to shelter them from the enemy's small shot. Mantelets are made of different materials, so as to render them musket proof; as of strong boards nailed together, and covered with tin; or of thick leather, or of layers of rope, &c, firmly bound together.

There are also other sorts of Mantelets, covered on the top, used by the miners in approaching the walls or works of an enemy. The double Mantelets form an

angle,

Plate XVII.

GEOGRAPHICAL MAPS.

Fig. 1.

Fig. 2.

Fig. 3.

Fig. 4.

OPERA *Glass*

Fig. 5.

PERAMBULATOR. *Fig. 6.*

angle, and ſtand ſquare, making two fronts. It appears from Vegetius, that Mantelets were in uſe among the Ancients, under the name of Vineæ.

MANTLE, or MANTLE-*tree*, is the lower part of the breaſt or front of a chimney. It was formerly a piece of timber that lay acroſs the jambs, ſupporting the breaſtwork ; but by a late act of parliament, chimney-breaſts are not to be ſupported by a wooden mantle-tree, or turning piece, but by an iron bar, or by an arch of brick or ſtone.

MAP, a plane figure repreſenting the ſurface of the earth, or ſome part of it ; being a projection of the globular ſurface of the earth, exhibiting countries, ſeas, rivers, mountains, cities, &c, in their due poſitions, or nearly ſo.

Maps are either Univerſal or Particular, that is Partial.

Univerſal MAPS are ſuch as exhibit the whole ſurface of the earth, or the two hemiſpheres.

Particular, or *Partial* MAPS, are thoſe that exhibit ſome particular region, or part of the earth.

Both kinds are uſually called Geographical, or Land-Maps, as diſtinguiſhed from Hydrographical, or Sea-Maps, which repreſent only the ſeas and ſea coaſts, and are properly called *Charts*.

Anaximander, the ſcholar of Thales, it is ſaid, about 400 years before Chriſt, firſt invented geographical tables, or Maps. The Pentingerian Tables, publiſhed by Cornelius Pentinger of Auſburgh, contain an itinerary of the whole Roman Empire ; all places, except ſeas, woods, and deſarts, being laid down according to their meaſured diſtances, but without any mention of latitude, longitude, or bearing.

The Maps publiſhed by Ptolomy of Alexandria, about the 144th year of Chriſt, have meridians and parallels, the better to define and determine the ſituation of places, and are great improvements on the conſtruction of Maps. Though Ptolomy himſelf owns that his Maps were copied from ſome that were made by Marinus, Tirus, &c, with the addition of ſome improvements of his own. But from his time till about the 14th century, during which, geography and moſt ſciences were neglected, no new Maps were publiſhed. Mercator was the firſt of note among the Moderns, and next to him Ortelius, who undertook to make a new ſet of Maps, with the modern diviſions of countries and names of places ; for want of which, thoſe of Ptolomy were become almoſt uſeleſs. After Mercator, many others publiſhed Maps, but for the moſt part they were mere copies of his. Towards the middle of the 17th century, Bleau in Holland, and Sanſon in France, publiſhed new ſets of Maps, with many improvements from the travellers of thoſe times, which were afterwards copied, with little variation, by the Engliſh, French, and Dutch ; the beſt of theſe being thoſe of Viſcher and De Witt. And later obſervations have furniſhed us with ſtill more accurate and copious ſets of Maps, by De Liſle, Robert, Wells, &c, &c. Concerning Maps, ſee Varenius's Geog. lib. 3, cap. 3, prop. 4 ; Fournier's Hydrog. lib. 4, c. 24 ; Wolfius's Elem. Hydrog. c. 9 ; John Newton's Idea of Navigation ; Mead's Conſtruction of Globes and Maps ; Wright's Conſtructions of Maps, &c, &c.

Conſtruction of MAPS. Maps are conſtructed by making a projection of the globe, either on the plane of ſome particular circle, or by the eye placed in ſome particular point, according to the rules of Perſpective, &c ; of which there are ſeveral methods.

Firſt, to conſtruct a Map of the World, or a general Map.

1ſt *Method*.—A map of the world muſt repreſent two hemiſpheres ; and they muſt both be drawn upon the plane of that circle which divides the two hemiſpheres. The firſt way is to project each hemiſphere upon the plane of ſome particular circle, by the rules of Orthographic projection, forming two hemiſpheres, upon one common baſe or circle. When the plane of projection is that of a meridian, the maps will be the eaſt and weſt hemiſpheres, the other meridians will be ellipſes, and the parallel circles will be right lines. Upon the plane of the equinoctial, the meridians will be right lines croſſing in the centre, which will repreſent the pole, and the parallels of latitude will be circles having that common centre, and the Maps will be the northern and ſouthern hemiſpheres. The fault of this way of drawing Maps, is, that near the outſide the circles are too near one another ; and therefore equal ſpaces on the earth are repreſented by very unequal ſpaces upon the Map.

2d *Method*.—Another way is to project the ſame hemiſpheres by the rules of Stereographic projection ; in which way, all the parallels will be repreſented by circles, and the meridians by circles or right lines. And here the contrary fault happens, viz, the circles towards the outſides are too far aſunder, and about the middle they are too near together.

3d *Method*.—To remedy the faults of the two former methods, proceed as follows. Firſt, for the eaſt and weſt hemiſpheres, deſcribe the circle PENQ for the meridian (pl. xvii, fig. 1), or plane of projection ; through the centre of which draw the equinoctial EQ, and axis PN perpendicular to it, making P and N the north and ſouth pole. Divide the quadrants PE, EN, NQ, and QP into 9 equal parts, each repreſenting 10 degrees, beginning at the equinoctial EQ : divide alſo CP and CN into 9 equal parts ; beginning at EQ ; and through the correſponding points draw the parallels of latitude. Again, divide CE and CQ into 9 equal parts ; and through the points of diviſion, and the two poles P and N, draw circles, or rather ellipſes, for the meridians. So ſhall the Map be prepared to receive the ſeveral places and countries of the earth.

Secondly, for the north or ſouth hemiſphere, draw AQBE, for the equinoctial (fig. 2), dividing it into the four quadrants EA, AQ, QB, and BE ; and each quadrant into 9 equal parts, repreſenting each 10 degrees of longitude ; and then, from the points of diviſion, draw lines to the centre C, for the circles of longitude. Divide any circle of longitude, as the firſt meridian EC, into 9 equal parts, and through theſe points deſcribe circles from the centre C, for the parallels of latitude ; numbering them as in the figure.

In this 3d method, equal ſpaces on the earth are repreſented by equal ſpaces on the Map, as near as any projection will bear ; for a ſpherical ſurface can no way be repreſented exactly upon a plane. Then the ſeveral countries of the world, ſeas, iſlands, ſea-coaſts, towns, &c,

MAP [78] MAP

&c, are to be entered in the Map, according to their latitudes and longitudes.

In filling up the Map, all places representing land are filled with such things as the countries contain; but the seas are left white; the shores adjoining to the sea being shaded. Rivers are marked by strong lines, or by double lines, drawn winding in form of the rivers they represent; and small rivers are expressed by small lines. Different countries are best distinguished by different colours, or at least the borders of them. Forests are represented by trees; and mountains shaded to make them appear. Sands are denoted by small points or specks; and rocks under water by a small cross. In any void space, draw the mariner's compass, with the 32 points or winds.

II. *To draw a Map of any particular Country.*

1st *Method.*—For this purpose its extent must be known, as to latitude and longitude; as suppose Spain, lying between the north latitudes 36 and 44, and extending from 10 to 23 degrees of longitude; so that its extent from north to south is 8 degrees, and from east to west 13 degrees.

Draw the line AB for a meridian passing through the middle of the country (fig. 3), on which set off 8 degrees from B to A, taken from any convenient scale; A being the north, and B the south point. Through A and B draw the perpendiculars CD, EF, for the extreme parallels of latitude. Divide AB into 8 parts, or degrees, through which draw the other parallels of latitude, parallel to the former.

For the meridians; divide any degree in AB into 60 equal parts, or geographical miles. Then, because the length of a degree in each parallel decreases towards the pole, from the table shewing this decrease, under the article DEGREE, take the number of miles answering to the latitude of B, which is $48\frac{1}{2}$ nearly, and set it from B, 7 times to E, and 6 times to F; so is EF divided into degrees. Again, from the same table take the number of miles of a degree in the latitude A, viz $43\frac{1}{2}$ nearly; which set off, from A, 7 times to C, and 6 times to D. Then from the points of division in the line CD, to the corresponding points in the line EF, draw so many right lines, for the meridians. Number the degrees of latitude up both sides of the Map, and the degrees of longitude on the top and bottom. Also, in some vacant place make a scale of miles; or of degrees, if the Map represent a large part of the earth; to serve for finding the distances of places upon the Map.

Then make the proper divisions and subdivisions of the country: and having the latitudes and longitudes of the principal places, it will be easy to set them down in the Map: for any town, &c, must be placed where the circles of its latitude and longitude intersect. For instance, Gibraltar, whose latitude is 36° 11′, and longitude 12° 27′, will be at G: and Madrid, whose lat. is 40° 10′, and long. 14° 44′, will be at M. In like manner the mouth of a river must be set down; but to describe the whole river, the latitude and longitude of every turning must be marked down, and the towns and bridges by which it passes. And so for woods, forests, mountains, lakes, castles, &c. The boundaries will be described, by setting down the re-

markable places on the sea coast, and drawing a continued line through them all. And this way is very proper for small countries.

2d *Method.*—Maps of particular places are but portions of the globe, and therefore may be drawn after the same manner as the whole is drawn. That is, such a Map may be drawn either by the orthographic or stereographic projection of the sphere, as in the last prob. But in partial Maps, an easier way is as follows. Having drawn the meridian AB (fig. 3), and divided it into equal parts as in the last method, through all the points of division draw lines perpendicular to AB, for the parallels of latitude; CD, EF being the extreme parallel. Then to divide these, set off the degrees in each parallel, diminished after the manner directed for the two extreme parallels CD, EF, in the last method: and through all the corresponding points draw the meridians, which will be curve lines; which were right lines in the last method; because only the extreme parallels were divided by the table. This method is proper for a large tract, as Europe, &c: in which case the parallels and meridians need only be drawn to every 5 or 10 degrees. This method is much used in drawing Maps; as all the parts are nearly of their due magnitude, but a little distorted towards the outside, from the oblique intersections of the meridians and parallels.

3d *Method.*—Draw PB of a convenient length, for a meridian; divide it into 9 equal parts, and through the points of division, describe as many circles for the parallels of latitude, from the centre P, which represents the pole. Suppose AB (fig. 4) the height of the Map; then CD will be the parallel passing through the greatest latitude, and EF will represent the equator. Divide the equator EF into equal parts, of the same size as those in AB, both ways, beginning at B. Divide also all the parallels into the same number of equal parts, but lesser, in proportion to the numbers for the several latitudes, as directed in the last method for the rectilineal parallels. Then through all the corresponding divisions, draw curve lines, which will represent the meridians, the extreme ones being EC and FD. Lastly, number the degrees of latitude and longitude, and place a scale of equal parts, either of miles or degrees, for measuring distances.—This is a very good way of drawing large Maps, and is called the globular projection; all the parts of the earth being represented nearly of their due magnitude, excepting that they are a little distorted on the outsides.

When the place is but small that a Map is to be made of, as if a county was to be exhibited; the meridians, as to sense, will be parallel to one another, and the whole will differ very little from a plane. Such a Map will be made more easily than by the preceding rules. It will here be sufficient to measure the distances of places in miles, and so lay them down in a plane rectangular map. But this belongs more properly to Surveying.

The Use of MAPS is obvious from their construction. The degrees of the meridians and parallels shew the latitudes and longitudes of places, and the scale of miles annexed, their distances; the situation of places, with regard to each other, as well as to the cardinal points, appears by inspection; the top of the map being always the north, the bottom the south, the right hand the east, eaft, and the left hand the weft; unlefs the compafs, ufually annexed, fhew the contrary.

MARALDI (JAMES PHILIP), a learned aftronomer and mathematician, was born in 1665 at Perinaldo in the county of Nice, a place already honoured by the birth of his maternal uncle the celebrated Caffini. Having made a confiderable progrefs in mathematics, at the age of 22 his uncle, who had been a long time fettled in France, invited him there, that he might himfelf cultivate the promifing genius of his nephew. Maraldi no fooner applied himfelf to the contemplation of the heavens, than he conceived the defign of forming a catalogue of the fixed ftars, the foundation of all the aftronomical edifice. In confequence of this defign, he applied himfelf to obferve them with the moft conftant attention; and he became by this means fo intimate with them, that on being fhewn any one of them, however fmall, he could immediately tell what conftellation it belonged to, and its place in that conftellation. He has been known to difcover thofe fmall comets, which aftronomers often take for the ftars of the conftellation in which they are feen, for want of knowing precifely what ftars the conftellation confifts of, when others, on the fpot, and with eyes directed equally to the fame part of the heavens, could not for a long time fee any thing of them.

In 1700 he was employed under Caffini in prolonging the French meridian to the northern extremity of France, and had no fmall fhare in completing it. He then fet out for Italy, where Clement the 11th invited him to affift at the affemblies of the Congregation then fitting in Rome to reform the calendar. Bianchini alfo availed himfelf of his affiftance to conftruct the great meridian of the Carthufian church in that city. In 1718 Maraldi, with three other academicians, prolonged the French meridian to the fouthern extremity of that country. He was admitted a member of the Academy of Sciences of Paris in 1699, in the department of Aftronomy, and communicated a great multitude of papers, which are printed in their memoirs, in almoft every year from 1699 to 1729, and ufually feveral papers in each of thofe years; for he was indefatigable in his obfervations of every thing that was curious and ufeful in the motions and phenomena of the heavenly bodies. As to the catalogue of the fixed ftars, it was not quite completed: juft as he had placed a mural quadrant on the terras of the obfervatory, to obferve fome ftars towards the north and the zenith, he fell fick, and died the 1ft of December 1729.

MARCH, the 3d month of the year, according to the common way of computing, and confifts of 31 days. The fun enters the fign Aries about the 20th or 21ft day of this month.

Among the Romans, March was the firft month; and in fome ecclefiaftical computations, that order is ftill preferved. In England, before the alteration of the ftile, March was the 1ft month in order, the year always commencing with the 25th day of the month.

It has been faid it was Romulus who firft divided the year into months; to the firft of which he gave the name of his fuppofed father Mars. It is obferved by Ovid, however, that the people of Italy had the month of March before the time of Romulus; but that they placed it differently; fome making it the third, fome the 4th, fome the 5th, and others the 10th month of the year.

MARINE BAROMETER. See BAROMETER.
MARINERS-COMPASS. See COMPASS.

MARIOTTE (EDME), an eminent French philofopher and mathematician, was born at Dijon, and admitted a member of the Academy of Sciences of Paris in 1666. His works however are better known than his life. He was a good mathematician, and the firft French philofopher who applied much to experimental phyfics. The law of the fhock or collifion of bodies, the theory of the preffure and motion of fluids, the nature of vifion, and of the air, particularly engaged his attention. He carried into his philofophical refearches, that fpirit of fcrutiny and inveftigation fo neceffary to thofe who would make any confiderable progrefs in it. He died in 1684.

He communicated a number of curious and valuable papers to the Academy of Sciences, which were printed in the collection of their Memoirs dated 1666, viz, from volume 1 to volume 10. And all his works were collected into 2 volumes in 4to, and printed at Leyden in 1717.

MARS, one of the feven primary planets now known, and the firft of the four fuperior ones, being placed immediately next above the earth. It is ufually denoted by this character ♂, being a mark rudely formed from a man holding a fpear protruded, reprefenting the god of war of the fame name.

The mean diftance of Mars from the fun, is 1524 of thofe parts, of which the diftance of the earth from the fun is 1000; his excentricity 141; and his real diftance 145 millions of miles. The inclination of his orbit to the plane of the ecliptic, is 1° 52'; the length of his year, or the period of one revolution about the fun, is 686⅔ of our days, or 667⅔ of his own days, which are 40 minutes longer than ours, the revolution on his axis being performed in 24 hours 40 minutes. His mean diameter is 4444 miles; and the fame feen from the fun is 11″: the inclination of the axis to his orbit 0° 0'; the inclination of his orbit to the ecliptic 1° 52'; place of the aphelion ♍ 0° 32; place of his afcending node ♉ 17° 17'; and his parallax, according to Dr. Hook and Mr. Flamfteed, is fcarce 30 feconds.

Dr. Hook, in 1665, obferved feveral fpots in Mars; which having a motion, he concluded the planet turned round its centre. In 1666, M. Caffini obferved feveral fpots in the two faces or hemifpheres of Mars, which he found made one revolution in 24hours 40minutes. Thefe obfervations were repeated in 1670, and confirmed by Miraldi in 1704, and 1719: whence both the motion and period, or natural day, of that planet, were determined.

In the Philof. Tranf. for 1781, Mr. Herfchel gave a feries of obfervations on the rotation of this planet about its axis, from which he concluded that one mean fidereal rotation was between 24 h. 39 m. 5 fec. and 24 h. 39 m. 22 fec.; and in the Philof. Tranf. for 1784, is given a paper by the fame gentleman, on the remarkable appearances at the polar regions of the planet Mars, the inclination of its axis, the pofition of its poles, and its fpheroidical figure; with a few hints relating to its real diameter and atmofphere, deduced from his

his obfervations taken from the year 1777 to 1783 inclufively. He obferved feveral remarkable bright fpots near both poles, which had fome fmall motion; and the refults of his obfervations are as follow; viz,

" Inclination of axis to the ecliptic, 59° 22'.

The node of the axis is in ♓ 17° 47'.

Obliquity of the planet's ecliptic 28° 42'.

The point Aries on Mars's ecliptic anfwers to our ♐ 19° 28'.

The figure of Mars is that of an oblate fpheroid, whofe equatorial diameter is to the polar one, as 1355 to 1272, or as 16 to 15 nearly.

The equatorial diameter of Mars, reduced to the mean diftance of the earth from the fun, is 9" 8'''.

And the planet has a confiderable, but moderate atmofphere, fo that its inhabitants probably enjoy a fituation in many refpects fimilar to ours."

Mars always appears with a ruddy troubled light; owing, it is fuppofed, to the nature of his atmofphere, through which the light paffes.

In the acronical rifing of this planet, or when in oppofition to the fun, it is five times nearer to us than when in conjunction with him; and fo appears much larger and brighter than at other times.

Mars, having his light from the fun, and revolving round it, has an increafe and decreafe like the moon: it may alfo be obferved almoft bifected, when in the quadratures, or in perigæon; but is never feen cornicular, as the inferior planets. All which fhews both that his orbit includes that of the earth within it, and that he fhines not by his own light.

MARTIN (Benjamin), was born in 1704, and became one of the moft celebrated mathematicians and opticians of his time. He firft taught a fchool in the country; but afterwards came up to London, where he read lectures on experimental philofophy for many years, and carried on a very extenfive trade as an optician and globe-maker in Fleet-ftreet, till the growing infirmities of old age compelled him to withdraw from the active part of bufinefs. Trufting too fatally to what he thought the integrity of others, he unfortunately, though with a capital more than fufficient to pay all his debts, became a bankrupt. The unhappy old man, in a moment of defperation from this unexpected ftroke, attempted to deftroy himfelf; and the wound, though not immediately mortal, haftened his death, which happened the 9th of February 1782, at 78 years of age.

He had a valuable collection of foffils and curiofities of almoft every fpecies; which after his death were almoft given away by public auction. He was indefatigable as an artift, and as a writer he had a very happy method of explaining his fubject, and wrote with clearnefs, and even confiderable elegance. He was chiefly eminent in the fcience of optics; but he was well fkilled in the whole circle of the mathematical and philofophical fciences, and wrote ufeful books on every one of them; though he was not diftinguifhed by any remarkable inventions or difcoveries of his own. His publications were very numerous, and generally ufeful: fome of the principal of them were as follow:

The Philofophical Grammar; being a View of the

8

prefent State of Experimental Phyfiology, or Natural Philofophy, 1735, 8vo.—A new, complete, and univerfal Syftem or Body of Decimal Arithmetic, 1735, 8vo.—The Young Student's Memorial Book, or Pocket Library, 1735, 8vo.—Defcription and Ufe of both the Globes, the Armillary Sphere and Orrery, Trigonometry, 1736, 2 vols. 8vo.—Syftem of the Newtonian Philofophy, 1759, 3 vols.—New Elements of Optics, 1759.—Mathematical Inftitutions, 1764, 2 vols. —Philologic and Philofophical Geography, 1759. —Lives of Philofophers, their inventions, &c. 1764. —Young Gentleman and Lady's Philofophy, 1764, 3 vols.—Mifcellaneous Correfpondence, 1764, 4 vols.— Inftitutions of Aftronomical Calculations, 3 parts, 1765. —Introduction to the Newtonian Philofophy, 1765.— Treatife of Logarithms.—Treatife on Navigation.— Defcription and Ufe of the Air-pump.—Defcription of the Torricellian Barometer.—Appendix to the Ufe of the Globes.—Philofophia Britannica, 3 vols.—Principles of Pump-work.—Theory of the Hydrometer.— Defcription and Ufe of a Cafe of Mathematical Inftruments.—Ditto of a Univerfal Sliding Rule.—Micrographia, on the Microfcope.—Principles of Perfpective. —Courfe of Lectures.—Optical Effays.—Effay on Electricity.—Effay on Vifual Glaffes or Spectacles.— Horologia Nova, or New Art of Dialling.—Theory of Comets.—Nature and Conftruction of Solar Eclipfes. —Venus in the Sun.—The Mariner's Mirror.—Thermometrum Magnum.—Survey of the Solar Syftem.— Effay on Ifland Chryftal.—Logarithmologia Nova, &c. &c.

MASCULINE *Signs*. Aftrologers divide the Signs, &c, into Mafculine and Feminine; by reafon of their qualities, which are either active, and hot, or cold, accounted Mafculine; or paffive, dry, and moift, which are feminine. On this principle they call the Sun, Jupiter, Saturn, and Mars, Mafculine; and the Moon and Venus, feminine. Mercury, they fuppofe, partakes of the two. Among the Signs, they account Aries, Libra, Gemini, Leo, Sagittarius, and Aquarius, Mafculine; but Cancer, Capricornus, Taurus, Virgo, Scorpio, and Pifces are feminine.

MASS, the quantity of matter in any body. This is rightly eftimated by its weight; whatever be its figure, or whether its bulk or magnitude be large or fmall.

MATERIAL, relating to Matter.

MATHEMATICAL, relating to Mathematics.

MATHEMATICAL *Sect*, is one of the two leading philofophical fects, which arofe about the beginning of the 17th century; the other being the Metaphyfical fect. The former directed its refearches by the principles of Gaffendi, and fought after truth by obfervation and experience. The difciples of this fect denied the poffibility of erecting on the bafis of metaphyfical and abftract truths, a regular and folid fyftem of philofophy, without the aid of affiduous obfervation and repeated experiments, which are the moft natural and effectual means of philofophical progrefs and improvement. The advancement and reputation of this fect, and of natural knowledge in general, were much owing to the plan of philofophizing propofed by lord Bacon, to the eftablifhment of the Royal Society in London,

London, to the genius and industry of Mr. Boyle, and to the unparalleled researches and discoveries of Sir Isaac Newton. Barrow, Wallis, Locke, and many other great luminaries in learning, adorned this sect.

MATHEMATICS, the science of quantity; or a science that considers magnitudes either as computable or measurable.

The word in its original, *μαθησις*, *mathesis*, signifies *discipline* or *science* in general; and, it seems, has been applied to the doctrine of quantity, either by way of eminence, or because, this having the start of all other sciences, the rest took their common name from it.

As to the origin of the Mathematics, Josephus dates it before the flood, and makes the sons of Seth observers of the course and order of the heavenly bodies: he adds, that to perpetuate their discoveries, and secure them from the injuries either of a deluge or a conflagration, they had them engraven on two pillars, the one of stone, the other of brick; the former of which, he says, was yet standing in Syria in his time.

Indeed it is pretty generally agreed that the first cultivators of Mathematics, after the flood, were the Assyrians and Chaldeans; from whom, Josephus adds, the science was carried by Abraham to the Egyptians; who proved such notable proficients, that Aristotle even fixes the first rise of Mathematics among them. From Egypt, 584 years before Christ, Mathematics passed into Greece, being carried thither by Thales; who having learned geometry of the Egyptian priests, taught it in his own country. After Thales, came Pythagoras; who, among other Mathematical arts, paid a particular regard to arithmetic; drawing the greatest part of his philosophy from numbers. He was the first, according to Laertius, who abstracted geometry from matter; and to him we owe the doctrine of incommensurable magnitude, and the five regular bodies, besides the first principles of music and astronomy. To Pythagoras succeeded Anaxagoras, Oenopides, Briso, Antipho, and Hippocrates of Scio; all of whom particularly applied themselves to the quadrature of the circle, the duplicature of the cube, &c; but the last with most success of any: he is also mentioned by Proclus, as the first who compiled elements of Mathematics.

Democritus excelled in Mathematics as well as physics; though none of his works in either kind are extant; the destruction of which is by some authors ascribed to Aristotle. The next in order is Plato, who not only improved geometry, but introduced it into physics, and so laid the foundation of a solid philosophy. From his school arose a crowd of mathematicians. Proclus mentions 13 of note; among whom was Leodamus, who improved the analysis first invented by Plato; Theætetus, who wrote Elements; and Archytas, who has the credit of being the first that applied Mathematics to use in life. These were succeeded by Neocles and Theon, the last of whom contributed to the elements. Eudoxus excelled in arithmetic and geometry, and was the first founder of a system of astronomy. Menechmus invented the conic sections, and Theudius and Hermotimus improved the elements.

For Aristotle, his works are so stored with Mathematics, that Blancanus compiled a whole book of them: out of his school came Eudemus and Theophrastus;

the first of whom wrote upon numbers, geometry, and invisible lines; and the latter composed a mathematical history. To Aristeus, Isidorus, and Hypsicles, we owe the books of Solids; which, with the other books of Elements, were improved, collected, and methodised by Euclid, who died 284 years before the birth of Christ.

A hundred years after Euclid, came Eratosthenes and Archimedes: and contemporary with the latter was Conon, a geometrician and astronomer. Soon after came Apollonias Pergæus; whose excellent conics are still extant. To him are also ascribed the 14th and 15th books of Euclid, and which, it is said, were contracted by Hypsicles. Hipparchus and Menelaus wrote on the subtenses of the arcs in a circle; and the latter also on spherical triangles. Theodosius's 3 books of Spherics are still extant. And all these, Menelaus excepted, lived before Christ.

Seventy years after Christ, was born Ptolomy of Alexandria; a good geometrician, and the prince of astronomers: to him succeeded the philosopher Plutarch, some of whose Mathematical problems are still extant. After him came Eutocius, who commented on Archimedes, and occasionally mentions the inventions of Philo, Diocles, Nicomedes, Sporus, and Heron, on the duplicature of the cube. To Ctesebes of Alexandria we are indebted for pumps; and Geminus, who lived soon after, is preferred by Proclus to Euclid himself.

Diophantus of Alexandria was a great master of numbers, and the first Greek writer on Algebra. Among others of the Ancients, Nicomachus is celebrated for his arithmetical, geometrical, and musical works: Serenus, for his books on the section of the cylinder; Proclus, for his commentaries on Euclid; and Theon has the credit among some, of being author of the books of elements ascribed to Euclid. The last to be named among the Ancients, is Pappus of Alexandria, who flourished about the year of Christ 400, and is justly celebrated for his books of Mathematical collections, still extant.

Mathematics are commonly distinguished into *Speculative* and *Practical*, *Pure* and *Mixed*.

Speculative MATHEMATICS, is that which barely contemplates the properties of things: and

Practical MATHEMATICS, that which applies the knowledge of those properties to some uses in life.

Pure MATHEMATICS is that branch which considers quantity abstractedly, and without any relation to matter or bodies.

Mixed MATHEMATICS considers quantity as subsisting in material being; for instance, length in a pole, depth in a river, height in a tower, &c.

Pure Mathematics, again, either considers quantity as discrete, and so computable, as arithmetic; or as concrete, and so measureable, as geometry.

Mixed Mathematics are very extensive, and are distinguished by various names, according to the different subjects it considers, and the different views in which it is taken; such as Astronomy, Geography, Optics, Hydrostatics, Navigation, &c, &c.

Pure Mathematics has one peculiar advantage, that it occasions no contests among wrangling disputants, as happens in other branches of knowledge: and the reason

reason is, because the definitions of the terms are premised, and every person that reads a proposition has the same idea of every part of it. Hence it is easy to put an end to all mathematical controversies, by shewing, either that our adversary has not stuck to his definitions, or has not laid down true premises, or else that he has drawn false conclusions from true principles; and in case we are not able to do either of these, we must acknowledge the truth of what he has proved.

It is true, that in mixed Mathematics, where we reason mathematically upon physical subjects, such just definitions cannot be given as in geometry: we must therefore be content with descriptions; which will be of the same use as definitions, provided we be consistent with ourselves, and always mean the same thing by those terms we have once explained.

Dr. Barrow gives a very elegant description of the excellence and usefulness of mathematical knowledge, in his inaugural oration, upon being appointed Professor of Mathematics at Cambridge. The Mathematics, he observes, effectually exercise, not vainly delude, nor vexatiously torment studious minds with obscure subtilties, but plainly demonstrate every thing within their reach, draw certain conclusions, instruct by profitable rules, and unfold pleasant questions. These disciplines likewise enure and corroborate the mind to a constant diligence in study; they wholly deliver us from a credulous simplicity, most strongly fortify us against the vanity of scepticism, effectually restrain us from a rash presumption, most easily incline us to a due assent, and perfectly subject us to the government of right reason. While the mind is abstracted and elevated from sensible matter, distinctly views pure forms, conceives the beauty of ideas, and investigates the harmony of proportions; the manners themselves are sensibly corrected and improved, the affections composed and rectified, the fancy calmed and settled, and the understanding raised and excited to more divine contemplations.

MATTER, an extended substance. Other properties of Matter are, that it resists, is solid, divisible, moveable, passive, &c; and it forms the principles of which all bodies are composed.

Matter and form, the two simple and original principles of all things, according to the Ancients, composing some simple natures, which they called Elements; from the various combinations of which all natural things were afterwards composed.

Dr. Woodward was of opinion, that Matter is originally and really various, being at first creation divided into several ranks, sets. or kinds of corpuscles, differing in substance, gravity, hardness, flexibility, figure, size, &c; from the various compositions and combinations of which, he thinks, arise all the varieties in bodies as to colour, hardness, gravity, tastes, &c. But it is Sir Isaac Newton's opinion, that all those differences result from the various arrangements of the same Matter; which he accounts homogeneous and uniform in all bodies.

The quantity of Matter in any body, is its measure arising from the joint consideration of the magnitude and density of the body: as if one body be twice as dense as another, and also occupy twice the space, then will it contain 4 times the Matter of the other. This quantity of Matter is best discovered by the weight or gravity of the body, to which it is always proportional.

Newton observes, that " it seems probable, God, in the beginning, formed Matter in solid, massy, hard, impenetrable, moveable particles, of such sizes, figures, and with such other properties, and in such proportion to space, as most conduced to the end for which he formed them; and that these primitive particles, being solid, are incomparably harder than any porous bodies compounded of them; even so very hard, as never to wear, and break in pieces: no ordinary power being able to divide what God himself made one in the first creation. While the particles continue entire, they may compose bodies of one and the same nature and texture in all ages; but should they wear away, or break in pieces, the nature of things depending on them would be changed. Water and earth, composed of old worn particles, would not be of the same nature and texture now with water and earth composed of entire particles in the beginning. And therefore, that nature may be lasting, the changes of corporeal things are to be placed only in the various separations and new associations and motions of these permanent particles; compound bodies being apt to break, not in the midst of solid particles, but where those particles are laid together, and touch in a few points. It seems farther, he continues, that these particles have not only a vis inertiæ, accompanied with such passive laws of motion as naturally result from that force, but also that they are moved by certain active principles, such as is that of gravity, and that which causeth fermentation, and the cohesion of bodies. These principles are to be considered not as occult qualities, supposed to result from the specific forms of things, but as general laws of nature, by which the things themselves are formed; their truth appearing to us by phenomena, though their causes are not yet discovered."

Hobbes, Spinoza, &c, maintain that all the beings in the universe are material, and that their differences arise from their different modifications, motions, &c. Thus they conceive that Matter extremely subtile, and in a brisk motion, may think; and so they exclude spirit out of the world.

Dr. Berkley, on the contrary, argues against the existence of Matter itself; and endeavours to prove that it is a mere *ens rationis*, and has no existence out of the mind.

Some late philosophers have advanced a new hypothesis concerning the nature and essential properties of Matter. The first of these who suggested, or at least published an account of this hypothesis, was M. Boscovich, in his Theoria Philosophiæ Naturalis. He supposes that Matter is not impenetrable, but that it consists of physical points only, endued with powers of attraction and repulsion, taking place at different distances, that is, surrounded with various spheres of attraction and repulsion; in the same manner as solid Matter is generally supposed to be. Provided therefore that any body move with a sufficient degree of velocity, or have sufficient momentum to overcome any power of repulsion that it may meet with, it will find no difficulty in making its way through any body whatever. If the velocity of such a body in motion be sufficiently great, Boscovich contends, that the particles of any body through which it passes, will not even be moved

moved out of their place by it. With a degree of velocity fomething lefs than this, they will be confiderably agitated, and ignition might perhaps be the confequence, though the progrefs of the body in motion would not be fenfibly interrupted; and with a ftill lefs momentum it might not pafs at all.

Mr. Michell, Dr. Prieftley, and fome others of our own country, are of the fame opinion. See Prieftley's Hiftory of Difcoveries relating to Light, pa. 390.— In conformity to this hypothefis, this author maintains, that Matter is not that inert fubftance that it has been fuppofed to be; that powers of attraction or repulfion are neceffary to its very being, and that no part of it appears to be impenetrable to other parts. Accordingly, he defines Matter to be a fubftance, poffeffed of the property of extenfion, and of powers of attraction or repulfion, which are not diftinct from Matter, and foreign to it, as it has been generally imagined, but abfolutely effential to its very nature and being: fo that when bodies are divefted of thefe powers, they become nothing at all. In another place, Dr. Prieftley has given a fomewhat different account of Matter; according to which it is only a number of centres of attraction and repulfion; or more properly of centres, not divifible, to which divine agency is directed; and as fenfation and thought are not incompatible with thefe powers, folidity, or impenetrability, and confequently a vis inertiæ only having been thought repugnant to them, he maintains, that we have no reafon to fuppofe that there are in man two fubftances abfolutely diftinct from each other. See Difquifitions on Matter and Spirit.

But Dr. Price, in a correfpondence with Dr. Prieftley, publifhed under the title of A Free Difcuffion of the Doctrines of Materialifm and Philofophical Neceffity, 1778, has fuggefted a variety of unanfwerable objections againft this hypothefis of the penetrability of Matter, and againft the conclufions that are drawn from it. The vis inertiæ of Matter, he fays, is the foundation of all that is demonftrated by natural philofophers concerning the laws of the collifion of bodies. This, in particular, is the foundation of Newton's philofophy, and efpecially of his three laws of motion. Solid Matter has the power of acting on other Matter by impulfe; but unfolid Matter cannot act at all by impulfe; and this is the only way in which it is capable of acting, by any action that is properly its own. If it be faid, that one particle of Matter can act upon another without contact and impulfe, or that Matter can, by its own proper agency, attract or repel other Matter which is at a diftance from it, then a maxim hitherto univerfally received muft be falfe, that " nothing can act where it is not." Newton, in his letters to Bentley, calls the notion, that Matter poffeffes an innate power of attraction, or that it can act upon Matter at a diftance, and attract and repel by its own agency, an abfurdity into which he thought no one could poffibly fall. And in another place he expreffly difclaims the notion of innate gravity, and has taken pains to fhew that he did not take it to be an effential property of bodies. By the fame kind of reafoning purfued, it muft appear, that Matter has not the power of attracting and repelling; that this power is the power of fome foreign caufe, acting upon Matter according to ftated laws; and confequently that attraction and repulfion, not being actions, much lefs inherent qualities of Matter, as fuch, it ought not to be defined by them. And if Matter has no other property, as Dr. Prieftley afferts, than the power of attracting and repelling, it muft be a non-entity; becaufe this is a property that cannot belong to it. Befides, all power is the power of fomething; and yet if Matter is nothing but this power, it muft be the power of nothing; and the very idea of it is a contradiction. If Matter be not folid extenfion, what can it be more than mere extenfion?

Farther, Matter that is not folid, is the fame with pore; and therefore it cannot poffefs what philofophers mean by the momentum or force of bodies, which is always in proportion to the quantity of Matter in bodies, void of pore.

MAUNDY Thursday, is the Thurfday in Paffion week; which was called *Maundy* or *Mandate Thurfday*, from the command which Chrift gave his apoftles to commemorate him in the Lord's Supper, which he this day inftituted; or from the new commandment which he gave them to love one another, after he had wafhed their feet as a token of his love to them.

MAUPERTUIS (Peter Louis Morceau de), a celebrated French mathematician and philofopher, was born at St Malo in 1698, and was there privately educated till he attained his 16th year, when he was placed under the celebrated profeffor of philofophy, M. le Blond, in the college of la Marche, at Paris; while M. Guifnée, of the Academy of Sciences, was his inftructor in mathematics. For this fcience he foon difcovered a ftrong inclination, and particularly for geometry. He likewife practifed inftrumental mufic in his early years with great fuccefs; but fixed on no profeffion till he was 20, when he entered into the army; in which he remained about 5 years, during which time he purfued his mathematical ftudies with great vigour; and it was foon remarked by M. Freret and other academicians, that nothing but mathematics could fatisfy his active foul and unbounded thirft for knowledge.

In the year 1723, he was received into the Royal Academy of Sciences, and read his firft performance, which was a memoir upon the conftruction and form of mufical inftruments. During the firft years of his admiffion, he did not wholly confine his attention to mathematics; he dipt into natural philofophy, and difcovered great knowledge and dexterity in obfervations and experiments upon animals.

If the cuftom of travelling into remote countries, like the fages of antiquity, in order to be initiated into the learned myfteries of thofe times, had ftill fubfifted, no one would have conformed to it with more eagernefs than Maupertuis. His firft gratification of this paffion was to vifit the country which had given birth to Newton; and during his refidence at London he became as zealous an admirer and follower of that philofopher as any one of his own countrymen. His next excurfion was to Bafil in Switzerland, where he formed a friendfhip with the celebrated John Bernoulli and his family, which continued till his death. At his return to Paris, he applied himfelf to his favourite ftudies with greater zeal than ever. And how well he fulfilled

filled the duties of an academician, may be seen by running over the Memoirs of the Academy from the year 1724 to 1744; where it appears that he was neither idle, nor occupied by objects of small importance. The most sublime questions in the mathematical sciences, received from his hand that elegance, clearness, and precision, so remarkable in all his writings.

In the year 1736, he was sent to the polar circle, to measure a degree of the meridian, in order to ascertain the figure of the earth; in which expedition he was accompanied by Meff. Clairault, Camus, Monnier, Outhier, and Celsus the celebrated professor of astronomy at Upsal. This business rendered him so famous, that on his return he was admitted a member of almost every academy in Europe.

In the year 1740, Maupertuis had an invitation from the king of Prussia to go to Berlin; which was too flattering to be refused. His rank among men of letters had not wholly effaced his love for his first profession, that of arms. He followed the king to the field, but at the battle of Molwitz was deprived of the pleasure of being present, when victory declared in favour of his royal patron, by a singular kind of adventure. His horse, during the heat of the action, running away with him, he fell into the hands of the enemy; and was at first but roughly treated by the Austrian Hussars, to whom he could not make himself known for want of language; but being carried prisoner to Vienna, he received such honours from the emperor as never were effaced from his memory. Maupertuis lamented very much the loss of a watch of Mr. Graham's, the celebrated English artist, which they had taken from him; the emperor, who happened to have another by the same artist, but enriched with diamonds, presented it to him, saying, " the Hussars meant only to jest with you, they have sent me your watch, and I return it to you."

He went soon after to Berlin; but as the reform of the academy which the king of Prussia then meditated was not yet mature, he repaired to Paris, where his affairs called him, and was chosen in 1742 director of the Academy of Sciences. In 1743 he was received into the French Academy; which was the first instance of the same person being a member of both the academies at Paris at the same time. Maupertuis again assumed the soldier at the siege of Fribourg, and was pitched upon by marshal Coigny and the count d'Argenson to carry the news to the French king of the surrender of that citadel.

Maupertuis returned to Berlin in the year 1744, when a marriage was negotiated and brought about, by the good offices of the queen mother, between our author and mademoiselle de Borck, a lady of great beauty and merit, and nearly related to M. de Borck at that time minister of state. This determined him to settle at Berlin, as he was extremely attached to his new spouse, and regarded this alliance as the most fortunate circumstance of his life.

In the year 1746, Maupertuis was declared, by the king of Prussia, President of the Royal Academy of Sciences at Berlin, and soon after by the same prince was honoured with the Order of Merit. However, all these accumulated honours and advantages, so far

from lessening his ardour for the sciences, seemed to furnish new allurements to labour and application. Not a day passed but he produced some new project or essay for the advancement of knowledge. Nor did he confine himself to mathematical studies only: metaphysics, chemistry, botany, polite literature, all shared his attention, and contributed to his fame. At the same time he had, it seems, a strange inquietude of spirit, with a dark atrabilaire humour, which rendered him miserable amidst honours and pleasures. Such a temperament did not promise a pacific life; and he was in fact engaged in several quarrels. One of these was with Koenig the professor of philosophy at Franeker, and another more terrible with Voltaire. Maupertuis had inserted in the volume of Memoirs of the Academy of Berlin for 1746, a discourse upon the laws of motion; which Koenig was not content with attacking, but attributed to Leibnitz. Maupertuis, stung with the imputation of plagiarism, engaged the academy of Berlin to call upon him for his proof; which Koenig failing to produce, his name was struck out of the academy, of which he was a member. Several pamphlets were the consequence of this measure; and Voltaire, for some reason or other, engaged in the quarrel against Maupertuis. We say, for some reason or other; because Maupertuis and Voltaire were apparently upon the most amicable terms; and the latter respected the former as his master in the mathematics. Voltaire upon this occasion exerted all his wit and satire against him; and upon the whole was so much transported beyond what was thought right, that he found it expedient in 1753 to quit the court of Prussia.

Our philosopher's constitution had long been considerably impaired by the great fatigues of various kinds in which his active mind had involved him; though from the amazing hardships he had undergone, in his northern expedition, most of his bodily sufferings may be traced. The intense sharpness of the air could only be supported by means of strong liquors; which helped but to lacerate his lungs, and bring on a spitting of blood, which began at least 12 years before he died. Yet still his mind seemed to enjoy the greatest vigour; for the best of his writings were produced, and most sublime ideas developed, during the time of his confinement by sickness, when he was unable to occupy his presidial chair at the academy. He took several journeys to St. Malo, during the last years of his life, for the recovery of his health: and though he always received benefit by breathing his native air, yet still, upon his return to Berlin, his disorder likewise returned with greater violence. His last journey into France was undertaken in the year 1757; when he was obliged, soon after his arrival there, to quit his favourite retreat at St. Malo, on account of the danger and confusion which that town was thrown into by the arrival of the English in its neighbourhood. From thence he went to Bourdeaux, hoping there to meet with a neutral ship to carry him to Hamburgh, in his way back to Berlin; but being disappointed in that hope, he went to Toulouse, where he remained seven months. He had then thoughts of going to Italy, in hopes a milder climate would restore him to health; but finding himself grow worse, he rather inclined towards Germany, and went no Neufchatel, where for three months

months he enjoyed the converfation of lord Marifchal, with whom he had formerly been much connected. At length he arrived at Bafil, October 16, 1758, where he was received by his friend Bernoulli and his family with the utmoft tendernefs and affection. He at firft found himfelf much better here than he had been at Neufchatel: but this amendment was of fhort duration; for as the winter approached, his diforder returned, accompanied by new and more alarming fymptoms. He languifhed here many months, during which he was attended by M. de la Condamine; and died in 1759, at 61 years of age.

The works which he publifhed were collected into 4 volumes 8vo, publifhed at Lyons in 1756, where alfo a new and elegant edition was printed in 1768. Thefe contain the following works: 1. Effay on Cofmology.—2. Difcourfe on the different Figures of the Stars.—3. Effay on Moral Philofophy.—4. Philofophical Reflections upon the Origin of Languages, and the Signification of Words—5. Animal Phyfics, concerning Generation &c.—6. Syftem of Nature, or the Formation of bodies—7. Letters on various fubjects.—8. On the Progrefs of the Sciences.—9. Elements of Geography—10. Account of the Expedition to the Polar Circle, for determining the Figure of the Earth; or the Meafure of the Earth at the Polar Circle.—11. Account of a Journey into the Heart of Lapland, to fearch for an Ancient Monument.—12. On the Comet of 1742.—13. Various Academical Difcourfes, pronounced in the French and Pruffian Academies.—14. Differtation upon Languages.—15. Agreement of the Different Laws of Nature, which have hitherto appeared incompatible.—16. Upon the Laws of Motion.—17. Upon the Laws of Reft.—18. Nautical Aftronomy.—19. On the Parallax of the Moon.—20. Operations for determining the Figure of the Earth, and the Variations of Gravity.—21. Meafure of a Degree of the Meridian at the Polar Circle.

Befide thefe works, Maupertuis was author of a great multitude of interefting papers, particularly thofe printed in the Memoirs of the Paris and Berlin Academies, far too numerous here to mention; viz, in the Memoirs of the Academy at Paris, from the year 1724, to 1749; and in thofe of the Academy of Berlin, from the year 1746, to 1756.

MAXIMUM, denotes the greateft ftate or quantity attainable in any given cafe, or the greateft value of a variable quantity. By which it ftands oppofed to Minimum, which is the leaft poffible quantity in any cafe.

As in the algebraical expreffion $a^2 - bx$, where a and b are conftant or invariable quantities, and x a variable one. Now it is evident that the value of this remainder or difference, $a^2 - bx$, will increafe as the term bx, or x, decreafes; and therefore that will be the greateft when this is the fmalleft; that is, $a^2 - bx$ is a maximum, when x is the leaft, or nothing at all.

Again, the expreffion or difference $a^2 - \dfrac{b}{x}$, evidently increafes as the fraction $\dfrac{b}{x}$ diminifhes; and this diminifhes as x increafes; therefore the given expreffion will be the greateft, or a maximum, when x is the greateft, or infinite.

I

Alfo, if along the diameter KZ *(the 3d fig. below)* of a circle, a perpendicular ordinate LM be conceived to move, from K towards Z; it is evident that, from K it increafes continually till it arrive at the centre, in the pofition NO, where it is at the greateft ftate; and from thence it continually decreafes again, as it moves along from N to Z, and quite vanifhes at the point Z. So that the maximum ftate of the ordinate is NO, equal to the radius of the circle.

Methodus de MAXIMIS *et* MINIMIS, a method of finding the greateft or leaft ftate or value of a variable quantity.

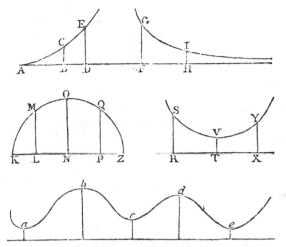

Some quantities continually increafe, and fo have no maximum but what is infinite; as the ordinates BC, DE of the parabola ACE: Some continually decreafe, and fo their leaft or minimum ftate is nothing; as the ordinates FG, HI, to the afymptotes of the hyperbola. Others increafe to a certain magnitude, which is their maximum, and then decreafe again; as the ordinates LM &c of the circle. And others again decreafe to a certain magnitude TV, which is their minimum, and then increafe again; as the ordinates of the curve SVY. While others admit of feveral maxima and minima; as the ordinates of the curve *abcde*, where at *b* and *d* they are maxima, and *a, c, e*, minima. And thus the maxima and minima of all other variable quantities may be conceived; expreffing thofe quantities by the ordinates of fome curves.

The firft maxima and minima are found in the Elements of Euclid, or flow immediately from them: thus, it appears, by the 5th prop. of book 2, that the greateft rectangle that can be made of the two parts of a given line, any how divided, is when the line is divided equally in the middle; prob. 7, book 3, fhews that the greateft line that can be drawn from a given point within a circle, is that which paffes through the centre; and that the leaft line that can be fo drawn, is the continuation of the fame to the other fide of the circle: prop. 8 ib. fhews the fame for lines drawn from a point without the circle: and thus other inftances might be pointed out in the Elements.—Other writers on the Maxima and Minima, are, Apollonius, in the whole 5th book of his Conic Sections;

and

and in the Preface or Dedication to that book, he says others had then also treated the subject, though in a slighter manner.—Archimedes; as in prop. 9 of his Treatise on the Sphere and Cylinder, where he demonstrates that, of all spherical segments under equal superficies, the hemisphere is the greatest.—Serenus, in his 2d book, or that on the Conic Sections.—Pappus, in many parts of his Mathematical Collections; as in lib. 3, prop. 28 &c, lib. 6, prop. 31 &c, where he treats of some curious cases of variable geometrical quantities, shewing how some increase and decrease both ways to infinity; while others proceed only one way, by increase or decrease, to infinity, and the other way to a certain magnitude; and others again both ways to a certain magnitude, giving a maximum and minimum; also lib. 7, prop. 13, 14, 165, 166, &c. And all these are the geometrical Maxima and Minima of the Ancients; to which may be added some others of the same kind, viz. Viviani De Maximis & Minimis Geometrica Divinatio in quintum Conicorum Apollonii Pergæi, in fol. at Flor. 1659; also an ingenious little tract in Thomas Simpson's Geometry, on the Maxima and Minima of Geometrical Quantities.

Other writings on the Maxima and Minima are chiefly treated in a more general way by the modern analysis; and first among these perhaps may be placed that of Fermat. This, and other methods, are best referred to, and explained by the ordinates of curves. For when the ordinate of a curve increases to a certain magnitude, where it is greatest, and afterwards decreases again, it is evident that two ordinates on the contrary sides of the greatest ordinate may be equal to each other; and the ordinates decrease to a certain point, where they are at the least, and afterwards increase again; there may also be two equal ordinates, one on each side of the least ordinate. Hence then an equal ordinate corresponds to two different abscisses, or for every value of an ordinate there are two values of abscisses. Now as the difference between the two abscisses is conceived to become less and less, it is evident that the two equal ordinates, corresponding to them, approach nearer and nearer together; and when the differences of the abscisses are infinitely little, or nothing, then the equal ordinates unite in one, which is either the maximum or minimum. The method hence derived then, is this: Find two values of an ordinate, expressed in terms of the abscisses: put those two values equal to each other, cancelling the parts that are common to both, and dividing all the remaining terms by the difference between the abscisses, which will be a common factor in them: next, supposing the abscisses to become equal, that the equal ordinates may concur in the maximum or minimum, that difference will vanish, as well as all the terms of the equation that include it; and therefore, striking those terms out of the equation, the remaining terms will give the value of the absciss corresponding to the maximum or minimum.

For example, suppose it were required to find the greatest ordinate in a circle KMQ. Put the diameter KZ $= a$, the abscis KL $= x$, the ordinate LM $= y$; hence the other part of the diameter is LZ $= a - x$, and consequently, by the nature of the circle KL $\times$ LZ being equal LM², $x \times \overline{a - x}$ or $ax - x^2 = y^2$.

Again, put another abscis KP $= x + d$, where d is the difference LP, the ordinate PQ, being equal to LM or y; here then again KP $\times$ PZ $=$ PQ², or $\overline{x + d} \times \overline{a - x - d} = ax - x^2 - 2dx + ad - d^2 = y^2$: put now these two values of y^2 equal to each other, so shall $ax - x^2 = ax - x^2 - 2dx + ad - d^2$; cancel the common terms ax and x^2, then $0 = -2dx + ad - d^2$, or $2dx + d^2 = ad$; divide all by d, so shall $2x + d = a$, a general equation derived from the equality of the two ordinates. Now, bringing the two equal ordinates together, or making the two abscisses equal, their difference d vanishes, and the last equation becomes barely $2x = a$, or $x = \frac{1}{2}a$, $=$ KN, the value of the abscis KN when the ordinate NO is a maximum, viz, the greatest ordinate bisects the diameter. And the operation and conclusion it is evident will be the same, to divide a given line into two parts, so that their rectangle shall be the greatest possible.

For a second example, let it be required to divide the given line AB into two such

¹———¹————¹
A   C   D   B

parts, that the one line drawn into the square of the other may be the greatest possible. Putting the given line AB $= a$, and one part AC $= x$; then the other part CB will be $a - x$, and therefore $x^2 \times \overline{a - x} = ax^2 - x^3$ is the product of one part by the square of the other. Again, let one part be AD $= x + d$, then the other part is $a - x - d$, and $\overline{x + d}^2 \times \overline{a - x - d} = ax^2 - x^3 - 3dx^2 + \overline{2ad - 3d^2} \cdot x + ad^2 - d^3$. Then, putting these two products equal to each other, cancelling the common terms $ax^2 - x^3$, and dividing the remainder by d, there results

$0 = -3x^2 + \overline{2a - 3d} \cdot x + ad - d^2$; hence, cancelling all the terms that contain d, there remains $0 = -3x^2 + 2ax$, or $3x = 2a$, and, $x = \frac{2}{3}a$; that is, the given line must be divided into two parts in the ratio of 3 to 2. See Fermat's Opera Varia, pa. 63, and his Letters to F. Mersenne.

The next method was that of John Hudde, given by Schooten among the additions to Des Cartes's Geometry, near the end of the 1st vol. of his edition. This method is also drawn from the property of an equation that has two equal roots. He there demonstrates that, having ranged the terms of an equation, that has two roots equal, according to the order of the exponents of the unknown quantity, taking all the terms over to one side, and so making them equal to nothing on the other side; if then the terms in that order be multiplied by the terms of any arithmetical progression, the resulting equation will still have one of its roots equal to one of the two equal roots of the former equation. Now since, by what has been said of the foregoing method, when the ordinate of a curve, admitting of a maximum or minimum, is expressed in terms of the abscissa, that abscissa, or the value of x, will be two-fold, because there are two ordinates of the same value; that is, the equation has at least two unequal roots or values of x: but when the ordinate becomes a maximum or minimum, the two abscisses unite in one, and the two roots, or values of x, are equal; therefore, from the above said property, the terms of this equation for the maximum or minimum being multiplied by the terms of any arithmetical progression, the root of the resulting equation
tion

tion will be one of the said equal roots, or the value of the absciss x when the ordinate is a maximum.

Although the terms of any arithmetic progression may be used for this purpose, some are more convenient than others; and Mr. Hudde directs to make use of that progression which is formed by the exponents of x, viz, to multiply each term by the exponent of its power, and putting all the resulting products equal to nothing; which, it is evident, is exactly the same process as taking the fluxions of all the terms, and putting them equal to nothing; being the common process now used for the same purpose.

Thus, in the former of the two foregoing examples, where $ax - x^2$, or y^2, is to be a maximum;
mult. by $\quad$ 1 $\quad$ 2
gives $\quad ax - 2x^2 = 0$; hence $2x = a$, and $x = \frac{1}{2}a$, as before.

And in the 2d example, where $ax^2 - x^3$, is to be a maximum; mult. by $\quad - \quad 2 \quad 3$
gives $\quad - \quad - \quad - 2ax^2 - 3x^3 = 0$;
hence $2a - 3x = 0$, or $3x = 2a$, and $x = \frac{2}{3}a$, as before.

The next general method, and which is now usually practised, is that of Newton, or the method of Fluxions, which proceeds upon a principle different from that of the two former methods of Fermat and Hudde. These proceed upon the idea of the two equal ordinates of a curve uniting into one, at the place of the maximum and minimum; but Newton's upon the principle, that the fluxion or increment of an ordinate is nothing, at the point of the maximum or minimum; a circumstance which immediately follows from the nature of that doctrine: for, since a quantity ceases to increase at the maximum, and to decrease at the minimum, at those points it neither increases nor decreases; and since the fluxion of a quantity is proportional to its increase or decrease, therefore the fluxion is nothing at the maximum or minimum. Hence this rule. Take the fluxion of the algebraical expression denoting the maximum or minimum, and put it equal to nothing; and that equation will determine the value of the unknown letter or quantity in question.

So in the first of the two foregoing examples, where it is required to determine x when $ax - x^2$ is a maximum: the fluxion of this is $a\dot{x} - 2x\dot{x} = 0$; divide by $\dot{x}$, so shall $a - 2x = 0$, or $a = 2x$, and $x = \frac{1}{2}a$.

Also, in the 2d example, where $ax^2 - x^3$ must be a maximum: here the fluxion is $2ax\dot{x} - 3x^2\dot{x} = 0$; hence $2a - 3x = 0$, or $2a = 3x$, and $x = \frac{2}{3}a$.

When a quantity becomes a maximum or minimum, and is expressed by two or more affirmative and negative terms, in which only one variable letter is contained; it is evident that the fluxion of the affirmative terms will be equal to the fluxion of the negative ones; since their difference is equal to nothing.

And when, in the expression for the fluxion of a maximum or minimum, there are two or more fluxionary letters, each contained in both affirmative and negative terms; the sum of the terms containing the fluxion of each letter, will be equal to nothing: For, in order that any expression be a maximum or minimum, which contains two or more variable quantities, it must produce a maximum or minimum, if but one of those quantities be supposed variable. So if $ax - 2xy + by$

denote a minimum; its fluxion is $a\dot{x} - 2y\dot{x} - 2x\dot{y} + b\dot{y}$; hence $a\dot{x} - 2y\dot{x} = 0$, and $b\dot{y} - 2x\dot{y} = 0$; from the former of these $y = \frac{1}{2}a$, and from the latter $x = \frac{1}{2}b$. Or, in such a case, take the fluxion of the whole expression, supposing only one quantity variable; then take the fluxion again, supposing another quantity only variable: and so on, for all the several variable quantities; which will give the same number of equations for determining those quantities. So, in the above example, $ax - 2xy + by$, the fluxion is $a\dot{x} - 2y\dot{x} = 0$, supposing only x variable; which gives $y = \frac{1}{2}a$: and the fluxion is $- 2x\dot{y} + b\dot{y} = 0$, when y only is variable; which gives $x = \frac{1}{2}b$; the same as before.

Farther, when any quantity is a maximum or minimum, all the powers or roots of it will be so too; as will also the result be, when it is increased or decreased, or multiplied, or divided by a given or constant quantity; and the logarithm of the same will be also a maximum or minimum.

To find whether a proposed algebraic quantity admits of a maximum or minimum.—Every algebraic expression does not admit of a maximum or minimum, properly so called; for it may either increase continually to infinity, or decrease continually to nothing; in both which cases there is neither a proper maximum nor minimum; for the true maximum is that value to which an expression increases, and after which it decreases again; and the minimum is that value to which the expression decreases, and after that it increases again. Therefore when the expression admits of a maximum, its fluxion is positive before that point, and negative after it; but when it admits of a minimum, its fluxion is negative before, and positive after it. Hence, take the fluxion of the expression a little before the fluxion is equal to nothing, and a little after it; if the first fluxion be positive, and the last negative, the middle state is a maximum; but if the first fluxion be negative, and the last positive, the middle state is a minimum. See Maclaurin's Fluxions, book 1, chap. 9, and book 2, chap. 5, art. 859.

MAY, *Maius*, the fifth month in the year, reckoning from our first or January; but the third, counting the year to begin with March, as the Romans did anciently. It was called Maius by Romulus, in respect to the senators and nobles of his city, who were named *majores;* as the following month was called *Junius*, in honour of the youth of Rome, *in honorem juniorum*, who served him in the war. Though some say it has been thus called from *Maia*, the mother of Mercury, to whom they offered sacrifice on the first day of this month: and Papias derives the name from *Madius, eo quod tunc terra madeat.*

In this month the sun enters the sign Gemini, and the plants of our hemisphere begin mostly to flower.

MAYER (Tobias), one of the greatest astronomers and mechanists of the 18th century, was born at Marpach, in the duchy of Wirtemberg, 1723. He taught himself mathematics, and at 14 years of age designed machines and instruments with the greatest dexterity and justness. These pursuits did not hinder him from cultivating the Belles Lettres. He acquired the Latin tongue, and wrote it with elegance. In 1750, the university of Gottingen chose him for their mathematical professor; and every year of his short life

6 $\qquad\qquad$ was

was thenceforward marked with fome confiderable difcoveries in geometry and aftronomy. He publifhed feveral works in this way, which are all accounted excellent of their kind; and fome papers are inferted in the fecond volume of the Memoirs of the Univerfity of Gottingen. He was very accurate and indefatigable in his aftronomical obfervations; indeed his labours feem to have very early exhaufted him; for he died worn out in 1762, at no more than 39 years of age.

His Table of Refractions, deduced from his aftronomical obfervations, very nicely agrees with that of Doctor Bradley; and his Theory of the Moon, and Aftronomical Tables and Precepts, were fo well efteemed, that they were rewarded by the Englifh Board of Longitude, with the premium of three thoufand pounds, which fum was paid to his widow after his death. Thefe tables and precepts were publifhed by the Board of Longitude in 1770.

MEAN, a middle ftate between two extremes : as a mean motion, mean diftance, arithmetical mean, geometrical mean, &c.

Arithmetical MEAN, is half the fum of the extremes. So, 4 is an arithmetical mean between 2 and 6, or between 3 and 5, or between 1 and 7; alfo an arithmetical mean between a and b is $\frac{a+b}{2}$ or $\frac{1}{2}a + \frac{1}{2}b$.

Geometrical MEAN, commonly called a mean proportional, is the fquare root of the product of the two extremes; fo that, to find a mean proportional between two given extremes, multiply thefe together, and extract the fquare root of the product. Thus, a mean proportional between 1 and 9, is $\sqrt{1 \times 9} = \sqrt{9} = 3$; a mean between 2 and $4\frac{1}{2}$ is $\sqrt{2 \times 4\frac{1}{2}} = \sqrt{9} = 3$ alfo; the mean between 4 and 6 is $\sqrt{4 \times 6} = \sqrt{24}$; and the mean between a and b is $\sqrt{ab}$.

The geometrical mean is always lefs than the arithmetical mean, between the fame two extremes. So the arithmetical mean between 2 and $4\frac{1}{2}$ is $3\frac{1}{4}$, but the geometrical mean is only 3. To prove this generally; let a and b be any two terms, a the greater, and b the lefs; then, univerfally, the arithmetical mean $\frac{a+b}{2}$ fhall be greater than the geometrical mean $\sqrt{ab}$, or $a+b$ greater than $2\sqrt{ab}$. For, by fquaring both, they are $a^2 + 2ab + b^2 > 4ab$; fubtr. $4ab$ from each, then $a^2 - 2ab + b^2 > 0$, that is - - - $(a-b)^2 > 0$.

To find a Mean Proportional Geometrically, between two given lines M and N. Join the two given lines together at C in one continued line AB; upon the diameter AB defcribe a femicircle, and erect the perpendicular CD; which will be the mean proportional between AC and CB; or M and N.

To find two Mean Proportionals between two given extremes. Multiply each extreme by the fquare of the other, viz, the greater extreme by the fquare of the lefs, and the lefs extreme by the fquare of the

greater; then extract the cube root out of each product, and the two roots will be the two mean proportionals fought. That is, $\sqrt[3]{a^2b}$ and $\sqrt[3]{ab^2}$ are the two means between a and b. So, between 2 and 16, the two mean proportionals are 4 and 8; for $\sqrt[3]{2^2 \times 16} = \sqrt[3]{64} = 4$, and $\sqrt[3]{2 \times 16^2} = \sqrt[3]{512} = 8$.

In a fimilar manner we proceed for three means, or four means, or five means, &c. From all which it appears that the feries of the feveral numbers of mean proportionals between a and b will be as follows : viz,

one mean, $\sqrt{ab}$;
two means, $\sqrt[3]{a^2b}$, $\sqrt[3]{ab^2}$;
three means, $\sqrt[4]{a^3b}$, $\sqrt[4]{a^2b^2}$, $\sqrt[4]{ab^3}$;
four means, $\sqrt[5]{a^4b}$, $\sqrt[5]{a^3b^2}$, $\sqrt[5]{a^2b^3}$; $\sqrt[5]{ab^4}$;
five means, $\sqrt[6]{a^5b}$, $\sqrt[6]{a^4b^2}$, $\sqrt[6]{a^3b^3}$, $\sqrt[6]{a^2b^4}$, $\sqrt[6]{ab^5}$;
&c, &c.

Harmonical MEAN, is double a fourth proportional to the fum of the extremes, and the two extremes themfelves a and b : thus, as $a+b : a :: 2b : \frac{2ab}{a+b} = m$ the harmonical mean between a and b. Or it is the reciprocal of the arithmetical mean between the reciprocals of the given extremes; that is, take the reciprocals of the extremes a and b, which will be $\frac{1}{a}$ and $\frac{1}{b}$; then take the arithmetical mean between thefe reciprocals, or half their fum, which will be $\frac{1}{2a} + \frac{1}{2b}$ or $\frac{a+b}{2ab}$; laftly, the reciprocal of this is $\frac{2ab}{a+b} = m$ the harmonical mean : for, arithmeticals and harmonicals are mutually reciprocals of each other;

fo that if a, m, b, &c be arithmeticals,
then fhall $\frac{1}{a}$, $\frac{1}{m}$, $\frac{1}{b}$, &c be harmonicals;
or if the former be harmonicals, the latter will be arithmeticals.

For example, to find a harmonical mean between 2 and 6; here $a = 2$, and $b = 6$; therefore $\frac{2ab}{a+b} = \frac{2 \times 2 \times 6}{2+6} = \frac{24}{8} = 3 = m$ the harmonical mean fought between 2 and 6.

In the 3d book of Pappus's Mathematical Collections we have a very good tract on all the three forts of mean proportionals, beginning at the 5th propofition. He obferves, that the Ancients could not refolve, in a geometrical way, the problem of finding two mean proportionals; and becaufe it is not eafy to defcribe the conic fections in plano, for that purpofe, they contrived eafy and convenient inftruments, by which they obtained good mechanical conftructions of that problem; as appears by their writings; as in the Mefolabe of Eratofthenes. of Philo, with the Mechanics and Catapultics of Hero. For thefe, rightly deeming the problem a folid one, effected the conftruction only by inftruments, and Apollonius Pergæus by means of the conic fections; which others again performed by the *loci folidi* of Ariftæus; alfo Nicomedes folved it by the conchoid, by means of which

which likewife he trifected an angle: and Pappus himfelf gave another folution of the fame problem.

Pappus adds definitions of the three foregoing different forts of means, with many problems and properties concerning them, and, among others, this curious fimilarity of them, viz, *a*, *m*, *b*, being three continued terms, either arithmeticals, geometricals, or harmonicals; then in the

Arithmeticals, $a : a :: a - m : m - b$;

Geometricals, $a : m :: a - m : m - b$;

Harmonicals, $a : b :: a - m : m - b$.

MEAN-*and Extreme Proportion*, or *Extreme-and-Mean Proportion*, is when a line, or any quantity is fo divided, that the lefs part is to the greater, as the greater is to the whole.

MEAN *Anomaly*, of a planet, is an angle which is always proportional to the time of the planet's motion from the aphelion, or perihelion, or proportional to the area defcribed by the radius vector; that is, as the whole periodic time in one revolution of the planet, is to the time paft the aphelion or perihelion, fo is 360° to the Mean anomaly. See ANOMALY.

MEAN *Axis*, in Optics. See AXIS.

MEAN *Conjunction* or *Oppofition*, is when the mean place of the fun is in conjunction, or oppofition, with the mean place of the moon in the ecliptic.

MEAN *Diameter*, in Gauging, is a Mean between the diameters at the head and bung of a calk.

MEAN *Diftance*, of a Planet from the Sun, is an arithmetical mean between the planet's greateft and leaft diftances; and this is equal to the femitranfverfe axis of the elliptic orbit in which it moves, or to the right line drawn from the fun or focus to the extremity of the conjugate axis of the fame.

MEAN *Motion*, is that by which a planet is fuppofed to move equably in its orbit; and it is always proportional to the time.

MEAN *Time*, or Equal time, is that which is meafured by an equable motion, as a clock; as diftinguifhed from apparent time, arifing from the unequal motion of the earth or fun.

MEASURE, denotes any quantity, affumed as unity, or one, to which the ratio of other homogeneous or like quantities may be expreffed.

MEASURE *of an Angle*, is an arc of a circle defcribed from the angular point as a centre, and intercepted between the legs or fides of the angle: and it is ufual to eftimate and exprefs the Meafure of the angle by the number of degrees and parts contained in that arc, of which 360 make up the whole circumference. So, the meafure of the angle BAC, is the arc BC to the radius AB, or the arc *bc* to the radius A*b*.

Hence, a right angle is meafured by a quadrant, or 90 degrees; and any angle, as BAC, is in proportion to a right angle, as the arc BC is to a quadrant, or as the degrees in BC are to 90 degrees.

Common MEASURE. See COMMON *Meafure*.

MEASURE *of a Figure*, or Plane Surface, is a fquare inch, or fquare foot, or fquare yard, &c, that is, a

fquare whofe fide is an inch, or a foot, or a yard, or fome other determinate length; and this fquare is called the *meafuring unit*.

MEASURE *of a Line*, is any right line taken at pleafure, and confidered as unity; as an inch, or a foot, or a yard, &c.

Line of MEASURES. See LINE *of Meafures*.

MEASURE *of a Mafs*, or *Quantity of Matter*, is its weight.

MEASURE *of a Number*, is any number that divides it, without leaving a remainder. So, 2 is a Meafure of 4, of 8, or of any even number; and 3 is a Meafure of 6, or of 9, or of 12, &c.

MEASURE *of a Ratio*, is its logarithm, in any fyftem of logarithms; or it is the exponent of the power to which the ratio is equal, the exponent of fome given ratio being affumed as unity. So, if the logarithm or Meafure of the ratio of 10 to 1, be affumed equal to 1; then the Meafure of the ratio of 100 to 1, will be 2, becaufe 100 is $= 10^2$, or becaufe 100 to 1 is in the duplicate ratio of 10 to 1; and the Meafure of the ratio of 1000 to 1, will be 3, becaufe 1000 is $= 10^3$, or becaufe 1000 to 1 is triplicate of the ratio of 10 to 1.

MEASURE *of a Solid*, is a cubic inch, or cubic foot, or cubic yard, &c; that is, a cube whofe fide is an inch, or a foot, or a yard, &c.

MEASURE *of a Superficies*, the fame as the Meafure of a figure.

MEASURE *of Velocity*, is the fpace uniformly paffed over by a moving body in a given time.

Univerfal or *Perpetual* MEASURE, is a kind of Meafure unalterable by time or place, to which the Meafures of different ages and nations might be reduced, and by which they may be compared and eftimated. Such a Meafure would be very ufeful, if it could be attained; fince, being ufed at all times, and in all places, a great deal of confufion and error would be avoided.

Huygens, in his Horol. Ofcil. propofes, for this purpofe, the length of a pendulum that fhould vibrate feconds, meafured from the point of fufpenfion to the point of ofcillation: the 3d part of fuch a pendulum to be called horary foot, and to ferve as a ftandard to which the Meafure of all other feet might be referred. Thus, for inftance, the proportion of the Paris foot to the horary foot, would be that of 864 to 881; becaufe the length of 3 Paris feet is 864 half lines, and the length of a pendulum, vibrating feconds, contains 881 half lines. But this Meafure, in order to its being univerfal, fuppofes that the action of gravity is the fame on every part of the earth's furface, which is contrary to fact: for which reafon it would really ferve only for places under the fame parallel of latitude: fo that, if every different latitude were to have its foot equal to the 3d part of the pendulum vibrating feconds there, any latitude would ftill have a different length of foot. And befides, the difficulty of meafuring exactly the diftance between the centres of motion and ofcillation are fuch, that hardly any two meafurers would make it the fame quantity.

M. Mouton, canon of Lyons, has alfo a treatife *De Menfura pofteris tranfmittenda*.

Since that time various other expedients have been propofed for eftablifhing an univerfal Meafure, but

hitherto

hitherto without the perfect effect. In 1779, a method was proposed to the Society of Arts, &c, by a Mr. Hatton, in consequence of a premium, which had been 4 years advertised by that institution, of a gold medal, or 100 guineas, ' for obtaining invariable standards for weights and Measures, communicable at all times and to all nations.' Mr. Hatton's plan consisted in the application of a moveable point of suspension to one and the same pendulum, in order to produce the full and absolute effect of two pendulums, the difference of whose lengths was the intended Measure. Mr. Whitehurst much improved upon this idea, by very curious and accurate machinery, in his tract published 1787, intitled ' An Attempt towards obtaining invariable Measures of Length, Capacity, and Weight, from the Mensuration of time, &c. Mr. Whitehurst's plan is, to obtain a Measure of the greatest length that conveniency will permit, from two pendulums whose vibrations are in the ratio of 2 to 1, and whose lengths coincide with the English standard in whole numbers. The numbers he has chosen shew great ingenuity. On a supposition that the length of a seconds pendulum, in the latitude of London, is 39·2 inches, the length of one vibrating 42 times in a minute, must be 80 inches ; and of another vibrating 84 times in a minute, must be 20 inches ; their difference, 60 inches or 5 feet, is his standard Measure. By his experiments, however, the difference in the lengths of the two pendulums was found to be 59·892 inches instead of 60, owing to the error in the assumed length of the seconds pendulum, 39·2 inches being greater than the truth. Mr. Whitehurst has fully accomplished his design, and shewn how an invariable standard may, at all times, be found for the same latitude. He has also ascertained a fact, as accurately as human powers seem capable of ascertaining it, of great consequence in natural philosophy. The difference between the lengths of the rods of two pendulums whose vibrations are known, is a datum from which may be derived the true length of pendulums, the spaces through which heavy bodies fall in a given time, with many other particulars relative to the doctrine of gravitation, the figure of the earth, &c, &c. The result deduced from this experiment is, that the length of a seconds pendulum, vibrating in a circular arc of 3° 20', is 39·119 inches very nearly ; but vibrating in the arc of a cycloid it would be 39·136 inches ; and hence, heavy bodies will fall, in the first second of their descent, 16·094 feet, or 16 feet 1⅛ inch, very nearly.

It is said, the French philosophers have a plan in contemplation, to take for a universal Measure, the length of a whole meridian circle of the earth, and take all other Measures from sub-divisions of that ; which will be a very good way.—Other projects have also been devised, but of little or no consideration.

MEASURE, in a legal, commercial, and popular sense, denotes a certain quantity or proportion of any thing, bought, sold, valued, or the like.

The regulation of weights and Measures ought to be universally the same throughout the nation, and indeed all nations ; and they should therefore be reduced to some fixed rule or standard.

Measures are various, according to the various kinds or dimensions of the things measured. Hence arise

Lineal or *Longitudinal* MEASURES, for lines or lengths :

Square MEASURES, for areas or superficies : and

Solid or *Cubic* MEASURES, for the solid contents and capacities of bodies.

The several Measures used in England, are as in the following Tables :

1. *English Long Measure.*

Barley
Corns

3 =	1 Inch				
36 =	12 =	1 Foot			
108 =	36 =	3 =	1 Yard		
594 =	198 =	16½ =	5½ =	1 Pole	
23760 =	7920 =	660 =	220 =	40 = 1 Furlong	
190082 =	63360 =	5280 =	1760 =	320 = 8 = 1 Mile	

Also,   4 Inches   = 1 Hand
      6 Feet, or 2 yds = 1 Fathom
      3 Miles   = 1 League
      60 Nautical or Geograph. Miles = 1 Degree
or   69½ Statute Miles = 1 Degree nearly
      360 Degrees, or 25000 Miles nearly = the Circumference of the Earth.

2. *Cloth Measure.*

Inches

2¼ =	1 Nail			
9 =	4 =	1 Quarter		
36 =	16 =	4 =	1 Yard	
27 =	12 =	3 =	1 Ell Flemish	
45 =	20 =	5 =	1 Ell English	
54 =	24 =	6 =	1 Ell French.	

3. *Square Measure.*

Inches

144 =	1 Foot			
1296 =	9 =	1 Yard		
39204 =	272¼ =	30¼ =	1 Pole	
1568160 =	10890 =	1210 =	40 = 1 Rood	
6272640 =	43560 =	4840 =	160 = 4 = 1 Acre.	

4. *Solid*, or *Cubical Measure.*

Inches

1728 =	1 Foot	
46656 =	27 =	1 Yard.

5. *Wine Measure.*

Pints

2 =	1 Quart						
8 =	4 =	1 Gallon = 231 Cubic Inches.					
336 =	168 =	42 =	1 Tierce				
504 =	252 =	63 =	1½ =	1 Hogshead			
672 =	336 =	84 =	2 =	1⅓ =	1 Puncheon		
1008 =	504 =	126 =	3 =	2 =	1½ =	1 Pipe	
2016 =	1008 =	252 =	6 =	4 =	3 =	2 =	1 Tun.

Also, 231 Cubic Inches = 1 Gallon
      10 Gallons   = 1 Anker
      18 Gallons   = 1 Runlet
      31½ Gallons   = 1 Barrel.

6. *Ale*

6. *Ale and Beer Measure.*

Pints.

2 =   1 Quart.
8 =   4 =   1 Gallon = 282 Cubic Inches.
72 =   36 =   9 =   1 Firkin.
144 =   72 =   18 =   2 = 1 Kilderkin.
288 = 144 =   36 =   4 = 2 = 1 Barrel.
432 = 216 =   54 =   6 = 3 = $1\frac{1}{2}$ = 1 Hogſhead.
576 = 288 =   72 =   8 = 4 = 2 = $1\frac{1}{3}$ = 1 Puncheon.
864 = 432 = 108 = 12 = 6 = 3 = 2 = $1\frac{1}{2}$ = 1 Butt.

Note, The Ale gallon contains 282 cubic inches.

7. *Dry Meaſure.*

Pints.

8 =   1 Gallon = $268\frac{4}{5}$ Cubic Inches.
16 =   2 =   1 Peck.
64 =   8 =   4 =   1 Buſhel.
256 =   32 =   16 =   4 =   1 Coom.
512 =   64 =   32 =   8 =   2 =   1 Quarter.
2560 = 320 = 160 = 40 = 10 =   5 = 1 Wey.
5120 = 640 = 320 = 80 = 20 = 10 = 2 = 1 Laſt.

Alſo, $268\frac{4}{5}$ Cubic Inches   = 1 Gallon.
and   36 Buſhels of Coals = 1 Chaldron.

8. *Proportions of the Long Meaſures of ſeveral Nations to the Engliſh Foot.*									
			Thouſandth Parts.	Inches.				Thouſandth Parts.	Inches.
Engliſh	-	-	foot	1000	12·000	Amſterdam	-	ell	2269	27·228
Paris	-	foot	$1065\frac{3}{4}$	12·792	Antwerp	-	-	ell	2273	27·276
Rynland, or Leyden	-	foot	1033	12·396	Rynland, or Leyden	-	ell	2260	27·120
Amſterdam	-	-	foot	942	11·304	Frankfort	-	-	ell	1826	21·912
Brill	-	-	foot	1103	13·236	Hamburgh	-	-	ell	1905	22·860
Antwerp	-	-	foot	946	11·352	Leipſic	-	-	ell	2260	27·120
Dort	-	-	foot	1184	14·208	Lubeck	-	-	ell	1908	22·896
Lorrain	-	-	foot	958	11·496	Noremburgh	-	ell	2227	26·724
Mechlin	-	-	foot	919	11·028	Bavaria	-	-	ell	954	11·448
Middleburgh	-	foot	991	11·892	Vienna	-	-	ell	1053	12·636
Straſburgh	-	-	foot	920	11·040	Bononia	-	-	ell	2147	25·764
Bremen	-	-	foot	964	11·568	Dantzic	-	-	ell	1903	22·836
Cologn	-	-	foot	954	11·448	Florence	-	Brace or ell	1913	22·956
Frankfort ad Mœnum	-	foot	948	11·376	Spaniſh, or Caſtile	-	palm	751	9·012
Spaniſh	-	-	foot	1001	12·012	Spaniſh	-	-	vare	3004	36·040
Toledo	-	-	foot	899	10·788	Liſbon	-	-	vare	2750	33·000
Roman	-	-	foot	967	11·604	Gibraltar	-	-	vare	2760	33·120
On the monument of Ceſtius Statilius }	-	foot	972	11·664	Toledo	-	-	vare	2685	32·220
Bononia	-	-	foot	1204	14·448			palm	861	10·332
Mantua	-	-	foot	1569	18·838	Naples	-	brace	2100	25·200
Venice	-	-	foot	1162	13·944			canna	6880	82·560
Dantzic	-	-	foot	944	11·328	Genoa	-	-	palm	830	9·960
Copenhagen	-	foot	965	11·580	Milan	-	-	calamus	6544	78·528
Prague	-	-	foot	1026	12·312	Parma	-	-	cubit	1866	22·392
Riga	-	-	foot	1831	21·972	China	-	-	cubit	1016	12·192
Turin	-	-	foot	1062	12·744	Cairo	-	-	cubit	1824	21·888
The Greek	-	foot	1007	12·084	Old Babylonian	-	cubit	1520	18·240
Old Roman	-	foot	970	11·640	Old Greek	-	cubit	1511	18·132
Lyons	-	-	ell	3967	47·604	Old Roman	-	cubit	1458	17·496
Bologna	-	-	ell	2076	24·912	Turkiſh	-	pike	2200	26·400
						Perſian	-	-	araſh	3197	38·364

MEASURING, the ſame as Mensuration, which ſee.

MECHANICS, a mixed mathematical ſcience, that treats of forces, motion, and moving powers, with their effects in machines, &c. The ſcience of Mechanics is diſtinguiſhed, by Sir Iſaac Newton, into Practical and Rational: the former treats of the Mechanical Powers, and of their various combinations; the latter, or Rational Mechanics, comprehends the whole theory and doctrine of forces, with the motions and effects produced by them.

That part of Mechanics, which treats of the weight, gravity,

gravity, and equilibrium of bodies and powers, is called Statics; as diftinguifhed from that part which confiders the Mechanical powers, and their application, which is properly called Mechanics.

Some of the principles of Statics were eftablifhed by Archimedes, in his Treatife on the Centre of Gravity of Plane Figures: befides which, little more upon Mechanics is to be found in the writings of the Ancients, except what is contained in the 8th book of Pappus's Mathematical Collections, concerning the five Mechanical Powers. Galileo laid the beft foundation of Mechanics, when he inveftigated the defcent of heavy bodies; and fince his time, by the affiftance of the new methods of computation, a great progrefs has been made, efpecially by Newton, in his Principia, which is a general treatife on Rational and Phyfical Mechanics, in its largeft extent. Other writers on this fcience, or fome branch of it, are, Guido Ubaldus, in his Liber Mechanicorum; Torricelli, Libri de Motu Gravium naturaliter Defcendentium & Projectorum; Bahanus, Tractatus de Motu naturali Gravium; Huygens, Horologium Ofcillatorium, and Tractatus de Motu Corporum ex Percuffione; Leibnitz, Refiftentia Solidorum in Acta Eruditor. an. 1684; Guldinus, De Centro Gravitatis; Wallis, Tractatus de Mechanica; Varignon, Projet d'une Nouvelle Mechanique, and his papers in the Memoir. Acad. an. 1702; Borelli, Tractatus De Vi-Percuffionis, De Motionibus Naturalibus a Gravitate pendentibus, and De Motu Animalium; De Chales, Treatife on Motion; Pardies, Difcourfe of Local Motion; Parent, Elements of Mechanics and Phyfics; Cafatus, Mechanica; Oughtred, Mechanical Inftitutions; Rohault, Tractatus de Mechanica; Lamy, Mechanique; Keill, Introduction to true Philofophy; De la Hire, Mechanique; Mariotte, Traité du Choc des Corps; Ditton, Laws of Motion; Herman, Phoronomia; Gravefande, Phyfics: Euler, Tractatus de Motu; Muffchenbroek, Phyfics; Boffu, Mechanique; Defaguliers, Mechanics; Rowning, Natural Philofophy; Emerfon, Mechanics; Parkinfon, Mechanics; La Grange, Mechanique Analytique; Nicholfon, Introduction to Natural Philofophy; Enfield, Inftitutes of Natural Philofophy, &c, &c. As to the Defcription of Machines, fee Strada, Zeifingius, Beffon, Auguftine de Ramellis, Boetler, Leopold, Sturmy, Perrault, Limberg, Emerfon, Royal Academy of Sciences, &c.

In treating of machines, we fhould confider the weight that is to be raifed, the power by which it is to be raifed, and the inftrument or engine by which this effect is to be produced. And, in treating of thefe, there are two principal problems that prefent themfelves: the firft is, to determine the proportion which the power and weight ought to have to each other, that they may juft be in equilibrio; the fecond is, to determine what ought to be the proportion between the power and weight, that a machine may produce the greateft effect in a given time. All writers on Mechanics treat on the firft of thefe problems, but few have confidered the fecond, though not lefs ufeful than the other.

As to the firft problem, this general rule holds in all

powers; namely, that when the power and weight are reciprocally proportional to the diftances of the directions in which they act, from the centre of motion; or when the product of the power by the diftance of its direction, is equal to the product of the weight by the diftance of its direction; this is the cafe in which the power and weight fuftain each other, and are in equilibrio; fo that the one would not prevail over the other, if the engine were at reft; and if it were in motion, it would continue to proceed uniformly, if it were not for the friction of its parts, and other refiftances. And, in general, the effect of any power, or force, is as the product of that force multiplied by the diftance of its direction from the centre of motion, or the product of the power and its velocity when in motion, fince this velocity is proportional to the diftance from that centre.

The fecond general problem in Mechanics, is, to determine the proportion between the power and weight, fo that when the power prevails, and the machine is in motion, the greateft effect poffible may be produced by it in a given time. It is manifeft, that this is an enquiry of the greateft importance, though few have treated of it. When the power is only a little greater than what is fufficient to fuftain the weight, the motion ufually is too flow; and though a greater weight be raifed in this cafe, it is not fufficient to compenfate for the lofs of time. On the other hand, when the power is much greater than what is fufficient to fuftain the weight, this is raifed in lefs time; but it may happen that this is not fufficient to compenfate for the lofs arifing from the fmallnefs of the load. It ought therefore to be determined when the product of the weight multiplied by its velocity, is the greateft poffible; for this product meafures the effect of the engine in a given time, which is always the greater in proportion both as the weight is greater, and as its velocity is greater. For fome calculations on this problem, fee Maclaurin's Account of Newton's Difcoveries, p. 171, &c; alfo his Fluxions, art. 908 &c. And, for the various properties in Mechanics, fee the feveral terms MOTION, FORCE, MECHANICAL POWERS, LEVER, &c.

MECHANIC, or MECHANICAL, fomething relating to Mechanics, or regulated by the nature and laws of motion.

MECHANICAL is alfo ufed in Mathematics, to fignify a conftruction or proof of fome problem, not done in an accurate and geometrical manner, but coarfely and unartfully, or by the affiftance of inftruments; as are moft problems relating to the duplicature of the cube, and the quadrature of the circle.

MECHANICAL *Affections*, fuch properties in matter, as refult from their figure, bulk, and motion.

MECHANICAL *Caufes*, are fuch as are founded on Mechanical Affections.

MECHANICAL *Curve*, called alfo *Tranfcendental*, is one whofe nature cannot be expreffed by a finite Algebraical equation.

MECHANICAL *Philofophy*, alfo called the *Corpufcular Philofophy*, is that which explains the phenomena of nature, and the operations of corporeal things, on the principles of Mechanics; viz, the motion, gravity, figure, arrangement, difpofition, greatnefs,

or

or fmallnefs of the parts which compofe natural bodies.

MECHANICAL *Solution*, of a Problem, is either when the thing is done by repeated trials, or when the lines ufed in the folution are not truly geometrical, or by organical conftruction.

MECHANICAL *Powers*, are certain fimple machines which are ufed for raifing greater weights, or overcoming greater refiftances than could be effected by the natural ftrength without them.

Thefe fimple machines are ufually accounted fix in number, viz, the Lever, the Wheel and Axle, or Axis in Peritrochio, the Pulley, the Inclined Plane, the Wedge, and the Screw. Of the various combinations of thefe fimple powers do all engines, or compound machines, confift : and in treating of them, fo as to fettle their theory and properties, they are confidered as mathematically exact, or void of weight and thicknefs, and moving without friction. See the properties and demonftrations of each of thefe under the feveral words LEVER, &c. To which may be added the following general obfervations on them all, in a connective way.

1. A *Lever*, the moft fimple of all the mechanic powers, is an engine chiefly ufed to raife large weights to fmall heights; fuch as a handfpike, when of wood ; and a crow, when of iron. In theory, a lever is confidered as an inflexible line, like the beam of a balance, and fubject to the fame proportions ; only that the power applied to it, is commonly an animal power ; and from the different ways of ufing it, or applying it, it is called a lever of the firft, fecond, or third kind : viz, of the 1ft kind, when the weight is on one fide of the prop, and the power on the other ; of the 2d kind, when the weight is between the prop and the power ; and of the 3d kind, when the power is between the prop and the weight.

Many of the inftruments in common ufe, are levers of one of the three kinds ; thus, pincers, fheers, forceps, fnuffers, and fuch like, are compounded of two levers of the firft kind ; for the joint about which they move, is the fulcrum, or centre of motion ; the power is applied to the handles, to prefs them together ; and the weight is the body which they pinch or cut. The cutting knives ufed by druggifts, patten-makers, blockmakers, and fome other trades, are levers of the 2d kind : for the knife is fixed by a ring at one end, which makes the fulcrum, or fixed point ; the other end is moved by the hand, or power ; and the body to be cut, or the refiftance to be overcome, is the weight. Doors are levers of the 2d kind ; the hinges being the centre of motion ; the hand applied to the lock is the power ; while the door or weight lies between them. A pair of bellows confifts of two levers of the 2d kind ; the centre of motion is where the ends of the boards are fixed near the pipe ; the power is applied at the handles ; and the air preffed out from between the boards, by its refiftance, acts againft the middle of the boards like a weight. The oars of a boat are levers of the 2d kind : the fixed point is the blade of the oar in the water ; the power is the hand acting at the other end ; and the weight to be moved is the boat. And the fame of the rudder of a veffel. Spring fheers and tongs

are levers of the 3d kind ; where the centre of motion is at the bow-fpring at one end ; the weight or refiftance is acted on by the other end ; and the hand or power is applied between the ends. A ladder reared by a man againft a wall, is a lever of the 3d kind : and fo are alfo almoft all the bones and mufcles of animals.

In all levers, the effect of any power or weight, is both proportional to tha power or weight, and alfo to its diftance from the centre of motion. And hence it is that, in raifing great weights by a lever, we chufe the longeft levers ; and alfo reft it upon a point as far from the hand or power, and as near to the weight, as poffible. Hence alfo there will be an equilibrium between the power and weight, when thofe two products are equal, viz, the power multiplied by its diftance, equal to the weight multiplied by its diftance ; when, alfo, the weight and power are to each other reciprocally as their diftances from the prop or fixed point.

2. The Axis in Peritrochio, or Wheel and Axle, is a fimple engine confifting of a wheel fixed upon the end of an axle, fo that they both turn round together in the fame time. This engine may be referred to the lever : for the centre of the axis, or wheel, is the fixed point ; the radius of the wheel is the diftance of the power, acting at the circumference of the wheel, from that point ; and the radius of the axle is the diftance of the weight from the fame point. Hence the effect of the power, independent of its own natural intenfity, is as the radius of the wheel ; and the effect of the weight is as the radius of the axle : fo that the two will be in equilibrio, when the two products are equal, which are made by multiplying each of thefe, the weight and power, by the radius, or diftance at which it acts ; and then alfo, the weight and power are reciprocally proportional to thofe radii.

In practice, the thicknefs of the rope, that winds upon the axle, and to which the weight is faftened, is to be confidered : which is done, by adding half its thicknefs to the radius of the axis, for its diftance from the fixed point, when there is only one fold of rope upon the axle ; or as many times the thicknefs as there are folds, wanting only one half when there are feveral folds of the rope, one over another : which is the reafon that more power muft be applied when the axis is thus thickened ; as often happens in drawing water from a deep and narrow well, over which a long axle cannot be placed.

If the rope to which the power is faftened, be fuccefffively applied to different wheels, whofe diameters are larger and larger ; the axis will be turned with ftill more and more eafe, unlefs the intenfity of the power be diminifhed in the fame proportion ; and if fo, the axis will always be drawn with the fame ftrength by a power continually diminifhing. This is practifed in fpring clocks and watches ; where the fpiral fpring, which is ftrongeft in its action when firft wound up, draws the fuzee, or continued axis in peritrochio, firft by the fmaller wheels, and as it unbends and becomes weak, draws at the larger wheels, in fuch manner that the watch work is always carried round with the fame force.

As a very fmall axis would be too weak for very great weights, or a large wheel would be expenfive as well

well as cumberfome, and take more room than perhaps can be fpared for it; therefore, that the action of the power may be increafed, without incurring either of thofe inconveniences, a compound Axis in Peritrochio is ufed, which is effected by combining wheels and axles by means of pinions, or fmall wheels, upon the axles, the teeth of which take hold of teeth made in the large wheels; as is feen in clocks, jacks, and other compound machines. And in fuch a combination of wheels and axles, the effect of the power is increafed in the ratio of the continual product of all the axles, or fmall wheels, to that of all the large ones. Thus, if there be two fmall wheels and an axle, turning three large wheels; the axle being 2 inches diameter, and each of the fmall wheels 4 inches, while the large ones are 2 feet or 24 inches diameter; then $2 \times 4 \times 4 = 32$ is the continual product of the fmall diameters, and $24 \times 24 \times 24 = 13824$ is that of the large ones; therefore 13824 to 32, or 432 to 1, is the ratio in which the power is increafed: and if the power be a man, whofe natural ftrength is equal, fuppofe, to 150 pounds weight, then $432 \times 150 = 64800$lb, or 28 ton 18 cwt 64lb, is the weight he would be able to balance, fufpended about the axle.

3. *A Single Pulley*, is a fmall wheel, moveable round an axis, called its centre pin; which of itfelf is not properly one of the mechanical powers, becaufe it produces no gain of power; for, as the weight hangs by one end of the cord that paffes over the pulley, and the power acts at the other end of the fame, thefe act at equal diftances from the centre or axis of motion, and confequently the power is equal to the weight when in equilibrio. So that the chief ufe of the fingle pulley is to change the direction of the power from upwards to downwards, &c, and to convey bodies to a great height or diftance, without a perfon moving from his place.

But by combining feveral fingle pulleys together, a confiderable gain of power is made, and that in proportion to the additional number of ropes made to pafs over them; and yet it enjoys at the fame time the properties of a fingle pulley, by changing the direction of the action in any manner.

4. *The Inclined Plane*, is made by planks, bars, or beams, laid aflope; by which, large and heavy bodies may be more eafily raifed or lowered, by fliding them up or down the plane; and the gain in power is in proportion as the length of the plane to its height, or as radius to the fine of the angle of inclination of the plane with the horizon.

In drawing a weight up an inclined plane, the power acts to the greateft advantage, when its direction is parallel to the plane.

5. *The Wedge*, which refembles a double inclined plane, is very ufeful to drive in below very heavy weights to raife them but a fmall height, alfo in cleaving and fplitting blocks of wood, and ftone &c; and the power gained, is in proportion of the flant fide to half the thicknefs of the back. So that, if the back of a wedge be 2 inches thick, and the fide 20 inches long, any weight preffing on the back will balance 20 times as much acting on the fide. But the great advantage of a wedge lies in its being urged, not

by preffure, but ufually by percuffion, as the blow of a hammer or mallet; by which means a wedge may be driven in below, and fo be made to lift, almoft any the greateft weight, as the largeft fhip, by a man ftriking the back of a wedge with a mallet.

To the wedge may be referred the axe or hatchet, the teeth of faws, the chifel, the augur, the fpade and fhovel, knives and fwords of all kinds, as alfo the bodkin and needle, and in a word all forts of inftruments which, beginning from edges or points, become gradually thicker as they lengthen; the manner in which the power is applied to fuch inftruments, being different according to their different fhapes, and the various ufes for which they have been contrived.

6. *The Screw*, is a kind of perpetual or endlefs Inclined Plane; the power of which is ftill farther affifted by the addition of a handle or lever, where the power acts; fo that the gain in power, is in the proportion of the circumference defcribed or paffed through by the power, to the diftance between thread and thread in the fcrew.

The ufes to which the fcrew is applied, are various; as, the preffing of bodies clofe together; fuch as the prefs for napkins, for bookbinders, for packers, hotpreffers, &c.

In the fcrew, and the wedge, the power has to overcome both the weight, and alfo a very great friction in thofe machines; fuch indeed as amounts fometimes to as much as the weight to be raifed, or more. But then this friction is of ufe in retaining the weight and machine in its place, even after the power is taken off.

If machines or engines could be made without friction, the leaft degree of power added to that which balances the weight, would be fufficient to raife it. In the lever, the friction is little or nothing; in the wheel and axle, it is but fmall; in pulleys, it is very confiderable; and in the inclined plane, wedge, and fcrew, it is very great.

It is a general property in all the Mechanic powers, that when the weight and power are regulated fo as to balance each other, in every one of thefe machines, if they be then put in motion, the power and weight will be to each other reciprocally as the velocities of their motion, or the power is to the weight as the velocity of the weight is to the velocity of the power; fo that their two momenta are equal, viz, the product of the power multiplied by its velocity, equal to the product of the weight multiplied by its velocity. And hence too, univerfally, what is gained in power, is loft in time; for the weight moves as much flower as the power is fmaller.

Hence alfo it is plain, that the force of the power is not at all increafed by engines; only the velocity of the weight, either in lifting or drawing, is fo diminifhed by the application of the inftrument, as that the momentum of the weight is not greater than the force of the power. Thus, for inftance, if any force can raife a pound weight with a given velocity, it is impoffible by any engine to raife 2 pound weight with the fame velocity: but by an engine it may be made to raife 2 pound weight with half the velocity, or even 1000 times the weight with the 1000th part of the velocity.

See

See Maclaurin's Account of Newton's Philof. Difcov. book 2, chap. 3; Hamilton's Philof. Eff. 1; Philof. Tranf. 53, pa. 116; or Landen's Memoirs, vol. 1, pa. 1.

MECHANISM, either the conftruction or the machinery employed in any thing; as the Mechanifm of the barometer, of the microfcope, &c.

MEDIUM, the fame as mean, either arithmetical, geometrical, or harmonical.

MEDIUM denotes alfo that fpace, or region, or fluid, &c, through which a body paffes in its motion towards any point. Thus, the air, or atmofphere, is the medium in which birds and beafts live and move, and in which a projectile moves; water is the medium in which fifhes move; and æther is a fuppofed fubtile Medium in which the planets move. Glafs is alfo called a Medium, being that through which the rays of light move and pafs.

Mediums refift the motion of bodies moving through them, in proportion to their denfity or fpecific gravity.

Subtile or *Ætherial* MEDIUM, is an univerfal one whofe exiftence is by Newton rendered probable. He makes it univerfal; and vaftly more rare, fubtile, elaftic, and active than air; and by that means freely permeating the pores and interftices of all other Mediums, and diffufing itfelf through the whole creation. By the intervention of this fubtile Medium he thinks it is that moft of the great phenomena of nature are effected. See ÆTHER.

This Medium it would feem he has recourfe to, as the firft and moft remote phyfical fpring, and the ultimate of all natural caufes. By the vibrations of this Medium, he fuppofes that heat is propagated from lucid bodies; as alfo the intenfenefs of heat increafed and preferved in hot bodies, and from them communicated to cold ones.

By this Medium, he fuppofes that light is reflected, inflected, refracted, and put alternately into fits of eafy reflection and tranfmiffion; which effects he alfo elfewhere afcribes to the power of attraction; fo that it would feem, this Medium is the fource and caufe even of attraction itfelf.

Again, this Medium being much rarer within the heavenly bodies, than in the heavenly fpaces, and growing denfer as it recedes farther from them, he fuppofes this is the caufe of the gravitation of thefe bodies towards each other, and of the parts towards the bodies.

Again, from the vibrations of this fame Medium, excited in the bottom of the eye by the rays of light, and thence propagated through the capillaments of the optic nerves into the fenforium, he fuppofes that vifion is performed: and fo likewife hearing, from the vibrations of this or fome other Medium, excited in the auditory nerves by the tremors of the air, and propagated through the capillaments of thofe nerves into the fenforium: and fo of the other fenfes.

And again, he conceives that mufcular motion is performed by the vibrations of the fame Medium, excited in the brain at the command of the will, and thence propagated through the capillaments of the nerves into the mufcles; and thus contracting and dilating them.

The elaftic force of this Medium, he fhews, muft be prodigioufly great. Light moves at the rate of confiderably more than 10 millions of miles in a minute; yet the vibrations and pulfations of this Medium, to caufe the fits of eafy reflection and tranfmiffion, muft be fwifter than light, which is yet 7 hundred thoufand times fwifter than found. The elaftic force of this Medium, therefore, in proportion to its denfity, muft be above 490000 million of times greater than the elaftic force of the air, in proportion to its denfity; the velocities and pulfes of the elaftic Mediums being in a fubduplicate ratio of the elafticities, and the rarities of the Mediums, taken together. And thus may it be conceived that the vibration of this Medium is the caufe alfo of the elafticity of bodies.

Farther, the particles of this Medium being fuppofed indefinitely fmall, even fmaller than thofe of light; if they be likewife fuppofed, like our air, endued with a repelling power, by which they recede from each other, the fmallnefs of the particles may exceedingly contribute to the increafe of the repelling power, and confequently to that of the elafticity and rarity of the Medium; by that means fitting it for the free tranfmiffion of light, and the free motions of the heavenly bodies. In this Medium may the planets and comets roll without any confiderable refiftance. If it be 700,000 times more elaftic, and as many times rarer, than air, its refiftance will be above 600 million times lefs than that of water; a refiftance that would caufe no fenfible alteration in the motion of the planets in ten thoufand years.

MEGAMETER. See MICROMETER.

MEIBOMIUS (MARCUS), a very learned perfon of the 17th century, of a family in Germany which had long been famous for learned men. He devoted himfelf to literature and criticifm, but particularly to the learning of the Ancients; as their mufic, the ftructure of their galleys, &c. In 1652 he publifhed a collection of feven Greek authors, who had written upon Ancient Mufic, to which he added a Latin verfion by himfelf. This work he dedicated to queen Chriftina of Sweden; in confequence of which he received an invitation to that Princefs's court, like feveral other learned men, which he accepted. The queen engaged him one day to fing an air of ancient mufic, while a perfon danced the Greek dances to the found of his voice; and the immoderate mirth which this occafioned in the fpectators, fo covered him with ridicule, and difgufted him fo vehemently, that he abruptly left the court of Sweden immediately, after heartily battering with his fifts the face of Bourdelot, the favourite phyfician and buffon to the queen, who had perfuaded her to exhibit that fpectacle.

Meibomius pretended that the Hebrew copy of the Bible was full of errors, and undertook to correct them by means of a metre, which he fancied he had difcovered in thofe ancient writings; but this it feems drew upon him no fmall raillery from the Learned. Neverthelefs, befides the work above mentioned, he produced feveral others, which fhewed him to be a good fcholar; witnefs his Notes upon Diogenes Laertius in Menage's edition; his *Liber de Fabrica Triremium*, 1671, in which he thinks he difcovered the

6   method

method in which the Ancients difpofed their banes of cars; his edition of the Ancient Greek Mythologifts; and his Dialogues on Proportions, a curious work, in which the interlocutors, or perfons reprefented as fpeaking, are Euclid, Archimedes, Apollonius, Pappus, Eutocius, Theo, and Hermotimus. This laft work was oppofed by Langius, and by Dr. Wallis, in a confiderable Tract, printed in the firft volume of his works.

MELODY, is the agreeable effect of different mufical founds, ranged or difpofed in a proper fucceffion, being the effect only of one fingle part, voice, or inftrument; by which it is diftinguifhed from harmony, which properly refults from the union of two or more mufical founds heard together.

MENISCUS, a lens or glafs, convex on one fide, and concave on the other. Sometimes alfo called a Lune or Lunula. See its figure under the article LENS.

To find the Focus of a Menifcus, the rule is, as the difference between the diameters of the convexity and concavity, is to either of them, fo is the other diameter, to the focal length, or diftance of the focus from the Menifcus. So that, having given the diameter of the convexity, it is eafy to find that of the concavity, fo as to remove the focus to any propofed diftance from the Menifcus. For, if D and *d* be the diameters of the two fides, and *f* the focal diftance; then fince,

by the rule $D - d : D :: d : f$,
therefore $d : D :: f - d : f$,
or $f - d : f :: d : D$.

Hence, if D the diameter of the concavity be double to *d* that of the convexity, *f* will be equal to D, or the focal diftance equal to the diameter; and therefore the Menifcus will be equivalent to a planoconvex lens.

Again, if $D = 3d$, or the diameter of the concavity triple to that of the convexity, then will $f = \frac{1}{2}D$, or the focal diftance equal to the radius of concavity; and therefore the Menifcus will be equivalent to a lens equally convex on either fide.

But if $D = 5d$, then will $f = \frac{1}{4}D$; and therefore the Menifcus will be equivalent to a fphere.

Laftly, if $D = d$, then will *f* be infinite; and therefore a ray falling parallel to the axis, will ftill continue parallel to it after refraction.

MENSTRUUM, SOLVENT, or DISSOLVENT, any fluid that will diffolve hard bodies, or feparate their parts. Sir Ifaac Newton accounts for the action of Menftruums from the acids with which they are impregnated; the particles of acids being endued with a ftrong attractive force, in which their activity confifts, and by virtue of which they diffolve bodies. By this attraction they gather together about the particles of bodies, whether metallic, ftony, or the like, and adhere very clofely to them, fo as fcarce to be feparated from them by diftillation, or fublimation. Thus ftrongly attracting, and gathering together on all fides, they raife, disjoin, and fhake afunder the particles of bodies, i. e. they diffolve them; and by the attractive power with which they rufh againft the particles of the bodies, they move the fluid, and fo excite heat, fhaking fome of the particles to that degree, as to convert them into air, and fo generating bubbles.

Dr. Keill has given the theory or foundation of the action of Menftruums. in feveral propofitions. See ATTRACTION. From thofe propofitions are perceived the reafons of the different effects of different Menftruums; why fome bodies, as metals, diffolve in a faline Menftruum; others again, as refins, in a fulphureous one; &c: particularly why filver diffolves in aqua fortis, and gold only in aqua regis; all the varieties of which are accountable for, from the different degrees of cohefion, or attraction in the parts of the body to be diffolved, the different diameters and figures of its pores, the different degrees of attraction in the Menftruum, and the different diameters and figures of its parts.

MENSURABILITY, the fitnefs of a body for being applied, or conformable to a certain meafure.

MENSURATION, the act, or art, of meafuring figured extenfion and bodies; or of finding the dimenfions, and contents of bodies; both fuperficial and folid.

Every different fpecies of Menfuration is eftimated and meafured by others of the fame kind; fo, the folid contents of bodies are meafured by cubes, as cubic inches, or cubic feet, &c; furfaces by fquares, as fquare inches, feet, &c; and lengths or diftances by other lines, as inches, feet, &c.

The contents of rectilinear figures, whether plane or folid, can be accurately determined, or expreffed; but of many curved ones, not. So the quadrature of the circle, and cubature of the fphere, are problems that have never yet been accurately folved. See the various kinds of Menfuration, as well as that of the different figures, under their refpective terms.

The firft writers on Geometry were chiefly writers on Menfuration; as Euclid, Archimedes, &c. See QUADRATURE; alfo the Preface to my Menfuration, for the moft ample information.

MERCATOR (GERARD), an eminent geographer and mathematician, was born in 1512, at Ruremonde in the Low Countries. He applied himfelf with fuch induftry to the fciences of geography and mathematics, that it has been faid he often forgot to eat and fleep. The emperor Charles the 5th encouraged him much in his labours, and the duke of Juliers made him his cofmographer. He compofed and publifhed a Chronology; a larger and fmaller Atlas; and fome Geographical Tables; befide other books in Philofophy and Divinity. He was alfo fo curious, as well as ingenious, that he engraved and coloured his maps himfelf. He made various maps, globes, and other mathematical inftruments for the ufe of the emperor; and gave the moft ample proofs of his uncommon fkill in what he profeffed. His method of laying down charts is ftill ufed, which bear the name of *Mercator's Charts*; alfo a part of navigation is from him called *Mercator's Sailing*.—He died at Duifbourg in 1594, at 82 years of age.—See MERCATOR'S *Chart*, below.

MERCATOR (*Nicholas*), an eminent mathematician and aftronomer, whofe name in High-Dutch was *Hauffman*, was born, about the year 1640, at Holftein in Denmark. From his works we learn, that he had an early and liberal education, fuitable to his diftinguifhed genius, by which he was enabled to extend his researches

researches into the mathematical sciences, and to make very considerable improvements : for it appears from his writings, as well as from the character given of him by other mathematicians, that his talent rather lay in improving, and adapting any discoveries and improvements to use, than invention. However, his genius for the mathematical sciences was very conspicuous, and introduced him to public regard and esteem in his own country, and facilitated a correspondence with such as were eminent in those sciences, in Denmark, Italy, and England. In consequence, some of his correspondents gave him an invitation to this country, which he some time after accepted, and he afterwards continued in England till his death. He had not been long here before he was admitted F. R. S. and gave frequent proofs of his close application to study, as well as of his eminent abilities in improving some branch or other of the sciences. But he is charged sometimes with borrowing the inventions of others, and adopting them as his own. And it appeared upon some occasions that he was not of an over liberal mind in scientific communications. Thus, it had some time before him been observed, that there was an analogy between a scale of logarithmic tangents and Wright's protraction of the nautical meridian line, which consisted of the sums of the secants ; though it does not appear by whom this analogy was first discovered. It appears however to have been first published, and introduced into the practice of navigation, by Henry Bond, who mentions this property in an edition of Norwood's Epitome of Navigation, printed about 1645 ; and he again treats of it more fully in an edition of Gunter's Works, printed in 1653, where he teaches, from this property, to resolve all the cases of Mercator's Sailing by the logarithmic tangents, independent of the table of meridional parts. This analogy had only been found to be nearly true by trials, but not demonstrated to be a mathematical property. Such demonstration seems to have been first discovered by Mercator, who, desirous of making the most advantage of this and another concealed invention of his in navigation, by a paper in the Philosophical Transactions for June 4, 1666, invites the public to enter into a wager with him on his ability to prove the truth or falsehood of the supposed analogy. This mercenary proposal it seems was not taken up by any one, and Mercator reserved his demonstration. Our author however distinguished himself by many valuable pieces on philosophical and mathematical subjects. His first attempt was, to reduce Astrology to rational principles, which proved a vain attempt. But his writings of more particular note, are as follow :

1. *Cosmographia, sive Descriptio Cæli & Terræ in Circulos, qua fundamentum sterniter sequentibus ordine Trigonometriæ Sphericorum Logarithmicæ, &c, a Nicolao Hauffman Holsato* ; printed at Dantzick, 1651, 12mo.

2. *Rationes Mathematicæ subductæ anno* 1653 ; Copenhagen, in 4to.

3. *De Emendatione annua Diatribæ duæ, quibus exponuntur & demonstrantur Cycli Solis & Lunæ, &c* ; in 4to.

4. *Hypothesis Astronomica nova, et Consensus ejus cum Observationibus* ; Lond. 1664, in folio.

5. *Logarithmotechnia, sive Methodus Construendi Logarithmos nova, accurata, et facilis ; scripto antehac communicata anno sc. 1667 nonis Augusti ; cui nunc accedit, Vera Quadratura Hyperbolæ, & Inventio summæ Logarithmorum. Auctore Nicolao Mercatore Holsato è Societate Regia. Huic etiam jungitur Michaelis Angeli Riccii Exercitatio Geometrica de Maximis et Minimis, hic ob argumenti præstantiam & exemplarium raritatem recusa :* Lond. 1668, in 4to.

6. *Institutionum Astronomicarum libri duo, de Motu Astrorum communi & proprio, secundum hypotheses veterum & recentiorum præcipuas ; deque Hypotheseon ex observatis constructione, cum tabulis Tychonianis, Solaribus, Lunaribus, Lunæ-solaribus, & Rudolphinis Solis, Fixarum & quinque Errantium, earumque usu præceptis et exemplis commonstrato. Quibus accedit Appendix de iis, quæ novissimis temporibus cælitus innotuerunt :* Lond. 1676, 8vo.

7. *Euclidis Elementa Geometrica, novo ordine ac methodo fere, demonstrata. Una cum Nic. Mercatoris in Geometriam Introductione brevi, qua Magnitudinum Ortus ex genuinis Principiis, & Ortarum Affectiones ex ipsa Genesi derivantur.* Lond. 1678, 12mo.

His papers in the Philosophical Transactions, are,

1. A Problem on some Points in Navigation : vol. 1, pa. 215.

2. Illustrations of the Logarithmo-technia : vol. 3, pa. 759.

3. Considerations concerning his Geometrical and Direct Method for finding the Apogees, Excentricities, and Anomalies of the Planets : vol. 5, pa. 1168.

Mercator died in 1594, about 54 years of age.

MERCATOR's *Chart*, or *Projection*, is a projection of the surface of the earth in plano, so called from Gerrard 'Mercator, a Flemish Geographer, who first published maps of this sort in the year 1556 ; though it was Edward Wright who first gave the true principles of such charts, with their application to Navigation, in 1599.

In this chart or projection, the meridians, parallels, and rhumbs, are all straight lines, the degrees of longitude being every where increased so as to be equal to one another, and having the degrees of latitude also increased in the same proportion ; namely, at every latitude or point on the globe, the degrees of latitude, and of longitude, or the parallels, are increased in the proportion of radius to the sine of the polar distance, or cosine of the latitude ; or, which is the same thing, in the proportion of the secant of the latitude to radius ; a proportion which has the effect of making all the parallel circles be represented by parallel and equal right lines, and all the meridians by parallel lines also, but increasing infinitely towards the poles.

From this proportion of the increase of the degrees of the meridian, viz, that they increase as the secant of the latitude, it is very evident that the length of an arch of the meridian, beginning at the equator, is proportional to the sum of all the secants of the latitude, i. e. that the increased meridian, is to the true arch of it, as the sum of all those secants, to as many times the radius. But it is not so evident that the same increased meridian is also analogous to a scale of the logarithmic tangents, which however it is. " It does not appear by whom, nor by what accident, was discovered the

O

analogy

analogy between a scale of logarithmic tangents and Wright's protraction of the nautical meridian line, which consisted of the sums of the secants. It appears however to have been first published, and introduced into the practice of navigation, by Mr. Henry Bond, who mentions this property in an edition of Norwood's Epitome of Navigation, printed about 1645; and he again treats of it more fully in an edition of Gunter's Works, printed in 1653, where he teaches, from this property, to resolve all the cases of Mercator's Sailing by the logarithmic tangents, independent of the table of meridional parts. This analogy had only been found however to be nearly true by trials, but not demonstrated to be a mathematical property. Such demonstration, it seems, was first discovered by Mr. Nicholas Mercator, which he offered a wager to disclose, but this not being accepted; Mercator reserved his demonstration; as mentioned in the account of his life in the foregoing page. The proposal however excited the attention of mathematicians to the subject, and demonstrations were not long wanting. The first was published about two years after, by James Gregory, in his Exercitationes Geometricæ; from hence, and other similar properties there demonstrated, he shews how the tables of logarithmic tangents and secants may easily be computed from the natural tangents and secants.

" The same analogy between the logarithmic tangents and the meridian line, as also other similar properties, were afterwards more elegantly demonstrated by Dr. Halley, in the Philos. Transf. for Feb. 1696, and various methods given for computing the same, by examining the nature of the spirals into which the rhumbs are transformed in the stereographic projection of the sphere on the plane of the equator: the doctrine of which was rendered still more easy and elegant by the ingenious Mr. Cotes, in his Logometria, first printed in the Philos. Transf. for 1714, and afterwards in the collection of his works published 1732, by his cousin Dr. Robert Smith, who succeeded him as Plumian professor of philosophy in the University of Cambridge."

The learned Dr. Isaac Barrow also, in his Lectiones Geometricæ, Lect. xi, Append. first published in 1672, delivers a similar property, namely, " that the sum of all the secants of any arc, is analogous to the logarithm of the ratio of $r + s$ to $r - s$, viz, radius plus sine to radius minus sine; or, which is the same thing, that the meridional parts answering to any degree of latitude, are as the logarithms of the ratios of the versed sines of the distances from the two poles." Preface to my Logarithms, pa. 100.

The meridian line in Mercator's Chart, is a scale of logarithmic tangents of the half colatitudes. The differences of longitude on any rhumb, are the logarithms of the same tangents, but of a different species; those species being to each other, as the tangents of the angles made with the meridian. Hence any scale of logarithmic tangents is a table of the differences of longitude, to several latitudes, upon some one determinate rhumb; and therefore, as the tangent of the angle of such a rhumb, is to the tangent of any other rhumb, so is the difference of the logarithms of any two tangents, to the difference of longitude

on the proposed rhumb, intercepted between the two latitudes, of whose half complements the logarithmic tangents were taken.

It was the great study of our predecessors to contrive such a chart in plano, with straight lines, on which all, or any parts of the world, might be truly set down, according to their longitudes and latitudes, bearings and distances. A method for this purpose was hinted by Ptolomy, near 2000 years since; and a general map, on such an idea, was made by Mercator; but the principles were not demonstrated, and a ready way shewn of describing the chart, till Wright explained how to enlarge the meridian line by the continual addition of secants; so that all degrees of longitude might be proportional to those of latitude, as on the globe: which renders this chart, in several respects, far more convenient for the navigator's use, than the globe itself; and which will truly shew the course and distance from place to place, in all cases of sailing.

MERCATOR'S *Sailing*, or more properly *Wright's* Sailing, is the method of computing the cases of sailing on the principles of Mercator's chart, which principles were laid down by Edward Wright in the beginning of the last century; or the art of finding on a plane the motion of a ship upon any assigned course, that shall be true as well in longitude and latitude, as distance; the meridians being all parallel, and the parallels of latitude straight lines.

In the right-angled triangle Abc, let Ab be the true difference of latitude between two places, the angle bAc the angle of the course sailed, and Ac the true distance sailed; then will bc be what is called the departure, as in plane sailing: produce Ab till AB be equal to the meridional difference of latitude, and draw BC parallel to bc; so shall BC be the difference of longitude.

Now from the similarity of the two triangles Abc, ABC, when three of the parts are given, the rest may be found; as in the following analogies: As

Radius : sin. course :: distance : departure;
Radius : cos. course :: distance : dif. lat.;
Radius : tan. course :: merid. dif. lat : dif. longitude.

And by means of these analogies may all the cases of Mercator's Sailing be resolved.

MERCURY, the smallest of the inferior planets, and the nearest to the sun, about which it is carried with a very rapid motion. Hence it was, that the Greeks called this planet after the name of the nimble messenger of the Gods, and represented it by the figure of a youth with wings at his head and feet; from whence is derived ☿, the character in present use for this planet.

The mean distance of Mercury from the sun, is to that of the earth from the sun, as 387 to 1000, and therefore his distance is about 36 millions of miles, or little more than one-third of the earth's distance from
the

the fun. Hence the fun's diameter will appear at Mercury, near 3 times as large as at the earth ; and hence alfo the fun's light and heat received there is about 7 times thofe at the earth ; a degree of heat fufficient to make water boil. Such a degree of heat therefore muft render Mercury not habitable to creatures of our conftitution : and if bodies on its furface be not inflamed, and fet on fire, it muft be becaufe their degree of denfity is proportionably greater than that of fuch bodies is with us.

The diameter of Mercury is alfo nearly one-third of the diameter of the earth, or about 2600 miles. Hence the furface of Mercury is nearly 1-9th, and his magnitude or bulk 1-27th of that of the earth.

The inclination of his orbit to the plane of the ecliptic, is 6° 54 ; his period of revolution round the fun, 87days 23hours ; his greateft elongation from the fun 28° ; the excentricity of his orbit ⅕ of his mean diftance, which is far greater than that of any of the other planets ; and he moves in his orbit about the fun at the amazing rate of 95000 miles an hour.

The place of his aphelion is ♐ 23° 8′ ; place of afcending node ♉ 14° 43′, and confequently that of the defcending node ♏ 14° 43′.

His Length of day, or rotation on his axis, Inclination of axis to his orbit, Gravity on his furface, Denfity, and Quantity of matter, are all unknown.

Mercury changes his phafes, like the moon, according to his various pofitions with regard to the earth and fun ; except only, that he never appears quite full, becaufe his enlightened fide is never turned directly towards us, unlefs when he is fo near the fun as to be loft to our fight in his beams. And as his enlightened fide is always towards the fun, it is plain that he fhines not by any light of his own ; for if he did, he would conftantly appear round.

The beft obfervations of this planet are thofe made when it is feen on the fun's difc, called its tranfit ; for in its lower conjunction, it fometimes paffes before the fun like a little fpot, eclipfing a fmall part of the fun's body, only obfervable with a telefcope. That node from which Mercury afcends northward above the ecliptic, is in the 15th degree of Taurus, and the oppofite in the 15th degree of Scorpio. The earth is in thofe parts on the 6th of November, and 4th of May, new ftyle ; and when Mercury comes to either of his nodes at his inferior conjunction about thefe times, he will appear in this manner to pafs over the difc of the fun. But in all other parts of his orbit, his conjunctions are invifible, becaufe he goes either above or below the fun. The firft obfervation of this kind was made by Gaffendi, in November 1631. Several following obfervations of the like tranfits are collected in Du Hamel's Hift. of the Royal Acad. of Sciences, pa. 470, ed. 2. And Mr. Whifton has given a lift of feveral periods at which Mercury may be feen on the fun's difc, viz, in 1782, Nov. 12, at 3h 44m afternoon ; in 1786, May 4th, at 6h 57m in the forenoon ; in 1789, Dec. 6th, at 3h 55m afternoon ; and in 1799, May 7th, at 2h 34m afternoon. There are alfo feveral intermediate tranfits, but none of them vifible at London See Dr. Halley's account of the Tranfits of Mercury and Venus, in the Philof. Tranf n 193.

MERIDIAN, in' Aftronomy, is a great circle of the celeftial fphere, paffing through the poles of the world, and both the zenith and nadir, croffing the equinoctial at right angles, and dividing the fphere into two equal parts, or hemifpheres, the one eaftern, and the other weftern. Or, the Meridian is a vertical circle paffing through the poles of the world.

It is called Meridian, from the Latin *meridies*, midday or noon, becaufe when the fun comes to the fouth part of this circle, it is noon to all thofe places fituated under it.

MERIDIAN, in Geography, is a great circle paffing through the poles of the earth, and any given place whofe Meridian it is ; and it lies exactly under, or in the plane of, the celeftial Meridian.

Thefe Meridians are various, and change according to the longitude of places ; fo that their number may be faid to be infinite, for that all places from eaft to weft have their feveral Meridians. Farther, as the Meridian invefts the whole earth, there are many places fituated under the fame Meridian. Alfo, as it is noon whenever the centre of the fun is in the celeftial Meridian ; and as the Meridian of the earth is in the plane of the former ; it follows, that it is noon at the fame time, in all places fituated under the fame Meridian.

Firft MERIDIAN, is that from which the reft are counted, reckoning both eaft and weft ; and is the beginning of longitude.

The fixing of the Firft Meridian is a matter merely arbitrary ; and hence different perfons, nations, and ages, have fixed it differently : from which circumftance fome confufion has arifen in geography. The rule among the Ancients was, to make it pafs through the place fartheft to the weft that was known. But the Moderns knowing that there is no fuch place on the earth as can be efteemed the moft wefterly, the way of computing the longitudes of places from one fixed point is much laid afide.

Ptolomy affumed the Meridian that paffes through the fartheft of the Canary Iflands, as his firft Meridian ; that being the moft weftern place of the world then known. After him, as more countries were difcovered in that quarter, the Firft Meridian was remov farther off. The Arabian geographers chofe to the Firft Meridian upon the utmoft fhore of the weftern ocean. Some fixed it to the ifland of St. Nicholas near the Cape Verd ; Hondius to the ifle of St. James ; others to the ifland of Del Corvo, one of the Azores ; becaufe on that ifland the magnetic needle at that time pointed directly north, without any variation : and it was not then known that the variation of the needle is itfelf fubject to variation. The lateft geographers, particularly the Dutch, have pitched on the Pike of Teneriffe ; others on the Ifle of Palm, another of the Canaries ; and laftly, the French, by order of the king, on the ifland of Fero, another of the Canaries.

But, without much regard to any of thefe rules, geographers and map-makers often affume the Meridian of the place where they live, or the capital of their country, or its chief obfervatory, for a Firft Meridian : and from thence reckon the longitudes of places, eaft and weft.

Aftronomers, in their calculations, ufually choofe the

the Meridian of the place where their observations are made, for their First Meridian; as Ptolomy at Alexandria; Tycho Brahe at Uranibourg; Riccioli at Bologna; Flamsteed at the Royal Observatory at Greenwich; and the French at the Observatory at Paris.

There is a suggestion in the Philos. Transf. that the Meridians vary in time. And it has been said that this is rendered probable, from the old Meridian line in the church of St. Petronio at Bologna, which is said to vary no less than 8 degrees from the true Meridian of the place at this time; and from the Meridian of Tycho at Uranibourg, which M. Picart observes, varies 18 minutes from the modern Meridian. If there be any thing of truth in this hint, Dr. Wallis says, the alteration must arise from a change of the terrestrial poles (here on earth, of the earth's diurnal motion), not of their pointing to this or that of the fixed stars: for if the poles of the diurnal motion remain fixed to the same place on the earth, the Meridians, which pass through these poles, must remain the same.

But the notion of the changes of the Meridian seems overthrown by an observation of M. Chazelles, of the French Academy of Sciences, who, when in Egypt, found that the four sides of a pyramid, built 3000 years ago, still looked very exactly to the four cardinal points. A position which cannot be considered as merely fortuitous.

MERIDIAN *of a Globe*, or *Sphere*, is the brazen circle, in which the globe hangs and turns.

It is divided into four 90's, or 360 degrees, beginning at the equinoctial: on it, each way, from the equinoctial, on the celestial globes, is counted the north and south declination of the sun, moon, or stars; and on the terrestrial globe, the latitude of places, north and south. There are two points on this circle called the poles; and a diameter, continued from thence through the centre of either globe, is called the axis of the earth, or heavens, on which it is supposed they turn round.

On the terrestrial globes there are usually drawn 36 Meridians, one through every 10th degree of the equator, or through every 10th degree of longitude.

The uses of this circle are, to set the globes in any particular latitude, to shew the sun's or a star's declination, right ascension, greatest altitude, &c.

MERIDIAN *Line*, an arch, or part, of the Meridian of the place, terminated each way by the horizon. Or, a Meridian line is the intersection of the plane of the Meridian of the place with the plane of the horizon, often called a north-and south line, because its direction is from north to south.

The Meridian line is of most essential use in astronomy, geography, dialling, &c; and the greatest pains are taken by astronomers to fix it at their observatories to the utmost precision. M. Cassini has distinguished himself by a Meridian line drawn on the pavement of the church of St. Petronio, at Bologna; being extended to 120 feet in length. In the roof of this church, 1000 inches above the pavement, is a small hole, through which the sun's image, when in the meridian, falling upon the line, marks his progress all the year. When finished, M. Cassini, by a public writing, quaintly informed the mathematicians of Europe, of a new oracle of Apollo, or the sun, established in a temple, which might be consulted, with entire confidence, as to all difficulties in astronomy. See GNOMON.

To draw a Meridian Line.—There are many ways of doing this; but some of the easiest and simplest are as follow:

1. On an horizontal plane describe several concentric circles AB, *ab*, &c, and on the common centre C erect a stile, or gnomon, perpendicular to the horizontal plane, of about a foot in length. About the 21st of June, between the hours of 9 and 11 in the morning, and between 1 and 3 in the afternoon, observe the points A, *a*, B, *b*, &c, in the circles, where the shadow of the stile terminates. Bisect the arches AB, *ab*, &c, in D, *d*, &c. If then the same right line DE bisect all these arches, it will be the Meridian line sought.

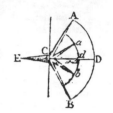

As it is not easy to determine precisely the extremity of the shadow, it will be best to make the stile flat at top, and to drill a small hole through it, noting the lucid point projected by it on the arches AB and *ab*, instead of marking the extremity of the shadow itself.

2. Another method is thus: Knowing the south quarter pretty nearly, observe the altitude FE of some star on the east side of it, and not far from the Meridian HZRN: then, keeping the quadrant firm on its axis, so as the plummet may still cut the same degree, direct it to the western side of the Meridian, and wait till you find the star has the same altitude as before, as *fe*. Lastly, bisect the angle EC*e*, formed by the intersection of the two planes in which the quadrant has been placed at the time of the two observations, by the right line HR, which will be the Meridian sought.

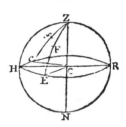

Many other methods are given by authors, of describing a Meridian line; as by the pole star, or by equal altitudes of the sun, &c; by Schooten in his Exercitationes Geometriæ; Grey, Derham, &c, in the Philos. Transf. and by Ferguson in his Lectures on Select Subjects.

From what has been said it is evident that whenever the shadow of the stile covers the Meridian line, the centre of the sun is in the Meridian, and therefore it is then noon. And hence the use of a Meridian line in adjusting the motion of clocks to the sun.

If another stile be erected perpendicularly on any other horizontal plane, and a signal be given when the shadow of the former stile covers the Meridian line drawn on another plane, noting the apex or extremity of the shadow projected by the second stile, a line drawn through that point and the foot of the stile will be a Meridian line at the 2d place.

Or, instead of the 2d stile, a plumb line may be hung up, and its shadow noted on a plane, upon a signal given that the shadow of another plummet, or

of

of a ftile, falls exactly in another Meridian line, at a little diftance; which fhadow will give the other Meridian line parallel to the former.

MERIDIAN *Line*, on a Dial, is a right line arifing from the interfection of the Meridian of the place with the plane of the dial. This is the line of noon, or 12 o'clock, and from hence the divifion of the hour-line begins.

MERIDIAN *Line*, on Gunter's fcale, is divided unequally towards 87 degrees, in fuch manner as the Meridian in Mercator's chart is divided and numbered.

This line is very ufeful in navigation. For, 1ft, It ferves to graduate a fea-chart according to the true projection. 2d, Being joined with a line of chords, it ferves for the protraction and refolution of fuch rectilineal triangles as are concerned in latitude, longitude, courfe, and diftance, in the practice of failing; as alfo in pricking the chart truly at fea.

Magnetical MERIDIAN, is a great circle paffing through or by the magnetical poles; to which Meridians the magnetical needle conforms itfelf.

Meridian Altitude, of the fun or ftars, is their altitude when in the meridian of the place where they are obferved.

MERIDIONAL *Diftance*, in Navigation, is the fame with the Departure, or eafting and wefting, or diftance between two meridians.

MERIDIONAL *Parts, Miles*, or *Minutes*, in Navigation, are the parts of the increafed or enlarged meridian, in the Mercator's chart. Tables of thefe parts are in moft books of navigation; and they ferve both for conftructing that fort of charts, and for working that kind of navigation.

Under the article MERCATOR's *Chart*, it is fhewn that the parts of the enlarged Meridian increafe in proportion as the cofine of the latitude to radius, or, which is the fame thing, as radius to the fecant of the latitude; and therefore it follows, that the whole length of the enlarged nautical Meridian, from the equator to any point, or latitude, will be proportional to the fum of all the fecants of the feveral latitudes up to that point of the Meridian. And on this principle was the firft Table of Meridional Parts conftructed, by the inventor of it, Mr. Edward Wright, and publifhed in 1599; viz, he took the Meridional parts

of 1′ = the fec. of 1′;
of 2′ = fec. of 1′ + fec. of 2′;
of 3′ = fecants of 1, 2, and 3 min.
of 4′ = fecants of 1, 2, 3, and 4 min.

and fo on by a conftant addition of the fecants.

The Tables of Meridional Parts, fo conftructed, are perhaps exact enough for ordinary practice in navigation; but they would be more accurate if the Meridian were divided into more or fmaller parts than fingle minutes; and the fmaller the parts, fo much the greater the accuracy. But, as a continual fubdivifion would greatly augment the labour of calculation, other ways of computing fuch a table have been devifed, and treated of, by Bond, Gregory, Oughtred, Sir Jonas Moor, Dr. Wallis, Dr. Halley, and others. See MERCATOR's *Chart*, and Robertfon's Navigation, vol. 2, book 8. The beft of thefe methods was derived from this property, viz, that the Meridian line, in a Mercator's chart, is analogous to a fcale of logarithmic tangents of half the complements of the latitudes; from which property alfo a method of computing the cafes of Mercator's Sailing has been deduced, by Dr. Halley. Vide ut fupra, alfo the Philof. Tranf. vol. 46, pa. 559.

To find the MERIDIONAL PARTS *to any Spheroid, with the fame exactnefs as in a Sphere.*

Let the femidiameter of the equator be to the diftance of the centre from the focus of the generating ellipfe, as *m* to 1. Let A reprefent the latitude for which the meridional parts are required, *s* the fine of the latitude, to the radius 1: Find the arc B, whofe fine is $\frac{s}{m}$; take the logarithmic tangent of half the complement of B, from the common tables; fubtract the log. tangent from 10·0000000, or the log. tangent of 45°; multiply the remainder by the number 7915·7044679, and divide the product by *m*; then the quotient fubtracted from the Meridional parts in the fphere, computed in the ufual manner for the latitude A, will give the Meridional parts, expreffed in minutes, for the fame latitude in the fpheroid, when it is the oblate one.

Example. If *mm* : 1 :: 1000 : 22, then the greateft difference of the Meridional parts in the fphere and fpheroid is 76·0929 minutes. In other cafes it is found by multiplying the remainder above mentioned by the number 1174·078.

When the fpheroid is oblong, the difference in the Meridional parts between the fphere and fpheroid, for the fame latitude, is then determined by a circular arc. See Philof. Tranf. no. 461, fect. 14. Alfo Maclaurin's Fluxions, art. 895, 899. And Murdoch's Mercator's Sailing &c.

MERLON, in Fortification, that part of the Parapet, which lies between two embrafures.

MERSENNE (MARTIN), a learned French author, was born at Bourg of Oyfe, in the province of Maine, 1588. He ftudied at La Fleche at the fame time with Des Cartes; with whom he contracted a ftrict friendfhip, which continued till death. He afterwards went to Paris, and ftudied at the Sorbonne; and in 1611 entered himfelf among the Minims. He became well fkilled in Hebrew, philofophy, and mathematics. From 1615 to 1619, he taught philofophy and theology in the convent of Nevers; and became the Superior of that convent. But being defirous of applying himfelf more freely and clofely to ftudy, he refigned all the pofts he enjoyed in his order, and retired to Paris, where he fpent the remainder of his life; excepting fome fhort excurfions which he occafionally made into Italy, Germany, and the Netherlands.

Study and literary converfation were afterwards his whole employment. He held a correfpondence with moft of the learned men of his time; being as it were the very centre of communication between literary men of all countries, by the mutual correfpondence which he managed between them; being in France what Mr. Collins was in England. He omitted no opportunity to engage them to publifh their works; and the world is obliged to him for feveral excellent difcoveries, which would probably have been loft, but for his encouragement; and on all accounts he had the reputation of being one of the beft men, as well as philofophers,

of

of his time. No perfon was more curious in penetrating into the fecrets of nature, and carrying all the arts and fciences to perfection. He was the chief friend and literary agent of Des Cartes at Paris; giving him advice and affiftance upon all occafions, and informing him of all that paffed at Paris and elfewhere. For, being a perfon of univerfal learning, but particularly excelling in phyfical and mathematical knowledge, Des Cartes fcarcely ever did any thing, or at leaft was not perfectly fatisfied with any thing he had done, without firft knowing what Merfenne thought of it. It is even faid, that when Merfenne gave out in Paris, that Des Cartes was erecting a new fyftem of phyfics upon the foundation of a vacuum, and found the public very indifferent to it on that very account, he immediately fent notice to Des Cartes, that a vacuum was not then the fafhion at Paris; upon which, that philofopher changed his fyftem, and adopted the old doctrine of a plenum.

Merfenne was a man of good invention alfo himfelf; and he had a peculiar talent in forming curious queftions, though he did not always fucceed in refolving them; however, he at leaft gave occafion to others to do it. It is faid he invented the Cycloid, otherwife called the Roulette. Prefently the chief geometricians of the age engaged in the contemplation of this new curve, among whom Merfenne himfelf held a diftinguifhed rank. After a very ftudious and ufeful life, he died at Paris in 1648, at 60 years of age.

Merfenne was author of many ufeful works, particularly the following:

1. *Queftiones celeberrimæ in Genefim.*
2. *Harmonicorum Libri.*
3. *De Sonorum Natura, Caufis, et Effectibus.*
4. *Cogitata Phyfico-Mathematica;* 2 vols. 4to.
5. *La Verité des Sciences.*
6. *Les Queftions inouies.*

Befides many letters in the works of Des Cartes, and other authors.

MESOLABE, or MESOLABIUM, a mathematical inftrument invented by the Ancients, for finding two mean proportionals mechanically, which they could not perform geometrically. It confifts of three parallelograms, moving in a groove to certain interfections. Its figure is defcribed by Eutocius, in his Commentary on Archimedes. See alfo Pappus, lib. 3.

MESO-LOGARITHM, a term ufed by Kepler to fignify the logarithms of the cofines and cotangents.

METO, or METON, the fon of Paufanias, a famous mathematician of Athens, who flourifhed 432 years before Chrift. In the firft year of the 87th Olympiad, he obferved the folftice at Athens: and publifhed his *Anneadecatoride,* that is, his *Cycle of* 19 *Years;* by which he endeavoured to adjuft the courfe of the fun to that of the moon, and to make the folar and lunar years begin at the fame point of time. See CYCLE.

METONIC CYCLE, called alfo the *Golden Number,* and *Lunar Cycle,* or *Cycle of the Moon,* that which was invented by Meton the Athenian; being a period of 19 years. See CYCLE.

METOPE, or METOPA, in Architecture, the fquare fpace between the triglyphs of the Doric Freeze; which among the Ancients ufed to be adorned with the heads of beafts, bafons, vafes, and other inftruments ufed in facrificing.

A *Demi-Metope* is a fpace fomewhat lefs than half a Metope, at the corner of the Doric Freeze.

MICHAELMAS, the feaft of St. Michael the archangel; held on the 29th of September.

MICROCOUSTICS, the fame with MICROPHONES.

MICROMETER, is an inftrument ufually fitted to a telefcope, in the focus of the object-glafs, for meafuring fmall angles or diftances; as the apparent diameters of the planets, &c.

There are feveral forts of thefe inftruments, upon different principles; the origin of which has been difputed. The general principle is, that the inftrument moves a fine wire parallel to itfelf, in the plane of the picture of an object, formed in the focus of a telefcope, and fo with great exactnefs to meafure its perpendicular diftance from a fixed wire in the fame plane: and thus are meafured fmall angles, fubtended by remote objects at the naked eye.

For example, Let a planet be viewed through the telefcope; and when the parallel wires are opened to fuch a diftance as to appear exactly to touch two oppofite points in the circumference of the planet, it is evident that the perpendicular diftance between the wires is then equal to the diameter of the picture of the planet, formed in the focus of the object-glafs. Let this diftance, whofe meafure is given by the mechanifm of the micrometer, be reprefented by the line

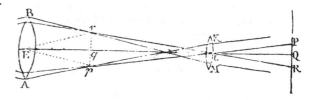

pq; then, fince the meafure of the focal diftance *q*L may be alfo known, the ratio of *q*L to *qp,* that is, of radius to the tangent of the angle *q*L*p,* will give the angle itfelf, by a table of fines and tangents; and this angle is equal to the oppofite angle PLQ, which the real diameter of the planet fubtends at L, or at the naked eye.

With refpect to the invention of the Micrometer; Meff. Azout and Picard have the credit of it in common fame, as being the firft who publifhed it, in the year 1666; but Mr. Townley, in the Philof. Tranf. reclaims it for one of our own countrymen, Mr. Gafcoigne. He relates that, from fome fcattered papers and letters of this gentleman, he had learnt that before our civil wars he had invented a Micrometer, of as much effect as that fince made by M. Azout, and had made ufe of it for fome years, not only in taking the diameters of the planets, and diftances upon land, but in determining other matters of nice importance in the heavens; as the moon's diftance, &c. Mr. Gafcoigne's inftrument alfo fell into the hands of Mr. Townley, who fays farther, that by the help of it he could make above 40,000 divifions in a foot. This inftrument being fhewn to Dr. Hook, he gave a drawing and defcription of it, and propofed feveral improvements in it; which may be feen in the Philof. Tranf. vol. 1, pa. 63, and Abr. vol. 1, pa. 217. Mr. Gafcoigne divided the image of an object, in the focus of the object-glafs, by the approach of

two

two pieces of metal, ground to a very fine edge; instead of which, Dr. Hook would substitute two fine hairs, stretched parallel to each other: and two other methods of Dr. Hook, different from this, are described in his posthumous works, pa. 497 &c. An account of several curious observations which Mr. Gascoigne made by the help of his Micrometer, particularly in measuring the diameter of the moon and other planets, may be seen in the Philos. Transf. vol. 48, pa. 190; where Dr. Bevis refers to an original letter of Mr. Gascoigne, to Mr. Oughtred, written in 1641, for an account given by the author of his own invention, &c.

Monsf. De la Hire, in a discourse on the æra of the inventions of the Micrometer, pendulum clock, and telescope, read before the Royal Academy of Sciences in 1717, makes M. Huygens the inventor of the Micrometer. That author, he observes, in his Observations on Saturn's Ring, &c, published in 1659, gives a method of finding the diameters of the planets by means of a telescope, viz, by putting an object, which he calls a virgula, of a size proper to take in the distance to be measured, in the focus of the convex object glass: in this case, says he, the smallest object will be seen very distinctly in that place of the glass. By such means, he adds, he measured the diameter of the planets, as he there delivers them. See Huygens's System of Saturn.

This Micrometer, M. De la Hire observes, is so very little different from that published by the marquis De Malvasia, in his Ephemerides, three years after, that they ought to be esteemed the same: and the Micrometer of the marquis differed yet less from that published four years after his, by Azout and Picard. Hence, De la Hire concludes, that it is to Huygens the world is indebted for the invention of the Micrometer; without taking any notice of the claim of our countryman Gascoigne, which however is many years prior to any of them.

De la Hire says, that there is no method more simple or commodious for observing the digits of an eclipse, than a net in the focus of the telescope. These, he says, were usually made of silken threads; and for this particular purpose six concentric circles had also been used, drawn upon oiled paper; but he advises to draw the circles on very thin pieces of glass, with the point of a diamond. He also gives some particular directions to assist persons in using them. In another memoir, he shews a method of making use of the same net for all eclipses, by using a telescope with two object-glasses, and placing them at different distances from each other. Mem. 1701 and 1717.

M. Cassini invented a very ingenious method of ascertaining the right ascensions and declinations of stars, by fixing four cross hairs in the focus of the telescope, and turning it about its axis, so as to make them move in a line parallel to one of them. But the later improved Micrometers will answer this purpose with greater exactness. Dr. Maskelyne has published directions for the use of it, extracted from Dr. Bradley's papers, in the Philos. Transf. vol. 62. See also Smith's Optics, vol. 2, pa. 343.

Wolfius describes a Micrometer of a very easy and simple structure, first contrived by Kirchius.

Dr. Derham tells us, that his Micrometer is not put into a tube, as is usual, but is contrived to measure the spectres of the sun on paper, of any radius, or to measure any part of them. By this means he can easily, and very exactly, with the help of a fine thread, take the declination of a solar spot at any time of the day; and, by his half seconds watch, measure the distance of the spot from either limb of the sun.

J. And. Segner proposed to enlarge the field of view in these Micrometers, by making them of a considerable extent, and having a moveable eye-glass, or several eye-glasses, placed opposite to different parts of it. He thought however, that two would be quite sufficient, and he gives particular directions how to make use of such Micrometers in astronomical observations. See Comm. Gotting. vol. 1, pa. 27.

A considerable improvement in the Micrometer was communicated to the Royal Society, in 1743, by Mr. S. Savary; an account of which, extracted from the minutes by Mr. Short, was published in the Philos. Transf. for 1753. The first hint of such a Micrometer was suggested by M. Roemer, in 1675: and M. Bouguer proposed a construction similar to that of M. Savary, in 1748; for which see HELIOMETER. The late Mr. Dollond made a farther improvement in this kind of Micrometer, an account of which was given to the Royal Society by Mr. Short, and published in the Philos. Transf. vol. 48. Instead of two object-glasses, he used only one, which he neatly cut into two semicircles, and fitted each semicircle in a metal frame, so that their diameters sliding in one another, by means of a screw, may have their centres so brought together as to appear like one glass, and so form one image; or by their centres receding, may form two images of the same object: it being a property of such glasses, for any segment to exhibit a perfect image of an object, although not so bright as the whole glass would give it. If proper scales are fitted to this instrument, shewing how far the centres recede, relative to the focal length of the glass, they will also shew how far the two parts of the same object are asunder, relative to its distance from the object glass; and consequently give the angle under which the distance of the parts of that object are seen. This divided object-glass Micrometer, which was applied by the late Mr. Dollond to the object end of a reflecting telescope, and has been with equal advantage adapted by his son to the end of an achromatic telescope, is of so easy use, and affords so large a scale, that it is generally looked upon by astronomers as the most convenient and exact instrument for measuring small distances in the heavens. However, the common Micrometer is peculiarly adapted for measuring differences of right ascension, and declination, of celestial objects, but less convenient and exact for measuring their absolute distances; whereas the object-glass Micrometer is peculiarly fitted for measuring distances, though generally supposed improper for the former purpose. But Dr. Maskelyne has found that this may be applied with very little trouble to that purpose also; and he has furnished the directions necessary to be followed when it is used in this manner. The addition requisite for this purpose, is a cell, containing two wires, intersecting each other at right angles, placed in the focus of the eye-glass of the telescope, and moveable round about, by the turning of a button. For the description of this apparatus, with the method of applying and using it, it, see Dr. Maskelyne's paper on the subject, in the Philos. Transf. vol. 61, pa. 536 &c.

After all, the use of the object-glass Micrometer is attended with difficulties, arising from the alterations in the focus of the eye, which are apt to cause it to give different measures of the same angle at different times. To obviate these difficulties, Dr. Maskelyne, in 1776, contrived a prismatic Micrometer, or a Micrometer consisting of two achromatic prisms, or wedges, applied between the object-glass and eye-glass of an achromatic telescope, by moving of which wedges nearer to or farther from the object-glass, the two images of an object produced by them appeared to approach to, or recede from, each other, so that the focal length of the object glass becomes a scale for measuring the angular distance of the two images. The rationale and use of this Micrometer are explained in the Philos. Transf. vol. 67, pa. 799, &c. And a similar invention by the abbé Rochon, and improved by the abbé Boscovich, was also communicated to the Royal Society, and published in the same volume of the Transactions, pa. 789 &c.

Mr. Ramsden has lately described two new Micrometers, which he has contrived for remedying the defects of the object-glass Micrometer. One of these is a catoptric Micrometer, which, besides the advantage it derives from the principle of reflection, of not being disturbed by the heterogeneity of light, avoids every defect of other Micrometers, and can have no aberration, nor any defect arising from the imperfection of materials, or of execution; as the great simplicity of its construction requires no additional mirrors or glasses, to those required for the telescope; and the separation of the image being effected by the inclination of the two specula, and not depending on the focus of lens or mirror, any alteration in the eye of an observer cannot affect the angle measured. It has peculiar to itself the advantages of an adjustment, to make the images coincide in a direction perpendicular to that of their motion; and also of measuring the diameter of a planet on both sides of the zero; which will appear no inconsiderable advantage to observers who know how much easier it is to ascertain the contact of the external edges of two images than their perfect coincidence.

The other Micrometer invented and described by Mr. Ramsden, is suited to the principle of refraction. This Micrometer is applied to the erect eye-tube of a refracting telescope, and is placed in the conjugate focus of the first eye-glass, as the image is considerably magnified before it comes to the Micrometer, any imperfection in its glass will be magnified only by the remaining eye-glasses, which in any telescope seldom exceeds 5 or 6 times; and besides, the size of the Micrometer glass will not be the 100th part of the area which would be required, if it were placed at the object-glass; and yet the same extent of scale is preserved, and the images are uniformly bright in every part of the field of the telescope. See the description and construction of these two Micrometers in the Philos. Transf. vol. 69, part 2, art. 27.

In vol. 72 of the Philos. Transf. for the year 1782, Dr. Herschel, after explaining the defects and imperfections of the parallel wire Micrometer, especially for measuring the apparent diameter of stars, and the distances between double and multiple stars, describes one,

8

for these purposes, which he calls a lamp Micrometer; one that is free from such defects, and has the advantage of a very enlarged scale. In speaking of the application of this instrument, he says, "It is well known to opticians and others, who have been in the habit of using optical instruments, that we can with one eye look into a microscope or telescope, and see an object much magnified, while the naked eye may see a scale upon which the magnified picture is thrown. In this manner I have generally determined the power of my telescopes; and any one who has acquired a facility of taking such observations, will very seldom mistake so much as one in 50 in determining the power of an instrument, and that degree of exactness is fully sufficient for the purpose.

" The Newtonian form is admirably adapted to the use of this Micrometer; for the observer stands always erect, and looks in a horizontal direction, notwithstanding the telescope should be elevated to the zenith. —The scale of the Micrometer at the convenient distance of 10 feet from the eye, with the power of 460, is above a quarter of an inch to a second; and by putting on my power of 932, I obtain a scale of more than half an inch to a second, without increasing the distance of the Micrometer; whereas the most perfect of my former Micrometers, with the same instrument, had a scale of less than the 2000th part of an inch to a second.

" The measures of this Micrometer are not confined to double stars only, but may be applied to any other objects that require the utmost accuracy, such as the diameters of the planets or their satellites, the mountains of the moon, the diameters of the fixed stars, &c."

The Micrometer has not only been applied to telescopes, and employed for astronomical purposes; but there have been various contrivances for adapting it to microscopical observations. Mr. Leeuwenhoek's method of estimating the size of small objects, was by comparing them with grains of sand, of which 100 in a line took up an inch. These grains he laid upon the same plate with his objects, and viewed them at the same time. Dr. Jurin's method was similar to this; for he found the diameter of a piece of fine silver wire, by wrapping it very close upon a pin, and observing how many rings made an inch: and he used this wire in the same manner as Leeuwenhoek used his sand. Dr. Hook used to look upon the magnified object with one eye, while at the same time he viewed other objects, placed at the same distance, with the other eye. In this manner he was able, by the help of a ruler, divided into inches and small parts, and laid on the pedestal of the microscope, as it were to cast the magnified appearance of the object upon the ruler, and thus exactly to measure the diameter which it appeared to have through the glass; which being compared with the diameter as it appeared to the naked eye, easily shewed the degree in which it was magnified. A little practice, says Mr. Baker, will render this method exceedingly easy and pleasant.

Mr. Martin, in his Optics, recommends such a Micrometer for a microscope as had been applied to telescopes; for he advises to draw a number of parallel lines on a piece of glass, with the fine point of a diamond, at the distance of one 40th of an inch from one another, and to place it in the focus of the eye-glass.

By

By this method, Dr. Smith contrived to take the exact draught of objects viewed by a double microscope; for he advises to get a lattice, made with small silver wires or squares, drawn upon a plain glass by the strokes of a diamond, and to put it into the place of the image formed by the object-glass. Then, by transferring the parts of the object, seen in the squares of the glass or lattice, upon similar corresponding squares drawn on paper, the picture may be exactly taken. Mr. Martin also introduced into compound microscopes another Micrometer, consisting of a screw. See both these methods described in his Optics, pa. 277.

A very accurate division of a scale is performed by Mr. Coventry, of Southwark. The Micrometers of his construction are parallel lines drawn on glass, ivory, or metal, from the 10th to the 10,000th part of an inch. These may be applied to microscopes, for measuring the size of minute objects, and the magnifying power of the glasses; and to telescopes, for measuring the size and distance of objects, and the magnifying power of the instrument. To measure the size of an object in a single microscope; lay it on a Micrometer, whose lines are seen magnified in the same proportion with it, and they give at one view the real size of the object. For measuring the magnifying power of the compound microscope, the best and readiest method is the following: On the stage in the focus of the object-glass, lay a Micrometer, consisting of an inch divided into 100 equal parts; count how many divisions of the Micrometer are taken into the field of view; then lay a two-foot rule parallel to the Micrometer: fix one eye on the edge of the field of light, and the other eye on the end of the rule, which move, till the edge of the field of light and the end of the rule correspond; then the distance from the end of the rule to the middle of the stage, will be half the diameter of the field: ex. gr. If the distance be 10 inches, the whole diameter will be 20, and the number of the divisions of the Micrometer contained in the diameter of the field, is the magnifying power of the microscope. For measuring the height and distance of objects by a Micrometer in the telescope, see TELESCOPE.

Mr. Adams has applied a Micrometer, that instantly shews the magnifying power of any telescope.

In the Philos. Transf. for 1791, a very simple scale Micrometer for measuring small angles with the telescope is described by Mr. Cavallo. This Micrometer consists of a thin and narrow slip of mother of-pearl finely divided, and placed in the focus of the eye-glass of a telescope, just where the image of the object is formed; whether the telescope is a reflector or a refractor, provided the eye-glass be a convex lens. This substance Mr. Cavallo, after many trials, found much more convenient than either glass, ivory, horn, or wood, as it is a very steady substance, the divisions very easy marked upon it, and when made as thin as common writing paper it has a very useful degree of transparency.

Upon this subject, see M. Azout's Tract on it, contained in *Divers Ouvrages de Mathematique & de Phisique; par Messieurs de l'Academie Royal des Sciences; M. de la Hire's Astronomicæ Tabulæ;* Mr. *Townley,* in the *Philos. Transf.* n°. 21; *Wolfius,* in his *Elem.*

VOL. II.

Astron. § 508; Dr. *Hook,* and many others, in the *Philos. Transf.* n°. 29 &c; *Hevelius,* in the *Acta Eruditorum, ann.* 1708; Mr. *Balshafer,* in his *Micrometria;* also several volumes of the *Paris Memoirs,* &c.

MICROPHONES, instruments contrived to magnify small sounds, as microscopes do small objects.

MICROSCOPE, an optical instrument, composed of lenses or mirrors, by means of which small objects are made to appear larger than they do to the naked eye.

MICROSCOPES are distinguished into simple and compound, or single and double.

Simple, or *Single* MICROSCOPES, are such as consist of a single lens, or a single spherule. And a

Compound MICROSCOPE consists of several lenses duly combined.—As optics have been improved, other varieties have been contrived in this instrument : Hence reflecting Microscopes, water Microscopes, &c.

It is not certainly known when, or by whom, Microscopes were first invented; although it is probable they would soon follow upon the use of telescopes, since a Microscope is like a telescope inverted. We are informed by Huygens, that one Drebell, a Dutchman, had the first Microscope, in the year 1621, and that he was reputed the inventor of it : though F. Fontana, a Neapolitan, in 1646, claims the invention to himself, and dates it from the year 1618. Be this as it may, it seems they were first used in Germany about 1621. According to Borelli, they were invented by Zacharias Jansen and his son, who presented the first Microscopes they had constructed to prince Maurice, and Albert arch-duke of Austria. William Borelli, who gives this account in a letter to his brother Peter, says, that when he was ambassador in England, in 1619, Cornelius Drebell shewed him a Microscope, which he said was the same that the arch-duke had given him, and had been made by Jansen himself. Borelli De vero Telescopii inventore, pa. 35. See LENS.

Theory and *Foundation* of MICROSCOPES.

If an object be placed in the focus of the convex lens of a single Microscope, and the eye be very near on the other side, the object will appear distinct in an erect situation, and magnified in the ratio of the focal distance of the lens, to the ordinary distance of distinct vision, viz, about 8 inches.

So, if the object AB be placed in the focus F, of a small glass sphere, and the eye behind it, as in the focus G, the object will appear distinct, and in an erect posture, increased as to diameter in the ratio of $\frac{3}{4}$ of the diameter EI to 8 inches. If, ex. gr. the diameter EI of the small sphere be $\frac{1}{10}$ of an inch; then $CE = \frac{1}{20}$, and $FE = \frac{1}{2}CE = \frac{1}{40}$, so that $CF = \frac{3}{40}$; then as $\frac{3}{40} : 8$, or as 3 : 320, or as 1 : $106\frac{2}{3}$ :: the natural size to the magnified appearance; that is, the object is magnified about 107 times.

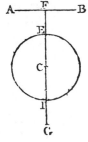

Hence the smaller the spherule or the lens is, so much the more is the object magnified. But then, so

much

much the lefs part is comprehended at one view, and fo much the lefs diftinct is the appearance of the object.

Equal appearances of the fame object, formed by different combinations, become obfcure in proportion as the number of rays conftituting each pencil decreafes, that is, in proportion to the fmallnefs of the object-glafs.

Wherefore, if the diameter of the object-glafs exceeds the diameter of the pupil, as many times as the diameter of the appearance exceeds the diameter of the object ; the appearance fhall be as clear and bright as the object itfelf.

The diameter of the object glafs cannot be fo much increafed, without increafing at the fame time the focal diftances of all the glaffes, and confequently the length of the inftrument : Otherwife the rays would fall too obliquely upon the eye-glafs, and the appearance become confufed and irregular.

There are feveral kinds of fingle Microfcopes ; of which the following is the moft fimple.

AB (Plate xviii, fig. 1) is a little tube, to one end of which BC, is fitted a plain glafs ; to which any object, as a gnat, the wing of an infect, or the like, is applied ; to the other end AD, at a proper diftance from the object, is applied a lens, convex on both fides, of about an inch in diameter : the plane glafs is turned to the fun, or the light of a candle, and the object is feen magnified. And if the tube be made to draw out, lenfes or fegments of different fpheres may be ufed.

Again, a lens, convex on both fides, is inclofed in a cell AC (fig. 2), and held there by the fcrew H. Through the ftem or pedeftal CD paffes a long fcrew EF, carrying a ftile or needle EG. In E is a fmall tube ; on which, and on the point G, the various objects are to be difpofed. Thus, lenfes of various fpheres may be applied.

A good fimple inftrument of this kind is Mr. Wilfon's pocket Microfcope, which has 9 different magnifying glaffes, 8 of which may be ufed with two different inftruments, for the better applying them to various objects. One of thefe inftruments is reprefented at AABB (fig. 3), which is made either of brafs or ivory. There are three thin brafs plates at E, and a fpiral fpring H of fteel wire within it : to one of the thin plates of brafs is fixed a piece of leather F, with a fmall furrow G, both in the leather, and brafs to which it is fixed : in one end of this inftrument there is a long fcrew D, with a convex glafs C, placed in the end of it : in the other end of the inftrument there is a hollow fcrew oo, in which any of the magnifying glaffes, M, are fcrewed, when they are to be made ufe of. The 9 different magnifying glaffes are all fet in ivory, 8 of which are fet in the manner expreffed at M. The greateft magnifier is marked upon the ivory, in which it is fet, number 1, the next number 2, and fo on to number 8 ; the 9th glafs is not marked, but is fet in the manner of a little barrel box of ivory, as at *b*. At *ee* is a flat piece of ivory, of which there are 8 belonging to this fort of Microfcopes (though any one who has a mind to keep a regifter of objects may have as many of them as he pleafes) ; in each of them there are 3 holes *fff*, in

which 3 or more objects are placed between two thin glaffes, or talcs, when they are to be ufed with the greater magnifiers.

The ufe of this inftrument AABB is this. A handle W, from fig. 4, being fcrewed upon the button S, take one of the flat pieces of ivory or fliders *ee*, and flide it between the two thin plates of brafs at E, through the body of the Microfcope, fo that the object to be viewed be juft in the middle ; remarking to put that fide of the plate *ee*, where the brafs rings are, fartheft from the end AA : then fcrew into the hollow fcrew, oo, the 3d, 4th, 5th, 6th, or 7th magnifying glafs M ; which being done, put the end AA clofe to your eye, and while looking at the object through the magnifying glafs, fcrew in or out the long fcrew D, which moving round upon the leather F, held tight to it by the fpiral wire H, will bring your object to the true diftance ; which may be known by feeing it clearly and diftinctly.

Thus may be viewed all tranfparent objects, dufts, liquids, cryftals of falts, fmall infects, fuch as fleas, mites, &c. If they are infects that will creep away, or fuch objects as are to be kept, they may be placed between the two regifter glaffes *ff*. For, by taking out the ring that keeps in the glaffes *ff*, where the object lies, they will fall out of themfelves ; fo the object may be laid between the two hollow fides of them, and the ring put in again as before ; but if the objects be dufts or liquids, a fmall drop of the liquid, or a little of the duft laid on the outfide of the glafs *ff*, and applied as before, will be feen very cafily.

As to the 1ft, 2d, and 3d magnifying glaffes, being marked with a + upon the ivory in which they are fet, they are only to be ufed with thofe plates or fliders that are alfo marked with a +, in which the objects are placed between two thin talcs ; becaufe the thicknefs of the glaffes in the other plates or fliders, hinders the object from approaching to the true diftance from thefe greater magnifiers. But the manner of ufing them is the fame with the former.

For viewing the circulation of the blood at the extremities of the arteries and veins, in the tranfparent parts of fifhes tails, &c, there are two glafs tubes, a larger and a fmaller, as expreffed at *gg*, into which the animal is put. When thefe tubes are to be ufed, unfcrew the end fcrew D in the body of the Microfcope, until the tube *gg* can be eafily received into that little cavity G of the brafs plate faftened to the leather F under the other two thin plates of brafs at E. When the tail of the fifh lies flat on the glafs tube, fet it oppofite to the magnifying glafs, and bringing it to the proper diftance by fcrewing in or out the end fcrew D, when the blood will be feen clearly circulating.

To view the blood circulating in the foot of a frog ; choofe fuch a frog as will juft go into the tube ; then with a little ftick expand its hinder foot, which apply clofe to the fide of the tube, obferving that no part of the frog hinders the light from coming on its foot ; and when it is brought to the proper diftance, by means of the fcrew D, the rapid motion of the blood will be feen in its veffels, which are very numerous, in the tranfparent thin membrane or web between the toes. For this object, the 4th and 5th magnifiers will do very
well ;

Plate XVIII.

MICROSCOPES

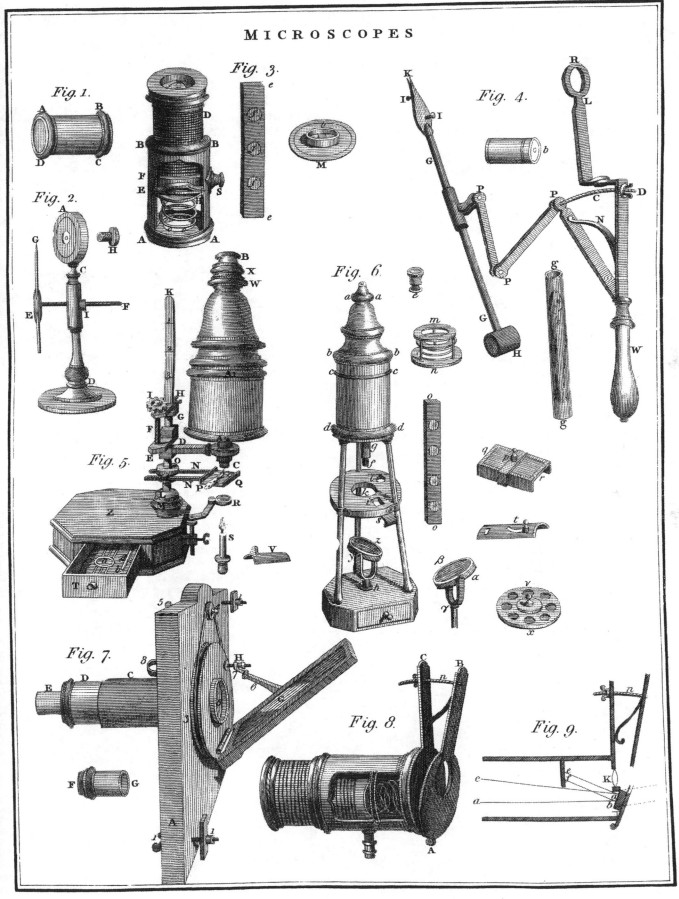

Fig. 1.

Fig. 2.

Fig. 3.

Fig. 4.

Fig. 5.

Fig. 6.

Fig. 7.

Fig. 8.

Fig. 9.

well; but the circulation may be seen in the tails of water-newts in the 6th and 7th glasses, because the globules of the blood of those newts are as large again as the globules of the blood of frogs or small fish, as has been remarked in number 280 of the Philos. Transf. pa. 1184.

The circulation cannot so well be seen by the 1st, 2d, and 3d magnifiers, because the thickness of the glass tube, containing the fish, hinders the approach of the object to the focus of the magnifying glass. Fig. 4 is another instrument for this purpose.

In viewing objects, one ought to be careful not to hinder the light from falling upon them by the hat, hair, or any other thing, especially in looking at opaque objects; for nothing can be seen with the best of glasses, unless the object be at a due distance, with a sufficient light. The best lights for the plates or sliders, when the object lies between the two glasses, is a clear sky-light, or where the sun shines on something white, or the reflection of the light from a looking glass. The light of a candle is also good for viewing very small objects, though it be a little uneasy to those who are not practised in the use of Microscopes.

To cast small Glass Spherules for MICROSCOPES.— There are several methods for this purpose. Hartsoeker first improved single Microscopes by using small globules of glass, melted in the flame of a candle; by which he discovered the animalculæ in semine masculino, and thereby laid the foundation of a new system of generation. Wolfius describes the following method of making such globules: A small piece of very fine glass, sticking to the wet point of a steel needle, is to be applied to the extreme bluish part of the flame of a lamp, or rather of spirits of wine, which will not black it; being there melted, and run into a small round drop, it is to be removed from the flame, on which it instantly ceases to be fluid. Then folding a thin plate of brass, and making very small smooth perforations, so as not to leave any roughness on the surfaces, and also smoothing them over to prevent any glaring, fit the spherule between the plates against the apertures, and put the whole in a frame, with objects convenient for observation.

Mr. Adams gives another method, thus: Take a piece of fine window-glass, and rase it, with a diamond, into as many lengths as you think needful, not more than 1-8th of an inch in breadth; then holding one of those lengths between the fore finger and thumb of each hand, over a very fine flame, till the glass begins to soften, draw it out till it be as fine as a hair, and break; then applying each of the ends into the purest part of the flame, you presently have two spheres, which may be made greater or less at pleasure: if they remain long in the flame, they will have spots; so they must be drawn out immediately after they are turned round. Break the stem off as near the globule as possible; and, lodging the remainder of the stem between the plates, by drilling the hole exactly round, all the protuberances are buried between the plates; and the Microscope performs to admiration.

Mr. Butterfield gave another manner of making these globules, in number 141 Philos. Transf.

In any of these ways may the spherules be made much smaller than any lens; so that the best single Mi-

croscopes, or such as magnify the most, are made of them Leeuwenhoeck and Musschenbroek have succeeded very well in spherical Microscopes, and their greatest magnifiers enlarged the diameter of an object about 160 times; Philos. Transf. vol. 7, pa. 129, and vol. 8, pa. 121. But the smallest globules, and consequently the highest magnifiers for Microscopes, were made by F. de Torre of Naples, who, in 1765, sent four of them to the Royal Society. The largest of them was only two Paris points in diameter, and magnified a line 640 times; the second was the size of one Paris point, and magnified 1280 times; and the 3d no more than half a Paris point, or the 144th part of an inch in diameter, and magnified 2560 times. But since the focus of a glass globule is at the distance of one-4th of its diameter, and therefore that of the 3d globule of de Torre, above mentioned, only the 576th part of an inch distant from the object, it must be with the utmost difficulty that globules so minute as those can be employed to any purpose; and Mr. Baker, to whose examination they were referred, considers them as matters of curiosity rather than of real use. Philos. Transf. vol. 55, pa. 246, vol. 56, pa. 67.

Water MICROSCOPE. Mr. S. Gray, and, after him, Wolfius and others, have contrived water Microscopes, consisting of spherules or lenses of water, instead of glass. But since the distance of the focus of a lens or sphere of water is greater than that in one of glass, the spheres of which they are segments being the same, consequently water Microscopes magnify less than those of glass, and therefore are less esteemed. Mr. Gray first observed, that a small drop or spherule of water, held to the eye by candle light or moon light, without any other apparatus, magnified the animalcules contained in it, vastly more than any other Microscope. The reason is, that the rays coming from the interior surface of the first hemisphere, are reflected so as to fall under the same angle on the surface of the hinder hemisphere, to which the eye is applied, as if they came from the focus of the spherule; whence they are propagated to the eye in the same manner as if the objects were placed without the spherule in its focus.

Hollow glass spheres of about half an inch diameter, filled with spirit of wine, are often used for Microscopes; but they do not magnify near so much.

Theory of Compound or Double MICROSCOPES.— Suppose an object-glass ED, the segment of a very small

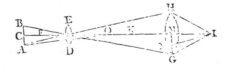

sphere, and the object AB placed without the focus F. Suppose an eye-glass GH, convex on both sides, and the segment of a sphere greater than that of DE, though not too great; and, the focus being at K, let it be so disposed behind the object,

that CF : CL :: CL : CK.
Lastly suppose LK : LM :: LM : LI.

If then O be the place where an object is seen distinct with the naked eye; the eye in this case, being placed in I, will see the object AB distinctly, in an inverted position, and magnified in the compound ratio of MK × LC to LK × CO ; as is proved by the laws of dioptrics; that is, the image is larger than the object, and we are able to view it distinctly at a less distance. For Examp.—If the image be 20 times larger than the object, and by the help of the eye-glass we are able to view it 5 times nearer than we could have done with the naked eye, it will, on both these accounts, be magnified 5 times 20, or 100 times.

Laws of Double MICROSCOPES.

1. The more an object is magnified by the Microscope, the less is its field, i. e. the less of it is taken in at one view.

2. To the same eye-glass may be successively applied object-glasses of various spheres, so as that both the entire objects, but less magnified, and their several parts, much more magnified, may be viewed through the same Microscope. In which case, on account of the different distance of the image, the tube in which the lenses are fitted, should be made to draw out.

3. Since it is proved, that the distance of the image LK, from the object-glass DE, will be greater, if another lens, concave on both sides, be placed before its focus; it follows, that the object will be magnified the more, if such a lens be here placed between the object-glass DE, and the eye-glass GH. Such a Microscope is much commended by Conradi, who used an object-lens, convex on both sides, whose radius was 2 digits, its aperture equal to a mustard seed; a lens, concave on both sides, from 12 to 16 digits; and an eye glass, convex on both sides, of 6 digits.

4. Since the image is projected to the greater distance, the nearer another lens, of a segment of a larger sphere, is brought to the object-glass; a Microscope may be composed of three lenses, which will magnify prodigiously.

5. From these considerations it follows, that the object will be magnified the more, as the eye-glass is the segment of a smaller sphere; but the field of vision will be the greater, as the same is a segment of a larger sphere. Therefore if two eye-glasses, the one a segment of a larger sphere, the other of a smaller one, be so combined, as that the object appearing very near through them, i. e. not farther distant than the focus of the first, be yet distinct; the object, at the same time, will be vastly magnified, and the field of vision much greater than if only one lens was used ; and the object will be still more magnified, and the field enlarged, if both the object-glass and eye-glass be double. But because an object appears dim when viewed through so many glasses, part of the rays being reflected in passing through each, it is not adviseable greatly to multiply glasses; so that, among compound Microscopes, the best are those which consist of one object-glass, and two eye glasses.

Dr. Hook, in the preface to his Micrography, tells us, that in most of his observations he used a Microscope of this kind, with a middle eye-glass of a considerable diameter, when he wanted to see much of the object at one view, and took it out when he would examine the small parts of an object more accurately : for the fewer refractions there are, the more light and clear the object appears.

For a Microscope of three lenses De Chales recommends an object glass of $\frac{1}{3}$ or $\frac{1}{4}$ of a digit ; and the first eye glass he makes 2 or $2\frac{1}{2}$ digits; and the distance between the object glass and eye-glass about 20 lines. Conradi had an excellent Microscope, whose object glass was half a digit, and the two eye-glasses (which were placed very near) 4 digits; but it answered best when, instead of the object-glass, he used two glasses, convex on both sides, their sphere about a digit and a half, and at most 2, and their convexities touching each other within the space of half a line. Eustachius de Divinis, instead of an object-glass convex on both sides, used two plano convex lenses, whose convexities touched. Grindelius did the same; only that the convexities did not quite touch. Zahnius made a binocular Microscope, with which both eyes were used. But the most commodious double Microscope, it is said, is that of our countryman Mr. Marshal ; though some improvement was made in it by Mr. Culpepper and Mr. Scarlet. These are exhibited in figures 5 and 6.

It is observed, that compound Microscopes sometimes exhibit a fallacious appearance, by representing convex objects concave, and vice versa. Philos. Trans. numb. 476, pa. 387.

To fit Microscopes, as well as Telescopes, to short-sighted eyes, the object-glass and the eye glass must be placed a little nearer together, so that the rays of each pencil may not emerge parallel, but may fall diverging upon the eye.

Reflecting MICROSCOPE, is that which magnifies by reflection, as the foregoing ones do by refraction. The inventor of this Microscope was Sir Isaac Newton.

The structure of such a Microscope may be conceived thus : near the focus of a concave speculum AB, place a minute object C, that its image may be formed larger than itself in D; to the speculum join a lens, convex on both sides, EF, so as the image D may be in its focus.

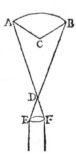

The eye will here see the image inverted, but distinct, and enlarged ; consequently the object will be larger than if viewed through the lens alone.

Any telescope is changed into a Microscope, by removing the object-glass to a greater distance from the eye-glass. And since the distance of the image is various, according to the distance of the object from the focus; and it is magnified the more, as its distance from the object-glass is greater ; the same telescope may be successively changed into Microscopes which magnify the object in different degrees. See some instruments of this sort described in Smith's Optics, Remarks, pa. 94.

Solar MICROSCOPE, called also the Camera Obscura Microscope, was invented by Mr. Lieberkhn in 1738 or 1739, and consists of a tube, a looking-glass, a convex lens, and a Wilson's Microscope. The tube (fig. 7) is brass, near 2 inches in diameter, fixed in a circular collar of mahogany, with a groove on the outside fide of its periphery, denoted by 2, 3, and connected by a cat-gut to the pulley 4 on the upper part; which turning round at pleafure, by the pin 5 within, in a fquare frame, may be eafily adjufted to a hole in the fhutter of a window, by the fcrews 1, 1, fo clofely that no light can enter the room but through the tube of the inftrument. The mirror G is faftened to the frame by hinges, on the fide that goes without the window: this glafs, by means of a jointed brafs wire, 6, 7, and the fcrew H 8, coming through the frame, may be moved either vertically or horizontally, to throw the fun's rays through the brafs tube into the darkened room. The end of the brafs tube without the fhutter has a convex lens, 5, to collect the rays thrown on it by the glafs G, and bring them to a focus in the other part, where D is a tube fliding in and out, to adjuft the object to a due diftance from the focus. And to the end G of another tube F, is fcrewed one of Wilfon's fimple pocket Microfcopes, containing the object to be magnified in a flider; and by tube F, fliding on the fmall end, E, of the other tube D, it is brought to a true focal diftance.

The Solar Microfcope has been introduced into the fmall and portable Camera Obfcura, as well as the large one: and if the image be received upon a piece of half-ground glafs, fhaded from the light of the fun, it will be fufficiently vifible. Mr. Lieberkuhn made confiderable improvements in his Solar Microfcope, particularly in adapting it to the viewing of opaque objects; and M. Aepinus, Nov. Com. Petrop. vol. 9, pa. 326, has contrived, by throwing the light upon the forefide of any object, before it is tranfmitted through the object lens, to reprefent all kinds of objects by it with equal advantage. In this improvement, the body of the common Solar Microfcope is retained, and only an addition made of two brafs plates, AB, AC, (fig 8), joined by a hinge, and held at a proper diftance by a fcrew. A fection of thefe plates, and of all the neceffary parts of the inftrument, may be feen in fig. 9 where a c reprefent rays of the fun converging from the illuminating lens, and falling upon the mirror bd, which is fixed to the nearer of the brafs plates. From this they are thrown upon the object at ef, and are thence tranfmitted through the object lens at K, and a perforation in the farther plate, upon a fcreen, as ufual. The ufe of the fcreen n is to vary the diftance of the two plates, and thereby to adjuft the mirror to the object with the greateft exactnefs. M. Euler alfo contrived a method of introducing vifion by reflected light into this Microfcope.

The MICROSCOPE *for Opaque Objects* was alfo invented by M Lieberkuhn, about the fame time with the former, and remedies the inconvenience of having the dark fide of an object next the eye; for by means of a concave fpeculum of filver, highly polifhed, having a magnifying lens placed in its centre, the object is fo ftrongly illuminated, that it may be examined with eafe. A convenient apparatus of this kind, with 4 different fpeculums and magnifiers of different powers, was brought to perfection by Mr. Cuff. Philof. Tranf. number 458, § 9.

MICROSCOPIC *Objects.* All things too minute to be viewed diftinctly by the naked eye, are proper objects for the Microfcope. Dr. Hook has diftinguifhed them into thefe three general kinds; viz, exceeding

5

fmall bodies, exceeeding fmall pores, or exceeding fmall motions. The fmall bodies may be feeds, infects, animalcules, fands, falts, &c: the pores may be the interftices between the folid parts of bodies, as in ftones, minerals, fhells, &c. or the mouths of minute veffels in vegetables, or the pores of the fkin, bones, and other parts of animals: the fmall motions, may be the movements of the feveral parts or members of minute animals, or the motion of the fluids, contained either in animal or vegetable bodies. Under one or other of thefe three general heads, almoft every thing about us affords matter of obfervation, and may conduce both to our amufement and inftruction.

Great caution is to be ufed in forming a judgment on what is feen by the Microfcope, if the objects are extended or contracted by force or drynefs.

Nothing can be determined about them, without making the proper allowances; and different lights and pofitions will often fhew the fame object as very different from itfelf. There is no advantage in any greater magnifier than fuch as is capable of fhewing the object in view diftinctly; and the lefs the glafs magnifies, the more pleafantly the object is always feen.

The colours of objects are very little to be depended on, as feen by the Microfcope; for their feveral component particles, being thus removed to great diftances from one another, may give reflections very different from what they would, if feen by the naked eye.

The motions of living creatures too, or of the fluids contained in their bodies, are by no means to be haftily judged of, from what we fee by the Microfcope, without due confideration; for as the moving body, and the fpace in which it moves, are magnified, the motion muft alfo be magnified; and therefore that rapidity with which the blood feems to pafs through the veffels of fmall animals, muft be judged of accordingly. Baker on the Microfcope, pa. 52, 62, &c. See alfo an elegant work on this fubject, lately publifhed by that ingenious optician Mr. George Adams.

MIDDLE *Latitude,* is half the fum of two given latitudes; or the arithmetical mean, or the middle between two parallels of latitude. Therefore,

If the latitudes be of the fame name, either both north or both fouth, add the one number to the other, and divide the fum by 2; the quotient is the middle latitudes, which is of the fame name with the two given latitudes. But

If the latitudes be of different names, the one north and the other fouth; fubtract the lefs from the greater, and divide the remainder by 2, fo fhall the quotient be the middle latitude, of the fame name with the greater of the two.

Ex. 1.   *Ex.* 2.
One lat. 35° 27′ N.   35° 27′ S.
the other 21 13 N.   21 13 N.

        2) 56 40   2) 14 14

Mid. lat. 28 20 N.   Mid. lat. 7   7 S.

MIDDLE *Latitude Sailing,* is a method of refolving the cafes of globular failing, by means of the Middle Latitude, on the principles of plane and parallel failing jointly.

This

This method is not quite accurate, yet often agrees pretty nearly with Mercator's Sailing, and is founded on the following principle, viz, That the departure is accounted a meridional distance in the middle latitude between the latitude sailed from and the latitude arrived at.

This artifice seems to have been invented, on account of the easy manner in which the several cases may be resolved by the Traverse Table, and to serve where a table of meridional parts is wanting. It is sufficiently near the truth either when the two parallels are near the equator, or not far distant from one another, in any latitude. It is performed by these two rules:

1. As the cosine of the middle latitude :
   Is to radius
   So is the departure   : :
   To the difference of longitude   .

2. As the cosine of the middle latitude :
   Is to the tangent of the course   : :
   So is the difference of latitude   :
   To the difference of longitude   .

Ex. A ship sails from latitude 37° north, steering constantly N. 33° 19 east, for 8 days, when she was found in latitude 51° 18 north; required her difference of longitude.

$$\begin{array}{cc} 51°\ 18' & 51°\ 18' \\ 37\ \ 00 & 37\ \ 00 \\ \hline 2\ )\ 88\ 18 & \text{Diff. lat. } 14\ \ 18 = 858\ m. \\ \hline \end{array}$$

As cos. mid. l.	44 09	-	0·14417
To tang. cour.	33 19	-	9·81776
So diff. lat.	858	-	2·93349
To diff. long.	786	-	2·89542

or 13° 6' diff. of long. sought.

MIDDLE *Region.* See REGION.

MID HEAVEN, *Medium Cœli,* is that point of the ecliptic which culminates, or is highest, or is in the meridian at any time.

MIDSUMMER-*Day*, is held on the 24th of June, the same day as the Nativity of St. John the Baptist is held.

MILE, a long measure, by which the English, Italians, and some other nations, use to express the distance between places: the same as the French use the word *League.*

The Mile is of different lengths in different countries. The geographical, or Italian Mile, contains 000 geometrical paces, *mille passus*, whence the term Mile is derived. The English Mile consists of 8 furlongs, each furlong of 40 poles, and each pole of 16½ feet: so that the Mile is = 8 furlongs = 320 poles = 1760 yards = 5280 feet.

The following table shews the length of the Mile, or league, in the principal nations of Europe, expressed in geometrical paces:

			Geomet. Paces.
Mile of Ruffa	-		750
of Italy	-	-	1000
of England	,	-	1200
of Scotland and Ireland			1500
Old League of France	-	-	1500
Small League, ibid.	;	-	2000
Mean League of France	-	-	2500
Great League, ibid.		-	3000
Mile of Poland	-	-	3000
of Spain	-	-	3428
of Germany	-	-	4000
of Sweden	-	-	5000
of Denmark	-	-	5000
of Hungary	-	-	6000

MILITARY *Architecture.* The same with Fortification.

MILKY WAY, *Via Lactea,* or *Galaxy,* a broad track or path, encompassing the whole heavens, distinguishable by its white appearance, whence it obtains the name. It extends itself in some parts by a double path, but for the most part it is single. Its course lies through the constellations Cassiopeia, Cygnus, Aquila, Perseus, Andromeda, part of Ophiucus and Gemini, in the northern hemisphere; and in the southern, it takes in part of Scorpio, Sagittarius, Centaurus, the Argonavis, and the Ara. There are some traces of the same kind of light about the south pole, but they are small in comparison of this: these are called by some, luminous spaces, and Magellanic clouds; but they seem to be of the same kind with the Milky way.

The Milky way has been ascribed to various causes. The Ancients fabled, that it proceeded from a stream of milk, spilt from the breast of Juno, when she pushed away the infant Hercules, whom Jupiter laid to her breast to render him immortal. Some again, as Aristotle, &c, imagined that this path consisted only of a certain exhalation hanging in the air; while Metrodorus, and some Pythagoreans, thought the sun had once gone in this track, instead of the ecliptic; and consequently that its whiteness proceeds from the remains of his light. But it is now well known, by the help of telescopes, that this track in the heavens consists of an immense multitude of stars, seemingly very close together, whose mingled light gives this appearance of whiteness; by Milton beautifully described as a path "powdered with stars."

MILL, properly denotes a machine for grinding corn, &c; but in a more general signification, is applied to all machines whose action depends on a circular motion. Of these there are several kinds, according to the various methods of applying the moving power; as water mills, wind mills, horse-mills, hand-mills, &c, and even steam-mills, or such as are worked by the force of steam; as that noble structure that was erected near Blackfriars Bridge, called the Albion Mills, but lately destroyed by fire.

The water acts both by its impulse and weight in an overshot water-mill, but only by its impulse in an undershot one; but here the velocity is greater, because the water is suffered to descend to a greater depth before it strikes the wheel. Mr. Ferguson observes, that where there is but a small quantity of water, and a fall great enough for the wheel to lie under it, the bucket or overshot wheel is always used: but where there is a large body of water, with a little fall, the breast or float-board wheel must take place: and where there is a large supply of water, as a river, or large stream or brook, with very little fall, then the undershot wheel is the easiest, cheapest, and most simple structure.

Dr.

Dr. Defaguliers, having had occasion to examine many undershot and overshot Mills, generally found that a well made overshot Mill ground as much corn, in the same time, as an undershot Mill does with ten times as much water; supposing the fall of water at the overshot to be 20 feet, and at the undershot about 6 or 7 feet: and he generally observed that the wheel of the overshot Mill was of 15 or 16 feet diameter, with a head of water of 4 or 5 feet, to drive the water into the buckets with some momentum.

In Water mills, some few have given the preference to the undershot wheel, but most writers prefer the overshot one. M. Belidor greatly preferred the undershot to any other construction. He had even concluded, that water applied in this way will do more than six times the work of an overshot wheel; while Dr. Defaguliers, in overthrowing Belidor's position, determined that an overshot wheel would do ten times the work of an undershot wheel with an equal quantity of water. So that between these two celebrated authors, there is a difference of no less than 60 to . In consequence of such monstrous disagreement, Mr. Smeaton began the course of experiments mentioned below.

In the Philos. Transf. vol. 51, for the year 1759, we have a large paper with experiments on Mills turned both by water and wind, by that ingenious and experienced engineer Mr. Smeaton. From those experiments it appears, pa. 129, that the effects obtained by the overshot wheel are generally 4 or 5 times as great as those with the undershot wheel, in the same time, with the same expence of water, descending from the same height above the bottom of the wheels; or that the former performs the same effect as the latter, in the same time, with an expence of only one-4th or one-5th of the water, from the same head or height. And this advantage seems to arise from the water lodging in the buckets, and so carrying the wheel about by their weight. But, in pa. 130, Mr. Smeaton reckons the effect of overshot only double to that of the undershot wheel. And hence he infers, in general, " that the higher the wheel is in proportion to the whole descent, the greater will be the effect; because it depends less upon the impulse of the head, and more upon the gravity of the water in the buckets. However, as every thing has its limits, so has this; for thus much is desirable, that the water should have somewhat greater velocity, than the circumference of the wheel, in coming thereon; otherwise the wheel will not only be retarded, by the buckets striking the water, but thereby dashing a part of it over, so much of the power is lost." He is farther of opinion, that the best velocity for an overshot wheel is when its circumference moves at the rate of about 3 feet in a second of time. See WIND MILL.

Considerable differences have also arisen as to the mathematical theory of the force of water striking the floats of a wheel in motion. M. Parent, Maclaurin, Defaguliers, &c, have determined, by calculation, that a wheel works to the greatest effect, when its velocity is equal to one-third of the velocity of the water which strikes it; or that the greatest velocity that the wheel acquires, is one-third of that of the water. And this determination, which has been followed by all mathematicians till very lately, necessarily results from a

position which they assume, viz, that the force of the water against the wheel, is proportional to the square of its relative velocity, or of the difference between the absolute velocity of the water and that of the wheel. And this position is itself an inference which they make from the force of water striking a body at rest, being as the square of the velocity, because the force of each particle is as the velocity it strikes with, and the number of particles or the whole quantity that strikes is also as the same velocity. But when the water strikes a body in motion, the quantity of it that strikes is still as the absolute velocity of the water, though the force of each particle be only as the relative velocity, or that with which it strikes. Hence it follows, that the whole force or effect is in the compound ratio of the absolute and relative velocities of the water; and therefore is greater than the before mentioned effect or force, in the ratio of the absolute to the relative velocity. The effect of this correction is, that the maximum velocity of the wheel becomes one-half the velocity of the water, instead of one-third of it only: a determination which nearly agrees with the best experiments, as those of Mr. Smeaton.

This correction has been lately made by Mr. W. Waring, in the 3d volume of the Transactions of the American Philosophical Society, pa. 144. This ingenious writer says, ' Being lately requested to make some calculations relative to Mills, particularly Dr. Barker's construction as improved by James Rumsey, I found more difficulty in the attempt than I at first expected. It appeared necessary to investigate new theorems for the purpose, as there are circumstances peculiar to this construction, which are not noticed, I believe, by any author; and the theory of Mills, as hitherto published, is very imperfect, which I take to be the reason it has been of so little use to practical mechanics.

' The first step, then, toward calculating the power of any water-mill (or wind-mill) or proportioning their parts and velocities to the greatest advantage, seems to be,

' *The Correction of an Essential Mistake adopted by Writers on the Theory of Mills.*

' This is attempted with all the deference due to eminent authors, whose ingenious labours have justly raised their reputation and advanced the sciences; but when any wrong principles are successively published by a series of such pens, they are the more implicitly received, and more particularly claim a public rectification; which must be pleasing, even to these candid writers themselves.'

A very ingenious writer in England, ' in his masterly treatise on the rectilinear motion and rotation of bodies, published so lately as 1784, continues this oversight, with its pernicious consequences, through his propositions and corollaries (pa. 275 to 284), although he knew the theory was suspected: for he observes (pa. 382) " Mr. Smeaton in his paper on mechanic " power (published in the Philosophical Transactions " for the year 1776) allows, that the theory usually " given will not correspond with matter of fact, when " compared with the motion of machines; and seems " to attribute this disagreement, rather to deficiency " in the theory, than to the obstacles which have prevented

4

" vented the application of it to the complicated mo-
" tion of engines, &c. In order to fatisfy himfelf con-
" cerning the reafon of this difagreement, he conftruct-
" ed a fet of experiments, which, from the known
" abilities and ingenuity of the author, certainly de-
" ferve great confideration and attention from every
" one who is interefted in thefe inquiries." ' And
notwithftanding the fame learned author fays, " The
evidence upon which the theory refts is fcarcely lefs
than mathematical ;" I am forry to find, in the prefent
ftate of the fciences, one of his abilities concluding
(pa. 380) " It is not probable that the theory of motion, however inconteftible its principles may be, can
afford much affiftance to the practical mechanic," although indeed his theory, compared with the above
cited experiments, might fuggeft fuch an inference.
But to come to the point, I would juft premife thefe

Definitions.

' If a ftream of water impinge againft a wheel in
motion, there are three different velocities to be confidered, appertaining thereto, viz,

First, the abfolute velocity of the water ;

Second, the abfolute velocity of the wheel ;

Third, the relative velocity of the water to that of
the wheel,

i. e. the difference of the abfolute velocities, or the velocity with which the water overtakes or ftrikes the
wheel.'

' Now the miftake confifts in fuppofing the momentum or force of the water againft the wheel, to be in
the *duplicate ratio of the relative velocity* : Whereas,

PROP. I.

' The force of an Invariable Stream, impinging
againft a Mill-wheel in Motion, is in the *Simple Direct
Proportion of the Relative Velocity*.'

' For, if the relative velocity of a fluid againft a fingle plane be varied, either by the motion of the plane,
or of the fluid from a given aperture, or both, then,
the number of particles acting on the plane in a given
time, and likewife the momentum of each particle,
being refpectively as the relative velocity, the force on
both thefe accounts, muft be in the *duplicate* ratio of
the relative velocity, agreeably to the common theory,
with refpect to this *fingle plane* : but, the number of
thefe planes, or parts of the wheel acted on in a given
time, will be as the velocity of the wheel, or *inverfely
as the relative velocity* ; therefore, the moving force of
the wheel muft be in the fimple direct ratio of the relative velocity. Q. E. D.

' Or the propofition is manifeft from this confideration ; that, while the ftream is invariable, whatever be
the velocity of the wheel, the fame number of particles
or quantity of the fluid, muft ftrike it fomewhere or
other in a given time ; confequently the variation of
force is *only* on account of the varied impingent velocity
of the fame body, occafioned by a change of motion
in the wheel ; that is, the momentum is as the relative
velocity.'

' Now, this true principle fubftituted for the erroneous one in ufe, will bring the theory to agree remarkably with the notable experiments of the ingenious

Smeaton, before mentioned, publifhed in the Philofophical Tranfactions of the Royal Society of London
for the year 1751, vol. 51, for which the honorary annual medal was adjudged by the fociety, and prefented
to the author by their prefident. An inftance or two
of the importance of this correction may be adduced as
below.'

PROP. II.

' The velocity of a wheel, moved by the impact of
a ftream, muft be half the velocity of the fluid, to produce the greateft poffible effect.—For let

$V =$ the velocity, $m =$ the momentum of the fluid ;

$v =$ the velocity, $p =$ the power of the wheel.

Then $V - v =$ the relative velocity, by def. 3d ;

and as $V : V - v :: m : \frac{m}{V} \times \overline{V - v} = p$ (prop. 1) ;

this multiplied by v, gives $pv = \frac{m}{V} \times \overline{Vv - v^2} = $ a

maximum ; hence $Vv - v^2 = $ a maximum, and its
fluxion (v being the variable quantity) is $V\dot{v} - 2v\dot{v} = 0$;
therefore $v = \frac{1}{2}V$, that is, the velocity of the wheel
$=$ half that of the fluid, at the place of impact, when
the effect is a maximum. Q. E. D.'

' The ufual theory gives $v = \frac{1}{3}V$; where the error
is not lefs than one third of the true velocity of the
wheel.'

' This propofition is applicable to underfhot wheels,
and correfponds with the accurate experiments before
cited, as appears from the author's conclufion (Philof.
Tranf. for 1776, pa. 457), viz, " The velocity of the
" wheel, which according to M. Parent's determina-
" tion, adopted by Defaguliers and Maclaurin, ought to
" be no more than one third of that of the water, varies
" at the maximum in the experiments of table 1, be-
" tween one third and one half ; but in all the cafes
" there related, in which the moft work is performed
" in proportion to the water expended, and which ap-
" proach the neareft to the circumftances of great
" works when properly executed, the maximum lies
" much nearer one half than one third, *one half feeming
" to be the true maximum*, if nothing were loft by the
" refiftance of the air, the fcattering of the water car-
" ried up by the wheel, &c." Thus he fully fhews
the common theory to have been very defective ; but,
I believe, none have fince pointed out wherein the deficiency lay, nor how to correct it ; and now we fee the
agreement of the true theory with the refult of his experiments.' For another problem,

PROB. III.

' Given, the momentum (m) and velocity (V) of the
fluid at I, the place of impact ; the radius ($R = IS$)
of the wheel ABC ; the radius ($r = DS$) of the fmall
wheel DEF on the fame axle or fhaft ; the weight (w)
or refiftance to be overcome at D, and the friction (f)
or force neceffary to move the wheel without the
weight ; required the velocity (v) of the wheel &c."

' Here we have $V : V - v :: m : m \times \dfrac{V - v}{V} = $

the acting force at I in the direction KI, as before
(prop. 2). Now $R : r :: w : \dfrac{rw}{R} = $ the power

at

at I neceſſary to counterpoiſe the weight w; hence $\frac{rw}{R} + f =$ the whole reſiſtance oppoſed to the action

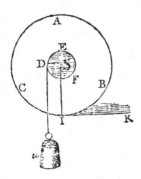

of the fluid at I; which deducted from the moving force, leaves $m \times \frac{V-v}{V} - \frac{rw}{R} - f =$ the accelerating force of the machine; which, when the motion becomes uniform, will be evaneſcent or $= o$; therefore $m \times \frac{V-v}{V} = \frac{rw}{R} + f$, which gives

$v = V \times \overline{1 - \frac{rw}{mR} - \frac{f}{m}} =$ the true velocity equired; or, if we reject the friction, then

$v = V \times \overline{1 - \frac{rw}{mR}}$ is the theorem for the velocity of the wheel. This, by the common theory, would be

$v = V \times \overline{1 - \sqrt{\frac{rw}{mR}}}$, which is too little by

$V \sqrt{\frac{rw}{mR}} - V \frac{rw}{mR}$. No wonder why we have hitherto derived ſo little advantage from the theory.'

'COROL. 1. If the weight (w) or reſiſtance be required, ſuch as juſt to admit of that velocity which would produce the greateſt effect; then, by ſubſtituting $\frac{1}{2}V$ for its equivalent v (by prop. 2), we have

$\frac{1}{2}V = V \times \overline{1 - \frac{rw}{mR} - \frac{f}{m}}$; hence $w = \frac{\frac{1}{2}m - f}{r} \times R$;

or, if $f = o$, $w = \frac{mR}{2r}$; but theoriſts make this $\frac{4mR}{9r}$,

where the error is $\frac{mR}{18r}$.'

'COROL. 2. We have alſo $r = \frac{\frac{1}{2}m - f}{w} \times R$; or, rejecting friction, $r = \frac{mR}{2w}$, when the greateſt effect is produced, inſtead of $r = \frac{4mR}{9w}$, as has been ſuppoſed: this is an important theorem in the conſtruction of mills.'

In the ſame volume of the American Tranſactions, pa. 185, is another ingenious paper, by the ſame author, on the power and machinery of Dr. Barker'ſ Mill, as improved by Mr. James Rumſey, with a deſcription of it. This is a Mill turned by the reſiſting force of a ſtream of water that iſſues from an orifice, the rotatory part, in which that orifice is, being impelled the contrary way by its reaction againſt the ſtream that iſſues from it.

Mr. Ferguſon has given the following directions for conſtructing water mills in the beſt manner; with a table of the ſeveral correſponding dimenſions proper to a great variety of perpendicular falls of the water.

When the float-boards of the water-wheel move with a 3d part of the velocity of the water that acts upon them, the water has the greateſt power to turn the Mill: and when the millſtone makes about 60 turns in a minute, it is found to perform its work the beſt: for, when it makes but about 40 or 50, it grinds too ſlowly; and when it makes more than 70, it heats the meal too much, and cuts the bran ſo ſmall that a great part of it mixes with the meal, and cannot be ſeparated from it by ſifting or boulting. Conſequently the utmoſt perfection of mill-work lies in making the train ſo as that the millſtone ſhall make about 60 turns in a minute when the water wheel moves with a 3d part of the velocity of the water. To have it ſo, obſerve the following rules:

1. Meaſure the perpendicular height of the fall of water, in feet, above the middle of the aperture, where it is let out to act by impulſe againſt the floatboards on the loweſt ſide of the underſhot wheel.

2. Multiply that height of the fall in feet by the conſtant number $64\frac{1}{3}$, and extract the ſquare root of the product, which will be the velocity of the water at the bottom of the fall, or the number of feet the water moves per ſecond.

3. Divide the velocity of the water by 3; and the quotient will be the velocity of the floats of the wheel in feet per ſecond.

4. Divide the circumference of the wheel in feet, by the velocity of its floats; and the quotient will be the number of ſeconds in one turn or revolution of the great water-wheel, on the axis of which is fixed the cogwheel that turns the trundle.

5. Divide 60 by the number of ſeconds in one turn of the water-wheel or cog-wheel; and the quotient will be the number of turns of either of theſe wheels in a minute.

6. Divide 60 (the number of turns the millſtone ought to have in a minute) by the aboveſaid number of turns; and the quotient will be the number of turns the millſtone ought to have for one turn of the water or cog-wheel. Then,

7. As the required number of turns of the millſtone in a minute is to the number of turns of the cogwheel in a minute, ſo muſt the number of cogs in the wheel be to the number of ſtaves or rounds in the trundle on the axis of the millſtone, in the neareſt whole number that can be found.

By theſe rules the following table is calculated; in which, the diameter of the water-wheel is ſuppoſed 18 feet, and conſequently its circumference $56\frac{4}{7}$ feet, and the diameter of the millſtone is 5 feet.

The MILL-WRIGHT's Table.

Perpendicular height of the fall of water.	Velocity of the water in feet per second.	Velocity of the wheel in feet per second.	Number of turns of the wheel in a minute.	Required no. of turns of the millstone for each turn of the wheel.	Nearest number of cogs and staves for that purpose.		Number of turns of the millstone for one turn of the wheel by these cogs and staves.	Number of turns of the millstone in a minute by these cogs and staves.
					Cogs.	Staves.		
1	8·02	2·67	2·83	21·20	127	6	21·17	59·9
2	11·40	3·78	4·00	15·00	105	7	15·00	60·00
3	13·89	4·63	4·91	12·22	98	8	12·25	60·14
4	16·04	5·35	5·67	10·58	95	9	10·56	59·87
5	17·93	5·98	6·34	9·46	85	9	9·44	59·84
6	19·64	6·55	6·94	8·64	78	9	8·66	60·10
7	21·21	7·07	7·50	8·00	72	9	8·00	60·00
8	22·68	7·56	8·02	7·48	67	9	7·44	59·67
9	24·05	8·02	8·51	7·05	70	10	7·00	59·57
10	25·35	8·45	8·97	6·69	67	10	6·70	60·09
11	26·59	8·86	9·40	6·38	64	10	6·40	60·16
12	27·77	9·26	9·82	6·11	61	10	6·10	59·90
13	28·91	9·64	10·22	5·87	59	10	5·80	60·18
14	30·00	10·00	10·60	5·66	56	10	5·60	59·36
15	31·05	10·35	10·99	5·46	55	10	5·40	60·48
16	32·07	10·69	11·34	5·29	53	10	5·30	60·10
17	33·06	11·02	11·70	5·13	51	10	5·10	59·67
18	34·02	11·34	12·02	4·99	50	10	5·00	60·10
19	34·95	11·65	12·37	4·85	49	10	4·80	60·61
20	35·86	11·92	12·68	4·73	47	10	4·70	59·59

For the theory and conſtruction of Wind-mills, ſee WIND-*mill.*

MILLION, the number of ten hundred thouſand, or a thouſand times a thouſand.

MINE, in Fortification &c, is a ſubterraneous canal or paſſage, dug under any place or work intended to be blown up by gunpowder. The paſſage of a mine leading to the powder is called the *Gallery;* and the extremity, or place where the powder is placed, is called the *Chamber.* The line drawn from the centre of the chamber perpendicular to the neareſt ſurface, is called the *Line of leaſt Reſiſtance;* and the pit or hole, made by the mine when ſprung, or blown up, is called the *Excavation.*

The Mines made by the beſiegers in the attack of a place, are called ſimply *Mines;* and thoſe made by the beſieged, *Counter-mines.*

The fire is conveyed to the Mine by a pipe or hoſe, made of coarſe cloth, of about an inch and half in diameter, called *Sauciſſon,* extending from the powder in the chamber to the beginning or entrance of the gallery, to the end of which is fixed a match, that the miner who ſets fire to it may have time to retire before it reaches the chamber.

It is found by experiments, that the figure of the excavation made by the exploſion of the powder, is nearly a paraboloid, having its focus in the centre of the powder, and its axis the line of leaſt reſiſtance; its diameter being more or leſs according to the quantity of the powder, to the ſame axis, or line of leaſt reſiſtance. Thus, M. Belidor lodged ſeven different quantities of powder in as many different mines, of the ſame depth, or line of leaſt reſiſtance 10 feet; the charges and greateſt diameters of the excavation, meaſured after the exploſion, were as follow:

	Powder.	Diam.
1ft	120lb	22⅔ feet
2d	160	26
3d	200	29
4th	240	31¼
5th	280	33½
6th	320	36
7th	360	38

From which experiments it appears that the excavation, or quantity of earth blown up, is in the ſame proportion with the quantity of powder; whence the charge of powder neceſſary to produce any other propoſed effect, will be had by the rule of Proportion.

MINE-*Dial,* is a box and needle, with a braſs ring divided into 360 degrees, with ſeveral dials graduated upon it, commonly made for the uſe of miners.

MINUTE, is the 60th part of a degree, or of an hour. The minutes of a degree are marked with the acute accent, thus ′; the ſeconds by two, ″; the thirds by three, ‴. The minutes, ſeconds, thirds, &c, in time, are ſometimes marked the ſame way; but, to avoid confuſion, the better way is, by the initials of the words; as minutes m, ſeconds s, thirds t, &c.

MINUTE, in Architecture, uſually denotes the 60th part of a module, but ſometimes only the 30th part.

3 MIRROR,

MIRROR, a speculum, looking-glass, or any polished body, whose use is to form the images of distinct objects by reflexion of the rays of light.

Mirrors are either plane, convex, or concave. The first sort reflects the rays of light in a direction exactly similar to that in which they fall upon it, and therefore represents bodies of their natural magnitude. But the convex ones make the rays diverge much more than before reflexion, and therefore greatly diminish the images of those objects which they exhibit: while the concave ones, by collecting the rays into a focus, not only magnify the objects they shew, but will also burn very fiercely when exposed to the rays of the sun; and hence they are commonly known by the name of *burning Mirrors*.

In ancient times the Mirrors were made of some kind of metal; and from a passage in the Mosaic writings we learn, that the Mirrors used by the Jewish women, were made of brass; a practice doubtless learned from the Egyptians.

Any kind of metal, when well polished, will reflect very powerfully; but of all others, silver reflects the most, though it has always been too expensive a material for common use. Gold is also very powerful; and all metals, or even wood, gilt and polished, will act very powerfully as burning Mirrors. Even polished ivory, or straw nicely plaited together, will form Mirrors capable of burning, if on a large scale.

Since the invention of glass, and the application of quicksilver to it, have become generally known, it has been universally employed for those plane Mirrors used as ornaments to houses; but in making reflecting telescopes they have been found much inferior to metallic ones. It does not appear however that the same superiority belongs to the metallic burning Mirrors, considered merely as burning speculums; since the Mirror with which Mr. Macquer melted platina, though only 22 inches diameter, and made of quicksilvered glass, produced much greater effects than M. Villette's metal speculum, which was of a much larger size. It is very probable, however, that M. Villette's Mirror was not so well polished as it ought to have been; as the art of preparing the metal for taking the finest polish, has but lately been discovered, and published in the Philos. Transactions, by Dr. Mudge of Plymouth, and, after him, by Mr. Edwards, Dr. Herschel, &c.

Some of the more remarkable laws and phenomena of plane Mirrors, are as follow:

1. A spectator will see his image of the same size, and erect, but reversed as to right and left, and as far beyond the speculum as he is before it. As he moves to or from the speculum, his image will, at the same time, move towards or from the speculum also on the other side. In like manner if, while the spectator is at rest, an object be in motion, its image behind the speculum will be seen to move at the same rate. Also when the spectator moves, the images of objects that are at rest will appear to approach or recede from him, after the same manner as when he moves towards real objects.

2. If several Mirrors, or several fragments or pieces of Mirrors, be all disposed in the same plane, they will only exhibit an object once.

3. If two plane Mirrors, or speculums, meet in any angle, the eye, placed within that angle, will see the image of an object placed within the same, as often repeated as there may be perpendiculars drawn determining the places of the images, and terminated without the angle. Hence, as the more perpendiculars, terminated without the angle, may be drawn as the angle is more acute; the acuter the angle, the more numerous the images. Thus, Z. Traber found, at an angle of one-3d of a circle, the image was represented twice, at ¼th thrice, at ⅙th five times, and at $\frac{1}{12}$th eleven times.

Farther, if the Mirrors be placed upright, and so contracted; or if you retire from them, or approach to them, till the images reflected by them coalesce, or run into one, they will appear monstrously distorted. Thus, if they be at an angle somewhat greater than a right one, the image of one's face will appear with only one eye; if the angle be less than a right one, you will see 3 eyes, 2 noses, 2 mouths, &c. At an angle still less, the body will have two heads. At an angle somewhat greater than a right one, at the distance of 4 feet, the body will be headless, &c. Again, if the Mirrors be placed, the one parallel to the horizon, the other inclined to it, or declined from it, it is easy to perceive that the images will be still more romantic. Thus, one being declined from the horizon to an angle of 144 degrees, and the other inclined to it, a man sees himself standing with his head to another's feet.

Hence it appears how Mirrors may be managed in gardens, &c, so as to convert the images of those near them into monsters of various kinds; and since glass Mirrors will reflect the image of a lucid object twice or thrice, if a candle, &c, be placed in the angle between two Mirrors, it will be multiplied a great number of times.

Laws of Convex MIRRORS.

1. In a spherical convex Mirror, the image is less than the object. And hence the use of such Mirrors in the art of painting, where objects are to be represented less than the life.

2. In a convex Mirror, the more remote the object, the less its image; also the smaller the Mirror, the less the image.

3. In a convex Mirror, the right hand is turned to the left, and the left to the right; and magnitudes perpendicular to the Mirror appear inverted.

4. The image of a right line, perpendicular to the Mirror, is a right line; but that of a right line oblique or parallel to the Mirror, is convex.

5. Rays reflected from a convex Mirror, diverge more than if reflected from a plane Mirror; and the smaller the sphere, the more the rays diverge.

Laws of Concave MIRRORS.

The effects of concave Mirrors are, in general, the reverse of those of convex ones; rays being made to converge more, or diverge less than in plane Mirrors; the image is magnified, and the more so as the sphere is smaller; &c, &c.

MITRE, in Architecture, is the workmen's term for an angle that is just 45 degrees, or half a right angle. And if the angle be the half of this, or a quarter of a right angle, they call it a *half-mitre*.

MIXT

MIXT *Angle*, or *Figure*, is one contained by both right and curved lines.

MIXT *Number*, is one that is partly an integer, and partly a fraction; as 3½.

MIXT *Ratio*, or *Proportion*, is when the sum of the antecedent and consequent is compared with the difference of the antecedent and consequent;

$$\text{as if} \begin{cases} 4:3::12:9 \\ a:b::c:d \end{cases}$$

$$\text{then} \begin{cases} 7 : 1 :: 21 : 3 \\ a+b.a-b::c+d:c-d. \end{cases}$$

MOAT, in Fortification, a deep trench dug round a town or fortress, to be defended, on the outside of the wall, or rampart.

The breadth and depth of a Moat often depend on the nature of the foil; according as it is marshy, rocky, or the like. The brink of the Moat next the rampart, is called the Scarp; and the opposite side, the Counterscarp.

Dry MOAT, is one that is without water; which ought to be deeper than one that has water, called a Wet Moat. A Dry Moat, or one that has a little water, has often a small notch or ditch run all along the middle of its bottom, called a Cuvette.

Flat-bottomed MOAT, is that which has no sloping, its corners being somewhat rounded.

Lined MOAT, is that whose scarp and counterscarp are cased with a wall of mason's work lying aslope.

MOBILE, *Primum*, in the Ancient Astronomy, was a 9th heaven, or sphere, conceived above those of the planets and fixed stars. It was supposed that this was the first mover, and carried all the lower spheres about with it; by its rapidity communicating to them a motion carrying them round in 24 hours. But the diurnal apparent revolution of the heavens is now better accounted for, by the rotation of the earth on its axis, without the assistance of any such Primum Mobile.

MOBILITY, an aptitude or facility to be moved.

The Mobility of Mercury is owing to the smallness and sphericity of its particles; and these also render its fixation so difficult.

The hypothesis of the Mobility of the earth is the most plausible, and is universally admitted by the later astronomers.

Pope Paul V. appointed commissioners to examine the opinion of Copernicus touching the Mobility of the earth. The result of their enquiry was, a prohibition to assert, not that the Mobility was possible, but that it was really true: that is, they allowed the Mobility of the earth to be held as an hypothesis, which gives an easy and sensible solution of the phenomena of the heavenly motions; but forbade the Mobility of the earth to be maintained as a thesis, or real effective thing; because they conceived it contrary to Scripture.

MODILLIONS, small inverted consoles under the soffit or bottom of the drip, or of the corniche, seeming to support the projecture of the larmier, in the Ionic, Composite, and Corinthian orders.

MODULE, or Little Measure, in Architecture, a certain measure, taken at pleasure, for regulating the proportions of columns, and the symmetry or distribu-tion of the whole building. Architects usually choose the diameter, or the semidiameter, of the bottom of the column, for their Module; which they subdivide into minutes; for estimating all the other parts of the building by.

MOINEAU, a flat bastion raised before a curtin when it is too long, and the bastions of the angles too remote to be able to defend one another. Sometimes the Moineau is joined to the curtin, and sometimes it is divided from it by a moat. Here musquetry are placed to fire each way.

MOLYNEUX (WILLIAM), an excellent mathematician and astronomer, was born at Dublin in 1656. After the usual grammar education, which he had at home, he was entered of the university of that city. Here he distinguished himself by the probity of his manners, as well as by the strength of his parts; and having made a remarkable progress in academical learning, and particularly in the new philosophy, as it was then called, after four years spent in this university, he was sent over to London, where he was admitted into the Middle Temple in 1675. Here he spent three years, in the study of the laws of his country. But the bent of his genius lay strongly toward mathematical and philosophical studies; and even at the university he conceived a dislike to scholastic learning, and fell into the methods of lord Bacon.

Returning to Ireland in 1678, he shortly after married Lucy the daughter of Sir William Domville, the king's attorney-general. Being master of an easy fortune, he continued to indulge himself in prosecuting such branches of natural and experimental philosophy as were most agreeable to his fancy; in which astronomy having the greatest share, he began, about 1681, a literary correspondence with Mr. Flamsteed, the king's astronomer, which he kept up for several years. In 1683 he formed a design of erecting a Philosophical Society at Dublin, in imitation of the Royal Society at London; and, by the countenance and encouragement of Sir William Petty, who accepted the office of president, began a weekly meeting that year, when our author was appointed their first secretary.

Mr. Molyneux's reputation for learning recommended him, in 1684, to the notice and favour of the first great duke of Ormond, then lord-lieutenant of Ireland; by whose influence chiefly he was appointed that year, jointly with sir William Robinson, surveyor-general of the king's buildings and works, and chief engineer.

In 1685, he was chosen fellow of the Royal Society at London; and that year he was sent by the government to view the most considerable fortresses in Flanders. Accordingly he travelled through that country and Holland, with part of Germany and France; and carrying with him letters of recommendation from Flamsteed to Cassini, he was introduced to him, and others, the most eminent astronomers in the several places through which he passed.

Soon after his return from abroad, he printed at Dublin, in 1686, his *Sciothericum Telescopium*, containing a Description of the Structure and Use of a Telescopic Dial, invented by him: another edition of which was published at London in 1700.

In 1688 the Philosophical Society of Dublin was broken up and dispersed by the confusion of the times.

Mr.

Mr. Molyneux had diftinguifhed himfelf as a Member of it from the beginning, and prefented feveral difcourfes upon curious fubjects; fome of which were tranfmitted to the Royal Society at London, and afterwards printed in the Philofophical Tranfactions. In 1689, among great numbers of other Proteftants, he withdrew from the difturbances in Ireland, occafioned by the feverities of Tyrconnel's government; and after a fhort ftay at London, he fixed himfelf with his family at Chefter. In this retirement, he employed himfelf in putting together the materials he had fome time before prepared for his *Dioptrics*, in which he was much affifted by Mr. Flamfteed; and in Auguft 1690, he went to London to put it to the prefs, where the fheets were revifed by Dr. Halley, who, at our author's requeft, gave leave for printing, in the appendix, his celebrated Theorem for finding the Foci of Optic Glaffes. Accordingly the book came out, 1692, in 4to, under the title of *Dioptrica Nova:* a Treatife of Dioptrics, in two parts; wherein the various effects and appearances of fpherical glaffes, both convex and concave, fingle and combined, in telefcopes and microfcopes, together with their ufefulnefs in many concerns of human life, are explained." He gave it the title of *Dioptrica Nova,* both becaufe it was almoft wholly new, very little being borrowed from other writers, and becaufe it was the firft book that appeared in Englifh upon the fubject. The work contains feveral of the moft generally ufeful propofitions for practice, demonftrated in a clear and eafy manner, for which reafon it was for many years ufed by the artificers: and the fecond part is very entertaining, efpecially in the hiftory which he gives of the feveral optical inftruments, and of the difcoveries made by them.

Before he left Chefter he loft his lady, who died foon after fhe had brought him a fon. Illnefs had deprived her of her eye-fight 12 years before, that is, foon after her marriage; from which time fhe had been very fickly, and afflicted with great pains in her head.

As foon as the public tranquillity was fettled in his native country, he returned home; and, upon the convening of a new parliament in 1692, was chofen one of the reprefentatives for the city of Dublin. In the next parliament, in 1695, he was chofen to reprefent the univerfity there, and continued to do fo to the end of his life; that learned body having lately conferred on him the degree of doctor of laws. He was likewife nominated by the lord-lieutenant one of the commiffioners for the forfeited eftates, to which employment was annexed a falary of 500l. a year; but looking upon it as an invidious office, he declined it!

In 1698, he publifhed " The Cafe of Ireland ftated, in regard to its being bound by Acts of Parliament made in England:" in which it is fuppofed he has delivered all, or moft, that can be faid upon this fubject, with great clearnefs and ftrength of reafoning.

Among many learned perfons with whom he maintained correfpondence and friendfhip, Mr. Locke was in a particular manner dear to him, as appears from their letters. In the above mentioned year, which was the laft of our author's life, he made a journey to England, on purpofe to pay a vifit to that great man; and not long after his return to Ireland, he was feized with a fit of the ftone, which terminated his exiftence.

I

Befides the three works already mentioned, viz, the *Sciothericum Telefcopium,* the *Dioptrica Nova,* and the *Cafe of Ireland ftated;* he publifhed a great number of pieces in the Philofophical Tranfactions, which are contained in the volumes 14, 15, 16, 18, 19, 20, 21, 22, 23, 26, 29, feveral papers commonly in each volume.

MOLYNEUX *(Samuel),* fon of the former, was born at Chefter in July 1689; and educated with great care by his father, according to the plan laid down by Locke on that fubject. When his father died, he fell under the management of his uncle, Dr. Thomas Molyneux, an excellent fcholar and phyfician at Dublin, and alfo an intimate friend of Mr. Locke, who executed his truft fo well, that Mr. Molyneux became afterwards a moft polite and accomplifhed gentleman, and was made fecretary to George the 3d when prince of Wales. Aftronomy and Optics being his favourite ftudies, as they had been his father's, he projected many fchemes for the advancement of them, and was particularly employed in the years 1723, 1724, and 1725, in perfecting the method of making telefcopes; one of which inftruments, of his own making, he had prefented to John the 5th, king of Portugal.

Being foon after appointed a commiffioner of the admiralty, he became fo engaged in public affairs, that he had not leifure to purfue thofe enquiries any farther, as he intended. He therefore gave his papers to Dr. Robert Smith, profeffor of aftronomy at Cambridge, whom he invited to make ufe of his houfe and apparatus of inftruments, in order to finifh what he had left imperfect. But Mr. Molyneux dying foon after, Dr. Smith loft the opportunity; he however fupplied what was wanting from M. Huygens and others, and publifhed the whole in his " Complete Treatife of Optics."

MOMENT, in Time, is fometimes taken for an extremely fmall part of duration; but, more properly, it is only an inftant or termination or limit in time, like a point in geometry. Maclaurin's Fluxions, vol. 1, pa. 245.

MOMENTS, in the new Doctrine of Infinites, denote the indefinitely fmall parts of quantity; or they are the fame with what are otherwife called infinitefimals, and differences, or increments and decrements; being the momentary increments or decrements of quantity confidered as in a continual flux.

Moments are the generative principles of magnitude: they have no determined magnitude of their own; but are only inceptive of magnitude.

Hence, as it is the fame thing, if, inftead of thefe Moments, the velocities of their increafes and decreafes be made ufe of, or the finite quantities that are proportional to fuch velocities; the method of proceeding which confiders the motions, changes, or fluxions of quantities, is denominated, by Sir Ifaac Newton, the Method of Fluxions.

Leibnitz, and moft foreigners, confidering thefe infinitely fmall parts, or infinitefimals, as the differences of two quantities; and thence endeavouring to find the differences of quantities, i. e. fome Moments, or quantities indefinitely fmall, which taken an infinite number of times fhall equal given quantities; call thefe Moments

ments, Differences; and the method of procedure, the Differential Calculus.

MOMENT, or *Momentum*, in Mechanics, is the same thing with Impetus, or the quantity of motion in a moving body.

In comparing the motions of bodies, the ratio of their Momenta is always compounded of the quantity of matter and the celerity of the moving body: so that the momentum of any such body, may be considered as the rectangle or product of the quantity of matter and the velocity of the motion. As, if b denote any body, or the quantity or mass of matter, and v the velocity of its motion; then bv will express, or be proportional to, its Momentum m. Also if B be another body, and V its velocity; then its Momentum M, is as BV. So that, in general, $M : m :: BV : bv$, i. e. the Momenta are as the products of the mass and velocity Hence, if the Momenta M and m be equal, then shall the two products BV and bv be equal also; and consequently $B : b :: v : V$, or the bodies will be to each other in the inverse or reciprocal ratio of their velocities; that is, either body is so much the greater as its velocity is less. And this force of Momentum is of a different kind from, and incomparably greater than, any mere dead weight, or pressure, whatever.

The Momentum also of any moving body, may be considered as the aggregate or sum of all the Momenta of the parts of that body; and therefore when the magnitudes and number of particles are the same, and also moved with the same celerity, then will the Momenta of the wholes be the same also.

MONADES. DIGITS.

MONOCEROS, the *Unicorn*, one of the new constellations of the northern hemisphere, or one of those which Hevelius has added to the 48 old asterisms, and formed out of the stellæ informes, or those which were not comprized within the outlines of any of the others. In Hevelius's catalogue, the Unicorn contains 19 stars, but in the Britannic catalogue 31.

MONOCHORD, a musical instrument with only one string, used by the Ancients to try the variety and proportion of sounds. It was formed of a rule, divided and subdivided into several parts, on which there is a moveable string stretched over two bridges at the extremes of it. In the interval between these is a sliding or moveable bridge, by means of which, in applying it to the different divisions of the line, the sounds are found to bear the same proportion to each other, as the division of the line cut by the bridge. This instrument is also called the *harmonical canon*, or the *canonical rule*, because it serves to measure the degrees of gravity or acuteness. Ptolomy examines his harmonical intervals by the Monochord. When the chord was divided into two equal parts, so that the parts were as 1 to 1, they called them *unisons*; but if they were as 2 to 1, they called them *octaves* or *diapasons*; when they were as 3 to 2, they called them *diapentes*, or *fifths*; if they were as 4 to 3, they called them *diatessarons*, or *fourths*; if the parts were as 5 to 4, they called them *diton*, or *major-third*; but if they were as 6 to 5, they were called a *demi-diton*, or *minor-third*; and lastly, if the parts were as 24 to 25, a *demitone*, or *dieze*.

The Monochord, being thus divided, was properly what they called a system, of which there were many kinds, according to the different divisions of the Monochord.

MONOCHORD is also used for any musical instrument consisting of only one chord or string. Such is the Trump-marine.

MONOMIAL, in Algebra, is a simple or single nomial, consisting of only one term; as a or ax, or a^2bx^3, &c.

MONOTRIGLYPH, a term in Architecture, denoting the space of one triglyph between two pilasters, or two columns.

MONSOON, a regular or periodical wind, that blows one way for 6 months together, and the contrary way the other 6 months of the year. These prevail in several parts of the eastern and southern oceans.

MONTH, the 12th part of the year, and is so called from the Moon, by whose motions it was regulated; being properly the time in which the moon runs through the zodiac. The lunar Month is either *illuminative*, *periodical*, or *synodical*.

Illuminative MONTH, is the interval between the first appearance of one new moon and that of the next following. As the moon appears sometimes sooner after one change than after another, the quantity of the Illuminative Month is not always the same. The Turks and Arabs reckon by this Month.

Lunar Periodical MONTH, is the time in which the moon runs through the zodiac, or returns to the same point again; the quantity of which is 27 days 7 hrs 43 m. 8 sec.

Lunar Synodical MONTH, called also a Lunation, is the time between two conjunctions of the moon with the sun, or between two new moons; the quantity of which is 29 days, 12 hours, 44 m. 3 sec. 11 thirds.

The ancient Romans used Lunar Months, and made them alternately of 29 and 30 days: They marked the days of each Month by three terms, viz, Calends, Nones, and Ides.

Solar MONTH, is the time in which the sun runs through one entire sign of the ecliptic, the mean quantity of which is 30 days 10 hours 29 min. 5 sec. being the 12th part of 365 ds. 5 hrs. 49 min. the mean solar year.

Astronomical or *Natural* MONTH, is that measured by some exact interval corresponding to the motion of the sun or moon. Such are the lunar and solar months above-mentioned.

Civil or *Common* MONTH, is an interval of a certain number of whole days, approaching nearly to the quantity of some astronomical month. These may be either lunar or solar. The

Civil Lunar MONTH, consists alternately of 29 and 30 days. Thus will two Civil Months be equal to astronomical ones, abating for the odd minutes; and so the new moon will be kept to the first day of such Civil Months for a long time together. This was the Month in Civil or common use among the Jews, Greeks, and Romans, till the time of Julius Cæsar. The

Civil Solar MONTH, consisted alternately of 30 and 31 days, excepting one Month of the twelve, which consisted only of 29 days, but every 4th year of 30 days. And this form of Civil Months was introduced by Julius Cæsar. Under Augustus, the 6th Month,

till

till then from its place called Sextilis, received the name Auguſtus, now Auguſt, in honour of that prince; and, to make the compliment ſtill the greater, a day was added to it; which made it conſiſt of 31 days, though till then it had only contained 30 days; to compenſate for which, a day was taken from February, making it conſiſt of 28 days, and 29 every 4th year. And ſuch are the Civil or Calendar Months now uſed through Europe.

MOON, *Luna*, ☽, one of the heavenly bodies, being a ſatellite, or ſecondary planet to the earth, conſidered as a primary planet, about which ſhe revolves in an elliptic orbit, or rather the earth and Moon revolve about a common centre of gravity, which is as much nearer to the earth's centre than to the Moon's, as the maſs of the former exceeds that of the latter.

The mean time of a revolution of the Moon about the earth, from one new moon to another, when ſhe overtakes the ſun again, is 29 d. 12 h. 44 m. 3 s. 11 th.; but ſhe moves once round her own orbit in 27 d. 7 h. 43 m. 8 s. moving about 2290 miles every hour; and turns once round her axis exactly in the time that ſhe goes round the earth, which is the reaſon that ſhe ſhews always the ſame ſide towards us; and that her day and night taken together are juſt as long as our lunar month.

The mean diſtance of the Moon from the earth is 60½ radii, or 30¼ diameters, of the earth; which is about 240,000 miles. The mean excentricity of her orbit is $\frac{55}{1000}$, or $\frac{1}{18}$th nearly of her mean diſtance, amounting to about 13,000 miles.

The Moon's diameter is to that of the earth, as 20 to 73, or nearly as 3 to 11, or 1 to 3⅔; and therefore it is equal to 2180 miles: her mean apparent diameter is 31′ 16″½, that of the ſun being 32′ 12″. The ſurface of the Moon is to the ſurface of the earth, as 1 to 13¼, or as 3 to 40; ſo that the earth reflects 13 times as much light upon tne Moon, as ſhe does upon the earth; and the ſolid content to that of the earth. as 3 to 146, or as 1 to 48⅔. The denſity of the Moon's body is to that of the earth, as 5 to 4; and therefore her quantity of matter to that of the earth, as 1 to 39 very nearly: the force of gravity on her ſurface, is to that on the earth, as 100 to 293. The Moon has little or no difference of ſeaſons; becauſe her axis is almoſt perpendicular to the ecliptic.

Phenomena and Phaſes of the MOON. The Moon being a dark, opaque, ſpherical body, only ſhining with the light ſhe receives from the ſun, hence only that half turned towards him, at any inſtant, can be illuminated, the oppoſite half remaining in its native darkneſs: then as the face of the Moon viſible on our earth, is that part of her body turned towards us; whence, according to the various poſitions of the Moon, with reſpect to the earth and ſun, we perceive different degrees of illumination; ſometimes a large and ſometimes a leſs portion of the enlightened ſurface being viſible: And hence the Moon appears ſometimes increaſing, then waning; ſometimes horned, then half round; ſometimes gibbous, then full and round. This may be eaſily illuſtrated by means of an ivory ball, which being before a candle in various poſitions, will preſent a greater or leſs portion of its illuminated hemiſphere to the view of the obſerver, according to its ſituation in moving it round the candle.

The ſame phaſes may be otherwiſe exhibited thus: Let S repreſent the ſun, T the earth, and ABCD &c the Moon's orbit. (Plate xv, fig. 3.) Now, when the Moon is at A, in conjunction with the ſun S, her dark ſide being entirely turned towards the earth, ſhe will be inviſible, as at *a*, and is then called the new Moon. When ſhe comes to her firſt octant at B, or has run through the 8th part of her orbit, a quarter of her enlightened hemiſphere will be turned towards the earth, and ſhe will then appear horned, as at *b*. When ſhe has run through the quarter of her orbit, and arrived at C, ſhe ſhews us the half of her enlightened hemiſphere, as at *c*, when it is ſaid ſhe is one half full. At D ſhe is in her 2d octant, and by ſhewing us more of her enlightened hemiſphere than at C, ſhe appears gibbous, as at *d*. At her oppoſition at E her whole enlightened ſide is turned towards the earth, when ſhe appears round, as at *e*, and ſhe is ſaid to be full; having increaſed all the way round from A to E. On the other ſide ſhe decreaſes again all the way from E to A: thus, in her 3d octant at F, part of her dark ſide being turned towards the earth, ſhe again appears gibbous, as at *f*. At G ſhe appears ſtill farther decreaſed, ſhewing again juſt one half of her illuminated ſide, as at *g*. But when ſhe comes to her 4th octant at H, ſhe preſents only a quarter of her enlightened hemiſphere, and ſhe again appears horned, as at *h*. And at A, having now completed her courſe, ſhe again diſappears, or becomes a new moon again, as at firſt. And the earth preſents all the very ſame phaſes to a ſpectator in the Moon, as ſhe does to us, but only in a contrary order, the one being full when the other changes, &c.

The Motions of the MOON are moſt of them very irregular, and very conſiderably ſo. The only equable motion ſhe has, is her revolution on her own axis, in the ſpace of a month, or time in which ſhe moves round the earth; which is the reaſon that ſhe always turns the ſame face towards us.

This expoſure of the ſame face is not ſo uniformly ſo however, but that ſhe turns ſometimes a little more of the one ſide, and ſometimes of the other, called the Moon's Libration; and alſo ſhews ſometimes a little more towards one pole, and ſometimes towards the other, by a motion like a kind of Wavering, or Vacillation. The former of theſe motions happens from this: the Moon's rotation on her axis is equable or uniform; while her motion in her orbit is unequal, being quickeſt when the Moon is in her perigee, and ſloweſt when in the apogee, like all other planetary motions; which cauſes that ſometimes more of one ſide is turned to the earth, and ſometimes of the other. And the other irregularity ariſes from this: that the axis of the Moon is not perpendicular, but a little inclined to the plane of her orbit: and as this axis maintains its parallelifm, in the Moon's motion round the earth; it muſt neceſſarily change its ſituation, in reſpect of an obſerver on the earth; whence it happens that ſometimes the one, and ſometimes the other pole of the Moon becomes viſible.

The very orbit of the Moon is changeable, and does not always perſevere in the ſame figure: for though her orbit be elliptical, or nearly ſo, having the earth in one focus, the excentricity of the ellipſe is varied, being ſometimes increaſed, and ſometimes diminiſhed; viz,

being

being greateſt when the line of the apſes coincides with that of the ſyzygies, and leaſt when theſe lines are at right angles to each other.

Nor is the apogee of the Moon without an irregularity; being found to move forward, when it coincides with the line of the ſyzygies; and backward, when it cuts that line at right angles. Neither is this progreſs or regreſs uniform; for in the conjunction or oppoſition, it goes briſkly forward; and in the quadratures, it either moves ſlowly forward, ſtands ſtill, or goes backward.

The motion of the nodes is alſo variable; being quicker and ſlower in different poſitions.

The Phyſical Cauſe of the Moon's *Motion*, about the earth, is the ſame as that of all the primary planets about the ſun, and of the ſatellites about their primaries, viz, the mutual attraction between the earth and Moon.

As for the particular irregularities in the Moon's motion, to which the earth and other planets are not ſubject, they ariſe from the ſun which acts on, and diſturbs her in her ordinary courſe through her orbit; and are all mechanically deducible from the ſame great law by which her general motion is directed, viz, the law of gravitation and attraction. The other ſecondary planets, as thoſe of Jupiter, Saturn, &c, are alſo ſubject to the like irregularities with the Moon; as they are expoſed to the ſame perturbating or diſturbing force of the ſun; but their diſtance ſecures them from being ſo greatly affected as the Moon is, and alſo from being ſo well obſerved by us.

For a familar idea of this matter, it muſt firſt be conſidered, that if the ſun acted equally on the earth and Moon, and always in parallel lines, this action would ſerve only to reſtrain them in their annual motions round the ſun, and no way affect their actions on each other, or their motions about their common centre of gravity. But becauſe the Moon is nearer the ſun, in one half of her orbit, than the earth is, but farther off in the other half of her orbit; and becauſe the power of gravity is always leſs at a greater diſtance; it follows, that in one half of her orbit the Moon is more attracted than the earth towards the ſun, and leſs attracted than the earth in the other half: and hence irregularities neceſſarily ariſe in the motions of the Moon; the exceſs of attraction in the firſt caſe, and the defect in the ſecond, becoming a force that diſturbs her motion: and beſides, the action of the ſun, on the earth and Moon, is not directed in parallel lines, but in lines that meet in the centre of the ſun; which makes the effect of the diſturbing force ſtill the more complex and embarraſſing. And hence, as well as from the various ſituations of the Moon, ariſe the numerous irregularities in her motions, and the equations, or corrections, employed in calculating her places, &c.

Newton, as well as others, has computed the quantities of theſe irregularities, from their cauſes. He finds that the force added to the gravity of the Moon in her quadratures, is to the gravity with which ſhe would revolve in a circle about the earth, at her preſent mean diſtance, if the ſun had no effect on her, as 1 to $178\frac{29}{40}$: he finds that the force ſubducted from her gravity in the conjunctions and oppoſitions, is

double of this quantity; and that the area deſcribed in a given time in the quarters, is to the area deſcribed in the ſame time in the conjunctions and oppoſitions, as 10973 to 11073: and he finds that, in ſuch an orbit, her diſtance from the earth in her quarters, would be to her diſtance in the conjunctions and oppoſitions, as 70 to 69. Upon theſe irregularities, ſee Maclaurin's Account of Newton's Diſcoveries, book 4, chap. 4; as alſo moſt books of aſtronomy. Other particulars relating to the Moon's motions, &c, have been ſtated as follow: The power of the Moon's influence, as to the tides, is to that of the ſun, as $6\frac{1}{3}$ to 1, according to Sir I. Newton; but different according to others.

As to the figure of the Moon, ſuppoſing her at firſt to have been a fluid, like the ſea, Newton calculates, that the earth's attraction would raiſe the water there near 90 feet high, as the attraction of the Moon raiſes our ſea 12 feet: whence the figure of the Moon muſt be a ſpheroid, whoſe greateſt diameter extended, will paſs through the centre of the earth; and will be longer than the other diameter, perpendicular to it, by 180 feet; and hence it comes to paſs, that we always ſee the ſame face of the Moon; for ſhe cannot reſt in any other poſition, but always endeavours to conform herſelf to this ſituation: Princip. lib. 3, prop. 38.

Newton eſtimates the mean apparent diameter of the Moon at 32′ 12′; as the ſun is 31′ 27′′.

The denſity of the Moon he concludes is to that of the earth, as 9 to 5 nearly; and that the maſs, or quantity of matter, in the Moon, is to that of the earth, as 1 to 26 nearly.

The plane of the Moon's orbit is inclined to that of the ecliptic, and makes with it an angle of about 5 degrees: but this inclination varies, being greateſt when ſhe is in the quarters, and leaſt when in her ſyzygies.

As to the inequality of the Moon's motion, ſhe moves ſwifter, and by the radius drawn from her to the earth deſcribes a greater area in proportion to the time, alſo has an orbit leſs curved, and by that means comes nearer to the earth, in her ſyzygies or conjunctions, than in the quadratures, unleſs the motion of her eccentricity hinders it: which eccentricity is the greateſt when the Moon's apogee falls in the conjunction, but leaſt when this falls in the quadratures: her motion is alſo ſwifter in the earth's aphelion, than in its perihelion. The apogee alſo goes forward ſwifter in the conjunction, and goes ſlower at the quadratures: but her nodes are at reſt in the conjunctions, and recede ſwifteſt of all in the quadratures.

The Moon alſo perpetually changes the figure of her orbit, or the ſpecies of the ellipſe ſhe moves in.

There are alſo ſome other inequalities in the motion of this planet, which it is very difficult to reduce to any certain rule: as the velocities or horary motions of the apogee and nodes, and their equations, with the difference between the greateſt eccentricity in the conjunctions, and the leaſt in the quadratures; and that inequality which is called the Variation of the Moon. All theſe do increaſe and decreaſe annually, in a triplicate ratio of the apparent diameter of the ſun: and this variation is increaſed and diminiſhed in a duplicate ratio of the time between the quadratures; as is proved by Newton in many parts of his Principia.

He

He also found that the apogees in the Moon's syzygies, go forward in respect of the fixed stars, at the rate of 23′ each day; and backwards in the quadratures 16′¾ per day: and therefore the mean annual motions he estimates at 40 degrees.

The gravity of the Moon towards the earth, is increased by the action of the sun, when the Moon is in the quadratures, and diminished in the syzygies: and, from the syzygies to the quadrature, the gravity of the Moon towards the earth is continually increased, and she is continually retarded in her motion: but from the quadrature to the syzygy, the Moon's motion is perpetually diminished, and the motion in her orbit is accelerated.

The Moon is less distant from the earth at the syzygies, and more at the quadratures.

As radius is to ⅔ of the sine of double the Moon's distance from the syzygy, so is the addition of gravity in the quadratures, to the force which accelerates or retards the Moon in her orbit.

And as radius is to the sum or difference of ½ the radius and ⅔ the cosine of double the distance of the Moon from the syzygy, so is the addition of gravity in the quadratures, to the decrease or increase of the gravity of the Moon at that distance.

The apses of the Moon go forward when she is in the syzygies, and backward in the quadratures. But, in a whole revolution of the Moon, the progress exceeds the regress.

In a whole revolution, the apses go forward the fastest of all when the line of the apses is in the nodes; and in the same case they go back the slowest of all in the same revolution.

When the line of the apses is in the quadratures, the apses are carried in consequentia, the least of all in the syzygies; but they return the swiftest in the quadratures; and in this case the regress exceeds the progress, in one entire revolution of the Moon.

The eccentricity of the orbit undergoes various changes every revolution. It is the greatest of all when the line of the apses is in the syzygies, and the least when that line is in the quadratures.

Considering one entire revolution of the Moon, cæteris paribus, the nodes move in antecedentia swiftest of all when she is in the syzygies; then slower and slower, till they are at rest, when she is in the quadratures.

The line of nodes acquires successively all possible situations in respect of the sun; and every year it goes twice through the syzygies, and twice through the quadratures.

In one whole revolution of the Moon, the nodes go back very fast when they are in the quadratures; then slower till they come to rest, when the line of nodes is in the syzygies.

The inclination of the plane of the orbit is changed by the same force with which the nodes are moved; being increased as the Moon recedes from the node, and diminished as she approaches it.

The inclination of the orbit is the least of all when the nodes are come to the syzygies. For in the motion of the nodes from the syzygies to the quadratures, and in one entire revolution of the Moon, the force which increases the inclination exceeds that which di-

minishes it; therefore the inclination is increased; and it is the greatest of all when the nodes are in the quadratures.

The Moon's motion being considered in general: her gravity towards the earth is diminished coming near the sun, and the periodical time is the greatest; as also the distance of the Moon, cæteris paribus, the greatest when the earth is in the perihelion.

All the errors in the Moon's motion are something greater in the conjunction than in the opposition.

All the disturbing forces are inversely as the cube of the distance of the sun from the earth; which when it remains the same, they are as the distance of the Moon from the earth. Considering all the disturbing forces together, the diminution of gravity prevails.

The figure of the Moon's *path,* about the earth, is, as has been said, nearly an ellipse; but her path, in moving, together with the earth about the sun, is made up of a series or repetition of epicycloids, and is in every point concave towards the earth. See Maclaurin's Account of Newton's Discov. pa. 336, 4to. Ferguson's Astron. pa. 129, &c; and Rowe's Flux. pa. 225, edit. 2.

Astronomy of the Moon.

To determine the Periodical and Synodical Months; or the period of the Moon's revolution about the earth, and the period between one opposition or conjunction and another.

In the middle of a lunar eclipse, the Moon is in opposition to the sun: compute therefore the time between two such eclipses, at some considerable distance of time from each other; and divide this by the number of lunations that have passed in the mean time; so shall the quotient be the quantity of the synodical month. Compute also the sun's mean motion during the time of this synodical month, which add to 360°. Then, as the sum is to 360°, so is the synodical to the periodical month.

For example, Copernicus observed two eclipses of the Moon, the one at Rome on November 6, 1500, at 12 at night, and the other at Cracow on August 1, 1523, at 4 h. 25 min. the dif. of meridians being 0 h. 29 min.: hence the quantity of the synodical month is thus determined:

2d Observ.	1523ʸ	237ᵈ	4ʰ	25ᵐ
1st Observ.	1500	310	0	29
Difference	22	292	3	56
Add intercalary days		5		
Exact interval	22	297	3	56

which divided by 282, the number of lunations in that time, gives the synodical month 29ᵈ 12ʰ 41ᵐ.

From two other observations of eclipses, the one at Cracow, the other at Babylon, the same author determines more accurately the quantity of the synodical month to be 29ᵈ 12ʰ 43ᵐ &c; and from other observations, probably more accurate still, the same is fixed at 29ᵈ 12ʰ 44ᵐ.

The sun's mean motion in that time 29° 6′ 24″ 18‴, added to 360°, gives the Moon's motion 389 6 24 18;

Therefore the periodical month is 27ᵈ 7ʰ 43ᵐ 5ˢ.

R
According

According to the obfervations of Kepler, the mean fynodical month is $29^d \ 12^h \ 44^m \ 3^s \ 2^{th}$, and the mean periodical month $27 \quad 7 \quad 43 \quad 8$

Hence, 1, the quantity of the periodical month being given, by the rule of three are found the Moon's diurnal or horary motion, &c : and thus may tables of the mean motion of the Moon be conftructed.

2. If the mean diurnal motion of the fun be fubtracted from that of the Moon, the remainder will give the Moon's diurnal motion from the fun : and thus may a table of this motion be conftructed.

3. Since the Moon is in the node at the time of a total eclipfe, if the fun's place be found for that time, and 6 figns be added to the fame, the fum will give the place of that node.

4. By comparing the ancient obfervations with the modern, it appears, that the nodes have a motion, and that they proceed in antecedentia, or backwards from Taurus to Aries, from Aries to Pifces, &c. Therefore if the diurnal motion of the nodes be added to the Moon's diurnal motion, the fum will be the motion of the Moon from the node ; and thence by the rule of three, may be found in what time the Moon goes 360° from the dragon's head, or afcending node, or in what time fhe goes from, and returns to it ; that is, the quantity of the Dracontic Month.

5. If the motion of the apogee be fubtracted from the mean motion of the Moon, the remainder will be the Moon's mean motion from the apogee ; and hence, by the rule of three, the quantity of the Anomaliftic Month is determined.

Thus, according to Kepler's obfervations,

The mean fynodical month is	29^d	$12^h \ 44^m$	3^s	2th
The periodical month	27	7	43	8
The place of the apogee for the year 1700 Jan. 1 old ftyle, was	11^s	8°	57′	1″
The place of the afcending node	4	27	39	17
Mean diurnal motion of the Moon	.	13	10	35
Diurnal motion of the apogee	.	.	6	41
Diurnal motion of the nodes	.	.	3	11
Theref. diurnal mot from the latter	.	13	13	46
And the diurnal motion from the apogee	.	13	3	54

Laftly, the eccentricity is 4362, of fuch parts as the femidiameter of the eccentric is 100,000.

To find nearly the MOON's *Age or Change.*

To the epact add the number and day of the month ; their fum, abating 30 if it be above, is the Moon's age; and her age taken from 30, fhews the day of the change.

The numbers of the months, or monthly epacts, are the Moon's age at the beginning of each month, when the folar and lunar years begin together ; and are thus :

0	2	2	3	4	5	6	8	8	10	10
Jan.	Feb.	Mar.	Ap.	Ma.	Jun.	Jul.	Aug.	Sep.	Oct.	Nov.	Dec.

For Ex. To find the Moon's age the 14th of Oct. 1783.

Here, the epact is 26
Number of the month 8
Day of the month 14
The fum is 48

Subtract or abate 30
Leaves Moon's age 18
Taken from 30
Days till the change 12
Anfwering to Oct. 26

To find nearly the MOON's *Southing,* or coming to the Meridian.

Take $\frac{4}{5}$ or $\frac{8}{10}$ of her age, for her fouthing nearly ; after noon, if it be lefs than 12 hours ; but if greater, the excefs is the time after laft midnight.

For Ex. Oct. 14, 1783 ;
The Moon's age is 18 days
$\frac{8}{10}$ of which is 14·4 or $14^h \ 24^m$
Subtract 12 00
Rem. Moon's fouthing 2 24 in the morning.

Mr. Ferguson, in his Select Exercifes, pa. 135 &c, has given very eafy tables and rules for finding the new and full Moons near enough the truth for any common almanac. But the Nautical Almanac, which is now always publifhed for feveral years before hand, in a great meafure fuperfedes the neceffity of thefe and other fuch contrivances.

Of the Spots and Mountains &c in the MOON.

The face of the Moon is greatly diverfified with inequalities, and parts of different colours, fome brighter and fome darker than the other parts of her difc. When viewed through a telefcope, her face is evidently diverfified with hills and valleys : and the fame is alfo fhewn by the edge or border of the Moon appearing jagged, when fo viewed, efpecially about the confines of the illuminated part when the Moon is either horned or gibbous.

The aftronomers Florenti, Langreni, Hevelius, Grimaldi, Riccioli, Caffini, and De la Hire, &c, have drawn the face of the Moon as viewed through telefcopes ; noting all the more fhining parts, and, for the better diftinction, marking them with fome proper name ; fome of thefe authors calling them after the names of philofophers, aftronomers, and other eminent men ; while others denominate them from the known names of the different countries, iflands, and feas on the earth. The names adopted by Riccioli however are moftly followed, as the names of Hipparchus, Tycho, Copernicus, &c. Fig. 4, plate xv, is a pretty exact reprefentation of the full Moon in her mean libration, with the numbers to the principal fpots according to Riccioli, Caffini, Mayer, &c, which denote the names as in the following Lift of them : alfo the afterifk refers to one of the volcanoes obferved by Herfchel.

✳	Herfchel's Volcano	12	Helicon
1	Grimaldi	13	Capuanus
2	Galileo	14	Bulliald
3	Ariftarchus	15	Eratofthenes
4	Kepler	16	Timocharis
5	Gaffendi	17	Plato
6	Schikard	18	Archimedes
7	Harpalus	19	Infula Sinus Medii
8	Heraclides	20	Pitatus
9	Lanfberg	21	Tycho
10	Reinhold	22	Eudoxus
11	Copernicus	23	Ariftotle

24 Ma

That the fpots in the Moon, which are taken for mountains and valleys, are really fuch, is evident from their fhadows. For in all fituations of the Moon, the elevated parts are conftantly found to caft a triangular fhadow in a direction from the fun; and, on the contrary, the cavities are always dark on the fide next the fun, and illuminated on the oppofite one; which is exactly conformable to what we obferve of hills and valleys on the earth. And as the tops of thefe mountains are confiderably elevated above the other parts of the furface; they are often illuminated when they are at a confiderable diftance from the confines of the enlightened hemifphere, and by this means afford us a method of determining their heights.

Thus, let ED be the Moon's diameter, ECD the boundary of light and darknefs; and A the top of a hill in the dark part beginning to be illuminated; with a telefcope take the proportion of AE to the diameter ED: then there are given the two fides AE, EC of a right-angled triangle ACE, the fquares of which being add-

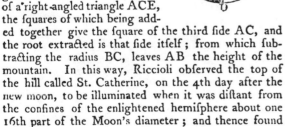

ed together give the fquare of the third fide AC, and the root extracted is that fide itfelf; from which fubtracting the radius BC, leaves AB the height of the mountain. In this way, Riccioli obferved the top of the hill called St. Catherine, on the 4th day after the new moon, to be illuminated when it was diftant from the confines of the enlightened hemifphere about one 16th part of the Moon's diameter; and thence found its height muft be near 9 miles.

It is probable however that this determination is too much. Indeed, Galileo makes AE to be only one 20th of ED, and Hevelius makes it only one 26th of ED; the former of thefe would give 5¼ miles, and the latter only 3¼ miles, for AB, the height of the mountain: and probably it fhould be ftill lefs than either of thefe.

Accordingly, they are greatly reduced by the obfervations of Herfchel, whofe method of meafuring them may be feen in the Philof. Tranf. an. 1780, pa. 507. This gentleman meafured the height of many of the lunar prominences, and draws at laft the following conclufions:—" From thefe obfervations I believe it is evident, that the height of the lunar mountains in general is greatly over-rated; and that, when we have excepted a few, the generality do not exceed half a

mile in their perpendicular elevation." And this is confirmed by the meafurement of feveral mountains, as may be feen in the place above quoted.

As the Moon has on her furface mountains and valleys in common with the earth, fome modern aftronomers have difcovered a ftill greater fimilarity, viz, that fome of thefe are really volcanoes, emitting fire as thofe on the earth do. An appearance of this kind was difcovered fome few years ago by Don Ulloa in an eclipfe of the fun. It was a fmall bright fpot like a ftar near the margin of the Moon, and which he at that time fuppofed to be a hole or valley with the fun's light fhining through it. Succeeding obfervations, however, have induced aftronomers to attribute appearances of this kind to the eruption of volcanic fire; and Mr. Herfchel has particularly obferved feveral eruptions of the lunar volcanos, the laft of which he gives an account of in the Philof. Tranf. for 1787. April 19, 10h. 36m. fidereal time, I perceived, fays he, three volcanos in different places of the dark part of the new Moon. Two of them are either already nearly extinct, or otherwife in a ftate of going to break out; which perhaps may be decided next lunation. The third fhews an actual eruption of fire or luminous matter: its light is much brighter than the nucleus of the comet which M. Mechain difcovered at Paris the 10th of this month." The following night he found it burnt with greater violence; and by meafurement he found that the fhining or burning matter muft be more than 3 miles in diameter; being of an irregular round figure, and very fharply defined on the edges. The other two volcanos refembled large faint nebulæ, that are gradually much brighter in the middle; but no well-defined luminous fpot was difcovered in them. He adds, " the appearance of what I have called the *actual fire*, or eruption of a volcano, exactly refembled a fmall piece of burning charcoal when it is covered by a very thin coat of white afhes, which frequently adhere to it when it has been fome time ignited; and it had a degree of brightnefs about as ftrong as that with which a coal would be feen to glow in faint day-light.

It has been difputed whether the Moon has any atmofphere or not. The following arguments have been urged by thofe who deny it.

1. The Moon, fay they, conftantly appears with the fame brightnefs when our atmofphere is clear; which could not be the cafe if fhe were furrounded with an atmofphere like ours, fo variable in its denfity, and fo often obfcured by clouds and vapours. 2. In an appulfe of the Moon to a ftar, when fhe comes fo near it that a part of her atmofphere comes between our eye and the ftar, refraction would caufe the latter to feem to change its place, fo that the Moon would appear to touch it later than by her own motion fhe would do. 3. Some philofophers are of opinion, that becaufe there are no feas or lakes in the Moon, there is therefore no atmofphere, as there is no water to be raifed up in vapours.

But all thefe arguments have been anfwered by other aftronomers in the following manner. It is denied that the Moon appears always with the fame brightnefs, even when our atmofphere appears equally clear. Hevelius relates, that he has feveral times found in

fkies

skies perfectly clear, when even stars of the 6th and 7th magnitude were visible, that at the same altitude of the Moon with the same elongation from the sun, and with the same telescope, the Moon and her maculæ do not appear equally lucid, clear, and conspicuous at all times; but are much brighter and more distinct at some times than at others. And hence it is inferred that the cause of this phenomenon is neither in our air, in the tube, in the Moon, nor in the spectator's eye; but must be looked for in something existing about the Moon. An additional argument is drawn from the different appearances of the Moon in total eclipses, which it is supposed are owing to the different constitutions of the lunar atmosphere.

To the 2d argument Dr. Long replies, that Newton has shewn (Princip. prop. 37, cor. 5), that the weight of any body upon the Moon is but a third part of what the weight of the same would be upon the earth: now the expansion of the air is reciprocally as the weight that compresses it; therefore the air surrounding the Moon, being pressed together by a weight of one-third, or being attracted towards the centre of the Moon by a force equal only to one-third of that which attracts our air towards the centre of the earth, it thence follows, that the lunar atmosphere is only one-third as dense as that of the earth, which is too little to produce any sensible refraction of the star's light. Other astronomers have contended, that such refraction was sometimes very apparent. Mr. Cassini says, that he often observed that Saturn, Jupiter, and the fixed stars, had their circular figures changed into an elliptical one, when they approached either to the Moon's dark or illuminated limb, though they own that, in other occultations, no such change could be observed. And, with regard to the fixed stars, it has been urged that, granting the Moon to have an atmosphere of the same nature and quantity as ours, no such effect as a gradual diminution of light ought to take place; at least none that we could be capable of perceiving. At the height of 44 miles, our atmosphere is so rare as to be incapable of refracting the rays of light: this height is the 180th part of the earth's diameter; but since clouds are never observed higher than 4 miles, it appears that the vapourous or obscure part is only the 1980th part. The mean apparent diameter of the Moon is 31' 29'', or 1889''; therefore the obscure parts of her atmosphere, when viewed from the earth, must subtend an angle of less than one second; which space is passed over by the Moon in less than two seconds of time. It can therefore hardly be expected that observation should generally determine whether the supposed obscuration takes place or not.

As to the 3d argument, it concludes nothing, because it is not known that there is no water in the Moon; nor, though this could be proved, would it follow that the lunar atmosphere answers no other purpose than the raising of water into vapour. There is however a strong argument in favour of the existence of a lunar atmosphere, taken from the appearance of a luminous circle round the Moon in the time of total solar eclipses; a circumstance that has been observed by many astronomers; especially in the total eclipse of the sun which happened May 1, 1706.

Of the Harvest Moon. It is remarkable that the Moon, during the week in which she is full about the time of harvest, rises sooner after sun-setting, than she does in any other full-moon week in the year. By this means she affords an immediate supply of light after sun-set, which is very beneficial for the harvest and gathering in the fruits of the earth: and hence this full Moon is distinguished from all the others in the year, by calling it the Harvest-Moon.

To conceive the reason of this phenomenon; it may first be considered, that the Moon is always opposite to the sun when she is full; that she is full in the signs Pisces and Aries in our harvest months, those being the signs opposite to Virgo and Libra, the signs occupied by the sun about the same season; and because those parts of the ecliptic rise in a shorter space of time than others, as may easily be shewn and illustrated by the celestial globe: consequently, when the Moon is about her full in harvest, she rises with less difference of time, or more immediately after sun-set, than when she is full at other seasons of the year.

In our winter, the Moon is in Pisces and Aries about the time of her first quarter, when she rises about noon; but her rising is not then noticed, because the sun is above the horizon.

In spring, the Moon is in Pisces and Aries about the time of her change; at which time, as she gives no light, and rises with the sun, her rising cannot be perceived.

In summer, the Moon is in Pisces and Aries about the time of her last quarter; and then, as she is on the decrease, and rises not till midnight, her rising usually passes unobserved.

But in autumn, the Moon is in Pisces and Aries at the time of her full, and rises soon after sun-set for several evenings successively; which makes her regular rising very conspicuous at that time of the year.

And this would always be the case, if the Moon's orbit lay in the plane of the ecliptic. But as her orbit makes an angle of 5° 18' with the ecliptic, and crosses it only in the two opposite points called the nodes, her rising when in Pisces and Aries will sometimes not differ above 1h. and 40 min. through the whole of 7 days; and at other times, in the same two signs she will differ 3 hours and a half in the time of her rising in a week, according to the different positions of the nodes with respect to these signs; which positions are constantly changing, because the nodes go backward through the whole ecliptic in 18 years 225 days.

This revolution of the nodes will cause the Harvest Moons to go through a whole course of the most and least beneficial states, with respect to the harvest, every 19 years. The following Table shews in what years the Harvest Moons are least beneficial as to the times of their rising, and in what years they are most beneficial, from the year 1790 to 1861; the column of years under the letter L, are those in which the Harvest-Moons are least of all beneficial, because they fall about the descending node; and those under the letter M are the most of all beneficial, because they fall about the ascending node.

Harvest

Harvest Moons.

L	M	L	M	L	M	L	M
1790	1798	1807	1816	1826	1835	1844	1843
1791	1799	1808	1817	1827	1836	1845	1854
1792	1800	1809	1818	1828	1837	1846	1855
1793	1801	1810	1819	1829	1838	1847	1856
1794	1802	1811	1820	1830	1839	1848	1857
1795	1803	1812	1821	1831	1840	1849	1858
1796	1804	1813	1822	1832	1841	1850	1859
1797	1805	1814	1823	1833	1842	1851	1860
	1806	1815	1824	1834	1843	1852	1861
			1825				

As to the Influence of the MOON, on the changes of the weather, and the constitution of the human body, it may be observed, that the vulgar doctrine concerning it is very ancient, and has also gained much credit among the Learned, though perhaps without sufficient examination. The common opinion is, that the Lunar Influence is chiefly exerted about the time of the full and change, but more especially the latter; and it would seem that long experience has in some degree established the fact: hence, persons observed at those times to be a little deranged in their intellects, are called Lunatics; and hence many persons anxiously look for the new Moon to bring a change in the weather. The Moon's Influence on the sea, in producing tides, being agreed upon on all hands, it is argued that she must also produce similar changes in the atmosphere, but in a much higher degree; which changes and commotions there, must, it is inferred, have a considerable influence on the weather, and on the human body.

Beside the observations of the Ancients, which tend to establish this doctrine, several among the Modern Philosophers have defended the same opinion, and that upon the strength of experience and observation; while others as strenuously deny the fact. The celebrated Dr. Mead was a believer in the Influence of the Sun and Moon on the human body, and published a book to this purpose, intitled, De Imperio Solis ac Lunæ in Corpore Humano. The existence of such influence is however opposed by Dr. Horsley, the present bishop of Rochester, in a learned paper upon this subject in the Philos. Transf. for the year 1775; where he gives a specimen of arranging tables of meteorological observations, so as to deduce from them facts, that may either confirm or refute this popular opinion; recommending it to the Learned, to collect a large series of such observations, as no conclusions can be drawn from one or two only. On the other hand professor Toaldo, and some French philosophers, take the opposite side of the question; and, from the authority of a long series of observations, pronounce decidedly in favour of the Lunar Influence.

Acceleration of the MOON. See ACCELERATION.

MOON-*Dial.* See DIAL.

Horizontal MOON. See *Apparent* MAGNITUDE.

MOORE (Sir JONAS), a very respectable mathematician, Fellow of the Royal Society, and Surveyor-general of the Ordnance, was born at Whitby in Yorkshire about the year 1620. After enjoying the advantages of a liberal education, he bent his studies principally to the mathematics, to which he had al-

ways a strong inclination. In the expeditions of King Charles the 1st into the northern parts of England, our author was introduced to him, as a person studious and learned in those sciences; when the king expressed much approbation of him, and promised him encouragement; which indeed laid the foundation of his fortune. He was afterwards appointed mathematical master to the king's second son James, to instruct him in arithmetic, geography, the use of the globes, &c. During Cromwell's government it seems he followed the profession of a public teacher of mathematics; for I find him styled, in the title-page of some of his publications, "professor of the mathematics." After the return of Charles the 2d, he found great favour and promotion, becoming at length surveyor-general of the king's ordnance. He was it seems a great favourite both with the king and the duke of York, who often consulted him, and were advised by him upon many occasions. And it must be owned that he often employed his interest with the court to the advancement of learning and the encouragement of merit. Thus, he got Flamsteed house built in 1675, as a public observatory, recommending Mr. Flamsteed to be the king's astronomer, to make the observations there: and being surveyor general of the ordnance himself, this was the reason why the salary of the astronomer royal was made payable out of the office of ordnance. Being a governor of Christ's hospital, it seems that by his interest the king founded the mathematical school there, allowing a handsome salary for a master to instruct a certain number of the boys in mathematics and navigation, to qualify them for the sea service. Here he soon found an opportunity of exerting his abilities in a manner somewhat answerable to his wishes, namely, that of serving the rising generation. And considering with himself the benefit the nation might receive from a mathematical school, if rightly conducted, he made it his utmost care to promote the improvement of it. The school was settled; but there still wanted a methodical institution from which the youths might receive such necessary helps as their studies required: a laborious work, from which his other great and assiduous employments might very well have exempted him, had not a predominant regard to a more general usefulness engaged him to devote all the leisure hours of his declining years to the improvement of so useful and important a seminary of learning.

Having thus engaged himself in the prosecution of this general design, he next sketched out the plan of a course or system of mathematics for the use of the school, and then drew up and printed several parts of it himself, when death put an end to his labours, before the work was completed. I have not found in what year this happened; but it must have been but little before 1681, the year in which the work was published by his sons-in-law, Mr. Hanway and Mr. Potinger. Of this work, the Arithmetic, Practical Geometry, Trigonometry, and Cosmography, were written by Sir Jonas himself, and printed before his death. The Algebra, Navigation, and the books of Euclid were supplied by Mr. Perkins, the then master of the mathematical school. And the Astronomy, or Doctrine of the Sphere, was written by Mr. Flamsteed, the astronomer royal.

The

The lift of Sir Jonas's works, as far as I have feen them, are the following:

1. The New Syftem of Mathematics; above mentioned, in 2 vols 4to, 1681.

2. Arithmetic in two books, viz, Vulgar Arithmetic and Algebra. To which are added two Treatifes, the one A new Contemplation Geometrical, upon the Oval Figure called the Ellipfis; the other, The two firft books of Mydorgius, his Conical Sections analized &c. 8vo, 1660.

3. A Mathematical Compendium; or Ufeful Practices in Arithmetic, Geometry, and Aftronomy, Geography and Navigation, &c, &c. 12mo, 4th edition in 1705.

4. A General Treatife of Artillery: or, Great Ordnance. Written in Italian by Tomafo Moretii of Brefcia. Tranflated into Englifh, with notes thereupon, and fome additions out of French for Sea-Gunners. By Sir Jonas Moore, Kt. 8vo, 1683.

MORTALITY. *Bills of Mortality,* are accounts or regifters fpecifying the numbers born, and buried, and fometimes married, in any town, parifh, or diftrict. Thefe are of great ufe, not only in the doctrine of Life Annuities, but in fhewing the degrees of healthinefs and prolificnefs, with the progrefs of population in the places where they are kept. It is therefore much to be wifhed that fuch accounts had always been correctly kept in every kingdom, and regularly publifhed at the end of every year. We fhould then have had under infpection the comparative ftrength of every kingdom, as far as it depends on the number of inhabitants, and its increafe or decreafe at different periods.

Such accounts are rendered ftill more ufeful, when they include the ages of the dead, and the diftempers of which they have died. In this cafe they convey fome of the moft important inftructions, by furnifhing the means of afcertaining the law which governs the wafte of human life, the values of annuities dependent on the continuance of any lives, or any furvivorfhips between them, and the favourablenefs or unfavourablenefs of different fituations to the duration of human life.

There are but few regifters of this kind; nor has this fubject, though fo interefting to mankind, ever engaged much attention till lately. Indeed, bills of Mortality for the feveral parifhes of the city of London have been kept from the year 1592, with little interruption; and a very ample account of them has been publifhed down to the year 1759, by Dr. Birch, in a large 4to vol. which is perhaps the fulleft work of the kind extant; containing befides the bills of Mortality, with the difeafes and cafualties, feveral other valuable tracts on the fubject of them, and on political arithmetic, by feveral other authors, as Capt. John Graunt, F R. S.; Sir William Petty, F. R. S.; Corbyn Morris, Efq. F R. S.; and J. P. Efq. F. R. S.; the whole forming a valuable repofitory of materials; and it would be well if a continuation were publifhed down to the prefent time, and fo continued from time to time.

Bills containing the ages of the dead, were long fince publifhed for the town of Breflaw in Silefia. It is well known what ufe has been made of thefe by Dr. Halley, and after him by Mr. De Moivre. A table of the probabilities of the duration of human life at every age, deduced from them by Dr. Halley, was publifhed in the Philof. Tranf. vol. 17, and has been inferted in this work under the article LIFE-*Annuities;* which is the firft table of this kind that has been publifhed. Since the publication of this table, fimilar bills have been eftablifhed in many other places, in England, Germany, Switzerland, France, Holland, &c, but moft efpecially in Sweden; the refults of fome of which may be feen in the large comparative table of the duration of life, under the article LIFE-*Annuities,* in this work.

MORTAR, or MORTAR-PIECE, a fhort piece of ordnance, thick and wide, proper for throwing bombfhells, carcafes, ftones, grape-fhot, &c.

It is thought that the ufe of Mortars is older than that of cannon: for they were employed in the wars of Italy, to throw bails of red hot iron, and ftones, long before the invention of fhells: and it is generally believed that the Germans were the firft inventors. The practice of throwing red-hot balls out of Mortars, was firft practifed at the fiege of Stralfund in 1675, by the elector of Brandenburg; though fome fay, in 1653, at the fiege of Bremen.

Mortars are made either of brafs or iron, and it is ufual to diftinguifh them by the diameter of the bore; as, the 13 inch, the 10 inch, or the 8 inch Mortar: there are fome of a fmaller fort, as Coehorns of 4.6 inches, and Royals of 5.8 inches in diameter. As to the larger fizes, as 18 inches, &c, they are now difufed by the Englifh, as well as moft other European nations. For the circumftances relating to Mortars, fee Muller's Artillery.

Coehorn MORTAR, a fmall kind of one, invented by the celebrated engineer baron Coehorn, to throw fmall fhells or grenades. Thefe Mortars are often fixed, to the number of a dozen, on a block of oak, at the elevation of 45°.

MOTION, or *Local* MOTION, is a continued and fucceffive change of place. Borelli defines it, the fucceffive paffage of a body from one place to another, in a determinate time, by becoming fucceffively contiguous to all the parts of the intermediate fpace.

Motion is confidered as of various kinds; as Natural, Violent, Abfolute and Relative, &c, &c.

Natural MOTION, is that which has its principle, or actuating force, within the moving body. Such is that of a ftone falling towards the earth. And

Violent MOTION, is that whofe principle is without, and againft which the moving body makes a refiftance. Such is that of a ftone thrown upwards, or of a ball fhot off from a gun, &c.

Motion is again divided into Abfolute and Relative.

Abfolute MOTION, is the change of abfolute place, in any moving body, confidered independently of any other motion; whofe celerity therefore will be meafured by the quantity of abfolute fpace which the moveable body runs through. And

Relative MOTION, is the change of the relative place of a moving body, or confidered with refpect to the motion of fome other body; and has its celerity eftimated by the quantity of relative fpace run through.

As to the Continuation of MOTION, or the caufe why a body once in Motion comes to perfevere in it: this has been

been much controverted among phyfical writers; and yet it follows very evidently from one of the grand Laws of Nature; viz, that all bodies perfevere in their prefent ftate, whether of reft or motion, unlefs difturbed by fome foreign powers. Motion therefore, once begun, would be continued in infinitum, were it to meet with no interruption from external caufes; as the power of gravity, the refiftance of the medium, &c.

Nor has the communication of motion, or how a moving body comes to affect another at reft, or how much of its motion is communicated by the firft to the laft, been lefs difputed. See the Laws of it under the word PERCUSSION.

Motion is the proper fubject of mechanics; and mechanics is the bafis of all natural philofophy; which hence becomes denominated Mechanical.

In effect, all the phenomena of nature, all the changes that happen in the fyftem of bodies, are owing to Motion; and are directed according to the laws of it. Hence the modern philofophers have applied themfelves with peculiar ardour to confider the doctrine of Motion; to inveftigate the properties and laws of it; by obfervation and experiment, joined to the ufe of geometry. And to this is owing the great advantage of the modern philofophy above that of the Ancients; who were extremely difregardful of the effects of Motion.

Among all the Ancients, there is nothing extant on Motion, excepting fome things in Archimedes's books, De Æquiponderantibus. To Galileo is owing a great part of the doctrine of Motion: he firft difcovered the general laws of it, and particularly of the defcent of heavy bodies, both perpendicularly and on inclined planes; the laws of the Motion of projectiles; the vibration of pendulums, and of ftretched cords, with the theory of refiftances, &c: things which the Ancients had little notion of.

Torricelli polifhed and improved the difcoveries of his matter, Galileo; and added many experiments concerning the force of percuffion, and the equilibrium of fluids. Huygens improved very confiderably on the doctrine of the pendulum; and both he and Borelli on the force of percuffion. Laftly, Newton, Leibnitz, Varignon, Mariotte, &c, have brought the doctrine of Motion ftill much nearer to perfection.

The general laws of Motion were firft brought into a fyftem, and analytically demonftrated together, by Dr. Wallis, Sir Chriftopher Wren, and M. Huygens, all much about the fame time; the firft in bodies not elaftic, and the two latter in elaftic bodies. Laftly, the whole doctrine of Motion, including all the difcoveries both of the Ancients and Moderns on that head, was given by Dr. Wallis in his Mechanica, five De Motu, publifhed in 1670.

Quantity of MOTION, is the fame as MOMENTUM, which fee. It is a principle maintained by the Cartefians, and fome others, that the Creator at the beginning impreffed a certain Quantity of Motion on bodies; and that under fuch laws, as that no part of it fhould be loft, but the fame portion of Motion fhould be conftantly preferved in matter: and hence they conclude, that if any moving body ftrike another body, the former lofes no more of its Motion than it communicates to the latter. This pofition however has been oppofed by other philofophers, and perhaps juftly, unlefs the prefervation

of Motion be underftood only of the quantity of it as eftimated always in the fame direction; for then it feems the principle will hold good. However, the reafoning ought to have proceeded in the contrary order; by firft obferving from experiment, or otherwife, that when two bodies act upon each other, the one gains exactly the Motion which is loft by the other, in the fame direction; and from hence made the inference, that there is therefore the fame Quantity of Motion preferved in the univerfe, as was created by God in the beginning; fince no body can act upon another, without being itfelf equally acted upon in the oppofite or contrary direction.

The Continuation of MOTION, or the caufe why a body once in Motion comes to perfevere in it, has been much controverted among phyfical writers; and yet it follows very evidently from one of the grand Laws of Nature; viz, that all bodies perfevere in their prefent ftate, whether of Motion or reft, unlefs they are difturbed by fome foreign powers. Motion therefore, once begun, would be continued for ever, were it to meet with no interruption from external caufes; as the power of gravity, the refiftance of the medium, &c.

The Communication of MOTION, or the manner in which a moving body comes to affect another at reft, or how much of its Motion is communicated by the firft to the laft, has alfo been the fubject of much difcuffion and controverfy. See the Laws of it under the word PERCUSSION.

MOTION may be confidered either as Equable, and Uniform; or as Accelerated, and Retarded. Equable Motion, again, may be confidered either as Simple, or as Compound; and Compound Motion either as Rectilinear, or as Curvilinear.

And all thefe again may be confidered either with regard to themfelves, or with regard to the manner of their production, and communication, by percuffion, &c.

Equable MOTION, is that by which the moving body proceeds with exactly the fame velocity or celerity; paffing always over equal fpaces in equal times.

The Laws of Uniform Motion, are thefe: 1. The fpaces defcribed, or paffed over, are in the compound ratio of the velocities, and the times of defcribing thofe fpaces. So that, if V and *v* be any two uniform velocities, S and *s* the fpaces defcribed or paffed over by them, in the refpective times T and *t* :

$$\text{then is } S : s :: TV : tv,$$
$$\text{or } 20 : 12 :: 4 \times 5 : 3 \times 4;$$
$$\text{taking } T = 4, t = 3, V = 5, \text{ and } v = 4.$$

2. In Uniform Motions, the time is as the fpace directly, and as the velocity reciprocally; or as the fpace divided by the velocity. So that

$$T : t :: \frac{S}{V} : \frac{s}{v} \text{ or } :: Sv : sV.$$

3. The velocity is as the fpace directly, and the time reciprocally; or as the fpace divided by the time.

$$\text{That is. } V : v :: \frac{S}{T} : \frac{s}{t} \text{ or } :: St : sT.$$

Accelerated MOTION, is that which continually receives frefh acceffions of velocity. And it is faid to be
uniformly

uniformly accelerated, when its acceffions of velocity are equal in equal times; fuch as that which is produced by the continual action of one and the fame force, like the force of gravity, &c.

Retarded Motion, is that whofe velocity continually decreafes. And it is faid to be uniformly Retarded, when its decreafe is continually proportional to the time, or by equal quantities in equal times; like that which is produced by the continual oppofition of one and the fame force; fuch as the force of gravity, in uniformly retarding the Motion of a body that is thrown upwards.

The Laws of Motion, uniformly accelerated or retarded, are thefe:

1. In uniformly varied motions, the fpace, S or *s*, is as the fquare of the time, or as the fquare of the greateft velocity, or as the rectangle or product of the time and velocity.

That is, $S : s :: T^2 : t^2 :: V^2 : v^2 :: TV : tv$.

2. The velocity is the time, or as the fpace divided by the the time, or as the fquare root of the fpace.

That is, $V : v :: T : t :: \dfrac{S}{T} : \dfrac{s}{t} :: \sqrt{S} : \sqrt{s}$.

3. The time is as the velocity, or as the fpace divided by the velocity, or as the fquare root of the fpace.

That is, $T : t :: V : v :: \dfrac{S}{V} : \dfrac{s}{v} :: \sqrt{S} : \sqrt{s}$.

4. When a fpace is defcribed, or paffed over, by an uniformly varied Motion, the velocity either beginning at nothing, and continually accelerated; or elfe beginning at fome determinate velocity, and continually retarded till the velocity be reduced to nothing; then the fpace, fo run over by the variable Motion, will be exactly equal to half the fpace that would be run over in the fame time by the greateft velocity if uniformly continued for that time. So, for inftance, if *g* denote the fpace run over in one fecond, or any other time, by fuch a variable Motion; then 2*g* would be the fpace that would be run over in one fecond, or the fame time, by the greateft velocity uniformly continued for the fame time; or 2*g* would be the greateft velocity per fecond which the moving body had. Confequently, if *t* be any other time, *s* the fpace run over in that time, and *v* the greateft velocity attained in it; then, from the foregoing articles, it will be

$$1'' : t'' :: 2g : 2gt = v \text{ the velocity,}$$
$$\text{and } 1^2 : t^2 :: g : gt^2 = s \text{ the fpace.}$$

And hence, for any fuch uniformly varied Motions, the relations among the feveral quantities concerned, will be expreffed by the following equations: viz,

$$s = gt^2 = \tfrac{1}{2}tv = \frac{v^2}{4g},$$

$$v = 2gt = \frac{2s}{t} = 2\sqrt{gs},$$

$$t = \frac{v}{2g} = \frac{2s}{v} = \sqrt{\frac{s}{g}},$$

$$g = \frac{v}{2t} = \frac{s}{t^2} = \frac{v^2}{4s}.$$

And thefe equations will hold good in the Motion either generated or deftroyed by the force of gravity, or by any other uniform force whatever. See alfo the articles Gravity, Acceleration, Retardation, &c. Again,

Simple Motion, is that which is produced by fome one power or force only, and is always rectilinear, or in one direction, whether the force be only momentary or continued. And

Compound Motion, is that which is produced by two or more powers acting in different directions See Compound, and Composition *of Motion*.

If a moving body be acted on by a double power; the one according to the direction AB, the other according to AC; with the Compound Motion, or that which is compounded of thefe two together, it will defcribe the diagonal AD of the parallelogram, whofe fides AB and AC it would have defcribed in the fame time with each of the refpective powers apart.

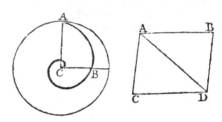

And if the radius of a circle be carried round upon the centre C, while a point in the radius fets off from A, and keeps moving along the radius towards the centre; then, by this Compound Motion, the path of the point will be a kind of a fpiral ABC.

For the Particular Laws of Motion, *arifing from the Collifion of bodies, both Elaftic and Non-elaftic, and that where the directions are both Perpendicular and Oblique,* see Percussion.

For Circular *Motion, and the Laws of* Projectiles, fee the refpective words.

For the Motion of Pendulums, and the Laws of Ofcillation, fee Pendulum.

Perpetual Motion, is a Motion which is fupplied and renewed from itfelf, without the intervention of any external caufe.

The celebrated problem of a Perpetual Motion, confifts in the inventing a machine, which has the principle of its Motion within itfelf; and is a problem that has employed the mathematicians for 2000 years; though none perhaps have profecuted it with attention and earneftnefs equal to thofe of the prefent age. Infinite are the fchemes, defigns, plans, engines, wheels, &c, to which this long defired Perpetual Motion has given birth.

But M. De la Hire has proved the impoffibility of any fuch machine, and finds that it amounts to this; viz, to find a body which is both heavier and lighter at the fame time; or to find a body which is heavier than itfelf. Indeed there feems but little in nature to countenance all this affiduity and expectation: among all the laws of matter and Motion, we know of none yet that feem likely to furnifh any principle or foundation for fuch an effect.

7 Action

Action and reaction it is allowed are always equal; and a body that gives any quantity of Motion to another, always loses just so much of its own; but under the present state of things, the resistance of the air, the friction of the parts of machines, &c, do necessarily retard every Motion.

To continue the Motion therefore either, first, there must be a supply from some foreign cause; which in a Perpetual Motion is excluded.

Or, 2dly, all resistance from the friction of the parts of matter must be removed; which necessarily implies a change in the nature of things.

Or, 3dly and lastly, there must be some method of gaining a force equivalent to what is lost, by the artful disposition and combination of mechanic powers; to which last point then all endeavours are to be directed: but how, or by what means, such force should be gained, is still a mystery.

The multiplication of powers or forces, it is certain, avails nothing; for what is gained in power is lost in time, so that the quantity of Motion still remains the same. This is an inviolable law of nature; by which nothing is left to art, but the choice of the several combinations that may produce the same effect.

There are various ways by which absolute force may be gained; but since there is always an equal gain in opposite directions, and no increase obtained in the same direction; in the circle of actions necessary to make a perpetual movement, this gain must be presently lost, and will not serve for the necessary expence of force employed in overcoming friction, and the resistance of the medium. And therefore, though it could be shewn, that in an infinite number of bodies, or in an infinite machine, there could be a gain of force for ever, and a Motion continued to infinity, it does not follow that a perpetual movement can be made. That which was proposed by M. Leibnitz in the Leipsic Acts of 1690, as a consequence of the common estimation of the forces of bodies in Motion, is of this kind, and for this and other reasons ought to be rejected. See PERPETUAL *Motion*; also ORFFYREUS's *Wheel*, &c.

Animal MOTION, is that by which the situation, figure, magnitude, &c, of the parts and members of animals are changed. Under these Motions, come all the animal functions; as respiration, circulation of the blood, excretion, walking, running, &c.

Animal Motions are usually divided into two species; viz, Natural and Spontaneous.

Natural MOTION, is that involuntary one which is effected without the command of the will, by the mere mechanism of the parts. Such as the Motion of the heart and pulse; the Peristaltic Motion of the intestines, &c. But

Spontaneous, or *Muscular* MOTION, is that which is performed by means of the muscles, at the command of the will; which is hence called Voluntary Motion. Borelli has a celebrated treatise on this subject, entitled De Motu Animalium.

Intestine MOTION, denotes an agitation of the particles of which a body consists.—Some philosophers will have every body, and every particle of a body, in continual Motion. As for fluids, it is the definition they give of them, that their parts are in continual Motion. And as to solids, they infer the like Motion

VOL. II.

from the effluvia continually emitted through their pores. Hence Intestine Motion is represented to be a Motion of the internal and smaller parts of matter, continually excited by some external, latent agent, which of itself is insensible, and only discovers itself by its effects; appointed by Nature to be the great instrument of the changes in bodies.

MOTION, in Astronomy, is peculiarly applied to the orderly courses of the heavenly bodies.

Mean MOTION. See MEAN.

The Motions of the celestial luminaries are of two kinds: Diurnal, or Common; and Secondary, or Proper.

Diurnal, or *Primary* MOTION, is that with which all the heavenly bodies, and the whole mundane sphere, appear to revolve every day round the earth, from east to west. This is also called the Motion of the Primum Mobile, and the Common Motion, to distinguish it from that rotation which is peculiar to each planet, &c.

Secondary, or *Proper* MOTION, is that with which a star, planet, or the like, advances a certain space every day from the west towards the east. See the several Motions of each luminary, with the irregularities, &c, of them, under the proper articles, EARTH, MOON, STAR, &c.

Angular MOTION, is that by which the angular position of any thing varies. See ANGULAR.

Horary MOTION, is the Motion during each hour. See HORARY.

Paracentric MOTION *of Impetus*. See PARACENTRIC.

MOTION *of Trepidation*, &c. See TREPIDATION and LIBRATION.

MOTIVE *Power* or *Force*, is the whole power or force acting upon any body, or quantity of matter, to move it; and is proportional to the momentum or quantity of motion it can produce in a given time. To distinguish it from the Accelerative force, which is considered as affecting the celerity only.

MOTRIX, something that has the power or faculty of moving. See *Vis Motrix*, and MOTION.

MOVEABLE, something susceptible of motion, or that is disposed to be moved. A sphere is the most Moveable of all bodies, or is the easiest to be moved on a plane. A door is Moveable on its hinges; the magnetic needle on a pin or pivot, &c. Moveable is often used in contradistinction to Fixed or Fixt.

MOVEABLE *Feasts*, are such as are not always held on the same day of the year or month; though they may be on the same day of the week. Thus, Easter is a Moveable Feast; being always held on the Sunday which falls upon or next after the first full moon following the 21st of March. See Philof. Transf. numb. 240, pa. 185. All the other Moveable Feasts follow Easter, keeping their constant distance from it; so that they are fixed with respect to it, though Moveable through the course of the year. Such are Septuagesima, Sexagesima, Ash-Wednesday, Ascension-Day, Pentecost, Trinity-Sunday, &c.

MOVEMENT, a term often used in the same sense with Automaton. The most usual Movements for keeping time, are Clocks and Watches: the latter are such as shew the parts of time by inspection, and are portable in the pocket; the former such as publish it by sounds, and are fixed as furniture.

S

MOVE-

MOVEMENT, in its popular ufe, fignifies all the inner works of a clock, watch, or other machine, that move, and by that motion carry on the defign of the inftrument. The Movement of a clock, or watch, is the infide; or that part which meafures the time, and ftrikes, &c; exclufive of the frame, cafe, dial plate, &c.

The parts common to both of thefe Movements are, the Main-fpring with its appurtenances, lying in the fpring box, and in the middle of it lapping about the fpring-arbor, to which one end of it is faftened. A-top of the fpring arbor is the Endlefs fcrew, and its wheel; but in fpring clocks this is a ratchet-wheel with its click, that ftops it. That which the main-fpring draws, and round which the chain or ftring is wrapped, is called the fufee: this is moftly taper; in large works, going with weights, it is cylindrical, and is called the barrel. The fmall teeth at the bottom of the fufee or barrel, which ftop it in winding up, is called the Ratchet; and that which ftops it when wound up, and is for that end driven up by the fpring, the Gardegut. The Wheels are various: the parts of a wheel are, the Hoop or Rim; the Teeth, the Crofs, and the Collet, or piece of brafs foldered on the arbor or fpindle on which the wheel is riveted. The little wheels, playing in the teeth of the larger, are called Pinions; and their teeth, which are 4, 5, 6, 8, &c, are called Leves; the ends of the fpindle are called Pivots; and the guttured wheel, with iron fpikes at bottom, in which the line of common clocks runs, the Pulley.

Theory of Calculating the Numbers for MOVEMENTS.

1. It is firft to be obferved, that a wheel, divided by its pinion, fhews how many turns the pinion has to one turn of the wheel.

2. That from the fufee to the balance the wheels drive the pinions, confequently the pinions run fafter, or make more revolutions, than the wheel; but it is the contrary from the great wheel to the dial-wheel.

3. That the wheels and pinions are written down either as vulgar fractions, or in the way of divifion in common arithmetic: for example, a wheel of 60, moving a pinion of 5, is fet down either thus $\frac{60}{5}$, or thus 5)60, which is better. And the number of turns the pinion has in one turn of the wheel, as a quotient, thus 5) 60 (12. A whole Movement may be written as follows:

$$
\begin{array}{r}
4)\ 36\ (9 \\
\hline
5)\ 55\ (11 \\
5)\ 45\ (9 \\
5)\ 40\ (8 \\
\hline
17
\end{array}
$$

where the uppermoft number expreffes the pinion of report 4, the dial-wheel 36, and the turns of the pinion 9: the fecond, the pinion and great wheel; the third, the fecond wheel &c; the fourth, the contrate wheel; and the laft, 17, the crown-wheel.

4. Hence, from the number of turns any pinion makes, in one turn of the wheel it works in, may be determined the number of turns a wheel or pinion has at any greater diftance, viz, by multiplying the quotients together; the product being the number of turns. Thus, fuppofe the wheels and pinions as in the cafe above; the quotient 11 multiplied by 9, gives 99, the number of turns in the fecond pinion 5 to one turn of the wheel 55, which runs concentrical, or on the fame fpindle, with the pinion 5. Again, 99 multiplied by 8, gives 792, the number of turns the laft pinion has to one turn of the firft wheel 5. Hence we proceed to find, not only the turns, but the number of beats of the balance, in the time of thofe turns. For, having found the number of turns the crown-wheel has in one turn of the wheel propofed, thofe turns multiplied by its notches, give half the number of beats in that one turn of the wheel. Suppofe, for example, the crown-wheel to have 720 turns, to one of the firft wheel; this number multiplied by 15, the notches in the crown-wheel, produces 10800, half the number of ftrokes of the balance in one turn of the firft wheel of 80 teeth.

The general divifion of a Movement is, into the clock, and watch parts.

MOULDINGS, in Architecture, are certain projections beyond the naked of a wall, column, wainfcot &c, the affemblage of which forms cornices, door-cafes, and other decorations of architecture.

MOULDINGS, are annexed to great guns by way of ornament, and perhaps in fome parts for ftrength; and probably are derived from the hoops or rings which bound the long iron bars together, anciently ufed in making cannon.

MOYNEAU. See MOINEAU.

MULLER (JOHN), commonly called REGIOMONTANUS, from Mons Regius, or Koningfberg, a town in Franconia, where he was born in 1436, and became the greateft aftronomer and mathematician of his time. He was indeed a very prodigy for genius and learning. Having firft acquired grammatical learning in his own country, he was admitted, while yet a boy, into the academy at Leipfic, where he formed a ftrong attachment to the mathematical fciences, arithmetic, geometry, aftronomy, &c. But not finding proper affiftance in thefe ftudies at this place, he removed, at only 15 years of age, to Vienna, to ftudy under the famous Purbach, the profeffor there, who read lectures in thofe fciences with the higheft reputation. A ftrong and affectionate friendfhip foon took place between thefe two, and our author made fuch rapid improvement in the fciences, that he was able to be affifting to his mafter, and to become his companion in all his labours. In this manner they fpent about ten years together; elucidating obfcurities, obferving the motions of the heavenly bodies, and comparing and correcting the tables of them; particularly thofe of Mars, which they found to difagree with the motions, fometimes as much as two degrees.

About this time there arrived at Vienna the cardinal Beffarion, who came to negociate fome affairs for the pope; who, being a lover of aftronomy, foon formed an acquaintance with Purbach and Regiomontanus. He had begun to form a Latin Verfion of Ptolomy's Almageft, or an Epitome of it; but not having time to go on with it himfelf, he requefted Purbach to complete the work, and for that purpofe to return with him into Italy, to make himfelf mafter of the Greek tongue, which he was as yet unacquainted with. To thefe propofals Purbach only affented, on condition that Regiomontanus would accompany him, and fhare in all the labours. They firft however, by

I means

means of an Arabic Version of Ptolomy, made some progress in the work; but this was soon interrupted by the death of Purbach, which happened in 1461, in the 39th year of his age. The whole task then devolved upon Regiomontanus, who finished the work, at the request of Purbach, made to him when on his death bed. This work our author afterwards revised and perfected at Rome, when he had learned the Greek language, and consulted the commentator Theon, &c.

Regiomontanus accompanied the cardinal Bessarion in his return to Rome, being then near 30 years of age. Here he applied himself diligently to the study of the Greek language; not neglecting however to make astronomical observations and compose various works in that science; as his Dialogue against the Theories of Cremonensis. The cardinal going to Greece soon after, Regiomontanus went to Ferrara, where he continued the study of the Greek language under Theodore Gaza; who explained to him the text of Ptolomy, with the commentaries of Theon; till at length he became so perfect in it, that he could compose verses, and read it like a critic.—In 1463 he went to Padua, where he became a member of the university; and, at the request of the students, explained Alfraganus, an Arabian philosopher.—In 1464 he removed to Venice, to meet and attend his patron Bessarion. Here he wrote, with great accuracy, his Treatise of Triangles, and a Refutation of the Quadrature of the Circle, which Cardinal Cusan pretended he had demonstrated. The same year he returned with Bessarion to Rome; where he made some stay, to procure the most curious books: those he could not purchase, he took the pains to transcribe, for he wrote with great facility and elegance; and others he got copied at a great expence. For as he was certain that none of these books could be had in Germany, he thought on his return thither, he would at his leisure translate and publish some of the best of them. During this time too he had a fierce contest with George Trabezonde, whom he had greatly offended by animadverting on some passages in his translation of Theon's Commentary.

Being now weary of rambling about, and having procured a great number of manuscripts, which was one great object of his travels, he returned to Vienna, and performed for some time the offices of his professorship, by reading of lectures &c. After being a while thus employed, he went to Buda, on the invitation of Matthias king of Hungary, who was a great lover of letters and the sciences, and had founded a rich and noble library there: for he had bought up all the Greek books that could be found on the sacking of Constantinople; also those that were brought from Athens, or wherever else they could be met with through the whole Turkish dominions, collecting them all together into a library at Buda. But a war breaking out in this country, he looked out for some other place to settle in, where he might pursue his studies, and for this purpose he retired to Noremberg. He tells us, that the reasons which induced him to desire to reside in this city the remainder of his life were, that the artists there were dextrous in fabricating his astronomical machines; and besides, he could from thence easily transmit his letters by the merchants into foreign countries. Being now well versed in all parts

of learning, and made the utmost proficiency in mathematics, he determined to occupy himself in publishing the best of the ancient authors, as well as his own lucubrations. For this purpose he set up a printing-house, and formed a nomenclature of the books he intended to publish, which still remains.

Here that excellent man, Bernard Walther, one of the principal citizens, who was well skilled in the sciences, especially astronomy, cultivated an intimacy with Regiomontanus; and as soon as he understood those laudable designs of his, he took upon himself the expence of constructing the astronomical instruments, and of erecting a printing-house. And first he ordered astronomical rules to be made of tin, for observing the altitudes of the sun, moon and planets. He next constructed a rectangular, or astronomical radius, for taking the distances of those luminaries. Then an armillary astrolabe, such as was used by Ptolomy and Hipparchus, for observing the places and motions of the stars. Lastly, he made other smaller instruments, as the torquet, and Ptolomy's meteoroscope, with some others which had more of curiosity than utility in them. From this apparatus it evidently appears, that Regiomontanus was a most diligent observer of the laws and motions of the celestial bodies, if there were not still stronger evidences of it in the accounts of the observations themselves which he made with them.

With regard to the printing-house, which was the other part of his design in settling at Noremberg, as soon as he had completed it, he put to press two works of his own, and two others. The latter were, The *New Theories* of his master Purbach, and the *Astronomicon* of Manilius. And his own were, the *New Calendar*, in which were given (as he says in the Index of the books which he intended to publish) the true conjunctions and oppositions of the luminaries, their eclipses, their true places every day, &c. His other work was his *Ephemerides*, of which he thus speaks in the said index: " The Ephemerides, which they vulgarly call an Almanac, for 30 years: where you may every day see the true motion of all the planets, of the moon's nodes, with the aspects of the moon to the sun and planets, the eclipses of the luminaries; and in the fronts of the pages are marked the latitudes." He published also most acute commentaries on Ptolomy's Almagest: a work which cardinal Bessarion so highly valued, that he scrupled not to esteem it worth a whole province. He prepared also new versions of Ptolomy's Cosmography; and at his leisure hours examined and explained works of another nature. He enquired how high the vapours are carried above the earth, which he fixed to be not more than 12 German miles. He set down observations of two comets that appeared in the years 1471 and 1472.

In 1474, pope Sixtus the 4th conceived a design of reforming the calendar; and sent for Regiomontanus to Rome, as the properest and ablest person to accomplish his purpose. Regiomontanus was very unwilling to interrupt the studies, and printing of books, he was engaged in at Noremberg; but receiving great promises from the pope, who also for the present named him bishop of Ratisbon, he at length consented to go. He arrived at Rome in 1475, but died there the year after, at only 40 years of age; not without a

suspicion

fufpicion of being poifoned by the fons of George Trabezonde, in revenge for the death of their father, which was faid to have been caufed by the grief he felt on account of the criticifms made by Regiomontanus on his tranflation of Ptolomy's Almageft.

Purbach firft of any reduced the trigonometrical tables of fines, from the old fexagefimal divifion of the radius, to the decimal fcale. He fuppofed the radius to be divided into 600000 equal parts, and computed the fines of the arcs to every ten minutes, in fuch equal parts of the radius, by the decimal notation. This project of Purbach was perfected by Regiomontanus; who not only extended the fines to every minute, the radius being 600000, as defigned by Purbach, but afterwards, difliking that fcheme, as evidently imperfect, he computed them likewife to the radius 1000000, for every minute of the quadrant. Regiomontanus alfo introduced the tangents into trigonometry, the canon of which he called *fœcundus*, becaufe of the many great advantages arifing from them. Befide thefe things, he enriched trigonometry with many theorems and precepts. Indeed, excepting for the ufe of logarithms, the trigonometry of Regiomontanus is but little inferior to that of our own time. His Treatife, on both Plane and Spherical Trigonometry, is in 5 books; it was written about the year 1464, and printed in folio at Noremberg in 1533. In the 5th book are various problems concerning rectilinear triangles, fome of which are refolved by means of algebra : a proof that this fcience was not wholly unknown in Europe before the treatife of Lucas de Burgo.

Regiomontanus was author of fome other works befide thofe before mentioned. Peter Ramus, in the account he gives of the admirable works attempted and performed by Regiomontanus, tells us, that in his workfhop at Noremberg there was an automaton in perpetual motion: that he made an artificial fly, which taking its flight from his hand, would fly round the room, and at laft, as if weary, would return to his mafter's hand : that he fabricated an eagle, which, on the emperor's approach to the city, he fent out, high in the air, a great way to meet him, and that it kept him company to the gates of the city. Let us no more wonder, adds Ramus, at the dove of Archytas, fince Noremberg can fhew a fly, and an eagle, armed with geometrical wings. Nor are thofe famous artificers, who were formerly in Greece, and Egypt, any longer of fuch account, fince Noremberg can boaft of her Regiomontanufes. For Wernerus firft, and then the Schoneri, father and fon, afterwards, revived the fpirit of Regiomontanus.

MULTANGULAR Figure, is one that has many angles, and confequently many fides alfo. Thefe are otherwife called polygons.

MULTILATERAL Figures, are fuch as have many fides, or more than four fides.

MULTINOMIAL, or Multinomial *Roots*, are fuch as are compofed of many names, parts, or members ; as, $a + b + c + d$ &c.

For the raifing an infinite Multinomial to any propofed power, or extracting any root out of fuch power, fee a method by Mr. De Moivre, in the Philof. Tranf. numb. 230. See alfo POLYNOMIAL.

MULTIPLE, Multiplex, a number which com-

prehends fome other number feveral times. Thus, 6 is a Multiple of 2, this being contained in 6 juft 3 times. Alfo 12 is a common Multiple of 6, 4, and 3 ; comprehending the firft twice, the fecond thrice, and the third four times.

Multiple *Ratio* or *Proportion*, is that which is between Multiple numbers &c. If the lefs term of a ratio be an aliquot part of the greater, the ratio of the greater to the lefs is called Multiple ; and that of the lefs to the greater Submultiple.

A Submultiple number, is that which is contained in the Multiple. Thus, the numbers 2, 3, and 4 are Submultiples of 12 and 24.

Duple, triple, &c ratios ; as alfo fubduples, fubtriples, &c, are fo many fpecies of Multiple and Submultiple ratios.

Multiple *Superparticular Proportion*, is when one number or quantity contains another more than once, and a certain aliquot part ; as 10 to 3, or $3\frac{1}{3}$ to 1.

Multiple *Superpartient Proportion*, is when one number or quantity contains another feveral times, and fome parts befides ; as 29 to 6, or $4\frac{5}{6}$ to 1.

MULTIPLICAND, is one of the two factors in the rule of multiplication, being that number given to be multiplied by the other, called the multiplicator, or multiplier.

MULTIPLICATION, is, in general, the taking or repeating of one number or quantity, called the Multiplicand, as often as there are units in another number, called the Multiplier, or Multiplicator ; and the number or quantity refulting from the Multiplication, is called the Product of the two foregoing numbers or factors.

Multiplication is a compendious addition ; performing at once, what in the ufual way of addition would require many operations : for the multiplicand is only added to itfelf, or repeated, as often as is exprefled by the units in the multiplier. Thus, if 6 were to be multiplied by 5, the product is 30, which is the fum arifing from the addition of the number 6 five times to itfelf.

In every Multiplication, 1 is in proportion to the multiplier, as the multiplicand is to the product.

Multiplication is of various kinds, in whole numbers, in fractions, decimals, algebra, &c.

1. MULTIPLICATION *of Whole Numbers*, is performed by the following rules : When the multiplier confifts of only one figure, fet it under the firft, or righthand figure, of the multiplicand ; then, drawing a line underneath, and beginning at the faid firft figure, multiply every figure of the multiplicand by the multiplier ; fetting down the feveral products below the line, proceeding orderly from right to left. But if any of thefe products amount to 10, or feveral 10's, either with or without fome overplus, then fet down only the overplus, or fet down 0 if there be no overplus ; and carry, to the next product, as many units as the former contained of tens. Thus, to multiply 35092 by 4.

Multiplicand	35092
Multiplier	4
Product	140368

When

When the multiplier confists of feveral figures; multiply the multiplicand by each figure of it, as before, and place the feveral lines of products underneath each other in fuch order, that the firft figure or cipher of each line may fall ftraight under its refpective multiplier, or multiplying figure; then add thefe feveral lines of products together, as they ftand, and the fum of them all will be the product of the whole multiplication. Thus, to multiply 63017 by 236:

Multiplicand - -	63017	
Multiplier	236	
Product of 63017 by 6 -	378102	
Product of 63017 by 30	189051	
Product of 63017 by 200	126034	
Whole product	14872012	

The feveral lines of products may be fet down in any order, or any of them firft, and any other of them fecond, &c; for the order of placing them can make no difference in the fum total. There are many abbreviations, and peculiar cafes, according to circumftances, which may be feen in moft books of arithmetic.

The mark or character now ufed for Multiplication, is either the × crofs or a fingle point · ; the former being introduced by Oughtred, and the latter I think by Leibnitz.

To Prove MULTIPLICATION. This may be done various ways; either by dividing the product by the multiplier, then the quotient will be equal to the multiplicand; or divide the fame product by the multiplicand, and the quotient will come out equal to the multiplier; or in general divide the product by either of the two factors, and the quotient will come out equal to the other factor, when the operations are all right. But the more ufual, and compendious way of proving Multiplication, is by what is called cafting out the nines; which is thus performed: Add the figures of the multiplicand all together, and as often as the fum amounts to 9, reject it always, and fet down the laft overplus as in the margin; this in the foregoing example is 8. Then do the fame by the multiplier, fetting down the laft overplus, which is 2, on the right of the former remainder 8. Next multiply thefe two remainders, 2 and 8, together, and from their product 16, caft out the 9, and there remains 7, which fet down over the two former. Laftly, add up, in the fame manner, all the figures of the whole product of the multiplication, viz 14872012, cafting out the 9's, and then there remains 7, to be fet down under the two firft remains. Then when the figure at top, is the fame as that at bottom, as they are here both 7's, the work it may be prefumed is right; but if thefe two figures fhould not be the fame, it is certainly wrong.

2. *To Multiply Money, or any other thing, confifting of different Denominations together, by any number, ufually called Compound Multiplication.* Beginning at the loweft, multiply the number of each denomination feparately by the multiplier, fetting down the products below them. But if any of thefe products amount to as much as 1 or more of the next higher denominations, carry fo many to the next product, and fet down only the overplus. *For Ex.* To find the amount of 9 things at 1 l 12 s 4½ d. each; or to multiply 1 l 12 s 4½ d by 9: fet the multiplier 9 under the given fum as in the margin, and multiply thus : 9 halfpence make 4d halfpenny, fet down ½ penny, and carry 4; then 9 times 4 are 36, and 4 to carry make 40 pence, which are 3 s and 4 d, fet down 4 and carry 3; next 9 times 12 are 108, and 3 to carry, make 111 fhillings, or 5 l 11 s, fet down 11, and carry 5; laftly 9 times 1 are 9, and 5 to carry, make 14, which fet down; and then the whole amount, or product, comes to 14 l 11 s 4½ d.

l	s	d
1	12	4¼
		9
14	11	4½

3. *To Multiply Vulgar Fractions.*—Multiply all the given numerators together for the numerator of the product, and all the denominators together for the denominator of the product fought.

Thus, $\frac{2}{3}$ multiplied by $\frac{4}{5}$, or $\frac{2}{3} \times \frac{4}{5}$ make $\frac{8}{15}$.

And $\frac{3}{5} \times \frac{2}{5} \times \frac{3}{7}$ make $\frac{18}{175}$.

And here it may be noted that, when there are any common numbers in the numerators and denominators, thefe may be omitted from both, which will make the operation fhorter, and bring out the whole product in a fraction much fimpler and in lower terms. Thus,

$\frac{2}{3} \times \frac{3}{4} \times \frac{5}{6}$, by leaving out the two 3's, become

$\frac{2 \times 5}{4 \times 6} = \frac{10}{24}$ or $\frac{5}{12}$

Alfo, when any numerators and denominators will both abbreviate or divide by one and the fame number, let them be divided, and the quotients ufed inftead of them. So, in the above example, after omitting the two 3's, let the 2 and 6 be both divided by 2, and ufe the quotients 1 and 3 inftead of them, fo fhall the expreffion become $\frac{1 \times 5}{4 \times 3} = \frac{5}{12}$, as before.

4. *To Multiply Decimals.*—Multiply the given numbers together the fame as if they were whole numbers, and point off as many decimals in the whole product as there are in both factors together; as in the annexed example, where the number of decimals is five, becaufe there are three in the multiplicand, and two in the multiplier.—When it happens that there are not fo many figures in the product as there muft be decimals, then prefix as many ciphers as will fupply the defect.

2·305
21·86
13830
18440
2305
4610
50·38730

5. *Crofs* MULTIPLICATION, otherwife called *Duodecimal Arithmetic*, is the multiplying of numbers together whofe fubdivifions proceed by 12's; as feet, inches, and parts, that is 12th parts, &c; a thing of very frequent ufe in fquaring, or multiplying toge-

ther the dimensions of the works of bricklayers, carpenters, and other artificers. *For Example.* To multiply 5 feet 3 inches by 2 feet 4 inches. Set them down as in the margin, and multiply all the parts of the multiplicand by each part of the multiplier; thus, 2 times 3 make 6 inches, and 2 times 5 make 10 feet ; then 4 times 3 make 12 parts, or 1 inch to carry ; and 4 times 5 make 20, and 1 to carry makes 21 inches, or 1 f. 9 inc. to set down below the former line :

F	I
5	3
2	4
10	6
1	9
12	3

Lastly adding the two lines together, the whole sum or product amounts to 12 f. 3 inc.

6. MULTIPLICATION *in Algebra.* This is performed, 1. When the quantities are simple, by only joining the letters together like a word ; and if the simple quantities have any coefficients or numbers joined with them, multiply the numbers together, and prefix the product of them to the letters so joined together. But, in algebra, we have not only to attend to the quantities themselves, but also to the signs of them ; and the general rule for the signs is this: When the signs are alike, or the same, either both + or both —, then the sign of the product will always be + ; but when the signs are different, or unlike, the one +, and the other —, then the sign of the product will be —. Hence these

EXAMPLES.

Mult.	$+a$	$-2a$	$+6x$	$-8x$	$-3ab$
By	$+b$	$-4b$	$-3a$	$+5a$	$-5ac$
Products	$+ab$	$+8ab$	$-18ax$	$-40ax$	$+15a^2bc$

2. In Compound quantities, multiply every term or part of the multiplicand by each term separately of the multiplier, and set down all the products with their signs, collecting always into one sum as many terms as are similar or like to one another.

EXAMPLES.

$a+b$	$a-b$	$a+b$
$a+b$	$a-b$	$a-b$
a^2+ab	a^2-ab	a^2+ab
$\quad +ab+b^2$	$\quad -ab+b^2$	$\quad -ab-b^2$
$a^2+2ab+b^2$	$a^2-2ab+b^2$	$a^2 \quad \cdot \quad -b^2$

$2a-3b$	$2a+4x$	a^2-ax
$4a+5b$	$2a-4x$	$2a+2x$
$8a^2-12ab$	$4a^2+8ax$	$2a^3-2a^2x$
$\quad +10ab-15b^2$	$\quad -8ax-16x^2$	$\quad +2a^2x-2ax^2$
$8a^2-2ab-15b^2$	$4a^2 \quad \cdot \quad -16x^2$	$2a^3 \quad \cdot \quad -2ax^3$

3. In Surd quantities, if the terms can be reduced to a common surd, the quantities under each may be multiplied together, and the mark of the same surd prefixed to the product ; but if not, then the different surds may be set down with some mark of multiplication between them, to denote their product.

EXAMPLES.

$7\sqrt{ax}$	$\sqrt{7}$	$\sqrt[3]{7ab}$	$\sqrt{12a}$	$6a\sqrt{2cx}$
$5\sqrt{cx}$	$\sqrt{5}$	$\sqrt[3]{4ac}$	$\sqrt{3a}$	$2b\sqrt{3ax}$
$35\sqrt{acx^2}$	$\sqrt{35}$	$\sqrt[3]{28a^2bc}$	$\sqrt{36a^2}=6a$	$12ab\sqrt{6acx^2}$

4. Powers or Roots of the same quantity are multiplied together, by adding their exponents.

Thus, $a^2 \times a^3 = a^5$; and $\overline{a+x}|^3 \times \overline{a+x}|^5 = \overline{a+x}|^8$: also $x^2 \times x^{\frac{1}{2}} = x^{\frac{5}{2}}$; and $a^{\frac{1}{2}} \times a^{\frac{1}{2}} = a^1$ or a.

To Multiply Numbers together by Logarithms.—This is performed by adding together the logarithms of the given numbers, and taking the number answering to that sum, which will be the product sought.

Des Cartes, at the beginning of his Geometry, performs Multiplication (and indeed all the other common arithmetical rules) in geometry, or by lines ; but this is no more than taking a 4th proportional to three given lines, of which the first represents unity, and the 2d and 3d the two factors or terms to be multiplied, the product being expressed by the 4th proportional ; because, in every multiplication, unity or 1 is to either of the two factors, as the other factor is to the product.

MULTIPLICATOR, is the number or quantity by which another is multiplied ; and is otherwise called the multiplier.

MULTIPLIER, or MULTIPLICATOR, is the number or quantity which multiplies another, called the multiplicand, in any operation of multiplication.

MUNSTER (SEBASTIAN), an eminent German divine and mathematician, was born at Ingelheim in 1489. At the age of 14 he was sent to Heidelberg to study. Two years after, he entered the convent of the Cordeliers ; where he assiduously studied divinity, mathematics, and geography. He was the first who published a Chaldee Grammar and Lexicon ; and he shortly after gave the world a Talmudic Dictionary. He afterwards became professor of the Hebrew language at Basil. He was one of the first who attached himself to Luther, and embraced Protestantism : yet behaved himself with great moderation ; never concerning himself with their disputes ; but shut himself up at home and pursued his favourite studies, which were mathematics, natural philosophy, with the Hebrew and other Oriental languages. He published a great number of books on these subjects ; particularly, a Latin version, from the Hebrew, of all the books of the Old Testament, with learned notes, printed at Basil in 1534 and 1546 ; Josephus's History of the Jews in Latin ; a Treatise of Dialling, in folio, 1536 ; Universal Cosmography, in 6 books folio, Basil 1550. For these works he was styled the German Strabo ; as he was the German Esdras, for his Oriental writings.

Munster was a meek-tempered, pacific, studious, retired man, who wrote a great number of books, but

never

never meddled in controverfy.—He died of the plague at Bafil, in 1552, at 63 years of age.

MURDERERS, a fmall fpecies of ordnance once ufed on fhipboard; but now out of ufe.

MUSIC, the fcience of found, confidered as capable of producing melody, or harmony.

Among the Ancients, Mufic was taken in a much more extenfive fenfe than among the Moderns: what we call the fcience of Mufic, was by the Ancients rather called Harmonica.

Mufic is one of the feven fciences called liberal, and comprehended alfo among the mathematical fciences, as having for its object difcrete quantity or number; not however confidering it in the abftract, like arithmetic; but in relation to time and found, with intent to conftitute a delightful harmony.

This fcience is alfo Theoretical and Practical. Theoretical, which examines the nature and properties of concords and difcords, explaining the proportions between them by numbers. And Practical, which teaches not only compofition, or the manner of compofing tunes, or airs; but alfo the art of finging with the voice, and playing on mufical inftruments.

It appears that Mufic was one of the moft ancient of the arts; and, of all others, Vocal Mufic muft doubtlefs have been the firft kind. For man had not only the various tones of his own voice to make his obfervations on, before any other art or inftrument was found out, but had the various natural ftrains of birds to give him occafion to improve his own voice, and the modulations of founds it was capable of. The firft invention of wind inftruments Lucretius afcribes to the obfervation of the winds whiftling in the hollow reeds. As for other kinds of inftruments, there were fo many occafions for cords or ftrings, that men could not be long in obferving their various founds; which might give rife to ftringed inftruments. And for the pulfative inftruments, as drums and cymbals, they might arife from the obfervation of the naturally hollow noife of concave bodies.

As to the inventors and improvers of Mufic, Plutarch, in one place, afcribes the firft invention of it to Apollo; and in another place to Amphion, the fon of Jupiter and Antiope. The latter indeed, it is pretty generally allowed, firft brought Mufic into Greece, and invented the lyre.

To him fucceeded Chiron, the demigod; then Democus; Hermes Trifmegiftus: Olympus; and Orpheus, whom fome make the firft introducer of Mufic into Greece, and the inventor of the lyre: to whom add Phemius, and Terpander, who was contemporary with Lycurgus, and fet his laws to Mufic; to whom alfo fome attribute the firft inftitution of mufical modes, and the invention of the lyre: laftly, Thales; and Thamyris, who, it has been faid, was the firft inventor of inftrumental Mufic without finging.

Thefe were the eminent muficians before Homer's time: others of a later date were, Lafus Hermionenfis, Melanippides, Philoxenus, Timotheus, Phrynnis, Epigonius, Lyfander, Simmicus, and Diodorus; who were all of them confiderable improvers of Mufic. Lafus, it is faid, was the firft author who wrote upon Mufic, in the time of Darius Hyftafpis; Epigonius invented an inftrument of 40 ftrings, called the Epigonium.

Simmicus alfo invented an inftrument of 35 ftrings, called a Simmicium; Diodorus improved the Tibia, by adding new holes; and Timotheus the Lyre, by adding a new ftring; for which he was fined by the Lacedemonians.

As the accounts we have of the inventors of mufical inftruments among the Ancients are very obfcure, fo alfo are the accounts of thofe inftruments themfelves; of moft of them indeed we know little more than the bare names.

The general divifion of inftruments is, into ftringed inftruments, wind inftruments, and thofe of the pulfatile kind. Of ftringed inftruments, mention is made of the ly raor cithara, the pfalterium, trigonum, fambuca, pectis, magas, barbiton, teftudo, epigonium, fimmicium, and panderon; which were all ftruck with the hand, or a plectrum. Of wind inftruments, were the tibia, fiftula, hydraulic organs, tubæ, cornua, and lituus. And the pulfatile inftruments were the tympanum, cymbalum, creptaculum, tintinnabulum, crotalum, and fiftrum.

Mufic has ever been in the higheft efteem in all ages, and among all people; nor could authors exprefs their opinion of it ftrongly enough, but by inculcating that it was ufed in heaven, and as one of the principal entertainments of the gods, and the fouls of the bleffed. The effects afcribed to it by the Ancients are almoft miraculous: by its means, it has been faid, difeafes have been cured, unchaftity corrected, feditions quelled, paffions raifed and calmed, and even madnefs occafioned. Athenæus affures us, that anciently all laws, divine and civil, exhortations to virtue, the knowledge of divine and human things, with the lives and actions of illuftrious men, were written in verfe, and publicly fung by a chorus to the found of inftruments; which was found the moft effectual means to imprefs morality on the minds of men, and a right fenfe of their duty.

Dr. Wallis has endeavoured to account for the furprifing effects attributed to the ancient Mufic; and afcribes them chiefly to the novelty of the art, and the hyperboles of the ancient writings: nor does he doubt, but the modern Mufic, in like cafes, would produce effects at leaft as confiderable as the ancient. The truth is, we can match moft of the ancient ftories of this kind in the modern hiftories. If Timotheus could excite Alexander's fury with the Phrygian mode, and footh him into indolence with the Lydian; a more modern mufician has driven Eric, king of Denmark, into fuch a rage, as to kill his beft fervants. Dr. Niewentyt fpeaks of an Italian who, by varying his Mufic from brifk to folemn, and the contrary, could fo move the foul, as to caufe diftraction and madnefs; and Dr. South has founded his poem, called Mufica Incantans, on an inftance he knew of the fame kind.

Mufic however is found not only to exert its force on the affections, but on the parts of the body alfo: witnefs the Gafcon knight, mentioned by Mr. Boyle, who could not contain his water at the playing of a bagpipe; and the woman, mentioned by the fame author, who would burft into tears at the hearing of a certain tune, with which other people were but a little affected. To fay nothing of the trite ftory of the Tarantula, we have an inftance, in the Hiftory of the Academy of Sciences, of a mufician being cured of a violent fever,

fever, by a little concert occasionally played in his room.

Nor are our minds and bodies alone affected with sounds, but even inanimate bodies are so. Kircher speaks of a large stone, that would tremble at the sound of one particular organ pipe; and Morhoff mentions one Petter, a Dutchman, who could break rummer-glasses with the tone of his voice. Merfenne also mentions a particular part of a pavement, that would shake and tremble, as if the earth would open, when the organs played. Mr. Boyle adds, that seats will tremble at the sound of organs; that he has felt his hat do so under his hand, at certain notes both of organs and discourse; and that he was well informed every well-built vault would thus answer to some determinate note.

It has been disputed among the Learned, whether the Ancients or Moderns best understood and practised Music. Some maintain that the ancient art of Music, by which such wonderful effects were performed, is quite lost; and others, that the true science of harmony is now arrived at much greater perfection than was known or practised among the Ancients. This point seems no other way to be determinable but by comparing the principles and practice of the one with those of the other. As to the theory or principles of harmonics, it is certain we understand it better than the Ancients; because we know all that they knew, and have improved considerably on their foundations. The great dispute then lies on the practice; with regard to which it may be observed, that among the Ancients, Music, in the most limited sense of the word, included Harmony, Rythmus, and Verse; and consisted of verses sung by one or more voices alternately, or in choirs, sometimes with the sound of instruments, and sometimes by voices only. Their musical faculties, we have just observed, were Melopœia, Rythmopœia, and Poesis; the first of which may be considered under two heads, Melody and Symphony. As to the latter, it seems to contain nothing but what relates to the conduct of a single voice, or making what we call Melody. It does not appear that the Ancients ever thought of the concert, or harmony of parts; which is a modern invention, for which we are beholden to Guido Aretine, a Benedictine friar.

Not that the Ancients never joined more voices or instruments than one together in the same symphony; but that they never joined several voices so as that each had a distinct and proper melody, which made among them a succession of various concords, and were not in every note unisons, or at the same distance from each other as octaves. This last indeed agrees to the general definition of the word Symphonia; yet it is plain that in such cases there is but one song, and all the voices perform the same individual melody. But when the parts differ, not by the tension of the whole, but by the different relations of the successive notes, this is the modern art, which requires so peculiar a genius, and on which account the modern Music seems to have much the advantage of the ancient. For farther satisfaction on this head, see Kircher, Perrault, Wallis, Malcolm, Cerceau, and others; who unanimously agree, that after all the pains they have taken to know the true state of the Music of the Ancients, they could not find

the least reason to think there was any such thing in their days as Music in parts.

The ancient musical notes are very mysterious and perplexed: Boethius and Gregory the Great first put them into a more easy and obvious method. In the year 1204, Guido Aretine, a Benedictine of Arezzo in Tuscany, first introduced the use of a staff with five lines, on which, with the spaces, he marked his notes by setting a point up and down upon them, to denote the rise and fall of the voice: though Kircher says this artifice was in use before Guido's time.

Another contrivance of Guido's was to apply the six musical syllables, *ut, re, mi, fa, sol, la*, which he took out of the Latin hymn,

UT queant laxis   REsonare fibris
MIra gestorum   FAmuli tuorum,
SOLve polluti   LAbii reatum,

O Pater Alme.

We find another application of them in the following lines.

UT RElevit MIserum FAtum, SOLitosque LAbores
Aevi, sit dulcis musica noster amor.

Besides his notes of Music, by which, according to Kircher, he distinguished the tones, or modes, and the seats of the semitones, he also invented the scale, and several musical instruments, called polyplectra, as spinets and harpsichords.

The next considerable improvement was in 1330, when Joannes Muria, or de Muris, doctor at Paris (or as Bayle and Gesner make him, an Englishman), invented the different figures of notes, which express the times or length of every note, at least their true relative proportions to one another, now called longs, breves, semi-breves, crotchets, quavers, &c.

The most ancient writer on Music was Lasus Hermionensis; but his works, as well as those of many others, both Greek and Roman, are lost. Aristoxenus, disciple of Aristotle, is the earliest author extant on the subject: after whom came Euclid, author of the Elements of Geometry; and Aristides Quintilianus wrote after Cicero's time. Alypius stands next; after him Gaudentius the philosopher, and Nicomachus the Pythagorean, and Bacchius. Of which seven Greek authors we have a fair copy, with a translation and notes, by Meibomius. Ptolomy, the celebrated astronomer, wrote in Greek on the principles of harmonics, about the time of the emperor Antoninus Pius. This author keeps a medium between the Pythagoreans and Aristoxenians. He was succeeded at a considerable distance by Manuel Bryennius.

Of the Latins, we have Boetius, who wrote in the time of Theodoric the Goth; and one Cassiodorus, about the same time; Martianus, and St. Augustine, not far remote.

And of the moderns are Zarlin, Salinas, Vincenzo Galileo, Doni, Kircher, Mersenne, Paran, De Caux, Perrault, Des Cartes, Wallis, Holder, Malcolm, Rousseau, &c.

MUSICAL *Numbers*, are the numbers 2, 3, and 5, together with their composites. They are so called, because all the intervals of music may be expressed by such numbers. This is now generally admitted by

musical

musical theorists. Mr. Euler seems to suppose, that 7 or other primes might be introduced; but he speaks of this as a doubtful and difficult matter. Here 2 corresponds to the octave, 3 to the fifth, or rather to the 12th, and 5 to the third major, or rather the seventeenth. From these three may all other intervals be found.

MUSICAL *Proportion*, or Harmonical Proportion, is when, of four terms, the first is to the 4th, as the difference of the 1st and 2d is to the difference of the 3d and 4th: as 2, 3, 4, and 8 are in Musical proportion, because 2 : 8 :: 1 : 4. And hence, if there be only three terms, the middle term supplying the place of both the 2d and 3d, the 1st is to the 3d, as the difference of the 1st and 2d, is to the difference of the 2d and 3d: as in these 2, 3, 6; where 2 : 6 :: 1 : 3. See HARMONICAL *Proportion*.

MUSSCHENBROEK (PETER), a very distinguished natural philosopher and mathematician, was born at Utrecht a little before 1700. He was first professor of these sciences in his own university, and afterwards invited to the chair at Leyden, where he died full of reputation and honours in 1761. He was a member of several academies, particularly the Academy of Sciences at Paris. He published several works in Latin, all of them shewing his great penetration and accuracy. As,

1. His Elements of Physico-Mathematics, in 1726.
2. Elements of Physics, in 1736.
3. Institutions of Physics; containing an abridgment of the new discoveries made by the Moderns; in 1748.
4. Introduction to Natural Philosophy; which he began to print in 1760; and which was completed and published at Leyden, in 1762, by M. Lulofs, after the death of the author. It was translated into French by M. Sigaud de la Fond, and published at Paris in 1769, in 3 vols 4to; under the title of A Course of Experimental and Mathematical Physics.

He had also several papers, chiefly on meteorology, printed in the volumes of Memoirs of the Academy of Sciences, viz, in those of the years 1734, 1735, 1736, 1753, 1756, and 1760.

MUTULE, a kind of square modillion in the Doric frize.

MYRIAD, the number of 10,000, or ten thousand.

N.

NAB

NABONASSAR, first king of the Chaldeans; memorable for the Jewish era which bears his name, which began on Wednesday February 26th in the 3967th year of the Julian period, or 747 years before Christ; the years of this epoch being Egyptian ones, of 365 days each. This is a remarkable era in chronology, because Ptolomy assures us there were astronomical observations made by the Chaldeans from Nabonassar to his time; also Ptolomy, and the other astronomers, account their years from that epoch.

Nabonassar was the first king of the Chaldeans or Babylonians. These having revolted from the Medes, who had overthrown the Assyrian monarchy, did, under Nabonassar, found a dominion, which was much increased under Nebuchadnezzar. It is probable this Nabonassar is that Baladan in the 2d Book of Kings, xx, 12, father of Merodach, who sent ambassadors to Hezekiah. See 2 Chron. xxii.

NADIR, that point of the heavens diametrically under our feet, or opposite to the zenith, which is directly over our heads. The zenith and Nadir are the two poles of the horizon, each being 90° distant from it.

The Sun's NADIR, is the axis of the cone projected by the shadow of the earth: so called, because that axis

NAP

being prolonged, gives a point in the ecliptic diametrically opposite to the sun.

NAKED, in Architecture, as the Naked of a wall, &c, is the surface, or plane, from whence the projectures arise; or which serves as a ground to the projectures.

NAPIER, or NEPER (JOHN), baron of Merchiston in Scotland, inventor of the logarithms, was the eldest son of Sir Archibald Napier of Merchiston, and born in the year 1550. Having given early indications of great natural parts, his father was careful to have them cultivated by a liberal education. After going through the ordinary course of education at the university of St. Andrew's, he made the tour of France, Italy, and Germany. On his return to his native country, his literature and other fine accomplishments soon rendered him conspicuous; he however retired from the world to pursue literary researches, in which he made an uncommon progress, as appears by the several useful discoveries with which he afterwards favoured mankind. He chiefly applied himself to the study of mathematics; without however neglecting that of the Scriptures; in both of which he discovered the most extensive knowledge and profound penetration. His Essay upon the book of the Apocalypse indicates the most acute investigation;

investigation; though time hath difcovered that his calculations concerning particular events had proceeded upon fallacious data. But what has chiefly rendered his name famous, was his great and fortunate difcovery of logarithms in trigonometry, by which the eafe and expedition in calculation have fo wonderfully affifted the fcience of aftronomy and the arts of practical geometry and navigation. Napier, having a great attachment to aftronomy, and fpherical trigonometry, had occafion to make many numeral calculations of fuch triangles, with fines, tangents, &c; and thefe being expreffed in large numbers, they hence occafioned a great deal of labour and trouble: To fpare themfelves part of this labour, Napier, and other authors about his time, fet themfelves to find out certain fhort modes of calculation, as is evident from many of their writings. To this neceffity, and thefe endeavours it is, that we owe feveral ingenious contrivances; particularly the computation by Napier's Rods, and feveral other curious and fhort methods that are given in his *Rabdologia;* and at length, after trials of many other means, the moft complete one of logarithms, in the actual conftruction of a large table of numbers in arithmetical progreffion, adapted to a fet of as many others in geometrical progreffion. The property of fuch numbers had been long known, viz, that the addition of the former anfwered to the multiplication of the latter, &c; but it wanted the neceffity of fuch very troublefome calculations as thofe above mentioned, joined to an ardent difpofition, to make fuch a ufe of that property. Perhaps alfo this difpofition was urged into action by certain attempts of this kind which it feems were made elfewhere; fuch as the following, related by Wood in his Athenæ Oxonienfes, under the article Briggs, on the authority of Oughtred and Wingate, viz, "That one Dr. Craig a Scotchman, coming out of Denmark into his own country, called upon John Neper baron of Marchefton near Edinburgh, and told him among other difcourfes of a new invention in Denmark (by Longomontanus as 'tis faid) to fave the tedious multiplication and divifion in aftronomical calculations. Neper being folicitous to know farther of him concerning this matter, he could give no other account of it, than that it was by proportionable numbers. Which hint Neper taking, he defired him at his return to call upon him again. Craig, after fome weeks had paffed, did fo, and Neper then fhewed him a rude draught of that he called *Canon Mirabilis Logarithmorum.* Which draught, with fome alterations, he printing in 1614, it came forthwith into the hands of our author Briggs, and into thofe of William Oughtred, from whom the relation of this matter came."

Whatever might be the inducement-however, Napier publifhed his invention in 1614, under the title of *Logarithmorum Canonis Defcriptio, &c,* containing the conftruction and canon of his logarithms, which are thofe of the kind that is called hyperbolic. This work coming prefently to the hands of Mr. Briggs, then Profeffor of Geometry at Grefham College in London, he immediately gave it the greateft encouragement, teaching the nature of the logarithms in his public lectures, and at the fame time recommending a change in the fcale of them, by which they might be advantageoufly altered to the kind which he afterwards

3

computed himfelf, which are thence called Briggs's Logarithms, and are thofe now in common ufe. Mr. Briggs alfo prefently wrote to lord Napier upon this propofed change, and made journeys to Scotland the two following years, to vifit Napier, and confult him about that alteration, before he fet about making it. Briggs, in a letter to archbifhop Ufher, March 10, 1615, writes thus: " Napier lord of Markinfton hath fet my head and hands at work with his new and admirable logarithms. I hope to fee him this fummer, if it pleafe God; for I never faw a book which pleafed me better, and made me more wonder." Briggs accordingly made him the vifit, and ftaid a month with him.

The following paffage, from the life of Lilly the aftrologer, contains a curious account of the meeting of thofe two illuftrious men. " I will acquaint you (fays Lilly) with one memorable ftory related unto me by John Marr, an excellent mathematician and geometrician, whom I conceive you remember. He was fervant to King James and Charles the Firft. At firft when the lord Napier, or Marchifton, made public his logarithms, Mr. Briggs, then reader of the aftronomy lectures at Grefham College in London, was fo furprifed with admiration of them, that he could have no quietnefs in himfelf until he had feen that noble perfon the lord Marchifton, whofe only invention they were: he acquaints John Marr herewith, who went into Scotland before Mr. Briggs, purpofely to be there when thefe two fo learned perfons fhould meet. Mr. Briggs appoints a certain day when to meet at Edinburgh; but failing thereof, the lord Napier was doubtful he would not come. It happened one day as John Marr and the lord Napier were fpeaking of Mr. Briggs; ' Ah, John (faid Marchifton), Mr. Briggs will not now come.' At the very inftant one knocks at the gate; John Marr hafted down, and it proved Mr. Briggs to his great contentment. He brings Mr. Briggs up into my lord's chamber, where almoft one quarter of an hour was fpent, each beholding other almoft with admiration before one word was fpoke. At laft Mr. Briggs began: 'My lord, I have undertaken this long journey purpofely to fee your perfon, and to know by what engine of wit or ingenuity you came firft to think of this moft excellent help into aftronomy, viz, the logarithms; but, my lord, being by you found out, I wonder no body elfe found it out before, when now known it is fo eafy.' He was nobly entertained by the lord Napier; and every fummer after that, during the lord's being alive, this venerable man Mr. Briggs went purpofely into Scotland to vifit him."

Napier made alfo confiderable improvements in fpherical trigonometry &c, particularly by his Catholic or Univerfal Rule, being a general theorem by which he refolves all the cafes of right-angled fpherical triangles in a manner very fimple, and eafy to be remembered, namely, by what he calls the Five Circular Parts. His Conftruction of Logarithms too, befide the labour of them, manifefts the greateft ingenuity. Kepler dedicated his Ephemerides to Napier, which were publifhed in the year 1617; and it appears from many paffages in his letter about this time, that he accounted Napier to be the greateft man of his age in the particular department to which he applied his abilities.

The

The laſt literary exertion of this eminent perſon was the publication of his *Rabdology and Promptuary*, in the year 1617; ſoon after which he died at Marchiſton, the 3d of April in the ſame year, in the 68th year of his age.—The liſt of his works is as follows:

1. A Plain Diſcovery of the Revelation of St. John; 1593.

2. *Logarithmorum Canonis Deſcriptio;* 1614.

3. *Mirifici Logarithmorum Canonis Conſtructio; et sorum ad Naturales ipſorum numeros habitudines; una cum appendice, de alia eaque præſtantiore Logarithmorum ſpecie condenda. Quibus acceſſere propoſitiones ad triangula ſphærica faciliore calculo reſolvenda. Una cum Annotationibus aliquot doctiſſimi D. Henrici Briggii in eas, & memoratam appendicem.* Publiſhed by the author's ſon in 1619.

4. *Rabdologia, ſeu Numerationis per Virgulas, libri duo;* 1617. This contains the deſcription and uſe of the Bones or Rods; with ſeveral other ſhort and ingenious modes of calculation.

5. His Letter to Anthony Bacon (the original of which is in the archbiſhop's library at Lambeth), intitled, Secret Inventions, Profitable and Neceſſary in theſe days for the Defence of this Iſland, and withſtanding Strangers Enemies to God's Truth and Religion; dated June 2, 1596.

NAPIER's *Bones*, or *Rods*, an inſtrument contrived by lord Napier, for the more eaſy performing of the arithmetical operations of multiplication, diviſion, &c. Theſe rods are five in number, made of Bone, ivory, horn, wood, or paſteboard, &c. Their faces are divided into nine little ſquares (fig. 7, pl. 16); each of which is parted into two triangles by diagonals. In theſe little ſquares are written the numbers of the multiplication-table; in ſuch manner as that the units, or right-hand figures, are found in the right-hand triangle: and the tens, or the left-hand figures, in the left-hand triangle; as in the figure.

To Multiply Numbers by NAPIER's *Bones.* Diſpoſe the rods in ſuch manner, as that the top figures may exhibit the multiplicand; and to theſe, on the left-hand, join the rod of units: in which ſeek the right-hand figure of the multiplier: and the numbers correſponding to it, in the ſquares of the other rods, write out, by adding the ſeveral numbers occurring in the ſame rhomb together, and their ſums. After the ſame manner write out the numbers correſponding to the other figures of the multiplier; diſpoſing them under one another as in the common multiplication; and laſtly add the ſeveral numbers into one ſum.

For example, ſuppoſe the multiplicand 5978, and the multiplier 937. From the outermoſt triangle on the right-hand (fig. 8, pl. 16) which correſponds to the right-hand figure of the multiplier 7, write out the figure 6, placing it under the line. In the next rhomb towards the left, add 9 and 5; their ſum being 14, write the right-hand figure 4, againſt 6; carrying the left-hand figure 1 to 4 and 3, which are found in the next rhomb: oin the ſum 8 to 46, already ſet down. After the ſame manner, in the laſt rhomb, add 6 and 5,

5978
937
——
41846
17934
53802
————
5601386
————

and the latter figure of the ſum 11, ſet down as before, and carry 1 to the 3 found in the left-hand triangle; the ſum 4 join as before on the left hand of 1846. Thus you will have 41846 for the product of 5978 by 7. And in the ſame manner are to be found the products for the other figures of the multiplier; after which the whole is to be added together as uſual.

To perform Diviſion by NAPIER's *Bones.* Diſpoſe the rods ſo, as that the uppermoſt figures may exhibit the diviſor; to theſe on the left-hand, join the rod of units. Deſcend under the diviſor, till you meet thoſe figures of the dividend in which it is firſt required how oft the diviſor is found, or at leaſt the next leſs number, which is to be ſubtracted from the dividend; then the number correſponding to this, in the place of units, ſet down for a quotient. And by determining the other parts of the quotient after the ſame manner, the diviſion will be completed.

For example; ſuppoſe the dividend 5601386, and the diviſor 5978; ſince it is firſt enquired how often 5978 is found in 56013, deſcend under the diviſor (fig. 8) till in the loweſt ſeries you find the number 53802, approaching neareſt to 56013; the former of which is to be ſubtracted from the latter, and the figure 9

5978)5601386(937
53802
——
22118
17934
——
41846
41846
——

correſponding to it in the rod of units ſet down for the quotient. To the remainder 2211 join the following figure 8 of the dividend; and the number 17934 being found as before for the next leſs number to it, the correſponding number 3 in the rod of units is to be ſet down for the next figure of the quotient. After the ſame manner the third and laſt figure of the quotient will be found to be 7; and the whole quotient 937.

NATIVITY, in Aſtrology, the ſcheme or figure of the heavens, and particularly the twelve houſes, at the moment when a perſon was born; called alſo the Horoſcope.

To Caſt the NATIVITY, is to calculate the poſition of the heavens, and erect the figure of them for the time of birth.

NATURAL *Day, Year, &c.* See DAY, YEAR, &c.

NATURAL *Horizon,* is the ſenſible or phyſical horizon.

NATURAL *Magic,* is that which only makes uſe of natural cauſes; ſuch as the Treatiſe of J. Bapt. Porta, Magia Naturalia.

NATURAL *Philoſophy,* otherwiſe called *Phyſics,* is that ſcience which conſiders the powers of nature, the properties of natural bodies, and their actions upon one another.

Laws of NATURE, are certain axioms, or general rules, of motion and reſt, obſerved by natural bodies in their actions upon one another. Of theſe Laws, Sir I. Newton has eſtabliſhed three:

1ſt LAW.—That every body perſeveres in the ſame ſtate, either of reſt, or uniform rectilinear motion; unleſs it is compelled to change that ſtate by the action of ſome foreign force or agent. Thus, projectiles perſevere in their motions, except ſo far as they are

retarded

retarded by the refiftance of the air, and the action of gravity : and thus a top, once fet up in motion, only ceafes to turn round, becaufe it is refifted by the air, and by the friction of the plane upon which it moves. Thus alfo the larger bodies of the planets and comets preferve their progreffive and circular motions a long time undiminifhed, in regions void of all fenfible refiftance.—As body is paffive in receiving its motion, and the direction of its motion, fo it retains them, or perfeveres in them, without any change, till it be acted upon by fomething external.

2d Law.—The Motion, or Change of Motion, is always proportional to the moving force by which it is produced, and in the direction of the right line in which that force is impreffed. If a certain force produce a certain motion, a double force will produce double the motion, a triple force triple the motion, and fo on. And this motion, fince it is always directed to the fame point with the generating force, if the body were in motion before, is either to be added to it, as where the motions confpire ; or fubtracted from it, as when they are oppofite ; or combined obliquely, when oblique : being always compounded with it according to the determination of each.

3d Law.—Re-action is always contrary, and equal to action ; or the actions of two bodies upon one another, are always mutually equal, and directed contrary ways ; and are to be eftimated always in the fame right line. Thus, whatever body preffes or draws another, is equally preffed or drawn by it. So, if I prefs a ftone with my finger, the finger is equally preffed by the ftone : if a horfe draw a weight forward by a rope, the horfe is equally oppofed or drawn back towards the weight ; the equal tenfion or ftretch of the rope hindering the progrefs of the one, as it promotes that of the other. Again, if any body, by ftriking on another, do in any manner change its motion, it will itfelf, by means of the other, undergo alfo an equal change in its own motion, by reafon of the equality of the preffure. When two bodies meet, each endeavours to perfevere in its ftate, and refifts any change : and becaufe the change which is produced in either may be equally meafured by the action which it excites upon the other, or by the refiftance which it meets with from it, it follows that the changes produced in the motions of each are equal, but are made in contrary directions : the one acquires no new force but what the other lofes in the fame direction ; nor does this laft lofe any force but what the other acquires ; and hence, though by their collifions, motion paffes from the one to the other, yet the fum of their motions, eftimated in a given direction, is preferved the fame, and is unalterable by their mutual actions upon each other. In thefe actions the changes are equal ; not thofe, we mean, of the velocities, but thofe of the motions, or momentums ; the bodies being fuppofed free from any other impediments. For the changes of velocities, which are likewife made contrary ways, inafmuch as the motions are equally changed, are reciprocally proportional to the bodies or maffes.

This law obtains alfo in attractions.

NAVIGATION, is the art of conducting a fhip at fea from one port or place to another.

This is perhaps the moft ufeful of all arts, and is of the higheft antiquity. It may be impoffible to fay who

5

were the inventors of it ; but it is probable that many people cultivated it, independent of each other, who inhabited the coafts of the fea, and had occafion, or found it convenient, to convey themfelves upon the water from place to place ; beginning from rafts and logs of wood, and gradually improving in the ftructure and management of their veffels, according to the length of time, and extent of their voyages. Writers however afcribe the invention of this art to different perfons, or nations, according to their different fources of information. Thus,

The poets refer the invention of Navigation to Neptune, fome to Bacchus, others to Hercules, to Jafon, or to Janus, who it is faid made the firft fhip. Hiftorians afcribe it to the Æginetes, the Phœnicians, Tyrians, and the ancient inhabitants of Britain. Some are of opinion that the firft hint was taken from the flight of the kite ; and fome, as Oppian (De Pifcibus, lib. 1) from the fifh called Nautilus ; while others afcribe it to accident ; and others again deriving the hint and invention from Noah's ark.

However, hiftory reprefents the Phœnicians, efpecially thofe of the capital Tyre, as the firft navigators that made any extenfive progrefs in the art, fo far as has come to our knowledge ; and indeed it muft have been this very art that made their city what it was. For this purpofe, Lebanon, and the other neighbouring mountains, furnifhing them with excellent wood for fhip-building, they were fpeedily mafters of a numerous fleet, with which conftantly hazarding new navigations, and fettling new trades, they foon arrived at an incredible pitch of opulence and populoufnefs ; fo as to be in a condition to fend out colonies, the principal of which was that of Carthage ; which, keeping up their Phœnician fpirit of commerce, in time far furpaffed Tyre itfelf ; fending their merchant fhips through Hercules's pillars, now the ftraits of Gibraltar, and thence along the weftern coafts of Africa and Europe ; and even, according to fome authors, to America itfelf. The city of Tyre being deftroyed by Alexander the Great, its Navigation and commerce were transferred by the conqueror to Alexandria, a new city, well fituated for thefe purpofes, and propofed for the capital of the empire of Afia, the conqueft of which Alexander then meditated. And thus arofe the Navigation of the Egyptians ; which was afterwards fo cultivated by the Ptolomies, that Tyre and Carthage were quite forgotten.

Egypt being reduced to a Roman province after the battle of Actium, its trade and Navigation fell into the hands of Auguftus ; in whofe time Alexandria was only inferior to Rome ; and the magazines of the capital of the world were wholly fupplied with merchandizes from the capital of Egypt.

At length, Alexandria itfelf underwent the fate of Tyre and Carthage ; being furprifed by the Saracens, who, in fpite of the emperor Heraclius, overfpread the northern coafts of Africa, &c ; whence the merchants being driven, Alexandria has ever fince been in a languifhing ftate, though ftill it has a confiderable part of the commerce of the chriftian merchants trading to the Levant.

The fall of Rome and its empire drew along with it not only that of learning and the polite arts, but that of

Naviga-

Navigation alfo; the barbarians, into whofe hands it fell, contenting themfelves with the fpoils of the induftry of their predeceffors.

But no fooner were the brave among thofe nations well fettled in their new provinces; fome in Gaul, as the Franks; others in Spain, as the Goths; and others in Italy, as the Lombards; but they began to learn the advantages of Navigation and commerce, with the methods of managing them, from the people they fub- dued; and this with fo much fuccefs, that in a little time fome of them became able to give new leffons, and fet on foot new inftitutions for its advantage. Thus it is to the Lombards we ufually afcribe the invention and ufe of banks, book-keeping, exchanges, rechanges, &c.

It does not appear which of the European people, after the fettlement of their new mafters, firft betook themfelves to Navigation and commerce.—Some think it began with the French; though the Italians feem to have the jufter title to it, and are ufually confidered as the reftorers of them, as well as of the polite arts, which had been banifhed together from the time the empire was torn afunder. It is the people of Italy then, and particularly thofe of Venice and Genoa, who have the glory of this reftoration; and it is to their advanta- geous fituation for Navigation that they in a great meafure owe their glory. From about the time of the 6th century, when the inhabitants of the iflands in the bottom of the Adriatic began to unite together, and by their union to form the Venetian ftate, their fleets of merchantmen were fent to all the parts of the Me- diterranean; and at laft to thofe of Egypt, particularly Cairo, a new city, built by the Saracen princes on the eaftern banks of the Nile, where they traded for their fpices and other products of the Indies. Thus they flourifhed, increafed their commerce, their Navi- gation, and their conquefts on the terra firma, till the league of Cambray in 1508, when a number of jealous princes confpired to their ruin; which was the more eafily effected by the diminution of their Eaft- India commerce, of which the Portuguefe had got one part, and the French another. Genoa too, which had cultivated Navigation at the fame time with Venice, and that with equal fuccefs, was a long time its dan- gerous rival, difputed with it the empire of the fea, and fhared with it the trade of Egypt; and other parts both of the eaft and weft.

Jealoufy foon began to break out; and the two re- publics coming to blows, there was almoft continual war for three centuries, before the fuperiority was afcertained; when, towards the end of the 14th century, the battle of Chioza ended the ftrife: the Genoefe, who till then had ufually the advantage, having now loft all; and the Venetians almoft become defperate, at one happy blow, beyond all expectation, fecured to themfelves the empire of the fea, and the fuperiority in commerce.

About the fame time that Navigation was retrieved in the fouthern parts of Europe, a new fociety of mer- chants was formed in the north, which not only car- ried commerce to the greateft perfection it was capa- ble of, till the difcovery of the Eaft and Weft Indies, but alfo formed a new fcheme of laws for the regula- tion of it, which ftill obtain under the name of, *Ufes and Cuftoms of the Sea.* This fociety is that ce- lebrated league of the Hanfe-towns, begun about the year 1164.

The art of Navigation has been greatly improved in modern times, both in refpect of the form of the veffels themfelves, and the methods of working or con- ducting them. The ufe of rowers is now entirely fuperceded by the improvements made in the fails, rigging, &c. It is alfo very probable, that the An- cients were neither fo well fkilled as the Moderns, in finding the latitudes, nor in fteering their veffels in places of difficult Navigation, as the Moderns. But the greateft advantage which thefe have over the Ancients, is from the mariner's compafs, by which they are enabled to find their way with as much fa- cility in the midft of an immeafurable ocean, as the Ancients could have done by creeping along the coaft, and never going out of fight of land. Some people indeed contend, that this is no new invention, but that the Ancients were acquainted with it. They fay, it was impoffible for Solomon's fhips to go to Ophir, Tarfhifh, and Parvaim, which laft they will have to be Peru, without this ufeful inftrument. They infift, that it was impoffible for the Ancients to be acquainted with the attractive virtue of the magnet, without knowing its polarity. They even affirm, that this property of the magnet is plainly mentioned in the book of Job, where the loadftone is called topaz, or the ftone that turns itfelf. But, not to mention that Mr. Bruce has lately made it appear highly probable that Solomon's fhips made no more than coafting voyages, it is certain that the Romans, who conquered Judea, were ignorant of this inftrument; and it is very probable, that fo ufeful an invention, if once it had been commonly known to a nation, would never have been forgotten, or perfectly concealed from fo prudent a people as the Romans, who were fo much interefted in the difcovery of it.

Among thofe who do agree that the mariner's com- pafs is a modern invention, it has been much difputed who was the inventor. Some give the honour of it to Flavio Gioia of Amalfi in Campania, about the begin- ning of the 14th century; while others fay that it came from the eaft, and was earlier known in Europe. But, at whatever time it was invented, it is certain, that the mariner's compafs was not commonly ufed in Navigation before the year 1420. In that year the fcience was confiderably improved under the aufpices of Henry duke of Vifco, brother to the king of Portugal. In the year 1485, Roderic and Jofeph, phyficians to king John the 2d of Portugal, together with one Mar- tin de Bohemia, a Portuguefe native of the ifland of Fayal, and pupil to Regiomontanus, calculated tables of the fun's declination for the ufe of failors, and re- commended the aftrolabe for taking obfervations at fea. The celebrated Columbus, it is faid, availed himfelf of Martin's inftructions, and improved the Spaniards in the knowledge of this art; for the farther progrefs of which, a lecture was afterwards founded at Seville by the emperor Charles the 5th.

The difcovery of the variation of the compafs, is claimed by Columbus, and by Sebaftian Cabot. The former certainly did obferve this variation without having heard of it from any other perfon, on the 14th of September 1492, and it is very probable that Cabot might do the fame. At that time it was found that there was no variation at the Azores, for which reafon.

fon fome geographers made that the firft meridian, though it has fince been difcovered that the variation alters in time. The ufe of the crofs-ftaff now began to be introduced among failors. This ancient inftrument is defcribed by John Werner of Nuremberg, in his annotations on the firft book of Ptolomy's Geography, printed in 1514: he recommends it for obferving the diftance between the moon and fome ftar, from which to determine the longitude.

At this time the art of Navigation was very imperfect, from the ufe of the plane chart, which was the only one then known, and which, by its grofs errors, muft have greatly mifled the mariner, efpecially in places far diftant from the equator; and alfo from the want of books of inftruction for feamen.

At length two Spanifh treatifes came out, the one by Pedro de Medina, in 1545; and the other by Martin Cortes, or Curtis as it is printed in Englifh, in 1556, though the author fays he compofed it at Cadiz in 1545, containing a complete fyftem of the art as far as it was then known. Medina, in his dedication to Philip prince of Spain, laments that multitudes of fhips daily perifhed at fea, becaufe there were neither teachers of the art, nor books by which it might be learned; and Cortes, in his dedication, boafts to the emperor, that he was the firft who had reduced Navigation into a compendium, valuing himfelf much on what he had performed. Medina defended the plane chart; but he was oppofed by Cortes, who fhewed its errors, and endeavoured to account for the variation of the compafs, by fuppofing the needle was influenced by a magnetic pole, different from that of the world, and which he called the *point attractive*: which notion has been farther profecuted by others. Medina's book was foon tranflated into Italian, French, and Flemifh, and ferved for a long time as a guide to foreign navigators. However, Cortes was the favourite author of the Englifh nation, and was tranflated in 1561, by Richard Eden, while Medina's work was much neglected, though tranflated alfo within a fhort time of the other. At that time a fyftem of Navigation confifted of materials fuch as the following: An account of the Ptolomaic hypothefis, and the circles of the fphere; of the roundnefs of the earth, the longitudes, latitudes, climates, &c, and eclipfes of the luminaries; a calendar; the method of finding the prime, epact, moon's age, and tides; a defcription of the compafs, an account of its variation, for the difcovering of which Cortes faid an inftrument might eafily be contrived; tables of the fun's declination for 4 years, in order to find the latitude from his meridian altitude; directions to find the fame by certain ftars: of the courfe of the fun and moon; the length of the days; of time and its divifions; the method of finding the hour of the day and night; and laftly, a defcription of the fea-chart, on which to difcover where the fhip is; they made ufe alfo of a fmall table, that fhewed, upon an alteration of one degree of the latitude, how many leagues were run on each rhumb, together with the departure from the meridian; which might be called a table of diftance and departure, as we have now a table of difference of latitude and departure. Befides, fome inftruments were defcribed, efpecially by Cortes; fuch as, one to find the place and declination of the fun, with the age and place of the moon; certain dials, the aftrolabe,

and crofs-ftaff; with a complex machine to difcover the hour and latitude at once.

About the fame time propofals were made for finding the longitude by obfervations of the moon. In 1530, Gemma Frifius advifed the keeping of the time by means of fmall clocks or watches, then newly invented, as he fays. He alfo contrived a new fort of crofs-ftaff, and an inftrument called the Nautical Quadrant; which laft was much praifed by William Cuningham, in his Cofmographical Glafs, printed in the year 1559.

In the year 1537 Pedro Nunez, or Nonius, publifhed a book in the Portuguefe language, to explain a difficulty in Navigation, propofed to him by the commander Don Martin Alphonfo de Sufa. In this work he expofes the errors of the plane chart, and gives the folution of feveral curious aftronomical problems; among which is that of determining the latitude from two obfervations of the fun's altitude and the intermediate azimuth being given. He obferved, that though the rhumbs are fpiral lines, yet the direct courfe of a fhip will always be in the arch of a great circle, by which the angle with the meridians will continually change: all that the fteerfman can here do for preferving the original rhumb, is to correct thefe deviations as foon as they appear fenfible. But thus the fhip will in reality defcribe a courfe without the rhumb-line intended; and therefore his calculations for affigning the latitude, where any rhumb-line croffes the feveral meridians, will be in fome meafure erroneous. He invented a method of dividing a quadrant by means of concentric circles, which, after being much improved by Dr. Halley, is ufed at prefent, and is called a Nonius.

In 1577, Mr William Bourne publifhed a treatife, in which, by confidering the irregularities in the moon's motion, he fhews the errors of the failors in finding her age by the epact, and alfo in determining the hour from obferving on what point of the compafs the fun and moon appeared. In failing towards high latitudes, he advifes to keep the reckoning by the globe, as the plane chart is moft erroneous in fuch fituations. He defpairs of our ever being able to find the longitude, unlefs the variation of the compafs fhould be occafioned by fome fuch attractive point as Cortes had imagined; of which however he doubts: but as he had fhewn how to find the variation at all times, he advifes to keep an account of the obfervations, as ufeful for finding the place of the fhip; which advice was profecuted at large by Simon Stevin in a treatife publifhed at Leyden in 1599; the fubftance of which was the fame year printed at London in Englifh by Mr. Edward Wright, intitled the *Haven-finding Art*. In the fame old tract alfo is defcribed the way by which our failors eftimate the rate of a fhip in her courfe, by the inftrument called the Log. The author of this contrivance is not known; neither was it farther noticed till 1607, when it is mentioned in an Eaft-India voyage publifhed by Purchas: but from this time it became common, and mentioned by all authors on Navigation; and it ftill continues to be ufed as at firft, though many attempts have been made to improve it, and contrivances propofed to fupply its place; fome of which have fucceeded in ftill water, but proved ufelefs in a ftormy fea.

In

In 1581 Michael Coignet, a native of Antwerp, published a Treatise, in which he animadverted on Medina. In this he shewed, that as the rhumbs are spirals, making endless revolutions about the poles, numerous errors must arise from their being represented by straight lines on the sea-charts; but though he hoped to find a remedy for these errors, he was of opinion that the proposals of Nonius were scarcely practicable, and therefore in a great measure useless. In treating of the sun's declination, he took notice of the gradual decrease in the obliquity of the ecliptic; he also described the Cross-Staff with three transverse pieces, as it was then in common use among the sailors. He likewise gave some instruments of his own invention; but all of them are now laid aside, excepting perhaps his Nocturnal. He constructed a sea-table, to be used by such as sailed beyond the 60th degree of latitude; and at the end of the book is delivered a Method of Sailing on a Parallel of Latitude, by means of a ring dial and a 24 hour glass.

In the same year Mr. Robert Norman published his Discovery of the Dipping-needle, in a pamphlet called the New Attractive; to which is always subjoined Mr. William Burroughs's Discourse of the Variation of the Compass.—In 1594, Capt. John Davis published a small treatise, entitled the Seaman's Secrets, which was much esteemed in its time.

The writers of this period complained much of the errors of the plane chart, which continued still in use, though they were unable to discover a proper remedy: till Gerrard Mercator contrived his Universal Map, which he published in 1569, without clearly understanding the principles of its construction: these were first discovered by Mr. Edward Wright, who sent an account of the true method of dividing the meridian from Cambridge, where he was a Fellow, to Mr. Blundeville, with a short table for that purpose, and a specimen of a chart so divided. These were published by Blundeville in 1594, among his Exercises; to the later editions of which was added his Discourse of Universal Maps, first printed in 1589. However, in 1599 Mr. Wright printed his Correction of certain Errors in Navigation, in which work he shews the reason of this division, the manner of constructing his table, and its uses in Navigation. A second edition of this treatise, with farther improvements, was printed in 1610, and a third edition by Mr. Moxon, in 1657.—The Method of Approximation, by what is called the middle latitude, now used by our sailors, occurs in Gunter's works, first printed in 1623.—About this time Logarithms began to be introduced, which were applied to Navigation in a variety of ways by Mr. Edmund Gunter; though the first application of the Logarithmic Tables to the Cases of Sailing, was by Mr. Thomas Addison, in his Arithmetical Navigation, printed in 1625.—In 1635 Mr. Henry Gellibrand printed a Discourse Mathematical on the Variation of the Magnetical Needle, containing his discovery of the changes to which the variation is subject.—In 1631, Mr. Richard Norwood published an excellent Treatise of Trigonometry, adapted to the invention of logarithms, particularly in applying Napier's general canons; and for the farther improvement of Navigation, he undertook the laborious work of measuring a degree of the

meridian, for examining the divisions of the log-line. He has given a full and clear account of this operation in his Seaman's Practice, first published in 1637; where he also describes his own excellent method of setting down and perfecting a sea-reckoning, &c. This treatise, and that of Trigonometry, were often reprinted, as the principal books for learning scientifically the art of Navigation. What he had delivered, especially in the latter of them, concerning this subject, was contracted as a manual for sailors in a very small piece, called his Epitome, which has gone through a great number of editions.—About the year 1645, Mr. Bond published, in Norwood's Epitome, a very great improvement in Wright's method, by a property in his meridian line, by which its divisions are more scientifically assigned than the author was able to effect; which he deduced from this theorem, that these divisions are analogous to the excesses of the logarithmic tangents of half the respective latitudes increased by 45 degrees, above the logarithm of the radius: this he afterwards explained more fully in the 3d edition of Gunter's works, printed in 1653; and the demonstration of the general theorem was supplied by Mr. James Gregory of Aberdeen, in his Exercitationes Geometricæ, printed at London in 1668, and afterwards by Dr. Halley, in the Philos. Trans. numb. 219, as also by Mr. Cotes, numb. 388.—In 1700, Mr. Bond, who imagined that he had discovered the longitude, by having discovered the true theory of the magnetic variation, published a general map, on which curve lines were drawn, expressing the paths or places where the magnetic needle had the same variation. The positions of these curves will indeed continually suffer alterations; and therefore they should be corrected from time to time, as they have already been for the years 1744, and 1756, by Mr. William Mountaine, and Mr. James Dodson.—The allowances proper to be made for lee-way, are very particularly set down by Mr. John Buckler, and published in a small tract first printed in 1702, intitled a New Compendium of the whole Art of Navigation, written by Mr. William Jones.

As it is now generally agreed that the earth is a spheroid, whose axis or polar diameter is shorter than the equatorial diameter, Dr. Murdoch published a tract in 1741, in which he adapted Wright's, or Mercator's sailing to such a figure; and in the same year Mr. Maclaurin also, in the Philos. Trans. numb. 461, for determining the meridional parts of a spheroid; and he has farther prosecuted the same speculation in his Fluxions, printed in 1742.

The method of finding the longitude at sea, by the observed distances of the moon from the sun and stars, commonly called the Lunar method, was proposed at an early stage in the Art of Navigation, and has now been happily carried into effectual execution by the encouragement of the Board of Longitude, which was established in England in the year 1714, for rewarding any successful endeavours to keep the longitude at sea. In the year 1767, this Board published a Nautical Almanac, which has been continued annually ever since, by the advice, and under the direction of the astronomer royal at Greenwich: this work is purposely adapted to the use of navigators in long voyages, and, among a great many useful articles, contains tables of the

lunar

lunar diftances accurately computed for every 3 hours in the year, for the purpofe of comparing the diftance thus known for any time, with the diftance obferved in an unknown place, from whence to compute the longitude of that place. Under the aufpices of this Board too, befides giving encouragement to the authors of many ufeful tables and other works, which would otherwife have been loft, time-keepers have been brought to a wonderful degree of perfection, by Mr. Harrifon, Mr. Arnold, and many other perfons, which have proved highly advantageous in keeping the time during long voyages at fea, and thence giving the longitude.

Some of the other principal writers on Navigation are Bartholomew Crefcenti, of Rome, in 1607; Willebrord Snell, at Leyden, in 1624, his Typhis Batavus; Geo. Fournier, at Paris, 1633; John Baptift Riccioli, at Bologna, in 1661; Dechales, in 1674 and 1677; the Sieur Blondel St. Aubin, in 1671 and 1673; M. Daffier, in 1683; M. Sauveur, in 1692; M. John Bouguer, in 1698; F. Pezenas, in 1733 and 1741; and M. Peter Bouguer, who, in 1753, publifhed a very elaborate treatife on this fubject, intitled, Nouveau Traité de Navigation; in which he gives a variation compafs of his own invention, and attempts to reform the Log, as he had before done in the Memoirs of the Academy of Sciences for 1747. He is alfo very particular in determining the lunations more accurately than by the common methods, and in defcribing the corrections of the dead reckoning. This book was abridged and improved by M. de la Caille, in 1760. To thefe may be added the Navigation of Don George Juan of Spain, in 1757. And, in our own nation, the feveral treatifes of Meffieurs Newhoufe, Seller, Hodgfon, Atkinfon, Harris, Patoun, Hauxley, Wilfon, Moore, Nicholfon, &c; but, over all, The Elements of Navigation, in 2 vols, by Mr. John Robertfon, firft printed about the year 1750, and fince often re-printed; which is the moft complete work of the kind extant; and to which work is prefixed a Differtation on the Rife and Progrefs of the modern Art of Navigation, by Dr. James Wilfon, containing a very learned and elaborate hiftory of the writings and improvements in this art.

For an account of the feveral inftruments ufed in this art, with the methods for the longitude, and the various kinds and methods of Navigation, &c, fee the refpective articles themfelves.

NAVIGATION is either Proper or Common.

NAVIGATION, *Common*, ufually called Coafting, in which the places are at no great diftance from one another, and the fhip fails ufually in fight of land, and moftly within foundings. In this, little elfe is required befides an acquaintance with the lands, the compafs, and founding-line; each of which, fee in its place.

NAVIGATION, *Proper*, is where the voyage is long, and purfued through the main ocean. And here, befides the requifites in the former cafe, are likewife required the ufe of Mercator's Chart, the azimuth and amplitude compaffes, the log-line, and other inftruments for celeftial obfervations; as foreftaffs, quadrants, and other fectors, &c.

Navigation turns chiefly upon four things; two of which being given or known, the reft are thence eafily

found out. Thefe four things are, the difference of latitude, difference of longitude, the reckoning or diftance run, and the courfe or rhumb failed on. The latitudes are eafily found, and that with fufficient accuracy: the courfe and diftance are had by the log-line, or dead reckoning, together with the compafs. Nor is there any thing wanting to the perfection of Navigation, but to determine the longitude. The mathematicians and aftronomers of many ages have applied themfelves, with great affiduity, to fupply this grand defideratum, but not altogether with the fuccefs that was defired, confidering the importance of the object, and the magnificent rewards offered by feveral ftates to the difcoverer. See LONGITUDE.

Sub-Marine NAVIGATION, or the art of failing under water, is mentioned by Mr. Boyle, as the defideratum of the art of Navigation. This, he fays, was fuccefsfully attempted, by Cornelius Drebbel; feveral perfons who were in the boat breathing freely all the time. See DIVING-*bell*.

Inland NAVIGATION, is that performed by fmall craft, upon canals &c, cut through a country.

NAVIGATOR, a perfon capable of conducting a fhip at fea to any place propofed.

NAUTICAL *Chart*, the fame as Sea-Chart.

NAUTICAL *Compafs*, the fame as Sea-Compafs.

NAUTICAL *Planifphere*, a projection or conftruction of the terreftrial globe upon a plane, for the ufe of mariners; fuch as the Plane Chart, and Mercator's Chart.

NEAP, or NEEP-*Tides*, are thofe that happen at equal diftances between the fpring tides. The Neap tides are the loweft, as the fpring tides are the higheft ones, being the oppofites to them. And as the higheft of the fpring tides happens about three days after the full or change of the moon, fo the loweft of the Neap tides fall about three days after the quarters, or four days before the full and change; when the feamen fay it is Deep Neap.

NEAPED. When a fhip wants water, fo that fhe cannot get out of the harbour, out of the dock, or off the ground, the feamen fay, fhe is Neaped, or Beneaped.

NEBULOUS, or Cloudy, a term applied to certain fixed ftars, which fhew a dim, hazy light; being lefs than thofe of the 6th magnitude, and therefore fcarcely vifible to the naked eye, to which at beft they only appear like little dufky fpecks or clouds.

Through a moderate telefcope, thefe Nebulous ftars plainly appear to be congeries or clufters of feveral little ftars. In the Nebulous ftar called Præfepe, in the breaft of Cancer, there are reckoned 36 little ftars, 3 of which Mr. Flamfteed fets down in his catalogue. In the Nebulous ftar of Orion, are reckoned 21. F. le Compte adds, that there are 40 in the Pleïades; 12 in the ftar in the middle of Orion's fword; 500 in the extent of two degrees of the fame conftellation; and 2500 in the whole conftellation. It may farther be obferved, that the galaxy, or milky-way, is a continued affemblage of Nebulæ, or vaft clufters of fmall ftars.

NEEDHAM (JOHN TUBERVILLE), a refpectable philofopher and catholic divine, was born at London December 10, 1713. His father poffeffed a confiderable

able patrimony at Hilston, in the county of Monmouth, being of the younger or catholic branch of the Needham family, and who died young, leaving but a small fortune to his four children. Our author, who was the eldest son, studied in the English college of Douai, where he took orders, taught rhetoric for several years, and surpassed all the other professors of that seminary in the knowledge of experimental philosophy.

In 1740, he was engaged by his superiors in the service of the English mission, and was entrusted with the direction of the school-erected at Twyford, near Winchester, for the education of the Roman Catholic youth. —In 1744 he was appointed professor of philosophy in the English college at Lisbon, where, on account of his bad health, he remained only 15 months. After his return, he passed several years at London and Paris, which were chiefly employed in microscopical observations, and in other branches of experimental philosophy. The results of these observations and experiments were published in the Philosophical Transactions of the Royal Society of London in the year 1749, and in a volume in 12mo at Paris in 1750; and an account of them was also given by M. Buffon, in the first volumes of his natural history. There was an intimate connection subsisted between Mr. Needham and this illustrious French naturalist: they made their experiments and observations together; though the results and systems which they deduced from the same objects and operations were totally different.

Mr. Needham was elected a member of the Royal Society of London in the year 1747, and of the Antiquarian Society some time after.—From the year 1751 to 1767 he was chiefly employed in finishing the education of several English and Irish noblemen, by attending them as tutor in their travels through France, Italy, and other countries. He then retired from this wandering life to the English seminary at Paris, and in 1768 was chosen by the Royal Academy of Sciences in that city a corresponding member.

When the regency of the Austrian Netherlands, for the revival of philosophy and literature in that country, formed the project of an Imperial Academy, which was preceded by the erection of a small literary society to prepare the way for its execution, Mr. Needham was invited to Brussels, and was appointed successively chief director of both these foundations; an appointment which he held, together with some ecclesiastical preferments in the Low Countries, till his death, which happened December the 30th 1781.

Mr. Needham's papers inserted in the Philosophical Transactions, were the following, viz:

1. Account of Chalky Tubulous Concretions, called Malm: vol. 42.

2. Microscopical Observations on Worms in Smutty Corn: vol. 42.

3. Electrical Experiments lately made at Paris: vol. 44.

4. Account of M. Buffon's Mirror, which burns at 66 feet: ib.

5. Observations upon the Generation, Composition, and Decomposition of Animal and Vegetable Substances: vol. 45.

6. On the Discovery of Asbestos in France: vol. 51.

Other works printed at Paris, in French, are,

1. New Microscopical Discoveries: 1745.

2. The same enlarged: 1750.

3. On Microscopical, and the Generation of Organized Bodies: 2 vols, 1769.

NEEDLE, *Magnetical*, denotes a Needle, or a slender piece of iron or steel, touched with a loadstone; which, when sustained on a pivot or centre, upon which it plays round at liberty, it settles at length in a certain direction, either duly, or nearly north-and-south, and called the magnetic meridian.

Magnetical Needles are of two kinds; Horizontal and Inclinatory.

Horizontal NEEDLES, are those equally balanced on each side of the pivot which sustains them; and which, playing horizontally, with their two extremes point out the north and south parts of the horizon.

Construction of a Horizontal NEEDLE. Having procured a thin light piece of pure steel, about 6 inches long, a perforation is made in the middle, over which a brass cap is soldered on, having its inner cavity conical, so as to play freely on the style or pivot, which has a fine steel point. To give the Needle its verticity, or directive faculty, it is rubbed or stroked leisurely on each pole of a magnet, from the south pole towards the north; first beginning with the northern end, and going back at each repeated stroke towards the south; being careful not to give a stroke in a contrary direction, which would take away the power again. Also the hand should not return directly back again the same way it came, but should return in a kind of oval figure, carrying the hand about 6 or 8 inches beyond the point where the touch ended, but not beyond on the side where the touch begins.

Before touching, the north end of the Needle, in our hemisphere, is made a little lighter than the other end; because the touch always destroys an exact balance, rendering the north end heavier than the south, and thus causing the Needle to dip. And if, after touching, the Needle be out of its equilibrium, something must be filed off from the heavier side, till it be found to balance evenly.

Needles may also acquire the magnetic virtue by means of artificial magnetic bars in the following manner: Lay two equal Needles parallel and about an inch asunder, with the north end of one and the south end of the other pointing the same way, and apply two conductors in contact with their ends: then, with two magnetic hard bars, one in each hand, and held as nearly horizontal as can be, with the upper ends, of contrary names, turned outwards to the right and left, let a Needle be stroked or rubbed from the middle to both ends at the same time, for ten or twelve times, the north end of a bar going over the south end of a Needle, and the south end of a bar going over the north end of a Needle: then, without moving from the place, change hands with the bars, or in the same hands turn the other ends downwards, and stroke the other Needle in like manner; so will they both be magnetical. But to make them still stronger, repeat the operation three or four times from Needle to Needle, and at laſt turn the lower ſide of each Needle upwards, and repeat the operations of ſtroking them, as on the former ſides.

The Needles that were formerly applied to the compaſs, on board merchant ſhips, were formed of two pieces of ſteel wire, each being bent in the middle, ſo as to form an obtuſe angle, while their ends, being applied together, made an acute one, ſo that the whole repreſented the form of a lozenge. Dr. Knight, who has ſo much improved the compaſs, found, by repeated experiments, that partly from the foregoing ſtructure, and partly from the unequal hardening of the ends, theſe Needles not only varied from the true direction, but from one another, and from themſelves.

Alſo the Needles formerly uſed on board the men of war, and ſome of the larger trading ſhips, were made of one piece of ſteel, of a ſpring temper, and broad towards the ends, but tapering towards the middle. Every Needle of this form is found to have ſix poles inſtead of two, one at each end, two where it becomes tapering, and two at the hole in the middle.

To remedy theſe errors and inconveniences, the Needle which Dr. Knight contrived for his compaſs, is a ſlender parallelopipedon, being quite ſtraight and ſquare at the ends, and ſo has only two poles, although the curves are a little confuſed about the hole in the middle ; though it is, upon the whole, the ſimpleſt and beſt.

Mr. Michell ſuggeſts, that it would be uſeful to increaſe the weight and length of magnetic Needles, which would render them both more accurate and permanent ; alſo to cover them with a coat of linſeed oil, or varniſh, to preſerve them from any ruſt.

A Needle on occaſion may be prepared without touching it on a loadſtone : for a fine ſteel ſewing Needle, gently laid on the water, or delicately ſuſpended in the air, will take the north-and-ſouth direction.— Thus alſo a Needle heated in the fire, and cooled again in the direction of the meridian, or only in an erect poſition, acquires the ſame faculty.

Declination or *Variation of the* NEEDLE, is the deviation of the horizontal Needle from the meridian ; or the angle it makes with the meridian, when freely ſuſpended in an horizontal plane.

A Needle is always changing the line of its direction, traverſing ſlowly to certain limits towards the eaſt and weſt ſides of the meridian. It was at firſt thought that the magnetic Needle pointed due north ; but it was obſerved by Cabot and Columbus that it had a deviation from the north, though they did not ſuſpect that this deviation had itſelf a variation, and was continually changing. This change in the Variation was firſt found out, according to Bond, by Mr. John Mair, ſecondly by Mr. Gunter, and thirdly by Mr. Gellibrand, by comparing together the obſervations made at different times near the ſame place by Mr. Burrowes, Mr. Gunter, and himſelf, and he publiſhed a Diſcourſe upon it in 1635. Soon after this, Mr. Bond ventured to deliver the rate at which the Variation changes for ſeveral years ; by which he foretold that at London in 1657 there would be no Variation of the compaſs, and from that time it would gradually increaſe the other way, or towards the weſt, making certain revolutions ; which happened ac-

cordingly : and upon this Variation he propoſed a method of finding the longitude, which has been farther improved by many others ſince his time, though with very little ſucceſs. See VARIATION.

The period or revolution of the Variation, Henry Philips made only 370 years, but according to Henry Bond it is 600 years, and their yearly motion 36 minutes. The firſt good obſervations of the Variation were by Burrowes, about the year 1580, when the Variation at London was 11° 15′ eaſt ; and ſince that time the Needle has been moving to the weſtward at that place ; alſo by the obſervations of different perſons, it has been found to point, at different times, as below :

Years.	Obſervers.		Variat. E. or W.
1580	Burrowes	-	11° 15′ Eaſt.
1622	Gunter	- -	5 56
1634	Gellibrand	- -	4 3
1640	Bond	- -	3 7
1657	Bond	- -	0 0
1665	Bond	- -	1 23 Weſt.
1666	Bond	- -	1 36
1672	-	- -	2 30
1683	-	- -	4 30
1692	-	- -	6 00
1723	Graham	- -	14 17
1747	-	- -	17 40
1774	Royal Society	-	21 16
1775	Royal Society	-	21 43
1776	Royal Society	-	21 47
1777	Royal Society	-	22 12
1778	Royal Society	-	22 20
1779	Royal Society	-	22 28
1780	Royal Society	-	22 41

By this Table it appears that, from the firſt obſervations in 1580 till 1657, the change in the Variation was 11° 15′ in 77 years, which is at the rate nearly of 9′ a year ; and from 1657 till 1780, or the ſpace of 123 years, it changed 22° 41′, which is at the rate of 11′ a year nearly ; which it may be preſumed is very near the truth.

The Variation and Dip of the Needle was for many years carefully obſerved by the Royal Society while they met at Crane Court ; and it is a pity that ſuch obſervations have not been continued ſince that time.

Dipping, or *Inclinatory* NEEDLE, is a Needle to ſhew the Dip of the Magnetic Needle, or how far it points below the horizon.

The Inclination or Dip of the Needle was firſt obſerved by Robert Norman, a compaſs-maker at Ratcliffe ; and according to him, the dip at that place, in the year 1576, was 71° 50′ ; and at the Royal Society it was obſerved for ſome years lately as follows :

viz in 1776	-	72° 30′
1778	-	72 25
1780	-	72 17.

Mr. Henry Bond makes the Variation and Dip of the Needle depend on the ſame motion of the magnetic poles in their revolution, and upon it he founded a method of diſcovering the longitude at ſea.

NEEP

NEEP *Tides.* See NEAP *Tides.*

NEGATIVE, in Algebra, something marked with the sign —, or minus, as being contrary to such as are positive, or marked with the sign plus, +. As Negative powers and roots, Negative quantities, &c. See POWER, ROOT, QUANTITY, &c.

NEGATIVE *Sign*, the sign of subtraction —, or that which denotes something in defect. Stifel is the first author I find who used this mark — for subtraction, or negation, before his time, the word minus itself was used, or else its initial *m*.

The use of the Negative sign in algebra, is attended with several consequences that at first fight are admitted with some difficulty, and has sometimes given occasion to notions that seem to have no real foundation. This sign implies, that the real value of the quantity represented by the letter to which it is prefixed, is to be subtracted; and it serves, with the positive sign, to keep in view what elements or parts enter into the composition of quantities, and in what manner, whether as increments or decrements, that is whether by addition or subtraction, which is of the greatest use in this art.

Hence it serves to express a quantity of an opposite quality to a positive; such as a line in a contrary position, a motion with opposite direction, or a centrifugal force in opposition to gravity; and thus it often saves the trouble of distinguishing, and demonstrating separately, the various cases of proportions, and preserves their analogy in view. But as the proportions of lines depend on their magnitude only, without regard to their position; and motions and forces are said to be equal or unequal, in any given ratio, without regard to their directions; and in general the proportion of quantities relates to their magnitude only, without determining whether they are to be considered as increments or decrements; so there is no ground to imagine any other proportion of $+ a$ and $- b$, than that of the real magnitudes of the quantities represented by a and b, whether these quantities are, in any particular case, to be added or subtracted.

As to the usual arithmetical operations of addition, subtraction, &c, the case is different, as the effect of the Negative sign is here to be carefully attended to, and is to be considered always as producing, in those operations, an effect just opposite to the positive sign. Thus, it is the same thing to subtract a decrement as to add an equal increment, or to subtract $- b$ from $a - b$, is to add $+ b$ to it: and because multiplying a quantity by a Negative number, implies only a repeated subtraction of it, the multiplying $- b$ by $- n$, is subtracting $- b$ as often as there are units in n, and is therefore equivalent to adding $+ b$ so many times, or the same as adding $+ nb$. But if we infer from this, that 1 is to $- n$ as $- b$ to nb, according to the rule, that unit is to one of the factors as the other factor is to the product, there is not ground to imagine that there is any mystery in this, or any other meaning than that the real quantities represented by 1, n, b, and nb are proportional. For that rule relates only to the magnitude of the factors and product, without determining whether any factor, or the product, is additive or subtractive. But this likewise must be determined in algebraic computations; and this is the proper use concerning the signs, without which the operation could not proceed. Because a quantity to be subtracted is never produced, in composition, by any repeated addition of a positive, or repeated subtraction of a Negative, a Negative square number is never produced by composition from a root. Hence the $\sqrt{- 1}$, or the square root of a Negative, implies an imaginary quantity, and in resolution is a mark or character of the impossible cases of a problem, unless it is compensated by another imaginary symbol or supposition, for then the whole expression may have a real signification. Thus $1 + \sqrt{- 1}$, and $1 - \sqrt{- 1}$, taken separately, are both imaginary, but yet their sum is the number 2: as the conditions that separately would render the solution of a problem impossible, in some cases destroy each others effect when conjoined. In the pursuit of general conclusions, and of simple forms for representing them, expressions of this kind must sometimes arise, where the imaginary symbol is compensated in a manner that is not always so obvious.

By proper substitutions, however, the expression may be transformed into another, wherein each particular term may have a real signification, as well as the whole expression.

The theorems that are sometimes briefly discovered by the use of this symbol, may be demonstrated without it by the inverse operation, or some other way; and though such symbols are of some use in the computations in the method of fluxions, &c, its evidence cannot be said to depend upon any arts of this kind. See Maclaurin's Fluxions, book 2, chap. 1.

Mr. Baron Maseres published a pretty large book in quarto, on the use of the Negative Sign in algebra.

For the rules or ways of using the Negative sign in the several rules of Algebra, see those rules severally, viz, ADDITION, SUBTRACTION, MULTIPLICATION, &c. And for the method of managing the roots of Negative quantities, see IMPOSSIBLES.

NEPER. See NAPIER.

NEWEL, the upright post that stairs turn about; being that part of the staircase which sustains the steps.

NEWTON (Dr. JOHN), an eminent English mathematician and divine, was the grandson of John Newton of Axmouth in Devonshire, and son of Humphrey Newton of Oundle in Northamptonshire, where he was born in 1622. After receiving the proper foundation of a grammar education, he was sent to Oxford, where he was entered a commoner of St. Edmund's Hall in 1637. He took the degree of bachelor of arts in 1641; and the year following he was created master, in precedence to many students of quality, on account of his distinguished talents in the great branches of literature. His genius leading him strongly to astronomy and mathematics, he applied himself diligently to those sciences, as well as to divinity, and made a great proficiency in them, which he found of some service to him during Cromwell's government.

After the restoration of Charles the 2d, he reaped the fruits of his loyalty: being created doctor of divinity at Oxford, Sept. 1661, he was made one of the king's chaplains, and rector of Rofs in Herefordshire, instead of Mr. John Toombes, ejected for nonconformity. He held this living till his death, which happened at Rofs on Christmas day 1678, at 56 years of age.

U 2

Mr

Mr. Wood gave him the character of a capricious and humourſome perſon. However that be, his writ ings are a proof of his great application to ſtudy, and a ſufficient monument of his genius and ſkill in the mathematical ſciences. Theſe are,

1. Aſtronomia Britannica, &c : in 4to, 1656.
2. Help to Calculation ; with Tables of Declination, &c : 4to, 1657.
3. Trigonometria Britannica, in two books ; the one compoſed by our author, and the other tranſlated from the Latin of Henry Gellibrand : folio, 1658.
4. Chiliades Centum Logarithmorum, printed with,
5. Geometrical Trigonometry : 1659.
6. Mathematical Elements, three parts : 4to, 1660.
7. A Perpetual Diary, or Almanac : 1662.
8. Deſcription of the Uſe of the Carpenter's Rule : 1667.
9. Ephemerides, ſhewing the intereſt and rate of money at 6 per cent. &c : 1667.
10. Chiliades Centum Logarithmorum et Tabula Partium Proportionalium : 1667.
11. The Rule of Intereſt, or the Caſe of Decimal Fractions, &c, part 2 : 8vo, 1668.
12. School-paſtimes for young children, &c : 8vo, 1669.
13. Art of Practical Gauging, &c : 1669.
14. Introduction to the art of Rhetoric : 1671.
15. The Art of Natural Arithmetic in Whole Numbers, and Fractions Vulgar and Decimal : 8vo, 1671.
16. The Engliſh Academy : 8vo, 1677.
17. Coſmography.
18. Introduction to Aſtronomy.
19. Introduction to Geography : 8vo, 1678.

NEWTON (Sir ISAAC), one of the greateſt philoſophers and mathematicians the world has produced, was born at Woolſtrop in Lincolnſhire on Chriſtmas day 1642. He was deſcended from the eldeſt branch of the family of Sir John Newton, bart. who were lords of the manor of Woolſtrop, and had been poſſeſſed of the eſtate for about two centuries before, to which they had removed from Weſtley in the ſame county, but originally they came from the town of Newton in Lancaſhire. Other accounts ſay, I think more truly, that he was the only child of Mr. John Newton of Coleſworth, near Grantham in Lincolnſhire, who had there an eſtate of about 120l. a year, which he kept in his own hands. His mother was of the ancient and opulent family of the Ayſcoughs, or Aſkews, of the ſame county. Our author loſing his father while he was very young, the care of his education devolved on his mother, who, though ſhe married again after his father's death, did not neglect to improve by a liberal education the promiſing genius that was obſerved in her ſon. At 12 years of age, by the advice of his maternal uncle, he was ſent to the grammar ſchool at Grantham, where he made a good proficiency in the languages, and laid the foundation of his future ſtudies. Even here was obſerved in him a ſtrong inclination to figures and philoſophical ſubjects. One trait of this early diſpoſition is told of him : he had then a rude method of meaſuring the force of the wind blowing againſt him, by obſerving how much farther he could leap in the direction of the wind, or blowing

on his back, than he could leap the contrary way, or oppoſed to the wind : an early mark of his original infantine genius.

After a few years ſpent here, his mother took him home ; intending, as ſhe had no other child, to have the pleaſure of his company ; and that, after the manner of his father before him, he ſhould occupy his own eſtate.

But inſtead of minding the markets, or the buſineſs of the farm, he was always ſtudying and poring over his books, even by ſtealth, from his mother's knowledge. On one of theſe occaſions his uncle diſcovered him one day in a hay-loft at Grantham, whither he had been ſent to the market, working a mathematical problem ; and having otherwiſe obſerved the boy's mind to be uncommonly bent upon learning, he prevailed upon his ſiſter to part with him ; and he was accordingly ſent, in 1660, to Trinity College in Cambridge, where his uncle, having himſelf been a member of it, had ſtill many friends. Iſaac was ſoon taken notice of by Dr. Barrow, who was ſoon after appointed the firſt Lucaſian profeſſor of mathematics ; and obſerving his bright genius, contracted a great friendſhip for him. At his outſetting here, Euclid was firſt put into his hands, as uſual, but that author was ſoon diſmiſſed ; ſeeming to him too plain and eaſy, and unworthy of taking up his time. He underſtood him almoſt before he read him ; and a caſt of his eye upon the contents of his theorems, was ſufficient to make him maſter of them : and as the analytical method of Des Cartes was then much in vogue, he particularly applied to it, and Kepler's Optics, &c, making ſeveral improvements on them, which he entered upon the margins of the books as he went on, as his cuſtom was in ſtudying any author.

Thus he was employed till the year 1664, when he opened a way into his new method of Fluxions and Infinite Series ; and the ſame year took the degree of bachelor of arts. In the mean time, obſerving that the mathematicians were much engaged in the buſineſs of improving teleſcopes, by grinding glaſſes into one of the figures made by the three ſections of a cone, upon the principle then generally entertained, that light was homogeneous, he ſet himſelf to grinding of optic glaſſes, of other figures than ſpherical, having as yet no diſtruſt of the homogeneous nature of light : but not hitting preſently upon any thing in this attempt to ſatisfy his mind, he procured a glaſs priſm, that he might try the celebrated phenomena of colours, diſcovered by Grimaldi not long before. He was much pleaſed at firſt with the vivid brightneſs of the colours produced by this experiment ; but after a while, conſidering them in a philoſophical way, with that circumſpection which was natural to him, he was ſurpriſed to ſee them in an oblong form, which, according to the received rule of refractions, ought to be circular. At firſt he thought the irregularity might poſſibly be no more than accidental ; but this was what he could not leave without further enquiry : accordingly, he ſoon invented an infallible method of deciding the queſtion, and the reſult was, his *New Theory of Light and Colours.*

However, the theory alone, unexpected and ſurpriſing as it was, did not ſatisfy him ; he rather conſidered the

the

the proper ufe that might be made of it for improving telefcopes, which was his firft defign. To this end, having now difcovered that light was not homogeneous, but an heterogeneous mixture of differently refrangible rays, he computed the errors arifing from this different refrangibility; and, finding them to exceed fome hundreds of times thofe occafioned by the circular figure of the glaffes, he threw afide his glafs works, and took reflections into confideration. He was now fenfible that optical inftruments might be brought to any degree of perfection defired, in cafe there could be found a reflecting fubftance which would polifh as finely as glafs, and reflect as much light as glafs tranfmits, and the art of giving it a parabolical figure he alfo attained: but thefe feemed to him very great difficulties; nay, he almoft thought them infuperable, when he further confidered, that every irregularity in a reflecting fuperficies makes the rays ftray five or fix times more from their due courfe, than the like irregularities in a refracting one.

Amidft thefe fpeculations, he was forced from Cambridge, in 1665, by the plague; and it was more than two years before he made any further progrefs in the fubject. However, he was far from paffing his time idly in the country; on the contrary, it was here, at this time, that he firft ftarted the hint that gave rife to the fyftem of the world, which is the main fubject of the Principia. In his retirement, he was fitting alone in a garden, when fome apples falling from a tree, led his thoughts upon the fubject of gravity; and, reflecting on the power of that principle, he began to confider, that, as this power is not found to be fenfibly diminifhed at the remoteft diftance from the centre of the earth to which we can rife, neither at the tops of the loftieft buildings, nor on the fummits of the higheft mountains, it appeared to him reafonable to conclude, that this power muft extend much farther than is ufually thought. "Why not as high as the moon? faid he to himfelf; and if fo, her motion muft be influenced by it; perhaps fhe is retained in her orbit by it: however, though the power of gravity is not fenfibly weakened in the little change of diftance at which we can place ourfelves from the centre of the earth, yet it is very poffible that, at the height of the moon, this power may differ in ftrength much from what it is here." To make an eftimate what might be the degree of this diminution, he confidered with himfelf, that if the moon be retained in her orbit by the force of gravity, no doubt the primary planets are carried about the fun by the like power; and, by comparing the periods of the feveral planets with their diftances from the fun, he found, that if any power like gravity held them in their courfes, its ftrength muft decreafe in the duplicate proportion of the increafe of diftance. This he concluded, by fuppofing them to move in perfect circles, concentric to the fun, from which the orbits of the greateft part of them do not much differ. Suppofing therefore the force of gravity, when extended to the moon, to decreafe in the fame manner, he computed whether that force would be fufficient to keep the moon in her orbit.

In this computation, being abfent from books, he took the common eftimate in ufe among the geographers and our feamen, before Norwood had meafured

the earth, namely that 60 miles make one degree of latitude; but as that is a very erroneous fuppofition, each degree containing about 69½ of our Englifh miles, his computation upon it did not make the power of gravity, decreafing in a duplicate proportion to the diftance, anfwerable to the power which retained the moon in her orbit: whence he concluded, that fome other caufe muft at leaft join with the action of the power of gravity on the moon. For this reafon he laid afide, for that time, any further thoughts upon the matter. Mr. Whifton (in his Memoirs, pa. 33) fays, he told him that he thought Des Cartes's vortices might concur with the action of gravity.

Nor did he refume this enquiry on his return to Cambridge, which was fhortly after. The truth is, his thoughts were now engaged upon his newly projected reflecting telefcope, of which he made a fmall fpecimen, with a metallic reflector fpherically concave. It was but a rude effay, chiefly defective by the want of a good polifh for the metal. This inftrument is now in the poffeffion of the Royal Society. In 1667 he was chofen Fellow of his college, and took the degree of mafter of arts. And in 1669 Dr. Barrow refigned to him the mathematical chair at Cambridge, the bufinefs of which appointment interrupted for a while his attention to the telefcope: however, as his thoughts had been for fome time chiefly employed upon optics, he made his difcoveries in that fcience the fubject of his lectures, for the firft three years after he was appointed Mathematical Profeffor: and having now brought his Theory of Light and Colours to a confiderable degree of perfection, and having been elected a Fellow of the Royal Society in Jan. 1672, he communicated it to that body, to have their judgment upon it; and it was afterwards publifhed in their Tranfactions, viz, of Feb. 19, 1672. This publication occafioned a difpute upon the truth of it, which gave him fo much uneafinefs, that he refolved not to publifh any thing further for a while upon the fubject; and in that refolution he laid up his Optical Lectures, although he had prepared them for the prefs. And the Analyfis by Infinite Series, which he had intended to fubjoin to them, unhappily for the world, underwent the fame fate, and for the fame reafon.

In this temper he refumed his telefcope; and obferving that there was no abfolute neceffity for the parabolic figure of the glaffes, fince, if metals could be ground truly fpherical, they would be able to bear as great apertures as men could give a polifh to, he completed another inftrument of the fame kind. This anfwering the purpofe fo well, as, though only half a foot in length, to fhew the planet Jupiter diftinctly round, with his four fatellites, and alfo Venus horned, he fent it to the Royal Society, at their requeft, together with a defcription of it, with further particulars; which were publifhed in the Philofophical Tranfactions for March 1672. Several attempts were alfo made by that fociety to bring it to perfection; but, for want of a proper compofition of metal, and a good polifh, nothing fucceeded, and the invention lay dormant, till Hadley made his Newtonian telefcope in 1723. At the requeft of Leibnitz, in 1676, he explained his invention of Infinite Series, and took notice how far he had improved it by his Method of Fluxions, which however
he

he still concealed, and particularly on this occasion, by a transposition of the letters that make up the two fundamental propositions of it, into an alphabetical order; the letters concerning which are inserted in Collins's Commercium Epistolicum, printed 1712. In the winter between the years 1676 and 1677, he found out the grand proposition, that, by a centripetal force acting reciprocally as the square of the distance, a planet must revolve in an ellipsis, about the centre of force placed in its lower focus, and, by a radius drawn to that centre, describe areas proportional to the times. In 1680 he made several astronomical observations upon the comet that then appeared ; which, for some considerable time, he took not to be one and the same, but two different comets; and upon this occasion several letters passed between him and Mr. Flamsteed.

He was still under this mistake, when he received a letter from Dr. Hook, explaining the nature of the line described by a falling body, supposed to be moved circularly by the diurnal motion of the earth, and perpendicularly by the power of gravity. This letter put him upon enquiring anew what was the real figure in which such a body moved ; and that enquiry, convincing him of another mistake which he had before fallen into concerning that figure, put him upon resuming his former thoughts with regard to the moon ; and Picart having not long before, viz, in 1679, measured a degree of the earth with sufficient accuracy, by using his measures, that planet appeared to be retained in her orbit by the sole power of gravity ; and consequently that this power decreases in the duplicate ratio of the distance ; as he had formerly conjectured. Upon this principle, he found the line described by a falling body to be an ellipsis, having one focus in the centre of the earth. And finding by this means, that the primary planets really moved in such orbits as Kepler had supposed, he had the satisfaction to see that this enquiry, which he had undertaken at first out of mere curiosity, could be applied to the greatest purposes. Hereupon he drew up about a dozen propositions, relating to the motion of the primary planets round the sun, which were communicated to the Royal Society in the latter end of 1683. This coming to be known to Dr. Halley, that gentleman, who had attempted the demonstration in vain, applied, in August 1684, to Newton, who assured him that he had absolutely completed the proof. This was also registered in the books of the Royal Society ; at whose earnest solicitation Newton finished the work, which was printed under the care of Dr. Halley, and came out about midsummer 1687, under the title of, *Philosophiæ naturalis Principia mathematica*, containing in the third book, the Cometic Astronomy, which had been lately discovered by him, and now made its first appearance in the world : a work which may be looked upon as the production of a celestial intelligence rather than of a man.

This work however, in which the great author has built a new system of natural philosophy upon the most sublime geometry, did not meet at first with all the applause it deserved, and was one day to receive. Two reasons concurred in producing this effect : Des Cartes had then got full possession of the world. His philosophy was indeed the creature of a fine imagination, gaily

dressed out : he had given her likewise some of nature's fine features, and painted the rest to a seeming likeness of her. On the other hand, Newton had with an unparalleled penetration, and force of genius, pursued nature up to her most secret abode, and was intent to demonstrate her residence to others, rather than anxious to describe particularly the way by which he arrived at it himself : he finished his piece in that elegant conciseness, which had justly gained the Ancients an universal esteem. In fact, the consequences flow with such rapidity from the principles, that the reader is often left to supply a long chain of reasoning to connect them : so that it required some time before the world could understand it. The best mathematicians were obliged to study it with care, before they could make themselves master of it ; and those of a lower rank durst not venture upon it, till encouraged by the testimonies of the more learned. But at last, when its value came to be sufficiently known, the approbation which had been so slowly gained, became universal, and nothing was to be heard from all quarters, but one general burst of admiration. " Does Mr. Newton eat, drink, or sleep like other men ?" says the marquis de l'Hospital, one of the greatest mathematicians of the age, to the English who visited him. " I represent him to myself as a celestial genius intirely disengaged from matter."

In the midst of these profound mathematical researches, just before his Principia went to the press in 1686, the privileges of the university being attacked by James the 2d, Newton appeared among its most strenuous defenders, and was on that occasion appointed one of their delegates to the high-commission court ; and they made such a defence, that James thought proper to drop the affair. Our author was also chosen one of their members for the Convention-Parliament in 1688, in which he sat till it was dissolved.

Newton's merit was well known to Mr. Montague, then chancellor of the exchequer, and afterwards earl of Halifax, who had been bred at the same college with him ; and when he undertook the great work of recoining the money, he fixed his eye upon Newton for an assistant in it ; and accordingly, in 1696, he was appointed warden of the mint, in which employment, he rendered very signal service to the nation. And three years after he was promoted to be master of the mint, a place worth 12 or 15 hundred pounds per annum, which he held till his death. Upon this promotion, he appointed Mr. Whitton his deputy in the mathematical professorship at Cambridge, giving him the full profits of the place, which appointment itself he also procured for him in 1703. The same year our author was chosen president of the Royal Society, in which chair he sat for 25 years, namely till the time of his death ; and he had been chosen a member of the Royal Academy of Sciences at Paris in 1699, as soon as the new regulation was made for admitting foreigners into that society.

Ever since the first discovery of the heterogeneous mixture of light, and the production of colours thence arising, he had employed a good part of his time in bringing the experiment, upon which the theory is founded, to a degree of exactness that might satisfy himself. The truth is, this seems to have been his favourite invention ; 30 years he had spent in this ardu-

ous

ous tafk, before he publifhed it in 1704. In infinite feries and fluxions, and in the power and rule of gravity in preferving the folar fyftem, there had been fome, though diftant hints, given by others before him: whereas in diffecting a ray of light into its primary conftituent particles, which then admitted of no further feparation; in the difcovery of the different refrangibility of thefe particles thus feparated; and that thefe conftituent rays had each its own peculiar colour inherent in it; that rays falling in the fame angle of incidence have alternate fits of reflection and refraction; that bodies are rendered transparent by the minutenefs of their pores, and become opaque by having them large; and that the moft tranfparent body, by having a great thinnefs, will become lefs pervious to the light: in all thefe, which make up his new theory of light and colours, he was abfolutely and entirely the firft ftarter; and as the fubject is of the moft fubtle and delicate nature, he thought it neceffary to be himfelf the laft finifher of it.

In fact, the affair that chiefly employed his refearches for fo many years, was far from being confined to the fubject of light alone. On the contrary, all that we know of natural bodies, feemed to be comprehended in it; he had found out, that there was a natural action at a diftance between light and other bodies, by which both the reflections and refractions, as well as inflections, of the former, were conftantly produced. To afcertain the force and extent of this principle of action, was what had all along engaged his thoughts, and what after all, by its extreme fubtlety, efcaped his moft penetrating fpirit. However, though he has not made fo full a difcovery of this principle, which directs the courfe of light, as he has in regard to the power by which the planets are kept in their courfes; yet he gave the beft directions poffible for fuch as fhould be difpofed to carry on the work, and furnifhed matter abundantly fufficient to animate them to the purfuit. He has indeed hereby opened a way of paffing from optics to an entire fyftem of phyfics; and, if we look upon his queries as containing the hiftory of a great man's firft thoughts, even in that view they muft be always at leaft entertaining and curious.

This fame year, and in the fame book with his Optics, he publifhed, for the firft time, his Method of Fluxions. It has been already obferved, that thefe two inventions were intended for the public fo long before as 1672; but were laid by then, in order to prevent his being engaged on that account in a difpute about them. And it is not a little remarkable, that even now this laft piece proved the occafion of another difpute, which continued for many years. Ever fince 1684, Leibnitz had been artfully working the world into an opinion, that he firft invented this method.— Newton faw his defign from the beginning, and had fufficiently obviated it in the firft edition of the Principia, in 1687 (viz, in the Scholium to the 2d lemma of the 2d book): and with the fame view, when he now publifhed that method, he took occafion to acquaint the world, that he invented it in the years 1665 and 1666. In the Acta Eruditorum of Leipfic, where an account is given of this book, the author of that account afcribed the invention to Leibnitz, intimating that Newton borrowed it from him. Dr. Keill, the

aftronomical profeffor at Oxford, undertook Newton's defence; and after feveral anfwers on both fides, Leibnitz complaining to the Royal Society, this body appointed a committee of their members to examine the merits of the cafe. Thefe, after confidering all the papers and letters relating to the point in controverfy, decided in favour of Newton and Keill; as is related at large in the life of this laft mentioned gentleman; and thefe papers themfelves were publifhed in 1712, under the title of Commercium Epiftolicum Johannis Collins, 8vo.

In 1705, the honour of knighthood was conferred upon our author by queen Anne, in confideration of his great merit. And in 1714 he was applied to by the Houfe of Commons, for his opinion upon a new method of difcovering the longitude at fea by fignals, which had been laid before them by Ditton and Whifton, in order to procure their encouragement; but the petition was thrown afide upon reading Newton's paper delivered to the committee.

The following year, 1715, Leibnitz, with the view of bringing the world more eafily into the belief that Newton had taken the method of fluxions from his Differential method, attempted to foil his mathematical fkill by the famous problem of the trajectories, which he therefore propofed to the Englifh by way of challenge; but the folution of this, though the moft difficult propofition he was able to devife, and what might pafs for an arduous affair to any other, yet was hardly any more than an amufement to Newton's penetrating genius: he received the problem at 4 o'clock in the afternoon, as he was returning from the Mint; and, though extremely fatigued with bufinefs, yet he finifhed the folution before he went to bed.

As Leibnitz was privy-counfellor of juftice to the elector of Hanover, fo when that prince was raifed to the Britifh throne, Newton came more under the notice of the court; and it was for the immediate fatisfaction of George the Firft, that he was prevailed on to put the laft hand to the difpute about the invention of Fluxions. In this court, Caroline princefs of Wales, afterwards queen confort to George the Second, happened to have a curiofity for philofophical enquiries; no fooner therefore was fhe informed of our author's attachment to the houfe of Hanover, than fhe engaged his converfation, which foon endeared him to her. Here fhe found in every difficulty that full fatisfaction, which fhe had in vain fought for elfewhere; and fhe was often heard to declare publicly, that fhe thought herfelf happy in coming into the world at a juncture of time, which put it in her power to converfe with him. It was at this princefs's folicitation, that he drew up an abftract of his Chronology; a copy of which was at her requeft communicated, about 1718, to fignior Conti, a Venetian nobleman, then in England, upon a promife to keep it fecret. But notwithftanding this promife, the abbe, who while here had alfo affected to fhew a particular friendfhip for Newton, though privately betraying him as much as lay in his power to Leibnitz, was no fooner got acrofs the water into France, than he difperfed copies of it, and procured an antiquary to tranflate it into French, as well as to write a confutation of it. This, being printed at Paris in 1725, was delivered as a prefent from the bookfeller that printed it to our author, that he might obtain, as was faid, his confent

to

to the publication; but though he exprefsly refufed fuch confent, yet the whole was publifhed the fame year Hereupon Newton found it neceffary to publifh a Defence of himfelf, which was inferted in the Philofophical Tranfactions. Thus he, who had fo much all his life long been ftudious to avoid difputes, was unavoidably all his life time, in a manner, involved in them; nor did this laft difpute even finifh at his death, which happened the year following. Newton's paper was republifhed in 1726 at Paris, in French, with a letter of the abbe Conti in anfwer to it; and the fame year fome differtations were printed there by father Souciet againft Newton's Chronological Index, an anfwer to which was inferted by Halley in the Philofophical Tranfactions, numb. 397.

Some time before this bufinefs, in his 80th year, our author was feized with an incontinence of urine, thought to proceed from the ftone in the bladder, and deemed to be incurable. However, by the help of a ftrict regimen and other precautions, which till then he never had occafion for, he procured confiderable intervals of eafe during the five remaining years of his life. Yet he was not free from fome fevere paroxyfms, which even forced out large drops of fweat that ran down his face. In thefe circumftances he was never obferved to utter the leaft complaint, nor exprefs the leaft impatience; and as foon as he had a moment's eafe, he would fmile and talk with his ufual chearfulnefs. He was now obliged to rely upon Mr. Conduit, who had married his niece, for the difcharge of his office in the Mint. Saturday morning March 18, 1727, he read the newfpapers, and difcourfed a long time with Dr. Mead his phyfician, having then the perfect ufe of all his fenfes and his underftanding; but that night he entirely loft them all, and, not recovering them afterwards, died the Monday following, March 20, in the 85th year of his age. His corpfe lay in ftate in the Jerufalemchamber, and on the 28th was conveyed into Weftminfter-abbey, the pall being fupported by the lord chancellor, the dukes of Montrofe and Roxburgh, and the earls of Pembroke, Suffex, and Macclesfield. He was interred near the entrance into the choir on the left hand, where a ftately monument is erected to his memory, with a moft elegant infcription upon it.

Newton's character has been attempted by M. Fontenelle and Dr. Pemberton, the fubftance of which is as follows. He was of a middle ftature, and fomewhat inclined to be fat in the latter part of his life. His countenance was pleafing and venerable at the fame time; efpecially when he took off his peruke, and fhewed his white hair, which was pretty thick. He never made ufe of fpectacles, and loft but one tooth during his whole life. Bifhop Atterbury fays, that, in the whole air of Sir Ifaac's face and make, there was nothing of that penetrating fagacity which appears in his compofitions; that he had fomething rather languid in his look and manner, which did not raife any great expectation in thofe who did not know him.

His temper it is faid was fo equal and mild, that no accident could difturb it. A remarkable inftance of which is related as follows. Sir Ifaac had a favourite little dog, which he called Diamond. Being one day called out of his ftudy into the next room, Diamond was left behind. When Sir Ifaac returned, having been ab-

fent but a few minutes, he had the mortification to find, that Diamond having overfet a lighted candle among fome papers, the nearly finifhed labour of many years was in flames, and almoft confumed to afhes. This lofs, as Sir Ifaac was then very far advanced in years, was irretrievable; yet, without once ftriking the dog, he only rebuked him with this exclamation, "Oh Diamond! Diamond! thou little knoweft the mifchief thou haft done!"

He was indeed of fo meek and gentle a difpofition, and fo great a lover of peace, that he would rather have chofen to remain in obfcurity, than to have the calm of life ruffled by thofe ftorms and difputes, which genius and learning always draw upon thofe that are the moft eminent for them.

From his love of peace, no doubt, arofe that unufual kind of horror which he felt for all difputes: a fteady unbroken attention, free from thofe frequent recoilings infeparably incident to others, was his peculiar felicity; he knew it, and he knew the value of it. No wonder then that controverfy was looked on as his bane. When fome objections, haftily made to his difcoveries concerning light and colours, induced him to lay afide the defign he had taken of publifhing his Optical Lectures, we find him reflecting on that difpute, into which he had been unavoidably drawn, in thefe terms: "I blamed my own imprudence for parting with fo real a bleffing as my quiet, to run after a fhadow." It is true this fhadow, as Fontenelle obferves, did not efcape him afterwards, nor did it coft him that quiet which he fo much valued, but proved as much a real happinefs to him as his quiet itfelf; yet this was a happinefs of his own making: he took a refolution from thefe difputes, not to publifh any more concerning that theory, till he had put it above the reach of controverfy, by the exacteft experiments, and the ftricteft demonftrations; and accordingly it has never been called in queftion fince. In the fame temper, after he had fent the manufcript to the Royal Society, with his confent to the printing of it by them; yet upon Hook's injurioufly infifting that he himfelf had demonftrated Kepler's problem before our author, he determined, rather than be involved again in a controverfy, to fupprefs the third book; and he was very hardly prevailed upon to alter that refolution. It is true, the public was thereby a gainer; that book, which is indeed no more than a corollary of fome propofitions in the firft, being originally drawn up in the popular way, with a defign to publifh it in that form; whereas he was now convinced that it would be beft not to let it go abroad without a ftrict demonftration.

In contemplating his genius, it prefently becomes a doubt, which of thefe endowments had the greateft fhare, fagacity, penetration, ftrength, or diligence; and, after all, the mark that feems moft to diftinguifh it is, that he himfelf made the jufteft eftimation of it, declaring, that if he had done the world any fervice, it was due to nothing but induftry and patient thought; that he kept the fubject of confideration conftantly before him, and waited till the firft dawning opened gradually, by little and little, into a full and clear light. It is faid, that when he had any mathematical problems or folutions in his mind, he would never quit the fubject on any account. And his fervant has

faid,

said, when he has been getting up in a morning, he has sometimes begun to dress, and with one leg in his breeches, sat down again on the bed, where he has remained for hours before he has got his clothes on: and that dinner has been often three hours ready for him before he could be brought to table. Upon this head several little anecdotes are related; among which is the following: Doctor Stukely coming in accidentally one day, when Newton's dinner was left for him upon the table, covered up, as usual, to keep it warm till he cou'd find it convenient to come to table; the doctor lifting the cover, found under it a chicken, which he presently ate, putting the bones in the dish, and replacing the cover. Some time after Newton came into the room, and after the usual compliments sat down to his dinner; but on taking up the cover, and seeing only the bones of the fowl left, he observed with some little surprise, "I thought I had not dined, but I now find that I have."

After all, notwithstanding his anxious care to avoid every occasion of breaking his intense application to study, he was at a great distance from being steeped in philosophy. On the contrary, he could lay aside his thoughts, though engaged in the most intricate researches, when his other affairs required his attention; and, as soon as he had leisure, resume the subject at the point where he had left off. This he seems to have done not so much by any extraordinary strength of memory, as by the force of his inventive faculty, to which every thing opened itself again with ease, if nothing intervened to ruffle him. The readiness of his invention made him not think of putting his memory much to the trial; but this was the offspring of a vigorous intenseness of thought, out of which he was but a common man. He spent therefore the prime of his age in those abstruse researches, when his situation in a college gave him leisure, and while study was his proper business. But as soon as he was removed to the mint, he applied himself chiefly to the duties of that office; and so far quitted mathematics and philosophy, as not to engage in any pursuits of either kind afterwards.

Dr. Pemberton observes, that though his memory was much decayed in the last years of his life, yet he perfectly understood his own writings, contrary to what I had formerly heard, says the doctor, in discourse from many persons. This opinion of theirs might arise perhaps from his not being always ready at speaking on these subjects, when it might be expected he should. But on this head it may be observed, that great geniuses are often liable to be absent, not only in relation to common life, but with regard to some of the parts of science that they are best informed of: inventors seem to treasure up in their minds what they have found out, after another manner, than those do the same things, who have not this inventive faculty. The former, when they have occasion to produce their knowledge, are in some measure obliged immediately to investigate part of what they want; and for this they are not equally fit at all times: from whence it has often happened, that such as retain things chiefly by means of a very strong memory, have appeared off-hand more expert than the discoverers themselves.

It was evidently owing to the same inventive faculty that Newton, as this writer found, had read fewer of

the modern mathematicians than one could have expected; his own prodigious invention readily supplying him with what he might have occasion for in the pursuit of any subject he undertook. However, he often censured the handling of geometrical subjects by algebraic calculations; and his book of algebra he called by the name of *Universal Arithmetic*, in opposition to the injudicious title of *Geometry* which Des Cartes had given to the treatise in which he shews how the geometrician may assist his invention by such kind of computations. He frequently praised Slusius, Barrow, and Huygens, for not being influenced by the false taste which then began to prevail. He used to commend the laudable attempt of Hugo d'Omerique to restore the ancient analysis; and very much esteemed Apollonius's book *De Sectione Rationis*, for giving us a clearer notion of that analysis than we had before. Dr. Barrow may be esteemed as having shewn a compass of invention equal, if not superior, to any of the Moderns, our author only excepted; but Newton particularly recommended Huygens's style and manner: he thought him the most elegant of any mathematical writer of modern times, and the truest imitator of the Ancients. Of their taste and mode of demonstration our author always professed himself a great admirer; and even censured himself for not following them yet more closely than he did; and spoke with regret of his mistake at the beginning of his mathematical studies, in applying himself to the works of Des Cartes, and other algebraic writers, before he had considered the Elements of Euclid with that attention which so excellent a writer deserves.

But if this was a fault, it is certain it was a fault to which we owe both his great inventions in speculative mathematics, and the doctrine of Fluxions and Infinite Series. And perhaps this might be one reason why his particular reverence for the Ancients is omitted by Fontenelle, who however certainly makes some amends by that just elogium which he makes of our author's modesty, which amiable quality he represents as standing foremost in the character of this great man's mind and manners. It was in reality greater than can be easily imagined, or will be readily believed: yet it always continued so without any alteration; though the whole world, says Fontenelle, conspired against it; let us add, though he was thereby robbed of his invention of Fluxions. Nicholas Mercator publishing his *Logarithmotechnia* in 1668, where he gave the quadrature of the hyperbola by an infinite series, which was the first appearance in the learned world of a series of this sort drawn from the particular nature of the curve, and that in a manner very new and abstracted; Dr. Barrow, then at Cambridge, where Mr. Newton, then about 26 years of age, resided, recollected, that he had met with the same thing in the writings of that young gentleman; and there not confined to the hyperbola only, but extended, by general forms, to all sorts of curves, even such as are mechanical; to their quadratures, their rectifications, and their centres of gravity; to the solids formed by their rotations, and to the superficies of those solids; so that, when their determinations were possible, the series stopped at a certain point, or at least their sums were given by stated rules: and if the absolute determinations were impossible, they could yet be infinitely approximated; which is the happiest and most
refined

refined method, fays Fontenelle, of fupplying the defects of human knowledge that man's imagination could poffibly invent. To be matter of fo fruitful and general a theory, was a mine of gold to a geometrician; but it was a greater glory to have been the difcoverer of fo furprifing and ingenious a fyftem. So that Newton, finding by Mercator's book, that he was in the way to it, and that others might follow in his track, fhould naturally have been forward to open his treafures, and fecure the property, which confifted in making the difcovery; but he contented himfelf with his treafure which he had found, without regarding the glory. What an idea does it give us of his unparalleled modefty, when we find him declaring, that he thought Mercator had entirely difcovered his fecret, or that others would, before he fhould become of a proper age for writing! His manufcript upon Infinite Series was communicated to none but Mr. John Collins and the lord Brounker, then Prefident of the Royal Society, who had alfo done fomething in this way himfelf; and even that had not been complied with, but for Dr. Barrow, who would not fuffer him to indulge his modefty fo much as he defired.

It is further obferved, concerning this part of his character, that he never talked either of himfelf or others, nor ever behaved in fuch a manner, as to give the moft malicious cenfurers the leaft occafion even to fufpect him of vanity. He was candid and affable, and always put himfelf upon a level with his company. He never thought either his merit or his reputation fufficient to excufe him from any of the common offices of focial life. No fingularities, either natural or affected, diftinguifhed him from other men. Though he was firmly attached to the church of England, he was averfe to the perfecution of the non-conformifts. He judged of men by their manners; and the true fchifmatics, in his opinion, were the vicious and the wicked. Not that he confined his principles to natural religion, for it is faid he was thoroughly perfuaded of the truth of Revelation; and amidft the great variety of books which he had conftantly before him, that which he ftudied with the greateft application was the Bible, at leaft in the latter years of his life: and he underftood the nature and force of moral certainty as well as he did that of a ftrict demonftration.

Sir Ifaac did not neglect the opportunities of doing good, when the revenues of his patrimony and a profitable employment, improved by a prudent œconomy, put it in his power. We have two remarkable inftances of his bounty and generofity; one to Mr. Maclaurin, extra profeffor of mathematics at Edinburgh, to encourage whofe appointment he offered 20 pounds a year to that office; and the other to his niece Barton, upon whom he had fettled an annuity of 100 pounds per annum. When decency upon any occafion required expence and fhew, he was magnificent without grudging it, and with a very good grace: at all other times, that pomp which feems great to low minds only, was utterly retrenched, and the expence referved for better ufes.

Newton never married; and it has been faid, that " perhaps he never had leifure to think of it; that, being immerfed in profound ftudies during the prime of his age, and afterwards engaged in an employment of great importance, and even quite taken up with the company which his merit drew to him, he was not fenfible of any vacancy in life, nor of the want of a companion at home." Thefe however do not appear to be any fufficient reafons for his never marrying, if he had had an inclination fo to do. It is much more likely that he had a conftitutional indifference to the ftate, and even to the fex in general; and it has even been faid of him, that he never once knew woman.—He left at his death, it feems, 32 thoufand pounds; but he made no will; which, Fontenelle tells us, was becaufe he thought a legacy was no gift.—As to his works, befides what were publifhed in his life-time, there were found after his death, among his papers, feveral difcourfes upon the fubjects of Antiquity, Hiftory, Divinity, Chemiftry, and Mathematics; feveral of which were publifhed at different times, as appears from the following catalogue of all his works; where they are ranked in the order of time in which thofe upon the fame fubject were publifhed.

1. Several Papers relating to his *Telefcope*, and his *Theory of Light and Colours*, printed in the Philofophical Tranfactions, numbs. 80, 81, 82, 83, 84, 85, 88, 96, 97, 110, 121, 123, 128; or vols 6, 7, 8, 9, 10, 11.

2. *Optics*, or a *Treatife of the Reflections, Refractions, and Inflections, and the Colours of Light*; 1704, 4to.— A Latin tranflation by Dr. Clarke; 1706, 4to.—And a French tranflation by Pet. Cofte, Amft. 1729, 2 vols 12mo.—Befide feveral Englifh editions in 8vo.

3. *Optical Lectures*; 1728, 8vo. Alfo in feveral Letters to Mr. Oldenburg, fecretary of the Royal Society, inferted in the General Dictionary, under our author's article.

4. *Lectiones Opticæ*; 1729, 4to.

5. *Naturalis Philofophiæ Principia Mathematica*; 1687, 4to.—A fecond edition in 1713, with a Preface, by Roger Cotes.—The 3d edition in 1726, under the direction of Dr. Pemberton.—An Englifh tranflation, by Motte, 1729, 2 volumes 8vo, printed in feveral editions of his works, in different nations, particularly an edition, with a large Commentary, by the two learned Jefuits, Le Seur and Jacquier, in 4 volumes 4to, in 1739, 1740, and 1742.

6. *A Syftem of the World*, tranflated from the Latin original; 1727, 8vo.—This, as has been already obferved, was at firft intended to make the third book of his Principia.—An Englifh tranflation by Motte, 1729, 8vo.

7. *Several Letters* to Mr. Flamfteed, Dr. Halley, and Mr. Oldenburg.—See our author's article in the General Dictionary.

8. *A Paper concerning the Longitude*; drawn up by order of the Houfe of Commons; ibid.

9. *Abregé de Chronologie*, &c; 1726, under the direction of the abbé Conti, together with fome Obfervations upon it.

10. *Remarks upon the Obfervations made upon a Chronological Index of Sir I. Newton, &c.* Philof. Tranf. vol. 33. See alfo the fame, vol. 34 and 35, by Dr. Halley.

11. The Chronology of Ancient Kingdoms amended, &c; 1728, 4to.

12. *Arithmetica Univerfalis, &c*; under the infpection

tion

tion of Mr. Whiston, Cantab. 1707, 8vo. Printed I think without the author's consent, and even against his will: an offence which it seems was never forgiven. There are also English editions of the same, particularly one by Wilder, with a Commentary, in 1769, 2 vols 8vo. And a Latin edition, with a Commentary, by Castilion, 2 vols 4to, Amst. &c.

13. *Analysis per Quantitatum Series, Fluxiones, et Differentias, cum Enumeratione Linearum Tertii Ordinis;* 1711, 4to; under the inspection of W. Jones, Esq. F. R. S.—The last tract had been published before, together with another on the *Quadrature of Curves*, by the Method of Fluxions, under the title of *Tractatus duo de Speciebus & Magnitudine Figurarum Curvilinearum;* subjoined to the first edition of his Optics in 1704; and other letters in the Appendix to Dr. Gregory's Catoptrics, &c, 1735, 8vo.—Under this head may be ranked *Newtoni Genesis Curvarum per Umbras;* Leyden, 1740.

14. *Several Letters relating to his Dispute with Leibnitz,* upon his Right to the Invention of Fluxions; printed in the *Commercium Epistolicum D. Johannis Collins & aliorum de Analysi Promota, jussu Societatis Regiæ editum;* 1712, 8vo.

15. Postscript and Letter of M. Leibnitz to the Abbé Conti, with Remarks, and a Letter of his own to that Abbé; 1717, 8vo. To which was added, Raphson's History of Fluxions, as a Supplement.

16. *The Method of Fluxions, and Analysis by Infinite Series,* translated into English from the original Latin; to which is added, a Perpetual Commentary, by the translator Mr. John Colson; 1736, 4to.

17. *Several Miscellaneous Pieces, and Letters,* as follow:—(1). A Letter to Mr. Boyle upon the subject of the Philosopher's Stone. Inserted in the General Dictionary, under the article BOYLE.—(2). A Letter to Mr. Afton, containing directions for his travels; ibid. under our author's article.—(3). An English Translation of a Latin Dissertation upon the Sacred Cubit of the Jews. Inserted among the miscellaneous works of Mr. John Greaves, vol. 2, published by Dr. Thomas Birch, in 1737, 2 vols 8vo. This Dissertation was found subjoined to a work of Sir Isaac's, not finished, intitled *Lexicon Propheticum.*—(4). Four Letters from Sir Isaac Newton to Dr. Bentley, containing some arguments in proof of a Deity; 1756, 8vo.—(5). Two Letters to Mr. Clarke, &c.

18. *Observations on the Prophecies of Daniel and the Apocalypse of St. John;* 1733, 4to.

19. *If. Newtoni Elementa Perspectivæ Universalis;* 1746, 8vo.

20. *Tables for purchasing College Leases;* 1742, 12mo.

21. Corollaries, by Whiston.

22. A Collection of several pieces of our author's, under the following title, *Newtoni If. Opuscula Mathematica Philos. & Philol.* collegit J. Castilioneus; Laus. 1744, 4to, 8 tomes.

23. *Two Treatises* of the Quadrature of Curves, and Analysis by Equations of an Infinite Number of Terms, explained: translated by John Stewart, with a large Commentary; 1745, 4to.

24. *Description of an Instrument* for observing the Moon's Distance from the Fixed Stars at Sea. Philos. Transf. vol. 42.

25. Newton also published *Barrow's Optical Lectures,* in 1699, 4to: and *Bern. Varenii Geographia, &c;* 1681, 8vo.

26. The whole works of Newton, published by Dr. Horsley; 1779, 4to, in 5 volumes.

The following is a list of the papers left by Newton at his death, as mentioned above.

A Catalogue of Sir Isaac Newton's Manuscripts and Papers, as annexed to a Bond, given by Mr. Conduit, to the Administrators of Sir Isaac; by which he obliges himself to account for any profit he shall make by publishing any of the papers.

Dr. Pellet, by agreement of the executors, entered into Acts of the Prerogative Court, being appointed to peruse all the papers, and judge which were proper for the press.

No.
1. Viaticum Nautarum; by Robert Wright.
2. Miscellanea; not in Sir Isaac's hand writing.
3. Miscellanea; part in Sir Isaac's hand.
4. Trigonometria; about 5 sheets.
5. Definitions.
6. Miscellanea; part in Sir Isaac's hand.
7. 40 sheets in 4to, relating to Church History.
8. 126 sheets written on one side, being foul draughts of the Prophetic Stile.
9. 88 sheets relating to Church History.
10. About 70 loose sheets in small 4to, of Chemical papers; some of which are not in Sir Isaac's hand.
11. About 62 ditto, in folio.
12. About 15 large sheets, doubled into 4to; Chemical.
13. About 8 sheets ditto, written on one side.
14. About 5 sheets of foul papers, relating to Chemistry.
15. 12 half-sheets of ditto.
16. 104 half-sheets, in 4to, ditto.
17. About 22 sheets in 4to, ditto.
18. 24 sheets, in 4to, upon the Prophecies.
19. 29 half-sheets; being an answer to Mr. Hook, on Sir Isaac's Theory of Colours.
20. 87 half-sheets relating to the Optics, some of which are not in Sir Isaac's hand.

From No. 1 to No. 20 examined on the 20th of May 1727, and judged not fit to be printed.

T. Pellet.
Witness, *Tho. Pilkington.*

21. 328 half-sheets in folio, and 63 in small 4to; being loose and foul papers relating to the Revelations and Prophecies.
22. 8 half-sheets in small 4to, relating to Church Matters.
23. 24 half-sheets in small 4to; being a discourse relating to the 2d of Kings.
24. 353 half-sheets in folio, and 57 in small 4to; being foul and loose papers relating to Figures and Mathematics.
25. 201 half-sheets in folio, and 21 in small 4to; loose and foul papers relating to the Commercium Epistolicum.

26.

26. 91 half-sheets in small 4to, in Latin, upon the Temple of Solomon.

27. 37 half-sheets in folio, upon the Host of Heaven the Sanctuary, and other Church Matters.

28. 44 half-sheets in folio, upon Ditto.

29. 25 half-sheets in folio; being a farther account of the Host of Heaven.

30. 51 half-sheets in folio; being an Historical Account of two notable Corruptions of Scripture.

31. 88 half-sheets in small 4to; being Extracts of Church History.

32. 116 half-sheets in folio; being Paradoxical Questions concerning Athanasius, of which several leaves in the beginning are very much damaged.

33. 56 half-sheets in folio, De Motu Corporum; the greatest part not in Sir Isaac's hand.

34. 61 half-sheets in small 4to; being various sections on the Apocalypse.

35. 25 half-sheets in folio, of the Working of the Mystery of Iniquity.

36. 20 half-sheets in folio, of the Theology of the Heathens.

37. 24 half-sheets in folio; being an Account of the Contest between the Host of Heaven, and the Transgressors of the Covenant.

38. 31 half-sheets in folio; being Paradoxical Questions concerning Athanasius.

39. 107 quarter-sheets in small 4to, upon the Revelations.

40. 174 half-sheets in folio; being loose papers relating to Church History.

May 22, 1727, examined from No. 21 to No. 40 inclusive, and judged them not fit to be printed; only No. 33 and No. 38 should be reconsidered.

T. Pellet.
Witness, *Tho. Pilkington.*

41. 167 half-sheets in folio; being loose and foul papers relating to the Commercium Epistolicum.

42. 21 half-sheets in folio; being the 3d letter upon Texts of Scripture, very much damaged.

43. 31 half-sheets in folio; being foul papers relating to Church Matters.

44. 495 half-sheets in folio; being loose and foul papers relating to Calculations and Mathematics.

45. 335 half-sheets in folio; being loose and foul papers relating to the Chronology.

46. 112 sheets in small 4to, relating to the Revelations and other Church Matters.

47. 126 half-sheets in folio; being loose papers relating to the Chronology, part in English and part in Latin.

48. 400 half-sheets in folio; being loose Mathematical papers.

49. 109 sheets in 4to, relating to the Prophecies, and Church Matters.

50. 127 half-sheets in folio, relating to the University; great part not in Sir Isaac's hand.

51. 18 sheets in 4to; being Chemical papers.

52. 255 quarter-sheets; being Chemical papers.

53. An Account of Corruptions of Scripture; not in Sir Isaac's hand.

54. 31 quarter-sheets; being Flammell's Explication of Hieroglyphical Figures.

55. About 350 half-sheets; being Miscellaneous papers.

56. 6 half-sheets; being An Account of the Empires &c represented by St. John.

57. 9 half-sheets folio, and 71 quarter-sheets 4to; being Mathematical papers.

58. 140 half-sheets, in 9 chapters, and 2 pieces in folio, titled, Concerning the Language of the Prophets.

59. 606 half-sheets folio, relating to the Chronology; 9 more in Latin.

60. 182 half-sheets folio; being loose papers relating to the Chronology and Prophecies.

61. 144 quarter sheets, and 95 half-sheets folio; being loose Mathematical papers.

62. 137 half-sheets folio; being loose papers relating to the Dispute with Leibnitz.

63. A folio Common-place book; part in Sir Isaac's hand.

64. A bundle of English Letters to Sir Isaac, relating to Mathematics.

65. 54 half-sheets; being loose papers found in the Principia.

66. A bundle of loose Mathematical Papers; not Sir Isaac's.

67. A bundle of French and Latin Letters to Sir Isaac.

68. 136 sheets folio, relating to Optics.

69. 22 half-sheets folio, De Rationibus Motuum &c; not in Sir Isaac's hand.

70. 70 half-sheets folio; being loose Mathematical Papers.

71. 38 half-sheets folio; being loose papers relating to Optics.

72. 47 half-sheets folio; being loose papers relating to Chronology and Prophecies.

73. 40 half-sheets folio; Processus Mysterii Magni Philosophicus, by Wm. Yworth; not in Sir Isaac's hand.

74. 5 half-sheets; being a letter from Rizzetto to Martine, in Sir Isaac's hand.

75. 41 half-sheets; being loose papers of several kinds, part in Sir Isaac's hand.

76. 40 half-sheets; being loose papers, foul and dirty, relating to Calculations.

77. 90 half-sheets folio; being loose Mathematical papers.

78. 176 half-sheets folio; being loose papers relating to Chronology.

79. 176 half-sheets folio; being loose papers relating to the Prophecies.

80. { 12 half-sheets folio; An Abstract of the Chronology.
{ 92 half-sheets, folio; The Chronology.

81. 40 half-sheets folio; The History of the Prophecies, in 10 chapters, and part of the 11th unfinished.

82. 5 small bound books in 12mo, the greatest part not in Sir Isaac's hand, being rough Calculations.

May

May 26th 1727, Examined from No. 41 to No. 82 inclufive, and judged not fit to be printed, except No. 80, which is agreed to be printed, and part of No. 61 and 81, which are to be reconfidered.

Th. Pellet.
Witnefs, *Tho. Pilkington.*

It is aftonifhing what care and induftry Sir Ifaac had employed about the papers relating to Chronology, Church Hiftory, &c; as, on examining the papers themfelves, which are in the poffeffion of the family of the earl of Portfmouth, it appears that many of them are copies over and over again, often with little or no variation; the whole number being upwards of 4000 fheets in folio, or 8 reams of folio paper; befide the bound books &c in this catalogue, of which the number of fheets is not mentioned. Of thefe there have been publifhed only the Chronology, and Obfervations on the Prophecies of Daniel and the Apocalypfe of St. John.

NEWTONIAN *Philofophy*, the doctrine of the univerfe, or the properties, laws, affections, actions, forces, motions, &c of bodies, both celeftial and terreftrial, as delivered by Newton.

This term however is differently applied; which has given occafion to fome confufed notions relating to it. For, fome authors, under this term, include all the corpufcular philofophy, confidered as it now ftands reformed and corrected by the difcoveries and improvements made in feveral parts of it by Newton. In which fenfe it is, that Gravefande calls his Elements of Phyfics, Introductio ad Philofophiam Newtonianam. And in this fenfe the Newtonian is the fame as the new philofophy; and ftands contradiftinguifhed from the Cartefian, the Peripatetic, and the ancient Corpufcular.

Others, by Newtonian Philofophy, mean the method or order ufed by Newton in philofophifing; viz, the reafoning and inferences drawn directly from phenomena, exclufive of all previous hypothefes; the beginning from fimple principles, and deducing the firft powers and laws of nature from a few felect phenomena, and then applying thofe laws &c to account for other things. In this fenfe, the Newtonian Philofophy is the fame with the Experimental Philofophy, or ftands oppofed to the ancient Corpufcular, and to all hypothetical and fanciful fyftems of Philofophy.

Others again, by this term, mean that Philofophy in which phyfical bodies are confidered mathematically, and where geometry and mechanics are applied to the folution of phenomena. In which fenfe, the Newtonian is the fame with the Mechanical and Mathematical Philofophy.

Others, by Newtonian Philofophy, underftand that part of phyfical knowledge which Newton has handled, improved, and demonftrated.

And laftly, others, by this Philofophy, mean the new principles which Newton has brought into Philofophy; with the new fyftem founded upon them, and the new folutions of phenomena thence deduced; or that which characterizes and diftinguifhes his Philofophy from all others. And this is the fenfe in which we fhall here chiefly confider it.

As to the hiftory of this Philofophy, confult the foregoing article. It was firft publifhed in the year 1687, the author being then profeffor of mathematics in the univerfity of Cambridge; a 2d edition, with confiderable additions and improvements, came out in 1713; and a 3d in 1726. An edition, with a very large Commentary, came out in 1739, by Le Seur and Jacquier; befides the complete edition of all Newton's works, with notes, by Dr. Horfley, in 1779 &c. Several authors have endeavoured to make it plainer; by fetting afide many of the more fublime mathematical refearches, and fubftituting either more obvious reafonings or experiments inftead of them; particularly Whifton, in his Praelect. Phyf. Mathem.; Gravefande, in Elem. & Inft.; Pemberton, in his View &c; and Maclaurin, in his Account of Newton's Philofophy.

The chief parts of the Newtonian Philofophy, as delivered by the author, except his Optical Difcoveries &c, are contained in his Principia, or Mathematical Principles of Natural Philofophy. He founds his fyftem on the following definitions.

1. *Quantity of Matter*, is the meafure of the fame, arifing from its denfity and bulk conjointly.—Thus, air of a double denfity, in the fame fpace, is double in quantity; in a double fpace, is quadruple in quantity; in a triple fpace, is fextuple in quantity, &c.

2. *Quantity of Motion*, is the meafure of the fame, arifing from the velocity and quantity of matter conjunctly.—This is evident, becaufe the motion of the whole is the motion of all its parts; and therefore in a body double in quantity, with equal velocity, the Motion is double, &c.

3. The *Vis Infita, Vis Inertiae*, or innate force of matter, is a power of refifting, by which every body, as much as in it lies, endeavours to perfevere in its prefent ftate, whether it be of reft, or moving uniformly forward in a right line.—This definition is proved to be juft by experience, from obferving the difficulty with which any body is moved out of its place, upwards, or obliquely, or even downwards when acted on by a body endeavouring to urge it quicker than the velocity given it by gravity; and any how to change its ftate of motion or reft. And therefore this force is the fame, whether the body have gravity or not; and a cannon ball, void of gravity, if it could be, being difcharged horizontally, will go the fame diftance in that direction, in the fame time, as if it were endued with gravity.

4. An *Impreffed Force*, is an action exerted upon a body, in order to change its ftate, whether of reft or motion.—This force confifts in the action only; and remains no longer in the body when the action is over. For a body maintains every new ftate it acquires, by its vis inertiae only.

5. A *Centripetal Force*, is that by which bodies are drawn, impelled, or any way tend towards a point, as to a centre.—This may be confidered of three kinds, abfolute, accelerative, and motive.

6. The *Abfolute quantity* of a centripetal force, is a meafure of the fame, proportional to the efficacy of the caufe that urges it to the centre.

7. The *Accelerative quantity* of a centripetal force, is the meafure of the fame, proportional to the velocity which it generates in a given time.

8. The

8. The Motive quantity of a centripetal force, is a measure of the same, proportional to the motion which it generates in a given time.—This is always known by the quantity of a force equal and contrary to it, that is just sufficient to hinder the descent of the body.

After these definitions, follow certain Scholia, treating of the nature and distinctions of Time, Space, Place, Motion, Absolute, Relative, Apparent, True, Real, &c. After which, the author proposes to shew how we are to collect the true motions from their causes, effects, and apparent differences; and vice versa, how, from the motions, either true or apparent, we may come to the knowledge of their causes and effects. In order to this, he lays down the following axioms or laws of motion.

1st LAW. Every body perseveres in its state of rest, or of uniform motion in a right line, unless it be compelled to change that state by forces impressed upon it.—Thus, " Projectiles persevere in their motions, so far as they are not retarded by the resistance of the air, or impelled downwards by the force of gravity. A top, whose parts, by their cohesion, are perpetually drawn aside from rectilinear motions, does not cease its rotation otherwise than as it is retarded by the air. The greater bodies of the planets and comets, meeting with less resistance in more free spaces, preserve their motions, both progressive and circular, for a much longer time."

2d LAW. The Alteration of motion is always proportional to the motive force impressed; and is made in the direction of the right line in which that force is impressed. Thus, if any force generate a certain quantity of motion, a double force will generate a double quantity, whether that force be impressed all at once, or in successive moments.

3d LAW. To every action there is always opposed an equal re-action: or the mutual actions of two bodies upon each other, are always equal, and directed to contrary parts. Thus, whatever draws or presses another, is as much drawn or pressed by that other. If you press a stone with your finger, the finger is also pressed by the stone: &c.

From this axiom, or law, Newton deduces the following corollaries.

1. A body by two forces conjoined will describe the diagonal of a parallelogram, in the same time that it would describe the sides by those forces apart.

2. Hence is explained the composition of any one direct force out of any two oblique ones, viz, by making the two oblique forces the sides of a parallelogram, and the diagonal the direct one.

3. The quantity of motion, which is collected by taking the sum of the motions directed towards the same parts, and the difference of those that are directed to contrary parts, suffers no change from the action of bodies among themselves; because the motion which one body loses, is communicated to another.

4. The common centre of gravity of two or more bodies does not alter its state of motion or rest by the actions of the bodies among themselves; and therefore the common centre of gravity of all bodies, acting upon each other, (excluding external actions and impe-

diments) is either at rest, or moves uniformly in a right line.

5. The motions of bodies included in a given space are the same among themselves, whether that space be at rest, or move uniformly forward in a right line without any circular motion. The truth of this is evident from the experiment of a ship; where all motions are just the same, whether the ship be at rest, or proceed uniformly forward in a straight line.

6. If bodies, any how moved among themselves, be urged in the direction of parallel lines by equal accelerative forces, they will all continue to move among themselves, after the same manner as if they had not been urged by such forces.

The mathematical part of the Newtonian Philosophy depends chiefly on the following lemmas; especially the first; containing the doctrine of prime and ultimate ratios.

LEM. 1. Quantities, and the ratios of quantities, which in any finite time converge continually to equality, and before the end of that time approach nearer the one to the other than by any given difference, become ultimately equal.

LEM. 2 shews, that in a space bounded by two right lines and a curve, if an infinite number of parallelograms be inscribed, all of equal breadth; then the ultimate ratio of the curve space and the sum of the parallelograms, will be a ratio of equality.

LEM. 3 shews, that the same thing is true when the breadths of the parallelograms are unequal.

In the succeeding lemmas it is shewn, in like manner, that the ultimate ratios of the sine, chord, and tangent of arcs infinitely diminished, are ratios of equality, and therefore that in all our reasonings about these, we may safely use the one for the other:—that the ultimate form of evanescent triangles, made by the arc, chord, or tangent, is that of similitude, and their ultimate ratio is that of equality; and hence, in reasonings about ultimate ratios, these triangles may safely be used one for another, whether they are made with the sine, the arc, or the tangent.—He then demonstrates some properties of the ordinates of curvilinear figures; and shews that the spaces which a body describes by any finite force urging it, whether that force is determined and immutable, or continually varied, are to each other, in the very beginning of the motion, in the duplicate ratio of the forces:—and lastly, having added some demonstrations concerning the evanescence of angles of contact, he proceeds to lay down the mathematical part of his system, which depends on the following theorems.

THEOR. 1. The areas which revolving bodies describe by radii drawn to an immoveable centre of force, lie in the same immoveable planes, and are proportional to the times in which they are described.—To this prop. are annexed several corollaries, respecting the velocities of bodies revolving by centripetal forces, the directions and proportions of those forces, &c; such as, that the velocity of such a revolving body, is reciprocally as the perpendicular let fall from the centre of force upon the line touching the orbit in the place of the body, &c.

THEOR. 2. Every body that moves in any curve line

line defcribed in a plane, and, by a radius drawn to a point either immoveable or moving forward with an uniform rectilinear motion, defcribes about that point areas proportional to the times, is urged by a centripetal force directed to that point.—With corollaries relating to fuch motions in refifting mediums, and to the direction of the forces when the areas are not proportional to the times.

THEOR. 3. Every body that, by a radius drawn to the centre of another body, any how moved, defcribes areas about that centre proportional to the times, is urged by a force compounded of the centripetal forces tending to that other body, and of the whole accelerative force by which that other body is impelled.—With feveral corollaries.

THEOR. 4. The centripetal forces of bodies, which by equal motions defcribe different circles, tend to the centres of the fame circles ; and are one to the other as the fquares of the arcs defcribed in equal times, applied to the radii of the circles.—With many corollaries, relating to the velocities, times, periodic forces, &c. And, in fcholium, the author farther adds, Moreover, by means of the foregoing propofition and its corollaries, we may difcover the proportion of a centripetal force to any other known force, fuch as that of gravity. For if a body by means of its gravity revolve in a circle, concentric to the earth, this gravity is the centripetal force of that body. But from the defcent of heavy bodies, the time of one entire revolution, as well as the arc defcribed in any given time, is given by a corol. to this prop. And by fuch propofitions, Mr. Huygens, in his excellent book De Horologio Ofcillatorio, has compared the force of gravity with the centrifugal forces of revolving bodies.

On thefe, and fuch-like principles, depends the Newtonian Mathematical Philofophy. The author farther fhews how to find the centre to which the forces impelling any body are directed, having the velocity of the body given : and finds that the centrifugal force is always as the verfed fine of the nafcent arc directly, and as the fquare of the time inverfely ; or directly as the fquare of the velocity, and inverfely as the chord of the nafcent arc. From thefe premifes, he deduces the method of finding the centripetal force directed to any given point when the body revolves in a circle ; and this whether the central point be near hand, or at immenfe diftance ; fo that all the lines drawn from it may be taken for parallels. And he fhews the fame thing with regard to bodies revolving in fpirals, ellipfes, hyperbolas, or parabolas. He fhews alfo, having the figures of the orbits given, how to find the velocities and moving powers ; and indeed refolves all the moft difficult problems relating to the celeftial bodies with a furprifing degree of mathematical fkill. Thefe problems and demonftrations are all contained in the firft book of the Principia : but an account of them here would neither be generally underftood, nor eafily comprized in the limits of this work.

In the fecond book, Newton treats of the properties and motion of fluids, and their powers of refiftance, with the motion of bodies through fuch refifting mediums, thofe refiftances being in the ratio of any powers of the velocities ; and the motions being either made in right lines or curves, or vibrating like pendulums.

And here he demonftrates fuch principles as entirely overthrow the doctrine of Des Cartes's vortices, which was the fafhionable fyftem in his time ; concluding the book with thefe words : " So that the hypothefis of vortices is utterly irreconcileable with aftronomical phenomena, and rather ferves to perplex than explain the heavenly motions. How thefe motions are performed in free fpaces without vortices, may be underftood by the firft book ; and I fhall now more fully treat of it in the following book Of the Syftem of the World."— In this fecond book he makes great ufe of the doctrine of Fluxions, then lately invented ; for which purpofe he lays down the principles of that doctrine in the 2d Lemma, in thefe words : " The moment of any Genitum is equal to the moments of each of the generating fides drawn into the indices of the powers of thofe fides, and into their coefficients continually :" which rule he demonftrates, and then adds the following fcholium concerning the invention of that doctrine : " In a letter of mine, fays he, to Mr. J. Collins, dated December 10, 1672, having defcribed a method of tangents, which I fufpected to be the fame with Slufius's method, which at that time was not made public ; I fubjoined thefe words : ' This is one particular, or rather a corollary, of a general method which extends itfelf, without any troublefome calculation, not only to the drawing of tangents to any curve lines, whether geometrical or mechanical, or any how refpecting right lines or other curves, but alfo to the refolving other abftrufer kinds of problems about the curvature, areas, lengths, centres of gravity of curves, &c ; nor is it (as Hudden's method de Maximis & Minimis) limited to equations which are free from furd quantities. This method I have interwoven with that other of working in equations, by reducing them to infinite feries.' So far that letter. And thefe laft words relate to a Treatife I compofed on that fubject in the year 1671." Which, at leaft, is therefore the date of the invention of the doctrine of Fluxions.

On entering upon the 3d book of the Principia, Newton briefly recapitulates the contents of the two former books in thefe words : " In the preceding books I have laid down the principles of philofophy ; principles not philofophical, but mathematical ; fuch, to wit, as we may build our reafonings upon in philofophical enquiries. Thefe principles are, the laws and conditions of certain motions, and powers or forces, which chiefly have refpect to philofophy. But left they fhould have appeared of themfelves dry and barren, I have illuftrated them here and there with fome philofophical fcholiums, giving an account of fuch things, as are of a more general nature, and which philofophy feems chiefly to be founded on ; fuch as the denfity and the refiftance of bodies, fpaces void of all matter, and the motion of light and founds. It remains, he adds, that from the fame principles I now demonftrate the frame of the fyftem of the world. Upon this fubject, I had indeed compofed the 3d book in a popular method, that it might be read by many. But afterwards confidering that fuch as had not fufficiently entered into the principles could not eafily difcern the ftrength of the confequences, nor lay afide the prejudices to which they had been many years accuftomed ; therefore to prevent the difputes which
might

might be raised upon such accounts, I chose to reduce the substance of that book into the form of propositions, in the mathematical way, which should be read by those only, who had first made themselves masters of the principles established in the preceding books."

As a necessary preliminary to this 3d part, Newton lays down the following rules for reasoning in natural philosophy :

1. We are to admit no more causes of natural things, than such as are both true and sufficient to explain their natural appearances.

2. Therefore to the same natural effects we must always assign, as far as possible, the same causes.

3. The qualities of bodies which admit neither intension nor remission of degrees, and which are found to belong to all bodies within the reach of our experiments, are to be esteemed the universal qualities of all bodies whatsoever.

4. In experimental philosophy, we are to look upon propositions collected by general induction from phenomena, as accurately or very nearly true, notwithstanding any contrary hypotheses that may be imagined, till such time as other phenomena occur, by which they may either be made more accurate, or liable to exceptions.

The phenomena first considered are, 1. That the satellites of Jupiter, by radii drawn to his centre, describe areas proportional to the times of description ; and that their periodic times, the fixed stars being at rest, are in the sesquiplicate ratio of their distances from that centre. 2. The same thing is likewise observed of the phenomena of Saturn. 3. The five primary planets, Mercury, Venus, Mars, Jupiter, and Saturn, with their several orbits, encompass the sun. 4. The fixed stars being supposed at rest, the periodic times of the said five primary planets, and of the earth, about the sun, are in the sesquiplicate proportion of their mean distances from the sun. 5. The primary planets, by radii drawn to the earth, describe areas no ways proportional to the times : but the areas which they describe by radii drawn to the sun are proportional to the times of description. 6. The moon, by a radius drawn to the centre of the earth, describes an area proportional to the time of description. All which phenomena are clearly evinced by astronomical observations. The mathematical demonstrations are next applied by Newton in the following propositions.

PROP. 1. The forces by which the satellites of Jupiter are continually drawn off from rectilinear motions, and retained in their proper orbits, tend to the centre of that-planet ; and are reciprocally as the squares of the distances of those satellites from that centre.

PROP. 2. The same thing is true of the primary planets, with respect to the sun's centre.

PROP. 3. The same thing is also true of the moon, in respect of the earth's centre.

PROP. 4. The moon gravitates towards the earth ; and by the force of gravity is continually drawn off from a rectilinear motion, and retained in her orbit.

PROP. 5. The same thing is true of all the other planets, both primary and secondary, each with respect to the centre of its motion.

PROP. 6. All bodies gravitate towards every planet ; and the weights of bodies towards any one and the same planet, at equal distances from its centre, are proportional to the quantities of matter they contain.

PROP. 7. There is a power of gravity tending to all bodies, proportional to the several quantities of matter which they contain.

PROP. 8. In two spheres mutually gravitating each towards the other, if the matter in places on all sides, round about and equidistant from the centres, be similar ; the weight of either sphere towards the other, will be reciprocally as the square of the distance between their centres.—Hence are compared together the weights of bodies towards different planets : hence also are discovered the quantities of matter in the several planets : and hence likewise are found the densities of the planets.

PROP. 9. The force of gravity, in parts downwards from the surface of the planets towards their centres, decreases nearly in the proportion of the distances from those centres.

These, and many other propositions and corollaries, are proved or illustrated by a great variety of experiments, in all the great points of physical astronomy ; such as, That the motions of the planets in the heavens may subsist an exceeding long time :—That the centre of the system of the world is immoveable :—That the common centre of gravity of the earth, the sun, and all the planets, is immoveable :—That the sun is agitated by a perpetual motion, but never recedes far from the common centre of gravity of all the planets :—That the planets move in ellipses which have their common focus in the centre of the sun ; and, by radii drawn to that centre, they describe areas proportional to the times of description :—The aphelions and nodes of the orbits of the planets are fixt :—To find the aphelions, eccentricities, and principal diameters of the orbits of the planets :—That the diurnal motions of the planets are uniform, and that the libration of the moon arises from her diurnal motion :—Of the proportion between the axes of the planets and the diameters perpendicular to those axes :—Of the weights of bodies in the different regions of our earth :—That the equinoctial points go backwards, and that the earth's axis, by a nutation in every annual revolution, twice vibrates towards the ecliptic, and as often returns to its former position :—That all the motions of the moon, and all the inequalities of those motions, follow from the principles above laid down :—Of the unequal motions of the satellites of Jupiter and Saturn :—Of the flux and reflux of the sea, as arising from the actions of the sun and moon :—Of the forces with which the sun disturbs the motions of the moon ; of the various motions of the moon, of her orbit, variation, inclinations of her orbit, and the several motions of her nodes :—Of the tides, with the forces of the sun and moon to produce them :—Of the figure of the moon's body :—Of the precession of the equinoxes :—And of the motions and trajectory of comets. The great author then concludes with a General Scholium, containing reflections on the principal parts of the great and beautiful system of the universe, and of the infinite, eternal Creator and Governor of it.

" The hypothesis of vortices, says he, is pressed

with

with many difficulties. That every planet by a radius drawn to the fun may defcribe areas proportional to the times of defcription, the periodic times of the feveral parts of the vortices fhould obferve the duplicate proportion of their diftances from the fun. But that the periodic times of the planets may obtain the fefquiplicate proportion of their diftances from the fun, the periodic times of the parts of the vortex ought to be in the fefquiplicate proportion of their diftances. That the fmaller vortices may maintain their leffer revolutions about Saturn, Jupiter, and other planets, and fwim quietly and undifturbed in the greater vortex of the fun, the periodic times of the parts of the fun's vortex fhould be equal. But the rotation of the fun and planets about their axes, which ought to correfpond with the motions of their vortices, recede far from all thefe proportions. The motions of the comets are exceeding regular, are governed by the fame laws with the motions of the planets, and can by no means be accounted for by the hypothefis of vortices. For comets are carried with very eccentric motions through all parts of the heavens indifferently, with a freedom that is incompatible with the notion of a vortex.

" Bodies, projected in our air, fuffer no refiftance but from the air. Withdraw the air, as is done in Mr. Boyle's vacuum, and the refiftance ceafes. For in this void a bit of fine down and a piece of folid gold defcend with equal velocity. And the parity of reafon muft take place in the celeftial fpaces above the earth's atmofphere; in which fpaces, where there is no air to refift their motions, all bodies will move with the greateft freedom; and the planets and comets will conftantly purfue their revolutions in orbits given in kind and pofition, according to the laws above explained. But though thefe bodies may indeed perfevere in their orbits by the mere laws of gravity, yet they could by no means have at firft derived the regular pofition of the orbits themfelves from thofe laws.

" The fix primary planets are revolved about the fun, in circles concentric with the fun, and with motions directed towards the fame parts, and almoft in the fame plane. Ten moons are revolved about the earth, Jupiter and Saturn, in circles concentric with them, with the fame direction of motion, and nearly in the planes of the orbits of thofe planets. But it is not to be conceived that mere mechanical caufes could give birth to fo many regular motions: fince the comets range over all parts of the heavens, in very eccentric orbits. For by that kind of motion they pafs eafily through the orbs of the planets, and with great rapidity; and in their aphelions, where they move the floweft, and are detained the longeft, they recede to the greateft diftances from each other, and thence fuffer the leaft difturbance from their mutual attractions. This moft beautiful fyftem of the fun, planets, and comets, could only proceed from the counfel and dominion of an intelligent and powerful Being. And if the fixed ftars are the centres of other like fyftems, thefe being formed by the like wife counfel, muft be all fubject to the dominion of one; efpecially, fince the light of the fixed ftars is of the fame nature with the light of the fun, and from every fyftem light paffes into all the other fyftems. And left the fyftem of the fixed ftars fhould, by their

gravity, fall on each other mutually, he hath placed thofe fyftems at immenfe diftances one from another."

Then, after a truly pious and philofophical defcant on the attributes of the Being who could give exiftence and continuance to fuch prodigious mechanifm, and with fo much beautiful order and regularity, the great author proceeds,

" Hitherto we have explained the phenomena of the heavens and of our fea, by the power of gravity, but have not yet affigned the caufe of this power. This is certain, that it muft proceed from a caufe that penetrates to the very centres of the fun and planets, without fuffering the leaft diminution of its force; that operates, not according to the quantity of the furfaces of the particles upon which it acts, (as mechanical caufes ufe to do,) but according to the quantity of the folid matter which they contain, and propagates its virtue on all fides, to immenfe diftances, decreafing always in the duplicate proportion of the diftances. Gravitation towards the fun, is made up out of the gravitations towards the feveral particles of which the body of the fun is compofed; and in receding from the fun, decreafes accurately in the duplicate proportion of the diftances, as far as the orb of Saturn, as evidently appears from the quiefcence of the aphelions of the planets; nay, and even to the remoteft aphelions of the comets, if thofe aphelions are alfo quiefcent. But hitherto I have not been able to difcover the caufe of thofe properties of gravity from phenomena, and I frame no hypothefes. For whatever is not deduced from the phenomena, is to be called an hypothefis; and hypothefes, whether metaphyfical or phyfical, whether of occult qualities or mechanical, have no place in experimental philofophy. In this philofophy particular propofitions are inferred from the phenomena, and afterwards rendered general by induction. Thus it was that the impenetrability, the mobility, and the impulfive force of bodies, and the laws of motion and of gravitation, were difcovered. And to us it is enough, that gravity does really exift, and act according to the laws which we have explained, and abundantly ferves to account for all the motions of the celeftial bodies, and of our fea.

" And now we might add fomething concerning a certain moft fubtle fpirit, which pervades and lies hid in all grofs bodies, by the force and action of which fpirit, the particles of bodies mutually attract one another at near diftances, and cohere, if contiguous, and electric bodies operate to greater diftances, as well repelling as attracting the neighbouring corpufcles; and light is emitted, reflected, refracted, inflected, and heats bodies; and all fenfation is excited, and the members of animal bodies move at the command of the will, namely, by the vibrations of this fpirit, mutually propagated along the folid filaments of the nerves, from the outward organs of fenfe to the brain, and from the brain into the mufcles. But thefe are things that cannot be explained in few words, nor are we furnifhed with that fufficiency of experiments which is required to an accurate determination and demonftration of the laws by which this electric and elaftic fpirit operates."

NICHE.

NICHE, a cavity, or hollow part, in the thickness of a wall, to place a figure or statue in.

NICOLE (FRANCIS), a very celebrated French mathematician, was born at Paris December the 23d, 1683. His early attachment to the mathematics induced M. Montmort to take the charge of his education : and he opened out to him the way to the higher geometry. He first became publicly remarkable by detecting the fallacy of a pretended quadrature of the circle. This quadrature a M. Mathulon so assuredly thought he had discovered, that he deposited, in the hands of a public notary at Lyons, the sum of 3000 livres, to be paid to any person who, in the judgment of the Academy of Sciences, should demonstrate the falsity of his solution. M. Nicole, piqued at this challenge, undertook the task, and exposing the paralogism, the Academy's judgment was, that Nicole had plainly proved that the rectilineal figure which Mathulon had given 'as equal to the circle, was not only unequal to it, but that it was even greater than the polygon of 32 sides circumscribed about the circle.—The prize of 3000 livres, Nicole presented to the public hospital of Lyons.

The Academy named Nicole, Eleve-Mechanician, March 12, 1707; Adjunct in 1716, Associate in 1718, and Pensioner in 1724; which he continued till his death, which happened the 18th of January 1758, at 75 years of age.

His works were all inserted in the different volumes of the Memoirs of the Academy of Sciences; and are as follow:

1. A General Method for determining the Nature of Curves formed by the Rolling of other Curves upon any Given Curve; in the volume for the year 1707.

2. A General Method for Rectifying all Roulets upon Right and Circular Bases; 1708.

3. General Method of determining the Nature of those Curves which cut an Infinity of other Curves given in Position, cutting them always in a Constant Angle; 1715.

4. Solution of a Problem proposed by M. de Lagny; 1716.

5. Treatise of the Calculus of Finite Differences; 1717.

6. Second Part of the Calculus of Finite Differences; 1723.

7. Second Section of ditto; 1723.

8. Addition to the two foregoing papers; 1724.

9. New Proposition in Elementary Geometry; 1725.

10. New Solution of a Problem proposed to the English Mathematicians, by the late M. Leibnitz; 1725.

11. Method of Summing an Infinity of New Series, which are not summable by any other known method; 1727.

12. Treatise of the Lines of the Third Order, or the Curves of the Second Kind; 1729.

13. Examination and Resolution of some Questions relating to Play; 1730.

14. Method of determining the Chances at Play.

15. Observations upon the Conic Sections; 1731.

16. Manner of generating in a Solid Body, all the Lines of the Third Order; 1731.

17. Manner of determining the Nature of Roulets formed upon the Convex Surface of a Sphere; and of determining which are Geometric, and which are Rectifiable; 1732.

18. Solution of a Problem in Geometry; 1732.

19. The Use of Series in resolving many Problems in the Inverse Method of Tangents; 1737.

20. Observations on the Irreducible Case in Cubic Equations; 1738.

21. Observations upon Cubic Equations; 1738.

22. On the Trisection of an Angle; 1740.

23. On the Irreducible Case in Cubic Equations; 1741.

24. Addition to ditto; 1743.

25. His Last Paper upon the same; 1744.

26. Determination, by Incommensurables and Decimals, the Values of the Sides and Areas of the Series in a Double Progression of Regular Polygons, inscribed in and circumscribed about a Circle; 1747.

NIEUWENTYT (BERNARD), an eminent Dutch philosopher and mathematician, was born on the 10th of August 1654, at Westgraafdyk in North Holland, where his father was minister. He discovered very early a good genius and a strong inclination for learning; which was carefully improved by a suitable education. He had also that prudence and sagacity, which led him to pursue literature by sure and proper steps, acquiring a kind of mastery in one science before he proceeded to another. His father had designed him for the ministry; but seeing his inclination did not lie that way, he prudently left him to pursue the bent of his genius. Accordingly young Nieuwentyt apprehending that nothing was more useful than fixing his imagination and forming his judgment well, applied himself early to logic, and the art of reasoning justly, in which he grounded himself upon the principles of Des Cartes, with whose philosophy he was greatly delighted. From thence he proceeded to the mathematics, in which he made a considerable proficiency; though the application he gave to that branch of learning did not hinder him from studying both law and physic. In fact he succeeded in all these sciences so well, as deservedly to acquire the character of a good philosopher, a great mathematician, an expert physician, and an able and just magistrate.

Although he was naturally of a grave and serious disposition, yet he was very affable and agreeable in conversation. His engaging manner procured the affection of every one; and by this means he often drew over to his opinion those who before differed very widely from him. Thus accomplished, he acquired a great esteem and credit in the council of the town of Puremerende, where he resided; as he did also in the states of that province, who respected him the more, inasmuch as he never engaged in any cabals or factions, in order to secure it; regarding in his conduct, an open, honest, upright behaviour, as the best source of satisfaction, and relying solely on his merit. In fact, he was more attentive to cultivate the sciences, than eager to obtain the honours of the government; contenting himself with being counsellor and burgomaster, without courting or accepting any other posts, which might interfere with his studies, and draw him too much out of his library.—Nieuwentyt died the 7th of May 1730, at 76 years

years of age—having been twice married.—He was author of several works, in the Latin, French, and Dutch languages, the principal of which are the following:

1. A Treatife in Dutch, *proving the Existence of God by the Wonders of Nature*; a much esteemed work, and went through many editions. It was translated also into several languages, as the French, and the English, under the title of, *The Religious Philosopher, &c.*

2. A Refutation of Spinoza, in the Dutch language.

3. *Analysis Infinitorum*; 1695, 4to.

4. *Considerationes secundæ circa Calculi Differentialis Principia*; 1696, 8vo.—In this work he attacked Leibnitz, and was answered by John Bernoulli and James Herman.

5. A Treatise on the New Use of the Tables of Sines and Tangents.

6. A Letter to Bothnia or Burmania, upon the Subject of Meteors.

NIGHT, that part of the natural day, during which the fun is below the horizon: though sometimes it is understood that the twilight is referred to the day, or time the fun is above the horizon; the remainder only being the Night.

Under the equator, the Nights, in the former sense, are always equal to the days; each being 12 hours long. But under the poles, the Night continues half a year. —The ancient Gauls and Germans divided their time not by days, but Nights; as appears from Cæsar and Tacitus; also the Arabs and the Icelanders do the same. The same may also be observed of our Saxon ancestors: whence our custom of saying, Sevennight, Fortnight, &c.

NOCTILUCA, a species of phosphorus, so called because it shines in the night, without any light being thrown on it: such is the phosphorus made of urine, By which it stands distinguished from some other species of phosphorus, which require to be exposed to the sunbeams before they will shine; as the Bononian-stone, &c. —Mr. Boyle has a particular Treatise on this subject.

NOCTURNAL *Arch*, is the arch of a circle described by the fun, or a star, in the night.

NOCTURNAL, or NOCTURLABIUM, denotes an instrument, chiefly used at sea, to take the altitude or depression of the pole star, and some other stars about the pole, for finding the latitude, and the hour of the night.

There are several kinds of this instrument; some of which are projections of the sphere; such as the hemispheres, or planispheres, on the plane of the equinoctial. The seamen commonly use two kinds; the one adapted to the pole star and the first of the guards of the Little Bear; the other to the pole star and the pointers of the Great Bear.

The Nocturnal consists of two circular plates (fig. 15, pl. xiii) applied over each other. The greater, which has a handle to hold the instrument, is about $2\frac{1}{2}$ inches diameter, and is divided into 12 parts, answering to the 12 months; also each month subdivided into every 5th day; and in such manner, that the middle of the handle corresponds to that day of the year in which the star here respected has the same right ascension with the fun. When the instrument is fitted for two stars, the han-

dle is made moveable. The upper circle is divided into 24 equal parts, for the 24 hours of the day, and each hour subdivided into quarters, as in the figure. These 24 hours are noted by 24 teeth; to be told in the night. In the centre of the two circular plates is adjusted a long index A, moveable upon the upper plate. And the three pieces, viz. the two circles and index, are joined by a rivet which is pierced through the centre, with a hole 2 inches in diameter, for the star to be observed through.

To Use the NOCTURNAL. Turn the upper plate till the longest tooth, marked 12, be against the day of the month on the under plate; and bringing the instrument near the eye, suspend it by the handle, with the plane nearly parallel to the equinoctial; then viewing the pole-star through the hole in the centre, turn the index about till, by the edge coming from the centre, you see the bright star or guard of the Little Bear, if the instrument be fitted to that star: then that tooth of the upper circle, under the edge of the index, is at the hour of the night on the edge of the hour-circle: which may be known without a flight, by counting the teeth from the longest, which is for the hour of 12.

NODATED *Hyperbola*, one, so called by Newton, which by turning round decussates or crosses itself: as in the 2d, and several other species, of his Enumeratio Linearum Tertii Ordinis.

NODES, the two opposite points where the orbit of a planet intersects the ecliptic. That, where the planet ascends from the south to the north side of the ecliptic, is called the Ascending Node, or the Dragon's Head, and marked thus ☊: and the opposite point, where the planet descends from the north to the south side of the ecliptic, is called the Descending Node, or Dragon's Tail, and is thus marked ☋. Also the right line drawn from the one Node to the other, is called the Line of the Nodes.

By observation it appears that, in all the planets, the Line of the Nodes continually changes its place, its motion being *in antecedentia*; i. e. contrary to the order of the signs, or from east to west; with a peculiar degree of motion for each planet. Thus, by a retrograde motion, the line of the moon's nodes completes its circuit in 18 years and 225 days, in which time the Node returns again to the same point of the ecliptic. Newton has not only shewn, that this motion arises from the action of the sun, but, from its cause, he has with great skill calculated all the elements and varieties in this motion. See his Princip. lib. 3, prop. 30, 31, &c.

The moon must be in or near one of the Nodes to make an eclipse either of the sun or moon.

NODUS, or *Node*, in Dialling, denotes a point or hole in the gnomon of a dial, by the shadow or light of which is shewn, either the hour of the day in dials without furniture, or the parallels of the sun's declination, and his place in the ecliptic, &c, in dials with furniture.

NOLLET (the Abbé JOHN ANTHONY), a considerable French philosopher, and a member of most of the philosophical societies and academies of Europe, was born at Pimpre, in the district of Noyon, the 19th of November 1700. From the profound retreat, in which the mediocrity of his fortune obliged him to live, his reputation continually increased from day to day.

M. Dufay

M. Dufay affociated him in his Electrical Refearches; and M. de Reaumur refigned to him his laboratory. It was under thefe mafters that he developed his talents. M. Dufay took him along with him in a journey he made into England; and Nollet profited fo well of this opportunity, as to inftitute a friendly and literary correfpondence with fome of the moft celebrated men in this country.

The king of Sardinia gave him an invitation to Turin, to perform a courfe of experimental philofophy to the duke of Savoy. From thence he travelled into Italy, where he collected fome good obfervations concerning the natural hiftory of the country.

In France he was mafter of philofophy and natural hiftory to the royal family; and profeffor royal of experimental philofophy to the college of Navarre, and to the fchools of artillery and engineers. The Academy of Sciences appointed him adjunct-mechanician in 1739, affociate in 1742, and penfioner in 1757. Nollet died the 24th of April 1770, regretted by all his friends, but efpecially by his relations, whom he always fuccoured with an affectionate attention. The works publifhed by Nollet, are the following:

1. Recueils de Lettres fur l'Electricité; 1753, 3 vols in 12mo.

2. Effai fur l'Electricité des Corps; 1 vol. in 12mo.

3. Recherches fur les Caufes particulieres des Phenomenes Electriques; 1 vol. in 12mo.

4. L'Art des Experiences; 1770, 3 vols in 12mo.

His papers printed in the different volumes of the Memoirs of the Academy of Sciences, are much too numerous to be particularized here; they are inferted in all or moft of the volumes from the year 1740 to the year 1767 inclufive, moftly feveral papers in each volume.

NONAGESIMAL, or NONAGESIMAL *Degree*, called alfo the Mid-heaven, is the higheft point, or 90th degree of the ecliptic, reckoned from its interfection with the horizon at any time; and its altitude is equal to the angle that the ecliptic makes with the horizon at their interfection, or equal to the diftance of the zenith from the pole of the ecliptic. It is much ufed in the calculation of folar eclipfes.

NONAGON, a figure having nine fides and angles. —In a regular Nonagon, or that whofe angles, and fides, are all equal, if each fide be 1, its area will be $6 \cdot 1818242 = \frac{9}{4}$ of the tangent of 70°, to the radius 1. See my Menfuration, p. 114, 2d edit.

NONES, in the Roman Calendar, the 5th day of the months January, February, April, June, Auguft, September, November, and December; and the 7th of the other months March, May, July, and October: thefe laft four months having 6 days before the Nones, and the others only four.—They had this name probably, becaufe they were always 9 days inclufively, from the firft of the Nones to the Ides, i. e. reckoning inclufively both thofe days.

NONIUS, or Nunez (Peter), a very eminent Portuguefe mathematician and phyfician, was born in 1497, at Alcazar in Portugal, anciently a remarkable city, known by the name of Salacia, from whence he was furnamed Salacienfis. He was profeffor of mathematics in the univerfity of Coimbra, where he publifhed

some pieces which procured him great reputation. He was mathematical preceptor to Don Henry, fon to king Emanuel of Portugal, and principal cofmographer to the king. Nonius was very ferviceable to the defigns, which this court entertained of carrying on their maritime expeditions into the Eaft, by the publication of his book *Of the Art of Navigation*, and various other works. He died in 1577, at 80 years of age.

Nonius was the author of feveral ingenious works and inventions, and juftly efteemed one of the moft eminent mathematicians of his age. Concerning his *Art of Navigation*, father Dechales fays, " In the year 1530, Peter Nonius, a celebrated Portuguefe mathematician, upon occafion of fome doubts propofed to him by Martinus Alphonfus Sofa, wrote a Treatife on Navigation, divided into two books; in the firft, he anfwers fome of thofe doubts, and explains the nature of Loxodromic lines. In the fecond book, he treats of rules and inftruments proper for navigation, particularly fea-charts, and inftruments ferving to find the elevation of the pole; but fays he is rather obfcure in his manner of writing."—Furetiere, in his Dictionary, takes notice that Peter Nonius was the firft who, in 1530, invented the angles which the Loxodromic curves make with each meridian, calling them in his language Rhumbs, and which he calculated by fpherical triangles.—Stevinus acknowledges, that Peter Nonius was fcarce inferior to the very beft mathematicians of the age. And Schottus fays, he explained a great many problems, and particularly the mechanical problem of Ariftotle on the motion of veffels by oars. His Notes upon Purbach's Theory of the Planets, are very much to be efteemed: he there explains feveral things, which had either not been noticed before, or not rightly underftood.

In 1542 he publifhed a Treatife on the Twilight, which he dedicated to John the 3d, king of Portugal; to which he added what Alhazen, an Arabian author, has compofed on the fame fubject. In this work he defcribes the method or inftrument called, from him, a Nonius, a particular account of which fee in the following article.—He corrected feveral mathematical miftakes of Orontius Finæus.—But the moft celebrated of all his works, or that at leaft he appeared moft to value, was his *Treatife of Algebra*, which he had compofed in Portuguefe, but tranflated it into the Caftilian tongue, when he refolved upon making it public, which he thought would render his book more ufeful, as this language was more generally known than the Portuguefe. The dedication, to his former pupil, prince Henry, was dated from Lifbon, Dec. 1, 1564. This work contains 341 pages in the Antwerp edition of 1567, in 8vo.

The catalogue of his works, chiefly in Latin, is as follows:

1. *De Arte Navigandi*, libri duo; 1530.

2. *De Crepufculis*; 1542.

3. *Annotationes in Ariftotelem*.

4. Problema Mechanicum de Motu Navigii ex Remis.

5. Annotationes in Planetarum Theorias Georgii Purbachii, &c.

6. Libro de Algebra en Arithmetica y Geometra; 1564.

NONIUS,

Nonius, is a name also erroneously given to the method of graduation now generally used in the division of the scales of various instruments, and which should be called Vernier, from its real inventor. The method of Nonius, so called from its inventor Pedro Nunez, or Nonius, and described in his treatise De Crepusculis, printed at Lisbon in 1542, consists in describing within the same quadrant, 45 concentric circles, dividing the outermost into 90 equal parts, the next within into 89, the next into 88, and so on, till the innermost was divided into 46 only. By this means, in most observations, the plumb-line or index must cross one or other of those circles in or very near a point of division: whence by calculation the degrees and minutes of the arch might easily be obtained. This method is also described by Nunez, in his treatise De Arte et Ratione Navigandi, lib. 2, cap. 6, where he imagines it was not unknown to Ptolomy. But as the degrees are thus divided unequally, and it is very difficult to attain exactness in the division, especially when the numbers, into which the arches are to be divided, are incomposite, of which there are no less than nine, the method of diagonals, first published by Thomas Digges, Esq. in his treatise Alæ seu Scalæ Mathematicæ, printed at Lond. in 1573, and said to be invented by one Richard Chanseler, a very skilful artist, was substituted in its stead. However, Nonius's method was improved at different times; but the admirable division now so much in use, is the most considerable improvement of it. See Vernier.

NORMAL, is used sometimes for a perpendicular.

NORTH Star, called also the Pole-star, is the last in the tail of the Little Bear.

Northern Signs, are those six that are in the north side of the equator; viz, Aries, Taurus, Gemini, Cancer, Leo, Virgo.

NORTHING, in Navigation, is the difference of latitude, which a ship makes in sailing northwards.

NOSTRADAMUS (Michel), an able physician and celebrated astrologer, was born at St. Remy in Provence in the diocese of Avignon, December 14, 1503. His father was a notary public, and his grandfather a physician, from whom he received some tincture of the mathematics. He afterwards completed his courses of languages and philosophy at Avignon. From hence, going to Montpelier, he there applied himself to physic; but being forced away by the plague, he travelled through different places till he came to Bourdeaux, undertaking all such patients as were willing to put themselves under his care. This course occupied him five years; after which he returned to Montpelier, and was created doctor of his faculty in 1529; after which he revisited the same places where he had practised physic before. At Agen he formed an acquaintance with Julius Cæsar Scaliger, and married his first wife; but having buried her, and two children which she brought him, he quitted Agen after a residence of about four years. He fixed next at Marseilles; but, his friends having provided an advantageous match for him at Salon, he repaired thither about the year 1544;

and married accordingly his second wife, by whom he had several children.

In 1546, Aix being afflicted with the plague, he went thither at the solicitation of the inhabitants, to whom he rendered great service, particularly by a powder of his own invention: so that the town, in gratitude, gave him a considerable pension for several years after the contagion ceased. In 1547 the city of Lyons, being visited with the same distemper, had recourse to our physician, who attended them also. Afterwards returning to Salon, he began a more retired course of life, and in this time of leisure applied himself closely to his studies. He had for a long time followed the trade of a conjurer occasionally; and now he began to fancy himself inspired, and miraculously illuminated with a prospect into futurity. As fast as these illuminations had discovered to him any future event, he entered it in writing, in simple prose, though in enigmatical sentences; but revising them afterwards, he thought the sentences would appear more respectable, and savour more of a prophetic spirit, if they were expressed in verse. This opinion determined him to throw them all into quatrains, and he afterward ranged them into centuries. For some time he could not venture to publish a work of this nature; but afterwards perceiving that the time of many events foretold in his quatrains was very near at hand, he resolved to print them, as he did, with a dedication addressed to his son Cæsar, an infant only some months old, and dated March 1, 1555. To this first edition, which comprises but seven centuries, he prefixed his name in Latin, but gave to his son Cæsar the name as it is pronounced in French, Notradame.

The public were divided in their sentiments of this work: many looked upon the author as a simple visionary; by others he was accused of magic or the black art, and treated as an impious person who held a commerce with the devil; while great numbers believed him to be really endued with the supernatural gift of prophecy. However, Henry the 2d, and queen Catharine of Medicis, his mother, were resolved to see our prophet, who receiving orders to that effect, he presently repaired to Paris. He was very graciously received at court, and received a present of 200 crowns. He was sent afterwards to Blois, to visit the king's children there, and report what he should be able to discover concerning their destinies. It is not known what his sentence was; however he returned to Salon loaded with honour, and good presents.

Animated with this success, he augmented his work to the number of 1000 quatrains, and published it with a dedication to the king in 1558. That prince dying the next year of a wound which he received at a tournament, our prophet's book was immediately consulted; and this unfortunate event was found in the 35th quatrain of the first century, which runs thus in the London edition of 1672:

Le Lion jeune le vieux surmontera,
En champ bellique, par singulier duelle,
Dans cage d'or l'œil il lui crevera,
Deux playes une, puis mourir mort cruelle.

In Englifh thus, from the fame edition :

> The young Lion fhall overcome the old one,
> In martial field by a fingle duel,
> In a golden cage he fhall put out his eye,
> Two wounds from one, then he fhall die a cruel death.

-So remarkable a prediction added new wings to his fame ; and he was honoured foon after with a vifit from Emanuel duke of Savoy, and the princefs Margaret of France, his confort. From this time Noftradamus found himfelf even overburdened with vifitors, and his fame made every day new acquifitions. Charles the 9th, coming to Salon, was eager above all things to have a fight of him : Noftradamus, who then was in waiting as one of the retinue of the magiftrates, being inftantly prefented to the king, complained of the little efteem his countrymen had for him ; upon which the monarch publicly declared that he would hold the enemies of Noftradamus to be his enemies, and defired to fee his children. Nor did that prince's favour ftop here ; in paffing, not long after, through the city of Arles, he fent for Noftradamus, and prefented him with a purfe of 200 crowns, together with a brevet, conftituting him his phyfician in ordinary, with the fame appointment as the reft. But our prophet enjoyed thefe honours only a fhort time, as he died 16 months after, viz, July 2, 1566, at Salon, being then in his grand climacteric, or 63d year.—He had publifhed feveral other pieces, chiefly relating to medicine.

He left three fons and three daughters. Cæfar the eldeft fon was born at Salon in 1555, and died in 1629 : he left a manufcript, giving an account of the moft remarkable events in the hiftory of Provence, from 1080 to 1494, in which he inferted the lives of the poets of that country. Thefe memoirs falling into the hands of his nephew Cæfar Noftradamus, gentleman to the duke of Guife, he undertook to complete the work ; and being encouraged by the eftates of the country, he carried the account up to the Celtic Gauls : the impreffion was finifhed at Lyons in 1614, and publifhed under the title of Chronique de l'Hiftoire de Provence. —The fecond fon, John, exercifed with reputation the bufinefs of a proctor in the parliament of Provence.— He wrote the Lives of the Ancient Provençal Poets, called Troubadours, and the work was printed at Lyons in 1575, 8vo.—The youngeft fon it is faid undertook the trade of peeping into futurity after his father.

NOTATION, is the reprefenting of numbers, or any other quantities, by Notes, characters, or marks.

The choice of arithmetical, and other, characters, is arbitrary ; and hence they are various in various nations : the figures 0, 1, 2, 3, &c, in common ufe, are derived from the Arabs and Indians, from whom they have their name, and the Notation by them, which forms the decimal or decuple fcale, is perhaps the moft convenient of any for arithmetical computations.

The Greeks, Hebrews, and other eaftern nations, as alfo the Romans, expreffed numbers by the letters of their common alphabet. See CHARACTER.

In Algebra, the quantities are reprefented moftly by the letters of the alphabet, &c ; and that as early as the time of Diophantus. See ALGEBRA.

NOTES, in Mufic, are characters which mark the tones, i. e. the elevations and fallings of the voice, or

4

found, and the fwiftnefs or flownefs of its motion, &c ; and thefe have undergone various alterations and improvements, before they arrived at their prefent ftate of perfection.

NOVEMBER, the eleventh month in the Julian year, but the ninth in the year of Romulus, beginning with March ; whence its name. In this month, which contains 30 days, the fun enters the fign ♐, viz, ufually about the 21ft day of the month.

NUCLEUS, the kernel, is ufed by Hevelius, and fome other aftronomers, for the body of a comet, which others call its head, as diftinguifhed from its tail, or beard.

NUCLEUS is alfo ufed by fome writers for the central parts of the earth, and other planets, which they fuppofe firmer, and as it were feparated from them, confidered as a cortex or fhell.

NUEL, the fame as NEWEL of a Staircafe.

NUMBER, a collection or affemblage of feveral units, or feveral things of the fame kind ; as 2, 3, 4, &c, exclufive of the number 1 : which is Euclid's definition of Number.—Stevinus defines Number as that by which the quantity of anything is expreffed : agreeably to which Newton conceives a Number to confift, not in a multitude of units, as Euclid defines it, but in the abftract ratio of a quantity of any kind to another quantity of the fame kind, which is accounted as unity: and in this fenfe, including all thefe three fpecies of Number, viz, Integers, Fractions, and Surds.

Wolfius defines Number to be fomething which refers to unity, as one right line refers to another. Thus, affuming a right line for unity, a Number may likewife be expreffed by a right line. And in this way alfo Des Cartes confiders numbers as expreffed by lines, where he treats of the arithmetical operations as performed by lines, in the beginning of his Geometry.

For the manner of characterizing NUMBERS, fee NOTATION. And

For reading and expreffing NUMBERS in combination, fee NUMERATION.

Mathematicians confider Number under a great many circumftances, and different relations, accidents, &c.

NUMBERS, *Abfolute, Abftract, Abundant, Amicable, Applicate, Binary, Cardinal, Circular, Compofite, Concrete, Defective, Fractional, Homogeneal, Irrational* or *Surd, Linear* or *Mixt, Ordinal, Polygonal, Prime, Pyramidal, Rational, Similar, &c,* fee the refpective adjectives.

Broken NUMBERS, or Fractions, are certain parts of unity, or of fome other Number.

Cubic NUMBER, is the product of a fquare Number multiplied by its root, or the continual product of a Number twice multiplied by itfelf;
as the Numbers - - 1, 8, 27, 64, 125, &c, which are the cubes of - 1, 2, 3, 4, 5, &c.

This feries of the cubes of the ordinal Numbers, may be raifed by addition only, viz, adding always the differences ; as was firft fhewn by Peletarius, at the end of his Algebra, firft printed in 1558, where he gives a table of the fquares and Cubes of the firft 140 numbers. See CUBE.

Every Cubic Number whofe root is lefs than 6, viz, the Cubic Numbers 1, 8, 27, 64, 125, being divided by 6, the remainder is the root itfelf :

Thus,

Thus,

$$\tfrac{1}{6} = 0\tfrac{1}{6}; \quad \tfrac{8}{6} = 1\tfrac{2}{6}; \quad \tfrac{27}{6} = 4\tfrac{3}{6}; \quad \tfrac{64}{6} = 10\tfrac{4}{6}; \quad \tfrac{125}{6} = 20\tfrac{5}{6};$$

where the remainders, or the numerators of the small fractions, are 0, 1, 2, 3, 4, 5, the same as the roots of the Cubes 0, 1, 8, 27, 64, 125. After these, the next six Cubic Numbers being divided by 6, the remainders will be respectively the same arithmetical series, viz - - - 0, 1, 2, 3, 4, 5; to each of which adding 6, gives 6, 7, 8, 9, 10, 11, for the roots of the next six cubes 216, 343, &c.

Then, again dividing the next set of six Cubic Numbers, viz, 1728, 2197, &c, by 6, the remainders are again } 0, 1, 2, 3, 4, 5, the same series, viz, to each of which adding 12, gives 12, 13, 14, 15, 16, 17, for the roots of the said next six cubes. And so on in infinitum, the series of remainders 0, 1, 2, 3, 4, 5, continually recurring, and to each set of these remainders the respective Numbers 0, 6, 12, 18, 24, &c, being added, the sums will be the whole series of roots, 0, 1, 2, 3, 4, 5, 6, &c.

M. de la Hire, from considering this property of the Number 6, with regard to Cubic Numbers, found that all other Numbers, raised to any power whatever, had each their divisor, which had the same effect with regard to them, that 6 has with regard to Cubes. And the general rule he has discovered is this: if the exponent of the power of a number be even, i. e. if that number be raised to the 2d, 4th, 6th. &c power, it must be divided by 2, then the remainder added to 2, or to a multiple of 2, gives the root of the Number corresponding to its power, i. e. the 2d, or 4th, &c, root. But if the exponent of the power of the Number be uneven, viz the 3d, 5th, 7th, &c power, the double of that exponent shall be the divisor, which shall have the property here required.

A Determinate NUMBER, is that which is referred to some given unit; as a ternary or three.

An *Even* NUMBER, is that which may be divided into two equal parts, without remainder or fraction, as the Numbers 2, 4, 6, 8, 10, &c.—The sums, differences, products, and powers of Even Numbers, are also Even Numbers.

An *Evenly-Even* NUMBER, is such as being divided by an even Number, the quotient is also an Even Number without a remainder: as 16, which divided by 8 gives 2 for the quotient.

An *Unevenly-Even* NUMBER, is such as being divided by an Even Number, the quotient is an Uneven one: as 20, which divided by 4, gives 5 for the quotient.

Figurate or *Figural* NUMBERS, are certain ranks of Numbers found by adding together first a rank of units, which is the first order, which gives the 2d order; then these added give the 3d order; and so on. Hence, the several orders of Figurate Numbers, are as follow:

First order	-	1 . 1 . 1 . 1 . 1 . &c.
2d order	-	1 . 2 . 3 . 4 . 5 . &c.
3d order	-	1 . 3 . 6 . 10 . 15 . &c.
4th order	-	1 . 4 . 10 . 20 . 35 . &c.
5th order	-	1 . 5 . 15 . 35 . 70 . &c.

The first order consists all of equals, and the 2d order of the natural arithmetical progression; the 3d order

is also called triangular Numbers, the 4th order pyramidals, &c.

See FIGURATE *Numbers*.

Heterogeneal NUMBERS, are such as are referred to different units. As three men and 4 trees.

Homogeneal NUMBERS, are such as are referred to the same unit. As 3 men and 4 men.

Imperfect NUMBERS, are those whose aliquot parts added together, make either more or less than the whole of the number itself; and are distinguished into Abundant and Defective.

Indeterminate NUMBER, is that which is referred to unity in the general; which is what we call Quantity.

Irrational or *Surd* NUMBER, is one that is not commensurable with unity; as $\sqrt{2}$, or $\sqrt[3]{4}$, &c

Perfect NUMBER, that which is just equal to the sum of its aliquot parts, added together. As, 6, 28, &c: for the aliquot parts of 6 are 1, 2, 3, whose sum is the same 6; and the aliquot parts of 28, are 1, 2, 4, 7, 14, whose sum is 28. See PERFECT *Number*.

Plane NUMBER, that which arises from the multiplication of two other Numbers: so 6 is a plane or rectangle, whose two sides are 2 and 3, for $2 \times 3 = 6$.

Square NUMBER, is a Number produced by multiplying any given Number by itself; as the

Square Numbers   1, 4, 9, 16, 25, &c,
produced from the roots   1, 2, 3, 4, 5, &c.

Every Square Number added to its root makes an even Number. See SQUARE.

Uneven NUMBER, or *Odd* NUMBER, that which differs from an even Number by one, or which cannot be divided into two equal integer parts; such as 1, 3, 5, 7, &c. The sums and differences of Uneven Numbers are even; but all the products and powers of them are Uneven Numbers. On the other hand, the sum or difference of an even and Uneven Number are both Uneven, but their product is even.

Whole NUMBER, or *Integer*, is unit, or a collection of units.

Golden NUMBER. See GOLDEN *Number* and CYCLE.

NUMBER of *Direction*, in Chronology, some one of the 35 Numbers between the Easter limits, or between the earliest and latest day on which it can fall, i. e. between March 22 and April 25, which are 35 days; being so called, because it serves as a Direction for finding Easter for any year; being indeed the Number that expresses how many days after March 21, Easter-day falls. Thus, Easter-day falling as in the first line below, the Number of Direction will be as on the lower line:

March   April
Easter-day, 22, 23, 24, 25, 26, 27, 28, 29, 30, 31, 1, 2, &c.
No of Dir. 1, 2, 3, 4, 5, 6, 7, 8, 9, 10, 11, 12, &c

and so on, till the Number of Direction on the lower line be 35, which will answer to April 25, being the latest that Easter can happen. Therefore add 21 to the Number of Direction, and the sum will be so many days in March for the Easter-day: if the sum exceed 31, the excess will be the day of April.

To find the NUMBER *of Direction.* Enter the following table (which is adapted to the New Style), with the Dominical Letter on the left hand, and the Golden Number at the top, then where the columns meet is
the

the Number of Direction for that year. See Ferguson's Aftron. pa. 381, ed. 8vo.

G. N.	1	2	3	4	5	6	7	8	9	10	11	12	13	14	15	16	17	18	19
Dom. Let.																			
A	29	19	5	26	11	33	19	12	26	19		5	26	12		26	12	32	19	12
B	27	13	6	27	13	34	20	13	27	20		6	27	13		20	13	34	20	6
C	8	4	7	21	14	35	21		7	28	21		7	28	14		7	21	14	28	21	7
D	19	15	8	22	15	29	22		8	29	15		8	29	15		1	22	15	29	22	8
E	30	16	2	23	16	30	23		9	30	16		9	2	16		2	23	9	30	23	9
F	24	17	3	24	10	31	24		10	31	17		10	2	17		3	24	10	31	17	10
G	25	18	4	25	11	32	18		11	32	18		11	2	18		4	25	11	32	18	11

Thus, for the year 1790, the Dominical Letter being C, and the Golden Number 5; on the line of C, and below 5, is 14 for the Number of Direction. To this add 21, the sum is 35 days from the 1st of March, which, deducting the 31 days of March, leaves 4 for the day of April, for Easter-day that year.

NUMERAL *Characters*. See CHARACTERS.

NUMERAL *Figures*. The antiquity of these in England has, for several reasons, been supposed as high as the eleventh century; in France about the middle of the tenth century; having been introduced into both countries from Spain, where they had been brought by the Moors or Saracens. See Wallis's Algebra, pa. 9 &c, and pa. 153 of additions at the end of the same. See also Philof. Tranf. numb. 439 and 475.

NUMERAL *Letters*, those letters of the alphabet that are commonly used for figures or numbers, as I, V, X, L, C, D, M.

NUMERATION, in Arithmetic, the art of estimating or pronouncing any number, or series of numbers.

Numbers are usually expressed by the ten following characters, 1, 2, 3, 4, 5, 6, 7, 8, 9, and 0; the first nine denoting respectively the first nine ordinal numbers; and the last, or cipher 0, joined to any of the others, denotes so many tens. In like manner, two ciphers joined to any one of the first nine significant figures, make it become so many hundreds, three ciphers make it thousands, and so on.

Weigelius indeed shews how to number, without going beyond a quaternary; i. e. by beginning to repeat at each fourth. And Leibnitz and De Lagny, in what they call their binary arithmetic, begin to repeat at every 2d place; using only the two figures 1 and 0. But these are rather matters of curiosity than any real use.

That the nine significant figures may express not only units, but also tens, hundreds, thousands, &c, they have a local value given them, as hinted above; so that, though when alone, or in the right-hand place, they denote only units or ones, yet in the 2d place they denote tens, in the 3d place hundreds, in the 4th place thousands, &c; as the number 5558 is, five thousand five hundred fifty and five.

Hence then, to express any written number, or assign the proper value to each character; beginning at the right hand, divide the proposed number into classes, of three characters to each class; and consider two classes as making up a period of six figures or places. Then every period, of six figures, has a name common to all the figures in it; the first being primes or units; the 2d is millions; the 3d is millions of-millions, or billions; the 4th is millions-of-millions-of-millions or trillions; and so on; also every class, or half-period, of three figures, is read separately by itself, so many hundreds, tens, and units; only, after the left-hand half of each period, the word thousands is added; and at the end of the 2d, 3d, 4th &c period, its common name millions, billions, &c, is expressed.

Thus the number 4,591, is 4 thousand 5 hundred and 91.

The number 210,463, is 2 hundred and 10 thousands, and 463.

The number 281,427,307, is 281 millions, 427 thousands, and 307.

NUMERATOR, of a Fraction, is the number which shews how many of those parts, which the integer is supposed to be divided into, are denoted by the fraction. And, in the notation the Numerator is set over the denominator, or number that shews into how many parts the integer is divided, in the fraction. So, ex. gr. ¾ denotes three-fourths, or 3 parts out of 4; where 3 is the numerator, and 4 the denominator.

NUMERICAL, NUMEROUS, or *Numeral*, something that relates to number.

NUMERAL *Algebra*, is that which makes use of numbers, in contradistinction from literal algebra, or that in which the letters of the alphabet are used.

O.

OBE

OBELISK, a kind of quadrangular pyramid, very tall and slender, raised as an ornament in some public place, or to serve as a memorial of some remarkable transaction.

OBJ

OBJECT, something presented to the mind, by sensation, or by imagination. Or something that affects us by its presence, that affects the eye, ear, or some other of the organs of sense.

The

The objects of the eye, or vision, are painted on the retina; though not there erect, but inverted, according to the laws of optics. This is easily shewn from Des Cartes's experiment, of laying bare the vitreous humour on the back part of the eye, and putting over it a bit of white paper, or the skin of an egg, and then placing the fore part of the eye to the hole of a darkened room. By this means there is obtained a pretty landscape of the external objects, painted invertedly on the back of the eye. In this case, how the Objects thus painted invertedly should be seen erect, is matter of controversy.

OBJECT is also used for the subject, or matter of an art or science; being that about which it is employed or concerned.

OBJECT-*Glass*, of a telescope or microscope, is the glass placed at the end of the tube which is next or towards the Object to be viewed.

To prove the goodness and regularity of an Objectglass; on a paper describe two concentric circles, the one having its diameter the same with the breadth of the Object-glass, and the other half that diameter; divide the smaller circumference into 6 equal parts, pricking the points of division through with a fine needle; cover one side of the glass with this paper, and, exposing it to the sun, receive the rays through these 6 holes upon a plane; then by moving the plane nearer to or farther from the glass, it will be found whether the six rays unite exactly together at any distance from the glass; if they do, it is a proof of the regularity and just form of the glass; and the said distance is also the focal distance of the glass.

A good way of proving the excellency of an Objectglass, is by placing it in a tube, and trying it with small eye-glasses, at several distant objects; for that Object-glass is always the best, which represents objects the brightest and most distinct, and which bears the greatest aperture, and the most convex and concave eyeglasses, without colouring or haziness.

A circular Object-glass is said to be truly centred, when the centre of its circumference falls exactly in the axis of the glass; and to be ill centred, when it falls out of the axis.

To prove whether Object-glasses be well centred, hold the glass at a due distance from the eye, and observe the two reflected images of a candle, varying the distance till the two images unite, which is the true centre point: then if this fall in the middle, or central point of the glass, it is known to be truly centred.

As Object-glasses are commonly included in cells that screw upon the end of the tube of a telescope, it may be proved whether they be well centred, by fixing the tube, and observing while the cell is unscrewed, whether the cross-hairs keep fixed upon the same lines of an object seen through the telescope.

For various methods of finding the true centre of an Object-glass, see Smith's Optics, book 3, chap. 3; also the Philos. Trans. vol. 48, pa. 177.

OBJECTIVE *Line*, in Perspective, is any line drawn on the geometrical plane, whose representation is sought for in a draught or picture.

OBJECTIVE *Plane*, in Perspective, is any plane situated in the horizontal plane, whose perspective representation is required.

OBLATE, flatted, or shortened; as an Oblate sphe-

roid, having its axis shorter than its middle diameter; being formed by the rotation of an ellipse about the shorter axis.

OBLATENESS, of the earth, the flatness about the poles, or the diminution of the polar axis in respect of the equatorial. The ratio of these two axes has been determined in various ways; sometimes by the measures of different degrees of latitude, and sometimes by the length of pendulums vibrating seconds in different latitudes, &c; the results of all which, as well as accounts of the means of determining them, see under the articles EARTH and DEGREE. To what is there said, may be added the following, from An Account of the Experiments made in Russia concerning the Length of a Pendulum which swings Seconds, by Mr. Krafft, contained in the 6th and 7th volumes of the New Petersburgh Transactions, for the years 1790 and 1793. These experiments were made at different times, and in various parts of the Russian empire: Mr. Krafft has collected and compared them, with a view to investigate the consequences that may be deduced from them. From the whole he concludes, that the length p of a pendulum, which swings seconds in any given latitude l, and in a temperature of 10 degrees of Reaumur's thermometer, may be determined by the following equation, in lines of a French foot: viz,

$$p = 439 \cdot 178 + 2 \cdot 321 \; \text{sine}^2 \, l.$$

This expression agrees, very nearly, not only with all the experiments made on the pendulum in Russia, but also with those of Mr. Graham, and those of Mr. Lyons in 79° 50 north latitude, where he found its length to be 441·38 lines.

It also shews the augmentation of gravity from the equator to the parallel of a given latitude l: for, putting g for the gravity under the equator, G for that under the pole, and z for that under the latitude l; Mr. Krafft finds $z = (1 + 0 \cdot 0052848 \; \text{sine}^2 \, l) \times g$; and consequently G $= 1 \cdot 0052848 g$:

From this proportion of Gravity under different latitudes, Mr. Krafft deduces, that on the hypothesis of the earth's being a homogeneous ellipsoid, its oblateness must be $\frac{1}{190}$; instead of $\frac{1}{230}$, which ought to be the result of this hypothesis: but on adopting the supposition that the earth is a heterogeneous ellipsoid, he finds its Oblateness, as deduced from these experiments, to be $\frac{1}{297}$; which agrees with that resulting from the measurement of degrees of the meridian.

This confirms an observation of M. De la Place, that, if the hypothesis of the earth's homogeneity be given up, then do theory, the measurement of degrees of latitude, and experiments with the pendulum, all agree in their result with respect to the Oblateness of the earth.

OBLIQUE, aslant, indirect, or deviating from the perpendicular. As,

OBLIQUE *Angle*, one that is not a right angle, but is either greater or less than this, being either obtuse or acute.

OBLIQUE-*angled Triangle*, that whose angles are all oblique.

OBLIQUE *Ascension*, is that point of the equinoctial which rises with the centre of the sun, or star, or any other point of the heavens, in an Oblique sphere.

OBLIQUE *Circle*, in the stereographic projection,

is any circle that is Oblique to the plane of projection.

OBLIQUE *Defcenfion*, that point of the equinoctial which fets with the centre of the fun, or ftar, or other point of the heavens in an Obliqne fphere.

OBLIQUE *Direction*, that which is not perpendicular to a line or plane.

OBLIQUE *Force*, or Percuffion, or *Power*, or Stroke, is that made in a direction Oblique to a body or plane. It is demonftrated that the effect of fuch Oblique force &c, upon the body, is to an equal perpendicular one, as the fine of the angle of incidence is to radius.

OBLIQUE *Line*, that which makes an Oblique angle with fome other line.

OBLIQUE *Planes*, in Dialling, are fuch as recline from the zenith, or incline towards the horizon.

OBLIQUE *Projection*, is that where a body is projected or impelled in a line of direction that makes an oblique angle with the horizontal line.

OBLIQUE *Sailing*, in Navigation, is that part which includes the application and calculation of Oblique-angled triangles.

OBLIQUE *Sphere*, in Geography, is that in which the axis is Oblique to the horizon of a place.—In this fphere, the equator and parallels of declination cut the horizon obliquely. And it is this obliquity that occafions the inequality of days and nights, and the variation of the feafons. See SPHERE.

OBLIQUITY, that which denotes a thing Oblique.

OBLIQUITY *of the Ecliptic*, is the angle which the ecliptic makes with the equator. See ECLIPTIC.

OBLONG, fometimes means any figure that is longer than it is broad ; but more properly it denotes a rectangle, or a right-angled parallelogram, whofe length exceeds its breadth.

OBLONG, is alfo ufed for the quality or fpecies of a figure that is longer than it is broad : as an Oblong fpheroid ; formed by an ellipfe revolved about its longer or tranfverfe axis ; in contradiftinction from the oblate fpheroid, or that which is flatted at its poles, being generated by the revolution of the ellipfe about its conjugate or fhorter axis.

OBSCURA *Camera*. See CAMERA *Obfcura*.

OBSCURA *Clara*. See CLARA *Obfcure*.

OBSERVATION, in Aftronomy and Navigation, is the obferving with an inftrument fome celeftial phenomenon ; as, the altitude of the fun, moon, or ftars, or their diftances afunder, &c. But by this term the feamen commonly mean only the taking the meridian altitudes, in order to find the latitude. And the finding the latitude from fuch obferved altitude, they call *working an obfervation*.

OBSERVATORY, a place deftined for obferving the heavenly bodies ; or a building, ufually in form of a tower, erected on fome eminence, and covered with a terrace, for making aftronomical obfervations.

Moft nations, at almoft all times, have had their obfervatories, either public or private ones, and in various degrees of perfection. A defcription of a great many of them may be feen in a differtation of Weidler's, De præfenti Specularum Aftronomicarum Statu, printed in 1727, and in different articles of his Hiftory of Aftronomy, printed in 1741, viz. pa. 86 &c ; as alfo in La Lande's Aftronomy, the preface pa. 34. The chief among thefe are the following :

I. The Greenwich Obfervatory, or Royal Obfervatory of England. This was built and endowed in the year 1676, by order of King Charles the 2d, at the inftance of Sir Jonas Moore, and Sir Chriftopher Wren : the former of thefe gentlemen being Surveyor General of the Ordnance, the office of Aftronomer Royal was placed under that department, in which it has continued ever fince.

This obfervatory was at firft furnifhed with feveral very accurate inftruments ; particularly a noble fextant of 7 feet radius, with telefcopic fights. And the firft Aftronomer Royal, or the perfon to whom the province of obferving was firft committed, was Mr. John Flamfteed ; a man who, as Dr. Halley expreffes it, feemed born for the employment. During 14 years he watched the motions of the planets with unwearied diligence, efpecially thofe of the moon, as was given him in charge ; that a new theory of that planet being found, fhewing all her irregularities, the longitude might thence be determined.

In the year 1690, having provided himfelf with a mural arch of near 7 feet radius, made by his Affiftant Mr. Abraham Sharp, and fixed in the plane of the meridian, he began to verify his catalogue of the fixed ftars, which had hitherto depended altogether on the diftances meafured with the fextant, after a new and very different manner, viz, by taking the meridian altitudes, and the moments of culmination, or in other words the right afcenfion and declination. And he was fo well pleafed with this inftrument, that he difcontinued almoft entirely the ufe of the fextant.

Thus, in the fpace of upwards of 40 years, the Aftronomer Royal collected an immenfe number of good obfervations ; which may be found in his Hiftoria Cœleftis Britannica, publifhed in 1725 ; the principal part of which is the Britannic catalogue of the fixed ftars.

Mr. Flamfteed, on his death in 1719, was fucceeded by Dr. Halley, and he by Dr. Bradley in 1742, and this laft by Mr. Blifs in 1762 ; but none of the obfervations of thefe gentlemen have yet been given to the public.

On the demife of Mr. Blifs, in 1765, he was fucceeded by Dr. Nevil Mafkelyne, the prefent worthy aftronomer royal, whofe valuable obfervations have been publifhed, from time to time, under the direction of the Royal Society, in feveral folio volumes.

The Greenwich Obfervatory is found, by very accurate obfervations, to lie in 51° 28′ 40″ north latitude, as fettled by Dr. Mafkelyne, from many of his own obfervations, as well as thofe of Dr. Bradley.

II. The Paris Obfervatory was built by Louis the 14th, in the fauxbourg St. Jaques, being begun in 1664, and finifhed in 1672. It is a fingular but magnificent building, of 80 feet in height, with a terrace at top ; and here M. De la Hire, M. Caffini, &c, the king's aftronomers, have made their obfervations. Its latitude is 48° 50′ 14″ north, and its longitude 9′ 20″ eaft of Greenwich Obfervatory.

In the Obfervatory of Paris is a cave, or pit, 170 feet deep, with fubterraneous paffages, for experiments that are to be made out of the reach of the fun, efpecially fuch as relate to congelations, refrigerations, &c. In this cave there is an old thermometer of M. De la Hire, which ftands always at the fame height ; thereby

2

fhewing

shewing that the temperature of the place remains always the same. From the top of the platform to the bottom of the cave is a perpendicular well or pit, used formerly for experiments on the fall of bodies; being also a kind of long telescopical tube, through which the stars are seen at mid-day.

III. Tycho Brahe's Observatory was in the little island Ween, or the Scarlet Island, between the coasts of Schonen and Zealand, in the Baltic sea. This Observatory was not well situated for some kinds of observations, particularly the risings and settings; as it lay too low, and was landlocked on all the points of the compass except three; and the land horizon being very rugged and uneven.

IV. Pekin Observatory. Father Le Compte describes a very magnificent Observatory, erected and furnished by the late emperor of China, in his capital, at the intercession of some Jesuit missionaries, chiefly father Verbiest, whom he appointed his chief observer. The instruments here are exceeding large; but the divisions are less accurate, and in some respects the contrivance is less commodious than in those of the Europeans. The chief are, an armillary zodiacal sphere, of 6 Paris feet diameter, an azimuthal horizon 6 feet diameter, a large quadrant 6 feet radius, a sextant 8 feet radius, and a celestial globe 6 feet diameter.

V. Bramins' Observatory at Benares, in the East Indies, which is still one of the principal seminaries of the Bramins or priests of the original Gentoos of Hindostan. This Observatory at Benares it is said was built about 200 years since, by order of the emperor Ackbar: for as this wise prince endeavoured to improve the arts, so he wished also to recover the sciences of Hindostan, and therefore ordered that three such places should be erected; one at Delhi, another at Agra, and the third at Benares.

Wanting the use of optical glasses, to magnify very distant or very small objects, these people directed their attention to the increasing the size of their instruments, for obtaining the greater accuracy and number of the divisions and subdivisions in their instruments. Accordingly, the Observatory contains several huge instruments, of stone, very nicely erected and divided, consisting of circles, columns, gnomons, dials, quadrants, &c, some of them of 20 feet radius, the circle divided first into 360 equal parts, and sometimes each of these into 20 other equal parts, each answering to 3, and of about two-tenths of an inch in extent. And although these wonderful instruments have been built upwards of 200 years, the graduations and divisions on the several arcs appear as well cut, and as accurately divided, as if they had been the performance of a modern artist. The execution, in the construction of these instruments, exhibits an extraordinary mathematical exactness in the fixing, bearing, fitting of the several parts, in the necessary and sufficient supports to the very large stones that compose them, and in the joining and fastening them into each other by means of lead and iron.

See a farther description, and drawing, of this Observatory, by Sir Robert Barker, in the Philos. Transf. vol. 67, pa. 598.

OBSERVATORY *Portable*. See EQUATORIAL.

OBTUSE *Angle*, one that is greater than a right-angle.

OBTUSE-*angled Triangle*, is a triangle that has one of its angles Obtuse: and it can have only one such.

OBTUSE *Cone*, or OBTUSE-*Angled Cone*, one whose angle at the vertex, by a section through the axis, is Obtuse.

OBTUSE *Hyperbola*, one whose asymptotes form an Obtuse angle.

OBTUSE-*angular Section of a Cone*, a name given to the hyperbola by the ancient geometricians, because they considered this section only in the Obtuse cone.

OCCIDENT, or OCCIDENTAL, west, or westward, in Astronomy; a planet is said to be Occident, when it sets after the sun.

OCCIDENT, in Geography, the westward quarter of the horizon, or that part of the horizon where the ecliptic, or the sun's place in it, descends into the lower hemisphere.

OCCIDENT *Equinoctial*, that point of the horizon where the sun sets, when he crosses the equinoctial, or enters the sign Aries or Libra.

OCCIDENT *Estival*, that point of the horizon where the sun sets at his entrance into the sign Cancer, or in our summer when the days are longest.

OCCIDENT *Hybernal*, that point of the horizon where the sun sets at midwinter, when entering the sign Capricorn.

OCCIDENTAL *Horizon*. See HORIZON.

OCCULT, in Geometry, is used for a line that is scarce perceivable, drawn with the point of the compasses, or a black-lead pencil. Occult or dry lines, are used in several operations; as the raising of plans, designs of building, pieces of perspective, &c. They are to be effaced or rubbed out when the work is finished.

OCCULTATION, the obscuration, or hiding from our sight, any star or planet, by the interposition of the body of the moon, or of some other planet.—The Occultation of a star by the moon, if observed in a place whose latitude and longitude are well determined, may be applied to the correction of the lunar tables; but if observed in a place whose latitude only is well known, may be applied to the determining the longitude of the place.

Circle of Perpetual OCCULTATION. See CIRCLE.

OCEAN, the vast collection of salt and navigable water, which encompasses most parts of the earth.

By computation it appears that the Ocean takes up considerably more of what we know of the terrestrial globe, than the dry land does. This is perhaps easiest known, by taking a good map of the world, and with a pair of scissars clipping out all the water from the land, and weighing the two parts separately: by which means it has been found, that the water occupies about two-thirds of the whole surface of the globe.

The great and universal Ocean is sometimes, by geographers, divided into three parts. As, 1st, the Atlantic and European Ocean, lying between part of Europe, Africa, and America; 2d, the Indian Ocean, lying between Africa, the East-Indian islands, and New Holland; 3d, the Pacific Ocean, or great south sea, which lies between the Philippine islands, China, Japan, and New Holland on the west, and the coast of America on the east. The Ocean also takes divers other names, according eording to the different countries it borders upon : as the British Ocean, German Ocean, &c. Also according to the position on the globe ; as the northern, southern, eastern, and western Oceans.

The Ocean, penetrating the land at several streights, quits its name of Ocean, and assumes that of sea or gulph ; as the Mediterranean sea, the Persian gulph, &c. In very narrow places, it is called a streight, &c

OCTAEDRON, or OCTAHEDRON, one of the five regular bodies ; contained under 8 equal and equilateral triangles.—It may be conceived as consisting of two quadrilateral pyramids joined together at their bases.

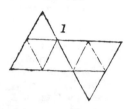 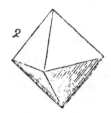

To form an Octaedron. Join together 8 equal and equilateral triangles, as in fig. 1 ; then cut the lines half through, and fold the figure up by these cut lines, till the extreme edges meet, and form the Octaedron, as in figure 2.

In an Octaedron, if

A be the linear edge or side,
B its whole surface,
C its solidity, or solid content,
R the radius of the circumscribed sphere, and
r the radius of the inscribed sphere : Then

$$A = r\sqrt{6} = R\sqrt{2} = \cdot \frac{B\sqrt{3}}{6} = \sqrt[3]{\frac{3C\sqrt{2}}{2}}.$$

$$B = 12r^2\sqrt{3} = 4R^2\sqrt{3} = 2A^2\sqrt{3} = 6\sqrt[3]{\frac{C^2\sqrt{3}}{2}}.$$

$$C = 4r^3\sqrt{3} = \frac{4}{3}R^3 = \frac{1}{3}A^3\sqrt{2} = \frac{B\sqrt{B\sqrt{3}}}{18}.$$

$$R = r\sqrt{3} = \frac{1}{2}A\sqrt{2} = \sqrt{\frac{B\sqrt{3}}{12}} = \sqrt[3]{\frac{3}{4}C}.$$

$$r = \frac{1}{3}R\sqrt{3} = \frac{1}{6}A\sqrt{6} = \frac{1}{6}\sqrt{B\sqrt{3}} = \sqrt[3]{\frac{C\sqrt{3}}{12}}.$$

See my *Mensuration*, pa. 251 &c, 2d edition.

OCTAGON, is a figure of 8 sides and angles ; which, when these are all equal, is also called a regular one, or may be inscribed in a circle.

If the side of a regular Octagon be *s* ; then

Its area $= 2s^2 \times \overline{1 + \sqrt{2}} = 4\cdot8284271s^2$; and

the Radius of its circumsc. circle $= \dfrac{s}{\sqrt{2-\sqrt{2}}}$.

OCTAGON, in Fortification, denotes a place that has 8 sides, or 8 bastions.

OCTANT, the 8th part of a circle.

OCTANT, or OCTILE, means also an aspect, or po-

5

sition of two planets, when their places are distant by the 8th part of a circle, or 45 degrees.

OCTAVE, or 8th, in Music, is an interval of 8 sounds ; every 8th note in the scale of the gamut being the same, as far as the compass of music requires.

Tones, or sounds, that are Octaves to each other, or at an Octave's distance, are alike, or the same nearly as the unison. In this case, the more acute of the two makes exactly two vibrations while the deeper or graver makes but one ; whence, they coincide at every two vibrations of the acuter, which, being more frequent, makes this concord more perfect than any other, and as it were an unison. Hence also, it happens, that two chords or strings, of the same matter, thickness, and tension, but the one double the length of the other, produce the Octave.

The Octave containing in it all the other simple concords, and the degrees being the differences of these concords ; it is evident, that the division of the Octave comprehends the division of all the rest.

By joining therefore all the simple concords to a common fundamental, we have the following series :

$$1 : \frac{5}{6} : \frac{4}{5} : \frac{3}{4} : \frac{2}{3} : \frac{3}{5} : \frac{1}{2}$$
Fund. 3d *l*, 3d *g*, 4th, 5th, 6th *l*, 6th *g*, 8ve.

Mr. Malcolm observes, that any wind instrument being over-blown, the sound will rise to an Octave, and no other concord ; which he ascribes to the perfection of the Octave, and its being next to unison.

Des Cartes, from an observation of the like kind, viz, that the sound of a whistle, or organ pipe, will rise to an Octave, if forcibly blown, concludes, that no sound is heard, but its acute Octave seems some way to echo or resound in the ear.

OCTILE. See OCTANT.

OCTOBER, the 8th month of the year, in Romulus's calendar ; but the tenth in that of Numa, Julius Cæsar, &c, after the addition of January and February. This month contains 31 days ; about the 22d of which, the sun enters the sign Scorpio ♏.

OCTOGON. See OCTAGON.

OCTOSTYLE, in Architecture, the face of a building adorned with 8 columns.

ODD, in Arithmetic, is said of a number that is not even. The series of Odd numbers is 1, 3, 5, 7, &c.

ODDLY-ODD. A number is said to be Oddly-Odd, when an Odd number measures it by an Odd number. So 15 is a number Oddly-odd, because the Odd number 3 measures it by the Odd number 5.

OFFING, or OFFIN, in Navigation, that part of the sea which is at a good distance from shore ; where there is deep water, and no need of a pilot to conduct the ship into port.

OFFSETS, in Surveying are the perpendiculars let fall, and measured from the station lines, to the corners or bends in the hedge, fence, or boundary of any ground.

OFFSET-*Staff*, a slender rod or staff, of 10 links, or other convenient length. Its use is for measuring the Offsets, and other short lines and distances.

OFFWARD, in Navigation, the same with from the shore, &c.

OGEE.

OGEE, or OG, an ornamental moulding in the shape of an S; consisting of two members, the one concave and the other convex.

OLDENBURG (HENRY), who wrote his name sometimes GRUBENDOL, reversing the letters, was a learned German gentleman, and born in the Duchy of Bremen in the Lower Saxony, about the year 1626, being descended from the counts of Aldenburg in Westphalia; whence his name. During the long English parliament in the time of Charles the 1st, he came to England as consul for his countrymen; in which capacity he remained at London in Cromwell's administration. But being discharged of that employment, he was engaged as tutor to the lord Henry Obryan, an Irish nobleman, whom he attended to the university of Oxford; and in 1656 he entered himself a student in that university, chiefly to have the benefit of consulting the Bodleian library. He was afterwards appointed tutor to lord William Cavendish, and became intimately acquainted with Milton the poet. During his residence at Oxford, he became also acquainted with the members of that society there, which gave birth to the Royal Society; and upon the foundation of this latter, he was elected a member of it: and when the Society found it necessary to have two secretaries, he was chosen assistant to Dr. Wilkins. He applied himself with extraordinary diligence to the duties of this office, and began the publication of the Philosophical Transactions with No. 1, in 1664. In order to discharge this task with more credit to himself and the Society, he held a correspondence with more than seventy learned persons, and others, upon a great variety of subjects, in different parts of the world. This fatigue would have been insupportable, had he not, as he told Dr. Lister, managed it so as to make one letter answer another; and that, to be always fresh, he never read a letter before he was ready immediately to answer it: so that the multitude of his letters did not clog him, nor ever lie upon his hands. Among others, he was a constant correspondent of Mr. Robert Boyle, and he translated many of that ingenious gentleman's works into Latin.

About the year 1674 he was drawn into a dispute with Mr. Hook, who complained, that the secretary had not done him justice, in the History of the Transactions, with respect to the invention of the spiral spring for pocket watches; the contest was carried on with some warmth on both sides, but was at length terminated to the honour of Mr. Oldenburg; for, pursuant to an open representation of the affair to the Royal Society, the council thought fit to declare, in behalf of their secretary, that they knew nothing of Mr. Hook having printed a book intitled *Lampas, &c*; but that the publisher of the Transactions had conducted himself faithfully and honestly in managing the intelligence of the Royal Society, and given no just cause for such reflections.

Mr. Oldenburg continued to publish the Transactions as before, to No. 136, June 25, 1677; after which the publication was discontinued till the January following; when they were again resumed by his successor in the secretary's office, Mr. Nehemiah Grew, who carried them on till the end of February 1678. Mr. Oldenburg died at his house at Charlton, between Greenwich and Woolwich, in Kent, August 1678, and was interred there, being 52 years of age.

He published, besides what has been already mentioned, 20 tracts, chiefly on theological and political subjects; in which he principally aimed at reconciling differences, and promoting peace.

OLYMPIAD, in Chronology, a revolution or period of four years, by which the Greeks reckoned their time: so called from the Olympic games, which were celebrated every fourth year, during 5 days, near the summer solstice, upon the banks of the river Alpheus, near Olympia, a town of Elis. As each Olympiad consisted of 4 years, these were called the 1st, 2d, 3d, and 4th year of each Olympiad; the first year commencing with the nearest new moon to the summer solstice.

The first Olympiad began the 3938 year of the Julian period, the 3208 of the creation, 776 years before the birth of Christ, and 24 years before the foundation of Rome. And the computation by these, ended with the 404th Olympiad, being the 440th year of the present vulgar Christian era.

OMBROMETER, a name given by Mr. Roger Pickering (Philos. Transf. No. 473, or Abridg. V, 456) to what is more commonly, though less properly, called a Pluviameter or Rain gage. See PLUVIAMETER.

OMPHALOPTER, or OMPHALOPTIC, in Optics, a glass that is convex on both sides, popularly called a Convex Lens.

OPACITY, a quality of bodies which renders them opake, or the contrary of transparency.

The Cartesians make opacity to consist in this; that the pores of the body are not all straight, or directly before each other; or rather not pervious every way.

This doctrine however is deficient: for though, to have a body transparent, its pores must be straight, or rather open every way; yet it is inconceivable how it should happen, that not only glass and diamonds, but even water, whose parts are so very moveable, should have all their pores open and pervious every way; while the finest paper, or the thinnest gold leaf, should exclude the light, for want of such pores. So that another cause of Opacity must be sought for.

Now all bodies have vastly more pores or vacuities than are necessary for an infinite number of rays to pass freely through them in right lines, without striking on any of the parts themselves. For since water is 19 times lighter or rarer than gold; and yet gold itself is so very rare, that magnetic effluvia pass freely through it, without any opposition; and quicksilver is readily received within its pores, and even water itself by compression; it must have much more pores than solid parts: consequently water must have at least 40 times as much vacuity as solidity.

The cause therefore, why some bodies are opake, does not consist in the want of rectilinear pores, pervious every way; but either in the unequal density of the parts, or in the magnitude of the pores; and to their being either empty, or filled with a different matter; by means of which, the rays of light, in their passage, iage, are arrested by innumerable refractions and reflections, till at length falling on some solid part, they become quite extinct, and are utterly absorbed.

Hence cork, paper, wood, &c, are opake; while glass, diamonds, &c, are pellucid. For in the confines or joining of parts alike in density, such as those of glass, water, diamonds, &c, among themselves, no refraction or reflection takes place, because of the equal attraction every way; so that such of the rays of light as enter the first surface, pass straight through the body, excepting such as are lost and absorbed, by striking on solid parts: but in the bordering of parts of unequal density, such as those of wood and paper, both with regard to themselves, and with regard to the air or empty space in their larger pores, the attraction being unequal, the reflections and refractions will be very great; and thus the rays will not be able to pass through such bodies, being continually driven about, till they become extinct.

That this interruption or discontinuity of parts is the chief cause of Opacity, Sir Isaac Newton argues, appears from hence; that all opake bodies immediately begin to be transparent, when their pores become filled with a substance of nearly equal density with their parts. Thus, paper dipped in water or oil, some stones steeped in water, linen cloth dipped in oil or vinegar, &c, become more transparent than before.

OPAKE, not translucent, nor transparent, or not admitting a free passage to the rays of light.

OPEN *Flank*, in Fortification, is that part of the flank which is covered by the orillon or shoulder.

OPENING *of the Trenches*, is the first breaking of ground by the besiegers, in order to carry on their approaches towards a place.

OPENING *of Gates*, in Astrology, is when one planet separates from another, and presently applies to a third, bearing rule in a sign opposite to that ruled by the planet with which it was before joined.

OPERA-*Glass*, in Optics, is so called from its use in play-houses, and sometimes a *Diagonal Perspective*, from its construction, which is as follows. ABCD (fig. 5, pl. xvii) represents a tube about 4 inches long; in each side of which there is a hole EF and GH, exactly against the middle of a plane mirror IK, which reflects the rays falling upon it to the convex glass LM; through which they are refracted to the concave eyeglass NO, whence they emerge parallel to the eye at the hole *rs*, in the end of the tube. Let P*a*Q be an object to be viewed, from which proceed the rays P*c*, *ab*, and Q*d*: these rays, being reflected by the plane mirror IK, will shew the object in the direction *cp*, *ba*, *dq*, in the image *pq*, equal to the object PQ, and as far behind the mirror as the object is before it: the mirror being placed so as to make an angle of 45 degrees with the sides of the tube. And as, in viewing near objects, it is not necessary to magnify them, the focal distances of both the glasses may be nearly equal; or if that of LM be 3 inches, and that of NO on e inch, the distance between them will be but 2 inches, and the object will be magnified 3 times, being sufficient for the purposes to which this glass is applied.

When the object is very near, as XY, it is viewed through a hole *xy*, at the other end of the tube AB, without an eye glass; the upper part of the mirror being polished for that purpose, as well as the under. The tube unscrews near the object-glass LM, for taking out and cleansing the glasses and mirror. The position of the object will be erect through the concave eye-glass.

The peculiar artifice of this glass is to view a person at a small distance, so that no one shall know who is observed; for the instrument points to a different object from that which is viewed; and as there is a hole on each side, it is impossible to know on which hand the object is situated, which you are viewing.

OPHIUCUS, a constellation of the northern hemisphere; called also Serpentarius.

OPPOSITE *Angles*, or Vertical Angles, are those opposite to each other, made by two intersecting lines; as *a* and *b*, or *c* and *d*.—The opposite angles are equal to each other.

OPPOSITE *Cones*, denote two similar cones vertically opposite, having the same common vertex and axis, and the same sides produced; as the cones A and B.

OPPOSITE *Sections*, or *Hyperbolas*, are those made by cutting the Opposite cones by the same plane; as the hyperbolas C and D.—These are always equal and similar, and have the same transverse axis EF, as also the same conjugate axis.

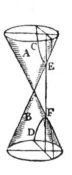

OPPOSITION, is that aspect or situation of two planets or stars, when they are diametrically opposite to each other; being 180°, or a semi-circle apart; and marked thus ☍.

The moon is in Opposition to the sun when she is at the full.

OPTIC, or OPTICAL, something that relates to vision, or the sense of seeing, or the science of optics.

OPTIC *Angle*. See ANGLE.

OPTIC *Axis*. See AXIS.

OPTIC *Chamber*. See CAMERA *Obscura*.

OPTIC *Glasses*, are glasses ground either concave or convex; so as either to collect or disperse the rays of light; by which means vision is improved, and the eye strengthened, preserved, &c.

Among these, the principal are spectacles, reading glasses, telescopes, microscopes, magic lanterns, &c.

OPTIC *Inequality*, in Astronomy, is an apparent irregularity in the motions of far distant bodies; so called, because it is not really in the moving bodies, but arising from the situation of the observer's eye. For if the eye were in the centre, it would always see the motions as they really are.

The

The Optic Inequality may be thus illustrated. Suppose a body revolving with a real uniform motion, in the periphery of a circle ABD &c; and suppose the eye in the plane of the same circle, but at a distance from it, viewing the motion of the body from O. Now when the body goes from A to B; its apparent motion is measured by the angle AOB or the arch or line HL, which it will seem to describe. But while it

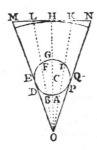

moves through the arch BD in an equal time, its apparent motion will be determined by the angle BOD, or the arch or line LM, which is less than the former LH. But it spends the same time in describing DE, as it does in AB or BD; during all which time of describing DE it appears stationary in the point M. When it really describes EFGIQ, it will appear to pass over MLHKN; so that it will seem to have gone retrograde. And lastly, from Q to P it will again appear stationary in the point N.

OPTIC *Nerves*, the second pair of nerves, springing from the crura of the medulla oblongata, and passing thence to the eye.

These are covered with two coats, which they take from the dura and pia mater; and which, by their expansions, form the two membranes of the eye, called the uvea and cornea. And the retina, which is a third membrane, and the immediate organ of sight, is only an expansion of the fibrous, or inner, and medullary part of these nerves.

OPTIC *Pencil*. See PENCIL *of Rays*.

OPTIC *Place*, of a star &c, is that point or part of its orbit, which is determined by our sight, when the star is seen there. This is either true or apparent; true, when the observer's eye is supposed to be at the centre of the motion; or apparent, when his eye is at the circumference of the earth. See also PLACE.

OPTIC *Pyramid*, in Perspective, is the pyramid ABCO, whose base is the visible object ABC, and the vertex is in the eye at O; being formed by rays

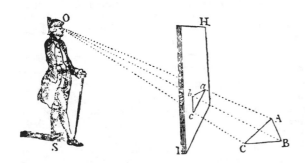

drawn from the several points of the perimeter to the eye.

Hence also may appear what is meant by Optic triangle.

OPTIC *Rays*, particularly means those by which an Optic pyramid, or Optic triangle, is terminated. As OA, OB, OC, &c.

OPTICS. the science of vision; including Catoptrics, and Dioptrics; and even Perspective; as also the whole doctrine of light and colours, and all the phenomena of visible objects.

Optics, in its more extensive acceptation, is a mixed mathematical science; which explains the manner in which vision is performed in the eye; treats of sight in general; gives the reasons of the several modifications or alterations, which the rays of light undergo in the eye; and shews why objects appear sometimes greater, sometimes smaller, sometimes more distinct, sometimes more confused, sometimes nearer and sometimes more remote. In this extensive signification it is considered by Newton, in his excellent work called Optics.

Indeed Optics makes a considerable branch of natural philosophy; both as it explains the laws of nature, according to which vision is performed; and as it accounts for abundance of physical phenomena, otherwise inexplicable.

The Principal Authors and Discoveries in Optics, are the following:

Euclid seems to be the earliest author on Optics that we have. He composed a treatise on the ancient Optics and catoptrics; dioptrics being less known to the Ancients; though it was not quite unnoticed by them, for among the phenomena, at the beginning of that work, Euclid remarks the effect of bringing an object into view, by refraction, in the bottom of a vessel, by pouring water into it, which could not be seen over the edge of the vessel, before the water was poured in; and other authors speak of the then known effects of glass globes &c, both as burning glasses, and as to bodies seen through them. Euclid's work however is chiefly on catoptrics, or reflected rays; in which he shews, in 31 propositions, the chief properties of them, both in plane, convex, and concave surfaces, in his usual geometrical manner; beginning with that concerning the equality of the angles of incidence and reflection, which he demonstrates; and in the last proposition, shewing the effect of a concave speculum, as a burning glass, when exposed to the rays of the sun.

The effects of burning glasses, both by refraction and reflection, are noticed by several others of the Ancients, and it is probable that the Romans had a method of lighting their sacred fire by some such means. Aristophanes, in one of his comedies, introduces a person as making use of a globe filled with water to cancel a bond that was against him, by thus melting the wax of the seal. And if we give but a small degree of credit to what some ancient historians are said to have written concerning the exploits of Archimedes, we shall be induced to think that he constructed some very powerful burning mirrors. It is even allowed that this eminent geometrician wrote a treatise on the subject of them, though it be not now extant; as also concerning the appearance of a ring or circle under water, and therefore could not have been ignorant of the common phenomena of refraction. We find many questions concerning such optical appearances in Aristotle. This author was also sensible that it is the reflection of light from the atmosphere which prevents total darkness after the sun sets, and in places where he does not shine in the day time. He was also of opinion, that rainbows, halos,

halos, and mock funs, were all occafioned by the reflection of the funbeams in different circumftances, by which an imperfect image of his body was produced, the colour only being exhibited, and not his proper figure.

The Ancients were not only acquainted with the more ordinary appearances of refraction, but knew alfo the production of colours by refracted light. Seneca fays, that when the light of the fun fhines through an angular piece of glafs, it fhews all the colours of the rainbow. Thefe colours however, he fays, are falfe, fuch as are feen in a pigeon's neck when it changes its pofition; and of the fame nature he fays is a fpeculum, which, without having any colour of its own, affumes that of any other body.

It appears alfo, that the Ancients were not unacquainted with the magnifying power of glafs globes filled with water, though it does not appear that they knew any thing of the reafon of this power: and it is fuppofed that the ancient engravers made ufe of a glafs globe filled with water to magnify their figures, that they might work to more advantage.

Ptolomy, about the middle of the fecond century, wrote a confiderable treatife on Optics. The work is loft; but from the accounts of others, it appears that he there treated of aftronomical refractions. The firft aftronomers were not aware that the intervals between ftars appear lefs when near the horizon than in the meridian; and on this account they muft have been much embarraffed in their obfervations: but it is evident that Ptolomy was aware of this circumftance by the caution which he gives to allow fomething for it, whenever recourfe is had to ancient obfervations. This philofopher alfo advances a very fenfible hypothefis to account for the remarkably great apparent fize of the fun and moon when feen near the horizon. The mind, he fays, judges of the fize of objects by means of a preconceived idea of their diftance from us: and this diftance is fancied to be greater when a number of objects are interpofed between the eye and the body we are viewing; which is the cafe when we fee the heavenly bodies near the horizon. In his Almageft, however, he afcribes this appearance to a refraction of the rays by vapours, which actually enlarge the angle under which the luminaries appear; juft as the angle is enlarged by which an object is feen from under water.

Alhazen, an Arabian writer, was the next author of confequence, who wrote about the year 1100. Alhazen made many experiments on refraction, at the furface between air and water, air and glafs, and water and glafs; and hence he deduced feveral properties of atmofpherical refraction; fuch as, that it increafes the altitudes of all objects in the heavens; and he firft advanced that the ftars are fometimes feen above the horizon by means of refraction, when they are really below it: which obfervation was confirmed by Vitello, Walther, and efpecially by the obfervations of Tycho Brahe. Alhazen obferved, that refraction contracts the diameters and diftances of the heavenly bodies, and that it is the caufe of the twinkling of the ftars. This refractive power he afcribed, not to the vapours contained in the air, but to its different degrees of tranfparency. And it was his opinion, that fo far from being the caufe of the heavenly bodies appearing larger near the horizon, that it would make them appear lefs; obferving that two ftars appear nearer together in the horizon, than near the meridian. This phenomenon he ranks among optical deceptions. We judge of diftance, he fays, by comparing the angle under which objects appear, with their fuppofed diftance; fo that if thefe angles be nearly equal, and the diftance of one object be conceived greater than that of the other, this will be imagined to be the larger. And he farther obferves, that the fky near the horizon is always imagined to be farther from us than any other part of the concave furface.

In the writings of Alhazen too, we find the firft diftinct account of the magnifying power of glaffes; and it is not improbable that his writings on this head gave rife to the ufeful invention of fpectacles: for he fays, that if an object be applied clofe to the bafe of the larger fegment of a fphere of glafs, it will appear magnified. He alfo treats of the appearance of an object through a globe, and fays that he was the firft who obferved the refraction of rays into it.

In 1270, Vitello, a native of Poland, publifhed a treatife on Optics, containing all that was valuable in Alhazen, and digefted in a better manner. He obferves, that light is always loft by refraction, which makes objects appear lefs luminous. He gave a table of the refults of his experiments on the refractive powers of air, water, and glafs, correfponding to different angles of incidence. He afcribes the twinkling of the ftars to the motion of the air in which the light is refracted; and he illuftrates this hypothefis, by obferving that they twinkle ftill more when viewed in water put in motion. He alfo fhews, that refraction is neceffary as well as reflection, to form the rainbow; becaufe the body which the rays fall upon is a tranfparent fubftance, at the furface of which one part of the light is always reflected, and another refracted. And he makes fome ingenious attempts to explain refraction, or to afcertain the law of it. He alfo confiders the foci of glafs fpheres, and the apparent fize of objects feen through them; though with but little accuracy.

To Vitello may be traced the idea of feeing images in the air. He endeavours to fhew, that it is poffible, by means of a cylindrical convex fpeculum, to fee the images of objects in the air, out of the fpeculum, when the objects themfelves cannot be feen.

The Optics of Alhazen and Vitello were publifhed at Bafil in 1572, by Fred. Rifner.

Contemporary with Vitello, was Roger Bacon, a man of very extenfive genius, who wrote upon almoft every branch of fcience; though it is thought his improvements in Optics were not carried far beyond thofe of Alhazen and Vitello. He even affents to the abfurd notion, held by all philofophers down to his time, that vifible rays proceed *from* the eye, inftead of *towards* it. From many ftories related of him however, it would feem, that he made greater improvements than appear in his writings. It is faid he had the ufe of fpectacles: that he had contrivances, by reflection from glaffes, to fee what was doing at a great diftance, as in an enemy's camp. And lord chancellor Bacon relates a ftory, of his having apparently walked in the air between two fteeples, and which he fuppofed was effected

by

by reflection from glasses while he walked upon the ground.

About 1279 was written a treatise on Optics by Peccam, archbishop of Canterbury.

One of the next who distinguished himself in this way, was Maurolycus, teacher of mathematics at Messina. In a treatise, De Lumine et Umbra, published in 1575, he demonstrates, that the crystalline humour of the eye is a lens that collects the rays of light issuing from the objects, and throws them upon the retina, where the focus of each pencil is. From this principle he discovered the reason why some people are short-sighted, and others long-sighted; also why the former are relieved by concave glasses, and the others by convex ones.

Contemporary with Maurolycus, was John Baptista Porta, of Naples. He discovered the Camera Obscura, which throws considerable light on the nature of vision. His house was the constant resort of all the ingenious persons at Naples, whom he formed into what he call-d An Academy of Secrets; each member being obliged to contribute something that was not generally known, and might be useful. By this means he was furnished with materials for his Magia Naturalis, which contains his account of the Camera Obscura, and the first edition of which was published, as he informs us, when he was not quite 15 years old. He also gave the first hint of the Magic Lantern; which Kircher afterwards followed and improved. His experiments with the camera obscura convinced him, that vision is performed by the intromission of something into the eye, and not by visual rays proceeding from it, as had been formerly imagined; and he was the first who fully satisfied himself and others upon this subject. He justly considered the eye as a camera obscura, and the pupil the hole in the window-shutter; but he was mistaken in supposing that the crystalline humour corresponds, to the wall which receives the images; nor was it discovered till the year 1604, that this office is performed by the retina. He made a variety of just remarks concerning vision; and particularly explained several cases in which we imagine things to be without the eye, when the appearances are occasioned by some affection of the eye itself, or by some motion within the eye. —He remarked also that, in certain circumstances, vision will be assisted by convex or concave glasses; and he seems even to have made some small advances towards the discovery of telescopes.

Other treatises on Optics, with various and gradual improvements, were afterwards successively published by several authors: as Aguilon, Opticorum libr. 6, Antv. 1613; L'Optique, Catoptrique, & Dioptrique of Herigone, in his Cursus Math. Paris 1637; the Dioptrics of Des Cartes, 1637; L'Optique & Catoptrique of Mersenne, Paris 1651: Scheiner, Optica, Lond. 1652: Manchini, Dioptrica Practica, Bologna, 1660: Barrow, Lectiones Opticæ, London 1663: James Gregory, Optica Promota, Lond. 1663: Grimaldi, Physico-mathesis de Lumine, Coloribus, & Iride, Bononia, 1665: Scaphusa, Cogitationes Physico-mechanicæ de Natura Visionis, Heidel. 1670: Kircher, Ars Magna Lucis & Umbræ, Rome 1671: Cherubin, Dioptrique Oculaire, Paris 1671: Leibnitz, Principe Generale de l'Optique, Leipsic Acts 1682:

Newton's Optics and Lectiones Opticæ, 4to and 8vo, 1704 &c: Molyneux, Dioptrics, Lond. 1692: Dr. Jurin's Theory of Distinct and Indistinct Vision.——There is also a large and excellent work on Optics, by Dr. Smith, 2 vols 4to; and an elaborate History of the Present State of Discoveries relating to Vision, Light, and Colours, by Dr. Priestley, 4to, 1772; with a multitude of other authors of inferior note; besides lesser and occasional tracts and papers in the Memoirs of the several learned Academies and Societies of Europe; with improvements by many other persons, among whom are the respectable names of Snell, Fermat, Kepler, Huygens, Hortensius, Boyle, Hook, De la Hire, Lowthorp, Cassini, Halley, Delisle, Euler, Dollond, Clairaut, D'Alembert, Zeiher, Bouguer, Buffon, Nollet, Baume; but the particular improvements by each author must be referred to the history of his life, under the article of their names; while the history and improvements of the several branches are to be found under the various particular articles, as, Light, Colours, Reflection, Refraction, Inflection, Transmission, &c, Spectacles, Telescope, Microscope, &c, &c.

ORB, a spherical shell, hollow sphere, or space contained between two concentric spherical surfaces.—The ancient astronomers conceived the heavens as consisting of several vast azure transparent Orbs or spheres, inclosing one another, and including the bodies of the planets.

The ORBIS *Magnus*, or *Great* ORB, is that in which the sun is supposed to revolve; or rather it is that in which the earth makes its annual circuit.

ORB, in Astrology, or ORB *of Light*, is a certain sphere or extent of light, which the astrologers allow a planet beyond its centre. They pretend that, provided the aspects do but fall within this Orb, they have almost the same effect as if they pointed directly against the centre of the planet.—The Orb of Saturn's light they make to be 10 degrees; that of Jupiter 12 degrees; that of Mars $7\frac{1}{2}$; that of the Sun 17 degrees; that of Venus 8 degrees; that of Mercury 7 degrees; and that of the Moon $12\frac{1}{2}$ degrees.

ORBIT, is the path of a planet or comet; being the curve line described by its centre, in its proper motion in the heavens. So the earth's Orbit, is the ecliptic, or the curve it describes in its annual revolution about the sun.

The ancient astronomers made the planets describe circular Orbits, with an uniform velocity. Copernicus himself could not believe they should do otherwise; being unable to disentangle himself entirely from the excentrics and epicycles to which they had recourse, to account for the inequalities in their motions.

But Kepler found, from observations, that the Orbit of the earth, and that of every primary planet, is an ellipsis, having the sun in one of its foci; and that they all move in these ellipses by this law, that a radius drawn from the centre of the sun to the centre of the planet, always describes equal areas in equal times; or, which is the same thing, in unequal times, it describes areas that are proportional to those times. And Newton has since demonstrated, from the nature of universal gravitation, and projectile motion, that the Orbits must of necessity be ellipses, and the motions observe that law, both of the primary and secondary planets; excepting in so far as their motions and paths are disturbed by their mutual actions upon one another; as the Orbit of the earth by that of the moon; or that of Saturn by the action of Jupiter; &c.

Of these elliptic Orbits, there have been two kinds assigned: the first that of Kepler and Newton, which is the common or conical ellipse; for which Seth Ward, though he himself keeps to it, thinks we might venture to substitute circular Orbits, by using two points, taken at equal distances from the centre, on one of the diameters, as is done in the foci of the ellipsis, and which is called his Circular Hypothesis. The second is that of Cassini, of this nature, viz, that the products of the two lines drawn from the two foci, to any point in the circumference, are every where equal to the same constant quantity; whereas, in the common ellipse, it is the sum of those two lines that is always a constant quantity.

The Orbits of the planets are not all in the same plane with the ecliptic, which is the earth's Orbit round the sun, but are variously inclined to it, and to each other: but still the plane of the ecliptic, or earth's Orbit, intersects the plane of the Orbit of every other planet, in a right line which passes through the sun, called the line of the nodes, and the points of intersection of the Orbits themselves are called the nodes.

The mean semidiameters of the several Orbits, or the mean distances of the planets from the sun, with the excentricities of the Orbits, their inclination to the ecliptic, and the places of their nodes, are as in the following table: where the 2d column contains the proportions of semidiameters of the Orbits, the true semidiameter of that of the earth being 95 millions of miles; and the 3d column shews what part of the semidiameters the excentricities are equal to.

	Propor. semid.	Excentr. pts. of semidiam.	Inclina. of Orbit.	Ascending Node, 1790.
Mercury	387	$\frac{4}{19}$	6° 54′	♉ 14° 43
Venus	723	$\frac{1}{138}$	3 20	♊ 13 59
Earth	1000	$\frac{1}{59}$	0 0	— —
Mars	1524	$\frac{1}{11}$	1 52	♉ 17 17
Jupiter	5201	$\frac{1}{21}$	1 20	♋ 7 29
Saturn	9539	$\frac{1}{18}$	2 30	♋ 21 13
Georgian	19034	$\frac{1}{21}$	0 48	♊ 12 54

The Orbits of the comets are also very excentric ellipses.

ORDER, in Architecture, a system of the several members, ornaments, and proportions of a column and pilaster.

There are five Orders of columns, of which three are Greek, viz, the Doric, Ionic, and Corinthian; and two Italic, viz, the Tuscan and Composite. The three Greek Orders represent the three different manners of building, viz, the solid, the delicate, and the middling: the two Italic ones are imperfect productions of these.

ORDER, in Astronomy. A planet is said to go according to the order of the signs, when it is direct; proceeding from Aries to Taurus, thence to Gemini, &c. As, on the contrary, it goes contrary to the Order of the signs, when it is retrograde, or goes backward, from Pisces to Aquarius, &c.

ORDER, in the Geometry of Curve Lines, is denominated from the rank or Order of the equation by which the geometrical line is expressed, so the simple equation, or 1st power, denotes the 1st Order of lines, which is the right line; the quadratic equation, or 2d power, defines the 2d Order of lines, which are the conic sections and circle; the cubic equation, or 3d power, defines the 3d Order of lines; and so on.

Or, the Orders of lines are denominated from the number of points in which they may be cut by a right line. Thus, the right line is of the 1st Order, because it can be cut only in one point by a right line; the circle and conic sections are of the 2d Order, because they can be cut in two points by a right line; while those of the 3d Order, are such as can be cut in 3 points by a right line; and so on.

It is to be observed, that the Order of curves is always one degree lower than the corresponding line; because the 1st Order, or right line, is no curve; and the circle and conic sections, which are the 2d Order of lines, are only the 1st Order of curves; &c.

See Newton's Enumeratio Linearum Tertii Ordinis.

ORDINATES, in the Geometry of Curve Lines, are right lines drawn parallel to each other, and cutting the curve in a certain number of points.

The parallel Ordinates are usually all cut by some other line, which is called the abscifs, and commonly the Ordinates are perpendicular to the abscissal line. When this line is a diameter of the curve, the property of the Ordinates is then the most remarkable; for, in the curves of the first kind, or the conic sections and circle, the Ordinates are all bisected by the diameter, making the part on one side of it equal to the part on the other side of it; and in the curves of the 2d order, which may be cut in three points by an Ordinate, then of the three parts of the Ordinate, lying between these three intersections of the curve and the intersection with the diameter, the part on one side the diameter is equal to both the two parts on the other side of it. And so for curves of any order, whatever the number of intersections may be, the sum of the parts of any Ordinate, on one side of the diameter, is equal to the sum of the parts on the other side of it.

The use of Ordinates in a curve, and their abscisses, is to define or express the nature of a curve, by means of the general relation or equation between them; and the greatest number of factors, or the dimensions of the highest term, in such equation, is always the same as the order of the line; that equation being a quadratic, or its highest term of two dimensions, in the lines of the 2d order, being the circle and conic sections; and a cubic equation, or its highest term containing 3 dimensions, in the lines of the 3d order; and so on.

Thus,

Thus, y denoting an Ordinate BC, and x its abfcifs AB ; alfo a, b, c, &c, given quantities : then $y^2 = ax^2 + bx + c$ is the general equation for the lines of the 2d order ; and $xy^2 - ey = ax^3 + bx^2 + cx + d$ is the equation for the lines of the 3d order ; and fo on.

ORDNANCE, are all forts of great guns, ufed in war ; fuch as cannon, mortars, howitzers, &c.

ORFFYREUS's *Wheel*, in Mechanics, is a machine fo called from its inventor, which he afferted to be a perpetual motion. This machine, according to the account given of it by Gravefande, in his Oeuvres Philofophiques, publifhed by Allemand, Amft. 1774, confifted externally of a large circular wheel, or rather drum, 12 feet in diameter, and 14 inches deep ; being very light, as it was formed of an affemblage of deals, having the intervals between them covered with waxed cloth, to conceal the interior parts of it. The two extremities of an iron axis, on which it turned, refted on two fupports. On giving a flight impulfe to the wheel, in either direction, its motion was gradually accelerated ; fo that after two or three revolutions it acquired fo great a velocity as to make 25 or 26 turns in a minute. This rapid motion it actually preferved during the fpace of 2 months, in a chamber of the landgrave of Heffe, the door of which was kept locked, and fealed with the landgrave's own feal. At the end of that time it was ftopped, to prevent the wear of the materials. The profeffor, who had been an eye-witnefs to thefe circumftances, examined all the external parts of it, and was convinced that there could not be any communication between it and any neighbouring room. Orffyreus however was fo incenfed, or pretended to be fo, that he broke the machine in pieces, and wrote on the wall, that it was the impertinent curiofity of profeffor Gravefande which made him take this ftep. The prince of Heffe, who had feen the interior parts of this wheel, but fworn to fecrefy, being afked by Gravefande, whether, after it had been in motion for fome time, there was any change obfervable in it, and whether it contained any pieces that indicated fraud or deception, anfwered both queftions in the negative, and declared that the machine was of a very fimple conftruction.

ORGANICAL *Defcription of Curves*, is the defcription of them upon a plane, by means of inftruments, and commonly by a continued motion. The moft fimple conftruction of this kind, is that of a circle by means of a pair of compaffes. The next is that of an ellipfe by means of a thread and two pins in the foci, or the ellipfe and hyperbola, by means of the elliptical and hyperbolic compaffes.

A great variety of defcriptions of this fort are to be found in Schooten De Organica Conic. Sect. in Piano Defcriptione ; in Newton's Arithmetica Univerfalis, De Curvarum Defcriptione Organica; Maclaurin's Geometria Organica ; Brackenridge's Defcriptio Linearum Curvarum ; &c.

ORGUES, or ORGANS, in Fortification, long and thick pieces of wood, fhod with pointed iron, and hung each by a feparate rope over the gate way of a town, ready on any furprife or attempt of the enemy to be let down to ftop up the gate. The ends of the feveral ropes are wound about a windlafs, fo as to be let down all together.

ORGUES is alfo ufed for a machine compofed of feveral harquebuffes or mufket-barrels, bound together ; fo as to make feveral explofions at the fame time. They are ufed to defend breaches and other places attacked.

ORIENT, the eaft, or the eaftern point of the horizon.

ORIENT *Equinoctial*, is ufed for that point of the horizon where the fun rifes when he is in the equinoctial, or when he enters the figns Aries and Libra.

ORIENT *Aestival*, is the point where the fun rifes in the middle of fummer, when the days are longeft.

ORIENT *Hybernal*, is the point where the fun rifes in the middle of winter, when the days are fhorteft.

ORIENTAL, fituated towards the eaft with regard to us : in oppofition to occidental or the weft.

ORIENTAL *Aftronomy, Philofophy*, &c. ufed for thofe of the eaft, or of the Arabians, Chaldeans, Perfians, Indians, &c.

ORILLON, in Fortification, a fmall rounding of earth, lined with a wall, raifed on the fhoulder of thofe baftions that have cafemates, to cover the cannon in the retired flank, and prevent their being difmounted by the enemy.

There are other forts of Orillons, properly called Epaulements, or Shoulderings, which are almoft of a fquare figure.

ORION, a conftellation of the fouthern hemifphere, with refpect to the ecliptic, but half in the northern, and half on the fouthern fide of the equinoctial, which runs acrofs the middle of his body.

The ftars in this conftellation are, 38 in Ptolomy's catalogue, 42 in Tycho's, 62 in Hevelius's, and 78 in Flamfteed's. But fome telefcopes have difcovered feveral thoufands of ftars in this conftellation.

Of thefe ftars, there are no lefs than two of the firft magnitude, and four of the fecond, befide a great many of the third and fourth. One of thofe two ftars of the firft magnitude is upon the middle of the left foot, and is called *Regel*; the other is on the right fhoulder, and called *Betelguefe*; of the four of the fecond magnitude, one is on the left fhoulder, and called *Bellatrix*, and the other three are in the belt, lying nearly in a right line and at equal diftances from each other, forming what is popularly called the *Yardwand*.

This conftellation is one of the 48 old afterifms, and one of the moft remarkable in the heavens. It is in the figure of a man, having a fword by his fide, and feems attacking the bull with a club in his right hand, his left bearing a fhield.

This conftellation is particularly mentioned by many of the ancient authors, and even in the Scriptures themfelves. The Greeks, according to their cuftom, give feveral fabulous accounts of him. One is, that this Orion was a fon of their fea-god Neptune by Euryale, the famous huntrefs. The fon poffeffed the difpofition of his mother, and became the greateft hunter in the world : and Neptune gave him the fingular privilege, that he fhould walk upon the furface of the fea as well as if it were on dry land. Another account of his origin is, that one Hyreius in Thebes, having entertained Jupiter and Mercury with great hospitality, requeſted of them the favour that he might have a ſon. The ſkin of the ox which he had ſacrificed to them, was buried in the ground, with certain ceremonies, and the ſon ſo much deſired was produced from it, a youth of promiſing ſpirit, and named Orion.

They farther tell us, that he viſited Chios when grown up, and raviſhed Penelope the daughter of Œnopron, for which the father put out his eyes, and baniſhed him the iſland: he thence went to Lemnos, where Vulcan received him, and gave him Cedalion for a companion. Afterwards, being reſtored to ſight by the ſun, he returned to Chios, and would have revenged himſelf on the king, but the people hid him. After this it ſeems he hunted with Diana, and was ſo exalted with his ſucceſs, that he uſed to ſay he would deſtroy every creature on the earth: the Earth, irritated at this, produced a Scorpion, which ſtung him to death, and both he and the reptile were taken up to the ſkies, the Scorpion making one of the twelve ſigns of the zodiac.

Others give a different account of his deſtruction: they tell us that he would have raviſhed the goddeſs of chaſtity Diana herſelf, and that ſhe killed him with her arrow. All the writers, however, are not agreed about this: they who make him the ſacrifice to the vengeance of the offended goddeſs, ſay, that herſelf afterwards placed his figure in the ſkies as a memorial of the attempt, and a terror to all ages. But there are ſome who ſay ſhe loved him ſo well that ſhe had thoughts of marrying him: theſe add, that Apollo could not bear ſo diſhonourable an alliance for his ſiſter, for which reaſon he killed him; and that Diana, after ſhedding ſhowers of tears over his corps, obtained of Jupiter a place for him in the heavens.

No conſtellation was ſo terrible to the mariners of the early periods, as this of Orion. He is mentioned in this way by all the Greek and Latin poets, and even by their hiſtorians; his riſing and ſetting being attended by ſtorms and tempeſts: and as the northern conſtellations are made the followers of the Pleiades; ſo are the ſouthern ones made the attendants of Orion.

The name of this conſtellation is alſo met with in Scripture ſeveral times, viz, in the books of Job, Amos, and Iſaiah. In Job it is aſked, "Canſt thou bind the ſweet influence of the Pleiades, or looſe the bands of Orion?" And Amos ſays, "Seek him that maketh the Seven Stars and Orion, and turneth the ſhadow of death into morning."

ORION's *River*, the ſame as the conſtellation Eridanus.

ORLE, ORLET, or ORLO, in Architecture, a fillet under the ovolo, or quarter-round of a capital — When it is at the top or bottom of the ſhaft, it is called the cincture.—Palladio alſo uſes Orlo for the plinth of the baſes of columns and pedeſtals.

ORRERY, an aſtronomical machine, for exhibiting the various motions and appearances of the ſun and planets; and hence often called a Planetarium.

The reaſon of the name Orrery was this: Mr. Rowley, a mathematical inſtrument-maker, having got one from Mr. George Graham, the original inventor, to be

ſent abroad with ſome of his own inſtruments, he copied it, and made the firſt for the earl of Orrery, Sir Richard Steel, who knew nothing of Mr. Graham's machine, thinking to do juſtice to the firſt encourager, as well as to the inventor of ſuch a curious inſtrument, called it an Orrery, and gave Rowley the praiſe due to Mr. Graham. Deſaguliers' Experim. Philoſ. vol. 1, pa. 430. The figure of this grand Orrery is exhibited at fig. 1, pl. 19. It is ſince made in various other figures.

ORTEIL, in Fortification. See BERME.

ORTELIUS (ABRAHAM), a celebrated geographer, was born at Antwerp, in 1527. He was well ſkilled in the languages and mathematics, and acquired ſuch reputation by his ſkill in geography, that he was ſurnamed the *Ptolomy of his time.* Juſtus Lipſius, and moſt of the great men of the 16th century, were our author's intimate friends. He paſſed ſome time at Oxford in the reign of Edward the 6th; and he viſited England a ſecond time in 1577.

His *Theatrum Orbis Terræ* was the completeſt work of the kind that had ever been publiſhed, and gained our author a reputation adequate to his immenſe-labour in compiling it. He wrote alſo ſeveral other excellent geographical works; the principal of which are, his *Theſaurus*, and his *Synonyma Geographica.*—The world is alſo obliged to him for the *Britannia*, which was undertaken by Cambden at his requeſt.—He died at Antwerp, 1598, at 71 years of age.

ORTHODROMICS, in Navigation, is Great-circle ſailing, or the art of ſailing in the arch of a great circle, which is the ſhorteſt courſe: For the arch of a great circle is Orthodromia, or the ſhorteſt diſtance between two points or places.

ORTHOGONIAL, in Geometry, is the ſame as rectangular, or right-angled.—When the term refers to a plane figure, it ſuppoſes one leg or ſide to ſtand perpendicular to the other: when ſpoken of ſolids, it ſuppoſes their axis to be perpendicular to the plane of the horizon.

ORTHOGRAPHIC or ORTHOGRAPHICAL *Projection of the Sphere*, is the projection of its ſurface or of the ſphere on a plane, paſſing through the middle of it, by an eye vertically at an infinite diſtance. See PROJECTION.

ORTHOGRAPHY, in Geometry, is the drawing or delineating the fore-right plan or ſide of any object, and of expreſſing the heights or elevations of every part. Being ſo called from its determining things by perpendicular right lines falling on the geometrical plan; or rather, becauſe all the horizontal lines are here ſtraight and parallel, and not oblique as in repreſentations of perſpective.

ORTHOGRAPHY, in Architecture, is the profile or elevation of a building, ſhewing all the parts in their true proportion. This is either external or internal.

External ORTHOGRAPHY, is a delineation of the outer face or front of a building; ſhewing the principal wall with its apertures, roof, ornaments, and every thing viſible to an eye placed before the building. And

Internal ORTHOGRAPHY, called alſo a Section, is a delineation or draught of a building, ſuch as it would appear if the external wall were removed.

ORTHOGRAPHY, in Fortification, is the profile, or

representation

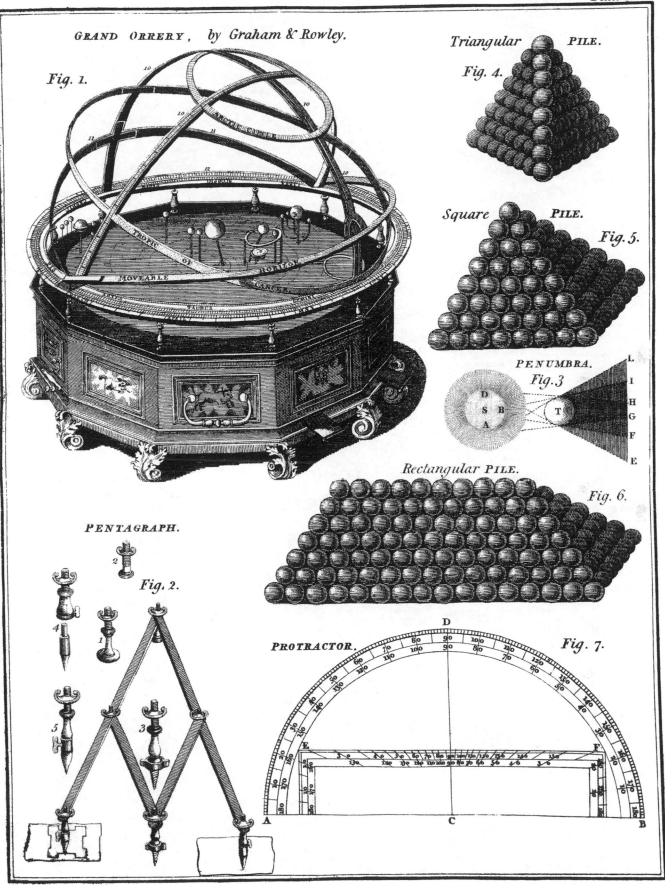

Plate XIX.

GRAND ORRERY, by Graham & Rowley.

Fig. 1.

Triangular PILE.

Fig. 4.

Square PILE.

Fig. 5.

PENUMBRA.

Fig. 3

Rectangular PILE.

Fig. 6.

PENTAGRAPH.

Fig. 2.

PROTRACTOR.

Fig. 7.

reprefentation of a work ; or a draught fo conducted, as that the length, breadth, height, and thicknefs of the feveral parts are expreffed, fuch as they would appear, if it were perpendicularly cut from top to bottom.

ORTIVE, or *Eaftern Amplitude*, in Aftronomy, is an arch of the horizon intercepted between the point where a ftar rifes, and the eaft point of the horizon.

OSCILLATION, in Mechanics, vibration, or the reciprocal afcent and defcent of a pendulum.

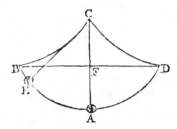

If a fimple pendulum be fufpended between two femicycloids BC, CD, that have the diameter CF of the generating circle equal to half the length of the ftring, fo that the ftring, as the body E Ofcillates, folds about them, then will the body Ofcillate in another cycloid BEAD, fimilar and equal to the former. And the time of the Ofcillation in any arc AE, meafured from the loweft point A, is always the fame conftant quantity, whether that arc be larger or fmaller. But the Ofcillations in a circle are unequal, thofe in the fmaller arcs being lefs than thofe in the larger; and fo always lefs and lefs as the arcs are fmaller, but ftill greater than the time of Ofcillation in a cycloidal arc ; till the circular arc becomes very fmall, and then the time of Ofcillation in it is very nearly equal to the time in the cycloid, becaufe the circle and cycloid have the fame curvature at the vertex, the length of the ftring being the common radius of curvature to them there.

· The time of one whole Ofcillation in the cycloid, or of an afcent and defcent in any arch of it, is to the time in which a heavy body would fall freely through CF or FA, the diameter of the generating circle, or through half the length of the pendulum ftring, as the circumference of a circle is to its diameter, that is as 3·1416 to 1. So that if l denote the length of the pendulum CA, and $g = 16\frac{1}{12}$ feet $= 193$ inches, the fpace a heavy body falls in the 1ft fecond of time, and $p = 3·1416$ the circumference of a circle whofe diameter is 1 : then by the laws of falling bodies,

it is $\sqrt{g} : \sqrt{\frac{1}{2}l} :: 1'' : \sqrt{\frac{l}{2g}}$, the time of falling through

CF or $\frac{1}{2}l$; therefore $1 : p :: \sqrt{\frac{l}{2g}} : p\sqrt{\frac{l}{2g}}$, which is the time of one vibration in any arch of the cycloid which has the diameter of its generating circle equal to $\frac{1}{2}l$. Or, by extracting the known numbers, the fame time of an Ofcillation becomes barely $\frac{4}{25}\sqrt{l}$ or $\frac{16}{100}\sqrt{l}$ very nearly, l being the length of the pendulum in inches. And therefore this is alfo very nearly the time of an Ofcillation in a fmall circular arc, whofe radius is l inches.

Hence the times of the Ofcillation of pendulums of different lengths, are directly in the fubduplicate ratio of their lengths, or as the fquare roots of their lengths.

The more exact time of Ofcillating in a circular arc, when this is of fome finite fmall length, is

$\frac{4}{25}\sqrt{l} \times (1 + \frac{b}{8l})$; where b is the height of the vibration, or the verfed fine of the fingle arc of afcent, or defcent, to the radius l.

The celebrated Huygens firft refolved the problem concerning the Ofcillations of pendulums, in his book De Horologio Ofcillatorio, reducing compound pendulums to fimple ones. And his doctrine is founded on this hypothefis, that the common centre of gravity of feveral bodies, connected together, muft afcend exactly to the fame height from which it fell, whether thofe bodies be united, or feparated from one another in afcending again, provided that each begin to afcend with the velocity acquired by its defcent.

This fuppofition was oppofed by feveral, and very much fufpected by others. And thofe even who believed the truth of it, yet thought it too daring to be admitted without proof into a fcience which demonftrates every thing.

At length Mr. James Bernoulli demonftrated it, from the nature of the lever; and publifhed his folution in the Mem. Acad. of Scienc. of Paris, for the year 1703. After his death, which happened in 1705, his brother John Bernoulli gave a more eafy and fimple folution of the fame problem, in the fame Memoirs for 1714; and about the fame time, Dr. Brook Taylor publifhed a fimilar folution in his Methodus Incrementorum : which gave occafion to a difpute between thefe two mathematicians, who accufed each other of having ftolen their folutions. The particulars of which difpute may be feen in the Leipfic Acts for 1716, and in Bernoulli's works, printed in 1743.

Axis of OSCILLATION, is a line parallel to the horizon, fuppofed to pafs through the centre or fixed point about which the pendulum ofcillates, and perpendicular to the plane in which the Ofcillation is made.

Centre of OSCILLATION, in a fufpended body, is a certain point in it, fuch that the Ofcillations of the body will be made in the fame time as if that point alone were fufpended at that diftance from the point of fufpenfion. Or it is the point into which if the whole weight of the body be collected, the feveral Ofcillations will be performed in the fame time as before : the Ofcillations being made only by the force of gravity of the ofcillating body. See CENTRE of *Ofcillation*.

OSCULATION, in Geometry, denotes the contact between any curve and its ofculatory circle, that is, the circle of the fame curvature with the given curve, at the point of contact or of Ofculation. If AC be the evolute of the involute curve AEF, and the tangent CE the radius of curvature at the point E, with which, and the centre C, if the circle BEG be defcribed ; this circle is faid to ofculate or kifs the curve AEF in the point E, which point E Mr. Huygens calls the point of Ofculation, or kiffing point.

The line CE is called the ofculatory radius, or the radius of curvature; and the circle BEG the ofculatory or kiffing circle.

The

The evolute AC is the locus of the centres of all the circles that osculate the involute curve AEF.

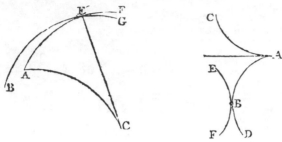

OSCULATION also means the point of concourse of two branches of a curve which touch each other. For example, if the equation of a curve be $y = \sqrt{x} + \sqrt[4]{x^3}$, it is easy to see that the curve has two branches touching one another at the point where $x = 0$, because the roots have each the signs $+$ and $-$.

The point of Osculation differs from the cusp or point of retrocession (which is also a kind of point of contact of two branches) in this, that in this latter case the two branches terminate, and pass no farther, but in the former the two branches exist on both sides of the point of Osculation. Thus, in the second figure above, the point B is the Osculation of the two branches ABD, EBF; but C, though it is also a tangent point, is a cusp or point of retrocession, of AC and AB, the branches not passing beyond the point A.

OSCULATORY *Circle*, or *Kissing Circle*, is the same as the circle of curvature; that is, the circle having the same curvature with any curve at a given point. See the foregoing article, Osculation, where BEG, in the last figure but one, is the Osculatory circle of the curve AEF at the point E; and CE the Osculatory radius, or the radius of curvature.

This circle is called Osculatory, or kissing, because that, of all the circles that can touch the curve in the same point, that one touches it the closest, in such manner that no other such tangent circle can be drawn between it and the curve; so that, in touching the curve, it embraces it as it were, both touching and cutting it at the same time, being on one side at the convex part of the curve, and on the other at the concave part of it.

In a circle, all the Osculatory radii are equal, being the common radius of the circle; the evolute of a circle being only a point, which is its centre. See some properties of the Osculatory circle in Maclaurin's Algebra, Appendix De Linearum Geometricarum Proprietatibus generalibus Tractatus, Theor. 2, § 15 &c, treated in a pure geometrical manner.

OSCULATORY *Parabola*. See PARABOLA.

OSCULATORY *Point*, the Osculation, or point of contact between a curve and its Osculatory circle.

OSTENSIVE *Demonstrations*, such as plainly and directly demonstrate the truth of any proposition. In which they stand distinguished from Apagogical ones, or reductions ad absurdum, or ad impossibile, which prove the truth proposed by demonstrating the absurdity or impossibility of the contrary

OTACOUSTIC, an instrument that aids or improves the sense of hearing. See ACOUSTICS.

OVAL, an oblong curvilinear figure, having two unequal diameters, and bounded by a curve line returning into itself. Or a figure contained by a single curve line, imperfectly round, its length being greater than its breadth, like an egg: whence its name.

The proper Oval, or egg-shape, is an irregular figure, being narrower at one end than the other; in which it differs from the ellipse, which is the mathematical Oval, and is equally broad at both ends.—The common people confound the two together: but geometricians call the Oval a False Ellipse.

The method of describing an Oval chiefly used among artificers, is by a cord or string, as FH*f*, whose length is equal to the greater diameter of the intended Oval, and which is fastened by its extremes to two points or pins, F and *f*, planted in its longer diameter; then, holding it always stretched out as at H, with a pin or pencil carried round the inside, the Oval is described: which will be so much the longer and narrower as the two fixed points are farther apart. This Oval so described is the true mathematical ellipse, the points F and *f* being the two foci.

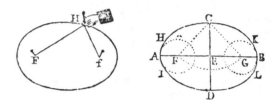

Another popular way to describe an Oval of a given length and breadth, is thus: Set the given length and breadth, AB and CD, to bisect each other perpendicularly at E; with the centre C, and radius AE, describe an arc to cross AB in F and G; then with these centres, F and G, and radii AF and BG, describe two little arcs HI and KL for the smaller ends of the Oval; and lastly, with the centres C and D, and radius CD, describe the arcs HK and IL, for the flatter or longer sides of the Oval.— Sometimes other points, instead of C and D, are to be taken by trial, as centres in the line CD, produced if necessary, so as to make the two last arcs join best with the two former ones.

OVAL denotes also certain roundish figures, of various and pleasant shapes, among curve lines of the higher kinds. These figures are expressed by equations of all dimensions above the 2d, and more especially the even dimensions, as the 4th, 6th, &c. Of this kind is the equation $a^2 y^2 = -x^4 + ax^3$, which denotes the

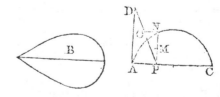

Oval B, in shape of the section of a pear through the middle, and is easily described by means of points. For, if

a circle

a circle be deſcribed whoſe diameter AC is = *a*, and AD be perpendicular and equal to AC ; then taking any point P in AC, joining DP, and drawing PN parallel to AD, and NO parallel to AC ; and laſtly taking PM = NO, the point M will be one point of the Oval ſought.

In like manner the equation

$$y^4 - 4y^2 = -ax^4 + bx^3 + cx^2 + dx + e$$

expreſſes ſeveral very pretty Ovals, among which the following 12 are ſome of the moſt remarkable. For when the equation

$$ax^4 = bx^3 + cx^2 + dx + e$$

has four real unequal roots, the given equation will denote the three following ſpecies, in fig. 1, 2, 3 :

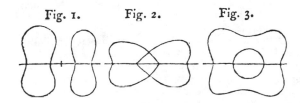

Fig. 1.   Fig. 2.   Fig. 3.

When the two leſs roots are equal, the three ſpecies will be expreſſed as in fig. 4, 5, 6, thus :

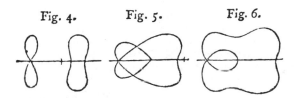

Fig. 4.   Fig. 5.   Fig. 6.

When the two leſs roots become imaginary, it will denote the three ſpecies as exhibited in fig. 7, 8, 9 :

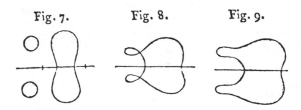

Fig. 7.   Fig. 8.   Fig. 9.

When the two middle roots are equal, the ſpecies will be as appears in fig. 10: when two roots are equal, and two more ſo, the ſpecies will be as in fig. 11 : and when the two middle roots become imaginary, the ſpecies will be as appears in fig. 12 :

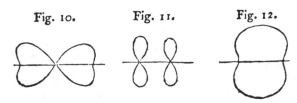

Fig. 10.   Fig. 11.   Fig. 12.

OUGHTRED (WILLIAM), an eminent Engliſh mathematician and divine, was born at Eton in Buckinghamſhire, 1573, and educated in the ſchool there ; whence he was elected to King's-college in Cambridge in 1592, where he continued about 12 years, and became a fellow ; employing his time in cloſe application to uſeful ſtudies, particularly the mathematical ſciences, which he contributed greatly, by his example and exhortation, to bring into vogue among his acquaintances there.

About 1603 he quitted the univerſity, and was preſented to the rectory of Aldbury, near Guildford in Surry, where he lived a long retired and ſtudious life, ſeldom travelling ſo far as London once a year ; his recreation being a diverſity of ſtudies : " as often, ſays he, as I was tired with the labours of my own profeſſion, I have allayed that tediouſneſs by walking in the pleaſant, and more than Elyſian Fields of the diverſe and various parts of human learning, and not of the mathematics only." About the year 1628 he was appointed by the earl of Arundel tutor to his ſon lord William Howard, in the mathematics, and his Clavis was drawn up for the uſe of that young nobleman. He always kept up a correſpondence by letters with ſome of the moſt eminent ſcholars of his time, upon mathematical ſubjects : the originals of which were preſerved, and communicated to the Royal Society, by William Jones, Eſq. The chief mathematicians of that age owed much of their ſkill to him ; and his houſe was always full of young gentlemen who came from all parts to receive his inſtruction : nor was he without invitations to ſettle in France, Italy, and Holland. " He was as facetious, ſays Mr. David Lloyd, in Greek and Latin, as ſolid in arithmetic, geometry, and the ſphere, of all meaſures, muſic, &c ; exact in his ſtyle as in his judgment ; handling his tube and other inſtruments at 80 as ſteadily as others did at 30 ; owing this, as he ſaid, to temperance and exerciſe ; principling his people with plain and ſolid truths, as he did the world with great and uſeful arts ; advancing new inventions in all things but religion, which he endeavoured to promote in its primitive purity, maintaining that prudence, meekneſs, and ſimplicity were the great ornaments of his life.

Notwithſtanding Oughtred's great merit, being a ſtrong royaliſt, he was in danger, in 1646, of a ſequeſtration by the committee for plundering miniſters ; ſeveral articles being depoſed and ſworn againſt him : but upon his day of hearing, William Lilly, the famous aſtrologer, applied to Sir Bulſtrode Whitlocke and all his old friends ; who appeared ſo numerous in his behalf, that though the chairman and many other Preſbyterian members were active againſt him, yet he was cleared by the majority. This is told us by Lilly himſelf, in the Hiſtory of his own Life, where he ſtyles Oughtred the moſt famous mathematician then of Europe.—He died in 1660, at 86 years of age, and was buried at Aldbury. It is ſaid he died of a ſudden ecſtaſy of joy, about the beginning of May, on hearing the news of the vote at Weſtminſter, which paſſed for the reſtoration of Charles the 2d —He left one ſon, whom he put apprentice to a watch-maker, and wrote a book of inſtructions in that art for his uſe.

He publiſhed ſeveral works in his life time ; the principal of which are the following :

1. *Arithmetica*

1. *Arithmeticæ in Numero & Speciebus Inſtitutio*, in 8vo, 1631. This treatiſe he intended ſhould ſerve as a general Key to the Mathematics. It was afterwards reprinted, with conſiderable alterations and additions, in 1648, under the title of *A Key to the Mathematics.* It was alſo publiſhed in Engliſh, with ſeveral additional tracts; viz, one on the Reſolution of all ſorts of Affected Equations in Numbers; a ſecond on Compound Intereſt; a third on the eaſy Art of Delineating all manner of Plain Sun-dials; alſo a Demonſtration of the Rule of Falſe-Poſition. A 3d edition of the ſame work was printed in 1652, in Latin, with the ſame additional tracts, together with ſome others, viz, On the Uſe of Logarithms; A Declaration of the 10th book of Euclid's Elements; a treatiſe of Regular Solids; and the Theorems contained in the books of Archimedes.

2. *The Circles of Proportion*, and a *Horizontal Inſtrument*; in 1633, 4to; publiſhed by his ſcholar Mr. William Foſter.

3. *Deſcription and Uſe of the Double Horizontal Dial;* 1636, 8vo.

4. *Trigonometria:* his treatiſe on Trigonometry, in Latin, in 4to, 1657: And another edition in Engliſh, together with Tables of Sines, Tangents, and Secants.

He left behind him a great number of papers upon mathematical ſubjects; and in moſt of his Greek and Latin mathematical books, there were found notes in his own hand writing, with an abridgment of almoſt every propoſition and demonſtration in the margin, which came into the muſeum of the late William Jones Eſq. F. R. S. Theſe books and manuſcripts then paſſed into the hands of his friend Sir Charles Scarborough the phyſician; the latter of which were carefully looked over, and all that were found fit for the preſs, printed at Oxford in 1676, in 8vo, under the title of

5. *Opuſcula Mathematica hactenus inedita.* This collection contains the following pieces: (1), Inſtitutiones Mechanicæ: (2), De Variis Corporum Generibus Gravitate & Magnitudine comparatis: (3), Automata: (4), Quæſtiones Diophanti Alexandrini, libri tres: (5), De Triangulis Planis Rectangulis: (6), De Diviſione Superficierum: (7), Muſicæ Elementa: (8) De Propugnaculorum Munitionibus: (9), Sectiones Angulares.

6. In 1660, Sir Jonas Moore annexed to his Arithmetic a treatiſe entitled, " *Conical Sections;* or, The ſeveral Sections of a Cone; being an Analyſis or Methodical Contraction of the two firſt books of Mydorgius, and whereby the nature of the Parabola, Hyperbola, and Ellipſis, is very clearly laid down. Tranſlated from the papers of the learned William Oughtred."

Oughtred, though undoubtedly a very great mathematician, was yet far from having the happieſt method of treating the ſubjects he wrote upon. His ſtyle and manner were very conciſe, obſcure, and dry; and his rules and precepts ſo involved in ſymbols and abbreviations, as rendered his mathematical writings very troubleſome to read, and difficult to be underſtood. Beſide the characters and abbreviations before made uſe of in Algebra, he introduced ſeveral others; as

× to denote multiplication;

: : for proportion or ſimilitude of ratios;

÷ for continued proportion;

⊐ ⊐ } for greater and leſs; &c.

OUNCE, a ſmall weight, being the 16th part of a pound avoirdupois; and the 12th part of a pound troy. —The avoirdupois Ounce is divided into 16 drachms or drams; alſo the Ounce troy into 24 pennyweights, and the pennyweight into 24 grains.

OVOLO, in Architecture, a round moulding, whoſe profile or ſweep, in the Ionic and Compoſite capital, is uſually a quadrant of a circle; whence it is alſo popularly called the Quarter round.

OUTWARD *Flanking Angle*, or the *Angle of the Tenaille*, is that comprehended by the two flanking lines of defence.

OUTWORKS, in Fortification, all thoſe works made on the outſide of the ditch of a fortified place, to cover and defend it.

Outworks, called alſo Advanced and Detached Works, are thoſe which not only ſerve to cover the body of the place, but alſo to keep the enemy at a diſtance, and prevent them from taking advantage of the cavities and elevations uſually found in the places about the counterſcarp; which might ſerve them either as lodgments, or as rideaux, to facilitate the carrying on their trenches, and planting their batteries againſt the place. Such are ravelins, tenailles, hornworks, queue d'arondes, envelopes, and crownworks. Of theſe, the moſt uſual are ravelins, or halfmoons, formed between the two baſtions, on the flanking angle of the counterſcarp, and before the curtain, to cover the gates and bridges.

It is a general rule in all Outworks, that if there be ſeveral of them, one before another, to cover one and the ſame tenaille of a place, the nearer ones muſt gradually, and one after another, command thoſe which are fartheſt advanced out into the campagne; that is, muſt have higher ramparts, that ſo they may overlook and fire upon the beſiegers, when they are maſters of the more outward works.

The gorges alſo of all Outworks ſhould be plain, and without parapets; leſt, when taken, they ſhould ſerve to ſecure the beſiegers againſt the fire of the retiring beſieged; whence the gorges of Outworks are only palliſadoed, to prevent a ſurprize.

OX-EYE, in Optics. See SCIOPTIC, and CAMERA *Obſcura.*

OXGANG, or OXGATE, of land, is uſually taken for 15 acres; being as much land as it is ſuppoſed one ox can plow in a year. In Lincolnſhire they ſtill corruptly call it Oſkin of land.—In Scotland, the term is uſed for a portion of arable land, containing 13 acres.

OXYGONE, in Geometry, is acute-angled, meaning a figure conſiſting wholly of acute angles, or ſuch as are leſs than 90 degrees each.—The term is chiefly applied to triangles, where the three angles are all acute.

OXYGONIAL, is acute-angular.

OZANAM (JAMES), an eminent French mathematician, was deſcended from a family of Jewiſh extraction, but which had long been converts to the Romiſh faith; and ſome of whom had held conſiderable places in the parliaments of Provence. He was born at Boligneux in Breſſia, in the year 1640; and being a younger ſon, fon, though his father had a good eftate, it was thought proper to breed him to the church, that he might enjoy fome fmall benefices which belonged to the family, to ferve as a provifion for him. Accordingly he ftudied divinity four years; but then, on the death of his father, he devoted himfelf entirely to the mathematics, to which he had always been ftrongly attached. Some mathematical books, which fell into his hands, firft excited his curiofity; and by his extraordinary genius, without the aid of a mafter, he made fo great a progrefs, that at the age of 15 he wrote a treatife of that kind.

For a maintenance, he firft went to Lyons to teach the mathematics; which anfwered very well there; and after fome time his generous difpofition procured him ftill better fuccefs elfewhere. Among his fcholars were two foreigners, who expreffing their uneafinefs to him, at being difappointed of fome bills of exchange for a journey to Paris; he afked them how much would do, and being told 50 piftoles, he lent them the money immediately, even without their note for it. Upon their arrival at Paris, mentioning this generous action to M. Dagueffeau, father of the chancellor, this magiftrate was touched with it; and engaged them to invite Ozanam to Paris, with a promife of his favour. The opportunity was eagerly embraced; and the bufinefs of teaching the mathematics here foon brought him in a confiderable income: but he wanted prudence for fome time to make the beft ufe of it. He was young, handfome, and fprightly; and much addicted both to gaming and gallantry, which continually drained his purfe. Among others, he had a love intrigue with a woman, who lodged in the fame houfe with him, and gave herfelf out for a perfon of condition. However, this expence in time led him to think of matrimony, and he foon after married a young woman without a fortune. She made amends for this defect however by her modefty, virtue, and fweet temper; fo that though the ftate of his purfe was not amended, yet he had more home-felt enjoyment than before, being indeed completely happy in her, as long as fhe lived. He had twelve children by her, who moftly all died young; and he was laftly rendered quite unhappy by the death of his wife alfo, which happened in 1701. Neither did this misfortune come fingle: for the war breaking out about the fame time, on account of the Spanifh fucceffion, it fwept away all his fcholars, who, being foreigners, were obliged to leave Paris. Thus he funk into a very melancholy ftate; under which however he received fome relief, and amufement, from the honour of being admitted this fame year an eleve of the Royal Academy of Sciences.

He feems to have had a pre-fentiment of his death, from fome lurking diforder within, of which no outward fymptoms appeared. In that perfuafion he refufed to engage with fome foreign noblemen, who offered to become his fcholars; alleging that he fhould not live long enough to carry them through their intended courfe. Accordingly he was feized foon after with an apoplexy, which terminated his exiftence in lefs than two hours, on the 3d of April 1717, at 77 years of age.

Ozanam was of a mild and calm difpofition, a chearful and pleafant temper, endeared by a generofity almoft unparalleled. His manners were irreproachable after marriage; and he was fincerely pious, and zea· loufly devout, though ftudioufly avoiding to meddle in theological queftions. He ufed to fay, that it was the bufinefs of the Sorbonne to difcufs, of the pope to decide, and of a mathematician to go ftraight to heaven in a perpendicular line. He wrote a great number of ufeful books; a lift of which is as follows:

1. A treatife of Practical Geometry; 12mo, 1684.

2. Tables of Sines, Tangents and Secants; with a treatife of Trigonometry; 8vo, 1685.

3. A treatife of Lines of the Firft Order; of the Conftruction of Equations; and of Geometric Lines, &c; 4to, 1687.

4. The Ufe of the Compaffes of Proportion, &c; with a treatife on the Divifion of Lands; 8vo, 1688.

5. An Univerfal Inftrument for readily refolving Geometrical Problems without calculation; 12mo, 1688.

6. A Mathematical Dictionary; 4to, 1690.

7. A General Method for drawing Dials, &c; 12mo, 1693.

8. A Courfe of Mathematics, in 5 volumes, 8vo, 1693.

9. A treatife on Fortification, Ancient and Modern; 4to, 1693.

10. Mathematical and Philofophical Recreations; 2 vols 8vo, 1694; and again with additions in 4 vols, 1724.

11. New Treatife on Trigonometry; 12mo, 1699.

12. Surveying, and meafuring all forts of Artificers Works; 12mo, 1699.

13. New Elements of Algebra; 2 vols 8vo, 1702.

14. Theory and Practice of Perfpective; 8vo, 1711.

15. Treatife of Cofmography and Geography; 8vo, 1711.

16. Euclid's Elements, by De Chales, corrected and enlarged; 12mo, 1709.

17. Boulanger's Practical Geometry enlarged, &c; 12mo, 1691.

- 18. Boulanger's treatife on the Sphere corrected and enlarged; 12mo.

Ozanam has alfo the following pieces in the *Journal des Sçavans*: viz, (1), Demonftration of this theorem, that neither the Sum nor the Difference of two Fourth Powers, can be a Fourth Power; Journal of May 1680. —(2), Anfwer to a Problem propofed by M. Comiers; Journal of Nov. 17, 1681.—(3), Demonftration of a Problem concerning Falfe and Imaginary Roots; Journal of April 2 and 9, 1685.—(4), Method of finding in Numbers the Cubic and Surfolid Roots of a Binomial, when it has one; Journal of April 9, 1691.

Alfo in the *Memoires de Trevoux*, of December 1703, he has this piece, viz, Anfwer to certain articles of Objection to the firft part of his Algebra.

And laftly, in the Memoirs of the Academy of Sciences, of 1707, he has Obfervations on a Problem of Spherical Trigonometry.

P.

PAGAN (BLAISE FRANÇOIS Comte de), an eminent French mathematician and engineer, was born at Avignon in Provence, 1604; and took to the profession of a soldier at 14 years of age. In 1620 he was employed at the siege of Caen, in the battle of Pont de Ce, and the reduction of the Navareins, and the rest of Béarn; where he signalized himself, and acquired a reputation far above his years. He was present, in 1621, at the siege of St. John d'Angeli, as also that of Clarac and Montauban, where he lost an eye by a musket-shot. After this time, there happened neither siege, battle, nor any other occasion, in which he did not signalize himself by some effort of courage and conduct. At the passage of the Alps, and the barricade of Suza, he put himself at the head of the Forlorn Hope, composed of the bravest youths among the guards; and undertook to arrive the first at the attack, by a private way which was extremely dangerous; when, having gained the top of a very steep mountain, he cried out to his followers, " There lies the way to glory!" Upon which, sliding along this mountain, they came first to the attack; when immediately commencing a furious onset, and the army coming to their assistance, they forced the barricades. When the king laid siege to Nancy in 1633, Pagan attended him, in drawing the lines and forts of circumvallation.—In 1642 he was sent to the service in Portugal, as field-marshal; and the same year he unfortunately lost the sight of his other eye by a distemper, and thus became totally blind.

But though he was thus prevented from serving his country with his conduct and courage in the field, he resumed the vigorous study of fortification and the mathematics; and in 1645 he gave the public a treatise on the former subject, which was esteemed the best extant.—In 1651 he published his *Geometrical Theorems*, which shewed an extensive and critical knowledge of his subject.—In 1655 he printed a *Paraphrase of the Account of the River of Amazons*, by father de Rennes; and, though blind, it is said he drew the chart of the river and the adjacent parts of the country, as in that work.—In 1657 he published *The Theory of the Planets*, cleared from that multiplicity of eccentric cycles and epicycles, which the astronomers had invented to explain their motions. This work distinguished him among astronomers as much as that of Fortification had among engineers. And in 1658 he printed his *Astronomical Tables*, which are plain and succinct.

Few great men are without some foible: Pagan's was that of a prejudice in favour of judicial astrology; and though he is more reserved than most others on that head, yet we cannot place what he did on that subject

among those productions which do honour to his understanding. He was beloved and respected by all persons illustrious for rank as well as science; and his house was the rendezvous of all the polite and learned both in city and court.—He died at Paris, universally regretted, Nov. 18, 1665.

Pagan had an universal genius; and, having turned his attention chiefly to the art of war, and particularly to the branch of Fortification, he made extraordinary progress and improvements in it. He understood mathematics not only better than is usual for a gentleman whose view is to push his fortune in the army, but even to a degree of perfection superior to that of the ordinary masters who teach that science. He had so particular a genius for this kind of learning, that he acquired it more readily by meditation than by reading authors upon it; and accordingly he spent less time in such books than he did in those of history and geography. He had also made morality and politics his particular study; so that he may be said to have drawn his own character in his *Homme Heroïque*, and to have been one of the completest gentlemen of his time.—Having never married, that branch of his family, which removed from Naples to France in 1552, became extinct in his person.

PALILICUM, the same as Aldebaran, a fixed star of the first magnitude, in the eye of the Bull, or sign Taurus.

PALISADES, or PALISADOES, in Fortification, stakes or small piles driven into the ground, in various situations, as some defence against the surprize of an enemy. They are usually about 6 or 7 inches square, and 9 or 10 feet long, driven about 3 feet into the ground, and 6 inches apart from each other, being braced together by pieces nailed across them near the tops; and secured by thick posts at the distance of every 4 or 5 yards.

PALISADES are placed in the covert-way, parallel to and at 3 feet distance from the parapet or ridge of the glacis, to secure it against a surprize. They are also used to fortify the avenues of open forts, gorges, half-moons, the bottoms of ditches, the parapets of covert-ways; and in general all places liable to surprize, and easy of access.

PALISADOES are usually planted perpendicularly; though some make an angle inclining out towards the enemy, that the ropes cast over them, to tear them up, may slip.

PALLADIO (ANDREW), a celebrated Italian architect in the 16th century, was a native of Vicenza in Lombardy, and the disciple of Triffin, a learned man, who was a Patrician, or Roman nobleman, of the same town

town of Vicenza. Palladio was one of those, who laboured particularly to restore the ancient beauties of architecture, and contributed greatly to revive a true taste in that art. Having learned the principles of it, he went to Rome; where, applying himself with great diligence to study the ancient monuments, he entered into the spirit of their architects, and possessed himself with all their beautiful ideas. This enabled him to restore their rules, which had been corrupted by the barbarous Goths. He made exact drawings of the principal works of antiquity which were to be met with at Rome; to which he added *Commentaries*, which went through several impressions, with the figures. This, though a very useful work, yet is greatly exceeded by the four books of architecture, which he published in 1570. The last book treats of the Roman temples, and is executed in such a manner, as gives him the preference to all his predecessors upon that subject. It was translated into French by Roland Friatt, and into English by several authors. Inigo Jones wrote some excellent remarks upon it, which were published in an edition of Palladio by Leoni, 1742, in 2 volumes folio.

PALLETS, in Clock and Watch Work, are those pieces or levers which are connected with the pendulum or balance, and receive the immediate impulse of the swing-wheel, or balance-wheel, so as to maintain the vibrations of the pendulum in clocks, and of the balance in watches.—The Pallets in all the ordinary constructions of clocks and watches, are formed on the verge or axis of the pendulum or balance, and are of various lengths and shapes, according to the construction of the piece, or the fancy of the artist.

PALLIFICATION, or PILING, in Architecture, denotes the piling of the ground-work, or the strengthening it with piles, or timber driven into the ground; which is practised when buildings are erected upon a moist or marshy soil.

PALLISADES. See PALISADES.

PALM, an ancient long measure, taken from the extent of the hand.

The Roman Palm was of two kinds: the great Palm, taken from the length of the hand, answered to our span, and contained 12 fingers, digits, or fingers breadths, or 9 Roman inches, equal to about $8\frac{1}{2}$ English inches. The small Palm, taken from the breadth of the hand, contained 4 digits or fingers, equal to about 3 English inches.

The Greek Palm, or Doron, was also of two kinds. The small contained 4 fingers, equal to little more than 3 inches. The great Palm contained 5 fingers. The Greek double Palm, called Dichas, contained also in proportion.

The Modern Palm is different in different places where it is used. It contains,

	Inc.	Lines
At Rome - - -	8	$3\frac{1}{2}$
At Naples, according to Riccioli,	8	0
Ditto, according to others, -	8	7
At Genoa - - -	9	9
At Morocco and Fez - -	7	2
Languedoc, and some other parts of France,	9	9
The English Palm is - -	3	0

PALM-SUNDAY, the last Sunday in Lent, or the Sunday next before Easter Day. So called, from the primitive days, on account of a pious ceremony then in use, of bearing Palms, in memory of the triumphant entry of Jesus Christ into Jerusalem, eight days before the feast of the passover.

PAPPUS, a very eminent Greek mathematician of Alexandria towards the latter part of the 4th century, particularly mentioned by Suidas, who says he flourished under the emperor Theodosius the Great, who reigned from the year 379 to 395 of Christ. His writings shew him to have been a consummate mathematician. Many of his works are lost, or at least have not yet been discovered. Suidas mentions several of his works, as also Vossius *de Scientiis Mathematicis*. The principal of these are, his *Mathematical Collections*, in 8 books, the first and part of the second being lost. He wrote also a *Commentary upon Ptolomy's Almagest; an Universal Chorography; A Description of the Rivers of Libya; A Treatise of Military Engines; Commentaries upon Aristarchus of Samos, concerning the Magnitude and Distance of the Sun and Moon;* &c. Of these, there have been published, The Mathematical Collections, in a Latin translation, with a large Commentary, by Commandine, in folio, 1588; and a second edition of the same in 1660. In 644, Mersenne exhibited a kind of abridgment of them in his Synopsis Mathematica, in 4to: but this contains only such propositions as could be understood without figures. In 1655, Meibomius gave some of the Lemmata of the 7th book, in his Dialogue upon Proportions. In 1688, Dr. Wall printed the last 12 propositions of the 2d book, at the end of his Aristarchus Samius. In 1703, Dr. David Gregory gave part of the preface of the 7th book, in the Prolegomena to his Euclid. And in 1706, Dr. Halley gave that Preface entire, in the beginning of his Apollonius.

As the contents of the principal work, the Mathematical Collections, are exceedingly curious, and no account of them having ever appeared in English, I shall here give a very brief analysis of those books, extracted from my notes upon this author.

Of the Third Book—The subjects of the third book consist chiefly of three principal problems; for the solution of which, a great many other problems are resolved, and theorems demonstrated. The first of these three problems is, To find Two Mean Proportionals between two given lines—The 2d problem is, To find, what are called, three Medietates in a semicircle; where, by a Medietas is meant a set of three lines in continued proportion, whether arithmetical, or geometrical, or harmonical; so that to find three medietates, is to find an arithmetical, a geometrical, and an harmonical set of three terms each. And the third problem is, From some points in the base of a triangle, to draw two lines to meet in a point within the triangle, so that their sum shall be greater than the sum of the other two sides which are without them. A great many curious properties are premised to each of these problems; then their solutions are given according to the methods of several ancient mathematicians, with an historical account of them, and his own demonstrations; and lastly, their applications to various matters of great importance. In his historical anecdotes, many curious things are preserved concerning mathematicians that were ancient

B b 2 even

even in his time, which we should otherwise have known nothing at all about.

In order to the solution of the first of the three problems above mentioned, he begins by premising four general theorems concerning proportions. Then follows a dissertation on the nature and division of problems by the Ancients, into Plane, Solid, and Linear, with examples of them, taken out of the writings of Eratosthenes, Philo, and Hero. A solution is then given to the problem concerning two mean proportionals, by four different ways, namely according to Eratosthenes, Nicomedes, Hero, and after a way of his own, in which he not only doubles the cube, but also finds another cube in any proportion whatever to a given cube.

For the solution of the second problem, he lays down very curious definitions and properties of *medietates* of all sorts, and shews how to find them all in a great variety of cases, both as to what the Ancients had done in them, and what was done by others whom he calls the Moderns. *Medietas* seems to have been a general term invented to express three lines, having either an arithmetical, or a geometrical, or an harmonical relation; for the words proportion (or ratio), and analogy (or similar proportions), are restricted to a geometrical relation only. But he shews how all the mediates may be expressed by analogies.

The solution of the 3d problem leads Pappus out into the consideration of a number of admirable and seemingly paradoxical problems, concerning the inflecting of lines to a point within triangles, quadrangles, and other figures, the sum of which shall exceed the sum of the surrounding exterior lines.

Finally, a number of other problems are added, concerning the inscription of all the regular bodies within a sphere. The whole being effected in a very general and pure mathematical way; making all together 58 propositions, viz, 44 problems and 14 theorems.

Of the 4th Book of Pappus.— In the 4th book are first premised a number of theorems relating to triangles, parallelograms, circles, with lines in and about circles, and the tangencies of various circles: all preparatory to this curious and general problem, viz, relative to an infinite series of circles inscribed in the space, called αρϐελον, arbelon, contained between the circumferences of two circles touching inwardly. Where it is shewn, that if the infinite series of circles be inscribed in the manner of this first figure, where three semicircles are described on the lines PR, PQ, QR, and

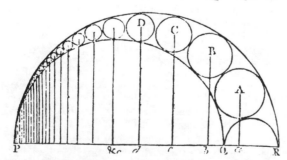

the perpendiculars A*a* B*b*, C*c*, &c, let fall from the centres of the series of inscribed circles; then the

property of these perpendiculars is this, viz, that the first perpendicular A*a* is equal to the diameter or double the radius of the circle A; the second perpendicular B*b* equal to double the diameter or 4 times the radius of the second circle B; the third perpendicular C*c* equal to 3 times the diameter or 6 times the radius of the third circle C; and so on, the series of perpendiculars being to the series of the diameters, as 1, 2, 3, 4, &c, to 1, or to the series of radii, as 2, 4, 6, 8, &c, to 1.

But if the several small circles be inscribed in the

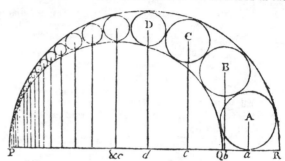

manner of this second circle, the first circle of the series touching the part of the line QR; then the series of perpendiculars A*a*, B*b*, C*c*, &c, will be 1, 3, 5, 7, &c, times the radii of the circles A, B, C, D, &c; viz, according to the series of odd numbers; the former proceeding by the series of even numbers.

He next treats of the Helix, or Spiral, proposed by Conon, and resolved by Archimedes, demonstrating its principal properties: in the demonstration of some of which, he makes use of the same principles as Cavallerius did lately, adding together an infinite number of infinitely short parallelograms and cylinders, which he imagines a triangle and cone to be composed of.—He next treats of the properties of the Conchoid which Nicomedes invented for doubling the cube: applying it to the solution of certain problems concerning Inclinations, with the finding of two mean proportionals, and cubes in any proportion whatever.——Then of the τιτραγωνιζουτα, or Quadratrix, so called from its use in squaring the circle, for which purpose it was invented and employed by Dinostratus, Nicomedes, and others: the use of which however he blames, as it requires postulates equally hard to be granted, as the problem itself to be demonstrated by it.—Next he treats of Spirals, described on planes, and on the convex surfaces of various bodies.—From another problem, concerning Inclinations, he there shews, how to trisect a given angle; to describe an hyperbola, to two given asymptotes, and passing through a given point; to divide a given arc or angle in any given ratio; to cut off arcs of equal lengths from unequal circles; to take arcs and angles in any proportion, and arcs equal to right lines; with parabolic and hyperbolic loci, which last is one of the inclinations of Archimedes.

Of the 5th Book of Pappus.—This book opens with reflections on the different natures of men and brutes, the former acting by reason and demonstration, the latter by instinct, yet some of them with a certain portion of reason or foresight, as bees, in the curious structure of their cells, which he observes are of such

a form

a form as to complete the space quite around a point, and yet require the least materials to build them, to contain the same quantity of honey. He shews that the triangle, square, and hexagon, are the only regular polygons capable of filling the whole space round a point; and remarks that the bees have chosen the fittest of these; proving afterwards, in the propositions, that of all regular figures of the same perimeter, that is of the largest capacity which has the greatest number of sides or angles, and consequently that the circle is the most capacious of all figures whatever.

And thus he finishes this curious book on Isoperimetrical figures, both plane and solid; in which many curious and important properties are strictly demonstrated, both of planes and solids, some of them being old in his time, and many new ones of his own. In fact, it seems he has here brought together into this book, all the properties relating to isoperimetrical figures then known, and their different degrees of capacity. In the last theorem of the book, he has a dissertation to shew, that there can be no more regular bodies beside the five Platonic ones, or, that only the regular triangles, squares, and pentagons, will form regular solid angles.

Of the 6th Book of Pappus.—In this book he treats of certain spherical properties, which had been either neglected, or improperly and imperfectly treated by some celebrated authors before his time.—— Such are some things in the 3d book of Theodosius's Spherics, and in his book on Days and Nights, as also some in Euclid's Phenomena. For the sake of these, he premises and intermixes many curious geometrical properties, especially of circles of the sphere, and spherical triangles. He adverts to some curious cases of variable quantities; shewing how some increase and decrease both ways to infinity; while others proceed only one way by increase or decrease, to infinity, and the other way to a certain magnitude; and others again both ways to a certain magnitude, giving a maximum and minimum.—Here are also some curious properties concerning the perspective of the circles of the sphere, and of other lines. Also the locus is determined of all the points from whence a circle may be viewed, so as to appear an ellipse, whose centre is a given point within the circle; which locus is shewn to be a semicircle passing through that point.

Of the 7th Book of Pappus.—In the introduction to this book, he describes very particularly the nature of the mathematical composition and resolution of the Ancients, distinguishing the particular process and uses of them, in the demonstration of theorems and solution of problems. He then enumerates all the analytical books of the Ancients, or those proceeding by resolution, which he does in the following order, viz, 1st, Euclid's Data, in one book: 2d, Apollonius's Section of a Ratio, 2 books: 3d, his Section of a Space, 2 books: 4th, his Tangencies, 2 books; 5th, Euclid's Porisms, 3 books: 6th, Apollonius's Inclinations, 2 books: 7th, his Plane Loci, 2 books: 8th, his Conics, 8 books: 9th, Aristæus's Solid Loci, 5 books: 10th, Euclid's Loci in Superficies, 2 books; and 11th, Eratosthenes's Medietates, 2 books. So that all the books are 31, the arguments or contents of which he exhibits, with the number of the Loci, determina-

tions, and cases, &c; with a multitude of lemmas and propositions laid down and demonstrated; the whole making 238 propositions, of the most curious geometrical principles and properties, relating to those books.

Of the 8th Book of Pappus.—The 8th book is altogether on Mechanics. It opens with a general oration on the subject of mechanics; defining the science, enumerating the different kinds and branches of it, and giving an account of the chief authors and writings on it. After an account of the centre of gravity, upon which the science of mechanics so greatly depends, he shews in the first proposition, that such a point really exists in all bodies. Some of the following propositions are also concerning the properties of the centre of gravity. He next comes to the Inclined Plane, and in prop. 9, shews what power will draw a given weight up a given inclined plane, when the power is given which can draw the weight along a horizontal plane. In the 10th prop. concerning the moving a given weight with a given power, he treats of what the Ancients called a Glossocomum, which is nothing more than a series of Wheels-and-axles, in any proportions, turning each other, till we arrive at the given power. In this proposition, as well as in several other places, he refers to some books that are now lost; as Archimedes on the Balance, and the Mechanics of Hero and of Philo. Then, from prop. 11 to prop. 19, treats on various miscellaneous things, as, the organical construction of solid problems; the diminution of an architectural column; to describe an ellipse through five given points; to find the axes of an ellipse organically; to find also organically, the inclination of one plane to another, the nearest point of a sphere to a plane, the points in a spherical surface cut by lines joining certain points, and to inscribe seven hexagons in a given circle. Prop. 20, 21, 22, 23, teach how to construct and adapt the *Tympani*, or wheels of the Glossocomum to one another, shewing the proportions of their diameters, the number of their teeth, &c. And prop. 24 shews how to construct the spiral threads of a screw.

He comes then to the *Five Mechanical Powers, by which a given weight is moved by a given power*. He here proposes briefly to shew what has been said of these powers by Hero and Philo, adding also some things of his own. Their names are, the Axis-in-peritrochio, the Lever, Pulley, Wedge and Screw; and he observes, those authors shewed how they are all reduced to one principle, though their figures be very different. He then treats of each of these powers separately, giving their figures and properties, their construction and uses.

He next describes the manner of drawing very heavy weights along the ground, by the machine Chelone, which is a kind of sledge placed upon two loose rollers, and drawn forward by any power whatever, a third roller being always laid under the fore part of the Chelone, as one of the other two is quitted and left behind by the motion of the Chelone. In fact this is the same machine as has always been employed upon many occasions in moving very great weights to moderate distances.

Finally, Pappus describes the manner of raising great weights to a height by the combination of mechanic

powers,

powers, as by cranes, and other machines; illustrating this, and the former parts, by drawings of the machines that are described.

PARABOLA, in Geometry, a figure arising from the section of a cone, when cut by a plane parallel to one of its sides, as the section ADE parallel to the side VB of the cone. See Conic *Sections*, where some general properties are given.

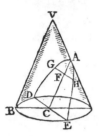 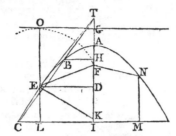

Some other Properties of the Parabola.

1. From the same point of a cone only one Parabola can be drawn; all the other sections between the Parabola and the parallel side of the cone being ellipses, and all without them hyperbolas. Also the Parabola has but one focus, through which the axis AC passes; all the other diameters being parallel to this, and infinite in length also.

2. The parameter of the axis is a third proportional to any abscifs and its ordinate; viz, AC : CD : : CD : p the parameter. And therefore if x denote any abscifs AC, and y the ordinate CD, it will be $x : y : : y : p = \frac{y^2}{x}$ the parameter; or, by multiplying extremes and means, $px = y^2$, which is the equation of the Parabola.

3. The focus F is the point in the axis where the double ordinate GH is equal to the parameter. Therefore, in the equation of the curve $px = y^2$, taking $p = 2y$, it becomes $2yx = y^2$, or $2x = y$, that is $2AF = FH$, or $AF = \frac{1}{2}FH$, or the focal distance from a vertex AF is equal to half the ordinate there, or $= \frac{1}{4}p$, one-fourth of the parameter.

4. The abscifses of a Parabola are to one another, as the squares of their corresponding ordinates. This is evident from the general equation of the curve $px = y^2$, where, p being constant, x is as y^2.

5. The line FE (*fig. 2 above*) drawn from the focus to any point of the curve, is equal to the sum of the focal distance and the abscifs of the ordinate to that point; that is FE = FA + AD = GD, taking AG = AF = $\frac{1}{4}p$. Or EF is always =, EO, drawn parallel to DG, to meet the perpendicular GO, called the Directrix.

6. If a line TBC cut the curve of a Parabola in two points, and the axis produced in T, and BH and CI be ordinates at those two points; then is AT a mean proportional between the abscifses AH and AI, or $AT^2 = AH : AI$.—And if TE touch the curve, then is AT = AD = the mean between AH and AI.

7. If FE be drawn from the focus to the point of contact of the tangent TE, and EK perpendicular to the same tangent; then is FT = FE = FK; and the subnormal DK equal to the constant quantity 2AF or $\frac{1}{2}p$.

8. The diameter EL being parallel to the axis AK, the perpendicular EK, to the curve or tangent at E, bisects the angle LEF. And therefore all rays of light LE, MN, &c, coming parallel to the axis, will be reflected into the point F, which is therefore called the focus, or burning point; for the angle of incidence LEK is = the angle of reflection KEF.

9. If IEK (*next fig. below*) be any line parallel to the axis, limited by the tangent TC and ordinate CKL to the point of contact; then shall IE : EK :: CK : KL. And the same thing holds true when CL is also in any oblique position.

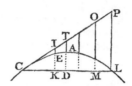 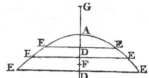

10. The external parts of the parallels IE, TA, ON, PL, &c, are always proportional to the squares of their intercepted parts of the tangent; that is, the external parts IE , TA , ON , PL , are proportional to CI², CT², CO², CP², or to the squares CK², CD², CM², CL².

And as this property is common to every position of the tangent, if the lines IE, TA, ON, &c, be appended to the points I, T, O, &c, of the tangent, and moveable about them, and of such lengths as that their extremities E, A, N, &c, be in the curve of a Parabola in any one position of the tangent; then making the tangent revolve about the point C, the extremities E, A, N, &c, will always form the curve of some Parabola, in every position of the tangent.

The same properties too that have been shewn of the axis, and its abscifses and ordinates, &c, are true of those of any other diameter. All which, besides many other curious properties of the Parabola, may be seen demonstrated in my Treatise on Conic Sections.

11. To Construct a Parabola by Points.

In the axis produced take AG = AF (*last fig. above*) the focal distance, and draw a number of lines EE, EE, &c, perpendicular to the axis AD; then with the distances GD, GD, &c, as radii, and the centre F, describe arcs crossing the parallel ordinates in E, E, &c. Then with a steady hand, or by the side of a slip of bent whale-bone, draw the curve through all the points E, E, E, &c.

12. To describe a Parabola by a continued Motion.

If the rule or the directrix BC be laid upon a plane, (*first fig. below*) with the square GDO, in such manner that one of its sides DG lies along the edge of that rule; and if the thread FMO equal in length to DO, the other side of the square, have one end fixed in the extremity of the rule at O, and the other end in some

point

point F: Then flide the fide of the fquare DG along the rule BC, and at the fame time keep the thread continually tight by means of the pin M, with its part MO clofe to the fide of the fquare DO; fo fhall the curve AMX, which the pin defcribes by this motion, be one part of a Parabola.

And if the fquare be turned over, and moved on the other fide of the fixed point F, the other part of the fame Parabola AMZ will be defcribed.

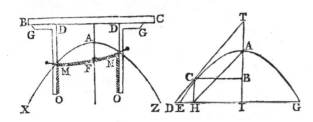

To draw Tangents to the Parabola.

13. If the point of contact C be given: *(laft fig. above)* draw the ordinate CB, and produce the axis till AT be = AB; then join TC, which will be the tangent.

14. Or if the point be given in the axis produced: Take AB = AT, and draw the ordinate BC, which will give C the point of contact; to which draw the line TC as before.

15. If D be any other point, neither in the curve nor in the axis produced, through which the tangent is to pafs: Draw DEG perpendicular to the axis, and take DH a mean proportional between DE and DG, and draw HC parallel to the axis, fo fhall C be the point of contact, through which and the given point D the tangent DCT is to be drawn.

16. When the tangent is to make a given angle with the ordinate at the point of contact: Take the abfcifs AI equal to half the parameter, or to double the focal diftance, and draw the ordinate IE: alfo draw AH to make with AI the angle HAI equal to the given angle; then draw HC parallel to the axis, and it will cut the curve in C the point of contact, where a line drawn to make the given angle with CB will be the tangent required.

17. *To find the Area of a Parabola.* Multiply the bafe EG by the perpendicular height AI, and $\frac{2}{3}$ of the product will be the area of the fpace AEGA; becaufe the Parabolic fpace is $\frac{2}{3}$ of its circumfcribing parallelogram.

18. *To find the Length of the Curve* AC, commencing at the vertex.—Let $y =$ the ordinate BC, $p =$ the parameter, $q = \frac{2y}{p}$, and $s = \sqrt{1 + q^2}$; then fhall $\frac{1}{2}p$ × $(qs + $ hyp. log. of $q + s)$ be the length of the curve AC.

See various other rules for the areas, and lengths of the curve, &c, in my Treatife on Menfuration, fec. 6, pa. 355, &c, 2d edition.

PARABOLAS *of the Higher Kinds*, are algebraic curves, defined by the general equation $a^{n-1} x = y^n$;

that is, either $a^2 x = y^3$, or $a^3 x = y^4$, or $a^4 x = y^5$, &c.

Some call thefe by the name of Paraboloids: and in particular, if $a^2 x = y^3$, they call it a Cubical Paraboloid; if $a^3 x = y^4$, they call it a Biquadratical Paraboloid, or a Surfolid Paraboloid. In refpect of thefe, the Parabola of the Firft Kind, above explained, they call the Apollonian, or Quadratic Parabola.

Thofe curves are alfo to be referred to Parabolas, that are expreffed by the general equation $a x^{n-1} = y^n$, where the indices of the quantities on each fide are equal, as before; and thefe are called Semi Parabolas: as $a x^2 = y^3$ the Semi-Cubical Parabola; or $a x^3 = y^4$ the Semi-Biquadratical Parabola; &c.

They are all comprehended under the more general equation $a^m x^n = y^{m+n}$, where the two indices on one fide are ftill equal to the index on the other fide of the equation; which include both the former kinds of equations, as well as fuch as thefe following ones, $a^2 x^2 = y^4$, or $a^2 x^3 = y^5$, or $a^4 x^3 = y^7$, &c.

Cartefian PARABOLA, is a curve of the 2d order expreffed by the equation

$$xy = a x^3 + b x^2 + c x + d,$$

containing four infinite legs, viz two hyperbolic ones

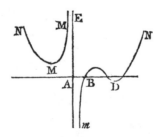

MM and B*m*, to the common afymptote AE, tending contrary ways, and two Parabolic legs MN and DN joining them, being Newton's 66th fpecies of lines of the 3d order, and called by him a Trident. It is made ufe of by Des Cartes in the 3d book of his Geometry, for finding the roots of equations of 6 dimenfions, by means of its interfections with a circle. Its moft fimple equation is $xy = x^3 + g^3$. And points through which it is to pafs may be eafily found by means of a common Parabola whofe abfcifs is $a x^2 + b x + c$, and an hyperbola whofe abfcifs is $\frac{d}{x}$; for y will be equal to the fum or difference of the correfponding ordinates of this Parabola and hyperbola.

Des Cartes, in the place abovementioned, fhews how to defcribe this curve by a continued motion. And Mr. Maclaurin does the fame thing in a different way, in his Organica Geometria.

Diverging PARABOLA, is a name given by Newton to a fpecies of five different lines of the 3d order, expreffed by the equation

$$y^2 = a x^3 + b x^2 + c x + d.$$

The

The first is a bell-form Parabola, with an oval at its head (*fig.* 1.); which is the cafe when the equation $0 = ax^3 + bx^2 + cx + d$, has three real and unequal roots; fo that one of the moft fimple equations of a curve of this kind is $py^2 = x^3 + ax^2 + a^2 x$.

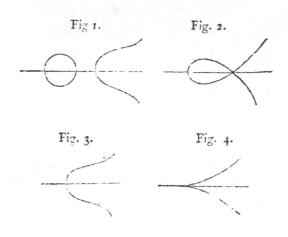

Fig 1.   Fig. 2.

Fig. 3.   Fig. 4.

The 2d is alfo a bell-form Parabola, with a conjugate point, or infinitely fmall oval, at the head (*fig.* 1.); being the cafe when the equation $0 = ax^3 + bx^2 + cx + d$ has its two lefs roots equal; the moft fimple equation of which is $py^2 = x^3 + ax^2$.

The third is a Parabola, with two diverging legs, crofling one another like a knot (*fig.* 2.); which happens when the equation $0 = ax^3 + bx^2 + cx + d$ has its two greater roots equal; the more fimple equation being $py^2 = x^3 + ax^2$.

The fourth a pure bell-form Parabola (*fig.* 3.); being the cafe when $0 = ax^3 + bx^2 + cx + d$ has two imaginary roots; and its moft fimple equation is $py^2 = x^3 + a^3$, or $py^2 = x^3 + a^2 x$.

The fifth a Parabola with two diverging legs, forming at their meeting a cufp or double point (*fig.* 4.); being the cafe when the equation $0 = ax^3 + bx^2 + cx + d$ has three equal roots; fo that $py^2 = x^3$ is the moft fimple equation of this curve, which indeed is the Semi-cubical, or Neilian Parabola.

If a folid generated by the rotation of a femi-cubical Parabola, about its axis, be cut by a plane, each of thefe five Parabolas will be exhibited by its fections. For, when the cutting plane is oblique to the axis, but falls below it, the fection is a diverging Parabola, with an oval at its head. When oblique to the axis, but paffes through the vertex, the fection is a diverging Parabola, having an infinitely fmall oval at its head. When the cutting is oblique to the axis, falls below it, and at the fame time touches the curve furface of the folid, as well as cuts it, the fection is a diverging Parabola, with a nodus or knot. When the cutting plane falls above the vertex, either parallel or oblique to the axis, the fection is a pure diverging Parabola. And laftly when the cutting plane paffes through the axis, the fection is the femi-cubical Parabola from which the folid was generated.

PARABOLIC *Afymptote*, is ufed for a Parabolic line approaching to a curve, fo that they never meet; yet by producing both indefinitely, their diftance from each other becomes lefs than any given line.

There may be as many different kinds of thefe Afymptotes as there are parabolas of different orders. When a curve has a common parabola for its Afymptote, the ratio of the fubtangent to the abfcifs approaches continually to the ratio of 2 to 1, when the axis of the parabola coincides with the bafe; but this ratio of the fubtangent to the abfcifs approaches to that of 1 to 2, when the axis is perpendicular to the bafe. And by obferving the limit to which the ratio of the fubtangent and abfcifs approaches, Parabolic Afymptotes of various kinds may be difcovered. See Maclaurin's Fluxions, art. 337.

PARABOLIC *Conoid*, is a folid generated by the rotation of a parabola about its axis.

This folid is equal to half its circumfcribed cylinder; and therefore if the bafe be multiplied by the height, half the product will be the folid content.

To find the Curve Surface of a Paraboloid.

Let BAD be the generating parabola, AC = AT, and BT a tangent at B. Put $p = 3 \cdot 1416$, $y = $ BC, $x = $ AC = AT, and $t = $ BT $= \sqrt{4x^2 + y^2}$; then is the curve furface $= \frac{2}{3}ay \times (y + \frac{tt}{t+y})$.

See various other rules and geometrical conftructions for the furfaces and folidities of Parabolic Conoids, in my Menfuration, part 3, fect. 6, 2d edition.

PARABOLIC *Pyramidoid*, is a folid figure thus named by Dr. Wallis, from its genefis, or formation, which is thus: Let all the fquares of the ordinates of a parabola be conceived to be fo placed, that the axis fhall pafs perpendicularly through all their centres; then the aggregate of all thefe planes will form the Parabolic Pyramidoid.

This figure is equal to half its circumfcribed parallelopipedon. And therefore the folid content is found by multiplying the bafe by the altitude, and taking half the product; or the one of thefe by half the other.

PARABOLIC *Space*, is the fpace or area included by the curve line and bafe or double ordinate of the parabola. The area of this fpace, it has been fhewn under the article Parabola, is $\frac{2}{3}$ of its circumfcribed parallelogram; which is its quadrature, and which was firft found out by Archimedes, though fome fay by Pythagoras.

PARABOLIC *Spindle*, is a folid figure conceived to be formed by the rotation of a parabola about its bafe or double ordinate.

This folid is equal to $\frac{8}{15}$ of its circumfcribed cylinder. See my Menfuration, prob. 15, pa. 390, &c, 2d edition.

PARABOLIC *Spiral*. See HELICOID *Parabola*.

PARABOLIFORM *Curves*, a name fometimes given to the parabolas of the higher orders.

PARABOLOIDES, Parabolas of the higher orders.——The equation for all curves of this kind being $a^{m-n} x^n = y^m$, the proportion of the area of any one to the complement of it to the circumfcribing parallelogram, will be as m to n.

PARA-

PARACENTRIC *Motion*, denotes the space by which a revolving planet approaches nearer to, or recedes farther from, the fun, or centre of attraction.

Thus, if a planet in A move towards B; then is SB — SA = *b*B the Paracentric motion of that planet: where S is the place of the fun.

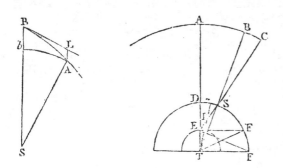

PARACENTRIC *Solicitation of Gravity*, is the same as the Vis Centripeta; and is expreffed by the line AL drawn from the point A, parallel to the ray SB (infinitely near SA), till it interfect the tangent BL.

PARALLACTIC *Angle*, called alfo fimply PARALLAX, is the angle EST (*laft fig. above*) made at the centre of a ftar, &c, by two lines, drawn, the one from the centre of the earth at T, and the other from its furface at E.—Or, which amounts to the fame thing, the Parallactic angle, is the difference of the two angles CEA and BTA, under which the real and apparent diftances from the zenith are feen.

The fines of the Parallactic angles ELT, EST, at the fame or equal diftances DS from the zenith, are in the reciprocal ratio of the diftances, TL, and TS, from the centre of the earth.

PARALLAX, is an arch of the heavens intercepted between the true place of a ftar, and its apparent place.

The true place of a ftar S, is that point of the heavens B, in which it would be feen by an eye placed in the centre of the earth at T. And the apparent place, is that point of the heavens C, where a ftar appears to an eye upon the furface of the earth at E.

This difference of places, is what is called abfolutely the Parallax, or the Parallax of Altitude; which Copernicus calls the Commutation; and which therefore is an angle formed by two vifual rays, drawn, the one from the centre, the other from the circumference of the earth, and traverfing the body of the ftar; being meafured by an arch of a great circle intercepted between the two points of true and apparent place, B and C.

The PARALLAX *of Altitude* CB is properly the difference between the true diftance from the zenith AB, and the apparent diftance AC. Hence the Parallax diminifhes the altitude of a ftar, or increafes its diftance from the zenith; and it has therefore a contrary effect to the refraction.

The Parallax is greateft in the horizon, called the Horizontal Parallax EFT. From hence it decreafes all the way to the zenith D or A, where it is nothing; the real and apparent places there coinciding.

The Horizontal Parallax is the fame, whether the ftar be in the true or apparent horizon.

The fixed ftars have no fenfible Parallax, by reafon of their immenfe diftance, to which the femidiameter of the earth is but a mere point.

Hence alfo, the nearer a ftar is to the earth, the greater is its Parallax; and on the contrary, the farther it is off, the lefs is the Parallax, at an equal elevation above the horizon. So the ftar at S has a lefs Parallax than the ftar at I. Saturn is fo high, that it is difficult to obferve in him any Parallax at all.

Parallax increafes the right and oblique afcenfion, and diminifhes the defcenfion; it diminifhes the northern declination and latitude in the eaftern part, and increafes them in the weftern; but it increafes the fouthern declination in the eaftern and weftern part; it diminifhes the longitude in the weftern part, and increafes it in the eaftern. Parallax therefore has juft oppofite effects to refraction.

The doctrine of Parallaxes is of the greateft importance, in aftronomy, for determining the diftances of the planets, comets, and other phenomena of the heavens; for the calculation of eclipfes, and for finding the longitude.

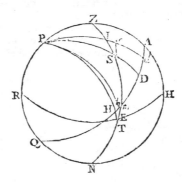

PARALLAX *of Right Afcenfion and Defcenfion*, is an arch of the Equinoctial D*d*, by which the Parallax of altitude increafes the afcenfion, and diminifhes the defcenfion.

PARALLAX *of Declination*, is an arch of a circle of declination *s*I, by which the Parallax of altitude increafes or diminifhes the declination of a ftar.

PARALLAX *of Latitude*, is an arch of a circle of latitude SI, by which the Parallax of altitude increafes or diminifhes the latitude.

Menftrual PARALLAX *of the Sun*, is an angle formed by two right lines; one drawn from the earth to the fun, and another from the fun to the moon, at either of their quadratures.

PARALLAX *of the Annual Orbit of the Earth*, is the difference between the heliocentric and geocentric place of a planet, or the angle at any planet, fubtended by the diftance between the earth and fun.

There are various methods for finding the Parallaxes of the celeftial bodies: fome of the principal and eafier of which are as follow:

To Obferve the PARALLAX *of a Celeftial Body.*—Obferve when the body is in the fame vertical with a fixed ftar which is near it, and in that pofition meafure its

apparent

apparent diftance from the ftar. Obferve again when the body and ftar are at equal altitudes from the horizon; and there meafure their diftance again. Then the difference of thefe diftances will be the Parallax very nearly.

To Obferve the Moon's PARALLAX.—Obferve very accurately the moon's meridian altitude, and note the moment of time. To this time, equated, compute her true latitude and longitude, and from thefe find her declination; alfo from her declination, and the elevation of the equator, find her true meridian altitude. Subtract the refraction from the obferved altitude: then the difference between the remainder and the true altitude, will be the Parallax fought. If the obferved altitude be not meridional, reduce it to the true altitude for the time of obfervation.

By this means, in 1583, Oct. 12 day 5 h. 19 m. from the moon's meridian altitude obferved at 13° 38′, Tycho found her Parallax to be 54 minutes.

To Obferve the Moon's PARALLAX *in an Eclipfe.*—In an eclipfe of the moon obferve when both horns are in the fame vertical circle, and at that moment take the altitudes of both horns; then half their fum will be nearly the apparent altitude of the moon's centre; from which fubtract the refraction, which gives the apparent altitude freed from refraction. But the true altitude is nearly equal to the altitude of the centre of the fhadow at that time: now the altitude of the centre of the fhadow is known, becaufe we know the fun's place in the ecliptic, and his depreffion below the horizon, which is equal to the altitude of the oppofite point of the ecliptic, in which the centre of the fhadow is. Having thus the true and apparent altitudes, their difference is the Parallax fought.

De la Hire makes the greateft horizontal Parallax 1° 1′ 25″, and the leaft 54′ 5″. M. le Monnier determined the mean Parallax of the Moon to be 57′ 12″. Others have made it 57′ 18″.

From the Moon's PARALLAX *EST, and altitude SF* (laft fig. but one); *to find her diftance from the Earth.* —From her apparent altitude given, there is given her apparent zenith diftance, i. e. the angle AES; or by her true altitude, the complement angle ATS. Wherefore, fince at the fame time, the Parallactic angle S is known, the 3d or fupplemental angle TES is alfo known. Then, confidering the earth's femidiameter TE as 1, in the triangle TES are given all the angles and the fide TE, to find ES the moon's diftance from the furface of the earth, or TS her diftance from the centre.

Thus Tycho, by the obfervation above mentioned, found the moon's diftance at that time from the earth, was 62 of the earth's femidiameters. According to De la Hire's determination, her diftance when in the perigee is near 56 femidiameters, but in her apogee near 63½; and therefore the mean nearly 59¾, or in round numbers 60 femidiameters.

Hence alfo, fince, from the moon's theory, there is given the ratio of her diftances from the earth in the feveral degrees of her anomaly; thofe diftances being found, by the rule of three, in femidiameters of the earth, the Parallax is thence determined to the feveral degrees of the true anomaly.

To Obferve the PARALLAX *of Mars.*—1. Suppofe Mars in the meridian and equator at H; and that the obferver, under the equator in A, obferves him culminating with fome fixed ftar. 2. If now the obferver were in the centre of the earth, he would fee Mars conftantly in the fame point of the heavens with the ftar; and therefore, together with it, in the plane of the horizon, or of the 6th horary: but fince Mars here has fome fenfible Parallax, and the fixed ftar has none, Mars will be feen in the horizon, when in P, the plane of the fenfible horizon; and the ftar, when in R, the plane of the true horizon: therefore obferve the time between the tranfit of Mars and of the ftar through the plane of the 6th hour.—3. Convert this time into minutes of the equator, at the rate of 15 degrees to the hour; by which means there will be obtained the arch PM, to which the angle PAM, and confequently the angle AMD, is nearly equal; which is the horizontal Parallax of Mars.

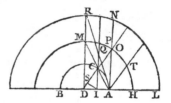

If the obferver be not under the equator, but in a parallel IQ, that difference will be a lefs arch QM: wherefore, fince the fmall arches QM and PM are nearly as their fines AD and ID; and fince ADG is equal to the diftance of the place from the equator, i. e. to the elevation of the pole, or the latitude; therefore AD to ID, as radius to the cofine of the latitude; fay, as the cofine of the latitude ID is to radius, fo is the Parallax obferved in I, to the Parallax under the equator.

Since Mars and the fixed ftar cannot be commodioufly obferved in the horizon; let them be obferved in the circle of the 3d hour: and fince the Parallax obferved there TO, is to the horizontal one PM, as IS to ID: fay, as the fine of the angle IDS, or 45° (fince the plane DO is in the middle between the meridian DH and the true horizon DM), is to radius, fo is the Parallax TO to the horizontal Parallax PM.

If Mars be likewife out of the plane of the equator, the Parallax found will be an arch of a parallel; which muft therefore be reduced, as above, to an arch of the equator.

Laftly, if Mars be not ftationary, but either direct or retrograde, by obfervations for feveral days find out what his motion is every hour, that his true place from the centre may be affigned for any given time.

By this method Caffini, who was the author of it, obferved the greateft horizontal Parallax of Mars to be 25″; but Mr. Flamfteed found it near 30″. Caffini obferved alfo the Parallax of Venus by the fame method.

To Find the Sun's PARALLAX.—The great diftance of the fun renders his Parallax too fmall to fall under even the niceft immediate obfervation. Many attempts have indeed been made, both by the ancients and moderns, and many methods invented for that purpofe. The firft was that of Hipparchus, which was followed by Ptolomy, &c, and was founded on the obfervation of lunar eclipfes.

eclipfes. The fecond was that of Ariftarchus, in which the angle fubtended by the femidiameter of the moon's orbit, feen from the fun, was fought from the lunar phafes. But thefe both proving deficient, aftronomers are now forced to have recourfe to the Parallaxes of the nearer planets, Mars and Venus. Now from the theory of the motions of the earth and planets, there is known at any time the proportion of the diftances of the fun and planets from us; and the horizontal Parallaxes being reciprocally proportional to thofe diftances; by knowing the Parallax of a planet, that of the fun may be thence found.

Thus Mars, when oppofite to the fun, is twice as near as the fun is, and therefore his Parallax will be twice as great as that of the fun. And Venus, when in her inferior conjunction with the fun, is fometimes nearer us than he is; and therefore her Parallax is greater in the fame proportion. Thus, from the Parallaxes of Mars and Venus, Caffini found the fun's Parallax to be 10"; from whence his diftance comes out 22000 femidiameters of the earth.

But the moft accurate method of determining the Parallaxes of thefe planets, and thence the Parallax of the fun, is that of obferving their tranfit. However, Mercury, though frequently to be feen on the fun, is not fit for this purpofe; becaufe he is fo near the fun, that the difference of their Parallaxes is always lefs than the folar Parallax required. But the Parallax of Venus, being almoft 4 times as great as the folar Parallax, will caufe very fenfible differences between the times in which fhe will feem to be paffing over the fun at different parts of the earth. With the view of engaging the attention of aftronomers to this method of determining the fun's Parallax, Dr. Halley communicated to the Royal Society, in 1691, a paper, containing an account of the feveral years in which fuch a tranfit may happen, computed from the tables which were then in ufe: thofe at the afcending node occur in the month of November O. S. in the years 918, 1161, 1396, 1631, 1639, 1874, 2109, 2117; and at the defcending node in May O. S. in the years 1048, 1283, 1291, 1518, 1526, 1761, 1769, 1996, 2004. Philof. Tranf. Abr. vol. 1, p. 435 &c.

Dr. Halley even then concluded, that if the interval of time between the two interior contacts of Venus with the fun, could be meafured to the exactnefs of a fecond, in two places properly fituated, the fun's Parallax might be determined within its 500dth part. And this conclufion was more fully explained in a fubfequent paper, concerning the tranfit of Venus in the year 1761, in the Philof. Tranf. numb. 348, or Abr. vol. 4, p. 213.

It does not appear that any of the preceding tranfits had been obferved; except that of 1639, by our ingenious countryman Mr. Horrox, and his friend Mr. Crabtree, of Manchefter. But Mr. Horrox died on the 3d of January, 1641, at the age of 25, juft after he had finifhed his treatife, *Venus in Sole vifa*, in which he difcovers a more accurate knowledge of the dimenfions of the folar fyftem, than his learned commentator Hevelius.

To give a general idea of this method of determining the horizontal Parallax of Venus, and from thence,

by analogy, the Parallax and diftance of the fun, and of all the planets from him; let DBA be the earth, V Venus, and TSR the eaftern limb of the fun. To an obferver at B, the point *t* of that limb will be on the meridian, its place referred to the heavens will be at E, and Venus will appear juft within it at S. But to an obferver at A, at the fame inftant, Venus is eaft of the fun, in the right line AVF; the point *t* of the fun's limb appears at *e* in the heavens, and if Venus were then vifible fhe would appear at F. The angle CVA is the horizontal Parallax of Venus; which is equal to the oppofite angle FVE, meafured by the arc FE. ASC is the fun's horizontal Parallax, equal to the oppofite angle *e*SE, meafured by the arc *e*E; and FA*e* or VA*e* is Venus's horizontal Parallax from the fun, which may be found by obferving how much later in abfolute time her total ingrefs on the fun is, as feen from A, than as feen from B, which is the time fhe takes to move from V to *v*, in her orbit OV*v*.

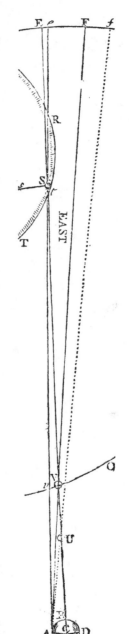

If Venus were nearer the earth, as at U, her horizontal Parallax from the fun would be the arch *fe*, which meafures the angle *fA e*; and this angle is greater than the angle FA*e*, by the difference of their meafures F*f*. So that as the diftance of the celeftial object from the earth is lefs, its Parallax is the greater.

Now it has been already obferved, that the horizontal Parallaxes of the planets are inverfely as their diftances from the earth's centre, therefore as the fun's diftance at the time of the tranfit is to Venus's diftance, fo is the Parallax of Venus to that of the fun: and as the fun's mean diftance from the earth's centre, is to his diftance on the day of the tranfit, fo is his horizontal Parallax on that day, to his horizontal Parallax at the time of his mean diftance from the earth's centre. Hence his true diftance in femidiameters of the earth may be obtained by the following analogy, viz, as the fine of the fun's Parallax is to radius, fo is unity or the earth's femidiameter, to the number of femidiameters of the earth in the fun's diftance from the centre; which number multiplied by the number of

miles

miles in the earth's femidiameter, will give the number of miles in the fun's diftance. Then from the proportional diftances of the planets, determined by the theory of gravity, their true diftances may be found. And from their apparent diameters at thefe known diftances, their real diameters and bulks may be found.

Mr. Short, with great labour, deduced the quantity of the fun's Parallax from the beft obfervations that were made of the tranfit of Venus, on the 6th of June, 1761, (for which fee Philof. Tranf. vol. 51 and 52) both in Britain and in foreign parts, and found it to have been 8″·52 on the day of the tranfit, when the fun was very nearly at his greateft diftance from the earth; and confequently 8″·65 when the fun is at his mean diftance from the earth. See Philof. Tranf. vol. 52, p. 611 &c. Whence,

As fin. 8″·65	- - -	log.	5·6219140
to radius	- - -		10·0000000
So is 1 femidiameter	- -		0·0000000
to 23882·84 femidiameters		-	4·3780860

that is, 23882 $\frac{84}{100}$ is the number of the earth's femidiameters contained in its diftance from the fun; and this number of femidiameters being multiplied by 3985, the number of Englifh miles contained in the earth's femidiameter, (though later obfervations make this femidiameter only 3956½ miles), there is obtained 95,173,127 miles for the earth's mean diftance from the fun. And hence, from the analogies under the article Distance, the mean diftances of all the reft of the planets from the fun, in miles, are found as follow, viz,

Mercury's diftance	- -	36,841,468
Venus's diftance	- -	68,891,486
Mars's diftance	- -	145,014,148
Jupiter's diftance	- -	494,990,976
Saturn's diftance	- -	907,956,130.

In another paper (Philof. Tranf. vol. 53, p. 169) Mr. Short ftates the mean horizontal Parallax of the fun at 8″·69. And Mr. Hornfby, from feveral obfervations of the tranfit of June 3d, 1769 (for which fee the Philof. Tranf. vol. 59) deduces the fun's Parallax for that day equal to 8·65, and the mean Parallax 8″·78; whence he makes the mean diftance of the earth from the fun to be 93,726,900 Englifh miles, and the diftances of the other planets thus:

Mercury's diftance	- -	36,281,700
Venus's diftance	- -	67,795,500
Mars's diftance	- -	142,818,000
Jupiter's diftance	- -	487,472,000
Saturn's diftance	- -	894,162,000

See the Philof. Tranf. vol. 61, p. 572.

But others, by taking the refults of thofe obfervations that are moft to be depended on, have made the fun's Parallax at his mean diftance from the earth to be 8·6045; and fome make it only 8·54. According to the former of thefe, the fun's mean diftance from the earth is 95,109,736 miles; and according to the latter it is 95,834,742 miles. Upon the whole there feems

reafon to conclude that the fun's horizontal Parallax may be ftated at 8″·6, and his diftance near 95 millions of miles. Hence, the following horizontal Parallaxes:

Mean Parallax of the fun	0′	8″·6
Moon's greateft	61	32
Moon's leaft	54	4
Moon's mean	57	48
Mars's	0	25

Of the Parallax *of the Fixed Stars.* As to the fixed ftars, their diftance is fo great, that it has never been found that they have any fenfible Parallax, neither with refpect to the earth's diameter, nor even with regard to the diameter of the earth's annual orbit round the fun, although this diameter be about 190 millions of miles. For, any of thofe ftars being obferved from oppofite ends of this diameter, or at the interval of half a year between the obfervations, when the earth is in oppofite points of her orbit, yet ftill the ftar appears in the fame place and fituation in the heavens, without any change that is fenfible, or meafurable with the very beft inftruments, not amounting to a fingle fecond of a degree. That is, the diameter of the earth's annual orbit, at the neareft of the fixed ftars, does not fubtend an angle of a fingle fecond; or, in comparifon of the diftance of the fixed ftars, the extent of 190 millions of miles is but as a point!

Parallax is alfo ufed, in Levelling, for the angle contained between the line of true level, and that of apparent level. And, in other branches of fcience, for the difference between the true and apparent places.

PARALLEL, in Geometry, is applied to lines, figures, and bodies, which are every where equidiftant from each other; or which, though infinitely produced, would never either approach nearer, or recede farther from, each other; their diftance being every where meafured by a perpendicular line between them. Hence,

Parallel *right lines* are thofe which, though infinitely produced ever fo far, would never meet: which is Euclid's definition of them.

Newton, in Lemma 22, book 1 of his Principia, defines Parallels to be fuch lines as tend to a point infinitely diftant.

Parallel Lines ftand oppofed to lines converging, and diverging.

Some define an inclining or converging line, to be that which will meet another at a finite diftance, and a Parallel line, that which will only meet at an infinite diftance.

As a perpendicular is by fome faid to be the fhorteft of all lines that can be drawn to another; fo a Parallel is faid to be the longeft.

It is demonftrated by geometricians, that two lines, AB and CD, that are both Parallel to one and the fame right line EF, are alfo Parallel to each other. And that if two Parallel lines AB and EF be cut by any other line GH; then 1ft, the alternate angles are equal; viz the angle $a = \angle b$, and $\angle c = \angle a$. 2d, The external angle is equal to the internal one on the fame fide of the cutting line; viz the $\angle e = \angle d$, and the $\angle f = \angle b$. 3d, That the two internal ones on the fame fide are, taken together, equal to two right

right angles; viz, $\angle a + \angle d = 180°$, or $\angle c + \angle b = 180°$.

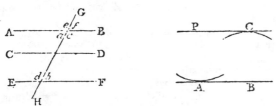

To draw a PARALLEL *Line.*—If the line to be Parallel to AB muſt paſs through a given point P: Take the neareſt diſtance between the point P and the given line AB, by ſetting one foot of the compaſſes in P, and with the other deſcribe an arc juſt to touch the line in A; then with that diſtance as a radius, and a centre B taken any where in the line, deſcribe another arc C; laſtly, through P draw a line PC juſt to touch the arc C, and that will be the Parallel ſought.

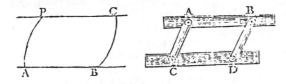

Otherwiſe.—With the centre P, and any radius, deſcribe an arc BC, cutting the given line in B. Next, with the ſame radius, and centre B, deſcribe another arc PA, cutting alſo the given line in A. Laſtly, take AP between the compaſſes, and apply it from B to C; and through P and C draw the Parallel PC required.

Or, draw the line with the Parallel Ruler, deſcribed below, by laying one edge of the ruler along AB, and extending the other to the given point or diſtance.

When the one line is to be at a given diſtance from the other; take that diſtance between the compaſſes as a radius, and with two centres taken any where in the given line, deſcribe two arcs; then lay a ruler juſt to touch the arcs, and by it draw the Parallel.

PARALLEL *Planes,* are every where equidiſtant, or have all the perpendiculars that are drawn between them, everywhere equal.

PARALLEL *Rays,* in Optics, are thoſe which keep always at an equal diſtance in reſpect to each other, from the viſual object to the eye, from which the object is ſuppoſed to be infinitely diſtant.

PARALLEL *Ruler,* is a mathematical inſtrument, conſiſting of two equal rulers, AB and CD, either of wood or metal, connected together by two ſlender croſs bars or blades AC and BD, moveable about the points or joints A, B, C, D.

There are other forms of this inſtrument, a little varied from the above; ſome having the two blades croſſing in the middle, and fixed only at one end of them, the other two ends ſliding in grooves along the two rulers; &c.

The uſe of this inſtrument is obvious. For the edge of one of the rulers being applied to any line, the other opened to any extent will be always parallel to the

former; and conſequently any Parallels to this may be drawn by the edge of the ruler, opened to any extent.

PARALLEL *Sailing,* in Navigation, is the ſailing on or under a Parallel of latitude, or Parallel to the equator. —Of this there are three caſes.

1. Given the Diſtance and Difference of Longitude; to find the Latitude.—Rule. As the difference of longitude is to the diſtance, ſo is radius to the coſine of the latitude.

2. Given the Latitude and Difference of Longitude; to find the Diſtance.—Rule. As radius is to the coſine of the latitude, ſo is the difference of longitude to the diſtance.

3. Given the Latitude and Diſtance; to find the difference of longitude.—Rule. As the coſine of latitude is to radius, ſo is the diſtance to the difference of longitude.

PARALLEL *Sphere,* is that ſituation of the ſphere where the equator coincides with the horizon, and the poles with the zenith and nadir.

In this ſphere all the Parallels of the equator become Parallels of the horizon; conſequently no ſtars ever riſe or ſet, but all turn round in circles Parallel to the horizon, as well as the ſun himſelf, which when in the equinoctial wheels round the horizon the whole day. Alſo, After the ſun riſes to the elevated pole, he never ſets for ſix months; and after his entering again on the other ſide of the line, he never riſes for ſix months longer.

This poſition of the ſphere is theirs only who live at the poles of the earth, if any ſuch there be. The greateſt height the ſun can riſe to them, is $23\frac{1}{2}$ degrees. They have but one day and one night, each being half a year long. See SPHERE.

PARALLELS, or *Places of Arms,* in a Siege, are deep trenches, 15 or 18 feet wide, joining the ſeveral attacks together; and ſerving to place the guard of the trenches in, to be at hand to ſupport the workmen when attacked.

There are uſually three in an attack: the firſt is about 600 yards from the covert-way, the ſecond between 3 and 400, and the third near or on the glacis. —It is ſaid they were firſt invented or uſed by Vauban.

PARALLELS *of Altitude,* or Almacantars, are circles Parallel to the horizon, conceived to paſs through every degree and minute of the meridian between the horizon and zenith; having their poles in the zenith.

PARALLELS, or PARALLEL *Circles,* called alſo Parallels of Latitude, and Circles of Latitude, are leſſer circles of the ſphere, Parallel to the equinoctial or equator.

PARALLELS *of Declination,* are leſſer circles Parallel to the equinoctial.

PARALLELS *of Latitude,* in Geography, are leſſer circles Parallel to the equator. But in Aſtronomy they are Parallel to the ecliptic.

PARALLELISM, the quality of a parallel, or that which denominates it ſuch. Or it is that by which two things, as lines, rays, or the like, become equidiſtant from one another.

PARALLELISM *of the Earth's Axis,* is that invariable ſituation of the axis, in the progreſs of the earth through the annual orbit, by which it always keeps parallel to itſelf; ſo that if a line be drawn parallel to its axis,

while

while in any one position; the axis, in all other positions or parts of the orbit, will always be parallel to the same line.

In consequence of this Parallelism, the axis of the earth points always, as to sense, to the same place or point in the heavens, viz to the poles. Because, though really the axis, in the annual motion, describes the surface of a cylinder, whose base is the circle of the earth's annual orbit, yet this whole circle is but as a point in comparison with the distance of the fixed stars; and therefore all the sides of the cylinder seem to tend to the same point, which is the celestial pole.—To this Parallelism is owing the change and variety of seasons; with the inequality of days and nights.

This Parallelism is the necessary consequence of the earth's double motion; the one round the sun, the other round its own axis. Nor is there any necessity to imagine a third motion, as some have done, to account for this Parallelism.

PARALLELISM *of Rows of Trees.* The eye placed at the end of an alley bounded by two rows of trees, planted in parallel lines, never sees them parallel, but always inclining to each other, towards the farther end.

Hence mathematicians have taken occasion to enquire, in what lines the trees must be disposed, to correct this effect of the perspective, and make the rows still appear parallel. And, to produce this effect, it is evident that the unequal intervals of any two opposite or corresponding trees may be seen under equal visual angles.

For this purpose, M. Fabry, Tacquet, and Varignon observe, that the rows must be opposite semi-hyperbolas. See the Mem. Acad. Sciences, an. 1717.

But notwithstanding the ingenuity of their speculations, it has been proved by D'Alembert, and Bouguer, that to produce the effect proposed, the trees are to be ranged merely in two diverging right lines.

PARALLELOGRAM, in Geometry, is a quadrilateral right-lined figure, whose opposite sides are parallel to each other.

A Parallelogram may be conceived as generated by the motion of a right line, along a plane, always parallel to itself.

Parallelograms have several particular denominations, and are of several species, according to certain particular circumstances, as follow:

When the angles of the Parallelogram are right ones, it is called a Rectangle.—When the angles are right, and all its sides equal, it is a square.—When the sides are equal, but the angles oblique ones, the figure is a Rhombus or Lozenge. And when both the sides and angles are unequal, it is a Rhomboides.

Every other quadrilateral whose opposite sides are neither parallel nor equal, is called a Trapezium.

Properties of the PARALLELOGRAM.—I. In every Parallelogram ABDC, the diagonal divides the figure into two equal triangles, ABD, ACD. Also the opposite angles and sides are equal, viz; the side AB = CD, and AC = BD, also the angle A = ∠ D, and the

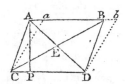

∠ B = ∠ C. And the sum of any two succeeding

angles, or next the same side, is equal to two right angles, or 180 degrees, as ∠ A + ∠ C = ∠ C + ∠ D = ∠ D + ∠ B = ∠ B + ∠ A = two right-angles.

2. All Parallelograms, as ABDC and *ab*DC, are equal, that are on the same base CD, and between the same parallels A*b*, CD; or that have either the same or equal bases and altitudes; and each is double a triangle of the same or equal base and altitude.

3. The areas of Parallelograms are to one another in the compound ratio of their bases and altitudes. If their bases be equal, the areas are as their altitudes; and if the altitudes be equal, the areas are as the bases. And when the angles of the one Parallelogram are equal to those of another, the areas are as the rectangles of the sides about the equal angles.

4. In every Parallelogram, the sum of the squares of the two diagonals, is equal to the sum of the squares of all the four sides of the figure, viz,

$$AD^2 + BC^2 = AB^2 + BD^2 + DC^2 + CA^2.$$

Also the two diagonals bisect each other; so that AE = ED, and BE = EC.

5. *To find the Area of a* PARALLELOGRAM.—Multiply any one side, as a base, by the height, or perpendicular let fall upon it from the opposite side. Or, multiply any two adjacent sides together, and the product by the sine of their contained angle, the radius being 1 : viz,

The area is = CD × AP = AC × CD × sin. ∠ C.

Complement of a PARALLELOGRAM. See COMPLEMENT.

Centre of Gravity of a PARALLELOGRAM. See CENTRE *of Gravity,* and CENTROBARIC *Method.*

PARALLELOGRAM, or PARALLELISM, or PENTAGRAPH, also denotes a machine used for the ready and exact reduction or copying of designs, schemes, plans, prints, &c, in any proportion. See PENTAGRAPH.

PARALLELOGRAM *of the Hyperbola,* is the Parallelogram formed by the two asymptotes of an hyperbola, and the parallels to them, drawn from any point of the curve. This term was first used by Huygens, at the end of his Dissertatio de Causa Gravitatis. This Parallelogram, so formed, is of an invariable magnitude in the same hyperbola; and the rectangle of its sides is equal to the power of the hyperbola.

This Parallelogram is also the modulus of the logarithmic system; and if it be taken as unity or 1, the hyperbolic sectors and segments will correspond to Napier's or the natural logarithms; for which reason these have been called the hyperbolic logarithms. If the Parallelogram be taken = ·43429448190 &c, these sectors and segments will represent Briggs's logarithms; in which case the two asymptotes of the hyperbola make between them an angle of 25° 44′ 25″½.

Newtonian or Analytic PARALLELOGRAM, a term used for an invention of Sir Isaac Newton, to find the first term of an infinite converging series. It is sometimes called the Method of the Parallelogram and Ruler; because a ruler or right line is also used in it:

This Analytical Parallelogram is formed by dividing any geometrical Parallelogram into equal small squares or Parallelograms, by lines drawn horizontally and perpendicularly

pendicularly

pendicularly through the equal divisions of the sides of the Parallelogram. The small cells, thus formed, are filled with the dimensions or powers of the species *x* and *y*, and their products.

For instance, the powers of *y*, as y^0 or 1, y, y^2, y^3, y^4, &c, being placed in the lowest horizontal range of cells; and the powers of *x*, as $x^0 = 1$, x, x^2, x^3, &c, in the vertical column to the left; or vice versa; these powers and their products will stand as in this figure:

A				D
x^4	x^4y	x^4y^2	x^4y^3	x^4y^4
x^3	x^3y	x^3y^2	x^3y^3	x^3y^4
x^2	x^2y	x^2y^2	x^2y^3	x^2y^4
x	xy	xy^2	xy^3	xy^4
1	y	y^2	y^3	y^4
B				C

Now when any literal equation is proposed, involving various powers of the two unknown quantities *x* and *y*, to find the value of one of these in an infinite series of the powers of the other; mark such of the cells as correspond to all its terms, or that contain the same powers and products of *x* and *y*; then let a ruler be applied to two, or perhaps more, of the Parallelograms so marked, of which let one be the lowest in the left hand column at AB, the other touching the ruler towards the right hand; and let all the rest, not touching the ruler, lie above it. Then select those terms of the equation which are represented by the cells that touch the ruler, and from them find the first term or quantity to be put in the quotient.

Of the application of this rule, Newton has given several examples in his Method of Fluxions and Infinite Series, p. 9 and 10, but without demonstration; which has been supplied by others. See Colson's Comment on that treatise, p. 192 & seq. Also Newton's Letter to Oldenburg, Oct. 24, 1676. Maclaurin's Algebra, p. 251. And especially Cramer's Analyses des Lignes Courbes, p. 148.—This author observes, that this invention, which is the true foundation of the method of series, was but imperfectly understood, and not valued as it deserved, for a long time. He thinks it however more convenient in practice to use the Analytical Triangle of the abbé de Gua, which takes in no more than the diagonal cells lying between A and C, and those which lie between them and B.

PARALLELOGRAM *Protractor*, a mathematical instrument, consisting of a semicircle of brass, with four rulers in form of a Parallelogram, made to move to any angle. One of these rulers is an index, which shews on the semicircle the quantity of any inward and outward angle.

PARALLELOPIPED, or PARALLELOPIPEDON, is a solid figure contained under six parallelograms, the opposites of which are equal and parallel. Or, it is a prism whose base is a parallelogram.

Properties of the PARALLELOPIPEDON.—All Parallelopipedons, whether right or oblique, that have their bases and altitudes equal, are equal; and each equal to triple a pyramid of an equal base and altitude.—A diagonal plane divides the Parallelopipedon into two equal triangular prisms.—See other properties under the general term PRISM, of which this is only a particular species.

To Measure the Surface and Solidity of a PARALLELOPIPEDON.—Find the areas of the three parallelograms AD, BE, and BG, which add into one sum; and double that sum will be the whole surface of the Parallelopipedon.

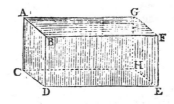

For the Solidity; multiply the base by the altitude; that is, any one face or side by its distance from the opposite side; as AD × DE, or AB × BE, or BG × BD.

PARAMETER, a certain constant right line in each of the three Conic Sections; otherwise called also Latus Rectum.

This line is called Parameter, or equal measurer, because it measures the conjugate axis by the same ratio which is between the two axes themselves; being indeed a third proportional to them; viz, a third proportional to the transverse and conjugate axes, in the ellipse and hyperbola; and, which is the same thing, a third proportional to any abscis and its ordinate in the parabola. So if *t* and *c* be the two axes in the ellipse and hyperbola, and *x* and *y* an abscis and its ordinate in the parabola;

then $t : c :: c : p = \dfrac{c^2}{t}$ the Param. in the former,

and $x : y :: y : p = \dfrac{y^2}{x}$ the Param. in the last.

The Parameter is equal to the double ordinate drawn through the focus of any of the three conic sections.

PARAPET, or *Breastwork*, in Fortification, is a defence or screen, on the extreme edge of a rampart, or other work, serving to cover the soldiers and the cannon from the enemy's fire.

The thickness of the Parapet is 18 or 20 feet, commonly lined with masonry; and 7 or 8 feet high, when the enemy has no command above the battery; otherwise, it should be raised higher, to cover the men while they

they load the guns. There are certain openings, called Embrasures, cut in the Parapet, from the top downwards, to within about 2½ or 3 feet of the bottom of it, for the cannon to fire through; the solid pieces of it between one embrasure and another, being called Merlons.

PARAPET is also a little breast-wall, raised on the brinks of bridges, quays, or high buildings; to serve as a stay, and prevent people from falling over.

PARDIES (IGNATIUS GASTON), an ingenious French mathematician and philosopher, was born at Pau, in the province of Gascony, in 1636; his father being a counsellor of the parliament of that city.— At the age of 16 he entered into the order of Jesuits, and made so great a proficiency in his studies, that he taught polite literature, and composed many pieces in prose and verse with a distinguished delicacy of thought and style, before he was well arrived at the age of manhood. Propriety and elegance of language appear to have been his first pursuits; for which purpose he studied the Belles Lettres, and other learned productions. But afterwards he devoted himself to mathematical and philosophical studies, and read, with due attention, the most valuable authors, ancient and modern, in those sciences: so that, in a short time he made himself master of the Peripatetic and Cartesian philosophy, and taught them both with great reputation. Notwithstanding he embraced Cartesianism, yet he affected to be rather an inventor in philosophy himself. In this spirit he sometimes advanced very bold opinions in natural philosophy, which met with opposers, who charged him with starting absurdities: but he was ingenious enough to give his notions a plausible turn, so as to clear them seemingly from contradictions. His reputation procured him a call to Paris, as Professor of Rhetoric in the College of Lewis the Great. He also taught the mathematics in that city, as he had before done in other places. He had from his youth a happy genius for that science, and made a great progress in it; and the glory which his writings acquired him, raised the highest expectations from his future labours; but these were all blasted by his early death, in 1673, at 37 years of age; falling a victim to his zeal, he having caught a contagious disorder by preaching to the prisoners in the Bicetre.

Pardies wrote with great neatness and elegance. His principal works are as follow:

1. Horologium Thaumaticum duplex; 1662, in 4to.

2. Dissertatio de Motu et Natura Cometarum; 1665, 8vo.

3. Discours du Mouvement Local; 1670, 12mo.

4. Elemens de Geometrie; 1670, 12mo.—This has been translated into several languages; in English by Dr. Harris, in 1711.

5. Discours de la Connoissance des Betes; 1672, 12mo.

6. Lettre d'un Philosophe à un Cartesien de ses amis; 1672, 12mo.

7. La Statique ou la Science des Forces Mouvantes; 1673, 12mo.

8. Description et Explication de deux Machines propres à faire des Cadrans avec une grande facilité; 1673, 12mo.

9. Remarques du Mouvement de la Lumiere.

10. Globi Cœlestis in tabula plana redacti Descriptio; 1675, folio.

Part of his works were printed together, at the Hague, 1691, in 12mo; and again at Lyons, 1725.— Pardies had a dispute also with Sir Isaac Newton, about his New Theory of Light and Colours, in 1672. His letters are inserted in the Philosophical Transactions for that year.

PARENT (ANTHONY), a respectable French mathematician, was born at Paris in 1666. He shewed an early propensity to the mathematics, eagerly perusing such books in that science as fell in his way. His custom was to write remarks in the margins of the books he read; and in this way he had filled a number of books with a kind of commentary by the time he was 13 years of age.

Soon after this he was put under a master, who taught rhetoric at Chartres. Here he happened to see a dodecaedron, upon every face of which was delineated a sun-dial, except the lowest on which it stood. Struck, as it were instantaneously with the curiosity of these dials, he attempted drawing one himself: but having only a book which taught the practical part, without the theory, it was not till after his master came to explain the doctrine of the sphere to him, that he began to understand how the projection of the circles of the sphere formed sun-dials. He then undertook to write a treatise upon gnomonics. To be sure the piece was rude and unpolished enough; however, it was entirely his own, and not borrowed. About the same time he wrote a book of geometry, in the same taste, at Beauvais.

His friends then sent for him to Paris to study the law; and in obedience to them he went through a course in that faculty: which was no sooner finished than, urged by his passion for mathematics, he shut himself up in the college of Dormans, that no avocation might take him from his beloved study: and, with an allowance of less than 200 livres a-year, he lived content in this retreat, from which he never stirred but to the Royal College, to hear the lectures of M. de la Hire or M. de Sauveur. When he thought himself capable of teaching others, he took pupils: and fortification being a branch of study which the war had brought into particular notice, he had often occasion to teach it: but after some time he began to entertain scruples about teaching a subject he had never seen, knowing it only by imagination. He imparted this scruple to M. Sauveur, who recommended him to the Marquis d'Aligre, who luckily at that time wanted to have a mathematician with him. M. Parent made two campaigns with the marquis, by which he instructed himself sufficiently in viewing fortified places; of which he drew a number of plans, though he had never learned the art of drawing.

From this period he spent his time in a continual application to the study of natural philosophy, and mathematics in all its branches, both speculative and practical; to which he joined anatomy, botany, and chemistry:—his genius joined with his indefatigable application overcoming every thing.

M. de Billettes being admitted into the Academy of Sciences at Paris in 1699, with the title of their mechanician, he named M. Parent for his eleve or disciple,

ciple, a branch of mathematics in which he chiefly excelled. It was foon difcovered in this fociety, that he engaged in all the different fubjects which were brought before them ; and indeed that he had a hand in every thing. But this extent of knowledge, joined to a natural warmth and impetuofity of temper, raifed a fpirit of contradiction in him, which he indulged on all occafions ; fometimes to a degree of precipitancy that was highly culpable, and often with but little regard to decency. Indeed the fame behaviour was returned to him, and the papers which he brought to the academy were often treated with much feverity. In his productions, he was charged with obfcurity ; a fault for which he was indeed fo notorious, that he perceived it himfelf, and could not avoid correcting it.

By a regulation of the academy in 1716, the clafs of eleves was fuppreffed, as that diftinction feemed to put too great an inequality between the members. M. Parent was made an adjunct or affiftant member for the clafs of geometry : though he enjoyed this promotion but a very fhort time ; being cut off by the fmall-pox the fame year, at 50 years of age.

M. Parent, befides leaving many pieces in manufcript, publifhed the following works :

1. Elemens de Mecanique & de Phyfique ; in 12mo, 1700.

2. Recherches de Mathematiques & de Phyfique ; 3 vols. 4to, 1714.

3. Arithmetique theorico-pratique ; in 8vo, 1714.

4. A great multitude of papers in the volumes of the Memoirs of the Academy of Sciences, from the year 1700 to 1714, feveral papers in almoft every volume, upon a variety of branches in the mathematics.

PARGETING, in Building, is ufed for the plaiftering of walls ; fometimes for plaifter itfelf.

PARHELION, or PARHELIUM, denotes a mock fun, or meteor, appearing as a very bright light by the fide of the fun ; being formed by the reflection of his beams in a cloud properly fituated.

Parhelia ufually accompany the coronæ, or luminous circles, and are placed in the fame circumference, and at the fame height. Their colours refemble thofe of the rainbow ; the red and yellow are on that fide towards the fun, and the blue and violet on the other. Though coronæ are fometimes feen entire, without any Parhelia ; and fometimes Parhelia without coronæ.

The apparent fize of Parhelia is the fame as that of the true fun ; but they are not always round, nor always fo bright as the fun ; and when feveral appear, fome are brighter than others. They are tinged externally with colours like the rainbow, and many of them have a long fiery tail oppofite to the fun, but paler towards the extremity. Some Parhelia have been obferved with two tails and others with three. Thefe tails moftly appear in a white horizontal circle, commonly paffing through all the Parhelia, and would go through the centre of the fun if it were entire. Sometimes there are arcs of leffer circles, concentric to this, touching thofe coloured circles which furround the fun : thefe are alfo tinged with colours, and contain other Parhelia.

Parhelia are generally fituated in the interfections of circles ; but Caffini fays, thofe which he faw in 1683, were on the outfide of the coloured circle, though the

tails were in the circle that was parallel to the horizon. M. Aepinus apprehends, that Parhelia with elliptical coronæ are more frequent in the northern regions, and thofe with circular ones in the fouthern. They have been vifible for one, two, three, or four hours together ; and it is faid that in North America they continue feveral days, and are vifible from fun-rife to fun-fet. When the Parhelia difappear, it fometimes rains, or there falls fnow in the form of oblong fpiculæ. And Mariotte accounts for the appearance of Parhelia from an infinity of fmall particles of ice floating in the air, which multiply the image of the fun, either by refracting or breaking his rays, and thus making him appear where he is not ; or by reflecting them, and ferving as mirrors.

Moft philofophers have written upon Parhelia ; as Ariftotle, Pliny, Scheiner, Gaffendi, Des Cartes, Huygens, Hevelius, De la Hire, Caffini, Grey, Halley, Maraldi, Muffchenbroek, &c. See Smith's Optics, book 1, chap. 11. Alfo Prieftley's Hift. of Light &c, p. 613. And Muffchenbroek's Introduction &c, vol. 2, p. 1038 quarto.

PARODICAL *Degrees*, in an equation, a term that has been fometimes ufed to denote the feveral regular terms in a quadratic, cubic, biquadratic, &c, equation, when the indices of the powers afcend or defcend orderly in an arithmetical progreffion. Thus, $x^3 + mx^2 + nx = p$ is a cubic equation where no term is wanting, but having all its Parodic Degrees ; the indices of the terms regularly defcending thus, 3, 2, 1, 0.

PART, *Aliquant, Aliquot, Circular Proportional, Similar*, &c. See the refpective adjectives.

PART *of Fortune*, in Judicial Aftrology, is the lunar horofcope ; or the point in which the moon is, at the time when the fun is in the afcending point of the eaft.

The fun in the afcendant is fuppofed, according to this fcience, to give life ; and the moon difpenfes the radical moifture, and is one of the caufes of fortune. In horofcopes the Part of Fortune is reprefented by a circle divided by a crofs.

PARTICLE, the minute part of a body, or an affemblage of feveral of the atoms of which natural bodies are compofed. Particle is fometimes confidered as fynonymous with atom, and corpufcle ; and fometimes they are diftinguifhed.

Particles are, as it were, the elements of bodies ; by the various arrangement and texture of which, with the difference of the cohefion, &c, are conftituted the feveral kinds of bodies, hard, foft, liquid, dry, heavy, light, &c. The fmalleft Particles or corpufcles cohere with the ftrongeft attractions, and always compofe larger Particles of weaker cohefion : and many of thefe, cohering, compofe ftill larger Particles, whofe vigour is ftill weaker ; and fo on for divers fucceffions, till the progreffion end in the largeft Particles, upon which the operations in chemiftry, and the colours of natural bodies, depend ; and which, by cohering, compofe bodies of fenfible magnitude.

PARTILE *Afpect*, in Aftrology, is when the planets are in the exact degree of any particular afpect. In contradiftinction to Platic Afpect, or when they do not regard each other with thofe very degrees. See ASPECT.

PARTY

PARTY *Arches*, in Architecture, are arches built between separate tenures, where the property is intermixed, and apartments over each other do not belong to the same estate.

PARTY *Walls*, are partitions of brick made between buildings in separate occupations, for preventing the spread of fire. These are made thicker than the external walls; and their thickness in London is regulated by act of parliament of the 14th of George the Third.

PASCAL (BLAISE), a respectable French mathematician and philosopher, and one of the greatest geniuses and best writers that country has produced. He was born at Clermont in Auvergne, in the year 1623. His father, Stephen Pascal, was president of the Court of Aids in his province: he was also a very learned man, an able mathematician, and a friend of Des Cartes. Having an extraordinary tenderness for this child, his only son, he quitted his office in his province, and settled at Paris in 1631, that he might be quite at leisure to attend to his son's education, which he conducted himself, and young Pascal never had any other master.

From his infancy Blaise gave proofs of a very extraordinary capacity. He was extremely inquisitive; desiring to know the reason of every thing; and when good reasons were not given him, he would seek for better; nor would he ever yield his assent but upon such as appeared to him well grounded. What is told of his manner of learning the mathematics, as well as the progress he quickly made in that science, seems almost miraculous. His father, perceiving in him an extraordinary inclination to reasoning, was afraid lest the knowledge of the mathematics might hinder his learning the languages, so necessary as a foundation to all sound learning. He therefore kept him as much as he could from all notions of geometry, locked up all his books of that kind, and refrained even from speaking of it in his presence. He could not however prevent his son from musing on that science; and one day in particular he surprised him at work with charcoal upon his chamber floor, and in the midst of figures. The father asked him what he was doing: I am searching, says Pascal, for such a thing; which was just the same as the 32d proposition of the 1st book of Euclid. He asked him then how he came to think of this: It was, says Blaise, because I found out such another thing; and so, going backward, and using the names of *bar* and *round*, he came at length to the definitions and axioms he had formed to himself. Does it not seem miraculous, that a boy should work his way into the heart of a mathematical book, without ever having seen that or any other book upon the subject, or knowing any thing of the terms? Yet we are assured of the truth of this by his sister, Madam Perier, and several other persons, the credit of whose testimony cannot reasonably be questioned.

From this time he had full liberty to indulge his genius in mathematical pursuits. He understood Euclid's Elements as soon as he cast his eyes upon them. At 16 years of age he wrote a treatise on Conic Sections, which was accounted a great effort of genius; and therefore it is no wonder that Des Cartes, who had been in Holland a long time, upon reading it, should choose to believe that M. Pascal the father was the

real author of it. At 19 he contrived an admirable arithmetical machine, which was esteemed a very wonderful thing, and would have done credit as an invention to any man versed in science, and much more to such a youth.

About this time his health became impaired, so that he was obliged to suspend his labours for the space of four years. After this, having seen Torricelli's experiment respecting a vacuum and the weight of the air, he turned his thoughts towards these objects, and undertook several new experiments, one of which was as follows: Having provided a glass tube, 46 feet in length, open at one end, and hermetically sealed at the other, he filled it with red wine, that he might distinguish the liquor from the tube, and stopped up the orifice; then having inverted it, and placed it in a vertical position, with the lower end immersed into a vessel of water one foot deep, he opened the lower end, and the wine descended to the distance of about 32 feet from the surface of the vessel, leaving a considerable vacuum at the upper part of the tube. He next inclined the tube gradually, till the upper end became only of 32 feet perpendicular height above the bottom, and he observed the liquor proportionally ascend up to the top of the tube. He made also a great many experiments with siphons, syringes, bellows, and all kinds of tubes, making use of different liquors, such as quicksilver, water, wine, oil, &c; and having published them in 1647, he dispersed his work through all countries.

All these experiments however only ascertained effects, without demonstrating the causes. Pascal knew that Torricelli conjectured that those phenomena which he had observed were occasioned by the weight of the air, though they had formerly been attributed to Nature's abhorrence of a vacuum; but if Torricelli's theory were true, he reasoned that the liquor in the barometer tube ought to stand higher at the bottom of a hill, than at the top of it. In order therefore to discover the truth of this theory, he made an experiment at the top and bottom of a mountain in Auvergne, called *le Puy de Dome*, the result of which gave him reason to conclude that the air was indeed heavy. Of this experiment he published an account, and sent copies of it to most of the learned men in Europe. He also renewed it at the top and bottom of several high towers, as those of Notre Dame at Paris, St. Jaques de la Boucherie, &c; and always remarked the same difference in the weight of the air, at different elevations. This fully convinced him of the general pressure of the atmosphere; and from this discovery he drew many useful and important inferences. He composed also a large treatise, in which he fully explained this subject, and replied to all the objections that had been started against it. As he afterwards thought this work rather too prolix, and being fond of brevity and precision, he divided it into two small treatises, one of which he intitled, A Dissertation on the Equilibrium of Fluids; and the other, An Essay on the Weight of the Atmosphere. These labours procured Pascal so much reputation, that the greatest mathematicians and philosophers of the age proposed various questions to him, and consulted him respecting such difficulties as they could not resolve. Upon one

3

of

of thefe occafions he difcovered the folution of a problem propofed by Merfenne, which had baffled the penetration of all that had attempted it. This problem was to determine the curve defcribed in the air by the nail of a coach-wheel, while the machine is in motion; which curve was thence called a rouliette, but now commonly known by the name of cycloid. Pafcal offered a reward of 40 piftoles to any one who fhould give a fatisfactory anfwer to it. No perfon having fucceeded, he publifhed his own at Paris; but as he began now to be difgufted with the fciences, he would not fet his real name to it, but fent it abroad under that of A. d'Ettonville.—This was the laft work which he publifhed in the mathematics; his infirmities, from a delicate conftitution, though ftill young, now increafing fo much, that he was under the neceffity of renouncing fevere ftudy, and of living fo reclufe, that he fcarcely admitted any perfon to fee him.—Another fubject on which Pafcal wrote very ingenioufly, and in which he has been fpoken of as an inventor, was what has been called his Arithmetical Triangle, being a fet of figurate numbers difpofed in that form. But fuch a table of numbers, and many properties of them, had been treated of more than a century before, by Cardan, Stifelius, and other arithmetical writers.

After having thus laboured abundantly in mathematical and philofophical difquifitions, he forfook thofe ftudies and all human learning at once, to devote himfelf to acts of devotion and penance. He was not 24 years of age, when the reading fome pious books had put him upon taking this refolution; and he became as great a devotee as any age has produced. He now gave himfelf up entirely to a ftate of prayer and mortification; and he had always in his thoughts thefe great maxims of renouncing all pleafure and all fuperfluity; and this he practifed with rigour even in his illneffes, to which he was frequently fubject, being of a very invalid habit of body.

Though Pafcal had thus abftracted himfelf from the world, yet he could not forbear paying fome attention to what was doing in it; and he even interefted himfelf in the conteft between the Jefuits and the Janfenifts. Taking the fide of the latter, he wrote his *Lettres Provinciales*, publifhed in 1656, under the name of *Louis de Montalte*, making the former the fubject of ridicule. " Thefe letters, fays Voltaire, may be confidered as a model of eloquence and humour. The beft comedies of Moliere have not more wit than the firft part of thefe letters; and the fublimity of the latter part of them, is equal to any thing in Boffuet. It is true indeed that the whole book was built upon a falfe foundation; for the extravagant notions of a few Spanifh and Flemifh Jefuits were artfully afcribed to the whole fociety. Many abfurdities might likewife have been difcovered among the Dominican and Francifcan cafuifts; but this would not have anfwered the purpofe; for the whole raillery was to be levelled only at the Jefuits. Thefe letters were intended to prove, that the Jefuits had formed a defign to corrupt mankind; a defign which no fect or fociety ever had, or can have." Voltaire calls Pafcal the firft of their fatirifts; for Defpréaux, fays he, muft be confidered as only the fecond. In another place, fpeaking of this work of Pafcal, he fays, that " Examples of all

the various fpecies of eloquence are to be found in it. Though it has now been written almoft 100 years, yet not a fingle word occurs in it, favouring of that viciffitude to which living languages are fo fubject. Here then we are to fix the epoch when our language may be faid to have affumed a fettled form. The bifhop of Lucon, fon of the celebrated Buffy, told me, that afking one day the bifhop of Meaux what work he would covet moft to be the author of, fuppofing his own performances fet afide, Boffu replied, The Provincial Letters," Thefe letters have been tranflated into all languages, and printed over and over again. Some have faid that there were decrees of formal condemnation againft them; and alfo that Pafcal himfelf, in his laft illnefs, detefted them, and repented of having been a Janfenift: but both thefe particulars are falfe and without foundation. It was fuppofed that Father Daniel was the anonymous author of a piece againft them, intitled *The Dialogues of Cleander and Eudoxus.*

Pafcal was but about 30 years of age when thefe letters were publifhed; yet he was extremely infirm, and his diforders increafing foon after fo much, that he conceived his end faft approaching, he gave up all farther thoughts of literary compofition. He refolved to fpend the remainder of his days in retirement and pious meditation; and with this view he broke off all his former connections, changed his habitation, and fpoke to no one, not even to his own fervants, and hardly ever even admitted them into his room. He made his own bed, fetched his dinner from the kitchen, and carried back the plates and difhes in the evening; fo that he employed his fervants only to cook for him, to go to town, and to do fuch other things as he could not abfolutely do himfelf. In his chamber nothing was to be feen but two or three chairs, a table, a bed, and a few books. It had no kind of ornament whatever; he had neither a carpet on the floor, nor curtains to his bed. But this did not prevent him from fometimes receiving vifits; and when his friends appeared furprifed to fee him thus without furniture, he replied, that he had what was neceffary, and that any thing elfe would be a fuperfluity, unworthy of a wife man. He employed his time in prayer, and in reading the Scriptures; writing down fuch thoughts as this exercife infpired. Though his continual infirmities obliged him to ufe very delicate food, and though his fervants employed the utmoft care to provide only what was excellent, he never relifhed what he ate, and feemed quite indifferent whether they brought him good or bad. His indifference in this refpect was fo great, that though his tafte was not vitiated, he forbad any fauce or ragout to be made for him which might excite his appetite.

Though Pafcal had now given up intenfe ftudy, and though he lived in the moft temperate manner, his health continued to decline rapidly; and his diforders had fo enfeebled his organs, that his reafon became in fome meafure affected. He always imagined that he faw a deep abyfs on one fide of him, and he never would fit down till a chair was placed there, to fecure him from the danger which he apprehended. At another time he pretended that he had a kind of vifion or ecftafy; a memorandum of which he preferved

during

during the remainder of his life on a bit of paper, put between the cloth and the lining of his coat, and which he always carried about him. After languishing for several years in this imbecile state of body and mind, M. Pascal died at Paris the 19th of August 1662, at 39 years of age.

In company, Pascal was distinguished by the amiableness of his behaviour; by great modesty; and by his easy, agreeable, and instructive conversation. He possessed a natural kind of eloquence, which was in a manner irresistible. The arguments he employed for the most part produced the effect which he proposed; and though his abilities intitled him to assume an air of superiority, he never displayed that haughty and imperious tone which may often be observed in men of shining talents. The philosophy of this extraordinary man consisted in renouncing all pleasure, and every superfluity. He not only denied himself the most common gratifications; but he took also without reluctance, and even with pleasure, either as nourishment or as medicine, whatever was disagreeable to the senses; and he every day retrenched some part of his dress, food, or other things, which he considered as not absolutely necessary. Towards the close of his life, he employed himself wholly in devout and moral reflections, writing down those which he deemed worthy of being preserved. The first bit of paper he could find was employed for this purpose; and he commonly set down only a few words of each sentence, as he wrote them merely for his own use. The scraps of paper upon which he had written these thoughts, were found after his death filed upon different pieces of string, without any order or connection; and being copied exactly as they were written, they were afterward arranged and published, under the title of *Pensées, &c,* or *Thoughts upon Religion and other Subjects;* being parts of a work he had intended against atheists and infidels, which has been much admired. After his death appeared also two other little tracts; the one intitled, *The Equilibrium of Fluids;* and the other, *The Weight of the Mass of Air.*

The works of Pascal were collected in 5 volumes 8vo, and published at the Hague, and at Paris, in 1779. This edition of Pascal's works may be considered as the first published; at least the greater part of them were not before collected into one body, and some of them had remained only in manuscript. For this collection, the public were indebted to the Abbé Bossu, and Pascal was deserving of such an editor. "This extraordinary man, says he, inherited from nature all the powers of genius. He was a mathematician of the first rank, a profound reasoner, and a sublime and elegant writer. If we reflect, that in a very short life, oppressed by continual infirmities, he invented a curious arithmetical machine, the elements of the calculation of chances, and a method of resolving various problems, respecting the cycloid; that he fixed in an irrevocable manner the wavering opinions of the learned concerning the weight of the air; that he wrote one of the completest works existing in the French language; and that in his *Thoughts* there are passages the depth and beauty of which are incomparable—we can hardly believe that a greater genius ever existed in any age or nation. All those who had oc-

casion to frequent his company in the ordinary commerce of the world, acknowledged his superiority; but it excited no envy against him, as he was never fond of shewing it. His conversation instructed, without making those who heard him sensible of their own inferiority; and he was remarkably indulgent towards the faults of others. It may be easily seen by his Provincial Letters, and by some of his other works, that he was born with a great fund of humour, which his infirmities could never entirely destroy. In company, he readily indulged in that harmless and delicate raillery which never gives offence, and which greatly tends to enliven conversation; but its principal object was generally of a moral nature. For example, ridiculing those authors who say, *My Book, my Commentary, my History,* they would do better (added he) to say, *Our book, our Commentary, our History;* since there is in them much more of other people's than their own."

The celebrated Baley too, speaking of this great man, says, a hundred volumes of sermons are not of so much avail as a simple account of the life of Pascal. His humanity and his devotion mortified the libertines more than if they had been attacked by a dozen of missionaries. In short, Bayle had so high an idea of this philosopher, that he calls him *a paradox in the human species.* "When we consider his character, says he, we are almost inclined to doubt whether he was born of a woman, like the man mentioned by Lucretius;

"*Ut vix humana videatur stirpe creatus.*"

PATE, in Fortification, a kind of platform, like what is called a Horse-shoe; not always regular, but commonly oval, encompassed only with a parapet, and having nothing to flank it. It is usually erected in marshy grounds, to cover a gate of a town, or the like.

PATH *of the Vertex,* a term frequently used by Mr. Flamsteed, in his Doctrine of the Sphere, denoting a circle, described by any point of the earth's surface, as the earth turns round its axis.

This point is considered as vertical to the earth's centre; and is the same with what is called the vertex or zenith in the Ptolomaic projection.

The semidiameter of this Path of the vertex, is always equal to the complement of the latitude of the point or place that describes it; that is, to the place's distance from the pole of the world.

PAVILION, in Architecture, is a kind of turret, or building usually insulated, and contained under a single roof; sometimes square and sometimes in form of a dome: thus called from the resemblance of its roof to a tent.

PAVO, *Peacock,* a new constellation, in the southern hemisphere, added by the modern astronomers. It contains 14 stars.

PAUSE, or REST, in Music, a character of silence and rest; called also by some a Mute Figure; because it shews that some part or person is to be silent, while the others continue the song.

PECK, a measure or vessel used in measuring grain, pulse, and the like dry substances.

The standard, or Winchester Peck, contains two gallons, or the 4th part of a bushel.

PEDESTAL,

PEDESTAL, in Architecture, the lowest part of an order of columns; being that which sustains the column, and serves it as a foot to stand upon. It is a square body or dye, with a cornice and base.

The proportions and ornaments of the Pedestal are different in the different orders. Vignola indeed, and most of the moderns, make the Pedestal, and its ornaments, in all the orders, one third of the height of the column, including the base and capital. But some deviate from this rule.

Perrault makes the proportions of the three constituent parts of Pedestals, the same in all the orders; viz, the base one fourth of the Pedestal; the cornice an eighth part; and the socle or plinth of the base, two thirds of the base itself. The height of the dye is what remains of the whole height of the Pedestal.

The *Tuscan* PEDESTAL is the simplest and lowest of all; from 3 to 5 modules high. It has only a plinth for its base, and an astragal crowned for its cornice.

The *Doric* PEDESTAL is made 4 or 5 modules in height, by the moderns; for no ancient columns, of this order, are found with any Pedestal, or even with any base.

The *Ionic* PEDESTAL is from 5 to 7 modules high.

The *Corinthian* PEDESTAL is the richest and most delicate of all, and is from 4 to 7 modules high.

The *Composite* PEDESTAL is of 6 or 7 modules in height.

Square PEDESTAL, is one whose breadth and height are equal.

Double PEDESTAL, is that which supports two columns, being broader than it is high.

Continued PEDESTAL, is that which supports a row of columns without any break or interruption.

PEDESTALS *of Statues*, are those serving to support figures or statues.

PEDIMENT, in Architecture, a kind of low pinnacle; serving to crown porticos, or finish a frontispiece; and placed as an ornament over gates, doors, windows, niches, altars, &c; being usually of a triangular form, but sometimes an arch of a circle. Its height is various, but it is thought most beautiful when the height is one fifth of the length of its base.

PEDOMETER, or PODOMETER, foot-measurer, or way-wiser; a mechanical instrument, in form of a watch, and consisting of various wheels and teeth; which, by means of a chain, or string, fastened to a man's foot, or to the wheel of a chariot, advance a notch each step, or each revolution of the wheel: by which it numbers the paces or revolutions, and so the distance from one place to another.

PEDOMETER is also sometimes used for the common surveying wheel, an instrument chiefly used in measuring roads; popularly called the way-wiser. See PERAMBULATOR.

PEER, in Building. See PIER.

PEGASUS, the Horse, a constellation of the northern hemisphere, figured in the form of a flying horse; being one of the 48 ancient constellations.

It is fabled, by the Greeks, to have been the offspring of an amour between Neptune and the Gorgon Medusa; and to have been that on which Bellerophon rode when he overcame the Chimera; and that flying from mount Helicon to heaven, he there became a constellation; having thrown his rider in the flight; and that the stroke of his hoof on the mount opened the sacred fountain Hippocrene.

The stars in this constellation, in Ptolomy's catalogue, are 20, in Tycho's 19, in Hevelius's 38, and in the Britannic catalogue 89.

PELECOIDES, or *Hatchet-form*, in Geometry, a figure in form of a hatchet. As the figure ABCDA, contained under the semicircle BCD and the two quadrantal arcs AB and AD.

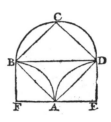

The area of the Pelecoides is equal to the square AC, and this again is equal to the rectangle BE. It is equal to the square, because the two segments AB and AD, which it wants of the square on the lower part, are compensated by the two equal segments BC and CD, by which it exceeds on the upper part. And the square is equal to the rectangle BE, because the triangle ABD, which is half the square, is also half the rectangle BE of the same base and height with it.

PELL (Dr. JOHN), an eminent English mathematician, descended from an ancient family in Lincolnshire, was born at Southwick in Suffex, March 1, 1610, where his father was minister. He received his grammar education at the free-school at Stenning in that county. At the age of 13 he was sent to Trinity College in Cambridge, being then as good a scholar as most masters of arts in that university; but though he was eminently skilled in the Greek and Hebrew languages, he never offered himself a candidate at the election of scholars or fellows of his college. His person was handsome; and being of a strong constitution, using little or no recreations, he prosecuted his studies with the more application and intensenefs.

In 1629 he drew up the " *Description and Use of the Quadrant, written for the Use of a Friend*," in two books; the original manuscript of which is still extant among his papers in the Royal Society. And the same year he held a correspondence with Mr. Briggs on the subject of logarithms.

In 1630 he wrote, *Modus supputandi Ephemerides Astronomicas, &c, ad an. 1630 accommodatus*; and, *A Key to unlock the meaning of Johannes Trithemius, in his Discourse of Steganography*: which Key he imparted to Mr. Samuel Hartlib and Mr. Jacob Homedæ. The same year he took the degree of Master of Arts at Cambridge. And the year following he was incorporated in the University of Oxford. June the 7th, he wrote *A Letter to Mr. Edmund Wingate on Logarithms*; and Oct. 5, 1631, *Commentationes in Cosmographiam Alstedii*.

In 1632 he married Ithamaria, second daughter of Mr. Henry Reginolles of London, by whom he had four sons and four daughters.—March 6, 1634, he finished his *Astronomical History of Observations of Heavenly Motions and Appearances*; and April the 10th, his *Ecliptica Prognostica, or Foreknower of the Eclipses, &c.*—In 1634 he translated *The Everlasting Tables of Heavenly*

Heavenly Motions, grounded upon the Obfervations of all Times, and agreeing with them all, by Philip Lanfberg, of Ghent in Flanders. And June the 12th, the fame year, he committed to writing, *The Manner of Deducing his Aftronomical Tables out of the Tables and Axioms of Philip Lanfberg.*—March the 9th, 1635, he wrote *A Letter of Remarks on Gellibrand's Mathematical Difcourfe on the Variation of the Magnetic Needle.* And the 3d of June following, another on the fame fubject.

His eminence in mathematical knowledge was now fo great, that he was thought worthy of a profeffor's chair in that fcience ; and, upon the vacancy of one at Amfterdam in 1639, Sir William Bofwell, the Englifh Refident with the States General, ufed his intereft, that he might fucceed in that profefforfhip : it was not filled up however till 1643, when Pell was chofen to it ; and he read with great applaufe public lectures upon Diophantus.—In 1644 he printed at Amfterdam, in two pages 4to, *A Refutation of Longomontanus's Difcourfe, De Vera Circuli Menfura.*

In 1646, on the invitation of the Prince of Orange, he removed to the new college at Breda, as Profeffor of Mathematics, with a falary of 1000 guilders a year.—His *Idea Mathefeos*, which he had addreffed to Mr. Hartlib, who in 1639 had fent it to Des Cartes and Merfenne, was printed 1650 at London, in 12mo, in Englifh, with the title of *An Idea of Mathematics*, at the end of Mr. John Durie's Reformed Library-keeper. It is alfo printed by Mr. Hook, in his Philofophical Collections, No. 5, p. 127 ; and is efteemed our author's principal work.

In 1652 Pell returned to England : and in 1654 he was fent by the protector Cromwell agent to the Proteftant Cantons in Switzerland ; where he continued till June 23, 1658, when he fet out for England, where he arrived about the time of Cromwell's death. His negociations abroad gave afterwards a general fatisfaction, as it appeared he had done no fmall fervice to the intereft of king Charles the Second, and of the church of England ; fo that he was encouraged to enter into holy orders ; and in the year 1661 he was inftituted to the rectory of Fobbing in Effex, given him by the king. In December that year, he brought into the upper houfe of convocation the calendar reformed by him, affifted by Sancroft, afterwards archbifhop of Canterbury.—In 1673 he was prefented by Sheldon, bifhop of London, to the rectory of Laingdon in Effex ; and, upon the promotion of that bifhop to the fee of Canterbury foon after, became one of his domeftic chaplains. He was then doctor of divinity, and expected to be made a dean ; but his improvement in the philofophical and mathematical fciences was fo much the bent of his genius, that he did not much purfue his private advantage. The truth is, he was a helplefs man, as to worldly affairs, and his tenants and relations impofed upon him, cozened him of the profits of his parfonage, and kept him fo indigent, that he wanted neceffaries, even ink and paper, to his dying day. He was for fome time confined to the King's-bench prifon for debt ; but, in March 1682, was invited by Dr. Whitler to live in the college of phyficians. Here he continued till June following ; when he was obliged,

by his ill ftate of health, to remove to the houfe of a grandchild of his in St. Margaret's Church-yard, Weftminfter. But he died at the houfe of Mr. Cothorne, reader of the church of St. Giles's in the Fields, December the 12th, 1685, in the 74th year of his age, and was interred at the expence of Dr. Bufby, mafter of Weftminfter fchool, and Mr. Sharp, rector of St. Giles's, in the rector's vault under that church.—Dr. Pell publifhed fome other things not yet mentioned, a lift of which is as follows : viz,

1. An Exercitation concerning Eafter ; 1644, in 4to.

2. A Table of 10,000 fquare numbers, &c ; 1672, folio.

3. An Inaugural Oration at his entering upon the Profefforfhip at Breda.

4. He made great alterations and additions to Rhonius's Algebra, printed at London 1668, 4to, under the title of, An Introduction to Algebra ; tranflated out of the High Dutch into Englifh by Thomas Branker, much altered and augmented by D. P. (Dr. Pell). Alfo a Table of Odd Numbers, lefs than 100,000, fhewing thofe that are incompofite, &c, fupputated by the fame Thomas Branker.

5. His Controverfy with Longomontanus concerning the Quadrature of the Circle ; Amfterdam, 1646, 4to.

He likewife wrote a Demonftration of the 2d and 10th books of Euclid ; which piece was in MS. in the library of lord Brereton in Chefhire : as alfo Archimedes's Arenarius, and the greateft part of Diophantus's 6 books of Arithmetic ; of which author he was preparing, Aug. 1644, a new edition, in which he intended to have corrected the tranflation, and made new illuftrations. He defigned likewife to publifh an edition of Apollonius, but laid it afide, in May, 1645, at the defire of Golius, who was engaged in an edition of that author from an Arabic manufcript given him at Aleppo 18 years before. Letters of Dr. Pell to Sir Charles Cavendifh, in the Royal Society.

Some of his manufcripts he left at Brereton in Chefhire, where he refided fome years, being the feat of William lord Brereton, who had been his pupil at Breda. A great many others came into the hands of Dr. Bufby ; which Mr. Hook was defired to ufe his endeavours to obtain for the Society. But they continued buried under duft, and mixed with the papers and pamphlets of Dr. Bufby, in four large boxes, till 1755 ; when Dr. Birch, fecretary to the Royal Society, procured them for that body, from the truftees of Dr. Bufby. The collection contains not only Pell's mathematical Papers, letters to him, and copies of thofe from him, &c, but alfo feveral manufcripts of Walter Warner, the mathematician and philofopher, who lived in the reigns of James the Firft and Charles the Firft.

Dr. Pell invented the method of ranging the feveral fteps of an algebraical calculus, in a proper order, in fo many diftinct lines, with the number affixed to each ftep, and a fhort defcription of the operation or procefs in the line. He alfo invented the character ÷ for divifion, ⊕ for involution, and ⊞ for evolution.

PENCIL *of Rays*, in Optics, is a double cone, or pyramid, of rays, joined together at the bafe ; as

BGSC :

BGSC: the one cone having its vertex in fome point of the object at B, and the cryftalline humour, or the glafs GLS for its bafe; and the other having its bafe on the fame glafs, or cryftalline, but its vertex in the point of convergence, as at C.

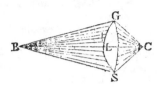

PENDULUM, in Mechanics, any heavy body, fo fufpended as that it may fwing backwards and forwards, about fome fixed point, by the force of gravity.

Thefe alternate afcents and defcents of the Pendulum, are called its Ofcillations, or Vibrations; each complete ofcillation being the defcent from the higheft point on one fide, down to the loweft point of the arch, and fo on up to the higheft point on the other fide. The point round which the Pendulum moves, or vibrates, is called its Centre of Motion, or Point of Sufpenfion; and a right line drawn through the centre of motion, parallel to the horizon, and perpendicular to the plane in which the Pendulum moves, is called the Axis of Ofcillation. There is alfo a certain point within every Pendulum, into which, if all the matter that compofes the Pendulum were collected, or condenfed as into a point, the times in which the vibrations would be performed, would not be altered by fuch condenfation; and this point is called Centre of Ofcillation. The length of the Pendulum is always eftimated by the diftance of this point below the centre of motion; being ufually near the bottom of the Pendulum; but in a cylinder, or any other uniform prifm or rod, it is at the diftance of one third from the bottom, or two-thirds from and below the centre of motion.

The length of a Pendulum, fo meafured to its centre of ofcillation, that it will perform each vibration in a fecond of time, thence called the fecond's Pendulum, has, in the latitude of London, been generally taken at $39\frac{2}{10}$ or $39\frac{1}{5}$ inches; but by fome very ingenious and accurate experiments, the late celebrated Mr. George Graham found the true length to be $39\frac{1.3.8}{1.0.0.0}$, inches, or $39\frac{1}{8}$ inches very nearly.

The length of the Pendulum vibrating feconds at Paris, was found by Varin, Des Hays, De Glos, and Godin, to be $440\frac{5}{9}$ lines; by Picard $440\frac{1}{2}$ lines; and by Mairan $440\frac{1.2}{1.2}$ lines.

Galileo was the firft who made ufe of a heavy body annexed to a thread, and fufpended by it, for meafuring time, in his experiments and obfervations. But according to Sturmius, it was Riccioli who firft obferved the ifochronifm of Pendulums, and made ufe of them in meafuring time. After him, Tycho, Langrene, Wendeline, Merfenne, Kircher, and others, obferved the fame thing; though, it is faid, without any intimation of what had been done by Riccioli. But it was the celebrated Huygens who firft demonftrated the principles and properties of Pendulums, and probably the firft who applied them to clocks. He demon-

ftrated, that if the centre of motion were perfectly fixed and immoveable, and all manner of friction, and refiftance of the air, &c, removed, then a Pendulum, once fet in motion, would for ever continue to vibrate without any decreafe of motion, and that all its vibrations would be perfectly ifochronal, or performed in the fame time. Hence the Pendulum has univerfally been confidered as the beft chronometer or meafurer of time. And as all Pendulums of the fame length perform their vibrations in the fame time, without regard to their different weights, it has been fuggefted, by means of them, to eftablifh an univerfal ftandard for all countries. On this principle Mouton, canon of Lyons, has a treatife, De Menfura pofteris tranfmittenda; and feveral others fince, as Whitchurft, &c. See Univerfal MEASURE.

Pendulums are either fimple or compound, and each of thefe may be confidered either in theory, or as in practical mechanics among artifans.

A Simple PENDULUM, in Theory, confifts of a fingle weight, as A, confidered as a point, and an inflexible right line AC, fuppofed void of gravity or weight, and fufpended from a fixed point or centre C, about which it moves.

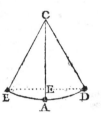

A Compound PENDULUM, in Theory, is a Pendulum confifting of feveral weights moveable about one common centre of motion, but connected together fo as to retain the fame diftance both from one another, and from the centre about which they vibrate.

The Doctrine and Laws of PENDULUMS.—1. A Pendulum raifed to B, through the arc of the circle AB, will fall, and rife again, through an equal arc, to a point equally high, as D; and thence will fall to A, and again rife to B; and thus continue rifing and falling perpetually. For it is the fame thing, whether the body fall down the infide of the curve BAD, by the force of gravity, or be retained in it by the action of the ftring; for they will both have the fame effect; and it is otherwife known, from the oblique defcents of bodies, that the body will defcend and afcend along the curve in the manner above defcribed.

Experience alfo confirms this theory, in any finite number of ofcillations. But if they be fuppofed infinitely continued, a difference will arife. For the refiftance of the air, and the friction and rigidity of the ftring about the centre C, will take off part of the force acquired in falling; whence it happens that it will not rife precifely to the fame point from whence it fell.

Thus, the afcent continually diminifhing the ofcillation, this will be at laft ftopped, and the Pendulum will hang at reft in its natural direction, which is perpendicular to the horizon.

Now as to the real time of ofcillation in a circular arc BAD: it is demonftrated by mathematicians, that if $p = 3.1416$, denote the circumference of a circle whofe diameter is 1; $g = 16\frac{1}{12}$ feet or 193 inches, the fpace a heavy body falls in the firft fecond of time; and $r = $ CA the length of the Pendulum; alfo $a = $ AE the height of the arch of vibration; then the

time

time of each oscillation in the arc BAD will be equal to $p\sqrt{\dfrac{r}{2g}} \times$ into the infinite series

$$1 + \frac{1^2 a}{2^2 d} + \frac{1^2 \cdot 3^2 a^2}{2^2 \cdot 4^2 d^2} + \frac{1^2 \cdot 3^2 \cdot 5^2 a^3}{2^2 \cdot 4^2 \cdot 6^2 d^3} \&c,$$

where $d = 2r$ is the diameter of the arc described, or twice the length of the Pendulum.

And here, when the arc is a small one, as in the case of the vibrating Pendulum of a clock, all the terms of this series after the 2d may be omitted, on account of their smallness; and then the time of a whole vibration will be nearly equal to $p\sqrt{\dfrac{r}{2g}} \times (1 + \dfrac{a}{8r})$.

So that the times of vibration of a Pendulum in different small arcs of the same circle, are as $8r + a$, or 8 times the radius, added to the versed sine of the semiarc.

And farther, if D denote the number of degrees in the semiarc AB, whose versed sine is a, then the quantity last mentioned, for the time of a whole vibration, is changed to $p\sqrt{\dfrac{r}{2g}} \times (1 + \dfrac{D^2}{52524})$. And therefore the times of vibration in different small arcs, are as $52524 + D^2$, or as the number 52524 added to the square of the number of degrees in the semiarc AB. See my Conic Sections and Select Exercises, p. 190.

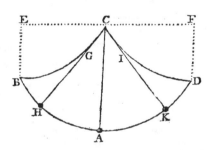

2. Let CB be a semicycloid, having its base EC parallel to the horizon, and its vertex B downwards; and let CD be the other half of the cycloid, in a similar position to the former. Suppose a Pendulum string, of the same length with the curve of each semicycloid BC, or CD, having its end fixed in C, and the thread applied all the way close to the cycloidal curve BC, and consequently the body or Pendulum weight coinciding with the point B. If now the body be let go from B, it will descend by its own gravity, and in descending it will unwind the string from off the arch BC, as at the position CGH; and the ball G will describe a semicycloid BHA, equal and similar to BGC, when it has arrived at the lowest point A; after which, it will continue its motion, and ascend, by another equal and similar semicycloid AKD, to the same height D, as it fell from at B, the string now wrapping itself upon the other arch CID. From D it will descend again, and pass along the whole cycloid DAB, to the point B; and thus perform continual successive oscillations between B and D, in the curve of a cycloid; as it before oscillated in the curve of a circle, in the former case.

This contrivance to make the Pendulum oscillate in the curve of a cycloid, is the invention of the celebrated Huygens, to make the Pendulum perform all its vibrations in equal times, whether the arch, or extent of the vibration be great or small; which is not the case in a circle, where the larger arcs take a longer time to run through them, than the smaller ones do, as is well known both from theory and practice.

The chief properties of the cycloidal Pendulum then, as demonstrated by Huygens, are the following. 1st, That the time of an oscillation in all arcs, whether larger or smaller, is always the same quantity, viz, whether the body begin to descend from the point B, and describe the semiarch BA; or that it begins at H, and describes the arch HA; or that it sets out from any other point; as it will still descend to the lowest point A in exactly the same time. And it is farther proved, that the time of a whole vibration through any double arc BAD, or HAK, &c, is in proportion to the time in which a heavy body will freely fall, by the force of gravity, through a space equal to $\frac{1}{2}$AC, half the length of the Pendulum, as the circumference of a circle is to its diameter. So that, if $g = 16\frac{1}{12}$ feet denote the space a heavy body falls in the first second of time, $p = 3\cdot1416$ the circumference of a circle whose diameter is 1, and $r = $ AC the length of the Pendulum; then, because, by the nature of descents by gravity, $\sqrt{g} : \sqrt{\frac{1}{2}r} :: 1^{\prime\prime} : \sqrt{\dfrac{r}{2g}}$ that is the time in which a body will fall through $\frac{1}{2}r$, or half the length of the Pendulum; therefore, by the above proportion, as $1 : p :: \sqrt{\dfrac{r}{2g}} : p\sqrt{\dfrac{r}{2g}}$, which is the time of an entire oscillation in the cycloid.

And this conclusion is abundantly confirmed by experience. For example, if we consider the time of a vibration as 1 second, to find the length of the Pendulum that will so oscillate in 1 second; this will give the equation $p\sqrt{\dfrac{r}{2g}} = 1$; which reduced, gives $r = \dfrac{2g}{p^2} = \dfrac{386,}{3\cdot1416^2}$ inches $= 39\cdot11$ or $39\frac{1}{9}$ inches, for the length of the second's Pendulum; which the best experiments shew to be about $39\frac{1}{8}$ inches.

3. Hence also, we have a method of determining, from the experimented length of a Pendulum, the space a heavy body will fall perpendicularly through in a given time: for, since $p\sqrt{\dfrac{r}{2g}} = 1$, therefore, by reduction, $g = \frac{1}{2}p^2 r$ is the space a body will fall through in the first second of time, when r denotes the length of the second's Pendulum; and as constant experience shews that this length is nearly $39\frac{1}{8}$ inches, in the latitude of London, in this case g or $\frac{1}{2}p^2 r$ becomes $\frac{1}{2} \times 3\cdot1416^2 \times 39\frac{1}{8} = 193\cdot07$ inches $= 16\frac{1}{12}$ feet, very nearly, for the space a body will fall in the first second of time, in the latitude of London: a fact which has been abundantly confirmed by experiments made there. And in the same manner, Mr. Huygens found the same space fallen through at Paris, to be 15 French feet.

The whole doctrine of Pendulums, oscillating between two semicycloids, both in theory and practice, was delivered by that author, in his Horologium Oscillatorium, five Demonstrationes de Motu Pendulorum. And every thing that regards the motion of Pendulums has since been demonstrated in different ways, and particularly by Newton, who has given an admirable theory on the subject, in his Principia, where he has extended to epicycloids the properties demonstrated by Huygens of the cycloids.

4. As the cycloid may be considered as coinciding, in A, with any small arc of a circle described from the centre C, passing through A, where it is known the two curves have the same radius and curvature; therefore the time in the small arc of such a circle, will be nearly equal to the time in the cycloid; so that the times in very small circular arcs are equal, because these small arcs may be considered as portions of the cycloid, as well as of the circle. And this is one great reason why the Pendulums of clocks are made to oscillate in as small arcs as possible, viz, that their oscillations may be the nearer to a constant equality.

This may also be deduced from a comparison of the times of vibration in the circle, and in the cycloid, as laid down in the foregoing articles. It has there been shewn, that the times of vibration in the circle and cycloid are thus, viz,

time in the circle nearly $p\sqrt{\frac{r}{2g}} \times (1 + \frac{a}{8r})$,

time in the cycloidal arc $p\sqrt{\frac{r}{2g}}$;

where it is evident, that the former always exceeds the latter in the ratio of $1 + \frac{a}{8r}$ to 1; but this ratio always approaches nearer to an equality, as the arc, or as its versed sine a, is smaller; till at length, when it is very small, the term $\frac{a}{8r}$ may be omitted, and then the times of vibration become both the same quantity, viz $p\sqrt{\frac{r}{2g}}$.

Farther, by the same comparison, it appears, that the time lost in each second, or in each vibration of the second's Pendulum, by vibrating in a circle, instead of a cycloid, is $\frac{a}{8r}$, or $\frac{D^2}{52524}$; and consequently the time lost in a whole day of 24 hours, is $\frac{5}{3}D^2$ nearly. In like manner, the seconds lost per day by vibrating in the arc of Δ degrees, is $\frac{5}{3}\Delta^2$. Therefore if the Pendulum keep true time in one of these arcs, the seconds lost or gained per day, by vibrating in the other, will be $\frac{5}{3}(D^2 - \Delta^2)$. So, for example, if a Pendulum measure true time in an arc of 3 degrees, on each side of the lowest point, it will lose $11\frac{2}{3}$ seconds a day by vibrating 4 degrees; and $26\frac{2}{3}$ seconds a day by vibrating 5 degrees; and so on.

5. The action of gravity is less in those parts of the earth where the oscillations of the same Pendulum are slower, and greater where these are swifter; for the time of oscillation is reciprocally proportional to $\sqrt{g}$. And it being found by experiment, that the oscillations of the same Pendulum are slower near the equator, than in places farther from it; it follows that the force of

gravity is less there; and consequently the parts about the equator are higher or farther from the centre, than the other parts; and the shape of the earth is not a true sphere, but somewhat like an oblate spheroid, flatted at the poles, and raised gradually towards the equator. And hence also the times of the vibration of the same Pendulum, in different latitudes, afford a method of determining the true figure of the earth, and the proportion between its axis and the equatorial diameter.

Thus, M. Richer found by an experiment made in the island Cayenna, about 4 degrees from the equator, where a Pendulum 3 feet $8\frac{3}{5}$ lines long, which at Paris vibrated seconds, required to be shortened a line and a quarter to make it vibrate seconds. And many other observations have confirmed the same principle. See Newton's Principia, lib. 3, prop. 20. By comparing the different observations of the French astronomers, Newton apprehends that 2 lines may be considered as the length a seconds Pendulum ought to be decreased at the equator.

From some observations made by Mr. Campbell, in 1731, in Black-river, in Jamaica, 18° north latitude, it is collected, that if the length of a simple Pendulum that swings seconds in London, be 39·126 English inches, the length of one at the equator would be 39·00, and at the poles 39·206. Philos. Transf. numb. 432; or Abr. vol. 8, part 1, pa. 238.

And hence Mr. Emerson has computed the following Table, shewing the length of a Pendulum that swings seconds at every 5th degree of latitude, as also the length of the degree of latitude there, in English miles.

Degrees of Lat.	Length of Pendulum.	Length of the Degree.
	inches.	miles.
0	39·027	68·723
5	39·029	68·730
10	39·032	68·750
15	39·036	68·783
20	39·044	68·830
25	39·057	68·882
30	39·070	68·950
35	39·084	69·020
40	39·097	69·097
45	39·111	69·176
50	39·126	69·256
55	39·142	69·330
60	39·158	69·401
65	39·168	69·467
70	39·177	69·522
75	39·185	69·568
80	39·191	69·601
85	39·195	69·620
90	39·197	69·628

6. If two Pendulums vibrate in similar arcs, the times of vibration are in the sub-duplicate ratio of their lengths. And the lengths of Pendulums vibrating in similar arcs, are in the duplicate ratio of the times

of

of a vibration directly; or in the reciprocal duplicate ratio of the number of oscillations made in any one and the same time. For, the time of vibration t being as $p\sqrt{\dfrac{r}{2g}}$, where p and g are constant or given, therefore t is as $\sqrt{r}$, and as t^2. Hence therefore the length of a half-second Pendulum will be $\frac{1}{4}r$ or $\dfrac{39\frac{1}{8}}{4} = 9.781$ inches; and the length of the quarter-second Pendulum will be $\frac{1}{16}r = \dfrac{39\frac{1}{8}}{16} = 2.445$ inches; and so of others.

7. The foregoing laws, &c, of the motion of Pendulums, cannot strictly hold good; unless the thread that sustains the ball be void of weight, and the gravity of the whole ball be collected into a point. In practice therefore, a very fine thread, and a small ball, but of a very heavy matter, are to be used. But a thick thread, and a bulky ball, disturb the motion very much; for in that case, the simple Pendulum becomes a compound one; it being much the same thing, as if several weights were applied to the same inflexible rod in several places.

8. M. Krafft in the new Petersburgh Memoirs, vols 6 and 7, has given the result of many experiments upon Pendulums, made in different parts of Russia, with deductions from them, from whence he derives this theorem: If x be the length of a Pendulum that swings seconds in any given latitude l, and in a temperature of 10 degrees of Reaumur's thermometer, then will the length of that Pendulum, for that latitude, be thus expressed, viz,

$$x = (439.178 + 2.321 \times \sin.^2 l) \text{ lines of a French foot.}$$

And this expression agrees very nearly, not only with all the experiments made on the Pendulum in Russia, but also, with those of Mr. Graham, and those of Mr. Lyons in 79° 50′ north latitude, where he found its length to be 441.38 lines. See OBLATENESS.

Simple PENDULUM, in Mechanics, an expression commonly used among artists, to distinguish such Pendulums as have no provision for correcting the effects of heat and cold, from those that have such provision. Also Simple Pendulum, and Detached Pendulum, are terms sometimes used to denote such Pendulums as are not connected with any clock, or clock-work.

Compound PENDULUM, in Mechanics, is a Pendulum whose rod is composed of two or more wires or bars of metal. These, by undergoing different degrees of expansion and contraction, when exposed to the same heat or cold, have the difference of expansion or contraction made to act in such manner as to preserve constantly the same distance between the point of suspension, and centre of oscillation, although exposed to very different and various degrees of heat or cold. There are a great variety of constructions for this purpose; but they may be all reduced to the Gridiron, the Mercurial, and the Lever Pendulum.

It may be just observed by the way, that the vulgar method of remedying the inconvenience arising from the extension and contraction of the rods of common Pendulums, is by applying the bob, or small ball, with a screw, at the lower end; by which means the Pendulum is at any time made longer or shorter, as the ball is screwed downwards or upwards, and thus the time of its vibration is kept continually the same.

The *Gridiron* PENDULUM was the invention of Mr. John Harrison, a very ingenious artist, and celebrated for his invention of the watch for finding the difference of longitude at sea, about the year 1725; and of several other time-keepers and watches since that time; for all which he received the parliamentary reward of between 20 and 30 thousand pounds. It consists of 5 rods of steel, and 4 of brass, placed in an alternate order, the middle rod being of steel, by which the Pendulum ball is suspended; these rods of brass and steel, thus placed in an alternate order, and so connected with each other at their ends, that while the expansion of the steel rods has a tendency to lengthen the Pendulum, the expansion of the brass rods, acting upwards, tends to shorten it. And thus, when the lengths of the brass and steel rods are duly proportioned, their expansions and contractions will exactly balance and correct each other, and so preserve the Pendulum invariably of the same length. The simplicity of this ingenious contrivance is much in its favour; and the difficulty of adjustment seems the only objection to it.

Mr. Harrison in his first machine for measuring time at sea, applied this combination of wires of brass and steel, to prevent any alterations by heat or cold; and in the machines or clocks he has made for this purpose, a like method of guarding against the irregularities arising from this cause is used.

The *Mercurial* PENDULUM was the invention of the ingenious Mr. Graham, in consequence of several experiments relating to the materials of which Pendulums might be formed, in 1715. Its rod is made of brass, and branched towards its lower end, so as as to embrace a cylindric glass vessel 13 or 14 inches long, and about 2 inches diameter; which being filled about 12 inches deep with mercury, forms the weight or ball of the Pendulum. If upon trial the expansion of the rod be found too great for that of the mercury, more mercury must be poured into the vessel: if the expansion of the mercury exceeds that of the rod, so as to occasion the clock to go fast with heat, some mercury must be taken out of the vessel, so as to shorten the column. And thus may the expansion and contraction of the quicksilver in the glass be made exactly to balance the expansion and contraction of the Pendulum rod, so as to preserve the distance of the centre of oscillation from the point of suspension invariably the same.

Mr. Graham made a clock of this sort, and compared it with one of the best of the common sort, for 3 years together; when he found the errors of his but about one-eighth part of those of the latter. Philos. Trans. numb. 392.

The *Lever* PENDULUM. From all that appears concerning this construction of a Pendulum, we are inclined to believe that the idea of making the difference of the expansion of different metals operate by means of a lever, originated with Mr. Graham, who in the year 1737 constructed a Pendulum, having its rod composed of one bar of steel between two of brass, which acted upon the short end of a lever, to the other end of which, the ball or weight of the Pendulum was suspended.

This

This Pendulum however was, upon trial, found to move by jerks; and therefore laid aside by the inventor, to make way for the mercurial Pendulum, just mentioned.

Mr. Short informs us in the Philof. Tranf. vol. 47, art. 88, that a Mr. Frotheringham, a quaker in Lincolnfhire, caufed a Pendulum of this kind to be made: it confifted of two bars, one of brafs, and the other of fteel, faftened together by fcrews, with levers to raife or let down the bulb; above which thefe levers were placed. M. Caffini too, in the Hiftory of the Royal Academy of Sciences at Paris, for 1741, defcribes two forts of Pendulums for clocks, compounded of bars of brafs and fteel, and in which he applies a lever to raife or let down the bulb of the Pendulum, by the expanfion or contraction of the bar of brafs.

Mr. John Ellicott alfo, in the year 1738, conftructed a Pendulum on the fame principle, but differing from Mr. Graham's in many particulars. The rod of Mr. Ellicott's Pendulum was compofed of two bars only; the one of brafs, and the other of fteel. It had two levers, each fuftaining its half of the ball or weight; with a fpring under the lower part of the ball to relieve the levers from a confiderable part of its weight, and fo to render their motion more fmooth and eafy. The one lever in Mr. Graham's conftruction was above the ball: whereas both the levers in Mr. Ellicott's were within the ball; and each lever had an adjufting fcrew, to lengthen or fhorten the lever, fo as to render the adjuftment the more perfect. See the Philof. Tranf. vol. 47, p. 479; where Mr. Ellicott's methods of conftruction are defcribed, and illuftrated by figures.

Notwithftanding the great ingenuity difplayed by thefe very eminent artifts on this conftruction, it muft farther be obferved, in the hiftory of improvements of this nature, that Mr. Cumming, another eminent artift, has given, in his Effays on the Principles of Clock and Watch-work, Lond. 1766, an ample defcription, with plates, of a conftruction of a Pendulum with levers, in which it feems he has united the properties of Mr. Graham's and Mr. Ellicott's, without being liable to any of the defects of either. The rod of this Pendulum is compofed of one flat bar of brafs, and two of fteel; he ufes three levers within the ball of the Pendulum; and, among many other ingenious contrivances, for the more accurate adjufting of this Pendulum to mean time, it is provided with a fmall ball and fcrew below the principal ball or weight, one entire revolution of which on its fcrew will only alter the rate of the clock's going one fecond per day; and its circumference is divided into 30, one of which divifions will therefore alter its rate of going one fecond in a month.

PENDULUM *Clock*, is a clock having its motion regulated by the vibration of a Pendulum.

It is controverted between Galileo and Huygens, which of the two firft applied the Pendulum to a clock. For the pretenfions of each, fee CLOCK.

After Huygens had difcovered, that the vibration made in arcs of a cycloid, however unequal they might be in extent, were all equal in time; he foon perceived, that a Pendulum applied to a clock, fo as to make it defcribe arcs of a cycloid, would rectify the otherwife unavoidable irregularities of the motion of the clock;

since, though the feveral caufes of thofe irregularities fhould occafion the Pendulum to make greater or fmaller vibrations, yet, by virtue of the cycloid, it would ftill make them perfectly equal in point of time; and the motion of the clock governed by it, would therefore be preferved perfectly equable. But the difficulty was, how to make the Pendulum defcribe arcs of a cycloid; for naturally the Pendulum, being tied to a fixed point, can only defcribe circular arcs about it.

Here M. Huygens contrived to fix the iron rod or wire, which bears the ball or weight, at the top to a filken thread, placed between two cycloidal cheeks, or two little arcs of a cycloid, made of metal. Hence the motion of vibration, applying fucceffively from one of thofe arcs to the other, the thread, which is extremely flexible, eafily affumes the figure of them, and by that means caufes the ball or weight at the bottom to defcribe a juft cycloidal arc.

This is doubtlefs one of the moft ingenious and ufeful inventions many ages have produced: by means of which it has been afferted there have been clocks that would not vary a fingle fecond in feveral days: and the fame invention alfo gave rife to the whole doctrine of involute and evolute curves, with the radius and degree of curvature, &c.

It is true, the Pendulum is ftill liable to its irregularities, how minute foever they may be. The filken thread by which it was fufpended, fhortens in moift weather, and lengthens in dry; by which means the length of the whole Pendulum, and confequently the times of the vibrations, are fomewhat varied.

To obviate this inconvenience, M. De la Hire, inftead of a filken thread, ufed a little fine fpring; which was not indeed fubject to fhorten and lengthen, from thofe caufes; yet he found it grew ftiffer in cold weather, and then made its vibrations fafter than in warm; to which alfo we may add its expanfion and contraction by heat and cold. He therefore had recourfe to a ftiff wire or rod, firm from one end to the other. Indeed by this means he renounced the advantages of the cycloid; but he found, as he fays, by experience, that the vibrations in circular arcs are performed in times as equal, provided they be not of too great extent, as thofe in cycloids. But the experiments of Sir Jonas Moore, and others, have demonftrated the contrary.

The ordinary caufes of the irregularities of Pendulums Dr. Derham afcribes to the alterations in the gravity and temperature of the air, which increafe and diminifh the weight of the ball, and by that means make the vibrations greater and lefs; an acceffion of weight in the ball being found by experiment to accelerate the motion of the Pendulum; for a weight of 6 pounds added to the ball, Dr. Derham found made his clock gain 13 feconds every day.

A general remedy againft the inconveniences of Pendulums, is to make them long, the ball heavy, and to vibrate but in fmall arcs. Thefe are the ufual means employed in England; the cycloidal cheeks being generally neglected. See the foregoing article.

Pendulum clocks refting againft the fame rail have been found to influence each other's motion. See the Philof. Tranf. numb. 453, fect. 5 and 6, where Mr. Ellicott has given a curious and exact account of this phenomenon.

PENDULUM

PENDULUM Royal, a name used among us for a clock, whose Pendulum swings seconds, and goes 8 days without winding up; shewing the hour, minute, and second. The numbers in such a piece are thus calculated. First cast up the seconds in 12 hours, which are the beats in one turn of the great wheel; and they will be found to be 43200 = 12 × 60 × 60. The swing wheel must be 30, to swing 60 seconds in one of its revolutions; now let the half of 43200, viz 21600, be divided by 30, and the quotient will be 720, which must be separated into quotients. The first of these must be 12, for the great wheel, which moves round once in 12 hours. Now 720 divided by 12, gives 60, which may also be conveniently broken into two quotients, as 10 and 6, or 12 and 5, or 8 and 7½; which last is most convenient: and if the pinions be all taken 8, the work will stand thus:

$$8) 96 (12$$
$$8) 64 (8$$
$$8) 60 (7\frac{1}{2}$$
$$30$$

According to this computation, the great wheel will go round once in 12 hours, to shew the hour; the next wheel once in an hour, to shew the minutes; and the swing-wheel once in a minute, to shew the seconds. See CLOCK-WORK.

Ballistic PENDULUM. See BALLISTIC Pendulum.
Level PENDULUM. See LEVEL.
PENDULUM Watch. See WATCH.

PENETRABILITY, capability of being penetrated. See IMPENETRABILITY.

PENETRATION, the act by which one thing enters another, or takes up the place already possessed by another.

The schoolmen define Penetration the co-existence of two or more bodies, so that one is present, or has its extension in the same place as the other.

Most philosophers hold the penetration of bodies absurd, i. e. that two bodies should be at the same time in the same place; and accordingly impenetrability is laid down as one of the essential properties of matter.

What is popularly called Penetration, only amounts to the matter of one body's being admitted into the vacuity of another. Such is the Penetration of water through the substance of gold.

PENINSULA, is a portion or extent of land which is almost surrounded with water, being joined to the continent only by an isthmus, or narrow neck. Such is Africa, the greatest Peninsula in the world, which is joined to Asia, by the neck at the end of the Red Sea; such also is Peloponnesus, or the Morea, joined to Greece: and Jutland, &c. Peninsula is the same with what is otherwise called Chersonesus.

PENNY, formerly a piece of silver coin, but now an imaginary sum, equal to two copper coins called a halfpenny.

The Penny was the first silver coin struck in England by our Saxon ancestors, being the 240th part of their pound, and its true weight was about 22½ grains Troy.

In Etheldred's time, the Penny was the 20th part of the Troy ounce, and equal in weight to our three pence; which value it retained till the time of Edward the Third.

Till the time of King Edward the First, the Penny was struck with a cross so deeply sunk in it, that it might, on occasion, be easily broken, and parted into two halves, thence called Halfpennies; or into four, thence called Fourthings, or Farthings. But that Prince coined it without the cross; instead of which he struck round Halfpence and Farthings. Though there are said to be instances of such round Halfpence having been made in the reign of Henry the First, if not also in that of the two Williams.

Edward the First also reduced the weight of the Penny to a standard; ordering that it should weigh 32 grains of wheat, taken out of the middle of the ear. This Penny was called the Penny Sterling; and 20 of them were to weigh an ounce; whence the Penny became a weight as well as a coin.

By the 9th of Edward the Third, it was diminished to the 26th part of the Troy ounce; by the 2d of Henry the Sixth it was the 32d part; by the 5th of Edward the Fourth, it became the 40th, and also by the 36th of Henry the Eighth, and afterwards, the 45th; but by the 2d of Elizabeth, 60 Pence were coined out of the ounce, and during her reign 62, which last proportion is still observed in our times.

The Penny Sterling is now disused as a coin; and scarce subsists, but as a money of account, containing two copper Halfpence, or the 12th part of a shilling, or the 240th part of a pound.

The French Penny, or Denier, is of two kinds; the Paris Penny, called Denier Parisis; and the Penny of Tours, called Denier Tournois.

The Dutch Penny, called Pennink, or Pening, is a real money, worth about one-fifth more than the French Penny Tournois. The Pennink is also used as a money of account, in keeping books by pounds, florins, and patards; 12 Penninks make the patard, and 20 patards the florin.

At Hamburg, Nuremberg, &c, the Penny or Pfennig of account is equal to the French Penny Tournois. Of these, 8 make the krieuk; and 60 the florin of those cities; also 90 the French crown, or 4s 6d sterling.

PENNY-Weight, a Troy weight, being the 20th part of an ounce, containing 24 grains; each grain weighing a grain of wheat gathered out of the middle of the ear, well dried. The name took its rise from its being actually the weight of one of our ancient silver Pennies. See the foregoing article.

PENTAGON, in Geometry, a plane figure consisting of five angles, and consequently five sides also. If the angles be all equal, it is a regular Pentagon.

It is a remarkable property of the Pentagon, that its side is equal in power to the sides of a hexagon and a decagon inscribed in the same circle; that is, the square of the side of the Pentagon, is equal to both the squares taken together of the sides of the other two figures; and consequently those three sides will consti-

6 tute

tute a right-angled triangle. Euclid, book 13, prop. 10.

Pappus has also demonstrated, that 12 regular Pentagons contain more than 20 triangles inscribed in the same circle; lib. 5, prop. 45.

The dodecahedron, which is the fourth regular body or solid, is contained under 12 equal and regular Pentagons.

To find the Area of a Regular PENTAGON. Multiply the square of its side by 1.7204774, or by $\frac{5}{4}$ of the tangent of 54°, or by $\frac{5}{4}\sqrt{1 + \frac{2}{5}\sqrt{5}}$. Hence if s denote the side of the Pentagon, its area will be $1.7204774 s^2 = \frac{5}{4}s^2 \sqrt{1 + \frac{2}{5}\sqrt{5}} = \frac{5}{4}s^2 \times \text{tang. } 54^\circ$.

PENTAGRAPH, otherwise called a Parallelogram, a mathematical instrument for copying designs, prints, plans, &c, in any proportion.

The common Pentagraph (Plate xix, fig. 2) consists of four rulers or bars, of metal or wood, two of them from 15 to 18 inches long, the other two half that length. At the ends, and in the middle, of the long rulers, as also at the ends of the shorter ones, are holes upon the exact fixing of which the perfection of the instrument chiefly depends. Those in the middle of the long rulers are to be at the same distance from those at the end of the long ones, and those of the short ones; so that, when put together, they may always make a parallelogram.

The instrument is fitted together for use, by several little pieces, particularly a little pillar, number 1, having at one end a nut and screw, joining the two long rulers together; and at the other end a small knot for the instrument to slide on. The piece numb. 2 is a rivet with a screw and nut by which each short ruler is fastened to the middle of each long one. The piece numb. 3 is a pillar, one end of which, being hollowed into a screw, has a nut fitted to it; and at the other end is a worm to screw into the table; when the instrument is to be used, it joins the ends of the two short rulers. The piece numb. 4 is a pen, or pencil, or portcrayon, screwed into a little pillar. Lastly, the piece numb. 5 is a brass point, moderately blunt, screwed likewise into a little pillar.

Use of the PENTAGRAPH.—1. To copy a design in the same size or scale as the original. Screw the worm numb. 3 into the table; lay a paper under the pencil numb. 4, and the design under the point numb. 5. This done, conducting the point over the several lines and parts of the design, the pencil will draw or repeat the same on the paper.

2. When the design is to be reduced — ex. gr. to half the scale; the worm must be placed at the end of the long ruler numb. 4, and the paper and pencil in the middle. In this situation conduct the brass point over the several lines of the design, as before; and the pencil at the same time will draw its copy in the proportion required; the pencil here only moving half the lengths that the point moves.

3. On the contrary, when the design is to be enlarged to a double size; the brass point, with the design, must be placed in the middle at numb. 3, the pencil and paper at the end of the long ruler, and the worm at the other end.

4. To reduce or enlarge in other proportions, there are holes drilled at equal distances on each ruler; viz, all along the short ones, and half way of the long ones, for placing the brass point, pencil, and worm, in a right line in them; i. e. if the piece carrying the point be put in the third hole, the other two pieces must be put each in its third hole; &c.

PENTANGLE, a plane figure of five angles, or the same as the PENTAGON.

PENUMBRA, in Astronomy, a faint or partial shade, in an eclipse, observed between the perfect shadow, and the full light.

The Penumbra arises from the magnitude of the sun's body: were he only a luminous point, the shadow would be all perfect; but by reason of the diameter of the sun it happens, that a place which is not illuminated by the whole body of the sun, does yet receive rays from some part of it.

Thus, suppose S the sun, and T the moon, and the shadow of the latter projected on a plane, as GH (Plate xix, fig. 3). The true proper shadow of T, viz GH, will be encompassed with an imperfect shadow, or Penumbra, HL and GE, each portion of which is illuminated by an entire hemisphere of the sun.

The degree of light or shade of the Penumbra, will be more or less in different parts, as those parts lie open to the rays of a greater or less part of the sun's body; thus from L to H, and from E to G, the light continually diminishes; and in the confines of G and H, the Penumbra is darkest, and becomes lost and confounded with the total shade: as near E and L it is thin and confounded with the total light.

A Penumbra must be found in all eclipses, whether of the sun, the moon, or the other planets, primary or secondary; but it is most considerable with us in eclipses of the sun; which is the case here referred to.

The Penumbra extends infinitely in length, and grows still wider and wider; two rays drawn from the two extremities of the earth's diameter, and which proceed always diverging, form its two edges; all that infinite diverging space, included between lines passing through E and L, is the Penumbra, except the cone of the shadow in the middle of it.

To determine how much of the surface of the earth can be involved in the Penumbra, let the apparent semidiameter of the sun be supposed the greatest, or about $16' 20''$, which is when the earth is in her perihelion; also let the moon be in her apogee, and therefore at her greatest distance from the earth, or about 64 of the earth's semidiameters. Let KNC be the earth, T the moon, and MKN the Penumbra, involving the part of the earth from K to

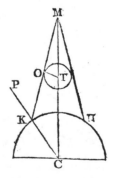

N, which it is required to find. Here then are given the angle $KMC = 16' 20''$, $TC = 64$, $KC = 1$, and $OT = \frac{11}{40}$ of KC. Hence, in the right-angled triangle OTM, as sin. OMT : radius :: OT : TM $= 210\frac{1}{2}$ OT $= 58$ KC nearly. Therefore MC $=$ MT + TC $= 58 + 64 = 122$ semidiameters of the earth. Then, in the triangle KMC, there are given KC $=$

$KC = 1$, and $MC = 122$, alfo the angle $KMC = 16' 20''$, to find the angle C; thus, as

$KC : MC :: \text{fin.} \angle KMC : \text{fin.} \angle MKP = 35° 25' 35''$; from this take the $\angle KMC$ - - 0 16 20, leaves the $\angle C$ - - - 35 9 11, the double of which is the arc KN 70 18 22, or nearly a fpace of 4866 miles in diameter.

PERAMBULATOR, an inftrument for meafuring diftances; called alfo Pedometer, Waywifer, and Surveying Wheel.

This wheel is contrived to meafure out a pole, or $16\frac{1}{2}$ feet, in making two revolutions; confequently its circumference is $8\frac{1}{4}$ feet, and its diameter $2\cdot 626$ feet, or 2 feet $5\frac{1}{4}$ inches and $\frac{12}{1000}$ parts, very nearly. It is either driven forward by two handles, by a perfon walking; or is drawn by a coach wheel, &c, to which it is attached by a pole. It contains various movements, by wheels, or clock-work, with indices on its face, which is like that of a clock, to point out the diftance paffed over, in miles, furlongs, poles, yards, &c.

Its advantages are its readinefs and expedition; being very ufeful for meafuring roads, and great diftances on level ground. See the fig. Plate xvii, fig. 6.

PERCH, in Surveying, a fquare meafure, being the 40th part of a rood, or the 160th part of an acre; that is, the fquare of a pole or rod, of the length of $5\frac{1}{2}$ yards, or $16\frac{1}{2}$ feet.

PERCH is by fome alfo made to mean a meafure of length; being the fame as the rod or pole of $5\frac{1}{2}$ yards or $16\frac{1}{2}$ feet long. But it is better, for preventing confufion, to diftinguifh them.

PERCUSSION, in Phyfics, the impreffion a body makes in falling or ftriking upon another; or the fhock or collifion of two bodies, which meeting alter each other's motion.

Percuffion is either Direct or Oblique. It is alfo either of Elaftic or Nonelaftic bodies, which have each their different laws. It is true, we know of no bodies in nature that are either perfectly elaftic or the contrary; but all partaking that property in different degrees; even the hardeft and the fofteft being not entirely divefted of it. But, for the fake of perfpicuity, it is ufual, and proper, to treat of thefe two feparately and apart.

Direct PERCUSSION is that in which the impulfe is made in the direction of a line perpendicular at the place of impact, and which alfo paffes through the common centre of gravity of the two ftriking bodies. As is the cafe in two fpheres, when the line of the direction of the ftroke paffes through the centres of both fpheres; for then the fame line, joining their centres, paffes perpendicularly through the point of impact. And

Oblique PERCUSSION, is that in which the impulfe is made in the direction of a line that does not pafs through the common centre of gravity of the ftriking bodies; whether that line of direction is perpendicular to the place of impact, or not.

The force of Percuffion is the fame as the momentum, or quantity of motion, and is reprefented by the product arifing from the mafs or quantity of matter moved, multiplied by the velocity of its motion; and that without any regard to the time or duration of action; for its action is confidered totally independent of time, or but as for an inftant, or an infinitely fmall time.

This confideration will enable us to refolve a queftion that has been greatly canvaffed among philofophers and mathematicians, viz, what is the relation between the force of Percuffion and mere preffure or weight? For we hence infer, that the former force is infinitely, or incomparably, greater than the latter. For, let M denote any mafs, body, or weight, having no motion or velocity, but fimply its preffure; then will that preffure or force be denoted by M itfelf, if it be confidered as acting for fome certain finite affignable time; but, confidered as a force of Percuffion, that is, as acting but for an infinitely fmall time, its velocity being 0, or nothing, its percuffive force will be $0 \times M$, that is 0, or nothing; and is therefore lefs than any the fmalleft percuffive force whatever. Again, let us confider the two forces, viz, of Percuffion and preffure, with refpect to the effects they produce: Now the intenfity of any force is very well meafured and eftimated by the effect it produces in a given time: But the effect of the preffure M, in 0 time, or an infinitely fmall time, is nothing at all; that is, it will not, in an infinitely fmall time, produce, for example, any motion, either in itfelf, or in any other body: its intenfity therefore, as its effect, is infinitely lefs than any the fmalleft force of Percuffion. It is true, indeed, that we fee motion and other confiderable effects produced by mere preffure, and to counteract which it will require the oppofition of fome confiderable percuffive force: but then it muft be obferved, that the former has been an infinitely longer time than the latter in producing its effect; and it is no wonder in mathematics that an infinite number of infinitely fmall quantities makes up a finite one. It has therefore only been for want of confidering the circumftance of *time*, that any queftion could have arifen on this head. Hence the two forces are related to each other, only as a furface is to a folid or body: by the motion of the furface through an infinite number of points, or through a finite right line, a folid or body is generated: and by the action of the preffure for an infinite number of moments, or for fome finite time, a quantity equal to a given percuffive force is generated: but the furface itfelf is infinitely lefs than any folid, and the preffure infinitely lefs than any percuffive force. This point may be eafily illuftrated by fome familiar inftances, which prove at leaft the enormous difproportion between the two forces, if not alfo their abfolute incomparability. And firft, the blow of a fmall hammer, upon the head of a nail, will drive the nail into a board; when it is hard to conceive any weight fo great as will produce a like effect, i. e. that will fink the nail as far into the board, at leaft unlefs it is left to act for a very confiderable time: and even after the greateft weight has been laid as a preffure on the head of the nail, and has funk it as far as it can as to fenfe, by remaining for a long time there without producing any farther fenfible effect; let the weight be removed from the head of the nail, and inftead of it, let it be ftruck a fmall blow with a hammer, and the nail will immediately fink farther into the wood. Again, it is alfo well known, that a fhip-carpenter, with a blow of his mallet, will drive a wedge in below the greateft fhip whatever, lying aground, and fo overcome her weight, and lift her up. Laftly, let us confider a man with a club to ftrike a fmall ball, upwards or in

7

any

any other direction; it is evident that the ball will acquire a certain determinate velocity by the blow, suppose that of 10 feet per second, or minute, or any other time whatever: now it is a law, universally allowed in the communication of motion, that when different bodies are struck with equal forces, the velocities communicated are reciprocally as the weights of the bodies that are struck; that is, that a double body, or weight, will acquire half the velocity from an equal blow; a body 10 times as great, one 10th of the velocity; a body 100 times as great, the 100th part of the velocity; a body a million times as great, the millionth part of the velocity; and so on without end: from whence it follows, that there is no body or weight, how great soever, but will acquire some finite degree of velocity, and be overcome, by any given small finite blow, or Percussion.

It appears that Des Cartes, first of any, had some ideas of the laws of Percussion; though it must be acknowledged, in some cases perhaps wide of the truth. The first who gave the true laws of motion in nonelastic bodies, was Doctor Wallis, in the Philos. Transf. numb. 43, where he also shews the true cause of reflections in other bodies, and proves that they proceed from their elasticity. Not long after, the celebrated Sir Christopher Wren and Mr. Huygens imparted to the Royal Society the laws that are observed by perfectly elastic bodies, and gave exactly the same construction, though each was ignorant of what the other had done. And all those laws, thus published in the Philos. Transf. without demonstration, were afterwards demonstrated by Dr. Keill, in his Philos. Lect. in 1700; and they have since been followed by a multitude of other authors.

In Percussion, we distinguish at least three several sorts of bodies; the perfectly hard, the perfectly soft, and the perfectly elastic. The two former are considered as utterly void of elasticity; having no force to separate them, or throw them off from each other again, after collision; and therefore either remaining at rest, or else proceeding uniformly forward together as one body or mass of matter.

The laws of Percussion therefore to be considered, are of two kinds: those for elastic, and those for nonelastic bodies.

The one only general principle, for determining the motions of bodies from Percussion, and which belongs equally to both the sorts of bodies, i.e. both the elastic and nonelastic. is this: viz, that there exists in the bodies the same momentum, or quantity of motion, estimated in any one and the same direction, both before the stroke and after it. And this principle is the immediate result of the third law of nature or motion, that reaction is equal to action, and in a contrary direction; from whence it happens, that whatever motion is communicated to one body by the action of another, exactly the same motion doth this latter lose in the same direction, or exactly the same does the former communicate to the latter in the contrary direction.

From this general principle too it results, that no alteration takes place in the common centre of gravity of bodies by their actions upon one another; but that the said common centre of gravity perseveres in the same state, whether of rest or of uniform motion, both before and after the shock of the bodies.

Now, from either of these two laws, viz, that of the preservation of the same quantity of motion, in one and the same direction, and that of the preservation of the same state of the centre of gravity, both before and after the shock, all the circumstances of the motions of both the kinds of bodies after collision may be made out; in conjunction with their own peculiar and separate constitutions, namely, that of the one sort being elastic, and the other nonelastic.

The effects of these different constitutions, here alluded to, are these; that nonelastic bodies, on their shock, will adhere together, and either remain at rest, or else move together as one mass with a common velocity; or if elastic, they will separate after the shock with the very same relative velocity with which they met and shocked. The former of these consequences is evident, viz, that nonelastic bodies keep together as one mass after they meet; because there exists no power to separate them; and without a cause there can be no effect. And the latter consequence results immediately from the very definition and essence of elasticity itself, being a power always equal to the force of compression, or shock; and which restoring force therefore, acting the contrary way, will generate the same relative velocity between the bodies, or the same quantity of matter, as before the shock, and the same motion also of their common centre of gravity.

A   B   C   b   D

To apply now the general principle to the determination of the motions of bodies after their shock; let B and b be any two bodies, and V and v their respective velocities, estimated in the direction AD; which quantities V and v will be both positive if the bodies both move towards D, but one of them as v will be negative if the body b move towards A, and v will be = 0 if the body b be at rest. Hence then BV is the momentum of B towards D, and bv is the momentum of b towards D, whose sum is BV + bv, which is the whole quantity of motion in the direction AD, and which momentum must also be preserved after the shock.

Now if the bodies have no elasticity, they will move together as one mass B + b after they meet, with some common velocity, which call y, in the direction AD; therefore the momentum in that direction after the shock, being the product of the mass and velocity, will be (B + b) × y. But the momenta, in the same direction, before and after the impact, are equal, that is BV + bv = (B + b) y; from which equation any one of the quantities may be determined, when the rest are given. So, if we would find the common velocity after the stroke, it will be $y = \frac{BV + bv}{B + b}$, equal to the sum of the momenta divided by the sum of the bodies; which is also equal to the velocity of the common centre of gravity of the two bodies, both before and after the collision. The signs of the terms, in this value of y, will be all positive, as above.

above, when the bodies move both the same way AD; but one term bv must be made negative when the motion of b is the contrary way; and that term will be absent or nothing, when b is at rest, before the shock.

Again, for the case of elastic bodies, which will separate after the stroke, with certain velocities, x and z, viz, x the velocity of B, and z the velocity of b after the collision, both estimated in the direction AD, which quantities will be either positive, or negative, or nothing, according to the circumstances of the masses B and b, with those of their celerities before the stroke. Hence then Bx and bz are the separate momenta after the shock, and $Bx + bz$ their sum, which must be equal to the sum $BV + bv$ in the same direction before the stroke: also $z - x$ is the relative velocity with which the bodies separate after the blow, and which must be equal to $V - v$ the same with which they meet; or, which is the same thing, that $V + x = v + z$; that is, the sum of the two velocities of the one body, is equal to the sum of the velocities of the other, taken before and after the stroke; which is another notable theorem. Hence then, for determining the two unknown quantities x and z, there are these two equations, or

$$viz, \quad BV + bv = Bx + bz,$$
$$and \quad V - v = z - x;$$
$$or \quad V + x = v + z;$$

the resolution of which equations gives those two velocities as below,

$$viz, \quad x = \frac{2bv + (B - b)V}{B + b},$$
$$and \quad z = \frac{2BV - (B - b)v}{B + b}.$$

From these general values of the velocities, which are to be understood in the direction AD, any particular cases may easily be drawn. As, if the two bodies B and b be equal, then $B - b = 0$, and $B + b = 2B$, and the two velocities in that case become, after impulse, $x = v$, and $z = V$, the very same as they were before, but changed to the contrary bodies, i. e. the bodies have taken each other's velocity that it had before, and with the same sign also. So that, if the equal bodies were before both moving the same way, or towards D, they will do the same after, but with interchanged velocities. But if they before moved contrary ways, B towards D, and b towards A, they will rebound contrary ways, B back towards A, and b towards D, each with the other's velocity. And, lastly, if one body, as b, were at rest before the stroke, then the other B will be at rest after it, and b will go on with the motion that B had before. And thus may any other particular cases be deduced from the first general values of and z.

We may now conclude this article with some remarks on these motions, and the mistakes of some authors concerning them. And first, we observe this striking difference between the motions that are communicated by elastic and by nonelastic bodies, viz, that a nonelastic body, by striking, communicates to the body it strikes, exactly its whole momentum; as is evident. But the stroke of an elastic body may either communicate its whole motion to the body it strikes, or it may communicate only a part of it, or it may even communicate more than it had. For, if the striking body remain at rest after the stroke, it has

just lost all its motion, and therefore has communicated all it had; but if it still move forward in the same direction, it has still some motion left in that direction, and therefore has only communicated a part of what motion it had; and if the striking body rebound back, and move in the contrary direction, the other body has received not only the whole of the motion that the first had, but also as much more as the first has acquired in the contrary direction.

It has been denied by some authors, and in the Encyclopédie, that the same quantity of motion remains after the shock, as before it; and hence they seize an opportunity to reprehend the Cartesians for making that assertion, which they do, not only with respect to the case of two bodies, but also of all the bodies in the whole universe. And yet nothing is more true, if the motion be considered as estimated always in one and the same direction, esteeming that as negative, which is in the contrary or opposite direction. For it is a general law of nature, that no motion, nor force, can be generated, nor destroyed, nor changed, but by some cause which must produce an equal quantity in the opposite direction. And this being the case in one body, or two bodies, it must necessarily be the case in all bodies, and in the whole solar system, since all bodies act upon one another. And hence also it is manifest, that the common centre of gravity of the whole solar system must always preserve its original condition, whether it be of rest or of uniform motion; since the state of that centre is not changed by the mutual actions of bodies upon one another, any more than their quantity of motion, in one and the same direction.

What may have led authors into the mistake above alluded to, which they bring no proof of, seems to be the discovery of M. Huygens, that the sums of the two products are equal, both before and after the shock, that are made by multiplying each body by the square of its velocity, viz, that $BV^2 + bv^2 = Bx^2 + bz^2$, where V and v are the velocities before the shock, and x and z the velocities after it. Such an expression, namely the product of the mass by the square of the velocity, is called the vis viva, or living force; and hence it has been inferred that the whole vis viva before the shock, or $BV^2 + bv^2$, is equal to that after the stroke, or $Bx^2 + bz^2$; which is indeed very true, as will be shewn presently. But when they hence infer, both that therefore the forces of bodies in motion are as the squares of the velocities, and that there is not the same quantity of motion between the two striking bodies, both before and after the shock, they are grossly mistaken, and thereby shew that they are ignorant of the true derivation of the equation $BV^2 + bv^2 = Bx^2 + bz^2$. For this equation is only a consequence of the very principle above laid down, and which is not acceded to by those authors, viz, that the quantity of motion is the same before and after the shock, or that $BV + bv = Bx + bz$, the truth of which last equation they deny, because they think the former one is true, never dreaming that they may be both true, and much less that the one is a consequence of the other, and derived from it; which however is now found to be the case, as is proved in this manner:

It has been shewn that the sum of the two momenta,

in

in the same direction, before and after the ſtroke, are equal, or that $BV + bv = Bx + bz$; and alſo that the ſum of the two velocities of the one body, is equal to the ſum of thoſe of the other, or that $V + x = v + z$; and it is now propoſed to ſhew that from theſe two equations there reſults the third equation $BV^2 + bv^2 = Bx^2 + bz^2$, or the equation of the living forces.

Now becauſe $BV + bx = Bx + bz$, by tranſpoſition it is $BV - Bz = bz - bv$; which ſhews that the difference between the two momenta of the one body, before and after the ſtroke, is equal to the difference between thoſe of the other body; which is another notable theorem. But now, to derive the equation of the vis viva, ſet down the two foregoing equations, and multiply them together, ſo ſhall the products give the ſaid equation required; thus

Mult. $BV - Bz = bz - bv$, the equat. of the momenta, by $V + x = z + v$, the equat. of the velocities,

produc. $\overline{BV^2 - Bx^2} = \overline{bz^2 - bv^2}$,

or $BV^2 + bv^2 = Bx^2 + bz^2$,

the very equation of the vis viva required. Which was to be proved.

When the elaſticity of the bodies is not perfect, but only partially ſo, as is the caſe with all the bodies we know of, the determination of the motions after colliſion may be determined in a ſimilar manner. See Keill's Lect. Philoſ. lect. 14, theor. 29, at the end. And for the geometrical determinations after impact, ſee the article COLLISION.

Centre of PERCUSSION, is the point in which the ſhock or impulſe of a body which ſtrikes another is the greateſt that it can be. See CENTRE.

The Centre of Percuſſion is the ſame as the centre of oſcillation, when the ſtriking body moves round a fixed axis. See OSCILLATION.

But if all the parts of the ſtriking body move with a parallel motion, and with the ſame velocity, then the Centre of Percuſſion is the ſame as the centre of gravity.

PERFECT NUMBER, is one that is equal to the ſum of all its aliquot parts, when added together. Eucl. lib. 7, def. 22. As the number 6, which is = $1 + 2 + 3$, the ſum of all its aliquot parts; alſo 28, for $28 = 1 + 2 + 4 + 7 + 14$, the ſum of all its aliquot parts.

It is proved by Euclid, in the laſt prop. of book the 9th, that if the common geometrical ſeries of numbers 1, 2, 4, 8, 16, 32, &c, be continued to ſuch a number of terms, as that the ſum of the ſaid ſeries of terms ſhall be a prime number, then the product of this ſum by the laſt term of the ſeries will be a perfect number.

This ſame rule may be otherwiſe expreſſed thus: If n denote the number of terms in the given ſeries 1, 2, 4, 8, &c; then it is well known that the ſum of all the terms of the ſeries is $2^n - 1$, and it is evident that the laſt term is 2^{n-1}: conſequently the rule becomes thus, viz. $2^{n-1} \times \overline{2^n - 1}$ = a perfect number, whenever $2^n - 1$ is a prime number.

Now the ſums of one, two, three, four, &c, terms of the ſeries 1, 2, 4, 8, &c, form the ſeries 1, 3, 7, 15, 31, &c; ſo that the number will be found perfect

whenever the correſponding term of this ſeries is a prime, as 1, 3, 7, 31, &c. Whence the table of perfect numbers may be found and exhibited as follows; where the 1ſt column ſhews the number of terms, or the value of n; the 2d column is the laſt term of the ſeries 1, 2, 4, 8, &c, and is expreſſed by 2^{n-1}; the 3d column contains the correſponding ſums of the ſaid ſeries, or the values of the quantity $2^n - 1$; which numbers in this 3d column are eaſily conſtructed by adding always the laſt number in this column to the next following number in the 2d column: and laſtly, the 4th column ſhews the correſpondent Perfect Numbers, or the values of $2^{n-1} \times \overline{2^n - 1}$, the product of the numbers in the 2d and 3d columns, when $2^n - 1$, or the number in the 3d column, is a prime number; the products in the other caſes being omitted, as not Perfect Numbers.

Values of n	Values of 2^{n-1}	Values of $2^n - 1$	Perf. Numbers, or $2^{n-1} \times (2^n - 1)$
1	1	1	1
2	2	3	6
3	4	7	28
4	8	15	.
5	16	31	496
6	32	63	.
7	64	127	8128

Hence the firſt four Perfect Numbers are found to be 6, 28, 496, 8128; and thus the table might be continued to find others, but the trouble would be very great, for want of a general method to diſtinguiſh which numbers are primes, as the caſe requires. Several learned mathematicians have endeavoured to facilitate this buſineſs, but hitherto with only a ſmall degree of perfection. After the foregoing four Perfect Numbers, there is a long interval before any more occur. The firſt eight are as follow, with the factors and products which produce them:

The firſt Perfect Numbers.		Their values.
6	- -	$= (2^2 - 1)\, 2$
28	- -	$= (2^3 - 1)\, 2^2$
496	- -	$= (2^5 - 1)\, 2^4$
8128	- -	$= (2^7 - 1)\, 2^6$
33550336	- -	$= (2^{13} - 1)\, 2^{12}$
8589869056	- -	$= (2^{17} - 1)\, 2^{16}$
137438691328	- -	$= (2^{19} - 1)\, 2^{18}$
2305843008139952128	- -	$= (2^{31} - 1)\, 2^{30}$

See ſeveral conſiderable tracts on the ſubject of Perfect Numbers in the Memoirs of the Peterſburgh Academy, vol. 2 of the new vols, and in ſeveral other volumes.

PERIÆCI. See PERIOECI.

PERIGÆUM, or PERIGEE, is that point of the orbit of the ſun or moon, which is the neareſt to the earth. In which ſenſe it ſtands oppoſed to Apogee, which is the moſt diſtant point from the earth.

PERIGEE.

PERIGEE, in the Ancient Aſtronomy, denotes a point in a planet's orbit, where the centre of its epicycle is at the leaſt diſtance from the earth.

PERIHELION, PERIHELIUM, that point in the orbit of a planet or comet which is neareſt to the ſun. In which ſenſe it ſtands oppoſed to Aphelion, or Aphelium, which is the higheſt or moſt diſtant point from the ſun.

Inſtead of this term, the Ancients uſed Perigeum; becauſe they placed the earth in the centre.

PERIMETER, in Geometry, the ambit, limit, or outer bounds of a figure; being the ſum of all the lines by which it is incloſed or formed.

In circular figures, &c, inſtead of this term, the word circumference or periphery is uſed.

PERIOD, in Aſtronomy, the time in which a ſtar or planet makes one revolution, or returns again to the ſame point in the heavens.

The ſun's, or properly the earth's tropical period, is 365 days 5 hours 48 minutes 45 ſeconds 30 thirds. That of the moon is 27 days 7 hours 43 minutes. That of the other planets as below.

There is a wonderful harmony between the diſtances of the planets from the ſun, and their Periods round him; the great law of which is, that the ſquares of the Periodic times are always proportional to the cubes of their mean diſtances from the ſun.

The Periods, both tropical and ſydereal, with the proportions of the mean diſtances of the ſeveral planets are as follow:

Planets	Tropical Periods	Sydereal Periods	Proport. Diſts.
Mercury	87ᵈ 23ʰ 14′	87ᵈ 23ʰ 16′	36710
Venus	224 16 42	224 16 49	72333
Earth	365 5 49	365 6 9	100000
Mars	686 22 18	686 23 31	152369
Jupiter	4330 8 58	4332 8 51	520110
Saturn	10749 7 22	10761 14 37	953800
Georgian or Herſchel	30456 1 41		1908180

As to the comets, the Periods of very few of them are known. There is one however of between 75 and 76 years, which appeared for the laſt time in 1759; another was ſuppoſed to have its Period of 129 years, which was expected to appear in 1789 or 1790, but it did not; and the comet which appeared in 1680 it is thought has its Period of 575 years.

PERIOD, in Chronology, denotes an epoch, or interval of time, by which the years are reckoned; or a ſeries of years by which time is meaſured, in different nations. Such are the Calippic and Metonic Periods, two different corrections of the Greek calendar, the Julian Period, invented by Joſeph Scaliger; the Victorian Period, &c.

Calippic PERIOD. See CALIPPIC *Period*.

Conſtantinopolitan PERIOD, is that uſed by the Greeks, and is the ſame as the *Julian* PERIOD, which ſee.

Chaldaic PERIOD. See SAROS.

Dionyſian PERIOD. See *Victorian* PERIOD.

Hipparchus's PERIOD, is a ſeries or cycle of 304 ſolar years, returning in a conſtant round, and reſtoring the new and full moons to the ſame day of the ſolar year; as Hipparchus thought.

This Period ariſes by multiplying the Calippic Period by 4. Hipparchus aſſumed the quantity of the ſolar year to be 365d. 5h. 55m. 12 ſec. and hence he concluded, that in 304 years Calippus's Period would err a whole day. He therefore multiplied the Period by 4, and from the product caſt away an entire day. But even this does not reſtore the new and full moons to the ſame day throughout the whole Period: but they are ſometimes anticipated 1d. 8h. 23 m. 29 ſec. 20 thirds.

Julian PERIOD, ſo called as being adapted to the Julian year, is a ſeries of 7980 Julian years; ariſing from the multiplications of the cycles of the ſun, moon, and indiction together, or the numbers 28, 19, 15; commencing on the 1ſt day of January in the 764th Julian year before the creation, and therefore is not yet completed. This comprehends all other cycles, Periods and epochs, with the times of all memorable actions and hiſtories; and therefore it is not only the moſt general, but the moſt uſeful of all Periods in Chronology.

As every year of the Julian Period has its particular ſolar, lunar, and indiction cycles, and no two years in it can have all theſe three cycles the ſame, every year of this Period becomes accurately diſtinguiſhed from another.

This Period was invented by Joſeph Scaliger, as containing all the other epochs, to facilitate the reduction of the years of one given epoch to thoſe of another. It agrees with the Conſtantinopolitan Period, uſed by the Greeks, except in this, that the cycles of the ſun, moon, and indiction, are reckoned differently; and alſo in that the firſt year of the Conſtantinopolitan Period differs from that of the Julian Period.

To find the year anſwering to any given year of the Julian Period, and vice verſa; ſee EPOCH.

Metonic PERIOD. See CYCLE *of the Moon*.

Victorian PERIOD, an interval of 532 Julian years; at the end of which, the new and full moons return again on the ſame day of the Julian year, according to the opinion of the inventor Victorinus, or Victorius, who lived in the time of pope Hilary.

Some aſcribe this Period to Dionyſius Exiguus, and hence they call it the Dionyſian Period: others again call it the Great Paſchal Cycle, becauſe it was invented for computing the time of Eaſter.

The Victorian Period is produced by multiplying the ſolar cycle 28 by the lunar cycle 19, the product being 532. But neither does this reſtore the new and full moons to the ſame day throughout its whole duration, by 1d. 16h. 58m 59s. 40 thirds.

PERIOD, in Arithmetic, is a diſtinction made by a point, or a comma, after every 6th place, or figure; and is uſed in numeration, for the readier diſtinguiſhing and naming the ſeveral figures or places, which are thus diſtinguiſhed into Periods of ſix figures each. See NUMERATION.

PERIOD is alſo uſed in Arithmetic, in the extraction of

of roots, to point off, or separate the figures of the given number into Periods, or parcels, of as many figures each as are expressed by the degree of the root to be extracted, viz, of two places each for the square root, three places for the cube root, and so on.

PERIODIC, or PERIODICAL, appertaining to Period, or going by periods. Thus, the Periodical motion of the moon, is that of her monthly period or course about the earth, called her Periodical month, containing 27 days 7 hours 45 minutes.

PERIODICAL *Month*. See MONTH.

PERIŒCI, or PERIOECIANS, in Geography, are such as live in opposite points of the same parallel of latitude. Hence they have the same seasons at the same time, with the same phenomena of the heavenly bodies; but their times of the day are opposite, or differ by 12 hours, being noon with the one when it is midnight with the other.

PERIPATETIC *Philosophy*, the system of philosophy taught and established by Aristotle, and maintained by his followers, the Peripatetics. See ARISTOTLE.

PERIPATETICS, the followers of Aristotle. Though some derive their establishment from Plato himself, the master of both Xenocrates and Aristotle.

PERIPHERY, in Geometry, is the circumference, or bounding line, of a circle, ellipse, or other regular curvilineal figure. See CIRCUMFERENCE, and CIRCLE.

PERISCII, or PERISCIANS, those inhabitants of the earth, whose shadows do, in one and the same day, turn quite round to all the points of the compass, without disappearing.

Such are the inhabitants of the two frozen zones, or who live within the compass of the arctic and antarctic circles; for, as the sun never sets to them, after he is once up, but moves quite round about, so do their shadows also.

PERISTYLE, in the ancient Architecture, a place or building encompassed with a row of columns on the inside; by which it is distinguished from the periptere, where the columns are disposed on the outside.

PERISTYLE is also used, by modern writers, for a range of columns, either within or without a building.

PERITROCHIUM, in Mechanics, is a wheel or circle, concentric with the base of a cylinder, and moveable together with it, about an axis. The axis, with the wheel, and levers fixed in it to move it, make that mechanical power, called AXIS *in Peritrochio*, which see.

PERMUTATIONS *of Quantities*, in Algebra, the ALTERNATIONS, CHANGES, or different COMBINATIONS of any number of things. See those terms.

PERPENDICULAR, in Geometry, or NORMAL. One line is Perpendicular to another, when the former meets the latter so as to make the angles on both sides of it equal to each other. And those angles are called right angles. And hence, to be Perpendicular to, or to make right-angles with, means one and the same

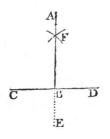

thing. So, when the angle ABC is equal to the angle ABD, the line AB is said to be Perpendicular, or normal, or at right angles to the line CD.

A line is Perpendicular to a curve, when it is perpendicular to the tangent of the curve at the point of contact.

A line is Perpendicular to a plane, when it is Perpendicular to every line drawn in the plane through the bottom of the Perpendicular. And one plane is Perpendicular to another, when a line in the one plane is Perpendicular to the other plane.

From the very principle and motion of a Perpendicular, it follows, 1. That the Perpendicularity is mutual, if the first AB is perpendicular to the second CD, then is the second Perpendicular to the first.—2. That only one Perpendicular can be drawn from one point in the same place.—3. That if a Perpendicular be continued through the line it was drawn Perpendicular to; the continuation BE will also be Perpendicular to the same.—4. That if there be two points, A and E, of a right line, each of which is at an equal distance from two points, C and D, of another right line; those lines are Perpendiculars.—5. That a line which is Perpendicular to another line, is also Perpendicular to all the parallels of the other.—6. That a Perpendicular is the shortest of all those lines which can be drawn from the same point to the same right line. Hence the distance of a point from a line or plane, is a line drawn from the point Perpendicular to the line or plane: and hence also the altitude of a figure is a Perpendicular let fall from the vertex to the base.

To Erect a Perpendicular from a given point in a line. —1. When the given point B is near the middle of the line; with any interval in the compasses take the two equal parts BC, BD: and from the two centres C and D, with any radius greater than BC or BD, strike two arcs intersecting in F; then draw BFA the Perpendicular required.

2. When the given point G is at or near the end of the line; with any centre I and radius IG describe an arc HGK through G; then a ruler laid by H and I will cut the arc in the point K, through which the Perpendicular GK must be drawn.

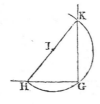

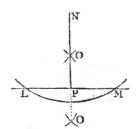

To let fall a Perpendicular upon a given line LM from a given point N. With the centre N, and a convenient radius, describe an arc cutting the given line in L and M; with these two centres, and any other convenient radius, strike

two

two other arcs interfecting in O, the point through which the Perpendicular NOP muſt be drawn.

Note, that Perpendiculars are beſt drawn, in practice, by means of a ſquare, laying one ſide of it along the given line, and the other to paſs through the given point.

PERPENDICULAR, in Gunnery, is a ſmall inſtrument uſed for finding the centre line of a piece, in the operation of pointing it to a given object. See *Pointing of a Gun*.

PERPETUAL *Motion*. See MOTION.

Circle of PERPETUAL *Occultation and Apparition*. See CIRCLE.

PERPETUAL, or *Endleſs Screw*. See SCREW.

PERPETUITY, in the Doctrine of Annuities, is the number of years in which the ſimple intereſt of any principal ſum will amount to the ſame as the principal itſelf. Or it is the quotient ariſing by dividing 100, or any other principal, by its intereſt for one year. Thus, the Perpetuity, at the rate of 5 per cent. intereſt, is $\frac{100}{5} = 20$; at 4 per cent. $\frac{100}{4} = 25$; &c.

PERRY (Captain JOHN), was a celebrated Engliſh engineer. After acquiring great reputation for his ſkill in this country, he reſided many years in Ruſſia, having been recommended to the czar Peter while in England, as a perſon capable of ſerving him on a variety of occaſions relating to his new deſign of eſtabliſhing a fleet, making his rivers navigable, &c. His ſalary in this ſervice was to be 300l. per annum, beſides travelling expences and ſubſiſtence money on whatever ſervice he ſhould be employed, with a farther reward to his ſatisfaction at the concluſion of any work he ſhould finiſh.

After ſome converſation with the czar himſelf, particularly reſpecting a communication between the rivers Volga and Don, he was employed on that work for three ſummers ſucceſſively; but not being well ſupplied with men, partly on account of the ill ſucceſs of Peter's arms againſt the Swedes at the battle of Narva, and partly by the diſcouragement of the governor of Aſtracan, he was ordered at the end of 1707 to ſtop, and next year was employed in refitting the ſhips at Veroniſe, and 1709 in making the river of that rane navigable. But after repeated diſappointments, and a variety of fruitleſs applications for his ſalary, he at length quitted the kingdom, under the protection of Mr. Whitworth, the Engliſh ambaſſador, in 1712. (See his Narrative in the Preface to *The State of Ruſſia*.)

In 1721 he was employed in ſtopping the breach at Dagenham, made in the bank of the river Thames, near the village of that name in Eſſex, and about 3 miles below Woolwich, in which he happily ſucceeded, after ſeveral other perſons had failed in that undertaking. He was alſo employed, the ſame year, about the harbour at Dublin, and publiſhed at that time an Anſwer to the objections made againſt it.—Beſide this piece, Captain Perry was author of, *The State of Ruſſia*, 1716, 8vo; and *An Account of the Stopping of Dagenham Breach*, 1721, 8vo.—He died February the 11th 1733.

PERSEUS, a conſtellation of the northern hemiſphere, being one of the 48 ancient aſteriſms.

The Greeks fabled that this is Perſeus, whom they make the ſon of Jupiter by Danae. The father of that lady had been told, that he ſhould be killed by his grandchild, and having only Danae to take care of, he locked her up ; but Jupiter found his way to her in a ſhower of gold, and Perſeus verified the oracle. He cut off alſo the head of the gorgon, and affixed it to his ſhield ; and after many other great exploits he reſcued Andromeda, the daughter of Caſſiopeia, whom the ſea nymphs, in revenge for that lady's boaſting of ſuperior beauty, had faſtened to a rock to be devoured by a monſter. Jupiter his father in honour of the exploit, they ſay, afterwards took up the hero, and the whole family with him, into the ſkies.

The number of ſtars in this conſtellation, in Ptolomy's catalogue, are 29 ; in Tycho's 29, in Hevelius's 46, and in the Britannic catalogue 59.

PERSIAN *Wheel*, in Mechanics, a machine for raiſing a quantity of water, to ſerve for various purpoſes. Such a wheel is repreſented in plate xx, fig. 1; with which water may be raiſed by means of a ſtream AB turning a wheel CDE, according to the order of the letters, with buckets *a, a, a, a*, &c, hung upon the wheel by ſtrong pins *b, b, b, b*, &c, fixed in the ſide of the rim ; which muſt be made as high as the water is intended to be raiſed above the level of that part of the ſtream in which the wheel is placed. As the wheel turns, the buckets on the right hand go down into the water, where they are filled, and return up full on the left hand, till they come to the top at K ; where they ſtrike againſt the end *n* of the fixed trough M, by which they are overſet, and ſo empty the water into the trough ; from whence it is to be conveyed in pipes to any place it is intended for : and as each bucket gets over the trough, it falls into a perpendicular poſition again, and ſo goes down empty till it comes to the water at A, where it is filled as before. On each bucket is a ſpring *r*, which going over the top or crown of the bar *m* (fixed to the trough M) raiſes the bottom of the bucket above the level of its mouth, and ſo cauſes it to empty all its water into the trough.

Sometimes this wheel is made to raiſe water no higher than its axis ; and then inſtead of buckets hung upon it, its ſpokes C, *d, e, f, g, h*, are made of a bent form, and hollow within ; theſe hollows opening into the holes C, D, E, F, in the outſide of the wheel, and alſo into thoſe at O in the box N upon the axis. So that, as the holes C, D, &c, dip into the water, it runs into them ; and as the wheel turns, the water riſes in the hollow ſpokes, *c, d*, &c, and runs out in a ſtream P from the holes at O, and falls into the trough Q, from whence it is conveyed by pipes.

PERSIAN, or PERSIC, in Architecture, a name common to all ſtatues of men ; ſerving inſtead of columns to ſupport entablatures.

PERSIAN *Era and Year*. See EPOCH and YEAR.

PERSPECTIVE, the art of delineating viſible objects on a plane ſurface, ſuch as they appear at a given diſtance, or height, upon a tranſparent plane, placed commonly perpendicular to the horizon, between the eye and the object. This is particularly called

Linear PERSPECTIVE, as regarding the poſition, magnitude, form, &c, of the ſeveral lines, or contours of objects, and expreſſing their diminution.

Some make this a branch of Optics ; others an art

and

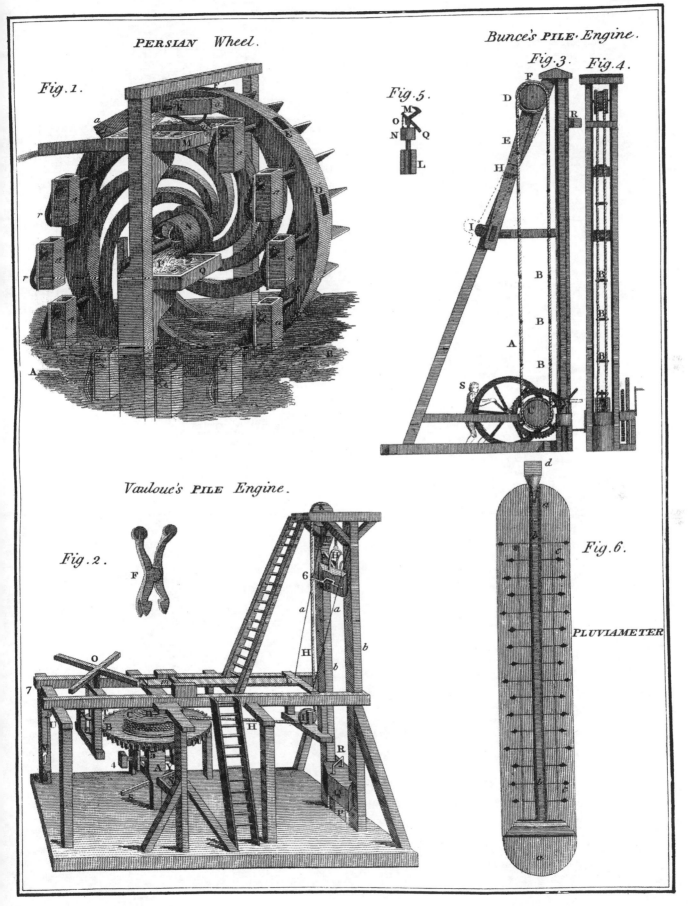

Plate XX.

PERSIAN *Wheel.*

Fig. 1.

Bunce's PILE *Engine.*

Fig. 3.   *Fig. 4.*

Fig. 5.

Vaulouc's PILE *Engine.*

Fig. 2.

Fig. 6.

PLUVIAMETER

and science derived from it : its operations however are all geometrical.

History of PERSPECTIVE. This art derives its origin from painting, and particularly from that branch of it which was employed in the decorations of the theatre, where landscapes were chiefly introduced. Vitruvius, in the proem to his 7th book, says that Agatharchus, at Athens, was the first author who wrote upon this subject, on occasion of a play exhibited by Æschylus, for which he prepared a tragic scene ; and that afterwards the principles of the art were more distinctly taught in the writings of Democritus and Anaxagoras, the disciples of Agatharchus, which are not now extant.

The Perspective of Euclid and of Heliodorus Larisseus contains only some general elements of optics, that are by no means adapted to any particular practice ; though they furnish some materials that might be of service even in the linear Perspective of painters.

Geminus, of Rhodes, a celebrated mathematician, in Cicero's time, also wrote upon this science.

It is also evident that the Roman artists were acquainted with the rules of Perspective, from the account which Pliny (Nat. Hist. lib. 35, cap. 4) gives of the representation on the scene of those plays given by Claudius Pulcher ; by the appearance of which the crows were so deceived, that they endeavoured to settle on the fictitious roofs. However, of the theory of this Art among the Ancients we know nothing; as none of their writings have escaped the general wreck of ancient literature in the dark ages of Europe. Doubtless this art must have been lost, when painting and sculpture no longer existed. However, there is reason to believe that it was practised much later in the Eastern empire.

John Tzetzes, in the 12th century, speaks of it as well acquainted with its importance in painting and statuary. And the Greek painters, who were employed by the Venetians and Florentines, in the 13th century, it seems brought some optical knowledge along with them into Italy : for the disciples of Giotto are commended for observing Perspective more regularly than any of their predecessors in the art had done ; and he lived in the beginning of the 14th century.

The Arabians were not ignorant of this art ; as may be presumed from the optical writings of Alhazen, about the year 1100. And Vitellus, a Pole, about the year 1270, wrote largely and learnedly on optics. And, of our own nation, friar Bacon, as well as John Peckham, archbishop of Canterbury, treated this subject with surprising accuracy; considering the times in which they lived.

The first authors who professedly laid down rules of Perspective, were Bartolomeo Bramantino, of Milan, whose book, Regole di Perspectiva, e Misure delle Antichita di Lombardia, is dated 1440 ; and Pietro del Borgo, likewise an Italian, who was the most ancient author met with by Ignatius Danti, and who it is supposed died in 1443. This last writer supposed objects placed beyond a transparent tablet, and so to trace the images, which rays of light, emitted from them, would make upon it. And Albert Durer constructed a machine upon the principles of Borgo, by which he could trace the Perspective appearance of objects.

Leon Battista Alberti, in 1450, wrote his treatise De Pictura, in which he treats chiefly of Perspective.

Balthazar Peruzzi, of Siena, who died in 1536, had diligently studied the writings of Borgo ; and his method of Perspective was published by Serlio in 1540. To him it is said we owe the discovery of points of distance, to which are drawn all lines that make an angle of 45° with the ground line.

Guido Ubaldi, another Italian, soon after discovered, that all lines that are parallel to one another, if they be inclined to the ground line, converge to some point in the horizontal line ; and that through this point also will pass a line drawn from the eye parallel to them. His Perspective was printed at Pisaro in 1600, and contained the first principles of the method afterwards discovered by Dr. Brook Taylor.

In 1583 was published the work of Giacomo Barozzi, of Vignola, commonly called Vignola, intitled The two Rules of Perspective, with a learned commentary by Ignatius Danti. In 1615 Marolois' work was printed at the Hague, and engraved and published by Hondius. And in 1625, Sirigatti published his treatise of Perspective, which is little more than an abstract of Vignola's.

Since that time the art of Perspective has been gradually improved by subsequent geometricians, particularly by professor Gravesande, and still more by Dr. Brook Taylor, whose principles are in a great measure new, and far more general than those of any of his predecessors. He did not confine his rules, as they had done, to the horizontal plane only, but made them general, so as to affect every species of lines and planes, whether they were parallel to the horizon or not ; and thus his principles were made universal. Besides, from the simplicity of his rules, the tedious progress of drawing out plans and elevations for any object, is rendered useless, and therefore avoided; for by this method, not only the fewest lines imaginable are required to produce any Perspective representation, but every figure thus drawn will bear the nicest mathematical examination. Farther, his system is the only one calculated for answering every purpose of those who are practitioners in the art of design ; for by it they may produce either the whole, or only so much of an object as is wanted ; and by fixing it in its proper place, its apparent magnitude may be determined in an instant. It explains also the Perspective of shadows, the reflection of objects from polished planes, and the inverse practice of Perspective.

His Linear Perspective was first published in 1715; and his New Principles of Linear Perspective in 1719, which he intended as an explanation of his first treatise. And his method has been chiefly followed by all others since.

In 1738 Mr. Hamilton published his Stereography, in 2 vols folio, after the manner of Dr. Taylor. But the neatest system of Perspective, both as to theory and practice, on the same principles, is that of Mr. Kirby. There are also good treatises on the subject, by Desargues, de Bosse, Albertus, Lamy, Niceron, Pozzo the Jesuit, Ware, Cowley, Priestley, Ferguson, Emerson, Malton, Henry Clarke, &c, &c.

Of the Principles of PERSPECTIVE. To give an idea

of

of the firſt principles and nature of this art ; ſuppoſe a tranſparent plane, as of glaſs &c, HI raiſed perpendicularly on a horizontal plane ; and the ſpectator S directing his eye O to the triangle ABC : if now we conceive the rays AO, BO, CO, &c, in their paſſage through the plane, to leave their traces or veſtiges in *a*, *b*, *c*, &c, on the plane ; there will appear the triangle *abc* ; which, as it ſtrikes the eye by the ſame rays *a*O, *b*O, *c*O, by which the reflected particles of light from the triangle are tranſmitted to the ſame, it will exhibit the true appearance of the triangle ABC, though the object ſhould be removed, the ſame diſtance and height of the eye being preſerved.

The buſineſs of Perſpective then, is to ſhew by what certain rules the points *a*, *b*, *c*, &c, may be found geometrically : and hence alſo we have a mechanical method of delineating any object very accurately.

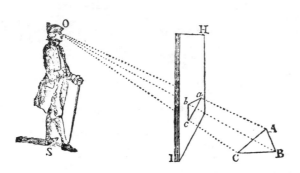

Hence it appears that *abc* is the ſection of the plane of the picture with the rays, which proceed from the original object to the eye : and therefore, when this is parallel to the picture, its repreſentation will be both parallel to the original, and ſimilar to it, though ſmaller in proportion as the original object is farther from the picture. When the original object is brought to coincide with the picture, the repreſentation is equal to the original ; but as the object is removed farther and farther from the picture, its image will become ſmaller and ſmaller, and alſo riſe higher and higher in the picture, till at laſt, when the object is ſuppoſed to be at an infinite diſtance, its image will vaniſh in an imaginary point, exactly as high above the bottom of the picture as the eye is above the ground plane, upon which the ſpectator, the picture, and the original object are ſuppoſed to ſtand.

This may be familiarly illuſtrated in the following manner : Suppoſe a perſon at a window looks through an upright pane of glaſs at any object beyond ; and, keeping his head ſteady, draws the figure of the object upon the glaſs, with a black-lead pencil, as if the point of the pencil touched the object itſelf ; he would then have a true repreſentation of the object in Perſpective, as it appears to his eye. For properly drawing upon the glaſs, it is neceſſary to lay it over with ſtrong gum water, which will be fit for drawing upon when dry, and will then retain the traces of the pencil. The perſon ſhould alſo look through a ſmall hole in a thin plate of metal, fixed about a foot from the glaſs, between it and his eye ; keeping his eye cloſe to the hole, other-

8

wiſe he might ſhift the poſition of his head, and ſo make a falſe delineation of the object.

Having traced out the figure of the object, he may go over it again, with pen and ink ; and when that is dry, cover it with a ſheet of paper, tracing the image upon this with a pencil ; then taking away the paper, and laying it upon a table, he may finiſh the picture, by giving it the colours, lights, and ſhades, as he ſees them in the object itſelf ; and thus he will have a true reſemblance of the object on the paper.

Of certain Definitions in PERSPECTIVE.

The *point of ſight*, in Perſpective, is the point E, where the ſpectator's eye ſhould be placed to view the

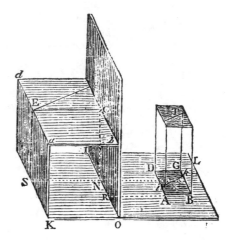

picture. And the *point of ſight*, in the picture, called alſo the *centre of the picture*, is the point C directly oppoſite to the eye, where a perpendicular from the eye at E meets the picture. Alſo this perpendicular EC is the *diſtance of the picture* : and if this diſtance be transferred to the horizontal line on each ſide of the point C, as is ſometimes done, the extremes are called the points of diſtance.

The *original plane*, or *geometrical plane*, is the plane KL upon which the real or original object ABGD is ſituated. The line OI, where the ground plane cuts the bottom of the picture, is called the *ſection* of the original plane, the *ground-line*, the *line of the baſe*, or the *fundamental line*.

If an original line AB be continued, ſo as to interſect the picture, the point of interſection R is called the interſection of that original line, or its *interſecting point*. The *horizontal plane* is the plane *abgd*, which paſſes through the eye, parallel to the horizon, and cuts the Perſpective plane or picture at right angles ; and the *horizontal line bg* is the common interſection of the horizontal plane with the picture.

The *vertical plane* is that which paſſes through the eye at right angles both to the ground plane and to the picture, as ECSN. And the *vertical line* is the common ſection of the vertical plane and the picture, as CN.

The *line of ſtation* SN is the common ſection of the vertical plane with the ground plane, and perpendicular to the ground line OI.

The

The *line of the height of the eye* is a perpendicular, as ES, let fall from the eye upon the ground plane.

The *vanishing line* of the original plane, is that line where a plane passing through the eye, parallel to the original plane, cuts the picture: thus *bg* is the vanishing line of ABGD, being the greatest height to which the image can rise, when the original object is infinitely distant.

The *vanishing point* of the original line, is that point where a line drawn from the eye, parallel to that original line, intersects the picture: thus C and *g* are the vanishing points of the lines AB and *ki*. All lines parallel to each other have the same vanishing point.

If from the point of sight a line be drawn perpendicular to any vanishing line, the point where that line intersects the vanishing line, is called the centre of that vanishing line: and the *distance of a vanishing line* is the length of the line which is drawn from the eye, perpendicular to the said line.

Measuring points are points from which any lines in the Perspective plane are measured, by laying a ruler from them to the divisions laid down upon the ground line. The measuring point of all lines parallel to the ground line, is either of the points of distance on the horizontal line, or point of sight. The measuring point of any line perpendicular to the ground line, is in the point of distance on the horizontal line; and the measuring point of a line oblique to the ground line is found by extending the compasses from the vanishing point of that line to the point of distance on the perpendicular, and setting off on the horizontal line.

Some general Maxims or Theorems in PERSPECTIVE.

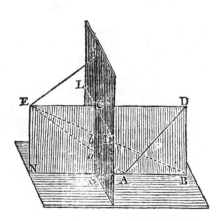

1. The representation *ab*, of a line AB, is part of a line SC, which passes through the intersecting point S, and the vanishing point C, of the original line AB.

2. If the original plane be parallel to the picture, it can have no vanishing line upon it; consequently the representation will be parallel. When the original is perpendicular to the ground line, as AB, then its vanishing point is in C, the centre of the picture, or point of sight; because EC is perpendicular to the picture, and therefore parallel to AB.

3. The image of a line bears a certain proportion to its original. And the image may be determined by transferring the length or distance of the given line to

the intersecting line; and the distance of the vanishing point to the horizontal line; i. e. by bringing both into the plane of the picture.

PROB. *To find the representation of an Objective point* A.
—Draw A1 and A2 at pleasure, intersecting the bot-

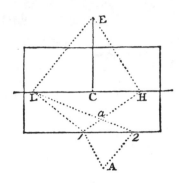

tom of the picture in 1 and 2; and from the eye E draw EH parallel to A1, and EL parallel to A2; then draw H1 and L2, which will intersect each other in *a*, the representation of the point A.

OTHERWISE. Let H be the given objective point.

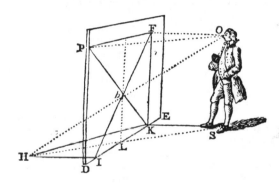

From which draw HI perpendicular to the fundamental line DE. From the fundamental line DE cut off IK = IH: through the point of sight F draw a horizontal line FP, and make FP equal to the distance of the eye SK: lastly, join FI and PK, and their intersection *h* will be the appearance of the given objective point H, as required.

And thus, by finding the representations of the two points, which are the extremes of a line, and connecting them together, there will be formed the representation of the line itself. In like manner, the representations of all the lines or sides of any figure or solid, determine those of the solid itself; which therefore are thus put into Perspective.

Aerial PERSPECTIVE, is the art of giving a due diminution or gradation to the strength of light, shade, and colours of objects, according to their different distances, the quantity of light which falls upon them, and the medium through which they are seen.

PERSPECTIVE *Machine*, is a machine for readily and easily making the Perspective drawing and appearance of any object, with little or no skill in the art. There have been invented various machines of this kind. One of which may even be seen in the works of Albert

Durer.

Durer. A very convenient one was invented by Dr. Bevis, and is deſcribed by Mr. Ferguſon, in his Perſpective, pa. 113. And another is deſcribed in Kirby's Perſpective, pa. 65.

PERSPECTIVE *Plan*, or *Plane*, is a glaſs or other tranſparent ſurface ſuppoſed to be placed between the eye and the object, and uſually perpendicular to the horizon.

Scenographic PERSPECTIVE. See SCENOGRAPHY.

PERSPECTIVE *of Shadows*. See SHADOW.

Specular PERSPECTIVE, is that which repreſents the objects in cylindrical, conical, ſpherical, or other mirrors.

PERTICA, a ſort of comet, being the ſame with VERU.

PETARD, a military engine, ſomewhat reſembling in ſhape a high-crowned hat ; ſerving formerly to break down gates, barricades, draw-bridges, or the like works intended to be ſurpriſed. It is about 8 or 9 inches wide, and weighs from 55 to 70 pounds. Its uſe was chiefly in a clandeſtine or private attack, to break down the gates &c. It has alſo been uſed in countermines, to break through the enemies galleries, and give vent to their mines : but the uſe of Petards is now diſcontinued.——Their invention is aſcribed to the French Hugonots in the year 1579. Their moſt ſignal exploit was the taking the city Cahors by means of them, as we are told by d'Aubigné.

PETIT (PETER), a conſiderable mathematician and philoſopher of France, was born at Montluçon in the dioceſe of Bourges, in the year 1589 according to ſome, but in 1600 according to others. — He firſt cultivated the mathematics and philoſophy in the place of his nativity ; but in 1633 he repaired to Paris, to which place his reputation had procured him an invitation. Here he became highly celebrated for his ingenious writings, and for his connections with Paſcal, Des Cartes, Merſenne, and the other great men of that time. He was employed on ſeveral occaſions by cardinal Richelieu ; he was commiſſioned by this miniſter to viſit the ſea-ports, with the title of the king's engineer ; and was alſo ſent into Italy upon the king's buſineſs. He was at Tours in 1640, where he married ; and was afterwards made intendant of the fortifications. Baillet, in his Life of Des Cartes, ſays, that Petit had a great genius for mathematics ; that he excelled particularly in aſtronomy ; and had a ſingular paſſion for experimental philoſophy. About 1637 he returned to Paris from Italy, when the Dioptrics of Des Cartes were much ſpoken of. He read them, and communicated his objections to Merſenne, with whom he was intimately acquainted. And yet he ſoon after embraced the principles of Des Cartes, becoming not only his friend, but his partiſan and defender alſo. He was intimately connected with Paſcal, with whom he made at Rouen the ſame experiments concerning the vacuum, which Torricelli had before made in Italy ; and was aſſured of their truth by frequent repetitions. This was in 1646 and 1647 ; and though there appears to be a long interval from this date to the time of his death, we meet with no other memoirs of his life. He died Auguſt the 20th 1667 at Lagny, near Paris, whither he had retired for ſome time before his deceaſe.

Petit was the author of ſeveral works upon phyſical and aſtronomical ſubjects ; the principal of which are,

1. Chronological Diſcourſe, &c, 1636, 4to. In defence of Scaliger.

2. Treatiſe on the Proportional Compaſſes.

3. On the Weight and Magnitude of Metals.

4. Conſtruction and Uſe of the Artillery Calipers.

5. On a Vacuum.

6. On Eclipſes.

7. On Remedies againſt the Inundations of the Seine at Paris.

8. On the Junction of the Ocean with the Mediterranean ſea, by means of the rivers Aude and Garonne.

9. On Comets.

10. On the proper Day for celebrating Eaſter.

11. On the Nature of Heat and Cold, &c.

PETTY (Sir WILLIAM), a ſingular inſtance of a univerſal genius, was the elder ſon of Anthony Petty, a clothier at Rumſey in Hampſhire, where he was born May the 16th, 1623. While a boy he took great delight in ſpending his time among the artificers there, whoſe trades he could work at when but 12 years of age. He then went to the grammar-ſchool in that place, where at 15 he became maſter of the Latin, Greek, and French languages, with arithmetic and thoſe parts of practical geometry and aſtronomy uſeful in navigation. Soon after, he went to the univerſity of Caen in Normandy ; and after ſome ſtay there he returned to England, where he was preferred in the king's navy. In 1643, when the civil war grew hot, and the times troubleſome, he went into the Netherlands and France for three years ; and having vigorouſly proſecuted his ſtudies, eſpecially in phyſic, at Utrecht, Leyden, Amſterdam, and Paris, he returned home to Rumſey. In 1647 he obtained a patent to teach the art of double writing for 17 years. In 1648 he publiſhed at London, " Advice to Mr. Samuel Hartlib, for the advancement of ſome particular parts of learning." At this time he adhered to the prevailing party of the nation ; and went to Oxford, where he taught anatomy and chemiſtry, and was created a doctor of phyſic, and grew into ſuch repute that the philoſophical meetings, which preceded and laid the foundation of the Royal Society, were firſt held at his houſe. In 1650 he was made profeſſor of anatomy there ; and ſoon after a member of the college of phyſicians in London, as alſo profeſſor of muſic at Greſham college London. In 1652 he was appointed phyſician to the army in Ireland ; as alſo to three lord lieutenants ſucceſſively, Lambert, Fleetwood, and Henry Cromwell. In Ireland he acquired a great fortune, but not without ſuſpicions and charges of unfair practices in his offices. After the rebellion was over in Ireland, he was appointed one of the commiſſioners for dividing the forfeited lands to the army who ſuppreſſed it. When Henry Cromwell became lieutenant of that kingdom, in 1655, he appointed Dr. Petty his ſecretary, and clerk of the council : he likewiſe procured him to be elected a burgeſs for Weſtloo in Cornwall, in Richard Cromwell's parliament, which met in January 1658. But, in March following, Sir Hierom Sankey, member for Woodſtock in Oxfordſhire, impeached him of high crimes and miſdemeanors in the execution of his office.

This

This gave the doctor a great deal of trouble, as he was summoned before the House of Commons; and notwithstanding the strenuous endeavours of his friends, in their recommendations of him to secretary Thurloe, and the defence he made before the house, his enemies procured his dismission from his public employments, in 1659. He then retired to Ireland, till the restoration of king Charles the Second; soon after which he came into England, where he was very graciously received by the king, resigned his professorship at Gresham college, and was appointed one of the commissioners of the Court of Claims. Likewise, April the 11th, 1661, he received the honour of knighthood, and the grant of a new patent, constituting him surveyor-general of Ireland, and was chosen a member of parliament there.

Upon the incorporating of the Royal Society, he was one of the first members, and of its first council. And though he had left off the practice of physic his name was continued as an honorary member of the college of physicians in 1663.

About this time he invented his double bottomed ship, to sail against wind and tide, and afterwards presented a model of this ship to the Royal Society; to whom also, in 1665, he communicated "A Discourse about the Building of Ships," containing some curious secrets in that art. But, upon trial, finding his ship failed in some respects, he at length gave up that project.

In 1666 Sir William drew up a treatise, called *Verbum Sapienti*, containing an account of the wealth and expences of England, and the method of raising taxes in the most equal manner.—The same year, 1666, he suffered a considerable loss by the fire of London.—The year following he married Elizabeth, daughter of Sir Hardresse Waller; and afterwards set up iron works and pilchard fishing, opened lead mines and a timber trade in Kerry, which turned to very good account. But all these concerns did not hinder him from the pursuit of both political and philosophical speculations, which he thought of public utility, publishing them either separately or by communication to the Royal Society, particularly on finances, taxes, political arithmetic, land carriage, guns, pumps, &c.

Upon the first meeting of the Philosophical Society at Dublin, upon the plan of that at London, every thing was submitted to his direction; and when it was formed into a regular society, he was chosen president in Nov. 1684. Upon this occasion he drew up a "Catalogue of mean, vulgar, cheap, and simple Experiments," proper for the infant state of the society, and presented it to them; as he did also his *Supellex Philosophica*, consisting of 45 instruments requisite to carry on the design of their institution. In 1685 he made his will; in which he declares, that being then about 60, his views were fixed upon improving his lands in Ireland, and to promote the trade of iron, lead, marble, fish, and timber, which his estate was capable of. And as for studies and experiments, " I think now, says he, to confine the same to the anatomy of the people, and political arithmetic; as also the improvement of ships, land-carriages, guns, and pumps, as of most use to mankind, not blaming the study of other men." But a few years after, all his pursuits were

determined by the effects of a gangrene in his foot, occasioned by the swelling of the gout, which put a period to his life, at his house in Piccadilly, Westminster, Dec. 16, 1687, in the 65th year of his age. His corpse was carried to Rumsey, and there interred, near those of his parents.

Sir William Petty died possessed of a very large fortune, as appears by his will; where he makes his real estate about 6,500l. per annum, his personal estate about 45,000l. his bad and desperate debts 30,000l. and the demonstrable improvements of his Irish estate, 4000l. per annum; in all, at 6 per cent. interest, 15,000l. per annum. This estate came to his family, which consisted of his widow and three children, Charles, Henry, and Anne: of whom Charles was created baron of Shelbourne, in the county of Waterford in Ireland, by king William the Third; but dying without issue, was succeeded by his younger brother Henry, who was created viscount Dunkeron, in the county of Kerry, and earl of Shelbourne Feb. 11, 1718. He married the lady Arabella Boyle, sister of Charles earl of Cork, who brought him several children. He was member of parliament for Great Marlow in Buckinghamshire, and a fellow of the Royal Society: he died April 17, 1751. Anne was married to Thomas Fitzmorris, baron of Kerry and Lixnaw, and died in Ireland in the year 1737.

The variety of pursuits, in which Sir William Petty was engaged, shews him to have had a genius capable of any thing to which he chose to apply it: and it is very extraordinary, that a man of so active and busy a spirit could find time to write so many things, as it appears he did, by the following catalogue.

1. Advice to Mr. S. Hartlib &c; 1648, 4to.—2. A Brief of Proceedings between Sir Hierom Sankey and the author &c; 1659, folio.—3. Reflections upon some persons and things in Ireland, &c; 1660, 8vo.—4. A Treatise of Taxes and Contribution, &c; 1662, 1667, 1685, 4to, all without the author's name. This last was re-published in 1690, with two other anonymous pieces, " The Privileges and Practice of Parliaments," and " The Politician Discovered;" with a new title-page, where it is said they were all written by Sir William, which, as to the first, is a mistake.—5. Apparatus to the History of the Common Practice of Dyeing;" printed in Sprat's History of the Royal Society, 1667, 4to.—6. A Discourse concerning the Use of Duplicate Proportion, together with a New Hypothesis of Springing or Elastic Motions; 1674, 12mo.—7. Colloquium Davidis cum Anima sua, &c; 1679, folio.—8. The Politician Discovered, &c; 1681, 4to.—9. An Essay in Political Arithmetic; 1682, 8vo.—10. Observations upon the Dublin Bills of Mortality in 1681, &c; 1683, 8vo.—11. An Account of some Experiments relating to Land-carriage, Philos. Transf. numb. 161.—12. Some Queries for examining Mineral Waters, ibid. numb. 166.—13. A Catalogue of Mean, Vulgar, Cheap, and Simple Experiments, &c; ibid. numb. 167.—14. Maps of Ireland, being an Actual Survey of the whole Kingdom, &c; 1685, folio.—15. An Essay concerning the Multiplication of Mankind; 1686, 8vo.—16. A further Assertion concerning the magnitude of London, vindicating it, &c; Philos. Transf. numb. 185.—17. Two Essays in Politi-

cal Arithmetic; 1687, 8vo.—18. Five Essays in Political Arithmetic; 1687, 8vo.—19. Observations upon London and Rome; 1687, 8vo.

His posthumous pieces are, (1), Political Arithmetic; 1690, 8vo, and 1755, with his life prefixed — (2), The Political Anatomy of Ireland, with Verbum Sapienti, 1691, 1719 —(3), A Treatise of Naval Philosophy; 1691, 12mo.—(4), What a complete Treatise of Navigation should contain; Philos. Transf. numb. 198.—(5), A Discourse of making Cloth with Sheep's Wool; in Birch's Hist. of the Roy. Soc.—(6), Supellex Philosophica; ibid.

PHÆNOMENON. See PHENOMENON.

PHARON, the name of a game of chance. See De Moivre's Doctrine of Chances, pa. 77 and 105.

PHASES, in Astronomy, the various appearances, or quantities of illumination of the moon, Venus, Mercury, and the other planets, by the sun. These Phases are very observable in the moon with the naked eye; by which she sometimes increases, sometimes wanes, is now bent into horns, and again appears a half circle; at other times she is gibbous, and again a full circular face. And by help of the telescope, the like variety of Phases is observed in Venus, Mars, &c.

Copernicus, a little before the use of telescopes, foretold, that after ages would find that Venus underwent all the changes of the moon; which prophecy was first fulfilled by Galileo, who, directing his telescope to Venus, observed her Phases to emulate those of the moon; being sometimes full, sometimes horned, and sometimes gibbous.

PHASES of an Eclipse. To determine these for any time: Find the moon's place in her visible way for that moment; and from that point as a centre, with the interval of the moon's semidiameter, describe a circle: In like manner find the sun's place in the ecliptic, from which, with the semidiameter of the sun, describe another circle: The intersection of the two circles shews the Phases of the eclipse, the quantity of obscuration, and the position of the cusps or horns.

PHENOMENON, or PHÆNOMENON, an appearance in physics, an extraordinary appearance in the heavens, or on earth; either discovered by observation of the celestial bodies, or by physical experiments, the cause of which is not obvious. Such are meteors, comets, uncommon appearance of stars and planets, earthquakes, &c. Such also are the effects of the magnet, phosphorus, &c.

PHILOLAUS, of Crotona, was a celebrated philosopher of the Ancients. He was of the school of Pythagoras, to whom that philosopher's Golden Verses have been ascribed. He made the heavens his chief object of contemplation; and has been said to be the author of that true system of the world which Copernicus afterwards revived; but erroneously, because there is undoubted evidence that Pythagoras learned that system in Egypt. On that erroneous supposition however it was, that Bulliald placed the name of Philolaus at the head of two works, written to illustrate and confirm that system.

"He was (says Dr. Enfield, in his History of Philosophy) a disciple of Archytas, and flourished in the time of Plato. It was from him that Plato purchased the written records of the Pythagorean system, contra-

ry to an express oath taken by the society of Pythagoreans, pledging themselves to keep secret the mysteries of their sect. It is probable that among these books were the writings of Timæus, upon which Plato formed the dialogue which bore his name. Plutarch relates, that Philolaus was one of the persons who escaped from the house which was burned by Cylon, during the life of Pythagoras; but this account cannot be correct. Philolaus was contemporary with Plato, and therefore certainly not with Pythagoras. Interfering in affairs of state, he fell a sacrifice to political jealousy.

"Philolaus treated the doctrine of nature with great subtlety, but at the same time with great obscurity; referring every thing that exists to mathematical principles. He taught, that reason, improved by mathematical learning, is alone capable of judging concerning the nature of things: that the whole world consists of infinite and finite; that number subsists by itself, and is the chain by which its power sustains the eternal frame of things; that the Monad is not the sole principle of things, but that the Binary is necessary to furnish materials from which all subsequent numbers may be produced; that the world is one whole, which has a fiery centre, about which the ten celestial spheres revolve, heaven, the sun, the planets, the earth, and the moon; that the sun has a vitreous surface, whence the fire diffused through the world is reflected, rendering the mirror from which it is reflected visible; that all things are preserved in harmony by the law of necessity; and that the world is liable to destruction both by fire and by water. From this summary of the doctrine of Philolaus it appears probable that, following Timæus, whose writings he possessed, he so far departed from the Pythagorean system as to conceive two independent principles in nature, God and matter, and that it was from the same source that Plato derived his doctrine upon this subject."

PHILOSOPHER, a person well versed in philosophy; or who makes a profession of, or applies himself to, the study of nature or of morality.

PHILOSOPHICAL TRANSACTIONS, those of the Royal Society. See TRANSACTIONS.

PHILOSOPHIZING, the act of considering some object of our knowledge, examining its properties, and the phenomena it exhibits, and enquiring into their causes or effects, and the laws of them; the whole conducted according to the nature and reason of things, and directed to the improvement of knowledge.

The Rules of PHILOSOPHIZING, as established by Sir Isaac Newton, are, 1. That no more causes of a natural effect be admitted than are true, and suffice to account for its phenomena. This agrees with the sentiments of most philosophers, who hold that nature does nothing in vain; and that it were vain to do that by many things, which might be done by fewer.

2. That natural effects of the same kind, proceed from the same causes. Thus, for instance, the cause of respiration is one and the same in man and brute; the cause of the descent of a stone, the same in Europe as in America; the cause of light, the same in the sun and in culinary fire; and the cause of reflection, the same in the planets as the earth.

3. Those qualities of bodies which are not capable of being heightened, and remitted, and which are found in

in all bodies on which experiments can be made, muft be confidered as univerfal qualities of all bodies. Thus, the extenfion of body is only perceived by our fenfes, nor is it perceivable in all bodies: but fince it is found in all that we have perception of, it may be affirmed of all. So we find that feveral bodies are hard; and argue that the hardnefs of the whole only arifes from the hardnefs of the parts: whence we infer that the particles, not only of thofe bodies which are fenfible, but of all others, are likewife hard. Laftly, if all the bodies about the earth gravitate towards the earth, and this according to the quantity of matter in each; and if the moon gravitate towards the earth alfo, according to its quantity of matter; and the fea again gravitate towards the moon; and all the planets and comets gravitate towards each other: it may be affirmed univerfally, that all bodies in the creation gravitate towards each other. This rule is the foundation of all natural philofophy.

PHILOSOPHY, the knowledge or ftudy of nature or morality, founded on reafon and experience. Literally and originally, the word fignified a love of wifdom. But by Philofophy is now meant the knowledge of the nature and reafons of things; as diftinguifhed from hiftory, which is the bare knowledge of facts; and from mathematics, which is the knowledge of the quantity and meafures of things.

Thefe three kinds of knowledge ought to be joined as much as poffible. Hiftory furnifhes matter, principles, and practical examinations; and mathematics completes the evidence.

Philofophy being the knowledge of the reafons of things, all arts muft have their peculiar Philofophy which conftitutes their theory: not only law and phyfic, but the loweft and moft abject arts are not without their reafons. It is to be obferved that the bare intelligence and memory of philofophical propofitions, without any ability to demonftrate them, is not Philofophy, but hiftory only. However, where fuch propofitions are determinate and true, they may be ufefully applied in practice, even by thofe who are ignorant of their demonftrations. Of this we fee daily inftances in the rules of arithmetic, practical geometry, and navigation; the reafons of which are often not underftood by thofe who practife them with fuccefs. And this fuccefs in the application produces a conviction of mind, which is a kind of medium between Philofophical or fcientific knowledge, and that which is hiftorical only.

If we confider the difference there is between natural philofophers, and other men, with regard to their knowledge of phenomena, we fhall find it confifts not in an exacter knowledge of the efficient caufe that produces them, for that can be no other than the will of the Deity; but only in a greater and more enlarged comprehenfion, by which analogies, harmonies, and agreements are defcribed in the works of nature, and the particular effects explained; that is, reduced to general rules, which rules grounded on the analogy and uniformnefs obferved in the production of natural effects, are more agreeable, and fought after by the mind; for that they extend our profpect beyond what is prefent, and near to us, and enable us to make very probable conjectures, touching things that may have happened

at very great diftances of time and place, as well as to predict things to come; which fort of endeavour towards omnifcience is much affected by the mind. Berkley, Princip. of Hum. Knowledge, fect. 104, 105.

From the firft broachers of new opinions, and the firft founders of fchools, Philofophy is become divided into feveral fects, fome ancient, others modern; fuch are the Platonifts, Peripatetics, Epicureans, Stoics, Pyrrhonians, and Academics; alfo the Cartefians, Newtonians, &c. See the particular articles for each.

Philofophy may be divided into two branches, or it may be confidered under two circumftances, theoretical and practical.

Theoretical or *Speculative* PHILOSOPHY, is employed in mere contemplation. Such is phyfics, which is a bare contemplation of nature, and natural things.

Theoretical Philofophy again is ufually fubdivided into three kinds, viz, pneumatics, phyfics or fomatics, and metaphyfics or ontology.

The firft confiders being, abftractedly from all matter: its objects are fpirits, their natures, properties, and effects. The fecond confiders matter, and material things: its objects are bodies, their properties, laws, &c.

The third extends to each indifferently: its objects are body or fpirit.

In the order of our difcovery, or arrival at the knowledge of them, phyfics is firft, then metaphyfics; the laft arifes from the two firft confidered together.

But in teaching, or laying down thefe feveral branches to others, we obferve a contrary order; beginning with the moft univerfal, and defcending to the more particular. And hence we fee why the Peripatetics call metaphyfics, and the Cartefians pneumatics, the *prima philofophia.*

Others prefer the diftribution of Philofophy into four parts, viz, 1. Pneumatics, which confiders and treats of fpirits. 2. Somatics, of bodies. 3. The third compounded of both, anthropology, which confiders man, in whom both body and fpirit are found. 4. Ontofophy, which treats of what is common to all the other three.

Again, Philofophy may be divided into three parts; intellectual, moral, and phyfical: the intellectual part comprifes logic and metaphyfics; the moral part contains the laws of nature and nations, ethics and politics; and laftly the phyfical part comprehends the doctrine of bodies, animate or inanimate: thefe, with their various fubdivifions, will comprize the whole of Philofophy.

Practical PHILOSOPHY, is that which lays down the rules of a virtuous and happy life; and excites us to the practice of them. Moft authors divide it into two kinds, anfwerable to the two forts of human actions to be directed by it; viz, Logic, which governs the operations of the underftanding; and Ethics, properly fo called, which direct thofe of the will.

For the feveral particular forts of Philofophy, fee the articles, Arabian, Ariftotelian, Atomical, Cartefian, Corpufcular, Epicurean, Experimental, Hermetical, Leibnitzian, Mechanical, Moral, Natural, Newtonian, Oriental, Platonic, Scholaftic, Socratic, &c. &c.

G g 2

PHOENIX,

PHOENIX, a conſtellation of the ſouthern hemi-ſphere. This is one of the new-added aſteriſms, un-known to the Ancients, and is not viſible in our north-ern parts of the globe. There are 13 ſtars in this conſtellation.

PHONICS, otherwiſe called ACOUSTICS, is the doctrine or ſcience of ſounds.

Phonics may be conſidered as an art analogous to Optics; and may be divided, like that, into Direct, Refracted, and Reflected. Theſe branches, the biſhop of Ferns, in alluſion to the parts of Optics, denomi-nates Phonics, Diaphonics, and Cataphonics. See ACOUSTICS.

PHOSPHORUS, a matter which ſhines, or even burns ſpontaneouſly, and without the application of any ſenſible fire.

Phoſphori are either natural or artificial.

Natural PHOSPHORI, are matters which become luminous at certain times, without the aſſiſtance of any art or preparation. Such are the glow-worms, fre-quent in our colder countries; lantern-flies, and other ſhining inſects, in hot countries; rotten-wood; the eyes, blood, ſcales, fleſh, ſweat, feathers, &c, of ſeveral animals; diamonds, when rubbed after a certain man-ner, or after having been expoſed to the ſun or light; ſu-gar and ſulphur, when pounded in a dark place; ſea water, and ſome mineral waters, when briſkly agitated; a cat's or horſe's back, duly rubbed with the hand, &c, in the dark; nay Dr. Croon tells us, that upon rubbing his own body briſkly with a well-warmed ſhirt, he has frequently made both to ſhine; and Dr. Sloane adds, that he knew a gentleman of Briſtol, and his ſon, both whoſe ſtockings would ſhine much after walking.

All natural Phoſphori have this in common, that they do not ſhine always, and that they never give any heat.

Of all the natural Phoſphori, that which has occa-ſioned the greateſt ſpeculation, is the

Barometrical or *Mercurial* PHOSPHORUS. M. Picard firſt obſerved, that the mercury of his barometer, when ſhaken in a dark place, emitted light. And many fanciful explanations have been given of this phenome-non, which however is now found to be a mere electrical effect.

Mr. Hawkſbee has ſeveral experiments on this ap-pearance. Paſſing air forcibly through the body of quickſilver, placed in an exhauſted receiver, the parts were violently driven againſt the ſide of the receiver, and gave all around the appearance of fire; conti-nuing thus till the receiver was half full again of air.

From other experiments he found, that though the appearance of light was not producible by agitating the mercury in the ſame manner in the common air, yet that a very fine medium, nearly approaching to a va-cuum, was not at all neceſſary. And laſtly, from other experiments he found that mercury incloſed in water, which communicated with the open air, by a violent ſhaking of the veſſel in which it was incloſed, emitted particles of light in great plenty, like little ſtars.

By including the veſſel of mercury, &c, in a receiver, and exhauſting the air, the phenomenon was changed; and upon ſhaking the veſſel, inſtead of ſparks of light,

the whole maſs appeared one continued circle of light.

Farther, if mercury be incloſed in a glaſs tube, cloſe ſtopped, that tube is found, on being rubbed, to give much more light, than when it had no mercury in it. When this tube has been rubbed, after raiſing ſucceſ-ſively its extremities, that the mercury might flow from one end to the other, a light is ſeen creeping in a ſer-pentine manner all along the tube, the mercury being all luminous. By making the mercury run along the tube afterwards without rubbing it, it emitted ſome light, though much leſs than before; this proves that the friction of the mercury againſt the glaſs, in running along, does in ſome meaſure electrify the glaſs, as the rubbing it with the hand does, only in a much leſs de-gree. This is more plainly proved by laying ſome very light down near the tube, for this will be attracted by the electricity raiſed by the running of the mercury, and will riſe to that part of the glaſs along which the mercury runs; from which it is plain, that what has been long known in the world under the name of the Phoſphorus of the barometer, is not a Phoſphorus, but merely a light raiſed by electricity, the mercury electrifying the tube. Philoſ. Tranſ. numb. 484.

Artificial PHOSPHORI, are ſuch as owe their luminous quality to ſome art or preparation. Some of theſe are made by the maceration of plants alone, and without any fire; ſuch as thread, linen cloth, but above all pa-per: the luminous appearance of this laſt, which it is now known is an electrical phenomenon, is greatly in-creaſed by heat. Almoſt all bodies, by a proper treat-ment, have that power of ſhining in the dark, which at firſt was ſuppoſed to be the property of one, and after-wards only of a few. See Philoſ. Tranſ. numb. 478, in vol. 44, pa. 83.

Of Artificial Phoſphori there are three principal kinds: the firſt *burning,* which conſumes every combuſ-tible it touches; the other two have no ſenſible heat, and are called the *Bononian* and *Hermetic* Phoſphorus; to which claſs others of a ſimilar kind may be re-ferred.

The Burning PHOSPHORUS, is a combination of phlo-giſton with a peculiar acid, and conſequently a ſpecies of ſulphur, tending to decompoſe itſelf, and ſo as to take fire on the acceſs of air only. This may be made of urine, blood, hairs, and generally of any part of an animal that yields an oil by diſtillation, and moſt eaſily of urine. It is of a yellowiſh colour, and of the conſiſt-ence of hard wax, in the condition it is left by the diſtil-lation; in which ſtate it is called *phoſphorus fulgurans,* from its corruſcations; and *phoſphorus ſmaragdinus,* be-cauſe its light is often green or blue, eſpecially in places that are not very dark; and from its conſiſtence it is called ſolid Phoſphorus. It diſſolves in all kinds of diſtilled oils, in which ſtate it is called liquid Phoſpho-rus. And it may be ground in all kinds of fat poma-tums, in which way it makes a luminous unguent.—So that theſe ſorts are all the ſame preparation, under dif-ferent circumſtances.

The diſcovery of this Phoſphorus was made in 1677, by one Brandt, a citizen of Hamburgh, in his re-ſearches for the philoſopher's ſtone. And the method was afterwards found out both by Kunckel, and Mr. Boyle, from only learning that urine was the chief ſub-

ſtance

ftance of it; fince then it has been called Kunckell's Phofphorus. It is prepared by firft evaporating the urine to a rob, or the confiftence of honey, and afterwards diftilling it in a very ftrong heat, &c. See Mem. Acad. Paris 1737; Philof. Tranf. numb. 196, or Abr. vol. 3, pa. 346; Mem. Acad. Berlin 1743.

Many curious and amufing experiments are made with Phofphorus; as by writing with it, when the letters will appear like flame in the dark, though in the light nothing appears but a dim fmoke; alfo a little bit of it rubbed between two papers, prefently takes fire, and burns vehemently; &c. By wafhing the face, or hands, &c, with liquid Phofphorus, they will fhine very confiderably in the dark, and the luftre will be communicated to adjacent objects, yet, without hurting the fkin: on bringing in the candle, the fhining difappears, and no change is perceivable.

Bolognian or *Bononian* PHOSPHORUS, is a preparation of a ftone called the Bononian ftone, from Bologna, a city in Italy, near which it is found. This Phofphorus has no fenfible heat, and only becomes luminous after being expofed to the fun or day light. For the method of preparing it, fee the Mem. Acad. Berlin 1749 and 1750.

The *Hermetic* PHOSPHORUS, or third kind, is a preparation of Englifh chalk, with aqua fortis, or fpirit of nitre, by the fire. It makes a body confiderably fofter than the Bolognian ftone, but having otherwife all the fame qualities. It is alfo called Baldwin's Phofphorus, from its inventor, a German chemift, called alfo Hermes in the fociety of the Naturæ Curioforum, whence its other name Hermetic: it was difcovered a little before the year 1677. See Acad. Par. 1693, pa. 271; and Grew's Muf. Reg. Soc. p. 353.

Ammoniacal PHOSPHORUS, firft difcovered by Homberg, is a combination of quick-lime with the acid of fal ammoniac, from which it receives its phlogifton. Mem. Acad. Par. 1693.

Antimonial PHOSPHORUS, is a kind difcovered by Mr. Geoffroy in his experiments on antimony. Mem. Acad. Par. 1736.

PHOSPHORUS *of the Berne-ftone*, a name given to a ftone from Berne, in Switzerland, where it is found, and which becomes a kind of Phofphorus when heated. Mem. Acad. Paris 1724.

Canton's PHOSPHORUS, a very good kind, prepared by Mr. Canton, an ingenious philofopher, from calcined oyfter fhells. Philof. Tranf. vol. 58, pa. 337.

PHOPSHORUS *Fæcalis*, a very good kind, exhibiting many wonderful phenomena, and prepared, by Mr. Homberg, from human dung mixed with alum. Mem. Acad. Par. 1711.

PHOSPHORUS *Metallorum*, a name given by fome chemifts to a preparation of a certain mineral fpar, found in the mines of Saxony, and other places where there is copper. Philof. Tranf. numb. 244, p. 365.

PHOSPHORUS *of Sulphur*, a new-difcovered fpecies, which readily takes fire on being expofed to the open air, and invented by M. Le Fevre. Mem. Acad. Par. 1728.

PHOSPHORUS, in Aftronomy, is the morning ftar, or the planet Venus, when fhe rifes before the fun. The Latins call it Lucifer, the French Etoile de berger, and the Greeks Phofphorus.

PHYSICAL, fomething belonging to nature, or exifting in it: Thus, we fay a Phyfical point, in oppofition to a mathematical one, which laft only exifts in the imagination. Or a Phyfical fubftance or body, in oppofition to fpirit, or metaphyfical fubftance, &c.

PHYSICAL, or *Senfible Horizon*. See HORIZON.

PHYSICO-*Mathematics*, or *Mixed Mathematics*, includes thofe branches of Phyfics which, uniting obfervation and experiment to mathematical calculation, explain mathematically the phenomena of nature.

PHYSICS, called alfo *Phyfiology*, and *Natural Philofophy*, is the doctrine of natural bodies, their phenomena, caufes, and effects, with their various affections, motions, operations, &c. So that the immediate and proper objects of Phyfics, are body, fpace, and motion.

The origin of Phyfics is referred, by the Greeks, to the Barbarians, viz, the brachmans, the magi, and the Hebrew and Egyptian priefts. From thefe it paffed to the Greek fages or fophi, particularly to Thales, who it is faid firft profeffed the ftudy of nature in Greece. Hence it defcended into the fchools of the Pythagoreans, the Platonifts, and the Peripatetics; from whence it paffed into Italy, and thence through the reft of Europe. Though the druids, bards, &c, had a kind of fyftem of Phyfics of their own.

Phyfics may be divided, with regard to the manner in which it has been handled, and the perfons by whom, into

Symbolical PHYSICS, or fuch as was couched under fymbols: fuch was that of the old Egyptians, Pythagoreans, and Platonifts; who delivered the properties of natural bodies under arithmetical and geometrical characters, and hieroglyphics.

Peripatetical PHYSICS, or that of the Ariftotelians, who explained the nature of things by matter, form, and privation, elementary and occult qualities, fympathies, antipathies, attractions, &c.

Experimental PHYSICS, which enquires into the reafons and natures of things from experiments: fuch as thofe in chemiftry, hydroftatics, pneumatics, optics, &c. And

Mechanical or *Corpufcular* PHYSICS, which explains the appearances of nature from the matter, motion, ftructure, and figure of bodies and their parts: all according to the fettled laws of nature and mechanics. See each of thefe articles under its own head.

PIASTER, a Spanifh money, more ufually called Piece of Eight, about the value of 4s. 6d.

PIAZZA, popularly called Piache, an Italian name for a portico, or covered walk, fupported by arches.

PICARD (JOHN), an able mathematician of France, and one of the moft learned aftronomers of the 17th century, was born at Fleche, and became prieft and prior of Rillie in Anjou. Coming afterwards to Paris, his fuperior talents for mathematics and aftronomy foon made him known and refpected. In 1666 he was appointed aftronomer in the Academy of Sciences. And five years after, he was fent, by order of the king, to the caftle of Uraniburgh, built by Tycho Brahe in Denmark, to make aftronomical obfervations there; and from thence he brought the original manufcripts, written by Tycho Brahe; which are the more valuable, as they differ in many places from the printed copies, and

contain

contain a book more than has yet appeared. These discoveries were followed by many others, particularly in astronomy: He was one of the first who applied the telescope to astronomical quadrants: he first executed the work called, *La Connoissance des Temps*, which he calculated from 1679 to 1683 inclusively: he first observed the light in the vacuum of the barometer, or the mercurial phosphorus: he also first of any went through several parts of France, to measure the degrees of the French meridian, and first gave a chart of the country, which the Cassinis afterwards carried to a great degree of perfection. He died in 1682 or 1683, leaving a name dear to his friends, and respectable to his contemporaries and to posterity. His works are,

1. A treatise on Levelling.
2. Practical Dialling by calculation.
3. Fragments of Dioptrics.
4. Experiments on Running Water.
5. Of Measurements.
6. Mensuration of Fluids and Solids.
7. Abridgment of the Measure of the Earth.
8. Journey to Uraniburgh, or Astronomical Observations made in Denmark.
9. Astronomical Observations made in divers parts of France.
10. La Connoissance des Temps, from 1679 to 1683.

All these, and some other of his works, which are much esteemed, are given in the 6th and 7th volumes of the Memoirs of the Academy of Sciences.

PICCOLOMINI (ALEXANDER), archbishop of Patras, and a native of Sienna, where he was born about the year 1508. He was of an illustrious and ancient family, which came originally from Rome, but afterwards settled at Sienna. He composed with success for the theatre; but he was not more distinguished by his genius, than by the purity of his manners, and his regard to virtue. His charity was great; and was chiefly exerted in relieving the necessities of men of letters. He was the first who made use of the Italian language in writing upon philosophical subjects. He died at Sienna the 12th of March 1578, at 70 years of age, leaving behind him a number of works in Italian, on a variety of subjects. A particular catalogue of them may be seen in the Typographical Dictionary; the principal of which are the following:

1. Various Dramatical pieces.
2. A treatise on the Sphere.
3. A Theory of the Planets.
4. Translation of Aristotle's Art of Rhetoric and Poetry.
5. A System of Morality, published at Venice, 1575, in 4to; translated into French by Peter de Larivey, and printed at Paris, 1581, in 4to.

These, with a variety of other works, prove his extensive knowledge in natural philosophy, mathematics, and theology.

PICCOLOMINI (Francis), of the same family with the foregoing, was born in 1520, and taught philosophy with success, for the space of 22 years, in the most celebrated universities of Italy, and afterwards retired to Sienna, where he died, in 1604, at 84 years of age. He was so much and so generally respected, that the city went into mourning on his death.

Piccolomini laboured to revive the doctrine of Plato,

2

and endeavoured also to imitate the manners of that philosopher. He had for his rival the famous James Zabar Alla, whom he excelled in facility of expression and neatness of diction; but to whom he was much inferior in point of argument, because he did not examine matters to the bottom as the other did; but passed too rapidly from one proposition to another.

PICKET, *Picquet*, or *Piquet*, in Fortification &c, a stake sharp at one end, and usually shod with iron; used in laying out ground, to mark the several bounds and angles of it. There are also larger Pickets, driven into the earth, to hold together fascines or faggots, in works that are thrown up in haste. As also various sorts of smaller Pickets for divers other uses.

PIECES, in Artillery, include all sorts of great guns and mortars; meaning Pieces of ordnance, or of artillery.

PIEDOUCHE, in Architecture, a little stand, or pedestal, either oblong or square, enriched with mouldings; serving to support a bust, or other little figure; and is more usually called a bracket pedestal.

PIEDROIT, in Architecture, a kind of square pillar, or pier, partly hid within a wall. Differing from the Pilaster by having no regular base nor capital.

PIEDROIT is also used for a part of the solid wall annexed to a door or window; comprehending the doorpost, chambranle, tableau, leaf, &c.

PIER, in Building, denotes a mass of stone, &c, opposed by way of fortress, against the force of the sea, or a great river, for the security of ships lying in any harbour or haven. Such are the Piers at Dover, or Ramsgate, or Yarmouth, &c.

PIERS are also used in Architecture for a kind of pilasters, or buttresses, raised for support, strength, and sometimes for ornament.

Circular PIERS, are called Massive Columns, and are either with or without caps. These are often seen in Saracenic architecture.

PIERS, of a Bridge, are the walls built to support the arches, and from which they spring as bases, to stand upon.

Piers should be built of large blocks of stone, solid throughout, and cramped together with iron, which will make the whole as one solid stone. Their extremities, or ends, from the bottom, or base, up to high-water mark, ought to project sharp out with a saliant angle, to divide the stream. Or perhaps the bottom part of the Pier should be built flat or square up to about half the height of low-water mark, to encourage a lodgment against it for the sand and mud, to cover the foundation; lest, being left bare, the water should in time undermine and ruin it. The best form of the projection for dividing the stream, is the triangle; and the longer it is, or the more acute the saliant angle, the better it will divide it, and the less will the force of the water be against the Pier; but it may be sufficient to make that angle a right one, as it will render the masonry stronger, and in that case the perpendicular projection will be equal to half the breadth or thickness of the Pier. In rivers where large heavy craft navigate, and pass the arches, it may perhaps be better to make the ends semicircular; for though this figure does not divide the water so well as the triangle, it will better turn off, and bear the shock of the craft.

The

The thickneſs of the Piers ought to be ſuch as will make them of weight, or ſtrength, ſufficient to ſupport their interjacent arch, independent of the aſſiſtance of any other arches. And then, if the middle of the Pier be run up to its full height, the centring may be ſtruck, to be uſed in another arch, before the hanches or ſpandrels are filled up. They ought alſo to be made with a broad bottom on the foundation, and gradually diminiſhed in thickneſs by offsets up to low-water mark.

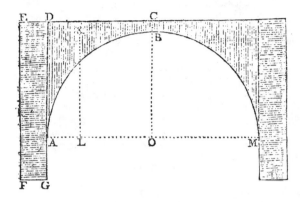

To find the thickneſs FG *of the Piers*, neceſſary to ſupport an arch ABM, this is a general rule. Let K be the centre of gravity of the half arch ADCB, A = its area; KL perpendicular to AM the ſpan of the arch, OB its height, and BC its thickneſs at the crown: then is the thickneſs of the pier

$$ FG = \sqrt{\frac{2GA \times AL}{EF \times KL}} \times A. $$

Some authors pretend to give numbers, in tables, for this purpoſe; but they are very erroneous. See my treatiſe on the Principles of Bridges, ſect. 3.

PIKE, an offenſive weapon, conſiſting of a ſhaft of wood, 12 or 14 feet long, headed with a flat-pointed ſteel, called the ſpear.

Pliny ſays the Lacedemonians were the inventors of the Pike. The Macedonian phalanx was evidently a battalion of Pikemen.

The Pike was long uſed by the infantry, to enable them to ſuſtain the attack of the cavalry; but it is now taken from them, and the bayonet, fixed to the muzzle of the firelock, is given inſtead of it.—It is ſtill uſed by ſome officers of infantry, under the name of ſpontoon.

Half PIKE is the weapon carried by an officer of foot; being only 8 or 9 feet long.

PILASTER, in Architecture, a ſquare column, ſometimes inſulated, but more frequently let within a wall, and only projecting by a 4th or 5th part of its thickneſs.

The Pilaſter is different in the different orders; borrowing the name of each order, and having the ſame proportions, and the ſame capitals, members, and ornaments, with the columns themſelves.

Demi PILASTER, called alſo *Membretto*, is a Pilaſter that ſupports an arch; and it generally ſtands againſt a pier or column.

PILES, in Building, are large ſtakes, or beams, ſharpened at the end, and ſhod with iron, to be driven into the ground, for a foundation to build upon in marſhy places.

Amſterdam, and ſome other cities, are wholly built upon Piles. The ſtoppage of Dagenham-breach was effected by dove-tail Piles, that is by Piles mortiſed into one another by a dovetail joint.

Piles are driven down by blows of a large iron weight, ram, or hammer, dropped continually upon them from a height, till the Pile is ſunk deep enough into the ground.

Notwithſtanding the momentum, or force of a body in motion, is as the weight multiplied by the velocity, or ſimply as its velocity, the weight being given, or conſtant; yet the effect of the blow will be nearly as the ſquare of that velocity, the effect being the quantity the Pile ſinks in the ground by the ſtroke. For the force of the blow, which is transferred to the Pile, being deſtroyed, in ſome certain definite time, by the friction of the part which is within the earth, which is nearly a conſtant quantity; and the ſpaces, in conſtant forces, being as the ſquares of the velocities; therefore the effects, which are thoſe ſpaces ſunk, are nearly as the ſquare of the velocities; or, which is the ſame thing, nearly as the heights fallen by the ram or hammer, to the head of the Pile. See, upon this ſubject, Leopold Belidor, alſo Deſaguliers's Exper. Philoſ. vol. 1, pa. 336, and vol. 2, pa. 417: and Philoſ. Tranſ. 1779, pa. 120.

There have been various contrivances for raiſing and dropping the hammer, for driving down the Piles; ſome ſimple and moved by ſtrength of men, and ſome complex and by machinery; but the completeſt Pile-Driver is eſteemed that which was employed in driving the Piles in the foundation of Weſtminſter bridge. This machine was the invention of a Mr. Vauloue, and the deſcription of it is as follows.

Deſcription of Vauloue's PILE-*Driver*. See fig. 2, pl. xx. A is the great upright ſhaft or axle, carrying the great wheel B and drum C, and turned by horſes attached to the bars S, S. The wheel B turns the trundle X, having a fly O at the top, to regulate the motion, and to act againſt the horſes, and keep them from falling when the heavy ram Q is diſengaged to drive the Pile P down into the mud &c. in the bottom of the river. The drum C is looſe upon the ſhaft A, but is locked to the wheel B by the bolt Y. On this drum the great rope HH is wound; one end of it being fixed to the drum, and the other to the follower G, paſſing over the pulleys I and K. In the follower G are contained the tongs F, which take hold of the ram Q, by the ſtaple R for drawing it up. D is a ſpiral or fuſee fixed to the drum, on which winds the ſmall rope T which goes over the pulley U, under the pulley V, and is faſtened to the top of the frame at 7. To the pulley-block V is hung the counterpoiſe W, which hinders the follower from accelerating as it goes down to take hold of the ram: for, as the follower tends to acquire velocity in its deſcent, the line T winds downwards upon the fuſee, on a larger and larger radius, by which means the counterpoiſe W acts ſtronger and ſtronger againſt it; and ſo allows it to come down with only a moderate and uniform velocity. The bolt Y locks the

drum

drum to the great wheel, being pushed upward by the small lever z, which goes through a mortise in the shaft A, turns upon a pin in the bar 3 fixed into the great wheel B, and has a weight 4, which always tends to push up the bolt Y through the wheel into the drum. L is the great lever turning on the axis m, and resting upon the forcing bar 5, 5, which goes down through a hollow in the shaft A, and bears upon the little lever 2.

By the horses going round, the great rope H is wound about the drum C, and the ram Q is drawn up by the tongs F in the follower G, till they come between the inclined planes E; which, by shutting the tongs at the top, open them below, and so discharge the ram, which falls down between the guide posts ib upon the Pile P, and drives it by a few strokes as far into the ground as it can go, or as is desired; after which, the top part is sawed off close to the mud, by an engine for that purpose. Immediately after the ram is discharged, the piece 6 upon the follower G takes hold of the ropes aa, which raise the end of the lever L, and cause its end N to descend and press down the forcing bar 5 upon the little lever 2, which, by drawing down the bolt Y, unlocks the drum C from the great wheel B; and then the follower, being at liberty, comes down by its own weight to the ram; and the lower ends of the tongs slip over the staple R, and the weight of their heads causes them to fall outward, and shuts upon it. Then the weight 4 pushes up the bolt Y into the drum, which locks it to the great wheel, and so the ram is drawn up as before.

As the follower comes down, it causes the drum to turn backward, and unwinds the rope from it, while the horses, the great wheel, trundle, and fly, go on with an uninterrupted motion: and as the drum is turning backward, the counterpoise W is drawn up, and its rope T wound upon the spiral fusee D.

There are several holes in the under side of the drum, and the bolt Y always takes the first one that it finds when the drum stops by the falling of the follower upon the ram; till which stoppage, the bolt has not time to slip into any of the holes.

The peculiar advantages of this engine are, that the weight, called the ram, or hammer, may be raised with the least force; that, when it is raised to a proper height, it readily disengages itself and falls with the utmost freedom; that the forceps or tongs are lowered down speedily, and instantly of themselves again lay hold of the ram, and lift it up; on which account this machine will drive the greatest number of piles in the least time, and with the fewest labourers.

This engine was placed upon a barge on the water, and so was easily conveyed to any place desired. The ram was a ton weight; and the guides b, b, by which it was let fall, were 30 feet high.

A new machine for driving piles has been invented lately by Mr. S. Bunce of Kirby-street, Hatton-street, London. This, it is said, will drive a greater number of Piles in a given time than any other; and that it can be constructed more simply to work by horses than Vauloue's engine above described.

Fig. 3 and 4, plate xx, represent a side and front section of the machine. The chief parts are, A, fig. 3, which are two endless ropes or chains, connected by cross pieces of iron B (fig. 4) corresponding with two

cross grooves cut diametrically opposite in the wheel C (fig. 3) into which they are received; and by which means the rope or chain A is carried round. FHK is a side-view of a strong wooden frame moveable on the axis H. D is a wheel, over which the chain passes and turns within at the top of the frame. It moves occasionally from F to G upon the centre H, and is kept in the position F by the weight I fixed to the end K. In fig. 5, L is the iron ram, which is connected with the cross pieces by the hook M. N is a cylindrical piece of wood suspended at the hook at O, which by sliding freely upon the bar that connects the hook to the ram, always brings the hook upright upon the chain when at the bottom of the machine, in the position of GP. See fig. 3.

When the man at S turns the usual crane-work, the ram being connected to the chain, and passing between the guides, is drawn up in a perpendicular direction; and when it is near the top of the machine, the projecting bar Q of the hook strikes against a cross piece of wood at R (fig. 3); and consequently discharges the ram, while the weight I of the moveable frame instantly draws the upper wheel into the position shewn at F, and keeps the chain free of the ram in its descent. The hook, while descending, is prevented from catching the chain by the wooden piece N: for that piece being specifically lighter than the iron weight below, and moving with a less degree of velocity, cannot come into contact with the iron, till it is at the bottom, and the ram stops. It then falls, and again connects the hook with the chain, which draws up the ram, as before.

Mr. Bunce has made a model of this machine, which performs perfectly well; and he observes, that, as the motion of the wheel C is uninterrupted, there appears to be the least possible time lost in the operation.

PILE is also used among Architects, for a mass or body of building.

PILE, in Artillery, denotes a collection or heap of shot or shells, piled up by horizontal courses into either a pyramidal or else a wedge-like form; the base being an equilateral triangle, a square, or a rectangle. In the triangle and square, the Pile terminates in a single ball or point, and forms a pyramid, as in plate xix, fig. 4 and 5, but with the rectangular base, it finishes at top in a row of balls, or an edge, forming a wedge, as in fig. 6.

In the triangular and square Piles, the number of horizontal rows, or courses, or the number counted on one of the angles from the bottom to the top, is always equal to the number counted on one side, in the bottom row. And in rectangular Piles, the number of rows, or courses, is equal to the number of balls in the breadth of the bottom row, or shorter side of the base: also, in this case, the number in the top row, or edge, is one more than the difference between the length and breadth of the base. All which is evident from the inspection of the figures, as above.

The courses in these Piles are figurate numbers.

In a triangular Pile, each horizontal course is a triangular number, produced by taking the successive sums of the ordinate numbers, viz,

$$1 = 1$$
$$1 + 2 = 3$$
$$1 + 2 + 3 = 6$$
$$1 + 2 + 3 + 4 = 10, \&c.$$

And

And the number of fhot in the triangular Pile, is the fum of all thefe triangular numbers, taken as far, or to as many terms, as the number in one fide of the bafe. And therefore, to find this fum, or the number of all the fhot in the Pile, multiply continually together, the number in one fide of the bafe row, and that number increafed by 1, and the fame number increafed by 2; then $\frac{1}{6}$ of the laft product will be the anfwer, or number of all the fhot in the Pile.

That is, $\frac{n \cdot n + 1 \cdot n + 2}{6}$ is the fum;

where n is the number in the bottom row.

Again, in Square Piles, each horizontal courfe is a fquare number, produced by taking the fquare of the number in its fide, or the fucceffive fums of the odd numbers, thus,

$$
\begin{aligned}
1. &= 1 \\
1 + 3 &= 4 \\
1 + 3 + 5 &= 9 \\
1 + 3 + 5 + 7 &= 16, \&c.
\end{aligned}
$$

And the number of fhot in the fquare Pile is the fum of all thefe fquare numbers, continued fo far, or to as many terms, as the number in one fide of the bafe. And therefore, to find this fum, multiply continually together, the number in one fide of the bottom courfe, and that number increafed by 1, and double the fame number increafed by 1; then $\frac{1}{6}$ of the laft product will be the fum or anfwer.

That is, $\frac{n \cdot n + 1 \cdot 2n + 1}{6}$ is the fum.

In a rectangular Pile, each horizontal courfe is a rectangle, whofe two fides have always the fame difference as thofe of the bafe courfe, and the breadth of the top row, or edge, being only 1: becaufe each courfe in afcending has its length and breadth always lefs by 1 than the courfe next below it. And thefe rectangular courfes are found by multiplying fucceffively the terms or breadths 1, 2, 3, 4, &c, by the fame terms added to the conftant difference of the two fides d; thus,

$$
\begin{aligned}
1 \cdot 1 + d &= 1 + d \\
2 \cdot 2 + d &= 4 + 2d \\
3 \cdot 3 + d &= 9 + 3d \\
4 \cdot 4 + d &= 16 + 4d, \&c.
\end{aligned}
$$

And the number of fhot in the rectangular Pile is the fum of all thefe rectangles, which, it is evident, confift of the fum of the fquares, together with the fum of an arithmetical progreffion, continued till the number of terms be the difference between the length and breadth of the bafe, and 1 lefs than the edge or top row. And therefore, to find this fum, multiply continually together, the number in the breadth of the bafe row, the fame number increafed by 1, and double the fame number increafed by 1, and alfo increafed by triple the difference between the length and breadth of the bafe; then $\frac{1}{6}$ of the laft product will be the anfwer.

That is, $\frac{b \cdot b + 1 \cdot 2b + 3d + 1}{6}$ is the fum.

where b is the breadth of the bafe, and d the difference between the length and breadth of the bottom courfe.

As to incomplete Piles, which are only fruftums,

VOL. II.

wanting a fimilar fmall Pile at the top; it is evident that the number in them will be found, by firft computing the number in the whole Pile, as if it were complete, and alfo the number in the fmall Pile wanting at top, both by their proper rule; and then fubtracting the one number from the other.

In piling of fhot, when room is an object, it may be obferved that the fquare Pile is the leaft eligible, of any, as it takes up more room, in proportion to the number of fhot contained in it, than either of the other two forms; and that the rectangular Pile is the moft eligible, as taking up the leaft room in proportion to the number it contains.

PILLAR, a kind of irregular column, round, and infulated, or detached from the wall. Pillars are not reftricted to any rules, their parts and proportions being arbitrary; fuch for example as thofe that fupport Saracenic vaults, and other buildings, &c.

PINION, in Mechanics, is an arbor, or fpindle, in the body of which are feveral notches, which are catched by the teeth of a wheel that ferves to turn it round. Or a Pinion is any leffer wheel that plays in the teeth of a larger.

In a watch, &c, the notches of a Pinion are called leaves, and not teeth, as in other wheels; and their number is commonly 4, 5, 6, 8, &c.

PINION of Report, is that Pinion, in a watch, commonly fixed on the arbor of a great wheel: and which ufed to have but four leaves in old watches; it drives the dial-wheel, and carries about the hand.

The number of turns to be laid upon the Pinion of report, is found by this proportion: as the beats in one turn of the great wheel, are to the beats in an hour, fo are the hours on the face of the clock (viz 12 or 24), to the quotient of the hour-wheel or dial-wheel divided by the Pinion of report, that is, by the number of turns which the Pinion of report hath in one turn of the dial-wheel. Which in numbers is 26928 : 20196 :: 12 : 9. —Or thus; as the hours of the watch's going, are to the numbers of the turns of the fufee, fo are the hours of the face, to the quotient of the Pinion of report. So, if the hours be 12, then as 16 : 12 :: 12 : 9; but if 24, then as 16 : 12 :: 24 : 18.

This rule may ferve to lay the Pinion of report on any other wheel, thus: as the beats in one turn of any wheel, are to the beats in an hour, fo are the hours of the face, or dial-plate, of the watch, to the quotient of the dial-wheel divided by the Pinion of report, fixed on the fpindle of the aforefaid wheel.

PINT, a meafure of capacity, being the 8th part of a gallon, both in ale and wine meafure, &c. The wine Pint of pure fpring water, weighs near 17 ounces avoirdupois, and the ale Pint a little above 20 ounces.

The Paris Pint contains about 2 pounds of common water. And the Scotch Pint contains $108\frac{2}{5}$ cubic inches, and therefore contains 3 Englifh Pints.

PISCES, the 12th fign or conftellation in the zodiac; in the form of two fifhes tied together by the tails.

The Greeks, who have fome fable to account for the origin of every conftellation, tell us, that when Venus and Cupid were one time on the banks of the Euphrates, there appeared before them that terrible giant Typhon, who was fo long a terror to all the Gods. Thefe deities immediately, they fay, threw themfelves

H h

into

into the water, and were there changed into thefe two fifhes, the Pifces, by which they efcaped the danger. But the Egyptians ufed the figns of the zodiac as part of their hieroglyphic language, and by the 12 they conveyed an idea of the proper employment during the 12 months of the year. The Ram and the Bull had, at that time, taken to the increafe of their flock, the young of thofe animals being then growing up; the maid Virgo, a reaper in the field, fpoke the approach of harveft; Sagittary declared autumn the time for hunting; and the Pifces, or fifhes tied together, in token of their being taken, reminded men that the approach of fpring was the time for fifhing.

The Ancients, as they gave one of the 12 months of the year to the patronage of each of the 12 fuperior deities, fo they alfo dedicated to, or put under the tutelage of each, one of the 12 figns of the zodiac. In this divifion, the fifhes naturally fell to the fhare of Neptune; and hence arifes that rule of the aftrologers, which throws every thing that regards the fate of fleets and merchandize, under the more immediate patronage and protection of this conftellation.

The ftars in the fign Pifces are, in Ptolomy's catalogue 38, in Tycho's 36, in Hevelius's 39, and in the Britannic catalogue 113.

PISCIS *Auftralis*, the Southern Fifh, is a conftellation of the fouthern hemifphere, being one of the old 48 conftellations mentioned by the Ancients.

The Greeks have here again the fable of Venus and her fon throwing themfelves into the fea, to efcape from the terrible Typhon. This fable is probably borrowed from the hieroglyphics of the Egyptians. With them, a fifh reprefented the fea, its element; and Typhon was probably a land flood, perhaps reprefented by the fign Aquarius, or water pourer, whofe ftream or river is reprefented as fwallowed up by this fifh, as the land floods and rivers are by the fea. And Venus was fome queen, perhaps Semiramis, otherwife called Hamamah, who took to the river or the fea with her fon, in a veffel, to avoid the flood, &c.

The remarkable ftar Fomahaut, of the 1ft magnitude, is juft in the mouth of this fifh. The ftars of this conftellation are, in Ptolomy's catalogue 18, and in Flamfteed's 24.

PISCIS *Volans*, the Flying Fifh, is a fmall conftellation of the fouthern hemifphere, unknown to the Ancients, but added by the Moderns. It is not vifible in our latitude, and contains only 8 ftars.

PISTOLE, a gold coin in Spain, Italy, Switzerland, &c, of the value of about 16s. 6d.

PISTON, a part or member in feveral machines, particularly pumps, air-pumps, fyringes, &c; called alfo the Embolus, and popularly the Sucker.

The Pifton of a pump is a fhort cylinder of wood or metal, fitted exactly to the cavity of the barrel, or body; and which, being worked up and down alternately, raifes the water; and when raifed, preffes it again, fo as to make it force up a valve with which it is furnifhed, and fo efcape through the fpout of the pump.

There are two forts of Piftons ufed in pumps; the one with a valve, called a bucket; and the other without a valve, called a forcer.

PLACE, in Philofophy, that part of infinite fpace which any body poffeffes.

Ariftotle and his followers divide Place into External and Internal.

Internal PLACE, is that fpace or room which the body contains. And

External PLACE, is that which includes or contains the body; and is by Ariftotle called the firft or concave and immoveable furface of the ambient body.

Newton better, and more intelligibly, diftinguifhes Place into Abfolute and Relative.

Abfolute and *Primary* PLACE, is that part of infinite and immoveable fpace which a body poffeffes. And

Relative, or *Secondary* PLACE, is the fpace it poffeffes confidered with regard to other adjacent objects.

Dr. Clark adds another kind of Relative Place, which he calls Relatively Common Place; and defines it, that part of any moveable or meafurable fpace which a body poffeffes; which Place moves together with the body.

PLACE, Mr. Locke obferves, is fometimes likewife taken for that portion of infinite fpace poffeffed by the material world; though this, he adds, were more properly called extenfion. The proper idea of Place, according to him, is the relative pofition of any thing, with regard to its diftance from certain fixed points; whence it is faid a thing has or has not changed Place, when its diftance is or is not altered with refpect to thofe bodies.

PLACE, in Optics, or *Optical* PLACE, is the point to which the eye refers an object.

Optic PLACE of a ftar, is a point in the furface of the mundane fphere in which a fpectator fees the centre of the ftar, &c.—This is divided into True and Apparent.

True, or *Real Optic* PLACE, is that point of the furface of the fphere, where a fpectator at the centre of the earth would fee the ftar, &c.

Apparent, or *Vifible Optic* PLACE, is that point of the furface of the fphere, where a fpectator at the furface of the earth fees the ftar, &c.

The diftance between thefe two optic Places makes what is called the Parallax.

PLACE *of the Sun*, or *Moon*, or *Star*, or *Planet*, in Aftronomy, fimply denotes the fign and degree of the zodiac which the luminary is in; and is ufually expreffed either by its latitude and longitude, or by its right afcenfion and declination.

PLACE *of Radiation*, in Optics, is the interval or fpace in a medium, or tranfparent body, through which any vifible object radiates.

PLACE, in Geometry, ufually called *Locus*, is a line ufed in the folution of problems, being that in which the determination of every cafe of the problem lies. See LOCUS, *Plane*, *Simple*, *Solid*, &c.

PLACE, in War and Fortification, a general name for all kinds of fortreffes, where a party may defend themfelves.

PLACE *of Arms*, a ftrong part where the arms &c are depofited, and where ufually the foldiers affemble and are drawn up.

PLAFOND, or PLATFOND, in Architecture, the cieling of a room.

PLAIN &c. See PLANE.

4

PLAN, a representation of something, drawn on a plane. Such as maps, charts, and ichnographies.

PLAN, in Architecture, is particularly used for a draught of a building; such as it appears, or is intended to appear, on the ground; shewing the extent, division, and distribution of its area into apartments, rooms, passages, &c. It is also called the Ground Plot, Platform, and Ichnography of the building; and is the first device or sketch the architect makes.

Geometrical PLAN, is that in which the solid and vacant parts are represented in their natural proportion.

Raised PLAN, is that where the elevation, or upright, is shewn upon the geometrical Plan, so as to hide the distribution.

Perspective PLAN, is that which is conducted and exhibited by degradations, or diminutions, according to the rules of Perspective.

PLANE, or PLAIN, in Geometry, denotes a Plane figure, or a surface lying evenly between its bounding lines. *Euclid.*

Some define a Plane, a surface, from every point of whose perimeter a right line may be drawn to every other point in the same, and always coinciding with it.

As the right line is the shortest extent from one point to another, so is a Plane the shortest extension between one line and another.

PLANES are much used in Astronomy, conic sections, spherics, &c, for imaginary surfaces, supposed to cut and pass through solid bodies.

When a Plane cuts a cone parallel to one side, it makes a parabola; when it cuts the cone obliquely, an ellipse or hyperbola; and when parallel to its base, a circle. Every section of a sphere is a circle.

The sphere is wholly explained by Planes, conceived to cut the celestial bodies, and to fill the areas or circumferences of the orbits. They are differently inclined to each other; and by us the inhabitants of the earth, the Plane of whose orbit is the Plane of the ecliptic, their inclination is estimated with regard to this Plane.

PLANE *of a Dial*, is the surface on which a dial is supposed to be described.

PLANE, in Mechanics. A *Horizontal* PLANE, is a Plane that is level, or parallel to the horizon.

Inclined PLANE, is one that makes an oblique angle with a horizontal Plane.

The doctrine of the motion of bodies on Inclined Planes, makes a very considerable article in mechanics, and has been fully explained under the articles, MECHANICAL *Powers*, and INCLINED *Plane*.

PLANE *of Gravity*, or *Gravitation*, is a Plane supposed to pass through the centre of gravity of the body, and in the direction of its tendency; that is, perpendicular to the horizon.

PLANE *of Reflection*, in Catoptrics, is a Plane which passes through the point of reflection; and is perpendicular to the Plane of the glass, or reflecting body.

PLANE *of Refraction*, is a Plane passing through the incident and refracted ray.

Perspective PLANE, is a Plane transparent surface, usually perpendicular to the horizon, and placed between the spectator's eye and the object he views; through which the optic rays, emitted from the several points of the object, are supposed to pass to the eye, and in their passage to leave marks that represent them on the said Plane.—Some call this the Table, or Picture, because the draught or Perspective of the object is supposed to be upon it. Others call it the Section, from its cutting the visual rays; and others again the Glass, from its supposed transparency.

Geometrical PLANE, in Perspective, is a Plane parallel to the horizon, upon which the object is supposed to be placed that is to be drawn.

Horizontal PLANE, in Perspective, is a Plane passing through the spectator's eye, parallel to the horizon.

Vertical PLANE, in Perspective, is a Plane passing through the spectator's eye, perpendicular to the geometrical Plane, and usually at right angles to the perspective Plane.

Objective PLANE, in Perspective, is any Plane situate in the horizontal Plane, of which the representation in perspective is required.

PLANE *of the Horopter*, in Optics, is a Plane passing through the horopter AB, and perpendicular to a Plane passing through the two optic axes CH and CI. See the fig. to the article HOROPTER.

PLANE *of the Projection*, is the Plane upon which the sphere is projected.

PLANE *Angle*, is an angle contained under two lines or surfaces.—It is so called in contradistinction to a solid angle, which is formed by three or more Planes.

PLANE *Triangle*, is a triangle formed by three right lines; in opposition to a spherical and a mixt triangle.

PLANE *Trigonometry* is the doctrine of Plane triangles, their measures, proportions, &c. See TRIGONOMETRY.

PLANE *Glass*, or *Mirror*, in Optics, is a glass or mirror having a flat or even surface.

PLANE *Chart*, in Navigation, is a sea-chart, having the meridians and parallels represented by parallel straight lines; and consequently having the degrees of longitude the same in every part. See CHART.

PLANE *Number*, is that which may be produced by the multiplication of two numbers the one by the other. Thus, 6 is a plane number, being produced by the multiplication of the two numbers 2 and 3; also 15 is a Plane number, being produced by the multiplication of the numbers 3 and 5. See NUMBER.

PLANE *Place*, *Locus Planus*, or *Locus ad Planum*, is a term used by the ancient geometricians, for a geometrical locus, when it was a right line or a circle, in opposition to a solid place, which was one of the conic sections.

These Plane Loci are distinguished by the Moderns into Loci ad Rectum, and Loci ad Circulum. See LOCUS.

PLANE *Problem*, is such a one as cannot be resolved geometrically, but by the intersection either of a right line and a circle, or of the circumferences of two circles. Such as this problem following: viz, Given the hypothenuse, and the sum of the other two sides, of a right-angled triangle; to find the triangle. Or this: Of four given lines to form a trapezium of a given area.

PLANE *Sailing*, in Navigation, is the art of working the several cases and varieties in a ship's motion on a Plane chart; or of navigating a ship upon principles

H h 2   deduced

deduced from the notion of the earth's being an extended Plane.

This principle, though notoriously falfe, yet places being laid down accordingly, and a long voyage broken into many fhort ones, the voyage may be performed tolerably well by it, efpecially near the fame meridian.

In Plain Sailing it is fuppofed that thefe three, the rhumb line, the meridian, and parallel of latitude, will always form a right-angled triangle ; and fo pofited, as that the perpendicular fide will reprefent part of the meridian, or north and fouth line, containing the difference of latitude ; the bafe of the triangle, the departure, or eaft-and weft line ; and the hypothenufe the diftance failed. The angle at the vertex is the courfe ; and the angle at the bafe, the complement of the courfe ; any two of which, befides the right angle, being given, the triangle may be protracted, and the other three parts found.

For the doctrine of Plane Sailing, fee SAILING.

PLANE *Scale*, is a thin ruler, upon which are graduated the lines of chords, fines, tangents, fecants, leagues, rhumbs, &c ; being of great ufe in moft parts of the mathematics, but efpecially in navigation. See its defcription and ufe under SCALE.

PLANE *Table*, an inftrument much ufed in land-furveying ; by which the draught, or plan, is taken upon the fpot, as the furvey or meafurement goes on, without any future protraction, or plotting.

This inftrument confifts of a Plane rectangular board, of any convenient fize, the centre of which, when ufed, is fixed by means of fcrews to a three-legged ftand, having a ball and focket, or univerfal joint, at the top, by means of which, when the legs are fixed on the ground, the table is inclined in any direction. To the table belongs,

1. A frame of wood, made to fit round its edges, for the purpofe of fixing a fheet of paper upon the table. The one fide of this frame is ufually divided into equal parts, by which to draw lines acrofs the table, parallel or perpendicular to the fides ; and the other fide of the frame is divided into 360 degrees, from a centre which is in the middle of the table ; by means of which the table is to be ufed as a theodolite, &c.

2. A magnetic needle and compafs fcrewed into the fide of the table, to point out directions and be a check upon the fights.

3. An index, which is a brafs two foot fcale, either with a fmall telefcope, or open fights erected perpendicularly upon the ends. Thefe fights and the fiducial edge of the index are parallel, or in the fame Plane.

General Ufe of the PLANE Table.

To ufe this inftrument properly, take a fheet of writing or drawing paper, and wet it to make it expand ; then fpread it flat upon the table, preffing down the frame upon the edges, to ftretch it, and keep it fixed there ; and when the paper is become dry, it will, by fhrinking again, ftretch itfelf fmooth and flat from any cramps or unevennefs. Upon this paper is to be drawn the plan or form of the thing meafured.

The general ufe of this inftrument, in land-furveying, is to begin by fetting up the table at any part of the ground you think the moft proper, and make a point upon a convenient part of the paper or table, to repre-

fent that point of the ground ; then fix in that point of the paper one leg of the compaffes, or a fine fteel pin, and apply to it the fiducial edge of the index, moving it round the table, clofe by the pin, till through the fights you perceive fome point defired, or remarkable object, as the corner of a field, or a picket fet up, &c ; and from the ftation point draw a dry or obfcure line along the fiducial edge of the index. Then turn the index to another object, and draw a line on the paper towards it. Do the fame by another ; and fo on till as many objects are fet as may be thought neceffary. Then meafure from your ftation towards as many of the objects as may be neceffary, and no more, taking the requifite off-fets to corners or crooks in the hedges, &c ; laying the meafured diftances, from a proper fcale, down upon the refpective lines on the paper. Then move the table to any of the proper places meafured to, for a fecond ftation, fixing it there in the original pofition, turning it about its centre for that purpofe, both till the magnetic needle point to the fame degree of the compafs as at firft, and alfo by laying the fiducial edge of the index along the line between the two ftations, and turning the table till through the index the former ftation can be feen ; and then fix the table there : from this new ftation repeat the fame operations as at the former ; fetting feveral objects, that is, drawing lines towards them, on the paper, by the edge of the index, meafuring and laying off the diftances. And thus proceed from ftation to ftation ; meafuring only fuch lines as are neceffary, and determining as many as you can by interfecting lines of direction drawn from different ftations.

Of Shifting the Paper on the PLANE *Table.* When one paper is full of the lines &c meafured, and the furvey is not yet completed ; draw a line in any manner through the fartheft point of the laft ftation line to which the work can be conveniently laid down ; then take the fheet off the table, and fix another fair fheet in its place, drawing a line upon it, in a part of it the moft convenient for the reft of the work, to reprefent the line drawn at the end of the work on the former paper. Then fold or cut the old fheet by the line drawn upon it ; apply it fo to the line on the new fheet, and, as they lie together in that pofition, continue or produce the laft ftation line of the old fheet upon the new one ; and place upon it the remainder of the meafurement of that line, beginning at where the work left off on the old fheet. And fo on, from one fheet to another, till the whole work is completed.

But it is to be noted, that if the faid joining lines, upon the old and new fheet, have not the fame inclination to the fide of the table, the needle will not refpect or point to the original degree of the compafs, when the table is rectified. But if the needle be required to refpect ftill the fame degree of the compafs, the eafieft way then of drawing the lines in the fame pofition, is to draw them both parallel to the fame fides of the table, by means of the equal parallel divifions marked on the other two fides of the frame.

When the work of furveying is done, and you would faften all the fheets together into one piece, or rough plan, the aforefaid lines are to be accurately joined together, in the fame manner as when the lines were tranfferred from the old fheets to the new ones.

See more full directions for the ufe of the Plane
Table,

Table, illuftrated with various examples, in my Treatife on Menfuration, 2d edit. pa. 509 &c.

PLANET, literally a wanderer, or a wandering ftar, in oppofition to a ftar, properly fo called, which remains fixed. It is a celeftial body, revolving around the fun, or fome other planet, as a centre, or at leaft as a focus, and with a moderate degree of excentricity, fo that it never is fo much farther from the fun at one time than at another, but that it can be feen as well from one part of its orbit as another; as diftinguifhed from the comets, which on the fartheft part of their trajectory go off to fuch vaft diftances, as to remain a long time invifible.

The Planets are ufually diftinguifhed into Primary and Secondary.

Primary PLANETS, called alfo fimply Planets, are thofe which move round the fun, as their centre, or focus of their orbit. Such as Mercury, Venus, the Earth, Mars, Jupiter, Saturn, the Georgian or Herfchel, and perhaps others. And the

Secondary PLANETS, are fuch as move round fome primary one, as their centre, in the fame manner as the primary ones do about the fun. Such as the moon, which moves round the earth, as a fecondary; and the three, Jupiter, Saturn, and Georgian, have each feveral fecondary Planets, or moons, moving round them.

Till very lately the number of the primary Planets was efteemed only fix, which it was thought conftituted the whole number of them in the folar fyftem; viz, Mercury, Venus, the Earth, Mars, Jupiter, and Saturn; all of which it appears were known to the aftronomers of all ages, who never dreamt of an increafe to their number. But a feventh has been lately difcovered, by Dr. Herfchel, viz, on March the 13th, 1781, lying beyond all the reft, and now called the Georgian, or Herfchel: and poffibly others may ftill remain undifcovered to this day.

The primary Planets are again diftinguifhed into Superior and Inferior.

The Superior Planets are thofe that are above the earth, or farther from the fun than the earth is; as, Mars, Jupiter, Saturn, and the Georgian or Herfchel. And

The Inferior Planets are thofe that are below the earth, or that are nearer the fun than the earth is; which are Venus and Mercury.

The Planets were reprefented by the fame characters as the chemifts ufe to reprefent their metals by, on account of fome fuppofed analogy between thofe celeftial and the fubterraneous bodies. Thus,

Mercury, the meffenger of the Gods, reprefented by ☿, the fame as that metal, imitating a man with wings on his head and feet, is a fmall bright planet, with a light tinct of blue, the fun's conftant attendant, from whofe fide it never departs above 28°, and by that means is ufually hid in his fplendor. It performs its courfe around him in about 3 months.

Venus, the goddefs of love, marked ♀, from the figure of a woman, the fame as denotes copper, from a flight tinge of that colour, or verging to a light ftraw colour. She is a very bright Planet, revolving next above Mercury, and never appears above 48 degrees from the fun, finifhing her courfe about him in about feven months. When this Planet goes before the fun;

6

or is a morning ftar, it has been called Phofphorus, and alfo Lucifer; and when following him, or when it fhines in the evening as an evening ftar, it is called Hefperus.

Tellus, the Earth, next above Venus, is denoted by ⊕, and performs its courfe about the fun in the fpace of a year.

Mars, the god of war, characterized ♂, a man holding out a fpear, the fame as iron, is a ruddy fierycoloured Planet, and finifhes his courfe about the fun in about 2 years.

Jupiter, the chief god, or thunderer, marked ♃, to reprefent the thunderbolts, denoting the fame as tin, from his pure white brightnefs. This Planet is next above Mars, and completes its courfe round the fun in about 12 years.

Saturn, the father of the Gods, is expreffed by ♄, to imitate an old man fupporting himfelf with a ftaff, and is the fame as denotes lead, from his feeble light and dufky colour. He revolves next above Jupiter, and performs his courfe in about 30 years.

Laftly, the Georgian, or Herfchel, is denoted by ♅, the initial of his name, with a crofs for the chriftian Planet, or that difcovered by the chriftians. This is the higheft, or outermoft, of the known Planets, and revolves around the fun in the fpace of about 90 years.

From thefe defcriptions a perfon may eafily diftinguifh all the Planets, except the laft, which requires the aid of a telefcope. For if after fun-fet he fees a Planet nearer the eaft than the weft, he may conclude it is neither Venus nor Mercury; and he may determine whether it is Saturn, Jupiter, or Mars, by the colour, light, and magnitude: by which alfo he may diftinguifh between Venus and Mercury.

It is probable that all the Planets are dark opake bodies, fimilar to the earth, and for the following reafons.

1. Becaufe, in Mercury, Venus, and Mars, only that part of the difk is found to fhine which is illuminated by the fun; and again, Venus and Mercury, when between the fun and the earth, appear like maculæ or dark fpots on the fun's face: from which it is evident, that thofe three Planets are opake bodies, illuminated by the borrowed light of the fun. And the fame appears of Jupiter, from his being void of light in that part to which the fhadow of his fatellites reaches as well as in that part turned from the fun: and that his fatellites are opake, and reflect the fun's light, like the moon, is abundantly fhewn. Moreover, fince Saturn, with his ring and fatellites, and alfo Herfchel, with his fatellites, only yield a faint light, confiderably fainter than that of the reft of the Planets, and than that of the fixed ftars, though thefe be vaftly more remote; it is paft a doubt that thefe Planets too, with their attendants, are opake bodies.

2. Since the fun's light is not tranfmitted through Mercury or Venus, when placed againft him, it is plain they are denfe opake bodies; which is likewife evident of Jupiter, from his hiding the fatellites in his fhadow; and therefore, by analogy, the fame may be concluded of Saturn and Herfchel.

3. From the variable fpots of Venus, Mars, and Jupiter, it is evident that thefe Planets have a changeable atmofphere; which fort of atmofphere may, by a like argument, be inferred of the fatellites of Jupiter; and therefore,

therefore, by fimilitude, the fame may be concluded of the other Planets.

4. In like manner, from the mountains obferved in the moon and Venus, the fame may be fuppofed in the other Planets.

5. Laftly, fince all thefe Planets are opake bodies, fhining with the fun's borrowed light, are furnifhed with mountains, and are encompaffed with a changeable atmofphere; they confequently have waters, feas &c, as well as dry land, and are bodies like the moon, and therefore like the earth. And hence, it feems alfo probable, that the other Planets have their animal inhabitants, as well as our earth has.

Of the Orbits of the PLANETS.

Though all the primary Planets revolve about the fun, their orbits are not circles, but ellipfes, having the fun in one of the foci. This circumftance was firft found out by Kepler, from the obfervations of Tycho Brahe: before that, all aftronomers took the planetary orbits for eccentric circles.

The Planes of thefe orbits do all interfect in the fun; and the line in which the plane of each orbit cuts that of the earth, is called the Line of the nodes; and the two points in which the orbits themfelves touch that plane, are the Nodes; alfo the angle in which each plane cuts that of the ecliptic, is called the Inclination of the plane or orbit.—The diftance between the centre of the fun, and the centre of each orbit, is called the excentricity of the Planet, or of its orbit.

The Motions of the PLANETS.

The motions of the primary Planets are very fimple and tolerably uniform, as being compounded only of a projectile motion, forward in a right line, which is a tangent to the orbit, and a gravitation towards the fun at the centre. Befides, being at fuch vaft diftances from each other, the effects of their mutual gravitation towards one another are in a confiderable degree, though not altogether, infenfible; for the action of Jupiter upon Saturn, for ex. is found to be $\frac{1}{20\frac{1}{3}}$ of the action of the fun upon Saturn, by comparing the matter of Jupiter with that of the fun, and the fquare of the diftance of each from Saturn. So that the elliptic orbit of Saturn will be found more juft, if its focus be fuppofed not in the centre of the fun, but in the common centre of gravity of the fun and Jupiter, or rather in the common centre of gravity of the fun and all the Planets below Saturn. And in like manner, the elliptic orbit of any other Planet will be found more accurate, by fuppofing its focus to be in the common centre of gravity of the fun and all the Planets that are below it. But the matter is far otherwife, in refpect of the fecondary Planets: for every one of thefe, though it chiefly gravitates towards its refpective primary one, as its centre, yet at equal diftances from the fun, it is alfo attracted towards him with an equally accelerated gravity, as the primary one is towards him; but at a greater diftance with lefs, and at a nearer diftance with greater: from which double tendency towards the fun, and towards their own primary Planets, it happens, that the motion of the fatellites, or fecondary Planets, comes to be very much compounded, and affected with various inequalities.

The motions even of the primary Planets, in their elliptic orbits, are not equable, becaufe the fun is not in their centre, but their focus. Hence they move; fometimes fafter, and fometimes flower, as they are nearer to or farther from the fun; but yet thefe irregularities are all certain, and follow according to an immutable law. Thus, the ellipfis PEA &c reprefenting the orbit of a Planet, and the focus S the fun's place: the axis of the ellipfe AP, is the line of the apfes; the point A, the higher apfis or aphelion; P the lower apfis or perihelion; CS the eccentricity; and ES the Planet's mean diftance from the fun. Now the motion of the Planet in its perihelion P is fwifteft, but in its aphelion A it is floweft; and at E the motion as well as the diftance is a

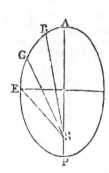

mean, being there fuch as would defcribe the whole orbit in the fame time it is really defcribed in. And the law by which the motion in every point is regulated, is this, that a line or radius drawn from the centre of the fun to the centre of the Planet, and thus carried along with an angular motion, does always defcribe an elliptic area proportional to the time; that is, the trilineal area ASB, is to the area ASG, as the time the Planet is in moving over AB, to the time it is in moving over AG. This law was firft found out by Kepler, from obfervations; and has fince been accounted for and demonftrated by Sir Ifaac Newton, from the general laws of attraction and projectile motion.

As to the periods and velocities of the Planets, or the times in which they perform their courfes, they are found to have a wonderful harmony with their diftances from the fun, and with one another: the nearer each Planet being to the fun, the quicker ftill is its motion, and its period the fhorter, according to this general and regular law; viz, that the fquares of their periodical times are as the cubes of their mean diftances from the fun or focus of their orbits. The knowledge of this law we owe alfo to the fagacity of Kepler, who found that it obtained in all the primary Planets; as aftronomers have fince found it alfo to hold good in the fecondary ones. Kepler indeed deduced this law merely from obfervation, by a comparifon of the feveral diftances of the Planets with their periods or times: the glory of inveftigating it from phyfical principles is due to Sir Ifaac Newton, who has demonftrated that, in the prefent ftate of nature, fuch a law was inevitable.

The phenomena of the Planets are, their Conjunctions, Oppofitions, Elongations, Stations, Retrogradations, Phafes, and Eclipfes; for which fee the refpective articles.

For a view of the comparative magnitudes of the Planets; and for a view of their feveral diftances, &c; fee the articles ORBIT and SOLAR SYSTEM, as alfo Plate xxi, fig. 1.

The following Table contains a fynopfis of the diftances, magnitudes, periods, &c, of the feveral Planets, according to the lateft obfervations and improvements.

TABLE

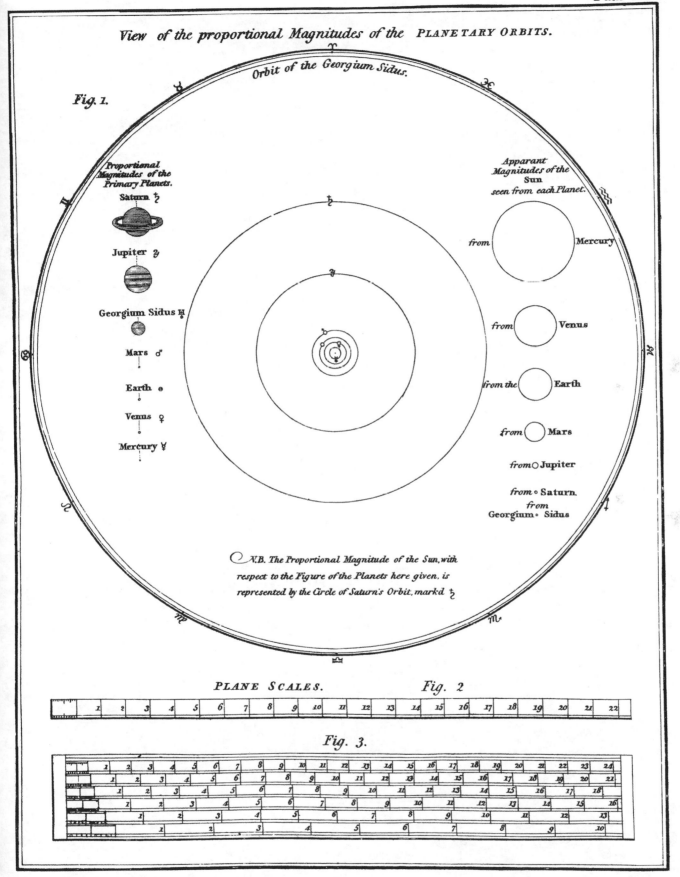

Plate XXI.

View of the proportional Magnitudes of the PLANETARY ORBITS.

Orbit of the Georgium Sidus.

Fig. 1.

Proportional
Magnitudes of the
Primary Planets.

Saturn ♄

Jupiter ♃

Georgium Sidus ♅

Mars ♂

Earth ⊕

Venus ♀

Mercury ☿

Apparant
Magnitudes of the
Sun
seen from each Planet.

from   Mercury

from   Venus

from the   Earth

from   Mars

from Jupiter

from Saturn
from
Georgium Sidus

N.B. The Proportional Magnitude of the Sun, with
respect to the Figure of the Planets here given, is
represented by the Circle of Saturn's Orbit, mark'd ♄

PLANE SCALES.   Fig. 2

Fig. 3.

TABLE *of the* PLANETARY MOTIONS, DISTANCES, &c.

Anno 1784.	MERCURY.	VENUS.	EARTH.	MARS.	JUPITER.	SATURN.	HERSCHEL, or GEORGIAN, 1782.
Greatest Elongation of Inferior, and Parallax of Superior Planets.	$28°\ 20'$	$47°\ 48'$	$*\ *$	$47°\ 24'$	$11°\ 51'$	$6°\ 29'$	$3°\ 4'\frac{1}{4}$
Periodical Revolutions round the Sun.	$87^{d}\ 23^{h}\ 15\frac{1}{2}^{m}$	$224^{d}\ 16^{h}\ 49\frac{1}{4}^{m}$	$365^{d}\ 6^{h}\ 9\frac{1}{4}^{m}$	$686^{d}\ 23^{h}\ 30\frac{3}{4}^{m}$	$4332^{d}\ 8^{h}\ 51\frac{1}{2}^{m}$	$10761^{d}\ 14^{h}\ 36\frac{3}{4}^{m}$	$30445^{d}\ 18^{h}$
Diurnal Rotations upon their Axes.	$*\ *\ *$	$23^{h}\ 22^{m}$	$23^{h}\ 56^{m}\ 4^{s}$	$24^{h}\ 39^{m}\ 22^{s}$	$9^{h}\ 56^{m}$	$*\ *$	$*\ *$
Inclinations of their Orbits to the Ecliptic.	$7°\ 0'$	$3°\ 23'\frac{1}{3}$	$*\ *$	$1°\ 51'$	$1°\ 19'\frac{1}{4}$	$2°\ 30'\frac{1}{3}$	$48'\ 0''$
Place of the Ascending Node.	$1^{s}\ 15°\ 46'\frac{3}{4}$	$2^{s}\ 14°\ 44'$	$*\ *\ *$	$1^{s}\ 17°\ 59'$	$3^{s}\ 8°\ 50'$	$3^{s}\ 21°\ 48'\frac{1}{4}$	$3^{s}\ 13°\ 1'$
Place of the Aphelion, or point farthest from the Sun.	$8^{s}\ 14°\ 13'$	$10^{s}\ 9°\ 38'$	$9^{s}\ 9°\ 15'\frac{1}{4}$	$5^{s}\ 2°\ 6'\frac{1}{4}$	$6^{s}\ 10°\ 57\frac{1}{2}'$	$9^{s}\ 0°\ 45'\frac{1}{2}$	$11^{s}\ 23°\ 23'$
Greatest Apparent Diameters, seen from the Earth.	$11''$	$58''$	$*$	$25''$	$46''$	$20''$	$4''$
Diameters in English Miles; that of the Sun being 883217.	3222	7687	7964	4189	89170	79042	35109
Proportional Mean Distances from the Sun.	38710	72333	100000	152369	520098	953937	1903421
Mean Distances from the Sun in Semidiameters of the Earth.	9210	17210	23799	36262	123778	227028	453000
Mean Distances from the Sun in English Miles.	37 millions	68 millions	95 millions	144 millions	490 millions	900 millions	1800 millions
Eccentricities or Distance of the Focus from the Centre.	7960	510	1680	14218	25277	53163	4759
Proportion of Light and Heat; that of the Earth being 100.	668	191	100	43	3·7	1·1	0·276
Proportion of Bulk; that of the Sun being 1380000.	$\frac{1}{15}$	$\frac{8}{9}$	1	$\frac{7}{24}$	$1\frac{2}{5}$	1000	90
Proportion of Density; that of the Sun being $\frac{1}{4}$.	2	$1\frac{1}{4}$	1	·7	·23	02	$*$

A Planet's motion, or diftance from its apogee, is called the mean anomaly of the Planet, and is meafured by the area it defcribes in the given time: when the Planet arrives at the middle of its orbit, or the point E, the area or time is called the true anomaly. When the Planet's motion is reckoned from the firft point of Aries, it is called its motion in longitude; which is either mean or true; viz, mean, which is fuch as it would have were it to move uniformly in a circle; and true, which is that with which the Planet actually defcribes its orbit, and is meafured by the arc of the ecliptic it defcribes. And hence may be found the Planet's place in its orbit for any given time after it has left the aphelion: for fuppofe the area of the ellipfis be fo divided by the line SG, that the whole elliptic area may have the fame proportion to the part ASG, as the whole periodical time in which the Planet defcribes its whole orbit, has to the given time; then will G be the Planet's place in its orbit fought.

PLANETARIUM, an aftronomical machine, contrived to reprefent the motions, orbits, &c, of the planets, as they really are in nature, or according to the Copernican fyftem. The larger fort of them are called Orreries. See ORRERY.

A very remarkable machine of this fort was invented by Huygens, and defcribed in his Opufc. Pofth. tom. 2. p. 157, edit. Amft. 1728. And it is ftill preferved among the curiofities of the univerfity at Leyden.

In this Planetarium, the five primary planets perform their revolutions about the fun, and the moon performs her revolution about the earth, in the fame time that they are really performed in the heavens. Alfo the orbits of the moon and planets are reprefented with their true proportions, eccentricity, pofition, and declination from the ecliptic or orbit of the earth. So that by this machine the fituation of the planets, with the conjunctions, oppofitions, &c, may be known, not only for the prefent time, but for any other time either paft or yet to come; as in a perpetual ephemeris.

There was exhibited in London, viz. in the year 1791, a ftill much more complete Planetarium of this fort; called " a Planetarium or aftronomical machine, which exhibits the moft remarkable phenomena, motions, and revolutions of the univerfe. Invented, and partly executed, by the celebrated M. Phil. Matthew Hahn, member of the academy of fciences at Erfurt. But finished and completed by Mr. Albert de Mylius." This is a moft ftupendous and elaborate machine; confifting of the folar fyftem in general, with all the orbits and planets in their due proportions and pofitions; as alfo the feveral particular planetary fyftems of fuch as have fatellites, as of the earth, Jupiter, &c; the whole kept in continual motion by a chronometer, or grand eight-day clock; by which all thefe fyftems are made perpetually to perform all their motions exactly as in nature, exhibiting at all times the true and real motions, pofitions, afpects, phenomena, &c, of all the celeftial bodies, even to the very diurnal rotation of the planets, and the unequal motions in their elliptic orbits. A defcription was publifhed of this moft fuperb machine; and it was purchafed and fent as one of the prefents to the emperor of China, in the embaffy of Lord Macartney, in the year 1793.

But the Planetariums or orreries now moft commonly ufed, do not reprefent the true times of the celeftial motions, but only their proportions; and are not kept in continual motion by a clock, but are only turned round occafionally with the hand, to help to give young beginners an idea only of the planetary fyftem; as alfo, if conftructed with fufficient accuracy, to refolve problems, in a coarfe way, relating to the motions of the planets, and of the earth and moon, &c.

Dr. Defaguliers (Exp Philof. vol. 1, p. 430.) defcribes a Planetarium of his own contrivance, which is one of the beft of the common fort. The machine is contrived to be rectified or fet to any latitude; and then by turning the handle of the Planetarium, all the planets perform their revolutions round the fun in proportion to their periodical times, and they carry indices which fhew the longitudes of the planets, by pointing to the divifions graduated on circles for that purpofe.

The Planetarium reprefented in fig. 1, plate xxii. is an inftrument contrived by Mr. Wm. Jones, of Holborn, London, mathematical inftrument maker, who has paid confiderable attention to fuch machines, to bring them to a great degree of fimplicity and perfection. It reprefents in a general manner, by various parts of its machinery, all the motions and phenomena of the planetary fyftem. This machine confifts of, the Sun in the centre, with the Planets in the order of their diftance from him, viz. Mercury, Venus, the Earth and Moon, Mars, Jupiter with his moons, and Saturn with his ring and moons; and to it is alfo occafionally applied an extra long arm for the Georgian Planet and his two moons. To the earth and moon is applied a frame CD, containing only four wheels and two pinions, which ferve to preferve the earth's axis in its due parallelifm in its motion round the fun, and to give the moon at the fame time her due revolution about the earth. Thefe wheels are connected with the wheelwork in the round box below, and the whole is fet in motion by the winch H. The arm M that carries round the moon, points out on the plate C her age and phafes for any fituation in her orbit, upon which they are engraved. In like manner the arm points out her place in the ecliptic B, in figns and degrees, called her geocentric place, that is, as feen from the earth. The moon's orbit is reprefented by the flat rim A; the two joints of it, upon which it turns, denoting her nodes; and the orbit being made to incline to any required angle. The terrella, or little earth, of this machine, is ufually made of a three inch globe papered, &c, for the purpofe; and by means of the terminating wire that goes over it, points out the changes of the feafons, and the different lengths of days and nights more confpicuoufly. By this machine are feen at once all the Planets in motion about the Sun, with the fame refpective velocities and periods of revolution which they have in the heavens; the wheelwork being calculated to a minute of time, from the lateft difcoveries. See Mr. Jones's Defcription of his new portable Orrery.

PLANETARY, fomething that relates to the planets. Thus, we fay Planetary worlds, Planetary inhabitants, Planetary motions, &c. Huygens and Fontenelle bring feveral probable arguments for the reality of Planetary worlds, animals, plants, men, &c.

PLANETARY Syftem, is the fyftem, or affemblage of the Planets, primary and fecondary, moving in their refpective

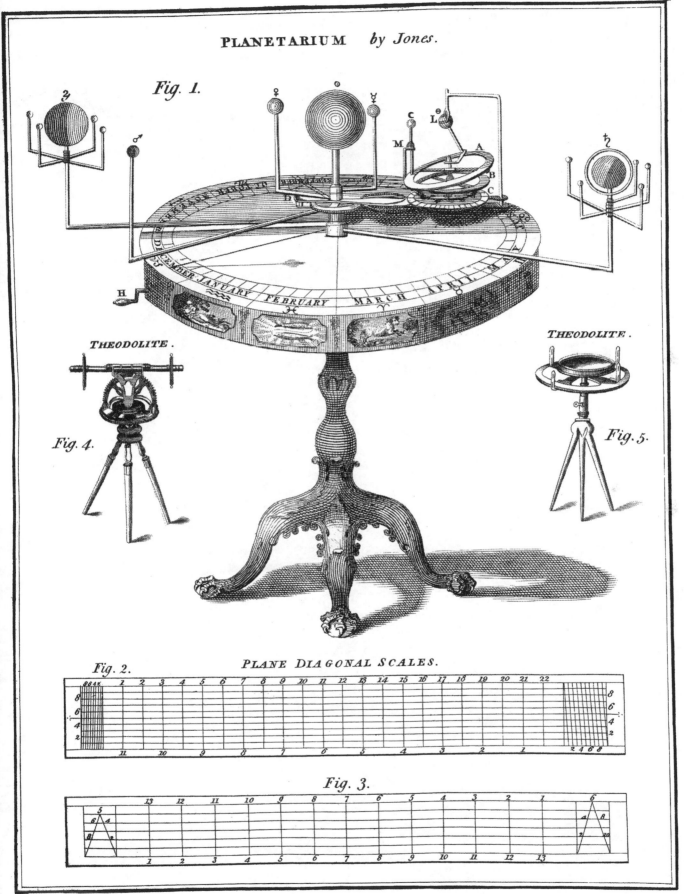

Plate XXII.

PLANETARIUM *by Jones.*

Fig. 1.

THEODOLITE.

Fig. 4.

THEODOLITE.

Fig. 5.

Fig. 2. PLANE DIAGONAL SCALES.

Fig. 3.

respective orbits, round their common centre the sun. See *Solar* SYSTEM.

PLANETARY *Days*. With the Ancients, the week was shared among the seven planets, each planet having its day. This we learn from Dion Cassius and Plutarch, Sympos. lib. 4. q. 7. Herodotus adds, that it was the Egyptians who first discovered what god, that is what planet, presides over each day; for that among this people the planets were directors. And hence it is, that in most European languages the days of the week are still denominated from the planets; as Sunday, Monday, &c.

PLANETARY *Dials*, are such as have the Planetary hours inscribed on them.

PLANETARY *Hours*, are the 12th parts of the artificial day and night. See *Planetary* HOUR.

PLANETARY *Squares*, are the squares of the seven numbers from 3 to 9, disposed magically. Cornelius Agrippa, in his book of magic, has given the construction of the seven Planetary squares. And M. Poignard, canon of Brussels, in his treatise on sublime squares, gives new, general, and easy methods, for making the seven Planetary squares, and all others to infinity, by numbers in all sorts of progressions. See MAGIC *square*.

PLANETARY *Years*, the periods of time in which the several planets make their revolutions round the sun, or earth.—As from the proper revolution of the earth, or the apparent revolution of the sun, the solar year takes its original; so from the proper revolutions of the rest of the planets about the earth, as many sorts of years do arise; viz. the Saturnian year, which is defined by 29 Egyptian years 174 days 58 minutes, equivalent in a round number to 30 solar years. The Jovial year, containing 11 years 317 days 14 hours 59 minutes. The Martial year, containing 1 year 321 days 23 hours 31 minutes. For Venus and Mercury, as their years, when judged of with regard to the earth, are almost equal to the solar year; they are more usually estimated from the sun, the true centre of their motions: in which case the former is equal to 224 days 16 hours 49 minutes; and the latter to 87 days 23 hours 16 minutes.

PLANIMETRY, that part of geometry which considers lines and plane figures, without any regard to heights or depths.—Planimetry is particularly restricted to the mensuration of planes and other surfaces; as contradistinguished from Stereometry, or the mensuration of solids, or capacities of length, breadth and depth.

Planimetry is performed by means of the squares of long measures, as square inches, square feet, square yards, &c; that is, by squares whose side is an inch, a foot, a yard, &c. So that the area or content of any surface is said to be found, when it is known how many such square inches, feet, yards, &c, it contains. See MENSURATION and SURVEYING.

PLANISPHERE, a projection of the sphere, and its various circles, on a plane; as upon paper or the like. In this sense, maps of the heavens and the earth, exhibiting the meridians and other circles of the sphere, may be called Planispheres.

Planisphere is sometimes also considered as an astronomical instrument, used in observing the motions of the heavenly bodies; being a projection of the celestial sphere upon a plane, representing the stars, constellations,

&c, in their proper situations, distances, &c. As the Astrolabe, which is a common name for all such projections.

In all Planispheres, the eye is supposed to be in a point, viewing all the circles of the sphere, and referring them to a plane beyond them, against which the sphere is as it were flattened: and this plane is called the Plane of Projection, which is always some one of the circles of the sphere itself, or parallel to some one.

Among the infinite number of Planispheres which may be furnished by the different planes of projection, and the different positions of the eye, there are two or three that have been preferred to the rest. Such as that of Ptolomy, where the plane of projection is parallel to the equator: that of Gemma Frisius, where the plane of projection is the colure, or solstitial meridian, and the eye the pole of the meridian, being a stereographical projection: or that of John de Royas, a Spaniard, whose plane of projection is a meridian, and the eye placed in the axis of that meridian, at an infinite distance; being an orthographical projection, and called the Analemma.

PLANO-*Concave* glass or lens, is that which is plane on one side, and concave on the other. And

PLANO-*Convex* glass or lens, is that which is plane on one side, and convex on the other. See LENS.

PLAT-BAND, in Architecture, is any flat square moulding, whose height much exceeds its projecture. Such are the faces of an architrave, and the Platbands of the modillions of a cornice.

PLATFORM, in Artillery and Gunnery, a small elevation, or a floor of wood, stone, or the like, on which cannon, &c, are placed, for more convenienty working and firing them.

PLATFORM, in Architecture, a row of beams that support the timber-work of a roof, lying on the top of the walls, where the entablature ought to be raised. Also a kind of flat walk, or plane floor, on the top of a building; from whence a fair view may be taken of the adjacent grounds. So, an edifice is said to be covered with a Platform, when it has no arched roof.

PLATO, one of the most celebrated among the ancient philosophers, being the founder of the sect of the Academics, was the son of Aristo, and born at Athens, about 429 years before Christ. He was of a royal and illustrious family, being descended by his father from Codrus, and by his mother from Solon. The name given him by his parents was *Aristocles;* but being of a robust make, and remarkably broad-shouldered, from this circumstance he was nick-named *Plato* by his wrestling-master, which name he retained ever after.

From his infancy, Plato distinguished himself by his lively and brilliant imagination. He eagerly imbibed the principles of poetry, music, and painting. The charms of philosophy however prevailing, drew him from those of the fine arts; and at the age of twenty he attached himself to Socrates only, who called him the *Swan of the Academy.* The disciple profited so well of his master's lessons, that at twenty-five years of age he had the reputation of a consummate sage. He lived with Socrates for eight years, in which time he committed to writing, according to the custom of the students, the purport of a great number of his master's excellent lectures, which he digested by way of philosophical conversations; but made so many judicious additions and improvements of his own, that Socrates, hearing him one day recite his Lysis, cried out, O Hercules! how many fine sentiments does this young man ascribe to me, that I never thought of! And Laertius assures us, that he composed several discourses which Socrates had no manner of hand in. At the time when Socrates was first arraigned, Plato was a junior senator, and he assumed the orator's chair to plead his master's cause, but was interrupted in that design, and the judges passed sentence of condemnation upon Socrates. Upon this occasion Plato begged him to accept from him a sum of money sufficient to purchase his enlargement, but Socrates peremptorily refused the generous offer, and suffered himself to be put to death.

The philosophers who were at Athens were so alarmed at the death of Socrates, that most of them fled, to avoid the cruelty and injustice of the government. Plato retired to Megara, where he was kindly entertained by Euclid the philosopher, who had been one of the first scholars of Socrates, till the storm should be over. Afterwards he determined to travel in pursuit of knowledge; and from Megara he went to Italy, where he conferred with Eurytus, Philolaus, and Archytas, the most celebrated of the Pythagoreans, from whom he learned all his natural philosophy, diving into the most profound and mysterious secrets of the Pythagoric doctrines. But perceiving other knowledge to be connected with them, he went to Cyrene, where he studied geometry and other branches of mathematics under Theodorus, a celebrated master.

Hence he travelled into Egypt, to learn the theology of their priests, with the sciences of arithmetic, astronomy, and the nicer parts of geometry. Having taken also a survey of the country, with the course of the Nile and the canals, he settled some time in the province of Sais, learning of the sages there their opinions concerning the universe, whether it had a beginning, whether it moved wholly or in part, &c; also concerning the immortality and transmigration of souls: and here it is also thought he had some communication with the books of Moses.

Plato's curiosity was not yet satisfied. He travelled into Persia, to consult the magi as to the religion of that country. He designed also to have penetrated into India, to learn of the Brachmans their manners and customs; but was prevented by the wars in Asia.

Afterwards, returning to Athens, he applied himself to the teaching of philosophy, opening his school in the Academia, a place of exercise in the suburbs of the city; from whence it was that his followers took the name of Academics.

Yet, settled as he was, he made several excursions abroad: one in particular to Sicily, to view the fiery ebullitions of Mount Etna. Dionysius the tyrant then reigned at Syracuse; a very bad man. Plato however went to visit him; but, instead of flattering him like a courtier, reproved him for the disorders of his court, and the injustice of his government. The tyrant, not used to disagreeable truths, grew enraged at Plato, and would have put him to death, if Dion and Aristomenes, formerly his scholars, and then favourites of that prince, had not powerfully interceded for him. Dionysius

however delivered him into the hands of an envoy of the Lacedemonians, who were then at war with the Athenians: and this envoy, touching upon the coast of Ægina, sold him for a slave to a merchant of Cyrene; who, as soon as he had bought him, liberated him, and sent him home to Athens.

Some time after, he made a second voyage into Sicily, in the reign of Dionysius the younger; who sent Dion, his minister and favourite, to invite him to court, that he might learn from him the art of governing his people well. Plato accepted the invitation, and went; but the intimacy between Dion and Plato raising jealousy in the tyrant, the former was disgraced, and the latter sent back to Athens. But Dion, being taken into favour again, persuaded Dionysius to recall Plato, who received him with all the marks of goodwill and friendship that a great prince could give. He sent out a fine galley to meet him, and went himself in a magnificent chariot, attended by all his court, to receive him. But this prince's uneven temper hurried him into new suspicions. It seems indeed that these apprehensions were not altogether groundless: for Ælian says, and Cicero was of the same opinion, that Plato taught Dion how to dispatch the tyrant, and to deliver the people from oppression. However this may be, Plato was offended and complained; and Dionysius, incensed at these complaints, resolved to put him to death: but Archytas, who had great interest with the tyrant, being informed of it by Dion, interceded for the philosopher, and obtained leave for him to retire.

The Athenians received him joyfully at his return, and offered him the administration of the government; but he declined that honour, choosing rather to live quietly in the Academy, in the peaceable contemplation and study of philosophy; being indeed so desirous of a private retirement that he never married. His fame drew disciples from all parts, when he would admit them, as well as invitations to come to reside in many of the other Grecian states; but the three that most distinguished themselves, were Spusippus his nephew, who continued the Academy after him, Xenocrates the Caledonian, and the celebrated Aristotle. It is said also that Theophrastus and Demosthenes were two of his disciples. He had it seems so great a respect for the science of geometry and the mathematics, that he had the following inscription painted in large letters over the door of his academy; Let no one enter here, unless he has a taste for Geometry and the Mathematics!

But as his great reputation gained him on the one hand many disciples and admirers, so on the other it raised him some emulators, especially among his fellow-disciples, the followers of Socrates. Xenophon and he were particularly disaffected to each other. Plato was of so quiet and even a temper of mind, even in his youth, that he never was known to express a pleasure with any greater emotion than that of a smile; and he had such a perfect command of his passions, that nothing could provoke his anger or resentment; from hence, and the subject and style of his writings, he acquired the appellation of the *Divine Plato*. But although he was naturally of a reserved and very pensive disposition; yet, according to Aristotle, he was affable, courteous, and perfectly good-natured; and sometimes would condescend fcend to crack little innocent jokes on his intimate acquaintances. Of his affability there needs no greater proof than his civil manner of conversing with the philosophers of his own times, when pride and envy were at their height. His behaviour to Diogenes is always mentioned in his history. This Cynic was greatly offended, it feems, at the politeness and fine taste of Plato, and used to catch all opportunities of snarling at him. Dining one day at his table with other company, when trampling upon the tapestry with his dirty feet, he uttered this brutish sarcasm, " I trample upon the pride of Plato :" to which the latter wisely and calmly replied, " with a greater pride."

This extraordinary man, being arrived at 81 years of age, died a very easy and peaceable death, in the midst of an entertainment, according to some ; but, according to Cicero, as he was writing. Both the life and death of this philosopher were calm and undisturbed ; and indeed he was finely composed for happiness. Beside the advantages of a noble birth, he had a large and comprehensive understanding, a vast fund of wit and good taste, great evenness and sweetness of temper, all cultivated and refined by education and travel ; so that it is no wonder he was honoured by his countrymen, esteemed by strangers, and adored by his scholars. Tully perfectly adored him : he tells us that he was justly called by Panætius, the divine, the most wise, the most sacred, the Homer of philosophers ; thinks, that if Jupiter had spoken Greek, he would have done it in Plato's style, &c. But, panegyric aside, Plato was certainly a very wonderful man, of a large and comprehensive mind, an imagination infinitely fertile, and of a most flowing and copious eloquence. However, the strength and heat of fancy prevailing over judgment in his composition, he was too apt to soar beyond the limits of earthly things, to range in the imaginary regions of general and abstracted ideas ; on which account, though there is always a greatness and sublimity in his manner, he did not philosophize so much according to truth and nature as Aristotle, though Cicero did not scruple to give him the preference.

The writings of Plato are all in the way of dialogue, where he seems to deliver nothing from himself, but every thing as the sentiments and opinions of others, of Socrates chiefly, of Timæus, &c. His style, as Aristotle observed, is between prose and verse : on which account some have not scrupled to rank him among the poets : and indeed, beside the elevation and grandeur of his style, his matter is frequently the offspring of imagination, instead of doctrines or truths deduced from nature. The first edition of Plato's works in Greek, was printed by Aldus at Venice in 1513 : but a Latin version of them by Marsilius Ficinus had been printed there in 1491. They were reprinted together at Lyons in 1588, and at Francfort in 1602. The famous printer Henry Stephens, in 1578, gave a beautiful and correct edition of Plato's works at Paris, with a new Latin version by Serranus, in three volumes folio.

PLATONIC, something that relates to Plato, his school, philosophy, opinions, or the like.

PLATONIC Bodies, so called from Plato who treated of them, are what are otherwise called the regular bodies. They are five in number ; the tetraedron, the hexaedron, the octaedron, the dodecaedron, and the icosaedron. See each of these articles, as also REGULAR BODIES.

PLATONIC Year, or the Great Year, is a period of time determined by the revolution of the equinoxes, or the time in which the stars and constellations return to their former places, in respect of the equinoxes.

The Platonic year, according to Tycho Brahe, is 25816 solar years, according to Riccioli 25920, and according to Cassini 24800 years.

This period being once accomplished, it was an opinion among the ancients, that the world was to begin anew, and the same series of things to return over again.

PLATONISM, the doctrine and sentiments of Plato and his followers, with regard to philosophy, &c. His disciples were called Academics, from Academia, the name of a villa in the suburbs of Athens where he opened his school. Among these were Xenocrates, Aristotle, Lycurgus, Demosthenes, and Isocrates. In physics, he chiefly followed Heraclitus ; in ethics and politics, Socrates ; and in metaphysics, Pythagoras.

After his death, two of the principal of his disciples, Xenocrates and Aristotle, continuing his office, and teaching, the one in the Academy, the other in the Lycæum, formed two sects, under different names, though in other respects the same ; the one retaining the denomination of ACADEMICS, the other assuming that of PERIPATETICS. See these two articles.

Afterwards, about the time of the first ages of Christianity, the followers of Plato quitted the title of Academists, and took that of Platonists. It is supposed to have been at Alexandria, in Egypt, that they first assumed this new title, after having restored the ancient academy, and re-established Plato's sentiments ; which had many of them been gradually dropped and laid aside. Porphyry, Plotin, Iamblichus, Proclus, and Plutarch, are those who acquired the chief reputation among the Greek Platonists ; Apuleius and Chalcidius, among the Latins ; and Philo Judæus, among the Hebrews. The modern Platonists own Plotin the founder, or at least the reformer, of their sect.

The Platonic philosophy appears very consistent with the Mosaic ; and many of the primitive fathers follow the opinions of that philosopher, as being favourable to Christianity. Justin is of opinion that there are many things in the works of Plato which this philosopher could not learn from mere natural reason ; but thinks he must have learnt them from the books of Moses, which he might have read when in Egypt. Hence Numenius the Pythagorean expressly calls Plato the Attic Moses, and upbraids him with plagiarism ; because he stole his doctrine concerning God and the world from the books of Moses. Theodoret says expressly, that he has nothing good and commendable concerning the Deity and his worship, but what he took from the Hebrew theology ; and Clemens Alexandrinus calls him the Hebrew Philosopher. Gale is very particular in his proof of the point, that Plato borrowed his philosophy from the Scriptures, either immediately, or by means of tradition ; and, beside the authority of the ancient writers, he brings some arguments from the thing itself. For example, Plato's confession, that the Greeks borrowed their knowledge of the one infinite God, from an ancient people, better and nearer to God than they; by which people, our author makes no doubt, he meant the Jews, from his account of the state of innocence; as, that man was born of the earth, that he was naked, that he enjoyed a truly happy state, that he conversed with brutes, &c. In fact, from an examination of all the parts of Plato's philosophy, physical, metaphysical, and ethical, this author finds, in every one, evident marks of its sacred original.

As to the manner of the creation, Plato teaches, that the world was made according to a certain exemplar, or idea, in the divine architect's mind. And all things in the universe, in like manner, he shews, do depend on the efficacy of internal ideas. This ideal world is thus explained by Didymus: ' Plato supposes certain patterns, or exemplars, of all sensible things, which he calls ideas; and as there may be various impressions taken off from the same seal, so he says are there a vast number of natures existing from each idea.' This idea he supposes to be an eternal essence, and to occasion the several things in nature to be such as itself is. And that most beautiful and perfect idea, which comprehends all the rest, he maintains to be the world.

Farther, Plato teaches that the universe is an intelligent animal, consisting of a body and a soul, which he calls *the generated God*, by way of distinction from what he calls the *immutable essence*, who was the cause of the generated God, or the universe.

According to Plato, there were two sorts of inferior and derivative gods; the mundane gods, all of which had a temporary generation with the world; and the supramundane eternal gods, which were all of them, one excepted, produced from that one, and dependent on it as their cause. Dr. Cudworth says, that Plato asserted a plurality of gods, meaning animated or intellectual beings, or dæmons, superior to men, to whom honour and worship are due; and applying the appellation to the sun, moon, and stars, and also to the earth. He asserts however, at the same time, that there was one supreme God, the self originated being, the maker of the heaven and earth, and of all those other gods. He also maintains, that the Psyche, or universal mundane soul, which is a self-moving principle, and the immediate cause of all the motion in the world, was neither eternal nor self-existent, but made or produced by God in time; and above this self-moving Psyche, but subordinate to the Supreme Being, and derived by emanation from him, he supposes an immoveable Nous or intellect, which was properly the Demiurgus, or framer of the world.

The first matter of which this body of the universe was formed, he observes, was a rude indigested heap, or chaos: Now, adds he, the creation was a mixed production; and the world is the result of a combination of necessity and understanding, that is, of matter, which he calls necessity, and the divine wisdom: yet so that mind rules over necessity; and to this necessity he ascribes the introduction and prevalence both of moral and natural evil.

The principles, or elements, which Plato lays down, are fire, air, water, and earth. He supposes two heavens, the Empyrean, which he takes to be of a fiery nature, and to be inhabited by angels, &c; and the Starry heaven, which he teaches is not adamantine, or solid, but liquid and spirable.

With regard to the human soul, Plato maintained its transmigration, and consequently its future immortality and pre-existence. He asserted, that human souls are here in a lapsed state, and that souls sinning should fall down into these earthly bodies. Eusebius expressly says, that Plato held the soul to be ungenerated, and to be derived by emanation from the first cause.

His physics, or doctrine *de corpore*, is chiefly laid down in his Timæus, where he argues on the properties of body in a geometrical manner; which Aristotle takes occasion to reprehend in him. His doctrine *de mente* is delivered in his 10th Book of Laws, and his Parmenides.

St. Augustine commends the Platonic philosophy; and even says, that the Platonists were not far from Christianity. It is also certain that most of the celebrated fathers were Platonists, and borrowed many of their explanations of scripture from the Platonic system. To account for this fact, it may be observed, that towards the end of the second century, a new sect of philosophers, called the modern, or later, Platonics, arose of a sudden, spread with amazing rapidity through the greatest part of the Roman empire, swallowed up almost all the other sects, and proved very detrimental to Christianity.

The school of Alexandria in Egypt, instituted by Ptolomy Philadelphus, renewed and reformed the Platonic philosophy. The votaries of this system distinguished themselves by the title of Platonics, because they thought that the sentiments of Plato concerning the Deity and invisible things, were much more rational and sublime than those of the other philosophers. This new species of Platonism was embraced by such of the Alexandrian Christians as were desirous to retain, with the profession of the gospel, the title, the dignity, and the habit of philosophers. Ammonius Saccas was its principal founder, who was succeeded by his disciple Plotinus, as this latter was by Porphyry, the chief of those formed in his school. From the time of Ammonius until the sixth century, this was almost the only system of philosophy publicly taught at Alexandria. It was brought into Greece by Plutarch, who renewed at Athens the celebrated Academy, from whence issued many illustrious philosophers. The general principle on which this sect was founded, was, that truth was to be pursued with the utmost liberty, and to be collected from all the different systems in which it lay dispersed. But none that were desirous of being ranked among these new Platonists, called in question the main doctrines; those, for example, which regarded the existence of one God, the fountain of all things; the eternity of the world; the dependance of matter upon the Supreme Being; the nature of souls; the plurality of gods, &c.

In the fourth century, under the reign of Valentinian, a dreadful storm of persecution arose against the Platonists; many of whom, being accused of magical practices, and other heinous crimes, were capitally convicted.

In the fifth century Proclus gave new life to the doctrine of Plato, and restored it to its former credit in Greece; with whom concurred many of the Christian doctors, who adopted the Platonic system. The

Platonic

Platonic philosophers were generally oppofers of Chriftianity ; but in the fixth century. Chalcidius gave the Pagan fyftem an evangelical afpect ; and thofe who, before it became the religion of the ftate, ranged themfelves under the ftandard of Plato, now repaired to that of Chrift, without any great change of their fyftem.

Under the emperor Juftinian. who iffued a particular edict, prohibiting the teaching of philofophy at Athens, which edict feems to have been levelled at modern Platonifm, all the celebrated philofophers of this fect took refuge among the Perfians, who were at that time the enemies of Rome ; and though they returned from their voluntary exile, when the peace was concluded between the Perfians and Romans, in 533, they could never recover their former credit, nor obtain the direction of the public fchools.

Platonifm however prevailed among the Greeks, and was by them, and particularly by Gemiftius Pletho, introduced into Italy, and eftablifhed, under the aufpices of Cofmo de Medicis, about the year 1439, who ordered Marfilius Ficinus to tranflate into Latin the works of the moft renowned Platonifts.

PLATONISTS, the followers of Plato ; otherwife called Academics, from Academia, the name of the place that philofopher chofe for his refidence at Athens.

PLEIADES, an affemblage of feven ftars in the neck of the conftellation Taurus, the bull ; although there are now only fix of them vifible to the naked eye. The largeft of thefe is of the third magnitude, and called Lucido Pleiadum.

The Greeks fabled, that the name Pleiades was given to thefe ftars from feven daughters of Atlas and Pleione one of the daughters of Oceanus, who having been the nurfes of Bacchus, were for their fervices taken up to heaven and placed there as ftars, where they ftill fhine. The meaning of which fable may be that Atlas firft obferved thefe ftars, and called them by the names of the daughters of his wife Pleione.

PLENILUNIUM, the full-moon.

PLENUM, in Phyfics, fignifies that ftate of things, in which every part of fpace, or extenfion, is fuppofed to be full of matter : in oppofition to a Vacuum, which is a fpace devoid of all matter.

The Cartefians held the doctrine of an abfolute Plenum ; namely on this principle, that the effence of matter confifts in extenfion ; and confequently, there being every where extenfion or fpace, there is every where matter : which is little better than begging the queftion.

PLINTH, in Architecture, a flat fquare member in form of a brick or tile ; ufed as the foot or foundation of columns and pillars, &c.

PLOT, in Surveying, the plan or draught of any parcel of ground ; as a field, farm, or manor, &c.

PLOTTING, in Surveying, the defcribing or laying down on paper, the feveral angles and lines, &c, of a tract of land, that has been furveyed and meafured.

Plotting is ufually performed by two inftruments, the protractor and Plotting-fcale ; the former ferving to lay off all the angles that have been meafured and fet down, and the latter all the meafured lines. See thefe two inftruments under their refpective names.

PLOTTING Scale, a mathematical inftrument chiefly ufed for the plotting of grounds in furveying, or fetting off the lengths of the lines. It is either 6, 9, or 12 inches in length, and about an inch and half broad ; being made either of box-wood, brafs, ivory, or filver ; thofe of ivory are the neateft.

This inftrument contains various fcales or divided lines, on both fides of it. On the one fide are a number of plane fcales, or fcales of equal divifions, each of a different number to the inch ; as alfo fcales of chords, for laying down angles ; and fometimes even the degrees of a circle marked on one edge, anfwering to a centre marked on the oppofite edge, by which means it ferves alfo as a protractor. On the other fide are feveral diagonal fcales, of different fizes, or different divifions to the inch ; ferving to take off lines expreffed by numbers to three dimenfions, as units, tens, and hundreds ; as alfo a fcale of divifions which are the 100th parts of a foot. But the moft ufeful of all the lines that can be laid upon this inftrument, though not always done, is a line or plane fcale upon the two oppofite edges, made thin for that purpofe. This is a very ufeful line in furveying ; for by laying the inftrument down upon the paper, with its divided edge along a line upon which are to be laid off feveral diftances, for the places of off-fets, &c ; thefe diftances are all transferred at once from the inftrument to the line on the paper, by making fmall marks or points againft the refpective divifions on the edge of the fcale. See fig. 2 & 3, plates xxi and xxii.

PLOTTING-Table, in Surveying, is ufed for a plane table, as improved by Mr. Beighton, who has obviated a good many inconveniencies attending the ufe of the common plane table. See Philof. Tranf. numb. 461, fect. 1.

PLOUGH, or PLOW, in Navigation, an ancient mathematical inftrument, made of box or pear-tree, and ufed to take the height of the fun or ftars, in order to find the latitude. This inftrument admits of the degrees to be very large, and has been much efteemed by many artifts ; though now quite out of ufe.

PLUMB-LINE, a term among artificers for a line perpendicular to the horizon.

PLUMMET, PLUMB-RULE, or PLUMB-LINE, an inftrument ufed by mafons, carpenters, &c, to draw perpendiculars ; in order to judge whether walls, &c, be upright, or planes horizontal, and the like.

PLUNGER, in Mechanics, a folid brafs cylinder, ufed as a forcer in forcing pumps.

PLUS. in Algebra, the affirmative or pofitive fign, +, fignifying more or addition, or that the quantity following it is either to be confidered as a pofitive or affirmative quantity, or that it is to be added to the other quantities ; fo 4 + 6 = 10, is read thus, 4 plus 6 is equal to 10. See AFFIRMATIVE Sign.

The more early writers of Algebra, as Lucas de Burgo, Cardan, Tartaglia, &c, wrote the word moftly at full length. Afterwards the word was contracted or abbreviated, ufing one or two of its firft letters ; which initial was, by the Germans I think, corrupted to the prefent character + ; which I find firft ufed by Stifelius, printed in his Arithmetic.

PLUVIAMETER, a machine for meafuring the quantity of rain that falls. There is defcribed in the Philof. Tranf. (numb. 473, or Abridg. x. 456), by Robert Pickering, under the name of an Ombrameter, an instrument of this kind. It consists of a tin funnel *d*, whose surface is an inch square (fig. 6, plate xx); a flat board *aa*; and a glass tube *bb*, set into the middle of it in a groove; and an index with divisions *cc*; the board and tube being of any length at pleasure. The bore of the tube is about half an inch, which Mr. Pickering says is the best size. The machine is fixed in some free and open place, as the top of the house, &c.

The Rain-gage employed at the house of the Royal Society is described by Mr. Cavendish, in the Philos. Transf. for 1776, p. 384. The veffel which receives the rain is a conical funnel, strengthened at the top by a brass ring, 12 inches in diameter. The sides of the funnel and inner lip of the brass ring are inclined to the horizon, in an angle of above 65°; and the outer lip in an angle of above 50°; which are such degrees of steepness, that there seems no probability either that any rain which falls within the funnel, or on the inner lip of the ring, shall dash out, or that any which falls

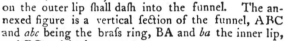

on the outer lip shall dash into the funnel. The annexed figure is a vertical section of the funnel, ABC and *abc* being the brass ring, BA and *ba* the inner lip, and BC and *bc* the outer.

Note, that in fixing Pluviameters care should be taken that the rain may have free access to them, without being impeded or overshaded by buildings, &c; and therefore the tops of houses are mostly to be preferred. Also when the quantities of rain collected in them, at different places, are compared together, the instruments ought to be fixed at the same height above the ground at both places; because at different heights the quantities are always different, even in the same place. And hence also, any register or account of rain in the Pluviameter, ought to be accompanied with a note of the height above the ground the instrument is placed at. See *Quantity of* RAIN.

PNEUMATICS, that branch of natural philosophy which treats of the weight, pressure, and elasticity of the air, or elastic fluids, with the effects arising from them. Wolfius, instead of Pneumatics, uses the term Aerometry.

This is a sifter science to Hydrostatics; the one considering the air in the same manner as the other does water. And some consider Pneumatics as a branch of mechanics; because it considers the air in motion, with the consequent effects.

For the nature and properties of air, see the article AIR, where they are pretty largely treated of. To which may be added the following, which respects more particularly the science of Pneumatics, as contained in a few propositions, and their corollaries.

PROP. I. *The Air is a heavy fluid body, which surrounds and gravitates upon all parts of the surface of the earth.*

These properties of air are proved by experience. That it is a fluid, is evident from its easily yielding to

any the least force impressed upon it, with little or no sensible resistance.

But when it is moved briskly, by any means, as by a fan, or a pair of bellows; or when any body is moved swiftly through it; in these cases we become sensible of it as a body, by the resistance it makes in such motions, and also by its impelling or blowing away any light substances. So that, being capable of resisting, or moving other bodies by its impulse, it must itself be a body, and be heavy, like all other bodies, in proportion to the matter it contains; and therefore it will press upon all bodies that are placed under it.

And being a fluid, it will spread itself all over upon the earth; also like other fluids it will gravitate upon, and press every where upon the earth's surface.

The gravity and pressure of the air is also evident from many experiments. Thus, for instance, if water, or quick-silver, be poured into the tube ACE, and the

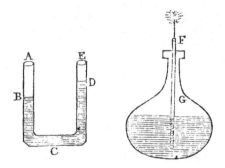

air be suffered to press upon it, in both ends of the tube; the fluid will rest at the same height in both the legs: but if the air be drawn out of one end as E, by any means; then the air pressing on the other end A, will press down the fluid in this leg at B, and raise it up in the other to D, as much higher than at B, as the pressure of the air is equal to. By which it appears, not only that the air does really press, but also what the quantity of that pressure is equal to. And this is the principle of the Barometer.

PROP. II. *The air is also an elastic fluid, being condensible and expansible. And the law it observes in this respect is this, namely, that its density is always proportional to the force by which it is compressed.*

This property of the air is proved by many experiments. Thus, if the handle of a syringe be pushed inwards, it will condense the inclosed air into a less space; by which it is shewn to be condensible. But the included air, thus condensed, will be felt to act strongly against the hand, and to resist the force compressing it more and more; and on withdrawing the hand, the handle is pushed back again to where it was at first. Which shews that the air is elastic.

Again, fill a strong bottle half full with water, and then insert a pipe into it, putting its lower end down near to the bottom, and cementing it very close round the mouth of the bottle. Then if air be strongly injected through the pipe, as by blowing with the mouth or otherwise, it will pass through the water from the lower end, and ascend up into the part before occupied

by

by the air at G, and the whole mafs of air become there condenfed becaufe the water is not eafily compreffed into a lefs fpace. But on removing the force which injected the air at F, the water will begin to rife from thence in a jet, being pufhed up the pipe by the increafed elafticity of the air G, by which it preffes on the furface of the water, and forces it through the pipe, till as much be expelled as there was air forced in; when the air at G will be reduced to the fame denfity as at firft, and, the balance being reftored, the jet will ceafe.

Likewife, if into a jar of water AB, be inverted an empty glafs tumbler C, or fuch like; the water will

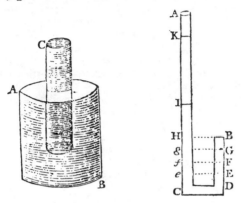

enter it, and partly fill it, but not near fo high as the water in the jar, compreffing and condenfing the air into a lefs fpace in the upper part C, and caufing the glafs to make a fenfible refiftance to the hand in pufhing it down. But on removing the hand, the elafticity of the internal condenfed air throws the glafs up again.—All thefe fhewing that the air is condenfible and elaftic.

Again, to fhew the rate or proportion of the elafticity to the condenfation; take a long flender glafs tube, open at the top A, bent near the bottom or clofe end B, and equally wide throughout, or at leaft in the part BD (2d fig. above). Pour in a little quickfilver at A, juft to cover the bottom to the bend at CD, and to ftop the communication between the external air and the air in BD. Then pour in more quickfilver, and obferve to mark the correfponding heights at which it ftands in the two legs: fo, when it rifes to H in the open leg AC, let it rife to E in the clofe one, reducing its included air from the natural bulk BD to the contracted fpace BE, by the preffure of the column H*e*; and when the quickfilver ftands at I and K, in the open leg, let it rife to F and G in the other, reducing the air to the refpective fpaces BF, BG, by the weights of the columns I*f*, K*g*. Then it is always found, that the condenfations and elafticities are as the compreffing weights, or columns of the quickfilver and the atmofphere together. So, if the natural bulk of the air BD be compreffed into the fpaces BE, BF, BG, or reduced by the fpaces DE, DF, DG, which are $\frac{1}{3}$, $\frac{1}{2}$, $\frac{2}{3}$ of BD, or as the numbers 1, 2, 3; then the atmofphere, together with the correfponding column H*e*, I*f*, K*g*, will alfo be found to be in the fame proportion. or as the numbers 1, 2, 3: and then the weights of the quickfilver are thus, viz, H*e* = $\frac{1}{3}$A, I*f* = A, and K*g* = 3A; where A denotes the weight of the atmofphere. Which fhews

that the condenfations are directly as the compreffing forces. And the elafticities are alfo in the fame proportion, fince the preffures in AC are fuftained by the elafticities in BD.

From the foregoing principles may be deduced many ufeful remarks, as in the following corollaries, viz:

Corol. 1. The fpace that any quantity of air is confined in, is reciprocally as the force that compreffes it. So, the forces which confine a quantity of air in the

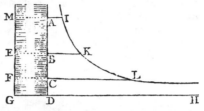

cylindrical fpaces AG, BG, CG, are reciprocally as the fame, or reciprocally as the heights AD, BD, CD. And therefore, if to the two perpendicular lines AD, DH, as afymptotes, the hyperbola IKL be defcribed, and the ordinates AI, BK, CL be drawn; then the forces which confine the air in the fpaces AG, BG, CG, will be as the correfponding ordinates AI, BK, CL, fince thefe are reciprocally as the abfciffes AD, BD, CD, by the nature of the hyperbola.

Corol. 2. All the air near the earth is in a ftate of compreffion, by the weight of the incumbent atmofphere.

Corol. 3. The air is denfer near the earth, than in high places; or denfer at the foot of a mountain, than at the top of it. And the higher above the earth, the rarer it is.

Corol. 4. The fpring or elafticity of the air, is equal to the weight of the atmofphere above it; and they will produce the fame effects; fince they are always fuftained and balanced by each other.

Corol. 5. If the denfity of the air be increafed, preferving the fame heat or temperature; its fpring or elafticity will likewife be increafed, and in the fame proportion.

Corol. 6. By the gravity and preffure of the atmofphere upon the furfaces of fluids, the fluids are made to rife in any pipes or veffels, when the fpring or preffure within is diminifhed or taken off.

PROP. III. *Heat increafes the elafticity of the air, and cold diminifhes it. Or heat expands, and cold contracts and condenfes the air.*

This property is alfo proved by experience.

Thus, tie a bladder very clofe, with fome air in it; and lay it before the fire; then as it warms, it will more and more diftend the bladder, and at laft burft it, if the heat be continued and increafed high enough. But if the bladder be removed from the fire; it will contract again to its former ftate by cooling.——It was upon this principle that the firft air-balloons were made by Montgolfier: for by heating the air within them, by a fire underneath, the hot air diftends them to a fize which occupies a fpace in the atmofphere whofe weight of common air exceeds that of the balloon.

Alfo, if a cup or glafs, with a little air in it, be inverted into a veffel of water; and the whole be heated

over

over the fire, or otherwise : the air in the top will expand till it fill the glass, and expel the water out of it ; and part of the air itself will follow, by continuing or increasing the heat.

Many other experiments to the same effect might be adduced, all proving the properties mentioned in the proposition.

Schol. Hence, when the force of the elasticity of the air is considered, regard must be had to its heat or temperature ; the same quantity of air being more or less elastic, as its heat is more or less. And it has been found by experiment that its elasticity is increased at the following rate, viz, by the 435th part, by each degree of heat expressed by Fahrenheit's thermometer, of which there are 180 between the freezing and boiling heat of water. It has also been found (Philos. Transf. 1777, pa. 560 &c), that water expands the 6666th part, with each degree of heat ; and mercury the 9600th part by each degree. Moreover, the relative or specific gravities of these three substances, are as follow : viz,

Air 1·232
Water 1000 } when the barom. is at 30,
Mercury 13600 } and the thermom. at 55.

Also these numbers are the weights of a cubic foot of each, in the same circumstances of the barometer and thermometer.

PROP. IV. *The weight or pressure of the atmosphere, upon any base at the surface of the earth, is equal to the weight of a column of quicksilver of the same base, and its height between 28 and 31 inches.*

This is proved by the barometer, an instrument which measures the pressure of the air ; the description of which see under its proper article. For at some seasons, and in some places, the air sustains and balances a column of mercury of about 28 inches ; but at others, it balances a column of 29, or 30, or near 31 inches high ; seldom in the extremes 28 or 31, but commonly about the means 29 or 30, and indeed mostly near 30. A variation which depends partly on the different degrees of heat in the air near the surface of the earth, and partly on the commotions and changes in the atmosphere, from winds and other causes, by which it is accumulated in some places, and depressed in others, being thereby rendered denser and heavier, or rarer and lighter ; which changes in its state are almost continually happening in any one place. But the medium state is from 29½ to 30 inches.

Corol. 1. Hence the pressure of the atmosphere upon every square inch at the earth's surface, at a medium, is very near 15 pounds avoirdupois. For, a cubic foot of mercury weighing nearly 13600 ounces, a cubic inch of it will weigh the 1728th part of it, or almost 8 ounces, or half a pound, which is the weight of the atmosphere for every inch of the barometer upon a base of a square inch ; and therefore 29¼ inches, the medium height of the barometer, weighs almost 15 pounds, or rather 14¾lb very nearly.

Corol. 2. Hence also the weight or pressure of the atmosphere, is equal to that of a column of water from 32 to 35 feet high, or on a medium 33 or 34 feet high. For water and quicksilver are in weight nearly as 1 to 13·6 ; so that the atmosphere will balance a column of water 13·6 times higher than one of quicksilver ; consequently 13·6 × 30 inches = 408 inches or 34 feet, is near the medium height of water, or it is more nearly 33³ feet. And hence it appears that a common sucking pump will not raise water higher than about 34 feet. And that a syphon will not run if the perpendicular height of the top of it be more than 33 or 34 feet.

Corol. 3. If the air were of the same uniform density, at every height, up to the top of the atmosphere, as at the surface of the earth ; its height would be about 5¼ miles at a medium. For the weights of the same volume of air and water, are nearly as 1·232 to 1000 ; therefore as 1·232 : 1000 : : 34 feet : 27600 feet, or 5¼ miles very nearly. And so high the atmosphere would be, if it were all of uniform density, like water. But, instead of that, from its expansive and elastic quality, it becomes continually more and more rare the farther above the earth, in a certain proportion which will be treated of below

Corol. 4. From this prop. and the last, it follows that the height is always the same, of an uniform atmosphere above any place, which shall be all of the uniform density with the air there, and of equal weight or pressure with the real height of the atmosphere above that place, whether it be at the same place at different times, or at any different places or heights above the earth ; and that height is always about 27600 feet, or 5¼ miles, as found above in the 3d corollary. For, as the density varies in exact proportion to the weight of the column, it therefore requires a column of the same height in all cases, to make the respective weights or pressures. Thus, if W and w be the weights of atmosphere above any places, D and d their densities, and H and h the heights of the uniform columns, of the same densities and weights : Then H × D = W, and $h × d = w$; therefore $\frac{W}{D}$ or H is equal to $\frac{w}{d}$ or h ; the temperature being the same.

PROP. V. *The density of the atmosphere, at different heights above the earth, decreases in such sort, that when the heights increase in arithmetical progression, the densities decrease in geometrical progression.*

Let the perpendicular line AP, erected on the earth, be conceived to be divided into a great number of very small parts A, B, C, D, &c, forming so many thin strata of air in the atmosphere, all of different density, gradually decreasing from the greatest at A : then the density of the several strata A, B, C, D, &c, will be in geometrical progression decreasing.

For, as the strata A, B, C, &c, are all of equal thickness, the quantity of matter in each of them, is as the density there ; but the density in any one, being as the compressing force, is as the weight or quantity of matter from that place upward to the top of the atmosphere ; therefore the quantity of matter in each stratum, is also as the whole quantity from that place upwards. Now if from the whole weight at any

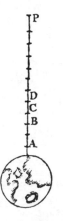

place

place as B, the weight or quantity in the ftratum B be fubtracted, the remainder will be the weight at the next higher ftratum C; that is, from each weight fubtracting a part which is proportional to itfelf, leaves the next weight; or, which is the fame thing, from each denfity fubtracting a part which is always proportional to itfelf, leaves the next denfity. But when any quantities are continually diminifhed by parts which are proportional to themfelves, the remainders then form a feries of continued proportionals; and confequently thefe denfities are in geometrical progreffion.

Thus, if the firft denfity be D, and from each there be taken its nth part; then there remains its $\frac{n-1}{n}$ part, or the $\frac{m}{n}$ part, putting m for $n-1$; and therefore the feries of denfities will be D, $\frac{m}{n}$D, $\frac{m^2}{n^2}$D, $\frac{m^3}{n^3}$D, &c, $\frac{m}{n}$ being the common ratio of the feries.

Schol. Becaufe the terms of an arithmetical feries, are proportional to the logarithms of the terms of a geometrical feries; therefore different altitudes above the earth's furface, are as the logarithms of the denfities, or weights of air, at thofe altitudes. So that,

if D denote the denfity at the altitude A,
and d the denfity at the altitude a;
then A being as the logarithm of D,
and a as the logarithm of d,
the dif. of altitude $A - a$ will be as

the log. of D − log. of d, or as log. of $\frac{D}{d}$.

And if A = o, or D the denfity at the furface of the earth, then any altitude above the furface a, is as the log. of $\frac{D}{d}$. Or, in general, the log. of $\frac{D}{d}$ is as the altitude of the one place above the other, whether the lower place be at the furface of the earth, or any where elfe.

And from this property is derived the method of determining the heights of mountains, and other eminences, by the barometer, which is an inftrument that meafures the weight or denfity of the air at any place. For by taking with this inftrument, the preffure or denfity at the foot of a hill for inftance, and again at the top of it, the difference of the logarithms of thefe two preffures, or the logarithms of their quotient, will be as the difference of altitude, or as the height of the hill; fuppofing the temperatures of the air to be the fame at both places, and the gravity of air not altered by the different diftances from the earth's centre.

But as this formula expreffes only the relations between different altitudes, with refpect to their denfities, recourfe muft be had to fome experiment, to obtain the real altitude which correfponds to any given denfity, or the denfity which correfponds to a given altitude. Now there are various experiments by which this may be done. The firft, and moft natural, is that which refults from the known fpecific gravity of air, with refpect to the whole preffure of the atmofphere on the furface of the earth. Now, as the altitude a is always as

log. $\frac{D}{d}$, affume h fo that a may be $= h \times$ log. $\frac{D}{d}$, where h will be of one conftant value for all altitudes; and to determine that value, let a cafe be taken in which we know the altitude a correfponding to a known denfity d: as for inftance take $a = 1$ foot, or 1 inch, or fome fuch fmall altitude; and becaufe the denfity D may be meafured by the preffure of the atmofphere, or the uniform column of 27600 feet, when the temperature is 55°; therefore 27600 feet will denote the denfity D at the lower place, and 27599 the lefs denfity d at one foot above it; confequently this equation arifes, viz, $1 = h \times$ log. of $\frac{27600}{27599}$, which, by the nature of logarithms, is nearly

$= h \times \frac{\cdot 43429448}{27600} = \frac{h}{63551}$ nearly; and hence $h = 63451$ feet; which gives for any altitude whatever, this general theorem, viz, $a =$

$63551 \times$ log. $\frac{D}{d}$, or $= 63551 \times$ log. $\frac{M}{m}$ feet, or

$10592 \times$ log. $\frac{M}{m}$ fathoms; where M is the column of mercury which is equal to the preffure or weight of the atmofphere at the bottom, and m that at the top of the altitude a; and where M and m may be taken in any meafure, either feet, or inches, &c.

Note, that this formula is adapted to the mean temperature of the air 55°. But for every degree of temperature different from this, in the medium between the temperatures at the top and bottom of the altitude a, that altitude will vary by its 435th part; which muft be added when the medium exceeds 55°, otherwife fubtracted.

Note alfo, that a column of 30 inches of mercury varies its length by about the 320th part of an inch for every degree of heat, or rather the 9600th part of the whole volume.

But the fame formula may be rendered much more convenient for ufe, by reducing the factor 10592 to 10000, by changing the temperature proportionably from 55°: thus, as the difference 592 is the 18th part of the whole factor 10592; and as 18 is the 24th part of 435; therefore the correfponding change of temperature is 24°, which reduces the 55° to 31°. So that the formula becomes $a = 10000 \times$ log. of $\frac{M}{m}$ fathoms when the temperature is 31 degrees; and for every degree above that, the refult muft be increafed by fo many times its 435th part.

See more on this head under the article BAROMETER, at the end.

By the weight and preffure of the atmofphere, the effect and operations of Pneumatic engines may be accounted for, and explained; fuch as fyphons, pumps, barometers, &c. See each of thefe articles, alfo AIR.

PNEUMATIC *Engine,* the fame as the AIR-PUMP.

POCKET *Electrical Apparatus.*—This is a contrivance of Mr. William Jones, in Holborn, the form of which is reprefented in plate xxiii, fig. 4.

This

This small machine is capable of a tolerably strong charge, or accumulation of electricity, and will give a small shock to one, two, three, or a greater number of persons.

A is the Leyden phial or jar that holds the charge. B is the discharger to discharge the jar when required without electrifying the person that holds it. C is a ribbon prepared in a peculiar manner so as to be excited, and communicate its electricity to the jar. D are two hair, &c, skin rubbers, which are to be placed on the first and middle fingers of the left hand.

To charge the Jar.

Place the two finger-caps D on the first and middle finger of the left hand; hold the jar A at the same time, at the joining of the red and black on the outside between the thumb and first finger of the same hand; then take the ribbon in your right hand, and steadily and gently draw it upwards between the two rubbers D, on the two fingers; taking care at the same time, the brass ball of the jar is kept nearly close to the ribbon, while it is passing through the fingers. By repeating this operation twelve or fourteen times, the electrical fire will pass into the jar which will become charged, and by placing the discharger C against it, as in the plate, you will see a sensible spark pass from the ball of the jar to that of the discharger. If the apparatus is dry and in good order, you will hear the crackling of the fire when the ribbon is passing through the fingers, and the jar will discharge at the distance represented in the figure.

To electrify a Person.

You must desire him to take the jar in one hand, and with the other touch the nob of it: or, if diversion is intended, desire the person to smell at the nob of it, in expectation of smelling the scent of a rose or a pink; this last mode has occasioned it to be sometimes called the Magic Smelling Bottle.

POETICAL *Numbers.* See NUMBERS.

POETICAL *Rising* and *Setting.* See RISING and SETTING.

The ancient poets, referring the rising and setting of the stars to that of the sun, make three kinds of rising and setting, viz, Cosmical, Acronical, and Heliacal. See each of these words in its place.

POINT, a term used in various arts and sciences.

POINT, in Architecture. Arches of the third Point, and Arches of the fourth Point. See ARCHES.

POINT, in Astronomy, is a term applied to certain parts or places marked in the heavens, and distinguished by proper epithets.

The four grand points or divisions of the horizon, viz, the east, west, north, and south, are called the Cardinal Points.—The zenith and nadir are the Vertical Points. —The Points where the orbits of the planets cut the plane of the ecliptic, are called the Nodes.—The Points where the ecliptic and equator intersect, are called the Equinoctial Points. In particular, that where the sun ascends towards the north pole is called the Vernal Point; and that where he descends towards the south, the Autumnal Point.—The highest and lowest Points of the ecliptic are called the Solstitial Points. Particularly, the former of them the Estival or Summer Point; the latter, the Brumal or Winter Point.

POINTS, in Electricity, are those acute terminations of bodies which facilitate the passage of the electrical fluid either *from* or *to* such bodies.

Mr. Jallabert was probably the first person who observed that a body pointed at one end, and round at the other, produced different appearances upon the same body, according as the pointed or round end was presented to it. But Dr. Franklin first observed and evinced the whole effect of pointed bodies, both in drawing and throwing off electricity at greater distances than other bodies could do it; though he candidly acknowledges, that the power of Points to throw off the electric fire was communicated to him by his friend Mr. Thomas Hopkinson.

Dr. Franklin electrified an iron shot, 3 or 4 inches in diameter, and observed that it would not attract a thread when the Point of a needle, communicating with the earth, was presented to it; and he found it even impossible to electrify an iron shot when a sharp needle lay upon it. This remarkable property, possessed by pointed bodies, of gradually and silently receiving or throwing off the electric fluid, has been evinced by a variety of other familiar experiments.

Thus, if one hand be applied to the outside coating of a large jar fully charged, and the Point of a needle held in the other, be directed towards the knob of the jar, and moved gradually near it, till the Point of the needle touch the knob or ball, the jar will be entirely discharged, so as to give no shock at all, or one that is hardly sensible. In this case the Point of the needle has gradually and silently drawn away the superabundant electricity from the electrified jar.

Farther, if the knob of a brass rod be held at such a distance from the prime conductor, that sparks may easily escape from the latter to the former, whilst the machine is in motion; then if the Point of a needle be presented, though at twice the distance of the rod from the conductor, no more sparks will be seen passing to the rod. When the needle is removed, the sparks will be seen; but upon presenting it again, they will again disappear. So that the Point of the needle draws off silently almost all the fluid, which is thrown by the cylinder or globe of the machine upon the prime conductor. This experiment may be varied, by fixing the needle upon the prime conductor with the point upward; and then, though the knob of a discharging rod, or the knuckle of the finger, be brought very near the prime conductor, and the excitation be very strong, little or no spark will be perceived.

The influence of points is also evinced in the amusing experiment, commonly called the electrical horse-race, and many others. See THUNDER-*house.*

The late Mr. Henly exhibited the efficacy of pointed bodies, by suspending a large bladder, well blown, and covered with gold, silver, or brass leaf, by means of gum-water, at the end of a silken thread 6 or 7 feet long, hanging from the cieling of a room, and electrifying the bladder by giving it a strong spark with the knob of a charged bottle: upon presenting to it the knob of a wire, it caused the bladder to move towards the knob, and when nearly in contact gave it a spark, thus discharging its electricity. By giving the bladder another charge, and presenting the Point of a needle to it, the bladder was not attracted by the Point, but rather

8

rather receded from it, especially when the needle was suddenly presented towards it.

But experiments evincing the efficacy of pointed bodies for silently receiving or throwing off the electric fluid, may be infinitely diversified, according to the fancy or convenience of the electrician.

It may be observed, that in the case of points throwing off or receiving electricity, a current of air is sensible at an electrified Point, which is always in the direction of the Point, whether the electricity be positive or negative. A fact which has been well ascertained by many electricians, and particularly by Dr. Priestley and Sig. Beccaria. The former contrived to exhibit the influence of this current on the flame of a candle, presented to a pointed wire, electrified negatively, as well as positively. The blast was in both cases alike, and so strong as to lay bare the greatest part of the wick, the flame being driven from the Point; and the effect was the same whether the electric fluid issued out of the Point or entered into it. He farther evinced this phenomenon by means of thin light vanes; and he found, as Mr. Wilson had before observed, that the vanes would not turn in vacuo, nor in a close unexhausted receiver where the air had no free circulation. And in much the same manner, Beccaria exhibited to sense the influence of the wind or current of air driven from points.

As to the *Theory* of the phenomena of Points, these are accounted for in a variety of ways, by different authors, though perhaps by none with perfect satisfaction. See Franklin's writings on Electricity; Lord Mahon's Principles of Electricity, 1779; Beccaria's Artificial Electricity, 1776, pa. 331; and Priestley's History of Electricity, vol. 2, pa. 191, edit. 1775.

As to the *Application* of the doctrine of Points; it may be observed that there is not a more important fact in the history of electricity, than the use to which the discovery of the efficacy of pointed bodies has been applied.

Dr. Franklin, having ascertained the identity of electricity and lightning, was presently led to propose a cheap and easy method of securing buildings from the damage of lightning, by fixing a pointed metal rod higher than any part of the building, and communicating with the ground, or with the nearest water. And this contrivance was actually executed in a variety of cases; and has usually been thought an excellent preservative against the terrible effects of lightning.

Some few instances however having occurred, in which buildings have been struck and damaged, though provided with these conductors; a controversy arose with regard to their expediency and utility. In this controversy Mr. Benjamin Wilson took the lead, and Dr. Musgrave, and some few other electricians, the least acquainted with the subject, concurred with him in their opposition to pointed elevated conductors. These alledge, that every Point, as such, solicits the lightning, and thus contributes not only to increase the quantity of every actual discharge, but also frequently to occasion a discharge when it might not otherwise have happened: whereas, say they, if instead of pointed conductors, those with blunted terminations were used, they would as effectually answer the purpose of conveying away the lightning safely, without the same tendency to increase or invite it. Accordingly, Mr. Wilson, in a

letter to the marquis of Rockingham (Philos. Trans. vol. 54, art. 44), expresses his opinion, that, in order to prevent lightning from doing mischief to high buildings, large magazines, and the like, instead of the elevated external conductors, that, on the inside of the highest part of such building, and within a foot or two of the top, it may be proper to fix a rounded bar of metal, and to continue it down along the side of the wall to any kind of moisture in the ground.

On the other hand, it is urged by the advocates for pointed conductors, that Points, instead of increasing an actual discharge, really prevent a discharge where it would otherwise happen, and that blunted conductors tend to invite the clouds charged with lightning. And it seems to be a certain fact, that though a sharp Point will draw off a charge of electricity silently at a much greater distance than a knob, yet a knob will be struck with a full explosion or shock, the charge being the same in both cases, at a greater distance than a sharp Point.

The efficacy of pointed bodies for preventing a stroke of lightning, is ingeniously explained by Dr. Franklin in the following manner:—An eye, he says, so situated as to view horizontally the underside of a thunder-cloud, will see it very ragged, with a number of separate fragments or small clouds one under another; the lowest sometimes not far from the earth. These, as so many stepping stones, assist in conducting a stroke between a cloud and a building. To represent these by an experiment, he directs to take two or three locks of fine loose cotton, and connect one of them with the prime conductor by a fine thread of 2 inches, another to that, and a third to the second, by like threads, which may be spun out of the same cotton. He then directs to turn the globe, and says we shall see these locks extending themselves towards the table, as the lower small clouds do towards the earth; but that, on presenting a sharp Point, erect under the lowest, it will shrink up to the second, the second up to the first, and all together to the prime conductor, where they will continue as long as the Point continues under them. May not, he adds, in like manner, the small electrified clouds, whose equilibrium with the earth is soon restored by the Point, rise up to the main body, and by that means occasion so large a vacancy, as that the grand cloud cannot strike in that place? Letters, pa. 121.

Mr. Henly too, as well as several other persons, with a view of determining the question, whether Points or knobs are to be preferred for the terminations of conductors, made several experiments, shewing in a variety of instances, the efficacy of Points in silently drawing off the electricity, and preventing strokes which would happen to knobs in the same situation. Philos. Trans. vol. 64, part 2, art. 18. See also THUNDER-Hous.

Indeed it has been universally allowed, that in cases where the quantity of electricity, with which thunder-clouds are charged, is small, or when they move slowly in their passage to and over a building, pointed conductors, which draw off the electrical fluid silently, within the distance at which rounded ends will explode, will gradually exhaust them, and thus contribute to prevent a stroke and preserve the buildings to which they are annexed.

But

But it has been said by those who are averse to the use of such conductors, that if clouds, of great extent, and highly electrified, should be driven directly over them with great velocity, or if a cloud hanging directly over buildings to which they are annexed, suddenly receives a charge by explosion from another cloud at a distance, so as to enable it instantly to strike into the earth, these pointed conductors must take the explosion; on account of their greater readiness to admit electricity at a much greater distance than those that are blunted, and in proportion to the difference of that striking distance, do mischief instead of good: and therefore, they add, that such pointed conductors, though they may be sometimes advantageous, are yet at other times prejudicial; and that, as the purpose for which conductors are fixed upon buildings, is not to protect them from one particular sort of clouds only, but if possible from all, it cannot be advisable to use that kind of conductors which, if they diminish danger on the one hand, will increase it on the other. Besides, it is alleged, that if pointed conductors are attended with any the slightest degree of danger, that danger must be considerably augmented by carrying them high up into the air, and by fixing them upon every angle of a building, and by making them project in every direction. Such is the reasoning of Dr. Musgrave: see his paper in the Philof. Tranf. vol. 68, part 2, art. 36.

Mr. Wilson too, dissenting from the report of a committee of the Royal Society, appointed to inspect the damage done by lightning to the house of the Board of Ordnance, at Purfleet, in 1777, was led to justify his dissent, and to disparage the use of pointed and elevated conductors, by means of a magnificent apparatus he constructed, with which he might produce effects similar to those that had happened in the case referred to the consideration and decision of the committee. With this view he procured a model of the Board-house at Purfleet, resembling it as nearly as possible in every essential appendage, and furnished with conductors of different lengths and terminations. And to construct a substitute for a cloud, he joined together the broad rims of 120 drums, forming together a cylinder of 155 feet in length, and above 16 inches in diameter; and this immense cylinder, of about 600 square feet of coated surface, was connected occasionally with one end of a wire 4800 feet long. As this bulky apparatus, representing the thunder-cloud, could not conveniently be put in motion, he contrived to accomplish the same end by moving the model of the building, with a velocity answering to that of the cloud, which he states, at a moderate computation, to be about 4 or 5 miles an hour. This apparatus was charged by a machine with one glass cylinder, about 10 or 11 feet from its nearest end; and the whole of the apparatus was disposed in the great room of the Pantheon, and applied to use in a variety of experiments. But it is impossible within the limits of this article to do justice to Mr. Wilson's experiments, or to the inferences which he deduces from them. Suffice it just to observe, that most of his experiments, in which the model of the house, which was passed swiftly under the artificial cloud, and having annexed to it either the pointed or blunt conductors at the same or different heights, were intended to shew, that pointed conductors are struck at a greater distance, and with a higher elevation, than the blunted ones: and from all his experiments made with pointed and rounded conductors, provided the circumstances be the same in both, he infers, that the rounded ones are much the safer of the two; whether the lightning proceeds from one cloud or from several; that those are still safer which rise little or nothing above the highest part of the building; and that this safety arises from the greatest resistance exerted at the larger surface. See Philof. Tranf. for 1778, pa. 232.

The committee of the Royal Society however, which was composed of nine of the most distinguished electricians in the kingdom, and to whom was referred the consideration of the most effectual method of securing the powder-magazines at Purfleet against the effects of lightning, express their united opinion, that elevated sharp rods, constructed and disposed in the manner which they direct, are preferable to low conductors terminated in rounded ends, knobs, or balls of metal; and that the experiments and reasonings, made and alleged to the contrary by Mr. Wilson, are inconclusive.

Mr. Nairne also, in order to obviate the objections of Mr. Wilson and others, and to vindicate the preference generally given to high and pointed conductors, constructed a much more simple apparatus than that of Mr Wilson, with which he made a number of well-designed and well-conducted experiments, which seem to prove the point as far as it is capable of being proved by an artificial electrical apparatus. From these last experiments it appears, that though the point was struck by means of a swift motion of the artificial cloud, yet a small ball of 3 tenths of an inch diameter was struck farther off than the Point, and a larger ball at a much greater distance than either, even with the swiftest motion. Upon the whole, Mr. Nairne seems to be justified in preferring elevated pointed conductors; next to them, those that are pointed, though they rise but little above the highest part of a building; and after them, those that are terminated in a ball, and placed even with the highest part of the building. See Philof. Tranf. 1778, pa. 823.

On the other part, Dr. Musgrave, not yet satisfied, gave in another paper, being " Reasons for dissenting from the Report of the Committee appointed to consider of Mr. Wilson's Experiments; including Remarks on some Experiments exhibited by Mr. Nairne;" which is inserted, by mistake, before Mr. Nairne's paper, being at pa. 801 of the same volume.

And farther, Mr. Wilson has another paper, on the same subject, at pa. 999 of the same vol. of Philof. Tranf. for 1778, entitled, " New Experiments upon the Leyden Phial, respecting the termination of conductors;" repeating and asserting his former objections and reasonings.

In the Philof. Tranf. too for 1779, pa. 454, Mr. William Swift has a paper, farther prosecuting this subject; making various experiments with simple and ingenious machinery, with models of houses and clouds, and with various sorts of conductors. From the experiments he infers in general, that " the whole current of these experiments tends to shew the preference of Points to balls, in order to diminish and draw off the electric matter when excited, or to prevent it from accumulating; and consequently the propriety or even necessity of terminating all conductors with Points, to make them useful to prevent damage to buildings from lightning. Nay the very construction of all electrical machines, in which it is necessary to round all the parts, and to avoid making edges and points which would hinder the matter from being excited, will, I imagine, on reflection, be another corroborating proof of the result of the experiments themselves."

There were other communications made to the Royal Society upon the important subject of conductors, some of which were received, and others rejected. Upon the whole, this contest turned out one of the most extraordinary that ever was agitated in the Society; producing the most remarkable disputes, differences, and strange consequences, that ever the Society experienced since it had existence; consequences which manifested themselves in various instances for many years after, and which continue to this very day. All which, with the various secret springs and astonishing intrigues, may probably be given to the public on some other occasion.

POINT, in Geometry, according to Euclid, is that which has no parts, or is indivisible; being void of all extension, both as to length, breadth, and depth.

This is what is otherwise called the Mathematical Point, being the intersection of two lines, and is only conceived by the imagination; yet it is in this that all magnitude begins and ends; the extremes of a line being Points; the extremes of a surface, Lines; and the extremes of a solid, Surfaces. And hence some define a Point, the inceptive of magnitude.

Proportion of Mathematical POINTS. It is a popular maxim, that all infinites are equal; yet is the maxim false, whether of quantities infinitely great, or infinitely little. Dr. Halley instances in several infinite quantities which are in a finite proportion to each other; and some that are infinitely greater than others. See INFINITE *Quantity*.

And the same is shewn by Mr. Robarts, of infinitely small quantities, or mathematical Points. He demonstrates, for instance, that the Points of contact between circles and their tangents, are in the subduplicate ratio of the diameters of the circles; that the Point of contact between a sphere and a plane is infinitely greater than between a circle and a line; and that the Points of contact in spheres of different magnitudes, are to each other as the diameters of the spheres. Philos. Transf. vol. 27, pa. 470.

Conjugate POINT, is used for that Point into which the conjugate oval, belonging to some kind of curves, vanishes. Maclaurin's Alg. pa. 308.

POINT of *Contrary Flexure*, &c. See INFLEXION, RETROGRADATION or RETROGRESSION, &c, of curves.

POINTS *of the Compass*, or *Horizon*, &c, in Geography and Navigation, are the Points of division when the whole circle, quite around, is divided into 32 equal parts. These Points are therefore at the distance of the 32d part of the circle, or $11°\ 15'$, from each other; hence $5°\ 37'\frac{1}{2}$ is the distance of the half points, and

$2°\ 48'\frac{1}{4}$ is the distance of the quarter Points. See COMPASS. The principal of these are the four cardinal Points, east, west, north and south.

Point is also used for a cape or headland, jutting out into the sea.——The seamen say two Points of land are one in another, when they are in a right line, the one behind the other.

POINT, in Optics. As the

POINT *of Concourse* or *Concurrence*, is that in which converging rays meet; and is usually called focus.

POINT of *Dispersion, Incidence, Reflection, Refraction,* and *Radiant* POINT. See these several articles.

POINT, in Perspective, is a term used for various parts or places, with regard to the perspective plane. As, the

POINT *of Sight*, or *of the eye*, called also the Principal Point, is the Point on a plane where a perpendicular from the eye meets it. See PERSPECTIVE.

Some authors, however, by the Point of Sight, or Vision, mean the Point where the eye is actually placed, and where all the rays terminate. See PERSPECTIVE.

POINT *of Distance*, is a Point in a horizontal line, at the same distance from the principal Point as the eye is from the same. See PERSPECTIVE.

Third POINT, is a Point taken at discretion in the line of distance, where all the diagonals meet that are drawn from the divisions of the geometrical plane.

Objective POINT, is a Point on a geometrical plane, whose representation on the perspective plane is required.

Accidental Point, and *Visual* POINT. See ACCIDENTAL and VISUAL.

POINT *of View*, with regard to Building, Painting, &c, is a Point at a certain distance from a building, or other object, where the eye has the most advantageous view or prospect of the same. And this Point is usually at a distance equal to the height of the building.

POINT, in Physics, is the smallest or least sensible object of sight, marked with a pen, or point of a compass, or the like. This is popularly called a Physical Point, and of such does all physical magnitude consist.

POINT-BLANC, *Point-Blank*, in Gunnery, denotes the horizontal or level position of a gun, or having its muzzle neither elevated nor depressed. And the Point-blanc range, is the distance the shot goes, before it strikes the level ground, when discharged in the horizontal or Point-blanc direction. O sometimes this means the distance the ball goes horizontally in a straight-lined direction.

POINTING, in Artillery and Gunnery, is the laying a piece of ordnance in any proposed direction, either horizontal, or elevated, or depressed, to any angle. This is usually effected by means of the gunner's quadrant, which, being applied to, or in, the muzzle of the piece, shews by a plummet the degree of elevation or depression.

POINTING, in Navigation, is the marking on the chart in what Point, or place, the vessel is.——This is done by means of the latitude and longitude, after these are known, or found by observation or computation. Thus, draw a line, with a pencil, across the chart according to the latitude; and another across the other way according to the longitude; then the intersection

section of these two lines, is the Point or place on the chart where the ship is; which is then marked black with a pen, and the pencil lines rubbed out. From the Point or place, thus found, the chart readily shews the direct distance and course run, as also yet to run to the intended port, &c.

POLAR, something that relates to the poles of the world: as polar virtue, polar tendency.

POLAR *Circles*, are two lesser circles of the sphere, or globe, one round each pole, and at the same distance from it as is equal to the sun's greatest declination or the obliquity of the ecliptic; that is, at present 23° 28'.—The space included within each polar circle, is the frigid zone; and to every part of this space, the sun never sets at some time of the year, and never rises at another time; each of these being a longer duration as the place is nearer the pole.

POLAR *Dials*, are such as have their planes parallel to some great circle passing through the poles, or to some one of the hour-circles; so that the pole is neither elevated above the plane, nor depressed below it.—This dial, therefore, can have no centre; and consequently its style, substyle, and hour-lines, are parallel.—This will therefore be an horizontal dial to those who live at the equator.

POLAR *Projection*, is a representation of the earth, or heavens, projected on the plane of one of the polar circles.

POLAR *Regions*, are those parts of the earth which lie near the north and south poles.

POLARITY, the quality of a thing having poles, or pointing to, or respecting some pole: as the magnetic needle, &c.

By heating an iron bar, and letting it cool again in a vertical position, it acquires a polarity, or magnetic virtue: the lower end becoming the north pole, and the upper end the south pole. But iron bars acquire a polarity by barely continuing a long time in an erect position, even without heating them. Thus, the upright iron bars of some windows, &c, are often found to have poles: Nay, an iron rod acquires a polarity, by the mere holding it erect; the lower end, in that case, attracting the south end of a magnetic needle; and the upper, the north end. But these poles are mutable, and shift with the situation of the rod.

Some modern writers, particularly Dr. Higgins, in his Philosophical Essay concerning Light, have maintained the polarity of the parts of matter, or that their simple attractions are more forcible in one direction, or axis of each atom, than in any other.

POLES, in Astronomy, the extremities of the axis upon which the whole sphere of the world revolves; or the points on the surface of the sphere through which the axis passes. These are on every side at the distance of a quadrant, or 90°, from every point of the equinoctial, and are called, by way of eminence, the poles of the world. That which is visible to us in Europe, or raised above our horizon, is called the Arctic or North Pole; and its opposite one, the Antarctic or South Pole.

POLES, in Geography, are the extremities of the earth's axis; or the points on the surface of the earth through which the axis passes. Of which, that elevated above our horizon is called the Arctic or North Pole; and the opposite one, the Antarctic or South Pole.

In consequence of the situation of the Poles, with the inclination of the earth's axis, and its parallelism during the annual motion of our globe round the sun, the Poles have only one day and one night throughout the year, each being half a year in length. And because of the obliquity with which the rays of the sun fall upon the polar regions, and the great length of the night in the winter season, it is commonly supposed the cold is so intense, that those parts of the globe which lie near the Poles have never been fully explored, though the attempt has been repeatedly made by the most celebrated navigators. And yet Dr. Halley was of opinion, that the solstitial day, at the Pole, is as hot as at the equator when the sun is in the zenith; because all the 24 hours of that day under the Pole the sun-beams are inclined to the horizon in an angle of 23° 28'; whereas at the equator, though the sun becomes vertical, yet he shines no more than 12 hours, being absent the other 12 hours: and besides, that during 3 hours 8 minutes of the 12 hours which he is above the horizon there, he is not so much elevated as at the Pole. Experience however seems to shew that this opinion and reasoning of Dr. Halley are not well founded: for in all the parts of the earth that we know, the middle of summer is always the less hot the farther the place is from the equator, or the nearer it is to the Pole.

The great object for which navigators have ventured themselves in the frozen seas about the north pole, was to find out a more quick and ready passage to the East Indies. And this has been attempted three several ways: one by coasting along the northern parts of Europe and Asia, called the north-east passage; another, by sailing round the northern part of the American continent, called the north-west passage; and the third, by sailing directly over the pole itself.

The possibility of succeeding in the north-east was for a long time believed; and in the last century many navigators, particularly the Hollanders, attempted it with great fortitude and perseverance. But it was always found impossible to surmount the obstacles which nature had thrown in the way; and subsequent attempts have in a manner demonstrated the impossibility of ever sailing eastward along the northern coast of Asia. The reason of this impossibility is, that in proportion to the extent of land, the cold is always greater in winter, and vice versa. This is the case even in temperate climates; but much more so in those frozen regions when the sun's influence, even in summer, is but small. Hence, as the continent of Asia extends a vast way from west to east, and has besides the continent of Europe joined to it on the west, it follows, that about the middle part of that tract of land the cold should be greater than any where else. Experience has determined this to be fact; and it now appears, that about the middle of the northern part of Asia, the ice never thaws; neither have even the hardy Russians and Siberians themselves been able to overcome the difficulties they meet with in that part of their voyages.

With regard to the north-west passage, the same difficulties occur as in the other. According to Captain Cook's voyage, it appears that if there is any strait which

which divides the continent of America into two, it muſt lie in a higher latitude than 70°, and conſequently be perpetually frozen up. And therefore if a north-weſt paſſage can be found, it muſt be by ſailing round the whole American continent, inſtead of ſeeking a paſſage through it, which ſome have ſuppoſed to exiſt in the bottom of Baffin's Bay. But the extent of the American continent to the northward is yet unknown; and there is a poſſibility of its being joined to that part of Aſia between the Piaſida and Chatanga, which has never yet been circumnavigated. Indeed a rumour has lately gone abroad of ſome remarkable inlet being obſerved on the weſtern coaſt of North America, which it is gueſſed may poſſibly lead to ſome communication with the eaſtern ſide, by the lakes, or a paſſage into Hudſon's Bay: but there ſeems little or no probability of any ſucceſs this way, in which many fruitleſs attempts have been made at various times. It remains therefore to conſider, whether there is any probability of attaining the wiſhed-for paſſage by ſailing directly north, between the eaſtern and weſtern continents.

The late celebrated mathematician, Mr. Maclaurin, was ſo fully perſuaded of the practicability of paſſing by this way to the South and Indian ſeas, that he uſed to ſay, if his other avocations would permit, he would undertake the voyage of trial, even at his own expence.

The practicability of this method, which would lead directly to the Pole itſelf, has alſo been ingeniouſly ſupported by Mr. Daines Barrington, in ſome tracts publiſhed in the years 1775 and 1776, in conſequence of the unſucceſsful attempt made by captain Phipps in the year 1773, to reach a higher northern latitude than 81°. Mr. Barrington inſtances a great number of navigators who have reached very high northern latitudes; nay, ſome who have been at the Pole itſelf, or gone beyond it. From all which he concludes, that if the voyage be attempted at a proper time of the year, there would not be any great difficulty in reaching the Pole. Thoſe vaſt pieces of ice which commonly obſtruct the navigators, he thinks, proceed from the mouths of the great Aſiatic rivers which run northward into the frozen ocean, and are driven eaſtward and weſtward by the currents. But, though we ſhould ſuppoſe them to come directly from the Pole, ſtill our author thinks that this affords an undeniable proof that the Pole itſelf is free from ice; becauſe, when the pieces leave it, and come to the ſouthward, it is impoſſible that they can at the ſame time accumulate at the Pole.

The Altitude or Elevation of the POLE, is an arch of the meridian intercepted between the Pole and the horizon of any places, and is equal to the latitude of the place.

To obſerve the Altitude of the POLE. With a quadrant, obſerve both the greateſt and leaſt meridian altitude of the Pole ſtar. Then half the ſum of the two altitudes, will be the height of the Pole, or the latitude of the place; and half the difference of the ſame will be the diſtance of the ſtar from the Pole. But, for accuracy, the obſerved altitudes ſhould be corrected on account of refraction, before their ſum or difference is taken. See REFRACTION.

POLE, in Spherics, or the Pole of a great circle, is a point upon the ſphere equally diſtant from every part of the circumference of the great circle; or a point 90° diſtant from the circumference in any part of it.—The zenith and nadir are the Poles of the horizon; and the Poles of the equator are the ſame with thoſe of the ſphere or globe.

POLES, in Magnetics, are two points in a loadſtone, correſponding to the Poles of the world; one pointing to the north, and the other to the ſouth.

If the ſtone be broken in ever ſo many pieces, every fragment will ſtill have its two Poles. And if a magnet be biſected by a plane perpendicular to the axis; the two points before joined will become oppoſite Poles, one in each ſegment.

To touch a needle, &c, with a magnet, that part intended for the north end is touched with the ſouth Pole of the magnet; and that intended for the ſouth end, with the north Pole; for the Poles of the needle become contrary to thoſe of the magnet.

A piece of iron acquires a polarity by only holding it upright; though its Poles are not fixed, but ſhift, and are inverted as the iron is. Fire deſtroys all fixed Poles; but it ſtrengthens the mutable ones.

Dr. Gilbert ſays, the end of a rod being heated, and left to cool pointing northward, it becomes a fixed north Pole; if ſouthward, a fixed ſouth Pole. When the end is cooled, held downward, it acquires rather more magnetiſm than if cooled horizontally towards the north. But the beſt way is to cool it a little inclined to the north. Repeating the operations of heating and cooling does not increaſe the effect.

Dr. Power ſays, if a rod be held northwards, and the north end be hammered in that poſition, it will become a fixed north Pole; and contrarily if the ſouth end be hammered. The heavier the blows are, cæteris paribus, the ſtronger will the magnetiſm be; and a few hard blows have as much effect as a great number. And what is ſaid of hammering, is to be likewiſe underſtood of filing, grinding, ſawing, &c; nay, a gentle rubbing, when long continued, will produce Poles.

Old punches and drills have all fixed north poles; becauſe they are almoſt conſtantly uſed downwards. New drills have either mutable Poles, or weak north ones. Drilling with ſuch a one ſouthward horizontally, it is a chance if you produce a fixed ſouth Pole; much leſs if you drill ſouth downwards; but by drilling ſouth upwards, you always make a fixed ſouth Pole.

Mr. Ballard ſays, that in 6 or 7 drills, made in his preſence, the bit of each became a north Pole, merely by hardening.

A weak fixed Pole may degenerate into a mutable one in a day, or even in a few minutes, by holding it in a poſition contrary to its pole. The loadſtone itſelf will not make a fixed Pole in every piece of iron: if the iron be thick, it is neceſſary that it have ſome conſiderable length.

POLE *of a Glaſs*, in Optics, is the thickeſt part of a convex glaſs, or the thinneſt part of a concave one; being the ſame as what is otherwiſe called the vertex of the glaſs; and which, when truly ground, is exactly in the middle of its ſurface.

POLE, or *Rod*, in Surveying, is a lineal meaſure containing $5\frac{1}{2}$ yards, or $16\frac{1}{2}$ feet.---The ſquare of it is called a ſquare Pole; but more uſually a perch, or a rod.

POLE-

POLE-STAR, is a star of the 2d magnitude near the north Pole, in the end of the tail of Ursa Minor, or the Little Bear. Its mean place in the heavens for the beginning of 1790, was as follows : viz,

Right Ascension	-	12°	31′	47″
Annual variat. in ditto	-	0	3	4
Declination	-	88	11	8
Annual variat. in ditto	-	0	0	$19\frac{6}{10}$

The nearness of this star to the Pole, on which account it is always above the horizon in these northern latitudes, makes it very useful in Navigation, &c, for determining the meridian line, the elevation of Pole, and consequently the latitude of the place, &c.

POLEMOSCOPE, in Optics, an oblique kind of prospective glass, contrived for the seeing of objects that do not lie directly before the eye. It was invented by Hevelius, in 1637, and is the same as OPERA Glass; which see.

POLITICAL *Arithmetic*, the application of arithmetical calculations to political uses and subjects; such as the public revenues, the number of people, the extent and value of lands, taxes, trade, commerce, or whatever relates to the power, strength, riches, &c, of a nation or commonwealth. Or, as Davenant concisely defines it, the art of reasoning by figures, upon things relating to government.

The chief authors who have attempted calculations of this kind, are, Sir William Petty, Major Graunt, Dr. Halley, Dr. Davenant, Mr. King, and Dr. Price.

Sir William Petty, among many other articles, states that, in his time, the people in England were about six millions, and their annual expence about 7l. each; that the rent of the lands was about eight millions, and the interests and profits of the personal estates as much; that the rent of the houses in England was four millions, and the profits of the labour of all the people twenty-six millions yearly; that the corn used in England, at 5s. the bushel for wheat, and 2s. 6d. for barley, amounts to ten millions per annum; that the navy of England required 36,000 men to man it, and the trade and other shipping about 48,000; that the whole people in England, Scotland, and Ireland, together, were about nine millions and a half; and those in France about thirteen millions and a half; and in the whole world about 350 millions; also that the whole cash of England, in current money, was then about six millions sterling. See his Political Arith. p. 74, &c.

Mr. Davenant gives some good reasons why many of Sir W. Petty's numbers are not to be entirely depended on; and advances others of his own, founded on the observations of Mr. Greg. King. Some of the particulars are, that the land of England is thirty-nine millions of acres; that the number of people in London was about 530,000, and in all England five millions and a half, increasing 9000 annually, or about the 600th part; the yearly rent of the lands ten millions, and that of the houses two millions; the produce of all kinds of grain 9 millions. Davenant's Essay upon the probable methods &c, in his works, vol. 6.

Major Graunt, in his observations on the bills of mortality, computes, that there are 39,000 square miles of land in England, or 25 million acres in England and Wales, and 4,600,000 persons, making about 5 acres and a half to each person; that the people of London were 640,000; and states the several numbers of persons living at the different ages.

Sir William Petty, in his discourse about duplicate proportion, farther states, that it is found by experience, that there are more persons living between 16 and 26 than of any other age; and from thence he infers, that the square roots of every number of men's ages under 16, whose root is 4, shew the proportion of the probability of such persons reaching the age of 70 years: thus, the probability of reaching that age by persons of the

ages of 16, 9, 4, and 1,
are as   4, 3, 2,   1, respectively.

Also that the probabilities of their order of dying, at ages above that, are as the square-roots of the ages: thus, the probabilities of the order of dying first,

of the ages   16, 25, 36, &c,
are as the roots 4, 5, 6, &c.

that is, the odds are 5 to 4 that a person of 25 dies before one of 16, and so on, declining up to 70 years of age.

Dr. Halley has made a very exact estimation of the degrees of mortality of mankind, from a curious table of the births and burials, at the city of Breslau, in Silesia; with an attempt to ascertain the price of annuities upon lives, and many other curious particulars. See the Philos. Transf. vol. 17, pa. 596. Another table of this kind is given by Mr. Simpson, for the city of London; and several by Dr. Price, for many different places.

Mr. Kerseboom, of Holland, has many and curious calculations and tables of the same kind. From his observations on the births of the people in England, it appears, that the number of males born, is in proportion to that of the females, as 18 to 17; and that the inhabitants living in Holland are in the same proportion.

Dr. Brackenridge has given an estimate of the number of people in England, formed both from the number of houses, and also from the quantity of bread consumed. Upon the former principle, he finds the number of houses in England and Wales to be about 900,000; and, allowing 6 persons to each house, the number of people near 5 millions and a half. And upon the latter principle, estimating the quantity of corn consumed at home at 2 millions of quarters, and 3 persons to every quarter of corn, makes the number of people 6 millions. See Philos. Transf. vol. 49, art. 45 and 113.

Dr. Derham, from a great number of registers of places, finds the proportions of the marriages to the births and burials; and Dr. Price has done the same for still more places; the mediums of all which are,

		Marriages to Births, as
Dr. Derham	-	1 to 4·7
Dr. Price	-	1 to 3·9

See Philos. Transf. Abr. vol. 7, part 4, pa. 46; also Dr. Price's Observations on Reversionary Payments; and the articles of this Dictionary, EXPECTATION *of Life*,

Life, LIFE-*Annuities*, MORTALITY, POPULATION, &c.

POLLUX, in Aſtronomy, the hind twin. or the poſterior part of the conſtellation Gemini.

POLLUX is alſo a fixed ſtar of the ſecond magnitude, in the conſtellation Gemini, or the Twins. See CASTOR and *Pollux*, alſo GEMINI.

POLYACOUSTICS, inſtruments contrived to multiply ſounds, as polyſcopes or multiplying glaſſes do the images of objects.

POLYEDRON. See POLYHEDRON.

POLYGON, in Geometry, a figure of many angles; and conſequently of many ſides alſo; for every figure has as many ſides as angles. If the angles be all equal among themſelves, the polygon is ſaid to be a regular one; otherwiſe, it is irregular. Polygons alſo take particular names according to the number of their ſides; thus a Polygon of

3 ſides is called a trigon,
4 ſides - a tetragon,
5 ſides - a pentagon,
6 ſides - a hexagon, &c.

and a circle may be conſidered as a Polygon of an infinite number of ſmall ſides, or as the limit of the Polygons.

Polygons have various properties, as below:

1. Every Polygon may be divided into as many triangles as it hath ſides.

2. The angles of any Polygon taken together, make twice as many right angles, wanting 4, as the figure hath ſides. Thus, if the Polygon has 5 ſides; the double of that is 10, from which ſubtracting 4, leaves 6 right angles, or 540 degrees, which is the ſum of the 5 angles of the pentagon. And this property, as well as the former, belongs to both regular and irregular Polygons.

3. Every regular Polygon may be either inſcribed in a circle, or deſcribed about it. But not ſo of the irregular ones, except the triangle, and another particular caſe as in the following property.

An equilateral figure inſcribed in a circle, is always equiangular.—But an equiangular figure inſcribed in a circle is not always equilateral; but only when the number of ſides is odd. For if the ſides be of an even number, then they may either be all equal; or elſe half of them may be equal, and the other half equal to each other, but different from the former half, the equals being placed alternately.

4. Every Polygon, circumſcribed about a circle, is equal to a right-angled triangle, of which one leg is the radius of the circle, and the other the perimeter or ſum of all the ſides of the Polygon. Or the Polygon is equal to half the rectangle under its perimeter and the radius of its inſcribed circle, or the perpendicular from its centre upon one ſide of the Polygon.

Hence, the area of a circle being leſs than that of its circumſcribing Polygon, and greater than that of its inſcribed Polygon, the circle is the limit of the inſcribed and circumſcribed Polygons: in like manner the circumference of the circle is the limit between the perimeters of the ſaid Polygons: conſequently the circle is equal to a right-angled triangle, having one leg

equal to the radius, and the other leg equal to the circumference; and therefore its area is found by multiplying half the circumference by half the diameter. In like manner, the area of any Polygon is found by multiplying half its perimeter by the perpendicular demitted from the centre upon one ſide.

5. The following Table exhibits the moſt remarkable particulars in all the Polygons, up to the dodecagon of 12 ſides; viz. the angle at the centre AOB, the angle of the Polygon C or CAB or double of OAB, and the area of the Polygon when each ſide AB is 1. (See the following figure.)

No. of ſides.	Name of Polygon.	Ang. O at cent.	Ang. C. of Polyg.	Area.
3	Trigon	120°	60°	0·4330127
4	Tetragon	90	90	1·0000000
5	Pentagon	72	108	1·7204774
6	Hexagon	60	120	2·5980762
7	Heptagon	51⅖	128⁴⁄₇	3·6339124
8	Octagon	45	135	4·8284271
9	Nonagon	40	140	6·1818242
10	Decagon	36	144	7·6942088
11	Undecagon	32¹	147⁷⁄₁₁	9·3656399
12	Dodecagon	30	150	11·1961524

By means of the numbers in this Table, any Polygons may be conſtructed, or their areas found: thus, (1ſt) *To inſcribe a Polygon in a given Circle.* At the centre make the angle O equal to the angle at the centre of the propoſed Polygon, found in the 3d column of the Table, the legs cutting the circle in A and B; and join A and B which will be one ſide of the Polygon. Then take AB between the compaſſes, and apply it continually round the circumference, to complete the Polygon.

(2d) *Upon the given Line AB to deſcribe a regular Polygon.* From the extremities draw the two lines AO and BO, making the angles A and B each equal to half the angle of the Polygon, found in the 4th column of the Table, and their interſection O will be the centre of the circumſcribed circle: then apply AB continually round the circumference as before.

(3d) *To deſcribe a Polygon about a given Circle.*— At the centre O make the angle of the centre as in the 1ſt art. its legs cutting the circle in *a* and *b*: join *ab*, and parallel to it draw AB to touch the circle: and meeting O*a* and O*b* produced in A and B: with the radius OA, or OB, deſcribe a circle, and around its circumference apply continual AB, which will complete the Polygon as before.

(4th) *To find the Area of any regular Polygon.*— Multiply the ſquare of its ſide by the tabular area, found on the line of its name in the laſt column of the Table, and the product will be the area. Thus to find

find the area of the trigon, or equilateral triangle, whose side is 20. The square of 20 being 400, multiply the tabular area ·4330127 by 400, as in the margin, and the product 173·20508 will be the area.

$$\begin{array}{r} 0\cdot4330127 \\ 400 \\ \hline 173\cdot2050800 \end{array}$$

6. There are several curious algebraical theorems for inscribing Polygons in circles, or finding the chord of any proposed part of the circumference, which is the same as angular sections. These kinds of sections, or parts and multiples of arcs, were first treated of by Vieta, as shewn in the Introduction to my Log. pa. 9, and since pursued by several other mathematicians, in whose works they are usually to be found. Many other particulars relating to Polygons may also be seen in my Mensuration, 2d edit. pa. 20, 21, 22, 23, 113, &c.

POLYGON, in Fortification, denotes the figure or perimeter of a fortress, or fortified place. This is either Exterior or Interior.

Exterior POLYGON is the perimeter or figure formed by lines connecting the points of the bastions to one another, quite round the work. And

Interior POLYGON, is the perimeter or figure formed by lines connecting the centres of the bastions, quite around.

Line of POLYGONS, is a line on some sectors, containing the homologous sides of the first nine regular Polygons inscribed in the same circle; viz, from an equilateral triangle to a dodecagon.

POLYGONAL *Numbers,* are the continual or successive sums of a rank of any arithmeticals beginning at 1, and regularly increasing; and therefore are the first order of figurate numbers; they are called Polygonals, because the number of points in them may be arranged in the form of the several Polygonal figures in geometry, as is illustrated under the article FIGURATE *Numbers,* which see.

The several sorts of Polygonal numbers, viz, the triangles, squares, pentagons, hexagons, &c, are formed from the addition of the terms of the arithmetical series, having respectively their common difference 1, 2, 3, 4, &c; viz, if the common difference of the arithmeticals be 1, the sums of their terms will form the triangles; if 2, the squares; if 3, the pentagons; if 4, the hexagons, &c. Thus:

Arith. Progres.	1,	2,	3,	4,	5,	6,	7.
Triang. Nos.	1,	3,	6,	10,	15,	21,	28.
Arith. Progres.	1,	3,	5,	7,	9,	11,	13.
Square Numbers	1,	4,	9,	16,	25,	36,	49.
Arith. Progres.	1,	4,	7,	10,	13,	16,	19.
Pentagonal Nos.	1,	5,	12,	22,	35,	51,	70.
Arith. Progres.	1,	5,	9,	13,	17,	21,	25.
Hexagonal Nos.	1,	6,	15,	28,	45,	66,	91.

The *Side* of a Polygonal number is the number of points in each side of the Polygonal figure when the points in the number are ranged in that form. And this is also the same as the number of terms of the arithmeticals that are added together in composing the Po-

lygonal number; or, in short, it is the number of the term from the beginning. So, in the 2d or squares,

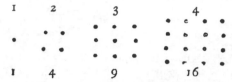

the side of the first (1) is 1, that of the second (4) is 2, that of the third (9) is 3, that of the fourth (16) is 4, and so on. And

The *Angles,* or Numbers of Angles, are the same as those of the figure from which the number takes its name. So the angles of the triangular numbers are 3, of the square ones 4, of the pentagonals 5, of the hexagonals 6, and so on. Hence, the angles are 2 more than the common difference of the arithmetical series from which any rank of Polygonals is formed: so the arithmetical series has for its common difference the number 1 or 2 or 3 &c as follows, viz, 1 in the triangles, 2 in the squares, 3 in the pentagons, &c; and, in general, if a be the number of angles in the Polygon, then $a - 2$ is $= d$ the common difference of the arithmetical series, or $d + 2 = a$ the number of angles.

PROB. 1. *To find any Polygonal Number proposed;* having given its side n and angles a. The Polygonal number being evidently the sum of the arithmetical progression whose number of terms is n and common difference $a - 2$, and the sum of an arithmetical progression being equal to half the product of the extremes by the number of terms, the extremes being 1 and $1 + d \cdot \overline{n - 1} = 1 + \overline{a - 2} \cdot n - 1$; therefore that number, or this sum, will be

$$\frac{n^2 \cdot l - n \cdot \overline{d - 2}}{2} \text{ or } \frac{n^2 \cdot \overline{a - 2} - n \cdot \overline{a - 4}}{2}, \text{ where}$$

d is the common difference of the arithmeticals that form the Polygonal number, and is always 2 less than the number of angles a.

Hence, for the several sorts of Polygons, any particular number whose side is n, will be found from either of these two formulæ, by using for d its values 1, 2, 3, 4, &c; which gives these following formulæ for the Polygonal number in each sort, viz, the

Triangular	-	$\dfrac{n^2 + n}{2},$
Square	-	$\dfrac{2n^2 - 0n}{2} = n^2,$
Pentagonal	-	$\dfrac{3n^2 - n}{2},$
Hexagonal	-	$\dfrac{4n^2 - 2n}{2},$
Heptagonal	-	$\dfrac{5n^2 - 3n}{2},$
&c.		

PROB. 2. *To find the Sum of any Number of Polygonal Numbers of any order.*—Let the angles of the Polygon be

be a, or the common difference of the arithmeticals that form the Polygonals, d; and n the number of terms in the Polygonal feries, whofe fum is fought: then is

$$\left(\frac{n^2-1}{6}d + \frac{n+1}{2}\right)n \text{ or } \left(\frac{n^2-1}{6}.a-2 + \frac{n+1}{2}\right)n$$

the fum of the n terms fought.

Hence, fubftituting fucceffively the numbers 1, 2, 3, 4, &c, for d, there is obtained the following particular cafes, or formulæ, for the fums of n terms of the feveral ranks of Polygonal numbers, viz, the fum of the

Triangulars - $\dfrac{n^2 + 3n + 2}{6}n$,

Squares - $\dfrac{2n^2 + 3n + 1}{6}n$,

Pentagonals - $\dfrac{3n^2 + 3n + 0}{6}n$,

Hexagonals - $\dfrac{4n^2 + 3n - 1}{6}n$,

Heptagonals - $\dfrac{5n^2 + 3n - 2}{6}n$,

&c

POLYGRAM, in Geometry, a figure confifting of many lines.

POLYHEDRON, or POLYEDRON, a body or folid contained by many rectilinear planes or fides.

When the fides of the Polyhedron are regular polygons, all fimilar and equal, then the Polyhedron becomes a regular body, and may be infcribed in a fphere; that is, a fphere may be defcribed about it, fo that its furface fhall touch all the angles or corners of the folid. There are but five of thefe regular bodies, viz, the tetraedron, the hexaedron or cube, the octaedron, the dodecaedron, and the icofaedron. See REGULAR *Body*, and each of thofe five bodies feverally.

Gnomonical POLYHEDRON, is a ftone with feveral faces, on which are projected various kinds of dials. Of this fort, that in the Privy-garden, London, now gone to ruin, was efteemed the fineft in the world.

POLYHEDRON, in Optics. See POLYSCOPE.

POLYHEDROUS *Figure*, in Geometry, a folid contained under many fides or planes. See POLYHEDRON.

POLYNOMIAL, in Algebra, a quantity of many names or terms, and is otherwife called a Multinomial. As $a + 3b - 2c + 4d$, &c. See MULTINOMIAL.

POLYOPTRUM, in Optics, a glafs through which objects appear multiplied, but diminifhed. The Polyoptrum differs both in ftructure and phenomena from the common multiplying glaffes called Polyhedra or Polyfcopes.

To conftruct the Polyoptrum.—From a glafs AB, plane on both fides, and about 3 fingers thick, cut out fpherical fegments, fcarce a 5th part of a digit in diameter.—If then the glafs be removed to fuch a diftance from the eye, that you can take in all the cavities at one view, you will fee the fame object, as if

through fo many feveral concave glaffes as there are cavities, and all exceeding fmall.—Fit this, as an objectglafs, in a tube ABCD, whofe aperture AB is equal to the diameter of the glafs, and the other CD is equal to that of an eye-glafs, as for inftance about a finger's breadth. The length of the tube AC is to be accommodated to the object-glafs and eye-glafs, by trial. In CD fit a convex eye-glafs, or in its ftead a menifcus having the diftance of its principal focus a little larger than the length of the tube; fo that the point from which the rays diverge after refraction in the objectglafs, may be in the focus. If then the eye be applied near the eye-glafs, a fingle object will be feen repeated as often as there are cavities in the object-glafs, but ftill diminifhed.

POLYSCOPE, or POLYHEDRON, in Optics, is a multiplying glafs, being a glafs or lens which reprefents a fingle object to the eye as if it were many. It confifts of feveral plane furfaces, difpofed into a convex form, through every one of which the object is feen.

Phenomena of the Polyfcope.—1. If feveral rays, as EF, AB, CD, fall parallel on the furface of a Poly-

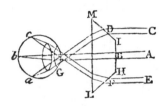

fcope, they will continue parallel after refraction. If then the Polyfcope be fuppofed regular, LH, HI, IM will be as tangents cutting the fpherical convex lens in F, B, and D; and confequently, rays falling on the points of contact, interfect the axis. Wherefore, fince the reft are parallel to thefe, they will alfo mutually interfect each other in G.

Hence, if the eye be placed where parallel rays decuffate, rays of the fame object will be propagated to it ftill parallel from the feveral fides of the glafs. Wherefore, fince the cryftalline humour, by its convexity, unites parallel rays, the rays will be united in as many different points of the retina, a, b, c, as the glafs has fides. Confequently the eye, through a Polyfcope, fees the object repeated as many times as there are fides. And hence, fince rays coming from very remote objects are parallel, a remote object is feen through a Polyfcope as often repeated as that has fides.

2. If rays AB, AC, AD, coming from a radiant

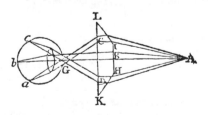

point A, fall on feveral fides of a regular Polyfcope; after

after refraction they will decussate in G, and proceed on a little diverging.

Hence, if the eye be placed where the rays decussate after coming from the several planes, the rays will be propagated to it from the several planes a little diverging, or as if they proceeded from different points. But since the crystalline humour, by its convexity, collects rays from several points into the same point ; the rays will be united in as many different points of the retina, *a*, *b*, *c*, as the glass has sides ; and consequently the eye, being placed in the focus G, will see even a near object through the Polyscope as often repeated as that has sides.

Thus may the images of objects be multiplied in a camera obscura, by placing a Polyscope at its aperture, and adding a convex lens at a due distance from it. And it makes a very pleasant appearance, if a prism be applied so that the coloured rays of the sun refracted from it be received on the Polyscope : for by this means they will be thrown on a paper or wall near at hand in little lucid specks, much exceeding the brightness of any precious stone ; and in the focus of the Polyscope, where the rays decussate (for in this experiment they are received on the convex side) will be a star of surprising lustre.

Farther, if images be painted in water-colours in the areolæ or little squares of a Polyscope, and the glass be applied to the aperture of a camera obscura ; the sun's rays, passing through it, will carry with them the images, and project them on the opposite wall.—This artifice bears a resemblance to that other, by which an image on paper is projected on the camera ; viz, by wetting the paper with oil, and straining it tight in a frame ; then applying it to the aperture of the camera obscura, so that the rays of a candle may pass through it upon the Polyscope.

To make an Anamorphosis, or Deformed Image, which shall appear regular and beautiful through a Polyscope, or Multiplying Glass.—At one end of a horizontal table erect another perpendicularly, upon which a figure may be designed ; and on the other end erect another, to serve as a fulcrum or support, moveable on the horizontal one. To the fulcrum apply a plano-convex Polyscope, consisting, for example, of 24 plane triangles ; and let the Polyscope be fitted in a draw-tube, of which that end towards the eye may have only a very small aperture, and a little farther off than the focus. Remove the fulcrum from the other perpendicular table, till it be out of the distance of the focus ; and the more so, as the image is to be greater. Before the little aperture place a lamp ; and trace the luminous areolæ projected from the sides of the Polyscope, with a black lead pencil, on the vertical plane, or a paper applied upon it.

In the several areolæ, design the different parts of an image, in such a manner as that, when joined together, they may make one whole, looking every now and then through the tube to guide and correct the colours, and to see that the several parts match and fit well together. As to the intermediate space, it may be filled up with any figures or designs at pleasure, contriving it so, as that to the naked eye the whole may exhibit some appearance very different from that intended to appear through the Polyscope.

The eye, now looking through the small aperture of the tube, will see the several parts and members dispersed among the areolæ to exhibit one continued image, all the intermediate parts disappearing. See ANAMORPHOSIS.

POLYSPASTON, in Mechanics, a machine so called by Vitruvius, consisting of an assemblage of several pullies, used for raising heavy weights.

PONTON, or PONTOON, a kind of flat-bottomed boat, whose carcass of wood is lined within and without with tin. Some nations line them on the outside only, and that with plates of copper, which is better. Our Pontoons are 21 feet long, nearly 5 feet broad, and 2 feet 1½ inch deep within. They are carried along with an army upon carriages, to make temporary bridges, called Pontoon-bridges. See the next article.

PONTOON-*Bridge*, a bridge made of Pontoons slipped into the water, and moored by anchors and otherwise fastened together by ropes, at small distances from one another ; then covered by beams of timber passing over them ; upon which is laid a flooring of boards. By this means, whole armies of infantry, cavalry, and artillery are quickly passed over rivers.—For want of Pontoons, &c, bridges are sometimes formed of empty powder casks, or powder barrels, which support the beams and flooring. Julius Cæsar and Aulus Gellius both mention Pontoons (pontones) ; but theirs were no more than a kind of square flat vessels, proper for carrying over horse &c.

PONT-VOLANT, or *Flying-bridge*, is a kind of bridge used in sieges, for surprising a post or outwork that has but narrow moats. It is made of two small bridges laid over each other, and so contrived that, by means of cords and pullies placed along the sides of the under bridge, the upper may be pushed forwards, till it join the place where it is designed to be fixed. The whole length of both ought not to be above 5 fathoms, lest it should break with the weight of the men.

PORES, are the small interstices between the particles of matter which compose bodies ; and are either empty, or filled with some insensible medium.

Condensation and rarefaction are only performed by closing and opening the Pores. Also the transparency of bodies is supposed to arise from their Pores being directly opposite to one another. And the matter of insensible perspiration is conveyed through the Pores of the cutis.

Mr. Boyle has a particular essay on the porosity of bodies, in which he proves that the most solid bodies have some kind of Pores : and indeed if they had not, all bodies would be alike specifically heavy.

Sir Isaac Newton shews, that bodies are much more rare and porous than is commonly believed. Water, for example, is 19 times lighter and rarer than gold ; and gold itself is so rare, as very readily, and without the least opposition, to transmit magnetic effluvia, and easily to admit even quicksilver into its pores, and to let water pass through it : for a concave sphere of gold hath, when filled with water, and soldered up, upon pressing it with a great force, suffered the water to squeeze through it, and stand all over its outside, in multitudes of small drops like dew, without bursting or cracking the gold. Whence it may be concluded, that

that gold has more pores than solid parts, and confequently that water has above 40 times more Pores than parts. Hence it is that the magnetic effluvia paffes freely through all cold bodies that are not magnetic; and that the rays of light pafs, in right lines, to the greateft diftances through pellucid bodies.

PORIME, *Porima*, in Geometry, a kind of eafy lemma, or theorem fo eafily demonftrated, that it is almoft felf-evident: fuch, for example, as that a chord is wholly within the circle.—Porime ftands oppofed to Aporime, which denotes a propofition fo difficult, as to be almoft impoffible to be demonftrated, or effected. Such as the quadrature of the circle, &c.

PORISM, *Porifma*, in Geometry, has by fome been defined a general theorem, or canon, deduced from a geometrical locus, and ferving for the folution of other general and difficult problems. Proclus derives the word from the Greek πορίζω, *I eftablifh*, and conclude from fomething already done and demonftrated: and accordingly he defines Porifma a theorem drawn occafionally from fome other theorem already proved: in which fenfe it agrees with what is otherwife called corollary.

Pappus fays, a Porifm is that in which fomething was propofed to be inveftigated.

Others derive it from πόρος, *a paffage*, and make it of the nature of a lemma, or a propofition neceffary for paffing to another more important one.

But Dr. Simfon, rejecting the erroneous accounts that have been given of a Porifm, defines it a propofition, either in the form of a problem or a theorem, in which it is propofed either to inveftigate, or demonftrate.

Euclid wrote three books of Porifms, being a curious collection of various things relating to the analyfis of the more difficult and general problems. Thofe books however are loft; and nothing remains in the works of the ancient geometricians concerning this fubject, befides what Pappus has preferved, in a very imperfect and obfcure ftate, in his Mathematical Collections, viz, in the introduction to the 7th book.

Several attempts have been made to reftore thefe writings in fome degree, befides that which Pappus has left upon the fubject. Thus, Fermat has given a few propofitions of this kind; which are to be found in the collection of his works, in folio, 1679, pa. 116. The like was done by Bullialdus, in his Exercitationes Geometricæ, 4to, 1657. Dr. Robert Simfon gave alfo a fpecimen, in two propofitions, in the Philof. Tranf. vol. 32, pa. 330; and befides left behind him a confiderable treatife on the fubject of Porifms, which has been printed in an edition of his works, at the expence of the earl of Stanhope, in 4to, 1776.

The whole three books of Euclid were alfo reftored by that ingenious mathematician Albert Girard, as appears by two notices that he gave, firft in his Trigonometry, printed in French, at the Hague, in 1629, and alfo in his edition of the works of Stevinus, printed at Leyden in 1634, pa. 459; but whether his intention of publifhing them was ever carried into execution, I have not been able to learn.

A learned paper on the fubject of Porifms, by the very ingenious Profeffor Playfair, has juft been inferted in the 3d volume of the *Tranfactions* of the Royal So-

ciety of Edinburgh. As this paper contains a number of curious obfervations on the geometry of the Ancients in general, as well as forms a complete treatife as it were on Porifm in particular, a pretty confiderable abftract of it cannot but be deemed in this place very ufeful and important.

" The reftoration of the ancient books of geometry (fays the learned profeffor) would have been impoffible, without the coincidence of two circumftances, of which, though the one is purely accidental, the other is effentially connected with the nature of the mathematical fciences. The firft of thefe circumftances is the prefervation of a fhort abftract of thofe books, drawn up by Pappus Alexandrinus, together with a feries of fuch lemmata, as he judged ufeful to facilitate the ftudy of them. The fecond is, the neceffary connection that takes place among the objects of every mathematical work, which, by excluding whatever is arbitrary, makes it poffible to determine the whole courfe of an inveftigation, when only a few points in it are known. From the union of thefe circumftances, mathematics has enjoyed an advantage of which no other branch of knowledge can partake; and while the critic or the hiftorian has only been able to lament the fate of thofe books of Livy and Tacitus which are loft, the geometer has had the high fatisfaction to behold the works of Euclid and Apollonius reviving under his hands.

" The firft reftorers of the ancient books were not, however, aware of the full extent of the work which they had undertaken. They thought it fufficient to demonftrate the propofitions, which they knew from Pappus, to have been contained in thofe books; but they did not follow the antient method of inveftigation, and few of them appear to have had any idea of the elegant and fimple analyfis by which thefe propofitions were originally difcovered, and by which the Greek Geometry was peculiarly diftinguifhed.

" Among thefe few, Fermat and Halley are to be particularly remarked. The former, one of the greateft mathematicians of the laft age, and a man in all refpects of fuperior abilities, had very juft notions of the geometrical analyfis, and appears often abundantly fkilful in the ufe of it; yet in his reftoration of the Loci Plani, it is remarkable, that in the moft difficult propofitions, he lays afide the analytical method, and contents himfelf with giving the fynthetical demonftration. The latter, among the great number and variety of his literary occupations, found time for a moft attentive ftudy of the ancient mathematicians, and was an inftance of, what experience fhews to be much rarer than might be expected, a man equally well acquainted with the ancient and the modern geometry, and equally difpofed to do juftice to the merit of both. He reftored the books of Apollonius, on the problem De Sectione Spatii, according to the true principles of the ancient analyfis.

" Thefe books, however, are but fhort, fo that the firft reftoration of confiderable extent that can be reckoned complete, is that of the Loci Plani by Dr. Simfon, publifhed in 1749, which, if it differs at all from the work it is intended to replace, feems to do fo only by its greater excellence. This much at leaft is certain, that the method of the ancient geometers does not appear to greater advantage in the moft entire of their writings

writings, than in the reftoration above mentioned ; and that Dr. Simfon has often facrificed the elegance to which his own analyfis would have led, in order to tread more exactly in what the lemmata of Pappus pointed out to him, as the track which Apollonius had pur-fued.

" There was another fubject, that of Porifms, the moft intricate and enigmatical of any thing in the an-cient geometry, which was ftill referved to exercife the genius of Dr Simfon, and to call forth that enthufiaftic admiration of antiquity, and that unwearied perfeve-rance in refearch, for which he was fo peculiarly diftin-guifhed. A treatife in three books, which Euclid had compofed on Porifms, was loft, and all that remained concerning them was an abftract of that treatife, inferted by Pappus Alexandrinus in his Mathematical Collec-tions, in which, had it been entire, the geometers of later times would doubtlefs have found wherewithal to confole themfelves for the lofs of the original work. But unfortunately it has fuffered fo much from the in-juries of time, that all which we can immediately learn from it is, that the Ancients put a high value on the propofitions which they called Porifms, and regard-ed them as a very important part of their analyfis. The Porifms of Euclid are faid to be, " Collectio artificio-" fiffima multarum rerum quæ fpectant ad analyfin dif-" ficiliorum et generalium problematum." The cu-riofity, however, which is excited by this encomium is quickly difappointed ; for when Pappus proceeds to ex-plain what a Porifm is, he lays down two definitions of it, one of which is rejected by him as imperfect, while the other, which is ftated as correct, is too vague and indefinite to convey any ufeful information.

" Thefe defects might neverthelefs have been fup-plied, if the enumeration which he next gives of Eu-clid's Propofitions had been entire ; but on account of the extreme brevity of his enunciations, and their refe-rence to a diagram which is loft, and for the conftruct-ing of which no directions are given, they are all, ex-cept one, perfectly unintelligible. For thefe reafons, the fragment in queftion is fo obfcure, that even to the learning and penetration of Dr. Halley, it feemed im-poffible that it could ever be explained ; and he there-fore concluded, after giving the Greek text with all poffible correctnefs, and adding the Latin tranflation, " Hactenus Porifmatum defcriptio nec mihi intellecta, " nec lectori profutura. Neque aliter fieri potuit, tam " ob defectum fchematis cujus fit mentio, quam ob " omiffa quædam et tranfpofita, vel aliter vitiata in pro-" pofitionis generalis expofitione, unde quid fibi velit " Pappus haud mihi datum eft conjicere. His adde " dictionis modum nimis contractum, ac in re difficili, " qualis hæc eft, minime ufurpandum."

" It is true, however, that before this time, Fermat had attempted to explain the nature of Porifms, and not altogether without fuccefs. Guiding his conjectures by the definition which Pappus cenfures as imperfect, becaufe it defined Porifms only " ab accidente," viz. " Porifma eft quod deficit hypothefi a, Theoremate Lo-" cali," he formed to himfelf a tolerably juft notion of thefe propofitions, and illuftrated his general defcription by examples that are in effect Porifms. But he was able to proceed no farther ; and he neither proved, that his notion of a Porifm was the fame with Euclid's, nor

attempted to reftore, or explain any one of Euclid's propofitions ; much lefs did he fuppofe, that they were to be inveftigated by an analyfis peculiar to themfelves. And fo imperfect indeed was this attempt, that the complete reftoration of the Porifms was neceffary to prove, that Fermat had even approximated to the truth.

" All this did not, however, deter Dr. Simfon from turning his thoughts to the fame fubject, which he ap-pears to have done very early, and long before the publication of the Loci Plani in 1749.

" The account he gives of his progrefs, and of the obftacles he encountered, will be always interefting to mathematicians. " Poftquam vero apud Pappum le-" geram, Porifmata Euclidis collectionem fuiffe artifi-" ciofiffimam multarum rerum, quæ fpectant ad analyfin " difficiliorum et generalium problematum, magno " defiderio tenebar, aliquid de iis cognofcendi ; quare " fæpius et multis variifque viis tum Pappi propofitio-" nem generalem, mancam et imperfectam, tum pri-" mum lib. i.

" Porifma, quod folum ex omnibus in tribus libris " integrum adhuc manet, intelligere et reftituere " conabar ; fruftra tamen, nihil enim proficiebam. " Cumque cogitationes de hac re multum mihi tempo-" ris confumpferint, atque moleftæ admodum evaferint, " firmiter animum induxi hæc nunquam in pofterum " inveftigare ; præfertim cum optimus geometra Hal-" leius fpem omnem de iis intelligendis abjeciffet. Un-" de quoties menti occurrebant, toties eas arcebam. " Poftea tamen accidit, ut improvidum et propofiti im-" memorem invaferint, meque detinuerint donec tan-" dem lux quædam effulferit, quæ fpem mihi faciebat " inveniendi faltem Pappi propofitionem generalem, " quam quidem multa inveftigatione tandem reftitui. " Hæc autem paulo poft una cum Porifmate primo " lib. i. impreffa eft inter Tranfactiones Phil. anni 1723, " num. 177."

" The propofitions mentioned, as inferted in the Phi-lofophical Tranfactions for 1723, are all that Dr. Sim-fon publifhed on the fubject of Porifms during his life, though he continued his inveftigations concerning them, and fucceeded in reftoring a great number of Euclid's propofitions, together with their analyfis. The propo-fitions thus reftored form a part of that valuable edition of the pofthumous works of this geometer which the mathematical world owes to the munificence of the late earl Stanhope

" The fubject of Porifms is not, however, exhaufted, nor is it yet placed in fo clear a light as to need no far-ther illuftration. It yet remains to enquire into the probable origin of thefe propofitions, that is to fay, in-to the fteps by which the ancient geometers appear to have been led to the difcovery of them.

" It remains alfo to point out the relations in which they ftand to the other claffes of geometrical truths ; to confider the fpecies of analyfis, whether geometrical or algebraical, that belongs to them ; and, if poffible, to affign the reafon why they have fo long efcaped the no-tice of modern mathematicians. It is to thefe points that the following obfervations are chiefly directed.

" I begin with defcribing the fteps that appear to have led the ancient geometers to the difcovery of Po-rifms ; and muft here fupply the want of exprefs teftimony by probable reafonings, fuch as are neceffary, whenever we would trace remote difcoveries to their fources, and which have more weight in mathematics than in any other of the fciences.

" It cannot be doubted, that it has been the folution of problems, which, in all ftates of the mathematical fciences, has led to the difcovery of moft geometrical truths. The firft mathematical enquiries, in particular, muft have occurred in the form of queftions, where fomething was given, and fomething required to be done; and by the reafonings neceffary to anfwer thefe queftions, or to difcover the relation between the things that were given, and thofe that were to be found, many truths were fuggefted, which came afterwards to be the fubjects of feparate demonftration. The number of thefe was the greater, that the ancient geometers always undertook the folution of problems with a fcrupulous and minute attention, which would fcarcely fuffer any of the collateral truths to efcape their obfervation. We know from the examples which they have left us, that they never confidered a problem as refolved, till they had diftinguifhed all its varieties, and evolved feparately every different cafe that could occur, carefully remarking whatever change might arife in the conftruction, from any change that was fuppofed to take place among the magnitudes which were given.

" Now as this cautious method of proceeding was not better calculated to avoid error, than to lay hold of every truth that was connected with the main object of enquiry, thefe geometers foon obferved, that there were many problems which, in certain circumftances, would admit of no folution whatever, and that the general conftruction by which they were refolved would fail, in confequence of a particular relation being fuppofed among the quantities which were given.

" Such problems were then faid to become impoffible; and it was readily perceived, that this always happened, when one of the conditions prefcribed was inconfiftent with the reft, fo that the fuppofition of their being united in the fame fubject, involved a contradiction. Thus, when it was required to divide a given line, fo that the rectangle under its fegment, fhould be equal to a given fpace, it was evident, that if this fpace was greater than the fquare of half the given line, the thing required could not poffibly be done; the two conditions, the one defining the magnitude of the line, and the other that of the rectangle under its fegments, being then inconfiftent with one another. Hence an infinity of beautiful propofitions concerning the maxima and the minima of quantities, or the limits of the poffible relations which quantities may ftand in to one another.

" Such cafes as thefe would occur even in the folution of the fimpleft problems; but when geometers proceeded to the analyfis of fuch as were more complicated, they muft have remarked, that their conftructions would fometimes fail, for a reafon directly contrary to that which has now been affigned. Inftances would be found where the lines that, by their interfection, were to determine the thing fought, inftead of interfecting one another, as they did in general, or of not meeting at all, as in the above-mentioned cafe of impoffibility, would coincide with one another entirely, and leave the queftion of confequence unrefolved. But

though this circumftance muft have created confiderable embarraffment to the geometers who firft obferved it, as being perhaps the only inftance in which the language of their own fcience had yet appeared to them ambiguous or obfcure, it would not probably be long till they found out the true interpretation to be put on it. After a little reflexion, they would conclude, that fince, in the general problem, the magnitude required was determined by the interfection of the two lines above mentioned, that is to fay, by the points common to them both; fo, in the cafe of their coincidence, as all their points were in common, every one of thefe points muft afford a folution; which folutions therefore muft be infinite in number; and alfo, though infinite in number, they muft all be related to one another, and to the things given, by certain laws, which the pofition of the two coinciding lines muft neceffarily determine.

" On enquiring farther into the peculiarity in the ftate of the data which had produced this unexpected refult, it might likewife be remarked, that the whole proceeded from one of the conditions of the problem involving another, or neceffarily including it; fo that they both together made in fact but one, and did not leave a fufficient number of independent conditions, to confine the problem to a fingle folution, or to any determinate number of folutions. It was not difficult afterwards to perceive, that thefe cafes of problems formed very curious propofitions, of an intermediate nature between problems and theorems, and that they admitted of being enunciated feparately, in a manner peculiarly elegant and concife. It was to fuch propofitions, fo enunciated, that the ancient geometers gave the name of Porifms.

" This deduction requires to be illuftrated by examples." Mr. Playfair then gives feveral problems by way of illuftration; one of which, which may here fuffice to fhew the method, is as follows:

" A triangle ABC being given, and alfo a point D, to draw through D a ftraight line DG, fuch, that, perpendiculars being drawn to it from the three angles of the triangle, viz. AE, BG, CF, the fum of the two perpendiculars on the fame fide of DG, fhall be equal to the remaining perpendicular: or, that AE and BG together, may be equal to CF.

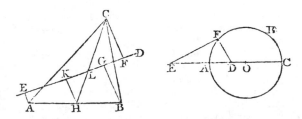

" Suppofe it done: Bifect AB in H, join CH, and draw HK perpendicular to DG.

" Becaufe AB is bifected in H, the two perpendiculars AE and BG are together double of HK; and as they are alfo equal to CF by hypothefis, CF muft be double of HK; and CL of LH. Now, GH is given in pofition, and magnitude; therefore the point L is
given;

given; and the point D being also given, the line DL is given in position, which was to be found.

"The construction was obvious. Bisect AB in H, join CH, and take HL equal to one third of CH; the straight line which joins the points D and L is the line required.

"Now, it is plain, that while the triangle ABC remains the same, the point L also remains the same, wherever the point D may be. The point D may therefore coincide with L; and when this happens, the position of the line to be drawn is left undetermined; that is to say, any line whatever drawn through L will satisfy the conditions of the problem. Here therefore we have another indefinite case of a problem, and of consequence another Porism, which may be thus enunciated: "A triangle being given in position, a point in it may be found, such, that any straight line whatever being drawn through that point, the perpendiculars drawn to this straight line from the two angles of the triangle which are on one side of it, will be together equal to the perpendicular that is drawn to the same line from the angle on the other side of it.

"This Porism may be made much more general; for if, instead of the angles of a triangle, we suppose ever so many points to be given in a plane, a point may be found such, that any straight line being drawn through it, the sum of all the perpendiculars that fall on that line from the given points on one side of it, is equal to the sum of the perpendiculars that fall on it from all the points on the other side of it.

"Or still more generally, any number of points being given not in the same plane, a point may be found, through which if any plane be supposed to pass, the sum of all the perpendiculars which fall on that plane from the points on one side of it, is equal to the sum of all the perpendiculars that fall on the same plane from the points on the other side of it. It is unnecessary to observe, that the point to be found in these propositions, is no other than the centre of gravity of the given points; and that therefore we have here an example of a Porism very well known to the modern geometers, though not distinguished by them from other theorems."

After some examples of other Porisms, and remarks upon them, the author then adds,

"From this account of the origin of Porisms, it follows, that a Porism may be defined, *A proposition affirming the possibility of finding such conditions as will render a certain problem indeterminate, or capable of innumerable solutions.*

"To this definition, the different characters which Pappus has given will apply without difficulty. The propositions described in it like those which he mentions, are, strictly speaking, neither theorems nor problems, but of an intermediate nature between both; for they neither simply enunciate a truth to be demonstrated, nor propose a question to be solved: but are affirmations of a truth, in which the determination of an unknown quantity is involved. In as far therefore as they assert, that a certain problem may become indeterminate, they are of the nature of theorems; and in as far as they seek to discover the conditions by which that is brought about, they are of the nature of problems.

"In the preceding definition also, and the instance from which it is deduced, we may trace that imperfect description of Porisms which Pappus ascribes to the later geometers, viz. "Porisma est quod deficit hypothesi a theoremate locali." Now, to understand this, it must be observed, that if we take the converse of one of the propositions called *Loci*, and make the construction of the figure a part of the hypothesis, we have what was called by the Ancients a Local Theorem. And again, if, in enunciating this theorem, that part of the hypothesis which contains the construction be suppressed, the proposition arising from thence will be a Porism; for it will enunciate a truth, and will also require, to the full understanding and investigation of that truth, that something should be found, viz. the circumstance in the construction, supposed to be omitted.

"Thus when we say; If from two given points E and D (2d fig. above), two lines EF and FD are in flected to a third point F, so as to be to one another in a given ratio, the point F is in the circumference of a circle given in position: we have a *Locus*.

"But when conversely it is said; If a circle ABC, of which the centre is O, be given in position, as also a point E, and if D be taken in the line EO, so that the rectangle EO OD be equal to the square of AO, the semidiameter of the circle; and if from E and D, the lines EF and DF be inflected to any point whatever in the circumference ABC; the ratio of EF to DF will be a given ratio, and the same with that of EA to AD we have a local theorem.

"And, lastly, when it is said; If a circle ABC be given in position, and also a point E, a point D may be found, such, that if the two lines EF and FD be inflected from E and D to any point whatever F, in the circumference, these lines shall have a given ratio to one another: the proposition becomes a Porism.

"Here it is evident, that the local theorem is changed into a Porism, by leaving out what relates to the determination of the point D, and of the given ratio. But though all propositions formed in this way, from the conversion of *Loci*, be Porisms, yet all Porisms are not formed from the conversion of *Loci*. The first and second of the preceding, for instance, cannot by conversion be changed into *Loci*; and therefore the definition which describes all Porisms as being so convertible, is not sufficiently comprehensive. Fermat's idea of Porisms, as has been already observed, was founded wholly on this definition, and therefore could not fail to be imperfect.

"It appears, therefore, that the definition of Porisms given above agrees with Pappus's idea of these propositions, as far at least as can be collected from the imperfect fragments which contain his general description of them. It agrees also with Dr. Simson's definition, which is this: "Porisma est propositio in qua proponitur demonstrare rem aliquam, vel plures datas esse, cui, vel quibus, ut et cuilibet ex rebus innumeris, non quidem datis, sed quæ ad ea quæ data sunt eandem habent relationem, convenire ostendendum est affectionem quandam communem in propositione descriptam.

"It cannot be denied, that there is a considerable degree of obscurity in this definition; notwithstanding of which it is certain, that every proposition to which it applies

applies muft contain a *problematical* part, viz " in qua proponitur demonftrare rem aliquam, vel plures datas effe," and alfo a *theoretical* part, which contains the property, or *communis affectio*, affirmed of certain things which have been previoufly defcribed.

" It is alfo evident, that the fubject of every fuch propofition, is the relation between magnitudes of three different kinds; determinate magnitudes which are given; determinate magnitudes which are to be found; and indeterminate magnitudes which, though unlimited in number, are connected with the others by fome common property. Now, thefe are exactly the conditions contained in the definitions that have been given here.

" To confirm the truth of this theory of the origin of Porifms, or at leaft the juftnefs of the notions founded on it, I muft add a quotation from an Effay on the fame fubject, by a member of this fociety, the extent and correctnefs of whofe views make every coincidence with his opinions peculiarly flattering. In a paper read feveral years ago before the Philofophical Society, Profeffor Dugald Stewart defined a Porifm to be " A propofition affirming the poffibility of finding one or more of the conditions of an indeterminate theorem." Where, by an indeterminate theorem, as he had previoufly explained it, is meant one which expreffes a relation between certain quantities that are indeterminate, both in magnitude and in number. The near agreement of this with the definition and explanations which have been given above, is too obvious to require to be pointed out; and I have only to obferve, that it was not long after the publication of Simfon's pofthumous works, when, being both of us occupied in fpeculations concerning Porifms, we were led feparately to the conclufions which I have now ftated.

" In an enquiry into the origin of Porifms, the etymology of the term ought not to be forgotten. The queftion indeed is not about the derivation of the word Πορισμα, for concerning that there is no doubt; but about the reafon why this term was applied to the clafs of propofitions above defcribed. Two opinions may be formed on this fubject, and each of them with confiderable probability: 1*mo*. One of the fignifications of πορίζω, is *to acquire or obtain*; and hence Πορισμα, *the thing obtained or gained.*

" Accordingly, Scapula fays, *Eft vox a geometris defumpta qui theorema aliquid ex demonftrativo fyllogifmo neceffario fequens inferentes, illud quafi lucrari dicuntur, quod non ex profeffo quidem theorematis hujus inftituta fit demonftratio, fed tamen ex demonftratis recte fequatur.* In this fenfe Euclid ufes the word in his Elements of Geometry, where he calls the corollaries of his propofition, *Porifmata.* This circumftance creates a prefumption, that when the word was applied to a particular clafs of propofitions, it was meant, in both cafes, to convey nearly the fame idea, as it is not at all probable, that fo correct a writer as Euclid, and fo fcrupulous in his ufe of words, fhould employ the fame term to exprefs two ideas which are perfectly different. May we not therefore conjecture, that thefe propofitions got the name of Porifms, entirely with a reference to their origin: According to the idea explained above, they would in general occur to mathematicians when engaged in the folution of the more difficult problems, and would arife

from thofe particular cafes, where one of the conditions of the data involved in it fome one of the reft. Thus a particular kind of theorem would be obtained, following as a corollary from the folution of the problem: and to this theorem the term Πορισμα might be very properly applied, fince, in the words of Scapula, already quoted, *Non ex profeffo theorematis hujus inftituta fit demonftratio, fed tamen ex demonftratis recte fequatur.*

" 2do. But though this interpretation agrees fo well with the fuppofed origin of Porifms, it is not free from difficulty. The verb πορίζω has another fignification, *to find out, to difcover, to devife*; and is ufed in this fenfe by Pappus, when he fays that the propofitions called Porifms, afford great delight, τοις δυναμενοις ορᾳν και πορίζμα, *to thofe who are able to underftand and inveftigate.* Hence comes πορισμος; *the act of finding it or difcovering*, and from πορισμος, in this fenfe, the fame author evidently confiders Πορισμα as being derived. His words are, Εφαςαν δι (οι αρχαιοι) Πορισμα εικαιτο προτεινομαιον εις Πορισμον αυτε προτεινομονη, *the Ancients faid, that a Porifm is fomething propofed for the* finding out, *or difcovering of the very thing propofed.* It feems fingular, however, that Porifms fhould have taken their name from a circumftance common to them with fo many other geometrical truths; and if this was really the cafe, it muft have been on account of the enigmatical form of their enunciations, which required, that in the analyfis of thefe propofitions, a fort of double difcovery fhould be made, not only of the Truth, but alfo of the Meaning *of the very thing which was propofed.* They may therefore have been called *Porifmata*, or *inveftigations*, by way of eminence.

" We might next proceed to confider the particular Porifms which Dr. Simfon has reftored, and to fhew, that every one of them is the indeterminate cafe of fome problem. But of this it is fo eafy for any one, who has attended to the preceding remarks, to fatisfy himfelf, by barely examining the enunciations of thofe propofitions, that the detail unto which it would lead feems to be unneceffary. I fhall therefore go on to make fome obfervations on that kind of analyfis which is particularly adapted to the inveftigation of Porifms.

" If the idea which we have given of thefe propofitions be juft, it follows, that they are always to be difcovered by confidering the cafes in which the conftruction of a problem fails in confequence of the lines which by their interfection, or the points which, by their pofition, were to determine the magnitude required, happening to coincide with one another—a Porifm may therefore be deduced from the problem it belongs to in the fame manner that the propofitions concerning the *ma ima* and *minima* of quantities are deduced from the problems of which they form the limitations; and fuch no doubt is the moft natural and moft obvious analyfis of which this clafs of propofitions will admit.

" It is not, however, the only one that they will admit of; and there are good reafons for wifhing to be provided with another, by means of which, a Porifm that is any how fufpected to exift, may be found out, independently of the general folution of the problem to which it belongs. Of thefe reafons, one is, that the Porifm may perhaps admit of being inveftigated more eafily than the general problem admits of being refolved;

and another is, that the former, in almost every case, helps to discover the simplest and most elegant solution that can be given of the latter.

" It is desirable to have a method of investigating Porisms, which does not require, that we should have previously resolved the problems they are connected with, and which may always serve to determine, whether to any given problem there be attached a Porism, or not. Dr. Simson's Analysis may be considered as answering to this description; for as that geometer did not regard these propositions at all in the light that is done here, nor in relation to their origin, an independent analysis of this kind, was the only one that could occur to him; and he has accordingly given one which is extremely ingenious, and by no means easy to be invented, but which he uses with great skilfulness and dexterity throughout the whole of his Restoration.

" It is not easy to ascertain whether this be the precise method used by the Ancients. Dr. Simson had here nothing to direct him but his genius, and has the full merit of the first inventor. It seems probable, however, that there is at least a great affinity between the methods, since the *lemmata* given by Pappus as necessary to Euclid's demonstrations, are subservient also to those of our modern geometer.

" It is, as we have seen, a general principle that a problem is converted into a Porism, when one, or when two, of the conditions of it, necessarily involve in them some one of the rest. Suppose then that two of the conditions are exactly in that state which determines the third; then, while they remain fixed or given, should that third one be supposed to vary, or differ, ever so little, from the state required by the other two, a contradiction will ensue. Therefore if, in the hypothesis of a problem, the conditions be so related to one another as to render it indeterminate, a Porism is produced; but if, of the conditions thus related to one another, some one be supposed to vary, while the others continue the same, an absurdity follows, and the problem becomes impossible. *Wherever therefore any problem admits both of an indeterminate, and an impossible case, it is certain, that these cases are nearly related to one another, and that some of the conditions by which they are produced, are common to both.*

" It is supposed above, that *two* of the conditions of a problem involve in them a third, and wherever that happens, the conclusion which has been deduced will invariably take place.

" But a Porism may sometimes be so simple, as to arise from the mere coincidence of *one* condition of a problem with another, though in no case whatever, any inconsistency can take place between them. Thus, in the second of the foregoing propositions, the coincidence of the point given in the problem with another point, viz, the centre of gravity of the given triangle, renders the problem indeterminate; but as there is no relation of distance, or position, between these points, that may not exist, so the problem has no impossible case belonging to it. There are, however, comparatively but few Porisms so simple in their origin as this, or that arise from problems in which the conditions are so little complicated; for it usually happens, that a problem which can become indefinite, may also become

impossible; and if so, the connection between these cases, which has been already explained, never fails to take place.

" Another species of impossibility may frequently arise from the porismatic case of a problem, which will very much affect the application of geometry to astronomy, or any of the sciences of experiment or observation. For when a problem is to be resolved by help of data furnished by experiment or observation, the first thing to be considered is, whether the data so obtained, be sufficient for determining the thing sought; and in this a very erroneous judgment may be formed, if we rest satisfied with a general view of the subject: For though the problem may in general be resolved from the data that we are provided with, yet these data may be so related to one another in the case before us, that the problem will become indeterminate, and instead of one solution, will admit of an infinite number.

" Suppose, for instance, that it were required to determine the position of a point F from knowing that it was situated in the circumference of a given circle ABC, and also from knowing the ratio of its distances from two given points E and D; it is certain that in general these data would be sufficient for determining the situation of F. But nevertheless, if E and D should be so situated, that they were in the same straight line with the centre of the given circle; and if the rectangle under their distances from that centre, were also equal to the square of the radius of the circle, then, the position of F could not be determined.

" This particular instance may not indeed occur in any of the practical applications of geometry; but there is one of the same kind which has actually occurred in astronomy: And as the history of it is not a little singular, affording besides an excellent illustration of the nature of Porisms, I hope to be excused for entering into the following detail concerning it.

" Sir Isaac Newton having demonstrated, that the trajectory of a comet is a parabola, reduced the actual determination of the orbit of any particular comet to the solution of a geometrical problem, depending on the properties of the parabola, but of such considerable difficulty, that it is necessary to take the assistance of a more elementary problem, in order to find, at least nearly, the distance of the comet from the earth, at the times when it was observed. The expedient for this purpose, suggested by Newton himself, was to consider a small part of the comet's path as rectilineal, and described with an uniform motion, so that four observations of the comet being made at moderate intervals of time from one another, four straight lines would be determined, viz, the four lines joining the places of the earth and the comet, at the times of observation, across which if a straight line were drawn, so as to be cut by them in three parts, in the same ratios with the intervals of time abovementioned; the line so drawn would nearly represent the comet's path, and by its intersection with the given lines, would determine, at least nearly, the distances of the comet from the earth at the time of observation.

" The geometrical problem here employed, of drawing a line to be divided by four other lines given in position, into parts having given ratios to one another, had been already resolved by Dr. Wallis and Sir Christopher

topher Wren, and to their solutions Sir Isaac Newton added three others of his own, in different parts of his works. Yet none of all these geometers observed that peculiarity in the problem which rendered it inapplicable to astronomy. This was first done by M. Boscovich, but not till after many trials, when, on its application to the motion of comets, it had never led to any satisfactory result. The errors it produced in some instances were so considerable, that Zanotti, seeking to determine by it the orbit of the comet of 1739, found, that his construction threw the comet on the side of the sun opposite to that on which he had actually observed it. This gave occasion to Boscovich, some years afterwards, to examine the different cases of the problem, and to remark that, in one of them, it became indeterminate, and that, by a curious coincidence, this happened in the only case which could be supposed applicable to the astronomical problem abovementioned; in other words, he found, that in the state of the data, which must there always take place, innumerable lines might be drawn, that would be all cut in the same ratio, by the four lines given in position. This he demonstrated in a dissertation published at Rome in 1749, and since that time in the third volume of his *Opuscula*. A demonstration of it, by the same author, is also inserted at the end of Castillon's Commentary on the *Arithmetica Universalis*, where it is deduced from a construction of the general problem, given by Mr. Thomas Simpson, at the end of his Elements of Geometry. The proposition, in Boscovich's words, is this: Problema quo quæritur recta linea quæ quatuor rectas positione datas ita secet, ut tria ejus segmenta sint invicem in ratione data, evadit aliquando indeterminatum, ita ut per quodvis punctum cujusvis ex iis quatuor rectis duci possit recta linea, quæ ei conditioni faciat satis.

" It is needless, I believe, to remark, that the proposition thus enunciated is a Porism, and that it was discovered by Boscovich, in the same way, in which I have supposed Porisms to have been first discovered by the geometers of antiquity.

" A question nearly connected with the origin of Porisms still remains to be solved, namely, from what cause has it arisen that propositions which are in themselves so important, and that actually occupied so considerable a place in the ancient geometry, have been so little remarked in the modern? It cannot indeed be said, that propositions of this kind were wholly unknown to the Moderns before the restoration of what Euclid had written concerning them; for besides M. Boscovich's proposition, of which so much has been already said, the theorem which asserts, that in every system of points there is a centre of gravity, has been shewn above to be a Porism; and we shall see hereafter, that many of the theorems in the higher geometry belong to the same class of propositions. We may add, that some of the elementary propositions of geometry want only the proper form of enunciation to be perfect Porisms. It is not therefore strictly true, that none of the propositions called Porisms have been known to the Moderns; but it is certain, that they have not met, from them, with the attention they met with from the Ancients, and that they have not been distinguished as a separate class of propositions. The cause of this difference is undoubtedly to be sought for in a comparison

of the methods employed for the solution of geometrical problems in ancient and modern times.

" In the solution of such problems, the geometers of antiquity proceeded with the utmost caution, and were careful to remark every particular case, that is to say, every change in the construction, which any change in the state of the data could produce. The different conditions from which the solutions were derived, were supposed to vary one by one, while the others remained the same; and all their possible combinations being thus enumerated, a separate solution was given, wherever any considerable change was observed to have taken place.

" This was so much the case, that the *Sectio Rationis*, a geometrical problem of no great difficulty, and one of which the solution would be dispatched, according to the methods of the modern geometry, in a single page, was made by Apollonius, the subject of a treatise consisting of two books: The first book has seven general divisions, and twenty four cases; the second, fourteen general divisions, and seventy-three cases, each of which cases is separately considered. Nothing, it is evident, that was any way connected with the problem, could escape a geometer, who proceeded with such minuteness of investigation.

" The same scrupulous exactness may be remarked in all the other mathematical researches of the Ancients: and the reason doubtless is, that the geometers of those ages, however expert they were in the use of their analysis, had not sufficient experience in its powers, to trust to the more general applications of it. That principle which we call the *law of continuity*, and which connects the whole system of mathematical truths by a chain of insensible gradations, was scarcely known to them, and has been unfolded to us, only by a more extensive knowledge of the mathematical sciences, and by that most perfect mode of expressing the relations of quantity, which forms the language of algebra; and it is this principle alone which has taught us, that though in the solution of a problem, it may be impossible to conduct the investigation without assuming the data in a *particular* state, yet the result may be perfectly *general*, and will accommodate itself to every case with such wonderful versatility, as is scarcely credible to the most experienced mathematician, and such as often forces him to stop, in the midst of his calculus, and look back, with a mixture of diffidence and admiration, on the unforeseen harmony of his conclusions. All this was unknown to the Ancients; and therefore they had no resource, but to apply their analysis separately to each particular case, with that extreme caution which has just been described; and in doing so, they were likely to remark many peculiarities, which more extensive views, and more expeditious methods of investigation, might perhaps have induced them to overlook.

" To rest satisfied, indeed, with too general results, and not to descend sufficiently into particular details, may be considered as a vice that naturally arises out of the excellence of the modern analysis. The effect which this has had, in concealing from us the class of propositions we are now considering, cannot be better illustrated than by the example of the Porism discovered by Boscovich, in the manner related above. Though the problem from which that Porism is derived, was

resolved

resolved by several mathematicians of the first eminence, among whom also was Sir Isaac Newton, yet the Porism which, as it happens is the most important case of it, was not observed by any of them. This is the more remarkable, that Sir Isaac Newton takes notice of the two most simple cases, in which the problem obviously admits of innumerable solutions, viz, when the lines given in position are either all parallel, or all meeting in a point, and these two hypotheses he therefore expressly excepts. Yet he did not remark, that there are other circumstances which may render the solution of the problem indeterminate as well as these; so that the porismatic case considered above, escaped his observation: and if it escaped the observation of one who was accustomed to penetrate so far into matters infinitely more obscure, it was because he satisfied himself with a general construction, without pursuing it into its particular cases. Had the solution been conducted after the manner of Euclid or Apollonius, the Porism in question must infallibly have been discovered."

PORISTIC *Method*, is that which determines when, by what means, and how many different ways, a problem may be resolved.

PORTA (JOHN BAPTISTA), called also in Italy *Giovan Batista de la Porta*, of Naples, lived about the end of the 16th century, and was famous for his skill in philosophy, mathematics, medicine, natural history, &c, as well as for his indefatigable endeavours to improve and propagate the knowledge of those sciences. With this view, he not only established private schools for particular sciences, but to the utmost of his power promoted public academies. He had no small share in establishing the academy at *Gli Ozioni*, at Naples, and had one in his own house, called *de Secreti*, into which none were admitted members, but such as had made some new discoveries in nature. He died at Pisa, in the kingdom of Naples, in the year 1615.

Porta gave the fullest proof of an extensive genius, and wrote a great many works; the principal of which are as follow:

1. His *Natural Magic;* a book abounding with curious experiments; but containing nothing of magic, the common acceptation of the word, as he pretends to nothing above the power of nature.
2. *Elements of Curve Lines.*
3. *A Treatise of Distillation.*
4. *A Treatise of Arithmetic.*
5. *Concerning Secret Letter-writing.*
6. *Of Optical Refractions.*
7. *A Treatise of Fortification.*
8. *A Treatise of Physiognomy.*

Beside some Plays and other pieces of less note.

PORTAIL, in Architecture, the face or frontispiece of a church, viewed on the side in which the great door is placed. It means also the great door or gate itself of a palace, castle, &c,

PORTAL, in Architecture, a term used for a little square corner of a room, cut off from the rest of the room by the wainscot; frequent in the ancient buildings, but now disused.

PORTAL is sometimes also used for a little gate, portella; where there are two gates, a large and a small one.

PORTAL is sometimes also used for a kind of arch of joiner's work before a door.

PORTCULLICE, called also *Herse*, and *Sarrasin*, in Fortification, an assemblage of several large pieces of wood laid or joined across one another, like a harrow, and each pointed at the bottom with iron. These were formerly used to be hung over the gateways of fortified places, to be ready to let down in case of a surprize, when the enemy should come so quick, as not to allow time to shut the gates. But the orgues are now more generally used, being found to answer the purpose better.

PORT-FIRE, in Gunnery, a paper tube, about 10 inches long, filled with a composition of meal-powder, sulphur, and nitre, rammed moderately hard; used to fire guns and mortars, instead of match.

PORTICO, in Architecture, is a kind of gallery, raised upon arches, under which people walk for shelter.

POSITION, or *Site*, or *Situation*, in Physics, is an affection of place, expressing the manner of a body's being in it.

POSITION, in Architecture, denotes the situation of a building, with respect to the points of the horizon. The best it is thought is when the four sides point directly to the four winds.

POSITION, in Astronomy, relates to the sphere. The position of the sphere is either right, parallel, or oblique; whence arise the inequality of days, the difference of seasons, &c.

Circles of POSITION, are circles passing through the common intersections of the horizon and meridian, and through any degree of the ecliptic, or the centre of any star, or other point in the heavens; used for finding out the position or situation of any star. These are usually counted six in number, cutting the equator into twelve equal parts, which the astrologers call the celestial houses.

POSITION, in Arithmetic, called also False Position, or Supposition, or Rule of False, is a rule so called, because it consists in calculating by false numbers supposed or taken at random, according to the process described in any question or problem proposed, as if they were the true numbers, and then from the results, compared with that given in the question, the true numbers are found. It is sometimes also called Trial-and-Error, because it proceeds by trials of false numbers, and thence finds out the true ones by a comparison of the errors.

Position is either Single or Double.

Single POSITION is when only one supposition is employed in the calculation. And

Double POSITION is that in which two suppositions are employed.

To the rule of Position properly belong such questions as cannot be resolved from a direct process by any of the other usual rules in arithmetic, and in which the required numbers do not ascend above the first power: such, for example, as most of the questions usually brought to exercise the reduction of simple equations in algebra. But it will not bring out true answers when the numbers sought ascend above the first power; for then the results are not proportional to the Positions,

or

or fuppofed numbers, as in the fingle rule ; nor yet the errors to the difference of the true number and each Pofition, as in the double rule. Yet in all fuch cafes, it is a very good approximation, and in exponential equations, as well as in many other things, it fucceeds better than perhaps any other method whatever.

Thofe queftions, in which the refults are proportional to their fuppofitions, belong to Single Pofition : fuch are thofe which require the multiplication or divifion of the number fought by any number ; or in which it is to be increafed or diminifhed by itfelf any number of times, or by any part or parts of it. But thofe in which the refults are not proportional to their pofitions, belong to the double rule : fuch are thofe, in which the numbers fought, or their multiples or parts, are increafed or diminifhed by fome given abfolute number, which is no known part of the number fought.

To work by *the Single Rule of* Position. Suppofe, take, or affume any number at pleafure, for the number fought, and proceed with it as if it were the true number, that is, perform the fame operations with it as, in the queftion, are defcribed to be performed with the number required : then if the refult of thofe operations be the fame with that mentioned or given in the queftion, the fuppofed number is the fame as the true one that was required ; but if it be not, make this proportion, viz, as your refult is to that in the queftion, fo is your fuppofed falfe number, to the true one required.

Example. Suppofe that a perfon, after fpending $\frac{1}{3}$ and $\frac{1}{4}$ of his money, has yet remaining 60l.; what fum had he at firft ?

Suppofe he had at firft 120l.

Now $\frac{1}{3}$ of 120 is	40
and $\frac{1}{4}$ of it is	30
their fum is	70
which taken from	120
leaves remaining	50, inftead of 60.

Therefore as 50 : 60 :: 120 : 144 the fum at firft.

Proof.

$\frac{1}{3}$ of 144 is	48
$\frac{1}{4}$ of it is	36
their fum	84
taken from	144
leaves juft	60 as per queft.

To work by the Double Rule of Position.

In this rule, make two different fuppofitions, or affumptions, and work or perform the operations with each, defcribed in the queftion, exactly as in the fingle rule : and if neither of the fuppofed numbers folve the queftion, that is, produce a refult agreeing with that in the queftion ; then obferve the errors, or how much each of the falfe refults differs from the true one, and alfo whether they are too great or too little ; marking them with + when too great, and with — when too little. Next multiply, croffwife, each pofition by the error of the other ; and if the errors be of the fame affection, that is both +, or both —, fubtract the one

product from the other, as alfo the one error from the other, and divide the former of thefe two remainders by the latter, for the anfwer, or number fought. But if the errors be unlike, that is, the one +, and the other —, add the two products together, and alfo the two errors together, and divide the former fum by the latter, for the anfwer.

And in this rule it is particularly ufeful to remember this part of the rule, viz. to fubtract when the errors are alike, both + or both —, but to add when unlike, or the one + and the other —.

Example. A fon afking his father how old he was, received this anfwer : Your age is now $\frac{1}{4}$ of mine ; but 5 years ago your age was only $\frac{1}{5}$ of mine at that time. What then were their ages ?

First, fuppofe the fon 15 ;

then $15 \times 4 = 60$ the father's ;	
alfo, 5 years ago the fon was	10,
and the father's muft be	55,
but ought to be 10×5 or	50,
therefore the error is	5—.

Again, fuppofe the fon 22 ;

then $22 \times 4 = 88$ is the father's ;	
alfo 5 years ago the fon was	17,
and the father's then	83,
but ought to be 17×5, or	85,
therefore the error is	2+.

And the errors, being unlike, muft be added, their fum being 7.

Then 15	22
2	5
30	110
	30

7) 140 (20 the fon's age, and confequently 80 the father's.

This rule of Pofition, or trial-and-error, is a good general way of approximating to the roots of the higher equations, to which it may be applied even before the equation is reduced to a final or fimple ftate, by which it often faves much trouble in fuch reductions. It is alfo eminently ufeful in refolving exponential equations, and equations involving arcs, or fines, &c, or logarithms, and in fhort in any equations that are very intricate and difficult. And even in the extraction of the higher roots of common numbers, it may be very ufefully applied. As for inftance, to extract the 3d or cubic root of the number 20.—Here it is evident that the root is greater than 2 and lefs than 3 ; making thefe two numbers therefore the fuppofitions, the procefs will be thus :

1ft fup.	$2^3 = 8$	2d fup.	$3^3 = 27$
given number	20	given number	20
1ft error	12—	2d error	7+
	3		2

$$\begin{array}{c} 12 \quad 36 \\ 7 \quad 14 \end{array} \Big\} \text{ add} \qquad 14$$

19) 50 (2·63 the firft approximation.

Again,

Again, as it thus appears the cube root of 20 is near 2·6 or 2·7, make suppoſition of theſe two, and repeat the proceſs with them, thus :

1ſt ſup. $2 \cdot 6^3 =$ 17·576	2d ſup. $2 \cdot 7^3 =$ 19·683
given number 20·	given number 20·
1ſt error 2·424—	2d error 0·317—
2·7	2·6
16968	1902
4848	634

$$\begin{array}{c} 2 \cdot 424 \quad 6 \cdot 5448 \\ \cdot 317 \quad \cdot 8242 \end{array} \Big\} \text{ſubtr.} \qquad \cdot 8242$$

$$2 \cdot 107) \; 5 \cdot 7206 \; (2 \cdot 714 \text{ root ſought.}$$

The rule of Poſition paſſed from the Moors into Europe, by Spain and Italy, along with their algebra, or method of equations, which was probably derived from the former.

POSITION, in Geometry, reſpects the ſituation, bearing, or direction of one thing, with regard to another. And Euclid ſays, " Points, lines, and angles, which have and keep always one and the ſame place and ſituation, are ſaid to be given by Poſition or ſituation." Data, def. 4.

POSITIVE *Quantities*, in Algebra, ſuch as are of a real, affirmative, or additive nature ; and which either have, or are ſuppoſed to have, the affirmative or poſitive ſign + before them ; as *a* or + *a*, or *bc*, &c. It is uſed in contradiſtinction from negative quantities, which are defective or ſubductive ones, and marked by the ſign — ; as — *a*, or —*ab*.

POSTERN, or *Sally-port*, in Fortification, a ſmall gate, uſually made in the angle of the flank of a baſtion, or in that of the curtain, or near the orillon, deſcending into the ditch ; by which the garriſon can march in and out, unperceived by the enemy, either to relieve the works, or to make private ſallies, &c.—It means alſo any private or back door.

POSTICUM, in Architecture, the poſtern gate, or back-door of any fabric.

POSTULATE, a demand, petition, or a problem of ſo obvious a nature, as to need neither demonſtration, nor explication, to render it either more plain or certain. This definition will nearly agree alſo to an axiom, which is a ſelf-evident theorem, as a Poſtulate is a ſelf-evident problem.

Euclid lays down theſe three Poſtulates in his Elements ; viz, 1ſt, That from one point to another a line can be drawn. 2d, That a right line can be produced out at pleaſure. 3d, That with any centre and radius a circle may be deſcribed.

As to axioms, he has a great number ; as, That two things which are equal to one and the ſame thing, are equal to each other, &c.

POUND, a certain weight ; which is of two kinds, viz, the pound troy, and the pound avoirdupois ; the former conſiſting of 12 ounces troy, and the latter of 16 ounces avoirdupois.—The pound troy is to the pound avoirdupois as 5760 to 6999½, or nearly 576 to 700.

POUND alſo is an imaginary money uſed in accounting, in ſeveral countries. Thus, in England there is the Pound ſterling, containing 20 ſhillings ; in France the Pound or livre Tournois and Pariſis ; in Holland and Flanders, a Pound or livre de gros, &c.—The term aroſe from hence, that the ancient pound ſterling, though it only contained 240 pence, as ours does ; yet each penny being equal to five of ours, the pound of ſilver weighed a Pound troy.

POUNDER, in Artillery, a term uſed to expreſs a certain weight of ſhot or ball, or how many pounds weight the proper ball is for any cannon : as a 24 pounder, a 12 pounder, &c.

POWDER, *Gun*. See GUNPOWDER.

POWDER-*Triers*. See EPROUVETTE.

POWER, in Mechanics, denotes ſome force which, being applied to a machine, tends to produce motion ; whether it does actually produce it or not. In the former caſe, it is called a moving Power ; in the latter, a ſuſtaining power.

POWER is alſo uſed in Mechanics, for any of the ſix ſimple machines, viz. the lever, the balance, the ſcrew, the wheel and axle, the wedge, and the pulley.

POWER *of a Glaſs*, in Optics, is by ſome uſed for the diſtance between the convexity and the ſolar focus.

POWER, in Arithmetic, the produce of a number, or other quantity, ariſing by multiplying it by itſelf, any number of times.

Any number is called the firſt power of itſelf. If it be multiplied once by itſelf, the product is the ſecond power, or ſquare ; if this be multiplied by the firſt power again, the product is the third power, or cube ; if this be multiplied by the firſt power again, the product is the fourth power, or biquadratic ; and ſo on ; the Power being always denominated from the number which exceeds the multiplications by one or unity, which number is called the index or exponent of the Power, and is now ſet at the upper corner towards the right of the given quantity or root, to denote or expreſs the Power. Thus,

3	or $3^1 =$	3 is the 1ſt power of 3,
3×3	or $3^2 =$	9 is the 2d power of 3,
$3^2 \times 3$	or $3^3 =$	27 is the 3d power of 3,
$3^3 \times 3$	or $3^4 =$	81 is the 4th power of 3,
&c.		&c.

Hence, to raiſe a quantity to a given Power or dignity, is the ſame as to find the product ariſing from its being multiplied by itſelf a certain number of times ; for example, to raiſe 2 to the 3d power, is the ſame thing as to find the factum, or product $8 = 2 \times 2 \times 2$. The operation of raiſing Powers, is called Involution.

Powers, of the ſame degree, are to one another in the ratio of the roots as manifold as their common exponent contains units : thus, ſquares are in a duplicate ratio of the roots ; cubes in a triplicate ratio ; 4th powers in a quadruplicate ratio.—And the Powers of proportional quantities are alſo proportional to one another : ſo, if $a : b :: c : d$, then, in any Powers alſo, $a^n : b^n :: c^n : d^n$.

The particular names of the ſeveral Powers, as introduced by the Arabians, were, ſquare, cube, quadrato-quadratum or biquadrate, ſurſolid, cube ſquared, ſecond ſurſolid, quadrato-quadrato-quadratum, cube of the cube,

cube, fquare of the furfolid, third furfolid, and fo on, according to the *products* of the indices.

And the names given by Diophantus, who is followed by Vieta and Oughtred, are, the fide or root, fquare, cube, quadrato-quadratum, quadrato-cubus, cubo-cubus, quadrato-quadrato-cubus, quadrato-cubo-cubus, cubo-cubo-cubus, &c. according to the *fums* of the indices.

But the moderns, after Des Cartes, are contented to diftinguifh moft of the Powers by the exponents; as 1ft, 2d, 3d, 4th, &c.

The characters by which the feveral Powers are denoted, both in the Arabic and Cartefian notation are as follow :

Arab.	1	R	q	c	bq	s	qc	Bſ	tq	lc
Cart.	a^0	a^1	a^2	a^3	a^4	a^5	a^6	a^7	a^8	a^9
	1	2	4	8	16	32	64	128	256	512

Hence, 1ft. The Powers of any quantity, form a feries of geometrical proportionals, and their exponents a feries of arithmetical proportionals, in fuch fort that the addition of the latter anfwers to the multiplication of the former, and the fubtraction of the latter anfwers to the divifion of the former, &c ; or in fhort, that the latter, or exponents, are as the logarithms of the former, or Powers.

Thus, $a^2 \times a^3 = a^5$, and $2 + 3 = 5$;
$$4 \times 8 = 32;$$
alfo $a^5 \div a^3 = a^2$, and $5 - 3 = 2$
$$32 \div 8 = 4.$$

2d. The o Power of any quantity, as a^0, is $= 1$.

3d. Powers of the fame quantity are multiplied, by adding their exponents : Thus,

Mult.	a^3	x^2	y^m	x^m	a^3
by	a^4	x^4	y^m	x^n	a^n
Prod.	a^7	x^6	y^{2m}	x^{m+n}	a^{3+n}

4th. Powers are divided by fubtracting their exponents. Thus,

Div.	a^7	x^6	y^{2m}	x^{m+n}	a^{3+n}
by	a^3	x^2	y^m	x^m	a^3
Quot.	a^4	x^4	y^m	x^n	a^n

5th. Powers are alfo confidered as negative ones, or having negative exponents, when they denote a divifor, or the denominator of a fraction. So $\frac{1}{a^3} = a^{-3}$, and $\frac{2}{a^2} = 2a^{-2}$, and $\frac{a^2}{x^4} = a^2 x^{-4}$, &c.

And hence any quantity may be changed from the denominator to the numerator, or from a divifor to a multiplier, or vice verfa, by changing the fign of its exponent ; and the whole feries of Powers proceeds indefinitely both ways from 1 or the o Power, pofitive on the one hand, and negative on the other. Thus,

&c $a^{-4} \ a^{-3} \ a^{-2} \ a^{-1} \ a^0 \ a^1 \ a^2 \ a^3 \ a^4$ &c,

or &c $\frac{1}{a^4} \ \frac{1}{a^3} \ \frac{1}{a^2} \ \frac{1}{a} \ 1 \ a \ a^2 \ a^3 \ a^4$ &c.

Powers are alfo denoted with fractional exponents, or

even with furd or irrational ones ; and then the numerator denotes the Power raifed to, and the denominator the exponent of fome root to be extracted: Thus,

$$\sqrt{a} = a^{\frac{1}{2}}, \text{ and } \sqrt{a^3} = a^{\frac{3}{2}}, \text{ and } \sqrt[3]{a^2} = a^{\frac{2}{3}}, \text{ &c.}$$

And thefe are fometimes called imperfect powers, or furds.

When the quantity to be raifed to any Power is pofitive, all its Powers muft be pofitive. And when the radical quantity is negative, yet all its even Powers muft be pofitive : becaufe $- \times -$ gives $+$: the odd Powers only being negative, or when their exponents are odd numbers : Thus, the Powers of $- a$,

are $+ 1, - a, + a^2, - a^3, + a^4, - a^5, + a^6$, &c.

where the even Powers a^2, a^4, a^6 are pofitive,
and the odd Powers a, a^3, a^5 are negative.

Hence, if a Power have a negative fign, no even root of it can be affigned ; fince no quantity multiplied by itfelf an even number of times, can give a negative product. Thus $\sqrt{- a^2}$, or the fquare or 2d root of $- a^2$, cannot be affigned ; and is called an impoffible root, or an imaginary quantity —Every Power has as many roots, real and imaginary, as there are units in the exponent.

M. De la Hire gives a very odd property common to all Powers. M. Carre had obferved with regard to the number 6, that all the natural cubic numbers, 8, 27, 64, 125, having their roots lefs than 6, being divided by 6, the remainder of the divifion is the root itfelf ; and if we go farther, 216, the cube of 6, being divided by 6, leaves no remainder ; but the divifor 6 is itfelf the root. Again, 343, the cube of 7, being divided by 6, leaves 1 ; which added to the divifor 6, makes the root 7, &c. M. De la Hire, on confidering this, has found that all numbers, raifed to any Power whatever, have divifors, which have the fame effect with regard to them, that 6 has with regard to cubic numbers. For finding thefe divifors, he difcovered the following general rule, viz, If the exponent of the Power of a number be even, i. e. if the number be raifed to the 2d, 4th, 6th, &c Power, it muft be divided by 2 ; the remainder of the divifion, when there is any, added to 2, or to a multiple of 2, gives the root of this number, correfponding to its Power, i. e. the 2d, 6th, &c root.

But if the exponent of the power be an uneven number, i. e. if the number be raifed to the 3d, 5th, 7th, &c Power ; the double of this exponent will be the divifor, which has the property abovementioned. Thus is it found in 6, the double of 3, the exponent of the Power of the cubes : fo alfo 10, the double of 5, is the divifor of all 5th Powers ; &c.

Any Power of the natural numbers 1, 2, 3, 4, 5, 6, &c, as the *n*th Power, has as many orders of differences as there are units in the common exponent of all the numbers ; and the laft of thofe differences is a conftant quantity, and equal to the continual product $1 \times 2 \times 3 \times 4 \times \cdots \times n$, continued till the laft factor, or the number of factors, be *n*, the exponent of the Powers. Thus,

3   the

the 1st Powers 1, 2, 3, 4, 5, &c, have but one order of differences 1 1 1 &c, and that difference is 1.

The 2d Pwrs. 1, 4, 9, 16, 25, &c, have two orders of differences 3 5 7 9
2 2 2

and the last of these is 2 = 1 × 2.

The 3d Pwrs. 1, 8, 27, 64, 125, &c, have three orders of differences 7 19 37 61
12 18 24
6 6

and the last of these is 6 = 1 × 2 × 3.

In like manner, the 4th or last differences of the 4th Powers, are each 24 = 1 × 2 × 3 × 4; and the 5th or last differences of the 5th Powers, are each 120 = 1 × 2 × 3 × 4 × 5. And so on. Which property was first noticed by Peletarius.

And the same is true of the Powers of any other arithmetical progression 1, 1 + d, 1 + 2d, 1 + 3d, &c,

viz, 1, $\overline{1 + d}^n$, $\overline{1 + 2d}^n$, $\overline{1 + 3d}^n$, &c,

the number of the orders of differences being still the same exponent n, and the last of those orders each equal to 1 × 2 × 3 ------ × nd^n, the same product of factors as before, multiplied by the same Power of the common difference d of the series of roots: as was shewn by Briggs.

And hence arises a very easy and general way of raising all the Powers of all the natural numbers, viz, by common addition only, beginning at the last differences, and adding them all continually, one after another, up to the Powers themselves. Thus, to generate the series of cubes, or 3d Powers, adding always 6, the common 3d difference gives the 2d differences 12, 18, 24, &c; and these added to the 1st of the 1st differences 7, gives the rest of the said 1st differences; and these again added to the 1st cube 1, gives the rest of the series of cubes, 8, 27, 64, &c, as below.

3dD.	2dD.	1stD.	Cubes.
			1
	12	7	
6		19	8
6	18		27
	24	37	
6		61	64
	30		125
		91	
			216
			&c.

Commensurable in Power, is said of quantities which, though not commensurable themselves, have their squares, or some other Power of them, commensurable. Euclid confines it to squares. Thus, the diagonal and side of a square are commensurable in Power, their squares being as 2 to 1, or commensurable; though they are not commensurable themselves, being as √2 to 1.

Power *of an Hyperbola*, is the square of the 4th part of the conjugate axis.

PRACTICAL Arithmetic, Geometry, Mathematics, &c, is the part that regards the practice, or application, as contradistinguished from the theoretical part.

PRACTICE, in Arithmetic, is a rule which expeditiously and compendiously answers questions in the golden rule, or rule-of-three, especially when the first term is 1. See rules for this purpose in all the books of practical arithmetic.

PRECESSION *of the Equinoxes,* is a very slow motion of them, by which they change their place, going from east to west, or backward, *in antecedentia,* as astronomers call it, or contrary to the order of the signs.

From the late improvements in astronomy it appears, that the pole, the solstices, the equinoxes, and all the other points of the ecliptic, have a retrograde motion, and are constantly moving from east to west, or from Aries towards Pisces, &c; by means of which, the equinoctial points are carried farther and farther back, among the preceding signs or stars, at the rate of about 50″¼ each year; which retrograde motion is called the Precession, Recession, or Retrocession of the Equinoxes.

Hence, as the stars remain immoveable, and the equinoxes go backward, the stars will seem to move more and more eastward with respect to them; for which reason the longitudes of all the stars, being reckoned from the first point of Aries, or the vernal equinox, are continually increasing.

From this cause it is, that the constellations seem all to have changed the places assigned to them by the ancient astronomers. In the time of Hipparchus, and the oldest astronomers, the equinoctial points were fixed to the first stars of Aries and Libra: but the signs do not now answer to the same points; and the stars which were then in conjunction with the sun when he was in the equinox, are now a whole sign, or 30 degrees, to the eastward of it: so, the first star of Aries is now in the portion of the ecliptic, called Taurus; and the stars of Taurus are now in Gemini; and those of Gemini in Cancer; and so on.

This seeming change of place in the stars was first observed by Hipparchus of Rhodes, who, 128 years before Christ, found that the longitudes of the stars in his time were greater than they had been before observed by Tymochares, and than they were in the sphere of Eudoxus, who wrote 380 years before Christ. Ptolomy also perceived the gradual change in the longitudes of the stars; but he stated the quantity at too little, making it but 1° in 100 years, which is at the rate of only 36″ per year. Y hang, a Chinese, in the year 721, stated the quantity of this change at 1° in 83 years, which is at the rate of 43″¼ per year. Other more modern astronomers have made this precession still more, but with some small differences from each other; and it is now usually taken at 50″¼ per year. All these rates are deduced from a comparison of the longitude of certain stars as observed by more ancient astronomers, with the later observations of the same stars; viz, by subtracting the former from the latter, and dividing the remainder by the number of years in the interval between the dates of the observations. Thus, by a medium of a great number of comparisons, the quantity of the annual change has been fixed at 50″¼, according to which rate it will require 25791 years for the equinoxes to make their revolution westward quite around the circle, and return to the same point again.

Thus

Thus, by taking the longitudes of the principal stars established by Tycho Brahe, in his book Astronomiæ Instauratæ Progymnasmata, pa. 208 and 232, for the beginning of 1586, and comparing them with the same as determined for the year 1750, by M. de la Caille, for that interval of 164 years, there will be obtained the following differences of longitude of several stars; viz,

γ Arietis	- -	2° 17′ 37″
Aldebaran	- -	2 17 45
μ Geminorum	- -	2 17 1
β Geminorum	- -	2 15 26
Regulus	- -	2 16 32
α Virginis	- -	2 18 18
α Aquilæ	- -	2 19 1
α Pegasi	- -	2 16 12
β Libræ	- -	2 17 52
Antares	- -	2 16 28
ε Tauri	- -	2 17 58
γ Geminorum	- -	2 18 38
γ Cancri	- -	2 19 12
γ Leonis	- -	2 19 38
γ Capricorni	- -	2 16 10
Medium of these 15 stars	-	2 17 35

which divided by 164, the interval of years, gives 50″·336, or nearly 50″⅓, or after the rate of 1° 23′ 53″⅓ in 100 years, or 25,748 years for the whole revolution, or circle of 360 degrees. And nearly the same conclusion results from the longitudes of the stars in the Britannic catalogue, compared with those of the still later catalogues. See De la Lande's Astronomie, in several places.

The Ancients, and even some of the Moderns, have taken the equinoxes to be immoveable; and ascribed that change in the distance of the stars from it, to a real motion of the orb of the fixed stars, which they supposed had a slow revolution about the poles of the ecliptic; so as that all the stars perform their circuits in the ecliptic, or its parallels, in the space of 25,791 years; after which they should all return again to their former places.

This period the Ancients called the Platonic, or great year; and imagined that at its completion every thing would begin as at first, and all things come round in the same order as they have done before.

The phenomena of this retrograde motion of the equinoxes, or intersections of the equinoctial with the ecliptic, and consequently of the conical motion of the earth's axis, by which the pole of the equator describes a small circle in the same period of time, may be understood and illustrated by a scheme, as follows: Let NZSVL be the earth, SONA its axis produced to the starry heavens, and terminating in A, the present north pole of the heavens, which is vertical to N, the north pole of the earth. Let EOQ be the equator, T♋Z the tropic of cancer, and VT♑ the tropic of capricorn; VOZ the ecliptic, and BO its axis, both of which are immoveable among the stars. But as the equinoctial points recede in the ecliptic, the earth's

axis SON is in motion upon the earth's centre O, in such a manner as to describe the double cone NO*n*

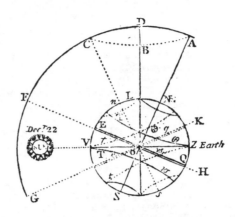

and SO*s*, round the axis of the ecliptic BO, in the time that the equinoctial points move round the ecliptic, which is 25,791 years; and in that length of time, the north pole of the earth's axis, produced, describes the circle ABCDA in the starry heavens, round the pole of the ecliptic, which keeps immoveable in the centre of that circle. The earth's axis being now 23° 28′ inclined to the axis of the ecliptic, the circle ABCDA, described by the north pole of the earth's axis produced to A, is 46° 56′ in diameter, or double the inclination of the earth's axis. In consequence of this, the point A, which is at present the north pole of the heavens, and near to a star of the 2d magnitude in the end of the Little Bear's tail, must be deserted by the earth's axis; which moving backwards 1 degree every 71¾ years nearly, will be directed towards the star or point B in 6447¾ years hence; and in double of that time, or 12,895½ years, it will be directed towards the star or point C; which will then be the north pole of the heavens, although it is at present 8½ degrees south of the zenith of London L. The present position of the equator EOQ will then be changed into *e*O*q*, the tropic of cancer T♋Z into V*t*♋, and the tropic of capricorn VT♑ into *tv*♑Z; as is evident by the figure. And the sun, in the same part of the heavens where he is now over the earthly tropic of capricorn, and makes the shortest days and longest nights in the northern hemisphere, will then be over the earthly tropic of cancer, and make the days longest and nights shortest. So that it will require 12,895½ years yet more, or from that time, to bring the north pole N quite round, so as to be directed toward that point of the heavens which is vertical to it at present. And then, and not till then, the same stars which at present describe the equator, tropics, and polar circles, &c, by the earth's diurnal motion, will describe them over again.

From this shifting of the equinoctial points, and with them all the signs of the ecliptic, it follows, that those stars which in the infancy of astronomy were in Aries, are now found in Taurus; those of Taurus in Gemini, &c. Hence likewise it is, that the stars which rose or set at any particular season of the year, in the times of Hesiod, Eudoxus, Virgil, Pliny, &c,

by

by no means anſwer at this time to their deſcriptions.

As to the phyſical cauſe of the Preceſſion of the equinoxes, Sir Iſaac Newton demonſtrates, that it ariſes from the broad or flat ſpheroidal figure of the earth; which itſelf ariſes from the earth's rotation about its axis: for as more matter has thus been accumulated all round the equatorial parts than any where elſe on the earth, the ſun and moon, when on either ſide of the equator, by attracting this redundant manner, bring the equator ſooner under them, in every return towards it, than if there was no ſuch accumulation.

Sir Iſaac Newton, in determining the quantity of the annual Preceſſion from the theory of gravity, on ſuppoſition that the equatorial diameter of the earth is to the polar diameter, as 230 to 229, finds the ſun's action ſufficient to produce a Preceſſion of $9''\frac{1}{8}$ only; and collecting from the tides the proportion between the ſun's force and the moon's to be as 1 to $4\frac{1}{2}$, he ſettles the mean Preceſſion reſulting from their joint actions, at $50''$; which, it muſt be owned, is nearly the ſame as it has ſince been found by the beſt obſervations; and yet ſeveral other mathematicians have ſince objected to the truth of Sir Iſaac Newton's computation.

Indeed, to determine the quantity of the Preceſſion ariſing from the action of the ſun, is a problem that has been much agitated among modern mathematicians; and although they ſeem to agree as to Newton's miſtake in the ſolution of it, they have yet generally diſagreed from one another. M. D'Alembert, in 1749, printed a treatiſe on this ſubject, and claims the honour of having been the firſt who rightly determined the method of reſolving problems of this kind. The ſubject has been alſo conſidered by Euler, Friſius, Silvabelle, Walmeſley, Simpſon, Emerſon, La Place, La Grange, Landen, Milner, and Vince.

M. Silvabelle, ſtating the ratio of the earth's axis to be that of 178 to 177, makes

the annual Preceſſion cauſed by the ſun 13″ 52‴,
and that of the moon - - 34 17;

making the ratio of the lunar force to the ſolar, to be that of 5 to 2; alſo the nutation of the earth's axis cauſed by the moon, during the time of a ſemirevolution of the pole of the moon's orbit, i. e. in $9\frac{1}{3}$ years, he makes 17″ 51‴.—M. Walmeſley, on the ſuppoſition that the ratio of the earth's diameters is that of 230 to 229, and the obliquity of the ecliptic to the equator 23° 28′ 30″, makes the annual Preceſſion, owing to the ſun's force, equal to 10″·583; but ſuppoſing the ratio of the diameters to be that of 178 to 177, that Preceſſion will be 13″·675.—Mr. Simpſon, by a different method of calculation, determines the whole annual preceſſion of the equinoxes cauſed by the ſun, at 21″ 6‴; and he has pointed out the errors of the computations propoſed by M. Silvabelle and M. Walmeſley.—Mr. Milner's deduction agrees with that of Mr. Simpſon, as well as Mr. Vince's; and their papers contain beſides ſeveral curious particulars relative to this ſubject. But for the various principles and reaſonings of theſe mathematicians, ſee Philoſ. Tranſ. vol. 48, pa. 385; vol. 49, pa. 704; vol. 69, pa. 505; and vol. 77. pa. 363; as alſo the writings of Simpſon, Emerſon, Landen, &c; alſo De la Lande's Aſtronomie, and the Memoirs of the Acad. Sci. in ſeveral places.

As to the effect of the planets upon the equinoctial points, M. De la Place, in his new reſearches on this article, finds that their action cauſes thoſe points to advance by $0''·2016$ in a year, along the equator, or $0''·1849$ along the ecliptic; from whence it follows that the quantity of the luni-ſolar Preceſſion muſt be $50''·4349$, ſince the total obſerved Preceſſion is $50''\frac{1}{4}$, or $50''·25$.

To find the Preceſſion in right aſcenſion and declination.

Put d = the declination of a ſtar,
and a = its right aſcenſion;

then their annual variations of Preceſſions will be nearly as follow:

viz, $20''·084 \times$ coſ. a = the annual preceſ. in declinat.
and $46''·0619 + 20''·084 \times$ ſin. $a \times$ tang. d = that of right aſcenſion. See the Connoiſſance des Temps for 1792, pa. 206, &c.

PRESS, in Mechanics, is a machine made of iron or wood, ſerving to compreſs or ſqueeze any body very cloſe, by means of ſcrews.

The common Preſſes conſiſt of ſix members, or pieces; viz, two flat and ſmooth planks; between which the things to be preſſed are laid; two ſcrews, or worms, faſtened to the lower plank, and paſſing through two holes in the upper; and two nuts, ſerving to drive the upper plank, which is moveable, againſt the lower, which is ſtable, and without motion.

PRESSION. See PRESSURE.

PRESSURE, is properly the action of a body which makes a continual effort or endeavour to move another; ſuch as the action of a heavy body ſupported by a horizontal table; in contradiſtinction from percuſſion, or a momentary force or action. Preſſure equally reſpects both bodies, that which preſſes, and that which is preſſed; from the mutual equality of action and reaction.

Preſſure, in the Carteſian Philoſophy, is an impulſive kind of motion, or rather an endeavour to move, impreſſed on a fluid medium, and propagated through it. In ſuch a preſſure the Carteſians ſuppoſe the action of light to conſiſt. And in the various modifications of this Preſſure, by the ſurfaces of bodies, on which that medium preſſes, they ſuppoſe the various colours to conſiſt, &c. But Newton ſhews, that if light conſiſted only in a Preſſure, propagated without actual motion, it could not agitate and warm ſuch bodies as reflect and refract it, as we actually find it does; and if it conſiſted in an inſtantaneous motion, or one propagated to all diſtances in an inſtant, as ſuch Preſſure ſuppoſes, there would be required an infinite force to produce that motion every moment, in every lucid particle. Farther, if light conſiſted either in Preſſure, or in motion propagated in a fluid medium, whether inſtantaneouſly, or in time, it muſt follow, that it would inflect itſelf *ad umbram*; for Preſſure, or motion, in a fluid medium, cannot be propagated in right lines, beyond any obſtacle which ſhall hinder any part of the motion; but will inflect and diffuſe itſelf, every way, into thoſe parts of the quieſcent medium which lie beyond the ſaid obſtacle.
Thus

Thus the force of gravity tends downward; but the Preſſure which ariſes from that force of gravity, tends every way with an equable force; and, with equal eaſe and force, is propagated in crooked lines, as in ſtraight ones. Waves on the ſurface of water, while they ſlide by the ſides of any large obſtacle, do inflect, dilate, and diffuſe themſelves gradually into the quieſcent water lying beyond the obſtacle. The waves, pulſes, or vibrations of the air, in which ſounds conſiſt, do manifeſtly inflect themſelves, though not ſo much as the waves of water; for the ſound of a bell, or of a cannon, can be heard over a hill, which intercepts the ſonorous object from our ſight; and ſounds are propagated as eaſily through crooked tubes, as through ſtraight ones. But light is never obſerved to go in curved lines, nor to inflect itſelf *ad umbram*; for the fixed ſtars do immediately diſappear on the interpoſition of any of the planets: as well as ſome parts of the ſun's body, by the interpoſition of the Moon, or Venus, or Mercury.

PRESSURE of *Air, Water, &c.* See AIR, WATER, &c.

The effects anciently aſcribed to the fuga vacui are now accounted for from the weight and Preſſure of the air.

The Preſſure of the air on the ſurface of the earth, is balanced by a column of water of the ſame baſe, and about 34 feet high; or of one of Mercury of near 30 inches high; and upon every ſquare inch at the earth's ſurface, that Preſſure amounts to about $14\frac{3}{4}$ pounds avoirdupois. The elaſticity of the air is equal to that Preſſure, and by means of that Preſſure, or elaſticity, the air would ruſh into a vacuum with a velocity of about 1370 feet per ſecond. At different heights above the earth's ſurface, the Preſſure of the air is as its denſity and elaſticity, and each decreaſes in ſuch ſort, that when the heights above the ſurface increaſe in arithmetical progreſſion, the Preſſure &c decreaſe in geometrical progreſſion: and hence if the axis BC of a logarithmic curve AD be erected perpendicular to the horizon, and if the ordinate AB denote the Preſſure or elaſticity, or denſity of the air, at the earth's ſurface, then will any other abſciſs

EF ⎫ denote the Preſſure &c ⎧ BE,
GH ⎬ at the altitude   ⎨ BG,
IK ⎭   ⎩ BI,

The Preſſure of water, as this fluid is every where of the ſame denſity, is as its depth at any place, and in all directions the ſame; and upon a ſquare foot of ſurface, every foot in height preſſes with the force of a weight of 1000 ounces or $62\frac{1}{2}$ lbs avoirdupois. And hence, if AB be the depth

of water in any veſſel, and BE denote its Preſſure at the depth B; by joining AE and drawing any other ordinates FG, HI; then ſhall theſe ordinates FG, HI, &c, denote the Preſſure at the correſponding depths AG, AI, &c; alſo the area of the triangle ABE will denote the whole Preſſure againſt the whole upright ſide AB, and which therefore is but half the Preſſure on the bottom of the ſame area as the ſide. Moreover, if a hole were opened in the bottom or ſide of the veſſel at B, the water, from the Preſſure of the ſuperincumbent fluid, would iſſue out with the velocity of $8\sqrt{AB}$ feet per ſecond nearly; AB being eſtimated in feet.

Centre of PRESSURE, in Hydroſtatics, is that point of any plane, to which, if the total Preſſure were applied, its effect upon the plane would be the ſame as when it was diſtributed unequally over the whole; or it is that point in which the whole Preſſure may be conceived to be united; or it is that point to which, if a force were applied equal to the total Preſſure, but with an oppoſite direction, it would exactly balance, or reſtrain the effect of the Preſſure, ſo that the body preſſed on will not incline to either ſide. Thus, if ABCD (2d fig. above) be a veſſel of water, and the ſide BC be preſſed upon with a force equivalent to 20 pounds of water, this force is unequally diſtributed over BC, for the parts near B are leſs preſſed than thoſe near C, which are at a greater depth; and therefore the efforts of all the particular Preſſures are united in ſome point E, which is nearer to C than to B; and that point E is called the centre of Preſſure: if to that point a force equivalent to 20 pounds will be applied, it will affect the plane BC in the ſame manner as by the Preſſure of the water diſtributed unequally over the whole; and if to the ſame point the ſame force be applied in a contrary direction to that of the Preſſure of the water, the force and the Preſſure will balance each other, and by oppoſite endeavours deſtroy each other's effects. Suppoſing a cord EFG fixed at E, and paſſing over the pulley F, has a weight of 20 pounds annexed to it, and that the part of the cord FE is perpendicular to BC; then the effort of the weight G is equal, and its direction contrary, to that of the Preſſure of the water. Now if E be the centre of Preſſure, theſe two powers will be in equilibrio, and mutually defeat each other's endeavours.

This point E, or the centre of Preſſure, is the ſame with the centre of percuſſion of the plane BC, the point of ſuſpenſion being B, the ſurface of the water. And if the plane be oblique, the caſe is ſtill the ſame, taking for the axis of ſuſpenſion, the interſection of that plane and the ſurface of the fluid, both produced if neceſſary. See Cotes's Lectures, pa. 40, &c.

The centre of Preſſure upon a plane parallel to the horizon, or upon any plane where the Preſſure is uniform, is the ſame as the centre of gravity of that plane. For the Preſſure acts upon every part in the ſame manner as gravity does.

PRIMARY *Planets*, are thoſe which revolve round the ſun as a centre. Such are the planets Mercury, Venus, Terra the Earth, Mars, Jupiter, Saturn, and Herſchel, and perhaps others. They are thus called, in contradiſtinction from the ſecondary planets, or ſatellites, which revolve about their reſpective Primaries. See PLANET.

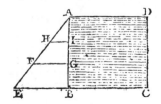
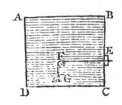

     PRIMES

PRIMES, denote the firſt diviſions into which ſome whole or integer is divided. As, a minute, or Prime minute, the 60th part of a degree; or the firſt place of decimals, being the 10th parts of units; or the firſt diviſion of inches in duodecimals, being the 12th parts of inches; &c.

PRIME *Numbers*, are thoſe which can only be meaſured by unity, or exactly divided without a remainder, 1 being the only aliquot part: as 2, 3, 5, 7, 11, 13, 17, &c. And they are otherwiſe called Simple, or Incompoſite numbers. No even number is a Prime, becauſe every even number is diviſible by 2. No number that ends with 0 or 5 is a Prime, the former being diviſible by 10, and the latter by 5. The following Table contains all the Prime numbers, and all the odd compoſite numbers, under 10,000, with the leaſt Prime diviſors of theſe; the deſcription, nature, and uſe of which, ſee immediately following the Table.

A Table of Prime and Compoſite Odd Numbers, under 10,000.

	c	1	2	3	4	5	6	7	8	9	10	11	12	13	14	15	16
01			3	7			3		3	17	7	3			3	19	
03			7	3	13		3	19	11	3	17		3		23	3	3	7
07			3	11	3			7	3		19		3	17		3	11	
09	3		11	3			3		3				3		7	3	
11		3			3	7	13	3			3	11		7	3	17		3
13			3		7	3		23	3	11		3	13	3	17		3	
17		3	7		3	11		3	19	7	3		3	13	37		3	
19		7	3	11			3			3	23		3	7		3		31
21	3	11	13	3			3	7		3		19	3		7	3		3
23		3		17	3		7	3	13		3			3		11	23	
27	3			3	7	17	3			3	13	7		3		3	43	
29		3		3	23	17	3			3			3		11	3		
31			3			3		17		3	7		3	11	3		7	
33	3	7		3		13	3		7	3	11	3	31		3	3	23	
37			3		19	3	7	11	3		17	3	7	3	29		3	
39	3			3		7	3			3		17	3	13		3	11	
41		3		11			3	29		3	7	17	3	11	23	3		
43		11	3	7		3			3	23	7	3	11	17	3		31	
47		3	13		3			3	7		3	31	29		3	7	3	
49	7		3			3	11	7	3	13		3		19	3			17
51	3			3	11	19	3		23	3			3	7		3	13	
53		3	11		3	7		3		3	7	3		23	31		3	
57	3			3		3			3	7	13	3	7	3		3		
59		3	7		3	13		3		7	3	19		3			3	
61		7	3	19			3	3	31		3	13		3		7	11	
63	3		3				3	23		13	3		29	3	7	3		
67			5			3	23	13	3		11		3	7	7	3		
69	3	13		3	7		3		11	3		7	3	37	13	3		
71			3		7	3		11	3	13		3		31	3			3
73		3		11	3			3	7	29	3	19		3	11		7	
77	7	3		13	3		3			3	11		3	7	19	3		
79			3			3	7	19	3	11	13	3		7	3			23
81	3			3	13	7	3	11		3	23		3			3	41	
83		3			3	11		3			3	7		3		3		7
87	3	11	7	3		3			3			3	19		3	7		
89		3	17		3	19	13		3	23	3	29		3	7		3	
91	7		3	17		3			7	3			3		13	3	37	19
93	3			3	17		3	13		3	19		3		7	3		
97		3		7	3	17		3			7	3		3		11	3	
99	3		13	3			3	17	29		3	7	11		3			3

		17	18	19	20	21	22	23	24	25	26	27	28	29	30	31	32	33
01		3				3	11	31		3		7	41		3	37		
03		13		3	11			3		7	3			19	3		3	29
07		3	13			3	7		3	29	23	3			7		3	31
09				3	23		7	3	47		3	13			3	53		3
11		29			3				3			3	7		3	41		
13		3	7		3			3	19		3	7	3		29	3	23	
17		17	23	3		29		3	7		3		11		3	11	3	
19		3	17	19		3	13		3	7	41	11		3		2		31
21				3	17	43	3		11		3		7	2	3			3
23				3		7		3	23		3	43	7	3	3			
27	11		3	41			3	17	13	3	7	37		3	11	3	53	7
29		7	31	3			3	17		7		3	11		3	29	13	3
31		3				3		23		3	11		3	19	3	7	31	3
33				3		19	3	7		3	17		3		7	3	13	53
37		3	11	13	3		7		43		3	7		3		3	43	41
39	37	3	7		3			2		7		3	17		3			3
41		7	3	13		3			3		19		3	17		3	7	13
43	3	19	29		3			3	7		3	41	17	7	3	17	
47		3	23	19	3			3		3	7	11		3	7	3	17	
49		3	43			3	7	13	3	31		7	3		47	3	17
51	17	3		7	3			3		11	3		13	3	23		3
53		17	3			3	13	11		3		3		43	3		7
57	7		3	19	11		3	37		3	7		3		3	7	3
59		11		3	29	17	3	7		3		31	3	11	7	3	
61		3		37	3		7	3	23	13		3	11		29		3
63	41	3	13		3	31	17	3	11		3			3		13	3
67	3		7	3	11		3	17			7	47		3		3	7
69	29	3	11			23	3	17	3	7	19		3			7	3
71	7		3	19	13	3		7			17	3		37	3		
73			3		41			3		31		3	47	13	7	19	3
77		3	31	7	3			3		11	3	13	17		3	29	11
79	3			3		43	3	37		3	7		3	11		3	31
81	13	3	7		3			3	29	7	3	43	11		3	17	3
83		7	3		37	3		13	3		11		3			7	17
87			3				7	3	13		3		29	3		19	3
89			3			11	3	19		3		3		7		3	11
91	3	31	11		3	7	29	3	47		3		7	3	11		3
93	11	3			7	3		3				3	11	41	3	31	37
97	3	7		3	13		3	11	7	3		3	19	23	3	43
99	3	7			3	11		3	23		3	13		3	7	3	43

A Table of Prime and Composite Odd Numbers, under 10,000.

	34	35	36	37	38	39	40	41	42	43	44	45	46	47	48	49	50		51	52	53	54	55	56	57	58	59	60	61	62	63	64	65	66	67
01	19	3	13		3	47		3		11	3	7	43	3		13	3			7	3	11		3			3	17		3		37	3	7	
03	41	31	3	7	3			11	3	13	7	3		17	3		3		3	11		3	13	3	7		3	17		3	19	7	3		
07		3		11	3			3	7	59	3		17	11	17	7	3			41	3		3	13		3		31	3		7	43	3		19
09	7	11	3		3	19		7	3	31		3		11	17	3			3		3	7	71	3	37	19	41		7	3	13	23	3		
11	3		23	3	37		3			3	11	13	3	7	17	3		11	19	3	47	7	3	31		3	23		3		3	17	11	3	
13		3		47	3	7		3	11	19	3		7	3	53		17	3	13	3	37	3	29		3	7		3	59	11	3	17		7	
17	3			3	11		3	23		3	7		3	53		3	29	17	7	3	13		3	41		3	61	11	3		3	7	3	3	
19	13	3	7		3		3		3	7	3		31	3	61		3	19		17	3		3	7	11	3	13	29	3	71	7	3		
21	11	7	3	61		3		13	3	29		3		3	7		3	23	17	3		7	3		31	3		3		3	11	3		3	11
23	3	13		3			3	7	41	3		3	7	3	47	3	11	3		59	3		19	3	7			61		3	7	
27	23		3	43	3		3		3		19	3	7	29	3	13	11	27	3		7	3	17	3		11	13	3		3	7	
29	3		19	3	7		3			3	43	7	3	11		3	47	23	3	73	61	3	13	17	3	7		3		3	7	3
31	47	3		7	3		29	3		61	3	23	11	3			3	7		3		3	11	7	3	37	3	13	59	3	19	53
33		3			3	37		3	7	11	3	41		3	7	8	11	43	3	19	17	3	23	3	7	47	3
37	7	3		37	3	31	11	3	19	3	13		3	7	3	11	11	3		7	3	13	3	17	3	41	3
39	19		3		11	3	7			3	23	3	7	3	11	13	19	3	29	3	3	7	17	3	47	13	3	23
41	3		11	3	23	7	3	41		3		19	3	11	47	3	71	53	3	7	3		3	13	7	3	79	17	3	31	29	3
43	11	3		19	3	13	3		43	3	7	3	29	3	7	37	7	3	23	3	3	13	17	3	7	11
47	3	7	3	3	11	31	3	3	3	47	37	3	7		3	13	3	19	3	3	23	11	3	17	3
49	3	41	23	3	11	3	3	3	13	7	3	19	29	3	31	3	3	23	11	3	7	3	61	17
51	7	53	3	11	3	17	3	19	3	3	3	59	3	7	3	11	3	7	3	3	43
53	3	11	13	3	59	3	3	61	29	3	7	23	3	31	3	53	7	3	11	3	13	3	3	3
57	3	13	7	3	3	67	3	13	3	7	11	3	3	7	3	47	3	11	79	3	29
59	3	3	17	37	3	3	7	47	3	43	3	7	3	23	53	3	13	3	59	73	3	11	3	7	3
61	3	7	3	17	31	3	7	3	59	3	11	3	13	3	43	67	3	7	3	11	61	3	7	3
63	7	3	53	3	17	23	3	3	11	3	7	61	3	19	31	3	7	3	11	67	3	7	3	23	3
67	13	19	3	7	3	17	11	3	13	3	31	3	23	3	7	19	3	73	3	3	29	3	59	67
69	43	3	53	3	13	11	3	17	41	3	7	19	3	37	3	11	7	3	3	47	3	31	3	3	7
71	3	3	7	11	3	43	3	17	7	3	13	3	11	3	41	3	53	29	3	13	3	23	3	7	3
73	23	3	7	3	29	3	3	17	3	11	3	7	3	13	3	23	7	3	3	3	13
77	3	7	3	41	3	7	3	11	23	3	17	3	31	3	19	3	7	53	3	43	59	3	7	3	11	3
79	7	3	13	3	23	3	11	29	3	19	3	7	13	3	3	3	37	3
81	59	3	19	3	7	37	3	13	3	31	7	3	17	3	3	13	3	7	11	3	3
83	3	29	11	7	3	47	3	3	19	3	13	71	3	7	3	3	31	7	3	61	13	3	29	41	3
87	11	17	3	7	13	3	61	53	3	41	7	3	43	3	17	3	7	3	37	11	3	7	3	23	3	13	7	3	11
89	3	37	7	3	3	59	3	67	13	3	3	17	11	3	7	3	53	3	19	3	11	3
91	3	17	3	13	3	7	3	3	67	7	3	29	11	3	17	3	43	3	41	3	7	3
93	7	3	17	3	7	3	23	3	13	3	11	3	67	3	7	3	71	13	3	11	7	3	43	19	3
97	13	3	3	7	17	3	3	7	3	59	19	3	3	23	29	3	11	3	7	73	3	37	7
99	3	59	3	29	7	3	13	3	53	11	3	37	3	7	3	11	41	3	17	3	67	3	13

A Table of Prime and Composite Odd Numbers, under 10,000.

	68	69	70	71	72	73	74	75	76	77	78	79	80	81	82	83
01	3	67			3	19	7	3	13	11		3	29		3	59
03		3	47		3		67	11	3				7	53	3	13
07	3		7	3					3			37		3	11	29
09	11		3	43		3		31		3	7	13	3	11		3
11	7			3	13		3		7	3	11	73	3			3
13	3	31			3		71	3	11	23	3	13	41	3	7	43
17	17			3	11	7		3		3			3		7	43
19	3	11			3		13	3	73	19	3	7		3	23	
21	19	3	7		3		41	3		3		7	3	89	13	3
23		7	3	17	31	3	13		3				3	71		3
27		3			3	17	7		3	29		3		23	3	7
29		13	3			3	17			3	59		3	7	11	3
31	3	29	79	3	7			3	17	13	3	41		3	47	
33		3	13	7	3			3	11		3		29	3	3	13
37	3	7	31	3		11	3		7	3	17		3	79		3
39	7	3		11	3	41	43	3		71	3	17		3	7	31
41		11	3	37	13	3	7		3			3	11	7	3	19
43	3	53		3		7	3	19		3	11	13	3	17		3
47	41		3	7		3	11		3	61	7	3	13		3	17
49	3		7	3	11		3			3	47		3	29	73	3
51	13	3	11		3			3	7	23	3		83	3	37	7
53	7	17	3	23		3	29	7	3			3	31	3		
57		3		17	3	7		3	13		3	73	7	3	23	61
59	19		3		7	3		3		29	3		41	3	13	
61	3		23	3	53	17	3		47	3	7	19	3		11	3
63		3	7	13	3	37	17	3	79	7	3		11	3		
67	3		37	3	13	53		3	7	11	3	31		3	7	3
69		3	3	67	3		7	3		17	3	13		3		3
71			3	7	11	3	31	67	3	19	17	3	7		3	11
73	3	19	11	3	7	73	3		3		3	7	11	3		
77	13		3		19	3		3	7		3	41	13	3		
79	3	7		3	29	47	3	11	7	3		79	3		17	3
81	7	3	73	43	3	11		3		31	3	23		3	7	17
83		3	11					7		3	43		3	59	7	3
87	71	3	19		3	83		3		13	3		7	3	3	83
89	83	29	3	7	37	3		3		3	7		19	3		
91	3		7	3	23	19	3		3	13	61	3				3
93	61	3	41		3		59	3	7	3		3		3		7
97	3		47	3		13		3	71	43	3	53	11	3	7	3
99		3	31	23	3	7		3		11		19	7	3	43	37

	84	85	86	87	88	89	90	91	92	93	94	95	96	97	98	99
01	31		3	7	13	3		19	3	71	7	3		89	3	
03	3	11	7	3		29	3			3		13	3	31		3
07	7	47		3		59	3	7	3	11	23	3	13	17	3	
09	3	67		3	23	59	3		3	97	37	3	7	17	3
11	13	3	79	31	3	7		3	19		3		7	3		11
13	47		3		7	3		13	3	67		3		11	3	23
17	19	3	7	23	3	37	71	3	13	7		3	31	59	3	47
19		7	3		3	29	11	3			3		3			7
21	3		37	3		11	3	7		3			3		7	3
23	3		11	3			7	3	23		3	89		3	11	3
27	3			3	7	79	3			3	11	7	3	71	31	
29	3		7	3			3	11	19	3	13		3	7		3
31	19	3			3	11	23	3	7		3			37	3
33	3	7	89	3	11		3		7	3		3		3		
37	11		3				3	7			3	23	7	3	19
39	3		53	3			3	13		3		3			3	
41	23	3			3		3			3	7	31	3	13	
43		3	7	37	3		41	3		7	3			3	61
47		3		3	23	83	3	7	13	3		11	3	43	7
49	7	83	3	13		3		7	3		11	3		3	
51	3	17	41	3	53		3	11	3	13		3	7		3
53	79	3	17		3	7	11	3	19	47	3	41	7	3	59	37
57	3	43	11	3	17	13	3		3	7	19	3	11	3
59	11	3	7	19	3	17		3	47	7	3	11	13	3	23
61		7	3		3	13		3	11		3		43	3	7
63	3		3		3	7	59	3	73	3	13	3	7	
67		13	3	11		3	89	3	17		3	7	3	3	
69	3	11		3	7	3	53	13	3	17	7	3	71	3
71	43	3	13	7	3		47	3	73		3	17	19	3	13
73	37		3	31	19	3	43		7		3		17	29	3
77	7	3		67	3	47	29	3		61	3		7	11
79	61	23	3		13	3	7	67	3	83		3	7	3	17
81	3			3	83	7	3		3	19	11	3		41	3
83	17	3	19		3	13	31		11	3	7	23	3		67
87	3	31	7	3		11	3		19	3	53		3	3
89	13	3		11	3	89	61	3	7	41	3	43	3	11	7
91	7	11	3	59	17	3		7	3		11	3		3	97
93	3	13		3		17	3	29		3	11	53	3	7	13	3
97	29		3	19	7	3	11	17	3		3	3		97	13
99	3			3	11		3		17	3	7	29	3	41	19	3

Qu

Out of the foregoing Table, are omitted all the odd numbers that end with 5, becaufe it is known that 5 is a divifor, or aliquot part of every fuch number.—— The difpofition of the Prime and compofite odd numbers in this Table, is along the top line, and down the firft or left-hand column; while their leaft Prime divifors are placed in the angles of meeting in the body of the page. Thus, the figures along the top line, viz, 0, 1, 2, 3, 4, &c, to 99, are fo many hundreds; and thofe down the firft column, from 1 to 99 alfo, are units or ones; and the former of thefe fet before the latter, make up the whole number, whether it be Prime or compofite; juft like the difpofition of the natural numbers in a table of logarithms. So the 16 in the top line, joined with the 19 in the firft column, makes the number 1619: the angle of their meeting, viz, of the column under 16, and of the line of 19, being blank, fhews that the number 1619 has no aliquot part or divifor, or that it is a Prime number. In like manner, all the other numbers are Primes that have no figure in their angle of meeting, as the numbers 41, 401, 919, &c. But when the two parts of any number have fome figure in their angle of meeting, that figure is the leaft divifor of the number, which is therefore not a Prime, but a compofite number: fo 301 has 7 for its leaft divifor, and 803 has 11 for its leaft divifor, and 1633 has 23 for its leaft divifor.

Hence, by the foregoing Table, are immediately known at fight all the Prime numbers up to 10,000; and hence alfo are readily found all the divifors or aliquot parts of the compofite numbers, namely in this manner: Find the leaft divifor of the given number in the Table, as above; divide the given number by this divifor, and confider the quotient as another or new number, of which find the leaft divifor alfo in the Table, dividing the faid quotient by this laft divifor; and fo on, dividing always the laft quotient by its leaft divifor found in the Table, till a quotient be found that is a Prime number: then are the faid divifors and the laft or Prime quotient, all the fimple or Prime divifors of the firft given number; and if thefe fimple divifors be multiplied together thus, viz, every two and every three, and every four, &c, of them together, the feveral products will make up the compound divifors or aliquot parts of the firft given number; noting, that if the given number be an even one, divide it by 2 till an odd number come out.

For example, to find all the divifors or component factors of the number 210. This being an even number, dividing it by 2, one of its divifors, gives 105; and this ending with 5, dividing it by 5, another of its factors, gives 21; and the leaft divifor of 21, by the Table is 3, the quotient from which is 7; therefore all the Prime or fimple factors of the given number are 2, 3, 5, 7. Set thefe therefore down in the firft line as in the margin; then multiply the 2 by the 3, and fet the product 6 below the 3; next multiply the 5 by all that precede it, viz, 2, 3, 6, and fet the products below the 5; laftly multiply the 7 by all the feven factors preceding it. and fet the products below the 7; fo fhall we have all the fac-

2	3	5	7
	6	10	14
		15	21
		30	42
			35
			70
			105
			210

tors or divifors of the given number 210, which are thefe, viz,

2, 3, 5, 6, 7, 10, 14, 15, 21, 30, 35, 42, 70, 105.

PRIME *Vertical*, is that vertical circle, or azimuth, which is perpendicular to the meridian, and paffes through the eaft and weft points of the horizon.

PRIME *Verticals*, in Dialling, or PRIME-*Vertical* Dials, are thofe that are projected on the plane of the Prime vertical circle, or on a plane parallel to it. Thefe are otherwife called direct, erect, north, or fouth dials.

PRIME *of the Moon*, is the new moon at her firft appearance, for about 3 days after her change. It means alfo the GOLDEN *Number*; which fee.

PRIMUM *Mobile*, in the Ptolomaic Aftronomy, is fuppofed to be a vaft fphere, whofe centre is that of the world, and in comparifon of which the earth is but a point. This they defcribe as including all other fpheres within it, and giving motion to them, turning itfelf and all the reft quite round in 24 hours.

PRINCIPAL, in Arithmetic, or in Commerce, is the fum lent upon intereft, either fimple or compound.

PRINCIPAL *Point*, in Perfpective, is a point in the perfpective plane, upon which falls the principal ray, or line from the eye perpendicular to the plane. This point is in the interfection of the horizontal and vertical planes; and is alfo called the *point of fight*, and *point of the eye*, or *centre of the picture*, or again the *point of concurrence*.

PRINCIPAL *Ray*, in Perfpective, is that which paffes from the fpectator's eye perpendicular to the picture or perfpective plane, and fo meeting it in the principal point.

PRINGLE (Sir JOHN), Baronet, the late worthy prefident of the Royal Society, was born at Stichelhoufe, in the county of Roxburgh, North Britain, April 10, 1707. His father was Sir John Pringle, of Stichel, Bart. and his mother Magdalen Elliott, was fifter to Sir Gilbert Elliott, of Stobs, Baronet. He was the youngeft of feveral fons, three of whom, befides himfelf, arrived to years of maturity. After receiving his grammatical education at home, he was fent to the univerfity of St. Andrews, where having ftaid fome years, he removed to Edinburgh in 1727, to ftudy phyfic, that being the profeffion which he now determined to follow. He ftaid however only one year at Edinburgh, being defirous of going to Leyden, which was then the moft celebrated fchool for medicine in Europe. Dr. Boerhaave, who had brought that univerfity into great reputation, was confiderably advanced in years, and Mr. Pringle was defirous of benefiting by that great man's lectures. After having gone through his proper courfe of ftudies at Leyden, he was admitted, in 1730, to his doctor of phyfic's degree; upon which occafion his inaugural differtation, *De Marcore Senili*, was printed. On quitting Leyden, Dr. Pringle returned and fettled at Edinburgh as a phyfician, where, in 1734, he was appointed, by the magiftrates and council of the city, to be joint profeffor of pneumatics and moral philofophy with Mr. Scott, during this gentleman's life, and fole profeffor after his deceafe; being alfo admitted at the fame time a member of the univerfity. In difcharging the duties of this new employment,

ment, his text-book was Puffendorff *de Officio Hominis et Civis*; agreeably to the method he pursued through life, of making fact and experiment the basis of science.

Dr. Pringle continued in the practice of Physic at Edinburgh, and in duly performing the office of professor, till 1742, when he was appointed physician to the earl of Stair, who then commanded the British army. By the interest of this nobleman, Dr. Pringle was constituted, the same year, physician to the military hospital in Flanders, with a salary of 20 shillings a-day, and the right to half-pay for life. On this occasion he was permitted to retain his professorship of moral philosophy; two gentlemen, Messrs. Muirhead and Cleghorn teaching in his absence, as long as he requested it. The great attention which Dr. Pringle paid to his duty as an army physician, is evident from every page of his *Treatise on the Diseases of the Army*, in the execution of which office he was sometimes exposed to very imminent dangers. He soon after also met with no small affliction in the retirement of his great friend the earl of Stair, from the army. He offered to resign with his noble patron, but was not permitted: he was therefore obliged to content himself with testifying his respect and gratitude to him, by accompanying the earl 40 miles on his return to England; after which he took leave of him with the utmost regret.

But though Dr. Pringle was thus deprived of the immediate protection of a nobleman who knew and esteemed his worth, his conduct in the duties of his station procured him effectual support. He attended the army in Flanders through the campaign of 1744, and so powerfully recommended himself to the duke of Cumberland, that in the spring following he had a commission, appointing him physician-general to the king's forces in the Low Countries, and parts beyond the seas; and on the next day he received a second commission from the duke, constituting him physician to the royal hospitals in those countries. In consequence of these promotions, he the same year resigned his professorship in the university of Edinburgh.

In 1745 he was also with the army in Flanders; but was recalled from that country in the latter end of the year, to attend the forces which were to be sent against the rebels in Scotland. At this time he had the honour of being chosen F. R. S. and the Society had good reason to be pleased with the addition of such a member. In the beginning of 1746, Dr. Pringle accompanied, in his official capacity, the duke of Cumberland in his expedition against the rebels; and remained with the forces, after the battle of Culloden, till their return to England the following summer. In 1747 and 1748, he again attended the army abroad; but in the autumn of 1748, he embarked with the forces for England, on the signing of the treaty of Aix-la-Chapelle.

From that time he mostly resided in London, where, from his known skill and experience, and the reputation he had acquired, he might reasonably expect to succeed as a physician. In 1749 he was appointed physician in ordinary to the duke of Cumberland. And in 1750 he published, in a letter to Dr. Mead, *Observations on the Gaol or Hospital Fever*: this piece, with some alterations, was afterwards included in his grand work on the *Diseases of the Army*.

In this, and the two following years Dr. Pringle communicated to the Royal Society his celebrated *Experiments upon Septic and Antiseptic Substances, with Remarks relating to their Use in the Theory of Medicine*; some of which were printed in the Philosophical Transactions, and the whole were subjoined, as an appendix, to his *Observations on the Diseases of the Army*. Those experiments procured for the ingenious author the honour of Sir Godfrey Copley's gold medal; besides gaining him a high and just reputation as an experimental philosopher. He gave also many other curious papers to the Royal Society: thus, in 1753, he presented, *An Account of several Persons seized with the Gaol Fever by working in Newgate; and of the Manner by which the Infection was communicated to one entire Family*; in the Philos. Trans. vol. 48. His next communication was, *A remarkable case of Fragility, Flexibility, and Dissolution of the Bones*; in the same vol.—In the 49th volume, are accounts which he gave of an Earthquake felt at Brussels; of another at Glasgow and Dunbarton; and of the Agitation of the Waters, Nov. 1, 1756, in Scotland and at Hamburgh.—The 50th volume contains his Observations on the Case of lord Walpole, of Woolerton; and a Relation of the Virtues of Soap, in Dissolving the Stone.—The next volume is enriched with two of the doctor's articles, of considerable length, as well as value. In the first, he hath collected, digested, and related, the different accounts that had been given of a very extraordinary Fiery Meteor, which appeared the 26th of November 1758; and in the second he hath made a variety of remarks upon the whole, displaying a great degree of philosophical sagacity.—Besides his communications in the Philosophical Transactions, he gave, in the 5th volume of the Edinburgh Medical Essays, an Account of the Success of the *Vitrum ceratum Antimonii*.

In 1752, Dr. Pringle married Charlotte, the second daughter of Dr. Oliver, an eminent physician at Bath: a connection which however did not last long, the lady dying in the space of a few years. And nearly about the time of his marriage, he gave to the public the first edition of his *Observations on the Diseases of the Army*; which afterwards went through many editions with improvements, was translated into the French, the German, and the Italian languages, and deservedly gained the author the highest credit and encomiums. The utility of this work however was of still greater importance than its reputation. From the time that the doctor was appointed a physician to the army, it seems to have been his grand object to lessen, as far as lay in his power, the calamities of war; nor was he without considerable success in his noble and benevolent design. The benefits which may be derived from our author's great work, are not solely confined to gentlemen of the medical profession. General Melville, a gentleman who unites with his military abilities the spirit of philosophy, and the feelings of humanity, was enabled, when governor of the Neutral Islands, to be singularly useful, in consequence of the instructions he had received from Dr. Pringle's book, and from personal conversation with him. By taking care to have his men always lodged in large, open, and airy apartments, and by never letting his force remain long enough in swampy places to be injured by the noxious air of such places, the general was the

happy

happy inftrument of faving the lives of feven hundred foldiers.

Though Dr. Pringle had not for fome years been called abroad, he ftill held his place of phyfician to the army; and in the war that began in 1755, he attended the camps in England during three feafons. In 1758, however, he entirely quitted the fervice of the army; and being now determined to fix wholly in London, he was the fame year admitted a licentiate of the college of phyficians.—After the acceffion of king George the 3d to the throne of Great Britain, Dr. Pringle was appointed, in 1761, phyfician to the queen's houfehold; and this honour was fucceeded, by his being conftituted, in 1763, phyfician extraordinary to the queen. The fame year he was chofen a member of the Academy of Sciences at Haarlem, and elected a fellow of the Royal College of Phyficians in London.—In 1764, on the deceafe of Dr. Wollafton, he was made phyfician in ordinary to the queen. In 1766 he was elected a foreign member, in the phyfical line, of the Royal Society of Sciences at Gottingen, and the fame year he was raifed to the dignity of a baronet of Great Britain. In 1768 he was appointed phyfician in ordinary to the late princefs dowager of Wales.

After having had the honour to be feveral times elected into the council of the Royal Society, Sir John Pringle was at length, viz. Nov. 30, 1772, in confequence of the death of James Weft Efq. elected prefident of that learned body. His election to this high ftation, though he had fo refpectable a character as the late Sir James Porter for his opponent, was carried by a very confiderable majority. Sir John Pringle's conduct in this honourable ftation fully juftified the choice the Society made of him as their prefident. By his equal, impartial, and encouraging behaviour, he fecured the good will and beft exertions of all for the general benefit of fcience, and true interefts of the Society, which in his time was raifed to the pinnacle of honour and credit. Inftead of fplitting the members into oppofite parties, by cruel, unjuft, and tyrannical conduct, as has fometimes been the cafe, to the ruin of the beft interefts of the Society, Sir John Pringle cherifhed and happily united the endeavours of all, collecting and directing the energy of every one to the common good of the whole. He happily alfo ftruck out a new way to diftinction and ufefulnefs, by the difcourfes which were delivered by him, on the annual affignment of Sir Godfrey Copley's medal. This gentleman had originally bequeathed five guineas, to be given at each anniverfary meeting of the Royal Society, by the determination of the prefident and council, to the perfon who fhould be the author of the beft paper of experimental obfervations for the year. In procefs of time, this pecuniary reward, which could never be an important confideration to a man of an enlarged and philofophical mind, however narrow his circumftances might be, was changed into the more liberal form of a gold medal; in which form it is become a truly honourable mark of diftinction, and a juft and laudable object of ambition. No doubt it was always ufual for the prefident, on the delivery of the medal, to pay fome compliment to the gentleman on whom it was beftowed; but the cuftom of making a fet fpeech on the occafion, and of entering into the hiftory of that part of philofophy to which the experiments, or the fubject of the

paper related, was firft introduced by Martin Folkes Efq. The difcourfes however which he and his fucceffors delivered, were very fhort, and were only inferted in the minute-books of the Society. None of them had ever been printed before Sir John Pringle was raifed to the chair. The firft fpeech that was made by him being much more elaborate and extended than ufual, the publication of it was defired; and with this requeft, it is faid, he was the more ready to comply, as an abfurd account of what he had delivered had appeared in a newspaper. Sir John was very happy in the fubject of his firft difcourfe. The difcoveries in magnetifm and electricity had been fucceeded by the inquiries into the various fpecies of air. In thefe enquiries, Dr. Prieftley, who had already greatly diftinguifhed himfelf by his electrical experiments, and his other philofophical purfuits and labours, took the principal lead. A paper of his, intitled, *Obfervations on different Kinds of Air*, having been read before the Society in March 1772, was adjudged to be deferving of the gold medal; and Sir John Pringle embraced with pleafure the occafion of celebrating the important communications of his friend, and of relating with accuracy and fidelity what had previoufly been difcovered upon the fubject.

It was not intended, we believe, when Sir John's firft fpeech was printed, that the example fhould be followed: but the fecond difcourfe was fo well received by the Society, that the publication of it was unanimoufly requefted. Both the difcourfe itfelf, and the fubject on which it was delivered, merited fuch a diftinction. The compofition of the fecond fpeech is evidently fuperior to that of the former one; Sir John having probably been animated by the favourable reception of his firft effort. His account of the Torpedo, and of Mr. Walfh's ingenious and admirable experiments relative to the electrical properties of that extraordinary fifh, is fingularly curious. The whole difcourfe abounds with ancient and modern learning, and exhibits the worthy prefident's knowledge in natural hiftory, as well as in medicine, to great advantage.

The third time that he was called upon to difplay his abilities at the delivery of the annual medal, was on a very beautiful and important occafion. This was no lefs than Mr. (now Dr.) Mafkelyne's fuccefsful attempt completely to eftablifh Newton's fyftem of the univerfe, by his *Obfervations made on the Mountain Schehallien, for finding its attraction.* Sir John laid hold of this opportunity to give a perfpicuous and accurate relation of the feveral hypothefes of the Ancients, with regard to the revolutions of the heavenly bodies, and of the noble difcoveries with which Copernicus enriched the aftronomical world. He then traces the progrefs of the grand principle of gravitation down to Sir Ifaac's illuftrious confirmation of it; to which he adds a concife account of Meffrs. Bouguer's and Condamine's experiment at Chimboraço, and of Mr. Mafkelyne's at Schehallien. If any doubts ftill remained with refpect to the truth of the Newtonian fyftem, they were now completely removed.

Sir John Pringle had reafon to be peculiarly fatisfied with the fubject of his fourth difcourfe; that fubject being perfectly congenial to his difpofition and ftudies. His own life had been much employed in pointing out the means which tended not only to cure, but to pre-

vent

vent the diseases of mankind; and it is probable, from his intimate friendship with captain Cook, that he might suggest to that sagacious commander some of the rules which he followed, in order to preserve the health of the crew of his ship, during his voyage round the world. Whether this was the case, or whether the method pursued by the captain to attain so salutary an end, was the result alone of his own reflections, the success of it was astonishing; and this celebrated voyager seemed well entitled to every honour which could be bestowed. To him the Society assigned their gold medal, but he was not present to receive the honour. He was gone out upon the voyage, from which he never returned. In this last voyage he continued equally successful in maintaining the health of his men.

The learned president, in his fifth annual dissertation, had an opportunity of displaying his knowledge in a way in which it had not hitherto appeared. The discourse took its rise from the adjudication of the prize medal to Mr. Mudge, then an eminent surgeon at Plymouth, on account of his valuable paper, containing *Directions for making the best Composition for the Metals of Reflecting Telescopes, together with a Description of the Process for Grinding, Polishing, and giving the Great Speculum the true Parabolic form.* Sir John hath accurately related a variety of particulars, concerning the invention of reflecting telescopes, the subsequent improvements of these instruments, and the state in which Mr. Mudge found them, when he first set about working them to a greater perfection, till he had truly realized the expectation of Newton, who, above an hundred years ago, presaged that the public would one day possess a parabolic speculum, not accomplished by mathematical rules, but by mechanical devices.

Sir John Pringle's sixth and last discourse, to which he was led by the assignment of the gold medal to myself, on account of my paper intitled, *The Force of fired Gunpowder, and the Initial Velocity of Cannon Balls, determined by Experiments,* was on the theory of gunnery. Though Sir John had so long attended the army, this was probably a subject to which he had heretofore paid very little attention. We cannot however help admiring with what perspicuity and judgment he hath stated the progress that was made, from time to time, in the knowledge of projectiles, and the scientific perfection to which it has been said to be carried in my paper. As Sir John Pringle was not one of those who delighted in war, and in the shedding of human blood, he was happy in being able to shew that even the study of artillery might be useful to mankind; and therefore this is a topic which he hath not forgotten to mention. Here ended our author's discourses upon the delivery of Sir Godfrey Copley's medal, and his presidency over the Royal Society at the same time, the delivering that medal into my hand being the last office he ever performed in that capacity; a ceremony which was attended by a greater number of the members, than had ever met together before upon any other occasion. Had he been permitted to preside longer in that chair, he would doubtless have found other occasions of displaying his acquaintance with the history of philosophy. But the opportunities which he had of signalizing himself in this respect were important in themselves, happily varied, and sufficient to gain him a solid and lasting reputation.

Several marks of literary distinction, as we have already seen, had been conferred upon Sir John Pringle, before he was raised to the president's chair. But after that event they were bestowed upon him in great abundance, having been elected a member of almost all the literary societies and institutions in Europe. He was also, in 1774, appointed physician extraordinary to the king.

It was at rather a late period of life when Sir John Pringle was chosen to be president of the Royal Society, being then 65 years of age. Considering therefore the great attention that was paid by him to the various and important duties of his office, and the great pains he took in the preparation of his discourses, it was natural to expect that the burthen of his honourable station should grow heavy upon him in a course of time. This burthen, though not increased by any great addition to his life, for he was only 6 years president, was somewhat augmented by the accident of a fall in the area in the back part of his house, from which he received some hurt. From these circumstances some persons have affected to account for his resigning the chair at the time when he did. But Sir John Pringle was naturally of a strong and robust frame and constitution, and had a fair prospect of being well able to discharge the duties of his situation for many years to come, had his spirits not been broken by the most cruel harassings and baitings in his office. His resolution to quit the chair arose from the disputes introduced into the Society, concerning the question, whether pointed or blunted electrical conductors are the most efficacious in preserving buildings from the pernicious effects of lightning, and from the cruel circumstances attending those disputes. These drove him from the chair. Such of those circumstances as were open and manifest to every one, were even of themselves perhaps quite sufficient to drive him to that resolution. But there were yet others of a more private nature, which operated still more powerfully and directly to produce that event; which may probably hereafter be laid before the public, when I shall give to them the history of the most material transactions of the Royal Society, especially those of the last 22 years, which I have from time to time composed and prepared with that view.

His intention of resigning however, was disagreeable to his friends, and the most distinguished members of the Society, who were many of them perhaps ignorant of the true motive for it. Accordingly, they earnestly solicited him to continue in the chair; but, his resolution being fixed, he resigned it at the anniversary meeting in 1778, immediately on delivering the medal, at the conclusion of his speech, as mentioned above.

Though Sir John Pringle thus quitted his particular relation to the Royal Society, and did not attend its meetings so constantly as he had formerly done, he still retained his literary connections in general. His house continued to be the resort of ingenious and philosophical men, whether of his own country, or from abroad; and he was frequent in his visits to his friends. He was held in particular esteem by eminent and learned foreigners, none of whom came to England without waiting upon him, and paying him the greatest respect. He treated them, in return, with distinguished civility and regard. When a number of gentlemen met at

his

his table, foreigners were usually a part of the company.

In 1780 Sir John spent the summer on a visit to Edinburgh; as he did also that of 1781; where he was treated with the greatest respect. In this last visit he presented to the Royal College of Physicians in that city, the result of many years labour, being ten folio volumes of *Medical and Physical Observations*, in manuscript, on condition that they should neither be published, nor lent out of the library of the college on any account whatever. He was at the same time preparing two other volumes, to be given to the university, containing the formulas referred to in his annotations. He returned again to London, and continued for some time his usual course of life, receiving and paying visits to the most eminent literary men, but languishing and declining in his health and spirits, till the 18th of January 1782, when he died, in the 75th year of his age; the account of his death being every where received in a manner which shewed the high sense that was entertained of his merit.

Sir John Pringle's eminent character as a practical physician, as well as a medical author, is so well known, and so universally acknowledged, that an enlargement upon it cannot be necessary. In the exercise of his profession he was not rapacious; being ready, on various occasions, to give his advice without pecuniary views. The turn of his mind led him chiefly to the love of science, which he built on the firm basis of fact. With regard to philosophy in general, he was as averse to theory, unsupported by experiments, as he was with respect to medicine in particular. Lord Bacon was his favourite author; and to the method of investigation recommended by that great man, he steadily adhered. Such being his intellectual character, it will not be thought surprising that he had a dislike to Plato. And to metaphysical disquisitions he lost all regard in the latter part of his life.

Sir John had no great fondness for poetry. He had not even any distinguished relish for the immortal Shakespeare: at least he seemed too highly sensible of the defects of that illustrious bard, to give him the proper degree of estimation. Sir John had not in his youth been neglectful of philological enquiries, nor did he desert them in the last stages of his life, but cultivated even to the last a knowledge of the Greek language. He paid a great attention to the French language; and it is said that he was fond of Voltaire's critical writings. Among all his other pursuits, he never forgot the study of the English language. This he regarded as a matter of so much consequence, that he took uncommon pains with regard to the style of his compositions; and it cannot be denied, that he excelled in perspicuity, correctness, and propriety of expression. His six discourses in particular, delivered at the annual meetings of the Royal Society, on occasion of the prize medals, have been universally admired as elegant compositions, as well as critical and learned dissertations. And this characteristic of them, seemed to increase and heighten, from year to year: a circumstance which argues rather an improvement of his faculties, than any decline of them, and that even after the accident which it was pretended occasioned his descent from the president's chair. So excellent indeed were these compositions esteem-

ed, that envy used to asperse his character with the imputation of borrowing the hand of another in those learned discourses. But how false such aspersion was, I, and I believe most of the other gentlemen who had the honour of receiving the annual medal from his hands, can fully testify. For myself in particular, I can witness for the last, and perhaps the best, that on the theory and improvements in gunnery, having been present or privy to his composition of every part of it.—Though our author was not fond of poetry, he had a great affection for the sister art music. Of this art he was not merely an admirer, but became so far a practitioner in it, as to be a performer on the violoncello, at a weekly concert given by a society of gentlemen at Edinburgh. Besides a close application to medical and philosophical science, during the latter part of his life, he devoted much time to the study of divinity: this being with him a very favourite and interesting object.

If, from the intellectual, we pass on to the moral character of Sir John Pringle, we shall find that the ruling feature of it was integrity. By this principle he was uniformly actuated in the whole of his conduct and behaviour. He was equally distinguished for his sobriety. I and other persons have heard him declare, that he had never once in his life been intoxicated with liquor. In his friendships, he was ardent and steady. The intimacies which were formed by him, in the early part of his life, continued unbroken to the decease of the gentlemen with whom they were made; and were kept up by a regular correspondence, and by all the good offices that lay in his power.

With regard to Sir John's external manner of deportment, he paid a very respectful attention to those who were honoured with his friendship and esteem, and to such strangers as came to him well recommended. Foreigners in particular had good reason to be satisfied with the uncommon pains which he took to shew them every mark of civility and regard. He had however at times somewhat of a dryness and reserve in his behaviour, which had the appearance of coldness; and this was the case when he was not perfectly pleased with the persons who were introduced to him, or who happened to be in his company. His sense of integrity and dignity would not permit him to adopt that false and superficial politeness, which treats all men alike, though ever so different in point of real estimation and merit, with the same shew of cordiality and kindness. He was above assuming the profession, without the reality of respect.

PRISM, in Geometry, is a body, or solid, whose two ends are any plane figures which are parallel, equal, and similar; and its sides, connecting those ends, are parallelograms.—Hence, every section parallel to the ends, is the same kind of equal and similar figure as the ends themselves are; and the Prism may be considered as generated by the parallel motion of this plane figure.

Prisms take their several particular names from the figure of their ends. Thus, when the end is a triangle, it is a Triangular Prism; when a square, a Square Prism; when a pentagon, a Pentagonal Prism; when a hexagon, a Hexagonal Prism; and so on. And hence the denomination Prism comprises also the cube and parallelopipedon, the former being a square Prism, and

the

PRO [284] PRO

the latter a rectangular one. And even a cylinder may be considered as a round Prism, or one that has an infinite number of sides. Also a Prism is said to be regular or irregular, according as the figure of its end is a regular or an irregular polygon.

The Axis of a Prism, is the line conceived to be drawn lengthways through the middle of it, connecting the centre of one end with that of the other end.

Prisms, again, are either right or oblique.

A *Right* Prism is that whose sides, and its axis, are perpendicular to its ends; like an upright tower. And

An *Oblique* Prism, is when the axis and sides are oblique to the ends; so that, when set upon one end, it inclines on one hand, like an inclined tower.

The principal properties of Prisms, are,

1. That all Prisms are to one another in the ratio compounded of their bases and heights.

2. Similar Prisms are to one another in the triplicate ratio of their like sides.

3. A Prism is triple of a pyramid of equal base and height; and the solid content of a Prism is found by multiplying the base by the perpendicular height.

4. The upright surface of a right Prism, is equal to a rectangle of the same height, and its breadth equal to the perimeter of the base or end. And therefore such upright surface of a right Prism, is found by multiplying the perimeter of the base by the perpendicular height. Also the upright surface of an oblique Prism is found by computing those of all its parallelogram sides separately, and adding them together.

And if to the upright surface be added the areas of the two ends, the sum will be the whole surface of the Prism.

PRISM, in Dioptrics, is a piece of glass in form of a triangular Prism: which is much used in experiments concerning the nature of light and colours.

The use and phenomena of the Prism arise from its sides not being parallel to each other; from whence it separates the rays of light in their passage through it, by coming through two sides of one and the same angle.

The more general of these phenomena are enumerated and illustrated under the article Colour; which are sufficient to prove, that colours do not either consist in the contorsion of the globules of light, as Des Cartes imagined; nor in the obliquity of the pulses of the etherial matter, as Hook fancied; nor in the constipation of light, and its greater or less concitation, as Dr. Barrow conjectured; but that they are original and unchangeable properties of light itself.

PRISMOID, is a solid, or body, somewhat resembling a prism, but that its ends are any dissimilar parallel plane figures of the same number of sides; the upright sides being trapezoids.——If the ends of the Prismoid be bounded by dissimilar curves, it is sometimes called a cylindroid.

PROBABILITY *of an Event*, in the Doctrine of Chances, is the ratio of the number of chances by which the event may happen, to the number by which it may both happen and fail. So that, if there be constituted a fraction, of which the numerator is the number of chances for the events happening, and the denominator the number for both happening and failing, that fraction

will properly express the value of the Probability of the event's happening. Thus, if an event have 3 chances for happening, and 2 for failing, the sum of which being 5, the fraction ⅗ will fitly represent the Probability of its happening, and may be taken to be the measure of it. The same thing may be said of the Probability of failing, which will likewise be measured by a fraction, whose numerator is the number of chances by which it may fail, and its denominator the whole number of chances both for its happening and failing: so the Probability of the failing of the above event, which has 2 chances to fail, and 3 to happen, will be expressed or measured by the fraction ⅖.

Hence, if there be added together the fractions which express the Probability for both happening and failing, their sum will always be equal to unity or 1; since the sum of their numerators will be equal to their common denominator. And since it is a certainty that an event will either happen or fail, it follows that a certainty, which may be considered as an infinitely great degree of Probability, is fitly represented by unity. See Simpson's or Demoivre's Doctrine of Chances; also Bernoulli's Ars Conjectandi; Monmort's Analyse des Jeux de Hasard; or M. De Parcieu's Essais sur les Probabilites de la Vie humaine. See also EXPECTATION, and GAMING.

PROBABILITY *of Life*. See EXPECTATION *of Life*, and LIFE-*Annuities*.

PROBLEM, in Geometry, is a proposition in which some operation or construction is required. As, to bisect a line, to make a triangle, to raise a perpendicular, to draw a circle through three points, &c.

A Problem, according to Wolfius, consists of three parts: The proposition, which expresses what is to be done; the resolution, or solution, in which are orderly rehearsed the several steps of the process or operation; and the demonstration, in which it is shewn, that by doing the several things prescribed in the resolution, the thing required is obtained.

PROBLEM, in Algebra, is a question or proposition which requires some unknown truth to be investigated or discovered; and the truth of the discovery demonstrated.

PROBLEM, *Kepler's*. See KEPLER'S *Problem*.

PROBLEM, *Determinate, Diophantine, Indeterminate, Limited, Linear, Local, Plane, Solid, Surfolid*, and *Unlimited*. See the adjectives.

Deliacal PROBLEM, in Geometry, is the doubling of a cube. This amounts to the same thing as the finding of two mean proportionals between two given lines: whence this also is called the Deliaeal Problem. See DUPLICATION.

PROCLUS, an eminent philosopher and mathematician among the later Platonists, was born at Constantinople in the year 410, of parents who were both able and willing to provide for his instruction in all the various branches of learning and knowledge. He was first sent to Xanthus, a city of Lycia, to learn grammar: from thence to Alexandria, where he was under the best masters in rhetoric, philosophy, and mathematics: and from Alexandria he removed to Athens, where he attended the younger Plutarch, and Syrian, both of them celebrated philosophers. He succeeded the latter in the

the government of the Platonic school at Athens; where he died in 485, at 75 years of age.

Marinus of Naples, who was his succeffor in the school, wrote his life; the firft perfect copy of which was publifhed, with a Latin verfion and notes, by Fabricius at Hamburgh, 1700, in 4to; and afterwards fubjoined to his *Bibliotheca Latina*, 1703, in 8vo.

Proclus wrote a great number of pieces, and upon many different fubjects; as, commentaries on philofophy, mathematics, and grammar; upon the whole works of Homer, Hefiod, and Plato's books of the republic: he wrote alfo on the conftruction of the Aftrolabe. Many of his pieces are loft; fome have been publifhed; and a few remain ftill in manufcript only. Of the publifhed, there are four very elegant hymns; one to the Sun, two to Venus, and one to the Mufes. There are commentaries upon feveral pieces of Plato; upon the four books of Ptolomy's work *de Judiciis Aftrorum*; upon the firft book of Euclid's Elements; and upon Hefiod's *Opera et Dies*. There are alfo works of Proclus upon philofophical and aftronomical fubjects; particularly the piece *De Sphæra*, which was publifhed, 1620, in 4to, by Bainbridge, the Savilian profeffor of aftronomy at Oxford. He wrote alfo 18 arguments againft the Chriftians, which are ftill extant, and in which he attacks them upon the queftion, whether the world be eternal? the affirmative of which he maintains.

The character of Proclus is the fame as that of all the later Platonifts, who it feems were not lefs enthufiafts and madmen, than the Chriftians their contemporaries, whom they reprefented in this light. Proclus was not reckoned quite orthodox by his own order: he did not adhere fo rigoroufly, as Julian and Porphyry, to the doctrines and principles of his mafter: " He had, fays Cudworth, fome peculiar fancies and whimfies of his own, and was indeed a confounder of the Platonic theology, and a mingler of much unintelligible ftuff with it."

PROCYON, in Aftronomy, a fixed ftar, of the fecond magnitude, in Canis Minor, or the Little Dog.

PRODUCING, in Geometry, denotes the continuing a line, or drawing it farther out, till it have an affigned length.

PRODUCT, in Arithmetic, or Algebra, is the factum of two numbers, or quantities, or the quantity arifing from, or produced by, the multiplication of, two or more numbers &c together. Thus, 48 is the product of 6 multiplied by 8.—In multiplication, unity is in proportion to one factor, as the other factor is to the product. So $1 : 6 :: 8 : 48$.

In Algebra, the product of fimple quantities is expreffed by joining the letters together like a word, and prefixing the product of the numeral coefficients: So the product of a and b is ab, of $3a$ and $4bc$ is $12abc$. But the product of compound factors or quantities is expreffed by fetting the fign of multiplication between them, and binding each compound factor in a vinculum: fo the product of $2a + 3b$ and $a - 4c$ is

$$\overline{2a + 3b} \times \overline{a - 4c}, \text{ or } (2a + 3b) \times (a - 4c).$$

In geometry, a rectangle anfwers to a product, its length and breadth being the two factors; becaufe the numbers expreffing the length and breadth being mul-

tiplied together, produce the content or area of the rectangle.

PROFILE, in Architecture, the figure or draught of a building, fortification, or the like; in which are expreffed the feveral heights, widths, and thickneffes, fuch as they would appear, were the building cut down perpendicularly from the roof to the foundation. Whence the Profile is alfo called the Section, and fometimes the Orthographical Section; and by Vitruvius the Sciography. In this fenfe, Profile amounts to the fame thing with Elevation; and fo ftands oppofed to a Plan or Ichnography.

PROFILE is alfo ufed for the contour, or outline of a figure, building, member of architecture, or the like; as a bafe, a cornice, &c.

PROGRESSION, an orderly advancing or proceeding in the fame manner, courfe, tenor, proportion, &c.

Progreffion is either Arithmetical, or Geometrical.

Arithmetical PROGRESSION, is a feries of quantities proceeding by continued equal differences, either increafing or decreafing. Thus,

increafing 1, 3, 5, 7, 9, &c, or
decreafing 21, 18, 15, 12, 9, &c;

where the former progreffion increafes continually by the common difference 2, and the latter feries or Progreffion decreafes continually by the common difference 3.

1. And hence, to conftruct an arithmetical Progreffion, from any given firft term, and with a given common difference; add the common difference to the firft term, to give the 2d; to the 2d, to give the 3d; to the 3d. to give the 4th; and fo on; when the feries is afcending or increafing: but fubtract the common difference continually, when the feries is a defcending one.

2. The chief property of an arithmetical Progreffion, and which arifes immediately from the nature of its conftruction, is this; that the fum of its extremes, or firft and laft terms, is equal to the fum of every pair of intermediate terms that are equidiftant from the extremes, or to the double of the middle term when there is an uneven number of the terms.

Thus, 1, 3, 5, 7, 9, 11, 13,
       13, 11, 9, 7, 5, 3, 1,
       ————————————————————
Sums   14, 14, 14, 14, 14, 14, 14,

where the fum of every pair of terms is the fame number 14.

Also, a, $a + d$, $a + 2d$, $a + 3d$, $a + 4d$,
      $a + 4d$, $a + 3d$, $a + 2d$, $a + d$, a
      ————————————————————————————————————
fums $2a + 4d$, $2a + 4d$, $2a + 4d$, $2a + 4d$, $2a + 4d$

3. And hence it follows, that double the fum of all the terms in the feries, is equal to the fum of the two extremes multiplied by the number of the terms; and confequently, that the fingle fum of all the terms of the feries, is equal to half the faid product. So the fum of the 7 terms

I,

$1, 3, 5, 7, 9, 11, 13$, is $\overline{1+13} \times \frac{7}{2} = \frac{14}{2} \times 7 = 49$. And the sum of the five terms

$a, a+d, a+2d, a+3d, a+4d$, is $\overline{a+4d} \times \frac{5}{2}$.

4. Hence also, if the first term of the Progression be 0, the sum of the series will be equal to half the product of the last term multiplied by the number of terms: i. e. the sum of

$0+d+2d+3d+4d \cdots n-1.d = \frac{1}{2}n.\overline{n-1}.d$, where n is the number of terms, supposing 0 to be one of them. That is, in other words, the sum of an arithmetical Progression, whether finite or infinite, whose first term is 0, is to the sum of as many times the greatest term, in the ratio of 1 to 2.

5. In like manner, the sum of the squares of the terms of such a series, beginning at 0, is to the sum of as many terms each equal to the greatest, in the ratio of 1 to 3. And

6. The sum of the cubes of the terms of such a series, is to the sum of as many times the greatest term, in the ratio of 1 to 4.

7. And universally, if every term of such a Progression be raised to the m power, then the sum of all those powers will be to the sum of as many terms equal to the greatest, in the ratio of $m+1$ to 1. That is,

the sum $0 + d + 2d + 3d \cdots l$,
is to $l^m + l^m + l^m + l^m \cdots l^m$,
in the ratio of 1 to $m+1$.

8. A synopsis of all the theorems, or relations, in an arithmetical Progression, between the extremes or first and last term, the sum of the series, the number of terms, and the common difference, is as follows; viz, if

a denote the least term,
z the greatest term,
d the common difference,
n the number of terms,
s the sum of the series;

then will each of these five quantities be expressed in terms of the others, as below:

$$a = z - \overline{n-1}.d = \frac{2s}{n} - z = \frac{s}{n} - \frac{n-1}{2}d = \sqrt{\overline{\tfrac{1}{2}d+z}^2 - 2ds} + \tfrac{1}{2}d.$$

$$z = a + \overline{n-1}.d = \frac{2s}{n} - a = \frac{s}{n} + \frac{n-1}{2}d = \sqrt{\overline{\tfrac{1}{2}d-a}^2 + 2ds} - \tfrac{1}{2}d.$$

$$d = \frac{z-a}{n-1} = \frac{s-na}{n-1}\cdot\frac{2}{n} = \frac{nz-s}{n-1}\cdot\frac{2}{n} = \frac{z+a.z-a}{2s-a-z}.$$

$$n = \frac{z-a}{d} + 1 = \frac{2s}{a+z} = \frac{\tfrac{1}{2}d-a+\sqrt{\overline{\tfrac{1}{2}d-a}^2+2ds}}{d} = \frac{\tfrac{1}{2}d+z-\sqrt{\overline{\tfrac{1}{2}d+z}^2-2ds}}{d}.$$

$$s = \frac{a+z}{2}n = \frac{a+z}{2}\cdot\frac{z-a+d}{d} = \frac{2a+\overline{n-1}.d}{2}n = \frac{2z-\overline{n-1}.d}{2}n.$$

And most of these expressions will become much simpler if the first term be 0 instead of a.

Geometrical PROGRESSION, is a series of quantities proceeding in the same continual ratio or proportion, either increasing or decreasing; or it is a series of quantities that are continually proportional; or which increase by one common multiplier, or decrease by one common divisor; which common multiplier or divisor is called the common ratio. As,

increasing, $1, 2, 4, 8, 16$, &c,
decreasing, $81, 27, 9, 3, 1$, &c;

where the former progression increases continually by the common multiplier 2, and the latter decreases by the common divisor 3.

Or ascending, a, ra, r^2a, r^3a, &c,

or descending, $a, \frac{a}{r}, \frac{a}{r^2}, \frac{a}{r^3}$, &c;

where the first term is a, and common ratio r.

1. Hence, the same principal properties obtain in a geometrical Progression, as have been remarked of the arithmetical one, using only multiplication in the geometricals for addition in the arithmeticals, and division in the former for subtraction in the latter. So that, to construct a geometrical Progression, from any given first term, and with a given common ratio; multiply the 1st term continually by the common ratio, for the rest of the terms when the series is an ascending one;

or divide continually by the common ratio, when it is a descending Progression.

2. In every geometrical Progression, the product of the extreme terms, is equal to the product of every pair of the intermediate terms that are equidistant from the extremes, and also equal to the square of the middle term when there is a middle one, or an uneven number of the terms.

Thus, $1, 2, 4, 8, 16$,
$16, 8, 4, 2, 1$
prod. $16, 16, 16, 16, 16$

Also a, ra, r^2a, r^3a, r^4a,
r^4a, r^3a, r^2a, ra, a
prod. $r^4a^2, r^4a^2, r^4a^2, r^4a^2, r^4a^2$

3. The last term of a geometrical Progression, is equal to the first term multiplied, or divided, by the ratio raised to the power whose exponent is less by 1 than the number of terms in the series; so $z = ar^{n-1}$ when the series is an ascending one, or $z = \frac{a}{r^{n-1}}$, when it is a descending Progression.

4. As the sum of all the antecedents, or all the terms except the least, is to the sum of all the consequents, or all the terms except the greatest, so is 1 to r the ratio. For,

3 if

if $\quad a + ra + r^2a + r^3a$ be all except the laſt,

then $ra + r^2a + r^3a + r^4a$ are all except the firſt;

where it is evident that the former is to the latter as 1 to r, or the former multiplied by r gives the latter. So that, z denoting the laſt term, a the firſt term, and r the ratio, alſo s the ſum of all the terms; then $s - z : s - a :: 1 : r$, or $s - a = \overline{s - z} . r$. And from this equation all the relations among the four quantities a, z, r, s, are eaſily derived; ſuch as, $s = \dfrac{rz - a}{r - 1}$; viz, multiply the greateſt term by the ratio, ſubtract the leaſt term from the product, then the remainder divided by 1 leſs than the ratio, will give the ſum of the ſeries. And if the leaſt term a be 0, which happens when the deſcending Progreſſion is infinitely continued, then the ſum is barely $\dfrac{rz}{r - 1}$. As in the infinite Progreſſion $1 + \dfrac{1}{2} + \dfrac{1}{4} + \dfrac{1}{8}$ &c, where $z = 1$, and $r = 2$, it is s or $\dfrac{rz}{r - 1} = \dfrac{2}{2 - 1} = \dfrac{2}{1} = 2$.

5. The firſt or leaſt term of a geometrical Progreſſion, is to the ſum of all the terms, as the ratio minus 1, to the n power of the ratio minus 1; that is $a : s :: r - 1 : r^n - 1$.

Other relations among the five quantities a, z, r, n, s, where

> a denotes the leaſt term,
> z the greateſt term,
> r the common ratio,
> n the number of terms,
> s the ſum of the Progreſſion,

are as below; 'viz,

$$a = \frac{z}{r^{n-1}} = zr - (r - 1)s = \frac{r - 1}{r^n - 1}s.$$

$$z = ar^{n-1} = \frac{a + (r - 1)s}{r} = \frac{r - 1}{r^n - 1}sr^{n-1}.$$

$$r = \frac{s - a}{s - z} = \sqrt[n-1]{\frac{z}{a}}.$$

$$n = \frac{\log . \frac{rz}{a}}{\log . r} = \frac{\log . \frac{a + (r-1)s}{a}}{\log . r} = \frac{\log . \frac{rz}{rz - (r-1)s}}{\log . r} = \frac{\log . \frac{s - az}{s - z.a}}{\log . \frac{s - a}{s - z}}.$$

$$s = \frac{rz - a}{r - 1} = \frac{r^n - 1}{r - 1}a = \frac{r^n - 1}{r - 1}\frac{z}{r^{n-1}} = \frac{\sqrt[n-1]{z^n} - \sqrt[n-1]{a^n}}{\sqrt[n-1]{z} - \sqrt[n-1]{a}}.$$

And the other values of a, z, and r are to be found from theſe equations, viz,

$$(s - z)^{n-1} z = (s - a)^{n-1} a,$$

$$r^n - \frac{s}{a} r = 1 - \frac{s}{a},$$

$$r^n - \frac{s}{s - z} r^{n-1} = \frac{z}{s - z}.$$

For other ſorts of Progreſſions, ſee SERIES.

PROJECTILE, or PROJECT, in Mechanics, is any body which, being put into a violent motion by an external force impreſſed upon it, is diſmiſſed from the agent, and left to purſue its courſe. Such as a ſtone thrown out of the hand or a ſling, an arrow from a bow, a ball from a gun, &c.

PROJECTILES, the ſcience of the motion, velocity, flight, range, &c, of a projectile put into violent motion by ſome external cauſe, as the force of gunpowder, &c. This is the foundation of gunnery, under which article may be found all that relates peculiarly to that branch.

All bodies, being indifferent as to motion or reſt, will neceſſarily continue the ſtate they are put into, except ſo far as they are hindered, and forced to change it by ſome new cauſe. Hence, a Projectile, put in motion, muſt continue eternally to move on in the ſame right line, and with the ſame uniform or conſtant velocity, were it to meet with no reſiſtance from the medium, nor had any force of gravity to encounter.

In the firſt caſe, the theory of Projectiles would be very ſimple indeed ; for there would be nothing more to do, than to compute the ſpace paſſed over in a given time by a given conſtant velocity; or either of theſe, from the other two being given.

But by the conſtant action of gravity, the Projectile is continually deflected more and more from its right-lined courſe, and that with an accelerated velocity ; which, being combined with its Projectile impulſe, cauſes the body to move in a curvilineal path, with a variable motion, which path is the curve of a parabola, as will be proved below ; and the determination of the range, time of flight, angle of projection, and variable velocity, conſtitutes what is uſually meant by the doctrine of Projectiles, in the common acceptation of the word.

What is ſaid above however, is to be underſtood of Projectiles moving in a non-reſiſting medium ; for when the reſiſtance of the air is alſo conſidered, which is enormouſly great, and which very much impedes the firſt Projectile velocity, the path deviates greatly from the parabola, and the determination of the circumſtances of its motion becomes one of the moſt complex and difficult problems in nature.

In the firſt place therefore it will be proper to conſider the common doctrine of Projectiles, or that on the parabolic theory, or as depending only on the nature of gravity and the Projectile motion, as abſtracted from the reſiſtance of the medium.

Little more than 200 years ago, philoſophers took the line deſcribed by a body projected horizontally, ſuch as a bullet out of a cannon, while the force of the powder greatly exceeded the weight of the bullet, to be a right line, after which they allowed it became a curve. Nicholas Tartaglia was the firſt who perceived the miſtake, maintaining that the path of the bullet was a curved line through the whole of its extent. But it was Galileo who firſt determined what particular curve it is that a Projectile deſcribes ; ſhewing that the path of a bullet projected horizontally from an eminence, was a parabola ; the vertex of which is the point where the bullet quits the cannon. And the ſame is proved generally, in the 2d ſection following, when the projection is made in any direction whatever, viz, that the

curve

curve is always a parabola, fuppofing the body moves in a non-refifting medium.

The Laws of the Motion of PROJECTILES.

I. If a heavy body be projected perpendicularly, it will continue to afcend or defcend perpendicularly; becaufe both the projecting and the gravitating force are found in the fame line of direction.

II If a body be projected in free fpace, either parallel to the horizon, or in any oblique direction; it will, by this motion, in conjunction with the action of gravity, defcribe the curve line of a parabola.

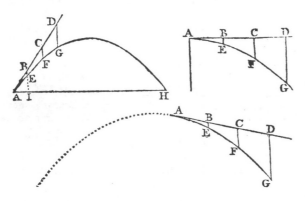

For let the body be projected from A, in the direction AD, with any uniform velocity; then in any equal portions of time it would, by that impulfe alone, defcribe the equal fpaces AB, BC, CD, &c, in the line AD, if it were not drawn continually down below that line by the action of gravity. Draw BE, CF, DG, &c, in the direction of gravity, or perpendicular to the horizon; and take BE, CF, DG, &c, equal to the fpaces through which the body would defcend by its gravity in the fame times in which it would uniformly pafs over the fpaces AB, AC, AD, &c, by the Projectile motion. Then, fince by thefe motions, the body is carried over the fpace AB in the fame time as the fpace BE, and the fpace AC in the fame time as the fpace CF, and the fpace AD in the fame time as the fpace DG, &c; therefore, by the compofition of motions, at the end of thofe times the body will be found refpectively in the points E, F, G, &c, and confequently the real path of the Projectile will be the curve line AEFG &c. But the fpaces AB, AC, AD, &c, being defcribed by uniform motion, are as the times of defcription; and the fpaces BE, CF, DG, &c, defcribed in the fame times by the accelerating force of gravity, are as the fquares of the times; confequently the perpendicular defcents are as the fquares of the fpaces in AD,

that is   BE, CF, DG, &c, are refpectively proportional to AB^2, AC^2, AD^2, &c,

which is the fame as the property of the parabola. Therefore the path of the Projectile is the parabolic line AEFG &c, to which AD is a tangent at the point A.

Hence, 1. The horizontal velocity of a Projectile is always the fame conftant quantity, in every point of the curve; becaufe the horizontal motion is in a con-

ftant ratio to the motion in AD, which is the uniform Projectile motion; viz, the conftant horizontal velocity being to the Projectile velocity, as radius to the cofine of the angle DAH, or angle of elevation or depreffion of the piece above or below the horizontal line AH.

2. The velocity of the Projectile in the direction of the curve, or of its tangent, at any point A, is as the fecant of its angle BAI of direction above the horizon. For the motion in the horizontal direction AI being conftant, and AI being to AB as radius to the fecant of the angle A; therefore the motion at A, in AB, is as the fecant of the angle A.

3. The velocity in the direction DG of gravity, or perpendicular to the horizon, at any point G of the curve, is to the firft uniform Projectile velocity at A, as 2GD to AD. For the times of defcribing AD and DG being equal, and the velocity acquired by freely defcending through DG being fuch as would carry the body uniformly over twice DG in an equal time, and the fpaces defcribed with uniform motions being as the velocities, it follows that the fpace AD is to the fpace 2DG, as the Projectile velocity at A is to the perpendicular velocity at G.

III. The velocity in the direction of the curve, at any point of it, as A, is equal to that which is generated by gravity in freely defcending through a fpace which is equal to one-fourth of the parameter of the diameter to the parabola at that point.

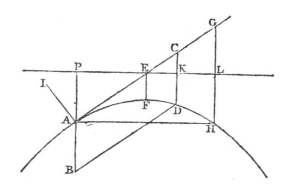

Let PA or AB be the height due to the velocity of the Projectile at any point A, in the direction of the curve or tangent AC, or the velocity acquired by falling through that height; and complete the parallelogram ACDB. Then is CD = AB or AP the height due to the velocity in the curve at A; and CD is alfo the height due to the perpendicular velocity at D, which will therefore be equal to the former: but, by the laft corollary, the velocity at A is to the perpendicular velocity at D, as AC to 2CD; and as thefe velocities are equal, therefore AC or BD is equal to 2CD or 2AB; and hence AB or AP is equal to ½BD or ¼ of the parameter of the diameter AB by the nature of the parabola.

Hence, 1. If through the point P, the line PL be drawn perpendicular to AP; then the velocity in the curve at every point, will be equal to the velocity acquired by falling through the perpendicular diftance
of

of the point from the said line PL ; that is, a body falling freely through

PA, acquires the velocity in the curve at **A**,
EF,   -   -   -   at F,
KD,   -   -   -   at D,
LH,   -   -   -   at H.

The reason of which is, that the line PL is what is called the Directrix of the parabola, the property of which is, that the perpendicular to it, from every point of the curve, is equal to one-fourth of the parameter of the diameter at that point, viz,

PA $= \frac{1}{4}$ the parameter of the diameter at A,
EF $=$ -   -   -   at F,
KD $=$ -   -   -   at D,
LH $=$ -   -   -   at H.

2. If a body, after falling through the height PA, which is equal to AB, and when it arrives at A if its course be changed, by reflection from a firm plane AI, or otherwise, into any direction AC, without altering the velocity ; and if AC be taken equal to 2AP or 2AB, and the parallelogram be completed; the body will describe the parabola passing through the point D.

3. Because AC $=$ 2AB, or 2CD or 2AP, therefore AC² $=$ 2AP.2CD or AP.4CD; and because all the perpendiculars EF, CD, GH are as AE², AC², AG²; therefore also AP.4EF $=$ AE², and AP.4GH $=$ AG², &c ; and because the rectangle of the extremes is equal to the rectangle of the means, of four proportionals, therefore it is always,

$$AP : AE :: AE : 4EF,$$
and $$AP : AC :: AC : 4CD,$$
and $$AP : AG :: AG : 4GH,$$
and so on.

IV. Having given the Direction of a Projectile, and the Impetus or Altitude due to the first velocity ; to determine the Greatest Height to which it will rise, and the Random or Horizontal Range.

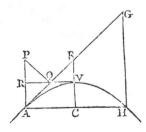

Let AP be the height due to the Projectile velocity at A, or the height which a body must fall to acquire the same velocity as the projectile has in the curve at A ; also AG the direction, and AH the horizon. Upon AG let fall the perpendicular PQ, and on AP the perpendicular QR; so shall AR be equal to the greatest altitude CV, and 4RQ equal to the horizontal range AH. Or, having drawn PQ perpendicular to AG, take AG $=$ 4AQ, and draw GH perpendicular to AH; then AH is the range.

For by the last cor. - - - AP : AG :: AG : 4GH,
and by sim. triangles, - - - AP : AG :: AQ : GH,
          or AP : AG :: 4AQ : 4GH;

therefore AG $=$ 4AQ; and, by similar triangles, AH $=$ 4RQ.

Also, if V be the vertex of the parabola, then AB or $\frac{1}{2}$AG $=$ 2AQ, or AQ $=$ QB; consequently AR $=$ BV which is $=$ CV by the nature of the parabola.

Hence, 1. Because the angle Q is a right angle, which is the angle in a semicircle, therefore if upon AP as a diameter a semicircle be described, it will pass through the point Q.

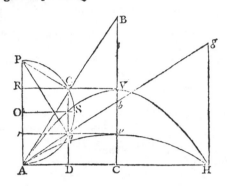

2. If the Horizontal Range and the Projectile Velocity be given, the Direction of the piece so as to hit the object H will be thus easily found : Take AD $=$ $\frac{1}{4}$AH, and draw DQ perpendicular to AH, meeting the semicircle described on the diameter AP in Q and q ; then either AQ or Aq will be the direction of the piece. And hence it appears, that there are two directions AB and Ab which, with the same Projectile velocity, give the very same horizontal range AH ; and these two directions make equal angles qAD and QAP with AH and AP, because the arc PQ is equal to the arc Aq.

3. Or if the Range AH and Direction AB be given ; to find the Altitude and Velocity or Impetus : Take AD $=$ $\frac{1}{4}$AH, and erect the perpendicular DQ meeting AB in Q ; so shall DQ be equal to the greatest altitude CV. Also erect AP perpendicular to AH, and QP to AQ ; so shall AP be the height due to the velocity.

4. When the body is projected with the same velocity, but in different directions ; the horizontal ranges AH will be as the sines of double the angles of elevation. Or, which is the same thing, as the rectangle of the sine and cosine of elevation. For AD or RQ, which is $\frac{1}{4}$AH, is the sine of the arc AQ, which measures double the angle QAD of elevation.

And when the direction is the same, but the velocities different, the horizontal ranges are as the square of the velocities, or as the height AP which is as the square of the velocity; for the sine AD or RQ, or $\frac{1}{4}$AH, is as the radius, or as the diameter AP.

Therefore, when both are different, the ranges are in the compound ratio of the squares of the velocities, and the sines of double the angles of elevation.

5. The greatest range is when the angle of elevation is half a right angle, or 45°. For the double of 45 is 90°, which has the greatest sine. Or the radius OS, which is $\frac{1}{4}$ of the range, is the greatest sine.

And hence the greatest range, or that at an elevation of 45°, is just double the altitude AP which is due to the

the

the velocity. Or equal to 4VC. And confequently, in that cafe, C is the focus of the parabola, and AH its parameter.

And the ranges are equal at angles equally above and below 45°.

6. When the elevation is 15°, the double of which, or 30°, having its fine equal to half the radius, confequently its range will be equal to AP, or half the greateft range at the elevation of 45°; that is, the range at 15° is equal to the impetus or height due to the projectile velocity.

7. The greateft altitude CV, being equal to AR, is as the verfed fine of double the angle of elevation, and alfo as AP or the fquare of the velocity. Or as the fquare of the fine of elevation, and the fquare of the velocity; for the fquare of the fine is as the verfed fine of the double angle.

8. The time of flight of the projectile, which is equal to the time of a body falling freely through GH or 4CV, 4 times the altitude, is therefore as the fquare root of the altitude, or as the projectile velocity and fine of the elevation.

9. And hence may be deduced the following fet of theorems, for finding all the circumftances relating to projectiles on horizontal planes, having any two of them given. Thus, let

s, c, t = fine, cofine, and tang. of elevation,

S, v = fine and verf. of double the elevation,

R the horizontal range, T the time of flight, V the projectile velocity, H the greateft height of the projectile, $g = 16\frac{1}{12}$ feet, and a = the impetus or the altitude due to the velocity V. Then,

$$R = 2aS = 4asc = \frac{SV^2}{2g} = \frac{scV^2}{g} = \frac{gcT^2}{s} = \frac{gT^2}{t} = \frac{4H}{t}$$

$$V = \sqrt{4ag} = \sqrt{\frac{2gR}{S}} = \sqrt{\frac{gR}{sc}} = \frac{gT}{s} = \frac{2\sqrt{gH}}{s}$$

$$T = \frac{sV}{g} = 2s\sqrt{\frac{a}{g}} = \sqrt{\frac{tR}{g}} = \sqrt{\frac{sR}{gc}} = 2\sqrt{\frac{H}{g}}$$

$$H = as^2 = \frac{1}{2}av = \frac{1}{4}tR = \frac{sR}{4c} = \frac{sV^2}{4g} = \frac{vV^2}{8g} = \frac{gT^2}{4}.$$

And from any of thefe, the angle of direction may be found.

V. To determine the Range on an oblique plane; having given the Impetus or the Velocity, and the Angle of Direction.

Let AE be the oblique plane, at a given angle above or below the horizontal plane AH; AG the direction of the piece; and AP the altitude due to the projectile velocity at A.

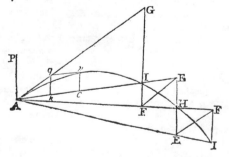

By the laft prop. find the horizontal range AH to the given velocity and direction; draw HE perpendicular to AH meeting the oblique plane in E; draw EF parallel to the direction AG, and FI parallel to HE; fo fhall the projectile pafs through I, and the range on the oblique plane will be AI. This is evident from prob. 17 of the Parabola in my treatife on Conic Sections, where it is proved, that if AH, AI be any two lines terminated at the curve, and IF, HE be parallel to the axis; then is EF parallel to the tangent AG.

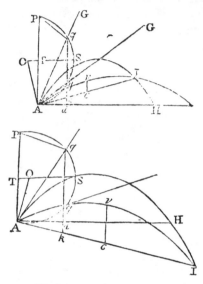

Hence, 1. If AO be drawn perpendicular to the plane AI, and AP be bifected by the perpendicular STO; then with the centre O defcribing a circle through A and P, the fame will alfo pafs through q, becaufe the angle GAI, formed by the tangent AG and AI, is equal to the angle APq, which will therefore ftand upon the fame arc Aq.

2. If there be given the Range and Velocity, or the Impetus, the Direction will then be eafily found thus: Take A$k = \frac{1}{4}$AI, draw kq perpendicular to AH, meeting the circle defcribed with the radius AO in two points q and q; then Aq or Aq will be the direction of the piece. And hence it appears that there are two directions, which, with the fame impetus, give the very fame range AI, on the oblique plane. And thefe two directions make equal angles with AI and AP, the plane and the perpendicular, becaufe the arc Pq = the arc Aq. They alfo make equal angles with a line drawn from A through S, becaufe the arc Sq = the arc Sq.

3. Or, if there be given the Range AI, and the Direction Aq; to find the Velocity or Impetus. Take A$k = \frac{1}{4}$AI, and erect kq perpendicular to AH meeting the line of direction in q; then draw qP making the angle AqP = the angle Akq; fo fhall AP be the impetus, or altitude due to the projectile velocity.

4. The range on an oblique plane, with a given elevation, is directly as the rectangle of the cofine of the direction of the piece above the horizon and the fine of the direction above the oblique plane, and reciprocally as the fquare of the cofine of the angle of the plane above or below the horizon.

For

For put s = fin. ∠ qAI or APq,

    c = cof. ∠ qAH or fin. PAq,

    C = cof. ∠ IAH or fin. Akd or Akq or AqP.

Then, in the tri. APq, - - - C : s :: AP : Aq,

and in the tri. Akq, - - - C : c :: Aq : Ak,

therefore by compof. - - - C² : cs :: AP : Ak = ¼AI.

So that the oblique range AI = $\frac{cs}{C^2}$ × 4AP.

Hence the range is the greateſt when Ak is the greateſt, that is when kq touches the circle in the middle point S, and then the line of direction paſſes through S, and biſects the angle formed by the oblique plane and the vertex. Alſo the ranges are equal at equal angles above and below this direction for the maximum.

5. The greateſt height cv or kq of the projectile, above the plane, is equal to $\frac{s^2}{c^2}$ × AP. And therefore it is as the impetus and ſquare of the fine of direction above the plane directly, and ſquare of the coſine of the plane's inclination reciprocally.

For C (fin. AqP) : s (fin. APq) :: AP : Aq,

and C (fin. Akq) : s (fin. kAq) :: Aq : kq,

therefore by comp. C² : s^2 :: AP : kq.

6. The time of flight in the curve AvI is = $\frac{2s}{C}\sqrt{\frac{AP}{g}}$, where g = 16 $\frac{1}{12}$ feet. And therefore it is as the velocity and fine of direction above the plane directly, and coſine of the plane's inclination reciprocally. For the time of deſcribing the curve, is equal to the time of falling freely through GI or 4kq or $\frac{4s^2}{C^2}$ × AP. Therefore, the time being as the ſquare root of the diſtance, $\sqrt{g} : \frac{2s}{C}\sqrt{AP} :: 1'' : \frac{2s}{C}\sqrt{\frac{AP}{g}}$ the time of flight.

7. From the foregoing corollaries may be collected the following ſet of theorems, relating to projects made on any given inclined planes, either above or below the horizontal plane. In which the letters denote as before, namely,

  c = cof. of direction above the horizon,

  C = cof. of inclination of the plane,

  s = fin. of direction above the plane,

  R the range on the oblique plane,

  T the time of flight,

  V the projectile velocity,

  H the greateſt height above the plane,

  a the impetus, or alt. due to the velocity V,

  g = 16 $\frac{1}{12}$ feet. Then

$$R = \frac{cs}{C^2} \times 4a = \frac{cs}{C^2 g}V^2 = \frac{gc}{s}T^2 = \frac{4c}{s}H.$$

$$H = \frac{s^2}{C^2}a = \frac{s^2 V^2}{4gC^2} = \frac{sR}{4c} = \frac{g}{4}T^2.$$

$$V = \sqrt{4ag} = C\sqrt{\frac{gR}{cs}} = \frac{gC}{s}T = \frac{2C}{s}\sqrt{gH}.$$

$$T = \frac{2s}{C}\sqrt{\frac{a}{g}} = \frac{sV}{gC} = \sqrt{\frac{sR}{gc}} = 2\sqrt{\frac{H}{g}}.$$

And from any of theſe, the angle of direction may be found.

Of the Path of PROJECTILES as depending on the Reſiſtance of the Air.

For a long time after Galileo, philoſophers ſeemed to be ſatiſfied with the parabolic theory of Projectiles, deeming the effect of the air's reſiſtance on the path as of no conſequence. In proceſs of time, however, as the true philoſophy began to dawn, they began to ſuſpect that the reſiſtance of the medium might have ſome effect upon the Projectile curve, and they ſet themſelves to conſider this ſubject with ſome attention.

Huygens, ſuppoſing that the reſiſtance of the air was proportional to the velocity of the moving body, concluded that the line deſcribed by it would be a kind of logarithmic curve.

But Newton, having clearly proved, that the reſiſtance to the body is not proportional to the velocity itſelf, but to the ſquare of it, ſhews, in his Principia, that the line a Projectile deſcribes, approaches nearer to an hyperbola than a parabola. Schol. prop. 10, lib. 2. Thus, if AGK be a curve of

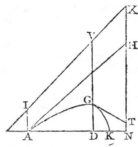

the hyperbolic kind, one of whoſe aſymptotes is NX, perpendicular to the horizon AK, and the other IX inclined to the ſame, where VG is reciprocally as DN^n, whoſe index is n: this curve will nearer repreſent the path of a Projectile thrown in the direction AH in the air, than a parabola. Newton indeed ſays, that theſe hyperbolas are not accurately the curves that a Projectile makes in the air; for the true ones are curves which about the vertex are more diſtant from the aſymptotes, and in the parts remote from the axis approach nearer to the aſymptotes than theſe hyperbolas; but that in practice theſe hyperbolas may be uſed inſtead of thoſe more compounded ones. And if a body be projected from A, in the right line AH, and AI be drawn parallel to the aſymptote NX, and GT a tangent to the curve at the vertex: Then the denſity of the medium in A will be reciprocally as the tangent AH, and the body's velocity will be as $\sqrt{\frac{AH^2}{AI}}$, and the reſiſtance of the medium will be to gravity, as AH to $\frac{2n^2 + 2n}{n + 2}$ × AI.

M. John Bernoulli conſtructed this curve by means of the quadrature of ſome tranſcendental curves, at the requeſt of Dr. Keil, who propoſed this problem to him in 1718. It was alſo reſolved by Dr. Taylor; and another ſolution of it may be found in Hermann's Phoronomia.

The commentators Le Sieur and Jacquier ſay, that the deſcription of the curve in which a Projectile moves, is ſo very perplexed, that it can ſcarcely be expected any deduction ſhould be made from it, either to philoſophical or mechanical purpoſes: vol. 2. pa. 118.

Dan. Bernoulli too proved, that the reſiſtance of the air has a very great effect on ſwift motions, ſuch as thoſe of cannon ſhot. He concludes from experiment, that a ball which aſcended only 7819 feet in the air, would have aſcended 58750 feet in vacuo, being near eight times as high. Comment. Acad. Petr. tom. 2.

M. Euler has farther inveſtigated the nature of this curve, and directed the calculation and uſe of a number of tables for the ſolution of all caſes that occur in gunnery, which may be accompliſhed with nearly as much expedition as by the common parabolic principles. Memoirs of the Academy of Berlin, for the year 1753.

But how raſh and erroneous the old opinion of the inconſiderable reſiſtance of the air is, will eaſily appear from the experiments of Mr. Robins, who has ſhewn that, in ſome caſes, this reſiſtance to a cannon ball, amounts to more than 20 times the weight of the ball; and I myſelf, having proſecuted this ſubject far beyond any former example, have ſometimes found this reſiſtance amount to near 100 times the weight of the ball, viz, when it moved with a velocity of 2000 feet per ſecond, which is a rate of almoſt 23 miles in a minute. What errors then may not be expected from an hypotheſis which neglects this force, as inconſiderable! Indeed it is eaſy to ſhew, that the path of ſuch Projectiles is neither a parabola nor nearly a parabola. For, by that theory, if the ball, in the inſtance laſt mentioned, flew in the curve of a parabola, its horizontal range, at 45° elevation, will be found to be almoſt 24 miles; whereas it often happens that the ball, with ſuch a velocity, ranges far ſhort of even one mile.

Indeed the falſeneſs of this hypotheſis almoſt appears at ſight, even in Projectiles ſlow enough to have their motion traced by the eye; for they are ſeen to deſcend through a curve manifeſtly ſhorter and more inclined to the horizon than that in which they aſcended, and the higheſt point of their flight, or the vertex of the curve, is much nearer to the place where they fall on the ground, than to that from whence they were at firſt diſcharged. Theſe things cannot for a moment be doubted of by any one, who in a proper ſituation views the flight of ſtones, arrows, or ſhells, thrown to any conſiderable diſtance.

Mr. Robins has not only detected the errors of the parabolic theory of gunnery, which takes no account of the reſiſtance of the air, but ſhews how to compute the real range of reſiſted bodies. But for the method which he propoſes, and the tables he has computed for this purpoſe, ſee his Tracts of Gunnery,

pa. 183, &c, vol. 1; and alſo Euler's Commentary on the ſame, tranſlated by Mr. Hugh Brown, in 1777.

There is an odd circumſtance which often takes place in the motion of bodies projected with conſiderable force, which ſhews the great complication and difficulty of this ſubject; namely, that bullets in their flight are not only depreſſed beneath their original direction by the action of gravity, but are alſo frequently driven to the right or left of that direction by the action of ſome other force.

Now if it were true that bullets varied their direction by the action of gravity only, then it ought to happen that the errors in their flight to the right or left of the mark they were aimed at, ſhould increaſe in the proportion of the diſtance of the mark from the piece only. But this is contrary to all experience; the ſame piece which will carry its bullet within an inch of the intended mark, at 10 yards diſtance, cannot be relied on to 10 inches in 100 yards, much leſs to 30 in 300 yards.

And this inequality can only ariſe from the track of the bullet being incurvated ſideways as well as downwards; for by this means the diſtance between the incurvated line and the line of direction, will increaſe in a much greater ratio than that of the diſtance; theſe lines coinciding at the mouth of the piece, and afterwards ſeparating in the manner of a curve from its tangent, if the mouth of the piece be conſidered as the point of contact.

This is put beyond a doubt from the experiments made by Mr. Robins; who found alſo that the direction of the ſhot in the perpendicular line was not leſs uncertain, falling ſometimes 200 yards ſhort of what it did at other times, although there was no viſible cauſe of difference in making the experiment. And I myſelf have often experienced a difference of one-fifth or one-ſixth of the whole range, both in the deflection to the right or left, and alſo in the extent of the range, of cannon ſhot.

If it be aſked, what can be the cauſe of a motion ſo different from what has been hitherto ſuppoſed? It may be anſwered, that the deflection in queſtion muſt be owing to ſome power acting obliquely to the progreſſive motion of the body, which power can be no other than the reſiſtance of the air. And this reſiſtance may perhaps act obliquely to the progreſſive motion of the body, from inequalities in the reſiſted ſurface; but its general cauſe is doubtleſs a whirling motion acquired by the bullet about an axis, by its friction againſt the ſides of the piece; for by this motion of rotation, combined with the progreſſive motion, each part of the ball's ſurface will ſtrike the air in a direction very different from what it would do if there was no ſuch whirl; and the obliquity of the action of the air, ariſing from this cauſe, will be greater, according as the rotatory motion of the bullet is greater in proportion to its progreſſive motion. Tracts, vol. 1, p. 149, &c.

M. Euler, on the contrary, attributes this deflection of the ball to its figure, and very little to its rotation: for if the ball was perfectly round, though its centre of gravity did not coincide, the deflection from the axis of the cylinder, or line of direction ſideways, would be very inconſiderable. But when it is not round, it will generally

generally go to the right or left of its direction, and so much the more, as its range is greater. From his reasoning on this subject he infers, that cannon shot, which are made of iron, and rounder and less susceptible of a change of figure in passing along the cylinder than those of lead, are more certain than musket shot. True Principles of Gunnery investigated, 1777, p. 304, &c.

PROJECTION, in Mechanics, the act of giving a projectile its motion.

If the direction of the force, by which the projectile is put in motion, be perpendicular to the horizon, the Projection is said to be perpendicular; if parallel to the apparent horizon, it is said to be an horizontal Projection; and if it make an oblique angle with the horizon, the Projection is oblique. In all cases, the angle which the line of direction makes with the horizontal line, is called the angle of Elevation of the projectile, or of Depression when the line of direction points below the horizontal line.

PROJECTION, in Perspective, denotes the appearance or representation of an object on the perspective plane. So, the Projection of a point, is a point, where the optic ray passes from the objective point through the plane to the eye; or it is the point where the plane cuts the optic ray. — And hence it is easy to conceive what is meant by the projection of a line, a plane, or a solid.

PROJECTION *of the Sphere in Plano*, is a representation of the several points or places of the surface of the sphere, and of the circles described upon it, upon a transparent plane placed between the eye and the sphere, or such as they appear to the eye placed at a given distance. For the laws of this Projection, see PERSPECTIVE; the Projection of the sphere being only a particular case of perspective.

The chief use of the Projection of the sphere, is in the construction of planispheres, maps, and charts; which are said to be of this or that Projection, according to the several situations of the eye, and the perspective plane, with regard to the meridians, parallels, and other points or places to be represented.

The most usual Projection of maps of the world, is that on the plane of the meridian, which exhibits a right sphere; the first meridian being the horizon. The next is that on the plane of the equator, which has the pole in the centre, and the meridians the radii of a circle, &c; and this represents a parallel sphere. See MAP.—The primitive circle is that great circle.

The Projection of the sphere is usually divided into Orthographic and Stereographic; to which may be added Gnomonic.

Orthographic PROJECTION, is that in which the surface of the sphere is drawn upon a plane, cutting it in the middle; the eye being placed at an infinite distance vertically to one of the hemispheres. And

Stereographic PROJECTION of the sphere, is that in which the surface and circles of the sphere are drawn upon the plane of a great circle, the eye being in the pole of that circle.

Gnomonical PROJECTION *of the Sphere*, is that in which

the surface of the sphere is drawn upon a plane without side of it, commonly touching it, the eye being at the centre of the sphere. See GNOMONICAL *Projection*.

Laws of the Orthographic Projection.

1. The rays coming from the eye, being at an infinite distance, and making the Projection, are parallel to each other, and perpendicular to the plane of Projection.

2. A right line perpendicular to the plane of Projection, is projected into a point, where that line meets the said plane.

3. A right line, as AB, or CD, not perpendicular, but either parallel or oblique to the plane of the Projection, is projected into a right line, as EF or GH, and is always comprehended between the extreme perpendiculars AE and BF, or CG and DH.

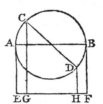

4. The Projection of the right line AB is the greatest, when AB is parallel to the plane of the Projection.

5. Hence it is evident, that a line parallel to the plane of the Projection, is projected into a right line equal to itself; but a line that is oblique to the plane of Projection, is projected into one that is less than itself.

6. A plane surface, as ACBD, perpendicular to the plane of the Projection, is projected into the right line, as AB, in which it cuts that plane —Hence it is evident, that the circle ACBD perpendicular to the plane of Projection, passing through its centre, is projected into that diameter AB in which it cuts the plane of the Projection. Also any arch as C*c* is projected into O*o*, equal to *ca*, the right sine of that arch; and the complemental arc *c*B is projected into *o*B, the versed sine of the same arc *c*B.

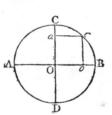

7. A circle parallel to the plane of the Projection, is projected into a circle equal to itself, having its centre the same with the centre of the Projection, and its radius equal to the cosine of its distance from the plane. And a circle oblique to the plane of the Projection, is projected into an ellipsis, whose greater axis is equal to the diameter of the circle, and its less axis equal to double the cosine of the obliquity of the circle, to a radius equal to half the greater axis.

Properties of the Stereographic Projection.

1. In this Projection a right circle, or one perpendicular to the plane of Projection, and passing through the eye, is projected into a line of half tangents.

2. The Projection of all other circles, not passing through the projecting point, whether parallel or oblique, are projected into circles.

Thus,

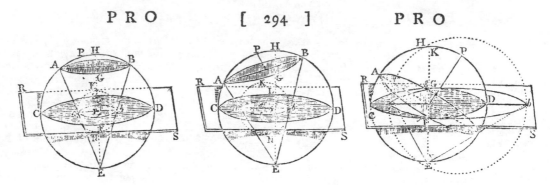

Thus, let ACEDB reprefent a fphere, cut by a plane RS, paffing through the centre I, perpendicular to the diameter EH, drawn from E the place of the eye; and let the fection of the fphere by the plane RS be the circle CFDL, whofe poles are H and E. Suppofe now AGB is a circle on the fphere to be projected, whofe pole moft remote from the eye is P; and the vifual rays from the circle AGB meeting in E, form the cone AGBE, of which the triangle AEB is a fection through the vertex E, and diameter of the bafe AB: then will the figure *aghf*, which is the Projection of the circle AGB, be itfelf a circle. Hence, the middle of the projected diameter is the centre of the projected circle, whether it be a great circle or a fmall one: Alfo the poles and centres of all circles, parallel to the plane of Projection, fall in the centre of the Projection: And all oblique great circles cut the primitive circle in two points diametrically oppofite.

2. The projected diameter of any circle fubtends an angle at the eye equal to the diftance of that circle from its neareft pole, taken on the fphere; and that angle is bifected by a right line joining the eye and that pole. Thus, let the plane RS cut the fphere HFEG through

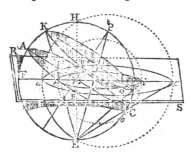

its centre I; and let ABC be any oblique great circle, whofe diameter AC is projected into *ac*; and KOL any fmall circle parallel to ABC, whofe diameter KL is projected in *kl*. The diftances of thofe circles from their pole P, being the arcs AHP, KHP; and the angles *aEc*, *kEl*, are the angles at the eye, fubtended by their projected diameters, *ac* and *kl*. Then is the angle *aEc* meafured by the arc AHP, and the angle *kEl* meafured by the arc KHP; and thofe angles are bifected by EP.

3. Any point of a fphere is projected at fuch a diftance from the centre of Projection, as is equal to the tangent of half the arc intercepted between that point and the pole oppofite to the eye, the femidiameter of the fphere being radius. Thus, let C*b*EB be a great circle of the fphere, whofe centre is *c*, GH the plane of Projection cutting the diameter of the fphere in *b*

and B; alfo E and C the poles of the fection by that plane; and *a* the projection of A. Then *ca* is equal

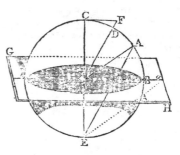

the tangent of half the arc AC, as is evident by drawing CF = the tangent of half that arc, and joining *c*F.

4. The angle made by two projected circles, is equal to the angle which thefe circles make on the fphere. For let IACE and ABL be two circles on a fphere

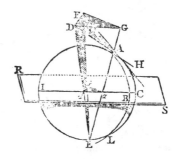

interfecting in A; E the projecting point; and RS the plane of Projection, in which the point A is projected in *a*, in the line IC, the diameter of the circle ACE. Alfo let DH and FA be tangents to the circles ACE and ABL. Then will the projected angle *daf* be equal to the fpherical angle BAC.

5. The diftance between the poles of the primitive circle and an oblique circle, is equal to the tangent of half the inclination of thofe circles; and the diftance of their centres, is equal to the tangent of their inclination; the femidiameter of the primitive being radius. For let AC be the diameter of a circle, whofe poles are P and Q, and inclined to the plane of Projection in the angle AIF; and let *a*, *c*, *p* be the Projections of the points A, C, P; alfo let H*a*E be the projected oblique circle, whofe centre is *q*. Now when the plane of Projection becomes the primitive circle, whofe pole is I; then is I*p* = tangent of half the angle AIF, or of half

8　the

the arch AF; and I*q* = tangent of AF, or of the angle FH*a* = AIF.

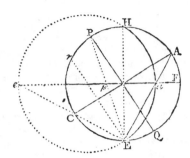

6. If through any given point in the primitive circle, an oblique circle be deſcribed; then the centres of all other oblique circles paſſing through that point, will be in a right line drawn through the centre of the firſt oblique circle, and perpendicular to a line paſſing through that centre, the given point, and the centre of the pri-

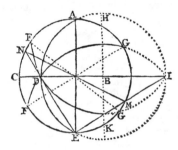

mitive circle. Thus, let GACE be the primitive circle, ADEI a great circle deſcribed through D, its centre being B. HK is a right line drawn through B perpendicular to a right line CI paſſing through D and B and the centre of the primitive circle. Then the centres of all other great circles, as FDG, paſſing through D, will fall in the line HK.

7. Equal arcs of any two great circles of the ſphere will be intercepted between two other circles drawn on the ſphere through the remoteſt poles of thoſe great circles. For let PBEA be a ſphere, on which AGB and

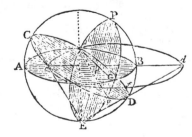

CFD are two great circles, whoſe remoteſt poles are E and P; and through theſe poles let the great circle PBEC and the ſmall circle PGE be drawn, cutting the great circles AGB and CFD in the points B, G, D, F.

Then are the intercepted arcs BG and DF equal to one another.

8. If lines be drawn from the projected pole of any great circle, cutting the peripherics of the projected circle and plane of Projection; the intercepted arcs of thoſe peripherics are equal; that is, the arc BG = *df*.

9. The radius of any leſſer circle, whoſe plane is perpendicular to that of the primitive circle, is equal to the tangent of that leſſer circle's diſtance from its pole; and the ſecant of that diſtance is equal to the diſtance of the centres of the primitive and leſſer circle. For let P be the pole, and AB the diameter of a leſſer circle, its plane being perpendicular to that of the primi-

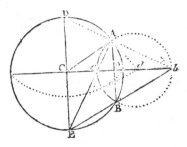

tive circle, whoſe centre is C: then *d* being the centre of the projected leſſer circle, *da* is equal to the tangent of the arc PA, and *d*C = the ſecant of PA. See STEREOGRAPHIC *Projection*.

Mercator's PROJECTION. See MERCATOR and CHART.

PROJECTION of Globes, &c. See GLOBE, &c.

Polar PROJECTION. See POLAR.

PROJECTION *of Shadows.* See SHADOW.

PROJECTION, or PROJECTURE, in Building, the outjetting or prominency which the mouldings and members have, beyond the plane or naked of the wall, column, &c.

Monſtrous PROJECTION. See ANAMORPHOSIS.

PROJECTIVE *Dialling*, a manner of drawing the hour lines, the furniture &c of dials, by a method of projection on any kind of ſurface whatever, without regard to the ſituation of thoſe ſurfaces, either as to declination, reclination, or inclination. See DIALLING.

PROLATE, or OBLONG *Spheroid*, is a ſpheroid produced by the revolution of a ſemiellipſis about its longer diameter; being longeſt in the direction of that axis, and reſembling an egg, or a lemon.

It is ſo called in oppoſition to the oblate or ſhort ſpheroid, which is formed by the rotation of a ſemiellipſis about its ſhorter axis; being therefore ſhorteſt in the direction of its axis, or flatted at the poles, and ſo reſembling an orange, or perhaps a turnip, according to the degree of flatneſs; and which is alſo the figure of the earth we inhabit, and perhaps of the planets alſo; having their equatorial diameter longer than the polar. See SPHEROID.

PROMONTORY, in Geography, is a rock or high point of land projecting out into the ſea. The extremity of which towards the ſea is uſually called a Cape, or Headland.

PROPORTION, in Arithmetic &c, the equality

or

or fimilitude of ratios. As the four numbers 4, 8, 15, 30 are proportionals, or in proportion, becaufe the ratio of 4 to 8 is equal or fimilar to the ratio of 15 to 30, both of them being the fame as the ratio of 1 to 2.

Euclid, in the 5th definition of the 5th book, gives a general definition of four proportionals, or when, of four terms, the firft has the fame ratio to the 2d, as the 3d has to the 4th, viz, when any equimultiples whatever of the firft and third being taken, and any equimultiples whatever of the 2d and 4th; if the multiple of the firft be lefs than that of the 2d, the multiple of the 3d is alfo lefs than that of the 4th; or if the multiple of the firft be equal to that of the 2d, the multiple of the 3d is alfo equal to that of the 4th; or if the multiple of the firft be greater than that of the 2d, the multiple of the 3d is alfo greater than that of the 4th. And this definition is general for all kinds of magnitudes or quantities whatever, though a very obfcure one.

Alfo, in the 7th book, Euclid gives another definition of proportionals, viz, when the firft is the fame equimultiple of the 2d, as the 3d is of the 4th, or the fame part or parts of it. But this definition appertains only to numbers and commenfurable quantities.

Proportion is often confounded with ratio; but they are quite different things. For, ratio is properly the relation of two magnitudes or quantities of one and the fame kind; as the ratio of 4 to 8, or of 15 to 30, or of 1 to 2; and fo implies or refpects only two terms or things. But Proportion refpects four terms or things, or two ratios which have each two terms. Though the middle term may be common to both ratios, and then the Proportion is expreffed by three terms only, as 4, 8, 64, where 4 is to 8 as 8 to 64.

Proportion is alfo fometimes confounded with progreffion. In fact, the two often coincide; the difference between them only confifting in this, that progreffion is a particular fpecies of Proportion, being indeed a continued Proportion, or fuch as has all the terms in the fame ratio, viz, the 1ft to the 2d, the 2d to the 3d, the 3d to the 4th, &c; as the terms 2, 4, 8, 16, &c; fo that progreffion is a feries or continuation of Proportions.

Proportion is either continual, or difcrete or interrupted.

The Proportion is continual when every two adjacent terms have the fame ratio, or when the confequent of each ratio is the antecedent of the next following ratio, and fo all the terms form a progreffion; as 2, 4, 8, 16, &c; where 2 is to 4 as 4 to 8, and as 8 to 16, &c.

Difcrete or interrupted Proportion, is when the confequent of the firft ratio is different from the antecedent of the 2d, &c; as 2, 4, and 3, 6.

Proportion is alfo either Direct or Inverfe.

Direct PROPORTION is when more requires more, or lefs requires lefs. As it will require more men to perform more work, or fewer men for lefs work, in the fame time.

Inverfe or *Reciprocal* PROPORTION, is when more requires lefs, or lefs requires more. As it will require more men to perform the fame work in lefs time, or fewer men in more time. Ex. If 6 men can perform a piece of work in 15 days, how many men can do the fame in 10 days. Then,

reciprocally - as $\frac{1}{15}$ to $\frac{1}{10}$ fo is 6 : 9 } the
or inverfely - as 10 to 15 fo is 6 : 9 } anfwer.

Proportion, again, is diftinguifhed into Arithmetical, Geometrical, and Harmonical.

Arithmetical PROPORTION is the equality of two arithmetical ratios, or differences. As in the numbers 12, 9, 6; where the difference between 12 and 9, is the fame as the difference between 9 and 6, viz 3.

And here the fum of the extreme terms is equal to the fum of the means, or to double the fingle mean when there is but one. As $12 + 6 = 9 + 9 = 18$.

Geometrical PROPORTION is the equality between two geometrical ratios, or between the quotients of the terms. As in the three 9, 6, 4, where 9 is to 6 as 6 is to 4, thus denoted $9 : 6 :: 6 : 4$; for $\frac{9}{6} = \frac{6}{4}$, being each equal $\frac{3}{2}$ or $1\frac{1}{2}$.

And in this Proportion, the rectangle or product of the extreme terms, is equal to that of the two means, or the fquare of the fingle mean when there is but one. For $9 \times 4 = 6 \times 6 = 36$.

Harmonical PROPORTION, is when the firft term is to the third, as the difference between the 1ft and 2d is to the difference between the 2d and 3d; or in four terms when the 1ft is to the 4th, as the difference between the 1ft and 2d is to the difference between the 3d and 4th; or the reciprocals of an arithmetical Proportion are in harmonical Proportion. As 6, 4, 3; becaufe $6 : 3 :: 6 - 4 = 2 : 4 - 3 = 1$; or becaufe $\frac{1}{6}, \frac{1}{4}, \frac{1}{3}$ are in arithmetical Proportion, making $\frac{1}{6} + \frac{1}{3} = \frac{1}{4} + \frac{1}{4} = \frac{1}{2}$. Alfo the four 24, 16, 12, 9 are in harmonical Proportion, becaufe $24 : 9 :: 8 : 3$.

See PROPORTIONALS.

Compafs of PROPORTION, a name by which the French, and fome Englifh authors, call the Sector.

Rule of PROPORTION, in Arithmetic, a rule by which a 4th term is found in Proportion to three given terms. And is popularly called the Golden Rule, or Rule of Three.

PROPORTIONAL, relating to Proportion. As, Proportional Compaffes, Parts, Scales, Spirals, &c. See the feveral terms.

PROPORTIONAL *Compaffes*, are compaffes with two pair of oppofite legs, like a St. Andrew's crofs, by which any fpace is enlarged or diminifhed in any proportion.

PROPORTIONAL *Part*, is a part of fome number that is analogous to fome other part or number; fuch as the Proportional parts in the logarithms, and other tables.

PROPORTIONAL *Scales*, called alfo Logarithmic Scales, are the logarithms, or artificial numbers, placed on lines, for the eafe and advantage of multiplying and dividing &c, by means of compaffes, or of fliding rulers. Thefe are in effect fo many lines of numbers, as they are called by Gunter, but made fingle, double, triple, or quadruple; beyond which they feldom go. See GUNTER's *Scale*, SCALE, &c.

PROPORTIONAL *Spiral*. See SPIRAL.

PROPORTIONALITY, the quality of Proportionals. This term is ufed by Gregory St. Vincent, for the proportion that is between the exponents of four ratios.

PROPORTIONALS, are the terms of a proportion; confifting of two extremes, which are the firft

and

and laſt terms of the ſet, and the means, which are the reſt of the terms. Theſe Proportionals may be either arithmeticals, geometricals, or harmonicals, and in any number above two, and alſo either continued or diſcontinued.

Pappus gives this beautiful and ſimple compariſon of the three kinds of Proportionals, arithmetical, geometrical, and harmonical, viz, a, b, c being the firſt, ſecond and third terms in any ſuch proportion, then

In the arithmeticals, a a ⎫
in the geometricals, $a \cdot b$ ⎬ $:: a - b : b - c$.
in the harmonicals, $a : c$ ⎭

See MEAN *Proportional*.

Continued Proportionals form what is called a progreſſion ; for the properties of which ſee PROGRESSION.

I. *Properties of Arithmetical* PROPORTIONALS.

(For what reſpects Progreſſions and Mean Proportionals of all ſorts, ſee MEAN, and PROGRESSION.)

1. Four Arithmetical Proportionals, as 2, 3, 4, 5, are ſtill Proportionals when inverſely,   5, 4, 3, 2;
or alternately, thus,   -   2, 4, 3, 5;
or inverſely and alternately, thus   - 5, 3, 4, 2.

2. If two Arithmeticals be added to the like terms of other two Arithmeticals, of the ſame difference or arithmetical ratio, the ſums will have double the ſame difference or arithmetical ratio.

So, to   3 and 5, whoſe difference is   2,
add   7 and 9, whoſe difference is alſo 2,
the ſums 10 and 14 have a double diff. viz   4.

And if to theſe ſums be added two other numbers alſo in the ſame difference, the next ſums will have a triple ratio or difference ; and ſo on. Alſo, whatever be the ratios of the terms that are added, whether the ſame or different, the ſums of the terms will have ſuch arithmetical ratio as is compoſed of the ſums of the others that are added.

So   3 ,   5, whoſe dif. is 2
and   7 , 10, whoſe dif. is 3
and 12   16, whoſe dif. is 4
——— ———
make 22 , 31, whoſe dif. is 9.

On the contrary, if from two Arithmeticals be ſubtracted others, the difference will have ſuch arithmetical ratio as is equal to the differences of thoſe.

So from 12 and 16, whoſe dif. is   4
take   7 and 10, whoſe dif. is   3
——— ———
leaves   5 and 6, whoſe dif. is   1

Alſo from 7 and 9, whoſe dif. is   2
take   3 and 5, whoſe dif. is   2
——— ———
leaves   4 and 4, whoſe dif. is   0

3. Hence, if Arithmetical Proportionals be multiplied or divided by the ſame number, their difference, or arithmetical ratio, is alſo multiplied or divided by the ſame number.

II. *Properties of Geometrical Proportionals.*

The properties relating to mean Proportionals are given under the term MEAN *Proportional ;* ſome are alſo given under the article Proportion ; and ſome additional ones are as below :

1. To find a 3d Proportional to two given numbers, or a 4th Proportional to three : In the former caſe, multiply the 2d term by itſelf, and divide the product by the 1ſt : and in the latter caſe, multiply the 2d term by the 3d, and divide the product by the 1ſt.

So 2 : 6 :: 6 : 18, the 3d prop. to 2 and 6:
and 2 : 6 :: 5 : 15, the 4th prop. to 2, 6, and 5.

2. If the terms of any geometrical ratio be augmented or diminiſhed by any others in the ſame ratio, or proportion, the ſums or differences will ſtill be in the ſame ratio or proportion.

So if $a : b :: c : d$,
then is $a : b :: a \pm c : b \pm d :: c : d$.

And if the terms of a ratio, or proportion, be multiplied or divided by any one and the ſame number, the products and quotients will ſtill be in the ſame ratio, or proportion.

Thus, $a : b :: na : nb :: \dfrac{a}{n} : \dfrac{b}{n}$.

3. If a ſet of continued Proportionals be either augmented or diminiſhed by the ſame part or parts of themſelves, the ſums or differences will alſo be Proportionals.

Thus if a, b, c, d, &c be Propors.

then are $a \pm \dfrac{a}{n}, b \pm \dfrac{b}{n}, c \pm \dfrac{c}{n}$, &c alſo Propors.

where the common ratio is $1 \pm \dfrac{1}{n}$.

And if any ſingle quantity be either augmented or diminiſhed by ſome part of itſelf, and the reſult be alſo increaſed or diminiſhed by the ſame part of itſelf, and this third quantity treated in the ſame manner, and ſo on ; then ſhall all theſe quantities be continued Proportionals. So, beginning with the quantity a, and taking always the nth part, then ſhall

$a, a \pm \dfrac{a}{n}, a \pm \dfrac{2a}{n} + \dfrac{a^2}{n^2}$, &c be Proportionals,

or $a, a \pm \dfrac{a}{n}, (a \pm \dfrac{a}{n})^2, (a \pm \dfrac{a}{n})^3$, &c Propors.

the common ratio being $1 \pm \dfrac{a}{n}$.

4. If one ſet of Proportionals be multiplied or divided by any other ſet of Proportionals, each term by each, the products or quotients will alſo be Proportionals.

Thus, if $a : na$ :: $b : nb$,
and $c : mc$ :: $d : md$;
then is $ac : mnac :: bd : mnbd$,
and $\dfrac{a}{c} : \dfrac{na}{mc} :: \dfrac{b}{d} : \dfrac{nb}{mb}$.

5. If there be ſeveral continued Proportionals, then whatever ratio the 1ſt has to the 2d, the 1ſt to the 3d ſhall

shall have the duplicate of the ratio, the 1st to the 4th the triplicate of it, and so on.

So in a, na, n^2a, n^3a, &c, the ratio being n; then $a : n^2a$, or 1 to n^2, the duplicate ratio, and $a : n^3a$, or 1 to n^3, the triplicate ratio, and so on.

6. In three continued Proportionals, the difference between the 1st and 2d term, is a mean Proportional between the 1st term and the second difference of all the terms.

Thus, in the three Propor. a, na, n^2a;

Terms	1st difs.	2d dif.
n^2a		
na	$n^2a - na$	
a	$na - a$	$n^2a - 2na + a$,

then $a : na - a :: na - a : n^2a - 2na + a$.

Or in the numbers 2, 6, 18;

18		
6	12	8 the 2d difference;
2	4	

then 2, 4, 8 are Proportionals.

7. When four quantities are in proportion, they are also in proportion by inversion, composition, division, &c; thus, a, na, b, nb being in proportion, viz,

1		$a : na :: b : nb$; then by
2. Inversion		$na : a :: nb : b$;
3. Alternation		$a : b :: na : nb$;
4. Composition		$a + na : na :: b + nb : nb$;
5. Conversion		$a + na : a :: b + nb : b$;
6. Division	$\begin{cases} a - na : a :: b - nb : b; \\ a - na : na :: b - nb : nb. \end{cases}$	

III. *Properties of Harmonical Proportionals.*

1. If three or four numbers in Harmonical Proportion, be either multiplied or divided by any number, the products or quotients will also be Harmonical Proportionals.

Thus, 6, 3, 2 being harmon. Propor.
then 12, 6, 4 are also harmon. Propor.
and $\frac{6}{2}$, $\frac{3}{2}$, $\frac{2}{2}$ are also harmon. Propor.

2. In the three Harmonical Proportionals a, b, c, when any two of these are given, the 3d can be found from the definition of them, viz, that $a : c :: a - b : b - c$; for hence

$$b = \frac{2ac}{a + c}$$ the harmonical mean, and.

$$c = \frac{ab}{2a - b}$$ the 3d harmon. to a and b.

3. And of the four Harmonicals, a, b, c, d, any three being given, the fourth can be found from the definition of them, viz, that $a : d :: a - b : c - d$; for thence the three b, c, d, will be thus found, viz,

$$b = \frac{2ad - ac}{d}; \quad c = \frac{2ad - bd}{a}; \quad d = \frac{ac}{2a - b}$$

4. If there be four numbers disposed in order, as 2, 3, 4, 6, of which one extreme and the two middle terms are in Arithmetical Proportion, and the other extreme and the same middle terms are in Harmonical Proportion; then are the four terms in Geometrical Proportion: so here

the three	2, 3, 4 are arithmeticals,
and the three	3, 4, 6 are harmonicals,
then the four	2, 3, 4, 6 are geometricals.

5. If between any two numbers, as 2 and 6, there be interposed an arithmetical mean 4, and also a harmonical mean 3, the four will then be geometricals, viz, $2 : 3 :: 4 : 6$.

6. Between the three kinds of proportion, there is this remarkable difference ; viz, that from any given number there can be raised a continued arithmetical series increasing ad infinitum, but not decreasing ; while the harmonical can be decreased ad infinitum, but not increased ; and the geometrical admits of both.

PROPOSITION, is either some truth advanced, and shewn to be such by demonstration ; or some operation proposed, and its solution shewn. In short, it is something proposed either to be demonstrated, or to be done or performed. The former is a theorem, and the latter is a problem.

PROSTHAPHERESIS, in Astronomy, the difference between the true and mean motion, or between the true and mean place, of a planet, or between the true and equated anomaly ; called also Equation of the Orbit, or Equation of the Centre, or simply the Equation ; and it is equal to the angle formed at the planet, and subtended by the excentricity of its orbit.

Thus, if S be the sun, and P the place of a planet in its orbit APB, whose centre is C.

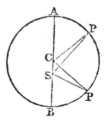

Then the mean anomaly is the $\angle$ ACP, and the true anomaly the $\angle$ ASP, the difference of which is the $\angle$ CPS,

which is the Prosthapheresis ; which is so called, because it is sometimes to be added to, and sometimes to be subtracted from the mean motion, to give the true one ; as is evident from the figure.

PROTRACTING, or PROTRACTION, in Surveying, the act of plotting or laying down the dimensions taken in the field, by means of a Protractor, &c : Protracting makes one part of surveying.

PROTRACTING-*Pin*, a fine pointed pin, or needle, fitted into a handle, used to prick off degrees and minutes from the limb of the Protractor.

PROTRACTOR, a mathematical instrument, chiefly used in surveying, for laying down angles upon paper, &c.

The simplest, and most natural Protractor consists of a semicircular limb ADB (fig. 7, pl. xix) commonly of metal, divided into 180°, and subtended by a diameter AB ; in the middle of which is a small notch C, called

called the centre of the Protractor. And for the convenience of reckoning both ways, the degrees are numbered from the left hand towards the right, and from the right hand towards the left.

But this instrument is made much more commodious by transferring the divisions from the circumference to the edge of a ruler, whose side EF is parallel to AB, which is easily done by laying a ruler on the centre C, and over the several divisions on the semicircumference ADB, and marking the intersections of that ruler on the line EF: so that a ruler with these divisions marked on one of its sides as above, and returned down the two ends, and numbered both ways as in the circular Protractor, the fourth or blank side representing the diameter of the circle, is both a more useful form than the circular Protractor, and better adapted for putting into a case.

To make any Angle with the Protractor.—Lay the diameter of the Protractor along the given line which is to be one side of the angle, and its centre at the given angular point; then make a mark opposite the given degree of the angle found on the limb of the instrument, and, removing the Protractor, by a plane ruler laid over that point and the centre, draw a line, which will form the angle sought.

In the same way is any given angle measured, to find the number of degrees it contains.

This Protractor is also very useful in drawing one line perpendicular to another; which is readily done by laying the Protractor across the given line, so that both its centre and the 90th degree on the opposite edge fall upon the line, also one of the edges passing over the given point, by which then let the perpendicular be drawn.

The Improved PROTRACTOR is an instrument much like the former, only furnished with a little more apparatus, by which an angle may be set off to a single minute.

The chief addition is an index attached to the centre, about which it is moveable, so as to play freely and steadily over the limb. Beyond the limb the index is divided, on both edges, into 60 equal parts of the portions of circles, intercepted by two other right lines drawn from the centre, so that each makes an angle of one degree with lines drawn to the assumed points from the centre.

To set off an angle of any number of degrees and minutes with this Protractor, move the index, so that one of the lines drawn on the limb, from one of the forementioned points, may fall upon the number of degrees given; and prick off as many of the equal parts on the proper edge of the index as there are minutes given; then drawing a line from the centre to that point so pricked off, the required angle is thus formed with the given line or diameter of the Protractor.

PROVING *of Gunpowder.* See EPROUVETTE, and GUNPOWDER.

PSEUDO-STELLA, any kind of meteor or phenomenon, appearing in the heavens, and resembling a star.

PTOLEMAIC, or PTOLOMAIC, something relating to Ptolomy; as the Ptolomaic System, the Ptolomaic Sphere, &c. See SYSTEM, SPHERE, &c.

PTOLEMY, or PTOLOMY, (CLAUDIUS), a very celebrated geographer, astronomer, and mathematician, among the Ancients, was born at Pelusium in Egypt, about the 70th year of the Christian era; and died, it has been said, in the 78th year of his age, and in the year of Christ 147. He taught astronomy at Alexandria in Egypt, where he made many astronomical observations, and composed his other works. It is certain that he flourished in the reigns of Marcus Antoninus and Adrian: for it is noted in his Canon, that Antoninus Pius reigned 23 years, which shews that he himself survived him; he also tells us in one place, that he made a great many observations upon the fixed stars at Alexandria, in the second year of Antoninus Pius; and in another, that he observed an eclipse of the moon, in the ninth year of Adrian; from which it is reasonable to conclude that this astronomer's observations upon the heavens were many of them made between the year 125 and 140.

Ptolomy has always been reckoned the prince of astronomers among the Ancients, and in his works has left us an entire body of that science. He has preserved and transmitted to us the observations and principal discoveries of the Ancients, and at the same time augmented and enriched them with his own. He corrected Hipparchus's catalogue of the fixed stars; and formed tables, by which the motions of the sun, moon, and planets, might be calculated and regulated. He was indeed the first who collected the scattered and detached observations of the Ancients, and digested them into a system; which he set forth in his Μεγαλη Συ.ταξις, five *Magna Constructio*, divided into 13 books. He adopts and exhibits here the ancient system of the world, which placed the earth in the centre of the universe; and this has been called from him, the Ptolomaic System, to distinguish it from those of Copernicus and Tycho Brahe.

About the year 827 this work was translated by the Arabians into their language, in which it was called *Almagestum*, by order of one of their kings; and from Arabic into Latin, about 1230, by the encouragement of the emperor Frederic the 2d. There were also other versions from the Arabic into Latin; and a manuscript of one, done by Girardus Cremonensis, who flourished about the middle of the 14th century, Fabricius says, is still extant in the library of All Souls College in Oxford. The Greek text of this work began to be read in Europe in the 15th century; and was first published by Simon Grynæus at Basil, 1538, in folio, with the eleven books of commentaries by Theon, who flourished at Alexandria in the reign of the elder Theodosius. In 1541 it was reprinted at Basil, with a Latin version by George Trapezond; and again at the same place in 1551, with the addition of other works of Ptolomy, and Latin versions by Camerarius. We learn from Kepler, that this last edition was used by Tycho.

Of this principal work of the ancient astronomers, it may not be improper to give here a more particular account. In general, it may be observed, that the work is founded upon the hypothesis of the earth's being at rest in the centre of the universe, and that the heavenly bodies, the stars and planets, all move around it in solid orbs, whose motions are all directed by one, which Pto-

lomy

lomy called the *Primum Mobile*, or Firſt Mover, of which he difcourſes at large. But, to be more particular, this great work is divided into 13 books.

In the firſt book, Ptolomy ſhews, that the earth is in the centre of thoſe orbs, and of the univerſe itſelf, as he underſtood it : he repreſents the earth as of a ſpherical figure, and but as a point in compariſon of the reſt of the heavenly bodies : he treats concerning the ſeveral circles of the earth, and their diſtances from the equator ; as alſo of the right and oblique aſcenſion of the heavenly bodies in a right ſphere.

In the 2d book, he treats of the habitable parts of the earth ; of the elevation of the pole in an oblique ſphere, and the various angles which the ſeveral circles make with the horizon, according to the different latitude of places ; alſo of the phenomena of the heavenly bodies depending on the ſame.

In the 3d book, he treats of the quantity of the year, and of the unequal motion of the ſun through the zodiac : he here gives the method of computing the mean motion of the ſun, with tables of the ſame ; and likewiſe treats of the inequality of days and nights.

In the 4th book, he treats of the lunar motions, and their various phenomena : he gives tables for finding the moon's mean motions, with her latitude and longitude : he diſcourſes largely concerning lunar epicycles ; and by comparing the times of a great number of eclipſes, mentioned by Hipparchus, Calippus, and others, he has computed the places of the ſun and moon, according to their mean motions, from the firſt year of Nabonazar, king of Egypt, to his own time.

In the 5th book, he treats of the inſtrument called the Aſtrolabe : he treats alſo of the eccentricity of the lunar orbit, and the inequality of the moon's motion, according to her diſtance from the ſun : he alſo gives tables, and an univerſal canon for the inequality of the lunar motions : he then treats of the different aſpects or phaſes of the moon, and gives a computation of the diameter of the ſun and moon, with the magnitude of the ſun, moon and earth compared together ; he ſtates alſo the different meaſures of the diſtance of the ſun and moon, according as they are determined by ancient mathematicians and philoſophers.

In the 6th book, he treats of the conjunctions and oppoſitions of the ſun and moon, with tables for computing the mean time when they happen ; of the boundaries of ſolar and lunar eclipſes ; of the tables and methods of computing the eclipſes of the ſun and moon, with many other particulars.

In the 7th book, he treats of the fixed ſtars ; and ſhews the methods of deſcribing them, in their various conſtellations, on the ſurface of an artificial ſphere or globe : he rectifies the places of the ſtars to his own time, and ſhews how different thoſe places were then, from what they had been in the times of Timocharis, Hipparchus, Ariſtillus, Calippus, and others : he then lays down a catalogue of the ſtars in each of the northern conſtellations, with their latitude, longitude, and magnitudes.

In the 8th book, he gives a like catalogue of the ſtars in the conſtellations of the ſouthern hemiſphere, and in the 12 ſigns or conſtellations of the zodiac. This is the firſt catalogue of the ſtars now extant, and forms the moſt valuable part of Ptolomy's works. He then treats of the galaxy, or milky-way ; alſo of the planetary aſpects, with the riſing and ſetting of the ſun, moon, and ſtars.

In the 9th book, he treats of the order of the ſun, moon, and planets, with the periodical revolutions of the five planets ; then he gives tables of the mean motions, beginning with the theory of Mercury, and ſhewing its various phenomena with reſpect to the earth.

The 10th book begins with the theory of the planet Venus, treating of its greateſt diſtance from the ſun ; of its epicycle, eccentricity, and periodical motions : it then treats of the ſame particulars in the planet Mars.

The 11th book treats of the ſame circumſtances in the theory of the planets Jupiter and Saturn. It alſo corrects all the planetary motions from obſervations made from the time of Nabonazar to his own.

The 12th book treats of the retrogreſſive motion of the ſeveral planets ; giving alſo tables of their ſtations, and of the greateſt diſtances of Venus and Mercury from the ſun.

The 13th book treats of the ſeveral hypotheſes of the latitude of the five planets ; of the greateſt latitude, or inclination of the orbits of the five planets, which are computed and diſpoſed in tables ; of the riſing and ſetting of the planets, with tables of them. Then follows a concluſion or winding up of the whole work.

This great work of Ptolomy will always be valuable on account of the obſervations he gives of the places of the ſtars and planets in former times, and according to ancient philoſophers and aſtronomers that were then extant ; but principally on account of the large and curious catalogue of the ſtars, which being compared with their places at preſent, we thence deduce the true quantity of their ſlow progreſſive motion according to the order of the ſigns, or of the preceſſion of the equinoxes.

Another great and important work of Ptolomy was, his *Geography*, in 7 books ; in which, with his uſual ſagacity, he ſearches out and marks the ſituation of places according to their latitudes and longitudes ; and he was the firſt that did ſo. Though this work muſt needs fall far ſhort of perfection, through the want of neceſſary obſervations, yet it is of conſiderable merit, and has been very uſeful to modern geographers. Cellarius indeed ſuſpects, and he was a very competent judge, that Ptolomy did not uſe all the care and application which the nature of his work required ; and his reaſon is, that the author delivers himſelf with the ſame fluency and appearance of certainty, concerning things and places at the remoteſt diſtance, which it was impoſſible he could know any thing of, that he does concerning thoſe which lay the neareſt to him, and fall the moſt under his cognizance. Salmaſius had before made ſome remarks to the ſame purpoſe upon this work of Ptolomy. The Greek text of this work was firſt publiſhed by itſelf at Baſil in 1533, in 4to : afterward with a Latin verſion and notes by Gerard Mercator at Amſterdam, 1605 ; which laſt edition was reprinted at the ſame place, 1618, in folio, with neat geographical tables, by Bertius.

Other

Other works of Ptolomy, though lefs confiderable than thefe two, are ftill extant. As, *Libri quatuor de Judiciis Aftrorum*, upon the firft two books of which Cardan wrote a commentary.—*Fructus Librorum fuorum*; a kind of fupplement to the former work.—*Recenfio Chronologica Regum* : this, with another work of Ptolomy, *De Hypothefibus Planetarum*, was publifhed in 1620, 4to, by John Bainbridge, the Savilian profeffor of Aftronomy at Oxford : And Scaliger, Petavius, Dodwell, and the other chronological writers, have made great ufe of it.—*Apparentiæ Stellarum Inerrantium* : this was publifhed at Paris by Petavius, with a Latin verfion, 1630, in folio ; but from a mutilated copy, the defects of which have fince been fupplied from a perfect one, which Sir Henry Saville had communicated to archbifhop Ufher, by Fabricius, in the 3d volume of his *Bibliotheca Græca*.—*Elementorum Harmonicorum libri tres ;* publifhed in Greek and Latin, with a commentary by Porphyry the philofopher, by Dr. Wallis at Oxford, 1682, in 4to ; and afterwards reprinted there, and inferted in the 3d volume of Wallis's works, 1699, in folio.

Mabillon exhibits, in his *German Travels*, an effigy of Ptolomy looking at the ftars through an optical tube ; which effigy, he fays, he found in a manufcript of the 13th century, made by Conradus a monk. Hence fome have fancied, that the ufe of the telefcope was known to Conradus. But this is only matter of mere conjecture, there being no facts or teftimonies, nor even probabilities, to fupport fuch an opinion.

It is rather likely that the tube was nothing more than a plain open one, employed to ftrengthen and defend the eye-fight, when looking at particular ftars, by excluding adventitious rays from other ftars and objects ; a contrivance which no obferver of the heavens can ever be fuppofed to have been without.

PULLEY, one of the five mechanical powers ; confifting of a little wheel, being a circular piece of wood or metal, turning on an axis, and having a channel around it, in its edge or circumference, in which a cord flides and fo raifes up weights.

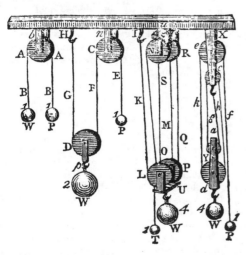

The Latins call it Trochlea ; and the feamen, when fitted with a rope, a Tackle. An affemblage of feveral

Pulleys is called a Syftem of Pulleys, or Polyfpafton : fome of which are in a block or cafe, which is fixed ; and others in a block which is moveable, and rifes with the weight. The wheel or rundle is called the Sheave or Shiver ; the axis on which it turns, the Gudgeon ; and the fixed piece of wood or iron, into which it is put, the Block.

Doctrine of the PULLEY.—1. If the equal weights P and W hang by the cord BB upon the pulley A, whofe block *b* is fixed to the beam HI, they will counterpoife each other, juft in the fame manner as if the cord were cut in the middle, and its two ends hung upon the hooks fixed in the Pulley at A and A, equally diftant from the centre.

Hence, a fingle Pulley, if the lines of direction of the power and the weight be tangents to the periphery, neither affifts nor impedes the power, but only changes its direction. The ufe of the Pulley therefore, is when the vertical direction of a power is to be changed into an horizontal one ; or an afcending direction into a defcending one ; &c. This is found a good provifion for the fafety of the workmen employed in drawing with the Pulley. And this change of direction by means of a Pulley has this farther advantage ; that if any power can exert more force in one direction than another, we are hence enabled to employ it with its greateft effect ; as for the convenience of a horfe to draw in a horizontal direction, or fuch like.

But the great ufe of the Pulley is in combining feveral of them together ; thus forming what Vitruvius and others call Polyfpafta ; the advantages of which are, that the machine takes up but little room, is eafily removed, and raifes a very great weight with a moderate force.

2. When a weight W hangs at the lower end of the moveable block *p* of the Pulley D, and the chord GF goes under the Pulley, it is plain that the part G of the cord bears one half of the weight W, and the part F the other half of it ; for they bear the whole between them ; therefore whatever holds the upper end of either rope, fuftains one half of the weight ; and thus the power P, which draws the cord F by means of the cord E, paffing over the fixed pulley C, will fuftain the weight W when its intenfity is only equal to the half of W ; that is, in the cafe of one moveable Pulley, the power gained is as 2 to 1, or as the number of ropes G and F to the one rope E.

In like manner, in the cafe of two moveable Pulleys P and L, each of thefe alfo doubles the power, and produces a gain of 4 to 1, or as the number of the ropes Q, M, S, K, fuftaining the weight W, to the 1 rope O fuftaining the power T ; that is, W is to T as 4 to 1. And fo on, for any number of moveable Pulleys, viz, 3 fuch Pulleys producing an increafe of power as 6 to 1 ; 4 Pulleys, as 8 to 1 ; &c ; each power adding 2 to the number. Alfo the effect is the fame, when the Pulleys are difpofed as in the fixed block X, and the other two as in the moveable block Y ; thefe in the lower block giving the fame advantage to the power, when they rife all together in one block with the weight.

But if the lower Pulleys do not rife all together in one block with the weight, but act upon one another, having the weight only faftened to the loweft of them, the force

force of the power is still more increased, each power doubling the former numbers, the gain of power in this case proceeding in the geometrical progression, 1, 2, 4, 8, 16, &c, according to the powers of 2; whereas in the former case, the gain was only in arithmetical progression, increasing by the addition of 2. Thus, a power whose intensity is equal to 8lb applied at *a* will, by means of the lower Pulley A, sustain 16lb; and a power equal to 4lb at *b*, by means of the Pulley, will sustain the power of 8lb acting at *a*, and consequently the weight of 16lb at W; also a third power equal to 2lb at *c*, by means of the Pulley C, will sustain the power of 4lb at *b*; and a 4th power of 1lb at *d*, by means of the Pulley D, will sustain the power 2 at *c*, and consequently the power 4 at B, and the power 8 at A, and the weight 16 at W.

3. It is to be noted however, that, in whatever proportion the power is gained, in that very same proportion is the length of time increased to produce the same effect. For when a power moves a weight by means of several Pulleys, the space passed over by the power is to the space passed over by the weight, as the weight is to the power. Hence, the smaller a force is that sustains a weight by means of Pulleys, the slower is the weight raised; so that what is saved or gained in force, is always spent or lost in time: which is the general property of all the mechanical powers.

The usual methods of arranging Pulleys in their blocks, may be reduced to two. The first consists in placing them one by the side of another, upon the same pin; the other, in placing them directly under one another, upon separate pins. Each of these methods however is liable to inconvenience; and Mr. Smeaton, to avoid the impediments to which these combinations are subject, proposes to combine these two methods in one. See the Philos. Transf. vol. 47, pa. 494.

Some instances of such combinations of Pulleys are exhibited in the following figures; beside which, there are also other varieties of forms.

A very considerable improvement in the construction of Pulleys has been made by Mr. James White, who has obtained a patent for his invention, and of which he gives the following description. The last of the three following figures shews the machine, consisting of two Pulleys Q and R, one fixed and the other moveable. Each of these has six concentric grooves, capable of having a line put round them, and thus acting like as many different Pulleys, having diameters equal to those of the grooves. Supposing then each of the grooves to be a distinct Pulley, and that all their diameters were equal, it is evident that if the weight 144 were to be raised by pulling at S till the Pulleys touch each other, the first Pulley must receive the length of line as many times as there are parts of the line hanging

between it and the lower Pulley. In the present case, there are 12 lines, *b*, *d*, *f*, &c, hanging between the

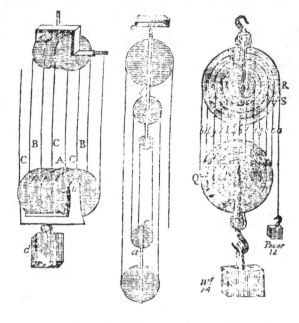

two pulleys, formed by its revolution about the six upper lower grooves. Hence as much line must pass over the uppermost Pulley as is equal to twelve times the distance of the two. But, from an inspection of the figure, it is plain, that the second Pulley cannot receive the full quantity of line by as much as is equal to the distance betwixt it and the first. In like manner, the third Pulley receives less than the first by as much as is the distance between the first and third; and so on to the last, which receives only one twelfth of the whole. For this receives its share of line *n* from a fixed point in the upper frame, which gives it nothing; while all the others in the same frame receives the line partly by turning to meet it, and partly by the line coming to meet them.

Supposing now these Pulleys to be equal in size, and to move freely as the line determines them, it appears evident, from the nature of the system, that the number of their revolutions, and consequently their velocities, must be in proportion to the number of suspending parts that are between the fixed point above mentioned and each Pulley respectively. Thus the outermost Pulley would go twelve times round in the time that the Pulley under which the part *n* of the line, if equal to it, would revolve only once; and the intermediate times and velocities would be a series of arithmetical proportionals, of which, if the first number were 1, the last would always be equal to the whole number of terms. Since then the revolutions of equal and distinct Pulleys are measured by their velocities, and that it is possible to find any proportion of velocity, on a single body running on a centre, viz, by finding proportionate distances from that centre; it follows, that if the diameters of certain grooves in the same substance be exactly adapted to the above series (the line itself being supposed inelastic, and of no magnitude) the necessity

of

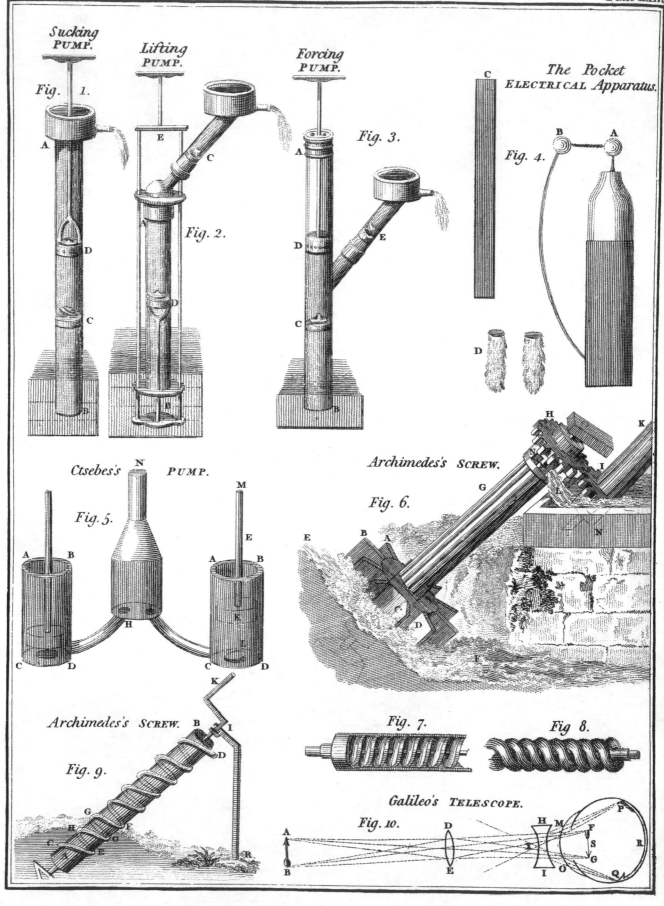

Plate XXIII

Sucking
PUMP.

Fig. 1.

Lifting
PUMP.

Forcing
PUMP.

Fig. 3.

The Pocket
ELECTRICAL *Apparatus.*

Fig. 4.

Fig. 2.

Ctsebes's PUMP.

Fig. 5.

Archimedes's SCREW.

Fig. 6.

Archimedes's SCREW.

Fig. 9.

Fig. 7.

Fig 8.

Galileo's TELESCOPE.

Fig. 10.

of ufing feveral Pulleys in each frame will be obviated, and with that fome of the inconveniencies to which the ufe of the Pulley is liable.

In the figure referred to, the coils of rope by which the weight is fupported, are reprefented by the lines *a, b, c* &c ; *a* is *the line of traction*, commonly called the fall, which paffes over and under the proper grooves, until it is faftened to the upper frame juft above *n*. In practice, however, the grooves are not arithmetical proportions, nor can they be fo ; for the diameter of the rope employed muft in all cafes be deducted from each term ; without which the fmaller grooves, to which the faid diameter bears a larger proportion than to the larger ones, will tend to rife and fall fafter than they, and thus introduce worfe defects than thofe which they were intended to obviate.

The principal advantage of this kind of Pulley is, that it deftroys lateral friction, and that kind of fhaking motion which is fo inconvenient in the common Pulley. And left (fays Mr. White) this circumftance fhould give the idea of weaknefs, I would obferve, that to have pins for the pulleys to run on, is not the only nor perhaps the beft method ; but that I fometimes ufe centres fixed to the Pulleys, and revolving on a very fhort bearing in the fide of the frame, by which ftrength is increafed, and friction very much diminifhed ; for to the laft moment the motion of the Pulley is perfectly circular : and this very circumftance is the caufe of its not wearing out in the centre as foon as it would, affifted by the ever increafing irregularities of a gullied bearing. Thefe Pulleys, when well executed, apply to jacks and other machines of that nature with peculiar advantage, both as to the time of going and their own durability ; and it is poffible to produce a fyftem of Pulleys of this kind of fix or eight parts only, and adapted to the pockets, which, by means of a fkain of fewing filk, or a clue of common thread, will raife upwards of an hundred weight.

As a fyftem of Pulleys has no great weight, and lies in a fmall compafs, it is eafily carried about, and can be applied for raifing weights in a great many cafes, where other engines cannot be ufed. But they are fubject to a great deal of friction, on the following accounts ; viz, 1ft, becaufe the diameters of their axes bear a very confiderable proportion to their own diameters ; 2d, becaufe in working they are apt to rub againft one another, or againft the fides of the block ; 3dly, becaufe of the ftiffnefs of the rope that goes over and under them. See Fergufon's Mech. pa. 37, 4to.

But the friction of the Pulley is now reduced to nothing as it were, by the ingenious Mr. Garnett's patent friction rollers, which produce a great faving of labour and expence, as well as in the wear of the machine, both when applied to Pulleys and to the axles of wheel carriages. His general principle is this ; between the axle and nave, or centre pin and box, a hollow fpace is left, to be filled up by folid equal rollers nearly touching each other. Thefe are furnifhed with axles inferted into a circular ring at each end, by which their relative diftances are preferved ; and they are kept parallel by means of wires faftened to the rings between the rollers, and which are rivetted to them.

The above contrivance is exhibited in the annexed figure ; where ABCD reprefents a piece of metal to

be inferted into the box or nave, of which E is the centre-pin or axle, and 1, 1, 1, &c, rollers of metal having

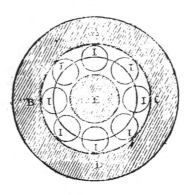

axes inferted in the brazen circle which paffes through their centres ; and both circles being rivetted together by means of bolts paffing between the rollers from one fide of the nave to the other ; and thus they are always kept feparate and parallel.

PUMP, in Hydraulics, a machine for raifing water, and other fluids.

Pumps are probably of very ancient ufe. Vitruvius afcribes the invention to Ctefebes of Athens, fome fay of Alexandria, about 120 years before Chrift. They are now of various kinds. As the Sucking Pump, the Lifting Pump, the Forcing Pump, Ship Pumps, Chain Pumps, &c. By means of the lifting and forcing Pumps, water may be raifed to any height, with a fufficient power, and an adequate apparatus : but by the fucking Pump the water may, by the general preffure of the atmofphere on the furface of the well, be raifed only about 33 or 34 feet ; though in practice it is feldom applied to the raifing it much above 28 ; becaufe, from the variations obferved in the barometer, it appears that the air may fometimes be lighter than 33 feet of water ; and whenever that happens, for want of the due counterpoife, this Pump may fail in its performance.

The Common Sucking PUMP.—This confifts of a pipe, of wood or metal, open at both ends, having a fixed valve in the lower part of it opening upwards, and a moveable valve or bucket by which the water is drawn or lifted up. This bucket is juft the fize of the bore of the Pump-pipe, in that part where it works, and leathered round fo as to fit it very clofe, that no air may pafs by the fides of it ; the valve hole being in the middle of the bucket. The bucket is commonly worked in the upper part of the barrel by a fhort rod, and another fixed valve placed juft below the defcent of the bucket. Thus, (fig. 1, pl. 23), AB is the Pump-pipe, C the lower fixed valve, opening upwards, and D is the bucket, or moving valve, alfo opening upwards.

In working the Pump ; draw up the bucket D, by means of the Pump rod, having any fort of a handle fixed to it : this draws up the water that is above it, or if not, the air ; in either cafe the water pufhes up the valve C, and enters to fupply the void left between C and D, being pufhed up by the preffure of the atmofphere on the furface of the water in the well below. Next, the bucket D is pufhed down, which fhuts the

valve

valve C, and prevents the return of the water downwards, which opens the valve D, by which the water afcends above it. And thus, by repeating the ftrokes of the Pump-rod handle, the valves alternately open and fhut, and the water is drawn up at every ftroke, and runs out at the nozle or fpout near the top.

The Lifting Pump differs from the fucking Pump only in the difpofition of it's valves and the form of its pifton frame. This kind of Pump is reprefented in fig. 2, pl. 23; where the lower valve D is moveable, being worked up and down with the Pump rod, which lifts the water up, and fo opens the upper valve C, which is fixed, and permits the water to iffue through it, and run out at top. Then as the pifton D defcends, the weight of the water above C fhuts that valve C, and fo prevents its return, till that valve be opened again by another lift of the pifton D. And fo alternately.

The Forcing Pump raifes the water through the fucker, or lower valve C (fig. 3, pl. 23), in the fame manner as the fucking Pump; but as the pifton or plunger D has no valve in it, the water cannot get above it when this is pufhed down again; inftead of which, a fide pipe is inferted between C and D, having a fixed valve at E opening upwards, through which the water is forced out of the Pump by pufhing down the plunger D.

Obfervations on Pumps.—The force required to work a Pump, is equal to the weight of water raifed at each ftroke; or equal to the weight of water filling the cavity of the pipe, and its height equal to the length of the ftroke made by the pifton. Hence if d denote the diameter of the pipe, and l the length of the ftroke, both in inches; then is $\cdot7854d^2l$ the content of the water raifed at a ftroke, in inches, or $\cdot0028d^2l$ in ale gallons; and the weight of it is $\frac{d^2l}{220}$ ounces or $\frac{d^2l}{3520}$ lb. But if the handle of the pump be a lever which gains in the power of p to 1, the force of the hand to work the Pump will be only $\frac{d^2l}{3520p}$ lb, or, when p is 5 for inftance, it will be $\frac{d^2l}{17600}$ lb. And all thefe over and above the friction of the moving parts of the Pump.

To the forcing Pump is fometimes adapted an air veffel, which, being compreffed by the water, by its elafticity acts upon the water again, and forces it out to a great diftance, and in a continued ftream, inftead of by jerks or jets. So, Mr. Newfham's water engine, for extinguifhing fire, confifts of two forcing Pumps, which alternately drive water into a clofe veffel of air, by which means the air in it is condenfed, and compreffes the water fo ftrongly, that it rufhes out with great impetuofity and force through a pipe that comes down into it, making a continued uniform ftream.

By means of forcing Pumps, water may be raifed to any height whatever above the level of a river or fpring; and machines may be contrived to work thefe Pumps, either by a running ftream, a fall of water, or by horfes.

Ctefebes's Pump, acts both by fuction and by preffion. Thus, a brafs cylinder ABCD (fig. 5, pl. 23) furnifhed with a valve at L, is placed in the water. In this is fitted the pifton KM, made of green wood, which will not fwell in the water, and is adjufted to the aperture of the cylinder with a covering of leather, but without any valve. Another tube NH is fitted on at H, with a valve I opening upwards.—Now the pifton being raifed, the water opens the valve I, and rifes into the cavity of the cylinder. When the pifton is depreffed again, the valve I is opened, and the water is driven up through the tube HN.

This was the Pump ufed among the Ancients, and that from which both the others have been deduced. Sir Samuel Morland has endeavoured to increafe its force by leffening the friction; which he has done to good effect, fo as to make it work with very little.

There are various kinds of Pumps ufed in fhips, for throwing the water out of the hold, and upon other occafions, as the Chain Pump, &c.

Air-Pump, in Pneumatics, is a machine, by means of which the air is emptied out of veffels, and a kind of vacuum produced in them. For the particulars of which, fee Air-*Pump*.

PUNCHEON, a meafure for liquids, containing $\frac{1}{3}$ of a tun, or a hogfhead and $\frac{1}{2}$, or 84 gallons.

PUNCHINS, or Punchions, in Building, fhort pieces of timber placed to support fome confiderable weight.

PUNCTATED *Hyperbola*, in the higher geometry, an hyperbola whofe conjugate oval is infinitely fmall, that is, a point.

PUNCTUM *ex Comparatione*, is either focus, in the ellipfe or hyperbola; fo called by Apollonius, becaufe the rectangle under two abfciffes made at the focus, is equal to one fourth part of what he calls the figure, which is the fquare of the conjugate axis, or the rectangle under the tranfverfe and the parameter.

PURBACH (George), a very eminent mathematician and aftronomer, was born at Purbach, a town upon the confines of Bavaria and Auftria, in 1423, and educated at Vienna. He afterwards vifited the moft celebrated univerfities in Germany, France, and Italy; and found a particular friend and patron in cardinal Cufa at Rome. Returning to Vienna, he was appointed mathematical profeffor, in which office he continued till his death, which happened in 1461, in the 39th year of his age only, to the great lofs of the learned world.

Purbach compofed a great number of pieces, upon mathematical and aftronomical fubjects; and his fame brought many ftudents to Vienna, and among them, the celebrated Regiomontanus, between whom and Purbach there fubfifted the ftricteft friendfhip and union of ftudies till the death of the latter. Thefe two laboured together to improve every branch of learning, by all the means in their power, though aftronomy feems to have been the favourite of both; and had not the immature death of Purbach prevented his further purfuits, there is no doubt but that, by their joint induftry, aftronomy would have been carried to very great perfection. That this is not merely furmife, may be learnt from thofe improvements which Purbach actually did make, to render the ftudy of it more eafy and practicable. His firft effay was, to amend the Latin tranflation of Ptolomy's Almageft, which had been made from the Arabic verfion: this he did, not by the help of the Greek text, for he was unacquainted with that language, but by drawing the moft probable conjectures from a ftrict attention to the fenfe of the author.

He

He then proceeded to other works, and among them, he wrote a tract, which he entitled, *An Introduction to Arithmetic* ; then a treatise on *Gnomonics*, or *Dialling*, with tables suited to the difference of climates or latitudes ; likewise a small tract concerning the *Altitudes of the Sun*, with a table ; also, *Astrolabic Canons*, with a table of the parallels, proportioned to every degree of the equinoctial.

After this, he constructed Solid Spheres, or Celestial Globes, and composed a new table of fixed stars, adding the longitude by which every star, since the time of Ptolomy, had increased. He likewise invented various other instruments, among which was the Gnomon, or Geometrical Square, with canons and a table for the use of it.

He not only collected the various tables of the Primum Mobile, but added new ones. He made very great improvements in Trigonometry, and by introducing the table of Sines, by a decimal division of the radius, he quite changed the appearance of that science : he supposed the radius to be divided into 600000 equal parts, and computed the sines of the arcs, for every ten minutes, in such equal parts of the radius, by the decimal notation, instead of the duodecimal one delivered by the Greeks, and preserved even by the Arabians till our author's time ; a project which was completed by his friend Regiomontanus, who computed the sines to every minute of the quadrant, in 1000000th parts of the radius.

Having prepared the tables of the fixed stars, he next undertook to reform those of the planets, and constructed some entirely new ones. Having finished his tables, he wrote a kind of Perpetual Almanac, but chiefly for the moon, answering to the periods of Meton and Calippus ; also an Almanac for the Planets, or, as Regiomontanus afterwards called it, an Ephemeris, for many years. But observing there were some planets in the heavens at a great distance from the places where they were described to be in the tables, particularly the sun and moon (the eclipses of which were observed frequently to happen very different from the times predicted), he applied himself to construct new tables, particularly adapted to eclipses ; which were long after famous for their exactness. To the same time may be referred his finishing that celebrated work, entitled, *A New Theory of the Planets*, which Regiomontanus afterwards published the first of all the works executed at his new printing-house.

PURE *Hyperbola*, is an Hyperbola without any oval, node, cusp, or conjugate point ; which happens through the impossibility of two of its roots.

PURE *Mathematics*, *Proposition*, *Quadratics*, &c. See the several articles.

PURLINES, in Architecture, those pieces of timber that lie across the rafters on the inside, to keep them from sinking in the middle of their length.

PYRAMID, a solid having any plane figure for its base, and its sides triangles whose vertices all meet in a point at the top, called the vertex of the pyramid ; the base of each triangle being the sides of the plane base of the Pyramid.—The number of triangles is equal to the number of the sides of the base ; and a cone is a round Pyramid, or one having an infinite number of sides. ——The Pyramid is also denominated from its base,

being triangular when the base is a triangle, quadrangular when a quadrangle, &c.

The *axis* of the Pyramid, is the line drawn from the vertex to the centre of the base. When this axis is perpendicular to the base, the Pyramid is said to be a *right* one ; otherwise it is *oblique*.

1. A Pyramid may be conceived to be generated by a line moved about the vertex, and so carried round the perimeter of the base.

2. All Pyramids having equal bases and altitudes, are equal to one another : though the figures of their bases should even be different.

3. Every Pyramid is equal to one-third of the circumscribed prism, or a prism of the same base and altitude ; and therefore the solid content of the Pyramid is found by multiplying the base by the perpendicular altitude, and taking $\frac{1}{3}$ of the product.

4. The upright surface of a Pyramid, is found by adding together the areas of all the triangles which form that surface.

5. If a Pyramid be cut by a plane parallel to the base, the section will be a plane figure similar to the base ; and these two figures will be in proportion to each other as the squares of their distances from the vertex of the Pyramid.

6. The centre of gravity of a Pyramid is distant from the vertex $\frac{3}{4}$ of the axis.

Frustum of a PYRAMID, is the part left at the bottom when the top is cut off by a plane parallel to the base.

The solid content of the Frustum of a Pyramid is found, by first adding into one sum the areas of the two ends and the mean proportional between them, the 3d part of which sum is a medium section, or is the base of an equal prism of the same altitude ; and therefore this medium area or section multiplied by the altitude gives the solid content. So, if A denote the area of one end, a the area of the other end, and h the height ; then $\frac{1}{3}(A + a + \sqrt{Aa})$ is the medium area or section, and $\frac{1}{3}(A + a + \sqrt{Aa}) \times h$ is the solid content.

PYRAMIDS *of Egypt*, are very numerous, counting both great and small ; but the most remarkable are the three Pyramids of Memphis, or, as they are now called, of Cheisa or Gize. They are square Pyramids, and the dimensions of the greatest of them, are 700 feet on each side of the base, and the oblique height or slant side the same ; its base covers, or stands upon, nearly 11 acres of ground. It is thought by some that these Pyramids were designed and used as gnomons, for astronomical purposes ; and it is remarkable that their four sides are accurately in the direction of the four cardinal points of the compass, east, west, north, and south.

PYRAMIDAL *Numbers*, are the sums of polygonal numbers, collected after the same manner as the polygonal numbers themselves are found from arithmetical progressions.

These are particularly called First Pyramidals. The sums of First Pyramidals are called Second Pyramidals ; and the sums of the 2d are 3d Pyramidals ; and so on. Particularly, those arising from triangular numbers, are called Prime Triangular Pyramidals ; those arising from pentagonal numbers, are called Prime Pentagonal Pyramidals; and so on.

The numbers 1, 4, 10, 20, 35, &c, formed by adding the triangulars } 1, 3, 6, 10, 15, &c,

are usually called simply by the name of Pyramidals; and the general formula for finding them is $n \times \frac{n-1}{2} \times \frac{n-2}{3}$; so the 4th Pyramidal is found by substituting 4 for n; the 5th by substituting 5 for v; &c. See FIGURATE *Numbers*, and POLYGONAL *Numbers*.

PYRAMIDOID, is sometimes used for the parabolic spindle, or the solid formed by the rotation of a semiparabola about its base or greatest ordinate. See PARABOLIC *Spindle*.

PYROMETER, or fire-measurer, a machine for measuring the expansion of solid bodies by heat.

Muschenbroek was the first inventor of this instrument; though it has since received several improvements by other philosophers. He has given a table of the expansions of the different metals, with various degrees of heat. Having prepared cylindric rods of iron, steel, copper, brass, tin, and lead, he exposed them first to a Pyrometer with one flame in the middle; then with two flames; then successively with three, four, and five flames. The effects were as in the following Table, where the degrees of expansion are marked in parts equal to the 12500th part of an inch.

Expansion of	Iron	Steel	Copp.	Brass	Tin	Lead
By 1 flame	80	85	89	110	153	155
By 2 flames placed close together	117	123	115	220		274
By 2 flames at 2½ inches distant	109	94	92	141	219	263
By 3 flames close together	142	168	193	275		
By 4 flames close together	211	270	270	361		
By 5 flames	230	310	310	377		

Tin easily melts when heated by two flames placed close together; and lead with three flames close together, when they burn long.

It hence appears that the expansions of any metal are in a less degree than the number of flames: so two flames give less than a double expansion, three flames less than a triple expansion, and so on, always more and more below the ratio of the number of flames. And the flames placed together cause a greater expansion, than with an interval between them.

For the construction of Muschenbroek's Pyrometer, with alterations and improvements upon it by Desaguliers, see Desag. Exper. Philos. vol. 1, pa. 421; see also Muschenbroek's translation of the Experiments of the Academy del Cimento, printed at Leyden in 1731;

and for a Pyrometer of a new construction, by which the expansions of metals in boiling fluids may be examined and compared with Fahrenheit's thermometer, see Muschenb. Introd. ad Philos. Nat. 4to, 1762, vol. 2, pa. 610.

But as it has been observed, that Muschenbroek's Pyrometer was liable to some objections, these have been removed in a good measure by Ellicott, who has given a description of his improved Pyrometer in the Philos. Trans. numb. 443; and the same may be seen in the Abridg. vol. 8, pa. 464. This instrument measures the expansions to the 7200th part of an inch; and by means of it, Mr. Ellicott found, upon a medium, that the expansions of bars of different metals, as nearly of the same dimensions as possible, by the same degree of heat, were as below:

Gold, Silver, Brass, Copper, Iron, Steel, Lead.
73   103   95   89   60   56   149

The great difference between the expansions of iron and brass, has been applied with good success to remove the irregularities in pendulums arising from heat. Philos. Trans. vol. 47, pa. 485.

Mr. Graham used to measure the minute expansions of metal bars, by advancing the point of a micrometer screw, till it sensibly stopped against the end of the bar to be measured. This screw, being small and very lightly hung, was capable of agreement within the 3000 or 4000th part of an inch. And on this general principle Mr. Smeaton contrived his Pyrometer, in which the measures are determined by the contact of a piece of metal with the point of a micrometer-screw. This instrument makes the expansions sensible to the 2345th part of an inch. And when it is used, both the instrument and the bar, to be measured, are immerged in a cistern of water, heated to any degree, up to boiling, by means of lamps placed under the cistern; and the water communicates the same degree of heat to the instrument and bar, and to a mercurial thermometer immerged in it, for ascertaining that degree.

With this Pyrometer Mr. Smeaton made several experiments, which are arranged in a table; and he remarks, that their result agrees very well with the proportions of expansions of several metals given by Mr. Ellicott. The following Table shews how much a foot in length of each metal expands by an increase of heat corresponding to 180 degrees of Fahrenheit's thermometer, or to the difference between the temperatures of freezing and boiling water, expressed in the 10000th part of an inch.

1. White glass barometer tube	100
2. Martial regulus of antimony	130
3. Blistered steel	138
4. Hard steel	147
5. Iron	151
6. Bismuth	167
7. Copper, hammered	204
8. Copper 8 parts, mixed with 1 part tin	218
9. Cast brass	225
10. Brass 16 parts, with 1 of tin	229
11. Brass wire	232
12. Speculum metal	232
13. Spelter solder, viz 2 parts brass and 1 zinc	247
14. Fine	

For a farther account of this inftrument, with its ufe, fee Philof. Tranf. vol. 48, pa. 598.

Mr. Fergufon has conftructed, and defcribed a Pyrometer (Lect. on Mechanics, Suppl. pa. 7, 4to), which makes the expanfion of metals by heat vifible to the 45000th part of an inch. And another plan of a Pyrometer has lately been invented by M. De Luc, in confequence of a hint fuggefted to him by Mr. Ramfden : for an account of which, with the principle of its conftruction and ufe, both in the comparative meafure of the expanfions of bodies by heat, and the meafure of their abfolute expanfion, as well as the experiments made with it, fee M. De Luc's elaborate effay on Pyrometry &c, in the Philof. Tranf. vol. 68, pa. 419—546.

Other very nice and ingenious contrivances, for the meafuring of expanfions by heat, have been made by Mr. Ramfden ; which he has fuccefsfully applied in the cafe of the meafuring rods and chains lately employed, by General Roy and Col. Williams, in meafuring the bafe on Hounflow Heath, &c ; which determine the expanfions, to great minutenefs, for each degree of the thermometer. See Philof. Tranf. 1785, &c.

PYROPHORUS, the name ufually given to that fubftance by fome called black phofphorus ; being a chemical preparation poffeffing the fingular property of kindling fpontaneoufly when expofed to the air ; which was accidentally difcovered by M. Homberg, who prepared it of alum and human fæces. Though it has fince been found, by the fon of M. Lemeri, that the fæces are not neceffary to it, but that honey, fugar, flour, and any animal or vegetable matter, may be ufed inftead of the fæces ; and M. De Suvigny has fhewn that moft vitriolic falts may be fubftituted for the alum. See Prieftley's Obferv. on Air, vol. 3, Append. p. 386, and vol. 4, Append. p. 479.

PYROTECHNY, the art of fire, or the fcience which teaches the application and management of fire in feveral operations.

Pyrotechny is of two kinds, military and chemical.

Military PYROTECHNY, is the fcience of artificial fire-works, and fire-arms, teaching the ftructure and ufe both of thofe employed in war, as gunpowder, cannon, fhells, carcaffes, mines, fufees, &c ; and of thofe made for amufement, as rockets, ftars, ferpents, &c.

Some call Pyrotechny by the name Artillery; though that word is ufually confined to the inftruments employed in war. Others choofe to call it Pyrobology, or rather Pyroballogy, or the art of miffile fires.

Wolfius has reduced Pyrotechny into a kind of mixt mathematical art. Indeed it will not allow of geometrical demonftrations ; but he brings it to tolerable rules and reafons ; whereas it had formerly been treated by authors at random, and without regard to any reafons at all. See the feveral articles CANNON, GUNPOWDER, ROCKET, SHELL, &c.

Chemical PYROTECHNY, is the art of managing and applying fire in diftillations, calcinations, and other operations of chemiftry.

Some reckon a third kind of Pyrotechny, viz, the art of fufing, refining, and preparing metals.

PYTHAGORAS, one of the greateft philofophers of antiquity, was born about the 47th Olympiad, or 590 years before Chrift. His father's principal refidence was at Samos, but being a travelling merchant, his fon Pythagoras was born at Sidon in Syria; but foon returning home again, our philofopher was brought up at Samos, where he was educated in a manner that was anfwerable to the great hopes that were conceived of him. He was called " the youth with a fine head of hair;" and from the great qualities that foon appeared in him, he was regarded as a good genius fent into the world for the benefit of mankind.

Samos however afforded no philofophers capable of fatisfying his thirft for knowledge; and therefore, at 18 years of age, he refolved to travel in queft of them elfewhere. The fame of Pherecydes drew him firft to the ifland of Syros : from hence he went to Miletus, where he converfed with Thales. He then travelled to Phœnicia, and ftayed fome time at Sidon, the place of his birth ; and from hence he paffed into Egypt, where Thales and Solon had been before him.

Having fpent 25 years in Egypt, to acquire all the learning and knowledge he could procure in that country, with the fame view he travelled through Chaldea, and vifited Babylon. Returning after fome time, he went to Crete ; and from hence to Sparta, to be inftructed in the laws of Minos and Lycurgus. He then returned to Samos; which, finding under the tyranny of Polycrates, he quitted again, and vifited the feveral countries of Greece. Paffing through Peloponnefus, he ftopped at Phlius, where Leo then reigned ; and in his converfation with that prince, he fpoke with fo much eloquence and wifdom, that Leo was at once ravifhed and furprifed.

From Peloponnefus he went into Italy, and paffed fome time at Heraclea, and at Tarentum, but made his chief refidence at Croton ; where, after reforming the manners of the citizens by preaching, and eftablifhing the city by wife and prudent counfels, he opened a fchool, to difplay the treafures of wifdom and learning he poffeffed. It is not to be wondered, that he was foon attended by a crowd of difciples, who repaired to him from different parts of Greece and Italy.

He gave his fcholars the rules of the Egyptian priefts, and made them pafs through the aufterities which he himfelf had endured. He at firft enjoined them a five years filence in the fchool, during which they were only to hear ; after which, leave was given them to ftart queftions, and to propofe doubts, under the caution however, to fay, " not a little in many words, but much in a few." Having gone through their probation, they were obliged, before they were admitted, to bring all their fortune into the common ftock, which was managed by perfons chofen on purpofe, and called œconomifts, and the whole community had all things in common.

The neceffity of concealing their myfteries induced the Egyptians to make ufe of three forts of ftyles, or ways of expreffing their thoughts; the fimple, the hieroglyphical,

hieroglyphical, and the symbolical. In the simple, they spoke plainly and intelligibly, as in common conversation; in the hieroglyphical, they concealed their thoughts under certain images and characters; and in the symbolical, they explained them by short expressions, which, under a sense plain and simple, included another wholly figurative. Pythagoras borrowed these three different ways from the Egyptians, in all the instructions he gave; but chiefly imitated the symbolical style, which he thought very proper to inculcate the greatest and most important truths: for a symbol, by its double sense, the proper and the figurative, teaches two things at once; and nothing pleases the mind more, than the double image it represents to our view.

In this manner Pythagoras delivered many excellent things concerning God and the human soul, and a great variety of precepts, relating to the conduct of life, political as well as civil; he made also some considerable discoveries and advances in the arts and sciences. Thus, among the works ascribed to him, there are not only books of physic, and books of morality, like that contained in what are called his *Golden Verses*, but treatises on politics and theology. All these works are lost: but the vastness of his mind appears from the wonderful things he performed. He delivered, as antiquity relates, several cities of Italy and Sicily from the yoke of slavery; he appeased seditions in others; and he softened the manners, and brought to temper the most savage and unruly spirits, of several people and tyrants. Phalaris, the tyrant of Sicily, it is said, was the only one who could withstand the remonstrances of Pythagoras; and he it seems was so enraged at his discourses, that he ordered him to be put to death. But though the lectures of the philosopher could make no impression on the tyrant, yet they were sufficient to re-animate the Sicilians, and to put them upon a bold action. In short, Phalaris was killed the same day that he had fixed for the death of the philosopher.

Pythagoras had a great veneration for marriage; and therefore himself married at Croton, a daughter of one of the chief men of that city, by whom he had two sons and a daughter: one of the sons succeeded his father in the school, and became the master of Empedocles: the daughter, named Damo, was distinguished both by her learning and her virtues, and wrote an excellent commentary upon Homer. It is related, that Pythagoras had given her some of his writings, with express commands not to impart them to any but those of his own family; to which Damo was so scrupulously obedient, that even when she was reduced to extreme poverty, she refused a great sum of money for them.

From the country in which Pythagoras thus settled and gave his instructions, his society of disciples was called the Italic sect of philosophers, and their reputation continued for some ages afterwards, when the Academy and the Lycæum united to obscure and swallow up the Italic sect. Pythagoras's disciples regarded the words of their master as the oracles of a god; his authority alone, though unsupported by reason, passed with them for reason itself: they looked on him as the most perfect image of God among men. His house was called the temple of Ceres, and his court yard the temple of the Muses: and when he went into towns, it was said he

2

went thither, " not to teach men, but to heal them."

Pythagoras however was persecuted by bad men in the last years of his life; and some say he was killed in a tumult raised by them against him; but according to others, he died a natural death, at 90 years of age, about 497 years before Christ.

Beside the high respect and veneration the world has always had for Pythagoras, on account of the excellence of his wisdom, his morality, his theology, and politics, he was renowned as learned in all the sciences, and a considerable inventor of many things in them; as arithmetic, geometry, astronomy, music, &c. In arithmetic, the common multiplication table is, to this day, still called Pythagoras's table. In geometry, it is said he invented many theorems, particularly these three; 1st, Only three polygons, or regular plane figures, can fill up the space about a point, viz, the equilateral triangle, the square, and the hexagon: 2d, The sum of the three angles of every triangle, is equal to two right angles: 3d, In any right-angled triangle, the square on the longest side, is equal to both the squares on the two shorter sides: for the discovery of this last theorem, some authors say he offered to the gods a hecatomb, or a sacrifice of a hundred oxen; Plutarch however says it was only one ox, and even that is questioned by Cicero, as inconsistent with his doctrine, which forbade bloody sacrifices: the more accurate therefore say, he sacrificed an ox made of flour, or of clay; and Plutarch even doubts whether such sacrifice, whatever it was, was made for the said theorem, or for the area of the parabola, which it was said Pythagoras also found out.

In astronomy his inventions were many and great. It is reported he discovered, or maintained the true system of the world, which places the sun in the centre, and makes all the planets revolve about him; from him it is to this day called the old or Pythagorean system; and is the same as that lately revived by Copernicus. He first discovered, that Lucifer and Hesperus were but one and the same, being the planet Venus, though formerly thought to be two different stars. The invention of the obliquity of the zodiac is likewise ascribed to him. He first gave to the world the name Κοσμ⊙, *Kosmos*, from the order and beauty of all things comprehended in it; asserting that it was made according to musical proportion: for as he held that the sun, by him and his followers termed the fiery globe of unity, was seated in the midst of the universe, and the earth and planets moving around him, so he held that the seven planets had an harmonious motion, and their distances from the sun corresponded to the musical intervals or divisions of the monochord.

Pythagoras and his followers held the transmigration of souls, making them successively occupy one body after another: on which account they abstained from flesh, and lived chiefly on vegetables.

PYTHAGORAS's *Table*, the same as the multiplication-table; which see.

PYTHAGOREAN, or PYTHAGORIC *System*, among the Ancients, was the same as the Copernican system among the Moderns. In this system, the sun is supposed at rest in the centre, with the earth and all the planets revolving about him, each in their orbits. See SYSTEM.

It

It was fo called, as having been maintained and cultivated by Pythagoras, and his followers; not that it was invented by him, for it was much older.

PYTHAGOREAN *Theorem*, is that in the 47th propofition of the 1ft book of Euclid's Elements; viz, that in a right-angled triangle, the fquare of the longeft fide, is equal to the fum of both the fquares of the two fhorter fides. It has been faid that Pythagoras offered a hecatomb, or facrifice of 100 oxen, to the gods, for infpiring him with the difcovery of fo remarkable a property.

PYTHAGOREANS, a fect of ancient philofophers, who followed the doctrines of Pythagoras. They were otherwife called the Italic fect, from the circumftance of his having fettled in Italy. Out of his fchool proceeded the greateft philofophers and legiflators, Zaleucus, Charondas, Archytas, &c. See the article PYTHAGORAS.

PYXIS *Nautica*, the feaman's compafs.

Q.

QUADRAGESIMA, a denomination given to the time of Lent, from its confifting of about 40 days; commencing on Afh Wednefday.

QUADRAGESIMA *Sunday*, is the firft Sunday in Lent, or the 1ft Sunday after Afh Wednefday.

QUADRANGLE, or QUADRANGULAR *figure*, in Geometry, is a plane figure having four angles; and confequently four fides alfo.

To the clafs of Quadrangles belong the fquare, parallelogram, trapezium, rhombus, and rhomboides.—A fquare is a regular Quadrangle; a trapezium an irregular one.

QUADRANT, in Geometry, is either the quarter or 4th part of a circle, or the 4th part of its circumference; the arch of which therefore contains 90 degrees.

QUADRANT alfo denotes a mathematical inftrument, of great ufe in aftronomy and navigation, for taking the altitudes of the fun and ftars, as alfo taking angles in furveying, heights-and-diftances, &c.

This inftrument is varioufly contrived, and furnifhed with different apparatus, according to the various ufes it is intended for; but they have all this in common, that they confift of the quarter of a circle, whofe limb or arch is divided into 90° &c. Some have a plummet fufpended from the centre, and are furnifhed either with plain fights, or a telefcope, to look through.

The principal and moft ufeful Quadrants, are the common Surveying Quadrant, the Aftronomical Quadrant, Adams's Quadrant, Cole's Quadrant, Collins's or Sutton's Quadrant, Davis's Quadrant, Gunter's Quadrant, Hadley's Quadrant, the Horodictical Quadrant, and the Sinical Quadrant, &c. Of thefe in their order.

1. *The Common, or Surveying* QUADRANT.—This inftrument ABC, fig. 1, pl. 24, is made of brafs, or wood, &c; the limb or arch of which BC is divided into 90°, and each of thefe farther divided into as many equal parts as the fpace will allow, either diagonally or otherwife. On one of the radii AC, are fitted two

moveable fights; and to the centre is fometimes alfo annexed a label, or moveable index AD, bearing two other fights; but inftead of thefe laft fights, there is fometimes fitted a telefcope. Alfo from the centre hangs a thread with a plummet; and on the under fide or face of the inftrument is fitted a ball and focket, by means of which it may be put into any pofition. The general ufe of it is for taking angles in a vertical plane, comprehended under right lines going from the centre of the inftrument, one of which is horizontal, and the other is directed to fome vifible point. But befides the parts above defcribed, there is often added on the face, near the centre, a kind of compartment EF, called a Quadrat, or Geometrical Square, which is a kind of feparate inftrument, and is particularly ufeful in Altimetry and Longimetry, or Heights-and-Diftances.

This Quadrant may be ufed in different fituations; in each of them, the plane of the inftrument muft be fet parallel to that of the eye and the objects whofe angular diftance is to be taken. Thus, for obferving heights or depths, its plane muft be difpofed vertically, or perpendicular to the horizon; but to take horizontal angles or diftances, its plane muft be difpofed parallel to the horizon.

Again, heights and diftances may be taken two ways, viz, by means of the fixed fights and plummet, or by the label; as alfo, either by the degrees on the limb, or by the Quadrat. Thus, fig. 2 pl. 24 fhews the manner of taking an angle of elevation with this Quadrant; the eye is applied at C, and the inftrument turned vertically about the centre A, till the object R be feen through the fights on the radius AC; then the angle of elevation RAH, made with the horizontal line KAH, is equal to the angle BAD, made by the plumb line and the other radius of the Quadrant, and the quantity of it is fhewn by the degrees in the arch BD cut off by the plumb line AD.

See the ufe of the inftrument in my Menfuration, under the fection of Heights-and-Diftances.

2. *The Aftronomical* QUADRANT, is a large one, ufually

ally made of brafs or iron bars ; having its limb EF (fig. 3 pl. 24) nicely divided, either diagonally or otherwife, into degrees, minutes, and feconds, if room will permit, and furnifhed either with two pair of plain fights or two telefcopes, one on the fide of the Quadrant at AB, and the other CD moveable about the centre by means of the fcrew G. The dented wheels I and H ferve to direct the inftrument to any object or phenomenon.

The application of this ufeful inftrument, in taking obfervations of the fun, planets, and fixed ftars, is obvious ; for being turned horizontally upon its axis, by means of the telefcope AB, till the object is feen through the moveable telefcope, then the degrees &c cut by the index, give the altitude &c required.

3. *Cole's* QUADRANT, is a very ufeful inftrument, invented by Mr. Benjamin Cole. It confifts of fix parts, viz, the ftaff AB (fig. 11 pl. 24) ; the quadrantal arch DE ; three vanes A, B, C ; and the vernier FG. The ftaff is a bar of wood about 2 feet long, an inch and a quarter broad, and of a fufficient thicknefs to prevent it from bending or warping. The quadrantal arch is alfo of wood ; and is divided into degrees and 3d parts of degrees, to a radius of about 9 inches ; and to its extremities are fitted two radii, which meet in the centre of the Quadrant by a pin, about which it eafily moves. The fight-vane A is a thin piece of brafs, near two inches in height, and one broad, fet perpendicularly on the end of the ftaff A, by means of two fcrews paffing through its foot. In the middle of this vane is drilled a fmall hole, through which the coincidence or meeting of the horizon and folar fpot is to be viewed. The horizon-vane B is about an inch broad, and 2 inches and a half high, having a flit cut through it of near an inch long, and a quarter of an inch broad ; this vane is fixed in the centre-pin of the inftrument, in a perpendicular pofition, by means of two fcrews paffing through its foot, by which its pofition with refpect to the fight-vane is always the fame, their angle of inclination being equal to 45 degrees. The fhade-vane C is compofed of two brafs plates. The one, which ferves as an arm, is about 4½ inches long, and ¾ of an inch broad, being pinned at one end to the upper limb of the Quadrant by a fcrew, about which it has a fmall motion ; the other end lies in the arch, and the lower edge of the arm is directed to the middle of the centre-pin : the other plate, which is properly the vane, is about 2 inches long, being fixed perpendicularly to the other plate, at about half an inch diftance from that end next the arch ; this vane may be ufed either by its fhade, or by the folar fpot caft by a convex lens placed in it. And becaufe the wood-work is often fubject to warp or twift, therefore this vane may be rectified by means of a fcrew, fo that the warping of the inftrument may occafion no error in the obfervation, which is performed in the following manner : Set the line G on the vernier againft a degree on the upper limb of the Quadrant, and turn the fcrew on the backfide of the limb forward or backward, till the hole in the fight-vane, the centre of the glafs, and the funk fpot in the horizon-vane, lie in a right line.

To find the Sun's Altitude by this inftrument. Turn your back to the fun, holding the ftaff of the inftru-

ment with the right hand, fo that it be in a vertical plane paffing through the fun ; apply one eye to the fight-vane, looking through that and the horizon-vane till the horizon be feen ; with the left hand flide the quadrantal arch upwards, till the folar fpot or fhade, caft by the fhade-vane, fall directly upon the fpot or flit in the horizon-vane ; then will that part of the quadrantal arch, which is raifed above G or S (according as the obfervation refpects either the folar fpot or fhade) fhew the altitude of the fun at that time. But for the meridian altitude, the obfervation muft be continued, and as the fun approaches the meridian, the fea will appear through the horizon-vane, which completes the obfervation ; and the degrees and minutes, counted as before, will give the fun's meridian altitude : or the degrees counted from the lower limb upwards will give the zenith diftance.

4. *Adams's* QUADRANT, differs only from Cole's Quadrant, juft defcribed, in having an horizontal vane, with the upper part of the limb lengthened ; fo that the glafs, which cafts the folar fpot on the horizon-vane, is at the fame diftance from the horizon-vane as the fight-vane at the end of the index.

5. *Collins's* or *Sutton's* QUADRANT, (fig. 8 pl. 24) is a ftereographic projection of one quarter of the fphere between the tropics, upon the plane of the ecliptic, the eye being in its north pole ; and fitted to the latitude of London. The lines running from right to left, are parallels of altitude ; and thofe croffing them are azimuths. The fmaller of the two circles, bounding the projection, is one quarter of the tropic of Capricorn ; and the greater is a quarter of the tropic of Cancer. The two ecliptics are drawn from a point on the left edge of the Quadrant, with the characters of the figns upon them ; and the two horizons are drawn from the fame point. The limb is divided both into degrees and time ; and by having the fun's altitude, the hour of the day may here be found to a minute. The quadrantal arches next the centre contain the calendar of months ; and under them, in another arch, is the fun's declination. On the projection are placed feveral of the moft remarkable fixed ftars between the tropics ; and the next below the projection is the Quadrant and line of fhadows.

To find the Time of the Sun's Rifing or Setting, his Amplitude, his Azimuth, Hour of the Day, &c. by this Quadrant. Lay the thread on the day and the month, and bring the bead to the proper ecliptic, either of fummer or winter, according to the feafon, which is called *rectifying ;* then by moving the thread bring the bead to the horizon, in which cafe the thread will cut the limb in the point of the time of the fun's rifing or fetting before or after 6 : and at the fame time the bead will cut the horizon in the degrees of the fun's amplitude.——Again, obferving the fun's altitude with the Quadrant, and fuppofing it found to be 45° on the 5th of May, lay the thread over the 5th of May ; then bring the bead to the fummer ecliptic, and carry it to the parallel of altitude 45° ; in which cafe the thread will cut the limb at 55° 15′, and the hour will be feen among the hour-lines to be either 41m. paft 9 in the morning, or 19m. paft 2 in the afternoon.——Laftly, the bead fhews among the azimuths the fun's diftance from the fouth 50° 41′.

7

But

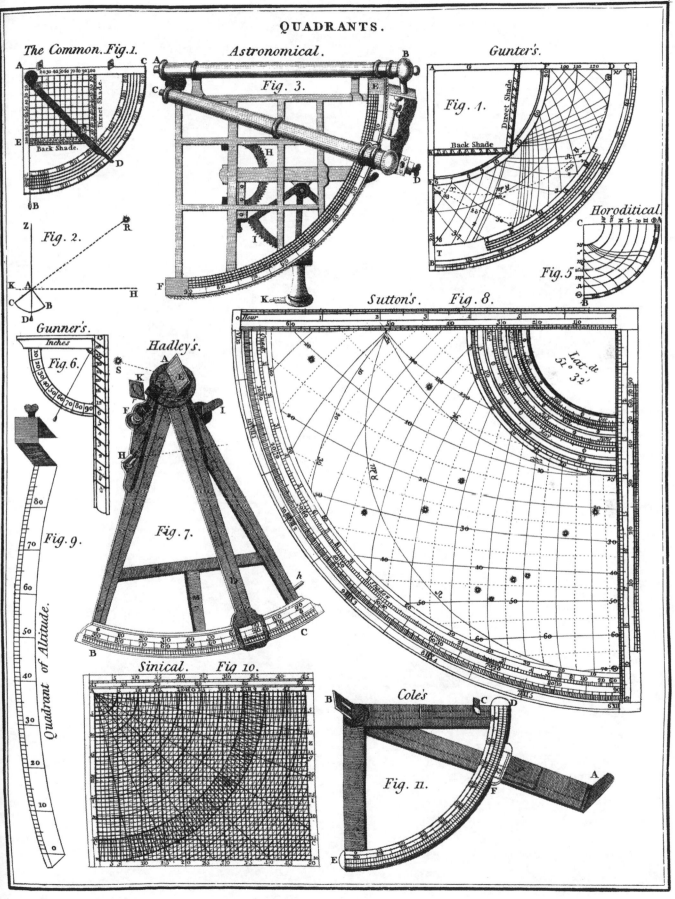

Plate XXIV.

QUADRANTS.

The Common. Fig.1.

Astronomical.

Fig. 3.

Gunter's.

Fig. 1.

Direct Shade.

Back Shade.

Fig. 2.

Horoditical.

Fig. 5.

Gunner's.

Inches

Fig. 6.

Hadley's.

Sutton's.   Fig. 8.

Hour

Lat. de 51° 32'

Fig. 9.

Quadrant of Altitude.

Fig. 7.

Sinical.   Fig 10.

Cole's

Fig. 11.

But note, that if the fun's altitude be lefs than what it is at 6 o'clock, the operation muft be performed among thofe parallels above the upper horizon; the bead being rectified to the winter ecliptic.

6. *Davis's* QUADRANT, the fame as the BACK-STAFF; which fee.

7. *Gunner's* QUADRANT, (fig. 6 pl. 24), fometimes called the *Gunner's Square*, is ufed for elevating and pointing cannon, mortars, &c, and confifts of two branches either of wood or brafs, between which is a quadrantal arch divided into 90°, and furnifhed with a thread and plummet.

The ufe of this inftrument is very eafy; for if the longer branch, or bar, be placed in the mouth of the piece, and it be elevated till the plummet cut the degree neceffary to hit a propofed object, the thing is done.

Sometimes on the fides of the longer bar, are noted the divifion of diameters and weights of iron balls, as alfo the bores of pieces.

8. *Gunter's* QUADRANT, fo called from its inventor Edmund Gunter, (fig. 4 pl. 24) befide the apparatus of other Quadrants, has a ftereographic projection of the fphere on the plane of the equinoctial; and alfo a calendar of the months, next to the divifions of the limb; by which, befide the common purpofes of other Quadrants, feveral ufeful queftions in aftronomy, &c, are eafily refolved.

Ufe of Gunter's Quadrant. — 1. To find the fun's meridian altitude for any given day, or converfely the day of the year anfwering to any given meridian altitude. Lay the thread to the day of the month in the fcale next the limb; then the degree it cuts in the limb is the fun's meridian altitude. And, contrariwife, the thread being fet to the meridian altitude, it fhews the day of the month.

2. To find the hour of the day. Having put the bead, which flides on the thread, to the fun's place in the ecliptic, obferve the fun's altitude by the Quadrant; then if the bead be laid over the fame in the limb, the bead will fall upon the hour required. On the contrary, laying the bead on a given hour, having firft rectified or fet it to the fun's place, the degree cut by the thread on the limb gives the altitude.

Note, the bead may be rectified otherwife, by bringing the thread to the day of the month, and the bead to the hour-line of 12.

3. To find the fun's declination from his place given; and the contrary. Bring the bead to the fun's place in the ecliptic, and move the thread to the line of declination E T, fo fhall the bead cut the degree of declination required. On the contrary, the bead being adjufted to a given declination, and the thread moved to the ecliptic, the bead will cut the fun's place.

4. The fun's place being given, to find the right afcenfion; or contrariwife. Lay the thread on the fun's place in the ecliptic, and the degree it cuts' on the limb is the right afcenfion fought. And the converfe.

5. The fun's altitude being given, to find his azimuth; and contrariwife. Rectify the bead for the time, as in the fecond article, and obferve the fun's altitude; bring the thread to the complement of that altitude; then the bead will give the azimuth fought, among the azimuth-lines.

.9. *Hadley's* QUADRANT, (fig. 7 pl. 24) fo called from its inventor John Hadley, Efq, is now univerfally ufed as the beft of any for nautical and other obfervations.

It feems the firft idea of this excellent inftrument was fuggefted by Dr. Hooke; for Dr. Sprat, in his Hiftory of the Royal Society, pa. 246, mentions the invention of a new inftrument for taking angles by reflection, by which means the eye at once fees the two objects both as touching the fame point, though diftant almoft to a femicircle; which is of great ufe for making exact obfervations at fea. This inftrument is defcribed and illuftrated by a figure in Hooke's Pofthumous works, pa. 503. But as it admitted of only one reflection, it would not anfwer the purpofe. The matter however was at laft effected by Sir Ifaac Newton, who communicated to Dr. Halley a paper of his own writing, containing the defcription of an inftrument with two reflections, which foon after the doctor's death was found among his papers by Mr. Jones, by whom it was communicated to the Royal Society, and it was publifhed in the Philof Tranf. for the year 1742. See alfo the Abridg. vol. 8, pa. 129. How it happened that Dr. Halley never mentioned this in his lifetime, is hard to fay; but it is very extraordinary; more efpecially as Mr. Hadley had defcribed, in the Tranfac. for 1731, his inftrument, which is conftructed on the fame principles. See alfo Abr. vol. 6, pa. 139. Mr. Hadley, who was well acquainted with Sir Ifaac Newton, might have heard him fay, that Dr. Hooke's propofal could be effected by means of a double reflection; and perhaps in confequence of this hint, he might apply himfelf, without any previous knowledge of what Newton had actually done, to the conftruction of his inftrument. Mr. Godfrey too, of Pennfylvania, had recourfe to a fimilar expedient; for which reafon fome gentlemen of that colony have afcribed the invention of this excellent inftrument to him. The truth may probably be, that each of thefe gentlemen difcovered the method independent of one another. See Abr. Philof. Tranf. vol. 8, pa. 366; alfo Tranf. of the American Society, vol. 1, pa. 21 Appendix.

This inftrument confifts of the following particulars: 1. An octant, or the 8th part of a circle, ABC. 2. An index D. 3. The fpeculum E. 4. Two horizontal glaffes, F, G. 5. Two fcreens, K and K. 6. Two fight-vanes, H and I.

The octant confifts of two radii, AB, AC, ftrengthened by the braces L, M, and the arch BC; which, though containing only 45°, is neverthelefs divided into 90 primary divifions, each of which ftands for degrees, and are numbered 0, 10, 20, 30, &c, to 90; beginning at each end of the arch for the convenience of numbering both ways, either for altitudes or zenith diftances: alfo each degree is fubdivided into minutes, by means of a vernier. But the number of thefe divifions varies with the fize of the inftrument.

The index D, is a flat bar, moveable about the centre of the inftrument; and that part of it which flides over the graduated arch, BC, is open in the middle, with Vernier's fcale on the lower part of it; and

and underneath is a screw, serving to fasten the index against any division.

The speculum E is a piece of flat glass, quicksilvered on one side, set in a brass box, and placed perpendicular to the plane of the instrument, the middle part of the former coinciding with the centre of the latter: and because the speculum is fixed to the index, the position of it will be altered by the moving of the index along the arch. The rays of an observed object are received on the speculum, and from thence reflected on one of the horizon glasses, F or G; which are two small pieces of looking glass placed on one of the limbs, their faces being turned obliquely to the speculum, from which they receive the reflected rays of objects. This glass F has only its lower part silvered, and set in brass-work; the upper part being left transparent to view the horizon. The glass G has in its middle a transparent slit, through which the horizon is to be seen. And because the warping of the materials, and other accidents, may distend them from their true situation, there are three screws passing through their feet, by which they may be easily replaced.

The screens are two pieces of coloured glass, set in two square brass frames K and K, which serve as screens to take off the glare of the sun's rays, which would otherwise be too strong for the eye; the one is tinged much deeper than the other; and as they both move on the same centre, they may be both or either of them used: in the situation they appear in the figure, they serve for the horizon-glass F; but when they are wanted for the horizon-glass G, they must be taken from their present situation, and placed on the Quadrant above G.

The sight-vanes are two pins, H and I, standing perpendicularly to the plane of the instrument: that at H has one hole in it, opposite to the transparent slit in the horizon-glass G; the other, at I, has two holes in it, the one opposite to the middle of the transparent part of the horizon-glass F, and the other rather lower than the quick-silvered part: this vane has a piece of brass on the back of it, which moves round a centre, and serves to cover either of the holes.

Of the Observations.—There are two sorts of observations to be made with this instrument: the one is when the back of the observer is turned towards the object, and therefore called the *back observation*; the other when the face of the observer is turned towards the object, which is called the *fore-observation.*

To Rectify the Instrument for the Fore-observation: Slacken the screw in the middle of the handle behind the glass F; bring the index close to the button *h*; hold the instrument in a vertical position, with the arch downwards; look through the right-hand hole in the vane I, and through the transparent part of the glass F, for the horizon; and if it lie in the same right line with the image of the horizon seen on the silvered part, the glass F is rightly adjusted; but if the two horizontal lines disagree, turn the screw which is at the end of the handle backward or forward, till those lines coincide; then fasten the middle screw of the handle, and the glass is rightly adjusted.

To take the Sun's Altitude by the Fore-observation. Having fixed the screens above the horizon-glass F, and suited them proportionally to the strength of the sun's rays, turn your face towards the sun, holding the instrument with your right hand, by the braces L and M, in a vertical position, with the arch downward; put your eye close to the right-hand hole in the vane I, and view the horizon through the transparent part of the horizon glass F, at the same time moving the index D with the left hand, till the reflex solar spot coincides with the line of the horizon; then the degrees counted from C, or that end next your body, will give the sun's altitude at that time, observing to add or subtract 16 minutes according as the upper or lower edge of the sun's reflex image is made use of.

But to get the sun's meridian altitude, which is the thing wanted for finding the latitude; the observations must be continued; and as the sun approaches the meridian the index D must be continually moved towards B, to maintain the coincidence between the reflex solar spot and the horizon; and consequently as long as this motion can maintain the same coincidence, the observation must be continued, till the sun has reached the meridian, and begins to descend, when the coincidence will require a retrograde motion of the index, or towards C; and then the observation is finished, and the degrees counted as before will give the sun's meridian altitude, or those from B will give the zenith distance; observing to add the semi-diameter, or 16′, when his lower edge is brought to the horizon; or to subtract 16′, when the horizon and upper edge coincide.

To take the Altitude of a Star by the Fore-observation. Through the vane H, and the transparent slit in the glass G, look directly to the star; and at the same time move the index, till the image of the horizon behind you, being reflected by the great speculum, be seen in the silvered part of G, and meet the star; then will the index shew the degrees of the star's altitude.

To Rectify the Instrument for the Back-observation. Slacken the screw in the middle of the handle, behind the glass G; turn the button *h* on one side, and bring the index as many degrees before o as is equal to double the dip of the horizon at your height above the water; hold the instrument vertical, with the arch downward; look through the hole of the vane H; and if the horizon, seen through the transparent slit in the glass G, coincide with the image of the horizon seen in the silvered part of the same glass, then the glass G is in its proper position; but if not, set it by the handle, and fasten the screw as before.

To take the Sun's Altitude by the Back-observation. Put the stem of the screens K and K into the hole *r*, and in proportion to the strength or faintness of the sun's rays, let either one or both or neither of the frames of those glasses be turned close to the face of the limb; hold the instrument in a vertical position, with the arch downward, by the braces L and M, with the left hand; turn your back to the sun, and put one eye close to the hole in the vane H, observing the horizon through the transparent slit in the horizon glass G; with the right hand move the index D, till the reflected image of the sun be seen in the silvered part of the glass G, and in a right line with the horizon; swing your body to and fro, and if the observation be well made, the sun's image will be observed to brush the horizon, and the degrees reckoned from C, or that part of the arch farthest from your body

body, will give the fun's altitude at the time of obfervation; obferving to add 16´ or the fun's femidiameter if the fun's upper edge be ufed, and fubtract the fame for the lower edge.

The directions juft given, for taking altitudes at fea, would be fufficient, but for two corrections that are neceffary to be made before the altitude can be accurately determined, viz, one on account of the obferver's eye being raifed above the level of the fea, and the other on account of the refraction of the atmofphere, efpecially in fmall altitudes.

The following tables therefore fhew the corrections to be made on both thefe accounts.

TABLE I. Dip of the Horizon of the Sea.		TABLE II. Refractions of the Stars &c in Altitude.			
Height of the Eye.	Dip of the Horizon.	Appar. Altitude in Deg.	Refraction.	Appar. Altitude in Deg.	Refraction.
Feet.	′ ″	°	′ ″	°	′ ″
1	0 57	0	33 0	11	4 47
2	1 21	¼	30 35	12	4 23
3	1 39	½	28 22	15	3 30
5	2 8	1	24 29	20	2 35
10	3 1	2	18 35	25	2 2
15	3 42	3	14 36	30	1 38
20	4 16	4	11 51	35	1 21
25	4 46	5	9 54	40	1 8
30	5 14	6	8 29	45	0 57
35	5 39	7	7 20	50	0 48
40	6 2	8	6 29	60	0 33
45	6 24	9	5 48	70	0 21
50	6 44	10	5 15	80	0 10

General Rules for thefe Corrections.

1. In the fore-obfervations, add the fum of both corrections to the obferved zenith diftance, for the true zenith diftance : or fubtract the faid fum from the obferved altitude, for the true one. 2. In the back-obfervation, add the dip and fubtract the refraction, for altitudes ; and for zenith diftances, do the contrary, viz, fubtract the dip, and add the refraction.

Example. By a back obfervation, the altitude of the fun's lower edge was found by Hadley's Quadrant to be 25° 12´ ; the eye being 30 feet above the horizon. By the tables, the dip on 30 feet is 5´ 14″, and the refraction on 25° 12´ is 2´ 1″. Hence

Appar. alt. lower limb	25°	12′	0″
Sun's femidiameter, fub.	0	16	0
Appar. alt. of centre	24	56	0
Dip. of horizon, add	0	5	14
	25	1	14
Refraction, fubtract	0	2	1
True alt. of centre	24	59	13

In the cafe of the moon, befides the two corrections above, another is to be made for her parallaxes. But

for all thefe particulars, fee the Requifite Tables for the Nautical Almanac, alfo Robertfon's Navigation, vol. 2, pa. 340 &c, edit. 1780.

10. *Horodictical* QUADRANT, a pretty commodious inftrument, and is fo called from its ufe in telling the hour of the day. Its conftruction is as follows. From the centre of the Quadrant C, (fig. 5 pl. 24), whofe limb AB is divided into 90°, defcribe feven concentric circles at any intervals ; and to thefe add the figns of the zodiac, in the order reprefented in the figure. Then, applying a ruler to the centre C and the limb AB, mark upon the feveral parallels the degrees correfponding to the altitude of the fun, when in them, for the given hours ; connect the points belonging to the fame hour with a curve line, to which add the number of the hour. To the radius CA fit a couple of fights, and to the centre of the Quadrant C tie a thread with a plummet, and upon the thread a bead to flide.

Now if the bead be brought to the parallel in which the fun is, and the Quadrant be directed to the fun, till a vifual ray pafs through the fights, the bead will fhew the hour. For the plummet, in this fituation, cuts all the parallels in the degrees correfponding to the fun's altitude. And fince the bead is in the parallel which the fun defcribes, and becaufe hour-lines pafs through the degrees of altitude to which the fun is elevated every hour, therefore the bead muft fhew the prefent hour.

11. *Sinical* QUADRANT, is one of fome ufe in Navigation. It confifts of feveral concentric quadrantal arches, divided into 8 equal parts by means of radii, with parallel right lines croffing each other at right angles. Now any one of the arches, as BC, (fig. 10 pl. 24) may reprefent a Quadrant of any great circle of the fphere, but is chiefly ufed for the horizon or meridian. If then BC be taken for a Quadrant of the horizon, either of the fides, as AB, may reprefent the meridian ; and the other fide, AC, will reprefent a parallel, or line of eaft-and-weft ; all the other lines, parallel to AB, will be alfo meridians ; and all thofe parallel to AC, eaft-and-weft lines, or parallels. Again, the eight fpaces into which the arches are divided by the radii, reprefent the eight points of the compafs in a quarter of the horizon ; each containing 11° 15′. The arch BC is likewife divided into 90°, and each degree fubdivided into 12, diagonalwife. To the centre is fixed a thread, which, being laid over any degree of the Quadrant, ferves to divide the horizon.

If the finical Quadrant be taken for a fourth part of the meridian, one fide of it, AB, may be taken for the common radius of the meridian and equator ; and then the other, AC, will be half the axis of the world. The degrees of the circumference, BC, will reprefent degrees of latitude ; and the parallels to the fide AB, affumed from every point of latitude to the axis AC, will be radii of the parallels of latitude, as likewife the cofine of thofe latitudes.

Hence, fuppofe it be required to find the degrees of longitude contained in 83 of the leffer leagues in the parallel of 48°: lay the thread over 48° of latitude on the circumference, and count thence the 83 leagues on AB, beginning at A ; this will terminate in H, allow-

S f ing

ing every fmall interval four leagues. Then tracing out the parallel HE, from the point H to the thread; the part AE of the thread fhews that 125 greater or equinoctial leagues make 6° 15′; and therefore that the 83 leffer leagues AH, which make the difference of longitude of the courfe, and are equal to the radius of the parallel HE, make 6° 15′ of the faid parallel.

When the fhip fails upon an oblique courfe, fuch courfe, befide the north and fouth greater leagues, gives leffer leagues eafterly and wefterly, to be reduced to degrees of longitude of the equator. But thefe leagues being made neither on the parallel of departure, nor on that of arrival, but in all the intermediate ones, there muft be found a mean proportional parallel between them. To find this, there is on the inftrument a fcale of crofs latitudes. Suppofe then it were required to find a mean parallel between the parallels of 40° and 60°; take with the compaffes the middle between the 40th and 60th degree on the fcale: this middle point will terminate againft the 51ft degree, which is the mean parallel fought.

The chief ufe of the finical Quadrant, is to form upon it triangles fimilar to thofe made by a fhip's way with the meridians and parallels; the fides of which triangles are meafured by the equal intervals between the concentric Quadrants and the lines N and S, E and W: and every 5th line and arch is made deeper than the reft. Now fuppofe a fhip has failed 150 leagues north-eaft-by-north, or making an angle of 33° 45′ with the north part of the meridian: here are given the courfe and diftance failed, by which a triangle may be formed on the inftrument fimilar to that made by the fhip's courfe; and hence the unknown parts of the triangle may be found. Thus, fuppofing the centre A to reprefent the place of departure; count, by means of the concentric circles along the point the fhip failed on, viz. AD, 150 leagues: then in the triangle AED, fimilar to that of the fhip's courfe, find AE = difference of latitude, and DE = difference of longitude, which muft be reduced according to the parallel of latitude come to.

Sutton's QUADRANT. See *Collins's* QUADRANT.

12. QUADRANT *of Altitude*, (fig. 9 pl. 24) is an appendix to the artificial globe, confifting of a thin flip of brafs, the length of a quarter part of one of the great circles of the globe, and graduated. At the end, where the divifion terminates, is a nut riveted on, and furnifhed with a fcrew, by means of which the inftrument is fitted on the meridian, and moveable round upon the rivet to all points of the horizon, as reprefented in the figure referred to.—Its ufe is to ferve as a fcale in meafuring of altitudes, amplitudes, azimuths, &c.

QUADRANTAL *Triangle*, is a fpherical triangle, which has one fide equal to a quadrant or quarter part of a circle.

QUADRAT, called alfo *Geometrical Square*, and *Line of Shadows*: it is often an additional member on the face of Gunter's and Sutton's quadrants; and is chiefly ufeful in taking heights or depths. See my Menfuration, the chap. on Altimetry and Longimetry, or Heights and Diftances.

QUADRAT, in Aftrology, is the fame as quartile, being an afpect of the heavenly bodies when they are

9

diftant from each other a quadrant, or 90°, or 3 figns, and is thus marked □.

QUADRATIC *Equations*, in Algebra, are thofe in which the unknown quantity is of two dimenfions, or raifed to the 2d power. See EQUATION.

Quadratic equations are either fimple, or affected, that is compound.

A Simple QUADRATIC *equation*, is that which contains the 2d power only of the unknown quantity, without any other power of it: as $x^2 = 25$, or $y^2 = ab$. And in this cafe, the value of the unknown quantity is found by barely extracting the fquare root on both fides of the equation: fo in the equations above, it will be $x = \pm 5$, and $y = \pm \sqrt{ab}$; where the fign of the root of the known quantity is to be taken either plus or minus, for either of thefe may be confidered as the fign of the value of the root x, fince either of thefe, when fquared, make the fame fquare, $\overline{+5}\vert^2 = 25$, and $\overline{-5}\vert^2 = 25$ alfo; and hence the root of every quadratic or fquare, has two values.

Compound or *Affected* QUADRATICS, are thofe which contain both the 1ft and 2d powers of the unknown quantity; as $x^2 + ax = b$, or $x^{2n} - ax^n = \pm b$, where n may be of any value, and then x^n is to be confidered as the root or unknown quantity.

Affected quadratics are ufually diftinguifhed into three forms, according to the figns of the terms of the equation:

Thus, 1ft form, $x^2 + ax = b$,
2d form, $x^2 - ax = b$,
3d form, $x^2 - ax = -b$.

But the method of extracting the root, or finding the value of the unknown quantity x, is the fame in all of them. And that method is ufually performed by what is called completing the fquare, which is done by taking half the coefficient of the 2d term or fingle power of the unknown quantity, then fquaring it, and adding that fquare to both fides of the equation, which makes the unknown fide a complete fquare. Thus, in the equation $x^2 + ax = b$, the coefficient of the 2d term being a, its half is $\frac{1}{2}a$, the fquare of which is $\frac{1}{4}a^2$, and this added to both fides of the equation, it becomes $x^2 + ax + \frac{1}{4}a^2 = \frac{1}{4}a^2 + b$, the former fide of which is now a complete fquare. The fquare being thus completed, its root is next to be extracted; in order to which, it is to be obferved that the root on the unknown fide confifts of two terms, the one of which is always x the fquare root of the firft term of the equation, and the other part is $\frac{1}{2}a$ or half the coefficient of the 2d term: thus then the root of $x^2 + ax + \frac{1}{4}a^2$ the firft fide of the completed equation being $x + \frac{1}{2}a$, and the root of the other fide $\frac{1}{4}a^2 + b$ being $\pm \sqrt{a^2 + b}$, it follows that $x + \frac{1}{2}a = \pm \sqrt{\frac{1}{4}a^2 + b}$, and hence, by transfpofing $\frac{1}{2}a$, it is $x = -\frac{1}{2}a \pm \sqrt{\frac{1}{4}a^2 + b}$, the two values of x, or roots of the given equation $x^2 + a = b$. And thus is found the root, or value of x, in the three forms of equations above mentioned: thus,

1ft form $x = -\frac{1}{2}a \pm \sqrt{\frac{1}{4}a^2 + b}$,
2d form $x = +\frac{1}{2}a \pm \sqrt{\frac{1}{4}a^2 + b}$,
3d form $x = +\frac{1}{2}a \pm \sqrt{\frac{1}{4}a^2 - b}$.

Where

Where it is obfervable that, becaufe of the double fign ±, every form has two roots : in the 1ft and 2d forms thofe roots are the one pofitive and the other negative, the pofitive root being the lefs of the two in the 1ft form, but the greater in the 2d form ; and in the 3d form the roots are both pofitive. Again, the two roots of the 1ft and 2d forms, are always both of them real ; but in the 3d form, the two roots are either both real or both imaginary, viz, both real when $\frac{1}{4}a^2$ is greater than b, or both imaginary when $\frac{1}{4}a^2$ is lefs than b, becaufe in this cafe $\frac{1}{4}a^2 - b$ will be a negative quantity, the root of which is impoffible, or an imaginary quantity.

Example of the 1ft form, let $x^2 + 6x = 7$. Here then $a = 6$, and $b = 7$; then $x = -\frac{1}{2}a \pm \sqrt{\frac{1}{4}a^2 + b} = -3 \pm \sqrt{16} = -3 \pm 4 = +1$ or -7.

Example of the 2d form, let $x^2 - 6x = 7$. Here alfo $a = 6$, and $b = 7$; then $x = +\frac{1}{2}a \pm \sqrt{\frac{1}{4}a^2 + b} = +3 \pm \sqrt{16} = +3 \pm 4 = +7$ or -1; the fame two roots as before, with the figns changed.

Example of the 3d form, let $x^2 - 6x = -7$. Here again $a = 6$, and $b = 7$; then $x = +\frac{1}{2}a \pm \sqrt{\frac{1}{4}a^2 - b} = +3 \pm \sqrt{2}$, the two roots both real.

But if $x^2 - 6x = -11$; then $a = 6$, and $b = 11$, which gives x or $+\frac{1}{2}a \pm \sqrt{\frac{1}{4}a^2 - b} = +3 \pm \sqrt{-2}$, the two roots both imaginary.

All equations whatever that have only two different powers of the unknown quantity, of which the index of the one is juft double to that of the other, are refolved like Quadratics, by completing the fquare. Thus, the equation $x^4 + ax^2 = b$, by completing the fquare becomes $x^4 + ax^2 + \frac{1}{4}a^2 = \frac{1}{4}a^2 + b$; whence, extracting the root on both fides, $x^2 + \frac{1}{2}a = \pm \sqrt{\frac{1}{4}a^2 + b}$; therefore $x^2 = -\frac{1}{2}a \pm \sqrt{\frac{1}{4}a^2 + b}$, and confequently $x = \pm \sqrt{-\frac{1}{2}a \pm \sqrt{\frac{1}{4}a^2 + b}}$, where the root x has four values, becaufe the given equation $x^4 + ax^2 = b$ rifes to the 4th power. See EQUATION.

QUADRATRIX, or QUADRATIX, in Geometry, is a mechanical line, by means of which, right lines are found equal to the circumference of circles, or other curves, and of the parts of the fame. Or, more accurately, the *Quadratrix of a curve*, is a tranfcendental curve defcribed on the fame axis, the ordinates of which being given, the quadrature of the correfpondent parts in the other curve is likewife given. See CURVE.—Thus, for example, the curve AND may be

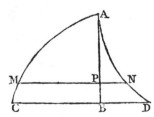

called the Quadratrix of the parabola AMC, when the area APMA bears fome fuch relation as the following to the abfcifs AP or ordinate PN, viz,

when APM = PN²,
or APM = AP × PN,
or APM = a × PN,

where a is fome given conftant quantity.

The moft diftinguifhed of thefe Quadratices are, thofe of Dinoftrates and of Tfchirnhaufen for the circle, and that of Mr. Perks for the hyperbola.

QUADRATRIX *of Dinoftrates*, is a curve AMD, by which the quadrature of the circle is effected, though not geometrically, but only mechanically. It is fo called from its inventor Dinoftrates ; and the genefis or defcription of it is as follows : Divide the quadrantal arc ANB into any number of equal parts, in the points N, n, n, &c ; and alfo the radius AC into the fame number of parts at the points P, p, p, &c. To the points of N, n, n, &c, draw the

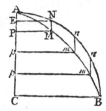

radii CN, Cn, &c ; and from the points P, p, &c, the parallels to CB, as PM, Pm, &c : then through all the points of interfection draw the curve AMmD, and it will be the Quadratrix of Dinoftrates.

Or the fame curve may be conceived as defcribed by a continued motion, thus : Conceive a radius CN to revolve with a uniform motion about the centre C, from the pofition AC to the pofition BC ; and at the fame time a ruler PM always moving uniformly parallel towards CB ; the two uniform motions being fo regulated that the radius and the ruler fhall arrive at the pofition BC at the fame time. For thus the continual interfection M, m, &c. of the revolving radius, and moving ruler, will defcribe the Quadratrix AMm &c. Hence,

1. For the *Equation of the Quadratrix* : Since, from the relation of the uniform motions, it is always, AB : AN :: AC : AP ; therefore if AB = a, AC = r, AP = x, and AN = z, it will be $a : z :: r : x$, or $ax = rz$, which is the equation of the curve.

Or, if s denote the fine NE of the arc AN, and y = PM the ordinate of the curve AM, its abfcifs AP being x ; then, by fimilar triangles, CE : CP :: EN : PM, that is, $\sqrt{r^2 - s^2} : r - x :: s : y$, and hence $y\sqrt{r^2 - s^2} = \overline{r - x} . s$, the equation of the curve. And when the relation between AB and AN is given, in terms of that between AC and AP, hence will be expreffed the relation between the fine EN and the radius CB, or s will be expreffed in terms of r and x ; and confequently the equation of the curve will be expreffed in terms of r, x, and y only.

2. The bafe of the Quadratrix CD is a third proportional to the quadrant AB and the radius AC or CB ; i. e. CD : CB :: CB : AB. Hence the rectification and quadrature of the circle.

3. A quadrantal arc DF defcribed with the centre C and radius CD, will be equal in length to the radius CA or CB.

4. CDF being a quadrant infcribed in the Quadratrix AMD, if the bafe CD be = 1, and the circular arc DG = x ; then in the area

$$CFMD = x - \frac{1}{9}x^3 - \frac{1}{225}x^5 - \frac{2}{6615}x^7$$

&c. See QUADRATURE.

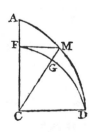

QUADRA-

QUADRATRIX of *Tſchirnhauſen*, is a tranſcendental curve AM*m*B by which the quadrature of the circle is likewiſe effected. This was invented by M. Tſchirnhauſen, and its geneſis, in imitation of that of Dinoſtrates, is as follows: Divide the quadrant ANC, and the radius AC, each into equal parts,

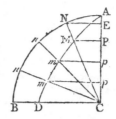

as before; and from the points P, *p*, &c, draw the lines PM, *pm*, &c, parallel to CB; alſo from the points N, *n*, &c, the lines NM, *nm*, &c, parallel to the other radius AC; ſo ſhall all the interſections M, *m*, &c, be in the curve of the Quadratrix AM*m*B.

Now for the Equation of this Quadratrix; it is, as before, AB : AN : : AC : AP,

or *a* : *z* : : *r* : *x*, or *ax* = *rz*.

Or, becauſe here *y* = PM = EN = *s*; therefore *s*, as before, expreſſed in terms of *r* and *x*, gives the equation of this Quadratrix in terms of *r*, *x*, and *y*, and that in a ſimpler form than the other. Thus, from the nature of the circle and the conſtruction of the Quadratrix, it is

$$y \text{ or } s = x + \frac{r^2 - x^2}{2 \cdot 3 r^2} A + \frac{3^2 r^2 - x^2}{4 \cdot 5 r^2} B + \frac{5^2 r^2 - x^2}{6 \cdot 7 r^2} C \text{ \&c,}$$

where A, B, C, &c, are the preceding terms; which is the equation of the curve or Quadratrix of Tſchirnhauſen.

By either Quadratix, it is evident that an arc or angle is eaſily divided into three, or any other number of equal parts; viz, by dividing the correſponding radius, or part of it, into the ſame number of equal parts: for AN is always the ſame part of AB, that AP is of AC.

QUADRATURE, in Aſtronomy, that aſpect or poſition of the moon when ſhe is 90° diſtant from the ſun. Or, the Quadratures or quarters are the two middle points of the moon's orbit between the points of conjunction and oppoſition; viz, the points of the 1ſt and 3d quarters; at which times the moon's face ſhews half full, being dichotomized or biſected.

The moon's orbit is more convex in the Quadratures than in the ſyzygies, and the greater axis of her orbit paſſes through the Quadratures, at which points alſo ſhe is moſt diſtant from the earth.—In the Quadratures, and within 35° of them, the apſes of the moon go backwards, or move in antecedentia; but in the ſyzygies the contrary.—When the nodes are in the Quadratures, the inclination of the moon's orbit is greateſt, but leaſt when they are in the ſyzygies.

QUADRATURE *Lines*, or *Lines of* QUADRATURE, are two lines often placed on Gunter's ſector. They are marked with the letter Q, and the figures 5, 6, 7, 8, 9, 10; of which Q denotes the ſide of a ſquare, and the figures denote the ſides of polygons of 5, 6, 7, &c ſides. Alſo S denotes the ſemidiameter of a circle, and 90 a line equal to the quadrant or 90° in circumference.

QUADRATURE, in Geometry, is the ſquaring of a figure, or reducing it to an equal ſquare, or finding a ſquare equal to the area of it.

The Quadrature of rectilineal figures falls under common geometry, or menſuration; as amounting to no more than the finding their areas, or ſuperficies; which are in effect their ſquares: which was fully effected by Euclid.

The QUADRATURE *of Curves*, that is, the meaſuring of their areas, or the finding a rectilineal ſpace equal to a propoſed curvilineal one, is a matter of much deeper ſpeculation; and makes a part of the ſublime or higher geometry. The lunes of Hypocrates are the firſt curves that were ſquared, as far as we know of The circle was attempted by Euclid and others before him: he ſhewed indeed the proportion of one circle to another, and gave a good method of approximating to the area of the circle, by deſcribing a polygon between any two concentric circles, however near their circumferences might be to each other. At this time the conic ſections were admitted in geometry, and Archimedes, perfectly, for the firſt time, ſquared the parabola, and he determined the relations of ſpheres, ſpheroids, and conoids, to cylinders and cones; and by purſuing the method of exhauſtions, or by means of inſcribed and circumſcribed polygons, he approximated to the periphery and area of the circle; ſhewing that the diameter is to the circumference nearly as 7 to 22, and the area of the circle to the ſquare of the diameter as 11 to 14 nearly. Archimedes likewiſe determined the relation between the circle and ellipſe, as well as that of their ſimilar parts: It is probable too that he attempted the hyperbola; but it is not likely that he met with any ſucceſs, ſince approximations to its area are all that can be given by the various methods that have ſince been invented. Beſide theſe figures, he left a treatiſe on a ſpiral curve; in which he determined the relation of its area to that of the circumſcribed circle; as alſo the relation of their ſectors.

Several other eminent men among the Ancients wrote upon this ſubject, both before and after Euclid and Archimedes; but their attempts were uſually confined to particular parts of it, and made according to methods not eſſentially different from theirs. Among theſe are to be reckoned Thales, Anaxagoras, Pythagoras, Bryſon, Antiphon, Hypocrates of Chios, Plato, Apollonius, Philo, and Ptolomy; moſt of whom wrote upon the Quadrature of the circle; and thoſe after Archimedes, by his method, uſually extended the approximation to a higher degree of accuracy.

Many of the Moderns have alſo proſecuted the ſame problem of the Quadrature of the circle, after the ſame methods, to ſtill greater lengths; ſuch are Vieta, and Metius; whoſe ratio between the diameter and the circumference, is that of 113 to 355, which is within about $\frac{3}{1000000}$ of the true ratio; but above all, Ludolph van Collen, or a Ceulen, who, with an amazing degree of induſtry and patience, by the ſame methods, extended the ratio to 36 places of figures, making the ratio to be that of

1 to 3·14159,26535,89793,23846,26433,83279 50288 + or 9 —. And the ſame was repeated and confirmed by his editor Snellius. See DIAMETER, and CIRCLE; alſo the Preface to my Menſuration.

Though the Quadrature, eſpecially of the circle, be a thing which many of the principal mathematicians, among the Ancients, were very ſolicitous about; yet nothing of this kind has been done ſo conſiderable, as about

about and fince the middle of the laft century; when, for example, in the year 1657, Sir Paul Neil, Lord Brouncker, and Sir Chriftopher Wren geometrically demonftrated the equality of fome curvilineal fpaces to rectilineal ones. Soon after this, other perfons did the like in other curves; and not long afterwards the thing was brought under an analytical calculus, the firft fpecimen of which ever publifhed, was given by Mercator in 1688, in a demonftration of Lord Brouncker's Quadrature of the hyperbola, by Dr. Wallis's method of reducing an algebraical fraction into an infinite feries by divifion.

Though, by the way, it appears that Sir Ifaac Newton had difcovered a method of attaining the area of all quadrable curves analytically, by his Method of Fluxions, before the year 1668. See his *Fluxions*, alfo his *Analyfis per Æquationes Numero Terminorum Infinitas*, and his *Introductio ad Quadruturam Curvarum;* where the Quadratures of Curves are given by general methods.

It is contefted, between Mr. Huygens and Sir Chriftopher Wren, which of the two firft found out the Quadrature of any determinate cycloidal fpace. Mr. Leibnitz afterwards difcovered that of another fpace; and Mr. Bernoulli, in 1699, found out the Quadrature of an infinity of cycloidal fpaces, both fegments and fectors &c.

As to the Quadrature of the Circle in particular, or the finding a fquare equal to a given circle, it is a problem that has employed the mathematicians of all ages, but ftill without the defired fuccefs. This depends on the ratio of the diameter to the circumference, which has never yet been determined in precife numbers. Many perfons have approached very near this ratio; for which fee CIRCLE.

Strict geometry here failing, mathematicians have had recourfe to other means, and particularly to a fort of curves called quadratices: but thefe being mechanical curves, inftead of geometrical ones, or rather tranfcendental inftead of algebraical ones, the problem cannot fairly be effected by them.

Hence recourfe has been had to analytics. And the problem has been attempted by three kinds of algebraical calculations. The firft of thefe gives a kind of tranfcendental Quadratures, by equations of indefinite degrees. The fecond by vulgar numbers, though irrationally fuch; or by the roots of common equations, which for the general Quadrature is impoffible. The third by means of certain feries, exhibiting the quantity of a circle by a progreffion of terms. See SERIES.

Thus, for example, the diameter of a circle being 1, it has been found that the quadrant, or one-fourth of the circumference, is equal to $\frac{1}{1} - \frac{1}{3} + \frac{1}{5} - \frac{1}{7} + \frac{1}{9}$ &c, making an infinite feries of fractions, whofe common numerator is 1, and denominators the natural feries of odd numbers; and all thefe terms alternately will be too great, and too little. This feries was difcovered by Leibnitz and Gregory. And the fame feries is alfo the area of the circle.

If the fum of this feries could be found, it would give the Quadrature of the circle: but this is not yet done; nor is it at all probable that it ever will be done;

though the impoffibility has never yet been demonftrated.

To this it may be added, that as the fame magnitude may be expreffed by feveral different feries, poffibly the circumference of the circle may be expreffed by fome other feries, whofe fum may be found. And there are many other feries, by which the quadrant, or area, to the diameter, has been expreffed; though it has never been found that any one of them is actually fummable.

Such as this feries, $1 - \frac{1}{6} - \frac{1}{40} - \frac{1}{112}$ &c, invented by Newton; with innumerable others.

But though a definite Quadrature of the whole circle was never yet given, nor of any aliquot part of it; yet certain other portions of it have been fquared. The firft partial Quadrature was given by Hippocrates of Chios; who fquared a portion called, from its figure, the *lune*, or *lunule;* but this Quadrature has no dependence on that of the circle. And fome modern geometricians have found out the Quadrature of any portion of the lune taken at pleafure, independently of the Quadrature of the circle; though ftill fubject to a certain reftriction, which prevents the Quadrature from being perfect, and what the geometricians call abfolute and indefinite. See LUNE. And for the Quadrature of the different kinds of curves, fee their feveral particular names.

QUADRATURES *by Fluxions.*—The moft general method of Quadratures yet difcovered, is that of Newton, by means of Fluxions, and is as follows. AC being any curve to be fquared, AB an abfcifs, and BC an ordinate perpendicular to it, alfo *bc* another ordinate indefinitely near to the former. Putting AB $= x$, and BC $= y$; 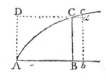 then is B*b* $= \dot{x}$ the fluxion of the abfcifs, and $y\dot{x} =$ C*b* the fluxion of the area ABC fought. Now let the value of the ordinate y be found in terms of the abfcifs x, or in a function of the abfcifs, and let that function be called X, that is $y = $ X; then fubftituting X for y in $y\dot{x}$, gives X$\dot{x}$ the fluxion of the area; and the fluent of this, being taken, gives the area or Quadrature of ABC as required, for any curve, whatever its nature may be.

Ex. Suppofe for example, AC to be a common parabola; then its equation is $px = y^2$, where p is the parameter; which gives $y = \sqrt{px}$, the value of y in a function of x, and is what is called X above; hence then $y\dot{x} = \dot{x}\sqrt{px} = p^{\frac{1}{2}}x^{\frac{1}{2}}\dot{x}$ is the fluxion of the area; and the fluent of this is $\frac{2}{3}p^{\frac{1}{2}}x^{\frac{3}{2}} = \frac{2}{3}x\sqrt{px} = \frac{2}{3}xy = \frac{2}{3}$ of the circumfcribing rectangle BD; which therefore is the Quadrature of the parabola.

Again, if AC be a circle whofe diameter is d; then its equation is $y^2 = dx - x^2$, which gives $y = \sqrt{dx - x^2}$, and the fluxion of the area $y\dot{x} = \dot{x}\sqrt{dx - x^2}$. But as the fluent of this cannot be found in finite terms, the quantity $\sqrt{dx - x^2}$ is thrown into a feries, and then the fluxion of the area is $y\dot{x} = \dot{x} \sqrt{\overline{dx - x^2}} = \dot{x} \sqrt{dx} \times (1 - \dfrac{x}{2d} - \dfrac{x^2}{2.4d^2} - \dfrac{1.3 x^3}{2.4.6 d^3}$

&c) ; and the fluent of this gives

$$x \sqrt{dx} \times (\dfrac{2}{3} - \dfrac{1}{5} \cdot \dfrac{x}{d} - \dfrac{1}{4.7} \cdot \dfrac{x^2}{d^2} - \dfrac{1.3}{4.6.9} \cdot \dfrac{x^3}{d^3} \&c)$$

for the general expreſſion of the area ABC. Now when the ſpace becomes a ſemicircle, x becomes $= d$, and

then the ſeries above becomes $d^2 (\dfrac{2}{3} - \dfrac{1}{5} - \dfrac{1}{4.7} - \dfrac{1.3}{4.6.9}$

&c) for the area of the ſemicircle whoſe diameter is d.

In ſpirals CAR, or curves referred to a centre C ; put $y =$ any radius CR, $x =$ BN the arc of a circle deſcribed about the centre C, at any diſtance CB $= a$, and Cnr another ray indefinitely near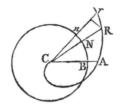

CNR: then $\frac{1}{2}$ CN . N$n =$

$\frac{1}{2} a\dot{x} =$ CNn, and by ſim. fig. CN² : CR² or $a^2 : y^2 ::$

CN$n : \dfrac{y^2 \dot{x}}{2a} =$ CRr the fluxion of the area deſcribed by the revolving ray CR ; then the fluent of this, for any particular caſe, will be the Quadrature of the ſpiral. So if, for inſtance, it be Archimedes's ſpiral, in which $x : y$ in a conſtant ratio ſuppoſe as $m : n$, or $my = nx$, and $y^2 = \dfrac{n^2 x^2}{m^2}$; hence then CR$r = \dfrac{y^2 \dot{x}}{2a} = \dfrac{n^2 x^2 \dot{x}}{2am^2}$ the fluxion of the area ; the fluent of which is $\dfrac{n^2 x^3}{6am^2} = \dfrac{xy^2}{6a}$ the general Quadrature of the ſpiral of Archimedes.

QUADRILATERAL, or QUADRILATERAL *Figure*, is a figure comprehended by four right lines ; and having conſequently alſo four angles, for which reaſon it is otherwiſe called a quadrangle.

The general term Quadrilateral comprehends theſe ſeveral particular ſpecies or figures, viz, the ſquare, parallelogram, rectangle, rhombus, rhomboides, and trapezium.

If the oppoſite ſides be parallel, the Quadrilateral is a parallelogram. If the parallelogram have its angles right ones, it is a rectangle ; if oblique, it is an oblique one. The rectangle having all its ſides equal, becomes a ſquare ; and the oblique parallelogram having all its ſides equal, is a rhombus, but if only the oppoſites be equal, it is a rhomboides. All other forms of the Quadrilateral, are trapeziums, including all the irregular ſhapes of it.

The ſum of all the four angles of any Quadrilateral, is equal to 4 right angles. Alſo, the two oppoſite angles of a Quadrilateral inſcribed in a circle taken together, are equal to two right angles. And in this caſe the rectangle of the two diagonals, is equal to the ſum of the two rectangles of the oppoſite ſides. For the properties of the particular ſpecies of Quadrilaterals, ſee their reſpective names, SQUARE, RECTANGLE, PARALLELOGRAM, RHOMBUS, RHOMBOIDES, and TRAPEZIUM.

QUADRIPARTITION, is the dividing by 4, or

into four equal parts.—Hence *quadripartite*, &c, the 4th part, or ſomething parted into four.

QUADRUPLE, is four-fold, or ſomething taken four times, or multiplied by 4 ; and ſo is the converſe of Quadripartition.

QUALITY, denotes generally the property or affection of ſome being, by which it affects our ſenſes in a certain way, &c.

Senſible Qualities are ſuch as are the more immediate object of the ſenſes : as figure, taſte, colour, ſmell, hardneſs, &c.

Occult Qualities, among the Ancients, were ſuch as did not admit of a rational ſolution in their way.

Dr. Keil demonſtrates, that every Quality which is propagated in orbem, ſuch as light, heat, cold, odour, &c, has its efficacy or intenſity either increaſed, or decreaſed, in a duplicate ratio of the diſtances from the centre of radiation inverſely. So at double the diſtance from the earth's centre, or from a luminous or hot body, the weight or light or heat, is but a 4th part ; and at 3 times the diſtance, they are 9 times leſs, or a 9th part, &c.

Sir Iſaac Newton lays it down as one of the rules of philoſophizing, that thoſe Qualities of bodies that are incapable of being intended and remitted, and which are found to obtain in all bodies upon which the experiment could ever be tried, are to be eſteemed univerſal Qualities of all bodies.

QUALITY *of Curvature*, in the higher geometry, is uſed to ſignify its form, as it is more or leſs inequable, or as it is varied more or leſs in its progreſs through different parts of the curve. Newton's Method of Fluxions, pa. 75 ; and Maclaurin's Fluxions, art. 369.

QUANTITY, denotes any thing capable of eſtimation, or menſuration ; or which being compared with another thing of the ſame kind, may be ſaid to be either greater or leſs, equal or unequal to it.

Mathematics is the doctrine or ſcience of Quantity.

Physical or *Natural* QUANTITY, is of two kinds : 1ſt, that which nature exhibits in matter, and its extenſion ; and 2dly, in the powers and properties of natural bodies ; as gravity, motion, light, heat, cold, denſity, &c.

Quantity is popularly diſtinguiſhed into continued and diſcrete.

Continued QUANTITY, is when the parts are connected together, and is commonly called magnitude ; which is the object of geometry.

Diſcrete QUANTITY, is when the parts, of which it conſiſts, exiſt diſtinctly, and unconnected ; which makes what is called multitude or number, the object of arithmetic.

The notion of continued Quantity, and its difference from diſcrete, appears to ſome without foundation. Mr. Machin conſiders all mathematical Quantity, or that for which any ſymbol is put, as nothing elſe but number, with regard to ſome meaſure which is conſidered as 1 ; for that we know nothing preciſely how much any thing is, but by means of number. The notion of continued Quantity, without regard to ſome meaſure, is indiſtinct and confuſed ; and though ſome ſpecies of ſuch Quantity, conſidered phyſically, may be deſcribed by motion, as lines by the motion of

points,

points, and furfaces by the motion of lines; yet the magnitudes, or mathematical Quantities, are not made by the motion, but by numbering according to a meafure. Philof. Tranf. numb. 447, pa. 228.

QUANTITY *of Action.* See ACTION.

QUANTITY *of Curvature* at any point of a curve is determined by the circle of curvature at that point, and is reciprocally proportional to the radius of curvature.

QUANTITY *of Matter* in any body, is its meafure arifing from the joint confideration of its magnitude and denfity, being expreffed by, or proportional to the product of the two. So,

if M and *m* denote the magnitude of two bodies, and D and *d* their denfities;
then DM and *dm* will be as their Quantities of matter.

The Quantity of matter of a body is beft difcovered by its abfolute weight, to which it is always proportional, and by which it is meafured.

QUANTITY *of Motion,* or the *Momentum,* of any body, is its meafure arifing from the joint confideration of its Quantity, and the velocity with which it moves. So,

if *q* denote the Quantity of matter,
and *v* the velocity of any body;
then *qv* will be its quantity of motion.

QUANTITIES, in Algebra, are the expreffions of indefinite numbers, that are ufually reprefented by letters. Quantities are properly the fubject of Algebra; which is wholly converfant in the computation of fuch Quantities.

Algebraic Quantities are either *given* and *known,* or elfe they are *unknown* and *fought.* The given or known Quantities are reprefented by the firft letters of the alphabet, as *a, b, c, d, e,* &c, and the unknown or required Quantities, by the laft letters, as *x, y, z, w,* &c.

Again, Algebraic Quantities are either pofitive or negative.

A pofitive or affirmative Quantity, is one that is to be added, and has the fign + or plus prefixed, or underftood; as *ab* or + *ab.* And a negative or privative Quantity, is one that is to be fubtracted, and has the fign — or minus prefixed; as — *ab.*

QUART, a meafure of capacity, being the quarter or 4th part of fome other meafure. The Englifh Quart is the 4th part of the gallon, and contains two pints. The Roman Quart, or quartarius, was the 4th part of their congius. The French, befides their Quart or pot of 2 pints, have various other Quarts, diftinguifhed by the whole of which they are Quarters; as Quart de muid, and Quart de boiffeau.

QUARTER, the 4th part of a whole, or one part of an integer, which is divided into four equal portions.

QUARTER, in weights, is the 4th part of the quintal, or hundred weight; and fo contains 28 pounds.

QUARTER is alfo a dry meafure, containing of corn 8 bufhels ftriked; and of coals the 4th part of a chaldron.

Quarter, in Aftronomy, the moon's period, or lunation, is divided into 4 ftages or Quarters; each containing between 7 and 8 days. The firft Quarter is from the new moon to the quadrature; the fecond is from thence to the full moon, and fo on.

QUARTER, in Navigation, is the Quarter or 4th part of a point, wind, or rhumb; or of the diftance between two points &c. The Quarter contains an arch of 2° 48′ 45″, being the 4th part of 11° 15′, which is one point.

QUARTER *Round,* in Architecture, is a term ufed by the workmen for any projecting moulding, whofe contour is a Quarter of a circle, or nearly fo.

QUARTILE, an afpect of the planets when they are at the diftance of 3 figns or 90° from each other: and is denoted by the character □.

QUEUE D'ARONDE, or *Swallow's Tail,* in Fortification, is a detached or outwork, whofe fides fpread or open towards the campaign, or draw narrower and clofer towards the gorge. Of this kind are either fingle or double tenailles, and fome horn-works, whofe fides are not parallel, but are narrow at the gorge, and open at the head, like the figure of a fwallow's tail.

On the contrary, when the fides are lefs than the gorge, the work is called *contre Queue d'aronde.*

QUEUE *d'aronde,* in Carpentry, a method of jointing, called alfo dove-tailing.

QUINCUNX, denotes $\frac{5}{12}$ths of any thing. So 10 is quincunx of 24, being $\frac{5}{12}$ of it.

QUINCUNX, in Aftronomy, is that pofition, or afpect, of the planets, when diftant from each other by $\frac{5}{12}$ths of the whole circle, or 5 figns out of the 12, that is 150 degrees. The Quincunx is marked Q, or Vc.

QUINDECAGON, is a plane figure of 15 angles, and confequently the fame number of fides. When thofe are all equal, it is a regular Quindecagon, otherwife not.

Euclid fhews how to infcribe this figure in a circle, prop. 16, lib. 4. And the fide of a regular Quindecagon, fo infcribed, is equal in power to the half difference between the fide of the equilateral triangle, and the fide of the pentagon; and alfo to the difference of the perpendiculars let fall on both fides, taken together.

QUINQUAGESIMA-*Sunday,* is the fame as Shrove-Sunday, and is fo called as being about the 50th day before Eafter, being indeed the 7th Sunday before it. Anciently the term Quinquagefima was ufed for Whitfunday, and for the 50 days between Eafter and Whitfunday; but to diftinguifh this Quinquagefima from that before Eafter, it was called the pafchal Quinquagefima.

QUINQUEANGLED, or Quinqueangular, confifting of 5 angles.

QUINTAL, the weight of a hundred pounds, in moft countries; but in England it is the hundred weight, or 112 pounds. Quintal was alfo formerly ufed for a weight of lead, iron, or other common metal, ufually equal to a hundred pounds, at 6 fcore to the hundred.

QUINTILE, in Aftronomy, an afpect of the planets when they are diftant the 5th part of the zodiac, or 72 degrees; and is marked thus, C, or O.

QUINTUPLE, is five-fold, or five times as much as another thing.

QUOIN, in Architecture, an angle or corner of ftone or brick walls. When thefe ftand out beyond the reft of the wall, their edges being chamferred off, they are called *ruftic Quoins.*

QUOIN, in Artillery, is a loofe wedge of wood, which

is put in below the breech of a cannon, to raife or deprefs it more or lefs.

QUOTIENT, in Arithmetic, is the refult of the operation of divifion, or the number that arifes by dividing the dividend by the divifor, fhewing how often the latter is contained in the former. Thus the Quotient of 12 divided by 3 is 4; which is ufually thus difpofed, or expreffed,

3) 12 (4 the quotient,

or thus $12 \div 3 = 4$ the Quotient, or thus $\frac{12}{3}$ like a vulgar fraction; all thefe meaning the fame thing. —In divifion, as the divifor is to the dividend, fo is unity or 1 to the Quotient; thus $3 : 12 :: 1 : 4$ the Quotient.

R.

RAD

RADIANT *Point*, or RADIATING *Point*, is any point from whence rays proceed.

Every Radiant point diffufes innumerable rays all around: but thofe rays only are vifible from which right lines can be drawn to the pupil of the eye; becaufe the rays are all in right lines. All the rays proceeding from the fame Radiant continually diverge; but the cryftalline collects or reunites them again.

RADIATION, is the cafting or fhooting forth of rays of light as from a centre.—Every vifible body is a radiating body; it being only by means of its rays that it affects the eye.—The furface of a radiating or vifible body, may be conceived as confifting of radiant points.

RADICAL *Sign*, in Algebra, the fign or character denoting the root of a quantity; and is this $\sqrt{}$. So $\sqrt{2}$ is the fquare root of 2, and $\sqrt[3]{2}$ is the cube root of 2, &c.

RADIOMETER, a name which fome writers give to the Radius Aftronomicus, or Jacob's Staff. See FORE-STAFF.

RADIUS, in Geometry, the femidiameter of a circle; or a right line drawn from the centre to the circumference.—It is implied in the definition of a circle, and it is apparent from its conftruction, that all the radii of the fame circle are equal.—The Radius is fometimes called, in Trigonometry, the Sinus Totus, or whole fine.

RADIUS, in the Higher Geometry. RADIUS *of the Evoluta,* RADIUS *Ofculi,* called alfo the *Radius of concavity,* and the *Radius of curvature,* is the right line CB, reprefenting a thread, by whofe evolution from off the curve AC, upon which it was wound, the curve AB is formed. Or it is the Radius of a circle having the fame curvature, in a given point of the curve at B, with that of the curve in that point. See CURVATURE and EVOLUTE, where the method of finding this Radius may be feen.

RAF

RADIUS *Aftronomicus,* an inftrument ufually called Jacob's Staff, the Crofs-ftaff, or Fore-ftaff.

RADIUS, in Mechanics, is applied to the fpokes of a wheel; becaufe iffuing like rays from its centre.

RADIUS, in Optics. See RAY.

RADIUS *Vector,* is ufed for a right line drawn from the centre of force in any curve in which a body is fuppofed to move by a centripetal force, to that point of the curve where the body is fuppofed to be.

RADIX, or *Root,* is a certain finite expreffion or function, which, being evolved or expanded according to the rules proper to its form, fhall produce a feries. That finite expreffion, or Radix, is alfo the value of the infinite feries. So $\frac{1}{3}$ is the radix of ·3333 &c, becaufe $\frac{1}{3}$ being evolved or expanded, by dividing 1 by 3, gives the infinite feries ·3333 &c. In like manner, the Radix

of $1 - r + r^2 - r^3 + r^4$ &c is $\dfrac{1}{1 + r}$,

of $1 - \dfrac{1}{2} + \dfrac{1}{4} - \dfrac{1}{8} + \dfrac{1}{16}$ &c is $\dfrac{1}{1 + \frac{1}{2}}$,

of $1 - 1 + 1 - 1 + 1$ &c is $\dfrac{1}{1 + 1}$,

of $1 - 2 + 4 - 8 + 16$ &c is $\dfrac{1}{1 + 2}$,

of $\dfrac{1}{2} - \dfrac{1}{4} + \dfrac{1}{8} - \dfrac{1}{16} + \dfrac{1}{32}$ &c is $\dfrac{1}{2 + 1}$,

of $1 + x + x^2 + x^3 + x^4$ &c is $\dfrac{1}{1 + x}$,

of $1 + 2x + 3x^2 + 4x^3$ &c is $\dfrac{1}{(1 - x)^2}$,

of $1 + \dfrac{x^2}{2} + \dfrac{3x^4}{8} + \dfrac{5x^6}{16}$ &c is $\sqrt{\dfrac{1}{1 - x^2}}$.

See my Tracts, vol. 1, pa. 9, and 31, &c.

RAFTERS, in Architecture, are pieces of timber which ftand by pairs on the raifing-piece, or wall plate, and meet in an angle at the top, forming the roof of a building.

5

RAIN,

RAIN, water that defcends from the atmofphere in the form of drops of a confiderable fize. Rain is apparently a precipitated cloud; as clouds are nothing but vapours raifed from moifture, waters, &c. By this circumftance it is diftinguifhed from dew and fog: in the former of which the drops are fo fmall that they are quite invifible; and in the latter, though their fize be larger, they feem to have very little more fpecific gravity than the atmofphere itfelf, and may therefore be reckoned hollow fpherules rather than drops.

It is univerfally agreed, that Rain is produced by the water previoufly abforbed by the heat of the fun, or otherwife, from the terraqueous globe, into the atmofphere, as vapours, or veficulæ. Thefe veficulæ, being fpecifically lighter than the atmofphere, are buoyed up by it, till they arrive at a region where the air is in a juft balance with them; and there they float, till by fome new agent they are converted into clouds, and thence either into Rain, fnow, hail, mift, or the like.

But the agent in this formation of the clouds into Rain, and even of the vapours into clouds, has been much controverted. Moft philofophers will have it, that the cold, which conftantly occupies the fuperior regions of the air, chills and condenfes the veficulæ, at their arrival from a warmer quarter; congregates them together, and occafions feveral of them to coalefce into little maffes: and thus their quantity of matter increafing in a higher proportion than their furface, they become an overload to the thin air, and fo defcend in Rain.

Dr. Derham accounts for the precipitation, hence; that the veficulæ being full of air, when they meet with a colder air than that they contain, this is contracted into a lefs fpace: and confequently the watry fhell or cafe becomes thicker, fo as to become heavier than the air, &c.

But this feparation cannot be afcribed to cold, fince Rain often takes place in very warm weather. And though we fhould fuppofe the condenfation owing to the cold of the higher regions, yet there is a remarkable fact which will not allow us to have recourfe to this fuppofition: for it is certain that the drops of Rain increafe in fize confiderably as they defcend. On the top of a hill for inftance, they will be fmall and inconfiderable, forming only a drizzling fhower; but half way down the hill it is much more confiderable; and at the bottom the drops will be very large, defcending in an impetuous Rain. Which fhews that the atmofphere condenfes the vapours as well where it is warm as where it is cold.

Others allow the cold only a part in the action, and bring in the winds as fharers with it: alledging, that a wind blowing againft a cloud will drive its veficulæ upon one another, by which means feveral of them, coalefcing as before, will be enabled to defcend; and that the effect will be ftill more confiderable, if two oppofite winds blow together towards the fame place: they add, that clouds already formed, happening to be aggregated by frefh acceffions of vapour continually afcending, may thence be enabled to defcend.

Yet the grand caufe, according to Rohault, is ftill behind. That author conceives it to be the heat of the air, which, after continuing for fome time near the earth, is at length carried up on high by a wind, and

there thawing the fnowy villi or flocks of the half frozen veficulæ, it reduces them into drops; which, coalefcing, defcend, and have their diffolution perfected in their progrefs through the lower and warmer ftages of the atmofphere.

Others, as Dr. Clarke, &c, afcribe this defcent of the clouds rather to an alteration of the atmofphere than of the veficulæ; and fuppofe it to arife from a diminution of the fpring or elaftic force of the air. This elafticity, which depends chiefly or wholly on the dry terrene exhalations, being weakened, the atmofphere finks under its burden; and the clouds fall, on the common principle of precipitation.

Now the fmall veficulæ, by thefe or any other caufes, being once upon the defcent, will continue to defcend notwithftanding the increafe of refiftance they every moment meet with in their progrefs through ftill denfer and denfer parts of the atmofphere. For as they all tend toward the fame point, viz, the centre of the earth, the farther they fall, the more coalitions will they make; and the more coalitions, the more matter will there be under the fame furface; the furface only increafing as the fquares, but the folidity as the cubes of the diameters: and the more matter under the fame furface, the lefs friction or refiftance there will be to the fame matter.

Thus then, if the caufes of rain happen to act early enough to precipitate the afcending veficulæ, before they are arrived at any confiderable height, the coalitions being few in fo fhort a defcent, the drops will be proportionably fmall; thus forming what is called dew. If the vapours prove more copious, and rife a little higher, there is produced a mift or fog. A little higher ftill, and they produce a fmall rain, &c. If they neither meet with cold nor wind enough to condenfe or diffipate them; they form a heavy, thick, dark fky, which lafts fometimes feveral days, or even weeks.

But later writers on this part of philofophical fcience have, with greater fhew of truth, confidered Rain as an electrical phenomenon. Signior Beccaria reckons Rain, hail, and fnow, among the effects of a moderate electricity in the atmofphere. Clouds that bring Rain, he thinks are produced in the fame manner as thunder clouds, only by a moderate electricity. He defcribes them at large; and the refemblance which all their phenomena bear to thofe of thunder clouds, is very ftriking. He notes feveral circumftances attending Rain without lightning, which render it probable that it is produced by the fame caufe as when it is accompanied with lightning. Light has been feen among the clouds by night in rainy weather; and even by day rainy clouds are fometimes feen to have a brightnefs evidently independent of the fun. The uniformity with which the clouds are fpread, and with which the Rain falls, he thinks are evidences of an uniform caufe like that of electricity. The intenfity alfo of electricity in his apparatus, ufually correfponded very nearly to the quantity of Rain that fell in the fame time. Sometimes all the phenomena of thunder, lightning, hail, Rain, fnow, and wind, have been obferved at one time; which fhews the connection they all have with fome common caufe. Signior Beccaria therefore fuppofes that, previous to Rain, a quantity of electric matter escapes out of the earth, in some place where there is a redundancy of it; and in its ascent to the higher regions of the air, collects and conducts into its path a great quantity of vapours. The same cause that collects, will condense them more and more; till, in the places of the nearest intervals, they come almost into contact, so as to form small drops; which, uniting with others as they fall, come down in the form of Rain. The Rain will be heavier in proportion as the electricity is more vigorous, and the cloud approaches more nearly to a thunder cloud: &c. See *Lettere dell Elettricismo*; and Priestley's Hist. &c of Electricity, vol. 1, pa. 427, &c, 8vo. And for farther accounts of the phenomena of Rain &c, see BAROMETER, EVAPORATION, OMBROMETER, PLUVIAMETER, VAPOUR, &c. See also the Theory of Rain, by Dr. James Hutton, art. 2 vol. 1 of Transactions of the Royal Society of Edinburgh.

Quantity of RAIN. As to the general quantity of Rain that falls, with its proportion in several places at the same time, and in the same place at different times, there are many observations, journals, &c, in the Philos. Transf. the Memoirs of the French Academy, &c. And upon measuring the rain that falls annually, its depth, on a medium, is found as in the following table:

Mean Annual Depth of Rain for several Places.

At	Observed by	Inch.
Townley, in Lancashire	Mr. Townley	$42\frac{1}{2}$
Upminster, in Essex	Dr. Derham	$19\frac{1}{4}$
Zurich, Switserland	Dr. Scheuchzer	$32\frac{1}{4}$
Pisa, in Italy	Dr. Mich. Ang. Tilli	$43\frac{1}{4}$
Paris, in France	M. De la Hire	19
Lisle, Flanders	M. De Vauban	24

Quantity of Rain fallen in several Years at Paris and Upminster.

At Paris.	Years.	At Upminster.
Inches 21·37	1700	19·03 Inches
27·77	1701	18·69
17·45	1702	20·38
18·51	1703	23·99
21·20	1704	15·80
14·82	1705	16·93
20·19	Mediums	19·14

Medium Quantity of Rain at London, for several Years, from the Philos. Transf.

Viz, in 1774	26·328 inches.
1775	24·083
1776	20·354
1777	25·371
1778	20·772
1779	26·785
1780	17·313
Medium of these 7 years	23·001

See also Philos. Transf. Abr. vol. 4, pt. 2, pa. 81, &c, and vol. 10 in many places; also the Meteorological Journal of the Royal Society, published annually in the Philos. Transf. and the article PLUVIAMETER or OMBROMETER.

It is reasonably to be expected, and all experience shews, that the most Rain falls in places near the sea coast, and less and less as the places are situated more inland. Some differences also arise from the circumstances of hills, valleys, &c. So when the quantity of Rain fallen in one year at London, is 20 inches, that on the western coast of England will often be twice as much, or 40 inches, or more. Those winds also bring most Rain, that blow from the quarter in which is the most and nearest sea; as our west and south-west winds.

It is also found, by the pluviameter or Rain-gage, that, in any one place, the more Rain is collected in the instrument, as it is placed nearer the ground; without any appearance of a difference, between two places, on account of their difference of level above the sea, provided the instrument is but as far from the ground at the one place, as it is from the ground at the other. These effects are remarked in the Philos. Transf for 1769 and 1771, the former by Dr. Heberden, and the latter by Mr. Daines Barrington. Dr. Heberden says, " A comparison having been made between the quantity of Rain, which fell in two places in London, about a mile distant from one another, it was found, that the Rain in one of them constantly exceeded that in the other, not only every month, but almost every time that it rained. The apparatus used in each of them was very exact, and both made by the same artist; and upon examining every probable cause, this unexpected variation did not appear to be owing to any mistake, but to the constant effect of some circumstance, which not being supposed to be of any moment, had never been attended to. The Rain-gage in one of these places was fixed so high, as to rise above all the neighbouring chimnies; the other was considerably below them; and there appeared reason to believe, that the difference of the quantity of Rain in these two places was owing to this difference in the placing of the vessel in which it was received. A funnel was therefore placed above the highest chimnies, and another upon the ground of the garden belonging to the same house, and there was found the same difference between these two, though placed so near one another, which there had been between them, when placed at similar heights in different parts of the town. After this fact was sufficiently ascertained, it was thought proper to try whether the difference would be greater at a much greater height; and a Rain gage was therefore placed upon the square part of the roof of Westminster Abbey. Here the quantity of Rain was observed for a twelvemonth, the Rain being measured at the end of every month, and care being taken that none should evaporate by passing a very long tube of the funnel into a bottle through a cork, to which it was exactly fitted. The tube went down very near to the bottom of the bottle, and therefore the Rain which fell into it would soon rise above the end of the tube, so that the water was no where open to the air except

for

for the fmall fpace of the area of the tube : and by trial it was found that there was no fenfible evaporation through the tube thus fitted up.

The following table fhews the refult of thefe obfertions.

From July the 7th 1766, to July the 7th 1767, there fell in a Rain-gage, fixed

1766.	Below the top of a houfe.	Upon the top of a houfe.	Upon Weftminfter Abbey.
From the 7th to	*Inches.*	*Inches.*	*Inches.*
the end of July	3·591	3·210	2·311
Auguft	0·558	0·479	} 0·508
September	0·421	0·344	
October	2·364	2·061	1·416
November	1·079	0·842	0·632
December	1·612	1·258	0·994
1767, January	2·071	1·455	1·035
February	2·864	2·494	1·335
March	1·807	1·303	0·587
April	1·437	1·213	0·994
May	2·432	1·745	1·142
June	1·997	1·426	} 1·145
July 7	0·395	0·309	
	22·608	18·139	12·099

By this table it appears, that there fell below the top of a houfe above a fifth part more Rain, than what fell in the fame fpace above the top of the fame houfe ; and that there fell upon Weftminfter Abbey not much above one half of what was found to fall in the fame fpace below the tops of the houfes. This experiment has been repeated in other places with the fame refult. What may be the caufe of this extraordinary difference, has not yet been difcovered ; but it may be ufeful to give notice of it, in order to prevent that error, which would frequently be committed in comparing the Rain of two places without attending to this circumftance.''

Such were the obfervations of Dr. Heberden on firft announcing this circumftance, viz, of different quantities of Rain falling at different heights above the ground. Two years afterward, Daines Barrington Efq. made the following experiments and obfervations, to fhew that this effect, with refpect to different places, refpected only the feveral heights of the inftrument above the ground at thofe places, without regard to any real difference of level in the ground at thofe places.

Mr. Barrington caufed two other Rain-gages, exactly like' thofe of Dr. Heberden, to be placed, the one upon mount Rennig, in Wales, and the other on the plane below, at about half a mile's diftance, the perpendicular height of the mountain being 450 yards, or 1350 feet ; each gage being at the fame height above the furface of the ground at the two ftations.

The refults of the Experiment are as below :

1770.	Bottom of the mountain.	Top of the mountain.
	Inches.	*Inches.*
From July 6 to 16	0·709	0·648
July 16 to 29	2·185	2·124
July 29 to Aug. 10.	0·610	0·656
Sept. 9 both bottles had run over.		
Sept. 9 to 30	3·234	2·464
Oct. 17. both bottles had run over.		
Oct. 17 to 22	0·747	0·885
Oct. 22 to 29	1·281	1·388
Nov. 20 both bottles were broken by the froft	8·766	8·165

'' The inference to be drawn from thefe experiments, Mr. Barrington obferves, feems to be, that the increafe of the quantity of Rain depends upon its nearer approximation to the earth, and fcarcely at all upon the height of places, provided the Rain-gages are fixed at about the fame diftance from the ground.

'' Poffibly alfo a much controverted point between the inhabitants of mountains and plains may receive a folution from thefe experiments ; as in an *adjacent valley, at leaft*, very nearly the fame quantity of Rain appears to fall within the fame period of time as upon the neighbouring mountains.''

Dr. Heberden alfo adds the following note. '' It may not be improper to fubjoin to the foregoing account, that, in places where it was firft obferved, a different quantity of Rain would be collected, according as the Rain gages were placed above or below the tops of the neighbouring buildings ; the Rain-gage below the top of the houfe, into which the greater quantity of Rain had for feveral years been found to fall, was above 15 feet above the level of the other Rain-gage, which in another part of London was placed above the top of the houfe, and into which the leffer quantity always fell. This difference therefore does not, as Mr. Barrington juftly remarks, depend upon the greater quantity of atmofphere, through which the Rain defcends : though this has been fuppofed by fome, who have thence concluded that this appearance might readily be folved by the accumulation of more drops, in a defcent through a great depth of atmofphere.''

RAINBOW, *Iris*, or fimply the *Bow*, is a meteor in form of a party-coloured arch, or femicircle, exhibited in a rainy fky, oppofite to the fun, by the refraction and reflection of his rays in the drops of falling rain. There is alfo a fecondary, or fainter bow, ufually feen invefting the former at fome diftance. Among naturalifts, we alfo read of lunar Rainbows, marine Rainbows, &c.

The Rainbow, Sir Ifaac Newton obferves, never appears but where it rains in the funfhine ; and it may be reprefented artificially, by contriving water to fall

in

in small drops, like rain, through which the sun shining, exhibits a bow to a spectator placed between the sun and the drops, especially if there be disposed beyond the drops some dark body, as a black cloth, or such like.

Some of the ancients, as appears by Aristotle's tract on Meteors, knew that the Rainbow was caused by the refraction of the sun's light in drops of falling rain. Long afterwards, one Fletcher of Breslaw, in a treatise which he published in 1571, endeavoured more particularly to account for the colours of the Rainbow by means of a double refraction, and one reflection. But he imagined that a ray of light, after entering a drop of rain, and suffering a refraction, both at its entrance and exit, was afterwards reflected from another drop, before it reached the eye of the spectator. It seems he overlooked the reflection at the farther side of the drop, or else he imagined that all the bendings of the light within the drop would not make a sufficient curvature, to bring the ray of the sun to the eye of the spectator. But Antonio de Dominis, bishop of Spalato, about the year 1590, whose treatise *De Radiis Visûs et Lucis* was published in 1611 by J. Bartolus, first advanced, that the double refraction of Fletcher, with an intervening reflection, was sufficient to produce the colours of the Rainbow, and also to bring the rays that formed them to the eye of the spectator, without any subsequent reflection. He distinctly describes the progress of a ray of light entering the upper part of the drop, where it suffers one refraction, and after being by that thrown upon the back part of the inner surface, is from thence reflected to the lower part of the drop; at which place undergoing a second refraction, it is thereby bent so as to come directly to the eye. To verify this hypothesis, he procured a small globe of solid glass, and viewing it when it was exposed to the rays of the sun, in the same manner in which he had supposed the drops of rain were situated with respect to them, he actually observed the same colours which he had seen in the true Rainbow, and in the same order. Thus this author shewed how the interior bow is formed in round drops of rain, viz, by two refractions of the sun's rays and one reflection between them; and he likewise shewed that the exterior bow is formed by two refractions and two sorts of reflections between them in each drop of water.

The theory of A. de Dominis was adopted, and in some degree improved with respect to the exterior bow, by Des Cartes, in his treatise on Meteors; and indeed he was the first who, by applying mathematics to the investigation of this surprising appearance, ever gave a tolerable theory of the Rainbow. Philosophers were however still at a loss when they endeavoured to assign reasons for all the particular colours, and for the order of them. Indeed nothing but the doctrine of the different refrangibility of the rays of light, a discovery which was reserved for the great Newton, could furnish a complete solution of this difficulty.

Dr. Barrow, in his Lectiones Opticæ, at Lect. 12, n. 14, says, that a friend of his (by whom we are to understand Mr. Newton) communicated to him a way of determining the angle of the Rainbow, which was hinted to Newton by Slusius, without making a table of the refractions, as Des Cartes did. The doctor shews

5

the method; as also several other matters, at n. 14, 15, 16, relating to the Rainbow, worthy the genius of those two eminent men. But the subject was given more perfectly by Newton afterwards, viz, in his Optics, prop. 9; where he makes the breadth of the interior bow to be nearly 2° 15′, that of the exterior 3° 40′, their distance 8° 25′, the greatest semidiameter of the interior bow 42° 17′, and the least of the exterior 50° 42′, when their colours appear strong and perfect.

The doctrine of the Rainbow may be illustrated and confirmed by experiment in several different ways. Thus, by hanging up a glass globe, full of water, in the sun shine, and viewing it in such a posture that the rays which come from the globe to the eye, may include an angle either of 42° or 50° with the sun's rays; for ex. if the angle be about 42°, the spectator will see a full red colour in that side of the globe opposite to the sun. And by varying the position so as to make that angle gradually less, the other colours, yellow, green, and blue, will appear successively, in the same side of the globe, and that very bright. But if the angle be made about 50°, suppose by raising the globe, there will appear a red colour in that side of the globe toward the sun, though somewhat faint; and if the angle be made greater, as by raising the globe still higher, this red will change successively to the other colours, yellow, green, and blue. And the same changes are observed by raising or depressing the eye, while the globe is at rest. Newton's Optics, pt. 2, prop. 9, prob. 4.

Again, a similar bow is often observed among the waves of the sea (called the *marine Rainbow*), the upper parts of the waves being blown about by the wind, and so falling in drops. This appearance is also seen by moon light (called the *lunar Rainbow*) though seldom vivid enough to render the colours distinguishable. Also it is sometimes seen on the ground, when the sun shines on a very thick dew. Cascades and fountains too, whose waters are in their fall divided into drops, exhibit Rainbows to a spectator, if properly situated during the time of the sun's shining; and even water blown violently out of the mouth of an observer, standing with his back to the sun, never fails to produce the same phenomenon. The artificial Rainbow may even be produced by candle light on the water which is ejected by a small fountain or jet d'eau. All these are of the same nature, and they depend upon the same causes; some account of which is as follows.

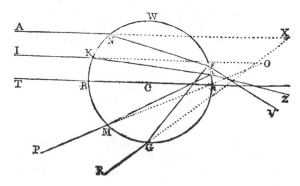

Let the circle WQGB reprefent a drop of water, or a globe, upon which a beam of parallel light falls, of which let TB reprefent a ray falling perpendicularly at B, and which confequently either paffes through without refraction, or is reflected directly back from Q. Suppofe another ray IK, incident at K, at a diftance from B, and it will be refracted according to a certain ratio of the fines of incidence and refraction to each other, which in rain water is as 529 to 396, to a point L, whence it will be in part tranfmitted in the direction L Z, and in part reflected to M, where it will again in part be reflected, and in part tranfmitted in the direction MP, being inclined to the line defcribed by the incident ray in the angle IOP. Another ray AN, ftill farther from B, and confequently incident under a greater angle, will be refracted to a point F, ftill farther from Q, whence it will be in part reflected to G, from which place it will in part emerge, forming an angle AXR with the incident AN, greater than that which was formed between the ray MP and its incident ray. And thus, while the angle of incidence, or diftance of the point of incidence from B, increafes, the diftance between the point of reflection and Q, and the angle formed between the incident and emergent reflected rays, will alfo increafe; that is, as far as it depends on the diftance from B: but as the refraction of the ray tends to carry the point of reflection towards Q, and to diminifh the angle formed between the incident and emergent reflected ray, and that the more the greater the diftance of the point of incidence from B, there will be a certain point of incidence between B and W, with which the greateft poffible diftance between the point of reflection and Q, and the greateft poffible angle between the incident and emergent reflected ray, will correfpond. So that a ray incident nearer to B fhall, at its emergence after reflection, form a lefs angle with the incident, by reafon of its more direct reflection from a point nearer to Q; and a ray incident nearer to W, fhall at its emergence form a lefs angle with the incident, by reafon of the greater quantity of the angles of refraction at its incidence and emergence. The rays which fall for a confiderable fpace in the vicinity of that point of incidence with which the greateft angle of emergence correfponds, will, after emerging, form an angle with the incident rays differing infenfibly from that greateft angle, and confequently will proceed nearly parallel to each other; and thofe rays which fall at a diftance from that point will emerge at various angles, and confequently will diverge. Now, to a fpectator, whofe back is turned towards the radiant body, and whofe eye is at a confiderable diftance from the globe or drop, the divergent light will be fcarcely, if at all, perceptible; but if the globe be fo fituated, that thofe rays that emerge parallel to each other, or at the greateft poffible angle with the incident, may arrive at the eye of the fpectator, he will, by means of thofe rays, behold it nearly with the fame fplendour at any diftance.

In like manner, thofe rays which fall parallel on a globe, and are emitted after two reflections, fuppofe at the points F and G, will emerge at H parallel to each oth   when the angle they make with the incident AN is the leaft poffible; and the globe muft be feen very refplendent when its pofition is fuch, that thofe parallel rays fall on the eye of the fpectator.

The quantities of thefe angles are determined by calculation, the proportion of the fines of incidence and refraction to each other being known. And this proportion being different in rays which produce different colours, the angles muft vary in each. Thus it is found, that the greateft angle in rain water for the leaft refrangible, or red rays, emitted parallel after one reflection, is 42° 2', and for the moft refrangible or violet rays, emitted parallel after one reflection, 40° 17'; likewife, after two reflections, the leaft refrangible, or red rays, will be emitted nearly parallel under an angle of 50° 57', and the moft refrangible, or violet, under an angle of 54° 7'; and the intermediate colours will be emitted nearly parallel at intermediate angles.

Suppofe now, that O is the fpectator's eye, and OP a line drawn parallel to the fun's rays, SE, SF, SG, and SH;

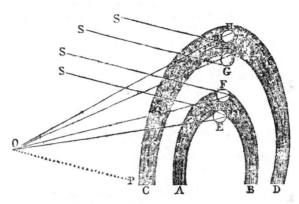

and let POE, POF, POG, POH be angles of 40° 17', 42° 2', 50° 57', and 54° 7' refpectively; then thefe angles turned about their common fide OP, will with their other fides OE, OF, OG, OH defcribe the verges of the two Rainbows, as in the figure. For, if E, F, G, H be drops placed any where in the conical fuperficies defcribed by OE, OF, OG, OH, and be illuminated by the fun's rays SE, SF, SG, SH; the angle SEO being equal to the angle POE, or 40° 17', will be the greateft angle in which the moft refrangible rays can, after one reflection, be refracted to the eye, and therefore all the drops in the line OE muft fend the moft refrangible rays moft copioufly to the eye, and fo ftrike the fenfe with the deepeft violet colour in that region. In like manner, the angle SFO being equal to the angle POF, or 42° 2', will be the greateft in which the leaft refrangible rays after one reflection can emerge out of the drops, and therefore thofe rays muft come moft copioufly to the eye from the drops in the line OF, and ftrike the fenfe with the deepeft red colour in that region. And, by the fame argument, the rays which have the intermediate degrees of refrangibility will come moft copioufly from drops between E and F, and ftrike the fenfes with the intermediate colours in the order which their degrees of refrangibility require; that is, in the

progrefs

progress from E to F, or from the infide of the bow to the outfide, in this order, violet, indigo, blue, green, yellow, orange, red. But the violet, by the mixture of the white light of the clouds, will appear faint, and inclined to purple.

Again, the angle SGO being equal to the angle POG, or 50° 57′, will be the leaft angle in which the leaft refrangible rays can, after two reflections, emerge out of the drops, and therefore the leaft refrangible rays muft come moft copioufly to the eye from the drops in the line OG, and ftrike the fenfe with the deepeft red in that region. And the angle SHO being equal to the angle POH, or 54° 7′, will be the leaft angle in which the moft refrangible rays, after two reflections, can emerge out of the drops, and therefore thofe rays muft come moft copioufly to the eye from the drops in the line OH, and ftrike the fenfe with the deepeft violet in that region. And, by the fame argument, the drops in the regions between G and H will ftrike the fenfe with the intermediate colours in the order which their degrees of refrangibility require; that is, in the progrefs from G to H, or from the infide of the bow to the outfide, in this order, red, orange, yellow, green, blue, indigo, and violet. And fince the four lines OE, OF, OG, OH may be fituated any where in the above-mentioned conical fuperficies, what is faid of the drops and colours in thefe lines, is to be underftood of the drops and colours every where in thofe fuperficies.

Thus there will be made two bows of colours, an interior and ftronger, by one reflection in the drops, and an exterior and fainter by two; for the light becomes fainter by every reflection; and their colours will lie in a contrary order to each other, the red of both bows bordering upon the fpace GF, which is between the bows. The breadth of the interior bow, EOF, meafured acrofs the colours, will be 1° 15′, and the breadth of the exterior GOH, will be 3° 10′, alfo the diftance between them GOF, will be 8° 55′, the greateft femidiameter of the innermoft, that is, the angle FOF, being 42° 2′, and the leaft femidiameter of the outermoft POG being 50° 57′. Thefe are the meafures of the bows as they would be, were the fun but a point; but by the breadth of his body, the breadth of the bows will be increafed by half a degree, and their diftance diminifhed by as much; fo that the breadth of the inner bow will be 2° 15′, that of the outer 3° 40′, their diftance 8° 25′; the greateft femidiameter of the interior bow 42° 17′, and the leaft of the exterior 50° 42′. And fuch are the dimenfions of the bows in the heavens found to be, very nearly, when their colours appear ftrong and perfect.

The light which comes through drops of rain by two refractions without any reflection, ought to appear ftrongeft at the diftance of about 26 degrees from the fun, and to decay gradually both ways as the diftance from the fun increafes and decreafes. And the fame is to be underftood of light tranfmitted through fpherical hailftones. If the hail be a little flatted, as it often is, the light tranfmitted may grow fo ftrong at a little lefs diftance than that of 26°, as to form a halo about the fun and moon; which halo, when the ftones are duly figured, may be coloured, and then it muft be

red within, by the leaft refrangible rays, and blue without, by the moft refrangible ones.

The light which paffes through a drop of rain after two refractions, and three or more reflections, is fcarce ftrong enough to caufe a fenfible bow.

As to the dimenfion of the Rainbow, Des Cartes firft determined its diameter by a tentative and indirect method; laying it down, that the magnitude of the bow depends on the degree of refraction of the fluid; and affuming the ratio of the fine of incidence to that of refraction, to be in water as 250 to 187. But Dr. Halley, in the Philof. Tranf. number 267, gave a fimple direct method of determining the diameter of the Rainbow from the ratio of the refraction of the fluid being given; or, vice verfa, the diameter of the Rainbow being given, to determine the refractive power of the fluid. And Dr. Halley's principles and conftruction were farther explained by Dr. Morgan, bifhop of Ely, in his Differtation on the Rainbow, among the notes upon Rohault's Syftem of Philofophy, part 3, chap. 17.

From the theory of the Rainbow, all the particular phenomena of it are eafily deducible. Hence we fee, 1ft, Why the iris is always of the fame breadth; becaufe the intermediate degrees of refrangibility of the rays between red and violet, which are its extreme colours, are always the fame.

2dly, Why the bow fhifts its fituation as the eye does; and, as the popular phrafe has it, flies from thofe who follow it, and follows thofe that fly from it; the coloured drops being difpofed under a certain angle, about the axis of vifion, which is different in different places: whence alfo it follows, that every different fpectator fees a different bow.

3dly, Why the bow is fometimes a larger portion of a circle, fometimes a lefs: its magnitude depending on the greater or lefs part of the furface of the cone, above the furface of the earth, at the time of its appearance; and the higher the fun, always the lefs the Rainbow.

4thly, Why the bow never appears when the fun is above a certain altitude; the furface of the cone, in which it fhould be feen, being loft in the ground at a little diftance from the eye, when the fun is above 42° high.

5thly, Why the bow never appears greater than a femicircle, on a plane; fince, be the fun never fo low, and even in the horizon, the centre of the bow is ftill in the line of afpect; which in this cafe runs along the earth, and is not at all raifed above the furface. Indeed if the fpectator be placed on a very confiderable eminence, and the fun in the horizon, the line of afpect, in which the centre of the bow is, will be confiderably raifed above the horizon. And if the eminence be very high, and the rain near, it is poffible the bow may be an entire circle.

6thly, How the bow may chance to appear inverted, or the concave fide turned upwards; viz, a cloud happening to intercept the rays, and prevent their fhining on the upper part of the arch: in which cafe, only the lower part appearing, the bow will feem as if turned upfide down; which has probably been the cafe in feveral prodigies of this kind, related by authors.

Lunar

Lunar RAINBOW. The moon fometimes alfo exhibits the phenomenon of an iris, by the refraction of her rays in the drops of rain in the night-time.

Ariftotle fays, he was the firft that ever obferved it; and adds, it is never feen but at the time of the full moon; her light at other times being too faint to affect the fight after two refractions and one reflection.

The lunar iris has all the colours of the folar, very diftinct and pleafant; only fainter, both from the different intenfity of the rays, and the different difpofition of the medium.

Marine RAINBOW. This is a phenomenon fometimes obferved in a much agitated fea; when the wind, fweeping part of the tops of the waves, carries them aloft; fo that the fun's rays, falling upon them, are refracted, &c. as in a common fhower, and there paint the colours of the bow. Thefe bows are lefs diftinguifhable and bright than the common bow: but then they exceed as to numbers, there being fometimes 20 or 30 feen together. They appear at noon day, and in a pofition oppofite to that of the common bow, the concave fide being turned upwards, as indeed it ought to be.

RAIN GAGE, an inftrument for meafuring the quantity of rain that falls. It is the fame as OMBROMETER, or PLUVIAMETER, which fee.

RAKED *Table*, or RAKING *Table*, in Architecture, a member hollowed in the fquare of a pedeftal, or elfewhere.

RAM, in Aftronomy. See ARIES.

RAM, *battering*. See BATTERING *Ram*.

RAMS HORNS, in Fortification, a name given by Belidor to the Tenailles.

RAMPART, or RAMPIER, in Fortification, a maffy bank or elevation of earth around a place, to cover it from the direct fire of an enemy, and of fufficient thicknefs to refift the efforts of their cannon for many days. It is formed into baftions, curtains, &c.

Upon the Rampart the foldiers continually keep guard, and the pieces of artillery are planted for defence. Alfo, to fhelter the men from the enemy's fhot, the outfide of the Rampart is built higher than the reft, i. e. a parapet is raifed upon it with a platform. It is encompaffed with a moat or ditch, out of which is dug the earth that forms the Rampart, which is raifed floping, that the earth may not flip down, and having a berme at bottom, or is otherwife fortified, being lined with a facing of brick or ftone.

The height of the Rampart need not be more than 3 fathoms, this being fufficient to cover the houfes from the battery of the cannon; neither need its thicknefs be more than 10 or 12, unlefs more earth come out of the ditch than can otherwife be beftowed.

The Ramparts of halfmoons are the better for being low, that the fmall fire of the defendants may the better reach the bottom of the ditch; but yet they muft be fo high as not to be commanded by the covertway.

RAMPART is alfo ufed, in civil architecture, for the void fpace left between the wall of a city and the houfes. This is what the Romans called Pomœrium, where it was forbidden to build, and where they planted rows of trees for the people to walk and amufe themfelves under.

RAMUS (PETER), a celebrated French mathematician and philofopher, was born in 1515, in a village of Vermandois in Picardy. He was defcended of a good family, which had been reduced to extreme poverty by the wars and other misfortunes. His own life too, fays Bayle, was the fport of fortune. In his infancy he was twice attacked by the plague. At 8 years of age, a thirft for learning urged him to go to Paris; but he was foon forced by poverty to leave that city. He returned to it again as foon as he could; but, being unable to fupport himfelf, he left it a fecond time: yet his paffion for ftudy was fo violent, that notwithftanding his bad fuccefs in the two former vifits, he ventured upon a third. He was maintained there fome months by one of his uncles; after which he was obliged to become a fervant in the college of Navarre. Here he fpent the day in waiting upon his mafters, and the greateft part of the night in ftudy.

After having finifhed claffical learning and rhetoric, he went through a courfe of philofophy, which took him up three years and a half in the fchools. The thefis, which he made for his mafter of arts degree, offended every one; for he maintained in it, that all that Ariftotle had advanced was falfe; and he gave very good anfwers to the objections of the profeffors. This fuccefs encouraged him to examine the doctrine of Ariftotle more clofely, and to combat it vigoroufly: but he confined himfelf chiefly to his logic. The two firft books he publifhed, the one entitled, *Inftitutiones Dialecticæ*, the other *Ariftotelicæ Animadverfiones*, occafioned great difturbances in the univerfity of Paris. The profeffors there, who were adorers of Ariftotle, ought to have refuted Ramus's books, if they could, by writings and lectures: but inftead of confining themfelves within the juft bounds of academical wars, they profecuted this anti-peripatetic before the civil magiftrate, as a man who was going to fap the foundations of religion. They raifed fuch clamours, that the caufe was carried before the parliament of Paris: but, perceiving that it would be examined equably, his enemies by their intrigues took it from that tribunal, to bring it before the king's council, in 1543. The king ordered, that Ramus and Anthony Govea, who was his principal adverfary, fhould choofe two judges each, to pronounce on the controverfy, after they fhould have ended their difputation; while he himfelf appointed a deputy. Ramus appeared before the five judges, though three of them were his declared enemies. The difpute lafted two days, and Govea had all the advantages he could defire; Ramus's books being prohibited in all parts of the kingdom, and their author fentenced not to teach philofophy any longer; upon which his enemies triumphed in the moft indecent manner.

The year after, the plague made great havoc in Paris, and forced moft of the ftudents in the college of Prefle to quit it; but Ramus, being prevailed upon to teach in it, foon drew together a great number of auditors. The Sorbonne attempted in vain to drive him from that college; for he held the headfhip of that houfe house by arrêt of parliament. Through the patronage and protection of the cardinal of Lorrain, he obtained from Henry the 2d, in 1547, the liberty of speaking and writing, and the regal professorship of philosophy and eloquence in 1551. The parliament of Paris had, before this, maintained him in the liberty of joining philosophical lectures to those of eloquence; and this arrêt or decree had put an end to several prosecutions, which Ramus and his pupils had suffered. As soon as he was made regius professor, he was fired with a new zeal for improving the sciences, notwithstanding the hatred of his enemies, who were never at rest.

Ramus bore at that time a part in a very singular affair. About the year 1550, the royal professors corrected among other abuses, that which had crept into the pronunciation of the Latin torgue. Some of the clergy followed this regulation; but the Sorbonnists were much offended at it as an innovation, and defended the old pronunciation with great zeal. Things at length were carried so far, that a minister, who had a good living, was very ill treated by them; and caused to be ejected from his benefice for having pronounced *quisquis, quanquam,* according to the new way, instead of *kiskis, kankam,* according to the old. The minister applied to the parliament; and the royal professors, with Ramus among them, fearing he would fall a victim to the credit and authority of the faculty of divines, for presuming to pronounce the Latin tongue according to their regulations, thought it incumbent on them to assist him. Accordingly, they went to the court of justice, and represented in such strong terms the indignity of the prosecution, that the minister was cleared, and every person had the liberty of pronouncing as he pleased.

Ramus was bred up in the Catholic religion, but afterwards deserted it. He began to discover his new principles by removing the images from the chapel of his college of Presle, in 1552. Hereupon such a persecution was raised against him by the Religionists, as well as Aristotelians, that he was driven out of his professorship, and obliged to conceal himself. For that purpose, with the king's leave he went to Fontainbleau; where, by the help of books in the king's library, he prosecuted geometrical and astronomical studies. As soon as his enemies found out his retreat, they renewed their persecutions; and he was forced to conceal himself in several other places. In the mean time, his curious and excellent collection of books in the college of Presle was plundered: but after a peace was concluded in 1563, between Charles the 9th and the Protestants, he again took possession of his employment, maintained himself in it with vigour, and was particularly zealous in promoting the study of the mathematics.

This continued till the second civil war in 1567, when he was forced to leave Paris, and shelter himself among the Hugonots, in whose army he was at the battle of St. Denys. Peace having been concluded some months after, he was restored to his professorship; but, foreseeing that the war would soon break out again, he did not care to venture himself in a fresh storm, and therefore obtained the king's leave to visit the universities of Germany. He accordingly undertook this journey in 1568, and received great honours wherever he came. He returned to France, after the third war in 1571; and lost his life miserably, in the massacre of St. Bartholomew's day, 1572, at 57 years of age. It is said, that he was concealed in a granary during the tumult; but discovered and dragged out by some peripatetic doctors who hated him; these, after stripping him of all his money under pretence of preserving his life, gave him up to the assassins, who, after cutting his throat and giving him many wounds, threw him out of the window; and his bowels gushing out in the fall, some Aristotelian scholars, encouraged by their masters, spread them about the streets; then dragged his body in a most ignominious manner, and threw it into the river.

Ramus was a great orator, a man of universal learning, and endowed with very fine qualities. He was sober, temperate, and chaste. He ate but little, and that of boiled meat; and drank no wine till the latter part of his life, when it was prescribed by the physicians. He lay upon straw; rose early, and studied hard all day; and led a single life with the utmost purity. He was zealous for the protestant religion, but at the same time a little obstinate, and given to contradiction. The protestant ministers did not love him much, for he made himself a kind of head of a party, to change the discipline of the protestant churches: his design was to introduce a democratical government in the church, but this design was traversed, and defeated in a national synod. His sect flourished however for some time afterwards, spreading pretty much in Scotland and England, and still more in Germany.

He published a great many books; but mathematics was chiefly obliged to him. Of this kind, his writings were principally these following:

1. *Scholarum Mathematicarum libri 31.*

2. *Arithmeticæ libri duo.—Algebræ libri duo.—Geometriæ libri 27.*

These were greatly enlarged and explained by Schoner, and published in 2 volumes 4to. There were several editions of them; mine is that of 1627, at Frankfort.—The Geometry, which is chiefly practical, was translated into English by William Bedwell, and published in 4to, at London, 1635.

RANDOM-Shot, is a shot discharged with the axis of the gun elevated above the horizontal or point-blank direction.

Random, of a shot, also sometimes means the range of it, or the distance to which it goes at the first graze, or where it strikes the ground. See Range.

RANGE, in Gunnery, sometimes means the path a shot flies in. But more usually,

Range now means the distance to which the shot flies when it strikes the ground or other object, called also the amplitude of the shot. But Range is the term in present use.

Were it not for the resistance of the air, the greatest Range, on a horizontal plane, would be when the shot is discharged at an angle of 45° above the horizon; and all other Ranges would be the less, the more the angle of elevation is above or below 45°; but so as that at equal distances above and below 45°, the two Ranges are equal to each other. But, on account of the resistance of the air, the Ranges are altered, and that in different proportions, both for the different

sizes

fizes of the fhot, and their different velocities: fo that the greateft Range, in practice, always lies below the elevation of 45°, and the more below it as the fhot is fmaller, and as its velocity is greater; fo as that the fmalleft balls, difcharged with the greateft velocity in practice, ranges the fartheft with an elevation of 30° or under, while the largeft fhot, with very fmall velocities, range fartheft with nearly 45° elevation; and at all the intermediate degrees in the other cafes. See Projectiles.

RARE, in Phyfics, is the quality of a body that is very porous, whofe parts are at a great diftance from one another, and which contains but little matter under a great magnitude. In which fenfe Rare ftands oppofed to denfe.

The corpufcular philofophers, viz, the Epicureans, Gaffendifts, Newtonians, &c, affert that bodies are rarer, fome than others, in virtue of a greater quantity of pores, or of vacuity lying between their parts or particles. The Cartefians hold, that a greater rarity only confifts in a greater quantity of materia fubtilis contained in the pores. And laftly, the Peripatetics contend, that rarity is a new quality fuperinduced upon a body, without any dependence on either vacuity or fubtile matter.

RAREFACTION, in Phyfics, the rendering a body rarer, that is bringing it to expand or occupy more room or fpace, without the acceffion of new matter: and it is oppofed to condenfation. The more accurate writers reftrict the term Rarefaction to that kind of expanfion which is effected by means of heat: and the expanfion from other caufes they term *dilatation*; if indeed there be other caufes; for though fome philofophers have attributed it to the action of a repulfive principle in the matter itfelf; yet from the many difcoveries concerning the nature and properties of the electric fluid and fire, there is great reafon to believe that this repulfive principle is no other than elementary fire.

The Cartefians deny any fuch thing as abfolute Rarefaction: extenfion, according to them, conftituting the effence of matter, they are obliged to hold all extenfion equally full. Hence they make Rarefaction to be no other than an acceffion of frefh, fubtile, and infenfible matter, which, entering the parts of bodies, fenfibly diftends them.

It is by Rarefaction that gunpowder has its effect; and to the fame principle alfo we owe eolipiles, thermometers, &c. As to the air, the degree to which it is rarefiable exceeds all imagination, experience having fhewn it to be far above 10,000 times more than the ufual ftate of the atmofphere; and as it is found to be about 1000 times denfer in gunpowder than the atmofphere, it follows that experience has found it differ by about 10 millions of times. Perhaps indeed its degree of expanfion is abfolutely beyond all limits.

Such immenfe Rarefaction, Newton obferves, is inconceivable on any other principle than that of a repelling force inherent in the air, by which its particles mutually fly from one another. This repelling force, he obferves, is much more confiderable in air than in other bodies, as being generated from the moft fixed bodies, and that with much difficulty, and fcarce without fermentation; thofe particles being always

Vol. II.

found to fly from each other with the greateft force, which, when in contact, cohere the moft firmly together. See Air.

Upon the Rarefaction of the air is founded the ufeful method of meafuring altitudes by the barometer, in all the cafes of which, the rarity of the air is found to be inverfely as the force that compreffes it, or inverfely as the weight of all the air above it at any place.

RARITY, thinnefs, fubtlety, or the contrary to denfity.

RATCH, or Rash, in Clock-Work, a fort of wheel having 12 fangs, which ferve to lift up the detents every hour, to make the clock ftrike.

RATCHETS, in a Watch, are the fmall teeth at the bottom of the fufee, or barrel, that ftop it in winding up.

RATIO, according to Euclid, is the habitude or relation of two magnitudes of the fame kind in refpect of quantity. So the ratio of 2 to 1 is double, that of 3 to 1 triple, &c. Several mathematicians have found fault with Euclid's definition of a Ratio, and others have as much defended it, efpecially Dr. Barrow, in his Mathematical Lectures, with great fkill and learning.

Ratio is fometimes confounded with proportion, but very improperly, as being quite different things; for proportion is the fimilitude or equality or identity of two Ratios. So the Ratio of 6 to 2 is the fame as that of 3 to 1, and the Ratio of 15 to 5 is that of 3 to 1 alfo; and therefore the Ratio of 6 to 2 is fimilar or equal or the fame with that of 15 to 5, which conftitutes proportion, which is thus expreffed, 6 is to 2 as 15 to 5, or thus 6 : 2 : : 15 : 5, which means the fame thing. So that Ratio exifts between two terms, but proportion between two Ratios or four terms.

The two quantities that are compared, are called the *terms* of the Ratio, as 6 and 2; the firft of thefe 6 being called the *antecedent*, and the latter 2 the *confequent*. Alfo the *index* or *exponent* of the Ratio, is the quotient of the two terms: fo the index of the Ratio of 6 to 2 is $\frac{6}{2}$ or 3, and which is therefore called a *triple Ratio*.

Wolfius diftinguifhes Ratios into *rational* and *irrational*.

Rational Ratio is that which can be expreffed between two rational numbers; as the Ratio of 6 to 2, or of $6\sqrt{3}$ to $2\sqrt{3}$, 3 to 1. And

Irrational Ratio is that which cannot be expreffed by that of one rational number to another; as the Ratio of $\sqrt{6}$ to $\sqrt{2}$, or of $\sqrt{3}$ to root $\sqrt{1}$, that is $\sqrt{3}$ to 1, which cannot be expreffed in rational numbers.

When the two terms of a Ratio are equal, the Ratio is faid to be that of *equality*; as of 3 to 3, whofe index is 1, denoting the fingle or equal Ratio. But when the terms are not equal, as of 6 to 2, it is a *Ratio of inequality*.

Farther, when the antecedent is the greater term, as in 6 to 2, it is faid to be the *Ratio of greater inequality*: but when the antecedent is the lefs term, as in the Ratio of 2 to 6, it is faid to be *the Ratio of*

U u *lefs*

less inequality. In the former cafe, if the lefs term be an aliquot part of the greater, the Ratio of greater inequality is faid to be *multiplex* or *multiple;* and the Ratio of the lefs inequality, *fub-multiple.* Particularly, in the firft cafe, if the exponent of the Ratio be 2, as in 6 to 3, the Ratio is called *duple* or *double;* if 3, as in 6 to 2, it is *triple;* and fo on. In the fecond cafe, if the Ratio be 2, as in 3 to 6, the Ratio is called *fub-duple;* if $\frac{1}{3}$, as in 2 to 6, it is *fubtriple;* and fo on.

If the greater term contain the lefs once, and one aliquot part of the fame over; the Ratio of the greater inequality is called *fuperparticular,* and the Ratio of the lefs *fubfuperparticular.* Particularly, in the firft cafe, if the exponent be $\frac{3}{2}$ or $1\frac{1}{2}$, it is called *fefquialterate;* if $\frac{4}{3}$ or $1\frac{1}{3}$, *fefquitertial;* &c. In the other cafe, if the exponent be $\frac{2}{3}$, the Ratio is called *fubfefquialterate;* if $\frac{3}{4}$, it is *fubfefquitertial.*

When the greater term contains the lefs once and feveral aliquot parts over, the Ratio of the greater inequality is called *fuperpartiens,* and that of the lefs inequality is *fubfuperpartiens.* Particularly, in the former cafe, if the exponent be $\frac{5}{3}$ or $1\frac{2}{3}$, the Ratio is called *fuperbipartiens tertias;* if the exponent be $\frac{7}{4}$ or $1\frac{3}{4}$, *fupertripartiens quartas;* if $\frac{11}{7}$ or $1\frac{4}{7}$, *fuperquadripartiens feptimas;* &c. In the latter cafe, if the exponent be the reciprocals of the former, or $\frac{3}{5}$, the Ratio is called *fubfuperbipartiens tertias;* if $\frac{4}{7}$, *fubfupertripartiens quartas;* if $\frac{7}{11}$, *fubfuperquadripartiens feptimas;* &c.

When the greater term contains the lefs feveral times, and fome one part over; the ratio of the greater inequality is called *multiplex fuperparticular;* and the Ratio of the lefs inequality is called *fubmultiplex fubfuperparticular.* Particularly, in the former cafe, if the exponent be $\frac{5}{2}$ or $2\frac{1}{2}$, the ratio is called *dupla fefquialtera;* if $\frac{13}{4}$ or $3\frac{1}{4}$, *tripla fefquiquarta,* &c. In the latter cafe, if the exponent be $\frac{2}{5}$, the Ratio is called *fubdupla fubfefquialtera;* if $\frac{4}{13}$, *fubtripla fubfefquiquarta,* &c. Laftly, when the greater term contains the lefs feveral times, and feveral aliquot parts over; the Ratio of the greater inequality is called *multiplex fuperpartiens;* that of the lefs inequality, *fubmultiplex fubfuperpartiens.* Particularly, in the former cafe, if the exponent be $\frac{8}{3}$ or $2\frac{2}{3}$, the Ratio is called *dupla fuperbipartiens tertias;* if $\frac{25}{7}$ or $3\frac{4}{7}$, *tripla fuperbiquadripartiens feptimas,* &c. In the latter cafe, if the exponent be $\frac{3}{8}$, the Ratio is called *fubdupla fubfuperbipartiens tertias;* if $\frac{7}{25}$, *fubtripla fubfuperquadripartiens feptimas;* &c.

Thefe are the various denominations of rational Ratios, names which are very neceffary to the reading of the ancient authors; though they occur but rarely among the modern writers, who ufe inftead of them the fmalleft numeral terms of the Ratios; fuch 2 to 1 for duple, and 3 to 2 for fefquialterate, &c.

Compound RATIO, is that which is made up of two or more other Ratios, viz, by multiplying the exponents together, and fo producing the compound Ratio of the product of all the antecedents to the product of all the confequents.

Thus the compound Ratio of 5 to 3,
and 7 to 4,
is the Ratio of - - - - 35 to 12.

Particularly, if a Ratio be compounded of two equal Ratios, it is called the *duplicate Ratio;* if of three equal Ratios, the *triplicate Ratio;* if of four equal Ratios, the *quadruplicate Ratio;* and fo on, according to the powers of the exponents, for all *multiplicate Ratios.* So the feveral multiplicate Ratios of

the fimple Ratio of - 3 to 2, are thus, viz.
the duplicate Ratio - 9 : 4,
the triplicate Ratio - 27 : 8,
the quadruplicate Ratio 81 : 16, &c.

Properties of RATIOS. Some of the more remarkable properties of Ratios are as follow:

1. The like multiples, or the like parts, of the terms of a Ratio, have the fame Ratio as the terms themfelves.

So $a : b$, and $na : nb$, and $\frac{a}{n} : \frac{b}{n}$ are all the fame Ratio.

2. If to, or from, the terms of any Ratio, be added or fubtracted either their like parts, or their like multiples, the fums or remainders will ftill have the fame Ratio.

So $a : b$, and $a \pm na : b \pm nb$, and $a \pm \frac{a}{n} : b \pm \frac{b}{n}$ are all the fame Ratio.

3. When there are feveral quantities in the fame continued Ratio, a, b, c, d, e, &c. whatever Ratio the firft has to the 2d,

the 1ft to the 3d has the duplicate of that Ratio,
the 1ft to the 4th has the triplicate of that Ratio,
the 1ft to the 5th has the quadruplicate of it,

and fo on. Thus, the terms of the continued Ratio being $1, r, r^2, r^3, r^4, r^5$, &c, where each term has to the following one the Ratio of 1 to r, the Ratio of the 1ft to the 2d; then $1 : r^2$ is the duplicate, $1 : r^3$ the triplicate, $1 : r^4$ the quadruplicate, and fo on, according to the powers of $1 : r$.

For other properties fee PROPORTION.

To approximate to a RATIO *in fmaller Terms.*—Dr. Wallis, in a fmall tract at the end of Horrox's works, treats of the nature and folution of this problem, but in a very tedious way; and he has profecuted the fame to a great length in his Algebra, chap. 10 and 11, where he particularly applies it to the Ratio of the diameter of a circle to its circumference. Mr. Huygens too has given a folution, with the reafons of it, in a much fhorter and more natural way, in his Defcrip. Autom. Planet. Opera Reliqua, vol 1, pa. 174.

So alfo has Mr. Cotes, at the beginning of his Harmon. Menfurarum. And feveral other perfons have done the fame thing, by the fame or fimilar methods. The problem is very ufeful, for expreffing a Ratio in fmall numbers, that fhall be near enough in practice, to any given Ratio in large numbers, fuch as that of the diameter of a circle to its circumference. The principle of all thefe methods, confifts in reducing the terms of the propofed Ratio into a feries of what are called continued fractions, by dividing the greater term by the lefs, and the lefs by the remainder, and fo on, always the laft divifor by the laft remainder, after the manner of finding the greateft common meafure of the two terms; then connecting all the quotients &c together in a feries of continued fractions; and laftly collecting gradually thefe fractions together one after another.

So if $\frac{b}{a}$ be any fraction, or exponent of any Ratio; then dividing thus,

$a) b (c$

a) $\dfrac{b}{\ \ }$ (c

d) $\dfrac{a}{\ \ }$ (e

f) $\dfrac{d}{\ \ }$ (g

b) $\dfrac{f}{\ \ }$ (i

k) $\dfrac{b}{\ \ }$ (l

&c.

gives c, e, g, i, &c, for the feveral quotient, **and thefe,** formed in the ufual way, give the approximate value of the given Ratio in a feries of continued fractions; thus,

$$\frac{b}{a} = c + \frac{1}{e} + \frac{1}{g} + \frac{1}{i} + \&c.$$

Then collecting the terms of this feries, one after another, fo many values of $\dfrac{b}{a}$ are obtained, always nearer and nearer; the firft value being c or $\dfrac{c}{1}$, the next

$$c + \frac{1}{e} = \frac{ce+1}{e} = \frac{A}{B},$$

the 3d value $c + \dfrac{1}{e} + \dfrac{1}{g} = c + \dfrac{1}{\dfrac{ge+1}{g}} = c + \dfrac{g}{ge+1} =$

$$\frac{cge + c + g}{ge + 1} = \frac{(ce+1)\,g + c}{ge + 1} = \frac{Ag + c}{Bg + 1} = \frac{C}{D};$$

in like manner,

the 4th value is $\dfrac{Ci + A}{Di + B} = \dfrac{E}{F}$;

the 5th value is $\dfrac{El + C}{Fl + D} = \dfrac{G}{H}$; &c.

From whence comes this general rule: Having found any two of thefe values, multiply the terms of the latter of them by the next quotient, and to the two products add the correfponding terms of the former value, and the fums will be the terms of the next value, &c.

For example, let it be required to find a feries of Ratios in leffer numbers, conftantly approaching to the Ratio of 100000 to 314159, or nearly the Ratio of the diameter of a circle to its circumference. Here firft dividing, thus,

100000) 314159 (3 = c

$d =$ 14159) 100000 (7 = e

$f =$ 887) 14159 (15 = g

$b =$ 854) 887 (1 = i, &c.

&c.

there are obtained the quotients 3, 7, 15, 1, 25, 1, 7, 4.

Hence 3 or $\dfrac{3}{1} = c$, the 1ft value;

$\dfrac{ce+1}{e} = \dfrac{3.7+1}{1.7} = \dfrac{22}{7} = \dfrac{A}{B}$, the 2d value;

$\dfrac{Ag+c}{Bg+1} = \dfrac{22.15+3}{7.15+1} = \dfrac{333}{106} = \dfrac{C}{D}$, the 3d value;

$\dfrac{Ci + A}{Di + B} = \dfrac{333.1 + 22}{106.1 + 7} = \dfrac{355}{113} = \dfrac{E}{F}$, the 4th value;

and fo on; where the fucceffive continual approximating values of the propofed Ratio are $\dfrac{3}{1}$, $\dfrac{22}{7}$, $\dfrac{333}{106}$, $\dfrac{355}{113}$, &c; the 2d of thefe, viz. $\dfrac{22}{7}$, being the approximation of Archimedes; and the 4th, viz $\dfrac{355}{133}$, is that of Metius, which is very near the truth, being equal to 3·1415929, the more accurate Ratio being - - 3·1415927.

The doctrine of Ratios and Proportions, as delivered by Euclid, in the fifth book of his Éléments, is confidered by moft perfons as very obfcure and objectionable, particularly the definition of proportionality; and feveral ingenious gentlemen have endeavoured to elucidate that fubject. Among thefe, the Rev. Mr. Abram Robertfon, of Chrift Church College, Oxford, lecturer in geometry in that univerfity, printed a neat little paper there in 1789, for the ufe of his claffes, being a demonftration of that definition, in 7 propofitions, the fubftance of which is as follows. He firft premifes this advertifement:

" As demonftrations depending upon proportionality pervade every branch of mathematical fcience, it is a matter of the higheft importance to eftablifh it upon clear and indifputable principles. Moft mathematicians, both ancient and modern, have been of opinion that Euclid has fallen fhort of his ufual perfpicuity in this particular. Some have queftioned the truth of the definition upon which he has founded it, and, almoft all who have admitted its truth and validity have objected to it *as a definition*. The author of the following propofitions ranks himfelf amongft objectors of the laft mentioned defcription. He thinks that Euclid muft have founded the definition in queftion upon the reafoning contained in the firft fix demonftrations here given, or upon a fimilar train of thinking; and in his opinion *a definition* ought to be as fimple, or as free from a multiplicity of conditions, as the fubject will admit."

He then lays down thefe four definitions:

" 1. Ratio is the relation which one magnitude has to another, of the fame kind, with refpect to quantity."

" 2. If the firft of four magnitudes be exactly as great when compared to the fecond, as the third is when compared to the fourth, the firft is faid to have to the fecond the fame Ratio that the third has to the fourth."

" 3. If the firft of four magnitudes be greater, when compared to the fecond, than the third is when compared to the fourth, the firft is faid to have to the fecond a greater Ratio than the third has to the fourth."

" 4. If the firft of four magnitudes be lefs, when compared to the fecond, than the third is when compared to the fourth, the firft is faid to have to the fecond a lefs Ratio than the third has to the fourth."

Mr. Robertfon then delivers the propofitions, which are the following:

" *Prop.* 1. If the firft of four magnitudes have to the fecond, the fame Ratio which the third has to the fourth; then,

then, if the firſt be equal to the ſecond, the third is equal to the fourth; if greater, greater; if leſs, leſs."

"*Prop.* 2. If the firſt of four magnitudes be to the ſecond as the third to the fourth, and if any equimultiples whatever of the firſt and third be taken, and alſo any equimultiples of the ſecond and fourth; the multiple of the firſt will be to the multiple of the ſecond as the multiple of the third to the multiple of the fourth."

"*Prop.* 3. If the firſt of four magnitudes be to the ſecond as the third to the fourth, and if any like aliquot parts whatever be taken of the firſt and third, and any like aliquot parts whatever of the ſecond and fourth, the part of the firſt will be to the part of the ſecond as the part of the third to the part of the fourth."

"*Prop.* 4. If the firſt of four magnitudes be to the ſecond as the third to the fourth, and if any equimultiples whatever be taken of the firſt and third, and any whatever of the ſecond and fourth; if the multiple of the firſt be equal to the multiple of the ſecond, the multiple of the third will be equal to the multiple of the fourth; if greater, greater; if leſs, leſs."

"*Prop.* 5. If the firſt of four magnitudes be to the ſecond as the third is to a magnitude leſs than the fourth, then it is poſſible to take certain equimultiples of the firſt and third, and certain equimultiples of the ſecond and fourth, ſuch, that the multiple of the firſt ſhall be greater than the multiple of the ſecond, but the multiple of the third not greater than the multiple of the fourth."

"*Prop.* 6. If the firſt of four magnitudes be to the ſecond as the third is to a magnitude greater than the fourth, then certain equimultiples can be taken of the firſt and third, and certain equimultiples of the ſecond and fourth, ſuch, that the multiple of the firſt ſhall be leſs than the multiple of the ſecond, but the multiple of the third not leſs than the multiple of the fourth."

"*Prop.* 7. If any equimultiples whatever be taken of the firſt and third of four magnitudes, and any equimultiples whatever of the ſecond and fourth; and if when the multiple of the firſt is leſs than that of the ſecond, the multiple of the third is alſo leſs than that of the fourth; or if when the multiple of the firſt is equal to that of the ſecond, the multiple of the third is alſo equal to that of the fourth; or if when the multiple of the firſt is greater than that of the ſecond, the multiple of the third is alſo greater than that of the fourth: then, the firſt of the four magnitudes ſhall be to the ſecond as the third to the fourth."

And all theſe propoſitions Mr. Robertſon demonſtrates by ſtrict mathematical reaſoning.

RATIONAL, in Arithmetic &c, the quality of numbers, fractions, quantities, &c, when they can be expreſſed by common numbers; in contradiſtinction to irrational or ſurd ones, which cannot be expreſſed in common numbers. Suppoſe any quantity to be 1; there are infinite other quantities, ſome of which are commenſurable to it, either ſimply, or in power: theſe Euclid calls *Rational quantities.* The reſt, that are incommenſurable to 1, he calls *irrational quantities,* or *ſurds.*

RATIONAL *Horizon,* or *True Horizon,* is that whoſe plane is conceived to paſs through the centre of the earth; and which therefore divides the globe into two equal portions or hemiſpheres. It is called the Rational horizon, becauſe only conceived by the underſtanding;

in oppoſition to the ſenſible or apparent horizon, or that which is viſible to the eye.

RAVELIN, in Fortification, was anciently a flat baſtion, placed in the middle of a curtain. But

RAVELIN is now a detached work, compoſed only of two faces, which form a ſalient angle uſually without flanks. Being a triangular work reſembling the point of a Baſtion with the flanks cut off. It raiſed before the curtain, on the counterſcarf of the place; and ſerving to cover it and the adjacent flanks from the direct fire of an enemy. It is alſo uſed to cover a bridge or a gate, and is always placed without the moat.

There are alſo double Ravelins, which ſerve to defend each other; being ſo called when they are joined by a curtain.

What the engineers call a Ravelin, the men uſually call a demilune, or halfmoon.

RAY, in Geometry, the ſame as RADIUS.

RAY, in Optics, a beam or line of light, propagated from a radiant point, through any medium.

If the parts of a Ray of light lie all in a ſtraight line between the radiant point and the eye, the Ray is ſaid to be *direct:* the laws and properties of which make the ſubject of Optics.—If any of them be turned out of that direction, or bent in their paſſage, the Ray is ſaid to be *refracted.*—If it ſtrike on the ſurface of any body, and be thrown off again, it is ſaid to be *reflected.*—In each caſe, the Ray, as it falls either directly on the eye, or on the point of reflection, or of refraction, is ſaid to be *incident.*

Again, if ſeveral Rays be propagated from the radiant object equidiſtantly from one another, they are called *parallel* Rays. If they come inclining towards each other, they are called *converging* Rays. And if they go continually receding from each other, they are called *diverging* Rays.

It is from the different circumſtances of Rays, that the ſeveral kinds of bodies are diſtinguiſhed in Optics. A body, for example, that diffuſes its own light, or emits Rays of its own, is called a *radiating* or *lucid* or *luminous* body. If it only reflect Rays which it receives from another, it is called an *illuminated* body. If it only tranſmit Rays, it is called a *tranſparent* or *tranſlucent* body. If it intercept the Rays, or refuſe them paſſage, it is called an *opaque* body.

It is by means of Rays reflected from the ſeveral points of illuminated objects to the eye, that they become viſible, and that viſion is performed; whence ſuch Rays are called *viſual* Rays.

The Rays of light are not homogeneous, or ſimilar, but differ in all the properties we know of; viz, refrangibility, reflexibility, and colour. It is probably from the different refrangibility that the other differences have their riſe; at leaſt it appears that thoſe Rays which agree or differ in this, do ſo in all the reſt. It is not however to be underſtood that the property or effect called colour, exiſts in the Rays of light themſelves; but from the different ſenſations the differently diſpoſed Rays excite in us, we call them *red Rays, yellow Rays,* &c. Each beam of light however, as it comes from the ſun, ſeems to be compounded of all the ſorts of Rays mixed together; and it is only by ſplitting or ſeparating the parts of it, that theſe different ſorts become obſervable; and this is done by tranſmitting the

beam

beam through a glaſs priſm, which refraçting it in the paſſage, and the parts that excite the different colours having different degrees of refrangibility, they are thus ſeparated from one another, and exhibited each apart, and appearing of the different colours.

Beſide refrangibility, and the other properties of the Rays of light already aſcertained by obſervation and experiment, Sir I. Newton ſuſpeçts they may have many more; particularly a power of being infleçted or bent by the açtion of diſtant bodies; and thoſe Rays which differ in refrangibility, he conceives likewiſe to differ in flexibility.

Theſe Rays he ſuſpeçts may be very ſmall bodies emitted from ſhining ſubſtances. Such bodies may have all the conditions of light: and there is that açtion and reaçtion between tranſparent bodies and light, which very much reſembles the attraçtive force between other bodies. Nothing more is required for the produçtion of all the various colours, and all the degrees of refrangibility, but that the Rays of light be bodies of different ſizes; the leaſt of which may make violet the weakeſt and darkeſt of the colours, and be the moſt eaſily diverted by refraçting ſurfaces from its reçtilinear courſe; and the reſt, as they are larger and larger, may make the ſtronger and more lucid colours, blue, green, yellow, and red. See COLOUR, LIGHT, REFRACTION, REFLECTION, INFLECTION, CONVERGING, DIVERGING, &c, &c.

Reflected RAYS, thoſe Rays of light which are reflected, or thrown back again, from the ſurfaces of bodies upon which they ſtrike. It is found that, in all the Rays of light, the angle of reflеçtion is equal to the angle of incidence.

Refracted RAYS, are thoſe Rays of light, which are bent or broken, in paſſing out of one medium into another.

Pencil of RAYS, a number of Rays iſſued from a point of an objeçt, and diverging in the form of a cone.

Principal RAY, in Perſpeçtive, is the perpendicular diſtance between the eye and the vertical plane or table, as ſome call it.

RAY *of Curvature.* See *Radius of* CURVATURE.

REAUMUR (RENE-ANTOINE-FERCHAULT, Sieur de), a reſpeçtable French philoſopher, was born at Rochelle in 1683. After the uſual courſe of ſchool education, he was ſent to Poitiers to ſtudy philoſophy, and, in 1699, to Bourges to ſtudy the law, the profeſſion for which he was intended. But philoſophy and mathematics having very early been his favourite purſuits, he quitted the law, and repaired to Paris in 1703, to purſue thoſe ſciences to the beſt advantage; and here his charaçter procured him a ſeat in the Academy in the year 1708; which he held till the time of his death, which happened the 18th of November 1757, at 74 years of age.

Reaumur ſoon juſtified the choice that was made of him by the Academy. He made innumerable obſervations, and wrote a great multitude of pieces upon the various branches of natural philoſophy. His Hiſtory of Inſeçts, in 6 vols. quarto, at Paris, is his principal work. Another edition was printed in Holland, in 12 vols. 12mo. He made alſo great and uſeful diſcoveries concerning iron; ſhewing how to change common wrought iron into ſteel, how to ſoften caſt iron, and to make works in caſt iron as fine as in wrought iron. His labours and diſcoveries concerning iron were rewarded by the duke of Orleans, regent of the kingdom, by a penſion of 12 thouſand livres, equal to about 500l. Sterling; which however he would not accept but on condition of its being put under the name of the Academy, who might enjoy it after his death. It was owing to Reaumur's endeavours that there were eſtabliſhed in France manufaçtures of tin-plates, of porcelain in imitation of china-ware, &c. They owe to him alſo a new thermometer, which bears his name, and is pretty generally uſed on the continent, while that of Fahrenheit is uſed in England, and ſome few other places. Reaumùr's thermometer is a ſpirit one, having the freezing point at 0, and the boiling point at 80.

Reaumur is eſteemed as an exaçt and clear writer; and there is an elegance in his ſtyle and manner, which is not commonly found among thoſe who have made only the ſciences their ſtudy. He is repreſented alſo as a man of a moſt amiable diſpoſition, and with qualities to make him beloved as well as admired. He left a great variety of papers and natural curioſities, which he bequeathed to the Academy of Sciences.

The works publiſhed by him, are the following.

1. The Art of changing Forged Iron into Steel; of Softening Caſt Iron; and of making works of Caſt Iron, as fine as of Wrought Iron. Paris, 1722, 1 vol. in 4to.

2. Natural Hiſtory of Inſeçts, 6 vols. in 4to.

His memoirs printed in the volumes of the Academy of Sciences, are very numerous, amounting to upwards of a hundred, and on various ſubjeçts, from the year 1708 to 1763, ſeveral papers in almoſt every volume.

RECEIVER, *of an Air Pump,* is part of its apparatus; being a glaſs veſſel placed on the top of the plate, out of which the air is to be exhauſted.

RECEPTION, in Aſtrology, is a dignity befalling two planets when they exchange houſes: for example, when the ſun arrives in Cancer, the houſe of the moon; and the moon, in her turn, arrives in the ſun's houſe.—The ſame term is alſo uſed when two planets exchange exaltation.

RECESSION *of the Equinoxes.* See PRECESSION *of the Equinoxes.*

RECIPROCAL, in Arithmetic, &c, is the quotient ariſing by dividing 1 by any number or quantity So, the Reciprocal of 2 is $\frac{1}{2}$; of 3 is $\frac{1}{3}$, and of a is $\frac{1}{a}$, &c. Hence, the Reciprocal of a vulgar fraçtion is found, by barely making the numerator and the denominator mutually change places: ſo the Reciprocal of $\frac{1}{2}$ is $\frac{2}{1}$ or 2; of $\frac{2}{3}$, is $\frac{3}{2}$; of $\frac{a}{b}$, is $\frac{b}{a}$, &c. Hence alſo, any quantity being multiplied by its Reciprocal, the product is always equal to unity or 1: ſo $\frac{1}{2} \times \frac{2}{1} = \frac{2}{2} = 1$, and $\frac{2}{3} \times \frac{3}{2} = \frac{6}{6} = 1$, and $\frac{a}{b} \times \frac{b}{a} = \frac{ab}{ab} = 1$.

TABLE

Table of RECIPROCALS.

No.	Recip.	No.	Recip.	No	Recip.	No.	Recip.	No.	Recip.	No.	Recip.
1	1	61	0163934	121	0082645	181	0055249	241	0041494	301	0033223
2	5	62	0161290	122	0081967	182	0054945	242	0041322	302	0033113
3	3333333	63	0158730	123	0081300	183	0054645	243	0041152	303	0033003
4	25	64	015625	124	0080645	184	0054348	244	0040984	304	0032895
5	2	65	0153846	125	008	185	0054054	245	0040816	305	0032787
6	1666666	66	0151515	126	0079365	186	0053763	246	004065	306	0032680
7	1428571	67	0149254	127	0078740	187	0053476	247	0040486	307	0032573
8	125	68	0147059	128	0078125	188	0053191	248	0040323	308	0032468
9	1111111	69	0144928	129	0077519	189	0052910	249	0040161	309	0032362
10	1	70	0142857	130	0076923	190	0052632	250	004	310	0032258
11	0909090	71	0140845	131	0076336	191	0052356	251	0039841	311	0032154
12	0833333	72	0138888	132	0075757	192	0052083	252	0039683	312	0032051
13	0769230	73	0136986	133	0075188	193	0051813	253	0039526	313	0031949
14	0714285	74	0135135	134	0074627	194	0051546	254	0039370	314	0031847
15	0666666	75	0133333	135	0074074	195	0051282	255	0039216	315	0031746
16	0625	76	0131579	136	0073529	196	0051020	256	0039063	316	0031646
17	0588235	77	0129870	137	0072993	197	0050761	257	0038911	317	0031546
18	0555555	78	0128205	138	0072464	198	0050505	258	0038760	318	0031447
19	0526316	79	0126582	139	0071942	199	0050251	259	0038610	319	0031348
20	05	80	0125	140	0071429	200	005	260	0038462	320	003125
21	0476190	81	0123457	141	0070922	201	0049751	261	0038314	321	0031153
22	0454545	82	0121950	142	0070423	202	0049504	262	0038168	322	0031056
23	0434783	83	0120482	143	0069930	203	0049261	263	0038023	323	0030960
24	0416666	84	0119048	144	0069444	204	0049020	264	0037878	324	0030846
25	04	85	0117647	145	0068966	205	0048750	265	0037736	325	0030769
26	0384615	86	0116279	146	0068493	206	0048544	266	0037594	326	0030675
27	0370370	87	0114943	147	0068027	207	0048309	267	0037453	327	0030581
28	0357143	88	0113636	148	0067567	208	0048077	268	0037313	328	0030488
29	0344828	89	0112360	149	0067114	209	0047847	269	0037175	329	0030395
30	0333333	90	0111111	150	0066666	210	0047619	270	0037037	330	0030303
31	0322581	91	0109890	151	0066225	211	0047393	271	0036900	331	0030211
32	03125	92	0108696	152	0065789	212	0047170	272	0036765	332	0030120
33	0303030	93	0107527	153	0065359	213	0046948	273	0036630	333	0030030
34	0294118	94	0106383	154	0064935	214	0046729	274	0036496	334	0029940
35	0285714	95	0105263	155	0064516	215	0046512	275	0036363	335	0029851
36	0277777	96	0104166	156	0064103	216	0046296	276	0036232	336	0029762
37	0270270	97	0103093	157	0063694	217	0046083	277	0036101	337	0029674
38	0263158	98	0102041	158	0063291	218	0045872	278	0035971	338	0029586
39	0256410	99	0101010	159	0062893	219	0045662	279	0035842	339	0029499
40	025	100	01	160	00625	220	0045454	280	0035714	340	0029412
41	0243902	101	0099009	161	0062112	221	0045249	281	0035587	341	0029326
42	0238095	102	0098039	162	0061728	222	0045045	282	0035461	342	0029240
43	0232558	103	0097087	163	0061350	223	0044843	283	0035336	343	0029155
44	0227272	104	0096154	164	0060975	224	0044643	284	0035211	344	0029070
45	0222222	105	0095238	165	0060606	225	0044444	285	0035088	345	0028986
46	0217391	106	0094340	166	0060241	226	0044248	286	0034965	346	0028902
47	0212766	107	0093458	167	0059880	227	0044053	287	0034843	347	0028818
48	0208333	108	0092592	168	0059524	228	0043860	288	0034722	348	0028736
49	0204082	109	0091743	169	0059172	229	0043668	289	0034602	349	0028653
50	02	110	0090909	170	0058824	230	0043478	290	0034483	350	0028571
51	0196078	111	0090090	171	0058480	231	0043290	291	0034364	351	0028490
52	0192308	112	0089286	172	0058141	232	0043103	292	0034246	352	0028409
53	0188679	113	0088496	173	0057803	233	0042918	293	0034130	353	0028329
54	0185185	114	0087719	174	0057471	234	0042735	294	0034014	354	0028248
55	0181818	115	0086957	175	0057143	235	0042553	295	0033898	355	0028169
56	0178571	116	0086207	176	0056818	236	0042373	296	0033783	356	0028070
57	0175439	117	0085470	177	0056497	237	0042194	297	0033670	357	0028011
58	0172414	118	0084745	178	0056180	238	0042017	298	0033557	358	0027932
59	0169490	119	0084034	179	0055866	239	0041841	299	0033445	359	0027855
60	0166666	120	0083333	180	0055555	240	0041666	300	0033332	360	0027777

Table of RECIPROCALS.

No.	Recip.	No.	Recip.	No.	Recip.	No.	Recip.	No.	Recip.	No.	Recip.
361	0027701	421	0023753	481	0020790	541	0018484	601	0016639	661	0015129
362	0027624	422	0023697	482	0020747	542	0018450	602	0016611	662	0015106
363	0027548	423	0023641	483	0020704	543	0018416	603	0016584	663	0015083
364	0027473	424	0023585	484	0020661	544	0018382	604	0016556	664	0015060
365	0027397	425	0023529	485	0020619	545	0018349	605	0016529	665	0015038
366	0027322	426	0023474	486	0020576	546	0018315	606	0016501	666	0015015
367	0027248	427	0023419	487	0020534	547	0018282	607	0016474	667	0014993
368	0027174	428	0023364	488	0020492	548	0018248	608	0016447	668	0014970
369	0027100	429	0023310	489	0020450	549	0018215	609	0016420	669	0014948
370	0027027	430	0023256	490	0020408	550	0018181	610	0016393	670	0014925
371	0026954	431	0023202	491	0020367	551	0018149	611	0016367	671	0014903
372	0026882	432	0023148	492	0020325	552	0018116	612	0016340	672	0014881
373	0026810	433	0023095	493	0020284	553	0018083	613	0016313	673	0014859
374	0026738	434	0023042	494	0020243	554	0018051	614	0016287	674	0014837
375	0026666	435	0022989	495	0020202	555	0018018	615	0016260	675	0014814
376	0026596	436	0022936	496	0020162	556	0017986	616	0016234	676	0014793
377	0026525	437	0022883	497	0020121	557	0017953	617	0016207	677	0014771
378	0026455	438	0022831	498	0020080	558	0017921	618	0016181	678	0014749
379	0026385	439	0022779	499	0020040	559	0017889	619	0016155	679	0014728
380	0026316	440	0022727	500	002	560	0017857	620	0016129	680	0014706
381	0026247	441	0022676	501	0019960	561	0017825	621	0016103	681	0014684
382	0026178	442	0022624	502	0019920	562	0017794	622	0016077	682	0014663
383	0026110	443	0022573	503	0019881	563	0017762	623	0016051	683	0014641
384	0026042	444	0022522	504	0019841	564	0017730	624	0016026	684	0014620
385	0025974	445	0022472	505	0019801	565	0017699	625	0016	685	0014599
386	0025907	446	0022422	506	0019763	566	0017668	626	0015974	686	0014577
387	0025840	447	0022371	507	0019724	567	0017637	627	0015949	687	0014556
388	0025773	448	0022321	508	0019685	568	0017606	628	0015924	688	0014535
389	0025707	449	0022272	509	0019646	569	0017575	629	0015898	689	0014514
390	0025641	450	0022222	510	0019608	570	0017544	630	0015873	690	0014493
391	0025575	451	0022173	511	0019569	571	0017513	631	0015848	691	0014472
392	0025510	452	0022124	512	0019531	572	0017483	632	0015823	692	0014451
393	0025445	453	0022075	513	0019493	573	0017452	633	0015798	693	0014430
394	0025381	454	0022026	514	0019455	574	0017422	634	0015773	694	0014409
395	0025316	455	0021978	515	0019417	575	0017391	635	0015748	695	0014388
396	0025252	456	0021930	516	0019380	576	0017361	636	0015723	696	0014368
397	0025189	457	0021882	517	0019342	577	0017331	637	0015699	697	0014347
398	0025126	458	0021834	518	0019305	578	0017301	638	0015674	698	0014327
399	0025063	459	0021786	519	0019268	579	0017271	639	0015649	699	0014306
400	0025	460	0021739	520	0019231	580	0017241	640	0015625	700	0014286
401	0024938	461	0021692	521	0019194	581	0017212	641	0015601	701	0014265
402	0024876	462	0021645	522	0019157	582	0017182	642	0015576	702	0014245
403	0024814	463	0021598	523	0019120	583	0017153	643	0015552	703	0014225
404	0024752	464	0021552	524	0019084	584	0017123	644	0015528	704	0014205
405	0024691	465	0021505	525	0019048	585	0017094	645	0015504	705	0014184
406	0024631	466	0021459	526	0019011	586	0017065	646	0015480	706	0014164
407	0024570	467	0021413	527	0018975	587	0017036	647	0015456	707	0014144
408	0024510	468	0021368	528	0018939	588	0017007	648	0015432	708	0014124
409	0024450	469	0021322	529	0018904	589	0016978	649	0015408	709	0014104
410	0024390	470	0021277	530	0018868	590	0016949	650	0015385	710	0014085
411	0024331	471	0021231	531	0018832	591	0016920	651	0015361	711	0014065
412	0024272	472	0021187	532	0018797	592	0016891	652	0015337	712	0014045
413	0024213	473	0021142	533	0018762	593	0016863	653	0015314	713	0014025
414	0024155	474	0021097	534	0018727	594	0016835	654	0015291	714	0014006
415	0024096	475	0021053	535	0018692	595	0016807	655	0015267	715	0013986
416	0024038	476	0021008	536	0018657	596	0016779	656	0015244	716	0013966
417	0023981	477	0020964	537	0018622	597	0016750	657	0015221	717	0013947
418	0023923	478	0020921	538	0018587	598	0016722	658	0015198	718	0013928
419	0023866	479	0020877	539	0018553	599	0016694	659	0015175	719	0013908
420	0023810	480	0020833	540	0018518	600	0016666	660	0015151	720	0013888

Table of RECIPROCALS.

No.	Recip.	No.	Recip.	No.	Recip.	No.	Recip.	No.	Recip.	No.	Recip.
721	0013870	768	0013021	815	0012270	862	0011601	909	0011001	956	0010460
722	0013850	769	0013004	816	0012255	863	0011587	910	0010989	957	0010449
723	0013831	770	0012987	817	0012240	864	0011574	911	0010977	958	0010438
724	0013812	771	0012970	818	0012225	865	0011561	912	0010965	959	0010428
725	0013793	772	0012953	819	0012210	866	0011547	913	0010953	960	0010416
726	0013774	773	0012937	820	0012195	867	0011534	914	0010941	961	0010406
727	0013755	774	0012920	821	0012180	868	0011521	915	0010929	962	0010395
728	0013736	775	0012903	822	0012165	869	0011507	916	0010917	963	0010384
729	0013717	776	0012887	823	0012151	870	0011494	917	0010905	964	0010373
730	0013699	777	0012870	824	0012136	871	0011481	918	0010893	965	0010363
731	0013680	778	0012853	825	0012121	872	0011468	919	0010881	966	0010352
732	0013661	779	0012837	826	0012106	873	0011455	920	0010870	967	0010341
733	0013643	780	0012821	827	0012092	874	0011442	921	0010858	968	0010331
734	0013624	781	0012804	828	0012077	875	0011429	922	0010846	969	0010320
735	0013605	782	0012788	829	0012063	876	0011416	923	0010834	970	0010309
736	0013587	783	0012771	830	0012048	877	0011403	924	0010823	971	0010299
737	0013569	784	0012755	831	0012034	878	0011390	925	0010810	972	0010288
738	0013550	785	0012739	832	0012019	879	0011377	926	0010799	973	0010277
739	0013532	786	0012723	833	0012005	880	0011363	927	0010787	974	0010267
740	0013513	787	0012706	834	0011990	881	0011351	928	0010776	975	0010256
741	0013495	788	0012690	835	0011976	882	0011338	929	0010764	976	0010246
742	0013477	789	0012674	836	0011962	883	0011325	930	0010753	977	0010235
743	0013459	790	0012658	837	0011947	884	0011312	931	0010741	978	0010225
744	0013441	791	0012642	838	0011933	885	0011299	932	0010730	979	0010215
745	0013423	792	0012626	839	0011919	886	0011287	933	0010718	980	0010204
746	0013405	793	0012610	840	0011905	887	0011274	934	0010707	981	0010194
747	0013387	794	0012594	841	0011891	888	0011261	935	0010695	982	0010183
748	0013369	795	0012579	842	0011876	889	0011249	936	0010684	983	0010173
749	0013351	796	0012563	843	0011862	890	0011236	937	0010672	984	0010163
750	0013333	797	0012547	844	0011848	891	0011223	938	0010661	985	0010152
751	0013316	798	0012531	845	0011834	892	0011211	939	0010650	986	0010142
752	0013298	799	0012516	846	0011820	893	0011198	940	0010638	987	0010132
753	0013280	800	00125	847	0011806	894	0011186	941	0010627	988	0010121
754	0013263	801	0012484	848	0011792	895	0011173	942	0010616	989	0010111
755	0013245	802	0012469	849	0011779	896	0011161	943	0010604	990	0010101
756	0013228	803	0012453	850	0011765	897	0011148	944	0010593	991	0010091
757	0013210	804	0012438	851	0011751	898	0011136	945	0010582	992	0010081
758	0013193	805	0012422	852	0011737	899	0011123	946	0010571	993	0010070
759	0013175	806	0012407	853	0011723	900	0011111	947	0010560	994	0010060
760	0013158	807	0012392	854	0011710	901	0011099	948	0010549	995	0010050
761	0013141	808	0012376	855	0011696	902	0011086	949	0010537	996	0010040
762	0013123	809	0012361	856	0011682	903	0011074	950	0010526	997	0010030
763	0013106	810	0012346	857	0011669	904	0011062	951	0010515	998	0010020
764	0013089	811	0012330	858	0011655	905	0011050	952	0010504	999	0010010
765	0013072	812	0012315	859	0011641	906	0011038	953	0010493	1000	001
766	0013055	813	0012300	860	0011628	907	0011025	954	0010482		
767	0013038	814	0012285	861	0011614	908	0011013	955	0010471		

Of the preceding Table, the use is evidently to shorten arithmetical calculations, and will appear eminently great to those mathematicians and others who are frequently concerned in such kinds of computations. The structure of the Table is evident; the first column contains the natural series of numbers from 1 to 1000, the 2d the Reciprocals. These Reciprocals (which are no other than the decimal values of the quotients resulting from the division of unity or 1 by each of the several numbers from 1 to 1000) are not only useful in shewing by inspection the quotient when the dividend is unity, but are also applied with much advantage in turning many divisions into multiplications, which are much easier performed, and are done by multiplying the Reciprocal of the divisor (as found in the Table) by the dividend, for the quotient; they will also apply to good purpose in summing the terms of many converging series.

The Reciprocals are carried on to 7 places of decimals (for the column of Reciprocals must be accounted all decimal figures, although they have not the decimal point placed before them, which is omitted to save room), each being set down to the nearest figure in the last place, that is, when the next figure beyond the last set down in the Table came out a 5 or more, the last figure

figure was increafed by 1, otherwife not; excepting in the repetends which occurred among the Reciprocals, where the real laft figure is always fet down; the Reciprocals, which in the Table confift of lefs than feven figures, are thofe which terminate, and are complete within that number; fuch as ·5 the Reciprocal of 2, ·25 the Reciprocal of 4, &c.

RECIPROCAL *Figures*, in Geometry, are fuch as have the antecedents and confequents of the fame ratio in both figures. So, in the two rectangles BE and BD, if AB : DC :: BC : AE, then thofe rectangles are reciprocal figures; and are alfo equal.

RECIPROCAL *Proportion*, is when, in four quantities, the two latter terms have the Reciprocal ratio of the two former, or are proportional to the Reciprocals of them. Thus, 24, 15, 5, 8 form a Reciprocal proportion, becaufe

$$\frac{1}{24} : \frac{1}{15} :: 5 : 8, \text{ or } 15 : 24 :: 5 : 8.$$

RECIPROCAL *Ratio*, of any quantity, is the ratio of the Reciprocal of the quantity.

RECIPROCALLY. One quantity is Reciprocally as another, when the one is greater in proportion as the other is lefs; or when the one is proportional to the Reciprocal of the other. So *a* is Reciprocally as *b*, when *a* is always proportional to $\frac{1}{b}$. Like as in the mechanic powers, to perform any effect, the lefs the power is, the greater muft be the time of performing it, or, as it is faid, what is gained in power, is loft in time. So that, if *p* denote any power or agent, and *t* the time of its performing any given fervice; then *p* is as $\frac{1}{t}$, and *t* is as $\frac{1}{p}$; that is, *p* and *t* are Reciprocally proportionals to each other.

RECKONING, in Navigation, is the eftimating the quantity of a fhip's way; or of the courfe and diftance run. Or, more generally, a fhip's Reckoning is that account, by which it may at any time be known where the fhip is, and confequently on what courfe or courfes fhe muft fteer to gain her intended port. The Reckoning is ufually performed by keeping an account of the courfes fteered, and the diftance run, with any accidental circumftances that occur. The courfes fteered are obferved by the compafs; and the diftances run are eftimated from the rate of running, and the time run upon each courfe. The rate of running is meafured by the log, from time to time; which however is liable to great irregularities. Anciently Vitruvius, for meafuring the rate of failing, advifed an axis to be paffed through the fides of the fhip, with two large heads protending out of the fhip, including wheels touching the water, by the revolution of which the fpace paffed over in a given time is meafured. And the fame has been fince recommended by Snellius.

RECKONING, *Dead*. See DEAD *Reckoning*.

RECLINATION *of a Plane*, in Dialling, is the angular quantity which a dial plane leans backwards, from an exactly upright or vertical plane, or from the zenith.

RECLINER, or RECLINING *Dial*, is a dial whofe plane eclines from the perpendicular, that is, leans backwards, or from you, when you ftand before it.

RECLINER, *Declining*, or *Declining* RECLINING *Dial*, is one which neither ftands perpendicularly, nor oppofite to one of the cardinal points.

RECOIL, or REBOUND, the refilition, or flying backward, of a body, efpecially a fire-arm. This is the motion by which, upon explofion, it ftarts or flies backwards; and the caufe of it is the refiftance of the ball and the impelling force of the powder, which acts equally on the gun and on the ball. It has been commonly faid by authors, that the momentum of the ball is equal to that of the gun with its carriage together; but this is a miftake; for the latter momentum is nearly equal to that of the ball and half the weight of the powder together, moving with the velocity of the ball. So that, if the gun, and the ball with half the powder, were of equal weight, the piece would recoil with the fame velocity as the ball is difcharged. But the heavier any body is, the lefs will its velocity be, to have the fame momentum, or force; and therefore fo many times as the cannon and carriage is heavier than the ball and half the powder, juft as many times will the velocity of the ball be greater than that of the gun; and in the fame ratio nearly is the length of the barrel before the charge, to the quantity the gun Recoils in the time the ball is paffing along the bore of the gun. So, if a 24 pounder of 10 feet long be 6400lb weight, and charged with 6lb of powder; then, when the ball quits the piece, the gun will have Recoiled $\frac{28}{6400} \times 10 = \frac{7}{160}$ of a foot, or nearly half an inch.

RECORDE (ROBERT), a learned phyfician and mathematician, was born of a good family in Wales, and flourifhed in the reigns of Henry the 8th, Edward the 6th, and Mary. There is no account of the exact time of his birth, though it muft have been early in the 16th century, as he was entered of the univerfity of Oxford about the year 1525, where he was elected fellow of Allfouls college in 1531. Making phyfic his profeffion, he went to Cambridge, where he was honoured with the degree of doctor in that faculty, in 1545, and highly efteemed by all that knew him for his great knowledge in feveral arts and fciences. He afterwards returned to Oxford, where, as he had done before he went to Cambridge, he publicly taught arithmetic, and other branches of the mathematics, with great applaufe. It feems he afterwards repaired to London, and it has been faid he was phyfician to Edward the 6th and Mary, to which princes he dedicates fome of his books; and yet he ended his days in the King's-bench prifon, Southwark, where he was confined for debt, in the year 1558, at a very immature age.

Recorde publifhed feveral mathematical books, which are moftly in dialogue, between the mafter and fcholar. They are as follow:

1. *The Pathway to Knowledge*, containing the firft Principles of Geometrie, as they may mofte aptly be applied unto practife, bothe for ufe of Inftrumentes Geometricall and Aftronomicall, and alfo for Projection of Plattes much neceffary for all fortes of men, Lond. 4to, 1551.

2. *The Ground of Arts*, teaching the perfect worke and practice of Arithmeticke, both in whole numbers and

and fractions, after a more easie and exact forme then in former time hath beene set forth, 8vo, 1552.—This work went through many editions, and was corrected and augmented by several other persons ; as first by the famous Dr. John Dee ; then by John Mellis, a schoolmaster, 1590 ; next by Robert Norton ; then by Robert Hartwell, practitioner in mathematics, in London ; and lastly by R. C. and printed in 8vo, 1623.

3. *The Castle of Knowledge*, containing the Explication of the Sphere bothe Celestiall and Materiall, and divers other things incident thereto. With sundry pleasaunt proofes and certaine newe demonstrations not written before in any vulgare woorkes. Lond. folio, 1556.

4. *The Whetstone of Witte*, which is the seconde part of Arithmetike : containing the Extraction of Rootes : the Cossike Practise, with the rules of Equation : and the woorkes of Surde Nombers. Lond. 4to, 1557.—For an analysis of this work on Algebra, with an account of what is new in it, see pa. 79 of vol. 1, under the article ALGEBRA.

Wood says he wrote also several pieces on physic, anatomy, politics, and divinity ; but I know not whether they were ever published. And Sherburne says that he published *Cosmographiæ Isagogen* ; also that he wrote a book, *De Arte faciendi Horologium* ; and another, *De Usu Globorum, & de Statu Temporum* ; which I have never seen.

RECTANGLE, in Geometry, is a right-angled parallelogram, or a right-angled quadrilateral figure.

If from any point O, lines be drawn to all the four

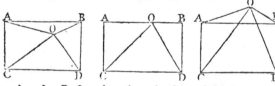

angles of a Rectangle ; then the sum of the squares of the lines drawn to the opposite corners will be equal, in whatever part of the plane the point O is situated ; viz, $OA^2 + OD^2 = OB^2 + OC^2$. For other properties of the Rectangle, see PARALLELOGRAM ; for the Rectangle being a species of the parallelogram, whatever properties belong to the latter, must equally hold in the former.

For the Area of a RECTANGLE. Multiply the length by the breadth or height.—*Otherwise* ; Multiply the product of the two diagonals by half the sine of their angle at the intersection.

That is, AB × AC, or AD × BC × ½ sin. ∠ P = area. A Rectangle, as of two lines AB and AC, is thus denoted, AB × AC, or AB.AC ; or else thus expressed, the Rectangle of, or under, AB and AC.

RECTANGLE, in Arithmetic, is the same with product or factum. So the Rectangle of 3 and 4, is 3 × 4 or 12 ; and of *a* and *b* is *a* × *b* or *ab*.

RECTANGLED, RIGHT-ANGLED, or RECTANGULAR, is applied to figures and solids that have at least one right angle, if not more. So a Right-angled triangle, has one right angle : a Right-angled parallelogram

is a rectangle, and has four right angles. Such also are squares, cubes, parallelopipedons.

Solids are also said to be Rectangular with respect to their situation, viz, when their axis is perpendicular to their base ; as right cones, pyramids, cylinders, &c.

The Ancients used the phrase *Rectangular section of a cone*, to denote a parabola ; that conic section, before Apollonius, being only considered in a cone having its vertex a right angle. And hence it was, that Archimedes entitled his books of the quadrature of the parabola, by the name of *Rectanguli Coni Sectio*.

RECTIFICATION, in Geometry, is the finding of a right line equal to a curve. The Rectification of curves is a branch of the higher geometry, a branch in which the use of the inverse method of Fluxions is especially useful. This is a problem to which all mathematicians, both ancient and modern, have paid the greatest attention, and particularly as to the Rectification of the circle, or finding the length of the circumference, or a right line equal to it ; but hitherto without the perfect effect : upon this also depends the quadrature of the circle, since it is demonstrated that the area of a circle is equal to a right-angled triangle, of which one of the sides about the right angle is the radius, and the other equal to the circumference : but it is much to be feared that neither the one nor the other will ever be accomplished. Innumerable approximations however have been made, from Archimedes, down to the mathematicians of the present day. See CIRCLE and CIRCUMFERENCE.

The first person who gave the Rectification of any curve, was Mr. Neal, son of Sir Paul Neal, as we find at the end of Dr. Wallis's treatise on the Cissoid ; where he says, that Mr. Neal's Rectification of the curve of the semicubical parabola, was published in July or August, 1657. Two years after, viz in 1659, Van Haureat, in Holland, also gave the Rectification of the same curve ; as may be seen in Schooten's Commentary on Des Cartes's Geometry.

The most comprehensive method of Rectification of curves, is by the inverse method of fluxions, which is thus : Let ACc be any curve line, AB an abscifs, and

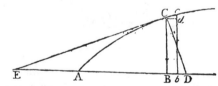

BC a perpendicular ordinate ; also *bc* another ordinate indefinitely near to BC ; and C*d* drawn parallel to the abscifs AB. Put the abscifs AB = *x*, the ordinate BC = *y*, and the curve AC = *z* : then is C*d* = B*b* = $\dot{x}$ the fluxion of the abscifs AB, and *cd* = $\dot{y}$ the fluxion of the ordinate BC, also C*c* = $\dot{z}$ the fluxion of the curve AB. Hence because C*cd* may be considered as a plane right-angled triangle, $Cc^2 = Cd^2 + cd^2$, or $\dot{z}^2 = \dot{x}^2 + \dot{y}^2$; and therefore $\dot{z} = \sqrt{\dot{x}^2 + \dot{y}^2}$; which is the fluxion of the length of any curve ; and consequently, out of this equation expelling either $\dot{x}$ or $\dot{y}$, by means of the particular equation expressing the nature of the curve in question, the fluents of the resulting equation, being then taken, will give the length of the curve, in finite terms when it is

rectifiable,

rectifiable, otherwise in an infinite series, or in a logarithmic or exponential &c expression, or by means of some other curve, &c.

Ex. 1. *To rectify the common parabola.*—In this case, the equation of the curve is $2ax = y^2$, where a is half the parameter. The fluxion of this equation is $2a\dot{x} = 2y\dot{y}$, and hence $\dot{x}^2 = \dfrac{y^2\dot{y}^2}{a^2}$; this being substituted in the general equation $\dot{z} = \sqrt{\dot{x}^2 + \dot{y}^2}$, it becomes $\dot{z} = \dfrac{\dot{y}\sqrt{aa + yy}}{a}$; the correct fluents of which give

$$z = \frac{y\sqrt{aa + yy}}{2a} + \tfrac{1}{2}a \times \text{hyp. log. of } \frac{y + \sqrt{aa + yy}}{a},$$

which is the length of the curve AC, when it is a parabola.

And the same might be expressed by an infinite series, by expanding the quantity $\sqrt{aa + yy}$. See my Mensuration, pa. 361, 2d edit.

Ex. 2. *To rectify the Circle.*—The equation of the circle may be expressed either in terms of the sine, or versed sine, or tangent, or secant, &c, and the radius. Let therefore the radius of the circle be DA or DC = r, the versed sine AB = x, the right sine BC = y, the tangent CE = t, and the secant DE = s; then, by the nature of the circle, we have these equations,

$$y^2 = 2rx - x^2 = \frac{r^2 t^2}{r^2 + t^2} = \frac{s^2 - r^2}{s^2}r^2 ;$$

and by means of the fluxions of these equations, with the general equation $\dot{z}^2 = \dot{x}^2 + \dot{y}^2$, are obtained the following fluxional forms for the fluxion of the curve, the fluent of any one of which will be the curve itself, viz,

$$\dot{z} = \frac{r\dot{x}}{\sqrt{2rx - xx}} = \frac{r\dot{y}}{\sqrt{rr - yy}} = \frac{r^2\dot{t}}{r^2 + t^2} = \frac{r^2\dot{s}}{\sqrt{s^2 - r^2}}.$$

Hence the value of the curve, from the fluent of each of these, gives the four following forms, in series, viz, the curve, putting $d = 2r$ the diameter, is z

$$= (1 + \frac{x}{2.3d} + \frac{3x^2}{2.4.5d^2} + \frac{3.5x^3}{2.4.6.7d^3} \text{ &c}) \sqrt{dv},$$

$$= (1 + \frac{y^2}{2.3r^2} + \frac{3y^4}{2.4.5r^4} + \frac{3.5y^6}{2.4.6.7r^6} \text{ &c}) y,$$

$$= (1 - \frac{t^2}{3r^2} + \frac{t^4}{5r^4} - \frac{t^6}{7r^6} + \frac{t^8}{9r^8} \text{ &c}) t,$$

$$= (\frac{s - r}{s} + \frac{s^3 - r^3}{2.3s^3} + \frac{3(s^5 - r^5)}{2.4.5s^5} \text{ &c}) r.$$

See my Mensur. 2d edit. pa. 118 &c, also most treatises on Fluxions.

It is evident that the simplest of these series is the third, or that which is expressed in terms of the tangent. It will therefore be the properest form to calculate an example by in numbers. And for this purpose it will be convenient to assume some arc whose tangent, or at least its square, is known to be some small finite number. Now the arc of 45° it is known has its tangent equal to the radius; and therefore, taking the radius $r = 1$, and consequently the tangent of 45° or $t = 1$ also, in this case the arc of 45° to the radius 1,

or the quadrant to the diameter 1, will be =
$$1 - \frac{1}{3} + \frac{1}{5} - \frac{1}{7} + \frac{1}{9} \text{ &c.}$$
But as this series converges very slowly, some smaller arch must be taken, that the series may converge faster; such as the arc of 30°, whose tangent is $= \sqrt{\dfrac{1}{3}} = 5773502$, or its square $t^2 = \dfrac{1}{3}$; and hence, after the first term, the succeeding terms will be found by dividing always by 3, and these quotients divided by the absolute numbers 3, 5, 7, 9, &c; and lastly adding every other term together into two sums, the one the sum of the positive terms, and the other the sum of the negative ones, then lastly the one sum taken from the other leaves the length of the arc of 30°, which is the 12th part of the whole circumference when the radius is 1, or the 6th part when the diameter is 1, and consequently 6 times that arc will be the length of the whole circumference to the diameter 1; therefore multiply the 1st term $\sqrt{\dfrac{1}{3}}$ by 6, and the product is $\sqrt{\dfrac{36}{3}}$ or $\sqrt{12} = 3.4641016$; hence the operation will be conveniently made as follows:

		+ Terms.	− Terms.
1)	3·4641016	(3·4641016	
3)	1·1547005	(	0·3849002
5)	3849002	(769800	
7)	1283001	(	183286
9)	427667	(47519	
11)	142556	(	12960
13)	47519	(3655	
15)	15840	(	1056
17)	5280	(311	
19)	1760	(	93
21)	587	(28	
23)	196	(	8
25)	65	(3	
27)	22	(	1

$$+ 3.5462332 - 0.4046406$$
$$- 0.4046406$$

$$3.1415926 \text{ the circumference.}$$

Various other series for the Rectification of the circle may be seen in different parts of my Mensuration, as at pa. 121, 122, 137, 138, 422, &c. See also my paper on this subject in the Philos. Trans. vol. 66, pa. 476.

RECTIFIER, in Navigation, is an instrument used for determining the variation of the compass, in order to rectify the ship's course. It consists of two circles, either laid upon, or let into one another, and so fastened together in their centres that they represent two compasses, the one fixed, and the other moveable. Each is divided into 32 points of the compass, and 360°, and numbered both ways, from the north and the south, ending at the east and west in 90°. The fixed compass represents the horizon, in which the north, and all the other points, are liable to variation. In the centre of

the

the moveable compafs is faftened a filk thread, long enough to reach the outfide of the fixed compafs: but when the inftrument is made of wood, an index is ufed inftead of the thread.

RECTIFYING *of Curves.* See RECTIFICATION.

RECTIFYING *of the Globe or Sphere*, is a previous adjuftment of it, to prepare it for the folution of problems. This ufually confifts in placing it in the fame pofition as the true fphere of the world has at fome certain time propofed; which is done firft by elevating the pole above the horizon as much as the latitude of the place is, then bringing the fun's place for the given day, found in the ecliptic, to the graduated fide of the brafs or general meridian, next move the hour-index to the upper hour of 12, fo fhall the globe be Rectified for noon of that day; and if the globe be turned about till the hour-index point at any propofed hour, then is the globe in the real pofition of the earth at that time, if the whole globe be fet in the north and fouth pofition by means of the compafs.

RECTILINEAL, RECTILINEAR, or *Right-lined*, is the quality or nature of figures that are bounded by right lines, or formed by right lines.

RECURRING *Series*, is a feries conftituted in fuch a manner, that having taken at pleafure any number of its terms, each following term fhall be related to the fame number of preceding terms according to a conftant law of relation. See *Recurring* SERIES.

RED, in Phyfics, or Optics, one of the fimple or primary colours of natural bodies, or rather of the rays of light.—The Red rays are the leaft refrangible of all the rays of light. And hence, as Newton fuppofes the different degrees of refrangibility to arife from the different magnitudes of the luminous particles of which the rays confift; therefore the Red rays, or Red light, is concluded to be that which confifts of the largeft particles. See COLOUR and LIGHT.

Authors diftinguifh three general kinds of Red: one bordering on the blue, as colombine, or dove-colour, purple, and crimfon; another bordering on yellow, as flame-colour and orange; and between thefe extremes is a medium, which is that which is properly called Red.

REDANS, or REDANT, or REDENS, in Fortification, is a kind of work indented like the teeth of a faw, with falient and re-entering angles; to the end that one part may flank or defend another. It is called alfo *faw work*, and *indented work*.

Redans are often ufed in fortifying of walls, where it is not neceffary to be at the expence of building baftions; as when they ftand on the fide of a river, or a marfh, or the fea, &c. But the fault of fuch fortification is, that the befiegers from one battery may ruin both the fides of the tenaille or front of a place, and make an affault without fear of being enfiladed, fince the defences are ruined.

The parapet of the corridor alfo is frequently Redented, or carried on by the way of Redans.

REDINTEGRATION, is the taking or finding the integral or fluent again, from the fluxion. See FLUXION and FLUENT.

REDOUBT, or REDOUTE, in Fortification, a fmall fort, without any defence but in front, ufed in trenches,

lines of circumvallation, contravallation, and approach; as alfo for the lodging of corps de garde, and to defend paffages.

A Detached REDOUBT, is a kind of work refembling a ravelin, with flanks, placed beyond the glacis.—It is made to occupy fome fpot of ground which might be advantageous to the befiegers; and alfo to oblige the enemy to open his trenches farther off than he would otherwife do

REDUCING *Scale*, or SURVEYING *Scale*, is a broad, thin flip of box, or ivory, having feveral lines and fcales of equal parts upon it; ufed by furveyors for turning chains and links into roods and acres, by infpection. They ufe it alfo to reduce maps and draughts from one dimenfion to another.

REDUCTION, in general, is the bringing or changing fome thing to a different form, ftate, or denomination.

REDUCTION, in *Arithmetic*, is commonly underftood of the changing of money, weights, or meafures, to other denominations, of the fame value; and it is of two kinds, *Reduction Defcending*, which is the changing a number to its equivalent value in a lower denomination; as pounds into fhillings or pence: and *Reduction Afcending*, which is the changing numbers to higher denominations; as pence to fhillings or pounds.

RULE. *To perform Reduction:* confider how many of the lefs denomination make one of the greater, as how many pence make a fhilling, or how many fhillings make a pound; and multiply by that number when the Reduction is defcending, but divide by it when it is afcending So to reduce 23l. into pence; and converfely thofe pence into pounds; multiply or divide by 12 and 20, as here below.

23 *l.*   12) 5520 *d.*
20

460 *sh.*   20) 460 *sh.*
12   23 *l.*

5520 *d.*

REDUCTION *of Fractions.* See FRACTION, and DECIMAL.

REDUCTION *of Equations*, in Algebra. See EQUATION.

REDUCTION *of Curves.* See CURVE.

REDUCTION *of a Figure, Defign, or Draught*, is the making a copy of it, either larger or fmaller than the original, but ftill preferving the form and proportion.

Figures and plans are reduced, and copied, in various ways; as by the Pentagraph, and Proportional compaffes. See PENTAGRAPH, and *Proportional* COMPASSES. The beft of the other methods of reducing are as below.

To reduce a Simple Rectilinear Figure by Lines.

Pitch upon a point P any where about the given figure ABCDE, either within it, or without it, or in one fide or angle; but near the middle is beft. From that point P draw lines through all the angles; upon one

of

of which take P*a* to PA in the proposed proportion of the scales, or linear dimensions; then draw *ab* parallel to AB, *bc* to BC, &c; so shall *abcde* be the reduced figure sought, either greater or smaller than the original.

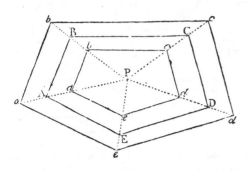

To Reduce a Figure by a Scale.—Measure all the sides, and diagonals, of the figure, as ABCDE, by a scale; and lay down the same measures respectively, from another scale, in the proportion required.

To Reduce a Map, Design, or Figure, by Squares.—Divide the original into a number of little squares; and divide a fresh paper, of the dimensions required, into the same number of other squares, either greater or smaller as required. This done, in every square of the second figure, draw what is found in the corresponding square of the first or original figure.

The cross lines forming these squares, may be drawn with a pencil, and these rubbed out again after the work is finished. But a more ready and convenient way, especially when such Reductions are often wanted, would be to keep always at hand frames of squares ready made, of several sizes; for by only just laying them down upon the papers, the corresponding parts may be readily copied. These frames may be made of four stiff or inflexible bars, strung across with horse hairs, or fine catgut.

REDUCTION *to the Ecliptic*, in Astronomy, is the difference between the argument of latitude, as NP, and an arc of the ecliptic NR, intercepted between the place of a planet, and the node.—To find this Reduction, or difference; in the right-angled spherical triangle NPR, are given the angle of inclination, and the argument of latitude NP; to find NR; then the difference between NP and NR is the Reduction sought.

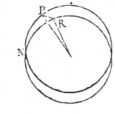

REDUNDANT *Hyperbola*, is a curve of the higher kind, so called because it exceeds the conical hyperbola in the number of legs; being a triple hyperbola, with 6 hyperbolic legs. See Newton's Enum. Lin. tertii Ordinis, nomina formarum, &c.

RE-ENTERING *Angle*, in Fortification, is an angle whose point is turned inwards, or towards the place.

REFLECTED *Ray*, or *Vision*, is that which is made by the reflection of light, or by light first received upon the surface of some body, and thence reflected again. See RAY, VISION, and REFLECTION.

REFLECTING, or REFLEXIVE, *Dial*, is a kind of dial which shews the hour by means of a thin piece of looking-glass plate, duly placed to throw the sun's rays to the top of a cieling, on which the hour-lines are drawn.

REFLECTION, or REFLEXION, in Mechanics, is the return, or regressive motion of a moveable body, occasioned by the resistance of another body, which hinders it from pursuing its former course of direction.

Reflection is conceived, by the latest and best authors, as a motion peculiar to elastic bodies, by which, after striking on others which they cannot remove, they recede, or turn back, or aside, by their elastic power.

On this principle it is asserted, that there may be, and is, a period of rest between the incidence and the reflection; since the reflected motion is not a continuation of the other, but a new motion, arising from a new cause or principle, viz, the power of elasticity.

It is one of the great laws of Reflection, that the angle of incidence is equal to the angle of Reflection; i. e. that the angle which the direction of motion of a striking body makes with the surface of the body struck, is equal to the angle made between the same surface and the direction of motion after the stroke. See INCIDENCE and PERCUSSION.

REFLECTION *of the Rays of Light*, like that of other bodies, is their motion after being reflected from the surfaces of bodies.

The Reflection of the rays of light from the surfaces of bodies, is the means by which those bodies become visible. And the disposition of bodies to reflect this, or that kind of rays most copiously, is the cause of their being of this or that colour. Also, the Reflection of light, from the surfaces of mirrors, makes the subject of catoptrics.

The Reflection of light, Newton has shewn, is not effected by the rays striking on the very parts of the bodies; but by some power of the body equally diffused throughout its whole surface, by which it acts upon the ray, attracting or repelling it without any real immediate contact. This power he also shews is the same by which, in other circumstances, the rays are refracted; and by which they are at first emitted from the lucid body.

Dr. Priestley says, it is not more probable, that the rays of light are transmitted from the sun, with an uniform disposition to be reflected or refracted, according to the circumstances of the bodies on which they impinge; and that the transmission of some of the rays, apparently under the same circumstances, with others that are reflected, is owing to the minute vibrations of the small parts of the surfaces of the mediums through which the rays pass; vibrations that are independent of action and reaction between the body and the particles of light at the time of their impinging, though probably excited by the action of preceding rays. Hist. of Light and Colours, pa. 309.

Newton concludes his account of the Reflection of light with observing, that if light be reflected not by impinging on the solid parts of bodies, but by some other principle, it is probable that as many of its

rays

rays as impinge on the folid parts of bodies are not reflected, but ftifled and loft in the bodies. Otherwife, he fays, we muft fuppofe two kinds of Reflection; for fhould all the rays be reflected which impinge on the internal parts of clear water or cryftal, thofe fubftances would rather have a cloudy colour, than a clear tranfparency. To make bodies look black, it is neceffary that many rays be ftopped, retained and loft in them; and it does not feem probable that any rays can be ftopped and ftifled in them, which do not impinge on their parts: and hence, he fays, we may underftand, that bodies are much more rare and porous than is commonly believed. However, M. Bouguer difputes the fact of light being ftifled or loft by impinging on the folid parts of bodies.

REFLECTION, in Catoptrics, is the return of a ray of light from the polifhed furface of a fpeculum or mirror, as driven thence by fome power refiding in it.

The ray thus returned is called a *reflex* or *reflected ray*, or a *ray of Reflection*; and the point of the fpeculum where the ray commences, is called the *point of Reflection.* Thus, the ray AB, proceeding from the radiant A, and ftriking on the point of the fpeculum B, being returned thence to C, BC reprefents the reflected ray, and B the point

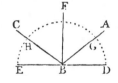

of Reflection; in refpect of which, AB reprefents the incident ray, or ray of incidence, and B the point of incidence; alfo the angle CBE is the angle of Reflection, and ABD the angle of incidence; where DE is the reflecting furface, or at leaft a tangent to it at the point B. Though fome count the angle of incidence and of Reflection from the perpendicular BF.

General Laws of REFLECTION.——I. *When a ray of light is reflected from a fpeculum of any form, the angle of incidence is always equal to the angle of Reflection.* This law obtains in the percuffions of all kinds of bodies; and confequently muft do fo in thofe of light; and the proof of it may be feen at the article INCIDENCE.

This law is confirmed alfo by experiments on all bodies; and on the rays of light in this manner: A ray from the fun falling on a mirror, in a dark room, through a fmall hole, you will have the pleafure to fee it rebound, fo as to make the angle of Reflection equal to the angle of incidence. And the fame may be fhewn in various other ways: thus ex. gr. placing a femicircle DFE on a mirror DE, its centre on B, and its limb or plane perpendicular to the fpeculum; and affuming equal arcs DG and E H; place an object in A, and the eye in C: then will the object be feen by a ray reflected from the point B. But by covering B, the object will ceafe to be feen.

II. *Every point of a fpeculum reflects rays falling on it, from every part of an object.*

III. *If the eye C and the radiant point A change places, the point will continue to radiate upon the eye, in the fame courfe or path as before.*

IV. *The plane of Reflection is perpendicular to the furface of the fpeculum; and it paffes through the centre in fpherical fpecula.*

REFLECTION *of the Moon,* is a term ufed by fome

authors for what is otherwife called *her variation*; being the 3d inequality in her motion, by which her true place out of the quadratures differs from her place twice equated.

REFLECTION is alfo ufed in the Copernican fyftem, for the diftance of the pole from the horizon of the difc; which is the fame thing as the fun's declination in the Ptolomaic fyftem.

REFLECTOIRE CURVE. See *Reflectoire* CURVE.

REFLEXIBILITY *of the rays of light,* is that property by which they are difpofed to be reflected. Or, it is their difpofition to be turned back into the fame medium, from any other medium on whofe furface they fall. Hence thofe rays are faid to be more or lefs reflexible, which are returned back more or lefs eafily under the fame incidence. Thus, if light pafs out of glafs into air, and by being inclined more and more to the common furface of the glafs and air, begins at length to be totally reflected by that furface, thofe forts of rays which at like incidences are reflected moft copioufly, or the rays which by being inclined begin fooneft to be totally reflected, are the moft reflexible rays.

That rays of light are of different colours, and endued with different degrees of reflexibility, was firft difcovered by Sir I. Newton; and it is fhewn by the following experiment. Applying a prifm DFE to

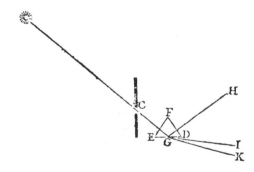

the aperture C of a darkened room in fuch manner that the light be reflected from the bafe in G; the violet rays are feen firft reflected into HG; the other rays continuing ftill refracted to I and K. After the violet, the blue are all reflected; then the green, &c.——Hence it appears, that the differently coloured rays differ in degree of Reflexibility. And from other experiments it appears, that thofe rays which are moft reflexible, are alfo moft refrangible.

REFLUX *of the Sea,* is the ebbing of the water, or its return from the fhore; being fo called, becaufe it is the oppofite motion to the flood or flux. See TIDE.

REFRACTED *Angle,* or *Angle of Refraction,* in Optics, is the angle which the refracted ray makes with the refracting furface; or fometimes it denotes the complement of that, or the angle it makes with the perpendicular to the faid furface.

REFRACTED *Dials,* or *Refracting Dials,* are fuch as fhew the hour by means of fome refracting tranfparent fluid.

REFRACTED *Ray,* or *Ray of* REFRACTION, is a ray

ray after it is broken or bent, at the common furface of two different mediums, where it paffes from the one into the other. See RAY, and REFRACTION.

REFRACTION, in Mechanics, is the deviation of a moving body from its direct courfe, by reafon of the different denfity of the medium it moves in; or a flexion and change of determination, occafioned by a body's paffing obliquely out of one medium into another of a different denfity.

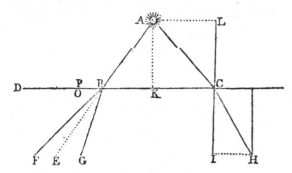

Thus a ball A, moving in the air in the line AB, and falling obliquely on the furface of the water CD, does not proceed ftraight in the fame direction, as to E, but deviates or is deflected to F. Again, if the ball move in water in the line AB, and fall obliquely on a furface of air CD; it will in this cafe alfo deviate from the fame continued direction BE, but now the contrary way, and will go to G, on the other fide of it. Now the deflection in either cafe is called the *Refraction*, the Refraction being towards the denfer furface BD in the former cafe, but from it in the latter.

Thefe Refractions are fuppofed to arife from hence; that the ball arriving at B, in the firft cafe finds more refiftance or oppofition on the one fide O, or from the fide of the water, than it did from the fide P, or that of the air; and in the latter more refiftance from the fide P, which is now the fide of the water, than the fide O, which is that of the air. And fo for any other different media: a vifible inftance of which is often perceived in the falling of fhot or fhells into the earth, as clay &c. when the perforation is found to rife a little upwards, toward the furface. However another reafon is affigned for the Refraction of the rays of light, whofe Refractions lie the contrary way to thofe above, as will be feen in what follows, viz, that water by its greater attraction accelerates the motion of the rays of light more than air does.

REFRACTION *of Light*, in Optics, is an inflection or deviation of the rays from their rectilinear courfe on paffing obliquely out of one medium into another, of a different denfity.

That a body may be refracted, it is neceffary that it fhould fall obliquely on the fecond medium: in perpendicular incidence there is no Refraction. Yet Vofcius and Snellius imagined they had obferved a perpendicular ray of light undergo a Refraction; a perpendicular object appearing in the water nearer than it really was but this was attributing that to a Refraction of the perpendicular rays, which was owing to

the divergency of the oblique rays after refraction, from a nearer point. Yet there is a manifeft Refraction even of perpendicular rays found in ifland cryftal.

Rohault adds, that though an oblique incidence be neceffary in all other mediums we know of, yet the obliquity muft not exceed a certain degree; if it do, the body will not penetrate the medium, but will be reflected inftead of being refracted. Thus, cannon-balls, in fea engagements, falling very obliquely on the furface of the water, are obferved to bound or rife from it, and to fweep the men from off the enemy's decks. And the fame thing happens to the little ftones with which children make their ducks and drakes along the furface of the water.

The ancients confounded Refraction with Reflection; and it was Newton who firft taught the true difference between them. He fhews however that there is a good deal of analogy between them, and particularly in the cafe of light.

The laws of Refraction of the rays of light in mediums differently terminated, i. e. whofe furfaces are plane, concave, and convex, make the fubject of Dioptrics.—By Refraction it is, that convex glaffes, or lenfes, collect the rays, magnify objects, burn, &c; and hence the foundation of microfcopes, telefcopes, &c.—And by Refraction it is, that all remote objects are feen out of their real places; particularly, that the heavenly bodies are apparently higher than they are in reality. The Refraction of the air has many times fo uncertain an influence on the places of celeftial objects, near the horizon, that wherever Refraction is concerned, the conclufions deduced from obfervations that are much affected by it, will always remain doubtful, and fometimes too precarious to be relied on. See Dr. Bradley in Philof. Tranf. number 485.

As to the caufe of Refraction, it does not appear that any perfon before Des Cartes attempted to explain it; this he undertook to do by the refolution of forces, on the principles of mechanics; in confequence of which, he was obliged to fuppofe that light paffes with more eafe through a denfe medium than a rare one: thus, the ray AC falling obliquely on a denfer medium at C is fuppofed to be acted on by two forces, one of them impelling it in the direction AL, and the other in AK, which alone can be affected by the change of medium: and fince, after the ray has entered the denfer medium, it approaches the perpendicular CI, it is plain that this force muft have received an increafe, whilft the other continued the fame.

The firft perfon who queftioned the truth of this explanation of the caufe of Refraction, was Fermat; he afferted, contrary to Des Cartes, that light fuffers greater refiftance in water than in air, and greater in glafs than in water; and he maintained that the refiftance of different mediums, with refpect to light, is in proportion to their denfities. Leibnitz alfo adopted the fame general idea; and they reafoned upon the fubject in the following manner. Nature, fay they, accomplifhes her ends by the fhorteft methods; and therefore light ought to pafs from one point to another, either by the fhorteft courfe, or by that in which the leaft time is required. But it is plain that the path in which light paffes, when it falls obliquely upon a denfer

fer medium, is not the moſt direct or the ſhorteſt; and therefore it muſt be that in which the leaſt time is ſpent. And whereas it is demonſtrable, that light falling obliquely upon a denſer medium (in order to take up the leaſt time poſſible, in paſſing from a point in one medium to a point in the other) muſt be refracted in ſuch a manner, that the ſine of the angles of incidence and Refraction muſt be to one another, as the different facilities with which light is tranſmitted in thoſe mediums; it follows that, ſince light approaches the perpendicular when it paſſes obliquely from air into water, the facility with which water ſuffers light to paſs through it, is leſs than that of the air; ſo that the light meets with greater reſiſtance in water than in air.

This method of arguing from final cauſes could not ſatisfy philoſophers. Dr. Smith obſerves, that it agrees only to the caſe of Refraction at a plane ſurface; and that the hypotheſis is altogether arbitrary.

Dechales, in explaining the law of Refraction, ſuppoſes that every ray of light is compoſed of ſeveral ſmaller rays, which adhere to one another; and that they are refracted towards the perpendicular, in paſſing into a denſer medium, becauſe one part of the ray meets with more reſiſtance than another part; ſo that the former traverſes a ſmaller ſpace than the latter; in conſequence of which the ray muſt neceſſarily bend a little towards the perpendicular. This hypotheſis was adopted by the celebrated Dr. Barrow, and indeed ſome ſay, he was the author of it. On this hypotheſis it is plain, that mediums of a greater refractive power, muſt give a greater reſiſtance to the paſſage of the rays of light, than mediums of a leſs refractive power; which is contrary to fact.

The Bernoullis, both father and ſon, have attempted to explain the cauſe of Refraction on mechanical principles; the former on the equilibrium of forces, and the latter on the ſame principles with the ſuppoſition of etherial vortices: but neither of theſe hypotheſes have gained much credit.

M. Mairan ſuppoſes a ſubtle fluid, filling the pores of all bodies, and extending, like an atmoſphere, to a ſmall diſtance beyond their ſurfaces; and then he ſuppoſes that the Refraction of light is nothing more than a neceſſary and mechanical effect of the incidence of a ſmall body in thoſe circumſtances. There is more, he ſays, of the refracting fluid, in water than in air, more in glaſs than in water, and in general more in a denſe medium than in one that is rarer.

Maupertuis ſuppoſes that the courſe which every ray takes, in paſſing out of one medium into another, is that which requires the leaſt quantity of action, which depends upon the velocity of the body and the ſpace it paſſes over; ſo that it is in proportion to the ſum of the products ariſing from the ſpaces multiplied by the velocities with which bodies paſs over them. From this principle he deduces the neceſſity of the ſine of the angle of incidence being in a conſtant proportion to that of Refraction; and alſo all the other laws relating to the propagation and reflection of light.

Dr. Smith (in his Optics, Remarks, p. 70) obſerves, that all other theories for explaining the reflection and Refraction of light, except that of Newton, ſuppoſe that it ſtrikes upon bodies and is reſiſted by

6

them; which has never been proved by any deduction from experience. On the contrary, it appears by various conſiderations, and might be ſhewn by the obſervations of Mr. Molyneux and Dr. Bradley on the parallax of the fixed ſtars, that their rays are not at all impeded by the rapid motion of the earth's atmoſphere, nor by the object glaſs of the teleſcope, through which they paſs. And by Newton's theory of Refraction, which is grounded on experience only, it appears that light is ſo far from being reſiſted and retarded by Refraction into any denſe medium, that it is ſwifter there than in vacuo in the ratio of the ſine of incidence in vacuo to the ſine of Refraction into the denſe medium. Prieſtley's Hiſt. of Light, &c, p. 102 and 333.

Newton ſhews that the Refraction of light is not performed by the rays falling on the very ſurface of bodies; but that it is effected, without any contact, by the action of ſome power belonging to bodies, and extending to a certain diſtance beyond their ſurfaces; by which ſame power, acting in other circumſtances, they are alſo emitted and reflected.

The manner in which Refraction is performed by mere attraction, without contact, may be thus accounted for: Suppoſe HI the boundary of two mediums, N and O; the firſt the rarer, ex. gr. air; the ſecond the denſer, ex. gr. glaſs; the attraction of the mediums here will be as their denſities. Suppoſe pS to be the diſtance to which the attracting force of the denſer medium exerts itſelf within the rarer. Now 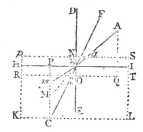 let a ray of light Aa fall obliquely on the ſurface which ſeparates the mediums, or rather on the ſurface pS, where the action of the ſecond and more reſiſting medium commences: as the ray arrives at a, it will begin to be turned out of its rectilinear courſe by a ſuperior force, with which it is attracted by the medium O, more than by the medium N; hence the ray is bent out of its right line in every point of its paſſage between pS and RT, within which diſtance the attraction acts; and therefore between theſe lines it deſcribes a curve aBb; but beyond RT, being out of the ſphere of attraction of the medium N, it will proceed uniformly in a right line, according to the direction of the curve in the point b.

Again, ſuppoſe N the denſer and more attracting medium, O the rarer, and HI the boundary as before; and let RT be the diſtance to which the denſer medium exerts its attractive force within the rarer: even when the ray has paſſed the point B, it will be within the ſphere of the ſuperior attraction of the denſer medium; but that attraction acting in lines perpendicular to its ſurface, the ray will be continually drawn from its ſtraight courſe BM perpendicularly towards HI: thus, having two forces or directions, it will have a compound motion, by which, inſtead of BM, it will deſcribe Bm, which Bm will in ſtrictneſs be a curve. Laſtly, after it has arrived at m, being out of the influence of the medium N, it will perſiſt uniformly, in a right line, in the direction in which the extremity of the

the curve leaves it.—Thus we see how Refraction is performed, both towards the perpendicular DE, and from it.

REFRACTION *in Dioptrics*, is the inflexion or bending of the rays of light, in passing the surfaces of glasses, lenses, and other transparent bodies of different densities. Thus, a ray, as AB, falling obliquely from the radiant A, upon a point B, in a diaphanous surface HI, rarer or denser than the medium along which it was propagated from the radiant, has its direction there altered by the action of the new medium; and instead of proceeding to M, it deviates, as for ex. to C.

This deviation is called the *Refraction of the ray*; BC the *refracted ray*, or *line of Refraction*; and B the *point of Refraction*.—The line AB is also called the *line of incidence*; and in respect of it, B is also called the *point of incidence*. The plane in which both the incident and refracted ray are found, is called the *plane of Refraction*; also a right line BE drawn in the refracting medium perpendicular to the refracting surface at the point of Refraction B, is called the *axis of Refraction*; and its continuation DB along the medium through which the ray falls, is called the *axis of incidence*.—Farther, the angle ABI, made by the incident ray and the refracting surface, is usually called the *angle of incidence*; and the angle ABD, between the incident ray and the axis of incidence, is the *angle of inclination*. Moreover, the angle MBC, between the refracted and incident rays, is called the *angle of Refraction*; and the angle CBE, between the refracted ray and the axis of Refraction, is the *refracted angle*. But it is also very common to call the angles ABD and CBE made by the perpendicular with the incident and refracted rays, the *angles of incidence and Refraction*.

General Laws of REFRACTION.—I. *A ray of light in its passage out of a rarer medium into a denser*, ex. gr. *out of air into water or into glass, is refracted towards the perpendicular*, i. e. *towards the axis of Refraction*. Hence, the refracted angle is less than the angle of inclination; and the angle of Refraction less than that of incidence; as they would be equal were the ray to proceed straight from A to M.

II. *The ratio of the sines of the angles* ABD, CBE, *made by the perpendicular with the incident and refracted rays, is a constant and fixed ratio*; whatever be the obliquity of the incident ray, the mediums remaining. Thus, the Refraction out of air, into water, is nearly as 4 to 3, and into glass it is nearly as 3 to 2. As to air in particular, it is shewn by Newton, that a ray of light, in traversing quite through the atmosphere, is refracted the same as it would be, were it to pass with the same obliquity out of a vacuum immediately into air of equal density with that in the lowest part of the atmosphere.

The true law of Refraction was first discovered by Willebrord Snell, professor of Mathematics at Leyden; who found by experiment that the cosecants of the angles of incidence and Refraction, are always in the same ratio. It was commonly attributed however to Des Cartes; who, having seen it in a MS. of Snell's, first published it in his Dioptrics, without naming Snellius, as Huygens asserts; Des Cartes having only

altered the form of the law, from the ratio of the cosecants, to that of the sines, which is the same thing.

It is to be observed however, that as the rays of light are not all of the same degree of refrangibility, this constant ratio must be different in different kinds: so that the ratio mentioned by authors, is to be understood of rays of the mean refrangibility, i. e. of green rays. The difference of Refraction between the least and most refrangible rays, that is, between violet and red rays, Newton shews, is about the $\frac{1}{27}$ of the whole Refraction of the mean refrangible; which difference, he allows, is so small, that it seldom needs to be regarded.

Different transparent substances have indeed very different degrees of Refraction, and those not according to any regular law; as appears by many experiments of Newton, Euler, Hawksbee, &c. See Newton's Optics, 3d edit. pa. 247; Hawksbee's Experim. pa. 292; Act. Berlin. 1762, pa. 302; Priestley's Hist. of Light &c, pa. 479.

Whence the different refractive powers in different fluids arise, has not been determined. Newton shews, that in many bodies, as glass, crystal, selenites, pseudo-topaz, &c, the refractive power is indeed proportionable to their densities; whilst in sulphureous bodies, as camphor, linseed, and olive oil, amber, spirit of turpentine, &c, the power is two or three times greater than in other bodies of equal density; and yet even these have the refractive power with respect to each other, nearly as their densities. Water has a refractive power in a medium degree between those two kinds of substances; whilst salts and vitriols have refractive powers in a middle degree between those of earthy substances and water, and accordingly are composed of those two sorts of matter. Spirit of wine has a refractive power in a middle degree between those of water and oily substances; and accordingly it seems to be composed of both, united by fermentation. It appears therefore, that all bodies seem to have their refractive powers nearly proportional to their densities, excepting so far as they partake more or less of sulphureous oily particles, by which those powers are altered.

Newton suspected that different degrees of heat might have some effect on the refractive power of bodies; but his method of determining the general Refraction was not sufficiently accurate to ascertain this circumstance. Euler's method however was well adapted to this purpose: from his experiments he infers, that the focal distance of a single lens of glass diminishes with the heat communicated to it; which diminution is owing to a change in the refractive power of the glass itself, which is probably increased by heat, and diminished by cold, as well probably as that of all other translucent substances.

From the law above laid down it follows, that one angle of inclination, and its corresponding refracted angle, being found by observation, the refracted angles corresponding to the several other angles of inclination are thence easily computed. Now, Zahnius and Kircher have found, that if the angle of inclination be 70°, the refracted angle, out of air into glass, will be 38° 50'; on which principle Zahnius has constructed a table of those Refractions for the several degrees of the angle

angle of inclination; a fpecimen of which here follows :

Angle of In-clination.	Refracted Angle.			Angle of Re-fraction.		
°	°	′	″	°	′	″
1	0	40	5	0	19	55
2	1	20	6	0	39	54
3	2	0	4	0	59	56
4	2	40	5	1	19	55
5	3	20	3	1	39	57
10	6	39	16	3	20	44
20	13	11	35	6	48	25
30	19	29	29	10	30	31
45	28	9	19	16	50	41
90	41	51	40	48	8	20

Hence it appears, that if the angle of inclination be
lefs than 20°, the angle of Refraction out of air into
glafs is almoft ⅓ of the angle of inclination; and therefore a ray is refracted to the axis of Refraction by almoft a third part of the quantity of its angle of inclination. And on this principle it is that Kepler, and
moft other dioptrical writers, demonftrate the Refractions in glaffes; though in eftimating the law of thefe
Refractions he followed the example of Alhazen and
Vitello, and fought to difcover it in the proportion of
the angles, and not in that of the fines, or cofecants,
as difcovered by Snellius, as mentioned above.

The refractive powers of feveral fubftances, as determined by different philofophers, may be feen in the following tables; in which the ray is fuppofed to pafs out
of air into each of the fubftances, and the annexed
numbers fhew the proportion to unity or 1, between the fines of the angles of incidence and Refraction.

1. By Sir Ifaac Newton's Obfervations.

Air - - -	0·9997
Rain water - - -	1·3358
Spirit of wine - -	1·3698
Oil of vitriol - -	1·4285
Alum - - -	1·4577
Oil olive - - -	1·4666
Borax - -	1·4667
Gum Arabic - -	1·4771
Linfeed oil - -	1·4814
Selenites - -	1·4878
Camphor - -	1·5000
Dantzick vitriol -	1·5000
Nitre - -	1·5238
Sal gem - -	1·5455
Glafs - - -	1·5500
Amber - -	1·5556
Rock cryftal -	1·5620
Spirit of turpentine -	1·5625
A yellow pfeudo-topaz -	1·6429
Ifland cryftal -	1·6666
Glafs of antimony -	1·8889
A Diamond -	2·4390

2. By Mr. Hawkfbee.

Water - -	1·3359
Spirit of honey - -	1·3359
Oil of amber - -	1·3377
Human urine - -	1·3419
White of an egg - -	1·3511
French brandy - -	1·3625
Spirit of wine - -	1·3721
Diftilled vinegar - -	1·3721
Gum ammoniac - -	1·3723
Aqua regia - -	1·3898
Aqua fortis - -	1·4044
Spirit of nitre - -	1·4076
Cryftalline humour of an ox's eye	1·4635
Oil of vitriol - -	1·4262
Oil of turpentine - -	1·4833
Oil of amber - -	1·5010
Oil of cloves - -	1·5136
Oil of cinnamon - -	1·5340

3. By Mr. Euler, junior.

Rain or diftilled water -	1·3358
Well water - -	1·3362
Diftilled vinegar -	1·3442
French wine - -	1·3458
A folution of gum arabic -	1·3467
French brandy -	1·3600
Ditto a ftronger kind -	1·3618
Spirit of wine rectified -	1.3583
Ditto more highly rectified -	1·3706
White of an egg - -	1·3685
Spirit of nitre - -	1·4025
Oil of Provence - -	1·4651
Oil of turpentine - -	1·4822

III. *When a ray paffes out of a denfer medium into a*
rarer, it is refracted from the perpendicular, or from the
axis of Refraction.

This is exactly the reverfe of the 2d law, and the
quantity of Refraction is equal in both cafes, or both
forwards and backwards; fo that a ray would take the
fame courfe back, by which another paffed forward,
viz. if a ray would pafs from A by B to C, another
would pafs from C by B to A. Hence, in this cafe,
the angle of Refraction is greater than the angle of
inclination. Hence alfo, if the angle of inclination be
lefs than 30°, MBC is nearly equal to ⅓ of MBE;
therefore MBC is ½ of CBE; confequently, if the
Refraction be out of glafs into air, and the angle of
inclination lefs than 30°, the ray is refracted from the
axis of Refraction by almoft the half of the angle of
inclination. And this is the other dioptrical principle
ufed by moft authors after Kepler, to demonftrate the
Refractions of glaffes.

If the Refraction be out of air into glafs, the ratio
of the fines of inclination and Refraction is as 3 to 2,
or more accurately as 17 to 11; if out of air into water
as 4 to 3; therefore if the courfe be the contrary way,
viz. out of glafs or water into air, the ratio of the
fines will be, in the former cafe as 2 to 3 or 11 to 17,
and in the latter as 3 to 4. So that, if the Refraction
be from water or glafs into air, and the angle of incidence

dence or inclination be greater than about 48½ degrees in water, or greater than about 40° in glass, the ray will not be refracted into air; but will be reflected into a line which makes the angle of reflection equal to the angle of incidence ; because the sines of 48½ and 40° are to the radius, as 3 to 4, and as 11 to 17′ nearly ; and therefore when the sine has a greater proportion to the radius than as above, the ray will not be refracted.

IV. *A ray falling on a curve surface, whether concave or convex, is refracted after the same manner as if it fell on a plane which is a tangent to the curve in the point of incidence.* Because the curve and its tangent have the point of contact common to both, where the ray is refracted.

Laws of REFRACTION in Plane Surfaces.

1. If parallel rays, AB and CD, be refracted out of one transparent medium into another of a different density, they will continue parallel after Refraction, as BE and DF. Hence a glass that is plane on both sides, being turned either directly or obliquely to the sun, &c, the light passing through it will be propagated in the same manner as if the glass were away.

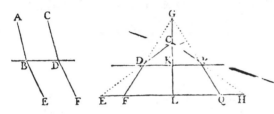

2. If two rays CD and CP, proceeding from the same radiant C, and falling on a plane surface of a different density, so that the points of Refraction D and P be equally distant from the perpendicular of incidence GK, the refracted rays DF and PQ have the same virtual focus, or the same point of dispersion G.—Hence, when refracted rays, falling on the eye placed out of the perpendicular of incidence, are either equally distant from the perpendicular, or very near each other, they will flow upon the eye as if they came to it from the point G ; consequently the point C will be seen by the refracted rays as in G. And hence also, if the eye be placed in a dense medium, objects in a rarer will appear more remote than they are ; and the place of the image, in any case, may be determined from the ratio of Refraction : Thus, to fishes swimming under water, objects out of the water must appear farther distant than in reality they are. But, on the contrary, if the eye at E be placed in a rarer medium, then an object G placed in a denser, appears, at C, nearer than it is ; and the place of the image may be determined in any given case by the ratio of Refraction : and thus the bottom of a vessel full of water is raised by Refraction a third part of its depth, with respect to an eye placed perpendicularly over the refracting surface ; and thus also fishes and other bodies, under water, appear nearer than they really are.

3. If the eye be placed in a rarer medium ; then an object seen in a denser, by a ray refracted in a plane surface, will appear larger than it really is. But if the eye be in a denser medium, and the object in a rarer,

the object will appear less than it is. And, in each case, the apparent magnitude FQ is to the real one EH, as the rectangle CK·GL to GK·CL, or in the compound ratio of the distance CK of the point to which the rays tend before Refraction, from the refracting surface DP, to the distance GK of the eye from the same, and of the distance GL of the object EH from the eye, to its distance CL from the point to which the rays tend before Refraction.—Hence, if the object be very remote, CL will be physically equal to GL ; and then the real magnitude EL is to the apparent magnitude FL, as GK to CK, or as the distance of the eye G from the refracting plane, to the distance of the point of convergence F from the same plane. And hence also, objects under water, to an eye in the air, appear larger than they are ; and to fishes under water, objects in the air appear less than they are.

Laws of REFRACTION in Spherical Surfaces, both concave and convex.

1. A ray of light DE, parallel to the axis, after a single refraction at E, meets the axis in the point F, beyond the centre C.

2. Also in that case, the semi-diameter CB or CE will be to the refracted ray EF, as the sine of the angle of refraction to the sine of the angle of inclination BCE. But the distance of the focus, or point of concurrence from the centre, CF, is to the refracted ray EF, as the sine of the refracted angle to the sine of the angle of inclination.

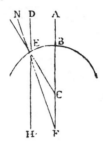

3. Hence also, in this case, the distance BF of the focus from the refracting surface, must be to CF its distance from the centre, in a ratio greater than that of the sine of the angle of inclination to the sine of the refracted angle. But those ratios will be nearly equal when the rays are very near the axis, and the angle of inclination BCE is only of a few degrees. And when the Refraction is out of air into glass, then

For rays near the axis,	For more distant rays,
BF : FC :: 3 : 2,	BF : FC > 3 : 2,
BC : BF :: 1 : 3.	BC : BF < 1 : 3.

But if the Refraction be out of air into water, then

For rays near the axis,	For more distant rays,
BF : FC :: 4 : 3,	BF : FC > 4 : 3,
BC : BF :: 1 : 4,	BC : BF < 1 : 4.

Hence, as the sun's rays are parallel as to sense, if they fall on the surface of a solid glass sphere, or of a sphere full of water, they will not meet the axis within the sphere : so that Vitello was mistaken when he imagined that the sun's rays, falling on the surface of a crystalline sphere, were refracted to the centre.

4. If a ray HE fall parallel to the axis FA, out of a rarer medium, on the concave spherical surface BE of a denser one ; the refracted ray EN will diverge from the point of the axis F, so that FE will be to FC, in the ratio of the sine of the angle of inclination, to the sine of the refracted angle. Consequently FB to FC is in a greater ratio than that ; unless when the rays are very near the axis, and the angle BCE is very small.

Y y 2

for

for then FB will be to FC nearly in that ratio. And hence, in the cases of Refraction out of air into water or glass, the ratios of BC, BF and CF, will be the same as specified in the last article.

5. If a ray DE, parallel to the axis FC, pass out of a denser into a rarer spherical convex medium, it will diverge from the axis after Refraction; and the distance FC of the point of dispersion, or of the virtual focus F, from the centre of the sphere, will be to its semidiameter CE or CB, as the sine of the refracted angle is to the sine of the angle of Refraction; but to the portion of the refracted ray drawn back, FE, it will be in the ratio of the sine of the refracted angle to the sine of the angle of inclination. Consequently FC will be to FB, in a greater ratio than this last one: unless when the rays DE fall very near the axis FC, for then FC to FB will be very nearly in that ratio.

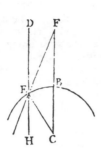

Hence, when the Refraction is out of glass into air; then,

For rays near the axis,	For more distant rays,
FC : FB :: 3 : 2,	FC : FB > 3 : 2,
BC : BF :: 1 : 2.	BC : BF > 1 : 2,

But when the Refraction is out of water into air; then,

For rays near the axis,	For more distant rays,
FC : FB :: 4 : 3,	FC : FB > 4 : 3,
BC : BF :: 1 : 3.	BC : BF > 1 : 3.

6. If the ray HE fall parallel to the axis CF, from a denser medium, upon the surface of a spherically concave rarer one; the refracted ray will meet with the axis in the point F, so that the distance CF from the centre, will be to the refracted ray FE, as the sine of the refracted angle, to the sine of the angle of inclination. Consequently FC will be to FB, in a greater ratio than that above mentioned: unless when the rays are very near the axis, for then FC is to FB very nearly in that ratio; and the three FB, FC, BC are, in the cases of air, water and glass, in the numeral ratios as specified at the end of the last article. See Wolfius, Elem. Mathes. tom. 3 p. 179 &c.

Refraction *in a Glass Prism.*

ABC being the transverse section of a prism; if a ray of light DE fall obliquely upon it out of the air; instead of proceeding straight on to F, being refracted

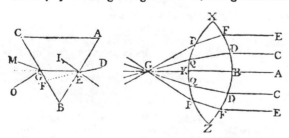

towards the perpendicular IE, it will decline to G; Again, since the ray EG, passing out of glass into air, falls obliquely on BC, it will be refracted to M, so as

to recede from the perpendicular GO. And hence arise the various phenomena of the prism. See COLOUR.

Refraction *in a Convex Lens.*

If parallel rays, AB, CD, EF, fall on the surface of a convex lens XBZ (the last fig. above); the perpendicular ray AB will pass unrefracted to K, where emerging, as before, perpendicularly, into air, it will proceed straight on to G. But the rays CD and EF, falling obliquely out of air into glass, at D and F, will be refracted towards the axis of Refraction, or towards the perpendiculars at D and F, and so decline to Q and P: where emerging again obliquely out of the glass into the surface of the air, they will be refracted from the perpendicular, and proceed in the directions QG and PG, meeting in G. And thus also will all the other rays be refracted so as to meet the rest near the place G. See FOCUS and LENS.——Hence the great property of convex glasses; viz, that they collect parallel rays, or make them converge into a point.

Refraction *in a Concave Lens.*

Parallel rays AB, CD, EF, falling on a concave lens GBHIMK, the ray AB falling perpendicularly on the glass at B, will pass unrefracted to M; where, being still perpendicular, it will pass into the air to L, with-

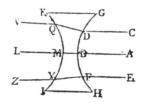 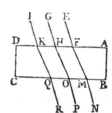

out Refraction. But the ray CD, falling obliquely on the surface of the glass, will be refracted towards the perpendicular at D, and proceed to Q; where again falling obliquely out of the glass upon the surface of air, it will be refracted from the perpendicular at Q, and proceed to V. After the same manner the ray EF is first refracted to Y, and thence to Z.——Hence the great property of concave glasses; viz, that they disperse parallel rays, or make them diverge. See LENS.

Refraction *in a Plane Glass.*

If parallel rays EF, GH, IK, (the last fig. above) fall obliquely on a plane glass ABCD; the obliquity being the same in all, by reason of their parallelism, they will be all equally refracted towards the perpendicular; and accordingly, being still parallel at M, O, and Q, they will pass out into the air equally refracted again from the perpendicular, and still parallel. Thus will the rays EF, GH, and IK, at their entering the glass, be inflected towards the right; and in their going out as much inflected to the left; so that the first Refraction is here undone by the second, thereby causing the rays on their emerging from the glass, to be parallel to their first direction before they entered it; though not so as that the object is seen in its true place; for the ray RQ, being produced back again, will not coincide with the ray IK, but will fall to the left of it; and this the more as the glass is thicker; however, as to the colour, the second Refraction does really undo the first. See Colour.

REFRACTION *in Aftronomy*, or REFRACTION *of the Stars*, is an inflexion of the rays of thofe luminaries, in paffing through our atmofphere ; by which the apparent altitudes of the heavenly bodies are increafed.

This Refraction arifes from hence, that the atmofphere is unequally denfe in different ftages or regions ; rareft of all at the top, and denfeft of all at the bottom ; which inequality in the fame medium, makes it equivalent to feveral unequal mediums, by which the courfe of the ray of light is continually bent into a continued curve line. See Atmosphere.——And Sir Ifaac Newton has fhewn, that a ray of light, in paffing from the higheft and rareft part of the atmofphere, down to the loweft and denfeft, undergoes the fame quantity of Refraction that it would do in paffing immediately, at the fame obliquity, out of a vacuum into air of equal denfity with that in the loweft part of the atmofphere.

The effect of this Refraction may be thus conceived. Suppofe ZV a quadrant of a vertical circle defcribed from the centre of the earth T, under which is AB a quadrant of a circle on the furface of the earth, and GH a quadrant of the furface of the atmofphere. Then fuppofe SE a ray of light emitted

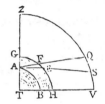

by a ftar at S, and falling on the atmofphere at E : this ray coming out of the ethereal medium, which is much rarer than our air, or perhaps out of a perfect vacuum, and falling on the furface of the atmofphere, will be refracted towards the perpendicular, or inclined down more towards the earth ; and fince the upper air is again rarer than that near the earth, and grows ftill denfer as it approaches the earth's furface, the ray in its progrefs will be continually refracted, fo as to arrive at the eye in the curve line EA. Then fuppofing the right line AF to be a tangent to the arch at A, the ray will enter the eye at A in the direction of AF ; and therefore the ftar will appear in the heavens at Q, inftead of S, higher or nearer the zenith than the ftar really is.

Hence arife the phenomena of the crepufculum or twilight ; and hence alfo it is that the moon is fometimes feen eclipfed, when fhe is really below the horizon, and the fun above it.

That there is a real Refraction of the ftars &c, is deduced not only from phyfical confiderations, and from arguments a priori, and a fimilitudine, but alfo from precife aftronomical obfervation : for there are numberlefs obfervations by which it appears that the fun, moon, and ftars rife much fooner, and appear higher, than they fhould do according to aftronomical calculations. Hence it is argued, that as light is propagated in right lines, no rays could reach the eye from a luminary below the horizon, unlefs they were deflected out of their courfe, at their entrance into the atmofphere ; and therefore it appears that the rays are refracted in paffing through the atmofphere.

Hence the ftars appear higher by Refraction than they really are ; fo that to bring the obferved or apparent altitudes to the true ones, the quantity of Refraction muft be fubtracted. And hence, the ancients,

as they were not acquainted with this Refraction, reckoned their altitudes too great, fo that it is no wonder they fometimes committed confiderable errors. Hence alfo, Refraction lengthens the day, and fhortens the night, by making the fun appear above the horizon a little before his rifing, and a little after his fetting. Refraction alfo makes the moon and ftars appear to rife fooner and fet later than they really do. The apparent diameter of the fun or moon is about 32′ ; the horizontal refraction is about 33′ ; whence the fun and moon appear *wholly* above the horizon when they are entirely below it. Alfo, from obfervations it appears that the Refractions are greater nearer the pole than at lefler latitudes, caufing the fun to appear fome days above the horizon, when he is really below it ; doubtlefs from the greater denfity of the atmofphere, and the greater obliquity of the incidence.

Stars in the zenith are not fubject to any Refraction : thofe in the horizon have the greateft of all : from the horizon, the Refraction continually decreafes to the zenith. All which follows from hence, that in the firft cafe, the rays are perpendicular to the medium ; in the fecond, their obliquity is the greateft, and they pafs through the largeft fpace of the lower and denfer part of the air, and through the thickeft vapours ; and in the third, the obliquity is continually decreafing.

The air is condenfed, and confequently Refraction is increafed, by cold ; for which reafon it is greater in cold countries than in hot ones. It is alfo greater in cold weather than in hot, in the fame country ; and the morning Refraction is greater than that of the evening, becaufe the air is rarefied by the heat of the fun in the day, and condenfed by the coldnefs of the night. Refraction is alfo fubject to fome fmall variation at the fame time of the day in the fineft weather.

At the fame altitudes the fun, moon, and ftars all undergo the fame Refraction : for at equal altitudes the incident rays have the fame inclinations ; and the fines of the refracted angles are as the fines of the angles of inclination, &c.

Indeed Tycho Brahe, who firft deduced the Refractions of the fun, moon, and ftars, from obfervation, and whofe table of the Refraction of the ftars is not much different from thofe of Flamfteed and Newton, except near the horizon, makes the folar Refractions about 4′ greater than thofe of the fixed ftars ; and the lunar Refractions alfo fometimes greater than thofe of the ftars, and fometimes lefs. But the theory of Refractions, found out by Snellius, was not fully underftood in his time.

The horizontal Refraction, being the greateft, is the caufe that the fun and moon appear of an oval form at their rifing and fetting ; for the lower edge of each being more refracted than the upper edge, the perpendicular diameter is fhortened, and the under edge appears more flatted alfo.——Hence alfo, if we take with an inftrument the diftance of two ftars when they are in the fame vertical and near the horizon, we fhall find it confiderably lefs than if we meafure it when they are both at fuch a height as to fuffer little or no Refraction ; becaufe the lower ftar is more elevated than the higher. There is alfo another alteration made by Refraction in the apparent diftance of ftars : when two ftars are in the fame almicantar, or parallel of declination, their ap-

parent

parent distance is less than the true; for since Refraction makes each of them higher in the azimuth or vertical in which they appear, it must bring them into parts of the vertical where they come nearer to each other; because all vertical circles converge and meet in the zenith. This contraction of distance, according to Dr. Halley (Philos. Transf. numb. 368) is at the rate of at least one second in a degree; so that, if the distance between two stars in a position parallel to the horizon measure 30°, it is at most to be reckoned only 29° 59' 30".

The quantity of the Refraction at every altitude, from the horizon, where it is greatest, to the zenith where it is nothing, has been determined by observation, by many astronomers; those of Dr. Bradley and Mr. Mayer are esteemed the most correct of any, being nearly alike, and are now used by most astronomers. Doctor Bradley, from his observations, deduced this very simple and general rule for the Refraction r at any altitude a whatever; viz,

as rad. 1 : cotang. $\overline{a + 3r}$ :: 57" : r" the Refraction in seconds.

This rule, of Dr. Bradley's, is adapted to these states of the barometer and thermometer, viz,

either 29·6 inc. barom. and 50° thermometer,
or 30 — barom. and 55 thermometer,
for both which states it answers equally the same. But for any other states of the barometer and thermometer, the Refraction above-found is to be corrected in this manner; viz, if b denote any other height of the barometer in inches, and t the degrees of the thermometer, r being the Refraction uncorrected, as found in the manner above. Then

as 29·6 : b :: r : R the Refraction corrected on account of the barometer,

and 400 : 450 t :: R : the Refraction corrected both on account of the barometer and thermometer; which final corrected Refraction is therefore $= \dfrac{450 - t}{11840} br$.

Or, to correct the same Refraction r by means of the latter state, viz, barom. 30 and therm. 55, it will be

as 30 : b :: r : R $= \dfrac{br}{30}$,

and 400 : 455 − t :: R : $\dfrac{455 - t}{400}$ R $= \dfrac{455 - t}{12000} br$ the correct Refraction.

From Dr. Bradley's rule, r = 57" × cot. $\overline{a + 3r}$, the following Table of the mean astronom. Refrac. is computed.

Mean Astronomical Refractions in Altitude.

Apparent Altitude	Refraction	Apparent Altitude	Refraction	Apparent Altitude	Refraction	Apparent Altitude	Refraction	Apparent Altitude	Refraction
0° 0	33 0"	3° 0	14 36"	8° 30	6' 8"	20° 0	2' 35"	54°	41"
0 5	32 10	3 5	14 20	8 40	6 1	20 30	2 31	55	40
0 10	31 22	3 10	14 4	8 50	5 55	21 0	2 27	56	38
0 15	30 35	3 15	13 49	9 0	5 48	21 30	2 24	57	37
0 20	29 50	3 20	13 34	9 10	5 42	22 0	2 20	58	35
0 25	29 6	3 25	13 20	9 20	5 36	23	2 14	59	34
0 30	28 22	3 30	13 6	9 30	5 31	24	2 7	60	33
0 35	27 41	3 40	12 40	9 40	5 25	25	2 2	61	31
0 40	27 0	3 50	12 15	9 50	5 20	26	1 56	62	30
0 45	26 20	4 0	11 51	10 0	5 15	27	1 51	63	29
0 50	25 42	4 10	11 29	10 15	5 7	28	1 47	64	28
0 55	25 5	4 20	11 8	10 30	5 0	29	1 42	65	26
1 0	24 29	4 30	10 48	10 45	4 53	30	1 38	66	25
1 5	23 54	4 40	10 29	11 0	4 47	31	1 35	67	24
1 10	23 20	4 50	10 11	11 15	4 40	32	1 31	68	23
1 15	22 47	5 0	9 54	11 30	4 34	33	1 28	69	22
1 20	22 15	5 10	9 38	11 45	4 29	34	1 24	70	21
1 25	21 44	5 20	9 23	12 0	4 23	35	1 21	71	19
1 30	21 15	5 30	9 8	12 20	4 16	36	1 18	72	18
1 35	20 46	5 40	8 54	12 40	4 9	37	1 16	73	17
1 40	20 18	5 50	8 41	13 0	4 3	38	1 13	74	16
1 45	19 51	6 0	8 28	13 20	3 57	39	1 10	75	15
1 50	19 25	6 10	8 15	13 40	3 51	40	1 8	76	14
1 55	19 0	6 20	8 3	14 0	3 45	41	1 5	77	13
2 0	18 35	6 30	7 51	14 20	3 40	42	1 3	78	12
2 5	18 11	6 40	7 40	14 40	3 35	43	1 1	79	11
2 10	17 48	6 50	7 30	15 0	3 30	44	0 59	80	10
2 15	17 26	7 0	7 20	15 30	3 24	45	0 57	81	9
2 20	17 4	7 10	7 11	16 0	3 17	46	0 55	82	8
2 25	16 44	7 20	7 2	16 30	3 10	47	0 53	83	7
2 30	16 24	7 30	6 53	17 0	3 4	48	0 51	84	6
2 35	16 4	7 40	6 45	17 30	2 59	49	0 49	85	5
2 40	15 45	7 50	6 37	18 0	2 54	50	0 48	86	4
2 45	15 27	8 0	6 29	18 30	2 49	51	0 46	87	3
2 50	15 9	8 10	6 22	19 0	2 45	52	0 44	88	2
2 55	14 52	8 20	6 15	19 30	2 39	53	0 43	89	1

Mr. Mayer fays his rule was deduced from theory, and, when reduced from French meafure and Reaumur's thermometer, to Englifh meafure and Fahrenheit's thermometer, it is this,

$$r = \frac{74.4b \times \text{cof. } a}{(1 + .00248t)^{\frac{1}{2}}} \left(\sqrt{14 \frac{17.14 \text{ fin. } a}{1 + .00248t}} - \frac{17.14 \text{ fin. } a}{(1 + .00248t)^{\frac{1}{2}}} \right)$$

or $r = \dfrac{74.4b \times \text{cof. } a \times \text{tang. } \frac{1}{2}A}{(1 + .00248t)^{\frac{3}{4}}}$ the Refraction in

feconds, correcked for both barometer and thermometer: where the letters denote the fame things as before, except A, which denotes the angle whofe tangent is

$$\frac{\sqrt{1 + .00248t}}{17.14 \text{ fin. } a}.$$

Mr. Simpfon too (Differt. pa. 46 &c) has ingenioufly determined by theory the aftronomical Refractions, from which he brings out this rule, viz, As 1 to .9986 or as radius to fine of 86° 58′ 30″, fo is the fine of any given zenith diftance, to the fine of an arc; then $\frac{2}{11}$ of the difference between this arc and the zenith diftance, is the Refraction fought for that zenith diftance. And by this rule Mr. Simpfon computed a Table of the mean Refractions, which are not much different from thofe of Dr. Bradley and Mr. Mayer, and are as in the following Table.

Mr. Simpfon's Table of Mean Refractions.

Appa-rent Alti-tude.	Refrac-tion.		Appa-rent Alti-tude.	Refrac-tion.		Appa-rent Alti-tude.	Refrac-tion.	
0°	33′	0″	17°	2′	50″	38°	1′	7″
1	23	50	18	2	40	40	1	2
2	17	43	19	2	31	42	0	58
3	13	44	20	2	23	44	0	54
4	11	5	21	2	16	46	0	50
5	9	10	22	2	9	48	0	47
6	7	49	23	2	3	50	0	44
7	6	48	24	1	57	52	0	41
8	5	59	25	1	52	54	0	38
9	5	21	26	1	47	56	0	35
10	4	50	27	1	42	58	0	32
11	4	24	28	1	38	60	0	30
12	4	2	29	1	34	65	0	24
13	3	43	30	1	30	70	0	19
14	3	27	32	1	23	75	0	14
15	3	13	34	1	17	80	0	9
16	3	1	36	1	12	85	0	$4\frac{1}{2}$

It is evident that all obferved altitudes of the heavenly bodies ought to be diminifhed by the numbers taken out of the foregoing Table. It is alfo evident that the Refraction diminifhes the right and oblique afcenfions of a ftar, and increafes the defcenfions: it increafes the northern declination and latitude, but decreafes the fouthern: in the eaftern part of the heavens it diminifhes the longitude of a ftar, but in the weftern part of the heavens it increafes the fame.

I

Refraction of *Altitude*, is an arc of a vertical circle, as AB, by which the altitude of a ftar AC is increafed by the Refraction.

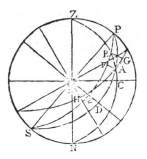

Refraction of *Afcenfion and Defcenfion*, is an arc DE of the equator, by which the afcenfion and defcenfion of a ftar, whether right or oblique, is increafed or diminifhed by the Refraction.

Refraction of *Declination*, is an arc BF of a circle of declination, by which the declination of a ftar DA or EF is increafed or diminifhed by Refraction.

Refraction of *Latitude* is an arc AG of a circle of latitude, by which the latitude of a ftar AH is increafed or diminifhed by the Refraction.

Refraction of *Longitude* is an arc IH of the ecliptic, by which the longitude of a ftar is increafed or diminifhed by means of the Refraction.

Terreftrial Refraction, is that by which terreftrial objects appear to be raifed higher than they really are, in obferving their altitudes. The quantity of this Refraction is eftimated by Dr. Mafkelyne at one-tenth of the diftance of the object obferved, expreffed in degrees of a great circle. So, if the diftance be 10000 fathoms, its 10th part 1000 fathoms, is the 60th part of a degree of a great circle on the earth, or 1′, which therefore is the Refraction in the altitude of the object at that diftance. (Requifite Tables, 1766, pa. 134).

But M. Le Gendre is induced, he fays, by feveral experiments, to allow only $\frac{1}{14}$th part of the diftance for the Refraction in altitude. So that, upon the diftance of 10000 fathoms, the 14th part of which is 714 fathoms, he allows only 44″ of terreftrial Refraction, fo many being contained in the 714 fathoms. See his Memoir concerning the Trigonometrical operations, &c.

Again, M. de Lambre, an ingenious French aftronomer, makes the quantity of the Terreftrial Refraction to be the 11th part of the arch of diftance. But the Englifh meafurers, Col. Edw. Williams, Capt. Mudge, and Mr. Dalby, from a multitude of exact obfervations made by them, determine the quantity of the medium Refraction to be the 12th part of the faid diftance.

The quantity of this Refraction, however, is found to vary confiderably, with the different ftates of the weather and atmofphere, from the 15th part of the diftance, to the 9th part of the fame; the medium of which is the 12th part, as above mentioned.

Some whimfical effects of this Refraction are alfo related, arifing from peculiar fituations and circumftances. Thus, it is faid, any perfon ftanding by the fide of the river Thames at Greenwich, when it is high-

water

water there, he can see the cattle grazing on the Isle of Dogs, which is the marshy meadow on the other side of the river at that place; but when it is low-water there, he cannot see any thing of them, as they are hid from his view by the land wall or bank on the other side, which is raised higher than the marsh, to keep out the waters of the river. This curious effect is probably owing to the moist and dense vapours, just above and rising from the surface of the water, being raised higher or lifted up with the surface of the water at the time of high tide, through which the rays pass, and are the more refracted.

Again, a similar instance is related in a letter to me, from an ingenious friend, Mr. Abr. Crocker of Frome in Somersetshire, dated January 12, 1795. "My Devonshire friend," says he, (whose seat is in the vicinity of the town of Modbury, 12 miles in a geographical line from Maker tower near Plymouth) "being on a pleasure spot in his garden, on the 4th of December 1793, with some friends, viewing the surrounding country, with an achromatic telescope, described an object like a perpendicular pole standing up in the chasm of a hedge which bounded their view at about 9 miles distance; which, from its direction, was conjectured to be the flagstaff on Maker tower.—Directing the glass, on the morning of the next day, to the same part of the horizon, a flag was perceived on the pole; which corroborated the conjecture of the preceding day. This day's view also discovered the pinnacles and part of the shaft of the tower.—Viewing the same spot at 8 in the morning on the 9th of January 1794, the whole tower and part of the roof of the church, with other remote objects not before noticed, became visible.

" It is necessary to give you the state of the weather there, on those days.

1793.	Barometer.	Thermom.	Wind	
Dec. 4	29·93, rising	36·0	N. E.	Frosty morning, a mist over the land below.
5	29·97, rising	35·2	W.	Ditto.
1794.				
Jan. 9	30·01, falling	29·8	W.	Hard white frost, a fog over the lowlands; clear in the surrounding country.

" The singularity of this phenomenon has occasioned repeated observations on it; from all which it appears that the summer season, and wet windy weather, are unfavourable to this refracted elevation; but that calm frosty weather, with the absence of the sun, are favourable to it.

" From hence a question arises; what is the principal or most general cause of atmospheric Refraction, which produces such extraordinary appearances?"

The following is also a copy of a letter to Mr. Crocker on this curious phenomenon, from his friend above mentioned, viz. Mr. John Andrews, of Traine, near Modbury, dated the 1st of February 1795.

" My good Friend,

" Finding, by your favour of last Sunday, the proceedings which are going on in respect to my observations on the phenomenon of *Looming*, I have thought it necessary to bestow about half this day in preparing, what I am obliged to call, in my way, *drawings*, illustrative of those observations.—I have endeavoured to distinguish, by different tints and shades, the grounds which lie nearer or more remote; but this will perhaps be better explained by the letters of reference, which I have inserted as they may be serviceable in future correspondence.—I believe the drawings, rough as they are, give a tolerably exact representation of the scenes: they may be properly copied to send to London by one of your ingenious sons.—I have been attentive in my observations, or rather in looking out for observations, during the late hard frosts, which you will be surprised to learn, have (except on one or two days) been very unpropitious to the phenomenon; but they have compensated for that disappointment, by a discovery, that a dry frost, though ever so intense, has no tendency to produce it. A hoar frost, or that kind of dewy vapour which, in a sufficient degree of cold, occasions a hoar frost, appears essentially necessary. This took place pretty favourably on the 6th of January, when the elevation was equal to that represented in the third drawing *(see plate 25, fig. 3)*, much like what it was on the 9th of January 1794, and confirmed me as to the certainty of some peculiar appearances, hinted at in my letter of the 14th of that month, but not there described. What I allude to, was a fluctuating appearance of two horizons, one above the other, with a complete vacancy between them, exactly like what may be often observed looking through an uneven pane of glass. Divers instances of this were seen by my brother and myself on the 6th of last January; continually varying and intermitting, but not rapidly, so that they were capable of distinct observation.—Till that day I had formed, as I thought, a plausible theory, to account for, as well this latter, as all the other phenomena; but now, unless my imagination deceives me, I am left in impenetrable darkness. The vacant line of separation, you will take notice, would often increase so much in breadth, as to efface entirely the upper of the two horizons; forming then a kind of dent or gap in the remaining horizon, which horizon at the places contiguous to the extremities of the vacancy, seemed of the same height as the upper horizon was, before effaced. This vacancy was several times seen to approach and take in the tower, and immediately to admit a view of the whole or most part of its body (like that in the third drawing) which was not the case before: exactly, to all appearance, as if it had opened a gap for that purpose in the intercepted ground.—It remains therefore to be determined by future observations, whether the separation is effected by an elevation of the upper, or depression of the lower horizon; and if the latter, why the vacancy does not cause the tower to disappear, as well as the intervening ground?—As an opportunity for this purpose may not soon occur, I hope you will not wait for it, in your communications to him who is, Dear Sir, yours very truly,

JOHN ANDREWS."

See the representations in plate 25, of the appearances, in three different states of the atmosphere, with the explanations of them.

REFRAN.

REFRANGIBILITY *of Light*, the difpofition of the rays to be refracted. And a greater or lefs Refrangibility, is a difpofition to be more or lefs refracted, in paffing at equal angles of incidence into the fame medium.

That the rays of light are differently refrangible, is the foundation of Newton's whole theory of light and colours; and the truth and circumftances of the principle he evinced from fuch experiments as the following.

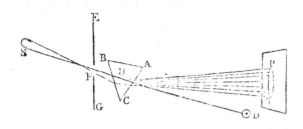

Let EG reprefent the window-fhutter of a dark room, and F a hole in it, through which the light paffes, from the luminous object S, to the glafs prifm ABC within the room, which refracts it towards the oppofite fide, or a fcreen, at PT, where it appears of an oblong form; its length being about five times the breadth, and exhibiting the various colours of the rainbow; whereas without the interpofition of the prifm, the ray of light would have proceeded on in its firft direction to D. Hence then it follows,

1. That the rays of light are refrangible. This appears by the ray being refracted from its original direction SHD, into another one, HP or HT, by paffing through a different medium.

2. That the ray SFH is a compound one, which, by means of the prifm, is decompounded or feparated into its parts, HP, HT, &c, which it hence appears are all endued with different degrees of Refrangibility, as they are tranfmitted to all the intermediate points from T to P, and there painting all the different colours.

From this, and a great variety of other experiments, Newton proved, that the blue rays are more refracted than the red ones, and that there is likewife unequal refraction in the intermediate rays; and upon the whole it appears that the fun's rays have not all the fame Refrangibility, and confequently are not of the fame nature. It is alfo obferved that thofe rays which are moft refrangible, are alfo moft reflexible. See REFLEXIBILITY; alfo Newton's Optics, pa. 22 &c, 3d edit.

The difference between Refrangibility and reflexibility was firft difcovered by Sir Ifaac Newton, in 1671-2, and communicated to the Royal Society, in a letter dated Feb. 6 of that year, which was publifhed in the Philof. Tranf. numb. 80, pa. 3075; and from that time it was vindicated by him, from the objections of feveral authors; particularly Pardies, Mariotte, Linus or Lin, and other gentlemen of the Englifh college at Liege; and at length it was more fully laid down, illuftrated, and confirmed, by a great variety of experiments, in his excellent treatife on Optics.

But farther, as not only thefe colours of light produced by refraction in a prifm, but alfo thofe

reflected from opaque bodies, have their different degrees of Refrangibility and reflexibility; and as a white light arifes from a mixture of the feveral coloured rays together, the fame great author concluded that all homogeneous light has its proper colour, correfponding to its degree of Refrangibility, and not capable of being changed by any reflexions, or any refractions; that the fun's light is compofed of all the primary colours; and that all compound colours arife from the mixture of the primary ones, &c.

The different degrees of Refrangibility, he conjectures to arife from the different magnitude of the particles compofing the different rays. Thus, the moft refrangible rays, that is the red ones, he fuppofes may confift of the largeft particles; the leaft refrangible, i. e. the violet rays, of the fmalleft particles; and the intermediate rays, yellow, green, and blue, of particles of intermediate fizes. See COLOUR.

For the method of correcting the effect of the different Refrangibility of the rays of light in glaffe, fee ABERRATION and TELESCOPE.

REGEL, or RIGEL, a fixed ftar of the firft magnitude, in the left foot of Orion.

REGIOMONTANUS. See *John* MULLER.

REGION, of the Air or Atmofphere. Authors divide the atmofphere into three ftages, called the upper, middle, and lower Regions.—The loweft Region is that in which we breathe, and is bounded by the reflexion of the fun's rays, that is, by the height to which they rebound from the earth.—The middle Region is that in which the clouds refide, and where meteors are formed, &c; extending from the extremity of the loweft, to the tops of the higheft mountains.—The upper Region commences from the tops of the mountains, and reaches to the utmoft limits of the atmofphere. In this Region there probably reigns a perpetual equable calmnefs, clearnefs, and ferenity.

Elementary REGION, according to the Ariftotelians, is a fphere terminated by the concavity of the moon's orb, comprehending the earth's atmofphere.

Ethereal REGION, is the whole extent of the univerfe, comprifing all the heavens with the orbs of the fixed ftars and other celeftial bodies.

REGION, in Geography, a country or particular divifion of the earth, or a tract of land inhabited by people of the fame nation.

REGIONS *of the Moon*: Modern aftronomers divide the moon into feveral Regions, or provinces, to each of which they give its proper name.

REGIONS *of the Sea*, are the two parts into which the whole depth of the fea is conceived to be divided. The upper of thefe extends from the furface of the water, down as low as the rays of the fun can pierce, and extend their influence; and the lower Region extends from thence to the bottom of the fea.

Subterranean REGIONS. Thefe are three, into which the earth is divided, at different depths below the furface, according to different degrees of cold or warmth; and it is imagined that the 2d or middlemoft of thefe Regions is the coldeft of the three.

REGIS (PETER SYLVAIN), a French philofopher, and great propagator of Cartefianifm, was born in Agenois 1632.

He ftudied the languages and philofophy under the

Jefuits

Jesuits at Cahors, and afterwards divinity in the university of that town, being designed for the church. His progress in learning was so uncommon, that at the end of four years he was offered a doctor's degree without the usual charges ; but he did not think it became him till he should study also in the Sorbonne at Paris. He accordingly repaired to the capital for that purpose ; but he soon became disgusted with theology ; and, as the philosophy of Des Cartes began at that time to make a noise through the lectures of Rohault, he conceived a taste for it, and gave himself up entirely to it.

Having, by attending those lectures, and by close study, become an adept in that philosophy, he went to Toulouse in 1665, where he set up lectures in it himself. Having a clear and fluent manner, and a happy way of explaining his subject, he drew all sorts of people to his discourses ; the magistrates, the literati, the ecclesiastics, and the very women, who all now affected to renounce the ancient philosophy.

In 1671, he received at Montpellier the same applauses for his lectures as at Toulouse. Finally, in 1680 he returned to Paris ; where the concourse about him was such, that the sticklers for Peripateticism began to be alarmed. These applying to the archbishop of Paris, he thought it expedient, in the name of the king, to put a stop to the lectures ; which accordingly were discontinued for several months. Afterwards his whole life was spent in propagating the new philosophy, both by lectures, and by publishing books. In defence of his system, he had disputes with Huet, Du Hamel, Malbranche, and others. His works, though abounding with ingenuity and learning, have been neglected in consequence of the great discoveries and advancement in philosophic knowledge that has been since made.—He was chosen a member of the Academy of Sciences in 1699 ; and died in 1707, at 75 years of age.

His works, which he published, are,

1. *A System of Philosophy* ; containing Logic, Metaphysics, and Morals ; in 1690, 3 vols in 4to. being a compilation of the different ideas of Des Cartes.—It was reprinted the year after at Amsterdam, with the addition of a Discourse upon Ancient and Modern Philosophy.

2. *The Use of Reason and of Faith.*

3. An Answer to Huet's Censures of the Cartesian Philosophy ; and an Answer to Du Hamel's Critical Reflections.

4. Some pieces against Malbranche, to shew that the apparent magnitude of an object depends solely on the magnitude of its image, traced on the retina.

5. A small piece upon the question, Whether Pleasure makes our present happiness ?

REGRESSION, or RETROGRADATION *of Curves*, &c. See RETROGRADATION.

REGULAR *Figure*, in Geometry, is a figure that is both equilateral and equiangular, or having all its sides and angles equal to one another.

For the dimensions, properties, &c, of regular figures, see POLYGON.

REGULAR *Body*, called also *Platonic Body*, is a body or solid comprehended by like, equal, and regular plane figures, and whose solid angles are all equal.

The plane figures by which the solid is contain-

ed, are the faces of the solid. And the sides of the plane figures are the edges, or linear sides of the solid.

There are only five Regular Solids, viz,

The tetraedron, or regular triangular pyramid, having 4 triangular faces ;

The hexaedron, or cube, having 6 square faces ;

The octaedron, having 8 triangular faces ;

The dodecaedron, having 12 pentagonal faces ;

The icosaedron, having 20 triangular faces.

Besides these five, there can be no other Regular Bodies in nature.

PROB. I. *To construct or form the Regular Solids.*—See the method of describing these figures under the article BODY.

2. *To find either the Surface or the Solid Content of any of the Regular Bodies.*—Multiply the proper tabular area or surface (taken from the following Table) by the square of the linear edge of the solid, for the superficies. And

Multiply the tabular solidity, in the last column of the Table, by the cube of the linear edge, for the solid content.

Surfaces and Solidities of Regular Bodies, the side being unity or 1.			
No. of sides.	Name.	Surface.	Solidity.
4	Tetraedron	1·7320508	0·1178513
6	Hexaedron	6·0000000	1·0000000
8	Octaedron	3·4641016	0·4714045
12	Dodecaedron	20·6457788	7·6631189
20	Icosaedron	8·6602540	2·1816950

3. The Diameter of a Sphere being given, to find the side of any of the Platonic bodies, that may be either inscribed in the sphere, or circumscribed about the sphere, or that is equal to the sphere.

Multiply the given diameter of the sphere by the proper or corresponding number, in the following Table, answering to the thing sought, and the product will be the side of the Platonic body required.

The diam. of a sphere being 1, the side of a	That may be inscribed in the sphere, is	That may be circumscribed about the sphere, is	That is equal to the sphere, is
Tetraedron	0·816497	2·44948	1·64417
Hexaedron	0·577350	1·00000	0·88610
Octaedron	0·707107	1·22474	1·03576
Dodecaedron	0·525731	0·66158	0·62153
Icosaedron	0·356822	0·44903	0·40883

4. The side of any of the five Platonic bodies being given, to find the diameter of a sphere, that may either be inscribed in that body, or circumscribed about it, or that is equal to it.—As the respective number in the Table above, under the title, *inscribed, circumscribed,* or *equal,* is to 1, so is the side of the given Platonic body,

body, to the diameter of its inscribed, circumscribed, or equal sphere.

5. The side of any one of the five Platonic bodies being given; to find the side of any of the other four bodies, that may be equal in solidity to that of the given body.—As the number under the title *equal* in the last column of the table above, against the given Platonic body, is to the number under the same title, against the body whose side is sought, so is the side of the given Platonic body, to the side of the body sought.

See demonstrations of many other properties of the Platonic bodies, in my Mensuration, part 3 sect. 2 pa. 249, &c, 2d edition.

REGULAR *Curve.* See CURVE.

REGULATOR *of a Watch*, is a small spring belonging to the balance, serving to adjust the going, and to make it go either faster or slower.

REGULUS, in Astronomy, a star of the first magnitude, in the constellation Leo; called also, from its situation, *Cor Leonis*, or the *Lion's Heart*; by the Arabs, *Alhabor*; and by the Chaldeans, *Kalbeleced*, or *Karbeleceid*; from an opinion of its influencing the affairs of the heavens; as Theon observes.

The longitude of Regulus, as fixed by Flamsteed, is 25° 31' 21'', and its latitude 0° 26' 38'' north. See LEO.

REINFORCE, in Gunnery, is that part of a gun next the breech, which is made stronger to resist the force of the powder. There are usually two Reinforces in each piece, called the first and second Reinforce. The second is somewhat smaller than the first, because the inflamed powder in that part is less strong.

REINFORCE *Rings* of a cannon, are flat mouldings, like iron hoops, placed at the breech end of the first and second Reinforce, projecting beyond the rest of the metal about a quarter of an inch.

REINHOLD (ERASMUS), an eminent astronomer and mathematician, was born at Salfeldt in Thuringia, a province in Upper Saxony, the 11th of October 1511. He studied mathematics under James Milichi at Wittemberg, in which university he afterwards became professor of those sciences, which he taught with great applause. After writing a number of useful and learned works, he died the 19th of February 1553, at 42 years of age only. His writings are chiefly the following:

1. *Theoriæ novæ Planetarum G. Purbachii*, augmented and illustrated with diagrams and Scholia in 8vo, 1542; and again in 1580.—In this work, among other things worthy of notice, he teaches (pa. 75 and 76) that the centre of the lunar epicycle describes an *oval figure* in each monthly period, and that the orbit of Mercury is also of the same oval figure.

2. *Ptolomy's Almagest*, the first book, in Greek, with a Latin version, and Scholia, explaining the more obscure passages; in 8vo, 1549.—At the end of pa. 123 he promises an edition of Theon's Commentaries, which are very useful for understanding Ptolomy's meaning; but his immature death prevented Reinhold from giving this and other works which he had projected.

3. *Prutenicæ Tabulæ Cælestium Motuum*, in 4to,

1551; again in 1571; and also in 1585.—Reinhold spent seven years labour upon this work, in which he was assisted by the munificence of Albert, duke of Prussia, from whence the tables had their name. Reinhold compared the observations of Copernicus with those of Ptolomy and Hipparchus, from whence he constructed these new tables, the uses of which he has fully explained in a great number of precepts and canons, forming a complete introduction to practical astronomy.

4. *Primus liber Tabularum Directionum*; to which are added, the *Canon Fæcundus*, or Table of Tangents, to every minute of the quadrant; and New Tables of Climates, Parallels and Shadows, with an Appendix containing the second Book of the Canon of Directions; in 4to, 1554.—Reinhold here supplies what was omitted by Regiomontanus in his Table of Directions, &c; shewing the finding of the sines, and the construction of the tangents, the sines being found to every minute of the quadrant, to the radius 10,000,000; and he produced the Oblique Ascensions from 60 degrees to the end of the quadrant. He teaches also the use of these tables in the solution of spherical problems.

Reinhold prepared likewise an edition of many other works, which are enumerated in the Emperor's Privilege, prefixed to the Prutenic Tables. Namely, Ephemerides for several years to come, computed from the new tables. Tables of the Rising and Setting of several Fixed Stars, for many different climates and times. The illustration and establishment of Chronology, by the eclipses of the luminaries, and the great conjunctions of the planets, and by the appearance of comets, &c. The Ecclesiastical Calendar. The History of Years, or Astronomical Calendar. *Isagoge Spherica*, or Elements of the Doctrine of the Primum Mobile. *Hypotyposes Orbium Cælestium*, or the Theory of Planets. Construction of a New Quadrant. The Doctrine of Plane and Spherical Triangles. Commentaries on the work of Copernicus. Also Commentaries on the 15 books of Euclid, on Ptolomy's Geography, and on the Optics of Alhazen the Arabian.—Reinhold also made Astronomical Observations, but with a wooden quadrant, which observations were seen by Tycho Brahe when he passed through Wittemberg in the year 1575, who wondered that so great a cultivator of astronomy was not furnished with better instruments.

Reinhold left a son, named also Erasmus after himself, an eminent mathematician and physician at Salfeldt: He wrote a small work in the German language, on Subterranean Geometry, printed in 4to at Erfurt 1575.—He wrote also concerning the New Star which appeared in Cassiopeia in the year 1572; with an Astrological Prognostication, published in 1574, in the German language.

RELAIS, in Fortification, a French term, the same with berme.

RELATION, in Mathematics, is the habitude or respect of quantities of the same kind to each other, with regard to their magnitude; more usually called *ratio.*—And the equality, identity, or sameness of two such Relations, is called proportion.

RELATION, *Inharmonical*, in Musical Composition,

Z z 2

is that whose extremes form a false or unnatural interval, incapable of being sung.—This is otherwise called a *false Relation*, and stands opposed to a just or true one.

RELATIVE *Gravity*, *Levity*, *Motion*, *Necessity*, *Place*, *Space*, *Time*, *Velocity*, &c. See the several substantives.

RELIEVO, in Architecture, denotes the sally or projecture of any ornament.

REMAINDER, is the difference between two quantities, or that which is left after subtracting one from the other.

RENDERING, in Building. See PARGETING.

REPELLING *Power*, in Physics, is a certain power or faculty, residing in the minute particles of natural bodies, by which, under certain circumstances, they mutually fly from each other. This is the reverse or opposite of the attractive power. Newton shews, from observation, that such a force does really exist; and he argues, that as in algebra, where positive quantities cease, there negative ones begin; so in physics, where the attractive force ceases, there a Repelling force must begin.

As the Repelling power seems to arise from the same principle as the attractive, only exercised under different circumstances, it is governed by the same laws. Now the attractive power we find is stronger in small bodies, than in great ones, in proportion to the masses; therefore the Repelling is so too. But the rays of light are the most minute bodies we know of; and therefore their Repelling force must be the greatest. It is computed by Newton, that the attractive force of the rays of light is above 10000000000000000, or one thousand million of millions of times stronger than the force of gravity on the surface of the earth: hence arises that inconceivable velocity with which light must move to reach from the sun to the earth in little more than 7 minutes of time. For the rays emitted from the body of the sun, by the vibrating motion of its parts, are no sooner got without the sphere of attraction of the sun, than they come within the action of the Repelling power.

The elasticity or springiness of bodies, or that property by which, after having their figure altered by an external force, they return to their former shape again, follows from the Repelling power. See REPULSION.

REPERCUSSION. See REFLECTION.

REPETEND, in Arithmetic, denotes that part of an infinite decimal fraction, which is continually repeated ad infinitum. Thus in the numbers 2·13 13 13 &c. the figures 13 are the Repetend, and marked thus 1̇3̇.

These Repetends chiefly arise in the reduction of vulgar fractions to decimals. Thus, $\frac{1}{3}$ = 0·333 &c = 0·3̇; and $\frac{1}{6}$ = 01666 &c = 0·16̇; and $\frac{1}{7}$ = 0·142857 142857 &c = 0·1̇42857̇. Where it is to be observed, that a point is set over the figure of a single Repetend, and a point over the first and last figure when there are several that repeat.

Repetends are either *single* or *compound*.

A *Single* REPETEND is that in which only one figure repeats; as 0·3̇, or 0·6̇, &c.

A *Compound* REPETEND, is that in which two or more figures are repeated; as ·1̇3̇, or ·2̇15̇, or ·1̇42857̇, &c.

Similar REPETENDS are such as begin at the same place, and consist of the same number of figures: as ·3̇ and ·6̇, or 1·341̇ and 2·156̇.

Dissimilar REPETENDS begin at different places, and consist of an unequal number of figures.

To find the finite *Value* of any *Repetend*, or to reduce it to a Vulgar Fraction. Take the given repeating figure or figures for the numerator; and for the denominator, take as many 9's as there are recurring figures or places in the given Repetend.

So ·3̇ = $\frac{3}{9}$ = $\frac{1}{3}$; and ·05̇ = ·0$\frac{5}{9}$ = $\frac{5}{90}$ = $\frac{1}{18}$;

and ·123̇ = $\frac{123}{999}$ = $\frac{41}{333}$; and 2·63̇ = 2$\frac{63}{99}$ = 2$\frac{7}{11}$;

and ·059̇4405̇ = $\frac{594405}{9999990}$ = $\frac{17}{286}$;

and ·769̇230̇ = $\frac{769230}{999999}$ = $\frac{10}{13}$.

Hence it follows, that every such infinite Repetend has a certain determinate and finite value, or can be expressed by a terminate vulgar fraction. And consequently, that an infinite decimal which does not repeat or circulate, cannot be completely expressed by a finite vulgar fraction.

It may farther be observed, that if the numerator of a vulgar fraction be 1, and the denominator any prime number, except 2 and 5, the decimal which shall be equal to that vulgar fraction, will always be a Repetend, beginning at the first place of decimals; and this Repetend must necessarily be a submultiple, or an aliquot part of a number expressed by as many 9's as the Repetend has figures; that is, if the Repetend have six figures, it will be a submultiple of 999999; if four figures, a submultiple of 9999 &c. From whence it follows, that if any prime number be called p, the series 9999 &c, produced as far as is necessary, will always be divisible by p, and the quotient will be the Repetend of the decimal fraction = $\frac{1}{p}$.

RESIDUAL *Figure*, in Geometry, the figure remaining after subtracting a less from a greater.

RESIDUAL *Root*, is a root composed of two parts or members, only connected together with the sign — or minus. Thus, $a-b$, or $5-3$, is a residual root; and is so called, because its true value is no more than the residue, or difference between the parts a and b, or 5 and 3, which in this case is 2.

RESIDUUM *of a Charge*, in Electricity, first discovered by Mr. Gralath, in Germany, in 1746, is that part of the charge that lay on the uncoated part of a Leyden phial, which does not part with all its electricity at once; so that it is afterwards gradually diffused to the coating.

RESISTANCE, or RESISTING *Force*, in Physics, any power which acts in opposition to another, so as to destroy or diminish its effect.

These

There are various kinds of Refiſtance, ariſing from the various natures and properties of the refiſting bodies, and governed by various laws: as, the Reſiſtance of ſolids, the Reſiſtance of fluids, the Reſiſtance of the air, &c. Of each of theſe in their order, as below.

RESISTANCE *of Solids*, in Mechanics, is the force with which the quieſcent parts of ſolid bodies oppoſe the motion of others contiguous to them.

Of theſe, there are two kinds. The firſt where the refiſting and the refiſted parts, i. e. the moving and quieſcent bodies, are only contiguous, and do not cohere; conſtituting ſeparate bodies or maſſes. This Reſiſtance is what Leibnitz calls *Refiſtance of the ſurface*, but which is more properly called *friction*: for the laws of which, ſee the article FRICTION.

The ſecond caſe of Reſiſtance, is where the refiſting and refiſted parts are not only contiguous, but cohere, being parts of the ſame continued body or maſs. This Reſiſtance was firſt conſidered by Galileo, and may properly be called *renitency*.

As to what regards the Reſiſtance of bodies when ſtruck by others in motion, ſee PERCUSSION, and COLLISION.

Theory of the Reſiſtance of the Fibres of Solid Bodies. —To conceive an idea of this Reſiſtance, or renitency of the parts, ſuppoſe a cylindrical body ſuſpended vertically by one end. Here all its parts, being heavy, tend downwards, and endeavour to ſeparate the two contiguous planes or ſurfaces where the body is the weakeſt; but all the parts of them reſiſt this ſeparation by the force with which they cohere, or are bound together. Here then are two oppoſite powers; viz, the weight of the cylinder, which tends to break it; and the force of coheſion of the parts, which reſiſts the fracture.

If now the baſe of the cylinder be increaſed, without increaſing its length; it is evident that both the Reſiſtance and the weight will be increaſed in the ſame ratio as the baſe; and hence it appears that all cylinders of the ſame matter and length, whatever their baſes be, have an equal Reſiſtance, when vertically ſuſpended.

But if the length of the cylinder be increaſed, without increaſing its baſe, its weight is increaſed, while the Reſiſtance or ſtrength continues unaltered; conſequently the lengthening has the effect of weakening it, or increaſes its tendency to break.

Hence to find the greateſt length a cylinder of any matter may have, when it juſt breaks with the addition of another given weight, we need only take any cylinder of the ſame matter, and faſten to it the leaſt weight that is juſt ſufficient to break it; and then conſider how much it muſt be lengthened, ſo that the weight of the part added, together with the given weight, may be juſt equal to that weight, and the thing is done. Thus, let l denote the firſt length of the cylinder, c its weight, g the given weight the lengthened cylinder is to bear, and w the leaſt weight that breaks the cylinder l, alſo x the length ſought;

then as $l : x :: c : \dfrac{cx}{l} = $ the weight of the longeſt cylinder ſought; and this, together with the given

weight g, muſt be equal to c together with the weight w; hence then

$\dfrac{cx}{l} + g = c + w$; therefore $x = \dfrac{c + w - g}{c} l = $ the whole length of the cylinder ſought. If the cylinder muſt juſt break with its own weight, then is $g = 0$, and in that caſe $x = \dfrac{c + w}{c} l$ is the whole length that juſt breaks by its own weight. By this means Galileo found that a copper wire, and of conſequence any other cylinder of copper, might be extended to 4801 braccios or fathoms of 6 feet each.

If the cylinder be fixed by one end into a wall, with the axis horizontally; the force to break it, and its Reſiſtance to fracture, will here be both different; as both the weight to cauſe the fracture, and the Reſiſtance of the fibres to oppoſe it, are combined with the effects of the lever; for the weight to cauſe the fracture, whether of the weight of the beam alone, or combined with an additional weight hung to it, is to be ſuppoſed collected into the centre of gravity, where it is conſidered as acting by a lever equal to the diſtance of that centre beyond the face of the wall where the cylinder or other priſm is fixed; and then the product of the ſaid whole weight and diſtance, will be the momentum or force to break the priſm. Again, the Reſiſtance of the fibres may be ſuppoſed collected into the centre of the tranſverſe ſection, and all acting there at the end of a lever equal to the vertical ſemidiameter of the ſection, the loweſt point of that diameter being immoveable, and about which the whole diameter turns when the priſm breaks; and hence the product of the adheſive force of the fibres multiplied by the ſaid ſemidiameter, will be the momentum of Reſiſtance, and muſt be equal to the former momentum when the priſm juſt breaks.

Hence, to find the length a priſm will bear, fixed ſo horizontally, before it breaks, either by its own weight, or by the addition of any adventitious weight; take any length of ſuch a priſm, and load it with weights till it juſt break. Then, put

$l = $ the length of this priſm,
$c = $ its weight,
$w = $ the weight that breaks it,
$a = $ diſtance of weight w,
$g = $ any given weight to be borne,
$d = $ its diſtance,
$x = $ the length required to break.

Then $l : x :: c : \dfrac{cx}{l}$ the weight of the priſm x,

and $\dfrac{cx}{l} \times \frac{1}{2} x = \dfrac{cx^2}{2l} = $ its momentum; alſo $dg = $ the momentum of the weight g; therefore $\dfrac{cx^2}{2l} + dg$ is the momentum of the priſm x and its added weight. In like manner $\frac{1}{2} cl + aw$ is that of the former or ſhort priſm and the weight that brake it; conſequently $\dfrac{cx^2}{2l} + dg = \frac{1}{2} cl + aw$, and $x = $

$\sqrt{\dfrac{aw + \frac{1}{2} cl - dg}{c}} \times 2l$ is the length ſought, that juſt

breaks

breaks with the weight g at the distance d. If this weight g be nothing, then $x = \sqrt{\dfrac{aw + \frac{1}{2}cl}{c}} \times 2l$ is the length of the prism that just breaks with its own weight.

If two prisms of the same matter, having their bases and lengths in the same proportion, be suspended horizontally; it is evident that the greater has more weight than the lesser, both on account of its length, and of its base; but it has less Resistance on account of its length, considered as a longer arm of a lever, and has only more Resistance on account of its base; therefore it exceeds the lesser in its momentum more than it does in its Resistance, and consequently it must break more easily.

Hence appears the reason why, in making small machines and models, people are apt to be mistaken as to the Resistance and strength of certain horizontal pieces, when they come to execute their designs in large, by observing the same proportions as in the small.

When the prism, fixed vertically, is just about to break, there is an equilibrium between its positive and relative weight; and consequently those two opposite powers are to each other reciprocally as the arms of the lever to which they are applied, that is, as half the diameter to half the axis of the prism. On the other hand, the Resistance of a body is always equal to the greatest weight which it will just sustain in a vertical position, that is, to its absolute weight. Therefore, substituting the absolute weight for the Resistance, it appears, that the absolute weight of a body, suspended horizontally, is to its relative weight, as the distance of its centre of gravity from the fixed point or axis of motion, is to the distance of the centre of gravity of its base from the same.

The discovery of this important truth, at least of an equivalent to it, and to which this is reducible, we owe to Galileo. On this system of Resistance of that author, Mariotte made an ingenious remark, which gave birth to a new system. Galileo supposes that where the body breaks, all the fibres break at once; so that the body always resists with its whole absolute force, or the whole force that all its fibres have in the place where it breaks. But Mariotte, finding that all bodies, even glass itself, bend before they break, shews that fibres are to be considered as so many little bent springs, which never exert their whole force, till stretched to a certain point, and never break till entirely unbent. Hence those nearest the fulcrum of the lever, or lowest point of the fracture, are stretched less than those farther off, and consequently employ a less part of their force, and break later.

This consideration only takes place in the horizontal situation of the body: in the vertical, the fibres of the base all break at once; so that the absolute weight of the body must exceed the united Resistance of all its fibres; a greater weight is therefore required here than in the horizontal situation, that is, a greater weight is required to overcome their united Resistance, than to overcome their several Resistances one after another.

Varignon has improved on the system of Mariotte, and shewn that to Galileo's system, it adds the consideration of the centre of percussion. In each system, the section, where the body breaks, moves on the axis of equilibrium, or line at the lower extremity of the same section; but in the second, the fibres of this section are continually stretching more and more, and that in the same ratio, as they are situated farther and farther from the axis of equilibrium, and consequently are still exerting a greater and greater part of their whole force.

These unequal extensions, like all other forces, must have some common centre where they are united, making equal efforts on each side of it; and as they are precisely in the same proportion as the velocities which the several points of a rod moved circularly would have to one another, the centre of extension of the section where the body breaks, must be the same as its centre of percussion. Galileo's hypothesis, where fibres stretch equally, and break all at once, corresponds to the case of a rod moving parallel to itself, where the centre of extension or percussion does not appear, as being confounded with the centre of gravity.

Hence it follows, that the Resistance of bodies in Mariotte's system, is to that in Galileo's, as the distance of the centre of percussion, taken on the vertical diameter of the fracture, is to the whole of that diameter. Hence also, the Resistance being less than what Galileo imagined, the relative weight must also be less, and in the ratio just mentioned. So that, after conceiving the relative weight of a body, and its Resistance equal to its absolute weight, as two contrary powers applied to the two arms of a lever, in the hypothesis of Galileo, there needs nothing to change it into that of Mariotte, but to imagine that the Resistance, or the absolute weight, is become less, in the ratio above mentioned, every thing else remaining the same.

One of the most curious, and perhaps the most useful questions in this research, is to find what figure a body must have, that its Resistance may be equal or proportional in every part to the force tending to break it. Now to this end, it is necessary, some part of it being conceived as cut off by a plane parallel to the fracture, that the momentum of the part retrenched be to its Resistance, in the same ratio as the momentum of the whole is to its Resistance; these four powers acting by arms of levers peculiar to themselves, and are proportional in the whole, and in each part, of a solid of equal Resistance. From this proportion, Varignon easily deduces two solids, which shall resist equally in all their parts, or be no more liable to break in one part than in another: Galileo had found one before. That discovered by Varignon is in the form of a trumpet, and is to be fixed into a wall at its greater end; so that its magnitude or weight is always diminished in proportion as its length, or the arm of the lever by which its weight acts, is increased. It is remarkable that, howsoever different the two systems may be, the solids of equal Resistance are the same in both.

For the Resistance of a solid supported at each end, as of a beam between two walls, see BEAM.

RESISTANCE of *Fluids*, is the force with which
bodies,

bodies, moving in fluid mediums, are impeded and retarded in their motion.

A body moving in a fluid is refifted from two caufes. The firft of thefe is the cohefion of the parts of the fluid. For a body, in its motion, feparating the parts of a fluid, muft overcome the force with which thofe parts cohere. The fecond is the inertia, or inactivity of matter, by which a certain force is required to move the particles from their places, in order to let the body pafs.

The retardation from the firft caufe is always the fame in the fame fpace, whatever the velocity be, the body remaining the fame ; that is, the Refiftance is as the fpace run through, in the fame time : but the velocity is alfo in the fame ratio of the fpace run over in the fame time : and therefore the Refiftance, from this caufe, is as the velocity itfelf.

The Refiftance from the fecond caufe, when a body moves through the fame fluid with different velocities, is as the fquare of the velocity. For, firft the Refiftance increafes according to the number of particles or quantity of the fluid ftruck in the fame time ; which number muft be as the fpace run through in that time, that is, as the velocity : but the Refiftance alfo increafes in proportion to the force with which the body ftrikes againft every part ; which force is alfo as the velocity of the body, fo as to be double with a double velocity, and triple with a triple one, &c : therefore, on both thefe accounts, the Refiftance is as the velocity multiplied by the velocity, or as the fquare of the velocity. Upon the whole therefore, on account of both caufes, viz, the tenacity and inertia of the fluid, the body is refifted partly as the velocity and partly as the fquare of the velocity.

But when the fame body moves through different fluids with the fame velocity, the Refiftance from the fecond caufe follows the proportion of the matter to be removed in the fame time, which is as the denfity of the fluid.

Hence therefore, if d denote the denfity of the fluid,
v the velocity of the body,
and a and b conftant coefficients :
then $adv^2 + bv$ will be proportional to the whole Refiftance to the fame body, moving with different velocities, in the fame direction, through fluids of different denfities, but of the fame tenacity.

But, to take in the confideration of different tenacities of fluids ; if t denote the tenacity, or the cohefion of the parts of the fluid, then $adv^2 + btv$ will be as the faid whole Refiftance.

Indeed the quantity of Refiftance from the cohefion of the parts of fluids, except in glutinous ones, is very fmall in refpect of the other Refiftance ; and it alfo increafes in a much lower degree, being only as the velocity, while the other increafes as the fquare of the velocity, and rather more. Hence then the term btv is very fmall in refpect of the other term adv^2 ; and confequently the Refiftance is nearly as this latter term ; or nearly as the fquare of the velocity. Which rule has been employed by moft authors, and is very near the truth in flow motions ; but in very rapid ones, it differs confiderably from the truth, as we fhall perceive below ; not indeed from the omiffion of the fmall term btv, due to the cohefion, but from the want of the full

counter preffure on the hinder part of the body, a vacuum, either perfect or partial, being left behind the body in its motion ; and alfo perhaps to fome compreffion or accumulation of the fluid againft the fore part of the body. Hence,

To conceive the Refiftance of fluids to a body moving in them, we muft diftinguifh between thofe fluids which, being greatly compreffed by fome incumbent weight, always clofe up the fpace behind the body in motion, without leaving any vacuity there ; and thofe fluids which, not being much compreffed, do not quickly fill up the fpace quitted by the body in motion, but leave a kind of vacuum behind it. Thefe differences, in the refifting fluids, will occafion very remarkable varieties in the laws of their Refiftance, and are abfolutely neceffary to be confidered in the determination of the action of the air on fhot and fhells ; for the air partakes of both thefe affections, according to the different velocities of the projected body.

In treating of thefe Refiftances too, the fluids may be confidered either as continued or difcontinued, that is, having their particles contiguous or elfe as feparated and unconnected ; and alfo either as elaftic or nonelaftic. If a fluid were fo conftituted, that all the particles compofing it were at fome diftance from each other, and having no action between them, then the Refiftance of a body moving in it would be eafily computed, from the quantity of motion communicated to thofe particles ; for inftance, if a cylinder moved in fuch a fluid in the direction of its axis, it would communicate to the particles it met with, a velocity equal to its own, and in its own direction, when neither the cylinder nor the parts of the fluid are elaftic : whence, if the velocity and diameter of the cylinder be known, and alfo the denfity of the fluid, there would thence be determined the quantity of motion communicated to the fluid, which (as action and reaction are equal) is the fame with the quantity loft by the cylinder, and confequently the Refiftance would thus be afcertained.

In this kind of difcontinued fluid, the particles being detached from each other, every one of them can purfue its own motion in any direction, at leaft for fome time, independent of the neighbouring ones ; fo that, inftead of a cylinder moving in the direction of its axis, if a body with a furface oblique to its direction be fuppofed to move in fuch a fluid, the motion which the parts of the fluid will hence acquire, will not be in the direction of the refifted body, but perpendicular to its oblique furface ; whence the Refiftance to fuch a body will not be eftimated from the whole motion communicated to the particles of the fluid, but from that part of it only which is in the direction of the refifted body. In fluids then, where the parts are thus difcontinued from each other, the different obliquities of that furface which goes foremoft, will occafion confiderable changes in the Refiftance ; although the tranfverfe fection of the folid fhould in all cafes be the fame : And Newton has particularly determined that, in a fluid thus conftituted, the Refiftance of a globe is but half the Refiftance of a cylinder of the fame diameter, moving, in the direction of its axis, with the fame velocity.

But though the hypothefis of a fluid thus conftituted
be

be of great use in explaining the nature of Resistances, yet we know of no such fluid existing in nature; and the fluids with which we are conversant being so formed, that their particles either lie contiguous to each other, or at least act on each other in the same manner as if they did: consequently, in these fluids, no one particle that is contiguous to the resisted body, can be moved, without moving at the same time a great number of others, some of which will be distant from it; and the motion thus communicated to a mass of the fluid, will not be in any one determined direction, but different in all the particles, according to the different positions in which they lie in contact with those from which they receive their impulse; whence, great numbers of the particles being diverted into oblique directions, the Resistance of the moving body, which will depend on the quantity of motion communicated to the fluid in its own direction, will be different in quantity from what it would be in the foregoing supposition, and its estimation becomes much more complicated and operose.

If the fluid be compressed by the incumbent weight of its upper parts (as all fluids are with us, except at their very surface), and if the velocity of the moving body be much less than that with which the parts of the fluid would rush into a void space, in consequence of their compression; it is evident, that in this case the space left by the moving body will be instantaneously filled up by the fluid; and the parts of the fluid against which the foremost part of the body presses in its motion, will, instead of being impelled forwards in the direction of the body, in some measure circulate towards the hinder part of the body, in order to restore the equilibrium, which the constant influx of the fluid behind the body would otherwise destroy; whence the progressive motion of the fluid, and consequently the Resistance of the body, which depends upon it, would in this instance be much less, than in the hypothesis where each particle is supposed to acquire, from the stroke of the resisting body, a velocity equal to that with which the body moved, and in the same direction. Newton has determined, that the Resistance of a cylinder, moving in the direction of its axis, in such a compressed fluid as we have here treated of, is but one-fourth part of the Resistance to the same cylinder, if it moved with the same velocity in a fluid constituted in the manner described in the first hypothesis, each fluid being supposed of the same density.

But again, it is not only in the quantity of their Resistance that these fluids differ, but also in the different manner in which they act upon solids of different forms moving in them. In the discontinued fluid, first described, the obliquity of the foremost surface of the moving body would diminish the Resistance; but the same thing does not hold true in compressed fluids, at least not in any considerable degree; for the chief Resistance in compressed fluids arises from the greater or less facility with which the fluid, impelled by the fore part of the body, can circulate towards its hinder part; and this being little, if at all, affected by the form of the moving body, whether it be cylindrical, conical, or spherical, it follows, that while the transverse section of the body is the same, and consequently the quan-

tity of impelled fluid also, the change of figure in the body will scarcely affect the quantity of its Resistance.

And this case, viz. the Resistance of a compressed fluid to a solid, moving in it with a velocity much less than what the parts of the fluid would acquire from their compression, has been very fully considered by Newton, who has ascertained the quantity of such a Resistance, according to the different magnitudes of the moving body, and the density of the fluid. But he expressly informs us that the rules he has laid down, are not generally true, but only upon a supposition that the compression of the fluid be increased in the greater velocities of the moving body: however, some unskilful writers, who have followed him, overlooking this caution, have applied his determination to bodies moving with all sorts of velocities, without attending to the different compressions of the fluids they are resisted by; and by this means they have accounted the Resistance, for instance, of the air to musket and cannon shot, to be but about one-third part of what it is found to be by experience.

It is indeed evident that the resisting power of the medium must be increased, when the resisted body moves so fast that the fluid cannot instantaneously press in behind it, and fill the deserted space; for when this happens, the body will be deprived of the pressure of the fluid behind it; which in some measure balanced its Resistance, or at least the fore pressure, and must support on its fore part the whole weight of a column of the fluid, over and above the motion it gives to the parts of the same; and besides, the motion in the particles driven before the body, is less affected in this case by the compression of the fluid, and consequently they are less deflected from the direction in which they are impelled by the resisted surface; whence it happens that this species of Resistance approaches more and more to that described in the first hypothesis, where each particle of the fluid being unconnected with the neighbouring ones, pursued its own motion, in its own direction, without being interrupted or deflected by their contiguity; and therefore, as the Resistance of a discontinued fluid to a cylinder, moving in the direction of its axis, is 4 times greater than the Resistance of a fluid sufficiently compressed of the same density, it follows that the Resistance of a fluid, when a vacuity is left behind the moving body, may be near 4 times greater than that of the same fluid, when no such vacuity is formed; for when a void space is thus left, the Resistance approaches in its nature to that of a discontinued fluid.

This then may probably be the case in a cylinder moving in the same compressed fluid, according to the different degrees of its velocity; so that if it set out with a great velocity, and moves in the fluid till that velocity be much diminished, the resisting power of the medium may be near 4 times greater in the beginning of its motion than in the end.

In a globe, the difference will not be so great, because, on account of its oblique surface, its Resistance in a discontinued medium is but about twice as much as in one properly compressed; for its oblique surface diminishes its Resistance in one case, and not in the other: however, as the compression of the medium,

even

even when a vacuity is left behind the moving body, may yet confine the oblique motion of the parts of the fluid, which are driven before the body, and as in an elaftic fluid, fuch as our air is, there will be fome degree of condenfation in thofe parts; it is highly probable that the Refiftance of a globe, moving in a compreffed fluid with a very great velocity, may greatly exceed the proportion of the Refiftance to flow motions.

And as this increafe of the refifting power of the medium will take place, when the velocity of the moving body is fo great, that a perfect vacuum is left behind it, fo fome degree of augmentation will be fenfible in velocities much fhort of this; for even when, by the compreffion of the fluid, the fpace left behind the body is inftantaneoufly filled up; yet, if the velocity with which the parts of the fluid rufh in behind, is not much greater than that with which the body moves, the fame reafons that have been urged above, in the cafe of an abfolute vacuity, will hold in a lefs degree in this inftance; and therefore it is not to be fuppofed that, in the increafed Refiftance which has been hitherto treated of, it immediately vanifhes when the compreffion of the fluid is juft fufficient to prevent a vacuum behind the refifted body; but we muft confider it as diminifhing only according as the velocity, with which the parts of the fluid follow the body, exceeds that with which the body moves.

Hence then it may be concluded, that if a globe fets out in a refifting medium, with a velocity much exceeding that with which the particles of the medium would rufh into a void fpace, in confequence of their compreffion, fo that a vacuum is neceffarily left behind the globe in its motion; the Refiftance of this medium to the globe will be much greater, in proportion to its velocity, than what we are fure, from Sir I. Newton, would take place in a flower motion. We may farther conclude, that the refifting power of the medium will gradually diminifh as the velocity of the globe decreafes, till at laft, when it moves with velocities which bear but a fmall proportion to that with which the particles of the medium follow it, the Refiftance becomes the fame with what is affigned by Newton in the cafe of a compreffed fluid.

And from this determination may be feen, how falfe that pofition is, which afferts the Refiftance of any medium to be always in the duplicate ratio of the velocity of the refifted body; for it plainly appears, by what has been faid, that this can only be confidered as nearly true in fmall variations of velocity, and can never be applied in comparing together the Refiftances to all velocities whatever, without incurring the moft enormous errors. See Robins's Gunnery, chap. 2 prop. 1, and my Select Exercifes pa. 235 &c. See alfo the articles RESISTANCE *of the Air*, PROJECTILE, and GUNNERY.

Refiftance and retardation are ufed indifferently for each other, as being both in the fame proportion, and the fame Refiftance always generating the fame retardation. But with regard to different bodies, the fame Refiftance frequently generates different retardations; the Refiftance being as the quantity of motion, and the retardation that of the celerity. For the difference and meafure of the two, fee RETARDATION.

The retardations from this Refiftance may be com-
VOL. II.

pared together, by comparing the Refiftance with the gravity or quantity of matter. It is demonftrated that the Refiftance of a cylinder, which moves in the direction of its axis, is equal to the weight of a column of the fluid, whofe bafe is equal to that of the cylinder, and its altitude equal to the height through which a body muft fall in vacuo, by the force of gravity, to acquire the velocity of the moving body. So that, if a denote the area of the face or end of the cylinder, or other prifm, v its velocity, and n the fpecific gravity of the fluid; then, the altitude due to the velocity v being $\frac{v^2}{4g}$, the whole Refiftance, or motive force m, will be $a \times n \times \frac{v^2}{4g} = \frac{anv^2}{4g}$; the quantity g being $= 16\frac{1}{12}$ feet, or the fpace a body falls, in vacuo, in the firft fecond of time. And the Refiftance to a globe of the fame diameter would be the half of this.—Let a ball, for inftance, of 3 inches diameter, be moved in water with a celerity of 16 feet per fecond of time: now from experiments on pendulums, and on falling bodies, it has been found, that this is the celerity which a body acquires in falling from the height of 4 feet; therefore the weight of a cylinder of water of 3 inches diameter, and 4 feet high, that is a weight of about 12 lb 4 oz, is equal to the Refiftance of the cylinder; and confequently the half of it, or 6 lb 2 oz is that of the ball. Or, the formula

$$\frac{anv^2}{4g} \text{ gives } \frac{\cdot 7854 \times 9 \times 1000 \times 16 \times 16}{144 \times 4 \times 16} = 196 \text{ oz,}$$

or 12 lb 4 oz, for the Refiftance of the cylinder, or 6 lb 2 oz for that of the ball, the fame as before.

Let now the Refiftance, fo difcovered, be divided by the weight of the body, and the quotient will fhew the ratio of the retardation to the force of gravity. So if the faid ball, of 3 inches diameter, be of caft iron, it will weigh nearly 61 ounces, or $3\frac{4}{5}$ lb; and the Refiftance being 6 lb 2 oz, or 98 ounces; therefore, the Refiftance being to the gravity as 98 to 61, the retardation, or retarding force, will be $\frac{98}{61}$ or $1\frac{3}{5}$, the force of gravity being 1. Or thus; becaufe a the area of a great circle of the ball, is $= pd^2$, where d is the diameter, and $p = \cdot 7854$, therefore the Refiftance to the ball is $m = \frac{pnd^2v^2}{8g}$; and becaufe its folid content is $w = \frac{1}{3}pd^3$, and its weight $\frac{1}{3}Npd^3$, where N denotes its fpecific gravity; therefore, dividing the Refiftance or motive force m by the weight w, gives $\frac{m}{w} = \frac{3nv^2}{16Ndg} = f$ the retardation, or retarding force, that of gravity being 1; which is therefore as the fquare of the velocity directly, and as the diameter inverfely: and this is the reafon why a large ball overcomes the Refiftance better than a fmall one, of the fame denfity. See my Select Exercifes, pa. 225 &c.

RESISTANCE *of Fluid Mediums to the Motion of Falling Bodies.*—A body freely defcending in a fluid, is accelerated by the relative gravity of the body, (that is, the difference between its own abfolute gravity and that of a like bulk of the fluid), which continually acts upon it, yet not equably, as in a vacuum: the Refiftance of the fluid occafions a retardation, or diminution

3 A
of

of acceleration, which diminution increases with the velocity of the body. Hence it happens, that there is a certain velocity, which is the greatest that a body can acquire by falling ; for if its velocity be such, that the Resistance arising from it becomes equal to the relative weight of the body, its motion can be no longer accelerated ; for the motion here continually generated by the relative gravity, will be destroyed by the Resistance, or the force of Resistance is equal to the relative gravity, and the body forced to go on equably ; for after the velocity is arrived at such a degree, that the resisting force is equal to the weight that urges it, it will increase no longer, and the globe must afterward continue to descend with that velocity uniformly. A body continually comes nearer and nearer to this greatest celerity, but can never attain accurately to it. Now, N and n being the specific gravities of the globe and fluid, $N - n$ will be the relative gravity of the globe in the fluid, and therefore $w = \frac{2}{3} p d^3 (N - n)$ is the weight by which it is urged downward; also $m = \frac{p n d^2 v^2}{8 g}$ is the Resistance, as above ; therefore these two must be equal when the velocity can be no farther increased, or $m = w$, that is $\frac{p n d^2 v^2}{8 g} = \frac{2}{3} p d^3 (N - n)$, or $n v^2 = \frac{16}{3} d g (N - n)$; and hence $v = \sqrt{\frac{16}{3} d g \times \frac{N - n}{n}}$ is the said uniform or greatest velocity to which the body may attain ; which is evidently the greater in the subduplicate proportion of v the diameter of the ball. But v is always $= \sqrt{4 g f s}$, the velocity generated by any accelerative force f in describing the space s ; which being compared with the former, it gives $s = \frac{4}{3} d$, when f is $= \frac{N - n}{n}$; that is, the greatest velocity is that which is generated by the accelerating force $\frac{N - n}{n}$ in passing over the space $\frac{4}{3} d$ or $\frac{4}{3}$ of the diameter of the ball, or it is equal to the velocity generated by gravity in describing the space $\frac{N - n}{n} \times \frac{4}{3} d$. For ex. if the ball be of lead, which is about $11\frac{1}{4}$ times the density of water; then $N = 11\frac{1}{4}, n = 1, N - n = \frac{N - n}{n} = 10\frac{1}{4} = \frac{41}{4}$, and $\frac{N - n}{n} \times \frac{4}{3} d = \frac{41}{3} d = 13\frac{2}{3} d$; that is, the uniform or greatest velocity of a ball of lead, descending in water, is equal to that which a heavy body acquired by falling in vacuo through a space equal to $13\frac{2}{3}$ of the diameter of the ball, which velocity is $v = 2 \sqrt{\frac{4}{3} d g \times \frac{N - n}{n}} = 2 \sqrt{13\frac{2}{3} d g} = 8 \sqrt{13\frac{2}{3} d}$ nearly, or 8 times the root of the same space.

Hence it appears, how soon small bodies come to their greatest or uniform velocity in descending in a fluid, as water, and how very small that velocity is : which explains the reason of the slow precipitation of mud, and small particles, in water, as also why, in precipitations, the larger and gross particles descend soonest, and the lowest.

Farther, where $N = n$, or the density of the fluid is equal to that of the body, then $N - n = 0$, consequently the velocity and distance descended are each nothing, and the body will just float in any part of the fluid.

Moreover, when the body is lighter than the fluid, then N is less than n, and $N - n$ becomes a negative quantity, or the force and motion tend the contrary way, that is, the ball will ascend up towards the top of the fluid by a motive force which is as $n - N$. In this case then, the body ascending by the action of the fluid, is moved exactly by the same laws as a heavier body falling in the fluid. Wherever the body is placed, it is sustained by the fluid, and carried up with a force equal to the difference of the weight of a quantity of the fluid of the same bulk as the body, from the weight of the body; there is therefore a force which continually acts equably upon the body ; by which not only the action of gravity of the body is counteracted, so as that it is not to be considered in this case; but the body is also carried upwards by a motion equably accelerated, in the same manner as a body heavier than a fluid descends by its relative gravity : but the equability of acceleration is destroyed in the same manner by the Resistance, in the ascent of a body lighter than the fluid, as it is destroyed in the descent of a body that is heavier.

For the circumstances of the correspondent velocity, space, and time, &c, of a body moving in a fluid in which it is projected with a given velocity, or descending by its own weight, &c, see my Select Exercises, prop. 29, 30, 31, and 32, pag. 221 &c.

RESISTANCE *of the Air*, in Pneumatics, is the force with which the motion of bodies, particularly of projectiles, is retarded by the opposition of the air or atmosphere. See GUNNERY, PROJECTILES, &c.

The air being a fluid, the general laws of the Resistance of fluids obtain in it ; subject only to some variations and irregularities from the different degrees of density in the different stations or regions of the atmosphere.

The Resistance of the air is chiefly of use in military projectiles, in order to allow for the differences caused in their flight and range by it. Before the time of Mr. Robins, it was thought that this Resistance to the motion of such heavy bodies as iron balls and shells, was too inconsiderable to be regarded, and that the rules and conclusions derived from the common parabolic theory, were sufficiently exact for the common practice of gunnery. But that gentleman shewed, in his New Principles of Gunnery, that, so far from being inconsiderable, it is in reality enormously great, and by no means to be rejected without incurring the grossest errors ; so much so, that balls or shells which range, at the most, in the air, to the distance of two or three miles, would in a vacuum range to 20 or 30 miles, or more. To determine the quantity of this Resistance, in the case of different velocities, Mr. Robins discharged musket balls, with various degrees of known velocity, against his ballistic pendulums, placed at several different distances, and so discovered by experiment the quantity of velocity lost, when passing through those distances

7

or

or fpaces of air, with the feveral known degrees of celerity. For having thus known, the velocity loft or deftroyed, in paffing over a certain fpace, in a certain time, (which time is very nearly equal to the quotient of the fpace divided by the medium velocity between the greateft and leaft, or between the velocity at the mouth of the gun and that at the pendulum); that is, knowing the velocity v, the fpace s, and time t, the refifting force is thence eafily known, being equal to $\dfrac{v b}{2 g t}$ or $\dfrac{v V b}{2 g s}$, where b denotes the weight of the ball, and V the medium velocity above-mentioned. The balls employed upon this occafion by Mr. Robins, were leaden ones, of $\frac{1}{12}$ of a pound weight, and $\frac{3}{4}$ of an inch diameter; and to the medium velocity of

1600 feet the Refiftance was 11 lb,
1065 feet - - - - - - - it was $2\frac{4}{5}$;

but by the theory of Newton, before laid down, the former of thefe fhould be only $4\frac{1}{2}$ lb, and the latter 2 lb: fo that, in the former cafe the real Refiftance is more than double of that by the theory, being increafed as 9 to 22; and in the leffer velocity the increafe is from 2 to $2\frac{4}{5}$, or as 5 to 7 only.

Mr. Robins alfo invented another machine, having a whirling or circular motion, by which he meafured the Refiftances to larger bodies, though with much fmaller velocities: it is defcribed, and a figure of it given, near the end of the 1ft vol. of his works.

That this refifting power of the air to fwift motions is very fenfibly increafed beyond what Newton's theory for flow motions makes it, feems hence to be evident. By other experiments it appears that the Refiftance is very fenfibly increafed, even in the velocity of 400 feet. However, this increafed power of Refiftance diminifhes as the velocity of the refifted body diminifhes, till at length, when the motion is fufficiently abated, the actual Refiftance coincides with that fuppofed in the theory nearly. For thefe varying Refiftances Mr. Robins has given a rule, extending to 1670 feet velocity.

Mr. Euler has fhewn, that the common doctrine of Refiftance anfwers pretty well when the motion is not very fwift, but in fwift motions it gives the Refiftance lefs than it ought to be, fon two accounts. 1. Becaufe in quick motions, the air does not fill up the fpace behind the body faft enough to prefs on the hinder parts, to counterbalance the weight of the atmofphere on the fore part. 2. The denfity of the air before the ball being increafed by the quick motion, will prefs more ftrongly on the fore part, and fo will refift more than lighter air in its natural ftate. He has fhewn that Mr. Robins has reftrained his rule to velocities not exceeding 1670 feet per fecond; whereas had he extended it to greater velocities, the refult muft have been erroneous; and he gives another formula himfelf, and deduces conclufions differing from thofe of Mr. Robins. See his Principles of Gunnery inveftigated, tranflated by Brown in 1777, pa. 224 &c.

Mr. Robins having proved that, in very great changes of velocity, the Refiftance does not accurately follow the duplicate ratio of the velocity, lays down two pofitions, which he thought might be of fome fervice in the practice of artillery, till a more complete and accurate theory of Refiftance, and the changes of its augmentation, may be obtained. The firft of thefe is, that till the velocity of the projectile furpafs 1100 or 1200 feet in a fecond, the Refiftance may be efteemed to be in the duplicate ratio of the velocity: and the fecond is, that when the velocity exceeds 1100 or 1200 feet, then the abfolute quantity of the Refiftance will be near 3 times as great as it fhould be by a comparifon with the fmaller velocities. Upon thefe principles he proceeds in approximating to the actual ranges of pieces with fmall angles of elevation, viz. fuch as do not exceed 8° or 10°, which he fets down in a table, compared with their correfponding potential ranges. See his Mathematical Tracts, vol. 1 pa. 179, &c. But we fhall fee prefently that thefe pofitions are both without foundation; that there is no fuch thing as a fudden or abrupt change in the law of Refiftance, from the fquare of the velocity to one that gives a quantity three times as much; but that the change is flow and gradual, continually from the fmalleft to the higheft velocities; and that the increafed real Refiftance no where rifes higher than to about double of that which Newton's theory gives it.

Mr. Glenie, in his Hiftory of Gunnery, 1776, pa. 49, obferves, in confequence of fome experiments with a rifled piece, properly fitted for experimental purpofes, that the Refiftance of the air to a velocity fomewhat lefs than that mentioned in the firft of the above propofitions, is confiderably greater than in the duplicate ratio of the velocity; and that, to a celerity fomewhat greater than that ftated in the fecond, the Refiftance is confiderably lefs than that which is treble the Refiftance in the faid ratio. Some of Robins's own experiments feem neceffarily to make it fo; fince, to a velocity no quicker than 400 feet in a fecond, he found the Refiftance to be fomewhat greater than in that ratio. But the true value of the ratio, and other circumftances of this Refiftance, will more fully appear from what follows.

The fubject of the Refiftance of the air, as begun by Robins, has been profecuted by myfelf, to a very great extent and variety, both with the whirling machine, and with cannon balls of all fizes, from 1 lb to 6 lb weight, as well as with figures of many other different fhapes, both on the fore part and hind part of them, and with planes fet at all varieties of angles of inclination to the path or motion of the fame; from all which I have obtained the real Refiftance to bodies for all velocities, from 1 up to 2000 feet per fecond; together with the law of the Refiftance to the fame body for all different velocities, and for different fizes with the fame velocity, and alfo for all angles of inclination; a full account of which would make a book of itfelf, and muft be referved for fome other occafion. In the mean time, fome general tables of conclufions may be taken as below.

Table I. *Resistances of different Bodies.*

Veloc per Sec.	Small Hemis. flat side	Large Hemis.		Cone		Cylinder	Whole globe	Resist. as the power of the veloc.
		flat side	round side	vertex	base			
feet	oz	oz	oz	oz	oz	oz	oz	oz
3	·028	·051	·020	·028	·064	·050	·027	
4	·048	·096	·039	·048	·109	·090	·047	
5	·072	·148	·065	·071	·162	·43	·068	
6	·103	·211	·092	·098	·225	·205	·094	
7	·141	·284	·123	·129	·298	·278	·125	
8	·184	·368	·160	·168	·382	·360	·162	
9	·233	·464	·199	·211	·478	·456	·205	
10	·287	·573	·242	·260	·587	·565	·255	
11	·349	·698	·292	·315	·712	·688	·310	2·052
12	·418	·836	·347	·376	·850	·826	·370	2·042
13	·492	·938	·409	·440	1·000	·979	·435	2·036
14	·573	1·154	·478	·512	1·166	1·145	·505	2·031
15	·661	1·336	·552	·589	1·346	1·327	·581	2·031
16	·754	1·538	·634	·673	1·546	1·526	·663	2·033
17	·853	1·757	·722	·762	1·763	1·745	·752	2·038
18	·959	1·998	·818	·858	2·002	1·986	·848	2·044
19	1·073	2·258	·922	·959	2·260	2·246	·949	2·047
20	1·196	2·542	1·033	1·069	2·540	2·528	1·057	2·051
Mean propor. Nos.	140	288	119	126	291	285	124	2·040
1	2	3	4	5	6	7	8	9

In this Table are contained the Resistances to several forms of bodies, when moved with several degrees of velocity, from 3 feet per second to 20. The names of the bodies are at the tops of the columns, as also which end went foremost through the air; the different velocities are in the first column, and the Resistances on the same line, in their several columns, in avoirdupois ounces and decimal parts. So on the first line are contained the Resistances when the bodies move with a velocity of 3 feet in a second, viz, in the 2d column for the small hemisphere, of $4\frac{3}{4}$ inches diameter, its Resistance ·028 oz when the flat side went foremost; in the 3d and 4th columns the Resistances to a larger hemisphere, first with the flat side, and next the round side foremost, the diameter of this, as well as all the following figures being $6\frac{3}{8}$ inches, and therefore the area of the great circle = 32 sq. inches, or $\frac{2}{9}$ of a sq. foot; then in the 5th and 6th columns are the Resistances to a cone, first its vertex and then its base foremost, the altitude of the cone being $6\frac{3}{8}$ inches, the same as the diameter of its base; in the 7th column the Resistance to the end of the cylinder, and in the 8th that against the whole globe or sphere. All the numbers shew the real weights which are equal to the Resistances; and at the bottoms of the columns are placed proportional numbers, which shew the mean proportions of the Resistances of all the figures to one another, with any velocity. Lastly, in the 9th column are placed the exponents of the power of the velocity which the Resistances in the 8th column bear to each other, viz, which that of the 10 feet velocity bears to each of the following ones, the medium of all of them being as the 2·04 power of the velocity, that is, very little above the square or second power of the velocity, so far as the velocities in this Table extend.

From this Table the following inferences are easily deduced.

1. That the Resistance is nearly in the same proportion as the surfaces; a small increase only taking place in the greater surfaces, and for the greater velocities. Thus, by comparing together the numbers in the 2d and 3d columns, for the bases of the two hemispheres, the areas of which bases are in the proportion of $17\frac{1}{4}$ to 32, or 5 to 9 very nearly, it appears that the numbers in those two columns, expressing the Resistances, are nearly as 1 to 2 or 5 to 10, as far as the velocity of 12 feet; but after that, the Resistances on the greater surface increase gradually more and more above that proportion.

2. The Resistance to the same surface, with different velocities, is, in these slow motions, nearly as the square of the velocity; but gradually increases more and more above that proportion as the velocity increases. This is manifest from all the columns; and the index of the power of the velocity is set down in the 9th column, for the Resistances in the 8th, the medium being 2·04; by which it appears that the Resistance to the same body is, in these slow motions, as the 2·04 power of the velocity, or nearly as the square of it.

3. The round ends, and sharp ends, of solids, suffer less Resistance than the flat or plane ends, of the same diameter; but the sharper end has not always the less Resistance. Thus, the cylinder, and the flat ends of the hemisphere and cone, have more Resistance, than the round or sharp ends of the same; but the round side of the hemisphere has less Resistance than the sharper end of the cone.

4. The Resistance on the base of the hemisphere, is to that on the round, or whole sphere, as $2\frac{1}{5}$ to 1, instead of 2 to 1, as the theory gives that relation. Also the experimented Resistance, on each of these, is nearly $\frac{1}{4}$ more than the quantity assigned by the theory.

5. The Resistance on the base of the cone, is to that on the vertex, nearly as $2\frac{3}{10}$ to 1; and in the same ratio is radius to the sine of the angle of inclination of the side of the cone to its path or axis. So that, in this instance, the Resistance is directly as the sine of the angle of incidence, the transverse section being the same.

6. When the hinder parts of bodies are of different forms, the Resistances are different, though the fore-parts be exactly alike and equal; owing probably to the different pressures of the air on the hinder parts. Thus, the Resistance to the fore part of the cylinder, is less than on the equal flat surface of the cone, or of the hemisphere; because the hinder part of the cylinder is more pressed or pushed, by the following air than those of the other two figures; also, for the same reason, the base of the hemisphere suffers a less Resistance than that of the cone, and the round side of the hemisphere less than the whole sphere.

TABLE

TABLE II. *Resistances both by Experiment and Theory, to a Globe of 1·965 Inches Diameter.*

Veloc. per sec. in feet.	Resist. by Exper. oz.	Resist. by Theory. oz.	Ratio of Exper. to Theory.	Resist. as the power of the veloc.
5	0·006	0·005	1·20	
10	0·024½	0·020	1·23	
15	0·055	0·044	1·25	
20	0·100	0·079	1·27	
25	0·157	0·123	1·28	2·022
30	0·23	0·177	1·30	2·055
40	0·42	0·314	1·33	2·068
50	0·67	0·491	1·36	2·075
100	2·72	1·964	1·38	2·059
200	11	7·9	1·40	2·041
300	25	18·7	1·41	2·039
400	45	31·4	1·43	2·039
500	72	49	1·47	2·044
600	107	71	1·51	2·051
700	151	96	1·57	2·059
800	205	126	1·63	2·067
900	271	159	1·70	2·077
1000	350	196	1·78	2·086
1100	442	238	1·86	2·095
1200	546	283	1·90	2·102
1300	661	332	1·99	2·107
1400	785	385	2·04	2·111
1500	916	442	2·07	2·113
1600	1051	503	2·09	2·113
1700	1186	568	2·08	2·111
1800	1319	636	2·07	2·108
1900	1447	709	2·04	2·104
2000	1569	786	2·00	2·098

In the first column of this Table are contained the several velocities, gradually from o up to the great velocity of 2000 feet per second, with which a ball or globe moved. In the 2d column are the experimented Resistances, in averdupois ounces. In the 3d column are the correspondent Resistances, as computed by the foregoing theory. In the 4th column are the ratios of these two Resistances, or the quotients of the former divided by the latter. And in the 5th or last, the indexes of the power of the velocity which is proportional to the experimented Resistance; which are found by comparing the Resistance of 20 feet velocity with each of the following ones.

From the 2d, 3d and 4th columns it appears, that at the beginning of the motion, the experimented Resistance is nearly equal to that computed by theory; but that, as the velocity increases, the experimented Resistance gradually exceeds the other more and more, till at the velocity of 1300 feet the former becomes just double the latter; after which the difference increases a little farther, till at the velocity of 1600 or 1700, where that excess is the greatest, and is rather less than 2 1/10; after this, the difference decreases gradually as the velocity increases, and at the velocity of 2000, the former Resistance again becomes just double the latter.

From the last column it appears that, near the begin-

ning, or in slow motions, the Resistances are nearly as the square of the velocities; but that the ratio gradually increases, with some small variation, till at the velocity of 1500 or 1600 feet it becomes as the 2⅛ power of the velocity nearly, which is its highest ascent; and after that it gradually decreases again, as the velocity goes higher. And similar conclusions have also been derived from experiments with larger balls or globes.

And hence we perceive that Mr. Robins's positions are erroneous on two accounts, viz, both in stating that the Resistance changes suddenly, or all at once, from being as the square of the velocity, so as then to become as some higher and constant power; and also when he states the Resistance as rising to the height of 3 times that which is given by the theory: since the ratio of the Resistance both increases gradually from the beginning, and yet never ascends higher than $2\frac{6}{100}$ of the theory.

TABLE III. *Resistance to a Plane, set at various Angles of Inclination to its Path.*

Angle with the Path.	Experim. Resistances. oz.	Resist. by this Formula. ·84 $s^{1·842}c$	Sines of the Angles to Radius ·840.
0°	·000	·000	·000
5	·015	·009	·073
10	·044	·035	·146
15	·082	·076	·217
20	·133	·131	·287
25	·200	·199	·355
30	·278	·278	·420
35	·362	·363	·482
40	·448	·450	·540
45	·534	·535	·594
50	·619	·613	·643
55	·684	·680	·688
60	·729	·736	·727
65	·770	·778	·761
70	·803	·808	·789
75	·823	·826	·811
80	·835	·836	·827
85	·839	·839	·838
90	·840	·840	·840

In the 2d column of this Table are contained the actual experimented Resistances, in ounces, to a plane of 32 square inches, or ⅔ of a square foot, moved through the air with a velocity of exactly 12 feet per second, when the plane was set so as to make, with the direction of its path, the corresponding angles in the first column.

And from these I have deduced this formula, or theorem, viz, $·84 s^{1·842}c$, which brings out very nearly the same numbers, and is a general theorem for every angle, for the same plane of ⅔ of a foot, and moved with the same velocity of 12 feet in a second of time; where s is the sine, and c the cosine of the angles of inclination in the first column.

If

If a theorem be defired for any other velocity v, and any other plane whofe area is a, it will be this : $\frac{1}{3 \frac{1}{3}} a v^2 s^{1.842c}$, or more nearly $\frac{1}{4 \frac{1}{2}} a v^{2.24} s^{1.842c}$; which denotes the Refiftance nearly to any plane furface whofe area is a, moved through the air with the velocity v, in a direction making with that plane an angle, whofe fine is s, and cofine c.

If it be water or any other fluid, different from air, this formula will be varied in proportion to the denfity of it.

By this theorem were computed the numbers in the 3d column ; which it is evident agree very nearly with the experiment Refiftances in the 2d column, excepting in two or three of the fmall numbers near the beginning, which are of the leaft confequence. In all other cafes, the theorem gives the true Refiftance very nearly. In the 4th or laft column are entered the fines of the angles of the firft column, to the radius ·84, in order to compare them with the Refiftances in the other columns. From whence it appears, that thofe Refiftances bear no fort of analogy to the fines of the angles, nor yet to the fquares of the fines, nor to any other power of them whatever. In the beginning of the columns, the fines much exceed the Refiftances all the way till the angle be between 55 and 60 degrees; after which the fines are lefs than the Refiftances all the way to the end, or till the angle become of 90 degrees.

Mr. James Bernoulli gave fome theorems for the Refiftances of different figures, in the Acta Erud. Lipf. for June 1693, pa. 252 &c. But as thefe are deduced from theory only, which we find to be fo different from experiment, they cannot be of much ufe. Meffieurs Euler, D'Alembert, Gravefande, and Simpfon, have alfo written pretty largely on the theory of Refiftances, befides what had been done by Newton.

Solid of Leaft RESISTANCE. Sir Ifaac Newton, from his general theory of Refiftance, deduces the figure of a folid which fhall have the leaft Refiftance of the fame bafe, height and content.

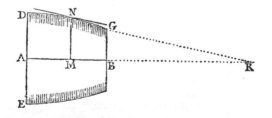

The figure is this. Suppofe DNG to be a curve of fuch a nature, that if from any point N the ordinate NM be drawn perpendicular to the axis AB ; and from a given point G there be drawn GR parallel to a tangent at N, and meeting the axis produced in R ; then if MN be to GR, as GR³ to 4BR × BG², a folid defcribed by the revolution of this figure about its axis AB, moving in a medium from A towards B, is lefs refifted than any other circular folid of the fame bafe, &c.

This theorem, which Newton gave without a demonftration, has been demonftrated by feveral mathe-

maticians, as Facio, Bernoulli, Hofpital, &c. See Maclaurin's Flux. fect. 606 and 607 ; alfo Hotflev's edit. of Newton, vol. 2, pag. 390. See alfo Act. Erud. 1699, pa. 514 ; and Mem. de l'Acad. &c ; alfo Robins's View of Newton's method for comparing the Refiftance of Solids, 8vo, 1734; and Simpfon's Fluxions, art. 413 ; or my Principles of Bridges, prop. 11 and 12.

M. Bouguer has refolved this problem in a very general manner ; not in fuppofing the folid to be formed by a revolution, of any figure whatever. The problem, as enunciated and refolved by M. Bouguer, is this : Any bafe being given, to find what kind of folid muft be formed upon it, fo that the impulfe upon it may be the leaft poffible. Properly however it ought to be the retardive force, or the impulfe divided by the weight or mafs of matter in the body, that ought to be the minimum.

RESOLUTION, in Phyfics, the reduction of a body into its original or natural ftate, by a diffolution or feparation of its aggregated parts. Thus, fnow and ice are faid to be refolved into water ; water refolves in vapour by heat ; and vapour is again refolved into water by cold ; alfo any compound is refolved into its ingredients, &c.

Some of the modern philofophers, particularly Boyle, Mariotte, Boerhaave, &c, maintain, that the natural ftate of water is to be congealed, or in ice ; in as much as a certain degree of heat, which is a foreign and violent agent, is required to make it fluid : fo that near the pole, where this foreign agent is wanting, it conftantly retains its fixed or icy ftate.

RESOLUTION, or SOLUTION, in Mathematics, is an orderly enumeration of feveral things to be done, to obtain what is required in a problem.

Wolfius makes a problem to confift of three parts : The *propofition* (or what is properly called the *problem*), the *Refolution*, and the *demonftration*.

As foon as a problem is demonftrated, it is converted into a theorem ; of which the Refolution is the hypothefis ; and the propofition the thefis.

For the procefs of a mathematical Refolution, fee the following article.

RESOLUTION in *Algebra*, or *Algebraical* RESOLUTION, is of two kinds ; the one practifed in numerical problems, the other in geometrical ones.

In Refolving a Numerical Problem Algebraically, the method is this. Firft, the given quantities are diftinguifhed from thofe that are fought ; and the former denoted by the initial letters of the alphabet, but the latter by the laft letters.—2. Then as many equations are formed as there are unknown quantities. If that cannot be done from the propofition or data, the problem is indeterminate ; and certain arbitrary affumptions muft be made, to fupply the defect, and which can fatisfy the queftion. When the equations are not contained in the problem itfelf, they are to be found by particular theorems concerning equations, ratios, proportions, &c.—Since, in an equation, the known and unknown quantities are mixed together, they muft be feparated in fuch a manner, that the unknown one remain alone on one fide, and the known ones on the other. This reduction, or feparation, is made by addition, fubtraction, multiplication, divifion, extraction

of

of roots, and raifing of powers ; refolving every kind of combination of the quantities, by their counter or reverfe ones, and performing the fame operation on all the quantities or terms, on both fides of the equation, that the equality may ftill be preferved.

To Refolve a Geometrical Problem Algebraically.— The fame fort of operations are to be performed, as in the former article ; befides feveral others, that depend upon the nature of the diagram, and geometrical properties. As 1ft, the thing required or propofed, muft be fuppofed done, the diagram being drawn or conftructed in all its parts, both known and unknown. 2. We muft then examine the geometrical relations which the lines of the figure have among themfelves, without regarding whether they are known or unknown, to find what equations arife from thofe relations, for finding the unknown quantities. 3. It is often neceffary to form fimilar triangles and rectangles, fometimes by producing of lines, or drawing parallels and perpendiculars, and forming equal angles, &c ; till equations can be formed, from them, including both the known and unknown quantities.

If we do not thus arrive at proper equations, the thing is to be tried in fome other way. And fometimes the thing itfelf, that is required, is not to be fought directly, but fome other thing, bearing certain relations to it, by means of which it may be found.

The final equation being at laft arrived at, the geometrical conftruction is to be deduced from it, which is performed in various ways according to the different kinds of equations.

RESOLUTION *of Forces,* or *of Motion,* is the refolving or dividing of any one force or motion, into feveral others, in other directions, but which, taken together, fhall have the fame effect as the fingle one ; and it is the reverfe of the compofition of forces or motions. See thefe articles.

Any fingle direct force AD, may be refolved into two oblique forces, whofe quantities and directions are AB, AC, having the fame effect, by defcribing any parallelogram ABDC, whofe diagonal is AD. And each of thefe may, in like manner, be refolved into two others ; and fo on, as far as we pleafe. And all thefe new forces, or motions, fo found, when acting together, will produce exactly the fame effect as the fingle original one. See alfo COLLISION, PERCUSSION, MOTION, &c.

REST, in Phyfics, the continuance of a body in the fame place ; or its continual application or contiguity to the fame parts of the ambient and contiguous bodies.—See SPACE.

Reft is either *abfolute* or *relative*, as place is.

Some define Reft to be the ftate of a thing without motion ; and hence again Reft becomes either abfolute or relative, as motion is.

Newton defines true or abfolute Reft to be the continuance of a body in the fame part of abfolute and immoveable fpace ; and relative Reft to be the continuance of a body in the fame part of relative fpace.

Thus, in a fhip under fail, relative Reft is the continuance of a body in the fame part of the fhip. But true or abfolute Reft is its continuance in the fame part of univerfal fpace in which the fhip itfelf is contained.

Hence, if the earth be really and abfolutely at Reft, the body relatively at Reft in the fhip will really and abfolutely move, and that with the fame velocity as the fhip itfelf. But if the earth do likewife move, there will then arife a real and abfolute motion of the body at Reft ; partly from the real motion of the earth in abfolute fpace, and partly from the relative motion of the fhip on the fea. Laftly, if the body be likewife relatively moved in the fhip, its real motion will arife partly from the real motion of the earth in immoveable fpace, and partly from the relative motion of the fhip on the fea, and of the body in the fhip.

It is an axiom in philofophy, that matter is indifferent as to Reft or motion. Hence Newton lays it down, as a law of nature, that every body perfeveres in its ftate, either of Reft or uniform motion, except fo far as it is difturbed by external caufes.

The Cartefians affert, that firmnefs, hardnefs, or folidity of bodies, confifts in this, that their parts are at Reft with regard to each other ; and this Reft they eftablifh as the great nexus, or principle of cohefion, by which the parts are connected together. On the other hand, they make fluidity to confift in a perpetual motion of the parts, &c. But the Newtonian philofophy furnifhes us with much better folutions.

Maupertuis afferts, that when bodies are in equilibrio, and any fmall motion is impreffed on them, the quantity of action refulting will be the leaft poffible. This he calls the law of *Reft* ; and from this law he deduces the fundamental propofition of ftatics. See Berlin Mem. tom. 2, pa. 294. And from the fame principle too he deduces the laws of percuffion.

RESTITUTION, in Phyfics, the returning of elaftic bodies, forcibly bent, to their natural ftate ; by fome called the *motion of Reftitution*.

RETARDATION, in Phyfics, the act of retarding, that is, of delaying the motion or progrefs of a body, or of diminifhing its velocity.

The Retardation of moving bodies arifes from two great caufes, the refiftance of the medium, and the force of gravity.

The RETARDATION *from the Refiftance* is often confounded with the refiftance itfelf ; becaufe, with refpect to the fame moving body, they are in the fame proportion.

But with refpect to different bodies, the fame refiftance often generates different Retardations. For if bodies of equal bulk, but different denfities, be moved through the fame fluid with equal velocity, the fluid will act equally on each ; fo that they will have equal refiftances, but different Retardations ; and the Retardations will be to each other, as the velocities which might be generated by the fame forces in the bodies propofed ; that is, they are inverfely as the quantities of matter in the bodies, or inverfely as the denfities.

Suppofe then bodies of equal denfity, but of unequal bulk, to move equally faft through the fame fluid ; then their refiftances increafe according to their fuperficies, that is as the fquares of their diameters ; but the quantities of matter are increafed according to their mafs or magnitude, that is as the cubes of their diameters : the refiftances are the quantities of motion ;
the

the Retardations are the celerities arising from them; and dividing the quantities of motion by the quantities of matter, we shall have the celerities; therefore the Retardations are directly as the squares of the diameters, and inversely as the cubes of the diameters, that is inversely as the diameters themselves.

If the bodies be of equal magnitude and density, and moved through different fluids, with equal celerity, their Retardations are as the densities of the fluids. And when equal bodies are carried through the same fluid with different velocities, the Retardations are as the squares of the velocities.

So that, if s denote the superficies of a body, w its weight, d its diameter, v the velocity, and n the density of the fluid medium, and N that of the body; then, in similar bodies, the resistance is as nsv^2 or as nd^2v^2, and the Retardation, or retarding force,

as $\dfrac{nsv^2}{w}$, or as $\dfrac{nd^2v^2}{Nd^3} = \dfrac{nv^2}{Nd}$.

The Retardation *from Gravity* is peculiar to bodies projected upwards. A body thrown upwards is retarded after the same manner as a falling body is accelerated; only in the one case the force of gravity conspires with the motion acquired, and in the other it acts contrary to it.

As the force of gravity is uniform, the Retardation from that cause will be equal in equal times. Hence, as it is the same force which generates motion in the falling body, and diminishes it in the rising one, a body rises till it lose all its motion; which it does in the same time in which a body falling would have acquired a velocity equal to that with which the body was thrown up.

Also, a body thrown up, will rise to the same height from which, in falling, it would acquire the same velocity with which it was thrown up: therefore the heights which bodies can rise to, when thrown up with different velocities, are to each other as the squares of the velocities.

Hence, the Retardations of motions may be compared together. For they are, first, as the squares of the velocities; 2dly, as the densities of the fluids through which the bodies are moved; 3dly, inversely as the diameters of those bodies; 4thly, inversely as the densities of the bodies themselves; as expressed by

the theorem above, viz. $\dfrac{nv^2}{Nd}$.

The Laws of Retardation, are the very same as those for acceleration; motion and velocity being destroyed in the one case, in the very same quantity and proportion as it is generated in the other.

RETICULA, or Reticule, in Astronomy, a contrivance for measuring very nicely the quantity of eclipses, &c.

This instrument, introduced some years since by the Paris Acad. of Sciences, is a little frame, consisting of 13 fine silken threads, parallel to, and equidistant from each other; placed in the focus of object-glasses of telescopes; that is, in the place where the image of the luminary is painted in its full extent. Consequently the diameter of the sun or moon is thus seen divided into 12 equal parts or digits: so that, to find the quantity of the eclipse, there is nothing to do but to number the parts that are dark, or that are luminous.

As a square Reticule is only proper for the diameter of the luminary, not for the circumference of it, it is sometimes made circular, by drawing 6 concentric equidistant circles; which represents the phases of the eclipse perfectly.

But it is evident that the Reticule, whether square or circular, ought to be perfectly equal to the diameter or circumference of the sun or star, such as it appears in the focus of the glass; otherwise the division cannot be just. Now this is no easy matter to effect, because the apparent diameter of the sun and moon differs in each eclipse; nay that of the moon differs from itself in the progress of the same eclipse.—Another imperfection in the Reticule is, that its magnitude is determined by that of the image in the focus; and of consequence it will only fit one certain magnitude.

But M. de la Hire has found a remedy for all these inconveniences, and contrived that the same Reticule shall serve for all telescopes, and all magnitudes of the luminary in the same eclipse. The principle upon which his invention is founded, is that two object-glasses applied against each other, having a common focus, and these forming an image of a certain magnitude, this image will increase in proportion as the distance between the two glasses is increased, as far as to a certain limit. If therefore a Reticule be taken of such a magnitude, as just to comprehend the greatest diameter the sun or moon can ever have in the common focus of two object-glasses applied to each other, there needs nothing but to remove them from each other, as the star comes to have a less diameter, to have the image still exactly comprehended in the same Reticule.

Farther, as the silken threads are subject to swerve from the parallelism, &c. by the different temperature of the air, another improvement is, to make the Reticule of a thin looking-glass, by drawing lines or circles upon it with the fine point of a diamond. See Micrometer.

RETIRED Flank, in Fortification. See Flank.

RETROCESSION *of Curves, &c.* See Retrogradation.

Retrocession *of the Equinox.* See Precession.

RETROGRADATION, or Retrogression, in Astronomy, is an apparent motion of the planets, by which they seem to go backwards in the ecliptic, and to move contrary to the order or succession of the signs.

When a planet moves in consequentia, or according to the order of the signs, as from Aries to Taurus, from Taurus to Gemini, &c, which is from west to east, it is said to be *direct.*—When it appears for some days in the same place, or point of the heavens, it is said to be *stationary.*—And when it goes in antecedentia, or backwards to the following signs, or contrary to the order of the signs, which is from east to west, it is said to be *retrograde.* All these different affections or circumstances, may happen in all the planets, except the sun and moon, which are seen to go direct only. But the times of the superior and inferior planets being retrograde are different; the former appearing so about their opposition, and the latter about their conjunction.

tion. The intervals of time also between two Retrogradations of the several planets, are very unequal:

In Saturn it is 1 year 13 days,
In Jupiter - - 1 - - 43
In Mars - - 2 - - 50
In Venus - - 1 - - 220
In Mercury - 0 - - 115

Again, Saturn continues retrograde 140 days, Jupiter 120, Mars 73, Venus 42, and Mercury 22; or nearly so; for the several Retrogradations of the same planet are not constantly equal.

These various circumstances however in the motions of the planets are not real, but only apparent; as the inequalities arise from the motion and position of the earth, from whence they are viewed; for when they are considered as seen from the sun, their motions appear always uniform and regular. These inequalities are thus explained:

Let S denote the sun; and ABCD &c the path or orbit of the earth, moving from west to east, and in that order; also GK &c the orbit of a superior planet, as Saturn for instance, moving the same way, or in the direction GKLG, but with a much less celerity than the earth's motion.

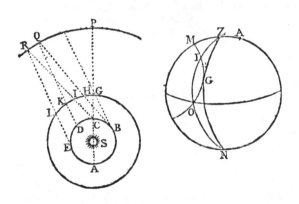

the earth's motion. Now when the earth is at the point A of its orbit, let Saturn be at G, in conjunction with the sun, when it will be seen at P in the zodiac, or among the stars; and when the earth has moved from A to B, let Saturn have moved from G to H in its orbit, when it will be seen in the line BHQ, and will appear to have moved from P to Q in the zodiac; also when the earth has got to C, let Saturn be arrived at I, but found at R in the zodiac, where being seen in the line CIR, it appears stationary, or without motion in the zodiac at R. But after this, Saturn will appear for some time in Retrogradation, viz, moving backwards, or the contrary way: for when the earth has moved to D, Saturn will have got to K, and, being seen in the line DKQ, will appear to have moved retrograde in the zodiac from R to Q; about which place the planet, ceasing to recede any farther, again becomes stationary, and afterward proceeds forward again; for while the earth moves from D to E, and Saturn from K to L, this latter, being now seen in the line ELR, appears to

have moved forward in the zodiac from Q to R. And so on; the superior planets always becoming retrograde a little before they are in opposition to the sun, and continuing so till some time after the opposition: the retrograde motion being swiftest when the planet is in the very opposition itself; and the direct motion swiftest when in the conjunction. The arch RQ which the planet describes while thus retrograde, is called the arch of Retrogradation. These arches are unequal in all the planets, being greatest in the most distant, and gradually less in the nearer ones.

In like manner may be shewn the circumstances of the Retrogradations of the inferior planets; by which it will appear, they become stationary a little before their inferior conjunction, and go retrograde till a little time after it; moving the quickest retrograde just at that conjunction, and the quickest direct just at the superior or further conjunction.

RETROGRADATION *of the Nodes of the Moon*, is a motion of the line of the nodes of her orbit, by which it continually shifts its situation from east to west, contrary to the order of the signs, completing its retrograde circulation in the period of about 19 years: after which time, either of the nodes, having receded from any point of the ecliptic, returns to the same again.—Newton has demonstrated, in his Principia, that the Retrogradation of the moon's nodes is caused by the action of the sun, which continually drawing this planet from her orbit, deflects this orbit from a plane, and causes its intersection with the ecliptic continually to vary; and his determinations on this point have been confirmed by observation.

RETROGRADATION *of the Sun*, a motion by which in some situations, in the torrid zone, he seems to move retrograde or backwards.

When the sun is in the torrid zone, and has his declination AM greater than the latitude of the place AZ, but either northern or southern as that is (last fig. above), the sun will appear to go retrograde, or backwards, both before and after noon. For draw the vertical circle ZGN to be a tangent to the sun's diurnal circle MGO in G, and another ZON through the sun's rising, at O: then it is evident, that all the intermediate vertical circles cut the sun's diurnal circle twice: first in the arc GO, and the second time in the arc GI. So that, as the sun ascends through the arc GO, he continually arrives at farther and farther verticals. But as he continues his ascent through the arc GI, he returns to his former verticals; and therefore is seen retrograde for some time before noon. And in like manner it may be shewn that he does the same thing for some time after noon. Hence, as the shadow always tends opposite to the sun, the shadow will be retrograde twice every day in all places of the torrid zone, where the sun's declination exceeds the latitude.

But the same thing can never happen without the tropics, in a natural way.

RETROGRADATION, or RETROGRESSION, in the Higher Geometry, is the same with what is otherwise called *contrary flexion* or *flexure*. See FLEXURE, and INFLEXION.

RETRO.

RETROGRADE, denotes backward, or contrary to the forward or natural direction. See RETROGRADATION.

RETROGRESSION, or RETROCESSION. The same with RETROGRADATION.

RETURNING *Stroke*, in Electricity, is an expression used by lord Mahon (now earl Stanhope) to denote the effect produced by the return of the electric fire into a body from which, in certain circumstances, it has been expelled.

To understand properly the meaning of these terms, it must be premised that, according to the noble author's experiments, an insulated smooth body, immerged within the electrical atmosphere, but beyond the striking distance of another body, charged positively, is at the same time in a state of threefold electricity. The end next to the charged body acquires negative electricity; the farther end is positively electrified; while a certain part of the body, somewhere between its two extremes, is in a natural, unelectrified, or neutral state; so that the two contrary electricities balance each other. It may farther be added, that if the body be not insulated, but have a communication with the earth, the whole of it will be in a negative state. Suppose then a brass ball, which may be called A, to be constantly placed at the striking distance of a prime conductor; so that the conductor, the instant when it becomes fully charged, explodes into it. Let another large or second conductor be suspended, in a perfectly insulated state, farther from the prime conductor than the striking distance, but within its electrical atmosphere: let a person standing on an insulated stool touch this second conductor very lightly with a finger of his right hand; while, with a finger of his left hand, he communicates with the earth, by touching very lightly a second brass ball fixed at the top of a metallic stand, on the floor, which may be called B. Now while the prime conductor is receiving its electricity, sparks pass (at least if the distance between the two conductors is not too great) from the second conductor to the right hand of the insulated person; while similar and simultaneous sparks pass out from the finger of his left hand into the second metallic ball B, communicating with the earth. At length however the prime conductor, having acquired its full charge, suddenly strikes into the ball A, of the first metallic stand, placed for that purpose at the striking distance. The explosion being made, and the prime conductor suddenly robbed of its elastic atmosphere, its pressure or action on the second conductor, and on the insulated person, as suddenly ceases; and the latter instantly feels a smart Returning Stroke, though he has no direct or visible communication (except by the floor) with either of the two bodies, and is placed at the distance of 5 or 6 feet from both of them. This Returning Stroke is evidently occasioned by the sudden re-entrance of the electric fire naturally belonging to his body and to the second conductor, which had before been expelled from them by the action of the charged prime conductor upon them; and which returns to its former place in the instant when that action or elastic pressure ceases. When the second conductor and the insulated person are placed in the densest part of the electrical atmo-

sphere of the prime conductor, or just beyond the striking distance, the effects are still more considerable; the Returning Stroke being extremely severe and pungent, and appearing considerably sharper than even the main stroke itself, received directly from the prime conductor. Lord Mahon observes, that persons and animals may be destroyed, and particular parts of buildings may be much damaged, by an electrical Returning Stroke, occasioned even by some very distant explosion from a thunder cloud; possibly at the distance of a mile or more. It is certainly not difficult to conceive that a charged extensive thunder cloud must be productive of effects similar to those produced by the prime conductor; but perhaps the effects are not so great, nor the danger so terrible, as it seems have been apprehended. If the quantity of electric fluid naturally contained, for example, in the body of a man, were immense or indefinite; then the estimate between the effects producible by a cloud, and those caused by a prime conductor, might be admitted; but surely no electrical cloud can expel from a body more than the natural quantity of electricity which this contains. On the sudden removal therefore of the pressure by which this natural quantity had been expelled, in consequence of the explosion of the cloud into the earth, no more (at the utmost) than his whole natural stock of electricity can re-enter his body, provided it be so situated, that the returning fire of other bodies must necessarily pass through his body. But perhaps we have no reason to suppose that this quantity is so great, as that its sudden re entrance into his body should destroy or injure him.

Allowing therefore the existence of the Returning Stroke, as sufficiently ascertained, and well illustrated, in a variety of circumstances, by the author's experiments, the magnitude and danger of it do not seem to be so alarming as he apprehends. See Lord Mahon's Principles of Electricity, &c. 4to. 1779, pa. 76, 113, and 131. Also Monthly Review, vol. 62, pa. 436.

REVERSION *of Series*, in Algebra, is the finding the value of the root, or unknown quantity, whose powers enter the terms of an infinite series, by means of another infinite series in which it is not contained. As, in the infinite series $z = ax + bx^2 + cx^3 + dx^4$ &c; then if there be found $x = Az + Bz^2 + Cz^3$ &c, that series is inverted, or its root x is found in an infinite series of other terms.

This was one of Newton's improvements in analysis, the first specimen of which was given in his Analysis per Æquationes Numero Terminorum Infinitas; and it is of great use in resolving many problems in various parts of the mathematics.

The most usual and general way of Reversion, is to assume a series, of a proper form, for the value of the required unknown quantity; then substitute the powers of this value, instead of those of that quantity into the given series; lastly compare the resulting terms with the said given series, and the values of the assumed coefficients will thus be obtained. So, to revert the series $z = ax + bx^2 + cx^3$, &c, or to find the value of x in terms of z; assume it thus, $x = Az + Bz^2 + Cz^3$ &c;

&c; then by involving this feries, for the feveral powers of x, and multiplying the correfponding powers by a, b, c, &c, there refults

$$z = aAz + dB x^3 + dC z^{3\cdots} + aD z^4, \quad \&c.$$
$$+ bA^2 z^2 + 2bA bz^{3\prime} + 2bACz^4$$
$$+ 2 b^3 z^4$$
$$+ cA^3 z^3 + 3cA^2 B z^4$$
$$+ dA^4 z^4$$

Then by comparing the correfponding terms of this laft feries, or making their coefficients equal, there are obtained thefe equations, viz,

$aA - 1$, and $aB + bA^2 = 0$, and $aC + 2bAB + cA^3 = 0$, &c, which give thefe values of the affumed coefficients, viz,

$$A = \frac{1}{a}; \quad B = -\frac{bA^2}{a} = -\frac{b}{a^3};$$

$$C = -\frac{2bAB + cA^3}{a} = \frac{2bb - a}{a^5} c;$$

$$D = -\frac{2bAC + bB^2 + 3cA^2B + dA^4}{a}$$

$$= -\frac{5abc - 5b^3 - a^2d}{a^7}; \quad \&c.$$

and confequently

$$x = \frac{1}{a} z - \frac{b}{a^3} z^2 + \frac{2bb - ac}{a^5} z^3 - \frac{5abc - 5b^3 - a^2d}{a^7} z^4$$

&c; which is therefore a general formula or theorem for every feries of the fame kind, as to the powers of the quantity x. Thus, for

Ex. Suppofe it were required to revert the feries $z = x - x^2 + x^3 - x^4$, &c.

Here $a = 1$, $b = -1$, $c = 1$, $d = -1$, &c; which values of thefe letters being fubftituted in the theorem, there refults $x = z + z^2 + z^3 + z^4$, &c, which is that feries reverted, or the value of x in it.

In the fame way it will be found that the theorem for reverting the feries $z = ax + bx^3 + cx^5 + dx^7$ &c, is

$$x = \frac{1}{a} z - \frac{b}{a^4} z^3 + \frac{3bb - ac}{a^7} z^5 - \frac{a^2d + 12b^3 - 8abc}{a^{10}},$$
&c.

And if $z = ax^m + bx^{m+n} + cx^{m+2n} + \&c$, then is

$$x = y^{\frac{1}{m}} - \frac{b}{ma} y^{\frac{1+n}{m}} + \frac{(1+2n+m) bb - 2mac}{2mmaa}$$

$$\times y^{\frac{1+2n}{m}} \&c. \text{ where } y \text{ is} = \frac{z}{a}.$$

Various methods of Reverfion may be feen as given by De Moivre in the Philof. Tranf. number 240; or Maclaurin's Algebra pa. 263; or Stuart's Explanation of Newton's Analyfis, &c. pa. 455; or Coulfon's Comment on Newton's Flux. pa. 219; or Horfley's ed. of Newton's works vol. 1, pa. 291; or Simpfon's Flux. vol. 2, pa. 302: or moft authors on Algebra.

REVETEMENT, in Fortification, a ftrong wall built on the outfide of the rampart and parapet, to fupport the earth, and prevent its rolling into the ditch.

REVOLUTION, in Geometry, the motion of rotation of a line about a fixed point or centre, or of any figure about a fixed axis, or upon any line or furface. Thus, the Revolution of a given line about a fixed centre, generates a circle; and that of a right-angled triangle about one fide, as an axis, generates a cone; and that of a femicircle about its diameter, generates a fphere or globe, &c.

REVOLUTION, in Aftronomy, is the period of a ftar, planet, or comet, &c; or its courfe from any point of its orbit, till it return to the fame again.

The planets have a twofold Revolution. The one about their own axis, ufually called their *diurnal rotation*, which conftitutes their day. The other about the fun, called their *annual Revolution*, or *period*, conftituting their year.

REYNEAU (CHARLES-RENE), commonly called Father Reyneau, a noted French mathematician, was born at Briffac in the province of Anjou, in the year 1656. At 20 years of age he entered himfelf among the Oratorians, a kind of religious order, in which the members lived in community without making any vows, and applied themfelves chiefly to the education of youth. He was foon after fent, by his fuperiors to teach philofophy at Pezenas, and then at Toulon. This requiring fome acquaintance with geometry; he contracted a great affection for this fcience, which he cultivated and improved to a great extent; in confequence he was called to Angers in 1683, to fill the mathematical chair; and the Academy of Angers elected him a member in 1694.

In this occupation Father Reyneau, not content with making himfelf mafter of every thing worth knowing, which the modern analyfis, fo fruitful in fublime fpeculations and ingenious difcoveries, had already produced, undertook to reduce into one body, for the ufe of his fcholars, the principal theories fcattered here and there in Newton, Defcartes, Leibnitz, Bernoulli, the Leipfic Acts, the Memoirs of the Paris Academy, and in other works; treafures which by being fo widely difperfed, proved much lefs ufeful than they otherwife might have been. The fruit of this undertaking, was his *Analyfe Demontrée*, or Analyfis Demonftrated, which he publifhed in 2 volumes 4to, 1708.

Father Reyneau called this ufeful work, Analyfis Demonftrated, becaufe he demonftrates in it feveral methods which had not been demonftrated by the authors of them, or at leaft not with fufficient perfpicuity and exactnefs; for it often happens that, in matters of this kind, a perfon is clear in a thing, without being able to demonftrate it. Some perfons too have been fo miftakingly fond of glory as to make a fecret of their demonftrations, in order to perplex thofe, whom it would become them much better to inftruct. This book of Reyneau's was fo well approved, that it foon became a maxim, at leaft in France, that to follow him was the beft, if not the only way, to make any extra-

3 B 2

ordinary

ordinary progress in the mathematics. This was considering him as the first master, as the Euclid of the sublime geometry.

Reyneau, after thus giving lessons to those who understood something of geometry, thought proper to draw up some for such as were utterly unacquainted with that science. This was in some measure a condescension in him, but his passion to be useful made it easy and agreeable. In 1714 he published a volume in 4to on calculation, under the title of *Science du Calcul des Grandeurs*, of which the then Censor Royal, a most intelligent and impartial judge, says, in his approbation of it, that " though several books had already appeared upon the same subject, such a treatise as that before him was still wanting, as in it every thing was handled in a manner sufficiently extensive, and at the same time with all possible exactness and perspicuity." In fact, though most branches of the mathematics had been well treated of before that period there were yet no good elements, even of practical geometry. Those who knew no more than what precisely such a book ought to contain, knew too little to complete a good one; and those who knew more, thought themselves probably above the task; whereas Reyneau possessed at once all the learning and modesty necessary to undertake and execute such a work.

As soon as the Royal Academy of Sciences at Paris, in consequence of a regulation made in the year 1716, opened its doors to other learned men, under the title of *Free Associates*, Father Reyneau was admitted of the number. The works however which we have already mentioned, besides a small piece upon *Logic*, are the only ones he ever published, or probably ever composed, except most of the materials for a second volume of his *Science du Calcul*, which he left behind him in manuscript. The last years of his life were attended with too much sickness to admit of any extraordinary application. He died in 1728, at 72 years of age, not more regretted on account of his great learning, than of his many virtues, which all conspired in an eminent degree to make that learning agreeable to those about him, and useful to the world. The first men in France deemed it an honour and a happiness to count him among their friends. Of this number were the chancellor of that kingdom, and Father Mallebranche, of whom Reyneau was a zealous and faithful disciple.

RHABDOLOGY, or RABDOLOGY, in Arithmetic, a name given by Napier to a method of performing some of the more difficult operations of numbers by means of certain square little rods. Upon these are inscribed the simple numbers; then by shifting them according to certain rules, those operations are performed by simply adding or subtracting of the numbers as they stand upon the rods. See Napier's Rabdologia, printed in 1617. See also the article NAPIER's *Bones*.

RHEO-STATICS, is used by some for the statics, or the science of the equilibrium of fluids.

RHETICUS (GEORGE JOACHIM), a noted German astronomer and mathematician, who was the colleague of Reinhold in the university of Wittemberg, being joint professors of mathematics there toge-

ther. He was born at Feldkirk in Tyrol the 15th of February 1514. After imbibing the elements of the mathematics at Tiguri with Oswald Mycone, he went to Wittemberg, where he diligently cultivated that science. Here he was made master of philosophy in 1535, and professor in 1537. He quitted this situation however two years after, and went to Fruenburg to put him under the assistance of the celebrated Copernicus, being induced to this step by his zeal for astronomical pursuits, and the great fame which Copernicus had then acquired. Rheticus assisted this astronomer for some years, and constantly exhorted him to perfect his work, *De Revolutionibus*, which he published after the death of Copernicus, viz, in 1543, folio, at Norimberg, together with an illustration of the same in a narration, dedicated to Schoner. Here too, to render astronomical calculations more accurate, he began his very elaborate canon of sines, tangents and secants, to 15 places of figures, and to every 10 seconds of the quadrant, a design which he did not live quite to complete. The canon of sines however, to that radius, for every 10 seconds, and for every single second in the first and last degree of the quadrant, computed by him, was published in folio at Francfort 1613 by Pitiscus, who himself added a few of the first sines computed to 22 places of figures. But the larger work, or canon of sines, tangents and secants, to every 10 seconds, was perfected and published after his death, viz, in 1596, by his disciple Valentine Otho, mathematician to the Electoral Prince Palatine; a particular account and analysis of which work may be seen in the Historical Introduction to my Logarithms, pa. 9.

After the death of Copernicus, Rheticus returned to Wittemberg, viz, in 1541 or 1542, and was again admitted to his office of professor of mathematics. The same year, by the recommendation of Melancthon, he went to Norimberg, where he found certain manuscripts of Werner and Regiomontanus. He afterwards taught mathematics at Leipsic. From Saxony he departed a second time, for what reason is not known, and went to Poland; and from thence to Cassovia in Hungary, where he died December the 4th, 1576, near 63 years of age.

His *Narratio de Libris Revolutionum Copernici*, was first published at Gedunum in 4to, 1540; and afterwards added to the editions of Copernicus's work. He also composed and published *Ephemerides*, according to the doctrine of Copernicus, till the year 1551.

Rheticus also projected other works, and partly executed them, though they were never published, of various kinds, astronomical, astrological, geographical, chemical, &c; as they are more particularly mentioned in his letter to Peter Ramus in the year 1568, which Adrian Romanus inserted in the preface to the first part of his Idea of Mathematics.

RHOMB SOLID, consists of two equal and right cones joined together at their bases.

RHOMBOID, or RHOMBOIDES, in Geometry, a quadrilateral figure, whose opposite sides and angles are equal; but which is neither equilateral nor equiangular.

RHOMBUS, is an oblique equilateral parallelogram;

or

or a quadrilateral figure, whose sides are equal and parallel, but the four angles not all equal, two of the opposite ones being obtuse, and the other two opposite ones acute.

The two diagonals of a Rhombus intersect at right angles; but not of a rhomboides.

As to the area of the Rhombus or rhomboides, it is found, like that of all other parallelograms, by multiplying the length or base by the perpendicular breadth.

RHOMBUS-*Solid*. See RHOMB-*Solid*.

RHUMB, RUMB, or RUM, in Navigation, a vertical circle of any given place; or the intersection of a part of such a circle with the horizon. Rhumbs therefore coincide with points of the world, or of the horizon. And hence mariners distinguish the Rhumbs by the same names as the points and winds. But we may observe, that the Rhumbs are denominated from the points of the compass in a different manner from the winds: thus, at sea, the north-east wind is that which blows from the north-east point of the horizon towards the ship in which we are; but we are said to sail upon the north-east Rhumb, when we go towards the north-east.

They usually reckon 32 Rhumbs, which are represented by the 32 lines in the rose or card of the compass.

Aubin defines a Rhumb to be a line on the terrestrial globe, or sea-compass, or sea-chart, representing one of the 32 winds which serve to conduct a vessel. So that the Rhumb a vessel pursues is conceived as its route, or course.

Rhumbs are divided and subdivided like points of the compass. Thus, the whole Rhumb answers to the cardinal point. The half Rhumb to a collateral point, or makes an angle of 45 degrees with the former. And the quarter Rhumb makes an angle of 22° 30' with it. Also the half-quarter Rhumb makes an angle of 11° 15' with the same.

For, a table of the Rhumbs, or points, and their distances from the meridian, see WIND.

RHUMB-LINE, *Loxodromia*, in Navigation, is a line prolonged from any point of the compass in a nautical chart, except the four cardinal points: or it is the line which a ship, keeping in the same collateral point, or rhumb, describes throughout its whole course.

The chief property of the Rhumb-line, or loxodromia, and that from which some authors define it, is, that it cuts all the meridians in the same angle.

This angle is called the *angle of the Rhumb*, or the *loxodromic* angle. And the angle which the Rhumbline makes with any parallel to the equator, is called the *complement of the Rhumb*.

An idea of the origin and properties of the Rhumbline, the great foundation of Navigation, may be conceived thus: a vessel beginning its course, the wind by which it is driven makes a certain angle with the meridian of the place; and as we shall suppose that the vessel runs exactly in the direction of the wind, it makes the same angle with the meridian which the wind makes. Supposing then the wind to continue the

same, as each point or instant of the progress may be esteemed the beginning, the vessel always makes the same angle with the meridian of the place where it is each moment, or in each point of its course which the wind makes.

Now a wind, for example, that is north-east, and which consequently makes an angle of 45 degrees with the meridian, is equally north-east wherever it blows, and makes the same angle of 45 degrees with all the meridians it meets. And therefore a vessel, driven by the same wind, always makes the same angle with all the meridians it meets with on the surface of the earth.

If the vessel sail north or south, it describes the great circle of a meridian. If it runs east or west, it cuts all the meridians at right angles, and describes either the circle of the equator, or else a circle parallel to it.

But if the vessel sails between the two, it does not then describe a circle; since a circle, drawn obliquely to a meridian, would cut all the meridians at unequal angles, which the vessel cannot do. It describes therefore another curve, the essential property of which is, that it cuts all the meridians in the same angle, and it is called the *loxodromy*, or *loxodromic curve*, or *Rhumbline*.

This curve, on the globe, is a kind of spiral, tending continually nearer and nearer to the pole, and making an infinite number of circumvolutions about it, but without ever arriving exactly at it. But the spiral Rhumbs on the globe become proportional spirals in the stereographic projection on the plane of the equator.

The length of a part of this Rhumb-line, or spiral, then, is the distance run by the ship while she keeps in the same course. But as such a spiral line would prove very perplexing in the calculation, it was necessary to have the ship's way in a right line; which right line however must have the essential properties of the curve line, viz, to cut all the meridians at right angles. The method of effecting which, see under the article CHART.

The arc of the Rhumb-line is not the shortest distance between any two places through which it passes; for the shortest distance, on the surface of the globe, is an arc of the great circle passing through those places; so that it would be a shorter course to sail on the arc of this great circle: but then the ship cannot be kept in the great circle, because the angle it makes with the meridians is continually varying, more or less.

Let P be the pole, RW the equator, ABCDEP a spiral Rhumb, divided into an indefinite number of equal parts at the points B, C, D, &c; through which are drawn the meridians, PS, PT, PV, &c, and the parallels FB, KC, LD, &c, also draw the parallel AN. Then, as a ship sails along the Rhumbline towards the pole, or in the direction ABCD &c, from A to E, the distance sailed AE

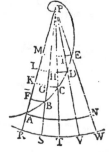

is made up of all the small equal parts of the Rhumb AB + BC + CD + DE ; and

the sum of all the small differences of latitude AF + BG + CH + DI make up the whole difference of latitude AM or EN ; and

the sum of all the small parallels FB + GC + HD + IE is what is called the departure in plane sailing ; and ME is the meridional distance, or distance between the first and last meridians, measured on the last parallel ; also RW is the difference of longitude, measured on the equator. So that these last three are all different, viz. the departure, the meridional distance, and the difference of longitude.

If the ship sail towards the equator, from E to A ; the departure, difference of latitude, and difference of longitude, will be all three the same as before ; but the meridional distance will now be AN, instead of ME ; the one of these AN being greater than the departure FB + GC + HD + IE, and the other ME is less than the same ; and indeed that departure is nearly a mean proportional between the two meridional distances ME, AN. Other properties are as below.

1. All the small elementary triangles ABF, BCG, CDH, &c. are mutually similar and equal in all their parts. For all the angles at A, B, C, D, &c are equal, being the angles which the Rhumb makes with the meridians, or the angles of the course ; also all the angles F, G, H, I, are equal, being right angles ; therefore the third angles are equal, and the triangles all similar. Also the hypotenuses AB, BC, CD, &c. are all equal by the hypothesis ; and consequently the triangles are both similar and equal.

2. As radius : distance run AE

:: sine of course ∠ A : departure FB + GC &c,
:: cosin. of course ∠ A : dif. of lat. AM.

For in any one ABF of the equal elementary triangles, which may be considered as small right-angled plane triangles, it is, as rad. or sin. ∠ F : sin. course A :: AB : FB :: (by composition) the sum of all the distances AB + BC + CD &c : the sum of all the departures FB + GC + HD &c.

And, in like manner, as radius : cos. course A :: AB : AF :: AB + BC &c : AF + BG &c.

Hence, of these four things, the course, the difference of latitude, the departure, and the distance run, having any two given, the other two are found by the proportions above in this article.

By means of the departure, the length of the Rhumb, or distance run, may be connected with the longitude and latitude, by the following two theorems.

3. As radius : half the sum of the cosines of both the latitudes, of A and E :: dif. of long. RW : departure.

Because RS : FB :: radius : sine of PA or cos. RA, and VW : IE :: radius : sine of PE or cos. EW.

4. As radius : cos. middle latitude :: dif. of longitude : departure.—Because cosine of middle latitude is nearly equal to half the sum of the cosines of the two extreme latitudes.

RICCIOLI (JOANNES BAPTISTA), a learned Ita-

lian astronomer, philosopher, and mathematician, was born in 1598, at Ferrara, a city in Italy, in the dominions of the Pope. At 16 years of age he was admitted into the society of the Jesuits. He was endowed with uncommon talents, which he cultivated with extraordinary application ; so that the progress he made in every branch of literature and science was surprising. He was first appointed to teach rhetoric, poetry, philosophy, and scholastic divinity, in the Jesuits' colleges at Parma and Bologna ; yet applied himself in the mean time to making observations in geography, chronology, and astronomy. This was his natural bent, and at length he obtained leave from his superiors to quit all other employment, that he might devote himself entirely to those sciences.

He projected a large work, to be divided into three parts, and to contain as it were a complete system of philosophical, mathematical, and astronomical knowledge. The first of these parts, which regards astronomy, came out at Bologna in 1651, 2 vols. folio, with this title, *J. B. Riccioli Almagestum Novum, Astronomiam veterem novamque complectens, observationibus aliorum et propriis, novisque theorematibus, problematibus ac tabulis promotam.* Riccioli imitated Ptolomy in this work, by collecting and digesting into proper order, with observations, every thing ancient and modern, which related to his subject ; so that Gassendus very justly called his work, " Promptuarium et thesaurum ingentem Astronomiæ."

In the first volume of this work, he treats of the sphere of the world, of the sun and moon, with their eclipses ; of the fixed stars, of the planets, of the comets and new stars, of the several mundane systems, and six sections of general problems serving to astronomy, &c. —In the second volume, he treats of trigonometry, or the doctrine of plane and spherical triangles.; proposes to give a treatise of astronomical instruments, and the optical part of astronomy (which part was never published) ; treats of geography, hydrography, with an epitome of chronology.—The third, comprehends observations of the sun, moon, eclipses, fixed stars and planets, with precepts and tables of the primary and secondary motions, and other astronomical tables.

Riccioli printed also, two other works, in folio, at Bologna, viz,

2. *Astronomia Reformata,* 1665 : the design of which was, that of considering the various hypotheses of several astronomers, and the difficulty thence arising of concluding any thing certain, by comparing together all the best observations, and examining what is most certain in them, thence to reform the principles of astronomy.

3. *Chronologia Reformata,* 1669.

Riccioli died in 1671, at 73 years of age.

RICOCHET Firing, in the Military Art, is a method of firing with small charges, and pieces elevated but in a small degree, as from 3 to 6 degrees. The word signifies duck-and drake, or rebounding, because the ball or shot, thus discharged, goes bounding and rolling along, and killing or destroying every thing in its way, like the bounding of a flat stone along the surface of water when thrown almost horizontally.

RIDEAU,

RIDEAU, in Fortification, a small elevation of earth, extending itself lengthways on a plain; serving to cover a camp, or give an advantage to a post.

RIDEAU is sometimes also used for a trench, the earth of which is thrown up on its side, to serve as a parapet for covering the men.

RIFLE GUNS, in the Military Art, are those whose barrels, instead of being smooth on the inside, are formed with a number of spiral channels, making each about a turn and a half in the length of the barrel. These carry their balls both farther and truer than the common pieces. For the nature and qualities of them, see Robins's Tracts, vol. 1 pa 328 &c.

RIGEL, in Astronomy. See REGEL.

RIGHT, in Geometry, something that lies evenly or equally, without inclining or bending one way or another. Thus, a Right-line is that whose small parts all tend the same way. In this sense, Right means the same as straight, as opposed to curved or crooked.

RIGHT-Angle, that which one line makes with another upon which it stands so as to incline neither to one side nor the other. And in this sense the word Right stands opposed to oblique.

RIGHT-angled, is said of a figure when its sides are at Right angles or perpendicular to each other.—This sometimes holds in all the angles of the figure, as in squares and rectangles; sometimes only in part, as in right-angled triangles.

RIGHT Cone, or Cylinder, or prism, or pyramid, one whose axis is at right-angles to the base.

RIGHT-lined Angle, one formed by Right lines.

RIGHT Sine, one that stands at Right-angles to the diameter; as opposed to versed sine.

RIGHT, Sphere, is that where the equator cuts the horizon at Right angles; or that which has the poles in the horizon, and the equinoctial in the zenith.

Such is the position of the sphere with regard to those who live at the equator, or under the equinoctial. The consequences of which are; that they have no latitude, nor elevation of the pole; they see both poles of the world, and all the stars rise, culminate and set; also the sun always rises and descends at Right angles, and makes their days and nights equal. In a Right sphere, the horizon is a meridian; and if the sphere be supposed to revolve, all the meridians successively become horizons, one after another.

RIGHT Ascension, Descension, Parallax, &c. See the respective Articles.

RIGHT Circle, in the Stereographic Projection of the Sphere, is a circle at Right angles to the plane of projection, or that is projected into a Right line.

RIGHT Sailing, is that in which a voyage is performed on some one of the four cardinal points, east, west, north, or south.

If the ship sail on a meridian, that is, north or south, she does not alter her longitude, but only changes the latitude, and that just as much as the number of degrees she has run.

But if she sail on the equator, directly east or west,

she varies not her latitude, but only changes the longitude, and that just as much as the number of degrees she has run.

And if she sail directly east or west upon any parallel, she again does not change her latitude, but only the longitude; yet not the same as the number of degrees of a great circle she hath sailed, as on the equator, but more, according as the parallel is remoter from the equinoctial towards the pole. For the less any parallel is, the greater is the difference of longitude answering to the distance run.

RIGIDITY, a brittle hardness; or that kind of hardness which is supposed to arise from the mutual indentation of the component particles within one another. Rigidity is opposed to ductility, malleability, &c.

RING, in Astronomy and Navigation, an instrument used for taking the sun's altitude &c. It is usually of brass, about 9 inches diameter, suspended by a little swivel, at the distance of 45° from the point of which is a perforation, which is the centre of a quadrant of 90° divided in the inner concave surface.

To use it, let it be held up by the swivel, and turned round to the sun, till his rays, falling through the hole, make a spot among the degrees, which marks the altitude required.

This instrument is preferred before the astrolabe, because the divisions are here larger than on that instrument.

RING, of Saturn, is a thin, broad, opaque circular arch encompassing the body of that planet, like the wooden horizon of an artificial globe, without touching it, and appearing double, when seen through a good telescope.

This Ring was first discovered by Huygens, who, after frequent observation of the planet, perceived two lucid points, like ansæ or handles, arising out from the body in a right line. Hence as in subsequent observations he always found the same appearance, he concluded that Saturn was encompassed with a permanent Ring; and accordingly produced his New System of Saturn, in 1659. However, Galileo first discovered that the figure of Saturn was not round.

Huygens makes the space between the globe of Saturn and the Ring equal to the breadth of the Ring, or rather more, being about 22000 miles broad; and the greatest diameter of the Ring, in proportion to that of the globe, as 9 to 4. But Mr. Pound, by an excellent micrometer applied to the Huygenian glass of 123 feet, determined this proportion, more exactly, to be as 7 to 3.

Observations have also determined, that the plane of the Ring is inclined to the plane of the ecliptic in an angle of 30 degrees; that the Ring probably turns, in the direction of its plane, round its axis, because when it is almost edgewise to us, it appears rather thicker on one side of the planet than on the other; and the thickest edge has been seen on different sides at different times: the sun shines almost 15 of our years together on one side of Saturn's Ring without setting, and as long on the other in its turn; so that the Ring is visible to the inhabitants of that planet for almost 15

of

of our years, and as long invisible, by turns, if its axis has no inclination to its Ring; but if the axis of the planet be inclined to the Ring, ex. gr. about 30 degrees, the Ring will appear and disappear once every natural day to all the inhabitants within 30 degrees of the equator, on both sides, frequently eclipsing the sun in a Saturnian day. Moreover, if Saturn's axis be so inclined to his Ring, it is perpendicular to his orbit; by which the inconvenience of different seasons to that planet is avoided.

This Ring, seen from Saturn, appears like a large luminous arch in the heavens, as if it did not belong to the planet.

When we see the Ring most open, its shadow upon the planet is broadest; and from that time the shadow grows narrower, as the Ring appears to do to us; until, by Saturn's annual motion, the sun comes to the plane of the Ring, or even with its edge; which, being then directed towards us, becomes invisible, on account of its thinness.

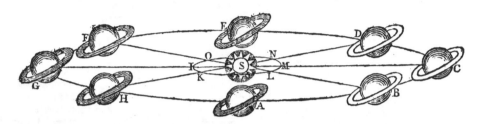

The phenomena of Saturn's Ring are illustrated by a view of this figure. Let S be the sun, ABCDEFGH Saturn's orbit, and IKLMNO the earth's orbit. Both Saturn and the earth move according to the order of the letters; and when Saturn is at A, his Ring is turned edgewise to the sun S, and he is then seen from the earth as if he had lost his Ring, let the earth be in any part of its orbit whatever, except between N and O; for whilst it describes that space, Saturn is apparently so near the sun as to be hid in his beams. As Saturn goes from A to C, his Ring appears more and more open to the earth; at C the Ring appears most open of all; and seems to grow narrower and narrower as Saturn goes from C to E; and when he comes to E, the Ring is again turned edgewise both to the sun and earth; and as neither of its sides is illuminated, it is invisible to us, because its edge is too thin to be perceptible; and Saturn appears again as if he had lost his Ring. But as he goes from E to G, his Ring opens more and more to our view on the under side; and seems just as open at G as it was at C, and may be seen in the night time from the earth in any part of its orbit, except about M, when the sun hides the planet from our view.

As Saturn goes from G to A, his Ring turns more and more edgewise to us, and, therefore, it seems to grow narrower and narrower; and at A it disappears as before.

Hence, while Saturn goes from A to E, the sun shines on the upper side of his Ring, and the under side is dark; and whilst he goes from E to A, the sun shines on the under side of his Ring, and the upper side is dark. The Ring disappears twice in every annual revolution of Saturn, viz, when he is in the 19th degree or Pisces and of Virgo, and when Saturn is in the middle between these points, or in the 19th degree either of Gemini or of Sagittarius, his Ring appears most open to us; and then its longest diameter is to its shortest, as 9 to 4. Ferguson's Astr. sect. 204.

There are various hypotheses concerning this Ring. Kepler, in his Epitom. Astron. Copern. and after him

Dr. Halley, in his Enquiry into the Causes of the Variation of the Needle, Phil. Transf. No 195, suppose our earth may be composed of several crusts or shells, one within another, and concentric to each other. If this be the case, it is possible the Ring of Saturn may be the fragment or remaining ruin of his formerly exterior shell, the rest of which is broken or fallen down upon the body of the planet. And some have supposed that the Ring may be a congeries or series of moons revolving about the planet.

Later observations have thrown much more light upon this curious phenomenon, especially respecting its dimensions, and rotation, and division into two or more parts. De la Lande and De la Place say, that Cassini saw the breadth of the Ring divided into two separate parts that are equal, or nearly so. Mr. Short assured M. De la Lande, that he had seen many divisions upon the Ring, with his 12 feet telescope. And Mr. Hadley, with an excellent $5\frac{1}{2}$ feet reflector, saw the Ring divided into two parts. Several excellent theories have been given in the French Memoirs, particularly by De la Place, contending for the division of the Ring into many parts. But finally the observations of Dr. Herschel, in several volumes of the Philof. Transf. seem to confirm the division into two concentric parts only. The dimensions of these two Rings, and the space between them, he states in the following proportion to each other.

	Miles.
Inner diam. of smaller Ring	146345
Outside diam. of ditto	184393
Inner diam. of larger Ring	190248
Outside diam. of ditto	204883
Breadth of the inner Ring	20000
Breadth of the outer Ring	7200
Breadth of the vacant space	2839
Ring revolves in its own plane, in $10^h 32' 1'' .4$	

So that the outside diameter of the larger Ring is almost 26 times the diameter of the earth.

Dr. Herschel adds, Some theories and observations,

of

of other perfons, " lead us to confider the queftion, whether the conftruction of this Ring is of a nature fo as permanently to remain in its prefent ftate? or whether it be liable to continual and frequent changes, in fuch a manner as in the courfe of not many years, to be feen fubdivided into narrow flips, and then again as united into one or two circular planes only. Now, without entering into a difcuffion, the mind feems to revolt, even at firft fight, againft an idea of the chaotic ftate in which fo large a mafs as the Ring of Saturn muft needs be, if phenomena like thefe can be admitted. Nor ought we to indulge a fufpicion of this being a reality, unlefs repeated and well-confirmed obfervations had proved, beyond a doubt, that this Ring was actually in fo fluctuating a condition " But from his own obfervations he concludes, " It does not appear to me that there is a fufficient ground for admitting the Ring of Saturn to be of a very changeable nature, and I guefs that its phenomena will hereafter be fo fully explained, as to reconcile all obfervations. In the mean while, we muft withhold a final judgment of its conftruction, till we can have more obfervations. Its divifion however into two very unequal parts, can admit of no doubt." See Philof. Tranf. vol. 80 pa. 4, 481 &c, and the vol. for 1792, pa. 1 &c. alfo Hift. de l'Acad. des Scienc. de Paris, 1787, pa. 249 &c.

RINGS *of Colours*, in Optics, a phenomenon firft obferved in thin plates of various fubftances, by Boyle, and Hook, but afterwards more fully explained by Newton.

Mr. Boyle having exhibited a variety of colours in colourlefs liquors, by fhaking them till they rofe in bubbles, as well as in bubbles of foap and water, and alfo in turpentine, procured glafs blown fo thin as to exhibit fimilar colours; and he obferves, that a feather of a proper fhape and fize, and alfo a black ribband, held at a proper diftance between his eye and the fun, fhewed a variety of little rainbows, as he calls them, with very vivid colours. Boyle's Works by Shaw, vol. 2, p. 70. Dr. Hook, about nine years after the publication of Mr. Boyle's Treatife on Colours, exhibited the coloured bubbles of foap and water, and obferved, that though at firft it appeared white and clear, yet as the film of water became thinner, there appeared upon it all the colours of the rainbow. He alfo defcribed the beautiful colours that appear in thin plates of Mufcovy glafs; which appeared, through the microfcope, to be ranged in Rings furrounding the white fpecks or flaws in them, and with the fame order of colours as thofe of the rainbow, and which were often repeated ten times. He alfo took two thin pieces of glafs, ground plane and polifhed, and putting them one upon another, preffed them till there began to appear a red coloured fpot in the middle; and preffing them clofer, he obferved feveral Rings of colours encompaffing the firft place, till, at laft, all the colours difappeared out of the middle of the circles, and the central fpot appeared white. The firft colour that appeared was red, then yellow, then green, then blue, then purple; then again red, yellow, green, blue, and purple; and again in the fame order; fo that he fometimes counted nine or ten of thefe circles, the red immediately next to the purple; and the laft colour that

appeared before the white was blue; fo that it began with red, and ended with purple. Thefe Rings, he fays, would change their places, by changing the pofition of the eye, fo that, the glaffes remaining the fame, that part which was red in one pofition of the eye, was blue in a fecond, green in the third, &c. Birch's Hift. of the Royal Society, vol. 3, pa. 54.

Newton, having demonftrated that every different colour confifts of rays which have a different and fpecific degree of refrangibility, and that natural bodies appear of this or that colour, according to their difpofition to reflect this or that fpecies of rays (fee COLOUR), purfued the hint fuggefted by the experiments of Dr. Hook, already recited, and cafually noticed by himfelf, with regard to thin tranfparent fubftances. Upon compreffing two prifms hard together, in order to make their fides touch one another, he obferved, that in the place of contact they were perfectly tranfparent, which appeared like a dark fpot, and when it was looked through, it feemed like a hole in that air, which was formed into a thin plate, by being impreffed between the glaffes. When this plate of air, by turning the prifms about their common axis, became fo little inclined to the incident rays, that fome of them began to be tranfmitted, there arofe in it many flender arcs of colours, which increafed, as the motion of the prifms was continued, and bended more and more about the tranfparent fpot, till they were completed into circles, or Rings, furrounding it; and afterwards they became continually more and more contracted.

By another experiment, with two object glaffes, he was enabled to obferve diftinctly the order and quality of the colours from the central fpot, to a very confiderable diftance. Next to the pellucid central fpot, made by the contact of the glaffes, fucceeded blue, white, yellow, and red. The next circuit immediately furrounding thefe, confifted of violet, blue, green, yellow, and red. The third circle of colours was purple, blue, green, yellow, and red. The fourth circle confifted of green and red. All the fucceeding colours became more and more imperfect and dilute, till, after three or four revolutions, they ended in perfect whitenefs.

When thefe Rings were examined in a darkened room, by the coloured light of a prifm caft on a fheet of white paper, they became more diftinct, and vifible to a far greater number than in the open air. He fometimes faw more than twenty of them, whereas in the open air he could not difcern above eight or nine.

From other curious obfervations on thefe Rings, made by different kinds of light thrown upon them, he inferred, that the thickneffes of the air between the glaffes, where the Rings are fucceffively made, by the limits of the feven colours, red, orange, yellow, green, blue, indigo, and violet, in order, are one to another as the cube roots of the fquares of the eight lengths of a chord, which found the notes in an octave, fol, la, fa, fol, la, mi, fa, fol; that is, as the cube roots of the fquares of the numbers 1, $\frac{8}{9}$, $\frac{5}{6}$, $\frac{3}{4}$, $\frac{2}{3}$, $\frac{3}{5}$, $\frac{9}{16}$, $\frac{1}{2}$. Thefe Rings appeared of that prifmatic colour, with which they were illuminated, and by projecting the prifmatic colours immediately upon the glaffes, he found that the light, which fell on the dark fpaces between the coloured Rings, was tranfmitted through the glaffes without any change of colour. From this circumftance he thought that the origin of thefe Rings is manifeft; becaufe the air between the glaffes is difpofed according to its various thicknefs, in fome places to reflect, and in others to tranfmit the light of any particular colour, and in the fame place to reflect that of one colour, where it tranfmits that of another.

In examining the phenomena of colours made by a denfer medium furrounded by a rarer, fuch as thofe which appear in plates of Mufcovy glafs, bubbles of foap and water, &c, the colours were found to be much more vivid than the others, which were made with a rarer medium furrounded by a denfer.

From the preceding phenomena it is an obvious deduction, that the tranfparent parts of bodies, according to their feveral feries, reflect rays of one colour and tranfmit thofe of another; on the fame account that thin plates, or bubbles, reflect or tranfmit thofe rays, and this Newton fuppofed to be the reafon of all their colours. Hence alfo he has inferred, that the fize of thofe component parts of natural bodies that affect the light, may be conjectured by their colours. See Colour and Reflection.

Newton, purfuing his difcoveries concerning the colours of thin fubftances, found that the fame were alfo produced by plates of a confiderable thicknefs, divifible into leffer thickneffes. The Rings formed in both cafes have the fame origin, with this difference, that thofe of the thin plates are made by the alternate reflexions and tranfmiffions of the rays at the fecond furface of the plate, after one paffage through it; but that, in the cafe of a glafs fpeculum, concave on one fide, and convex on the other, and quickfilvered over on the convex fide, the rays go through the plate and return before they are alternately reflected and tranfmitted. Newton's Optics, p. 169, &c. or Newton's Opera, Horfley's edit. vol. 4, p. 121, &c. p. 184, &c.

The abbé Mazeas, in his experiments on the Rings of colours that appear in thin plates, has difcovered feveral important circumftances attending them, which were overlooked by the fagacious Newton, and which tend to invalidate his theory for explaining them. In rubbing the flat fide of an object-glafs againft another piece of flat and fmooth glafs, he found that they adhered very firmly together after this friction, and that the fame colours were exhibited between thefe plane glaffes, which Newton had obferved between the convex object glafs of a telefcope, and another that was plane; and that the colours were in proportion to their adhefion. When the furfaces of pieces of glafs, that are tranfparent and well polifhed, are equally preffed, a refiftance will be perceived; and wherever this is felt, two or three very fine curve lines will be difcovered, fome of a pale red, and others of a faint green. If the friction be continued, the red and green lines increafe in number at the place of contact; the colours being fometimes mixed without any order, and fometimes difpofed in a regular manner; in which cafe the coloured lines are generally concentric circles, or ovals, more or lefs elongated, as the furfaces are more or lefs united.

When the colours are formed, the glaffes adhere with considerable force; but if the glaffes be feparated fuddenly, the colours will appear immediately upon their being put together, without the leaft friction. Beginning with the flighteft touch, and increafing the preffure by infenfible degrees, there firft appears an oval plate of a faint red, and in the centre of it a fpot of light green, which enlarges by the preffure, and becomes a green oval, with a red fpot in the centre; and this enlarging in its turn, difcovers a green fpot in its centre. Thus the red and green fucceed one another in turns, affuming different fhades, and having other colours mixed with them. The greateft difference between thefe colours exhibited between plane furfaces, and thofe by curve ones, is, that, in the former cafe, preffure alone will not produce them, except in the cafe above mentioned.

In rubbing together two prifms, with very fmall refracting angles, which were joined fo as to form a parallelopiped, the colours appeared with a furprifing luftre at the places of contact, and differently coloured ovals appeared.

In the centre there was a black fpot, bordered by a deep purple; next to this appeared violet, blue, orange, red tinged with purple, light green, and faint purple.

The other Rings appeared to the naked eye to confift of nothing but faint reds and greens. When thefe coloured glaffes were fufpended over the flame of a candle, the colours difappeared fuddenly, though they ftill adhered; but being fuffered to cool, the colours returned to their former places, in the fame order as before. At firft the abbe Mazeas had no doubt but that thefe colours were owing to a thin plate of air between the glaffes, to which Newton has afcribed them; but the remarkable difference in the circumftances attending thofe produced by the flat plates and thofe produced by the object glaffes of Newton, convinced him that the air was not the caufe of this appearance. The colours of the flat plates vanifhed at the approach of flame, but thofe of the object glaffes did not. Nor was this difference owing to the plane glaffes being lefs compreffed than the convex ones; for though the former were compreffed ever fo much by a pair of forceps, it did not in the leaft hinder the effect of the flame. Afterwards he put both the plane glaffes and the convex ones into the receiver of an air-pump, fufpending the former by a thread, and keeping the latter compreffed by two ftrings; but he obferved no change in the colours of either of them, in the moft perfect vacuum that he could make. Sufpecting ftill that the air adhered to the furface of the glaffes, fo as not to be feparated from them by the force of the pump, he had recourfe to other experiments, which rendered it ftill more improbable that the air fhould be the caufe of thefe colours. Having laid the coloured plates, after warming them gradually, on burning coals; and thus, when they were nearly red, rubbing them together, he obferved the fame coloured circles and ovals as before. When he ceafed to prefs upon them, the colours feemed to vanifh; but they returned, as he renewed the friction. In order to determine whether the colours were owing to the thicknefs of fome matter interpofed between the glaffes, he rubbed them together

2

ther with fuet and other foft fubftances between them ; yet his endeavour to produce the colours had no effect. However by continuing the friction with fome degree of violence, he obferved, that a candle appeared through them encompaffed with two or three concentric greens, and with a lively red inclining to yellow, and a green like that of an emerald, and at length the Rings affumed the colours of blue, yellow, and violet. The abbe was confirmed in his opinion that there muft be fome error in Newton's hypothefis, by confidering that, according to his meafures, the colours of the plates varied with the difference of a millionth part of an inch; whereas he was fatisfied that there muft have been much greater differences in the diftance between his glaffes, when the colours remained unchanged. From other experiments he concluded, that the plate of water introduced between the glaffes was not the caufe of their colours, as Newton apprehended, and that the coloured Rings could not be owing to the compreffion of the glaffes. After all, he adds, that the theory of light, thus reflected from thin plates, is too delicate a fubject to be completely afcertained by a fmall number of obfervations. Berlin Mem. for 1752, or Memoires Prefentes, vol. 2, pa. 28—43. M. du Tour repeated the experiments of the abbe Mazeas, and added fome obfervations of his own. See Mem. Pref. vol. 4, pa. 283.

Muffchenbroeck is alfo of opinion, that the colours of thin plates do not depend upon the air; but as to the caufe of them, he acknowledges that he could not fatisfy himfelf about it. Introd. ad Phil. Nat. vol. 2, p. 738.

See on this fubject Prieftley's Hift. of Light, &c. per. 6, fect. 5, pa. 498, &c.

For an account of the Rings of colours produced by electrical explofions, fee *Colours* of *natural bodies*, CIRCULAR *fpots*, and FAIRY *circles*.

RISING, in Aftronomy, the appearance of the fun, or a ftar, or other luminary, above the horizon, which before was hid beneath it.

By reafon of the refraction of the atmofphere, the heavenly bodies always appear to rife before their time; that is, they are feen above the horizon, while they are really below it, by about $33\frac{1}{3}$ of a degree.

There are three poetical kinds of Rifing of the ftars. See ACRONICAL, COSMICAL, and HELIACAL.

RIVER, in Geography, a ftream or current of frefh water, flowing in a bed or channel, from a fource or fpring, into the fea.

When the ftream is not large enough to bear boats, or fmall veffels, loaden, it is properly called by the diminutive, *rivulet* or *brook*; but when it is confiderable enough to carry larger veffels, it is called by the general name River.

Rivulets have their rife fometimes from great rains, or great quantities of thawed fnow, efpecially in mountainous places; but they more ufually arife from fprings.

Rivers themfelves all arife either from the confluence of feveral rivulets, or from lakes.

RIVER, in Phyfics, denotes a ftream of water running by its own gravity, from the more elevated parts of the earth towards the lower parts, in a natural bed or channel open above.

When the channel is artificial, or cut by art, it is called a canal; of which there are two kinds, viz, that whofe channel is every where open, without fluices, called an artificial River, and that whofe water is kept up and let off by means of fluices, which is properly a *canal*.

Modern philofophers endeavour to reduce the motion and flux of Rivers to precife laws; and with this view they have applied geometry and mechanics to this fubject; fo that the doctrine of Rivers is become a part of the new philofophy.

The authors who have moft diftinguifhed themfelves in this branch, are the Italians, the French, and the Dutch, but efpecially the firft, and among them more efpecially Gulielmini, and Ximenes.

Rivers, fays Gulielmini, ufually have their fources in mountains or elevated grounds; in the defcent from which it is moftly that they acquire the velocity, or acceleration, which maintains their future current. In proportion as they advance farther, this velocity diminifhes, on account of the continual friction of the water againft the bottom and fides of the channel; as well as from the various obftacles they meet with in their progrefs, and from their arriving at length in plains where the defcent is lefs, and confequently their inclination to the horizon greater. Thus the Reno, a River in Italy, which he fays gave occafion, in fome meafure, to his fpeculations, is found to have near its mouth a declivity of fcarce 52 feconds.

When the acquired velocity is quite fpent, through the many obftacles, fo that the current becomes horizontal, there will then nothing remain to propagate the motion, and continue the ftream, but the depth, or the perpendicular preffure of the water, which is always proportional to the depth. And, happily for us, this refource increafes, as the occafion for it increafes; for in proportion as the water lofes of the velocity acquired by the defcent, it rifes and increafes in its depth.

It appears from the laws of motion pertaining to bodies moved on inclined planes, that when water flows freely upon an inclined bed, it acquires a velocity, which is always as the fquare root of the quantity of defcent of the bed. But in an horizontal bed, opened by fluices or otherwife, at one or both ends, the water flows out by its gravity alone.

The upper parts of the water of a River, and thofe at a diftance from the banks, may continue to flow, from the fimple caufe or principle of declivity, how fmall foever it be; for not being detained by any obftacle, the minuteft difference of level will have its effect; but the lower parts, which roll along the bottom, will fcarce be fenfible of fo fmall a declivity; and will only have what motion they receive from the preffure of the fuperincumbent waters.

The greateft velocity of a River is about the middle of its depth and breadth, or that point which is the fartheft poffible from the furface of the water, and from the bottom and fides of the bed or channel. Whereas, on the contrary, the leaft velocity of the water is at the bottom and fides of the bed, becaufe there the refiftance arifing from friction is the greateft, which is communicated to the other parts of the fection of the

River inversely as the distances from the bottom and sides.

To find whether the water of a River, almost horizontal, flows by means of the velocity acquired in its descent, or by the pressure of its depth ; set up an obstacle perpendicular to it : then if the water rise and swell immediately against the obstacle, it runs by virtue of its fall ; but if it first stop a little while, in virtue of its pressure.

Rivers, according to this author, almost always make their own beds. If the bottom have originally been a large declivity, the water, hence falling with a great force, will have swept away the most elevated parts of the soil, and carrying them lower down, will gradually render the bottom more nearly horizontal.

The water having made its bed horizontal, becomes so itself, and consequently takes with the less force against the bottom, till at length that force becomes only equal to the resistance of the bottom, which is now arrived at a state of permanency, at least for a considerable time ; and the longer according to the quality of the soil, clay and chalk resisting longer than sand or mud.

On the other hand, the water is continually wearing away the brims of its channel, and this with the more force, as, by the direction of its stream, it impinges more directly against them. By this means it has a continual tendency to render them parallel to its own course. At the same time that it has thus rectified its edges, it has widened its own bed, and thence becoming less deep, it loses part of its force and pressure : this it continues to do till there is an equilibrium between the force of the water and the resistance of its banks, and then they will remain without farther change. And it appears by experience that these equilibriums are all real, as we find that Rivers only dig and widen to a certain pitch.

The very reverse of all these things does also on some occasions happen. Rivers, whose waters are thick and muddy, raise their bed, by depositing part of the heterogeneous matters contained in them : they also contract their banks, by a continual opposition of the same matter, in brushing over them. This matter, being thrown aside far from the stream of water, might even serve, by reason of the dullness of the motion, to form new banks.

If these various causes of resistance to the motion of flowing waters did not exist, viz, the attraction and continual friction of the bottom and sides, the inequalities in both, the windings and angles that occur in their course, and the diminution of their declivity the farther they recede from their springs, the velocity of their currents would be accelerated to 10, 15, or even 20 times more than it is at present in the same Rivers, by which they would become absolutely unnavigable.

The union of two Rivers into one, makes the whole flow the swifter, because, instead of the friction of four shores, they have only two to overcome, and one bottom instead of two ; also the stream, being farther distant from the banks, goes on with the less interruption, besides, that a greater quantity of water, moving with a greater velocity, digs deeper in the bed, and of

course retrenches of its former width. Hence also it is, that Rivers, by being united, take up less space on the surface of the earth, and are more advantageous to low grounds, which drain their superfluous moisture into them, and have also less occasion for dykes to prevent their overflowing.

A very good and simple method of measuring the velocity of the current of a River, or canal, is the following. Take a cylindrical piece of dry, light wood, and of a length something less than the depth of the water in the River ; about one end of it let there be suspended as many small weights, as may keep the cylinder in a vertical or upright position, with its head just above water. To the centre of this end fix a small straight rod, precisely in the direction of the cylinder's axis ; to the end that, when the instrument is suspended in the water, the deviations of the rod from a perpendicularity to the surface of it, may indicate which end of the cylinder goes foremost, by which may be discovered the different velocities of the water at different depths ; for when the rod inclines forward, according to the direction of the current, it is a proof that the surface of the water has the greatest velocity ; but when it reclines backward, it shews that the swiftest current is at the bottom ; and when it remains perpendicular, it is a sign that the velocities at the top and bottom are equal.

This instrument, being placed in the current of a River or canal, receives all the percussions of the water throughout the whole depth, and will have an equal velocity with that of the whole current from the surface to the bottom at the place where it is put in, and by that means may be found, both with exactness and ease, the mean velocity of that part of the River for any determinate distance and time.

But to obtain the mean velocity of the whole section of the River, the instrument must be put successively both in the middle and towards the sides, because the velocities at those places are often very different from each other. Having by this means found the several velocities, from the spaces run over in certain times, the arithmetical mean proportional of all these trials, which is found by dividing the common sum of them all by the number of the trials, will be the mean velocity of the River or canal. And if this medium velocity be multiplied by the area of the transverse section of the waters at any place, the product will be the quantity running through that place in a second of time.

If it be required to find the velocity of the current only at the surface, or at the middle, or at the bottom, a sphere of wood loaded, or a common bottle corked with a little water in it, of such a weight as will remain suspended in equilibrium with the water at the surface or depth which we want to measure, will be better for the purpose than the cylinder, because it is only affected by the water of that sole part of the current where it remains suspended.

It follows from what has been said in the former part of this article, that the deeper the waters are in their bed in proportion to its breadth, the more their motion is accelerated ; so that their velocity increases in the inverse ratio of the breadth of the bed, and also

of

of the magnitude of the section; whence, in order to augment the velocity of water in a River or canal, without augmenting the declivity of the bed, we must increase the depth of the channel, and diminish its breadth. And these principles are agreeable to observation; as it is well known, that the velocity of flowing waters depends much more on the quantity and depth of the water, and on the compress on of the upper parts on the lower, than on the declivity of the bed; and therefore the declivity of a River must be made much greater in the beginning than toward the end of its course, where it should be almost insensible. If the depth or volume of water in a River or canal be considerable, it will suffice, in the part next the mouth to allow one foot of declivity through 6000, or 8000, or even (according to Dechales, De Fontibus et Fluviis, prop. 49) 10000 feet in horizontal extent; at most it need not be above 1 in 6 or 7 thousand. From hence the quantity of declivity in equal spaces must slowly and gradually increase as far as the current is to be made fit for navigation; but in such a manner, as that at this upper end there may not be above one foot of perpendicular declivity in 4000 feet of horizontal extent.

To conclude this article, M. de Buffon observes, that people accustomed to Rivers can easily foretell when there is going to be a sudden increase of water in the bed from floods produced by sudden falls of rain in the higher countries through which the Rivers pass. This they perceive by a particular motion in the water, which they express by saying, that the River's bottom moves, that is, the water at the bottom of the channel runs off faster than usual; and this increase of motion at the bottom of a River always announces a sudden increase of water coming down the stream. Nor, says he, is their opinion ill grounded; because the motion and weight of the waters coming down, though not yet arrived, must act upon the waters in the lower parts of the River, and communicate by impulsion part of their motion to them, within a certain distance.

On the subject of this article, see an elaborate treatise on Rivers and canals, in the Philos. Transf. vol. 69, pa. 555.&c, by Mr. Mann, who has availed himself of the observations of Gulielmini, and most other writers.

RIXDOLLAR, a silver coin, struck in several states and free cities in Germany, as also in Flanders, Poland, Denmark, Sweden, &c.

There is but little difference between the Rixdollar and the dollar, another silver coin struck in Germany, each being nearly equal to the French crown of three livres, or the Spanish piece of eight, or 4s. 6d. sterling.

ROBERVAL (GILES-PERSONNE), an eminent French mathematician, was born in 1602, at Roberval, a parish in the diocese of Beauvais. He was first professor of mathematics at the College of Maitre-Gervais, and afterwards at the College-royal. A similarity of taste connected him with Gassendi and Morin; the latter of whom he succeeded in the mathematical chair at the Royal College, without quitting however that of Ramus.

Roberval made experiments on the Torricellian vacuum: he invented two new kinds of balance, one of which was proper for weighing air; and made many other curious experiments. He was one of the first members of the ancient Academy of Sciences of 1666; but died in 1675, at 73 years of age. His principal works are,

I. A treatise on Mechanics.
II. A work entitled Aristarchus Samos.

He had several memoirs inserted in the volumes of the Academy of Sciences of 1666, viz,

1. Experiments concerning the Pressure of the Air.

2. Observations on the Composition of Motion, and on the Tangents of Curve Lines.

3. The Recognition of Equations.

4. The Geometrical Resolution of Plane and Cubic Equations.

5. Treatise on Indivisibles.

6. On the Trochoid, or Cycloid.

7. A Letter to Father Mersenne.

8. Two Letters from Torricelli.

9. A new kind of Balance.

ROBERVALLIAN Lines, a name given to certain lines, used for the transformation of figures: thus called from their inventor Roberval.

These lines bound spaces that are infinitely extended in length, which are nevertheless equal to other spaces that are terminated on all sides.

The abbot Gallois, in the Memoirs of the Royal Academy, anno 1693, observes, that the method of transforming figures, explained at the latter end of Roberval's treatise of Indivisibles, was the same with that afterwards published by James Gregory, in his Geometria Universalis, and also by Barrow in his Lectiones Geometricæ; and that, by a letter of Torricelli, it appears, that Roberval was the inventor of this manner of transforming figures, by means of certain lines, which Torricelli therefore called Robervallian Lines.

He adds, that it is highly probable, that J. Gregory first learned the method in the journey he made to Padua in 1668, the method itself having been known in Italy from the year 1646, though the book was not published till the year 1692.

This account David Gregory has endeavoured to refute, in vindication of his uncle James. His answer is inserted in the Philos. Transf. of 1694, and the abbot rejoined in the French Memoirs of the Academy of 1703.

ROBINS (BENJAMIN), an English mathematician and philosopher of great genius and eminence, was born at Bath in Somersetshire, 1707. His parents were of low condition, and Quakers; and consequently neither able from their circumstances, nor willing from their religious profession, to have him much instructed in that kind of learning which they are taught to despise as human. Nevertheless, he made an early and surprising progress in various branches of science and literature, particularly in the mathematics; and his friends being desirous that he might continue his pursuits, and that his merit might not be buried in obscurity, wished that he could be properly recommended

to

to teach that science in London. Accordingly, a specimen of his abilities in this way was sent up thither, and shewn to Dr. Pemberton, the author of the " View of Sir Isaac Newton's Philosophy ;" who, thence conceiving a good opinion of the writer, for a farther trial of his skill sent him some problems, which Robins resolved very much to his satisfaction. He then came to London, where he confirmed the opinion which had been preconceived of his abilities and knowledge.

But though Robins was possessed of much more skill than is usually required in a common teacher; yet being very young, it was thought proper that he should employ some time in perusing the best writers upon the sublimer parts of the mathematics, before he should undertake publicly the instruction of others. In this interval, besides improving himself in the modern languages, he had opportunities of reading in particular the works of Archimedes, Apollonius, Fermat, Huygens, De Witt, Slusius, Gregory, Barrow, Newton, Taylor, and Cotes. These authors he readily understood without any assistance, of which he gave frequent proofs to his friends: one was, a demonstration of the last proposition of Newton's treatise on Quadratures, which was thought not undeserving a place in the Philosophical Transactions for 1727.

Not long after, an opportunity offered him of exhibiting to the public a specimen also of his knowledge in Natural Philosophy. The Royal Academy of Sciences at Paris had proposed, among their prize questions in 1724 and 1726, to demonstrate the laws of motion in bodies impinging on one another. John Bernoulli here condescended to be a candidate; and as his dissertation lost the reward, he appealed to the learned world by printing it in 1727. In this piece he endeavoured to establish Leibnitz's opinion of the force of bodies in motion from the effects of their striking against springy materials; as Poleni had before attempted to evince the same thing from experiments of bodies falling on soft and yielding substances. But as the insufficiency of Poleni's arguments had been demonstrated in the Philosophical Transactions, for 1722 ; so Robins published in the Present State of the Republic of Letters, for May 1728, a Confutation of Bernoulli's performance, which was allowed to be unanswerable.

Robins now began to take scholars; and about this time he quitted the garb and profession of a Quaker; for, having neither enthusiasm nor superstition in his nature, as became a mathematician, he soon shook off the prejudices of such early habits. But though he professed to teach the mathematics only, he would frequently assist particular friends in other matters; for he was a man of universal knowledge: and the confinement of this way of life not suiting his disposition, which was active, he gradually declined it, and went into other courses, that required more exercise. Hence he tried many laborious experiments in gunnery; believing that the resistance of the air had a much greater effect on swift projectiles, than was generally supposed. And hence he was led to consider those mechanic arts that depend upon mathematical principles, in which he might employ his invention: as, the constructing of mills, the building of bridges, draining of fens, rendering of rivers navigable, and making of harbours. Among other arts of this kind, fortification very much engaged his attention ; in which he met with opportunities of perfecting himself, by a view of the principal strong places of Flanders, in some journeys he made abroad with persons of distinction.

On his return home from one of these excursions, he found the learned here amused with Dr. Berkeley's treatise, printed in 1734, entitled, " The Analyst ;" in which an examination was made into the grounds of the doctrine of Fluxions, and occasion thence taken to explode that method. Robins was therefore advised to clear up this affair, by giving a full and distinct account of Newton's doctrines, in such a manner, as to obviate all the objections, without naming them, which had been advanced by Berkeley ; and accordingly he published, in 1735, *A Discourse concerning the Nature and Certainty of Sir Isaac Newton's Method of Fluxions, and of Prime and Ultimate Ratios*. This is a very clear, neat, and elegant performance : and yet some persons, even among those who had written against The Analyst, taking exception at Robins's manner of defending Newton's doctrine, he afterwards wrote two or three additional discourses.

In 1738, he defended Newton against an objection, contained in a note at the end of a Latin piece, called " Matho, sive Cosmotheoria puerilis," written by Baxter, author of the " Inquiry into the Nature of the Human Soul :" and the year after he printed *Remarks on Euler's Treatise of Motion*, on *Smith's System of Optics*. and on *Jurin's Discourse of Distinct and Indistinct Vision*, annexed to Dr. Smith's work.

In the mean time Robins's performances were not confined to mathematical subjects: for, in 1739, there came out three pamphlets upon political affairs, which did him great honour. The first was entitled, *Observations on the present Convention with Spain*: the second, *A Narrative of what passed in the Common Hall of the Citizens of London, assembled for the Election of a Lord Mayor*: the third, *An Address to the Electors and other free Subjects of Great Britain, occasioned by the late Succession; in which is contained a Particular Account of all our Negociations with Spain, and their Treatment of us for above ten years past*. These were all published without our author's name ; and the first and last were so universally esteemed, that they were generally reputed to have been the production of the great man himself, who was at the head of the opposition to Sir Robert Walpole. They proved of such consequence to Mr. Robins, as to occasion his being employed in a very honourable post ; for, the patriots at length gaining ground against Sir Robert, and a committee of the House of Commons being appointed to examine into his past conduct, Robins was chosen their secretary. But after the committee had presented two reports of their proceedings, a sudden stop was put to their farther progress, by a compromise between the contending parties.

In 1742, being again at leisure, he published a small treatise, entitled, *New Principles of Gunnery;* containing the result of many experiments he had made, by which are discovered the force of gunpowder, and the difference in the resisting power of the air to swift and

slow

flow motions. To this treatife was prefixed a full and learned account of the progrefs which modern fortification had made from its firft rife; as alfo of the invention of gunpowder, and of what had already been performed in the theory of gunnery. It feems that the occafion of this publication, was the difappointment of a fituation at the Royal Military Academy at Woolwich. On the new modelling and eftablifhing of that Academy, in 1741, our author and the late Mr. Muller were competitors for the place of profeffor of fortification and gunnery. Mr. Muller held then fome poft in the Tower of London, under the Board of Ordnance, fo that, notwithftanding the great knowledge and abilities of our author, the intereft which Mr. Muller had with the Board of Ordnance carried the election in his favour. Upon this difappointment Mr. Robins, indignant at the affront, determined to fhew them, and the world, by his military publications, what fort of a man he was that they had rejected.

Upon a difcourfe containing certain experiments being publifhed in the Philofophical Tranfactions, with a view to invalidate fome of Robins's opinions, he thought proper, in an account he gave of his book in the fame Tranfactions, to take notice of thofe experiments: and in confequence of this, feveral differtations of his on the refiftance of the air were read, and the experiments exhibited before the Royal Society, in 1746 and 1747; for which he was prefented with the annual gold medal by that Society.

In 1748 came out Anfon's Voyage round the World; which, though it bears Walter's name in the title-page, was in reality written by Robins. Of this voyage the public had for fome time been in expectation of feeing an account, compofed under that commander's own infpection: for which purpofe the reverend Richard Walter was employed, as having been chaplain on board the Centurion the greateft part of the expedition. Walter had accordingly almoft finifhed his tafk, having brought it down to his own departure from Macao for England; when he propofed to print his work by fubfcription. It was thought proper however that an able judge fhould firft review and correct it, and Robins was appointed; when, upon examination, it was refolved, that the whole fhould be written entirely by Robins, and that what Walter had done, being moftly taken verbatim from the journals, fhould ferve as materials only. Hence it was that the whole of the introduction, and many differtations in the body of the work, were compofed by Robins, without receiving the leaft hint from Walter's manufcript; and what he had tranfcribed from it regarded chiefly the wind and weather, the currents, courfes, bearings, diftances, offings, foundings, moorings, the qualities of the ground they anchored on, and fuch particulars as ufually fill up a feaman's account. No production of this kind ever met with a more favourable reception, four large impreffions having been fold off within a year: it was alfo tranflated into moft of the European languages; and it ftill fupports its reputation, having been repeatedly reprinted in various fizes. The fifth edition at London in 1749 was revifed and corrected by Robins himfelf; and the 9th edition was printed there in 1761.

Thus becoming famous for his elegant talents in

writing, he was requefted to compofe an apology for the unfortunate affair at Preftonpans in Scotland. This was added as a preface to the Report of the Proceedings and Opinion of the Board of General Officers on their Examination into the Conduct of Lieutenant General Sir John Cope, &c, printed at London in 1749; and this preface was efteemed a mafter-piece in its kind.

Robins had afterwards, by the favour of lord Anfon, opportunities of making farther experiments in Gunnery; which have been publifhed fince his death, in the edition of his works by his friend Dr. Wilfon. He alfo not a little contributed to the improvements made in the Royal Obfervatory at Greenwich, by procuring for it, through the intereft of the fame noble perfon, a fecond mural quadrant, and other inftruments; by which it became perhaps the completeft of any obfervatory in the world.

His reputation being now arrived at its full height, he was offered the choice of two very confiderable employments. The firft was to go to Paris, as one of the commiffaries for adjufting the limits in Acadia; the other, to be engineer general to the Eaft India Company, whofe forts, being in a moft ruinous condition, wanted an able perfon to put them into a proper ftate of defence. He accepted the latter, as it was fuitable to his genius, and as the Company's terms were both advantageous and honourable. He defigned, if he had remained in England, to have written a fecond part of the Voyage round the World; as appears by a letter from lord Anfon to him, dated Bath, Oct. 22, 1749, as follows.

"Dear Sir, when I laft faw you in town, I forgot to afk you, whether you intended to publifh the fecond volume of my Voyage before you leave us; which I confefs I am very forry for. If you fhould have laid afide all thoughts of favouring the world with more of your works, it will be much difappointed, and no one in it more than your very obliged humble fervant,

"ANSON."

Robins was alfo preparing an enlarged edition of his New Principles of Gunnery: but, having provided himfelf with a complete fet of aftronomical and other inftruments, for making obfervations and experiments in the Indies, he departed hence at Chriftmas in 1749; and after a voyage, in which the fhip was near being caft away, he arrived at India in July following. There he immediately fet about his proper bufinefs with the greateft diligence, and formed complete plans for Fort St. David and Madras: but he did not live to put them into execution. For the great difference of the climate from that of England being beyond his conftitution to fupport, he was attacked by a fever in September the fame year; and though he recovered out of this, yet about eight months after he fell into a languifhing condition, in which he continued till his death, which happened the 29th of July 1751, at only 44 years of age.

By his laft will, Mr. Robins left the publifhing of his Mathematical Works to his honoured and intimate friend Martin Folkes, Efq. prefident of the Royal Society, and to Dr. James Wilfon; but the former of thefe gentlemen

gentlemen being incapacitated by a paralytic diforder, for fome time before his death, they were afterwards publifhed by the latter, in 2 volumes 8vo, 1761. To this collection, which contains his mathematical and philofophical pieces only, Dr. Wilfon has prefixed an account of Mr Robins, from which this memoir is chiefly extracted. He added alfo a large appendix at the end of the fecond volume, containing a great many curious and critical matters in various interefting parts of the mathematics. As to Mr. Robins's own papers in thefe two volumes, they are as follow: viz, in vol. I,

1. New Principles of Gunnery. Firft printed in 1742.

2. An Account of that book. Read before the Royal Society, April the 14th and 21ft 1743.

3. Of the Refiftance of the Air. Read the 12th of June 1746.

4. Of the Refiftance of the Air; together with the Method of computing the Motions of Bodies projected in that Medium. Read June 19, 1746.

5. Account of Experiments relating to the Refiftance of the Air. Read the 4th of June 1747.

6. Of the Force of Gunpowder, with the Computation of the Velocities thereby communicated to military projectiles. Read the 25th of June 1747.

7. A Comparifon of the Experimental Ranges of Cannon and Mortars, with the Theory contained in the preceding papers. Read the 27th of June 1751.

8. Practical Maxims relating to the Effects and Management of Artillery, and the Flight of Shells and Shot.

9. A Propofal for increafing the Strength of the Britifh Navy. Read the 2d of April 1747.

10. A Letter to Martin Folkes, Efq. Prefident of the Royal Society. Read the 7th of January 1748.

11. A Letter to Lord Anfon. Read the 26th of October 1749.

12. On Pointing, or Directing of Cannon to ftrike diftant objects.

13. Obfervations on the Height to which Rockets afcend. Read the 4th of May 1749.

14. An Account of fome Experiments on Rockets, by Mr. Ellicott.

15. Of the Nature and Advantage of Rifled Barrel Pieces, by Mr. Robins. Read the 2d of July 1747.

In volume II are,

16. A Difcourfe concerning the Nature and Certainty of Sir Ifaac Newton's Methods of Fluxions, and of Prime and Ultimate Ratios.

17. An Account of the preceding Difcourfe.

18. A Review of fome of the principal Objections, that have been made to the Doctrine of Fluxions and Ultimate Proportions, with fome Remarks on the different Methods, that have been taken to obviate them.

19. A Differtation fhewing, that the Account of the Doctrines of Fluxions and of Prime and Ultimate Ratios, delivered in Mr. Robins's Difcourfe, is agreeable to the real Meaning of their great Inventor.

20. A Demonftration of the Eleventh Propofition of Sir Ifaac Newton's Treatife of Quadratures.

21. Remarks on Bernoulli's Difcourfe upon the Laws of the Communication of Motion.

22. An Examination of a Note concerning the Sun's Parallax, publifhed at the end of Baxter's Matho.

23. Remarks on Euler's Treatife of Motion; Dr. Smith's Syftem of Optics; and Dr. Jurin's Effay on Diftinct and Indiftinct Vifion.

24. Appendix by the Publifher.

It is but juftice to fay, that Mr. Robins was one of the moft accurate and elegant mathematical writers that our language can boaft of; and that he made more real improvements in Artillery, the flight and the refiftance of projectiles, than all the preceding writers on that fubject. His New Principles of Gunnery were tranflated into feveral other languages, and commented upon by feveral eminent writers. The celebrated Euler tranflated the work into the German language, accompanied with a large and critical commentary; and this work of Euler's was again tranflated into Englifh in 1714, by Mr. Hugh Brown, with Notes, in one volume 4to.

ROBINS, or ROBYNS (JOHN), an Englifh mathematician, was born in Staffordfhire about the clofe of the 15th century, as he was entered a ftudent at Oxford in 1516, where he was educated for the church. But the bent of his genius lay to the fciences, and he foon made fuch a progrefs, fays Wood, in "the pleafant ftudies of mathematics and aftrology, that he became the ableft perfon in his time for thofe ftudies, not excepted his friend Record, whofe learning was more general. At length, taking the degree of bachelor of divinity in 1531, he was the year following made by king Henry the VIIIth (to whom he was chaplain) one of the canons of his college in Oxon, and in December 1543 canon of Windfor, and in fine chaplain to Queen Mary, who had him in great veneration for his learning. Among feveral things that he hath written relating to aftrology (or aftronomy) I find thefe following:

" De Culminatione Fixarum Stellarum, &c.
De Ortu & Occafu Stellarum Fixarum, &c.
Annotationes Aftrologicæ, &c. lib. 3.
Annotationes Edwardo VI.
Tractatus de Prognofticatione per Eclipfin.

" All which books, that are in MS, were fome time in the choice library of Mr. Thomas Allen of Gloceffter Hall. After his death, coming into the hands of Sir Kenelm Digby, they were by him given to the Bodleian library, where they yet remain. It is alfo faid, that he the faid Robyns hath written a book intitled, De Portentofis Cometis, but fuch a thing I have not yet feen, nor do I know any thing elfe of the author, only that paying his laft debt to nature the 25th of Auguft 1558, he was buried in the chappel of St. George at Windfore"

ROCKET, in Pyrotechny, an artificial firework, ufually confifting of a cylindrical cafe of paper, filled with a compofition of certain combuftible ingredients; which being tied to a rod, mounts into the air to a confiderable height, and there burfts. Thefe are called Sky Rockets. Befide which, there are others called Water Rockets, from their acting in water.

The

The composition with which Rockets are filled, confists of the three following ingredients, viz, faltpetre, charcoal, and sulphur, all well ground; and in the smaller sizes, gunpowder duft is also added. But the proportions of all the ingredients vary with the weight of the Rocket, as in the following Table.

Compositions for Rockets of Various Sizes.

The General Composition for Rockets is,

Saltpetre	4 lb.
Sulphur	1 lb.
Charcoal	1 lb.

But for large Rockets,

Saltpetre	4 lb.
Sulphur	1 lb.
Mealpowder	1 lb.

For Rockets of a Middle Size,

Saltpetre	3 lb.
Sulphur	2 lb.
Mealpowder	1 lb.
Charcoal	1 lb.

When Rockets are intended to mount upwards, they have a long flender rod fixed to the lower end, to direct their motion.

Theory of the Flight of Rockets.—Mariotte takes the rife of Rockets to be owing to the impulfe or refiftance of the air against the flame. Defaguliers accounts for it thus.

Conceive the Rocket to have no vent at the choke, and to be fet on fire in the conical bore; the confequence would be, either that the Rocket would burft in the weakeft place, or that, if all parts were equally ftrong, and able to fuftain the impulfe of the flame, the Rocket would burn out immoveable. Now, as the force of the flame is equable, fuppofe its action downwards, or that upwards, fufficient to lift 40 pounds; as thefe forces are equal, but their directions contrary, they will deftroy each other's action.

Imagine then the Rocket opened at the choke; by this means the action of the flame downwards is taken away, and there remains a force equal to 40 pounds acting upwards, to carry up the Rocket, and the ftick or rod it is tied to. Accordingly we find that if the compofition of the Rocket be very weak, fo as not to give an impulfe greater than the weight of the Rocket and ftick, it does not rife at all; or if the compofition be flow, fo that a fmall part of it only kindles at firft, the Rocket will not rife.

The ftick ferves to keep it perpendicular; for if the Rocket fhould begin to tumble, moving round a point in the choke, as being the common centre of gravity of Rocket and ftick, there would be fo much friction against the air, by the ftick between the centre and the point, and the point would beat against the air with fo much velocity, that the reaction of the medium would reftore it to its perpendicularity. When the compofition is burnt out, and the impulfe upwards has ceafed, the common centre of gravity is brought lower towards the middle of the ftick; by which means the velocity of the point of the ftick is decreafed, and that of the

point of the Rocket is increafed; fo that the whole will tumble down, with the Rocket end foremoft.

All the while the Rocket burns, the common centre of gravity is fhifting and getting downwards, and ftill the fafter and the lower as the ftick is lighter; fo that it fometimes begins to tumble before it is quite burnt out: but when the ftick is too heavy, the common centre of gravity will not get fo low, but that the Rocket will rife ftraight, though not fo faft.

From the experiments of Mr. Robins, and other gentlemen, it appears that the Rockets of 2, 3, or 4 inches diameter, rife the higheft; and they found them rife to all heights in the air, from 400 to 1254 yards, which is about three quarters of a mile. See Robins's Tracts, vol. 2, pa. 317, and the Philof. Tranf. vol. 46, pa. 578.

ROD, or *Pole*, is a long meafure, of 16½ feet, or 5¼ yards, or the 4th part of a Gunter's chain, for landmeafuring.

ROEMER (Olaus), a noted Danifh aftronomer and mathematician, was born at Arhufen in Jutland, 1644; and at 18 years of age was fent to the univerfity of Copenhagen. He applied affiduoufly to the ftudy of the mathematics and aftronomy, and became fo expert in thofe fciences, that when Picard was fent by Lewis the XIVth in 1671, to make obfervations in the north, he was greatly furprifed and pleafed with him. He engaged him to return with him to France, and had him prefented to the king, who honoured him with the dauphin as a pupil in mathematics, and fettled a penfion upon him. He was joined with Picard and Caffini, in making aftronomical obfervations; and in 1672 he was admitted a member of the academy of fciences.

During the ten years he refided at Paris, he gained great reputation by his difcoveries; yet it is faid he complained afterwards, that his coadjutors ran away with the honour of many things which belonged to him. Here it was that Roemer, firft of any one, found out the velocity with which light moves, by means of the eclipfes of Jupiter's fatellites. He had obferved for many years that, when Jupiter was at his greateft diftance from the earth, where he could be obferved, the emerfions of his firft fatellite happened conftantly 15 or 16 minutes later than the calculation gave them. Hence he concluded that the light reflected by Jupiter took up this time in running over the excefs of diftance, and confequently that it took up 16 or 18 minutes in running over the diameter of the earth's orbit; and 8 or 9 in coming from the fun to us, provided its velocity was nearly uniform. This difcovery had at firft many oppofers; but it was afterwards confirmed by Dr. Bradley in the moft ingenious and beautiful manner.

In 1681 Roemer was recalled back to his own country by Chriftian the Vth, king of Denmark, who made him profeffor of aftronomy at Copenhagen. The king employed him alfo in reforming the coin and the architecture, in regulating the weights and meafures, and in meafuring and laying out the high roads throughout the kingdom; offices which he difcharged with the greateft credit and fatisfaction. In confequence he was honoured by the king with the appointment of chancellor of the exchequer and other dignities. Finally he became counfellor of ftate and burgomafter of Copen-

hagen,

hagen, under Frederic the IVth, the fucceffor of Chriftian. Roemer was preparing to publifh the refult of
his obfervations, when he died the 19th of September
1710, at 66 years of age: but this lofs was fupplied by
Horrebow, his difciple, then profeffor of aftronomy at
Copenhagen, who publifhed, in 4to, 1753, various
obfervations of Roemer, with his method of obferving,
under the title of *Bafis Aftronomiæ*.—He had alfo
printed various aftronomical obfervations and pieces,
in feveral volumes of the Memoirs of the Royal Academy of Sciences at Paris, of the inftitution of 1666,
particularly vol. 1 and 10 of that collection.

ROHAULT (JAMES), a French philofopher, was
the fon of a rich merchant at Amiens, where he was
born in 1620. He cultivated the languages and belles
lettres in his own country, and then was fent to Paris
to ftudy philofophy. He feems to have been a great
lover of truth, at leaft what he thought fo, and to
have fought it with much impartiality. He read the
ancient and modern philofophers; but Des Cartes was
the author who moft engaged his notice. Accordingly
he became a zealous follower of that great man, and
drew up an abridgment and explanation of his philofophy with great clearnefs and method. In the preface
to his *Phyfics*, for fo his work is called, he makes no
fcruple to fay, that "the abilities and accomplifhments
of this philofopher muft oblige the whole world to confefs, that France is at leaft as capable of producing and
raifing men verfed in all arts and branches of knowledge, as ancient Greece." Clerfelier, well known
for his tranflation of many pieces of Des Cartes, conceived fuch an affection for Rohault, on account of his
attachment to this philofopher, that he gave him his
daughter in marriage againft all the remonftrances of
his family.

Rohault's Phyfics were written in French, but have
been tranflated into Latin by Dr. Samuel Clarke,
with notes, in which the Cartefian errors are corrected
upon the Newtonian fyftem. The fourth and beft edition of *Rohault's Phyfica*, by Clarke, is that of 1718,
in 8vo. He wrote alfo,

> *Elemens de Mathematiques*,
> *Traite de Mechanique*, and
> *Entretiens fur la Philofophie*.

But thefe dialogues are founded and carried on upon
the principles of the Cartefian philofophy, which has
now little other merit, than that of having corrected
the errors of the Ancients. Rohault died in 1675,
and left behind him the character of an amiable, as
well as a learned and philofophic man.

His pofthumous works were collected and printed in
two neat little volumes, firft at Paris, and then at the
Hague in 1690. The contents of them are, 1. The
firft 6 books of Euclid. 2. Trigonometry. 3. Practical Geometry. 4. Fortification. 5. Mechanics.
6. Perfpective. 7. Spherical Trigonometry. 8. Arithmetic.

ROLLE (MICHEL), a French mathematician, was
born at Ambert, a fmall town in Auvergne, the 21ft of
April 1652. His firft ftudies and employments were
under notaries and attorneys; occupations but little
fuited to his genius. He went to Paris in 1675, with
the only refource of fine penmanfhip, and fubfifted by

giving leffons in writing. But as his inclination for
the mathematics had drawn him to that city, he attended the mafters in this fcience, and foon became
one himfelf. Ozanam propofed a queftion in arithmetic to him, to which Rolle gave fo clear and good a
folution, that the minifter Colbert made him a handfome gratuity, which at laft grew into a fixed penfion.
He then abandoned penmanfhip, and gave himfelf up
entirely to algebra and other branches of the mathematics. His conduct in life gained him many friends; in
which his fcientific merit, his peaceable and regular behaviour, with an exact and fcrupulous probity of manners, were his only folicitors.

Rolle was chofen a member of the Ancient Academy
of Sciences in 1685, and named fecond geometrical-
penfionary on its renewal in 1699; which he enjoyed
till his death, which happened the 5th of July 1719,
at 67 years of age.

The works publifhed by Rolle, were,

I. A Treatife of Algebra; in 4to, 1690.

II. A method of refolving Indeterminate Queftions
in Algebra; in 1699. Befides a great many curious
pieces inferted in the Memoirs of the Academy of
Sciences, as follow:

1. A Rule for the Approximation of Irrational
Cubes: an. 1666, vol. 10.

2. A Method of Refolving Equations of all Degrees
which are expreffed in General Terms: an. 1666,
vol. 10.

3. Remarks upon Geometric Lines: 1702 and
1703.

4. On the New Syftem of Infinity: 1703, pa.
312.

5. On the Inverfe Method of Tangents: 1705, pa.
25, 171, 222.

6. Method of finding the Foci of Geometric Lines
of all kinds: 1705, pa. 284.

7. On Curves, both Geometrical and Mechanical,
with their Radii of Curvature: 1707, pa. 370.

8. On the Conftruction of Equations: 1708, and
1709.

9. On the Extermination of the Unknown Quantities in the Geometrical Analyfis: 1709, pa. 419.

10. Rules and Remarks for the Conftruction of
Equations: 1711, pa. 86.

11. On the Application of Diophantine Rules to
Geometry: 1712.

12. On a Paradox in Geometric Effections: 1713,
pa. 243.

13. On Geometric Conftructions: 1713, pa. 261,
and 1714, pa. 5.

ROLLING, or *Rotation*, in Mechanics, a kind of
circular motion, by which the moveable body turns
round its own axis, or centre, and continually applies
new parts of its furface to the body it moves upon.
Such is that of a wheel, a fphere, a garden roller, or
the like.

The motion of Rolling is oppofed to that of fliding;
in which latter motion the fame furface is continually
applied to the plane it moves along.

In a wheel, it is only the circumference that properly and fimply rolls; the reft of the wheel proceeds
in a compound angular kind of motion, and partly
rolls,

rolls, partly slides. The want of distinguishing between which two motions, occasioned the difficulty of that celebrated problem of Aristotle's Wheel.

The friction of a body in rolling, is much less than the friction in sliding. And hence arises the great use of wheels, rolls, &c, in machines; as much of the action as possible being laid upon it, to make the resistance the less.

ROMAN *Order*, in Architecture, is the same as the composite. It was invented by the Romans, in the time of Augustus; and it is made up of the Ionic and Corinthian orders, being more ornamental than either.

RONDEL, in Fortification, a round tower, sometimes erected at the foot of a bastion.

ROOD, a square measure, being a quantity of land just equal to the 4th part of an acre, or, equal to 40 perches or square poles.

ROOF, in Architecture, the uppermost part of a building; being that which forms the covering of the whole. In this sense, the Roof comprises the timber work, together with its furniture, of slate, or tile, or lead, or whatever else serves for a covering: though the carpenters usually restrain Roof to the timberwork only.

The form of a Roof is various: viz, 1. *Pointed*, when the ridge, or angle formed by the two sides, is an acute angle.—2. *Square*, when the pitch or angle of the ridge is a right angle, called the true pitch.—3. *Flat* or pediment Roof, being only pediment pitch, or the angle very obtuse. There are also various other forms, as hip Roofs, valley Roofs, hopper Roofs, double ridges, platforms, round, &c.—In the true pitch, when the sides form a square or right angle, the girt over both sides of the Roof, is accounted equal to the breadth of the building and the half of the same.

ROOKE (LAWRENCE), an English astronomer and geometrician, was born at Deptford in Kent, 1623, and educated at Eton school. From hence he removed to King's College, Cambridge, in 1639. After taking the degree of master of arts in 1647, he retired into the country. But in the year 1650 he went to Oxford, and settled in Wadham College, that he might have the company of, and receive improvement from Dr. Wilkins, and Mr. Seth Ward the Astronomy Professor; and that he might also accompany Mr. Boyle in his chemical operations.

After the death of Mr. Foster, he was chosen Astronomy Professor in Gresham College, London, in the year 1652. He made some observations upon the comet at Oxford, which appeared in the month of December that year; which were printed by Mr. Seth Ward the year following. And, in 1655, Dr. Wallis publishing his treatise on Conic Sections, he dedicated that work to those two gentlemen.

In 1657, Mr. Rooke was permitted to exchange the astronomy professorship for that of geometry. This step might seem strange, as astronomy still continued to be his favourite study; but it was thought to have been from the convenience of the lodgings, which opened behind the reading hall, and therefore were proper for the reception of those gentlemen after the lectures, who in the year 1660 formed the Royal Society there.

Mr. Rooke having thus successively enjoyed those two places some years before the restoration in 1658, most of those gentlemen who had been accustomed to assemble with him at Oxford, coming to London, joined with other philosophical gentlemen, and usually met at Gresham College to hear Mr. Rooke's lectures, and afterwards withdrew into his apartment; till their meetings were interrupted by the quartering of soldiers in the college that year. And after the Royal Society came to be formed and settled into a regular body, Mr. Rooke was very zealous and serviceable in promoting that great and useful institution; though he did not live till it received its establishment by the Royal charter.

The Marquis of Dorchester, who was not only a patron of learning, but learned himself, used to entertain Mr. Rooke at his seat at Highgate after the restoration, and bring him every Wednesday in his coach to the Royal Society, which then met on that day of the week at Gresham College. But the last time Mr. Rooke was at Highgate, he walked from thence; and it being in the summer, he overheated himself, and taking cold after it, he was thrown into a fever, which cost him his life. He died at his apartments at Gresham College the 27th of June 1662, in the 40th year of his age.

One other very unfortunate circumstance attended his death, which was, that it happened the very night that he had for some years expected to finish his accurate observations on the satellites of Jupiter. When he found his illness prevented him from making that observation, Dr. Pope says, he sent to the Society his request, that some other person, properly qualified, might be appointed for that purpose; so intent was he to the last on making those curious and useful discoveries, in which he had been so long engaged.

Mr. Rooke made a nuncupatory will, leaving what he had to Dr. Ward, then lately made bishop of Exeter: whom he permitted to receive what was due upon bond, if the debtors offered payment willingly, otherwise he would not have the bonds put in suit: " for, said he, as I never was in law, nor had any contention with any man, in my life-time; neither would I be so after my death."

Few persons have left behind them a more agreeable character than Mr. Rooke, from every person that was acquainted with him, or with his qualifications; and in nothing more than for his veracity: for what he asserted positively, might be fully relied on: but if his opinion was asked concerning any thing that was dubious, his usual answer was, " I have no opinion." Mr. Hook has given this copious, though concise character of him: " I never was acquainted with any person who knew more, and spoke less, being indeed eminent for the knowledge and improvement of astronomy." Dr. Wren and Dr. Seth Ward describe him, as a man of profound judgment, a vast comprehension, prodigious memory, and solid experience. His skill in the mathematics was reverenced by all the lovers of those studies, and his perfection in many other forts of learning deserves no less admiration; but above all, as another writer characterizes him, his extensive knowledge

ledge

ledge had a right influence on the temper of his mind, which had all the humility, goodnefs, calmnefs, ftrength, and fincerity, of a found and unaffected philofopher. Thefe accounts give us his picture only in miniature; but his fucceffor, Dr. Ifaac Barrow, has drawn it in full proportion, in his oration at Grefham College; which is too long to be inferted in this place.

His writings were chiefly;

1. *Obfervations on the Comet* of Dec. 1652. This was printed by Dr. Seth Ward, in his Lectures on Comets, 4to, 1653.

2. *Directions for Seamen going to the Eaft and Weft Indies.* Publifhed in the Philofophical Tranfactions for Jan. 1665.

3. *A Method of Obferving the Eclipfes of the Moon &c.* In the Philof. Tranf. for Feb. 1666.

4. *A Difcourfe concerning the Obfervations of the Eclipfes of the Satellites of Jupiter.* In the Hiftory of the Royal Society, pa. 183.

5. *An Account of an Experiment made with Oil in a long Tube.* Read to the Royal Soc. April 23, 1667.—By this experiment it was found, that the oil funk when the fun fhone out, and rofe when he was clouded; the proportions of which are fet down in the account.

ROOT, in Arithmetic and Algebra, denotes a quantity which being multiplied by itfelf produces fome higher power; or a quantity confidered as the bafis or foundation of a higher power, out of which this arifes and grows, like as a plant from its Root.

In the involution of powers, from a given Root, the Root is alfo called the firft power; when this is once multiplied by itfelf, it produces the fquare or fecond power; this multiplied by the Root again, makes the cube or 3d power; and fo on. And hence the Roots alfo come to be denominated the fquare Root, or cube-Root, or 2d Root, or 3d Root, &c, according as the given power or quantity is confidered as the fquare, or cube, or 2d power, or 3d power, &c. Thus, 2 is the fquare-Root or 2d Root of 4, and the cube-Root or 3d Root of 8, and the 4th Root of 16, &c.

Hence, the fquare-Root is the mean proportional between 1 and the fquare or given power; and the cube-Root is the firft of two mean proportionals between 1 and the given cube; and fo on.

To Extract the Root of a given number or power. This is the fame thing as to find a number or quantity, which being multiplied the proper number of times, will produce the given number or power. So, to find the cube Root of 8, is finding the number 2, which multiplied twice by itfelf produces the given number 8.

For the ufual methods of extracting the Roots of Numbers, fee the common treatifes on Arithmetic.

A Root, of any power, that confifts of two parts, is called a binomial Root; as 12 or 10 + 2. If it confift of three parts, it is a trinomial Root; as 126 or 100 + 20 + 6. And fo on.

The extraction of the Roots of algebraic quantities, is alfo performed after the fame manner as that of numbers; as may be feen in any treatife on algebra. See alfo the article EXTRACTION of Roots.

A general method for all Roots, is alfo by Newton's binomial theorem. See BINOMIAL *Theorem.*

Finite approximating rules for the extraction of Roots have alfo been given by feveral authors, as Raphfon, De Lagney, Halley, &c. See the articles APPROXIMATION and EXTRACTION. See alfo Newton's Univerfal Arith. the Appendix; Philof. Tranf. numb. 210, or Abridg. vol. 1, pa. 81; Maclaurin's Alg. pa. 242; Simpfon's Alg. pa. 155; or his Effays, pa. 82, or his Differtations, pa. 102, or his Select Exerc. pa. 215: where various general theorems for approximating to the Roots of pure powers are given. See alfo EQUATION and REDUCTION of Equations, APPROXIMATION, and CONVERGING.

But the moft commodious and general rule of any, for fuch approximations, I believe, is that which has been invented by myfelf, and explained in my Tracts, vol. 1, pa. 49: which theorem is this;

$$\frac{\overline{n+1}.N + \overline{n-1}.a^n}{\overline{n-1}.N + \overline{n+1}.a^n}\, a = \sqrt[n]{N}.$$ That is, having to extract the nth Root of the given number N; take a^n the nearest rational power to that given quantity N, whether greater or lefs, its Root of the fame kind being a; then the required Root, or $\sqrt{N}$, will be as is exprefled in this formula above; or the fame exprefled in a proportion will be thus:

$$\overline{n-1}.N + \overline{n+1}.a^n : \overline{n+1}.N + \overline{n-1}.a^n :: a : \sqrt[n]{N}$$

the Root fought very nearly. Which rule includes all the particular rational formulas of De Lagney, and Halley, which were feparately inveftigated by them; and yet this general formula is perfectly fimple and eafy to apply, and more eafily kept in mind than any one of the faid particular formulas.

Ex. Suppofe it be required to double the cube, or to extract the cube Root of the number 2.

Here N = 2, n = 3, the neareft Root a = 1, alfo a^3 = 1; hence, for the cube Root the formula becomes $\frac{4N + 2a^3}{2N + 4a^3}\, a$ or $\frac{2N + a^3}{N + 2a^3}\, a = \sqrt[3]{N}$.

But N + $2a^3$ = 4, and 2N + a^3 = 5; therefore as 4 : 5 :: 1 : $\frac{5}{4}$ = 1·25 = the Root nearly by a firft approximation.

Again, for a fecond approximation, take $a = \frac{5}{4}$, and confequently $a^3 = \frac{125}{64}$;

hence 2N + a^3 = 4 + $\frac{125}{64}$ = $\frac{381}{64}$,

and N + $2a^3$ = 2 + $\frac{250}{64}$ = $\frac{378}{64}$;

therefore as 378 : 381, or as 126 : 127 :: $\frac{5}{4}$: $\frac{635}{504}$ = 1·259921 &c, for the required cube Root of 2, which is true even in the laft place of decimals.

ROOT *of an Equation*, denotes the value of the unknown quantity in an equation; which is fuch a quantity,

quantity, as being fubftituted inftead of that unknown letter, into the equation, fhall make all the terms to vanifh, or both fides equal to each other. Thus, of the equation $3x + 5 = 14$, the Root or value of x is 3, becaufe fubftituting 3 for x, makes it become $9 + 5 = 14$. And the Root of the equation $2x^2 = 32$ is 4, becaufe $2 \times 4^2 = 32$. Alfo the Root of the equation $x^2 = a^2 + c^2$ is $x = \sqrt{a^2 + c^2}$.

For the Nature of Roots, and for extracting the feveral Roots of equations, fee EQUATION.

Every equation has as many Roots, or values of the unknown quantity, as are the dimenfions or higheft power in it. As a fimple equation one Root, a quadratic two, a cubic three, and fo on.

Roots are pofitive or negative, real or imaginary, rational or radical, &c. See EQUATION.

Cubic Root. This is threefold, even for a fimple cubic. So the cube Root of a^3, is either

a, or $\dfrac{-1 + \sqrt{-3}}{2}a$, or $\dfrac{-1 - \sqrt{-3}}{2}a$.

And even the cube Root of 1 itfelf is either

1, or $\dfrac{-1 + \sqrt{-3}}{2}$, or $\dfrac{-1 - \sqrt{-3}}{2}$.

Real and Imaginary Roots. The odd Roots, as the 3d, 5th, 7th, &c Roots, of all real quantities, whether pofitive or negative, are real, and are refpectively pofitive or negative. So the cube Root of a^3 is a, and of $-a^3$ is $-a$.

But the even Roots, as the 2d, 4th, 6th, &c, are only real when the quantity is pofitive; being imaginary or impoffible when the quantity is negative. So the fquare Root of a^2 is a, which is real; but the fquare Root of $-a^2$, that is, $\sqrt{-a^2}$, is imaginary or impoffible; becaufe there is no quantity, neither $+a$ nor $-a$, which by fquaring will make the given negative fquare $-a^2$.

Table *of* Roots, &c.

THE following Table of Roots, Squares, and Cubes, is very ufeful in many calculations, and will ferve to find fquare-Roots and cube Roots, as well as fquare and cubic powers. The Table confifts of three columns: in the firft column are the feries of common numbers, or Roots, 1, 2, 3, 4, 5, 6, &c; in the fecond column are the fquares, and in the third column the cubes of the fame. For example, to find the fquare or the cube of the number or Root 49. Finding this number 49 in the firft column; upon the fame line with it, ftands its fquare 2401 in the fecond column, and its cube 117649 in the third column.

Again, to find the fquare Root of the number 700. Near the beginning of the Table, it appears that the next lefs and greater tabular fquares are 676 and 729, whofe Roots are 26 and 27, and therefore the fquare Root of 700 is between 26 and 27. But a little further on, viz, among the hundreds, it appears that the required Root lies between 26·4 and 26·5, the tabular fquares of thefe being 696·96 and 702·25, cutting off the proper part of the figures for

decimals. Take the difference between the lefs fquare 696·96 and the given number 700, which gives 3·04, and divide the half of it, viz 1·52, by the lefs given tabular Root, viz 26·4, and the quotient 575 gives as many more figures of the Root, to be joined to the firft three, and thus making the Root equal to 26·4575, which is true in all its places.

Alfo to find the cube Root of the number 7000; near the beginning of the Table, among the tens, it appears that the cube Root of this number is between 19 and 20; but farther on, among the hundreds, it appears that it lies between 19·1 and 19·2, allowing for the proper number of integers. But if more figures are required; from the given number 7000 take the next lefs tabular one, or the cube of 19·1, viz 6967871, and there remains 32·129, the 3d part of which, or 10·730, divide by the fquare of 19·1, viz 364·81, found on the fame line, and the quotient 293 is the next three figures of the Root, and therefore the whole cubic Root is 19·1293, which is true in all its figures.—The Table follows.

TABLE of Square and Cubic Roots.

Root.	Square.	Cube.	Root.	Square.	Cube.	Root.	Square.	Cube.	Root.	Square.	Cube.
1	1	1	64	4096	262144	127	16129	2048383	190	36100	6859000
2	4	8	65	4225	274625	128	16384	2097152	191	36481	6967871
3	9	27	66	4356	287496	129	16641	2146689	192	36864	7077888
4	16	64	67	4489	300763	130	16900	2197000	193	37249	7189057
5	25	125	68	4624	314432	131	17161	2248091	194	37636	7301384
6	36	216	69	4761	328509	132	17424	2299968	195	38025	7414875
7	49	343	70	4900	343000	133	17689	2352637	196	38416	7529536
8	64	512	71	5041	357911	134	17956	2406104	197	38809	7645373
9	81	729	72	5184	373248	135	18225	2460375	198	39204	7762392
10	100	1000	73	5329	389017	136	18496	2515456	199	39601	7880599
11	121	1331	74	5476	405224	137	18769	2571353	200	40000	8000000
12	144	1728	75	5625	421875	138	19044	2628072	201	40401	8120601
13	169	2197	76	5776	438976	139	19321	2685619	202	40804	8242408
14	196	2744	77	5929	456533	140	19600	2744000	203	41209	8365427
15	225	3375	78	6084	474552	141	19881	2803221	204	41616	8489664
16	256	4096	79	6241	493039	142	20164	2863288	205	42025	8615125
17	289	4913	80	6400	512000	143	20449	2924207	206	42436	8741816
18	324	5832	81	6561	531441	144	20736	2985984	207	42849	8869743
19	361	6859	82	6724	551368	145	21025	3048625	208	43264	8998912
20	400	8000	83	6889	571787	146	21316	3112136	209	43681	9123329
21	441	9261	84	7056	592704	147	21609	3176523	210	44100	9261000
22	484	10648	85	7225	614125	148	21904	3241792	211	44521	9393931
23	529	12167	86	7396	636056	149	22201	3307949	212	44944	9528128
24	576	13824	87	7569	658503	150	22500	3375000	213	45369	9663597
25	625	15625	88	7744	681472	151	22801	3442951	214	45796	9800344
26	676	17576	89	7921	704969	152	23104	3511808	215	46225	9938375
27	729	19683	90	8100	729000	153	23409	3581577	216	46656	10077696
28	784	21952	91	8281	753571	154	23716	3652264	217	47089	10218313
29	841	24389	92	8464	778688	155	24025	3723875	218	47524	10360282
30	900	27000	93	8649	804357	156	24336	3796416	219	47961	10503459
31	961	29791	94	8836	830584	157	24649	3869893	220	48400	10648000
32	1024	32768	95	9025	857375	158	24964	3944312	221	48841	10793861
33	1089	35937	96	9216	884736	159	25281	4019679	222	49284	10941048
34	1156	39304	97	9409	912673	160	25600	4096000	223	49729	11089567
35	1225	42875	98	9604	941192	161	25921	4173281	224	50176	11239424
36	1296	46656	99	9801	970299	162	26244	4251528	225	50625	11390625
37	1369	50653	100	10000	1000000	163	26569	4330747	226	51076	11543176
38	1444	54872	101	10201	1030301	164	26896	4410944	227	51529	11697083
39	1521	59319	102	10404	1061208	165	27225	4492125	228	51984	11852352
40	1600	64000	103	10609	1092727	166	27556	4574296	229	52441	12008989
41	1681	68921	104	10816	1124864	167	27889	4657463	230	52900	12167000
42	1764	74088	105	11025	1157625	168	28224	4741632	231	53361	12326391
43	1849	79507	106	11236	1191016	169	28561	4826809	232	53824	12487168
44	1936	85184	107	11449	1225043	170	28900	4913000	233	54289	12649337
45	2025	91125	108	11664	1259712	171	29241	5000211	234	54756	12812904
46	2116	97336	109	11881	1295029	172	29584	5088448	235	55225	12977875
47	2209	103823	110	12100	1331000	173	29929	5177717	236	55696	13144256
48	2304	110592	111	12321	1367631	174	30276	5268024	237	56169	13312053
49	2401	117649	112	12544	1404928	175	30625	5359375	238	56644	13481272
50	2500	125000	113	12769	1442897	176	30976	5451776	239	57121	13651919
51	2601	132651	114	12996	1481544	177	31329	5545233	240	57600	13824000
52	2704	140608	115	13225	1520875	178	31684	5639752	241	58081	13997521
53	2809	148877	116	13456	1560896	179	32041	5735339	242	58564	14172488
54	2916	157464	117	13689	1601613	180	32400	5832000	243	59049	14348907
55	3025	166375	118	13924	1643032	181	32761	5929741	244	59536	14526784
56	3136	175616	119	14161	1685159	182	33124	6028568	245	60025	14706125
57	3249	185193	120	14400	1728000	183	33489	6128487	246	60516	14886936
58	3364	195112	121	14641	1771561	184	33856	6229504	247	61009	15069223
59	3481	205379	122	14884	1815848	185	34225	6331625	248	61504	15252992
60	3600	216000	123	15129	1860867	186	34596	6434856	249	62001	15438249
61	3721	226981	124	15376	1906624	187	34969	6539203	250	62500	15625000
62	3844	238328	125	15625	1953125	188	35344	6644672	251	63001	15813251
63	3969	250047	126	15876	2000376	189	35721	6751269	252	63504	16003008

Table of Square and Cubic Roots.

Root.	Square.	Cube.	Root.	Square.	Cube.	Root.	Square.	Cube.	Root.	Square.	Cube.
253	64009	16194277	316	99856	31554496	379	143641	54439939	442	195364	86350888
254	64516	16387064	317	100489	31855013	380	144400	54872000	443	196249	86938307
255	65025	16581375	318	101124	32157432	381	145161	55306341	444	197136	87528384
256	65536	16777216	319	101761	32461759	382	145924	55742968	445	198025	88121125
257	66049	16974593	320	102400	32768000	383	146689	56181887	446	198916	88716536
258	66564	17173512	321	103041	33076161	384	147456	56623104	447	199809	89314623
259	67081	17373979	322	103684	33386248	385	148225	57066625	448	200704	89915392
260	67600	17576000	323	104329	33698267	386	148996	57512456	449	201601	90518849
261	68121	17779581	324	104976	34012224	387	149769	57960603	450	202500	91125000
262	68644	17984728	325	105625	34328125	388	150544	58411072	451	203401	91733851
263	69169	18191447	326	106276	34645976	389	151321	58863869	452	204304	92345408
264	69696	18399744	327	106929	34965783	390	152100	59319000	453	205209	92959677
265	70225	18609625	328	107584	35287552	391	152881	59776471	454	206116	93576664
266	70756	18821096	329	108241	35611289	392	153664	60236288	455	207025	94196375
267	71289	19034163	330	108900	35937000	393	154449	60698457	456	207936	94818816
268	71824	19248832	331	109561	36264691	394	155236	61162984	457	208849	95443993
269	72361	19465109	332	110224	36594368	395	156025	61629875	458	209764	96071912
270	72900	19683000	333	110889	36926037	396	156816	62099136	459	210681	96702579
271	73441	19902511	334	111556	37259704	397	157609	62570773	460	211600	97336000
272	73984	20123648	335	112225	37595375	398	158404	63044792	461	212521	97972181
273	74529	20346417	336	112896	37933056	399	159201	63521199	462	213444	98611128
274	75076	20570824	337	113569	38272753	400	160000	64000000	463	214369	99252847
275	75625	20796875	338	114244	38614472	401	160801	64481201	464	215296	99897344
276	76176	21024576	339	114921	38958219	402	161604	64964808	465	216225	100544625
277	76729	21253933	340	115600	39304000	403	162409	65450827	466	217156	101194696
278	77284	21484952	341	116281	39651821	404	163216	65939264	467	218089	101847563
279	77841	21717639	342	116964	40001688	405	164025	66430125	468	219024	102503232
280	78400	21952000	343	117649	40353607	406	164836	66923416	469	219961	103161709
281	78961	22188041	344	118336	40707584	407	165649	67419143	470	220900	103823000
282	79524	22425768	345	119025	41063625	408	166464	67917312	471	221841	104487111
283	80089	22665187	346	119716	41421736	409	167281	68417929	472	222784	105154048
284	80656	22906304	347	120409	41781923	410	168100	68921000	473	223729	105823817
285	81225	23149125	348	121104	42144192	411	168921	69426531	474	224676	106496424
286	81796	23393656	349	121801	42508549	412	169744	69934528	475	225625	107171875
287	82369	23639903	350	122500	42875000	413	170569	70444997	476	226576	107850176
288	82944	23887872	351	123201	43243551	414	171396	70957944	477	227529	108531333
289	83521	24137569	352	123904	43614208	415	172225	71473375	478	228484	109215352
290	84100	24389000	353	124609	43986977	416	173056	71991296	479	229441	109902239
291	84681	24642171	354	125316	44361864	417	173889	72511713	480	230400	110592000
292	85264	24897088	355	126025	44738875	418	174724	73034632	481	231361	111284641
293	85849	25153757	356	126736	45118016	419	175561	73560059	482	232324	111980168
294	86436	25412184	357	127449	45499293	420	176400	74088000	483	233289	112678587
295	87025	25672375	358	128164	45882712	421	177241	74618461	484	234256	113379904
296	87616	25934336	359	128881	46268279	422	178084	75151448	485	235225	114084125
297	88209	26198073	360	129600	46656000	423	178929	75686967	486	236196	114791256
298	88804	26463592	361	130321	47045881	424	179776	76225024	487	237169	115501303
299	89401	26730899	362	131044	47437928	425	180625	76765625	488	238144	116214272
300	90000	27000000	363	131769	47832147	426	181476	77308776	489	239121	116930169
301	90601	27270901	364	132496	48228544	427	182329	77854483	490	240100	117649000
302	91204	27543608	365	133225	48627125	428	183184	78402752	491	241081	118370771
303	91809	27818127	366	133956	49027896	429	184041	78953589	492	242064	119095488
304	92416	28094464	367	134689	49430863	430	184900	79507000	493	243049	119823157
305	93025	28372625	368	135424	49836032	431	185761	80062991	494	244036	120553784
306	93636	28652616	369	136161	50243409	432	186624	80621568	495	245025	121287375
307	94249	28934443	370	136900	50653000	433	187489	81182737	496	246016	122023936
308	94864	29218112	371	137641	51064811	434	188356	81746504	497	247009	122763473
309	95481	29503629	372	138384	51478848	435	189225	82312875	498	248004	123505992
310	96100	29791000	373	139129	51895117	436	190096	82881856	499	249001	124251499
311	96721	30080231	374	139876	52313624	437	190969	83453453	500	250000	125000000
312	97344	30371328	375	140625	52734375	438	191844	84027672	501	251001	125751501
313	97969	30664297	376	141376	53157376	439	192721	84604519	502	252004	126506008
314	98596	30959144	377	142129	53582633	440	193600	85184000	503	253009	127263527
315	99225	31255875	378	142884	54010152	441	194481	85766121	504	254016	128024064

Table of Square and Cube Roots.

Root.	Square.	Cube.	Root.	Square.	Cube.	Root.	Square.	Cube.	Root.	Square.	Cube.
505	255025	128787625	568	322624	183250432	631	398161	251239591	694	481636	334255384
506	256036	129554216	569	323761	184220009	632	399424	252435968	695	483025	335702375
507	257049	130323843	570	324900	185193000	633	400689	253636137	696	484416	337153536
508	258064	131096512	571	326041	186169411	634	401956	254840104	697	485809	338608873
509	259081	131872229	572	327184	187149248	635	403225	256047875	698	487204	340068392
510	260100	132651000	573	328329	188132517	636	404496	257259456	699	488601	341532099
511	261121	133432831	574	329476	189119224	637	405769	258474853	700	490000	343000000
512	262144	134217728	575	330625	190109375	638	407044	259694072	701	491401	344472101
513	263169	135005697	576	331776	191102976	639	408321	260917119	702	492804	345948008
514	264196	135796744	577	332929	192100033	640	409600	262144000	703	494209	347428927
515	265225	136590875	578	334084	193100552	641	410881	263374721	704	495616	348913664
516	266256	137388096	579	335241	194104539	642	412164	264609288	705	497025	350402625
517	267289	138188413	580	336400	195112000	643	413449	265847707	706	498436	351895816
518	268324	138991832	581	337561	196122941	644	414736	267089984	707	499849	353393243
519	269361	139798359	582	338724	197137368	645	416025	268336125	708	501264	354894912
520	270400	140608000	583	339889	198155287	646	417316	269586136	709	502681	356400829
521	271441	141420761	584	341056	199176704	647	418609	270840023	710	504100	357911000
522	272484	142236648	585	342225	200201625	648	419904	272097792	711	505521	359425431
523	273529	143055667	586	343396	201230056	649	421201	273359449	712	506944	360944128
524	274576	143877824	587	344569	202262003	650	422500	274625000	713	508369	362467097
525	275625	144703125	588	345744	203297472	651	423801	275894451	714	509796	363994344
526	276676	145531576	589	346921	204336469	652	425104	277167808	715	511225	365525875
527	277729	146363183	590	348100	205379000	653	426409	278445077	716	512656	367061696
528	278784	147197952	591	349281	206425071	654	427716	279726264	717	514089	368601813
529	279841	148035889	592	350464	207474688	655	429025	281011375	718	515524	370146232
530	280900	148877000	593	351649	208527857	656	430336	282300416	719	516961	371694959
531	281961	149721291	594	352836	209584584	657	431649	283593393	720	518400	373248000
532	283024	150568768	595	354025	210644875	658	432964	284890312	721	519841	374805361
533	284089	151419437	596	355216	211708736	659	434281	286191179	722	521284	376367048
534	285156	152273304	597	356409	212776173	660	435600	287496000	723	522729	377933067
535	286225	153130375	598	357604	213847192	661	436921	288804781	724	524176	379503424
536	287296	153990656	599	358801	214921799	662	438244	290117528	725	525625	381078125
537	288369	154854153	600	360000	216000000	663	439569	291434247	726	527076	382657176
538	289444	155720872	601	361201	217081801	664	440896	292754944	727	528529	384240583
539	290521	156590819	602	362404	218167208	665	442225	294079625	728	529984	385828352
540	291600	157464000	603	363609	219256227	666	443556	295408296	729	531441	387420489
541	292681	158340421	604	364816	220348864	667	444889	296740963	730	532900	389017000
542	293764	159220088	605	366025	221445125	668	446224	298077632	731	534361	390617891
543	294849	160103007	606	367236	222545016	669	447561	299418309	732	535824	392223168
544	295936	160989184	607	368449	223648543	670	448900	300763000	733	537289	393832837
545	297025	161878625	608	369664	224755712	671	450241	302111711	734	538756	395446904
546	298116	162771336	609	370881	225866529	672	451584	303464448	735	540225	397065375
547	299209	163667323	610	372100	226981000	673	452929	304821217	736	541696	398688256
548	300304	164566592	611	373321	228099131	674	454276	306182024	737	543169	400315553
549	301401	165469149	612	374544	229220928	675	455625	307546875	738	544644	401947272
550	302500	166375000	613	375769	230346397	676	456976	308915776	739	546121	403583419
551	303601	167284151	614	376996	231475544	677	458329	310288733	740	547600	405224000
552	304704	168196608	615	378225	232608375	678	459684	311665752	741	549081	406869021
553	305809	169112377	616	379456	233744896	679	461041	313046839	742	550564	408518488
554	306916	170031464	617	380689	234885113	680	462400	314432000	743	552049	410172407
555	308025	170953875	618	381924	236029032	681	463761	315821241	744	553536	411830784
556	309136	171879616	619	383161	237176659	682	465124	317214568	745	555025	413493625
557	310249	172808693	620	384400	238328000	683	466489	318611987	746	556516	415160936
558	311364	173741112	621	385641	239483061	684	467856	320013504	747	558009	416832723
559	312481	174676879	622	386884	240641848	685	469225	321419125	748	559504	418508992
560	313600	175616000	623	388129	241804367	686	470596	322828856	749	561001	420189749
561	314721	176558481	624	389376	242970624	687	471969	324242703	750	562500	421875000
562	315844	177504328	625	390625	244140625	688	473344	325660672	751	564001	423564751
563	316969	178453547	626	391876	245314376	689	474721	327082769	752	565504	425259008
564	318096	179406144	627	393129	246491883	690	476100	328509000	753	567009	426957777
565	319225	180362125	628	394384	247673152	691	477481	329939371	754	568516	428661064
566	320356	181321496	629	395641	248858189	692	478864	331373888	755	570025	430368875
567	321489	182284263	630	396900	250047000	693	480249	332812557	756	571536	432081216

Table of Square and Cubic Roots.

Root	Square	Cube	Root	Square	Cube	Root	Square	Cube	Root	Square	Cube
757	573049	433798093	820	672400	551368000	883	779689	688465387	946	894916	846590536
758	574564	435519512	821	674041	553387661	884	781456	690807104	947	896809	849378123
759	576081	437245479	822	675684	555412248	885	783225	693154125	948	898704	851971392
760	577600	438976000	823	677329	557441767	886	784996	695506456	949	900601	854670349
761	579121	440711081	824	678976	559476224	887	786769	697864103	950	902500	857375000
762	580644	442450728	825	680625	561515625	888	788544	700227072	951	904401	860085351
763	582169	444194947	826	682276	563559976	889	790321	702595369	952	906304	862801408
764	583696	445943744	827	683929	565609283	890	792100	704969000	953	908209	865523177
765	585225	447697125	828	685584	567663552	891	793881	707347971	954	910116	868250664
766	586756	449455096	829	687241	569722789	892	795664	709732288	955	912025	870983875
767	588289	451217663	830	688900	571787000	893	797449	712121957	956	913936	873722816
768	589824	452984832	831	690561	573856191	894	799236	714516984	957	915849	876467493
769	591361	454756609	832	692224	575930368	895	801025	716917375	958	917764	879217912
770	592900	456533000	833	693889	578009537	896	802816	719323136	959	919681	881974079
771	594441	458314011	834	695556	580093704	897	804609	721734273	960	921600	884736000
772	595984	460099648	835	697225	582182875	898	806404	724150792	961	923521	887503681
773	597529	461889917	836	698896	584277056	899	808201	726572699	962	925444	890277128
774	599076	463684824	837	700569	586376253	900	810000	729000000	963	927369	893056347
775	600625	465484375	838	702244	588480472	901	811801	731432701	964	929296	895841344
776	602176	467288576	839	703921	590589719	902	813604	733870808	965	931225	898632125
777	603729	469097433	840	705600	592704000	903	815429	736314327	966	933156	901428696
778	605284	470910952	841	707281	594823321	904	817216	738763264	967	935089	904231063
779	606841	472729139	842	708964	596947688	905	819025	741217625	968	937024	907039232
780	608400	474552000	843	710649	599077107	906	820836	743677416	969	938961	909853209
781	609961	476379541	844	712336	601211584	907	822649	746142643	970	940900	912673000
782	611524	478211768	845	714025	603351125	908	824464	748613312	971	942841	915498611
783	613089	480048687	846	715716	605495736	909	826281	751089429	972	944784	918330048
784	614656	481890304	847	717409	607645423	910	828100	753571000	973	946729	921167317
785	616225	483736625	848	719104	609800192	911	829921	756058031	974	948676	924010424
786	617796	485587656	849	720801	611960049	912	831744	758550528	975	950625	926859375
787	619369	487443403	850	722500	614125000	913	833569	761048497	976	952576	929714176
788	620944	489303872	851	724201	616295051	914	835396	763551944	977	954529	932574833
789	622521	491169069	852	725904	618470208	915	837225	766060875	978	956484	935441352
790	624100	493039000	853	727609	620650477	916	839056	768575296	979	958441	938313739
791	625681	494913671	854	729316	622835864	917	840889	771095213	980	960400	941192001
792	627264	496793088	855	731025	625026375	918	842724	773620632	981	962361	944076141
793	628849	498677257	856	732736	627222016	919	844561	776151559	982	964324	946966168
794	630436	500566184	857	734449	629422793	920	846400	778688000	983	966289	949862087
795	632025	502459875	858	736164	631628712	921	848241	781229961	984	968256	952763904
796	633616	504358336	859	737881	633839779	922	850084	783777448	985	970225	955671625
797	635209	506261573	860	739600	636050000	923	851929	786330467	986	972196	958585256
798	636804	508169592	861	741321	638277381	924	853776	788889024	987	974169	961504803
799	638401	510082399	862	743044	640503928	925	855625	791453125	988	976144	964430272
800	640000	512000000	863	744769	642735647	926	857476	794022776	989	978121	967361669
801	641601	513922401	864	746496	644972544	927	859329	796597983	990	980100	970299000
802	643204	515849608	865	748225	647214625	928	861184	799178752	991	982081	973242271
803	644809	517781627	866	749956	649461896	929	863041	801765089	992	984064	976191488
804	646416	519718464	867	751689	651714363	930	864900	804357000	993	986049	979146657
805	648025	521660125	868	753424	653972032	931	866761	806954491	994	988036	982107784
806	649636	523606616	869	755161	656234909	932	868624	809557368	995	990025	985074875
807	651249	525557943	870	756900	658503000	933	870489	812166237	996	992016	988047936
808	652864	527514112	871	758641	660776311	934	872356	814780504	997	994009	991026973
809	654481	529475129	872	760384	663054848	935	874225	817400375	998	996004	994011992
810	656100	531441000	873	762129	665338617	936	876096	820025856	999	998001	997002999
811	657721	533411731	874	763876	667627624	937	877969	822656953	1000	1000000	1000000000
812	659344	535387328	875	765625	669921875	938	879844	825293672	1001	1002001	1003003001
813	660969	537366797	876	767376	672221376	939	881721	827936019	1002	1004004	1006012008
814	662596	539353144	877	769129	674526133	940	883600	830584000	1003	1006009	1009027027
815	664225	541343375	878	770884	676836152	941	885481	833237621	1004	1008016	1012048064
816	665856	543338496	879	772641	679151439	942	887364	835896888	1005	1010025	1015075125
817	667489	545338513	880	774400	681472000	943	889249	838561807	1006	1012036	1018108216
818	669124	547343432	881	776161	683797841	944	891136	841232384	1007	1014049	1021147343
819	670761	549353259	882	777924	686128968	945	893025	843908625	1008	1016064	1024192512

The following is another Table of the Square Roots of the first 1000 Numbers to 10 places of decimal figures beside the integers, which needs no farther explanation, as Numbers stand always in the first column, and their Square Roots in the next.

Table of Square Roots *to ten Decimal Places.*

No.	Square Root.	No.	Square Root.	No.	Square Root.	No.	Square Root.
1	1·0000000000	64	8·0000000000	127	11·2694276696	190	13·7840487521
2	1·4142135624	65	8·0622577483	128	11·3137084990	191	13·8202749611
3	1·7320508076	66	8·1240384046	129	11·3578166916	192	13·8564064606
4	2·0000000000	67	8·1853527719	130	11·4017542510	193	13·8924439894
5	2·2360679775	68	8·2462112512	131	11·4455231423	194	13·9283882772
6	2·4494897428	69	8·3066238629	132	11·4891252931	195	13·9642400438
7	2·6457513111	70	8·3666002653	133	11·5325625947	196	14·0000000000
8	2·8284271247	71	8·4261497732	134	11·5758369028	197	14·0356688441
9	3·0000000000	72	8·4852813742	135	11·6189500386	198	14·0712472795
10	3·1622776602	73	8·5440037453	136	11·6619037897	199	14·1067359797
11	3·3166247904	74	8·6023252670	137	11·7046999111	200	14·1421356237
12	3·4641016151	75	8·6602540378	138	11·7473443808	201	14·1774468788
13	3·6055512755	76	8·7177978871	139	11·7898261226	202	14·2126704036
14	3·7416573868	77	8·7749643874	140	11·8321595662	203	14·2478068488
15	3·8729833462	78	8·8317608663	141	11·8743420870	204	14·2828568571
16	4·0000000000	79	8·8881944173	142	11·9163752878	205	14·3178210633
17	4·1231056256	80	8·9442719100	143	11·9582607431	206	14·3527000944
18	4·2426406871	81	9·0000000000	144	12·0000000000	207	14·3874945699
19	4·3588989435	82	9·0553851381	145	12·0415945788	208	14·4222051019
20	4·4721359550	83	9·1104335791	146	12·0830459736	209	14·4568322948
21	4·5825756950	84	9·1651513899	147	12·1243556530	210	14·4913767462
22	4·6904157598	85	9·2195444573	148	12·1655250606	211	14·5258390463
23	4·7958315233	86	9·2736184955	149	12·2065556153	212	14·5602197786
24	4·8989794856	87	9·3273790531	150	12·2474487139	213	14·5945195193
25	5·0000000000	88	9·3808315196	151	12·2882057274	214	14·6287388383
26	5·0990195136	89	9·4339811321	152	12·3288280059	215	14·6628782986
27	5·1961524227	90	9·4868329805	153	12·3693168769	216	14·6969384567
28	5·2915026221	91	9·5393920142	154	12·4096736460	217	14·7309198627
29	5·3851648071	92	9·5916630466	155	12·4498995980	218	14·7648230602
30	5·4772255751	93	9·6436507610	156	12·4899959968	219	14·7986485869
31	5·5677643628	94	9·6953597148	157	12·5299640861	220	14·8323969742
32	5·6568542495	95	9·7467943448	158	12·5698050900	221	14·8660687473
33	5·7445626465	96	9·7979589711	159	12·6095202129	222	14·8996644258
34	5·8309518948	97	9·8488578018	160	12·6491106407	223	14·9331845231
35	5·9160797831	98	9·8994949366	161	12·6885775404	224	14·9666295471
36	6·0000000000	99	9·9498743711	162	12·7279220614	225	15·0000000000
37	6·0827625303	100	10·0000000000	163	12·7671453348	226	15·0332963784
38	6·1644140030	101	10·0498756211	164	12·8062484749	227	15·0665191733
39	6·2449979984	102	10·0995049384	165	12·8452325787	228	15·0996688705
40	6·3245553203	103	10·1488915651	166	12·8840987267	229	15·1327459504
41	6·4031242374	104	10·1980390272	167	12·9228479833	230	15·1657508881
42	6·4807406984	105	10·2469507660	168	12·9614813968	231	15·1986841536
43	6·5574385243	106	10·2956301410	169	13·0000000000	232	15·2315462117
44	6·6332495807	107	10·3440804328	170	13·0384048104	233	15·2643375225
45	6·7082039325	108	10·3923048454	171	13·0766968306	234	15·2970585408
46	6·7823299831	109	10·4403065089	172	13·1148770486	235	15·3297097168
47	6·8556546004	110	10·4880884817	173	13·1529464380	236	15·3622914957
48	6·9282032303	111	10·5356537529	174	13·1909059583	237	15·3948043183
49	7·0000000000	112	10·5830052443	175	13·2287565553	238	15·4272486209
50	7·0710678119	113	10·6301458127	176	13·2664991614	239	15·4596248337
51	7·1414284285	114	10·6770782520	177	13·3041346957	240	15·4919333848
52	7·2111025509	115	10·7238052948	178	13·3416640641	241	15·5241746963
53	7·2801098893	116	10·7703296143	179	13·3790881603	242	15·5563491861
54	7·3484692283	117	10·8166538264	180	13·4164078650	243	15·5884572681
55	7·4161984871	118	10·8627804912	181	13·4536240471	244	15·6204993518
56	7·4833147735	119	10·9087121146	182	13·4907375632	245	15·6524758425
57	7·5498344353	120	10·9544511501	183	13·5277492585	246	15·6843871414
58	7·6157731059	121	11·0000000000	184	13·5646599663	247	15·7162336455
59	7·6811457479	122	11·0453610172	185	13·6014705087	248	15·7480157480
60	7·7459666924	123	11·0905365064	186	13·6381816970	249	15·7797338381
61	7·8102496759	124	11·1355287257	187	13·6747913312	250	15·8113883008
62	7·8740078740	125	11·1803398875	188	13·7113092008	251	15·8429795178
63	7·9372539332	126	11·2249721603	189	13·7477270849	252	15·8745078664

Table of Square Roots.

No.	Square Root.	No.	Square Root.	No.	Square Root.	No.	Square Root.
253	15·9059737206	316	17·7763888346	379	19·4679223339	442	21·0237960416
254	15·9373774505	317	17·8044938148	380	19·4935886896	443	21·0475651798
255	15·9687194227	318	17·8325545001	381	19·5192212959	444	21·0713075057
256	16·0000000000	319	17·8605710995	382	19·5448202857	445	21·0950231097
257	16·0312195419	320	17·8885438200	383	19·5703857908	446	21·1187120819
258	16·0623784042	321	17·9164728672	384	19·5959179423	447	21·1423745119
259	16·0934769394	322	17·9443584449	385	19·6214168703	448	21·1660104885
260	16·1245154966	323	17·9722007556	386	19·6468827044	449	21·1896201004
261	16·1554944214	324	18·0000000000	387	19·6723155729	450	21·2132034356
262	16·1864140562	325	18·0277563773	388	19·6977156036	451	21·2367605816
263	16·2172747402	326	18·0554700853	389	19·7230829231	452	21·2602916255
264	16·2480768092	327	18·0831413200	390	19·7484176581	453	21·2837966538
265	16·2788205961	328	18·1107702763	391	19·7737199333	454	21·3072757527
266	16·3095054303	329	18·1383571472	392	19·7989898732	455	21·3307290077
267	16·3401346384	330	18·1659021246	393	19·8242276016	456	21·3541565041
268	16·3707055437	331	18·1934053987	394	19·8494332413	457	21·3775583264
269	16·4012194669	332	18·2208671583	395	19·8746069144	458	21·4009345590
270	16·4316767252	333	18·2482875909	396	19·8997487421	459	21·4242852856
271	16·4620776332	334	18·2756668825	397	19·9248588452	460	21·4476105895
272	16·4924225025	335	18·3030052177	398	19·9499373433	461	21·4709105536
273	16·5227116419	336	18·3303027798	399	19·9749843554	462	21·4941852579
274	16·5529453569	337	18·3575597507	400	20·0000000000	463	21·5174347914
275	16·5831239518	338	18·3847763169	401	20·0249843945	464	21·5406592285
276	16·6132477258	339	18·4119526395	402	20·0499376558	465	21·5638586528
277	16·6433169771	340	18·4390889146	403	20·0748598999	466	21·5870331449
278	16·6733320005	341	18·4661853126	404	20·0997512422	467	21·6101827850
279	16·7032930885	342	18·4932420089	405	20·1246117975	468	21·6333076528
280	16·7332005307	343	18·5202591775	406	20·1494416796	469	21·6564078277
281	16·7630546142	344	18·5472369910	407	20·1742410018	470	21·6794833887
282	16·7928556237	345	18·5741756210	408	20·1990098767	471	21·7025344142
283	16·8226038413	346	18·6010752377	409	20·2237484162	472	21·7255609824
284	16·8522995464	347	18·6279360102	410	20·2484567313	473	21·7485631709
285	16·8819430161	348	18·6547581062	411	20·2731349327	474	21·7715410571
286	16·9115345253	349	18·6815416923	412	20·2977831302	475	21·7944947177
287	16·9410743461	350	18·7082869339	413	20·3224014329	476	21·8174242293
288	16·9705627485	351	18·7349939952	414	20·3469899494	477	21·8403296678
289	17·0000000000	352	18·7616630393	415	20·3715487875	478	21·8632111091
290	17·0293863659	353	18·7882942281	416	20·3960780544	479	21·8860686282
291	17·0587221092	354	18·8148877222	417	20·4205778567	480	21·9089023002
292	17·0880074906	355	18·8414436814	418	20·4450483003	481	21·9317121995
293	17·1172427686	356	18·8679622641	419	20·4694894905	482	21·9544984024
294	17·1464281995	357	18·8944436277	420	20·4939015319	483	21·9772609758
295	17·1755640373	358	18·9208879284	421	20·5182845287	484	22·0000000000
296	17·2046505341	359	18·9472953215	422	20·5426385842	485	22·0227155455
297	17·2336879396	360	18·9736659610	423	20·5669638012	486	22·0454076850
298	17·2626765016	361	19·0000000000	424	20·5912602820	487	22·0680764907
299	17·2916164658	362	19·0262975904	425	20·6155281281	488	22·0907220344
300	17·3205080757	363	19·0525588833	426	20·6397674406	489	22·1133443875
301	17·3493515729	364	19·0787840283	427	20·6639783198	490	22·1359436212
302	17·3781471969	365	19·1049731745	428	20·6881608656	491	22·1585198062
303	17·4068951855	366	19·1311264697	429	20·7123151772	492	22·1810730128
304	17·4355957742	367	19·1572440607	430	20·7364413533	493	22·2036033112
305	17·4642491966	368	19·1833260933	431	20·7605394920	494	22·2261107709
306	17·4928556845	369	19·2093727123	432	20·7846096908	495	22·2485954613
307	17·5214154679	370	19·2353840617	433	20·8086520467	496	22·2710574513
308	17·5499287748	371	19·2613602843	434	20·8326666560	497	22·2934968096
309	17·5783958312	372	19·2873015220	435	20·8566536146	498	22·3159136044
310	17·6068168617	373	19·3132079158	436	20·8806130178	499	22·3383079039
311	17·6351920885	374	19·3390537514	437	20·9045449604	500	22·3606797750
312	17·6635217327	375	19·3649167310	438	20·9284495365	501	22·3830292856
313	17·6918060130	376	19·3907194297	439	20·9523268398	502	22·4053565024
314	17·7200451467	377	19·4164878389	440	20·9761769634	503	22·4276614920
315	17·7482393493	378	19·4422220952	441	21·0000000000	504	22·4499441206

Table of Square Roots.

No.	Square Root.	No.	Square Root.	No.	Square Root.	No.	Square Root.
505	22·4722050542	568	23·8327505756	631	25·1197133742	694	26·3438797446
506	22·4944437584	569	23·8537208838	632	25·1396101800	695	26·3628526529
507	22·5166604984	570	23·8746727726	633	25·1594912508	696	26·3818119165
508	22·5388553392	571	23·8956062907	634	25·1793566201	697	26·4007575649
509	22·5610283454	572	23·9165214862	635	25·1992063367	698	26·4196896272
510	22·5831795813	573	23·9374184072	636	25·2190404258	699	26·4386081328
511	22·6053091109	574	23·9582971014	637	25·2388589282	700	26·4575131106
512	22·6274169980	575	23·9791576166	638	25·2586618806	701	26·4764045897
513	22·6495033058	576	24·0000000000	639	25·2784493195	702	26·4952825990
514	22·6715680975	577	24·0208242989	640	25·2982212813	703	26·5141471671
515	22·6936114358	578	24·0416305603	641	25·3179778023	704	26·5329983228
516	22·7156333832	579	24·0624188310	642	25·3377189186	705	26·5518360947
517	22·7376340018	580	24·0831683962	643	25·3574446662	706	26·5706605112
518	22·7596133535	581	24·1039415864	644	25·3771550809	707	26·5894716006
519	22·7815714998	582	24·1246761636	645	25·3968501984	708	26·6082693913
520	22·8035085020	583	24·1453929353	646	25·4165300543	709	26·6270539114
521	22·8254244210	584	24·1660919472	647	25·4361946849	710	26·6458251889
522	22·8473193176	585	24·1867732449	648	25·4558441227	711	26·6645832519
523	22·8691932521	586	24·2074368736	649	25·4754784057	712	26·6833281283
524	22·8910462845	587	24·2280828792	650	25·4950975680	713	26·7020598456
525	22·9128784748	588	24·2487113060	651	25·5147016443	714	26·7207784318
526	22·9346898824	589	24·2693221990	652	25·5342906696	715	26·7394839142
527	22·9564805665	590	24·2899156030	653	25·5538646784	716	26·7581763205
528	22·9782505862	591	24·3104915623	654	25·5734237051	717	26·7768556780
529	23·0000000000	592	24·3310501212	655	25·5929677841	718	26·7955220139
530	23·0217288664	593	24·3515913238	656	25·6124969497	719	26·8141753556
531	23·0434372436	594	24·3721152139	657	25·6320112360	720	26·8328157300
532	23·0651251893	595	24·3926218353	658	25·6515106768	721	26·8514431642
533	23·0867927612	596	24·4131112315	659	25·6709953060	722	26·8700576851
534	23·1084400166	597	24·4335834457	660	25·6904651573	723	26·8886593195
535	23·1300670124	598	24·4540385213	661	25·7099202544	724	26·9072480941
536	23·1516738056	599	24·4744765010	662	25·7293606605	725	26·9258240357
537	23·1732604525	600	24·4948974278	663	25·7487863792	726	26·9443871706
538	23·1948270095	601	24·5153013443	664	25·7681974535	727	26·9629375254
539	23·2163735325	602	24·5356882928	665	25·7875939165	728	26·9814751265
540	23·2379000772	603	24·5560583156	666	25·8069758011	729	27·0000000000
541	23·2594066992	604	24·5764114549	667	25·8263431403	730	27·0185121722
542	23·2808934536	605	24·5967477525	668	25·8456959666	731	27·0370116692
543	23·3023603955	606	24·6170672502	669	25·8650343128	732	27·0554985169
544	23·3238075794	607	24·6373699895	670	25·8843582111	733	27·0739727414
545	23·3452350599	608	24·6576560119	671	25·9036676940	734	27·0924343683
546	23·3666428911	609	24·6779253585	672	25·9229627936	735	27·1108834235
547	23·3880311271	610	24·6981780705	673	25·9422435421	736	27·1293199325
548	23·4093998214	611	24·7184141836	674	25·9615099715	737	27·1477439210
549	23·4307490277	612	24·7386337537	675	25·9807621135	738	27·1661554144
550	23·4520787991	613	24·7588368063	676	26·0000000000	739	27·1845544381
551	23·4733891886	614	24·7790233867	677	26·0192236625	740	27·2029410175
552	23·4946802489	615	24·7991935353	678	26·0384331326	741	27·2213151776
553	23·5159520326	616	24·8193472920	679	26·0576284416	742	27·2396769438
554	23·5372045919	617	24·8394846967	680	26·0768096208	743	27·2580263409
555	23·5584379788	618	24·8596057893	681	26·0959767014	744	27·2763633940
556	23·5796522451	619	24·8797106092	682	26·1151297144	745	27·2946881279
557	23·6008474424	620	24·8997991960	683	26·1342686907	746	27·3130005671
558	23·6220236220	621	24·9198715888	684	26·1533936612	747	27·3313007374
559	23·6431808351	622	24·9399278267	685	26·1725046566	748	27·3495886624
560	23·6643191324	623	24·9599679487	686	26·1916017074	749	27·3678643668
561	23·6854385647	624	24·9799919936	687	26·2106848442	750	27·3861278753
562	23·7065391823	625	25·0000000000	688	26·2297540972	751	27·4043792121
563	23·7276210354	626	25·0199920064	689	26·2488094968	752	27·4226184016
564	23·7486841741	627	25·0399680511	690	26·2678510731	753	27·4408454680
565	23·7697286480	628	25·0599281723	691	26·2868788562	754	27·4590604355
566	23·7907545067	629	25·0798724080	692	26·3058928759	755	27·4772633281
567	23·8117617996	630	25·0998007960	693	26·3248931622	756	27·4954541697

Table of Square Roots.

No.	Square Root.	No.	Square Root.	No.	Square Root.	No.	Square Root.
757	27·5136329844	818	28·6006992922	879	29·6479324743	940	30·6594194335
758	27·5317997959	819	28·6181760425	880	29·6647939484	941	30·6757233004
759	27·5499546279	820	28·6356421266	881	29·6816441593	942	30·6920185064
760	27·5680975042	821	28·6530975638	882	29·6984848098	943	30·7083050656
761	27·5862284483	822	28·6705423737	883	29·7153159162	944	30·7245829915
762	27·6043474837	823	28·6879765756	884	29·7321374946	945	30·7408522979
763	27·6224546339	824	28·7054001888	885	29·7489495613	946	30·7571129985
764	27·6405499222	825	28·7228132327	886	29·7657521323	947	30·7733651069
765	27·6586333719	826	28·7402157264	887	29·7825452237	948	30·7896086367
766	27·6767050062	827	28·7576076891	888	29·7993288515	949	30·8058436015
767	27·6947648483	828	28·7749891399	889	29·8161030318	950	30·8220700148
768	27·7128129211	829	28·7923600978	890	29·8328677804	951	30·8382878902
769	27·7308492477	830	28·8097205818	891	29·8496231132	952	30·8544972417
770	27·7488738510	831	28·8270706108	892	29·8663690461	953	30·8706980809
771	27·7668867538	832	28·8444102037	893	29·8831055950	954	30·8868904230
772	27·7848879789	833	28·8617393793	894	29·8998327755	955	30·9030742807
773	27·8028775489	834	28·8790581564	895	29·9165506033	956	30·9192496675
774	27·8208554865	835	28·8963665536	896	29·9332590942	957	30·9354165965
775	27·8388218142	836	28·9136645896	897	29·9499582637	958	30·9515750811
776	27·8567765544	837	28·9309522830	898	29·9666481275	959	30·9677251344
777	27·8747197295	838	28·9482296523	899	29·9833287011	960	30·9838667697
778	27·8926513620	839	28·9654967159	900	30·0000000000	961	31·0000000000
779	27·9105714739	840	28·9827534924	901	30·0166620396	962	31·0161248385
780	27·9284800875	841	29·0000000000	902	30·0333148354	963	31·0322412984
781	27·9463772250	842	29·0172362571	903	30·0499584026	964	31·0483493925
782	27·9642629082	843	29·0344622819	904	30·0665927567	965	31·0644491340
783	27·9821371593	844	29·0516780927	905	30·0832179130	966	31·0805405358
784	28·0000000000	845	29·0688837075	906	30·0998338866	967	31·0966236109
785	28·0178514522	846	29·0860791445	907	30·1164406928	968	31·1126983722
786	28·0356915378	847	29·1032644217	908	30·1330383466	969	31·1287648325
787	28·0535202782	848	29·1204395571	909	30·1496268634	970	31·1448230048
788	28·0713376881	849	29·1376015687	910	30·1662062580	971	31·1608729018
789	28·0891438104	850	29·1547594742	911	30·1827765456	972	31·1769145362
790	28·1069386451	851	29·1719042916	912	30·1993377411	973	31·1929479210
791	28·1247222209	852	29·1890390387	913	30·2158898595	974	31·2089730687
792	28·1424945589	853	29·2061637330	914	30·2324329157	975	31·2249899920
793	28·1602556807	854	29·2232783924	915	30·2489669245	976	31·2409987036
794	28·1780056072	855	29·2403830344	916	30·2654919008	977	31·2569992162
795	28·1957443597	856	29·2574776767	917	30·2820078595	978	31·2729915422
796	28·2134719593	857	29·2745623366	918	30·2985148151	979	31·2889756943
797	28·2311884270	858	29·2916370318	919	30·3150127824	980	31·3049516850
798	28·2488937837	859	29·3087017795	920	30·3315017762	981	31·3209195267
799	28·2665880502	860	29·3257565972	921	30·3479818110	982	31·3368792320
800	28·2842712475	861	29·3428015022	922	30·3644529014	983	31·3528308132
801	28·3019433962	862	29·3598365118	923	30·3809150619	984	31·3687742827
802	28·3196045170	863	29·3768616431	924	30·3973683071	985	31·3847096530
803	28·3372546306	864	29·3938769134	925	30·4138126515	986	31·4006369362
804	28·3548937575	865	29·4108823397	926	30·4302481094	987	31·4165561448
805	28·3725219182	866	29·4278779391	927	30·4466746953	988	31·4324672910
806	28·3901391332	867	29·4448637287	928	30·4630924235	989	31·4483703870
807	28·4077454227	868	29·4618397253	929	30·4795013083	990	31·4642654451
808	28·4253408071	869	29·4788059460	930	30·4959013640	991	31·4801524774
809	28·4429253067	870	29·4957624075	931	30·5122926048	992	31·4960314960
810	28·4604989415	871	29·5127091267	932	30·5286750449	993	31·5119025132
811	28·4780617318	872	29·5296461205	933	30·5450486986	994	31·5277655409
812	28·4956136976	873	29·5465734054	934	30·5614135799	995	31·5436205912
813	28·5131548588	874	29·5634909982	935	30·5777697028	996	31·5594676761
814	28·5306852354	875	29·5803989155	936	30·5941170816	997	31·5753068077
815	28·5482048472	876	29·5972971739	937	30·6104557300	998	31·5911379979
816	28·5657137142	877	29·6141857899	938	30·6267856622	999	31·6069612586
817	28·5832118559	878	29·6310647801	939	30·6431068921	1000	31·6227766017

ROTA, in Mechanics. See WHEEL.

ROTA *Ariſtotelica*, or *Ariſtotle's Wheel*, denotes a celebrated problem in mechanics, concerning the motion or rotation of a wheel about its axis ; ſo called becauſe firſt noticed by Ariſtotle.

The difficulty is this. While a circle makes a revolution on its centre, advancing at the ſame time in a right line along a plane, it deſcribes, on that plane, a right line which is equal to its circumference. Now if this circle, which may be called the deferent, carry with it another ſmaller circle, concentric with it, like the nave of a coach wheel ; then this little circle, or nave, will deſcribe a line in the time of the revolution, which ſhall be equal to that of the large wheel or circumference itſelf; becauſe its centre advances in a right line as faſt as that of the wheel does, being in reality the ſame with it.

The ſolution given by Ariſtotle, is no more than a good explication of the difficulty.

Galileo, who next attempted it, has recourſe to an infinite number of infinitely little vacuities in the right line deſcribed by the two circles ; and imagines that the little circle never applies its circumference to thoſe vacuities ; but in reality only applies it to a line equal to its own circumference ; though it appears to have applied it to a much larger. But all this is nothing to the purpoſe.

Tacquet will have it, that the little circle, making its rotation more ſlowly than the great one, does on that account deſcribe a line longer than its own circumference ; yet without applying any point of its circumference to more than one point of its baſe. But this is no more ſatisfactory than the former.

After the fruitleſs attempts of ſo many great men, M. Dortous de Meyran, a French gentleman, had the good fortune to hit upon a ſolution, which he ſent to the Academy of Sciences ; where being examined by Meſſ. de Lonville and Soulmon, appointed for that purpoſe, they made their report that it was ſatisfactory. The ſolution is to this effect :

The wheel of a coach is only acted on, or drawn in a right line ; its rotation or circular motion ariſes purely from the reſiſtance of the ground upon which it is applied. Now this reſiſtance is equal to the force which draws the wheel in the right line, inaſmuch as it defeats that direction ; of conſequence the cauſes of the two motions, the one right and the other circular, are equal. And hence the wheel deſcribes a right line on the ground equal to its circumference.

As for the nave of the wheel, the caſe is otherwiſe. It is drawn in a right line by the ſame force as the wheel ; but it only turns round becauſe the wheel does ſo, and can only turn in the ſame time with it. Hence it follows, that its circular velocity is leſs than that of the wheel, in the ratio of the two circumferences ; and therefore its circular motion is leſs than the rectilinear one. Since then it neceſſarily deſcribes a right line equal to that of the wheel, it can only do it partly by ſliding, and partly by revolving, the ſliding part being more or leſs as the nave itſelf is ſmaller or larger. See CYCLOID.

ROTATION, *Rolling*, in Mechanics. See ROLLING.

ROTATION, in Geometry, the circumvolution of a

ſurface round an immoveable line, called the *axis of Rotation*. By ſuch Rotation of planes, the figures of certain regular ſolids are formed or generated. Such as, a cylinder by the Rotation of a rectangle, a cone by the Rotation of a triangle, a ſphere or globe by the Rotation of a ſemicircle, &c.

The method of cubing ſolids that are generated by ſuch Rotation, is laid down by Mr. Demoivre, in his ſpecimen of the uſe of the doctrine of fluxions, Philoſ. Tranſ. numb. 216 ; and indeed by moſt of the writers on Fluxions. In every ſuch ſolid, all the ſections perpendicular to the axis are circles, and therefore the fluxion of the ſolid, at any ſection, is equal to that circle multiplied by the fluxion of the axis. So that, if x denote an abſciſs of that axis, and y an ordinate to it in the revolving plane, which will alſo be the radius of that circle ; then, n being put for 3·1416, the area of the circle is ny^2, and conſequently the fluxion of the ſolid is $ny^2\dot{x}$; the fluent of which will be the content of the ſolid.

Such ſolid may alſo be expreſſed in terms of the generating plane and its centre of gravity ; for the ſolid is always equal to the product ariſing from the generating plane multiplied by the path of its centre of gravity, or by the line deſcribed by that centre in the Rotation of the plane. And this theorem is general, by whatever kind of motion the plane is moved, in deſcribing a ſolid.

ROTATION, *Revolution*, in Aſtronomy. See REVOLUTION.

Diurnal ROTATION. See DIURNAL, and EARTH.

ROTONDO, or ROTUNDO, in Architecture, a popular term for any building that is round both within and withoutſide, whether it be a church, hall, a ſaloon, a veſtibule, or the like.

ROUND, ROUNDNESS, ROTUNDITY, the property of a circle and ſphere or globe &c.

ROWNING (JOHN), an ingenious Engliſh mathematician and philoſopher, was fellow of Magdalen College, Cambridge, and afterwards Rector of Anderby in Lincolnſhire, in the gift of that ſociety. He was a conſtant attendant at the meetings of the Spalding Society, and was a man of a great philoſophical habit and turn of mind, though of a cheerful and companionable diſpoſition. He had a good genius for mechanical contrivances in particular. In 1738 he printed at Cambridge, in 8vo, *A Compendious Syſtem of Natural Philoſophy*, in 2 vols 8vo ; a very ingenious work, which has gone through ſeveral editions. He had alſo two pieces inſerted in the Philoſophical Tranſactions, viz, 1. *A Deſcription of a Barometer wherein the Scale of Variation may be increaſed at pleaſure* ; vol. 38, pa. 39. And 2. *Direction for making a Machine for finding the Roots of Equations univerſally, with the Manner of uſing it* ; vol. 60, pa. 240.—Mr. Rowning died at his lodgings in Carey-ſtreet near Lincoln's-Inn Fields, the latter end of November 1771, at 72 years of age.

Though a very ingenious and pleaſant man, he had but an unpromiſing and forbidding appearance : he was tall, ſtooping in the ſhoulders, and of a ſallow downlooking countenance.

ROYAL *Oak*, *Robur Carolinum*, in Aſtronomy, one of the new ſouthern conſtellations, the ſtars of which,

which, according to Sharp's catalogue, annexed to the Britannic, are 12.

ROYAL *Society of England*, is an academy or body of persons, supposed to be eminent for their learning, instituted by king Charles the IId, for promoting natural knowledge.

This once illustrious body originated from an assembly of ingenious men, residing in London, who, being inquisitive into natural knowledge, and the new and experimental philosophy, agreed, about the year 1645, to meet weekly on a certain day, to discourse upon such subjects. These meetings, it is said, were suggested by Mr. Theodore Haak, a native of the Palatinate in Germany; and they were held sometimes at Dr. Goddard's lodgings in Wood-street, sometimes at a convenient place in Cheapside, and sometimes in or near Gresham College. This assembly seems to be that mentioned under the title of the *Invisible, or Philosophical College*, by Mr. Boyle, in some letters written in 1646 and 1647. About the years 1648 and 1649, the company which formed these meetings, began to be divided, some of the gentlemen removing to Oxford, as Dr. Wallis, and Dr. Goddard, where, in conjunction with other gentlemen, they held meetings also, and brought the study of natural and experimental philosophy into fashion there; meeting first in Dr. Petty's lodgings, afterwards at Dr. Wilkins's apartments in Wadham College, and, upon his removal, in the lodgings of Mr. Robert Boyle; while those gentlemen who remained in London continued their meetings as before. The greater part of the Oxford Society coming to London about the year 1659, they met once or twice a week in Term-time at Gresham College, till they were dispersed by the public distractions of that year, and the place of their meeting was made a quarter for soldiers. Upon the restoration, in 1660, their meetings were revived, and attended by many gentlemen, eminent for their character and learning.

They were at length noticed by the government, and the king granted them a charter, first the 15th of July 1662, then a more ample one the 22d of April 1663, and thirdly the 8th of April 1669; by which they were erected into a corporation, *consisting of a president, council, and fellows, for promoting natural knowledge*, and endued with various privileges and authorities.

Their manner of electing members is by balloting; and two-thirds of the members present are necessary to carry the election in favour of the candidate. The council consists of 21 members, including the president, vice-president, treasurer, and two secretaries; ten of which go out annually, and ten new members are elected instead of them, all chosen on St. Andrew's day. They had formerly also two curators, whose business it was to perform experiments before the society.

Each member, at his admission, subscribes an engagement, that he will endeavour to promote the good of the society; from which he may be freed at any time, by signifying to the president that he desires to withdraw.

The charges are five guineas paid to the treasurer at admission; and one shilling per week, or 52s. per year, as long as the person continues a member; or, in lieu of the annual subscription, a composition of 25 guineas in one payment.

The ordinary meetings of the society, are once a week, from November till the end of Trinity term the next summer. At first, the meeting was from 3 o'clock till 6 afternoon. Afterwards, their meeting was from 6 till 7 in the evening, to allow more time for dinner, which continued for a long series of years, till the hour of meeting was removed, by the present president, to between 8 and 9 at night, that gentlemen of fashion, as was alleged, might have the opportunity of coming to attend the meetings after dinner.

Their design is to " make faithful records of all the " works of nature or art, which come within their " reach; so that the present, as well as after ages, " may be enabled to put a mark on errors which have " been strengthened by long prescription; to restore " truths that have been long neglected; to push those " already known to more various uses; to make the " way more passable to what remains unrevealed, " &c."

To this purpose they have made a great number of experiments and observations on most of the works of nature; as eclipses, comets, planets, meteors, mines, plants, earthquakes, inundations, springs, damps, fires, tides, currents, the magnet, &c: their motto being *Nullius in Verba*. They have registered experiments, histories, relations, observations, &c, and reduced them into one common stock. They have, from time to time, published some of the most useful of these, under the title of Philosophical Transactions, &c. usually one volume each year, which were, till lately, very respectable, both for the extent or magnitude of them, and for the excellent quality of their contents. The rest, that are not printed, they lay up in their registers.

They have a good library of books, which has been formed, and continually augmenting, by numerous donations. They had also a museum of curiosities in nature, kept in one of the rooms of their own house in Crane Court Fleet-street, where they held their meetings, with the greatest reputation, for many years, keeping registers of the weather, and making other experiments; for all which purposes those apartments were well adapted. But, disposing of these apartments, in order to remove into those allotted them in Somerset Place, where having neither room nor convenience for such purposes, the museum was obliged to be disposed of, and their useful meteorological registers discontinued for many years.

Sir Godfrey Copley, bart. left 5 guineas to be given annually to the person who should write the best paper in the year, under the head of experimental philosophy: this reward, which is now changed to a gold medal, is the highest honour the society can bestow; and it is conferred on St. Andrew's day: but the communications of late years have been thought of so little importance, that the prize medal remains sometimes for years undisposed of.

Indeed this once very respectable society, now consisting of a great proportion of honorary members, who do not usually communicate papers; and many scientific members being discouraged from making their

usual

ufual communications, by what is deemed the prefent arbitrary government of the fociety; the annual volumes have in confequence become of much lefs importance, both in refpect of their bulk and the quality of their contents.

ROYAL *Society of Scotland*. See SOCIETY.

RUDOLPHINE *Tables*, a fet of aftronomical tables that were publifhed by the celebrated Kepler, and fo called from the emperor Rudolph or Rudolphus.

RULE, *The Carpenters*, a folding ruler generally ufed by carpenters and other artificers; and is otherwife called the fliding Rule.

This inftrument confifts of two equal pieces of boxwood, each one foot in length, connected together by a folding joint. One fide or face, of the Rule, is divided into inches, and half-quarters, or eighths. On the fame face alfo are feveral plane fcales, divided into 12th parts by diagonal lines; which are ufed in planning dimenfions that are taken in feet and inches. The edge of the Rule is commonly divided decimally, or into 10ths; viz, each foot into 10 equal parts, and each of thefe into 10 parts again, or 100dth parts of the foot: fo that by means of this laft fcale, dimenfions are taken in feet and tenths and hundreds, and multiplied together as common decimal numbers, which is the beft way.

On the one part of the other face are four lines, marked A, B, C, D, the two middle ones B and C being on a flider, which runs in a groove made in the ftock. The fame numbers ferve for both thefe two middle lines, the one line being above the numbers, and the other below them.

Thefe four lines are logarithmic ones, and the three A, B, C, which are all equal to one another, are double lines, as they proceed twice over from 1 to 10. The loweft line D is a fingle one, proceeding from 4 to 40. It is alfo called the girt line, from its ufe in cafting up the contents of trees and timber: and upon it are marked WG at 17·15, and AG at 18·95, the wine and ale gauge points, to make this inftrument ferve the purpofe of a gauging rule.

Upon the other part of this face is a table of the value of a load, or 50 cubic feet, of timber, at all prices, from 6 pence to 2s. a foot.

When 1 at the beginning of any line is accounted only 1, then the 1 in the middle is 10, and the 10 at the end 100; and when the 1 at the beginning is accounted 10, then 1 in the middle is 100, and the 10 at the end 1000; and fo on. All the fmaller divifions being alfo altered proportionally.

By means of this Rule all the ufual operations of arithmetic may be eafily and quickly performed, as multiplication, divifion, involution, evolution, finding mean proportionals, 3d and 4th proportionals, or the Rule of-three, &c. For all which, fee my Menfuration, part 5, fect. 3, 2d edition.

RULES *of Philofophizing*. See PHILOSOPHIZING.

RULE, in Arithmetic, denotes a certain mode of operation with figures to find fums or numbers unknown, and to facilitate computations.

Each Rule in arithmetic has its particular name, according to the ufe for which it is intended. The firft four, which ferve as a foundation of the whole art, are

called *addition, fubtraction, multiplication,* and *divifion*.

From thefe arife numerous other Rules, which are indeed only applications of thefe to particular purpofes and occafions; as the Rule-of-three, or Golden Rule, or Rule of Proportion; alfo the Rules of Fellowfhip, Intereft, Exchanges, Pofition, Progreffions, &c, &c. For which, fee each article feverally.

RULE-*of-Three*, or *Rule of Proportion*, commonly called the *Golden Rule* from its great ufe, is a Rule that teaches how to find a 4th proportional number to three others that are given.

As, if 3 degrees of the equator contain 208 miles, how many are contained in 360 degrees, or the whole circumference of the earth?

The Rule is this: State, or fet the three given terms down in the form of the firft three terms of a proportion, ftating them proportionally, thus:

$$
\begin{array}{cc}
\text{deg. mil.} & \text{deg. miles.} \\
\text{as} \quad 3 : 208 :: & 360 : 24960 \\
& 360
\end{array}
$$

$$
\begin{array}{r}
12480 \\
624 \\
\hline
3\,)74880 \\
\hline
24960
\end{array}
$$

Then multiply the 2d and 3d terms together, and divide the product by the 1ft term, fo fhall the quotient be the 4th term in proportion, or the anfwer to the queftion, which in this example is 24960 or nearly 25 thoufand miles, for the circumference of the earth.

This rule is often confidered as of two kinds, viz. *Direct*, and *Inverfe*.

Rule-of-Three Direct, is that in which more requires more, or lefs requires lefs. As in this; if 3 men mow 21 yards of grafs in a certain time, how much will 6 men mow in the fame time? Here more requires more, that is, 6 men, which are more than 3 men, will alfo perform more work, in the fame time. Or if it were thus; if 6 men mow 42 yards, how much will 3 men mow in the fame time? here then lefs requires lefs, or 3 men will perform proportionally lefs work, in the fame time. In both thefe cafes then, the Rule, or the proportion, is direct; and the ftating muft be

thus, as 3 : 21 :: 6 : 42,

or thus, as 6 : 42 :: 3 : 21.

Rule-of-Three Inverfe, is when more requires lefs, or lefs requires more. As in this; if 3 men mow a certain quantity of grafs in 14 hours, in how many hours will 6 men mow the like quantity? Here it is evident that 6 men, being more than 3, will perform the fame work in lefs time, or fewer hours; hence then more requires lefs, and the Rule or queftion is inverfe, and muft be ftated by making the number of men change places, thus, as 6 : 14 :: 3 : 7 hours, the time in which 6 men will perform the work; ftill multiplying the 2d and 3d terms together, and dividing by the 1ft.

For various abbreviations, and other particulars relating

5

lating to thefe Rules, fee any of the common books of arithmetic.

Rule-*of-Five*, or *Compound Rule-of-Three*, is where two Rules-of three are required to be wrought, or to be combined together, to find out the number fought.

This Rule may be performed, either by working the two ftatings or proportions feparately, making the refult or 4th term of the 1ft operation to be the 2d term of the laft proportion; or elfe by reducing the two ftatings into one, by multiplying the two firft terms together, and the two third terms together, and ufing the products as the 1ft and 3d terms of the compound ftating. As, if the queftion be this: If 100l. in 2 years yield 9l. intereft, how much will 500l. yield in 6 years. Here, the two ftatings are,

$$\left.\begin{matrix}100\\2\end{matrix}\right\} : 9 :: \left\{\begin{matrix}500\\6\end{matrix}\right.$$

Then, to work the two ftatings feparately,

as 100 : 9 :: 500 : 45l.
and 2 : 45 :: 6 : 135l.

fo that 135l. is the intereft or anfwer fought. But to work by one ftating, it will be thus,

$$\begin{matrix} 100 & & 500 \\ 2 & & 6 \\ \hline 200 : 9 :: 3000 : 135l. \text{ the anfwer.} \end{matrix}$$

2'00) 270.00 (135l.

See the books of arithmetic for more particulars.
Central Rule. See Central *Rule.*
Parallel Ruler. See Parallel *Ruler.*
RUMB, or Rum. See Rhumb.
Rumb-*Line*, or *Loxodromic.* See Rhumb-*Line.*
RUSTIC, in Architecture, denotes a manner of building in imitation of fimple or rude nature, rather than according to the rules of art.
Rustic *Quoins.* See Quoin.
Rustic *Work* is where the ftones in the face &c of a building, inftead of being fmooth, are hatched or picked with the point of an inftrument.

Regular Rustics, are thofe in which the ftones are chamfered off at the edges, and form angular or fquare receffes of about an inch deep at their jointings, or beds, and ends.

Rustic *Order*, is an order decorated with ruftic quoins, or ruftic work, &c.

RUTHERFORD (Thomas, D. D.), an ingenious Englifh philofopher, was the fon of the Rev. Thomas Rutherford, rector of Papworth Everard in the county of Cambridge, who had made large collections for the hiftory of that county.

Our author was born the 13th of October 1712. He ftudied at Cambridge, and became fellow of St. John's college, and regius profeffor of divinity, in that univerfity; afterwards rector of Shenfield in Effex, and of Barley in Hertfordfhire, and archdeacon of Effex. He died the 5th of October 1771, at 59 years of age.

Dr. Rutherford, befides a number of theological writings, publifhed, at Cambridge,

1. *Ordo Inftitutionum Phyficarum,* 1743, in 4to.
2. *A Syftem of Natural Philofophy,* in 2 vols, 4to, 1748. A work which has been much efteemed.
3. He communicated alfo to the Gentleman's Society at Spalding, a curious correction of Plutarch's defcription of the inftrument ufed to renew the Veftal fire, as relating to the triangle with which the inftrument was formed. It was nothing elfe, it feems, but a concave fpeculum, whofe principal focus, which collected the rays, is not in the centre of concavity, but at the diftance of half a diameter from its furface. But fome of the Ancients thought otherwife, as appears from prop. 31 of Euclid's Catoptrics.

The writer of his epitaph fays, " He was eminent no lefs for his piety and integrity, than his extenfive learning; and filled every public ftation in which he was placed with general approbation. In private life, his behaviour was truly amiable. He was efteemed, beloved, and honoured by his family and friends; and his death was fincerely lamented by all who had ever heard of his well deferved character."

S.

SAI

S, IN books of Navigation, &c, denotes fouth. So alfo S. E. is fouth eaft; S. W. fouth-weft; and S. S. E. fouth-fouth-eaft, &c. See COMPASS.

SAGITTA, in Aftronomy, the *Arrow* or *Dart*, a conftellation of the northern hemifphere near the eagle, and one of the 48 old afterifms. The Greeks fay that this conftellation owes its origin to one of the arrows of Hercules, with which he killed the eagle or vulture that gnawed the liver of Prometheus.

The ftars in this conftellation, in the catalogues of Ptolomy, Tycho, and Hevelius, are only 5, but in Flamfteed's they are extended to 18.

SAGITTA, in Geometry, is a term ufed by fome writers for the abfcifs of a curve.

SAGITTA, in Trigonometry &c, is the fame as the verfed fine of an arch; being fo called becaufe it is like a dart or arrow, ftanding on the chord of the arch.

SAGITTARIUS, SAGITTARY, the *Archer*, one of the figns of the zodiac, being the 9th in order, and marked with the character ♐ of a dart or arrow. This conftellation is drawn in the figure of a Centaur, or an animal half man and half horfe, in the act of fhooting an arrow from a bow. This figure the Greeks feign to be Crotus, the fon of Eupheme, the nurfe of the mufes. Among more ancient nations the figure was probably meant for a hunter, to denote the hunting feafon, when the fun enters this fign.

The ftars in this conftellation are, in Ptolomy's catalogue 31, in Tycho's 14, in Hevelius's 22, and in the Britannic catalogue 69.

SAILING, in a general fenfe, denotes the movement by which a veffel is wafted along the furface of the water, by the action of the wind upon her fails.

Sailing is alfo ufed for the art or act of navigating; or of determining all the cafes of a fhip's motion, by means of fea charts &c. Thefe charts are conftructed either on the fuppofition that the earth is a large extended flat furface, whence we obtain thofe that are called plane charts; or on the fuppofition that the earth is a fphere, whence are derived globular charts. Accordingly, Sailing may be diftinguifhed into two general kinds, viz, *plane Sailing*, and *globular Sailing*. Sometimes indeed a third fort is added, viz, *fpheroidical Sailing*, which proceeds upon the fuppofition of the fpheroidical figure of the earth.

Plane SAILING is that which is performed by means of a plane chart; in which cafe the meridians are confidered as parallel lines, the parallels of latitude are at right angles to the meridians, the lengths of the degrees on the meridians, equator, and parallels of latitude, are every where equal.

SAI

In Plane Sailing, the principal terms and circumftances made ufe of, are, courfe, diftance, departure, difference of latitude, rhumb, &c; for as to longitude, that has no place in plane Sailing, but belongs properly to globular or fpherical failing. For the explanation of all which terms, fee the refpective articles.

If a fhip fails either due north or fouth, fhe fails on a meridian, her diftance and difference of latitude are the fame, and fhe makes no departure; but where the fhip fails either due eaft or weft fhe runs on a parallel of latitude, making no difference of latitude, and her departure and diftance are the fame. It may farther be obferved, that the departure and difference of latitude always make the legs of a right-angled triangle, whofe hypotenufe is the diftance the fhip has failed; and the angles are the courfe, its complement, and the right angle; therefore among thefe four things, courfe, diftance, difference of latitude, and departure, any two of them being given, the reft may be found by plane trigonometry.

Thus, in the annexed figure, fuppofe the circle FHFH to reprefent the horizon of the place A, from whence a fhip fails; AC the rhumb fhe fails upon, and C the place arrived at: then HH reprefents the parallel of latitude fhe failed from, and CC the parallel of the latitude arrived in: fo that

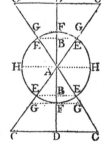

AD becomes the difference of latitude.
DC the departure,
AC the diftance failed,
∠DAC is the courfe, and
∠DCA the comp. of the courfe.

And all thefe particulars will be alike reprefented, whether the fhip fails in the NE, or NW, or SE, or SW quarter of the horizon.

From the fame figure, in which
AE or AF or AH reprefents the rad. of the tables,
EB the fine of the courfe,
AB the cofine of the courfe,

we may eafily deduce all the proportions or canons, as they are ufually called by mariners, that can arife in Plane Sailing; becaufe the triangles ADC and ABE and AFG are evidently fimilar. Thefe proportions are exhibited in the following Table, which confifts of 6 cafes, according to the varieties of the two parts that can be given.

Cafe

Cafe.	Given.	Required.	Solutions.
1	∠A and AC, i. e. courfe and diftance.	AD and DC, i. e. difference of latitude and departure.	AE : AB :: AC : AD, i. e. rad. : f. courfe :: dift. : dif. lat. AE : EB :: AC . DC, i. e. rad. : cof. courfe :: dift. : depart.
2	∠A and AD, i. e. courfe and difference of latitude.	AC and DC, i. e. diftance and departure.	AB : AE :: AD : AC, i. e. cof. cour. : rad. :: dif. lat. : dift. AB : BE :: AD : DC, i. e. cof. cour. : f. cour. :: dif. lat. : dep.
3	∠A and DC, i. e. courfe and departure.	AC and AD, i. e. diftance and difference of latitude.	BE : AE :: DC : AC, i. e. f. cour. : rad. :: depart. : dift. BE : AB :: DC : AD, i. e. f. cour. : cof. cour. :: oep. : dif. lat.
4	AC and AD, i. e. diftance and difference of latitude.	∠A and DC, i. e. courfe and departure.	AC : AD :: AE : AB, i. e. dift. : dif. lat. :: rad. : cof. courfe. AE : EB :: AC : DC, i. e. rad. : f. courfe :: dift. : depart.
5	AC and DC, i. e. diftance and departure.	∠A and AD, i. e. courfe and difference of latitude.	AC : DC :: AE : EB, i. e. dift. : dep. :: rad. : f. courfe. AE : AB :: AC : AD, i. e. rad. : cof. cour. :: dift. : dif. lat.
6	AD and DC, i. e. difference of latitude and departure.	∠A and AC, i. e. courfe and diftance.	AD : DC :: AF : FG, i. e. dif. lat. : dep. :: rad. : tang. courfe. BE : AE :: DC : AC, i. e. f. cour. : rad. :: dep. : dift.

For the ready working of any fingle courfe, there is a table, called a *Traverfe Table*, ufually annexed to treatifes of navigation; which is fo contrived, that by finding the given courfe in it, and a diftance not exceeding ∞ or 120 miles, the ufual extent of the table; then the difference of latitude and the departure are had by infpection. And the fame table will ferve for greater diftances, by doubling, or trebling, or quadrupling, &c; or taking proportional parts. See TRAVERSE *Table.*

An ex. to the firft cafe may fuffice to fhew the method. Thus, A fhip from the latitude 47° 30 N, has failed SW by S 98 miles; required the departure made, and the latitude arrived in.

1. *By the Traverfe Table.* In the column of the courfe, viz 3 points, againft the diftance 98, ftands the number 54 45 miles for the departure, and 81·5 miles for the diff. of lat.; which is 1° 21′½; and this being taken from the given lat. 47° 30′, leaves 46° 8′½ for the lat. come to.

2. *By Conftruction.* Draw the meridian AD; and drawing an arc, with the chord of 60, make PQ or angle A equal to 3 points; through Q draw the diftance AQE = 98 miles, and through E the departure ED perp. to AD. Then, by meafuring, the diff. of lat. AD meafures about 81½ miles, and the departure DE about 54¼ miles.

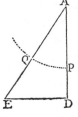

3. *By Computation.*

Firft, as radius - - - - - -	10·00000
to fin. courfe 33° 45′ - - -	9·74474
fo dift. 98 - - - - - - - -	1·99123
to depart. 54·45 - - - -	1·73597
Again, as radius - - - - -	10·00000
to cof. courfe - - - - - -	9·91985
fo dift. 98 - - - - - - - -	1·99123
to diff. of lat. 81·48 - - -	1·91108

4. *By Gunter's Scale.* The extent from radius, or 8 points, to 3 points, on the line of fine rhumbs, applied to the line of numbers, will reach from 98 to 54⅔ the departure. And the extent from 8 points to 5 points, of the rhumbs, reaches from 98 to 81½ on the line of numbers, for the difference of latitude.

And in like manner for other cafes.

Traverfe SAILING, or *Compound Courfes,* is the uniting of feveral cafes of plane failing together into one; as when a fhip fails in a zigzag manner, certain diftances upon feveral different courfes, to find the whole difference of latitude and departure made good on all of them This is done by working all the cafes feparately, by means of the traverfe table, and conftructing the figure as in this example.

3 E 2   Ex. A

Ex. A ſhip ſailing from a place in latitude 24° 32′ N, has run five different courſes and diſtances, as ſet down in the 1ſt and 2d columns of the following traverſe table; required her preſent latitude, with the departure, and the direct courſe and diſtance, between the place ſailed from, and the place come to.

			Traverſe Table.		
Courſes.	Diſt.	N	S	E	W
SW b S	45		25·0		37·4
ESE	50		19·1	46·2	
SW	30		21·2		21·2
SE b E	60		33·3	49·9	
SW b S ¼ W	63		50·6		37·5
			149·2	96·1	96·1

Here, by finding, in the general traverſe table, the difference of latitude and departure anſwering to each courſe and diſtance, they are ſet down on the ſame lines with each courſe, and in their proper columns of northing, ſouthing, eaſting, or weſting, according to the quarter of the compaſs the ſhip ſails in, at each courſe. As here, there is no northing, the differences of latitude are all ſouthward, alſo two departures are eaſtward, and three are weſtward. Then, adding up the numbers in each column, the ſum of the eaſtings appears to be exactly equal to the ſum of the weſtings, conſequently the ſhip is arrived in the ſame meridian, without making any departure; and the ſouthings, or difference of latitude being 149·2 miles or minutes,

that is - - - - - - - - 2° 29′,

which taken from - - 24 32 , the latitude dep. from,
                        ‾‾‾‾‾‾
leaves - - - - - - - - - 22 3 N, the latitude come to.

To Conſtruct this Traverſe.
With the chord of 60 degrees deſcribe the circle N 135 S &c, and quarter it by the two perpendicular diameters; then from S ſet upon it the ſeveral courſes, to the points marked 1, 2, 3, 4, 5, through which points draw lines from the centre A, or conceive them to be drawn; laſtly, upon the firſt line lay off the firſt diſtance 45 from A to B, alſo draw BC = 50 and parallel to A 2, and CD = 30 parallel to A 3, and DE = 60 parallel to A 4, and

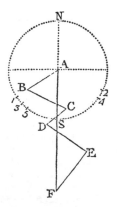

EF = 63, parallel to A 5; then it is found that the point F falls exactly upon the meridian NAF produced, thereby ſhewing that there is no departure; and by meaſuring AF, it gives 149 miles for the difference of latitude.

Oblique SAILING, is the reſolution of certain caſes and problems in Sailing by oblique triangles, or in which oblique triangles are concerned.

2

In this kind of Sailing, it may be obſerved, that *to ſet an object*, means to obſerve what rhumb or point of the nautical compaſs is directed to it. And the *bearing* of an object is the rhumb on which it is ſeen; alſo the bearing of one place from another, is reckoned by the name of the rhumb paſſing through thoſe two places.

In every figure relating to any caſe of plane Sailing, the bearing of a line, not running from the centre of the circle or horizon, is found by drawing a line parallel to it, from the centre, and towards the ſame quarter.

Ex. A ſhip ſailing at ſea, obſerved a point of land to bear E by S; and then after ſailing NE 12 miles, its bearing was found to be SE by E. Required the place of that point, and its diſtance from the ſhip at the laſt obſervation.

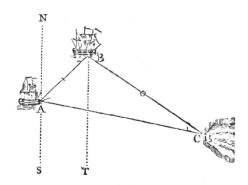

Conſtruction. Draw the meridian line NAS, and, aſſuming A for the firſt place of the ſhip, draw AC the E by S rhumb, and AB the NE one, upon which lay off 12 miles from A to B; then draw the meridian BT parallel to NS, from which ſet off the SE by E point BC, and the point C will be the place of the land required; then the diſtance BC meaſures 26 miles.

By Computation. Here are given the ſide AB, and the two angles A and B, viz, the ∠A = 5 points or 56° 15′, and the ∠B = 9 points or 101° 15′; conſequently the ∠C = 2 points or 22° 30 . Then, by plane trigonometry,

As ſin. ∠C 22° 30′ - - - - 9·58284
To ſin. ∠B 56 15 - - - - 9·91985
So is AB 12 miles - - - - 1·07918
                              ‾‾‾‾‾‾‾
To BC 26·073 miles - - - - 1·41619

SAILING *to Windward*, is working the ſhip towards that quarter of the compaſs from whence the wind blows.

For rightly underſtanding this part of navigation, it will be neceſſary to explain the terms that occur in it, though moſt of them may be ſeen in their proper places in this work.

When the wind is directly, or partly, againſt a ſhip's direct courſe for the place ſhe is bound to, ſhe reaches her port by a kind of zigzag or z like courſe; which is made by ſailing with the wind firſt on one ſide of the ſhip, and then on the other ſide.

In a ſhip, when you look towards the head, *Starboard* denotes the right hand ſide.

Larboard

Larboard the left hand fide.

Forwards, or *afore,* is towards the head.

Aft, or *abaft,* is towards the ftern.

The *beam* fignifies athwart or acrofs the middle of the fhip.

When a fhip fails the fame way that the wind blows, fhe is faid to fail or run before the wind; and the wind is faid to be *right aft,* or *right aftern;* and her courfe is then 16 points, or the fartheft poffible, from the wind, that is from the point the wind blows from.—When the fhip fails with the wind blowing directly acrofs her, fhe is faid to have the *wind on the beam;* and her courfe is 8 points from the wind.—When the wind blows obliquely acrofs the fhip, the wind is faid to be *abaft the beam* when it purfues her, or blows more on the hinder part, but *before the beam* when it meets or oppofes her courfe, her courfe being more than 8 points from the wind in the former cafe, but lefs than 8 points in the latter cafe.—When a fhip endeavours to fail towards that point of the compafs from which the wind blows, fhe is faid to *fail on a wind,* or to *ply to windward.*—And a veffel failing as near as fhe can to the point from which the wind blows, fhe is faid to be *clofe hauled.* Moft fhips will lie within about 6 points of the wind; but floops, and fome other veffels, will lie much nearer. To know how near the wind a fhip will lie; obferve the courfe fhe goes on each tack, when fhe is clofe hauled; then half the number of points between the two courfes, will fhew how near the wind the fhip will lie.

The *windward,* or *weather fide,* is that fide of the fhip on which the wind blows; and the other fide is called the *leeward,* or *lee fide.*—*Tacks* and *fheets* are large ropes faftened to the lower corners of the fore and main fails; by which either of thefe corners is hauled fore or aft.—When a fhip fails on a wind, the windward tacks are always hauled forwards, and the leeward fheets aft.—The *ftarboard tacks are aboard,* when the ftarboard fide is to windward, and the larboard fide to leeward. And the *larboard tacks are aboard,* when the larboard fide is to windward, and the ftarboard to leeward.

The moft common cafes in turning to windward may be conftructed by the following precepts. Having drawn a circle with the chord of 60°, for the compafs, or the horizon of the place, quarter it by drawing the meridian and parallel of latitude perpendicular to each other, and both through the centre; mark the place of the wind in the circumference; draw the rhumb paffing through the place bound to, and lay on it, from the centre, the diftance of that place. On each fide of the wind lay off, in the circumference, the points or degrees fhewing how near the wind the fhip can lie; and draw thefe rhumbs.—Now the firft courfe will be on one of thefe rhumbs, according to the tack the fhip leads with. Draw a line through the place bound to, parallel to the other rhumb, and meeting the firft; and this will fhew the courfe and diftance on the other tack.

Ex. The wind being at north, and a fhip bound to a port 25 miles directly to windward; beginning with the ftarboard tacks, what muft be the courfe and diftance on each of two tacks to reach the port?

Conftruction. Having drawn the circle &c, as above defcribed, where A is the port, AP and AQ the two rhumbs, each within 6 points of AN; in NA

produced take AB = 25 miles, then B is the place of the fhip; draw BC parallel to AP, and meeting QA produced in C; fo fhall BC and CA be the diftances on the two tacks; the former being WNW, and the latter ENE.

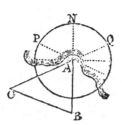

Computation.

$$\text{Here } \angle B = NAP = 6 \text{ points,}$$
$$\text{and } \angle A = NAQ = 6 \text{ points,}$$
$$\text{theref. } \angle C = 4 \text{ points.}$$

So that all the angles are given, and the fide AB, to find the other two fides AC and BC, which are equal to each other, becaufe their oppofite angles A and B are equal. Hence

as fin. C : AB :: fin. A : BC,

i. e. s. 45° : 25 :: s. 67° 30 : $32\frac{2}{3}$ = BC or AC, the diftance to be run on each tack.

SAILING *in Currents,* is the method of determining the true courfe and diftance of a fhip when her own motion is affected and combined with that of a current

A *current* or *tide* is a progreffive motion of the water caufing all floating bodies to move that way towards which the ftream is directed.—The *fetting* of a tide, or current, is that point of the compafs towards which the waters run; and the *drift* of the current is the rate at which it runs per hour.

The drift and fetting of the moft remarkable tides and currents, are pretty well known; but for unknown currents, the ufual way to find the drift and fetting, is thus: Let three or four men take a boat a little way from the fhip; and by a rope, faftened to the boat's ftem, let down a heavy iron pot, or loaded kettle, into the fea, to the depth of 80 or 100 fathoms, when it can be done: by which means the boat will ride almoft as fteady as at anchor. Then heave the log, and the number of knots run out in half a minute will give the current's rate, or the miles which it runs per hour; and the bearing of the log fhews the fetting of the current.

A body moving in a current, may be confidered in three cafes: viz,

1. Moving with the current, or the fame way it fets.

2. Moving againft it, or the contrary way it fets.

3. Moving obliquely to the current's motion.

In the 1ft cafe, or when a fhip fails with a current, its velocity will be equal to the fum of its proper motion, and the current's drift. But in the 2d cafe, or when a fhip fails againft a current, its velocity will be equal to the difference of her own motion and the drift of the current: fo that if the current drives ftronger than the wind, the fhip will drive aftern, or lofe way. In the 3d cafe, when the current fets oblique to the courfe of the fhip, her real courfe, or that made good, will be fomewhere between that in which the fhip endeavours to go, and the track in which the current tries to drive her; and indeed it will always be along the diagonal of a parallelogram, of which one fide reprefents the fhip's courfe fet, and the other adjoining fide is the current's drift.

Thus;

Thus, if ABDC be a parallelogram. Now if the wind alone would drive the ship from A to B in the same time as the current alone would drive her from A to C: then, as the wind neither helps nor hinders the ship from coming towards the line CD, the current will bring her there in the same time as if the wind did not act. And as the current neither helps nor hinders the ship from coming towards the line BD, the wind will bring her there in the same time as if the current did not act. Therefore the ship must, at the end of that time, be found in both those lines, that is, in their meeting D. Consequently the ship must have passed from A to D in the diagonal AD.

Hence, drawing the rhumbs for the proper course of the ship and of the current, and setting the distances off upon them, according to the quantity run by each in the given time; then forming a parallelogram of these two, and drawing its diagonal, this will be the real course and distance made good by the ship.

Ex. 1. A ship sails E. 5 miles an hour, in a tide setting the same way 4 miles an hour: required the ship's course, and the distance made good.

The ship's motion is   5m. E.
The current's motion is   4m. E.

Theref. the ship's run is 9m. E.

Ex. 2. A ship sails SSW. with a brisk gale, at the rate of 9 miles an hour, in a current setting NNE. 2 miles an hour: required the ship's course, and the distance made good.

The ship's motion is   SSW. 9m.
The current's motion is NNE. 2m.

Theref. ship's true run is SSW. 7m.

Ex. 3. A ship running south at the rate of 5 miles an hour, in 10 hours crosses a current, which all that time was setting east at the rate of 3 miles an hour: required the ship's true course and distance sailed.

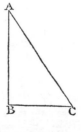

Here the ship is first supposed to be at A, her imaginary course is along the line AB, which is drawn south; and equal to 50 miles, the run in 10 hours; then draw BC east, and equal to 30 miles, the run of the current in 10 hours. Then the ship is found at C, and her true path is in the line AC = 58·31 her distance, and her course is the angle at A = 30° 58′ from the south towards the east.

Globular SAILING is the estimating the ship's motion and run upon principles derived from the globular figure of the earth, viz. her course, distance, and difference of latitude and longitude.

The principles of this method are explained under the articles RHUMB-*line*, *Mercator's* CHART, and MERIDIONAL *Parts*; which see.

Globular Sailing, in the extensive sense here applied

6

to the term, comprehends *Parallel Sailing, Middle-latitude Sailing*, and *Mercator's Sailing*; to which may be added *Circular Sailing*, or *Great-circle Sailing*. Of each of which it may be proper here to give a brief account.

Parallel SAILING is the art of finding what distance a ship should run due east or west, in sailing from the meridian of one place to that of another place, in any parallel of latitude.

The computations in parallel sailing depend on the following rule:

As radius,
To cosine of the lat. of any parallel;
So are the miles of long. between any two meridians,
To the dist. of these meridians in that parallel.

Also, for any two latitudes,
As the cosine of one latitude,
Is to the cosine of another latitude;
So is a given meridional dist. in the 1st parallel,
To the like meridional dist. in the 2d parallel.

Hence, counting 60 nautical miles to each degree of longitude, or on the equator; then, by the first rule the number of miles in each degree on the other parallels, will come out as in the following table.

Table of Meridional Distances.					
Lat.	Miles.	Lat.	Miles.	Lat.	Miles.
1	59·99	31	51·43	61	29·09
2	59·96	32	50·88	62	28·17
3	59·92	33	50·32	63	27·24
4	59·85	34	49·74	64	26·30
5	59·77	35	49·15	65	25·36
6	59·67	36	48·54	66	24·41
7	59·56	37	47·92	67	23·44
8	59·42	38	47·28	68	22·48
9	59·26	39	46·63	69	21·50
10	59·09	40	45·96	70	20·52
11	58·89	41	45·28	71	19·53
12	58·69	42	44·59	72	18·54
13	58·46	43	43·88	73	17·54
14	58·22	44	43·16	74	16·54
15	57·95	45	42·43	75	15·53
16	57·67	46	41·68	76	14·51
17	57·38	47	40·92	77	13·50
18	57·06	48	40·15	78	12·48
19	56·73	49	39·36	79	11·45
20	56·38	50	38·57	80	10·42
21	56·01	51	37·76	81	9·38
22	55·63	52	36·94	82	8·35
23	55·23	53	36·11	83	7·32
24	54·81	54	35·27	84	6·28
25	54·38	55	34·41	85	5·23
26	53·93	56	33·55	86	4·18
27	53·46	57	32·68	87	3·14
28	52·97	58	31·79	88	2·09
29	52·47	59	30·90	89	1·05
30	51·96	60	30·00	90	0·00

See

See another table of this kind, allowing 69$\frac{1}{15}$ Englifh miles to one degree, under the article DE-GREE.

To find the meridional diftance to any number of minutes between any of the whole degrees in the table, as for inftance in the parallel of 48° 26′; take out the tabular diftances for the two whole degrees between which the parallel or the odd minutes lie, as for 48° and 49°; fubtract the one from the other, and take the proportional part of the remainder for the odd minutes, by multiplying it by thofe minutes, and dividing by 60; and laftly, fubtract this proportional part from the greater tabular number. Thus,

Lat. 48° - - 40·15
Lat. 49. - - 39·36

As 60′ : 26′ : :   0·79 rem. : 0·34
                26

            474
            158

60) ·20·54

          0·34   pro. part
Taken from - 40·15   for lat. 48

Leaves merid. dift. 39·81   for lat. 48° 26′

And, in like manner, by the counter operation, to find what latitude anfwers to a given meridional diftance. As, for ex. in what latitude 46·08 miles anfwer to a degree of longitude.

From 46·63 for 39° | from 46·63 for 39°
Take 45·96 for 40° | take 46·08 given number.

Then as 0·67 : 60′ : :   0·55 : 49′
                  60

67) 3300

              49′ pro. part.

Therefore the latitude fought is 39° 49′.

Ex. 3 Given the latitude and meridional diftance; to find the correfponding difference of longitude. As, if a fhip, in latitude 53° 36′, and longitude 10° 18 eaft, fail due weft 236 miles; required her prefent longitude.

Here, by the firft rule,
As cof. lat. 53° 36′ comp.   0·22664
To radius - 90 00 - -   10·00000
So merid. dift.   236 m.   -   2·37291

To diff. long. 397·7 - -   2·59955

Its 60th gives   6°   38′ W. diff. long.
Taken from   10   18 E. long. from

Leaves - - 3   40 E. long. come to.

By the table; the length of a degree on the parallel of 53° 36 is 35·6.

Then as 35·6 : 60 : : 236 : 397·7, the diff. of long. the fame as before.

Middle-latitude SAILING, is a method of refolving the cafes of globular Sailing by means of the middle latitude between the latitude departed from, and that come to. This method is not quite accurate, being only an approximation to the truth, and it makes ufe of the principles of plane Sailing and parallel Sailing conjointly.

The method is founded on the fuppofition that the departure is reckoned as a meridional diftance in that latitude which is a middle parallel between the latitude failed from, and the latitude come to. And the method is not quite accurate, becaufe the arithmetical mean, or half fum of the cofines of two diftant latitudes, is not exactly the cofine of the middle latitude, or half the fum of thofe latitudes; nor is the departure between two places, on an oblique rhumb, equal to the meridional diftance in the middle latitude; as is prefumed in this method. Yet when the parallels are near the equator, or near to each other, in any latitude, the error is not confiderable.

This method feems to have been invented on account of the eafy manner in which the feveral cafes may be refolved by the traverfe table, and when a table of meridional parts is wanting. The computations depend on the following rules:

1. Take half the fum, or the arithmetical mean, of the two given latitudes, for the middle latitude. Then,

2. As cofine of middle latitude,
Is to the radius;
So is the departure,
To the diff. of longitude. And,

3. As cofine of middle latitude,
Is to the tangent of the courfe;
So is the difference of latitude,
To the difference of longitude.

Mercator's SAILING, is the art of refolving the feveral cafes of globular Sailing, by plane trigonometry, with the affiftance of a table of meridional parts, or of logarithmic tangents. And the computations are performed by the following rules:

1. As meridional diff lat.
To diff. of longitude;
So is the radius;
To tangent of the courfe.

2. As the proper diff lat.
To the departure;
So is merid. diff. lat.
To diff. of longitude.

3. As diff. log. tang. half colatitudes,
To tang. of 51° 38′ 09″;
So is a given diff. longitude,
To tangent of the courfe.

The manner of working with the meridional parts and logarithmic tangents, will appear from the two following cafes.

1. Given the latitudes of two places; to find their meridional difference of latitude.

By the Merid. Parts. When the places are both on the

the fame fide of the equator, take the difference of the meridional parts anfwering to each latitude ; but when the places are on oppofite fides of the equator, take the fum of the fame parts, for the meridional difference of latitude fought,

By the Log. Tangents. In the former cafe, take the difference of the long. tangents of the half colatitudes ; but in the latter cafe, take the fum of the fame ; then the faid difference or fum divided by 12·63, will give the meridional difference of latitude fought.

2. Given the latitude of one place, and the meridional difference of latitude between that and another place ; to find the latitude of this latter place.

By the Merid. Parts. When the places have like names, take the fum of the merid. parts of the given lat. and the given diff. ; but take the difference between the fame when they have unlike names ; then the refult, being found in the table of meridional parts, will give the latitude fought.

By the Log. Tangents. Multiply the given meridional diff. of lat. by 12·63 ; then in the former cafe fubtract the product from the log. tangent of the given half colatitude, but in the latter cafe add them ; then feek the degrees and minutes anfwering to the refult among the log. tangents, and thefe degrees, &c. doubled will be the colatitude fought.

Circular SAILING, or *Great-circle* SAILING, is the art of finding what places a fhip muft go through, and what courfes to fteer, that her track may be in the arc of a great circle on the globe, or nearly fo, paffing through the place failed from and the place bound to.

This method of Sailing has been propofed, becaufe the fhorteft diftance between two places on the fphere, is an arc of a great circle intercepted between them, and not the fpiral rhumb paffing through them, unlefs when that rhumb coincides with a great circle, which can only be on a meridian, or on the equator.

As the folutions of the cafes in Mercator's Sailing are performed by plane triangles, in this method of Sailing they are refolved by means of fpherical triangles. A great variety of cafes might be here propofed, but thofe that are the moft ufeful, and moft commonly occur, pertain to the following problem.

Problem I. Given the latitudes and longitudes of two places on the earth ; to find their neareft diftance on the furface, together with the angles of pofition from either place to the other.

This problem comprehends 6 cafes.

Cafe 1. When the two places lie under the fame meridian ; then their difference of latitude will give their diftance, and the pofition of one from the other will be directly north and fouth.

Cafe 2. When the two places lie under the equator ; their diftance is equal to their difference of longitude, and the angle of pofition is a right angle, or the courfe from one to the other is due eaft or weft.

Cafe 3. When both places are in the fame parallel of latitude. Ex. gr. The places both in 37° north, but the longitude of the one 25° weft, and of the other 76° 23′ weft.

Let P denote the north pole, and A and B the two places on the fame parallel BDA, alfo BIA their diftance afunder, or the arc of a great circle

passing through them. Then is the angle A or B that of pofition, and the angle BPA = 51° 23′ the difference of longitude, and the fide PA or PB = 53° the colatitude.

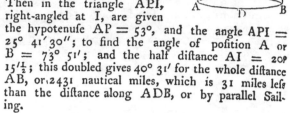

Draw PI perp. to AB, or bifecting the angle at P. Then in the triangle API, right-angled at I, are given the hypotenufe AP = 53°, and the angle API = 25° 41′ 30″ ; to find the angle of pofition A or B = 73° 51′ ; and the half diftance AI = 20° 15½′ ; this doubled gives 40° 31′ for the whole diftance AB, or 2431 nautical miles, which is 31 miles lefs than the diftance along ADB, or by parallel Sailing.

Cafe 4. When one place has latitude, and the other has none, or is under the equator. For example, fuppofe the Ifland of St. Thomas, lat. 0°, and long. 1° 0′ eaft, and Port St. Julian, in lat. 48° 51′ fouth, and long. 65° 10′ weft.

Port St. Julian, lat. 48° 51′ S.	- long.	65° 10′W.
Ifle St. Thomas - 0 00	- - - - -	1 00 E
Julian's colat. 41 09	Diff. long.	66 10

Hence, if S denote the fouth pole, A the Ifle St. Thomas at the equator, and B St. Julian ; then in the triangle are given SA a quadrant or 90°, BS = 41° 9′ the colat. of St. Julian, and the ∠S = 66° 10′ the dif. of longitude ; to find AB = 74° 35′ = 4475 miles, which is lefs by 57 miles than the diftance found by Mercator's Sailing ; alfo the angle of pofition at A = 51° 22′, and the angle of pofition B = 108° 24′.

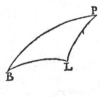

Cafe 5. When the two given places are both on the fame fide of the equator ; for example the Lizard, and the ifland of Bermudas.

The Lizard, lat. 49° 57′N.	- long.	5° 21′ W.
Bermudas, 32 35 N.	-	63 32 W.
		58 11

Here, if P be the north pole, L the Lizard, and B Bermudas ; there are given, PL = 40° 03′ colat. of the Lizard, PB = 57 25 colat. of Bermudas, ∠P = 58 11 diff. of longitude ; to find BL = 45° 44 = 2744 miles the diftance, and ∠ of pofition B = 49° 27′, alfo ∠ of pofition L = 90° 31′.

Cafe 6. When the given places lie on different fides of the equator ; as fuppofe St. Helena and Bermudas. Here

PB

PB = 57° 25′ polar diſt. Bermudas,

PH = 105 55 polar diſt. St. Helena,

∠P = 57 43 diff. long.

To find BH = 73°.26 = 4406 miles, the diſtance, alſo the angle of poſition H = 48° 0′, and the angle of poſition B = 121° 59′.

From the ſolutions of the foregoing caſes it appears, that to ſail on the arc of a great circle, the ſhip muſt continually alter her courſe ; but as this is a difficulty too great to be admitted into the practice of navigation, it has been thought ſufficiently exact to effect this buſineſs by a kind of approximation, that is, by a method which nearly approaches to the ſailing on a great circle : namely, upon this principle, that in ſmall arcs, the difference between the arc and its chord or tangent is ſo ſmall, that they may be taken for one another in any nautical operations : and accordingly it is ſuppoſed that the great circles on the earth are made up of ſhort right lines, each of which is a ſegment of a rhumb line. On this ſuppoſition the ſolution of the following problem is deduced.

Problem II. Having given the latitudes and longitudes of the places ſailed from and bound to ; to find the ſucceſſive latitudes on the arc of a great circle in thoſe places where the alteration in longitude ſhall be a given quantity ; together with the courſes and diſtances between thoſe places.

1. Find the angle of poſition at each place, and their diſtance, by one of the preceding caſes.

2. Find the greateſt latitude the great circle runs through, i. e. find the perpendicular from the pole to that circle ; and alſo find the ſeveral angles at the pole, made by the given alterations of longitude between this perpendicular and the ſucceſſive meridians come to.

3. With this perpendicular and the polar angles ſeverally, find as many correſponding latitudes, by ſaying, as radius : tang. greateſt lat. : : coſ. 1ſt polar angle : tang. 1ſt lat. : : coſ. 2d polar angle : tang. of 2d lat. &c.

4. Having now the ſeveral latitudes paſſed through, and the difference of longitude between each, then by Mercator's Sailing find the courſes and diſtances between thoſe latitudes. And theſe are the ſeveral courſes and diſtances the ſhip muſt run, to keep nearly on the arc of a great circle.

The ſmaller the alterations in longitude are taken, the nearer will this method approach to the truth ; but it is ſufficient to compute to every 5 degrees of difference of longitude ; as the length of an arc of 5 degrees differs from its chord, or tangent, only by 0·002.

The track of a ſhip, when thus directed nearly in the arc of a great circle, may be delineated on the Mercator's chart, by marking on it, by help of the latitudes and longitudes, the ſucceſſive places where the ſhip is to alter her courſe ; then thoſe places or points, being joined by right lines, will ſhew the path along which the ſhip is to ſail, under the propoſed circumſtances.

' On the ſubject of theſe articles, ſee Robertſon's Elements of Navigation, vol. 2.

Spheroidical SAILING, is computing the caſes of navigation on the ſuppoſition or principles of the ſpheroidical figure of the earth. See Robertſon's Navigation, vol. 2, b. 8. ſect. 8.

SAILING, *in a more confined ſenſe*, is the art of conducting a ſhip from place to place, by the working or handling of her ſails and rudder.

To bring Sailing to certain rules, M. Renau computes the force of the water, againſt the ſhip's rudder, ſtem, and ſide ; and the force of the wind againſt her ſails. In order to this, he firſt conſiders all fluid bodies, as the air, water, &c, as compoſed of little particles, which when they act upon, or move againſt any ſurface, do all move parallel to one another, or ſtrike againſt the ſurface after the ſame manner. Secondly, that the motion of any body, with regard to the ſurface it ſtrikes, muſt be either perpendicular, parallel, or oblique.

From theſe principles he computes, that the force of the air or water, ſtriking perpendicularly upon a ſail or rudder, is to the force of the ſame ſtriking obliquely, in the duplicate ratio of radius to the ſine of the angle of incidence : and conſequently that all oblique forces of the wind againſt the ſails, or of the water againſt the rudder, will be to one another in the duplicate ratio of the ſines of the angles of incidence.

Such are the concluſions from theory ; but it is very different in real practice, or experiments, as appears from the tables of experiments inſerted at the article RESISTANCE.

Farther, when the different degrees of velocity are conſidered, it is alſo found that the forces are as the ſquares of the velocities of the moving air or water nearly ; that is, a wind that blows twice as ſwift, as another, will have 4 times the force upon the ſail ; and when 3 times as ſwift, 9 times the force, &c. And it being alſo indifferent, whether we conſider the motion of a ſolid in a fluid at reſt, or of the fluid againſt the ſolid at reſt ; the reciprocal impreſſions being always the ſame ; if a ſolid be moved with different velocities in the ſame fluid matter, as water, the different reſiſtances which it will receive from that water, will be in the ſame proportion as the ſquares of the velocities of the moving body.

He then applies theſe principles to the motions of a ſhip, both forwards and ſideways, through the water, when the wind, with certain velocities, ſtrikes the ſails in various poſitions. After this, the author proceeds to demonſtrate, that the beſt poſition or ſituation of a ſhip, ſo as ſhe may make the leaſt lee-way, or ſide motion, but go to windward as much as poſſible, is this : that, let the ſail have what ſituation it will, the ſhip be always in a line biſecting the complement of the wind's angle of incidence upon the ſail. That is, ſuppoſing the ſail in the poſition BC, and the wind blowing from A to B, and conſequently the angle of the wind's incidence on the ſail is ABC, the complement of which is CBE : then muſt the ſhip be put in the poſition BK, or move in the line BL, biſecting the ∠ CBE.

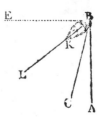

He

He shews farther, that the angle which the sail ought to make with the wind, i. e. the angle ABC, ought to be but 24 degrees; that being the most advantageous situation to go to windward the most possible.

To this might be added many curious particulars from Borelli de Vi Percussionis, concerning the different directions given to a vessel by the rudder, when sailing with a wind, or floating without sails in a current: in the former case, the head of the ship always coming to the rudder, and in the latter always flying off from it; as also from Euler, Bouguer, and Juan, who have all written learnedly on this subject.

SALIANT, in Fortification, is said of an angle that projects its point outwards; in opposition to a re-entering angle, which has its point turned inwards. Instances of both kinds of these we have in tenailles and star-works.

SALON, or SALOON, in Architecture, a grand, lofty, spacious sort of hall, vaulted at top, and usually comprehending two stories, with two ranges of windows. It is sometimes built square, sometimes round or oval, sometimes octagonal, as at Marly, and sometimes in other forms.

SAP, or SAPP, in Building, as to sap a wall, &c, is to dig out the ground from beneath it, so as to bring it down all at once for want of support.

SAP, in the Military Art, denotes a work carried on under cover of gabions and fascines on the flank, and mantlets or stuffed gabions on the front, to gain the descent of a ditch, or the like.

It is performed by digging a deep trench, descending by steps from top to bottom, under a corridor, carrying it as far as the bottom of the ditch, when that is dry; or as far as the surface of the water, when wet.

SAROS, in Chronology, a period of 223 lunar months. The etymology of the word is said to be Chaldean, signifying restitution, or return of eclipses; that is, conjunctions of the sun and moon in nearly the same place of the ecliptic. The Saros was a cycle like to that of Meto.

SARRASIN, or SARRAZIN, in Fortification, a kind of port-cullis, otherwise called a herse, which is hung with ropes over the gate of a town or fortress, to be let fall in case of a surprise.

SATELLITES, in Astronomy, are certain secondary planets, moving round the other planets, as the moon does round the earth. They are so called because always found attending them, from rising to setting, and making the tour about the sun together with them.

The words moon and Satellite are sometimes used indifferently: thus we say, either Jupiter's moons, or Jupiter's Satellites; but usually we distinguish, restraining the term moon to the earth's attendant, and applying the term Satellite to the little moons more recently discovered about Jupiter, Saturn, and the Georgian planet, by the assistance of the telescope, which is necessary to render them visible.

The Satellites move round their primary planets, as their centres, by the same laws as those primary ones do round their centre the sun; viz, in such manner that, in the Satellites of the same planet, the squares of the periodic times are proportional to the cubes of their distances from the primary planet. For the physical cause of their motions, see GRAVITY. See also PLANETS.

We know not of any Satellites beside those above mentioned, what other discoveries may be made by farther improvements in telescopes, time only can bring to light.

SATELLITES of Jupiter. There are are four little moons, or secondary planets now known performing their evolutions about Jupiter, as that planet does about the Sun.

Simon Marius, mathematician of the elector of Brandenburg, about the end of November 1609, observed three little stars moving round Jupiter's body, and proceeding along with him; and in January 1610, he found a 4th. In January 1610 Galileo also observed the same in Italy, and in the same year published his observations. These Satellites were also observed in the same month of January 1710, by Thomas Harriot, the celebrated author of a work upon algebra, and who made constant observations of these Satellites, from that time till the 26th of February 1612; as appears by his curious astronomical papers, lately discovered by Dr. Zach, at the seat of the earl of Egremont, at Petworth in Sussex.

One Antony Maria Schyrlæus di Reita, a capuchin of Cologne, imagined that, besides the four known Satellites of Jupiter, he had discovered five more, on December 29, 1642. But the observation being communicated to Gassendus, who had observed Jupiter on the same day, he soon perceived that the monk had mistaken five fixed stars, in the effusion of the water of Aquarius, marked in Tycho's catalogue 24, 25, 26, 27, 28, for Satellites of Jupiter.

When Jupiter comes into a line between any of his Satellites and the sun, the Satellite disappears, being then *eclipsed*, or involved in his shadow.—When the Satellite goes behind the body of Jupiter, with respect to an observer on the earth, it is then said to be *occulted*, being hid from our sight by his body, whether in his shadow or not.—And when the Satellite comes into a position between Jupiter and the Sun, it casts a shadow upon the face of that planet, which we see as an obscure round spot.—And lastly, when the Satellite comes into a line between Jupiter and us, it is said to *transit* the disc of the planet, upon which it appears as a round black spot.

The periods or revolutions of Jupiter's Satellites, are found out from their conjunctions with that planet; after the same manner, as those of the primary planets are discovered from their oppositions to the sun. And their distances from the body of Jupiter, are measured by a micrometer, and estimated in semidiameters of that planet, and thence in miles.

By the latest and most exact observations, the periodical times and distances of these Satellites, and the angles under which their orbits are seen from the earth, at its mean distance from Jupiter, are as below:

SATEL-

Satellites *of* Jupiter.				
Satel-lites.	Periodic Times.	Diftances in		Angles of Orbit.
		Semidia-meters.	Miles.	
1	$1^d\ 18^h\ 27'\ 34''$	$5\frac{2}{3}$	266,000	$3'\ 55''$
2	3 13 13 42	$9\frac{1}{39}$	423,000	6 14
3	7 3 42 36	$14\frac{5}{15}$	676,000	9 58
4	16 16 32 9	$25\frac{3}{10}$	1,189,000	17 30

The eclipfes of the Satellites, efpecially of thofe of Jupiter, are of very great ufe in aftronomy. Firft, in determining pretty exactly the diftance of Jupiter from the earth. A fecond advantage ftill more confiderable, which is drawn from thefe eclipfes, is the proof which they give of the progreffive motion of light. It is demonftrated by thefe eclipfes, that light does not come to us in an inftant, as the Cartefians pretended, although its motion is extremely rapid. For if the motion of light were infinite, or came to us in an inftant, it is evident that we fhould fee the commencement of an eclipfe of a Satellite at the fame moment, at whatever diftance we might be from it; but, on the contrary, if light move progreffively, then it is as evident, that the farther we are from a planet, the later we fhall be in feeing the moment of its eclipfe, becaufe the light will take up a longer time in arriving at us; and fo it is found in fact to happen, the eclipfes of thefe Satellites appearing always later and later than the true computed times, as the earth removes farther and farther from the planet. When Jupiter and the earth are at their neareft diftance, being in conjunction both on the fame fide of the fun, then the eclipfes are feen to happen the fooneft; and when the fun is directly between Jupiter and the earth, they are at their greateft diftance afunder, the diftance being more than before by the whole diameter of the earth's annual orbit, or by double the earth's diftance from the fun, then the eclipfes are feen to happen the lateft of any, and later than before by about a quarter of an hour. Hence therefore it follows, that light takes up a quarter of an hour in travelling acrofs the orbit of the earth, or near 8 minutes in paffing from the fun to the earth; which gives us about 12 millions of miles per minute, or 200,000 miles per fecond, for the velocity of light. A difcovery that was firft made by M. Roemer.

The third and greateft advantage derived from the eclipfes of the Satellites, is the knowledge of the longitudes of places on the earth. Suppofe two obfervers of an eclipfe, the one, for example, at London, the other at the Canaries; it is certain that the eclipfe will appear at the fame moment to both obfervers; but as they are fituated under different meridians, they count different hours, being perhaps 9 o'clock to the one, when it is only 8 to the other; by which obfervations of the true time of the eclipfe, on communication, they find the difference of their longitudes to be one hour in time, which anfwers to 15 degrees of longitude.

Satellites *of Saturn*, are 7 little fecondary planets revolving about him.

One of them, which till lately was reckoned the 4th in order from Saturn, was difcovered by Huygens, the 25th of March 1655, by means of a telefcope 12 feet long; and the 1ft, 2d, 3d, and 5th, at different times, by Caffini; viz, the 5th in October 1671, by a telefcope of 17 feet; the 3d in December 1672, by a telefcope of Campani's, 35 feet long; and the firft and fecond in March 1684, by help of Campani's glaffes, of 100 and 136 feet. Finally, the 6th and 7th Satellites have lately been difcovered by Dr. Herfchel, with his 40 feet reflecting telefcope, viz, the 6th on the 19th of Auguft 1787, and the 7th on the 17th of September 1788. Thefe two he has called the 6th and 7th Satellites, though they are nearer to the planet Saturn than any of the former five, that the names or numbers of thefe might not be miftaken or confounded, with regard to former obfervations of them.

Moreover, the great diftance between the 4th and 5th Satellite, gave occafion to Huygens to fufpect that there might be fome intermediate one, or elfe that the 5th might have fome other Satellite moving round it, as its centre. Dr. Halley, in the Philof. Tranf. (numb. 145, or Abr. vol. 1. pa. 371) gives a correction of the theory of the motions of the 4th or Huygenian Satellite. Its true period he makes $11^d\ 22^h\ 41'\ 6''$.

The periodical revolutions, and diftances of thefe Satellites from the body of Saturn, expreffed in femidiameters of that planet, and in miles, are as follow.

Satellites *of* Saturn.				
Satel-lites.	Periods.	Diftances in		Diam. of Orbit.
		Semidi-ameters.	Miles.	
1	$1^d\ 21^h\ 18'\ 27''$	$4\frac{3}{8}$	170,000	$1'\ 27''$
2	2 17 41 22	$5\frac{1}{2}$	217,000	1 52
3	4 12 25 12	8	303,000	2 36
4	15 22 41 13	18	704,000	6 18
5	79 7 48 0	54	2,050,000	17 4
6	1 8 53 9	$3\frac{5}{9}$	135,000	1 14
7	0 22 40 46	$2\frac{5}{8}$	107,000	0 57

The four firft defcribe ellipfes like to thofe of the ring, and are in the fame plane. Their inclination to the ecliptic is from 30 to 31 degrees. The 5th defcribes an orbit inclined from 17 to 18 degrees with the orbit of Saturn; his plane lying between the ecliptic and thofe of the other Satellites, &c. Dr. Herfchel obferves that the 5th Satellite turns once round its axis exactly in the time in which it revolves about the planet Saturn; in which refpect it refembles our moon, which does the fame thing. And he makes the angle of its diftance from Saturn, at his mean diftance, 17′ 2″. Philof. Tranf. 1792, pa. 22. See a long account of obfervations of thefe Satellites, with tables of their mean motions, by Dr. Herfchel, Philof. Tranf. 1790, pa. 427 &c.

Satellites *of the Georgian Planet*, or *Herfchel*, are two little moons that revolve about him, like thofe of

Jupiter and Saturn. Thefe Satellites were difcovered by Dr. Herfchel, in the month of January 1787 who gave an account of them in the Philof. Tranf. of that year, pa. 125 &c; and a ftill farther account of them in the vol. for 1788, pa. 364 &c; from which it appears that their fynodical periods, and angular diftances from their primary, are as follow:

Satellite.	Periods.	Dift.
1	8ᵈ 17ʰ 1′ 19″	0′ 33″
2	13 11 5 1½	0 44⅖

The orbits of thefe Satellites are nearly perpendicular to the ecliptic; and in magnitude they are probably not lefs than thofe of Jupiter.

SATELLITE of *Venus.* Caffini thought he faw one, and Mr. Short and other aftronomers have fufpected the fame thing. (Hift. de l'Acad. 1741, Philof. Tranf. numb. 459). But the many fruitlefs fearches that have been fince made to difcover it, leave room to fufpect that it has been only an optical illufion, formed by the glaffes of telefcopes; as appears to be the opinion of F. Hell, at the end of his Ephemeris for 1766, and Bofcovich, in his 5th Optical Differtation.

Neither has it been difcovered that either of the other planets Mars and Mercury have any Satellites revolving about them.

SATURDAY, the 7th or laft day of the week, fo called, as fome have fuppofed, from the idol Seater, worfhipped on this day by the ancient Saxons, and thought to be the fame as the Saturn of the Latins. In aftronomy, every day of the week is denoted by fome one of the planets, and this day is marked with the planet ♄ Saturn. Saturday anfwers to the Jewifh fabbath.

SATURN, one of the primary planets, being the 6th in order of diftance from the fun, and the outermoft of all, except the Georgian planet, or Herfchel, lately difcovered; and is marked with the character ♄, denoting an old man fupporting himfelf with a ftaff, reprefenting the ancient god Saturn.

Saturn fhines with but a feeble light, partly on account of his great diftance, and partly from its dull red colour. This planet is perhaps one of the moft engaging objects that aftronomy offers to our view; it is furrounded with a double ring, one without the other, and beyond thefe by 7 Satellites, all in the plane of the rings; the rings and planets being all dark and denfe bodies; like Saturn himfelf, thefe bodies cafting their fhadows mutually one upon another; though the reflected light of the rings is ufually brighter than that of the planet itfelf.

Saturn has alfo certain obfcure zones, or belts, appearing at times acrofs his difc, like thofe of Jupiter, which are changeable, and are probably obfcurations in his atmofphere. Dr Herfchel, Philof. Tranf. 1790, fhews that Saturn has a denfe atmofphere; that he revolves about an axis, which is perpendicular to the plane of the rings; that his figure is, like the other planets, the oblate fpheroid, being flatted at the poles, the polar diameter being to the equatorial one

as 10 to 11; that his ring has a motion of rotation in its own plane, its axis of motion being the fame as that of Saturn himfelf, and its periodical time equal to 10ʰ 32′ 15″·4. See alfo RING, and SATELLITE.

Concerning the difcovery of the ring and figure of Saturn; we find that Galileo firft perceived that his figure is not round: but Huygens fhewed, in his Syftema Saturniana 1659, that this was owing to the pofitions of his ring; for his fpheroidical form could only be feen by Herfchel's telefcope; though indeed Caffini, in an obfervation made June 19, 1692, faw the oval figure of Saturn's fhadow upon his ring.

Mr. Bugge determines (Philof. Tranf. 1787, pa. 42) the heliocentric longitude of Saturn's defcending node to be 9ˢ 21° 5′ 8″½; and that the planet was in that node Auguft 21, 1784, at 18ʰ 20′ 10″, time at Copenhagen.

The annual period of Saturn about the fun, is 10759 days 7 hours, or almoft 30 years; and his diameter is about 67000 miles, or near 8½ times the diameter of the earth; alfo his diftance is about 9½ times that of the earth. Hence fome have concluded that his light and heat are entirely unfit for rational inhabitants. But that their light is not fo weak as we imagine, is evident from their brightnefs in the night time. Befides, allowing the fun's light to be 45000 times as ftrong, with refpect to us, as the light of the moon when full, the fun will afford 500 times as much light to Saturn as the full moon does to us, and 1600 times as much to Jupiter. So that thefe two planets, even without any moon, would be much more enlightened than we at firft imagine; and by having fo many, they may be very comfortable places of refidence. Their heat, fo far as it depends on the force of the fun's rays, is certainly much lefs than ours; to which no doubt the bodies of their inhabitants are as well adapted as ours are to the feafons we enjoy. And if it be confidered that Jupiter never has any winter, even at his poles, which probably is alfo the cafe with Saturn, the cold cannot be fo intenfe on thefe two planets as is generally imagined. To this may be added, that there may be fomething in the nature of their mould warmer than in that of our earth; and we find that all our heat does not depend on the rays of the fun; for if it did, we fhould always have the fame months equally hot or cold at their annual return, which is very far from being the cafe.

See the articles PLANET, PERIOD, RING, SATELLITE.

SAUCISSE, in Artillery, a long train of powder inclofed in a roll or pipe of pitched cloth, and fometimes of leather, about 2 inches in diameter; ferving to fet fire to mines or caiffons. It is ufually placed in a wooden pipe, called an auget, to prevent its growing damp.

SAUCISSON, in Fortification, a kind of faggot, made of thick branches of trees, or of the trunks of fhrubs, bound together; for the purpofe of covering the men, and to ferve as epaulements; and alfo to repair breaches, ftop paffages, make traverfes over a wet ditch, &c.

The Sauciffon differs from the fafcine, which is only made of fmall branches; and by its being bound at both ends, and in the middle.

SAVILLE (Sir HENRY), a very learned Englifhman,

man, the second son of Henry Saville, Efq. was born at Bradley, near Halifax, in Yorkfhire, November the 30th, 1549. He was entered of Merton-college, Oxford, in 1561, where he took the degrees in arts, and was chofen fellow. When he proceeded 'mafter of arts in 1570, he read for that degree on the Almageft of Ptolomy, which procured him the reputation of a man eminently fkilled in mathematics and the Greek language; in the former of which he voluntarily read a public lecture in the univerfity for fome time.

In 1578 he travelled into France and other countries; where, diligently improving himfelf in all ufeful learning, in languages, and the knowledge of the world, he became a moft accomplifhed gentleman. At his return, he was made tutor in the Greek tongue to queen Elizabeth, who had a great efteem and liking for him.

In 1585 he was made warden of Merton college, which he governed fix-and-thirty years with great honour, and improved it by all the means in his power — In 1596 he was chofen provoft of Eton-college; which he filled with many learned men —James the Firft, upon his acceffion to the crown of England, expreffed a great regard for him, and would have preferred him either in church or ftate; but Saville declined it, and only accepted the ceremony of knighthood from the king at Windfor in 1604. His only fon Henry dying about that time, he thenceforth devoted his fortune to the promoting of learning. Among other things, in 1619, he founded, in the univerfity of Oxford, two lectures, or profefforfhips, one in geometry, the other in aftronomy; which he endowed with a falary of 160l. a year each, befides a legacy of 600l. to purchafe more lands for the fame ufe. He alfo furnifhed a library with mathematical books near the mathematical fchool, for the ufe of his profeffors; and gave 100l. to the mathematical cheft of his own appointing: adding afterwards a legacy of 40l. a year to the fame cheft, to the univerfity, and to his profeffors jointly. He likewife gave 120l. towards the new-building of the fchools, befide feveral rare manufcripts and printed books to the Bodleian library; and a good quantity of Greek types to the printing-prefs at Oxford.

After a life thus fpent in the encouragement and promotion of fcience and literature in general, he died at Eton-college the 19th of February 1622, in the 73d year of his age, and was buried in the chapel there. On this occafion, the univerfity of Oxford paid him the greateft honours, by having a public fpeech and verfes made in his praife, which were publifhed foon after in 4to, under the title of *Ultima Linea Savilii.*

As to the character of Saville, the higheft encomiums are beftowed on him by all the learned of his time: by Cafaubon, Mercerus, Meibomius, Jofeph Scaliger, and efpecially the learned bifhop Montague; who, in his *Diatriba* upon Selden's Hiftory of Tythes, ftyles him, " that magazine of learning whofe memory fhall be honourable amongft not only the learned, but the righteous for ever."

Several noble inftances of his munificence to the republic of letters have already been mentioned: in the account of his publications many more, and even greater, will appear. Thefe are,

1. *Four Books of the Hiftories of Cornelius Tacitus, and*

the *Life of Agricola;* with Notes upon them, in folio, dedicated to Queen Elizabeth, 1581.

2. *A View of certain Military Matters,* or Commentaries concerning Roman Warfare, 1598.

3. *Rerum Anglicarum Scriptores poft Bedam,* &c 1596. This is a collection of the beft writers of our Englifh hiftory; to which he added chronological tables at the end, from Julius Cæfar to William the Conqueror.

4. *The Works of St. Chryfoftom,* in Greek, in 8 vols. folio, 1613. This is a very fine edition, and compofed with great coft and labour. In the preface he fays, " that having himfelf vifited, about 12 years before, all the public and private libraries in Britain, and copied out thence whatever he thought ufeful to this defign, he then fent fome learned men into France, Germany, Italy, and the Eaft, to tranfcribe fuch parts as he had not already, and to collate the others with the beft manufcripts." At the fame time, he makes his acknowledgments to feveral eminent men for their affiftance; as Thuanus, Velferus, Schottus, Cafaubon, Ducæus, Gruter, Höefchelius, &c. In the 8th volume are inferted Sir Henry Saville's own notes, with thofe of other learned men. The whole charge of this edition, including the feveral fums paid to learned men, at home and abroad, employed in finding out, tranfcribing, and collating the beft manufcripts, is faid to have amounted to no lefs than 8000l. Several editions of this work were afterwards publifhed at Paris.

5. In 1618 he publifhed a Latin work, written by Thomas Bradwardin, abp. of Canterbury, againft Pelagius, intitled, *De Caufa Dei contra Pelagium, et de virtute caufarum;* to which he prefixed the life of Bradwardin.

6. In 1621 he publifhed a collection of his own Mathematical Lectures on Euclid's Elements; in 4to.

7. *Oratio coram Elizabetha Regina Oxoniæ habita,* anno 1592. Printed at Oxford in 1658, in 4to.

8. He tranflated into Latin king James's *Apology for the Oath of Allegiance.* He alfo left feveral manufcripts behind him, written by order of king James; all which are in the Bodleian library. He wrote notes likewife upon the margin of many books in his library, particularly Eufebius's *Ecclefiaftical Hiftory;* which were afterwards ufed by Valefius, in his edition of that work in 1659.—Four of his letters to Camden are publifhed by Smith, among *Camden's Letters,* 1691, 4to

Sir Henry Saville had a younger brother, *Thomas* SAVILLE, who was admitted probationer fellow of Merton-college, Oxford, in 1580. He afterwards travelled abroad into feveral countries. Upon his return he was chofen fellow of Eton-college; but he died at London in 1593. Thomas Saville was alfo a man of great learning, and an intimate friend of Camden; among whofe letters, juft mentioned, there are 15 of Mr. Saville's to him.

SAUNDERSON (Dr. Nicholas), an illuftrious profeffor of mathematics in the univerfity of Cambridge, and a fellow of the Royal Society, was born at Thurlfton in Yorkfhire in 1682. When he was but twelve months old, he loft not only his eye fight, but his very eye-balls, by the fmall-pox; fo that he could retain no more ideas of vifion than if he had been born blind. At an early age, however, being of very promifing mising parts, he was sent to the free-school at Penniston, and there laid the foundation of that knowledge of the Greek and Latin languages, which he afterwards improved so far, by his own application to the classic authors, as to hear the works of Euclid, Archimedes, and Diophantus read in their original Greek.

Having acquired a grammatical education, his father, who was in the excise, instructed him in the common rules of arithmetic. And here it was that his excellent mathematical genius first appeared: for he very soon became able to work the common questions, to make very long calculations by the strength of his memory, and to form new rules to himself for the better resolving of such questions as are often proposed to learners as trials of skill.

At the age of 18, our author was introduced to the acquaintance of Richard West, of Underbank, Esq. a lover of mathematics, who, observing Mr. Saunderson's uncommon capacity, took the pains to instruct him in the principles of algebra and geometry, and gave him every encouragement in his power to the prosecution of these studies. Soon after this he became acquainted also with Dr. Nettleton, who took the same pains with him. And it was to these two gentlemen that Mr. Saunderson owed his first institution in the mathematical sciences: they furnished him with books, and often read and expounded them to him. But he soon surpassed his masters, and became fitter to teach, than to learn any thing from them.

His father, otherwise burdened with a numerous family, finding a difficulty in supporting him, his friends began to think of providing both for his education and maintenance. His own inclination led him strongly to Cambridge, and it was at length determined he should try his fortune there, not as a scholar, but as a master: or, if this design should not succeed, they promised themselves success in opening a school for him at London. Accordingly he went to Cambridge in 1707, being then 25 years of age, and his fame in a short time filled the university. Newton's Principia, Optics, and Universal Arithmetic, were the foundations of his lectures, and afforded him a noble field for the displaying of his genius; and great numbers came to hear a blind man give lectures on optics, discourse on the nature of light and colours, explain the theory of vision, the effect of glasses, the phenomenon of the rainbow, and other objects of sight.

As he instructed youth in the principles of the Newtonian philosophy, he soon became acquainted with its incomparable author, though he had several years before left the university; and frequently conversed with him on the most difficult parts of his works: he also held a friendly communication with the other eminent mathematicians of the age, as Halley, Cotes, Demoivre, &c.

Mr. Whiston was all this time in the mathematical professor's chair, and read lectures in the manner proposed by Mr. Saunderson on his settling at Cambridge; so that an attempt of this kind looked like an encroachment on the privilege of his office; but, as a good-natured man, and an encourager of learning, he readily consented to the application of friends made in behalf of so uncommon a person.

Upon the removal of Mr. Whiston from his profes-

sorship, Mr. Saunderson's merit was thought so much superior to that of any other competitor, that an extraordinary step was taken in his favour, to qualify him with a degree, which the statute requires: in consequence he was chosen in 1711, Mr. Whiston's successor in the Lucasian professorship of mathematics, Sir Isaac Newton interesting himself greatly in his favour. His first performance, after he was seated in the chair, was an inaugural speech made in very elegant latin, and a style truly Ciceronian; for he was well versed in the writings of Tully, who was his favourite in prose, as Virgil and Horace were in verse. From this time he applied himself closely to the reading of lectures, and gave up his whole time to his pupils. He continued to reside among the gentlemen of Christ-college till the year 1723, when he took a house in Cambridge, and soon after married a daughter of Mr. Dickens, rector of Boxworth in Cambridgeshire, by whom he had a son and a daughter.

In the year 1728, when king George visited the university, he expressed a desire of seeing so remarkable a person; and accordingly our professor attended the king in the senate, and by his favour was there created doctor of laws.

Dr. Saunderson was naturally of a strong healthy constitution; but being too sedentary, and constantly confining himself to the house, he became a valetudinarian: and in the spring of the year 1739 he complained of a numbness in his limbs, which ended in a mortification in his foot, of which he died the 19th of April that year, in the 57th year of his age.

There was scarcely any part of the mathematics on which Dr. Saunderson had not composed something for the use of his pupils. But he discovered no intention of publishing any thing till, by the persuasion of his friends, he prepared his Elements of Algebra for the press, which after his death were published by subscription in 2 vols 4to, 1740.

He left many other writings, though none perhaps prepared for the press. Among these were some valuable comments on Newton's Principia, which not only explain the more difficult parts, but often improve upon the doctrines. These are published in Latin at the end of his posthumous Treatise on Fluxions, a valuable work, published in 8vo, 1756.—His manuscript lectures too, on most parts of natural philosophy, which I have seen, might make a considerable volume, and prove an acceptable present to the public if printed.

Dr. Saunderson, as to his character, was a man of much wit and vivacity in conversation, and esteemed an excellent companion. He was endued with a great regard to truth; and was such an enemy to disguise, that he thought it his duty to speak his thoughts at all times with unrestrained freedom. Hence his sentiments on men and opinions, his friendship or disregard, were expressed without reserve; a sincerity which raised him many enemies.

A blind man, moving in the sphere of a mathematician, seems a phenomenon difficult to be accounted for, and has excited the admiration of every age in which it has appeared. Tully mentions it as a thing scarce credible in his own master in philosophy, Diodotus; that he exercised himself in it with more assiduity

2

duity

duity after he became blind; and, what he thought next to impossible to be done without sight, that he professed geometry, describing his diagrams so exactly to his scholars, that they could draw every line in its proper direction. St. Jerome relates a still more remarkable instance in Didymus of Alexandria, who, though blind from his infancy, and therefore ignorant of the very letters, not only learned logic, but geometry also to very great perfection, which seems most of all to require sight. But, if we consider that the ideas of extended quantity, which are the chief objects of mathematics, may as well be acquired by the sense of feeling as that of sight, that a fixed and steady attention is the principal qualification for this study, and that the blind are by necessity more abstracted than others (for which reason it is said that Democritus put out his eyes, that he might think more intensely), we shall perhaps find reason to suppose that there is no branch of science so much adapted to their circumstances.

At first, Dr. Saunderson acquired most of his ideas by the sense of feeling; and this, as is commonly the case with the blind, he enjoyed in great perfection. Yet he could not, as some are said to have done, distinguish colours by that sense; for, after having made repeated trials, he used to say, it was pretending to impossibilities. But he could with great nicety and exactness observe the smallest degree of roughness or defect of polish in a surface. Thus, in a set of Roman medals, he distinguished the genuine from the false, though they had been counterfeited with such exactness as to deceive a connoisseur who had judged by the eye. By the sense of feeling also, he distinguished the least variation; and he has been seen in a garden, when observations have been making on the sun, to take notice of every cloud that interrupted the observation almost as justly as they who could see it. He could also tell when any thing was held near his face, or when he passed by a tree at no great distance, merely by the different impulse of the air on his face.

His ear was also equally exact. He could readily distinguish the 5th part of a note. By the quickness of this sense he could judge of the size of a room, and of his distance from the wall. And if ever he walked over a pavement, in courts or piazzas which reflected a sound, and was afterwards conducted thither again, he could tell in what part of the walk he stood, merely by the note it sounded.

Dr. Saunderson had a peculiar method of performing arithmetical calculations, by an ingenious machine and method which has been called his Palpable Arithmetic, and is particularly described in a piece prefixed to the first volume of his Algebra. That he was able to make long and intricate calculations, both arithmetical and algebraical, is a thing as certain as it is wonderful. He had contrived for his own use, a commodious notation for any large numbers, which he could express on his abacus, or calculating table, and with which he could readily perform any arithmetical operations, by the sense of feeling only, for which reason it was called his Palpable Arithmetic.

His calculating table was a smooth thin board, a little more than a foot square, raised upon a small frame so as to lie hollow; which board was divided into a great number of little squares, by lines intersecting one another perpendicularly, and parallel to the sides of the table, and the parallel ones only one-tenth of an inch from each other; so that every square inch of the table was thus divided into 100 little squares. At every point of intersection the board was perforated by small holes, capable of receiving a pin; for it was by the help of pins, stuck up to the head through these holes, that he expressed his numbers. He used two sorts of pins, a larger and a smaller sort; at least their heads were different, and might easily be distinguished by feeling. Of these pins he had a large quantity in two boxes, with their points cut off, which always stood ready before him when he calculated. The writer of that account describes particularly the whole process of using the machine, and concludes, " He could place and displace his pins with incredible nimbleness and facility, much to the pleasure and surprize of all the beholders. He could even break off in the middle of a calculation, and resume it when he pleased, and could presently know the condition of it, by only drawing his fingers gently over the table."

SAURIN (JOSEPH), an ingenious French mathematician, was born in 1659, at Courtaison, in the principality of Orange. His father, minister at Grenoble, was a man of a very studious disposition, and was the first preceptor or instructor to our author; who made a rapid progress in his studies, and at a very early age was admitted a minister at Eure in Dauphiny. But preaching an offensive sermon, he was obliged to quit France in 1683. On this occasion he retired to Geneva; from whence he went into the State of Berne, and was appointed to a living at Yverdun. He was no sooner established in this his post, than certain theologians raised a storm against him. Saurin, disgusted with the controversy, and still more with the Swiss, where his talents were buried, passed into Holland, and from thence into France, where he put himself under the protection of the celebrated Bossu, to whom he made his abjuration in 1690, as it is suspected, that he might find protection, and have an opportunity of cultivating the sciences at Paris. And he was not disappointed: he met with many flattering encouragements; was even much noticed by the king, had a pension from the court, and was admitted of the Academy of Sciences in 1707, in the quality of geometrician. This science was now his chief study and delight; with many writings upon which he enriched the volumes of the Journal des Savans, and the Memoirs of the Academy of Sciences. These were the only works of this kind that he published: he was author of several other pieces of a controversial nature, against the celebrated Rousseau, and other antagonists, over whom with the assistance of government he was enabled to triumph. The latter part of his life was spent in more peace, and in cultivating the mathematical sciences; and he died the 29th of December 1737, of a lethargic fever, at 78 years of age.

The character of Saurin was lively and impetuous, endued with a considerable degree of that noble independence and loftiness of manner, which is apt to be mistaken for haughtiness or insolence; in consequence of which, his memory was attacked after his death, as his reputation had been during his life; and it was even

said

said he had been guilty of crimes, by his own confession, that ought to ha e been punished with death.

Saurin's mathematical and philosophical papers, printed in the Memoirs of the Academy of Sciences, which are pretty numerous, are to be found in the volumes for the years following; viz, 1709, 1710, 1713, 1716, 1718, 1720, 1722, 1723, 1725, 1727.

SAUVEUR (JOSEPH), an eminent French mathematician, was born at La Fleche the 24th of March 1653. He was absolutely dumb till he was seven years of age; and then the organs of speech did not disengage so effectually, but that he was ever after obliged to speak very slowly and with difficulty. He very early discovered a great turn for mechanics, and was always inventing and constructing something or other in that way.

He was sent to the college of the Jesuits to learn polite literature, but made very little progress in poetry and eloquence. Virgil and Cicero had no charms for him; but he read with greediness books of arithmetic and geometry. However, he was prevailed on to go to Paris in 1670, and, being intended for the church, there he applied himself for a time to the study of philosophy and theology; but still succeeded no better. In short, mathematics was the only study he had any passion or relish for, and this he cultivated with extraordinary success; for during his course of philosophy, he learned the first six books of Euclid in the space of a month, without the help of a master.

As he had an impediment in his voice, though otherwise endued with extraordinary abilities, he was advised by M. Bossuet, to give up all designs upon the church, and to apply himself to the study of physic: but this being utterly against the inclination of his uncle, from whom he drew his principal resources, Sauveur determined to devote himself to his favourite study, and to perfect himself in it, so as to teach it for his support; and in effect he soon became the fashionable preceptor in mathematics, so that at 22 years of age he had prince Eugene for his scholar.—He had not yet read the geometry of Des Cartes; but a foreigner of the first quality desiring to be taught it, he made himself master of it in an inconceivably small space of time. —Basset being a fashionable game at that time, the marquis of Dangeau asked him for some calculations relating to it, which gave such satisfaction, that Sauveur had the honour to explain them to the king and queen.

In 1681 he was sent with M. Mariotte to Chantilli, to make some experiments upon the waters there, which he did with much applause. The frequent visits he made to this place inspired him with the design of writing a treatise on fortification; and, in order to join practice with theory, he went to the siege of Mons in 1691, where he continued all the while in the trenches. With the same view also he visited all the towns of Flanders; and on his return he became the mathematician in ordinary at the court, with a pension for life.— In 1680 he had been chosen to teach mathematics to the pages of the Dauphiness. In 1686 he was appointed mathematical professor in the Royal College. And in 1696 admitted a member of the Academy of Sciences, where he was in high esteem with the members of that society.—He became also particularly acquainted with

the prince of Condé, from whom he received many marks of favour and affection. Finally, M. Vauban having been made marshal of France, in 1703, he proposed Sauveur to the king as his successor in the office of examiner of the engineers; to which the king agreed, and honoured him with a pension, which our author enjoyed till his death, which happened the 9th of July 1716, in the 64th year of his age.

Sauveur, in his character, was of a kind obliging disposition, of a sweet, uniform, and unaffected temper and although his fame was pretty generally spread abroad. it did not alter his humble deportment, and the simplicity of his manners. He used to say, that what one man could accomplish in mathematics, another might do also, if he chose it.

He was twice married The first time he took a very singular precaution; so he would not meet the lady till he had been with a notary to have the conditions, he intended to insist on, reduced into a written form; for fear the sight of her should not leave him enough master of himself. This was acting very wisely, and like a true mathematician; who always proceeds by rule and line, and makes his calculations when his head is cool —He had children by both his wives; and by the latter a son who, like himself, was dumb for the first seven years of his life

An extraordinary part of Sauveur's character is, that though he had neither a musical voice nor ear, yet he studied no science more than music, of which he composed an entire new system. And though he was obliged to borrow other people's voice and ears, yet he amply repaid them with such demonstrations as were unknown to former musicians. He also introduced a new diction in music, more appropriate and extensive. He invented a new doctrine of sounds. And he was the first that discovered, by theory and experiment, the velocity of musical strings, and the spaces they describe in their vibrations, under all circumstances of tension and dimensions. It was he also who first invented for this purpose the monochord and the echometer. In short, he pursued his researches even to the music of the ancient Greeks and Romans, to the Arabs, and to the very Turks and Persians themselves; so jealous was he, lest any thing should escape him in the science of sounds.

Sauveur s writings, which consist of pieces rather than of set works, are all inserted in the volumes of the Memoirs of the Academy of Sciences, from the year 1700 to the year 1716, on various geometrical, mathematical, philosophical, and musical subjects.

SCALE, a mathematical instrument, consisting of certain lines drawn on wood, metal, or other matter, divided into various parts, either equal or unequal. It is of great use in laying down distances in proportion, or in measuring distances already laid down.

There are Scales of various kinds, accommodated to the several uses: the principal are the *plane Scale*, the *diagonal Scale*, *Gunter's Scale*, and the *plotting Scale*.

Plane or Plain SCALE, a mathematical instrument of very extensive use and application; which is commonly made of 2 feet in length; and the lines usually drawn upon it are the following, viz,

1. Lines

1	Lines of	Equal parts, and marked	E. P.
2	- -	Chords - - -	Cho.
3	- -	Rhumbs - - -	Ru.
4	- -	Sines - - -	Sin.
5	- -	Tangents - -	Tan.
6	- -	Secants - -	Sec.
7	- -	Semitangents - -	S. T.
8	- -	Longitude - -	Long.
9	- -	Latitude - -	Lat.

1. The lines of equal parts are of two kinds, viz, simply divided, and diagonally divided. The first of these are formed by drawing three lines parallel to one another, and dividing them into any equal parts by short lines drawn across them, and in like manner subdividing the first division or part into 10 other equal small parts; by which numbers or dimensions of two figures may be taken off. Upon some rulers, several of these scales of equal parts are ranged parallel to each other, with figures set to them to shew into how many equal parts they divide the inch; as 20, 25, 30, 35, 40, 45, &c. The 2d or diagonal divisions are formed by drawing eleven long parallel and equidistant lines, which are divided into equal parts, and crossed

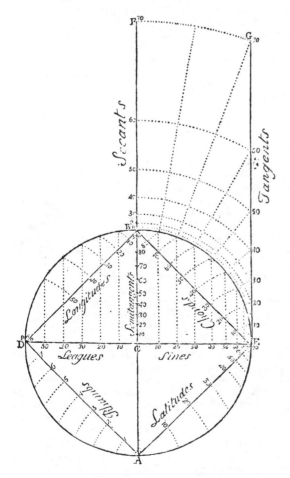

by other short lines, as the former; then the first of the equal parts have the two outermost of the eleven pa-

rallels divided into 10 equal parts, and the points of division being connected by lines drawn diagonally, the whole scale is thus divided into dimensions or numbers of three places of figures.

The other lines upon the scales are such as are commonly used in trigonometry, navigation, astronomy, dialling, projection of the sphere, &c, &c; and their constructions are mostly taken from the divisions of a circle, as follow:

Describe a circle with any convenient radius, and quarter it by drawing the diameters AB and DE at right angles to each other; continue the diameter AB out towards F, and draw the tangent line EG parallel to it; also draw the chords AD, DB, BE, EA. Then,

2. For the line of chords, divide a quadrant BE into 90 equal parts; on E as a centre, with the compasses transfer these divisions to the chord line EB, which mark with the corresponding numbers, and it will become a line of chords, to be transferred to the ruler.

3. For the line of rhumbs, divide the quadrant AD into 8 equal parts; then with the centre A transfer the divisions to the chord AD, for the line of rhumbs.

4. For the line of sines, through each of the divisions of the arc BE, draw right lines parallel to the radius BC, which will divide the radius CE into the sines, or versed sines, numbering it from C to E for the sines, and from E to C for the versed sines.

5. For the line of tangents, lay a ruler on C, and the several divisions of the arc BE, and it will intersect the line EG, which will become a line of tangents, and numbered from E to G with 10, 20, 30, 40, &c.

6. For the line of secants, transfer the distances between the centre C and the divisions on the line of tangents to the line BF, from the centre C, and these will give the divisions of the line of secants, which must be numbered from B towards F, with 10, 20, 30, &c.

7. For the line of semitangents, lay a ruler on D and the several divisions of the arc EB, which will intersect the radius CB in the divisions of the semitangents, which are to be marked with the corresponding figures of the arc EB.

The chief uses of the sines, tangents, secants, and semitangents, are to find the poles and centres of the several circles represented in the projections of the sphere.

8. For the line of longitude, divide the radius CD into 60 equal parts; through each of these, parallels to the radius BC will intersect the arc BD in as many points: from D as a centre the divisions of the arc BD being transferred to the chord BD, will give the divisions of the line of longitude.

If this line be laid upon the scale close to the line of chords, both inverted, so that 60° in the scale of longitude be against 0° in the chords, &c; and any degree of latitude be counted on the chords, there will stand opposite to it, in the line of longitude, the miles contained in one degree of longitude, in that latitude; the measure of 1 degree under the equator being 60 geographical miles.

9. For

9. For the line of latitude, lay a ruler on B, and the several divisions on the lines on CE, and it will interfect the arc AE in as many points; on A as a centre transfer the interfections of the arc AE to the chord AE, for the line of latitude.

See also Robertfon's Defcription and ufe of Mathematical Inftruments.

Decimal, or *Gunter's*, or *Plotting*, or *Proportional*, or *Reducing* SCALE. See the feveral articles.

SCALE, in Architecture and Geography, a line divided into equal parts, placed at the bottom of a map or draught, to ferve as a common meafure to all the parts of the building, or all the diftances and places of the map.

In maps of large tracts, as kingdoms and provinces, &c, the Scale ufually confifts of miles; whence it is denominated a Scale of miles.—In more particular maps, as thofe of manors, &c, the Scale is ufually of chains &c.—The Scales ufed in draughts of buildings moftly confift of modules, feet, inches, palms, fathoms, or the like.

To find the diftance between two towns &c, in a map, the interval is taken in the compaffes, and fet off in the fcale; and the number of divifions it includes gives the diftance. The fame method ferves to find the height of a ftory, or other part in a defign.

Front SCALE, in Perfpective, is a right line in the draught, parallel to the horizontal line; divided into equal parts, reprefenting feet, inches, &c.

Flying SCALE, is a right line in the draught, tending to the point of view, and divided into unequal parts, reprefenting feet, inches, &c.

Differential SCALE, is ufed for the fcale of relation fubtracted from unity. See SERIES.

SCALE *of Relation*, in Algebra, an expreffion denoting the relation of the terms of recurring feries to each other. See SERIES.

Hour SCALE. See HOUR.

SCALE, in Mufic, is a denomination given to the arrangement of the fix fyllables, invented by Guido Aratino, *ut re mi fa fol la*; called alfo gammut. It is called Scale, or ladder, becaufe it reprefents a kind of ladder, by means of which the voice rifes to acute, or finks to grave; each of the fix fyllables being as it were one ftep of the ladder.

SCALE is alfo ufed for a feries of founds rifing or falling towards acutenefs or gravity, from any given pitch of tune, to the greateft diftance that is fit or practicable, through fuch intermediate degrees as to make the fucceffion moft agreeable and perfect, and in which we have all the harmonical intervals moft commodioufly divided.

The fcale is otherwife called an *univerfal fyftem*, as including all the particular fyftems belonging to mufic. See SYSTEM.

There were three different Scales in ufe among the Ancients, which had their denominations from the three feveral forts of mufic, viz, the *diatonic*, *chromatic*, and *inharmonic*. Which fee.

SCALENE, or SCALENOUS *triangle*, is a triangle whofe fides and angles are all unequal.—A cylinder or cone, whofe axis is oblique or inclined to its bafe, is alfo faid to be fcalenous: though more frequently it is called oblique.

SCALIGER (JOSEPH JUSTUS), a celebrated French chronologer and critic, was the fon of Julius Cæfar Scaliger, and born at Agen in France, in 1540. He ftudied in the college of Bourdeaux; after which his father took him under his own care, and employed him in tranfcribing his poems; by which means he obtained fuch a tafte for poetry, that before he was 17 years old, he wrote a tragedy upon the fubject of Oedipus, in which he introduced all the poetical ornaments of ftyle and fentiment.

His father dying in 1558, he went to Paris the year following, with a defign to apply himfelf to the Greek tongue. For this purpofe he for two months attended the lectures of Turnebus; but finding that in the ufual courfe he fhould be a long time in gaining his point, he fhut himfelf up in his clofet, and by conftant application for two years gained a perfect knowledge of the Greek language. After which he applied himfelf to the Hebrew, which he learned by himfelf with great facility. And in like manner he ran through many other languages, till he could fpeak it is faid no lefs than 13 ancient and modern ones. He made no lefs progrefs in the fciences; and his writings procured him the reputation of one of the greateft men of that or any other age. He embraced the reformed religion at 22 years of age. In 1563, he attached himfelf to Lewis Cafteignier de la Roch Pazay, whom he attended in feveral journies. And, in 1593, the curators of the univerfity of Leyden invited him to an honorary profefforfhip in that univerfity, where he lived 16 years, and where he died of a dropfy in 1609, at 69 years of age.

Scaliger was a man of great temperance; was never married; and was fo clofe a ftudent, that he often fpent whole days in his ftudy without eating: and though his circumftances were always very narrow, he conftantly refufed the prefents that were offered him.

He was author of many ingenious works on various fubjects. His elaborate work, *De Emendatione Temporum*; his exquifite animadverfions on Eufebius; with his *Canon Ifagogicus Chronologiæ*; and his accurate comment upon Manilius's *Aftronomicon*, fufficiently evince his knowledge in the aftronomy, and other branches of learning, among the Ancients, and who, according to the opinion of the celebrated Vieta, was far fuperior to any of that age. And he had no lefs a character given him by the learned Cafaubon.—He wrote *Cyclometrica et Diatriba de Equinoctiorum Anticipatione*. He wrote alfo notes upon Seneca, Varro, and Aufonius's Poems. But that which above all things renders the name of Scaliger memorable to pofterity, is the invention of the Julian period, which confifts of 7980 years, being the continued product of the three cycles, of the fun 28, the moon 19, and Roman indiction 15. This period had its beginning fixed to the 764th year before the creation, and is not yet completed, and comprehends all other cycles, periods, and epochas, with the times of all memorable actions and hiftories.—The collections intitled *Scaligeriana*, were collected from his converfations by one of his friends; and being ranged in alphabetical order, were publifhed by Ifaac Voffius.

SCANTLING, a meafure, fize, or ftandard, by which the dimenfions &c of things are to be determined.
The

The term is particularly applied to the dimensions of any piece of timber, with regard to its breadth and thickness.

SCAPEMENT, in Clock-work, a general term for the manner of communicating the impulse of the wheels to the pendulum. The ordinary Scapements consist of the swing-wheel and pallets only; but modern improvements have added other levers or detents, chiefly for the purposes of diminishing friction, or for detaching the pendulum from the pressure of the wheels during part of the time of its vibration. Notwithstanding the very great importance of the Scapement to the performance of clocks, no material improvement was made in it from the first application of the pendulum to clocks to the days of Mr. George Graham; nothing more was attempted before his time, than to apply the impulse of the swing-wheel in such manner as was attended with the least friction, and would give the greatest motion to the pendulum. Dr. Halley discovered, by some experiments made at the Royal Observatory at Greenwich, that by adding more weight to the pendulum, it was made to vibrate larger arcs, and the clock went faster; by diminishing the weight of the pendulum, the vibrations became shorter, and the clock went slower; the result of these experiments being diametrically opposite to what ought to be expected from the theory of the pendulum, probably first roused the attention of Mr. Graham, and led him to such farther trials as convinced him, that this seeming paradox was occasioned by the retrograde motion, which was given to the swing-wheel by every construction of Scapement that was at that time in use; and his great sagacity soon produced a remedy for this defect, by constructing a Scapement which prevented all recoil of the wheels, and restored to the clock pendulum, wholly in theory, and nearly in practice, all its natural properties in its detached simple state; this Scapement was named by its celebrated inventor the *dead beat*, and its great superiority was so universally acknowledged, that it was soon introduced into general use, and still continues in universal esteem. The importance of the Scapement to the accurate going of clocks, was by this improvement rendered so unquestionable, that artists of the first rate all over Europe, were forward in producing each his particular construction, as may be seen in the works of Thiout l'ainé, M. J. A. Lepante, M. le Roy, M. Ferdinand Bertoud, and Mr. Cummings' Elements of Clock and Watchwork, in which we have a minute description of several new and ingenious constructions of Seapements, with an investigation of the principles on which their claim to merit is founded; and a comparative view of the advantages or defects of the several constructions. Besides the Scapements described in the above works, many curious constructions have been produced by eminent artists, who have not published any account of them, nor of the motives which have induced each to prefer his favourite construction: Mr. Harrison, Mr. Hindley of York, Mr. Ellicot, Mr. Mudge, Mr. Arnold, Mr. Whitehurst, and many other ingenious artists of this country, have made Scapements of new and peculiar constructions, of which we are unable, for the above reason, to give any farther account than that those of Mr. Harrison and Mr. Hindley had scarce

any friction, with a certain mode and quantity of recoil; those of all the other gentlemen, we believe, have been on the principle of the dead beat, with such other improvements as they severally judged most conducive to a good performance.

Count Bruhl has just published (in 1794) a small pamphlet, " On the Investigation of Astronomical Circles," to which he has annexed, " a Description of the Scapement in Mr. Mudge's first Timekeeper, drawn up in August 1771." Before entering upon the Description, the Count premises a few observations, in one of which he recognizes a hint concerning the nature of Mr. Mudge's Scapement, thrown out by this artist in a small tract printed by him in the year 1763, which is this: " The force derived from the mainspring should be made as equal as possible, by making the mainspring wind up another smaller spring at a less distance from the balance, at short intervals of time. *I think it would not be impracticable to make it wind up at every vibration, a small spring similar to the pendulum spring, that should immediately act on the balance, by which the whole force acting on the balance would be reduced to the greatest simplicity, with this advantage, that the force would increase in proportion to the arch.*" From this hint, Count Bruhl is surprised that no other artist have taken up Mr. Mudge's invention He then gives the Description of that invention as follows: " Mr. Mudge's Timekeeper has five wheels, with numbers high enough to admit pinions of twelve, and yet to go eight days. The Scapement consists of a wheel almost like that of a common crown wheel, and acts on pallets, each of which has a separate axis lying in the same line. To each pallet a spring is fixed in the shape of a pendulum spring; these springs are wound up alternately by the action of the last wheel upon the pallets, which is performed in the following manner:— Whenever one of the pallets (for instance the upper one) is set in motion by a tooth of the wheel sliding upon it, and then resting against a hook, or, rather a bearing at its end, the balance is entirely detached from it, being then employed in carrying the other pallet the contrary way. When the balance returns from that vibration (partly by the force of the pendulum spring, and partly by that of one of the two small springs which it had bent by the motion of that pallet which it had carried along with itself) it lays hold of the upper pallet and carries it round in the same manner as it did before the lower one, and, of course, in the same direction which the upper pallet had received from the power of the mainspring at the time that it was quite unconnected with the balance. The communication of motion from the balance to the pallets, and vice versa, is effected by two pins fixed to a crank, which in following the balance, hit each its proper pallet alternately. By what has been said, it is evident that whatever inequality there may be in the power derived from the mainspring (provided the latter be sufficient to wind up those little pallet springs) it can never interfere with the regularity of the balance's motion, but at the instant of unlocking the pallets, which is so instantaneous an operation, and the resistance so exceedingly small, that it cannot possibly amount to any sensible error. The removal of this great obstacle was certainly never so effectually done by any other contri-

I

vance

vance, and deserves the highest commendation, as a probable means to perfect a portable machine that will measure time correctly But this is not the only, nor indeed the principal advantage which this timekeeper will possess over any other; for, as it is impossible to reduce friction to so small a quantity as not to affect the motion of a balance, the consequence of which is, that it describes sometimes greater and sometimes smaller arcs, it became necessary to think of some method by which the balance might be brought to describe those different arcs in the same time. If a balance could be made to vibrate without friction or resistance from the medium in which it moves, the mere expanding and contracting of the pendulum spring, would probably produce the so much wished-for effect, as its force is supposed to be proportional to the arcs described; but as there is no machine void of friction, and as from that cause, the velocity of every balance decreases more rapidly than the spaces gone through decrease, this inequality could only be removed by a force acting on the balance, which assuming different ratios in its different stages, could counterbalance that inequality. This very material and important remedy, ·Mr. Mudge has effected by the construction of his Scapement; for his pallet springs having a force capable of being increased almost at pleasure, at the commencement of every vibration, the proportion in their different degrees of tension may be altered till it answers the intended purpose. This shews how effectually Mr. Mudge's Scapement removes the two greatest difficulties that have hitherto baffled the attempts of every other artist, namely, the inequalities of the power derived from the main spring, and the irregularities arising from friction, and the variable resistance of the medium in which the balance moves. Although at the time I am writing this account of his invention, the machine is not yet finished; I am not the less confident that whenever it is, it will be found to be one of the most useful of any which has as yet appeared."

SCARP, in Fortification, the interior slope of the ditch of a place; that is, the slope of that side of a ditch which is next to the place, or on the outside of the rampart at its foot, facing the champaign or open country. The slope on the outer side of the ditch is called the *counterscarp*.

SCENOGRAPHY, in Perspective, the perspective representation of a body on a plane; or a description and view of it in all its parts and dimensions, such as it appears to the eye in any oblique view.

This differs essentially from the ichnography and the orthography. The ichnography of a building, &c, represents the plan or ground work of the building, or section parallel to it; and the orthography the elevation, or front, or one side, also in its natural dimensions; but the Scenography exhibits the whole of the building that appears to the eye, front, sides, height, and all, not in their real dimensions or extent, but raised on the geometrical plan in perspective.

In architecture and fortification, Scenography is the manner of delineating the several parts of a building or fortress, as they are represented in perspective.

To exhibit the SCENOGRAPHY *of any body*. 1. Lay down the basis, ground plot, or plan, of the body, in the perspective ichnography, that is, draw the perspec-

tive appearance of the plan or basement, by the proper rules of perspective. 2. Upon the several points of the said perspective plan, raise the perspective heights, and connect the tops of them by the proper slope or oblique lines. So will the Scenography of the body be completed, when a proper shade is added. See PERSPECTIVE.

SCHEINER (CHRISTOPHER), a considerable German mathematician and astronomer, was born at Mundeilhein in Schwaben in 1575. He entered into the society of the Jesuits at 20 years of age; and afterwards taught the Hebrew tongue and the mathematics at Ingolstadt, Friburg, Brisac, and Rome. At length he became confessor to the archduke Charles, and rector of the college of the Jesuits at Neisse in Silesia, where he died in 1650, at 75 years of age.

Scheiner was chiefly remarkable for being one of the first who observed the spots in the sun with he telescope, though not the very first; for his observations of those spots were first made, at Ingolstadt, in the latter part of the year 1611, whereas Galileo and Harriot both observed them in the latter part of the year before, or 1610. Scheiner continued his observations on the solar phenomena for many years afterwards at Rome, with great assiduity and accuracy, constantly making drawings of them on paper, describing their places, figures, magnitude, revolutions and periods, so that Riccioli delivered it as his opinion that there was little reason to hope for any better observations of those spots. Des Cartes and Hevelius also say, that in their judgment, nothing can be expected of that kind more satisfactory. These observations were published in one volume folio, 1630, under the title of *Rosa Ursina*, &c; almost every page of which is adorned with an image of the sun with the spots. He wrote also several smaller pieces relating to mathematics and philosophy, the principal of which are,

2. *Oculus, sive Fundamentum Opticum, &c*; which was reprinted at London, in 1652, in 4to.

3. *Sol Eclipticus, Disquisitiones Mathematicæ*.

4. *De Controversiis et Novitatibus Astronomicis*.

SCHEME, a draught or representation of any geometrical or astronomical figure, or problem, by lines sensible to the eye; or of the celestial bodies in their proper places for any moment; otherwise called a diagram.

SCHEME *Arches*. See ARCH.

SCHOLIUM, a note, remark, or annotation, occasionally made on some passage, proposition, or the like.

The term is much used in geometry, and other parts of the mathematics; where, after demonstrating a proposition, it is used to point out how it might be done some other way; or to give some advice or precaution, in order to prevent mistakes; or to add some particular use or application of it.

Wolfius has given abundance of curious and useful arts and methods, and a good part of the modern philosophy, with the description of mathematical instruments, &c; all by way of Scholia to the respective propositions in his Elementa Matheseos.

SCHONER (JOHN), a noted German philosopher and mathematician, was born at Carolostadt in the year 1477, and died in 1547, at 70 years of age.—His early propensity to those sciences may be deemed a just prognostication of the great progress which

which he afterwards made in them. So that from his uncommon acquirements, he was chosen mathematical professor at Nuremburg when he was but a young man. He wrote a great many works, and was particularly famous for his astronomical tables, which he published after the manner of those of Regiomontanus, and to which he gave the title of *Resoluta*, on account of their clearness. But notwithstanding his great knowledge, he was, after the fashion of the times, much addicted to judicial astrology, which he took great pains to improve. The list of his writings is chiefly as follows:

1. Three Books of Judicial Astrology.
2. The Astronomical Tables named *Resolutæ*.
3. *De Usu Globi Stelliferi ; De Compositione Globi Cælestis ; De Usu Globi Terrestris, et de Compositione ejusdem.*
4. *Æquatorium Astronomicum.*
5. *Libellus de Distantiis Locorum per Instrumentum et Numeros Investigandis.*
6. *De Compositione Torqueti.*
7. *In Constructionem et Usum Rectanguli sive Radii Astronomici Annotationes.*
8. *Horarii Cylindri Canones.*
9. *Planisphærium, seu Meteoriscopium.*
10. *Organum Uranicum.*
11. *Instrumentum Impedimentorum Lunæ.*

All printed at Nuremburg, in folio, 1551. Of these, the large treatise of dialling rendered him more known in the learned world than all his other works besides; in which he discovers a surprising genius and fund of learning of that kind.

SCHOOL, a place where the languages, humanities, or arts and sciences, &c, are taught.

School is also used for a whole faculty, university, or sect; as Plato's school, the school of Epicurus, the school of Paris, &c.——The school of Tiberias was famous among the ancient Jews ; and it is to this we owe the Massora, and Massoretes.

School *Philosophy*, &c. the same with *scholastic.*

SCIAGRAPHY, or Sciography, the profile or vertical section of a building; used to shew the inside of it.

Sciagraphy, in Astronomy &c, is a term used by some authors for the art of finding the hour of the day or night, by the shadow of the sun, moon, stars, &c. See Dial.

SCIENCE, a clear and certain knowledge of any thing, founded on demonstration, or on self evident principles.—In this sense, *doubting* is opposed to science ; and *opinion* is the middle between the two.

Science is more particularly used for a formed system of any branch of knowledge, comprehending the doctrine, reason, and theory of the thing, without any immediate application of it to any uses or offices of life. And in this sense, the word is used in opposition to *art.*

Science may be divided into these three sorts : First, the knowledge of things, their constitutions, properties, and operations, whether material or immaterial. And this, in a little more enlarged sense of the word, may be called physics, or natural philosophy. Secondly the skill of rightly applying our own powers and actions for the attainment of good and useful things, as *Ethics.* Thirdly, the doctrine of signs ; as words, logic, &c.

SCIENTIFIC, or Scientifical, something relating to the pure and sublimer sciences; or that abounds in science, or knowledge.

A work, or method, &c, is said to be scientifical, when it is founded on the pure reason of things, or conducted wholly on the principles of them. In which sense the word stands opposed to narrative, arbitrary, opinionative, positive, tentative, &c.

SCIOPTIC, or Scioptric *Ball*, a sphere or globe of wood, with a circular hole or perforation, where a lens is placed. It is so fitted that, like the eye of an animal, it may be turned round every way, to be used in making experiments of the darkened room.

SCIOPTRICS. See Camera Obscura.

SCIOTHERICUM *Telescopium*, is an horizontal dial, adapted with a telescope for observing the true time both by day and night, to regulate and adjust pendulum clocks, watches, and other time-keepers. It was invented by Mr. Molyneux, who published a book with this title, which contains an accurate description of this instrument, with all its uses and applications.

SCLEROTICA, one of the common membranes of the eye, on its hinder part. It is a large, thick, firm, hard, opaque membrane, extended from the external circumference of the cornea to the optic nerve, and forms much the greater part of the external globe of the eye. The Sclerotica and the cornea compose the case in which all the internal coats of the eye and its humours are contained.

SCONCES, small forts, built for the defence of some pass, river, or other place. Some Sconces are made regular, of four, five, or six bastions ; others are of smaller dimensions, fit for passes, or rivers; and others for the field.

SCORE, in Music, denotes partition, or the original draught of the whole composition, in which the several parts, viz the treble, second treble, bass, &c are distinctly scored, and marked.

SCORPIO, the *Scorpion*, the 8th sign of the zodiac, denoted by the character ♏, being a rude design of the animal of that name.

The Greeks, who would be supposed the inventors of astronomy, and who have, with that intent, fathered some story or other of their own upon every one of the constellations, give a very singular account of the origin of this one. They tell us that this is the creature which killed Orion. The story goes, that the famous hunter of that name boasted to Diana and Latona, that he would destroy every animal that was upon the earth; the earth, they say, enraged at this, sent forth the poisonous reptile the Scorpion, which insignificant creature stung him, that he died. Jupiter, they say, raised the Scorpion to the heavens, giving him this place among the constellations; and that afterwards Diana requested of him to do the same honour to Orion, which he at last consented to, but placed him in such a situation, that when the Scorpion rises, he sets.

But the Egyptians, or whatever early nation it was that framed the zodiac, probably placed this poisonous reptile in that part of the heavens to denote that when the sun arrived at it, fevers and sicknesses, the maladies of autumn, would begin to rage. This they represented by an animal whose sting was of power to occasion some

of

of them; and it was thus they formed all the constellations.

The ancients allotted one of the twelve principal among their deities to be the guardian for each of the 12 signs of the zodiac. The Scorpion, as their history of it made it a fierce and fatal animal that had killed the great Orion, fell naturally to the protection of the god of war; Mars is therefore its tutelary deity; and to this single circumstance is owing all that jargon of the astrologers, who tell us that there is a great analogy between the planet Mars and the constellation Scorpio. To this also is owing the doctrine of the alchymists, that iron, which they call Mars, is also under the dominion of the same constellation, and that the transmutation of that metal into gold can only be performed when the sun is in this sign.

The stars in Scorpio, in Ptolomy's catalogue, are 24; in that of Tycho 10, in that of Hevelius 20, but in that of Flamsteed and Sharp 44.

SCORPION is also the name of an ancient military engine, used chiefly in the defence of walls, &c.

Marcellinus describes the Scorpion, as consisting of two beams bound together by ropes. From the middle of the two, rose a third beam, so disposed, as to be pulled up and let down at pleasure: and on the top of this were fastened iron hooks, where a sling was hung, either of iron or hemp; and under the third beam lay a piece of hair-cloth full of chaff, tied with cords. It had its name Scorpio, because when the long beam or tiller was erected, it had a sharp top in manner of a sting.

To use the engine, a round stone was put into the sling, and four persons on each side, loosening the beams bound by the ropes, drew back the erect beam to the hook; then the engineer, standing on an eminence, gave a stroke with a hammer on the chord to which the beam was fastened with its hook, which set it at liberty; so that hitting against the soft hair cloth, it struck out the stone with a great force.

SCOTIA, in Architecture, a semicircular cavity or channel between the tores, in the bases of columns; and sometimes under the larmier or drip, in the cornice of the Doric order. The workmen often call it the Casement, and it is also otherwise called the Trochilus.

SCREW, or SCRUE, one of the six mechanical powers; chiefly used in pressing or squeezing bodies close, though sometimes also in raising weights.

The Screw is a spiral thread or groove cut round a cylinder, and everywhere making the same angle with the length of it. So that, if the surface of the cylinder, with this spiral thread upon it, were unfolded and stretched into a plane, the spiral thread would form a straight inclined plane, whose length would be to its height, as the circumference of the cylinder is to the distance between two threads of the Screw; as is evident by considering, that in making one round, the spiral rises along the cylinder the distance between the two threads.

Hence the threads of a Screw may be traced upon the smooth surface of a cylinder thus: Cut a sheet of paper into the form of a right-angled triangle, having its base to its height in the above proportion, viz, as the circumference of the cylinder of the Screw is to the intended distance between two threads; then wrap this paper triangle about the cylinder, and the hypothenuse of it will trace out the line of the spiral thread.

When the spiral thread is upon the outside of a cylinder, the Screw is said to be a *male* one. But if the thread be cut along the inner surface of a hollow cylinder, or a round perforation, it is said to be *female*. And this latter is also sometimes called the *box* or *nut*.

When motion is to be given to something, the male and female Screw are necessarily conjoined; that is, whenever the screw is to be used as a simple engine, or mechanical power. But when joined with an axis in peritrochio, there is no occasion for a female; but in that case it becomes part of a compound engine.

The Screw cannot properly be called a simple machine, because it is never used without the application of a lever, or winch, to assist in turning it.

Of the Force and Power of the Screw.

1. The force of a power applied to turn a Screw round, is to the force with which it presses upwards or downwards, setting aside the friction, as the distance between two threads is to the circumference where the power is applied.

For, the Screw being only an inclined plane, or half wedge, whose height is the distance between two threads, and its base the said circumference; and the force in the horizontal direction being to that in the vertical one as the lines perpendicular to them, viz, as the height of the plane, or distance of the two threads, is to the base of the plane, or circumference at the place where the power is applied; therefore the power is to the pressure, as the distance of two threads, is to that circumference.

2. Hence, when the Screw is put in motion; then the power is to the weight which would keep it in equilibrio, as the velocity of the latter is to that of the former. And hence their two momenta are equal, which are produced by multiplying each weight or power by its own velocity. Two different forms of Screw presses, are as below.

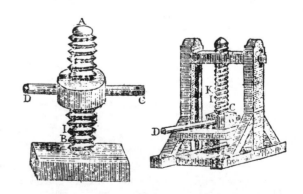

3. Hence we can easily compute the force of any machine turned by a Screw. Let the annexed figure represent a press driven by a Screw, whose threads are each a quarter of an inch asunder; and let the Screw be turned by a handle of 4 feet long from C to D; then if the natural force of a man, by which he can lift, or

pull,

pull, or draw, be 150 pounds; and it be required to determine with what force the Screw will prefs on the board, when the man turns the handle at C and D with his whole force. The diameter CD of the power being 4 feet, or 48 inches, its circumference is 48 × 3·1416 or 150⁴ nearly; and the diftance of the threads being ¼ of an inch; therefore the power is to the preffure, as ¼ to 150⅘ or as 1 to 603½: but the power is equal to 150lb; therefore as 1 : 603⅓ :: 150 : 90,480; and confequently the preffure at the bottom of the Screw, is equal to a weight of 90,480 pounds, independent of friction.

But the power has to overcome, not only the weight, or other refiftance, but alfo the friction of the Screw, which in this machine is very great, in fome cafes equal to the weight itfelf, fince it is fometimes fufficient to fuftain the weight, when the power is taken off.

Mr. Hunter has defcribed a new method of applying the Screw with advantage in particular cafes, in the Philof Tranf. vol. 71, pa. 58 &c.

The Endlefs SCREW, or Perpetual SCREW, is one which works in, and turns, a dented wheel DF, without a concave or female Screw; being fo called becaufe it may be turned for ever, without coming to an end. From the following fchemes it is evident, that while the Screw turns once round, the wheel only advances the diftance of one tooth.

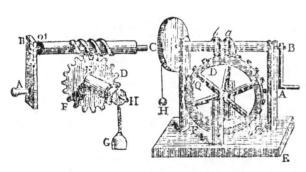

1. If the power applied to the lever, or handle, of an endlefs Screw AB, be to the weight, in a ratio compounded of the periphery of the axis of the wheel EH, to the periphery defcribed by the power in turning the handle, and of the revolutions of the wheel DF to the revolutions of the Screw CB, the power will balance the weight. Hence,

2. As the motion of the wheel is very flow, a fmall power may raife a very great weight, by means of an endlefs Screw. And therefore the chief ufe of fuch a Screw is, either where a great weight is to be raifed through a little fpace; or where only a flow gentle motion is wanted. For which reafon it is very ufeful in clocks and watches.

3 Having given the number of teeth, the diftance of the power from the centre of the Screw B, the radius of the axis HE, and the power; to find the weight it will raife. Multiply the diftance of the power from the centre of the Screw AB, by the number of the teeth, and the product will be the fpace paffed through by the power, while the weight paffes through a fpace equal to the periphery of the axis: then fay, as the radius of

the axis is to the fpace of the power juft found, fo is the power to a 4th proportional, which will be the weight the power is able to fuftain. Thus, if AB = 3, the radius of the axis HE = 1, the power 150 pounds, and the number of teeth of the wheel DF 48; then the weight will be found = 21,600 = 3 × 150 × 48. Whence it appears that the endlefs Screw exceeds all others in increafing the force of a power.

4. A machine for fhewing the power of the Screw may be contrived in the following manner. Let the wheel C (laft fig) have a Screw a b on its axis, working in the teeth of the wheel D, which fuppofe to be 48 in number. It is plain that for every revolution of the wheel C, and Screw ab, by the winch A, the wheel D will be moved one tooth by the Screw; and therefore in 48 revolutions of the winch, the wheel D will be turned once round. Then if the circumference of a circle, defcribed by the handle of the winch, be equal to the circumference of a groove e round the wheel D, the velocity of the handle will be 48 times as great as the velocity of any given point in the groove. Confequently when a line G goes round the groove e, and has a weight of 48lb hung to it below the pedeftal EF a power equal to one pound at the handle will balance and fupport the weight.

Archimedes's SCREW, is a fpiral pump, being a machine for raifing water, firft invented by Archimedes.

Its ftructure and ufe will be underftood by the following defcription of it.

ABCD (Pl. xxiii, fig. 6) is a wheel, which is turned round, according to the order of thofe letters, by the fall of water EF, which need not be more than 3 feet. The axis G of the wheel is raifed fo as to make an angle of about 44° with the horizon; and on the top of that axle is a wheel H, which turns fuch another wheel I of the fame number of teeth; the axle K of this laft wheel being parallel to the axle G of the two former wheels. The axle G is cut into a double threaded Screw, as in the annexed figure (fig. 7), exactly refembling the Screw on the axis of the fly of a common jack, which muft be what is called a right-handed Screw, if the firft wheel turns in the direction ABCD; but a left-handed Screw, if the ftream turns the wheel the contrary way; and the Screw on the axle G muft be cut in a contrary way to that on the axle K, becaufe thefe axes turn in contrary directions. Thefe Screws muft be covered clofe over with boards, like thofe of a cylindrical cafk; and then they will be fpiral tubes. Or they may be made of tubes or pipes of lead, and wrapt round the axles in fhallow grooves cut in it, like the figure 8. The lower end of the axle G turns conftantly in the ftream that turns the wheel, and the lower ends of the fpiral tubes are open into the water. So that, as the wheel and axle are turned round, the water rifes in the fpiral tubes, and runs out at L through the holes M, N, as they come about below the axle. Thefe holes, of which there may be any number, as 4 or 6, are in a broad clofe ring on the top of the axle, into which ring the water is delivered from the upper open ends of the Screw tubes, and falls into the open box N. The lower end of the axle K turns on a gudgeon in the water in N; and the fpiral tubes in that axle take up the water from N, and deliver it into another fuch box under the top of K: on which there may be fuch another

wheel

wheel as I, to turn a third axle by such a wheel upon it. And in this manner may water be raised to any proposed height, when there is a stream sufficient for that purpose to act on the broad float boards of the first wheel. Archimedes's Screw, or a still simpler form of it, is also represented in fig. 9.

SCROLLS, or SCROWLS, or *Volutes*, a term in Architecture. See VOLUTES.

SCRUE. See SCREW.

SCRUPLE, the least of the weights used by the ancients. Among the Romans it was the 24th part of an ounce, or the third part of a drachm.

SCRUPLE is still a small weight among us, equal to 20 grains, or the 3d part of a drachm. Among gold-smiths the scruple is 24 grains.

SCRUPLE, in Chronology, a small portion of time much used by the Chaldeans, Jews, Arabs, and other eastern people, in computations of time. It is the 1080th part of an hour, and by the Hebrews called *helakim*.

SCRUPLES, in Astronomy. As

SCRUPLES *Eclipsed*, denote that part of the moon's diameter which enters the shadow, expressed in the same measure in which the apparent diameter of the moon is expressed. See DIGIT.

SCRUPLES *of Half Duration*, an arch of the moon's orbit, which the moon's centre describes from the beginning of an eclipse to its middle.

SCRUPLES *of Immersion*, or *Incidence*, an arch of the moon's orbit, which her centre describes from the beginning of the eclipse, to the time when the centre falls into the shadow. See IMMERSION.

SCRUPLES *of Emersion*, an arch of the moon's orbit, which her centre describes in the time from the first emersion of the moon's limb, to the end of the eclipse.

SCYTALA, in Mechanics, a term which some writers use for a kind of radius, or spoke, standing out from the axis of a machine, as a lever or handle, to turn it round, and work it by.

SEA, in Geography, is frequently used for that vast tract of water encompassing the whole earth, more properly called ocean. But

SEA is more properly used for a particular part or division of the ocean, denominated from the countries it washes, or from other circumstances. Thus we say, the Irish sea, the Mediterranean sea, the Baltic sea, the Red sea, &c.

SEA among sailors is variously applied, to a single wave, or to the agitation produced by a multitude of waves in a tempest, or to their particular progress and direction. Thus they say, a heavy sea broke over our quarter, or we shipped a heavy sea; there is a great sea in the offing; the sea sets to the southward. Hence a ship is said to head the sea, when her course is opposed to the setting or direction of the surges. A *Long Sea* implies a steady and uniform motion of long and extensive waves. On the contrary, a *Short Sea* is when they run irregularly, broken, and interrupted, so as frequently to burst over a vessel's side or quarter.

Properties and Affections of the SEA.

1. *General Motion of the Sea.* M. Daffie of Paris, in a work long since published, has been at great pains

to prove that the Sea has a general motion, independent of winds and tides, and of more consequence in navigation than is usually supposed. He affirms that this motion is from east to west, inclining toward the north when the sun is on the north side of the equinoctial, but toward the south when he is on the south side of it. Philos. Trans. No. 135.

2. *Bason or Bottom of the* SEA, or *Fundus Maris*, a term used to express the bed or bottom of the sea in general. Mr. Boyle has published a treatise on this subject, in which he has given an account of its irregularities and various depths founded on the observations communicated to him by mariners.

Count Marsigli has, since his time, given a much fuller account of this part of the globe. The materials which compose the bottom of the Sea, may reasonably be supposed, in some degree, to influence the taste of its waters; and this author has made many experiments to prove that fossil coal, and other bituminous substances, which are found in plenty at the bottom of the Sea, may communicate in great part its bitterness to it.

It is a general rule among sailors, and is found to hold true in many instances, that the more the shores of any place are steep and high, forming perpendicular cliffs, the deeper the Sea is below; and that on the contrary, level shores denote shallow Seas. Thus the deepest part of the Mediterranean is generally allowed to be under the height of Malta. And the observation of the strata of earth and other fossils, on and near the shores, may serve to form a good judgment as to the materials to be found in its bottom. For the veins of salt and of bitumen doubtless run on the same, and in the same order, as we see them at land; and the strata of rocks that serve to support the earth of hills and elevated places on shore, serve also, in the same continued chain, to support the immense quantity of water in the bason of the Sea.

The coral fisheries have given occasion to observe that there are many, and those very large caverns or hollows in the bottom of the Sea, especially where it is rocky; and that the like caverns are sometimes found in the perpendicular rocks which form the steep sides of those fisheries. These caverns are often of great depth, as well as extent, and have sometimes wide mouths, and sometimes only narrow entrances into large and spacious hollows.

The bottom of the Sea is covered with a variety of matters, such as could not be imagined by any but those who have examined into it, especially in deep water, where the surface only is disturbed by tides and storms, the lower part, and consequently its bed at the bottom, remaining for ages perhaps undisturbed. The soundings, when the plummet first touches the ground on approaching the shores, give some idea of this. The bottom of the plummet is hollowed, and in that hollow there is placed a lump of tallow; which being the part that first touches the ground, the soft nature of the fat receives into it some part of those substances which it meets with at the bottom: this matter, thus brought up, is sometimes pure sand, sometimes a kind of sand made of the fragment of shells, beaten to a sort of powder, sometimes it is made of a like powder of the several sorts of corals, and sometimes it is composed of

o

of fragments of rocks; but beside these appearances, which are natural enough, and are what might well be expected, it brings up substances which are of the most beautiful colours. Marsigli Hist. Phys. de la Mer.

Dr. Donati, in an Italian work, containing an essay towards a natural history of the Adriatic Sea, printed at Venice in 1750, has related many curious observations on this subject, and which confirm the observations of Marsigli: having carefully examined the soil and productions of the various countries that surround the Adriatic Sea, and compared them with those which he took up from the bottom of the Sea, he found that there is very little difference between the former and the latter. At the bottom of the water there are mountains, plains, vallies, and caverns, similar to those upon land. The soil consists of different strata placed one upon another, and mostly parallel and correspondent to those of the rocks, islands, and neighbouring continents. They contain stones of different sorts, minerals, metals, various putrefied bodies, pumice stones, and lavas formed by volcanos.

One of the objects which most excited his attention, was a crust, which he discovered under the water, composed of crustaceous and testaceous bodies, and beds of polypes of different kinds, confusedly blended with earth, sand, and gravel; the different marine bodies which form this crust, are found at the depth of a foot or more, entirely petrified and reduced into marble; these he supposes are naturally placed under the Sea when it covers them, and not by means of volcanos and earthquakes, as some have conjectured. On this account he imagines that the bottom of the Sea is constantly rising higher and higher, with which other obvious causes of increase concur; and from this rising of the bottom of the Sea, that of its level or surface naturally results; in proof of which this writer recites a great number of facts. Philos. Transf. vol. 49, pa. 585.

3. *Luminousness of the* SEA. This is a phenomenon that has been noticed by many nautical and philosophical writers. Mr. Boyle ascribes it to some cosmical law or custom of the terrestrial globe, or at least of the planetary vortex.

Father Bourzes, in his voyage to the Indies, in 1704, took particular notice of this phenomenon, and very minutely describes it, without assigning the true cause.

The Abbe Nollet was long of opinion, that the light of the Sea proceeded from electricity; and others have had recourse to the same principle, and shewn that the luminous points in the surface of the Sea are produced merely by friction.

There are however two other hypotheses, which have more generally divided between them the solution of this phenomenon; the one of these ascribes it to the shining of luminous insects or animalcules, and the other to the light proceeding from the putrefaction of animal substances. The Abbé Nollet, who at first considered this luminousness as an electrical phenomenon, having had an opportunity of observing the circumstances of it, when he was at Venice in 1749, relinquished his former opinion, and concluded that it was occasioned either by the luminous aspect, or by

some liquor or effluvia of an insect which he particularly describes, though he does not altogether exclude other causes, and especially the spawn or fry of fish.

The same hypothesis had also occurred to M. Vianelli; and both he and Grizellini, a physician in Venice, have given drawings of the insects from which they imagined this light to proceed.

A similar conjecture is proposed by a correspondent of Dr. Franklin, in a letter read at the Royal Society in 1756; the writer of which apprehends, that this appearance may be caused by a great number of little animals, floating on the surface of the Sea. And Mr. Forster, in his account of a voyage round the world with captain Cook, in the years 1772, 3, 4, and 5, describes this phenomenon as a kind of blaze of the Sea; and, having attentively examined some of the shining water, expresses his conviction that the appearance was occasioned by innumerable minute animals of a round shape, moving through the water in all directions, which show separately as so many luminous sparks when taken up on the hand: he imagines that these small gelatinous luminous specks may be the young fry of certain species of some medusæ, or blubber. And M. Dagelat and M. Rigaud observed several times, and in different parts of the ocean, such luminous appearances by vast masses of different animalcules; and a few days after the Sea was covered, near the coasts, with whole banks of small fish in innumerable multitudes, which they supposed had proceeded from the shining animalcules.

But M. le Roi, after giving much attention to this phenomenon, concludes that it is not occasioned by any shining insects, especially as, after carefully examining with a microscope some of the luminous points, he found them to have no appearance of an animal; and he also found that the mixture of a little spirit of wine with water just drawn from the Sea, would give the appearance of a great number of little sparks, which would continue visible longer than those in the ocean: the same effect was produced by all the acids, and various other liquors. M. le Roi is far from asserting that there are no luminous insects in the Sea; for he allows that several gentlemen have found them; but he is satisfied that the Sea is luminous chiefly on some other account, though he does not so much as offer a conjecture with respect to the true cause.

Other authors, equally dissatisfied with the hypothesis of luminous insects, for explaining the phenomenon which is the subject of this article, have ascribed it to some substance of the phosphoric kind, arising from putrefaction. The observations of F. Bourzes, above referred to, render it very probable, that the luminousness of the Sea arises from slimy and other putrescent matter, with which it abounds, though he does not mention the tendency to putrefaction, as a circumstance of any consequence to the appearance. But the experiments of Mr. Canton, which have the advantage of being easily made, seem to leave no room to doubt that the luminousness of the Sea is chiefly owing to putrefaction. And his experiments confirm an observation of Sir John Pringle's, that the quantity of salt contained in Sea water hastens putrefaction; but since that precise quantity of salt which promotes putre-

faction

faction the moſt, is leſs than that which is found in Sea-water, it is probable, Mr. Canton obſerves, that if the Sea were leſs ſalt, it would be more luminous. See Philoſ. Tranſ. vol. 59, pa. 446, and Franklin's Exper. and Obſerv. pa. 274.

Of the Depth of the Sea, its Surface, &c.

What proportion the ſuperficies of the Sea bears to that of the land, is not accurately known, though it is ſaid to be ſomewhat more than two to one. This proportion of the ſurface of the Sea to the land, has been found by experiment thus: takin the printed paper map or covering of a terreſtrial globe, with a pair of ſciſſors clip out the parts that are land, and thoſe that are water; then weighing theſe parcels ſeparately in a pair of fine ſcales, the land is found to be near ⅓, and the water rather more than ⅔ of the whole.

With regard to the profundity or depth of the Sea, Varenius affirms, that it is in ſome places unfathomable, and in others very various, being in certain places from ¹⁄₂₀th of a mile to 4½ miles in depth, in other places deeper, but much leſs in bays than in oceans. In general, the depths of the Sea bear a great analogy to the height of mountains on the land, ſo far as is hitherto diſcovered.

There are two ſpecial reaſons why the Sea does not increaſe by means of rivers, &c, running every where into it. The firſt is, becauſe waters return from the Sea by ſubterranean cavities and aqueducts, through various parts of the earth. Secondly, becauſe the quantity of vapours raiſed from the Sea, and falling in rain upon the land, only cauſe a circulation of the water, but no increaſe of it. It has been found by experiment and calculation, that in a ſummer's day, there may be raiſed in vapours from the ſurface of the Mediterranean Sea, 528 millions of tuns of water; and yet this Sea receiveth not, from all its nine great rivers, above 183 millions of tuns per day, which is but about a third part of what is exhauſted in vapours; and this defect in the ſupply by the rivers, may ſerve to account for the continual influx of a current by the mouth or ſtraits at Gibraltar. Indeed it is rather probable, that the waters of the Sea ſuffer a continual ſlow decreaſe as to their quantity, by ſinking always deeper into the earth, by filtering through the fiſſures in the ſtrata and component parts.

SEASONS, certain portions or quarters of the year, diſtinguiſhed by the ſigns which the ſun then enters. Upon them depend the different temperatures of the air, different works in tillage, &c.

The year is divided into four Seaſons, ſpring, ſummer, autumn, winter, which take their beginnings when the ſun enters the firſt point of the ſigns Aries, Cancer, Libra, Capricorn.

The Seaſons are very well illuſtrated by fig. 1, plate viii; where the candle at I repreſents the ſun in the centre, about which the earth moves in the ecliptic ABCD, which cuts the equinoctial *abcd* in the two equinoxes E and G. When the earth is in theſe two points, it is evident that the ſun equally illuminates both the poles, and makes the days and nights equal all over the earth. But while the earth moves from G by C to ♑, the upper or north pole becomes more and more enlightened, the days become longer, and the nights ſhorter; ſo that when the earth is at ♑, or the ſun at ♋, our days are at the longeſt, as at midſummer. While the earth moves from ♑ by D to E, our days continually decreaſe, by the north pole gradually declining from the ſun, till at E or autumn they become equal to the nights, or 12 hours long. Again, while the earth moves from E by A to F, the north pole becomes always more and more involved in darkneſs, and the days grow always ſhorter, till at F or ♋, when it is midwinter to the inhabitants of the northern hemiſphere. Laſtly, while the earth moves from ♋ by B to G, the north parts come more and more out of darkneſs, and the days grow continually longer, till at G the two poles are equally enlightened, and the days equal to the nights again. And ſo on continually year after year.

SECANT, in Geometry, a line that cuts another, whether right or curved; Thus the line PA or PB, &c, is a Secant of the circle ABD, becauſe cutting it in the point F, or G, &c. Properties of ſuch Secants to the circle are as follow:

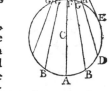

1. Of ſeveral Secants PA, PB, PD, &c, drawn from the ſame point P, that which paſſes through the centre C is the greateſt; and from thence they decreaſe more and more as they recede farther from the centre; viz. PB leſs than PA, and PD leſs than PB, and ſo on, till they arrive at the tangent ac E, which is the limit of all the Secants.

2. Of theſe Secants, the external parts PF, PG, PH, &c, are in the reverſe order, increaſing continually from F to E, the greater Secant having the leſs external part, and in ſuch ſort, that any Secant and its external part are in reciprocal proportion, or the whole is reciprocally as its external part, and conſequently that the rectangle of every Secant and its external part is equal to a conſtant quantity, viz, the ſquare of the tangent. That is,

$$PA : \frac{1}{PF} :: PB : \frac{1}{PG} :: PD : \frac{1}{PH} \ \&c,$$

or $PA \times PF = PB \times PG = PD \times PH = PE^2.$

3. The tangent PE is a mean proportional between any Secant and its external part; as between PA and PF, or PB and PG, or PD and PH, &c.

4. The angle DPB, formed by two Secants, is meaſured by half the difference of its intercepted arcs DB and GH.

SECANT, in Trigonometry, denotes a right line drawn from the centre of a circle, and, cutting the circumference, proceeds till it meets with a tangent to the ſame circle.

Thus, the line CD, drawn from the centre C, till it meets the tangent BD, is called a Secant; and particularly the Secant of the arc BE, to which BD is a tangent. In like manner, by producing DC to meet the tangent A*d* in *d*, then C*d*, equal to CD, is the Secant of the arch AE which is the ſupplement of the arch BE.

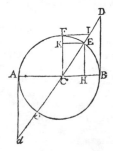

So that an arch and its supplement have their Secants equal, only the latter one is negative to the former, being drawn the contrary way. And thus the Secants in the 2d and 3d quadrant are negative, while those in the 1st and 4th quadrants are positive.

The Secant CI of the arc EF, which is the complement of the former arch BE, is called the *cosecant* of BE, or the Secant of its complement. The cosecants in the 1st and 2d quadrants are affirmative, but in the 3d and 4th negative.

The Secant of an arc is reciprocally as the cosine, and the cosecant reciprocally as the sine; or the rectangle of the Secant and cosine, and the rectangle of the cosecant and sine, are each equal to the square of the radius.

For CD : CE : : CB : CH, or $f : r : : r : c$,
and CI : CE : : CF : CK, or $\sigma : r : : r : s$;

and consequently $r^2 = cf = s\sigma$; where r denotes the radius, s the sine, c the cosine, f the Secant, and σ the cosecant.

An arc a, to the radius r, being given, the Secant f, and cosecant σ, and their logarithms, or the logarithmic Secant and cosecant, may be expressed in infinite series, as follows, viz,

$$f = r + \frac{a^2}{2r} + \frac{5a^4}{24r^3} + \frac{61a^6}{720r^5} + \frac{277a^8}{8064r^7} \ \&c.$$

$$\sigma = \frac{r^2}{a} + \frac{a}{6} + \frac{7a^3}{360r^2} + \frac{31a^5}{15120r^4} + \frac{127a^7}{604800r^6} \ \&c.$$

$$log. \ f = m \times \left(\frac{a^2}{2} + \frac{a^4}{12} + \frac{a^6}{45} + \frac{17a^8}{2520} \ \&c. \right)$$

$$log. \sigma = -log. a + m \times \left(\frac{a^2}{6} + \frac{a^4}{180} + \frac{a^6}{2835} + \frac{a^8}{37800} \&c. \right)$$

where m is the modulus of the system of logarithms.

SECANTS, *Figure of.* See FIGURE *of Secants.*

SECANTS, *Line of.* See SECTOR, and SCALE.

SECOND, in Geometry, or Astronomy, &c, the 60th part of a prime or minute: either in the division of circles, or in the measure of time. A degree, or an hour, are each divided into 60 minutes, marked thus '; a minute is subdivided into 60 Seconds, marked thus "; a Second into 60 thirds, marked thus '''; &c.

We sometimes say a *Second minute*, a *third minute*, &c, but more usually only *Second, third,* &c.

The Seconds pendulum, or pendulum that vibrates Seconds, in the latitude of London, is $39\frac{1}{8}$ inches long.

SECONDARY *Circles of the Ecliptic*, are circles of longitude of the stars; or circles which, passing through the poles of the ecliptic, are at right angles to the ecliptic.

By means of these Secondary circles, all points in the heavens are referred to the ecliptic; that is, any star, planet, or other phenomenon, is understood to be in that point of the ecliptic, which is cut by the Secondary circle that passes through such star, &c.

If two stars be thus referred to the same point of the ecliptic, they are said to be in conjunction; if in opposite points, they are in opposition; if they are referred to two points at a quadrant's distance, they are said to be in a quartile aspect, if the points differ a 6th part of the ecliptic, they are in sextile aspect, &c.

In general, all circles that intersect one of the six greater circles of the sphere at right angles, may be called Secondary circles. As the azimuth or vertical circles in respect of the horizon, &c; the meridian in respect of the equator, &c.

SECONDARY *Planets*, or *Satellites*, are those moving round other planets as the centres of their motion, and along with them round the sun.

SECTION, in Geometry, denotes a side or surface appearing of a body, or figure, cut by another; or the place where lines, planes, &c, cut each other.

The common Section of two planes is always a right line; being the line supposed to be drawn by one plane in its cutting or entering the other. If a sphere be cut in any manner by a plane, the figure of the Section will be a circle; also the common intersection of the surfaces of two spheres, is the circumference of a circle; and the two common Sections of the surfaces of a right cone and a sphere, are the circumferences of circles if the axis of the cone pass through the centre of the sphere, otherwise not; moreover, of the two common Sections of a sphere and a cone, whether right or oblique, if the one be a circle the other will be a circle also, otherwise not. See my Tracts, tract 7, prop. 7, 8, 9.

The Sections of a cone by a plane, are five; viz, a triangle, circle, ellipse, hyperbola, and parabola. See each of these terms, as also CONIC SECTION.

Sections of Buildings and Bodies, &c, are either vertical, or horizontal, &c. The

Vertical SECTION, or simply the SECTION, of a building, denotes its profile, or a delineation of its heights and depths raised on the plan; as if the fabric had been cut asunder by a vertical plane, to discover the inside. And

Horizontal SECTION is the ichnography or ground plan, or a Section parallel to the horizon.

SECTOR, of a Circle, is a portion of the circle comprehended between two radii and their included arc. Thus, the mixt triangle ABC, contained between the two radii AC and BC, and the arc AB, is a Sector of the circle.

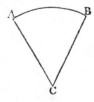

The Sector of a circle, as ABC, is equal to a triangle, whose base is the arc AB, and its altitude the radius AC or BC. And therefore the radius being drawn into the arc, half the product gives the area.

Similar SECTORS, are those which have equal angles included between their radii. These are to each other as the squares of their bounding arcs, or as their whole circles.

SECTOR also denotes a mathematical instrument, which is of great use in geometry, trigonometry, surveying, &c, in measuring and laying down and finding proportional quantities of the same kind: as between lines and lines, surfaces and surfaces, &c: whence the French call it the *compass of proportion.*

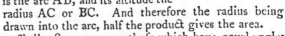

The

The great advantage of the Sector above the common scales, &c, is, that it is contrived so as to suit all radii, and all scales. By the lines of chords, sines, &c, on the Sector, we have lines of chords, sines, &c, to any radius between the length and breadth of the Sector when open.

The Sector is founded on the 4th proposition of the 6th book of Euclid; where it is demonstrated, that similar triangles have their like sides proportional. An idea of the theory of its construction may be conceived thus. Let the lines AB, AC represent the legs of the Sector; and AD, AE, two equal sections from the centre: then if the points BC and DE be connected, the lines BC and DE will be parallel; therefore the triangles ABC, ADE will be similar, and consequently the sides AB, BC, AD, DE proportional, that is, as AB : BC : : AD : DE; so that if AD be the half, 3d, or 4th part of AB, then DE will be a half, 3d, or 4th part of BC : and the same holds of all the rest. Hence, if DE be the chord, sine or tangent, of any arc, or of any number of degrees, to the radius AD, then BC will be the same to the radius AB.

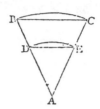

The Sector, it is supposed, was the invention of Guido Baldo or Ubaldo, about the year 1568. The first printed account of it was in 1584, by Gaspar Mordente at Antwerp, who indeed says that his brother Fabricius Mordente invented it, in the year 1554. It was next treated of by Daniel Speckle, at Strasburgh, in 1589; after that by Dr. Thomas Hood, at London, in 1598: and afterwards by many other writers on practical geometry, in all the nations of Europe.

Description of the Sector. This instrument consists of two rules or legs, the longer the better, made of box, or ivory, or brass, &c, representing the radii, moveable round an axis or joint, the middle of which represents the centre; from whence several scales are drawn on the faces. See the fig. 1, plate xxvi.

The scales usually set upon Sectors, may be distinguished into single and double. The single scales are such as are set upon plane scales: the double scales are those which proceed from the centre; each of these being laid twice on the same face of the instrument, viz, once on each leg. From these scales, dimensions or distances are to be taken, when the legs of the instrument are set in an angular position.

The scales set upon the best Sectors are

Single		A line of		marked
1		Inches, each divided into 8 and 10 parts.		
2		Decimals, containing 100 parts.		
3		Chords		Cho
4		Sines		Sin.
5		Tangents		Tang
6		Rhumbs		Rhum.
7		Latitude		Lat.
8		Hours		Hou.
9		Longitude		Lon.
10		Inclin. Merid.		In. mer.
11		the	Numbers	Num.
12		logarithms	Sines	Sin.
13		of	Versed Sines	V. Sin.
14			Tangents	Tan.

Double		a line of		marked
1		Lines, or equal parts		Lin.
2		Chords		Cho.
3		Sines		Sin.
4		Tangents to 45°		Tan.
5		Secants		Sec.
6		Tangents to above 45°		Tan.
7		Polygons		Pol.

The manner in which these scales are disposed on the Sector, is best seen in the figure.

The scales of lines, chords, sines, tangents, rhumbs, latitudes, hours, longitude, incl. merid. may be used, with the instrument either shut or open, each of these scales being contained on one of the legs only. The scales of inches, decimals, log. numbers, log. sines, log. versed sines, and log. tangents, are to be used with the Sector quite open, with the two rulers or legs stretched out in the same direction, part of each scale lying on both legs.

The double scales of lines, chords, sines, and lower tangents, or tangents under 45°, are all of the same radius or length; they begin at the centre of the instrument, and are terminated near the other extremity of each leg; viz, the lines at the division 10, the chords at 60, the sines at 90, and the tangents at 45; the remainder of the tangents, or those above 45°, are on other scales beginning at ¼ of the length of the former, counted from the centre, where they are marked with 45, and run to about 76 degrees.

The secants also begin at the same distance from the centre, where they are marked with 10, and are from thence continued to as many degrees as the length of the Sector will allow, which is about 75°.

The angles made by the double scales of lines, of chords, of sines, and of tangents to 45 degrees, are always equal. And the angles made by the scales of upper tangents, and of secants, are also equal.

The scales of polygons are set near the inner edge of the legs; and where these scales begin, they are marked with 4, and from thence are figured backwards, or towards the centre, to 12.

From this disposition of the double scales, it is plain, that those angles that are equal to each other while the legs of the Sector were close, will still continue to be equal, although the Sector be opened to any distance.

The scale of inches is laid close to the edge of the Sector, and sometimes on the edge; it contains as many inches as the instrument will receive when opened; each inch being usually divided into 8, and also into 10 equal parts. The decimal scale lies next to this: it is of the length of the Sector when opened, and is divided into 10 equal parts, or primary divisions, and each of these into 10 other equal parts; so that the whole is divided into 100 equal parts: and by this decimal scale, all the other scales, that are taken from tables, may be laid down. The scales of chords, rhumbs, sines, tangents, hours, &c, are such as are described under Plane Scale.

The scale of logarithmic or artificial numbers, called Gunter's scale, or Gunter's line, is a scale expressing the logarithms of common numbers, taken in their natural order.

The construction of the double scale will be evident by inspecting the instrument. As to the scale of polygons,

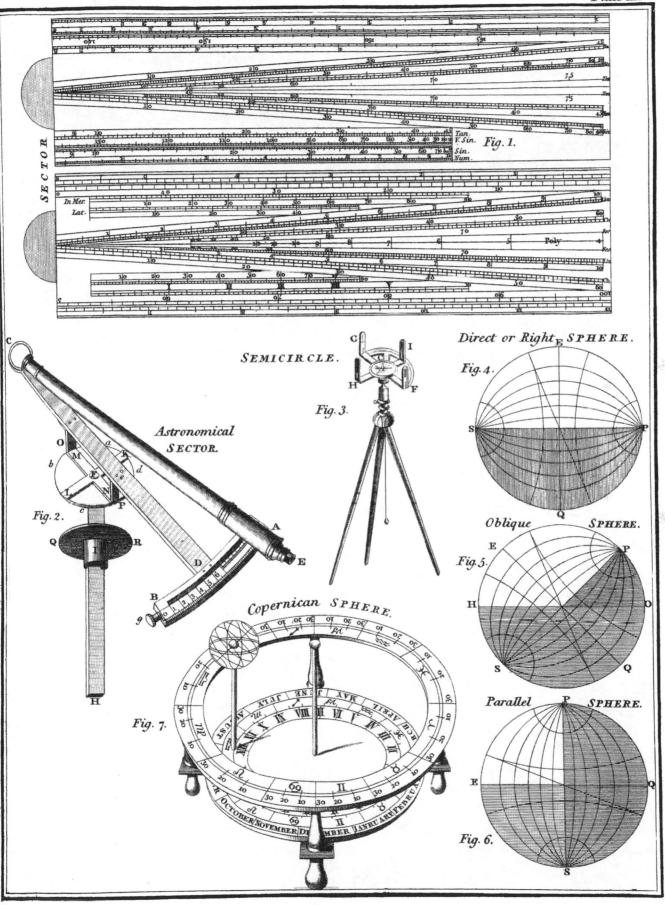

Plate XXVI

SECTOR

Tan.
V. Sin. Fig. 1.
Sin.
Num.

In Mer.
Lat.

Poly

SEMICIRCLE.

Fig. 3.

Direct or Right SPHERE.

Fig. 4.

Astronomical
SECTOR.

Fig. 2.

Oblique SPHERE.

Fig. 5.

Copernican SPHERE.

Fig. 7.

Parallel SPHERE.

Fig. 6.

gons, it usually comprehends the sides of the polygons from 6 to 12 sides inclusive : the divisions are laid down by taking the lengths of the chords of the angles at the centre of each polygon, and laying them down from the centre of the instrument. When the polygons of 4 and 5 sides are also introduced, this line is constructed from a scale of chords, where the length of 90° is equal to that of 60° of the double scale of chords on the Sector.

In describing the use of the Sector, the terms *lateral distance* and *transverse distance* often occur. By the former is meant the distance taken with the compasses on one of the scales only, beginning at the centre of the Sector ; and by the latter, the distance taken between any two corresponding divisions of the scales of the same name, the legs of the Sector being in an angular position.

Uses of the SECTOR.

Of the Line of Lines. This is useful, to divide a given line into any number of equal parts, or in any proportion, or to make scales of equal parts, or to find 3d and 4th proportionals, or mean proportionals, or to increase or decrease a given line in any proportion. Ex. 1. To divide a given line into any number of equal parts, as suppose 9 : make the length of the given line a transverse distance to 9 and 9, the number of parts proposed ; then will the transverse distance of 1 and 1 be one of the equal parts, or the 9th part of the whole ; and the transverse distance of 2 and 2 will be 2 of the equal parts, or ⅖ of the whole line ; and so on. 2. Again, to divide a given line into any number of parts that shall be in any assigned proportion, as suppose three parts, in the proportion of 2, 3, and 4. Make the given line a transverse distance to 9, the sum of the proposed numbers 2, 3, 4 ; then the transverse distances of these numbers severally will be the parts required.

Of the Scale of Chords. 1. To open the Sector to any angle, as suppose 50 degrees : Take the distance from the joint to 50 on the chords, the number of degrees proposed ; then open the Sector till the transverse distance from 60 to 60, on each leg, be equal to the said lateral distance of 50 ; so shall the scale of chords make the proposed angle of 50 degrees.—By the converse of this operation, may be known the angle the Sector is opened to ; viz, taking the transverse distance of 60, and applying it laterally from the joint.

2. To protract or lay down an angle of any given number of degrees. At any opening of the Sector, take the transverse distance of 60°, with which extent describe an arc ; then take the transverse distance of the number of degrees proposed, and apply it to that arc ; and through the extremities of this distance on the arc draw two lines from the centre, and they will form the angle as proposed. When the angle exceeds 60°, lay it off at twice or thrice.—By the converse operation any angle may be measured ; viz, With any radius describe an arc from the angular point ; set that radius transversely from 60 to 60 ; then take the distance of the intercepted arc and apply it transversely to the chords, which will shew the degrees in the given angle.

Of the Line of Polygons. 1. In a given circle to inscribe a regular polygon, for example an octagon. Open the legs of the Sector till the transverse distance from 6 to 6 be equal to the radius of the circle ; then will the transverse distance of 8 and 8 be the side of the inscribed octagon. 2. Upon a line given to describe a regular polygon. Make the given line a transverse dist. to 5 and 5 ; and at that opening of the Sector take the transverse distance of 6 and 6 ; with which as a radius, from the extremities of the given line describe arcs to intersect each other, which intersection will be the centre of a circle in which the proposed polygon may be inscribed ; then from that centre describe the said circle through the extremities of the given line, and apply this line continually round the circumference, for the several angular points of the polygon.—3. On a given right line as a base, to describe an isosceles triangle, having the angles at the base double the angle at the vertex. Open the Sector till the length of the given line fall transversely on 10 and 10 on each leg ; then take the transverse distance to 6 and 6, and it will be the length of each of the equal sides of the triangle.

Of the Sines, Tangents, and Secants. By the several lines disposed on the sector, we have scales of several radii. So that, 1. Having a length or radius given not exceeding the length of the Sector when opened, we can find the chord, sine, &c, to the same : for ex. suppose the chord, sine, or tangent of 20 degrees to a radius of 3 inches be required. Make 3 inches the opening or transverse distance to 60 and 60 on the chords ; then will the same extent reach from 45 to 45 on the tangents, and from 90 to 90 on the sines ; so that to whatever radius the line of chords is set, to the same are all the others set also. In this disposition therefore, if the transverse distance between 20 and 20 on the chords be taken with the compasses, it will give the chord of 20 degrees ; and if the transverse of 20 and 20 be in like manner taken on the sines, it will be the sine of 20 degrees ; and lastly, if the transverse distance of 20 and 20 be taken on the tangents, it will be the tangent of 20 degrees, to the same radius.—2. If the chord or tangent of 70 degrees were required. For the chord, the transverse distance of half the arc, viz 35, must be taken, as before ; which distance taken twice gives the chord of 70 degrees. To find the tangent of 70 degrees, to the same radius, the scale of upper tangents must be used, the under one only reaching to 45 : making therefore 3 inches the transverse distance to 45 and 45 at the beginning of that scale, the extent between 70 and 70 degrees on the same, will be the tangent of 70 degrees to 3 inches radius.—3. To find the secant of an arc ; make the given radius the transverse distance between 0 and 0 on the secants ; then will the transverse distance of 20 and 20, or 70 and 70, give the secant of 20 or 70 degrees.—4. If the radius, and any line representing a sine, tangent, or secant, be given, the degrees corresponding to that line may be found by setting the Sector to the given radius, according as a sine, tangent, or secant is concerned ; then taking the given line between the compasses, and applying the two feet transversely to the proper scale, and sliding the feet along till they both rest on like divisions on both legs ; then the divisions will shew the degrees and parts corresponding to the given line.

Use

Use of the Sector in Trigonometry, or in working any other proportions.

By means of the double scales, which are the parts more peculiar to the Sector, all proportions are worked by the property of similar triangles, making the sides proportional to the bases, that is, on the Sector, the lateral distances proportional to the transverse ones; thus, taking the distance of the first term, and applying it to the 2d, then the distance of the 3d term, properly applied, will give the 4th term: observing that the sides of triangles are taken off the line of numbers laterally, and the angles are taken transversely, off the sines or tangents or secants, according to the nature of the proportion. For example, in a plane triangle ABC, given two sides and an angle opposite to one of them, to find the rest; viz, given AB = 56, AC = 64, and ∠B = 46° 30', to find BC and the angles A and C. In this case, the sides are proportional to the sines of their opposite angles; hence these proportions,

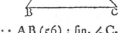

as AC (64) : sin. ∠B (46° 30') : : AB (56) : sin. ∠C, and as sin. B : AC : : sin. A : BC.

Therefore, to work these proportions by the Sector, take the lateral distance of 64 = AC from the lines, and open the Sector to make this a transverse distance of 46° 30' = ∠B, on the sines; then take the lateral distance of 56 = AB on the lines, and apply it transversely on the sines, which will give 39° 24' = ∠C. Hence, the sum of the angles B and C, which is 85° 54', taken from 180°, leaves 94° 6' = ∠A. Then, to work the 2d proportion, the Sector being set at the same opening as before, take the transverse distance of 94° 6' = ∠A, on the sines, or, which is the same thing, the transverse distance of its supplement 85° 54'; then this applied laterally to the lines, gives 88 = the side BC sought.

For the complete history of the Sector, with its more ample and particular construction and uses, see Robertson's *Treatise of such Mathematical Instruments, as are usually put into a Portable Case,* the Introduction.

SECTOR *of a Sphere,* is the solid generated by the revolution of the Sector of a circle about one of its radii; the other radius describing the surface of a cone, and the circular arc a circular portion of the surface of the sphere of the same radius. So that the spherical Sector consists of a right cone, and of a segment of the sphere having the same common base with the cone. And hence the solid content of it will be found by multiplying the base or spherical surface by the radius of the sphere, and taking a 3d part of the product.

SECTOR *of an ellipse,* or *of an hyperbola,* &c, is a part resembling the circular Sector, being contained by three lines, two of which are radii, or lines drawn from the centre of the figure to the curve, and the intercepted arc or part of that curve.

Astronomical SECTOR, an instrument invented by Mr. George Graham, for finding the difference in right ascension and declination between two objects, whose distance is too great to be observed through a fixed

I

telescope, by means of a micrometer. This instrument (fig. 2, pl. 26,) consists of a brass plate, called the Sector, formed like a T, having the shank CD, as a radius, about 2½ feet long, and 2 inches broad at the end D, and an inch and a half at C; and the cross-piece AB, as an arch, about 6 inches long, and one and a half broad; upon which, with a radius of 30 inches, is described an arch of 10 degrees, each degree being divided in as many parts as are convenient. Round a small cylinder C, containing the centre of this arch, and fixed in the shank, moves a plate of brass, to which is fixed a telescope CE, having its line of collimation parallel to the plane of the Sector, and passing over the centre C of the arch AB, and the index of a Vernier's dividing plate, whose length, being equal to 16 quarters of a degree, is divided into 15 equal parts, fixed to the eye end of the telescope, and made to slide along the arch; which motion is performed by a long screw, G, at the back of the arch, communicating with the Vernier through a slit cut in the brass, parallel to the divided arch. Round the centre F of a circular brass plate *abc,* of 5 inches diameter, moves a brass cross KLMN, having the opposite ends O and P of one bar turned up perpendicularly about 3 inches, to serve as supporters to the Sector, and screwed to the back of its radius; so that the plane of the Sector is parallel to the plane of the circular plate, and can revolve round the centre of that plate in this parallel position. A square iron axis HIF, 18 inches long, is screwed flat to the back of the circular plate along one of its diameters, so that the axis is parallel to the plane of the Sector. The whole instrument is supported on a proper pedestal, so that the said axis shall be parallel to the earth's axis, and proper contrivances are annexed to fix it in any position. The instrument, thus supported, can revolve round its axis HI, parallel to the earth's axis, with a motion like that of the stars, the plane of the Sector being always parallel to the plane of some hour circle, and consequently every point of the telescope describing a parallel of declination; and if the Sector be turned round the joint F of the circular plate, its graduated arch may be brought parallel to an hour-circle; and consequently any two stars, whose difference of declination does not exceed the degrees in that arch, will pass over it.

To observe their passage, direct the telescope to the preceding star, and fix the plane of the Sector a little to the westward of it; move the telescope by the screw G, and observe at the transit of each over the cross wires the time shewn by the clock, and also the division upon the arch AB, shewn by the index; then is the difference of the arches the difference of the declination; and that of the times shews the difference of the right ascension of those stars. For a more particular description of this instrument, see Smith's Optics, book iii, chap. 9.

SECULAR *Year,* the same with Jubilee.

SECUNDANS, an infinite series of numbers, beginning from nothing, and proceeding according to the squares of numbers in arithmetical progression, as 0, 1, 4, 9, 16, 25, 36, 49, 64, &c.

SEEING, the act of perceiving objects by the organ of sight; or the sense we have of external objects by means of the eye.

For the apparatus, or difpofition of the parts necef-
fary to Seeing, fee EYE. And for the manner in which
Seeing is performed, and the laws of it, fee VI-
SION.

Our beft anatomifts differ greatly as to the caufe
why we do not fee double with the two eyes? Galen,
and others after him, afcribe it to a coalition, or decuffation, of the optic nerve, behind the os fphenoides.
But whether they decuffate or coalefce, or only barely
touch one another, is not well agreed upon.

The Bartholines and Vefalius fay exprefsly, they are
united by a perfect confufion of their fubftance; Dr.
Gibfon allows them to be united by the clofeft conjunction, but not by a confufion of their fibres.

Alhazen, an Arabian philofopher of the 12th century, accounts for fingle vifion by two eyes, by fuppofing that when two correfponding parts of the
retina are affected, the mind perceives but one
image.

Des Cartes and others account for the effect another way; viz, by fuppofing that the fibrillæ conftituting the medullary part of thofe nerves, being fpread in
the retina of each eye, have each of them correfponding parts in the brain, fo that when any of thofe fibrillæ are ftruck by any part of an image, the correfponding parts of the brain are affected by it. Somewhat
like which is the opinion of Dr. Briggs, who takes the
optic nerves of each eye to confift of homologous fibres, having their rife in the thalamus nervorum opticorum, and being thence continued to both the retinæ,
which are compofed of them; and farther, that thofe
fibrillæ have the fame parallelifm, tenfion, &c, in both
eyes; confequently when an image is painted on the
fame correfponding fympathizing parts of each retina,
the fame effects are produced, the fame notice carried
to the thalamus, and fo imparted to the foul. Hence
it is, that double vifion enfues upon an interruption of
the parallelifm of the eyes; as when one eye is depreffed
by the finger, or their fymphony is interrupted by
difeafe: but Dr. Briggs maintains, that it is but in
few fubjects there is any decuffation; and in none any
conjunction more than mere contact; though his notion is by no means confonant to facts, fnd it is attended with many improbable circumftances.

It was the opinion of Sir Ifaac Newton, and of
many others, that objects appear fingle, becaufe the
two optic nerves unite before they reach the brain.
But Dr. Porterfield fhews, from the obfervation of feveral anatomifts, that the optic nerves do not mix or
confound their fubftance, being only united by a clofe
cohefion; and objects have appeared fingle, where the
optic nerves were found to be disjoined. To account
for this phenomenon, this ingenious writer fuppofes,
that, by an original law in our natures, we imagine an
object to be fituated fomewhere in a right line drawn
from the picture of it upon the retina, through the
centre of the pupil; confequently the fame object appearing to both eyes to be in the fame place, we cannot
diftinguifh it into two. In anfwer to an objection to this
hypothefis, from objects appearing double when one eye
is diftorted, he fays, the mind miftakes the pofition of
the eye, imagining, that it had moved in a manner
correfponding to the other, in which cafe the conclufion would have been juft: in this he feems to have re-

courfe to the power of habit, though he difclaims that
hypothefis. This principle however has been thought
fufficient to account for this appearance.

'Originally, every object making two pictures, one
in each eye, is imagined to be double; but, by degrees, we find that when two correfponding parts of
the retina are impreffed, the object is but one; but if
thofe correfponding parts be changed by the diftortion
of one of the eyes, the object muft again appear double as at the firft. This feems to be verified by Mr.
Chefelden, who informs us, that a gentleman, who,
from a blow on his head, had one eye diftorted, found
every object to appear double, but by degrees the moft
familiar ones came to appear fingle again, and in time
all objects did fo without amendment of the diftortion.
A fimilar cafe is mentioned by Dr. Smith.

On the other hand, Dr. Reid is of opinion, that the
correfpondence of the centres of two eyes, on which
fingle vifion depends, does not arife from cuftom, but
from fome natural conftitution of the eye, and of the
mind.

M. du Tour adopts an opinion, long before fuggefted
by Gaffendi, that the mind attends to no more than
the image made in one eye at a time; in fupport of
which, he produces feveral curious experiments; but
as M. Buffon obferves, it is a fufficient anfwer to this
hypothefis, that we fee more diftinctly with two eyes
than with one; and that when a round object is near
us, we plainly fee more of the furface in one cafe than
in the other.

With refpect to fingle vifion with two eyes, Dr.
Hartley obferves, that it deferves particular attention,
that the optic nerves of man, and fuch other animals as
look the fame way with both eyes, unite in the fella
turrica in a ganglion, or little brain, as it may be called,
peculiar to themfelves, and that the affociations between fynchronous impreffions on the two retinas, muft
be made fooner and cemented ftronger on this account;
alfo that they ought to have a much greater power over
one another's image, than in any other part of the
body. And thus an impreffion made on the right eye
alone by a fingle object, propagates itfelf into the left,
and there raifes up an image almoft equal in vividnefs to
itfelf; and, confequently, when we fee with one eye
only, we may however have pictures in both eyes.

It is a common obfervation, fays Dr. Smith, that
objects feen with both eyes appear more vivid and
ftronger than they do to a fingle eye, efpecially when
both of them are equally good. Porterfield on the
Eye, vol. ii, pa. 285, 315. Smith's Optics, Remarks
pa. 31. Reid's Inquiry, pa. 267. Mem. Préfentes,
pa. 514. Acad. Par. 1747. Mem. Pr. 334. Hartley on
Man, vol. i, pa. 207. Prieftley's Hift. of Light and
Colours, pa. 663, &c.

Whence it is that we fee objects erect, when it is
certain, that the images thereof are painted invertedly
on the retina, is another difficulty in the theory of Seeing. Des Cartes accounts for it hence, that the notice which the foul takes of the object, does not depend on any image, nor any action coming from the
object, but merely on the fituation of the minute parts
of the brain, whence the nerves arife. Ex. gr. the fituation of a capillament brain, which occafions the
foul

foul to fee all thofe places lying in a right line with it.

But Mr. Molyneux gives another account of this matter. The eye, he obferves, is only the organ, or inftrument; it is the foul that fees. To enquire then, how the foul perceives the object erect by an inverted image, is to enquire into the foul's faculties. Again, imagine that the eye receives an impulfe on its lower part, by a ray from the upper part of an object; muft not the vifive faculty be hereby directed to confider this ftroke as coming from the top, rather than the bottom of the object, and confequently be determined to conclude it the reprefentation of the top?

Upon thefe principles, we are to confider, that inverted is only a relative term, and that there is a very great difference between the real object, and the means or image by which we perceive it. When all the parts of a diftant profpect are painted upon the retina (fuppofing that to be the feat of vifion), they are all right with refpect to one another, as well as the parts of the profpect itfelf; and we can only judge of an object being inverted, when it is turned reverfe to its natural pofition with refpect to other objects which we fee and compare it with.

The eye or vifive faculty (fays Molyneux) takes no notice of the internal furface of its own parts, but ufes them as an inftrument only, contrived by nature for the exercife of fuch a faculty. If we lay hold of an upright ftick in the dark, we can tell which is the upper or lower part of it, by moving our hand upward or downward; and very well know that we cannot feel the upper end by moving our hand downward. Juft fo, we find by experience and habit, that by directing our eyes towards a tall object, we cannot fee its top by turning our eyes downward, nor its foot by turning our eyes upward; but muft trace the object the fame way by the eye to fee it from head to foot, as we do by the hand to feel it; and as the judgement is informed by the motion of the hand in one cafe, fo it is alfo by the motion of the eye in the other.

Molyneux's Dioptr. pa. 105, &c. Muffchenbroek's Int. ad Phil. Nat. vol. ii, pa. 762. Ferguson's Lectures, pa. 132. See SIGHT, VISIBLE, &c.

SEGMENT, in Geometry, is a part cut off the top of a figure by a line or plane; and the part remaining at the bottom, after the Segment is cut off, is called a *fruftum*, or a *zone*. So, a

SEGMENT *of a Circle*, is a part of the circle cut off by a chord, or a portion comprehended by an arch and its chord; and may be either greater or lefs than a femicircle. Thus, the portion ABCA is a Segment lefs than a femicircle; and ADCA a Segment greater.

The angle formed by lines drawn from the extremities of a chord to meet in any point of the arc, is called an angle *in* the Segment. So the angle ABC is an angle *in* the Segment ABCA; and the angle ADC, an angle *in* the Segment ADCA.

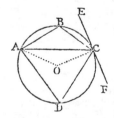

Alfo the angle B is faid to be the angle *upon* the

Segment ADC, and D the angle *on* the Segment ABC.

The angle which the chord AC makes with a tangent EF, is called the angle *of* a Segment; and it is equal to the angle in the alternate or fupplemental Segment, or equal to the fupplement of the angle in the fame Segment. So the angle ACE is the angle *of* the Segment ABC, and is equal to the angle ADC, or to the fupplement of the angle B; alfo the angle ACF is the angle *of* the Segment ADC, and is equal to the angle B, or to the fupplement of the angle D.

The area of a Segment ABC, is evidently equal to the difference between the fector OABC- of the fame arc, and the triangle OAC on the fame chord; the triangle being fubtracted from the fector, to give the Segment, when lefs than a femicircle; but to be added when greater. See more rules for the Segment in my Menfuration, pa. 122 &c, 2d edition.

Similar SEGMENTS, are thofe that have their chords directly proportional to their radii or diameters, or that have fimilar arcs, or fuch as contain the fame number of degrees.

SEGMENT *of a Sphere*, is a part cut off by a plane.

The bafe of a Segment is always a circle. And the convex furfaces of different Segments, are to each other as their altitudes, or verfed fines. And as the whole convex furface of the fphere is equal to 4 of its great circles, or 4 circles of the fame diameter; fo the furface of any Segment, is equal to 4 circles on a diameter equal to the chord of half the arc of the Segment. So that if d denote the diameter of the fphere, or the chord of half the circumference, and c the chord of half the arc of any other Segment, alfo a the altitude or verfed fine of the fame; then,

$3\cdot1416d^2$ is the furface of the whole fphere, and $3\cdot1416c^2$, or $3\cdot1416ad$, the furface of the Segment.

For the folid content of a Segment, there are two rules ufually given; viz. 1. To 3 times the fquare of the radius of its bafe, add the fquare of its height; multiply the fum by the height, and the product by 5236. Or, 2dly, From 3 times the diameter of the fphere, fubtract twice the height of the fruftum; multiply the remainder by the fquare of the height, and the product by ·5236. That is, in fymbols, the folid content is either

$$= \cdot5236a \times \overline{3r^2 + a^2}, \text{ or } = \cdot5236a^2 \times \overline{3d - 2a};$$

where a is the altitude of the Segment, r the radius of its bafe, and d the diameter of the whole fphere.

Line of SEGMENTS, are two particular lines, fo called, on Gunter's fector. They lie between the lines of fines and fuperficies, and are numbered with 5, 6, 7, 8, 9, 10. They reprefent the diameter of a circle, fo divided into 100 parts, as that a right line drawn through thofe parts, and perpendicular to the diameter, fhall cut the circle into two Segments, the greater of which fhall have the fame proportion to the whole circle, as the parts cut off have to 100.

SELENOGRAPHY, the defcription and reprefentation of the moon, with all the parts and appearances of her difc or face; like as geography does thofe of the earth.

7

Since the invention of the telescope, Selenography is very much improved. We have now distinct names for most of the regions, seas, lakes, mountains, &c, visible in the moon's body. Hevelius, a celebrated astronomer of Dantzic, and who published the first Selenography, named the several places of the moon from those of the earth. But Riccioli afterwards called them after the names of the most celebrated astronomers and philosophers. Thus, what the one calls *mons Porphyrites*, the other calls *Aristarchus* ; what the one calls *Ætna, Sinai, Athos, Apenninus*, &c, the other calls, *Copernicus, Posidonius, Tycho, Gassendus*, &c.

M. Cassini has published a work called *Instructions Seleniques*, and has published the best map of the moon.

SELEUCIDÆ, in Chronology, the era of the Seleucidæ, or the Syro-Macedonian era, which is a computation of time, commencing from the establishment of the Seleucidæ, a race of Greek kings, who reigned as successors of Alexander the Great, in Syria, as the Ptolomies did in Egypt. According to the best accounts, the first year of this era falls in the year 311 before Christ, which was 12 years after the death of Alexander.

SELL, in Building, is of two kinds, viz, *Ground-Sell*, which denotes the lowest piece of timber in a wooden building, and that upon which the whole superstructure is raised. And Sell of a window, or of a door, which is the bottom piece in the frame of them, upon which they rest.

SEMICIRCLE, in Geometry, is half a circle, or a figure comprehended between the diameter of a circle, and half the circumference.

SEMICIRCLE is also an instrument in Surveying, sometimes called the *graphometer*.

It consists of a semicircular limb or arch, as FIG (fig. 3, pl. 26) divided into 180 degrees, and sometimes subdivided diagonally or otherwise into minutes. This limb is subtended by a diameter FG, having two sights erected at its extremities. In the centre of the Semicircle, or the middle of the diameter, is fixed a box and needle ; and on the same centre is fitted an alidade, or moveable index, carrying two other sights, as H, I : the whole being mounted on a staff, with a ball and socket &c.

Hence it appears, that the Semicircle is nothing but half a theodolite ; with this only difference, that whereas the limb of the theodolite, being an entire circle, takes in all the 360° successively ; while in the Semicircle the degrees only going from 1 to 180, it is usual to have the remaining 180°, or those from 180° to 360°, graduated in another line on the limb within the former.

To take an Angle with a Semicircle.—Place the instrument in such manner, as that the radius CG may hang over one leg of the angle to be measured, with the centre C over the vertex of the same. The first is done by looking through the sights F and G, at the extremities of the diameter, to a mark fixed up in one extremity of the leg ; and the latter is had by letting fall a plummet from the centre of the instrument. This done, turn the moveable index HI on its centre towards the other leg of the angle, till through the sights fixed in it, you see a mark in the extremity of the leg. Then the degree which the index cuts on the limb, is the quantity or measure of the angle.

Other uses are the same as in the theodolite.

SEMICUBICAL PARABOLA, a curve of the 2d order, of such a nature that the cubes of the ordinates are proportional to the squares of the abscisses. Its equation is $ay^2 = x^3$. This curve, AM*m*, is one of

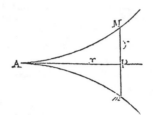

Newton's five diverging parabolas, being his 70th species ; having a cusp at its vertex at A. It is otherwise named the Neilian parabola, from the name of the author who first treated of it.

The area of the space APM, is $= \frac{4}{5}xy = \frac{4}{5}AP \times PM$, or $\frac{4}{5}$ of the circumscribing rectangle.

The content of the solid generated by the revolution of the space APM about the axis AP, is $\frac{1}{4}pxy^2 = .7854AP \times PM^2$, or $\frac{1}{4}$ of the circumscribing cylinder. And a circle equal to the surface of that solid may be found from the quadrature of an hyperbolic space.

Also the length of any arc AM of the curve may be easily obtained from the quadrature of a space contained under part of the curve of the common parabola, two semiordinates to the axis, and the part of the axis contained between them.

This curve may be described by a continued motion, viz, by fastening the angle of a square in the vertex of a common parabola ; and then carrying the intersection of one side of this square and a long ruler (which ruler always moves perpendicularly to the axis of the parabola) along the curve of that parabola. For the intersection of the ruler, and the other side of the square will describe a Semicubical parabola. Maclaurin performs this without a common parabola, in his Geometria Organica.

SEMIDIAMETER, or *Radius*, of a circle or sphere, is a line drawn from the centre to the circumference. And in any curve that has diameters and a centre, it is the radius, or half diameter, or a line drawn from the centre to some point in the curve.

The distances, diameters, &c, of the heavenly bodies, are usually estimated by astronomers in Semidiameters of the earth ; the number of which terrestrial Semidiameters, contained in that of each of those planets, is as below.

The Earth - -	1 Semidiam.
The Sun - -	111¼
The Moon - -	0·27
Mercury - -	0·38
Venus - -	1·15
Mars - -	0·65
Jupiter - -	11·81
Saturn - -	9·77
Herschel - -	4·32.

SEMIDIAPENTE, in Music, a defective or imperfect fifth, called usually by the Italians, *falsa quinta*, and by us a false fifth.

SEMIDIAPASON, in Music, a defective or imperfect octave ; or an octave diminished by a lesser semitone, or 4 commas.

SEMIDIATESSARON, in Music, a defective fourth, called also a false fourth.

SEMIDITONE, in Music, is the lesser third, having its terms as 6 to 5.

SEMIORDINATES, in Geometry, the halves of the ordinates or applicates, being the lines applied between the abscis and the curve.

SEMIPARABOLA, &c, in Geometry, the half of the whole parabola, &c.

SEMIQUADRATE, or SEMIQUARTILE, is an aspect of the planets, when distant from each other one sign and a half, or 45 degrees.

SEMIQUAVER, in Music, the half of a quaver.

SEMIQUINTILE, is an aspect of the planets when distant from each other the half of a 5th of the circle, or by 36 degrees.

SEMISEXTILE, an aspect of two planets, when they are distant from each other 30 degrees, or the half of a sextile, which is 2 signs or 60°. The Semisextile is marked s. s.

SEMITONE, in Music, a half tone or half note, one of the degrees or intervals of concords.

There are three degrees, or less intervals, by which a sound can move upwards and downwards, successively from one extreme of any concord to the other, and yet produce true melody. These degrees are the greater tone, the less tone, and the semitone. The ratios defining these intervals are these, viz, the greater tone 8 to 9, the less tone 9 to 10, and the Semitone 15 to 16. Its compass is 5 commas, and it has its name from being nearly half a whole, though it is really somewhat more.

There are several species of Semitones ; but those that usually occur in practice are of two kinds, distinguished by the addition of greater and less. The first is expressed by the ratio of 16 to 15, or $\frac{16}{15}$; and the second by 25 to 24, or $\frac{25}{24}$. The octave contains 10 Semitones major, and 2 dieses, nearly, or 17 Semitones minor, nearly ; for the measure of the octave being expressed by the logarithm   1,00000, the Semitone major will be measured by   0,09311, and the Semitone minor by   0,05889. These two differ by a whole enharmonic diesis ; which is an interval practicable by the voice. It was much in use among the Ancients, and is not unknown among modern practitioners. Euler Tent. Nov. Theor. Mus. pa. 107. See INTERVAL.

These Semitones are called *fictitious notes* ; and, with respect to the natural ones, they are expressed by characters called *flats* and *sharps*. The use of them is to remedy the defects of instruments, which, having their sounds fixed, cannot always be made to answer to the diatonic scale. By means of these, we have a new kind of scale, called the

SEMITONIC *Scale*, or the *Scale of Semitones*, which is a scale or system of music, consisting of 12 degrees, or 13 notes, in the octave, being an improvement on the natural or diatonic scale, by inserting between each two notes of it, another note, which divides the interval or tone into two unequal parts, called Semitones.

The use of this scale is for instruments that have fixed sounds, as the organ, harpsichord, &c, which are exceedingly defective on the foot of the natural or diatonic scale. For the degrees of the scale being unequal, from every note to its octave there is a different order of degrees ; so that from any note we cannot find every interval in a series of fixed sounds ; which yet is necessary, that all the notes of a piece of music, carried through several keys, may be found in their just tune, or that the same song may be begun indifferently at any note, as may be necessary for accommodating some instrument to others, or to the voice, when they are to accompany each other in unison.

The diatonic scale, beginning at the lowest note, being first settled on an instrument, and the notes of it distinguished by their names *a*, *b*, *c*, *d*, *e*, *f*, *g* ; the inserted notes, or Semitones, are called fictitious notes, and take the name or letter below with a ✳, as *c* ✳ called *c* sharp ; signifying that it is a Semitone higher than the sound of *c* in the natural series ; or this mark ♭, called a flat, with the name of the note above signifying it to be a Semitone lower.

Now $\frac{15}{16}$ and $\frac{128}{135}$ being the two Semitones the greater tone is divided into, and $\frac{15}{16}$ and $\frac{24}{25}$, the Semitones the less tone is divided into, the whole octave will stand as in the following scheme, where the ratios of each term to the next are written fraction-wise between them below.

Scale of Semitones.

c.	*c* ✳	*d.*	*d* ✳	*e.*	*f.*	*f* ✳	*g.*	*g* ✳	*a* ♭	*b.*	*cc.*
$\frac{15}{16}$	$\frac{128}{135}$	$\frac{15}{16}$	$\frac{24}{25}$	$\frac{15}{16}$	$\frac{15}{16}$	$\frac{128}{135}$	$\frac{15}{16}$	$\frac{15}{16}$	$\frac{24}{25}$	$\frac{15}{16}$	$\frac{128}{135}$

for the names of the intervals in this scale, it may be considered, that as the notes added to the natural scale are not designed to alter the species of melody, but leave it still diatonic, and only correct certain defects arising from something foreign to the office of the scale of music, viz, the fixing and limiting the sounds ; we see the reason why the names of the natural scale are continued, only making a distinction of each into a greater and less. Thus an interval of one Semitone, is called a less second ; of two Semitones, a greater second ; of three Semitones, a less third ; of four, a greater third, &c.

A second kind of Semitonic scale we have from another division of the octave into Semitones, which is performed by taking an harmonical mean between the extremes of the greater and less tone of the natural scale, which divides it into two Semitones nearly equal. Thus, the greater tone 8 to 9 is divided into two Semitones, which are 16 to 17, and 17 to 18 ; where 16, 17, 18, is an arithmetical division, the numbers representing the lengths of the chords ; but if they represent the vibration, the lengths of the chords are reciprocal ; viz as 1, $\frac{16}{17}$, $\frac{8}{9}$; which puts the greater Semitone

mitone $\frac{16}{17}$ next the lower part of the tone, and the leffer $\frac{17}{18}$ next the upper, which is the property of the harmonical divifion. And after the fame manner the lefs tone 9 to 10 is divid d into two Semitones, 18 to 19, and 19 to 20; and the whole octave ftands thus:

c. c✳. d. d✳. e. f. f✳. g. g✳. a. ♭. b. c.

$\frac{16}{17}$ $\frac{17}{18}$ $\frac{18}{19}$ $\frac{19}{20}$ $\frac{15}{16}$ $\frac{16}{17}$ $\frac{17}{18}$ $\frac{18}{19}$ $\frac{19}{20}$ $\frac{15}{16}$ $\frac{16}{17}$ $\frac{17}{18}$ $\frac{18}{19}$.

This fcale, Mr. Salmon tells us, in the Philofophical Tranfactions, he made an experiment of before the Royal Society, on chords, exactly in thefe proportions, which yielded a perfect concert with other inftruments, touched by the beft hands. Mr. Malcolm adds, that, having calculated the ratios of them, for his own fatiffaction, he found more of them falfe than in the preceding fcale, but then their errors were confiderably lefs, which made amends. Malcolm's Mufic, chap. 10. § 2.

SENSIBLE *Horizon*, or *Point*, or *Quality*, &c. See the fubftantives.

SEPTUAGESIMA, in the Calendar, is the 9th Sunday before Eafter, fo called, as fome have fuppofed, becaufe it is near 70 days, though in reality it is only 63 days, before it.

SERIES, in Algebra, denotes a rank or progreffion of quantities or terms, which ufually proceed according to fome certain law.

As the Series $1 + \frac{1}{2} + \frac{1}{4} + \frac{1}{8} + \frac{1}{16}$ &c,

or the Series, $1 + \frac{1}{2} + \frac{1}{3} + \frac{1}{4} + \frac{1}{5}$ &c.

where the former is a geometrical Series, proceeding by the conftant divifion by 2, or the denominators multiplied by 2; and the latter is an harmonical Series, being the reciprocals of the arithmetical Series 1, 2, 3, 4, &c, or the denominators being continually increafed by 1.

The doctrine and ufe of Series, one of the greateft improvements of the prefent age, we owe to Nicholas Mercator; though it feems he took the firft hint of it from Dr. Wallis's Arithmetic of Infinites; but the genius of Newton firft gave it a body and a form.

It is chiefly ufeful in the quadrature of curves; where, as we often meet with quantities which cannot be expreffed by any precife definite numbers, fuch as is the ratio of the diameter of a circle to the circumference, we are glad to exprefs them by a Series, which, infinitely continued, is the value of the quantity fought, and which is called an Infinite Series.

The Nature, Origin, &c, of SERIES.

Infinite Series commonly arife, either from a continued divifion, as was practifed by Mercator, or the extraction of roots, as firft performed by Newton, who alfo explained other general ways for the expanding of quantities into infinite Series, as by the binomial theorem. Thus, to divide 1 by 3, or to expand the fraction $\frac{1}{3}$ into an infinite Series; by divifion in decimals in the ordinary way, the feries is 0·3333 &c, or $\frac{3}{10} + \frac{3}{100} + \frac{3}{1000} + \frac{3}{10000}$ &c, where the law of

continuation is manifeft. Or, if the fame fraction $\frac{1}{3}$ be fet in this form $\frac{1}{2+1}$, and divifion be performed in the algebraic manner, the quotient will be

$$\frac{1}{3} = \frac{1}{2+1} = \frac{1}{2} - \frac{1}{4} + \frac{1}{8} - \frac{1}{16} + \frac{1}{32}\ \&c.$$

Or, if it be expreffed in this form $\frac{1}{3} = \frac{1}{4-1}$, by a like divifion there will arife the Series,

$$\frac{1}{3} = \frac{1}{4} + \frac{1}{16} + \frac{1}{64}\ \&c = \frac{1}{4} + \frac{1}{4^2} + \frac{1}{4^3}\ \&c.$$

And, thus, by dividing 1 by 5 — 2, or 6 — 3, or 7 — 4, &c, the Series anfwering to the fraction $\frac{1}{3}$, may be found in an endlefs variety of infinite Series; and the finite quantity $\frac{1}{3}$ is called the value or radix of the Series, or alfo its fum, being the number or fum to which the Series would amount, or the limit to which it would tend or approximate, by fumming up its terms, or by collecting them together one after another.

In like manner, by dividing 1 by the algebraic fum $a + c$, or by $a - c$, the quotient will be in thefe two cafes, as below, viz,

$$\frac{1}{a+c} = \frac{1}{a} - \frac{c}{a^2} + \frac{c^2}{a^3} - \frac{c^3}{a^4}\ \&c,$$

$$\frac{1}{a-c} = \frac{1}{a} + \frac{c}{a^2} + \frac{c^2}{a^3} + \frac{c^3}{a^4}\ \&c.$$

where the terms of each Series are the fame, and they differ only in this, that the figns are alternately pofitive and negative in the former, but all pofitive in the latter.

And hence, by expounding a and c by any numbers whatever, we obtain an endlefs variety of infinite Series, whofe fums or values are known. So, by taking a or c equal to 1 or 2 or 3 or 4, &c, we obtain thefe Series, and their values;

$$\frac{1}{1+1} = \frac{1}{2} = 1 - 1 + 1 - 1 + 1 - 1\ \&c,$$

$$\frac{1}{3-1} = \frac{1}{2} = \frac{1}{3} + \frac{1}{3^2} + \frac{1}{3^3} + \frac{1}{3^4}\ \&c,$$

$$\frac{1}{2+1} = \frac{1}{3} = \frac{1}{2} - \frac{1}{2^2} + \frac{1}{2^3} - \frac{1}{2^4}\ \&c,$$

$$\frac{1}{1+2} = \frac{1}{3} = 1 - 2 + 2^2 - 2^3\ \&c,$$

$$\frac{1}{3+1} = \frac{1}{4} = \frac{1}{3} - \frac{1}{3^2} + \frac{1}{3^3} - \frac{1}{3^4}\ \&c.$$

And hence it appears, that the fame quantity or radix may be expreffed by a great variety of infinite Series, or that many different Series may have the fame radix or fum.

Another way in which an infinite Series arifes, is by the extraction of roots. Thus, by extracting the fquare root of the number 3 in the common way, we obtain its value in a feries as follows, viz, $\sqrt{3} = 1·73205$ &c $= 1 + \frac{7}{10} + \frac{3}{100} + \frac{2}{1000} + \frac{5}{100000}$ &c;

in which way of refolution the law of the progreffion

of

of the Series is not vifible, as it is when found by divifion. And the fquare root of the algebraic quantity $a^2 + c^2$ gives

$$\sqrt{a^2 + c^2} = a + \frac{c^2}{2a} - \frac{c^4}{8a^3} + \frac{c^6}{16a^5} \ \&c.$$

And a 3d way is by Newton's binomial theorem, which is a univerfal method, that ferves for all forts of quantities, whether fractional or radical ones: and by this means the fame root of the laft given quantity becomes $\sqrt{a^2 + c^2} =$

$$= a + \frac{c^2}{2a} - \frac{1.c^4}{2.4a^3} + \frac{1.3c^6}{2.4.6a^5} + \frac{1.3.5c^8}{2.4.6.8a^7} \ \&c.$$

where the law of continuation is vifible.

See EXTRACTION *of Roots*, and BINOMIAL *Theorem*.

From the fpecimens above given, it appears that the figns of the terms may be either all plus, or alternately plus and minus. Though they may be varied in many other ways. It alfo appears that the terms may be either continually fmaller and fmaller, or larger and larger, or elfe all equal. In the firft cafe therefore the Series is faid to be a *decreafing* one, in the 2d cafe an *increafing* one, and in the 3d cafe an *equal* one. Alfo the firft Series is called a *converging* one, becaufe that by collecting its terms fucceffively, taking in always one term more, the fucceffive fums approximate or converge to the value or fum of the whole infinite Series. So, in the Series

$$\frac{1}{3 - 1} = \frac{1}{2} = \frac{1}{3} + \frac{1}{9} + \frac{1}{27} + \frac{1}{81}, \ \&c,$$

the firft term $\frac{1}{3}$ is too little, or below $\frac{1}{2}$ which is the value or fum of the whole infinite Series propofed; the fum of the firft two terms $\frac{1}{3} + \frac{1}{9}$ is $\frac{4}{9} = \cdot 4444 \ \&c,$ is alfo too little, but nearer to $\frac{1}{2}$ or $\cdot 5$ than the former;

and the fum of three terms $\frac{1}{3} + \frac{1}{9} + \frac{1}{27}$ is $\frac{13}{27} = \cdot 481481 \ \&c,$ is nearer than the laft, but ftill too little; and the fum of four terms

$$\frac{1}{3} + \frac{1}{9} + \frac{1}{27} + \frac{1}{81}, \text{ is } \frac{40}{81} = \cdot 493827 \ \&c.$$

which is again nearer than the former, but ftill too little; which is always the cafe when the terms are all pofitive. But when the converging Series has its terms alternately pofitive and negative, then the fucceffive fums are alternately too great and too little, though ftill approaching nearer and nearer to the final fum or value. Thus in the Series

$$\frac{1}{3 + 1} = \frac{1}{4} = 0\cdot 25 = \frac{1}{3} - \frac{1}{9} + \frac{1}{27} - \frac{1}{81} \ \&c,$$

the 1ft term $\frac{1}{3} = \cdot 333 \ \&c,$ is too great,

two terms $\frac{1}{3} - \frac{1}{9} = \cdot 222 \ \&c,$ are too little,

three terms $\frac{1}{3} - \frac{1}{9} + \frac{1}{27} = \cdot 259259 \ \&c,$ are too great,

four terms $\frac{1}{3} - \frac{1}{9} + \frac{1}{27} - \frac{1}{81} = \cdot 246913 \ \&c,$ are too great, and fo on, alternately too great and too fmall, but every fucceeding fum ftill nearer than the former, or converging.

In the fecond cafe, or when the terms grow larger and larger, the Series is called a *diverging* one, becaufe that by collecting the terms continually, the fucceffive fums diverge, or go always farther and farther from the true value or radix of the Series; being all too great when the terms are all pofitive, but alternately too great and too little when they are alternately pofitive and negative. Thus, in the Series

$$\frac{1}{1 + 2} = \frac{1}{3} = 1 - 2 + 4 - 8 \ \&c.$$

the firft term $+ 1$ is too great,
two terms $1 - 2 = - 1$ are too little,
three terms $1 - 2 + 4 = + 3$ are too great,
four terms $1 - 2 + 4 - 8 = - 5$ are too little,
and fo on continually, after the 2d term, diverging more and more from the true value or radix $\frac{1}{3}$, but alternately too great and too little, or pofitive and negative. But the alternate fums would be always more and more too great if the terms were all pofitive, and always too little if negative.

But in the third cafe, or when the terms are all equal, the Series of equals, with alternate figns, is called a *neutral* one, becaufe the fucceffive fums, found by a continual collection of the terms, are always at the fame diftance from the true value or radix, but alternately pofitive and negative, or too great and too little. Thus, in the Series

$$\frac{1}{1 + 1} = \frac{1}{2} = 1 - 1 + 1 - 1 + 1 - 1 \ \&c,$$

the firft term 1 is too great,
two terms $1 - 1 = 0$ are too little,
three terms $1 - 1 + 1 = 1$ too great,
four terms $1 - 1 + 1 - 1 = 0$ too little,
and fo on continually, the fucceffive fums being alternately 1 and 0, which are equally different from the true value or radix $\frac{1}{2}$, the one as much above it, as the other below it.

A Series may be terminated and rendered finite, and accurately equal to the fum or value, by affuming the fupplement, after any particular term, and combining it with the foregoing terms. So, in the Series $\frac{1}{2} - \frac{1}{4} + \frac{1}{8} - \frac{1}{16} \ \&c,$ which is equal to $\frac{1}{3}$, and found by dividing 1 by $2 + 1$, after the firft term, $\frac{1}{2}$, of the quotient, the remainder is $- \frac{1}{2}$, which divided by $2 + 1$, or 3, gives $- \frac{1}{6}$ for the fupplement, which

combined

combined with the firſt term $\frac{1}{2}$, gives $\frac{1}{2} - \frac{1}{6} = \frac{1}{3}$ the true ſum of the Series. Again, after the firſt two terms $\frac{1}{2} - \frac{1}{4}$, the remainder is $+ \frac{1}{4}$, which divided by the ſame diviſor 3, gives $\frac{1}{12}$ for the ſupplement, and this combined with thoſe two terms $\frac{1}{2} - \frac{1}{4}$, makes $\frac{1}{2} - \frac{1}{4} + \frac{1}{12} = \frac{1}{4} + \frac{1}{2} = \frac{4}{12}$ or $\frac{1}{3}$ the ſame ſum or value as before. And in general, by dividing 1 by $a + c$, there is obtained

$$\frac{1}{a+c} = \frac{1}{a} - \frac{c}{a^2} + \frac{c^2}{a^3} \cdots \pm \frac{c^n}{a^{n+1}} \mp \frac{c^{n+1}}{a^{n+1}(a+c)};$$

where, ſtopping the diviſion at any term as $\frac{c^n}{a^{n+1}}$, the remainder after this term is $\frac{c^{n+1}}{a^{n+1}}$, which being divided by the ſame diviſor $a + c$, gives $\frac{c^{n+1}}{a^{n+1}(a+c)}$ for the ſupplement as above.

The Law of Continuation.—A Series being propoſed, one of the chief queſtions concerning it, is to find the law of its continuation. Indeed, no univerſal rule can be given for this; but it often happens that the terms of the Series, taken two and two, or three and three, or in greater numbers, have an obvious and ſimple relation, by which the Series may be determined and produced indefinitely. Thus, if 1 be divided by $1 - x$, the quotient will be a geometrical progreſſion, viz, $1 + x + x^2 + x^3$ &c, where the ſucceeding terms are produced by the continual multiplication by x. In like manner, in other caſes of diviſion, other progreſſions are produced.

But in moſt caſes the relation of the terms of a Series is not conſtant, as it is in thoſe that ariſe by diviſion. Yet their relation often varies according to a certain law, which is ſometimes obvious on inſpection, and ſometimes it is found by dividing the ſucceſſive terms one by another, &c. Thus, in the Series

$1 + \frac{2}{3}x + \frac{8}{15}x^2 + \frac{16}{35}x^3 + \frac{128}{315}x^4$ &c, by dividing the 2d term by the 1ſt, the 3d by the 2d, the 4th by the 3d, and ſo on, the quotients will be

$$\frac{2}{3}x, \quad \frac{4}{5}x, \quad \frac{6}{7}x, \quad \frac{8}{9}x, \&c;$$

and therefore the terms may be continued indefinitely by the ſucceſſive multiplication by theſe fractions. Alſo in the following Series

$1 + \frac{1}{6}x + \frac{3}{40}x^2 + \frac{5}{128}x^3 + \frac{35}{1152}x^4$ &c, by dividing the adjacent terms ſucceſſively by each other, the Series of quotients is

$$\frac{1}{6}x, \quad \frac{9}{20}x, \quad \frac{25}{42}x, \quad \frac{49}{72}x, \&c, \text{ or}$$

$$\frac{1 \cdot 1}{2 \cdot 3}x, \quad \frac{3 \cdot 3}{4 \cdot 5}x, \quad \frac{5 \cdot 5}{6 \cdot 7}x, \quad \frac{7 \cdot 7}{8 \cdot 9}x, \&c;$$

and therefore the terms of the Series may be continued by the multiplication of theſe fractions.

Another method of expreſſing the law of a Series, is one that defines the Series itſelf, by its *general term*, ſhewing the relation of the terms generally by their diſtances from the beginning, or by differential equations. To do this, Mr. Stirling conceives the terms of the Series to be placed as ſo many ordinates on a right line given by poſition, taking unity as the common interval between theſe ordinates. The terms of the Series he denotes by the initial letters of the alphabet, A, B, C, D, &c; A being the firſt, B the 2d, C the 3d, &c: and he denotes any term in general by the letter T, and the reſt following it in order by T', T'', T''', T'''', &c; alſo the diſtance of the term T from any given term, or from any given intermediate point between two terms, he denotes by the indeterminate quantity z: ſo that the diſtances of the terms T', T'', T''', &c, from the ſaid term or point, will be $z + 1$, $z + 2$, $z + 3$, &c; becauſe the increment of the abſciſs is the common interval of the ordinates, or terms of the Series, applied to the abſciſs.

Theſe things being premiſed, let this Series be propoſed, viz,

$$1, \quad \frac{1}{2}x, \quad \frac{3}{8}x^2, \quad \frac{5}{16}x^3, \quad \frac{35}{128}x^4, \quad \frac{63}{256}x^5, \&c;$$

in which it is found, by dividing the terms by each other, that the relations of the terms are,

$$B = \frac{1}{2}Ax, \; C = \frac{3}{4}Bx, \; D = \frac{5}{6}Cx, \; E = \frac{7}{8}Dx, \&c:$$

then the relation in general will be defined by the equation $T' = \frac{2z+1}{2z+2}Tx$ or $\frac{z+\frac{1}{2}}{z+1}Tx$, where z denotes the diſtance of T from the firſt term of the Series. For by ſubſtituting 0, 1, 2, 3, 4, &c, ſucceſſively inſtead of z, the ſame relations will ariſe as in the propoſed Series above. In like manner, if z be the diſtance of T from the 2d term of the Series, the equation will be $T = \frac{2z+3}{2z+4}Tx$ or $\frac{z+\frac{3}{2}}{z+2}Tx$, as will appear by ſubſtituting the numbers -1, 0, 1, 2, 3, &c, ſucceſſively for z. Or, if z denote the place or number of the term T in the Series, its ſucceſſive values will be 1, 2, 3, 4, &c, and the equation or general term will be $T' = \frac{2z-1}{2z}Tx$.

It appears therefore, that innumerable differential equations may define one and the ſame Series, according to the different points from whence the origin of the abſciſs z is taken. And, on the contrary, the ſame equation defines innumerable different Series, by taking different ſucceſſive values of z. For in the equation $T' = \frac{2z-1}{2z}Tx$ which defines the foregoing Series

when

when 1, 2, 3, 4, &c are the fucceffive values of the abfciffes; if $1\frac{1}{2}$, $2\frac{1}{2}$, $3\frac{1}{2}$, $4\frac{1}{2}$, &c, be fucceffively fubftituted for z, the relations of the terms arifing will be,

$$B = \frac{2}{3} A x, \quad C = \frac{4}{5} B x, \quad D = \frac{6}{7} C x, \quad \&c, \text{ from}$$

whence will arife the Series

$$A, \quad \frac{2}{3} A x, \quad \frac{8}{15} A x^2, \quad \frac{16}{35} A x^3, \quad \frac{128}{315} A x^4, \quad \&c,$$

which is different from the former.

And thus the equation w.ll always determine the Series from the given values of the abfcifs and of the firft term, when the equation includes but two terms of the Series, as in the laft example, where the firft term being given, all the reft will be given.

But when the equation includes three terms, then two muft be given; and three muft be given, when it includes four; and fo on. So, if there be propofed the

Series $x, \quad \frac{1}{6} x^3, \quad \frac{3}{40} x^5, \quad \frac{5}{128} x^7, \quad \frac{35}{1152} x^9, \quad \&c,$

where the relations of the terms are,

$$B = \frac{1 \cdot 1}{2 \cdot 3} A x^2, \quad C = \frac{3 \cdot 3}{4 \cdot 5} B x^2, \quad D = \frac{5 \cdot 5}{6 \cdot 7} C x^2, \quad \&c.$$

the equation defining this Series will be

$$T = \frac{2z - 1 \cdot 2z - 1}{2z \cdot 2z + 1} T x^2 = \frac{4zz - 4z + 1}{4zz + 2z} T x^2,$$

where the fucceffive values of z are 1, 2, 3, 4, &c. See Stirling's Methodus Differentialis, in the introduction.

This may fuffice to give a notion of thefe differential equations, defining the nature of Series. But as to the application of thefe equations in interpolations, and finding the fums of Series, it would require a treatife to explain it. We muft therefore refer to that excellent one juft quoted, as alfo to De Moivre's Mifcellanea Analytica; and feveral curious papers by Euler in the Acta Petropolitana.

A Series often converges fo flowly, as to be of no ufe in practice. Thus, if it were required to find the fum of the Series

$$\frac{1}{1 \cdot 2} + \frac{1}{3 \cdot 4} + \frac{1}{5 \cdot 6} + \frac{1}{7 \cdot 8} + \frac{1}{9 \cdot 10} \&c,$$

which Lord Brouncker found for the quadrature of the hyperbola, true to 9 figures, by the mere addition of the terms of the Series; Mr. Stirling computes that it would be neceffary to add a thoufand millions of terms for that purpofe; for which the life of man would be too fhort. But by that gentleman's method, the fum of the Series may be found by a very moderate computation. See Method. Differ. pa. 26.

Series are of various kinds or defcriptions. So,

An *Afcending* SERIES, is one in which the powers of the indeterminate quantity increafe; as

$$1 + a x + b x^2 + c x^3 \&c. \quad \text{And a}$$

Defcending SERIES, is one in which the powers decreafe, or elfe increafe in the denominators, which is the fame thing; as

$$1 + a x^{-1} + b x^{-2} + c x^{-3} \&c, \text{ or } 1 + \frac{a}{x} + \frac{b}{x^2} + \frac{c}{x^3} \&c.$$

A Circular SERIES, which denotes a Series whofe

4

fum depends on the quadrature of the circle. Such is the Series $1 \quad \frac{1}{3} + \frac{1}{5} - \frac{1}{7} + \frac{1}{9} \&c$: See Demoivre Mifcel. Analyt. pa. 111, or my Menfur. pa. 119. Or the fum of the Series $1 + \frac{1}{4} + \frac{1}{9} + \frac{1}{16} + \frac{1}{25} \&c$, continued ad infinitum, according to Euler's difcovery.

Continued Fraction or *Series*, is a fraction of this kind, to infinity,

$$b + \cfrac{a}{d + \cfrac{c}{f + \cfrac{e}{k \cfrac{g}{} \&c.}}}$$

The firft Series of this kind was given by Lord Brouncker, firft prefident of the Royal Society, for the quadrature of the circle, as related by Dr. Wallis, in his Algebra, pa. 317. His feries is

$$1 + \cfrac{1}{2 + \cfrac{9}{2 + \cfrac{25}{2 + \cfrac{49}{2 + \cfrac{81}{2 + \&c,}}}}}$$

which denotes the ratio of the fquare of the diameter of a circle to its area. Mr. Euler has treated on this kind of Series, in the Peterfburgh Commentaries, vol. 11, and in his Analyf. Infinit. vol. 1, pa. 295, where he fhews various ufes of it, and how to transform ordinary fractions and common Series into continued fractions. A common fraction is transformed into a continued one, after the manner of feeking the greateft common meafure of the numerator and denominator, by dividing the greater by the lefs, and the laft divifor always by the laft remainder. Thus to change $\frac{1461}{59}$ to a continued fraction.

59) 1461 (24
   118
   281
   236
   45) 59 (1
     45
     14) 45 (3
      42
      3) 14 (4
       12
       2) 3 (1
        2
        1) 2 (2
         2
         0

Therer. $\dfrac{1461}{59} = 24 + \cfrac{1}{1 + \cfrac{1}{3 + \cfrac{1}{4 + \cfrac{1}{1 + \frac{1}{2}}}}}$

Alfo $\sqrt{2} = 1 + \cfrac{1}{2 + \cfrac{1}{2 + \cfrac{1}{2 + \frac{1}{2} \&c.}}}$

Converging

Converging SERIES, is a Series whose terms continually decrease, or the successive sums of whose terms approximate or converge always nearer to the ultimate sum of the whole Series. And, on the contrary, a

Diverging SERIES, is one whose terms continually increase, or that has the successive sums of its terms diverging, or going off always the farther, from the sum or value of the Series.

Determinate SERIES, is a Series whose terms proceed by the powers of a determinate quantity; as

$$1 + \frac{1}{2} + \frac{1}{2^2} + \frac{1}{2^3} + \&c.$$ If that determinate

quantity be unity, the Series is said to be determined by unity. De Moivre, Miscel. Analyt. pa. 111. And an

Indeterminate SERIES is one whose terms proceed by the powers of an indeterminate quantity x; as

$$x + \frac{1}{2} x^2 + \frac{1}{3} x^3 + \frac{1}{4} x^4 \&c.$$; or sometimes also with

indeterminate exponents, or indeterminate coefficients.

The *Form of a* SERIES, is used for that affection of an indeterminate Series, such as

$$ax^n + bx^{n+r} + cx^{n+2r} + dx^{n+3r} \&c,$$ which arises from the different values of the indices of x. Thus,

If $n = 1$, and $r = 1$, the Series will take the form
$$ax + bx^2 + cx^3 + dx^4 \&c.$$

If $n = 1$, and $r = 2$, the form will be
$$ax + bx^3 + cx^5 + dx^7 \&c.$$

If $n = \frac{1}{2}$, and $r = 1$, the form is
$$ax^{\frac{1}{2}} + bx^{\frac{3}{2}} + cx^{\frac{5}{2}} + dx^{\frac{7}{2}} \&c.$$ And

If $n = 0$, and $r = -1$, the form will be
$$a + bx^{-1} + cx^{-2} + dx^{-3} \&c.$$

When the value of a quantity cannot be found exactly, it is of use in algebra, as well as in common arithmetic, to seek an approximate value of that quantity, which may be useful in practice. Thus, in arithmetic, as the true value of the square root of 2 cannot be assigned, a decimal fraction is found to a sufficient degree of exactness in any particular case; which decimal fraction is in reality, no more than an infinite series of fractions converging or approximating to the true value of the root sought. For the expression $\sqrt{2} = 1 \cdot 414213$

&c, is equivalent to this $\sqrt{2} = 1 + \frac{4}{10} + \frac{1}{100} + \frac{4}{1000}$

&c; or supposing $x = 10$, to this

$$\sqrt{2} = 1 + \frac{4}{x} + \frac{1}{x^2} + \frac{4}{x^3} + \frac{2}{x^4} \&c.$$

or $= 1 + 4x^{-1} + x^{-2} + 4x^{-3} + 2x^{-4} \&c,$ which last Series is a particular case of the more general indeterminate Series $ax^n + bx^{n+r} + cx^{n+2r} \&c,$ viz, when $n = 0$, $r = -1$, and the coefficients $a = 1$, $b = 4$, $c = 1$, $d = 4$, &c.

But the application of the notion of approximations in numbers, to species, or to algebra, is not so obvious. Newton, with his usual sagacity, took the hint,

and prosecuted it; by which were discovered general methods in the doctrine of infinite Series, which had before been treated only in a particular manner, though with great acuteness, by Wallis and a few others. See Newton's Method of Fluxions and Infinite Series, with Colson's Comment; as also the Analysis per Æquationes Numero Terminorum Infinitas, published by Jones in 1711, and since translated and explained by Stewart, together with Newton's Tract on Quadratures, in 1745. To these may be added Maclaurin's Algebra, part 2, chap. 10, pa. 244; and Cramer's Analyse des Lignes Courbes Algebraiques, chap. 7, pa. 148; and many other authors.

Among the various methods for determining the value of a quantity by a converging Series, that seems preferable to the rest, which consists in assuming an indeterminate Series as equal to the quantity whose value is sought, and afterwards determining the values of the terms of this assumed Series. For instance, suppose a logarithm were given, to find the natural number answering to it. Suppose the logarithm to be z, and the corresponding number sought $1 + x$: then by the nature of logarithms and fluxions, $\dot{z} = \frac{\dot{x}}{1 + x}$, or

$\dot{z} + x\dot{z} = \dot{x}$. Now assume a Series for the value of the unknown quantity x, and substitute it and its fluxion instead of x and $\dot{x}$ in the last equation, then determine the assumed coefficients by comparing or equating the like terms of the equation. Thus,

assume $x = az + bz^2 + cz^3 + dz^4 \&c,$
then $\dot{x} = a\dot{z} + 2bz\dot{z} + 3cz^2\dot{z} + 4dz^3\dot{z} \&c;$
and $\dot{x} = (\dot{z} + x\dot{z}) = \dot{z} + az\dot{z} + bz^2\dot{z} + cz^3\dot{z} \&c;$

hence, comparing the like terms of these two values of $\dot{x}$,

there arises $a = 1$, $b = \frac{1}{2}$, $c = \frac{1}{6}$, $d = \frac{1}{24}$, &c;

which values being substituted for a, b, c, &c, in the assumed Series $ax + bx^2 + cx^3 \&c,$ it gives

$$x = z + \frac{1}{2} z^2 + \frac{1}{6} z^3 + \frac{1}{24} z^4 + \frac{1}{120} z^5, \&c, \text{ or}$$

$$x = x + \frac{1}{1 \cdot 2} z^2 + \frac{1}{1 \cdot 2 \cdot 3} z^3 + \frac{1}{1 \cdot 2 \cdot 3 \cdot 4} z^4 + \frac{1}{1 \cdot 2 \cdot 3 \cdot 4 \cdot 5} z^5$$

&c; and consequently the number sought will be

$$1 + x = 1 + z + \frac{1}{1 \cdot 2} z^2 + \frac{1}{1 \cdot 2 \cdot 3} z^3 \&c.$$

But the indeterminate Series $az + bz^2 + cz^3 \&c,$ was here assumed arbitrarily, with regard to its exponents 1, 2, 3, &c, and will not succeed in all cases, because some quantities require other forms for the exponents. For instance, if from an arc given, it were required to find the tangent. Let $x =$ the tangent, and z the arc, the radius being $= 1$. Then, from the nature of the circle we shall have $\frac{\dot{x}}{1 + x^2} = \dot{z}$, or

$\dot{x} = \dot{z} + x^2\dot{z}$. Now if, to find the value of x, we suppose $x = az + bz^2 + cz^3 \&c,$ and proceed as before, we shall find all the alternate coefficients $b, d, f,$ &c, or those of the even powers of z, to be each $= 0$; and therefore the Series assumed is not of a proper form. But

But supposing $x = az + bz^3 + cz^5 + dz^7$, &c, then we find $a = 1, b = \frac{1}{3}, c = \frac{2}{15}, d = \frac{17}{315}$, &c, and consequently $x = z + \frac{1}{3} z^3 + \frac{2}{15} z^5 + \frac{17}{315} z^7$ &c.

And other quantities require other forms of Series.

Now to find a proper indeterminate Series in all cases, tentatively, would often be very laborious, and even impracticable. Mathematicians have therefore endeavoured to find out a general rule for this purpose; though till lately the method has been but imperfectly understood and delivered. Most authors indeed have explained the manner of finding the coefficients a, b, c, d, &c, of the indeterminate Series $ax^n + bx^{n+r} + cx^{n+2r}$ &c, which is easy enough; but the values of n and r, in which the chief difficulty lies, have been assigned by many in a manner as if they were self-evident, or at least discoverable by an easy trial or two, as in the last example.

As to the number n, Newton himself has shewn the method of determining it, by his rule for finding the first term of a converging Series, by the application of his parallelogram and ruler. For the particulars of this method, see the authors above cited; see also PARALLELOGRAM.

Taylor, in his Methodus Incrementorum, investigates the number r; but Stirling observes that his rule sometimes fails. Lineæ Tert. Ordin. Newton. pa. 28. Mr. Stirling gives a correction of Taylor's rule, but says he cannot affirm it to be universal, having only found it by chance. And again Gravesande observes, that though he thinks Stirling's rule never leads into an error, yet that it is not perfect. See Gravesande, De Determin. Form. Seriei Infinit. printed at the end of his Matheseos Universalis Elementa. This learned professor has endeavoured to rectify the rule. But Cramer has shewn that it is still defective in several respects; and he himself, to avoid the inconveniences to which the methods of former authors are subject, has had recourse to the first principles of the method of infinite Series, and has entered into a more exact and instructive detail of the whole method, than is to be met with elsewhere; for which reason, and many others, his treatise deserves to be particularly recommended to beginners.

But it is to be observed, that in determining the value of a quantity by a converging Series, it is not always necessary to have recourse to an indeterminate Series: for it is often better to find it by division, or by extraction of roots. See Newton's Meth. of Flux. and Inf. Series, above cited. Thus, if it were required to find the arc of a circle from its given tangent, that is, to find the value of z in the given fluxional equation, $\dot{z} = \frac{\dot{x}}{1 + xx}$, by an infinite Series: dividing $\dot{x}$ by $1 + xx$, the quotient will be the Series $\dot{x} - x^2 \dot{x} + x^4 \dot{x} - x^6 \dot{x}$ &c $= \dot{z}$; and taking the fluents of the terms, there results $z = x - \frac{1}{3} x^3 + \frac{1}{5} x^5 - \frac{1}{7} x^7$ &c, which is the Series often used for the quadrature of the circle. If $x = 1$, or the tangent of $45°$, then

will $z = 1 - \frac{1}{3} + \frac{1}{5} - \frac{1}{7}$ &c $=$ the length of an arc of $45°$, or $\frac{1}{8}$ of the circumference, to the radius 1, or $\frac{1}{4}$ of the circumference to the diameter 1. Consequently, if 1 be the diameter, then $1 - \frac{1}{3} + \frac{1}{5} - \frac{1}{7}$ &c will be the area of the circle, because $\frac{1}{4}$ of the circumference multiplied by the diameter, gives the area of the circle. And this Series was first given by Leibnitz and James Gregory.

See the form of the Series for the binomial theorem, determined, both as to the coefficients and exponents, in my Tracts, vol. 1, pa. 79.

Harmonical SERIES, the reciprocal of arithmeticals. See HARMONICAL.

Hyperbolic SERIES, is used for a Series whose sum depends upon the quadrature of the hyperbola. Such is the Series $\frac{1}{1} + \frac{1}{2} + \frac{1}{3} + \frac{1}{4}$ &c. De Moivre's Miscel. Analyt. pa. 111.

Interpolation of SERIES, the inserting of some terms between others, &c. See INTERPOLATION.

Interscendent SERIES. See INTERSCENDENT.

Mixt SERIES, one whose sum depends partly on the quadrature of the circle, and partly on that of the hyperbola. De Moivre, Miscel. Analyt. pa. 111.

Recurring SERIES, is used for a Series which is so constituted, that having taken at pleasure any number of its terms, each following term shall be related to the same number of preceding terms according to a constant law of relation. Thus, in the following Series,

$$a \quad b \quad c \quad d \quad e \quad f$$
$$1 + 2x + 3x^2 + 10x^3 + 34x^4 + 97x^5 \text{ \&c,}$$

in which the terms being respectively represented by the letters a, b, c, &c, set over them, we shall have

$$d = 3cx - 2bx^2 + 5ax^3,$$
$$e = 3dx - 2cx^2 + 5bx^3,$$
$$f = 3ex - 2dx^2 + 5cx^3,$$
&c, &c,

where it is evident that the law of relation between d and e, is the same as between e and f, each being formed in the same manner from the three terms which precede it in the Series.

The quantities $3x - 2x^2 + 5x^3$, taken together and connected by their proper signs, form what De Moivre calls the *index*, or the *scale of relation*; though sometimes the bare coefficients $3 - 2 + 5$ are called the scale of relation. And the scale of relation subtracted from unity, is called the *differential scale*. On the subject of Recurring Series, see De Moivre's Miscel. Analyt. pa. 27 and 72, and his Doctrine of Chances, 3d edit. pa. 220; also Euler's Analys. Infinit. tom. 1, pa. 175.

Having given a recurring Series, with its scale of relation, the sum of the whole infinite Series will also be given. For instance, suppose a Series

$a +$

$a + bx + cx^2 + dx^3$ &c, where the relation between the coefficient of any term and the coefficients of any two preceding terms may be expressed by $f - g$; that is, $e = fd - gc$, and $d = fc - gb$, &c; then will the sum of the Series, infinitely continued, be

$$\frac{a + (b - fa)\,x}{1 - fx + g.x^2}.$$

Thus, for example, assume 2 and 5 for the coefficients of the first two terms of a recurring Series; and suppose f and g to be respectively 2 and 1; then the recurring Series will be

$$2 + 5x + 8x^2 + 11x^3 + 14x^4 + 18x^5 \text{ \&c},$$

and its sum $= \dfrac{2 + 5x - 4x}{1 - 2x + xx} = \dfrac{2 + x}{(1 - x)^2}$. For the proof of which divide $2 + x$ by $(1 - x)^2$, and there arises the said Series $2 + 5x + 8x^2 + 11x^3$ &c. And similar rules might be derived for more complex case

De Moivre's general rule is this: 1. Take as many terms of the Series as there are parts in the scale of relation. 2. Subtract the scale of relation from unity, and the Remainder is the differential scale. 3. Multiply the terms taken in the Series by the differential scale, beginning at unity, and so proceeding orderly, remembering to leave out what would naturally be extended beyond the last of the terms taken. Then will the product be the numerator, and the differential scale will be the denominator of the fraction expressing the sum required.

But it must here be observed, that when the sum of a recurring Series extended to infinity, is found by De Moivre's rule, it ought to be supposed that the Series converges indefinitely, that is, that the terms may become less than any assigned quantity. For if the Series diverge, that is, if its terms continually increase, the rule does not give the true sum. For the sum in such case is infinite, or greater than any given quantity, whereas the sum exhibited by the rule, will often be finite. The rule therefore in this case only gives a fraction expressing the radix of the Series, by the expansion of which the Series is produced. Thus $\dfrac{1}{(1 - x)^2}$ by expansion becomes the recurring Series $1 + 2x + 3x^2$ &c, whose scale of relation is $2 - 1$, and its sum by the rule will be $\dfrac{a + bx - fax}{1 - fx + gxx} = \dfrac{1 + 2x - 2x}{1 - 2x + xx} = \dfrac{1}{(1 - x)^2}$, the quantity from which the Series arose. But this quantity cannot in all cases be deemed equal to the infinite Series $1 + 2x + 3x^2$ &c: for stop where you will, there will always want a supplement to make the product of the quotient by the divisor equal to the dividend. Indeed when the Series converges infinitely, the supplement, diminishing continually, becomes less than any assigned quantity, or equal to nothing; but in a diverging Series, this supplement becomes infinitely great, and the Series deviates indefinitely from the truth. See Colson's Comment on Newton's Method of Fluxions and Infinite Series, pa. 152: Stirling's Method. Differ. pa. 36; Bernoulli de Serieb. Infin. pa. 249; and Cramer's Analyse des Lignes Courbes, pa. 174.

A recurring Series being given, the sum of any

finite number of the terms of that Series may be found. This is prob. 3, pa. 73, De Moivre's Miscel. Analyt. and prob. 5, pa. 2:3 of his Doctrine of Chances. The solution is effected, by taking the difference between the sums of two infinite Series, differing by the terms answering to the given number; viz, from the sum of the whole infinite Series, commencing from the beginning, subtract the sum of another infinite number of terms of the same Series, commencing after so many of the first terms whose sum is required; and the difference will evidently be the sum of that number of terms of the Series. For example, to find the sum of n terms of the infinite geometrical Series $a + ax + ax^2 + ax^3$ &c. Here are two infinite Series; the one beginning with a, and the other with ax^n, which is the next term after the first n terms of the original Series. By the rule, the sum of the first infinite progression will be $\dfrac{a}{1 - x}$, and the sum of the second $\dfrac{ax^n}{1 - x}$; the difference of which is $\dfrac{a - ax^n}{1 - x}$, which is therefore the sum of the first n terms of the Series. This quantity $\dfrac{a - ax^n}{1 - x}$ is equal to $\dfrac{ax^n - a}{x - 1}$ which last expression, putting $ax^{n-1} = l$, will be equivalent to this, $\dfrac{lx - a}{x - 1}$, which is the common rule for finding the sum of any geometric progression, having given the first term a, the last term l, and the ratio x. See Miscel. Analyt. pa. 167, 168.

In a recurring Series, any term may be obtained whose place is assigned. For after having taken so many terms of the Series as there are terms in the scale of relation, the Series may be protracted till it reach the place assigned. But when that place is very distant from the beginning of the Series, the continuing the terms is very laborious; and therefore other methods have been contrived. See Miscel. Analyt. pa. 33; and Doctrine of Chances, pa. 224.

These questions have been resolved in many cases, besides those of recurring Series. But as there is no universal method for the quadrature of curves, neither is there one for the summation of Series; indeed there is a great analogy between these things, and similar difficulties arising in both. See the authors above cited.

The investigation of Daniel Bernoulli's method for finding the roots of algebraic equations, which is inserted in the Petersburgh Acts, tom. 3, pa. 92, depends upon the doctrine of recurring Series. See Euler's Analysis Infinitorum, tom. 1, pa. 276.

Reversion of SERIES. See REVERSION *of Series.*

Summable SERIES, is one whose sum can be accurately found. Such is the Series $\dfrac{1}{2} + \dfrac{1}{4} + \dfrac{1}{8}$ &c, the sum of which is said to be unity, or, to speak more accurately, the limit of its sum is unity or 1.

An indefinite number of summable infinite Series may

may be assigned : such are, for instance, all infinite recurring converging Series, and many others, for which, consult De Moivre, Bernoulli, Stirling, Euler, and Maclaurin ; viz, Miscel. Analyt. pa. 110 ; De Serieb. Infinit. passim ; Method. Different. pa. 34 ; Acta Petrop. passim ; Fluxions, art. 350.

The obtaining the sums of infinite Serieses of fractions has been one of the principal objects of the modern method of computation ; and these sums may often be found, and often not. Thus the sums of the two following Series of geometrical progressionals are easily found to be 1 and $\frac{1}{2}$,

viz, $1 = \frac{1}{2} + \frac{1}{4} + \frac{1}{8} + \frac{1}{16}$ &c,

and $\frac{1}{2} = \frac{1}{3} + \frac{1}{9} + \frac{1}{27} + \frac{1}{81}$ &c.

But the Serieses of fractions that occur in the solution of problems, can seldom be reduced to geometric progressions ; nor can any general rule, in cases so infinitely various, be given. The art here, as in most other cases, is only to be acquired by examples, and by a careful observation of the arts used by great authors in the investigation of such Series of fractions as they have considered. And the general methods of infinite Series, which have been carried so far by De Moivre, Stirling, Euler, &c, are often found necessary to determine the sum of a very simple Series of fractions. See the quotations above.

The sum of a Series of fractions, though decreasing continually, is not always finite. This is the case of the Series $\frac{1}{1} + \frac{1}{2} + \frac{1}{3} + \frac{1}{4} + \frac{1}{5}$ &c, which is the harmonic Series, consisting of the reciprocals of arithmeticals, the sum of which exceeds any given number whatever ; and this is shewn from the analogy between this progression and the space comprehended by the common hyperbola and its asymptote ; though the same may be shewn also from the nature of progressions. See James Bernoulli, de Seriebus Infin. But, what is curious, the square of it is finite, for if the same terms of the harmonic Series, $\frac{1}{1} + \frac{1}{2} + \frac{1}{3}$ &c, be squared, forming the Series $\frac{1}{1} + \frac{1}{4} + \frac{1}{9}$ &c, being the reciprocals of the squares of the natural Series of numbers ; the sum of this Series of fractions will not only be limited, but it is remarkable that this sum will be precisely equal to the 6th part of the number which expresses the ratio of the square of the circumference of a circle to the square of its diameter. That is, if c denote $3\cdot14159$ &c, the ratio of the circumference to the diameter, then is $\frac{1}{6} c^2 = \frac{1}{1} + \frac{1}{4} + \frac{1}{9} + \frac{1}{16} + \frac{1}{25}$

&c. This property was first discovered by Euler ; and his investigation may be seen in the Acta Petrop. vol. 7. And Maclaurin has since observed, that this may easily be deduced from his Fluxions, art. 822. Philos. Trans. numb. 469.

It would require a whole treatise to enumerate the various kinds of Series of fractions which may or may not be summed. Sometimes the sum cannot be assigned, either because it is infinite, as in the harmonic Series $\frac{1}{1} + \frac{1}{2} + \frac{1}{3} + \frac{1}{4}$ &c, or although its sum be finite (as in the Series $\frac{1}{1} + \frac{1}{4} + \frac{1}{9}$ &c), yet its sum cannot be assigned in finite terms, or by the quadrature of the circle or hyperbola, which was the case of this Series before Euler's discovery ; but yet the sum of any given numb of terms of the Series may be expeditiously found, and the whole sum may be assigned by approximation, independent of the circle. See Stirling's Method. Different. and De Moivre's Miscel. Analyt.

Besides the Serieses of fractions, the sums of which converge to a certain quantity, there sometimes occur others, which converge by a continued multiplication. Of this kind is the Series found by Wallis, for the quadrature of the circle, which he expresses thus,

$$\square = \frac{3 \times 3 \times 5 \times 5 \times 7 \times 7 \times 9 \times 9 \times \text{&c}}{2 \times 4 \times 4 \times 6 \times 6 \times 8 \times 8 \times 10 \times \text{&c}},$$

where the character $\square$ denotes the ratio of the square of the diameter to the area of the circle. Hence the denominator of this fraction, is to its numerator, both infinitely continued, as the circle is to the square of the diameter. It may farther be observed that this Series is equivalent to

$$\frac{9}{8} \times \frac{25}{24} \times \frac{49}{48} \times \text{&c, or to } \frac{3^2}{3^2-1} \times \frac{5^2}{5^2-1} \times \frac{7^2}{7^2-1} \times$$

&c, that is, the product of the squares of all the odd numbers 3, 5, 7, 9, &c, is to the product of the same squares severally diminished by unity, as the square of the diameter is to the area of the circle. See Arithmet. Infinit. prop. 191, Oper. vol. 1, pa. 469. Id. Oper. vol. 2, pa. 819. And these products of fractions, and the like quantities arising from the continued multiplication of certain factors, have been particularly considered by Euler, in his Analysis Infinit. vol. 1, chap. 15, pa. 221.

For an easy and general method of summing all alternate Series, such as $a - b + c - d$ &c, see my Tracts, vol. 1, pa. 11 ; and in the same vol. may be seen many other curious tracts on infinite Series.

Summation of Infinite SERIES, is the finding the value of them, or the radix from which they may be raised. For which, consult all the authors upon this science.

To find an infinite Series by extracting of roots ; and to find an infinite Series by a presupposed Series ; see QUADRATURE *of the Circle.*

To extract the roots of an infinite Series, see EXTRACTION *of Roots.*

To raise an infinite Series to any power, see INVOLUTION, and POWER.

Transcendental SERIES. See TRANSCENDENTAL.

There are many other important writings upon the subject of Infinite Series, besides those above quoted. A very good elementary tract on this science is that of James Bernoulli, intituled, *Tractatus de Seriebus Infinitis, his*, and annexed to his *Ars Conjectandi*, published in 4to, 1713.

SERPENS, in Astronomy, a constellation in the northern hemisphere, being one of the 48 old constellations mentioned by all the Ancients, and is called more particularly *Serpens Ophiuchi*, being grasped in the hands of the constellation Ophiuchus. The Greeks, in their fables, have ascribed it sometimes to one of Triptolemus's dragons, killed by Carnabos; and sometimes to the serpent of the river Segaris, destroyed by Hercules. This is by some supposed to be the same as the author of the book of Job calls the *Crooked Serpent*; but this expression more probably meant the constellation Draco, near the north pole.

The stars in the constellation Serpens, in Ptolomy's catalogue are 18, in Tycho's 13, in Hevelius's 22, and in the Britannic catalogue 64.

SERPENTARIUS, a constellation of the northern hemisphere, being one of the 48 old constellations mentioned by all the Ancients. It is called also Ophiuchus, and anciently Æsculapius. It is in the figure of a man grasping the serpent.

The Greeks had different fables about this, and other constellations, because they were ignorant of the true meaning of them. Some of them say, it represents Carnabos, who killed one of the dragons of Triptolemus. Others say, it was Hercules, killing the serpent at the river Segaris. And others again say, it represents the celebrated physician Æsculapius, to denote his skill in medicine to cure the bite of the serpent.

The stars in the constellation Serpentarius, in Ptolomy's catalogue are 29, in Tycho's 15, in Hevelius's 40, and in the Britannic catalogue they are 74.

SERPENTINE *Line*, the same with spiral.

SESQUI, an expression of a certain ratio, viz, the second ratio of inequality, called also *superparticular* ratio; being that in which the greater term contains the less once, and some certain part over; as 3 to 2, where the first term contains the second once, and unity over, which is a quota part of 2. Now if this part remaining be just half the less term, the ratio is called *sesquialtera*; if the remaining part be a 3d part of the less term, as 4 to 3, the ratio is called *sesquitertia*, or *sesquiterza*; if a 4th part, as 5 to 4, the ratio is called *sesquiquarta*; and so on continually, still adding to Sesqui the ordinal number of the smaller term.

In English we sometimes say, *sesquialteral*, or *sesquialterate*, *sesquithird*, *sesquifourth*, &c.

As to the kinds of triples expressed by the particle *sesqui*, they are these:

SESQUIALTERATE, *the greater perfect*, which is a triple, where the breve is three measures, or semibreves.

Sesquialterate, *greater imperfect*, which is where the breve, when pointed, contains three measures, and without any point, two.

Sesquialterate, *less imperfect*, a triple, where the semibreve with a point contains three measures, and two without.

Sesquialterate, in Arithmetic and Geometry, is a ratio between two numbers, or lines, &c, where the greater is equal to once and a half of the less. Thus 6 and 9 are in a Sesquialterate ratio, as also 20 and 30.

SESQUIDITONE, in Music, a concord resulting from the sounds of two strings whose vibrations, in equal times, are to each other in the ratio of 5 to 6.

SESQUIDUPLICATE *Ratio*, is that in which the greater term contains the less, twice and a half; as the ratio of 15 to 6, or 50 to 20.

SESQUIQUADRATE, an aspect or position of the planets, when they are distant by 4 signs and a half, or 135 degrees.

SESQUIQUINTILE, is an aspect of the planets when they are distant ⅕ of the circle and a half, or 108 degrees.

SESQUITERTIONAL *Proportion*, is that in which the greater term contains the less once and one third; as 4 to 3, or 12 to 9.

SETTING, in Astronomy, the withdrawing of a star or planet, or its sinking below the horizon.

Astronomers and poets count three different kinds of Setting of the stars, viz, Achronical, Cosmical, and Heliacal. See these terms respectively.

Setting, in Navigation, Surveying, &c, denotes the observing the bearing or situation of any distant object by the compass, &c, to discover the angle it makes with the nearest meridian, or with some other line. See Bearing.

Thus, to *set the land*, or *the sun*, by the compass, is to observe how the land bears on any point of the compass, or on what point of the compass the sun is. Also, when two ships come in sight of each other, to mark on what point the chace bears, is termed *Setting the chace by the compass*.

Setting also denotes the direction of the wind, current, or sea, particularly of the two latter.

SEVEN Stars, a common denomination given to the cluster of stars in the neck of the sign Taurus, the bull, properly called the pleiades. They are so called from their number Seven which appear to the naked eye, though some eyes can discover only 6 of them; but by the help of telescopes there appears to be a great multitude of them.

SEVENTH, *Septima*, an interval in Music, called by the Greeks *heptachordon*.

SEXAGENARY, something relating to the number 60.

Sexagenary *Arithmetic*. See Sexagesimal.

Sexagenary *Tables*, are tables of proportional parts, shewing the product of two Sexagenaries that are to be multiplied, or the quotient of two that are to be divided.

SEXAGESIMA, the eighth Sunday before Easter; being so called because near 60 days before it.

SEXAGESIMAL or Sexagenary *Arithmetic*, a method of computation proceeding by 60ths. Such is that used in the division of a degree into 60 minutes, of the minute into 60 seconds, of the second into 60 thirds, &c.

SEXAGESIMALS, or Sexagesimal *Fractions*, are fractions whose denominators proceed in a sexagecuple ratio; that is, a prime, or the first minute $= \frac{1}{60}$, a second $= \frac{1}{3600}$, a third $= \frac{1}{216000}$.

Anciently there were no other than Sexagesimals used in astronomical operations, for which reason they are sometimes called *astronomical fractions*, and they are still retained in many cases, as in the divisions of time and of a circle; but decimal arithmetic is now much used ufed in the calculations. Sexagefimals were probably firft ufed for the divifions of a circle, 360, or 6 times 60 making up the whole circumference, on account that 360 days made up the year of the Ancients, in which time the fun was fuppofed to complete his courfe in the circle of the ecliptic.

In thefe fractions, the denominator being always 60, or a multiple of it, it is ufually omitted, and the numerator only written down: thus, $3°\ 45'\ 24''\ 40'''$ &c, is to be read, 3 degrees, 45 minutes, 24 feconds, 40 thirds, &c.

SEXANGLE, in Geometry, a figure having 6 angles, and confequently 6 fides alfo.

SEXTANS, a fixth part of certain things.

The Romans divided their *as*, which was a pound of brafs, into 12 ounces, called *uncia*, from *unum*; and the quantity of 2 ounces was called *fextans*, as being the 6th part of the pound.

SEXTANS was alfo a meafure, which contained 2 ounces of liquor, or 2 cyathi.

SEXTANS, the Sextant, in Aftronomy, a new conftellation, placed acrofs the equator, but on the fouth fide of the ecliptic, and by Hevelius made up of fome unformed ftars, or fuch as were not included in any of the 48 old conftellations. In Hevelius's catalogue it contains 11 ftars, but in the Britannic catalogue 41.

SEXTANT, denotes the 6th part of a circle, or an arch containing 60 degrees.

SEXTANT is more particularly ufed for an aftronomical inftrument. It is made like a quadrant, excepting that its limb only contains 60 degrees. Its ufe and application are the fame with thofe of the QUADRANT; which fee.

SEXTARIUS, an ancient Roman meafure, containing 2 cotylæ, or 2 heminæ.

SEXTILE, the afpect or pofition of two planets, when they are diftant the 6th part of the circle, viz, 2 figns or 60 degrees; and it is marked thus ✳.

SEXTUPLE, denotes 6 fold in general. But in mufic it denotes a mixed fort of triple time, which is beaten in double time.

SHADOW, *Shade*, in Optics, a certain fpace deprived of light, or where the light is weakened by the interpofition of fome opaque body before the luminary.

The doctrine of Shadows makes a confiderable article in optics, aftronomy, and geography; and is the general foundation of dialling.

As nothing is feen but by light, a mere fhadow is invifible; and therefore when we fay we fee a fhadow, we mean, partly that we fee bodies placed in the Shadow, and illuminated by light reflected from collateral boies, and partly that we fee the confines of the light.

When the opaque body, that projects the Shadow, is perpendicular to the horizon, and the plane it is projected on is horizontal, the Shadow is called a *right* one: fuch as the Shadows of men, trees, buildings, mountains, &c. But when the body is placed parallel to the horizon, it is called a *verfed Shadow*; as the arms of a man when ftretched out, &c.

Laws of the Projection of Shadows.

1. Every opaque body projects a Shadow in the

same direction with the rays of light; that is, towards the part oppofite to the light. Hence, as either the luminary or the body changes place, the Shadow likewife changes its place.

2. Every opaque body projects as many Shadows as there are luminaries to enlighten it.

3. As the light of the luminary is more intenfe, the fhadow is the deeper. Hence, the intenfity of the Shadow is meafured by the degrees of light that fpace is deprived of. In reality, the Shadow itfelf is not deeper; but it appears fo, becaufe the furrounding bodies are more intenfely illuminated.

4. When the luminous body and opaque one are equal, the Shadow is always of the fame breadth with the opaque body. But when the luminous body is the larger, the Shadow grows always lefs and lefs, the farther from the body. And when the luminous body is the fmaller of the two, the Shadow increafes always the wider, the farther from the body. Hence, the Shadow of an opaque globe is, in the firft cafe a cylinder, in the fecond cafe it is a cone verging to a point, and in the third cafe a truncated cone that enlarges ftill the more the farther from the body. Alfo, in all thefe cafes, a tranfverfe Section of the Shadow, by a plane, is a circle, refpectively, in the three cafes, equal, lefs, or greater than a great circle of the globe.

5. To find the length of the Shadow, or the axis of the fhady cone, projected by a fphere, when it is illuminated by a larger one; the diameters and diftance of the two fpheres being known. Let C and D be the

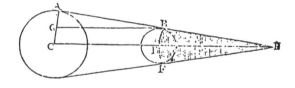

centres of the two fpheres, CA the femidiameter of the larger, and DB that of the fmaller, both perpendicular to the fide of the conical Shadow BEF, whofe axis is DE, continued to C; and draw BG parallel to the fame axis. Then, the two triangles AGB and BDE being fimilar, it will be AG : GB or CD :: BD : DE, that is, as the difference of the femidiameters is to the diftance of the centres, fo is the femidiameter of the opaque fphere to the axis of the Shadow, or the diftance of its vertex from the faid opaque fphere.

Ex. gr. If BD = 1 be the femidiameter of the earth, and AC = 101 the mean femidiameter of the fun, alfo their diftance CD or GB = 24000; then as 100 : 24000 :: 1 : 240 = DE, which is the mean height of the earth's Shadow, in femidiameters of the bafe.

6. To find the length of the fhadow AC projected by an opaque body AB; having given the altitude of the luminary, for ex. of the fun, above the horizon, viz, the angle C, and the height of the object AB. Here the proportion is, as tang. ∠ C : radius :: AB : AC.

Or, if the length of the Shadow AC be given, to find the height AB, it will be,

as radius : tang. ∠ C :: AC : AB.

Or,

Or, if the length of the Shadow AC, and of the object AB, be given, to find the fun's altitude above the horizon, or the angle at C. It will be, as AC : AB : : radius : tang. ∠ C fought.

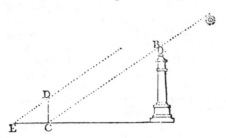

7. To meafure the height of any object, ex. gr. a tower AB, by means of its fhadow projected on an horizontal plane.—At the extremity of the fhadow, at C, erect a ftick or pole CD, and meafure the length of its fhadow CE ; alfo meafure the length of the Shadow AC of the tower. Then, by fimilar triangles, it will be, as EC : CD : : CA : AB. So if EC = 10 feet, CD = 6 feet, and CA = 95 feet; then as 10 : 6 : : 95 : 57 feet = AB, the height of the tower fought.

SHADOW, in Geography. The inhabitants of the earth are divided, with refpect to their fhadows, into ASCII, AMPHISCII, HETEROSCII, and PERISCII. See thefe terms in their places.

SHADOW, in Perfpective, is of great ufe in this art. —Having given the appearance of an opaque body, and a luminous one, whofe rays diverge, as a candle, or lamp, &c ; to find the juft appearance of the Shadow, according to the laws of perfpective. The method is this : From the luminous body, which is here confidered as a point, let fall a perpendicular to the perfpective plane or table ; and from the feveral angles, or raifed points of the body, let fall perpendiculars to the fame plane ; then connect the points on which thefe latter perpendiculars fall, by right lines, with the point on which the firft falls ; continuing thefe lines beyond the fide oppofite to the luminary, till they meet with as many other lines drawn from the centre of the luminary through the faid angles or raifed points ; fo fhall the points of interfection of thefe lines be the extremes or bounds of the Shadow.

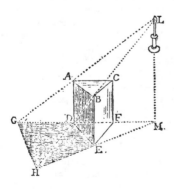

For Example, to project the appearance of the Shadow of a prifm ABCDEF, fcenographically delineated. Here M is the place of the perpendicular of the light L, and D, E, F thofe of the raifed points A, B, C, of the prifm ; therefore, draw MEH, MDG, &c, and LBH, LAG, &c, which will give DEGH &c for the appearance of the Shadow.

As for thofe Shadows that are intercepted by other objects, it may be obferved, that when the Shadow of a line falls upon any object, it muft neceffarily take the form of that object. If it fall upon another plane, it will be a right line ; if upon a globe, it will be circular ; and if upon a cylinder or cone, it will be circular, or oval, &c. If the body intercepting it be a plane, whatever be the fituation of it, the fhadow falling upon it might be found by producing that plane till it intercepted the perpendicular let fall upon it from the luminous body ; for then a line drawn from that point would determine the Shadow, juft as if no other plane had been concerned. But the appearance of all thefe Shadows may be drawn with lefs trouble, by firft drawing it through thefe intercepted objects, as if they had not been in the way, and then making the Shadow to afcend perpendicularly up every perpendicular plane, and obliquely on thofe that are fituated obliquely, in the manner defcribed by Dr. Prieftley, in his Perfpective, pa. 73 &c.

Here we may obferve in general, that fince the Shadows of all objects which are caft upon the ground, will vanifh into the horizontal line ; fo, for the fame reafon, the vanifhing points of all Shadows, which are caft upon any inclined or other plane, will be fomewhere in the vanifhing line of that plane.

When objects are not fuppofed to be viewed by the light of the fun, or of a candle, &c, but only in the light of a cloudy day, or in a room into which the fun does not fhine, there is no fenfible Shadow of the upper part of the object, and the lower part only makes the neighbouring parts of the ground, on which it ftands, a little darker than the reft. This imperfect obfcure kind of Shadow is eafily made, being nothing more than a fhade on the ground, oppofite to the fide on which the light comes ; and it may be continued to a greater or lefs diftance, according to the fuppofed brightnefs of the light by which it is made. It is in this manner (in order to fave trouble, and fometimes to prevent confufion) that the Shadows in moft drawings are made. On this fubject, fee Prieftley's Perfpect. above quoted ; alfo Kirby's Perfp. book 2, ch. 4.

SHAFT *of a Column*, in Building, is the body of it ; thus called from its ftraightnefs : but by architects more commonly the Euft.

SHAFT is alfo ufed for the fpire of a church fteeple ; and for the fhank or tunnel of a chimney.

SHARP (ABRAHAM), an eminent mathematician, mechanift, and aftronomer, was defcended from an ancient family at Little-Horton, near Bradford, in the Weft Riding of Yorkfhire, where he was born about the year 1651. At a proper age he was put apprentice to a merchant at Manchefter ; but his genius led him fo ftrongly to the ftudy of mathematics, both theoretical and practical, that he foon became uneafy in that fituation of life. By the mutual confent therefore of his mafter and himfelf, though not altogether with that of his father, he quitted the bufinefs of a merchant. Upon this he removed to Liverpool, where he

give

gave himself up wholly to the study of mathematics, astronomy, &c; and where, for a subsistance, he opened a school, and taught writing and accounts, &c.

He had not been long at Liverpool when he accidentally fell in company with a merchant or tradesman visiting that town from London, in whose house it seems the astronomer Mr. Flamsteed then lodged. With the view therefore of becoming acquainted with this eminent man, Mr. Sharp engaged himself with the merchant as a book-keeper. In consequence he soon contracted an intimate acquaintance and friendship with Mr. Flamsteed, by whose interest and recommendation he obtained a more profitable employment in the dock-yard at Chatham; where he continued till his friend and patron, knowing his great merit in astronomy and mechanics, called him to his assistance, in contriving, adapting, and fitting up the astronomical apparatus in the Royal Observatory at Greenwich, which had been lately built, namely about the year 1676; Mr. Flamsteed being then 30 years of age, and Mr. Sharp 25.

In this situation he continued to assist Mr. Flamsteed in making observations (with the mural arch, of 80 inches radius, and 140 degrees on the limb, contrived and graduated by Mr. Sharp) on the meridional zenith distances of the fixed stars, sun, moon, and planets, with the times of their transits over the meridian; also the diameters of the sun and moon, and their eclipses, with those of Jupiter's satellites, the variation of the compass, &c. He assisted him also in making a catalogue of near 3000 fixed stars, as to their longitudes and magnitudes, their right ascensions and polar distances, with the variations of the same while they change their longitude by one degree.

But from the fatigue of continually observing the stars at night, in a cold thin air, joined to a weakly constitution, he was reduced to a bad state of health; for the recovery of which he desired leave to retire to his house at Horton; where, as soon as he found himself on the recovery, he began to fit up an observatory of his own; having first made an elegant and curious engine for turning all kinds of work in wood or brass, with a maundril for turning irregular figures, as ovals, roses, wreathed pillars, &c. Beside these, he made himself most of the tools used by joiners, clockmakers, opticians, mathematical instrument-makers, &c. The limbs or arcs of his large equatorial instrument, sextant, quadrant, &c, he graduated with the nicest accuracy, by diagonal divisions into degrees and minutes. The telescopes he made use of were all of his own making, and the lenses ground, figured, and adjusted with his own hands.

It was at this time that he assisted Mr. Flamsteed in calculating most of the tables in the second volume of his *Historia Cælestis*, as appears by their letters, to be seen in the hands of Mr. Sharp's friends at Horton. Likewise the curious drawings of the charts of all the constellations visible in our hemisphere, with the still more excellent drawings of the planispheres both of the northern and southern constellations. And though these drawings of the constellations were sent to be engraved at Amsterdam by a masterly hand, yet the originals far exceeded the engravings in point of beauty and elegance: these were published by Mr. Flamsteed, and both copies may be seen at Horton.

5

The mathematician meets with something extraordinary in Sharp's elaborate treatise of *Geometry Improved* (in 4to 1717, signed A. S Philomath.), 1st, by a large and accurate table of segments of circles, its construction and various uses in the solution of several difficult problems, with compendious tables for finding a true proportional part; and their use in these or any other tables exemplified in making logarithms, or their natural numbers, to 60 places of figures; there being a table of them for all primes to 1100, true to 61 figures. 2d, His concise treatise of Polyedra, or solid bodies of many bases, both the regular ones and others: to which are added twelve new ones, with various methods of forming them, and their exact dimensions in surds, or species, and in numbers: illustrated with a variety of copper-plates, neatly engraved by his own hands. Also the models of these polyedra he cut out in box wood with amazing neatness and accuracy. Indeed few or none of the mathematical instrument-makers could exceed him in exactly graduating or neatly engraving any mathematical or astronomical instrument, as may be seen in the equatorial instrument above mentioned, or in his sextant, quadrants and dials of various sorts; also in a curious armillary sphere, which, beside the common properties, has moveable circles &c, for exhibiting and resolving all spherical triangles; also his double sector, with many other instruments, all contrived, graduated and finished, in a most elegant manner, by himself. In short, he possessed at once a remarkably clear head for contriving, and an extraordinary hand for executing, any thing, not only in mechanics, but likewise in drawing, writing, and making the most exact and beautiful schemes or figures in all his calculations and geometrical constructions.

The quadrature of the circle was undertaken by him for his own private amusement in the year 1699, deduced from two different series, by which the truth of it was proved to 72 places of figures; as may be seen in the introduction to Sherwin's tables of logarithms; that is, if the diameter of a circle be 1, the circumference will be found equal to 3·14159265358979323 84626433832795028841971693993751058209749 44592307816405, &c. In the same book of Sherwin's may also be seen his ingenious improvements on the making of logarithms, and the constructing of the natural sines, tangents, and secants.

He also calculated the natural and logarithmic sines, tangents, and secants, to every second in the first minute of the quadrant: the laborious investigation of which may probably be seen in the archives of the Royal Society, as they were presented to Mr. Patrick Murdoch for that purpose; exhibiting his very neat and accurate manner of writing and arranging his figures, not to be equalled perhaps by the best penman now living.

The late ingenious Mr. Smeaton says (Philos. Transf. an. 1786, pa. 5, &c):

"In the year 1689, Mr. Flamsteed completed his mural arc at Greenwich; and, in the Prolegomena to his *Historia Cœlestis*, he makes an ample acknowledgment of the particular assistance, care, and industry of Mr. Abraham Sharp; whom, in the month of August 1688, he brought into the observatory, as his amanuensis; and being as Mr. Flamsteed tells us, not only

only a very skilful mathematician, but exceedingly expert in mechanical operations, he was principally employed in the construction of the mural arc ; which in the compass of 14 months he finished, so greatly to the satisfaction of Mr. Flamsteed, that he speaks of him in the highest terms of praise.

" This celebrated instrument, of which he also gives the figure at the end of the Prolegomena, was of the radius of 6 feet 7¼ inches ; and, in like manner as the sextant, was furnished both with screw and diagonal divisions, all performed by the accurate hand of Mr. Sharp. But yet, whoever compares the different parts of the table for conversion of the revolutions and parts of the screw belonging to the mural arc into degrees, minutes, and seconds, with each other, at the same distance from the zenith on different sides ; and with their halves, quarters, &c, will find as notable a disagreement of the screw-work from the hand divisions, as had appeared before in the work of Mr. Tompion : and hence we may conclude, that the method of Dr. Hook, being executed by two such masterly hands as Tompion and Sharp, and found defective, is in reality not to be depended upon in nice matters.

" From the account of Mr. Flamsteed it appears also, that Mr. Sharp obtained the zenith point of the instrument, or line of collimation, by observation of the zenith stars, with the face of the instrument on the east and on the west side of the wall : and that having made the index stronger (to prevent flexure) than that of the sextant, and thereby heavier, he contrived, by means of pulleys and balancing weights, to relieve the hand that was to move it from a great part of its gravity. Mr. Sharp continued in strict correspondence with Mr. Flamsteed as long as he lived, as appeared by letters of Mr. Flamsteed's found after Mr. Sharp's death ; many of which I have seen.

" I have been the more particular relating to Mr. Sharp, in the business of constructing this mural arc ; not only because we may suppose it the first good and valid instrument of the kind, but because I look upon Mr. Sharp to have been the first person that cut accurate and delicate divisions upon astronomical instruments ; of which, independent of Mr. Flamsteed's testimony, there still remain considerable proofs : for, after leaving Mr. Flamsteed, and quitting the department above-mentioned, he retired into Yorkshire, to the village of Little Horton, near Bradford, where he ended his days about the year 1743 (should be, in 1742) ; and where I have seen not only a large and very fine collection of mechanical tools, the principal ones being made with his own hands, but also a great variety of scales and instruments made with them, both in wood and brass, the divisions of which were so exquisite, as would not discredit the first artists of the present times : and I believe there is now remaining a quadrant, of 4 or 5 feet radius, framed of wood, but the limb covered with a brass plate ; the subdivisions being done by diagonals, the lines of which are as finely cut as those upon the quadrants at Greenwich. The delicacy of Mr. Sharp's hand will indeed permanently appear from the copper-plates in a quarto book, published in the year 1718, intituled *Geometry Improved* by A. Sharp, Philomath." (or rather 1717, by A. S. Philomath.)

' whereof not only the geometrical lines upon the plates,

but the whole of the engraving of letters and figures, were done by himself, as I was told by a person in the mathematical line, who very frequently attended Mr. Sharp in the latter part of his life. I therefore look upon Mr. Sharp as the first person that brought the affair of hand division to any degree of perfection."

Mr. Sharp kept up a correspondence by letters with most of the eminent mathematicians and astronomers of his time, as Mr. Flamsteed, Sir Isaac Newton, Dr. Halley, Dr. Wallis, Mr. Hodgson, Mr. Sherwin, &c, the answers to which letters are all written upon the backs, or empty spaces, of the letters he received, in a short-hand of his own contrivance. From a great variety of letters (of which a large chest full remain with his friends) from these and many other celebrated mathematicians, it is evident that Mr. Sharp spared neither pains nor time to promote real science. Indeed, being one of the most accurate and indefatigable computers that ever existed, he was for many years the common resource for Mr. Flamsteed, Sir Jonas Moore, Dr. Halley, and others, in all sorts of troublesome and delicate calculations.

Mr. Sharp continued all his life a bachelor, and spent his time as recluse as a hermit. He was of a middle stature, but very thin, being of a weakly constitution ; he was remarkably feeble the last three or four years before he died, which was on the 18th of July 1742, in the 91st year of his age.

In his retirement at Little Horton, he employed four or five rooms or apartments in his house for different purposes, into which none of his family could possibly enter at any time without his permission. He was seldom visited by any persons, except two gentlemen of Bradford, the one a mathematician, and the other an ingenious apothecary : these were admitted, when he chose to be seen by them, by the signal of rubbing a stone against a certain part of the outside wall of the house. He duly attended the dissenting chapel at Bradford, of which he was a member, every Sunday ; at which time he took care to be provided with plenty of halfpence, which he very charitably suffered to be taken singly out of his hand, held behind him during his walk to the chapel, by a number of poor people who followed him, without his ever looking back, or asking a single question.

Mr. Sharp was very irregular as to his meals, and remarkably sparing in his diet, which he frequently took in the following manner. A little square hole, something like a window, made a communication between the room where he was usually employed in calculations, and another chamber or room in the house where a servant could enter ; and before this hole he had contrived a sliding board : the servant always placed his victuals in this hole, without speaking or making any the least noise ; and when he had a little leisure he visited his cupboard to see what it afforded to satisfy his hunger or thirst. But it often happened, that the breakfast, dinner, and supper have remained untouched by him, when the servant has gone to remove what was left—so deeply engaged had he been in calculations.—Cavities might easily be perceived in an old English oak table where he sat to write, by the frequent rubbing and wearing of his elbows.—*Gutta cavat lapidem, &c.*

By

By Mr. Sharp's epitaph it appears that he was related to archbishop Sharp. And Mr. Sharp the eminent surgeon, who it seems has lately retired from business, is the nephew of our author. Another nephew was the father of Mr. Ramsden, the present celebrated instrument maker, who says that his grand uncle Abraham, our author, was some time in his younger days an exciseman; which occupation he quitted on coming to a patrimonial estate of about 200l. a year.

SHARP, in Music, a kind of artificial note, or character, thus formed ✳: this being prefixed to any note, shews that it is to be sung or played a semitone or half note higher than the natural note is. When a Sharp is placed at the beginning of a stave or movement, it shews that all notes that are found on the same line, or space, throughout, are to be raised half a tone above their natural pitch, unless a natural intervene. When a Sharp occurs accidentally, it only affects as many notes as follow it on the same line or space, without a natural, in the compass of a bar.

SHEAVE, in Mechanics, a solid cylindrical wheel, fixed in a channel, and moveable about an axis, as being used to raise or increase the mechanical powers applied to remove any body.

SHEERS, aboard a ship, an engine used to hoist or displace the lower masts of a ship.

SHEKEL, or SHEKLE, an ancient Hebrew coin and weight, equal to 4 Attic drachmas, or 4 Roman denarii, or 2s. 9½d. sterling. According to father Mersenne, the Hebrew Shekel weighs 268 grains, and is composed of 20 oboli, each obolus weighing 16 grains of wheat.

SHILLING, an English silver coin, equal to 12 pence, or the 20th part of a pound sterling.

This was a Saxon coin, being the 48th part of their pound weight. Its value at first was 5 pence; but it was reduced to 4 pence about a century before the conquest. After the conquest, the French solidus of 12 pence, which was in use among the Normans, was called by the English name of Shilling; and the Saxon Shilling of 4 pence took a Norman name, and was called the *groat*, or *great* coin, because it was the largest English coin then known in England. From this time, the Shilling underwent many alterations.

Many other nations have also their Shillings. The English Shilling is worth about 23 French sols; those of Holland and Germany about half as much, or 11½ sols; those of Flanders about 9. The Dutch Shillings are also called *sols de gros*, because equal to 12 grofs. The Danes have copper Shillings, worth about one fourth of a farthing sterling.

In the time of Edward the 1st, the pound troy was the same as the pound sterling of silver, consisting of 20 Shillings; so that the Shilling weighed the 20th part of a pound, or more than half an ounce troy. But some are of opinion, there were no coins of this denomination, till Henry the 7th, in the year 1504, first coined silver pieces of 12 pence value, which we call Shillings. Since the reign of Elizabeth, a Shilling weighs the 62nd part of a pound troy, or 3 dwts. 20⅗ grs. the pound weight of silver making 62 Shillings. And hence the ounce of silver is worth 5s. 2d. or 5⅙ Shillings.

SHIVERS, in a ship, the seamen's term for those little round wheels, in which the rope of a pully or block runs. They turn with the rope, and have pieces of brass in their centres, into which the pin of the block goes, and on which they turn.

SHORT-SIGHTEDNESS *myopia*, a defect in the conformation of the eye, when the crystalline &c being too convex, the rays that enter the eye are refracted too much, and made to converge too fast, so as to unite before they reach the retina, by which means vision is rendered dim and confused.

It is commonly thought that Short-sightedness wears off in old age, on account of the eye becoming flatter; but Dr. Smith questions whether this be matter of fact, or only hypothesis. It is remarkable that Short-sighted persons commonly write a small hand, and love a small print, because they can see more of it at one view. That it is customary with them not to look at the person they converse with, because they cannot well see the motion of his eyes and features, and are therefore attentive to his words only. That they see more distinctly, and somewhat farther off, by a strong light, than by a weak one; because a strong light causes a contraction of the pupil, and consequently of the pencils, both here and at the retina, which lessens their mixture, and consequently the apparent confusion; and therefore, to see more distinctly, they almost close their eye-lids, for which reason they were anciently called *myopes*. Smith's Optics, vol. 2, Rem. p. 10.

Dr. Jurin observes, that persons who are much and long accustomed to view objects at small distances, as students in general, watchmakers, engravers, painters in miniature, &c, see better at small distances, and worse at great distances, than other people. And he gives the reasons, from the mechanical effect of habit in the eye. Essay on Dist. and Indist. Vision.

The ordinary remedy for Short-sightedness is a concave lens, held before the eye; for this causing the rays to diverge, or at least diminishing much of their convergency, it makes a compensation for the too great convexity of the crystalline. Dr. Hook suggests another remedy; which is to employ a convex glass, in a position between the object and the eye, by means of which, the object may be made to appear at any distance from the eye, and so the eye be made to contemplate the picture in the same manner as if the object itself were in its place. But here unfortunately the image will appear inverted: for this however he has some whimsical expedients; viz, in reading to turn the book upside down, and to learn to write upside down. As to distant objects, the Doctor asserts, from his own experience, that with a little practice in contemplating inverted objects, one gets as good an idea of them as if seen in their natural posture.

SHOT, in the Military Art, includes all sorts of balls or bullets for fire arms, from the cannon to the pistol. As to those for mortars, they are usually called shells.

Shot are mostly of a round form, though there are other shapes. Those for cannon are of iron; but those for muskets and pistols are of lead.

Cannon shot and shells are usually set up in piles, or heaps, tapering from the base towards the top; the base being either a triangle, a square, or a rectangle;

from

from which the number in the pile is eafily computed. See PILE.

The weight and dimenfions of balls may be found, the one from the other, whether they are of iron or of lead. Thus,

The weight of an iron ball of 4 inches diameter, is 9lb, and becaufe the weight is as the cube of the diameter, therefore as $4^3 : 9 :: d^3 : \frac{9}{64}$, $d^3 = w$, the weight of the iron ball whofe diameter is d; that is, $\frac{9}{64}$ of the cube of its diameter. And, converfely, if the weight be given, to find the diameter, it will be $\sqrt[3]{\frac{64}{9} w} = d$; that is, take $\frac{64}{9}$ or $7\frac{1}{9}$ of the weight, and the cube root of that will be the diameter of the iron ball.

For leaden balls; one of $4\frac{1}{4}$ inches diameter weighs 17 pounds; therefore as the cube of $4\frac{1}{4}$ is to 17, or nearly as $9 : 2 : : d^3 : \frac{2}{9} d^3 = w$, the weight of the leaden ball whofe diameter is d, that is, $\frac{2}{9}$ of the cube of the diameter. On the contrary, if the weight be given, to find the diameter, it will be $\sqrt[3]{\frac{9}{2} w} = d$; that is, $\frac{9}{2}$ or $4\frac{1}{2}$ of the weight, and the cube root of the product. See my Conic Sections and Select Exercifes, pa. 141.

SHOULDER *of a Baftion*, in Fortification, is the angle where the face and the flank meet.

SHOULDERING, in Fortification. See *Epaulement*.

SHWAN-*pan*, a Chinefe inftrument, compofed of a number of wires, with beads upon them, which they move backwards and forwards, and which ferves to affift them in their computations. See ABACUS.

SIDE, *latus*, in Geometry. The fide of a figure is a line making part of the periphery of any fuperficial figure, viz, a part between two fucceffive angles.

In triangles, the fides are alfo called *legs*. In a right-angled triangle, the two fides that include the right angle, are called *catheti*, or fometimes the *bafe* and *perpendicular*; and the third fide, the *hypothenufe*.

SIDE *of a Polygonal Number*, is the number of terms in the arithmetical progreffion that are fummed up to form the number.

SIDE *of a Power*, is what is ufually called the root or radix.

SIDES of *Horn-works, Crown-works, Double-tenailles*, &c, are the ramparts and parapets which inclofe them on the right and left, from the gorge to the head.

SIDEREAL, or SIDERIAL, fomething relating to the ftars. As Sidereal year, day, &c, being thofe marked out by the ftars.

SIDEREAL *Year*. See YEAR.

SIDEREAL *Day*, is the time in which any ftar appears to revolve from the meridian to the meridian again; which is 23 hours 56′ 4″ 6‴ of mean folar time; there being 366 Sidereal days in a year, or in the time of 365 diurnal revolutions of the fun; that is, exactly, if the equinoctial points were at reft in the heavens. But the equinoctial points go backward, with refpect to the ftars, at the rate of 50″ of a degree in a Julian year; which caufeth the ftars to have an apparent pro-

greffive motion eaftward 50″ in that time. And as the fun's mean motion in the ecliptic is only 11 figns 29° 45′ 40″ 15‴ in 365 days, it follows, that at the end of that time he will be 14′ 19″ 45‴ fhort of that point of the ecliptic from which he fet out at the beginning; and the ftars will be advanced 50″ of a degree with refpect to that point.

Confequently, if the fun's centre be on the meridian with any ftar on any given day of the year, that ftar will be 14′ 19″ 45‴ + 50″ or 15′ 9″ 45‴ eaft of the fun's centre, on the 365th day afterward, when the fun's centre is on the meridian; and therefore that ftar will not come to the meridian on that day till the fun's centre has paffed it by 1′ 0″ 38‴ 57⁗ of mean folar time; for the fun takes fo much time to go through an arc of 15′ 9″ 45‴; and then, in 365ᵈᵃ 0ʰ 1′ 0″ 38‴ 57⁗ the ftar will have juft completed its 366th revolution to the meridian.

In the following table, of Sidereal revolutions, the firft column contains the number of revolutions of the ftars; the others next it fhew the times in which thefe revolutions are made, as fhewn by a well regulated clock; and thofe on the right hand fhew the daily accelerations of the ftars, that is, how much any ftar gains upon the time fhewn by fuch a clock, in the corresponding revolutions.

Revol. of the ftars.	Times in which the revolutions are made.					Accelerations of the ftars.				
	da.	ho.	m.	fec.	th.	fo.	ho.	m.	fec.	th.	fo.
1	0	23	56	4	6	0	0	3	55	54	0
2	1	23	52	8	12	1	0	7	51	47	59
3	2	23	48	12	18	1	0	11	47	41	59
4	3	23	44	16	24	2	0	15	43	35	58
5	4	23	40	20	30	2	0	19	39	29	58
6	5	23	36	24	36	3	0	23	35	23	57
7	6	23	32	28	42	3	0	27	31	17	57
8	7	23	28	32	48	4	0	31	27	11	56
9	8	23	24	36	54	4	0	35	23	5	56
10	9	23	20	41	0	5	0	39	18	59	55
11	10	23	16	45	6	5	0	43	14	53	55
12	11	23	12	49	12	6	0	47	10	47	54
13	12	23	8	53	18	6	0	51	6	41	54
14	13	23	4	57	24	7	0	55	2	35	53
15	14	23	1	1	30	7	0	58	58	29	53
16	15	22	57	5	36	8	1	2	54	23	52
17	16	22	53	9	42	8	1	6	50	17	52
18	17	22	49	13	48	9	1	10	46	11	51
19	18	22	45	17	54	9	1	14	42	5	51
20	19	22	41	22	0	10	1	18	37	59	50
21	20	22	37	26	6	10	1	22	33	53	50
22	21	22	33	30	12	11	1	26	29	47	49
23	22	22	29	34	18	11	1	30	25	41	49
24	23	22	25	38	24	12	1	34	21	35	48
25	24	22	21	42	30	12	1	38	17	29	48
26	25	22	17	46	36	13	1	42	13	23	47
27	26	22	13	50	42	13	1	46	9	17	47
28	27	22	9	54	48	14	1	50	5	11	46
29	28	22	5	58	54	14	1	54	1	5	46
30	29	22	2	3	0	15	1	57	56	59	45
40	39	21	22	44	0	19	2	37	15	59	41
50	49	20	43	25	0	24	3	16	34	59	36
100	99	17	26	50	0	48	6	33	9	59	12
200	199	10	53	40	1	37	13	6	19	58	23
300	299	4	20	30	2	25	19	39	29	57	35
360	359	0	24	36	2	54	23	35	23	57	6
365	364	0	4	56	32	56	23	55	3	27	4
366	365	0	1	0	38	57	23	58	59	21	3

3 M

This

This table will not differ the 2799360000000th part of a second of time from the truth in a whole year. It was calculated by Mr. Ferguson; and it is the only table of the kind in which the recession of the equinoctial points has been taken into the calculation.

SIDUS *Georgium*, a new primary planet, discovered by Dr. Herschel at Bath, in the night of March 13, 1781. It is sometimes also called the *Georgian Planet*, and the *New Planet*, from its having been newly or lately discovered, also *Herschel's Planet*, from the name of its discoverer, and the *Planet Herschel*, or simply *Herschel*, by which name it is distinguished by the astronomers of almost all foreign nations. The planet is denoted by this character ♅, a Roman H as the initial of the name, the horizontal bar being crossed by a perpendicular line, forming a kind of cross, the emblem of Christianity, meaning thereby perhaps that its discovery was made by a Christian, or since the birth of Christ, as all the other planets were discovered long before that period

This planet is the remotest of all those that are yet known, though not the largest, being in point of magnitude less than Saturn and Jupiter. Its light, says Dr. Herschel, of a blueish-white colour, and its brilliancy between that of Venus and the moon. With a telescope that magnifies about 300 times, it appears to have a very well defined visible disk; but with instruments of a small power, it can hardly be distinguished from a fixed star of between the 6th and 7th magnitude. In a very fine clear night, when the moon is absent, a good eye will perceive it without a telescope.

From the observations and calculations of Dr. Herschel and other astronomers, the elements and dimensions &c of this planet, have been collected as below.

Place of the node - - - - - 2ˢ 11° 49′ 30″
Place of the aphelion in 1795 11   23   33   55
Inclination of the orbit - - ·   ·   43   35
Time of the perihelion passage Sep. 7, 1799.
Eccentricity of the orbit - - ·8203
Half the greater axis - - - - 19·0818 of Earth's dist.
Revolution - - - - - - - - 83⅓ sidereal years
Diameter of the planet - - - 34217 miles
Propor. of diam. to the earth's 4·3177 to 1
Its bulk to the earth's - - - 80·4926 to 1
Its density as - - - - - - - - ·2204 to 1
Its quantity of matter - - - 17·7406 to 1

And heavy bodies fall on its surface 18 feet 8 inches in one second of time.

SIGN, in Algebra, a symbol or CHARACTER.

SIGNS, *like, positive, negative, radical*, &c. See the adjectives.

SIGN, in Astronomy, a 12th part of the ecliptic, or zodiac; or a portion containing 30 degrees of the same.

The ancients divided the zodiac into 12 segments, called Signs; commencing at the point where the ecliptic and equinoctial intersect, and so counting forward from west to east, according to the course of the sun: these Signs they named from the 12 constellations which possessed those segments in the time of Hipparchus. But the constellations have since so changed their places, by the precession of the equinox, that Aries is now found in the sign called Taurus, and Taurus in that of Gemini, &c.

The names, and characters, of the 12 Signs, and their order, are as follow: Aries ♈, Taurus ♉, Gemini ♊, Cancer ♋, Leo ♌, Virgo ♍, Libra ♎, Scorpio ♏, Sagittarius ♐, Capricornus ♑, Aquarius ♒, Pisces ♓; each of which, with the stars in them, see under its proper article, ARIES, TAURUS, &c.

The Signs are distinguished, with regard to the season of the year when the sun is in them, into vernal, æstival, autumnal, and brumal.

Vernal or *Spring* SIGNS, are Aries, Taurus, Gemini.
Æstival or *Summer* SIGNS, are Cancer, Leo, Virgo.
Autumnal SIGNS, are Libra, Scorpio, Sagittary.
Brumal or *Winter* SIGNS, are Capricorn, Aquarius, Pisces.

The vernal and summer Signs are also called *northern* Signs, because they are on the north side of the equinoctial; and the autumnal and winter Signs are called *southern* ones, because they are on the south side of the same.

The Signs are also distinguished into *ascending* and *descending*, according as they are ascending toward the north, or descending toward the south. Thus, the

Ascending SIGNS, are the winter and spring signs, or those six from the winter solstice to the summer solstice, viz, the Signs Capricorn, Aquarius, Pisces, Aries, Taurus, Gemini. And the

Descending SIGNS are the summer and autumn Signs, or the Signs Cancer, Leo, Virgo, Libra, Scorpio, Sagittary.

SIGNS, *Fixed, Masculine*, &c; see the adjectives.

SILLON, in Fortification, an elevation of earth, made in the middle of the moat, to fortify it, when too broad. It is more usually called the Envelope.

SIMILAR, in Arithmetic and Geometry, the same with like. Similar things have the same disposition or conformation of parts, and differ in nothing but as to their quantity or magnitude; as two squares, or two circles, &c.

In Mathematics, Similar parts, as A, *a*, have the same ratio to their wholes B, *b*; and if the wholes have the same ratio to the parts, the parts are Similar.

SIMILAR *angles*, are also equal angles.

SIMILAR *arcs*, of circles, are such as are like parts of their whole peripheries. And, in general, similar arcs of any like curves, are the like parts of the wholes.

SIMILAR *bodies*, in Natural Philosophy, are such as have their particles of the same kind and nature one with another.

SIMILAR *Curves*. Two segments of two curves are said to be Similar when, any right-lined figure being inscribed within one of them, we can inscribe always a Similar rectilineal figure in the other.

SIMILAR *Conic Sections*, are such as are of the same kind, and have their principal axes and parameters proportional. So, two ellipses are figures of the same kind, but they are not Similar unless the axes of the one have the same ratio as the axes of the other. And the same of two hyperbolas, or two parabolas. And generally, those curves are Similar, that are of the same kind, and have their corresponding dimensions in the same ratio.—All circles are Similar figures.

SIMILAR *Diameters of Conic Sections*, are such as make equal angles with their ordinates.

SIMILAR *Figures*, or plane figures, are such as have all their angles equal respectively, each to each, and their

their fides about the equal angles proportional. And the fame of Similar polygons.—Similar plane figures have their areas or contents, in the duplicate ratio of their like fides, or as the fquares of thofe fides.

SIMILAR *Plane Numbers*, are fuch as may be ranged into the form of Similar rectangles ; that is, into rectangles whofe fides are proportional. Such are 12 and 48 ; for the fides of 12 are 6 and 2, and the fides of 48 are 12 and 4, which are in the fame proportion, viz, 6 : 2 :: 12 : 4.

SIMILAR *Polygons*, are polygons of the fame number of angles, and the angles in the one equal feverally to the angles in the other, alfo the fides about thofe angles proportional.

SIMILAR *Rectangles*, are thofe that have their fides about the like angles proportional.—All fquares are Similar.

SIMILAR *Segments of circles*, are fuch as contain equal angles.

SIMILAR *Solids*, are fuch as are contained under the fame number of Similar planes, alike fituated.—Similar folids are to each other as the cubes of their like linear dimenfions.

SIMILAR *Solid Numbers*, are thofe whofe little cubes may be fo ranged, as to form Similar parallelopipedons.

SIMILAR *Triangles*, are fuch as are eq iangular ones, or have all their three angles refpectively equal in each triangle For it is fufficient for triangles to be fimilar, that they be equiangular, becaufe that being equiangular, they neceffarily have their fides proportional, which is a condition of Similarity in all figures. As to other figures, having more fides than three, they may be equiangular, without having their fides proportional, and therefore without being fimilar.—Similar triangles are as the fquares of their like fides.

SIMILITUDE, in Arithmetic and Geometry, denotes the relation of things that are fimilar to each other.

Euclid and, after him, moft other authors, demonftrate every thing in geometry from the principle of congruity. Wolfius, inftead of it, fubftitutes that of Similitude ; which, he fays, was communicated to him by Leibnitz, and which he finds of very confiderable ufe in geometry, as ferving to demonftrate many things directly, which are only demonftrable from the principle of congruity in a very tedious manner.

SIMPLE, fomething not mixed, or not compounded ; in which fenfe it ftands oppofed to compound.

The elements are Simple bodies, from the compofition of which there refult all forts of mixed bodies.

SIMPLE *Equation*, *Fraction*, and *Surd*. See the fubftantives.

SIMPLE *Quantities*, in Algebra, are thofe that confift of one term only ; as *a*, or — *ab*, or 3 *abc*. In oppofition to compound quantities, which confift of two or more terms ; as *a* + *b*, or *a* + 2 *b* — 3 *ac*.

SIMPLE *Flank*, and *Tenaille*, in Fortification. See the fubftantives.

SIMPLE *Machine*, *Motion*, *Pendulum*, and *Wheel*, in Mechanics. See the fubftantives.

The fimpleft machines are always the moft efteemed. And in geometry, the moft fimple demonftrations are the beft.

SIMPLE *Problem*, in Mathematics. See LINEAR *Problem*.

SIMPLE *Vifion*, in Optics. See VISION.

SIMPSON (THOMAS), F. R. S. a very eminent mathematician, and profeffor of Mathematics in the Royal Military Academy at Woolwich, was born at Market Bofworth, in the county of Leicefter, the 20th of Auguft 1710. His father was a ftuff weaver in that town ; and though in tolerable circumftances, yet, intending to bring up his fon Thomas to his own bufinefs, he took fo little care of his education, that he was only taught to read Englifh. But nature had furnifhed him with talents and a genius for far other purfuits ; which led him afterwards to the higheft rank in the mathematical and philofophical fciences.

Young Simpfon very foon gave indications of his turn for ftudy in general, by eagerly reading all books he could meet with, teaching himfelf to write, and embracing every opportunity he could find of deriving knowledge from other perfons. His father obferving him thus to neglect his bufinefs, by fpending his time in reading what he thought ufelefs books, and following other fuch like purfuits, ufed all his endeavours to check fuch proceedings, and to induce him to follow his profeffion with fteadinefs and better effect. But after many ftruggles for this purpofe, the differences thus produced between them at length rofe to fuch a height, that our author quitted his father's houfe entirely.

Upon this occafion he repaired to Nuneaton, a town at a fmall diftance from Bofworth, where he went to lodge at the houfe of a taylor's widow, of the name of Swinfield, who had been left with two children, a daughter and a fon, by her hufband, of whom the fon, who was the younger, being but about two years older than Simpfon, had become his intimate friend and companion. And here he continued fome time, working at his trade, and improving his knowledge by reading fuch books as he could procure.

Among feveral other circumftances which, long before this, gave occafion to fhew our author's early thirft for knowledge, as well as proving a frefh incitement to acquire it, was that of a large folar eclipfe, which took place on the 11th day of May, 1724. This phenomenon, fo awful to many who are ignorant of the caufe of it, ftruck the mind of young Simpfon with a ftrong curiofity to difcover the reafon of it, and to be able to predict the like furprifing events. It was however feveral years before he could obtain his defire, which at length was gratified by the following accident. After he had been fome time at Mrs. Swinfield's, at Nuneaton, a travelling pedlar came that way, and took a lodging at the fame houfe, according to his ufual cuftom. This man, to his profeffion of an itinerant merchant, had joined the more profitable one of a fortune-teller, which he performed by means of judicial aftrology. Every one knows with what regard perfons of fuch a caft are treated by the inhabitants of country villages ; it cannot be furprifing therefore that an untutored lad of nineteen fhould look upon this man as a prodigy, and, regarding him in this light, fhould endeavour to ingratiate himfelf into his favour ; in which he fucceeded fo well, that the fage was no lefs taken with the quick natural parts and genius of his new acquaintance. The pedlar, intending a journey to Briftol fair, left in the hands of young Simpfon an old edition of Cocker's Arithmetic, to which was fubjoined a fhort Appendix on Algebra, and a book upon Genitures, by Partridge the almanac maker. Thefe books he had perufed to fo good purpofe, during the abfence of his friend, as to excite his amazement upon his return; in consequence of which he set himself about erecting a genethliacal figure, in order to a presage of Thomas's future fortune.

This position of the heavens having been maturely considered *secundum artem*, the wizard, with great confidence, pronounced, that, " within two years time Simpson would turn out a greater man than himself!"

In fact, our author profited so well by the encouragement and assistance of the pedlar, afforded him from time to time when he occasionally came to Nuneaton, that, by the advice of his friend, he at length made an open profession of casting nativities himself; from which, together with teaching an evening school, he derived a pretty pittance, so that he greatly neglected his weaving, to which indeed he had never manifested any great attachment, and soon became the oracle of Nuneaton, Bosworth, and the environs. Scarce a courtship advanced to a match, or a bargain to a sale, without previously consulting the infallible Simpson about the consequences. But as to helping people to stolen goods, he always declared that above his skill; and over life and death he declared he had no power: all those called *lawful questions* he readily resolved, provided the persons were certain as to the horary *data* of the horoscope: and, he has often declared, with such success, that if from very cogent reasons he had not been thoroughly convinced of the vain foundation and fallaciousness of his art, he never should have dropt it, as he afterwards found himself in conscience bound to do.

About this time he married the widow Swinfield, in whose house he lodged, though she was then almost old enough to be his grandmother, being upwards of fifty years of age. After this the family lived comfortably enough together for some short time, Simpson occasionally working at his business of a weaver in the daytime, and teaching an evening school or telling fortunes at night; the family being also farther assisted by the labours of young Swinfield, who had been brought up in the profession of his father.

But this tranquillity was soon interrupted, and our author driven at once from his home and the profession of astrology, by the following accident. A young woman in the neighbourhood had long wished to hear or know something of her lover, who had been gone to sea; but Simpson had put her off from time to time, till the girl grew at last so importunate, that he could deny her no longer. He asked her if she would be afraid if he should raise the devil, thinking to deter her; but she declared she feared neither ghost nor devil: so he was obliged to comply. The scene of action pitched upon was a barn, and young Swinfield was to act the devil or ghost; who being concealed under some straw in a corner of the barn, was, at a signal given, to rise slowly out from among the straw, with his face marked so that the girl might not know him. Every thing being in order, the girl came at the time appointed; when Simpson, after cautioning her not to be afraid, began muttering some mystical words, and chalking round about them, till, on the signal given, up rises the taylor slow and solemn, to the great terror of the poor girl, who, before she had seen half his shoulders, fell into violent fits, crying out it was the very image of her lover; and the effect upon her was so dreadful,

that it was thought either death or madness must be the consequence. So that poor Simpson was obliged immediately to abandon at once both his home and the profession of a conjuror.

Upon this occasion it would seem he fled to Derby, where he remained some two or three years, viz, from 1733 till 1735 or 1736; instructing pupils in an evening school, and working at his trade by day.

It would seem that Simpson had an early turn for versifying, both from the circumstance of a song written here in favour of the Cavendish family, on occasion of the parliamentary election at that place, in the year 1733; and from his first two mathematical questions that were published in the Ladies Diary, which were both in a set of verses, not ill written for the occasion. These were printed in the Diary for 1736, and therefore must at latest have been written in the year 1735. These two questions, being at that time pretty difficult ones, shew the great progress he had even then made in the mathematics; and from an expression in the first of them, viz, where he mentions his residence as being in latitude 52°, it appears he was not then come up to London, though he must have done so very soon after.

Together with his astrology, he had soon furnished himself with arithmetic, algebra, and geometry sufficient to be qualified for looking into the Ladies Diary (of which he had afterwards for several years the direction), by which he came to understand that there was a still higher branch of the mathematical knowledge than any he had yet been acquainted with; and this was the method of *Fluxions*. But our young analyst was quite at a loss to discover any English author who had written on the subject, except Mr. Hayes; and his work being a folio, and then pretty scarce, exceeded his ability of purchasing: however an acquaintance lent him Mr. Stone's *Fluxions*, which is a translation of the *Marquis de l'Hospital's Analyse des Infiniment Petits*: by this one book, and his own penetrating talents, he was, as we shall see presently, enabled in a very few years to compose a much more accurate treatise on this subject than any that had before appeared in our language.

After he had quitted astrology and its emoluments, he was driven to hardships for the subsistence of his family, while at Derby, notwithstanding his other industrious endeavours in his own trade by day, and teaching pupils at evenings. This determined him to repair to London, which he did in 1735 or 1736.

On his first coming to London, Mr. Simpson wrought for some time at his business in Spitalfields, and taught mathematics at evenings, or any spare hours. His industry turned to so good account, that he returned down into the country, and brought up his wife and three children, she having produced her first child to him in his absence. The number of his scholars increasing, and his abilities becoming in some measure known to the public, he was encouraged to make proposals for publishing by subscription, A new Treatise of Fluxions: wherein the Direct and Inverse Methods are demonstrated after a new, clear, and concise Manner, with their Application to Physics and Astronomy: also the Doctrine of Infinite Series and Reverting Series universally, are amply explained, Fluxionary and Exponential Equations solved: together with a variety of new and curious Problems.

When

When Mr. Simpson firſt propoſed his intentions of publiſhing ſuch a work, he did not know of any Engliſh book, founded on the true principles of Fluxions, that contained any thing material, eſpecially the practical part; and though there had been ſome very curious things done by ſeveral learned and ingenious gentlemen, the principles were nevertheleſs left obſcure and defective, and all that had been done by any of them in *infinite ſeries*, very inconſiderable.

The book was publiſhed in 4to, in the year 1737, although the author had been frequently interrupted from furniſhing the preſs ſo faſt as he could have wiſhed, through his unavoidable attention to his pupils for his immediate ſupport. The principles of fluxions treated of in this work, are demonſtrated in a method accurately true and genuine, not eſſentially different from that of their great inventor, being entirely expounded by finite quantities.

In 1740, Mr. Simpſon publiſhed a Treatiſe on The Nature and Laws of Chance, in 4to. To which are annexed, Full and clear Inveſtigations of two important Problems added in the 2d Edition of Mr. De Moivre's Book on Chances, as alſo two New Methods for the Summation of Series.

Our author's next publication was a 4to volume of Eſſays on ſeveral curious and intereſting Subjects in Speculative and Mixed Mathhmatics; printed in the ſame year 1740: dedicated to Francis Blake, Eſq. ſince Fellow of the Royal Society, and our author's good friend and patron.—Soon after the publication of this book, he was choſen a member of the Royal Academy at Stockholm.

Our author's next work was, The Doctrine of Annuities and Reverſions, deduced from general and evident Principles: with uſeful Tables, ſhewing the Values of Single and Joint Lives, &c. in 8vo, 1742. This was followed in 1743, by an Appendix containing ſome Remarks on a late book on the ſame Subject (by Mr. Abr. De Moivre, F. R. S.) with Anſwers to ſome perſonal and malignant Repreſentations in the Preface thereof. To this anſwer Mr. De Moivre never thought fit to reply. A new edition of this work has lately been publiſhed, augmented with the tract upon the ſame ſubject that was printed in our author's Select Exerciſes.

In 1743 alſo was publiſhed his Mathematical Diſſertations on a variety of Phyſical and Analytical Subjects, in 4to; containing, among other particulars,

A Demonſtration of the true Figure which the Earth, or any Planet, muſt acquire from its rotation about an Axis. A general Inveſtigation of the Attraction at the Surfaces of Bodies nearly ſpherical. A Determination of the Meridional Parts, and the Lengths of the ſeveral Degrees of the Meridian, according to the true Figure of the Earth. An Inveſtigation of the Height of the Tides in the Ocean. A new Theory of Aſtronomical Refractions, with exact Tables deduced from the ſame. A new and very exact Method for approximating the Roots of Equations in Numbers; which quintuples the number of Places at each Operation. Several new Methods for the Summation of Series. Some new and very uſeful Improvements in the Inverſe Method of Fluxions. The work being dedicated to Martin Folkes, Eſq. Preſident of the Royal Society.

His next book was A Treatiſe of Algebra, wherein the fundamental Principles are demonſtrated, and applied to the Solution of a variety of Problems. To which he added, The Conſtruction of a great Number of Geometrical Problems, with the Method of reſolving them numerically.

This work, which was deſigned for the uſe of young beginners, was inſcribed to William Jones, Eſq. F. R. S. and printed in 8vo, 1745. And a new edition appeared in 1755, with additions and improvements; among which was a new and general method of reſolving all Biquadratic Equations, that are complete, or having all their terms. This edition was dedicated to James Earl of Morton, F. R. S. Mr. Jones being then dead. The work has gone through ſeveral other editions ſince that time: the 6th, or laſt, was in 1790.

His next work was, " Elements of Geometry, with their Application to the Menſuration of Superficies and Solids, to the Determination of Maxima and Minima, and to the Conſtruction of a great Variety of geometrical Problems:" firſt publiſhed in 1747, in 8vo. And a ſecond edition of the ſame came out in 1760, with great alterations and additions, being in a manner a new work, deſigned for young beginners, particularly for the gentlemen educated at the Royal Military Academy at Woolwich, and dedicated to Charles Frederick, Eſq. Surveyor General of the Ordnance. And other editions have appeared ſince.

Mr. Simpſon met with ſome trouble and vexation in conſequence of the firſt edition of his Geometry. Firſt, from ſome reflections made upon it, as to the accuracy of certain parts of it, by Dr. Robert Simſon, the learned profeſſor of mathematicks in the univerſity of Glaſgow, in the notes ſubjoined to his edition of Euclid's Elements. This brought an anſwer to thoſe remarks from Mr. Simpſon, in the notes added to the 2d edition as above; to ſome parts of which Dr. Simſon again replied in his notes on the next edition of the ſaid Elements of Euclid.

The ſecond was by an illiberal charge of having ſtolen his Elements from Mr. Muller, the profeſſor of fortification and artillery at the ſame academy at Woolwich, where our author was profeſſor of geometry and mathematics. This charge was made at the end of the preface to Mr. Muller's Elements of Mathematics, in two volumes, printed in 1748; which was fully refuted by Mr. Simpſon in the preface to the 2d edition of his Geometry.

In 1748 came out Mr. Simpſon's Trigonometry, Plane and Spherical, with the Conſtruction and Application of Logarithms, 8vo. This little book contains ſeveral things new and uſeful.

In 1750 came out, in two volumes, 8vo, The Doctrine and Application of Fluxions, containing, beſides what is common on the Subject, a Number of new Improvements in the Theory, and the Solution of a Variety of new and very intereſting Problems in different Branches of the Mathematics.—In the preface the author offers this to the world as a new book, rather than a ſecond edition of that which was publiſhed in 1737, in which he acknowledges, that, beſides errors of the preſs, there are ſeveral obſcurities and defects, for want of experience, and the many diſadvantages he then laboured under, in his firſt ſally.

The

The idea and explanation here given of the first principles of Fluxions, are not essentially different from what they are in his former treatise, though expressed in other terms. The consideration of *time* introduced into the general definition, will, he says, perhaps be disliked by those who would have fluxions to be *mere velocities:* but the advantage of considering them otherwise, viz, not as the velocities themselves, but as magnitudes they would uniformly generate in a given time, appears to obviate any objection on that head. By taking fluxions as mere velocities, the imagination is confined as it were to a point, and without proper care insensibly involved in metaphysical difficulties. But according to this other mode of explaining the matter, less caution in the learner is necessary, and the higher orders of fluxions are rendered much more easy and intelligible. Besides, though Sir Isaac Newton defines fluxions to be the velocities of motions, yet he has recourse to the increments or moments generated in equal particles of time, in order to determine those velocities; which he afterwards teaches to expound by finite magnitudes of other kinds. This work was dedicated to George earl of Macclesfield.

In 1752 appeared, in 8vo, the *Select Exercises for young Proficients in the Mathematics.* This neat volume contains, "A great Variety of algebraical Problems, with their Solutions. A select Number of Geometrical Problems, with their Solutions, both algebraical and geometrical. The Theory of Gunnery, independent of the Conic Sections. A new and very comprehensive Method for finding the Roots of Equations in Numbers. A short Account of the first Principles of Fluxions. Also the Valuation of Annuities for single and joint Lives, with a Set of new Tables, far more extensive than any extant. This last part was designed as a supplement to his Doctrine of Annuities and Reversions; but being thought too small to be published alone, it was inserted here at the end of the Select Exercises; from whence however it has been removed in the last editions, and referred to its proper place, the end of the Annuities, as before mentioned. The examples that are given to each problem in this last piece, are according to the London bills of mortality; but the solutions are general, and may be applied with equal facility and advantage to any other table of observations. The volume is dedicated to John Bacon, Esq. F. R. S.

Mr. Simpson's Miscellaneous Tracts, printed in 4to, 1757, were his last legacy to the public: a most valuable bequest, whether we consider the dignity and importance of the subjects, or his sublime and accurate manner of treating them.

The first of these papers is concerned in determining the Precession of the Equinox, and the different Motions of the Earth's Axis, arising from the Attraction of the Sun and Moon. It was drawn up about the year 1752, in consequence of another on the same subject, by M. de Sylvabelle, a French gentleman. Though this gentleman had gone through one part of the subject with success and perspicuity, and his conclusions were perfectly conformable to Dr. Bradley's observations; it nevertheless appeared to Mr. Simpson, that he had greatly failed in a very material part, and that indeed the only very difficult one; that is, in the determination of the momentary alteration of the po-

sition of the earth's axis, caused by the forces of the sun and moon; of which forces, the quantities, but not the effects, are truly investigated. The second paper contains the Investigation of a very exact Method or Rule for finding the Place of a Planet in its Orbit, from a Correction of Bishop Ward's circular Hypothesis, by Means of certain Equations applied to the Motion about the upper Focus of the Ellipse. By this Method the Result, even in the Orbit of Mercury, may be found within a Second of the Truth, and that without repeating the Operation. The third shews the Manner of transferring the Motion of a Comet from a parabolic Orbit, to an elliptic one; being of great Use, when the observed Places of a (new) Comet are found to differ sensibly from those computed on the Hypothesis of a parabolic Orbit. The fourth is an Attempt to shew, from mathematical Principles, the Advantage arising from taking the Mean of a Number of Observations, in practical Astronomy; wherein the Odds that the Result in this Way, is more exact than from one single Observation, is evinced, and the Utility of the Method in Practice clearly made appear. The fifth contains the Determination of certain Fluents, and the Resolution of some very useful Equations, in the higher Orders of Fluxions, by Means of the Measures of Angles and Ratios, and the right and versed Sines of circular Arcs. The 6th treats of the Resolution of algebraical Equations, by the Method of Surd-divisors; in which the Grounds of that Method, as laid down by Sir Isaac Newton, are investigated and explained. The 7th exhibits the Investigation of a general Rule for the Resolution of Isoperimetrical Problems of all Orders, with some Examples of the Use and Application of the said Rule. The 8th, or last part, comprehends the Resolution of some general and very important Problems in Mechanics and Physical Astronomy; in which, among other Things, the principal Parts of the 3d and 9th Sections of the first Book of Newton's Principia are demonstrated in a new and concise Manner. But what may perhaps best recommend this excellent tract, is the application of the general equations, thus derived, to the determination of the Lunar Orbit.

According to what Mr. Simpson had intimated at the conclusion of his Doctrine of Fluxions, the greatest part of this arduous undertaking was drawn up in the year 1750. About that time M. Clairaut, a very eminent mathematician of the French Academy, had started an objection against Newton's general law of gravitation. This was a motive to induce Mr. Simpson (among some others) to endeavour to discover whether the motion of the moon's apogee, on which that objection had its whole weight and foundation, could not be truly accounted for, without supposing a change in the received law of gravitation, from the inverse ratio of the squares of the distances. The success answered his hopes, and induced him to look farther into other parts of the theory of the moon's motion, than he had at first intended: but before he had completed his design, M. Clairaut arrived in England, and made Mr. Simpson a visit; from whom he learnt, that he had a little before printed a piece on that subject, a copy of which Mr. Simpson afterwards received as a present, and found in it the same things demonstrated, to which he himself had directed his enquiry, besides several others.

The

The facility of the method Mr. Simpson fell upon, and the extensiveness of it, will in some measure appear from this, that it not only determines the motion of the apogee, in the same manner, and with the same ease, as the other equations, but utterly excludes all that dangerous kind of terms that had embarrassed the greatest mathematicians, and would, after a great number of revolutions, entirely change the figure of the moon's orbit. From whence this important consequence is derived, that the moon's mean motion, and the greatest quantities of the several equations, will remain unchanged, unless disturbed by the intervention of some foreign or accidental cause. These tracts are inscribed to the Earl of Macclesfield, President of the Royal Society.

Besides the foregoing, which are the whole of the regular books or treatises that were published by Mr. Simpson, he wrote and composed several other papers and fugitive pieces, as follow:

Several papers of his were read at the meetings of the Royal Society, and printed in their Transactions: But as most, if not all of them, were afterwards inserted, with alterations or additions, in his printed volumes, it is needless to take any farther notice of them here.

He proposed, and resolved many questions in the Ladies Diaries, &c; sometimes under his own name, as in the years 1735 and 1736; and sometimes under feigned or fictitious names; such as, it is thought, Hurlothrumbo, Kubernetes, Patrick O'Cavenah, Marmaduke Hodgson, Anthony Shallow, Esq, and probably several others; see the Diaries for the years 1735, 1736, 42, 43, 53, 54, 55, 56, 57, 58, 59, and 60. Mr. Simpson was also the editor or compiler of the Diaries from the year 1754 till the year 1760, both inclusive, during which time he raised that work to the highest degree of respect. He was succeeded in the Editorship by Mr. Edw. Rollinson. See my Diarian Miscellany, vol. 3.

It has also been commonly supposed that he was the real editor of, or had a principal share in, two other periodical works of a miscellaneous mathematical nature; viz, the Mathematician, and Turner's Mathematical Exercises, two volumes, in 8vo, which came out in periodical numbers, in the years 1750 and 1751, &c. The latter of these seems especially to have been set on foot to afford a proper place for exposing the errors and absurdities of Mr. Robert Heath, the then conductor of the Ladies Diary and the Palladium; and which controversy between them ended in the disgrace of Mr. Heath, and expulsion from his office of editor to the Ladies Diary, and the substitution of Mr. Simpson in his stead, in the year 1753.

In the year 1760, when the plans proposed for erecting a new bridge at Blackfriars were in agitation, Mr. Simpson, among other gentlemen, was consulted upon the best form for the arches, by the New-bridge Committee. Upon this occasion he gave a preference to the semicircular form; and, besides his report to the Committee, some letters also appeared, by himself and others, on the same subject, in the public newspapers, particularly in the Daily Advertiser, and in Lloyd's Evening Post. The same were also collected in the Gentleman's Magazine for that year, page 143 and 144.

It is probable that this reference to him, gave occasion to the turning his thoughts more seriously to this subject, so as to form the design of composing a regular treatise upon it: for his family have often informed me, that he laboured hard upon this work for some time before his death, and was very anxious to have completed it, frequently remarking to them, that this work, when published, would procure him more credit than any of his former publications. But he lived not to put the finishing hand to it. Whatever he wrote upon this subject, probably fell, together with all his other remaining papers, into the hands of major Henry Watson, of the engineers, in the service of the India Company, being in all a large chest full of papers. This gentleman had been a pupil of Mr. Simpson's, and had lodged in his house. After Mr. Simpson's death, Mr. Watson prevailed upon the widow to let him have the papers, promising either to give her a sum of money for them, or else to print and publish them for her benefit. But neither of these was ever done; this gentleman always declaring, when urged on this point by myself and others, that no use could be made of any of the papers, owing to the very imperfect state in which he said they were left. And yet he persisted in his refusal to give them up again.

From Mr. Simpson's writings, I now return to himself. Through the interest and solicitations of the beforementioned William Jones, Esq, he was, in 1743, appointed professor of mathematics, then vacant by the death of Mr. Derham, in the Royal Academy at Woolwich; his warrant bearing date August 25th. And in 1745 he was admitted a fellow of the Royal Society, having been proposed as a candidate by Martin Folkes, Esq. President, William Jones, Esq. Mr. George Graham, and Mr. John Machin, Secretary; all very eminent mathematicians. The president and council, in consideration of his very moderate circumstances, were pleased to excuse his admission fees, and likewise his giving bond for the settled future payments.

At the academy he exerted his faculties to the utmost, in instructing the pupils who were the immediate objects of his duty, as well as others, whom the superior officers of the ordnance permitted to be boarded and lodged in his house. In his manner of teaching, he had a peculiar and happy address; a certain dignity and perspicuity, tempered with such a degree of mildness, as engaged both the attention, esteem and friendship of his scholars; of which the good of the service, as well as of the community, was a necessary consequence.

It must be acknowledged however, that his mildness and easiness of temper, united with a more inactive state of mind, in the latter years of his life, rendered his services less useful; and the same very easy disposition, with an innocent, unsuspecting simplicity, and playfulness of mind, rendered him often the dupe of the little tricks of his pupils. Having discovered that he was fond of listening to little amusing stories, they took care to furnish themselves with a stock; so that, having neglected to learn their lessons perfect, they would get round him in a crowd, and, instead of demonstrating a proposition, would amuse him with some comical story, at which he would laugh and shake very heartily, especially if it were tinctured with somewhat of the

ludicrous

ludicrous or smutty; by which device they would contrive imperceptibly to wear out the hours allotted for instruction, and so avoid the trouble of learning and repeating their lesson. They tell also of various tricks that were practised upon him in consequence of the loss of his memory in a great degree, in the latter stage of his life.

It has been said that Mr. Simpson frequented low company, with whom he used to guzzle porter and gin: but it must be observed that the misconduct of his family put it out of his power to keep the company of of gentlemen, as well as to procure better liquor.

In the latter stage of his existence, when his life was in danger, exercise and a proper regimen were prescribed him, but to little purpose; for he sunk gradually into such a lowness of spirits, as often in a manner deprived him of his mental faculties, and at last rendered him incapable of performing his duty, or even of reading the letters of his friends; and so trifling an accident as the dropping of a tea-cup would flurry him as much as if a house had tumbled down.

The physicians advised his native air for his recovery; and in February, 1761, he set out, with much reluctance (believing he should never return) for Bosworth, along with some relations. The journey fatigued him to such a degree, that upon his arrival he betook himself to his chamber, where he grew continually worse and worse, to the day of his death, which happened the 14th of May, in the fifty-first year of his age.

SINE, or *Right* SINE, of an arc, in Trigonometry, a right line drawn from one extremity of the arc, perpendicular to the radius drawn to the other extremity of it: Or, it is half the chord of double the arc. Thus the line DE is the sine of the arc BD; either because it is drawn from one end D of that arc, perpendicular to CB the radius drawn to the other end B of the arc; or also because it is half the chord DF of double the arc DBF. For the

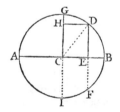

same reason also DE is the Sine of the arc AD, which is the supplement of BD to a semicircle or 180 degrees; that is, every Sine is common to two arcs, which are supplements to each other, or whose sum make up a semicircle, or 180 degrees.

Hence the Sines increase always from nothing at B till they become the radius CG, which is the greatest, being the Sine of the quadrant BG. From hence they decrease all the way along the second quadrant from G to A, till they quite vanish at the point A, thereby shewing that the Sine of the semicircle BGA, or 180 degrees, is nothing. After this they are negative all the way along the next semicircle, or 3d and 4th quadrants AFB, being drawn on the opposite side, or downwards from the diameter AB.

Whole SINE, or *Sinus Totus*, is the Sine of the quadrant BG, or of 90 degrees; that is, the Whole Sine is the same with the radius CG.

SINE-*Complement*, or *Cosine*, is the sine of an arc DG, which is the complement of another arc BD, to a quadrant. That is, the line DH is the Cosine of the arc BD; because it is the sine of DG which is the

complement of BD And for the same reason DE is the Cosine of DG. Hence the sine and Cosine and radius, of any arc, form a right-angled triangle CDE or CDH, of which the radius CD is the hypotenuse; and therefore the square of the radius is equal to the sum of the squares of the sine and Cosine of any arc, that is, $CD^2 = CE^2 + ED^2$ or $= CH^2 + DH^2$.

It is evident that the Cosine of 0 or nothing, is the whole radius CB. From B, where this Cosine is greatest, the Cosine decreases as the arc increases from B along the quadrant BDG, till it become 0 for the complete quadrant BG. After this, the Cosines, decreasing, become negative more and more all the way to the complete semicircle at A. Then the Cosines increase again all the way from A through I to B; at I the negation is destroyed, and the Cosine is equal to 0 or nothing; from I to B it is positive, and at B it is again become equal to the radius. So that, in general, the Cosines in the 1st and 4th quadrants are positive, but in the 2d and 3d negative.

Versed-SINE, is the part of the diameter between the sine and the arc. So BE is the Versed Sine of the arc BD, and AE the Versed Sine of AD, also GH the Versed-Sine of DG, &c. All Versed Sines are affirmative. The sum of the Versed Sine and cosine, of any arc or angle, is equal to the radius, that is, $BE + EC = AC$.—The sine, cosine, and Versed Sine, of an arc, are also the same of an angle, or the number of degrees &c, which it measures.

The Sines &c, of every degree and minute in a quadrant, are calculated to the radius 1, and ranged in tables for use. But because operations with these natural Sines require much labour in multiplying and dividing by them, the logarithms of them are taken, and ranged in tables also; and these logarithmic Sines are commonly used in practice, instead of the natural ones, as they require only additions and subtractions, instead of the multiplications and divisions. For the method of constructing the scales of Sines &c, see the article SCALE.

The Sines were introduced into trigonometry by the Arabians. And for the etymology of the word *Sine* see Introduction to my Logarithms, pa. 17 &c. And the various ways of calculating tables of the Sines, may be seen in the same place, pa. 13 &c.

Theorems for the Sines, Cosines, &c, one from another. From the definitions of them, and the common property of right-angled triangles, with that of the circle, viz, that $DE^2 = CD^2 - CE^2 = AE \times EB$, are easily deduced these following values of the Sines, &c, viz, putting

$$s = \text{the sine DE,}$$
$$c = \text{the cosine CE,}$$
$$v = \text{versed sine BE,}$$
$$\mathrm{v} = \text{suppl. versed sine AE,}$$
$$r = \text{radius AC or CB,}$$
$$a = \text{arc BD; then}$$

$$s = \sqrt{r^2 - c^2} = \sqrt{\mathrm{v}v} = \sqrt{2rv - vv} = \sqrt{2r\mathrm{v} - \mathrm{v}\mathrm{v}}$$
$$c = \sqrt{r^2 - s^2} = r - v = \mathrm{v} - r = \tfrac{1}{2}\mathrm{v} - \tfrac{1}{2}v.$$
$$v = r - c = 2r - \mathrm{v} = r - \sqrt{r^2 - s^2} = \mathrm{v} - 2c.$$
$$\mathrm{v} = r + c = 2r - v = r + \sqrt{r^2 - s^2} = v + 2c.$$

The

The tangent $= \dfrac{rs}{\sqrt{r^2 - s^2}}$. And Cotang. $= \dfrac{r\sqrt{r^2 - s^2}}{s}$.

The Secant $= \dfrac{rr}{\sqrt{r^2 - s^2}}$. And Cosec. $= \dfrac{rr}{s}$.

$$s = a - \frac{a^3}{2.3\,r^2} + \frac{a^5}{2.3.4.5\,r^4} - \frac{a^7}{2.3.4.5.6.7\,r^6} \ \&c.$$

$$a = s + \frac{s^3}{2.3\,r^2} + \frac{1.3\,s^5}{2.4.5\,r^4} + \frac{1.3.5\,s^7}{2.4.6.7\,r^6} \ \&c.$$

Log. $s = $ log. $a - M \left(\dfrac{n^2}{6} + \dfrac{a^4}{180} + \dfrac{a^6}{2835} + \dfrac{a^8}{37800} \ \&c \right)$

or Log. $s = -\frac{1}{2} M (c^2 + \frac{1}{2} c^4 + \frac{1}{3} c^6 + \frac{1}{4} c^8 \ \&c)$

or Log. $s = -2 M (z + \frac{1}{3} z^3 + \frac{1}{5} z^5 + \frac{1}{7} z^7 \ \&c.)$

when $z = \dfrac{1-s}{1+s}$, radius 1, and $M = \cdot 43429448$ &c.

If A be any other arc, S its sine, and C its cosine. Then

Sin. $\overline{A+a} = \dfrac{Sc + sC}{r}$ Cof. $\overline{A+a} = \dfrac{Cc - Ss}{r}$.

Sin. $\overline{A-a} = \dfrac{Sc - sC}{r}$.   Cof. $\overline{A-a} = \dfrac{Cc + Ss}{r}$.

Sin. $A \times$ cof. $a = \frac{1}{2}$ fin. $\overline{A-a} + \frac{1}{2}$ fin. $\overline{A+a}$.

Sin. $A \times$ fin. $a = \frac{1}{2}$ cof. $\overline{A-a} - \frac{1}{2}$ cof. $\overline{A+a}$.

Cof. $A \times$ cof. $a = $ cof. $\dfrac{A-a}{2} + $ cof. $\dfrac{A+a}{2}$

If $b = 2\cdot718281828$ &c, the number whose hyp. log. is 1; then

$$\text{Sin. } a = s = \frac{b^{a\sqrt{-1}} - b^{-a\sqrt{-1}}}{2\sqrt{-1}}.$$

$$\text{Cof. } a = c = \frac{b^{a\sqrt{-1}} + b^{-a\sqrt{-1}}}{2}.$$

See many other curious expressions of this kind in Bougainville's Calcul Integral, and in Bertrand's Mathematics.

And, in general,

$$\text{Sin. } na = nsc^{n-1} - \frac{n.\overline{n-1}.\overline{n-2}}{1.2.3} s^3 c^{n-3} + \frac{n.\overline{n-1}.\overline{n-2}.\overline{n-3}.\overline{n-4}}{1.2.3.4.5} s^5 c^{n-5} \ \&c.$$

$$\text{or Sin. } na = ns - \frac{n.\overline{n^2-1^2}}{2.3} s^3 + \frac{n.\overline{n^2-1^2}.\overline{n^2-3^2}}{2.3.4.5} s^5 \ \&c.$$

$$\text{Cof. } na = c^n - \frac{n.\overline{n-1}}{2} s^2 c^{n-2} + \frac{n.\overline{n-1}.\overline{n-2}.\overline{n-3}}{2.3.4} s^4 c^{n-4} \ \&c.$$

$$\text{or Cof. } na = 1 - \frac{n^2}{2} s^2 + \frac{n.\overline{n^2-2^2}}{2.3.4} s^4 - \frac{n^2.\overline{n^2-2^2}.\overline{n^2-4^2}}{2.3.4.5.6} s^6 \ \&c.$$

Sin. $\frac{1}{2} a = r \sqrt{\dfrac{r-c}{2r}}$. And cof. $\frac{1}{2} a = r \sqrt{\dfrac{r+c}{2r}}$. Radius being r.

Of the Tables of Sines, &c.

In estimating the quantity of the Sines &c, we assume radius for unity; and then compute the quantity

From some of the foregoing theorems the Sines of a great variety of angles, or number of degrees, may be computed. Ex. gr. as below.

Angles.	Sines.
90°	r
75	$\frac{1}{2}r\sqrt{2 + \sqrt{3}} = r \times \dfrac{\sqrt{6} + \sqrt{2}}{4}$
72	$\frac{1}{2}r\sqrt{\dfrac{5 + \sqrt{5}}{2}}$
67½	$\frac{1}{2}r\sqrt{2 + \sqrt{2}}$
60	$\frac{1}{2}r\sqrt{3}$
54	$\frac{1}{2}r\sqrt{\dfrac{3 + \sqrt{5}}{2}} = r \times \dfrac{\sqrt{5} + 1}{4}$
45	$\frac{1}{2}r\sqrt{2}$
36	$\frac{1}{2}r\sqrt{\dfrac{5 - \sqrt{5}}{2}}$
30	$\frac{1}{2}r$
22½	$\frac{1}{2}r\sqrt{2 - \sqrt{2}}$
18	$\frac{1}{2}r\sqrt{\dfrac{3 - \sqrt{5}}{2}} = r \times \dfrac{\sqrt{5} - 1}{4}$
15	$\frac{1}{2}r\sqrt{2 - \sqrt{3}} = r \times \dfrac{\sqrt{6} - \sqrt{2}}{4}$

Radius being 1. Then for multiple arcs:

the Sin. $\overline{n+1}.a = 2c \times$ fin. $na - $ fin. $\overline{n-1}.a$,

and Cof. $\overline{n+1}.a = 2c \times$ cof. $na - $ cof. $\overline{n-1}.a$;

That is, multiplying any Sine or cosine by $2c$, and the next preceding Sine or cosine subtracted from it, it gives the next following Sine or cosine. Hence

fin. $0a = 0$.	cof. $0a = 1$ or radius.
fin. $a = s$.	cof. $a = c$.
fin. $2a = 2sc$.	cof. $2a = c^2 - s^2$.
fin. $3a = 3sc^2 - s^3$.	cof. $3a = c^3 - 3cs^2$.
fin. $4a = 4sc^3 - 4s^3c$.	cof. $4a = c^4 - 6c^2s^2 + s^4$.
fin. $5a = 5sc^4 - 10s^3c^2 + s^5$.	cof. $5a = c^5 - 10c^3s^2 + 5cs^4$.
&c.	&c.

of the Sines, tangents, and secants, in fractions of it. From Ptolomy's Almagest we learn, that the ancients divided the radius into 60 parts, which they called degrees, and thence determined the chords in minutes,

nutes, feconds, and thirds; that is, in fexagefimal fractions of the radius, which they likewife ufed in the refolution of triangles. As to the Sines, tangents and fecants, they are modern inventions; the Sines being introduced by the Moors or Saracens, and the tangents and fecants afterwards by the Europeans. See Introd. to my Logs. pa. 1 to 19.

Regiomontanus, at firft, with the ancients, divided the radius into 60 degrees; and determined the Sines of the feveral degrees in decimal fractions of it. But he afterwards found it would be more convenient to affume 1 for radius, or 1 with any number of cyphers, and take the Sines in decimal parts of it; and thus he introduced the prefent method in trigonometry. In this way, different authors have divided the radius into more or fewer decimal parts; but in the common tables of Sines and tangents, the radius is conceived as divided into 10000000 parts; by which all the Sines are eftimated.

An idea of fome of the modes of conftructing the tables of Sines, may be conceived from what here follows: Firft, by common geometry the fides of fome of the regular polygons inferibed in the circle are computed, from the given radius, which will be the chords of certain portions of the circumference, denoted by the number of the fides; viz, the fide of the triangle the chord of the 3d part, or 120 degrees; the fide of the pentagon the chord of the 5th part, or 72 degrees; the fide of the hexagon the chord of the 6th part, or 60 degrees; the fide of the octagon the chord of the 8th part, or 45 degrees; and fo on. By this means there are obtained the chords of feveral of fuch arcs; and the halves of thefe chords will be the Sines of the halves of the fame arcs. Then the theorem $c = \sqrt{1 - s^2}$ will give the cofines of the fame half arcs. Next, by bifecting thefe arcs continually, there will be found the Sines and cofines of a continued feries as far as we pleafe by thefe two theorems,

$$\text{Sin. } \tfrac{1}{2} a = \frac{\sqrt{1-c}}{2}; \text{ and cof. } \tfrac{1}{2} a = \sqrt{\frac{1+c}{2}}.$$

Then, by the theorems for the fums and differences of arcs, from the foregoing feries, will be derived the Sines and cofines of various other arcs, till we arrive at length at the arc of 1′, or 1″, &c, whofe Sine and cofine thus become known.

Or, rather, the fine of 1 minute will be much more eafily found from the feries

$$s = a - \frac{a^3}{6} + \frac{a^5}{120} \text{ &c,}$$

becaufe the arc is equal to its Sine in fmall arcs; whence $s = a$ only in fuch fmall arcs. But the length of the arc of 180° or 10800 is known to be 3·14159265, &c; therefore, by proportion, as 10800′ : 1′ :: 3·14159265 : 0·000.908882 = a the arc or s the fine of 1′, which number is true to the laft place of decimals. Then, for the cofine of 1′, it is $c = \sqrt{1 - s^2}$ = 0·9999999577 the cofine of the fame 1′.

Hence we fhall readily obtain the Sines and cofines of all the multiples of 1′, as of 2′, 3′ 4′ 5′, &c, by the application of thefe two theorems,

Sin. $\overline{n + 1}.a = 2c \times$ fin. $na -$ fin. $\overline{n - 1}.a,$

Cof. $\overline{n + 1}.a = 2c \times$ cof. $na -$ cof. $\overline{n - 1}.a$;

for fuppofing $a =$ the arc of 1, then $c = 0.9999999577$, and taking n fuccefively equal to 1, 2, 3, 4, &c, the theorems for the Sines and cofines give feverally the Sines and cofines of 1′, 2′, 3′, 4′, &c; viz, the Sines thus:

fin. 1′ = s - - - - - - - - - = ·0002908882
fin. 2′ = 2c × fin. 1′ — fin. 0′ = ·0005817764
fin. 3′ = 2c × fin. 2′ — fin. 1′ = ·0008726645
fin. 4′ = 2c × fin. 3′ — fin. 2′ = ·0011635526
fin. 5′ = 2c × fin. 4′ — fin. 3′ = ·0014544406
&c.

And the Cofines thus,

cof. 1′ = c - - - - - - - - - - = ·9999999577
cof. 2′ = 2c × cof. 1′ — cof. 0′ = ·9999998308
cof. 3′ = 2c × cof. 2′ — cof. 1′ = ·9999996192
cof. 4′ = 2c × cof. 3′ — cof. 2′ = ·9999993231
cof. 5′ = 2c × cof. 4′ — cof. 3′ = ·9999989423
&c.

In this manner then all the Sines and cofines are made, by only one conftant multiplication and a fubtraction, up to 30 degrees, forming thus the Sines of the firft and laft 30 degrees of the quadrant, or from 0 to 30° and from 60° to 90°; or, which will be much the fame thing, the Sines only may be thus computed all the way up to 60°.

Then the Sines of the remaining 30°, from 60 to 90, will be found by one addition only for each of them, by means of this theorem, viz,

$$\text{Sin. } \overline{60 + a} = \text{fin. } \overline{60 - a} + \text{fin. } a;$$

that is, to the fine of any arc below 60°, add the Sine of its defect below 60, and the fum will be the Sine of another arc which is juft as much above 60.

The Sines of all arcs being thus found, they give alfo very eafily the verfed fines, the tangents, and the fecants. The verfed fines are only the arithmetical complements to 1, that is, each cofine taken from the radius 1.

The tangents are found by thefe three theorems:

1. As cofine to fine, fo is radius to tangent.

2. Radius is a mean proportional between the tangent and cotangent.

3. Half the difference between the tangent and cotangent, is equal to the tangent of the difference between the arc and its complement. Or, the fum arifing from the addition of double the tangent of an arc with the tangent of half its complement, is equal to the tangent of the fum of that arc and the faid half complement.

By the 1ft and 2d of thefe theorems, the tangents are to be found for one half of the quadrant: then the other half of them will be found by one fingle addition, or fubtraction, for each, by the 3d theorem.

This done, the fecants will be all found by addition or fubtraction only, by thefe two theorems: 1ft. The fecant of an arc, is equal to the fum of its tangent and the tangent of half its complement. 2nd. The fecant of of an arc, is equal to the difference between the tangent of that arc and the tangent of the arc added to half its complement.

Artificial SINES, are the logarithmic Sines, or the logarithms of the Sines.

Curve or *Figure of the* SINES. See FIGURE *of the Sines, &c.* To what is there said of the figure of the Sines, may be here added as follows, from a property just given above, viz. if *x* denote the abscifs of this curve, or the corresponding circular arc, and *y* its ordinate, or the Sine of that arc; then the equation of the curve will be this,

$$y = \text{fin. } x = \frac{b^{x\sqrt{-1}} - b^{-x\sqrt{-1}}}{2\sqrt{-1}} ;$$

where $b = 2.718281828$, &c, the number whose hyp. log. is 1.

Line of SINES, is a line on the sector, or Gunter's scale, &c, divided according to the Sines, or expressing the Sines. See those articles.

SINE *of Incidence*, or *of Reflection*, or *of Refraction*, is used for the Sine of the *angle* of incidence, &c.

SINICAL *Quadrant*, is a quadrant, made of wood or metal, with lines drawn from each side intersecting one another, with an index, divided by fines, also with 90 degrees on the limb, and two sights at the edge. Its use is to take the altitude of the sun. Instead of the fines, it is sometimes divided all into equal parts; and then it is used by seamen to resolve, by inspection, any problem of plane sailing.

SIPHON, or SYPHON, in Hydraulics, a crooked pipe or tube used in the raising of fluids, emptying of vessels, and in various hydrostatical experiments. It is otherwise called a crane.

Wolfius describes two vessels under the name of Siphons; the one cylindrical in the middle and conical at the two extremes; the other globular in the middle, with two narrow tubes fitted to it axis-wise; both serving to take up a quantity of liquid, and to retain it when up.

But the most usual Syphon is that which is here represented; where ABC is any crooked tube, having two legs of unequal lengths; but such however that, in any position, the perpendicular altitude BD, of B above A, when AB is filled with any fluid, the weight of that fluid may not be more than about 15 lb. upon every square inch of the base, or equal to the pressure of the atmosphere, because the pressure of

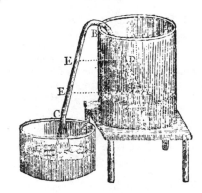

the atmosphere will raise or suspend the fluid so high, when the tube is exhausted of air. This height is about 30 inches when the fluid is quicksilver, and about 34 feet when it is water; and so on for other fluids, according to the rarity of them.

To use the Siphon, in drawing off any fluid; immerse the shorter end A into the fluid, then suck or draw the air out by the other or lower end C, and the fluid will presently follow, and run out by the Siphon, from the vessel at A to the vessel at C; till such time as the surface of the fluid sink as low as the orifice at A,

when the decanting will cease, and the Siphon will empty itself of the fluid, the whole of that which is in it running out at C. The principle upon which the Siphon acts, is this: when the tube is exhausted of air, the pressure of the atmosphere upon the surface of the fluid at D, forces it into the tube by the orifice at A, as in the barometer tube, and down the leg BC, if B is not above the surface at D more than 34 feet for water, or 30 inches for quicksilver, &c. Here, if the external leg of the Siphon terminate at E, on a horizontal level with the immersed end at A, or rather on a level with the water at D, the perpendicular pressures of the fluid in each leg, and of the external air, against each orifice, being alike in both, the fluid will be at rest in the Siphon, completely filling it, but without running or preponderating either way. But if the external end be the lower, terminating at C, then the fluid in this end being the heavier, or having more pressure, will preponderate and run out by the orifice at C, this would leave a vacuum at B but for the continual pressure of the atmosphere at D, which forces the fluid up by A to B, and so producing a continued motion of it through the tube, and a discharge or stream at C.

Instead of sucking out the air at C, another method is, first to fill the tube completely with the fluid, in an inverted position with the angle B downward; and, stopping the two orifices with the fingers, revert the tube again, and immerge the end A in the fluid; then take off the fingers, and immediately the stream commences from the end C.

Either of the two foregoing methods can be conveniently practised when the Siphon is small, and easily managed by the hand; as in decanting off liquors from casks, &c. But when the Siphon is very large, and many feet in height, as in exhausting water from a valley or pit, the following method is then recommended: Stop the orifice C, and, by means of an opening made in the top at B, fill the tube completely with water; then stop the opening at B with a plug, and open that at C; upon which the water will presently flow out at C, and

fo continue till that at A is exhaufted. And this method of conveying water over a hill, from one valley to another, is defcribed by Hero, the chief author of any confequence upon this fubject among the Ancients. But in this experiment it muft be noted, that the effect will not be produced when the hill at B is more than 33 or 34 feet above the furface of the water at A.

In an experiment of this kind, it is even faid the water in the legs, unlefs it be purged of its air, will not reft at a height of quite 30 feet above the water in the veffels; becaufe air will extricate itfelf out of the water, and getting above the water in the legs, prefs it downward, fo that its height will be lefs to balance the preffure of the atmofphere. But with very fine, or capillary tubes, the experiment will fucceed to a height fomewhat greater; becaufe the attraction of the matter of the very fine tube will attract the fluid, and fupport it at fome certain height, independent of the preffure of the atmofphere. For which reafon alfo it is, that the experiment fucceeds for fmall heights in the exhaufted receiver; as has been tried both with water and mercury, by Defaguliers and many other philofophers. Exper. Philof. vol. 2, pa. 168.

The figure of the veffel may be varied at pleafure, provided the orifice C be but below the level of the furface of the water to be drawn up, but ftill the farther it is below it, the quicker will the fluid run off. And if, in the courfe of the efflux, the orifice A be drawn out of the fluid; all the liquor in the Siphon will iffue out at the lower orifice C; that in the leg BC dragging, as it were, that in the fhorter leg AB after it.

But if a filled Siphon be fo difpofed, as that both orifices, A and C, be in the fame horizontal line; the fluid will remain pendant in each leg, how unequal foever the length of the legs may be. So that fluids in Siphons feem, as it were, to form one continued body; the heavier part defcending like a chain, and drawing the lighter after it.

The *Wirtemberg* SIPHON, is a very extraordinary

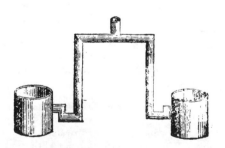

machine, performing feveral things which the common Siphon will not reach. This Siphon was projected by Jordan Pelletier, and executed at the expence of prince Frederic Charles, adminiftrator of Wirtemberg, by his mathematician Shahackard, who made each branch 20 feet long, and fet them 18 feet apart; and the defcription of it was publifhed by Reifelius, the duke's phyfician. This gave occafion to Papin to invent another, which performed the fame things, and is defcribed in

the Philof. Tranf. vol. 14, or Abr. vol. 1. Reifelius, in another paper in the fame volume, ingenuoufly owns that this is the fame with the Wirtemberg Siphon.

In this engine, though the legs be on the fame level, yet the water rifes up the one, and defcends through the other: The water rifes even through the aperture if the lefs leg be only half immerged in water: The Siphon has its effect after continuing dry a long time: Either of the apertures being opened, the other remaining fhut for a whole day, and then opened, the water flows out as ufual: Laftly, the water rifes and falls indifferently through either leg.

Muffchenbroek, in accounting for the operation of this Siphon, obferves that no difcharge could be made by it, unlefs the water applied to either leg caufe the one to be fhorter, and the other longer by its own weight. Introd. ad Phil. Nat. tom. 2, pa. 853, ed. 4to. 1762.

SIRIUS, the *Dog-ftar*; a very bright ftar of the firft magnitude, in the mouth of the conftellation *Canis Major*, or the Great Dog.

This is the brighteft of all the ftars in our firmament, and therefore probably, fays Dr. Mafkelyne, the aftronomer royal, the neareft to us of them all, in a paper recommending the difcovery of its parallax, Philof. Tranf vol. 2, pa. 889, Some however fuppofe Arcturus to be the neareft.

The Arabs call it *Afchere, Elfchecre, Scera*; the Greeks, *Sirius*; and the Latins, *Canicula*, or *Canis candens*. See CANICULA.

This is one of the earlieft named ftars in the whole heavens. Hefiod and Homer mention only four or five conftellations, or ftars, and this is one of them. Sirius and Orion, the Hyades, Pleiades, and Arcturus are almoft the whole of the old poetical aftronomy. The three laft the Greeks formed of their own obfervation, as appears by the names; the two others were Egyptian. Sirius was fo called from the Nile, one of the names of that river being Siris; and the Egyptians, feeing that river begin to fwell at the time of a particular rifing of this ftar, paid divine honours to the ftar, and called it by a name derived from that of the river, expreffing the ftar of the Nile.

SITUS, in Algebra and Geometry, denotes the fituation of lines, furfaces, &c. Wolfius delivers fome things in geometry, which are not deduced from the common analyfis, particularly matters depending on the *Situs* of lines and figures. Leibnitz has even founded a particular kind of analyfis upon it, called *Calculus Situs*.

SKY, the blue expanfe of the air or atmofphere.

The azure colour of the fky is attributed, by Newton, to vapours beginning to condenfe, having attained confiftence enough to reflect the moft reflexible rays, viz, the violet ones; but not enough to reflect any of the lefs reflexible ones.

De la Hire attributes it to our viewing a black object, viz the dark fpace beyond the regions of the atmofphere, through a white or lucid one, viz the air illuminated by the fun; a mixture of black and white always appearing blue. But this hypothefis is not originally his; being as old as Leonardo da Vinci.

SLIDING, in Mechanics, is when the fame point of a body, moving along a furface, defcribes a line on
that

that furface. Such is the motion of a parallelopipedon moved along a plane.

From Sliding arifes friction.

SLIDING *Rule*, a mathematical inftrument ferving to perform computations in gauging, meafuring, &c, without the ufe of compaffes ; merely by the fliding of the parts of the inftrument one by another, the lines and divifions of which give the anfwer or amount by in-fpection.

This inftrument is varioufly contrived and applied by different authors, particularly Gunter, Partridge, Hunt, Everard, and Coggefhall ; but the moft ufual and ufe-ful ones are thofe of the two latter.

Everard's SLIDING *Rule* is chiefly ufed in cafk gauging. It is commonly made of box, 12 inches long, 1 inch broad, and $\frac{6}{10}$ of an inch thick. It confifts of three parts ; viz, the ftock juft mentioned, and two thin flips, of the fame length, fliding in fmall grooves in two oppofite fides of the ftock : confequently, when both thefe pieces are drawn out to their full extent, the in-ftrument is 3 feet long.

On the firft broad face of the inftrument are four lo-garithmic lines of numbers ; for the properties &c, of which, fee GUNTER's *Line*. The firft, marked A, confifting of two radii numbered 1, 2, 3, 4, 5, 6, 7, 8, 9, 1 ; and then 2, 3, 4, 5, &c, to 10. On this line are four brafs centre-pins, two in each radius ; one in each of them being marked MB, for malt bufhel, is fet at 2150·42 the number of cubic inches in a malt-bufhel ; the other two are marked with A, for ale-gal-lon, at 282, the number of cubic inches in an ale gal-lon. The 2d and 3d lines of numbers are on the flid-ing pieces,. and are exactly the fame with the firft ; but they are diftinguifhed by the letter B. In the firft ra-dius is a dot, marked Si, at ·707, the fide of a fquare infcribed in a circle whofe diameter is 1. Another dot, marked Se. ftands at ·886, the fide of a fquare equal to the area of the fame circle. A third dot, marked W, is at 231, the cubic inches in a wine gallon. And a fourth, marked C, at 3·14, the circumference of the fame circle whofe diameter is 1. The fourth line of numbers, marked MD, to fignify malt-depth, is a broken line of two radii, numbered 2, 10, 9, 8, 7, 6, 5, 4, 3, 2, 1, 9, 8, 7, &c ; the number 1 being fet directly againft MB on the firft radius.

On the fecond broad face, marked *cd*, are feveral lines: as 1ft, a line marked D, and numbered 1, 2, 3, &c, to 10. On this line are four centre pins : the firft, marked WG, for wine-gauge, is at 17·15, the gauge-point for wine gallons, being the diameter of a cylin-der whofe height is. one inch, and content 31 cubic inches, or a wine gallon : the fecond centre-pin, marked AG, for ale-gauge, is at 18·95, the like diameter for an ale gallon: the 3d, marked MS, for malt fquare, is at 46·3, the fquare root of 2150·42, or the fide of a fquare whofe content is equal to the number of inches in a folid bufhel : and the fourth, marked MR, for malt-round, is at 52·32, the diameter of a cylinder, or bufhel, the area of whofe bafe is the fame 2150·42, the inches in a bufhel. 2dly, Two lines of numbers on the fliding piece, on the other fide, marked C. On thefe are two dots ; the one, marked *c*, at ·0795, the area of a circle whofe circumference is 1 ; and the other, marked *d*, at ·785, the area of the circle whofe diame-

ter is 1. 3dly, Two lines of fegments, each numbered 1, 2, 3, to 100; the firft for finding the ullage of a cafk, taken as the middle fruftum of a fpheroid, lying with its axis parallel to the horizon ; and the other for finding the ullage of a cafk ftanding.

Again, on one of the narrow fides, noted *c*, are, 1ft, a line of inches, numbered 1, 2, 3, &c to 12, each fubdivided into 10 equal parts. 2dly, A line by which,. with that of inches,. we find a mean diameter for a cafk, in the figure of the middle fruftum of a fpheroid : it is marked *Spheroid*, and numbered 1, 2, 3, &c to 7. 3dly, A line for finding the mean diameter of a cafk, in the form of the middle fruftum of a parabolic fpindle, which gaugers call the fecond variety of cafks ; it is therefore marked *Second Variety*, and is numbered 1, 2, 3, &c.

4thly, A line by which is found the mean diameter of a cafk of the third variety, confifting of the fruftums of two parabolic conoids, abutting on a common bafe ; it is therefore marked *Third Variety*, and is. numbered 1, 2, 3, &c.

On the other narrow face, marked *f*, are 1ft, a line of a foot divided into 100 equal parts,. marked FM. 2dly, A line of inches, like that before mentioned, marked IM. 3dly, A line for finding the mean diame-ter of the fourth variety of cafks, which is formed of the fruftums of two cones, abutting on a common bafe. It is numbered 1, 2, 3, &c ; and marked FC, for fruf-tum of a cone.

On the backfide of the two fliding pieces is a line of inches, from 12 to 36, for the whole extent of the 3 feet, when the pieces are put endwife ; and againft that, the correfpondent gallons, and 100th parts, that any fmall tub, or the like open veffel, will contain at 1 inch deep.

For the various ufes of this inftrument, fee the authors mentioned above, and moft other writers on Gauging.

Coggefhall's SLIDING *Rule* is chiefly ufed in meafuring the fuperficies and folidity of timber, mafonry, brick-work, &c.

This confifts of two rulers, each a foot long, which are united together in various ways. Sometimes they are made to flide by one another, like glaziers' rules : fometimes a groove is made in the fideof a common two-foot joint rule, and a thin fliding piece in one fide, and Coggefhall's lines added on that fide ; thus forming the common or Carpenter's rule : and fometimes one of the two rulers is made to flide in a groove made in the fide of the other.

On the Sliding fide of the rule are four lines of num-bers, three of which are double, that is, are lines to two radii, and the fourth is a fingle broken line of numbers. The firft three, marked A, B, C, are fi-gured 1, 2, 3, &c to 9 ; then 1, 2, 3, &c to 10 ; the conftruction and ufe of them being the fame as thofe on Everard's Sliding rule. The fingle line, called the *girt line*, and marked D, whofe radius is equal to the two radii of any of the other lines, is broken for the eafier meafuring of timber, and figured 4, 5, 6, 7, 8, 9, 10, 20, 30, &c. From 4 to 5 it is divided into 10 parts, and each 10th fubdivided into 2 ; and fo on from 5 to 10, &c.

On the backfide of the rule are, 1ft, a line of inch meafure, from 1 to 12 ; each inch being divided and fubdivided. 2dly, A line of foot meafure, confifting

of

of one foot divided into 100 equal parts, and figured 10, 20, 30, &c.

The backside of the sliding piece is divided into inches, halves, &c, and figured from 12 to 24; so that when the slide is out, there may be a measure of 2 feet.

In the Carpenter's rule, the inch measure is on one side, continued all the way from 1 to 24, when the rule is unfolded, and subdivided into 8ths or half-quarters: on this side are also some diagonal scales of equal parts. And upon the edge, the whole length of 2 feet is divided into 200 equal parts, or 100ths of a foot.

SLING, a string instrument, serving for the casting of stones &c with the greater violence.

Pliny, lib. 76, chap. 5, attributes the invention of the Sling to the Phœnicians; but Vegetius ascribes it to the inhabitants of the Balearic islands, who were celebrated in antiquity for the dextrous management of it. Florus and Strabo say, those people bore three kinds of Slings; some longer, others shorter, which they used according as their enemies were more remote or nearer hand. Diodorus adds, that the first served them for a head-band, the 2d for a girdle, and that the third they constantly carried with them in the hand. But it must be impossible to tell who were the first inventors of the Sling, as the instrument is so simple, and has been in general use by almost all nations. The instrument is much spoken of in the wars and history of the Israelites. David was so expert a slinger, that he ventured to go out, with one in his hand, against the giant and champion Goliath, and at a distance struck him on the forehead with the stone. And there were a number of left-handed men of one of the tribes of Israel, who it is said could Sling a stone at an hair's breadth.

The motion of a stone discharged from a Sling arises from its centrifugal force, when whirled round in a circle. The velocity with which it is discharged, is the same as that which it had in the circle, and is much greater than what can be given to it by the hand alone. And the direction in which it is discharged, is that of the tangent to the circle at the point of discharge. Whence its motion and effect may be computed as a projectile.

SLUSE, or SLUSIUS (René Francis Walter) of Vise, a small town in the county of Liege, where he enjoyed honours and preferment. He then became abbé of Amas, canon, councellor and chancellor of Liege, and made his name famous for his knowledge in theology, physics, and mathematics. The Royal Society of London elected him one of their members, and inserted several of his compositions in their Transactions. This very ingenious and learned man died at Liege in 1683, at 63 years of age.

Of Slusius's works there have been published, some learned letters, and a work intitled, *Mesolabum et Problemata solida*; beside the following pieces in the Philosophical Transactions, viz,

1. Short and Easy Method of drawing Tangents to all Geometrical Curves; vol. 7, pa. 5143.

2. Demonstration of the same; vol. 8, pa. 6059, 6119.

3. On the Optic Angle of Alhazen; vol. 8, pa. 6139.

SMEATON (JOHN), F. R. S. and a very cele-

brated civil engineer, was born the 28th of May 1724, at Austhorpe, near Leeds, in a house built by his grandfather, where the family have resided ever since, and where our author died the 28th of October 1792, in the 65th year of his age.

Mr. Smeaton seems to have been born an engineer. The originality of his genius and the strength of his understanding appeared at a very early age. His playthings were not those of children, but the tools men work with; and he had always more amusement in observing artificers work, and asking them questions, than in any thing else. Having watched some mill-wrights at work, he was one day, soon after, seen (to the distress of his family) on the top of his father's barn, fixing up something like a windmill. Another time, attending some men who were fixing a pump at a neighbouring village, and observing them cut off a piece of bored pipe, he contrived to procure it, of which he made a working pump that actually raised water. These anecdotes refer to circumstances that happened when he was hardly out of petticoats, and probably before he had reached the 6th year of his age. About his 14th or 15th year, he had made for himself an engine to turn rose-work; and he made several presents to his friends of boxes in ivory and wood, turned by him in that way.

His friend and partner in the Deptford Waterworks, Mr. John Holmes, an eminent clock and watch maker in the Strand, says, he visited Mr. Smeaton and spent a month with him at his father's house, in the year 1742, when consequently our author was about 18 years of age. Mr. Holmes could not but view young Smeaton's works with astonishment: he forged his own iron and steel, and melted his own metals; he had tools of every sort, for working in wood, ivory, and metals: he had made a lathe, by which he had cut a perpetual screw in brass, a thing very little known at that day.

Thus had Mr. Smeaton, by the strength of his genius, and indefatigable industry, acquired, at 18 years of age, an extensive set of tools, and the art of working in most of the mechanical trades, without the assistance of any master, and which he continued to do a part of every day when at the place where his tools were: and few men could work better.

Mr. Smeaton's father was an attorney, and was desirous of bringing him up to the same profession. He therefore came up to London in 1742, and for some time attended the courts in Westminster Hall. But finding that the profession of the law did not suit *the bent of his genius*, as his usual expression was, he wrote a strong memorial to his father on the subject, whose good sense from that moment left Mr. Smeaton to pursue the bent of his genius in his own way.

Mr. Smeaton after this continued to reside in London, and about 1750 he commenced philosophical instrument maker, which he continued for some time, and became acquainted with most of the ingenious men of that time; and this same year he made his first communication to the Royal Society, being an account of Dr. Knight's improvements of the mariner's compass. Continuing his very useful labours, and making experiments, he communicated to that learned body, the two following years, a number of other ingenious improvements,

ments, as will be enumerated in the lift of his writings, at the end of this account of him.

In 1751 he began a courfe of experiments, to try a machine of his invention, for meafuring a fhip's way at fea; and alfo made two voyages in company with Dr. Knight to try it, as well as a compafs of his own invention.

In 1753 he was elected a member of the Royal Society; and in 1759 he was honoured with their gold medal, for his paper concerning the natural powers of water and wind to turn mills, and other machines depending on a circular motion. This paper, he fays, was the refult of experiments made on working models in the years 1752 and 1753, but not communicated to the Society till 1759, having in the interval found opportunities of putting the refult of thefe experiments into real practice, in a variety of cafes, and for various purpofes, fo as to affure the Society he had found them to anfwer.

In 1754 his great thirft after experimental knowledge led him to undertake a voyage to Holland and the Low Countries, where he made himself acquainted with moft of the curious works of art fo frequent in thofe places.

In December 1755, the Edyftone lighthoufe was burnt down, and the proprietors, being defirous of rebuilding it in the moft fubftantial manner, enquired of the earl of Macclesfield, then prefident of the Royal Society, who he thought might be the fitteft perfon to rebuild it, when he immediately recommended our author. Mr. Smeaton accordingly undertook the work, which he completed with ftone in the fummer of 1759. Of this work he gives an ample defcription in a folio volume, with plates, publifhed in 1791. A work which contains, in a great meafure, the hiftory of four years of his life, in which the originality of his genius is fully difplayed, as well as his activity, induftry, and perfeverance.

Though Mr. Smeaton completed the building of the Edyftone lighthoufe in 1759, yet it feems he did not foon get into full bufinefs as a civil engineer; for in 1764, while in Yorkfhire, he offered himfelf a candidate for one of the receivers of the Derwentwater eftate; in which he fucceeded, though two other perfons, ftrongly recommended and powerfully fupported, were candidates for the employment. In this appointment he was very happy, by the affiftance and abilities of his partner Mr. Walton the younger, of Farnacres near Newcaftle, one of the prefent receivers, who, taking upon himfelf the management and the accounts, left Mr. Smeaton leifure and opportunity to exert his abilities on public works, as well as to make many improvements in the mills, and in the eftates of Greenwich hofpital.

By the year 1775, he had fo much bufinefs, as a civil engineer, that he was defirous of refigning the appointment for that hofpital, and would have done it then, had not his friends prevailed upon him to continue in the office about two years longer.

Mr. Smeaton having thus got into full bufinefs as a civil engineer, it would be an endlefs tafk to enumerate all the variety of concerns he was engaged in. A very few of them however may be juft mentioned in this place. —He made the river Calder navigable: a work that required great fkill and judgment; owing to the very impetuous floods in that river.—He planned and attended the execution of the great canal in Scotland, for conveying the trade of the country, either to the Atlantic or German ocean; and having brought it to a conclufion, he declined a handfome yearly falary, that he might not be prevented from attending to the multiplicity of his other bufinefs.

On opening the great arch at London bridge, the excavation around and under the fterlings was fo confiderable, that it was thought the bridge was in great danger of falling; the apprehenfions of the people on this head being fo great, that few would pafs over or under it. He was then in Yorkfhire, where he was fent for by exprefs, and he arrived in town with the greateft expedition. He applied himfelf immediately to examine it, and to found about the fterlings as minutely as he could. The committee being called together, adopted his advice, which was, to repurchafe the ftones that had been taken from the middle pier, then lying in Moorfields, and to throw them into the river to guard the fterlings, a practice he had before adopted on other occafions. Nothing fhews the apprehenfions of the bridge falling, more than the alacrity with which his advice was purfued: the ftones were repurchafed that day; horfes, carts, and barges were got ready, and the work inftantly begun though it was Sunday morning. Thus Mr. Smeaton, in all human probability, faved London bridge from falling, and fecured it till more effectual methods could be taken.

In 1771, he became, jointly with his friend Mr. Holmes above mentioned, proprietor of the works for fupplying Deptford and Greenwich with water; which by their united endeavours they brought to be of general ufe to thofe they were made for, and moderately beneficial to themfelves.

About the year 1785, Mr. Smeaton's health began to decline; in confequence he then took the refolution to endeavour to avoid any new undertakings in bufinefs as much as he could, that he might thereby alfo have the more leifure to publifh fome account of his inventions and works. Of this plan however he got no more executed than the account of the Edyftone lighthoufe, and fome preparations for his intended treatife on mills; for he could not refift the folicitations of his friends in various works; and Mr. Aubert, whom he greatly loved and refpected, being chofen chairman of Ramfgate harbour, prevailed upon him to accept the office of engineer to that harbour; and to their joint efforts the public are chiefly indebted for the improvements that have been made there within thefe few years; which fully appears in a report that Mr. Smeaton gave in to the board of truftees in 1791, which they immediately publifhed.

It had for many years been the practice of Mr. Smeaton to fpend part of the year in town, and the remainder in the country, at his houfe at Aufthorpe; on one of thefe excurfions in the country, while walking in his garden, on the 16th of September 1792, he was ftruck with the palfy, which put an end to his ufeful life the 28th of October following, to the great regret of a numerous fet of friends and acquaintances.

The great variety of mills conftructed by Mr. Smeaton, fo much to the fatisfaction and advantage of the owners, will fhew the great ufe he made of his experiments

ments in 1752 and 1753. Indeed he scarcely trusted to theory in any case where he could have an opportunity to investigate it by experiment; and for this purpose he built a steam-engine at Austhorpe, that he might make experiments expressly to ascertain the power of Newcomen's steam-engine, which he improved and brought to a much greater degree of certainty, both in its construction and powers, than it was before.

During many years of his life, Mr Smeaton was a constant attendant on parliament, his opinion being continually called for. And here his natural strength of judgment and perspicuity of expression had their full display. It was his constant practice, when applied to, to plan or support any measure, to make himself fully acquainted with it, and be convinced of its merits, before he would be concerned in it. By this caution, joined to the clearness of his description, and the integrity of his heart, he seldom failed having the bill he supported carried into an act of parliament. No person was heard with more attention, nor had any one ever more confidence placed in his testimony. In the courts of law he had several compliments paid to him from the bench, by the late lord Mansfield and others, on account of the new light he threw upon difficult subjects.

As a civil engineer, he was perhaps unrivalled, certainly not excelled by any one, either of the present or former times. His building the Edystone lighthouse, were there no other monument of his fame, would establish his character. The Edystone rocks have obtained their name from the great variety of contrary *sets* of the tide or current in their vicinity. They are situated nearly S. S. W. from the middle of Plymouth Sound. Their distance from the port of Plymouth is about 14 miles. They are almost in the line which joins the Start and the Lizard points; and as they lie nearly in the direction of vessels coasting up and down the channel, they were unavoidably, before the establishment of a light-house there, very dangerous, and often fatal to ships. Their situation with regard to the Bay of Biscay and the Atlantic is such, that they lie open to the swells of the bay and ocean, from all the south-western points of the compass; so that all the heavy seas from the south-west quarter come uncontroled upon the Edystone rocks, and break upon them with the utmost fury. Sometimes, when the sea is to all appearance smooth and even, and its surface unruffled by the slightest breeze, the *ground swell* meeting the slope of the rocks, the sea beats upon them in a frightful manner, so as not only to obstruct any work being done on the rock, or even landing upon it, when, figuratively speaking, you might go to sea in a walnut-shell. That circumstances fraught with danger surrounding it should lead mariners to wish for a light-house, is not wonderful; but the danger attending the erection leads us to wonder that any one could be found hardy enough to undertake it. Such a man was first found in the person of Mr. H. Winstanley, who, in the year 1696, was furnished by the Trinity house with the necessary powers. In 1700 it was finished; but in the great storm of November 1703, it was destroyed, and the projector perished in the ruins. In 1709 another, upon a different construction, was erected by a Mr. Rudyerd, which, in 1755, was unfortunately consumed by fire. The next

building was under the direction of Mr. Smeaton, who, having considered the errors of the former constructions, has judiciously guarded against them, and erected a building, the demolition of which seems little to be dreaded, unless the rock on which it is erected should perish with it.——Of his works, in constructing bridges, harbours, mills, engines, &c, &c, it were endless to speak. Of his inventions and improvements of philosophical instruments, as of the air-pump, the pyrometer, hygrometer, &c, &c, some idea may be formed from the list of his writings inserted below.

In his person, Mr. Smeaton was of a middle stature, but broad and strong made, and possessed of an excellent constitution. He had a great simplicity and plainness in his manners: he had a warmth of expression that might appear, to those who did not know him well, to border on harshness; but such as were more closely acquainted with him, knew it arose from the intense application of his mind, which was always in the pursuit of truth, or engaged in the investigation of difficult subjects. He would sometimes break out hastily, when any thing was said that was contrary to his ideas of the subject; and he would not give up any thing he argued for, till his mind was convinced by sound reasoning.

In all the social duties of life, Mr. Smeaton was exemplary; he was a most affectionate husband, a good father, a warm, zealous and sincere friend, always ready to assist those he respected, and often before it was pointed out to him in what way he could serve them. He was a lover and an encourager of merit wherever he found it; and many persons now living are in a great measure indebted for their present situation to his assistance and advice. As a companion, he was always entertaining and instructive, and none could spend their time in his company without improvement.

As to the list of his writings; beside the large work abovementioned, being the History of Edystone Lighthouse, and numbers of reports and memorials, many of which were printed, his communications to the Royal Society, and inserted in their Transactions, are as follow:

1. An Account of Dr. Knight's Improvements of the Mariner's Compass; an. 1750, pa. 513.

2. Some improvements in the Air-pump; an. 1752, pa. 413.

3. An Engine for raising Water by Fire; being an improvement on Savary's construction, to render it capable of working itself: invented by M. de Moura, of Portugal. Ib. pa. 436.

4. Description of a new Tackle, or Combination of Pulleys. Ib. 494.

5. Experiments upon a machine for measuring the Way of a Ship at Sea. An. 1754, pa. 532.

6. Description of a new Pyrometer. Ib. pa. 598.

7. Effects of Lightning on the Steeple and Church of Lestwithial in Cornwall. An. 1757, pa. 198.

8. Remarks on the different Temperature of the Air at Edystone Light-house, and at Plymouth. An. 1758, pa. 488.

9. Experimental enquiry concerning the natural powers of Water and Wind to turn mills and other machines depending on a circular motion. An. 1759, pa. 100.

10. On

10. On the Menstrual Parallax arising from the mutual gravitation of the earth and moon, its influence on the observation of the sun and planets, with a method of observing it. An. 1768, pa. 156.

11. Description of a new method of Observing the heavenly bodies out of the meridian. An. 1768, pa. 170.

12. Observations on a Solar Eclipse. An. 1769, pa. 286.

13. Description of a new Hygrometer. An. 1771, pa. 198.

14. An Experimental Examination of the quantity and proportion of Mechanical Power, necessary to be employed in giving different degrees of velocity to heavy bodies from a state of rest. An. 1776, pa. 450.

SMOKE, or *Smoak*, a humid matter exhaled in form of vapour by the action of heat, either external or internal; or Smoke consists of palpable particles, elevated by means of the rarefying heat, or by the force of the ascending current of air, from certain bodies exposed to heat; which particles vary much in their properties, according to the substances from which they are produced.

Sir Isaac Newton observes, that Smoke ascends in the chimney by the impulse of the air it floats in: for that air, being rarefied by the heat of the fire underneath, has its specific gravity diminished; and thus, being disposed to ascend itself, it carries up the Smoke along with it. The tail of a comet, the same author supposes, ascends from the nucleus after the same manner.

Smoke of fat unctuous woods, as fir, beech, &c, makes what is called lamp-black.

There are various inventions for preventing and curing smoky chimneys: as the æolipiles of Vitruvius, the ventiducts of Cardan, the windmills of Bernard, the capitals of Serlio, the little drums of Paduanus, and several artifices of De Lorme. See also the philosophical works of Dr. Franklin. Pans, resembling sugar pans, placed over the tops of chimneys, are useful to make them draw better; and the fire-grates called register-stoves, are always a sure remedy.

In the Philosophical Transactions is the description of an engine, invented by M. Dalesme, which consumes the Smoke of all sorts of wood so effectually, that the eye cannot discover it in the room, nor the nose distinguish the smell of it, though the fire be made in the middle of the room. It consists of several iron hoops, 4 or 5 inches in diameter, which shut into one another, and is placed on a trevet.

The late invention called Argand's lamp, also consumes the Smoke, and gives a very strong light. Its principle is a thin broad cotton wick, rolled into the form of a hollow cylinder; the air passes up the hollow of it, and the Smoke is almost all consumed.

SMOKE *Jack*, is a jack for turning a spit, turned by the Smoke of the kitchen fire, by means of thin iron sails set obliquely on an axis in the flue of the chimney. See JACK.

SNELL (RODOLPH), a respectable Dutch philosopher, was born at Oudenwater in 1546. He was some time professor of Hebrew and mathematics at Leyden, where he died in 1613, at 67 years of age. He was author of several works on geometry, and on all parts of

Vol. II.

the philosophy of his time; but I have not obtained a particular list of them.

SNELL (*Willebrord*), son of Rodolph above mentioned, an excellent mathematician, was born at Leyden in 1591, where he succeeded his father in the mathematical chair in 1613, and where he died in 1626, at only 35 years of age.

Willebrord Snell was author of several ingenious works and discoveries. Thus, it was he who first discovered the true law of the refraction of the rays of light; a discovery which he made before it was announced by Des Cartes, as Huygens assures us. Though the work which Snell prepared upon this subject, and upon optics in general, was never published, yet the discovery was very well known to belong to him, by several authors about his time, who had seen it in his manuscripts.——He undertook also to measure the earth. This he effected by measuring a space between Alcmaer and Bergen op-zoom, the difference of latitude between these places being 1° 11′ 30″. He also measured another distance between the parallels of Alcmaer and Leyden; and from the mean of both these measurements, he made a degree to consist of 55021 French toises or fathoms. These measures were afterwards repeated and corrected by Musschenbroek, who found the degree to contain 57033 toises.——He was author of a great many learned mathematical works, the principal of which are,

1. *Apollonius Batavus*; being the restoration of some lost pieces of Apollonius, concerning Determinate Section, with the Section of a Ratio and Space: in 4to, 1608, published in his 17th year.

2. *Eratosthenes Batavus*; in 4to, 1617. Being the work in which he gives an account of his operations in measuring the earth.

3. A translation out of the Dutch language, into Latin, of Ludolph van Collen's book *De Circulo & Adscriptis*, &c; in 4to, 1619.

4. *Cyclometricus, De Circuli Dimensione* &c; 4to, 1621. In this work, the author gives several ingenious approximations to the measure of the circle, both arithmetical and geometrical.

5. *Tiphis Batavus*; being a treatise on Navigation and Naval Affairs; in 4to, 1624.

6. A posthumous treatise, being four books *Doctrinæ Triangulorum Canonicæ*; in 8vo, 1627. In which are contained the canon of secants; and in which the construction of sines, tangents, and secants, with the dimension or calculation of triangles, both plane and spherical, are briefly and clearly treated.

7. Hessian and Bohemian Observations; with his own notes.

8. *Libra Astronomica & Philosophica*; in which he undertakes the examination of the principles of Galileo concerning comets.

9. Concerning the Comet which appeared in 1618, &c.

SNOW, a well known meteor, formed by the freezing of the vapours in the atmosphere. It differs from hail and hoar-frost in being as it were crystallized, which they are not. This appears on examination of a flake of Snow by a magnifying glass; when the whole of it appears to be composed of fine shining spicula diverging like rays from a centre. As the flakes descend

3 O

turough

through the atmosphere, they are continually joined by more of these radiated spicula, and thus increase in bulk like the drops of rain or hailstones; so that it seems as if the whole body of Snow were an infinite mass of icicles irregularly figured.

The lightness of Snow, although it is firm ice, is owing to the excess of its surface, in comparison to the matter contained under it; as even gold itself may be extended in surface, till it will float upon the least breath of air.

According to Beccaria, clouds of Snow differ in nothing from clouds of rain, but in the circumstance of cold that freezes them. Both the regular diffusion of the Snow, and the regularity of the structure of its parts, shew that clouds of Snow are acted upon by some uniform cause like electricity; and he endeavours to shew how electricity is capable of forming these figures. He was confirmed in his conjectures by observing, that his apparatus for shewing the electricity of the atmosphere, never failed to be electrified by Snow as well as by rain. Professor Wintrop sometimes found his apparatus electrified by Snow when driven about by the wind, though it had not been affected by it when the Snow itself was falling. A more intense electricity, according to Beccaria, unites the particles of hail more closely than the more moderate electricity does those of Snow, in the same manner as we see that the drops of rain which fall from the thunder-clouds, are larger than those which fall from others, though the former descend through a less space.

In the northern countries, the ground is covered with snow for several months; which proves exceedingly favourable for vegetation, by preserving the plants from those intense frosts which are common in such countries, and which would certainly destroy them. Bartholin ascribes great virtues to Snow water, but experience does not seem to warrant his assertions. Snow-water, or ice-water, is always deprived of its fixed air: and those nations who live among the Alps, and use it for their constant drink, are subject to affections of the throat, which it is thought are occasioned by it.

From some late experiments on the quantity of water yielded by Snow, it appears that the latter gives only about one-tenth of its bulk in water.

SOCIETY, an assemblage or union of several learned persons, for their mutual assistance, improvement, or information, and for the promotion of philosophical or other knowledge. There are various philosophical Societies instituted in different parts of the world. See ROYAL *Society*.

American Philosophical SOCIETY, was established at Philadelphia in the year 1769, for promoting useful knowledge, under the direction of a patron, a president, three vice-presidents, a treasurer, four secretaries, and three curators. The first volume of their Transactions comprehends a period of two years, viz, from Jan. 1, 1769, to Jan. 1, 1771. Their labours seem to have been interrupted during the troubles in America, which commenced soon after; but since their termination, some more volumes have been published, containing a number of very ingenious and useful memoirs.

American Academy of Arts and Sciences, was established by a law of the Commonwealth of Massachusetts in North America, in the year 1780.

Boston Academy of Arts and Sciences. This is a Society similar to the former, which has lately been established at Boston in New England, under the title of the Academy of Arts and Sciences &c.

Berlin SOCIETY. The Society of Natural Historians at Berlin, was founded by Dr. Martini. There is also a Philosophical Society in the same place.

Brussels SOCIETY. The Imperial and Royal Academy of Sciences and Belles Lettres of Brussels was founded in 1773. Several volumes of their Transactions have now been published.

Dublin SOCIETY. This is an Experimental Society, for promoting natural knowledge, which was instituted in 1777: the members meet once a week, and distribute three honorary gold medals annually for the most approved discovery, invention, or essay, on any mathematical or philosophical subject. The Society is under the direction of a president, two vice presidents, and a secretary.

Edinburgh Philosophical SOCIETY, succeeded the Medical Society, and was formed upon the plan of including all the different branches of natural knowledge and the antiquities of Scotland. The meetings of this Society, interrupted in 1745, were revived in 1752; and in 1754 the first volume of their collection was published, under the title of Essays or Observations Physical and Literary, which has been succeeded by other volumes. This Society has been lately incorporated by royal charter, under the name of the Royal Society of Scotland, instituted for the advancement of learning and useful knowledge. The members are divided into two classes, physical and literary; and those who are near enough to Edinburgh to attend the meetings, pay a guinea on admission, and the same sum annually. The first meeting was held on the first Monday of August 1783; when there were chosen, a president, two vice-presidents, a secretary, treasurer, and a council of 12 persons. Three of the volumes of their Transactions have been published, which are very respectable both for their magnitude and contents.

In *France* there have been several institutions of this kind for the improvement of science, besides those recounted under the word ACADEMY: As, the Royal Academy at Soissons, founded in 1674; at Villefranche, Beaujolois, in 1679; at Nismes, in 1682; at Angers, in 1685; the Royal Society at Montpelier, in 1706, which is so intimately connected with the Royal Academy of Sciences of Paris, as to form with it, in some respects, one body; the literary productions of this Society are published in the memoirs of the academy: the Royal Academy of Sciences and Belles Lettres at Lyons, in 1700; at Bourdeaux, in 1703; at Marseilles, in 1726; at Rochelle, in 1734; at Dijon, in 1740; at Pau in Bern, in 1721; at Beziers, in 1723; at Montauban, in 1744; at Rouen, in 1744; at Amiens, in 1750; at Toulouse, in 1750; at Besançon, in 1752; at Metz, in 1760; at Arras, in 1773; and at Chalons sur Maine, in 1775. For other institutions of a similar nature, and their literary productions, see the articles ACADEMY, JOURNAL, and TRANSACTIONS.

Manchester Literary and Philosophical SOCIETY, is of considerable reputation, and has been lately established there, under the direction of two presidents, four vice-presidents, and two secretaries. The number of of members is limited to 50; befides thefe there are feveral honorary members, all of whom are elected by ballot; and the officers are chofen annually in April. Several valuable effays have been already read at the meetings of this Society.

*Newcaſtle-upon-Tyne Literary and Philoſophical So-*CIETY. This Society was inſtituted the 7th of February 1793, under the direction of a preſident four vice-preſidents, two ſecretaries, a treaſurer, which together with four of the ordinary members form a committee, all annually elected at a general meeting. The ſubjects propoſed for the confideration and improvement of this Society, comprehend the mathematics, natural philoſophy and hiſtory, chemiſtry, polite literature, antiquities, civil hiſtory, biography, queſtions of general law and policy, commerce, and the arts. From ſuch ample ſcope in the objects of the Society, with the known reſpectability, zeal, and talents of the members, the greateſt improvements and diſcoveries may be expected to be made in thoſe important branches of uſeful knowledge.

SOCRATES, the chief of the ancient philoſophers, was born at Alopece, a ſmall village of Attica, in the 4th year of the 77th olympiad, or about 467 years before Chriſt. Sophroniſcus, his father, being a ſtatuary or carver of images in ſtone, our author followed the ſame profeſſion for ſome time; for a ſubſiſtence. But being naturally averſe to this profeſſion; he only followed it when neceſſity compelled him; and upon getting a little before hand would for a while lay it afide. Theſe intermiſſions of his trade were beſtowed upon philoſophy, to which he was naturally addicted; and this being obſerved by Crito, a rich philoſopher of Athens, Socrates was at length taken from his ſhop, and put into a condition of philoſophiſing at his eaſe and leiſure.

He had various inſtructors in the ſciences, as Anaxagoras, Archylaus, Damon, Prodicus, to whom may be added the two learned women Diotyma and Aſpaſia, of the laſt of whom he learned rhetoric: of Euenus he learned poetry; of Ichomachus, huſbandry; and of Theodorus, geometry.

At length he began vhimſelf to teach; and was ſo eloquent, that he could lead the mind to approve or diſapprove whatever he pleaſed; but never uſed this talent for any other purpoſe than to conduct his fellow-citizens into the path of virtue. The academy of the Lycæum, and a pleaſant meadow without the city on the ſide of the river Ilyſſus, were places where he chiefly delivered his inſtructions, though it ſeems he was never out of his way in that reſpect, as he made uſe of all times and places for that purpoſe.

He is repreſented by Xenophon as excellent in all kinds of learning, and particularly inſtances arithmetic, geometry, and aſtrology or aſtronomy. Plato mentions natural philoſophy; Idomeneus, rhetoric; Laertius, medicine. Cicero affirms, that by the teſtimony of all the learned, and the judgment of all Greece, he was, as well in wiſdom, acuteneſs, politeneſs, and ſubtlety, as in eloquence, variety, and richneſs, in whatever he applied himſelf to, without exception, the prince of all.

It has been obſerved by many, that Socrates little

affected travel; his life being wholly ſpent at home, excepting when he went out upon military ſervices. In the Peloponneſian war he was thrice perſonally engaged, upon which occaſions it is ſaid he outwent all the ſoldiers in hardineſs: and if at any time, ſaith Alcibiades, as it often happens in war, the proviſions failed, there were none who could bear the want of meat and drink like Socrates; yet, on the other hand, in times of feaſting, he alone ſeemed to enjoy them; and though of himſelf he would not drink, yet being invited he far outdrank every one, though he was never ſeen intoxicated.

To this great philoſopher Greece was principally indebted for her glory and ſplendor. He formed the manners of the moſt celebrated perſons of Greece, as Alcibiades, Xenophon, Plato, &c. But his great ſervices and the excellent qualities of his mind could not ſecure him from envy, perſecution, and calumny. The thirty tyrants forbad his inſtructing youth; and as he derided the plurality of the Pagan deities, he was accuſed of impiety. The day of trial being come, Socrates made his own defence, without procuring an advocate, as the cuſtom was, to plead for him. He did not defend himſelf with the tone and language of a ſuppliant or guilty perſon, but, as if he were maſter of the judges themſelves, with freedom, firmneſs, and ſome degree of contumacy. Many of his friends alſo ſpoke in his behalf; and laſtly, Plato went up into the chair, and began a ſpeech in theſe words: "Though I, Athenians, am the youngeſt of thoſe that come up into this place"—but they ſtopped him, crying out, " of thoſe that go down," which he was thereupon conſtrained to do; and then proceeding to vote, they condemned Socrates to death, which was effected by means of poiſon, when he was 70 years of age. Plato gives an affecting account of his impriſonment and death, and concludes, " This was the end of the beſt, the wiſeſt, and the juſteſt of men." And that account of it by Plato, Tully profeſſes, he could never read without tears.

As to the perſon of Socrates, he is repreſented as very homely; he was bald, had a dark complexion, a flat noſe, eyes ſticking out, and a ſevere downcaſt look. But the defects of his perſon were amply compenſated by the virtues and accompliſhments of his mind. Socrates was indeed a man of all virtues; and ſo remarkably frugal, that how little ſoever he had, it was always enough. When he was amidſt a great variety of rich and expenſive objects, he would often ſay to himſelf, " How many things are there which I do not want!"

Socrates had two wives, one of which was the noted Xantippe; whom Aulus Gellius deſcribes as an accurſed froward woman, always chiding and ſcolding, by day and by night, and whom it was ſaid he made choice of as a trial and exerciſe of his temper. Several inſtances are recorded of her impatience and his forbearance. One day, before ſome of his friends, ſhe fell into the uſual extravagances of her paſſion; when he, without anſwering a word, went abroad with them: but on his going out of the door, ſhe ran up into the chamber, and threw down water upon his head; upon which, turning to his friends, " Did not I tell you (fays he), that after fo much thunder we fhould have rain?" Another time fhe pulled his cloak from his fhoulders in the open forum; and fome of his friends advifing him to beat her, "Yes (fays he), that while we two fight, you may all ftand by, and cry, Well done, Socrates; to him, Xantippe."

They who affirm that Socrates wrote nothing, mean only in refpect to his philofophy; for it is attefted and allowed, that he affifted Euripides in compofing tragedies, and was the author of fome pieces of poetry. Dialogues alfo and epiftles are afcribed to him: but his philofophical difputations were committed to writing only by his fcholars; and that chiefly by Plato and Xenophon. The latter fet the example to the reft in doing it firft, and alfo with the greateft punctuality; as Plato did it with the moft liberty, intermixing fo much of his own, that it is hardly poffible to know what part belongs to each. Hence Socrates, hearing him recite his Lyfis, cried out, "How many things doth this young man feign of me!" Accordingly, the greateft part of his philofophy is to be found in the writings of Plato. To Socrates is afcribed the firft introduction of moral philofophy. Man having a twofold relation to things divine and human, his doctrines were with regard to the former metaphyfical, to the latter moral. His metaphyfical opinions were chiefly, that, There are three principles of all things, God, matter, and idea. God is the univerfal intellect; matter the fubject of generation and corruption; idea, an incorporeal fubftance, the intellect of God; God the intellect of the world. God is one, perfect in himfelf, giving the being and well-being of every creature.—That God, not chance, made the world and all creatures, is demonftrable from the reafonable difpofition of their parts, as well for ufe as defence; from their care to preferve themfelves, and continue their fpecies.—That he particularly regards man in his body, appears from his noble upright form, and from the gift of fpeech; in his foul, from the excellency of it above others.—That God takes care of all creatures, is demonftrable from the benefit he gives them of light, water, fire, and fruits of the earth in due feafon. That he hath a particular regard of man, from the deftination of all plants and creatures for his fervice; from their fubjection to man, though they may exceed him ever fo much in ftrength; from the variety of man's fenfe, accommodated to the variety of objects, for neceffity, ufe, and pleafure; from reafon, by which he difcourfeth through reminifcence from fenfible objects; from fpeech, by which he communicates all he knows, gives laws, and governs ftates. Finally, that God, though invifible himfelf, at once fees all, hears all, is every where, and orders all.

As to the other great object of metaphyfical refearch, the foul, Socrates taught, that it is pre-exiftent to the body, endued with the knowledge of eternal ideas, which in its union to the body it lofeth, as ftupefied, until awakened by difcourfe from fenfible objects; on which account, all its learning is only reminifcence, a recovery of its firft knowledge. That the body, being compounded, is diffolved by death; but that the foul, being fimple, paffeth into another life, incapable of corruption. That the fouls of men are divine. That the fouls of the good after death are in a happy ftate, united to God in a bleffed inacceffible

place; that the bad in convenient places fuffer condign punifhment.

All the Grecian fects of philofophers refer their origin to the difcipline of Socrates; particularly the Platonics, Peripatetics, Academics, Cyrenaics, Stoics, &c.

SOL, in Aftrology, &c, fignifies the fun.

SOLAR, fomething relating to the fun. Thus, we fay Solar fire in contradiftinction to culinary fire.

SOLAR *Civil Month.* See MONTH.

SOLAR *Cycle.* See CYCLE.

SOLAR *Comet.* See DISCUS.

SOLAR *Eclipfe,* is a privation of the light of the fun, by the interpofition of the opake body of the moon. See ECLIPSE.

SOLAR *Month, Rifing, Spots.* See the fubftantives.

SOLAR *Syftem,* the order and difpofition of the feveral heavenly bodies, which revolve round the fun as the centre of their motion; viz, the planets, primary and fecondary, and the comets. See SYSTEM.

SOLAR *Year.* See YEAR.

SOLID, in Phyfics, a body whofe minute parts are connected together, fo as not to give way, or flip from each other, on the fmalleft impreffion. The word is ufed in this fenfe, in contradiftinction to fluid.

SOLID, in Geometry, is a magnitude extended in every poffible direction, quite around. Though it is commonly faid to be endued with three dimenfions only, length, breadth, and depth or thicknefs.

Hence, as all bodies have thefe three dimenfions, and nothing but bodies, Solid and body are often ufed indifcriminately.

The extremes of Solids are furfaces. That is, Solids are terminated either by one furface, as a globe, or by feveral, either plane or curved. And from the circumftances of thefe, Solids are diftinguifhed into regular and irregular.

Regular SOLIDS, are thofe that are terminated by regular and equal planes. Thefe are the tetraedron, hexaedron, or cube, octaedron, dodecaedron, and icofaedron; nor can there poffibly be more than thefe five regular Solids or bodies, unlefs perhaps the fphere or globe be confidered as one of an infinite number of fides. See thefe articles feverally, alfo the article *Regular* BODY.

Irregular SOLIDS, are all fuch as do not come under the definition of regular ones: fuch as cylinder, cone, prifm, pyramid, &c.

Similar Solids are to one another in the triplicate ratio of their like fides, or as the cubes of the fame. And all forts of prifms, as alfo pyramids, are to one another in the compound ratio of their bafes and altitudes.

SOLID *Angle,* is that formed by three or more plane angles meeting in a point; like an angle of a die, or the point of a diamond well cut.

The fum of all the plane angles forming a Solid angle, is always lefs than 360°; otherwife they would conftitute the plane of a circle, and not a Solid.

Atmofphere of SOLIDS. See ATMOSPHERE.

SOLID *Baftion.* See BASTION.

Cubature of SOLIDS. See CUBATURE and SOLIDITY.

Meafure

Measure of a Solid. See Measure.

Solid Foot. See Foot.

Solid *Numbers*, are those which arise from the multiplication of a plane number, by any other number whatever. Thus, 18 is a Solid number, produced from the plane number 6 and 3, or from 9 and 2.

Solid *Place*. See Locus.

Solid *Problem*, is one which cannot be constructed geometrically; but by the intersection of a circle and a conic section, or by the intersection of two conic sections. Thus, to describe an isosceles triangle on a given base, so that either angle at the base shall be triple of that at the vertex, is a Solid problem, resolved by the intersection of a parabola and circle, and it serves to inscribe a regular heptagon in a given circle.

In like manner, to describe an isosceles triangle having its angles at the base each equal to 4 times that at the vertex, is a Solid problem, effected by the intersection of an hyperbola and a parabola, and serves to inscribe a regular nonagon in a given circle.

And such a problem as this has four solutions, and no more; because two conic sections can intersect but in 4 points.

How all such problems are constructed, is shewn by Dr. Halley, in the Philos. Transf. num. 188.

Solid of *Least Resistance*. See Resistance.

Surfaces of Solids. See Area and Superficies.

Solid *Theorem*. See Theorem.

SOLIDITY, in Physics, a property of matter or body, by which it excludes every other body from that place which is possessed by itself.

Solidity in this sense is a property common to all bodies, whether solid or fluid. It is usually called *impenetrability*; but Solidity expresses it better, as carrying with it somewhat more of positive than the other, which is a negative idea.

The idea of Solidity, Mr. Locke observes, arises from the resistance we find one body makes to the entrance of another into its own place. Solidity, he adds, seems the most extensive property of body, as being that by which we conceive it to fill space; it is distinguished from mere space, by this latter not being capable of resistance or motion.

It is distinguished from hardness, which is only a firm cohesion of the solid parts.

The difficulty of changing situation gives no more Solidity to the hardest body than to the softest; nor is the hardest diamond properly a jot more solid than water. By this we distinguish the idea of the extension of body, from that of the extension of space: that of body is the continuity or cohesion of solid, separable, moveable parts; that of space the continuity of unsolid, inseparable, immoveable parts.

The Cartesians however will, by all means, deduce Solidity, or as they call it impenetrability, from the nature of extension; they contend, that the idea of the former is contained in that of the latter; and hence they argue against a vacuum. Thus, say they, one cubic foot of extension cannot be added to another without having two cubic feet of extension; for each has in itself all that is required to constitute that magnitude. And hence they conclude, that every part of space is solid, or impenetrable, as of its own nature it excludes all others. But the conclusion is false, and the instance they give follows from this, that the parts of space are immoveable, not from their being impenetrable or solid. See Matter.

Solidity is also used for hardness, or firmness; as opposed to fluidity; viz, when body is considered either as fluid or solid, or hard or firm.

Solidity, in Geometry, denotes the quantity of space contained in a solid body, or occupied by it; called also the *solid content*, or the *cubical content*; for all solids are measured by cubes, whose sides are inches, or feet, or yards, &c; and hence the Solidity of a body is said to be so many cubic inches, feet, yards, &c, as will fill its capacity or space, or another of an equal magnitude.

The Solidity of a cube, parallelopipedon, cylinder, or any other prismatic body, i. e. one whose parallel sections are all equal and similar throughout, is found by multiplying the base by the height or perpendicular altitude. And of any cone or other pyramid, the Solidity is equal to one-third part of the same prism, because any pyramid is equal to the 3d part of its circumscribing prism. Also, because a sphere or globe may be considered as made up of an infinite number of pyramids, whose bases form the surface of the globe, and their vertices all meet in the centre, or having their common altitude equal to the radius of the globe; therefore the solid content of it is equal to one-third part of the product of its radius and surface. For the Solidity of other figures, see each figure separately.

The foregoing rules are such as are derived from common geometry. But there are in nature numberless other forms, which require the aid of other methods and principles, as follows.

Of the Solidity *of Bodies formed by a Plane revolving about any Axis, either within or without the Body.*—Concerning such bodies, there is a remarkable property or relation between their Solidity and the path or line described by the centre of gravity of the revolving plane; viz, the Solidity of the body generated, whether by a whole revolution, or only a part of one, is always equal to the product arising from the generating plane drawn into the path or line described by its centre of gravity, during its motion in describing the body. And this rule holds true for figures generated by all sorts of motion whatever, whether rotatory, or direct or parallel, or irregularly zigzag, &c, provided the generating plane vary not, but continue the same throughout. And the same law holds true also for all surfaces any how generated by the motion of a right line. This is called the Centrobaric method. See my Mensuration, sect. 3. part 4, pa. 501, 2d edit.

Of the Solidity *of Bodies by the Method of Fluxions.*—This method applies very advantageously in all cases also in which a body is conceived to be generated by the revolution of a plane figure about an axis, or, which is much the same thing, by the parallel motion of a circle, gradually expanding and contracting itself, according to the nature of the generating plane. And this method is particularly useful for the solids generated by any curvilineal plane figures. Thus, let the plane AED revolve about the axis AD; then it will generate the solid ABFEC. But as every ordinate DE, per-

pendicular

pendicular to the axis AD, describes a circle BCEF in the revolution, therefore the same solid may be conceived as generated by a circle BCEF, gradually expanding itself larger and larger, and moving perpendicularly along the axis AD. Consequently the area of that circle being drawn into the fluxion of the axis, will produce the fluxion of the solid; and therefore the fluent, when taken, will give the Solidity of that body. That is, AD × circle BCF, (whose radius is DE, or diameter BE) is the fluxion of the Solidity.

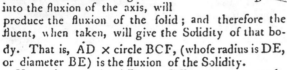

Hence then, putting AD $= x$, DE $= y$, $c = 3 \cdot 1416$; because cy^2 is equal to the area of the circle BCF; therefore cy^2 is the fluxion of the solid. Consequently if the value of either y^2 or x be found in terms of each other, from the given equation expressing the nature of the curve, and that value be substituted for it in the fluxional expression $cy^2\dot{x}$, the fluent of the resulting quantity, being taken, will be the required Solidity of the body.

For Ex. Suppose the figure of a parabolic conoid, generated by the rotation of the common parabola ADE about its axis AD. In this case, the equation of the curve of the parabola is $px = y^2$, where p denotes the parameter of the axis. Substituting therefore px instead of y^2, in the fluxion $cy^2 x$, it becomes $cpx\dot{x}$, and the fluent of this is $\frac{1}{2}cpx^2 = \frac{1}{2}cxy^2$ for the Solidity; that is, half the product of the base of the solid drawn into its altitude; for cy^2 is the area of the circular base BCF, and x is the altitude. And so on for other such figures. See the content of each solid under its proper article.

For the SOLIDITY of Irregular Solids, or such as cannot be considered as generated by some regular motion or description; they must either be considered as cut or divided into several parts of known forms, as prisms, or pyramids, or wedges, &c, and the contents of these parts found separately. Or, in the case of the smaller bodies, of forms so irregular as not to be easily divided in that way, put them into some hollow regular vessel, as a hollow cylinder or parallelopipedon, &c; then pour in water or sand so as it may fill the vessel just up to the top of the inclosed irregular body, noting the height it rises to; then take out the body, and note the height the fluid again stands at; the difference of these two heights is to be considered as the altitude of a prism of the same base and form as the hollow vessel; and consequently the product of that altitude and base will be the accurate Solidity of the immerged body, be it ever so irregular.

SOLSTICE, in Astronomy, is the time when the sun is in one of the solstitial points, that is, when he is at the greatest distance from the equator, which is now nearly 23° 28′ on either side of it. It is so called, because the sun then seems to stand still, and not to change his place, as to declination, either way.

There are two Solstices, in each year, when the sun is at the greatest distance on the north and south sides of the ecliptic; viz, the *estival* or *summer solstice*, and the *hyemal* or *winter solstice*.

The *Summer Solstice* is when the sun is in the tropic of Cancer; which is about the 21st of June, when he makes the longest day. And

The *Winter Solstice* is when he enters the first degree of Capricorn; which is about the 22d day of December, when he makes the shortest day.

This is to be understood, as in our northern hemisphere; for in the southern, the sun's entrance into Capricorn makes their summer Solstice, and that into Cancer the winter one. So that it is more precise and determinate, to say the northern and southern Solstice.

SOLSTITIAL *Points*, are those points of the ecliptic the sun is in at the times of the two Solstices, being the first points of Cancer and Capricorn, which are diametrically opposite to each other.

SOLSTITIAL *Colure*, is that which passes through the Solstitial points.

SOLUTION, in Mathematics, is the answering or resolving of a question or problem that is proposed. See RESOLUTION, and REDUCTION *of Equations*.

SOLUTION, in Physics, is the reduction of a solid or firm body, into a fluid state, by means of some menstruum.—Solution is often confounded with what is called dissolution, though there is a difference.

SOSIGENES, was an Egyptian mathematician, whose principal studies were chronology and the mathematics in general, and who flourished in the time of Julius Cæsar. He is represented as well versed in the mathematics and astronomy of the Ancients; particularly of those celebrated mathematicians, Thales, Archimedes, Hipparchus, Calippus, and many others, who had undertaken to determine the quantity of the solar year; which they had ascertained much nearer the truth than one can well imagine they should, with instruments so very imperfect; as may appear by reference to Ptolomy's Almagest.

It seems Sosigenes made great improvements, and gave proofs of his being able to demonstrate the certainty of his discoveries; by which means he became popular, and obtained repute with those who had a genius to understand and relish such enquiries. Hence he was sent for by Julius Cæsar, who being convinced of his capacity, employed him in reforming the calendar; and it was he who formed the Julian year which begins 45 years before the birth of Christ. His other works are lost since that period.

SOUND, in Geography, denotes a strait or inlet of the sea, between two capes or head-lands.

The SOUND is used, by way of eminence, for that celebrated strait which connects the German sea to the Baltic. It is situated between the island of Zealand and the coast of Schonen. It is about 16 leagues in length, and in general about 5 in breadth, except near the castle of Cronenberg, where it is but one; so that there is no passage for vessels but under the cannon of the fortress.

SOUND, in Physics, a perception of the mind, communicated by means of the ear; being an effect of the collision of bodies, and their consequent tremulous motion, communicated to the ambient fluid, and so propagated through it to the organs of hearing.

To illustrate the cause of Sound, it is to be observed, 1st, That a motion is necessary in the sonorous body for the production of sound. 2dly, That this motion exists first in the small and insensible parts of the sonorous bodies

bodies, and is excited in them by their mutual collision against each other, which produces the tremulous motion so observable in bodies that have a clear sound, as bells, musical chords, &c. 3dly, That this motion is communicated to, or produces a like motion in the air, or such parts of it as are fit to receive and propagate it. Lastly, That this motion must be communicated to those parts that are the proper and immediate instruments of hearing.

Now that motion of a sonorous body, which is the immediate cause of Sound, may be owing to two different causes ; either the percussion between it and other hard bodies, as in drums, bells, chords, &c ; or the beating and dashing of the sonorous body and the air immediately against each other, as in flutes, trumpets, &c.

But in both these cases, the motion, which is the consequence of the mutual action, as well as the immediate cause of the sonorous motion which the air conveys to the ear, is supposed to be an invisible, tremulous or undulating motion, in the small and insensible parts of the body. Perrault adds, that the visible motion of the grosser parts contributes no otherwise to Sound, than as it causes the invisible motion of the smaller parts, which he calls particles, to distinguish them from the sensible ones, which he calls parts, and from the smallest of all, which are called corpuscles.

The sonorous body having made its impression on the contiguous air, that impression is propagated from one particle to another, according to the laws of pneumatics.

A few particles, for instance, driven from the surface of the body, push or press their adjacent particles into a less space ; and the medium, as it is thus rarefied in one place, becomes condensed in the other ; but the air thus compressed in the second place, is, by its elasticity, returned back again, both to its former place and its former state ; and the air contiguous to that is compressed ; and the like obtains when the air less compressed, expanding itself, a new compression is generated. Therefore from each agitation of the air there arises a motion in it, analogous to the motion of a wave on the surface of the water ; which is called a *wave* or *undulation* of air.

In each wave, the particles go and return back again, through very short equal spaces ; the motion of each particle being analogous to the motion of a vibrating pendulum while it performs two oscillations ; and most of the laws of the pendulum, with very little alteration, being applicable to the former.

Sounds are as various as are the means that concur in producing them. The chief varieties result from the figure, constitution, quantity, &c. of the sonorous body ; the manner of percussion, with the velocity &c, of the consequent vibrations ; the state and constitution of the medium ; the disposition, distance, &c, of the organ ; the obstacles between the organ and the sonorous object and the adjacent bodies. The most notable distinction of Sounds, arising from the various degrees and combinations of the conditions above mentioned, are into *loud* and *low* (or strong and weak) ; into *grave* and *acute* (or sharp and flat, or high and low) ; and into *long* and *short*. The management of which is the office of music.

Euler is of opinion, that no Sound making fewer vibrations than 30 in a second, or more than 7520, is distinguishable by the human ear. According to this doctrine, the limit of our hearing, as to acute and grave, is an interval of 8 octaves. *Tentam. Nov. Theor. Mus. cap.* 1, *sect.* 13.

The velocity of Sound is the same with that of the aerial waves, and does not vary much, whether it go with the wind or against it. By the wind indeed a certain quantity of air is carried from one place to another ; and the Sound is accelerated while its waves move through that part of the air, if their direction be the same as that of the wind. But as Sound moves vastly swifter than the wind, the acceleration it will hereby receive is but inconsiderable ; and the chief effect we can perceive from the wind is, that it increases and diminishes the space of the waves, so that by help of it the Sound may be heard to a greater distance than otherwise it would.

That the air is the usual medium of Sound, appears from various experiments in rarefied and condensed air. In an unexhausted receiver, a small bell may be heard to some distance ; but when exhausted, it can scarce be heard at the smallest distance. When the air is condensed, the Sound is louder in proportion to the condensation, or quantity of air crowded in ; of which there are many instances in Hauksbee's experiments, in Dr. Priestley's, and others.

Besides, sounding bodies communicate tremors to distant bodies ; for example, the vibrating motion of a musical string puts others in motion, whose tension and quantity of matter dispose their vibrations to keep time with the pulses of air, propagated from the string that was struck. Galileo explains this phenomenon by observing, that a heavy pendulum may be put in motion by the least breath of the mouth, provided the blasts be often repeated, and keep time exactly with the vibrations of the pendulum ; and also by the like art in raising a large bell.

It is not air alone that is capable of the impressions of Sound, but water also ; as is manifest by striking a bell under water, the Sound of which may plainly enough be heard, only not so loud, and also a fourth deeper, according to good judges in musical notes. And Mersenne says, a Sound made under water is of the same tone or note, as if made in air, and heard under the water.

The velocity of Sound, or the space through which it is propagated in a given time, has been very differently estimated by authors who have written concerning this subject. Roberval states it at the rate of 560 feet in a second ; Gassendus at 1473 ; Mersenne at 1474 ; Duhamel, in the History of the Academy of Sciences at Paris, at 1338 ; Newton at 968 ; Derham, in whose measure Flamsteed and Halley acquiesce, at 1142.

The reason of this variety is ascribed by Derham, partly to some of those gentlemen using strings and plummets instead of regular pendulums ; and partly to the too small distance between the sonorous body and the place of observation ; and partly to no regard being had to the winds.

But by the accounts since published by M. Cassini de Thury, in the Memoirs of the Royal Acad. of Scien-

ccs

...es at Paris, 1738, where cannon were fired at various as well as great distances, under many varieties of weather, wind, and other circumstances, and where the measures of the different places had been settled with the utmost exactness, it was found that Sound was propagated, on a medium, at the rate of 1038 French feet in a second of time. But the French foot is in proportion to the English as 15 to 16; and consequently 1038 French feet are equal to 1107 English feet. Therefore the difference of the measures of Derham and Cassini is 35 English feet, or 33 French feet, in a second. The medium velocity of Sound therefore is nearly at the rate of a mile, or 5280 feet, in $4\frac{4}{5}$ seconds, or a league in 14 seconds, or 13 miles in a minute. But sea miles are to land miles nearly as 7 to 6; and therefore Sound moves over a sea mile in $5\frac{1}{4}$ seconds nearly, or a sea league in 16 seconds.

Farther, it is a common observation, that persons in good health have about 75 pulsations, or beats of the artery at the wrist, in a minute; consequently in 75 pulsations, Sound flies about 13 land miles, or $11\frac{1}{4}$ sea miles, which is about 1 land mile in 6 pulses, or one sea mile in 7 pulses, or a league in 20 pulses.

And hence the distance of objects may be found, by knowing the time employed by Sound in moving from those objects to an observer. For Ex. On seeing the flash of a gun at sea, if 54 beats of the pulse at the wrist were counted before the report was heard; the distance of the gun will easily be found by dividing 54 by 20, which gives 2·7 leagues, or about 8 miles.

Upon the nature, production, and propagation of Sound, see the article PHONICS and ECHO; also the Memoirs of the Acad. and the Philof. Tranf. in many places; Newton, Principia; Kircher, Mefurgia Univerfalis; Merfenne; Borelli, Del Suono; Priestley, Exper. and Obferv. vol. 5; Hales, Sonorum Doctrina rationalis et experimentalis; 4to 1778. See also an ingenious treatife published 1790, by Mr. Geo. Saunders, on Theatres; in which he relates many experiments made by himfelf, on the nature and propagation of Sound. In this work, he shews the great effect of water, and some other bodies, in conducting of Sound, probably by rendering the air more denfe near them. Some of his conclusions and observations are as follow:

Earth may be supposed to have a twofold property with respect to Sound. Being very porous, it absorbs Sound, which is counteracted by its property of conducting Sound, and occasions it to pass on a plane, in an equal proportion to its progress in air, unencumbered by any body.

If a Sound be sufficiently intense to imprefs the earth in its tremulous quality, it will be carried to a confiderable di...ce, as when the earth is struck with any thing hard, as by the motion of a carriage, horfes feet, &c.

Platter is proportionally better than loose earth for conducting Sound, as it is more compact.

Clothes of every kind, particularly woollen cloths, are very prejudicial to Sound: their absorption of Sound, may be compared to that of water, which they greedily imbibe.

A number of people feated before others, as in the pit or gallery of a theatre, do considerably prevent the voice reaching those behind; and hence it is, that

we hear fo much better in the front of the galleries, or of any situation, than behind others, though we may be nearer to the speaker. Our feats, rifing fo little above each other, occasion this defect, which would be remedied, could we have the feats to rife their whole height above each other, as in the ancient theatres.

Paint has generally been thought unfavourable to Sound, from its being fo to mufical instruments, whose effects it quite deftroys.

Mufical instruments moftly depend on the vibrative or tremulous property of the material, which a body of colour hardened in oil muft very much alter; but we should diftinguish that this regards the formation of Sound, which may not altogether be the cafe in the progrefs of it.

Water has been little noticed, with refpect to its conducting Sound; but it will be found to be of the greateft confequence. I had often perceived in newly-finished houfes, that while they were yet damp, they produced echoes; but that the echoing abated as they dried.

Exp. When I made the following experiment there was a gentle wind; consequently the water was proportionally agitated. I chofe a quiet part of the river Thames, near Chelfea Hofpital, and with two boats tried the diftance the voice would reach. On the water we could diftinctly hear a perfon read at the diftance of 140 feet, on land at that of 76. It should be obferved, that on land no noife intervened; but on the river fome noife was occafioned by the flowing of the water against the boats; fo that the difference on land and on water muft be much more.

Watermen obferve, that when the water is still, and the weather quite calm, if no noife intervene, a whifper may be heard acrofs the river; and that with the current it will be carried to a much greater diftance, and vice verfa against the current.

Mariners well know the difference of Sound on fea and land.

When a canal of water was laid under the pit floor of the theatre of Argentino, at Rome, a furprising difference was obferved; the voice has fince been heard at the end very diftinctly, where it was before fcarce diftinguishable. It is obfervable that, in this part, the canal is covered with a brick arch, over which there is a quantity of earth, and the timber floor over all.

The villa Simonetta near Milan, fo remarkable for its echoes, is entirely over arcades of water.

Another villa near Rouen, remarkable for its echo, is built over fubterraneous cavities of water.

A refervoir of water domed over, near Stanmore, has a strong echo.

I do not remember ever being under the arches of a ftone bridge that did not echo; which is not always the cafe with fimilar ftructures on land.

A houfe in Lambeth Marfh, inhabited by Mr. Turtle, is very damp during winter, when it yields an echo which abates as the houfe becomes dry in summer.

Kircher obferves, that echoes repeat more by night than during the day: he makes the difference to be double.

Dr Plott fays, the echo in Woodftock park repeated 17 times by day, and 20 by night. And Addifon's experi-

experiment at the Villa Simonetta was in a fog, when it produced 56 repetitions.

After all thefe inftances, I think little doubt can remain of the influence water has on Sound; and I conclude that it conducts Sound more than any other body whatever.

After water, ftone may be reckoned the beft conductor of Sound. To what caufe it may be attributed, I leave to future enquiries: I have confined myfelf to fpeak of facts only as they appear.

Stone is fonorous, but gives a harfh difagreeable tone, unfavourable to mufic.

Brick, in refpect to Sound, has nearly the fame properties as ftone. Part of the garden wall of the late W. Pitt, Efq. of Kingfton in Dorfetfhire, conveys a whifper to the diftancé of near 200 feet.

Wood is fonorous, conductive, and vibrative; of all materials it produces a tone the moft agreeable and melodious; and it is therefore the fitteft for mufical inftruments, and for lining of rooms and theatres.

The common notion that whifpering at one end of a long piece of timber would be heard at the other end, I found by experiment to be erroneous. A ftick of timber 65 feet long being flightly ftruck at one end, a found was heard at the other, and the tremor very perceptible: which is eafily accounted for, when we confider the number or length of the fibres that compofe it, each of which may be compared to a ftring of catgut.

For the Reflection, Refraction, &c, of Sound; *fee* Echo, *and* Phonics.

Articulate Sound. See Articulate.

Sound, in Mufic, denotes a quality in the feveral agitations of the air, fo as to make mufic or harmony.

Sound is the object of mufic; which is nothing but the art of applying Sounds, under fuch circumftances of tone and time, as to raife agreeable fenfations. The principal affection of Sound, by which it becomes fitted to have this end, is that by which it is diftinguifhed into acute and grave. This difference depends on the nature of the fonorous body; the particular figure and quantity of it; and even in fome cafes, on the part of the body where it is ftruck: and it is this that conftitutes what are called *different tones.*

The caufe of this difference appears to be no other than the different velocities of the vibrations of the founding body. Indeed the tone of a Sound is found, by numerous experiments, to depend on the nature of thofe vibrations, whofe differences we can conceive no otherwife than as having different velocities: and fince it is proved that the fmall vibrations of the fame chord are all performed in equal times, and that the tone of a Sound, which continues for fome time after the ftroke, is the fame from firft to laft, it follows, that the tone is neceffarily connected with a certain quantity of time in making each vibration, or each wave; or that a certain number of vibrations or waves, made in a given time, conftitute a certain and determinate tone. From this principle are all the phænomena of tune deduced.

If the vibrations be ifochronous, or performed in the

fame time, the Sound is called mufical, and is faid to continue at the fame pitch; and it is alfo accounted acuter, fharper, or higher than any other Sound, whofe vibrations are flower, and therefore graver, flatter, or lower, than any other whofe vibrations are quicker. See Unison.

From the fame principle arife what are called *concords,* &c; which refult from the frequent unions and coincidences of the vibrations of two fonorous bodies, and confequently of the pulfes or the waves of the air occafioned by them.

On the contrary, the refult of lefs frequent coincidences of thofe vibrations, is what is called *difcord.*

Another confiderable diftinction of mufical Sounds, is that by which they are called *long* and *fhort,* owing to the continuation of the impulfe of the efficient caufe on the fonorous body for a longer or fhorter time, as in the notes of a violin &c, which are made longer or fhorter by ftrokes of different length or quicknefs. This continuity is properly a fucceffion of feveral Sounds, or the effect of feveral diftinct ftrokes, or repeated impulfes, on the fonorous body, fo quick, that we judge it one continued Sound, efpecially where it is continued in the fame degree of ftrength; and hence arifes the doctrine of *meafure* and *time.*

Mufical Sounds are alfo divided into *fimple* and *compound;* and that in two different ways. In the firft, a Sound is faid to be compound, when a number of fucceffive vibrations of the fonorous body, and the air, come fo faft upon the ear, that we judge them the fame continued Sound; like as in the phenomenon of the circle of fire, caufed by putting the fired end of a ftick in a quick circular motion; where fuppofing the end of the ftick in any point of the circle, the idea we receive of it there continues till the impreffion is renewed by a fudden return.

A *Simple* Sound then, with regard to this compofition, fhould be the effect of a fingle vibration, or of as many vibrations as are neceffary to raife in us the idea of Sound.

In the fecond fenfe of compofition, a fimple Sound is the product of one voice, or one inftrument, &c.

A *Compound* Sound confifts of the Sounds of feveral diftinct voices or inftruments all united in the fame individual time, and meafure of duration, that is, all ftriking the ear together, whatever their other differences may be. But in this fenfe again, there is a twofold compofition; a natural and an artificial one.

The natural compofition is that proceeding from the manifold reflections of the firft Sound from adjacent bodies, where the reflections are not fo fudden as to occafion echoes, but are all in the fame tune with the firft note.

The artificial compofition, which alone comes under the mufician's province, is that mixture of feveral Sounds, which being made by art, the ingredient Sounds are feparable, and diftinguifhable from one another. In this fenfe the diftinct Sounds of feveral voices or inftruments, or feveral notes of the fame inftrument, are called fimple Sounds, in contradiftinction to the compound ones, in which, to anfwer the end

of

of mufic, the fimples muft have fuch an agreement in all relations, chiefly as to acutenefs and gravity, as that the ear may receive the mixture with pleafure.

Another diftinction of Sounds, with regard to mufic, is that by which they are faid to be *fmooth* or *even*, and *rough* or *harfh*, alfo *clear* and *hoarfe :* the caufe of which difference depends on the difpofition and ftate of the fonorous body, or the circumftances of the place ; but the ideas of the differences muft be fought from obfervation.

Smooth and *Rough* Sounds depend chiefly on the founding body ; of which we have a notable inftance in ftrings that are uneven, and not of the fame dimenfion and conftitution throughout.

As to *clear* and *hoarfe* Sounds, they depend on circumftances that are accidental to the fonorous body. Thus, a voice or inftrument will be hollow and hoarfe if founded within an empty hogfhead, that yet is clear and bright out of it : the effect is owing to the mixture of different Sounds, raifed by reflections, which corrupt and change the fpecies of the primitive Sound.

For Sounds to be fit to obtain the end of mufic, they ought to be fmooth and clear, efpecially the firft ; fince, without this, they cannot have one certain and difcernible tone, capable of being compared to others, in a certain relation of acutenefs, which the ear may judge of. So that, with Malcolm, we call that an harmonic or mufical Sound which, being clear and even, is agreeable to the ear, and gives a certain and difcernible tune (hence called tunable Sound), which is the fubject of the whole theory of harmony.

Wood has a particular vibrating quality, owing to its elafticity ; and all mufical inftruments made of this matter, are of a thicknefs proportioned to the fuperficies of the wood, and the tone they are to produce.

Metals are fonorous and vibrative, producing a harfh tone, very ferviceable to fome parts of mufic. Moft wind inftruments are made of metal, which is acted upon in its elaftic and tremulous quality, being capable of being reduced very thin for that purpofe. Inftruments of this kind are fuch as horns, trumpets, &c. Some inftruments however depend more on the form than the material ; as flutes, for inftance, which, if their lengths and bore be the fame, have very little difference in their Sounds, whatever the matter of them may be. See HARMONICAL.

SOUND-BOARD, the principal part of an organ, and that which makes the whole machine play. This Sound-board, or fummer, is a refervoir into which the wind, drawn in by the bellows, is conducted by a portvent, and thence diftributed into the pipes placed over the holes of its upper part. This wind enters them by valves, which open by preffing upon the ftops or keys, after drawing the regifters, which prevent the air from going into any of the other pipes befide thofe it is required in.

SOUND-*board* denotes alfo a thin broad board placed over the head of a public fpeaker, to enlarge and extend or ftrengthen his voice.

Sound-boards, in theatres, are found by experience to be of no fervice ; their diftance from the fpeaker

being too great, to be impreffed with fufficient force. But Sound-boards immediately over a pulpit have often a good effect, when the cafe is made of a juft thicknefs, and according to certain principles.

SOUND-*Poft*, is a poft placed withinfide of a violin, &c, as a prop between the back and the belly of the inftrument, and nearly under the bridge.

SOUNDING, in Navigation, the act of trying the depth of the water, and the quality of the bottom, by a line and plummet, or other artifice.

At fea, there are two plummets ufed for this purpofe, both fhaped like the fruftum of a cone or pyramid. One of thefe is called the hand-lead, weighing about 8 or 9lb ; and the other the deep-fea-lead, weighing from 25 to 30lb. The former is ufed in fhallow waters, and the latter at a great diftance from the fhore. The line of the hand-lead, is about 25 fathoms in length, and marked at every 2 or 3 fathoms, in this manner, viz, at 2 and 3 fathoms from the lead there are marks of black leather ; at 5 fathoms a white rag, at 7 a red rag, at 10 and at 13 black leather, at 15 a white rag, and at 17 a red one.

Sounding with the hand-lead, which the feamen call heaving the lead, is generally performed by a man who ftands in the main-chains to windward. Having the line all ready to run out, without interruption, he holds it nearly at the diftance of a fathom from the plummet, and having fwung the latter backwards and forwards three or four times, in order to acquire the greater velocity, he fwings it round his head, and thence as far forward as is neceffary ; fo that, by the lead's finking whilft the fhip advances, the line may be almoft perpendicular when it reaches the bottom. The perfon founding then proclaims the depth of the water in a kind of fong refembling the cries of hawkers in a city ; thus, if the mark of 5 be clofe to the furface of the water, he calls, ' by the mark 5,' and as there is no mark at 4, 6, 8, &c, he eftimates thofe numbers, and calls, ' by the dip four, &c.' If he judges it to be a quarter or a half more than any particular number, he calls, ' and a quarter 5,' ' and a half 4' &c. If he conceives the depth to be three quarters more than a particular number, he calls it a quarter lefs than the next : thus, at 4 fathom $\frac{3}{4}$, he calls, ' a quarter lefs 5,' and fo on.

The deep-fea-lead line is marked with 2 knots at 20 fathom, 3 at 30, 4 at 40, &c to the end. It is alfo marked with a fingle knot at the middle of each interval, as at 25, 35, 45 fathoms, &c. To ufe this lead more effectually at fea, or in deep water on the fea-coaft, it is ufual previoufly to bring-to the fhip, in order to retard her courfe : the lead is then thrown as far as poffible from the fhip on the line of her drift, fo that, as it finks, the fhip drives more perpendicularly over it. The pilot feeling the lead ftrike the bottom, readily difcovers the depth of the water by the mark on the line neareft its furface. The bottom of the lead, which is a little hollowed there for the purpofe, being alfo well rubbed over with tallow, retains the diftinguifhing marks of the bottom, as fhells, ooze, gravel, &c, which naturally adhere to it.

The depth of the water, and the nature of the ground, which are called the Soundings, are carefully marked in the log-book, as well to determine the diftance of the

the place from the shore, as to correct the observations of former pilots. *Falconer.*

For a machine to measure unfathomable depths of the sea, see ALTITUDE.

SOUNDING *the pump*, at sea, is done by letting fall a small line, with some weight at the end, down into the pump, to know what depth of water there is in it.

SOUTH, one of the four cardinal points of the wind, or compass, being that which is directly opposite to the north.

SOUTH *Direct Dials.* See PRIME *Verticals.*

SOUTHERN *Hemisphere, Signs, &c,* those in the south side of the equator.

SOUTHING, in Navigation, the difference of latitude made by a ship in sailing to the southward.

SPACE, denotes room, place, distance, capacity, extension, duration, &c.

When Space is considered barely in length between any two bodies, it gives the same idea as that of distance. When it is considered in length, breadth, and thickness, it is properly called capacity. And when considered between the extremities of matter, which fills the capacity of Space with something solid, tangible, and moveable, it is then called extension.

So that extension is an idea belonging to body only; but Space may be considered without it. Therefore Space, in the general signification, is the same thing with distance considered every way, whether there be any matter in it or not.

Space is usually divided into *absolute* and *relative.*

Absolute SPACE is that which is considered in its own nature, without regard to any thing external, which always remains the same, and is infinite and immoveable.

Relative SPACE is that moveable dimension, or measure of the former, which our senses define by its positions to bodies within it; and this the vulgar use for immoveable Space.

Relative Space, in magnitude and figure, is always the same with absolute; but it is not necessary it should be so numerically. Thus, when a ship is perfectly at rest, then the places of all things within her are the same both absolutely and relatively, and nothing changes its place: but, on the contrary, when the ship is under sail, or in motion, she continually passes through new parts of absolute Space; though all things on board, considered relatively, in respect to the ship, may yet be in the same places, or have the same situation and position, in regard to one another.

The Cartesians, who make extension the essence of matter, assert, that the Space any body takes up, is the same thing with the body itself; and that there is no such thing in the universe as mere Space, void of all matter; thus making Space or extension a substance. See this disproved under VACUUM.

Among those too who admit a vacuum, and consequently an essential difference between Space and matter, there are some who assert that Space is a substance. Among these we find Gravesande, *Introd. ad Philos.* sect. 19.

Others again put Space into the same class of beings as time and number; thus making it to be no more than a notion of the mind. So that according to these authors, absolute Space, of which the Newtonians

speak, is a mere chimera. See the writings of the late bishop Berkley.

Space and time, according to Dr. Clarke, are attributes of the Deity; and the impossibility of annihilating these, even in idea, is the same with that of the necessary existence of the Deity.

SPACE, in Geometry, denotes the area of any figure; or that which fills the interval or distance between the lines that terminate or bound it. Thus,

The Parabolic Space is that included in the whole parabola. The conchoidal Space, or the cissoidal Space, is what is included within the cavity of the conchoid or cissoid. And the asymptotic Space, is what is included between an hyperbolic curve and its asymptote. By the new methods now introduced, of applying algebra to geometry, it is demonstrated that the conchoidal and cissoidal Spaces, though infinitely extended in length, are yet only finite magnitudes or Spaces.

SPACE, in Mechanics, is the line a moveable body, considered as a point, is conceived to describe by its motion.

SPANDREL, with Builders, is the space included between the curve of an arch and the straight or right lines which inclose it; as the space *a*, or *b*.

SPEAKING *Trumpet.* See *Speaking* TRUMPET.

SPECIES, in Algebra, are the letters, symbols, marks, or characters, which represent the quantities in any operation or equation.

This short and advantageous way of notation was chiefly introduced by Vieta, about the year 1590; and by means of which he made many discoveries in algebra, not before taken notice of.

The reason why Vietà gave this name of Species to the letters of the alphabet used in algebra, and hence called Arithmetica Speciosa, seems to have been in imitation of the Civilians, who call cases in law that are put abstractedly, between John a Nokes and Tom a Stiles, between A and B; supposing those letters to stand for any persons indefinitely. Such cases they call Species: whence, as the letters of the alphabet will also as well represent quantities, as persons, and that also indefinitely, one quantity as well as another, they are properly enough called Species; that is general symbols, marks, or characters. From whence the literal algebra hath since been often called Specious Arithmetic, or Algebra in Species.

SPECIES, in Optics, the image painted on the retina by the rays of light reflected from the several points of the surface of an object, received in by the pupil, and collected in their passage through the crystalline, &c.

Philosophers have been in great doubt, whether the Species of objects, which give the soul an occasion of seeing, are an effusion of the substance of the body; or a mere impression which they make on all ambient bodies, and which these all reflect, when in a proper disposition and distance; or lastly, whether they are not some other more subtile body, as light, which receives all these impressions from bodies, and is continually sent and returning from one to another, with the different impressions and figures it has taken. But the moderns have decided this point by their invention of

artificial

artificial eyes, inwhich the Species of objects are received on a paper, in the same manner as they are received in the natural eye.

SPECIFIC, in Philofophy, that which is proper and peculiar to any thing; or that characterifes it, and diftinguifhes it from every other thing. Thus, the attracting of iron is Specific to the loadftone, or is a Specific property of it.

A juft definition fhould contain the Specific notion of the thing defined, or that which fpecifies and diftinguifhes it from every thing elfe.

SPECIFIC *Gravity*, in Hydroftatics, is the relative proportion of the weight of bodies of the fame bulk. See *Specific* GRAVITY.

SPECIFIC *Gravity of living men.* Mr. John Robertfon, late librarian to the Royal Society, in order to determine the Specific gravity of men, prepared a ciftern 78 inches long, 30 inches wide, 30 inches deep; and having procured 10 men for his purpofe, the height of each was taken and his weight; and afterwards they plunged fuccefively into the ciftern. A ruler or fcale, graduated to inches and decimal parts, was fixed to one end of the ciftern, and the height of the water fhown by it was noted before each man went in, and to what height it rofe when he immerfed himfelf under its furface. The following table contains the feveral refults of his experiments:

No. of men.	Height. Ft. In.	Weight. lbs.	Water raifed. Inches.	Solidity. Feet.	Wt. of water. lbs.	Specific gravity. (Wat. 1)
1	6 2	161	1·90	2·573	160·8	1·001
2	5 10⅜	147	1·91	2·586	161·6	0·901
3	5 9½	156	1·85	2·505	156·6	0·991
4	5 6¾	140	2·04	2·763	172·6	0·801
5	5 5⅞	158	2·08	2·817	176·0	0·900
6	5 5½	158	2·17	2·939	183·7	0·849
7	5 4⅜	140	2·01	2·722	170·1	0 823
8	5 4⅛	121	1·79	2·424	151·5	0·800
9	5 3¼	146	1·73	2·343	146·4	0·997
10	5 3⅛	132	1·85	2·505	156·6	0·843
medium of all.	5 6⅔	146	1·933	2·618	163·6	0·891

One of the reafons, Mr. Robertfon fays, that induced him to make thefe experiments, was a defire of knowing what quantity of timber would be fufficient to keep a man afloat in water, thinking that moft men were fpecifically heavier than river or common frefh water; but the contrary appears from the trials above recited; for, except the firft, every man was lighter than an equal bulk of frefh water, and much more fo than that of feawater. So that, if perfons who fall into water had prefence of mind enough to avoid the fright ufual on fuch occafions, many might be preferved from drowning; and a piece of wood not larger than an oar, would buoy a man partly above water as long as he had ftrength or fpirits to keep his hold. Philof. Tranf. vol. 50, art. 5.

From the laft line of the table appears the medium of all the circumftances of height, weight, &c; particu-

larly the mean Specific Gravity, 0·891, which is about ⅑ lefs than common water.

SPECTACLES, an optical machine, confifting of two lenfes fet in a frame, and applied on the nofe, to affift in defects of the organ of fight.

Old people, and all prefbytæ, ufe Spectacles of convex lenfes, to make amends for the flatnefs of the eye, which does not make the rays converge enough to have them meet in the retina.

Short-fighted people, or myopes, ufe concave lenfes, to prevent the rays from converging fo faft, on account of the greater roundnefs of the eye, or fmallnefs of the fphere, which is fuch as to make them meet before they reach the retina.

F. Cherubin, a capuchin, defcribes a kind of Spectacle telefcopes, for viewing remote objects with both eyes; and hence called *binoculi:* Though F. Rheita had mentioned the fame before him, in his Oculus Enoch et Eliæ. See BINOCLE. The fame author invented a kind of Spectacles, with three or four glaffes, which performed very well.

The invention of Spectacles has been much difputed. They were certainly not known to the ancients. Francifco Redi, in a learned treatife on Spectacles, contends that they were firft invented between the years 1280 and 1311, probably about 1290; and adds, that Alexander de Spina, a monk of the order of Predicants of St. Catharine, at Pifa, firft communicated the fecret, which was of his own invention, upon learning that another perfon had it as well as himfelf.

The author tells us, that in an old manufcript ftill preferved in his library, compofed in 1299, Spectacles are mentioned as a thing invented about that time: and that a celebrated Jacobin, one Jourdon de Rivalto, in a treatife compofed in 1305, fays exprefsly, that it was not yet 20 years fince the invention of Spectacles. He likewife quotes Bernard Gordon in his Lilium Medicinæ, written the fame year, where he fpeaks of a collyrium, good to enable an old man to read without Spectacles.

Muffchenbroek obferves, (Introd. vol. 2, pa. 786) that it is infcribed on the tomb of Salvinus Armatus, a nobleman of Florence, who died in 1317, that he was the inventor of Spectacles.

Du-Cange, however, carries the invention of Spectacles farther back; affuring us, that there is a Greek poem in manufcript in the French king's library, which fhews that Spectacles were in ufe in the year 1150; however the dictionary of the Academy Della Crufca, under the word *occhiale,* inclines to Redi's fide; and quotes a paffage from Jourdon's fermons, which fays that Spectacles had not been 20 years in ufe; and Salvati has obferved that thofe fermons were compofed between the years 1330 and 1336.

It is probable that the firft hint of the conftruction and ufe of Spectacles, was derived from the writings either of Alhazen, who lived in the 12th century, or of our own countryman Roger Bacon, who was born in 1214, and died in 1292, or 1294. The following remarkable paffage occurs in Bacon's Opus Majus by Jebb, p. 352. Si vero homo afpiciat literas et alias res minutas per medium cryftalli, vel vitri, vel alterius perfpicui fuppofiti literis, et fit portio minor fphæræ, cujus convexitas fit verfus oculum et oculus fit in aëre,

longe

longe melius videbit literas, et apparebunt ei majores.—
Et ideo hoc inſtrumentum eſt utile ſenibus et habentibus
oculos debiles : nam literam quantumcunque parvam
poſſunt videre in ſufficienti magnitudine. Hence, and
from other paſſages in his writings, much to the ſame
purpoſe, Molyneux, Plott, and others, have attributed
to him the invention of reading-glaſſes. Dr. Smith
indeed, obſerving that there are ſome miſtakes in his
reaſoning on this ſubject, has diſputed his claim. See
Molyneux's Dioptr. p. 256. Smith's Optics, Rem.
86—89.

SPECULATIVE *Geometry, Mathematics, Muſic,*
and *Philoſophy.* See the SUBSTANTIVES.

SPECULUM, or *Mirror,* in Optics, any poliſhed
body, impervious to the rays of light : ſuch as poliſhed
metals, and glaſſes lined with quickſilver, or any other
opake matter, popularly called Looking-glaſſes ; or
even the ſurface of mercury or of water, &c.

For the ſeveral kinds and forms of Specula, plane,
concave, and convex, with their theory and phenomena,
ſee MIRROR. And for their laws and effects, ſee
REFLECTION and BURNING-*Glaſs.*

As for the Specula of reflecting teleſcopes, it may
here be obſerved, that the perfection of the metal of
which they ſhould be made, conſiſts in its hardneſs,
whiteneſs, and compactneſs ; for upon theſe properties
the reflective powers and durability of the Specula depend. There are various compoſitions recommended
for theſe Specula, in Smith's Optics, book 3, ch. 2,
ſect. 787 ; alſo by Mr. Mudge in the Philoſ. Tranſ.
vol. 67 ; and in various other places, as by Mr. Edwards, in the Naut. Alm. for 1787, whoſe metal is the
whiteſt and beſt of any that I have ſeen.—For the method of grinding, ſee GRINDING.

Mr. Hearne's method of cleaning a tarniſhed Speculum was this : Get a little of the ſtrongeſt ſoap ley
from the ſoap-makers, and having laid the Speculum
on a table with its face upwards, put on as much of
the ley as it will hold, and let it remain about an hour :
then rub it ſoftly with a ſilk or muſlin, till the ley is all
gone ; then put on ſome ſpirit of wine, and rub it dry
with another part of the ſilk or muſlin. If the Speculum will not perform well after this, it muſt be new
poliſhed. A few faint ſpots of tarniſh may be rubbed
off with ſpirit of wine only, without the ley. Smith's
Optics, Rem. p. 107.

SPHERE, in Geometry, a ſolid body contained
under one ſingle uniform ſurface, every point of which
is equally diſtant from a certain point in the middle
called its centre.

The Sphere may be ſuppoſed
to be generated by the revolution of a ſemicircle ABD about
its diameter AB, which is alſo
called the *axis* of the Sphere,
and the extreme points of the
axis, A and B, the *poles* of the
Sphere ; alſo the middle of the
axis C is the *centre,* and half the axis, AC, the *radius.*

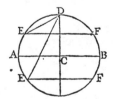

Properties of the SPHERE, are as follow.

1. A Sphere may be conſidered as made up of an
infinite number of pyramids, whoſe common altitude

6

is equal to the radius of the Sphere, and all their baſes
form the ſurface of the Sphere. And therefore the
ſolid content of the Sphere is equal to that of a pyramid whoſe altitude is the radius, and its baſe is equal
to the ſurface of the Sphere, that is, the ſolid content is equal to $\frac{1}{3}$ of the product of its radius and
ſurface.

2. A Sphere is equal to $\frac{2}{3}$ of its circumſcribing cylinder, or of the cylinder of the ſame height and diameter, and therefore equal to the cube of the diameter
multiplied by ·5236, or $\frac{2}{3}$ of ·7854 ; or equal to double a cone of the ſame baſe and height. Hence alſo different Spheres are to one another as the cubes of their
diameters. And their ſurfaces as the ſquares of the
ſame diameters.

3. The ſurface or ſuperficies of any Sphere, is equal
to 4 times the area of its great circle, or of a circle of
the ſame diameter as the Sphere. Or

4. The ſurface of the whole Sphere is equal to the
area of a circle whoſe radius is equal to the diameter of
the Sphere. And, in like manner, the curve ſurface
of any ſegment EDF, whether greater or leſs than a
hemiſphere, is equal to a circle whoſe radius is the
chord line DE, drawn from the vertex D of the ſegment to the circumference of its baſe, or the chord of
half its arc.

5. The curve ſurface of any ſegment or zone of a
Sphere, is alſo equal to the curve ſurface of a cylinder of
the ſame height with that portion, and of the ſame
diameter with the Sphere. Alſo the ſurface of the
whole Sphere, or of an hemiſphere, is equal to the
curve ſurface of its circumſcribing cylinder. And the
curve ſurfaces of their correſponding parts are equal,
that are contained between any two places parallel to
the baſe. And conſequently the ſurface of any ſegment or zone of a Sphere, is as its height or altitude.

Moſt of theſe properties are contained in Archimedes's
treatiſe on the Sphere and cylinder. And many other
rules for the ſurfaces and ſolidities of Spheres, their
ſegments, zones, fruſtums, &c, may be ſeen in my
Menſuration, part 3, ſect. 1, prob. 10, &c.

Hence, if d denote the diameter or axis of a Sphere,
s its curve ſurface, c its ſolid content, and $a = \cdot7854$
the area of a circle whoſe diam. is 1 ; then we ſhall,
from the foregoing properties, have theſe following
general values or equations, viz,

$$s = 4 . a d^2 = \frac{6c}{d} = 6 \sqrt[3]{\tfrac{2}{3} a c^2}.$$

$$c = \tfrac{1}{6} d s = \tfrac{2}{3} a d^3 = \tfrac{1}{12} \sqrt{\frac{s^3}{a}}.$$

$$d = \frac{6c}{s} = \sqrt{\frac{s}{4a}} = \sqrt[3]{\frac{3c}{2a}}.$$

Doctrine of the SPHERE. See SPHERICS.

Projection of the SPHERE. See PROJECTION.

SPHERE *of Activity,* of any body, is that determinate
ſpace or extent all around it, to which, and no farther,
the effluvia or the virtue of that body reaches, and in
which it operates according to the nature of the body.
See ACTIVITY.

SPHERE, in Aſtronomy, that concave orb or ex-
panſe which inveſts our globe, and in which the heavenly

venly bodies, the sun, moon, stars, planets, and comets, appear to be fixed at an equal distance from the eye. This is also called the Sphere of the world; and it is the subject of spherical astronomy.

This Sphere, as it includes the fixed stars, from whence it is sometimes called the *Sphere of the fixed stars*, is immensely great. So much so; that the diameter of the earth's orbit is vastly small in respect of it; and consequently the centre of the Sphere is not sensibly changed by any alteration of the spectator's place in the several parts of the orbit: but still in all points of the earth's surface, and at all times, the inhabitants have the same appearance of the Sphere; that is, the fixed stars seem to possess the same points in the surface of the Sphere. For, our way of judging of the places &c of the stars, is to conceive right lines drawn from the eye, or from the centre of the earth, through the centres of the stars, and thence continued till they cut the Sphere; and the points where these lines so meet the Sphere, are the apparent places of those stars.

The better to determine the places of the heavenly bodies in the Sphere, several circles are conceived to be drawn in the surface of it, which are called circles of the Sphere.

SPHERE, in Geography, &c, denotes a certain disposition of the circles on the surface of the earth, with regard to one another, which varies in the different parts of it.

The circles originally conceived on the surface of the Sphere of the world, are almost all transferred, by analogy, to the surface of the earth, where they are conceived to be drawn directly underneath those of the Sphere, or in the same positions with them; so that, if the planes of those of the earth were continued to the Sphere of the stars, they would coincide with the respective circles on it. Thus, we have an horizon, meridian, equator, &c, on the earth. And as the equinoctial, or equator, in the heavens, divides the Sphere into two equal parts, the one north and the other south, so does the equator on the surface of the earth divide its globe in the same manner. And as the meridians in the heavens pass through the poles of the equinoctial, so do those on the earth, &c. With regard then to the position of some of these circles in respect of others, we have a *right*, an *oblique*, and a *parallel* Sphere.

A Right or Direct SPHERE, (fig. 4, plate 26), is that which has the poles of the world PS in its horizon, and the equator EQ in the zenith and nadir. The inhabitants of this Sphere live exactly at the equator of the earth, or under the line. They have therefore no latitude, nor no elevation of the pole. They can see both poles of the world; all the stars do rise, culminate, and set to them; and the sun always rises at right-angles to their horizon, making their days and nights always of equal length, because the horizon bisects the circle of the diurnal revolution.

An Oblique SPHERE, (fig. 5, plate 26), is that in which the equator EQ, as also the axis PS, cuts the horizon HO obliquely. In this Sphere, one pole P is above the horizon, and the other below it; and therefore the inhabitants of it see always the former pole, but never the latter; the sun and stars &c all rise and set obliquely; and the days and nights are always varying, and growing alternately longer and shorter.

A Parallel SPHERE, (fig. 6, plate 26), is that which has the equator in or parallel to the horizon, as well as all the sun's parallels of declination. Hence, the poles are in the zenith and nadir; the sun and stars move always quite around parallel to the horizon, the inhabitants, if any, being just at the two poles, having 6 months continual day, and 6 months night, in each year; and the greatest height to which the sun rises to them, is 23° 28′, or equal to his greatest declination.

Armillary or Artificial SPHERE, is an astronomical instrument, representing the several circles of the Sphere in their natural order; serving to give an idea of the office and position of each of them, and to resolve various problems relating to them.

It is thus called, as consisting of a number of rings of brass, or other matter, called by the Latins *armillæ*, from their resembling of bracelets or rings for the arm.

By this, it is distinguished from the globe, which, though it has all the circles of the Sphere on its surface; yet is not cut into armillæ or rings, to represent the circles simply and alone; but exhibits also the intermediate spaces between the circles.

Armillary Spheres are of different kinds, with regard to the position of the earth in them; whence they become distinguished into Ptolomaic and Copernican Spheres: in the first of which, the earth is in the centre, and in the latter near the circumference, according to the position which that planet obtains in those systems.

The Ptolomaic SPHERE, is that commonly in use, and is represented in fig. 6, plate 2, vol. 1, with the names of the several circles, lines, &c of the Sphere inscribed upon it. In the middle, upon the axis of the Sphere, is a ball T, representing the earth, on the surface of which are the circles &c of the earth. The Sphere is made to revolve about the said axis, which remains at rest; by which means the sun's diurnal and annual courses about the earth are represented according to the Ptolomaic hypothesis: and even by means of this, all problems relating to the phenomena of the sun and earth are resolved, as upon the celestial globe, and after the same manner; which see described under GLOBE.

Copernican SPHERE, fig. 7, plate 26, is very different from the Ptolomaic, both in its constitution and use; and is more intricate in both. Indeed the instrument is in the hands of so few people, and its use so inconsiderable, except what we have in the other more common instruments, particularly the globe and the Ptolomaic Sphere, that any farther account of it is unnecessary.

Dr. Long had an Armillary Sphere of glass, of a very large size, which is described and represented in his Astronomy. And Mr. Ferguson constructed a similar one of brass, which is exhibited in his Lectures, p. 194 &c

SPHERICAL, something relating to the sphere. As,

SPHERICAL *Angle*, is the angle formed on the surface of a Sphere or globe by the circumferences of two

two great circles. This angle, formed by the circumferences, is equal to that formed by the planes of the fame circles, or equal to the inclination of thofe two planes; or equal to the angle made by their tangents at the angular point. Thus, the inclination of the two

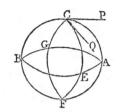

planes CAF, CEF, forms the Spherical Angle ACE, equal to the tangential angle PCQ.

The meafure of a Spherical Angle, ACE, is an arc of a great circle AE, defcribed from the vertex C, as from a pole, and intercepted between the legs CA and CE.

Hence, 1ft, Since the inclination of the plane CEF to the plane CAF, is every where the fame, the angles in the oppofite interfections, C and F, are equal.—2d, Hence the meafure of a Spherical Angle ACE, is an arc defcribed at the interval of a quadrant CA or CE, from the vertex C between the legs CA, CE.—3d, If a circle of the fphere CEFG cut another AEBG, the adjacent angles AEC and BEC are together equal to two right angles; and the vertical angles AEC, BEF are equal to one another. Alfo all the angles formed at the fame point, on the fame fide of a circle, are equal to two right angles, and all thofe quite around any point equal to four right angles.

Spherical *Triangle*, is a triangle formed upon the furface of a fphere, by the interfecting arcs of three great circles; as the triangle ACE.

Spherical Triangles are either *right-angled*, *oblique*, *equilateral*, *ifofceles*, or *fcalene*, in the fame manner as plane triangles. They are alfo faid to be *quadrantal*, when they have one fide a quadrant. Two fides or two angles are faid to be of the *fame affection*, when they are at the fame time either both greater, or both lefs than a quadrant or a right angle or 90°; and of *different affections*, when one is greater and the other lefs than 90 degrees.

Properties of Spherical *Triangles*.

1. Spherical Triangles have many properties in common with plane ones: Such as; That, in a triangle, equal fides fubtend equal angles, and equal angles are fubtended by equal fides: That the greater angles are fubtended by the greater fides, and the lefs angles by the lefs fides.

2. In every Spherical Triangle, each fide is lefs than a femicircle: any two fides taken together are greater than the third fide: and all the three fides taken together are lefs than the whole circumference of a circle.

3. In every Spherical Triangle, any angle is lefs than 2 right angles; and the fum of all the three angles taken together, is greater than 2, but lefs than 6, right angles.

4. In an oblique Spherical Triangle, if the angles at the bafe be of the fame affection, the perpendicular from the other angle falls within the triangle; but if they be of different affections, the perpendicular falls without the triangle.

Dr. Mafkelyne's remarks on the properties of Spherical Triangles, are as follow: (See the Introd. to my Logs. pa. 160, 2d edition.)

5. " A Spherical Triangle is equilateral, ifofcelar, or fcalene, according as it has its three angles all equal, or two of them equal, or all three unequal; and vice verfa.

6. The greateft fide is always oppofite the greateft angle, and the fmalleft fide oppofite the fmalleft angle.

7. Any two fides taken together are greater than the third.

8. If the three angles are all acute, or all right, or all obtufe; the three fides will be, accordingly, all lefs than 90°, or equal to 90°, or greater than 90°; and vice verfa.

9. If from the three angles A, B, C, of a triangle ABC, as poles, there be defcribed, upon the furface of the fphere, three arches of a great circle DE, DF, FE, forming by their interfections a new Spherical Triangle DEF; each fide of the new triangle will be the fupplement of the angle at its pole; and each angle of the fame triangle, will be the fupplement of the fide oppofite to it in the triangle ABC.

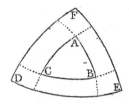 

10. In any triangle GHI or G*b*I, right angled in G, 1ft, The angles at the hypotenufe are always of the fame kind as their oppofite fides; 2dly, The hypotenufe is lefs or greater than a quadrant, according as the fides including the right angle, are of the fame or different kinds; that is to fay, according as thefe fame fides are either both acute, or both obtufe, or as one is acute and the other obtufe. And, vice verfa, 1ft, The fides including the right angle, are always of the fame kind as their oppofite angles; 2dly, The fides including the right angle will be of the fame or different kinds, according as the hypotenufe is lefs or more than 90°; but one at leaft of them will be of 90°, if the hypotenufe is fo."

Of the Area of a Spherical *Triangle*. The menfuration of Spherical Triangles and polygons was firft found out by Albert Girard, about the year 1600, and is given at large in his *Invention Nouvelle en l'Algebre*, pa. 50, &c; 4to, Amft. 1629. In any Spherical Triangle, the area, or furface inclofed by its three fides upon the furface of the globe, will be found by this proportion:

As 8 right angles or 720°,
Is to the whole furface of the fphere;
Or, as 2 right angles or 180°,
To one great circle of the fphere;
So is the excefs of the 3 angles above 2 right angles,
To the area of the Spherical Triangle.

Hence, if *a* denote ·7854,

$$d = \text{diam. of the globe, and}$$
$$s = \text{fum of the 3 angles of the triangle;}$$

then

then *add* $\times \dfrac{s-180}{180} =$ area of the Spherical Triangle.

Hence also, if r denote the radius of the sphere, and c its circumference; then the area of the triangle will thus be variously expressed; viz, Area $=$

$$ad^2 \times \dfrac{s-180}{180} = cd \times \dfrac{s-180}{720} = cr \times \dfrac{s-180}{360};$$

or barely $= r \times s-180°$, in square degrees, when the radius r is estimated in degrees; for then the circumference c is $= 360°$.

Farther, because the radius r, of any circle, when estimated in degrees, is, $= \dfrac{180}{3\cdot14159 \text{ \&c.}} = 57\cdot2957795$, the last rule $r \times s - 180$, for expressing the area A of the Spherical Triangle, in square degrees, will be barely

$$A = 57\cdot2957795 \ s - 10313\cdot24 =$$
$$= 57\tfrac{50}{169}s - 10313\tfrac{1}{4} \text{ very nearly.}$$

Hence may be found the sums of the three angles in any Spherical Triangle, having its area A known; for the last equation gives the sum

$$s = \dfrac{A}{r} + 180 = \dfrac{A}{57\cdot29 \text{ \&c.}} + 180 = \dfrac{169A}{9683} + 180.$$

So that, for a Triangle on the surface of the earth, whose three sides are known; if it be but small, as of a few miles extent, its area may be found from the known lengths of its sides, considering it as a plane Triangle, which gives the value of the quantity A; and then the last rule above will give the value of s, the sum of the three angles; which will serve to prove whether those angles are nearly exact, that have been taken with a very nice instrument, as in large and extensive measurements on the surface of the earth.

Resolution of SPHERICAL *Triangles.* See TRIANGLE, and TRIGONOMETRY.

SPHERICAL *Polygon*, is a figure of more than three sides, formed on the surface of a globe by the intersecting arcs of great circles.

The area of any Spherical Polygon will be found by the following proportion; viz,

As 8 right-angles or 720°,
To the whole surface of the sphere;
Or, as 2 right angles or 180°,
To a great circle of the sphere;
So is the excess of all the angles above the product of 180 and 2 less than the number of angles,
To the area of the spherical polygon.

That is, putting $n =$ the number of angles,
$s =$ sum of all the angles,
$d =$ diam. of the sphere,
$a = \cdot78539$ &c,

Then $A = aa^2 \times \dfrac{s-(n-2)180}{180} =$ the area of the Spherical Polygon.

Hence other rules might be found, similar to those for the area of the Spherical Triangle.

Hence also, the sum s of all the angles of any Spherical Polygon, is always less than $180n$, but greater than $180 (n-2)$, that is less than n times 2 right angles, but greater than $n-2$ times 2 right angles.

SPHERICAL *Astronomy*, that part of astronomy which considers the universe such as it appears to the eye. See ASTRONOMY.

Under Spherical Astronomy, then, come all the phenomena and appearances of the heavens and heavenly bodies, such as we perceive them, without any enquiry into the reason, the theory, or truth of them. By which it is distinguished from theorical astronomy, which considers the real structure of the universe, and the causes of those phenomena.

In the Spherical Astronomy, the world is conceived to be a concave Spherical surface, in whose centre is the earth, or rather the eye, about which the visible frame revolves, with stars and planets fixed in the circumference of it. And on this supposition all the other phenomena are determined.

The theorical astronomy teaches us, from the laws of optics, &c, to correct this Scheme and reduce the whole to a juster system.

SPHERICAL *Compasses.* See COMPASSES.

SPHERICAL *Geometry*, the doctrine of the sphere; particularly of the circles described on its surface, with the method of projecting the same on a plane; and measuring their arches and angles when projected.

SPHERICAL *Numbers.* See CIRCULAR *Numbers.*

SPHERICAL *Trigonometry.* See *Spherical* TRIGONOMETRY.

SPHERICITY, the quality of a sphere; or that by which a thing becomes spherical or round.

SPHERICS, the Doctrine of the sphere, particularly of the several circles described on its surface; with the method of projecting the same on a plane. See PROJECTION *of the Sphere.*

A circle of the sphere is that which is made by a plane cutting it. If the plane pass through the centre, it is a *great* circle: if not, it is a *little* circle.

The *pole* of a circle, is a point on the surface of the sphere equidistant from every point of the circumference of the circle. Hence every circle has two poles, which are diametrically opposite to each other; and all circles that are parallel to each other have the same poles.

Properties of the Circles of the Sphere.

1. If a sphere be cut in any manner by a plane, the section will be a circle. And a great circle when the section passes through the centre, otherwise it is a *little* circle. Hence, all great circles are equal to each other: and the line of section of two great circles of the sphere, is a diameter of the sphere: and therefore two great circles intersect each other in points diametrically opposite; and make equal angles at those points; and divide each other into two equal parts; also any great circle divides the whole sphere into two equal parts.

2. If a great circle be perpendicular to any other circle, it passes through its poles. And if a great circle

pass

pass through the pole of any other circle, it cuts it at right angles, and into two equal parts.

3. The distance between the poles of two circles, is equal to the angle of their inclination.

4. Two great circles passing through the poles of another great circle, cut all the parallels to this latter into similar arcs. Hence, an angle made by two great circles of the sphere, is equal to the angle of inclination of the planes of these great circles. And hence also the lengths of those parallels are to one another as the sines of their distances from their common pole, or as the cosines of their distances from their parallel great circle. Consequently, as radius is to the cosine of the latitude of any point on the globe, so is the length of a degree at the equator, to the length of a degree in that latitude.

5. If a great circle pass through the poles of another; this latter also passes through the poles of the former; and the two cut each other perpendicularly.

6. If two or more great circles intersect each other in the poles of another great circle; this latter will pass through the poles of all the former.

7. All circles of the sphere that are equally distant from the centre, are equal; and the farther they are distant from the centre, the less they are.

8. The shortest distance on the surface of a sphere, between any two points on that surface, is the arc of a great circle passing through those points. And the smaller the circle is that passes through the same points, the longer is the arc of distance between them. Hence the proper measure, or distance, of two places on the surface of the globe, is an arc of a great circle intercepted between the same. See Theodosius and other writers on Spherics.

SPHEROID, a solid body approaching to the figure of a sphere, though not exactly round, but having one of its diameters longer than the other.

This solid is usually considered as generated by the rotation of an oval plane figure about one of its axes. If that be the longer or transverse axis, the solid so generated is called an *oblong* Spheroid, and sometimes *prolate*, which resembles an egg, or a lemon; but if the oval revolve about its shorter axis, the solid will be an *oblate* Spheroid, which resembles an orange, and in this shape also is the figure of the earth, and the other planets.

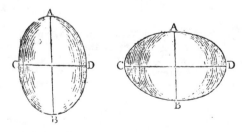

The axis about which the oval revolves, is called the *fixed* axis, as AB; and the other CD is the *revolving* axis: whichever of them happens to be the longer.

When the revolving oval is a perfect ellipse, the so-

lid generated by the revolution is properly called an *ellipsoid*, as distinguished from the Spheroid, which is generated from the revolution of any oval whatever, whether it be an ellipse or not. But generally speaking, in common acceptation, the term Spheroid is used for an ellipsoid; and therefore, in what follows, they are considered as one and the same thing.

Any section of a Spheroid, by a plane, is an ellipse (except the sections perpendicular to the fixed axe, which are circles); and all parallel sections are similar ellipses, or having their transverse and conjugate axes in the same constant ratio; and the sections parallel to the fixed axe are similar to the ellipse from which the solid was generated. See my Mensuration pa. 267 &c, 2d edit.

For the Surface of a Spheroid, whether it be oblong or oblate. Let f denote the fixed axe,

r the revolving axe,

$$a = \cdot 7854, \text{ and } q = \frac{ff \text{ or } rr}{ff} - ;$$

then will the surface s be expressed by the following series, using the upper signs for the oblong spheroid, and the under signs for the oblate one; viz,

$$s = 4\,arf \times \left(1 \mp \frac{1}{2.3}q - \frac{1}{2.4.5}q^2 \mp \frac{3}{2.4.6.7}q^3 \&c\right);$$

where the signs of the terms, after the first, are all negative for the oblong Spheroid, but alternately positive and negative for the oblate one.

Hence, because the factor $4arf$ is equal to 4 times the area of the generating ellipse, it appears that the surface of the oblong Spheroid is less than 4 times the generating ellipse, but the surface of the oblate Spheroid is greater than 4 times the same: while the surface of the sphere falls in between the two, being just equal to 4 times its generating circle.

Huygens, in his Horolog. Oscillat. prop. 9, has given two elegant constructions for describing a circle equal to the superficies of an oblong and an oblate Spheroid, which he says he found out towards the latter end of the year 1657. As he gave no demonstrations of these, I have demonstrated them, and also rendered them more general, by extending and adapting them to the surface of any segment or zone of the Spheroid. See my Mensuration, pa. 308 &c, 2d ed. where also are several other rules and constructions for the surfaces of Spheroids, besides those of their segments, and frustums.

Of the Solidity of a Spheroid. Every Spheroid, whether oblong or oblate, is, like a sphere, exactly equal to two-thirds of its circumscribing cylinder. So that, if f denote the fixed axe, r the revolving axe, and $a = \cdot 7854$; then $\frac{2}{3}afr^2$ denotes the solid content of either Spheroid. Or, which comes to the same thing, if t denote the transverse, and c the conjugate axe of the generating ellipse;

then $\frac{2}{3}ac^2t$ is the content of the oblong Spheroid,
and $\frac{2}{3}act^2$ is the content of the oblate Spheroid.

Consequently, the proportion of the former solid to the latter, is as c to t, or as the less axis to the greater.

Farther, if about the two axes of an ellipse be generated

nerated two fpheres and two fpheroids, the four folids will be continued proportionals, and the common ratio will be that of the two axes of the ellipfe; that is, as the greater fphere, or the fphere upon the greater axe, is to the oblate Spheroid, fo is the oblate Spheroid to the oblong Spheroid, and fo is the oblong Spheroid to the lefs fphere, and fo is the tranfverfe axis to the conjugate. See my Menfuration, pa. 327 &c, 2d ed. where may be feen many other rules for the folid contents of Spheroids, and their various parts. See alfo Archimedes on Spheroids and Conoids.

Dr. Halley has demonftrated, that in a fphere, Mercator's nautical meridian line is a fcale of logarithmic tangents of the half complements of the latitudes. But as it has been found that the fhape of the earth is fpheroidal, this figure will make fome alteration in the numbers refulting from Dr. Halley's theorem. Maclaurin has therefore given a rule, by which the meridional parts to any Spheroid may be found with the fame exactnefs as in a fphere. There is alfo an ingenious tract by Mr. Murdoch on the fame fubject. See Philof. Tranf. No. 219. Mr. Cotes has alfo demonftrated the fame propofition, Harm. Menf. pa. 20, 21. See MERIDIONAL *Parts*.

Univerfal SPHEROID, a name given to the folid generated by the rotation of an ellipfe about fome other diameter, which is neither the tranfverfe nor conjugate axis. This produces a figure refembling a heart. See my Menfuration, pa. 352, 2d ed.

SPINDLE, in Geometry, a folid body generated by the revolution of fome curve line about its bafe or double ordinate AB; in oppofition to a conoid, which is generated by the rotation of the curve about its axis or abfcifs, perpendicular to its ordinate.

A   B

The Spindle is denominated circular, elliptic, hyperbolic, or parabolic, &c, according to the figure of its generating curve. See my Menfur. in feveral places.

SPINDLE, in Mechanics, fometimes denotes the axis of a wheel, or roller, &c; and its ends are the pivots. See alfo *Double* CONE.

SPIRAL, in Geometry, a curve line of the circular kind, which, in its progrefs, recedes always more and more from a point within, called its centre; as in winding from the vertex of a cone down to its bafe.

The firft treatife on a Spiral is by Archimedes, who thus defcribes it: Divide the circumference of a circle App &c into any number of equal parts, by a continual bifection at the points pp &c. Divide alfo the radius AC into the fame number of equal parts, and make Cm, Cm, Cm, &c, equal to 1, 2, 3, &c of thefe equal parts; then a line drawn, with a fteady hand, drawn through all the points m, m, m, &c, will trace out the Spiral.

This is more particularly called the *firft* Spiral, when it has made one complete revolution to the point A; and the fpace included between the Spiral and the radius CA, is the *Spiral fpace*.

The firft Spiral may be continued to a *fecond*, by defcribing another circle with double the radius of the

firft; and the fecond may be continued to a *third*, by a third circle; and fo on.

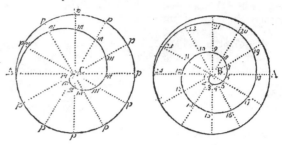

Hence it follows, that the parts of the circumference Ap are as the parts of the radii Cm; or Ap is to the whole circumference, as Cm is to the whole radius. Confequently, if c denote the circumference, r the radius, $x = Cm$, and $y = Ap$; then there arifes this proportion $r : c :: x : y$, which gives $ry = cx$ for the equation of this Spiral; and which therefore it has in common with the quadratrix of Dinoftrates, and that of Tfchirnhaufen: fo that $r^n y^m = c^n x^m$ will ferve for infinite Spirals and quadratrices. See QUADRATRIX.

The Spiral may alfo be conceived to be thus generated, by a continued uniform motion. If a right line, as AB (*laft fig. above*) having one end moveable about a fixed point at B, be uniformly turned round, fo as the other end A may defcribe the circumference of a circle; and at the fame time a point be conceived to move uniformly forward from B towards A, in the right line or radius AB, fo that the point may defcribe that line, while the line generates the circle; then will the point, with its two motions, defcribe the curve B, 1, 2, 3, 4, 5, &c, of the fame Spiral as before.

Again, if the point B be conceived to move twice as flow as the line AB, fo that it fhall get but half way along BA, when that line fhall have formed the circle; and if then you imagine a new revolution to be made of the line carrying the point, fo that they fhall end their motion at laft together, there will be formed a *double* Spiral line, as in the laft figure. From the manner of this defcription may eafily be drawn thefe corollaries:

1. That the lines B12, B11, B10, &c, making equal angles with the firft and fecond Spiral (as alfo B12, B10, B8), &c, are in arithmetical progreffion.

2. The lines B7, B10, &c, drawn any how to the firft Spiral, are to one another as the arcs of the circle intercepted between BA and thofe lines; becaufe whatever parts of the circumference the point A defcribes, as fuppofe 7, the point B will alfo have run over 7 parts of the line AB.

3. Any lines drawn from B to the fecond Spiral, as B18, B22, &c, are to each other as the aforefaid arcs, together with the whole circumference added on both fides: for at the fame time that the point A runs over 12, or the whole circumference, or perhaps 7 parts more, fhall the point B have run over 12, and 7 parts of the line AB, which is now fuppofed to be divided into 24 equal parts.

4. The

4. The firſt Spiral line is equal to half the circumference of the firſt circle; for the radii of. the ſectors, and conſequently of the arcs, are in a ſimple arithmetic progreſſion, while the circumference of the circle contains as many arcs equal to the greateſt; therefore the circumference is in proportion to all thoſe Spiral arcs, as 2 to 1.

5. The firſt Spiral ſpace is equal to ⅓ of the firſt or circumſcribing circle. That is, the area CABDE of the Spiral, is equal to ⅓ part of the circle deſcribed with the radius CE. In like manner, the whole Spiral area, generated by the ray drawn from the point C to the curve, when it makes two revolutions, is ⅔ of the circle deſcribed with the radius 2CE.

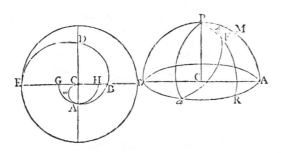

And, generally, the whole area generated by the ray from the beginning of the motion, till after any number _n_ of revolutions, is equal to $\frac{n}{3}$ of the circle whoſe radius is _n_ × CE, that is equal to the 3d part of the ſpace which is the ſame multiple of the circle deſcribed with the greateſt ray, as the number of revolutions is of unity.

In like manner alſo, any ſector or portion of the area of the Spiral, terminated by the curve C_m_A and the right line CA, is equal to ⅓ of the circular ſector CAG terminated by the right lines CA and CG, this latter being the ſituation of the revolving ray when the point that deſcribes the curve ſets out from C. See Maclaurin's Flux. Introd. pa. 30, 31. Se alſo QUADRATURE of the Spiral of Archimedes.

SPIRAL, _Logiſtic_, or _Logarithmic_. See LOGISTIC and QUADRATURE.

SPIRAL _of Pappus_, a Spiral formed on the ſurface of a ſphere, by a motion ſimilar to that by which the Spiral of Archimedes is deſcribed on a plane. This Spiral is ſo called from its inventor Pappus. Collect. Mathem. lib. 4 prop. 30. Thus, if C be the centre of the ſphere, ARBA a great circle, P its pole; and while the quadrant PMA revolves about the pole P with an uniform motion, if a point proceeding from P move with a given velocity along the quadrant, it will trace upon the ſpherical ſurface the Spiral PZF_a_.

Now if we ſuppoſe the quadrant PMA to make a complete revolution in the ſame time that the point, which traces the Spiral on the ſurface of the ſphere, deſcribes the quadrant, which is the caſe conſidered by Pappus; then the portion of the ſpherical ſurface terminated by the whole Spiral, and the circle ARBA, and the quadrant PMA, will be equal to the ſquare of the diameter AB. In any other caſe, the area PMA_a_ FZP is to the ſquare of that diameter AB, as

the arc A_a_ is to the whole circumference ARBA. And this area is always to the ſpherical triangle PA_a_, as a ſquare is to its circumſcribing circle, or as the diameter of a circle is to half its circumference, or as 2 is to 3·14159 &c. See Maclaurin's Fluxions, Introd. pa. 31—33.

The portion of the ſpherical ſurface, terminated by the quadrant PMA, with the arches AR, FR, and the ſpiral PZF, admits of a perfect quadrature. when the ratio of the arch A_a_ to the whole circumference can be aſſigned. See Maclaurin, ibid. pa. 33.

Parabolic SPIRAL. See HELICOID.

Proportional SPIRAL, is generated by ſuppoſing the radius to revolve uniformly, and a point from the circumference to move towards the centre with a motion decreaſing in geometrical progreſſion. See LOGISTIC.

From the nature of a decreaſing geometrical progreſſion, it is eaſy to conceive that the radius CA may be continually divided; and although each ſucceſſive diviſion becomes ſhorter than the next preceding one, yet there muſt be an infinite number of diviſions or terms before the laſt of them become of no finite magnitude. Whence it follows, that this Spiral winds continually round the centre, without ever falling into it in any finite number of revolutions.

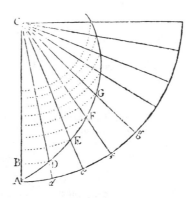

It is alſo evident that any Proportional Spiral cuts the intercepted radii at equal angles: for if the diviſions A_d_, _de_, _ef_, _fg_, &c. of the circumference be very ſmall, the ſeveral radii will be ſo cloſe to one another, that the intercepted parts AD, DE, EF, FG, &c. of the Spiral may be taken as right lines; and the triangles CAD, CDE, CEF, &c. will be ſimilar, having equal angles at the point C, and the ſides about thoſe angles proportional; therefore the angles at A, D, E, F, &c. are equal, that is, the ſpiral cuts the radii at equal angles. Robertſon's Elem. of Navig. book 2, pa. 87.

Proportional Spirals are ſuch Spiral lines as the rhumb lines on the terraqueous globe; which, becauſe they make equal angles with every meridian, muſt alſo make equal angles with the meridians in the ſtereographic projection on the plane of the equator, and therefore will be, as Dr. Halley obſerves, Proportional Spirals about the polar point. From whence he demonſtrates, that the meridian line is a ſcale of log. tangents of the half complements of the latitudes. See RHUMB, LOXODROMY, and MERIDIONAL *Parts*.

SPIRAL *Pump*. See *Archimedes*'s SCREW.

SPIRAL, in Architecture and Sculpture, denotes a curve that ascends, winding about a cone, or spire, so that all the points of it continually approach the axis.

By this it is distinguished from the Helix, which winds in the same manner about a cylinder.

SPORADES, in Astronomy, a name by which the ancients distinguished such stars as were not included in any constellation. These the moderns more usually call *unformed*, or *extraconstellary* stars.

Many of the Sporades of the ancients have been since formed into new constellations: thus, of those between Ursa Major and Leo, Hevelius has formed a constellation named Leo Minor; and of those between Ursa Minor and Auriga, he also formed the Lynx; and of those under the tail of Ursa Minor, another called Canis Venaticus; &c.

SPOTS, in Astronomy, are dark places observed on the disks or faces of the sun, moon, and planets.

The Spots on the sun are seldom if ever visible, except through a telescope. I have indeed met with persons whose eyes were so good that they have declared they could distinguish the solar Spots; and it is mentioned in Josephus a Costa's Natural and Moral History of the West Indies, book 1, ch. 2, before the use of telescopes, that in Peru there are Spots to be seen in the sun, which are not to be seen in Europe. See a memoir by Dr. Zach, in the Astronomical Ephemeris of the Acad. of Berlin for 1788, relating to the discoveries and unpublished papers of Thomas Harriot the celebrated algebraist. In that memoir it is shewn, for the first time, that Harriot was also an excellent astronomer, both theoretical and practical; that he made innumerable observations with telescopes from the year 1610, and, amongst them, 199 observations of the solar Spots, with their drawings, calculations, and the determinations of the sun's revolution round his axis. These Spots were also discovered near about the same time by Galileo and Scheiner. See Joh. Fabricius Phrysius De Maculis in Sole observatis & apparente eorum cum sole conversione narratio, 1611; also Galileo's Istoria e Demonstrazioni intorne alle Machie Solare e loro accidenti, 1613.

Some distinguish the Spots into Maculæ, or dark Spots; and Faculæ, or bright Spots; but there seems but little foundation for any such division. They are very changeable as to number, form, &c; and are sometimes in a multitude, and sometimes none at all. Some imagine they may become so numerous, as to hide the whole face of the Sun, or at least the greater part of it; and to this they ascribe what Plutarch mentions, viz, that in the first year of the reign of Augustus, the sun's light was so faint and obscure, that one might look steadily at it with the naked eye. To which Kepler adds, that in 1547, the Sun appeared reddish, as when viewed through a thick mist; and hence he conjectures that the Spots in the sun are a kind of dark smoke, or clouds, floating on his surface.

Some again will have them stars, or planets, passing over the body of the sun: but others, with more probability, think they are opake bodies, in manner

of crusts, formed like the scums on the surface of liquors.

Dr. Derham, from a variety of particulars, which he has recited, concerning the solar Spots, and their congruity to what we observe in our own globe, infers, that they are caused by the eruption of some new volcano in the sun, which pouring out at first a prodigious quantity of smoke and other opake matter, causeth the Spots: and as that fuliginous matter decays and spends itself, and the volcano at last becomes more torrid and flaming, so the Spots decay and become umbræ, and at last faculæ: which faculæ he supposes to be no other than more flaming lighter parts than any other parts of the sun. Philos. Transf. vol. 23, p. 1504, or Abr. vol. 4, p. 235.

Dr. Franklin (in his Exper. and Observ p. 266.) suggests a conjecture, that the parts of the Sun's sulphur separated by fire, rise into the atmosphere, and there being freed from the immediate action of the fire, they collect into cloudy masses, and gradually becoming too heavy to be longer supported, they descend to the sun, and are burnt over again. Hence, he says, the Spots appearing on his face, which are observed to diminish daily in size, their consuming edges being of particular brightness.

For another solution of these phenomena, see MACULÆ. Various other accounts and hypotheses of these Spots may be seen in many of the other volumes of the Philos. Transf. In one of these, viz, vol. 57, pa. 398, Dr. Horsley attempts to determine the height of the sun's atmosphere from the height of the solar Spots above his surface.

By means of the observations of these Spots, has been determined the period of the sun's rotation about his axis, viz, by observing their periodical return.

The lunar Spots are fixed: and astronomers reckon about 48 of them on the moon's face; to each of which they have given names. The 21st, called *Tycho*, is one of the most considerable.

Circular SPOTS, in Electricity. See CIRCULAR Spots and COLOURS.

Lucid SPOTS, *in the heavens*, are several little whitish Spots, that appear magnified, and more luminous when seen through telescopes; and yet without any stars in them. One of these is in Andromeda's girdle, and was first observed in 1612, by Simon Marius: it has some whitish rays near its middle, is liable to several changes, and is sometimes invisible. Another is near the ecliptic, between the head and bow of Sagittarius; it is small, but very luminous. A third is in the back of the Centaur, which is too far south to be seen in Britain. A fourth, of a smaller size, is before Antinous's right foot, having a star in it, which makes it appear more bright. A fifth is in the constellation Hercules, between the stars ϵ and η, which is visible to the naked eye, though it is but small, when the sky is clear and the moon absent. It is probable that with more powerful telescopes these lucid Spots will be found to be congeries of very minute fixed stars.

Planetary SPOTS, are those of the planets. Astronomers find that the planets are not without their spots. Jupiter, Mars, and Venus, when viewed through a telescope, shew several very remarkable ones: and it is

by

by the motion of these Spots, that the rotation of the planets about their axes is concluded, in the same manner as that of the sun is deduced from the apparent motion of his maculæ.

SPOUT, or *Water* Spout, an extraordinary meteor, or appearance, consisting of a moving column or pillar of water; called by the Latins *typho*, and *sipho*; and by the French *trompe*, from its shape, which resembles a speaking trumpet, the widest end uppermost.

Its first appearance is in form of a deep cloud, the upper part of which is white, and the lower black. From the lower part of this cloud there hangs, or rather falls down, what is properly called the Spout, in manner of a conical tube, largest at top. Under this tube is always a great boiling and flying up of the water of the sea, as in a jet d'eau. For some yards above the surface of the sea, the water stands as a column, or pillar; from the extremity of which it spreads, and goes off, as in a kind of smoke. Frequently the cone descends so low as to the middle of this column, and continues for some time contiguous to it; though sometimes it only points to it at some distance, either in a perpendicular, or in an oblique line.

Frequently it can scarce be distinguished, whether the cone or the column appear the first, both appearing all of a sudden against each other But sometimes the water boils up from the sea to a great height, without any appearance of a Spout pointing to it, either perpendicularly or obliquely. Indeed, generally, the boiling or flying up of the water has the priority, this always preceding its being formed into a column. For the most part the cone does not appear hollow till towards the end, when the sea water is violently thrown up along its middle, as smoke up a chimney: soon after this, the Spout or canal breaks and disappears; the boiling up of the water, and even the pillar, continuing to the last, and for some time afterwards; sometimes till the Spout form itself again, and appear anew, which it will do several times in a quarter of an hour. See a description of several Water-Spouts by Mr. Gordon, and by Dr. Stuart, in Phil. Transf. Abr. vol. iv, pa. 103 &c.

M. de la Pryme, from a near observation of two or three Spouts in Yorkshire, described in the Philosophical Transactions, num. 281, or Abr. vol. iv, pa. 106, concludes, that the Water Spout is nothing but a gyration of clouds by contrary winds meeting in a point, or centre; and there, where the greatest condensation and gravitation is, falling down into a pipe, or great tube, somewhat like Archimedes's spiral screw; and, in its working and whirling motion, absorbing and raising the water, in the same manner as the spiral screw does; and thus destroying ships &c.

Thus, June the 21st, he observed the clouds mightily agitated above, and driven together; upon which they became very black, and were hurried round; whence proceeded a most audible whirling noise like that usually heard in a mill. Soon after there issued a long tube, or Spout, from the centre of the congregated clouds, in which he observed a spiral motion, like that of a screw, by which the water was raised up.

Again, August 15, 1687, the wind blowing at the same time out of the several quarters, created a great vortex and whirling among the clouds, the centre of which every now and then dropt down, in shape of a long thin black pipe, in which he could distinctly behold a motion like that of a screw, continually drawing upwards, and screwing up, as it were, wherever it touched.

In its progress it moved slowly over a grove of trees, which bent under it like wands, in a circular motion. Proceeding, it tore off the thatch from a barn, bent a huge oak tree, broke one of its greatest limbs, and threw it to a great distance. He adds, that whereas it is commonly said, the water works and rises in a column, before the tube comes to touch it, this is doubtless a mistake, owing to the fineness and transparency of the tubes, which do most certainly touch the surface of the sea, before any considerable motion can be raised in it; but which do not become opake and visible, till after they have imbibed a considerable quantity of water.

The dissolution of Water-Spouts he ascribes to the great quantity of water they have glutted: which, by its weight, impeding their motion, upon which their force, and even existence depends, they break, and let go their contents; which use to prove fatal to whatever is found underneath.

A notable instance of this may be seen in the Philosophical Transactions (num. 363, or Abr. vol. iv. pa 108) related by Dr. Richardson. A Spout, in 1718, breaking on Emmotmoor, nigh Coln, in Lancashire, the country was immediately overflowed; a brook, in a few minutes, rose six feet perpendicularly high; and the ground upon which the Spout fell, which was 66 feet over, was torn up to the very rock, which was no less than 7 feet deep; and a deep gulf was made for above half a mile, the earth being raised in vast heaps on each side. See a description and figure of a Water-Spout, with an attempt to account for it in Franklin's Exp. and Obs. pa. 226, &c.

Signor Beccaria has taken pains to show that Water-Spouts have an electrical origin. To make this more evident, he first describes the circumstances attending their appearance, which are the following.

They generally appear in calm weather. The sea seems to boil, and to send up a smoke under them, rising in a hill towards the Spout. At the same time, persons who have been near them have heard a rumbling noise. The form of a Water-Spout is that of a speaking trumpet, the wider end being in the clouds, and the narrower end towards the sea.

The size is various, even in the same Spout. The colour is sometimes inclining to white, and sometimes to black. Their position is sometimes perpendicular to the sea, sometimes oblique; and sometimes the Spout itself is in the form of a curve. Their continuance is very various, some disappearing as soon as formed, and some continuing a considerable time. One that he had heard of continued a whole hour. But they often vanish, and presently appear again in the same place. The very same things that Water-Spouts are at sea, are some kinds of whirlwinds and hurricanes by land. They have been known to tear up trees, to throw down buildings, and make caverns in the earth; and in all these cases, to scatter earth, bricks, stones, timber, &c,

to

to a great diftance in every direction. Great quantities of water have been left, or raifed by them, fo as to make a kind of deluge; and they have always been attended by a prodigious rumbling noife.

That thefe phenomena depend upon electricity cannot but appear very probable from the nature of feveral of them; but the conjecture is made more probable from the following additional circumftances. They generally appear in months peculiarly fubject to thunder-ftorms, and are commonly preceded, accompanied, or followed by lightning, rain, or hail, the previous ftate of the air being fimilar. Whitifh or yellowifh flafhes of light have fometimes been feen moving with prodigious fwiftnefs about them. And laftly, the manner in which they terminate exactly refembles what might be expected from the prolongation of one of the uniform protuberances of electrified clouds, mentioned before, towards the fea; the water and the cloud mutually attracting one another: for they fuddenly contract themfelves, and difperfe almoft at once; the cloud rifing, and the water of the fea under it falling to its level. But the moft remarkable circumftance, and the moft favourable to the fuppofition of their depending on electricity, is, that they have been difperfed by prefenting to them fharp pointed knives or fwords. This, at leaft, is the conftant practice of mariners, in many parts of the world, where thefe Water-Spouts abound, and he was affured by feveral of them, that the method has often been undoubtedly effectual.

The analogy between the phenomena of Water Spouts and electricity, he fays, may be made vifible, by hanging a drop of water to a wire communicating with the prime conductor, and placing a veffel of water under it. In thefe circumftances, the drop affumes all the various appearances of a Water Spout, both in its rife, form, and manner of difappearing. Nothing is wanting but the fmoke, which may require a great force of electricity to become vifible.

Mr. Wilcke alfo confiders the Water-Spout as a kind of great electrical cone, raifed between the cloud ftrongly electrified, and the fea or the earth, and he relates a very remarkable appearance which occurred to himfelf, and which ftrongly confirms his fuppofition. On the 20th of July 1758, at three o'clock in the afternoon, he obferved a great quantity of duft rifing from the ground, and covering a field, and part of the town in which he then was. There was no wind, and the duft moved gently towards the eaft, where appeared a great black cloud, which, when it was near its zenith, electrified his apparatus pofitively, and to as great a degree as ever he had obferved it to be done by natural electricity. This cloud paffed his zenith, and went gradually towards the weft, the duft then following it, and continuing to rife higher and higher till it compofed a thick pillar, in the form of a fugar-loaf, and at length feemed to be in contact with the cloud. At fome diftance from this, there came, in the fame path, another great cloud, together with a long ftream of fmaller clouds, moving fafter than the preceding. Thefe clouds electrified his apparatus negatively, and when they came near the pofitive cloud, a flafh of lightning was feen to dart through the cloud of duft, the pofitive cloud, the large negative cloud, and, as far as the eye could diftinguifh, the whole train of fmaller negative clouds

4

which followed it. Upon this, the negative clouds fpread very much, and diffolved in rain, and the air was prefently clear of all the duft. The whole appearance lafted not above half an hour. See Prieftley's Electr. vol. 1, pa. 438, &c.

This theory of Water-Spouts has been farther confirmed by the account which Mr. Forfter gives of one of them, in his Voyage Round the World, vol. 1, pa. 191, &c. On the coaft of New Zealand he had an opportunity of feeing feveral, one of which he has particularly defcribed. The water, he fays, in a fpace of fifty or fixty fathoms, moved towards the centre, and there rifing into vapour, by the force of the whirling motion, afcended in a fpiral form towards the clouds. Directly over the whirlpool, or agitated fpot in the fea, a cloud gradually tapered into a long flender tube, which feemed to defcend to meet the rifing fpiral, and foon united with it into a ftraight column of a cylindrical form. The water was whirled upwards with the greateft violence in a fpiral, and appeared to leave a hollow fpace in the centre; fo that the water feemed to form a hollow tube, inftead of a folid column; and that this was the cafe, was rendered ftill more probable by the colour, which was exactly like that of a hollow glafs tube. After fome time, this laft column was incurvated, and broke like the others; and the appearance of a flafh of lightning which attended its disjunction, as well as the hail ftones which fell at the time, feemed plainly to indicate, that Water-Spouts either owe their formation to the electric matter, or, at leaft, that they have fome connection with it.

In Pliny's time, the feamen ufed to pour vinegar into the fea, to affuage and lay the Spout when it approached them: our modern feamen think to keep it off, by making a noife with filing and fcratching violently on the deck; or by difcharging great guns to difperfe it.

See the figure of a Water-Spout, fig. 1, plate 27.

SPRING, in Natural Hiftory, a fountain or fource of water, rifing out of the ground.

The moft general and probable opinion among philofophers, on the formation of Springs, is, that they are owing to rain. The rain-water penetrates the earth till fuch time as it meets a clayey foil, or ftratum; which proving a bottom fufficiently folid to fuftain and ftop its defcent, it glides along it that way to which the earth declines, till, meeting with a place or aperture on the furface, through which it may efcape, it forms a Spring, and perhaps the head of a ftream or brook.

Now, that the rain is fufficient for this effect, appears from hence, that upon calculating the quantity of rain and fnow which falls yearly on the tract of ground that is to furnifh, for inftance, the water of the Seine, it is found that this river does not take up above onefixth part of it.

Springs commonly rife at the bottom of mountains; the reafon is, that mountains collect the moft waters, and give them the greateft defcent the fame way. And if we fometimes fee Springs on high grounds, and even on the tops of mountains, they muft come from other remoter places, confiderably higher, along beds of clay, or clayey ground, as in their natural channels. So that if there happen to be a valley between a mountain on whofe top is a Spring, and the mountain which is to

furnifh

furnish it with water, the Spring must be considered as water conducted from a reservoir of a certain height, through a subterraneous channel, to make a jet of an almost equal height.

As to the manner in which this water is collected, so as to form reservoirs for the different kinds of Springs, it seems to be this: the tops of mountains usually abound with cavities and subterraneous caverns, formed by nature to serve as reservoirs; and their pointed summits, which seem to pierce the clouds, stop those vapours which float in the atmosphere; which being thus condensed, they precipitate in water, and by their gravity and fluidity easily penetrate through beds of sand and the lighter earth, till they become stopped in their descent by the denser strata, such as beds of clay, stone, &c, where they form a bason or cavern, and working a passage horizontally, or a little declining, they issue out at the sides of the mountains. Many of these Springs discharge water, which running down between the ridges of hills, unite their streams, and form rivulets or brooks, and many of these uniting again on the plain, become a river.

The perpetuity of divers Springs, always yielding the same quantity of water, equally when the least rain or vapour is afforded as when they are the greatest, furnish, in the opinion of some, considerable objections to the universality or sufficiency of the theory above. Dr. Derham mentions a Spring in his own parish of Upminster, which he could never perceive by his eye was diminished in the greatest droughts, even when all the ponds in the country, as well as an adjoining brook, had been dry for several months together; nor ever to be increased in the most rainy seasons, excepting perhaps for a few hours, or at most for a day, from sudden and violent rains. Had this Spring, he thought, derived its origin from rain or vapours, there would be found an increase and decrease of its water corresponding to those of its causes; as we actually find in such temporary Springs, as have undoubtedly their rise from rain and vapour.

Some naturalists therefore have recourse to the sea, and derive the origin of Springs immediately from thence. But how the sea-water should be raised up to the surface of the earth, and even to the tops of the mountains, is a difficulty, in the solution of which they cannot agree. Some fancy a kind of hollow subterranean rocks to receive the watery vapours raised from channels communicating with the sea, by means of an internal fire, and to act the part of alembics, in freeing them from their saline particles, as well as condensing and converting them into water. This kind of subterranean laboratory, serving for the distillation of sea-water, was the invention of Des Cartes: see his Princip. part 4, § 64. Others, as De la Hire &c (Mem. de l'Acad. 1703) set aside the alembics, and think it enough that there be large subterranean reservoirs of water at the height of the sea, from whence the warmth of the bottom of the earth, &c, may raise vapours; which pervade not only the intervals and fissures of the strata, but the bodies of the strata themselves, and at length arrive near the surface; where, being condensed by the cold, they glide along on the first bed of clay they meet with, till they issue forth by some aperture in the ground. De la Hire adds, that the salts of stones and minerals may contribute to the de-

taining and fixing the vapours, and converting them into water. Farther, it is urged by some, that there is a still more natural and easy way of exhibiting the rise of the sea-water up into mountains &c, viz, by putting a little heap of sand, or ashes, or the like, into a bason of water; in which case the sand &c will represent the dry land, or an island; and the bason of water, the sea about it. Here, say they, the water in the bason will rise to the top of the heap, or nearly so, in the same manner, and from the same principle, as the waters of the sea, lakes, &c, rise in the hills. The principle of ascent in both is accordingly supposed to be the same with that of the ascent of liquids in capillary tubes, or between contiguous planes, or in a tube filled with ashes; all which are now generally accounted for by the doctrine of attraction.

Against this last theory, Perrault and others have urged several unanswerable objections. It supposes a variety of subterranean passages and caverns, communicating with the sea, and a complicated apparatus of alembics, with heat and cold, &c, of the existence of all which we have no sort of proof. Besides, the water that is supposed to ascend from the depths of the sea, or from subterranean canals proceeding from it, through the porous parts of the earth, as it rises in capillary tubes, ascends to no great height, and in much too small a quantity to furnish springs with water, as Perrault has sufficiently shewn. And though the sand and earth through which the water ascends may acquire some saline particles from it, they are nevertheless incapable of rendering it so fresh as the water of our fountains is generally found to be. Not to add, that in process of time the saline particles of which the water is deprived, either by subterranean distillation or filtration, must clog and obstruct those canals and alembics, by which it is supposed to be conveyed to our Springs, and the sea must likewise gradually lose a considerable quantity of its salt.

Different sorts of SPRINGS. Springs are either such as run continually, called perennial; or such as run only for a time, and at certain seasons of the year, and therefore called *temporary* Springs. Others again are called *intermitting* Springs, because they flow and then stop, and flow and stop again; and *reciprocating* Springs, whose waters rise and fall, or flow and ebb, by regular intervals.

In order to account for these differences in Springs, let ABCDE (fig. 2, pl. 27) represent the declivity of a hill, along which the rain descends; passing through the fissures or channels BF, CG, DH, and LK, into the cavity or reservoir FGHKMI; from this cavity let there be a narrow drain or duct KE, which discharges the water at E. As the capacity of the reservoir is supposed to be large in proportion to that of the drain, it will furnish a constant supply of water to the spring at E. But if the reservoir FGHKMI be small, and the drain large, the water contained in the former, unless it is supplied by rain, will be wholly discharged by the latter, and the Spring will become dry: and so it will continue, even though it rains, till the water has had time to penetrate through the earth, or to pass through the channels into the reservoir; and the time necessary for furnishing a new supply to the drain KE will depend on the size of the fissures, the nature

ture

ture of the foil, and the depth of the cavity with which it communicates. Hence it may happen, that the Spring at E may remain dry for a confiderable time, and even while it rains; but when the water has found its way into the cavity of the hill, the Spring will begin to run. Springs of this kind, it is evident, may be dry in wet weather, efpecially if the duct KE be not exactly level with the bottom of the cavity in the hill, and difcharge water in dry weather; and the intermiffions of the Spring may continue feveral days. But if we fuppofe XOP to reprefent another cavity, fupplied with water by the channel NO, as well as by fiffures and clefts in the rock, and by the draining of the adjacent earth; and another channel STV, communicating with the bottom of it at S, afcending to T, and terminating on the furface at V, in the form of a fiphon; this difpofition of the internal cavities of the earth, which we may reafonably fuppofe that nature has formed in a variety of places, will ferve to explain the principle of reciprocating Springs; for it is plain, that the cavity XOP muft be fupplied with water to the height QPT, before it can pafs over the bend of the channel at T, and then it will flow through the longer leg of the fiphon TV, and be difcharged at the end V, which is lower than S. Now if the channel STV be confiderably larger than NO, by which the water is principally conveyed into the refervoir XOP, the refervoir will be emptied of its water by the fiphon; and when the water defcends below its orifice S, the air will drive the remaining water out of the channel STV, and the Spring will ceafe to flow. But in time the water in the refervoir will again rife to the height QPT, and be difcharged at V as before. It is eafy to conceive, that the diameters of the channels NO and STV may be fo proportioned to one another, as to afford an intermiffion and renewal of the Spring V at regular intervals. Thus, if NO communicates with a well fupplied by the tide, during the time of flow, the quantity of water conveyed by it into the cavity XOP may be fufficient to fill it up to QPT; and STV may be of fuch a fize as to empty it, during the time of ebb. It is eafy to apply this reafoning to more complicated cafes, where feveral refervoirs and fiphons communicating with each other, may fupply Springs with circumftances of greater variety. See Muffchenbroek's Introd. ad Phil. Nat. tom. ii. pa. 1010. Defagu. Exp Phil. vol. ii, pa. 173, &c.

We fhall here obferve, that Defaguliers calls thofe *reciprocating* Springs which flow conftantly, but with a ftream fubject to increafe and decreafe; and thus he diftinguifhes them from *intermitting* Springs, which flow or ftop alternately.

It is faid that in the diocefe of Paderborn, in Weftphalia, there is a Spring which difappears after twenty-four hours, and always returns at the end of fix hours with a great noife, and with fo much force, as to turn three mills, not far from its fource. It is called the Bolderborn, or boifterous Spring. Phil. Tranf. num. 7, pa. 127.

There are many Springs of an extraordinary nature in our own country, which it is needlefs to recite, as they are explicable by the general principles already illuftrated.

SPRING, *Ver*, in Aftronomy and Cofmography, denotes one of the feafons of the year; commencing, in the northern parts of the earth, on the day the fun enters the firft degree of Aries, which is about the 21ft day of March, and ending when the fun enters Cancer, at the fummer folftice, about the 21ft of June; Spring ending when the fummer begins.

Or, more ftrictly and generally, for any part of the earth, or on either fide of the equator, the Spring feafon begins when the meridian altitude of the fun, being on the increafe, is at a medium between the greateft and leaft; and ends when the meridian altitude is at the greateft. Or the Spring is the feafon, or time, from the moment of the fun's croffing the equator till he rife to the greateft height above it.

Elater SPRING, in Phyfics, denotes a natural faculty, or endeavour, of certain bodies, to return to their firft ftate, after having been violently put out of the fame by compreffing, or bending them, or the like.

This faculty is ufually called by philofophers, *elaftic force*, or *elafticity*.

SPRING, in Mechanics, is ufed to fignify a body of any fhape, perfectly elaftic.

Elafticity of a SPRING. See ELASTICITY.

Length of a SPRING, may, from its etymology, fignify the length of any elaftic body; but it is particularly ufed by Dr. Jurin to fignify the greateft length to which a Spring can be forced inwards, or drawn outwards, without prejudice to its elafticity. He obferves, this would be the whole length, were the Spring confidered as a mathematical line; but in a material Spring, it is the difference between the whole length, when the Spring is in its natural fituation, or the fituation it will reft in when not difturbed by any external force, and the length or fpace it takes up when wholly compreffed and clofed, or when drawn out.

Strength or Force of a SPRING, is ufed for the force or weight which, when the Spring is wholly compreffed or clofed, will juft prevent it from unbending itfelf. Alfo the Force of a Spring partly bent or clofed, is the force or weight which is juft fufficient to keep the Spring in that ftate, by preventing it from unbending itfelf any farther.

The theory of Springs is founded on this principle, *ut intenfio, fic vis*; that is, the intenfity is as the compreffing force; or if a Spring be any way forced or put out of its natural fituation, its refiftance is proportional to the fpace by which it is removed from that fituation. This principle has been verified by the experiments of Dr. Hook, and fince him by thofe of others, particularly by the accurate hand of Mr. George Graham. Lectures De Potentia Reftitutiva, 1678.

For elucidating this principle, on which the whole theory of Springs depends, fuppofe a Spring CL, refting at L againft any immoveable fupport, but otherwife lying in its natural fituation, and at full liberty. Then if this Spring be preffed inwards by any force p, or from C towards L, through the fpace of one inch, and can be there detained by that force p, the refiftance of the Spring and the force p, exactly counterbalancing each other; then will the double force $2p$ bend the Spring through the fpace of 2 inches, and the triple force $3p$ through 3 inches, and the quadruple force $4p$ through 4 inches, and fo on. The fpace CL through which the Spring is bent, or by which its end C is removed from its natural fituation, being always

ways proportional to the force which will bend it fo far, and will juft detain it when fo bent. On the other hand, if the end C be drawn outwards to any place λ, and be there detained from returning back by any force p, the fpace Cλ, through which it is fo drawn outwards, will be alfo proportional to the force p, which is juft able to retain it in that fituation.

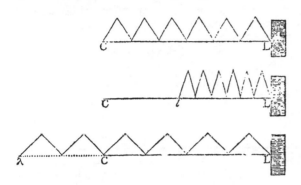

It may here be obferved, that the Spring of the air, or its elaftic-force, is a power of a different nature, and governed by different laws, from that of a palpable rigid Spring. For fuppofing the line LC to reprefent a cylindrical volume of air, which by compreffion is reduced to Ll, or by dilatation is extended to Lλ, its elaftic force will be reciprocally as Ll or Lλ; whereas the force or refiftance of a Spring is directly as Cl or Cλ.

This principle being premifed, Dr. Jurin lays down a general theorem concerning the action of a body ftriking on one end of a Spring, while the other end is fuppofed to reft againft an immoveable fupport.

Thus, if a Spring of the ftrength P, and the length CL, lying at full liberty upon an horizontal plane, reft with one end L againft an immoveable fupport; and a body of the weight M, moving with the velocity V, in the direction of the axis of the Spring, ftrike directly on the other end C, and fo force the Spring inwards, or bend it through any fpace CB; and if a mean proportional CG

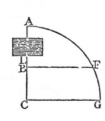

be taken between $\frac{M}{P} \times CL$ and $2a$, where a denotes the height to which a body would afcend in vacuo with the velocity V; and farther, if upon the radius R = CG be defcribed the quadrant of a circle GFA: then,

1. When the Spring is bent through the right fine CB of any arc GF, the velocity v of the body M is to the original velocity V, as the cofine BF is to the radius CG; that is $v : V :: BF : CG$, or $v = \frac{BF}{R} \times V$.

2. The time t of bending the Spring through the fame fine CB, is to T, the time of a heavy body's afcending in vacuo with the velocity V, as the corre-

fponding arc is to $2a$; that is $t : T :: GF : 2a$, or $t = \frac{GF}{2a} \times T$.

The doctor gives a demonftration of this theorem, and deduces a great many curious corollaries from it. Thefe he divides into three claffes. The firft contains fuch corollaries as are of more particular ufe when the Spring is wholly clofed before the motion of the body ceafes: the fecond comprehends thofe relating to the cafe, when the motion of the body ceafes before the Spring is wholly clofed: and the third when the motion of the body ceafes at the inftant that the Spring is wholly clofed.

We fhall here mention fome of the laft clafs, as being the moft fimple; having firft premifed, that P = the ftrength of the Spring, L = its length, V = the initial velocity of the body clofing the Spring, M = its mafs, t = time fpent by the body in clofing the Spring, A = height from which a heavy body will fall in vacuo in a fecond of time, a = the height to which a body would afcend in vacuo with the velocity V, C = the velocity gained by the fall, m = the circumference of a circle, whofe diameter is 1. Then, the motion of the ftriking body ceafing when the Spring is wholly clofed, it will be,

1. $V = C \sqrt{\dfrac{PL}{2MA}}$.

2. $Vt = \dfrac{mCL}{4A} \times 1''$.

3. $MV = C \sqrt{\dfrac{PLM}{2A}}$ the firft momentum.

4. If a quantity of motion MV bend a Spring through its whole length, and be deftroyed by it; no other quantity of motion equal to the former, as $nM \times \dfrac{V}{n}$, will clofe the fame Spring, and be wholly deftroyed by it.

5. But a quantity of motion, greater or lefs than MV, in any given ratio, may clofe the fame Spring, and be wholly deftroyed in clofing it; and the time fpent in clofing the Spring will be refpectively greater or lefs, in the fame given ratio.

6. The initial vis viva, or MV^2 is $= \dfrac{C^2 PL}{2A}$; and $2aM = PL$; alfo the initial vis viva is as the rectangle under the length and ftrength of the Spring, that is, MV^2 is as PL.

7. If the vis viva MV^2 bend a Spring through its whole length, and be deftroyed in clofing it; any other vis viva, equal to the former, as $n^2M \times \dfrac{V^2}{n^2}$, will clofe the fame Spring, and be deftroyed by it.

8. But the time of clofing the Spring by the vis viva $n^2M \times \dfrac{V^2}{n^2}$, will be to the time of clofing it by the vis viva MV^2, as n to 1.

9. If the vis viva MV^2 be wholly confumed in clofing a Spring, of the length L, and ftrength P; then the

vis viva n^2MV^2 will be sufficient to close, 1st, Either a Spring of the length L and strength n^2P. 2d, Or a Spring of the length nL and strength nP. 3d, Or of the length n^2L and strength P. 4th, Or, if n be a whole number, the number n^2 of Springs, each of the length L and strength P.—It may be added, that it appears from hence, that the number of similar and equal Springs a given body in motion can wholly close, is always proportional to the squares of the velocity of that body. And it is from this principle that the chief argument, to prove that the force of a body in motion is as the square of its velocity, is deduced. See FORCE.

The theorem given above, and its corollaries, will equally hold good, if the Spring be supposed to have been at first bent through a certain space, and by unbending itself to press upon a body at rest, and thus to drive that body before it, during the time of its expansion: only V, instead of being the initial velocity with which the body struck the Spring, will now be the final velocity with which the body parts from the Spring when totally expanded.

It may also be observed, that the theorem, &c, will equally hold good, if the Spring, instead of being pressed inward, be drawn outward by the action of the body. The like may be said, if the Spring be supposed to have been already drawn outward to a certain length, and in restoring itself draw the body after it. And lastly, the theorem extends to a Spring of any form whatever, provided L be the greatest length it can be extended to from its natural situation, and P the force which will confine it to that length. See Philos. Trans. num. 472, sect. 10, or vol. 43, art. 10.

SPRING is more particularly used, in the Mechanic Arts, for a piece of tempered steel, put into various machines to give them motion, by the endeavour it makes to unbend itself.

In watches, it is a fine piece of well beaten steel, coiled up in a cylindrical case, or frame; which by stretching itself forth, gives motion to the wheels, &c.

SPRING *Arbor*, in a Watch, is that part in the middle of the Spring-box, about which the Spring is wound or turned, and to which it is hooked at one end.

SPRING *Box*, in a Watch, is the cylindrical case, or frame, containing within it the Spring of the watch.

SPRING-*Compasses*. See COMPASSES.

SPRING *of the Air*, or its elastic force. See AIR, and ELASTICITY.

SPRING-*Tides*, are the higher tides, about the times of the new and full moon. See TIDE.

SPRINGY, or *Elastic Body*. See ELASTIC *Body*.

SQUARE, in Geometry, a quadrilateral figure, whose angles are right, and sides equal. Or it is an equilateral rectangle. Or an equilateral rectangular parallelogram.

A Square, and indeed any other parallelogram, is bisected by its diagonal. And the side of a Square is incommensurable to its diagonal, being in the ratio of 1 to $\sqrt{2}$.

To find the Area of a SQUARE. Multiply the side by itself, and the product is the area. So, if the side be 10, the area is 100; and if the side be 12, the area is 144.

SQUARE *Foot*, is a Square each side of which is equal to a foot, or 12 inches; and the area, or Square foot is equal to 144 square inches.

Geometrical SQUARE, a compartment often added on the face of a quadrant, called also *Line of* SHADOWS, and QUADRANT.

Gunner's SQUARE. See QUADRANT.

Magic SQUARE. See MAGIC *Square*.

SQUARE *Measures*, the Squares of the lineal measures; as in the following Table of Square Measures:

Squa. Inches.	Sq. Feet.	Sq. Yards.	Sq. Poles.	S. Chs.	Acres.	S. Miles.
144	1					
1296	9	1				
39204	272¼	30¼	1			
627264	4356	484	16	1		
6272640	43560	4840	160	10	1	
4014489600	27878400	3097600	102400	6400	640	1

Normal SQUARE, is an instrument, made of wood or metal, serving to describe and measure right angles;

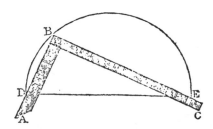

such is ABC. It consists of two rulers or branches fastened together perpendicularly. When the two legs are moveable on a joint, it is called a bevel.

To examine whether the Square is exact or not. Describe a semicircle DBE, with any radius at pleasure; in the circumference of which apply the angle of the Square to any point as B, and the edge of one leg to one end of the diameter as D, then if the other leg pass just by the other extremity at E, the Square is true; otherwise not.

SQUARE *Number*, is the product arising from a number multiplied by itself. Thus, 4 is the Square of 2, and 16 the Square of 4.

The series of Square integers, is 1, 4, 9, 16, 25, 36, &c; which are the Squares of - - 1, 2, 3, 4, 5, 6, &c.

Or the Square fractions - - $\frac{1}{4}$, $\frac{4}{9}$, $\frac{9}{16}$, $\frac{16}{25}$, $\frac{25}{36}$, $\frac{36}{49}$, &c, which are the Squares of - - $\frac{1}{2}$, $\frac{2}{3}$, $\frac{3}{4}$, $\frac{4}{5}$, $\frac{5}{6}$, $\frac{6}{7}$, &c.

A Square number is so called, either because it denotes the area of a Square, whose side is expressed by the root of the Square number; as in the annexed Square, which

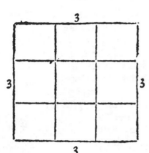

which confifts of 9 little fquares, the fide being equal to 3; or elfe, which is much the fame thing, becaufe the points in the number may be ranged in the form of a Square, by making the root, or factor, the fide of the Square.

Some properties of Squares are as follow : 1. Of the

Natural feries of Squares, 1^2, 2^2, 3^2, 4^2, &c, which are equal to 1, 4, 9, 16, &c;

The mean proportional mn between any two of thefe Squares m^2 and n^2, is equal to the lefs fquare *plus* its root multiplied by the difference of the roots ; or alfo equal to the greater fquare *minus* its root multiplied by the faid difference of the roots. That is,

$$mn = m^2 + dm = n^2 - dn ;$$

where $d = n - m$ is the difference of their roots.

2. An arithmetical mean between any two Squares m^2 and n^2, exceeds their geometrical mean, by half the Square of the difference of their roots.

That is $\frac{1}{2}m^2 + \frac{1}{2}n^2 = mn + \frac{1}{2}d^2$.

3. Of three equidiftant Squares in the Series, the geometrical mean between the extremes, is lefs than the middle Square by the Square of their common diftance in the Series, or of the common difference of their roots.

That is, $mp = n^2 - d^2$;

where m, n, p, are in arithmetical progreffion, the common difference being d.

4. The difference between the two adjacent Squares m^2, and n^2, is $n^2 - m^2 = 2m + 1$; in like manner, $p^2 - n^2 = 2n + 1$, the difference between the next two adjacent Squares n^2 and p^2 ; and fo on, for the next following Squares. Hence the difference of thefe differences, or the fecond difference of the Squares, is $2n - 2m = 2 \times \overline{n - m} = 2$ only, becaufe $n - m = 1$; that is, the fecond differences of the Squares are each the fame conftant number 2 : therefore the firft differences will be found by the continual addition of the number 2 ; and then the Squares themfelves will be found by the continual addition of the firft differences ; and thus the whole feries of Squares is conftructed by addition only, as here below :

2d Diff.		2	2	2	2	2	2	&c.
1ft Diff.	1	3	5	7	9	11	13	&c.
Squares.	1	4	9	16	25	36	49	&c.

And this method of conftructing the table of Square numbers I find firft noticed by Peletarius, in his Algebra.

5. Another curious property, alfo noted by the fame author, is, that the fum of any number of the cubes of the natural feries 1, 2, 3, 4, &c, taken from the beginning, always makes a Square number; and that the feries of Squares, fo formed, have for their roots the numbers 1, 3, 6, 10, 15, 21, &c, the diffs. of which are 1, 2, 3, 4, 5, 6, &c, viz, $1^3 = 1^2$,

$1^3 + 2^3 = 3^2$,

$1^3 + 2^3 + 3^3 = 6^2$,

$1^3 + 2^3 + 3^3 + 4^3 = 10^2$; and in general

$1^3 + 2^3 + 3^3 + n^3 = (1 + 2 + 3 + n)^2 = \overline{\frac{1}{2}n \cdot n + 1}$;

where n is the number of the terms or cubes.

SQUARE *Root*, a number confidered as the root of a fecond power or Square number : or a number which multiplied by itfelf, produces the given number. See EXTRACTION *of Roots*, and alfo the article ROOT, where tables of Squares and roots are inferted.

T. SQUARE, or *Tee* SQUARE, an inftrument ufed in drawing, fo called from its refemblance to the capital letter T.

This inftrument confifts of two ftraight rulers AB and CD, fixed at right angles to each other. To which is fometimes added a third EF, moveable about the pin C, to fet it to make any angle with CD.—It is very ufeful for drawing parallel and perpendicular lines, on the face of a fmooth drawing-board.

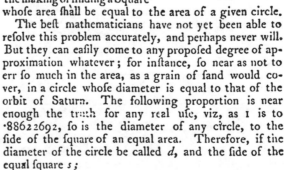

SQUARED · *fquare*, SQUARED-*cube*, &c. See POWER.

SQUARING. See QUADRATURE.

SQUARING *the Circle*, is the making or finding a Square whofe area fhall be equal to the area of a given circle.

The beft mathematicians have not yet been able to refolve this problem accurately, and perhaps never will. But they can eafily come to any propofed degree of approximation whatever ; for inftance, fo near as not to err fo much in the area, as a grain of fand would cover, in a circle whofe diameter is equal to that of the orbit of Saturn. The following proportion is near enough the truth for any real ufe, viz, as 1 is to ·88622692, fo is the diameter of any circle, to the fide of the fquare of an equal area. Therefore, if the diameter of the circle be called d, and the fide of the equal fquare s ;

then is $s = ·88622692d = \frac{3 \cdot 3}{4 \cdot 3}d$ nearly,

and $d = \dfrac{s}{·88622692} = \frac{4 \cdot 4}{3 \cdot 9}s$ nearly.

See CIRCLE, DIAMETER, and QUADRATURE.

3 R 2   STADIUM.

STADIUM, an ancient Greek long measure, containing 125 geometrical paces, or 625 Roman feet; corresponding to our furlong.

Eight Stadia make a geometrical or Roman mile; and 20, according to Dacier, a French league: but according to others, 800 Stadia make $41\frac{2}{3}$ leagues.

Guilletiere observes, that the Stadium was only 600 Athenian feet, which amount to 625 Roman, or 566 French, or 604 English feet: so that the Stadium should have been only 113 geometrical paces. It must be observed however, that the Stadium was different at different times and places.

STAFF, *Almucantar's, Augural, Back, Cross, Fore, Offset, &c.* See these several articles.

STAR, Stella, in Astronomy, a general name for all the heavenly bodies.

The Stars are distinguished, from the phenomena, &c, into *fixed* and *erratic* or *wandering*.

Erratic or *Wandering* Stars, are those which are continually changing their places and distances, with regard to each other. These are what are properly called *planets*. Though to the same class may likewise be referred comets or blazing Stars.

Fixed Stars, called also barely *Stars*, by way of eminence, are those which have usually been observed to keep the same distance, with regard to each other.

The chief circumstances observable in the fixed Stars, are their *distance, magnitude, number, nature,* and *motion*. Of each of which in their order.

Distance of the Fixed Stars. The fixed Stars are so extremely remote from us, that we have no distances in the planetary system to compare to them. Their immense distance appears from hence, that they have no sensible parallax; that is, that the diameter of the earth's annual orbit, which is nearly 190 millions of miles, bears no sensible proportion to their distance.

Mr. Huygens (Cosmotheor. lib. 4) attempts to determine the distance of the Stars, by making the aperture of a telescope so small, as that the sun through it appears no larger than Sirius; which he found to be only as 1 to 27664 of his diameter, when seen with the naked eye. So that, were the sun's distance 27664 times as much as it is, it would then be seen of the same diameter with Sirius. And hence, supposing Sirius to be a sun of the same magnitude with our sun, the distance of Sirius will be found to be 27664 times the distance of the sun, or 345 million times the earth's diameter.

Dr. David Gregory investigated the distance of Sirius, by supposing it of the same magnitude with the sun, and of the same apparent diameter with Jupiter in opposition: as may be seen at large in his Astronomy, lib. 3, prop. 47.

Cassini (Mem. Acad. 1717), by comparing Jupiter and Sirius, when viewed through the same telescope, inferred, that the diameter of that planet was 10 times as great as that of the Star; and the diameter of Jupiter being 50″, he concluded that the diameter of Sirius was about 5″; supposing then that the real magnitude of Sirius is equal to that of the sun, and the distance of the sun from us 12000 diameters of the earth, and the apparent diameter of Sirius being to that of the sun as 1 to 384, the distance of Sirius becomes equal to 4608000 diameters of the earth.

Thefe methods of Huygens, Gregory, and Cassini, are conjectural and precarious; both because the sun and Sirius are supposed of equal magnitude, and also because it is supposed the diameter of Sirius is determined with sufficient exactness.

Mr. Michell has proposed an enquiry into the probable parallax and magnitude of the fixed Stars, from the quantity of light which they afford us, and the peculiar circumstances of their situation. With this view he supposes, that they are, on a medium, equal in magnitude and natural brightness to the sun; and then proceeds to inquire, what would be the parallax of the sun, if he were to be removed so far from us, as to make the quantity of the light, which we should then receive from him, no more than equal to that of the fixed Stars. Accordingly, he assumes Saturn in opposition, as equal, or nearly equal in light to the brightest fixed Star. As the mean distance of Saturn from the sun is equal to about 2082 of the sun's semidiameters, the density of the sun's light at Saturn will consequently be less than at his own surface, in the ratio of the square of 2082 or 4334724 to 1: If Saturn therefore reflected all the light that falls upon him, he would be less luminous in that same proportion. And besides, his apparent diameter, in the opposition, being but about the 105th part of that of the sun, the quantity of light which we receive from him must be again diminished in the ratio of the square of 105 or 11025 to 1. Consequently, by multiplying these two numbers together, we shall have the whole of the light of the sun to that of Saturn, as the square nearly of 220,000 or 48,400,000,000 to 1. Hence, removing the sun to 220,000 times his present distance, he would still appear at least as bright as Saturn, and his whole parallax upon the diameter of the earth's orbit would be less than 2 seconds: and this must be assumed for the parallax of the brightest of the fixed Stars, upon the supposition that their light does not exceed that of Saturn.

By a like computation it may be found, that the distance, at which the sun would afford us as much light as we receive from Jupiter, is not less than 46,000 times his present distance, and his whole parallax in that case, upon the diameter of the earth's orbit, would not be more than 9 seconds; the light of Jupiter and Saturn, as seen from the earth, being in the ratio of about 22 to 1, when they are both in opposition, and supposing them to reflect equally in proportion to the whole of the light that falls upon them. But if Jupiter and Saturn, instead of reflecting the whole of the light that falls upon them, should really reflect only a part of it, as a 4th, or a 6th, which may be the case, the above distances must be increased in the ratio of 2 or $2\frac{1}{2}$ to 1, to make the sun's light no more than equal to theirs; and his parallax would be less in the same proportion. Supposing then that the fixed Stars are of the same magnitude and brightness with the sun, it is no wonder that their parallax should hitherto have escaped observation; since in this case it could hardly amount to 2 seconds, and probably not more than one in Sirius himself, though he had been placed in the pole of the ecliptic; and in those that appear much less luminous, as γ Draconis, which is only of the 3d magnitude, it could hardly be expected to be sensible with such instruments as have hitherto been used. However, Mr. Michell

chell fuggefts, that it is not impracticable to conftruct inftruments capable of diftinguifhing even to the 20th part of a fecond provided the air will admit of that degree of exactnefs. This ingenious writer apprehends that the quantity of light which we receive from Sirius, does not exceed the light we receive from the leaft fixed Star of the 6th magnitude, in a greater ratio than that of 1000 to 1, nor lefs than that of 400 to 1; and the fmaller Stars of the 2d magnitude feem to be about a mean proportional between the other two. Hence the whole parallax of the leaft fixed Stars of the 6th magnitude, fuppofing them of the fame fize and native brightnefs with the fun, fhould be from about 2''' to 3''', and their diftance from about 8 to 12 million times that of the fun : and the parallax of the fmaller Stars of the 2d magnitude, upon the fame fuppofition, fhould be about 12''', and their diftance about 2 million times that of the fun.

This author farther fuggefts, that, from the apparent fituation of the Stars in the heavens, there is the greateft probability that the Stars are collected together in clufters in fome places, where they form a kind of fyftems, whilft in others there are either few or none of them ; whether this difpofition be owing to their mutual gravitation, or to fome other law or appointment of the Creator. Hence it may be inferred, that fuch double Stars, &c. as appear to confift of two or more Stars placed very near together, do really confift of Stars placed near together, and under the influence of fome general law: and he proceeds to inquire whether, if the Stars be collected into fyftems, the fun does not likewife make one of fome fyftem, and which fixed Stars thofe are that belong to the fame fyftem with him.

Thofe Stars, he apprehends, which are found in clufters, and furrounded by many others at a fmall diftance from them, belong probably to other fyftems, and not to ours. And thofe Stars, which are furrounded with nebulæ, are probably only very large Stars which, on account of their fuperior magnitude, are fingly vifible, while the others, which compofe the remaining parts of the fame fyftem, are fo fmall as to efcape our fight. And thofe nebulæ in which we can difcover either none or only a few Stars, even with the affiftance of the beft telefcopes, are probably fyftems that are ftill more diftant than the reft. For other particulars of this inquiry, fee Philof. Tranf. vol. 57, pa. 234 &c.

As the diftance of the fixed Stars is beft determined by their parallax, various methods have been purfued, though hitherto without fuccefs, for inveftigating it ; the refult of the moft accurate obfervations having given us little more than a diftant approximation ; from which however we may conclude, that the neareft of the fixed Stars cannot be lefs than 40 thoufand diameters of the whole annual orbit of the earth diftant from us.

The method pointed out by Galileo, and attempted by Hook, Flamfteed, Molyneux, and Bradley, of taking the diftances of fuch Stars from the zenith as pafs very near it, has given us a much jufter idea of the immenfe diftance of the ftars, and furnifhed an approximation to their parallax, much nearer the truth, than any we had before.

Dr. Bradley affures us (Philof. Tranf. num. 406, or Abr. vol. 6, pa. 162), that had the parallax amounted to a fingle fecond, or two at moft, he fhould have perceived it in the great number of obfervations which he made, efpecially upon γ Draconis ; and that it feemed to him very probable, that the annual parallax of this Star does not amount to a fingle fecond, and confequently that it is above 400 thoufand times farther from us than the fun.

But Dr. Herfchel, to whofe induftry and ingenuity, in exploring the heavens, aftronomy is already much indebted, remarks, that the inftrument ufed on this occafion, being the fame with the prefent zenith fectors, can hardly be allowed capable of fhewing an angle of one or even two feconds, with accuracy : and befides, the Star on which the obfervations were made, is only a bright Star of the 3d magnitude, or a fmall Star of the 2d ; and that therefore its parallax is probably much lefs than that of a Star of the firft magnitude. So that we are not warranted in inferring, that the parallax of the Stars in general does not exceed 1 '', whereas thofe of the firft magnitude may have, notwithftanding the refult of Dr. Bradley's obfervations, a parallax of feveral feconds.

As to the method of zenith diftances, it is liable to confiderable errors, on account of refraction, the change of pofition of the earth's axis, arifing from nutation, preceffion of the equinoxes, or other caufes, and the aberration of light.

Dr. Herfchel has propofed another method, by means of double Stars, which is free from thefe errors, and of fuch a nature, that the annual parallax, even if it fhould not exceed the 10th part of a fecond, may ftill become vifible, and be afcertained at leaft much nearer than heretofore. This method, which was firft propofed in an imperfect manner by Galileo, and has been alfo mentioned by other authors, is capable of every improvement which the telefcope and mechanifm of micrometers can furnifh. To give a general idea of it, let O and E be two oppofite points of the annual orbit, taken in the fame plane with two ftars A, B, of unequal magnitudes. Let the angle AOB be obferved when the earth is at O, and AEB be obferved when the earth is at E. From the difference of thefe angles, when there is any, the parallax of the Stars may be computed, according to the theory fubjoined. Thefe two Stars ought to be as near as poffible to each other, and alfo to differ as much in magnitude as we can find them.

This theory of the annual parallax of double Stars, with the method of computing from thence what is ufually called the parallax of the fixed Stars, or of fingle Stars of the firft magnitude, fuch as are neareft to us, fuppofes 1ft, that the Stars are all about the fize of the fun ; and 2dly, that the difference in their apparent magnitudes, is owing to their different diftances, fo as that a Star of the 2d, 3d, or 4th magnitude, is 2, 3, or 4 times as far off as one of the firft. Thefe principles, which Dr. Herfchel premifes as poftulata, have fo great a probability in their favour, that they will

fcarcely

scarcely be objected to by those who are in the least acquainted with the doctrine of chances. See Mr. Michell's Inquiry, &c. already cited. And Philof. Tranf. vol. 57, pa. 234 - - - - 240. Also Dr. Halley, on the Number, Order, and Light of the fixed Stars, in the Philof. Tranf. vol. 31, or Abr. vol. 6, pa. 148.

Therefore, let EO be the whole diameter of the earth's annual orbit; and let A, B, C be three Stars fituated in the ecliptic, in fuch a manner, that they may appear all in one line OABC when the earth is at O. Now if OA, AB, BC be equal to each other, A will be a Star of the firft magnitude, B of the fecond, and C of the third. Let us next fuppofe the angle OAE, or parallax of the whole orbit of the earth, to be 1′ of a degree; then, becaufe very fmall angles, having the fame fubtenfe EO, may be taken to be in the inverfe ratio of the lines OA, OB, OC, &c, we fhall have EBO = $\frac{1}{2}''$, and ECO = $\frac{1}{3}''$, &c, alfo becaufe EA = AB nearly, the angle AEB = ABE = $\frac{1}{2}''$; and becaufe BC = $\frac{1}{2}$ BO = $\frac{1}{2}$ BE nearly, the angle BEC = $\frac{1}{2}$ BCE = $\frac{1}{6}''$, and hence AEC = $\frac{1}{2}$ + $\frac{1}{6}$ = $\frac{2}{3}''$; from all which it follows that, when the earth is at E,
the Stars A and B appear at $\frac{1}{2}''$ diftant from one another,
the Stars A and C at $\frac{2}{3}''$ diftant, and
the Stars B and C only $\frac{1}{6}''$ diftant. In like manner may be deduced a general expreffion for the parallax that will become vifible in the change of diftance between the two Stars, by the removal of the earth from one extreme of her orbit to the other. Let P denote the total parallax of a fixed Star of the magnitude of the M order, and m the number of the order of a fmaller Star, p denoting the partial parallax to be obferved by the change in the diftance of a double Star;

then is $p = \frac{m-M}{mM}P$, or $P = \frac{mMp}{m-M}$, which gives

P, when p is found by obfervation.

For Ex. Suppofe a Star of the 1ft magnitude fhould have a fmall Star of the 12th magnitude near it; then will the partial parallax we are to expect to fee be

$\frac{12-1}{12 \times 1}P = \frac{11}{12}P$, or $\frac{11}{12}$ of the total parallax of the

larger Star; and if we fhould, by obfervation, find the partial parallax between two fuch Stars to amount to 1″, then will the total parallax $P = \frac{12}{11}p = 1''\frac{1}{11}$. Again, if the Stars be of the 3d and 24th magnitude,

the total parallax will be $P = \frac{24 \times 3}{24-3}p = \frac{72}{21}p = \frac{24}{7}p$;

fo that if by obfervation p be found to be $\frac{1}{10}$ of a fecond, the whole parallax P will come out $\frac{24}{70}'' = 0.3428''$.

Farther, the Stars being ftill in the ecliptic, fuppofe

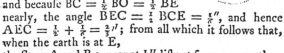

they fhould appear in one line, when the earth is in fome other part of her orbit between E and O; then will the parallax be ftill expreffed by the fame algebraic formula, and one of the maxima will ftill lie at E, the other at O; but the whole effect will be divided into two parts, which will be in proportion to each other, as radius — fine to radius + fine of the Star's diftance from the neareft conjunction or oppofition.

When the Stars are any where out of the ecliptic, fituated fo as to appear in one line OABC perpendicular to EO, the maximum of parallax will ftill be ex-

preffed by $\frac{m-M}{mM}P$; but there will arife another ad-

ditional parallax in the conjunction and oppofition, which will be to that which is found 90° before or after the fun, as the fine (s) of the latitude of the Stars feen at O, is to radius (1); and the effect of this parallax will be divided into two parts; half of it lying on one fide of the large Star, the other half on the other fide of it. This latter parallax will alfo be compounded with the former, fo that the diftance of the Stars in the conjunction and oppofition will then be reprefented by the diagonal of a parallelogram, whofe fides are the two femiparallaxes; a general expreffion for which will be

$$\frac{m-M}{2mM}P\sqrt{1+s^2} \text{ or } \frac{1}{2}p\sqrt{1+s^2}.$$

When the Stars are in the pole of the ecliptic, s will be = 1, and the laft formula becomes $\frac{1}{2}p\sqrt{2} = .7071p$.

Again, let the Stars be at fome diftance, as 5″, from each other, and let them be both in the ecliptic. This cafe is refolvable into the firft; for imagine the Star A to ftand at I; then the angle AEI may be accounted equal to AOI; and as the foregoing formula,

$p = \frac{m-M}{mM}P$, gives us the angles AEB, AEC,

we are to add AEI or 5″ to AEB, which will give IEB. In general, let the diftance of the Stars be d, and let the obferved diftance at E be D; then will $D = d + p$, and therefore the whole parallax of the

annual orbit will be expreffed by $\frac{D-d}{m-M}Dd = P$.

Suppofe now the Stars to differ only in latitude, one being in the ecliptic, the other at fome diftance as 5″ north, when feen at O. This cafe may alfo be refolved by the former; for imagine the Stars B and C to be elevated at right angles above the plane of the figure, fo that AOB, or AOC, may make an angle of 5″ at O; then inftead of the lines OABC, EA, EB, EC, imagine them all to be planes at right angles to the figure; and it will appear that the parallax of the Stars in longitude, muft be the fame as if the fmall Star had been without latitude. And fince the Stars B, C, by the motion of the earth from O to E, will not change their latitude, we fhall have the following conftruction for finding the diftance of the Stars AB and AC at E, and from thence the parallax P.

Let

Let the triangle $ab\beta$ reprefent the fituation of the Stars; ab is the fubtenfe of $5''$, the angle under which they are fuppofed to be feen at O. The quantity $b\beta$ by the for-mer theorem is found $= \frac{m-M}{mM}P,$

which is the partial parallax, that would have been feen by the earth's moving from O to E, if both Stars had been in the ecliptic; but, on account of the difference in latitude, it will now be reprefented by $a\beta$, the hypotenufe of the triangle $ab\beta$: therefore in general, putting $ab = d$, $a\beta = D$, we have $\frac{mM}{m-M}\sqrt{D^2-d^2} = P.$ Hence, D being found by obfervation, and the three d, m, M given, the total parallax is obtained.

When the Stars differ in longitude as well as latitude, this cafe may be refolved in the following manner. Let the triangle $ab\beta$ reprefent the fituation of the Stars, $ab = d$ being their diftance feen at O, $a\beta = D$ their diftance feen at E. That the change $b\beta$, which is pro-duced by the earth's motion, will be truly expreffed by $\frac{m-M}{mM}P$, may be proved as

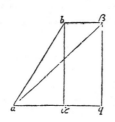

before, by fuppofing the Star a to have been placed at α. Now let the angle of pofition $ba\alpha$ be taken by a micro-meter, or by any other method fufficiently exact; then, by refolving the triangle $ab\alpha$, we obtain the longitudi-nal and latitudinal differences $a\alpha$ and $b\alpha$ of the two ftars. Put $a\alpha = x$, $ba' = y$, and it will be $x + b\beta = aq$, whence

$$D=\sqrt{(x+\frac{m-M}{mM}P)^2+y^2}; \text{ and } P=\frac{\sqrt{D^2-y^2}-x}{m-M}mM.$$

If neither of the Stars fhould be in the ecliptic, nor have the fame longitude or latitude, the laft theorem will ftill ferve to calculate the total parallax, whofe maxi-mum will lie in E. There will alfo arife another pa-rallax, whofe maximum will be in the conjunction and oppofition, which will be divided, and lie on different fides of the large Star; but as the whole parallax is ex-tremely fmall, it is not neceffary to inveftigate every particular cafe of this kind; for by reafon of the divi-fion of the parallax, which renders obfervations taken at any other time, except where it is greateft, very un-favourable, the formulæ would be of little ufe.

Dr. Herfchel clofes his account of this theory, with a general obfervation on the time and place where the maxima of parallax will happen. Thus, when two un-equal Stars are both in the ecliptic, or, not being in the ecliptic, have equal latitudes, north or fouth, and the larger Star has moft longitude, the maximum of the apparent diftance will be when the fun's longi-tude is $90°$ more than the Star's, or when obferved in the morning: and the minimum, when the longitude of the fun is $90°$ lefs than that of the Star, or when obferved in the evening. But when the fmall Star has moft

longitude, the maximum and minimum, as well as the time of obfervation, will be the reverfe of the former. And when the Stars differ in latitude, this makes no alteration in the place ot the maximum or minimum, nor in the time of obfervation; that is, it is immaterial which of the two Stars has the greater latitude. Philof. Tranf. vol. 72, art. 11.

The diftance of the Star γ Draconis appears, by Bradley's obfervations, already recited, to be at leaft 400,000 times that of the fun, and the diftance of the neareft fixed Star, not lefs than 40,000 diameters of the earth's annual orbit: that is, the diftance from the earth, of the former at leaft 38,000,000,000,000 miles, and the latter not lefs than 7,600,000,000,000 miles.

As thefe diftances are immenfely great, it may both be amufing, and help to a clearer and more familiar idea, to compare them with the velocity of fome moving body, by which they may be meafured.

The fwifteft motion we know of, is that of light, which paffes from the fun to the earth in about 8 mi-nutes; and yet this would be above 6 years traverfing the firft fpace, and near a year and a quarter in paffing from the neareft fixed Star to the earth. But a cannon-ball, moving on a medium at the rate of about 20 miles in a minute, would be 3 million 8 hundred thou-fand years in paffing from γ Draconis to the earth, and 760 thoufand years paffing from the neareft fixed Star. Sound, which moves at the rate of about 13 miles in a minute, would be 5 million 600 thoufand years tra-verfing the former diftance, and 1 million 128 thou-fand, in paffing through the latter.

The celebrated Huygens purfued fpeculations of this kind fo far, as to believe it not impoffible, that there may be Stars at fuch inconceivable diftances, that their light has not yet reached the earth fince its creation.

Dr. Halley has alfo advanced, what he fays feems to be a metaphyfical paradox (Philof. Tranf. number 364, or Abr. vol. 6, pa. 148), viz, that the number of fixed Stars muft be more than finite, and fome of them more than at a finite diftance from others: and Addifon has juftly obferved, that this thought is far from being ex-travagant, when we confider that the univerfe is the work of infinite power, prompted by infinite good-nefs, and having an infinite fpace to exert itfelf in; fo that our imagination can fet no bounds to it.

Magnitude of the fixed STARS. The magnitudes of the Stars appear to be very different from one ano-ther; which difference may probably arife, partly from a diverfity in their real magnitude, but principally from their diftances, which are different.

To the bare eye, the Stars appear of fome fenfible magnitude, owing to the glare of light arifing from the numberlefs reflections from the aërial particles &c. about the eye: this makes us imagine the Stars to be much larger than they would appear, if we faw them only by the few rays which come directly from them, fo as to enter our eyes without being intermixed with others.

Any perfon may be fenfible of this, by look-ing at a Star of the firft magnitude through a long narrow tube; which, though it takes in as much of the fky as would hold a thoufand fuch ftars, fcarce ren-ders that one vifible.

The

The Stars, on account of their apparently various fizes, have been diftributed into feveral claffes, called *magnitudes*. The 1ft clafs, or Stars of the firft magnitude, are thofe that appear largeft, and may probably be neareft to us. Next to thefe, are thofe of the 2d magnitude; and fo on to the 6th, which comprehends the fmalleft Stars vifible to the naked eye. All beyond thefe, that can be perceived by the help of telefcopes, are called *telefcopic* ftars. Not that all the Stars of each clafs appear juftly of the fame magnitude; there being great latitude in this refpect; and thofe of the firft magnitude appearing almoft all different in luftre and fize. There are alfo other Stars, of intermediate magnitudes, which aftronomers cannot refer to one clafs rather than another, and therefore they place them between the two. Procyon, for inftance, which Ptolomy makes of the firft magnitude, and Tycho of the 2d, Flamfteed lays down as between the 1ft and 2d. So that, inftead of 6 magnitudes, we may fay there are almoft as many orders of Stars, as there are Stars; fo great variations being obfervable in the magnitude, colour, and brightnefs of them.

There feems to be little chance of difcovering with certainty the real fize of any of the fixed Stars; we muft therefore be content with an approximation, deduced from their parallax, if this fhould ever be found; and the quantity of light they afford us, compared with that of the fun. And to this purpofe, Dr. Herfchel informs us, that with a magnifying power of 6450, and by means of his new micrometer, he found the apparent diameter of α Lyræ to be 0″·355.

The Stars are alfo diftinguifhed, with regard to their fituation, into *afterifms*, or *conftellations*; which are nothing but affemblages of feveral neighbouring Stars, confidered as conftituting fome determinate figure, as of an animal, &c, from which it is therefore denominated: a divifion as ancient as the book of Job, in which mention is made of Orion, the Pleiades, &c.

Befides the Stars thus diftinguifhed into magnitudes and conftellations, there are others not reduced to either. Thofe not reduced into conftellations, are called *informes*, or *unformed* Stars; of which kind feveral, fo left at large by the ancients, have fince been formed into new conftellations by the modern aftronomers, and efpecially by Hevelius.

Thofe not reduced to claffes or magnitudes, are called nebulous Stars; but fuch as only appear faintly in clufters, in form of little lucid fpots, nebulæ, or clouds.

Ptolomy fets down five of fuch nebulæ, viz, one at the extremity of the right hand of Perfeus, which appears through the telefcope, thick fet with Stars; one in the middle of the crab, called *Præfepe*, or the Manger, in which Galileo counted above 40 Stars; one unformed near the fting of the Scorpion; another in the eye of Sagittarius, in which two Stars may be feen in a clear fky with the naked eye, and feveral more with the telefcope; and the fifth in the head of Orion, in which Galileo counted 21 Stars.

Flamfteed obferved a cloudy Star before the bow of Sagittarius, which confifts of a great number of fmall Stars; and the Star *d* above the right fhoulder of this conftellation is encompaffed with feveral more. Flamfteed and Caffini alfo difcovered one between the great and little dog, which is very full of Stars, that are vifible only by the telefcope.

But the moft remarkable of all the cloudy Stars, is that in the middle of Orion's fword, in which Huygens and Dr. Long obferved 12 Stars, 7 of which (3 of them, now known to be 4, being very clofe together) feem to fhine through a cloud, very lucid near the middle, but faint and ill defined about the edges. But the greateft difcoveries of nebulæ and clufters of Stars, we owe to the powerful telefcopes of Dr. Herfchel, who has given accounts of fome thoufands of fuch nebulæ, in many of which the Stars feem to be innumerable, like grains of fand. See Philof. Tranf. 1784, 1785, 1786, 1789. See Galaxy, and Magellanic *clouds*, and *lucid* Spots.

Caffini is of opinion, that the brightnefs of thefe proceeds from Stars fo minute, as not to be diftinguifhed by the beft glaffes: and this opinion is fully confirmed by the obfervations of Dr. Herfchel, whofe powerful telefcopes fhew thofe lucid fpecks to be compofed entirely of maffes of fmall Stars, like heaps of fand.

There are alfo many Stars which, though they appear fingle to the naked eye, are yet difcovered by the telefcope to be double, triple, &c. Of thefe, feveral have been obferved by Caffini, Hooke, Long, Mafkelyne, Hornfby, Pigott, Mayer, &c; but Dr. Herfchel has been much the moft fuccefsful in obfervations of this kind; and his fuccefs has been chiefly owing to the very extraordinary magnifying powers of the Newtonian 7 feet reflector which he has ufed, and the advantage of an excellent micrometer of his own conftruction. The powers which he has ufed, have been 146, 227, 278, 460, 754, 932, 1159, 1536, 2010, 3168, and even 6450. He has already formed a catalogue, containing 269 double Stars, 227 of which, as far as he knows, have not been noticed by any other perfon. Among thefe there are alfo fome Stars that are treble, double-double, quadruple, double-treble, and multiple. His catalogue comprehends the names of the Stars, and the number in Flamfteed's catalogue, or fuch a defcription of thofe that are not contained in it, as will be found fufficient to diftinguifh them; alfo the comparative fize of the Stars; their colours as they appeared to his view; their diftances determined in feveral different ways; their angle of pofition with regard to the parallel of declination; and the dates when he firft perceived the Stars to be double, treble, &c. His obfervations appear to commence with the year 1776, but almoft all of them were made in the years 1779, 1780, 1781.

Dr. Herfchel has diftributed the double Stars contained in his catalogue, into 6 different claffes. In the firft he has placed all thofe which require a very fuperior telefcope, with the utmoft clearnefs of air, and every other favourable circumftance, to be feen at all, or well enough to judge of them; and there are 24 of thefe. To the 2d clafs belong all thofe double Stars that are proper for eftimations by the eye, and very delicate meafures by the micrometer; the number being 38. The 3d clafs comprehends all thofe double Stars, that are between 5″ and 15″ afunder; the number of them being 46. The 4th, 5th, and 6th claffes contain double

double Stars that are from 15″ to 30″, and from 30″ to 1′, and from 1′ to 2′ or more afunder; of which there are 44 in the 4th clafs. 51 in the 5th clafs, and 66 in the 6th clafs: the laft of this clafs is α Tauri, number 87 of Flamfteed, whofe apparent diameter, upon the meridian meafured with a power of 460 at a mean of two obfervations 1″ 46‴, and with a power of 932 at a mean of two obfervations 1″ 12‴. See the lift at large, Philofoph. Tranf, vol. 72, art. 12.

The Stars are alfo diftinguifhed, in each conftellation, by numbers, or by the letters of the alphabet. This fort of diftinction was introduced by John Bayer, in his Uranometria, 1654; where he denotes the Stars, in each conftellation, by the letters of the Greek alphabet, α, β, γ, δ, ε, &c, viz, the moft remarkable Star of each by α, the 2d by β, the 3d by γ, &c; and when there are more Stars in a conftellation than the characters in the Greek alphabet, he denotes the reft, in their order, by the Roman letters A, b, c, d, &c. But as the number of the Stars, that have been obferved and regiftered in catalogues, fince Bayer's time, is greatly increafed, as by Flamfteed and others, the additional ones have been marked by the ordinal numbers 1, 2, 3, 4, 5, &c.

The *Number of* STARS. The number of the Stars appears to be immenfely great, almoft infinite; yet have aftronomers long fince afcertained the number of fuch as are vifible to the eye, which are much fewer than at firft fight could be imagined. See CATALOGUE *of the Stars.*

Of the 3000 contained in Flamfteed's catalogue, there are many that are only vifible through a telefcope; and a good eye fcarce ever fees more than a thoufand at the fame time in the cleareft heaven; the appearance of innumerable more, that are frequent in clear winter nights, arifing from our fight's being deceived by their twinkling, and from our viewing them confufedly, and not reducing them to any order. But neverthelefs we cannot but think the Stars are almoft, if not altogether, infinite. See Halley, on the number, order, and light of the fixed Stars, Philof. Tranf. number 364, or Abr. vol. 6, pa. 148.

Riccioli, in his New Almageft, affirms, that a man who fhall fay there are above 20 thoufand times 20 thoufand, would fay nothing improbable. For a good telefcope, directed indifferently to almoft any point of the heavens, difcovers multitudes that are loft to the naked eye; particularly in the milky way, which fome take to be an affemblage of Stars, too remote to be feen fingly, but fo clofely difpofed as to give a luminous appearance to that part of the heavens where they are. And this fact has been confirmed by Herfchel's obfervations: though it is difputed by others, who contend that the milky way muft be owing to fome other caufe.

In the fingle conftellation of the Pleiades, inftead of 6, 7, or 8 Stars feen by the beft eye; Dr. Hook, with a telefcope 12 feet long, told 78, and with larger glaffes many more, of different magnitudes. And F. de Rheita affirms, that he has obferved above 2000 Stars in the fingle conftellation of Orion. The fame author found above 188 in the Pleiades. And Huygens, looking at the Star in the middle of Orion's

sword, inftead of one, found it to be 12. Galileo found 80 in the fpace of the belt of Orion's fword, 21 in the nebulous Star of his head, and above 500 in another part of him, within the compafs of one or two degrees fpace, and more than 40 in the nebulous Star Præfepe.

The Changes that have happened in the STARS are very confiderable. The firft change that is upon record, was about 120 years before Chrift; when Hipparchus, difcovering a new Star to appear, was firft induced to make a catalogue of the Stars, that pofterity might perceive any future changes of the like nature.

In the year 1572, Cornelius Gemma and Tycho Brahe obferved another new Star in the conftellation Caffiopeia, which was likewife the occafion of Tycho's making a new catalogue. At firft its magnitude and brightnefs exceeded the largeft of the Stars, Sirius and Lyra; and even equalled the planet Venus when neareft the earth, and was feen in fair day-light. It continued 16 months; towards the latter end of which it began to dwindle, and at length, in March 1574, it totally difappeared, without any change of place in all that time.

Leovicius tells us of another Star appearing in the fame conftellation, about the year 945, which refembled that of 1572; and he quotes another ancient obfervation, by which it appears that a new Star was feen about the fame place in 1264. Dr. Keil thinks thefe were all the fame Star; and indeed the periodical intervals, or diftance of time between thefe appearances, were nearly equal, being from 318 to 319 years; and if fo, its next appearance may be expected about 1890.

Fabricius, in 1596, difcovered another new Star, called the *ftella mira*, or *wonderful Star*, in the neck of the whale, which has fince been found to appear and difappear periodically, 7 times in 6 years, continuing in its greateft luftre for 15 days together; and is never quite extinguifhed. Its courfe and motion are defcribed by Bulliald, in a treatife printed at Paris in 1667. Dr. Herfchel has lately, viz, in the years 1777, 1778, 1779, and 1780, made feveral obfervations on this Star, an account of which may be feen in the Philof. Tranf. vol. 70, art. 21.

In the year 1600, William Janfen difcovered a changeable Star in the neck of the Swan, which gradually decreafed till it became fo fmall as to be thought to difappear entirely, till the years 1657, 1658, and 1659, when it regained its former luftre and magnitude; but foon decayed again, and is now of the fmalleft fize.

In the year 1604, a new Star was feen by Kepler, and feveral of his friends, near the heel of the right foot of Serpentarius, which was particularly bright and fparkling; and it was obferved to be every moment changing into fome of the colours of the rainbow, except when it is near the horizon, at which time it was generally white. It furpaffed Jupiter in magnitude, but was eafily diftinguifhed from him, by the fteady light of the planet. It difappeared about the end of the year 1605, and has not been feen fince that time.

Simon Marius difcovered another in Andromeda's

                                          girdle,

girdle, in 1612 and 1613; though Bulliald says it had been seen before, in the 15th century.

In July 1670, Hevelius discovered a second changeable Star in the Swan, which was so diminished in October as to be scarce perceptible. In April following it regained its former lustre, but wholly disappeared in August. In March 1672 it was seen again, but appeared very small, and has not been visible since.

In 1686 a third changeable Star was discovered by Kirchius in the Swan, viz, the Star χ of that constellation, which returned periodically in about 405 days.

In 1672 Cassini saw a Star in the neck of the Bull, which he thought was not visible in Tycho's time, nor when Bayer made his figures.

It is certain, from the old catalogues, that many of the ancient Stars are not now visible. This has been particularly remarked with regard to the Pleiades.

M. Montanari, in his letter to the Royal Society in 1070, observes that there are now wanting in the heavens two Stars of the 2d magnitude, in the stern of the ship Argo, and its yard, which had been seen till the year 1664. When they first disappeared is not known; but he assures us there was not the least glimpse of them in 1668. He adds, he has observed many more changes in the fixed Stars, even to the number of a hundred. And many other changes of the Stars have been noticed by Cassini, Maraldi, and other observers. See Gregory's Astron. lib. 2, prop. 30.

But the greatest numbers of variable Stars have been observed of late years, and the most accurate observations made on their periods, &c, by Herschel, Goodricke, Pigott, &c, in the late volumes of the Philos. Trans. particularly in the vol. for 1786, where the last of these gentlemen has given a catalogue of all that have been hitherto observed, with accounts of the observations that have been made upon them.

Various hypotheses have been devised to account for such changes and appearances in the Stars. It is not probable they could be comets, as they had no parallax, even when largest and brightest. It has been supposed that the periodical Stars have vast dark spots, or dark sides, and very slow rotations on their axes, by which means they must disappear when the darker side is turned towards us. And as for those which break out suddenly with such lustre, these may perhaps be suns who e fuel is almost spent, and again supplied by some of their comets falling upon them, and occasioning an uncommon blaze and splendor for some time; which it is conjectured may be one use of the cometary part of our system.

Maupertuis, in his Dissertation on the figures of the Celestial Bodies (pa. 61—63), is of opinion that some Stars, by their prodigious swift rotation on their axes, may not only assume the figures of oblate spheroids, but that by the great centrifugal force arising from such rotations, they may become of the figures of mill-stones, or be reduced to flat circular planes, so thin as to be quite invisible when their edges are turned towards us, as Saturn's ring is in such position. But when very eccentric planets or comets go round any flat Star in orbits much inclined to its equator, the attraction of

the planets or comets in their perihelions must alter the inclination of the axis of that Star; on which account it will appear more or less large and luminous, as its broad side is more or less turned towards us. And thus he imagines we may account for the apparent changes of magnitude and lustre of those Stars, and also for their appearing and disappearing.

Hevelius apprehends (Cometograph. pa. 380), that the Sun and Stars are surrounded with atmospheres, and that by whirling round their axes with great rapidity, they throw off great quantities of matter into those atmospheres, and so cause great changes in them; and that thus it may come to pass that a Star, which, when its atmosphere is clear, shines out with great lustre, may at another time, when it is full of clouds and thick vapours, appear greatly diminished in brightness and magnitude, or even become quite invisible.

Nature of the fixed STARS. The immense distance of the Stars leaves us greatly at a loss about the nature of them. What we can gather for certain from their phenomena, is as follows:

1st, That the fixed Stars are greater than our earth: because if that was not the case, they could not be visible at such an immense distance.

2nd, The fixed Stars are farther distant from the earth than the farthest of the planets. For we frequently find the fixed Stars hid behind the body of the planets: and besides, they have no parallax, which the planets have.

3rd, The fixed Stars shine with their own light; for they are much farther from the Sun than Saturn, and appear much smaller than Saturn; but since, notwithstanding this, they are found to shine much brighter than that planet, it is evident they cannot borrow their light from the same source as Saturn does, viz, the Sun; but since we know of no other luminous body beside the Sun, whence they might derive their light, it follows that they shine with their own native light.

Besides, it is known, that the more a telescope magnifies, the less is the aperture through which the Star is seen; and consequently, the fewer rays it admits into the eye. Now since the Stars appear less in a telescope which magnifies two hundred times, than they do to the naked eye, insomuch that they seem to be only indivisible points, it proves at once that the Stars are at immense distances from us, and that they shine by their own proper light. If they shone by borrowed light, they would be as invisible without telescopes as the satellites of Jupiter are; for the satellites appear larger when viewed with a good telescope than the largest fixed Stars do.

Hence,

1. We deduce, that the fixed Stars are so many suns; for they have all the characters of suns

2. That in all probability the Stars are not smaller than our sun.

3. That it is highly probable each Star is the centre of a system, and has planets or earths revolving round it, in the same manner as round our sun, i. e. it has opake bodies illuminated, warmed, and cherished by its light and heat. As we have incomparably more light from the moon than from all the Stars together, it is absurd to imagine that the Stars were made for no other purpose than to cast a faint light upon the earth;

especially

especially since many more require the assistance of a good telescope to find them out, than are visible without that instrument. Our sun is surrounded by a system of planets and comets, all which would be invisible from the nearest fixed Star; and from what we already know of the immense distance of the Stars, it is easy to prove, that the sun, seen from such a distance, would appear no larger than a Star of the first magnitude.

From all this it is highly probable, that each Star is a sun to a system of worlds moving round it, though unseen by us; especially as the doctrine of a plurality of worlds is rational, and greatly manifests the power, the wisdom, and the goodness of the great creator.

How immense, then, does the universe appear! Indeed, it must either be infinite, or infinitely near it.

Kepler, it is true, denies, that each Star can have its system of planets as ours has; and takes them all to be fixed in the same surface or sphere; urging, that were one twice or thrice as remote as another, it would be twice or thrice as small, supposing their real magnitudes equal; whereas there is no difference in their apparent magnitudes, justly observed, at all. But to this it is opposed, that Huygens has not only shewn, that fires and flames are visible at distances where other bodies, comprehended under equal angles, disappear; but it should likewise seem, that the optic theorem about the apparent diameters of objects, being reciprocally proportional to their distances from the eye, does only hold while the object has some sensible ratio to its distance.

As for periodical Stars, &c. see CHANGES, &c of *Stars*, supra.

Motion of the STARS. The fixed Stars have two kinds of apparent motion; one called the *first*, *common*, or *diurnal motion*, arising from the earth's motion round its axis: by this they seem to be carried along with the sphere or firmament, in which they appear fixed, round the earth, from east to west, in the space of 24 hours.

The other, called the *second*, or *proper motion*, is that by which they appear to go backwards from west to east, round the poles of the ecliptic, with an exceeding slow motion, as describing a degree of their circle only in the space of 71½ years, or 50⅓ seconds in a year. This apparent motion is owing to the recession of the equinoctial points, which is 50⅓ seconds of a degree in a year backward, or contrary to the order of the signs of the zodiac.

In consequence of this second motion, the longitude of the Stars will be always increasing. Thus, for example, the longitude of Cor Leonis was found at different periods, to be as follows: viz,

	Year.	Long.
By Ptolomy, in - - -	138	to be 2° 30′
By the Persians, in -	1115	- - - 17 30
By Alphonsus, in - -	1364	- - - 20 40
By Prince of Hesse, in	1586	- - - 24 11
By Tycho, in - - -	1601	- - - 24 17
By Flamsteed, in - -	1690	- - - 25 31⅓

Whence the proper motion of the Stars, according to the order of the signs, in circles parallel to the ecliptic, is easily inferred.

It was Hipparchus who first suspected this motion, upon comparing his own observations with those of Timocharis and Aristyllus. Ptolomy, who lived three

centuries after Hipparchus, demonstrated the same by undeniable arguments.

The increase of longitude in a century, as stated by different astronomers, is as follows:

By Tycho Brahe - - - -	1° 25′	0″
Copernicus - - - - -	1 23	40½
Flamsteed and Riccioli	1 23	20
Bulliald - - - - - - -	1 21	54½
Hevelius - - - - - -	1 24	46½
Dr. Bradley, &c. - -	1 23	55

which is at the rate of 50⅓ seconds per year.

From these data, the increase in the longitude of a Star for any given time, is easily had, and thence its longitude at any time: ex. gr. the longitude of Sirius, in Flamsteed's tables, for the year 1690, being 9° 49′ 1″, its longitude for the year 1800, is found by multiplying the interval of time, viz, 110 years, by 50⅓, the product 5537″, or 1° 32′ 17″, added to the given longitude - - - - 9 49 1

gives the longitude - - 11 21 18 for the year 1800.

The chief phenomena of the fixed Stars, arising from their common and proper motion, besides their longitude, are their altitudes, right ascensions, declinations, occultations, culminations, risings, and settings.

Some have supposed that the latitudes of the Stars are invariable. But this supposition is founded on two assumptions, which are both controverted among astronomers. The one of these is, that the orbit of the earth continues unalterable in the same plane, and consequently that the ecliptic is invariable; the contrary of which is now very generally allowed.

The other assumption is, that the Stars are so fixed as to keep their places immoveably. Ptolomy, Tycho, and others, comparing their observations with those of the ancient astronomers, have adopted this opinion. But from the result of the comparison of our best modern observations, with such as were formerly made with any tolerable degree of exactness, there appears to have been a real change in the position of some of the fixed Stars, with respect to each other; and several Stars of the first magnitude have already been observed, and others suspected to have a proper motion of their own.

Dr. Halley (Philos. Transf. number 355, or Abr. vol. 4, p. 225) has observed, that the three following Stars, the Bull's eye, Sirius, and Arcturus, are now found to be above half a degree more southerly than the ancients reckoned them: that this difference cannot arise from the errors of the transcribers, because the declinations of the Stars, set down by Ptolomy, as observed by Timocharis, Hipparchus, and himself, shew their latitudes given by him are such as those authors intended: and it is scarce to be believed that those three observers could be deceived in so plain a matter. To this he adds, that the bright Star in the shoulder of Orion has, in Ptolomy, almost a whole degree more southerly latitude than at present: that an ancient observation, made at Athens in the year 509, as Bulliald supposes, of an appulse of the moon to the Bull's eye, shews that Star to have had less latitude at that time than it now has: that as to Sirius, it appears by Tycho's observations, that he found him 4½′ more northerly

than

than he is at this time. All thefe obfervations, compared together, feem to favour an opinion, that fome of the Stars have a proper motion of their own, which changes their places in the fphere of heaven: this change of place, as Dr. Halley obferves, may fhew itfelf in fo long a time as 1800 years, though it be entirely imperceptible in the fpace of one fingle century; and it is likely to be foonest difcovered in fuch Stars as thofe juft now mentioned; becaufe they are all of the first magnitude, and may, therefore, probably be fome of the neareft to our folar Syftem. Arcturus, in particular, affords a strong proof of this: for if its prefent declination be compared with its place, as determined either by Tycho or Flamfteed, the difference will be found to be much greater than what can be fufpected to arife from the uncertainty of their obfervations. See ARCTURUS, and Mr Hornfby's enquiry into the quantity and direction of the proper motion of Arcturus, Phil. Tranf. vol. 63, part 1, pa. 93, &c.

For an account of Dr. Bradley's obfervations, fee the fequel of this article.

Dr. Herfchel has alfo lately obferved, that the diftance of the two Stars forming the double Star γ Draconis, is 54″ 48‴, and their pofition 44° 19′ N. preceding. Whereas, from the right afcenfion and declination of thefe Stars in Flamfteed's catalogue, their diftance, in his time, appears to have been 1′ 11″ ·418, and their pofition 44° 23′ N. preceding. Hence he infers, that as the difference in the diftance of thefe two Stars is fo confiderable, we can hardly account for it, otherwife than by admitting a proper motion in one or the other of the Stars, or in our folar fyftem: moft probably he fays, neither of the three is at reft. He alfo fufpects a proper motion in one of the double Stars, in Cauda Lyncis Media, and in o Ceti. Phil. Tranf. vol. 72, part 1, p. 117, 143, 150.

It is reafonable to expect, that other inftances of the like kind muft alfo occur among the great number of vifible Stars, becaufe their relative pofitions may be altered by various means. For if our own folar fyftem be conceived to change its place with refpect to abfolute fpace, this might, in procefs of time, occafion an apparent change in the angular diftances of the fixed Stars; and in fuch a cafe, the places of the neareft Stars being more affected than of thofe that are very remote, their relative pofition might feem to alter, though the Stars themfelves were really immoveable; and vice verfa, we may furmife, from the obferved motion of the Stars, that our fun, with all its planets and comets, may have a motion towards fome particular part of the heavens, on account of a greater quantity of matter collected in a number of Stars and their furrounding planets there fituated, which may perhaps occafion a gravitation of our whole folar fyftem towards it. If this furmife fhould have any foundation, as Dr. Herfchel obferves, ubi fupra, p. 103, it will fhew itfelf in a feries of fome years; fince from that motion there will arife another kind of hitherto unknown parallax (fuggefted by Mr. Michell, Philof. Tranf. vol. 57, p. 252), the inveftigation of which may account for fome part of the motions already obferved in fome of the principal Stars; and for the purpofe of determining the direction and quantity of fuch a motion, accurate obfervations of the diftance of Stars, that are near enough to be meafured

with a micrometer, and a very high power of telefcopes, may be of confiderable ufe, as they will undoubtedly give us the relative places of thofe Stars to a much greater degree of accuracy than they can be had by inftruments or fectors, and thereby much fooner enable us to difcover any apparent change in their fituation, occafioned by this new kind of fecular or fyftematical parallax, if we may fo exprefs the change arifing from the motion of the whole folar fyftem.

And, on the other hand, if our fyftem be at reft, and any of the Stars really in motion, this might likewife vary their apparent pofitions; and the more fo, the nearer they are to us, or the fwifter their motions are; or the more proper the direction of the motion is to be rendered perceptible by us. Since then the relative places of the Stars may be changed from fuch a variety of caufes, confidering the amazing diftance at which it is certain fome of them are placed, it may require the obfervations of many ages to determine the laws of the apparent changes, even of a fingle Star; much more difficult, therefore, muft it be to fettle the laws relating to all the moft remarkable Stars.

When the caufes which affect the places of all the Stars in general are known; fuch as the preceffion, aberration, and nutation, it may be of fingular ufe to examine nicely the relative fituations of particular Stars, and efpecially of thofe of the greateft luftre, which, it may be prefumed, lie neareft to us, and may therefore be fubject to more fenfible changes, either from their own motion, or from that of our fyftem. And if, at the fame time the brighter Stars are compared with each other, we likewife determine the relative pofitions of fome of the fmalleft that appear near them, whofe places can be afcertained with fufficient exactnefs, we may perhaps be able to judge to what caufe the change, if any be obfervable, is owing. The uncertainty that we are at prefent under, with refpect to the degree of accuracy with which former aftronomers could obferve, makes us unable to determine feveral things relating to this fubject; but the improvements, which have of late years been made in the methods of taking the places of the heavenly bodies, are fo great, that a few years may hereafter be fufficient to fettle fome points, which cannot now be fettled; by comparing even the earlieft obfervations with thofe of the prefent age.

Dr. Hook communicated feveral obfervations on the apparent motions of the fixed Stars; and as this was a matter of great importance in aftronomy, feveral of the learned were defirous of verifying and confirming his obfervations. An inftrument was accordingly contrived by Mr. George Graham, and executed with furprifing exactnefs.

With this inftrument the Star γ, in the conftellation Draco, was frequently obferved by Meffrs. Molyneux, Bradley, and Graham, in the years 1725, 1726; and the obfervations were afterwards repeated by Dr. Bradley with an inftrument contrived by the fame ingenious perfon, Mr. Graham, and fo exact, that it might be depended on to half a fecond. The refult of thefe obfervations was, that the Star did not always appear in the fame place, but that its diftance from the zenith varied, and that the difference of the apparent places amounted to 21 or 22 feconds. Similar obfervations were made on other Stars, and a like apparent motion

was

was found in them, proportional to the latitude of the Star. This motion was by no means such as was to have been expected, as the effect of a parallax, and it was some time before any way could be found of accounting for this new phenomenon. At length Dr. Bradley resolved all its variety, in a satisfactory manner, by the motion of light and the motion of the earth compounded together. See LIGHT, and Phil. Transf. No. 406, p. 364, or Abr. vol. vi, p. 149, &c.

Our excellent astronomer, Dr. Bradley, had no sooner discovered the cause, and settled the laws of aberration of the fixed Stars, than his attention was again excited by another new phenomenon, viz, an annual change of declination in some of the fixed Stars, which appeared to be sensibly greater than a precession of the equinoctial points of 50″ in a year, the mean quantity now usually allowed by astronomers, would have occasioned.

This apparent change of declination was observed in the Stars near the equinoctial colure, and there appearing at the same time an effect of a quite contrary nature, in some Stars near the solstitial colure, which seemed to alter their declination less than a precession of 50″ required, Dr. Bradley was thereby convinced, that all the phenomena in the different Stars could not be accounted for merely by supposing that he had assumed a wrong quantity for the precession of the equinoctial points. He had also, after many trials, sufficient reason to conclude, that these second unexpected deviations of the Stars were not owing to any imperfection of his instruments. At length, from repeated observations he began to guess at the real cause of these phenomena.

It appeared from the Doctor's observations, during his residence at Wansted, from the year 1727 to 1732, that some of the Stars near the solstitial colure had changed their declinations 9″ or 10″ less than a precession of 50″ would have produced; and, at the same time, that others near the equinoctial colure had altered theirs about the same quantity more than a like precession would have occasioned: the north pole of the equator seeming to have approached the Stars, which come to the meridian with the sun about the vernal equinox, and the winter solstice; and to have receded from those, which come to the meridian with the sun about the autumnal equinox and the summer solstice.

From the consideration of these circumstances, and the situation of the ascending node of the moon's orbit when he first began to make his observations, he suspected that the moon's action upon the equatorial parts of the earth might produce these effects.

For if the precession of the equinox be, according to Sir Isaac Newton's principles, caused by the actions of the sun and moon upon those parts; the plane of the moon's orbit being, at one time, above 10 degrees more inclined to the plane of the equator than at another, it was reasonable to conclude, that the part of the whole annual precession, which arises from her action, would, in different years, be varied in its quantity; whereas the plane of the ecliptic, in which the sun appears, keeping always nearly the same inclination to the equator, that part of the precession, which is owing to the sun's action, may be the same every year; and from hence it would follow, that although the mean annual precession, proceeding from the joint actions of the sun and moon, were 50″; yet the apparent annual precession might sometimes exceed, and sometimes fall short of that mean quantity, according to the various situations of the nodes of the moon's orbit.

In the year 1727, the moon's ascending node was near the beginning of Aries, and consequently her orbit was as much inclined to the equator as it can at any time be; and then the apparent annual precession was found, by the Doctor's first year's observations, to be greater than the mean; which proved, that the Stars near the equinoctial colure, whose declinations are most of all affected by the precession, had changed theirs, above a tenth part more than a precession of 50″ would have caused. The succeeding year's observations proved the same thing; and, in three or four years' time, the difference became so considerable as to leave no room to suspect it was owing to any imperfection either of the instrument or observation.

But some of the Stars, that were near the solstitial colure, having appeared to move, during the same time, in a manner contrary to what they ought to have done, by an increase of the precession; and the deviations in them being as remarkable as in the others, it was evident that something more than a mere change in the quantity of the precession would be requisite to solve this part of the phenomenon. Upon comparing the observations of Stars near the solstitial colure, that were almost opposite to each other in right ascension, they were found to be equally affected by this cause. For whilst γ Draconis appeared to have moved northward the small Star, which is the 35th Camelopardali Hevelii, in the British catalogue, seemed to have gone as much towards the south; which shewed, that this apparent motion in both those Stars might proceed from a nutation of the earth's axis; whereas the comparison of the Doctor's observations of the same Stars formerly enabled him to draw a different conclusion, with respect to the cause of the annual aberrations arising from the motion of light. For the apparent alteration in γ Draconis, from that cause, being as large again as in the other small Star, proved, that that did not proceed from a nutation of the earth's axis; as, on the contrary, this may.

Upon making the like comparison between the observations of other Stars, that lie nearly opposite in right ascension, whatever their situations were with respect to the cardinal points of the equator, it appeared, that their change of declination was nearly equal, but contrary; and such as a nutation or motion of the earth's axis would effect.

The moon's ascending node being got back towards the beginning of Capricorn in the year 1732, the Stars near the equinoctial colure appeared about that time to change their declinations no more than a precession of 50″ required; whilst some of those near the solstitial colure altered theirs above 2″ in a year less than they ought. Soon after the annual change of declination of the former was perceived to be diminished, so as to become less than 50″ of precession would cause; and it continued to diminish till the year 1736, when the moon's ascending node was about the beginning of Libra, and her orbit had the least inclination to the equator. But by this time, some of the Stars near the solstitial colure had altered their declinations 18″

less

lefs fince the year 1727, than they ought to have done from a precession of 50'. For γ Draconis, which in thofe 9 years would have gone about 8" more foutherly, was obferved, in 1736, to appear 10" more northerly than it did in the year 1727.

As this appearance in γ Draconis indicated a diminution of the inclination of the earth's axis to the plane of the ecliptic, and as feveral aftronomers have fuppofed that inclination to diminifh regularly; if this phenomenon depend upon fuch a caufe and amounted to 18" in 9 years, the obliquity of the ecliptic would, at that rate, alter a whole minute in 30 years; which is much fafter than any obfervations before made would allow. The Doctor had therefore reafon to think, that fome part of this motion at leaft, if not the whole, was owing to the moon's action on the equatorial parts of the earth, which he conceived might caufe a libratory motion of the earth's axis. But as he was unable to judge, from only 9 years obfervation, whether the axis would entirely recover the fame pofition that it had in the year 1727, he found it neceffary to continue his obfervations through a whole period of the moon's nodes; at the end of which he had the fatisfaction to fee, that the Stars returned into the fame pofitions again, as if there had been no alteration at all in the inclination of the earth's axis; which fully convinced him, that he had guelfed rightly as to the caufe of the phenomenon. This circumftance proves likewife, that if there be a gradual diminution of the obliquity of the ecliptic, it does not arife only from an alteration in the pofition of the earth's axis, but rather from fome change in the plane of the ecliptic itfelf; becaufe the Stars, at the end of the period of the moon's nodes, appeared in the fame places, with refpect to the equator, as they ought to have done if the earth's axis had retained the fame inclination to an invariable plane.

The Doctor having communicated thefe obfervations, and his fufpicion of their caufe, to the late Mr. Machin, that excellent geometrician foon after fent him a table, containing the quantity of the annual preceffion in the various pofitions of the moon's nodes, as alfo the correfponding nutations of the earth's axis; which was computed upon the fuppofition that the mean annual preceffion is 50", and that the whole is governed by the pole of the moon's orbit only; and therefore Mr. Machin imagined, that the numbers in the table would be too large, as, in fact, they were found to be. But it appeared that the changes which Dr. Bradley had obferved, both in the annual preceffion and nutation, kept the fame law, as to increafing and decreafing, with the numbers of Mr. Machin's table. Thofe were calculated on the fuppofition, that the pole of the equator, during a period of the moon's nodes, moved round in the periphery of a little circle, whofe centre was 23° 29' diftant from the pole of the ecliptic; having itfelf alfo an angular motion of 50" in a year about the fame pole. The north pole of the equator was conceived to be in that part of the fmall circle which is fartheft from the north pole of the ecliptic at the fame time when the moon's afcending node is in the beginning of Aries; and in the oppofite point of it, when the fame node is in Libra.

If the diameter of the little circle, in which the pole of the equator moves, be fuppofed equal to 18", which is the whole quantity of the nutation, as collected from Dr. Bradley's obfervations of the Star γ Draconis, then all the phenomena of the feveral Stars which he obferved will be very nearly folved by this hypothefis. But for the particulars of his folution, and the application of his theory to the practice of aftronomy, we muft refer to the excellent author himfelf; our intention being only to give the hiftory of the invention.

The corrections arifing from the aberration of light, and from the nutation of the earth's axis, muft not be neglected in aftronomical obfervations; fince fuch neglects might produce errors of near a minute in the polar diftance of fome Stars.

As to the allowance to be made for the aberration of light, Dr. Bradley affures us, that having again examined thofe of his own obfervations, which were moft proper to determine the tranfverfe axis of the ellipfis, which each Star feems to defcribe, he found it to be neareft to 40"; and this is the number he makes ufe of in his computations relating to the nutation.

Dr. Bradley fays, in general, that experience has taught him, that the obfervations of fuch Stars as lie neareft the zenith, generally agree beft with one another, and are therefore fitteft to prove the truth of any hypothefis. Phil. Tranf. N°. 485, vol. 45, p. 1, &c.

Monfieur d'Alembert has publifhed a treatife, entitled, Recherches fur la Preceffion des Equinoxes, et fur la Nutation de la Terre dans le Syfteme Newtonien, 4to. Paris, 1749. The calculations of this learned gentleman agree in general with Dr. Bradley's obfervations. But Monfieur d'Alembert finds, that the pole of the equator defcribes an ellipfis in the heavens, the ratio of whofe axes is that of 4 to 3; whereas, according to Dr. Bradley, the curve defcribed is either a circle or an ellipfis, the ratio of whofe axes is as 9 to 8.

The feveral Stars in each conftellation, as in Taurus, Bootes, Hercules, &c, fee under the proper article of each conftellation, TAURUS, BOOTES, HERCULES, &c.

To learn to know the feveral fixed Stars by the globe, fee GLOBE.

The parallax and diftance of the fixed Stars, fee under PARALLAX and DISTANCE.

Circumpolar STARS. See CIRCUMPOLAR.
Morning STAR. See MORNING.
Place of a STAR. See PLACE.
Pole STAR. See POLE.
Twinkling of the STARS. See TWINKLING.
Unformed STARS. See INFORMES.

The following two catalogues of Stars are taken from Dr. Zach's Tabulæ Motuum Solis &c, and are adapted to the beginning of the year 1800. The former contains 381 Stars, fhewing their names and Bayer's mark, their magnitude, declination, and right afcenfion, both in time and, in arcs or degrees of a great circle, with the annual variations of the fame. And the latter contains 162 principal Stars, fhewing their declinations to feconds of a degree, with their annual variations. The explanations are fufficiently clear from the titles of the columns.

A CATA-

A CATALOGUE *of the most remarkable* FIXED STARS, *with their Magnitudes, Right Ascensions, Declinations and Annual Variations, for the Beginning of the Year* 1800.

No. of Stars.	Names and Characters of the Stars.	Magnitude.	Right Ascens. in time.	Annual Variat. in ditto.	Right Ascension in degrees &c.	Annual Variat. in ditto.	Declination North and South.	Annual Variat. in ditto
			h. m. s. $\frac{1}{100}$	s $\frac{1}{1000}$	° ' " $\frac{1}{100}$	" $\frac{1}{100}$	° ' "	
				+		+		
1	γ Pegasi	2	0 2 56·79	3·063	0 44 11·85	45·95	14 4 N	
2	ι Ceti	3	0 5 13·51	3·059	2 18 22·66	45·89	9 57 S	
3	χ Cassiopeæ	4	0 21 45·12	3·301	5 26 16·75	49·51	61 50 N	
4	ζ Cassiopeæ	4	0 25 53·93	3·262	6 28 29·01	48·93	52 49 N	
5	δ Andromedæ	3	0 28 39·02	3·161	7 9 45·31	47·42	29 45 N	
6	α Cassiopeæ	3	0 29 14·40	3·311	7 18 35·95	49·66	55 26 N	
7	β Ceti	2 3	0 33 31·83	3·001	8 22 57·40	45·01	19 5 S	
8	η Cassiopeæ	4	0 37 1·44	3·389	9 15 21·64	50·83	56 46 N	
9	δ Piscium	4	0 38 19·08	3·093	9 34 46·15	46·39	6 30 N	
10	γ Cassiopeæ	3	0 44 44·75	3·505	11 11 11·29	52·58	59 38 N	
11	ε Piscium	4	0 52 33·95	3·103	13 8 29·20	46·55	6 49 N	
12	β Andromedæ	2	0 58 34·23	3·297	14 38 33·38	49·46	34 33 N	
13	θ Cassiopeæ	4	0 59 0·22	3·531	14 45 3·33	52·96	53 35 N	
14	ζ Piscium	4	1 3 16·99	3·109	15 49 14·80	46·63	6 31 N	
15	δ Cassiopeæ	3	1 12 50·58	3·761	18 12 38·70	56·42	59 11 N	
16	μ Piscium	5	1 19 42·07	3·108	19 55 31·07	46·62	5 7 N	
17	π Piscium	5	1 30 30·70	3·164	22 37 40·56	47·46	11 7 N	
18	ι Piscium	4 5	1 31 1·77	3·107	22 45 26·62	46·61	4 28 N	
19	ο Piscium	4 5	1 34 50·72	3·144	23 42 40·84	47·76	8 9 N	
20	ε Cassiopeæ	3	1 40 10·01	4·155	25 2 30·13	62·33	62 41 N	
21	ζ Ceti	3	1 41 36·69	2·953	25 24 10·33	44·30	11 20 S	
22	α Triang. Bor.	3 4	1 41 42·74	3·379	25 25 41·15	50·68	28 36 N	
23	γ Arietis	4	1 42 34·52	3·258	25 38 37·73	48·87	18 19 N	
24	β Arietis	3	1 43 36·77	3·277	25 54 11·48	49·15	19 50 N	
25	λ Arietis	5	1 46 48·86	3·315	26 42 12·83	49·73	22 37 N	
26	γ Andromedæ	2	1 51 41·05	3·615	27 55 15·76	54·23	41 22 N	
27	✳ preced. α ♈	∙	1 50 26·15	∙ ∙ ∙	27 36 32·25	∙ ∙ ∙	∙ ∙ ∙ ∙	
28	α Arietis	2	1 55 55·27	3·335	28 58 49·05	50·02	22 31 N	
29	✳ seq. α ♈	∙	1 59 38·13	∙ ∙ ∙	29 54 31·95	∙ ∙ ∙	∙ ∙ ∙ ∙	
30	51 Arietis	5 6	2 7 1·64	3·308	31 45 24·55	49·62	18 58 N	
31	ο Ceti (Variab.)	2	2 9 14·70	3·019	32 18 40·50	45·29	3 54 S	
32	π⁰ Arietis	6	2 38 9·29	3·321	39 32 19·32	49·81	16 38 N	
33	σ Arietis	6	2 40 28·09	3·285	40 7 1·39	49·28	14 15 N	
34	δ Ceti	3	2 29 14·17	3·060	37 18 32·54	45·90	0 33 S	
35	ε Ceti	3	2 29 53·42	2·884	37 28 21·37	43·27	12 44 S	
36	γ Ceti	3	2 32 57·18	·3·102	38 14 17·76	46·53	2 23 N	
37	π Ceti	3	2 34 36·02	2·849	38 39 0·29	42·74	14 43 S	
38	γ Lilii Bor.	4	2 35 57·70	·3·521	38 59 25·49	52·81	28 25 N	
39	γ Lilii Aust.	4	2 38 14·43	3·489	39 33 36·49	52·34	26 26 N	
40	ρ² Arietis	6	2 44 35·65	3·344	41 8 54·72	50·16	17 31 N	
41	ρ³ Arietis	5 6	2 45 9·35	3·340	41 17 20·19	50·10	17 13 N	
42	η Eridani	3	2 46 39·72	2·917	41 39 55·78	43·75	9 42 S	
43	ε Arietis	5	2 47 48·01	3·401	41 57 11·80	51·01	20 32 N	
44	γ Persei	3	2 50 24·42	4·250	42 36 6·25	63·75	52 43 N	
45	α Ceti	2	2 51 50·07	3·119	42 57 31·06	46·66	3 18 N	

Catalogue of the principal Fixed Stars for the Beginning of the Year 1800.

No. of Stars.	Names and Characters of the Stars.	Magnitude.	Right Ascens. in time. (h. m. s. 1/100)	Annual Variat. in ditto. (s 1/1000)	Right Ascens. in degrees. &c. (° ' " 1/100)	Annual Variat. in ditto. (" 1/100)	Declination North and South. (° ' ")	Annual Variat. in ditto.
				+		+		
46	* feq. α Ceti	·	2 51 54·61	· · ·	42 58 39·15	· · ·	· · ·	
47	β Perfei	2 .3	2 55 12 ⊃7	3·846	43 48 1·04	57·69	40 11 N	
48	δ Arietis	4	3 0 12·71	3·393	45 3 10·59	50·89	18 58 N	
49	ζ Arietis	5	3 3 25·98	3·422	45 51 29·77	51·33	20 18 N	
50	ξ Eridani	3	3 6 7·48	2·904	46 31 52·20	43·56	9 34 S	
51	τ¹ Arietis	7	3 9 42·36	3·433	47 25 35·39	51·49	20 25 N	
52	α Perfei	2	3 10 6·85	4·203	47 31 42·77	63·05	49 8 N	
53	τ² Arietis	6	3 11 16·37	3·428	47 49 5·48	51·42	20 1 N	
54	65 Arietis	7	3 12 55·52	3·430	48 13 52·74	51·45	20 5 N	
55	f Tauri	5	3 19 54·65	3·289	49 57 39·78	49·33	12 15 N	
56	ε Eridani	3 4	3 23 31·52	2·883	50 52 52·84	43·24	10 9 S	
57	δ Perfei	3	3 28 44·98	4·203	52 11 13·91	63·05	47 8 N	
58	η Lucida Plei.	3	3 35 37·17	3·535	53 54 17·56	53·03	23 29 N	
59	ζ Perfei	3	3 41 35·20	3·734	55 23 48·00	56·01	31 17 N	
60	ε Perfei	3	3 44 29·19	3·977	56 7 17·87	59·66	39 25 N	
61	γ Eridani	2 3	3 48 42·25	2·786	57 10 33·77	41·79	14 5 S	
62	A Tauri	5	3 52 53·46	3·515	58 13 21·83	52·72	21 32 N	
63	γ Tauri	3	4 8 25·23	3·387	62 6 18·48	50·80	15 8 N	
64	δ¹ Tauri	3 4	4 11 24·68	3·432	62 51 10·26	51·48	17 4 N	
65	δ² Tauri	4	4 12 34·93	3·431	63 8 43·89	51·46	16 58 N	
66	κ¹ Tauri	5	4 13 27·66	3·545	63 21 54·93	53·17	21 50 N	
67	κ² Tauri	4 5	4 13 31·13	3·543	63 22 47·02	53·14	21 44 N	
68	ε Tauri	3 4	4 16 56·98	3·475	64 14 14·76	52·12	18 44 N	
69	δ¹ Tauri	5	4 17 9·30	3·401	64 17 19·53	51·02	15 31 N	
70	δ² Tauri	5	4 17 19·53	3·399	64 19 52·91	50·99	15 25 N	
71	* præced. α ♉	·	4 22 12·56	· · ·	65 33 8·40	· · ·	· · ·	
72	Aldebaran	1	4 24 27·29	3·421	66 6 49·38	51·31	16 6 N	
73	* fequ. α ♉	·	4 26 43·44	· · ·	66 40 51·60	· · ·	· · ·	
74	σ² Tauri	6	4 27 44·78	3·406	66 56 11·77	51·09	15 24 N	
75	υ² Eridani	3 4	4 27 47·26	2·329	66 56 48·83	34·94	30 59 S	
76	σ² Tauri	6	4 27 50·67	3·409	66 57 40·87	51·13	15 31 N	
77	Eridani	3 4	4 31 43·15	2·615	67 55 47·21	39·23	20 4 S	
78	ι Tauri	4	4 51 9·38	3·565	72 47 20·75	53·47	21 18 N	
79	β Eridani	3	4 58 2·38	2·948	74 30 35·74	44·22	5 21 S	
80	λ Eridani	4	4 59 34·73	2·863	74 53 40·95	42·95	9 1 S	
81	* præc. α Aurig.	·	5 1 39·44	· · ·	75 24 51·60	· · ·	· · ·	
82	Capella	1	5 1 56·16	4·414	75 29 2·40	66·21	45 47 N	
83	* feq. α Aurig.	·	5 3 14·28	· · ·	75 48 34·20	· · ·	· · ·	
84	* præc. β Orio.	·	5 3 56·39	· · ·	75 59 5·85	· · ·	· · ·	
85	Rigel	1	5 4 55 54	2·867	76 13 53·10	43·01	8 27 S	
86	* feq. β Orionis	·	5 8 24·57	· · ·	77 6 8·55	· · ·	· · ·	
87	β Tauri	2	5 13 39·38	3·778	78 24 50·70	56·67	28 26 N	
88	γ Orionis	2	5 14 24·54	3·209	78 36 8·16	48·13	6 9 N	
89	β Leporis	3 4	5 19 41·09	2·565	79 55 16·40	38·47	20 56 S	
90	δ Orionis	2	5 21 47·38	3·057	80 26 53·69	45·86	0 28 S	

Catalogue of the principal Fixed Stars for the Beginning of the Year 1800.

No. of Stars.	Names and Characters of the Stars.	Magnitude.	Right Ascenf. in time.	Annual Variat. in ditto.	Right Ascension in degrees, &c.	Annual Variat. in ditto	Declination North and South.	Annual Variat. in ditto.
			h. m. s. $\frac{1}{100}$	s. $\frac{1}{1000}$	° ′ ″ $\frac{1}{100}$	″ $\frac{1}{100}$	° ′ ″	
				+		+		
91	α Leporis	3	5 23 54·94	2·639	80 58 44·13	39·59	17 59 S	
92	ƺ Tauri	3	5 25 42·41	3·575	81 25 36·14	53·62	21 0 N	
93	ε Orionis	2	5 26 4·15	3·037	81 31 2·19	45·55	1 20 S	
94	ƺ Orionis	2	5 30 40·57	3·020	82 40 8·57	45·30	2 4 S	
95	α Columbæ	2	5 32 25·03	2·167	83 6 15·45	32·50	34 11 S	
96	γ Leporis	3 4	5 36 9·02	2·517	84 2 15·34	37·75	22 31 S	
97	κ Orionis	4	5 38 16·28	2·839	84 34 4·21	42·59	9 45 S	
98	✳ præc. α Orio.	·	5 41 27·74	· · ·	85 21 56·10	· · ·	· ·	
99	α Orionis	I	5 44 20·57	3·239	86 5 8·55	48·59	7 21 N	
100	✳ feq. α Orionis	·	5 47 35·59	· · ·	86 58 23·85		· ·	
101	β Aurigæ	2 3	5 44 51·68	4·398	86 12 55·22	65·97	44 55 N	
102	H Gemin. (prop.)	4 5	5 51 57·72	3·642	87 59 25·77	54·63	23 16 N	
103	η Geminorum	3 4	6 2 48·29	3·623	90 42 4·34	54·34	22 33 N	
104	μ Geminorum	3	6 10 51·44	3·624	92 42 51·64	54·36	22 36 N	
105	ƺ Canis majoris	3	6 12 39·08	2·298	93 9 46·24	34·47	29 59 S	
106	β Canis majoris	2 3	6 13 53·76	2·638	93 28 26·33	39·57	17 52 S	
107	ν Geminorum	4	6 17 5·48	3·562	94 16 22·22	53·43	20 20 N	
108	γ Geminorum	2 3	6 26 9·35	3·463	96 32 20·32	51·95	16 34 N	
109	ι Geminorum	3	6 31 37·34	3 695	97 54 20·05	55·42	25 19 N	
110	✳ præc. α Can. maj.	·	6 29 41·80	· · ·	97 25 27·00	· · ·	· ·	
111	Sirius	I	6 36 19·91	2·647	99 4 58·65	39·71	16 26 S	
112	✳ feq. α Can. maj.	·	6 41 26·68	· · ·	100 21 40·20		· ·	
113	ε Canis majoris	2 3	6 50 46·21	2·354	102 41 33·20	35·31	28 43 S	
114	ƺ Geminorum	3 4	6 52 14·55	3·567	103 3 38·24	53·47	20 51 N	
115	δ Canis majoris	2 3	7 0 15·39	2·436	105 3 53·85	36 54	26 5 S	
116	δ Geminorum	3	7 8 10·06	3·594	107 2 30·94	52·91	22 20 N	
117	β Canis minoris	3	7 16 18·01	3·261	109 4 30·21	48·92	8 41 N	
118	✳ præc. α Gemin.	·	7 16 13·66	· · ·	109 3 24·90	· · ·	· ·	
119	Caſtor	I 2	7 21 48·81	3·855	110 27 12·15	57·83	32 19 N	
120	✳ feq. α Gemin.	·	7 27 4·84	· · ·	111 46 12·60	· · ·	· ·	
121	υ Geminorum	4 5	7 23 34·54	3·715	110 53 38·04	55·72	27 21 N	
122	✳ præc. α Can. min.	·	7 26 40·27	· · ·	111 40 4·05		· ·	
123	Procyon	I 2	7 28 49·10	3·137	112 12 16·50	47·06	5 44 N	
124	✳ feq. α Can. min.	·	7 30 27·12	· · ·	112 36 46·80	· · ·	· ·	
125	Pollux	2	7 33 3·18	3·687	113 15 47·70	55·31	28 30 N	
126	✳ feq. β Gemin.	·	7 35 28·78	· · ·	113 52 11·70	· · ·	· ·	
127	μ² Cancri	5	7 55 57·64	3·545	118 59 24·55	53·18	22 9 N	
128	ψ² Cancri	4	7 58 23·25	3·639	119 35 48·73	54·58	26 7 N	
129	β Cancri	3 4	8 5 39·37	3·266	121 24 50·61	48·99	9 48 N	
130	θ Cancri	5 6	8 20 10·35	3·441	125 2 35·31	51·61	18 46 N	
131	η Cancri	6 7	8 21 7·79	3·491	125 16 56·87	52·36	21 6 N	
132	δ Hydræ	4	8 27 3·04	3·189	126 45 45·53	47·83	6 23 N	
133	γ Cancri	4	8 31 41·86	3·499	127 55 27·85	52·49	22 10 N	
134	δ Cancri	4	8 33 18·11	3·428	128 19 31·70	51·42	18 53 N	
135	ε Hydræ	4	8 36 10·14	3·199	129 2 32·03	47·98	7 8 N	

Catalogue of the principal Fixed Stars for the Beginning of the Year 1800.

No. of Stars.	Names and Characters of the Stars.	Magnitude.	Right Ascens. in time.	Annual Variat. in ditto.	Right Ascension in degrees, &c.	Annual Variat. in ditto.	Declination North and South.		Annual Variat. in ditto.
			h. m. s. $\frac{1}{100}$	s. $\frac{1}{1000}$	° ′ ″ $\frac{1}{100}$	″ $\frac{1}{100}$	° ′ ″		
				+		+			
136	ζ Hydræ	4 5	8 44 48·86	3·187	131 12 12·84	47·81	6 42	N	
137	α¹ Cancri	4 5	8 44 59·39	3·290	131 14 50·81	49·35	12 24	N	
138	α² Cancri	3 4	8 47 31·82	3·202	131 52 57·26	49·38	12 37	N	
139	κ Cancri	4 5	8 56 54·33	3·263	134 13 34·92	48·95	11 28	N	
140	ξ¹ Cancri	5 6	8 58 30·37	3·472	134 27 35·48	52·08	22 51	N	
141	ϑ Hydræ	4	9 3 54·80	3·120	135 58 42·03	46·80	3 10	N	
142	κ Leonis	4	9 12 58·32	3·524	138 14 34·77	52·86	27 2	N	
143	Alphard	2	9 17 44·97	2·935	139 26 14·55	44·03	7 48	S	
144	✳ seq. α Hydræ	·	9 23 9·19	· · ·	140 47 17·85	· · ·	· · ·	·	
145	ξ Leonis	4	9 21 9·26	3 253	140 17 18·97	48 80	12 11	N	
146	o Leonis	4	9 30 27·65	3 224	142 36 54·82	48·36	10 48	N	
147	ε Leonis	3	9 34 28·29	3 434	143 37 4·31	51·51	24 41	N	
148	μ Leonis	3	9 41 21 84	3·457	145 20 27·54	51·85	26 57	N	
149	ν Leonis	4	9 47 26·92	3·243	146 51 43·76	48·65	13 24	N	
150	π Leonis	4	9 49 37·99	3·183	147 24 29·89	47 75	9 0	N	
151	η Leonis	3 4	9 52 24·60	3·289	149 6 9·04	49 33	17 44	N	
152	Regulus	1	9 57 42 02	3·204	149 25 30·30	48·06	12 56	N	
153	✳ seq. α Leonis	·	10 4 28·58	· · ·	151 7 8·70	· · ·	· · ·	·	
154	ζ Leonis	3	10 5 32·34	3·361	151 23 5·16	50·42	24 25	N	
155	γ² Leonis	2 3	10 8 55·22	3·306	152 13 48·23	49·60	20 51	N	
156	μ Ursæ majoris	3	10 10 21·35	3·635	152 35 23·32	54·52	42 30	N	
157	ρ Leonis	4	10 22 15·77	3·170	155 33 56·49	47·55	10 20	N	
158	β Ursæ majoris	2	10 49 39·53	3·709	162 24 54·93	55·63	57 27	N	
159	α Crateris	4	10 50 4·55	2·943	162 31 8·25	44·14	17 14	S	
160	α Ursæ majoris	1 2	10 51 15·84	3 847	162 48 57·61	57·70	62 50	N	
161	β Crateris	3 4	11 1 50 06	2·933	165 27 30·97	44·02	31 44	S	
162	δ Leonis	2 3	11 3 26·39	3·199	165 51 35·91	47·98	21 37	N	
163	ϑ Leonis	3	11 3 44·23	3·165	165 56 3·49	47·48	16 31	N	
164	λ Crateris	5 6	11 13 28·15	2·981	168 22 2·21	44·72	17 17	S	
165	ι Leonis	4	11 13 28·32	3·125	168 22 4·85	46·87	11 38	N	
166	τ Leonis	4	11 17 39·49	3 085	169 24 52·24	46·28	3 57	N	
167	υ Leonis	4	11 26 42·82	3·069	171 40 42·29	46·04	0 17	N	
168	ν Virginis	5	11 35 34·11	3·087	173 53 51·70	46·31	7 39	N	
169	✳ præc. β Leonis	·	11 38 19·46	· · ·	174 34 51·90	· · ·	· · ·	·	
170	Denebola	1 2	11 38 50·49	3·062	174 42 37·35	45·93	15 41	N	
171	β Virginis	3	11 40 16·38	3·122	175 4 5·70	46·83	2 54	N	
172	γ Ursæ majoris	2	11 43 14·22	3·212	175 48 33·33	48·18	54 43	N	
173	α Corvi	4	11 58 6·94	3·062	179 31 44·10	45·93	23 37	S	
174	ε Corvi	4	11 59 51·63	3·067	179 57 54·47	46·00	21 30	S	
175	δ Ursæ majoris	3	12 5 27·23	3·021	181 21 48·42	45·32	58 9	N	
176	γ Corvi	3	12 5 32·31	3·077	181 23 4·58	46·16	16 26	S	
177	η Virginis	3	12 9 40·74	3·067	182 25 11·13	46·01	0 27	N	
178	β Corvi	3	12 23 54·39	3·124	185 58 35·92	46·86	22 17	S	
179	κ Draconis	3	12 24 47·65	2·661	186 11 54·72	39·91	70 53	N	
180	γ¹ Virginis	3	12 31 33·85	3·069	187 53 27·72	46·03	0 21	S	

Catalogue of the principal Fixed Stars for the Beginning of the Year 1800.

No. of Stars.	Names and Characters of the Stars.	Mag. ni-tude.	Right Ascens. in time. (h. m. s. 1/100)	Annual Variat. in ditto. (s. 1/1000) +	Right Ascens. in degrees. &c. (o ' " 1/100)	Annual Variat. in ditto. (" 1/100) +	Declination North and South. (o ' ")	Annual Variat. in ditto.
181	ε Ursæ majoris	2 3	12 45 12.58	2.746	191 18 8.67	41.19	57 3 N	
182	δ Virginis	3	12 45 33.66	3.047	191 23 24.93	45.71	4 29 N	
183	ε Virginis	3	12 52 13.33	3.004	193 3 20.01	45.06	12 2 N	
184	θ Virginis	3 4	12 59 36.46	3.095	194 54 6.97	46.42	4 28 S	
185	γ Hydræ	3	13 8 4.31	3.225	197 1 4.64	48.38	22 7 S	
186	✳ præc. a Virg.		13 9 13.00	...	197 18 15.00	...	• • •	
187	Spica	1	13 14 40.11	3.137	198 40 1.66	47.06	10 7 S	
188	ζ Ursæ majoris	3	13 15 49.62	2.425	198 57 24.26	36.37	55 59 N	
189	ι Virginis	4	13 16 10.79	3.129	199 2 41.83	46.93	11 40 S	
190	ζ Virginis	3	13 24 30.65	3.064	201 7 39.68	45.96	0 26 N	
191	τ Bootis	4	13 37 46.36	2.884	204 26 35.35	43.26	18 27 N	
192	η Ursæ majoris	2 3	13 39 38.85	2.355	204 54 42.80	35.88	50 19 N	
193	η Bootis	3	13 45 69.21	2.860	206 17 18.12	42.90	19 25 N	
194	α Draconis	2 3	13 58 58.88	1.628	209 44 43.24	24.42	65 20 N	
195	x Virginis	4	14 2 14.87	3.179	210 33 43.04	47.68	9 20 S	
196	Arcturus	1	14 5 32.21	2.722	211 38 3.16	40.83	20 15 N	
197	✳ seq. α Bootis	•	14 6 36.46	...	211 39 6.90	...	• • •	
198	λ Virginis	4	14 8 19.14	3.223	212 4 47.16	48.35	12 27 S	
199	γ Bootis	3	14 24 1.50	2.428	216 0 22.54	36.42	39 11 N	
200	ζ Bootis	3	14 31 35.56	2.854	217 53 53.42	42.81	14 36 N	
201	ε Bootis	3	14 36 14.99	2.612	219 3 44.80	39.33	27 56 N	
202	μ Libræ	5	14 38 22.95	3.268	219 35 44.22	49.02	13 18 S	
203	α¹ Libræ	6	14 39 38.74	3.299	219 54 41.10	49.49	15 9 S	
204	✳ præc. α² ♎	•	14 39 38.77	...	219 54 41.55	...	• • •	
205	α² Libræ	2 3	14 39 49.97	3.289	219 57 29.55	49.34	15 12 S	
206	β Ursæ minoris	3	14 51 27.55	-0.329	222 51 53.19	-4.94	74 59 N	
207	γ Scorpii	3	14 52 24.35	3.482	223 6 5.22	52.23	24 29 S	
208	β Bootis	3	14 54 24.99	2.262	223 36 14.85	33.93	41 11 N	
209	ψ Bootis	5	14 55 52.50	2.580	223 58 7.57	38.70	27 44 N	
210	β Libræ	2 3	15 6 15.61	3.215	226 33 54.21	48.22	8 38 S	
211	δ Bootis	3	15 7 26.62	2.409	226 51 39.28	36.13	34 4 N	
212	o Coron. bor.	6	15 11 51.97	2.487	227 57 59.55	37.30	30 21 N	
213	η Coron. bor.	5	15 14 56.05	2.465	228 44 0.78	36.97	31 1 N	
214	β Coron. bor.	4	15 19 34.88	2.483	229 53 43.13	37.24	29 48 N	
215	γ² Ursæ minoris	2 3	15 21 11.76	-0.209	230 17 56.39	-3.14	72 33 N	
216	ζ⁴ Libræ	3 4	15 21 38.42	3.365	230 24 36.26	50.48	16 10 S	
217	γ Libræ	4	15 24 21.22	3.328	231 5 18.14	49.92	14 7 S	
218	δ Serpentis	3	15 25 15.84	2.861	231 18 57.61	42.91	11 13 N	
219	Gemma	2	15 26 13.29	2.543	231 33 19.35	38.15	27 24 N	
220	x Libræ	4	15 30 27.12	3.433	232 30 46.77	51.49	19 1 S	
221	α Serpentis	2	15 34 25.21	2.936	233 36 18.00	44.04	7 4 N	
222	✳ seq. α Serpentis	•	15 36 23.53	...	234 5 53.05	...	• • •	
223	β Serpentis	3	15 36 57.70	2.756	234 14 25.47	41.34	16 4 N	
224	μ Serpentis	4	15 39 10.30	3.023	234 47 34.55	45.35	2 48 S	
225	ε Serpentis	3 4	15 40 50.97	2.969	235 12 44.51	44.54	5 6 N	

Catalogue of the principal Fixed Stars for the Beginning of the Year 1800.

No. of Stars.	Names and Characters of the Stars.	Magnitude.	Right Ascens. in time.	Annual Variat. in ditto.	Right Ascension in degrees, &c.	Annual Variat. in ditto.	Declination North and South.	Annual Variat. in ditto.
			h. m. s. 1/100	s. 1/1000	o ' " 1/100	" 1/10	o ' "	
				+		+		
226	δ Coron. bor.	4	15 41 12·45	2·515	235 18 6·71	37·73	26 42 N	
227	λ Libræ	4	15 41 44·86	3·457	235 26 12·93	51·86	19 33 S	
228	℮ Scorpii	3 4	15 44 33·34	3·671	236 8 20·06	55·06	28 37 S	
229	π Scorpii	3	15 46 46·43	3·600	236 41 36·48	54·00	25 31 S	
230	↓ Libræ	4	15 47 0·91	3·339	236 45 13·63	50·09	13 41 S	
231	γ Serpentis	3	15 47 12·93	2·740	236 48 13·98	41·10	16 21 N	
232	δ Scorpii	3	15 48 31·89	3·521	237 7 58·38	52·82	22 2 S	
233	ε Coron. bor.	4 5	15 49 18·54	2·483	237 19 38·04	37·24	27 28 N	
234	π Serpentis	4	15 53 41·18	2·576	2 8 25 17·72	38·64	23 21 N	
235	β Scorpii	2	15 53 49·71	3·465	238 27 25·65	51·97	19 15 S	
236	ϑ Draconis	3 4	15 58 8·28	1·142	239 32 4·27	17·13	59 6 N	
237	ν Scorpii	4	16 0 23·31	3·465	240 5 49·60	51·96	18 56 S	
238	δ Ophiuchi	3	16 3 52·80	3·132	240 58 11·95	46·98	3 10 S	
239	ε Ophiuchi	3 4	16 7 45·09	3·154	241 56 16·31	47·30	4 12 S	
240	γ Herculis	3	16 13 5·83	2·642	243 16 27·41	39·63	19 38 N	
241	Antares	1	16 17 9·69	3·645	244 17 25·35	54·68	25 58 S	
242	✱ α Scorpii	·	16 19 6·66	· · ·	244 46 39·90	· · ·	· · · ·	
243	φ Ophiuchi	4 5	16 19 43·03	3·418	244 55 45·42	51·27	16 10 S	
244	η Draconis	3 4	16 21 18·33	0·785	245 19 34·92	11·78	61 58 N	
245	β Herculis	3	16 21 37·87	2·579	245 24 28·08	38·68	21 56 N	
246	τ Scorpii	4	16 23 26·96	3·709	245 51 44·36	55·64	27 47 S	
247	ζ Ophiuchi	2 3	16 26 9·55	3·287	246 32 23·24	49·30	10 9 S	
248	ζ Herculis	3 4	16 33 45·64	2·292	248 26 24·67	34·38	32 1 N	
249	η Herculis	3 4	16 36 3·14	2·046	249 0 47·15	30·69	39 19 N	
250	ε Herculis	3	16 52 38·68	2·2·2	253 9 40·16	34·38	31 16 N	
251	η Ophiuchi	2 3	16 58 55·15	3·424	254 43 47·26	51·36	15 28 S	
252	✱ præc. α Herc.	·	17 5 12·70	· · ·	256 18 10·50	· · ·	· · ·	
253	α Herculis	2 3	17 5 31·76	2·726	256 22 56·40	40·89	14 38 N	
254	δ Herculis	3 4	17 6 49·41	2·459	256 42 21·17	36·88	25 5 N	
255	ϑ Ophiuchi	3	17 9 44·25	3·669	257 26 3·68	55·04	24 47 S	
256	λ Scorpii	3	17 20 2·64	4·057	260 0 39·58	60·85	36 57 S	
257	✱ præc. α Ophi.	·	17 24 45·90	· · ·	261 11 28·50	· · ·	· · ·	
258	α Ophiuchi	2	17 25 38·97	2·768	261 24 44·55	41·52	17 43 N	
259	✱ seq. α Ophi.	·	17 29 11·02	· · ·	262 17 45·32	· · ·	· · ·	
260	β Draconis	3	17 25 55·99	1·348	261 28 59·82	20·22	52 27 N	
261	β Ophiuchi	3	17 33 35·77	2·959	263 23 56·54	44·39	4 40 N	
262	γ Ophiuchi	3	17 37 52·04	3·003	264 28 0·56	45·05	2 48 N	
263	ζ Serpentis	3 4	17 49 54·59	3·153	267 28 38·87	47·30	3 40 S	
264	o Ophiuchi	4	17 50 37·47	2·999	267 39 21·99	44·99	2 57 S	
265	γ Draconis	2 3	17 51 57·79	1·389	267 59 26·85	20·83	51 31 N	
266	γ Sagittarii	3 4	17 52 58·05	3·851	268 14 30·76	57·77	30 25 S	
267	b Taur. Poniat.	·	18 0 41·10	2·993	270 10 16·50	44·90	3 19 N	
268	μ¹ Sagittarii	4	18 1 48·37	3·584	270 27 5·63	53·76	21 6 S	
269	μ² Sagittarii	4 6	18 3 16·90	3·575	270 49 13·57	53·62	20 46 S	
270	ε Sagittarii	2 3	18 10 53·67	3·984	272 43 25·11	59·76	34 28 S	

Catalogue of the principal Fixed Stars for the Beginning of the Year 1800.

No of Stars.	Names and Characters of the Stars.	Magnitude.	Right Ascens. in time.	Annual Variat. in ditto.	Right Ascens. in degrees, &c.	Annual Variat. in ditto.	Declination North and South.	Annual Variat. in ditto.
			h. m. s. $\frac{1}{100}$	s.' $\frac{1}{1000}$	o ' '' $\frac{1}{100}$	'' $\frac{1}{100}$	o ' ''	
				+		+		
271	λ Sagittarii	4	18 15 37·66	3·705	273 54 24·92	55·57	25 31 S	
272	✳ præc. α Lyræ	·	18 28 40·12	...	277 10 1·88			
273	Wega	1	18 30 9·89	1·994	277 32 28·35	29·91	38 36 N	
274	✳ seq. α Lyræ	·	18 31 40·00	...	277 54 50·00			
275	φ Sagittarii	3 4	18 33 9·39	3·747	278 17 20·8.	6·21	27 11 S	
276	ι Lyræ	5	18 37 42·87	1·983	279 25 43·03	29·74	39 28 N	
277	,¹ Sagittarii	4 5	18 42 5·41	3·625	280 31 21·22	54·38	22 59 S	
278	β Lyræ	3	18 42 41·86	2·211	280 40 27·89	33·16	33 9 N	
279	σ Sagittarii	3	18 42 51·40	3·724	280 42 50·99	55·86	26 32 S	
280	,² Sagittarii	4 5	18 43 1·13	3·623	280 45 16·99	54·35	22 54 S	
281	θ Serpentis	3	18 46 16·82 / 18·32	2·977	281 34 12·35 / 34·84	44·66	3 57 N	
282	δ Lyræ	3 4	18 47 31·16	2·095	281 52 47·43	31·42	36 39 N	
283	o Draconis	4	18 48 14·15	0·880	282 3 32·20	13 21	59 9 N	
284	γ Lyræ	3	18 51 27·04	2·241	282 51 45·55	33·61	32 26 N	
285	o Sagittarii	4	18 52 41·18	3·595	283 10 17·72	53·92	22 1 S	
286	τ Sagittarii	4	18 54 26·45	3·758	283 36 36·82	56·37	27 57· S	
287	λ Antinoi	3 4	18 55 38·10	3·186	283 54 31·55	47·79	5 10 S	
288	ζ Aquilæ	3	18 56 12·69	2·755	284 3 10·37	41·33	13 35 N	
289	π Sagittarii	3 4	18 57 51·37	3 574	284 27 50·57	53·61	21 20 S	
290	↓ Sagittarii	4 5	19 3 15·27	3·685	285 48 49·04	55·27	25 35 S	
291	d Sagittarii	4 6	19 5 55·86	3·517	286 28 57·90	52·76	19 18 S	
292	δ Draconis	3	19 12 27·95	0·033	288 6 59·21	0 49	67 19 N	
293	χ Cygni	4	19 12 28·28	1·383	88 7 4·19	20·73	52 58 N	
294	δ Aquilæ	3	19 15 24·12	3·008	288 51 1·79	45·12	2 44 N	
295	β Cygni	3	19 22 38·53	2·415	290 39 37·97	36·23	27 33 N	
296	ι Cygni	4 6	19 24 39·61	1·511	291 9 54·19	22·67	51 19 N	
297	ι Antinoi	3 4	19 26 22·16	·106	291 35 32·37	46·59	1 43 S	
298	θ Cygni	4	19 31 5·16	1·645	292 46 17·40	24·68	49 46 N	
299	α Sagittæ	4	19 31 9·25	2·678	292 47 18·74	40·17	17 34 N	
300	f Sagittarii	6	19 34 41·43	3·520	293 40 21·39	52·80	20 14 S	
301	✳ præc. γ Aquilæ	·	19 35 12·50	...	293 48 7·50	...	· · N	
302	γ Aquilæ	3	19 36 44·50	2·837	294 11 7·50	42·59	10 8 N	
303	✳ seq. γ Aquilæ	·	19 39 0·81	...	294 45 12·16		· · N	
304	δ Cygni	3	19 38 43·09	1·869	294 40 46·34	28·02	44 39 N	
305	✳ præc. α Aquilæ	·	19 38 35·87	...	294 38 58·05		·	
306	Atair	1 2	19 41 1·02	2·918	295 15 15·30	43·78	8 21 N	
307	✳ seq. α Aquilæ	·	19 42 52·37	...	295 43 5·55		·	
308	η Antinoi	3 4	19 42 17·14	3·058	295 34 17 08	45·87	0 30 S	
309	b Sagittarii	4 5	19 44 39·56	3·699	296 9 53·46	55·48	27 41 S	
310	β Aquilæ	3 4	19 45 28·97	2·939	296 22 14·55	44·08	5 55 N	
311	θ Aquilæ	3	20 0 58·52	3·097	300 14 37·75	46·45	1 24 S	
312	α¹ Capricorni	3 4	20 6 32·79	3·330	301 38 11·88	49·95	13 7 S	
313	✳ præc. α² Capri.	·	20 5 17·48	...	301 19 22·20	...	·	
314	α² Capricorni	3	20 6 56·48	3·331	301 44 7·10	49·96	13 9 S	
315	✳ seq. α² Capri.	·	20 9 33·32	...	302 23 19·80		·	

Catalogue of the principal Fixed Stars for the Beginning of the Year 1800.

No. of Stars.	Names and Characters of the Stars.	Magnitude	Right Ascens. in time.	Annual Variat. in ditto.	Right Ascens. in degrees, &c.	Annual Variat. in ditto.	Declination North and South.	Annual Variat. n ditto.
			h. m. s. $\frac{1}{100}$	s. $\frac{1}{1000}$	° ' " $\frac{1}{100}$	" $\frac{1}{100}$	° ' "	
				+				
316	β Capricorni	•	20 9 31·35	2 380	302 22 50·19	50·70	15 24 S	
317	ν Capricorni	6	20 9 33·69	3·337	302 23 25·39	50·06	13 23 S	
318	β Capricorni	3	20 9 45·50	3·380	302 26 22·54	50·70	15 24 S	
319	γ Cygni	3	20 15 2·63	2·148	303 45 39·39	32·22	39 38 N	
320	ρ Capricorni	6	20 17 26·45	3·438	304 21 36·72	51·57	18 28 S	
321	ζ Delphini	4 5	20 25 57·50	2·801	306 29 22·44	42·01	14 0 N	
322	β Delphini	3	20 28 10·32	2·804	307 2 34·74	42·06	13 55 N	
323	α Delphini	3	20 30 20·76	2·780	307 35 11·37	41·70	15 13 N	
324	Deneb.	1 2	20 34 36·68	2·034	308 39 10·20	30·51	44 34 N	
325	✱ seq. α Cygni	•	20 40 28·55	· · ·	310 7 8·25	· · ·	• • •	
326	ε Aquarii	4 5	20 36 50·38	3·255	309 12 35·72	48·83	10 13 S	
327	✱ præc. γ Delphini	•	20 37 21·94	· · ·	309 20 29·08	· · ·	• • •	
328	γ Delphini	3	20 37 22·96	2·783	309 20 44·34	41·75	15 25 N	
329	ε Cygni	3	20 38 6·70	2·393	309 31 40·56	35·89	33 13 N	
330	μ Aquarii	4 5	20 41 51·17	3·233	310 27 47·61	48·65	9 43 S	
331	Aquarii	6	20 46 4·59	3·255	311 31 8 80	48·82	10 28 S	
332	9 Capricorni	5 4	20 54 39·75	3·384	313 39 56·25	50·76	18 1 S	
333	ν Aquarii	5	20 58 41·24	3·274	314 40 18·64	49·11	12 10 S	
334	α Equulei	4	21 5 48·78	2·997	316 27 11·73	44·96	4 26 N	
335	ι Capricorni	5	21 11 5·67	3·355	317 46 25·00	50·33	17 41 S	
336	β Equulei	6	21 12 57·71	2·981	318 14 25·62	44 72	5 58 N	
337	Aquarii	6	21 13 14·77	3·286	318 18 41·53	49·29	13 44 S	
338	α Cephei	3	21 13 46·85	1·427	318 26 42·69	21·40	61 45 N	
339	β Aquarii	3	21 21 1·12	3·165	320 15 16·74	47·48	6 27 S	
340	ε Capricorni	4	21 25 52·29	3·379	321 28 4·34	50·68	20 21 S	
341	β Cephei	3	21 26 1·18	0·821	321 30 17·76	12·32	69 41 N	
342	γ Capricorni	3 4	21 28 59·14	3·329	322 14 47·15	49·93	17 33 S	
343	χ Capricorni	5	21 31 27·98	3·360	322 51 59·74	50·40	19 46 S	
344	ε Pegasi	3	21 34 21·53	2·943	323 35 22·99	44·15	8 58 N	
345	π¹ Cygni	4	21 34 59·63	2·116	323 44 54·38	31·74	50 17 N	
346	δ Capricorni	3	21 35 58·68	3·310	323 59 40·2,	49·65	17 1 S	
347	✱ præc. α Aquarii	•	21 55 8·21	· · ·	328 47 3·15	· · ·	• • •	
348	α Aquarii	3	21 55 29·75	3·067	328 52 26·25	46·00	1 17 S	
349	γ Aquarii	3	22 11 18·89	3·094	332 49 43·39	46·41	2 23 S	
350	π Aquarii	4 5	22 15 4·00	3·065	333 45 59·98	45 97	0 22 N	
351	ζ Aquarii	4	22 18 31·68	3·079	334 37 55·26	46·18	1 2 S	
352	σ Aquarii	5	22 20 2·99	3·186	335 0 44·84	47·79	11 42 S	
353	Lacertæ	4	22 23 7·51	2·431	335 46 54·14	36·46	49 16 N	
354	η Aquarii	4	22 25 4·59	3·079	336 16 8·78	46·19	1 9 S	
355	κ Aquarii	5	22 27 23·38	3·117	336 50 50·67	46·76	5 14 S	
356	ζ Pegasi	3	22 31 29·06	2·961	337 52 15·89	44·72	9 48 N	
357	η Pegasi	3	22 33 37·87	3·792	338 24 27·91	41·88	29 11 N	
358	τ¹ Aquarii	5	22 37 4·70	3·197	339 16 10·57	47·96	15 7 S	
359	τ² Aquarii	5 6	22 38 59·29	3·190	339 44 49·41	47·85	14 39 S	
360	λ Aquarii	4	22 42 10·52	3·137	340 32 37·87	47·05	8 38 S	

Catalogue of the principal Fixed Stars for the Beginning of the Year 1800.

No. of Stars.	Names and Characters of the Stars.	Magnitude.	Right Ascens. in time.	Annual Variat. in ditto.	Right Ascens. in degrees, &c.	Annual Variat. in ditto.	Declination North and South.	Annual Variat. in ditto.
			h. m. s. $\frac{1}{100}$	s. $\frac{1}{1000}$	° ' " $\frac{1}{100}$	" $\frac{1}{100}$	° ' "	
				+		+		
361	ι Cephei	4	22 42 35·33	2·109	340 38 49·95	31·63	65 9 N	
362	δ Aquarii	3	22 44 1·80	3·201	341 0 27·06	48·07	16 53 S	
363	✳ præc. αPisc.auſt.	·	22 40 17·35	· · ·	340 4 20·25	· · ·	· · S	
364	Fomalhaut	1	22 46 33·60	3·330	341 38 24·00	49·95	30 41 S	
365	✳ ſeq. αPiſc. auſt.	·	22 48 38·04	· · ·	342 9 30·60	· · ·	· ·	
366	β Pegaſi	2	22 54 5·50	2·874	343 31 22·47	43·11	27 0 N	
367	Markab	2	22 54 47·99	2·964	343 41 59·85	44·46	14 8 N	
368	✳ ſeq. α Pegaſi	·	22 55 35·64	· · ·	343 53 54·60	· · ·	· · ·	
369	φ Aquarii	4 5	23 3 57·39	3·109	345 59 20·79	46·64	7 7 S	
370	ψ¹ Aquarii	5	23 5 23·05	3·125	346 20 45 68	46·88	10 10 S	
371	γ Piſcium	5	23 6 46·42	3·057	346 41 36·37	45·85	2 12 N	
372	ψ³ Aquarii	3	23 8 32·63	3·125	347 8 9·48	46·88	10 42 S	
373	Piſcium	6	23 26 11·40	3·065	351 32 50·96	45·97	1 0 N	
374	λ Piſcium	5	23 31 51·01	3·066	352 57 45·08	45·99	0 40 N	
375	Piſcium	5 6	23 36 10·88	3·062	354 2 43·18	45·93	2 23 N	
376	ω Piſcium	5	23 49 2·88	3·061	357 15 43·16	45·92	5·46 N	
377	✳ præc. αAndrom.	·	23 55 45·47	· · ·	358 56 22·05	· · ·	·	
378	✳ præc. α Androm.	··	23 56 15·28	3·060	359 3 49·19	45·90	· ·	
379	α Andromedæ	2	23 58 4·32	3·065	359 31 4·95	45·97	27 59 N	
380	✳ ſeq. αAndrom.	·	0 1 32·98	· · ·	0 23 14·70	· · ·	· ·	
381	β Caſſiopeiæ	2 3	23 58 34·32	3·051	359 38 34·75	45·76	58 3 N	

Another CATALOGUE of 162 PRINCIPAL STARS, ſhewing their Mean Declinations to Beginning of the Year 1800.

No.	Stars Names.	Mean Declin. north.	Annual Variation.	No.	Stars Names.	Mean Declin. north.	Annual Variation.
		° ' "	"			° ' "	"
1	Polaris	88 14 25	} + 19·57	11	α Lyræ	38 36 15	} + 2·59
2	Polaris	88 14 26		12	α Lyræ	38 36 10	
3	η Urſæ majoris	50 19 4	− 18·20	13	ζ Herculis	32 58 19	− 7·40
4	α Perſei	49 8 10	+ 13·59	14	Caſtor	32 18 54	} − 6 95
5	ι Urſæ majoris	48 49 9	− 13·21	15	Caſtor	32 18 41	
6	δ Perſei	47 8 14	+ 12·35	16	Pollux	28 29 47	− 7·46
7	Capella	45 46 50	+ 5·09	17	β Tauri	28 25 25	} + 4·08
8	α Cygni	41 34 20	} + 12·52	18	β Tauri	28 25 30	
9	α Cygni	44 34 19		19	ε Bootis	27 55 32	− 15·59
10	β Bootis	41 11 11	− 14·54	20	α Andromedæ	27 59 15	+ 20·25

The Mean Declinations of 162 principal Stars for the Beginning of the Year 1800.

No.	Stars Names.	Mean Declin. north.			Annual Variation.	No.	Stars Names.	Mean Declin. north.			Annual Variation.
		o	′	″	″			o	′	″	″
21	α Andromedæ	27	59	11	+ 20·25	66	γ Geminorum	16	33	28	− 2·22
22	β Cygni	27	32	51	+ 7·04	67	η Serpentis	16	19	40	− 11·01
23	Gemma	27	23	48	} − 12·50	68	β Serpentis	16	3	21	− 11·75
24	Gemma	27	23	49		69	Aldebaran	16	5	43	} + 8·16
25	μ Leonis	26	56	37	− 16·46	70	Aldebaran	16	5	45	
26	β Pegasi	26	59	58	+ 19·21	71	β Leonis	15	41	30	} − 19·96
27	ε Geminorum	25	18	55	} − 2·72	72	β Leonis	15	41	27	
28	ε Geminorum	25	18	56		73	γ Delphini	15	24	40	+ 12·68
29	δ Herculis	25	5	4	− 4·56	74	α Delphini	15	12	48	+ 12·21
30	ε Leonis	24	41	17	− 16 10	75	γ Tauri	15	8	1	+ 9·42
31	ζ Leonis	24	24	27	− 17·56	76	ζ Bootis	14	35	34	− 15·85
32	Alcione	23	28	34	+ 11·88	77	α Herculis	14	37	38	− 4·75
33	Electra	23	28	27	+ 12·04	78	α Pegasi	14	7	49	} + 19·22
34	Atlas	23	25	54	+ 11·74	79	α Pegasi	14	7	57	
35	Propus	23	15	40	+ 0·75	80	γ Pegasi	14	4	16	+ 20·04
36	τ Pegasi	22	38	51	+ 19·57	81	γ Pegasi	14	4	15	+ 20·04
37	μ Geminorum	22	36	12	− 0·89	82	β Delphini	13	54	31	+ 12·05
38	η Geminorum	22	32	59	− 0·19	83	ζ Aquilæ	13	34	32	+ 4·83
39	η Geminorum	22	33	5	− 0·19	84	Regulus	12	56	23	} − 17·24
40	α Arietis	22	30	37	+ 17·55	85	Regulus	12	56	20	
41	α Arietis	22	30	40	+ 17·55	86	α Cancri	12	37	30	− 13·18
42	δ Geminorum	22	20	19	− 5·83	87	α Ophiuchi	12	42	55	} − 3·05
43	γ Cancri	22	10	43	− 12·28	88	α Ophiuchi	12	43	7	
44	μ Cancri	22	9	9	− 9·67	89	ε Virginis	12	2	11	− 19·54
45	β Herculis	21	56	2	− 8·38	90	δ Serpentis	11	12	56	− 12·57
46	δ Leonis	21	37	1	− 19·43	91	ο Leonis	10	47	40	− 15·94
47	ζ Tauri	21	0	32	+ 3·05	92	ε Delphini	10	37	50	+ 11·73
48	γ Leonis	20	50	54	− 17·72	93	ς Leonis	19	19	52	− 18·24
49	ζ Geminorum	20	51	0	} − 4·48	94	γ Aquilæ	10	8	6	} + 8·17
50	ζ Geminorum	20	51	5		95	γ Aquilæ	10	8	10	
51	ν Geminorum	20	19	34	− 1·44	96	ε Pegasi	8	57	43	+ 16·10
52	Arcturus	20	13	45	− 19·10	97	β Canis minoris	8	40	51	− 6·51
53	Arcturus	20	13	45	*− 19·10	98	α Aquilæ	8	20	58	} + 8·51
54	γ Herculis	19	37	52	− 9·05	99	α Aquilæ	8	20	48	
55	η Bootis	19	24	19	− 18·00	100	α Orionis	7	21	27	+ 1·42
56	δ Cancri	18	52	52	− 12·40	101	α Orionis	7	21	27	+ 1·42
57	ε Pegasi	18	57	14	+ 14·91	102	ε Hydræ	7	8	37	− 12 60
58	β Arietis	18	49	30	+ 18·03	103	α Serpentis	7	3	55	} − 11·94
59	γ Arietis	18	18	29	+ 18·09	104	α Serpentis	7	3	50	
60	δ Sagittæ	18	3	15	+ 7·73	105	δ Hydræ	6	23	22	− 11·97
61	η Leonis	17	43	57	− 17·18	106	β Aquilæ	5	55	4	+ 8·86
62	α Sagittæ	17	33	48	+ 7·73	107	β Aquilæ	5	55	19	+ 8·86
63	δ¹ Tauri	17	3	38	+ 9·19	108	Procyon	5	44	11	− 7·51
64	θ Leonis	16	31	13	− 19·43	109	β Ophiuchi	4	39	41	− 2·35
65	γ Geminorum	16	33	27	− 2·22	110	δ Virginis	4	29	13	− 19·66

The Mean Declinations of 162 PRINCIPAL STARS for the Beginning of the Year 1800.

No	Stars Names.	Mean Declin. north.			Annual Variation.	No.	Stars Names.	Mean Declin. fouth.			Annual Variation.
		o	′	″	″			o	′	″	″
111	ϑ Serpentis	3	57	12	+ 3·97	156	γ Eridani	14	5	3	− 10·90
112	α Ceti	3	17	49	} + 14·70	157	α Libræ	15	12	0	} + 15·40
113	α Ceti	3	18	0		158	α Libræ	15	12	0	
114	β Virginis	2	53	35	} − 19·97	159	δ Corvi	15	24	5	+ 19·98
115	β Virginis	2	53	38		160	β Capricorni	15	24	10	− 10·71·
116	γ Ophiuchi	2	47	42	− 1·97	161	γ Canis majoris	15	21	6	+ 4·69
117	δ Aquilæ	2	43	36	+ 6·44	162	η Ophiuchi	15	27	58	+ 5·33
118	γ Ceti	2	23	17	+ 15·77	163	ι Aquarii	15	48	54	− 17·14
119	α Piſcium	1	47	40	+ 17·73	164	γ Corvi	16	25	47	+ 20·04
120	η Antinoi	0	30	12	+ 8·63	165	Sirius	16	27	7	+ 4·43
121	δ Orionis	0	27	17	− 3·38	166	Sirius	16	27	5	+ 4·33
122	ζ Virginis	0	25	48	− 18·72	167	δ Aquarii	16	52	59	−·18·85
		South Decl.									
123	ι Hydræ	0	8	18	− 15·86	168	δ Capricorni	17	1	38	− 16·19
124	γ Virginis	0	21	4	+ 19·86	169	α Crateris	17	14	11	+ 19·11
125	δ Ceti	0	32	18	− 15·97	170	ι Capricorni	17	33	22	− 15·82
126	α Aquarii	1	17	7	} − 17·15	171	γ Capricorni	17	39	58	− 14·97
127	α Aquarii	1	17	7		172	β Canis majoris	17	51	55	+ 1·18
128	ι Orionis	1	20	24	− 3·02	173	α Leporis	17	58	22	− 3·18
129	ϑ Antinoi	1	24	10	− 10·05	174	ϑ Capricorni	18	1	3	− 13·81
130	ζ Orionis	2	3	33	− 2·60	175	ι Scorpii	18	55	46	+ 10·03
131	γ Aquarii	2	23	31	− 17·81	176	β Ceti	19	5	9	− 19·84
132	δ Ophiuchi	3	10	8	+ 9·77	177	β Scorpii	19	14	46	+ 10·52
133	ζ Serpentis	3	39	32	− 0·93	178	μ Sagittarii	21	5	50	− 0·09
134	ι Ophiuchi	4	11	37	+ 9·47	179	π Sagittarii	21	19	45	− 4·95
135	ϑ Virginis	4	28	4	+ 19·39	180	ε Corvi	21	30	32	+ 20·05
136	β Eridani	5	21	16	− 5·41	181	δ Scorpii	22	2	30	+ 10·92
137	ι Orionis	6	3	0	− 3·04	182	o Sagittarii	22	1	21	− 4·51
138	β Aquarii	6	26	37	− 15·39	183	β Corvi	22	17	17	+ 19·94
139	φ Aquarii	7	7	25	− 19·44	184	γ Leporis	22	31	15	− 2·11
140	α Hydræ	7	47	56	+ 15·21	185	α Corvi	23	36	50	+ 20·04
141	α Hydræ	7	47	53	+ 15·21	186	γ Scorpii	24	29	10	+ 14·67
142	Rigel	8	26	35	− 4·81	187	ϑ Ophiuchi	24	47	17	+ 4·43
143	β Libræ	8	38	14	+ 13·82	188	σ Scorpii	25	6	8	+ 9·38
144	λ Aquarii	8	38	29	− 18·89	189	π Scorpii	25	31	36	+ 11·06
145	α Spica	10	6	46	+ 19·01	190	Antares	25	58	38	+ 8·75
146	Spicæ	10	6	45	+ 19·01	191	Antares	25	58	23	+ 8·75
147	ζ Ophiuchi	10	9	1	+ 8·02	192	δ Canis majoris	26	5	10	+ 5·18
148	δ Eridani	10	27	5	− 11·99	193	ε Canis majoris	28	42	29	+ 4·38
149	μ Ceti	11	22	48	+ 15·71	194	ζ Canis majoris	29	58	50	+ 1·07
150	λ Virginis	12	26	26	+ 17·01	195	Fomalhaut	30	40	38	− 19·01
151	α¹ Capricorni	13	6	58	} − 10·47						
152	α¹ Capricorni	13	7	0							
153	α² Capricorni	13	9	15	} − 10·50						
154	α² Capricorni	13	9	17							
155	γ Libræ	14	6	42	+ 12·63						

STAR, in Electricity, denotes the appearance of the electric matter on a point into which it enters. Beccaria supposes that the Star is occasioned by the difficulty with which the electric fluid is extricated from the air, which is an electric substance. See BRUSH.

STAR, in Fortification, denotes a small fort, having 5 or more points, or saliant and re-entering angles, flanking one another, and their faces 90 or 100 feet long.

STAR, in Pyrotechny, a composition of combustible matters; which being borne, or thrown aloft into the air, exhibits the appearance of a real Star.—Stars are chiefly used as appendages to rockets, a number of them being usually inclosed in a conical cap, or cover, at the head of the rocket, and carried up with it to its utmost height, where the Stars, taking fire, are spread around, and exhibit an agreeable spectacle.

To make Stars.—Mix 3lbs of saltpetre, 11 ounces of sulphur, one of antimony, and 3 of gunpowder dust: or, 12 ounces of sulphur, 6 of saltpetre, 5½ of gunpowder dust, 4 of olibanum, one of mastic, camphor, sublimate of mercury, and half an ounce of antimony and orpiment. Moisten the mass with gumwater, and make it into little balls, of the size of a chesnut; which dry either in the sun, or in the oven. These being set on fire in the air, will represent Stars.

STAR-*Board* denotes the right hand side of a ship, when a person on board stands with the face looking forward towards the head or fore part of the ship. In contradistinction from *Larboard*, which denotes the left hand side of the ship in the same circumstances.— They say, *Starboard the helm*, or *helm a Starboard*, when the man at the helm should put the helm to the right hand side of the ship.

Falling STAR, or *Shooting* STAR, a luminous meteor darting rapidly through the air, and resembling a Star falling.—The explication of this phenomenon has puzzled all philosophers, till the modern discoveries in electricity have led to the most probable account of it. Signior Beccaria makes it pretty evident, that it is an electrical appearance, and recites the following fact in proof of it. About an hour after sunset, he and some friends that were with him, observed a falling Star directing its course towards them, and apparently growing larger and larger, but it disappeared not far from them; when it left their faces, hands, and clothes, with the earth, and all the neighbouring objects, suddenly illuminated with a diffused and lambent light, not attended with any noise at all. During their surprize at this appearance, a servant informed them that he had seen a light shine suddenly in the garden, and especially upon the streams which he was throwing to water it. All these appearances were evidently electrical; and Beccaria was confirmed in his conjecture, that electricity was the cause of them, by the quantity of electric matter which he had seen gradually advancing towards his kite, which had very much the appearance of a falling Star. Sometimes also he saw a kind of glory round the kite, which followed it when it changed its place, but left some light, for a small space of time, in the place it had quitted. Priestley's Elect. vol. 1, pa. 434, 8vo. See IGNIS *Fatuus.*

STAR-*fort*, or *Redoubt*, in Fortification. See STAR, REDOUBT, and FORT.

STARLINGS, or STERLINGS, or *Jettees*, a kind of case made about a pier of stilts, &c, to secure it. See STILTS.

STATICS, a branch of mathematics which considers weight or gravity, and the motion of bodies resulting from it.

Those who define mechanics, the science of motion, make Statics a part of it; viz, that part which considers the motion of bodies arising from gravity.

Others make them two distinct doctrines; restraining mechanics to the doctrine of motion and weight, as depending on, or connected with, the power of machines; and Statics to the doctrine of motion, considered merely as arising from the weight of bodies, without any immediate respect to machines. In this way, Statics should be the doctrine or theory of motion; and mechanics, the application of it to machines.

For the laws of Statics, see GRAVITY, DESCENT, &c.

STATION, or STATIONARY, in Astronomy, the position or appearance of a planet in the same point of the zodiac, for several days. This happens from the observer being situated on the earth, which is far out of the centre of their orbits, by which they seem to proceed irregularly; being sometimes seen to go forwards, or from west to east, which is their natural *direction;* sometimes to go backwards, or from east to west, which is their *retrogradation;* and between these two states there must be an intermediate one, where the planet appears neither to go forwards nor backwards, but to stand still, and keep the same place in the heavens, which is called her *Station*, and the planet is then said to be *Stationary.*

Apollonius Pergæus has shewn how to find the Stationary point of a planet, according to the old theory of the planets, which supposes them to move in epicycles; which was followed by Ptolomy in his Almag. lib. 12, cap. 1, and others, till the time of Copernicus. Concerning this, see Regiomontanus in Epitome Almagesti, lib. 12, prop. 1; Copernicus's Revolutiones Cœlest. lib. 5, cap. 35 and 36; Kepler in Tabulis Rudolphinis, cap. 24; Riccioli's Almag. lib. 7, sect. 5, cap. 2: Harman in Miscellan. Berolinens, pa. 197. Dr. Halley, Mr. Facio, Mr. De Moivre, Dr. Keil, and others have treated on this subject. See also the articles RETROGRADE and STATIONARY in this Dictionary.

STATION, in Practical Geometry &c, is a place pitched upon to make an observation, or take an angle, or such like, as in surveying, measuring heights and distances, levelling, &c.

An accessible height is taken from one Station; but an inaccessible height or distance is only to be taken by making two Stations, from two places whose distance asunder is known. In making maps of counties, provinces, &c, Stations are fixed upon certain eminencies &c of the country, and angles taken from thence to the several towns, villages, &c.—In surveying, the instrument is to be adjusted by the needle, or otherwise, to answer the points of the horizon at every Station; the distance from hence to the last Station is to be measured, and an angle is to be taken to the next Station; which process repeated includes the chief practice of surveying.

furveying.—In levelling, the inftrument is rectified, or placed level at each Station, and obfervations made forwards and backwards.

There is a method of meafuring diftances at one Station, in the Philof. Tranf. numb 7, by means of a telefcope. I have heard of another, by Mr. Ramfden; and have feen a third ingenious way by Mr. Green of Deptford, not yet publifhed; this confifts of a permanent fcale of divifions, placed at any point whofe diftance is required; then the number of divifions feen through the telefcope, gives the diftance fought.

STATION *Line*, in Surveying, and *Line of Station*, in Perfpective. See LINE.

STATIONARY, in Aftronomy, the ftate of a planet when, to an obferver on the earth, it appears for fome time to ftand ftill, or remain immoveable in the fame place in the heavens. For as the planets, to fuch an obferver, have fometimes a progreffive motion, and fometimes a retrograde one, there muft be fome point between the two where they muft appear Stationary. Now a planet will be feen Stationary, when the line that joins the centres of the earth and planet is conftantly directed to the fame point in the heavens, which is when it keeps parallel to itfelf. For all right lines drawn from any point of the earth's orbit, parallel to one another, do all point to the fame ftar; the diftance of thefe lines being infenfible, in comparifon of that of the fixed ftars.

The planet Herfchel is feen Stationary at the diftance of   from the fun; Saturn at fomewhat more than 90°; Jupiter at the diftance of 52°; and Mars at a much greater diftance; Venus at 47°, and Mercury at 28°.

Herfchel is Stationary   days, Saturn 8, Jupiter 4, Mars 2, Venus $1\frac{1}{2}$, and Mercury $\frac{1}{2}$ a day: though the feveral ftations are not always equal; becaufe the orbits of the planets are not circles which have the fun in their centre.

STEAM, the fmoke or vapour arifing from water, or any other liquid or moift body, when confiderably heated. Subterranean Steams often affect the furface of the earth in a remarkable manner, and promote or prevent vegetation more than any thing elfe. It has been imagined that Steams may be the generative caufe of both minerals and metals, and of all the peculiarities of fprings. See Philof. Tranf. vol. 5, pa. 1154, or Abr. vol. 2, pa. 833.—Of the ufe of the air to elevate the Steams of bodies, fee pa. 2048 and 297 ib.— Concerning the warm and fertilizing temperature and Steams of the earth, fee Phil. Tranf. vol. 10, pa. 307 and 357. See alfo Dr. Hamilton " On the Afcent of Vapours."

The Steam raifed from hot water is an elaftic fluid, which, like elaftic air, has its elafticity proportional to its denfity when the heat is the fame, or proportional to the heat when the denfity is the fame. The Steam raifed with the ordinary heat of boiling water, is almoft 3000 times rarer than water, or about $3\frac{1}{2}$ times rarer than air, and has its elafticity about equal to that of the common air of the atmofphere. And by great heat it has been found that the Steam may be expanded into 14000 times the fpace of water, or may be made about 5 times ftronger than the atmofphere. But from fome accidents that have happened,

it appears that Steam, fuddenly raifed from water, or moift fubftances, by the immediate application of ftrong heat, is vaftly ftronger than the atmofphere, or even than gunpowder itfelf. Witnefs the accident that happened to a foundery of cannon at Moorfields, when upon the hot metal firft running into the mould in the earth, fome fmall quantity of water in the bottom of it was fuddenly changed into Steam, which by its exploion, blew the foundery all to pieces. I remember another fuch accident at a foundery at Newcaftle; the founder having purchafed, among fome old brafs, a hollow brafs ball that had been ufed for many years as a valve in a pump, withinfide of which it would feem fome water had got infinuated; and having put it into his fire to melt, when it had become very hot, it fuddenly burft with a prodigious noife, and blew the adjacent parts of the furnace in pieces.

Steam may be applied to many purpofes ufeful in life, but one of its chief ufes is in the Steam-engine defcribed in the following article.

‡STEAM *Engine*, an engine for raifing water by the force of Steam produced from boiling water; and often called the *Fire-engine*, on account of the fire employed in boiling the water to produce the Steam. This is one of the moft curious and ufeful machines, which modern art can boaft, for raifing water from ponds, wells, or pits, for draining mines, &c. Were it not for the ufe of this moft important invention, it is probable we fhould not now have the benefit of coal fires in England; as our forefathers had, before the prefent century, excavated all the mines of coal as deep as it could be worked, without the benefit of this engine to draw the water from greater depths.

This engine is commonly a forcing pump, having its rod fixed to one end of a lever, which is worked by the weight or preffure of the atmofphere upon a pifton, at the other end, a temporary vacuum being made below it, by fuddenly condenfing the Steam, that had been let into the cylinder in which this pifton works, by a jet of cold water thrown into it. A partial vacuum being thus made, the weight of the atmofphere preffes down the pifton, and raifes the other end of the ftraight lever with the water from the well &c. Then immediately a hole is uncovered in the bottom of the cylinder, by which a frefh fill of hot Steam rufhes in from a boiler of water below it, which proves a counterbalance for the atmofphere above the pifton, upon which the weight of the pump rods at the other end of the lever carries that end down, and raifes the pifton of the Steam cylinder. Immediately the Steam hole is fhut, and the cock opened for injecting the cold water into the cylinder of Steam, which condenfes it to water again, and thus making another vacuum below the pifton, the atmofphere above it preffes it down, and raifes the pump rods with another lift of water; and fo on continually. This is the common principle: but there are alfo other modes of applying the force of the Steam, as we fhall fee in the following fhort hiftory of this invention and its various improvements.

The earlieft account to be met with of the invention of this engine, is in the marquis of Worcefter's fmall book intitled a Century of Inventions (being a defcription of 100 notable difcoveries), publifhed in the year 1663, where he propofed the raifing of great quantities of water by the force of Steam, raifed from water by means of fire; and he mentions an engine of that kind, of his own contrivance, which could raife a continual ftream like a fountain 40 feet high, by means of two cocks which were alternately and fucceffively turned by a man to admit the Steam, and to re-fill the veffel with cold water, the fire being continually kept up.

However, this invention not meeting with encouragement, probably owing to the confufed ftate of public affairs at that time, it was neglected, and lay dormant feveral years, until one Captain Thomas Savery, having read the marquis of Worcefter's books, feveral years afterwards, tried many experiments upon the force and power of Steam; and at laft hit upon a method of applying it to raife water. He then bought up and deftroyed all the marquis's books that could be got, and claimed the honour of the invention to himfelf, and obtained a patent for it, pretending that he had difcovered this fecret of nature by accident. He contrived an engine which, after many experiments, he brought to fome degree of perfection, fo as to raife water in fmall quantities: but he could not fucceed in raifing it to any great height, or in large quantities, for the draining of mines; to effect which by his method, the Steam was required to be fo ftrong as would have burft all his veffels; fo that he was obliged to limit himfelf to raifing the water only to a fmall height, or in fmall quantities. The largeft engine he erected, was for the York-buildings Company in London, for fupplying the inhabitants in the Strand and that neighbourhood with water.

The principle of this machine was as follows: H (fig. 3, pl. 27) reprefents a copper boiler placed on a furnace. E is a ftrong iron veffel, communicating with the boiler by means of a pipe at top, and with the main pipe AB by means of a pipe I at bottom; AB is the main pipe immerfed in the water at B; D and C are two fixed valves, both opening upwards, one being placed above, and the other below the pipe of communication I. Laftly, at G is a cock that ferves occafionally to wet and cool the veffel E, by water from the main pipe, and F is a cock in the pipe of communication between the veffel E and the boiler.

The engine is fet to work, by filling the copper in part with water, and alfo the upper part of the main pipe above the valve C, the fire in the furnace being lighted at the fame time. When the water boils ftrongly, the cock F is opened, the Steam rufhes into the veffel E, and expels the air from thence through the valve C. The veffel E thus filled, and violently heated by the Steam, is fuddenly cooled by the water which falls upon it by turning the cock C; the cock F being at the fame time fhut, to prevent any frefh acceffion of Steam from the boiler. Hence, the Steam in E becoming condenfed, it leaves the cavity within almoft intirely a vacuum; and therefore the preffure of the atmofphere at B forces the water through the valve D till the veffel E is nearly filled. The condenfing cock G is then fhut, and the Steam cock F again opened; hence the Steam, rufhing into E, expels the water through the valve C, as it before did the air. Thus E becomes again filled with hot Steam, which is again cooled and condenfed by the water from G, the fupply of Steam being cut off by fhutting F, as in the former

operation: the water confequently rufhes through D; by the preffure of the atmofphere at B, and E is again filled. This water is forced up the main pipe through C, by opening F and fhutting G, as before. And thus it is eafy to conceive, that by this alternate opening and fhutting the cocks, water will be continually raifed, as long as the boiler continues to fupply the Steam.

For the fake of perfpicuity, the drawing is divefted of the apparatus that ferves to turn the two cocks at once, and of the contrivances for filling the copper to the proper quantity. But it may be found complete, with a full account of its ufes and application, in Mr. Savery's book intituled the *Miner's Friend*. The engines of this conftruction were ufually made to work with two receivers or Steam veffels, one to receive the Steam, while the other was raifing water by the condenfation. This engine has been fince improved, by admitting the end of the condenfing pipe G into the veffel E, by which means the Steam is more fuddenly and effectually condenfed than by water on the outfide of the veffel.

The advantages of this engine are, that it may be erected in almoft any fituation, that it requires but little room, and is fubject to very little friction in its parts.—Its difadvantages are, that great part of the Steam is condenfed and lofes its force upon coming into contact with the water in the veffel E, and that the heat and elafticity of the Steam muft be increafed in proportion to the height that the water is required to be raifed to. On both thefe accounts a large fire is required, and the copper muft be very ftrong, when the height is confiderable, otherwife there is danger of its burfting.

While captain Savery was employed in perfecting his engine, Dr. Papin of Marburg was contriving one on the fame principles, which he defcribes in a fmall book publifhed in 1707, intitled *Ars Nova ad Aquam Ignis adminiculo efficaciffimè elevandam.* Capt. Savery's engine however was much completer than that propofed by Dr. Papin.

About the fame time alfo one Monf. Amontons of Paris was engaged in the fame purfuit: but his method of applying the force of Steam was different from thofe before-mentioned; for he intended it to drive or turn a wheel, which he called a *fire-mill*, which was to work pumps for raifing water; but he never brought it to perfection. Each of thefe three gentlemen claimed the originality of the invention; but it is moft probable they all took the hint from the book publifhed by the marquis of Worcefter, as before-mentioned.

In this imperfect ftate it continued, without farther improvements, till the year 1705, when Mr. Newcomen, an iron-monger, and Mr. John Cowley, a glazier, both of Dartmouth, contrived another way to raife water by Steam, bringing the engine to work with a beam and pifton, and where the Steam, even at the greateft depths of mines, is not required to be greater than the preffure of the atmofphere: and this is the ftructure of the engine as it has fince been chiefly ufed. Thefe gentlemen obtained a patent for the fole ufe of this invention, for 14 years. The firft propofal they made for draining of mines by this engine, was in the year 1711; but they were very coldly received by

many

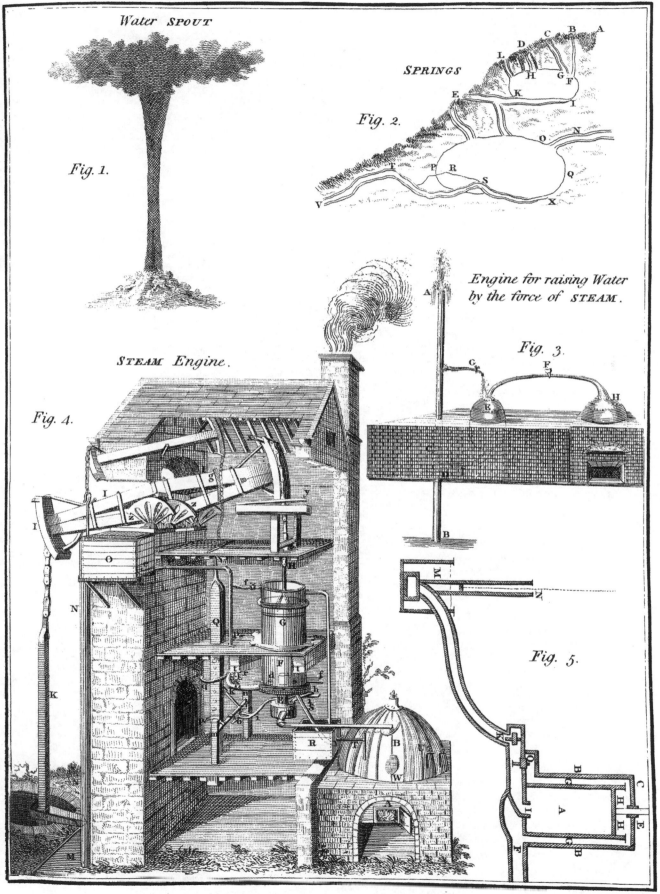

Plate XXVII.

Water SPOUT

Fig. 1.

SPRINGS

Fig. 2.

Engine for raising Water
by the force of STEAM.

Fig. 3.

STEAM Engine.

Fig. 4.

Fig. 5.

many perfons in the fouth of England, who did not underftand the nature of it. In 1712 they came to an agreement with the owners of a colliery at Griff in Warwickfhire, where they erected an engine with a cylinder of 22 inches diameter. At firft they were under great difficulties in many things; but by the affiftance of fome good workmen they got all the parts put together in fuch a manner, as to anfwer their intention tolerably well: and this was the firft engine of the kind erected in England. There was at firft one man to attend the Steam-cock, and another to attend the injection cock; but they afterwards contrived a method of opening and fhutting them by fome fmall machinery connected with the working beam. The next engine erected by thefe patentees, was at a colliery in the county of Durham, about the year 1718, where was concerned, as an agent, Mr. Henry Beighton, F. R. S. and conductor of the Ladies' Diary from the year 1714 to the year 1744: this gentleman, not approving of the intricate manner of opening and fhutting the cocks by ftrings and catches, as in the former engine, fubftituted the hanging beam for that purpofe as at prefent ufed, and likewife made improvements in the pipes, valves, and fome other parts of the engine.

In a few years afterwards, thefe engines came to be better underftood than they had been; and their advantages, efpecially in draining of mines, became more apparent: and from the great number of them erected, they received additional improvements from different perfons, till they arrived at their prefent degree of perfection: as will appear in the fequel, after we have a little confidered the general principles of this engine, which are as follow.

Principles of the Steam Engine.

The principles on which this engine acts, are truly philofophical; and when all the parts of the machine are proportioned to each other according to thefe principles, it never fails to anfwer the intention of the engineer.

1. It has been proved in pneumatics, that the preffure of the atmofphere upon a fquare inch at the earth's furface, is about $14\frac{3}{4}$lb avoirdupois at a medium, or $11\frac{1}{2}$lb on a circular inch, that is on a circle of an inch diameter. And,

2. If a vacuum be made by any means in a cylinder, which has a moveable pifton fufpended at one end of a lever equally divided, the air will endeavour to rufh in, and will prefs down the pifton, with a force proportionable to the area of the furface, and will raife an equal weight at the other end of the lever.

3. Water may be rarefied near 14000 times by being reduced into Steam, and violently heated: the particles of it are fo ftrongly repellent, as to drive away air of the common denfity, only by a heat fufficient to keep the water in a boiling ftate, when the Steam is almoft 3000 times rarer than water, or $3\frac{1}{2}$ times rarer than air, as appears by an experiment of Mr. Beighton's: by increafing the heat, the Steam may be rendered much ftronger; but this requires great ftrength in the veffels. This Steam may be again condenfed into its former ftate by a jet of cold water difperfed through it; fo that 14000 cubic inches of Steam admitted into a cy-

linder, may be reduced into the fpace of one cubic inch of water only, by which means a partial vacuum is obtained.

4. Though the preffure of the atmofphere be about $14\frac{3}{4}$ pounds upon every fquare inch, or $11\frac{1}{2}$ pounds upon a circular inch; yet, on account of the friction of the feveral parts, the refiftance from fome air which is unavoidably admitted with the jet of cold water, and from fome remainder of Steam in the cylinder, the vacuum is very imperfect, and the pifton does not defcend with a force exceeding 8 or 9 pounds upon every fquare inch of its furface.

5. The gallon of water of 282 cubic inches weighs $10\frac{1}{5}$ pounds avoirdupois, or a cubic foot $62\frac{1}{2}$ pounds, or 1000 ounces. The pifton being preffed by the atmofphere with a force proportional to its area in inches, multiplied by about 8 or 9 pounds, depreffes that end of the lever, and raifes a column of water in the pumps of equal weight at the other end, by means of the pump-rods fufpended to it. When the Steam is again admitted, the pump-rods fink by their fuperior weight, and the pifton rifes; and when that Steam is condenfed, the pifton defcends, and the pump-rods lift; and fo on alternately as long as the pifton works.

It has been obferved above, that the pifton does not defcend with a force exceeding 8 or 9 pounds upon every fquare inch of its furface; but by reafon of accidental frictions, and alterations in the denfity of the air, it will be fafeft, in calculating the power of the cylinder, to allow fomething lefs than 8 pounds for the preffure of the atmofphere, upon every fquare inch, viz 7lb. 10 oz. $= 7\frac{6}{4}$lb, or juft 6lb. upon every circular inch; and it being allowed that the gallon of water, of 282 cubic inches, weighs $10\frac{1}{5}$lb, from thefe premifes the dimenfions of the cylinder, pumps, &c, for any Steam-engine, may be deduced as follows:—Suppofe

$c =$ the cylinder's diameter in inches,
$p =$ the pump's ditto,
$f =$ the depth of the pit in fathoms,
$g =$ gallons drawn by a ftroke of 6 feet or a fathom,
$h =$ the hogfheads drawn per hour,
$s —$ the number of ftrokes per minute.

Then c^2 is the area of the cylinder in circular inches, theref. $6c^2$ is the power of the cylinder in pounds.

And $\dfrac{p^2 \times \cdot 7854 \times 72}{282}$ or $\frac{1}{5}p^2$ is $= g$ the gallons

contained in one fathom or 72 inches of any pump; which multiplied by f fathoms, gives $\frac{1}{5}p^2f$ for the gallons contained in f fathoms of any pump whofe diameter is p.

Hence $\frac{1}{5}p^2f \times 10\frac{1}{5}$lb. gives $2p^2f$ nearly, for the weight in pounds of the column of water which is to be equal to the power of the cylinder, which was before found equal to $6c^2$. Hence then we have the cd equation,

viz, $6c^2 = 2p^2f$, or $3c^2 = p^2f$;
the firft equation being $\frac{1}{5}p^2 = g$, or $p^2 = 5g$.

From which two equations, any particular circumftance may be determined.

Or if, inftead of 6lb, for the preffure of the air on each circular inch of the cylinder, that force be fuppofed

pofed any number as a pounds; then will the power of the cyclinder be ac^2, and the 2d equation becomes $ac^2 = 2p^2f = 10fg$, by fubftituting $5g$ inftead of p^2.

And farther, $63h = 60gs$, or $21h = 20gs$.

From a comparifon of thefe equations, the following theorems are derived, which will determine the fize of the cylinder and pumps of any Steam-engine capable of drawing a certain quantity of water from any affigned depth, with the preffure of the atmofphere on each circular inch of the cylinder's area.

Thefe theorems are more particularly adapted to one pump in a pit. But it often happens in practice, that an engine has to draw feveral pumps of different diameters from different depths; and in this cafe, the fquare of the diameter of each pump muft be multiplied by its depth, and double the fum of all the products will be the weight of water drawn at each ftrok, which is to he ufed inftead of $2p^2f$ for the power of the cylinder.

The following is a Table, calculated from the foregoing theorems, of the powers of cylinders from 30 to 70 inches diameter; and the diameter and lengths of pumps which thofe cylinders are capable of working, from a 6 inch bore to that of 20 inches, together with the quantity of water drawn per ftroke and per hour, allowing the engine to make 12 ftrokes of 6 feet per minute, and the preffure of the atmofphere at the rate of 7 lb 10 oz per fquare inch, or 6 lb per circular inch.

	A TABLE of THEOREMS *for the readier computing the Powers of a* STEAM-ENGINE.
1	$a = \dfrac{2fp^2}{c^2} = \dfrac{10fg}{c^2} = \dfrac{21fh}{2c^2s}$
2	$c = \sqrt{\dfrac{2fp^2}{a}} = \sqrt{\dfrac{10fg}{a}} = \sqrt{\dfrac{21fh}{2as}}$
3	$f = \dfrac{ac^2}{2p^2} = \dfrac{ac^2}{10g} = \dfrac{2ac^2s}{21h}$
4	$g = \dfrac{p^2}{5} = \dfrac{ac^2}{10f} = \dfrac{21h}{20s}$
5	$h = \dfrac{4p^2s}{21} = \dfrac{20gs}{21} = \dfrac{2ac^2s}{21f}$
6	$p = \sqrt{5g} = \sqrt{\dfrac{ac^2}{2g}} = \sqrt{\dfrac{21h}{4s}}$
7	$s = \dfrac{21h}{4p^2} = \dfrac{21h}{20g} = \dfrac{21fh}{2ac^2}$

TABLE of the Power and Effects of STEAM-ENGINES, allowing 12 Strokes, of 6 Feet long each, per Minute, and the pressure of the Air 7lb. 10oz per Square Inch, or 6lb per Circular Inch.

The Diameters of the Cylinders in Inches.	The Diameters of the Pumps in Inches.															Power of the cylinders and weight of water in pounds.
	6	7	8	9	10	11	12	13	14	15	16	17	18	19	20	
30	75	55	42	33	27	22	19	16	14	12	10	·	·	·	·	5400
31	80	58	45	35	29	24	20	17	15	13	11	10	·	·	·	5766
32	83	61	47	37	30	25	21	18	16	13	12	10	·	·	·	6144
33	90	67	51	40	3	27	22	19	17	14	13	11	10	·	·	6534
34	94	-0	53	42	34	28	23	20	18	15	14	12	10	·	·	693
35	102	75	57	45	37	30	26	22	19	16	14	13	11	·	·	7350
36	·	79	61	48	39	32	27	23	20	17	15	14	12	10	·	7776
37	·	84	64	51	41	34	29	24	21	18	16	14	12	11	10	8214
38	·	88	68	53	43	35	30	26	22	19	17	15	13	12	10	8664
39	·	93	71	56	45	37	32	27	23	20	18	16	14	12	11	9126
40	·	98	75	59	48	39	34	28	24	21	19	17	15	13	12	9600
42	·	108	83	65	53	43	38	31	27	23	21	18	16	14	13	10584
44	·	·	90	71	58	48	41	34	30	26	23	20	18	16	14	11616
46	·	·	99	78	63	52	45	37	33	29	25	21	19	17	16	12696
48	·	·	·	85	69	57	49	41	35	31	27	24	21	19	17	13824
50	·	·	·	92	75	62	53	44	38	34	29	26	23	21	19	15000
52	·	·	·	100	81	67	57	48	41	36	31	28	25	22	20	16224
54	·	·	·	·	87	72	61	52	44	38	34	30	27	24	22	17496
56	·	·	·	·	94	78	66	56	48	42	37	32	29	26	23	18816
58	·	·	·	·	101	83	70	59	51	44	39	34	31	28	25	20184
60	·	·	·	·	·	89	75	63	55	48	42	37	33	30	27	21600
62	·	·	·	·	·	95	80	68	58	51	45	39	35	32	28	23064
64	·	·	·	·	·	·	85	72	62	54	48	42	38	34	30	24546
66	·	·	·	·	·	·	90	77	66	57	51	45	40	36	32	26676
68	·	·	·	·	·	·	96	82	70	61	54	48	42	38	34	27744
70	·	·	·	·	·	·	·	86	75	64	57	50	45	40	36	29400
Quan.drawn at one stroke in gallons.	7·2	10	13	16·2	20	24·2	28·8	33·8	39·2	45	51·2	57·8	64·8	72·2	80	
Quan.drawn in one hour in hogsheads.	82	114	148	184	228	276	328	385	447	513	583	659	738	823	912	
Diameter of pumps.	6	7	8	9	10	11	12	13	14	15	16	17	18	19	20	

Lct.

Let us now defcribe the feveral parts of an engine, and exemplify the application of the foregoing principles, in the conftruction of one of the completeft of the modern engines. See fig. 4. pl. 27.

A reprefents the fire-place under the boiler, for the boiling of the water, and the afh-hole below it.

B, the boiler, filled with water about three feet above the bottom, made of iron plates.

C, the Steam pipe, through which the Steam paffes from the boiler into the receiver.

D, the receiver, a clofe iron veffel, in which is the regulator or Steam-cock. which opens and fhuts the hole of communication at each ftroke.

E, the communication pipe between the receiver and the cylinder; it rifes 5 or 6 inches up, in the infide of the cylinder bottom, to prevent the injected water from defcending into the receiver.

F, the cylinder, of caft iron, about 10 feet long, bored fmooth in the infide; it has a broad flanch in the middle on the outfide, by which it is fupported when hung in the cylinder-beams.

G, the pifton, made to fit the cylinder exactly: it has a flanch rifing 4 or 5 inches upon its upper furface, between which and the fide of the cylinder a quantity of junk or oakum is ftuffed, and kept down by weights, to prevent the entrance of air or water and the efcaping of Steam.

H, the chain and pifton fhank, by which it is connected to the working beam.

I I, the working-beam or lever: it is made of two or more large logs of timber, bent together at each end, and kept at the diftance of 8 or 9 inches from each other in the middle by the gudgeon, as reprefented in the Plate. The arch-heads, II, at the ends, are for giving a perpendicular direction to the chains of the pifton and pump-rods.

K, the pump-rod which works in the fucking pump.

L, and draws the water from the bottom of the pit to the furface.

M, a ciftern, into which the water drawn out of the pit is conducted by a trough, fo as to keep it always full: and the fuperfluous water is carried off by another trough.

N, the jack head pump, which is a fucking-pump wrought by a fmall lever or working-beam, by means of a chain connected to the great beam or lever near the arch g at the inner end, and the pump rod at the outer end. This pump commonly ftands near the corner of the front of the houfe, and raifes the column of water up to the ciftern O, into which it is conducted by a trough.

O, the jack head ciftern for fupplying the injection, which is always kept full by the pump N: it is fixed fo high as to give the jet a fufficient velocity into the cylinder when the cock is opened. This ciftern has a pipe on the oppofite fide for conveying away the fuperfluous water.

P P, the injection-pipe, of 3 or 4 inches diameter, which turns up in a curve at the lower end, and enters the cylinder bottom: it has a thin plate of iron upon the end a, with 3 or 4 adjutage holes in it, to prevent the jet of cold water of the jack-head ciftern

from flying up againft the pifton, and yet to condenfe the Steam each ftroke, when the injection-cock is open.

e, a valve upon the upper end of the injection pipe within the ciftern, which is fhut when the engine is not working, to prevent any wafte of the water.

f, a fmall pipe which branches off from the injection-pipe, and has a fmall cock to fupply the pifton with a little water to keep it air-tight.

Q, the working plug, fufpended by a chain to the arch g of the working beam. It is ufually a heavy piece of timber, with a flit vertically down its middle, and holes bored horizontally through it, to receive pins for the purpofe of opening and fhutting the injection and Steam cocks, as it afcends and defcends by the motion of the working beam.

h, the handle of the fteam-cock or regulator. It fixed to the regulator by a fpindle which comes up through the top of the receiver. The regulator is a circular plate of brafs or caft iron, which is moved horizontally by the handle h, and opens or fhuts the communication at the lower end of the pip E within the receiver. It is reprefented in the plate by a circular dotted line.

i i, the fpanner, which is a long rod or plate of iron for communicating motion to the handle of the regulator: to which it is fixed by means of a flit in the latter, and fome pins put through to faften it.

k l, the vibrating lever, called the Y, having the weight k at one end and two legs at the other end. It is fixed to an horizontal axis, moveable about its centre-pins or pivots m n, by means of the two fhanks o p fixed to the fame axis, which are alternately thrown backwards and forwards by means of two pins in the working plug; one pin on the outfide depreffing the fhank o, throws the loaded end k of the Y from the cylinder into the pofition reprefented in the plate, and caufes the leg l to ftrike againft the end of the fpanner; which forcing back the handle of the regulator or fteam cock, opens the communication, and permits the fteam to fly into the cylinder. The pifton immediately rifing by the admiffion of the Steam, the working beam I I rifes; which alfo raifes the working-plug, and another pin which goes through the flit raifes the fhank p, which throws the end k of the Z towards the cylinder, and, ftriking the end of the fpanner, forces it forward, and fhuts the regulator Steam-cock.

q r, the lever for opening and fhutting the injection cock, called the F. It has two toes from its centre, which take between them the key of the injection cock. When the working-plug has afcended nearly to its greateft height, and fhut the regulator, a pin catches the end q of the F and raifes it up, which opens the injection-cock, admits a jet of cold water to fly into the cylinder, and, condenfing the Steam, makes a vacuum; then the preffure of the atmofphere bringing down the pifton in the cylinder, and alfo the plug-frame, another pin fixed in it catches the end of the lever in its defcent, and, by preffing it down, fhuts the injection-cock, at the fame time the regulator is opened to admit Steam, and fo on alternately; when the regulator is fhut the injection is open, and when the former is open the latter is fhut.

R, the

4

R, the hot-well, a small cistern made of planks, which receives all the waste water from the cylinder.

S, the sink-pit to convey away the water which is injected into the cylinder at each stroke. Its upper end is even with the inside of the cylinder bottom, its lower end has a lid or cover moveable on a hinge which serves as a valve to let out the injected water, and shuts close each stroke of the engine, to prevent the water being forced up again when the vacuum is made.

T, the feeding pipe, to supply the boiler with water from the hot-well. It has a cock to let in a large or small quantity of water as occasion requires, to make up for what is evaporated; it goes nearly down to the boiler bottom.

U, two gage cocks, the one larger than the other, to try when a proper quantity of water is in the boiler: upon opening the cocks, if one give Steam and the other water, it is right; if they both give Steam, there is too little water in the boiler; and if they both give water, there is too much.

W, a plate which is screwed on to a hole on the side of the boiler, to allow a passage into the boiler for the convenience of cleaning or repairing it.

X, the Steam-clack or puppet valve, which is a brass valve on the top of a pipe opening into the boiler, to let off the Steam when it is too strong. It is loaded with lead, at the rate of one pound to an inch square; and when the Steam is nearly strong enough to keep it open, it will do for the working of the engine.

f, the snifting valve, by which the air is discharged from the cylinder each stroke, which was admitted with the injection, and would otherwise obstruct the due operation of the engine.

t t, the cylinder-beams; which are strong joists going through the house for supporting the cylinder.

v, the cylinder cap of lead, soldered on the top of the cylinder, to prevent the water upon the piston from flashing over when it rises too high.

w, the waste-pipe, which conducts the superfluous water from the top of the cylinder to the hot-well.

xx, iron bars, called the catch-pins, fixed horizontally through each arch head, to prevent the beam descending too low in case the chain should break.

yy, two strong wooden springs, to weaken the blow given by the catch pins when the stroke is too long.

zz, two friction-wheels, on which the gudgeon or centre of the great beam is hung; they are the third or fourth part of a circle, and move a little each way as the beam vibrates. Their use is to diminish the friction of the axis, which, in so heavy a lever, would otherwise be very great.

When this engine is to be set to work, the boiler must be filled about three or four feet deep with water, and a large fire made under it; and when the Steam is found to be of a sufficient strength by the puppet-clack, then by thrusting back the spanner, which opens the regulator or Steam-cock, the Steam is admitted into the cylinder, which raises the piston to the top of the cylinder, and forces out all the air at the snifting valve; then by turning the key of the injection-cock, a jet of cold water is admitted into the cylinder which condenses the Steam and makes a vacuum; and the atmosphere then pressing upon the piston, forces it down to the lower part of the cylinder, and makes a stroke by raising the column of water at the other end of the beam. After two or three strokes are made in this manner, by a man opening and shutting the cocks to try if they be right, then the pins may be put into the pin-holes in the working plug, and the engine left to turn the cocks of itself; which it will do with greater exactness than any man can do.

There are in some engines, methods of shutting and opening the cocks different from the one above described, but perhaps none better adapted to the purpose; and as the principles on which they all act are originally the same, any difference in the mechanical construction of the small machinery will have no influence of consequence upon the total effect of the grand machine.

The furnace or fire-place should not have the bars so close as to prevent the free admission of fresh air to the fire, nor so open as to permit the coals to fall through them; for which purpose two inches or thereabouts is sufficient for the distance betwixt the bars. The size of the furnace depends upon the size of the boiler; but in every case the ash-hole ought to be capacious to admit the air, and the greater its height the better. If the flame is conducted in a flue or chimney round the outside of the boiler, or in a pipe round the inside of it, it ought to be gradually diminished from the entrance at the furnace to its egress at the chimney; and the section of the chimney at that place should not exceed the section of the flue or pipe, and should also be somewhat less at the chimney-top.

The boiler or vessel in which the water is rarefied by the force of fire, may be made of iron plates, or cast iron, or such other materials as can withstand the effects of the fire, and the elastic force of the Steam. It may be considered as consisting of two parts; the upper part which is exposed to the Steam, and the under part which is exposed to the fire. The form of the latter should be such as to receive the full force of the fire in the most advantageous manner, so that a certain quantity of fuel may have the greatest possible effect in heating and evaporating the water; which is best done by making the sides cylindrical, and the bottom a little concave, and then conducting the flame by an iron flue or pipe round the inside of the boiler beneath the surface of the water, before it reach the chimney. For, by this means, after the fire in the furnace has heated the water by its effect on the bottom, the flame heats it again by the pipe being wholly included in the water, and having every part of its surface in contact with it; which is preferable to carrying it in a flue or chimney round the outside of the boiler, as a third or a half of the surface of the flame only could be in contact with the boiler, the other being spent upon the brick-work. This cylindric lower part may be less in its diameter than the upper part, and may contain from four to six feet perpendicular height of water in it.

The upper part of the boiler is beſt made hemiſpherical, for reſiſting the elaſticity of the Steam ; yet any other form may do, provided it be of ſufficient ſtrength for the purpoſe. The quick going of the engine depends much on the capaciouſneſs of the boiler-top ; for if it be too ſmall, it requires the Steam to be heated to a great degree, to increaſe its elaſtic force ſo much as to work the engine. If the top is ſo capacious as to contain eight or ten times the quantity of Steam uſed each ſtroke, it will require no more fire to preſerve its elaſticity than is ſufficient to keep the water in a proper ſtate of boiling ; this, therefore, is the beſt ſize for a boiler top. If the diameter of the cylinder be *c*, and works a ſix-foot ſtroke, and the diameter of the boiler be ſuppoſed *b*, then

$$200c^2 = b^3, \text{ or } b = \sqrt[3]{200c^2}.$$

The effect of the injection in condenſing the Steam in the cylinder, depends upon the height of the reſervoir and the diameter of the adjutage. If the engine makes a 6 feet ſtroke, then the jackhead ciſtern ſhould be 12 feet perpendicular above the bottom of the cylinder or the adjutage. The ſize of the adjutage may be from 1 to 2 inches in diameter ; or if the cylinder be very large, it is proper to have three or four holes rather than one large one, in order that the jet may be diſperſed the more effectually over the whole area of the cylinder. The injection pipe, or pipe of conduct, ſhould be ſo large as to ſupply the injection freely with water ; if the diameter of the injection pipe be called *p*, and the diameter of the adjutage, *a*, then $4a^2 = p^2$, and $a^2 = \frac{1}{4}p^2$, or $a = \frac{1}{2}p$.

For a further account of theſe engines, ſee Deſaguliers's Exp. Philoſ. vol. 2, ſect. 14, pa. 465, &c. ; or for an abſtract, Martin's Phil. Brit. number 461, or Nicholſon's Nat. Philoſ. p. 83 &c. And for an account of the improvement made in the fire-engine by Mr. Payne, ſee Philoſ. Tranſ. number 461, or Martin's Philoſ. Brit. p. 87 &c.

Mr. Blakey communicated to the Royal Society, in 1752, remarks on the beſt proportions for Steam-engine cylinders of a given content : and Mr. Smeaton deſcribes an engine of this kind, invented by Mr. De Moura of Portugal, being an improvement of Savery's conſtruction, to render it capable of working itſelf : for both which accounts, ſee Philoſ. Tranſ. vol. 47 art. 29 and 72.

We are informed in the new edit. of the Biograph. Brit. in the article Brindley, that in 1756 this gentleman, ſo well known for his concern in our inland navigations, undertook to erect a Steam engine near Newcaſtle-under-Line, upon a new plan. The boiler of it was made with brick and ſtone, inſtead of iron plates, and the water was heated by iron flues of a peculiar conſtruction ; by which contrivances the conſumption of fuel, neceſſary for working a Steam engine, was reduced one half. He introduced alſo in his engine, wooden cylinders, made in the manner of cooper's ware, inſtead of iron ones ; the former being both cheaper and more eaſily managed in the ſhafts : and he likewiſe ſubſtituted wood for iron in the chains which worked at the end of the beam. He had formed deſigns of introducing other improvements into the conſtruction of this uſeful engine ; but was diſcouraged by obſtacles that were thrown in his way.

Mr. Blakey, ſome years ago, obtained a patent for his improvement of Savery's Steam-engine, by which it is excellently adapted for raiſing water out of ponds, rivers, wells, &c, and for forcing it up to any height wanted for ſupplying houſes, gardens, and other places ; though it has not power ſufficient to drain off the water from a deep mine. The principles of his conſtruction are explained by Mr. Ferguſon, in the Supplement to his Lectures, pa. 19 ; and a more particular deſcription of it, accompanied with a drawing, is given by the patentee himſelf in the Gentleman's Magazine for 1769, p. 392.

Mr. Blakey, it is ſaid, is the firſt perſon who ever thought of making uſe of air as an intermediate body between Steam and water ; by which means the Steam is always kept from touching the water, and conſequently from being condenſed by it : and on this new principle he has obtained a patent. The engine may be built at a trifling expence, in compariſon of the common fire-engine now in uſe ; it will ſeldom need repairs, and will not conſume half ſo much fuel. And as it has no pumps with piſtons, it is clear of all their friction ; and the effect is equal to the whole ſtrength or compreſſive force of the Steam ; which the effect of the common fire-engine never is, on account of the great friction of the piſtons in their pumps.

Ever ſince Mr. Newcomen's invention of the Steam fire engine, the great conſumption of fuel with which it is attended, has been complained of as an immenſe drawback upon the profits of our mines. It is a known fact, that every fire-engine of conſiderable ſize conſumes to the amount of three thouſand pounds worth of coals in every year. Hence many of our engineers have endeavoured, in the conſtruction of theſe engines, to ſave fuel. For this purpoſe, the fire-place has been diminiſhed, the flame has been carried round from the bottom of the boiler in a ſpiral direction, and conveyed through the body of the water in a tube before its arrival at the chimney ; ſome have uſed a double boiler, ſo that fire might act in every poſſible point of contact ; and ſome have built a moor-ſtone boiler, heated by three tubes of flame paſſing through it. But the moſt important improvements which have been made in the Steam-engine for more than thirty years paſt, we owe to the ſkill of Mr. James Watt ; of which we ſhall give ſome account : premiſing, that the internal ſtructure of his new engines ſo much reſembles that of the common ones, that thoſe who are acquainted with them will not fail to underſtand the mechaniſm of his from the following deſcription : he has contrived to obſerve an uniform heat in the cylinder of his engines, by ſuffering no cold water to touch it, and by protecting it from the air, or other cold bodies, by a ſurrounding caſe filled with Steam, or with hot air or water, and by coating it over with ſubſtances that tranſmit heat ſlowly. He makes his vacuum to approach nearly to that of the barometer, by condenſing the Steam in a ſeparate veſſel, called the condenſer, which may be cooled at pleaſure without cooling the cylinder, either by an injection of cold water, or by ſurrounding the

the condenſer with it, and generally by both. He extracts the injection water, and detached air, from the cylinder or condenſer by pumps, which are wrought by the engine itſelf, or blows them out by the Steam. As the entrance of air into the cylinder would ſtop the operation of the engines, and as it is hardly to be expected that ſuch enormous piſtons as thoſe of Steam-engines can move up and down, and yet be abſolutely tight in the common engines; a ſtream of water is kept always running upon the piſton, which prevents the entry of the air: but this mode of ſecuring the piſton, though not hurtful in the common ones, would be highly prejudicial to the new engines. Their piſton is therefore made more accurately; and the outer cylinder, having a lid, covers it, the Steam is introduced above the piſton; and when a vacuum is produced under it, acts upon it by its elaſticity, as the atmoſphere does upon common engines by its gravity. This way of working effectually excludes the air from the inner cylinder, and gives the advantage of adding to the power, by increaſing the elaſticity of the Steam.

In Mr. Watt's engines, the cylinder, the great beams, the pumps, &c, ſtand in their uſual poſitions. The cylinder is ſmaller than uſual, in proportion to the load, and is very accurately bored.

In the moſt complete engines, it is ſurrounded at a ſmall diſtance, with another cylinder, furniſhed with a bottom and a lid. The interſtice between the cylinders communicates with the boilers by a large pipe, open at both ends: ſo that it is always filled with Steam, and thereby maintains the inner cylinder always of the ſame heat with the Steam, and prevents any condenſation within it, which would be more detrimental than an equal condenſation in the outer one. The inner cylinder has a bottom and piſton as uſual: and as it does not reach up quite to the lid of the outer cylinder, the Steam in the interſtice has always free acceſs to the upper ſide of the piſton. The lid of the outer cylinder has a hole in its middle; and the piſton rod, which is truly cylindrical, moves up and down through that hole, which is kept Steam-tight by a collar of oakum ſcrewed down upon it. At the bottom of the inner cylinder, there are two regulating valves, one of which admits the Steam to paſs from the interſtice into the inner cylinder below the piſton, or ſhuts it out at pleaſure: the other opens or ſhuts the end of a pipe, which leads to the condenſer. The condenſer conſiſts of one or more pumps furniſhed with clacks and buckets (nearly the ſame as in common pumps) which are wrought by chains faſtened to the great working beam of the engine. The pipe, which comes from the cylinder, is joined to the bottom of theſe pumps, and the whole condenſer ſtands immerſed in a ciſtern of cold water ſupplied by the engine. The place of this ciſtern is either within the houſe or under the floor, between the cylinder and the lever wall; or without the houſe between that wall and the engine ſhaft, as conveniency may require. The condenſer being exhauſted of air by blowing, and both the cylinders being filled with Steam, the regulating valve which admits the Steam into the inner cylinder is ſhut, and the other regulator which communicates with the condenſer is opened, and the Steam ruſhes into the vacuum of the condenſer with

violence: but there it comes into contact with the cold ſides of the pumps and pipes, and meets a jet of cold water, which was opened at the ſame time with the exhauſtion regulator; theſe inſtantly deprive it of its heat, and reduce it to water; and the vacuum remaining perfect, more Steam continues to ruſh in, and be condenſed until the inner cylinder be exhauſted. Then the Steam which is above the piſton, ceaſing to be counteracted by that which was below it, acts upon the piſton with its whole elaſticity, and forces it to deſcend to the bottom of the cylinder, and ſo raiſes the buckets of the pumps which are hung to the other end of the beam. The exhauſtion regulator is now ſhut, and the Steam one opened again, which, by letting in the Steam, allows the piſton to be pulled up by the ſuperior weight of the pump rods; and ſo the engine is ready for another ſtroke.

But the nature of Mr. Watt's improvement will be perhaps better underſtood from the following deſcription of it as referred to a figure.—The cylinder or Steam veſſel A, of this engine (fig. 5, pl. 27), is ſhut at bottom and opened at top as uſual; and is included in an outer cylinder or caſe BB, of wood or metal, covered with materials which tranſmit heat ſlowly. This caſe is at a ſmall diſtance from the cylinder, and cloſe at both ends. The cover C has a hole in it, through which the piſton rod E ſlides; and near the bottom is another hole F, by which the Steam from the boiler has always free entrance into this caſe or outer cylinder, and by the interſtice GG between the two cylinders has acceſs to the upper ſide of the piſton HH. To the bottom of the inner cylinder A is joined a pipe I, with a cock or valve K, which is opened and ſhut when neceſſary, and forms a paſſage to another veſſel L called a *Condenſer*, made of thin metal. This veſſel is immerſed in a ciſtern M full of cold water, and it is contrived ſo as to expoſe a very great ſurface externally to the water, and internally to the Steam. It is alſo made air-tight, and has pumps N wrought by the engine, which keep it always exhauſted of air and water.

Both the cylinders A and BB being filled with Steam, the paſſage K is opened from the inner one to the condenſer L, into which the Steam violently ruſhes by its elaſticity, becauſe that veſſel is exhauſted; but as ſoon as it enters it, coming into contact with the cold matter of the condenſer, it is reduced to water, and, the vacuum ſtill remaining, the Steam continues to ruſh in till the inner cylinder A below the piſton is left empty. The Steam which is above the piſton, ceaſing to be counteracted by that which is below it, acts upon the piſton HH, and forces it to deſcend to the bottom of the cylinder, and ſo raiſes the bucket of the pump by means of the lever. The paſſage K between the inner cylinder and the condenſer is then ſhut, and another paſſage O is opened, which permits the Steam to paſs from the outer cylinder, or from the boiler into the inner cylinder under the piſton; and then the ſuperior weight of the bucket and pump rods pulls down the outer end of the lever or great beam, and raiſes the piſton, which is ſuſpended to the inner end of the ſame beam.

The advantages that accrue from this conſtruction are, firſt, that the cylinder being ſurrounded with the Steam from the boiler, it is kept always uniformly as hot as the Steam itſelf, and is therefore incapable of deſtroy-

ing

ing any part of the Steam, which should fill it, as the common engines do. Secondly, the condenser being kept always as cold as water can be procured, and colder than the point at which it boils in vacuo, the Steam is perfectly condensed, and does not oppose the descent of the piston; which is therefore forced down by the full power of the Steam from the boiler, which is somewhat greater than that of the atmosphere.

In the common fire-engines, when they are loaded to 7 pounds upon the inch, and are of a middle size, the quantity of Steam which is condensed in restoring to the cylinder the heat which it had been deprived of by the former injection of cold water, is about one full of the cylinder, besides what it really required to fill that vessel; so that twice the full of the cylinder is employed to make it raise a column of water equal to about 7 pounds for each square inch of the piston: or, to take it more simply, a cubic foot of Steam raises a cubic foot of water about 8 feet high, besides overcoming the friction of the engine, and the resistance of the water to motion.

In the improved engine, about one full and a fourth of the cylinder is required to fill it, because the Steam is one-fourth more dense than in the common engine. This engine raises a load equal to 12 pounds and a half upon the square inch of the piston; and each cubic foot of Steam of the density of the atmosphere, raises one cubic foot of water 22 feet high.

The working of these engines is more regular and steady than the common ones, and from what has been said, their other advantages seem to be very considerable.

It is said, that the savings amount at least to two thirds of the fuel, which is an important object, especially where coals are dear. The new engines will raise from twenty thousand to twenty-four thousand cubic feet of water, to the height of twenty-four feet by one hundred weight of good pit coal: and Mr. Watt has proposed to produce engines upon the same principles, though somewhat differing in construction, which will require still much less fuel, and be more convenient for the purposes of mining, than any kind of engine yet used. Mr. Watt has also contrived a kind of mill wheel, which turns round by the power of Steam exerted within it.

The improvements above recited were invented by Mr. James Watt, at Glasgow, in Scotland, in 1764: he obtained the king's letters patent for the sole use of his invention in 1768; but meeting with difficulties in the execution of a large machine, and being otherwise employed, he laid aside the undertaking till the year 1774, when, in conjunction with Mr. Boulton near Birmingham, he completed both a reciprocating and rotative or wheel engine. He then applied to parliament for a prolongation of the term of his patent, which was granted by an act passed in 1775. Since that time, Mr. Watt and Mr. Boulton have erected several engines in Staffordshire, Shropshire, and Warwickshire, and a small one at Stratford near London. They have also lately finished another at Hawkesbury colliery near Coventry, which is justly supposed to be the most powerful engine in England. It has a cylinder 58 inches in diameter, which works a pump 14 inches in diameter, 65 fathoms high, and makes regularly twelve strokes, each 8 feet long, in a minute. They have also erected several engines in Cornwall; one of which has a cylinder 30 inches in diameter, that works a pump 6½ inches in diameter in two shafts, by flat rods with great friction, 300 feet distant from each other, 45 fathoms high in each shaft, equal in all to 90 fathoms, and can make 14 strokes, 8 feet long, in a minute, with a consumption of coals less than 20 bushels in 24 hours. The terms they offer to the public are, to take in lieu of all profits, one third part of the annual savings in fuel, which their engine makes when compared with a common engine of the same dimensions in the neighbourhood. The engines are built at the expence of those who use them, and Messrs. Boulton and Watt furnish such drawings, directions, and attendance, as may be necessary to enable a resident engineer to complete the machine. See the appendix to Pryce's Mineralogia, &c, 1778.

It has been said that some useful improvements have been made in the Steam engine by Mr. William Powel, who had lately the direction and care of an engine of this kind at a colliery near Swansea, in Glamorganshire.

It is hardly necessary to add, that Dr. Falck, in 1776, published an account and description of an improved Steam-engine, which, as he says, will, with the same quantity of fuel, and in an equal space of time, raise above double the quantity of water raised by any lever engine of the same dimensions; as he does not seem to have constructed even a working model of his proposed engine. The principal improvement, however, which he suggests, is to use two cylinders; into which the Steam is let alternately to ascend, by a common regulator, which always opens the communication of the Steam to one, whilst it shuts up the opening of the other: the piston rods are kept (by means of a wheel fixed to an arbour) in a continual ascending and descending motion, by which they move the common arbour, to which is affixed another wheel, moving the pump rods, in the same alternate direction as the piston rods, by which continual motion the pumps are kept in constant action.

STEELYARD, or STILYARD, in Mechanics, a kind of balance, called also, *Statera Romana*, or the *Roman Balance*, by means of which the weights of different bodies are discovered by using one single weight only.

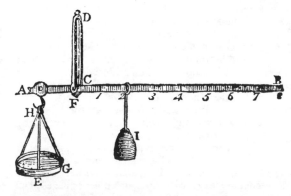

The common Steelyard consists of an iron beam AB,

in

in which is aſſumed a point at pleaſure, as C, on which is raiſed a perpendicular CD. On the ſhorter arm AC is hung a ſcale or baſon to receive the bodies weighed: the moveable weight I is ſhifted backward and forward on the beam, till it be a counterbalance to 1, 2, 3, 4, &c pounds placed in the ſcale; and the points are noted where the conſtant weight I weighs, as 1, 2, 3, 4, &c pounds. From this conſtruction of the Steelyard, the manner of uſing it is evident. But the inſtrument is very liable to deceit, and therefore is not much uſed in ordinary commerce.

Chineſe Steelyard. The Chineſe carry this Statera about them to weigh their gems, and other things of value. The beam or yard is a ſmall rod of wood or ivory, about a foot in length: upon this are three rules of meaſure, made of a fine ſilver-ſtudded work; they all begin from the end of the beam, whence the firſt is extended 8 inches, the ſecond $6\frac{1}{2}$, the third $8\frac{1}{2}$. The firſt is the European meaſure, the other two ſeem to be Chineſe meaſures. At the other end of the yard hangs a round ſcale, and at three ſeveral diſtances from this end are faſtened ſo many ſlender ſtrings, as different points of ſuſpenſion. The firſt diſtance makes $1\frac{3}{5}$ or $\frac{8}{5}$ of an inch, the ſecond $3\frac{1}{5}$ or double the firſt, and the third $4\frac{4}{5}$ or triple of the firſt. When they weigh any thing, they hold up the yard by ſome one of theſe ſtrings, and hang a ſealed weight, of about $1\frac{1}{4}$oz troy weight, upon the reſpective diviſions of the rule, as the thing requires. Grew's Muſeum, pa. 369.

Spring Steelyard, is a kind of portable balance, ſerving to weigh any matter, from 1 to about 40 pounds.

It is compoſed of a braſs or iron tube, into which goes a rod, and about that is wound a ſpring of tempered ſteel in a ſpiral form. On this rod are the diviſions of pounds and parts of pounds, which are made by ſucceſſively hanging on, to a hook faſtened to the other end, 1, 2, 3, 4, &c, pounds.

Now the ſpring being faſtened by a ſcrew to the bottom of the rod; the greater the weight is that is hung upon the hook, the more will the ſpring be contracted, and conſequently a greater part of the rod will come out of the tube; the proportions or quantities of which greater weights are indicated by the figures appearing againſt the extremity of the tube.

Steelyard-*Swing*. In the Philoſ. Tranſ. (no. 462, ſect. 5) is given an account of a Steelyard ſwing, propoſed as a mechanical method for aſſiſting children labouring under deformities, owing to the contraction of the muſcles on one ſide of the body. The crooked perſon is ſuſpended with cords under his arm, and theſe are placed at equal diſtances from the centre of the beam. It is ſuppoſed that the gravity of the body will affect the contracted ſide, ſo as to put the muſcles upon the ſtretch; and hence by degrees the defect may be remedied.

STEEPLE, an appendage uſually raiſed on the weſtern end of a church to contain the bells.—Steeples are denominated from their form, either *ſpires*, or *towers*. The firſt are ſuch as riſe continually diminiſhing like a cone or other pyramid. The latter are mere parallelopipedons, or ſome other priſm, and are covered at top platform like.—In each kind there is uſually a ſort of windows, or loop-holes, to let out the ſound, and ſo contrived as to throw it downward.

Maſius, in his treatiſe on bells, treats likewiſe of Steeples. The moſt remarkable in the world, it is ſaid, is that at Piſa, which leans ſo much to one ſide, that you fear every moment it will fall; yet is in no danger. This odd diſpoſition, he obſerves, is not owing to a ſhock of an earthquake, as is generally imagined; but was contrived ſo at firſt by the architect; as is evident from the cielings, windows, doors, &c, which are all in the bevel.

STEERAGE, in a ſhip, that part next below the quarter-deck, before the bulk-head of the great cabin, where the ſteerſman ſtands in moſt ſhips of war. In large ſhips of war it is uſed as a hall, through which it is neceſſary to paſs to or from the great cabin. In merchant ſhips it is moſtly the habitation of the lower officers and ſhip's crew.

Steerage, in Sea-language, is alſo uſed to expreſs the effort of the helm: and hence

Steerage-*way* is that degree of progreſſive motion communicated to a ſhip, by which ſhe becomes ſuſceptible of the effect of the helm to govern her courſe.

STEERING, in Navigation, the art of directing the ſhip's way by the movements of the helm; or of applying its efforts to regulate her courſe when ſhe advances.

The perfection of Steering conſiſts in a vigilant attention to the motion of the ſhip's head, ſo as to check every deviation from the line of her courſe in the firſt inſtant of its motion; and in applying as little of the power of the helm as poſſible. By this means ſhe will run more uniformly in a ſtraight path, as declining leſs to the right and left; whereas, if a greater effort of the helm be employed, it will produce a greater declination from the courſe, and not only increaſe the difficulty of Steering, but alſo make a crooked and irregular path through the water.

The helmſman, or ſteerſman, ſhould diligently watch the movements of the head by the land, clouds, moon, or ſtars; becauſe, although the courſe is in general regulated by the compaſs, yet the vibrations of the needle are not ſo quickly perceived, as the ſallies of the ſhip's head to the right or left, which, if not immediately reſtrained, will acquire additional velocity in every inſtant of their motion, and require a more powerful impulſe of the helm to reduce them; the application of which will operate to turn her head as far on the contrary ſide of her courſe.

The phraſes uſed in Steering a ſhip, vary according to the relation of the wind to her courſe. Thus, when the wind is large or fair, the phraſes uſed by the pilot or officer who ſuperintends the Steerage, are *port*, *ſtarboard*, and *ſteady*: the firſt of which is intended to direct the ſhip's courſe farther to the right; the ſecond to the left; and the laſt is deſigned to keep her exactly in the line on which ſhe advances, according to the intended courſe. The exceſs of the firſt and ſecond movement is called *hard-a-port*, and *hard-a-ſtarboard*; the former of which gives her the greateſt poſſible inclination to the right, and the latter an equal tendency to the left.—If, on the contrary, the wind be ſcant or foul, the phraſes are *luff*, *thus*, and *no nearer*: the firſt of which is the order to keep her cloſe to the wind; the ſecond, to retain her

her in her present situation; and the third, to keep her sails full.

STELLA. See STAR.

STENTOROPHONIC *Tube*, a *Speaking Trumpet*, or tube employed to speak to a person at a great distance. It has been so called from Stentor, a person mentioned in the 5th book of the Iliad, who, as Homer tells us, could call out louder than 50 men. The Stentorophonic horn of Alexander the Great is famous; with this it is said he could give orders to his army at the distance of 100 stadia, which is about 12 English miles.

The present speaking trumpet it is said was invented by Sir Samuel Moreland. But Derham, in his Physico-Theology, lib. 4, ch. 3, says, that Kircher found out this instrument 20 years before Moreland, and published it in his Mesurgia; and it is farther said that Gaspar Schottus had seen one at the Jesuits' College at Rome. Also one Conyers, in the Philos. Transf. number 141, gives a description of an instrument of this kind, different from those commonly made. Gravesande, in his Philosophy, disapproves of the usual figures of these instruments; he would have them to be parabolic conoids, with the focus of one of its parabolic sections at the mouth.—Concerning this instrument, see Sturmy's Collegium Curiosum, Pt. 2, Tentam. 8; also Philos. Transf. vol. 6, pa. 3056, vol. 12, pa. 1027, or Abridg. vol. 1, pa. 505.

STEREOGRAPHIC *Projection of the Sphere*, is that in which the eye is supposed to be placed in the surface of the sphere. Or it is the projection of the circles of the sphere on the plane of some one great circle, when the eye, or a luminous point, is placed in the pole of that circle.—For the fundamental principles and chief properties of this kind of projection, see PROJECTION.

STEREOGRAPHY, is the art of drawing the forms of solids upon a plane.

STEVIN, STEVINUS (SIMON), a Flemish mathematician of Bruges, who died in 1633. He was master of mathematics to prince Maurice of Nassau, and inspector of the dykes in Holland. It is said he was the inventor of the sailing chariots, sometimes made use of in Holland. He was a good practical mathematician and mechanist, and was author of several useful works: as, treatises on Arithmetic, Algebra, Geometry, Statics, Optics, Trigonometry, Geography, Astronomy, Fortification, and many others, in the Dutch language, which were translated into Latin, by Snellius, and printed in 2 volumes folio. There are also two editions in the French language, in folio, both printed at Leyden, the one in 1608, and the other in 1634, with curious notes and additions, by Albert Girard.—For a particular account of Stevin's inventions and improvements in Algebra, which were many and ingenious, see our article Algebra, vol. 1, pa. 82 and 83.

STEWART (the Rev. Dr. MATTHEW), late professor of mathematics in the university of Edinburgh, was the son of the reverend Mr. Dugald Stewart, minister of Rothsay in the Isle of Bute, and was born at that place in the year 1717. After having finished his course at the grammar school, being intended by his father for the church, he was sent to the university of Glasgow, and was entered there as a student in 1734.

His academical studies were prosecuted with diligence and success; and he was so happy as to be particularly distinguished by the friendship of Dr. Hutcheson, and Dr. Simson the celebrated geometrician, under whom he made great progress in that science.

Mr. Stewart's views made it necessary for him to attend the lectures in the university of Edinburgh in 1741; and that his mathematical studies might suffer no interruption, he was introduced by Dr. Simson to Mr. Maclaurin, who was then teaching with so much success, both the geometry and the philosophy of Newton, and under whom Mr. Stewart made that proficiency which was to be expected from the abilities of such a pupil, directed by those of so great a master. But the modern analysis, even when thus powerfully recommended, was not able to withdraw his attention from the relish of the ancient geometry, which he had imbibed under Dr. Simson. He still kept up a regular correspondence with this gentleman, giving him an account of his progress, and of his discoveries in geometry, which were now both numerous and important, and receiving in return many curious communications with respect to the *Loci Plani*, and the Porisms of Euclid. Mr. Stewart pursued this latter subject in a different, and new direction. In doing so, he was led to the discovery of those curious and interesting propositions, which were published, under the title of *General Theorems*, in 1746. They were given without the demonstrations; but they did not fail to place their discoverer at once among the geometricians of the first rank. They are, for the most part, Porisms, though Mr. Stewart, careful not to anticipate the discoveries of his friend, gave them only the name of Theorems. They are among the most beautiful, as well as most general propositions, known in the whole compass of geometry, and are perhaps only equalled by the remarkable locus to the circle in the second book of Apollonius, or by the celebrated theorem of Mr. Cotes.

Such is the history of the invention of these propositions; and the occasion of the publication of them was as follows. Mr. Stewart, while engaged in them, had entered into the church, and become minister of Roseneath. It was in that retired and romantic situation, that he discovered the greater part of those theorems. In the summer of 1746, the mathematical chair in the university of Edinburgh became vacant, by the death of Mr. Maclaurin. The General Theorems had not yet appeared; Mr. Stewart was known only to his friends; and the eyes of the public were naturally turned on Mr. Stirling, who then resided at Leadhills, and who was well known in the mathematical world. He however declined appearing as a candidate for the vacant chair; and several others were named, among whom was Mr. Stewart. Upon this occasion he printed the *General Theorems*, which gave their author a decided superiority above all the other candidates. He was accordingly elected professor of mathematics in the university of Edinburgh, in September 1747.

The duties of this office gave a turn somewhat different to his mathematical pursuits, and led him to think of the most simple and elegant means of explaining those difficult propositions, which were hitherto only accessible to men deeply versed in the modern analysis. In doing this, he was pursuing the object which,

of

of all others, he moft ardently wifhed to attain, viz, the application of geometry to such problems as the algebraic calculus alone had been thought able to refolve. His folution of Kepler's problem was the firft fpecimen of this kind which he gave to the world ; and it was perhaps impoffible to have produced one more to the credit of the method he followed, or of the abilities with which he applied it. Among the excellent folutions hitherto given of this famous problem, there were none of them at once direct in its method, and fimple in its principles. Mr. Stewart was fo happy as to attain both thefe objects. He founds his folution on a general property of curves, which, though very fimple, had perhaps never been obferved ; and by a moft ingenious application of that property, he fhows how the approximation may be continued to any degree of accuracy, in a feries of refults which converge with great rapidity.

This folution appeared in the fecond volume of the Effays of the Philofophical Society of Edinburgh, for the year 1756. In the firft volume of the fame collection, there are fome other propofitions of Mr. Stewart's, which are an extenfion of a curious theorem in the 4th book of Pappus. They have a relation to the fubject of Porifms, and one of them forms the 91ft of Dr. Simfon's Reftoration.

It has been already mentioned, that Mr. Stewart had formed the plan of introducing into the higher parts of mixed mathematics, the ftrict and fimple form of ancient demonftration. The profecution of this plan produced the *Tracts Phyfical and Mathematical*, which were publifhed in 1761. In the firft of thefe, Mr. Stewart lays down the doctrine of centripetal forces, in a feries of propofitions, demonftrated (if we admit the quadrature of curves) with the utmoft rigour, and requiring no previous knowledge of the mathematics, except the elements of plane Geometry, and of Conic Sections. The good order of thefe propofitions, added to the clearnefs and fimplicity of the demonftrations, renders this Tract perhaps the beft elementary treatife of Phyfical Aftronomy that is any where to be found.

In the three remaining Tracts, our author had it in view to determine, by the fame rigorous method, the effect of thofe forces which difturb the motions of a fecondary planet. From this he propofed to deduce, not only a theory of the moon, but a determination of the fun's diftance from the earth. The former, it is well known, is the moft difficult fubject to which mathematics have been applied, and the refolution required and merited all the clearnefs and fimplicity which our author poffeffed in fo eminent a degree. It muft be regretted therefore, that the decline of Dr. Stewart's health, which began foon after the publication of the Tracts, did not permit him to purfue this inveftigation.

The other object of the Tracts was, to determine the diftance of the fun, from his effect in difturbing the motions of the moon ; and his enquiries into the lunar irregularities had furnifhed him with the means of accomplifhing it.

The theory of the compofition and refolution of forces enables us to determine what part of the folar force is employed in difturbing the motions of the moon ; and therefore, could we meafure the inftanta-

neous effect of that force, or the number of feet by which it accelerates or retards the moon's motion in a fecond, we fhould be able to determine how many feet the whole force of the fun would make a body, at the diftance of the moon, or of the earth, defcend in a fecond of time, and confequently how much the earth is, in every inftant, turned out of its rectilineal courfe. Thus the curvature of the earth's orbit, or, which is the fame thing, the radius of that orbit, that is, the diftance of the fun from the earth, would be determined. But the fact is, that the inftantaneous effects of the fun's difturbing force are too minute to be meafured ; and that it is only the effect of that force, continued for an entire revolution, or fome confiderable portion of a revolution, which aftronomers are able to obferve.

There is yet a greater difficulty which embarraffes the folution of this problem. For as it is only by the difference of the forces exerted by the fun on the earth and on the moon, that the motions of the latter are difturbed, the farther off the fun is fuppofed, the lefs will be the force by which he difturbs the moon's motions ; yet that force will not diminifh beyond a fixed limit, and a certain difturbance would obtain, even if the diftance of the fun were infinite. Now the fun is actually placed at fo great a diftance, that all the difturbances, which he produces on the lunar motions, are very near to this limit, and therefore a fmall miftake in eftimating their quantity, or in reafoning about them, may give the diftance of the fun infinite, or even impoffible. But all this did not deter Dr. Stewart from undertaking the folution of the problem, with no other affiftance than that which geometry could afford. Indeed the idea of fuch a problem had firft occurred to Mr. Machin, who, in his book on the laws of the moon's motion, has juft mentioned it, and given the refult of a rude calculation (the method of which he does not explain), which affigns 8″ for the parallax of the fun. He made ufe of the motion of the nodes ; but Dr. Stewart confidered the motion of the apogee, or of the longer axis of the moon's orbit, as the irregularity beft adapted to his purpofe. It is well known that the orbit of the moon is not immoveable ; but that, in confequence of the difturbing force of the fun, the longer axis of that orbit has an angular motion, by which it goes back about 3 degrees in every lunation, and completes an entire revolution in 9 years nearly. This motion, though very remarkable and eafily determined, has the fame fault, in refpect of the prefent problem, that was afcribed to the other irregularities of the moon : for a very fmall part of it only depends on the parallax of the fun ; and of this Dr. Stewart feems not to have been perfectly aware.

The propofitions however which defined the relation between the fun's diftance and the mean motion of the apogee, were publifhed among the Tracts, in 1761. The tranfit of Venus happened in that fame year : the aftronomers returned, who had viewed that curious phenomenon, from the moft diftant ftations ; and no very fatisfactory refult was obtained from a comparifon of their obfervations. Dr. Stewart then refolved to apply the principles he had already laid down ; and, in 1763, he publifhed his effay on the Sun's Diftance, where the computation being actually made, the parallax of the fun was found to be no more than 6″·9,

and

and confequently his diftance almoft 29875 femidiameters of the earth, or nearly 119 millions of miles.

A determination of the fun's diftance, that fo far exceeded all former eftimations of it, was received with furprife, and the reafoning on which it was founded was likely to undergo a fevere examination. But, even among aftronomers, it was not every one who could judge in a matter of fuch difficult difcuffion. Accordingly, it was not till about 5 years after the publication of the fun's diftance, that there appeared a pamphlet, under the title of *Four Propofitions*, intended to point out certain errors in Dr. Stewart's inveftigation, which had given a refult much greater than the truth. From his defire of fimplifying, and of employing only the geometrical method of reafoning, he was reduced to the neceffity of rejecting quantities, which were confiderable enough to have a great effect on the laft refult. An error was thus introduced, which, had it not been for certain compenfations, would have become immediately obvious, by giving the fun's diftance near three times as great as that which has been mentioned.

The author of the pamphlet, referred to above, was the firft who remarked the dangerous nature of thefe fimplifications, and who attempted to eftimate the error to which they had given rife. This author remarked what produced the compenfation above mentioned, viz, the immenfe variation of the fun's diftance, which correfponds to a very fmall variation of the motion of the moon's apogee. And it is but juftice to acknowledge that, befides being juft in the points already mentioned, they are very ingenious, and written with much modefty and good temper. The author, who at firft concealed his name, but has now confented to its being made public, was Mr. Dawfon, a furgeon at Sudbury in Yorkfhire, and one of the moft ingenious mathematicians and philofophers this country now poffeffes.

A fecond attack was foon after this made on the Sun's Diftance, by Mr. Landen; but by no means with the fame good temper which has been remarked in the former. He fancied to himfelf errors in Dr. Stewart's inveftigation, which have no exiftence; he exaggerated thofe that were real, and feemed to triumph in the difcovery of them with unbecoming exultation. If there are any fubjects on which men may be expected to reafon difpaffionately, they are certainly the properties of number and extenfion; and whatever pretexts moralifts or divines may have for abufing one another, mathematicians can lay claim to no fuch indulgence. The afperity of Mr. Landen's animadverfions ought not therefore to pafs uncenfured, though it be united with found reafoning and accurate difcuffion. The error into which Dr. Stewart had fallen, though firft taken notice of by Mr. Dawfon, whofe pamphlet was fent by me to Mr. Landen as foon as it was printed (for I had the care of the edition of it) yet this gentleman extended his remarks upon it to greater exactnefs. But Mr. Landen, in the zeal of correction, brings many other charges againft Dr. Stewart, the greater part of which feem to have no good foundation. Such are his objections to the fecond part of the inveftigation, where Dr. Stewart finds the relation between the difturbing force of the fun, and the motion of the apfes of the lunar orbit. For this part, inftead of being liable to objection, is deferving of the greateft praife,

fince it refolves, by geometry alone, a problem which had eluded the efforts of fome of the ableft mathematicians, even when they availed themfelves of the utmoft refources of the integral calculus, Sir Ifaac Newton, though he affumed the difturbing force very near the truth, computed the motion of the apfes from thence only at one half of what it really amounts to; fo that, had he been required, like Dr. Stewart, to invert the problem, he would have committed an error, not merely of a few thoufandth parts, as the latter is alleged to have done, but would have brought out a refult double of the truth. (*Princip. Math. lib.* 3, *prop.* 3.) Machin and Callendrini, when commenting on this part of the Principia, found a like inconfiftency between their theory and obfervation. Three other celebrated mathematicians, Clairaut, D'Alembert, and Euler, feverally experienced the fame difficulties, and were led into an error of the fame magnitude. It is true, that, on refuming their computations, they found that they had not carried their approximations to a fufficient length, which when they had at laft accomplifhed, their refults agreed exactly with obfervation. Mr. Walmfley and Dr. Stewart were, I think, the firft mathematicians who, employing in the folution of this difficult problem, the one the algebraic calculus, and the other the geometrical method, were led immediately to the truth; a circumftance fo much for the honour of both, that it ought not to be forgotten. It was the bufinefs of an impartial critic, while he examined our author's reafonings, to have remarked and to have weighed thefe confiderations.

The *Sun's Diftance* was the laft work which Dr. Stewart publifhed; and though he lived to fee the animadverfions made on it, that have been taken notice of above, he declined entering into any controverfy. His difpofition was far from polemical; and he knew the value of that quiet, which a literary man fhould rarely fuffer his antagonifts to interrupt. He ufed to fay, that the decifion of the point in queftion was now before the public; that if his inveftigation was right, it would never be overturned, and that if it was wrong, it ought not to be defended.

A few months before he publifhed the Effay juft mentioned, he gave to the world another work, entitled, *Propofitiones More Veterum Demonftratæ*. It confifts of a feries of geometrical theorems, moftly new; inveftigated, firft by an analyfis, and afterwards fynthetically demonftrated by the inverfion of the fame analyfis. This method made an important part in the analyfis of the ancient geometricians; but few examples of it have been preferved in their writings, and thofe in the *Propofitiones Geometricæ* are therefore the more valuable.

Doctor Stewart's conftant ufe of the geometrical analyfis had put him in poffeffion of many valuable propofitions, which did not enter into the plan of any of the works that have been enumerated. Of thefe, not a few have found a place in the writings of Dr. Simfon, where they will for ever remain, to mark the friendfhip of thefe two mathematicians, and to evince the efteem which Dr. Simfon entertained for the abilities of his pupil. Many of thefe are in the work upon the Porifms, and others in the Conic Sections, viz, marked with the letter *x*; alfo a theorem in the edition of Euclid's Data,

Soon

Soon after the publication of the *Sun's Distance*, Dr. Stewart's health began to decline, and the duties of his office became burdenfome to him. In the year 1772, he retired to the country, where he afterwards fpent the greater part of his life, and never refumed his labours in the univerfity. He was however fo fortunate as to have a fon to whom, though very young, he could commit the care of them with the greateft confidence. Mr. Dugald Stewart, having begun to give lectures for his father from the period above mentioned, was elected joint profeffor with him in 1775, and gave an early fpecimen of thofe abilities, which have not been confined to a fingle fcience.

After mathematical ftudies (on account of the bad ftate of health into which Dr. Stewart was falling) had ceafed to be his bufinefs, they continued to be his amufement. The analogy between the circle and hyperbola had been an early object of his admiration. The extenfive views which that analogy is continually opening; the alternate appearance and difappearance of refemblance in the midft of fo much diffimilitude, make it an object that aftonifhes the experienced, as well as the young geometrician. To the confideration of this analogy therefore the mind of Dr. Stewart very naturally returned, when difengaged from other fpeculations. His ufual fuccefs ftill attended his inveftigations; and he has left among his papers fome curious approximations to the areas, both of the circle and hyperbola. For fome years toward the end of his life, his health fcarcely allowed him to profecute ftudy even as an amufement. He died the 23d of January 1785, at 68 years of age.

The habits of ftudy, in a man of original genius, are objects of curiofity, and deferve to be remembered. Concerning thofe of Dr. Stewart, his writings have made it unneceffary to remark, that from his youth he had been accuftomed to the moft intenfe and continued application. In confequence of this application, added to the natural vigour of his mind, he retained the memory of his difcoveries in a manner that will hardly be believed. He feldom wrote down any of his inveftigations, till it became neceffary to do fo for the purpofe of publication. When he difcovered any propofition, he would fet down the enunciation with great accuracy, and on the fame piece of paper would conftruct very neatly the figure to which it referred. To thefe he trufted for recalling to his mind, at any future period, the demonftration, or the analyfis, however complicated it m ght be. Experience had taught him that he might place this confidence in himfelf without any danger of difappointment; and for this fingular power, he was probably more indebted to the activity of his invention, than to the mere tenacioufnefs of his memory.

Though Dr. Stewart was extremely ftudious, he read but few books, and thus verified the obfervation of D'Alembert, that, of all the men of letters, mathematicians read leaft of the writings of one another. Our author's own inveftigations occupied him fufficiently; and indeed the world would have had reafon to regret the mifapplication of his talents, had he employed, in the mere acquifition of knowledge, that time which he could dedicate to works of invention.

It was Dr. Stewart's cuftom to fpend the fummer at

a delightful retreat in Ayrfhire, where, after the academical labours of the winter were ended, he found the leifure neceffary for the profecution of his refearches. In his way thither, he often made a vifit to Dr. Simfon of Glafgow, with whom he had lived from his youth in the moft cordial and uninterrupted friendfhip. It was pleafing to obferve, in thefe two excellent mathematicians, the moft perfect efteem and affection for each other, and the moft entire abfence of jealoufy, though no two men ever trode more nearly in the fame path. The fimilitude of their purfuits ferved only to endear them to each other, as it will ever do with men fuperior to envy. Their fentiments and views of the fcience they cultivated, were nearly the fame: they were both profound geometricians; they equally admired the ancient mathematicians, and were equally verfed in their methods of inveftigation; and they were both apprehenfive that the beauty of their favourite fcience would be forgotten, for the lefs elegant methods of algebraic computation. This innovation they endeavoured to oppofe; the one, by reviving thofe books of the ancient geometry which were loft; the other, by extending that geometry to the moft difficult enquiries of the moderns. Dr. Stewart, in particular, had remarked the intricacies, in which many of the greateft of the modern mathematicians had involved themfelves in the application of the calculus, which a little attention to the ancient geometry would certainly have enabled them to avoid. He had obferved too the elegant fynthetical demonftrations that, on many occafions, may be given of the moft difficult propofitions, inveftigated by the inverfe method of fluxions. Thefe circumftances had perhaps made a ftronger impreffion than they ought, on a mind already filled with admiration of the ancient geometry, and produced too unfavourable an opinion of the modern analyfis. But if it be confeffed that Dr. Stewart rated, in any refpect too high, the merit of the former of thefe fciences, this may well be excufed in the man whom it had conducted to the difcovery of the *General Theorems*, to the *folution of Kepler's Problem*, and to an *accurate* determination of the *Sun's difturbing force*. His great modefty made him afcribe to the method, he ufed, that fuccefs which he owed to his own abilities.

The foregoing account of Dr. Stewart and his writings, is chiefly extracted from the learned hiftory of them, by Mr. Playfair, in the 1ft volume of the Edinburgh Philofophical Tranfactions, pa. 57, &c.

STIFELS, STIFELIUS (MICHAEL), a Proteftant minifter, and very fkilful mathematician, was born at Eflingen a town in Germany; and died at Jena In Thuringia, in the year 1567, at 58 years of age according to Voffius, but fome others fay 80. Stifels was one of the beft mathematicians of his time. He publifhed, in the German language, a treatife on Algebra, and another on the Calendar or Ecclefiaftical computation. But his chief work, is the *Arithmetica Integra*, a complete and excellent treatife, in Latin, on Arithmetic and Algebra, printed in 4to at Norimberg 1544. In this work there are a number of ingenious inventions, both in common arithmetic and in algebra; of which, thofe relating to the latter are amply explained under the article *Algebra* in this dictionary, vol. 1, pa. 77 &c; to which may be added fome particulars

ticulars concerning the arithmetic, from my volume of *Tracts* printed in 1786, pa. 68. In this original work are contained many curious things, some of which have mistakenly been ascribed to a much later date. He here treats pretty fully and ably, of progressional and figurate numbers, and in particular of the triangular table, for constructing both them and the coefficients of the terms of all powers of a binomial; which has been so often used since his time for these and other purposes, and which more than a century after was, by Pascal, otherwise called the Arithmetical Triangle, and who only mentioned some additional properties of the table. Stifels shews, that the horizontal lines of the table furnish the coefficients of the terms of the corresponding powers of a binomial; and teaches how to make use of them in the extraction of roots of all powers whatever. Cardan seems to ascribe the invention of that table to Stifelius; but I apprehend that is only to be understood of its application to the extraction of roots.

It is remarkable too, how our author, at p. 35 &c of the same book, treats of the nature and use of logarithms; not under that name indeed, but under the idea of a series of arithmeticals, adapted to a series of geometricals. He there explains all their uses; such as, that the addition of them answers to the multiplication of their geometricals; subtraction to division; multiplication of exponents, to involution; and dividing of exponents to evolution. He also exemplifies the use of them in cases of the Rule-of-three, and in finding mean proportionals between given terms, and such like, exactly as is done in logarithms. So that he seems to have been in the full possession of the idea of logarithms, and wanted only the necessity of troublesome calculations to induce him to make a table of such numbers.

Stifels was a zealous, though weak disciple of Luther. He took it into his head to become a prophet, and he predicted that the end of the world would happen on a certain day in the year 1553, by which he terrified many people. When the proposed day arrived, he repaired early, with multitudes of his followers, to a particular place in the open air, spending the whole day in the most fervent prayers and praises, in vain looking for the coming of the Lord, and the universal conflagration of the elements, &c.

STILE. See STYLE.

STILYARD. See STEELYARD.

STOFLER (JOHN), a German mathematician, was born at Justingen in Suabia, in 1452, and died in 1531, at 79 years of age. He taught mathematics at Tubinga, where he acquired a great reputation, which however he in a great measure lost again, by intermeddling with the prediction of future events. He announced a great deluge, which he said would happen in the year 1524, a prediction with which he terrified all Germany, where many persons prepared vessels proper to escape with from the floods. But happily the prediction failing, it enraged the astrologer, though it served to convince him of the vanity of his prognostications.—He was author of several works in mathematics, and astrology, full of foolish and chimerical ideas; such as,

1. Elucidatio Fabric. Ususque Astrolabii; fol. 1513.
2. Procli Sphæram Comment. fol. 154.

3. Cosmographicæ aliquot Descriptiones; 4to, 1537.

STONE, (EDMUND), a good Scotch mathematician, who was author of several ingenious works. I know not the particular place or date of his birth, but it was probably in the shire of Argyle, and about the beginning of the present century, or conclusion of the last. Nor have we any memoirs of his life, except a letter from the Chevalier de Ramsay, author of the Travels of Cyrus, in a letter to father Castel, a Jesuit at Paris, and published in the Memoires de Trevoux, p. 109, as follows : " True genius overcomes all the disadvantages of birth, fortune, and education; of which Mr. Stone is a rare example. Born a son of a gardener of the duke of Argyle, he arrived at 8 years of age before he learnt to read.—By chance a servant having taught young Stone the letters of the alphabet, there needed nothing more to discover and expand his genius. He applied himself to study, and he arrived at the knowledge of the most sublime geometry and analysis, without a master, without a conductor, without any other guide but pure genius."

" At 18 years of age he had made these considerable advances without being known, and without knowing himself the prodigies of his acquisitions. The duke of Argyle, who joined to his military talents, a general knowledge of every science that adorns the mind of a man of his rank, walking one day in his garden, saw lying on the grass a Latin copy of Sir Isaac Newton's celebrated Principia. He called some one to him to take and carry it back to his library. Our young gardener told him that the book belonged to him. *To you?* replied the Duke. *Do you understand geometry, Latin, Newton?* I know a little of them, replied the young man with an air of simplicity arising from a profound ignorance of his own knowledge and talents. The Duke was surprised; and having a taste for the sciences, he entered into conversation with the young mathematician : he asked him several questions, and was astonished at the force, the accuracy, and the candour of his answers. *But how*, said the Duke, *came you by the knowledge of all these things ?* Stone replied, *A servant taught me, ten years since, to read : does one need to know any thing more than the 24 letters in order to learn every thing else that one wishes ?* The Duke's curiosity redoubled—he sat down upon a bank, and requested a detail of all his proceedings in becoming so learned."

" *I first learned to read*, said Stone: *the masons were then at work upon your house : I went near them one day, and I saw that the architect used a rule, compasses, and that he made calculations. I enquired what might be the meaning of and use of these things ; and I was informed that there was a science called Arithmetic ; I purchased a book of arithmetic, and I learned it.—I was told there was another science called Geometry : I bought the books, and I learnt geometry. By reading I found that there were good books in these two sciences in Latin : I bought a dictionary, and I learned Latin. I understood also that there were good books of the same kind in French : I bought a dictionary, and I learned French. And this, my lord, is what I have done : it seems to me that we may learn every thing when we know the 24 letters of the alphabet.*

This account charmed the Duke. He drew this wonderful genius out of his obscurity; and he provided him with an employment which left him plenty

of

of time to apply himself to the sciences. He discovered in him also the same genius for music, for painting, for architecture, for all the sciences which depend on calculations and proportions."

"I have seen Mr. Stone. He is a man of great simplicity. He is at present sensible of his own knowledge: but he is not puffed up with it. He is possessed with a pure and disinterested love for the mathematics; though he is not solicitous to pass for a mathematician; vanity having no part in the great labour he sustains to excel in that science. He despises fortune also: and he has solicited me twenty times to request the duke to give him less employment, which may not be worth the half of that he now has, in order to be more retired, and less taken off from his favourite studies. He discovers sometimes, by methods of his own, truths which others have discovered before him. He is charmed to find on these occasions that he is not a first inventor, and that others have made a greater progress than he thought. Far from being a plagiary, he attributes ingenious solutions, which he gives to certain problems, to the hints he has found in others, although the conrection is but very distant," &c.

Mr. Stone was author and translator of several useful works; viz.

1. A New Mathematical Dictionary, in 1 vol. 8vo, first printed in 1726.

2. Fluxions, in 1 vol. 8vo, 1730. The Direct Method is a translation from the French, of Hospital's Analyse des Infiniments Petits; and the Inverse Method was supplied by Stone himself.

3. The Elements of Euclid, in 2 vols. 8vo, 1731. A neat and useful edition of those Elements, with an account of the life and writings of Euclid, and a defence of his elements against modern objectors.

Beside other smaller works.

Stone was a fellow of the Royal Society, and had inserted in the Philos. Transactions (vol. 41, pa. 218) an "Account of two species of lines of the 3d order, not mentioned by Sir Isaac Newton, or Mr. Stirling."

STRABO, a celebrated Greek geographer, philosopher, and historian, was born at Amasia, and was descended from a family settled at Gnossus in Crete. He was the disciple of Xenarchus, a Peripatetic philosopher, but at length attached himself to the Stoics. He contracted a strict friendship with Cornelius Gallus, governor of Egypt; and travelled into several countries, to observe the situation of places, and the customs of nations.

Strabo flourished under Augustus; and died under Tiberius about the year 25, in a very advanced age.— He composed several works; all of which are lost, except his *Geography*, in 17 books; which are justly esteemed very precious remains of antiquity. The first two books are employed in showing, that the study of geography is not only worthy of a philosopher, but even necessary to him; the 3d describes Spain; the 4th, Gaul and the Britannic isles; the 5th and 6th, Italy and the adjacent isles; the 7th, which is imperfect at the end, Germany, the countries of the Getæ and Illyrii, Taurica, Chersonesus, and Epirus; the 8th, 9th, and 10th, Greece with the neighbouring isles; the four following, Asia within Mount Taurus; the 15th and 16th, Asia without Taurus, India, Persia,

Syria, Arabia; and the 17th, Egypt, Ethiopia, Carthage, and other parts of Africa.

Strabo's work was published with a Latin version by Xylander, and notes by Isaac Casaubon, at Paris 1620, in folio; but the best edition is that of Amsterdam in 1707, in 2 volumes folio, by the learned Theodore Janson of Almelooveen, with the entire notes of Xylander, Casaubon, Meursius, Cluver, Holsten, Salmasius, Bochart, Ez. Spanheim, Cellar, and others. To this edition is subjoined the *Chrestomathiæ*, or Epitome of Strabo; which, according to Mr. Dodswell, who has written a very elaborate and learned dissertation about it, was made by some unknown person, between the years of Christ 676 and 996. It has been found of some use, not only in helping to correct the original, but in supplying in some measure the defect in the 7th book. Mr. Dodswell's dissertation is prefixed to this edition.

STRAIT, or STRAIGHT, or STREIGHT, in Hydrography, is a narrow channel or arm of the sea, shut up between lands on either side, and usually affording a passage out of one great sea into another. As the Straits of Magellan, of Le Maire, of Gibraltar, &c.

STRAIT is also sometimes used, in Geography, for an isthmus, or neck of land between two seas, preventing their communication.

STRENGTH, *vis*, force, power.

Some authors make the Strength of animals, of the same kind, to depend on the quantity of blood; but most on the size of the bones, joints, and muscles; though we find by daily experience, that the animal spirits contribute greatly to Strength at different times.

Emerson has most particularly treated of the Strength of bodies depending on their dimensions and weight. In the General Scholium after his propositions on this subject, he adds; If a certain beam of timber be able to support a given weight; another beam, of the same timber, similar to the former, may be taken so great, as to be able but just to bear its own weight: while any larger beam cannot support itself, but must break by its own weight; but any less beam will bear something more. For the Strength being as the cube of the depth; and the stress, being as the length and quantity of matter, is as the 4th power of the depth; it is plain therefore, that the stress increases in a greater ratio than the Strength. Whence it follows, that a beam may be taken so large, that the stress may far exceed the Strength: and that, of all similar beams, there is but one that will just support itself, and nothing more. And the like holds true in all machines, and in all animal bodies. And hence there is a certain limit, in regard to magnitude, not only in all machines and artificial structures, but also in natural ones, which neither art nor nature can go beyond; supposing them made of the same matter, and in the same proportion of parts.

Hence it is impossible that mechanic engines can be increased to any magnitude at pleasure. For when they arrive at a particular size, their several parts will break and fall asunder by their weight. Neither can any buildings of vast magnitudes be made to stand, but must fall to pieces by their great weight, and go to ruin.

It

It is likewife impoffible for nature to produce animals of any vaft fize at pleafure : except fome fort of matter can be found, to make the bones of, which may be fo much harder and ftronger than any hitherto known : or elfe that the proportion of the parts be fo much altered, and the bones and mufcles made thicker in proportion ; which will make the animal diftorted, and of a monftrous figure, and not capable of performing any proper actions. And being made fimilar and of common matter, they will not be able to ftand or move : but, being burthened with their own weight, muft fall down. Thus, it is impoffible that there can be any animal fo large as to carry a caftle upon his back ; or any man fo ftrong as to remove a mountain, or pull up a large oak by the roots : nature will not admit of thefe things ; and it is impoffible that there can be animals of any fort beyond a determinate fize.

Fifh may indeed be produced to a larger fize than land animals ; becaufe their weight is fupported by the water. But yet even thefe cannot be increafed to immenfity, becaufe the internal parts will prefs upon one another by their weight, and deftroy their fabric.

On the contrary, when the fize of animals is diminifhed, their Strength is not diminifhed in the fame proportion as the weight. For which reafon a fmall animal will carry far more than a weight equal to its own, whilft a great one cannot carry fo much as its weight. And hence it is that fmall animals are more active, will run fafter, jump farther, or perform any motion quicker, for their weight, than large animals : for the lefs the animal, the greater the proportion of the Strength to the ftrefs. And nature feems to know no bounds as to the fmallnefs of animals, at leaft in regard to their weight.

Neither can any two unequal and fimilar machines refift any violence alike, or in the fame proportion ; but the greater will be more hurt than the lefs. And the fame is true of animals ; for large animals by falling break their bones, while leffer ones, falling higher, receive no damage. Thus a cat may fall two or three yards high, and be no worfe, and an ant from the top of a tower.

It is likewife impoffible in the nature of things, that there can be any trees of immenfe fize ; if there were any fuch, their limbs, boughs, and branches, muft break off and fall down by their own weight. Thus it is impoffible there can be an oak a quarter of a mile high ; fuch a tree cannot grow or ftand, but its limbs will drop off by their weight. And hence alfo fmaller plants can better fuftain themfelves than large ones.

As to the due proportion of Strength in feveral bodies, according to their particular pofitions, and the weights they are to bear ; he farther obferves that, If a piece of timber is to be pierced with a mortife-hole, the beam will be ftronger when it is taken out of the middle, than when taken out of either fide. And in a beam fupported at both ends, it is ftronger when the hole is made in the upper fide than when made in the under, provided a piece of wood is driven hard in to fill up the hole.

If a piece is to be fpliced upon the end of a beam, to be fupported at both ends ; it will be the ftronger

5

when fpliced on the under fide of a beam : but if the piece is fupported only at one end, to bear a weight on the other ; it is ftronger when fpliced on the upper fide.

When a fmall lever, &c, is nailed to a body, to move it or fufpend it by ; the ftrain is greater upon the nail neareft the hand, or point where the power is applied.

If a beam be fupported at both ends ; and the two ends reach over the props, and be fixed down immoveable ; it will bear twice as much weight, as when the ends only lie loofe or free upon the fupporters.

When a flender cylinder is to be fupported by two pieces ; the diftance of the pins ought to be nearly ⅘ of the length of the cylinder, and the pins equidiftant from its ends ; and then the cylinder will endure the leaft bending or ftrain by its weight.

A beam fixed at one end, and bearing a weight at the other ; if it be cut in the form of a wedge, and placed with its parallel fides parallel to the horizon ; it will be equally ftrong every where ; and no fooner break in one place than another.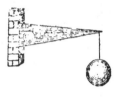

When a beam has all its fides cut in form of a concave parabola, having the vertex at the end, and its abfcifs perpendicular to the axis of the folid, and the bafe a fquare, or a circle, or any regular polygon ; fuch a beam 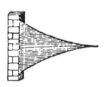 fixed horizontally, at one end, is equally ftrong throughout for fupporting its own weight.

Alfo when a wall faces the wind, and if the vertical fection of it be a right-angled triangle ; or if the fore part next the wind &c be perpendicular to the horizon, and the back part a floping plane ; fuch a wall will be equally ftrong in all its parts to refift the wind ; if the parts of the wall cohere ftrongly together ; but when it is built of loofe materials, it is better to be convex on the back part in form of a parabola.

When a wall is to fupport a bank of earth or any fluid body, it ought to be built concave in form of a femicubical parabola, whofe vertex is at the top of the wall, provided the parts of the wall adhere firmly together. But if the parts be loofe, then a right line or floping plane ought to be its figure. Such walls will be equally ftrong throughout.

All fpires of churches in the form of cones or pyramids, are equally ftrong in all parts to refift the wind. But when the parts do not cohere together, then they ought to be parabolic conoids, to be equally ftrong throughout.

Likewife if there be a pillar erected in form of the logarithmic curve, the afymptote being the axis ; it cannot be crufhed to pieces in one part fooner than in another, by its own weight. And if fuch a pillar be turned upfide down, and fufpended by the thick end, it will not be more liable to feparate in one part than another, by its own weight.

Moreover,

S T R   [533]   S T R

Moreover, if AE be a beam in form of a triangular prifm; and if AD = $\frac{1}{9}$AB, and AI = $\frac{1}{5}$ AC, and the edge or fmall fimilar prifm ADIF be cut away parallel to the bafe; the remaining beam DIBEF will bear a greater weight P, than the whole

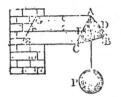

ABCEG, or the part will be ftronger than the whole; which is a paradox in Mechanics.

As to the Strength of feveral forts of wood, drawn from experiments, he fays, On a medium, a piece of good oak, an inch fquare, and a yard long, fupported at both ends, will bear in the middle, for a very fhort time, about 330lb averdupois, but will break with more than that weight. But fuch a piece of wood fhould not, in, practice, be trufted for any length of time, with more than a third or a fourth part of that weight. And the proportion of the Strength of feveral forts of wood, he found to be as follows:

Box, oak, plumbtree, yew	11
Afh, elm	8$\frac{1}{2}$
Thorn, walnut	7$\frac{1}{2}$
Apple tree, elder, red fir, holly, plane	7
Beech, cherry, hazle	6$\frac{2}{3}$
Alder, afp, Birch, white-fir, willow	6
Iron	107
Brafs	50
Bone	22
Lead	6$\frac{1}{2}$
Fine free ftone	1

As to the Strength of bodies in direction of the fibres, he obferves, A cylindric rod of good clean fir, of an inch circumference, drawn in length, will bear at extremity 400lb; and a fpear of fir 2 inches diameter, will bear about 7 ton.——A rod of good iron, of an inch circumference, will bear near 3 ton weight. And a good hempen rope of an inch circumference, will bear 1000lb. at extremity.

All this fuppofes thefe bodies to be found and good throughout; but none of them fhould be put to bear more than a third or a fourth part of that weight, especially for any length of time. From what has been faid; if a fpear of fir, or a rope, or a fpear of iron, of d inches diameter, were to lift $\frac{1}{4}$ the extreme weight; then

The fir would bear 8$\frac{2}{3}dd$ hundred weight.
The rope would bear 22dd hundred weight.
The iron would bear 6$\frac{3}{4}dd$ ton weight.

As to Animals; Men may apply their Strength feveral ways, in working a machine. A man of ordinary Strength turning a roller by the handle, can act for a whole day againft a refiftance equal to 30lb. weight; and if he works 10 hours a day; he will raife a weight of 30lb. through 3$\frac{1}{2}$ feet in a fecond of time; or if the weight be greater, he will raife it fo much lefs in proportion. But a man may act, for a fmall time, againft a refiftance of 50lb. or more.

If two men work at a windlafs, or roller, they can more eafily draw up 70lb, than one man can 30lb, provided the elbow of one of the handles be at right angles

to that of the other. And with a fly, or heavy wheel, applied to it, a man may do $\frac{1}{3}$ part more work; and for a little while he can act with a force, or overcome a continual refiftance, of 80l; and work a whole day when the refiftance is but 40lb.

Men ufed to bear loads, fuch as porters, will carry, fome 150lb, others 200 or 250lb. according to their Strength.

A man can draw but about 70 or 80lb. horizontally; for he can but apply about half his weight.

If the weight of a man be 140lb, he can act with no greater a force in thrufting horizontally, at the height of his fhoulders, than 27lb.

As to Horfes: A horfe is, generally fpeaking, as ftrong as 5 men. A horfe will carry 240 or 270lb. A horfe draws to greateft advantage, when the line of direction is a little elevated above the horizon, and the power acts againft his breaft: and he can draw 200lb. for 8 hours a day, at 2$\frac{1}{2}$ miles an hour. If he draw 240lb, he can work but 6 hours, and not go quite fo faft. And in both cafes, if he carries fome weight, he will draw the better for it. And this is the weight a horfe is fuppofed to be able to draw over a pulley out of a well. But in a cart, a horfe may draw 1000lb, or even double that weight, or a ton weight, or more.

As the moft force a horfe can exert, is when he draws a little above the horizontal pofition: fo the worft way of applying the ftrength of a horfe, is to make him carry or draw uphill: And three men in a fteep hill, carrying each 100lb, will climb up fafter than a horfe with 300l. Alfo, though a horfe may draw in a round walk of 18 feet diameter; yet fuch a walk fhould not be lefs than 25 or 30 feet diameter. Emerfon's Mechan. pa. 111 and 177.

STRIKE, or STRYKE, a meafure containing 4 bufhels or half a quarter.

STRIKING-*wheel*, in a clock, the fame as that by fome called the *pin-wheel*, becaufe of the pins which are placed on the round or rim, the number of which is the quotient of the pinion divided by the pinion of the detent-wheel. In fixteen-day clocks, the firft or great wheel is ufually the pin-wheel; but in fuch as go 8 days, the fecond wheel is the pin-wheel, or ftriking-wheel.

STRING, in Mufic. See CHORD.

If two Strings or chords of a mufical inftrument only differ in length; their tones, or the number of vibrations they make in the fame time, are in the inverfe ratio of their lengths. If they differ only in thicknefs, their tones are in the inverfe ratio of their diameters

As to the tenfion of Strings, to meafure it regularly, they muft be conceived ftretched or drawn by weights; and then, cæteris paribus, the tones of two Strings are in a direct ratio of the fquare roots of the weights that ftretch them; that is, ex. gr. the tone of a String ftretched by a weight 4, is an octave above the tone of a String ftretched by the weight 1.

It is an obfervation of very old ftanding, that if a viol or lute-ftring be touched with the bow, or the hand, another String on the fame inftrument, or even on another, not far from it, if in unifon with it, or in octave, or the like, will at the fame time tremble of

ii3

its own accord. But it is now found, that it is not the whole of that other String that thus trembles, but only the parts, feverally, according as they are unifons to the whole, or the parts, of the String fo ftruck. Thus, fuppofing AB to be an upper octave to *ab*, and therefore an unifon to each half of it, ftopped at *c*; if while *ab* is open, AB be

A————————B
a————————b
      d   c   e

ftruck, the two halves of this other, that is, *ac*, and *cb*, will both tremble; but the middle point will be at reft; as will be eafily perceived, by wrapping a bit of paper lightly about the ftring *ab*, and moving it fucceffively from one end of the ftring to the other. In like manner, if AB were an upper 12th to *ab*, and confequently an unifon to its three parts *ad, de, eb*; then, *ab* being open, if AB be ftruck, the three parts of the other, *ad, de, eb* will feverally tremble; but the points *d* and *e* remain at reft.

This, Dr. Wallis tells us, was firft difcovered by Mr. William Noble of Merton college; and after him by Mr. T. Pigot of Wadham college, without knowing that Mr. Noble had obferved it before. To which may be added, that M. Sauveur, long afterwards, propofed it to the Royal Academy at Paris, as his own difcovery, which in reality it might be; but upon his being informed, by fome of the members then prefent, that Dr. Wallis had publifhed it before, he immediately refigned all the honour of it. Philof. Tranf. Abridg. vol. I, pa. 606.

STURM, STURMIUS (JOHN CHRISTOPHER), a noted German mathematician and philofpher, was born at Hippolftein in 1635. He became profeffor of philofophy and mathematics at Altdorf, where he died in 1703, at 68 years of age.

He was author of feveral ufeful works on mathematics and philofophy, the moft efteemed of which are,

1. His *Mathefis enucleata*, in 1 vol. 8vo.

2. *Mathefis Juvenilis*, in 2 large volumes 8vo.

3. *Collegium Experimentale, five Curiofum, in quo primaria Seculi fuperioris Inventa & Experimenta Phyfico-Mathematica, Speciatim Campanæ Urinatoriæ, Cameræ obfcuræ, Tubi Torricelliani, feu Barofcopii, Antliæ Pneumaticæ, Thermometrorum Phænomena & Effecta; partim ac aliis jampridem exhibita, partim noviter iftis fuperaddita, &c.* in one large vol. 4to, Norimberg, 1701.

This is a very curious work, containing a multitude of interefting experiments, neatly illuftrated by copperplate figures printed upon almoft every page, by the fide of the letter-prefs. Of thefe, the 10th experiment is an improvement on father Lana's project for navigating a fmall veffel fufpended in the atmofphere by feveral globes exhaufted of air.

STYLE, in Chronology, a particular manner of counting time; as the *Old Style*, the *New Style*. See CALENDAR.

Old STYLE, is the Julian manner of computing, as inftituted by Julius Cæfar, in which the mean year confifts of 365¼ days.

New STYLE, is the Gregorian manner of computation, inftituted by pope Gregory the 13th, in the year 1582, and is ufed by moft catholic countries, and many other ftates of Europe.

The Gregorian, or new Style, agrees with the true folar year, which contains only 365 days 5 hours 49 minutes. In the year of Chrift 200, there was no difference of Styles. In the year 1582, when the new Style was firft introduced, there was a difference of 10 days. At prefent there is 11 days difference, and accordingly at the diet of Ratifbon, in the year 1700, it was decreed by the body of proteftants of the empire, that 11 days fhould be retrenched from the old Style, to accommodate it for the future to the new. And the fame regulation has fince paffed into Sweden, Denmark, and into England, where it was eftablifhed in the year 1752, when it was enacted, that in all dominions belonging to the crown of Great Britain, the fupputation, according to which the year of our lord begins on the 25th day of March, fhall not be ufed from and after the laft day of December 1751; and that from thenceforth, the 1ft day of January every year fhall be reckoned to be the firft day of the year: and that the natural day next immediately following the 2d day of September 1752, fhall be accounted the 14th day of September, omitting the 11 intermediate nominal days of the common calendar. It is farther enacted, that all kinds of writings, &c, fhall bear date according to the new method of computation, and that all courts and meetings &c, feafts, fafts, &c, fhall be held and obferved accordingly. And for preferving the calendar in the fame regular courfe for the future, it is enacted, that the feveral years of our lord 1800, 1900, 2100, 2200, 2300, &c, except only every 400th year, of which the year 2000 fhall be the firft, fhall be common years of 365 days, and that the years 2000, 2400, 2800, &c, and every other 400th year from the year 2000 inclufive, fhall be leap years, confifting of 366 days. See BISSEXTILE and CALENDAR.

The following table fhews by what number of days the new ftyle differs from the old, from 5900 years before the birth of Chrift, to 5900 years after it. The days under the fign — (viz from 6000 years before to 200 years after Chrift) are to be fubtracted from the old Style, to reduce it to the new; and the days under the fign + (viz from 200 to 5900 years after Chrift) are to be added to the old Style, to reduce it to the new.—N.B. All the years mentioned in the table are leap years in the old Style; but thofe only that are marked with an L are leap years in the new.

Years before Christ. New Style.	Days diff. —	Years after Christ. New Style.	Days diff. ∓
5900	46	L 0	−2
5800	45	100	−1
5700	44	200	0
L 5600	44	300	+1
5500	43	L 400	1
5400	42	500	2
5300	41	600	3
L 5200	41	700	4
5100	40	L 800	4
5000	39	900	5
4900	38	1000	6
L 4800	38	1100	7
4700	37	L 1200	7
4600	36	1300	8
4500	35	1400	9
L 4400	35	1500	10
4300	34	L 1600	10
4200	33	1700	11
4100	32	1800	12
L 4000	32	1900	13
3900	31	L 2000	13
3800	30	2100	14
3700	29	2200	15
L 3600	29	2300	16
3500	28	L 2400	16
3400	27	2500	17
3300	26	2600	18
L 3200	26	2700	19
3100	25	L 2800	19
3000	24	2900	20
2900	23	3000	21
L 2800	23	3100	22
2700	22	L 3200	22
2600	21	3300	23
2500	20	3400	24
L 2400	20	3500	25
2300	19	L 3600	25
2200	18	3700	26
2100	17	3800	27
L 2000	17	3900	28
1900	16	L 4000	28
1800	15	4100	29
1700	14	4200	30
L 1600	14	4300	31
1500	13	L 4400	31
1400	12	4500	32
1300	11	4600	33
L 1200	11	4700	34
1100	10	L 4800	34
1000	9	4900	35
900	8	5000	36
L 800	8	5100	37
700	7	L 5200	37
600	6	5300	38
500	5	5400	39
L 400	5	5500	40
300	4	L 5600	40
200	3	5700	41
100	2	5800	42
L 0	2	5900	43

The French nation has lately commenced another new Style, or computation of time, viz, in the year 1792; according to which, the year commences usually on our 22d of September. The year is divided into 12 months of 30 days each; and each month into 3 decades of 10 days each. For the names and computations of which, see the article CALENDAR.

STYLE, in Dialling, denotes the cock or gnomon, raised above the plane of the dial, to project a Shadow. —The edge of the Style, which by its shadow marks the hours on the face of the dial, is to be set according to the latitude, always parallel to the axis of the world.

STYLOBATA, or STYLOBATON, in Architecture, the same with the pedestal of a column. It is sometimes taken for the trunk of the pedestal, between the cornice and the base, and is then called *truncus*. It is also otherwise named *abacus*.

SUBCONTRARY position, in Geometry, is when two equiangular triangles, as VAB and VCD are so placed as to have one common angle V at the vertex, and yet their bases not parallel. Consequently the angles at the bases are equal, but on the contrary sides; viz, the ∠A = ∠C, and the ∠B = ∠D.

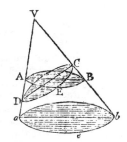

If the oblique cone VAB or V*ab*, having the circular base AEB, or *aeb*, be so cut by a plane DEC, that the angle D be = the ∠B, or the ∠C = ∠A, then the cone is said to be cut, by this plane, in a Subcontrary position to the base AEB, or *aeb*; and in this case the section DEC is always a circle, as well as the base AEB or *aeb*.

SUBDUCTION, in Arithmetic, the same as Subtraction.

SUBDUPLE *Ratio*, is when any number or quantity is the half of another, or contained twice in it. Thus, 3 is said to be subduple of 6, as 3 is the half of 6, or is twice contained in it.

SUBDUPLICATE *Ratio*, of any two quantities, is the ratio of their square roots, being the opposite to duplicate ratio, which is the ratio of the squares. Thus, of the quantities, a and b, the subduplicate ratio is that of $\sqrt{a}$ to $\sqrt{b}$ or $a^{\frac{1}{2}}$ to $b^{\frac{1}{2}}$, as the duplicate ratio is that of a^2 to b^2.

SUBLIME *Geometry*, the higher geometry, or that of curve lines. See GEOMETRY.

SUBLUNARY, is said of all things below the moon; as all things on the earth, or in its atmosphere, &c.

SUBMULTIPLE, the contrary of a multiple, being a number or quantity which is contained exactly a certain number of times in another of the same kind; or it is the same as an aliquot part of it. Thus, 3 is a Submultiple of 21, or an aliquot part of it, because 21 is a multiple of 3.

SUBMULTIPLE *Ratio*, is the ratio of a Submultiple or aliquot part, to its multiple; as the ratio of 3 to 21.

SUBNORMAL, in Geometry, is the subperpendicular AC, or line under the perpendicular to the curve BC, a term used in curve lines to denote the distance AC in the axis, between the ordinate AB, and the perpendicular

pendicular BC to the curve or to the tangent. And the said perpendicular BC is the normal.

In all curves, the Subnormal AC is a 3d proportional to the subtangent TA and the ordinate AB; and in the parabola, it is equal to half the parameter of the axis.

SUBSTITUTION, in Algebra, is the putting and using, in an equation, one quantity instead of another which is equal to it, but expressed after another manner. See REDUCTION of Equations.

SUBSTRACTION, or SUBTRACTION, in Arithmetic, is the taking of one number or quantity from another, to find the remainder or difference between them; and is usually made the second rule in arithmetic.

The greater number or quantity is called the *minuend*, the less is the *subtrahend*, and the *remainder* is the *difference*. Also the sign of Subtraction is —, or minus.

SUBTRACTION of *Whole Numbers*, is performed by setting the less number below the greater, as in addition, units under units, tens under tens, &c; and then, proceeding from the right hand towards the left, subtract or take each lower figure from that just above, and set down the several remainders or differences underneath; and these will compose the whole remainder or difference of the two given numbers. But when any one of the figures of the under number is greater than that of the upper; from which it is to be taken, you must add 10 (in your mind) to that upper figure, then take the under one from this sum, and set the difference underneath, carrying or adding 1 to the next under figure to be subtracted. Thus, for example, to subtract 2904821 from 37409732

Minuend	37409732
Subtrahend	2904821
Difference	34504911
Proof	37409732

To prove Subtraction: Add the remainder or difference to the less number, and the sum will be equal to the greater when the work is right.

SUBTRACTION *of Decimals*, is performed in the same manner as in whole numbers, by observing only to set the figures or places of the same kind under each other. Thus:

	From	351·04	·479	27
	Take	72·71	·0573	0·936
	Diff.	278·33	·4217	26·064

To Subtract Vulgar Fractions. Reduce the two fractions to a common denominator, if they have different ones; then take the less numerator from the greater, and set the remainder over the common denominator, for the difference sought.—N. B. It is best to set the less fraction after the greater, with the sign (—) of subtraction between them, and the mark of equality (=) after them.

Thus, $\frac{7}{9} - \frac{2}{9} = \frac{5}{9}$.

And $\frac{3}{5} - \frac{4}{7} = \frac{21}{35} - \frac{20}{35} = \frac{1}{35}$.

SUBTRACTION, *in Algebra*, is performed by changing the signs of all the terms of the subtrahend, to their contrary signs, viz, + into —, and — into +; and then uniting the terms with those of the minuend after the manner of addition of Algebra.

Ex. From	$+ 6a$			
Take	$+ 2a$			
Rem.	$6a - 2a = 4a.$			

From	$+ 6a$			
Take	$- 2a$			
Rem.	$6a + 2a = 8a.$			

From	$- 6a$			
Take	$+ 2a$			
Rem.	$- 6a - 2a = - 8a.$			

From	$- 6a$			
Take	$- 4a$			
Rem.	$- 6a + 4a = - 2a.$			

From	$2a$	$- 3$	$+ 5z$	$- 6$
Take	$6a$	$+ 4x$	$+ 5z$	$+ 4$
Rem.	$-4a$	$- 7x$	0	$- 10$

SUBSTILE, or SUBSTILAR *Line*, in Dialling, a right line upon which the stile or gnomon of a dial is erected, being the common section of the face of the dial and a plane perpendicular to it passing through the stile.

The angle included between this line and the stile, is called the elevation or height of the stile.

In polar, horizontal, meridional, and northern dials, the Substilar line is the meridional line, or line of 12 o'clock; or the intersection of the plane of the dial with that of the meridian.—In all declining dials, the Substile makes an angle with the hour line of 12, and this angle is called the distance of the Substile from the meridian.—In easterly and westerly dials, the substilar line is the line of 6 o'clock, or the intersection of the dial plane with the prime vertical.

SUBSUPERPARTICULAR. } See RATIO.
SUBSUPERPARTICUS. }

SUBTANGENT *of a curve*, is the line TA in the axis below the tangent TB, or limited between the tangent and ordinate to the point of contact. (See the last figure above).

The tangent, subtangent, and ordinate, make a right-angled triangle.

In all paraboliform and hyperboliform figures, the Subtangent is equal to the abscifs multiplied by the exponent of the power of the ordinate in the equation of the curve. Thus, in the common parabola, whose property or equation is $px = y^2$, the Subtangent is equal to $2x$, double the abscifs. And if $ax^2 = y^3$, or

$px =$

$px = y^{\frac{3}{2}}$, then the Subtangent is $= \frac{2}{3}x$. Also if

$a^m x^n = y^{m+n}$, or $px = y^{\frac{m+x}{n}}$, the Subtangent is $= \frac{m+n}{n}x$. See *Method of* TANGENTS.

SUBTENSE, in Geometry, of an arc, is the same as the chord of the arc; but of an angle, it is a line drawn across from the one leg of the angle to the other, or between the two extremes of the arc that measures the angle.

SUBTRACTION. See SUBSTRACTION.

SUBTRIPLE, is when one quantity is the 3d part of another; as 2 is Subtriple of 6 And SUBTRIPLE *Ratio*, is the ratio of 1 to 3.

SUBTRIPLICATE *Ratio*, is the ratio of the cube roots. So the Subtriplicate ratio of *a* to *b* is the ratio of $\sqrt[3]{a}$ to $\sqrt[3]{b}$, or of $a^{\frac{1}{3}}$ to $b^{\frac{1}{3}}$.

SUCCESSION *of Signs*, in Astronomy, is the order in which they are reckoned, or follow one another, and according to which the sun enters them; called also *consequentia*. As Aries, Taurus, Gemini, Cancer, &c.

When a planet goes according to the order and succession of the signs, or *in consequentia*, it is said to be direct; but retrograde when contrary to the succession of the signs, or *in antecedentia*, as from Gemini to Taurus, then to Aries, &c.

SUCCULA, in Mechanics, a bare axis or cylinder with staves in it to move it round; but without any tympanum, or peritrochium.

SUCKER, in Mechanics, a name by which sometimes is called the piston or bucket, in a sucking pump; and sometimes the pump itself is so called.

SUCKING-*Pump*, the common pump, working by two valves opening upwards. See PUMP.

SUMMER, the name of one of the seasons of the year, being one of the quarters when the year is divided into 4 quarters, or one half when the year is divided only into two, Summer and winter. In the former case, Summer is the quarter during which, in northern climates, the sun is passing through the three signs Cancer, Leo, Virgo, or from the time of the greatest declination, till the sun come to the equinoctial again, or have no declination; which is from about the 21st of June, till about the 22d of September. In the latter case, Summer contains the 6 warmer months, while the sun is on one side of the equinoctial; and winter the other 6 months, when the sun is on the other side of it.

It is said, that a frosty winter produces a dry Summer; and a mild winter, a wet Summer. See Philos. Transf. no. 458, sect. 10.

SUMMER *Solstice*, the time or point when the sun comes to his greatest declination, and nearest the zenith of the place. See SOLSTICE.

SUM, the quantity produced by addition, or by adding two or more numbers or quantities together. So the Sum of 6 and 4 is 10, and the Sum of *a* and *b* is $a + b$.

SUN, SOL, ☉, in Astronomy, the great luminary
VOL. II.

which enlightens the world, and by his presence constitutes day.

The Sun, which was reckoned among the planets in the infancy of astronomy, should rather be counted among the fixed stars. He only appears brighter and larger than they do, because we keep constantly near the Sun; whereas we are immensely farther from the stars. But a spectator, placed as near to any star as we are to the Sun, would probably see that star a body as large and as bright as the Sun appears to us; and, on the other hand, a spectator as far distant from the Sun as we are from the stars, would see the Sun as small as we see a star, divested of all his circumvolving planets; and he would reckon it one of the stars in numbering them.

According to the Pythagorean and Copernican hypothesis, which is now generally received, and has been demonstrated to be the true system, the Sun is the common centre of all the planetary and cometary system; around which all the planets and comets, and our earth among the rest, revolve, in different periods, according to their different distances from the Sun.

But the Sun, though thus eased of that prodigious motion by which the Ancients imagined he revolved daily round our earth, yet is he not a perfectly quiescent body. For, from the phenomena of his maculæ or spots, it evidently appears, that he has a rotation round his axis, like that of the earth by which our natural day is measured, but only slower. For, some of these spots have made their first appearance near the edge or margin of the Sun, from thence they have seemed gradually to pass over the Sun's face to the opposite edge, then disappear; and hence, after an absence of about 14 days, they have reappeared in their first place, and have taken the same course over again; finishing their entire circuit in 27 days 12^h 20^m; which is hence inferred to be the period of the Sun's rotation round his axis: and therefore the periodical time of the Sun's revolution to a fixed star is 25^d 15^h 16^m; because in 27^d 12^h 20^m of the month of May, when the observations were made, the earth describes an angle about the Sun's centre of 26° $22'$, and therefore as the angular motion

$$360^\circ + 26^\circ 22' : 360^\circ :: 27^d 12^h 20^m : 25^d 15^h 16^m.$$

This motion of the spots is from west to east: whence we conclude the motion of the Sun, to which the other is owing, to be from east to west.

Beside this motion round his axis, the Sun, on account of the various attractions of the surrounding planets, is agitated by a small motion round the centre of gravity of the system.—Whether the Sun and stars have any proper motion of their own in the immensity of space, however small, is not absolutely certain. Though some very accurate observers have intimated conjectures of this kind, and have made such a general motion not improbable. See STARS.

As for the apparent annual motion of the SUN *round the earth*; it is easily shewn, by astronomers, that the real annual motion of the earth, about the Sun, will cause such an appearance. A spectator in the Sun would see the earth move from west to east, for the same reason as we see the Sun move from east to west: and all the phenomena resulting from this annual motion in whichsoever of the bodies it be, will appear the same

from

from either. And hence arifes that apparent motion of the Sun, by which he is feen to advance infenfibly towards the eaftern ftars; in fo much that, if any ftar, near the ecliptic, rife at any time with the Sun; after a few days the Sun will be got more to the eaft of the ftar, and the ftar will rife and fet before him.

Nature, Properties, Figure, &c, of the SUN.

Thofe who have maintained that the fubftance of the Sun is fire, argue in the following manner: The Sun fhines, and his rays, collected by concave mirrors, or convex lenfes, do burn, confume, and melt the moft folid bodies, or elfe convert them into afhes, or glafs: therefore, as the force of the folar rays is diminifhed, by their divergency, in a duplicate ratio of the diftances reciprocally taken; it is evident that their force and effect are the fame, when collected by a burning lens, or mirror, as if we were at fuch diftance from the fun, where they were equally denfe. The Sun's rays therefore, in the neighbourhood of the Sun, produce the fame effects, as might be expected from the moft vehement fire: confequently the Sun is of a fiery fubftance.

Hence it follows, that its furface is probably every where fluid; that being the condition of flame. Indeed, whether the whole body of the Sun be fluid, as fome think; or folid, as others; they do not prefume to determine: but as there are no other marks, by which to diftinguifh fire from other bodies, but light, heat, a power of burning, confuming, melting, calcining, and vitrifying; they do not fee what fhould hinder but that the Sun may be a globe of fire, like our fires, invefted with flame: and, fuppofing that the maculæ are formed out of the folar exhalations, they infer that the Sun is not pure fire; but that there are heterogeneous parts mixed along with it.

Philofophers have been much divided in opinion with refpect to the nature of fire, light, and heat, and the caufes that produce them: and they have given very different accounts of the agency of the Sun, with which, whether we confider them as fubftances or qualities, they are intimately connected, and on which they feem primarily to depend. Some, among whom we may reckon Sir Ifaac Newton, confider the rays of light as compofed of fmall particles, which are emitted from fhining bodies, and move with uniform velocities in uniform mediums, but with variable velocities in mediums of variable denfities. Thefe particles, fay they, act upon the minute conftituent parts of bodies, not by impact, but at fome indefinitely fmall diftance; they attract and are attracted; and in being reflected or refracted, they excite a vibratory motion in the component particles. This motion increafes the diftance between the particles, and thus occafions an augmentation of bulk, or an expanfion in every dimenfion, which is the moft certain characteriftic of fire. This expanfion, which is the beginning of a difunion of the parts, being increafed by the increafing magnitude of the vibrations proceeding from the continued agency of light, it may eafily be apprehended, that the particles will at length vibrate beyond their fphere of mutual attraction, and thus the texture of the body will be altered or deftroyed; from folid it may become fluid, as in melted gold; or

from being fluid, it may be difperfed in vapour, as in boiling water.

Others, as Boerhaave, reprefent fire as a fubftance *fui generis*, unalterable in its nature, and incapable of being produced or deftroyed; naturally exifting in equal quantities in all places, imperceptible to our fenfes, and only difcoverable by its effects, when, by various caufes, it is collected for a time into a lefs fpace than that which it would otherwife occupy. The matter of this fire is not in any wife fuppofed to be derived from the Sun: the folar rays, whether direct or reflected, are of ufe only as they impel the particles of fire in parallel directions: that parallelifm being deftroyed, by intercepting the folar rays, the fire inftantly affumes its natural ftate of uniform diffufion. According to this explication, which attributes heat to the matter of fire, when driven in parallel directions, a much greater degree muft be given it when the quantity, fo collected, is amaffed into a focus; and yet the focus of the largeft fpeculum does not heat the air or medium in which it is is found, but only bodies of denfities different from that medium.

M. de Luc (Lettres Phyfiques, is of opinion, that the folar rays are the principal caufe of heat; but that they heat fuch bodies only as do not allow them a free paffage. In this remark he agrees with Newton; but then he differs totally from him, as well as from Boerhaave, concerning the nature of the rays of the Sun. He does not admit the emanation of any luminous corpufcles from the Sun, or other felf-fhining fubftances, but fuppofes all fpace to be filled with an ether of great elafticity and fmall denfity, and that light confifts in the vibrations of this ether, as found confifts in the vibrations of the air. "Upon Newton's fuppofition, fays an excellent writer, the caufe by which the particles of light, and the corpufcles conftituting other bodies are mutually attracted and repelled, is uncertain. The reafon of the uniform diffufion of fire, of its vibration, and repercuffion, as ftated in Boerhaave's opinion, is equally inexplicable. And in the laft mentioned hypothefis, we may add to the other difficulties attending the fuppofition of an univerfal ether, the want of a firft mover to make the Sun vibrate. Of thefe feveral opinions concerning elementary fire, it may be faid, as Cicero remarked upon the opinions of philofophers concerning the nature of the foul: *Harum fententiarum quæ vera fit, Deus aliquis viderit; quæ verifimillima, magna queftio eft.*" Watfon's Chem. Eff. vol. 1, pa. 164.

As to the Figure of the SUN; this, like the planets, is not perfectly globular, but fpheroidical, being higher about the equator than at the poles. The reafon of which is this: the Sun has a motion about his own axis; and therefore the folar matter will have an endeavour to recede from the axis, and that with the greater force as their diftances from it, or the circles they move in, are greater: but the equator is the greateft circle; and the reft, towards the poles, continually decreafe; therefore the folar matter, though at firft in a fpherical form, will endeavour to recede from the centre of the equator farther than from the centres of the parallels. Confequently, fince the gravity, by which it is retained in its place, is fuppofed to be uniform throughout the whole Sun, it will really recede from the centre more at

the

the equator, than at any of the parallels; and hence the Sun's diameter will be greater through the equator, than through the poles; that is, the Sun's figure is not perfectly fpherical, but fpheroidical.

Several particulars of the Sun, related by Newton, in his Principia, are as follow:

1. That the denfity of the Sun's heat, which is proportional to his light, is 7 times as great at Mercury as with us; and therefore our water there would be all carried off, and boil away: for he found by experiments of the thermometer, that a heat but 7 times greater than that of the Sun beams in fummer, will ferve to make water boil.

2. That the quantity of matter in the Sun is to that in Jupiter, nearly as 1100 to 1; and that the diftance of that planet from the Sun, is in the fame ratio to the Sun's femidiameter.

3. That the matter in the Sun is to that in Saturn, as 2360 to 1; and the diftance of Saturn from the Sun is in a ratio but little lefs than that of the Sun's femidiameter. And hence, that the common centre of gravity of the Sun and Jupiter is nearly in the fuperficies of the Sun; of the Sun and Saturn, a little within it.

4. And by the fame mode of calculation it will be found, that the common centre of gravity of all the planets, cannot be more than the length of the folar diameter diftant from the centre of the Sun. This common centre of gravity he proves is at reft; and therefore though the Sun, by reafon of the various pofitions of the planets, may be moved every way, yet it cannot recede far from the common centre of gravity, and this, he thinks, ought to be accounted the centre of our world. Book 3, prop. 12.

5. By means of the folar fpots it hath been difcovered, that the Sun revolves round his own axis, without moving confiderably out of his place, in about 25 days, and that the axis of this motion is inclined to the ecliptic in an angle of 87° 30′ nearly. The Sun's apparent diameter being fenfibly longer in December than in June, the Sun muft be proportionably nearer to the earth in winter than in Summer; in the former of which feafons therefore will be the perihelion, in the latter the aphelion: and this is alfo confirmed by the earth's motion being quicker in December than in June, as it is, by about $\frac{1}{15}$ part. For fince the earth always defcribes equal areas in equal times, whenever it moves fwifter, it muft needs be nearer to the Sun: and for this reafon there are about 8 days more from the fun's vernal equinox to the autumnal, than from the autumnal to the vernal.

6. That the Sun's diameter is equal to 100 diameters of the earth; and therefore the body of the Sun muft be 1000000 times greater than that of the earth.—Mr. Azout affures us, that he obferved, by a very exact method, the Sun's diameter to be no lefs than 21′ 45″ in his apogee, and not greater than 32′ 45″ in his perigee.

7. According to Newton, in his theory of the moon, the mean apparent diameter of the Sun is 32′ 12″.— The Sun's horizontal parallax is now fixed at 8″ $\frac{5}{10}$.

8. If you divide 360 degrees (the whole ecliptic) by the quantity of the folar year, it will give 59′ 8″ &c, which therefore is the medium quantity of the Sun's daily motion; and if this 59′ 8″ be divided by 24, you

have the Sun's horary motion equal to 2′ 28″: and if this laft be divided by 60, it will give his motion in a minute, &c. And in this way are the tables of the Sun's mean motion conftructed, as placed in books of Aftronomical tables and calculations.

SUNDAY, the firft day of the week; thus called by our idolatrous anceftors, becaufe fet apart for the worfhip of the fun.

It is fometimes called the *Lord's Day*, becaufe kept as a feaft in memory of our Lord's refurrection on this day: and alfo *Sabbath-day*, becaufe fubftituted under the new law inftead of the Sabbath in the old law.

It was Conftantine the Great who firft made a law for the proper obfervation of Sunday; and who, according to Eufebius, appointed that it fhould be regularly celebrated throughout the Roman empire.

SUNDAY *Letter*. See DOMINICAL *Letter*.

SUPERFICIAL, relating to Superficies.

SUPERFICIES, or SURFACE, in Geometry, the outfide or exterior face of any body. This is confidered as having the two dimenfions of length and breadth only, but no thicknefs; and therefore it makes no part of the fubftance or folid content or matter of the body.

The terms or bounds or extremities of a Superficies, are lines; and Superficies may be confidered as generated by the motions of lines.

Superficies are either rectilinear, curvilinear, plane, concave, or convex. A

Rectilinear SUPERFICIES, is that which is bounded by right lines.

Curvilinear SUPERFICIES, is bounded by curve lines.

Plane SUPERFICIES is that which has no inequality in it, nor rifings, nor finkings, but lies evenly and ftraight throughout, fo that a right line may wholly coincide with it in all parts and directions.

Convex SUPERFICIES, is that which is curved and rifes outwards.

Concave SUPERFICIES, is curved and finks inward.

The meafure or quantity of a Surface, is called the *area* of it. And the finding of this meafure or area, is fometimes called the *quadrature* of it, meaning the reducing it to an equal fquare, or to a certain number of fmaller fquares. For all plane figures, and the Surfaces of all bodies, are meafured by fquares; as fquare inches, or fquare feet, or fquare yards, &c; that is, fquares whofe fides are inches, or feet, or yards, &c. Our leaft fuperficial meafure is the fquare inch, and other fquares are taken from it according to the proportion in the following Table of fuperficial or fquare meafure:

Table of Superficial or Square Meafure.

144 fquare inches	= 1 fquare foot
9 fquare feet	= 1 fquare yard
30¼ fquare yards	= 1 fquare pole
16 fquare poles	= 1 fquare chain
10 fquare chains	= 1 acre
640 acres	= 1 fquare mile.

The Superficial meafure of all bodies and figures depends entirely on that of a rectangle; and this is found by drawing or multiplying the length by the breadth of

it ; as is proved from plane geometry only, in my Menfuration, pt. 2, fect. 1, prob. 1. From the area of the
rectangle we obtain that of any oblique parallelogram,
which, by geometry, is equal to a rectangle of equal
bafe and altitude ; thence a triangle, which is the half
of fuch a parallelogram or rectangle ; and hence, by
compofition, we obtain the Superficies of all other figures whatever, as thefe may be confidered as made up
of triangles only.

Befide this way of deriving the Superficies of all figures, which is the moft fimple and natural, as proceeding on common geometry alone, there are certain other
methods ; fuch as the methods of exhauftions, of fluxions,
&c. See thefe articles in their places, as alfo QUADRA-
TURES.

Line of SUPERFICIES, a line ufually found on the
fector, and Gunter's fcale. The defcription and ufe
of which, fee under SECTOR and GUNTER's *Scale.*

SUPERPARTICULAR *Proportion,* or *Ratio,* is
that in which the greater term exceeds the lefs by unit
or 1. As the ratio of 1 to 2, or 2 to 3, or 3 to 4, &c.

SUPERPARTIENT *Proportion,* or *Ratio,* is when
the greater term contains the lefs term, once, and leaves
fome number greater than 1 remaining. As the ratio

of 3 to 5, which is equal to that of 1 to $1\frac{2}{3}$;
of 7 to 10, which is equal to that of 1 to $1\frac{3}{7}$; &c.

SUPPLEMENT, of an arch, or angle, in Geometry or Trigonometry, is what it wants of a femicircle,
or of 180 degrees ; as the *complement* is what it wants of
a quadrant, or of 90 degrees. So, the Supplement of
50° is 130° ; as the complement of it is 40°.

SURD, in Arithmetic, denotes a number or quantity
that is incommenfurate to unity ; or that is inexpreffi-
ble in rational numbers by any known way of notation,
otherwife than by its radical fign or index.—This is
otherwife called an *irrational* or *incommenfurable number,*
as alfo an *imperfect power.*

Thefe Surds arife in this manner : when it is propofed to extract a certain root of fome number or quantity, which is not a complete power or a true figurate
number of that kind ; as, if its fquare root be demanded,
and it is not a true fquare ; or if its cube root be required, and it is not a true cube, &c ; then it is impoffible to affign, either in whole numbers, or in fractions,
the exact root of fuch propofed number. And whenever this happens, it is ufual to denote the root by fetting
before it the proper mark of radicality, which is √, and
placing above this radical fign the number that fhews
what kind of root is required. Thus, $\sqrt[3]{2}$ or $\sqrt{2}$ fignifies the fquare root of 2, and $\sqrt[3]{10}$ fignifies the cube
root of 10 ; which roots, becaufe it is impoffible to exprefs them in numbers exactly, are properly called
Surd roots.

There is alfo another way of notation, now much in
ufe, by which roots are expreffed by fractional indices,
without the radical fign : thus, like as x^2, x^3, x^4, &c,
denote the fquare, cube, 4th power, &c, of x ; fo
$x^{\frac{1}{2}}$, $x^{\frac{1}{3}}$, $x^{\frac{1}{4}}$, &c, denote the fquare root, cube root, 4th
root, &c, of the fame quantity x.—The reafon of this
is plain enough ; for fince $\sqrt{x}$ is a geometrical mean
proportional between 1 and x, fo $\frac{1}{2}$ is an arithmetical
mean between 0 and 1 ; and therefore, as 2 is the index

of the fquare of x, $\frac{1}{2}$ will be the proper index of its
fquare root, &c.

It may be obferved that, for convenience, or the fake
of brevity, quantities which are not naturally Surds, are
often expreffed in the form of Surd roots. Thus $\sqrt{4}$,
$\sqrt[3]{\frac{8}{1}}$, $\sqrt{27}$, are the fame as 2, $\frac{8}{1}$, 3.

Surds are either *fimple* or *compound.*

Simple SURDS, are fuch as are expreffed by one fingle
term ; as $\sqrt{2}$, or $\sqrt[3]{a}$, &c.

Compound SURDS, are fuch as confift of two or more
fimple Surds connected together by the figns + or — ; as
$\sqrt{3} + \sqrt{2}$, or $\sqrt{3} - \sqrt{2}$, or $\sqrt[3]{5 + \sqrt{2}}$: which
laft is called an *univerfal* root, and denotes the cubic
root of the fum arifing by adding 5 and the root of 2
together.

Of certain Operations by Surds.

1. Such Surds as $\sqrt{2}$, $\sqrt{3}$, $\sqrt{5}$, &c. though they
are themfelves incommenfurable with unity, according
to the definition, are commenfurable in power with it,
becaufe their powers are integers, which are multiples
of unity. They may alfo be fometimes commenfurable
with one another ; as $\sqrt{8}$ and $\sqrt{2}$, which are to one
another as 2 to 1, as is found by dividing them by their
greateft common meafure, which is $\sqrt{2}$, for then thofe
two become $\sqrt{4} = 2$, and 1 the ratio.

2. *To reduce Rational Quantities to the form of any
propofed Surd Roots.*—Involve the rational quantity
according to the index of the power of the Surd, and
then prefix before that power the propofed radical
fign.

Thus, $a = \sqrt{a^2} = \sqrt[3]{a^3} = \sqrt[4]{a^4} = \sqrt[n]{a^n}$, &c.
and $4 = \sqrt{16} = \sqrt[3]{64} = \sqrt[4]{256} = \sqrt[n]{4^n}$, &c.

And in this way may a fimple Surd fraction, whofe
radical fign refers to only one of its terms, be changed
into another, which fhall include both numerator and

denominator. Thus, $\frac{\sqrt{2}}{5}$ is reduced to $\sqrt{\frac{2}{25}}$, and
$\frac{5}{\sqrt[3]{4}}$ to $\sqrt[3]{\frac{125}{4}}$: thus alfo the quantity a reduced to

the form of $x^{\frac{1}{n}}$ or $\sqrt[n]{x}$, is $\overline{a^n}^{\frac{1}{n}}$ or $\sqrt[n]{a^n}$. And thus
may roots with rational coefficients be reduced fo
as to be wholly affected by the radical fign ; as
$a\sqrt[n]{x} = \sqrt[n]{a^n x}$.

3. *To reduce Simple Surds, having different radical
figns (which are called hetcrogeneal Surds) to others that
may have one common radical fign, or which are homogeneal :
Or to reduce roots of different names to roots of the fame
name.*—Involve the powers reciprocally, each according
to the index of the other, for new powers ; and multiply their indices together, for the common index.
Otherwife, as Surds may be confidered as powers with
fractional exponents, reduce thefe fractional exponents
to fractions having the fame value and a common
denominator.

Thus, by the 1ft way,
$\sqrt[n]{a}$ and $\sqrt[m]{x}$ become $\sqrt[mn]{a^m}$ and $\sqrt[mn]{x^n}$;

and,

and, by the 2d way,

$a^{\frac{1}{n}}$ and $x^{\frac{1}{m}}$ become $\overline{a^n}^{\frac{1}{mn}}$ and $\overline{x^n}^{\frac{1}{mn}}$.

Also $\sqrt{3}$ and $\sqrt[3]{2}$ are reduced to $\sqrt[6]{27}$ and $\sqrt[6]{4}$, which are equal to them, and have a common radical sign.

4. *To reduce Surds to their most simple expressions, or to the lowest terms possible.*—Divide the Surd by the greatest power, of the same name with that of the root, which you can discover is contained in it, and which will measure or divide it without a remainder; then extract the root of that power, and place it before the quotient or Surd so divided; this will produce a new Surd of the same value with the former, but in more simple terms. Thus, $\sqrt{16a^2 x}$, by dividing by $16a^2$, and prefixing its root $4a$, before the quotient $\sqrt{x}$, becomes $4a\sqrt{x}$; in like manner, $\sqrt{12}$ or $\sqrt{4 \times 3}$, becomes $2\sqrt{3}$;

And $\sqrt[3]{ab^3 x}$ reduces to $b\sqrt[3]{ax}$.

Also $\sqrt[3]{81} = \sqrt[3]{27 \times 3} = \sqrt[3]{3^3 \times 3} = 3\sqrt[3]{3}$.

And $\sqrt{288} = \sqrt[3]{144 \times 2} = 12\sqrt{2}$.

5. *To Add and Subtract Surds.*—When they are reduced to their lowest terms, if they have the same irrational part, add or subtract their rational coefficients, and to the sum or difference subjoin the common irrational part.

Thus, $\sqrt{75} + \sqrt{48} = 5\sqrt{3} + 4\sqrt{3} = 9\sqrt{3}$;
and $\sqrt{150} - \sqrt{54} = 5\sqrt{6} - 3\sqrt{6} = 2\sqrt{6}$;
also $\sqrt{a^2 x} + \sqrt{c^2 x} = a\sqrt{x} + c\sqrt{x} = \overline{a+c}\cdot\sqrt{x}$.

Or such Surds may be added and subtracted, by first squaring them (by uniting the square of each part with double their product), and then extracting the root universal of the whole. Thus, for the first example above,

$\sqrt{75} + \sqrt{48} = \sqrt{75 + 48 + 2\sqrt{75 \times 48}} =$
$\sqrt{123 + 2\sqrt{3600}} = \sqrt{123 + 120} = \sqrt{243} =$
$9\sqrt{3}$.

If the quantities cannot be reduced to the same irrational part, they may just be connected by the signs $+$ or $-$.

6. *To Multiply and Divide Surds.*—If the terms have the same radical, they will be multiplied and divided like powers, viz, by adding their indices for multiplication, and subtracting them for division.

Thus,

$\sqrt{a} \times \sqrt[3]{a} = a^{\frac{1}{2}} \times a^{\frac{1}{3}} = a^{\frac{3}{6}} \times a^{\frac{2}{6}} = a^{\frac{5}{6}} = \sqrt[6]{a^5}$;

and $\sqrt{2} \times \sqrt[3]{2} = 2^{\frac{5}{6}} = \sqrt[6]{2^5} = \sqrt[6]{32}$;

also $\sqrt{a} \div \sqrt[3]{a} = a^{\frac{1}{2}} \div a^{\frac{1}{3}} = a^{\frac{1}{6}} = \sqrt[6]{a}$;

and $\sqrt{2} \div \sqrt[3]{2} = 2^{\frac{1}{6}} = \sqrt[6]{2}$.

If the quantities be different, but under the same radical sign; multiply or divide the quantities, and place the radical sign to the product or quotient.

Thus, $\sqrt{2} \times \sqrt{5} = \sqrt{10}$;
and $\sqrt[3]{a^2} \times \sqrt[3]{c} = \sqrt[3]{a^2 c}$;
also $\sqrt[3]{20} \div \sqrt[3]{4} = \sqrt[3]{5}$.

But if the Surds have not the same radical sign, reduce them to such as shall have the same radical sign, and proceed as before.

Thus, $\sqrt[m]{a} \times \sqrt[n]{b} = \sqrt[mn]{a^n} \times \sqrt[mn]{b^m} = \sqrt[mn]{a^n b^m}$;
and $\sqrt{2} \times \sqrt[3]{4} = \sqrt[6]{2^3} \times \sqrt[6]{4^2} = \sqrt[6]{8 \times 16} = \sqrt[6]{128}$.

If the Surds have any rational coefficients, their product or quotient must be prefixed.

Thus, $5\sqrt{6} \times 2\sqrt{3} = 10\sqrt{18} = 30\sqrt{2}$;
and $8\sqrt{5} \div 2\sqrt{6} = 4\sqrt{\frac{5}{6}}$.

7. *Involution and Evolution of Surds.*——Surds are involved, or raised to any power, by multiplying their indices by the index of the power; and they are evolved or extracted, by dividing their indices by the index of the root.

Thus, the square of $\sqrt[3]{2}$ or of $2^{\frac{1}{3}}$ is $2^{\frac{2}{3}} = \sqrt[3]{4}$;

and the cube of $\sqrt{5}$ or of $5^{\frac{1}{2}}$, is $5^{\frac{3}{2}} = \sqrt{125}$;

also the square root of $\sqrt[3]{4}$ or $4^{\frac{1}{3}}$, is $4^{\frac{1}{6}} = 2^{\frac{1}{3}} = \sqrt[3]{2}$.

Or thus : involve or extract the quantity under the radical sign according to the power or root required, continuing the same radical sign.

So the square of $\sqrt[3]{2}$ is $\sqrt[3]{4}$;
and the square root of $\sqrt[3]{4}$, is $\sqrt[3]{2}$.

Unless the index of the power is equal to the name of the Surd, or a multiple of it, for in that case the power of the Surd becomes rational. Thus, the square of $\sqrt{3}$ is 3, and the cube of $\sqrt[3]{a^2}$ is a^2.

Simple Surds are commensurable in power, and by being multiplied by themselves give, at length, rational quantities : but compound Surds, multiplied by themselves, commonly give irrational products. Yet, in this case, when any compound Surd is proposed, there is another compound Surd, which, multiplied by it, gives a rational product.

Thus, $\sqrt{a} + \sqrt{b}$ multiplied by $\sqrt{a} - \sqrt{b}$ gives $a - b$;
and $\sqrt[3]{a} - \sqrt[3]{b}$ mult. by $\sqrt[3]{a^2} + \sqrt[3]{ab} + \sqrt[3]{b^2}$ gives $a-b$.
The finding of such a Surd as multiplying the proposed Surd gives a rational product, is made easy by three theorems, delivered by Maclaurin, in his Algebra, pa. 109 &c.

This operation is of use in reducing Surd expressions to more simple forms. Thus, suppose a binomial Surd divided by another, as $\sqrt{20} + \sqrt{12}$ by $\sqrt{5} - \sqrt{3}$, the quotient might be expressed by

$\dfrac{\sqrt{20} + \sqrt{12}}{\sqrt{5} - \sqrt{3}} = \dfrac{2\sqrt{5} + 2\sqrt{3}}{\sqrt{5} - \sqrt{3}}$; but this will be expressed in a more simple form, by multiplying both numerator and denominator by such a Surd as makes the product of the denominator become a rational quantity: thus, multiplying them by $\sqrt{5} + \sqrt{3}$, the fraction or quotient becomes

$2 \times \dfrac{\sqrt{5} + \sqrt{3}}{\sqrt{5} - \sqrt{3}} \times \dfrac{\sqrt{5} + \sqrt{3}}{\sqrt{5} + \sqrt{3}} = 2 \times \dfrac{\overline{\sqrt{5} + \sqrt{3}}|^2}{5 - 3 = 2} =$
$\overline{\sqrt{5} + \sqrt{3}}|^2 = 8 + 2\sqrt{15}$.

To do this generally, see Maclaurin's Alg. p. 113.

When the square root of a Surd is required, it may be found nearly, by extracting the root of a rational quantity that approximates to its value. Thus, to find the
square

square root of $3 + 2\sqrt{2}$; first calculate $\sqrt{2} = 1\cdot41421$; hence $3 + 2\sqrt{2} = 5\cdot82842$, the root of which is nearly $2\cdot41421$.

In like manner we may proceed with any other proposed root. And if the index of the root be very high, a table of logarithms may be used to advantage: thus, to extract the root $\sqrt[7]{5 + \sqrt[13]{17}}$; take the logarithm of 17, divide it by 13, find the number answering to the quotient, add this number to 5, find the log. of the sum, and divide it by 7, and the number answering to this quotient will be nearly equal to $\sqrt[7]{5 + \sqrt[13]{17}}$.

But it is sometimes requisite to express the roots of Surds exactly by other Surds. Thus, in the first example, the square root of $3 + 2\sqrt{2}$ is $1 + \sqrt{2}$, for $\overline{1 + \sqrt{2}}^{\,2} = 1 + 2\sqrt{2} + 2 = 3 + \sqrt{2}$. For the method of performing this, the curious may consult Maclaurin's Algeb. p. 115, where also rules for trinomials &c may be found. See also the article BINOMIAL *Roots*, in this Dictionary.

For extracting the higher roots of a binomial, whose two members when squared are commensurable numbers, we have a rule in Newton's Arith. pa. 59, but without demonstration. This is supplied by Maclaurin, in his Alg. p. 120: as also by Gravesande, in his Matheseos Univerf. Elem. p. 211.

It sometimes happens, in the resolution of cubic equations, that binomials of this form $a \pm b\sqrt{-1}$ occur, the cube roots of which must be found; and to these Newton's rule cannot always be applied, because of the impossible or imaginary factor $\sqrt{-1}$; yet if the root be expressible in rational numbers, the rule will often yield to it in a short way, not merely tentative, the trials being confined to known limits. See Maclaurin's Alg. p. 127. It may be farther observed, that such roots, whether expressible in rational numbers or not, may be found by evolving the quantity $a + b\sqrt{-1}$ by Newton's binomial theorem, and summing up the alternate terms. Maclaurin, p. 130.

Those who are desirous of a general and elegant solution of the problem, *to extract any root of an impossible binomial* $a + b\sqrt{-1}$, *or of a possible binomial* $a + \sqrt{b}$, may have recourse to the appendix to Saunderson's Algebra, and to the Philof. Transf. number 451, or Abridg. vol. 8, p. 1. On the management of Surds, see also the numerous authors upon Algebra.

SURDESOLID. See SURSOLID.

SURFACE, in Geometry. See SUPERFICIES.

A *mathematical* SURFACE is the mere exterior face of a body, but is not any part of it, being of no thickness, but only the bare figure or termination of the body.

A *Physical* SURFACE is considered as of some very small thickness.

SURSOLID, or SURDESOLID, in Arithmetic, the 5th power of a number, considered as a root. The number 2, for instance, considered as a root, produces the powers thus:

$$2 \quad . \quad = 2 \text{ the root or 1st power,}$$
$$2 \times 2 = 4 \text{ the square or 2d power,}$$
$$2 \times 4 = 8 \text{ the cube or 3d power,}$$
$$2 \times 8 = 16 \text{ the biquadratic or 4th power,}$$
$$2 \times 16 = 32 \text{ the Sursolid or 5th power.}$$

SURSOLID PROBLEM, is that which cannot be resolved but by curves of a higher kind than the conic sections.

SURVEYING, the art, or act, of measuring land. This comprises the three following parts; viz, taking the dimensions of any tract or piece of ground; the delineating or laying the same down in a map or draught; and finding the superficial content or area of the same; beside the dividing and laying out of lands.

The first of these is what is properly called *Surveying*; the second is called *plotting*, or *protracting*, or *mapping*; and the third *casting up*, or *computing the contents*.

The first again consists of two parts, the making of observations for the angles, and the taking of lineal measures for the distances.

The former of these is performed by some of the following instruments; the theodolite, circumferentor, semicircle, plain-table, or compass, or even by the chain itself: the latter is performed by means either of the chain, or the perambulator. The description and manner of using each of these, see under its respective article or name.

It is useful in Surveying, to take the angles which the bounding lines form with the magnetic needle, in order to check the angles of the figure, and to plot them conveniently afterwards. But, as the difference between the true and magnetic meridian perpetually varies in all places, and at all times; it is impossible to compare two surveys of the same place, taken at distant times, by magnetic instruments, without making due allowance for this variation. See observations on this subject, by Mr. Molineux, Philof. Transf. number 230, p. 625, or Abr. vol. 1, p. 125.

The second branch of Surveying is performed by means of the protractor, and plotting scale. The description of which, see under their proper names.

If the lands in the survey are hilly, and not in any one plane, the measured lines cannot be truly laid down on paper, till they are reduced to one plane, which must be the horizontal one, because angles are taken in that plane. And in this case, when observing distant objects, for their elevation or depression, the following table shews the links or parts to be subtracted from each chain in the hypothenufal line, when the angle is the corresponding number of degrees.

A TABLE *of the links to be subtracted out of every chain in hypothenufal lines, of several degrees of altitude or depression, for reducing them to horizontal.*

		links			links
4°	3′	1	19°	57′	6
5	44	1½	21	34	7
7	1	1¾	23	4	8
8	7	1	24	30	9
11	29	2	24	50	10
14	4	3	27	8	11
16	16	4	28	22	12
18	12	5	29	32	13

For

For example, if a station line measure 1250 links, or 12¼ chains, on an ascent, or a descent, of 11°; here it is after the rate of almost two links per chain, and it will be exact enough to take only the 12 chains at that rate, which make 24 links in all, to be deducted from 1250, which leaves 1226 links, for the length to be laid down.

Practical surveyors say, it is best to make this deduction at the end of every chain-length while measuring, by drawing the chain forward every time as much as the deduction is; viz, in the present instance, drawing the chain on 2 links at each chain-length.

The third branch of Surveying, namely computing or casting-up, is performed by reducing the several inclosures and divisions into triangles, trapeziums, and parallelograms, but especially the two former; then finding the areas or contents of these several figures, and adding them all together.

The Practice of Surveying.

1. Land is measured with a chain, called Gunter's chain, of 4 poles or 22 yards in length, which consists of 100 equal links, each link being $\frac{25}{100}$ of a yard, or $\frac{66}{100}$ of a foot, or 7·92 inches long, that is nearly 8 inches or ⅔ of a foot.

An acre of land is equal to 10 square chains, that is, 10 chains in length and 1 chain in breadth.

Or it is 40 × 4 or 160 square poles.
Or it is 220 × 22 or 4840 square yards.
Or it is 1000 × 100 or 100000 square links.
These being all the same quantity.

Also, an acre is divided into 4 parts called roods, and a rood into 40 parts called perches, which are square poles, or the square of a pole of 5½ yards long, or the square of ¼ of a chain, or of 25 links, which is 625 square links. So that the divisions of land measure will be thus:

$$625 \text{ sq. links} = 1 \text{ pole or perch}$$
$$40 \text{ perches} = 1 \text{ rood}$$
$$4 \text{ roods} = 1 \text{ acre}$$

The length of lines, measured with a chain, are set down in links as integers, every chain in length being 100 links; and not in chains and decimals. Therefore, after the content is found, it will be in square links; then cut off five of the figures on the right-hand for decimals, and the rest will be acres. Those decimals are then multiplied by 4 for roods, and the decimals of these again by 40 for perches.

Ex. Suppose the length of a rectangular piece of ground be 792 links, and its breadth 385: to find the area in acres, roods, and perches.

```
          792
          385
        ------
         3960
         6336
         2376
        ------              ac. ro. p.
        3·04920    Ans. 3   0   7
             4
        ------
         ·19680
            40
        ------
        7·87200
```

2. Among the various instruments for surveying, the plain-table is the easiest and most generally useful, especially in crooked difficult places, as in a town among houses, &c. But although the plain-table be the most generally useful instrument, it is not *always* so; there being many cases in which sometimes one instrument is the properest, and sometimes another; nor is that surveyor master of his business who cannot in any case distinguish which is the fittest instrument or method, and use it accordingly: nay, sometimes no instrument at all, but barely the chain itself, is the best method, particularly in regular open fields lying together; and even when you are using the plain-table, it is often of advantage to measure such large open parts with the chain only, and from those measures lay them down upon the table.

The perambulator is used for measuring roads, and other great distances on level ground, and by the sides of rivers. It has a wheel of 8¼ feet, or half a pole, in circumference, upon which the machine turns; and the distance measured is pointed out by an index, which is moved round by clock work.

Levels, with telescopic or other sights, are used to find the level between place and place, or how much one place is higher or lower than another.

An offset-staff is a very useful and necessary instrument, for measuring the offsets and other short distances. It is 10 links in length, being divided and marked at each of the 10 links.

Ten small arrows, or rods of iron or wood, are used to mark the end of every chain length, in measuring lines. And sometimes pickets, or staves with flags, are set up as marks or objects of direction.

Various scales are also used in protracting and measuring on the plan or paper; such as plane scales, line of chords, protractor, compasses, reducing scales, parallel and perpendicular rulers, &c. Of plane scales, there should be several sizes, as a chain in 1 inch, a chain in ¾ of an inch, a chain in ½ of an inch, &c. And of these, the best for use are those that are laid on the very edges of the ivory scale, to prick off distances by, without compasses.

3. *The Field Book.*

In surveying with the plain-table, a field-book is not used, as every thing is drawn on the table immediately when it is measured. But in surveying with the theodolite, or any other instrument, some sort of a field-book must be used, to write down in it a register or account of all that is done and occurs relative to the survey in hand.

This book every one contrives and rules as he thinks fittest for himself. The following is a specimen of a form that has formerly been much used. It is ruled into 3 columns: the middle, or principal column, is for the stations, angles, bearings, distances measured, &c; and those on the right and left are for the offsets on the right and left; which are set against their corresponding distances in the middle column; as also for such remarks as may occur, and be proper to note for drawing the plan, &c.

Here ⊙ 1 is the first station, where the angle or bearing is 10,° 25′ On the left, at 73 links in the distance

distance or principal line, is an offset of 92; and at 610 an offset of 24 to a cross hedge. On the right, at 0, or the beginning, an offset 25 to the corner of the field; at 248 Brown's boundary hedge commences; at 610 an offset 35; and at 954, the end of the first line, the 0 denotes its terminating in the hedge. And so on for the other stations.

Draw a line under the work, at the end of every station line, to prevent confusion.

Offsets and Remarks on the left.	Stations, Bearings, and Distances.	Offsets and Remarks on the right.
	☉ 1	
	105°25′	
	00	25 corner
92	73	
	248	Brown's hedge
cross a hedge 24	610	35
	954.	00 his.
	☉ 2	
	53°10′	
	00	00
house corner 51	25	21
	120	29 a tree
34	734	40 a stile
	☉ 3	
	67°20′	
	61	35
a brook 30	248	
	639	16 a spring
foot path 16	810	
cross hedge 18	973	20 a pond

But a few skilful surveyors now make use of a different method for the field book, namely, beginning at the bottom of the page and writing upwards; by which they sketch a neat boundary on either hand, as they pass it; an example of which will be given below, with the plan of the ground to accompany it.

In smaller surveys and measurements, a very good way of setting down the work, is, to draw, by the eye, on a piece of paper, a figure resembling that which is to be measured; and so write the dimensions, as they are found, against the corresponding parts of the figure. And this method may be practised to a considerable extent, even in the larger surveys.

4. *To measure a line on the ground with the chain:* Having provided a chain, with 10 small arrows, or rods, to stick one into the ground, as a mark, at the end of every chain; two persons take hold of the chain, one at each end of it, and all the 10 arrows are taken by one of them who goes foremost, and is called the leader; the other being called the follower, for distinction's sake.

A picket, or station staff, being set up in the direction of the line to be measured, if there do not appear some marks naturally in that direction; the follower stands at the beginning of the line, holding the ring at the end of the chain in his hand, while the leader drags forward the chain by the other end of it, till it is stretched straight, and laid or held level, and the leader directed, by the follower waving his hand, to the right or left, till the follower see him exactly in a line with the mark or direction to be measured to; then both of them stretching the chain straight, and stooping and holding it level, the leader having the head of one of his arrows in the same hand by which he holds the end of the chain, he there sticks one of them down with it, while he holds the chain stretched. This done, he leaves the arrow in the ground, as a mark for the follower to come to, and advances another chain forward, being directed in his position by the follower standing at the arrow, as before; as also by himself now, and at every succeeding chain's length, by moving himself from side to side, till he brings the follower and the back mark into a line. Having then stretched the chain, and stuck down an arrow, as before, the follower takes up his arrow, and they advance again in the same manner another chain-length. And thus they proceed till all the 10 arrows are employed, and are in the hands of the follower; and the leader, without an arrow, is arrived at the end of the 11th chain length. The follower then sends or brings the 10 arrows to the leader, who puts one of them down at the end of his chain, and advances with the chain as before. And thus the arrows are changed from the one to the other at every 10 chains' length, till the whole line is finished; the number of changes of the arrows shews the number of tens, to which the follower adds the arrows he holds in his hand, and the number of links of another chain over to the mark or end of the line. So if there have been 3 changes of the arrows, and the follower hold 6 arrows, and the end of the line cut off 45 links more, the whole length of the line is set down in links thus, 3645.

5. *To take Angles and Bearings.*

Let B and C be two objects, or two pickets set up perpendicular; and let it be required to take their bearings, or the angle formed between them at any station A.

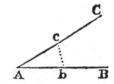

1st. *With the Plain Table.* The table being covered with a paper, and fixed on its stand; plant it at the station A, and fix a fine pin, or a point of the compasses, in a proper part of the paper, to represent the point A: Close by the side of this pin lay the fiducial edge of the index, and turn it about, still touching the pin, till one object B can be seen through the sights: then by the fiducial edge of the index draw a line. In the very same manner draw another line in the direction of the other object C. And it is done.

2d. *With the Theodolite, &c.* Direct the fixed sights along one of the lines, as AB, by turning the instrument about till you see the mark B through these sights; and there screw the instrument fast. Then

turn

turn the moveable index about till, through its fights, you fee the other mark C. Then the degrees cut by the index, upon the graduated limb or ring of the inftrument, fhew the quantity of the angle.

3d. *With the Magnetic Needle and Compafs.* Turn the inftrument, or compafs, fo, that the north end of the needle point to the flower-de-luce. Then direct the fights to one mark, as B, and note the degrees cut by the needle. Then direct the fights to the other mark C, and note again the degrees cut by the needle. Then their fum or difference, as the cafe is, will give the quantity of the angle BAC.

4th. *By Meafurement with the Chain, &c.* Meafure one chain length, or any other length, along both directions, as to b and c. Then meafure the diftance b c, and it is done.—This is eafily transferred to paper, by making a triangle A b c with thefe three lengths, and then meafuring the angle A as in Practical Geometry.

6. *To Meafure the Offsets.*

A h i k l m n being a crooked hedge, or river, &c. From A meafure in a ftraight direction along the fide of it to B. And in meafuring along this line AB obferve when you are directly oppofite any bends or corners of the hedge, as at c d, e, &c; and from thence meafure the perpendicular offsets, ch, di, &c, with the offset-ftaff, if they are not very large, otherwife with the chain itfelf; and the work is done. And the regifter, or field-book, may be as follows:

Offs. left.	Bafe line AB	
	0	⊙ A
ch 62	45	Ac
di 84	220	Ad
e k 70	340	Ae
f l 98	510	Af
gm 57	634	Ag
B n 91	785	AB

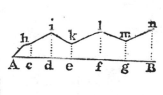

7. *To Survey a triangular Field* ABC.

1ft. *By the Chain.*

AP	794
AB	1321
PC	826

Having fet up marks at the corners, which is to be done in all cafes where there are not marks naturally; meafure with the chain from A to P, where a perpendicular would fall from the angle C, and there meafure from P to C; then complete the diftance AB by meafuring from P to B; fetting down each of thefe meafured diftances. And thus, having the bafe and perpendicular, the area from them is eafily found. Or having the place P of the perpendicular, the triangle is eafily conftructed.

Or, meafure all the three fides with the chain, and note them down. From which the content is eafily found, or the figure conftructed.

2d. *By taking one or more of the Angles.*

Meafure two fides AB, AC, and the angle A between them. Or meafure one fide AB, and the two adjacent angles A and B. From either of thefe ways the figure is eafily planned: then by meafuring the perpendicular CP on the plan, and multiplying it by half AB, you have the content.

8. *To meafure a Four-fided Field.*

1ft. *By the Chain.*

AE	214	210	DE
AF	362	306	BF
AC	592		

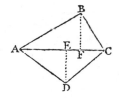

Meafure along either of the diagonals, as AC; and either the two perpendiculars DE, BF, as in the laft problem; or elfe the fides AB, BC, CD, DA. From either of which the figure may be planned and computed as before directed.

Otherwife by the Chain.

AP	110	352	PC
AQ	74	595	QD
AB	1110		

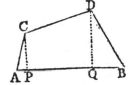

Meafure on the longeft fide, the diftances AP, AQ, AB; and the perpendiculars PC, QD.

2d. *By taking one or more of the Angles.*

Meafure the diagonal AC (fee the firft fig. above), and the angles CAB, CAD, ACB, ACD.—Or meafure the four fides, and any one of the angles as BAD.

Thus		Or thus	
AC	591	AB	486
CAB	37°20	BC	394
CAD	41 15	CD	410
ACB	72 25	DA	462
ACD	54 40	BAD	78°35'

9. *To Survey any Field by the Chain only.*

Having fet up marks at the corners, where neceffary, of the propofed field ABCDEFG. Walk over the ground, and confider how it can beft be divided into triangles and trapeziums; and meafure them feparately as in the laft two problems. And in this way it will be proper to divide it into as few feparate triangles, and as many trapeziums as may be, by drawing diago-

nals

nals from corner to corner: and fo, as that all the perpendiculars may fall within the figure. Thus, the following figure is divided into the two trapeziums ABCG, GDEF, and the triangle GCD. Then, in the firft, beginning at A, meafure the diagonal AC, and the two perpendiculars Gm, Bn. Then the bafe GC and the perpendicular Dq. Laftly the diagonal DF, and the two perpendiculars pE, oG. All which meafures write againft the correfponding parts of a rough figure drawn to refemble the figure to be furveyed, or fet them down in any other form you choofe.

Am	135	130	mG
An	410	180	nB
Ac	550		
Cq	152	230	qD
CG	440		
FO	206	120	oG
FP	288	80	pE
FD	520		

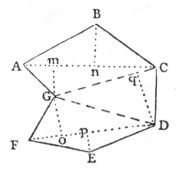

Or thus:

Meafure all the fides AB, BC, CD, DE, EF, FG, and GA; and the diagonals AC, CG, GD, DF.

Otherwife,

Many pieces of land may be very well furveyed, by meafuring any bafe line, either within or without them, together with the perpendiculars let fall upon it from every corner of them. For they are by thefe means divided into feveral triangles and trapezoids, all whofe parallel fides are perpendicular to the bafe line; and the fum of thefe triangles and trapeziums will be equal to the figure propofed if the bafe line fall within it; if not, the fum of the parts which are without being taken from the fum of the whole which are both within and without, will leave the area of the figure propofed.

In pieces that are not very large, it will be fufficiently exact to find the points, in the bafe line, where the feveral perpendiculars will fall, by means of the *crofs*, and from thence meafuring to the corners for the lengths of the perpendiculars.—And it will be moft convenient to draw the line fo as that all the perpendiculars may fall within the figure.

Thus, in the following figure, beginning at A, and meafuring along the line AG, the diftances and perpendiculars, on the right and left, are as below.

Ab	315	350	bB
Ac	440	70	cC
Ad	585	320	dD
Ae	610	50	eE
Af	990	470	fF
AG	1020	0	

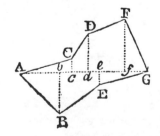

10. *To Survey any Field with the Plain Table.*

1ft. *From one Station.*

Plant the table at any angle, as C, from whence all the other angles, or marks fet up, can be feen; and turn the table about till the needle point to the flower-de-luce: and there fcrew it faft. Make a point for C on the paper on the table, and lay the edge of the index to C, turning it about there till through the fights you fee the mark D; and by the edge of the index draw a dry or obfcure line: then meafure the diftance CD, and lay that diftance down on the line CD. Then turn the index about the point C, till the mark E be feen through the fights, by which draw a line, and meafure the diftance to E, laying it on the line from C to E. In like manner determine the pofitions of CA and CB, by turning the fights fucceffively to A and B; and lay the lengths of thofe lines down. Then connect the points with the boundaries of the field, by drawing the black lines CD, DE, EA, AB, BC.

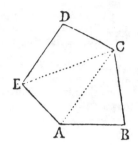

2d. *From a Station within the Field.*

When all the other parts cannot be feen from one angle, choofe fome place O within; or even without, if more convenient, from whence the other parts can be feen. Plant the table at O, then fix it with the needle north, and mark the point O upon it. Apply the index fucceffively to O, turning it round with the fights to each angle A, B, C, D, E, drawing dry lines to them by the edge of the index, then meafuring the diftances OA, OB, &c, and laying them down upon thofe lines. Laftly draw the boundaries AB, BC, CD, DE, EA.

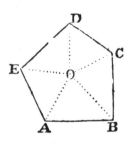

3d. *By going round the Figure.*

When the figure is a wood or water, or from fome other obftruction you cannot meafure lines acrofs it; begin at any point A, and meafure round it, either within or without the figure, and draw the directions of all the fides thus: Plant the table at A, turn it with the needle to the north or flower-de-luce, fix it and mark the point A. Apply the index to A, turning it till you can fee the point E, there draw a line; and then the point B, and there draw a line: then meafure thefe lines, and lay them down from A to E and B. Next move the table to B, lay the index along the line AB, and turn the table about till you can fee the mark A, and fcrew faft the table; in which pofition alfo the needle will again point to the flower-de-luce, as it will

do

do indeed at every station when the table is in the right position. Here turn the index about B till through the fights you fee the mark C; there draw a line, meafure BC, and lay the diftance upon that line after you have fet down the table at C. Turn it then again into its proper pofition, and in like manner find the next line CD. And fo on quite round by E to A again. Then the proof of the work will be the joining at A: for if the work is all right, the laft direction EA on the ground, will pafs exactly through the point A on the paper; and the meafured diftance will alfo reach exactly to A. If thefe do not coincide, or nearly fo, fome error has been committed, and the work muft be examined over again.

11. To Survey a Field with the Theodolite, &c.

1ft. From one Point or Station.

When all the angles can be feen from one point, as the angle C (laft fig. but one), place the inftrument at C, and turn it about till, through the fixed fights, you fee the mark B, and there fix it. Then turn the moveable index about till the mark A is feen through the fights, and note the degrees cut on the inftrument. Next turn the index fucceffively to E and D, noting the degrees cut off at each; which gives all the angles BCA, BCE, BCD. Laftly, meafure the lines CB, CA, CE, CD; and enter the meafures in a field-book, or rather againft the correfponding parts of a rough figure drawn by guefs to refemble the field.

2d. From a Point within or without.

Plant the inftrument at O, (laft fig.) and turn it about till the fixed fights point to any object, as A; and there fcrew it faft. Then turn the moveable index round till the fights point fucceffively to the other points E, D, C, B, noting the degrees cut off at each of them; which gives all the angles round the point O. Laftly, meafure the diftances OA, OB, OC, OD, OE, noting them down as before, and the work is done.

3d. By going round the Field.

By meafuring round, either within or without the field, proceed thus. Having fet up marks at B, C, &c. near the corners as ufual, plant the inftrument at any point A, and turn it till the fixed index be in the direction AB, and there fcrew it faft:

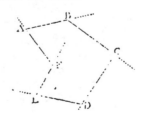

then turn the moveable index to the direction AF; and the degrees cut off will be the angle A. Meafure the line AB, and plant the inftrument at B, and there in the fame manner obferve the angle A. Then meafure BC, and obferve the angle C. Then meafure the diftance CD, and take the angle D. Then meafure DE, and take the angle E. Then meafure EF, and take the angle F. And laftly meafure the diftance FA.

To prove the work; add all the inward angles, A, B, C, &c, together, and when the work is right, their fum will be equal to twice as many right angles as the figure has fides, wanting 4 right angles. And when there is an angle, as F, that bends inwards, and

you meafure the external angle, which is lefs than two right angles, fubtract it from 4 right angles, or 360 degrees, to give the internal angle greater than a femi-circle or 180 degrees.

Otherwife. Inftead of obferving the internal angles, you may take the external angles, formed without the figure by producing the fides further out. And in this cafe, when the work is right, their fum altogether will be equal to 360 degrees. But when one of them, as F, runs inwards, fubtract it from the fum of the reft, to leave 360 degrees.

12. To Survey a Field with crooked Hedges, &c.

With any of the inftruments meafure the lengths and pofitions of imaginary lines running as near the fides of the field as you can; and in going along them meafure the offsets in the manner before taught; and you will have the plan on the paper in ufing the plain table, drawing the crooked hedges through the ends of the offsets; but in furveying with the theodolite, or other inftrument, fet down the meafures properly in a field-book, or memorandum-book, and plan them after returning from the field, by laying down all the lines and angles.

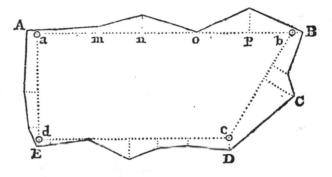

So, in furveying the piece ABCDE, fet up marks a, b, c, d, dividing it into as few fides as may be. Then begin at any ftation a, and meafure the lines ab, bc, cd, da, and take their pofitions, or the angles a, b, c, d; and in going along the lines meafure all the offsets, as at m, n, o, p, &c, along every ftation line.

And this is done either within the field, or without, as may be moft convenient. When there are obftructions within, as wood, water, hills, &c; then meafure without, as in the figure here below.

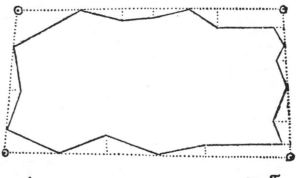

13. *To Survey a Field or any other Thing by Two Stations.*

This is performed by choosing two stations, from whence all the marks and objects can be seen, then measuring the distance between the stations, and at each station taking the angles formed by every object, from the station line or distance.

The two stations may be taken either within the bounds, or in one of the sides, or in the direction of two of the objects, or quite at a distance, and without the bounds of the objects, or part to be surveyed.

In this manner, not only grounds may be surveyed, without even entering them, but a map may be taken of the principal parts of a country, or the chief places of a town, or any part of a river or coast surveyed, or any other inaccessible objects; by taking two stations, on two towers, or two hills, or such like.

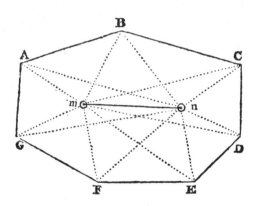

When the plain table is used; plant it at one station m, draw a line m n on it, along which lay the edge of the index, and turn the table about till the sights point directly to the other station; and there screw it fast. Then turn the sights round m successively to all the objects ABC, &c, drawing a dry line by the edge of the index at each, as mA, mB, mC, &c. Then measure the distance to the other station, there plant the table, and lay that distance down on the station line from m to n. Next lay the index by the line nm, and turn the table about till the sights point to the other station m, and there screw it fast. Then direct the sights successively to all the objects A, B, C, &c, as before, drawing lines each time, as nA, nB, nC, &c: and their intersection with the former lines will give the places of all the objects, or corners, A, B, C, &c.

When the theodolite, or any other instrument for taking angles, is used; proceed in the same way, measuring the station distance mn, planting the instrument first at one station, and then at another; then placing the fixed sights in the direction mn, and directing the moveable sights to every object, noting the degrees cut off at each time. Then, these observations being planned, the intersections of the lines will give the objects as before.

When all the objects, to be surveyed, cannot be seen from two stations; then three stations may be used, or four, or as many as is necessary; measuring always the distance from one station to another; placing the instrument in the same position at every station, by means described before; and from each station observing or setting every object that can be seen from it, by taking its direction or angular position, till every object be determined by the intersection of two or more lines of direction, the more the better. And thus may very extensive surveys be taken, as of large commons, rivers, coasts, countries, hilly grounds, and such like.

14. *To Survey a Large Estate.*

If the estate be very large, and contain a great number of fields, it cannot well be done by surveying all the fields singly, and then putting them together; nor can it be done by taking all the angles and boundaries that inclose it. For in these cases, any small errors will be so multiplied, as to render it very much distorted.

1st. Walk over the estate two or three times, in order to get a perfect idea of it, and till you can carry the map of it tolerably in your head. And to help your memory, draw an eye draught of it on paper, or at least, of the principal parts of it to guide you.

2d. Choose two or more eminent places in the estate, for your stations, from whence you can see all the principal parts of it: and let these stations be as far distant from one another as possible; as the fewer stations you have to command the whole, the more exact your work will be: and they will be fitter for your purpose, if these station lines be in or near the boundaries of the ground, and especially if two or more lines proceed from one station

3d. Take angles, between the stations, such as you think necessary, and measure the distances from station to station, always in a right line: these things must be done, till you get as many angles and lines as are sufficient for determining all your points of station. And in measuring any of these station distances, mark accurately where these lines meet with any hedges, ditches, roads, lanes, paths, rivulets, &c, and where any remarkable object is placed, by measuring its distance from the station line, and where a perpendicular from it cuts that line; and always mind, in any of these observations, that you be in a right line, which you will know by taking backsight and foresight, along your station line. And thus as you go along any main station line, take offsets to the ends of all hedges, and to any pond, house, mill, bridge, &c, omitting nothing that is remarkable. And all these things must be noted down; for these are your data, by which the places of such objects are to be determined upon your plan. And be sure to set marks up at the intersections of all hedges with the station line, that you may know where to measure from, when you come to survey these particular fields, which must immediately be done, as soon as you have measured that station line, whilst they are fresh in memory. In this way all your station lines are to be measured, and the situation of all places adjoining to them determined, which is the first grand point to be obtained. It will be proper for you to lay down your work upon paper every night, when you go home, that you may see how you go on.

4th. As to the inner parts of the estate, they must be determined

determined in like manner, by new ftation lines : for, after the main ftations are determined, and every thing adjoining to them, then the eftate muft be fubdivided into two or three parts by new ftation lines ; taking inner ftations at proper places, where you can have the beft view. Meafure thefe ftation lines as you did the firft, and all their interfections with hedges, and all offsets to fuch objects as appear. Then you may proceed to furvey the adjoining fields, by taking the angles that the fides make with the ftation line, at the interfections, and meafuring the diftances to each corner, from the interfections. For every ftation line will be a bafis to all the future operations; the fituation of all parts being entirely dependant upon them ; and therefore they fhould be taken of as great a length as poffible; and it is beft for them to run along fome of the hedges or boundaries of one or more fields, or to pafs through fome of their angles. All things being determined for thefe ftations, you muft take more inner ones, and fo continue to divide and fubdivide, till at laft you come to fingle fields; repeating the fame work for the inner ftations, as for the outer ones, till all be done : and clofe the work as often as you can, and in as few lines as poffible. And that you may choofe ftations the moft conveniently, fo as to caufe the leaft labour, let the ftation lines run as far as you can along fome hedges, and through as many corners of the fields, and other remarkable points, as you can. And take notice how one field lies by another; that you may not mifplace them in the draught.

5th. An eftate may be fo fituated, that the whole cannot be furveyed together; becaufe one part of the eftate cannot be feen from another. In this cafe, you may divide it into three or four parts, and furvey the parts feparately, as if they were lands belonging to different perfons; and at laft join them together.

6th. As it is neceffary to protract or lay down your work as you proceed in it, you muft have a fcale of a due length to do it by. To get fuch a fcale, you muft meafure the whole length of the eftate in chains; then you muft confider how many inches in length the map is to be ; and from thefe you will know how many chains you muft have in an inch; then make your fcale, or choofe one already made, accordingly.

7th. The trees in every hedge row muft be placed in their proper fituation, which is foon done by the plain table ; but may be done by the eye without an inftrument ; and being thus taken by guefs, in a rough draught, they will be exact enough, being only to look at ; except it be fuch as are at any remarkable places, as at the ends of hedges, at ftiles, gates, &c, and thefe muft be meafured. But all this need not be done till the draught is finifhed. And obferve in all the hedges, what fide the gutter or ditch is on, and confequently to whom the fences belong.

8th. When you have long ftations, you ought to have a good inftrument to take angles with ; and the plain table may very properly be made ufe of, to take the feveral fmall internal parts, and fuch as cannot be taken from the main ftations, as it is a very quick and ready inftrument.

15. Inftead of the foregoing method, an ingenious friend (Mr. Abraham Crocker), after mentioning the new and improved method of keeping the field book by writing from bottom to top of the pages, obferves that " In the former method of meafuring a large eftate, the accuracy of it depends on the correctnefs of the inftruments ufed in taking the angles. To avoid the errors incident to fuch a multitude of angles, other methods have of late years been ufed by fome few fkilful furveyors ; the moft practical, expeditious, and correct, feems to be the following.

" As was advifed in the foregoing method, fo in this, choofe two or more eminences, as grand ftations, and meafure a principal bafe line from one ftation to the other, noting every hedge, brook, or other remarkable object as you pafs by it ; meafuring alfo fuch fhort perpendicular lines to fuch bends of hedges as may be near at hand. From the extremities of this bafe line, or from any convenient parts of the fame, go off with other lines to fome remarkable object fituated towards the fides of the eftate, without regarding the angles they make with the bafe line or with one another ; ftill remembering to note every hedge, brook or other object that you pafs by. Thefe lines, when laid down by interfections, will with the bafe line form a grand triangle upon the eftate ; feveral of which, if need be, being thus laid down, you may proceed to form other fmaller triangles and trapezoids on the fides of the former : and fo on, until you finifh with the enclofures individually.

" To illuftrate this excellent method, let us take AB (in the plan of an eftate, fig. 1, pl. 28) for the principal bafe line. From B go off to the tree at C ; noting down, in the field-book, every crofs hedge, as you meafure on ; and from C meafure back to the firft ftation at A, noting down every thing as before directed.

" This grand triangle being completed, and laid down on the rough-plan paper, the parts, exterior as well as interior, are to be completed by fmaller triangles and trapezoids.

" When the whole plan is laid down on paper, the contents of each field might be calculated by the methods laid down below, at article 20.

" In countries where the lands are enclofed with high hedges, and where many lanes pafs through an eftate, a theodolite may be ufed to advantage, in meafuring the angles of fuch lands ; by which means, a kind of fkeletor of the eftate may be obtained, and the lane-lines ferve as the bafes of fuch triangles and trapezoids as are neceffary to fill up the interior parts."

The method of meafuring the other crofs lines, offfets and interior parts and enclofures, appears in the plan, fig. 1, laft referred to.

16. Another ingenious correfpondent (Mr. John Rodham of Richmond, Yorkfhire) has alfo communicated the following example of the new method of furveying, accompanied by the field-book, and its correfponding plan. His account of the method is as follows.

The field-book is ruled into three columns. In the middle one are fet down the diftances on the chain line at which any mark, offset, or other obfervation is made; and in the right and left hand columns are entered; the offsets and obfervations made on the right and left hand refpectively of the chain line.

It is of great advantage, both for brevity and perfpicuity,

fpicuity, to begin at the bottom of the leaf and write upwards ; denoting the croffing of fences, by lines drawn acrofs the middle column, or only a part of fuch a line on the right and left oppofite the figures, to avoid confufion, and the corners of fields, and other remarkable turns in the fences where offsets are taken to, by lines joining in the manner the fences do, as will be beft feen by comparing the book with the plan annexed, fig. 2, pl. 28.

The marks called, *a*, *b*, *c*, &c, are beft made in the fields, by making a fmall hole with a fpade, and a chip or fmall bit of wood, with the particular letter upon it, may be put in, to prevent one mark being taken for another, on any return to it. But in general, the name of a mark is very eafily had by referring in the book to the line it was made in. After the fmall alphabet is gone through, the capitals may be next, the print letters afterwards, and fo on, which anfwer the purpofe of fo many different letters ; or the marks may be numbered.

The letter in the left hand corner at the beginning of every line, is the mark or place measured *from* ; and, that at the right hand corner at the end, is the mark measured *to* : But when it is not convenient to go exactly from a mark, the place measured from, is described *fuch a diftance* from *one mark* towards *another* ; and where a mark is not measured to, the exact place is ascertained by faying, turn to the right or left hand, *fuch a diftance* to *fuch a mark*, it being always understood that thofe diftances are taken in the chain line.

The characters ufed, are Γ for *turn to the right hand*, ⊣ for turn to the left hand, and ∧ placed over an offset, to fhew that it is not taken at right angles with the chain line, but in the line with fome ftraight fence ; being chiefly ufed when croffing their directions, and is a better way of obtaining their true places than by offsets at right angles.

When a line is measured whofe pofition is determined, either by former work (as in the cafe of producing a given line or measuring from one known place or mark to another) or by itfelf (as in the third fide of a triangle) it is called a *faft line*, and a double line acrofs the book is drawn at the conclufion of it ; but if its pofition is not determined (as in the fecond fide of a triangle) it is called a *loofe line*, and a fingle line is drawn acrofs the book. When a line becomes determined in pofition, and is afterwards continued, a double line half through the book is drawn.

When a loofe line is measured, it becomes abfolutely neceffary to measure fome line that will determine its pofition. Thus, the firft line *ab*, being the bafe of a triangle, is always determined ; but the pofition of the fecond fide *hj*, does not become determined, till the third fide *jb* is measured; then the triangle may be constructed, and the pofition of both is determined.

At the beginning of a line, to fix a loofe line to the mark or place measured from, the fign of turning to the right or left hand muft be added (as at *j* in the third line) ; otherwife a ftranger, when laying down the work may as eafily construct the triangle *hjb* on the wrong fide of the line *ah*, as on the right one : but this error cannot be fallen into, if the fign above named be carefully obferved.

In choofing a line to fix a loofe one, care muft be taken that it does not make a very acute or obtufe angle ; as in the triangle *pBr*, by the angle at B being very obtufe, a fmall deviation from truth, even the breadth of a point at *p* or *r*, would make the error at B when conftructed very confiderable ; but by conftructing the triangle *pBq*, fuch a deviation is of no confequence.

Where the words *leave off* are written in the fieldbook, it is to fignify that the taking of offsets is from thence difcontinued ; and of courfe fomething is wanting between that and the next offset.

The field-book above referred to, is engraved on plate 29, in parts, reprefenting fo many pages, each of which is fuppofed to begin at the bottom, and end at top. And the map or plan belonging to it, in fig. 2, pl. 28.

17. *To Survey a County, or Large Tract of Land.*

1ft. Choofe two, three, or four eminent places for ftations ; fuch as the tops of high hills or mountains, towers, or church fteeples, which may be feen from one another ; and from which moft of the towns, and other places of note, may alfo be feen. And let them be as far diftant from one another as poffible. Upon thefe place raife beacons, or long poles, with flags of different colours flying at them ; fo as to be vifible from all the other ftations.

2d. At all the places, which you would fet down in the map, plant long poles with flags at them of feveral colours, to diftinguifh the places from one another ; fixing them upon the tops of church fteeples, or the tops of houfes, or in the centres of leffer towns.

But you need not have thefe marks at many places at once, as fuppofe half a fcore at a time. For when the angles have been taken, at the two ftations, to all thefe places, the marks may be moved to new ones ; and fo fucceffively to all the places you want. Thefe marks then being fet up at a convenient number of places, and fuch as may be feen from both ftations; go to one of thefe ftations, and with an inftrument to take angles, ftanding at that ftation, take all the angles between the other ftation, and each of thefe marks, obferving which is blue, which red, &c, and which hand they lie on ; and fet all down with their colours. Then go to the other ftation, and take all the angles between the firft ftation, and each of the former marks, and fet them down with the others, each againft his fellow with the fame colour. You may, if you can, alfo take the angles at fome third ftation, which may ferve to prove the work, if the three lines interfect in that point, where any mark ftands. The marks muft ftand till the obfervations are finifhed at both ftations ; and then they muft be taken down, and fet up at frefh places. And the fame operations muft be performed, at both ftations, for thefe frefh places ; and the like for others Your inftrument for taking angles muft be an exceeding good one, made on purpofe with telefcopic fights ; and of three, four, or five feet radius. A circumferentor is reckoned a good inftrument for this purpofe.

3d. And though it is not abfolutely neceffary to measure any diftance, becaufe a ftationary line being laid down from any fcale, all the other lines will be
proportional

(1)

	1794	to l
	1464	22
	1050	
~190	920	32
	650	60
	350	48
a	0	14 ....

	3074	to l
	2494	l
	2100	
0	2072	
54	1730	
80	1530	
	1420	k
56+30	1170	
52	620	
j	32	280	40

	2574	J
	2494	
	2000	44
	1880	30
	1840	
50	1794	i
34+50	1464	
76	1328	
96	1240	
52+34	1130	
34	860	
h 66	190	

	1450	h
	3570	g
	2620	f
	2590	
	2210	
	2080	e
	1574	d
	1550	
	1510	c
	990	b
a	806	

(2)

30	1480	x
0	1320	
30	1110	
	1080	
	990	w
u	750	50

	4440	36
	4420	v
	3884	u
	3380	60
	2992	90
	2692	t
120	2624	
	2592	
	2500	s
	2070	56 leave off
	1900	
	1840	r
60	1770	
	1320	q
	808	p
leave off 40	650	
80	360	
O 20	170	

	220	O
h produce by i	190	46

	1310	56 to l
	836	56
n	681	50

	1480	90 to g
	960	24
	930	n
	700	48
m	400	30

	1430	to i
	1290	40
	1004	36
	980	m
	610	24
k	280	32

(3)

D	488	32

	2280	
	2270	E
	2230	
20	2050	
56	2030	

	1940	180 to w
	1552	180
	1380	96
	950	110
	860	

	768	to A
	526	70
	496	
	460	
70	124	
D 40	100	

	455	D
	400	76
C	48	10

	600	to r
	432	C
B 56	160	
44	36	

B	152	to q
	480	B
r 24	160	
0	1750	

	1600	44 to s
	1028	
	940	4
44	666	
70	310	3
d 60	236	

	2148	480 to b
	1950	y
	1836	
120	1724	
60	1600	

(4)

	580	to v
10	500	
76	300	
F 16	100	

	360	to F
J 20	150	

	954	J
15	850	

	730	to E
30	490	
	340	60
0	280	
I 20	170	50

	744	to H
	672	0
a 70	450	0

	1160	to y
32	1000	
	890	
	780	32
	590	40
	570	I
	530	40
	376	H
	256	150
	190	64
G	144	130 leave off

	1676	G
	1676	30
	896	24
	632	
	620	50
180 from u towards r	588	F

	1068	to x
0	1032	
28	850	
70 from z towards A 44	528	
	644	to f

proportional to it; yet it is better to measure some of the lines, to ascertain the distances of places in miles; and to know how many geometrical miles there are in any length; and from thence to make a scale to measure any distance in miles. In measuring any distance, it will not be exact enough to go along the high roads; by reason of their turnings and windings, and hardly ever lying in a right line between the stations, which would cause endless reductions, and create trouble to make it a right line; for which reason it can never be exact. But a better way is to measure in a right line with a chain, between station and station, over hills and dales or level fields, and all obstacles. Only in case of water, woods, towns, rocks, banks, &c, where one cannot pass, such parts of the line must be measured by the methods of inaccessible distances; and besides, allowing for ascents and descents, when we meet with them. And a good compass that shews the bearing of the two stations, will always direct you to go straight, when you do not see the two stations; and in your progress, if you can go straight, you may take offsets to any remarkable places, likewise noting the intersection of the stationary line with all roads, rivers, &c.

4th. And from all the stations, and in the whole progress, be very particular in observing sea coasts, river mouths, towns, castles, houses, churches, windmills, watermills, trees, rocks, sands, roads, bridges, fords, ferries, woods, hills, mountains, rills, brooks, parks, beacons, sluices, floodgates, locks, &c; and in general all things that are remarkable.

5th. After you have done with the first and main station lines, which command the whole county; you must then take inner stations, at some places already determined; which will divide the whole into several partitions: and from these stations you must determine the places of as many of the remaining towns as you can. And if any remain in that part, you must take more stations, at some places already determined; from which you may determine the rest. And thus proceed through all the parts of the country, taking station after station, till we have determined all we want. And in general the station distances must always pass through such remarkable points as have been determined before, by the former stations.

6th. Lastly, the position of the station line you measure, or the point of the compass it lies on, must be determined by astronomical observation. Hang up a thread and plummet in the sun, over some part of the station line, and observe when the shadow runs along that line, and at that moment take the sun's altitude; then having his declination, and the latitude, the azimuth will be found by spherical trigonometry. And the azimuth is the angle the station line makes with the meridian; and therefore a meridian may easily be drawn through the map: Or a meridian may be drawn through it by hanging up two threads in a line with the pole star, when he is just north, which may be known from astronomical tables. Or thus; observe the star Alioth, or that in the rump of the great bear, being that next the square; or else Cassiopeia's hip; I say, observe by a line and plummet when either of these stars and the pole star come into a perpendicular; and at that time they are due north. There-

fore two perpendicular lines being fixed at that moment, towards these two stars, will give the position of the meridian.

18. *To Survey a Town or City.*

This may be done with any of the instruments for taking angles, but best of all with the plain table, where every minute part is drawn while in sight. It is proper also to have a chain of 50 feet long, divided into 50 links, and an offset-staff of 10 feet long.

Begin at the meeting of two or more of the principal streets, through which you can have the longest prospects, to get the longest station lines. There having fixed the instrument, draw lines of direction along those streets, using two men as marks, or poles set in wooden pedestals, or perhaps some remarkable places in the houses at the farther ends, as windows, doors, corners, &c. Measure these lines with the chain, taking offsets with the staff, at all corners of streets, bendings, or windings, and to all remarkable things, as churches, markets, halls, colleges, eminent houses, &c. Then remove the instrument to another station along one of these lines; and there repeat the same process as before. And so on till the whole is finished.

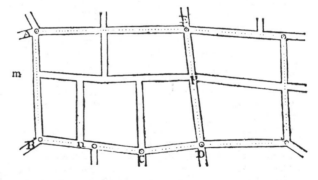

Thus, fix the instrument at A, and draw lines in the direction of all the streets meeting there; and measure AB, noting the street on the left at m. At the second station B, draw the directions of the streets meeting there; measure from B to C, noting the places of the streets at n and o as you pass by them. At the 3d station C, take the direction of all the streets meeting there, and measure CD. At D do the same, and measure DE, noting the place of the cross streets at p. And in this manner go through all the principal streets. This done, proceed to the smaller and intermediate streets; and lastly to the lanes, alleys, courts, yards, and every part that it may be thought proper to represent.

Of Planning, Computing, and Dividing.

19. *To Lay down the Plan of any Survey.*

If the survey was taken with a plain table, you have a rough plan of it already on the paper which covered the

the table. But if the furvey was with any other inftrument, a plan of it is to be drawn from the meafures that were taken in the furvey, and firft of all a rough plan upon paper.

To do this, you muft have a fet of proper inftruments, for laying down both lines and angles, &c ; as fcales of various fizes, the more of them, and the more accurate, the better ; fcales of chords, protraĉtors, perpendicular and parallel rulers, &c. Diagonal fcales are beft for the lines, becaufe they extend to three figures, or chains and links, which are hundredth parts of chains. And in ufing the diagonal fcale, a pair of compaffes muft be employed to take off the lengths of the principal lines very accurately. But a fcale with a thin edge divided, is much readier for laying down the perpendicular offsets to crooked hedges, and for marking the places of thofe offsets upon the ftation line ; which is done at only one application of the edge of the fcale to that line, and then pricking off all at once the diftances along it. Angles are to be laid down either with a good fcale of chords, which is perhaps the moft accurate way ; or with a large protraĉtor, which is much readier when many angles are to be laid down at one point, as they are pricked off all at once round the edge of the protraĉtor.

Very particular direĉtions for laying down all forts of figures cannot be neceffary in this place, to any perfon who has learned praĉtical geometry, or the conftruĉtion of figures, and the ufe of his inftruments. It may therefore be fufficient to obferve, that all lines and angles muft be laid down on the plan in the fame order in which they were meafured in the field, and in which they are written in the field-book ; laying down firft the angles for the pofition of lines, then the lengths of the lines, with the places of the offsets, and then the lengths of the offsets themfelves, all with dry or obfcure lines ; then a black line drawn through the extremities of all the offsets, will be the hedge or bounding line of the field, &c. After the principal bounds and lines are laid down, and made to fit or clofe properly, proceed next to the fmaller objeĉts, till you have entered every thing that ought to appear in the plan, as houfes, brooks, trees, hills, gates, ftiles, roads, lanes, mills, bridges, woodlands, &c, &c.

The north fide of a map or plan is commonly placed uppermoft, and a meridian fomewhere drawn, with the compafs or flower-de-luce pointing north. Alfo, in a vacant place, a fcale of equal parts or chains is drawn, and the title of the map in confpicuous charaĉters, and embellifhed with a compartment. All hills muft be fhadowed, to diftinguifh them in the map. Colour the hedges with different colours ; reprefent hilly grounds by broken hills and valleys ; draw fingle dotted lines for foot-paths, and double ones for horfe or carriage roads. Write the name of each field and remarkable place within it, and, if you choofe, its content in acres, roods, and perches.

In a very large eftate, or a county, draw vertical and horizontal lines through the map, denoting the fpaces between them by letters, placed at the top, and bottom, and fides, for readily finding any field or other objeĉt, mentioned in a table.

In mapping counties, and eftates that have uneven

6

grounds of hills and valleys, reduce all oblique line meafured up hill and down hill, to horizontal ftraight lines, if that was not done during the furvey, before the were entered in the field-book, by making a proper a lowance to fhorten them. For which purpofe, there commonly a fmall table engraven on fome of the inftruments for Surveying.

20. *To Compute the Contents of Fields.*

1ft. Compute the contents of the figures, whethe triangles, or trapeziums, &c, by the proper rules fo the feveral figures laid down in meafuring ; multiply ing the lengths by the breadths, both in links ; th produĉt is acres after you have cut off five figures on the right, for decimals ; then bring thefe decimals to roods and perches, by multiplying firft by 4, and ther by 40. An example of which was given in the defcrip tion of the chain, art. 1.

2d. In fmall and feparate pieces, it is ufual to caf up their contents from the meafures of the lines taken in furveying them, without making a correĉt plan o them.

Thus, in the triangle in art. 7, where we hac AP = 794, and AB = 1321
$$PC = 826$$

$$7926$$
$$2642$$
$$10568$$

2) 10·91146
5·45573   ac r p
4 Anf. 32 1 33 nearl

1·82292
40

32·91680

Or the firft example to art. 8, thus :

AE 214	210 ED
AF 362	306 FB
AC 592	

516 fum of perp.
592 AC

1032
4644
2580

3·05472   ac r p
4   Anf. 3 0 8

·21888
40

8·75520

On.

Or the 2d example to the fame article, thus:

AP ·	110		352	PC
AQ	745		595	QD
AB	1110			

| PC | 352 | | PC | 352 | | QD | 595 |
| AP | 110 | | QD | 595 | | QB | 365 |

2 APC	38720		fum	947			2975
			PQ	635			3570
							1785
				4735			
				2841		217175	= 2QDB
				5682		601345	= 2PCDQ
						38720	= 2APC

2PCDQ 601345

2) 8·57240 = dou. the whole
4·2862
4

          ac r p
Anf. 4 1 5

1·1448
40

5·7920

3d. In pieces bounded by very crooked and winding hedges, meafured by offsets, all the parts between the offsets are moft accurately meafured feparately as fmall trapezoids. Thus, for the example to art. 6, where

Ac	45		62	ch
Ad	220		84	di
Ae	340		70	ek
Af	510		98	fl
Ag	634		57	gm
AB	785		91	Bn

Then

Ac	45	ch	62	di	84	ek	70	fl	98	gm	57
ch	62	di	84	ek	70	fl	98	gm	57	Bn	91
	90		146		154		168		145		148
	270	cd	175	de	120	ef	170	fg	124	gB	151
	2790		730		18480		11760		580		148
			1022				168		290		740
			146						145		148
							28560				
	25550								17980		22348

2790
25550
18480
28560
17980
22348

2) 1·15708
·57854

                   ac r p
Content 0 2 12

4

2·31416
40

12·56640

4th. Sometimes fuch pieces as that above, are computed by finding a mean breadth, by dividing the fum of the offsets by the number of them, accounting that for one of them where the boundary meets the ftation line, as at A; then multiply the length AB by that mean breadth.

Thus:

00		785	AB
62		66	mean breadth
84			
70		4710	ac r p
98		4710	Content 0 2 2 by this method,
57			which is 10 perches too little.
91		·51810	

7) 462
66

2·07240
40

2·89600

But this method is always erroneous, except when the offsets ftand at equal diftances from one another.

5th. But in larger pieces, and whole eftates, confifting of many fields, it is the common practice to make a rough plan of the whole, and from it compute the contents quite independent of the meafures of the lines and angles that were taken in Surveying. For then new lines are drawn in the fields in the plan, fo as to divide them into trapeziums and triangles, the bafes and perpendiculars of which are meafured on the plan by means of the fcale from which it was drawn, and fo multiplied together for the contents. In this way the work is very expeditioufly done, and fufficiently correct; for fuch dimenfions are taken, as afford the moft eafy method of calculation; and, among a number of parts, thus taken and applied to a fcale, it is likely that fome of the parts will be taken a fmall matter too little, and others too great; fo that they will, upon the whole, in all probability, very nearly balance one another. After all the fields, and particular parts, are thus computed feparately, and added all together into one fum, calculate the whole eftate independent of the fields, by dividing it into large and arbitrary triangles and trapeziums, and add thefe alfo together. Then if this fum be equal to the former, or nearly fo, the work is right; but if the fums have any confiderable difference, it is wrong, and they muft be examined, and recomputed, till they nearly agree.

A fpecimen of dividing into one triangle, or one trapezium, which will do for moft fingle fields, may be feen in the examples to the laft article; and a fpecimen of dividing a large tract into feveral fuch trapeziums and triangles, in article 9, where a piece is fo divided, and its dimenfions taken and fet down; and again in articles 15, 16.

6th. But the chief fecret in cafting up, confifts in finding the contents of pieces bounded by curved, or very irregular lines, or in reducing fuch crooked fides of fields or boundaries to ftraight lines, that fhall inclofe the fame or equal area with thofe crooked fides, and fo obtain the area of the curved figure by means of the right-lined one, which will commonly be a trape-

4 B   zium.

zium. Now this reducing the crooked sides to straight ones, is very easily and accurately performed thus : Apply the straight edge of a thin, clear piece of lanthorn-horn to the crooked line, which is to be reduced, in such a manner, that the small parts cut off from the crooked figure by it, may be equal to those which are taken in : which equality of the parts included and excluded, you will presently be able to judge of very nicely by a little practice : then with a pencil draw a line by the straight edge of the horn. Do the same by the other sides of the field or figure. So shall you have a straight-sided figure equal to the curved one ; the contents of which, being computed as before directed, will be the content of the curved figure proposed.

Or, instead of the straight edge of the horn, a horsehair may be applied across the crooked sides in the same manner ; and the easiest way of using the hair, is to string a small slender bow with it, either of wire, or cane, or whale-bone, or such like slender springy matter ; for, the bow keeping it always stretched, it can be easily and neatly applied with one hand, while the other is at liberty to make two marks by the side of it, to draw the straight line by.

Ex. Thus, let it be required to find the contents of the same figure as in art. 12, to a scale of 4 chains to an inch.

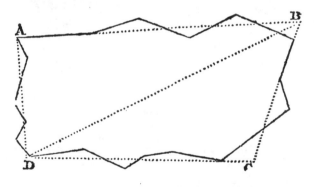

Draw the four dotted straight lines AB, BC, CD, DA, cutting off equal quantities on both sides of them, which they do as near as the eye can judge : so is the crooked figure reduced to an equivalent right-lined one of four sides ABCD. Then draw the diagonal BD, which, by applying a proper scale to it, measures 1256. Also the perpendicular, or nearest distance, from A to this diagonal, measures 456 ; and the distance of C from it, is 428. Then

$$
\begin{array}{rl}
456 & 2\ )\ 11{\cdot}10304 \\
428 & \overline{5{\cdot}55152} \\
\overline{884} & 4 \\
1256 & \overline{2{\cdot}20608} \\
\overline{5024} & 40 \\
10048 & \overline{8{\cdot}24320} \\
10048 & \\
\overline{1110304} &
\end{array}
$$

And thus the content of the trapezium, and consequently of the irregular figure, to which it is equal, is easily found to be 5 acres, 2 roods, 8 perches.

21. *To Transfer a Plan to another Paper, &c.*

After the rough plan is completed, and a fair one is wanted ; this may be done, either on paper or vellum, by any of the following Methods.

First Method.—Lay the rough plan upon the clean paper, and keep them always pressed flat and close together, by weights laid upon them. Then, with the point of a fine pin or pricker, prick through all the corners of the plan to be copied. Take them asunder, and connect the pricked points on the clean paper, with lines ; and it is done. This method is only to be practised in plans of such figures as are small and tolerably regular, or bounded by right lines.

Second Method.—Rub the back of the rough plan over with black lead powder ; and lay the said black part upon the clean paper, upon which the plan is to be copied, and in the proper position. Then, with the blunt point of some hard substance, as brass, or such like, trace over the lines of the whole plan ; pressing the tracer so much as that the black lead under the lines may be transferred to the clean paper ; after which take off the rough plan, and trace over the leaden marks with common ink, or with Indian ink, &c.—Or, instead of blacking the rough plan, you may keep constantly a blacked paper to lay between the plans.

Third Method.—Another way of copying plans, is by means of squares. This is performed by dividing both ends and sides of the plan, which is to be copied, into any convenient number of equal parts, and connecting the corresponding points of division with lines ; which will divide the plan into a number of small squares. Then divide the paper, upon which the plan is to be copied, into the same number of squares, each equal to the former when the plan is to be copied of the same size, but greater or less than the others, in the proportion in which the plan is to be increased or diminished, when of a different size. Lastly, copy into the clean squares, the parts contained in the corresponding squares of the old plan ; and you will have the copy either of the same size, or greater or less in any proportion.

Fourth Method.—A fourth way is by the instrument called a pentagraph, which also copies the plan in any size required.

Fifth Method.—But the neatest method of any is this. Procure a copying frame or glass, made in this manner ; namely, a large square of the best window glass, set in a broad frame of wood, which can be raised up to any angle, when the lower side of it rests on a table. Set this frame up to any angle before you, facing a strong light ; fix the old plan and clean paper together with several pins quite around, to keep them together, the clean paper being laid uppermost, and upon

upon the face of the plan to be copied. Lay them, with the back of the old plan, upon the glafs, namely, that part which you intend to begin at to copy firft ; and, by means of the light fhining through the papers, you will very diftinctly perceive every line of the plan through the clean paper. In this ftate then trace all the lines on the paper with a pencil. Having drawn that part which covers the glafs, flide another part over the glafs, and copy it in the fame manner. And then another part. And fo on till the whole be copied.

Then, take them afunder, and trace all the pencil-lines over with a fine pen and Indian ink, or with common ink.

And thus you may copy the fineft plan, without in-juring it in the leaft.

When the lines, &c, are copied upon the clean paper or vellum, the next bufinefs is to write fuch names, remarks, or explanations as may be judged neceffary ; laying down the fcale for taking the lengths of any parts, a flower-de-luce to point out the direction, and the proper title ornamented with a compartment ; and illuftrating or colouring every part in fuch manner as fhall feem moft natural, fuch as fhading rivers or brooks with crooked lines, drawing the reprefentations of trees, bufhes, hills, woods, hedges, houfes, gates, roads, &c, in their proper places ; running a fingle dotted line for a foot path, and a double one for a carriage road ; and either reprefenting the bafes or the elevations of buildings, &c.

22. Of the Divifion of Lands.

In the divifion of commons, after the whole is furveyed and caft up, and the proper quantities to be allowed for roads, &c, deducted, divide the net quantity remaining among the feveral proprietors, by the rule of Fellowfhip, in proportion to the real value of their eftates, and you will thereby obtain their proportional quantities of the land. But as this divifion fuppofes the land, which is to be divided, to be all of an equal goodnefs, you muft obferve that if the part in which any one's fhare is to be marked off, be better or worfe than the general mean quality of the land, then you muft diminifh or augment the quantity of his fhare in the fame proportion.

Or, which comes to the fame thing, divide the ground among the claimants in the direct ratio of the value of their claims, and the inverfe ratio of the quality of the ground allotted to each ; that is, in proportion to the quotients arifing from the divifion of the value of each perfon's eftate, by the number which expreffes the quality of the ground in his fhare.

But thefe regular methods cannot always be put in practice ; fo that, in the divifion of commons, the ufual way is, to meafure feparately all the land that is of different values, and add into two fums the contents and the values ; then, the value of every claimant's fhare is found, by dividing the whole value among them in proportion to their eftates ; and, laftly, by the 24th

article, a quantity is laid out for each perfon, that fhall be of the value of his fhare before found.

23. *It is required to divide any given Quantity of Ground, or its Value, into any given Number of Parts, and in Proportion as any given Numbers.*

Divide the given piece, or its value, as in the rule of Fellowfhip, by dividing the whole content or value by the fum of the numbers expreffing the proportions of the feveral fhares, and multiplying the quotient feverally by the faid proportional numbers for the refpective fhares required, when the land is all of the fame quality. But if the fhares be of different qualities, then divide the numbers expreffing the proportions or values of the fhares, by the numbers which exprefs the qualities of the land in each fhare ; and ufe the quotients inftead of the former proportional numbers.

Ex. 1. If the total value of a common be 2500 pounds, it is required to determine the values of the fhares of the three claimants A, B, C, whofe eftates are of thefe values, 10000, and 15000, and 25000 pounds.

The eftates being in proportion as the numbers 2, 3, 5, whofe fum is 10, we fhall have $2500 \div 10 = 250$; which being feverally multiplied by 2, 3, 5, the products 500, 750, 1250, are the values of the fhares required.

Ex. 2. It is required to divide 300 acres of land among A, B, C, D, E, F, G, and H, whofe claims upon it are refpectively in proportion as the numbers 1, 2, 3, 5, 8, 10, 15, 20.

The fum of thefe proportional numbers is 64, by which dividing 300, the quotient is 4 ac. 2 r. 30 p. which being multiplied by each of the numbers, 1, 2, 3, 5, &c, we obtain for the feveral fhares as below :

	Ac.	R.	P.
A =	4	2	30
B =	9	1	20
C =	14	0	10
D =	23	1	30
E =	37	2	00
F =	46	3	20
G =	70	1	10
H =	93	3	00
Sum =	300	0	00

Ex. 3. It is required to divide 780 acres among A, B, and C, whofe eftates are 1000, 3000, and 4000 pounds a year ; the ground in their fhares being worth 5, 8, and 10 fhillings the acre refpectively.

Here their claims are as 1, 3, 4 ; and the qualities of their land are as 5, 8, 10 ; therefore their quantities muft be as $\frac{1}{5}, \frac{3}{8}, \frac{4}{10}$, or, by reduction, as 8, 15, 16. Now the fum of thefe numbers is 39 ; by which dividing the 780 acres, the quotient is 20 ; which being multiplied feverally by the three numbers 8, 15, 16, the three products are 160, 300, 320, for the fhares of A, B, C, refpectively.

24. *To Cut off from a Plan a Given Number of Acres,
&c, by a Line drawn from any Point in the Side
of it.*

Let A be the given point in the annexed plan, from
which a line is to be drawn cutting off suppose 5 ac.
2 r. 14 p.

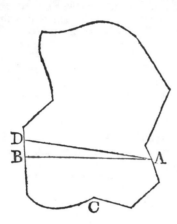

Draw AB cutting off the part ABC as near as can
be judged equal to the quantity proposed; and let the
true quantity of ABC, when calculated, be only 4 ac.
3 r. 20 p. which is less than 5 ac. 2 r. 14 p. the true quantity, by 0 ac. 2 r. 34 p. or 71250 square links. Then
measure AB, which suppose = 1234 links, and divide
71250, by 617 the half of it, and the quotient 115 links
will be the altitude of the triangle to be added, and
whose base is AB. Therefore if upon the centre B,
with the radius 115, an arc be described; and a line be
drawn parallel to AB, touching the arc, and cutting
BD in D; and if AD be drawn, it will be the line
cutting off the required quantity ADCA.

NOTE. If the first piece had been too much, then D
must have been set below B.

In this manner the several shares of commons, to be
divided, may be laid down upon the plan, and transferred
from thence to the ground itself.

Also for the greater ease and perfection in this business, the following problems may be added.

25. *From an Angle in a Given Triangle, to draw Lines
to the opposite Side, dividing the Triangle into any Number of Parts, which shall be in any assigned Proportion to each other.*

Divide the base into the same number of parts, and
in the same proportion, by article 22; then from the
several points of division draw lines to the proposed angle, and they will divide the triangle as required.—
For, the several parts are triangles of the same altitude,
and which therefore are as their bases, which bases are
taken in the assigned proportion.

Ex. Let the triangle ABC, of 20 acres, be divided
into five parts, which shall be in proportion to the numbers 1, 2, 3, 5, 9; the lines of division to be drawn
from A to CB, whose length is 1600 links.

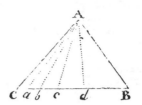

Here $1 + 2 + 3 + 5 + 9 = 20$, and $1600 \div 20 = 80$;
which being multiplied by each of the proportional
numbers, we have 80, 160, 240, 400, and 720. Therefore make $Ca = 80$, $ab = 160$, $bc = 240$, $cd =
400$, and $dB = 720$; then by drawing the lines Aa,
Ab, Ac, Ad, the triangle is divided as required.

26. *From any Point in one side of a Given Triangle, to
draw Lines to the other two Sides, dividing the Triangle into any Number of Parts which shall be in any assigned Ratio.*

From the given point D, draw DB to the angle opposite the side AC in which the point is taken; then divide the same side AC into as many parts AE, EF, FG,
GC, and in the same proportion with the required
parts of the triangle, like as was done in the last problem; and from the points of division draw lines EK,
FI, GH, parallel to the line BD, and meeting the
other sides of the triangle in K, I, H; lastly, draw
KD, ID, HD, so shall ADK, KDI, IDHB, HDC
be the parts required.—The example to this will be
done exactly as the last.

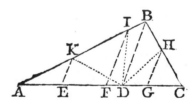

For, the triangles ADK, KDI, IDB, being of the
same height, are as their bases AK, KI, IB; which,
by means of the parallels EK, FI, DB, are as AE,
EF, FD; in like manner, the triangles CDH, HDB,
are to each other as CG, GD: but the two triangles
IDB, BDH, having the same base BD, are to each
other as the distances of I and H from BD, or as FD
to DG; consequently the parts DAK, DKI, DIBH,
DHC, are to each other as AE, EF, FG, GC.

Surveying of Harbours.

The method of Surveying harbours, and of forming
maps of them, as also of the adjacent coasts, sands,
&c, depends on the same principles, and is chiefly conducted like that of common Surveying. The operation is indeed more complicated and laborious; as it is
necessary to erect a number of signals, and to mark a
variety of objects along the coast, with different bearings from one another, and the several parts of the harbour; and likewise to measure a great number of angles
at

at different stations, whether on the land or the water. For this purpose, the best instrument is Hadley's quadrant, as all these operations may be performed by it, not only with greater ease, but also with much more precision, than can be hoped for by any other means, as it is the only instrument in use, in which neither the exactness of the observations, nor the ease with which they may be made, are sensibly affected by the motion of a vessel: and hence a single observer, in a boat, may generally determine the situation of any place at pleasure, with a sufficient degree of exactness, by taking the angles subtended by several pairs of objects properly chosen upon shores round about him; but it will be still better to have two observers, or the same observer at different stations, to take the like angles to the several objects, and also to the stations. By this means, two angles and one side are given, in every triangle, from whence the situation of every part of them will be known. By such observations, when carefully made with good instruments, the situation of places may be easily determined to 20 or 30 feet, or less, upon every 3 or 4 miles. See Philos. Transf. vol. 55, pa. 70; also Mackenzie's Maritime Surveying.

SURVEYING *Cross*. See CROSS.

SURVEYING *Quadrant*. See QUADRANT.

SURVEYING *Scale*, the same with Reducing Scale.

SURVEYING *Wheel*. See PERAMBULATOR.

SURVIVORSHIP, the doctrine of reversionary payments that depend upon certain contingencies, or contingent circumstances.

Payments which are not to be made till some future period, are termed *reversions*, to distinguish them from payments that are to be made immediately.

Reversions are either *certain* or *contingent*. Of the former sort, are all sums or annuities, payable certainly or absolutely at the expiration of any terms, or on the extinction of any lives. And of the latter sort, are all such reversions as depend on any contingency; and particularly the Survivorship of any lives beyond or after other lives. An account of the former may be found under the articles Assurance, Annuities, and Life annuities. But the latter form the most intricate and difficult part of the doctrine of reversions and life-annuities; and the books in which this subject is treated most at large, and at the same time with the most precision, are Mr. Simpson's Select Exercises; Dr. Price's Reversionary Payments; and Mr. Morgan's Annuities and Assurances on Lives and Survivorships. The whole likewise of the 3d volume of Dodson's Mathematical Repository is on this subject; but his investigations are founded on De Moivre's false hypothesis, viz of an equal decrement of life through all its stages, and which is explained under Life-annuities: but as this hypothesis does not agree near enough to fact and experience, the rules deduced from it cannot be sufficiently correct. For this reason, Dr. Price, and also the ingenious Mr. Maseres, cursitor baron of the exchequer (in two volumes lately published, entitled the Principles of the Doctrine of Life Annuities), have discarded the valuations of lives grounded upon it; and the former in particular, in order to obviate all occasion for using them, has substituted in their stead, a great variety of new tables of the probabilities and values of lives, at every age and in every situation; calculated, not upon any hypothesis, but in strict conformity to the best observations. These tables, added to other new tables of the same kind, in Mr. Baron Maseres's work just mentioned, form a complete set of tables, by which all questions relating to annuities on lives and Survivorships, may be answered with as much correctness as the nature of the subject allows.

Rules for calculating correctly, in most cases, the values of reversions depending on Survivorships, may be found in the three treatises just mentioned. Mr. Morgan, in particular, has gone a good way towards exhausting this subject, as far as any questions can include in them any Survivorships between two or three lives, either for terms, or the whole duration of the lives.

There is, however, one circumstance necessary to be attended to in calculating such values, to which no regard could be paid till lately. This circumstance is the shorter duration of the lives of males than of females; and the consequent advantage in favour of females in all cases of Survivorship. In the 4th edition of Dr. Price's Treatise on Reversionary Payments, this fact is not only ascertained, but separate tables of the duration and values of lives are given for males and females.

SUSPENSION, in Mechanics, as in a balance, are those points in the axis or beam where the weights are applied, or from which they are suspended.

SUTTON's *Quadrant*. See QUADRANT.

SWAN, in Astronomy. See CYGNUS.

SWALLOW's-TAIL, in Fortification, is a single Tenaille, which is narrower towards the place than towards the country.

SWING-*Wheel*, in a royal pendulum, is that wheel which drives the pendulum. In a watch, or balance clock, it is called the *crown-wheel*.

SYDEREAL *Day*, or *Year*. See SIDEREAL.

SYMMETRY, the relation of parity, both in respect of length, breadth, and height, of the parts necessary to compose a beautiful whole.

Symmetry arises from that proportion which the Greeks call *analogy*, which is the relation of conformity of all the parts of a building, and of the whole, to some certain measure; upon which depends the nature of Symmetry.

According to Vitruvius, Symmetry consists in the union and conformity of the several members of a work to their whole, and of the beauty of each of the separate parts to that of the intire work; regard being had to some certain measure: so the body, for instance, is framed with Symmetry, by the due relation which the arm, elbow, hand, fingers, &c, have to each other, and to their whole.

SYMPHONY, is a consonance or concert of several sounds agreeable to the ear; whether they be vocal or instrumental, or both; called also *harmony*.

The Symphony of the Ancients went no farther than to two or more voices or instruments set to unison; for they had no such thing as music in parts; as is very well proved by Perrault: at least, if ever they knew such a thing, it must have been early lost.

It is to Guido Aretine, about the year 1022, that most writers agree in ascribing the invention of composition: it was he, they say, who first joined in one harmony several distinct melodies; and brought it even to the

the length of 4 parts, viz. bass, tenor, counter-tenor, and treble.

The term Symphony is now applied to instrumental music, both that of pieces designed only for instruments, as sonatas and concertos, and that in which the instruments are accompanied with the voice, as in operas, &c.

A piece is said to be in grand Symphony, when, besides the bass and treble, it has also two other instrumental parts, viz, tenor and 5th of the violin.

SYNCHRONISM, the being or happening of several things together, at or in the same time.

The happening or performing of several things in equal times, as the vibrations of pendulums, &c, is more properly called *isochronism*: though some authors confound the two.

SYNCOPATION, in Music, denotes a striking or breaking of the time; by which the distinctness of the several times or parts of the measure is interrupted.

SYNCOPATION, or SYNCOPE, is more particularly used for the connecting the last note of one measure or bar with the first of the following measure; so as to make only one note of both.

SYNCOPATION is also used when a note of one part ends on the middle of a note of the other part. This is otherwise called *binding*.

SYNODICAL *Month*, is the period or interval of time in which the moon passes from one conjunction with the sun to another. This period is also called a *Lunation*, since in this period the moon puts on all her phases, or appearances, as to increase and decrease. —Kepler found the quantity of the mean Synodical month to be 29 days, 12 hrs, 44 min. 3 sec. 11 thirds.

SYNTHESIS denotes a method of composition, as opposed to analysis.

In the Synthesis, or synthetic method, we pursue the truth by reasons drawn from principles before established, or assumed, and propositions formerly proved; thus proceeding by a regular chain till we come to the conclusion; and hence called also the *direct* method, and *composition*, in opposition to analysis or resolution.

Such is the method in Euclid's Elements, and most demonstrations of the ancient mathematicians, which proceed from definitions and axioms, to prove theorems &c, and from those theorems proved, to demonstrate others. See ANALYSIS.

SYNTHETICAL *Method*, the method by Synthesis, or composition, or the direct method. See SYNTHESIS.

SYPHON. See SIPHON.

SYRINGE, in Hydraulics, a small simple machine, serving first to imbibe or suck in a quantity of water, or other fluid, and then to squirt or expel the same with violence in a small jet.

The Syringe is just a small single sucking pump, without a valve, the water ascending in it on the same principle. It consists, like the pump, of a small cylinder, with an embolus or sucker, moving up and down in it by means of a handle, and fitting it very close within. At the lower end is either a small hole, or a smaller tube fixed to it than the body of the instrument, through which the fluid or the water is drawn up, and squirted out again.

Thus, the embolus being first pushed close down, introduce the lower end of the pipe into the fluid, then draw up, by the handle, the sucker, and the fluid will immediately follow, so as to fill the whole tube of the Syringe, and will remain there, even when the pipe is taken out of the fluid; but by thrusting forward the embolus, it will drive the water before it; and, being partly impeded by the smallness of the hole, or pipe, it will hence be expelled in a smart jet or squirt, and to the greater distance, as the sucker is pushed down with the greater force, or the greater velocity.

This ascent of the water the Ancients, who supposed a plenum, attributed to Nature's abhorrence of a vacuum; but the Moderns, more reasonably, as well as more intelligibly, attribute it to the pressure of the atmosphere on the exterior surface of the fluid. For, by drawing up the embolus, the cavity of the cylinder would become a vacuum, or the air left there extremely rarefied; so that being no longer a counterbalance to the air incumbent on the surface of the fluid, this prevails, and forces the water through the little tube, or hole, up into the body of the Syringe.

SYSTEM, in a general Sense, denotes an assemblage or chain of principles and conclusions: or the whole of any doctrine, the several parts of which are bound together, and follow or depend on each other. As a System of astronomy, a System of planets, a System of philosophy, a System of motion, &c.

SYSTEM, in Astronomy, denotes an hypothesis or a supposition of a certain order and arrangement of the several parts of the universe; by which astronomers explain all the phenomena or appearances of the heavenly bodies, their motions, changes, &c.

This is more peculiarly called the *System of the world*, and sometimes the *Solar System*.

System and hypothesis have much the same signification; unless perhaps hypothesis be a more particular System, and System a more general hypothesis.

Some late authors indeed make another distinction: an hypothesis, say they, is a mere supposition or fiction, founded rather on imagination than reason; while a System is built on the firmest ground, and raised by the severest rules; it is founded on astronomical observations, and physical causes, and confirmed by geometrical demonstrations.

The most celebrated Systems of the world, are the Ptolomaic, the Copernican or Pythagorean, and the Tychonic: the economy of each of which is as follows.

Ptolomaic SYSTEM is so called from the celebrated astronomer Ptolomy. In this System, the earth is placed at rest, in the centre of the universe, while the heavens are considered as revolving about it, from east to west, and carrying along with them all the heavenly bodies, the stars and planets, in the space of 24 hours.

The principal assertors of this System, are Aristotle, Hipparchus, Ptolomy, and many of the old philosophers, followed by the whole world, for a great number of ages, and long adhered to in many universities, and other places. But the late improvements in philosophy and reasoning, have utterly exploded this erroneous System from the place it so long held in the minds of men.

Copernican SYSTEM, is that System of the world which

Plate XXXII

UNIVERSAL SOLAR SYSTEM.

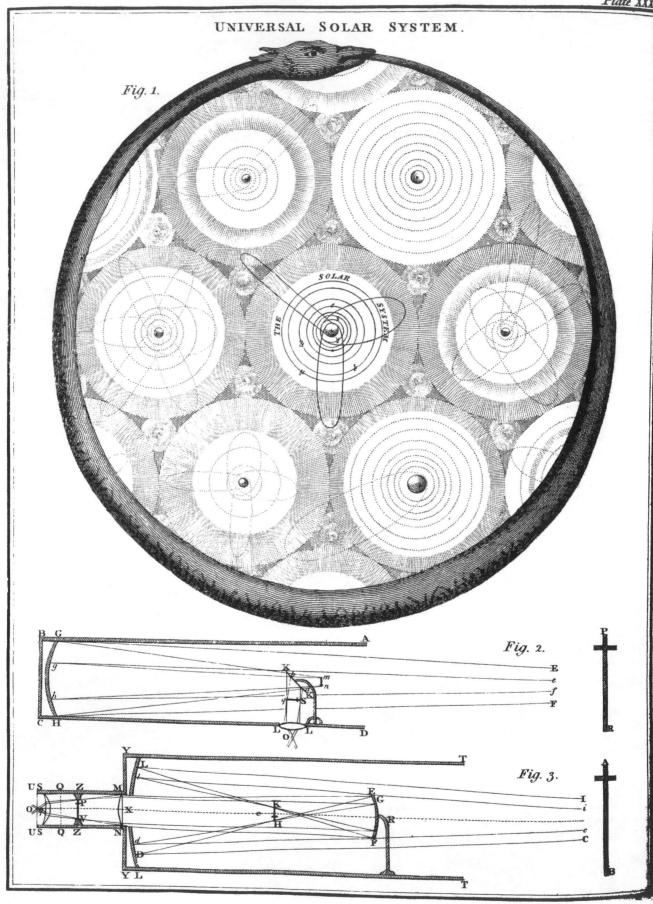

Fig. 1.

THE SOLAR SYSTEM

Fig. 2.

Fig. 3.

which places the Sun at reſt, in the centre of the world, and the earth and planets all revolving round him, in their ſeveral orbits. See this more particularly explained under the article COPERNICAN *Syſtem*.

Solar or *Planetary* SYSTEM, is uſually confined to narrower bounds; the ſtars, by their immenſe diſtance, and the little relation they ſeem to bear to us, being accounted no part of it. It is highly probable that each fixed ſtar is itſelf a Sun, and the centre of a particular Syſtem, ſurrounded with a company of planets &c, which, in different periods, and at different diſtances, perform their courſes round their reſpective ſun, which enlightens, warms, and cheriſhes them. Hence we have a very magnificent idea of the world, and the immenſity of it. Hence alſo ariſes a kind of Syſtem of Syſtems.

The Planetary Syſtem, deſcribed under the article COPERNICAN, is the moſt ancient in the world. It was firſt of all, as far as we know, introduced into Greece and Italy by Pythagoras; from whom it was called the Pythagorean Syſtem. It was followed by Philolaus, Plato, Archimedes, &c: but it was loſt under the reign of the Peripatetic philoſophy; till happily retrieved about the year 1500 by Nic. Copernicus.

Tychonic SYSTEM, was taught by Tycho, a Dane; who was born An. Dom. 1546. It ſuppoſes that the earth is fixed in the centre of the univerſe or firmament of ſtars, and that all the ſtars and planets revolve round the earth in 24 hours; but it differs from the Ptolomaic Syſtem, as it not only allows a menſtrual motion to the moon round the earth, and that of the ſatellites about Jupiter and Saturn, in their proper periods, but it makes the ſun to be the centre of the orbits of the primary planets Mercury, Venus, Mars, Jupiter, &c, in which they are carried round the ſun in their reſpective years, as the ſun revolves round the earth in a ſolar year; and all theſe planets, together with the ſun, are ſuppoſed to revolve round the earth in 24 hours. This hypotheſis was ſo embarraſſed and perplexed, that very few perſons embraced it. It was afterwards altered by Longomontanus and others, who allowed the diurnal motion of the earth on its own axis, but denied its annual motion round the ſun. This hypotheſis, partly true and partly falſe, is called the *Semi-Tychonic Syſtem.* See the figure and economy of theſe Syſtems, in plates 30, 31, 32, 33.

SYSTEM, in Muſic, denotes a compound interval; or an interval compoſed, or conceived to be compoſed of ſeveral leſs intervals. Such is the octave, &c.

SYSTYLE, in Architecture, the manner of placing columns, where the ſpace between the two fuſts conſiſts of 2 diameters, or 4 modules.

SYZYGY, a term equally uſed for the conjunction and oppoſition of a planet with the ſun.

On the phenomena and circumſtances of the Syzygies, a great part of the lunar theory depends. See MOON. For,

1. It is ſhewn in the phyſical aſtronomy, that the force which diminiſhes the gravity of the moon in the Syzygies, is double that which increaſes it in the quadratures; ſo that, in the Syzygies, the gravity of the moon is diminiſhed by a part which is to the whole gravity, as 1 to 89·36; for in the quadratures, the addition of gravity is to the whole gravity, as 1 to 178·73.

2. In the Syzygies, the diſturbing force is directly as the diſtance of the moon from the earth, and inverſely as the cube of the diſtance of the earth from the ſun. And at the Syzygies, the gravity of the moon towards the earth receding from its centre, is more diminiſhed than according to the inverſe ratio of the ſquare of the diſtance from that centre.—Hence, in the moon's motion from the Syzygies to the quadratures, the gravity of the moon towards the earth is continually increaſed, and the moon is continually retarded in her motion; but in the moon's motion from the quadratures to the Syzygies, her gravity is continually diminiſhed, and the motion in her orbit is accelerated.

3. Farther, in the Syzygies, the moon's orbit, or circuit round the earth, is more convex than in the quadratures; for which reaſon ſhe is leſs diſtant from the earth at the former than the latter.—Alſo, when the moon is in the Syzygies, her apſes go backward, or are retrograde.—Moreover, when the moon is in the Syzygies, the nodes move in antecedentia faſteſt; then ſlower and ſlower, till they become at reſt when the moon is in the quadratures.—Laſtly, when the nodes are come to the Syzygies, the inclination of the plane of the orbit is the leaſt of all.

However, theſe ſeveral irregularities are not equal in each Syzygy, being all ſomewhat greater in the conjunction than in the oppoſition.

T.

TABLE, in Architecture, a ſmooth, ſimple member or ornament, of various forms, but moſt commonly in that of a parallelogram.

TABLE, in Perſpective, is ſometimes uſed for the perſpective plane, or the tranſparent plane upon which the objects are formed in their reſpective appearance.

TABLE *of Pythagoras*, is the ſame as the MULTIPLICATION

CATION Table; which fee; as alfo PYTHAGORAS's Table.

TABLES of *Houfes*, among aftrologers, are certain Tables, ready drawn up, for the affiftance of practitioners in that art, for the erecting or drawing of figures or fchemes. See HOUSE.

TABLES, in Mathematics, are fyftems or feries of numbers, calculated to be ready at hand for expediting any fort of calculations in the various branches of mathematics.

Aftronomical TABLES; are computations of the motions, places, and other phenomena of the planets, both primary and fecondary.

The oldeft aftronomical Tables, now extant, are thofe of Ptolomy, found in his Almageft. Thefe however are not now of much ufe, as they no longer agree with the motions of the heavens.

In 1252, Alphonfo XI, king of Caftile, undertook the correcting of them, chiefly by the affiftance of Ifaac Hazen, a learned Jew; and fpent 400,000 crowns on the bufinefs. Thus arofe the *Alphonfine Tables*, to which that prince himfelf prefixed a preface. But the deficiency of thefe alfo was foon perceived by Purbach and Muller, or Regiomontanus; upon which the latter, and after him Walther Warner, applied themfelves to celeftial obfervations, for farther improving them; but death, or various difficulties, prevented the effect of thefe good defigns.

Copernicus, in his books of the celeftial revolutions, gives other Tables, calculated by himfelf, partly from his own obfervations, and partly from the Alphonfine Tables.

From Copernicus's obfervations and theorems, Erafmus Reinhold afterwards compiled the *Prutenic Tables*, which have been printed feveral times, and in feveral places.

Tycho Brahe, even in his youth, became fenfible of the deficiency of the Prutenic Tables: which determined him to apply himfelf with fo much vigour to celeftial obfervations. From thefe he adjufted the motions of the fun and moon; and Longomontanus, from the fame obfervations, made out Tables of the motions of the planets, which he added to the Theories of the fame, publifhed in his Aftronomia Danica; thofe being called the *Danifh Tables*. And Kepler alfo, from the fame obfervations, publifhed in 1627 his *Rudolphine Tables*, which are much efteemed.

Thefe were afterwards, viz in 1650, changed into another form, by Maria Cunitia, whofe Aftronomical Tables, comprehending the effect of Kepler's phyfical hypothefis, are very eafy, fatisfying all the phenomena without any mention of logarithms, and with little or no trouble of calculation. So that the Rudolphine calculus is here greatly improved.

Merca or made a like attempt in his Aftronomical Inftitution, publifhed in 1676. And the like did J. Bap. Morini, whofe abridgment of the Rudolphine Tables was prefix 1 to a Latin verfion of Street's Aftronomia Carolina, publifhed in 1705.

Lanfbergius indeed endeavoured to difcredit the Rudolphine Tables, and framed *Perpetual Tables*, as he calls them, of the heavenly motions. But his attempt was never much regarded by the aftronomers; and our countryman Horrox warmly attacked him, in his defence of the Keplerian aftronomy.

Since the Rudolphine Tables, many others have been framed, and publifhed: as the *Philolaic Tables* of Bulliald; the *Britannic Tables* of Vincent Wing, calculated on Bulliald's hypothefis; the *Britannic Tables* of John Newton; the French ones of the Count Pagan; the *Caroline Tables* of Street, all calculated on Ward's hypothefis; and the *Novalmajeftic Tables* of Riccioli. Among thefe, however, the Philolaic and Caroline Tables are efteemed the beft; infomuch that Mr. Whifton, by the advice of Mr. Flamfteed, thought fit to fubjoin the Caroline Tables to his aftronomical lectures.

The *Ludovician Tables*, publifhed in 1702, by De la Hire, were conftructed wholly from his own obfervations, and without the affiftance of any hypothefis; which, before the invention of the micrometer telefcope and the pendulum clock, was held impoffible.

Dr. Halley alfo long laboured to perfect another fet of Tables; which were printed in 1719, but not publifhed till 1752.

M. Monnier, in 1746, publifhed, in his Inftitutions Aftronomiques, Tables of the motions of the fun and moon, with the fatellites, as alfo of refractions, and the places of the fixed ftars. La Hire alfo publifhed Tables of the planets, and La Caille Tables of the fun: Gael Morris publifhed Tables of the fun and moon, and Mayer conftructed Tables of the moon, which were publifhed by the Board of Longitude. Tables of the fame have alfo been computed by Charles Mafon, from the principles of the Newtonian philofophy, which are found to be very accurate, and are employed in computing the Nautical Ephemeris. Many other fets of aftronomical Tables have alfo been publifhed by various perfons and academies; and divers fets of them may be found in the modern books of aftronomy, navigation, &c, of which thofe are efteemed the beft and moft complete, that are printed in Lalande's Aftronomy. For an account of feveral, and efpecially of thofe publifhed annually under the direction of the Commiffioners of Longitude, fee ALMANAC, EPHEMERIS, and LONGITUDE.

For TABLES *of the Stars*, fee CATALOGUE.

TABLES of *Sines*, *Tangents*, and *Secants*, ufed in trigonometry, &c, are ufually called CANONS. See SINE.

TABLES of *Logarithms*, *Rhumbs*, &c, ufed in geometry, navigation, &c, fee LOGARITHM, and RHUMB.

TABLES, *Loxodromic*, and *of Difference of Latitude and Departure*, are Tables ufed in computing the way and reckoning of a fhip on a voyage, and are publifhed in moft books of navigation.

TACQUET (ANDREW), a Jefuit of Antwerp, who died in 1660. He was a moft laborious and voluminous writer in mathematics. His works were collected, and printed at Antwerp in one large volume in folio, 1669.

TACTION, in Geometry, the fame as tangency, or touching. See ANGENT.

TALUS, or TALUD, in Architecture, the inclination or flope of a work; as of the outfide of a wall, when its thicknefs is diminifhed by degrees, as it rifes in height, to make it the firmer.

TALUS,

TALUS, in Fortification, means also the slope of a work, whether of earth or masonry.

The *Exterior Talus* of a work, is its slope on the side outwards or towards the country ; which is always made as little as possible, to prevent the enemy's escalade, unless the earth be bad, for then it is necessary to allow a considerable Talus for its parapet, and sometimes to support the earth with a slight wall, called a revetement.

The *Interior Talus* of a work, is its slope on the inside, towards the place. This is larger than the former, and it has, at the angles of the gorge, and sometimes in the middle of the curtains, ramps, or sloping roads for mounting upon the terreplain of the rampart.

Superior TALUS *of the Parapet*, is a slope on the top of the parapet, that allows of the soldiers defending the covert-way with small-shot, which they could not do if it were level.

TAMBOUR, in Architecture, a term applied to the Corinthian and Composite capitals, as bearing some resemblance to a tambour or drum.

TAMUZ, in Chronology, the 4th month of the Jewish ecclesiastical year, answering to part of our June and July. The 17th day of this month is observed by the Jews as a fast, in memory of the destruction of Jerusalem by Nebuchadnezzar, in the 11th year of Zedekiah, and the 588th before Christ.

TANGENT, in Geometry, is a line that touches a curve, &c, that is, which meets it in a point without cutting it there, though it be produced both ways ; as the Tangent AB of the circle BD. The point B, where the Tangent touches the curve, is called the *point of contact*.

The direction of a curve at the point of contact, is the same as the direction of the Tangent.

It is demonstrated in Geometry ;

1. That a Tangent to a circle, as AB, is perpendicular to the radius BC drawn to the point of contact.

2. The Tangent AB is a mean proportional between AF and AE, the whole secant and the external part of it ; and the same for any other secant drawn from the same point A.

3. The two Tangents AB and AD, drawn from the same point A, are always equal to one another. And therefore also, if a number of Tangents be drawn to different points of the curve quite around, and an equal length BA be set off upon each of them from the points of contact, the locus of all the points A will be a circle having the same centre C.

4. The angle of contact ABE, formed at the point of contact, between the Tangent AB and the arc BE, is less than any rectilineal angle.

5. The Tangent of an arc is the right line that limits the position of all the secants that can pass through the point of contact ; though strictly speaking it is not one of the secants, but only the limit of them.

6. As a right line is the Tangent of a circle, when it touches the circle so closely, that no right line can be drawn through the point of contact between it and

the arc, or within the angle of contact that is formed by them ; so, in general, when any right line touches an arc of any curve, in such a manner, that no right line can be drawn through the point of contact, between the right line and the arc, or within the angle of contact that is formed by them, then is that line the Tangent of the curve at the said point ; as AB.

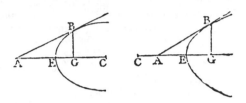

7. In all the conic sections ; if C be the centre of the figure, and BG an ordinate drawn from the point of contact and perpendicular to the axis ; then is CG : CE :: CE : CA, or the semiaxis CE is a mean proportional between CG and CA.

TANGENT, in Trigonometry. A TANGENT *of an arc*, is a right line drawn touching one extremity of the arc, and limited by a secant or line drawn through the centre and the other extremity of the arc.

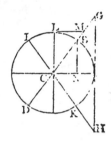

So, AG is the Tangent of the arc AB, or of the arc ABD ; and AH is the Tangent of the arc AI, or of the arc AIDK.

The same are also the Tangents of the angles that are subtended or measured by the arcs.

Hence, 1. The Tangents in the 1st and 3d quadrants are positive, in the 2d and 4th negative, or drawn the contrary way. But of 0 or 180° the semicircle, the Tangent is 0 or nothing ; while those of 90° or a quadrant, and 270° or 3 quadrants, are both infinite ; the former infinitely positive, and the latter infinitely negative. That is,

Between 0 and 90°, or bet. 180° and 270°, the Tangents are positive.
Bet. 90° and 180°, or bet. 270° and 360°, the Tangents are negative.

2. The Tangent of an arc and the Tangent of its supplement, are equal, but of contrary affections, the one being positive, and the other negative ;

as of a and 180° — a, where a is any arc.

Also 180° + a } have the same Tangent, and of the
and a } same affection.

Or 180° + a } have the same Tangent, but of
and 180° — a } different affections.

3. The Tangent of an arc is a 4th proportional to the cosine the sine and the radius ; that is, CN : NB :: CA : AG. Hence, a canon of sines being made or given, the canon of Tangents is easily constructed from them.

*Co-*TANGENT, contracted from complement-tangent, is the Tangent of the complement of the arc or angle, or of what it wants of a quadrant or 90°. So LM is the Cotangent of the arc AB, being the Tangent of its complement BL.

The Tangent is reciprocally as the cotangent ; or the

Tangent and cotangent are reciprocally proportional with the radius. That is Tang. is as $\frac{1}{\text{cotan.}}$, or Tang. : radius :: radius : cotan. And the rectangle of the Tangent and cotangent is equal to the square of the radius; that is, Tan. × cot. = radius².

Artificial TANGENTS, or *logarithmic* TANGENTS, are the logarithms of the tangents of arcs; so called, in contradistinction from the natural Tangents, or the Tangents expressed by the natural numbers.

Line of TANGENTS, is a line usually placed on the sector, and Gunter's scale; the description and uses of which see under the article SECTOR.

Sub-TANGENT, a line lying beneath the Tangent, being the part of the axis intercepted by the Tangent and the ordinate to the point of contact; as the line AG in the 2d and 3d figures above.

Method of TANGENTS, is a method of determining the quantity of the Tangent and subtangent of any algebraic curve; the equation of the curve being given.

This method is one of the great results of the doctrine of fluxions. It is of great use in Geometry; because that in determining the Tangents of curves, we determine at the same time the quadrature of the curvilinear spaces: on which account it deserves to be here particularly treated on.

To Draw the Tangent, or to find the Subtangent, of a curve.

If AE be any curve, and E any point in it, to which it is required to draw a Tangent TE. Draw the ordinate DE: then if we can determine the subtangent TD, by joining the points T and E, the line TE will be the Tangent sought.

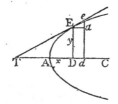

Let *dae* be another ordinate indefinitely near to DE, meeting the curve, or Tangent produced, in *e*; and let E*a* be parallel to the axis AD. Then is the elementary triangle E*ae* similar to the triangle TDE;

and therefore - *ea* : *a*E :: ED : DS ;
but ------- *ea* : *a*E :: flux. ED : flux. AD ;
therefore ---- flux. ED : flux. AD :: DE : DT ;

that is ------ $\dot{y} : \dot{x} :: y : \frac{y\dot{x}}{\dot{y}} = DT$;

which is therefore the value of the subtangent sought; where *x* is the abscifs AD, and *y* the ordinate DE.

Hence we have this general rule : By means of the given equation of the curve, find the value either of $\dot{x}$ or $\dot{y}$, or of $\frac{\dot{x}}{\dot{y}}$, which value substitute for it in the expression DT = $\frac{y\dot{x}}{\dot{y}}$, and, when reduced to its simplest terms, it will be the value of the subtangent sought. This we may illustrate in the following examples.

Ex. 1. The equation defining a circle is $2ax - xx = y^2$, where *a* is the radius; and the fluxion of this is

$2ax - 2x\dot{x} = 2y\dot{y}$; hence $\frac{\dot{x}}{\dot{y}} = \frac{y}{a-x}$; this multi-

plied by *y*, gives $\frac{y\dot{x}}{\dot{y}} = \frac{y^2}{a-x} = \frac{DE^2}{CD} = $ the subtangent TD, or CD : DE :: DE : TD, which is a property of the circle we also know from common geometry.

Ex. 2. The equation defining the common parabola is $ax = y^2$, *a* being the parameter, and *x* and *y* the abscifs and ordinate in all cases. The fluxion of this is $a\dot{x} = 2y\dot{y}$; hence $\frac{\dot{x}}{\dot{y}} = \frac{2y}{a}$; consq. $\frac{y\dot{x}}{\dot{y}} = \frac{2y^2}{a} = \frac{2ax}{a} = 2x = TD$; that is, the subtangent TD is double the abscifs AD, or TA is = AD, which is a well-known property of the parabola.

Ex. 3. The equation defining an ellipsis is $c^2 . \overline{2ax - x^2} = a^2 y^2$, where *a* and *c* are the semiaxes. The fluxion of it is $c^2 . \overline{2a\dot{x} - 2x\dot{x}} = 2a^2 y\dot{y}$; hence

$\frac{y\dot{x}}{\dot{y}} = \frac{a^2 y^2}{c^2 (a-x)} = \frac{c^2 (2ax - x^2)}{c^2 (a-x)} = \frac{2a-x}{a-x} x = TD$

the subtangent; or by adding CD which is = *a*−*x*, it becomes CT = $\frac{2ax - x^2}{a-x} + a - x = \frac{a^2}{a-x} = \frac{CA^2}{CD}$; or CD : CA :: CA : CT, a well-known property of the ellipse.

Ex. 4. The equation defining the hyperbola is $c^2 . \overline{2ax + x^2} = a^2 y^2$, which is similar to that for the ellipse, having only + x^2 for − x^2; hence the conclusion is exactly similar also, viz,

$\frac{2a + x}{a + x} x$ or $\frac{2ax + xx}{a + x} = TD$, which taken from

CD or *a* + *x*, gives CT = $\frac{CA^2}{CD}$, or CD : CA :: CA : CT.

And so on, for the Tangents to other curves.

The Inverse Method of TANGENTS. This is the reverse of the foregoing, and consists in finding the nature of the curve that has a given subtangent. The method of solution is to put the given subtangent equal to the general expression $\frac{y\dot{x}}{\dot{y}}$, which serves for all sorts of curves; then the equation reduced, and the fluents taken, will give the fluential equation of the curve sought.

Ex. 1. To find the curve line whose subtangent is = $\frac{2y^2}{a}$. Here $\frac{2y^2}{a} = \frac{y\dot{x}}{\dot{y}}$; hence $2y\dot{y} = a\dot{x}$, and the fluents of this give $y^2 = ax$, the equation to a parabola, which therefore is the curve sought.

Ex. 2. To find the curve whose subtangent is = $\frac{yy}{2a - x}$, or a third proportional to 2*a* − *x* and *y*.

Here $\frac{yy}{2a - x} = \frac{y\dot{x}}{\dot{y}}$; hence $y\dot{y} = 2a\dot{x} - x\dot{x}$, the fluents of which give $y^2 = ax - x^2$, the equation to a circle, which therefore is the curve sought.

TAN.

TANTALUS's *Cup*, in Hydraulics, is a cup, as A, with a hole in the bottom, and the longer leg of a syphon BCED cemented into the hole ; so that the end D of the shorter leg DE may always touch the bottom of the cup within. Then, if water be poured into this cup, it will rise in the shorter leg by its upward pressure, extruding the air before it through the longer leg, and when the cup is filled above the bend of the syphon at E, the pressure of the water in the cup will force it over the bend; from whence it will descend in the longer leg EB,

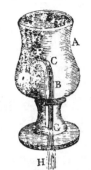

and through the bottom at G, till the cup be quite emptied. The legs of this syphon are almost close together, and it is sometimes concealed by a small hollow statue, or figure of a man placed over it ; the bend E being within the neck of the figure as high as the chin. So that poor thirsty Tantalus stands up to the chin in water, according to the fable, imagining it will rise a little higher, as more water is poured in, and he may drink ; but instead of that, when the water comes up to his chin, it immediately begins to descend, and therefore, as he cannot stoop to follow it, he is left as much tormented with thirst as ever. Ferguson's Lect. p. 72, 4to.

TARRANTIUS (LUCIUS), surnamed *Firmanus*, because he was a native of Firmum, a town in Italy, flourished at the same time with Cicero, and was one of his friends. He was a mathematical philosopher, and therefore was thought to have great skill in judicial astrology. He was particularly famous by two horoscopes which he drew, the one the horoscope of Romulus, and the other of Rome. Plutarch says, " Varro, who was the most learned of the Romans in history, had a particular friend named Tarrantius, who, out of curiosity, applied himself to draw horoscopes, by means of astronomical tables, and was esteemed the most eminent in his time:" Historians controvert some particular circumstances of his calculations ; but all agree in conferring on him the honorary title *Prince of astrologers*.

TARTAGLIA, or TARTALEA (NICHOLAS), a noted mathematician who was born at Brescia in Italy, probably towards the conclusion of the 15th century, as we find he was a considerable master or preceptor in mathematics in the year 1521, when the first of his collection of questions and answers was written, which he afterwards published in the year 1546, under the title of *Quesiti et Inventioni diverse*, at Venice, where he then resided as a public lecturer on mathematics, he having removed to this place about the year 1534. This work consists of 9 chapters, containing answers to a number of questions on all the different branches of mathematics and philosophy then in vogue. The last or 9th of these, contains the questions in Algebra, among which are those celebrated letters and communications between Tartalea and Cardan, by which our author put the latter in possession of the rules for cubic equations, which he first discovered in the year 1530.

But the first work of Tartalea's that was published, was his *Nova Scientia inventa*, in 4to, at Venice in

1537. This is a treatise on the theory and practice of gunnery, and the first of the kind, he being the first writer on the flight and path of balls and shells. This work was translated into English, by Lucar, and printed at London in 1588, in folio, with many notes and additions by the translator.

Tartalea published at Venice, in folio, 1543, the whole books of Euclid, accompanied with many curious notes and commentaries.

But the last and chief work of Tartalea, was his *Trattato di Numeri et Misure*, in folio, 1556 and 1560. This is an universal treatise on arithmetic, algebra, geometry, mensuration, &c. It contains many other curious particulars of the disputes between our author and Cardan, which ended only with the death of Tartalea, before the last part of this work was published, or about the year 1558.

For many other circumstances concerning Tartalea and his writings, see the article ALGEBRA, vol. 1, pa. 73.

TATIUS (ACHILLES), an ancient Greek writer of Alexandria ; but the age he lived in is uncertain. According to Suidas, who calls him Statius, he was at first a Heathen, then a Christian, and afterwards a bishop. He wrote a book upon the Sphere, which seems to have been nothing more than a commentary upon Aratus. Part of it is extant, and was translated into Latin by father Petavius, under the title of *Isagoges in Phænomena Arati*. He wrote also, *Of the Loves of Cliptophon and Leucippe*, in 8 books. He is well spoken of by Photius.

TAURUS, *the Bull*, in Astronomy, one of the 12 signs in the zodiac, and the second in order.

The Greeks fabled that this was the bull which carried Europa safe across the seas to Crete ; and that Jupiter, in reward for so signal a service, placed the creature, whose form he had assumed on that occasion, among the stars, and that this is the constellation formed of it. But it is probable that the Egyptians, or Babylonians, or whoever invented the constellations of the zodiac, placed this figure in that part of it which the sun entered about the time of the bringing forth of calves ; like as they placed the ram in the first part of spring, as the lambs appear before them, and the two kids (for that was the original figure of the sign Gemini), afterward, to denote the time of the goats bringing forth their young.

In the constellation Taurus there are some remarkable stars that have names ; as Aldebaran in the south or right eye of the bull. the cluster called the Pleiades in the neck, and the cluster called Hyades in the face.

The stars in the constellation Taurus, in Ptolomy's catalogue are 44, in Tycho's catalogue 43, in Hevelius's catalogue 51, and in the Britannic catalogue 141.

TEBET, or THEVET, the 4th month of the civil year of the Hebrews, and the 10th of their ecclesiastical year. It answered to part of our December and January, and had only 29 days.

TEETH, of various sorts of machines, as of mill wheels, &c. These are often called cogs by the workmen ; and by working in the pinions, rounds, or trundles, the wheels are made to turn one another.

Mr. Emerson (in his Mechanics, prop. 25), treats of the theory of Teeth, and shews that they ought to have

have the figure of epicycloids, for properly working in one another. Camus too (in 'his Cours de Mathematique, tom. 2, p. 349, &c, Edit. 1767) treats more fully on the same subject; and demonstrates that the Teeth of the two wheels should have the figures of epicycloids, but that the generating circles of these epicycloids should have their diameters only the half of what Mr. Emerson makes them.

Mr. Emerson observes, that the Teeth ought not to act upon one another before they arrive at the line which joins their centres. And though the inner or under sides of the Teeth may be of any form; yet it is better to make them both sides alike, which will serve to make the wheels turn backwards. Also a part may be cut away on the back of every Tooth, to make way for those of the other wheel. And the more Teeth that work together, the better; at least one Tooth should always begin before the other hath done working. The Teeth ought to be disposed in such manner as not to trouble or hinder one another, before they begin to work; and there should be a convenient length, depth and thickness given to them, as well for strength, as that they may more easily disengage themselves.

TELEGRAPH, a machine, brought into use by the French nation, in the year 1793, contrived to communicate words or signals from one person to another at a great distance, in a very small space of time.

The Telegraph it seems was originally the invention of William Amontons, an ingenious philosopher, born in Normandy in the year 1663. See his life in this Dictionary, vol. 1, pa. 105; where it is related that he pointed out a method to acquaint people at a great distance, and in a very little time, with whatever one pleased. This method was as follows: let persons be placed in several stations, at such distances from each other, that, by the help of a telescope, a man in one station may see a signal made by the next before him: this person immediately repeats the same signal to the third man; and this again to a fourth, and so on through all the stations to the last.

This, with considerable improvements, it seems has lately been brought into use by the French, and called a Telegraph. It is said they have availed themselves of this contrivance to good purpose, in the present war; and from the utility of the invention, it has also just been brought into use in this country.

The following account of this curious instrument is copied from Barrere's report in the sitting of the French Convention of August 15, 1794.—" The new-invented telegraphic language of signals is an artful contrivance to transmit thoughts, in a peculiar language, from one distance to another, by means of machines, which are placed at different distances, of from 12 to 15 miles from one another, so that the expression reaches a very distant place in the space of a few minutes. Last year an experiment of this invention was tried in the presence of several Commissioners of the Convention. From the favourable report which the latter made of the efficacy of the contrivance, the Committee of Public Welfare tried every effort to establish, by this means, a correspondence between Paris and the frontier places, beginning with Lisle. Almost a whole twelvemonth has been spent in collecting the necessary instruments for the machines, and to teach the people employed how to use them. At present,

the telegraphic language of signals is prepared in such a manner, that a correspondence may be conducted with Lisle upon every subject, and that every thing, nay even proper names, may be expressed; an answer may be received, and the correspondence thus be renewed several times a day. The machines are the invention of Citizen Chappe, and were constructed under his own eye; he also directs their establishment at Paris. They have the advantage of resisting the changes in the atmosphere, and the inclemencies of the seasons. The only thing which can interrupt their effect is, if the weather is so very bad and turbid that the objects and signals cannot be distinguished. By this invention, remoteness and distance almost disappear; and all the communications of correspondence are effected with the rapidity of the twinkling of an eye. The operations of Government can be very much facilitated by this contrivance, and the unity of the Republic can be the more consolidated by the speedy communication with all its parts. The greatest advantage which can be derived from this correspondence is, that, if one chooses, its object shall only be known to certain individuals, or to one individual alone, or to the extremities of any distance; so that the Committee of Public Welfare may now correspond with the Representative of the People at Lisle without any other persons getting acquainted with the object of the correspondence. Hence it follows that, were Lisle even besieged, we should know every thing at Paris that might happen in that place, and could send thither the Decrees of the Convention without the enemy's being able to discover or to prevent it."—The description and figure of the French machine, as given in some English prints, are as follow.

Explanation of the Machine (Telegraph) placed on the Mountain of Bellville, near Paris, for the purpose of communicating Intelligence.

AA is a beam or mast of wood, placed upright on a rising ground (fig. 3, pl. 28) which is about 15 or 16 feet high. BB is a beam or balance, moving upon the centre AA. This balance-beam may be placed vertically, or horizontally, or any how inclined, by means of strong cords, which are fixed to the wheel D, on the edge of which is a double groove, to receive the two chords. This balance is about 11 or 12 feet long, and 9 inches broad, having at the ends two pieces of wood CC, which likewise turn upon angles by means of four other cords that pass through the axis of the main balance, otherwise the balance would derange the cords; the pieces C are each about 3 feet long, and may be placed either to the right or left, straight or square with the balance-beam. By means of these three, the combination of movement is said to be very extensive, remarkably simple, and easy to perform. Below is a small wooden gouge or hut, in which a person is employed to observe the movements of the machine. In the mountain nearest to this, another person is to repeat these movements, and a third to write them down. The time taken up for each movement is 20 seconds; of which the motion alone is 4 seconds, the other 16 the machine is stationary. The stations of this machine are about 3 or 4 leagues distance; and there is an observatory near the Committee of Public Safety.

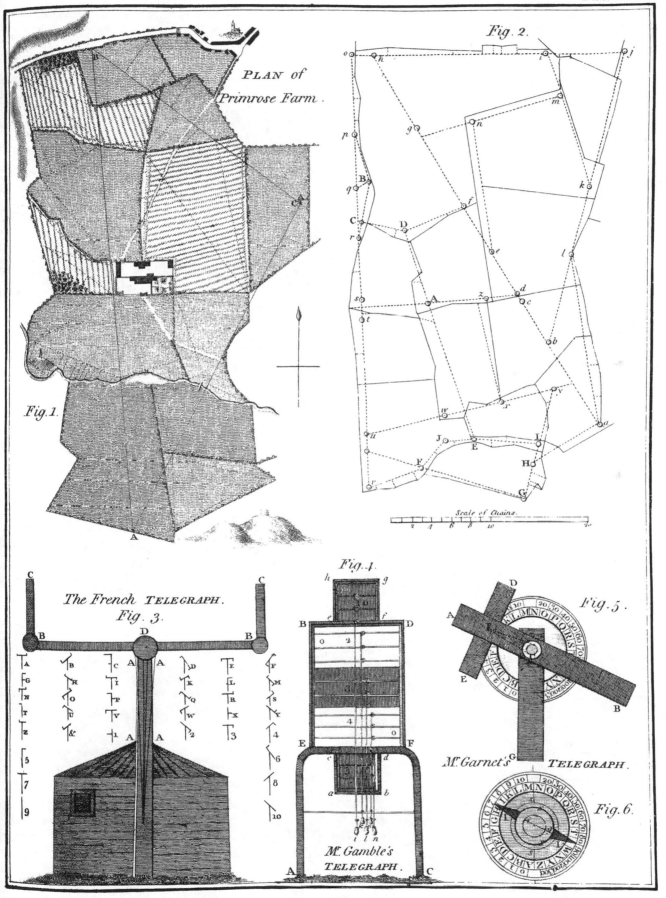

Plate XXVIII.

PLAN of
Primrose Farm.

Fig.1.

Fig. 2.

Scale of Chains.

The French TELEGRAPH.
Fig. 3.

Fig. 4.

Fig. 5.

Mr. Gamble's
TELEGRAPH.

Mr. Garnet's TELEGRAPH.

Fig. 6.

Safety to obferve the motions of the laft, which is at Bellville. The figns are fometimes made in words, and fometimes in letters; when in words, a fmall flag is hoifted, and, as the alphabet may be changed at pleafure, it is only the correfponding perfon who knows the meaning of the figns. In general, news are given every day, about 11 or 12 o'clock; but the people in the wooden gouge obferve from time to time, and, as foon as a certain fignal is given and anfwered, they begin, from one end to the other, to move the machine. It is painted of a dark brown colour.

Such is the account given of the French invention. Various improved contrivances have been fince made in England, and a pamphlet has lately been publifhed, giving an account of fome of them, by the Rev. J. Gamble, under the title of, *Obfervations and Telegraphic Experiments*, from whence the following remarks are extracted.

The object propofed is, to obtain an intelligible figurative language, which may be diftinguifhed at a diftance, and by which the obvious delay in the difpatch of orders or information by meffenger may be avoided.

On firft reflection we find the practical modes of fuch diftant communication muft be confined to Sound and Vifion. Each of which is in a great degree fubject to the ftate of the atmofphere: as, independent of the wind's direction, it is known that the air is fometimes fo far deprived of its elafticity, or whatever other quality the conveyance of found depends on, that the heavieft ordnance is fcarce heard farther than the fhot flies; it is alfo well known, that in thick hazy weather the largeft objects become totally obfcured at a fhort diftance. No inftrument therefore defigned for the purpofe can be perfect. We can only endeavour to diminifh thefe irremediable defects as much as may be.

It feems the Romans had a method in their walled cities, either by a hollow formed in the mafonry, or by tubes affixed to it, fo to confine and augment found as to convey information to any part they wifhed; and in lofty houfes it is now fometimes the cuftom to have a pipe, by way of fpeaking trumpet, to give orders from the upper apartments to the lower: by this mode of confining found its volume may be carried to a very great diftance; but beyond a certain extent the found, lofing articulation, would only convey alarm, not give directions.

Every city among the antients had its watch-towers; and the caftra ftativa of the Romans, had always fome fpot, elevated either by nature or art, from whence fignals were given to the troops cantoned or foraging in the neighbourhood. But I believe they had not arrived to greater refinement than that on feeing a certain fignal they were immediately to repair to their appointed ftations.

A beacon or bonfire made of the firft inflammable materials that offered, as the moft obvious, is perhaps the moft antient mode of general alarm; and by being previoufly concerted, the number or point where the fires appeared might have its particular intelligence affixed. The fame obfervations may be referred to the throwing up of rockets, whofe number or point from whence thrown may have its affixed fignification.

Flags or enfigns with their various devices are of earlieft invention, efpecially at fea; where, from the firft idea, which moft probably was that of a vane to fhew the direction of the wind, they have been long adopted as the diftinguifhing mark of nations, and are now fo neatly combined by the ingenuity of a great naval commander, that by his fyftem every requifite order and queftion is received and anfwered by the moft diftant fhips of a fleet.

To the adopting this or a fimilar mode in land fervice, the following are objections: That in the latter cafe, the variety of matter neceffary to be conveyed, is fo infinitely greater, that the combinations would become too complicated. And if the perfon for whom the information is intended fhould be in the direction of the wind, the flag would then prefent a ftraight line only, and at a little diftance be fcarce vifible. The Romans were fo well aware of this inconvenience of flags, that many of their ftandards were folid, and the name manipulus denotes the rudeft of their modes, which was a trufs of hay fixed on a pole.

The principle of water always keeping its own level has been fuggefted, as a mode of conveying intelligence, by Mr. Daniel Brent, of Rotherhithe, and put in practice on a fmall fcale. As for example, fuppofe a pipe AB to reach from London to Dover, and to have a per-

pendicular tube connected to each extremity, as AC and BD. Then, if the pipe be conftantly filled with water to a certain height, as AE, it will alfo rife to its level in the oppofite perpendicular tube BF; and if one inch of water be added in the tube AC, it will almoft inftantly produce a fimilar elevation of the tube BD; fo that by correfponding letters being adapted to the tubes AC and BD, at different heights, intelligence might be conveyed. But the method is liable to fuch objections, that it is not likely it can ever be adopted to facilitate the object of very diftant communication.

Full as many, if not greater objections, will perhaps operate againft every mode of electricity being ufed as the vehicle of information.—And the requifite magnitude of painted or illuminated letters offers an unfurmountable obftacle; befides, in them one object would be loft, that of the language being figurative.

As to the French machine, it is evident that to every angular change of the greater beam or of the leffer end arms, a different letter or figure may be annexed. But where the whole difference confifts in the variation of the angle of the greater or leffer pieces, much error may be expected, from the inaccuracy either of the operator or the obferver: befides other inconveniences arifing from the great magnitude of the machine.

Another idea is perfectly numerical; which is to raife and deprefs a flag or curtain a certain number of times for each letter, according to a previoufly concerted fyftem: as, fuppofe one elevation to mean A, two to mean B, and fo on through the alphabet. But in this cafe, the leaft inaccuracy in giving or noting the number

number changes the letter; and besides, the last letters of the alphabet would be a tedious operation.

Another method that has been proposed, is an ingenious combination of the magnetical experiment of Comus, and the telescopic micrometer. But as this is only an imperfect idea of Mr. Garnet's very ingenious machine, described in the latter part of this article, no farther notice need be taken of it here.

Mr. Gamble then proposes one on a new idea of his own. The principle of it is simply that of a Venetian blind, or rather what are called the lever boards of a brewhouse, which, when horizontal, present so small a surface to the distant observer, as to be lost to his view, but are capable of being in an instant converted into a screen of a magnitude adapted to the required distance of vision.—Let AB and CD (fig. 4. pl. 28), two upright posts fixed in the ground, and joined by the braces BD and EF, be considered as the frame work for 9 lever boards working upon centres in EB and DF, and opening in three divisions by iron rods connected with each three of the lever boards. Let abcd and efgh be two lesser frames fixed to the great one, having also three lever boards in each, and moving by iron rods, in the same manner as the others. If all these rods be brought so near the ground as to be in the management of the operator, he will then have five, of what may be called, keys to play on. Now as each of the handles iklmn commands three lever boards, by raising any one of them, and fixing it in its place by a catch or hook, it will give a different appearance to the machine; and by the proper variation of these five movements, there will be more than 25 of what may be called mutations, in each of which the machine exhibits a different appearance, and to which any letter or figure may be annexed at pleasure.

Should it be required to give intelligence in more than one direction, the whole machine may be easily made to turn to different points on a strong centre, after the manner of a single-post windmill.—To use this machine by night, another frame must be connected with the back part of the Telegraph, for raising five lamps, of different colours, behind the openings of the lever boards; these lamps by night answer for the openings by day.

M. Gamble gives also particular directions for placing and using the machine, and for writing down the several figures or movements.

I shall now conclude this article with a short idea of Mr. John Garnet's most simple and ingenious contrivance. This is merely a bar or plank turning upon a centre, like the sail of a windmill, and being moved into any position, the distant observer turns the tube of a telescope into the same position, by bringing a fixed wire within it to coincide with or parallel to the bar, which is a thing extremely easy to do. The centre of motion of the bar has a small circle about it, with letters and figures around the circumference, and an index moving round with the bar, pointing to any letter or mark that the operator wishes to set the bar to, or to communicate to the observer. The eye end of the telescope without has a like index and circle, with the corresponding letters or other marks. The consequence is obvious: the telescope being turned round till its wire cover or become parallel to the bar,

the index of the former necessarily points out the same letter or mark in its circle, as that of the latter, and the communication of sentiment is immediate and perfect. The use of this machine is so easy, that I have seen it put into the hands of two common labouring men, who had never seen it before, and they have immediately held a quick and distant conversation together.

The more particular description and figure of this machine, take as follows. ABDE (fig. 5, pl. 28), is the Telegraph, on whose centre of gravity C, about which it revolves, is a fixed pin, which goes through a hole or socket in the firm upright post G, and on the opposite side of which is fixed an index CI. Concentric to C, on the same post, is fixed a wooden or brass circle, of 6 or 8 inches diameter, divided into 48 equal parts, 24 of which represent the letters of the alphabet, and between the letters, numbers. So that the index, by means of the arm AB, may be moved to any letter or number. The length of the arm should be $2\frac{1}{2}$ or 3 feet for every mile of distance. Two revolving lamps of different colours suspended occasionally at A and B, the ends of the arm, would serve equally at night.

Let ss (fig. 6, pl. 28) represent the section of the outward tube of a telescope perpendicular to its axis, and xx the like section of the sliding or adjusting tube, on which is fixed an index II. On the part of the outward tube next to the observer, there is fixed a circle of letters and numbers, similarly divided and situated to the circle in figure 3; then the index II, by means of the sliding or adjusting tube, may be turned to any letter or number.—Now there being a cross hair, or fine silver wire fg, fixed in the focus of the eye glass, in the same direction as the index II; so that when the arm AB (fig. 5) of the Telegraph is viewed at a distance through the telescope, the cross hair may be turned, by means of the sliding tube, to the same direction of the arm AB; then the index II (fig. 6) will point to the same letter or number on its own circle, as the index I (fig. 5) points to on the Telegraphic circle.

If, instead of using the letters and numbers to form words at length, they be used as signals, three motions of the arm will give above a hundred thousand different signals.

TELESCOPE, an optical instrument which serves for discovering and viewing distant objects, either directly by glasses, or by reflection, by means of specula, or mirrors. Accordingly,

Telescopes are either refracting or reflecting; the former consisting of different lenses, through which the objects are seen by rays refracted through them to the eye; and the latter of specula, from which the rays are reflected and passed to the eye. The lens or glass turned towards the object, is called the *object-glass*; and that next the eye, the *eye-glass*; and when the Telescope consists of more than two lenses, all but that next the object are called *eye-glasses*.

The invention of the Telescope is one of the noblest and most useful these ages have to boast of: by means of it, the wonders of the heavens are discovered to us, and astronomy is brought to a degree of perfection which former ages could have no idea of.

The

The difcovery indeed was owing rather to chance than defign; fo that it is the good fortune of the difcoverer, rather than his fkill or ability, we are indebted to: on this account it concerns us the lefs to know, who it was that firft hit upon this admirable invention. Be that as it may, it is certain it muft have been cafual, fince the theory it depends upon was not then known.

John Baptifta Porta, a Neapolitan, according to Wolfius, firft made a Telefcope, which he infers from this paffage in the *Magia Naturalis* of that author, printed in 1560: " If you do but know how to join " the two (viz, the concave and convex glaffes) rightly " together, you will fee both remote and near objects, " much larger than they otherwife appear, and withal " very diftinct. In this we have been of good help " to many of our friends, who either faw remote " things dimly, or near ones confufedly; and have " made them fee every thing perfectly."

But it is certain, that Porta did not underftand his own invention, and therefore neither troubled himfelf to bring it to a greater perfection, nor ever applied it to celeftial obfervation. Befides, the account given by Porta of his concave and convex lenfes, is fo dark and indiftinct, that Kepler, who examined it by defire of the emperor Rudolph, declared to that prince, that it was perfectly unintelligible.

Thirty years afterwards, or in 1590, a Telefcope 16 inches long was made, and prefented to prince Maurice of Naffau, by a fpectacle maker of Middleburg: but authors are divided about his name. Sirturus, in a treatife on the Telefcope, printed in 1618, will have it to be John Lipperfheim: and Borelli, in a volume exprefsly on the inventor of the Telefcope, publifhed in 1655, fhews that it was Zacharias Janfen, or, as Wolfius writes it, Hanfen.

Now the invention of Lipperfheim is fixed by fome in the year 1609, and by others in 1605: Fontana, in his *Nova Obfervationes Cælestium et Terreftrium Rerum*, printed at Naples in 1646, claims the invention in the year 1608. But Borelli's account of the difcovery of Telefcopes is fo circumftantial, and fo well authenticated, as to render it very probable that Janfen was the original inventor.

In 1620, James Metius of Alcmaer, brother of Adrian Metius who was profeffor of mathematics at Franeker, came with Drebel to Middleburg, and there bought Telefcopes of Janfen's children, who had made them public; and yet this Adr. Metius has given his brother the honour of the invention, in which too he is miftakenly followed by Defcartes.

But none of thefe artificers made Telefcopes of above a foot and a half: Simon Marius in Germany, and Galileo in Italy, it is faid, firft made long ones fit for celeftial obfervations; though, from the recently difcovered aftronomical papers of the celebrated Harriot, author of the Algebra, it appears that he muft have made ufe of Telefcopes in viewing the folar maculæ, which he did quite as early as they were obferved by Galileo. Whether Harriot made his own Telefcopes, or whether he had them from Holland, does not appear: it feems however that Galileo's were made by himfelf; for Le Roffi relates, that Galileo, being then at Venice, was told of a fort of optic glafs made in Holland, which brought objects nearer: upon which, fetting himfelf to think how it fhould be, he ground two pieces of glafs into form as well as he could, and fitted them to the two ends of an organpipe; and with thefe he fhewed at once all the wonders of the invention to the Venetians, on the top of the tower of St. Mark. The fame author adds, that from this time Galileo devoted himfelf wholly to the improving and perfecting the Telefcope; and that he hence almoft deferved all the honour ufually done him, of being reputed the inventor of the inftrument, and of its being from him called *Galileo's tube*. Galileo himfelf, in his *Nuncius Sidereus*, publifhed in 1610, acknowledges that he firft heard of the inftrument from a German; and that, being merely informed of its effects, firft by common report, and a few days after by letter from a French gentleman, James Badovere, at Paris, he himfelf difcovered the conftruction by confidering the nature of refraction. He adds in his *Saggiatore*, that he was at Venice when he heard of the effects of prince Maurice's inftrument, but nothing of its conftruction; that the firft night after his return to Padua, he folved the problem, and made his inftrument the next day, and foon after prefented it to the Doge of Venice, who, in honour of his grand invention, gave him the ducal letters, which fettled him for life in his lecturefhip, at Padua, and doubled his falary, which then became treble of what any of his predeceffors had enjoyed before. And thus Galileo may be confidered as an inventor of the Telefcope, though not the firft inventor.

F. Mabillon indeed relates, in his travels through Italy, that in a monaftery of his own order, he faw a manufcript copy of the works of Commeftor, written by one Conradus, who lived in the 13th century; in the 3d page of which was feen a portrait of Ptolomy, viewing the ftars through a tube of 4 joints or draws: but that father does not fay that the tube had glaffes in it. Indeed it is more than probable, that fuch tubes were then ufed for no other purpofe but to defend and direct the fight, or to render it more diftinct, by fingling out the particular object looked at, and fhutting out all the foreign rays reflected from others, whofe proximity might have rendered the image lefs precife. And this conjecture is verified by experience; for we have often obferved that without a tube, by only looking through the hand, or even the fingers, or a pinhole in a paper, the objects appear more clear and diftinct than otherwife.

Be this as it may, it is certain that the optical principles, upon which Telefcopes are founded, are contained in Euclid, and were well known to the ancient geometricians; and it has been for want of attention to them, that the world was fo long without that admirable invention; as doubtlefs there are many others lying hid in the fame principles, only waiting for reflection or accident to bring them forth.

To the foregoing abftract of the hiftory of the invention of the Telefcope, it may be proper to add fome particulars relating to the claims of our own celebrated countryman, friar Bacon, who died in 1294. Mr. W. Molyneux, in his Dioptrica Nova, pa. 256, declares his opinion, that Bacon did perfectly well underftand all forts of optic glaffes, and knew likewife the way

way of combining them, so as to compose some such instrument as our Telescope: and his son, Samuel Molyneux, asserts more positively, that the invention of Telescopes, in its first original, was certainly put in practice by an Englishman, friar Bacon; although its first application to astronomical purposes may probably be ascribed to Galileo. The passages to which Mr. Molyneux refers, in support of Bacon's claims, occur in his Opus Majus, pa. 348 and 357 of Jebb's edit. 1773. The first is as follows: *Si vero non sint corpora plana, per quæ visus videt, sed sphæria, tunc est magna diversitas; nam vel concavitas corporis est versus oculum vel convexitas:* whence it is inferred, that he knew what a concave and a convex glass was. The second is comprised in a whole chapter, where he says, *De visione fracta majora sunt; nam de facili patet per canones supra dictos, quod maxima possunt apparere minima, et e contra, et longe distantia videbuntur propinquissime, et e converso. Nam possumus sic figurare perspicua, et taliter ea ordinare respectu nostri visus et rerum, quod frangentur radii, et flectentur quorsumcunque voluerimus, ut sub quocunque angulo voluerimus, videbimus rem prope vel longe, &c. Sic etiam faceremus solem et lunam et stellas descendere secundum apparentiam hic inferius, &c:* that is, Greater things than these may be performed by refracted vision; for it is easy to understand by the canons above mentioned, that the greatest things may appear exceeding small, and the contrary; also that the most remote objects may appear just at hand, and the converse; for we can give such figures to transparent bodies, and dispose them in such order with respect to the eye and the objects, that the rays shall be refracted and bent towards any place we please; so that we shall see the object near at hand or at a distance, under any angle we please, &c. So that thus the sun, moon, and stars may be made to descend hither in appearance, &c. Mr. Molyneux has also cited another passage out of Bacon's Epistle ad Parisiensem, of the Secrets of Art and Nature, cap. 5, to this purpose, *Possunt etiam sic figurari perspicua, ut longissime posita appareant propinqua, et è contrario; ita quod ex incredibili distantia legeremus literas minutissimas, et numeraremus res quantumquo parvas, et stellas faceremus apparere quo vellemus:* that is, Glasses, or diaphanous bodies may be so formed, that the most remote objects may appear just at hand, and the contrary; so that we may read the smallest letters at an incredible distance, and may number things though never so small, and may make the stars appear as near as we please.

Moreover, Doctor Jebb, in the dedication of his edition of the Opus Majus, produces a passage from a manuscript, to shew that Bacon actually applied Telescopes to astronomical purposes: *Sed longe magis quam hæc,* says he, *oporteret homines haberi, qui bene, immo optime, scirent perspectivam et instrumenta ejus—quia instrumenta astronomia non vadunt nisi per visionem secundum leges istius scientiæ.*

From these passages, it is not unreasonable to conclude, that Bacon had actually combined glasses so as to have produced the effects which he mentions, though he did not complete the construction of Telescopes. Dr. Smith, however, to whose judgment particular deference is due, is of opinion that the celebrated friar wrote hypothetically, without having made any actual

trial of the things he mentions: to which purpose he observes, that this author does not assert one single trial or observation upon the sun or moon, or any thing else, though he mentions them both: on the other hand, he imagines some effects of Telescopes that cannot possibly be performed by them. He adds, that persons unexperienced in looking through Telescopes expect, in viewing any object, as for instance the face of a man, at the distance of one hundred yards, through a Telescope that magnifies one hundred times, that it will appear much larger than when they are close to it: this he is satisfied was Bacon's notion of the matter; and hence he concludes that he had never looked through a Telescope.

It is remarkable that there is a passage in Thomas Digges's Stratioticos, pa. 359, where he affirms that his father, Leonard Digges, among other curious practices, had a method of discovering, by perspective glasses set at due angles, all objects pretty far distant that the sun shone upon, which lay in the country round about; and that this was by the help of a manuscript book of Roger Bacon of Oxford, who he conceived was the only man besides his father (since Archimedes) who knew it. This is the more remarkable, because the Stratioticos was first printed in 1579, more than 30 years before Metius or Galileo made their discovery of those glasses; and therefore it has hence been thought that Roger Bacon was the first inventor of Telescopes, and Leonard Digges the next reviver of them. But from what Thomas Digges says of this matter, it would seem that the instrument of Bacon, and of his father, was something of the nature of a camera obscura, or, if it were a Telescope, that it was of the reflecting kind; although the term *perspective* glass seems to favour a contrary opinion.

There is also another passage to the same effect in the preface to the Pantometria of Leonard Digges, but published by his son Thomas Digges, some time before the Stratioticos, and a second time in the year 1591. The passage runs thus: *My father by his continuall painfull practises, assisted with demonstrations mathematical, was able, and sundrie times hath by Proportional Glasses duely situate in convenient angles, not only discovered things farre off, read letters, numbered peeces of money with the very coyne and superscription thereof, cast by some of his freends of purpose upon downes in open fields, but also seven myles off declared what hath beene doone at that instant in private places: He hath also sundrie times by the sunne beames fixed (should be fired) powder, and dischargde ordinance halfe a mile and more distante,* &c.

But to whomsoever we ascribe the honour of first inventing the Telescope, the rationale of this admirable instrument, depending on the refraction of light in passing through mediums of different forms, was first explained by the celebrated Kepler, who also pointed out methods of constructing others, of superior powers, and more commodious application, than that first used: though something of the same kind, it is said, was also done by Maurolycus, whose treatise De Lumine et Umbra was published in 1575.

The Principal Effects of TELESCOPES, depend upon this plain maxim, viz, that objects appear larger in proportion to the angles which they subtend at the

eye;

Plate XX

Telescopic Appearances, of the Western Horizon, in three different States of the Atmosphere; taken from the Laurel Mount, at Traine, in Modbury, Devonshire.

THE SCALE — Minutes of a Degree.
0   10   20   30   40   50   60

Ordinary or common Appearance.

Appearance somewhat elevated by Refraction.

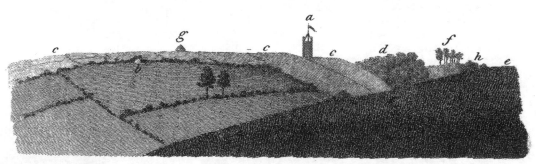

Appearance when more considerably elevated.

a. *Maker tower, about 12¾ miles distant, in a straight line.* b. *Gate place (at present appearing like an Arch) on a Hill about 3¾ miles.* cc. *Ground about 9¾ miles.* d. *A wood in Mount Edgcumbe park, about 12¼ miles.* ee. *A hill about 3 miles.* f. *Trees in the Park.* g. *A mow, seen by the Refraction on the Ground.* cc. h. *Another set of Trees in the Park.*

eye; and the effect is the same, whether the pencils of rays, by which objects are visible to us, come directly from the objects themselves, or from any place nearer to the eye, where they may have been united, so as to form an image of the object; because they issue again from those points in certain directions, in the same manner as they did from the corresponding points in the objects themselves. In fact therefore, all that is effected by a Telescope, is first to make such an image of a distant object, by means of a lens or mirror, and then to give the eye some assistance for viewing that image as near as possible; so that the angle, which it shall subtend at the eye, may be very large, compared with the angle which the object itself would subtend in the same situation. This is done by means of an eye-glass, which so refracts the pencils of rays, as that they may afterwards be brought to their several foci, by the natural humours of the eye. But if the eye had been so formed as to be able to see the image, with sufficient distinctness, at the same distance, without an eye-glass, it would appear to him as much magnified, as it does to another person who makes use of a glass for that purpose, though he would not in all cases have so large a field of view.

Although no image be actually formed by the foci of the pencil without the eye, yet if, by the help of an eye-glass, the pencils of rays shall enter the pupil, just as they would have done from any place without the eye, the visual angle will be the same as if an image had been actually formed in that place. Priestley's History of Light &c, pa. 69, &c.

As to the Grinding of Telescopic Glasses, the first persons who distinguished themselves in that way, were two Italians, Eustachio Divini at Rome, and Campani at Bologna, whose fame was much superior to that of Divini, or that of any other person of his time; though Divini himself pretended, that in all the trials that were made with their glasses, his of a great focal distance performed better than those of Campani, and that his rival was not willing to try them fairly, viz, with equal eye-glasses. It is however generally supposed, that Campani really excelled Divini, both in the goodness and the focal length of his object-glasses.

It was with Campani's Telescopes that Cassini discovered the nearest satellites of Saturn. They were made at the express desire of Lewis XIV, and were of 86, 100, and 136, Paris feet focal length.

Campani's laboratory was purchased, after his death, by pope Benedict XIV, who made a present of it to the academy at Bologna called the Institute; and by the account which Fougeroux has given, we learn that (except a machine which Campani constructed, to work the basons on which he ground his glasses) the goodness of his lenses depended upon the clearness of his glass, his Venetian tripoli, the paper with which he polished his glasses, and his great skill and address as a workman. It does not appear that he made many lenses of a very great focal distance. Accordingly Dr. Hook, who probably speaks with the partiality of an Englishman, says that some glasses, made by Divini and Campani, of 36 and 50 feet focal distance, did not excel Telescopes of 12 or 15 feet made in England. He adds, that Sir Paul Neilli made Telescopes of 36 feet, pretty good; and one of 50, but not of proportionable goodness.

Afterwards, Mr. Reive first, and then Mr. Cox, who were the most celebrated in England, as grinders of optic glasses, made some good Telescopes of 50 and 60 feet focal distance; and Mr. Cox made one of 100, but how good Dr. Hook could not assert. Borelli also in Italy, made object glasses of a great focal length, one of which he presented to the Royal Society. But, with respect to the focal length of Telescopes, these and all others were far exceeded by those of Auzout, who made one object-glass of 600 feet focus; but he was never able to manage it, so as to make any use of it. And Hartsoeker, it is said, made some of a still greater focal length.. Philos. Transf. Abr. vol. i, p. 193. Hook's Exper. by Derham, p. 261. Priestley as above, p. 211. See GRINDING.

Telescopes are of several kinds, distinguished by the number and form of their lenses, or glasses, and denominated from their particular uses &c: such are the *terrestrial* or *land Telescope,* the *celestial* or *astronomical Telescope;* to which may be added, the *Galilean* or *Dutch Telescope,* the *reflecting Telescope,* the *refracting Telescope,* the *aerial Telescope,* achromatic *Telescope, &c.*

Galileo's, or the *Dutch Telescope,* is one consisting of a convex object-glass, and a concave eye-glass.

This is the most ancient form of any, being the only kind made by the inventors, Galileo, &c. or known, before Huygens. The first Telescope, constructed by Galileo, magnified only 3 times; but he soon made another, which magnified 18 times: and afterwards, with great trouble and expence, he constructed one that magnified 33 times; with which he discovered the satellites of Jupiter, and the spots of the sun. The construction, properties. &c, of it, are as follow:

Construction of Galileo's, or the *Dutch* TELESCOPE. In a tube prepared for the purpose, at one end is fitted a convex object lens, either a plain convex, or convex on both sides, but a segment of a very large sphere: at the other end is fitted an eye-glass, concave on both sides, and the segment of a less sphere, so disposed as to be at the distance of the virtual focus before the image of the convex lens.

Let AB (fig. 10, pl. 23) be a distant object, from every point of which pencils of rays issue, and falling upon the convex glass DE, tend to their foci at FSG. But a concave lens HI (the focus of which is at FG) being interposed, the converging rays of each pencil are made parallel when they reach the pupil; so that by the refractive humours of the eye, they can easily be brought to a focus on the retina at PRQ. Also the pencils themselves diverging, as if they came from X, MXO is the angle under which the image will appear, which is much larger than the angle under which the object itself would have appeared. Such then is the Telescope that was at first discovered and used by philosophers: the great inconvenience of which is, that the field of view, which depends, not on the breadth of the eye-glass, as in the astronomical Telescope, but upon the breadth of the pupil of the eye, is exceedingly small: for since the pencils of the rays enter the eye very much diverging from one another, but few of them can be intercepted by the pupil; and this inconvenience increases with the magnifying power of the Telescope, so that philosophers may now well wonder

at

at the patience and addrefs with which Galileo and others, with fuch an inftrument, made the difcoveries they did. And yet no other Telefcope was thought of for many years after the difcovery. Defcartes, who wrote 30 years after, mentions no other as actually conftructed, though Kepler had fuggefted fome. Hence,

1. In an inftrument thus framed, all people, except myopes, or fhort-fighted perfons, muft fee objects diftinctly in an erect fituation, and increafed in the ratio of the diftance of the virtual focus of the eye-glafs, to the diftance of the focus of the object glafs.

2. But for myopes to fee objects diftinctly through fuch an inftrument, the eye-glafs muft be fet nearer the object-glafs, fo that the rays of each pencil may not emerge parallel, but may fall diverging upon the eye; in which cafe the apparent magnitude will be altered a little, though fcarce fenfibly.

3. Since the focus of a plano-convex object lens, and the vertical focus of a plano-concave eye-lens, are at the diftance of the diameter; and the focus of an object-glafs convex on both fides, and the vertical focus of an eye-glafs concave on both fides, are at the diftance of a femidiameter; if the object-glafs be plano-convex, and the eye glafs plano-concave, the Telefcope will increafe the diameter of the object, in the ratio of the diameter of the concavity to that of the convexity: if the object-glafs be convex on both fides, and the eye glafs concave on both fides, it will magnify in the ratio of the femidiameter of the concavity to that of the convexity: if the object-glafs be plano-convex, and the eye-glafs concave on both fides, the femidiameter of the object will be increafed in the ratio of the diameter of the convexity to the femidiameter of the concavity: and laftly, if the object-glafs be convex on both fides, and the eye-glafs plano-concave, the increafe will be in the ratio of the diameter of the concavity to the femidiameter of the convexity.

4. Since the ratio of the femidiameters is the fame as that of the diameters, Telefcopes magnify the object in the fame manner, whether the object-glafs be plano-convex, and the eye-glafs plano-concave; or whether the one be convex on both fides, and the other concave on both.

5. Since the femidiameter of the concavity has a lefs ratio to the diameter of the convexity than its diameter has, a Telefcope magnifies more if the object-glafs be plano-convex, than if it be convex on both fides. The cafe is the fame if the eye-glafs be concave on both fides, and not plano-concave.

6. The greater the diameter of the object-glafs, and the lefs that of the eye-glafs, the lefs ratio has the diameter of the object, viewed with the naked eye, to its femidiameter when viewed with a Telefcope, and confequently the more is the object magnified by it.

7. Since a Telefcope exhibits fo much a lefs part of the object, as it increafes its diameter more, for this reafon, mathematicians were determined to look out for another Telefcope, after having clearly found the imperfection of the firft, which was difcovered by chance. Nor were their endeavours vain, as appears from the aftronomical Telefcope defcribed below.

If the femidiameter of the eye-glafs have too fmall a ratio to that of the object-glafs, an object through the Telefcope will not appear fufficiently clear, becaufe the great divergency of the rays will occafion the feveral pencils reprefenting the feveral points of the object on the retina, to confift of too few rays.

It is alfo found that equal object-lenfes will not bear the fame eye-lenfes, if they be differently tranfparent, or if there be a difference in their polifh; a lefs tranfparent object-glafs, or one lefs accurately ground, requiring a more fpherical eye-glafs than another more tranfparent, &c.

Hevelius recommends an object-glafs convex on both fides, whofe diameter is 4 feet; and an eye-glafs concave on both fides, whofe diameter is $4\frac{1}{2}$ tenths of a foot. An object-glafs, equally convex on both fides, whofe diameter is 5 feet, he obferves, will require an eye-glafs of $5\frac{1}{2}$ tenths; and adds, that the fame eye-glafs will alfo ferve an object-glafs of 8 or 10 feet.

Hence, as the diftance between the object-glafs and eye-glafs is the difference between the diftance of the vertical focus of the eye-glafs, and the diftance of the focus of the object glafs; the length of the telefcope is had by fubtracting that from this. That is, the length of the Telefcope is the difference between the diameters of the object-glafs and eye-glafs, if the former be plano-convex, and the latter plano-concave; or the difference between the femidiameters of the object-glafs and eye-glafs, if the former be convex on both fides, and the latter concave on both; or the difference between the femidiameter of the object-glafs and the diameter of the eye-glafs, if the former be convex on both fides, and the latter plano-concave; or laftly the difference between the diameter of the object-glafs and the femidiameter of the eye-glafs, if the former be plano-convex, and the latter concave on both fides. Thus, for inftance, if the diameter of an object-glafs, convex on both fides, be 4 feet, and that of an eye-glafs, concave on both fides, be $4\frac{1}{2}$ tenths of a foot; then the length of the Telefcope will be 1 foot and $7\frac{1}{2}$ tenths.

Aftronomical TELESCOPE; this is one that confifts of an object-glafs, and an eye-glafs, both convex. It is fo called from being wholly ufed in aftronomical obfervations.

It was Kepler who firft fuggefted the idea of this Telefcope; having explained the rationale, and pointed out the advantages of it in his Catoptrics, in 1611. But the firft perfon who actually made an inftrument of this conftruction, was father Scheiner, who has given a defcription of it in his Rofa Urfina, publifhed in 1630. To this purpofe he fays, If you infert two fimilar convex lenfes in a tube, and place your eye at a convenient diftance, you will fee all terreftrial objects, inverted indeed, but magnified and very diftinct, with a confiderable extent of view. He afterwards fubjoined an account of a Telefcope of a different conftruction, with two convex eye-glaffes, which again reverfes the images, and makes them appear in their natural pofition. Father Reita however foon after propofed a better conftruction, ufing three eye-glaffes inftead of two.

Conftruction of the Aftronomical TELESCOPE. The tube being prepared, an object-glafs, either plano-convex,

vex, or convex on both fides, but a fegment of a large fphere, is fitted in at one end ; and an eye-glafs, convex on both fides, which is the fegment of a fmall fphere, is fitted into the other end ; at the common diftance of the foci.

Thus the rays of each pencil iffuing from every point of the object ABC, (fig. 3 pl. 30) paffing through the object-glafs DEF, become converging, and meet in their foci at IHG, where an image of the object will be formed. If then another convex lens KM, of a fhorter focal length, be fo placed, as that its focus fhall be in IHG, the rays of each pencil, after paffing through it, will become nearly parallel, fo as to meet upon the retina, and form an enlarged image of the object at RST. If the procefs of the rays be traced, it will prefently be perceived that this image muft be inverted. For the pencil that iffues from A, has its focus in G, and again in R, on the fame fide with A. But as there is always one inverfion in fimple vifion, this want of inverfion produces juft the reverfe of the natural appearance. The field of view in this Telefcope will be large, becaufe all the pencils that can be received on the furface of the lens KM, being converging after paffing through it, are thrown into the pupil of the eye, placed in the common interfection of the pencils at P.

Theory of the Aftronomical TELESCOPE.—An eye placed near the focus of the eye-glafs, of fuch a Telefcope, will fee objects diftinctly, but inverted, and magnified in the ratio of the diftance of the focus of the eye-glafs to the diftance of the focus of the object-glafs.

If the fphere of concavity in the eye-glafs of the Galilean Telefcope, be equal to the fphere of convexity in the eye-glafs of another Telefcope, their magnifying power will be the fame. The concave glafs however being placed between the object-glafs and its focus, the Galilean Telefcope will be fhorter than the other, by twice the focal length of the eye-glafs. Confequently, if the length of the Telefcopes be the fame ; the Galilean will have the greater magnifying power. Vifion is alfo more diftinct in thefe Telefcopes, owing in part perhaps to there being no intermediate image between the eye and the object. Befides, the eye-glafs being very thin in the centre, the rays will be lefs liable to be diftorted by irregularities in the fub-ftance of the glafs. Whatever be the caufe, we can fometimes fee Jupiter's fatellites very clearly in a Galilean Telefcope, of 20 inches or 2 feet long, when one of 4 or 5 feet, of the common fort, will hardly make them vifible.

As the aftronomical Telefcope exhibits objects inverted, it ferves commodioufly enough for obferving the ftars, as it is not material whether *they* be feen erect or inverted ; but for terreftrial objects it is much lefs proper, as the inverting often prevents them from being known. But if a plane well-polifhed metal fpeculum, of an oval figure, and about an inch long, and inclined to the axis in an angle of 45°, be placed behind the eye-glafs ; then the eye, conveniently placed, will fee the image, hence reflected, in the fame magnitude as before, but in an erect fituation ; and therefore, by the addition of fuch a fpeculum, the

aftronomical Telefcope is thus rendered fit to obferve terreftrial objects.

Since the focus of the glafs convex on both fides is diftant from the glafs itfelf a femidiameter, and that of a plano-convex glafs, a diameter ; if the object-glafs be convex on both fides, the Telefcope will magnify the femidiameter of the object, in the ratio of the diameter of the eye-glafs to the diameter of the object-glafs ; but if the object-glafs be a plano-convex, in the ratio of the femidiameter of the eye-glafs to the diameter of the object glafs. And therefore a Telefcope magnifies more if the object-glafs be a plano-convex, than if convex on both fides. And for the fame reafon, a Telefcope magnifies more when the eye-glafs is convex on both fides, than when it is plano-convex.

A Telefcope magnifies the more, as the object glafs is a fegment of a great fphere, and the eye-glafs of a lefs one. And yet the eye-glafs muft not be too fmall in refpect of the object-glafs ; for if it be, it will not refract rays enough to the eye from each point of the object ; nor will it feparate fufficiently thofe that come from different points ; by which means the vifion will be rendered obfcure and confufed.—De Chales obferves, that an object-lens of 2½ feet will require an eye-glafs of 1½ tenth of a foot ; and an object-glafs of 8 or 10 feet, an eye-glafs of 4 tenths ; in which he is confirmed by Euftachio Divini.

To fhorten the Aftronomical TELESCOPE ; that is, to conftruct a Telefcope fo, as that, though fhorter than the common one, it fhall magnify as much.

Having provided a drawing tube, fit in it an object-lens EO which is a fegment of a moderate fphere ;

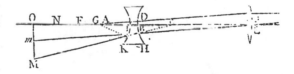

let the firft eye-glafs BD be concave on both fides, and fo placed in the tube, as that the focus of the object-glafs A may be behind it, but nearer to the centre of the concavity G : then will the image be thrown in Q, fo as that GA : GI : : AB : QI. Laftly, fit in another object-glafs, convex on both fides, and a fegment of a fmaller fphere, fo as that its focus may be in Q.

This Telefcope will magnify the diameter of the object more than if the object-glafs were to reprefent its image at the fame diftance EQ ; and confequently a fhorter Telefcope, conftructed this way, is equivalent to a longer in the common way. See Wolfius Elem. Math. vol. 3, p. 245.

Sir Ifaac Newton furnifhes us with another method of conftructing the Telefcope, in his catoptrical or reflecting Telefcope, the conftruction of which is given below. See *Achromatic* TELESCOPE.

Aërial TELESCOPE, a kind of aftronomical Telefcope, the lenfes of which are ufed without a tube. In ftrictnefs however, the aërial Telefcope is rather a particular manner of mounting and managing long

4 D 2 Telefcopes

Telescopes for celestial observation in the night-time, by which the trouble of long unwieldy tubes is saved, than a particular kind of Telescope; and the contrivance was one of Huygens's. This invention was successfully practised by the inventor himself and others, particularly with us by Dr. Pound and Dr. Bradley, with an object-glass of 123 feet focal distance, and an apparatus belonging to it, made and presented by Huygens to the Royal Society, and described in his Astroscopia Compendiaria Tubi Optici Molimine L berata, printed at the Hague in 1684.

The principal parts of this Telescope may be comprehended from a view of fig. 4, pl. 30, where AB is a long pole, or a mast, or a high tree, &c, in a groove of which slides a piece that carries a small tube LK in which is fixed an object glass; which tube is connected by a fine line, with another small tube OQ, which contains the eye-glass, &c.

La Hire contrived a little machine for managing the object-glass which is described Mem. de l'Acad. 1715. See Smith's Optics, book 3, chap. 10.

Hartsoeker, who made Telescopes of a very considerable focal length, contrived a method of using them without a tube, by fixing them to the top of a tree, a high wall, or the roof of a house. Miscel. Berol. vol. 1, p. 261.

Huygens's great Telescope, with which Saturn's true face, and one of his satellites were first discovered, consists of an object-glass of 12 feet, and an eye glass of a little more than 3 inches; though he frequently used a Telescope of 23 feet long, with two eye-glasses joined together, each $1\frac{1}{2}$ inch diameter; so that the two were equal to one of 3 inches.

The same author observes, that an object-glass of 30 feet requires an eye-glass of $3\frac{3}{10}$ inches; and has given a table of proportions for constructing astronomical Telescopes, an abridgment of which is as follows:

Dist. of Foc. of Obj. Glass.	Diameter of Apert.	Dist. of Foc. of Eye-glass.	Power or Magnitude of Diam.
Feet.	Inches and Decim.	Inches and Decim.	
1	0·55	0·61	20
2	0·77	0·85	28
3	0·95	1·05	34
4	1·09	1·20	40
5	1·23	1·35	44
6	1·34	1·47	49
7	1·45	1·60	53
8	1·55	1·71	56
9	1·64	1·80	60
10	1·73	1·90	63
15	2·12	2·33	77
20	2·45	2·70	89
25	2·74	3·01	100
30	3·00	3·30	109
40	3·46	3·81	120

Dist. of Foc. of Object-glass.	Diameter of Apert.	Dist. of Foc. of Eye-glass.	Power or Magnitude of Diam.
Feet.	Inches and Decim.	Inches and Decim.	
50	3·87	4·26	141
60	4·24	4·66	154
70	4·58	5·04	166
80	4·90	5·39	178
90	5·20	5·72	189
100	5·49	6·03	200
120	6·00	6·60	218
140	6·48	7·12	235
160	6·93	7·62	252
180	7·35	8·09	267
200	7·75	8·53	281
220	8·12	8·93	295
240	8·48	9·33	308
260	8·83	9·71	321
280	9·16	10·08	333
300	9·49	10·44	345
400	10·95	12·05	400
500	12·25	13·47	445
600	13·42	14·76	488

Dr. Smith (Rem. p. 78) observes, that the magnifying powers of this table are not so great as Huygens himself intended, or as the best object-glasses now made will admit of. For the author, in his Astroscopia Compendiaria, mentions an object-glass of 34 feet focal distance, which, in astronomical observations, bore an eye-glass of $2\frac{1}{5}$ inches focal distance, and consequently magnified 163 times. According to this standard, a Telescope of 35 feet ought to magnify 166 times, and of 1 foot 28 times; whereas the table allows but 118 times to the former, and but 20 to the latter. Now $\frac{166}{118}$ or $\frac{28}{20} = 1·4$; by which if we multiply the numbers in the given column of magnifying powers, we shall gain a new column, shewing how much those object-glasses ought to magnify if wrought up to the perfection of this standard.

The new apertures and eye glasses must also be taken in the same proportions to one another, as the old ones have in the table; or the eye-glasses may be found by dividing the length of each Telescope by its magnifying power. And thus a new table may be easily made for this or any other more perfect standard when offered.

The rule for computing this table depends on the following theorem, viz, that in refracting Telescopes of different lengths, a given object will appear equally bright and equally distinct, when their linear apertures and the focal distances of their eye-glasses are severally in a subduplicate ratio of their lengths, or focal distances of their object-glasses; and then also the breadth of their apertures will be in the subduplicate ratio of their lengths.

The rule is this: Multiply the number of feet in the focal distance of any proposed object-glass by 3000, and the square-root of the product will give the breadth of its aperture in centesms, or 100th parts of an inch;

that

that is, $\sqrt{3000\text{F}}$ is the breadth of the aperture in centefms of an inch, where F is the focal diftance of the object-glafs in feet. Alfo, the fame breadth of the aperture increafed by the 10th part of itfelf, gives the focal diftance of the eye-glafs in centefms of an inch. And the magnifying powers are as the breadths of the apertures.

If, in different Telefcopes, the ratio between the object-glafs and eye-glafs be the fame, the object will be magnified the fame in both. Hence fome may conclude the making of large Telefcopes a needlefs trouble. But it muft be remembered, that an eye-glafs may be in a lefs ratio to a greater object-glafs than to a fmaller: thus, for example, in Huygens's Telefcope of 25 feet, the eye-glafs is 3 inches: now, keeping this proportion in a Telefcope of 50 feet, the eye-glafs fhould be 6 inches; but the table fhews that $4\frac{1}{2}$ are fufficient. Hence, from the fame table it appears, that a Telefcope of 50 feet magnifies in the ratio of 1 to 141; whereas that of 25 feet only magnifies in the ratio of 1 to 100.

Since the diftance of the lens is equal to the aggregate of the diftances of the foci of the object and eye-glaffes; and fince the focus of a glafs convex on each fide is a femidiameter's diftance from the lens, and that of a plano-convex at a diameter's diftance from the fame; the length of a Telefcope is equal to the aggregate of the femidiameters of the lenfes, if the object-glafs be convex on both fides; and to the fum of the femidiameter of the eye-glafs and the whole diameter of the object glafs, if the object-glafs be a plano-convex.

But as the diameter of the eye-glafs is very fmall in refpect of that of the object-glafs, the length of the Telefcope is ufually eftimated from the diftance of the object-glafs; i. e. from its femidiameter if it be convex on both fides, or its whole diameter if plano-convex. Thus, a Telefcope is faid to be 12 feet, if the femidiameter of the object-glafs, convex on both fides, be 12 feet, &c.

Since myopes fee near objects beft; for them, the eye-glafs is to be removed nearer to the object-glafs, that the rays refracted through it may be the more diverging.

To take in the larger field at one view, fome make ufe of two eye-glaffes, the foremoft of which is a fegment of a larger fphere than that behind; to this it muft be added, that if two lenfes be joined immediately together, fo as the one may touch the other, the focus is removed to double the diftance which that of one of them would be at.

Land TELESCOPE, or *Day* TELESCOPE, is one adapted for viewing objects in the day-time, on or about the earth. This contains more than two lenfes, ufually it has a convex object-glafs, and three convex eye-glaffes; exhibiting objects erect, yet different from that of Galileo.

In this Telefcope, after the rays have paffed the firft eye glafs HI (fig. 2, pl. 30), as in the former conftruction, inftead of being there received by the eye, they pafs on to another equally convex lens, fituated at twice its focal diftance from the other, fo that the rays of each pencil, being parallel in that whole interval, thofe pencils crofs one another in the common

focus, and the rays conftituting them are tranfmitted parallel to the fecond eye-glafs LM; after which the rays of each pencil converge to other foci at NO, where a fecond image of the object is formed, but inverted with refpect to the former image in EF. This image then being viewed by a third eye-glafs QR, is printed upon the retina at XYZ, exactly as before, only in a contrary pofition.

Father Reita was the author of this conftruction; which is effected by fitting in at one end of a tube an object-glafs, which is either convex on both fides, or plano-convex, and a fegment of a large fphere; to this add three eye-glaffes, all convex on both fides, and fegments of equal fpheres; difpofing them in fuch a manner as that the diftance between any two may be the aggregate of the diftances of their foci. Then will an eye applied to the laft lens, at the diftance of its focus, fee objects very diftinctly, erect, and magnified in the ratio of the diftance of the focus of one eye-glafs, to the diftance of the focus of the object-glafs.

Hence, 1. An aftronomical Telefcope is eafily converted into a Land Telefcope, by ufing three eye-glaffes for one; and the Land Telefcope, on the contrary, into an aftronomical one, by taking away two eye-glaffes, the faculty of magnifying ftill remaining the fame.

2. Since the diftance of the eye-glaffes is very fmall, the length of the Telefcope is much the fame as if you only ufed one.

3. The length of the Telefcope is found by adding five times the femidiamer of the eye-glaffes, to the diameter of the object-glafs when this is a plano-convex, or to its femidiameter when convex on both fides.

Huygens firft obferved, both in the aftronomical and Land Telefcope, that it contributes confiderably to the perfection of the inftrument, to have a ring of wood or metal, with an aperture a little lefs than the breadth of the eye-glafs, fixed in the place where the image is found to radiate upon the lens next the eye: by means of which, the colours, which are apt to difturb the clearnefs and diftinctnefs of the object, are prevented, and the whole compafs taken in at one view, perfectly defined.

Some make Land Telefcopes of three lenfes, which yet reprefent objects erect, and magnified as much as the former. But fuch Telefcopes are fubject to very great inconveniences, both as the objects in them are tinged with falfe colours, and as they are diftorted about the margin.

Some again ufe five lenfes, and even more; but as fome parts of the rays are intercepted in paffing every lens, objects are thus exhibited dim and feeble.

Telefcopes of this kind, longer than 20 feet, will be of hardly any ufe in obferving terreftrial objects, on account of the continual motion of the particles of the atmofphere, which thefe powerful Telefcopes render vifible, and give a tremulous motion to the objects themfelves.

The great length of dioptric Telefcopes, adapted to any important aftronomical purpofe, rendered them extremely inconvenient for ufe; as it was neceffary to increafe their length in no lefs a proportion than the

duplicate

duplicate of the increase of their magnifying power: so that, in order to magnify twice as much as before, with the same light and distinctness, the Telescope required to be lengthened 4 times; and to magnify thrice as much, 9 times the length, and so on. This unwieldiness of refracting Telescopes, possessing any considerable magnifying power, was one cause, why the attention of astronomers, &c, was directed to the discovery and construction of reflecting Telescopes. And indeed a refracting Telescope, even of 1000 feet focus, supposing it possible to make use of such an instrument, could not be made to magnify with distinctness more than 1000 times; whereas a reflecting Telescope, of 9 or 10 feet, will magnify 12 hundred times. The perfection of refracting Telescopes, it is well known, is very much limited by the aberration of the rays of light from the geometrical focus: and this arises from two different causes, viz, from the different degrees of refrangibility of light, and from the figure of the sphere, which is not of a proper curvature for collecting the rays in a single point. Till the time of Newton, no optician had imagined that the object glasses of Telescopes were subject to any other error beside that which arose from their spherical figure, and therefore all their efforts were directed to the construction of them, with other kinds of curvature: but that author had no sooner demonstrated the different refrangibility of the rays of light, than he discovered in this circumstance a new and a much greater cause of error in Telescopes. Thus, since the pencils of each kind of light have their foci in different places, some nearer and some farther from the lens, it is evident that the whole beam cannot be brought into any one point, but that it will be drawn the nearest to a point in the middle place between the focus of the most and least refrangible rays; so that the focus will be a circular space of a considerable diameter. Newton shews that this space is about the 55th part of the aperture of the Telescope, and that the focus of the most refrangible rays is nearer to the object-glass than the focus of the least refrangible ones, by about the $27\frac{1}{2}$ part of the distance between the object-glass, and the focus of the mean refrangible rays. But he says, that if the rays flow from a lucid point, as far from the lens on one side, as their foci are on the other, the focus of the most refrangible rays will be nearer to the lens than that of the least refrangible, by more than the 14th part of the whole distance. Hence, he concludes, that if all the rays of light were equally refrangible, the error in Telescopes, arising from the spherical figure of the glass, would be many hundred times less than it now is: because the error arising from the spherical figure of the glass, is to that arising from the different refrangibility of the rays of light, as 1 to 5449. See ABERRATION.

Upon the whole he observes, that it is a wonder that Telescopes represent objects so distinctly as they do. The reason of which is, that the dispersed rays are not scattered uniformly over all the circular space above mentioned, but are infinitely more dense in the centre than in any other part of the circle; and that in the way from the centre to the circumference they grow continually rarer and rarer, till at the circumference they become infinitely rare: for which reason,

these dispersed rays are not copious enough to be visible, except about the centre of the circle. He also mentions another argument to prove, that the different refrangibility of the rays of light is the true cause of the imperfection of Telescopes: For the dispersions of the rays arising from the spherical figures of object-glasses, are as the cubes of their apertures; and therefore, to cause Telescopes of different lengths to magnify with equal distinctness, the apertures of the object-glasses, and the charges or magnifying powers ought to be as the cubes of the square roots of their lengths, which does not answer to experience. But the errors of the rays, arising from the different refrangibility, are as the apertures of the object glasses; and thence, to make Telescopes of different lengths to magnify with equal distinctness, their apertures and charges ought to be as the square roots of their lengths; and this answers to experience.

Were it not for this different refrangibility of the rays, Telescopes might be brought to a sufficient degree of perfection, by composing the object-glass of two glasses with water between them. For by this means, the refractions on the concave sides of the glasses will very much correct the errors of the refractions on the convex sides, so far as they arise from their spherical figure: but on account of the different refrangibility of different kinds of rays, Newton did not see any other means of improving Telescopes by refraction only, except by increasing their length. Newton's Optics, pa. 73, 83, 89, 3d edition.

This important desideratum in the construction of dioptric Telescopes, has been since discovered by the ingenious Mr. Dollond; an account of which is given below.

Achromatic TELESCOPE, is a name given to the refracting Telescope, invented by Mr. John Dollond, and so contrived as to remedy the aberration arising from colours, or the different refrangibility of the rays of light. See ACHROMATIC.

The principles of Mr. Dollond's discovery and construction, have been already explained under the articles ABERRATION, and ACHROMATIC. The improvement made by Mr. Dollond in his Telescopes, by making two object-glasses of crown-glass, and one of flint, which was tried with success when concave eye glasses were used, was completed by his son Peter Dollond; who, conceiving that the same method might be practised with success with convex eye-glasses, found, after a few trials, that it might be done. Accordingly he finished an object-glass of 5 feet focal length, with an aperture of $3\frac{3}{4}$ inches, composed of two convex lenses of crown-glass, and one concave of white flint glass. But apprehending afterward that the apertures might be admitted still larger, he completed one of $3\frac{1}{2}$ feet focal length, with the same aperture of $3\frac{3}{4}$ inches. Philos. Transf. vol. 55, p. 56.

But beside the obligation we are under to Mr. Dollond, for correcting the aberration of the rays of light in the focus of object-glasses, arising from their different refrangibility, he made another considerable improvement in Telescopes, viz, by correcting. in a great measure, both this kind of aberration, and also that which arises from the spherical form of lenses, by an expedient of a very different nature, viz, increasing

the

the number of eye-glasses. If any person, says he, would have the visual angle of a Telescope to contain 20 degrees, the extreme pencils of the field must be bent or refracted in an angle of 10 degrees; which, if it be performed by one eye-glass, will cause an aberration from the figure, in proportion to the cube of that angle: but if two glasses be so proportioned and situated as that the refraction may be equally divided between them, they will each of them produce a refraction equal to half the required angle; and therefore, the aberration being in this case proportional to double the cube of half the angle, will be but a 4th part of that which is in proportion to the cube of the whole angle; because twice the cube of 1 is but ¼ of the cube of 2: so that the aberration from the figure, where two eye-glasses are rightly proportioned, is but a 4th part of what it must unavoidably be, where the whole is performed by a single eye-glass. By the same way of reasoning, when the refraction is divided among three glasses, the aberration will be found to be but the 9th part of what would be produced from a single glass; because 3 times the cube of 1 is but the 9th part of the cube of 3. Whence it appears, that by increasing the number of eye-glasses, the indistinctness, near the borders of the field of a Telescope, may be very much diminished, though not entirely taken away.

The method of correcting the errors arising from the different refrangibility of light, is of a different consideration from the former: for, whereas the errors from the figure can only be diminished in a certain proportion to the number of glasses, in this they may be entirely corrected, by the addition of only one glass; as we find in the astronomical Telescope, that two eye-glasses, rightly proportioned, will cause the edges of objects to appear free from colours quite to the borders of the field. Also, in the day telescope, where no more than two eye glasses are absolutely necessary for erecting the object, we find, by the addition of a third rightly situated, that the colours, which would otherwise confuse the image, are entirely removed: but this must be understood with some limitation; for though the different colours, which the extreme pencils must necessarily be divided into by the edges of the eye-glasses, may in this manner be brought to the eye in a direction parallel to each other, so as, by its humours, to be converged to a point in the retina, yet if the glasses exceed a certain length, the colours may be spread too wide to be capable of being admitted through the pupil or aperture of the eye; which is the reason that, in long Telescopes, constructed in the common way, with three eye-glasses, the field is always very much contracted.

These considerations first set Mr. Dollond upon contriving how to enlarge the field, by increasing the number of eye-glasses, without any hindrance to the distinctness or brightness of the image: and though others had been about the same work before, yet observing that the five-glass Telescopes, sold in the shops, would admit of a farther improvement, he endeavoured to construct one with the same number of glasses in a better manner; which so far answered his expectations, as to be allowed by the best judges to be a considerable improvement on the former. Encouraged by this success, he resolved to try if he could not make

some farther enlargement of the field, by the addition of another glass, and by placing and proportioning the glasses in such a manner, as to correct the aberrations as much as possible, without any detriment to the distinctness: and at last he obtained as large a field as is convenient or necessary, and that even in the longest Telescopes that can be made. These Telescopes, with 6 glasses, having been well received both at home and abroad, the author has settled the date of the invention in a letter addressed to Mr. Short, and read at the Royal Society, March 1, 1753. Philos. Trans. vol. 48, art. 14.

Of the Achromatic Telescopes, invented by Mr. Dollond, there are several different sizes, from one foot to 8 feet in length, made and sold by his sons P. and J. Dollond. In the 17-inch improved Achromatic Telescope, the object glass is composed of three glasses, viz, two convex of crown-glass, and one concave of white flint-glass: the focal distance of this combined object-glass is about 17 inches, and the diameter of the aperture 2 inches. There are 4 eye-glasses contained in the tube, to be used for land objects; the magnifying power with these is near 50 times; and they are adjusted to different sights, and to different distances of the object, by turning a finger screw at the end of the outer tube. There is another tube, containing two eye-glasses that magnify about 70 times, for astronomical purposes. The Telescope may be directed to any object by turning two screws in the stand on which it is fixed, the one giving a vertical motion, and the other a horizontal one. The stand may be inclosed in the inside of the brass tube.

The object-glass of the 2¼ and 3½ feet Telescopes is composed of two glasses, one convex of crown glass, and the other concave of white flint glass; and the diameters of their apertures are 2 inches and 2⅜ inches. Each of them is furnished with two tubes; one for land objects, containing four eye-glasses, and another with two eye glasses for astronomical uses. They are adjusted by buttons on the outside of the wooden tube; and the vertical and horizontal motions are given by joints in the stands. The magnifying power of the least of these Telescopes, with the eye-glass for land objects, is near 50 times, and with those for astronomical purposes, 80 times; and that of the greatest for land objects is near 70 times, but for astronomical observations 80 and 130 times: for this has two tubes, either of which may be used as occasion requires. This Telescope is also moved by a screw and rackwork, and the screw is turned by means of a Hook's joint.

These opticians also construct an Achromatic pocket perspective glass, or Galilean Telescope; so contrived, that all the different parts are put together and contained in one piece 4¼ inches long. This small Telescope is furnished with 4 concave eye-glasses, the magnifying powers of which are 6, 12, 18, and 28 times. With the greatest power of this Telescope, the satellites of Jupiter and the ring of Saturn may be easily seen. They have also contrived an Achromatic Telescope, the sliding tubes of which are made of very thin brass, which pass through springs or tubes; the outside tube being either of mahogany or brass. These Telescopes, which from their convenience for gentlemen in the army are called military Telescopes, have 4 convex eye-glasses

glasses, whose surfaces and focal lengths are so proportioned, as to render the field of view very large. They are of 4 different lengths and sizes, usually called one foot, 2, 3, and 4 feet: the first is 14 inches when in use, and 5 inches when shut up, having the aperture of the object-glass $1\frac{1}{10}$ inch, and magnifying 22 times: the second 28 inches for use, 9 inches shut up, the aperture $1\frac{5}{10}$ inch, and magnifying 35 times; the third 40 inches, and 10 inches shut, with the aperture 2 inches, and magnifying 45 times; and the fourth 52 inches, and 14 inches shut, with the aperture $2\frac{3}{4}$ inches, and magnifying 55 times.

Mr. Euler, who, in a memoir of the Academy of Berlin for the year 1757, p. 323, had calculated the effects of all possible combinations of lenses in Telescopes and microscopes, published another long memoir on the subject of these Telescopes, shewing with precision of what advantages they are naturally capable. See Miscel. Taurin. vol. 3, par. 2, pag. 92.

Mr. Caleb Smith, having paid much attention to the subject of shortening and improving Telescopes, thought he had found it possible to rectify the errors which arise from the different degrees of refrangibility, on the principle that the sines of refraction of rays differently refrangible, are to one another in a given proportion, when their sines of incidence are equal; and the method he proposed for this purpose, was to make the specula of glass, instead of metal, the two surfaces having different degrees of concavity. But it does not appear that this scheme was ever carried into practice. See Philos. Transf. number 456, pa. 326, or Abr. vol. 8, pa. 113.

The ingenious Mr. Ramsden has lately described a new construction of eye-glasses for such Telescopes as may be applied to mathematical instruments. The construction which he proposes, is that of two plano-convex lenses, both of them placed between the eye and the observed image formed by the object-glass of the instrument, and thus correcting not only the aberration arising from the spherical figure of the lenses, but also that arising from the different refrangibility of light. For a more particular account of this construction, its principle, and its effects, see Philos. Transf. vol. 73, art. 5.

A construction, similar at least in its principle to that above, is ascribed, in the Synopsis Optica Honorati Fabri, to Eustachio Divini, who placed two equal narrow plano-convex lenses, instead of one eye lens, to his Telescopes, which touched at their vertices; the focus of the object-glass coinciding with the centre of the plano-convex lens next it. And this, it is said, was done at once both to make the rays that come parallel from the object fall parallel upon the eye, to exclude the colours of the rainbow from it, to augment the angle of sight, the field of view, the brightness of the object, &c. This was also known to Huygens, who sometimes made use of the same construction, and gives the theory of it in his Dioptrics. See Hugenii Opera Varia, vol. 4, ed. 1728.

TELESCOPE, *Reflecting*, or *Catoptric*, or *Catadioptric*, is a Telescope which, instead of lenses, consists chiefly of mirrors, and exhibits remote objects by reflection instead of refraction.

A brief account of the history of the invention of this important and useful Telescope, is as follows. The ingenious Mr. James Gregory, of Aberdeen, has been commonly considered as the first inventor of this Telescope —But it seems the first thought of a reflector had been suggested by Mersenne, about 20 years before the date of Gregory's invention: a hint to this purpose occurs in the 7th proposition of his Catoptrics, which was printed in 1651: and it appears from the 3d and 29th letters of Descartes, in vol. 2 of his Letters, which it is said were written in 1639, though they were not published till the year 1666, that Mersenne proposed a Telescope with specula to Descartes in that correspondence; though indeed in a manner so very unsatisfactory, that Descartes, who had given particular attention to the improvement of the Telescope, was so far from approving the proposal, that he endeavoured to convince Mersenne of its fallacy. This point has been largely discussed by Le Roi in the Encyclopedia, art. Telescope, and by Montucla in his Hist. des Mathem. tom. 2, p. 643.

Whether Gregory had seen Mersenne's treatise on optics and catoptrics, and whether he availed himself of the hint there suggested, or not, perhaps cannot now be determined. He was led however to the invention by seeking to correct two imperfections in the common Telescope: the first of these was its too great length, which made it troublesome to manage; and the second was the incorrectness of the image. It had been already demonstrated, that a pencil of rays could not be collected in a single point by a spherical lens; and also, that the image transmitted by such a lens would be in some degree incurvated. These inconveniences he thought might be obviated by substituting for the object-glass a metallic speculum, of a parabolical figure, to receive the image, and to reflect it towards a small speculum of the same metal; this again was to return the image to an eye-glass placed behind the great speculum, which was, for that purpose, to be perforated in its centre. This construction he published in 1663, in his Optica Promota. But as Gregory, according to his own account, possessed no mechanical skill, and could not find a workman capable of realizing his invention, after some fruitless trials, he was obliged to give up the thoughts of bringing Telescopes of this kind into use.

Sir Isaac Newton however interposed, to save this excellent invention from perishing, and to bring it forward to maturity. Having applied himself to the improvement of the Telescope, and imagining that Gregory's specula were neither very necessary, nor likely to be executed, he began with prosecuting the views of Descartes; who aimed at making a more perfect image of an object, by grinding lenses, not to the figure of a sphere, but to that formed from one of the conic sections. But, in the year 1666, having discovered the different refrangibility of the rays of light, and finding that the errors of Telescopes, arising from that cause alone, were much more considerable than such as were occasioned by the spherical figure of lenses, he was constrained to turn his thoughts to reflectors. The plague however interrupted his progress in this business; so that it was towards the end of 1668, or in the beginning of 1669, when, despairing of perfecting Telescopes by means of refracted light,

and

and recurring to the construction of reflectors, he set about making his own specula, and early in the year 1672 completed two small reflecting Telescopes. In these he ground the large speculum into a spherical concave, being unable to accomplish the parabolic form proposed by Gregory; but though he then despaired of performing that work by geometrical rules, yet (as he writes in a letter that accompanied one of these instruments, which he presented to the Royal Society) he doubted not but that the thing might in some measure be accomplished by mechanical devices. With a perseverance equal to his ingenuity, he, in a great measure, overcame another difficulty, which was to find a metallic substance that would be of a proper hardness, have the fewest pores, and receive the smoothest polish: this difficulty he deemed almost insurmountable, when he considered that every irregularity in a reflecting surface would make the rays of light deviate 5 or 6 times more out of their due course, than the like irregularities in a refracting surface. After repeated trials, he at last found a composition that answered in some degree, leaving it to those who should come after him to find a better. These difficulties have accordingly been since obviated by other artists, particularly by Dr. Mudge, the rev. Mr. Edwards, and Dr. Herschel, &c. Newton having succeeded so far, he communicated to the Royal Society a full and satisfactory account of the construction and performance of his Telescope. The Society, by their secretary Mr. Oldenburgh, transmitted an account of the discovery to Mr. Huygens, celebrated as a distinguished improver of the refractor; who not only replied to the Society in terms expressing his high approbation of the invention, but drew up a favourable account of the new Telescope, which he caused to be published in the Journal des Scavans of the year 1672, and by this mode of communication it was soon known over Europe. See Huygenii Opera Varia, tom. 4.

Notwithstanding the excellence and utility of this contrivance, and the honourable manner in which it was announced to the world, it seems to have been greatly neglected for nearly half a century. Indeed when Newton had published an account of his Telescopes in the Philos. Transf. M. Cassegrain, a Frenchman, in the Journal des Scavans of 1672, claimed the honour of a similar invention, and said, that, before he heard of Newton's improvement, he had hit upon a better construction, by using a small convex mirror instead of the reflecting prism. This Telescope, which was the Gregorian one disguised, the large mirror being perforated, and which it is said was never executed by the author, is much shorter than the Newtonian; and the convex mirror, by dispersing the rays, serves greatly to increase the image made by the large concave mirror.

Newton made many objections to Cassegrain's construction, but several of them equally affect that of Gregory, which has been found to answer remarkably well in the hands of good artists.

Dr. Smith took the pains to make many calculations of the magnifying power, both of Newton's and Cassegrain's Telescopes, in order to their farther improvement, which may be seen in his Optics, Rem. p. 97.

Mr. Short, it is also said, made several Telescopes on the plan of Cassegrain.

Dr. Hook constructed a Reflecting Telescope (mentioned by Dr. Birch in his Hist. of the Royal Soc. vol. 3, p. 122) in which the great mirror was perforated, so that the spectator looked directly towards the object, and it was produced before the Royal Society in 1674. On this occasion it was said that this construction was first proposed by Mersenne, and afterwards repeated by Gregory, but that it never had been actually executed before it was done by Hook. A description of this instrument may be seen in Hook's Experiments, by Derham, p. 269.

The Society also made an unsuccessful attempt, by employing an artificer to imitate the Newtonian construction; however, about half a century after the invention of Newton, a Reflecting Telescope was produced to the world, of the Newtonian construction, which the venerable author, ere yet he had finished his very distinguished course, had the satisfaction to find executed in such a manner, as left no room to fear that the invention would longer continue in obscurity. This effectual service to science was accomplished by Mr. John Hadley, who, in the year 1723, presented to the Royal Society a Telescope, which he had constructed upon Newton's plan. The two Telescopes which Newton had made, were but 6 inches long, were held in the hand for viewing objects, and in power were compared to a 6-feet refractor: but the radius of the sphere, to which the principal speculum of Hadley's was ground, was 10 feet $5\frac{1}{4}$ inches, and consequently its focal length was $62\frac{5}{8}$ inches. In the Philos. Transf. Abr. vol. 6, p. 133, may be seen a drawing and description of this Telescope, and also of a very ingenious but complex apparatus, by which it was managed. One of these Telescopes, in which the focal length of the large mirror was not quite $5\frac{1}{4}$ feet, was compared with the celebrated Huygenian Telescope, which had the focal length of its object-glass 123 feet; and it was found that the former would bear such a charge, as to make it magnify the object as many times as the latter with its due charge; and that it represented objects as distinctly, though not altogether so clear and bright. With this Reflecting Telescope might be seen whatever had been hitherto discovered by the Huygenian, particularly the transits of Jupiter's satellites, and their shades over the disk of Jupiter, the black list in Saturn's ring, and the edge of the shade of Saturn cast upon his ring. Five satellites of Saturn were also observed with this Telescope, and it afforded other observations on Jupiter and Saturn, which confirmed the good opinion which had been conceived of it by Pound and Bradley.

Mr. Hadley, after finishing two Telescopes of the Newtonian construction, applied himself to make them in the Gregorian form, in which the large mirror is perforated. This scheme he completed in the year 1726.

Dr. Smith prefers the Newtonian construction to that of Gregory; but if long experience be admitted as a final judge in such matters, the superiority must be adjudged to the latter; as it is now, and has been for many years past, the only instrument in request.

Mr.

Mr. Hadley spared no pains, after having completed his construction, to instruct Mr. Molyneux and Dr. Bradley; and when these gentlemen had made a good proficiency in the art, being desirous that these Telescopes should become more public, they liberally communicated to some of the chief instrument-makers of London, the knowledge they had acquired from him : and thus, as it is reasonable to imagine, reflectors were completed by other and better methods than even those in which they had been instructed. Mr. James Short in particular signalized himself as early as the year 1734, by his work in this way. He at first made his specula of glass; but finding that the light reflected from the best glass specula was much less than the light reflected from metallic ones, and that glass was very liable to change its form by its own weight, he applied himself to improve metallic specula; and, by giving particular attention to the curvature of them, he was able to give them greater apertures than other workmen could do; and by a more accurate adjustment of the specula, &c, he greatly improved the whole instrument. By some which he made, in which the larger mirror was 15 inches focal distance, he and some other persons were able to read in the Philos. Transf. at the distance of 500 feet; and they several times saw the five satellites of Saturn together, which greatly surprised Mr. Maclaurin, who gave this account of it, till he found that Cassini had sometimes seen them all with a 17 feet refractor. Short's Telescopes were all of the Gregorian construction. It is supposed that he discovered a method of giving the parabolic figure to his great speculum; a degree of perfection which Gregory and Newton despaired of attaining, and which Hadley it seems had never attempted in either of his Telescopes. However, the secret of working that configuration, whatever it was, it seems died with that ingenious artist. Though lately in some degree discovered by Dr. Mudge and others.

On the History of Reflecting Telescopes, see Dr. David Gregory's Elem. of Catopt. and Dioptr. Appendix by Desaguliers: Smith's Optics, book 3, c. 2, Rem. on art. 489: and Sir John Pringle's excellent Discourse on the Invention &c of the Reflecting Telescope.

Construction of the Reflecting-Telescope of the Newtonian form.—Let ABCD (fig. 2, pl. 32) be a large tube, open at AD, and closed at BC, and its length at least equal to the distance of the focus from the metallic spherical concave speculum GH placed at the end BC. The rays EG, FH, &c, proceeding from a remote object PR, intersect one another somewhere before they enter the tube, so that EG and *eg* are those that come from the lower part of the object, and *fh* FH from its upper part: these rays, after falling on the speculum GH, will be reflected so as to converge and meet in *mn*, where they will form a perfect image of the object. But as this image cannot be seen by the spectator, they are intercepted by a small plane metallic speculum KK, intersecting the axis at an angle of 45°, by which the rays tending to *m*, *n*, will be reflected towards a hole LL in the side of the tube, and the image of the object will be thus formed in *q*S; which image will be less distinct, because some of the rays which would otherwise fall on the concave speculum

GH, are intercepted by the plane speculum : it will nevertheless appear pretty distinct, because the aperture AD of the tube, and the speculum GH, are large. In the lateral hole LL is fixed a convex lens, whose focus is at S*q*; and therefore this lens will refract the rays that proceed from any point of the image, so as at their exit they will appear parallel, and those that proceed from the extreme points S, *q*, will converge after refraction, and form an angle at O, where the eye is placed; which will see the image S*q*, as if it were an object, through the lens LL: consequently the object will appear enlarged, inverted, bright, and distinct. In LL may be placed lenses of different convexities, which, by being moved nearer to the image and farther from it, will represent the object more or less magnified, if the surface of the speculum GH be of a figure truly spherical. If, instead of one lens LL, three lenses be disposed in the same manner with the three eye glasses of the refracting Telescope, the object will appear erect, but less distinct than when it is observed with one lens. On account of the position of the eye in this Telescope, it is extremely difficult to direct the instrument towards any object: Huygens therefore first thought of adding to it a small refracting Telescope, having its axis parallel to that of the reflector: this is called a *finder* or *director*. The Newtonian Telescope is also furnished with a suitable apparatus for the commodious use of it.

To determine the magnifying power of this Telescope, it is to be considered that the plane speculum KK is of no use in this respect: let us then suppose that one ray proceeding from the object coincides with the axis GLIA of the lens and speculum; let *bb* be

another ray proceeding from the lower extremity of the object, and passing through the focus I of the speculum KH; this will be reflected in the direction *bid*, parallel to the axis GLA, and falling on the lens *dLd*, will be refracted to G, so that GL will be equal to LI, and *d*G = *d*I. To the naked eye the object would appear under the angle I*bi* = *b*IA; but by means of the Telescope it appears under the angle *d*GL = *d*IL = I*di*: and the angle I*di* is to the angle I*bi* as *b*I to I*d*; consequently the apparent magnitude by the Telescope, is to that with the naked eye, as the distance of the focus of the speculum from the speculum, to the distance of the focus of the lens from the lens.

Construction of the Gregorian Reflecting Telescope.—Let TYYT (fig. 3, pl. 32) be a brass tube, in which L*ld*D is a metallic concave speculum, perforated in the middle at X; and EF a less concave mirror, so fixed by the arm or strong wire RT, which is moveable by means of a long screw on the outside of the tube, as to be moved nearer to, or farther from the larger speculum L*ld*D; its axis being kept in the same line with that of the great one. Let AB represent a very remote object, from each part of which issue pencils of rays, as *cd*, CD, from A the upper extremity of the object,

object, and IL, *il*, from the lower part B; the rays IL, CD, from the extremities, crossing one another before they enter the tube. These rays, falling upon the larger mirror LD, are reflected from it into the focus KH, where they form an inverted image of the object AB, as in the Newtonian Telescope. From this image the rays, issuing as from an object, fall upon the small mirror EF, the centre of which is at *e*, so that after reflection they would meet in their foci at QQ, and there form an erect image. But since an eye at that place could see but a small part of an object, in order to bring rays from more distant parts of it into the pupil, they are intercepted by the plano convex lens MN, by which means a smaller erect image is formed at PV, which is viewed through the meniscus SS, by an eye at O. This meniscus both makes the rays of each pencil parallel, and magnifies the image PV. At the place of this image all the foreign rays are intercepted by the perforated partition ZZ. For the same reason the hole near the eye O is very narrow. When nearer objects are viewed by this Telescope, the small speculum EF is removed to a greater distance from the larger LD, so that the second image may be always formed in PV : and this distance is to be adjusted (by means of the screw on the outside of the great tube) according to the form of the eye of the spectator. It is also necessary that the axis of the Telescope should pass through the middle of the speculum EF, and its centre, the centre of the speculum LL, and the middle of the hole X, the centres of the lenses MN, SS, and the hole near O. As the hole X in the speculum LL can reflect none of the rays issuing from the object, that part of the image which corresponds to the middle of the object, must appear to the observer more dark and confused than the extreme parts of it. Besides, the speculum EF will also intercept many rays proceeding from the object; and therefore, unless the aperture TT be large, the object must appear in some degree obscure.

The magnifying power of this Telescope is estimated in the following manner. Let LD be the larger mirror (fig. 3, pl. 31), having its focus at G, and aperture in A; and FF the small mirror with the focus of parallel rays in I, and the axis of both the specula and lenses MN, SS, be in the right line DIGAOK. Let *bb* be a ray of light coming from the lower extremity of a very distant visible object, passing through the focus G, and falling upon the point *b* of the speculum LD; which, after being reflected from *b* to F in a direction parallel to the axis of the mirror DAK, is reflected by the speculum F so as to pass through the focus I in the direction FIN to N, at the extremity of the lens MN, by which it would have been refracted to K; but by the interposition of another lens SS is brought to O, so that the eye in O sees half the object under the angle TOS. The angle G*b*F, or AG*b*, under which the object is viewed by the naked eye, is to SOT under which it is viewed by the Telescope, in the ratio of G*b*F to IF*i* = *n*IN, of *n*IN to NK*n*, and of NK*n* to SOT.

But G*b*F : IF*i* :: DI : GA,
and *n*IN : *n*KN :: *n*K : *n*I,
and *n*KN : SOT :: TO : TK;

theref. G*b*F : SOT :: DI × *n*K × TO : GA × *n*I × TK. Muschenbroek's Introd. vol. 2, p. 819.

In Reflecting Telescopes of different lengths, a given object will appear equally bright and equally distinct, when their linear apertures, and also their linear breadths, are as the 4th roots of the cubes of their lengths; and consequently when the focal distances of their eye-glasses are also as the 4th roots of their lengths. See the demonstration of this proposition in Smith's Optics, art. 361.

Hence he has deduced a rule, by which he has computed the following table for Telescopes of different lengths, taking, for a standard, the middle eye-glass and aperture of Hadley's Reflecting Telescope, described in Philos. Transf. number 376 and 378 : the focal distances and linear apertures being given in 1000th parts of an inch.

Table for Telescopes of different Lengths.			
Length of the Tel. or focal dist. of the conc.	Focal dist. of the Eye-glass.	Linear amplifying or magnifying power.	Linear aperture of the concave metal.
feet	inches	- - -	inches
½	0·167	36	0·864
1	0·199	60	1·440
2	0·236	102	2·448
3	0·261	138	3·312
4	0·281	171	4·104
5	0·297	202	4·843
6	0·311	232	5·568
7	0·323	260	6·240
8	0·334	287	6·888
9	0·344	314	7·536
10	0·353	340	8·160
11	0·362	365	8·760
12	0·367	390	9·360
13	0·377	414	9·936
14	0·384	437	10·488
15	0·391	460	11·040
16	0·397	483	11·592
17	0·403	506	12·143

Mr. Hadley's Telescope, above mentioned, magnified 228 or 230 times; but we are informed that an object-metal of 3¼ feet focal distance was wrought by Mr. Haukesbee to so great a perfection, as to magnify 226 times, and therefore it was scarcely inferior to Hadley's of 5½ feet. If Haukesbee's Telescope be taken for a new standard, it follows that a speculum of one foot focal distance ought to magnify 93 times, whereas the above table allows it but 60. Now 9²/₆₀ = 1·55, and the given column of magnifying powers multiplied by this number, gives a new column, shewing how much the object-metals ought to magnify if wrought up to the perfection of Haukesbee's. And thus a new table may be easily made for this or any other more perfect standard, taking also the new eye-glasses and apertures in the same ratio to one another as the old ones have in this table. Smith's Optics, Rem. p. 79.

     The

The magnifying power of any Telescope may be easily found by experiment, viz, by looking with one eye through the Telescope upon an object of known dimensions, and at a given distance, and throwing the image upon another object seen with the naked eye. Dr. Smith has given a particular account of the process, Rem. p. 79.

But the easiest method of all, is to measure the diameter of the aperture of the object-glass, and that of the little image of it, which is formed at the place of the eye. For the proportion between these gives the ratio of the magnifying power, provided no part of the original pencil be intercepted by the bad construction of the Telescope. For in all cases the magnifying power of Telescopes, or microscopes, is measured by the proportion of the diameter of the original pencil, to that of the pencil which enters the eye. Priestley's Hist. of Light, p. 747.

But the most considerable, and indeed truly astonishing magnifying powers, that have ever been used, are those of Dr. Herschel's Reflecting Telescopes. Some account of these, and of the discoveries made by them, has been already introduced under the article Star. For his method of ascertaining them, see Philos. Transf. vol. 72, pa. 173 &c. See also several of the other late volumes of the Philos. Transf.

Dr. Herschel observes, that though opticians have proved, that two eye-glasses will give a more correct image than one, he has always (from experience) persisted in refusing the assistance of a second glass, which is sure to introduce errors greater than those he would correct. "Let us resign, says he, the double eye-glass to those who view objects merely for entertainment, and who must have an exorbitant field of view. To a philosopher, this is an unpardonable indulgence. I have tried both the single and double eye-glass of equal powers, and always found that the single eye-glass had much the superiority in point of light and distinctness. With the double eye-glass I could not see the belts in Saturn, which I very plainly saw with the single one. I would however except all those cases where a large field is absolutely necessary, and where power joined to distinctness is not the sole object of our view." Philos. Transf. vol. 72, p. 95.

Mr. Green of Deptford has lately added both to the reflecting and refracting Telescope an apparatus, which fits it for the purposes of surveying, levelling, measuring angles and distances, &c. See his Description and Use of the improved Reflecting and Refracting Telescopes, and Scale of Surveying &c, 1778.— Mr. Ramsden too has lately adapted Telescopes to the like purpose of measuring distances from one station, &c.

Meridian Telescope, is one that is fixed at right angles to an axis, and turned about it in the plane of the meridian; and is otherwise called a *transit instrument*.—The common use of it is to correct the motion of a clock or watch, by daily observing the exact time when the sun or a star comes to the meridian. It serves also for a variety of other uses. The transverse axis is placed horizontal by a spirit level. For the farther description and method of fixing this instrument by means of its levels &c, see Smith's Optics, p. 321. See also Transit *Instrument*.

, TELESCOPICAL *Stars*, are such as are not visible to the naked eye, being only discernible by means of a telescope. See Star.

All stars less than those of the 6th magnitude, are Telescopic to an ordinary eye.

TEMPERAMENT, in Music, usually denotes a rectifying or amending the false or imperfect concords, by transferring to them part of the beauty of the perfect ones.

TENACITY, in Natural Philosophy, is that quality of bodies by which they sustain a considerable pressure or force without breaking; and is the opposite quality to fragility or brittleness. Mem. Acad. Berlin. 1745, p. 47.

TENAILLE, in Fortification, a kind of outwork, consisting of two parallel sides, with a front, having a re-entering angle. In fact; that angle, and the faces which compose it, are the Tenaille.

The Tenaille is of two kinds, *simple* and *double*.

Simple or *Single* Tenaille, is a large outwork, consisting of two faces or sides, including a re-entering angle.

Double, or *Flanked* Tenaille, is a large outwork, consisting of two simple Tenailles, or three saliant and two re-entering angles.

The great defects of Tenailles are, that they take up too much room, and on that account are advantageous to the enemy; that the re-entering angle is not defended; the height of the parapet preventing the seeing down into it, so that the enemy can lodge there under cover; and the sides are not sufficiently flanked. For these reasons, Tenailles are now mostly excluded out of fortification by the best engineers, and never made but where time does not serve to form a hornwork.

Tenaille *of the Place*, is the front of the place, comprehended between the points of two neighbouring bastions; including the curtain, the two flanks raised on the curtain, and the two sides of the bastions which face one another. So that the Tenaille, in this sense, is the same with what is otherwise called the *face of a fortress*.

Tenaille *of the Ditch*, is a low work raised before the curtain, in the middle of the foss or ditch; the parapet of which is only 2 or 3 feet higher than the level ground of the ravelin.

The use of Tenailles in general, is to defend the bottom of the ditch by a grazing fire, and likewise the level ground of the ravelin, which cannot be so conveniently defended from any other place. The first sort do not defend the ditch so well as the others, because they are too oblique a defence; but as they are not subject to be enfiladed, Vauban has generally preferred them in the fortifying of places. Those of the second sort defend the ditch much better than the first, and add a low flank to those of the bastions; but as these flanks are liable to be enfiladed, they have not been much used. This defect however might be remedied, by making them so as to be covered by the extremities of the parapets of the opposite ravelins, or by some other work. And the same thing may be said of the third sort as of the second.

The *Ram's-horn* is a curved Tenaille, raised in the foss before the flanks, and presenting its convexity to the

the covered way. This work seems preferable to either of the other Tenailles, both on account of its simplicity, and the defence for which it is constructed.

TENAILLONS, in Fortification, are works constructed on each side of the ravelin, much like the lunettes. They differ, as one of the faces of a Tenaillon is in the direction of the ravelin, whereas that of the lunette is perpendicular to it.

TENOR, in Music, the first mean or middle part, or that which is the ordinary pitch, or Tenor, of the voice, when not either raised to the treble, or lowered to the bass.

TENSION, the state of a thing tight, or stretched. Thus, animals sustain and move themselves by the Tension of their muscles and nerves. A chord, or string, gives an acuter or deeper sound, as it is in a greater or less degree of Tension, that is, more or less stretched or tightened.

TERM, in Geometry, is the extreme of any magnitude, or that which bounds and limits its extent. So the Terms of a line, are points; of a superficies, lines; of a solid, superficies.

TERMS, of an equation, or of any quantity, in Algebra, are the several names or members of which it is composed, separated from one another by the signs $+$ or $-$. So, the quantity $ax + 2bc - 3ax^2$, consists of the three Terms ax and $2bc$ and $3ax^2$.

In an equation, the Terms are the parts which contain the several powers of the same unknown letter or quantity: for if the same unknown quantity be found in several members in the same degree or power, they shall pass but for one Term, which is called a compound one, in distinction from a simple or single Term. Thus, in the equation $x^3 + \overline{a - 3b} \cdot x^2 - acx = b^3$, the four terms are x^3 and $\overline{a - 3b} \cdot x^2$ and acx and b^3; of which the second Term $\overline{a - 3b} \cdot x^2$ is compound, and the other three are simple Terms.

TERMS, of a Product, or of a Fraction, or of a Ratio, or of a Proportion, &c, are the several quantities employed in forming or composing them. Thus, the Terms

of the product ab, are a and b;
of the fraction $\frac{5}{8}$, are 5 and 8;
of the ratio 6 to 7, are 6 and 7;
of the proportion $a : b :: 5 : 9$, are a, b, 5, 9.

TERMS, are also used for the several times or seasons of the year in which the public colleges or universities, or courts of law, are open, or sit. Such are the Oxford and Cambridge Terms; also the Terms for the courts of King's-Bench, Common Pleas, and the Exchequer, which are the high courts of common law. But the high court of Parliament, the Chancery, and inferior courts, do not observe the Terms.—The rest of the year, out of Term-time, is called *vacation*.

There are four law Terms in the year; viz,

Hilary-Term, which, at London, begins the 23d day of January, and ends the 12th of February.

Easter-Term, which begins the 3d Wednesday after Easter-day, and ends on the Monday next after Ascension-day.

Trinity-Term, which begins the Friday next after Trinity-Sunday, and ends the 4th Wednesday after Trinity-Sunday.

Michaelmas-Term, which begins the 6th of November, and ends the 28th of November.

All these terms have also their returns, the days of which are expressed in the following table or synopsis.

Table of the Law Terms, and their Returns.							
Term	Begin.	1st Return	2d Return	3d Return	4th Return	5th Return	End.
Hilary	January 23	January 20	January 27	February 3	February 9	- - - -	February 12
Easter	3 Wed. af. East.	2 Wks. af. East.	3 Wks. af. East.	4 Wks. af. East.	5 Wks. af. East.	Ascenf. day	Mond. af. Ascens.
Trinity	Frid. af. Trin. S.	Trinity Mond.	1 Wk. af. Trin.	2 Wks. af. Trin.	3 Wks. af. Trin.	- - - -	4th Wed. af. Trin. S.
Mich.	November 6	November 3	November 12	November 18	November 25	- - - -	November 28

N. B. When the beginning or ending of any of these Terms happens on a Sunday, it is held on the Monday after.

Oxford TERMS. These are four; which begin and end as below:

Terms	Begin.	End.
Lent Term	January 14	Sat. bef. Palm-Sund.
Easter Term	Wed. af. Low-Sund.	Thurs. bef. Whitsun.
Trinity Term	Wed. af. Trin. Sund	Sat. after the Act
Michaelmas T.	October 10	December 17.

N. B. The *Act* is 1st Monday after the 6th of July. —When the day of the beginning or ending happens on a Sunday, the Terms begin or end the day after.

Cambridge-TERMS. These are three, as below:

Terms	Begin.	End.
Lent Term	January 13	Frid. bef. Palm-Sund.
Easter Term	Wed. aft. Low-Sund.	Frid. aft. Commence.
Michaelmas T.	October 10	December 16

N. B. The *Commencement* is the 1st Tuesday in July. —There is no difference on account of the beginning or ending being Sunday.

Scottish

Scottish TERMS. In Scotland, *Candlemas Term* begins January 23d, and ends February the 12th. *Whitsuntide-Term* begins May 25th, and ends June 15th. *Lammas-Term* begins July the 20th, and ends August the 8th. *Martinmas-Term* begins November the 3d, and ends November the 29th.

Irish TERMS. In Ireland the Terms are the same as at London, except *Michaelmas-Term*, which begins October the 13th, and adjourns to November the 3d, and thence to the 6th.

TERMINATOR, in Astronomy, a name sometimes given to the circle of illumination, from its property of terminating the boundaries of light and darkness.

TERRA, in Geography. See EARTH.

TERRA-*firma*, in Geography, is sometimes used for a continent, in contradistinction to islands. Thus, Asia, the Indies, and South America, are usually distinguished into Terra-firmas and islands.

TERRAQUEOUS, in Geography, an epithet given to our globe or earth, considered as consisting of land and water, which together constitute one mass.

TERRE-PLEIN, or TERRE-PLAIN, in Fortification, the top, platform, or horizontal surface of the rampart, upon which the cannon are placed, and where the defenders perform their office. It is so called, because it lies level, having only a little slope outwardly to counteract the recoil of the cannon. Its breadth is from 24 to 30 feet; being terminated by the parapet on the outer side, and inwardly by the inner talus.

TERRELLA, or little earth, is a magnet turned of a spherical figure, and placed so as that its poles, equator, &c, do exactly correspond with those of the world. It was so first called by Gilbert, as being a just representation of the great magnetic globe we inhabit. Such a Terrella, it was supposed, if nicely poised, and hung in a meridian like a globe, would be turned round like the earth in 24 hours by the magnetic particles pervading it; but experience has shewn that this is a mistake.

TERRESTRIAL, something relating to the earth. As Terrestrial globe, Terrestrial line, &c.

TERTIAN; denotes an old measure, containing 84 gallons, so called because it is the 3d part of a tun.

TERTIATE, in Gunnery. To Tertiate a great gun, is to examine the thickness of the metal at the muzzle, by which to judge of the strength of the piece, and whether it be sufficiently fortified or not.

TETRACHORD, in Music, called by the moderns a *fourth*, is a concord or interval of four tones.—The Tetrachord of the ancients, was a rank of four strings, accounting the Tetrachord for one tone, as it is often taken in music.

TETRADIAPASON, or *quadruple diapason*, is a musical chord, otherwise called a quadruple eighth, or a nine and-twentieth.

TETRAEDRON, or TETRAHEDRON, in Geometry, is one of the five Platonic or regular bodies or solids, comprehended under four equilateral and equal triangles. Or it is a triangular pyramid of four equal and equilateral faces.

It is demonstrated in geometry, that the side of a Tetraedron is to the diameter of its circumscribing sphere, as $\sqrt{2}$ to $\sqrt{3}$; consequently they are incommensurable.

If *a* denote the linear edge or side of a Tetraedron, *b* its whole superficies, *c* its solidity, *r* the radius of its inscribed sphere, and R the radius of its circumscribing sphere; then the general relation among all these is expressed by the following equations, viz,

$$a = 2r\sqrt{6} = \tfrac{2}{3}R\sqrt{6} = \sqrt{\tfrac{1}{3}b\sqrt{3}} = \sqrt[3]{6c\sqrt{2}}.$$
$$b = 24r^2\sqrt{3} = \tfrac{8}{3}R^2\sqrt{3} = a^2\sqrt{3} = 6\sqrt[3]{c^2\sqrt{3}}.$$
$$c = 8r^3\sqrt{3} = \tfrac{8}{27}R^3\sqrt{3} = \tfrac{1}{12}a^3\sqrt{2} = \tfrac{1}{3c}b\sqrt{2b\sqrt{3}}.$$
$$R = 3r = \tfrac{1}{4}a\sqrt{6} = \tfrac{1}{4}\sqrt{2b\sqrt{3}} = \tfrac{3}{2}\sqrt[3]{\tfrac{1}{3}c\sqrt{3}}.$$
$$r = \tfrac{1}{3}R = \tfrac{1}{12}a\sqrt{6} = \tfrac{1}{12}\sqrt{2b\sqrt{3}} = \tfrac{1}{2}\sqrt[3]{\tfrac{1}{3}c\sqrt{3}}.$$

See my Mensuration, pa. 248 &c, 2d ed. See also the articles REGULAR and BODIES.

TETRAGON, in Geometry, a quadrangle, or a figure having 4 angles. Such as a square, a parallelogram, a rhombus, and a trapezium. It sometimes also means peculiarly a square.

TETRAGON, in Astrology, denotes an aspect of two planets with regard to the earth, when they are distant from each other a 4th part of a circle, or 90 degrees. The Tetragon is expressed by the character □, and is otherwise called a square or quartile aspect.

TETRAGONIAS, a meteor, whose head is of a quadrangular figure, and its tail or train is long, thick, and uniform. It does not differ much from the meteor called *Trabs* or beam.

TETRAGONISM, a term which some authors use to express the quadrature of the circle, because the quadrature is the finding a square equal to it.

TETRASPASTON, in Mechanics, a machine in which are four pulleys.

TETRASTYLE, in the Ancient Architecture, a building, and particularly a temple, with four columns in front.

THALES, a celebrated Greek philosopher, and the first of the seven wisemen of Greece, was born at Miletum, about 640 years before Christ. After acquiring the usual learning of his own country, he travelled into Egypt and several parts of Asia, to learn astronomy, geometry, mystical divinity, natural knowledge or philosophy, &c. In Egypt he met for some time great favour from the king, Amasis; but he lost it again, by the freedom of his remarks on the conduct of kings, which it is said occasioned his return to his own country, where he communicated the knowledge he had acquired to many disciples, among the principal of whom were Anaximander, Anaximenes, and Pythagoras, and was the author of the Ionian sect of philosophers. He always however lived very retired, and re fused the proffered favours of many great men. He was often visited by Solon; and it is said he took great pleasure in the conversation of Thrasybulus, whose excellent wit made him forget that he was Tyrant of Miletum.

Laertius, and several other writers, agree, that he was the father of the Greek philosophy; being the first that made any researches into natural knowledge, and mathematics. His doctrine was, that water was the principle of which all the bodies in the universe are composed; that the world was the work of God; and that God sees the most secret thoughts in the heart of man. He said, that in order to live well, we ought to abstain from what we find fault with in others: that

bodily

Plate XXX

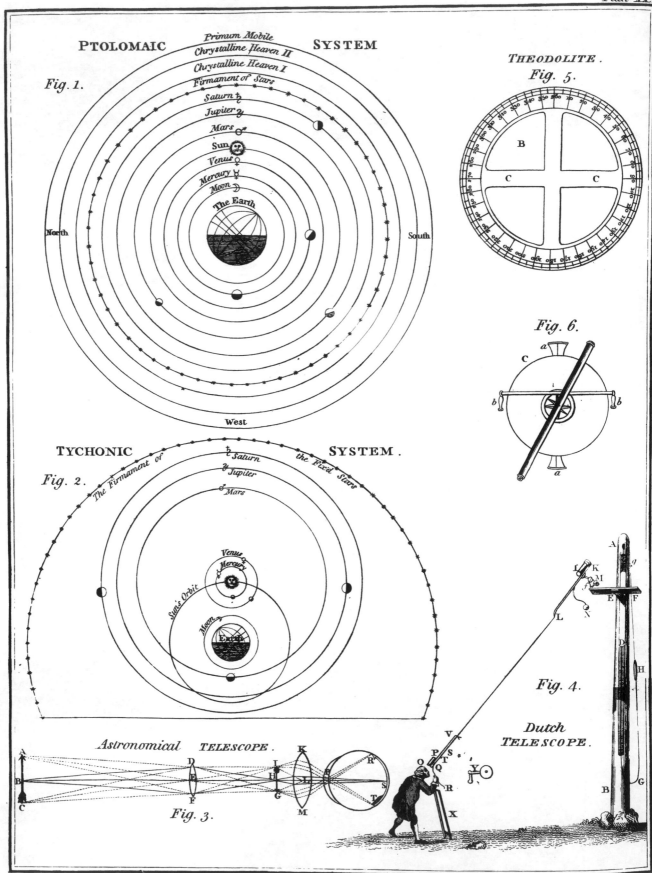

PTOLOMAIC **SYSTEM**

Fig. 1.

Primum Mobile
Chrystalline Heaven II
Chrystalline Heaven I
Firmament of Stars
Saturn ♄
Jupiter ♃
Mars ♂
Sun ☉
Venus ♀
Mercury ☿
Moon ☽
The Earth

North

South

West

THEODOLITE.
Fig. 5.

B

C   C

Fig. 6.

a

C

b   b

a

TYCHONIC **SYSTEM.**

Fig. 2. The Firmament of   the Fix'd Stars

♄ Saturn
♃ Jupiter
♂ Mars

Venus
Mercury

Sun's Orbit

Moon

Earth

A

Fig. 4.

Dutch
TELESCOPE.

Astronomical **TELESCOPE.**

Fig. 3.

bodily felicity confifts in health; and that of the mind in knowledge. That the moft ancient of beings is God, becaufe he is uncreated: that nothing is more beautiful than the world, becaufe it is the work of God; nothing more extenfive than fpace, quicker than fpirit, ftronger than neceffity, wifer than time. He ufed to obferve, that we ought never to fay that to any one which may be turned to our prejudice; and that we fhould live with our friends as with perfons that may become our enemies.

In Geometry, it has been faid, he was a confiderable inventor, as well as an improver; particularly in triangles. And all the writers agree, that he was the firft, even in Egypt, who took the height of the pyramids by the fhadow.

His knowledge and improvements in aftronomy were very confiderable. He divided the celeftial fphere into five circles or zones, the arctic and antarctic circles, the two tropical circles, and the equator. He obferved the apparent diameter of the fun, which he made equal to half a degree; and formed the conftellation of the Little Bear. He obferved the nature and courfe of eclipfes, and calculated them exactly; one in particular, memorably recorded by Herodotus, as it happened on a day of battle between the Medes and Lydians, which, Laertius fays, he had foretold to the Ionians. And the fame author informs us, that he divided the year into 365 days. Plutarch not only confirms his general knowledge of eclipfes, but that his doctrine was, that an eclipfe of the fun is occafioned by the intervention of the moon, and that an eclipfe of the moon is caufed by the intervention of the earth.

His morals were as juft, as his mathematics well grounded, and his judgment in civil affairs equal to either. He was very averfe to tyranny, and efteemed monarchy little better in any fhape.—Diogenes Laertius relates, that walking to contemplate the ftars, he fell into a ditch; on which a good old woman, that attended him, exclaimed, "How canft thou know what is doing in the heavens, when thou feeft not what is at thy feet?"—He went to vifit Crœfus, who was marching a powerful army into Cappadocia, and enabled him to pafs the river Halys without making a bridge. Thales died foon after, at above 90 years of age, it is faid, at the Olympic games, where, oppreffed with heat, thirft, and a load of years, he, in public view, funk into the arms of his friends.

Concerning his writings, it remains doubtful whether he left any behind him; at leaft none have come down to us. Auguftine mentions fome books of Natural Philofophy; Simplicius, fome written on Nautic Aftrology; Laertius, two treatifes on the Tropics and Equinoxes; and Suidas, a treatife on Meteors, written in verfe.

THAMMUZ, in Chronology, the 10th month of the year of the Jews, containing 29 days, and anfwering to our June.

THEMIS, in Aftronomy, a name given by fome to the 3d fatellite of Jupiter.

THEODOLITE, an inftrument much ufed in furveying, for taking angles, diftances, altitudes, &c.

This inftrument is varioufly made; different perfons having their feveral ways of contriving it, each attempting to make it more fimple and portable, more accurate

and expeditious, than others. It ufually confifts of a brafs circle, about a foot diameter, cut in form of fig. 5, pl. 31; having its limb divided into 360 degrees, and each degree fubdivided either diagonally, or otherwife, into minutes. Underneath, at cc, are fixed two little pillars bb (fig. 6), which fupport an axis, bearing a telefcope, for viewing remote objects.

On the centre of the circle moves the index C, which is a circular plate, having a compafs in the middle, the meridian line of which anfwers to the fiducial line aa; at bb are fixed two pillars to fupport an axis, bearing a telefcope like the former, whofe line of collimation anfwers to the fiducial line aa. At each end of either telefcope is, or may be, fixed a plain fight, for the viewing of nearer objects.

The ends of the index aa are cut circularly, to fit the divifions of the limb B; and when that limb is diagonally divided, the fiducial line at one end of the index fhews the degrees and minutes upon the limb. It is alfo furnifhed with crofs fpirit levels, for fetting the plane of the circle truly horizontal; and a vertical arch, divided into degrees, for taking angles of elevation and depreffion. The whole inftrument is mounted with a ball and focket, upon a three-legged ftaff.

Many Theodolites however have no telefcopes, but only four plain fights, two of them faftened on the limb, and two on the ends of the index. Two different ones, mounted on their ftand, are reprefented in fig. 2 and 3, plate 33.

The ufe of the Theodolite is abundantly fhewn in that of the femicircle, which is only half a Theodolite. And the index and compafs of the Theodolite ferve alfo for a circumferentor, and are ufed as fuch.

The ingenious Mr. Ramfden has lately made a moft excellent Theodolite, for the ufe of the military furvey now carrying on in England.

THEODOSIUS a celebrated mathematician, who flourifhed in the times of Cicero and Pompey; but the time and place of his death are unknown. This Theodofius, the Tripolite, as mentioned by Suidas, is probably the fame with Theodofius the philofopher of Bythinia, who Strabo fays excelled in the mathematical fciences, as alfo his fons; for the fame perfon might have travelled from the one of thofe places to the other, and fpent part of his life in each of them; like as Hipparchus was called by Strabo the Bythinian; but by Ptolomy and others the Rhodian.

Theodofius chiefly cultivated that part of geometry which relates to the doctrine of the fphere, concerning which he publifhed three books. The firft of thefe contains 22 propofitions; the fecond 23; and the third 14; all demonftrated in the pure geometrical manner of the Ancients. Ptolomy made great ufe of thefe propofitions, as well as all fucceeding writers. Thefe books were tranflated by the Arabians, out of the original Greek, into their own language. From the Arabic, the work was again tranflated into Latin, and printed at Venice. But the Arabic verfion being very defective, a more complete edition was publifhed in Greek and Latin, at Paris 1558, by John Pena, Regius Profeffor of Aftronomy. And Vitello acquired reputation by tranflating Theodofius into Latin. This author's works were alfo commented on and illuftrated by Clavius, Heleganius, and Guarinus, and laftly by
De

De Chales, in his Curfus Mathematicus. But that edition of Theodofius's Spherics which is now moft in ufe, was tranflated, and publifhed, by our countryman the learned Dr. Barrow, in the year 1675, illuftrated and demonftrated in a new and concife method. By this author's account, Theodofius appears not only to be a great mafter in this more difficult part of geometry, but the firft confiderable author of antiquity who has written on that fubject.

Theodofius too wrote concerning the Celeftial Houfes; alfo of Days and Nights; copies of which, in Greek, are in the king's library at Paris. Of which there was a Latin edition, publifhed by Peter Dafypody, in the year 1572.

THEON, of Alexandria, a celebrated Greek philofopher and mathematician, who flourifhed in the 4th century, about the year 380, in the time of Theodofius the Great; but the time and manner of his death are unknown. His genius and difpofition for the ftudy of philofophy were very early improved by a clofe application to ftudy; fo that he acquired fuch a proficiency in the fciences, as to render his name venerable in hiftory; and to procure him the honour of being prefident of the famous Alexandrian fchool. One of his pupils was the admirable Hypatia, his daughter, who fucceeded him in the prefidency of the fchool; a truft, which, like himfelf, fhe difcharged with the greateft honour and ufefulnefs. [See her life in its place in the firft volume of this Dictionary.]

The ftudy of nature led Theon to many juft conceptions concerning God, and to many ufeful reflections in the fcience of moral philofophy; hence, it is faid, he wrote with great accuracy on divine providence. And he feems to have made it his ftanding rule, to judge the truth of certain principles, or fentiments, from their natural or neceffary tendency. Thus, he fays, that a full perfuafion, that the Deity fees every thing we do, is the ftrongeft incentive to virtue; for he infifts, that the moft profligate have power to refrain their hands, and hold their tongues, when they think they are obferved, or overheard, by fome perfon whom they fear or refpect. With how much more reafon then, fays he, fhould the apprehenfion and belief, that God fees all things, reftrain men from fin, and conftantly excite them to their duty? He alfo reprefents this belief, concerning the Deity, as productive of the greateft pleafure imaginable, efpecially to the virtuous, who might depend with greater confidence on the favour and protection of Providence. For this reafon, he recommends nothing fo much as meditation on the prefence of God: and he recommended it to the civil magiftrate, as a reftraint on fuch as were profane and wicked, to have the following infcription written, in large characters, at the corner of every ftreet; GOD SEES THEE, O SINNER.

Theon wrote notes and commentaries on fome of the ancient mathematicians. He compofed alfo a book, entitled *Progymnafmata*, a rhetorical work, written with great judgment and elegance; in which he criticifed on the writings of fome illuftrious orators and hiftorians; pointing out, with great propriety and judgment, their beauties and imperfections; and laying down proper rules for propriety of ftyle. He recommends concifenefs of expreffion, and perfpicuity, as the principal orna-

ments. This book was printed at Bafle, in the year 1541; but the beft edition is that of Leyden, in 1626, in 8vo.

THEOPHRASTUS, a celebrated Greek philofopher, was the fon of Melanthus, and was born at Eretus in Bœotia. He was at firft the difciple of Lucippus, then of Plato, and laftly of Ariftotle; whom he fucceeded in his fchool, about the 322d year before the Chriftian era, and taught philofophy at Athens with great applaufe.

He faid of an orator without judgment, "that he was a horfe without a bridle." He ufed alfo to fay, "There is nothing fo valuable as time, and thofe who lavifh it are the moft inexcufable of all prodigals."— He died at about 100 years of age.

Theophraftus wrote many works, the principal of which are the following —1. An excellent moral treatife entitled, *Characters*, which, he fays in the preface, he compofed at 99 years of age. Ifaac Cafaubon has written learned commentaries on this fmall treatife. It has been tranflated from the Greek into French, by Bruyere; and it has alfo been tranflated into Englifh.— 2. A curious treatife on Plants.—3. A treatife on foffils or ftones; of which Dr. Hill has given a good edition, with an Englifh tranflation, and learned notes, in 8vo.

THEOREM, a propofition which terminates in theory, and which confiders the properties of things already made or done. Or, a Theorem is a fpeculative propofition, deduced from feveral definitions compared together. Thus, if a triangle be compared with a parallelogram ftanding on the fame bafe, and of the fame altitude, and partly from their immediate definitions, and partly from other of their properties already determined, it is inferred that the parallelogram is double the triangle; that propofition is a Theorem.

Theorem ftands contradiftinguifhed from problem, which denotes fomething to be done or conftructed, as a Theorem propofes fomething to be proved or demonftrated.

There are two things to be chiefly regarded in every Theorem, viz, the propofition, and the demonftration. In the firft is expreffed what agrees to fome certain thing, under certain conditions, and what does not. In the latter, the reafons are laid down by which the underftanding comes to conceive that it does or does not agree to it.

Theorems are of various kinds: as,

Univerfal THEOREM, is that which extends to any quantity without reftriction, univerfally. As this, that the rectangle or product of the fum and difference of any two quantities, is equal to the difference of their fquares.

Particular THEOREM, is that which extends only to a particular quantity. As this, in an equilateral rectilinear triangle, each angle is equal to 60 degrees.

Negative THEOREM, is that which expreffes the impoffibility of any affertion. As, that the fum of two biquadrate numbers cannot make a fquare number.

Local THEOREM is that which relates to a furface. As, that triangles of the fame bafe and altitude are equal.

Plane THEOREM, is that which relates to a furface that is either rectilinear or bounded by the circumference of a circle. As, that all angles in the fame fegment of a circle are equal.

Solid THEOREM, is that which confiders a fpace terminated

THERMOMETERS

Fig. 3.

Fig. 1.

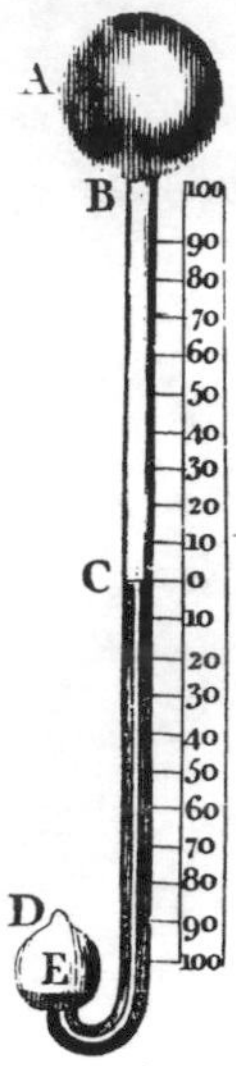

Fig. 2.

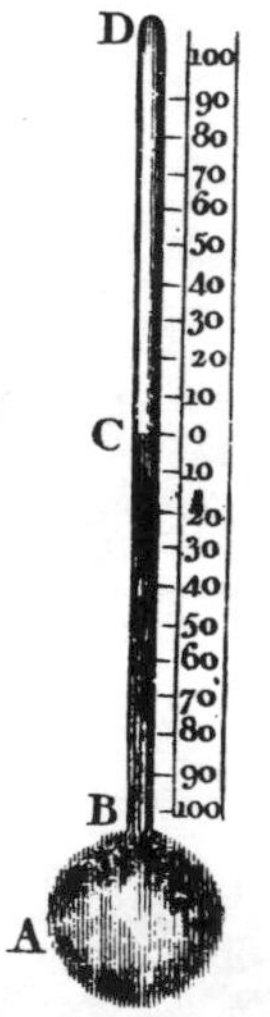

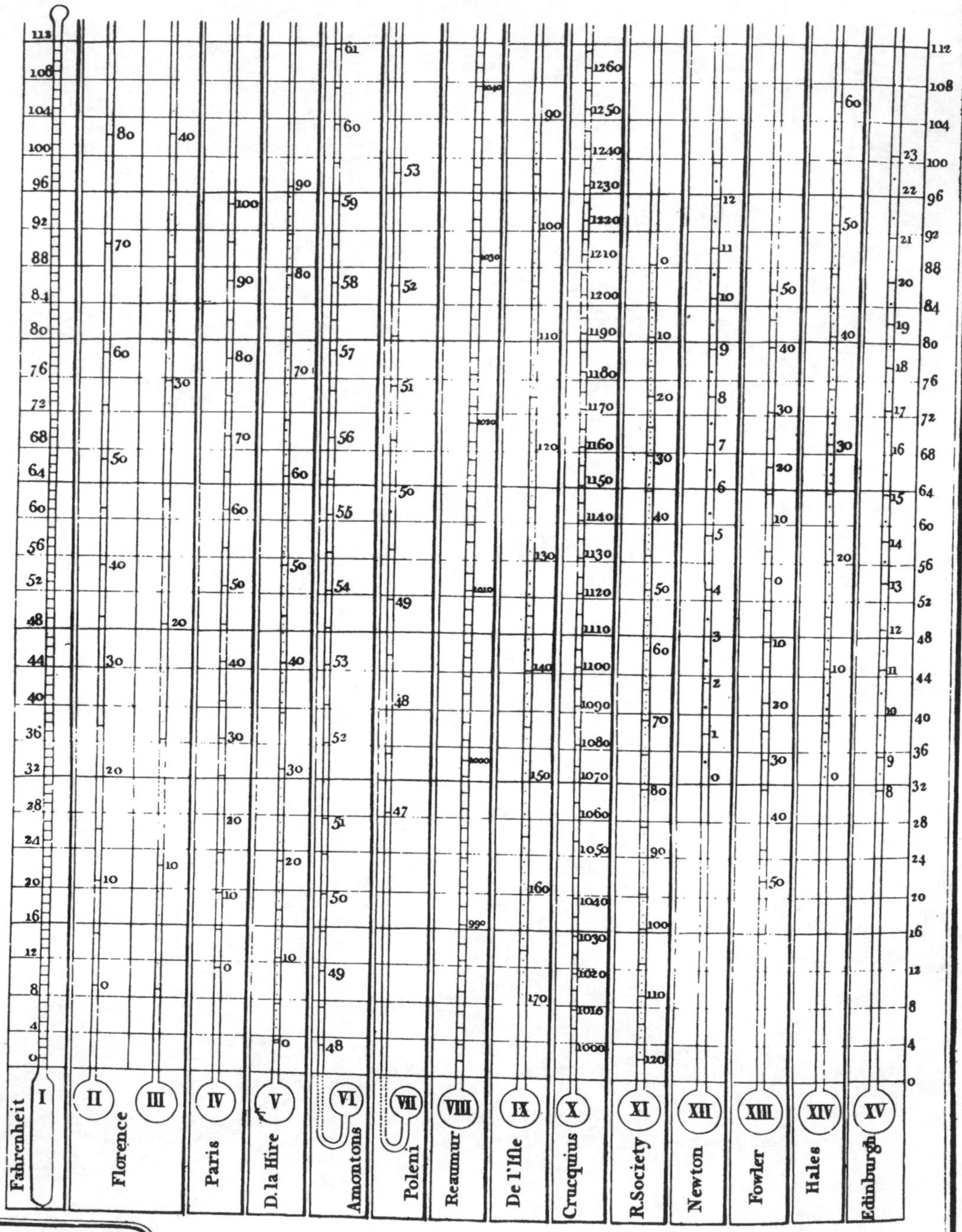

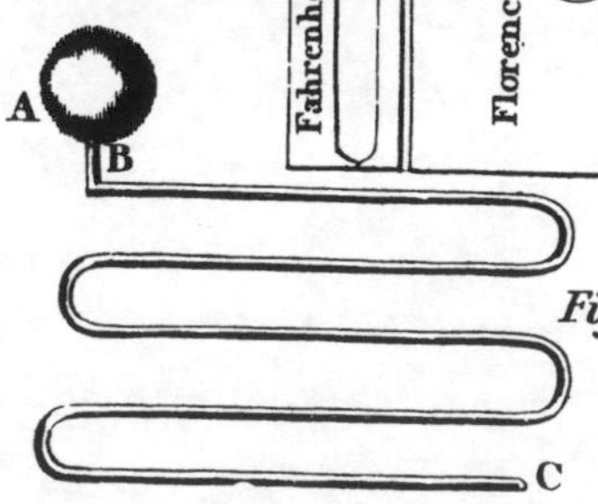

Fig. 4.

minated by a folid line ; that is, by any of the three conic fections. As this, that if a right line cut two afymptotic parabolas, its two parts terminated by them fhall be equal.

Reciprocal THEOREM, is one whofe converfe is true. As, that if a triangle have two fides equal, it has alfo two angles equal : the converfe of which is likewife true, viz, that if the triangle have two angles equal, it has alfo two fides equal.

THEORY, a doctrine which terminates in the fole fpeculation or confideration of its object, without any view to the practice or application of it.

To be learned in an art, &c, the Theory is fufficient; to be a mafter of it, both the Theory and practice are requifite.—Machines often promife very well in Theory, but fail in the practice.

We fay Theory of the moon, Theory of the rainbow, of the microfcope, of the camera obfcura, &c.

THEORIES *of the Planets*, &c, are fyftems or hypothefes, according to which the aftronomers explain the reafons of the phenomena or appearances of them.

THERMOMETER, an inftrument for measuring the temperature of the air, &c, as to heat and cold.

The Thermometer and thermofcope are ufually accounted the fame thing. But Wolfius makes a difference ; and he alfo fhews that what we call Thermometers, are really no more than thermofcopes.

The invention of the Thermometer is attributed to feveral perfons by different authors, viz, to Sanctorio, Galileo, father Paul, and to Drebbel. Thus, the invention is afcribed to Cornelius Drebbel of Alcmar, about the beginning of the 17th century, by his countrymen Boerhaave (Chem. 1, pa. 152, 156), and Muffchenbroeck (Introd. ad Phil. Nat. vol. 2, pa. 625).—Fulgenzio, in his Life of father Paul, gives him the honour of the firft difcovery.—Vincenzio Viviani (Vit. de l'Galil. pa. 67 ; alfo Oper. di Galil. pref. pa. 47) fpeaks of Galileo as the inventor of Thermometers.—But Sanctorino (Com. in Galen. Art. Med. pa. 736, 842, Com. in Avicen. Can. Fen. 1, pa. 22, 78, 219) exprefsly affumes to himfelf this invention : and Borelli (De Mot. Animal. 2, prop. 175) and Malpighi (Oper. Pofth. pa. 30) afcribe it to him without referve. Upon which Dr. Martine remarks, that thefe Florentine academicians are not to be fufpected of partiality in favour of one of the Patavinian fchool.

But whoever was the firft inventor of this inftrument, it was at firft very rude and imperfect; and as the various degrees of heat were indicated by the different contraction or expanfion of air, it was afterwards found to be an uncertain and fometimes a deceiving meafure of heat, becaufe the bulk of the air was affected, not only by the difference of heat, but alfo by the variable weight of the atmofphere.

There are various kinds of Thermometers, the conftruction, defects, theory, &c, of which, are as follow.

The Air THERMOMETER.—This inftrument depends on the rarefaction of the air. It confifts of a glafs tube BE (fig. 1, pl. 34) connected at one end with a large glafs ball A, and at the other end immerfed in an open veffel, or terminating in a ball DE, with a narrow orifice at D ; which veffel, or ball,

contains any coloured liquor that will not eafily freeze. Aquafortis tinged of a fine blue colour with folution of vitriol or copper, or fpirit of wine tinged with cochineal, will anfwer this purpofe. But the ball A muft be firft moderately warmed, fo that a part of the air contained in it may be expelled through the orifice D ; and then the liquor preffed by the weight of the atmofphere, will enter the ball DE, and rife, for example, to the middle of the tube at C, at a mean temperature of the weather ; and in this ftate the liquor by its weight, and the air included in the ball and tube ABC, by its elafticity, will counterbalance the weight of the atmofphere. As the furrounding air becomes warmer, the air in the ball and the upper part of the tube, expanding by heat, will drive the liquor into the lower ball, and confequently its furface will defcend ; on the contrary, as the ambient air becomes colder, that in the ball is condenfed, and the liquor, preffed by the weight of the atmofphere, will afcend : fo that the liquor in the tube will afcend or defcend more or lefs, according to the ftate of the air contiguous to the inftrument. To the tube is affixed a fcale of the fame length, divided upwards and downwards, from the middle C, into 100 equal parts, by means of which may be obferved the afcent and defcent of the liquor in the tube, and confequently the variations alfo in the temperature of the atmofphere.

A fimilar Thermometer may be conftructed by putting a fmall quantity of mercury, not exceeding the bulk of a pea, into the tube BC (fig. 4, pl. 33), bent into wreaths, that taking up the lefs height, it may be the more manageable, and lefs liable to harm ; divide this tube into any number of equal parts to ferve for a fcale. Here the approaches of the mercury towards the ball A will fhew the increafe of the degree of heat. The reafon of which is the fame as in the former.

The defect of both thefe inftruments confifts in this, that they are liable to be acted on by a double caufe : for, not only a decreafe of heat, but alfo an increafe of weight of the atmofphere, will make the liquor rife in the one, and the mercury in the other; and, on the contrary, either an increafe of heat, or decreafe of the weight of the atmofphere, will caufe them to defcend.

For thefe, and other reafons, Thermometers of this kind have been long difufed. However, M. Amontons, in 1702, with a view of perfecting the aërial Thermometer, contrived his *Univerfal Thermometer*. Finding that the changes produced by heat and cold in the bulk of the air were fubject to invincible irregularities, he fubftituted for thefe the variations produced by heat in the elaftic force of this fluid. This Thermometer confifted of a long tube of glafs (fig. 3, pl. 34) open at one end, and recurved at the other end, which terminated in a ball. A certain quantity of air was compreffed into this ball by the weight of a column of mercury, and alfo by the weight of the atmofphere. The effect of heat on this included air was to make it fuftain a greater or lefs weight; and this effect was meafured by the variation of the column of mercury in the tube, corrected by that of the barometer, with refpect to the changes of the weight of the external air. This inftrument, though much more perfect than the former, is neverthelefs fubject to very confiderable defects and incon-

inconveniences. Its length of 4 feet renders it unfit for a variety of experiments, and its construction is difficult and complex : it is extremely inconvenient for carriage, as a very small inclination of the tube would suffer the included air to escape : and the friction of the mercury in the tube, and the compressibility of the air, contribute to render the indications of this instrument extremely uncertain. Besides, the dilatation of the air is not so regularly proportional to its heat, nor is its dilatation by a given heat nearly so uniform as he supposed. This depends much on its moisture ; for dry air does not expand near so much by a given heat, as air stored with watery particles. For these, and other reasons, enumerated by De Luc (Recherches sur les Mod. de l'Atmo. tom. 1; pa. 278 &c), this instrument was imitated by very few, and never came into general use.

Of the Florentine THERMOMETER.—The academists del Cimento, about the middle of the 17th century, considering the inconveniencies of the air Thermometers above described, attempted another, that should measure heat and cold by the rarefaction and condensation of spirit of wine ; though much less than those of air, and consequently the alterations in the degree of heat likely to be much less sensible.

The spirit of wine coloured, was included in a very fine and cylindrical glass tube (fig. 2, pl. 34), exhausted of its air, having a hollow ball at one end A, and hermetically sealed at the other end D. The ball and tube are filled with rectified spirit of wine to a convenient height, as to C, when the weather is of a mean temperature, which may be done by inverting the tube into a vessel of stagnant coloured spirit, under a receiver of the air-pump, or in any other way. When the thermometer is properly filled, the end D is heated red hot by a lamp, and then hermetically sealed, leaving the included air of about ⅓ of its natural density, to prevent the air which is in the spirit from dividing it in its expansion. To the tube is applied a scale, divided from the middle, into 100 equal parts, upwards and downwards.

Now spirit of wine rarefying and condensing very considerably ; as the heat of the ambient atmosphere increases, the spirit will dilate, and so ascend in the tube ; and as the heat decreases, the spirit will descend ; and the degree or quantity of the motion will be shewn by the attached scale.

These Thermometers could not be subject to any inconvenience by an evaporation of the liquor, or a variable gravity of the incumbent atmosphere. Instruments of this kind were first introduced into England by Mr. Boyle, and they soon came into general use among philosophers in other countries. They are however subject to considerable inconveniences, from the weight of the liquor itself, and from the elasticity of the air above it in the tube, both which prevent the freedom of its ascent ; besides, the rarefactions are not exactly proportional to the surrounding heat. Moreover spirit of wine is incapable of bearing very great heat or very great cold : it boils sooner than any other liquor ; and therefore the degrees of heat of boiling fluids cannot be determined by this Thermometer. And though it retains its fluidit in pretty severe cold, yet it seems not to condense very regularly in them : and at

Torneao, near the polar circle, the winter cold was so severe, as Maupertuis informs us, that the spirits were frozen in all their Thermometers. So that the degrees of heat and cold, which spirit of wine is capable of indicating, is much too limited to be of very great or general use.

Another great defect of these, and other Thermometers, is, that their degrees cannot be compared with each other. It is true they mark the variations of heat and cold ; but each marks for itself, and after its own manner ; because they do not proceed from any point of temperature that is common to all of them.

From these and various other imperfections in these Thermometers, it happens, that the comparisons of them become so precarious and defective : and yet the most curious and interesting use of them, is what ought to arise from such comparison. It is by this we should know the heat or cold of another season, of another year, another climate, &c ; and what is the greatest degree of heat or cold that men and other animals can subsist in.

Reaumur contrived a new Thermometer, in which the inconveniences of the former are proposed to be remedied. He took a large ball and tube, the content or dimensions of which are known in every part ; he graduated the tube, so that the space from one division to another might contain a 1000th part of the liquor, which liquor would contain 1000 parts when it stood at the freezing point : then putting the ball of his Thermometer and part of the tube into boiling water, he observed whether it rose 80 divisions : if it exceeded these, he changed his liquor, and by adding water lowered it, till upon trial it should just rise 80 divisions ; or if the liquor, being too low, fell short of 80 divisions, he raised it by adding rectified spirit to it. The liquor thus prepared suited his purpose, and served for making a Thermometer of any size, whose scale would agree with his standard. Such liquor, or spirits, being about the strength of common brandy, may easily be had any where, or made of a proper degree of density by raising or lowering it.

The abbé Nollet made many excellent Thermometers upon Reaumur's principle. Dr. Martine however expresses his apprehensions that Thermometers of this kind cannot admit of such accuracy as might be wished. The balls or bulbs, being large, as 3 or 4 inches in diameter, are neither heated nor cooled soon enough to shew the variations of heat. Small bulbs and small tubes, he says, are much more convenient, and may be constructed with sufficient accuracy. Though it must be allowed that Reaumur, by his excellent scale, and by depriving the spirit of its air and expelling the air by means of heat from the ball and tube of his Thermometer, has brought it to as much perfection as may be ; yet it is liable to some of the inconveniences of spirit Thermometers, and is much inferior to mercurial ones. These two kinds do not agree together in indicating the same degrees of intense cold ; for when the mercury has stood at 22° below 0, the spirit indicated only 18°, and when the mercury stood at 28° or 37° below 0, the spirit rested at 25° or 29°. See the description of Reaumur's Thermometer at large in Mem. de l'Acad. des Scienc. an. 1730, pa. 645, Hist. pa. 15. Ib. an. 1731, pa. 354, Hist. pa. 7.

Mercurial

Mercurial THERMOMETER.——It is a most important circumstance in the construction of Thermometers, to procure a fluid that measures equal variations of heat by corresponding equal variations in its own bulk : and the fluid which possesses this essential requisite in the most perfect degree, is mercury : the variations in its bulk approaching nearer to a proportion with the corresponding variations of its heat, than any other fluid. Besides, it is the most easy to purge of its air ; and is also the most proper for measuring very considerable variations of heat and cold, as it will bear more cold before freezing, and more heat before boiling, than any other fluid. Mercury is also more sensible than any other fluid, air excepted, or conforms more speedily to the several variations of heat. Moreover, as mercury is an homogeneous fluid, it will in every Thermometer exhibit the same dilatation or condensation by the same variations of heat.

Dr. Halley, though apprized only of some of the remarkable properties of mercury above recited, seems to have been the first who suggested the application of this fluid to the construction of Thermometers. Philos. Transf. Abr. vol. 2, pa. 34.

Boerhaave (Chem. 1, pa. 720) says, these mercurial Thermometers were first contrived by Olaus Roemer ; but the claims of Fahrenheit of Amsterdam, who gave an account of his invention to the Royal Society in 1724, (Philos. Transf. num. 381, or Abr. vol. 7, pa. 49) have been generally allowed. And though Prius and others, in England, Holland, France, and other countries, have made this instrument as well as Fahrenheit, most of the mercurial Thermometers are graduated according to his scale, and are called *Fahrenheit's Thermometers.*

The cone or cylinder, which these Thermometers are often made with, instead of the ball, is made of glass of a moderate thickness, left, when the exhausted tube is hermetically sealed, its internal capacity should be diminished by the weight of the ambient atmosphere. When the mercury is thoroughly purged of its air and moisture by boiling, the Thermometer is filled with a sufficient quantity of it ; and before the tube is hermetically sealed, the air is wholly expelled from it by heating the mercury, so that it may be rarefied and ascend to the top of the tube. To the side of the tube is annexed a scale (fig. 3, pl. 34), which Fahrenheit divided into 600 parts, beginning with that of the severe cold which he had observed in Iceland in 1709, or that produced by surrounding the bulb of the Thermometer with a mixture of snow or beaten ice and sal ammoniac or sea salt. This he apprehended to be the greatest degree of cold, and accordingly he marked this, as the beginning of his scale, with 0 ; the point at which mercury begins to boil, he conceived to shew the greatest degree of heat, and this he made the limit of his scale. The distance between these two points he divided into 600 equal parts or degrees ; and by trials he found at the freezing point, when water just begins to freeze, or snow or ice just begins to thaw, that the mercury stood at 32 of these divisions, therefore called the degree of the freezing point ; and when the tube was immersed in boiling water, the mercury rose to 212, which therefore is the boiling point, and is just 180 degrees above the former or freezing point.

But the present method of making the scale of these Thermometers, which is the sort in most common use, is first to immerge the bulb of the Thermometer in ice or snow just beginning to thaw, and mark the place where the mercury stands with a 32 ; then immerge it in boiling water, and again mark the place where the mercury stands in the tube, which mark with the num. 212, exceeding the former by 180 ; dividing therefore the intermediate space into 180 equal parts, will give the scale of the Thermometer, and which may afterwards be continued upwards and downwards at pleasure.

Other Thermometers of a similar construction have been accommodated to common use, having but a portion of the above scale. They have been made of a small size and portable form, and adapted with appendages to particular purposes ; and the tube with its annexed scale has often been enclosed in another thicker glass tube, also hermetically sealed, to preserve the Thermometer from injury. And all these are called *Fahrenheit's Thermometers.*

In 1733, M. De l'Isle of Petersburgh constructed a mercurial Thermometer (see fig. 3, pl. 34), on the principles of Reaumur's spirit Thermometer. In his Thermometer, the whole bulk of quicksilver, when immerged in boiling water, is conceived to be divided into 100,000 parts ; and from this one fixed point the various degrees of heat, either above or below it, are marked in these parts on the tube or scale, by the various expansion or contraction of the quicksilver in all imaginable varieties of heat.—Dr. Martine apprehends it would have been better if De l'Isle had made the integer 100,000 parts, or fixed point, at freezing water, and from thence computed the dilatations or condensations of the quicksilver in those parts ; as all the common observations of the weather, &c, would have been expressed by numbers increasing as the heat increased, instead of decreasing, or counting the contrary way. However, in practice it will not be very easy to determine exactly all the divisions from the alteration of the bulk of the contained fluid. And besides, as glass itself is dilated by heat, though in a less proportion than quicksilver, it is only the excess of the dilatation of the contained fluid above that of the glass that is observed ; and therefore if different kinds of glass be differently affected by a given degree of heat, this will make a seeming difference in the dilatations of the quicksilver in the Thermometers constructed in the Newtonian method, either by Reaumur's rules or De l'Isle's. Accordingly it has been found, that the quicksilver in De l'Isle's Thermometers, has stood at different degrees of the scale when immerged in thawing snow : having stood in some at 154°, while in others it has been at 156 or even 158°.

Metallic THERMOMETER.—This is a name given to a machine composed of two metals, which, whilst it indicates the variations of heat, serves to correct the errors hence resulting in the going of pendulum clocks and watches. Instruments of this kind have been contrived by Graham, Le Roy, Ellicot, Harrison, and other eminent artificers. See the Philos. Transf. vol. 44, pa. 689, and vol. 45, pa. 129, and vol. 51, pa. 823, where the particular descriptions &c may be seen.

M. De Luc has likewise described two Thermometers

    of

of metal, which he uses for correcting the effects of heat upon a barometer, and an hygrometer of his construction connected with them. See Philos. Transf. vol. 68, p. 437.

Oil THERMOMETERS.—To this class belongs Newton's Thermometer, constructed in 1701, with linseed oil, instead of spirit of wine. This fluid has the advantage of being sufficiently homogeneous, and capable of a considerable rarefaction, not less than 15 times greater than that of spirit of wine. It has not been observed to freeze even in very great colds; and it sustains a great heat, about 4 times that of water, before it boils. With these advantages it was made use of by Sir I. Newton, who discovered by it the comparative degree of heat for boiling water, melting wax, boiling spirit of wine, and melting tin; beyond which it does not appear that this Thermometer was applied. The method he used for adjusting the scale of this oil Thermometer, was as follows: supposing the bulb, when immersed in thawing snow, to contain 0,000 parts, he found the oil expanded by the heat of the human body so as to take up a 39th more space, or 10256 such parts; and by the heat of water boiling strongly 10725; and by the heat of melting tin 11516. So that, reckoning the freezing point as a common limit between heat and cold, he began his scale there, marking it 0, and the heat of the human body he made 12°; and consequently, the degrees of heat being proportional to the degrees of rarefaction, or $256 : 725 :: 12 : 34$, this number 34 will express the heat of boiling water; and, by the same rule, 72 that of melting tin. Philos. Transf. number 270, or Abridg. vol. 4, par. 2, p. 3.

There is an insuperable inconvenience attending all Thermometers made with oil, or any other viscid fluid, viz, that such liquor adheres too much to the sides of the tube, and so inevitably disturbs the regularity and uniformity of the Thermometer.

Of the fixed points of THERMOMETERS.—Various methods have been proposed by different authors, for finding a fixed, point or degree of heat, from which to reckon the other degrees, and adjust the scale; so that different observations and instruments might be compared together. Mr. Boyle was very sensible of the inconveniences arising from the want of a universal scale and mode of graduation; and he proposed either the freezing of the essential oil of aniseeds, or of distilled water, as a term to begin the numbers at, and from thence to graduate them according to the proportional dilatations or contractions of the included spirits.

Dr. Halley (Philos. Transf. Abr. vol. 2, p. 36) seems to have been fully apprized of the bad effects of the indefinite method of constructing Thermometers, and wished to have them adjusted to some determined points. What he seems to prefer, for this purpose, is the degree of temperature found in subterranean places, where the heat in summer or cold in winter appears to have no influence. But this degree of temperature, Dr. Martine shews, is a term for the universal construction of Thermometers, both inconvenient and precarious, as it cannot be easily ascertained, and as the difference of soils and depths may occasion a considerable variation. Another term of heat, which he thought might be of use in a general graduation of Thermometers, is that of boiling spirit of wine that has been highly rectified.

The first trace that occurs of the method of actually applying fixed points or terms to the Thermometer, and of graduating it, so that the unequal divisions of it might correspond to equal degrees of heat, is the project of Renaldini, professor at Padua, in 1694: it is thus described in the Acta Erud. Lipf. "Take a slender tube, about 4 palms long, with a ball fastened to the same; pour into it spirit of wine, enough just to fill the ball, when surrounded with ice, and not a drop over: in this state seal the orifice of the tube hermetically, and provide 12 vessels, each capable of containing a pound of water, and somewhat more; and into the first pour 11 ounces of cold water, into the second 10 ounces, into the third 9, &c; this done, immerge the Thermometer in the first vessel, and pour into it one ounce of hot water, observing how high the spirit rises in the tube, and noting the point with unity, then remove the Thermometer into the second vessel, into which are to be poured 2 ounces of hot water, and note the place the spirit rises to with 2: by thus proceeding till the whole pound of water is spent, the instrument will be found divided into 12 parts, denoting so many terms or degrees of heat; so that at 2 the heat is double to that at 1, at 3 triple, &c."

But this method, though plausible, Wolfius shews, is deceitful, and built on false suppositions; for it takes for granted, that we have one degree of heat, by adding one ounce of hot water to 11 of cold; two degrees by adding 2 ounces to 10, &c: it supposes also, that a single degree of heat acts on the spirit of wine, in the ball, with a single force; a double with a double force, &c: lastly it supposes, that if the effect be produced in the Thermometer by the heat of the ambient air, which is here produced by the hot water, the air has the same degree of heat with the water.

Soon after this project of Renaldini, viz, in 1701, Newton constructed his oil Thermometer, and placed the base or lowest fixed point of his scale at the temperature of thawing snow, and 12 at that of the human body, &c, as above explained.—De Luc observes, that the 2d term of this scale should have been at a greater distance from the first, and that the heat of boiling water would have answered the purpose better than that of the human body.

In 1702, Amontons contrived his universal Thermometer, the scale of which was graduated in the foling manner. He chose for the first term, the weight that counterbalanced the air included in his Thermometer, when it was heated by boiling water: and in this state he so adjusted the quantity of mercury contained in it, till the sum of its height in the tube, and of its height in the barometer at the moment of observation, was equal to 73 inches. Fixing this number at the point to which the mercury in the tube rose by plunging it in boiling water, it is evident that if the barometer at this time was at 28 inches, the height of the column of mercury in the Thermometer, above the level of that in the ball, was 45 inches; but if the height of the barometer was less by a certain quantity, the column of the Thermometer ought to be greater by the same quantity, and reciprocally. He formed his scale on the supposition, that the weight of the atmosphere was always equal to that of a column of mercury of 28 inches, and he divided it into inches

from

from the point 73 downward, marking the divisions with 72, 71, 70, &c, and subdividing the inches into lines. But as the weight of the atmosphere is variable, the barometer must be observed at the same time with the Thermometer, that the number indicated by this last instrument may be properly corrected, by adding or subtracting the quantity which the mercury is below or above 28 inches in the barometer. In this scale then, the freezing point is at $51\frac{1}{2}$ inches, corresponding to 32 degrees of Fahrenheit, and the heat of boiling water at 73 inches, answering to 212 of Fahrenheit's; and thus they may be easily compared together.

The fixed points of Fahrenheit's Thermometer, as has been already observed, are the congelation produced by sal ammoniac and the heat of boiling water. The interval between these points is divided into 212 equal parts; the former of these points being marked 0, and the other 212.

Reaumur in his Thermometer, the construction of which he published in 1730, begins his scale at an artificial congelation of water in warm weather, which, as he uses large bulbs for his glasses, gives the freezing point much higher than it should be, and at boiling water he marks 80 degrees, which point Dr. Martine thinks is more vague and uncertain than his freezing point. In order to determine the correspondence of his scale with that of Fahrenheit, it is to be considered that his boiling water heat, is really only the boiling heat of weakened spirit of wine, coinciding nearly, as Dr. Martine apprehends, with Fahrenheit's 180 degrees. And as his $10\frac{1}{4}$ degrees is the constant heat of the cave of the observatory at Paris, or Fahrenheit's 53°, he thence finds his freezing point, instead of answering just to 32°, to be somewhat above 34°.

De l'Isle's Thermometer, an account of which he presented to the Petersburgh Academy in 1733, has only one fixed point, which is the heat of boiling water, and, contrary to the common order, the several degrees are marked from this point downward, according to the condensations of the contained quicksilver, and consequently by numbers increasing as the heat decreases. The freezing point of De l'Isle's scale, Dr. Martine makes near to his 150°, corresponding to Fahrenheit's 32, by means of which they may be compared; but Ducrest says, that this point ought to be marked at least at 154°.

Ducrest, in his spirit Thermometer, constructed in 1740, made use of two fixed points; the first, or 0, indicated the temperature of the earth, and was marked on his scale in the cave of the Paris Observatory; and the other was the heat of boiling water, which that spirit in his Thermometer was made to endure, by leaving the upper part of the tube full of air. He divided the interval between these points into 100 equal parts; calling the divisions upward, degrees of heat, and those below 0, degrees of cold.—It is said that he has since regulated his Thermometer by the degree of cold indicated by melting ice, which he found to be $10\frac{2}{5}$.

The Florentine Thermometers were of two sorts. In one sort the freezing point, determined by the

degree at which the spirit stood in the ordinary cold of ice or snow (probably in a thawing state) and coinciding with 32° of Fahrenheit, fell at 20°; and in the other sort at $13\frac{1}{2}$. And the natural heat of the viscera of cows and deer, &c, raised the spirit in the latter, or less sort, to about 40°, coinciding with their summer heat, and nearly with 102° in Fahrenheit's; and in their other or long Thermometer, the spirit, when exposed to the great midsummer heat in their country, rose to the point at which they marked 80°.

In the Thermometer of the Paris Observatory, made of spirit of wine by De la Hire, the spirit always stands at 48° in the cave of the observatory, corresponding to 53 degrees in Fahrenheit's; and his 28° corresponded with 51 inches 6 lines in Amontons' Thermometer, and consequently with the freezing point, or 32° of Fahrenheit's.

In Poleni's Thermometer, made after the manner of Amontons', but with less mercury, 47 inches corresponded, according to Dr. Martine, with 51 in that of Amontons, and 53 with $59\frac{1}{2}$.

In the standard Thermometer of the Royal Society of London, according to which Thermometers were for a long time constructed in England, Dr. Martine found that $34\frac{1}{2}$ degrees answered to 64° in Fahrenheit, and 0 to 89.

In the Thermometers graduated for adjusting the degrees of heat proper for exotic plants, &c, in stoves and greenhouses, the middle temperature of the air is marked at 0, and the degrees of heat and cold are numbered both above and below. Many of these are made on no regular and fixed principles. But in that formerly much used, called Fowler's regulator, the spirit fell, in melting snow, to about 34° under 0; and Dr. Martine found that his 16° above 0, nearly coincided with 64° of Fahrenheit.

Dr. Hales (Statical Essays, vol. 1, p. 58), in his Thermometer, made with spirit of wine, and used in experiments on vegetation, began his scale with the lowest degree of freezing, or 32° of Fahrenheit, and carried it up to 100°, which he marked where the spirit stood when the ball was heated in hot water, upon which some wax floating first began to coagulate, and this point Dr. Martine found to correspond with 142° of Fahrenheit. But by experience it is found that Hales's 100 falls considerably above our 142.

In the Edinburgh Thermometer, made with spirit of wine, and used in the meteorological observations published in the Medical Essays, the scale is divided into inches and tenths. In melting snow the spirit stood at $8\frac{2}{10}$, and the heat of the human skin raised it to $22\frac{7}{10}$. Dr. Martine found that the heat of the person who graduated it, was 97 of Fahrenheit.

As it is often of use to compare different Thermometers, in order to judge of the result of former observations, I have annexed from Dr. Martine's Essays, the table by which he compared 15 different thermometers. See Plate 34, fig. 3.

There is a Thermometer which has often been used in London, called the Thermometer of Lyons, because

M. Criftin

9

M. Criſtin brought it there into uſe, which is made of mercury: the freezing point is marked o, and the interval from that point to the heat of boiling water is divided into 100 equal degrees.

From the abſtract of the hiſtory of the conſtruction of Thermometers it appears, that freezing and boiling water have furniſhed the diſtinguiſhing points that have been marked upon almoſt all Thermometers. The inferior fixed point is that of freezing, which ſome have determined by the freezing of water, and others by the melting of ice, plunging the ball of the Thermometer into the water and ice, while melting, which is the beſt way. The ſuperior fixed point of almoſt all Thermometers, is the heat of boiling water. But this point cannot be conſidered as fixed and certain, unleſs the heat be produced by the ſame degree of boiling, and under the ſame weight of the atmoſphere; for it is found that the higher the barometer, or the heavier the atmoſphere, the greater is the heat when the water boils. It is now agreed therefore that the operation of plunging the ball of the Thermometer in the boiling water, or ſuſpending it in the ſteam of the ſame in an incloſed veſſel, be performed when the water boils violently, and when the barometer ſtands at 30 Engliſh inches, in a temperature of 55° of the atmoſphere, marking the height of the Thermometer then for the degree of 212 of Fahrenheit; the point of melting ice being 32 of the ſame; thus having 180 degrees between thoſe two fixed points, ſo determined. This was Mr. Bird's method, who it is apprehended firſt attended to the ſtate of the barometer, in the making of Thermometers. But theſe inſtruments may be made equally true under any preſſure of the atmoſphere, by making a proper allowance for the difference in the height of the barometer from 30 inches. M. De Luc, in his Recherches ſur les Mod. de l'Atmoſphere, from a ſeries of experiments, has given an equation for the allowance on account of this difference, in Paris meaſure, which has been verified by Sir George Schuckburgh, Philoſ. Tranſ. 1775 and 1778; alſo Dr. Horſley, Dr. Maſkelyne, and Sir George Shuckburgh have adapted the equation and rules, to Engliſh meaſures, and have reduced the allowances into tables for the uſe of the artiſt. Dr. Horſley's rule, deduced from De Luc's, is this:

$$\frac{99}{8990000} \log. z - 92.804 = h,$$

where h denotes the height of a Thermometer plunged in boiling water, above the point of melting ice, in degrees of Bird's Fahrenheit, and z the height of the barometer in 10ths of an inch. From this rule he has computed the following table, for finding the heights, to which a good Bird's Fahrenheit will riſe, when plunged in boiling water, in all ſtates of the barometer, from 27 to 31 Engliſh inches; which will ſerve, among other uſes, to direct inſtrument makers in making a true allowance for the effect of the variation of the barometer, if they ſhould be obliged to finiſh a Thermometer at a time when the barometer is above or below 30 inches; though it is beſt to fix the boiling point when the barometer is at that height.

Equation of the Boiling Point.

Barometer.	Equation.	Difference.
31·0	+ 1·57	
		0·78
30·5	+ 0·79	
		0·79
30·0	0·00	
		0·80
29·5	— 0·80	
		9·82
29·0	— 1·62	
		0·83
28·5	— 2·45	
		0·85
28·0	— 3·31	
		0·86
27·5	— 4·16	
		0·88
27·0	— 5·04	

The numbers in the firſt column of this table expreſs heights of the quickſilver in the barometer in Engliſh inches and decimal parts: the 2d column ſhews the equation to be applied, according to the ſign prefixed, to 212° of Bird's Fahrenheit, to find the true boiling point for every ſuch ſtate of the barometer. The boiling point for all intermediate ſtates of the barometer may be had with ſufficient accuracy by taking proportional parts, by means of the 3d column of differences of the equations. See Philoſ. Tranſ. vol. 64, art. 30; alſo Dr. Maſkelyne's paper, vol. 64, art. 20.

Sir Geo. Shuckburgh (Philoſ. Tranſ. vol. 69, pa. 362) has alſo given ſeveral tables and rules relating to the boiling point, both from his own obſervations and De Luc's, form whence is extracted the following table, for the uſe of artiſts in conſtructing the Thermometer.

Height of the Barometer.	Corr. of the Boil. Point.	Differences.	Correct. accord. to De Luc.	Differences.
26·0	— 7·09		— 6·83	
		0·91		0·90
26·5	— 6·18		— 5·93	
		0·91		0·89
27·0	— 5·27		— 5·04	
		0·90		0·88
27·5	— 4·37		— 4·16	
		0·89		0·87
28·0	— 3·48		— 3·31	
		0·89		0·86
28·5	— 2·59		— 2·45	
		0·87		0·83
29·0	— 1·72		— 1·62	
		0·87		0·82
29·5	— 0·85		— 0·80	
		0·85		0·80
30·0	0·00		0·00	
		0·85		0·79
30·5	+ 0·85		+ 0·79	
		0·84		0·78
31·0	+ 1·60		+ 1·57	

The Royal Society too, fully ſenſible of the importance of adjuſting the fixed points of Thermometers, appointed a committee of ſeven gentlemen to conſider of the beſt method for this purpoſe; and their report may be ſeen in the Philoſ. Tranſ. vol. 67, art. 37.

They obſerve, that although the boiling point be placed ſo much higher on ſome of the Thermometers now made, than on others, yet this does not produce any conſiderable error in the obſervations of the weather, at leaſt in this climate; for an error of 1½ degree in the poſition of the boiling point, will make an error only of half a degree in the poſition of 92°, and of not more than

than a quarter of a degree in the point of 62°. It is only in nice experiments, or in trying the heat of hot liquors, that this error in the boiling point can be of much signification.

In adjusting the freezing, as well as the boiling point, the quicksilver in the tube ought to be kept of the same heat as that in the ball. When the freezing point is placed at a considerable distance from the ball, the pounded ice should be piled up very near to it; if it be not so piled, then the observed point, to be very accurate, should be corrected, according to the following table.

Heat of the Air.	Correction.
42°	·00087
52	·00174
62	·00261
72	·00348
82	·00435

The correction in this table is expressed in 1000th parts of the distance between the freezing point and the surface of the ice: ex. gr. if the freezing point stand 6 inches above the surface of the ice, and the heat of the room be 62, then the point of 32 should be placed 6 × 00261, or ·01566 of an inch lower down than the observed point.

The committee farther observe, that in trying the heat of liquors, care should be taken that the quicksilver in the tube of the Thermometer be heated to the same degree as that in the ball; or if this cannot be done conveniently, the observed heat should be corrected on that account; for the manner of doing which, and a table calculated for that purpose, see Philos. Transf. vol. 67, art. 37.

It was for some time thought, especially from the experiments at Petersburgh, that quicksilver suffered a cold of several hundred degrees below 0 before it congealed and became fixed and malleable; but later experiments have shewn that this persuasion was merely owing to a deception in the experiments, and later ones have made it appear that its point of congelation is no lower than − 40°, or rather − 39°, of Fahrenheit's scale. But that it will bear however to be cooled a few degrees below that point, to which it leaps up again on beginning to congeal; and that its rapid descent in a Thermometer, through many hundred degrees, when it has once passed the above-mentioned limit, proceeds merely from its great contraction in the act of freezing. See Philos. Transf. vol. 73, art. *20, 20, 21.

Miscellaneous Observations.

It is absolutely necessary that those who would derive any advantage from these instruments, should agree in using the same liquor, and in determining, according to the same method, the two fundamental points. If they agree in these fixed points, it is of no great importance whether they divide the interval between them into a greater or a less number of equal parts. The scale of Fahrenheit, in which the fundamental interval between 212°, the point of boiling water,

and 32° that of melting ice, is divided into 180 parts, should be retained in the northern countries, where Fahrenheit's Thermometer is used: and the scale in which the fundamental interval is divided into 80 parts, will serve for those countries where Reaumur's Thermometer is adopted. But no inconvenience is to be apprehended from varying the scale for particular uses, provided care be taken to signify into what number of parts the fundamental interval is divided, and the point where 0 is placed.

With regard to the choice of tubes, it is best to have them exactly cylindrical through their whole length. The capillary tubes are preferable to others, because they require smaller bulbs, and they are also more sensible, and less brittle. The most convenient size for common experiments has the internal diameter about the 40th or 50th of an inch, about 9 inches long, and made of thin glass, that the rise and fall of the mercury may be better seen.

For the whole process of filling, marking, and graduating, see De Luc's Recherches &c, tom. 1, p. 393, &c.

Experiments with THERMOMETERS.

The following is a table of some observations made with Fahrenheit's Thermometer, the barometer standing at 29 inches, or little higher.

At 600° Mercury boils.

546	Oil of vitriol boils.
242	Spirit of nitre boils.
240	Lixivium of tartar boils.
213	Cow's milk boils.
212	Water boils.
206	Human urine boils.
190	Brandy boils.
175	Alcohol boils.
156	Serum of blood and white of eggs harden.
146	Kills animals in a few minutes.
108 to 99,	Hens hatch eggs.
107 103	Heat of skin in ducks, geese, hens, pigeons, partridges, and swallows.
106	Heat of skin in a common ague and fever.
103 100	Heat of skin in dogs, cats, sheep, oxen, swine, and most other quadrupeds.
99 to 92,	Heat of the human skin in health.
97	Heat of a swarm of bees.
96	A perch died in 3 minutes in water so warm.
80	Heat of air in the shade, in very hot weather.
74	Butter begins to melt.
64	Heat of air in the shade, in warm weather.
55	Mean temperature of air in England.
43	Oil of olives begins to stiffen and grow opake.
32	Water just freezing, or snow and ice just melting.
30	Milk freezes.
28	Urine and common vinegar freezes.
25	Blood out of the body freezes.
20	Burgundy, Claret, and Madeira freeze.
5	Greatest cold in Pennsylvania in 1731-2, lat. 40°.
4	Greatest cold at Utrecht in 1728-9.
0	A mixture of snow and salt, which can freeze oil of tartar per deliquium, but not brandy.
39	Mercury freezes.

Martine's Essays, p. 284, &c.

On

On the general subject of Thermometers also see Martine's Essays, Medical and Philosophical. Desaguliers's Exp. Phil. vol. 2, p. 289. Musschenbroeck's Int. ad Phil. Nat. vol. 2, p. 625, ed. 1762. De Luc's Recherches sur les Modif. &c, tom. 1, part 2, ch. 2. Nollet's Leçons de Physique, tom. 4, p. 375.

THERMOMETERS *for particular uses*—In 1757, lord Cavendish presented to the Royal Society an account of a curious construction of Thermometers, of two different forms; one contrived to shew the greatest degree of heat, and the other the greatest cold, that may happen at any time in a person's absence. Philos. Transf. vol. 50, p. 300.

Since the publication of Mr. Canton's discovery of the compressibility of spirits of wine and other fluids, there are two corrections necessary to be made in the result given by lord Cavendish's Thermometer. For in estimating, for instance, the temperature of the sea at any depth, the Thermometer will appear to have been colder than it really was: and besides, the expansion of spirits of wine by any given number of degrees of Fahrenheit's Thermometer, is greater in the higher degrees than in the lower. For the method of making these two corrections by Mr. Cavendish, see Phipps's Voyage to the North Pole, p. 145.

Instruments of this kind, for determining the degree of heat or cold in the absence of the observer, have been invented and described by others. Van Swinden (Diss. sur la Comparaison du Therm. p. 253 &c) describes one, which he says was the first of the kind, made on a plan communicated by Bernoulli to Leibnitz. Mr. Kraft, he also tells us, made one nearly like it. Mr. Six has lately, viz, in 1782, proposed another construction of a Thermometer of the same kind, described in the Philos. Transf. vol. 72, p. 72 &c

Mr. De Luc has described the best method of constructing a Thermometer, fit for determining the temperature of the air, in the measuring of heights by the barometer. He has also shewn how to divide the scale of a Thermometer, so as to adapt it for astronomical purposes in the observation of refractions. See Recherches &c, tom. 2, p. 35 and 265.

Mr. Cavallo, in 1781, proposed the construction of a *Thermometrical Barometer*, which, by means of boiling water, might indicate the various gravity of the atmosphere, or the height of the barometer. This Thermometer, he says, with its apparatus, might be packed up into a small portable box, and serve for determining the heights of mountains &c, with greater facility, than with the common portable barometer. The instrument, in its present state, consists of a cylindrical tin vessel, about 2 inches in diameter, and 5 inches high, in which vessel the water is contained, which may be made to boil by the flame of a large wax-candle. The Thermometer is fastened to the tin vessel in such a manner, as that its bulb may be about an inch above the bottom. The scale of this Thermometer, which is of brass, exhibits on one side of the glass tube a few degrees of Fahrenheit's scale, viz, from 200° to 216°. On the other side of the tube are marked the various barometrical heights, at which the boiling water shews those particular degrees of heat which are set down in Sir Geo. Shuckburgh's table. With this instrument the barometrical height is shewn within one

10th of an inch. The degrees of this Thermometer are rather longer than one 9th of an inch, and therefore may be divided into many parts, especially by a Nonius. But a considerable imperfection arises from the smallness of the tin vessel, which does not admit a sufficient quantity of water; but when the quantity of water shall be sufficiently large, as for instance 10 or 12 ounces, and is kept boiling in a proper vessel, its degree of heat under the same pressure of the atmosphere is very settled; whereas when a Thermometer is kept in a small quantity of boiling water, the mercury in its stem does not stand very steady, sometimes rising or falling so much as half a degree. Mr. Cavallo proposes a farther improvement of this instrument, in the Philos. Transf. vol 71, p. 524.

The ingenious Mr. Wedgwood, so well known for his various improvements in the different sorts of pottery ware, has contrived to make a Thermometer for measuring the higher degrees of heat, by means of a distinguishing property of argillaceous bodies, viz, the diminution of their bulk by fire. This diminution commences in a low red-heat, and proceeds regularly, as the heat increases, till the clay becomes vitrified. The total contraction of some good clays which he has examined in the strongest of his own fires, is considerably more than one-fourth part in every dimension. By measuring the contraction of such substances then, Mr. Wedgwood contrived to measure the most intense heats of ovens, furnaces, &c. For the curious particulars of which, see Philos. Transf. vol. 72, p. 305 &c.

THERMOSCOPE, an instrument shewing the changes happening in the air with respect to heat and cold.

The word Thermoscope is often used indifferently with that of thermometer. There is some difference however in the literal import of the two; the first signifying an instrument that shews or exhibits the changes of heat &c to the eye; and the latter an instrument that measures those changes; so that a thermometer should be a more accurate Thermoscope.

THIR, in Chronology, the name of the 5th month of the Ethiopians, which corresponds, according to Ludolf, to the month of January.

THIRD, in Music, a concord resulting from a mixture of two sounds containing an interval of 2 degrees: being called a third, because containing 3 terms, or sounds, between the extremes.

There is a greater and a less Third. The former takes its form from the sesquiquarta ratio, 4 to 5. The logarithm or measure of the octave $\frac{2}{1}$ being 1·00000, the measure of the greater Third $\frac{5}{4}$ will be 0·32193.— The *greater Third* is by practitioners often taken for the third part of an octave; which is an error, since three greater Thirds fall short of the octave by a diesis; for $\frac{5}{4} \times \frac{5}{4} \times \frac{5}{4} \times \frac{128}{125} = \frac{2}{1}$.

The *lesser Third* takes its form from the sesquiquinta ratio 5 to 6; the measure or logarithm of this lesser Third $\frac{6}{5}$, being 0·26303, that of the octave $\frac{2}{1}$ being 1·00000.

Both these Thirds are of great use in melody; making as it were the foundation and life of harmony.

THIRD *Point*, or *Tierce-point*, in Architecture, the point of section in the vertex of an equilateral triangle. —Arches or vaults of the Third Point, are those consisting fifting of two arches of a circle, meeting in an angle at top.

THREE-*legged-ftaff*, an inftrument confifting of three wooden legs, made with joints, fo as to fhut all together, and to take off in the middle for the better carriage. It has ufually a ball and focket on the top; and its ufe is to fupport and adjuft inftruments for aftronomy, furveying, &c.

THUNDER, a noife in the lower region of the air, excited by a fudden explofion of electrical clouds; which are therefore called Thunder clouds.

The phenomenon of Thunder is varioufly accounted for. Seneca, Rohault, and fome other authors, both ancient and modern, account for Thunder, by fuppofing two clouds impending over one another, the upper and rarer of which, becoming condenfed by a frefh acceffion of air raifed by warmth from the lower parts of the atmofphere, or driven upon it by the wind, immediately falls forcibly down upon the lower and denfer cloud; by which fall, the air interpofed between the two being compreffed, that next the extremities of the two clouds is fqueezed out, and leaves room for the extremity of the upper cloud to clofe tight upon the under; thus a great quantity of the air is enclofed, which at length efcaping through fome winding irregular vent or paffage, occafions the noife called Thunder.

But this lame device could only reach at moft to the cafe of Thunder heard without lightning; and therefore recourfe has been had to other modes of folution. Thus, it has been faid that Thunder is not occafioned by the falling of clouds, but by the kindling of fulphurous exhalations, in the fame manner as the noife of the aurum fulminans. "There are fulphurous exhalations, fays Sir I. Newton, always afcending into the air when the earth is dry; there they ferment with the nitrous acids, and, fometimes taking fire, generate Thunder, lightning, &c."

The effects of Thunder are fo like thofe of fired gunpowder, that Dr. Wallis thinks we need not fcruple to afcribe them to the fame caufe; and the principal ingredients in gunpowder, we know, are nitre and fulphur; charcoal only ferving to keep the parts feparate, for their better kindling. Hence, if we conceive in the air a convenient mixture of nitrous and fulphurous particles; and thofe, by any caufe, to be fet on fire, fuch explofion may well follow, and with fuch noife and light as attend the firing of gunpowder; and being once kindled, it will run from place to place, different ways, as the exhalations happen to lead it; much as is found in a train of gunpowder.

But a third, and moft probable opinion is, that Thunder is the report or noife produced by an electrical explofion in the clouds. Ever fince the year 1752, in which the identity of the matter of lightning and of the electrical fluid has been afcertained, philofophers have generally agreed in confidering Thunder as a concuffion produced in the air by an explofion of electricity. For the illuftration and proof of this theory, fee ELECTRICITY, and LIGHTNING.

It may here be obferved, that Mr. Henry Eeles, in a letter written in 1751, and read before the Royal Society in 1752, confiders the electrical fire as the caufe of Thunder, and accounts for it on this hypothefis; and he tells us, that he did not know of any other

person's having made the fame conjecture. Philof. Tranf. vol. 47, p. 524 &c.

That rattling in the noife of Thunder, which makes it feem as if it paffed through arches, or were varioufly broken, is probably owing to the found being excited among clouds hanging over one another, and the agitated air paffing irregularly between them.

The explofion, if high in the air, and remote from us, will do no mifchief; but when near, it may deftroy trees, animals, &c.

This proximity, or fmall diftance, may be eftimated nearly by the interval of time between feeing the flafh of lightning, and hearing the report of the Thunder, eftimating the diftance, after the rate of 1142 feet per fecond of time, or 3⅖ feconds to the mile. Dr. Wallis obferves, that commonly the difference between the two is about 7 feconds, which, at the rate above mentioned, gives the diftance almoft 2 miles. But fometimes it comes in a fecond or two, which argues the explofion very near us, and even among us. And in fuch cafes, the doctor affures us, he has fometimes foretold the mifchiefs that happened.

The noife of Thunder, and the flame of lightning, are eafily made by art. If a mixture of oil or fpirit of vitriol be made with water, and fome filings of fteel added to it, there will immediately arife a thick fmoke, or vapour, out of the mouth of the veffel; and if a lighted candle be applied to this, it will take fire, and the flame will immediately defcend into the veffel, which will be burft to pieces with a noife like that of a cannon.

This is fo far analogous to Thunder and lightning, that a great explofion and fire are occafioned by it; but in this they differ, that this matter when once fired is deftroyed, and can give no more explofions; whereas, in the heavens, one clap of Thunder ufually follows another, and there is a continued fucceffion of them for a long time. Mr. Homberg explained this by the lightnefs of the air above us, in comparifon of that near, which therefore would not fuffer all the matter fo kindled to be diffipated at once, but keeps it for feveral returns.

THUNDERBOLT. When lightning acts with extraordinary violence, and breaks or fhatters any thing, it is called a *Thunderbolt*, which the vulgar, to fit it for fuch effects, fuppofe to be a hard body, and even a ftone.—But that we need not have recourfe to a hard folid body to account for the effects commonly attributed to the Thunderbolt, will be evident to any one, who confiders thofe of the pulvis fulminans, and of gunpowder; but more efpecially the aftonifhing powers of electricity, when only collected and employed by human art, and much more when directed and exercifed in the courfe of nature.

When we confider the known effects of electrical explofions, and thofe produced by lightning, we fhall be at no lofs to account for the extraordinary operations vulgarly afcribed to Thunderbolts. As ftones and bricks ftruck by lightning are often found in a vitrified ftate, we may reafonably fuppofe, with Beccaria, that fome ftones in the earth, having been ftruck in this manner, gave occafion to the vulgar opinion of the Thunderbolt.

THUNDER-*clouds*, in Phyfiology, are thofe clouds which

which are in a state fit for producing lightning and thunder.

From Beccaria's exact and circumstantial account of the external appearances of Thunder-clouds, the following particulars are extracted.

The first appearance of a Thunder storm, which usually happens when there is little or no wind, is one dense cloud, or more, increasing very fast in size, and rising into the higher regions of the air. The lower surface is black and nearly level; but the upper finely arched, and well defined. Many of these clouds often seem piled upon one another, all arched in the same manner; but they are continually uniting, swelling, and extending their arches.

At the time of the rising of this cloud, the atmosphere is commonly full of a great many separate clouds, that are motionless, and of odd whimsical shapes. All these, upon the appearance of the Thunder-cloud, draw towards it, and become more uniform in their shapes as they approach; till, coming very near the Thunder-cloud, their limbs mutually stretch toward one another, and they immediately coalesce into one uniform mass. These he calls adscititious clouds, from their coming in, to enlarge the size of the Thunder-cloud. But sometimes the Thunder-cloud will swell, and increase very fast, without the conjunction of any adscititious clouds; the vapours in the atmosphere forming themselves into clouds wherever it passes. Some of the adscititious clouds appear like white fringes, at the skirts of the Thunder-cloud, or under the body of it, but they keep continually growing darker and darker, as they approach to unite with it.

When the Thunder-cloud is grown to a great size, its lower surface is often ragged, particular parts being detached towards the earth, but still connected with the rest. Sometimes the lower surface swells into various large protuberances bending uniformly downward; and sometimes one whole side of the cloud will have an inclination to the earth, and the extremity of it nearly touch the ground. When the eye is under the Thunder-cloud, after it is grown larger, and well formed, it is seen to sink lower, and to darken prodigiously; at the same time that a number of small adscititious clouds (the origin of which can never be perceived) are seen in a rapid motion, driving about in very uncertain directions under it. While these clouds are agitated with the most rapid motions, the rain commonly falls in the greatest plenty, and if the agitation be exceedingly great, it commonly hails.

While the Thunder-cloud is swelling, and extending its branches over a large tract of country, the lightning is seen to dart from one part of it to another, and often to illuminate its whole mass. When the cloud has acquired a sufficient extent, the lightning strikes between the cloud and the earth, in two opposite places, the path of the lightning lying through the whole body of the cloud and its branches. The longer this lightning continues, the less dense does the cloud become, and the less dark its appearance; till at length it breaks in different places, and shews a clear sky.

These Thunder-clouds were sometimes in a positive as well as a negative state of electricity. The electricity continued longer of the same kind, in proportion as the Thunder-cloud was simple, and uniform in its di-

rection: but when the lightning changed its place, there commonly happened a change in the electricity of the apparatus, over which the clouds passed. It would change suddenly after a very violent flash of lightning, but the change would be gradual when the lightning was moderate, and the progress of the Thunder-cloud slow. Beccar. Lettere dell'Elettricismo pa. 107; or Priestley's Hist. Elec. vol. 1, p. 397. See also Lightning.

Thunder-House, in Electricity, is an instrument invented by Dr. James Lind, for illustrating the manner in which buildings receive damage from lightning, and to evince the utility of metallic conductors in preserving them from it.

A (fig. 1, pl. 35), is a board about ¾ of an inch thick, and shaped like the gable end of a house. This board is fixed perpendicularly upon the bottom board B, upon which the perpendicular glass pillar CD is also fixed in a hole about 8 inches distant from the basis of the board A. A square hole ILMK, about a quarter of an inch deep, and nearly one inch wide, is made in the board A, and is filled with a square piece of wood, nearly of the same dimensions. It is nearly of the same dimensions, because it must go so easily into the hole, that it may drop off, by the least shaking of the instrument. A wire LK is fastened diagonally to this square piece of wood. Another wire HH of the same thickness, having a brass ball H, screwed on its pointed extremity, is fastened upon the board A: so also is the wire MN, which is shaped in a ring at O. From the upper extremity of the glass pillar CD, a crooked wire proceeds, having a spring socket F, through which a double knobbed wire slips perpendicularly, the lower knob G of which falls just above the knob H. The glass pillar DC must not be made very fast into the bottom board; but it must be fixed so that it may be pretty easily moved round its own axis, by which means the brass ball G may be brought nearer to or farther from the ball H, without touching the part EFG. Now when the square piece of wood LMIK (which may represent the shutter of a window or the like) is fixed into the hole so that the wire LK stands in the dotted representation IM, then the metallic communication from H to O is complete, and the instrument represents a house furnished with a proper metallic conductor; but if the square piece of wood LMIK be fixed so that the wire LK stands in the direction LK, as represented in the figure, then the metallic conductor HO, from the top of the house to its bottom, is interrupted at LM, in which case the house is not properly secured.

Fix the piece of wood LMIK, so that its wire may be as represented in the figure, in which case the metallic conductor HO is discontinued. Let the ball G be fixed at about half an inch perpendicular distance from the ball H; then, by turning the glass pillar DC, remove the former ball from the latter; by a wire or chain connect the wire EF with the wire Q of the jar P; and let another wire or chain, fastened to the hook O, touch the outside coating of the jar. Connect the wire Q with the prime conductor, and charge the jar; then, by turning the glass pillar DC, let the ball G come gradually near the ball H, and when they are arrived sufficiently near one another, you will observe, that the jar explodes and the piece of wood LMIK is

pushed

2

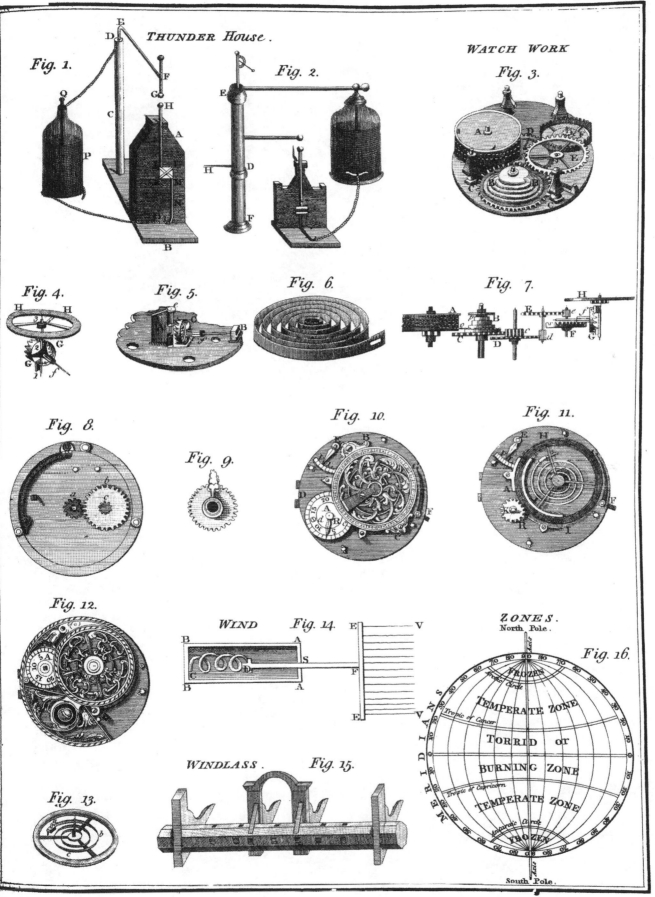

Plate XXXV.

THUNDER House.

WATCH WORK

Fig. 1.

Fig. 2.

Fig. 3.

Fig. 4.

Fig. 5.

Fig. 6.

Fig. 7.

Fig. 8.

Fig. 9.

Fig. 10.

Fig. 11.

Fig. 12.

WIND

Fig. 14.

ZONES.

North Pole.

Fig. 16.

MERIDIANS

FROZEN

TEMPERATE ZONE

Tropic of Cancer

TORRID or

BURNING ZONE

Tropic of Capricorn

TEMPERATE ZONE

FROZEN

South Pole.

WINDLASS.

Fig. 15.

Fig. 13.

pushed out of the hole to a considerable distance from the Thunder house.

Now the ball G, in this experiment, represents an electrified cloud, which, when it is arrived sufficiently near the top of the house A, the electricity strikes it; and as this house is not secured with a proper conductor, the explosion breaks part of it, i. e. knocks off the piece of wood IM.

Repeat the experiment with only this variation, viz, that this piece of wood IM be situated so that the wire LK may stand in the situation IM; in which case the conductor HO is not discontinued; and you will observe that the explosion will have no effect upon the piece of wood LM; this remaining in the hole unmoved; which shews the usefulness of the metallic conductor.

Farther, unscrew the brass ball H from the wire HI, so that this may remain pointed, and with this difference only in the apparatus repeat both the above experiments, and you will find that the piece of wood IM is in neither case moved from its place, nor will any explosion be heard; which not only demonstrates the preference of conductors with pointed terminations to those with blunted ones, but also shews that a house, furnished with sharp terminations, although not furnished with a regular conductor, is almost sufficiently guarded against the effects of lightning.

Mr. Henley, having connected a jar containing 509 square inches of coated surface with his prime conductor, observed that if it was so charged as to raise the index of his electrometer to 60°, by bringing the ball on the wire of the Thunder-house, to the distance of half an inch from that connected with the prime conductor, the jar would be discharged, and the piece in the Thunder-house thrown out to a considerable distance. Using a pointed wire for a conductor to the Thunder-house, instead of the knob, the charge being the same as before, the jar was discharged silently, though suddenly; and the piece was not thrown out of the Thunder-house. In another experiment, having made a double circuit to the Thunder-house, the first by the knob, the second by a sharp-pointed wire, at an inch and a quarter distance from each other, but of exactly the same height (as in fig. 2) the charge being the same; although the knob was brought first under that connected with the prime conductor, which was raised half an inch above it, and followed by the point, yet no explosion could fall upon the knob; the point drew off the whole charge silently, and the piece in the Thunder-house remained unmoved.

Phil. Transf. vol. 64, p. 136. See POINTS in Electricity.

THURSDAY, the 5th day of the Christian's week, but the 6th of the Jews. The name is from Thor, one of the Saxon Gods.

THUS, in Sea-Language, a word used by the pilot in directing the helmsman or steersman to keep the ship in her present situation when sailing with a scant wind, so that she may not approach too near the direction of the wind, which would shiver her sails, nor fall to leeward, and run farther out of her course.

TIDES, two periodical motions of the waters of the sea; called also the *flux* and *reflux*, or the *ebb* and *flow*.

The Tides are found to follow periodically the course of the sun and moon, both as to time and quantity. And hence it has been suspected, in all ages, that the Tides were somehow produced by the influence of these luminaries. Thus, several of the ancients, and among others, Pliny, Ptolomy, and Macrobius, were acquainted with the influence of the sun and moon upon the Tides; and Pliny says expressly, that the cause of the ebb and flow is in the sun, which attracts the waters of the ocean; and adds, that the waters rise in proportion to the proximity of the moon to the earth. It is indeed now well known, from the discoveries of Sir Isaac Newton, that the Tides are caused by the gravitation of the earth towards the sun and moon. Indeed the sagacious Kepler, long ago, conjectured this to be the cause of the Tides: " If, says he, the earth ceased to attract its waters towards itself, all the water in the ocean would rise and flow into the moon: the sphere of the moon's attraction extends to our earth, and draws up the water." Thus thought Kepler, in his Introd ad Theor. Mart. This surmise, for it was then no more, is now abundantly verified in the theory laid down by Newton, and by Halley, from his principles.

As to the Phenomena of the TIDES: 1. The sea is observed to flow, for about 6 hours, from south towards north; the sea gradually swelling; so that, entering the mouths of rivers, it drives back the river-waters towards their heads, or springs. After a continual flux of 6 hours, the sea seems to rest for about a quarter of an hour; after which it begins to ebb, or retire back again, from north to south, for 6 hours more; in which time, the water sinking, the rivers resume their natural course. Then, after a seeming pause of a quarter of an hour, the sea again begins to flow, as before: and so on alternately.

2. Hence, the sea ebbs and flows twice a day, but falling every day gradually later and later, by about 48 minutes, the period of a flux and reflux being on an average about 12 hours 24 minutes, and the double of each 24 hours 48 minutes; which is the period of a lunar day, or the time between the moon's passing a meridian, and coming to it again. So that the sea flows as often as the moon passes the meridian, both the arch above the horizon, and that below it; and ebbs as often as she passes the horizon, both on the eastern and western side.

Other phenomena of the Tides are as below; and the reasons of them will be noticed in the Theory of the Tides that follows.

3. The elevation towards the moon a little exceeds the opposite one. And the quantity of the ascent of the water is diminished from the equator towards the poles.

4. From the sun, every natural day, the sea is twice elevated, and twice depressed, the same as for the moon. But the solar Times are much less than the lunar ones, on account of the immense distance of the sun; yet they are both subject to the same laws.

5. The Tides which depend upon the actions of the sun and moon, are not distinguished, but compounded, and so forming as to sense one united Tide, increasing and decreasing, and thus making neap and spring Tides: for, by the action of the sun, the

lunar

lunar Tide is only changed; which change varies every day, by reason of the inequality between the natural and lunar day.

6. In the fyzygies the elevations from the action of both luminaries concur, and the fea is more elevated. But the fea afcends lefs in the quadratures; for where the water is elevated by the action of the moon, it is deprefled by the action of the fun; and vice. verfa. Therefore, while the moon paffes from the fyzygy to the quadrature, the daily elevations are continually diminifhed: on the contrary, they are increafed while the moon moves from the quadrature to the fyzygy. At a new moon alfo, *cæteris paribus*, the elevations are greater; and thofe that follow one another the fame day, are more different than at full moon.

7. The greateft elevations and depreffions are not obferved till the 2d or 3d day after the new or full moon. And if we confider the luminaries receding from the plane of the equator, we fhall perceive that the agitation is diminifhed, and becomes lefs, according as the declination of the luminaries becomes greater.

8. In the fyzygies, and near the equinoxes, the Tides are obferved to be the greateft, both luminaries being in or near the equator.

9. The actions of the fun and moon are greater, the nearer thofe bodies are to the earth; and the lefs, as they are farther off. Alfo the greateft Tides happen near the equinoxes, or rather when the fun is a little to the fouth of the equator, that is, a little before the vernal, and after the autumnal equinox. But yet this does not happen regularly every year, becaufe fome variation may arife from the fituation of the moon's orbit, and the diftance of the fyzygy from the equinox.

10. All thefe phenomena obtain, as defcribed, in the open fea, where the ocean is extended enough to be fubject to thefe motions. But the particular fituations of places, as to fhores, capes, ftraits, &c, difturb thefe general rules. Yet it is plain, from the moft common and univerfal obfervations, that the Tides follow the laws above laid down.

11. The mean force of the moon to move the fea, is to that of the fun, nearly as 4¼ to 1. And therefore, if the action of the fun alone produce a Tide of 2 feet, which it has been ftated to do, that of the moon will be 9 feet; from which it follows, that the fpring Tides will be 11 feet, and the neap Tides 7 feet high. But as to fuch elevations as far exceed thefe, they happen from the motion of the waters againft fome obftacles, and from the fea violently entering into ftraits or gulphs where the force is not broken till the water rifes higher.

Theory of the TIDES.

1. If the earth were entirely fluid, and quiefcent, it is evident that its particles, by their mutual gravity towards each other, would form the whole mafs into the figure of an exact fphere. Then fuppofe fome power to act on all the particles of this fphere with an equal force, and in parallel directions; by such a power the whole mafs will be moved together, but its figure will fuffer no alteration by it, being ftill the fame perfect fphere, whofe centre will have the fame motion as each particle.

Upon this fuppofition, if the motion of the earth round the common centre of gravity of the earth and moon were deftroyed, and the earth left to the influence of its gravitation towards the moon, as the acting power above mentioned; then the earth would fall or move ftraight towards the moon, but ftill retaining its true fpherical figure.

But the fact is, that the effects of the moon's action, as well as the action itfelf, on different parts of the earth, are not equal: thofe parts, by the general law of gravity, being moft attracted that are neareft the moon, and thofe being leaft attracted that are fartheft from her, while the parts that are at a middle diftance are attracted by a mean degree of force: befides, all the parts are not acted on in parallel lines, but in lines directed towards the centre of the moon: on both which accounts the fpherical figure of the fluid earth muft fuffer fome change from the action of the moon. So that, in falling, as above, the nearer parts, being moft attracted, would fall quickeft; the farther parts, being leaft attracted, would fall floweft; and the fluid mafs would be lengthened out, and take a kind of fpheroidical form.

Hence it appears, and what muft be carefully obferved, that it is not the action of the moon itfelf, but the inequalities in that action, that caufe any variation from the fpherical figure; and that, if this action were the fame in all the particles as in the central parts, and operating in the fame direction, no fuch change would enfue.

Let us now admit the parts of the earth to gravitate toward its centre: then, as this gravitation far exceeds the action of the moon, and much more exceeds the differences of her actions on different parts of the earth, the effect that refults from the inequalities of thefe actions of the moon, will be only a fmall diminution of the gravity of thofe parts of the earth which it endeavoured in the former fuppofition to feparate from its centre; that is, thofe parts of the earth which are neareft to the moon, and thofe that are fartheft from her, will have their gravity toward the earth fomewhat abated; to fay nothing of the lateral parts. So that fuppofing the earth fluid, the columns from the centre to the neareft, and to the fartheft parts, muft rife, till by their greater height they be able to balance the other columns, whofe gravity is lefs altered by the inequalities of the moon's action. And thus the figure of the earth muft ftill be an oblong fpheroid.

Let us now confider the earth, inftead of falling toward the moon by its gravity, as projected in any direction, fo as to move round the centre of gravity of the earth and moon: it is evident that in this cafe, the feveral parts of the fluid earth will ftill preferve their relative pofitions; and the figure of the earth will remain the fame as if it fell freely toward the moon; that is, the earth will ftill affume a fpheroidal form, having its longeft diameter directed toward the moon.

From

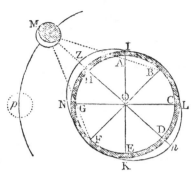

From the above reafoning it appears, that the parts of the earth directly under the moon, as at H, and alfo the oppofite parts at D, will have the flood or highwater at the fame time; while the parts, at B and F, at 90° diftance, or where the moon appears in the horizon, will have the ebbs or loweft waters at that time.

Hence, as the earth turns round its axis from the moon to the moon again in 24 hours 48 minutes, this oval of water muft fhift with it; and thus there will be two Tides of flood and two of ebb in that time.

But it is further evident that, by the motion of the earth on her axis, the moft elevated part of the water is carried beyond the moon in the direction of the rotation. So that the water continues to rife after it has paffed directly under the moon, though the immediate action of the moon there begins to decreafe, and comes not to its greateft elevation till it has got about half a quadrant farther. It continues alfo to defcend after it has paffed at 90° diftance from the point below the moon, to a like diftance of about half a quadrant. The greateft elevation therefore is not in the line drawn through the centres of the earth and moon, nor the loweft points where the moon appears in the horizon, but all thefe about half a quadrant removed eaftward from thefe points, in the direction of the motion of rotation. Thus in open feas, where the water flows freely, the moon M is generally paft the north and fouth meridian, as at p, when the high water is at Z and at n: the reafon of which is plain, becaufe the moon acts with the fame force after fhe has paffed the meridian, and thus adds to the libratory or waving motion, which the water acquired when fhe was in the meridian; and therefore the time of high water is not precifely at the time of her coming to the meridian, but fome time after, &c.

Befides, the Tides anfwer not always to the fame diftance of the moon, from the meridian, at the fame places; but are varioufly affected by the action of the fun, which brings them on fooner when the moon is in her firft and third quarters, and keeps them back later when fhe is in her 2d and 4th; becaufe, in the former cafe the Tide raifed by the fun alone would be earlier than the Tide raifed by the moon, and in the latter cafe later.

2. We have hitherto adverted only to the action of the moon in producing Tides; but it is manifeft that, for the fame reafons, the inequality of the fun's action on different parts of the earth, would produce a like

effect, and a like variation from the exact fpherical figure of a fluid earth. So that in reality there are two Tides every natural day from the action of the fun, as there are in the lunar day from that of the moon, fubject to the fame laws; and the lunar Tide, as we have obferved, is fomewhat changed by the action of the fun, and the change varies every day on account of the inequality between the natural and the lunar day. Indeed the effect of the fun in producing Tides, becaufe of his immenfe diftance, muft be confiderably lefs than that of the moon, though the gravity toward the fun be much greater: for it is not the action of the fun or moon itfelf, but the inequalities in that action, that have any effect: the fun's diftance is fo great, that the diameter of the earth is but as a point in comparifon with it, and therefore the difference between the fun's actions on the neareft and fartheft parts, becomes vaftly lefs than it would be if the fun were as near as the moon. However the immenfe bulk of the fun makes the effect ftill fenfible, even at fo great a diftance; and therefore, though the action of the moon has the greateft fhare in producing the Tides, the action of the fun adds fenfibly to it when they confpire together, as in the full and change of the moon, when they are nearly in the fame line with the centre of the earth, and therefore unite their forces: confequently, in the fyzygies, or at new and full moon, the Tides are the greateft, being what are called the *Spring-Tides*. But the action of the fun diminifhes the effect of the moon's action in the quarters, becaufe the one raifes the water in that cafe where the other depreffes it; therefore the Tides are the leaft in the quadratures, and are called *Neap-Tides*.

Newton has calculated the effects of the fun and moon refpectively upon the Tides, from their attractive powers. The former he finds to be to the force of gravity, as 1 to 12868200, and to the centrifugal force at the equator as 1 to 44527. The elevation of the waters by this force is confidered by Newton as an effect fimilar to the elevation of the equatorial parts above the polar parts of the earth, arifing from the centrifugal force at the equator; and as it is 44527 times lefs, he finds it to be $24\frac{1}{2}$ inches, or 2 feet and $\frac{1}{4}$ an inch.

To find the force of the moon upon the water, Newton compares the fpring tides at the mouth of the river Avon, below Briftol, with the neap-tides there, and finds the proportion as 9 to 5; whence, after feveral neceffary corrections, he concludes that the force of the moon to that of the fun, in raifing the waters of the ocean, is as 4·4815 to 1; fo that the force of the moon is able of itfelf to produce an elevation of 9 feet $1\frac{3}{4}$ inch, and the fun and moon together may produce an elevation of about 11 feet 2 inches, when at their mean diftances from the earth, or an elevation of about $12\frac{3}{4}$ feet, when the moon is neareft the earth. The height to which the water is found to rife, upon coafts of the open and deep ocean, is agreeable enough to this computation.

Dr. Horfley eftimates the force of the moon to that of the fun, as 5·0469 to 1, in his edit. of Newton's Princip. See the Princip. lib. 3, fect. 3, pr. 36 and 37; alfo Maclaurin's Differt. de Caufa Phyfica Fluxus et Refluxus Maris apud Phil. Nat. Princ. Math. Comment.

I

ment. le Seur & Jacquier, tom. 3, p. 272. And other calculators make the proportion still more different.

3. It must be observed, that the spring-tides do not happen precisely at new and full moon, nor the neap-tides at the quarters, but a day or two after; because, as in other cases, so in this, the effect is not greatest or least when the immediate influence of the cause is greatest or least. As, for example, the greatest heat is not on the day of the solstice, when the immediate action of the sun is greatest, but some time after; so likewise, if the actions of the sun and moon should suddenly cease, yet the Tides would continue to have their course for some time; and like also as the waves of the sea continue after a storm.

4. The different distances of the moon from the earth produce a sensible variation in the Tides. When the moon approaches toward the earth, her action on every part increases, and the differences of that action, on which the Tides depend, also increase; and as the moon approaches, her action on the nearest parts increases more quickly than that on the remote parts, so that the Tides increase in a higher proportion as the moon's distances decrease. In fact, it is shewn by Newton, that the Tides increase in proportion as the cubes of the distances decrease; so that the moon at half her distance would produce a Tide 8 times greater.

The moon describes an oval about the earth, and at her nearest distance produces a Tide sensibly greater than at her greatest distance from the earth : and hence it is that two great spring tides never succeed each other immediately; for if the moon be at her least distance from the earth at the change, she must be at her greatest distance at the full, having made half a revolution in the intervening time, and therefore the spring-tide then will be much less than that at the last change was; and for the same reason, if a great spring-tide happen at the time of full moon, the Tide at the ensuing change will be less.

5. The spring-tides are highest, and the neap-tides lowest, about the time of the equinoxes, or the latter end of March and September; and, on the contrary, the spring-tides are the lowest, and the neap-tides the highest, at the solstices, or about the latter end of June and December: so that the difference between the spring and neap Tides, is much more considerable about the equinoctial than the solstitial seasons of the year. To illustrate and evince the truth of this observation, let us consider the effect of the luminaries upon the Tides, when in and out of the plane of the equator. Now it is manifest, that if either the sun or moon were in the pole, they could not have any effect on the Tides; for their action would raise all the water at the equator, or at any parallel, quite around, to a uniform height; and therefore any place of the earth, in describing its parallel to the equator, would not meet, in its course, with any part of the water more elevated than another; so that there could be no Tide in any place, that is, no alteration in the height of the waters.

On the other hand, the effect of the sun or moon is greatest when in the equinoctial; for then the axis of the spheroidal figure, arising from their action, moves in the greatest circle, and the water is put into the greatest agitation; and hence it is that the spring-tides produced when the sun and moon are both in the equinoctial, are the greatest of any, and the neaptides the least of any about that time. And when the luminary is any where between the equinoctial and the pole, the Tides are the smaller.

6. The highest spring tides are after the autumnal and before the vernal equinox : the reason of which is, because the sun is nearer the earth in winter than in summer.

7. Since the greatest of the two Tides happening in every diurnal revolution of the moon, is that in which the moon is nearest the zenith, or nadyr : for this reason, while the sun is in the northern signs, the greater of the two diurnal Tides in our climates, is that arising from the moon above the horizon; when the sun is in the southern signs, the greatest is that arising from the moon below the horizon. Thus it is found by observation that the evening Tides in the summer exceed the morning Tides, and in winter the morning Tides exceed the evening Tides. The difference is found at Bristol to amount to 15 inches, and at Plymouth to 12. It would be still greater, but that a fluid always retains an impressed motion for some time; so that the preceding Tides affect always those that follow them. Upon the whole, while the moon has a north declination, the greatest Tides in the northern hemisphere are when she is above the horizon, and the reverse while her declination is south.

8. Such would the Tides regularly be, if the earth were all over covered with the sea very deep, so that the water might freely follow the influence of the sun and moon; but, by reason of the shoalness of some places, and the narrowness of the straits in others, through which the Tides are propagated, there arises a great diversity in the effect according to the various circumstances of the places. Thus, a very slow and imperceptible motion of the whole body of water, where it is very deep, as 2 miles for instance, will suffice to raise its surface 10 or 12 feet in a Tide's time : whereas, if the same quantity of water were to be conveyed through a channel of 40 fathoms deep, it would require a very rapid stream to effect it in so large inlets as are the English channel, and the German ocean; whence the Tide is found to set strongest in those places where the sea grows narrowest, the same quantity of water being in that case to pass through a smaller passage. This is particularly observable in the straits between Portland and Cape la Hogue in Normandy, where the Tide runs like a sluice : and would be yet more so between Dover and Calais, if the Tide coming round the island did not check it.

This force, when once impressed, continues to carry the water above the ordinary height in the ocean, especially where the water meets a direct obstacle, as it does in St. Maloes; and where it enters into a long channel which, running far into the land, grows very strait at its extremity, as it does into the Severn sea at Chepstow and Bristol.

This shoalness of the sea, and the intercurrent continents, are the reasons that in the open ocean the Tides rise but to very small heights in proportion to what they do in wide-mouthed rivers, opening in the direction

tion of the stream of the Tide; and that high water is not soon after the moon's appulse to the meridian, but some hours after it, as it is observed upon all the western coast of Europe and Africa, from Ireland to the Cape of Good Hope; in all which a south-west moon makes high water; and the same it is said is the case on the western side of America. So that Tides happen to different places at all distances of the moon from the meridian, and consequently at all hours of the day.

To allow the Tides their full motion, the ocean in which they are produced, ought to be extended from east to west 90 degrees at least; because that is the distance between the places where the water is most raised and depressed by the moon. Hence it appears that it is only in the great oceans that such Tides can be produced, and why in the larger Pacific ocean they exceed those in the Atlantic ocean. Hence also it is obvious, why the Tides are not so great in the torrid zone, between Africa and America, where the ocean is narrower, as in the temperate zones on either side; and hence we may also understand why the Tides are so small in islands that are very far distant from the shores. It is farther manifest that, in the Atlantic ocean, the water cannot rise on one shore but by descending on the other; so that at the intermediate islands it must continue at a mean height between its elevations on those two shores. But when Tides pass over shoals, and through straits into bays of the sea, their motion becomes more various, and their height depends on many circumstances.

To be more particular. The Tide that is produced on the western coasts of Europe, in the Atlantic, corresponds to the situation of the moon already described. Thus it is high water on the western coasts of Ireland, Portugal and Spain, about the 3d hour after the moon has passed the meridian: from thence it flows into the adjacent channels, as it finds the easiest passage. One current from it, for instance, runs up by the south of England, and another comes in by the north of Scotland; they take a considerable time to move all this way, making always high water sooner in the places to which they first come; and it begins to fall at these places while the currents are still going on to others that are farther distant in their course. As they return, they are not able to raise the Tide, because the water runs faster off than it returns, till, by a new Tide propagated from the open ocean, the return of the current is stopped, and the water begins to rise again. The Tide propagated by the moon in the German ocean, when she is 3 hours past the meridian, takes 12 hours to come from thence to London bridge; so that when it is high water there, a new Tide is already come to its height in the ocean; and in some intermediate place it must be low water at the same time. Consequently when the moon has north declination, and we should expect the Tide at London to be the greatest when the moon is above the horizon, we find it is least: and the contrary when she has south declination.

At several places it is high water 3 hours before the moon comes to the meridian; but that Tide, which the moon pushes as it were before her, is only the Tide opposite to that which was raised by her when she was 9 hours past the opposite meridian.

It would be endless to recount all the particular solutions, which are easy consequences from this doctrine: as, why the lakes and seas, such as the Caspian sea and the Mediterranean sea, the Black sea and the Baltic, have little or no sensible Tides: for lakes are usually so small, that when the moon is vertical she attracts every part of them alike, so that no part of the water can be raised higher than another: and having no communication with the ocean, it can neither increase nor diminish their water, to make it rise and fall; and seas that communicate by such narrow inlets, and are of so immense an extent, cannot speedily receive and empty water enough to raise or sink their surface any thing sensibly.

In general; when the time of high water at any place is mentioned, it is to be understood on the days of new and full moons.—Among pilots, it is customary to reckon the time of flood, or high water, by the point of the compass the moon bears on, at that time, allowing $\frac{3}{4}$ of an hour for each point. Thus, on the full and change days, in places where it is flood at noon, the Tide is said to flow north and south, or at 12 o'clock: in other places, on the same days, where the moon bears 1, 2, 3, 4, or more points to the east or west of the meridian, when it is high water, the Tide is said to flow on such point; thus, if the moon bears SE, at flood, it is said to flow SE and NW, or 3 hours before the meridian, that is, at 9 o'clock; if it bears SW, it flows SW and NE, or at 3 hours after the meridian; and in like manner for the other points of the moon's bearing.

The times of high water in any place fall about the same hours after a period of about 15 days, or between one spring Tide and another; but during that period, the times of high water fall each day later by about 48 minutes.

On the subject of this article, see Newton Princ. Math. lib. 3, prop. 24, and De System. Mundi sect. 38, &c. Apud Opera edit. Horsley, tom. 3, pa. 52 &c. p. 203 &c. Maclaurin's Account of Newton's Discoveries, book 4, ch. 7. Ferguson's Astron. ch. 17. Robertson's Navig. book 6, sect. 7, 8, 9. Lalande's Astron. vol. 4.

TIDE *Dial*, an instrument contrived by Mr. Ferguson, for exhibiting and determining the state of the Tides. For the construction and use of which see his Astron. p. 297.

TIDE *Tables*, are tables commonly exhibiting the times of high water at sundry places, as they fall on the days of the full and change of the moon, and sometimes the height of them also. These are common in most books on Navigation, particularly Robertson's, and the 2d ed. of Tables requisite to be used with the Nautical Almanac. See one at HIGH-*water*.

TIERCE, or TEIRCE, a liquid measure, as of wine, oil, &c, containing 42 gallons, or the 3d part of a pipe; whence its name.

TIME, a succession of phenomena in the universe; or a mode of duration, marked by certain periods and measures; chiefly indeed by the motion and

and revolution of the luminaries, and particularly of the fun.

The idea of Time in general, Locke obferves, we acquire by confidering any part of infinite duration, as fet out by periodical meafures: the idea of any particular Time, or length of duration, as a day, an hour, &c, we acquire firft by obferving certain appearances at regular and feemingly equidiftant periods. Now, by being able to repeat thefe lengths or meafures of Time as often as we will, we can imagine duration, where nothing really endures or exifts; and thus we imagine tomorrow, or next year, &c.

Some of the later fchool-philofophers define Time to be the duration of a thing whofe exiftence is neither without beginning nor end: by this, Time is diftinguifhed from eternity.

Ariftotle and the Peripatetics define it, *numerus motus fecundum prius & pofterius*, or a multitude of tranfient parts of motion, fucceeding each other, in a continual flux, in the relation of priority and pofteriority. Hence it fhould follow that Time is motion itfelf, or at leaft the duration of motion, confidered as having feveral parts, fome of which are continually fucceeding to others. But on this principle, Time or temporal duration would not agree to bodies at reft, which yet nobody will deny to exift in Time, or to endure for a Time.

To avoid this inconvenience, the Epicureans and Corpufcularians made Time to be a fort of flux different from motion, confifting of infinite parts, continually and immediately fucceeding each other, and this from eternity to eternity. But others directly explode this notion, as eftablifhing an eternal being, independent of God. For how fhould there be a flux before any thing exifted to flow? and what fhould that flux be, a fubftance, or an accident? According to the philofophic poet,

"Time of itfelf is nothing, but from thought
Receives its rife; by labouring fancy wrought
From things confider'd, whilft we think on fome
As prefent, fome as paft, or yet to come.
No thought can think on Time, that's ftill confeft,
But thinks on things in motion or at reft."

And fo on. Vide Lucretius, book i.

Time may be diftinguifhed, like place, into *abfolute* and *relative*.

Abfolute TIME, is Time confidered in itfelf, and without any relation to bodies, or their motions.

Relative or *Apparent* TIME, is the fenfible meafure of any duration by means of motion.

Some authors diftinguifh Time into *aftronomical* and *civil*.

Aftronomical TIME, is that which is taken purely from the motion of the heavenly bodies, without any other regard.

Civil TIME, is the former Time accommodated to civil ufes, and formed or diftinguifhed into years, months, days, &c.

Time makes the fubject of chronology.

TIME, in mufic, is an affection of found, by which it is faid to be long or fhort, with regard to its continuance in the fame tone or degree of tune.

Mufical Time is diftinguifhed into *common* or *duple* Time, and *triple* Time.

Double, *duple*, or *common Time*, is when the notes are in a duple duration of each other, viz, a femibreve equal to 2 minims, a minim to 2 crotchets, a crotchet to 2 quavers, &c.

Common or double Time is of two kinds. The firft when every bar or meafure is equal to a femibreve, or its value in any combination of notes of a lefs quantity. The fecond is where every bar is equal to a minim, or its value in lefs notes. The movements of this kind of meafure are various, but there are three common diftinctions; the firft *flow*, denoted at the beginning of the line by the mark C; the 2d *brifk*, marked thus ₵; and the 3d *very brifk*, thus marked.

Triple Time is when the durations of the notes are triple of each other, that is, when the femibreve is equal to 3 minims, the minim to 3 crotchets, &c. and it is marked T.

TIME-*keepers*, in a general fenfe, denote inftruments adapted for meafuring time. See CHRONOMETER.

In a more peculiar and definite fenfe, Time-keeper is a term firft applied by Mr. John Harrifon to his watches, conftructed and ufed for determining the longitude at fea, and for which he received, at different times, the parliamentary reward of 20 thoufand pounds. And feveral other artifts have fince received alfo confiderable fums for their improvements of Time-keepers; as Arnold, Mudge, &c. See LONGITUDE.

This appellation is now become common among artifts, to diftinguifh fuch watches as are made with extraordinary care and accuracy for nautical or aftronomical obfervations.

The principles of Mr. Harrifon's Time-keeper, as they were communicated by himfelf, to the commiffioners appointed to receive and publifh the fame in the year 1765, are as below:

"In this Time-keeper there is the greateft care taken to avoid friction, as much as can be, by the wheel moving on fmall pivots, and in ruby-holes, and high numbers in the wheels and pinions.

"The part which meafures time goes but the eighth part of a minute without winding up; fo that part is very fimple, as this winding-up is performed at the wheel next to the balance-wheel; by which means there is always an equal force acting at that wheel, and all the reft of the work has no more to do in the meafuring of time than the perfon that winds up once a day.

"There is a fpring in the infide of the fufee, which I will call a fecondary main fpring. This fpring is always kept ftretched to a certain tenfion by the main fpring; and during the time of winding-up the Time keeper, at which time the main-fpring is not fuffered to act, this fecondary-fpring fupplies its place.

"In common watches in general, the wheels have about one-third the dominion over the balance, that the balance-fpring has; that is, if the power which the balance-fpring has over the balance be called three,
that

that from the wheel is one : but in this my Time-keeper, the wheels have only about one-eightieth part of the power over the balance that the balance spring has ; and it must be allowed, the less the wheels have to do with the balance, the better. The wheels in a common watch having this great dominion over the balance, they can, when the watch is wound up, and the balance at rest, set the watch a-going ; but when my Time-keeper's balance is at rest, and the spring is wound up, the force of the wheels can no more set it a-going, than the wheels of a common regulator can, when the weight is wound-up, set the pendulum a-vibrating ; nor will the force from the wheels move the balance when at rest, to a greater angle in proportion to the vibration that it is to fetch, than the force of the wheels of a common regulator can move the pendulum from the perpendicular, when it is at rest.

" My Time-keeper's balance is more than three times the weight of a large sized common watch balance, and three times its diameter ; and a common watch balance goes through about six inches of space in a second, but mine goes through about twenty-four inches in that time : so that had my Time-keeper only these advantages over a common watch, a good performance might be expected from it. But my Time-keeper is not affected by the different degrees of heat and cold, nor agitation of the ship ; and the force from the wheels is applied to the balance in such a manner, together with the shape of the balance-spring, and (if I may be allowed the term) an artificial cycloid, which acts at this spring ; so that from these contrivances, let the balance vibrate more or less, all its vibrations are performed in the same time ; and therefore if it go at all, it must go true. So that it is plain from this, that such a Time-keeper goes entirely from principle, and not from chance."

We must refer those who may desire to see a minute account of the construction of Mr. Harrison's Time-keeper, to the publication by order of the commissioners of longitude.

We shall here subjoin a short view of the improvements in Mr. Harrison's watch, from the account presented to the board of longitude by Mr. Ludlam, one of the gentlemen to whom, by order of the commissioners, Mr. Harrison discovered and explained the principle upon which his Time-keeper is constructed. The defects in common watches which Mr. Harrison proposes to remedy, are chiefly these : 1. That the main spring acts not constantly with the same force upon the wheels, and through them upon the balance : 2. That the balance, either urged with an unequal force, or meeting with a different resistance from the air, or the oil, or the friction, vibrates through a greater or less arch : 3. That these unequal vibrations are not performed in equal times : and, 4. That the force of the balance-spring is altered by a change of heat.

To remedy the first defect, Mr. Harrison has contrived that his watch shall be moved by a very tender spring, which never unrolls itself more than one-eighth part of a turn, and acts upon the balance through one wheel only. But such a spring cannot keep the watch in motion a long time. He has, therefore, joined another, whose office is to wind up the first

spring eight times in every minute, and which is itself wound up but once a day. To remedy the second defect, he uses a much stronger balance spring than in a common watch. For if the force of this spring upon the balance remains the same, whilst the force of the other varies, the errors arising from that variation will be the less, as the fixed force is the greater. But a stronger spring will require either a heavier or a larger balance. A heavier balance would have a greater friction. Mr. Harrison, therefore, increases the diameter of it. In a common watch it is under an inch, but in Mr. Harrison's two inches and two tenths. However, the methods already described only lessening the errors, and not removing them, Mr. Harrison uses two ways to make the times of the vibrations equal, though the arches may be unequal : one is to place a pin, so that the balance-spring pressing against it, has its force increased, but increased less when the variations are larger : the other to give the pallets such a shape, that the wheels press them with less advantage, when the vibrations are larger. To remedy the last defect, Mr. Harrison uses a bar compounded of two thin plates of brass and steel, about two inches in length, riveted in several places together, fastened at one end and having two pins at the other, between which the balance spring passes. If this bar be straight in temperate weather (brass changing its length by heat more than steel) the brass side becomes convex when it is heated, and the steel side when it is cold : and thus the pins lay hold of a different part of the spring in different degrees of heat, and lengthen or shorten it as the regulator does in a common watch.

The principles, on which Mr. Arnold's Time-keeper is constructed, are these : The balance is unconnected with the wheel work, except at the time it receives the impulse to make it continue its motion, which is only whilst it vibrates 10° out of 380° which is the whole vibration ; and during this small interval it has little or no friction, but what is on the pivots, which work in ruby holes on diamonds. It has but one pallet, which is a plane surface formed out of a ruby, and has no oil on it. Watches of this construction, says Mr. Lyons, go whilst they are wound up ; they keep the same rate of going in every position, and are not affected by the different forces of the spring ; and the compensation for heat and cold is absolutely adjustable. Phipps's Voyage to the North Pole, p. 230. See Longitude.

TISRI, or Tizri, in chronology, the first Hebrew month of the civil year, and the 7th of the ecclesiastical or sacred year. It answered to part of our September and October.

TOD *of wool*, is mentioned in the statute 12 Carol. II. c. 32, as a weight containing 2 stone, or 28 pounds.

TOISE, a French measure, containing 6 of their feet, similar to our fathom.

TONDIN, or Tandino, in Architecture. See Tore.

TONE, or Tune, in Music, a property of sound, by which it comes under the relation of grave and acute ; or the degree of elevation any sound has, from the degree of swiftness of the vibrations of the parts of the sonorous body.

For the cause, measure, degree, difference, &c, of Tones, see TUNE.

The word Tone is taken in four different senses among the ancients. 1, For any sound. 2, For a certain interval; as when it is said the difference between the diapente and diatessaron is a Tone. 3, For a certain locus or compass of the voice; in which sense they used the Dorian, Phrygian, Lydian Tones. 4. For tension; as when they speak of an acute, a grave, or a middle Tone. Wallis's Append. Ptolom. Harm. p. 172.

TONE is more particularly used, in music, for a certain degree or interval of tune, by which a sound may be either raised or lowered from one extreme of a concord to the other, so as still to produce true melody.

In tempered scales of music, the Tones are made equal, but in a true and accurate practice of singing they are not so. Pepusch, in Philos. Transf. No. 481, p. 274.

Beside the concords, or harmonical intervals, musicians admit three less kinds of intervals, which are the measures and component parts of the greater, and are called *degrees*.

Of these degrees, two are called Tones, and the third a semitone. Their ratios in numbers are 8 to 9, called a *greater Tone*; 9 to 10, called a *lesser Tone*; and 15 to 16, a *semitone*.

The Tones arise out of the simple concords, and are equal to their differences. Thus the greater Tone, 8 : 9, is the difference of a 5th and a 4th; the less Tone 9 : 10, the difference of a less 3d and a 4th, or of a 5th and a greater 6th; and the semitone 15 : 16, the difference of a greater 3d and a 4th.

Of these Tones and semitones every concord is compounded, and consequently every one is resolvable into a certain number of them. Thus the less 3d consists of one greater Tone and one semitone: the greater 3d, of one greater Tone and one less Tone: the 4th, of one greater Tone, one less Tone, and one semitone: and the 5th, of two greater Tones, one less Tone, and one semitone.

TONSTALL (CUTHBERT), a learned English divine and mathematician, was born in the year 1476. He entered a student at the university of Oxford about the year 1491; but afterwards, being driven from thence by the plague, he went to Cambridge, and shortly after to the university of Padua in Italy, which was then in a flourishing state of literature, where his genius and learning acquired him great respect from every one, particularly for his knowledge in mathematics, philosophy, and jurisprudence.

Upon his return home, he met with great favours from the government, obtaining several church preferments, and the office of secretary to the cabinet of the king, Henry the 8th. This prince, having also employed him on several foreign embassies, was so well satisfied with his conduct, that he first gave him the bishopric of London in 1522, and afterwards that of Durham in 1530.

Tonstall approved at first of the dissolution of the marriage of his benefactor with Catherine of Spain, and even wrote a book in favour of that dissolution; but he afterwards condemned that work, and experi-

enced a great reverse of fortune. He was ejected from the see of Durham for his religion in the time of Edward the 6th, to which however he was restored again by queen Mary in the beginning of her reign, but was again expelled in 1559 when queen Elizabeth was settled in her throne, and he died in a prison a few months after, in the 84th year of his age.

Tonstall was doubtless one of the most learned men of his time. " He was, says Wood, a very good Grecian and Ebritian, an eloquent rhetorician, a skilful mathematician, a noted civilian and canonist, and a profound divine. But that which maketh for his greatest commendation, is, that Erasmus was his friend, and he a fast friend to Erasmus, in an epistle to whom from Sir Thomas More, I find this character of Tonstall, that, " As there was no man more adorned with knowledge and good literature, no man more severe and of greater integrity for his life and manners; so there was no man a more sweet and pleasant companion, with whom a man would rather choose to converse. "

His writings that were published, were chiefly the following :

1. *In Laudem Matrimonii*, Lond. 1518, 4to.—But that for which he is chiefly entitled to a place in this work, was his book upon arithmetic, viz,

2. *De Arte Supputandi*, Lond 1522, 4to, dedicated to Sir Thomas More. This was afterwards several times printed abroad.

3. A Sermon on Palm Sunday before king Henry the 8th, &c. Lond. 1539 and 1633, 4to.

4. *De Veritate Corporis & Sanguinis Domini in Eucharistia.* Lutet. 1554, 4to.

5. *Compendium in decem Libros Ethicorum Aristotelis.* Par. 1554, in 8vo.

6. *Contra impi Blasphematores Dei predestinationis opera.* Antw. 1555, 4to.

7. Godly and devout Prayers in English and Latin. 1558, in 8vo.

TOPOGRAPHY, is a description or draught of some particular place, or small tract of land; as that of a city or town, manor or tenement, field, garden, house, castle, or the like; such as surveyors set out in their plots, or make draughts of, for the information and satisfaction of the proprietors.

Topography differs from Chorography, as a particular from a more general.

TORNADO, a sudden and violent gust of wind arising suddenly from the shore, and afterwards veering round all points of the compass like a hurricane; very frequent on the coast of Guinea.

TORRENT, in Hydrography, a temporary stream of water, falling suddenly from mountains, &c, where there have been great rains, or an extraordinary thaw of snow; sometimes making great ravages in the plains.

TORRICELLI (EVANGELISTE), an illustrious mathematician and philosopher of Italy, was born at Faenza in 1608, and trained up in Greek and Latin literature by an uncle, who was a monk. Natural inclination led him to cultivate mathematical knowledge, which he pursued some time without a master; but at about 20 years of age, he went to Rome, where he continued

continued the purfuit of it under father Benedict Caftelli. Caftelli had been a fcholar of the great Galileo, and had been appointed by the pope profeffor of mathematics at Rome. Torricelli made fuch progrefs under this mafter, that having read Galileo's *Dialogues*, he compofed a *Treatife concerning motion* upon his principles. Caftelli, furprifed at the performance, carried it and read it to Galileo, who heard it with great pleafure, and conceived a high efteem and friendfhip for the author. Upon this, Caftelli propofed to Galileo, that Torricelli fhould come and live with him; recommending him as the moft proper perfon he could have, fince he was the moft capable of comprehending thofe fublime fpeculations, which his own great age, infirmities, and want of fight, prevented him from giving to the world. Galileo accepted the propofal, and Torricelli the employment, as things of all others the moft advantageous to both. Galileo was at Florence, at which place Torricelli arrived in 1641, and began to take down what Galileo dictated, to regulate his papers, and to act in every refpect according to his directions. But he did not long enjoy, the advantages of this fituation, as Galileo died at the end of only three months.

Torricelli was then about returning to Rome; but the Grand Duke engaged him to continue at Florence, making him his own mathematician for the prefent, and promifing him the profeffor's chair as foon as it fhould be vacant.

Here he applied himfelf intenfely to the ftudy of mathematics, phyfics, and aftronomy, making many improvements and fome difcoveries. Among others, he greatly improved the art of making microfcopes and telefcopes; and it is generally acknowledged that he firft found out the method of afcertaining the weight of the atmofphere by a proportionate column of quickfilver, the barometer being called from him the *Torricellian tube*, and *Torricellian experiment*. In fhort, great things were expected from him, and great things would probably have been farther performed by him, if he had lived: but he died, after a few days illnefs, in 1647, when he was but juft entered the 40th year of his age.

Torricelli publifhed at Florence in 1644, a volume of ingenious pieces, intitled, *Opera Geometrica*, in 4to. There was alfo publifhed at the fame place, in 1715, *Lezzioni Accademiche*, confifting of 96 pages in 4to. Thefe are difcourfes that had been pronounced by him upon different occafions. The firft of them was to the academy of La Crufca, by way of thanks for admitting him into their body. The reft are upon fubjects of mathematics and phyfics. Prefixed to the whole is a long life of Torricelli by Thomas Buonaventuri, a Florentine gentleman.

TORRICELLIAN, a term very frequent among phyfical writers, ufed in the phrafes, *Torricellian tube*, or *Torricellian experiment*, on account of the inventor Torricelli, a difciple of the great Galileo.

TORRICELLIAN *Tube*, is the barometer tube, being a glafs tube, open at one end, and hermetically fealed at the other, about 3 feet long, and $\frac{1}{10}$ of an inch in diameter.

TORRICELLIAN *Experiment*, or the filling the baro-meter tube, is performed by filling the Torricellian tube with mercury, then ftopping the open orifice with the finger, inverting the tube, and plunging that orifice into a veffel of ftagnant mercury. This done, the finger is removed, and the tube fuftained perpendicular to the furface of the mercury in the veffel.

The confequence is, that part of the mercury falls out of the tube into the veffel, and there remains only enough in the tube to fill about 30 inches of its capacity, above the furface of the ftagnant mercury in the veffel; thefe being fuftained in the tube by the preffure of the atmofphere on the furface of the ftagnant mercury; and according as the atmofphere is more or lefs heavy, or as the winds, blowing upward or downward, heave up or deprefs the air, and fo increafe or diminifh its weight and fpring, more or lefs mercury is fuftained, from 28 to 31 inches.

The Torricellian Experiment conftitutes what we now call the *Barometer*.

TORRICELLIAN *Vacuum*, is the vacuum produced by filling a tube with mercury, and when inverted allowing it to defcend to fuch a height as is counterbalanced by the preffure of the atmofphere, as in the Torricellian Experiment and Barometer, the vacuum being that part of the tube above the furface of the mercury.

TORRID *Zone*, is that round the middle of the earth, extending to $23\frac{1}{2}$ degrees on both fides of the equator.

TORUS, or TORE, in Architecture, is a large round moulding in the bafes of the columns.

TOUCAN, or *American Goofe*, is one of the modern conftellations of the fouthern hemifphere, confifting of 9 fmall ftars.

TRACTION, or *Drawing*, is the act of a moving power, by which the moveable is brought nearer to the mover, called alfo attraction.

TRACTRIX, in Geometry, a curve line called alfo Catenaria; which fee.

TRAJECTORY, a term often ufed generally for the path of any body moving either in a void, or in a medium that refifts its motion; or even for any curve paffing through a given number of points. Thus Newton, Princip. lib. 1, prob. 22, propofes to defcribe a Trajectory that fhall pafs through five given points.

TRAJECTORY *of a Comet*, is its path or orbit, or the line it defcribes in its motion. This path, Hevelius, in his Cometographia, will have to be very nearly a right line; but Dr. Halley concludes it to be, as it really is, a very excentric ellipfis; though its place may often be well computed on the fuppofition of its being a parabola.—Newton, in prop. 41 of his 3d book, fhews how to determine the Trajectory of a comet from three obfervations; and in his laft prop. how to correct a Trajectory graphically defcribed.

TRAMMELS, in Mechanics, an inftrument ufed by artificers for drawing ovals upon boards, &c. One part of it confifts of a crofs with two grooves at right angles; the other is a beam carrying two pins which flide in thofe grooves, and alfo the defcribing pencil. All the engines for turning ovals are conftructed on the fame principles with the Trammels: the only difference is, that in the Trammels the board is at reft, and the pen-

cil

cil moves upon it : in the turning engine, the tool, which supplies the place of the pencil, is at rest, and the board moves against it. See a demonstration of the chief properties of these instruments by Mr. Ludlam, in the Philof. Tranf. vol. 70, pa. 378 &c.

TRANSACTIONS, *Philosophical,* are a collection of the principal papers and matters read before certain philosophical societies, as the Royal Society of London, and the Royal Society of Edinburgh. These Transactions contain the several discoveries and histories of nature and art, either made by the members of those societies, or communicated by them from their correspondents, with the various experiments, observations, &c, made by them, or transmitted to them, &c.

The Philof. Tranf. of the Royal Society of London were set on foot in 1665, by Mr. Oldenburg, the then secretary of that Society, and were continued by him till the year 1677. They were then discontinued upon his death, till January 1678, when Dr. Grew resumed the publication of them, and continued it for the months of December 1678, and January and February 1679, after which they were intermitted till January 1683. During this last interval their want was in some measure supplied by Dr. Hook's Philosophical Collections. They were also interrupted for 3 years, from December 1687 to January 1691, beside other smaller interruptions amounting to near a year and a half more, before October 1695, since which time the Transactions have been carried on regularly to the present day, with various degrees of credit and merit.

Till the year 1752 these Transactions were published in numbers quarterly, and the printing of them was always the single act of the respective secretaries till that time ; but then the society thought fit that a committee should be appointed to consider the papers read before them, and to select out of them such as they should judge most proper for publication in the future Transactions. For this purpose the members of the council for the time being, constitute a standing committee : they meet on the first Thursday of every month, and no less than seven of the members of the committee (of which number the president, or in his absence a vice president, is always to be one) are allowed to be a *quorum,* capable of acting in relation to such papers ; and the question with regard to the publication of any paper, is always decided by the majority of votes taken by ballot.

They are published annually in two parts, at the expence of the society ; and each fellow, or member, is entitled to receive one copy *gratis* of every part published after his admission into the society. For many years past, the collection, in two parts, has made one volume in each year ; and in the year 1793 the number of the volumes was 83, being 10 less than the number of the year in the century. They were formerly much respected for the great number of excellent papers and discoveries contained in them ; but within the last dozen years there has been a great falling off, and the volumes are now considered as of very inferior merit, as well as quantity.

There is also a very useful Abridgment, of those volumes of the Transactions that were published before the year 1752, when the society began to publish the Transactions on their own account. Those to the end of the year 1700 were abridged, in 3 volumes, by Mr. John Lowthorp : those from the year 1700 to 1720 were abridged, in 2 volumes, by Mr. Henry Jones : and those from 1719 to 1733 were abridged, in 2 volumes, by Mr. John Eames and Mr. John Martyn ; Mr. Martyn also continued the abridgment of those from 1732 to 1744 in 2 volumes, and of those from 1744 to 1750 in 2 volumes ; making in all 11 volumes, of very curious and useful matters in all the arts and sciences.

The Royal Society of Edinburgh, instituted in 1783, have also published 3 volumes of their Philosophical Transactions ; which are deservedly held in the highest respect for the importance of their contents.

TRANSCENDENTAL *Quantities,* among Geometricians, are indeterminate ones ; or such as cannot be expressed or fixed to any constant equation : such is a transcendental curve, or the like.

M. Leibnitz has a dissertation in the Acta Erud. Lipf. in which he endeavours to shew the origin of such quantities ; viz, why some problems are neither plain, solid, nor surfolid, nor of any certain degree, but do *transcend* all algebraic equations.

He also shews how it may be demonstrated without calculus, that an algebraic quadratrix for the circle or hyperbola is impossible : for if such a quadratrix could be found, it would follow, that by means of it any angle, ratio, or logarithm, might be divided in a given proportion of one right line to another, and this by one universal construction : and consequently the problem of the section of an angle, or the invention of any number of mean proportionals, would be of a certain finite degree. Whereas the different degrees of algebraic equations, and therefore the problem understood in general of any number of parts of an angle, or mean proportionals, is of an indefinite degree, and *transcends* all algebraical equations.

Others define Transcendental equations, to be such fluxional equations as do not admit of fluents in common finite algebraical equations, but as expressed by means of some curve, or by logarithms, or by infinite series ; thus the expression $\dot{y} = \dfrac{\dot{x}}{\sqrt{aa - xx}}$ is a Transcendental equation, because the fluents cannot both be expressed in finite terms. And the equation which expresses the relation between an arc of a circle and its sine is a Transcendental equation ; for Newton has demonstrated that this relation cannot be expressed by any finite algebraic equation, and therefore it can only be by an infinite or a Transcendental equation.

It is also usual to rank exponential equations among Transcendental ones ; because such equations, although expressed in finite terms, have variable exponents, which cannot be expunged but by putting the equation into fluxions, or logarithms, &c. Thus, the exponential equation

equation $y = a^x$, gives $\dot{x} \times$ log. $a =$ log. y, or $\dot{x} \times$ log. $a = \dfrac{\dot{y}}{y}$.

TRANSCENDENTAL *Curve*, in the Higher Geometry, is such a one as cannot be defined by an algebraic equation; or of which, when it is expressed by an equation, one of the terms is a variable quantity, or a curve line. And when such curve line is a geometrical one, or one of the first degree or kind, then the Transcendental curve is said to be of the second degree or kind, &c.

These curves are the same with what Des Cartes, and others after him, call mechanical curves, and which they would have excluded out of geometry; contrary however to the opinion of Newton and Leibnitz; for as much as, in the construction of geometrical problems, one curve is not to be preferred to another as it is defined by a more simple equation, but as it is more easily described than that other: besides, some of these Transcendental, or mechanical curves, are found of greater use than almost all the algebraical ones.

M. Leibnitz, in the Acta Erudit. Lipf. has given a kind of Transcendental equations, by which these Transcendental curves are actually defined, and which are of an indefinite degree, or are not always the same in every point of the curve. Now whereas algebraists use to assume some general letters or numbers for the quantities sought, in these Transcendental problems Leibnitz assumes general or indefinite equations for the lines sought; thus, for example, putting x and y for the abscifs and ordinate, the equation he uses for a line required, is $a + bx + cy + cxy + fxx + gyy$ &c. $= 0$: by the help of which indefinite equation, he seeks for the tangent; and comparing that which results with the given property of tangents, he finds the value of the assumed letters a, b, c, &c. and thus defines the equation of the line sought.

If the comparison abovementioned do not succeed, he pronounces the line sought not to be an algebraical, but a Transcendental one.

This supposed, he proceeds to find the species of Transcendency: for some Transcendentals depend on the general division or section of a ratio, or upon logarithms, others upon circular arcs, &c.

Here then, beside the symbols x and y, he assumes a third, as v, to denote the Transcendental quantity; and of these three he forms a general equation of the line sought, from which he finds the tangent according to the differential method, which succeeds even in Transcendental quantities. This found, he compares it with the given properties of the tangents, and so discovers not only the values of a, b, c, &c. but also the particular nature of the Transcendental quantity.

Transcendental problems are very well managed by the method of fluxions. Thus, for the relation of a circular arc and right line, let a denote the arc, and x the versed sine, to the radius 1; then is $a =$ fluent of $\dfrac{\dot{x}}{\sqrt{2x - xx}}$; and if the ordinate of a cycloid be y; then is $y = \sqrt{2x - xx} +$ fluent of $\dfrac{\dot{x}}{\sqrt{2x - xx}}$.

Thus is the analytical calculus extended to those lines which have hitherto been excluded, for no other cause but that they were thought incapable of it.

TRANSFORMATION, in Geometry, is the changing or reducing of a figure, or of a body, into another of the same area, or the same solidity, but of a different form. As, to Transform or reduce a triangle to a square, or a pyramid to a parallelopipedon.

TRANSFORMATION *of Equations*, in Algebra, is the changing equations into others of a different form, but of equal value. This operation is often necessary, to prepare equations for a more easy solution, some of the principal cases of which are as follow.—1. The signs of the roots of an equation are changed, viz, the positive roots into negative, and the negative roots into positive ones, by only changing the signs of the 2d, 4th, and all the other even terms of the equation. Thus, the roots of the equation

$x^4 - x^3 - 19x^2 + 49x - 30 = 0$, are $+1, +2, +3, -5$;

whereas the roots of the same equation having only the signs of the 2d and 4th terms changed, viz, of

$x^4 + x^3 - 19x^2 - 49x - 30 = 0$, are $-1, -2, -3, +5$.

2. To Transform an equation into another that shall have its roots greater or less than the roots of the proposed equation by some given difference, proceed as follows. Let the proposed equation be the cubic $x^3 - ax^2 + bx - c = 0$; and let it be required to Transform it into another, whose roots shall be less than the roots of this equation by some given difference d; if the root y of the new equation must be the less, take it $y = x - d$, and hence $x = y + d$; then instead of x and its powers substitute $y + d$ and its powers, and there will arise this new equation

$$(A) \quad \left. \begin{array}{rrr} y^3 + 3dy^2 + 3d^2y + & d^3 \\ - \ ay^2 - 2ady - & ad^2 \\ + & by + & bd \\ - & c \end{array} \right\} = 0,$$

whose roots are less than the roots of the former equation by the difference d. If the roots of the new equation had been required to be greater than those of the old one, we must then have substituted $y = x + d$, or $x = y - d$, &c.

3. To take away the 2d or any other particular term out of an equation; or to Transform an equation, so as the new equation may want its 2d, or 3d, or 4th, &c term of the given equation $x^3 - ax^2 + bx - c = 0$, which is transformed into the equation (A) in the last article. Now to make any term of this equation (A) vanish, is only to make the coefficient of that term $= 0$, which will form an equation that will give the value of the assumed quantity d, so as to produce the desired effect, viz, to make that term vanish. So, to take away the 2d term, make $3d - a = 0$, which makes the assumed quantity $d = \frac{1}{3}a$. To take away the 3d term, we must put the sum of the coefficients of that term $= 0$, that is $3d^2 - 2ad + b = 0$, or $3d^2 - 2ad = -b$; then by resolving this quadratic equation, there is found the assumed quantity $d = \frac{1}{3}a \pm \frac{1}{3}\sqrt{a^2 - 3c}$, by the substitution of which for d, the 3d term will be taken away out of the equation.

In like manner, to take away the 4th term, we must make the sum of its coefficients $d^3 - ad^2 + bd - c = 0$; and

and so on for any other term whatever. And in the same manner we must also proceed when the proposed equation is not a cubic, but of any height whatever, as

$$x^n - ax^{n-1} + bx^{n-2} - cx^{n-3} \ \&c = 0:$$

this is first, by substituting $y + d$ for x, to be Transformed to this new equation

$$\left.\begin{array}{l} y^n + ndy^{n-1} + \frac{1}{2}n.\overline{n-1}.d^2y^{n-2} \ \&c \\ \quad - ay^{n-1} - a.\overline{n-1}.dy^{n-2} \ \&c \\ \qquad + by^{n-2} \ \&c \end{array}\right\} = 0;$$

then, to take away the 2d term, we must make $nd - a = 0$, or $d = \frac{a}{n}$; to take away the 3d term, we must make $\frac{1}{2}n.\overline{n-1}.d^2 - a.\overline{n-1}d + b = 0$, or $d^2 - \frac{2a}{n}d = -\frac{2b}{n(n-1)}$; and so on.

From whence it appears that, to take away the 2d term of an equation, we must resolve a simple equation; for the 3d term, a quadratic equation; for the 4th term, a cubic equation, and so on.

4. To multiply or divide the roots of an equation by any quantity; or to Transform a given equation to another, that shall have its roots equal to any multiple or submultiple of those of the proposed equation. This is done by substituting, for x and its powers, $\frac{y}{m}$ or py, and their powers, viz. $\frac{y}{m}$ for x, to multiply the roots by m; and py for x, to divide the roots by p.

Thus, to multiply the roots by m, substituting $\frac{y}{m}$ for x in the proposed equation

$$x^n - ax^{n-1} + bx^{n-2} \ \&c = 0, \text{ and it becomes}$$

$$\frac{y^n}{m^n} - \frac{ay^{n-1}}{m^{n-1}} + \frac{by^{n-2}}{m^{n-2}} \ \&c = 0;$$

or multiply all by m^n, then is

$$y^n - amy^{n-1} + bm^2y^{n-2} - cm^3y^{n-3} \ \&c = 0,$$

an equation that hath its roots equal to m times the roots of the proposed equation.

In like manner, substituting py for x, in the proposed equation, &c, it becomes

$$y^n - \frac{ay^{n-1}}{p} + \frac{by^{n-2}}{p^2} - \frac{cy^{n-3}}{p^3} \ \&c = 0,$$

an equation that hath its roots equal to those of the proposed equation divided by p.

From whence it appears, that to multiply the roots of an equation by any quantity m, we must multiply its terms, beginning at the 2d term, respectively by the terms of the geometrical series, m, m^2, m^3, m^4, &c. And to divide the roots of an equation by any quantity p, that we must divide its terms, beginning at the 2d, by the corresponding terms of this series p, p^2, p^3, p^4, &c.

5. And sometimes, by these Transformations, equations are cleared of fractions, or even of surds. Thus the equation

$$x^3 - ax^2\sqrt{p} + bx - c\sqrt{p} = 0, \text{ by putting } y = x\sqrt{p},$$

or multiplying the terms, from the 2d, by the geometricals $\sqrt{p}$, p, $p\sqrt{p}$, is Transformed to

$$y^3 - apy^2 + bpy - cp^2 = 0.$$

6. An equation, as $x^3 - ax^2 + bx - c = 0$, may be Transformed into another, whose roots shall be the reciprocals of the roots of the given equation, by substituting $\frac{1}{y}$ for x; by which it becomes

$$\frac{1}{y^3} - \frac{a}{y^2} + \frac{b}{y} - c = 0, \text{ or, multiplying all by } y^3, \text{ the}$$

same becomes $cy^3 - by^2 + ay - 1 = 0$.

On this subject, see Newton's Alg. on the Transmutation of Equations; Maclaurin's Algeb. pt. 2, chap. 3 and 4. Saunderson's Algebra, vol. 2, pa. 687, &c.

TRANSIT, in Astronomy, denotes the passage of any planet, just before or over another planet or star; or the passing of a star or planet over the meridian, or before an astronomical instrument.

Venus and Mercury, in their Transits over the sun, appear like dark specks.

Doctor Halley computed the times of a number of these visible Transits, for the last and present century, and published m the Philos. Transf. numb. 193. See also Abridg. vol. 1, pa. 427 &c. A Synopsis of these Transits is as follows, those of Mercury happening in the months of April and October, and those of Venus in May and November, both old-style; and if 11 days be added to the dates below, the sums will give the times for the new-style. First for Mercury, and then for Venus.

A Series of the Moments when Mercury is seen in Conjunction with the Sun, and within his Disc, with the Distances of the same Planet from the Sun's Centre.

In April, Old-Style.

Years.	Times of Mercury's Conjunction.			Distances from the the Sun's Centre.		
	d.	h.	min.	′	″	
1615	22	21	38 ✳	7	20	N
1628	25	5	15 ✳	9	35	S
1661	23	4	52 ✳	4	27	N
1674	26	12	29	12	28	S
1707	24	12	6	1	34	N
1720	26	19	43 ✳	15	21	S
1740	21	11	43	15	36	N
1758	24	19	20 ✳	1	19	S
1786	22	18	57 ✳	12	43	N
1799	26	2	34 ✳	4	12	S

Years.	Times of Mercury's Conjunction.			Distances from the Sun's Centre.		
	d.	h.	m.	′	″	
1605	22	8	29	12	48	S
1618	25	2	4*	4	45	S
1631	27	19	37*	3	18	N
1644	30	13	11	11	21	N
1651	23	13	20	11	26	S
1664	25	6	54*	3	23	S
1677	28	0	28**	4	40	N
1690	30	18	2*	12	43	N
1697	23	18	11*	10	4	S
1710	26	11	45	2	1	S
1723	29	5	19*	6	2	N
1730	22	5	28	16	45	S
1736	30	22	53**	13	5	N
1743	24	23	2**	8	42	S
1756	26	16	36	0	38	S
1769	29	10	10	7	24	N
1776	22	10	19	15	23	S

In October, Old-Style.

November

1782	1	3	44*	15	27	N

October

1789	25	3	53*	7	20	S

April.

Distance in Min.	Half duration.	
′	h.	m.
0	4	0½
1	4	0
2	3	58½
3	3	56
4	3	53
5	3	48¼
6	3	43
7	3	36
8	3	28
9	3	18½
10	3	7
11	2	54
12	2	38
13	2	19
14	1	55
15	1	21½
15½	0	56

October.

Distance in Min.	Half duration.	
′	h.	m.
0	2	44½
1	2	44
2	2	43
3	2	41½
4	2	39½
5	2	36½
6	2	33
7	2	28½
8	2	23
9	2	17
10	2	10
11	2	1
12	1	51
13	1	39
14	1	24
15	1	4
15½	0	50
16	0	30

" Those Transits which have the mark *, are but partly visible at London; but those which are marked **, are totally visible.

" Now it is to be observed, that at the ascending node of Mercury in the month of October, the diameter of the sun takes up 32′ 34″, and therefore the greatest duration of a central Transit is 5h 29m. But in the month of April the diameter of the sun is 31′ 54″, whence by reason of the slower motion of the planet, there arises the greatest duration 8h 1m. Now if Mercury approaches obliquely, these durations become shorter on account of the distance from the centre of the sun. And that the calculation may be more perfect, I have added the following Tables, in which are exhibited the half durations of these eclipses, to every minute of the distance seen from the centre of the sun. These added to or subtracted from the moment of conjunction found by the foregoing Table, give the beginning and end of the whole phenomenon."

Of the Visible Conjunction of Venus with the Sun.

" Though Venus is the most beautiful of all the stars, yet (says Dr. Halley) like the rest of her sex, she does not care to appear in sight without her borrowed ornaments, and her assumed splendor. For the confined laws of motion envy this spectacle to the mortals of a whole age, like the secular games of the Ancients; though it be far the most noble among all those that astronomy can pretend to shew. Now it shall be declared hereafter, that by this one observation alone, the distance of the sun from the earth may be determined with the greatest certainty which hitherto has been included within wide limits, because of the parallax which is otherwise insensible. But as to the periods, they cannot be described so accurately as those of Mercury, since Venus has been observed within the sun's disk but once since the beginning of the world, and that by our Horrox." Dr. Halley then exhibits the principles of calculating these Transits, from whence he infers that,

" After 18 years Venus returns to the sun, taking away 2d 10h 52½m, from the moment of the foregoing Transit; and the planet proceeds in a path which is 24′ 41″ more to the south than the former.

" After 235 years adding 2d 10h 9m, Venus may again enter the sun, but in a more northern path by 11′ 33″. But if the foregoing year is bissextile, 3d 10 9m must be added.

" After 243 years Venus may also pass the sun, only taking away 0h 43m from the time of the former; but

but proceeds more foutherly by 13′ 8″. Now if the foregoing year be biffextile, add 23ʰ 17ᵐ.

" And in all thefe appulfes of Venus to the fun, in the month of November, the angle of her path with the ecliptic is 9° 5′, and her horary motion within the fun is 4′ 7″. And fince the femidiameter of the fun is 16′ 21″, the greateft duration of the Tranfit of the centre of Venus comes out 7ʰ 56ᵐ.

" Then let the fun and Venus be in conjunction at the defcending node in the month of May ; and by the fame numbers the fame intervals may be computed. After 8 years let there be taken away 2ᵈ 6ʰ 55′. And Venus will make her Tranfit in a more northern path by 19′ 58″.

" After 235 years add 2ᵈ 8ʰ 18ᵐ, or if the foregoing year be biffextile 3ᵈ 8ʰ 18ᵐ, and you will have Venus more to the South by 9′ 21″.

" Laftly, after 243 years add 0ᵈ 1ʰ 23ᵐ, or if the foregoing year be biffextile 1ᵈ 1ʰ 23ᵐ, and Venus will be found in conjunction with the fun, but in a more northerly path by 10′ 37″.

" In every Tranfit within the fun at this node, the angle of Venus's path with the ecliptic is 8° 28′, and her horary motion is 4′ 0″; and the femidiameter of the fun fubtending 15′ 51″, the greateft duration of the central Tranfit comes out alfo 7ʰ 56ᵐ, exactly the fame as at the other node.

" As to the epochs, from that only ingrefs which Horrox obferved, the fun being then juft ready to fet, it is concluded, that Venus was in conjunction with the fun at London in the year 1639, Nov. 24ᵈ 6ʰ 37ᵐ, and that fhe declined towards the fouth 8′ 30″. But in the month of May no mortal has feen her as yet within the fun. But from my numbers, which I judge to be not very different from the heavens, it appears that Venus for the next time will enter the fun in 1761, May 25ᵈ 17ʰ 55ᵐ, that being the middle of the eclipfe, and then will be diftant from his centre 4′ 15″, towards the fouth. Hence and from the foregoing revolutions all the phenomena of this kind will be eafily exhibited for a whole millennium, as I have computed them in the following Table.

In November.

Years.	Times of Conjunction.			Diftance from the Sun's Centre.		
	d.	h.	m.	′	″	
918	20	21	53	6	12	N
1161	20	21	10	6	55	S
1396	23	7	20	4	38	N
1631	26	17	29	16	11	N
1639	24	6	37	8	30	S
1874	26	16	46	3	3	N
2109	29	2	57	14	36	N
2117	26	16	3	10	5	S

In May.

Years.	Times of Conjunction.			Diftance from the Sun's Centre.		
	d.	h.	m.	′	″	
1048	24	13	45	3	50	N
1283	23	8	14	5	31	S
1291	25	15	9	14	27	N
1518	25	16	32	14	52	S
1526	23	9	37	5	6	N
1761	25	17	55	4	15	S
1769	23	11	0	15	43	N
1996	28	2	13	13	36	S
2004	25	19	18	6	22	N

" As for the durations of thefe eclipfes of Venus, they may be computed after the fame manner as thofe of Mercury in refpect of the centre. But fince Venus's diameter is pretty large, and fince the parallaxes alfo may bring a very notable difference as to time, a particular calculation muft neceffarily be made for every place.

" Now the diameter of Venus is fo great, that while fhe adheres to the fun's limb almoft 20 minutes of time will be elapfed, that is, when fhe applies directly to the fun. But when fhe is incident obliquely, fhe continues longer in the limb. Now that diameter, according to Horrox's obfervation, takes up 1′ 18″, when fhe is in conjunction with the fun at the afcending node, and 1′ 12″ at the other node.

" Now the chief ufe of thefe conjunctions is accurately to determine the fun's diftance from the earth, or his parallax, which aftronomers have in vain attempted to find by various other methods ; for the minutenefs of the angles required eafily eludes the niceft inftruments. But in obferving the ingrefs of Venus into the fun, and her egrefs from the fame, the fpace of time between the moments of the internal contacts, obferved to a fecond of time, that is, to $\frac{1}{15}$ of a fecond or 4‴ of an arch, may be obtained by the affiftance of a moderate telefcope and a pendulum clock, that is confiftent with itfelf exactly for 6 or 8 hours. Now from two fuch obfervations rightly made in proper places, the diftance of the fun within a 500th part may be certainly concluded, &c." See PARALLAX.

TRANSIT *Inftrument*, in Aftronomy, is a telefcope fixed at right angles to a horizontal axis ; this axis being fo fupported that the line of collimation may move in the plane of the meridian.

The axis, to the middle of which the telefcope is fixed, fhould gradually taper toward its ends, and terminate in cylinders well turned and fmoothed ; and a proper weight or balance is put on the tube, fo that it may ftand at any elevation when the axis refts on the fupporters. Two upright pofts of wood or ftone, firmly fixed at a proper diftance, are to fuftain the fupporters to this inftrument ; thefe fupporters are two thick brafs

plates,

plates, having well fmoothed angular notches in their upper ends to receive the cylindrical arms of the axis; each of the notched plates is contrived to be moveable by a fcrew, which flides them upon the furfaces of two other plates immoveably fixed to the two upright pofts; one plate moving in a vertical direction, and the other horizontally, they adjuft the telefcope to the planes of the horizon and meridian; to the plane of the horizon, by a fpirit level hung in a pofition parallel to the axis, and to the plane of the meridian in the following manner. Obferve the times by the clock when a circumpolar ftar, feen through this inftrument, Tranfits both above and below the pole; then if the times of defcribing the eaftern and weftern parts of its circuit be equal, the telefcope is then in the plane of the meridian; otherwife the notched plates muft be gently moved till the time of the ftar's revolution is bifected by both the upper and lower Tranfits, taking care at the fame time that the axis keeps its horizontal pofition.

When the telefcope is thus adjufted, a mark muft be fet up, or made, at a confiderable diftance (the greater the better) in the horizontal direction of the interfection of the crofs wires, and in a place where it can be illuminated in the night-time by a lanthorn near it, which mark, being on a fixed object, will ferve at all times afterwards to examine the pofition of the telefcope, by firft adjufting the tranverfe axis by the level.

To adjuft a clock by the fun's Tranfit over the meridian, note the times by the clock, when the preceding and following edges of the fun's limb touch the crofs wires: the difference between the middle time and 12 hours, fhews how much the mean, or clock time, is fafter and flower than the apparent or folar time, for that day; to which the equation of time being applied, it will fhew the time of mean noon for that day, by which the clock may be adjufted.

TRANSMISSION, in Optics, &c, denotes the property of a tranfparent or tranflucent body, by which it admits the rays of light to pafs through its fubftance; in which fenfe, the word ftands oppofed to reflection.

For the caufe of Tranfmiffion, or the reafon why fome bodies Tranfmit the rays, and others reflect them, fee TRANSPARENCY and OPACITY.

The rays of light, Newton obferves, are fubject to fits of eafy Tranfmiffion and reflection. See LIGHT, and REFLECTION.

TRANSMUTATION, or TRANSFORMATION, in Geometry, denotes the reduction or change of one figure or body into another of the fame area or folidity; as a triangle into a fquare, a pyramid into a cube, &c.

TRANSMUTATION, in the Higher Geometry, has been ufed for the converting of a figure into another of the fame kind and order, whofe refpective parts rife to the fame dimenfions in an equation, and admit the fame tangents, &c.

If a rectilineal figure be to be Tranfmuted into another, it is fufficient that the interfections of the lines which compofe it be transferred, and lines drawn through the fame in the new figure. But if the figure to be Tranfmuted be curvilinear, the points, tangents, and

other right lines, by means of which the curve line is to be defined, muft be transferred.

TRANSOM, among Builders, the piece that is framed acrofs a double light window.

TRANSOM, among Mathematicians, denotes the vane of a crofs-ftaff; being a wooden member fixed acrofs it, with a fquare upon which it flides, &c.

TRANSPARENCY, or TRANSLUCENCY, in Phyfics, a quality in certain bodies, by which they give paffage to the rays of light.

The Tranfparency of natural bodies, as glafs, water, air, &c, is afcribed by fome, to the great number and fize of the pores or interftices between the particles of thofe bodies. But this account is very defective; for the moft folid and opaque body in nature, that we know of, contains a great deal more pores than it does matter; furely a great deal more than is neceffary for the paffage of fo very fine and fubtle a body as light.

Ariftotle, Des Cartes, &c, make Tranfparency to confift in ftraightnefs or rectilineal direction of the pores; by means of which, fay they, the rays can make their way through, without ftriking againft the folid parts, and fo being reflected back again. But this account, Newton fhews, is imperfect; the quantity of pores in all bodies being fufficient to tranfmit all the rays that fall upon them, however thofe pores be fituated with refpect to each other.

The reafon then why all bodies are not Tranfparent, is not to be afcribed to their want of rectilineal pores; but either to the unequal denfity of the parts, or to the pores being filled with fome foreign matters, or to their being quite empty, by means of which the rays, in paffing through, undergoing a great variety of reflections and refractions, are perpetually diverted different ways, till at length falling on fome of the folid parts of the body, they are extinguifhed and abforbed.

Thus cork, paper, wood, &c, are opake; while glafs, diamonds, &c, are Tranfparent; and the reafon is, that in the neighbourhood of parts equal in denfity with refpect to each other, as thefe latter bodies, the attraction being equal on every fide, no reflection or refraction enfues: but the rays which entered the firft furface of the body proceed quite through it without interruption, thofe few only excepted that chance to meet with the folid parts: but in the neighbourhood of parts that differ much in denfity, fuch as the parts of wood and paper are, both in refpect of themfelves and of the air, or the empty fpace in their pores; as the attraction is very unequal, the reflections and refractions muft be very great; and therefore the rays will not be able to make their way through fuch bodies, but will be varioufly deflected, and at length quite ftopped. See OPACITY.

TRANSPOSITION, in Algebra, is the bringing any term of an equation over to the other fide of it. Thus, if $a + x = c$, and you make $x = c - a$, then a is faid to be Tranfpofed.

This operation is to be performed in order to bring all the known terms to one fide of the equation, and all thofe that are unknown to the other fide of it; and every term thus Tranfpofed muft always have its fign

changed,

changed, from + to —, or from — to + ; which in fact is no more than subtracting or adding such term on both sides of the equation. See REDUCTION of Equations.

TRANSVERSE-*Axis*, or *Diameter*, in the Conic Sections, is the first or principal diameter, or axis. See AXIS, DIAMETER, and LATUS TRANSVERSUM.

In an ellipse the Transverse is the longest of all the diameters ; but the shortest of all in the hyperbola ; and in the parabola the diameters are all equal, or at least in a ratio of equality.

TRAPEZIUM, in Geometry, a plane figure contained under four right lines, of which both the opposite pairs are not parallel.—When this figure has two of its sides parallel to each other, it is sometimes called a *trapezoid*.—The chief properties of the Trapezium are as follow :

1. Any three sides of a Trapezium taken together, are greater than the third side.

2. The two diagonals of any Trapezium divide it into four proportional triangles, *a*, *b*, *c*, *d*. That is, the triangle $a : b :: c : d$.

3. The sum of all the four inward angles, A, B, C, D, taken together, is equal to 4 right angles, or 360°.

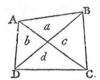 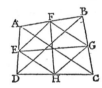

4. In a Trapezium ABCD, if all the sides be bisected, in the points E, F, G, H, the figure EFGH formed by joining the points of bisection will be a parallelogram, having its opposite sides parallel to the corresponding diagonals of the Trapezium, and the area of the said inscribed parallelogram is just equal to half the area of the Trapezium.

5. The sum of the squares of the diagonals of the Trapezium, is equal to twice the sum of the squares of the diagonals of the parallelogram, or of the two lines drawn to bisect the opposite sides of the Trapezium. That is, $AC^2 + BD^2 = 2EG^2 + 2FH^2$.

6. In any Trapezium, the sum of the squares of all the four sides, is equal to the sum of the squares of the two diagonals together with 4 times the square of the line KI joining their middle points. That is, (first fig. below)
$$AB^2 + BC^2 + CD^2 + DA^2 = AC^2 + BD^2 + 4IK^2.$$

 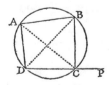

7. In any Trapezium, the sum of the two diago-

nals, is less than the sum of any four lines that can be drawn, to the four angles, from any point within the figure, beside the intersection of the diagonals.

8. The area of any Trapezium, is equal to half the rectangle or product under either diagonal and the sum of the two perpendiculars drawn upon it from the two opposite angles.

9. The area of any Trapezium may also be found thus : Multiply the two diagonals together, then that product, multiplied by the sine of their angle of intersection, to the radius 1, will be the area. That is,

$$AC \times BD \times \text{sin.} \angle L = \text{area.}$$

10. The same area will be otherwise found thus : Square each side of a Trapezium, add the squares of each pair of opposite sides together, subtract the less sum from the greater, multiply the remainder by the tangent of the angle of intersection of the diagonals (to radius 1), and ¼ of the product will be the area. That is, $(\overline{AB^2 + DC^2} - \overline{AD^2 + BC^2}) \times \frac{1}{4} \text{tang.} \angle L = \text{area.}$

11. The area of a Trapezoid, or one that has two sides parallel, is equal to the rectangle or product under the sum of the two parallel sides and the perpendicular distance between them.

12. If a Trapezium be inscribed in a circle ; the sum of any two opposite angles is equal to two right angles ; and if the sum of two opposite angles be equal to two right angles, the sum of the other two will also be equal to two right angles, and a circle may be described about it ; and farther, if one side, as DC, be produced out, the external angle will be equal to the interior opposite angle. That is, (last fig. above)

$$\angle A + \angle C = \angle B + \angle D = 2 \text{ right angles,}$$
and $\angle A = \angle BCP$.

13. If a Trapezium be inscribed in a circle ; the rectangle of the two diagonals, is equal to the sum of the two rectangles contained under the opposite sides. That is,

$$AC \times BD = AB \times DC + AD \times BC.$$

14. If a Trapezium be inscribed in a circle ; its area may be found thus : Multiply any two adjacent sides together, and the other two sides together ; then add these two products together, and multiply the sum by the sine of the angle included by either of the pairs of sides that are multiplied together, and half this last product will be the area. That is, the area is equal either

to $(AB \times AD + CB \times CD) \times \frac{1}{2} \text{sin.} \angle A$ or $\angle C$, or $(AB \times BC + AD \times DC) \times \frac{1}{2} \text{sin.} \angle B$ or $\angle D$.

15. Or, when the Trapezium can be inscribed in a circle, the area may be otherwise found thus : Add all the four sides together, and take half the sum ; then from this half subtract each side severally ; multiply the four remainders continually together, and the square root of the last product will be the area.

16. Lastly, the area of the Trapezium inscribed in a circle may be otherwise found thus :

Put

Put $m = AB \times BC + AD \times DC$,

$\quad\; n = BA \times AD + BC \times CD$,

$\quad\; p = AB \times DC + AD \times BC$,

$\quad\; r$ = radius of the circumscribing circle,

then $\dfrac{\sqrt{mnp}}{4r}$ = the area of the Trapezium.

TRAPEZOID, sometimes denotes a trapezium that has two of its sides parallel to each other; and sometimes an irregular solid figure, having four sides not parallel to each other.

TRAVERSE, in Gunnery, is the turning a piece of ordnance about, as upon a centre, to make it point in any particular direction.

TRAVERSE, in Fortification, denotes a trench with a little parapet, sometimes two, one on each side, to serve as a cover from the enemy that might come in flank.

TRAVERSE, in a wet fofs, is a fort of gallery, made by throwing faucissons, joists, fafcines, stones, earth, &c, into the fofs, opposite the place where the miner is to be put, in order to fill up the ditch, and make a passage over it.

TRAVERSE also denotes a wall of earth, or stone, raised across a work, to stop the shot from rolling along it.

TRAVERSE also sometimes signifies any retrenchment, or line fortified with fafcines, barrels or bags of earth, or gabions.

TRAVERSE, in Navigation, is the variation or alteration of a ship's course, occasioned by the shifting of the winds, or currents, &c; or a Traverse is a compound course, consisting of several different courses and distances.

TRAVERSE *Sailing*, is the method of working, or calculating, Traverses or compound courses, so as to bring them into one, &c.

Traverse Sailing is used when a ship, having failed from one port towards another, whose course and distance from the former is known, is by reason of contrary winds, or other accidents, forced to shift and fail upon several courses, which are to be brought into one course, to learn, after so many turnings and windings, the true course and distance made good, or the true point the ship is arrived at; and so to know what must be the new course and distance to the intended port.

To Construct a Traverse. Assume a convenient point or centre, to begin at, to represent the place failed from. From that point as a centre, with the chord of 60°, describe a circle, which quarter with two perpendicular lines intersecting in the centre, one to represent the meridian, or north-and south line, and the other the east-and-west line. From the intersections of these lines with the circle, set off upon the circumference, the arcs or degrees, taken from the chords, for the several courses that have been failed upon, marking the points they reach to in the circumference with the figures for the order or number of the courses, 1, 2, 3, 4, &c; and from the centre draw lines to these several points in the circumference, or conceive them to be drawn. Upon the first of these lines lay off the first distance failed; from the extremity of this distance draw a line parallel to the second radius, or line drawn in the circle, upon which lay off the 2d distance; through

the end of this 2d distance draw a line parallel to the 3d radius, for the direction of the 3d course, and upon it lay off the 3d distance; and so on, through all the courses and distances. This done, draw a line from the centre to the end of the last distance, which will be the whole distance made good, and it will cut the circle in a point shewing the course made good. Lastly, draw a line from the end of the last distance to the point representing the port bound to, and it will shew the distance and course yet to be failed, to gain that port.

To work a Traverse, or to compute it by the Traverse Table of Difference of Latitude and Departure.

Make a little tablet with 6 columns; the 1st for the courses, the 2d for the distances, the 3d for the northing, the 4th for the southing, the 5th for the easting, and the 6th for the westing; first entering the several courses and distances, in so many lines, in the 1st and 2d columns. Then, from the Traverse table, take out the quantity of the northings or southings, and eastings or westings, answering to the several given courses and distances, entering them on their corresponding lines, and in the proper columns of easting, westing, northing, and southing. This done, add up into one sum the numbers in each of these last four columns, which will give four sums shewing the whole quantity of easting, westing, northing, and southing made good; then take the difference between the whole easting and westing, and also between the northing and southing, so shall these shew the spaces made good in these two directions, viz, east or west, and north or south; which being compared with the given difference of latitude and departure, will shew those yet to be made good in failing to the desired port, and thence the course and distance to it.

Example. A ship from the latitude 28° 32′ north, bound to a port distant 100 miles, and bearing NE by N, has run the following courses and distances, viz, 1st, NW by N dist. 20 miles; 2d, SW 40 miles; 3d, NE by E 60 miles; 4th, SE 55 miles; 5th, W by S 41 miles; 6th, ENE 66 miles. Required her present latitude, with the direct course and distance made good, and those for the port bound to.

The numbers being taken out of the Traverse table, and entered opposite the several courses and distances, the tablet will be as here follows:

Courses.	Dift.	North.	South.	Eaft.	Weft.
NW by N	20	16·6	·	·	11·1
SW	40	·	28·3	·	28·3
NE by E	60	33·3	·	49·9	·
SE	55	·	38·9	38·9	·
W by S	41	·	8·0	·	40·2
ENE	66	25·3	·	61·0	·
		75·2	75·2	149·8	79·6
		75·2		79·6	
		0		70·2	Dep.

where the fums of the northings and fouthings, being both alike, 75·2, fhews that the fhip is come to the fame parallel of latitude fhe fet out from. And the difference between the fums of the eaftings and weftings, fhews that the fhip is 70·2 miles more to the eaftward, that being the greater. Confequently the courfe made good is due eaft, and the diftance is 70·2 miles.

But, by the Traverfe table, the northing and eafting to the propofed courfe NE by N, and diftance 100, are thus, viz, northing 83·1 and eafting 55·6

diff. from made good   0   and eafting 70·2

give - - northing 83·1 and wefting 14·6

yet to be made good to arrive at the intended port; and therefore, by finding thefe in tne Traverfe table, anfwering to them are the intended courfe and diftance, viz, diftance 85, and courfe N 10° W.

The geometrical conftruction, according to the method before defcribed, gives the figure as below: where A is the port fet out from, B is the port bound to,

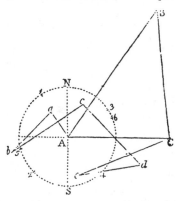

C is the place come to, by failing the feveral courfes and diftances Aa, ab, bc, cd, de, and eC; then CB is the diftance to be failed to arrive at the port B, and its courfe, or direction with the meridian, is nearly 10°, or the angle ACB, made with the eaft-and-weft line, nearly 80°.—Note, the radii from the centre to the feveral points in the circumference, are omitted, to prevent a confufion in the figure.

TRAVERSE-*Board*, in a fhip, a fmall round board, hanging up in the fteerage, and pierced full of holes in lines fhewing the points of the compafs: upon which, by moving a fmall peg from hole to hole, the fteerfman keeps an account how many glaffes, that is half hours, the fhip fteers upon any point.

TRAVERSE-*Table*, in Navigation, is the fame with a table of difference of latitude and departure; being the difference of latitude and departure ready calculated to every point, half point, quarter point, degree, &c, of the quadrant; and for every diftance, up to 50 or 100 or 120, &c. Though it may ferve for any greater diftance whatever, by adding two or more together; or by taking their halves, thirds, fourths, &c, and then doubling, tripling, quadrupling, &c, the difference of latitude and departure found to thofe parts of the diftance.

This table is one of the moft neceffary and ufeful things a navigator has occafion for; for by it he can readily reduce all his courfes and diftances, run in the fpace of 24 hours, into one courfe and diftance; whence he finds the latitude he is in, and the departure from the meridian.

One of the beft tables of this kind is in Robertfon's Navigation, at the end of book 7, vol. 1. The diftances are there carried to 120, for the fake of more eafy fubdivifions; and it is divided into two parts; the firft containing the courfes for every quarter point of the compafs, and the 2d adapted to every 15′, or quarter of a degree, in the quadrant. See TRAVERSE *Sailing*.

A fpecimen of fuch a Traverfe Table is the following, otherwife called a Table of Difference of Latitude and Departure. The diftances are placed at top and bottom of the columns, from 1 to 10; but may be extended to any quantity by multiplying the parts, and taking out at feveral times. The courfes, or angles of a rightangled triangle, are in a column, on both fides, each in two parts, the one containing the even points and quarter points, and the other whole degrees, as far as to 45°, or half the quadrant, on the left-hand fide, and the other half quadrant, from 45° to 90°, returned upwards from bottom to top on the right-hand fide. The correfponding Difference of Latitude and Departure are in two columns below or above the diftances, viz, below them when the courfe or angle is within 45°, or found on the left-hand fide; but above them when between 45 and 90°, or found on the right-hand fide.

The fame table ferves alfo to work all cafes of rightangled triangles, for any other purpofes. For example, Suppofe a given courfe be 15°, and diftance 35 miles, to find the correfponding difference of latitude and the departure: Or, in a right-angled triangle, given the hypotenufe 35, and one angle 15°, to find the two legs.

Here, the diftance 3 in the table muft be accounted 30, moving the decimal point proportionally or one place in the other numbers; and thofe numbers taken out at twice, viz, once from the columns under 3 for the 30, and the other from the columns under the diftance 5. Thus, on the line of 15°, and under the

	Dift.	Lat.	Dep.
	30 are	28·978 and	7·765
	5 are	4·830 and	1·294
theref. for	35 are	33·808 and	9·059

So that the other two legs of the triangle are 33·808 and 9·059.—If the courfe had been 75°, or the complement of the former, which is only the other angle of the fame triangle, and which is found on the fame line of the table, but on the right-hand fide of it, then the numbers in the columns will be the fame as before, and will give the fame fums for the two legs of the triangle, only with the contrary names, as to Latitude and Departure, which change places.

A TABLE

A Table of the Difference of Latitude and Departure, for Degrees and Quarter Points.

Course		Dist. 1		Dist. 2		Dist. 3		Dist. 4		Dist. 5		Course	
Pts.	D.	Lat.	Dep.	Lat.	Dep.	Lat.	Dep.	Lat.	Dep.	Lat.	Dep.	D.	Pts.
	1	0·9998	0·0175	1·9997	0·0349	2·9995	0·0524	3·9994	0·0698	4·9992	0·0873	89	
	2	0·9994	0·0349	1·9988	0·0698	2·9982	0·1047	3·9976	0·1396	4·9970	0·1745	88	
0¼		0·9988	0·0491	1·9976	0·0981	2·9964	0·1472	3·9952	0·1963	4·9940	0·2453		7¾
	3	0·9986	0·0523	1·9973	0·1047	2·9959	0·1570	3·9945	0·2093	4·9931	0·2617	87	
	4	0·9976	0·0698	1·9951	0·1395	2·9927	0·2093	3·9903	0·2790	4·9878	0·3488	86	
	5	0·9962	0·0872	1·9924	0·1743	2·9886	0·2615	3·9848	0·3486	4·9810	0·4358	85	
0½		0·9952	0·0980	1·9904	0·1960	2·9856	0·2940	3·9807	0·3921	4·9759	0·4901		7½
	6	0·9945	0·1045	1·9890	0·2091	2·9836	0·3136	3·9781	0·4181	4·9726	0·5226	84	
	7	0·9925	0·1219	1·9851	0·2437	2·9776	0·3656	3·9702	0·4875	4·9627	0·6093	83	
	8	0·9903	0·1392	1·9805	0·2783	2·9708	0·4175	3·9611	0·5567	4·9513	0·6959	82	
0¾		0·9892	0·1467	1·9784	0·2935	2·9675	0·4402	3·9567	0·5869	4·9459	0·7337		7¼
	9	0·9877	0·1564	1·9754	0·3129	2·9631	0·4693	3·9508	0·6257	4·9384	0·7822	81	
	10	0·9848	0·1736	1·9696	0·3473	2·9544	0·5209	3·9392	0·6946	4·9240	0·8682	80	
	11	0·9816	0·1908	1·9633	0·3816	2·9449	0·5724	3·9265	0·7632	4·9081	0·9540	79	
1		0·9808	0·1951	1·9616	0·3902	2·9424	0·5853	3·9231	0·7804	4·9039	0·9754		7
	12	0·9781	0·2079	1·9563	0·4158	2·9344	0·6237	3·9126	0·8316	4·8907	1·0396	78	
	13	0·9744	0·2250	1·9487	0·4499	2·9231	0·6749	3·8975	0·8998	4·8718	1·1248	77	
	14	0·9703	0·2419	1·9406	0·4838	2·9108	0·7258	3·8812	0·9677	4·8515	1·2096	76	
1¼		0·9700	0·2430	1·9401	0·4860	2·9101	0·7289	3·8801	0·9719	4·8502	1·2149		6¾
	15	0·9659	0·2588	1·9319	0·5176	2·8978	0·7765	3·8637	1·0353	4·8296	1·2941	75	
	16	0·9613	0·2756	1·9225	0·5513	2·8838	0·8269	3·8450	1·1025	4·8063	1·3782	74	
1½		0·9569	0·2903	1·9139	0·5806	2·8708	0·8209	3·8278	1·1611	4·7847	1·4514		6½
	17	0·9563	0·2924	1·9126	0·5847	2·8689	0·8771	3·8252	1·1695	4·7815	1·4619	73	
	18	0·9511	0·3090	1·9021	0·6180	2·8532	0·9271	3·8042	1·2361	4·7553	1·5451	72	
	19	0·9455	0·3256	1·8910	0·6511	2·8366	0·9767	3·7821	1·3023	4·7276	1·6278	71	
1¾		0·9415	0·3369	1·8831	0·6738	2·8246	1·0107	3·7662	1·3476	4·7077	1·6844		6¼
	20	0·9397	0·3420	1·8794	0·6840	2·8191	1·0261	3·7588	1·3681	4·6985	1·7101	70	
	21	0·9336	0·3584	1·8672	0·7167	2·8007	1·0751	3·7343	1·4335	4·6679	1·7918	69	
	22	0·9272	0·3746	1·8544	0·7492	2·7816	1·1238	3·7087	1·4984	4·6359	1·8730	68	
2		0·9239	0·3827	1·8478	0·7654	2·7716	1·1480	3·6955	1·5307	4·6194	1·9134		6
	23	0·9205	0·3907	1·8410	0·7815	2·7615	1·1722	3·6820	1·5629	4·6025	1·9537	67	
	24	0·9135	0·4067	1·8270	0·8135	2·7406	1·2202	3·6542	1·6269	4·5677	2·0337	66	
	25	0·9063	0·4226	1·8126	0·8452	2·7189	1·2679	3·6252	1·6905	4·5315	2·1131	65	
2¼		0·9040	0·4276	1·8080	0·8551	2·7120	1·2827	3·6160	1·7102	4·5199	2·1378		5¼
	26	0·8988	0·4384	1·7976	0·8767	2·6964	1·3151	3·5952	1·7535	4·4940	2·1919	64	
	27	0·8910	0·4540	1·7820	0·9080	2·6730	1·3620	3·5640	1·8160	4·4550	2·2699	63	
	28	0·8829	0·4695	1·7659	0·9389	2·6488	1·4084	3·5318	1·8779	4·4147	2·3474	62	
2½		0·8819	0·4714	1·7638	0·9428	2·6458	1·4142	3·5277	1·8850	4·4096	2·3570		5½
	29	0·8746	0·4848	1·7492	0·9696	2·6239	1·4544	3·4985	1·9392	4·3731	2·4240	61	
	30	0·8660	0·5000	1·7320	1·0000	2·5981	1·5000	3·4641	2·0000	4·3301	2·5000	60	
2¾		0·8577	0·5141	1·7155	1·0282	2·5732	1·5423	3·4309	2·0564	4·2886	2·5705		5¼
	31	0·8572	0·5150	1·7143	1·0301	2·5715	1·5451	3·4287	2·0602	4·2858	2·5752	59	
	32	0·8480	0·5299	1·6961	1·0598	2·5441	1·5896	3·3922	2·1197	4·2402	2·6496	58	
	33	0·8387	0·5446	1·6773	1·0893	2·5160	1·6339	3·3547	2·1786	4·1934	2·7232	57	
3		0·8315	0·5556	1·6629	1·1111	2·4944	1·6667	3·3259	2·2223	4·1573	2·7778		5
	34	0·8290	0·5592	1·6581	1·1184	2·4871	1·6776	3·3162	2·2368	4·1452	2·7960	56	
	35	0·8192	0·5736	1·6383	1·1472	2·4575	1·7207	3·2766	2·2943	4·0958	2·8679	55	
	36	0·8090	0·5878	1·6180	1·1756	2·4271	1·7634	3·2361	2·3511	4·0451	2·9389	54	
3¼		0·8032	0·5957	1·6064	1·1914	2·4096	1·7871	3·2128	2·3828	4·0160	2·9785		4¾
	37	0·7986	0·6018	1·5973	1·2036	2·3959	1·8054	3·1945	2·4073	3·9932	3·0091	53	
	38	0·7880	0·6157	1·5760	1·2313	2·3640	1·8470	3·1520	2·4626	3·9401	3·0783	52	
	39	0·7771	0·6293	1·5543	1·2586	2·3314	1·8880	3·1086	2·5173	3·8857	3·1466	51	
3½		0·7730	0·6344	1·5460	1·2688	2·3190	1·9032	3·0920	2·5376	3·8650	3·1720		4½
	40	0·7660	0·6428	1·5321	1·2856	2·2981	1·9284	3·0642	2·5712	3·8302	3·2139	50	
	41	0·7547	0·6561	1·5094	1·3121	2·2641	1·9682	3·0188	2·6242	3·7736	3·2803	49	
	42	0·7431	0·6691	1·4863	1·3383	2·2294	2·0074	2·9726	2·6765	3·7157	3·3457	48	
3¾		0·7410	0·6716	1·4819	1·3431	2·2229	2·0147	2·9638	2·6862	3·7048	3·3578		4¼
	43	0·7314	0·6820	1·4628	1·3640	2·1941	2·0460	2·9254	2·7280	3·6568	3·4100	47	
	44	0·7193	0·6947	1·4387	1·3894	2·1580	2·0840	2·8774	2·7786	3·5967	3·4733	46	
4	45	0·7071	0·7071	1·4142	1·4142	2·1213	2·1213	2·8284	2·8284	3·5355	3·5355	45	4
Pts.	Deg.	Dep.	Lat.	Dep.	Lat.	Dep.	Lat.	Dep.	Lat.	Dep.	Lat.	Deg.	Pts.
		Dist. 1		Dist. 2		Dist. 3		Dist. 4		Dist. 5			

TABLE *of the Difference of Latitude and Departure, for Degrees and Quarter Points.*

Course		Dist. 6		Dist. 7		Dist. 8		Dist. 9		Dist. 10		Course	
Pts.	D.	Lat.	Dep.	Lat.	Dep.	Lat.	Dep.	Lat.	Dep.	Lat.	Dep.	D.	Pts.
	1	5·9991	0·1047	6·9989	0·1222	7·9988	0·1396	8·9986	0·1571	9·9985	0·1745	89	
	2	5·9963	0·2094	6·9957	0·2443	7·9951	0·2792	8·9945	0·3141	9·9939	0·3490	88	
0 ¼		5·9928	0·2944	6·9916	0·3435	7·9904	0·3925	8·9892	0·4416	9·9880	0·4907		7 ¾
	3	5·9918	0·3140	6·9904	0·3664	7·9890	0·4187	8·9877	0·4710	9·9863	0·5234	87	
	4	5·9854	0·4185	6·9829	0·4883	7·9805	0·5580	8·9781	0·6278	9·9756	0·6975	86	
	5	5·9772	0·5229	6·9734	0·6101	7·9696	0·6972	8·9658	0·7844	9·9619	0·8716	85	
0 ½		5·9711	0·5881	6·9663	0·6861	7·9615	0·7841	8·9567	0·8822	9·9518	0·9802		7 ½
	6	5·9671	0·6272	6·9617	0·7317	7·9562	0·8362	8·9507	0·9408	9·9452	1·0453	84	
	7	5·9553	0·7312	6·9478	0·8531	7·9404	0·9750	8·9329	1·0968	9·9255	1·2187	83	
	8	5·9416	0·8350	6·9319	0·9742	7·9221	1·1134	8·9124	1·2526	9·9027	1·3917	82	
0 ¾		5·9351	0·8804	6·9242	1·0271	7·9134	1·1738	8·9026	1·3206	9·8918	1·4674		7 ¼
	9	5·9261	0·9386	6·9138	1·0950	7·9015	1·2515	8·8892	1·4075	9·8760	1·5643	81	
	10	5·9088	1·0419	6·8937	1·2155	7·8785	1·3892	8·8633	1·5628	9·8481	1·7365	80	
	11	5·8898	1·1449	6·8714	1·3357	7·8530	1·5265	8·8346	1·7173	9·8165	1·9081	79	
1		5·8847	1·1705	6·8655	1·3656	7·8463	1·5607	8·8271	1·7558	9·8079	1·9509		7
	12	5·8689	1·2475	6·8470	1·4554	7·8252	1·6633	8·8033	1·8712	9·7815	2·0791	78	
	13	5·8462	1·3497	6·8206	1·5746	7·7950	1·7996	8·7693	2·0245	9·7437	2·2495	77	
	14	5·8218	1·4515	6·7921	1·6935	7·7624	1·9354	8·7327	2·1773	9·7030	2·4192	76	
1 ¼		5·8202	1·4579	6·7902	1·7009	7·7602	1·9438	8·7303	2·1868	9·7003	2·4298		6 ¾
	15	5·7956	1·5529	6·7615	1·8117	7·7274	2·0706	8·6933	2·3204	9·6593	2·5882	75	
	16	5·7676	1·6538	6·7288	1·9295	7·6901	2·2051	8·6513	2·4807	9·6120	2·7562	74	
1 ½		5·7416	1·7417	6·6986	2·0320	7·6555	2·3223	8·6125	2·6126	9·5694	2·9028		6 ½
	17	5·7378	1·7542	6·6941	2·0466	7·6504	2·3390	8·6067	2·6313	9·5630	2·9237	73	
	18	5·7063	1·8541	6·6574	2·1631	7·6084	2·4721	8·5595	2·7812	9·5106	3·0902	72	
	19	5·6731	1·9534	6·6186	2·2790	7·5642	2·6045	8·5097	2·9301	9·4552	3·2557	71	
1 ¾		5·6493	2·0213	6·5908	2·3582	7·5324	2·6951	8·4739	3·0320	9·4154	3·3689		6 ¼
	20	5·6382	2·0521	6·5779	2·3941	7·5175	2·7362	8·4572	3·0782	9·3969	3·4202	70	
	21	5·6015	2·1502	6·5351	2·5086	7·4686	2·8669	8·4022	3·2253	9·3358	3·5837	69	
	22	5·5631	2·2476	6·4903	2·6222	7·4175	2·9969	8·3447	3·3715	9·2718	3·7461	68	
2		5·5433	2·2961	6·4672	2·6788	7·3910	3·0615	8·3149	3·4441	9·2388	3·8268		6
	23	5·5230	2·3444	6·4435	2·7351	7·3640	3·1258	8·2845	3·5166	9·2050	3·9075	67	
	24	5·4813	2·4404	6·3948	2·8472	7·3084	3·2539	8·2219	3·6606	9·1355	4·0674	66	
	25	5·4378	2·5357	6·3442	2·9583	7·2505	3·3809	8·1568	3·8036	9·0631	4·2262	65	
2 ¼		5·4239	2·5653	6·3279	2·9929	7·2319	3·4204	8·1359	3·8480	9·0399	4·2756		5 ¾
	26	5·3928	2·6302	6·2916	3·0686	7·1904	3·5070	8·0891	3·9453	8·9879	4·3837	64	
	27	5·3460	2·7239	6·2370	3·1779	7·1280	3·6319	8·0191	4·0859	8·9101	4·5399	63	
	28	5·2977	2·8168	6·1806	3·2863	7·0636	3·7558	7·9465	4·2252	8·8295	4·6947	62	
2 ½		5·2915	2·8284	6·1734	3·2998	7·0554	3·7712	7·9373	4·2426	8·8192	4·7140		5 ½
	29	5·2477	2·9089	6·1223	3·3937	6·9970	3·8785	7·8716	4·3633	8·7462	4·8481	61	
	30	5·1961	3·0000	6·0622	3·5000	6·9282	4·0000	7·7942	4·5000	8·6603	5·0000	60	
2 ¾		5·1464	3·0846	6·0041	3·5987	6·8618	4·1128	7·7196	4·6269	8·5773	5·1410		5 ¼
	31	5·1430	3·0902	6·0002	3·6052	6·8573	4·1203	7·7145	4·6353	8·5717	5·1504	59	
	32	5·0883	3·1795	5·9363	3·7094	6·7843	4·2394	7·6324	4·7693	8·4805	5·2992	58	
	33	5·0320	3·2678	5·8707	3·8125	6·7094	4·3571	7·5480	4·9018	8·3867	5·4464	57	
3		4·9888	3·3334	5·8203	3·8890	6·6518	4·4446	7·4832	5·0001	8·3147	5·5557		5
	34	4·9742	3·3552	5·8033	3·9144	6·6323	4·4735	7·4613	5·0327	8·2904	5·5919	56	
	35	4·9149	3·4415	5·7341	4·0150	6·5532	4·5880	7·3724	5·1622	8·1915	5·7358	55	
	36	4·8541	3·5267	5·6631	4·1145	6·4721	4·7023	7·2812	5·2901	8·0902	5·8779	54	
3 ¼		4·8192	3·5742	5·6224	4·1699	6·4257	4·7656	7·2289	5·3613	8·0321	5·9570		4 ¾
	37	4·7918	3·6109	5·5904	4·2127	6·3891	4·8145	7·1877	5·4163	7·9864	6·0182	53	
	38	4·7281	3·6940	5·5161	4·3096	6·3041	4·9253	7·0921	5·5409	7·8801	6·1560	52	
	39	4·6629	3·7759	5·4400	4·4052	6·2172	5·0346	6·9943	5·6639	7·7715	6·2932	51	
3 ½		4·6381	3·8064	5·4111	4·4408	6·1841	5·0751	6·9571	5·7095	7·7301	6·3439		4 ½
	40	4·5963	3·8567	5·3623	4·4995	6·1284	5·1423	6·8944	5·7851	7·6604	6·4279	50	
	41	4·5283	3·9363	5·2830	4·5924	6·0377	5·2485	6·7924	5·9045	7·5471	6·5606	49	
	42	4·4589	4·0148	5·2020	4·6839	5·9452	5·3530	6·6883	6·0222	7·4314	6·6913	48	
3 ¾		4·4457	4·0294	5·1867	4·7009	5·9276	5·3725	6·6680	6·0440	7·4095	6·7156		4 ¼
	43	4·3881	4·0920	5·1195	4·7740	5·8508	5·4560	6·5822	6·1380	7·3135	6·8200	47	
	44	4·3160	4·1679	5·0354	4·8626	5·7547	5·5573	6·4741	6·2519	7·1934	6·9466	46	
4	45	4·2426	4·2426	4·9497	4·9497	5·6509	5·6569	6·3640	6·3640	7·0711	7·0711	45	4
Pts.	Deg.	Dep.	Lat.	Dep.	Lat.	Dep.	Lat.	Dep.	Lat.	Dep.	Lat.	Deg.	Pts.
		Dist. 6		Dist. 7		Dist. 8		Dist. 9		Dist. 10			

TREBLE, in Mufic, the higheft or acuteft of the four parts in fymphony, or that which is heard the cleareft and fhrilleft in a concert. In the like fenfe we fay, a Treble violin, Treble hautboy, &c.

In vocal mufic, the Treble is ufually committed to boys and girls; their proper part being the Treble.

The Treble is divided into firft or higheft Treble, and fecond or bafs Treble. The half Treble is the fame with the counter-tenor.

TRENCHES, in Fortification, are ditches which the befiegers cut to approach more fecurely to the place attacked; whence they are called *lines of approach.* Their breadth is 8 or 10 feet, and depth 6 or 7.

They fay, *mount the Trenches,* that is, go upon duty in them. To *relieve the Trenches,* is to relieve fuch as have been upon duty there. The enemy is faid to have *cleared the Trenches,* when he has driven away or killed the foldiers who guarded them.

Tail of the TRENCH, is the place where it was begun. And the *Head* is the place where it ends.

Opening of the TRENCHES, is when the befiegers firft begin to work upon them, or to make them; which is ufually done in the night.

TREPIDATION, in the Ancient Aftronomy, denotes what they call a libration of the 8th fphere; or a motion which the Ptolomaic fyftem attributed to the firmament, to account for certain almoft infenfible changes and motions obferved in the axis of the world; by means of which the latitudes of the fixed ftars come to be gradually changed, and the ecliptic feems to approach reciprocally, firft towards one pole, then towards the other.

This motion is alfo called the *motion of the firft libration.*

TRET, in Commerce, is an allowance made for the wafte, or the duft, that may be mixed with any commodity; which is always 4 pounds on every 104 pounds weight. See TARE.

TRIANGLE, in Geometry, a figure bounded or contained by three lines or fides, and which confequently has three angles, from whence the figure takes its name.

Triangles are either plane or fpherical or curvilinear. Plane when the three fides of the Triangle are right lines; but fpherical when fome or all of them are arcs of great circles on the fphere.

Plane Triangles take feveral denominations, both from the relation of their angles, and of their fides, as below. And 1ft with regard to the fides.

An *Equilateral Triangle,* is that which has all its three fides equal to one another; as A.

An *Ifofceles* or *Equicrural Triangle,* is that which has two fides equal; as B.

A *Scalene Triangle* has all its fides unequal; as C.

6

Again, with refpect to the Angles.

A *Rectangular* or *Right-angled Triangle,* is that which has one right angle; as D.

An *Oblique Triangle* is that which has no right angle, but all oblique ones; as E or F.

An *Acutangular* or *Oxygone Triangle,* is that which has three acute angles; as E.

An *Obtufangular* or *Amblygone Triangle,* is that which has an obtufe angle; as F.

A *Curvilinear* or *Curvilineal Triangle,* is one that has all its three fides curve lines.

A *Mixtilinear Triangle* is one that has its fides fome of them curves, and fome right lines.

A *Spherical Triangle* is one that has its fides, or at leaft fome of them, arcs of great circles of the fphere.

Similar Triangles are fuch as have the angles in the one equal to the angles in the other, each to each.

The *Bafe* of a Triangle, is any fide on which a perpendicular is drawn from the oppofite angle, called the *vertex;* and the two fides about the perpendicular, or the vertex, are called the *legs.*

The Chief Properties of Plane Triangles, are as follow, viz, In any plane Triangle,

1. The greateft fide is oppofite to the greateft angle, and the leaft fide to the leaft angle, &c. Alfo, if two fides be equal, their oppofite angles are equal; and if the Triangle be equilateral, or have all its fides equal, it will alfo be equiangular, or have all its angles equal to one another.

2. Any fide of a Triangle is lefs than the fum, but greater than the difference, of the other two fides.

3. The fum of all the three angles, taken together, is equal to two right angles.

4. If one fide of a Triangle be produced out, the external angle, made by it and the adjacent fide, is equal to the fum of the two oppofite internal angles.

5. A line drawn parallel to one fide of a Triangle, cuts the other two fides proportionally, the corresponding fegments being proportional, each to each, and to the whole fides; and the Triangle cut off is fimilar to the whole Triangle.

If a perpendicular be let fall from any angle of a Triangle, as a vertical angle, upon the oppofite fide as a bafe; then

6. The rectangle of the fum and difference of the fides, is equal to twice the rectangle of the bafe and the diftance of the perpendicular from the middle of the bafe.—Or, which is the fame thing in other words,

7. The difference of the fquares of the fides, is equal to the difference of the fquares of the fegments of the bafe. Or, as the bafe is to the fum of the fides, fo is the difference of the fides, to the difference of the fegments of the bafe.

8. The rectangle of the legs or fides, is equal to the rectangle of the perpendicular and the diameter of the circumfcribing circle.

If a line be drawn bifecting any angle, to the bafe or oppofite fide; then,

9. The

9. The segments of the bafe, made by the line bi-fecting the oppofite angle, are proportional to the fides adjacent to them.

10. The fquare of the line bifecting the angle, is equal to the difference between the rectangle of the fides and the rectangle of the fegments of the bafe.

If a line be drawn from any angle to the middle of the oppofite fide, or bifecting the bafe; then

11. The fum of the fquares of the fides, is equal to twice the fum of the fquares of half the bafe and the line bifecting the bafe.

12. The angle made by the perpendicular from any angle and the line drawn from the fame angle to the middle of the bafe, is equal to half the difference of the angles at the bafe.

13. If through any point D, within a Triangle ABC, three lines EF, GH, 1K, be drawn parallel to the three fides of the Triangle; the continual products or folids made by the alternate fegments of thefe lines will be equal; viz,

$$DE \times DK \times DH = DG \times DF \times DI.$$

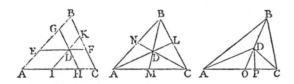

14. If three lines AL, BM, CN, be drawn from the three angles through any point D within a Triangle, to the oppofite fides; the folid products of the alternate fegments of the fides are equal; viz,
AN × BL × CM = AM × CL × BN, (2d fig. above).

15. Three lines drawn from the three angles of a Triangle to bifect the oppofite fides, or to the middle of the oppofite fides, do all interfect one another in the fame point D, and that point is the centre of gravity of the Triangle, and the diftance AD of that point from any angle as D$_r$ is equal to double the diftance DL from the oppofite fide; or one fegment of any of thefe lines is double the other fegment: moreover the fum of the fquares of the three bifecting lines, is ¾ of the fum of the fquares of the three fides of the Triangle.

16. Three perpendiculars bifecting the three fides of a Triangle, all interfect in one point, and that point is the centre of the circumfcribing circle.

17. Three lines bifecting the three angles of a Triangle, all interfect in one point, and that point is the centre of the infcribed circle.

18. Three perpendiculars drawn from the three angles of a Triangle, upon the oppofite fides, all interfect in one point.

19. If the three angles of a Triangle be bifected by the lines AD, BD, CD (3d fig. above), and any one as BD be continued to the oppofite fide at O, and DP be drawn perp. to that fide; then is
∠ADO = ∠CDP, or ∠ADP = ∠CDO.

20. Any Triangle may have a circle circumfcribed about it, or touching all its angles, and a circle infcribed within it, or touching all its fides.

21. The fquare of the fide of an equilateral Triangle, is equal to 3 times the fquare of the radius of its cir-cumfcribing circle.

22. If the three angles of one Triangle be equal to the three angles of another Triangle, each to each; then thofe two Triangles are fimilar, and their like fides are proportional to one another, and the areas of the two Triangles are to each other as the fquares of their like fides.

23. If two Triangles have any three parts of the one (except the three angles), equal to three corre-fponding parts of the other, each to each; thofe two Triangles are not only fimilar, but alfo identical, or having all their fix correfponding parts equal, and their areas equal.

24. Triangles ftanding upon the fame bafe, and between the fame parallels, are equal; and Triangles upon equal bafes, and having equal altitudes, are equal.

25. Triangles on equal bafes, are to one another as their altitudes: and Triangles of equal altitudes, are to one another as their bafes; alfo equal Triangles have their bafes and altitudes reciprocally proportional.

26. Any Triangle is equal to half its circumfcribing parallelogram, or half the parallelogram on the fame or an equal bafe, and of the fame or equal altitude.

27. Therefore the area of any Triangle is found, by multiplying the bafe by the altitude, and taking half the product.

28. The area is alfo found thus: Multiply any two fides together, and multiply the product by the fine of their included angle, to radius 1, and divided by 2.

29. The area is alfo otherwife found thus, when the three fides are given: Add the three fides together, and take half their fum; then from this half fum fub-tract each fide feverally, and multiply the three re-mainders and the half fum continually together; then the fquare root of the laft product will be the area of the Triangle.

30. In a right-angled Triangle, if a perpendicular be let fall from the right angle upon the hypothenufe, it will divide it into two other Triangles fimilar to one another, and to the whole Triangle.

31. In a right-angled Triangle, the fquare of the hypothenufe is equal to the fum of the fquares of the two fides; and, in general, any figure defcribed upon the hypothenufe, is equal to the fum of two fimilar figures defcribed upon the two fides.

32. In an ifofceles Triangle, if a line be drawn from the vertex to any point in the bafe; the fquare of that line together with the rectangle of the fegments of the bafe, is equal to the fquare of the fide.

33. If one angle of a Triangle be equal to 120°; the fquare of the bafe will be equal to the fquares of both the fides, together with the rectangle of thofe fides; and if thofe fides be equal to each other, then the fquare of the bafe will be equal to three times the fquare of one fide, or equal to 12 times the fquare of the perpendicular from the angle upon the bafe.

34. In the fame Triangle, viz, having one angle equal to 120°; the difference of the cubes of the fides, about that angle, is equal to a folid contained by the difference of the fides and the fquare of the bafe; and the fum of the cubes of the fides, is equal to a folid contained by the fum of the fides and the difference
between

between the square of the base and twice the rectangle of the sides.

There are many other properties of Triangles to be found among the geometrical writers ; so Gregory St. Vincent has written a folio volume upon Triangles ; there are also several in his Quadrature of the circle. See also other properties under the article TRIGONOMETRY.

For the properties of spherical Triangles, see SPHE-RICAL *Triangles*.

Solution of TRIANGLES. See TRIGONOMETRY.

TRIANGLE, in Astronomy, one of the 48 ancient Constellations, situated in the northern hemisphere. There is also the *Southern Triangle* in the southern hemisphere, which is a modern constellation. The stars in the Northern Triangle are, in Ptolomy's catalogue 4, in Tycho's 4, in Hevelius's 12, and in the British catalogue 16.

The stars in the Southern Triangle are, in Sharp's catalogue, 5.

Arithmetical TRIANGLE, a kind of numeral Triangle, or Triangle of numbers, being a table of certain numbers disposed in form of a Triangle. It was so called by Pascal ; but he was not the inventor of this table, as some writers have imagined, its properties having been treated of by other authors, some centuries before him, as is shewn in my Mathematical Tracts, vol. 1, pa. 69 &c.

The form of the Triangle is as follows :

```
1
1    1
1    2    1
1    3    3    1
1    4    6    4    1
1    5   10   10    5    1
1    6   15   20   &c
1    7   21   &c
1    8   &c
1    9
1
```

And it is constructed by adding always the last two numbers of the next two preceding columns together, to give the next succeeding column of numbers.

The first vertical column consists of units ; the 2d a series of the natural numbers 1, 2, 3, 4, 5, &c ; the 3d a series of Triangular numbers 1, 3, 6, 10, &c ; the 4th a series of pyramidal numbers, &c. The oblique diagonal rows, descending from left to right, are also the same as the vertical columns. And the numbers taken on the horizontal lines are the co-efficients of the different powers of a binomial. Many other properties and uses of these numbers have been delivered by various authors, as may be seen in the Introduction to my Mathematical Tables, pages 7, 8, 75, 76, 77, 89, 2d edition.

After these, Pascal wrote a treatise on the Arithmetical Triangle, which is contained in the 5th volume of his works, published at Paris and the Hague in 1779, in 5 volumes, 8vo.

In this publication is also a description, taken from the 1st volume of the French Encyclopedie, art. *Arithmetique Machine,* of that admirable machine in-

VOL. II.

vented by Pascal at the age of 19, furnishing an easy and expeditious method of making all sorts of arithmetical calculations without any other assistance than the eye and the hand.

TRIANGULAR, relating to a triangle ; as

TRIANGULAR *Canon,* tables relating to trigonometry ; as of sines, tangents, secants, &c.

TRIANGULAR *Compasses,* are such as have three legs or feet, by which any triangle, or three points, may be taken off at once. These are very useful in the construction of maps, globes, &c.

TRIANGULAR *Numbers,* are a kind of polygonal numbers ; being the sums of arithmetical progressions, which have 1 for the common difference of their terms.

Thus, from these arithmeticals 1 2 3 4 5 6, are formed the Triang. Numb. 1 3 6 10 15 21, or the 3d column of the arithmetical triangle abovementioned.

The sum of any number n of the terms of the Triangular numbers, 1, 3, 6, 10, &c, is =

$$\frac{n^3}{6} + \frac{n^2}{2} + \frac{n}{3}, \text{ or } \frac{n}{1} \times \frac{n+1}{2} \times \frac{n+2}{3}$$

which is also equal to the number of shot in a triangular pile of balls, the number of rows, or the number in each side of the base, being n.

The sum of the reciprocals of the Triangular series, infinitely continued, is equal to 2 ; viz,

$$1 + \tfrac{1}{3} + \tfrac{1}{6} + \tfrac{1}{10} + \tfrac{1}{15} \&c = 2.$$

For the rationale and management of these numbers, see Malcolm's Arith. book 5, ch. 2 ; and Simpson's Algeb. sec. 15.

TRIANGULAR *Quadrant,* is a sector furnished with a loose piece, by which it forms an equilateral triangle. Upon it is graduated and marked the calendar, with the sun's place, and other useful lines ; and by the help of a string and a plummet, with the divisions graduated on the loose piece, it may be made to serve for a quadrant.

TRIBOMETER, in Mechanics, a term applied by Muschenbroek to an instrument invented by him for measuring the friction of metals. It consists of an axis formed of hard steel, passing through a cylindrical piece of wood : the ends of the axis, which are highly polished, are made to rest on the polished semicircular cheeks of various metals, and the degree of friction is estimated by means of a weight suspended by a fine silken string or ribband over the wooden cylinder. For a farther description and the figure of this instrument, with the results of various experiments performed with it, see Muschenb. Introd. ad Phil. Nat. vol. 1, p. 151.

TRIDENT, is a particular kind of parabola, used by Descartes in constructing equations of 6 dimensions. See the article *Cartesian* PARABOLA.

TRIGLYPH, in Architecture, is a member of the Doric Frize, placed directly over each column, and at equal distances in the intercolumnation, having two entire glyphs or channels engraven in it, meeting in an angle, and separated by three legs from the two demichannels of the sides.

TRIGON,

TRIGON, a figure of three angles, or a triangle.

TRIGON, in Aſtrology. See TRIPLICITY.

TRIGON, in Aſtronomy, denotes an aſpect of two planets when they are 120 degrees diſtant from each other; called alſo a Trine, being the 3d part of 360 degrees.—The Trigons of Mars and Saturn are by aſtrologers held malific or malignant aſpects.

TRIGON, in Dialling, is an inſtrument of a triangular form.

TRIGON, in Muſic, denoted a muſical inſtrument, uſed among the ancients. It was a kind of triangular lyre, or harp, invented by Ibycus; and was uſed at feaſts, being played on by women, who ſtruck it either with a quill, or beat it with ſmall rods of different lengths and weights, to occaſion a diverſity in the ſounds.

TRIGONAL Numbers. See TRIANGULAR Numbers.

TRIGONOMETER Armillary. See ARMILLARY Trigonometer.

TRIGONOMETRY, the art of meaſuring the ſides and angles of triangles, either plane or ſpherical, from whence it is accordingly called either Plane Trigonometry, or Spherical Trigonometry.

Every triangle has 6 parts, 3 ſides, and 3 angles; and it is neceſſary that three of theſe parts be given, to find the other three. In ſpherical Trigonometry, the three parts that are given, may be of any kind, either all ſides, or all angles, or part the one and part the other. But in plane Trigonometry, it is neceſſary that one of the three parts at leaſt be a ſide, ſince from three angles can only be found the proportions of the ſides, but not the real quantities of them.

Trigonometry is an art of the greateſt uſe in the mathematical ſciences, eſpecially in aſtronomy, navigation, ſurveying, dialling, geography, &c. &c. By it, we come to know the magnitude of the earth, the planets and ſtars, their diſtances, motions, eclipſes, and almoſt all other uſeful arts and ſciences. Accordingly we find this art has been cultivated from the earlieſt ages of mathematical knowledge.

Trigonometry, or the reſolution of triangles, is founded on the mutual proportions which ſubſiſt between the ſides and angles of triangles; which proportions are known by finding the relations between the radius of a circle and certain other lines drawn in and about the circle, called chords, ſines, tangents, and ſecants. The ancients Menelaus, Hipparchus, Ptolomy, &c, performed their Trigonometry, by means of the chords. As to the ſines, and the common theorems relating to them, they were introduced into Trigonometry by the Moors or Arabians, from whom this art paſſed into Europe, with ſeveral other branches of ſcience. The Europeans have introduced, ſince the 15th century, the tangents and ſecants, with the theorems relating to them. See the hiſtory and improvements at large, in the Introduction to my Mathematical Tables.

The proportion of the ſines, tangents, &c, to their radius, is ſometimes expreſſed in common or natural numbers; which conſtitute what we call the tables of natural ſines, tangents, and ſecants. Sometimes it is expreſſed in logarithms, being the logarithms of the ſaid natural ſines, tangents, &c; and theſe conſtitute the table of artificial ſines, &c. Laſtly, ſometimes the proportion is not expreſſed in numbers; but the ſeveral ſines, tangents, &c, are actually laid down upon lines of ſcales; whence the line of ſines, of tangents, &c. See SCALE.

In Trigonometry, as angles are meaſured by arcs of a circle deſcribed about the angular point, ſo the whole circumference of the circle is divided into a great number of parts, as 360 degrees, and each degree into 60 minutes, and each minute into 60 ſeconds, &c; and then any angle is ſaid to conſiſt of ſo many degrees, minutes and ſeconds, as are contained in the arc that meaſures the angle, or that is intercepted between the legs or ſides of the angle.

Now the ſine, tangent, and ſecant, &c, of every degree and minute, &c, of a quadrant, are calculated to the radius 1, and ranged in tables for uſe; as alſo the logarithms of the ſame; forming the triangular canon. And theſe numbers, ſo arranged in tables, form every ſpecies of right-angled triangles, ſo that no ſuch triangle can be propoſed, but one ſimilar to it may be there found, by compariſon with which, the propoſed one may be computed by analogy or proportion.

As to the ſcales of chords, ſines, tangents, &c, uſually placed on inſtruments, the method of conſtructing them is exhibited in the ſcheme annexed to the article SCALE; which, having the names added to each, needs no farther explanation.

There are uſually three methods of reſolving triangles, or the caſes of Trigonometry; viz, geometrical conſtruction, arithmetical computation, and inſtrumental operation. In the 1ſt method, the triangle in queſtion is conſtructed by drawing and laying down the ſeveral parts of their magnitudes given, viz, the ſides from a ſcale of equal parts, and the angles from a ſcale of chords, or other inſtrument; then the unknown parts are meaſured by the ſame ſcales, and ſo they become known.

In the 2d method, having ſtated the terms of the proportion according to rule, which terms conſiſt partly of the numbers of the given ſides, and partly of the ſines, &c, of angles taken from the tables, the proportion is then reſolved like all other proportions, in which a 4th term is to be found from three given terms, by multiplying the 2d and 3d together, and dividing the product by the firſt. Or, in working with the logarithms, adding the log. of the 2d and 3d terms together, and from the ſum ſubtracting the log. of the 1ſt term, then the number anſwering to the remainder is the 4th term ſought.

To work a caſe inſtrumentally, as ſuppoſe by the log. lines on one ſide of the two-foot ſcales: Extend the compaſſes from the 1ſt term to the 2d, or 3d, which happens to be of the ſame kind with it; then that extent will reach from the other term to the 4th. In this operation, for the ſides of triangles, is uſed the line of numbers (marked Num.); and for the angles, the line of ſines or tangents (marked ſin. and tan.) according as the proportion reſpects ſines or tangents.

In every caſe of triangles, as has been hinted before, there

there muſt be three parts, one at leaſt of which muſt be a ſide. And then the different circumſtances, as to the three parts that may be given, admit of three caſes or varieties only ; viz,

1ſt. When two of the three parts given, are a ſide and its oppoſite angle.—2d, When there are given two ſides and their contained angle.—3d, And thirdly, when the three ſides are given.

To each of theſe caſes there is a particular rule, or proportion, adapted, for reſolving it by.

1ſt. *The Rule for the ſt Caſe*, or that in which, of the three parts that are given, an angle and its oppo-ſite ſide are two of them, is this, viz, That the ſides are proportional to the ſines of their oppoſite angles,

That is,

As one ſide given   :
To the ſine of its oppoſite angle : :
So is another ſide given   :
To the ſine of its oppoſite angle.

Or,

As the ſine of an angle given   :
To its oppoſite ſide   : :
So is the ſine of another angle given :
To its oppoſite ſide.

So that, to find an angle, we muſt begin the pro-portion with a given ſide that is oppoſite to a given angle ; and to find a ſide, we muſt begin with an angle oppoſite to a given ſide.

Ex. Suppoſe, in the triangle ABC, there be given

AB = 365 feet,
AC = 154·33 f.
∠C = 98° 3′
to find the other ſide,
and the angles.

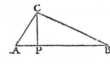

1. *Geometrically, by Conſtruction.*

Draw AC = 154·33 from a ſcale of equal parts : Make the angle C = 98° 3′, producing CB indefinitely : With centre A, and radius 365 feet, croſs CB in B : Then join AB, and the figure is conſtructed. Then, by meaſuring the unknown angles and ſide, the former by the line of cords or otherwiſe, and the ſide by the line of equal parts, they will be found, as near as they can be meaſured, as below, viz,
BC = 310; the ∠A = 57°¼; and ∠B = 24°¾.

2. *Arithmetically, by Tables of Logs.*

As AB - = 365 -   log. 2·5622929
To AC - = 154·33 - - 2·1884504
So ſin. ∠C = 98° 3′ or 81° 57′ 9·9956993

To ſin. ∠ B = 24° 45′ - -   9·6218568

the ſum   -   122 48
taken from -   180 00

leaves ∠A   57 12

Then, again,

As ſin. ∠C = 98° 3′ · log. 9·9956993
To AB - = 365 - - - 2·5622929
So ſin. ∠A = 57° 12′ - - 9·9245721

To BC - = 309·86 - - 2·4911657

3. *Inſtrumentally, by Gunter's Lines.*

In the firſt proportion, Extend the compaſſes from 365 to 154¼ on the line of numbers ; and that extent will reach, upon the line of ſines, from 82° to 24¾, which gives the angle B. And, in the ſecond propor-tion, Extend from 98° to 57¼ on the ſines ; and that extent will reach, upon the numbers, from 365 to 310, or the ſide BC nearly.

2d Caſe, when there are Given two Sides and their contained angle, to find the reſt, the rule is this :

As the ſum of the two given ſides :
Is to the difference of the ſides : :
So is the tang. of half the ſum of the two oppo-
     ſite angles, or cotangent of half the given angle :
To tang. of half the diff. of thoſe angles.

Then the half diff. added to the half ſum, gives the greater of the two unknown angles ; and ſubtracted, leaves the leſs of the two angles.

Hence, the angles being now all known, the remain-ing 3d ſide will be found by the former caſe.

Ex. Suppoſe, in the triangle ABC, there be given
the ſide   AC = 154·33
the ſide   BC = 309·86
the included ∠C = 98° 3′
to find the other ſide and the angles.

1. *Geometrically.*—Draw two indefinite lines making the angle C = 98° 3′ : upon theſe lines ſet off CA = 154¼, and CB = 310 : Join the points A and B, and the figure is made. Then, by meaſurement, as before, we find the
∠A = 57¼ ; ∠B·24¾ ; and ſide AB = 365.

2. *By Logarithms.*

As CB + CA = 464·19 -   log. 2·6666958
To CB − CA = 155·53 - - 2·1918142
So tan. ½A + ½B = 40° 58½′ - - 9·9387803

To tan. ½A − ½B = 16 13¼ - - 9·4638987

ſum gives ∠A 57 12
diff. gives ∠B 24 45

Then,

As ſin. ∠B = 24° 45′ -   log. 9·6218612
To ſide AC = 154·33 -   2·1884504
So ſin. ∠C = 98° 3′. or 81°57 9·9956993

To ſide AB = 365   - - 2·5622885

3. *Inſtrumentally.*—Extend the compaſſes from 464 to 155¼ upon the line of numbers ; then that extent will reach, upon the line of tangents, from 41° to 16°¼. Then, in the 2d proportion, extend the com-paſſes from 24°¼ to 82° on the ſines ; and that extent

will reach, upon the numbers, from $154\frac{1}{3}$ to 365, which is the third side.

3d *Case*, is when the three sides are given, to find the three angles; and the method of resolving this case is, to let a perpendicular fall from the greatest angle, upon the opposite side or base, dividing it into two segments, and the whole triangle into two smaller right-angled triangles: then it will be,

As the base, or sum of the two segments :
Is to the sum of the other two sides : :
So is the difference of those sides :
To the difference of the segments of the base.

Then half this difference of the two segments added to the half sum, or half the base, gives the greater segments, and subtracted, gives the less. Hence, in each of the two right-angled triangles, there are given the hypotenuse, and the base, besides the right angle, to find the other angles by the 1st case.

Ex. In the trangle ABC, suppose there are given the three sides, to find the three angles, viz,

$$\left.\begin{array}{l} AB = 365 \\ AC = 154{\cdot}33 \\ BC = 309{\cdot}86 \end{array}\right\} \text{to find the angles.}$$

1. *Geometrically.*—Draw the base AB = 365: with the radius $154\frac{1}{3}$ and centre A describe an arc; and with the radius 310 and centre B describe another arc, cutting the former in C; then join AC and BC, and the triangle is constructed. And by measuring the angles, they are found, viz.

$$\angle A = 57^\circ\tfrac{1}{4}\text{;}\quad \angle B = 24^\circ\tfrac{3}{4}\text{;}\quad \angle C = 98^\circ \text{ nearly.}$$

2. *Arithmetically.*—Having let fall the perpendicular CP, dividing the base into the two segments AP, PB, and the given triangle ABC into the two right-angled triangles ACP, BCP. Then,

As AB	= 365	- - log.	2·5622929
To CB + CA	= 464·19	- - -	2·6666958
So CB − CA	= 155·53	- - -	2·1918142
To BP − PA	= 197·80	- - -	2·2962171
its half	= 98·90		
½AB	= 182·50		
sum BP	= 281·40		
dif. AP	= 83·60		

Then, in the triangle APC, right-angled at P,

As AC	= 154·33	- - log.	2·1884504
To sin. ∠P	= 90°	- - - -	10·0000000
So AP	= 83·6	- - -	1·9222063
To sin. ∠ACP	= 32° 48′	- - -	9·7337559
its comp. ∠A	= 57° 12		

And in the triangle BPC, right angled at P,

As BC	= 309·86	- - log.	2·4911655
To sin. ∠P	= 90°	- -	10·0000000
So BP	= 281·4	- -	2·4493241
To sin. ∠BCP	= 65° 15′	- -	9·9581586
its comp. ∠B	= 24 45		
Also to ∠ACP	= 32 48		
add ∠BCP	= 65 15		
makes ∠ACB	= 98 3		

3. *Instrumentally.*—In the 1st proportion, Extend the compasses from 365 to 464 on the line of numbers, and that extent will reach, on the same line, from $155\frac{1}{2}$ to 197·8 nearly.—In the 2d proportion, Extend the compasses from $154\frac{1}{3}$ to 83·6 on the line of numbers, and that extent will reach, on the sines, from 90° to $32^\circ\frac{3}{4}$ nearly.—In the 3d proportion, Extend the compasses from 310 to $281\frac{1}{2}$ on the line of numbers; then that extent will reach, on the sines, from 90° to $65^\circ\frac{1}{4}$.

The foregoing three cases include all the varieties of plane triangles that can happen, both of right and oblique-angled triangles. But beside these, there are some other theorems that are useful upon many occasions, or suited to some particular forms of triangles, which are often more expeditious in use than the foregoing general ones; one of which, for right-angled triangles, as the case for which it serves so often occurs, may be here inserted, and is as follows.

Case 4. When, in a right-angled triangle, there are given the angles and one leg, to find the other leg, or the hypothenuse. Then it will,

As radius :
To given leg AB : :
So tang. adjacent ∠A :
To the opp. leg BC, and : :
So sec. of same ∠A :
To hypot. AC .

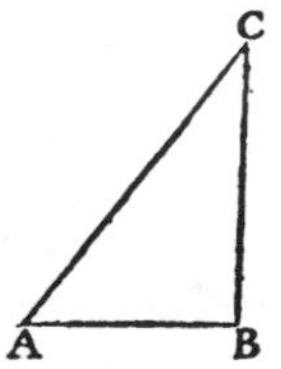

Ex. In the triangle ABC, right-angled at B,

$$\left.\begin{array}{l} \text{Given the leg } AB = 162 \\ \text{and the } \angle A = 53^\circ\ 7'\ 48'' \\ \text{conseq. } \angle C = 36\ 52\ \ 12 \end{array}\right\} \begin{array}{l}\text{to find BC} \\ \text{and AC.}\end{array}$$

1. *Geometrically.*—Draw the leg AB = 162: Erect the indefinite perpendicular BC: Make the angle $A = 53^\circ\frac{1}{8}$, and the side AC will cut BC in C, and form the triangle ABC. Then, by measuring, there will be found AC = 270, and BC = 216.

2. *Arithmetically.*

As radius	= 10	- -	log. 10·0000000
To AB	= 162	- - -	2·2095150
So tan. ∠A	= 53° 7′ 48″		10·1249372
To BC	= 216	- -	2·3344522
So sec. ∠A	= 53° 7′ 48″	- -	10·2218477
To AC	- = 270	- - -	2 4313627

3. *Instrumentally.*

3. *Inſtrumentally.*—Extend the compaſſes from 45° at the end of the tangents (the radius) to the tangent of 53°⅛; then that extent will reach, on the line of numbers, from 162 to 216, for BC. Again, extend the compaſſes from 36° 52′ to 90 on the ſines; then that extent will reach, on the line of numbers, from 162 to 270 for AC.

Note, another method, by making every ſide radius, is often added by the authors on Trigonometry, which is thus: The given right-angled triangle being ABC, make firſt the hypotenuſe AC radius, that is, with the extent of AC as a radius, and each of the centres A and C, deſcribe arcs CD and AE; then it is evi-

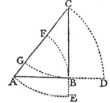

dent that each leg will repreſent the ſine of its oppoſite angle, viz, the leg BC the ſine of the arc CD or of the angle A, and the leg AB the ſine of the arc AE or of the angle C. Again, making either leg radius, the other leg will repreſent the tangent of its oppoſite angle, and the hypotenuſe the ſecant of the ſame angle; thus, with radius AB and centre A deſcribing the arc BF, BC repreſents the tangent of that arc, or of the angle A, and the hypotenuſe AC the ſecant of the ſame; or with the radius BC and centre C deſcribing the arc BG, the other leg AB is the tangent of that arc BG, or of the angle C, and the hypotenuſe CA the ſecant of the ſame.

And then the general rule for all theſe caſes is this, viz, that the ſides bear to each other the ſame proportions as the parts or things which they repreſent. And this is called making every ſide radius.

Spherical Trigonometry, is the reſolution and calculation of the ſides and angles of ſpherical triangles, which are made by three interſecting arcs of great circles on a ſphere. Here, any three of the ſix parts being given, even the three angles, the reſt can be found; and the ſides are meaſured or eſtimated by degrees, minutes, and ſeconds, as well as the angles.

Spherical Trigonometry is divided into right-angled and oblique-angled, or the reſolution of right and oblique-angled ſpherical triangles. When the ſpherical triangle has a right angle, it is called a right-angled triangle, as well as in plane triangles; and when a triangle has one of its ſides equal to a quadrant of a circle, it is called a quadrantal triangle.

For the reſolution of ſpherical Triangles, there are various theorems and proportions, which are ſimilar to thoſe in plane Trigonometry, by ſubſtituting the ſines of ſides inſtead of the ſides themſelves, when the proportion reſpects ſines; or tangents of the ſides for the ſides, when the proportion reſpects tangents, &c; ſome of the principal of which theorems are as follow:

Theor. 1. In any ſpherical triangle, the ſines of the ſides are proportional to the ſines of their oppoſite angles.

Theor 2. In any right-angled triangle,

As radius :
To ſine of one ſide : :
So tang. of the adjacent angle :
To tang. of the oppoſite ſide.

Theor. 3. If a perpendicular be let fall from any angl upon the baſe or oppoſite ſide of a ſpherical triangle it will be,

As the ſine of the ſum of the two ſides :
To the ſine of their difference : :
So cotan. ½ ſum angles at the vertex :
To tang. of half their difference.

Theor. 4.

As tang. half ſum of the ſides :
To tang. half their difference : :
So tang. ½ ſum ∠s at the baſe :
To tang. half their difference.

Theor. 5.

As cotan. ½ ſum of ∠s at the baſe :
To tang. half their difference : :
So tang. ½ ſum of ∠s at the vertex :
To tang. half their difference.

Theor. 6.

As tang. ½ ſum ſegments of baſe :
To tang. half ſum of the ſides : :
So tang. half difference of the ſides :
To tang. ½ diff. ſegments of baſe.

Theor. 7.

As ſin. ſum of ∠s at the baſe :
To ſine of their difference : :
So tang. ½ ſum ſegments of baſe :
To tang. of half their difference.

Theor. 8.

As ſin. ſum of ſegments of baſe :
To ſine of their difference : :
So ſin. ſum of angles at the vertex :
To ſine of their difference.

Theor. 9.

As ſine of the baſe :
To ſine of the vertical angle : :
So ſin. of diff. ſegments of the baſe :
To ſin. diff. ∠s at vertex, when the perp. falls within : :

Or ſo ſin. ſum ſegments of baſe :
To ſin. ſum vertical ∠s, where the perp. falls without.

Theor. 10.

As coſin. half ſum of the two ſides :
To coſine of half their difference : :
So cotang. of half the included angle :
To tang. half ſum of oppoſite angles.

Theor. 11.

As ſin. of half ſum of two ſides :
To ſine of half their difference : :
So cotang. half the included angle :
To tang. ½ diff. of the oppoſ. angles.

Theor.

Theor. 12.

As cosin. half sum of two angles :
To cosine of half their difference : :
So tang. of half the included side :
To tang ½ sum of the opposite sides.

Theor. 13.

As sin. half sum of two angles :
To sine of half their difference : :
So tang. half the included side :
To tang. ½ diff. of the opposite sides.

Theor. 14. In a right-angled triangle,

As sin. sum of hypot. and one side :
To sin. of their difference : .
So radius squared :
To square of tang. ½ contained angle.

Theor. 15. In any spherical triangle;
The product of the sines of two sides and of the cosine of the included angle, added to the product of the cosines of those sides, is equal to the cosine of the third side; the radius being 1.

Theor. 16. In any spherical triangle;
The product of the sines of two angles and of the cosine of the included side, minus the product of the cosines of those angles, is equal to the cosine of the third angle; the radius being 1.

By some or other of these theorems may all the cases of spherical triangles be resolved, both right angled and oblique: viz, the cases of right-angled triangles by the 1st and 2d theorems, and the oblique triangles by some of the other theorems.

In treatises on Trigonometry are to be found many other theorems, as well as synopses or tables of all the cases, with the theorem that is peculiar or proper to each. See the Introduction to my Mathematical Tables, p. 155 &c; or Robertson's Navigation, vol. 1, p. 162. See also Napier's Catholic or Universal Rule, in this Dictionary.

To the foregoing Theorems may be added the following synopsis of rules for resolving all the cases of plane and spherical triangles, under the title of

Trigonometrical Rules.

1. In a right-lined triangle, whose sides are A, B, C, and their opposite angles *a*, *b*, *c*; having given any three of these, of which one is a side; to find the rest.

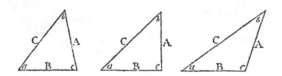

Put s for the sine, s the cosine, t the tangent, and t the cotangent of an arch or angle, to the radius *r*; also L for a logarithm, and L' its arithmetical complement. Then

Case 1. When three sides A, B, C, are given.
Put $P = \frac{1}{2}. \overline{A+B+C}$ or semiperimeter.

Then s. $\frac{1}{2}c = r \sqrt{\frac{\overline{P-A} \times \overline{P-B}}{A \times B}}$.

And s' $\frac{1}{2}c = r \sqrt{\frac{P \times \overline{P-C}}{A \times B}}$.

L. s. $\frac{1}{2}c = \frac{1}{2} (L. \overline{P-A} + L. \overline{P-B} + L'A + L'B)$,
L'. s. $\frac{1}{2}c = \frac{1}{2} (L. P + L. \overline{P-C} + L'A + L'B)$.

Note, When A = B, then

s. $\frac{1}{2}c = \frac{C}{A} \times \frac{r}{2}$. And s' $\frac{1}{2}c = r \sqrt{\frac{A^2 - \frac{1}{4}C^2}{A^2}}$.

Case 2. Given two sides A, B, and their included angle *c*.

Put $s = 90° - \frac{1}{2}c$, and t. $d = \frac{A-B}{A+B} \times$ t. s;
then $a = s + d$; and $b = s - d$. And

$$C = \sqrt{\frac{4AB + s^2 \frac{1}{2}c}{rr} + \overline{A-B}^2}.$$

Or in logarithms, putting L. Q = 2L. $\overline{A-B}$, and L. R = L. 2A + L. 2B + 2L. s. $\frac{1}{2}c - 20$, we shall have L. C = ½ L. $\overline{Q+R}$.

If the angle *c* be right, or = 90°; then

$$t. a = \frac{A}{B}r; \quad t. b = \frac{B}{A}r;$$

$$C = \frac{r}{s.a}A, \text{ or } = \frac{r}{s.b}B, \text{ or } = \sqrt{A^2 + B^2}.$$

If A = B; we shall have
$a = b = 90° - \frac{1}{2}c$, and $\Big\}$ $C = \frac{s. \frac{1}{2}c}{r} \times 2A$.

Case 3. When a side and its opposite angle are among the terms given.

Then $\frac{A}{s. a} = \frac{B}{s. b} = \frac{C}{s. c}$; from which equations any term wanted may be found.

When an angle, as *a*, is 90°, and A and C are given, then

$$B = \sqrt{A^2 - C^2} = \sqrt{\overline{A+C} \times \overline{A-C}}.$$
And L. B = ½ (L. $\overline{A+C}$ + L. $\overline{A-C}$).

Note, When two sides A, B, and an angle *a* opposite to one of them, are given; if A be less than B, then *b*, *c*, C have each two values; otherwise, only one value.

II. In a spherical triangle, whose three sides are A, B, C, and their opposite angles *a*, *b*, *c*; any three of these six terms being given, to find the rest.

Case

Case 1.   Given the three fides A, B, C.

Calling 2P the perim. or $P = \frac{1}{2} . \overline{A + B + C}$.

Then $s . \frac{1}{2} c = r \sqrt{\dfrac{s . \overline{P - A} \times s . \overline{P - B}}{s . A \times s . B}}$

And $s' \frac{1}{2} c = r \sqrt{\dfrac{s . P \times s . \overline{P - C}}{s . A \times s . B}}$.

$L . s . \frac{1}{2} c = \frac{1}{2} (L . s . \overline{P - A} + L . s . \overline{P - s} + L' s . A + L' s . B)$

$L . s' c = \frac{1}{2} (L . s . P + L . s . \overline{P - C} + L' s . A + L' s . B)$.

And the fame for the other angles.

Case 2.   Given the three angles.

Put $2p = a + b + c$.   Then

$s . \frac{1}{2} C = r \sqrt{\dfrac{s' p \times s' \overline{p - c}}{s . a \times s . b}}$.   And

$s' \frac{1}{2} C = r \sqrt{\dfrac{s' \overline{p - a} \times s' \overline{p - b}}{s . a \times s . b}}$.

$L . s . \frac{1}{2} C = \frac{1}{2} (L . s' p + L s' \overline{p - c} + L' s . a + L' s . b)$

$L . s' \frac{1}{2} C = \frac{1}{2} (L . s' \overline{p - a} + L . s' \overline{p - b} + L' s . a + L s' . b)$

And the fame for the other fides.

Note. The fign $>$ fignifies greater than, and $<$ lefs; alfo $\backsim$ the difference.

Case 3.   Given A, B, and included angle c.

To find an angle a oppofite the fide A,

let $r : s' c :: t . A : t . M$, like or unlike A, as c is $>$ or $< 90°$; alfo $N = B \backsim M$: then $s . N : s . M :: t . c : t . a$, like or unlike c as M is $>$ or $< B$.

Or let $s' \frac{1}{2} . \overline{A + B} : s . \frac{1}{2} . \overline{A \backsim B} :: t' \frac{1}{2} c : t . M$, which is $>$ or $< 90°$ as A + B is $>$ or $< 180°$; and $s . \frac{1}{2} . \overline{A + B} : s . \overline{A \backsim B} :: t' \frac{1}{2} c : t . N, > 90°$. then $a = M + N$; and $b = M - N$.

Again let $r : s' c :: t . A : t . M$, like or unlike A as c is $>$ or $< 90°$; and $N = B \backsim M$.

Then $s' M : s' N :: s' A , s' C$, like or unlike N as c is $>$ or $< 90°$. Or,

$s . \frac{1}{2} C = \sqrt{\dfrac{s . A \times s . B \times s^2 \frac{1}{2} c}{r^n}} + s^2 \frac{1}{2} . A \backsim B$.

In logarithms, put $L . Q = 2 L . s . \frac{1}{2} A \backsim B$; and $L . R = L . s . A + L . s . B + 2 L . s . \frac{1}{2} c - 20$; then $L . s . \frac{1}{2} C = \frac{1}{2} L . \overline{Q + R}$.

Case 4.   Given a, b, and included fide C.

Firft, let $r : s' C :: t . a : t' m$, like or unlike a as C is $>$ or $< 90°$; alfo $n = b \backsim m$.

Then $s' n : s' m :: t . C : t . A$, like or unlike n as a is $>$ or $< 90°$.

Or, let $s' \frac{1}{2} \overline{a + b} : s . \frac{1}{2} . \overline{a \backsim b} :: t . \frac{1}{2} C : t . M, >$ or $< 90°$ as $a + b$ is $>$ or $< 180°$;

and $s . \frac{1}{2} \overline{a + b} : s . \frac{1}{2} \overline{a \backsim b} :: t . \frac{1}{2} C : t N, > 90°$; then $A = M \pm N$; and $B = M \mp N$.

Again, let $r : s' C :: t . a : t' m$, like or unlike a as C is $>$ or $< 90°$; and $n = b \backsim m$:

then $s . m :: s . n :: s' a : s' c$, like or unlike a as m is $>$ or $< b$.

Case 5.   Given A, B, and an oppofite angle a.

1ft. $s . A : s . a :: s . B : s . b, >$ or $< 90°$.

2nd. Let $r : s' B :: t . a : t' m$, like or unlike s as a is $>$ or $< 90°$;

and $t . A : t . B :: s' m : s' n$, like or unlike A as a is $>$ or $> 90°$: then $c = m \pm n$, two values alfo.

3dly. Let $r : s' a :: t . B : t . M$, like or unlike B as a is $>$ or $< 90°$;

and $s' B : s' A :: s' M : s' N$, like or unlike A as a is $>$ or $< 90°$: then $C = M \pm N$, two values alfo.

But if A be equal to B, or to its fupplement, or between B and its fupplement; then is b like to B: alfo c is $= m \mp n$, and $C = M \pm N$, as B is like or unlike a.

Case 6.   Given a, b, and an oppofite fide A.

1ft. $s . a : s . A :: s . b : s . B, >$ or $< 90°$.

2nd. Let $r : s' b :: t . A : to t M$, like or unlike b as A is $>$ or $< 90°$;

and $t . a : t . b :: s . M : s . N, >$ or $< 90°$: then $C = M \mp N$, as a is like or unlike b.

3dly. Let $r : s' A :: t . b : t' m$, like or unlike b as A $>$ or $< 90°$;

and $s' b : s . a :: s . m : s . n, >$ or $< 90°$: then $c = m \pm n$, as a is like or unlike b.

But if A be equal to B, or to its fupplement, or between B and its fupplement; then B is unlike b, and only the lefs values of N, n, are poffible.

Note. When two fides A, B, and their oppofite angles a, b, are known; the third fide C, and its oppofite angle c, are readily found thus:

$s . \frac{1}{2} \overline{a \backsim b} : s . \frac{1}{2} . \overline{a + b} :: t . \frac{1}{2} \overline{A \backsim B} : t . \frac{1}{2} C$.

$s . \frac{1}{2} . \overline{A \backsim B} : s . \frac{1}{2} \overline{A + B} :: t . \frac{1}{2} . \overline{a \backsim b} : t . \frac{1}{2} c$.

III. In a right-angled fpheric triangle, where H is the hypotenufe, or fide oppofite the right angle, B, P the other two fides, and b, p their oppofite angles; any two of thefe five terms being given, to find the reft; the cafes, with their folutions, are as in the following Table.

2   The

The fame Table will alfo ferve for the quadrantal triangle, or that which has one fide = 90°, H being the angle oppofite that fide, B, P the other two angles, and *b*, *p* their oppofite fides : obferving, inftead of H, to take its fupplement : and mutually change the terms *like* and *unlike* for each other where H is concerned.

Cafe	Given	Reqᵈ	SOLUTIONS.
1	H B	*b* *p* P	s. H. : r :: s B : s*b*, and is like B r : t'H :: t. B : s'*p* } ▷ or ◁90° as H is like or unlike B s'B : r :: s'H : s'P }
2	H *b*	B P *p*	r : s'H :: s.*b* : s.B, like *b* r : s'*b* :: t.H : t.P } ▷ or ◁ 90° as H is like or unlike *b* r : s'H :: t.*b* : t'*p* }
3	B *b*	H P *p*	s.*b* : r :: s.B : s.H } r : t.B :: t'*b* : s.P } each ▷ or ◁ 90°; both values true s'B : r :: s'*b* : s.*p* }
4	B *p*	H *b* *p*	r : t'B :: s'*p* : t'H, ▷ or ◁ 90 as B is like or unlike *p* r : s'B :: s.*p* : s'*b*, like B r : s.B :: t.*p* : t.P, like *p*
5	B P	H *b* *p*	r : s'B :: s'P : s'H, ▷ or ◁ 90° as B is like or unlike P r : s.P :: t'B : t'*b*, like B r : s.B :: t'P : t'*p*, like P
6	*b* *p*	H B P	r : t'*b* :: t'*p* : s'H, ▷ or ◁ 90° as *b* is like or unlike *p* s.*p* : r :: s'*b* : s'B like *b* s.*b* : r :: s'*p* : s'P like *p*

The following Propofitions and Remarks, concerning Spherical Triangles, (felected and communicated by the reverend Nevil Mafkelyne, D. D. Aftronomer Royal, F. R. S.) will alfo render the calculation of them perfpicuous, and free from ambiguity.

" 1. A fpherical triangle is equilateral, ifofcelar, or fcalene, according as it has its three angles all equal, or two of them equal, or all three unequal ; and *vice verfa.*

2. The greateft fide is always oppofite the greateft angle, and the fmalleft fide oppofite the fmalleft angle.

3. Any two fides taken together, are greater than the third.

4. If the three angles are all acute, or all right, or all obtufe ; the three fides will be, accordingly, all lefs than 90°, or equal to 90°, or greater than 90°; and *vice verfa.*

5. If from the three angles A, B, C, of a triangle ABC, as poles, there be defcribed, upon the furface of the fphere, three arches of a great circle DE, DF, FE, forming by their interfections a new fpherical triangle DEF ; each fide of the new triangle will be the fupplement of the angle at its pole ; and each angle of the fame triangle, will be the fupplement of the fide oppofite to it in the triangle ABC.

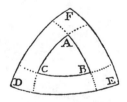

6. In any triangle ABC, or A*b*C, right angled in A, 1ft, The angles at the hypotenufe are always of the fame kind as their oppofite fides ; 2dly, The hypotenufe is lefs or greater than a quadrant, according as the fides including the right

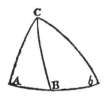

angle are of the fame or different kinds ; that is to fay, according as thefe fame fides are either both acute or both obtufe, or as one is acute and the other obtufe. And, *vice verfa,* 1ft, The fides including the right angle, are always of the fame kind as their oppofite angles : 2dly, The fides including the right angle will be of the fame or different kinds, according as the hypotenufe is lefs or more than 90°; but one at leaft of them will be of 90°, if the hypotenufe is fo."

TRILATERAL, three fided, a term applied to all figures of three fides, or triangles.

TRILLION, in Arithmetic, the number of a million of billions, or a million of million of millions.

TRIMMERS, in Architecture, pieces of timber framed at right-angles to the joifts, againft the ways for chimneys to fupport the hearths, and the well-holes for ftairs.

TRINE *Dimenfion,* or *threefold dimenfion,* includes length, breadth, and thicknefs. The Trine dimenfion is peculiar to bodies or folids.

TRINE, in Aftrology, is the afpect or fituation of one planet with refpect to another, when they are diftant

tant

tant ⅓ part of the circle, or 4 figns, or 120 degrees. It is alfo called trigon, and is denoted by the character △.

TRINITY *Sunday*, is the next after Whitfunday; fo called, becaufe on that day was anciently held a feftival (as it ftill continues to be in the Romifh Church) in honour of the Holy Trinity.—The obfervance of this feftival was firft enjoined by the 6th canon of the council of Arles, in 1260; and John the 22d, who diftinguifhed himfelf fo much by his opinion concerning the beatific vifion, it is faid, fixed the office for this feftival in 1334.

TRINODA, or TRINODIA *Terræ*, in fome ancient writers, denotes the quantity of 3 perches of land.

TRINOMIAL, in Algebra, is a quantity, or a root, confifting of three parts or terms, connected together by the figns $+$ or $-$: as $a + b - c$, or $x + y + z$.

TRIO, in Mufic, a part of a concert in which three perfons fing; or rather a mufical compofition confifting of 3 parts.—Trios are the fineft kind of mufical compofition, and pleafe moft in concerts.

TRIOCTILE, in Aftrology, an afpect or fituation of two planets, with regard to the earth, when they are 3 octants, or ⅜ of a circle, which is 135°, diftant from each other.—This afpect, which fome call the *fefquiquadrans*, is one of the new afpects added to the old ones by Kepler.

TRIONES, in Aftronomy, a fort of conftellation, or affemblage of 7 ftars in the Urfa Major, popularly called *Charles's Wain*.—From the *Septem Triones* the north pole takes the denomination *Septentrio*.

TRIPARTITION, is a divifion by 3, or the taking of the 3d part of any number or quantity.

TRIPLE, threefold. See RATIO and SUBTRIPLE.

TRIPLE, in Mufic, is one of the fpecies of meafure or time, and is taken from hence, that the whole, or half meafure, is divifible into 3 equal parts, and is beaten accordingly.

TRIPLICATE *Ratio*, is the ratio which cubes, or any fimilar folids, bear to each other; and is the cube of the fimple ratio, or this twice multiplied by itfelf. Thus 1 to 8 is the Triplicate ratio of 1 to 2, and 1 to 27 Triplicate of 1 to 3.

TRIPLICITY, or TRIGON, with Aftrologers, is a divifion of the 12 figns, according to the number of the 4 elements, earth, water, air, fire; each divifion confifting of 3 figns, making the earthly Triplicity, the watery Triplicity, the airy Triplicity, and the fiery Triplicity.

Triplicity is fometimes confounded with trine afpect; though they are, ftrictly fpeaking, very different things; as Triplicity is only ufed with regard to the figns, and trine with regard to the planets. The figns of Triplicity are thofe which are of the fame nature, and not thofe that are in trine afpect: thus Aries, Leo, and Sagittary are figns of Triplicity, becaufe thofe figns are, by thefe writers, all fuppofed fiery.

The figns in each of the four Triplicities, are as follow:

Earthly.	Watery.	Airy.	Fiery.
♉ Taurus.	♋ Cancer.	♊ Gemini.	♈ Aries.
♍ Virgo.	♏ Scorpio.	♎ Libra.	♌ Leo.
♑ Capricorn.	♓ Pifces.	♒ Aquarius.	♐ Sagittary.

TRIS-DIAPASON, or *Triple Diapafon Chord*, in Mufic, is what is otherwife called a *triple eighth*.

TRISECTION, the dividing a thing into three equal parts. The term is chiefly ufed in Geometry, for the divifion of an angle into three equal parts. The *Trifection of an angle* geometrically, is one of thofe great problems whofe folution has been fo much fought, for by mathematicians, for 2000 years paft; being, in this refpect, on a footing with the famous quadrature of the circle, and the duplicature of the cube.

The Ancients Trifected an angle by means of the conic fections, and the book of Inclinations; and Pappus enumerates feveral ways of doing it, in the 4th book of his Mathematical Collections, prop. 31, 32, 33, 34, 35, &c. He farther obferves, that the problem of Trifecting an angle, is a folid problem, or a problem of the 3d degree, being expreffed by the refolution of a cubic equation, in which way it has been refolved by Vieta, and others of the Moderns. See his Angular Sections, with thofe of other authors, and the Trifection in particular by cubic equations, as in Guifne's Application of Algebra to Geometry, in l'Hofpital's Conic Sections, and in Emerfon's Trigonometry, book 1, fec. 4. The cubic equation by which the problem of Trifection is refolved, is as follows: Let *c* denote the chord of a given arc, or angle, and *x* the cord of the 3d part of the fame, to the radius 1; then is

by the refolution of which cubic equation is found the value of *x*, or the chord of the 3d part of the given arc or angle, whofe chord is *c*; and the refolution of this equation, by Cardan's rule, gives the chord

$$x = \sqrt[3]{\frac{-c + \sqrt{c^2 - 4}}{2}} + \frac{1}{\sqrt[3]{\frac{-c + \sqrt{c^2 - 4}}{2}}},$$

$$\text{or } x = \sqrt[3]{\frac{-c + \sqrt{c^2 - 4}}{2}} + \sqrt[3]{\frac{-c - \sqrt{c^2 - 4}}{2}}$$

TRISPAST, or TRISPASTON, in Mechanics, a machine with 3 pulleys, or an affemblage of 3 pulleys, for raifing great weights; being a lower fpecies of the polyfpafton.

TRITE, in Mufic, the 3d mufical chord in the fyftem of the Ancients.

TRITONE, in Mufic, a falfe concord, confifting of three tones, or a greater third, and a greater tone. Its ratio or proportion in numbers, is that of 45 to 32.

TROCHILE, in Architecture, is that hollow ring, or cavity, which runs round a column next to the tore.

TROCHLEA, in Mechanics, one of the mechanic powers, more ufually called the pulley.

TRO-

TROCHOID, in the Higher Geometry, a curve described by a point in any part of the radius of a wheel, during its rotatory and progressive motions. This is the same curve as what is more usually called the *Cycloid*, where the construction and properties of it are shewn.

TRONE *Weight*, was the same with what we now call *Troy Weight*.

TRONE *Pound*, in Scotland, contains 20 Scotch ounces. Or because it is usual to allow one to the score, the Trone-pound is commonly 21 ounces.

TRONE-*Stone*, in Scotland, according to Sir John Skene, contains 19½ pounds.

TROPHY, in Architecture, an ornament which represents the trunk of a tree, charged or encompassed all around with arms or military weapons, both offensive and defensive.

TROPICAL, something relating to the Tropics. As, TROPICAL-*Winds*. See WIND, and TRADE-*Winds*.

TROPICAL *Year*, the space of time during which the sun passes round from a tropic, till his return to it again. See YEAR.

TROPICS, in Astronomy, two fixed circles of the sphere, drawn parallel to the equator, through the solstitial points, or at such distance from the equator, as is equal to the sun's greatest recess or declination, or to the obliquity of the ecliptic

Of the two Tropics, that on the north side of the equator, passes through the first point of Cancer, and is therefore called the *Tropic of Cancer*. And the other on the south side, passing through the first point of Capricorn, is called the *Tropic of Capricorn*.

To determine the distance between the two Tropics, and thence the sun's greatest declination, or the obliquity of the ecliptic ; observe the sun's meridian altitude, both in the summer and winter solstice, and subtract the latter from the former, so shall the remainder be the distance between the two Tropics ; and the half of this will be the quantity of the greatest declination, or the obliquity of the ecliptic ; the medium of which is now 23° 28″ nearly.

TROPICS, in Geography, are two lesser circles of the globe, drawn parallel to the equator through the beginnings of Cancer and Capricorn, being in the planes of the celestial Tropics, and consequently at 23° 28′ distance either way from the equator.

TROY-*Weight*, anciently called *Trone weight*, is supposed to be taken from a weight of the same name in France, and that from the name of the town of Troyes there.

The original of all weights used in England, was a corn or grain of wheat gathered out of the middle of the ear, and, when well dried, 32 of them were to make one pennyweight, 20 pennyweights 1 ounce, and 12 ounces 1 pound Troy. Vide Statutes of 51 Hen. III ; 31 Ed. I. and 12 Hen. VII.

But afterward it was thought sufficient to divide the said pennyweight into 24 equal parts, called grains, being the least weight now in common use ; so that the divisions of Troy weight now are these :

24 grains	= 1 pennyweight	*dwt.*
20 pennyweights	= 1 ounce	*oz.*
12 ounces	= 1 pound	*lb.*

By Troy-weight are weighed jewels, gold, silver, and all liquors.

TRUCKS, among Gunners, are the small wooden wheels fixed on the axletrees of gun carriages, especially those for ship service, to move them about by.

TRUE *Conjunction*, in Astronomy. See *True* CONJUNCTION.

TRUE *Place of a Planet or Star*, is a point in the heavens shewn by a right line drawn from the centre of the earth, through the centre of the star or planet.

TRUMPET, *Listening or Hearing*, is an instrument invented by Joseph Landini, to assist the hearing of persons dull of that faculty, or to assist us to hear persons who speak at a great distance.

Instruments of this kind are formed of tubes, with a wide mouth, and terminating in a small canal, which is applied to the ear. The form of these instruments evidently shews how they conduce to assist the hearing ; for the greater quantity of the weak and languid pulses of the air being received and collected by the large end of the tube, are reflected to the small end, where they are collected and condensed ; thence entering the ear in this condensed state, they strike the tympanum with a greater force than they could naturally have done from the ear alone.

' Hence it appears, that a speaking Trumpet may be applied to the purpose of a hearing Trumpet, by turning the wide end towards the sound, and the narrow end to the ear.

' *Speaking* TRUMPET, is a tube of a considerable length, from 6 to 15 feet, used for speaking with to make the voice be heard to a greater distance.

This tube, which is made of tin, is straight throughout its length, but opening to a large aperture outwards, and the other end terminating in a proper shape and size to receive both the lips in the act of speaking, the speaker pushing his voice or the sound outwards, by which means it may be heard at the distance of a mile or more.

The invention of this Trumpet is held to be modern, and has been ascribed to Sir Samuel Moreland, who called it the *tuba stentorophonica*, and in a work of the same name, published at London in 1671, that author gave an account of it, and of several experiments made with it. With one of these instruments, of 5½ feet long, 21 inches diameter at the greater end, and 2 inches at the smaller, tried at Deal-Castle, the speaker was heard to the distance of 3 miles, the wind blowing from the shore.

But it seems that Kircher has a better title to the invention ; for it is certain that he had such an instrument before ever Moreland thought of his. That author, in his Phonurgia Nova, published in 1673, says, that the tromba, published last year in England, he invented 24 years before, and published in his Mesurgia. He adds, that Jac. Albanus Ghibbisius and Fr. Eschinardus ascribe it to him ; and that G. Schottus testifies of him, that he had such an instrument in his chamber in the

Roman

Roman college, with which he could call to, and receive answers from the porter.

But, considering how famous the tube or horn of Alexander the Great was, it is rather strange that the Moderns should pretend to the invention. With his stentorophonic horn or tube he used to speak to his army, and make himself be distinctly heard, it is said, 100 stadia or furlongs. A figure of this tube is preserved in the Vatican; and it is nearly the same as that now in use. See STENTOROPHONIC.

The principle of this instrument is obvious; for as sound is stronger in proportion to the density of the air, it follows that the voice in passing through a tube, or trumpet, must be greatly augmented by the constant reflection and agitation of the air through the length of the tube, by which it is condensed, and its action on the external air greatly increased at its exit from the tube.

It has been found, that a man speaking through a tube of 4 feet long, may be understood at the distance of 500 geometrical paces; with a tube 16⅔ feet, at the distance of 1800 paces; and with a tube 24 feet long, at more than 2500 paces.

Although some advantage in heightening the sound, both in speaking and hearing; be derived from the shape of the tube, and the width of the outer end, yet the effect depends chiefly upon its length. As to the form of it, some have asserted that the best figure is that which is formed by the revolution of a parabola about its axis; the mouth-piece being placed in the focus of the parabola, and consequently the sonorous rays reflected parallel to the axis of the tube. But Mr. Martin observes, that this parallel reflection is by no means essential to increasing the sound: on the contrary, it prevents the infinite number of reflections and reciprocations of sound, in which, according to Newton, its augmentation chiefly consists; the augmentation of the impetus of the pulses of air being proportional to the number of repercussions from the sides of the tube, and therefore to its length, and to such a figure as is most productive of them. Hence he infers, that the parabolic Trumpet is the most unfit of any for this purpose; and he endeavours to shew, that the logarithmic or logistic curve gives the best form, viz, by a revolution about its axis. Martin's Philos. Brit. vol. 2, pa. 248, 3d edit.

But Cassegrain is of opinion that an hyperbola, having the axis of the tube for an asymptote, is the best figure for this instrument. Musschenb. Intr. ad Phil. Nat. tom. 2, pa. 926, 4to.

For other constructions of Speaking Trumpets, by Mr. Conyers, see Philos. Transf. numb. 141, for 1678.

TRUNCATED *Pyramid* or *Cone*, is the frustum of one, being the part remaining at the bottom, after the top is cut off by a plane parallel to the base. See FRUSTUM.

TRUNNIONS, of a piece of ordnance, are those knobs or short cylinders of metal on the sides, by which it rests on the cheeks of the carriage.

TRUNNION-*Ring*, is the ring about cannon, next before the Trunnions.

TSCHIRNHAUSEN (ERNFROY WALTER), an ingenious mathematician, lord of Killingswald and of Stolzenberg in Lusatia, where he was born in 1651. After having served as a volunteer in the army of Holland in 1672, he travelled into most parts of Europe, as England, Germany, Italy, France, &c. He went to Paris for the third time in 1682; where he communicated to the Academy of Sciences, the discovery of the curves called, from him, Tschirnhausen's Caustics; and the Academy in consequence elected the inventor one of its foreign members. On returning to Italy, he was desirous of perfecting the science of optics; for which purpose he established two glass-works, from whence resulted many new improvements in dioptrics and physics, particularly the noted burning-glass which he presented to the regent.—It was to him too that Saxony owed its porcelane manufactory.

Content with the enjoyment of literary fame, Tschirnhausen refused all other honours that were offered him. Learning was his sole delight. He searched out men of talents, and gave them encouragement. He was often at the expence of printing the useful works of other men, for the benefit of the public; and died, beloved and regretted, the 11th of September 1708.

Tschirnhausen wrote, *De Medicina Mentis & Corporis*, printed at Amsterdam in 1687. And the following memoirs were printed in the volumes of the Academy of Sciences.

1. Observations on Burning Glasses of 3 or 4 feet diameter: vol. 1699.

2. Observations on the Glass of a Telescope, convex on both sides, of 32 feet focal distance; 1700.

3. On the Radii of Curvature, with the finding the Tangents, Quadratures, and Rectifications of many curves; 1701.

4. On the Tangents of Mechanical Curves; 1702.

5. On a method of Quadratures; 1702.

TUBE, a pipe, conduit, or canal; being a hollow cylinder, either of metal, wood, glass, or other matter, for the conveyance of air, or water, &c.

The term is chiefly applied to those used in physics, astronomy, anatomy, &c. On other ordinary occasions, we more usually say *pipe*.

In the memoirs of the French Academy of Sciences, Varignon has given a treatise on the proportions for the diameters of tubes, to give any particular quantities of water. The result of his paper gives these two analogies, viz, that the diminutions of the velocity of water, occasioned by its friction against the sides of Tubes, are as the diameters; the Tubes being supposed equally long: and the quantities of water issuing out at the Tubes, are as the square roots of their diameters, deducting out of them the quantity that each is diminished.

TUBE, in Astronomy, is sometimes used for telescope; but more properly for that part of it into which the lenses are fitted, and by which they are directed and used.

TUESDAY,

TUESDAY, the 3d day of the week, so called from Tuesco, one of the Saxon Gods, similar to Mars; for which reason the astronomical mark for this day of the week, is ♂.

TUMBREL, is a kind of carriage with two wheels, used either in Husbandry for dung, or in Artillery to carry the tools of the pioneers, &c, and sometimes likewise the money of an army.

TUN, is a measure for liquids, as wine, oil, &c.

The English Tun contains 2 pipes, or 4 hogsheads, or 252 gallons.

TUNE, or TONE, in Music, is that property of sounds by which they come under the relation of acute and grave.

If two or more sounds be compared together in this relation, they are either equal or unequal in the degree of Tune: such as are equal, are called *unisons*. The unequal constitute what are called *intervals*, which are the differences of Tone between sounds.

Sonorous bodies are found to differ in Tone: 1st, According to the different kinds of matter; thus the sound of a piece of gold, is much graver than that of a piece of silver of the same shape and dimensions. 2d, According to the different quantities of the same matter in bodies of the same figure; as a solid sphere of brass of 1 foot diameter, sounds acuter than a sphere of brass of 2 feet diameter.

But the measures of Tone are only to be sought in the relations of the motions that are the cause of sound, which are most discernible in the vibration of chords. Now, in general, we find that in two chords all things being equal, excepting the tension, the thickness, or the length, the Tones are different; which difference can only be in the velocity of their vibratory motions, by which they perform a different number of vibrations in the same time; as it is known that all the small vibrations of the same chord are performed in equal times. Now the frequenter or quicker those vibrations are, the more acute is the Tone; and the slower and fewer they are in the same space of time, by so much the more grave is the Tone. So that any given note of a Tune is made by one certain measure of velocity of vibrations, that is, such a certain number of vibrations of a chord or string, in such a certain space of time, constitutes a determinate Tone.

This theory is strongly supported by the best and latest writers on music, Holder, Malcolm, Smith, &c, both from reason and experience. Dr. Wallis, who owns it very reasonable, adds, that it is evident the degrees of acuteness are reciprocally as the lengths of the chords; though, he says, he will not positively affirm that the degrees of acuteness answer the number of vibrations, as their only true cause: but his diffidence arises from hence, that he doubts whether the thing has been sufficiently confirmed by experiment.

TUNNAGE. See TONNAGE.

TURN, is used for a circular motion; in which sense it agrees with revolution.

TURN, in Clock or Watch-work, particularly denotes the revolution of a wheel or pinion.

In calculation, the number of Turns which the pinion hath, is denoted in common arithmetic thus, 5) 60 (12, where the pinion 5, playing in a wheel of 60, moves round 12 times in one Turns of the wheel. Now by knowing the number of Turns which any pinion hath, in one Turn of the wheel it works in, you may easily find how many Turns a wheel or pinion has at a greater distance; as the contrat-wheel, crown-wheel, &c, by multiplying together the quotients, and the number produced is the number of Turns, as in the example here annexed: the first of

5) 55 (11

5) 45 (9

5) 40 (8

these three numbers has 11 Turns, the next 9, and the last 8; if you multiply 11 by 9, it produces 99; that is, in one Turn of the wheel 55, there are 99 Turns of the second pinion 5, or the wheel 40, which runs concentrical or on the same arbor with the second pinion 5: and if you again multiply 99 by the last quotient 8, it produces 792, which is the number of Turns the third pinion 5 hath. See CLOCK-*work*, and PINION.

TURNING *to windward*, in Sea Language, denotes that operation in sailing when a ship endeavours to make a progress against the direction of the wind, by a compound course, inclined to the place of her destination.—This method of navigation is otherwise called *plying to windward*.

TUSCAN *Order*, in Architecture, is the first, the simplest, and the strongest or most massive of any. Its column has 7 diameters in height; and its capital, base, and entablature, have no ornaments, and but few mouldings.

TWELFTH-*Day*. the festival of the Epiphany, or the manifestation of Christ to the Gentiles, so called, as being the Twelfth day, exclusive, from the nativity or Christmas-day; of course it falls always on the 6th day of January.

TWILIGHT, in Astronomy, is that faint light which is perceived before the sun-rising, and after sun-setting. The Twilight is occasioned by the earth's atmosphere refracting the rays of the sun, and reflecting them among its particles.

The depression of the sun below the horizon, at the beginning of the morning, and end of the evening Twilight, has been variously stated, at different seasons, and by different observers: by Alhazen it was observed to be 19°; by Tycho 17°; by Rothman 24°; by Stevinus 18°; by Cassini 15°; by Riccioli, at the time of the equinox in the morning 16°, in the evening 20°½; in the summer solstice in the morning 21° 25', and in the winter 17° 15'. Whence it appears that the cause of the Twilight is variable; but, on a medium, about 18° of the sun's depression will serve tolerably well for our latitude, for the beginning and end of Twilight, and according to which Dr. Long, (in his Astronomy, vol. 1, pa. 258) gives the following Table, of the duration of Twilight, in different latitudes, and for several different declinations of the sun.

Latitude.

Latitude.	0		10		20		30		40		45		50		52½		55		60		65		70		75		80		85		90	
	h	m	h	m	h	m	h	m	h	m	h	m	h	m	h	m	h	m	h	m	h	m	h	m	h	m	h	m	h	m	h	m
☉ Enters ♋	1	18	1	21	1	28	1	41	2	8	2	39	w	n	w	n	w	n	w	n	w	n	c	d	c	d	c	d	c	d	c	d
♊ ♌	1	16	1	19	1	25	1	36	1	58	2	19	3	3	w	n	w	n	w	n	w	n	c	d	c	d	c	d	c	d	c	d
♉ ♍	1	13	1	15	1	20	1	28	1	43	1	55	2	12	2	25	2	41	3	55	w	n	w	n	w	n	c	d	c	d	c	d
♈ ♎	1	12	1	13	1	17	1	24	1	35	1	44	1	55	2	2	2	10	2	33	3	8	4	18	w	n	w	n	w	n	w	n
♓ ♏	1	13	1	14	1	18	1	24	1	35	1	43	1	54	2	0	2	8	2	27	2	56	8	41	5	2	17	32	w	n	w	n
♒ ♐	1	16	1	17	1	21	1	28	1	40	1	49	2	1	2	8	2	18	2	43	3	26	11	38	11	14	10	32	8	38	c	n
♑	1	18	1	19	1	23	1	30	1	43	1	53	2	6	2	15	2	26	2	57	4	4	10	24	9	30	7	46	c	n	c	n

Where *c d* fignify that it is then continual day, *c n* continual night, and *w n* that the Twilight lafts the whole night.

Prob.—To find the Beginning or End of Twilight.

In this problem, there are given the fides of an oblique fpherical triangle, to find an angle; viz, given the fide ZP the colatitude of the place; P☉ the codeclination, or polar diftance; and Z☉ the zenith diftance, which is always equal to 108°, viz, 90° from the zenith to the horizon, and 18° more for the fun's diftance below the horizon. For example, fuppofe the place London in latitude 51° 32', and the time the 1ft of May, when the fun's declination is 15° 12' north. Here then ZP = 38° 28' the complement of 51° 32', and P☉ = 74° 48', the complement of 15° 12'. Then the calculation is as follows.

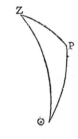

$$P☉ = 74° \ 48'$$
$$PZ = 38 \ \ 28$$
$$P☉ - PZ = 36 \ \ 20 = D$$
$$Z☉ = 108 \ \ 00$$

$$Z☉ + D = 144 \ \ 20 \quad | \quad 72° \ 10' = \tfrac{1}{2} \ \overline{Z☉ + D}$$
$$2) \underline{\qquad\qquad}$$
$$Z☉ - D = 71 \ \ 40 \quad | \quad 35 \ \ 50 = \tfrac{1}{2} \ \overline{Z☉ - D}$$

Then;

Co-ar. fin. polar dift.	= 74° 48'	- 0·01547
Co-ar. fin. colat. -	- = 38 28	- 0·20617
Sine ½ $\overline{Z☉ + D}$	- = 72 10	- 9·97861
Sine ½ $\overline{Z☉ - D}$	- = 35 50	- 9·76747

Sum of thefe four logs. - - 19·96772
Half fum gives 74° 28½' - 9·98386

Which doubled gives 148 57 for the angle ZP☉.

This 148° 57' reduced to time, at the rate of 15° per hour, gives 9ʰ 55ᵐ 48ˢ, either before or after noon; that is, the twilight begins at 2ʰ 4ᵐ 12ˢ in the morning, and ends at 9ʰ 55ᵐ 48ˢ in the evening on the given day at London.

TWINKLING *of the Stars*, denotes that tremulous motion which is obferved in the light proceeding from the fixed ftars.

This Twinkling in the ftars has been varioufly accounted for. Alhazen, a Moorifh philofopher of the 12th century, confiders refraction as the caufe of this phenomenon.

Vitello, in his Optics, (compofed before the year 1270) pá. 449, afcribes the Twinkling of the ftars to the motion of the air, in which the light is refracted; and he obferves, in confirmation of this hypothefis, that they Twinkle ftill more when they are viewed in water put into motion.

Dr. Hook (Microgr. pa. 231, &c) afcribes this phenomenon to the inconftant and unequal refraction of the rays of light, occafioned by the trembling motion of the air and interfperfed vapours, in confequence of variable degrees of heat and cold in the air, producing correfponding variations in its denfity, and alfo of the action of the wind, which muft caufe the fucceffive rays to fall upon the eye in different directions, and confequently upon different parts of the retina at different times, and alfo to hit and mifs the pupil alternately; and this alfo is the reafon, he fays, why the limbs of the fun, moon, and planets appear to wave or dance.

Thefe tremors of the air are manifeft to the eye by the tremulous motion of fhadows caft from gh towers; and by looking at objects through the fmoke of a chimney, or through fteams of hot water, or at objects fituated beyond hot fands, efpecially if the air be moved tranfverfely over them. But when ftars are feen through telefcopes that have large apertures; they Twinkle but little, and fometimes not at all. For, as Newton has obferved, (Opt. pa. 98) the rays of light which pafs through different parts of the aperture, tremble each of them apart, and by means of their various, and contrary tremors, fall at one and the fame time

time upon different points in the bottom of the eye, and their trembling motions are too quick and confused to be separately perceived. And all these illuminated points constitute one broad lucid point, composed of those many trembling points confusedly and insensibly mixed with one another by very short and swift tremors, and so cause the star to appear broader than it is, and without any trembling of the whole.

Dr. Jurin, in his Essay upon Distinct and Indistinct Vision, has recourse to Newton's hypothesis of fits of easy refraction and reflection for explaining the Twinkling of the stars: thus, he says, if the middle part of the image of a star be changed from light to dark, and the adjacent ring at the same time be changed from dark to light, as must happen from the least motion of the eye towards or from the star, this will occasion such an appearance as Twinkling.

Mr. Michell (Philos. Transf. vol. 57, pa. 262) supposes that the arrival of fewer or more rays at one time, especially from the smaller or more remote fixed stars, may make such an unequal impression on the eye, as may at least have some share in producing this effect: since it may be supposed that even a single particle of light is sufficient to make a sensible impression on the organs of sight; so that very few particles arriving at the eye in a second of time, perhaps not more than three or four, may be sufficient to make an object constantly visible. See LIGHT.

Hence, he says, it is not improbable that the number of the particles of light which enter the eye in a second of time, even from Syrius himself, may not exceed 3 or 4 thousand, and from stars of the 2d magnitude they may probably not exceed 100. Now the apparent increase and diminution of the light, which we observe in the Twinkling of the stars, seem to be repeated at intervals not very unequal, perhaps about 4 or 5 times in a second. He therefore thought it reasonable to suppose, that the inequalities which will naturally arise from the chance of the rays coming sometimes a little denser, and sometimes a little rarer, in so small a number of them, as must fall upon the eye in the 4th or 5th part of a second, may be sufficient to account for this appearance.

Since these observations were published however, Mr. Michell (as we are informed by Dr. Priestley in his Hist. of Light, pa. 495) has entertained some suspicion, that the unequal density of light does not contribute to this effect in so great a degree as he had imagined; especially as he has observed that even Venus does sometimes Twinkle. This he once observed her to do remarkably when she was about 6 degrees high, though Jupiter, which was then about 16 degrees high, and was sensibly less luminous, did not Twinkle at all. If, notwithstanding the great number of rays which doubtless come to the eye from such a surface as this planet presents, its appearance be liable to be affected in this manner, it must be owing to such undulations in the atmosphere, as will probably render the effect of every other cause altogether insensible.

Musschenbroek suspects (Introd. ad Phil. Nat. vol. 2, sect. 1741, pa. 707) that the Twinkling of the stars arises from some affection of the eye, as well as the state of the atmosphere. For, says he, in Holland, when the weather is frosty, and the sky very clear, the stars Twinkle most manifestly to the naked eye, though not in telescopes; and since he does not suppose there is any great exhalation, or dancing of the vapour, at that time, he questions whether the vivacity of the light, affecting the eye, may not be concerned in the phenomenon.

But this philosopher might have satisfied himself with respect to this hypothesis, by looking at the stars near the zenith, when the light traverses but a small part of the atmosphere, and therefore might be expected to affect the eye most sensibly. For he would have found that they do not Twinkle near so much as they do near the horizon, when much more of their light is intercepted by the atmosphere.

Some astronomers have lately endeavoured to explain the Twinkling of the fixed stars, by the extreme minuteness of their apparent diameter; so that they suppose the light of them is intercepted by every mote that floats in the air. To this purpose Dr. Long observes (Altron. vol. 1, pa. 170) that our air near the earth is so full of various kinds of particles, which are in continual motion, that some one or other of them is perpetually passing between us and any star we look at, which makes us every moment alternately see it and lose sight of it: and this Twinkling of the stars, he says, is greatest in those that are nearest the horizon, because they are viewed through a great quantity of thick air, where the intercepting particles are most numerous; whereas stars that are near the zenith do not Twinkle so much, because we do not look at them through so much thick air, and therefore the intercepting particles, being fewer, come less frequently before them. With respect to the planets, it is observed that, because they are much nearer to us than the stars, they have a sensible apparent magnitude, so that they are not covered by the small particles floating in the atmosphere, and therefore do not Twinkle, but shine with a steady light.

The fallacy of this hypothesis appears from the observation of Mr. Michell, that no object can hide a star from us that is not large enough to exceed the apparent diameter of the star, by the diameter of the pupil of the eye; so that if a star were even a mathematical point, or of no diameter, the interposing object must still be equal in size to the pupil of the eye; and indeed it must be large enough to hide the star from both eyes at the same time.

The principal cause therefore of the Twinkling of the stars, is now acknowledged to be the unequal refraction of light, in consequence of inequalities and undulations in the atmosphere.

Besides a variation in the quantity of light, it may here be added, that a momentary change of colour has likewise been observed in some of the fixed stars. Mr. Melville (Edinb. Essays, vol. 2, pa. 81) says, that when one looks steadfastly at Sirius, or any bright star, not much elevated above the horizon, its colour seems not to be constantly white, but appears tinctured, at every Twinkling, with red and blue. Mr. Melville could not entirely satisfy himself as to the cause of this phenomenon;

nomenon; obferving that the feparation of the colours by the refractive power of the atmofphere, is probably too fmall to be perceived. Mr. Michell's hypothefis above mentioned, though not adequate to the explication of the Twinkling of the ftars, may pretty well account for this circumftance. For the red and blue rays being much fewer than thofe of the intermediate colours, and therefore much more liable to inequalities from the common effect of chance, a fmall excefs or defect in either of them will make a very fenfible difference in the colour of the ftars.

TYCHONIC *Syftem*, or *Hypothefis*, is an order or arrangement of the heavenly bodies, of an intermediate nature between the Copernican and Ptolomaic; and is fo called from its inventor Tycho Brahe. See SYSTEM.

TYMPAN, or TYMPANUM, in Architecture, is the area of a pediment, being that part which is on a level with the naked of the frize. Or it is the fpace included between the three cornices of a triangular pediment, or the two cornices of a circular one.

TYMPAN is alfo ufed for that part of a pedeftal called the *trunk* or *dye*.

TYMPAN, among Joiners, is alfo applied to the pannels of doors.

TYMPAN *of an Arch*, is a triangular fpace or table in the corners of fides of an arch, ufually hollowed and enriched, fometimes with branches of laurel, olive-tree, or oak; or with trophies, &c; fometimes with flying figures, as fame, &c; or fitting figures, as the cardinal virtues.

TYMPAN, in Mechanics, is a kind of wheel placed round an axis, or cylindrical beam, on the top of which are two levers, or fixed ftaves, for more eafily turning the axis about, in order to raife a weight. The Tympanum is much the fame with the peritrochium; but that the cylinder of the axis of the peritrochium is much fhorter and lefs than the cylinder of the Tympanum.

TYMPANUM of a machine, is alfo ufed for a hollow wheel, in which people or animals walk, to turn it; fuch as that of fome cranes, calenders, &c.

TYR, in the Ethiopian Calendar, the name of the 5th month of the Ethiopian year. It commences on the 25th of December of the Julian year.

TYSHAS, among the Ethiopians, the name of the 4th month of their year, commencing the 27th of November in the Julian year.

U AND V.

V A C

V Is a numeral letter, in the Roman numeration, denoting 5 or five. And with a dafh over the top thus $\overline{V}$, it denoted 5000.

VACUUM, in Phyfics, a fpace empty or devoid of all matter.

Whether there be any fuch thing in nature as an abfolute Vacuum; or whether the univerfe be completely full, and there be an abfolute plenum; is a queftion that has been agitated by the philofophers of all ages.

The Ancients, in their controverfies, diftinguifhed two kinds; a *Vacuum coacervatum*, and a *Vacuum interfperfum*, or *diffeminatum*.

VACUUM *Coacervatum*, is conceived as a confiderably large fpace deftitute of matter; fuch, for inftance, as there would be, fhould God annihilate all the air, and other bodies, within the walls of a chamber.

The exiftence of fuch a Vacuum is maintained by the Pythagoreans, Epicureans, and the Atomifts or Corpufcularians; moft of whom affert, that fuch a Vacuum actually exifts without the limits of the fenfible world. But the modern Corpufcularians, who hold a *Vacuum coacervatum*, deny that appellation; as conceiving that

V A C

fuch a Vacuum muft be infinite, eternal, and uncreated.

According then to the later philofophers, there is no Vacuum coacervatum without the bounds of the fenfible world; nor would there be any other Vacuum, provided God fhould annihilate divers contiguous bodies, than what amounts to a mere privation, or nothing; the dimenfions of fuch a fpace, which the Ancients held to be real, being by thefe held to be mere negations; that is, in fuch a place there is fo much length, breadth, and depth wanting, as a body muft have to fill it. To fuppofe then, that when all the matter in a chamber is annihilated, there fhould yet be real dimenfions, is to fuppofe, fay they, corporeal dimenfions without body; which is abfurd.

The Cartefians however deny any *Vacuum coacervatum* at all, and affert that if God fhould immediately annihilate all the matter, for example in a chamber, and prevent the ingrefs of any other matter, the confequence would be, that the walls would become contiguous, and include no fpace at all. They add, that if there be no matter in a chamber, the walls cannot be conceived otherwife than as contiguous; thofe things
being

being said to be contiguous, between which there is not any thing intermediate: but if there be no body between, there is, say they, no extension between; extension and body being the same thing: and if there be no extension between, then the walls are contiguous; and where is the Vacuum?——But this reasoning, or rather quibbling, is founded on the mistake, that body and extension are the same thing.

Vacuum *Disseminatum*, or *Interspersum*, is that supposed to be naturally interspersed in and among bodies, in the interstices between different bodies, and in the pores of the same body.

It is this kind of Vacuum which is chiefly contested among the modern philosophers; the Corpuscularians strenuously asserting it; and the Peripatetics and Cartesians as tenaciously denying it. See Cartesian and Leibnitzian.

The great argument urged by the Peripatetics against a Vacuum interspersum, is, that there are divers bodies frequently seen to move contrary to their own nature and inclination; and that for no other apparent reason, but to avoid a Vacuum: whence they conclude, that nature abhors a Vacuum; and give us a new class of motions ascribed to the *fuga vacui* or nature's flying a Vacuum. Such, they say, is the rise of water in a syringe, upon the drawing up of the piston; and such is the ascent of water in pumps, and the swelling of the flesh in a cupping glass, &c.—But since the weight, elasticity, &c, of the air have been ascertained by sure experiments, those motions and effects are universally, and justly, ascribed to the gravity and pressure of the atmosphere.

The Cartesians deny, not only the actual existence, but even the possibility of a Vacuum; and that on this principle, that extension being the essence of matter, or body, wherever extension is, there is matter; but mere space, or vacuity, is supposed to be extended; therefore it is material. Whoever asserts an empty space, say they, conceives dimensions in that space, i. e. he conceives an extended substance in it; and therefore he denies a Vacuum, at the same time that he admits it.—But Descartes, if we may believe some accounts, rejected a Vacuum from a complaisance to the taste which prevailed in his time, against his own first sentiments; and among his familiar friends he used to call his system his philosophical romance.

On the other hand, the corpuscular authors prove, not only the possibility, but the actual existence, of a Vacuum, from divers considerations; particularly from the consideration of motion in general; and that of the planets, comets, &c, in particular; as also from the fall of bodies; from the vibration of pendulums; from rarefaction and condensation; from the different specific gravities of bodies; and from the divisibility of matter into parts.

1. First, there could be no linear or progressive motion without a Vacuum; for if all space were full of matter, no body could be moved out of its place, for want of another place unoccupied, to move into. And this argument was stated even by Lucretius.

2. The motions of the planets and comets also prove a Vacuum. Thus, Newton argues, " that there is no such fluid medium as æther," (to fill up the porous parts of all sensible bodies, and so make a plenum);

seems probable; because the planets and comets proceed with so regular and lasting a motion, through the celestial spaces; for hence it appears that those celestial spaces are void of all sensible resistance, and consequently of all sensible matter. Consequently if the celestial regions were as dense as water, or as quicksilver, they would resist almost as much as water or quicksilver; but if they were perfectly dense, without any intersperced vacuity, though the matter were ever so fluid and subtle, they would resist more than quicksilver does: a perfectly solid globe, in such a medium, would lose above half its motion, in moving 3 lengths of its diameter; and a globe not perfectly solid, such as the bodies of the planets and comets are, would be stopped still sooner. Therefore, that the motion of the planets and comets may be regular, and lasting, it is necessary that the celestial spaces be void of all matter; except perhaps some few and much rarefied effluvia of the planets and comets, and the passing rays of light."

3. The same great author also deduces a Vacuum from the consideration of the weights of bodies; thus: " All bodies about the earth gravitate towards it; and the weights of all bodies, equally distant from the earth's centre, are as the quantities of matter in those bodies. If the æther therefore, or any other subtile matter, were altogether destitute of gravity, or did gravitate less than in proportion to the quantity of its matter; because (as Aristotle, Descartes, and others, argue) it differs from other bodies only in the form of matter; the same body might, by the change of its form, gradually be converted into a body of the same constitution with those which gravitate most in proportion to the quantity of matter: and, on the other hand, the heaviest bodies might gradually lose their gravity, by gradually changing their form; and so the weights would depend upon the forms of bodies, and might be changed with them; which is contrary to all experiment."

4. The descent of bodies proves, that all space is not equally full; for the same author goes on, " If all spaces were equally full, the specific gravity of that fluid with which the region of the air would, in that case, be filled, would not be less than the specific gravity of quicksilver or gold, or any other the most dense body; and therefore neither gold, nor any other body, could descend in it. For bodies do not descend in a fluid, unless that fluid be specifically lighter than the body. But by the air-pump we can exhaust a vessel, till even a feather shall fall with a velocity equal to that of gold in the open air; and therefore the medium through which this feather falls, must be much rarer than that through which the gold falls in the other case. The quantity of matter therefore in a given space may be diminished by rarefaction: and why may it not be diminished ad infinitum? Add, that we conceive the solid particles of all bodies to be of the same density; and that they are only rarefiable by means of their pores; and hence a Vacuum evidently follows."

5. " That there is a Vacuum, is evident too from the vibrations of pendulums: for since those bodies, in places out of which the air is exhausted, meet with no resistance to retard their motion, or shorten their vibrations; it is evident that there is no sensible matter in those spaces, or in the occult pores of those bodies."

6. That

6. That there are interfperfed vacuities, appears from matter's being actually divided into parts, and from the figures of thofe parts; for, on fuppofition of an abfolute plenum, we do not conceive how any part of matter could be actually divided from that next adjoining, any more than it is poffible to divide actually the parts of abfolute fpace from one another: for by the actual divifion of the parts of a continuum from one another, we conceive nothing elfe underftood, but the placing of thofe parts at a diftance from one another, which in the continuum were at no diftance from one another: but fuch divifions between the parts of matter muft imply vacuities between them.

7. As for the figures of the parts of bodies, upon the fuppofition of a plenum, they muft either be all rectilinear, or all concavo-convex; otherwife they would not adequately fill fpace; which we do not find to be true in fact.

8. The denying a Vacuum fuppofes what it is impoffible for any one to prove to be true, viz, that the material world has no limits.

However, we are told by fome, that it is impoffible to conceive a Vacuum. But this furely muft proceed from their having imbibed Defcartes's doctrine, that the effence of body is conftituted by extenfion; as it would be contradictory to fuppofe fpace without extenfion. To fuppofe that there are fluids penetrating all bodies and replenifhing fpace, which neither refift nor act upon bodies, merely in order to avoid admitting a Vacuum, is feigning two forts of matter, without any neceffity or foundation; or is tacitly giving up the queftion.

Since then the effence of matter does not confift in extenfion, but in folidity, or impenetrability, the univerfe may be faid to confift of folid bodies moving in a Vacuum: nor need we at all fear, left the phenomena of nature, moft of which are plaufibly accounted for from a plenum, fhould become inexplicable when the plenitude is fet afide. The principal ones, fuch as the tides; the fufpenfion of the mercury in the barometer; the motion of the heavenly bodies, and of light, &c, are more eafily and fatisfactorily accounted for from other principles.

VACUUM *Boileanum*, is ufed to exprefs that approach to a real Vacuum, which we arrive at by means of the air-pump. Thus, any thing put in a receiver fo exhaufted, is faid to be put *in vacuo*: and thus moft of the experiments with the air-pump are faid to be performed *in vacuo*, or *in vacuo Boileano*.

Some of the principal phenomena obferved of bodies in vacuo, are; that the heavieft and lighteft bodies, as a guinea and a feather, fall here with equal velocity:—that fruits, as grapes, cherries, peaches, apples, &c, kept for any time in vacuo, retain their nature, frefhnefs, colour, &c, and thofe withered in the open air recover their plumpnefs in vacuo:—all light and fire become immediately extinct in vacuo:—little or no found is heard from a bell rung in vacuo:—a bladder half full of air, will diftend the bladder, and lift up 40 pound weight in vacuo:—moft animals foon expire in vacuo.

By experiments made in 1704, Dr. Derham found that animals which have two ventricals, and no foramen ovale, as birds, dogs, cats, mice, &c, die in lefs than half a minute; counting from the firft exfuction: a

mole died in one minute; a bat lived 7 or 8. Infects, as wafps, bees, grafshoppers, &c, feemed dead in two minutes; but after being left in vacuo 24 hours, they came to life again in the open air: fnails continued 24 hours in vacuo, without appearing much incommoded. —Seeds planted in vacuo do not grow: Small beer dies, and lofes all its tafte, in vacuo: And air rufhing through mercury into a Vacuum, throws the mercury in a kind of fhower upon the receiver, and produces a great light in a dark room.

The air-pump can never produce a perfect Vacuum; as is evident from its ftructure, and the manner of its working: in effect, every exfuction only takes away a part of the air; fo that there is ftill fome left after any finite number of exfuctions. For the air-pump has no longer any effect but while the fpring of the air remaining in the receiver is able to lift up the valves; and when the rarefaction is come to that degree, you can come no nearer to a Vacuum; unlefs perhaps the air valves can be opened mechanically, independent of the fpring of the air, as it is faid they are in fome new improved air-pumps.

Torricellian VACUUM, is that made in the barometer tube, between the upper end and the top of the mercury. This is perhaps never a perfect and entire Vacuum; as all fluids are found to yield or to rife in elaftic vapours, on the removal of the preffure of the atmofphere. See TORRICELLIAN, and BAROMETER.

VALVE, in Hydraulics, Pneumatics, &c, is a kind of lid or cover to a tube or veffel, contrived to open one way; but which, the more forcibly it is preffed the other way, the clofer it fhuts the aperture: fo that it either admits the entrance of a fluid into the tube, or veffel, and prevents its return; or permits it to efcape, and prevents its re-entrance.

Valves are of great ufe in the air-pump, and other wind machines; in which they are ufually made of pieces of bladder. In hydraulic engines, as the emboli of pumps, they are moftly of ftrong leather, of a round figure, and fitted to fhut the apertures of the barrels or pipes. Sometimes they are made of two round pieces of leather enclofed between two others of brafs; having divers perforations, which are covered with another piece of brafs, moveable upwards and downwards, on a kind of axis, which goes through the middle of them all. Sometimes they are made of brafs, covered over with leather, and furnifhed with a fine fpring, which gives way upon a force applied againft it; but upon the ceafing of that, returns the Valve over the aperture. See PUMP. See alfo Defaguliers' Exper. Philof. vol. 2, p. 156, and p. 180.

VANE, in a fhip, &c, a thin flip of fome kind of matter, placed on high in the open air, turning eafily round on an axis or fpindle, and veered about by the wind, to fhew its direction or courfe.

VANES, in Mathematical or Philofophical Inftruments, are fights made to flide and move upon crofsftaves, fore-ftaves, quadrants, &c.

VAPOUR, in Meteorology, a watery exhalation raifed up either by the heat of the fun, or any other heat, as fire, &c. Vapour is confidered as a thin veficle of water, or other humid matter, filled or inflated with air; which, being rarefied to a certain degree by the action of heat, afcends to fome height in the atmofphere, where it is fufpended, till it returns in form of rain, fnow, or the like. An affemblage of a number of particles or veficles of vapour, conftitutes what is called a cloud.

Some ufe the term Vapour indifferently, for all fumes emitted, either from moift bodies, as fluids of any kind; or from dry bodies, as fulphur, &c. But Newton, and other authors, better diftinguifh between humid and dry fumes, calling the latter *exhalations*.

For the manner in which Vapours are raifed, and again precipitated, fee CLOUD, DEW, RAIN, BAROMETER, and particularly EVAPORATION.

It may here be added, with refpect to the principles of folution adopted to account for evaporation, and largely illuftrated under that article, that Dr. Halley, about the beginning of the prefent century, feems to have been acquainted with the folvent power of air on water; for he fays, that fuppofing the earth to be covered with water, and the fun to move diurnally round it, the air would of itfelf imbibe a certain quantity of aqueous Vapours, and retain them like falts diffolved in water; and that the air warmed by the fun would fuftain a greater proportion of Vapours, as warm water will hold more diffolved falts; which would be difcharged in dews, fimilar to the precipitation of falts on the cooling of liquors. Philof. Tranf. Abr. vol. 2, p. 127.

Mr. Eeles, in 1755, endeavoured to account for the afcent of Vapour and exhalation, and their fufpenfion in the atmofphere, by means of the electric fire. The fun, he acknowledges, is the great agent in detaching Vapour and exhalations from their maffes, whether he acts immediately by himfelf, or by his rendering the electric fire more active in its vibrations: but their fubfequent afcent he attributes entirely to their being rendered fpecifically lighter than the lower air, by their conjunction with electrical fire: each particle of Vapour, with the electrical fluid that furrounds it, occupying a greater fpace than the fame weight of air. Mr. Eeles alfo endeavours to fhew, that the afcent and defcent of Vapour, attended by this fire, are the caufe of all the winds, and that they furnifh a fatisfactory folution of the general phenomena of the weather and barometer. Philof. Tranf. vol. 49, pa. 124.

Dr. Darwin, in 1757, publifhed remarks on the theory of Mr. Eeles, with a view of confuting it; and attempting to account for the afcent of Vapours, by confidering the power of expanfion which the conftituent parts of fome bodies acquire by heat, and alfo that fome bodies have a greater affinity to heat, or acquire it fooner, and retain it longer, than others. On thefe principles, he thinks, it is eafily underftood how water, whofe parts appear from the æolipile to be capable of immeafurable expanfion, fhould by heat alone become fpecifically lighter than the common atmofphere. A fmall degree of heat is fufficient to detach or raife the Vapour of water from the mafs to which it belongs; and the rays of the fun communicate heat only to thofe bodies by which they are refracted, reflected, or obftructed, whence, by their impulfe, a motion or vibration is caufed in the parts of fuch bodies. Hence he infers, that the fphericles of Vapour will, by refracting the folar rays, acquire a conftant heat,

though the furrounding atmofphere remain cold. If it be afked, how clouds are fupported in the abfence of the fun? It muft be remembered, that large maffes of Vapour muft for a confiderable time retain much of the heat they have acquired in the day; at the fame time reflecting how fmall a quantity of heat was neceffary to raife them, and that doubtlefs even a lefs will be fufficient to fupport them; as from the diminifhed preffure of the atmofphere at a given height, a lefs power may be able to continue them in their prefent ftate of rarefaction; and laftly, that clouds of particular fhapes will be fuftained or elevated by the motion they acquire from winds. Philof. Tranf. vol. 50, p. 246.

For the Effect of Vapour in the Formation of Springs, &c, fee SPRING, and RIVER.

The quantity of Vapour raifed from the fea by the warmth of the fun, muft be far greater than is commonly imagined. Dr. Halley has attempted to eftimate it. For the refult of his calculations, fee EVAPORATION.

VARIABLE, in Geometry and Analytics, is a term applied by mathematicians, to fuch quantities as are confidered in a Variable or changeable ftate, either increafing or decreafing. Thus, the abfciffes and ordinates of an ellipfis, or other curve line, are Variable quantities; becaufe thefe vary or change their magnitude together, the one at the fame time with the other. But fome quantities may be Variable by themfelves alone, or while thofe connected with them are conftant: as the abfciffes of a parallelogram, whofe ordinates may be confidered as all equal, and therefore conftant. Alfo the diameter of a circle, and the parameter of a conic fection, are *conftant*, while their abfciffes are *Variable*.

Variable quantities are ufually denoted by the laft letters of the alphabet, z, y, x, &c; while the conftant ones are denoted by the leading letters, a, b, c, &c.

Some authors, inftead of *Variable* and *conftant* quantities, ufe the terms *fluent* and *ftable* quantities.

The indefinitely fmall quantity by which a Variable quantity is continually increafed or decreafed, in very fmall portions of time, is called the *differential*, or *increment* or *decrement*. And the rate of its increafe or decreafe at any point, is called its *fluxion*; while the Variable quantity itfelf is called the *fluent*. And the calculation of thefe, is the fubject of the new *Methodus Differentialis*, or *Doctrine of Fluxions*.

VARENIUS (BERNARD), a learned Dutch geographer and phyfician, of the laft century, who was author of the beft mathematical treatife on Geography, intitled, *Geographia Univerfalis*, in qua affectiones generalis Telluris explicantur. This excellent work has been tranflated into all languages, and was honoured by an edition, with improvements, by Sir Ifaac Newton, for the ufe of his academical ftudents at Cambridge.

VARIATION, of *Quantities*, in Algebra. See CHANGES, and COMBINATION.

VARIATION, in Aftronomy.—*The Variation of the Moon*, called by Bulliald, the *Reflection of her Light*, is the third inequality obferved in the moon's motion; by which, when out of the quadratures, her true place differs from her place twice equated. See PLACE, EQUATION, &c.

Newton makes the moon's variation to arife partly from the form of her orbit, which is an ellipfis; and

partly

partly from the inequality of the fpaces, which the moon defcribes in equal times, by a radius drawn to the earth.

To find the Greateft Variation. Obferve the moon's longitude in the octants; and to the time of obfervation compute the moon's place twice equated; then the difference between the computed and obferved place, is the greateft Variation.

Tycho makes the greateft Variation 40' 30''; and Kepler makes it 51' 49''.—But Newton makes the greateft Variation, at a mean diftance between the fun and the earth, to be 35' 10'': at the other diftances, the greateft Variation is in a ratio compounded of the duplicate ratio of the times of the moon's fynodical revolution directly, and the triplicate ratio of the diftance of the fun from the earth inverfely. And therefore in the fun's apogee, the greateft Variation is 33' 14'', and in his perigee 37' 11''; provided that the eccentricity of the fun be to the tranfverfe femidiameter of the orbis magnus, as 16⅕⁵⁄₈ to 1000. Or, taking the mean motions of the moon from the fun, as they are ftated in Dr. Halley's tables, then the greateft Variation at the mean diftance of the earth from the fun will be 35' 7'', in the apogee of the fun 33' 27'', and in his perigee 36' 51''. Philof. Nat. Princ. pr. 29, lib. 3.

VARIATION, in Geography, Navigation, &c, a term applied to the deviation of the magnetic needle, or compafs, from the true north point, either towards the eaft or weft; called alfo the *declination.* Or the Variation of the compafs is properly defined, the angle which a magnetic needle, fufpended at liberty, makes with the meridian line on an horizontal plane; or an arch of the horizon, comprehended between the true and the magnetic meridians.

In the fea-language, the Variation is ufually called *north-eafting,* or *north-wefting.*

All magnetic bodies are found to range themfelves, in fome fort, according to the meridian; but they feldom agree precifely with it: in one place they decline, from the north toward the eaft, in another toward the weft; and that too differently at different times.

The Variation of the compafs could not long remain a fecret, after the invention of the compafs itfelf: accordingly Ferdinand, the fon of Columbus, in his life written in Spanifh, and printed in Italian at Venice in 1571, afferts, that his father obferved it on the 14th of September 1492: though others feem to attribute the difcovery of it to Sebaftian Cabat, a Venetian, employed in the fervice of our king Henry VII, about the year 1500.—It now appears however, that this Variation or declination of the needle was known even fome centuries earlier, though it does not appear that the ufe of the needle itfelf in navigation was then known. For it feems there is in the library of the univerfity of Leyden, a fmall manufcript tract on the Magnet, in Latin, written by one Peter Adfiger, bearing date the 8th of Auguft 1269; in which the declination of the needle is particularly mentioned. Mr. Cavallo has printed the chief part of this letter in the Supplement to his Treatife on Magnetifm, with a tranflation; and I think it is to be wifhed he had printed the whole of fo curious a paper. The curiofity of this letter, fays Mr. Cavallo, confifts in its containing almoft all that

is at prefent known of the fubject, at leaft the moft remarkable parts of it, mixed however with a good deal of abfurdity. The laws of magnetic attraction, and of the communication of that power to iron, the directive property of the natural magnet, as well as of the iron that has been touched by it, and even the declination of the magnetic needle, are clearly and unequivocally mentioned in it.

As this Variation differs in different places, Gonzales d'Oviedi found there was none at the Azores; from whence fome geographers thought fit in their maps to make the firft meridian pafs through one of thefe iflands: it not being then known that the Variation altered in time. See MAGNET; alfo Gilbert De Magnete, Lond. 1600, p. 4 and 5; or Purchas's Pilgrims, Lond. 1625, book 2, fect. 1.

Various are the hypothefes that have been framed to account for this extraordinary phenomenon: we fhall only notice fome of the latter, and more probable: juft premifing, that Robert Norman, the inventor of the Dipping needle, difputes againft Cortes's notion, that the Variation was caufed by a point in the heavens; contending that it fhould be fought for in the earth, and propofes how to difcover its place.

The firft is that of Gilbert (De Magnete, lib. 4, p. 151 &c), which is followed by Cabeus, &c. This notion is, that it is the earth, or land, that draws the needle out of its meridian direction: and hence they argue, that the needle varied more or lefs, as it was more or lefs diftant from any great continent; and confequently that if it were placed in the middle of an ocean, equally diftant from equal tracts of land on each fide, eaftward and weftward, it would not decline either to the one or the other, but point exactly north and fouth. Thus, fay they, in the Azores iflands, which are equally diftant from Africa on the eaft, and America on the weft, there is no Variation: but as you fail from thence towards Africa, the needle begins to decline toward the eaft, and that ftill more and more till you reach the fhore. If you proceed ftill farther eaftward, the declination gradually diminifhes again, by reafon of the land left behind on the weft, which continues to draw the needle. The fame holds till you arrive at a place where the tracts of land on each fide are equal; and there again the Variation will be nothing. But the misfortune is, the law does not hold univerfally; for multitudes of obfervations of the Variation, in different parts, made and collected by Dr. Halley, overturn the whole theory.

Others therefore have recourfe to the frame and compages of the earth, confidered as interfperfed with rocks and fhelves, which being generally found to run towards the polar regions, the needle comes to have a general tendency that way; but it feldom happens that their direction is exactly in the meridian, and the needle has confequently, for the moft part, fome Variation.

Others hold that divers parts of the earth have different degrees of the magnetic virtue, as fome are more intermixed with heterogeneous matters, which prevent the free action or effect of it, than others are.

Others again afcribe all to magnetic rocks and iron mines, which, affording more of the magnetic matter than other parts, draw the needle more.

Laftly,

Lastly, others imagine that earthquakes, or high tides, have disturbed and dislocated several considerable parts of the earth, and so changed the magnetic axis of the globe, which was originally the same with the axis of the earth itself.

But none of these theories can be the true one; for still that great phenomenon, the *Variation of the Variation*, i. e. the continual change of the declination, in one and the same place, is not accountable for, on any of these foundations, nor is it even consistent with them.

Doctor Hook communicated to the Royal Society, in 1674, a theory of the Variation; the substance of which is, that the magnet has its peculiar pole, distant 10 degrees from the pole of the earth, about which it moves, so as to make a revolution in 370 years: whence the Variation, he says, has altered of late about 10 or 11 minutes every year, and will probably

so continue to do for some time, when it will begin to proceed slower and slower, till at length it become stationary and retrograde, and so return back again. Birch's Hist. of the Royal Society, vol. 3, p. 131.

Dr. Halley has given a new system, the result of numerous observations, and even of a number of voyages made at the public expence on this account. The light which this author has thrown upon this obscure part of natural history, is very great, and of important consequence in navigation, &c. In this system he has reduced the several Variations in divers places to a precise rule, or order, which before appeared all precarious and arbitrary.

His theory will therefore deserve a more ample detail. The observations it is built upon, as laid down in the Philos. Transf. number 148, or Abr. vol. 2, p. 610, are as follow:

Observed Variations of the Needle in divers places, and at divers times.

Places observed at.	Longitude from London.		Latitude		Year of Observation.	Variation observed.		Places observed at.	Longitude from London.		Latitude		Year of Observation.	Variation observed.	
	o	′	o	′		o	′		o	′	o	′		o	′
London - - -	0	0	51 31 n		1580	11	15 e	Baldivia - -	73	0 w	40	0 s	1670	8	10 e
					1622	6	0 e	Cape Aguillas -	16	30 e	34	50 s	1622	2	0 w
					1634	4	5 e						1675	8	0 w
					1672	2	30 w	At Sea - - -	1	0 e	34	30 s	1675	0	0
					1683	4	30 w	At Sea - - -	20	0 w	34	0 s	1675	10	30 e
Paris - - - -	2	25 e	48 51 n		1640	3	0 e	At Sea - - -	32	0 w	24	0 s	1675	10	30 e
					1666	0	0	St. Helena - -	6	30 w	16	0 s	1677	0	40 e
					1681	2	30 w	Isle Ascension -	14	30 w	7	50 s	1678	1	0 e
Uraniburg - -	13	0 e	55 54 n		1672	2	35 w	Johanna - - -	44	0 e	12	15 s	1675	19	30 w
Copenhagen -	12	53 e	55 41 n		1649	1	53 e	Mombasa - - -	40	0 e	4	0 s	1675	16	0 w
					1672	3	45 w	Zocatra - - -	56	0 e	12	30 n	1674	17	0 w
Dantzick - -	19	0 e	54 23 n	1679	7	0 w	Aden, Mouth of Red Sea	47	30 e	13	0 n	1674	15	0 w
Montpelier - -	4	0 e	43 37 n	1674	1	10 w								
Brest - - -	4	25 w	48 23 n	1680	1	45 w	Diego Roiz - -	61	0 e	20	0 s	1676	20	30 w
Rome - - -	13	0 e	41 50 n	1681	5	0 w	At Sea - - -	64	30 e	0	0	1676	15	30 w
Bayonne - - -	1	20 w	43 30 n	1680	1	20 w	At Sea - - -	55	0 e	27	0 s	1676	24	0 w
Hudson's Bay -	79	40 w	51 0 n	1668	19	15 w	Bombay - - -	72	30 e	19	0 n	1676	12	0 w
In Hudson's Straits -	57	0 w	61 0 n	1668	29	30 w	Cape Comorin -	76	0 e	8	15 n	1680	8	48 w
							Ballasore - -	87	0 e	21	30 n	1680	8	10 w
Beffin's Bay, Sir T. Smith's Sound - -	80	0 w	78 0 n	1616	57	0 w	Fort St. George	80	0 e	13	15 n	1680	8	10 w
							West Point of Java - -	104	0 e	6	40 s	1676	3	10 w
At Sea - - -	57	0 w	38 40 n	1682	7	30 w	At Sea - - -	58	0 e	39	0 s	1677	27	30 w
At Sea - - -	31	30 w	43 50 n	1682	5	30 w	I. St. Paul - -	72	0 e	38	0 s	1677	23	30 w
At Sea - - -	42	0 w	21 0 n	1678	0	40 e	At Van Diemen's	142	0 e	42	25 s	1642	0	0
Cape St. Augustine - -	35	30 w	28 0 s	1670	5	30 e	At New Zealand - -	170	0 e	40	50 s	1642	9	0 e
Off the mouth of River Plate	53	0 w	39 30 s	1670	20	30 e	Three - kings Isle in ditto	169	30 e	34	35 s	1642	8	40 e
Cape Frio - -	41	10 w	22 40 s	1670	12	10 e	I. Rotterdam in the South Sea	184	0 e	20	15 s	1642	6	20 e
Entrance of Magellan's Straits - -	68	0 w	52 30 s	1670	17	0 e	Coast of New Guinea - -	149	0 e	4	30 s	1643	8	45 e
West Entrance of ditto -	75	0 w	53 0 s	1670	14	10 e	West Point of ditto - -	126	0 e	0	26 s	1643	5	30 e

Upon

Upon thefe obferved Variations Dr. Halley makes feveral remarks, as to the Variation in different parts of the world at the time of his writing, eaftward and weftward, and the fituation and direction of the lines or places of no Variation : from the whole he deduces the following theory.

Dr. *Halley's Theory of the Variation of the Needle.* That the whole globe of the earth is one great magnet, having four magnetical poles, or points of attraction ; near each pole of the equator two ; and that in thofe parts of the world which lie nearly adjacent to any one of thefe magnetic poles, the needle is governed by it ; the neareft pole being always predominant over the more remote.

The pole which at prefent is neareft to us, he conjectures to lie in or near the meridian of the Land's-end of England, and not above 7° from the north pole : by this pole, the Variations in all Europe and Tartary, and the North Sea, are chiefly governed ; though ftill with fome regard to the other northern pole, whofe fituation is in the meridian paffing about the middle of California, and about 15° from the north pole of the world, to which the needle has chiefly refpect in all North America, and in the two oceans on either fide of it, from the Azores weftward to Japan, and farther.

The two fouthern magnetic poles, he imagines, are rather more diftant from the fouth pole of the world ; the one being about 16° from it, on a meridian 20° to the weftward of the Magellanic Streights, or 95° weft from London : this pole commands the needle in all South America, in the Pacific Ocean, and the greateft part of the Ethiopic Ocean. The other magnetic pole feems to have the greateft power, and the largeft dominion of all, as it is the moft remote from the pole of the world, being little lefs than 20° diftant from it, in the meridian which paffes through New Holland, and the ifland Celebes, about 120° eaft from London : this pole is predominant in the fouth part of Africa, in Arabia, and the Red Sea, in Perfia, India, and its iflands, and all over the Indian fea, from the Cape of Good Hope eaftward, to the middle of the Great South Sea that divides Afia from America.

Such, he obferves, feems to be the prefent difpofition of the magnetic virtue thoughout the whole globe of the earth. It is then fhewn how this hypothefis accounts for all the Variations that have been obferved of late, and how it anfwers to the feveral remarks drawn from the table.

It is there inferred that from the whole it appears, that the direction of the needle, in the temperate and frigid zones, depends chiefly upon the counterpoife of the forces of two magnetic poles of the fame nature : as alfo why, under the fame meridian, the Variation fhould be in one place $29\frac{1}{2}$ degrees weft, and in another $20\frac{1}{2}$ degrees eaft.

In the torrid zone, and particularly about the equator, refpect muft be had to all the four poles, and their pofitions muft be well confidered, otherwife it will not be eafy to determine what the Variation fhould be, the neareft pole being always ftrongeft ; yet fo however as to be fometimes counterbalanced by the united forces of two more remote ones. Thus, in failing from St. Helena, by the ifle of Afcenfion, to the equator, on the north-weft courfe, the Variation is very little eafterly, and unalterable in that whole track ; becaufe the South-American pole (which is much the neareft in the aforefaid places), requiring a great eafterly variation, is counterpoifed by the contrary attraction of the North-American and the Afiatic fouth poles ; each of which fingly is, in thefe parts, weaker than the American fouth pole ; and upon the north-weft courfe the diftance from this latter is very little varied ; and as you recede from the Afiatic fouth pole, the balance is ftill preferved by an accefs towards the North-American pole. In this cafe no notice is taken of the European north pole ; its meridian being a little removed from thofe of thefe places, and of itfelf requiring the fame Variations which are here found.

After the fame manner may the Variations in other places about the equator be accounted for, upon Dr. Halley's hypothefis.

To obferve the Variation of the Needle. Draw a meridian line, as directed under MERIDIAN ; then a ftile being erected in the middle of it, place a needle upon it, and draw the right line which it hangs over. Thus will the quantity of the Variation appear.

Or thus : As the former method of finding the Variation cannot be applied at fea, others have been devifed, the principal of which are as follow. Sufpend a thread and plummet over the compafs, till the fhadow pafs through the centre of the card ; obferve the rhumb, or point of the compafs which the fhadow touches when it is the fhorteft. For the fhadow is then a meridian line ; and confequently the Variation is fhewn.

Or thus : Obferve the point of the compafs upon which the fun, or fome ftar, rifes and fets ; bifect the arch intercepted between the rifing and fetting, and the line of bifection will be the meridian line ; confequently the Variation is had as before. The fame may alfo be obtained from two equal altitudes of the fame ftar, obferved either by day or night. Or thus : Obferve the rhumb upon which the fun or ftar rifes and fets ; and from the latitude of the place find the eaftern or weftern amplitude : for the difference between the amplitude, and the diftance of the rhumb obferved, from the eaftern rhumb of the card, is the Variation fought.

Or thus : Obferve the altitude of the fun, or fome ftar S, whofe declination is known ; and note the rhumb in the compafs to which it then correfponds. Then in the triangle ZPS, are known three fides, viz. PZ the colatitude, PS the codeclination,  and ZS the coaltitude ; the angle PZS is thence found by fpherical trigonometry ; the fupplement to which, viz. AZS, is the azimuth from the fouth. Then the difference between the azimuth and the obferved diftance of the rhumb from the fouth, is the Variation fought. See *Azimuth* COMPASS.

The ufe of the Variation is to correct the courfes a fhip has fteered by the compafs, which muft always be done before they are worked, or calculated.

VARIATION

VARIATION *of the Variation*, is a gradual and continual change in the Variation, obferved in any place, by which the quantity of the Variation is found to be different at different times.

This Variation, according to Henry Bond (in his *Longitude found*, Lond. 1670, pa. 6) " was firft found to decreafe by Mr. John Mair; 2dly, by Mr. Edmund Gunter: 3dly, by Mr. Henry Gellibrand; 4thly, by myfelf (Henry Bond) in 1640; and laftly, by Dr. Robert Hook, and others, in 1665;" which they found out by comparing together obfervations made at the fame place, at different times. The difcovery was foon known abroad; for Kircher, in his treatife intitled Magnes, firft printed at Rome in 1641, fays that our countryman Mr. John Greaves had informed him of it, and then he gives a letter of Merfenne's, containing a diftinct account of it.

This continual change in the Variation, is gradual and univerfal, as appears by numerous obfervations. Thus, the Variation was,

At Paris, according to Orontius Finæus,

in 1550 -	8° 0′E.
in 1640 -	3 0 E.
in 1660 -	0 0
in 1681 -	2 2W.
in 1759 -	18 10W.
in 1760 -	18 20W.

M. De la Lande (Expofition du Calcul Aftronomique) obferves, that the Variation has changed, at Paris, 26° 20′ in the fpace of 150 years, allowing that in 1610 the Variation was 8° E: and fince 1740 the needle, which was always ufed by Maraldi, is more than 3° advanced toward the weft, beyond what it was at that period; which is a change after the rate nearly of 9′½ per year.

At Cape d'Agulhas, in 1600, it had no Variation; (whence the Portuguefe gave it that name);

in 1622 it was	2°W.
in 1675 -	8 W.
in 1692 -	11 W.

which is a change of nearly 8′ per year.

At St. Helena, the Variation, in 1600 was 8° 0′E.

in 1623 -	6 0 E.
in 1677 -	0 40 E.
in 1692 -	1 0 W.

which is a change of nearly 5′½ per year.

At Cape Comorin, the Variation,

in 1620 was	14° 20′W.
in 1680 -	8 44 W.
in 1688 -	7 30 W.

which is a change of nearly 6′½ per year.

At London, the Variation, in 1580 was 11° 15′E.

in 1622 -	6 0 E.
in 1634 -	4 5 E.
in 1657 -	0 0
in 1672 -	2 30 W.
in 1692 -	6 0 W.
in 1723 -	14 17 W.
in 1747 -	17 40 W.
in 1780 -	22 41 W.

which is a change after the rate of 10′ per year, upon a courfe of exactly 200 years. See Philof. Tranf. No. 148 and No. 383, or Abr. vol. 2, p. 615, and vol. 7, p. 290; and Philof. Tranf. vol. 45, p. 280, and vol. 66, p. 393. On the fubject of the Variation, fee alfo Norman's New Attractive 1614; Burrows's Difcovery of the Variation 1581; Bond's Longitude found, 1676; &c.

Mr. Thomas Harding, in the Tranfactions of the Royal Irifh Academy, vol. 4, has given obfervations on the Variation of the magnetic needle, at Dublin, which are rather extraordinary. He fays the change in the Variation at that place is *uniform*. That from the year 1657, in which the Variation was nothing (the fame as at London in that year), it has been going on at the medium rate of 12′ 20″ annually, and was in May 1791, 27° 23′ weft: exceeding that at London now by 3 or 4 degrees. He brings proof of his affertion of the uniformity of the Variation, from different authentic records, and ftates the operations by which it is calculated. He concludes with recommending accuracy in marking the exifting Variation when maps are made, as not only conducing to the exact definition of boundaries, but as laying the beft foundation for a difcovery of the longitude by fea or land.

Theory of the Variation of the Variation. According to Dr. Halley's theory, this change in the Variation of the compafs, is fuppofed owing to the difference of velocity in the motions of the internal and external parts of the globe. From the obfervations that have been cited, it feems to follow, that all the magnetical poles have a motion weftward, but yet not exactly round the axis of the earth, for then the Variations would continue the fame in the fame parallel of latitude, contrary to experience.

From the difagreement of fuch a fuppofition with experiments therefore, the learned author of the theory invented the following hypothefis: The external parts of the globe he confiders as the fhell, and the internal as a nucleus, or inner globe; and between the two he conceives a fluid medium. That inner earth having the fame common centre and axis of diurnal rotation, may revolve with our earth every 24 hours: Only the outer fphere having its turbinating motion fomewhat fwifter or flower than the internal ball; and a very minute difference in length of time, by many repetitions, becoming fenfible; the internal parts will gradually recede from the external, and they will appear to move, either eaftward or weftward, by the difference of their motions.

Now, fuppofing fuch an internal fphere, having fuch a motion, the two great difficulties in the former hypothefis are eafily folved; for if this exterior fhell of earth be a magnet, having its pole at a diftance from the poles of diurnal rotation; and if the internal nucleus be likewife a magnet, having its poles in two other places, diftant alfo from the axis; and thefe latter, by a flow gradual motion, change their place in refpect of the external, a reafonable account may then be given of the four magnetical poles before mentioned, and alfo of the changes of the needle's Variation.

The author thinks that two of thefe poles are fixed, and the other two moveable; viz, that the fixed poles are the poles of the external cortex or fhell of the earth;

earth ; and the other the poles of the magnetical nucleus, included and moveable within the former. From the obfervations he infers, that the motion is weftwards, and confequently that the nucleus has not precifely attained the fame velocity with the exterior parts in their diurnal rotation ; but fo very nearly equals it, that in 365 revolutions the difference is fcarcely fenfible.

That there is any difference of this kind, arifes from hence, that the impulfe by which the diurnal motion was impreffed on the earth, was given to the external parts, and from thence in time communicated to the internal ; but fo as not yet perfectly to equal the velocity of the firft motion impreffed on the fuperficial parts of the globe, and ftill preferved by them.

As to the precife period, obfervations are wanting to determine it, though the author thinks we may reafonably conjecture that the American pole has moved weftward 46° in 90 years, and that its whole period is performed in about 700 years.

Mr. Whifton, in his New Laws of Magnetifm, raifes feveral objections againft this theory. See MAGNETISM.

M. Euler, too, the fon of the celebrated mathematician of that name, has controverted and cenfured Dr. Halley's theory. He thinks, that two magnetic poles, placed on the furface of the earth, will fufficiently account for the Variation : and he then endeavours to fhew, how we may determine the declination of the needle, at any time, and on every part of the globe, from this hypothefis. For the particulars of this reafoning, fee the Hiftoire de l'Academie des Sciences & Belles Lettres of Berlin, for 1757; alfo Mr. Cavallo's Treatife on Magnetifm, p. 117.

Variation of the Needle by Heat and Cold.—There is a fmall Variation of the Variation of the magnetic needle, amounting only to a few minutes of a degree in the fame place, at different hours of the fame day, which is only difcoverable by nice obfervations. Mr. George Graham made feveral obfervations of this kind in the years 1722 and 1723, profeffing himfelf altogether ignorant of the caufe of the phenomena he obferved. Philof. Tranf. No. 383, or Abr. vol. 7, p. 290.

About the year 1750, Mr. Wargentin, fecretary of the Swedifh Academy of Sciences, took notice both of the regular diurnal Variation of the needle, and alfo of its being difturbed at the time of the aurora borealis, as recorded in the Philof. Tranf. vol. 47, p. 126.

About the year 1756, Mr. Canton commenced a feries of obfervations, amounting to near 4000, with an excellent Variation-compafs, of about 9 inches diameter. The number of days on which thefe obfervations were made, was 603, and the Diurnal Variation on 574 of them was regular, fo as that the abfolute Variation of the needle weftward was increafing from about 8 or 9 o'clock in the morning, till about 1 or 2 in the afternoon, when the needle became ftationary for fome time ; after that, the abfolute Variation weftward was decreafing, and the needle came back again to its former fituation, or nearly fo, in the night, or by the next morning. The Diurnal Variation is irregular when the needle moves flowly eaftward in the latter part of the morning, or weftward in the latter

part of the afternoon ; alfo when it moves much either way after night, or fuddenly both ways in a fhort time. Thefe irregularities feldom happen more than once or twice in a month, and are always accompanied, as far as Mr. Canton obferved, with an aurora borealis.

Mr. Canton lays down and evinces, by experiment, the following principle, viz. that the attractive power of the magnet (whether natural or artificial) will decreafe while the magnet is heating, and increafe while it is cooling. He then proceeds to account for both the regular and irregular Variation. It is evident, he fays, that the magnetic parts of the earth in the north, on the eaft fide, and on the weft fide of the magnetic meridian, equally attract the north end of the needle. If then the eaftern magnetic parts be heated fafter by the fun in the morning, than the weftern parts, the needle will move weftward, and the abfolute Variation will increafe : when the attracting parts of the earth on each fide of the magnetic meridian have their heat increafing equally, the needle will be ftationary, and the abfolute Variation will then be greateft ; but when the weftern magnetic parts are either heating fafter, or cooling flower, than the eaftern, the needle will move eaftward, or the abfolute Variation will decreafe ; and when the eaftern and weftern magnetic parts are cooling equally faft, the needle will again be ftationary, and the abfolute Variation will then be leaft.

By this theory, the Diurnal Variation in the fummer ought to exceed that in winter ; and accordingly it is found by obfervation, that the Diurnal Variation in the months of June and July is almoft double of that in December and January.

The irregular Diurnal Variation muft arife from fome other caufe than that of heat communicated by the fun ; and here Mr. Canton has recourfe to fubterranean heat, which is generated without any regularity as to time, and which will, when it happens in the north, affect the attractive power of the magnetic parts of the earth on the north end of the needle. That the air neareft the earth will be moft warmed by the heat of it, is obvious ; and this has been often noticed in the morning, before day, by means of thermometers at different diftances from the ground. Philof. Tranf. vol. 48, pa. 526.

Mr. Canton has annexed to his paper on this fubject, a complete year's obfervations ; from which it appears, that the Diurnal Variation increafes from January to June, and decreafes from June to December. Philof. Tranf. vol. 51, pa. 398.

It has alfo been obferved, that different needles, efpecially if touched with different loadftones, will differ a few minutes in their Variation. See Poleni Epift. Phil. Tranf. num. 421.

Dr. Lorimer (in the Supp. to Cavallo's Magnetifm) adduces fome ingenious obfervations on this fubject. It muft be allowed, fays he, according to the obfervations of feveral ingenious gentlemen, that the collective magnetifm of this earth arifes from the magnetifm of all the ferruginous bodies contained in it, and that the magnetic poles fhould therefore be confidered as the centres of the powers of thofe magnetic fubftances. Thefe poles muft therefore change their places according as the magnetifm of fuch fubftances is affected, and if

with.

with Mr. Canton we allow, that the general cause of the Diurnal Variation arises from the sun's heat in the forenoon and afternoon of the same day, it will naturally occur, that the same cause, being continued, may be sufficient to produce the general Variation of the magnetic needle for any number of years. For we must consider, that ever since any attentive observations have been made on this subject, the natural direction of the magnetic needle in Europe has been constantly moving, from west to east, and that in other parts of the world it has continued its motion with equal constancy.

As we must therefore admit, says Dr. Lorimer, that the heat in the different seasons depends chiefly on the sun, and that the months of July and August are commonly the hottest, while January and February are the coldest months of the year; and that the temperature of the other months falls into the respective intermediate degrees; so we must consider the influence of heat upon magnetism to operate in the like manner, viz, that for a short time it scarcely manifests itself; yet in the course of a century, the constancy and regularity of it becomes sufficiently apparent. It would therefore be idle to suppose, that such an influence could be derived from an uncertain or fortuitous cause. But if it be allowed to depend upon the constancy of the sun's motion, and this appears to be a cause sufficient to explain the phenomena, we should (agreeably to Newton's first law of philosophizing) look no farther.

As we therefore consider, says he, the magnetic powers of the earth to be concentrated in the magnetic poles, and that there is a diurnal Variation of the magnetic needle, these poles must perform a small diurnal revolution proportional to such Variation, and return again to the same point nearly. Suppose then that the sun in his diurnal revolution passes along the northern tropic, or along any parallel of latitude between it and the equator, when he comes to that meridian in which the magnetic pole is situated, he will be much nearer to it, than in any other; and in the opposite meridian he will of course be the farthest from it. As the influence of the sun's heat will therefore act most powerfully at the least, and less forcibly at the greatest distance, the magnetic pole will consequently describe a figure something of the elliptical kind; and as it is well known that the greatest heat of the day is some time after the sun has passed the meridian, the longest axis of this elliptical figure will lie north easterly in the northern, and south-easterly in the southern hemisphere. Again, as the influence of the sun's heat will not from those quarters have so much power, the magnetic poles cannot be moved back to the very same point, from which they set out; but to one which will be a little more northerly and easterly, or more southerly and easterly, according to the hemispheres in which they are situated. The figures therefore which they describe, may more properly be termed elliptoidal spirals.

In this manner the Variation of the magnetic needle in the northern hemisphere may be accounted for. But with respect to the southern hemisphere we must recollect, that though the lines of declination in the northern hemisphere have constantly moved from west to east, yet in the southern hemisphere, it is equally certain that they have moved from east to west, ever since any observations have been made on the subject. Hence

4

then the lines of magnetic declination, or Halleyan curves, as they are now commonly called, appear to have a contrary motion in the southern hemisphere, to what they have in the northern; though both the magnetic poles of the earth move in the same direction, that is from west to east.

In the northern hemisphere there was a line of no Variation, which had east Variation on its eastern side, and west Variation on its western side. This line evidently moved from west to east during the two last centuries; the lines of east Variation moving before it, while the lines of west Variation followed it with a proportional pace. These lines first passed the Azores or Western Islands, then the meridian of London, and after a certain number of years still later, they passed the meridian of Paris. But in the southern hemisphere there was another line of no Variation, which had east Variation on its western side, and west Variation on its eastern; the lines of east Variation moving before it, while those of the west Variation followed it. This line of no Variation first passed the Cape des Aiguilles, and then the Cape of Good Hope; the lines of 5°, 10°, 15°, and 29° west Variation following it, the same as was the case in the northern hemisphere, but in the contrary direction.

We may just farther mention the idea of Dr. Gowin Knight, which was, that this earth had originally received its magnetism, or rather that its magnetical powers had been brought into action, by a shock, which entered near the southern tropic, and passed out at the northern one. His meaning appears to have been, that this was the course of the magnetic fluid, and that the magnetic poles were at first diametrically opposite to each other. Though, according to Mr. Canton's doctrine, they would not have long continued so; for from the intense heat of the sun in the torrid zone, according to the principles already explained, the north pole must have soon retired to the north-eastward, and the south pole to the south-eastward. It is also curious to observe, that on account of the southern hemisphere being colder upon the whole than the northern hemisphere, the magnetic poles would have moved with unequal pace: that is, the north magnetic pole would have moved farther in any given time to the north-east, than the south magnetic pole could have moved to the south-east. And, according to the opinions of the most ingenious authors on this subject, it is generally allowed, that at this time the north magnetic pole is considerably nearer to the north pole of the earth, than the south magnetic pole is to the south pole of the earth.

It may farther be added, that several ingenious sea officers are of opinion, that in the western parts of the English Channel the Variation of the magnetic needle has already begun to decrease; having in no part of it ever amounted to 25°. There are however other persons who assert that the Variation is still increasing in the Channel, and as far westward as the 15th degree of longitude and 51° of latitude, at which place they say that it amounts to about 30°.

Of the Variation Chart. Doctor Halley having collected a multitude of observations made on the Variation of the needle in many parts of the world, was hence enabled to draw, on a Mercator's chart, certain lines, shewing the Variation of the compass in all those places

places over which they paſſed, in the year 1700, when he publiſhed the firſt chart of this kind, called the *Variation Chart*.

From the conſtruction of this chart it appears, that the longitude of any of thoſe places may be found by it, when the latitude and the Variation in that place are known. Thus, having found the Variation of the compaſs, draw a parallel of latitude on the chart through the latitude found by obſervation; and the point where it cuts the curved line, whoſe Variation is the ſame with that obſerved, will be the ſhip's place. A ſimilar project of thus finding the longitude, from the known latitude and inclination or dip of the needle, was before propoſed by Henry Bond, in his treatiſe intitled, The Longitude Found, printed in 1676.

This method however is attended with two conſiderable inconveniences: 1ſt, That wherever the Variation lines run eaſt and weſt, or nearly ſo, this way of finding the longitude becomes imperfect, as their interſection with the parallel of latitude muſt be very indefinite: and among all the trading parts of the world, this imperfection is at preſent found chiefly on the weſtern coaſts of Europe, between the latitudes of 45° and 53°; and on the eaſtern ſhores of North America, with ſome parts of the Weſtern Ocean and Hudſon's Bay, lying between the ſaid ſhores: but for the other parts of the world, a Variation Chart may be attended with conſiderable benefit. However, the Variation curves, when they run eaſt and weſt, may ſometimes be applied to good purpoſe in correcting the latitude, when meridian obſervations cannot be had, as it often happens on the northern coaſts of America, in the Weſtern Ocean, and about Newfoundland; for if the Variation can be obtained exactly, then the eaſt and weſt curve, anſwering to the Variation in the chart, will ſhew the latitude.

2dly, As the deviation of the magnetical meridian, from the true one, is ſubject to continual alteration, therefore a chart to which the Variation lines are fitted for any year, muſt in time become uſeleſs, unleſs new lines, ſhewing the ſtate of the Variation at that time, be drawn on the chart: but as the change in the Variation is very ſlow, therefore new Variation Charts publiſhed every 7 or 8 years, will anſwer the purpoſe tolerably well. And thus it has happened that Halley's Variation Chart has become uſeleſs, for want of encouragement to renew it from time to time.

However, in the year 1744, Mr. William Mountaine and Mr. James Dodſon publiſhed a new Variation Chart, adapted for that year, which was well received; and ſeveral inſtances of its great utility having been communicated to them, they fitted the Variation lines anew for the year 1756, and in the following year publiſhed the 3d Variation Chart, and alſo preſented to the Royal Society a curious paper concerning the Variation of the magnetic needle, with a ſet of tables annexed, containing the reſult of upwards of 50 thouſand obſervations, in ſix periodical reviews, from the year 1700 to 1756 incluſive, and adapted to every 5 degrees of latitude and longitude in the more frequented oceans; which paper and tables were printed in the Tranſactions for the year 1757.

From theſe tables of obſervations, ſuch extraordi-

nary and whimſical irregularities occur in the Variation, that we cannot think it wholly under the direction of one general and uniform law; but rather conclude, with Dr. Gowen, in the 87th prop. of his Treatiſe upon Attraction and Repulſion, that it is influenced by various and different magnetic attractions, perhaps occaſioned by the heterogeneous compoſitions in the great magnet, the earth.

Many other obſervations on the Variation of the magnetic needle, are to be found in ſeveral volumes of the Philoſ. Tranſ. See particularly vol. 48, p. 875; vol. 50, p. 329; vol. 56, p. 220; and vol. 61, p. 422.

VARIATION *Compaſs*. See COMPASS.

VARIATION *of Curvature*, in Geometry, is uſed for that inequality or change which takes place in the curvature of all curves except the circle, by which their curvature is more or leſs in different parts of them. And this Variation conſtitutes the quality of the curvature of any line.

Newton makes the index of the inequality, or Variation of Curvature, to be the ratio of the fluxion of the radius of curvature to the fluxion of the curve itſelf: and Maclaurin, to avoid the perplexity that different notions, connected with the ſame terms, occaſion to learners, has adopted the ſame definition: but he ſuggeſts, that this ratio gives rather the Variation of the ray of curvature, and that it might have been proper to have meaſured the Variation of Curvature rather by the ratio of the fluxion of the curvature itſelf to the fluxion of the curve; ſo that, the curvature being inverſely as the radius of curvature, and conſequently its fluxion as the fluxion of the radius itſelf directly, and the ſquare of the radius inverſely, its Variation would have been directly as the meaſure of it according to Newton's definition, and inverſely as the ſquare of the radius of curvature.

According to this notion, it would have been meaſured by the angle of contact contained by the curve and circle of curvature, in the ſame manner as the curvature itſelf is meaſured by the angle of contact contained by the curve and tangent. The reaſon of this remark may appear from this example: The Variation of curvature, according to Newton's explication, is uniform in the logarithmic ſpiral, the fluxion of the radius of curvature in this figure being always in the ſame ratio to the fluxion of the curve; and yet, while the ſpiral is produced, though its curvature decreaſes, it never vaniſhes; which muſt appear a ſtrange paradox to thoſe who do not attend to the import of Newton's definition. Newton's Method of Fluxions and Inf. Series, pa. 76. Maclaurin's Flux. art. 386. Philoſ. Tranſ. num. 468, pa. 342.

The Variation of curvature at any point of a conic ſection, is always as the tangent of the angle contained by the diameter that paſſes through the point of contact, and the perpendicular to the curve at the ſame point, or to the angle formed by the diameter of the ſection, and of the circle of curvature. Hence the Variation of curvature vaniſhes at the extremities of either axis, and is greateſt when the acute angle, contained by the diameter, paſſing through the point of contact and the tangent, is leaſt.

When the conic ſection is a parabola, the Variation is

as the tangent of the angle, contained by the right line drawn from the point of contact to the focus, and the perpendicular to the curve. See CURVATURE.

From Newton's definition may be derived practical rules for the Variation of curvature, as follows:

1. Find the radius of curvature, or rather its fluxion; then divide this fluxion by the fluxion of the curve, and the quotient will give the Variation of curvature; exterminating the fluxions when necessary, by the equation of the curve, or perhaps by expressing their ratio by help of the tangent, or ordinate, or subnormal, &c.

2. Since $\frac{\dot z^3}{-\dot x \dot y}$, or $\frac{\dot z^3}{-\ddot y}$ (putting $\dot x = 1$) denotes the radius of curvature of any curve z, whose absciss is x, and ordinate y; if the fluxion of this be divided by $\dot z$, and $\dot z$ and z be exterminated, the general value of the Variation will come out $\frac{-3\ddot y^2 + \dddot y(1 + \dot y^2)}{\ddot y^3}$; then substituting the values of $\dot y$, $\ddot y$, $\dddot y$ (found from the equation of the curve) into this quantity, it will give the Variation sought.

Ex. Let the curve be the parabola, whose equation is $ax = y^2$. Here then $2y\dot y = a\dot x = a$, and $\dot y = \frac{a}{2y}$; hence $\ddot y = \frac{-a\dot y}{2yy} = \frac{-aa}{4y^3}$, and $\dddot y = \frac{-3aa\dot y}{2y^4} = \frac{3a^3}{8y^5}$.

Therefore $\frac{-3\ddot y^2 + \dddot y(1 + \dot y^2)}{\ddot y^2} = -3\dot y + \dddot y \times \frac{1 + \dot y^2}{\ddot y^2} = \frac{-3a}{2y} + \frac{3a^3}{8y^5} \times (1 + \frac{aa}{4yy}) \times \frac{16y^6}{a^4} = \frac{6y}{a}$, the Variation sought. Emerson's Flux. pa. 228.

VARIGNON (PETER), a celebrated French mathematician and priest, was born at Caen in 1654, and died suddenly in 1722, at 68 years of age. He was the son of an architect in middling circumstances, but had a college education, being intended for the church. An accident threw a copy of Euclid's Elements in his way, which gave him a strong turn to that kind of learning. The study of geometry led him to the works of Des Cartes on the same science, and there he was struck with that new light which has, from thence, spread over the world.

He abridged himself of the necessaries of life to purchase books of this kind, or rather considered them of that number, as indeed they ought to be. What contributed to heighten this passion in him was, that he studied in private: for his relations observing that the books he studied were not such as were commonly used by others, strongly opposed his application to them. As there was a necessity for his being an ecclesiastic, he continued his theological studies, yet not entirely sacrificing his favourite subject to them.

At this time the Abbé St. Pierre, who studied philosophy in the same college, became acquainted with him. A taste in common for rational subjects, whether physics or metaphysics, and continual disputations, formed the bonds of their friendship. They were mutually serviceable to each other in their studies. The Abbé, to enjoy Varignon's company with greater ease, lodged him with himself; thus, growing still more

sensible of his merit, he resolved to give him a fortune, that he might fully pursue his genius, and improve his talents; and, out of only 18 hundred livres a year, which he had himself, he conferred 300 of them upon Varignon.

The Abbé, persuaded that he could not do better than go to Paris to study philosophy, settled there in 1686, with M. Varignon, in the suburbs of St. Jacques. There each studied in his own way; the Abbé applying himself to the study of men, manners, and the principles of government; whilst Varignon was wholly occupied with the mathematics.

I, says Fontenelle, who was their countryman, often went to see them, sometimes spending two or three days with them. They had also room for a couple of visitors, who came from the same province. We joined together with the greatest pleasure. We were young, full of the first ardour for knowledge, strongly united, and, what we were not then perhaps disposed to think so great a happiness, little known. Varignon, who had a strong constitution, at least in his youth, spent whole days in study, without any amusement or recreation, except walking sometimes in fine weather. I have heard him say, that in studying after supper, as he usually did, he was often surprised to hear the clock strike two in the morning; and was much pleased that four hours rest were sufficient to refresh him. He did not leave his studies with that heaviness which they usually create, nor with that weariness that a long application might occasion. He left off gay and lively, filled with pleasure, and impatient to renew it. In speaking of mathematics, he would laugh so freely, that it seemed as if he had studied for diversion. No condition was so much to be envied as his; his life was a continual enjoyment, delighting in quietness.

In the solitary suburb of St. Jacques, he formed however a connection with many other learned men; as Du Hamel, Du Verney, De la Hire, &c. Du Verney often asked his assistance in those parts of anatomy connected with mechanics: they examined together the positions of the muscles, and their directions; hence Varignon learned a good deal of anatomy from Du Verney, which he repaid by the application of mathematical reasoning to that subject.

At length, in 1687, Varignon made himself known to the public by a Treatise on New Mechanics, dedicated to the Academy of Sciences. His thoughts on this subject were, in effect, quite new. He discovered truths, and laid open their sources. In this work, he demonstrated the necessity of an equilibrium, in such cases as it happens in, though the cause of it is not exactly known. This discovery Varignon made by the theory of compound motions, and is what this essay turns upon.

This new Treatise on Mechanics was greatly admired by the mathematicians, and procured the author two considerable places, the one of Geometrician in the Academy of Sciences, the other of Professor of Mathematics in the College of Mazarine, to which he was the first person raised.

Varignon catched eagerly at the Science of Infinitesimals as soon as it appeared in the world, and became one of its most early cultivators. When that sublime and beautiful method was attacked in the Academy itself
self

felf (for it could not efcape the fate of all innovations) he became one of its moft zealous defenders, and in its favour he put a violence upon his natural character, which abhorred all contention. He fometimes lamented, that this difpute had interrupted him in his enquiries into the Integral Calculation fo far, that it would be difficult for him to refume his difquifition where he had left it off. He facrificed Infinitefimals to the intereft of Infinitefimals, and gave up the pleafure and glory of making a farther progrefs in them when called upon by duty to undertake their defence.

All the printed volumes of the Academy bear witnefs to his application and induftry. His works are never detached pieces, but complete theories of the laws of motion, central forces, and the refiftance of mediums to motion. In thefe he makes fuch ufe of his rules, that nothing efcapes him that has any connection with the fubject he treats.

Geometrical certainty is by no means incompatible with obfcurity and confufion, and thofe are fometimes fo great, that it is furprifing a mathematician fhould not mifs his way in fo dark and perplexing a labyrinth. The works of M. Varignon never occafion this difagreeable furprife, he makes it his chief care to place every thing in the cleareft light; he does not, as fome great men do, confult his eafe by declining to take the trouble of being methodical, a trouble much greater than that of compofition itfelf; he does not endeavour to acquire a reputation for profoundnefs, by leaving a great deal to be gueffed by the reader.

He was perfectly acquainted with the hiftory of mathematics. He learned it not merely out of curiofity, but becaufe he was defirous of acquiring knowledge from every quarter. This hiftorical knowledge is doubtlefs an ornament in a mathematician, but it is an ornament which is by no means without its utility. Indeed it may be laid down as a maxim, the more different ways the mind is occupied in, upon a fubject, the more it improves.

Though Varignon's conftitution did not feem eafy to be impaired, affiduity and conftant application brought upon him a fevere difeafe in 1705. Great abilities are generally dangerous to the poffeffors. He was fix months in danger, and three years in a languid ftate, which proceeded from his fpirits being almoft entirely exhaufted. He faid that fometimes when delirious with a fever, he thought himfelf in the midft of a foreft, where all the leaves of the trees were covered with algebraical calculations. Condemned by his phyficians, his friends, and himfelf, to lay afide all ftudy, he could not, when alone in his chamber, avoid taking up a book of mathematics, which he hid as foon as he heard any perfon coming. He again refumed the attitude and behaviour of a fick man, and feldom had occafion to counterfeit.

In regard to his character, Fontenelle obferves, that it was at this time that a writing of his appeared, in which he cenfured Dr. Wallis for having advanced that there are certain fpaces more than infinite, which that great geometrician afcribes to hyperbolas. He maintained, on the contrary, that they were finite. The criticifm was foftened with all the politenefs and refpect imaginable; but a criticifm it was, though he had written it only for himfelf. He let M. Carre fee it,

when he was in a ftate that rendered him indifferent about things of that kind; and that gentleman, influenced only by the intereft of the fciences, caufed it to be printed in the memoirs of the Academy of Sciences, unknown to the author, who thus made an attack againft his inclination.

He recovered from his difeafe; but the remembrance of what he had fuffered did not make him more prudent for the future. The whole impreffion of his *Project for a New Syftem of Mechanics*, having been fold off, he formed a defign to publifh a fecond edition of it, or rather a work entirely new, though upon the fame plan, but more extended. It muft be eafy to perceive how much learning he muft have acquired in the interval; but he often complained, that he wanted time, though he was by no means difpofed to lofe any. Frequent vifits, either of French or of foreigners, fome of whom went to fee him that they might have it to fay that they had feen him; and others to confult him and improve by his converfation: works of mathematics, which the authority of fome, or the friendfhip he had for others, engaged him to examine, and which he thought himfelf obliged to give the moft exact account of; a literary correfpondence with all the chief mathematicians of Europe; all thefe obftructed the book he had undertaken to write. Thus a man acquires reputation by having a great deal of leifure time, and he lofes this precious leifure as foon as he has acquired reputation. Add to this, that his beft fcholars, whether in the College of Mazarine or the Royal College (for he had a profeffor's chair in both), fometimes requefted private lectures of him, which he could not refufe. He fighed fer his two or three months of vacation, for that was all the leifure time he had in the year; no fooner were they come but he retired into the country, where his time was entirely his own, and the days feemed always quickly ended.

Notwithftanding his great defire of peace, in the latter part of his life he was involved in a difpute. An Italian monk, well verfed in mathematics, attacked him upon the fubject of tangents and the angle of contact in curves, fuch as they are conceived in the arithmetic of infinites; he anfwered by the laft memoir he ever gave to the Academy, and the only one which turned upon a difpute.

In the laft two years of his life he was attacked with an afthmatic complaint. This diforder increafed every day, and all remedies were ineffectual. He did not however ceafe from any of his cuftomary bufinefs; fo that, after having finifhed his lecture at the College of Mazarine, on the 22d of December 1722, he died fuddenly the following night.

His character, fays Fontenelle, was as fimple as his fuperior underftanding could require. He was not apt to be jealous of the fame of others: indeed he was at the head of the French mathematicians, and one of the beft in Europe. It muft be owned however, that when a new idea was offered to him, he was too hafty to object. The fire of his genius, the various infights into every fubject, made too impetuous an oppofition to thofe that were offered; fo that it was not eafy to obtain from him a favourable attention.

His works that were publifhed feparately, were,

1. Projet d'une Nouvelle Mechanique; 4to, Paris 1687.

2. Des

2. Des Nouvelles Conjectures sur la Pesanteur.

3. Nouvelle Mechanique ou Statique, 2 tom. 4to, 1725.

As to his memoirs in the volumes of the Academy of Sciences, they are far too numerous to be here particularized; they extend through almost all the volumes, down to his death in 1722.

VASA *Concordiæ*, in Hydraulics, are two vessels, so constructed, as that one of them, though full of wine, will not run a drop, unless the other, being full of water, do run also. Their structure and apparatus may be seen in Wolfius, Element. Mathes. tom. 3, Hydraul.

VAULT, in Architecture, an arched roof, so contrived, as that the several stones of which it consists, by their disposition into the form of a curve, mutually sustain each other; as the arches of bridges, &c.

Vaults are to be preferred, on many occasions, to soffits, or flat ceilings, as they give a greater rise and elevation, and are also more firm and durable.

The Ancients, Salmasius observes, had only three kinds of vaults: the first the *fornix*, made cradlewise; the 2d, the *testudo*, tortoise-wise, or oven-wise; the 3d, the *concha*, made shell-wise.

But the Moderns subdivide these three sorts into a great many more, to which they give different names, according to their figures and use: some are circular, others elliptical, &c.

Again, the sweeps of some are larger, and others less portions of a sphere: all above hemispheres are called *high*, or *surmounted Vaults*; all that are less than hemispheres, are *low*, or *surbased Vaults*, &c.

In some the height is greater than the diameter; in others it is less: there are others again quite flat, only made with haunses; others oven-like, and others growing wider as they lengthen, like a trumpet.

Of Vaults, some are *single*, others *double*, *cross*, *diagonal*, *horizontal*, *ascending*, *descending*, *angular*, *oblique*, *pendent*, &c, &c. There are also *Gothic* Vaults, with *pendentives*, &c.

Master VAULTS, are those which cover the principal parts of buildings; in contradistinction from the *less*, or subordinate Vaults, which only cover some small part; as a passage, a gate, &c.

Double VAULT, is such a one as, being built over another, to make the exterior decoration range with the interior, leaves a space between the convexity of the one, and the concavity of the other: as in the dome of St. Paul's at London, and that of St. Peter's at Rome.

VAULTS *with Compartiments*, are such whose sweep, or inner face, is enriched with pannels of sculpture, separated by platbands. These compartiments, which are of different figures, according to the Vaults, and are usually gilt on a white ground, are made with stucco, on brick Vaults; as in the church of St. Peter's at Rome; and with plaster, on timber Vaults.

Theory of VAULTS.—In a semicircular Vault, or arch, being a hollow cylinder cut by a plane through its axis, standing on two imposts, and all the stones that compose it, being cut and placed in such a manner, as that their joints, or beds, being prolonged, do all meet in the centre of the vault; it is evident that all the stones must be cut wedge-wise, or wider at top and above,

than below; by virtue of which, they sustain each other, and mutually oppose the effort of their weight, which determines them to fall.

The stone in the middle of the Vault, which is perpendicular to the horizon, and is called the *key of the Vault*, is sustained on each side by the two contiguous stones, as by two inclined planes.

The second stone, which is on the right or left of the key-stone, is sustained by a third; which, by virtue of the figure of the Vault, is necessarily more inclined to the second, than the second is to the first; and consequently the second, in the effort it makes to fall, employs a less part of its weight than the first.

For the same reason, all the stones, reckoning from the keystone, employ still a less and less part of their weight to the last; which, resting on the horizontal plane, employs no part of its weight, or makes no effort to fall, as being entirely supported by the impost.

Now a great point to be aimed at in Vaults, is, that all the several stones make an equal effort to fall: to effect this, it is evident that as each stone, reckoning from the key to the impost, employs a still less and less part of its whole weight; the first only employing, for example, one-half; the 2d, one-third; the 3d, one-fourth; &c; there is no other way to make these different parts equal, but by a proportionable augmentation of the whole; that is, the second stone must be heavier than the first, the third heavier than the second, and so on to the last, which should be vastly heavier.

La Hire demonstrates what that proportion is, in which the weights of the stones of a semicircular arch must be increased, to be in equilibrio, or to tend with equal forces to fall; which gives the firmest disposition that a vault can have. Before him, the architects had no certain rule to conduct themselves by; but did all at random. Reckoning the degrees of the quadrant of the circle, from the keystone to the impost; the length or weight of each stone must be so much greater, as it is farther from the key. La Hire's rule is, to augment the weight of each stone above that of the key stone, as much as the tangent of the arch to the stone exceeds the tangent of the arch of half the key. Now the tangent of the last stone becomes infinite, and consequently the weight should be so too; but as infinity has no place in practice, the rule amounts to this, that the last stone be loaded as much as possible, and the others in proportion, that they may the better resist the effort which the Vault makes to separate them; which is called the *shoot* or *drift* of the Vault.

M. Parent, and other authors, have since determined the curve, or figure, which the extrados or outside of a Vault, whose intrados or inside is spherical, ought to have, that all the stones may be in equilibrio.

The above rule of La Hire's has since been found not accurate. See ARCH, and BRIDGE. See also my Treatise on the Principles of Bridges, and Emerson's Construction of Arches.

Key of a VAULT. See KEY, and VOUSSOIR.

Reins or *fillings up of a* VAULT, are the sides which sustain it.

Pendentive of a VAULT. See PENDENTIVE.

Impost of a VAULT, is the stone upon which is laid the first voussoir, or arch-stone of the Vault.

VEADAR,

VEADAR, in Chronology, the 13th month of the Jewish ecclefiaftical year, anfwering commonly to our March; this month is intercalated, to prevent the beginning of Nifan from being removed to the end of February.

VECTIS, in Mechanics, one of the fimple mechanical powers, more ufually called the LEVER.

VECTOR, or *Rádius Vector*, in Aftronomy, is a line fuppofed to be drawn from any planet moving round a centre, or the focus of an ellipfe, to that centre, or focus. It is fo called, becaufe it is that line by which the planet feems to be carried round its centre; and with which it defcribes areas proportional to the times.

VELOCITY, or *Swiftnefs*, in Mechanics, is that affection of motion, by which a moving body paffes over a certain fpace in a certain time. It is alfo called celerity; and it is always proportional to the fpace moved over in a given time, when the Velocity is uniform, or always the fame during that time.

Velocity is either *uniform* or *variable*. Uniform, or equal *Velocity*, is that with which a body paffes always over equal fpaces in equal times. And it is *variable*, or *unequal*, when the fpaces paffed over in equal times are unequal; in which cafe it is either *accelerated* or *retarded* Velocity; and this acceleration, or retardation, may alfo be equal or unequal, i. e. uniform or variable, &c. See ACCELERATION, and MOTION.

Velocity is alfo either *abfolute* or *relative*. *Abfolute Velocity* is that we have hitherto been confidering, in which the Velocity of a body is confidered fimply in itfelf, or as paffing over a certain fpaces in a certain time. But *relative* or *refpective Velocity*, is that with which bodies approach to, or recede from one another, whether they both move, or one of them be at reft. Thus, if one body move with the abfolute Velocity of 2 feet per fecond, and another with that of 6 feet per fecond; then if they move directly towards each other, the relative velocity with which they approach is that of 8 feet per fecond; but if they move both the fame way, fo that the latter overtake the former, then the relative Velocity with which that overtakes it, is only that of 4 feet per fecond. or only half of the former; and confequently it will take double the time of the former before they come in contact together.

VELOCITY *in a Right Line.*—When a body moves with a uniform Velocity, the fpaces paffed over by it, in different times, are proportional to the times; alfo the fpaces defcribed by two different uniform Velocities, in the fame time, are proportional to the Velocities; and confequently, when both times and Velocities are unequal, the fpaces defcribed are in the compound ratio of the times and Velocities. That is, $S \propto TV$, and $s \propto tv$; or $S : s :: TV : tv$. Hence alfo, $V : v :: \frac{S}{T} : \frac{s}{t}$, or the Velocity is as the fpace directly and the time reciprocally.

But in uniformly accelerated motions; the laft degree of Velocity uniformly gained by a body in beginning from reft, is proportional to the time; and the fpace defcribed from the beginning of the motion, is as the product of the time and Velocity, or as the fquare of the Velocity, or as the fquare of the time. That is,

in uniformly accelerated motions, $v \propto t$, and $s \propto tv$ or $\propto v^2$ or $\propto t^2$. And, in fluxions, $s = vt$.

VELOCITY *of Bodies moving in Curves.*—According to Galileo's fyftem of the fall of heavy bodies, which is now univerfally admitted among philofophers, the Velocities of a body falling vertically are, at each moment of its fall, as the fquare roots of the heights from whence it has fallen; reckoning from the beginning of the defcent. And hence he inferred, that if a body defcend along an inclined plane, the Velocities it has, at the different times, will be in the fame ratio: for fince its Velocity is all owing to its fall, and it only falls as much as there is perpendicular height in the inclined plane, the Velocity fhould be ftill meafured by that height, the fame as if the fall were vertical.

The fame principle led him alfo to conclude, that if a body fall through feveral contiguous inclined planes, making any angles with each other, much like a ftick when broken, the Velocity would ftill be regulated after the fame manner, by the vertical heights of the different planes taken together, confidering the laft Velocity as the fame that the body would acquire by a fall through the fame perpendicular height.

This conclufion it feems continued to be acquiefced in, till the year 1672, when it was demonftrated to be falfe, by James Gregory, in a fmall piece of his intitled *Tentamina quædam Geometrica de Motu Penduli & Projectorum.* This piece has been very little known, becaufe it was only added to the end of an obfcure and pfeudonymous piece of his, then publifhed, to expofe the errors and vanity of Mr. Sinclair, profeffor of natural philofophy at Glafgow. This little jeu d'efprit of Gregory is intitled, *The great and new Art of Weighing Vanity: or a difcovery of the Ignorance and Arrogance of the great and new Artift, in his Pfeudo-Philofophical writings: by M. Patrick Mathers, Arch-Bedal to the Univerfity of S. Andrews.* In the *Tentamina*, Gregory fhews what the real Velocity is, which a body acquires by defcending down two contiguous inclined planes, forming an obtufe angle, and that it is different from the Velocity a body acquires by defcending perpendicularly through the fame height; alfo that the Velocity in quitting the firft plane, is to that with which it enters the fecond, and in this latter direction, as radius to the cofine of the angle of inclination between the two planes.

This conclufion however, Gregory obferves, does not apply to the motions of defcent down any curve lines, becaufe the contiguous parts of curve lines do not form any angle between them, and confequently no part of the Velocity is loft by paffing from one part of the curve to the other; and hence he infers, that the Velocities acquired in defcending down a continued curve line, are the fame as by falling perpendicularly through the fame height. This principle is then applied, by the author, to the motion of pendulums and projectiles.

Varignon too, in the year 1693, followed in the fame track, fhewing that the Velocity loft in paffing from one right lined direction to another, becomes indefinitely fmall in the courfe of a curve line; and that therefore the doctrine of Galileo holds good for the defcent of bodies down a curve line; viz, that the Velocity

6   acquired

acquired at any point of the curve, is equal to that which would be acquired by a fall through the same perpendicular altitude.

The nature of every curve is abundantly determined by the ratio of the ordinates to the corresponding abscisses; and the essence of curves in general may be conceived as consisting in this ratio, which may be varied in a thousand different ways. But this same ratio will be also that of two simple Velocities, by whose joint effect a body may describe the curve in question; and consequently the essence of all curves, in general, is the same thing as the concourse or combination of all the forces which, taken two by two, may move the same body. Thus we have a most simple and general equation of all possible curves, and of all possible Velocities. By means of this equation, as soon as the two simple Velocities of a body are known, the curve resulting from them is immediately determined.

It may be observed, in particular, according to this equation, that an uniform Velocity, combined with a Velocity that always varies as the square roots of the heights, the two produce the particular curve of a parabola, independent of the angle made by the directions of the two forces that give the Velocities; and consequently a cannon ball, shot either horizontally or obliquely to the horizon, must always describe a parabola, were it not for the resistance of the air.

Circular VELOCITY. See CIRCULAR.

Initial VELOCITY, in Gunnery, denotes the Velocity with which military projectiles issue from the mouth of the piece by which they are discharged. This, it is now known, is much more considerable than was formerly apprehended. For the method of estimating it,

and the result of a variety of experiments, by Mr. Robins, and myself, &c, see the articles GUN, GUNNERY, PROJECTILE, and RESISTANCE.

Mr. Robins had hinted in his New Principles of of Gunnery, at another method of measuring the Initial Velocities of military projectiles, viz, from the arc of vibration of the gun itself, in the act of expulsion, when it is suspended by an axis like a pendulum. And Mr. Thompson, in his experiments (Philos. Transf. vol. 71, p. 229) has pursued the same idea at considerable length, in a number of experiments, from whence he deduces a rule for computing the Velocity, which is somewhat different from that of Mr. Robins, but which agrees very well with his own experiments.

This rule however being drawn only from the experiments with a musket barrel, and with a small charge of powder, and besides being different from that in the theory as proposed by Robins; it was suspected that it would not hold good when applied to cannon, or other large pieces of ordnance, of different and various lengths, and to larger charges of powder. For this reason, a great multitude of experiments, as related in my Tracts, vol. 1, were instituted with cannon of various lengths and charged with many different quantities of powder; and the Initial Velocities of the shot were computed both from the vibration of a ballistic pendulum, and from the vibration of the gun itself; but the consequence was, that these two hardly ever agreed together, and in many cases they differed by almost 400 feet per second in the Velocity. A brief abstract for a comparison between these two methods, is contained in the following tablet, viz.

Comparison of the Velocities by the Gun and Pendulum.

Gun No.	2 Ounces.			4 Ounces.			8 Ounces.			16 Ounces.		
	Velocity by		Diff	Velocity by		Diff.	Velocity by		Diff.	Velocity by		Diff.
	Gun	Pend.		Gun	Pend.		Gun	Pend.		Gun	Pend.	
1	830	780	50	1135	1100	35	1445	1430	15	1345	1377	— 32
2	863	835	28	1203	1180	23	1521	1580	—59	1485	1656	—171
3	919	920	— 1	1294	1300	—6	1631	1790	—159	1680	1998	—318
4	929	970	—41	1317	1370	—53	1669	1940	—271	1730	2106	—376

In this table, the first column shews the number of the gun, as they were of different lengths; viz, the length of number 1 was 30⅛ inches, number 2 was 40⅜ inches, number 3 was 60 inches, and number 4 was 83 inches, nearly. After the first column, the rest of the table is divided into four spaces, for the four charges, 2, 4, 8, 16 ounces of powder: and each of these is divided into three columns: in the first of the three is the Velocity of the ball as determined from the vibration of the gun; in the second is the Velocity as determined from the vibration of the pendulum; and in the third is the difference between the two, being so many feet per second, which is marked with the nega-

tive sign, or —, when the former Velocity is too little, otherwise it is positive.

From the comparison contained in this table, it appears, in general, that the Velocities, determined by the two different ways, do not agree together; and that therefore the method of determining the Velocity of the ball from the recoil of the gun, is not generally true, although Mr. Robins and Mr. Thompson had suspected it to be so: and consequently that the effect of the inflamed powder on the recoil of the gun, is not exactly the same when it is fired without a ball, as when it is fired with one. It also appears, that this difference is no ways regular, neither in the different

guns

guns with the fame charge of powder, nor in the fame gun with different charges : That with very fmall charges, the Velocity by the gun is greater than that by the pendulum; but that the latter always gains upon the former, as the charge is increafed, and foon becomes equal to it ; and afterwards goes on to exceed it more and more : That the particular charge, at which the two Velocities become equal, is different in the different guns ; and that this charge is lefs, or the equality fooner takes place, as the gun is longer. And all this, whether we ufe the actual Velocity with which the ball ftrikes the pendulum, or the fame increafed by the Velocity loft by the refiftance of the air, in its flight from the gun to the pendulum.

VENTILATOR, a machine by which the noxious air of any clofe place, as an hofpital, gaol, fhip, chamber, &c, may be difcharged and changed for frefh air.

The noxious qualities of bad air have been long known ; and Dr. Hales and others have taken great pains to point out the mifchiefs arifing from foul air, and to prevent or remedy them. That philofopher propofed an eafy and effectual one, by the ufe of his Ventilators ; the account of which was read before the Royal Society in May 1741 ; and a farther account of it may be feen in his Defcription of Ventilators, printed at London in 8vo, 1743 ; and ftill farther in part 2, p. 32, printed in 1758 ; where the ufes and applications of them are pointed out for fhips, and prifons, &c. For what is faid of the foul air of fhips may be applied to that of gaols, mines, workhoufes, hofpitals, barracks, &c. In mines, Ventilators may guard againft the fuffocations, and other terrible accidents arifing from damps. The air of gaols has often proved infectious ; and we had a fatal proof of this by the accident that happened fome years fince at the Old Bailey feffions. After that, Ventilators were ufed in the prifon, which were worked by a fmall windmill, placed on the top of Newgate ; and the prifon became more healthy.

Dr. Hales farther fuggefts, that Ventilators might be of ufe in making falt ; for which purpofe there fhould be a ftream of water to work them ; or they might be worked by a windmill, and the brine fhould be in long narrow canals, covered with boards of canvas, about a foot above the furface of the brine, to confine the ftream of air, fo as to make it act upon the furface of the brine, and carry off the water in vapours. Thus it might be reduced to a dry falt, with a faving of fuel, in winter and fummer, or in rainy weather, or any ftate of the air whatever. Ventilators, he apprehends, might alfo ferve for drying linen hung in low, long, narrow galleries, efpecially in damp or rainy weather, and alfo in drying woollen cloths, after they are fulled or dyed ; and in this cafe, the Ventilators might be worked by the fulling water-mill. Ventilators might alfo be an ufeful appendage to malt and hop kilns ; and the fame author is farther of opinion, that a ventilation of warm dry air from the adjoining ftove, with a cautious hand, might be of fervice to trees and plants in green-houfes ; where it is well known that air full of the rancid vapours which perfpire from the plants, is very unkindly to them, as well as the vapours from human bodies are to men : for frefh air is as neceffary

to the healthy ftate of vegetables, as of animals.—Ventilators are alfo of excellent ufe for drying corn, hops, and malt.—Gunpowder may be thoroughly dried, by blowing air up through it by means of Ventilators ; which is of great advantage to the ftrength of it. Thefe Ventilators, even the fmaller ones, will alfo ferve to purify moft eafily, and effectually, the bad air of a fhip's well, before a perfon is fent down into it, by blowing air through a trunk, reaching near the bottom of it. And in a fimilar manner may ftinking water, and ill-tafted milk, &c, be fweetened, viz, by paffing a current of air through them, from bottom to top, which will carry the offenfive particles along with it.

For thefe and other ufes to which they might be applied, as well as for a particular account of the conftruction and difpofition of Ventilators in fhips, hofpitals, prifons, &c, and the benefits attending them. fee Hales's Treatife on Ventilators, part 2 paffim ; and the Philof. Tranf. vol. 49, p. 332.

The method of drawing off air from fhips by means of fire-pipes, which fome have preferred to Ventilators, was publifhed by Sir Robert Moray in the Philof. Tranf. for 1665. Thefe are metal pipes, about $2\frac{1}{2}$ inches diameter, one of which reaches from the fireplace to the well of the fhip, and other three branches go to other parts of the fhip ; the ftove hole and afh hole being clofed up, the fire is fupplied with air through thefe pipes. The defects of thefe, compared with Ventilators, are particularly examined by Dr. Hales, ubi fupra, p. 113.

In the latter part of the year 1741, M. Triewald, military architect to the king of Sweden, informed the fecretary to the Royal Society, that he had in the preceding fpring invented a machine for the ufe of fhips of war, to draw out the foul air from under their decks, which exhaufted 36172 cubic feet of air in an hour, or at the rate of 21732 tuns in 24 hours. In 1742 he fent one of thefe to France, which was approved of by the Academy of Sciences at Paris, and the navy of France was ordered to be furnifhed with the like Ventilators.

Mr. Erafmus King propofed to have Ventilators worked by the fire engines, in mines. And Mr. Fitzgerald has fuggefted an improved method of doing this, which he has alfo illuftrated by figures. See Philof. Tranf. vol. 50, p. 727.

There are various ways of Ventilation, or changing the air of rooms. Mr. Tidd contrived to admit frefh air into a room, by taking out the middle upper fafh pane of glafs, and fixing in its place a frame box, with a round hole in its middle, about 6 or 7 inches diameter ; in which hole are fixed, behind each other, a fet of fails of very thin broad copper-plates, which fpread over and cover the circular hole, fo as to make the air which enters the room, and turning round thefe fails, to fpread round in thin fheets fideways ; and fo not to incommode perfons, by blowing directly upon them, as it would do if it were not hindered by the fails.

This method however is very unfeemly and difagreeable in good rooms : and therefore, inftead of it, the late ingenious Mr. John Whitehurft fubftituted another ; which was, to open a fmall fquare or rectangular hole in the party wall of the room, in the upper part near the cieling, at a corner or part diftant from the

fire

fire ; and before it he placed a thin piece of metal or pasteboard &c, attached to the wall in its lower part just below the hole, but declining from it upwards, so as to give the air, that enters by the hole, a direction upwards against the cieling, along which it sweeps and disperses itself through the room, without blowing in a current against any person. This method is very useful to cure smoky chimneys, by thus admitting conveniently fresh air. A picture placed before the hole prevents the sight of it from disfiguring the room. This, and many other methods of Ventilating, he meant to have published, and was occupied upon, when death put an end to his useful labours. These have since been published, viz in 1794, 4to, by Dr. Willan.

VENUS, in Astronomy, one of the inferior planets, but the brightest and to appearance the largest of all the planets ; and is designed by the mark ♀, supposed to be a rude representation of a female figure, with her trailing robe.

Venus is easily distinguished from all the other planets, by her whiteness and brightness, in which she exceeds all the rest, even Jupiter himself, and which is so considerable, that in a dusky place she causes an object to project a sensible shadow, and she is often visible in the day-time. Her place in the system is the second from the sun, viz, between Mercury and the earth, and in magnitude is about equal to the earth, or rather a little larger according to Dr. Herschel's observations.

As Venus moves round the sun, in a circle beneath that of the earth, she is never seen in opposition to him, nor indeed very far from him ; but seems to move backward and forward, passing him from side to side, to the distance of about 47 or 48 degrees, both ways, which is her greatest elongation.

When she appears west of the sun, which is from her inferior conjunction to her superior, she rises before him, or is a morning star, and is called *Phosphorus*, or *Lucifer*, or the *Morning Star* ; and when she is eastwards from the sun, which is from her superior conjunction to her inferior, she sets after him, or is an evening star, and is called *Hesperus*, or *Vesper*, or the *Evening star* : being each of those in its turn for 290 days.

The real diameter of Venus is nearly equal to that of the earth, being about 7900 miles ; her apparent mean diameter seen from the earth 59″, seen from the sun, or her horizontal parallax, 30″ ; but as seen from the earth 18″·79 according to Dr. Herschel : her distance from the sun 70 million of miles ; her eccentricity $\frac{7}{1000}$ths of the same, or 490,000 miles ; the inclination of her orbit to the plane of the ecliptic 3° 23′ ; the points of their intersection or nodes are 14° of ♊ and ♐ ; the place of her aphelion ♒ 4° 20′ ; her axis inclined to her orbit 75° 0′ ; her periodical course round the sun 224 days 17 hours ; the diurnal rotation round her axis very uncertain, being according to Cassini only 23 hours, but according to the observations of Bianchini it is in 24 days 8 hours ; though Dr. Herschel thinks it cannot be so much. See also PLANETS.

Venus, when viewed through a telescope, is rarely seen to shine with a full face, but has phases and changes just like those of the moon, being increasing, decreasing, horned, gibbous, &c : her illuminated part being constantly turned toward the sun, or directed toward the east when she is a morning star, and toward the west when an evening star.

These different phases of Venus were first discovered by Galileo ; who thus fulfilled the prediction of Copernicus : for when this excellent astronomer revived the ancient Pythagorean system, asserting that the earth and planets move round the sun, it was objected that in such a case the phases of Venus should resemble those of the moon ; to which Copernicus replied, that some time or other that resemblance would be found out. Galileo sent an account of the first discovery of these phases in a letter, written from Florence in 1611, to William de Medici, the duke of Tuscany's ambassador at Prague ; desiring him to communicate it to Kepler. The letter is extant in the preface to Kepler's Dioptrics, and a translation of it in Smith's Optics, p. 416. Having recited the observations he had made, he adds, " We have hence the most certain, sensible decision and demonstration of two grand questions, which to this day have been doubtful and disputed among the greatest masters of reason in the world. One is, that the planets in their own nature are opake bodies, attributing to Mercury what we have seen in Venus : and the other is, that Venus necessarily moves round the sun ; as also Mercury and the other planets ; a thing well believed indeed by Pythagoras, Copernicus, Kepler, and myself, but never yet proved, as now it is, by ocular inspection upon Venus."

Cassini and Campani, in the years 1665 and 1666, discovered spots in the face of Venus : from the appearances of which the former ascertained her motion round her axis ; concluding that this revolution was performed in less than a day ; or at least that the bright spot which he observed, finished its period either by revolution or libration in about 23 hours. And de la Hire, in 1700, through a telescope of 16 feet, discovered spots in Venus ; which he found to be larger than those in the moon.

The next observations of the same kind that occur, are those of signior Binanchini at Rome, in 1726, 1727, 1728, who, with Campani's glasses, discovered several dark spots in the disc of Venus, of which he gave an account and a representation in his book entitled Hesperi et Phosphori Nova Phenomena, published at Rome in 1728. From several successive observations Bianchini concludes, that a rotation of Venus about her axis was not completed in 23 hours, as Cassini imagined, but in 24½ days ; that the north pole of this rotation faced the 20th degree of Aquarius, and was elevated 15° above the plane of the ecliptic, and that the axis kept parallel to itself, during the planet's revolution about the sun. Cassini the son, though he admits the accuracy of Bianchini's observations, disputes the conclusion drawn from them, and finally observes, that if we suppose the period of the rotation of Venus to be 23 h. 20 min. it agrees equally well with the observations both of his father and Bianchini ; but if she revolve in 24 d. 8 h. then his father's observations must be rejected as of no consequence.

In the Philos. Transf. 1792, are published the results of a course of observations on the planet Venus, begun in the year 1780, by Mr. Schroeter, of Lilienthal, Bremen. From these observations, the author infers,

that

that Venus has an atmosphere in some respects similar to that of our earth, but far exceeding that of the moon in density, or power to weaken the rays of the sun: that the diurnal period of this planet is probably much longer than that of other planets: that the moon also has an atmosphere, though less dense and high than that of Venus: and that the mountains of this planet are 5 or 6 times as high as those on the earth.

Dr. Herschel too, between the years 1777 and 1793, has made a long series of observations on this planet, accounts of which are given in the Philos. Transf. for 1793. The results of these observations are: that the planet revolves about its axis, but the time of it is uncertain: that the position of its axis is also very uncertain: that the planet's atmosphere is very considerable: that the planet has probably hills and inequalities on its surface, but he has not been able to see much of them, owing perhaps to the great density of its atmosphere; as to the mountains of Venus, no eye, he says, which is not considerably better than his, or assisted by much better instruments, will ever get a sight of them: and that the apparent diameter of Venus, at the mean distance from the earth, is 18″·79; from whence it may be inferred, that this planet is somewhat larger than the earth, instead of being less, as former astronomers have imagined.

Sometimes Venus is seen in the disc of the sun, in form of a dark round spot. These appearances, called Transits, happen but seldom, viz, when the earth is about her nodes at the time of her inferior conjunction. One of these transits was seen in England in 1639, by Mr. Horrox and Mr. Crabtree; and two in the present century, viz, the one June 6, 1761, and the other in June 1769. There will not happen another of them till the year 1874. See PARALLAX.

Except such transits as these, Venus exhibits the same appearances to us regularly every 8 years; her conjunctions, elongations, and times of rising and setting, being very nearly the same, on the same days, as before.

In 1672 and 1686, Cassini, with a telescope of 34 feet, thought he saw a satellite move round this planet, at the distance of about ⅗ of Venus's diameter. It had the same phases as Venus, but without any well defined form; and its diameter scarce exceeded ¼ of the diameter of Venus. Dr. Gregory (Astron. lib. 6, prop. 3) thinks it more than probable that this was a satellite; and supposes that the reason why it is not more frequently seen, is the unfitness of its surface to reflect the rays of the sun's light; as is the case of the spots in the moon; for if the whole disc of the moon were composed of such, he thinks she could not be seen so far as to Venus.

Mr. Short, in 1740, with a reflecting telescope of 16½ inches focus, perceived a small star near Venus: with another telescope of the same focus, magnifying 50 or 60 times, and fitted with a micrometer, he found its distance from Venus about 10′; and with a magnifying power of 240, he observed the star assume the same phases with Venus; its diameter seemed to be about ⅓, or somewhat less, of the diameter of Venus; its light not so bright and vivid, but exceeding sharp and well defined. He viewed it for the space of an hour; but never had the good fortune to see it after the

first morning. Philos. Transf. number 459, p. 646, or Abr. vol. 8, p. 208.

M. Montaign, of Limoges in France, preparing for observing the transit of 1761, discovered in the preceding month of May a small star, about the distance of 20′ from Venus, the diameter of it being about ¼ of that of the planet. Others have also thought they saw a like appearance. And indeed it must be acknowledged, that Venus may have a satellite, though it is difficult for us to see it. Its enlightened side can never be fully turned towards us, but when Venus is beyond the sun; in which case Venus herself appears little bigger than an ordinary star, and therefore her satellite may be too small to be perceived at such a distance. When she is between us and the sun, her moon has its dark side turned towards us; and when Venus is at her greatest elongation, there is but half the enlightened side of the moon turned toward us, and even then it may be too far distant to be seen by us. But it was presumed, that the two transits of 1761, and 1769, would afford opportunity for determining this point; and yet we do not find, although many observers directed their attention to this object, that any satellite was then seen in the sun's disc; unless we except two persons, viz, an anonymous writer in the London Chronicle of May 18, who says that he saw the satellite of Venus on the sun the day of the transit, at St. Neot's in Huntingdonshire; that it moved in a track parallel to that of Venus, but nearer the ecliptic; that Venus quitted the sun's disc at 31 minutes after 8, and the satellite at 6 minutes after 9; and M. Montaign at Limoges, whose account of his observations is in the Memoirs of the Academy of Paris, from whence the following certificate is extracted:—CERTIFICATE. " We having examined, by order of the Academy, the remarks of M. Baudouin on a new observation of the satellite of Venus, made at Limoges the 11th of May by M. Montaign. This fourth observation, of great importance for the theory of the satellite, has shewn that its revolution must be longer than appeared by the first three observations. M. Baudouin believes it may be fixed at 12 days; as to its distance, it appears to him to be 50 semidiameters of Venus; whence he infers that the mass of Venus is equal to that of the earth. This mass of Venus is a very essential element to astronomy, as it enters into many computations, and produces different phenomena: &c.

Signed   L' Abbé De La Caille,
De La Lande."

VERBERATION, in Physics, a term used to express the cause of sound, which arises from a Verberation of the air, when struck, in divers manners, by the several parts of the sonorous body first put into a vibratory motion.

VERNAL, something belonging to the spring season: as vernal signs, vernal equinox, &c.

VERNIER, is a scale, or a division, well adapted for the graduation of mathematical instruments, so called from its inventor Peter Vernier, a gentleman of Franche Comté, who communicated the discovery to the world in a small tract, entitled La Construction, l'Usage, et les Proprietez du Quadrant Nouveau de Mathematique &c, printed at Brussels in 1631. This was an improvement on the method of divifion propofed by Jacobus Curtius, printed by Tycho in Clavius's Aftrolabe, in 1593. Vernier's method of divifion, or dividing plate, has been very commonly, though erroneoufly, called by the name of Nonius; the method of Nonius being very different from that of Vernier, and much lefs convenient.

When the relative unit of any line is fo divided into many fmall equal parts, thofe parts may be too numerous to be introduced, or if introduced, they may be too clofe to one another to be readily counted or eftimated; for which reafon there have been various methods contrived for eftimating the aliquot parts of the fmall divifions, into which the relative unit of a line may be commodioufly divided; and among thofe methods, Vernier's has been moft juftly preferred to all others. For the hiftory of this, and other inventions of a fimilar nature, fee Robins's Math. Tracts, vol. 2, p 265, &c.

Vernier's fcale is a fmall moveable arch, or fcale, fliding along the limb of a quadrant, or any other graduated fcale, and divided into equal parts, that are one lefs in number than the divifions of the portion of the limb correfponding to it. So, if we want to fubdivide the graduations on any fcale into for ex. 10 equal parts; we muft make the Vernier equal in length to 11 of thofe graduations of the fcale, but dividing the fame length of the Vernier itfelf only into 10 equal parts; for then it is evident that each divifion on the Vernier will be $\frac{1}{10}$th part longer than the gradations on the inftrument, or that the divifion of the former is equal to $1\frac{1}{10}$ of the degree on the latter, as that gains 1 in 10 upon this.

Thus let AB be a part of the upper end of a barometer tube, the quickfilver ftanding at the point C; from 28 to 31 is a part of the fcale of inches, viz, from 28 inches to 31 inches, divided into 10ths of inches; and the middle piece, from 1 to 10, is the Vernier, that flides up and down in a groove, and having 10 of its divifions equal to 11 tenths of the inches, for the purpofe of fubdividing every 10th of the inch into 10 parts, or, the inches into centefms or 100th parts. In practice, the method of counting is by obferving when the Vernier is fet with its index at top pointing exactly againft the upper furface of the mercury in the tube) which divifion of the Vernier it is that exactly, or neareft, coincides with a divifion in the fcale of 10ths of inches, for that will fhew the number of 100ths, over the 10ths of inches next below the index at top. So, in the annexed figure, the top of the Vernier is between 2 and 3 tenths above the 30 inches of the barometer; and becaufe the 8th divifion of the Vernier is feen to coincide with a divifion of the fcale, this fhews that it is 8 centefms more: fo that the height of the quickfilver altogether, is 30·28, that is, 30

inches, and 28 hundredths, or 2 tenths and 8 hundredths.

If the fcale were not inches and 10ths, but degrees of a quadrant, &c, then the 8 would be $\frac{8}{10}$ of a degree, or 48′; or if every divifion on the fcale be 10 minutes, then the Vernier will fubdivide it into fingle minutes, and the 8 will then be 8 minutes. And fo for any other cafe.

By altering the number of divifions, either in the degrees or in the Vernier, or in both, an angle can be obferved to many different degrees of accuracy. Thus, if a degree on a quadrant be divided into 12 parts, each being 5 minutes, and the length of the Vernier be 21 fuch parts, or $1°\frac{3}{4}$, and divided into 20 parts, then

$$\frac{1}{12} \times \frac{1}{20} = \frac{1°}{240} = \frac{1'}{4} = 15'',$$

is the fmalleft divifion the Vernier will meafure to: Or, if the length of the Vernier be $2°\frac{7}{12}$, and divided into 30 parts, then

$$\frac{1}{12} \times \frac{1}{30} = \frac{1°}{360} = \frac{1'}{6} = 10'',$$

is the fmalleft part in this cafe: Alfo

$$\frac{1}{12} \times \frac{1}{50} = \frac{1°}{600} = \frac{1'}{10} = 6'',$$

is the fmalleft part when the Vernier extends $4°\frac{1}{4}$. See Robertfon's Navigation, book 5, p. 279.

For the method of applying the Vernier to a quadrant, fee *Hadley's* QUADRANT. And for the application of it to a telefcope, and the principles of its conftruction, fee Smith's Optics, book 3, fect. 861.

VERSED-*Sine*, of an arch, is the part of the diameter intercepted between the fine and the commencement of the arc; and it is equal to the difference between the radius and the cofine. See *Verfed*-SINE. And for *coverfed fine*, fee COVERSED-*Sine*.

VERTEX *of an Angle*, is the angular point, or the point where the legs or fides of the angle meet.

VERTEX *of a Figure*, is the uppermoft point, or the vertex of the angle oppofite to the bafe.

VERTEX *of a Curve*, is the extremity of the axis or diameter, or it is the point where the diameter meets the curve; which is alfo the vertex of the diameter.

VERTEX *of a Glafs*, in Optics, the fame as its pole.

VERTEX is alfo ufed, in Aftronomy, for the point of the heavens vertically or perpendicularly over our heads, alfo called the zenith.

VERTEX, *Path of the*. See PATH.

VERTICAL, fomething relating to the vertex or higheft point. As,

VERTICAL *Point*, in Aftronomy, is the fame with vertex, or zenith.—Hence a ftar is faid to be Vertical, when it happens to be in that point which is perpendicularly over any place.

VERTICAL *Circle*, is a great circle of the fphere, paffing through the zenith and nadir of a place.—The Vertical circles are alfo called *azimuths*. The meridian of any place is a Vertical circle, viz, that particular one which paffes through the north or fouth point of the horizon.—All the Vertical circles interfect one another in the zenith and nadir.

2. The

The ufe of the Vertical circles is to eftimate or meafure the height of the ftars &c, and their diftances from the zenith, which is reckoned on thefe circles; and to find their eaftern and weftern amplitude, by obferving how many degrees the Vertical, in which the ftar rifes or fets, is diftant from the meridian.

Prime VERTICAL, is that Vertical circle, or azimuth, which paffes through the poles of the meridian; or which is perpendicular to the meridian, and paffes through the equinoctial points.

Prime VERTICALS, in Dialling. See PRIME *Verticals*.

VERTICAL *of the Sun*, is the Vertical which paffes through the centre of the fun at any moment of time. —Its ufe is, in Dialling, to find the declination of the plane on which the dial is to be drawn, which is done by obferving how many degrees that Vertical is diftant from the meridian, after marking the point or line of the fhadow upon the plane at any times.

VERTICAL *Dial.* See *Vertical* DIAL.

VERTICAL *Line*, in Dialling, is a line in any plane perpendicular to the horizon.——This is beft found and drawn on an erect and reclining plane, by fteadily holding up a ftring and plummet, and then marking two points of the fhadow of the thread on the plane, a good diftance from one another: and drawing a line through thefe marks.

VERTICAL *Line*, in Conics, is a line drawn on the Vertical plane, and through the vertex of the cone.

VERTICAL *Line*, in Perfpective. See *Vertical* LINE.

VERTICAL *Plane*, in Conics, is a plane paffing through the vertex of a cone, and parallel to any conic fection.

VERTICAL *Plane*, in Perfpective. See PLANE and PERSPECTIVE.

VERTICAL *Angles*, or *Oppofite Angles*, in Geometry, are fuch as have their legs or fides continuations of each other, and which confequently have the fame vertex or angular point. So the angles *a* and *b* are Vertical angles; as alfo the angles *c* and *d*.

VERTICITY, is that property of the magnet or loadftone, or of a needle &c touched with it, by which it turns or directs itfelf to fome peculiar point, as to its pole.—The attraction of the magnet was known long before its Verticity.

VERU, a comet, according to fome writers, refembling a fpit, being nearly the fame as the lonchites, only its head is rounder, and its train longer and fharper pointed.

VESPER, in Aftronomy, called alfo *Hefperus*, and the *Evening Star*, is the planet Venus, when fhe is eaftward of the fun, and confequently fets after him, and fhines as an evening ftar.

VESPERTINE, in Aftronomy, is when a planet is defcending to the weft after fun fet, or fhines as an evening ftar.

VIA LACTEA, in Aftronomy, is the milky way, or Galaxy. See GALAXY.

VIA SOLIS, or *fun's way*, is ufed among aftronomers, for the ecliptic line, or path in which the fun feems always to move.

VIBRATION, in Mechanics, a regular reciprocal motion of a body, as, for example, a pendulum, which being freely fufpended, fwings or vibrates from fide to fide.

Mechanical authors, inftead of Vibration, often ufe the term *ofcillation*, efpecially when fpeaking of a body that thus fwings by means of its own gravity or weight.

The Vibrations of the fame pendulum are all ifochronal; that is, they are performed in an equal time, at leaft in the fame latitude; for in lower latitudes they are found to be flower than in higher ones. See PENDULUM. In our latitude, a pendulum $39\frac{1}{8}$ inches long, vibrates feconds, making 60 Vibrations in a minute.

The Vibrations of a longer pendulum take up more time than thofe of a fhorter one, and that in the fubduplicate ratio of the lengths, or the ratio of the fquare roots of the lengths. Thus, if one pendulum be 40 inches long, and another only 10 inches long, the former will be double the time of the latter in performing a Vibration; for $\sqrt{40} : \sqrt{10} :: \sqrt{4} : \sqrt{1}$, that is as 2 to 1. And becaufe the number of Vibrations, made in any given time, is reciprocally as the duration of one Vibration, therefore the number of fuch Vibrations is in the reciprocal fubduplicate ratio of the lengths of the pendulums.

M. Mouton, a prieft of Lyons, wrote a treatife, exprefsly to fhew, that by means of the number of Vibrations of a given pendulum, in a certain time, may be eftablifhed an univerfal meafure throughout the whole world; and may fix the feveral meafures that are in ufe among us, in fuch a manner, as that they might be recovered again, if at any time they fhould chance to be loft, as is the cafe of moft of the ancient meafures, which we now only know by conjecture.

The VIBRATIONS *of a Stretched Chord*, or *String*, arife from its elafticity; which power being in this cafe fimilar to gravity, as acting uniformly, the Vibrations of a chord follow the fame laws as thofe of pendulums. Confequently the Vibrations of the fame chord equally ftretched, though they be of unequal lengths, are ifochronal, or are performed in equal times; and the fquares of the times of Vibration are to one another inverfely as their tenfions, or powers by which they are ftretched.

The Vibrations of a fpring too are proportional to the powers by which it is bent. Thefe follow the fame laws as thofe of the chord and pendulum; and confequently are ifochronal; which is the foundation of fpring watches.

VIBRATIONS are alfo ufed in *Phyfics*, &c, and for feveral other regular alternate motions. Senfation, for inftance, is fuppofed to be performed by means of the vibratory motion of the contents of the nerves, begun by external objects, and propagated to the brain.

This doctrine has been particularly illuftrated by Dr. Hartley, who has extended it farther than any other writer, in eftablifhing a new theory of our mental operations.

The fame ingenious author alfo applies the doctrine of Vibrations to the explanation of mufcular motion, which he thinks is performed in the fame general manner as fenfation and the perception of ideas. For a particular account of his theory, and the arguments by which it is fupported, fee his Obfervations on Man, vol. 1.

The

The feveral forts and rays of light, Newton conceives to make Vibrations of divers magnitudes; which, according to thofe magnitudes, excite fenfations of feveral colours; much after the fame manner as Vibrations of air, according to their feveral magnitudes, excite fenfations of feveral founds. See the article Co-LOUR.

Heat, according to the fame author, is only an accident of light, occafioned by the rays putting a fine, fubtile, ethereal medium, which pervades all bodies, into a vibrative motion, which gives us that fenfation. See HEAT.

From the Vibrations or pulfes of the fame medium, he accounts for the alternate fits of eafy reflexion and eafy tranfmiffion of the rays.

In the Philofophical Tranfactions it is obferved, that the butterfly, into which the filk-worm is transformed, makes 130 Vibrations or motions of its wings, in one coition.

VIETA (FRANCIS), a very celebrated French mathematician, was born in 1540 at Fontenai, or Fontenaile-Comté, in Lower Poitou, a province of France. He was Mafter of Requefts at Paris, where he died in 1603, being the 63d year of his age. Among other branches of learning in which he excelled, he was one of the moft refpectable mathematicians of the 16th century, or indeed of any age. His writings abound with marks of great originality, and the fineft genius, as well as intenfe application. His application was fuch, that he has fometimes remained in his ftudy for three days together, without eating or fleeping. His inventions and improvements in all parts of the mathematics were very confiderable. He was in a manner the inventor and introducer of Specious Algebra, in which letters are ufed inftead of numbers, as well as of many beautiful theorems in that fcience, a full explanation of which may be feen under the article AL-GEBRA. He made alfo confiderable improvements in geometry and trigonometry. His angular fections are a very ingenious and mafterly performance: by thefe he was enabled to refolve the problem of Adrian Roman, propofed to all mathematicians, amounting to an equation of the 45th degree. Romanus was fo ftruck with his fagacity, that he immediately quitted his refidence of Wirtzbourg in Franconia, and came to France to vifit him, and folicit his friendfhip. His Apollonius Gallus, being a reftoration of Apollonius's tract on Tangencies, and many other geometrical pieces to be found in his works, fhew the fineft tafte and genius for true geometrical fpeculations.—He gave fome mafterly tracts on Trigonometry, both plane and fpherical, which may be found in the collection of his works, publifhed at Leyden in 1646, by Schooten, befides another large and feparate volume in folio, publifhed in the author's life-time at Paris in 1579, containing extenfive trigonometrical tables, with the conftruction and ufe of the fame, which are particularly defcribed in the introduction to my Logarithms, pa. 4 &c. To this complete treatife on Trigonometry, plane and fpherical, are fubjoined feveral mifcellaneous problems and obfervations, fuch as, the quadrature of the circle, the duplication of the cube, &c. Computations are here given of the ratio of the diameter of a circle to the circumfeence, and of the length of the fine of 1 minute, both

to a great many places of figures; by which he found that the fine of 1 minute is

between   2908881959
and   2908882056;

alfo the diameter of a circle being 1000 &c, that the perimeter of the infcribed and circumfcribed polygon of 393216 fides, will be as follows, viz, the

perim. of the infcribed polygon   3.1415926535
perim. of the circumfcribed polygon   3.1415926537

and that therefore the circumference of the circle lies between thofe two numbers.

Vieta having obferved that there were many faults in the Gregorian Calendar, as it then exifted, he compofed a new form of it, to which he added perpetual canons, and an explication of it, with remarks and objections againft Clavius, whom he accufed of having deformed the true Lelian reformation, by not rightly underftanding it.

Befides thofe, it feems a work greatly efteemed, and the lofs of which cannot be fufficiently deplored, was his *Harmonicon Cœlefte*, which, being communicated to father Merfenne, was, by fome perfidious acquaintance of that honeft-minded perfon, furreptitioufly taken from him, and irrecoverably loft, or fuppreffed, to the great detriment of the learned world. There were alfo, it is faid, other works of an aftronomical kind, that have been buried in the ruins of time.

Vieta was alfo a profound decipherer, an accomplifhment that proved very ufeful to his country. As the different parts of the Spanifh monarchy lay very diftant from one another, when they had occafion to communicate any fecret defigns, they wrote them in ciphers and unknown characters, during the diforders of the league: the cipher was compofed of more than 500 different characters, which yielded their hidden contents to the penetrating genius of Vieta alone. His fkill fo difconcerted the Spanifh councils for two years, that they publifhed it at Rome, and other parts of Europe, that the French king had only difcovered their ciphers by means of magic.

VINCULUM, in Algebra, a mark or character, either drawn over, or including, or fome other way accompanying, a factor, divifor, dividend, &c, when it is compounded of feveral letters, quantities, or terms, to connect them together as one quantity, and fhew that they are to be multiplied, or divided, &c, together.

Vieta, I think, firft ufed the bar or line over the quantities, for a Vinculum, thus $\overline{a+b}$; and Albert Girard the parenthefis thus $(a+b)$; the former way being now chiefly ufed by the Englifh, and the latter by moft other Europeans. Thus $\overline{a+b} \times c$, or $(a+b) \times c$, denotes the product of c and the fum $a+b$ confidered as one quantity. Alfo $\sqrt{a+b}$, or $\sqrt{(a+b)}$, denotes the fquare root of the fum $a+b$. Sometimes the mark : is fet before a compound factor, as a Vinculum, efpecially when it is very long, or an infinite feries; thus $3a \times : 1 - 2x + 3x^2 - 4x^3 + 5x^5$ &c.

VINDEMIATRIX, or VINDEMIATOR, a fixed ftar of the third magnitude, in the northern wing of the conftellation Virgo.

VIRGO,

VIRGO, in Aftronomy, one of the figns or conftellations of the zodiac, which the fun enters about the 21ft or 22d of Auguft; being one of the 48 old conftellations, and is mentioned by the aftronomers of all ages and nations, whofe works have reached us. Anciently the figure was that of a girl, almoft naked, with an ear of corn in her hand, evidently to denote the time of harveft among the people who invented this fign, whoever they were. But the Greeks much altered the figure, with clothes, wings, &c, and varioufly explained the origin of it by their own fables: thus, they tell us that the virgin, now exalted into the fkies, was, while on earth, that Juftitia, the daughter of Aftræus and Ancora, who lived in the golden age, and taught mankind their duty; but who, when their crimes increafed, was obliged to leave the earth, and take her place in the heavens. Again, Hefiod gives the celeftial maid another origin, and fays fhe was the daughter of Jupiter and Themis. There are alfo others who depart from both thefe accounts, and make her to have been Erigone, the daughter of Icarius: while others make her Parthene, the daughter of Apollo, who placed her there; and others, from the ear of corn, make it a reprefentation of Ceres; and others, from the obfcurity of her head, of Fortune.

The ancients, as they gave each of the 12 months of the year to the care of fome one of the 12 principal deities, fo they alfo threw into the protection of each of thefe one of the 12 figns of the zodiac. Hence Virgo, from the ear of corn in her hand, naturally fell to the lot of Ceres, and we accordingly find it called Signum Cereris.

The ftars in the conftellation Virgo, in Ptolomy's catalogue, are 32; in Tycho's 33; in Hevelius's 50; and in the Britannic 110.

VIRTUAL *Focus*, in Optics, is a point in the axis of a glafs, where the continuation of a refracted ray meets it. Thus, let D be the centre, and DBE

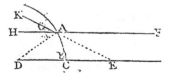

the axis of the glafs AB; upon which falls the ray FA. Now this ray will not proceed ftraight forward, as AH, after paffing the glafs, but will take the courfe AK, being deflected from the perpendicular AD. If then the refracted ray KA be produced, by AE, to the axis at E, this point E Mr. Molineux calls the *Virtual focus*, or *point of divergence*.

VIS, a Latin word, fignifying force or power; adopted by writers on phyfics, to exprefs divers kinds of natural powers or faculties.

The term Vis is either active or paffive: the *Vis activa* is the power of producing motion; the *Vis paffiva* is that of receiving or lofing it. The *Vis activa* is again fubdivided into *Vis viva* and *Vis mortua*.

Vis *Abfoluta*, or *abfolute force*, is that kind of centripetal force which is meafured by the motion that would be generated by it in a given body, at a given diftance, and depends on the efficacy of the caufe producing it.

Vis *Acceleratrix*, or *accelerating force*, is that centripetal force which produces an accelerated motion, and is proportional to the velocity which it generates in a given time; or it is as the motive or abfolute force directly, and as the quantity of matter moved inverfely.

Vis *Impreffa* is defined by Newton to be the action exercifed on any body to change its ftate, either of reft or moving uniformly in a right line.

This force confifts altogether in the action; and has no place in the body after the action is ceafed: for the body perfeveres in every new ftate by the *Vis inertiæ* alone.

This Vis impreffa may arife from various caufes; as from percuffion, preffion, and centripetal force.

Vis *Inertiæ*, or power of inactivity, is defined by Newton to be a power implanted in all matter, by which it refifts any change endeavoured to be made in its ftate, that is, by which it becomes difficult to alter its ftate, either of reft or motion.

This power then agrees with the *Vis refiftendi*, or power of refifting; by which every body endeavours, as much as it can, to perfevere in its own ftate, whether of reft or uniform rectilinear motion; which power is ftill proportional to the body, or to the quantity of matter in it, the fame as the weight or gravity of the body; and yet it is quite different from, and even independent of the force of gravity, and would be and act juft the fame if the body were devoid of gravity. Thus, a body by this force refifts the fame in all directions, upwards or downwards or obliquely; whereas gravity acts only downwards.

Bodies only exert this power in changes brought on their ftate by fome *Vis impreffa*, force impreffed on them. And the exercife of this power is, in different refpects, both refiftance and impetus; refiftance, as the body oppofes a force impreffed on it to change its ftate; and impetus, as the fame body endeavours to change the ftate of the refifting obftacle. Phil. Nat. Princ. Math. lib. 1.

The *Vis inertiæ*, the fame great author elfewhere obferves, is a paffive principle, by which bodies perfift in their motion or reft, and receive motion, in proportion to the force impreffing it, and refift as much as they are refifted. See **Resistance**.

Vis *Infita*, or *innate force* of matter, is a power of refifting, by which every body, as much as in it lies, endeavours to perfevere in its prefent ftate, whether of reft or of moving uniformly forward in a right line. This force is always proportional to the quantity of matter in the body, and differs in nothing from the Vis inertiæ, but in our manner of conceiving it.

Vis *Centripeta*. See **Centripetal** *Force*.

Vis *Motrix*, or *moving force* of a centripetal body, is the tendency of the whole body towards the centre, refulting from the tendency of all the parts, and is proportional to the motion which it generates in a given time; fo that the Vis motrix is to the Vis acceleratrix, as the motion is to the celerity: and as the quantity of motion in a body is eftimated by the product of the velocity into the quantity of matter, fo the Vis motrix ariles from the Vis acceleratrix multiplied by the quantity of matter.

The followers of Leibnitz use the term *Vis motrix* for the force of a body in motion, in the same sense as the Newtonians use the term *Vis inertiæ*; this latter they allow to be inherent in a body at rest; but the former, or Vis motrix, a force inherent in the same body only whilst in motion, which actually carries it from place to place, by acting upon it always with the same intensity in every physical part of the line which it describes.

Vis *Mortua*, and Vis *Viva*, in Mechanics, are terms used by Leibnitz and his followers for force, which they distinguish into two kinds, *Vis mortua*, and *Vis viva*; understanding by the former any kind of pressure, or an endeavour to move, not sufficient to produce actual motion, unless its action on a body be continued for some time; and by the latter, that force or power of acting which resides in a body in motion.

VISIBLE, something that is an object of vision or sight, or the property of a thing seen.

The Cartesians say that light alone is the proper object of vision. But according to Newton, colour alone is the proper object of sight; colour being that property of light by which the light itself is Visible, and by which the images of opake bodies are painted on the retina.

As to the Situation and Place of Visible Objects:

Philosophers in general had formerly taken for granted, that the place to which the eye refers any Visible object, seen by reflection or refraction, is that in which the visual ray meets a perpendicular from the object upon the reflecting or the refracting plane. That this is the case with respect to plane mirrors is universally acknowledged; and some experiments with mirrors of other forms seem to favour the same conclusion, and thus afford reason for extending the analogy to all cases of vision. If a right line be held perpendicularly over a convex or concave mirror, its image seems to make one line with it. The same is the case with a right line held perpendicularly within water; for the part which is within the water seems to be a continuation of that which is without. But Dr. Barrow called in question this method of judging of the place of an object, and so opened a new field of inquiry and debate in this branch of science. This, with other optical investigations, he published in his Optical Lectures, first printed in 1674. According to him, we refer every point of an object to the place from which the pencils of light issue, or from which they would have issued, if no reflecting or refracting substance intervened. Pursuing this principle, Dr. Barrow proceeded to investigate the place in which the rays issuing from each of the points of an object, and that reach the eye after one reflection or refraction, meet; and he found that when the refracting surface was plane, and the refraction was made from a denser medium into a rarer, those rays would always meet in a place between the eye and a perpendicular to the point of incidence. If a convex mirror be used, the case will be the same; but if the mirror be plane, the rays will meet in the perpendicular, and beyond it if it be concave. He also determined, ac-

cording to these principles, what form the image of a right line will take when it is presented in different manners to a spherical mirror, or when it is seen through a refracting medium.

Dr. Barrow however notices an objection against the maxim above mentioned, concerning the supposed place of visible objects, and candidly owns that he was not able to give a satisfactory solution of it. The objection is this: Let an object be placed beyond the focus of a convex lens, and if the eye be close to the lens, it will appear confused, but very near to its true place. If the eye be a little withdrawn, the confusion will increase, and the object will seem to come nearer; and when the eye is very near the focus, the confusion will be very great, and the object will seem to be close to the eye. But in this experiment the eye receives no rays but those that are converging; and the point from which they issue is so far from being nearer than the object, that it is beyond it; notwithstanding which the object is conceived to be much nearer than it is, though no very distinct idea can be formed of its precise distance.

The first person who took much notice of Dr. Barrow's hypothesis, and the difficulty attending it, was Dr. Berkeley, who (in his Essay on a New Theory of Vision, p. 30) observes, that the circle formed upon the retina, by the rays which do not come to a focus, produce the same confusion in the eye, whether they cross one another before they reach the retina, or tend to it afterwards; and therefore that the judgment concerning distance will be the same in both the cases, without any regard to the place from which the rays originally issued; so that in this case, by receding from the lens, as the confusion increases, which always accompanies the nearness of an object, the mind will judge that the object comes nearer. See *Apparent* Distance.

M. Bouguer (in his Traité d'Optique, p. 104) adopts Barrow's general maxim, in supposing that we refer objects to the place from which the pencils of rays seemingly converge at their entrance into the pupil. But when rays issue from below the surface of a vessel of water, or any other refracting medium, he finds that there are always two different places of this seeming convergence: one of them of the rays that issue from it in the same vertical circle, and therefore fall with different degrees of obliquity upon the surface of the refracting medium; and another of those that fall upon the surface with the same degree of obliquity, entering the eye laterally with respect to one another. He says, sometimes one of these images is attended to by the mind, and sometimes the other; and different images may be observed by different persons. And he adds, that an object plunged in water affords an example of this duplicity of images.

G. W. Krafft has ably supported Barrow's opinion, that the place of any point seen by reflection from the surface of any medium, is that in which rays issuing from it, infinitely near to one another, would meet; and considering the case of a distant object viewed in a concave mirror, by an eye very near it, when the image, according to Euclid and other writers, would be between the eye and the object, and Barrow's rule cannot be applied, he says that in this case the speculum may

b

be confidered as a plane, the effect being the fame, only that the image is more obfcure. Com. Petrepol. vol. 12, p. 252, 256. See Prieftley's Hift. of Light &c, p. 89, 688.

From the principle above illuftrated feveral remarkable phenomena of vifion may be accounted for: as—That if the diftance between two Vifible objects be an angle that is infenfible, the diftant bodies will appear as if contiguous: whence, a continuous body being the refult of feveral contiguous ones, if the diftances between feveral Vifibles fubtend infenfible angles, they will appear one continuous body; which gives a pretty illuftration of the notion of a continuum.—Hence alfo parallel lines, and long viftas, confifting of parallel rows of trees, feem to converge more and more the farther they are extended from the eye; and the roofs and floors of long extended alleys feen, the former to defcend, and the latter to afcend, and approach each other; becaufe the apparent magnitudes of their perpendicular intervals are perpetually diminifhing, while at the fame time we miftake their diftance.

As to the Different Diftances of Vifible Objects:

The mind perceives the diftance of Vifible objects, ift, From the different configurations of the eye, and the manner in which the rays ftrike the eye, and in which the image is impreffed upon it. For the eye difpofes itfelf differently, according to the different diftances it is to fee; viz, for remote objects the pupil is dilated, and the cryftalline brought nearer the retina, and the whole eye is made more globous; on the contrary, for near objects, the pupil is contracted, the cryftalline thruft forwards, and the eye lengthened. The mode of performing this however, has greatly divided the opinions of philofophers. See Prieftley's Hift. of Light &c, p. 638—652, where the feveral opinions of Defcartes, Kepler, La Hire, are Le Roi, Porterfield, Jurin, Muffchenbroek, &c, ftated and examined.

Again, the diftance of Vifible objects is judged of by the angle the object makes; from the diftinct or confufed reprefentation of the objects; and from the brifknefs or feeblenefs, or the rarity or denfity of the rays.

To this it is owing, ift, That objects which appear obfcure or confufed, are judged to be more remote; a principle which the painters make ufe of to caufe fome of their figures to appear farther diftant than others on the fame plane. 2d, To this it is likewife owing, that rooms whofe walls are whitened appear the fmaller; that fields covered with fnow, or white flowers, fhew lefs than when clothed with grafs; that mountains covered with fnow, in the night time, appear the nearer, and that opake bodies appear the more remote in the twilight.

The Magnitude of Vifible Objects.

The quantity or magnitude of Vifible objects, is known chiefly by the angle contained between two rays drawn from the two extremes of the object to the centre of the eye. An object appears fo large as is the angle it fubtends; or bodies feen under a greater angle

appear greater; and thofe under a lefs angle lefs, &c. Hence the fame thing appears greater or lefs as it is nearer the eye or farther off. And this is called the apparent magnitude.

But to judge of the real magnitude of an object, we muft confider the diftance; for fince a near and a remote object may appear under equal angles, though the magnitudes be different, the diftance muft neceffarily be eftimated, becaufe the magnitude is great or fmall according as the diftance is. So that the real magnitude is in the compound ratio of the diftance and the apparent magnitude; at leaft when the fubtended angle, or apparent magnitude, is very fmall; otherwife, the real magnitude will be in a ratio compounded of the diftance and the fine of the apparent magnitude, nearly, or nearer ftill its tangent.

Hence, objects feen under the fame angle, have their magnitudes in the fame ratio as their diftances. The chord of an arc of a circle appears of equal magnitude from every point in the circumference, though one point be vaftly nearer than another. Or if the eye be fixed in any point in the circumference, and a right line be moved round fo as its extremes be always in the periphery, it will appear of the fame magnitude in every pofition. And the reafon is, becaufe the angle it fubtends is always of the fame magnitude. And hence alfo, the eye being placed in any angle of a regular polygon, the fides of it will all appear of equal magnitude; being all equal chords of a circle defcribed about it.

If the magnitude of an object directly oppofite to the eye be equal to its diftance from the eye, the whole object will be diftinctly feen, or taken in by the eye, but nothing more. And the nearer you approach an object, the lefs part you fee of it.

The leaft angle under which an ordinary object becomes vifible, is about one minute of a degree.

Of the Figure of Vifible Objects.

This is eftimated chiefly from our opinion of the fituation of the feveral parts of the object. This opinion of the fituation, &c, enables the mind to apprehend an external object under this or that figure, more juftly than any fimilitude of the images in the retina, with the object can; the images being often elliptical, oblong, &c, when the objects they exhibit to the mind are circles, or fquares, &c.

The laws of vifion, with regard to the figures of Vifible objects, are,

1. That if the centre of the eye be exactly in the direction of a right line, the line will appear only as a point.

2. If the eye be placed in the direction of a furface, it will appear only as a line.

3. If a body be oppofed directly towards the eye, fo as only one plane of the furface can radiate on it, the body will appear as a furface.

4. A remote arch, viewed by an eye in the fame plane with it, will appear as a right line.

5. A fphere, viewed at a diftance, appears a circle.

6. Angular figures, at a diftance, appear round.

7. If the eye look obliquely on the centre of a regular figure, or a circle, the true figure will not be feen; but the circle will appear oval, &c.

VISIBLE

VISIBLE *Horizon, Place, &c.* See the substantives

VISION, is the act of seeing, or of perceiving external objects by the organ of sight.

When an object is so disposed, that the rays of light, coming from all parts of it, enter the pupil of the eye, and present its image on the retina, that object is then seen. This is proved by experiment; for if the eye of any animal be taken out, and the skin and fat be carefully stripped off from the back part of it, till only the thin membrane, which is called the retina, remains to terminate it behind, and any object be placed before the front of the eye, the picture of that object will be seen figured as with a pencil on that membrane. There are thousands of experiments which prove that this is the mechanical effect of Vision, or seeing, but none of them all appear so conveniently as this, which is made with the very eye itself of an animal; an eye of an ox newly killed shews this happily, and with very little trouble. It will indeed appear singular in this, that the object is inverted, in the picture thus drawn of it, in the eye; and the case is the same in the eye of a living person.

Various other opinions however have been held concerning the means of Vision among philosophers.

The Platonists and Stoics held Vision to be effected by the emission of rays out of the eyes; conceiving that there was a sort of light thus darted out; which, with the light of the external air, taking hold as it were of the objects, rendered them visible; and thus returning back again to the eye, altered and new modified by the contact of the object, made an impression on the pupil, which gave the sensation of the object.

Our own countryman, Roger Bacon, distinguished as he was in many respects, also assents to the opinion that visual rays proceed from the eye; giving this reason for it, that every thing in nature is qualified to discharge its proper functions by its own powers, in the same manner as the sun, and other celestial bodies. Opus Majus, pa. 289.

The Epicureans held, that Vision is performed by the emanation of corporeal species or images from objects; or a sort of atomical effluvia continually flying off from the intimate parts of objects, to the eye.

The Peripatetics hold, with Epicurus, that Vision is performed by the reception of species; but they differ from him in the circumstances; for they will have the species (which they call *intentionales*) to be incorporeal. It is true, Aristotle's doctrine of Vision, delivered in his chapter De Aspectu, amounts to no more than this, that objects must have some intermediate body, that by this they may move the organ of sight. To which he adds, in another place, that when we perceive bodies, it is their species, not their matter, that we receive; as a seal makes an impression on wax, without the wax receiving any thing of the seal.

But this vague and obscure account the Peripatetics have thought fit to improve. Accordingly, what their master calls species, the disciples, understanding of real proper species, assert, that every visible object expresses a perfect image of itself in the air contiguous to it; and this image another, somewhat less, in the next air; and the third another; and so on till the last image arrives at the crystalline, which they hold for the chief organ of sight, or that which immediately moves the soul. These images they call *intentional species*.

The modern philosophers, as the Cartesians and Newtonians, give a better account of Vision. They all agree, that it is performed by rays of light reflected from the several points of objects received in at the pupil, refracted and collected in their passage, through the coats and humours, to the retina; and this striking, or making an impression, on so many points of it; which impression is conveyed, by the correspondent capillaments of the optic nerve, to the brain, &c.

Baptista Porta's experiments with the camera obscura, about the middle of the 16th century, convinced him that vision is performed by the intermission of something into the eye, and not by visual rays proceeding from the eye, as had been the general opinion before his time; and he was the first who fully satisfied himself and others upon this subject; though several philosophers still adhered to the old opinion.

As for the Peripatetic series or chain of images, it is a mere chimera; and Aristotle's meaning is better understood without than with them. In fact, setting these aside, the Aristotelian, Cartesian, and Newtonian doctrines of Vision, are very consistent with one another; for Newton imagines that Vision is performed chiefly by the vibrations of a fine medium (which penetrates all bodies) excited in the bottom of the eye by the rays of light, and propagated through the capillaments of the optic nerves, to the sensorium. And Des Cartes maintains, that the sun pressing the materia subtilis, with which the whole universe is every where filled, the vibrations and pulses of that matter reflected from objects, are communicated to the eye, and thence to the sensory: so that the action or vibration of a medium is equally supposed in all.

It is generally concluded then, that the images of objects are represented on the retina; which is only an expansion of the fine capillaments of the optic nerve, and from whence the optic nerve is continued into the brain. Now any motion or vibration, impressed on one extremity of the nerve, will be propagated to the other: hence the impulse of the several rays, sent from the several points of the object, will be propagated as they are on the retina (that is, in their proper colours, &c, or in particular vibrations, or modes of pressure, corresponding to them) to the place where those capillaments are interwoven into the substance of the brain. And thus is Vision brought to the common case of sensation.

Experience teaches us that the eye is capable of viewing objects at a certain distance, without any mental exertion. Beyond this distance, no mental exertion can be of any avail: but, within it, the eye possesses a power of adapting itself to the various occasions that occur, the exercise of which depends on the volition of the mind. How this is effected, is a problem that has very much engaged the attention of optical writers: but it is doubted whether it has yet been satisfactorily explained. The first theory for the solution of this problem is that of Kepler. He supposes that the ciliary processes contract the diameter of the eye, and lengthen its axis by a muscular power. But Mr. Thomas Young (in some ingenious Observations on Vision in the Philos. Transf. 1793) observes, ferves, that thefe proceffes neither appear to contain any mufcular fibres, nor have any attachment by which they can be capable of performing this action.

Des Cartes afcribed this contraction and elongation to a mufcularity of the cryftalline, of which he fuppofed the ciliary proceffes to be the tendons: but he neither demonftrated this mufcularity, nor fufficiently confidered the connection with the ciliary proceffes.

De la Hire allows of no change in the eye, except the contraction and dilatation of the pupil: this opinion he founds on an experiment which Dr. Smith has fhewn to be fallacious. Haller adopted his hypothefis, notwithftanding its inconfiftency with the principles of optics and conftant experience.

Dr. Pemberton fuppofes that the cryftalline contains mufcular fibres, by which one of its furfaces is flattened, while the other is made convex: but he has not demonftrated the exiftence of thefe fibres; and Dr. Jurin has proved that fuch a change as this is inadequate to the effect.

Dr. Porterfield conceives that the ciliary proceffes draw the cryftalline forward, and make the cornea more convex. But the ciliary proceffes are incapable of this action; and it appears from Dr. Jurin's calculations, that a fufficient motion of this kind requires a very vifible increafe in the length of the axis of the eye; an increafe which has never yet been obferved.

Dr. Jurin maintains that the uvea, at its attachment to the cornea, is mufcular; and that the contraction of this ring makes the cornea more convex. But this hypothefis is not fufficiently confirmed by obfervation.

Muffchenbroek conjectures that the relaxation of this ciliary zone, which is nothing but the capfule of the vitreous humour where it receives the impreffion of the ciliary proceffes, permits the coats of the eye to pufh forward the cryftalline and cornea. Such a voluntary relaxation however, Mr. Young obferves, is wholly without example in the animal economy: befides, if it actually occurred, the coats of the eye could not act as he conceives; nor could they act in this manner without being obferved. He adds, that the contraction of the ciliary zone is equally inadequate and unneceffary.

Mr. Young, having examined thefe theories, and fome others of lefs moment, proceeds to inveftigate a more probable folution of this optical difficulty.—Adverting to the obfervation of Dr. Porterfield, that thofe who have been couched have not the power of accommodating the eye to different diftances; and to the reflections of other writers on this fubject; he was led to conclude that the rays of light, emitted by objects at a fmall diftance, could only be brought to foci on the retina by a nearer approach of the cryftalline to a fpherical form; and he imagined that no other power was capable of producing this change, befide a mufcularity of part, or of the whole of its capfule:—but, on clofely examining firft with the naked eye and then with a magnifier, the cryftalline of an ox's eye turned out of its capfule, he difcovered a ftructure which feemed to remove the difficulties that have long embarraffed this branch of optics.

" The cryftalline of the ox, fays he, is compofed of various fimilar coats, each of which confifts of fix mufcles, intermixed with a gelatinous fubftance, and attached to fix membranous tendons. Three of the tendons

are anterior, three pofterior; their length is about two-thirds of the femidiameter of the coat; their arrangement is that of three equal and equidiftant rays, meeting in the axis of the cryftalline: one of the anterior is directed towards the outer angle of the eye, and one of the pofterior towards the inner angle, fo that the pofterior are placed oppofite to the middle of the interftices of the anterior: and planes paffing through each of the fix, and through the axis, would mark on either furface fix regular equidiftant rays. The mufcular fibres arife from both fides of each tendon; they diverge till they reach the greateft circumference of the coat; and, having paffed it, they again converge, till they are attached refpectively to the fides of the neareft tendons of the oppofite furface. The anterior or pofterior portion of the fix, viewed together, exhibits the appearance of three penniform-radiated mufcles. The anterior tendons of all the coats are fituated in the fame planes, and the pofterior ones in the continuations of thefe planes beyond the axis. Such an arrangement of fibres can be accounted for on no other fuppofition than that of mufcularity. This mafs is inclofed in a ftrong membranous capfule, to which it is loofely connected by minute veffels and nerves; and the connection is more obfervable near its greateft circumference. Between the mafs and its capfule is found a confiderable quantity of an aqueous fluid, the liquid of the cryftalline.

" When the will is exerted to view an object at a fmall diftance, the influence of the mind is conveyed through the lenticular ganglion, formed from branches of the third and fifth pair of nerves by the filaments perforating the fclerotica, to the orbiculus ciliaris, which may be confidered as an annular plexus of nerves and veffels; and thence by the ciliary proceffes to the mufcle of the cryftalline, which, by the contraction of its fibres, becomes more convex, and collects the diverging rays to a focus on the retina. The difpofition of fibres in each coat is admirably adapted to produce this change; for, fince the leaft furface that can contain a given bulk is that of a fphere (Simpfon's Fluxions, pa. 486) the contraction of any furface muft bring its contents nearer to a fpherical form. The liquid of the cryftalline feems to ferve as a fynovia in facilitating the motion, and to admit a fufficient change of the mufcular part, with a fmaller motion of the capfule.

Mr. Young proceeds to enquire whether thefe fibres can produce an alteration in the form of the lens fufficiently great to account for the known effects; and he finds, by calculation, that, fuppofing the cryftalline to affume a fpherical form, its diameter will be 642 thoufandths of an inch, and its focal diftance in the eye ·926. Then, difregarding the thicknefs of the cornea, we find (by Smith, art. 370) that fuch an eye will collect thofe rays on the retina, which diverge from a point at the diftance of 12 inches and 8 tenths. This is a greater change than is neceffary for an ox's eye; for if it be fuppofed capable of diftinct Vifion at a diftance fomewhat lefs than 12 inches, yet it is probably far fhort of being able to collect parallel rays. The human cryftalline is fufceptible of a much greater change of form. The ciliary zone may admit of as much extenfion as this diminution of the diameter of the cryftal-

line

line will require ; and its elasticity will assist the cellular texture of the vitreous humour, and perhaps the gelatinous part of the crystalline, in restoring the indolent form.—Mr. Young apprehends that the sole office of the optic nerve is to convey sensation to the brain ; and that the retina does not contribute to supply the lens with nerves.—As the human crystalline resembles that of the ox, it may reasonably be presumed that the action of both organs depends on the same general principles.

This theory of Mr. Young's however is strongly opposed by Dr. Hosack, (Philos. Transf. 1794, part 2, pa. 196). He contests the existence of the muscles, which Mr. Young has described, for, several reasons. First, from the transparency they must possess : otherwise there would be some irregularity in the refraction of those rays which pass through the several parts, differing both in shape and density. Another circumstance is the number of these muscles. Mr. Young describes 6 in each lamina ; and as Leuwenhoek makes 2000 laminæ in all, therefore the number of muscles must amount to 12 thousand, the action of which, Dr. Hosack apprehends, must exceed comprehension. But the existence of these muscles is still more doubtful, if the accuracy of Dr. Hosack's observations be admitted. With the assistance of the best glasses, and with the greatest attention, he could not discover the structure of the crystalline described by Mr. Young, but found it to be perfectly transparent. He first observed the lens in its viscid state, and then exposed different lenses to a moderate degree of heat, so that they became opaque and dry ; and it was easy to separate the distinct layers described by Mr. Young. These were so numerous as not to admit of having, each of them, 6 muscles. Another consideration, which seems to prove that these layers possess no distinct muscles, is that, in this opaque state, they are not visible, but consist of an almost infinite number of concentric fibres, not divided into particular bundles, but similar to as many of the finest hairs of equal thickness, arranged in similar order. This regular structure of layers, composed of concentric fibres, Dr. Hosack thinks is much better adapted to the transmission of the rays of light than the irregular structure of muscles. Besides, it ought to be considered that the crystalline lens is not the most essential organ in viewing objects at different distances ; and if this be the case, the power of the eye cannot be owing to any changes in this lens. It is a fact, says Dr. Hosack, that we can, in a great degree, do without it ; as is the case after couching or extraction, by which operation all its parts must be destroyed. Dr. Porterfield, however, and Mr. Young, on his authority, maintain that patients, after the operation of couching, have not the power of accommodating the eye to different distances of objects. On the whole, Dr. Hosack concludes that no such muscles, as Mr. Young has described, exist, and that he must have been deceived by some other appearances that resembled muscles : neither will he allow the effects ascribed to the ciliary processes in changing the shape or situation of the lens.

Dr. Hosack then proceeds to illustrate the structure and use of the external muscles of the eye ; which are 6 in number, 4 called recti or straight, and 2 oblique, and by means of which he thinks the business is effect-ed. The common purposes to which these muscles are subservient are well known : but beside these, Dr. Hosack suggests that it is not inconsistent with the general laws of nature, nor even with the animal œconomy, to imagine that, from their combination, they should have a different action and an additional use. In describing the precise action of these muscles, he supposes an object to be seen distinctly first at the distance of 6 feet ; in which case the picture of it falls exactly on the retina. He then directs his attention to another object at the distance of 6 inches, as nearly as possible in the same line. While he is viewing this, he loses sight of the first object, though the rays proceeding from it still fall on the eye ; and hence he infers that the eye must have undergone some change ; so that the rays meet either before or behind the retina. But, as rays from a more distant object concur sooner than those from a nearer one, the picture of the more remote object must fall before the retina, while the others form a distinct image upon it. But yet the eye continued in the same place ; and therefore the retina must, by some means, have been removed to a greater distance from the fore-part of the eye, so as to receive the picture of the nearer object. This object, he contends, could not be seen distinctly, unless the retina were removed to a greater distance, or the refracting power of the media through which the rays passed were augmented :—but as the lens is the chief refracting medium, if we admit that this has no power of changing itself, we are under the necessity of adopting the first of these two suppositions.

The next object of inquiry is, how the external muscles are capable of producing these changes. The recti are strong, broad, and flat, and arise from the back part of the orbit of the eye ; and, passing over the ball as over a pulley, they are inserted by broad flat tendons at the anterior part of the eye. The oblique are inserted towards the posterior part by similar tendons. When these different muscles act jointly, the eye being in the horizontal position, and every muscle in action contracting itself, the four recti by their combination must compress the various parts of the eye and lengthen its axis, while the oblique muscles serve to keep the eye in its proper direction and situation. The convexity of the cornea, by means of its great elasticity, is also increased in proportion to the degree of pressure, and thus the rays of light passing through it are necessarily more converged. The elongation of the eye serves also to lengthen the media, in the aqueous, crystalline, and vitreous humours through which the rays pass, so that their powers of refraction are proportionably increased. This is the general effect of the contraction of the external muscles, according to Dr. Hosack's statement of it : to which it may be added, that we possess the same power of relaxing them in proportion to the greater distance of the object, till we arrive at the utmost extent of indolent Vision. Dr. Hosack also illustrates this hypothesis by some experiments.

The misrepresentations of Vision often depend upon the distance of the object. Thus, if an opake globe be placed at a moderate distance from the eye, the picture of it upon the retina will be a circle properly diversified with light and shade, so that it will excite in the mind the sensation of a sphere or globe ; but, if the

the globe be placed at a great diftance from the eye, the diftance between thofe lights and fhades, which form the picture of a globe, will be imperceptible, and the globe will appear no otherwife than as a circular plane. In a luminous globe, diftance is not neceffary in order to take off the reprefentation of prominent and flat; an iron bullet, heated very red hot, and held but a few yards diftance from the eye, appears a plane, not a prominent body; it has not the look of a globe, but of a circular plane. It is owing to this mifreprefentation of Vifion that we fee the fun and moon flat by the naked eye, and the planets alfo, through telefcopes flat. It is in this light that aftronomers, when they fpeak of the fun, moon, and planets, as they appear to our view, call them the difcs of the fun, moon, and planets, which we fee.

The nearer a globe is to the eye, the fmaller fegment of it is vifible, the farther off the greater, and at a due diftance the half; and, on the fame principle, the nearer the globe is to the eye, the greater is its apparent diameter, that is, under the greater angle it will appear, the farther off the globe is placed, the lefs is its apparent diameter. This is a propofition of importance, for, on this principle, we know that the fame globe, when it appears larger, is nearer to our eye, and, when fmaller, is farther off from it. Therefore, as we know that the globes of the fun and moon continue always of the fame fize, yet appear fometimes larger, and fometimes fmaller, to us, it is evident, that they are fometimes nearer, and fometimes farther off from the place whence we view them. Two globes, of different magnitude, may be made to appear of exactly the fame diameter, if they be placed at different diftances, and thofe diftances be exactly proportioned to their diameters. To this it is owing, that we fee the fun and moon nearly of the fame diameter; they are, indeed, vaftly different in real bulk, but, as the moon is placed greatly nearer to our eyes, the apparent magnitude of that little globe is nearly the fame with that of the greater.

In this inftance of the fun and moon (for there cannot be a more ftriking one) we fee the mifreprefentation of Vifion in two or three feveral ways. The apparent diameters of thefe globes are fo nearly equal, that, in their feveral changes of place, they do, at times, appear to us abfolutely equal, or mutually greater than one another. This is often to be feen, but it is at no time fo obvious, and fo perfectly evinced, as in eclipfes of the fun, which are total. In thefe we fee the apparent magnitudes of the two globes vary fo much according to their diftances, that fometimes the moon is large enough exactly to cover the difc of the fun, fometimes it is larger, and a part of it every where extends beyond the difc of the fun; and, on the contrary, fometimes it is fmaller, and, though the eclipfe be abfolutely central, yet it is annular, or a part of the fun's difc is feen in the middle of the eclipfed part, enlightened, and furrounding the opake body of the moon in form of a lucid ring.

When an object, which is feen above, without other objects of comparifon, is of a known magnitude, we judge of its diftance by its apparent magnitude; and cuftom teaches us to do this with tolerable accuracy. This is a practical ufe of the mifreprefentation of Vifion, and in the fame manner, knowing that we fee

things, which are near us, diftinctly, and thofe which are diftant, confufedly, we judge of the diftance of an object by the clearnefs, or confufion, in which we fee it. We alfo judge yet more eafily and truly of the diftance of an object by comparing it to another feen at the fame time, the diftance of which is better known, and yet more by comparing it with feveral others, the diftances of which are more or lefs known, or more or lefs eafily judged of. Thefe are the circumftances which affift us, even by the mifreprefentation of Vifion, to judge of diftance; but, without one or more of thefe, the eye does not, in reality, enable us to judge concerning the diftance of objects.

This mifreprefentation, although it ferves us on fome occafions, yet is very limited in its effects. Thus, though it helps us greatly in diftinguifhing the diftance of objects that are about us, both with refpect to ourfelves and them, and with refpect to themfelves with one another, yet it can do nothing with the very remote. We fee that immenfe concave circle, in which we fuppofe the fixed ftars to be placed, at all this vaft remove from us, and no change of place that we could make to get nearer to it, would be of any avail for determining the diftance of the ftars from one another. If we look at three or four churches from a diftance of as many miles, we fee them ftand in a certain pofition with regard to one another. If we advance a great deal nearer to them, we fee that pofition differ, but, if we move forward only eight or ten feet, the difference is not feen.

VISION, in Optics. The laws of Vifion, brought under mathematical demonftrations, make the fubject of Optics, taken in the greateft latitude of that word: for, among mathematical writers, optics is generally taken, in a more reftricted fignification, for the doctrine of *direct Vifion*; catoptrics, for the doctrine of *reflected Vifion*; and dioptrics, for that of *refracted Vifion*.

Direct or *Simple* VISION, is that which is performed by means of direct rays; that is, of rays paffing directly, or in right lines, from the radiant point to the eye. Such is that explained in the preceding article VISION.

Reflected VISION, is that which is performed by rays reflected from fpeculums, or mirrors. The laws of which, fee under REFLECTION, and MIRROR.

Refracted VISION, is that which is performed by means of rays refracted, or turned out of their way, by paffing through mediums of different denfity; chiefly through glaffes and lenfes. The laws of this, fee under the article REFRACTION.

Arch of VISION. See ARCH.

Diftinct VISION, is that by which an object is feen diftinctly. An object is faid to be feen diftinctly, when its outlines appear clear and well defined, and the feveral parts of it, if not too fmall, are plainly diftinguifhable, fo that they can eafily be compared one with another, in refpect to their figure, fize, and colour.

In order to fuch Diftinct Vifion, it had commonly been thought that all the rays of a pencil, flowing from a phyfical point of an object, muft be exactly united in a phyfical, or at leaft in a fenfible point of the retina. But Dr. Jurin has made it appear from experiments, that fuch an exact union of rays is not always neceffary to Diftinct Vifion. He fhews that objects may be feen with fufficient diftinctnefs, though the pencils of rays iffuing from the points of them do not unite precifely in the fame point on the retina ; but that fince, in this cafe, pencils from every point either meet before they reach the retina, or tend to meet beyond it, the light that comes from them muft cover a circular fpot upon it, and will therefore paint the image larger than perfect Vifion would reprefent it. Whence it follows, that every object placed either too near or too remote for perfect Vifion, will appear larger than it is by a pen-umbra of light, caufed by the circular fpaces, which are illuminated by pencils of rays proceeding from the extremities of the object.

The fmalleft diftance of perfect Vifion, or that in which the rays of a fingle pencil are collected into a phyfical point on the retina in the generality of eyes, Dr. Jurin, from a number of obfervations, ftates at 5, 6, or 7 inches. The greateft diftance of diftinct and perfect Vifion he found was more difficult to deter-mine ; but by confidering the proportion of all the parts of the eye, and the refractive power of each, with the interval that may be difcerned between two ftars, the diftance of which is known, he fixes it, in fome cafes, at 14 feet 5 inches ; though Dr. Porterfield had reftricted it to 27 inches only, with refpect to his own eye.

For other obfervations on this fubject, fee Jurin's Effay on Diftinct and Indiftinct Vifion, at the end of Smith's Optics ; and Robins's Remarks on the fame, in his Math. Tracts, vol. 2, pa. 278 &c. See alfo an in-genious paper on Vifion in the Philof. Tranf. 1793, pa. 169, by Mr. Thomas Young.

Field of VISION. See FIELD.

VISUAL, relating to fight, or feeing.

VISUAL *Angle*, is the angle under which an object is feen, or which it fubtends. See ANGLE.

VISUAL *Line*. See LINE.

VISUAL *Point*, in Perfpective, is a point in the ho-rizontal line, where all the ocular rays unite. Thus, a perfon ftanding in a ftraight long gallery, and looking forward ; the fides, floor, and cieling feem to meet and touch one another in this point, or common centre.

VISUAL *Rays*, are lines of light, conceived to come from an object to the eye.

VITELLIO, or VITELLO, a Polifh mathematician of the 13th century, as he flourifhed about 1254. We have of his a large *Treatife on Optics*, the beft edition of which is that of 1572. Vitello was the firft optical writer of any confequence among the modern Europeans. He collected all that was given by Euclid, Archimedes, Ptolomy, and Alhazen ; though his work is of but lit-tle ufe now.

VITREOUS *Humour*, or *Vitreus Humor*, denotes the third or glaffy humour of the eye ; thus called from its refemblance to melted glafs. It lies under the cryftalline ; by the impreffion of which, its fore-part is rendered concave. It greatly exceeds in quanti-ty both the aqueous and cryftalline humours taken to-gether, and confequently occupies much the greateft part of the cavity of the globe of the eye. Scheiner fays, that the refractive power of this humour is a me-dium between thofe of the aqueous, which does not

differ much from water, and of the cryftalline, which is nearly the fame with glafs. Hawkfbee makes its re-fractive power the fame with that of water ; and, ac-cording to Robertfon, its fpecific gravity agrees nearly with that of water.

VITRUVIUS (MARCUS VITRUVIUS POLLIO), a celebrated Roman architect, of whom however nothing is known, but what is to be collected from his ten books *De Architectura*, ftill extant. In the preface to the fixth book he writes, that he was carefully educated by his parents, and inftructed in the whole circle of arts and fciences ; a circumftance which he fpeaks of with much gratitude, laying it down as certain, that no man can be a complete architect, without fome knowledge and fkill in every one of them. And in the preface to the firft book he informs us, that he was known to Ju-lius Cæfar ; that he was afterwards recommended by Octavia to her brother Auguftus Cæfar ; and that he was fo favoured and provided for by this emperor, as to be out of all fear of poverty as long as he might live.

It is fuppofed that Vitruvius was born either at Rome or Verona ; but it is not known which. His books of architecture are addreffed to Auguftus Cæfar, and not only fhew confummate fkill in that particular fcience, but alfo very uncommon genius and natural abilities. Cardan, in his 16th book *De Subtilitate*, ranks Vitru-vius as one of the 12 perfons, whom he fuppofes to have excelled all men in the force of genius and inven-tion ; and would not have fcrupled to have given him the firft place, if it could be imagined that he had de-livered nothing but his own difcoveries. Thofe 12 per-fons were, Euclid, Archimedes, Apollonius Pergæus, Ariftotle, Archytas of Tarentum, Vitruvius, Achin-dus, Mahomet Ibn Mofes the inventor or improver of Algebra, Duns Scotus, John Suiffet furnamed the Cal-culator, Galen, and Heber of Spain.

The architecture of Vitruvius has been often print-ed ; but the beft edition is that of Amfterdam in 1649. Perrault alfo, the noted French architect, gave an ex-cellent French tranflation of the fame, and added notes and figures : the firft edition of which was publifhed at Paris in 1673, and the fecond much improved, in 1684. —Mr. William Newton too, an ingenious architect, and late Surveyor to the works at Greenwich Hofpital, publifhed in 1780 &c, curious commentaries on Vitru-vius, illuftrated with figures ; to which is added a de-fcription, with figures, of the Military Machines ufed by the Ancients.

VIVIANI (VINCENTIO), a celebrated Italian ma-thematician, was born at Florence in 1621, fon fay 1622. He was a difciple of the illuftrious Galileo, and lived with him from the 17th to the 20th year of his age. After the death of his great mafter, he paffed two or three years more in profecuting geometrical ftu-dies without interruption, and in this time it was that he formed the defign of his Reftoration of Arifteus. This ancient geometrician, who was contemporary with Euclid, had compofed five books of problems *De Locis Solidis*, the bare propofitions of which were collected by Pappus, but the books are entirely loft ; which Viviani undertook to reftore by the force of his genius.

He broke this work off before it was finifhed, in or-der to apply himfelf to another of the fame kind, and that

6

that was, to reftore the 5th book of Apollonius's Conic Sections. While he was engaged in this, the famous Borelli found, in the library of the Grand Duke of Tufcany, an Arabic manufcript, with a Latin infcription, which imported, that it contained the eight books of Apollonius's Conic Sections: of which the 8th however was not found to be there. He carried this manufcript to Rome, in order to tranflate it, with the affiftance of a profeffor of the Oriental languages. Viviani, very unwilling to lofe the fruits of his labours, procured a certificate that he did not underftand the Arabic language, and knew nothing of that manufcript: he was fo jealous on this head, that he would not even fuffer Borelli to fend him an account of any thing relating to it. At length he finifhed his book, and publifhed it, 1659, in folio, with this title, *De Maximis & Minimis Geometrica Divinatio in quintum Conicorum Apollonii Pergæi*. It was found that he had more than divined; as he feemed fuperior to Apollonius himfelf.

After this he was obliged to interrupt his ftudies for the fervice of his prince, in an affair of great importance, which was, to prevent the inundations of the Tiber, in which Caffini and he were employed for fome time, though nothing was entirely executed.

In 1664 he had the honour of a penfion from Louis the 14th, a prince to whom he was not fubject, nor could be ufeful. In confequence he refolved to finifh his Divination upon Arifteus, with a view to dedicate it to that prince; but he was interrupted in this tafk again by public works, and fome negotiations which his mafter entrufted to him.—In 1666 he was honoured by the Grand Duke with the title of his firft mathematician.—He refolved three problems, which had been propofed to all the mathematicians of Europe, and dedicated the work to the memory of Mr. Chapelain, under the title of *Enodatio Problematum* &c.—He propofed the problem of the quadrable arc, of which Leibnitz and l'Hofpital gave folutions by the Calculus Differentialis.—In 1669, he was chofen to fill, in the Royal Academy of Sciences, a place among the eight foreign affociates. This new favour reanimated his zeal; and he publifhed three books of his Divination upon Arifteus, at Florence in 1701, which he dedicated to the King of France. It is a thin folio, intitled, *De Locis Solidis fecunda Divinatio Geometrica*, &c. This was a fecond edition enlarged; the firft having been printed at Florence in 1673.—Viviani laid out the fortune, which he had raifed by the bounties of his prince, in building a magnificent houfe at Florence; in which he placed a buft of Galileo, with feveral infcriptions in honour of that great man; and died in 1703, at 81 years of age.

Viviani had, fays Fontenelle, that innocence and fimplicity of manners which perfons commonly preferve, who have lefs commerce with men than with books; without that roughnefs and a certain favage fiercenefs which thofe often acquire who have only to deal with books, not with men. He was affable, modeft, a faft and faithful friend, and, what includes many virtues in one, he was grateful in the higheft degree for favours.

ULLAGE, *of a Cafk*, in Gauging, is fo much as it wants of being full.

ULTERIOR, in Geography, is applied to fome part of a country or province, which, with regard to the reft of that country, is fituate on the farther fide of a river, or mountain, or other boundary, which divides the country into two parts.

ULTRAMUNDANE, beyond the world, is that part of the univerfe fuppofed to be without or beyond the limits of our world or fyftem.

UMBILICUS, and **UMBILICAL** *Point*, in Geometry, the fame with focus.

UMBRA, a Shadow. See LIGHT, SHADOW, PENUMBRA, &c.

UNCIA, a term generally ufed for the 12th part of a thing. In which fenfe it occurs in Latin writers, both for a weight, called by us an *ounce*, and a meafure called an *inch*.

UNCIÆ, in Algebra, firft ufed by Vieta, are the numbers prefixed to the letters in the terms of any power of a binomial; now more ufually, and generally, called *coefficients*. Thus, in the 4th power of $a + b$, viz,

$$a^4 + 4a^3b + 6a^2b^2 + 4ab^3 + b^4,$$

the Unciæ are 1, 4, 6, 4, 1.

Briggs firft fhewed how to find thefe Unciæ, one from another, in any power, independent of the foregoing powers. They are now ufually found by what is called Newton's binomial theorem, which is the fame rule as Briggs's in another form. See BINOMIAL.

UNDECAGON, is a polygon of eleven fides.

If the fide of a regular Undecagon be 1, its area will be $9.3656399 = \frac{11}{4} \times$ tang. of $73\frac{7}{11}$ degrees; and therefore if this number be multiplied by the fquare of the fide of any other regular Undecagon, the product will be the area of that Undecagon. See my Menfuration, pa. 114 &c, 2d edit.

UNDETERMINED, is fometimes ufed for INDETERMINATE.

UNDULATORY *Motion*, is applied to a motion in the air, by which its parts are agitated like the waves of the fea; as is fuppofed to be the cafe when the ftring of a mufical inftrument is ftruck. This Undulatory motion of the air is fuppofed the matter or caufe of found.—Inftead of the Undulatory, fome authors choofe to call this a *vibratory* motion.

UNEVEN *Number*, the fame as odd number, or fuch as cannot be divided by 2 without leaving 1 remaining. The feries of Uneven Numbers are 1, 3, 5, 7, 9, &c. See NUMBER, and ODD *Number*.

UNGULA, in Geometry, is a part cut off a cylinder, cone, &c, by a plane paffing obliquely through the bafe, and part of the curve furface; fo called from its refemblance to the (ungula) hoof of a horfe &c. For the contents &c of fuch Ungulas, fee my Menfuration, pa. 218—246, 2d edition.

UNICORN, in Aftronomy. See MONOCEROS.

UNIFORM, or *Equable Motion*, is that by which a body paffes always with the fame celerity, or over equal fpaces in equal times. See MOTION.

In Uniform motions, the fpaces defcribed or paffed over, are in the compound ratio of the times and velocities; but the fpaces are fimply as the times, when the velocity is given; and as the velocities, when the time is given.

UNIFORM

UNIFORM *Matter*, in Natural Philofophy, is that which is all of the fame kind and texture.

UNISON, in Mufic, is when two founds are exactly alike, or the fame note, or tone.

What conftitutes a Unifon, is the equality of the number of vibrations, made in the fame time, by the two fonorous bodies.

It is a noted phenomenon in mufic, that an intenfe found being raifed, either with the voice, or a fonorous body, another fonorous body near it, whofe tone is either Unifon, or octave to that tone, will found its proper note Unifon, or octave, to the given note. The experiment is eafily tried with the ftrings of two inftruments ; or with a voice and harpfichord ; or a bell, or even a drinking glafs.

This phenomenon is thus accounted for : one ftring being ftruck, and the air put into a vibratory motion by it ; every other ftring, within the reach of that motion, will receive fome impreffion from it : but each ftring can only move with a determinate velocity of recourfes or vibrations ; and all Unifons proceed from equal vibrations ; and other concords from other proportions of vibration. The Unifon ftring then, keeping equal pace with the founding ftring, as having the fame meafure of vibrations, muft have its motion continued, and ftill improved, till at length its motion become fenfible, and it give a diftinct found. Other concording ftrings have their motions propagated in different degrees, according to the frequency of the coincidence of their vibrations with thofe of the founded ftring : the octave therefore moft fenfibly ; then the 5th ; after which, the croffing of the motions prevents any effect.

This is illuftrated, as Galileo firft fuggefted, by the pendulum, which being fet a-moving, the motion may be continued and augmented, by making frequent, light, coincident impulfes ; as blowing on it when the vibration is juft finifhed : but if it be touched by any crofs or oppofite motion, and that frequently too, the motion will be interrupted, and ceafe altogether. So, of two Unifon ftrings, if the one be forcibly ftruck, it communicates motion, by the air, to the other ; and both performing their vibrations together, the motion of that other will be improved and heightened by the frequent impulfes received from the vibrations of the firft, becaufe given precifely when the other has finifhed its vibration, and is ready to return : but if the vibrations of the chords be unequal in duration, there will be a croffing of motions, more or lefs, according to the proportion of the inequality ; by which the motion of the untouched ftring will be fo checked, as never to be fenfible. And this we find to be the cafe in all confonarces, except Unifon, octave, and the fifth.

UNIT, UNITE, or UNITY, in Arithmetic, the number one, or one fingle individual part of difcrete quantity. See NUMBER.—The place of units, is the firft place on the right hand in integer numbers.

According to Euclid, Unity is not a number, for he defines number to be a multitude of Units.

UNITY, the abftract or quality which conftitutes or denominates a thing *one*.

UNIVERSE, a collective name, fignifying the affemblage of heaven and earth, with all things in them.

The Ancients, and after them the Cartefians, ima-

gine the Univerfe to be infinite ; and the reafon they give is, that it implies a contradiction to fuppofe it finite or bounded ; fince it is impoffible not to conceive fpace beyond any limits that can be affigned it ; which fpace, according to the Cartefians, is body, and confequently part of the Univerfe.

UNLIKE *Quantities*, in Algebra, are fuch as are expreffed by different letters, or by different powers of the fame letter. Thus, a, and b, and a^2, and ab are all Unlike quantities.

UNLIKE *Signs*, are the different figns $+$ and $-$.

UNLIMITED or *Indeterminate Problem*, is fuch a one as admits of many, or even of infinite anfwers. As, to divide a given triangle into two equal parts ; or to defcribe a circle through two given points. See DIOPHANTINE, and INDETERMINATE.

VOID *Space*, in Phyfics. See VACUUM.

VOLUTE, in Architecture, a kind of fpiral fcroll, and ufed in the Ionic and Compofite capitals ; of which it makes the principal characteriftic and ornament.

VORTEX, or *Whirlwind*, in Meteorology, a fudden, rapid, violent motion of the air, in circular whirling directions.

VORTEX is alfo ufed for an eddy or whirlpool, or a body of water, in certain feas and rivers, which runs rapidly round, forming a fort of cavity in the middle.

VORTEX, in the Cartefian Philofophy, is a fyftem or collection of particles of matter moving the fame way, and round the fame axis.

Such Vortices are the grand machines by which thefe philofophers folve moft of the motions and other phenomena of the heavenly bodies. And accordingly, the doctrine of thefe Vortices makes a great part of the Cartefian philofophy.

The matter of the world they hold to have been divided at the beginning into innumerable little equal particles, each endowed with an equal degree of motion, both about its own centre, and feparately, fo as to conftitute a fluid.

Several fyftems, or collections of this matter, they farther hold to have been endowed with a common motion about certain points, as common centres, placed at equal diftances, and that the matters, moving round thefe, compofed fo many Vortices.

Then, the primitive particles of the matter they fuppofe, by thefe inteftine motions, to become, as it were, ground into fpherical figures, and fo to compofe globules of divers magnitudes ; which they call the matter of the fecond element : and the particles rubbed, or ground off them, to bring them to that form, they call the matter of the firft element.

And fince there would be more of the firft element than would fuffice to fill all the vacuities between, the globules of the fecond, they fuppofe the remaining part to be driven towards the centre of the Vortex, by the circular motion of the globules ; and that being there amaffed into a fphere, it would produce a body like the fun.

This fun being thus formed, and moving about its own axis with the common matter of the Vortex, would neceffarily throw out fome parts of its matter, through the vacuities of the globules of the fecond element conftituting the Vortex ; and this efpecially at fuch places as are fartheft from its poles ; receiving, at the fame time,

in,

in, by thefe poles, as much as it lofes in its equatorial parts. And, by this means, it would be able to carry round with it thofe globules that are nearelt, with the greater velocity; and the remoter, with lefs. And by this means, thofe globules, which are nearelt the centre of the fun, mult be fmalleft; becaufe, were they greater, or equal, they would, by reafon of their velocity, have a greater centrifugal force, and recede from the centre. If it fhould happen, that any of thefe funlike bodies, in the centres of the feveral Vortices, fhould be fo incruftated, and weakened, as to be carried about in the Vortex of the true fun; if it were of lefs folidity, or had lefs motion, than the globules towards the extremity of the folar Vortex, it would defcend towards the fun, till it met with globules of the fame folidity, and fufceptible of the fame degree of motion with itfelf; and thus, being fixed there, it would be for ever after carried about by the motion of the Vortex, without either approaching any nearer to the fun, or receding from it; and fo would become a planet.

Suppofing then all this; we are next to imagine, that our fyftem was at firft divided into feveral Vortices, in the centre of each of which was a lucid fpherical body; and that fome of thefe, being gradually incruftated, were fwallowed up by others which were larger, and more powerful, till at length they were all deftroyed, and fwallowed up by the largeft folar Vortex; except fome few which were thrown off in right lines from one Vortex to another, and fo become comets.

But this doctrine of Vortices is, at beft, merely hypothetical. It does not pretend to fhew by what laws and means the celeftial motions are effected, fo much as by what means they poffibly might, in cafe it fhould have fo pleafed the Creator. But we have another principle which accounts for the fame phenomena as well, nay, better than that of Vortices; and which we plainly find has an actual exiftence in the nature of things: and this is gravity, or the weight of bodies.

The Vortices, then, fhould be thrown out of philofophy, were it only that two different adequate caufes of the fame phenomena are inconfiftent.

But there are other objections againft them. For, 1°, if the bodies of the planets and comets be carried round the fun in Vortices, the bodies with the parts of the Vortex immediately invefting them, muft move with the fame velocity, and in the fame direction; and befides, they muft have the fame denfity, or the fame vis inertiæ. But it is evident, that the planets and comets move in the very fame parts of the heavens with different velocity, and in different directions. It follows, therefore, that thofe parts of the Vortex muft revolve at the fame time, in different directions, and with different velocities; fince one velocity, and direction, will be required for the paffage of the planets, and another for that of the comets.

2°, If it were granted, that feveral Vortices are contained in the fame fpace, and do penetrate each other, and revolve with divers motions; fince thefe motions muft be conformable to thofe of the bodies, which are perfectly regular, and performed in conic fections; it may be afked, How they fhould have been preferved entire fo many ages, and not difturbed and confounded

by the adverfe actions and fhocks of fo much matter as they muft meet withal?

3°, The number of comets is very great, and their motions are perfectly regular, obferving the fame laws with the planets, and moving in orbits, that are exceedingly eccentric. Accordingly, they move every way, and towards all parts of the heavens, freely pervading the planetary regions, and going frequently contrary to the order of the figns; which would be impoffible unlefs thefe Vortices were away.

4°, If the planets move round the fun in Vortices, thofe parts of the Vortices next the planets, we have already obferved, would be equally denfe with the planets themfelves: confequently the vortical matter, contiguous to the perimeter of the earth's orbit, would be as denfe as the earth itfelf: and that between the orbits of the earth and Saturn, muft be as denfe, or denfer. For a Vortex cannot maintain itfelf, unlefs the more denfe parts be in the centre, and the lefs denfe towards the circumference: and fince the periodical times of the planets are in fefquialterate ratio of their diftances from the fun, the parts of the Vortex muft be in the fame ratio. Whence it follows, that the centrifugal forces of the parts will be reciprocally as the fquares of the diftances. Such, therefore, as are at a greater diftance from the centre, will endeavour to recede from it with the lefs force. Accordingly, if they be lefs denfe, they muft give way to the greater force, by which the parts nearer the centre endeavour to rife. Thus, the more denfe will rife, and the lefs denfe defcend; and thus there will be a change of places, till the whole fluid matter of the Vortex be fo adjufted as that it may reft in equilibrio.

Thus will the greateft part of the Vortex without the earth's orbit, have a degree of denfity and inactivity, not lefs than that of the earth itfelf. Whence the comets muft meet with a very great refiftance, which is contrary to all appearances. Cotes, Præf. ad Newt. Princip. The doctrine of Vortices, Newton obferves, labours under many difficulties: for a planet to defcribe areas proportional to the times, the periodical times of a Vortex fhould be in a duplicate ratio of their diftances from the fun; and for the periodical time of the planets, to be in a fefquiplicate proportion of their diftances from the fun, the periodical times of the parts of the Vortex fhould be in the fame proportion of their diftances: and, laftly, for the lefs Vortices about Jupiter, Saturn, and the other planets, to be preferved, and fwim fecurely in the fun's Vortex, the periodical times of the fun's Vortex fhould be equal. None of which proportions are found to obtain in the revolutions of the fun and planets round their axes. Phil. Nat. Princ. Math. apud Schol. Gen. in Calce.

Befides, the planets, according to this hypothefis, being carried about the fun in ellipfes, and having the fun in the focus of each figure, by lines drawn from themfelves to the fun, they always defcribe areas proportionable to the times of their revolutions, which that author fhews the parts of no Vortex can do. Schol. prop. ult. lib. ii. Princip.

Again, Dr. Keill proves, in his Examination of Burnet's Theory, that if the earth were carried in a Vortex, it would move fafter in the proportion of three to

two,

two, when it is in Virgo than when it is in Pifces; which all experience proves to be falfe.

There is, in the Philofophical Tranfactions, a Phyfico-mathematical demonftration of the impoffibility and infufficiency of Vortices to account for the Celeftial Phenomena; by Monf. de Sigorne. See Num. 457. Sect. vi. pa. 409 et feq.

This author endeavours to fhew, that the mechanical generation of a Vortex is impoffible; and that it has only an axifugal force, and not a centrifugal and centripetal one; that it is not fufficient for explaining gravity and its properties; that it deftroys Kepler's aftronomical laws; and therefore he concludes, with Newton, that the hypothefis of Vortices is fitter to difturb than explain the celeftial motions. We muft refer to the differtation itfelf for the proof of thefe affertions. See CARTESIAN PHILOSOPHY.

VOSSIUS (GERARD JOHN), one of the moft learned and laborious writers of the 17th century, was of a confiderable family in the Netherlands: and was born in 1577, in the Palatinate near Heidelberg, at a place where his father, John Voffius, was minifter. He firft learned Latin, Greek, and Philofophy at Dort, where his father had fettled, and died. In 1595 he went to Leyden, where he farther purfued thefe ftudies, joining mathematics to them, in which fcience he made a confiderable progrefs. He became Mafter of Arts and Doctor in Philofophy in 1598; and foon after, Director of the College at Dort; then, in 1614, Director of the Theological College juft founded at Leyden; and, in 1618, Profeffor of Eloquence and Chronology in the Academy there, the fame year in which appeared his Hiftory of the Pelagian Controverfy. This hiftory procured him much odium and difgrace on the continent, but an ample reward in England, where archbifhop Laud obtained leave of king Charles the 1ft for Voffius to hold a prebendary in the church of Canterbury, while he refided at Leyden: this was in 1629, when he came over to be inftalled, took a Doctor of Laws degree at Oxford, and then returned.—In 1633 he was called to Amfterdam to fill the chair of a Profeffor of Hiftory; where he died in 1649, at 72 years of age; after having written and publifhed as many works as, when they came to be collected and printed at Amfterdam in 1695 &c, made 6 volumes folio, works which will long continue to be read with pleafure and profit. The principal of thefe are, —1. *Etymologicon Linguæ Latinæ.*—2. *De Origine & Progreffu Idololatriæ.*—3. *De Hiftoricis Græcis*—4. *De Hiftoricis Latinis.*—5. *De Arte Grammatica.*—6. *De Vitiis Sermonis & Gloffematis Latino-Barbaris.*—7. *Inftitutiones Oratoriæ.*—8. *Inftitutiones Poeticæ.*—9. *Ars Hiftorica.*—10. *De quatuor Artibus popularibus, Grammatice, Gymnaftice, Muficæ, & Graphice.*—11. *De Philologia.*—12. *De Univerfa Mathefeos Natura & Conftitutione.*—13. *De Philofophia.*—14. *De Philofophorum Sectis.*—15. *De Veterum Poetarum Temporibus.*

VOSSIUS (*Denis*), fon of the foregoing Gerard John, died at 22 years of age, a prodigy of learning, whofe inceffant ftudies brought on him fo immature a death. There are of his, among other fmaller pieces, Notes upon Cæfar's Commentaries, and upon Maimonides on Idolatry.

VOSSIUS (*Francis*), brother of Denis and fon of Gerard John, died in 1645, after having publifhed a Latin poem in 1640, on a naval victory gained by the celebrated Van Tromp.

VOSSIUS (*Gerard*), brother of Denis and Francis, and fon of Gerard John, wrote Notes upon Paterculus, which were printed in 1639. He was one of the moft learned critics of the 17th century; but died in 1640, like his two brothers, at a very early age, and before their father.

VOSSIUS (*Ifaac*), was the youngeft fon of Gerard John, and the only one that furvived him. He was born at Leyden in 1618, and was a man of great talents and learning. His father was his only preceptor, and his whole time was fpent in ftudying. His merit recommended him to a correfpondence with queen Chriftina of Sweden, who employed him in fome literary commiffions. At her requeft, he made feveral journeys into Sweden, where he had the honour to teach her the Greek language; though fhe afterwards difcarded him on hearing that he intended to write againft Salmafius, for whom fhe had a particular regard. In 1663 he received a handfome prefent of money from Louis the 14th of France, accompanied with a complimentary letter from the minifter Colbert.—In 1670 he came over to England, when he was created Doctor of Laws at Oxford, and king Charles the 2d made him Canon of Windfor; though he knew his character well enough to fay, there was nothing that Voffius refufed to believe, excepting the Bible. He appears indeed, by his publications, which are neither fo numerous nor fo ufeful as his father's, to have been a moft credulous man, while he afforded many circumftances to bring his religious faith in queftion. He died at his lodgings in Windfor Caftle, in 1688; leaving behind him the beft private library, as it was then fuppofed, in the world; which, to the fhame and reproach of England, was fuffered to be purchafed and carried away by the univerfity of Leyden. His publications chiefly were:—1. *Periplus Scylacis Caryandenfis, &c,* 1639.—2. Juftin, with Notes, 1640.—3. *Ignatii Epiftolæ, & Barnabæ Epiftola,* 1646.—4. *Pomponius Mela de Situ Orbis,* 1648.—5. *Differtatio de vera Ætate Mundi, &c,* 1659.—6. *De Septuaginta Interpretibus, &c,* 1661.—7. *De Luce,* 1662.—8. *De Motu Marium & Ventorum.*—9. *De Nili & aliorum Fluminum Origine.*—10. *De Poematum Cantu &, Viribus Rythmi,* 1673.—11. *De Sybillinis aliifque, quæ Chrifti natalem præceffere,* 1679.—12. *Catullus, & in eum Ifaaci Voffii Obfervationes,* 1684.—13. *Variarum Obfervationum liber,* 1685, in which are contained the following pieces: viz, *De Antiquæ Romæ & aliarum quarundam Urbium Magnitudine; De Artibus & Scientiis Sinarum; De Origine & Progreffu Pulveris Bellici apud Europæos; De Triremium & Liburnicarum Conftructione; De Emendatione Longitudinum; De patefacienda per Septentrionem ad Japonenfes & Indos Navigatione; De apparentibus in Luna circulis; Diurna Telluris converfione omnia gravia ad medium tendere.*

VOUSSOIRS, vault-ftones, are the ftones which immediately form the arch of a bridge, &c, being cut fomewhat in the manner of a truncated pyramid, their under fides conftituting the intrados, to which their

joints

joints or ends should be every where in a perpendicular direction.

The length of the middle Vouffoir, or key-stone, and which is the least of all, should be about $\frac{1}{12}$th or $\frac{1}{18}$th of the span of the arch; from hence these stones should be made larger and larger, all the way down to the impost; that they may the better sustain the great weight which rests upon them, without being crushed or broken, and that they may also bind the firmer together.

To find the just length of the Vouffoirs, or the figure of the extrados, when that of the intrados is given; see my Principles of Bridges, or Emerson's Construction of Arches, in his volume of Miscellanies.

URANIBURGH, or celestial town, the name of a celebrated observatory, in a castle in the little island Weenen, in the Sound; built by the celebrated Danish astronomer, Tycho Brahe, who furnished it with instruments for observing the course and motions of the heavenly bodies.

This observatory, which was finished about the year 1580, had not subsisted above 17 years when Tycho, who little thought to have erected an edifice of so short a duration, and who had even published the figure and position of the heavens, which he had chosen for the moment to lay the first stone in, was obliged to abandon his country.

Soon after this, the persons to whom the property of the island was given, demolished the building: part of the ruins was dispersed into divers places: the rest served to build Tycho a handsome seat upon his ancient estate, which to this day bears the name of Uraniburgh; and it was here that Tycho composed his catalogue of the stars. Its latitude is 55° 54′ north, and longitude 12° 47′ east of Greenwich.

M. Picart, making a voyage to Uraniburgh, found that Tycho's meridian line, there drawn, deviated from the meridian of the world; which seems to confirm the conjecture of some persons, that the position of the meridian line may vary.

URSA, in Astronomy, the Bear, a name common to two constellations of the northern hemisphere, near the pole, distinguished by *Major* and *Minor*.

URSA *Major*, or the *Great Bear*, one of the 48 old constellations, and perhaps more ancient than many of the others; being familiarly known and alluded to by the oldest writers, and is mentioned by Homer as observed by navigators. It is supposed that this constellation is that mentioned in the book of Job, under the name of *Chesil*, which our translation has rendered Orion, where it is said, " Canst thou loose the bands of Chesil (Orion)?" It is farther said that the Ancients represented each of these two constellations under the form of a waggon drawn by a team of horses, and the Greeks originally called them waggons and two bears; they are to this day popularly called the wains, or waggons, and the greater of them Charles's Wain. Hence is remarked the propriety of the expression, " loose the bands &c," the binding and loosing being terms very applicable to a harness, &c.

Perhaps the Egyptians, or whoever else were the people that invented the constellations, placed those stars, which are near the pole, in the figure of a bear, as being an animal inhabiting towards the north pole, and making neither long journeys, nor swift motions.

But the Greeks, in their usual way, have adapted some of their fables to it. They say this bear was Callisto, daughter of Lycaon, king of Arcadia; that being debauched by Jupiter, he afterwards placed her in the heavens, as well as her son Arcturus.

The Greeks called this constellation Arctos and Helice, from its turning round the pole. The Latins from the name of the nymph, as variously written, Callisto, Megisto, and Flemisto, and from the Arabians, sometimes Feretrum Majus, the Great Bier. And the Ursa Minor, they called Feretrum Minus, the Little Bier. The Italians have followed the same custom, and call them Cataletto. They spoke also of the Phenicians being guided by the Lesser Bear, but the Greeks by the Greater.

There are two remarkable stars in this constellation, viz, those in the middle of his body, considered as the two hindermost of the wain, and called the pointers, because they always point nearly in a direction towards the north pole star, and so are useful in finding this star out.

The stars in Ursa Major, are, according to Ptolomy's catalogue, 35; in Tycho's 56; in Hevelius's 73; but in the Britannic catalogue 87.

URSA *Minor*, the *Little Bear*, called also *Arctos Minor*, *Phœnice*, and *Cynosura*, one of the 48 old constellations, and near the north pole, the large star in the tip of its tale being very near to it, and thence called the pole-star.

The Phenicians guided their navigations by this constellation, for which reason it was called Phenice, or the Phenician constellation. It was also called Cynosura by the Greeks, because, according to some, that was one of the dogs of the huntress Callisto, or the Great Bear; but according to others Cynosura was one of the Idæan nymphs that nursed the infant Jupiter; and some say that Callisto was another of them, and that, for their care, they were taken up together to the skies.

Ptolomy places in this constellation 8 stars, Tycho 7, Hevelius 12, and Flamsteed 24.

URSUS (NICHOLAS RAIMARUS), a very extraordinary person, and distinguished in the science of astronomy, was born at Henstedt in Dithmarsen, in the duchy of Holstein, about the year 1550 He was a swineherd in his youth, and did not begin to read till he was 18 years of age; but then he employed all the hours he could spare from his daily labour, in learning to read and write. He afterwards applied himself to learn the languages; and, having a strong genius, made a rapid progress in Greek and Latin. He quickly learned also the French language, the mathematics, astronomy, and philosophy; and most of them without the assistance of a master.

Having left his native country, he gained a maintenance by teaching; which he did in Denmark in 1584, and on the frontiers of Pomerania and Poland in 1585. It was in this place that he invented a new system of astronomy, very little different from that of Tycho Brahe. This he communicated, in 1586, to the landgrave of Hesse, which gave rise to a terrible dispute between him and Tycho. This celebrated astronomer charged him with being a plagiary: who, as he related, happening to come with his master into his study, saw there, drawn on a piece of paper, the figure of his

his fyftem; and afterwards infolently boafted that he himfelf was the inventor of it. Urfus, upon this accufation, wrote furioufly againft Tycho, called the honour of his invention in queftion, afcribing the fyftem to Apollonius Pergæus; and in fhort abufed him in fo brutal a manner, that he was going to be profecuted for it.

Urfus was afterwards invited by the emperor to teach the mathematics in Prague; from which city, to avoid the prefence of Tycho, he withdrew filently in 1589, and died foon after.

He made fome improvements in trigonometry, and wrote feveral books, which difcover the marks of his hafty ftudies; his erudition being indigefted, and his ftyle incorrect, as is almoft always to be obferved of perfons that are late-learned.

VULPECULA et ANSER, the *Fox and Goofe*, in Aftronomy, one of the new conftellations of the northern hemifphere, made out of the unformed ftars by Hevelius, in which he reckons 27 ftars; but Flamfteed counts 35.

W.

WAL

WAL

WAD, or WADDING, in Gunnery, a ftopple of paper, hay, ftraw, old rope-yarn, or tow, rolled firmly up like a ball, or a fhort cylinder, and forced into a gun upon the powder, to keep it clofe in the chamber; or put up clofe to the fhot, to keep it from rolling out, as well as, according to fome, to prevent the inflamed powder from dilating round the fides of the ball, by its windage, as it paffes along the chace, which it was thought would much diminifh the effort of the powder. But, from the accurate experiments lately made at Woolwich, it has not been found to have any fuch effect.

WADHOOK, or WORM, a long pole with a ferew at the end, to draw out the wad, or the charge, or paper &c from a gun.

WAGGONER, in Aftronomy, is the conftellation Urfa Major, or the Great Bear, called alfo vulgarly Charles's Wain.

WAGGONER is alfo ufed for a routier, or book of charts, defcribing the feas, their coafts, &c.

WALLIS (Dr. JOHN), an eminent Englifh mathematician, was the fon of a clergyman, and born at Afhford in Kent, Nov. 23. 1616. After being inftructed, at different fchools, in grammar learning, in Latin, Greek, and Hebrew, with the rudiments of logic, mufic, and the French language, he was placed in Emanuel college, Cambridge. About 1640 he entered into orders, and was chofen fellow of Queen's college. He kept his fellowfhip till it was vacated by his marriage, but quitted his college to be chaplain to Sir Richard Darley; after a year fpent in this fituation, he fpent two more as chaplain to lady Vere. While he lived in this family, he cultivated the art of deciphering, which proved very ufeful to him on feveral occafions: he met with rewards and preferment from the government at home for deciphering letters for them; and it is faid, that the elector of Brandenburg fent him a gold chain and medal, for explaining for him fome letters written in ciphers.

5

In 1643 he publifhed *Truth Tryd*, or Animadverfions on lord Brooke's treatife, called The Nature of Truth &c; ftyling himfelf " a minifter in London," probably of St. Gabriel Fenchurch, the fequeftration of which had been granted to him.—In 1644 he was chofen one of the fcribes or fecretaries to the affembly of divines at Weftminfter.

Academical ftudies being much interrupted by the civil wars in both the univerfities, many learned men from them reforted to London, and formed affemblies there. Wallis belonged to one of thefe, the members of which met once a week, to difcourfe on philofophical matters; and this fociety was the rife and beginning of that which was afterwards incorporated by the name of the Royal Society, of which Wallis was one of the moft early members.

The Savilian profeffor of geometry at Oxford being ejected by the parliamentary vifitors, in 1649, Wallis was appointed to fucceed him, and he opened his lectures there the fame year. In 1650 he publifhed fome Animadverfions on a book of Mr. Baxter's, intitled, " Aphorifms of Juftification and the Covenant." And in 1653, in Latin, a Grammar of the Englifh tongue, for the ufe of foreigners; to which was added, a tract *De Loquela feu Sonorum formatione, &c,* in which he confiders philofophically the formation of all founds ufed in articulate fpeech, and fhews how the organs being put into certain pofitions, and the breath pufhed out from the lungs, the perfon will thus be made to fpeak, whether he hear himfelf or not. Purfuing thefe reflections, he was led to think it poffible, that a deaf perfon might be taught to fpeak, by being directed fo to apply the organs of fpeech, as the found of each letter required, which children learn by imitation and frequent attempts, rather than by art. He made a trial or two with fuccefs; and particularly upon one Popham, which involved him in a difpute with Dr. Holder, of which fome account has already been given in the life of that gentleman.

In

In 1654 he took the degree of Doctor in Divinity; and the year after became engaged in a long controversy with Mr. Hobbes. This philosopher having, in 1655, printed his treatise *De Corpore Philosophico*, Dr. Wallis the same year wrote a confutation of it in Latin, under the title of *Elenchus Geometriæ Hobbianæ*; which so provoked Hobbes, that in 1656 he published it in English, with the addition of what he called, " Six Lessons to the Professors of Mathematics in Oxford." Upon this Dr. Wallis wrote an answer in English, intitled, " Due Correction for Mr. Hobbes; or School discipline for not saying his Lessons right," 1656: to which Mr. Hobbes replied in a pamphlet called " ΣΤΙΓΜΑΙ, &c, or Marks of the absurd Geometry, Rural Language, Scottish Church politics, and Barbarisms, of John Wallis, 1657." This was immediately rejoined to by Dr. Wallis, in *Hobbiani Puncti Dispunctio*, 1657. And here this controversy seems to have ended, at this time: but in 1661 Mr. Hobbes printed *Examinatio & Emendatio Mathematicorum Hodiernorum in sex Dialogis*; which occasioned Dr. Wallis to·publish the next year, *Hobbius Heautontimorumenos*, addressed to Mr. Boyle.

In 1657 he collected and published his mathematical works, in two parts, entitled, *Mathesis Universalis*, in 4to; and in 1658, *Commercium Epistolicum de Quæstionibus quibusdam Mathematicis nuper habitum*, in 4to; which was a collection of letters written by many learned men, as Lord Brounker, Sir Kenelm Digby, Fermat, Schooten, Wallis, and others.

He was this year chosen *Custos Archivorum* of the university. Upon this occasion Mr. Stubbe, who, on account of his friend Mr. Hobbes, had before waged war against Wallis, published a pamphlet, intitled, " The Savilian Professor's Case Stated," 1658. Dr. Wallis replied to this; and Mr. Stubbe republished his case, with enlargements, and a vindication against the exceptions of Dr. Wallis.

Upon the Restoration he met with great respect; the king thinking favourably of him on account of some services he had done both to himself and his father Charles the first. He was therefore confirmed in his places, also admitted one of the king's chaplains in ordinary, and appointed one of the divines empowered to revise the book of Common Prayer. He complied with the terms of the act of uniformity, and continued a steady conformist till his death. He was a very useful member of the Royal Society; and kept up a literary correspondence with many learned men. In 1670 he published his *Mechanica; sive de Motu*, 4to. In 1676 he gave an edition of *Archimedis Syracusani Arenarius & Dimensio Circuli*; and in 1682 he published from the manuscripts, *Claudii Ptolomæi Opus Harmonicum*, in Greek, with a Latin version and notes; to which he afterwards added, *Appendix de veterum Harmonica ad hodiernam comparata, &c*. In 1685 he published some theological pieces; and, about 1690, was engaged in a dispute with the Unitarians; also, in 1692, in another dispute about the Sabbath. Indeed his books upon subjects of divinity are very numerous, but nothing near so important as his mathematical works.

In 1685 he published his History and Practice of Algebra, in folio; a work that is full of learned and useful matter. Besides the works above mentioned, he

published many others, particularly his *Arithmetic of Infinites*, a book of genius and good invention, and perhaps almost his only work that is so, for he was much more distinguished for his industry and judgment, than for his genius. Also a multitude of papers in the Philos. Transf. in almost every volume, from the 1st to the 25th volume. In 1697, the curators of the University press at Oxford thought it for the honour of the university to collect the doctor's mathematical works, which had been printed separately, some in Latin, some in English, and published them all together in the Latin tongue, in 3 vols folio, 1699.

Dr. Wallis died at Oxford the 28th of October 1703, in the 88th year of his age, leaving behind him one son and two daughters. We are told that he was of a vigorous constitution, and of a mind which was strong, calm, serene, and not easily ruffled or discomposed. He speaks of himself, in his letter to Mr. Smith, in a strain which shews him to have been a very cautious and prudent man, whatever his secret opinions and attachments might be: he concludes, " It hath been my endeavour all along to act by moderate principles, being willing, whatever side was uppermost, to promote any good design, for the true interest of religion, of learning, and of the public good."

WARD (Dr. SETH), an English prelate, chiefly famous for his knowledge in mathematics and astronomy, was the son of an attorney, and born at Buntingford, Hertfordshire, in 1617 or 1618. From hence he was removed and placed a student in Sidney college, Cambridge, in 1632. Here he applied with great vigour to his studies, particularly to the mathematics, and was chosen fellow of his college. In 1640 he was pitched upon by the Vice-chancellor to be Prævaricator, which at Oxford is called Terræ-filius; whose office it was to make a witty speech, and to laugh at any thing or any body: a privilege which he exercised so freely, that the Vice-chancellor actually suspended him from his degree; though he reversed the censure the day following.

The civil war breaking out, Ward was involved not a little in the consequences of it. He was ejected from his fellowship for refusing the Covenant; against which he soon after joined with several others, in drawing up that noted treatise, which was afterwards printed. Being now obliged to leave Cambridge, he resided for some time with certain friends about London, and at other times at Aldbury in Surry, with the noted mathematician Oughtred, where he prosecuted his mathematical studies. He afterwards lived for the most part, till 1649, with Mr. Ralph Freeman at Aspenden in Hertfordshire, whose sons he instructed as their preceptor; after which he resided some months with lord Wenman, of Thame Park, in Oxfordshire.

He had not been long in this family before the visitation of the university of Oxford began; the effect of which was, that many learned and eminent persons were turned out, and among them Mr. Greaves, the Savilian professor of Astronomy: this gentleman laboured to procure Ward for his successor, whose abilities in his way were universally known and acknowledged; and effected it; Dr. Wallis succeeding to the Geometry professorship at the same time. Mr. Ward then entered himself of Wadham college, for the sake of

Dr.

Dr. Wilkins, who was the warden; and he presently applied himself to bring the astronomy lectures, which had long been neglected and disused, into repute again; and for this purpose he read them very constantly, never missing one reading day, all the while he held the lecture.

In 1654, both the Savilian professors did their exercises, in order to proceed doctors in divinity; and when they were to be presented, Wallis claimed precedency. This occasioned a dispute; which being decided in favour of Ward, who was really the senior, Wallis went out grand compounder, and so obtained the precedency. In 1659, Ward was chosen president of Trinity college; but was obliged at the Restoration to resign that place. He had amends made him, however, by being presented in 1660 to the rectory of St. Laurence Jewry. The same year he was also installed precentor of the church of Exeter. In 1661 he became fellow of the Royal Society, and dean of Exeter; and the year following he was advanced to the bishopric of the same church. In 1667 he was translated to the see of Salisbury; and in 1671 was made chancellor of the order of the garter; an honour which he procured to be permanently annexed to the see of Salisbury, after it had been held by laymen for above 150 years.

Dr. Ward was one of those unhappy persons who have the misfortune to survive their senses, which happened in consequence of a fever ill cured: he lived till the Revolution, but without knowing any thing of the matter; and died in January 1689, about 71 years of age. He was the author of several Latin works in astronomy and different parts of the mathematics, which were thought excellent in their day; but their use has been superseded by later improvements and the Newtonian philosophy. Some of these were,

1. A Philosophical Essay towards an Eviction of the Being and Attributes of God, &c. 1652.

2. De Cometis, &c; 4to, 1653.

3. In Ismaelis Bullialdi Astronomia Inquisitio; 4to, 1653.

4. Idea Trigonometriæ demonstratæ; 4to, 1654.

5. Astronomia Geometrica; 8vo, 1656. In this work, a method is proposed, by which the astronomy of the planets is geometrically resolved, either upon the Elliptical or Circular motion; it being in the third or last part of this work that he proposes and explains what is called Ward's Circular Hypothesis.

6. Exercitatio epistolica in Thomæ Hobbii Philosophiam, ad D. Joannem Wilkins; 1656, 8vo.

But that by which he hath chiefly signalized himself, as to astronomical invention, is his celebrated approximation to the true place of a planet, from a given mean anomaly, founded upon an hypothesis, that the motion of a planet, though it be really performed in an elliptic orbit, may yet be considered as equable as to angular velocity, or with an uniform circular motion round the upper focus of the ellipse, or that next the aphelion, as a centre. By this means he rendered the praxis of calculation much easier than any that could be used in resolving what has been commonly called Kepler's problem, in which the coequate anomaly was to be immediately investigated from the mean elliptic one. His hypothesis agrees very well with those orbits which are elliptical but in a very small degree, as that of the Earth and Venus: but in others, that are more elliptical, as those of Mercury, Mars, &c, this approximation stood in need of a correction, which was made by Bulliald. Both the method, and the correction, are very well explained and demonstrated, by Keill, in his Astronomy, lecture 24.

WARGENTIN (Peter), an ingenious Swedish mathematician and astronomer, was born Sept. 22, 1717, and died Dec. 13, 1783. He became secretary to the Academy at Stockholm in 1749, when he was only 32 years of age; and he became successively a member of most of the literary academies in Europe, as London, Paris, Petersburg, Gottingen, Upsal, Copenhagen, Drontheim, &c. In this country he is probably most known on account of his tables for computing the eclipses of Jupiter's satellites, which are annexed to the Nautical Almanac of 1779. I know not that he has published any separate work; but his communications were very numerous to several of those Academies of which he was a member; as the Academy of Stockholm, in which are 52 of his memoirs; in the Philosophical Transactions, the Upsal Acts, the Paris Memoirs, &c.

WATCH, a small portable machine, or movement, for measuring time; having its motion commonly regulated by a spiral spring. Perhaps, strictly speaking, watches are all such movements as *shew* the parts of time; as clocks are such as *publish* them, by striking on a bell, &c. But commonly, the term Watch is appropriated to such as are carried in the pocket; and clock to the large movements, whether they strike the hour or not.

Spring or *Pendulum* WATCHES stand pretty much on the same principle with pendulum clocks. For if a pendulum, describing small circular arcs, make vibrations of unequal lengths, in equal times, it is because it describes the greater arc with a greater velocity; so a spring put in motion, and making greater and less vibrations, as it is more or less stiff, and as it has a greater or less degree of motion given it, performs them nearly in equal times. Hence, as the vibrations of the pendulum had been applied to large clocks, to rectify the inequality of their motions; so, to correct the unequal motions of the balance in Watches, a spring is added, by the isochronism of whose vibrations the correction is to be affected. The spring is usually wound into a spiral; that, in the little compass allotted it, it may be as long as possible; and may have strength enough not to be mastered, and dragged about, by the inequalities of the balance it is to regulate. The vibrations of the two parts, viz, the spring and the balance, should be of the same length; but so adjusted, as that the spring, being more regular in the length of its vibrations than the balance, may occasionally communicate its regularity to the latter.

The Invention of Spring or Pocket Watches, is due to the last age. It is true, it is said, in the history of Charles the 5th, that a Watch was presented to that prince: but this was probably no more than a kind of clock to be set on a table: some resemblance of which we have still remaining in the ancient pieces made before the year 1670. Some accounts also say, the first Watches were made at Nuremberg in 1500, by Peter

Hell,

Hell, and were called Nuremberg eggs, on account of their oval form. And farther, that the same year George Purbach, a mathematician of Vienna, employed a watch that pointed to seconds, for astronomical observations, which was probably a kind of clock. In effect, it is between Hook and Huygens that the glory of this excellent invention lies: but to which of them it properly belongs, has been greatly disputed; the English ascribing it to the former, and the French, Dutch, &c, to the latter. Derham, in his Artificial Clockmaker, says roundly, that Dr. Hook was the inventor; and adds, that he contrived various ways of regulation: one way was with a loadstone: another with a tender straight spring, one end of which played backward and forward with the balance; so that the balance was to the spring as the ball of a pendulum, and the spring as the rod of the same: a third method was with two balances, of which there were divers sorts; some having a spiral spring to the balance for a regulator, and others without. But the way that prevailed, and which still continues in mode, was with one balance, and one spring running round the upper part of the verge of it: though this has a disadvantage, which those with two springs &c were free from; in that, a sudden jerk, or confused shake will alter its vibrations, and flurry it very much.

The time of these inventions was about the year 1658; as appears, among other evidences, from an inscription on one of the double-balance Watches presented to king Charles the second, viz, Rob. Hook inven. 1658. T. Tompion fecit, 1675. The invention soon came into repute both at home and abroad; and two of the machines were sent for by the Dauphin of France. Soon after this, M. Huygens's Watch with a spiral spring got abroad, and made a great noise in England, as if the longitude could be found by it. It is certain however, that this invention was later than the year 1673, when his book De Horol. Oscillat. was published; in which there is no mention of this, though he speaks of several other contrivances in the same way.

One of these the lord Brounker sent for out of France, where M. Huygens had got a patent for them. This Watch agreed with Dr. Hook's, in the application of the spring to the balance; only that of Huygens had a longer spiral spring, and its pulses and beats were much slower; also the balance, instead of turning quite round, as Dr. Hook's, turned several times every vibration. Huygens also invented divers other kinds of Watches, some of them without any string or chain at all, which he called pendulum Watches.

Mr. Derham suggests that he suspects Huygens's fancy was first set to work by some intelligence he might have of Hook's invention from Mr. Oldenburg, or some other of his correspondents in England; though Mr. Oldenburg vindicates himself against that charge, in the Philos. Transf. numbers 118 and 129.

Watches, since their first invention, have gone on in a continued course of improvement, and they have lately been brought to great perfection, both in England and in France, but more especially the former, particularly owing to the great encouragement that has been given to them by the Board of Longitude. Some of the chief writers and improvers of Watches, are,

Le Roy, Cummins, Harrison, Mudge, Emery, and Arnold, whose Watches are now in very high repute, and in frequent use in the navy and India ships, for keeping the longitude. See Derham's Artificial Clockmaker; Cummins's Principles of Clock and Watch work; Mudge's Thoughts on the Means of improving Watches, &c.

Striking WATCHES, are such as, besides the proper Watch part, for measuring time, have a clock part, for striking the hours, &c. These are real clocks; only moved by a spring instead of a weight; and are properly called pocket clocks.

Repeating WATCHES, are such as, by pulling a string, &c, repeat the hour, quarter, or minute, at any time of the day or night.—This repetition was the invention of Mr. Barlow, being first put in practice by him in larger movements or clocks, about the year 1676. The contrivance immediately set the other artists to work, who soon contrived divers ways of effecting the same. But its application to pocket Watches was not known before K. James the second's reign; when the ingenious inventor above mentioned was soliciting a patent for it. The talk of a patent engaged Mr. Quare to resume the thoughts of a like contrivance, which he had in view some years before: he now effected it; and being pressed to endeavour to prevent Mr. Barlow's patent, a Watch of each kind was produced before the king and council; upon trial of which, the preference was given to Mr. Quare's. The difference between them was, that Barlow's was made to repeat by pushing in two pieces on each side the Watch-box; one of which repeated the hour, and the other the quarter: whereas Quare's was made to repeat by a pin that stuck out near the pendant, which being thrust in (as now is done by thrusting in the pendant itself) repeated both the hour and quarter with the same thrust.

Of the Mechanism of a WATCH.

Watches, as well as clocks, are composed of wheels and pinions, with a regulator to direct the quickness or slowness of the wheels, and of a spring which communicates motion to the whole machine. But the regulator and spring of a Watch are vastly inferior to the weight and pendulum of a clock, neither of which can be employed in Watches. Instead of a pendulum, therefore, they are obliged to use a balance. (Pl. 34, fig. 4) to regulate the motion of a Watch; and of a spring (fig. 6), which serves instead of a weight, to give motion to the wheels and balance.

The wheels of a Watch, like those of a clock, are placed in a frame, formed of two plates and four pillars. Fig. 3 represents the inside of a Watch, after the plate (Fig. 5) is taken off. A is the barrel which contains the spring (fig. 6); the chain is rolled about the barrel, with one end of it fixed to the barrel A, and the other to the fusee B.

When a Watch is wound up, the chain which was upon the barrel winds about the fusee, and by this means the spring is stretched; for the interior end of the spring is fixed by a spring to the immoveable axis, about which the barrel revolves; the exterior end of the spring is fixed to the inside of the barrel, which turns upon an axis. It is there easy to perceive how the spring extends itself, and how its elasticity forces

the

the barrel to turn round, and confequently obliges the chain which is upon the fufee to unfold and turn the fufee; the motion of the fufee is communicated to the wheel CC; then by means of the teeth, to the pinion c, which carries the wheel D; then to the pinion d, which carries the wheel E; then to the pinion e, which carries the wheel F; then to the pinion f, upon which is the balance-wheel G, whofe pivot runs in the piece A, called the potance, and B called a follower, which are fixed on the plate fig. 5. This plate, of which only a part is reprefented, is applied to that of fig. 3, in fuch a manner, that the pivots of the wheels enter into holes made in the plate fig. 3. Thus the impreffed force of the fpring is communicated to the wheels: and the pinion f being then connected to the wheel F, obliges it to turn (fig. 7). This wheel acts upon the pallats of the verge 1, 2. (fig. 4) the axis of which carries the balance HH (fig. 4). The pivot I, in the end of the verge, enters into the hole G in the potance A (fig. 5). In this figure the pallats are reprefented; but the balance is on the other fide of the plate, as may be feen in fig. 11. The pivot 3 of the balance enters into a hole of the cock BC (fig. 10), a perfpective view of which is reprefented in fig. 12. Thus the balance, turns between the cock and the potance c (fig. 5), as in a kind of cage. The action of the balance-wheel upon the pallats 1, 2, (fig. 4) is the fame with that of the fame wheel in the clock; i. e. in a Watch the balance-wheel obliges the balance to vibrate backwards and forwards like a pendulum.

At each vibration of the balance a pallat allows a tooth of the balance wheel to efcape; fo that the quicknefs of the motion of the wheels is entirely determined by the quicknefs of the vibrations of the balance, and thefe vibrations of the balance and motion of the wheels are produced by the action of the fpring.

But the quicknefs or flownefs of the vibrations of the balance depends not folely upon the action of the great fpring, but chiefly upon the action of the fpring abc, called the fpiral fpring (fig. 13) fituated under the balance H, and reprefented in perfpective (fig. 11); the exterior end of the fpiral is fixed to the pin a (fig. 13). This pin is applied near the plate in a (fig. 11); the interior end of the fpiral is fixed by a peg to the centre of the balance. Hence if the balance be turned upon itfelf, the plates remaining immoveable, the fpring will extend itfelf, and make the balance perform one revolution. Now, after the fpiral is thus extended, if the balance be left to itfelf, the elafticity of the fpiral will bring back the balance, and in this manner the alternate vibrations of the balance are produced. In fig. 7 all the wheels above defcribed are reprefented in fuch a manner, that we may eafily perceive at firft fight how the motion is communicated from the barrel to the balance.

In fig. 8 are reprefented the wheels under the dial-plate, by which the hands are moved. The pinion a is adjufted to the force of the prolonged pivot of the wheel D (fig. 7), and is called a cannon pinion. This wheel revolves in an hour. The end of the axis of the pinion a, upon which the minute hand is fixed, is fquare; the pinion (fig. 8) is indented into the wheel b, which is carried by the pinion a. Fig. 9 is a wheel fixed upon a barrel, into the cavity of which the pinion

a enters, and upon which it turns freely. This wheel d revolves in 12 hours, and carries along with it the hour-hand.

WATER, in Phyfiology, a clear, infipid, and colourlefs fluid, coagulable into a tranfparent folid fubftance, called ice, when placed in a temperature of 32° of Fahrenheit's thermometer, or lower, but volatile and fluid in every degree of heat above that; and when pure, or freed from heterogeneous particles, is reckoned one of the four elements.

By fome late experiments of Meffrs. Lavoifier, Watt, Cavendifh, Prieftley, Kirwan, &c, it appears, that Water confifts of dephlogifticated air, and inflammable air or phlogifton intimately united; or, as Mr. Watt conceives, of thofe two principles deprived of part of their latent heat. And in fome inftances it appears that air and Water are mutually convertible into each other. Thus, Mr. Cavendifh (Philof. Tranf. vol. 74, p. 128) recites feveral experiments, in which he changed common air into pure Water, by decompofing it in conjunction with inflammable air. Dr. Prieftley likewife, having decompofed dephlogifticated and inflammable air, by firing them together by the electric explofion, found a manifeft decompofition of Water, which, as nearly as he could judge, was equal in weight to that of the decompofed air. He alfo made a number of other curious experiments, which feemed to favour the idea of a converfion of Water into air, without abfolutely proving it. The difficulty which M. De Luc and others have found in expelling all air from Water, is beft accounted for on the fuppofition of the generation of air from Water; and admitting that the converfion of Water into air is effected by the intimate union of what is called the principle of heat with the Water, it appears fufficiently analogous to other changes, or rather combinations, of fubftances. Is not, fays Dr. Prieftley, the acid of nitre, and alfo that of vitriol, a thing as unlike to air as Water is, their properties being as remarkably different? And yet it is demonftrable that the acid of nitre is convertible into the pureft refpirable air, and probably by the union of the fame principle of heat. Philof. Tranf. vol. 73, p. 414 &c.

Indeed there feems to be Water in all bodies, and particles of almoft all kinds of matter in Water; fo that it is hardly ever fufficiently pure to be confidered as an element. Water, if it could be had alone, and pure, Boerhaave argues, would have all the requifites of an element, and be as fimple as fire; but there is no expedient hitherto difcovered for procuring it fo pure. Rain Water, which feems the pureft of all thofe we know of, is replete with infinite exhalations of all kinds which it imbibes from the air: fo that if filtered and diftilled a thoufand times, there ftill remain fæces. Befides this, and the numberlefs impurities it acquires after it is raifed, by mixing with all forts of effluvia in the atmofphere, and by falling upon and running over the earth, houfes, and other places. There is alfo fire contained in all Water; as appears from its fluidity, which is owing to fire alone. Nor can any kinds of filtering through fand, ftone, &c, free it entirely from falts &c. Nor have all the experiments that have been invented by the philofophers, ever been able to derive Water perfectly pure. Hence Boerhaave fays, that he is convinced nobody ever faw a drop of pure Water;

that

that the utmoſt of its purity known, only amounts to its being free from this or that ſort of matter; and that it can never, for inſtance, be quite deprived of ſalt; ſince air will always accompany Water, and air always contains ſalt.

Water ſeems to be diffuſed everywhere, and to be preſent in all ſpace wherever there is matter. There are hardly any bodies in nature but what will yield Water: it is even aſſerted that fire itſelf is not without it. A ſingle grain of the fiery ſalt, which in a moment's time will penetrate through a man's hand, readily imbibes half its weight of Water, and melts even in the drieſt air imaginable. Among innumerable inſtances, hartſhorn, kept 40 years, and turned as hard and dry as any metal, ſo that it will yield ſparks of fire when ſtruck againſt a flint, yet being put into a glaſs veſſel, and diſtilled, will afford $\frac{1}{8}$th part of its quantity of Water. Bones dead and dried 25 years, and thus become almoſt as hard as iron, yet by diſtillation have yielded half their weight of Water. And the hardeſt ſtones, ground and diſtilled, always diſcover a portion of it. But hitherto no experiment ſhews, that Water enters as a principle into the combination of metallic matters, or even into that of vitreſcible ſtones.

From ſuch conſiderations, philoſophers have been led to hold the opinion, that all things were made of Water. Baſil Valentine, Paracelſus, Van Helmont, and others have maintained, that Water is the elemental matter or ſtamen of all things, and ſuffices alone for the production of all the viſible creation. Thus too Newton: "All birds, beaſts, and fiſhes, inſects, trees, and vegetables, with their ſeveral parts, do grow out of Water, and watery tinctures, and ſalts; and by putrefaction they all return again to watery ſubſtances." And the ſame doctrine is held, and confirmed by experiments, by Van Helmont, Boyle, and others.

But Dr. Woodward endeavours to ſhew that the whole is a miſtake —Water containing extraneous corpuſcles, ſome of which, according to him, are the proper matter of nutrition; the Water being ſtill found to afford ſo much the leſs nouriſhment, the more it is purified by diſtillation. So that Water, as ſuch, does not ſeem to be the proper nutriment of vegetables; but only the vehicle which contains the nutritious particles, and carries them along with it, through all the parts of the plant.

Helmont however carries his ſyſtem ſtill farther, and imagines that all bodies may be reconverted into Water. His alkaheſt, he affirms, adequately reſolves plants, animals, and minerals, into one liquor, or more, according to their ſeveral internal differences of parts; and the alkaheſt, being abſtracted again from theſe liquors, in the ſame weight, and with the ſame virtues, as when it diſſolved them, the liquors may, by frequent cohobations from chalk, or ſome other proper matter, be totally deprived of their ſeminal endowments, and at laſt return to their firſt matter; which is inſipid Water.

Spirit of wine, of all other ſpirits, ſeems freeſt from Water: yet Helmont affirms, it may be ſo united with Water, as to become Water itſelf. He adds, that it is material Water, only under a ſulphureous diſguiſe. And the ſame thing he obſerves of all ſalts, and of oils, which may be almoſt wholly changed into Water.

No ſtandard for the Weight and Purity of WATER.—Water ſcarce ever continues two moments exactly of the ſame weight; by reaſon of the air and fire contained in it. The expanſion of Water in boiling ſhews what effect the different degrees of fire have on the gravity of Water. This makes it difficult to fix the ſpecific gravity of Water, in order to ſettle its degree of purity. However, the pureſt Water we can obtain, according to the experiments of Mr. Hawſkbee, is 850 times heavier than air: or according to the experiments of Mr. Cavendiſh, the thermometer being at 50° and the barometer at $29\frac{3}{4}$, about 800 times as heavy as air: and according to the experiments of Sir Geo. Shuckburgh, when the barometer is at 29·27 and the thermometer at 53°, Water is 836 times heavier than air; whence alſo may be deduced this general proportion, which may be accounted a ſtandard, viz, that, when the barometer is at 30° and the thermometer at 55°, then Water is 820 times heavier than air; alſo that in ſuch a ſtate the cubic foot of Water weighs 1000 ounces avoirdupois, and that of air 1·222, or $1\frac{2}{9}$ nearly, alſo that of mercury 13600 ounces; and for other ſtates of the thermometer and barometer, the allowance is after this rate, viz, that the column of mercury in the barometer varies its length by the 10 thouſandth part of itſelf for a change of each ſingle degree of temperature, and Water changes by $\frac{3}{20000}$ part of its height or magnitude by each degree of the ſame. However, we have not any very exact ſtandard in air; for Water being ſo much heavier than air, the more Water there is contained in the air, the heavier of courſe muſt the air be; as indeed a conſiderable part of the weight of the atmoſphere ſeems to ariſe from the Water that is in it.

Properties and Effects of WATER.——Water is a very volatile body. It is entirely reduced into vapours and diſſipated, when expoſed to the fire and unconfined.

Water heated in an open veſſel, acquires no more than a certain determinate degree of heat, whatever be the intenſity of the fire to which it is expoſed; which greateſt degree of heat is when it boils violently.

It has been found that the degree of heat neceſſary to make Water boil, is variable, according to the purity of the Water and the weight of the atmoſphere. The following table ſhews the degree of heat at which Water boils, at various heights of the barometer, being a medium between thoſe reſulting from the experiments of Sir Geo. Shuckburgh and M. De Luc:

Height of the Barometer.	Heat of Boiling Water.
Inches.	°
26	205
$26\frac{1}{2}$	206
27	206·9
$27\frac{1}{2}$	207·7
28	208·5
$28\frac{1}{2}$	209·4
29	210·3
$29\frac{1}{2}$	211·2
30	212·0
$30\frac{1}{2}$	212·8
31	213·6

Water

Water is found the moſt penetrative of all bodies, after fire, and the moſt difficult to confine; paſſing through leather, bladders, &c, which will confine air; making its way gradually through woods; and is only retainable in glaſs and metals; nay it was found by experiment at Florence, that when ſhut up in a ſpherical veſſel of gold, which was preſſed with a great force, it made its way through the pores even of the gold itſelf.

Water, by this penetrative quality alone, may be inferred to enter the compoſition of all bodies, both vegetable, animal, foſſil, and even mineral; with this particular circumſtance, that it is eaſily, and with a gentle heat, ſeparable again from bodies it had united with.

And yet the ſame Water, as little coheſive as it is, and as eaſily ſeparated from moſt bodies, will cohere firmly with ſome others, and bind them together in the moſt ſolid maſſes; as in the tempering of earth, or aſhes, clay, or powdered bones, &c, with Water, and then dried and burnt, when the maſſes become hard as ſtones, though without the Water they would be mere duſt or powder. Indeed it appears wonderful that Water, which is otherwiſe an almoſt univerſal diſſolvent, ſhould nevertheleſs be a great coagulator.

Some have imagined that Water is incompreſſible, and therefore nonelaſtic; founding their opinion on the celebrated Florentine experiment above mentioned, with the globe of gold; when the Water being, as they ſay, incapable of condenſation, rather than yield, tranſuded through the pores of the metal, ſo that the ball was found wet all over the outſide; till at length making a cleft in the gold, it ſpun out with great vehemence. But the truth of the concluſions drawn from this Florentine experiment has been very juſtly queſtioned; Mr. Canton having proved by accurate experiments, that Water is actually compreſſed even by the weight of the atmoſphere. See COMPRESSION.

Beſides, the diminution of ſize which Water ſuffers when it paſſes to a leſs degree of heat, ſufficiently ſhews that the particles of this fluid are, like thoſe of all other known ſubſtances, capable of approaching nearer together.

Ditch WATER, is often uſed as an object for the microſcope, and ſeldom fails to afford a great variety of animalcules; often appearing of a greeniſh, reddiſh, or yellowiſh colour, from the great multitudes of them. And to the ſame cauſe is to be aſcribed the green ſkim on the ſurface of ſuch Water. Dunghill Water is alſo full of an immenſe crowd of animalcules.

Freſh WATER, is ſaid of that which is inſipid, or without ſalt, and inodorous; being the natural and pure ſtate of the element.

Hard WATER, or _Crude_ WATER, is that in which ſoap does not diſſolve completely or uniformly, but is curdled. The diſſolving power of hard Water is leſs than that of ſoft; and hence its unfitneſs for waſhing, bleaching, dyeing, boiling kitchen vegetables, &c.

The hardneſs of Water may ariſe either from ſalts, or from gas. That which ariſes from ſalts, may be diſcovered and remedied by adding ſome drops of a ſolution of fixed alkali; but the latter by boiling, or expoſure to the open air.

Spring Waters are often hard; but river Water ſoft. Hard Waters are remarkably indiſpoſed to corrupt;

they even preſerve putreſcible ſubſtances for a conſiderable length of time: hence they ſeem to be beſt fitted for keeping at ſea, eſpecially as they are ſo eaſily ſoftened by a little alkaline ſalt.

Putrid WATER, is that which has acquired an offenſive ſmell and taſte by the putreſcence of animal or vegetable ſubſtances contained in it. This kind of Water is in the higheſt degree pernicious to the human frame, and capable of bringing on mortal diſeaſes even by its ſmell. Quicklime put into water is uſeful to preſerve it longer ſweet; or even expoſure to the air in broad ſhallow veſſels. And putrid Water may be in a great meaſure ſweetened, by paſſing a current of freſh air through it, from bottom to top.

Rain WATER may be conſidered as the pureſt diſtilled Water, but impregnated during its paſſage through the air with a conſiderable quantity of phlogiſtic and putreſcent matter; whence it is ſuperior to any other in fertilizing the earth. Hence alſo it is inferior for domeſtic purpoſes to ſpring or river Water, even if it could be readily procured: but ſuch as is gotten from ſpouts placed below the roofs of houſes, the common way of procuring it in this country, is evidently very impure, and becomes putrid in a ſhort time.

River or _Running_ WATER, is next in purity to ſnow or diſtilled water; and for domeſtic purpoſes ſuperior to both, in having leſs putreſcent matter, and more fixed air. That however is much the pureſt that runs over a clean rocky or ſtony bottom.

River Waters generally putrefy ſooner than thoſe of ſprings. During the putrefaction, they throw off a part of their heterogeneous matter, and at length become ſweet again, and purer than at firſt; after which they commonly preſerve a long time: this is remarkably the caſe with the Thames Water, taken up about London; which is commonly uſed by ſeamen, in their voyages.

Salt WATER, ſuch as has much ſalt in it, ſo as to be ſenſible to the taſte.

i _Sea_ WATER, or Water of the ſea, is an aſſemblage of bodies, in which Water can ſcarce be ſaid to have the principal part: it is an univerſal colluvies of all the bodies in nature, ſuſtained and kept ſwimming in Water as a vehicle: being a ſolution of common ſalt, ſal catharticus amarus, a ſelenitic ſubſtance, and a compound of muriatic acid with magneſia, mixed together in various proportions. It may be freſhened by ſimple diſtillation without any addition, and thus it has ſometimes been uſeful in long voyages at ſea. Sea Water by itſelf has a purgative quality, owing to the ſalts it contains; and has been greatly recommended in ſcrophulous diſorders.

Sea Water is about 3 parts in 100 heavier than common Water; and its temperature at great depths is from 34 to 40 degrees; but near the ſurface it follows more nearly the temperature of the air.

Snow WATER, is the pureſt of all the common Waters, when the ſnow has been collected pure. Kept in a warm place, in clean glaſs veſſels, not cloſely ſtopped, but covered from duſt, &c, ſnow water becomes in time putrid; though in well-ſtopped bottles it remains unaltered for ſeveral years. But diſtilled Water ſuffers no alteration in either circumſtance

Spring WATER is commonly impregnated with a
ſmall

small portion of imperfect neutral salt, extracted from the different strata through which it percolates. Some contain a vast quantity of stony matter, which they deposit as they run along, and thus form masses of stone; sometimes incrustating various animal and vegetable matters, which they are therefore said to petrify. Spring-Water is much used for domestic purposes, and on account of its coolness is an agreeable drink; but on account of its being usually somewhat hard, is inferior to that which has run for a considerable way in a channel.

Spring water arises from the rain, and from the mists and moisture in the atmosphere. These falling upon hills and other parts of the earth, soak into the ground, and pass along till they find a vent out again, in the form of a spring.

WATER-*Bellows*, in Mechanics, are bellows, for blowing air into furnaces, that are worked by the force of water.

WATER-*Clock*. See CLEPSYDRA.

WATER-*Engine*, an engine for extinguishing fires; or any engine to raise water; or any engine moved by the force of Water. See ENGINE, and STEAM-*Engine*.

WATER-*Gage*, an instrument for measuring the depth or quantity of any water. See GAGE.

WATER-*Level*, is the true level which the surface of still Water takes, and is the truest of any.

WATER-*Logged*, in Sea-Language, denotes the state of a ship when, by receiving a great quantity of Water into her hold, by leaking, &c, she has become heavy and inactive upon the sea, so as to yield without resistance to the effort of every wave rushing over her deck.

WATER-*Machine*. See MACHINE.

WATER *Measure*. Salt, sea-coal, &c, while on board vessels in the pool, or river, are measured with the corn-bushel heaped up; or else 5 striked pecks are allowed to the bushel. This is called Water-measure; and it exceeds Winchester-measure by about 3 gallons in the bushel.

WATER-*Microscope*. See MICROSCOPE.

WATER-*Mill*. See MILL.

Motion of WATER, in Hydraulics. The theory of the motion of running Water is one of the principal objects of hydraulics, and to which many eminent mathematicians have paid their attention. But it were to be wished that their theories were more consistent with each other, and with experience. The inquisitive reader may consult Newton's Principia, lib. 2, pr. 36, with the comment. Dan. Bernoulli's Hydrodynamica. J. Bernoulli, Hydraulica, Oper. tom. 4, pa. 389. Dr. Jurin, in the Philos. Transf. num. 452, or Abridg. vol. 8, pa. 282. Gravesande, Physic. Elem. Mathem. lib. 3, par. 2. Maclaurin's Flux. art. 537. Poleni de Castellis, Ximenes, D'Alembert, Bossu, Buat, and many others.

But notwithstanding the labours of all these eminent authors, this intricate subject still remains in a great measure obscure and uncertain. Even the simple case of the motion of running water, when it issues from a hole in the bottom of a vessel, has never yet been determined, so as to give universal satisfaction to the learned. On this head, it is now pretty generally allowed,

that the velocity of the issuing stream, is equal to that which a heavy body acquires by falling through the height of the fluid above the hole, as may be demonstrated by theory: but in practice, the quantity of the effluent Water is much less than what is given by this theory; owing to the obstruction to the motion in the hole, partly from the sides of it, and partly from the different directions of the parts of the Water in entering it, which thence obstruct each other's motion. And this obstruction, and the diminution in the quantity of Water run out, is still the more in proportion as the hole is the smaller; in such sort, that when the hole is very small, the quantity is diminished in the ratio of $\sqrt{2}$ to 1 very nearly, which is the ratio of the greatest diminution; and for larger holes, the diminution is always less and less. This fact is ascertained, or admitted by Newton, and all the other philosophers abovementioned, with some small variations.

That the velocity of the Water in the hole, or at least some part of it, as that for example in the middle of the stream, is equal to that abovementioned, is even evinced by experiment, by directing the stream either sideways, or upwards: for in the former case, it is found to range upon an horizontal plane, a distance that just answers to that velocity, by the nature of projectiles; and in the latter case, the jet rises nearly to the height of the Water in the vessel; which it could not do, if its velocity were not equal to that acquired by the free descent of a body through that height. Hence it is evident then, that the particles of the Water, which are in the hole at the same moment of time, do not all burst out with the same velocity; and, in fact, the velocity is found to decrease all the way from the middle of the hole, where it is greatest, towards the side or edge, where it is the least.

At a small distance from the hole, the diameter of the vein of Water is much less than that of the hole. Thus, if the diameter of the hole be 1, the diameter of the vein of Water just without it, will be $\frac{21}{25}$, or 0.84, according to Newton's measure, who first observed this phenomenon; and according to Poleni's measure 0.78 nearly.

By the experiments of Buat (Principes d'Hydraulique), the quantity by theory is to that by experiment, for a small hole made in the thin side of a reservoir, as 8 to 5. When a short pipe is added to the hole outwards, of the length of two or three times its diameter, that ratio is as 16 to 13. And when the short pipe is all within side the vessel, as in the margin, the same ratio becomes that of 3 to 2. Poleni also found that the quantity of Water flowing through a pipe or tube, was much greater than that through a hole of the same diameter in the thin side or bottom of the vessel, the height of the head of Water above each being the same. See also many other curious circumstances in Buat's Principes above mentioned.

Some authors give this rule for finding the height due to the velocity in a flat orifice, or a medium among all the parts of it, such that this medium velocity being drawn into the area of the hole, shall give the quantity per second that runs through: viz, let *A* denote the area

area of the furface of the Water in the veffel, a the area of the orifice by which the Water iffues, and H the height of the Water above the orifice; then, $2A - a : A :: H : h$, the height due to the medium velocity, or the height from which a body muft freely defcend, by the force of gravity, to acquire that mean velocity.

Authors are not yet agreed as to the force with which a vein of Water, fpouting from a round hole in the fide of a veffel, preffes upon a plane directly oppofed to the motion of the vein. Moft authors agree, that the preffure of this vein, flowing uniformly, ought to be equal to the weight of a cylinder of Water, whofe bafe is equal to the hole through which the Water flows, and its height equal to the height of the Water in the veffel above the hole. The experiments made by Mariotte, and others, feem to countenance this opinion. But Dan. Bernoulli rejects it, and eftimates this preffure by the weight of a column of the fluid, whofe diameter is equal to the contracted vein (according to Newton's obfervation abovementioned), and the height of which is equal to double the altitude due to the real velocity of the fpouting Water; and this preffure is alfo equal to the force of repulfion, arifing from the reaction of the fpouting Water upon the veffel. The ingenious author remarks that he fpeaks only of fingle veins of Water, the whole of which are received by the planes upon which they prefs; for as to the preffures exerted by fluids furrounding the bodies they prefs upon, as the wind, or a river, the cafe is different, though confounded with the former by writers on this fubject. Hydrodynamica, pa. 289.

Another rule however had been adopted by the Academicians of Paris, who made a number of experiments to confirm or eftablifh it. Hift. Acad. Paris, ann. 1679, fect. 3, cap. 5.

D. Bernoulli, on the other hand, thinks his own theory fufficiently eftablifhed by the experiments he relates; for the particulars of which fee the Acta Petropolitana, vol. 8, pa. 122.

This ingenious author is of opinion that his theory of the quantity of the force of repulfion, exerted by a vein of fpouting Water, might be ufefully applied to move fhips by pumping; and he thinks the motion produced by this repulfive force would fall little, if at all, fhort of that produced by rowing. He has given his reafons and computations at length in his Hydrodynamica, pa. 293 &c.

This fcience of the preffures exerted by Water or other fluids in motion, is what Bernoulli calls *Hydraulico-ftatica*. This fcience differs from hydroftatics, which confiders only the preffure of Water and other fluids at reft; whereas hydraulico-ftatics confiders the preffure of Water in motion. Thus the preffure exerted by Water moving through pipes, upon the fides of thofe pipes, is an hydraulico-ftatical confideration, and has been erroneoufly determined by many, who have given no other rules in thefe cafes, but fuch as are applicable only to the preffure of fluids at reft. See Hydrodynam. pa. 256 &c.

WATER-*Poife*. See HYDROMETER, and AREOMETER.

Dr. Hook contrived a Water-poife, which may be of good fervice in examining the purity &c of Water. It confifts of a round glafs ball, like a bolt head, about 3 inches diameter, with a narrow ftem or neck, the 24th of an inch in diameter; which being poifed with red lead, fo as to make it but little heavier than pure fweet Water, and thus fitted to one end of a fine balance, with a counterpoife at the other end; upon the leaft addition of even the 2000th part of falt to a quantity of Water, half an inch of the neck will emerge above the water. Philof. Tranf. num. 197.

Raifing of WATER, in Hydraulics. The great ufe of raifing Water by engines for the various purpofes of life, is well known. Machines have in all ages been contrived with this view; a detail of the beft of which, with the theory of their conftruction, would be very curious and inftructive. M. Belidor has executed this in part in his Architecture Hydraulique. Dr. Defaguliers has alfo given a defcription of feveral engines to raife Water, in his Courfe of Experimental Philofophy, vol. 2, and there are feveral other fmaller works of the fame kind.

Engines for raifing Water are either fuch as throw it up with a great velocity, as in jets; or fuch as raife it from one place to another by a gentle motion. For the general theory of thefe engines, fee Bernoulli's Hydrodynamica.

Defaguliers has fettled the maximum of engines for raifing water, thus: a man with the beft Water engine cannot raife above one hogfhead of Water in a minute, 10 feet high, to hold it all day; but he can do almoft twice as much for a minute or two.

WATER-*Spout*. See SPOUT.

WATER-*Wheel*, an engine for raifing Water in great quantity out of a deep well, &c. See PERSIAN-*Wheel*.

WATER *Works*. See *Raifing of* WATER.

WAVE, in Phyfics, a volume of water elevated by the action of the wind &c, upon its furface, into a ftate of fluctuation, and accompanied by a cavity. The extent from the bottom or loweft point of one cavity, and acrofs the elevation, to the bottom of the next cavity, is the breadth of the Wave.

Waves are confidered as of two kinds, which may be diftinguifhed from one another by the names of natural and accidental Waves. The natural Waves are thofe which are regularly proportioned in fize to the ftrength of the wind which produces them. The accidental Waves are thofe occafioned by the wind's reacting upon itfelf by repercuffion from hills or high fhores, and by the dafhing of the Waves themfelves, otherwife of the natural kind, againft rocks and fhoals; by which means thefe Waves acquire an elevation much above what they can have in their natural ftate.

Mr. Boyle proved, by numerous experiments, that the moft violent wind never penetrates deeper than 6 feet into the water; and it feems a natural confequence of this, that the water moved by it can only be elevated to the fame height of 6 feet from the level of the furface in a calm; and thefe 6 feet of elevation being added to the 6 of excavation, in the part from whence that water fo elevated was raifed, fhould give 12 feet for the utmoft elevation of a Wave. This is a calculation that does great honour to its author; as many experiments and observations obfervations have proved that it is very nearly true in deep feas, where the Waves are purely natural, and have no accidental caufes to render them larger than their juft proportion.

It is not to be underftood however, that no Wave of the fea can rife more than 6 feet above its natural level in open and deep water; for Waves vaftly higher than thefe are formed in violent tempefts in the great feas. Thefe however are not to be accounted Waves in their natural ftate, but as compound Waves formed by the union of many others; for in thefe wide plains of water, when one Wave is raifed by the wind, and would elevate itfelf up to the exact height of 6 feet, and no more, the motion of the water is fo great, and the fucceffion of Waves fo quick, that while this is rifing, it receives into it feveral other Waves, each of which would have been at the fame height with itfelf; thefe run into the firft Wave one after another, as it is rifing; by which means its rife is continued much longer than it naturally would have been, and it becomes accumulated to an enormous fize. A number of thefe complicated Waves rifing together, and being continued in a long fucceffion by the continuation of the ftorm, make the Waves fo dangerous to fhips, which the failors in their phrafe call mountains high.

Different Waves do not difturb one another when they move in different directions. The reafon is, that whatever figure the furface of the water has acquired by the motion of the Waves, there may in that be an elevation and depreffion; as alfo fuch a motion as is required in the motion of a Wave.

Waves are often produced by the motion of a tremulous body, which alfo expand themfelves circularly, though the body goes and returns in a right line; for the water which is raifed by the agitation, defcending, forms a cavity, which is every where furrounded with a rifing.

The Motion of the WAVES, makes an article in the Newtonian philofophy; that author having explained their motions, and calculated their velocity from mathematical principles, fimilar to the motion of a pendulum, and to the reciprocation of water in the two legs of a bent and inverted fyphon or tube.

His propofition concerning fuch canal or tube is the 44th of the 2d book of his Principia, and is this: "If water afcend and defcend alternately in the erected legs of a canal or pipe; and a pendulum be conftructed, whofe length between the point of fufpenfion and the centre of ofcillation, is equal to half the length of the water in the canal; then the water will afcend and defcend in the fame times in which the pendulum ofcillates." The author hence infers, in prop. 45, that the velocity of Waves is in the fubduplicate ratio of their breadths; and in prop. 46, he proceeds "To find the velocity of Waves," as follows: "Let a pendulum be conftructed, whofe length between the point of fufpenfion and the centre of ofcillation is equal to the breadth of the Waves; and in the time that the pendulum will perform one fingle ofcillation, the Waves will advance forward nearly a fpace equal to their breadth. That which I call the breadth of the Waves, is the tranfverfe meafure lying between the deepeft part of the hollows, or between the tops of the ridges.

Let ABCDEF reprefent the furface of ftagnant water afcending and defcending in fucceffive Waves; alfo let

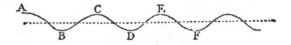

A, C, E, &c, be the tops of the Waves; and B, D, F, &c, the intermediate hollows. Becaufe the motion of the Waves is carried on by the fucceffive afcent and defcent of the water, fo that the parts of it, as A, C, E, &c, which are higheft at one time, become loweft immediately after; and becaufe the motive force, by which the higheft parts defcend and the loweft afcend, is the weight of the elevated water, that alternate afcent and defcent will be analogous to the reciprocal motion of the water in the canal, and obferve the fame laws as to the times of its afcent and defcent; and therefore (by prob. 44, above mentioned) if the diftances between the higheft places of the Waves A, C, E, and the loweft B, D, F, be equal to twice the length of any pendulum, the higheft parts A, C, E, will become the loweft in the time of one ofcillation, and in the time of another ofcillation will afcend again. Therefore between the paffage of each Wave, the time of two ofcillations will intervene; that is, the Wave will defcribe its breadth in the time that the pendulum will ofcillate twice; but a pendulum of 4 times that length, and which therefore is equal to the breadth of the Waves, will juft ofcillate once in that time. *Q. E. I.*

" *Corol.* 1. Therefore Waves, whofe breadth is equal to $39\frac{1}{8}$ inches, or $3\frac{2}{7}\frac{5}{8}$ feet, will advance through a fpace equal to their breadth in one fecond of time; and therefore in one minute they will go over a fpace of $195\frac{5}{8}$ feet; and in an hour a fpace of 11737 feet, nearly, or 2 miles and almoft a quarter.

" *Corol.* 2. And the velocity of greater or lefs Waves, will be augmented or diminifhed in the fubduplicate ratio of their breadth.

" Thefe things (Newton adds) are true upon the fuppofition, that the parts of water afcend or defcend in a right line; but in fact, that afcent and defcent is rather performed in a circle; and therefore I propofe the time defined by this propofition as only near the truth."

Stilling WAVES *by means of Oil.* This wonderful property, though well known to the Ancients, as appears from the writings of Pliny, was for many ages either quite unnoticed, or treated as fabulous by fucceeding philofophers. Of late it has, by means of Dr. Franklin, again attracted the attention of the learned; though it appears, from fome anecdotes, that feafaring people have always been acquainted with it. In Martin's defcription of the Weftern Iflands of Scotland, we have the following paffage: " The fteward of Kilda, who lives in Pabbay, is accuftomed, in time of a ftorm, to tie a bundle of puddings, made of the fat of feafowl, to the end of his cable, and lets it fall into the fea behind his rudder. This, he fays, hinders the Waves from breaking, and calms the fea." Mr. Pennant, in his Britifh Zoology, vol. iv, under the article

Seal, takes notice, that when these animals are devouring a very oily fish, which they always do under water, the Waves above are remarkably smooth; and by this mark the fishermen know where to find them. Sir Gilbert Lawson, who served long in the army at Gibraltar, assured Dr. Franklin, that the fishermen in that place are accustomed to pour a little oil on the sea, in order to still its motion, that they may be enabled to see the oysters lying at its bottom, which are there very large, and which they take up with a proper instrument. A similar practice obtains among fishermen in various other parts, and Dr. Franklin was informed by an old sea-captain, that the fishermen of Lisbon, when about to return into the river, if they saw too great a surf upon the bar, would empty a bottle or two of oil into the sea, which would suppress the breakers, and allow them to pass freely.

The Doctor having revolved in his mind all these pieces of information, became impatient to try the experiment himself. At last having an opportunity of observing a large pond very rough with the wind, he dropped a small quantity of oil upon it. But having at first applied it on the lee side, the oil was driven back again upon the shore. He then went to the windward side, and poured on about a tea-spoon full of oil; this produced an instant calm over a space several yards square, which spread amazingly, and extended itself gradually till it came to the lee-side; making all that quarter of the pond, perhaps half an acre, as smooth as glass. This experiment was often repeated in different places, and always with success. Our author accounts for it in the following manner:

" There seems to be no natural repulsion between water and air, to keep them from coming into contact with each other. Hence we find a quantity of air in water; and if we extract it by means of the air pump, the same water again exposed to the air will soon imbibe an equal quantity.—Therefore air in motion, which is wind, in passing over the smooth surface of water, may rub as it were upon that surface, and raise it into wrinkles; which, if the wind continues, are the elements of future Waves. The smallest Wave once raised does not immediately subside and leave the neighbouring water quiet; but in subsiding raises nearly as much of the water next to it, the friction of the parts making little difference. Thus a stone dropped into a pool raises first a single Wave round itself, and leaves it, by sinking to the bottom; but that first Wave subsiding raises a second, the second a third, and so on in circles to a great extent.

" A small power continually operating, will produce a great action. A finger applied to a weighty suspended bell, can at first move it but little; if repeatedly applied, though with no greater strength, the motion increases till the bell swings to its utmost height, and with a force that cannot be resisted by the whole strength of the arm and body. Thus the small first raised Waves being continually acted upon by the wind, are, though the wind does not increase in strength, continually increased in magnitude, rising higher and extending their bases, so as to include a vast mass of water in each Wave, which in its motion acts with great violence. But if there be a mutual repulsion between the particles

of oil, and no attraction between oil and water, oil dropped on water will not be held together by adhesion to the spot whereon it falls; it will not be imbibed by the water; it will be at liberty to expand itself; and it will spread on a surface that, besides being smooth to the most perfect degree of polish, prevents, perhaps by repelling the oil, all immediate contact, keeping it at a minute distance from itself; and the expansion will continue, till the mutual repulsion between the particles of the oil is weakened and reduced to nothing by their distance.

" Now I imagine that the wind blowing over water thus covered with a film of oil cannot easily catch upon it, so as to raise the first wrinkles, but slides over it, and leaves it smooth as it finds it. It moves the oil a little indeed, which being between it and the water, serves it to slide with, and prevents friction, as oil does between those parts of a machine that would otherwise rub hard together. Hence the oil dropped on the windward side of a pond proceeds gradually to leeward, as may be seen by the smoothness it carries with it quite to the opposite side. For the wind being thus prevented from raising the first wrinkles that I call the elements of Waves, cannot produce Waves, which are to be made by continually acting upon and enlarging those elements; and thus the whole pond is calmed.

" Totally therefore we might suppress the Waves in any required place, if we could come at the windward place where they take their rise. This in the ocean can seldom if ever be done. But perhaps something may be done on particular occasions to moderate the violence of the Waves when we are in the midst of them, and prevent their breaking when that would be inconvenient. For when the wind blows fresh, there are continually rising on the back of every great Wave a number of small ones, which roughen its surface, and give the wind hold, as it were, to push it with greater force. This hold is diminished by preventing the generation of those small ones. And possibly too, when a Wave's surface is oiled, the wind, in passing over it, may rather in some degree press it down, and contribute to prevent its rising again, instead of promoting it.

" This, as mere conjecture, would have little weight, if the apparent effects of pouring oil into the midst of Waves were not considerable, and as yet not otherwise accounted for.

" When the wind blows so fresh, as that the Waves are not sufficiently quick in obeying its impulse, their tops being thinner and lighter, are pushed forward, broken, and turned over in a white foam. Common Waves lift a vessel without entering it; but these, when large, sometimes break above and pour over it, doing great damage.

" That this effect might in any degree be prevented, or the height and violence of Waves in the sea moderated, we had no certain account; Pliny's authority for the practice of seamen in his time being slighted. But discoursing lately on this subject with his excellency Count Bentinck of Holland, his son the honourable Captain Bentinck, and the learned professor Allemand (to all whom I showed the experiment of smoothing in a windy day the large piece of water at the head of the green

green park), a letter was mentioned which had been received by the Count from Batavia, relative to the saving of a Dutch ship in a storm by pouring oil into the sea."

WAY *of a Ship*, is sometimes used for her wake or track. But more commonly the term is understood of the course or progress which she makes on the water under sail: thus, when she begins her motion, she is said to be *under Way;* when that motion increases, she is said to have *fresh Way* through the water; when she goes apace, they say *she has a good Way;* and the account of her rate of sailing by the log, they call, *keeping an account of her Way.* And because most ships are apt to fall a little to the leeward of their true course; it is customary, in casting up the log-board, to allow something for her *leeward Way,* or *leeway.* Hence also a ship is said to have *head Way,* and *stern-Way.*

WAYWISER, an instrument for measuring the road, or distance travelled; called also PERAMBULATOR, and PEDOMETER. See these two articles.

Mr. Lovell Edgworth communicated to the Society of Arts, &c, an account of a Way-wiser of his invention; for which he obtained a silver medal. This machine consists of a nave, formed of two round flat pieces of wood, 1 inch thick and 8 inches in diameter. In each of the pieces there are cut eleven grooves, ⅝ of an inch wide, and ⅜ deep; and when the two pieces are screwed together, they enclose eleven spokes, forming a wheel of spokes, without a rim: the circumference of the wheel is exactly one pole; and the instrument may be easily taken to pieces, and put up in a small compass. On each of the spokes there is driven a ferril, to prevent them from wearing out; and in the centre of the nave, there is a square hole to receive an axle. Into this hole is inserted an iron or brass rod, which has the thread of a very fine screw worked upon it from one end to the other; upon this screw hangs a nut which, as the rod turns round with the wheel, advances towards the nave of the wheel or recedes from it. The nut does this, because it is prevented from turning round with the axle, by having its centre of gravity placed at some distance below the rod, so as always to hang perpendicularly like a plummet. Two sides of this screw are filed away flat, and have figures engraved upon them, to shew by the progressive motion of the nut, how many circumvolutions of the wheel and its axle have been made: on one side the divisions of miles, furlongs, and poles are in a direct order, and on the other side the same divisions are placed in a retrograde order.

If the person who uses this machine places it at his right hand side, holding the axle loosely in his hands, and walks forward, the wheel will revolve, and the nut advance from the extremity of the rod towards the nave of the wheel. When two miles have been measured, it will have come close to the wheel. But to continue this measurement, nothing more is necessary than to place the wheel at the left hand of the operator; and the nut will, as he continues the course, recede from the axletree, till another space of two miles is measured.

It appears from the construction of this machine, that it operates like circular compasses; and does not, like the common wheel Way-wiser, measure the surface of every stone and molehill, &c, but passes over most of

the obstacles it meets with, and measures the chords only, instead of the arcs of any curved surfaces upon which it rolls.

WEATHER, denotes the state or disposition of the atmosphere, with regard to heat and cold, drought and moisture, fair or foul, wind, rain, hail, frost, snow, fog, &c. See ATMOSPHERE, HAIL, HEAT, FROST, RAIN, &c.

There does not seem in all philosophy any thing of more immediate concernment to us, than the state of the Weather; as it is in, and by means of the atmosphere, that all plants are nourished, and all animals live and breathe; and as any alterations in the density, heat, purity, &c, of that, must necessarily be attended with proportionable ones in the state of these.

The great, but regular alterations, a little change of Weather makes in many parts of inanimate matter, every person knows, in the common instance of barometers, thermometers, hygrometers, &c; and it is owing partly to our inattention, and partly to our unequal and intemperate course of life, that we also, like many other animals, do not feel as great and as regular ones in the tubes, chords, and fibres of our own bodies.

To establish a proper theory of the Weather, it would be necessary to have registers carefully kept in divers parts of the globe, for a long series of years; from whence we might be enabled to determine the directions, breadth, and bounds of the winds, and of the weather they bring with them; with the correspondence between the Weather of divers places, and the difference between one sort and another at the same place. We might thus in time learn to foretell many great emergencies; as, extraordinary heats, rains, frosts, droughts, dearths, and even plagues, and other epidemical diseases, &c.

It is however but very few, and partial registers or accounts of the Weather, that have been kept. The Royal Society, the French Academy, and a few particular philosophers, have at times kept such registers as their fancies have dictated, but at no time a regular and correspondent series in many different places, at the same time, followed with particular comparisons and deductions from the whole, &c. The most of what has been done in this way, is as follows: The volumes of the Philosophical Transactions from year to year; the same, for instructions and examples pertaining to the subject, vol. 65, part 2, art. 16; Eras. Bartholin has observations of the Weather for every day in the year 1671: Mr. W. Merle made the like at Oxford, for 7 years: Dr. Plot did the same at the same place, for the year 1684: Mr. Hillier, at Cape Corse, for the years 1686 and 1687: Mr. Hunt and others at Gresham College, for the years 1695 and 1696: Dr. Derham at Upminster in Essex, for the years 1691, 1692, 1697, 1698, 1699, 1703, 1704, 1705: Mr. Townley, in Lancashire, in 1697, 1698: Mr. Cunningham, at Emin in China, for the years 1698, 1699, 1700, 1701: Mr. Locke, at Oats in Essex, 1692: Dr. Scheuchzer, at Zurich, 1708; and Dr. Tilly, at Pisa, the same year: Professor Toaldo, at Padua, for many years: Mr. T. Barker, at Lyndon, in Rutland, for many years in the Philos. Transf.: Mr. Dalton for Kendal, and Mr. Crosthwaite for Keswick, in the years 1788,

1788, 1789, 1790, 1791, 1792, &c; and several others. The register now kept, for many years, in the Philof. Tranf. contains an account, two times every day, of the thermometer, barometer, hygrometer, quantity of rain, direction and strength of the wind, and appearance of the atmosphere, as to fair, cloudy, foggy, rainy, &c. And if similar registers were kept in many other parts of the globe, and printed in such-like public Transactions, they might readily be consulted, and a proper use made of them, for establishing this science on the true basis of experiment.

From many experiments, some general observations have been made, as follow: That barometers generally rise and fall together, even at very distant places, and a consequent conformity and similarity of Weather; but this is the more uniformly so, as the places are nearer together, as might be expected. That the variations of the barometer are greater, as the places are nearer the pole; thus, for instance, the mercury at London has a greater range by 2 or 3 lines than at Paris; and at Paris, a greater than at Zurich; and at some places near the equator, there is scarce any variation at all. That the rain in Switzerland and Italy is much greater in quantity, for the whole year, than in Effex; and yet the rains are more frequent, or there are more rainy days, in Effex, than at either of those places. That cold contributes greatly to rain; and this apparently by condensing the suspended vapours, and so making them descend: thus, very cold months, or seasons, are commonly followed immediately by very rainy ones; and cold summers are always wet ones. That high ridges of mountains, as the Alps, and the snows with which they are covered, not only affect the neighbouring places by the colds, rains, vapours, &c, which they produce; but even distant countries, as England, often partake of their effects. See a collection of ingenious and meteorological observations and conjectures, by Dr. Franklin, in his Experiments, &c, pa. 182, &c. Also a Meteorological Register kept at Mansfield Woodhouse, from 1784 to 1794, Nottingham 1795, 8vo; and Kirwin's ingenious papers on this subject in the Transactions of the Irish Academy, vol. 5. See also the articles EVAPORATION, RAIN, and WIND.

Other Prognostics and Observations, are as follow:

That a thick dark sky, lasting for some time, without either sun or rain, always becomes first fair, and then foul, i. e. it changes to a fair clear sky, before it turns to rain. And the reason is obvious: the atmosphere is replete with vapours which, though sufficient to reflect and intercept the sun's rays from us, yet want density to descend; and while the vapours continue in the same state, the Weather will do so too: accordingly, such Weather is commonly attended with moderate warmth, and with little or no wind to disturb the vapours, and a heavy atmosphere to sustain them; the barometer being commonly high: but when the cold approaches, and by condensing the vapours drives them into clouds or drops, then way is made for the sun beams; till the same vapours, by farther condensation, be formed into rain, and fall down in drops.

That a change in the warmth of the Weather is followed by a change in the wind. Thus, the northerly and southerly winds, though commonly accounted the *causes* of cold and warm Weather, are really the *effects* of the cold or warmth of the atmosphere; of which Dr. Derham assures us he had so many confirmations, that he makes no doubt of it. Thus, it is common to see a warm southerly wind suddenly changed to the north, by the fall of snow or hail; or to see the wind, in a cold frosty morning, north, when the sun has well warmed the air, wheel towards the south; and again turn northerly or easterly in the cold evening.

That most vegetables expand their flowers and down in sunshiny Weather: and towards the evening, and against rain, close them again; especially at the beginning of their flowering, when their seeds are tender and sensible. This is visible enough in the down of Dandelion, and other downs: and eminently so in the flowers of pimpernel; the opening and shutting of which make what is called the countryman's *Weatherwiser*, by which he foretels the Weather of the following day. The rule is, when the flowers are close shut up, it betokens rain, and foul Weather; but when they are spread abroad, fair Weather.

The stalk of trefoil, lord Bacon observes, swells against rain, and grows more upright: and the like may be observed, though less sensibly, in the stalks of most other plants. He adds, that in the stubble fields there is found a small red flower, called by the country people pimpernel, which opening in a morning, is a sure indication of a fine day.

It is very conceivable that vegetables should be affected by the same causes as the Weather, as they may be considered as so many hygrometers and thermometers, consisting of an infinite number of tracheæ, or air-vessels; by which they have an immediate communication with the air, and partake of its moisture, heat, &c.

Hence it is, that all wood, even the hardest and most solid, swells in moist Weather; the vapours easily insinuating into the pores, especially of the lighter and drier kinds. And hence is derived a very extraordinary use of wood, viz, for breaking rocks or millstones. The method at the quarries is this: Having cut a rock into the form of a cylinder, the workmen divide it into several thinner cylinders, of horizontal courses, by making holes at proper distances round the great one; into these holes they drive pieces of sallow wood, dried in an oven; these in moist Weather, imbibing the humidity from the air, swell, and acting like wedges they break or cleave the rock into several flat stones. And, in like manner, to separate large blocks of stone in the quarry, they wedge such pieces of wood into holes, forming the block into the intended shape, and then pour water upon the wedges, to produce the effect more immediately.

WEATHER-*Glasses*, are instruments contrived to shew the state of the atmosphere, as to heat, cold, moisture, weight, &c; and so to measure the changes that take place in those respects; by which means we are enabled to predict the alteration of Weather, as to rain, wind, frost, &c.

Under the class of Weather-glasses, are comprehended barometers, thermometers, hygrometers, manometers, and anemometers.

WEDGE,

WEDGE, in Geometry, is a solid having a rectangular base, and two of its oppofite fides ending in an acies or edge. Thus, AB is the rectangular base ; and DC the edge ; a perpendicular CE, from the edge to the base, is the height of the Wedge. When the length of the edge DC is equal to the length of the base BF, which is the most common form of it, the Wedge is equal to half a rectangular prism of the fame base AB and height EC ; or it is then a whole triangular prism, having the triangle BCG for its base, and AG or DC for its height. If the edge be more or lefs than AG, its folid content will be more or lefs. But, in all cafes of the Wedge, the following is a general rule for finding the content of it, viz,

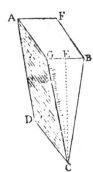

To twice the length of the base add the length of the edge, multiply the fum by the breadth of the base, and the product by the height of the Wedge; then ⅓ of the laft product will be the folid content.

That is, $\overline{2AG + DC} \times AF \times \frac{1}{3} EC$ = the content. See this rule demonftrated, and illuftrated with examples, in my Menfuration, p. 191, 2d edition.

WEDGE, in Mechanics, one of the five mechanical powers, or fimple engines; being a geometrical Wedge, or very acute triangular prism, applied to the fplitting of wood, or rocks, or raifing great weights.

The Wedge is made of iron, or fome other hard matter, and applied to the raifing of vaft weights, or feparating large or very firm blocks of wood or ftone, by introducing the thin edge of the Wedge, and driving it in by blows ftruck upon the back by hammers or mallets.

The Wedge is the moft powerful of all the fimple machines, having an almoft unlimited and double advantage over all the other fimple mechanical powers ; both as it may be made vaftly thin, in proportion to its height ; in which confifts its own natural power ; and as it is urged by the force of percuffion, or of fmart blows, which is a force incomparably greater than any mere dead weight or preffure, fuch as is employed upon other machines. And accordingly we find it produces effects vaftly fuperior to thofe of any other power whatever ; fuch as the fplitting and raifing the largeft and hardeft rocks; or even the raifing and lifting the largeft fhip, by driving a Wedge below it ; which a man can do by the blow of a mallet: and thus the fmall blow of a hammer, on the back of a Wedge, appears to be incomparably greater than any mere preffure, and will overcome it.

To the Wedge may be referred all edge-tools, and tools that have a fharp point, in order to cut, cleave, flit, fplit, chop, pierce, bore, or the like ; as knives, hatchets, fwords, bodkins, &c.

In the Wedge, the friction againft the fides is very great, at leaft equal to the force to be overcome ; becaufe the Wedge retains any pofition to which it is driven ; and therefore the refiftance is at leaft doubled by the friction.

Authors have been of various opinions concerning the principle from whence the Wedge derives its power. Ariftotle confiders it as two levers of the firft kind, inclined towards each other, and acting oppofite ways. Guido Ubaldi, Merfenne, &c, will have them to be levers of the fecond kind. But De Lanis fhews, that the Wedge cannot be reduced to any lever at all. Others refer the Wedge to the inclined plane. And others again, with De Stair, will hardly allow the Wedge to have any force at all in itfelf; afcribing much the greateft part to the mallet which drives it.

The doctrine of the force of the Wedge, according to fome writers, is contained in this propofition : " If a power directly applied to the head of a Wedge, be to the refiftance to be overcome, as the breadth of the back GB, is to the height EC ; then the power will be equal to the refiftance ; and if increafed, it will overcome it."

But Defaguliers has proved that, when the refiftance acts perpendicularly againft the fides of the Wedge, the power is to the whole refiftance, as the thicknefs of the back is to the length of both the fides taken together. And the fame proportion is adopted by Wallis (Op. Math. vol. 1, p. 1016), Keill (Intr. ad Ver. Phyf), Gravefande (Elem. Math. Lib. 1, cap. 14), and by almoft all the modern mathematicians. Gravefande indeed diftinguifhes the mode in which the Wedge acts, into two cafes, one in which the parts of a block of wood, &c, are feparated farther than the edge has penetrated to, and the other in which they have not feparated farther : In his Scholium de Ligno findendo (ubi fupra), he obferves, that when the parts of the wood are feparated before the Wedge, the equilibrium will be when the force by which it is pufhed in, is to the refiftance of the wood, as the line DE drawn from the middle of the bafe to the fide of the Wedge but perpendicular to the feparated fide of the wood continued FG, is to the height of the Wedge DC; but when the parts of the wood are feparated no farther than the Wedge is driven in, the equilibrium will be, when the power is to the refiftance, as the half bafe AD, is to its fide AC.

Mr. Fergufon, in eftimating the proportion of equilibrium in the two cafes laft mentioned by Gravefande, agrees with this author, and other modern philofophers, in the latter cafe; but in the former he contends, that when the wood cleaves to any diftance before the Wedge, as it generally does, then the power impelling the Wedge, will be to the refiftance of the wood, as half its thicknefs, is to the length of either fide of the cleft, eftimated from the top or acting part of the Wedge : for, fuppofing the Wedge to be lengthened down to the bottom of the cleft, the power will be to the refiftance, as half the thicknefs of the Wedge is to the length of either of its fides. See Fergufon's Lect. p. 40, &c, 4to. See alfo Defagu. Exp. Phil. vol. 1, p. 107 ; and Ludlam's Effay on the Power of the Wedge, printed in 7,0 ; &c.

The generally acknowledged property of the Wedge, and the fimpleft way of demonftrating it, feem to be the following : When a Wedge is kept in equilibrio, the power acting againft the back, is to the force acting

prepe

Perpendicularly against either side, as the breadth of the back AB, is to the length of the side AC or BC. —*Demonstra.* For any three forces which sustain one another in equilibrio, are as the corresponding sides of a triangle that are drawn perpendicular to the directions in which the forces act. But AB is perpendicular to the force acting on the back, to drive the Wedge forward; and the sides AC and BC are perpendicular to the forces acting upon them; therefore the three forces are as the said lines AB, AC, BC.

Hence, the thinner a Wedge is, the greater is its effect, in splitting any body, or in overcoming any resistance against the side of the Wedge.

WEDNESDAY, the 4th day of the week, formerly consecrated by the inhabitants of the northern nations to Woden or Oden; who, being reputed the author of magic and inventor of all the arts, was thought to answer to the Mercury of the Greeks and Romans, in honour of whom the same day was by them called *dies Mercurii*; and hence it is denoted by astronomers by the character of Mercury ☿.

WEEK, a division of time that comprises seven days.

The origin of this division of Weeks, or of computing time by sevenths, is much controverted. It has often been thought to have taken its rise from the four quarters or intervals of the moon, between her changes of phases, which, being about 7 days distant, gave occasion to the division: but others more probably from the seven planets.

Be this however as it may, the division is certainly very ancient. The Syrians, Egyptians, and most of the oriental nations, appear to have used it from the earliest ages: though it did not get footing in the west till brought in by christianity. The Romans reckoned their days not by sevenths, but by ninths; and the ancient Greeks by decads, or tenths; in imitation of which the new French calendar seems to have been framed.

The Jews divided their time by Weeks, of 7 days each, as prescribed by the law of Moses; in which they were appointed to work 6 days, and to rest the 7th, in commemoration of the creation, which being effected in 6 days, God rested on the 7th.

Some authors will even have the use of Weeks, among the other eastern nations, to have proceeded from the Jews; but with little appearance of probability. It is with better reason that others suppose the use of Weeks, among the eastern nations, to be a remnant of the tradition of the creation, which they had still retained with divers others; or else from the number of the planets.

The Jews denominated the days of the Week, the first, second, third, fourth, and fifth; and the sixth day they named the preparation of the sabbath, or 7th day, which answered to our Saturday. And the like method is still kept up by the christian Arabs, Persians, Ethiopians, &c.

The ancient heathens denominated the days of the Week from the seven planets; which names are still mostly retained among the christians of the west: thus, the first day was called *dies solis, sun-day*; the 2d *dies lunæ, moon day*; &c; a practice the more natural on Dion's principle, that the Egyptians took the division of the Week itself from the seven planets.

In fact, the true reason for these denominations seems to be founded in astrology. For the astrologers distributing the government and direction of all the hours in the Week among the seven planets, ♄ ♃ ♂ ☉ ♀ ☿ ☽, so as that the government of the first hour of the first day fell to Saturn, that of the second day to Jupiter, &c, they gave each day the name of the planet which, according to their doctrine, presided over the first hour of it, and that according to the order above stated. So that the order of the planets in the Week, bears little relation to that in which they follow in the heavens: the former being founded on an imaginary power each planet has, in its turn, on the first hour of each day.

Dion Cassius gives another reason for the denomination, drawn from the celestial harmony. For it being observed, that the harmony of the diatessaron, which consists in the ratio of 4 to 3, is of great force and effect in music; it was judged meet to proceed directly from Saturn to the Sun; because, according to the old system, there are three planets between Saturn and the Sun, and 4 from the Sun to the Moon.

Our Saxon ancestors, before their conversion to Christianity, named the seven days of the Week from the Sun and Moon and some of their deified heroes, to whom they were peculiarly consecrated, and representing the ancient gods or planets; which names we received and still retain: Thus, Sunday was devoted to the Sun; Monday to the Moon; Tuesday to Tuisco; Wednesday to Woden; Thursday to Thor, the thunderer; Friday to Friga or Friya or Fræa, the wife of Thor; and Saturday to Seater. And nearly according to this order, the modern astronomers express the days of the Week by the seven planets as below:

☉ Sunday
☾ Monday
♂ Tuesday
☿ Wednesday
♃ Thursday
♀ Friday
♄ Saturday.

In the same order and number also do these obtain in the Hindoo days of the Week. See Kindersley's Specimens of Hindoo Literature, just published, 8vo.

WEIGH, WAY, or WEY, a weight of cheese, wool, &c, containing 256 pounds avoirdupois. Of corn, the Weigh contains 40 bushels; of barley or malt, 6 quarters.

WEIGHT, or *Gravity*, in Physics, a quality in natural bodies, by which they tend downwards toward the centre of the earth. See GRAVITY.

Weight, like gravity, may be distinguished into *absolute*, *specific*, and *relative*.

Newton demonstrates, 1. That the Weights of all bodies, at equal distances from the centre of the earth, are directly proportional to the quantities of matter that each contains: Whence it follows, that the Weights of bodies have no dependence on their shapes or textures; and that all spaces are not equally full of matter.

2. On different parts of the earth's surface, the Weight of the same body is different; owing to the spheroidal figure of the earth, which causes the body on the surface to be nearer the centre in going from the equator toward the poles: and the increase in the Weight is nearly

nearly in proportion to the verfed fine of double the latitude; or, which is the fame thing, to the fquare of the right fine of the latitude : the Weight at the equator to that at the pole, being as 229 to 230; or the whole increafe of Weight from the equator to the pole, is the 229th part of the former.

3. That the Weights of the fame body, at different diftances above the earth, are inverfely as the fquares of the diftances from the centre. So that, a body at the diftance of the moon, which is 60 femidiameters from the earth's centre, would weigh only the 3600th part of what it weighs at the earth's furface.

4. That at different diftances within the earth, or below the furface, the weights of the fame body are directly as the diftances from the earth's centre : fo that, at half way toward the centre, a body would weigh but half as much, and at the very centre it would be no Weight at all.

5. A body immerfed in a fluid, which is fpecifically lighter than itfelf, lofes fo much of its Weight, as is equal to the Weight of a quantity of the fluid of the fame bulk with itfelf. Hence, a body lofes more of its weight in a heavier fluid than in a lighter one; and therefore it weighs more in a lighter fluid than in a heavier one.

The Weight of a cubic foot of pure water, is 1000 ounces, or 62½ pounds, avoirdupois. And the Weights of the cubic foot of other bodies, are as fet down under the article *Specific* GRAVITY.

In the Philof. Tranf. (number 458, p. 457 &c) is contained fome account of the analogy between English Weights and meafures, by Mr. Barlow. He ftates, that anciently the cubic foot of water was affumed as a general ftandard for liquids. This cubic foot, of 62½ lb, multiplied by 32, gives 2000, the weight of a ton : and hence 8 cubic feet of water made a hogfhead, and 4 hogfheads a tun, or ton, in capacity and denomination, as well as Weight.

Dry meafures were raifed on the fame model. A bufhel of wheat, affumed as a general ftandard for all forts of grain, alfo weighed 62½ lb. Eight of thefe bufhels make a quarter, and 4 quarters, or 32 bufhels, a ton Weight. Coals were fold by the chaldron, fuppofed to weigh a ton, or 2000 pounds; though in reality it weighs perhaps upwards of 3000 pounds.

Hence a ton in Weight is the common ftandard for liquids, wheat, and coals. Had this analogy been adhered to, the confufion now complained of would have been avoided.—It may reafonably be fuppofed that corn and other commodities, both dry and liquid, were firft fold by Weight; and that meafures, for convenience, were afterwards introduced, as bearing fome analogy to the Weights before ufed.

WEIGHT, *Pondus*, in Mechanics, denotes any thing to be raifed, fuftained, or moved by a machine; or any thing that in any manner refifts the motion to be produced.

In all machines, there is a natural, and fixed ratio between the Weight and the moving power : and if they be fuch as to balance each other in equilibrio, and then the machine be put in motion by any other force; the Weight and power will always be reciprocally as the velocities of them, or of their centres of gravity; or their momentums will be equal, that is, the pro-

duct of the Weight multiplied by its velocity, will be equal to the product of the power multiplied by its velocity.

WEIGHT, in Commerce, denotes a body of a known Weight, appointed to be put into a balance againft other bodies, whofe Weight is required to be known. Thefe Weights are ufually of lead, iron, or brafs; though in feveral parts of the Eaft Indies common flints are ufed; and in fome places a fort of little beans.

The diverfity of Weights, in all nations, and at all times, makes one of the moft perplexing circumftances in commerce, &c. And it would be a very great convenience if all nations could agree upon a univerfal ftandard, and fyftem, both of Weights and meafures.

Weights may be diftinguifhed into *ancient* and *modern, foreign* and *domeftic.*

Modern WEIGHTS, *ufed in the feveral parts of Europe, and the Levant.*

Englifh WEIGHTS. By the 27th chapter of Magna Charta, the Weights are to be the fame all over England : but for different commodities there are two different forts, viz, *troy Weight*, and *averdupois Weight*.

The origin from which both of thefe are raifed, is the grain of wheat, gathered in the middle of the ear :

32 of thefe, well dried, made one pennyweight,
20 pennyweights - - - - - one ounce, and
12 ounces - - - - - - - - one pound troy;
by Stat. 51 Hen. III; 31 Edw. I; 12 Henry VII.

A learned writer has fhewn that, by the laws of affize, from William the Conqueror to the reign of Henry VII, the legal pound Weight contained a pound of 12 ounces, raifed from 32 grains of wheat; and the legal gallon meafure contained 8 of thofe pounds of wheat, 8 gallons making the bufhel, and 8 bufhels the quarter.

Henry VII. altered the old Englifh Weight, and introduced the troy pound in its ftead, being 3 quarters of an ounce only heavier than the old Saxon pound, or 1-16th heavier. The firft ftatute that directs the ufe of the averdupois Weight, is that of 24 Henry VIII; and the particular ufe to which this Weight is thus directed, is fimply for weighing butcher's meat in the market; though it is now ufed for weighing all forts of coarfe and large articles. This pound contains 7000 troy grains; while the troy pound itfelf contains only 5760 grains, and the old Saxon pound Weight but 5400 grains. Philof. Tranf. vol. 65, art. 3.

Hence there are now in common ufe in England, two different Weights, viz, troy Weight, and averdupois Weight, the former being employed in weighing fuch fine articles as jewels, gold, filver, filk, liquors, &c; and the latter for coarfe and heavy articles, as bread, corn, flefh, butter, cheefe, tallow, pitch, tar, iron, copper, tin, &c. and all grocery wares. And Mr. Ward fuppofes that it was brought into ufe from this circumftance, viz, as it was cuftomary to allow larger Weight, of fuch coarfe articles, than the law had expreffly enjoined, and this he obferves happened to be a 6th part more. Apothecaries buy their drugs by averdupois Weight, but they compound them by troy Weight, though under fome little variation of name and divifions.

The

The troy or trone pound Weight in Scotland, which by statute is to be the same as the French pound, is commonly supposed equal to $15\frac{3}{4}$ English troy ounces, or 7560 grains; but by a mean of the standards kept by the dean of gild of Edinburgh, it weighs $7599\frac{1}{8}$ or 7600 grains nearly.

The following tables shew the divisions of the troy and averdupois Weights.

Table of Troy Weight, as used,

1. By the Goldsmiths, &c.

Grains	Pennywt.		
24 =	1 dwt.		
		Ounce	
480 =	20 =	1 oz.	
			Pound
5760 =	240 =	12 =	1 lb.

2. By the Apothecaries.

Grains	Scruples			
20 =	1 Ə			
		Drams		
60 =	3 =	1 ʒ		
			Ounces	
480 =	24 =	8 =	1 ℥	
				Pound
5760 =	288 =	96 =	12 =	1 lb.

Table of Averdupois Weight.

Drams	Ounces				
16 =	1				
		Pounds			
256 =	16 =	1			
			Quarters		
7168 =	448 =	28 =	1		
				Hund. wt.	
28672 =	1792 =	112 =	4 =	1	
					Ton
573440 =	35840 =	2240 =	80 =	20 =	1

Mr. Ferguson (Lect. on Mech. p. 100, 4to) gives the following comparison between troy and averdupois Weight.

175 troy pounds are equal to 144 averdup pounds.
175 troy ounces are equal to 192 averdup. ounces.
1 troy pound contains 5760 grains.
1 averdupois pound contains 7000 grains.
1 averdupois ounce contains $437\frac{1}{2}$ grains.
1 averdupois dram contains 27.34375 grains.
1 troy pound contains 13 oz. 2.651428576 drams averdupois
1 averdup. lb. contains 1 lb 2 oz. 11 dwts 16 gr troy

The moneyers, jewellers, &c, have a particular class of Weights, for gold and precious stones, viz, *carat* and *grain*; and for silver, the *pennyweight* and *grain*. The moneyers have also a peculiar subdivision of the troy grain: thus, dividing

the grain into 20 mites
the mite into 24 droits
the droit into 20 periots
the periot into 24 blanks.

The dealers in wool have likewise a particular set of Weights; viz, the *sack, weigh, tod, stone,* and *clove,* the proportions of which are as below: viz,

the sack containing   2   weighs
the weigh - - - - $6\frac{1}{2}$ tods
the tod - - - - - - 2 stones
the stone - - - - - 2 cloves
the clove - - - - - 7 pounds.

Also 12 sacks make a last or 4368 pounds.

Farther,
56 lb of old hay, or 60 lb new hay, make a truss.
40 lb of straw make a truss.
36 trusses make a load, of hay or straw.
14 lb make a stone.
5 lb of glass a stone.

French WEIGHTS. The common or Paris pound Weight, is to the English troy pound, as 21 to 16, and to the averdupois pound as 27 to 25; it therefore contains 7560 troy grains; and it is divided into 16 ounces like the pound averdupois, but more particularly thus: the pound into 2 *marcs*; the marc into 8 *ounces*; the ounce into 8 *gros*, or *drams*; the gross or dram into 3 deniers, Paris scruples or pennyweights; and the pennyweight into 24 *grains*; the grain being an equivalent to a grain of wheat. So that the Paris ounce contains $472\frac{1}{4}$ troy grains, and therefore it is to the English troy ounce as 63 to 64. But in several of the French provinces, the pound is of other different Weights. A *quintal* is equal to 100 pounds.

The Weights above enumerated under the two articles of English and French Weights, are the same as are used throughout the greatest part of Europe; only under somewhat different names, divisions, and proportions. And besides, particular nations have also certain Weights peculiar to themselves, of too little consequence here to be enumerated. But to shew the proportion of these several Weights to one another, there may be here added a reduction of the divers pounds in use throughout Europe, by which the other Weights are estimated, to one standard pound, viz, the pound of Amsterdam, Paris, and Bourdeaux; as they were accurately calculated by M. Ricard, and published in the new edition of his Traité de Commerce, in 1722.

Proportion of the WEIGHTS *of the chief Cities in Europe to that of Amsterdam.*

100 pounds of Amsterdam are equal to

108 lbs of	Alicant	100 lbs of	Bilboa
105	Antwerp	105	Bois le Duc
120	Archangel, or	151	Bologna
	3 poedes	100	Bourdeaux
105	Arschot	104	Bourg en Bresse
120	Avignon	103	Bremen
98	Basil	125	Breslaw
100	Bayonne	105	Bruges
166	Bergamo	105	Brussels
97	Berg. op Zoom	105	Cadiz
$95\frac{1}{4}$	Bergen, Norw.	105	Cologne
111	Bern	$107\frac{1}{2}$	Copenhagen
100	Besancon	87	Constantinople

5

WEIGHTS

WEIGHTS *continued.*

100 pounds of Amsterdam are equal to

$113\frac{1}{2}$ lbs	of Dantzic	154 lbs	of Messina
100	Dort	168	Milan
97	Dublin	120	Montpelier
97	Edinburgh	125	Muscovy
143	Florence	100	Nantes
98	Franckfort, sur Maine	100	Nancy
		169	Naples
105	Gaunt	98	Nuremberg
89	Geneva	100	Paris
163	Genoa	$112\frac{1}{2}$	Revel
102	Hamburgh	109	Riga
125	Koningsberg	100	Rochel
105	Leipsic	146	Rome
106	Leyden	100	Rotterdam
143	Leghorn	96	Rouen
$105\frac{1}{2}$	Liege	100	S. Malo
106	Lisbon	100	S. Sebastian
114	Lisle	$158\frac{1}{7}$	Saragosa
109	London, averdupois	100	Seville
		114	Smyrna
105	Louvain	110	Stetin
105	Lubeck	81	Stockholm
$141\frac{1}{2}$	Lucca	118	Tholouse
116	Lyons	151	Turin
114	Madrid	$158\frac{1}{2}$	Valencia
105	Malines	182	Venice.
$123\frac{1}{2}$	Marseilles		

Ancient WEIGHTS.

1. The Weights of the ancient Jews, reduced to the English troy Weights, will stand as below:

	lb	oz	dwt	gr
Shekel - - - -	0	0	9	$2\frac{4}{7}$
Manch - - - -	2	3	6	$10\frac{2}{7}$
Talent - - - -	113	10	1	$10\frac{2}{7}$

2. Grecian and Roman Weights, reduced to English troy Weight, are as in the following table:

	lb	oz	dwt	gr
Lentes - - - -	0	0	0	$0\frac{35}{112}$
Siliquæ - - - -	0	0	0	$3\frac{1}{28}$
Obolus - - - -	0	0	0	$9\frac{3}{28}$
Scriptulum - -	0	0	0	$18\frac{3}{14}$
Drachma - - -	0	0	2	$6\frac{5}{14}$
Sextula - - - -	0	0	3	$0\frac{6}{7}$
Sicilicus - - - -	0	0	4	$13\frac{2}{7}$
Duella - - - - -	0	0	6	$1\frac{5}{7}$
Uncia - - - - -	0	0	18	$5\frac{1}{7}$
Libra - - - - -	0	10	18	$13\frac{5}{7}$

The Roman ounce is the English averdupois ounce, which they divide into 7 denarii, as well as 8 drachms: and as they reckoned their denarius equal to the Attic drachm, this will make the Attic Weights one-eighth heavier than the correspondent Roman Weights. Arbuth.

Regulation of WEIGHTS *and Measures.* This is a branch of the king's prerogative. For the public convenience, these ought to be universally the same throughout the nation, the better to reduce the prices of articles to equivalent values. But as Weight and measure are things in their nature arbitrary and uncertain, it is necessary that they be reduced to some fixed rule or standard. It is however impossible to fix such a standard by any written law or oral proclamation; as no person can, by words only, give to another an adequate idea of a pound Weight, or foot rule. It is therefore expedient to have recourse to some visible, palpable, material standard; by forming a comparison with which, all Weights and measures may be reduced to one uniform size. Such a standard was anciently kept at Winchester: and we find in the laws of king Edgar, near a century before the conquest, an injunction that that measure should be observed throughout the realm.

Most nations have regulated the standard of measures of length from some parts of the human body; as the palm, the hand, the span, the foot, the cubit, the ell (*ulna* or arm), the pace, and the fathom. But as these are of different dimensions in men of different proportions, ancient historians inform us, that a new standard of length was fixed by our king Henry the first; who commanded that the *ulna* or ancient ell, which answers to the modern yard, should be made of the exact length of his own arm.

A standard of long measure being once gained, all others are easily derived from it; those of greater length by multiplying that original standard, those of less by dividing it. Thus, by the statute called *compositio ulnarum et perticarum*, $5\frac{1}{2}$ yards make a perch; and the yard is subdivided into 3 feet, and each foot into 12 inches; which inches will be each of the length of 3 barley corns. But some, on the contrary, derive all measures, by composition, from the barley corn.

Superficial measures are derived by squaring those of length; and measures of capacity by cubing them.

The standard of Weights was originally taken from grains or corns of wheat, whence our lowest denomination of Weights is still called a *grain*; 32 of which are directed, by the statute called *compositio mensurarum*, to compose a pennyweight. 20 of which make an ounce, and 12 ounces a pound, &c.

Under king Richard the first it was ordained, that there should be only one Weight and one measure throughout the nation, and that the custody of the assize or standard of Weights and measures, should be committed to certain persons in every city and borough; from whence the ancient office of the king's ulnager seems to have been derived. These original standards were called *pondus regis*, and *mensura domini regis*, and are directed by a variety of subsequent statutes to be kept in the exchequer chamber, by an officer called the *clerk of the market*, except the wine gallon, which is committed to the city of London, and kept in Guildhall.

The Scottish standards are distributed among the oldest boroughs. The elwand is kept at Edinburgh, the pint at Stirling, the pound at Lanark, and the firlot at Linlithgow.

The two principal Weights established in Great Britain, are troy Weight, and avoirdupois Weight,

 as

as before mentioned. Under the head of the former it may farther be added, that

A carat is a Weight of 4 grains; but when the term is applied to gold, it denotes the degree of fineness. Any quantity of gold is supposed divided into 24 parts. If the whole mass be pure gold, it is said to be 24 carats fine; if there be 23 parts of pure gold, and one part of alloy or base metal, it is said to be 23 carats fine, and so on.

Pure gold is too soft to be used for coin. The standard coin of this kingdom is 22 carats fine. A pound of standard gold is coined into $44\frac{1}{2}$ guineas, and therefore every guinea should weigh 5 dwts $9\frac{38}{89}$ grains.

A pound of silver for coin contains 11 oz 2 dwts pure silver, and 18 dwts alloy: and standard silver-plate, 11 ounces pure silver, with 1 ounce alloy. A pound of standard silver is coined into 62 shillings; and therefore the Weight of a shilling should be 3 dwts $20\frac{48}{31}$ grains.

Universal Standard for WEIGHTS and Measures.

Philosophers, from their habits of generalizing, have often made speculations for forming a general standard for Weights and measures through the whole world. These have been devised chiefly of a philosophical nature, as best adapted to universality. After the invention of pendulum clocks, it first occurred that the length of a pendulum which should vibrate seconds, would be proper to be made a universal standard for lengths; whether it should be called a yard, or any thing else. But it was found, that it would be difficult in practice, to measure and determine the true length of such a pendulum, that is the distance between the point of suspension and the point of oscillation. Another cause of inaccuracy was afterwards discovered, when it was found that the seconds pendulum was of different lengths in all the different latitudes, owing to the spheroidal figure of the earth, which causes that all places in different latitudes are at different distances from the centre, and consequently the pendulums are acted upon by different forces of gravity, and therefore require to be of different lengths. In the latitude of London this is found to be $39\frac{1}{8}$ inches.

The Society of Arts in London, among their many laudable and patriotic endeavours, offered a handsome premium for the discovery of a proper standard for Weights and measures. This brought them many frivolous expedients, as well as one which was an improvement on the method of the pendulum, by one Hatton. This consisted in measuring the difference of the lengths of two pendulums of different times of vibration; which could be performed more easily and accurately than that of the length of one single pendulum. This method was put in practice, and fully explained and illustrated, by the late Mr. Whitehurst, in his attempt to ascertain an Universal Standard of Weights and Measures. But still the same kind of inaccuracy of measurement &c, obtains in this way, as in the single pendulum, though in a smaller degree.

Another method that has been proposed for this purpose, is the space that a heavy body falls freely through in 1 second of time. But this is an experiment more difficult than the former to be made with accuracy; on which account, different persons will all make the space fallen to be of different quantities, which would give as many different standards of length. Add to this, that the spheroidal form of the earth here again introduces a diversity in the space, owing to the different distances from the centre, and the consequent diversity in the force of gravity by which the body falls. This space has been found to be 193 inches, or $16\frac{1}{12}$ feet, in the latitude of London; but it will be a different quantity in other latitudes.

Many other inferior expedients have also been proposed for the purpose of universal measures, and Weights; but there is another which now has the best prospect of success, and is at present under particular experiments, by the philosophers both of this and the French nation. This method is by the measure of the degrees of latitude; which would give a large quantity, and admit of more accurate measures, by subdivision, than what could be obtained by beginning from a small quantity, or measure, and thence to proceed increasing by multiples. This measure might be taken either from the extent of the whole compass of the earth, or of all the 360 degrees, or a medium degree among them all, or from the measure of a degree in the medium latitude of 45 degrees. It will also be most convenient to make the subdivisions of this measure, when found, to proceed decimally, or continually by 10ths.

The universal standard for lengths being once established, those of Weights, &c, would easily follow. For instance, a vessel, of certain dimensions, being filled with distilled water, or some other homogeneous matter, the Weight of that may be considered as a standard for Weights.

WEIGHT of the Air, Water, &c. See those articles severally. See also SPECIFIC GRAVITY.

WERST, a Russian, measure of length, equal to 3500 English feet.

WEST, one of the cardinal points of the horizon, or of the compass, diametrically opposite to the east, or lying on the left hand when we face the north. Or West is strictly the intersection of the prime vertical with the horizon, on that side where the sun sets.

WEST *Wind*, is also called *Zephyrus*, and *Favonius*.

WEST *Dial*. See DIAL.

WESTERN *Amplitude, Horizon, Ocean.* See the several articles.

WESTING, in Navigation, is the quantity of departure made good to the westward from the meridian.

WEY. See WEIGH.

WHALE, in Astronomy, one of the constellations. See CETUS.

WHEEL, in Mechanics, a simple machine, consisting of a circular piece of wood, metal, or other matter, that revolves on an axis. This is otherwise called *Wheel and Axle*, or AXIS *in Peritrochio*, as a mechanical power, being one of the most frequent and useful of any. In this capacity of it, the Wheel is a kind of perpetual lever, and the axis another lesser one: or the radius of the Wheel and that of its axis may be considered as the longer and shorter arms of a lever, the centre of the Wheel being the fulcrum or point of suspen *suspension.* Whence it is, that the power of this machine is estimated by this rule, as the radius of the axis is to the radius of the Wheel or of the circumference, so is any given power, to the weight it will sustain.

Wheels, as well as their axes, are frequently dented, or cut into teeth, and are then of use upon innumerable occasions; as in jacks, clocks, mill-work, &c; by which means they are capable of moving and acting on one another and of being combined together to any extent; the teeth either of the axis or circumference working in those of other Wheels or axles; and thus, by multiplying the power to any extent, an amazing great effect is produced.

To compute the power of a combination of Wheels; the teeth of the axis of every Wheel acting on those in the circumference of the next following. Multiply continually together the radii of all the axes, as also the radii of all the Wheels; then it will be, as the former product is to the latter product, so is a given power applied to the circumference, to the weight it can sustain. Thus, for example, in a combination of five Wheels and axles, to find the weight a man can sustain, or raise, whose force is equal to 150 pounds, the radii of the Wheels being 30 inches, and those of the axes 3 inches. Here $3 \times 3 \times 3 \times 3 \times 3 = 243$,

and $30 \times 30 \times 30 \times 30 \times 30 = 24300000$,
therefore as $243 : 24300000 :: 150 : 15000000$ lb, the weight he can sustain, which is more than 6696 tons weight. So prodigious is the increase of power in a combination of Wheels!

But it is to be observed, that in this, as well as every other mechanical engine, whatever is gained in power, is lost in time; that is, the weight will move as much slower than the power, as the force is increased or multiplied, which in the example above is 100000 times slower.

Hence, having given any power, and the weight to be raised, with the proportion between the Wheels and axles necessary to that effect; to find the number of the Wheels and axles. Or, having the number of the Wheels and axles given, to find the ratio of the radii of the Wheels and axles. Here, putting

$p =$ the power acting on the last wheel,
$w =$ the weight to be raised,
$r =$ the radius of the axles,
$R =$ the radius of the wheels,
$n =$ the number of the wheels and axles;

then, by the general proportion, as $r^n : R^n :: p : w$; therefore $pR^n = wr^n$ is a general theorem, from whence may be found any one of these five letters or quantities, when the other four are given. Thus, to find n the number of Wheels: we have first

$$\frac{R^n}{r^n} = \frac{w}{p}, \text{ then } n = \frac{\log. w - \log. p}{\log. R - \log. r}.$$

And to find $\frac{R}{r}$, the ratio of the Wheel to the axle; it is

$$\frac{R}{r} = \sqrt[n]{\frac{w}{p}}.$$

WHEELS *of a Clock, &c,* are, the crown wheel, contrat wheel, great wheel, second wheel, third wheel, striking wheel, detent wheel, &c.

WHEELS *of Coaches, Carts, Waggons, &c.* With respect to Wheels of carriages, the following particulars are collected from the experiments and observations of Desaguliers, Beighton, Camus, Ferguson, Jacob, &c.

1. The use of Wheels, in carriages, is twofold; viz, that of diminishing or more easily overcoming the resistance or friction from the carriage; and that of more easily overcoming obstacles in the road. In the first case the friction on the ground is transferred in some degree from the outer surface of the Wheel to its nave and axle; and in the latter, they serve easily to raise the carriage over obstacles and asperities met with on the roads. In both these cases, the height of the Wheel is of material consideration, as the spokes act as levers, the top of an obstacle being the fulcrum, their length enables the carriage more easily to surmount them; and the greater proportion of the Wheel to the axle serves more easily to diminish or to overcome the friction of the axle. See Jacob's Observations on Wheel Carriages, p. 23 &c.

2. The Wheels should be exactly round; and the fellies at right angles to the naves, according to the inclination of the spokes.

3. It is the most general opinion, that the spokes be somewhat inclined to the naves, so that the Wheels may be dishing or concave. Indeed if the Wheels were always to roll upon smooth and level ground, it would be best to make the spokes perpendicular to the naves, or to the axles; because they would then bear the weight of the load perpendicularly. But because the ground is commonly uneven, one Wheel often falls into a cavity or rut, when the other does not, and then it bears much more of the weight than the other does; in which case it is best for the Wheels to be dished, because the spokes become perpendicular in the rut, and therefore have the greatest strength when the obliquity of the road throws most of the weight upon them; whilst those on the high ground have less weight to bear, and therefore need not be at their full strength.

4. The axles of the Wheels should be quite straight, and perpendicular to the shafts, or to the pole. When the axles are straight, the rims of the Wheels will be parallel to each other, in which case they will move the easiest, because they will be at liberty to proceed straight forwards. But in the usual way of practice, the ends of the axles are bent downwards; which always keeps the sides of the Wheels that are next the ground nearer to one another than their upper sides are; and this not only makes the Wheels drag sideways as they go along, and gives the load a much greater power of crushing them than when they are parallel to each other, but also endangers the overturning the carriage when a Wheel falls into a hole or rut, or when the carriage goes on a road that has one side lower than the other, as along the side of a hill. Mr. Beighton however has offered several reasons to prove that the axles of Wheels ought not to be straight; for which see Desaguliers's Exp. Phil. vol. 2, Appendix.

5. Large Wheels are found more advantageous for rolling than small ones, both with regard to their power as a longer lever, and to the degree of friction, and to the advantage in getting over holes, rubs, and

hence,

ftones, &c. If we confider Wheels with regard to the friction upon their axles, it is evident that fmall Wheels, by turning oftener round, and fwifter about the axles, than large ones, mult have much more friction. Again, if we confider Wheels as they fink into holes or foft earth, the large Wheels, by finking lefs, mult be much eafier drawn out of them, as well as more eafily over ftones and obftacles, from their greater length of lever or fpokes. Defaguliers has brought this matter to a mathematical calculation, in his Experim. Philof. vol. 1, p. 171, &c. See alfo Jacob's Obferv. p. 63.

From hence it appears then, that Wheels are the more advantageous as they are larger, provided they are not more than 5 or 6 feet diameter; for when they exceed thefe dimenfions, they become too heavy; or if they are made light, their ftrength is proportionably diminifhed, and the length of the fpokes renders them more liable to break: befides, horfes applied to fuch Wheels would not be capable of exerting their utmoft ftrength, by having the axles higher than their breafts, fo that they would draw downwards; which is even a greater difadvantage than fmall Wheels have in occafioning the horfes to d aw upwards.

6. Carriages with 4 Wheels, as waggons or coaches, are much more advantageous than carriages with 2 Wheels, as carts and chaifes; for with 2 wheels it is plain the tiller horfe carries part of the weight, in one way or other: in going down hill, the weight bears upon the horfe; and in going up hill, the weight falls the other way, and lifts the horfe, which is ftill worfe. Befides, as the Wheels fink into the holes in the roads, fometimes on one fide, fometimes on the other, the fhafts ftrike againft the tiller's fides, which deftroys many horfes: moreover, when one of the Wheels finks into a hole or rut, half the weight falls that way, which endangers the overturning of the carriage.

7. It would be much more advantageous to make the 4 Wheels of a coach or waggon large, and nearly of a height, than to make the fore Wheels of only half the diameter of the hind Wheels, as is ufual in many places. The fore Wheels have commonly been made of a lefs fize than the hind ones, both on account of turning fhort, and to avoid cutting the braces. Crane-necks have alfo been invented for turning yet fhorter, and the fore Wheels have been lowered, fo as to go quite under the bend of the crane-neck.

It is held, that it is a great difadvantage in fmall Wheels, that as their axle is below the bow of the horfes breafts, the horfes not only have the loaded carriage to draw along, but alfo part of its weight to bear, which tires them foon, and makes them grow much ftiffer in their hams, than they would be if they drew on a level with the fore axle.

But Mr. Beighton difputes the propriety of fixing the line of traction on a level with the breaft of a horfe, and fays it is contrary to reafon and experience. Horfes, he fays, have little or no power to draw but what they derive from their weight; without which they could not take hold of the ground, and then they muft flip, and draw nothing. Common experience alfo teaches, that a horfe muft have a certain weight on his back or fhoulders, that he may draw the better. And

when a horfe draws hard, it is obferved that he bends forward, and brings his breaft near the ground; and then if the Wheels are high, he is pulling the carriage againft the ground. A horfe tackled in a waggon will draw two or three ton, becaufe the point or line of traction is below his breaft, by the lownefs of the Wheels. It is alfo common to fee, when one horfe is drawing a heavy load, efpecially up hill, his fore feet will rife from the ground; in which cafe it is ufual to add a weight on his back, to keep his fore part down, by a perfon mounting on his back or fhoulders, which will enable him to draw that load, which he could not move before. The greateft ftrefs, or main bufinefs of drawing, fays this ingenious writer, is to overcome obftacles; for on level plains the drawing is but little, and then the horfe's back need be preffed but with a fmall weight.

8. The utility of broad Wheels, in amending and preferving the roads, has been fo long and generally acknowledged, as to have occafioned the legiflature to enforce their ufe. At the same time, the proprietors and drivers of carriages feem to be convinced by experience, that a narrow-wheeled carriage is more eafily and fpeedily drawn by the fame number of horfes, than a broad-wheeled one of the fame burthen: probably becaufe they are much lighter, and have lefs friction on the axle.

On the fubject of this article, fee Jacob's Obferv. &c. on Wheel-Carriages, 1773, p. 81. Defagul. Exper. Phil. vol. 1, p. 201. Ferguson's Lect. 4to, p. 56. Martin's Phil. Brit. vol. 1, p. 229.

Blowing WHEEL, is a machine contrived by Defaguliers, for drawing the foul air out of any place, or for forcing in frefh, or doing both fucceffively, without opening doors or windows. See Philof. Tranf. number 437. The intention of this machine is the fame as that of Hales's ventilator, but not fo effectual, nor fo convenient. See Defag. Exper. Philof. vol. 2, p. 563, 568.—This Wheel is alfo called a *centrifugal Wheel*, becaufe it drives the air with a centrifugal force.

Water WHEEL, of a Mill, that which receives the impulfe of the ftream by means of ladle-boards or floatboards. M. Parent, of the Academy of Sciences, has determined that the greateft effect of an underfhot Wheel, is when its velocity is equal to the 3d part of the velocity of the water that drives it; but it ought to be the half of that velocity, as is fully fhewn in the article Mill, pa. 111. In fixing an underfhot Wheel, it ought to be confidered whether the water can run clear off, fo as to caufe no back-water to ftop its motion. Concerning this article, fee Defagul. Exp. Philof. vol. 2, p. 422. Alfo a variety of experiments and obfervations relating to underfhot and overfhot Wheels, by Mr. Smeaton, in the Philof. Tranf. vol. 51, p. 100.

Ariftotle's WHEEL. See ROTA *Ariftotelica.*
Meafuring WHEEL. See PERAMBULATOR.
Orffyreus's WHEEL. See ORFFYREUS.
Perfian WHEEL. See PERSIAN.
WHEEL-*Barometer.* See BAROMETER.
WHIRL-POOL, an eddy, vortex, or gulph, where the water is continually turning round.
WHIRLING-TABLE, a machine contrived for
repre-

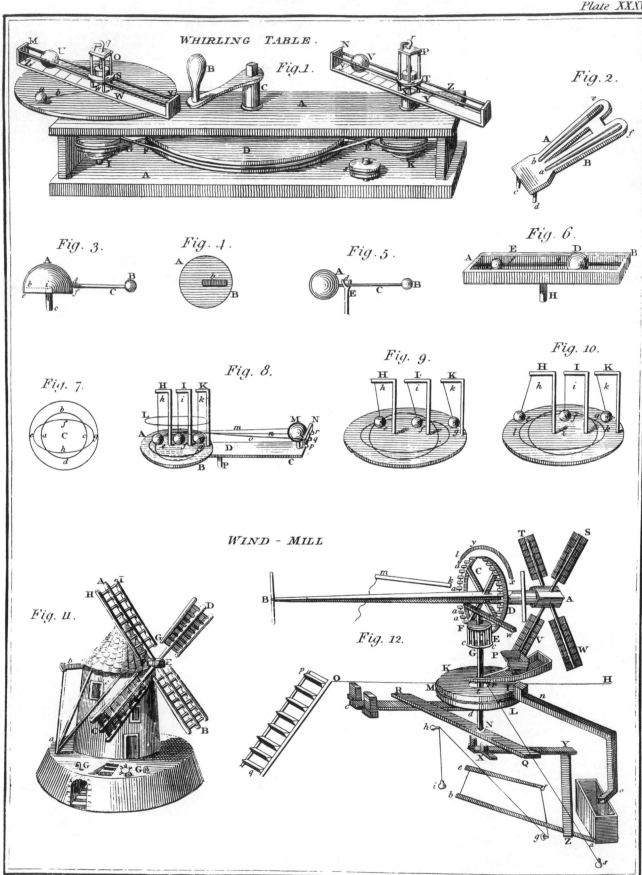

Plate XXXVI.

WHIRLING TABLE.

Fig. 1.

Fig. 2.

Fig. 3.

Fig. 4.

Fig. 5.

Fig. 6.

Fig. 7.

Fig. 8.

Fig. 9.

Fig. 10.

WIND - MILL

Fig. 11.

Fig. 12.

representing several phenomena in philosophy, and nature; as, the principal laws of gravitation, and of the planetary motions in curvilinear orbits.

The figure of this instrument is exhibited fig. 1, pl. 35: where AA is a strong frame of wood; B a winch fixed on the axis C of the wheel D, round which is the catgut string F, which also goes round the small wheels G and K, crossing between them and the great wheel D On the upper end of the axis of the wheel G, above the frame, is fixed the round board d, to which may be occasionally fixed the bearer MSX. On the axis of the wheel H is fixed the bearer NTZ, and when the winch B is turned, the wheels and bearers are put into a Whirling motion. Each bearer has two wires W, X, and Y, Z, fixed and screwed tight into them at the ends by nuts on the outside; and when the nuts are unscrewed, the wires may be drawn out in order to change the balls U, V, which slide upon the wires by means of brass loops fixed into the balls, and preventing their touching the wood below them. Through each ball there passes a silk line, which is fixed to it at any length from the centre of the bearer to its end, by a nut-screw at the top of the ball; the shank of the screw going into the centre of the ball, and pressing the line against the under side of the whole which it goes through. The line goes from the ball, and under a small pulley fixed in the middle of the bearer; then up through a socket in the round plate (S and T) in the middle of each bearer; then through a slit in the middle of the square top (O and P) of each tower, and going over a small pulley on the top comes down again the same way, and is at last fastened to the upper end of the socket fixed in the middle of the round plate above mentioned. Each of these plates S and T has four round holes near their edges, by which they slide up and down upon the wires which make the corner of each tower. The balls and plates being thus connected, each by its particular line, it is plain that if the balls be drawn outward, or towards the end M and N of their respective bearers, the round plates S and T will be drawn up to the top of their respective towers O and P.

There are several brass weights, some of two, some of three, and others of four ounces, to be occasionally put within the towers O and P, upon the round plates S and T: each weight having a round hole in the middle of it, for going upon the sockets or axes of the plates, and being slit from the edge to the hole, that it may slip over the line which comes from each ball to its respective plate.

For a specimen of the experiments which may be made with this machine, may be subjoined the following.

1. Removing the bearer MX, put the loop of the line b to which the ivory ball a is fastened over a pin in the centre of the board d, and turn the winch B; and the ball will not immediately begin to move with the board, but, on account of its inactivity, endeavour to remain in its state of rest. But when the ball has acquired the same velocity with the board, it will remain upon the same part of the board, having no relative motion upon it. However, if the board be suddenly stopped, the ball will continue to revolve upon

it, until the friction thereof stops its motion: so that matter resists every change of state, from that of rest to that of motion, and *vice versa*.

2. Put a longer cord to this ball; let it down through the hollow axis of the bearer MX and wheel G, and fix a weight to the end of the cord below the machine; and this weight, if left at liberty, will draw the ball from the edge of the Whirling board to its centre. Draw off the ball a little from the centre, and turn the winch; then the ball will go round and round with the board, and gradually fly farther from the centre, raising up the weight below the machine. And thus it appears that all bodies, revolving in circles, have a tendency to fly off from those circles, and must be retained in them by some power proceeding from or tending to the centre of motion. Stop the machine, and the ball will continue to revolve for some time upon the board; but as the friction gradually stops its motion, the weight acting upon it will bring it nearer and nearer to the centre in every revolution, till it brings it quite thither. Hence it appears, that if the planets met with any resistance in going round the sun, its attractive power would bring them nearer and nearer to it in every revolution, till they would fall into it.

3. Take hold of the cord below the machine with one hand, and with the other throw the ball upon the round board as it were at right angles to the cord, and it will revolve upon the board. Then, observing the velocity of its motion, pull the cord below the machine, and thus bring the ball nearer the centre of the board, and the ball will be seen to revolve with an increasing velocity, as it approaches the centre: and thus the planets which are nearest the sun perform quicker revolutions than those which are more remote, and move with greater velocity in every part of their respective circles.

4. Remove the ball a, and apply the bearer MX, whose centre of motion is in its middle at w, directly over the centre of the Whirling board d. Then put two balls (V and U) of equal weight upon their bearing wires, and having fixed them at equal distances from their respective centres of motion w and x upon their silk cords, by the screw nuts, put equal weights in the towers O and P. Lastly, put the catgut strings E and F upon the grooves G and H of the small wheels, which, being of equal diameters, will give equal velocities to the bearers above, when the winch B is turned; and the balls U and V will fly off toward M and N, and raise the weights in the towers at the same instant. This shews, that when bodies of equal quantities of matter revolve in equal circles with equal velocities, their centrifugal forces are equal.

5. Take away these equal balls, and put a ball of 6 ounces into the bearer MX, at a 6th part of the distance wz from the centre, and put a ball of one ounce into the opposite bearer, at the whole distance xy = wz; and fix the balls at these distances on their cords, by the screw nuts at the top: then the ball U, which is 6 times as heavy as the ball V, will be at only a 6th part of the distance from its centre of motion; and consequently will revolve in a circle of only a 6th part of the circumference of the circle in which V revolves. Let equal weights be put into the towers, and the winch be turned; which (as the catgut string

:s on equal wheels below, will caufe the balls to revolve in equal times : but V will move 6 times as faft as U, becaufe it revolves in a circle of 6 times its radius, and both the weights in the towers will rife at once. Hence it appears, that the centrifugal forces of revolving bodies are in direct proportion to their quantities of matter multiplied into their refpective velocities, or into their diftance from the centres of their refpective circles.

If thefe two balls be fixed at equal diftances from their refpective centres of motion, they will move with equal velocities ; and if the tower O has 6 times as much weight put into it as the tower P has, the balls will raife their weights exactly at the fame moment : i. e. the ball U, being 6 times as heavy as the ball V, has 6 times as much centrifugal force in defcribing an equal circle with an equal velocity.

6. Let two balls, U and V, of equal weights, be fixed on their cords at equal diftances from their refpective centres of motion w and x ; and let the catgut ftring E be put round the wheel K (whofe circumference is only half that of the wheel H or G) and over the pulley s to keep it tight, and let 4 times as much weight be put into the tower P as in the tower O. Then turn the winch B, and the ball V will revolve twice as faft as the ball U in a circle of the fame diameter, becaufe they are equidiftant from the centres of the circles in which they revolve ; and the weights in the towers will both rife at the fame inftant ; which fhews that a double velocity in the fame circle will exactly balance a quadruple power of attraction in the centre of the circle : for the weights in the towers may be confidered as the attractive forces in the centres, acting upon the revolving balls ; which moving in equal circles, are as if they both moved in the fame circle. Whence it appears that, if bodies of equal weights revolve in equal circles with unequal velocities, their centrifugal forces are as the fquares of the velocities.

7. The catgut ftring remaining as before, let the diftance of the ball V from the centre x be equal to 2 of the divifions on its bearer ; and the diftance of the ball U from the centre w be 3 and a 6th part ; the balls themfelves being equally heavy, and V making two revolutions by turning the winch, whilft U makes one ; fo that if we fuppofe the ball V to revolve in one moment, the ball U will revolve in 2 moments, the fquares of which are 1 and 4 : therefore, the fquare of the period of V is contained 4 times in the fquare of the period of U. But the diftance of V is 2, the cube of which is 8, and the diftance of U is $3\frac{1}{6}$, the cube of which is 32 very nearly, in which 8 is contained 4 times : and therefore, the fquares of the periods V and U are to one another as the cubes of their diftances from x and w, the centres of their refpective circles. And if the weight in the tower O be 4 ounces, or equal to the fquare of 2, which is the diftance of V from the centre x ; and the weight in the tower P be 10 ounces, nearly equal to the fquare of $3\frac{1}{6}$, the diftance of U from w ; it will be found upon turning the machine by the winch, that the balls U and V will raife their refpective weights at very nearly the fame inftant of time. This experiment confirms the famous propofition of Kepler, viz, that the fquares of the periodical times of the planets round the fun are in propor-

tion as the cubes of their diftances from him ; and that the fun's attraction is inverfely as the fquare of the diftance from his centre.

8. Take off the ftring E from the wheels D and H, and let the ftring F remain upon the wheels D and G ; take away alfo the bearer MX from the Whirlingboard d, and inftead of it put on the machine AB (fig. 2), fixing it to the centre of the board by the pins c and d, fo that the end ef may rife above the board to an angle of 30 or 40 degrees. On the upper part of this machine, there are two glafs tubes a and b, clofe ftopped at both ends, each tube being about three quarters full of water. In the tube a is a little quickfilver, which naturally falls down to the end a in the water ; and in the tube b is a fmall cork, floating on the top of the water, and fmall enough to rife or fall in the tube. While the board b with this machine upon it continues at reft, the quickfilver lies at the bottom of the tube a, and the cork floats on the water near the top of the tube b. But, upon turning the winch and moving the machine, the contents of each tube fly off towards the uppermoft ends, which are fartheft from the centre of motion ; the heavieft with the greateft force. Confequently, the quickfilver in the tube a will fly off quite to the end f, occupying its bulk of fpace, and excluding the water, which is lighter than itfelf : but the water in the tube b, flying off to its higher end e, will exclude the cork from that place, and caufe it to defcend toward the loweft end of the tube ; for the heavier body, having the greater centrifugal force, will poffefs the upper part of the tube, and the lighter body will keep between the heavier and the lower part.

This experiment demonftrates the abfurdity of the Cartefian doctrine of vortices ; for, if a planet be more denfe or heavy than its bulk of the vortex, it will fly off in it farther and farther from the fun ; if lefs denfe, it will come down to the loweft part of the vortex, at the fun : and the whole vortex itfelf, unlefs prevented by fome obftacle, would fly quite off, together with the planets.

9. If a body be fo placed upon the Whirling-board of the machine (fig. 1.) that the centre of gravity of the body be directly over the centre of the board, and the board be moved ever fo rapidly by the winch B, the body will turn round with the board, without removing from its middle ; for, as all parts of the body are in equilibrio round its centre of gravity, and the centre of gravity is at reft in the centre of motion, the centrifugal force of all parts of the body will be equal at equal diftances from its centre of motion, and therefore the body will remain in its place. But if the centre of gravity be placed ever fo little out of the centre of motion, and the machine be turned fwiftly round, the body will fly off towards that fide of the board on which its centre of gravity lies. Then if the wire C (fig. 3) with its little ball B be taken away from the femi-globe A, and the flat fide f of the femiglobe be laid upon the Whirling-board, fo that their centres may coincide ; if then the board be turned ever fo quickly by the winch, the femi-globe will remain where it was placed : but if the wire C be fcrewed into the femi-globe at d, the whole becomes one body, whofe centre of gravity is at or near d. Fix the pin c

in the centre of the Whirling-board, and let the deep groove *b* cut in the flat fide of the femi-globe be put upon the pin, fo that the pin may be in the centre of A (fee fig. 4) where the groove is to be reprefented at *b*, and let the board be turned by the winch, which will carry the little ball B (fig. 3) with its wire C, and the femi-globe A, round the centre-pin *c i* ; and then, the centrifugal force of the little ball B, weighing one ounce, will be fo great as to draw off the femi-globe A, weighing two pounds, until the end of the groove at *c* ſtrikes againſt the pin *c*, and fo prevents A from going any farther : otherwife, the centrifugal force of B would have been great enough to have carried A quite off the whirling-board. Hence we fee that, if the fun were placed in the centre of the orbits of the planets, it could not poffibly remain there ; for the centrifugal forces of the planets would carry them quite off, and the fun with them ; efpecially when feveral of them happened to be in one quarter of the heavens. For the fun and planets are as much connected by the mutual attraction fubfifting between them, as the bodies A and B are by the wire C fixed into them both. And even if there were but one planet in the whole heavens to go round ever fo large a fun in the centre of its orbit, its centrifugal force would foon carry off both itfelf and the fun : for the greateſt body placed in any part of free fpace could be eafily moved ; becaufe, if there were no other body to attract it, it would have no weight or gravity of itfelf, and confequently, though it could have no tendency of itfelf to remove from that part of fpace, yet it might be very eafily moved by any other fubftance.

10. As the centrifugal force of the light body B will not allow the heavy body A to remain in the centre of motion, even though it be 24 times as heavy as B ; let the ball A (fig. 5) weighing 6 ounces be con-nected by the wire C with the ball B, weighing one ounce, and let the fork E be fixed into the centre of the Whirling-board ; then, hang the balls upon the fork by the wire C in fuch a manner that they may exactly balance each other, which will be when the centre of gravity between them, in the wire at *d*, is fupported by the fork. And this centre of gravity is as much nearer to the centre of the ball A than to the centre B, as A is heavier than B ; allow-ing for the weight of the wire on each fide of the fork. Then, let the machine be moved, and the balls A and B will go round their common centre of gra-vity *d*, keeping their balance, becaufe either will not allow the other to fly off with it. For, fuppofing the ball B to be only one ounce in weight, and the ball A to be fix ounces ; then, if the wire C were equally heavy on each fide of the fork, the centre of gravity *d* would be 6 times as far from the centre of B as from the centre of A, and confequently B will revolve with a velocity 6 times as great as A does ; which will give B 6 times as much centrifugal force as any fingle ounce of A has ; but then as B is only one ounce, and A fix ounces, the whole centrifugal force of A will ex-actly balance that of B ; and therefore, each body will detain the other, fo as to make it keep in its circle.

Hence it appears, that the fun and planets muſt all move round the common centre of gravity of the whole

fyſtem, in order to preferve that juſt balance which takes place among them.

11. Take away the forks and balls from the Whirl-ing-board, and place the trough AB (fig. 6) thereon, fixing its centre to that of the board by the pin H. In this trough are two balls D and E of unequal weights, connected by a wire *f*, and made to flide eafily upon the wire ſtretched from end to end of the trough, and made faſt by nut fcrews on the outfide of the ends. Place thefe balls on the wire *c*, fo that their common centre of gravity *g*, may be directly over the centre of the Whirling-board. Then turn the machine by the winch ever fo fwiftly, and the trough and balls will go round their centre of gravity, fo as neither of them will fly off ; becaufe, on account of the equilibrium, each ball detains the other with an equal force acting againſt it. But if the ball E be drawn a little more towards the end of the trough at A, it will remove the centre of gravity towards that end from the centre of mo-tion ; and then, upon turning the machine, the little ball E will fly off, and ſtrike with a confiderable force againſt the end A, and draw the great ball B into the middle of the trough. Or, if the great ball D be drawn towards the end B of the trough, fo that the centre of gravity may be a little towards that end from the centre of motion ; and the machine be turned by the winch, the great ball D will fly off, and ſtrike violently againſt the end B of the trough, and will bring the little ball E into the middle of it. If the trough be not made very ſtrong, the ball D will break through it.

12. Mr. Fergufon has explained the reafon why the tides rife at the fame time on oppofite fides of the earth, and confequently in oppofite directions, by the following new experiment on the Whirling-table. For this purpofe, let *a b c d* (fig. 7) reprefent the earth, with its fide *c* turned toward the moon, which will then attract the water fo as to raife them from *c* to *g* : and in order to ſhew that they will rife as high at the fame time on the oppofite fide from *a* to *e* ; let a plate AB (fig. 8) be fixed upon one end of the flat bar DC, with fuch a circle drawn upon it as *a b c d* (fig. 7) to reprefent the round figure of the earth and fea ; and an ellipfe as *e f g h* to reprefent the fwelling of the tide at *e* and *g*, occafioned by the influence of the moon. Over this plate AB fufpend the three ivory balls *e*, *f*, *g*, by the filk lines *h*, *i*, *k*, faſtened to the tops of the wires H, I, K, fo that the ball at *e* may hang freely over the fide of the circle *e*, which is fartheſt from the moon M at the other end of the bar ; the ball at *f* over the centre, and the ball at *g* over the fide of the circle *g*, which is neareſt the moon. The ball *f* may reprefent the centre of the earth, the ball *g* water on the fide next the moon, and the ball *e* water on the oppofite fide. On the back of the moon M is fixed a ſhort bar N parallel to the horizon, and there are three holes in it above the little weights *p*, *q*, *r*. A filken thread *o* is tied to the line *k* clofe above the ball *g*, and paffing by one fide of the moon M goes through a hole in the bar N, and has the weight *p* hung to it. Such another thread *m* is tied to the line *i*, clofe above the ball *f*, and, paffing through the cen-tre of the moon M and middle of the bar N, has the weight *q* hung to it which is lighter than the weight *p*. A third thread *m* is tied to the line *h*. clofe

above

above the ball *e*, and, passing by the other side, of the moon M through the bar N, has the weight *r* hung. to it, which is lighter than the weight *q*. The use of these three unequal weights is to represent the moon's unequal attraction at different distances from her; so that if they are left at liberty, they will draw all the three balls towards the moon with different degrees of force, and cause them to appear as in fig. 9, in which case they are evidently farther from each other than if they hung freely by the perpendicular lines *h, i, k*. Hence it appears, that as the moon attracts the side of the earth which is nearest her with a greater degree of force than she does the centre of the earth, she will draw the water on that side more than the centre, and cause it to rise on that side: and as she draws the centre more than the opposite side, the centre will recede farther from the surface of the water on that opposite side, and leave it as high there as she raised it on the side next her. For, as the centre will be in the middle between the tops of the opposite elevations, they must of course be equally high on both sides at the same time.

However, upon this supposition, the earth and moon would soon come together; and this would be the case if they had not a motion round their common centre of gravity, to produce a degree of centrifugal force, sufficient to balance their mutual attraction. Such motion they have; for as the moon revolves in her orbit every month, at the distance of 240000 miles from the earth's centre, and of 234000 miles from the centre of gravity of the earth and moon, the earth also goes round the same centre of gravity every month at the distance of 6000 miles from it, i. e. from it to the centre of the earth. But the diameter of the earth being, in round numbers, 8000 miles, its side next the moon is only 2000 miles from the common centre of gravity of the earth and moon, its centre 6000 miles from it, and its farthest side from the moon 10000 miles. Consequently the centrifugal forces of these parts are as 2000, 6000, and 10000; i. e. the centrifugal force of any side of the earth, when it is turned from the moon, is five times as great as when it is turned toward the moon. And as the moon's attraction, expressed by the number 6000 at the earth's centre, keeps the earth from flying out of this monthly circle, it must be greater than the centrifugal force of the waters on the side next her; and consequently, her greater degree of attraction on that side is sufficient to raise them; but as her attraction on the opposite side is less than the centrifugal force of the water there, the excess of this force is sufficient to raise the water just as high on the opposite side.

To prove this experimentally, let the bar DC with its furniture be fixed on the Whirling-board of the machine (fig. 1.) by pushing the pin P into the centre of the board; which pin is in the centre of gravity of the whole bar with its three balls, *e, f, g*, and moon M. Now if the Whirling-board and bar be turned slowly round by the winch, till the ball *f* hangs over the centre of the circle, as in fig. 10, the ball *g* will be kept towards the moon by the heaviest weight *p* (fig. 8), and the ball *e*, on account of its greater centrifugal force, and the less weight *r*, will fly off as far to the other side, as in fig. 10. And thus, whilst the

machine is kept turning, the balls *e* and *g* will hang over the ends of the ellipse *l f k*. So that the centrifugal force of the ball *e* will exceed the moon's attraction just as much as her attraction exceeds the centrifugal force of the ball *g*, whilst her attraction just balances the centrifugal force of the ball *f*, and makes it keep in its circle. Hence it is evident, that the tides must rise to equal heights at the same time on opposite sides of the earth. See Ferguson's Lectures on Mechanics, lect. 2, and Desag. Ex. Phil. vol. 1, lect. 5.

WHIRLWIND, a wind that rises suddenly, is exceedingly rapid and impetuous, in a Whirling direction, and often progressively also; but it is commonly soon spent.

Dr. Franklin, in his Physical and Meteorological Observations, read to the Royal Society in 1756, supposes a Whirlwind and a waterspout to proceed from the same cause: their only difference being, that the latter passes over the water, and the former over the land. This opinion is corroborated by the observations of M. de la Pryme, and many others, who have remarked the appearances and effects of both to be the same. They have both a progressive as well as a circular motion; they usually rise after calms and great heats, and mostly happen in the warmer latitudes: the wind blows every way from a large surrounding space, both to the waterspout and whirlwind; and a waterspout has, by its progressive motion, passed from the sea to the land, and produced all the phenomena and effects of a Whirlwind: so that there is no reason to doubt that they are meteors arising from the same general cause, and explicable upon the same principles, furnished by electrical experiments and discoveries. See HURRICANE, and WATERSPOUT. For Dr. Franklin's ingenious method of accounting for both these phenomena, see his Letters and Papers, &c, vol. p. 191, 216, &c.

WHISPERING-*Places*, are places where a Whisper, or other small noise, may be heard from one part to another, to a great distance. They depend on a principle, that the voice, &c, being applied to one end of an arch, easily passes by repeated reflections to the other. Thus,

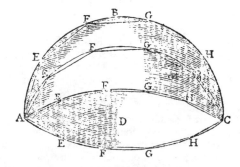

let ABC represent the segment of a sphere; and suppose a low voice uttered at A, the vibrations extending themselves every way, some of them will impinge upon the points E, E, &c; and thence will be reflected to the points F, F, &c; thence to G, G, &c; till at last they meet in C; where by their union they cause a much stronger sound than in any part of the segment whatever,

whatever, even louder than at the point from whence they set out. Accordingly, all the contrivance in a Whispering-place is, that near the person who Whispers, there be a smooth wall, arched either cylindrically, or elliptically, &c. A circular arch will do, but not so well.

Some of the most remarkable places for Whispering, are the following : viz, The prison of Dionysius at Syracuse, which increased a soft Whisper to a loud noise ; or a clap of the hand to the report of a cannon, &c. The aqueducts of Claudius, which carried a voice 16 miles : beside divers others mentioned by Kircher in his Phonurgia. In England, the most considerable Whispering places are, the dome of St. Paul's church, London, where the ticking of a watch may be heard from side to side, and a very soft Whisper may be sent all round the dome : this Dr. Derham found to hold not only in the gallery below, but above upon the scaffold, where a Whisper would be carried over a person's head round the top of the arch, though there be a large opening in the middle of it into the upper part of the dome. And the celebrated Whispering-place in Gloucester cathedral, which is only a gallery above the east end of the choir, leading from one side of it to the other. See Birch's Hist. of the Royal Soc. vol. 1, pa. 120.

WHISTON (WILLIAM), an English divine, philosopher, and mathematician, of uncommon parts, learning, and extraordinary character, was born the 9th of December 1667, at Norton in the county of Leicester, where his father was rector. He was educated under his father till he was 17 years of age, when he was sent to Tamworth school, and two years after admitted of Clare-hall, Cambridge, where he pursued his studies, and particularly the mathematics, with great diligence. During this time he became afflicted with a great weakness of sight, owing to close study in a whitened room ; which was in a good measure relieved by a little relaxation from study, and taking off the strong glare of light by hanging the place opposite his seat with green.

In 1693 he was made master of arts and fellow of the college, and soon after commenced one of the tutors ; but his ill state of health soon after obliged him to relinquish this profession. Having entered into orders, in 1694 he became chaplain to Dr. More, bishop of Norwich ; and while in this station he published his first work, intitled, *A New Theory of the Earth &c* ; in which he undertook to prove that the Mosaic doctrine of the earth was perfectly agreeable to reason and philosophy : which work, having much ingenuity, though it was written against by Mr. John Keill, brought considerable reputation to the author.

In the year 1698, bishop More gave him the living of Lowestoff in Suffolk, where he immediately went to reside, and devoted himself with great diligence to the discharge of that trust.—In the beginning of this century he was made Sir Isaac Newton's deputy, and afterwards his successor in the Lucasian professorship of mathematics ; when he resigned his living at Lowestoff, and went to reside at Cambridge. From this time his publications became very frequent, both in theology and mathematics. Thus, in 1702 he published, A Short View of the Chronology of the Old Testament,

and of the Harmony of the four Evangelists.—In 1707, *Prælectiones Astronomicæ* ; beside eight Sermons on the Accomplishment of the Scripture Prophecies, preached at Boyle's lecture ; and Newton's Arithmetica Universalis.—In 1708, Tacquet's Euclid, with select Theorems of Archimedes ; the former of which had accidentally been his first introduction to the study of the mathematics.—In the same year he drew up an Essay upon the Apostolical Constitutions, which the Vice-chancellor refused his licence for printing. The author tells us, he had read over the two first centuries of the church, and found that the Eusebian or Arian doctrine was chiefly the doctrine of those ages, which, though deemed heterodox, he thought it his duty to discover. —In 1709, he published a volume of Sermons and Essays on various subjects.—In 1710, Prælectiones Physico Mathematicæ, which, with the Prælectiones Astronomicæ, were translated and published in English. And it may be said, with no small honour to the memory of Mr. Whiston, that he was one of the first who explained the Newtonian philosophy in a popular way, so as to be intelligible to the generality of readers.— Among other things also, he translated the Apostolical Constitutions into English, which favoured the doctrine of the supremacy of the father and subordination of the son, vulgarly called the Arian heresy : Upon which his friends began to be alarmed for him ; and the consequence shewed it was not groundless ; for, Oct. 30, 1710, he was deprived of his professorship, and expelled the university of Cambridge, after he had been formally convened and interrogated for some days together.—At the conclusion of this year, he wrote his Historical preface, afterwards prefixed to his Primitive Christianity Revived, containing the reasons for his dissent from the commonly received notions of the Trinity, which work he published the next year, in 4 volumes 8vo, for which the Convocation fell upon him most vehemently.

In 1713, he and Mr. Ditton composed their scheme for finding the longitude, which they published the year following, a method which consisted in measuring distances by means of the velocity of sound ; some more particulars of which are related in the life of Mr. Ditton.—In 1719, he published an ironical Letter of Thanks to doctor Robinson, bishop of London, for his late Letter to his clergy against the use of New Forms of Doxology. And, the same year, a Letter to the earl of Nottingham, Concerning the Eternity of the Son of God, and his Holy Spirit.—In 1720, he was proposed by Sir Hans Sloane and Dr. Halley to the Royal Society as a member; but was refused admittance by Sir Isaac Newton the president.

On Mr. Whiston's expulsion from Cambridge, he went to London, where he conferred with Doctors Clarke, Hoadly, and other learned men, who endeavoured to moderate his zeal, which however he would not suffer to be tainted or corrupted, and many were not much satisfied with the authority of these constitutions, but approved his integrity. Mr. Whiston now settled in London with his family ; where, without suffering his zeal to be intimidated, he continued to write, and to propagate his Primitive Christianity with as much ardour as if he had been in the most flourishing circumstances ; which however were so bad, that, in

1721, a fubfcription was made for the fupport of his family, which amounted to 470l. For though he drew fome profits from reading aftronomical and philofophical lectures, and alfo from his publications, which were very numerous, yet thefe of themfelves were very infufficient : nor, when joined with the benevolence and charity of thofe who loved and efteemed him for his learning, integrity, and piety, did they prevent his being frequently in great diftrefs.—In 1722 he publifhed an Effay towards reftoring the true text of the Old Teftament.—In 1724, The Literal Accomplifhment of Scripture Prophecies.—Alfo, The Calculation of Solar Eclipfes without Parallaxes.—In 1726, Of the Thundering Legion &c.—In 1727, A Collection of Authentic Records belonging to the Old and New Teftament.—In 1730, Memoirs of the Life of Dr. Samuel Clarke.—In 1732, A Vindication of the Teftimony of Phlegon, or an Account of the Great Darknefs and Earthquake at our Saviour's Paffion, defcribed by Phlegon.—In 1736, Athanafian Forgeries, &c. And the Primitive Eucharift revived.—In 1737, The Aftronomical Year, particularly of the Comet foretold by Sir Ifaac Newton.—Alfo the Genuine Works of Flavius Jofephus.—In 1739, Mr. Whifton put in his claim to the mathematical profefforfhip at Cambridge, then vacant by the death of Dr. Saunderfon, in a letter to Dr. Afhton, the mafter of Jefus-college; but no regard was paid to it.—In 1745, he publifhed his Primitive New Teftament in Englifh.—In 1748, his Sacred Hiftory of the Old and New Teftament. Alfo, Memoirs of his own Life and Writings, which are very curious.

Whifton continued many years a member of the eftablifhed church ; but at length forfook it, on account of the reading of the Athanafian Creed, and went over to the Baptifts; which happened while he was at the houfe of Samuel Barker, Efq. at Lindon in Rutlandfhire, who had married his daughter; where he died, after a week's illnefs, the 22d of Auguft 1752, at upwards of 84 years of age.—We have mentioned the principal of his writings in the foregoing memoir ; to which may be added, Chronological Tables, publifhed in 1750.

The character of this confcientious and worthy man has been attempted by two very able perfonages, who were well acquainted with him, namely, bifhop Hare and Mr. Collins, who unite in giving him the higheft applaufes, for his integrity, piety, &c.—Mr. Whifton left fome children behind him ; among them, Mr. John Whifton, who was for many years a very confiderable bookfeller in London.

WHITE, one of the colours of bodies. Though White cannot properly be faid to be one colour, but rather a compofition of all the colours together: for Newton has demonftrated that bodies only appear White by reflecting all the kinds of coloured rays alike ; and that even the light of the fun is only White, becaufe it confifts of all colours mixed together.

This may be fhewn mechanically in the following manner: Take feven parcels of coloured fine powders, the fame as the primary colours of the rainbow, taking fuch quantities of thefe as fhall be proportional to the refpective breadths of thefe colours in the rainbow, which are of red 45 parts, orange 27, yellow 48, green 60, blue 60, indigo 40, and of violet 80 ; then mix intimately together thefe feven parcels of powders, and the mixture will be a pretty White colour : and this is only fimilar to the uniting the prifmatic colours together again, to form a White ray or pencil of light of the whole of them. The fame thing is done conveniently thus : Let the flat upper furface of a top be divided into 360 equal parts, all around its edge ; then divide the fame furface into feven fectors in the proportion of the numbers above, by feven radii or lines drawn from the centre ; next let the refpective colours be painted in a lively manner on thefe fpaces, but fo as the edge of each colour may be made nearly like the colour next adjoining, that the feparation may not be well diftinguifhed by the eye ; then if the top be made to fpin, the colours will thus feem to be mixed all together, and the whole furface will appear of a uniform whitenefs: and if a large round black fpot be painted in the middle, fo as there may be only a broad flat ring of colours around it, the experiment will fucceed the better. See Newton's Optics, prop. 6, book 1 ; and Fergufon's Tracts, pa. 296.

White bodies are found to take heat flower than black ones ; becaufe the latter abforb or imbibe rays of all kinds and colours, and the former reflect them. Hence it is that black paper is fooner put in flame, by a burning glafs, than White ; and hence alfo black clothes, hung up in the fun by the dyers, dry fooner than white ones.

WHITEHURST (JOHN), an ingenious Englifh philofopher, was born at Congleton in the county of Chefhire, the 10th of April 1713, being the fon of a clock and watch-maker there. Of the early part of his life but little is known ; he who dies at an advanced age, leaving few behind him to communicate anecdotes of his youth. On his quitting fchool, where it feems the education he received was very defective, he was bred by his father to his own profeffion, in which he foon gave hopes of his future eminence.

It was very early in life that, from his vicinity to the many ftupendous phenomena in Derbyfhire, which were conftantly prefented to his obfervation, his attention was excited to enquire into the various caufes of them.

At about the age of 21, his eagernefs after new ideas carried him to Dublin, having heard of an ingenious piece of mechanifm in that city, being a clock with certain curious appendages, which he was very defirous of feeing, and no lefs fo of converfing with the maker On his arrival however, he could neither procure a fight of the former, nor draw the leaft hint from the latter concerning it. Thus difappointed, he fell upon an expedient for accomplifhing his defign ; and accordingly took up his refidence in the houfe of the mechanic, paying the more liberally for his board, as he had hopes from thence of more readily obtaining the indulgence wifhed for. He was accommodated with a room directly over that in which the favourite piece was kept carefully locked up : and he had not long to wait for his gratification : for the artift, while one day employed in examining his machine, was fuddenly called down ftairs ; which the young enquirer happening to overhear, foftly flipped into the room, infpected the machine, and, prefently fatisfying himfelf as to the fecret, efcaped undif-

6
covered

covered to his own apartment. His end thus compaffed, he fhortly after bid the artift farewell, and returned to his father in England.

About two or three years after his return from Ireland, he left Congleton, and entered into bufinefs for himfelf at Derby, where he foon got into great employment, and diftinguifhed himfelf very much by feveral ingenious pieces of mechanifm, both in his own regular line of bufinefs, and in various other refpects, as in the conftruction of curious thermometers, barometers, and other philofophical inftruments, as well as in ingenious contrivances for water-works, and the erection of various larger machines: being confulted in almoft all the undertakings in Derbyfhire, and in the neighbouring counties, where the aid of fuperior fkill, in mechanics, pneumatics, and hydraulics, was requifite.

In this manner his time was fully and ufefully employed in the country, till, in 1775, when the act paffed for the better regulation of the gold coin, he was appointed ftamper of the money-weights; an office conferred upon him, altogether unexpectedly, and without folicitation. Upon this occafion he removed to London, where he fpent the remainder of his days, in the conftant habits of cultivating fome ufeful parts of philofophy and mechanifm. And here too his houfe became the conftant refort of the ingenious and fcientific at large, of whatever nation or rank, and this to fuch a degree, as very often to impede him in the regular profecution of his own fpeculations.

In 1778, Mr. Whitehurft publifhed his Inquiry into the Original State and Formation of the Earth; of which a fecond edition appeared in 1786, confiderably enlarged and improved; and a third in 1792. This was the labour of many years; and the numerous inveftigations neceffary to its completion, were in themfelves alfo of fo untoward a nature, as at times, though he was naturally of a ftrong conftitution, not a little to prejudice his health. When he firft entered upon this fpecies of refearch, it was not altogether with a view to inveftigate the formation of the earth, but in part to obtain fuch a competent knowledge of fubterraneous geography as might become fubfervient to the purpofes of human life, by leading mankind to the difcovery of many valuable fubftances which lie concealed in the lower regions of the earth.

May the 13th, 1779, he was elected and admitted a Fellow of the Royal Society. He was alfo a member of fome other philofophical focieties, which admitted him of their refpective bodies, without his previous knowledge; but fo remote was he from any thing that might favour of oftentation, that this circumftance was known only to a very few of his moft confidential friends. Before he was admitted a member of the Royal Society, three feveral papers of his had been inferted in the Philofophical Tranfactions, viz, Thermometrical Obfervations at Derby, in vol. 57; An Account of a Machine for raifing Water, at Oulton, in Chefhire, in vol. 65; and Experiments on Ignited Subftances, vol. 66: which three papers were printed afterwards in the collection of his works in 1792.

In 1783 he made a fecond vifit to Ireland, with a view to examine the Giant's Caufeway, and other northern parts of that ifland, which he found to be chiefly compofed of volcanic matter: an account and reprefenta-

tions of which are inferted in the latter editions of his Inquiry. During this excurfion, he erected an engine, for raifing water from a well, to the fummit of a hill, in a bleaching ground, at Tullidoi, in the county of Tyrone: it is worked by a current of water, and for its utility is perhaps unequalled in any country.

In 1787 he publifhed, An Attempt toward obtaining Invariable Meafures of Length, Capacity, and Weight, from the Menfuration of Time. His plan is, to obtain a meafure of the greateft length that conveniency will permit, from two pendulums whofe vibrations are in the ratio of 2 to 1, and whofe lengths coincide nearly with the Englifh ftandard in whole numbers. The numbers which he has chofen fhew much ingenuity. On a fuppofition that the length of a feconds pendulum, in the latitude of London, is $39\frac{1}{5}$ inches, the length of one vibrating 42 times in a minute, muft be 80 inches; and of another vibrating 84 times in a minute muft be 20 inches; and their difference, 60 inches, or 5 feet, is his ftandard meafure. By the experiments however, the difference between the lengths of the two pendulum rods, was found to be only 59·892 inches, inftead of 60, owing to the error in the affumed length of the feconds pendulum, $39\frac{1}{5}$ inches being greater than the truth, which ought to be $39\frac{1}{8}$ very nearly. By this experiment, Mr. Whitehurft obtained a fact, as accurately as may be in a thing of this nature, viz, the difference between the lengths of two pendulum rods whofe vibrations are known: a datum from whence may be obtained, by calculation, the true lengths of pendulums, the fpaces through which heavy bodies fall in a given time, and many other particulars relating to the doctrine of gravitation, the figure of the earth, &c, &c.

Mr. Whitehurft had been at times fubject to flight attacks of the gout, and he had for feveral years felt himfelf gradually declining. By an attack of that difeafe in his ftomach, after a ftruggle of two or three months, it put an end to his laborious and ufeful life, on the 18th of February 1788, in the 75th year of his age, at his houfe in Bolt-court, Fleet-ftreet, being the fame houfe where another eminent felf-taught philofopher, Mr. James Ferguson, had immediately before him lived and died.

For feveral years before his death, Mr. Whitehurft had been at times occupied in arranging and completing fome papers, for a treatife on Chimneys, Ventilation, and Garden-ftoves; which have fince been collected and given to the public, by Dr. Willan, in 1794.

However refpectable Mr. Whitehurft may have been in mechanics, and thofe parts of natural fcience which he more immediately cultivated, he was of ftill higher account with his acquaintance and friends on the fcore of his moral qualities. To fay nothing of the uprightnefs and punctuality of his dealings in all tranfactions relative to bufinefs; few men have been known to poffefs more benevolent affections than he, or, being poffeffed of fuch, to direct them more judiciously to their proper ends. As to his perfon, he was above the middle ftature, rather thin than otherwife, and of a countenance expreffive at once of penetration and mildnefs. His fine gray locks, unpolluted by art, gave a venerable air to his whole appearance. In drefs he was plain, in diet temperate, in his general intercourfe with
mankind

mankind eafy and obliging. In company he was cheerful or grave alike, according to the dictate of the occafion; with now and then a peculiar fpecies of humour about him, delivered with fuch gravity of manner and utterance, that thofe who knew him but flightly were apt to underftand him as ferious, when he was merely playful. But where any defire of information on fubjects in which he was converfant was expreffed, he omitted no opportunity of imparting it.

WHITSUNDAY, the 50th day or feventh funday from Eafter.—The feafon properly called Pentecoft, is popularly called *Whitfuntide*; becaufe, it is faid, in the primitive church, the newly baptized perfons came to church between Eafter and Pentecoft in *white* garments.

WILKINS (Dr. John), a very ingenious and learned Englifh bifhop and mathematician, was the fon of a goldfmith at Oxford, and born in 1614. After being educated in Greek and Latin, in which he made a very quick progrefs, he was entered a ftudent of New-Inn in that univerfity, when he was but 13 years of age; but after a fhort ftay there, he was removed to Magdalen Hall; where he took his degrees. Having entered into holy orders, he firft became chaplain to William Lord Say, and afterwards to Charles Count Palatine of the Rhine, with whom he continued fome time. Adhering to the Parliament during the civil wars, they made him warden of Wadham college about the year 1648. In 1656 he married the fifter of Oliver Cromwell, then lord protector of England, who granted him a difpenfation to hold his wardenfhip, notwithftanding his marriage. In 1659, he was by Richard Cromwell made mafter of Trinity college in Cambridge; but ejected the year following, upon the reftoration. He was then chofen preacher to the fociety of Gray's Inn, and rector of St. Lawrence Jewry, London, upon the promotion of Dr. Seth Ward to the bifhoprick of Exeter. About this time he became a member of the Royal Society, was chofen of their council, and proved one of their moft eminent members. He was afterwards made dean of Rippon, and in 1668 bifhop of Chefter; but died of the ftone in 1672, at 58 years of age.

Bifhop Wilkins was a man who thought it prudent to fubmit to the powers in being; he therefore fubfcribed to the folemn league and covenant, while it was enforced; and was equally ready to fwear allegiance to king Charles when he was reftored: this, with his moderate fpirit towards diffenters, rendered him not very agreeable to the churchmen; and yet feveral of them could not but give him one of the beft of characters. Burnet writes, that " he was a man of as great a mind, as true a judgment, as eminent virtues, and of as good a foul, as any he ever knew: that though he married Cromwell's fifter, yet he made no other ufe of that alliance, but to do good offices, and to cover the univerfity of Oxford from the fournefs of Owen and Goodwin. At Cambridge, he joined with thofe who ftudied to propagate better thoughts, to take men off from being in parties. or from narrow notions, from fuperftitious conceits, and fiercenefs about opinions. He was alfo a great obferver and promoter of experimental philofophy, which was then a new thing, and much looked after. He was naturally ambitious, but was the wifeft clergyman I ever knew. He was a lover of

mankind, and had a delight in doing good." The fame hiftorian mentions afterwards another quality which Wilkins poffeffed in a fupreme degree, and which it was well for him he did, fince he had great occafion for the ufe of it; and that was, fays he, " a courage, which could ftand againft a current, and againft all the reproaches with which ill-natured clergymen ftudied to load him."

Of his publications, which are all of them very ingenious and learned, and many of them particularly curious and entertaining, the firft was in 1638, when he was only 24 years of age, viz. The Difcovery of a New World; or, A Difcourfe to prove, that it is probable there may be another Habitable World in the Moon; with a Difcourfe concerning the Poffibility of a Paffage thither.—In 1640, A Difcourfe concerning a New Planet, tending to prove that it is probable our earth is one of the Planets.—In 1641, Mercury; or, the Secret and Swift Meffenger; fhewing, how a man may with Privacy and Speed communicate his Thoughts to a Friend at any Diftance, 8vo.—In 1648, Mathematical Magic; or, the Wonders that may be performed by Mathematical Geometry, 8vo. All thefe pieces were publifhed entire in one volume 8vo, in 1708, under the title of, The Mathematical and Philofophical Works of the right rev. John Wilkins, &c; with a print of the author and general title page handfomely engraven, and an account of his life and writings. To this collection is alfo fubjoined an abftract of a larger work, printed in 1668, folio, intitled, An Effay towards a Real Character and a Philofophical Language. Thefe were all his mathematical and philofophical works; befide which, he wrote feveral tracts in theology, natural religion, and civil polity, which were much efteemed for their piety and moderation, and went through feveral-editions.

WINCH, a popular term for a windlafs. Alfo the bent handle for turning round wheels, grind-ftones, &c.

WIND, a current, or ftream of air, efpecially when it is moved by fome natural caufe.

Winds are denominated from the point of the compafs or horizon they blow from; as the eaft Wind, north Wind, fouth Wind, &c.

Winds are alfo divided into feveral kinds; as *general, particular, perennial, ftated, variable, &c.*

Conftant or *Perennial* WINDS, are thofe that always blow the fame way; fuch as the remarkable one between the two tropics, blowing conftantly from eaft to weft, called alfo the *general trade-Wind.*

Stated or *Periodical* WINDS, are thofe that conftantly return at certain times. Such are the fea and land breezes, blowing from land to fea in the morning, and from fea to land in the evening. Such alfo are the fhifting or particular trade Winds, which blow one way during certain months of the year, and the contrary way the reft of the year.

Variable or *Erratic* WINDS, are fuch as blow without any regularity either as to time, place, or direction. Such as the Winds that blow in the interior parts of England, &c: though fome of thefe claim their certain times of the day; as, the north-Wind is moft frequent in the morning, the weft-Wind about noon, and the fouth-Wind in the night.

General

General WIND, is such as blows at the same time the same way, over a very large tract of ground, most part of the year; as the general trade Wind.

Particular WINDS, include all others, excepting the general trade Winds.

Those peculiar to one little canton or province, are called *topical* or *provincial Winds*. The Winds are also divided, with respect to the points of the compass or of the horizon, into *cardinal* and *collateral*.

Cardinal WINDS, are those blowing from the four cardinal points, east, west, north, and south.

Collateral WINDS, are the intermediate Winds between any two cardinal Winds, and take their names from the point of the compass or horizon they blow from.

In Navigation, when the Wind blows gently, it is called a *breeze*; when it blows harder, it is called a *gale*, or a *stiff gale*; and when it blows very hard, a *storm*.

For a particular account of the trade-Winds, monsoons, &c, see Philos. Transf. number 183, or Abridg. vol. 2, p. 133. Also Robertson's Navigation book 5, sect. 6.

A Wind blowing from the sea, is always moist; as bringing with it the copious evaporation and exhalations from the waters: also, in summer, it is cool; and in winter warm. On the contrary, a Wind from the continent, is always dry; warm in summer, and cold in winter. Our northerly and southerly Winds however, which are usually accounted the causes of cold and warm weather, Dr. Derham observes, are really rather the effect of the cold or warmth of the atmosphere. Hence it is that we often find a warm southerly Wind suddenly change to the north, by the fall of snow or hail; and in a cold frosty morning, we find the Wind north; which afterward wheels about to the southerly quarter, when the sun has well warmed the air; and again in the cold evening, turns northerly, or easterly.

Physical Cause of WINDS. Some philosophers, as Descartes, Rohault, &c, account for the general Wind, from the diurnal rotation of the earth; and from this general Wind they derive all the particular ones. Thus, as the earth turns eastward, the particles of the air near the equator, being very light, are left behind; so that, in respect of the earth's surface, they move westwards, and become a constant easterly wind, as they are found between the tropics, in those parallels of latitude where the diurnal motion is swiftest. But yet, against this hypothesis, it is urged, that the air, being kept close to the earth by the principle of gravity, would in time acquire the same degree of velocity that the earth's surface moves with, as well in respect of the diurnal rotation, as of the annual revolution about the sun, which is about 30 times swifter.

Dr. Halley therefore substitutes another cause, capable of producing a like constant effect, not liable to the same objections, but more agreeable to the known properties of the elements of air and water, and the laws of the motion of fluid bodies. And that is the action of the sun's beams, as he passes every day over the air, earth, and water, combined with the situation of the adjoining continents. Thus, the air which is less rarefied or expanded by heat, must have a motion towards those parts which are more rarefied, and less ponderous, to bring the whole to an equilibrium; and as the sun keeps continually shifting to the westward, the tendency of the whole body of the lower air is that way. Thus a general easterly Wind is formed, which being impressed upon the air of a vast ocean, the parts impel one another, and so keep moving till the next return of the sun, by which so much of the motion as was lost, is again restored; and thus the easterly Wind is made perpetual. But as the air towards the north and south is less rarefied than in the middle, it follows that from both sides it ought to tend towards the equator.

This motion, compounded with the former easterly Wind, accounts for all the phenomena of the general trade-Winds, which, if the whole surface of the globe were sea, would blow quite round the world, as they are found to do in the Atlantic and the Ethiopic oceans. But the large continents of land in this middle tract, being excessively heated, communicate their heat to the air above them, by which it is exceedingly rarefied, which makes it necessary that the cooler and denser air should rush in towards it, to restore the equilibrium. This is supposed to be the cause why, near the coast of Guinea, the wind always sets in upon the land, blowing westerly instead of easterly.

From the same cause it happens, that there are such constant calms in that part of the ocean called the *rains*; for this tract being placed in the middle, between the westerly Winds blowing on the coast of Guinea, and the easterly trade-Winds blowing to the westward of it; the tendency of the air here is indifferent to either, and so stands in equilibrio between both; and the weight of the incumbent atmosphere being diminished by the continual contrary Winds blowing from hence, is the reason that the air here retains not the copious vapour it receives, but lets it fall in so frequent rains.

It is also to be considered, that to the northward of the Indian ocean there is every where land, within the usual limits of the latitude of 30°, viz, Arabia, Persia, India, &c, which are subject to excessive heats when the sun is to the north, passing nearly vertical; but which are temperate enough when the sun is removed towards the other tropic, because of a ridge of mountains at some distance within the land, said to be often in winter covered with snow, over which the air as it passes must needs be much chilled. Hence it happens that the air coming, according to the general rule, out of the north-east, to the Indian sea, is sometimes hotter, sometimes colder, than that which, by a circulation of one current over another, is returned out of the south-west; and consequently sometimes the under current, or Wind, is from the north-east, sometimes from the south-west.

That this has no other cause, appears from the times when these Winds set, viz, in April: when the sun begins to warm these countries to the north, the southwest monsoons begin, and blow during the heats till October, when the sun being retired, and all things growing cooler northward, but the heat increasing to the south, the north east Winds enter, and blow all the winter, till April again. And it is doubtless from the same principle, that to the southward of the equator, in part of the Indian ocean, the north-west Winds succeed the south-east, when the sun draws near the tropic

tropic of Capricorn. Philof. Tranf. num. 183; or Abridg. vol. 2, pa. 193.

But fome philofophers, not fatisfied with Dr. Halley's theory above recited, or thinking it not fufficient for explaining the various phenomena of the Wind, have had recourfe to another caufe, viz, the gravitation of the earth and its atmofphere towards the fun and moon, to which the tides are confeffedly owing. They allege that, though we cannot difcover aërial tides, of ebb or flow, by means of the barometer, becaufe columns of air of unequal height, but different denfity, may have the fame preffure or weight; yet the protuberance in the atmofphere, which is continually following the moon, muft, fay they, occafion a motion in all parts, and fo produce a Wind more or lefs to every place, which confpiring with, or being counteracted by the Winds arifing from other caufes, makes them greater or lefs. Several differtations to this purpofe were publifhed, on occafion of the fubject propofed by the Academy of Sciences at Berlin, for the year 1746. But Muffchenbroek will not allow that the attraction of the moon is the caufe of the general Wind; becaufe the eaft Wind does not follow the motion of the moon about the earth; for in that cafe there would be more than 24 changes, to which it would be fubject in the courfe of a year, inftead of two. Introd. ad Phil. Nat. vol. 2, pa. 1102.

And Mr. Henry Eeles, conceiving that the rarefaction of the air by the fun cannot fimply be the caufe of all the regular and irregular motions which we find in the atmofphere, afcribes them to another caufe, viz, the afcent and defcent of vapour and exhalation, attended by the electrical fire or fluid; and on this principle he has endeavoured to explain at large the general phenomena of the weather and barometer. Philof. Tranf. vol. 49, pa. 124.

Laws of the Production of Wind.

The chief laws concerning the production of Wind, may be collected under the following heads.

1. If the fpring of the air be weakened in any place more than in the adjoining places, a Wind will blow through the place where the diminution is; becaufe the lefs elaftic or forcible will give way to that which is more fo, and thence induce a current of air into that place, or a Wind. Hence, becaufe the fpring of the air increafes, as the compreffing weight increafes, and compreffed air is denfer than that which is lefs compreffed; all Winds blow into rarer air, out of a place filled with a denfer.

2. Therefore, becaufe a denfer air is fpecifically heavier than a rarer; an extraordinary lightnefs of the air in any place muft be attended with extraordinary Winds, or ftorms. Now, an extraordinary fall of the mercury in the barometer fhewing an extraordinary lightnefs of the atmofphere, it is no wonder if that foretels ftorms of Wind and rain.

3. If the air be fuddenly condenfed in any place, its fpring will be fuddenly diminifhed: and hence, if this diminution be great enough to affect the barometer, a Wind will blow through the condenfed air. But fince the air cannot be fuddenly condenfed, unlefs it has before been much rarefied, a Wind will blow through the air, as it cools, after having been violently heated.

4. In like manner, if air be fuddenly rarefied, its fpring is fuddenly increafed; and it will therefore flow through the air not acted on by the rarefying force. Hence a Wind will blow out of a place, in which the air is fuddenly rarefied; and on this principle probably it is, that the fun, by rarefying the air, muft have a great influence on the production of Winds:

5. Moft caves are found to emit Wind, either more or lefs. Muffchenbroek has enumerated a variety of caufes that produce Winds, exifting in the bowels of the earth, on its furface, in the atmofphere, and above it. See Introd. ad Phil. Nat. vol. 2, pa. 1116.

6. The rifing and changing of the Winds are determined by weathercocks, placed on the tops of high buildings, &c. But thefe only indicate what paffes about their own height, or near the furface of the earth. And Wolfius affures us, from obfervations of feveral years, that the higher Winds, which drive the clouds, are different from the lower ones, which move the weathercocks. Indeed it is no uncommon thing to fee one tier of clouds driven one way by a Wind, and another tier juft over the former driven the contrary way, by another current of air, and that often with very different velocities. And the late experiments with air balloons have proved the frequent exiftence of counter Winds, or currents of air, even when it was not otherwife vifible, nor at all expected; by which they have been found to take very different and unexpected courfes, as they have afcended higher and higher in the atmofphere.

Laws of the Force and Velocity of the Wind.

Wind being only air in motion, and the motion of a fluid againft a body at reft, creating the fame refiftance as when the body moves with the fame velocity through the fluid at reft; it follows, that the force of the Wind, and the laws of its action upon bodies, may be referred to thofe of their refiftance when moved through it; and as thefe circumftances have been treated pretty fully under the article RESISTANCE of the Air, there is no occafion here to make a repetition of them. We there laid down both the quantity and laws of fuch a force, upon bodies of different fhapes and fizes, moving with all degrees of velocity up to 2000 feet per fecond, and alfo for planes fet at all degrees of obliquity, or inclination to the direction of motion; all which circumftances having, for the firft time, been determined by real experiments.

As to the Velocity of the Wind: philofophers have made ufe of various methods for determining it. The method employed by Dr. Derham, was by letting light downy feathers fly in the air, and nicely obferving the diftance to which they were carried in any number of half feconds. He fays that he thus meafured the velocity of the Wind in the great ftorm of Auguft 1705, which he found moved at the rate of 33 feet in half a fecond, or 45 miles per hour: whence he concludes, that the moft vehement Wind does not fly at the rate of above 50 or 60 miles an hour; and that at a medium the velocity of Wind is at the rate of 12 or 15 miles per hour. Philof. Tranf. number 313, or Abridg. vol. 4, p. 411.

Mr. Brice obferves however, that experiments with feathers are liable to much uncertainty; as they hardly

ever

ever go forward in a ftraight direction, but fpirally, or elfe irregularly from fide to fide, or up and down.

He therefore confiders the motion of a cloud, by means of its fhadow over the furface of the earth, as a much more accurate meafure of the velocity of the Wind. In this way he found that the Wind, in a confiderable ftorm, moved at the rate of near 63 miles an hour; and when it blew a frefh gale, at the rate of 21 miles per hour; and in a fmall breeze it was near 10 miles an hour. Philof. Tranf. vol. 56, p. 226.

The velocity and force of the Wind are alfo determined experimentally by various machines, called *anemometers*, *wind-meafurers*, or *wind-gages*; the defcription of which fee under thefe articles.

In the Philof. Tranf. for 1759, p. 165, Mr. Smeaton has given a table, communicated to him by a Mr. Roufe, for fhewing the force of the Wind, with feveral different velocities, which I fhall infert below, as I find the numbers nearly agree with my own experiments made on the refiftance of the air, when the refifting furfaces are reduced to the fame fize, by a due proportion for the refiftance, which is in a higher degree than that of the furfaces.

N. B. The table of my refults is printed in pa. 111, vol. 1, under the article ANEMOMETER; where it is to be noted, that the numbers in the third column of that table, for the velocity of the Wind per hour, are all erroneoufly printed, only the 4th part of what each of them ought to be; fo that thofe numbers muft be all multiplied by 4.

A Table of the different Velocities and Forces of the Wind, according to their common appellations.

Velocity of the Wind		Perpendicular force on one fq. foot, in averdupois pounds.	Common appellations of the Winds.
Miles in one hour.	= feet in one fecond.		
1	1·47	·005	Hardly perceptible.
2	2·93	·020	} Juft perceptible.
3	4·40	·044	
4	5·87	·079	} Gentle pleafant wind.
5	7·33	·123	
10	14·67	·492	} Pleafant brifk gale.
15	22·00	1·107	
20	29·34	1·968	} Very brifk.
25	36·67	3·075	
30	44·01	4·429	} High Winds.
35	51·34	6·027	
40	58·68	7·873	} Very high.
45	66·01	9·963	
50	73·35	12·300	A ftorm or tempeft.
60	88·02	17·715	A great ftorm.
80	117·36	31·490	A hurricane.
100	146·70	49·200	{ A hurricane that tears up trees, and carries buildings &c before it.

The force of the Wind is nearly as the fquare of the velocity, or but little above it, in thefe velocities. But the force is much more than in the fimple ratio of the

furfaces, with the fame velocity, and this increafe of the ratio is the more, as the velocity is the more. By accurate experiments with two planes, the one of 17¼ fquare inches, the other of 32, which are nearly in the ratio of 5 to 9, I found their refiftances, with a velocity of 20 feet per fecond, to be, the one 1·196 ounces, and the other 2·542 ounces; which are in the ratio of 8 to 17, being an increafe of between ⅓ and ⅛ part more than the ratio of the furfaces.

WIND-*Gage*, in Pneumatics, an inftrument ferving to determine the velocity and force of the Wind. See ANEMOMETER, ANEMOSCOPE, and the article juft above concerning the Force and Velocity of the Wind.

Dr. Hales had various contrivances for this purpofe. He found (Statical Effays, vol. 2, p. 326) that the air rufhed out of a fmith's bellows, at the rate of 68¾ feet in a fecond of time, when compreffed with a force of half a pound upon every fquare inch lying on the whole upper furface of the bellows. The velocity of the air, as it paffed out of the trunk of his ventilators, was found to be at the rate of 3000 feet in a minute, which is at the rate of 34 miles an hour. The fame author fays, that the velocity with which impelled air paffes out at any orifice, may be determined by hanging a light valve over the nofe of a bellows, by pliant leathern hinges, which will be much agitated and lifted up from a perpendicular to a more than horizontal pofition by the force of the rufhing air. There is alfo another more accurate way, he fays, of eftimating the velocity of air, viz, by holding the orifice of an inverted glafs fiphon full of water, oppofite to the ftream of air, by which the water will be depreffed in one leg, and raifed in the other, in proportion to the force with which the water is impelled by the air. Defcrip. of Ventilators, 1743, p. 12. And this perhaps gave Dr. Lind the idea of his Wind-gage, defcribed below.

M. Bouguer contrived a fimple inftrument, by which may be immediately difcovered the force which the Wind exerts on a given furface. This is a hollow tube AABB (fig. 14, pl. 30), in which a fpiral fpring CD is fixed, that may be more or lefs compreffed by a rod FSD, paffing through a hole within the tube at AA. Then having obferved to what degree different forces or given weights are capable of compreffing the fpiral, mark divifions on the rod in fuch a manner, that the mark at S may indicate the weight requifite to force the fpring into the fituation CD: afterwards join at right angles to this rod at F, a plane furface CFE of any given area at pleafure; then let this inftrument be oppofed to the Wind, fo that it may ftrike the furface perpendicularly, or parallel to the rod; then will the mark at S fhew the weight to which the force of the Wind is equivalent.

Dr. Lind has alfo contrived a fimple and eafy apparatus of this kind, nearly upon the laft idea of Dr. Hales mentioned above. This inftrument is fully explained at the article ANEMOMETER, vol. 1, pa. 111, and a figure of it given, pl. 3, fig. 4.

Mr. Benjamin Martin, from a hint firft fuggefted by Dr. Burton, contrived an anemofcope, or Wind-gage, of a conftruction like a Wind-mill, with four fails; but the axis which the fails turn, is not cylindrical, but conical, like the fufee of a watch; about this fufee winds a cord, having a weight at the end, which is

      wound

wound always, by the force of the Wind, upon the sails, till the weight just balances that force, which will be at a thicker part of the fusee when the Wind is strong, and at a smaller part of it when it is weaker. But although this instrument shews when a Wind is stronger or weaker, it will neither shew what is the actual velocity of the Wind, nor yet its force upon a square foot of direct surface; because the sails are set at an uncertain oblique angle to the Wind, and this acts at different distances from the axis or centre of motion. Martin's Phil. Brit. vol. 2, p. 211. See the fig. 5, plate 3, vol. 1.

WIND-*Gun*, the same as AIR-*Gun*; which see.

WIND-*Mill*, a kind of mill which receives its motion from the impulse of the Wind.

The internal structure of the Windmill is much the same with that of watermills: the difference between them lying chiefly in an external apparatus, for the application of the power. This apparatus consists of an axis EF (fig. 11, pl. 36), through which pass perpendicular to it, and to each other, two arms or yards, AB and CD, usually about 32 feet long: on these yards are formed a kind of sails, vanes, or flights, in a trapezoid form, with parallel ends; the greater of which HI is about 6 feet, and the less FG are determined by radii drawn from the centre E, to I and H.

These sails are to be capable of being always turned to the wind, to receive its impulse: for which purpose there are two different contrivances, which constitute the two different kinds of Windmills in common use.

In the one, the whole machine is supported upon a moveable arbor, or axis, fixed upright on a stand or foot; and turned round occasionally to suit the wind, by means of a lever.

In the other, only the cover or roof of the machine, with the axis and sails, in like manner turns round with a parallel or horizontal motion. For this purpose, the cover is built turret-wise, and encompassed with a wooden ring, having a groove, at the bottom of which are placed, at certain distances, a number of brass truckles; and within the groove is another ring, upon which the whole turret stands. To the moveable ring are connected beams *ab* and *fe*; and to the beam *ab* is fastened a rope at *b*, having its other end fitted to a windlass, or axis-in-peritrochio: this rope being drawn through the iron hook G, and the windlass turned, the sails are moved round, and set fronting the wind, or with the axis pointing straight against the wind.

The internal mechanism of a Windmill is exhibited in fig. 12; where AHO is the upper room, and H*o*Z the lower one; AB the axle-tree passing through the mill; STVW the sails covered with canvas, set obliquely to the wind, and turning round in the order of the letters; CD the cogwheel, having about 48 cogs or teeth, *a, a, a*, &c, which carry round the lantern EF, having 8 or 9 trundles or rounds *c, c, c*, &c, together with its upright axis GN; IK is the upper mill-stone, and LM the lower one; QR is the bridge, supporting the axis or spindle GN; this bridge is supported by the beams *cd*, XY, wedged up at *c, d* and X; ZY is the lifting tree, which stands upright; *ab* and *ef* are levers, whose centres of motion are Z and *e*; *fghi* is a cord, with a stone *i*, going about the pins *g* and *h*, and serving as a balance or counterpoise. The spindle *t*N

5

is fixed to the upper millstone IK, by a piece of iron called the rynd, and fixed in the lower side of the stone, which is the only one that turns about, and its whole weight rests upon a hard stone, fixed in the bridge QR at N. The trundle EF, and its axis G*t*, may be taken away; for it rests by its lower part at *t* by a square socket, and the top runs in the edge of the beam *w*. By bearing down the end *f* of the lever *fe*, *b* is raised, which raises ZY, and this raises YX, which lifts up the bridge QR, with the axis NG, and the upper stone IK; and thus the stones are set at any distance. The lower or immoveable stone is fixed upon strong beams, and is broader than the upper one: the flour is conveyed through the tunnel *no* into a chest; P is the hopper, into which is put the corn, which runs through the spout *r* into the hole *t*, and so falls between the stones, where it is ground to meal. The axis G*t* is square, which shaking the spout *r*, as it goes round, makes the corn run out; *rs* is a string going about the pin *s*, and serving to move the spout nearer to the axis or farther from it, so as to make the corn run faster or slower, according to the velocity and force of the wind. And when the wind is strong, the sails are only covered in part, or on one side, or perhaps only one half of two opposite sails. Toward the end B of the axletree is placed another cogwheel, trundle, and millstones, with an apparatus like that just described; so that the same axis moves two stones at once; and when only one pair is to grind, one of the trundles and its spindle are taken out: *xyl* is a girth of pliable wood, fixed at the end *x*; the other end *l* being tied to the lever *km*, moveable about *k*; and the end *m* being put down, draws the girth *xyl* close to the cogwheel, which gently and gradually stops the motion of the mill, when required: *pq* is a ladder for ascending to the higher part of the mill; and the corn is drawn up by means of a rope, rolled about the axis AB, when the mill is at work. See MILL.

Theory of the WINDMILL, *Position of the Sails, &c.*

Were the sails set square upon their arms or yards, and perpendicular to the axletree, or to the wind, no motion would ensue, because the direct wind would keep them in an exact balance. But by setting them obliquely to the common axis, like the sails of a smoke-jack, or inclined like the rudder of a ship, the wind, by striking the surface of them obliquely, turns them about. Now this angle which the sails are to make with their common axis, or the degree of *weathering*, as the mill-wrights call it, so as that the wind may have the greatest effect, is a matter of nice enquiry, and has much occupied the thoughts of the mathematician and the artist.

In examining the compound motions of the rudder of a ship, we find that the more it approaches to the direction of the keel, or to the course of the water, the more weakly this strikes it; but, on the other hand, the greater is the power of the lever to turn the vessel about. The obliquity of the rudder therefore has, at the same time, both an advantage and a disadvantage. It has been a point of inquiry therefore to find the position of the rudder when the ratio of the advantage over the disadvantage is the greatest. And M. Renau, in his

his theory of the working of ships, has found, that the best situation of the rudder is when it makes an angle of about 55 degrees with the keel.

The obliquity of the sails, with regard to their axis, has precisely the same advantage, and disadvantage, with the obliquity of the rudder to the keel. And M. Parent, seeking by the new analysis the most advantageous situation of the sails on the axis, finds it the same angle of about 55 degrees. This obliquity has been determined by many other mathematicians, and found to be more accurately 54° 44 . See Maclaurin's Fluxions, p. 733; Simpson's Fluxions, prob. 17, p. 521; Martin's Philof. Britan. vol. 1, p. 220, vol. 2, p. 212; &c.

This angle, however, is only that which gives the wind the greatest force to put the sail in motion, but not the angle which gives the force of the wind a maximum upon the sail when in motion: for when the sail has a certain degree of velocity, it yields to the wind; and then that angle must be increased. to give the wind its full effect. Maclaurin, in his Fluxions, p. 734, has shewn how to determine this angle.

It may be observed, that the increase of this angle should be different according to the different velocities from the axletree to the further extremity of the sail. At the beginning, or axis, it should be 54° 44'; and thence continually increasing, giving the vane a twist, and so causing all the ribs of the vane to lie in different planes.

It is farther observed, that the ribs of the vane or sail ought to decrease in length from the axis to the extremity, giving the vane a curvilinear form; so that no part of the force of any one rib be spent upon the rest, but all move on independent of each other. The twist above mentioned, and the diminution of the ribs, are exemplified in the wings of birds.

As the ends of the sail nearest the axis cannot move with the same velocity which the tips or farthest ends have, although the wind acts equally strong upon them both, Mr. Ferguson (Lect. on Mech. pa. 52) suggests, that perhaps a better position than that of stretching them along the arms directly from the centre of motion, might be, to have them set perpendicularly across the farther ends of the arms, and there adjusted lengthwise to the proper angle: for in that case both ends of the sails would move with the same velocity; and being farther from the centre of motion they would have so much the more power, and then there would be no occasion for having them so large as they are generally made; which would render them lighter, and consequently there would be so much the less friction on the thick neck of the axle, when it turns in the wall.

Mr. Smeaton (Philof. Transf. 1759), from his experiments with Windmill sails, deduces several practical maxims: as,

1. That when the wind falls upon a concave surface, it is an advantage to the power of the whole, though every part, taken separately, should not be disposed to the best advantage. By several trials he has found that the curved form and position of the sails will be best regulated by the numbers in the following table.

6th Parts of the radius or sail.	Angle with the axis.	Angle with the plane of motion.
1	72°	18°
2	71	19
3	72	18 middle.
4	74	16
5	77½	12½
6	83	7 end.

2. That a broader sail requires a greater angle; and that when the sail is broader at the extremity, than near the centre, this shape is more advantageous than that of a parallelogram.

3. When the sails, made like sectors of circles, joining at the centre or axis, filled up about 7 8ths of the whole circular space, the effect was the greatest.

4. The velocity of Windmill sails, whether unloaded, or loaded so as to produce a maximum of effect, is nearly as the velocity of the Wind; their shape and position being the same.

5. The load at the maximum is nearly, but somewhat less than, as the square of the velocity of the wind.

6. The effects of the same sails at a maximum, are nearly, but somewhat less than, as the cubes of the velocity of the wind.

7. In sails of a similar figure and position, the number of turns in a given time, are reciprocally as the radius or length of the sail.

8. The effects of sails of similar figure and position, are as the square of their length.

9. The velocity of the extremities of Dutch mills, as well as of the enlarged sails, in all their usual positions, is considerably greater than the velocity of the wind.

M. Parent, in considering what figure the sails of a Windmill should have, to receive the greatest impulse from the wind, finds it to be a sector of an ellipsis, whose centre is that of the axletree of the mill; and the less semiaxis the height of 32 feet; as for the greater, it follows necessarily from the rule that directs the sail to be inclined to the axis in the angle of 55 degrees.

On this foundation he assumes four such sails, each being a quarter of an ellipse; which he shews will receive all the wind, and lose none, as the common ones do. These 4 surfaces, multiplied by the lever, with which the wind acts on one of them, express the whole power the wind has to move the machine, or the whole power the machine has when in motion.

A Windmill with 6 elliptical sails, he shews, would still have more power than one with only four. It would only have the same surface with the four; since the 4 contain the whole space of the ellipsis, as well as the 6. But the force of the 6 would be greater than that of the 4, in the ratio of 245 to 231. If it were desired to have only two sails, each being a semiellipsis, the surface would be still the same; but the power would be diminished by near 1-3d of that with 6 sails; because the greatness of the sectors would much shorten the lever with which the wind acts.

The same author has also considered which form, among the rectangular sails, will be most advantageous;

i. e.

i. e. that which shall have the product of the surface by the lever of the wind, the greatest. The result of this enquiry is, that the width of the rectangular sail should be nearly double its length; whereas usually the length is made almost 5 times the width.

The power of the mill, with four of these new rectangular sails, M. Parent shews, will be to the power of four elliptic sails, nearly as 13 to 23; which leaves a considerable advantage on the side of the elliptic ones; and yet the force of the new rectangular sails will still be considerably greater than that of the common ones.

M. Parent also considers what number of the new sails will be most advantageous: and finds that the fewer the sails, the more surface there will be, but the power the less. Farther, the power of a Windmill with 6 sails is denoted by 14, that of another with 4 will be as 13, and another with 2 sails will be denoted by 9. That as to the common Windmill, its power still diminishes as the breadth of the sails is smaller, in proportion to the length: and therefore the usual proportion of 5 to 1 is exceedingly disadvantageous.

WINDWARD, in Sea Language, denotes any thing towards that point from whence the wind blows, in respect of a ship.

Sailing to WINDWARD. See SAILING.

WINDWARD *Tide*, denotes a tide that runs against the wind.

WINDAGE *of a Gun*, is the difference between the diameter of the bore of the gun and the diameter of the ball.

Heretofore the Windage appointed in the English service, viz, 1-20th of the diameter of the ball, which has been used almost from the beginning, has been far too much, owing perhaps to the first want of roundness in the ball, or to rust, foulness, or irregularities in the bore of the gun. But lately a beginning has been made to diminish the Windage, which cannot fail to be of very great advantage; as the shot will both go much truer, and have less room to bounce about from side to side, to the great damage of the gun; and besides much less powder will serve for the same effect, as in some cases $\frac{1}{3}$ or $\frac{1}{2}$ the inflamed powder escapes by the Windage. The French allowance of Windage is 1-25th of the diameter of the ball.

WINDLASS, or WINDLACE, a particular machine used for raising heavy weights, as guns, stones, anchors, &c.

This is a very simple machine, consisting only of an axis or roller, supported horizontally at the two ends by two pieces of wood and a pulley: the two pieces of wood meet at top, being placed diagonally so as to prop each other; and the axis or roller goes through the two pieces, and turns in them. The pulley is fastened at top, where the pieces join. Lastly, there are two staves or hand spikes which go through the roller, to turn it by; and the rope, which comes over the pulley, is wound off and on the same.

WINDLASS, in a Ship, is an instrument in small ships, placed upon the deck, just abaft the foremast. It is made of a long and thick piece of timber, either cylindrical, or octagonal, &c, in form of an axletree, placed horizontally across the ship, a foot or more above the deck; and it is turned about by the help of handspikes put into holes made for that purpose.

This machine will purchase or raise much more than a capstan, and that without any danger to those that heave; for if in heaving the Windlass about, any of the handspikes should happen to slip or break, the Windlass will stop of itself, as it does at the end of every pull or heave of the men, being prevented from returning by means of a catch that falls into notches. See fig. 15, pl. 35.

WINDOW, q. d. *wind-door*, an aperture or opening in the wall of a house, to admit the air and light.

Before the use of glass became general, which was not till towards the end of the 12th century, the Windows in England seem generally to have been composed of paper, oiled, both to defend it against the weather, and to make it more transparent; as now is sometimes used in workshops and unfinished buildings. Some of the better sort were furnished with lattices of wood or sheets of linen. These it seems were fixed in frames, called *capsamenta*, and hence our *casements* still so common in some of the counties.

The chief rules with regard to Windows are, 1. That they be as few in number, and as moderate in dimensions, as may be consistent with other respects; inasmuch as all openings are weakenings.

2. That they be placed at a convenient distance from the angles or corners of the buildings: both for strength and beauty.

3. That they be made all equal one with another, in their rank and order; so that those on the right hand may answer to those on the left, and those above be right over those below: both for strength and beauty.

As to their dimensions, care is to be taken, to give them neither more nor less than is needful; regard being had to the size of the rooms, and of the building. The apertures of Windows in middle sized houses, may be from 4 to 5 feet; in the smaller ones less; and in large buildings more. And the height may be double their width at the least: but in lofty rooms, or large buildings, the height may be a 4th, or 3d, or half their breadth more than the double.

to Such are the proportions for Windows of the first story; and the breadth must be the same in the upper stories; but as to the height, the second story may be a 3d part lower than the first, and the third story a 4th part lower than the second.

WINTER, one of the four seasons or quarters of the year.

Winter properly commences on the day when the sun's distance from the zenith of the place is the greatest, or when his declination is the greatest on the contrary side of the equator; and it ends on the day when that distance is a mean between the greatest and least, or when he next crosses the equinoctial.

At and near the equator, the Winter, as well as the other seasons, return twice every year; but all other places have only one Winter in the year; which in the northern hemisphere begins when the sun is in the tropic of Capricorn, and in the southern hemisphere when he is in the tropic of Cancer: so that all places in the same hemisphere have their Winter at the same time.

Notwithstanding the coldness of this season it is proved in astronomy, that the sun is really nearer to the earth in our Winter than in summer: the reason of the

defect

defect of heat being owing to the lowness of the fun, or to the obliquity of his rays.

WOLFF, Wolfius, (Christian), baron of the Roman empire, privy counsellor to the king of Pruffia, and chancellor to the univerfity of Halle in Saxony, as well as member of many of the literary academies in Europe, was born at Breflau in 1679. After ftudying philofophy and mathematics at Breflau and Jena, he obtained permiffion to give lectures at Leipfic; which, in 1703, he opened with a differtation called *Philofophia Practica Univerfalis, Methodo Mathematica confcripta*, which ferved greatly to enhance the reputation of his talents. He publifhed two other differtations the fame year; the firft *De Rotis Dentatis*, the other *De Algorithmo Infinitefimali Differentiali*; which obtained him the honourable appellation of Affiftant to the Faculty of Philofophy at Leipfic.

He now accepted the profefforfhip of mathematics at Halle, and was elected into the fociety at Leipfic, at that time engaged in publifhing the *Acta Eruditorum*. After having inferted in this work many important pieces relating to mathematics and phyfics, he undertook, in 1709, to teach all the various branches of philofophy, beginning with a fmall Logical treatife in Latin, being Thoughts on the Powers of the Human Underftanding. He carried himfelf through thefe great purfuits with amazing affiduity and ardour: the king of Pruffia rewarded him with the office of counfellor to the court in 1721, and augmented the profits of that poft by very confiderable appointments: he was also chofen a member of the Royal Society of London and of Pruffia

In the midft of all this profperity however, Wolff raifed an ecclefiaftical ftorm againft himfelf, by a Latin oration he delivered in praife of the Chinefe philofophy: every pulpit immediately refounded againft his tenets; and the faculty of theology, who entered into a ftrict examination of his productions, refolving that the doctrine he taught was dangerous to the laft degree, an order was obtained in 1723 for difplacing him, and commanding him to leave Halle in 24 hours.

Wolff now retired to Caffel, where he obtained the profefforfhip of mathematics and philofophy in the univerfity of Marbourg, with the title of Counfellor to the Landgrave of Heffe; to which a profitable penfion was annexed. Here he renewed his labours with redoubled ardour; and it was in this retreat that he publifhed the greateft part of his numerous works.

In 1725, he was declared an honorary profeffor of the academy of fciences at Peterfburg, and in 1733 was admitted into that of Paris. The king of Sweden alfo declared him one of the council of regency; but the pleafing fituation of his new abode, and the multitude of honours which he had received, were too alluring to permit him to accept of many advantageous offers; among which was the office of prefident of the academy at Peterfburg.

The king of Pruffia too, who was now recovered from the prejudices he had been made to conceive againft Wolff, wanted to re-eftablifh him in the univerfity of Halle in 1733, and made another attempt to effect it in 1739; which Wolff for a time thought fit to decline, but at laft fubmitted: he returned therefore in 1741, invefted with the characters of privy counfellor, vice

chancellor, and profeffor of the law of nature and of nations. The king afterwards, upon a vacancy, raifed him to the dignity of chancellor of the univerfity; and the elector of Bavaria created him a baron of the empire. He died at Halle in Saxony, of the gout in his ftomach, in 1754, in the 76th year of his age, after a life filled up with a train of actions as wife and fyftematical as his writings, of which he compofed in Latin and German more than 60 diftinct pieces. The chief of his mathematical compofitions, is his *Elementa Mathefeos Univerfæ*, the beft edition of which is that of 1732 at Geneva, in 5 vols 4to; which does not however comprife his Mathematical Dictionary in the German language, in 1 vol. 8vo, nor many other diftinct works on different branches of the mathematics, nor his Syftem of Philofophy, in 23 vols. in 4to.

WORKING *to Windward*, in Sea Language, is the operation by which a fhip endeavours to make a progrefs againft the wind.

WREN (Sir Christopher), a great philofopher and mathematician, and one of the moft learned and eminent architects of his age, was the fon of the rev. Chriftopher Wren, dean of Windfor, and was born at Knoyle in Wiltfhire in 1632. He ftudied at Wadham college, Oxford; where he took the degree of mafter of arts in 1653, and was chofen fellow of Allfouls college there. Soon after, he became one of that ingenious and learned fociety, who then met at Oxford for the improvement of natural and experimental philofophy, and which at length produced the Royal Society.

When very young, he difcovered a furprifing genius for the mathematics, in which fcience he made great advances before he was 16 years of age.—In 1657 he was made profeffor of aftronomy in Grefham college, London; and his lectures, which were much frequented, tended greatly to the promotion of real knowledge: in his inaugural oration, among other things, he propofed feveral methods by which to account for the fhadows returning backward 10 degrees on the dial of king Ahaz, by the laws of nature. One fubject of his lectures was upon telefcopes, to the improvement of which he had greatly contributed; another was on certain properties of the air, and the barometer. In the year 1658 he read a defcription of the body and different phafes of the planet Saturn; which fubject he propofed to inveftigate while his colleague, Mr. Rooke, then profeffor of geometry, was carrying on his obfervations upon the fatellites of Jupiter. The fame year he communicated fome demonftrations concerning cycloids to Dr. Wallis, which were afterwards publifhed by the doctor at the end of his treatife upon that fubject. About that time alfo, he refolved the problem propofed by Pafcal, under the feigned name of John de Montford, to all the Englifh mathematicians; and returned another to the mathematicians in France, formerly propofed by Kepler, and then refolved likewife by himfelf, to which they never gave any folution.—In 1660, he invented a method for the conftruction of folar eclipfes: and in the latter part of the fame year, he with ten other gentlemen formed themfelves into a fociety, to meet weekly, for the improvement of natural and experimental philofophy; being the foundation of the Royal Society.—In the beginning of 1661, he was chofen Savilian profeffor of aftronomy at Oxford,

in

in the room of Dr. Seth Ward; where he was the same year created Doctor of Laws.

Among his other accomplishments, Dr. Wren had gained so considerable a skill in architecture, that he was sent for the same year, from Oxford, by order of king Charles the 2d, to assist Sir John Denham, surveyor general of the works.—In 1663, he was chosen fellow of the Royal Society; being one of those who were first appointed by the Council after the grant of their charter. Not long after, it being expected that the king would make the society a visit, the lord Brounker, then president, by a letter requested the advice of Dr. Wren, concerning the experiments which might be most proper on that occasion: to whom the doctor recommended principally the Torricellian experiment, and the weather needle, as being not mere amusements, but useful, and also neat in their operation. Indeed upon many occasions Dr. Wren did great honour to that illustrious body, by many curious and useful discoveries, in astronomy, natural philosophy, and other sciences, related in the History of the Royal Society, where Dr. Sprat has inserted them from the registers and other books of the society to 1665. Among others of his productions there enumerated, is a lunar globe; representing the spots and various degrees of whiteness upon the moon's surface, with the hills, eminences and cavities: the whole contrived so, that by turning it round to the light, it shews all the lunar phases, with the various appearances that happen from the shadows of the mountains and valleys, &c: this lunar model was placed in the king's cabinet. Another of these productions, is a tract on the Doctrine of Motion that arises from the impact between two bodies, illustrated by experiments. And a third is, The History of the Seasons, as to the temperature, weather, productions, diseases, &c, &c. For which purpose he contrived many curious machines, several of which kept their own registers, tracing out the lines of variations, so that a person might know what changes the weather had undergone in his absence: as wind-gages, thermometers, barometers, hygrometers, rain-gages, &c.—He made also great additions to the new discoveries on pendulums; and among other things shewed, that there may be produced a natural standard for measure from the pendulum for common use.—He invented many ways to make astronomical observations more easy and accurate: He fitted and hung quadrants, sextants, and radii more commodiously than formerly: he made two telescopes to open with a joint like a sector, by which observers may infallibly take a distance to half minutes, &c. He made many sorts of retes, screws, and other devices, for improving telescopes to take small distances, and apparent diameters, to seconds: He made apertures for taking in more or less light, as the observer pleases, by opening and shutting, the better to fit glasses for crepusculine observations.—He added much to the theory of dioptrics; much to the manufacture of grinding good glasses: He attempted, and not without success, the making of glasses of other forms than spherical. He exactly measured and delineated the spheres of the humours of the eye, the proportions of which to one another were only guessed at before: a discussion shewing the reasons why we see objects erect, and that reflection conduces

as much to vision as refraction. He displayed a natural and easy theory of refractions, which exactly answered every experiment. He fully demonstrated all dioptrics in a few propositions, shewing not only, as in Kepler's Dioptrics, the common properties of glasses, but the proportions by which the individual rays cut the axis, and each other, upon which the charges of the telescopes, or the proportion of the eye glasses and apertures, are demonstrably discovered.—He made constant observations on Saturn, and a true theory of that planet, before the printed discourse by Huygens, on that subject, appeared.—He made maps of the Pleiades and other telescopic stars: and proposed methods to determine the great question as to the earth's motion or rest, by the small stars about the pole to be seen in large telescopes—In navigation he made many improvements. He framed a magnetical terella, which he placed in the midst of a plane board with a hole, into which the terella is half immersed, till it be like a globe with the poles in the horizon: the plane is then dusted over with steel filings from a sieve: the dust, by the magnetical virtue, becomes immediately figured into furrows that bend like a sort of helix, proceeding as it were out at one pole, and returning in by the other; the whole plane becoming figured like the circles of a planisphere.—It being a question in his time among the problems of navigation, to what mechanical powers sailing against the wind was reducible; he shewed it to be a wedge: and he demonstrated, how a transient force upon an oblique plane would cause the motion of the plane against the first mover: and he made an instrument mechanically producing the same effect, and shewed the reason of sailing on all winds. The geometrical mechanism of rowing, he shewed to be a lever on a moving or cedent fulcrum: for this end, he made instruments and experiments, to find the resistance to motion in a liquid medium; with other things that are the necessary elements for laying down the geometry of sailing, swimming, rowing, flying, and constructing of ships— He invented a very speedy and curious way of etching. He started many things towards the emendation of water-works. He likewise made some instruments for respiration, and for straining the breath from fuliginous vapours, to try whether the same breath, so purified, will serve again.—He was the first inventor of drawing pictures by microscopical glasses. He found out perpetual, or at least longlived lamps, for keeping a perpetual regular heat, in order to various uses, as hatching of eggs and insects, production of plants, chemical preparations, imitating nature in producing fossils and minerals, keeping the motion of watches equal, for the longitude and astronomical uses.—He was the first author of the anatomical experiment of injecting liquor into the veins of animals. By this operation, divers creatures were immediately purged, vomited, intoxicated, killed, or revived, according to the quality of the liquor injected. Hence arose many other new experiments, particularly that of transfusing blood, which has been prosecuted in sundry curious instances. This is a short account of the principal discoveries which Dr. Wren presented, or suggested, to the Royal Society, or were improved by him.

As to his architectural works: It has before been
observed

observed that he had been sent for to assist Sir John Denham. In 1665 he travelled into France, to examine the most beautiful edifices and curious mechanical works there, when he made many useful observations. Upon his return home, he was appointed architect, and one of the commissioners for repairing St. Paul's cathedral. Within a few days after the fire of London, 1666, he drew a plan for a new city, and presented it to the king; but it was not approved of by the parliament. In this model, the chief streets were to cross each other at right angles, with lesser streets between them; the churches, public buildings, &c, so disposed as not to interfere with the streets, and four piazzas placed at proper distances.—Upon the death of Sir John Denham, in 1668, he succeeded him in the office of surveyor-general of the king's works; and from this time he had the direction of a great many public edifices, by which he acquired the highest reputation. He built the magnificent theatre at Oxford, St. Paul's cathedral, the Monument, the modern part of Hampton Court, Chelsea-college, one of the wings of Greenwich hospital, the churches of St. Stephen Walbrook, and St. Mary-le-bow, with upwards of 60 other churches and public works, which that dreadful fire made necessary. In the management of which business, he was assisted in the measurements, and laying out of private property, by the ingenious Dr. Robert Hook. The variety of business in which he was by this means engaged, requiring his constant attendance and concern, he resigned his Savilian professorship at Oxford in 1673; and the year following he received from the king the honour of knighthood.—He was one of the commissioners who, on the motion of Sir Jonas Moore, surveyor-general of the ordnance, had been appointed to find out a proper place for erecting an observatory; and he proposed Greenwich, which was approved of; the foundation stone of which was laid the 10th of August 1675, and the building was presently finished under the direction of Sir Jonas, with the advice and assistance of Sir Christopher.

In 1680 he was chosen president of the Royal Society; afterwards appointed architect and commissioner of Chelsea-college; and in 1684, principal officer or comptroller of the works in Windsor-castle. Sir Christopher sat twice in Parliament, as a representative for two different boroughs. While he continued surveyor-general, his residence was in Scotland-yard; but after his removal from that office, in 1718, he lived in St. James's-street, Westminster. He died the 25th of February 1723, at 91 years of age; and he was interred with great solemnity in St. Paul's cathedral, in the vault under the south wing of the choir, near the east end.

As to his person, Sir Christopher Wren was of a low stature, and thin frame of body; but by temperance and skilful management he enjoyed a good state of health, to a very unusual length of life. He was modest, devout, strictly virtuous, and very communicative of his knowledge. Besides his peculiar eminence as an architect, his learning and knowledge were very extensive in all the arts and sciences, and especially in the mathematics.

Sir Christopher never printed any thing himself, but several of his works have been published by others: some in the Philosophical Transactions, and some by Dr. Wallis and other friends.—His posthumous works and draughts were published by his son.

WRIGHT (EDWARD), a noted English mathematician, who flourished in the latter part of the 16th century, and beginning of the 17th; dying in the year 1615. He was contemporary with Mr. Briggs, and much concerned with him in the business of the logarithms, the short time they were published before his death. He also contributed greatly to the improvement of navigation and astronomy. The following memoirs of him are translated from a Latin paper in the annals of Gonvile and Caius college in Cambridge, viz, " This year (1615) died at London, Edward Wright of Garveston in Norfolk, formerly a fellow of this college; a man respected by all for the integrity and simplicity of his manners, and also famous for his skill in the mathematical sciences: so that he was not undeservedly styled a most excellent mathematician by Richard Hackluyt, the author of an original treatise of our English navigations. What knowledge he had acquired in the science of mechanics, and how usefully he employed that knowledge to the public as well as private advantage, abundantly appear both from the writings he published, and from the many mechanical operations still extant, which are standing monuments of his great industry and ingenuity. He was the first undertaker of that difficult but useful work, by which a little river is brought from the town of Ware in a new canal, to supply the city of London with water; but by the tricks of others he was hindered from completing the work he had begun. He was excellent both in contrivance and execution, nor was he inferior to the most ingenious mechanic in the making of instruments, either of brass or any other matter. To his invention is owing whatever advantage Hondius's geographical charts have above others; for it was Wright who taught Jodocus Hondius the method of constructing them, which was till then unknown; but the ungrateful Hondius concealed the name of the true author, and arrogated the glory of the invention to himself. Of this fraudulent practice the good man could not help complaining, and justly enough, in the preface to his treatise of the Correction of Errors in the Art of Navigation; which he composed with excellent judgment, and after long experience, to the great advancement of naval affairs. For the improvement of this art he was appointed mathematical lecturer by the East-India Company, and read lectures in the house of that worthy knight Sir Thomas Smith, for which he had a yearly salary of 50 pounds. This office he discharged with great reputation, and much to the satisfaction of his hearers. He published in English a book on the doctrine of the sphere, and another concerning the construction of sun-dials. He also prefixed an ingenious preface to the learned Gilbert's book on the loadstone. By these and other his writings, he has transmitted his fame to latest posterity. While he was yet a fellow of this college, he could not be concealed in his private study, but was called forth to the public business of the nation, by the queen, about the year 1593. [Other accounts say 1589.] He was ordered to attend the earl of Cumberland in some fome maritime expeditions. One of thefe he has given a faithful account of, in the manner of a journal or ephemeris, to which he has prefixed an elegant hydrographical chart of his own contrivance. A little before his death he employed himfelf about an Englifh tranflation of the book of logarithms, then lately difcovered by lord Napier, a Scotchman, who had a great affection for him. This pofthumous work of his was publifhed foon after, by his only fon Samuel Wright, who was alfo a fcholar of this college. He had formed many other ufeful defigns, but was hindered by death from bringing them to perfection. Of him it may truly be faid, that he ftudied more to ferve the public than himfelf; and though he was rich in fame, and in the promifes of the great, yet he died poor, to the fcandal of an ungrateful age." So far the memoir; other particulars concerning him, are as follow.

Mr. Wright firft difcovered the true way of dividing the meridian line, according to which the Mercator's charts are conftructed, and upon which Mercator's failing is founded. An account of this he fent from Caius college, Cambridge, where he was then a fellow, to his friend Mr. Blondeville, containing a fhort table for that purpofe, with a fpecimen of a chart fo divided, together with the manner of dividing it. All which Blondeville publifhed, in 1594, among his Exercifes. And, in 1597, the reverend Mr. William Barlowe, in his Navigator's Supply, gave a demonftration of this divifion as communicated by a friend.

At length, in 1599, Mr. Wright himfelf printed his celebrated treatife, intitled, *The Correction of certain Errors in Navigation*, which had been written many years before; where he fhews the reafon of this divifion of the meridian, the manner of conftructing his table, and its ufes in navigation, with other improvements. In 1610 a fecond edition of Mr. Wright's book was publifhed, and dedicated to his royal pupil, prince Henry; in which the author inferted farther improvements; particularly he propofed an excellent way of determining the magnitude of the earth; at the fame time recommending very judicioufly, the making our common meafures in fome certain proportion to that of a degree on its furface, that they might

not depend on the uncertain length of a barley-corn. Some of his other improvements were; The Table of Latitudes for dividing the meridian, computed as far as to minutes: An inftrument, he calls the Sea-rings, by which the variation of the compafs, the altitude of the fun, and the time of the day, may be readily determined at once in any place, provided the latitude be known: The correcting of the errors arifing from the eccentricity of the eye in obferving by the crofs-ftaff: A total amendment in the Tables of the declinations and places of the fun and ftars, from his own obfervations, made with a fix-foot quadrant, in the years 1594, 95, 96, 97: A fea-quadrant, to take altitudes by a forward or backward obfervation; having alfo a contrivance for the ready finding the latitude by the height of the pole-ftar, when not upon the meridian. And that this book might be the better underftood by beginners, to this edition is fubjoined a tranflation of Zamorano's Compendium; and added a large table of the variation of the compafs as obferved in very different parts of the world, to fhew it is not occafioned by any magnetical pole. The work has gone through feveral other editions fince. And, befide the books above mentioned, he wrote another on navigation, intitled, *The Haven-finding Art.* Other accounts of him fay alfo, that it was in the year 1589 that he firft began to attend the earl of Cumberland in his voyages. It is alfo faid that he made, for his pupil, prince Henry, a large fphere with curious movements, which, by the help of fpring-work, not only reprefented the motions of the whole celeftial fphere, but fhewed likewife the particular fyftems of the fun and moon, and their circular motions, together with their places and poffibilities of eclipfing each other: there is in it a work for a motion of 17100 years, if it fhould not be ftopt, or the materials fail. This fphere, though thus made at a great expence of money and ingenious induftry, was afterwards in the time of the civil wars caft afide, among duft and rubbifh, where it was found, in the year 1646, by Sir Jonas Moore, who at his own expence reftored it to its firft ftate of perfection, and depofited it at his own houfe in the Tower, among his other mathematical inftruments and curiofities.

X.

XENOCRATES, an eminent philofopher among the ancient Greeks, was born at Chalcedon, and died 314 years before Chrift, at about 90 years of age. He became early a difciple of Plato, ftudying under this great mafter at the fame time with Ariftotle, though he was not poffeffed of equal talents; the for-

mer wanting a fpur, and the latter a bridle. He was fond of the mathematics; and permitted none of his fcholars to be ignorant of them. There was fomething flovenly in the behaviour of Xenocrates; for which reafon Plato frequently exhorted him to facrifice to the graces. Serioufnefs and feverity were always feen in his deport-

deportment: yet notwithstanding this severe cast of mind, he was very compassionate. There was something extraordinary in the rectitude of his morals: he was absolute master of his passions; and was not fond of pleasure, riches, or applause. Indeed, so great was his reputation for sincerity and probity, that he was the only person whom the magistrates of Athens dispensed from confirming his testimony with an oath. And yet he was so ill treated by them, as to be sold because he could not pay the poll-tax laid upon foreigners. Demetrius Phalereus bought Xenocrates, paid the debt to the Athenians, and immediately gave him his liberty. At Alexander's request, he composed a treatise on the Art of Reigning; 6 books on Nature; 6 books on Philosophy; one on Riches, &c; but none of them have come down to these times:—His theology it seems was but poor stuff: Cicero refutes him in the first book of the Nature of the Gods.

XENOPHANES, a Greek philosopher, born in Colophon, was, according to some authors, the disciple of Archelaus; in which case he must have been contemporary with Socrates. Others relate, that he taught himself all he knew, and that he lived at the same time with Anaximander: according to which account he must have flourished before Socrates, and about the 60th Olympiad, as Diogenes Laertius affirms. He founded the Eleatic sect; and wrote several poems on philosophical subjects; as also a great many on the foundation of Colophon, and on that of the colony of Elea. He wrote also against Homer and Hesiod. He was banished from his country, withdrew to Sicily, and lived in Zanche and Catana. His opinion with regard to the nature of God differs not much from that of Spinoza.—When he saw the Egyptians pour forth lamentations during their festivals, he thus advised them: "If the objects of your worship are Gods, do not weep: if they are men, offer not sacrifices to them." The answer he made to a man with whom he refused to play at dice, is highly worthy of a philosopher: This man calling him a coward, "Yes, replied he, I am excessively so with regard to all shameful actions."

XENOPHON, a celebrated Greek general, philosopher, and historian, was born at Athens, and became early a disciple of Socrates, who, says Strabo, saved his life in battle. About the 50th year of his age he engaged in the expedition of Cyrus, and accomplished his immortal retreat in the space of 15 months. The jealousy of the Athenians banished him from his native city, for engaging in the service of Sparta and Cyrus. On his return therefore he retired to Scillus, a town of Elis, where he built a temple to Diana, which he mentions in his epistles, and devoted his leisure to philosophy and rural sports. But commotions arising in that country, he removed to Corinth, where it seems he wrote his Grecian History, and died at the age of 90, in the year 360 before Christ.

By his wife Philesia he had two sons, Diodorus and Gryllus. The latter rendered himself immortal by killing Epaminondas in the famous battle of Mantinea, but perished in that exploit, which his father lived to record.

The best editions of his works are those of Franckfort in 1674, and of Oxford, in Greek and Latin, in 1703, 5 vols. 8vo. Separately have been published his *Cyropædia*, Oxon. 1727, 4to, and 1736, 8vo. *Cyri Anabasis*, Oxon. 1735, 4to, and 1747, 8vo. *Memorabilia Socratis*, Oxon. 1741, 8vo.—His *Cyropædia* has been admirably translated into English by Spelman.

XIPHIAS, in Astronomy, is the Dorado or Swordfish, a constellation of the southern hemisphere; being one of the new constellations added by modern astronomers; and consisting of 6 stars only. See DORADO.

Y.

YEA

YARD, a lineal measure, or measure of length, used in England and Spain chiefly to measure cloth, stuffs, &c. The Yard was settled by Henry the 1st, from the length of his own arm.

The English Yard contains 3 feet; and it is equal
    to 4-5ths of the English ell,
    to 7-9ths of the Paris ell,
    to 4-3ds of the Flemish ell,
    to 56-51sts of the Spanish vara or Yard.

YARD, or *Golden* YARD, is also a popular name given to the 3 stars which compose the belt of Orion.

YEAR, in the full extent of the word, is a system or cycle of several months, usually 12. Others define Year, in the general, a period or space of time, measured out by the revolution of some celestial body in

YEA

its orbit. Thus, the time in which the fixed stars make a revolution, is called the *great Year*; and the times in which Jupiter, Saturn, the Sun, Moon, &c, complete their courses, and return to the same point of the zodiac, are respectively called the Years of Jupiter, and Saturn, and the Solar, and Lunar Years, &c.

As Year denoted originally a revolution, and was not limited to that of the sun; accordingly we find by the oldest accounts, that people have, at different times, expressed other revolutions by it, particularly that of the moon: and consequently that the Years of some accounts, are to be reckoned only months, and sometimes periods of 2, or 3, or 4 months. This will help us greatly in understanding the accounts that certain nations give of their own antiquity, and perhaps of the age of men. We read exprefsly, in feveral of the old Greek writers, that the Egyptian Year, at one period, was only a month; and we are farther told that at other periods it was 3 months, or 4 months: and it is probable that the children of Ifrael followed the Egyptian account of their Years. The Egyptians talked, almoft 2000 years ago, of having accounts of events 48 thoufand Years diftance. A great deal muft be allowed to fallacy, on the above account; but befide this, the Egyptians had, in the time of the Greeks, the fame ambition which the Chinefe have at prefent, and wanted to pafs themfelves upon that people, as thefe others do upon us, for the oldeft inhabitants of the earth. They had recourfe alfo to the fame means, and both the prefent and the early impoftors have pretended to ancient obfervations of the heavenly bodies, and recounted eclipfes in particular, to vouch for the truth of their accounts. Since the time in which the folar Year, or period of the earth's revolution round the fun, has been received, we may account with certainty; but for thofe remote ages, in which we do not know of a certainty what is meant by the term Year, it is impoffible to form any conjecture of the duration of time in the accounts. The Babylonians pretend to an antiquity of the fame romantic kind; they talk of 47 thoufand Years in which they had kept obfervations; but we may judge of thefe as of the others, and of the obfervations as of the Years. The Egyptians fpeak of the ftars having four times altered their courfes in that period which they claim for their hiftory, and that the fun fet twice in the eaft. They were not fuch perfect aftronomers, but, after a round-about voyage, they might perhaps miftake the eaft for the weft when they came in again.

YEAR, or SOLAR YEAR, properly, and by way of eminence fo called, is the fpace of time in which the fun moves through the 12 figns of the ecliptic. This, by the obfervations of the beft modern aftronomers, contains 365 days, 5 hours, 48 min. 48 feconds: the quantity affumed by the authors of the Gregorian calendar is 365 days, 5 hours, 49 min. But in the civil or popular account, this Year only contains 365 days; except every 4th Year, which contains 366.

The viciffitude of feafons feems to have given occafion to the firft inftitution of the Year. Man, naturally curious to know the caufe of that diverfity, foon found it was the proximity and diftance of the fun; and therefore gave the name Year to the fpace of time in which that luminary performed his whole courfe, by returning to the fame point of his orbit. According to the accuracy in their obfervations, the Year of fome nations was more perfect than that of others, but none of them quite exact, nor whofe parts did not fhift with regard to the parts of the fun's courfe.

According to Herodotus, it was the Egyptians who firft formed the Year, making it to contain 360 days, which they fubdivided into 12 months, of 30 days each. Mercury Trifmegiftus added 5 days more to the account. And on this footing it is faid that Thales inftituted the Year among the Greeks; though that form of the Year did not hold throughout all Greece. Alfo, the Jewifh, Syrian, Roman, Perfian, Ethiopic, Arabic, &c Years, were all different. In fact, confidering the imperfect ftate of aftronomy in thofe ages,

9

it is no wonder that different people fhould difagree in the calculation of the fun's courfe. We are even affured by Diod. Siculus, lib. 1. Plutarch, in Numa, and Pliny, lib. 7, cap. 48, that the Egyptian Year itfelf was at firft very different from that now reprefented.

The folar Year is either *aftronomical* or *civil*.

The *Aftronomical Solar* YEAR, is that which is determined precifely by aftronomical obfervations; and is of two kinds, *tropical*, and *fidereal* or *aftral*.

Tropical, or *Natural* YEAR, is the time the fun takes in paffing through the zodiac; which, as before obferved, is 365 d. 5 h. 48 m. 48 fec.; or 365 d. 5 h. 49 min. This is the only proper or natural Year, becaufe it always keeps the fame feafons to the fame months.

Sidereal or *Aftral* YEAR, is the fpace of time the fun takes in paffing from any fixed ftar, till his return to it again. This confifts of 365 d. 6 h. 9 m. 17 fec.; being 20 m. 29 fec. longer than the true folar year.

Lunar YEAR, is the fpace of 12 lunar months. Hence, from the two kinds of fynodical lunar months, there arife two kinds of lunar Years; the one *aftronomical*, the other *civil*.

Lunar Aftronomical YEAR, confifts of 12 lunar fynodical months; and therefore contains 354d. 8h. 48m. 38fec. and is therefore 10d. 21h. 0m. 10 L fhorter than the folar Year. A difference which is the foundation of the Epact.

Lunar Civil YEAR, is either common or embolifmic.

The *Common Lunar* YEAR confifts of 12 lunar civil months; and therefore contains 354 days. And

The *Embolifmic* or *Intercalary Lunar* YEAR, confifts of 13 lunar civil months, and therefore contains 384 days.

Thus far we have confidered Years and months, with regard to aftronomical principles, upon which the divifion is founded. By this, the various forms of civil Years that have formerly obtained, or that do ftill obtain, in divers nations, are to be examined.

Civil YEAR, is that form of Year which every nation has contrived or adopted, for computing their time by. Or the civil is the tropical Year, confidered as only confifting of a certain number of whole days: the odd hours and minutes being fet afide, to render the computation of time, in the common occafions of life, more eafy. As the tropical Year is 365 d. 5 h. 49 m. or almoft 365 d. 6 h. which is 365 days and a quarter; therefore if the civil Year be made 365 days, every 4th year it muft be 366 days, to keep nearly to the courfe of the fun. And hence the civil Year is either *common* or *biffextile*. The

Common Civil YEAR, is that confifting of 365 days; having feven months of 31 days each, four of 30 days, and one of 28 days; as indicated by the following well known memorial verfes:

Thirty days hath September,
April, June, and November;
February twenty-eight alone,
And all the reft have thirty one.

Biffextile or *Leap* YEAR, confifts of 366 days; having one day extraordinary; called the intercalary, or biffextile day; and takes place every 4th Year. This additional day to every 4th Year, was firft introduced

by

by Julius Cæsar; who, to make the civil Years keep pace with the tropical ones, contrived that the 6 hours which the latter exceeded the former, should make one day in 4 years, and be added between the 24th and 23d of February, which was their 6th of the calends of March; and as they then counted this day twice over, or had *bis sexto calendas*, hence the Year itself came to be called *bis sextus*, and *bissextile*.

However, among us, the intercalary day is not introduced by counting the 23d of February twice over, but by adding a day at the end of that month, which therefore in that Year contains 29 days.

A farther reformation was made in this year by Pope Gregory. See *Gregorian* YEAR, CALENDAR, BISSEXTILE, and LEAP-*Year*.

The Civil or Legal Year, in England, formerly commenced on the day of the Annunciation, or 25th of March; though the historical Year began on the day of the Circumcision, or 1st of January; on which day the German and Italian Year also begins. The part of the Year between these two terms was usually expressed both ways: as 1745-6, or 1745⁶. But by the act for altering the stile, the civil Year now commences with the 1st of January.

Ancient Roman YEAR. This was the lunar Year, which, as first settled by Romulus, contained only ten months, of unequal numbers of days in the following order: viz,

March 31; April 30; May 31; June 30; Quintilis 31; Sextilis 30; September 30; October 31; November 30; December 30; in all 304 days; which came short of the true lunar Year by 50 days; and of the solar by 61 days. Hence, the beginning of Romulus's Year was vague, and unfixed to any precise season; to remove which inconvenience, that prince ordered so many days to be added yearly as would make the state of the heavens correspond to the first month, without calling them by the name of any month.

Numa Pompilius corrected this irregular constitution of the Year, composing two new months, January and February, of the days that were used to be added to the former Year. Thus Numa's year consisted of 12 months, of different days, as follow; viz,

January - 29; February 28; March - - 31;
April - - 29; May - 31; June - - - 29;
Quintilis 31; Sextilis 29; September 29;
October - 31; November 29; December 29;

in all 355 days; therefore exceeding the quantity of a lunar civil Year by one day; that of a lunar astronomical Year by 15ʰ 11ᵐ 22ˢ; but falling short of the common solar Year by 10 days; so that its beginning was still vague and unfixed.

Numa, however, desiring to have it begin at the winter solstice, ordered 22 days to be intercalated in February every 2d Year, 23 every 4th, 22 every 6th, and 23 every 8th Year.

But this rule failing to keep matters even, recourse was had to a new way of intercalating; and instead of 23 days every 8th Year, only 15 were to be added. The care of the whole was committed to the pontifex maximus; who however, neglecting the trust, let things run to great confusion. And thus the Roman Year stood till Julius Cæsar reformed it. See CALEN-

DAR. And for the manner of reckoning the days of the Roman months, see CALENDS, NONES, and IDES.

Julian YEAR. This is in effect a solar Year, commonly containing 365 days; though every 4th Year, called Bissextile, it contains 366. The months of the Julian Year, with the number of their days, stood thus:

January - 31; February - 28; March - 31;
April - - 30; May - - 31; June - - 30;
July - - 31; August - - 31; September 30;
October - 31; November 30; December 31.

But every Bissextile Year had a day added in February, making it then to contain 29 days.

The mean quantity therefore of the Julian Year is 365¼ days, or 365ᵈ 6ʰ; exceeding the true solar Year by somewhat more than 11 minutes; an excess which amounts to a whole day in almost 131 years. Hence the times of the equinoxes go backward, and fall earlier by one day in about 130 or 131 Years. And thus the Roman Year stood, till it was farther corrected by pope Gregory.

For settling this Year, Julius Cæsar brought over from Egypt, Sosigenes, a celebrated mathematician; who, to supply the defect of 67 days, which had been lost through the neglect of the priests, and to bring the beginning of the Year to the winter solstice, made one Year to consist of 15 months, or 445 days; on which account that Year was used to be called *annus confusionis*, the *Year of confusion*. See *Julian* CALENDAR.

Gregorian YEAR. This is the Julian Year corrected by this rule, viz, that instead of every secular or 100th Year being a bissextile, as it would be in the former way, in the new way three of them are common Years, and only the 4th is bissextile.

The error of 11 minutes in the Julian Year, by continual repetition, had accumulated to an error of 13 days from the time when Cæsar made his correction; by which means the equinoxes were greatly disturbed. In the Year 1582, the equinoxes were fallen back 10 days, and the full moons 4 days, more backward than they were in the time of the Nicene council, which was in the Year 325; viz, the former from the 20th of March to the 10th, and the latter from the 5th to the 1st of April. To remedy this increasing irregularity, pope Gregory the 13th, in the year 1582, called together the chief astronomers of his time, and concerted this correction, throwing out the 10 days above mentioned. He exchanged the lunar cycle for that of the epacts, and made the 4th of October of that Year to be the 15th; by that means restoring the vernal equinox to the 21st of March. It was also provided, by the omission of 3 intercalary days in 400 Years, to make the civil Year keep pace nearly with the solar Year, for the time to come. See CALENDAR.

In the Year 1700, the error of 10 days was grown to 11; upon which, the protestant states of Germany, to prevent farther confusion, adopted the Gregorian correction. And the same was accepted also in England in the year 1752, when 11 days were thrown out after the 2d of September that Year, by accounting the 3d to be the 14th day of the month: calling this the new stile, and the former the old stile. And the Gregorian, or new ſtile, is now in like manner uſed in moſt countries of Europe.

Yet this laſt correction is ſtill not quite perfect; for as it has been ſhewn that in 4 centuries, the Julian Year gains 3ᵈ 2ʰ 40ᵐ; and as it is only the 3 days that are kept out in the Gregorian Year; there is ſtill an exceſs of 2ʰ 40ᵐ in 4 centuries, which amounts to a whole day in 36 centuries, or in 3600 Years. See CALENDAR, *New* or *Gregorian* STILE, &c.

Egyptian YEAR, called alſo the *Year of Nabonaſſar*, on account of the epoch of Nabonaſſar, is the ſolar Year of 365 days, divided into 12 months, of 30 days each, beſide 5 intercalary days, added at the end. The order and names of theſe months are as follow :

1. Thoth ;	2. Paophi ;	3. Athyr ;
4. Chojac ;	5. Tybi ;	6. Mecheir ;
7. Phamenoth ;	8. Pharmuthi ;	9. Pachon ;
10. Pauni ;	11. Epiphi ;	12. Meſori.

As the Egyptian Year, by neglecting the 6 hours, in every 4 Years loſes a whole day of the Julian Year, its beginning runs through every part of the Julian Year in the ſpace of 1460 Years; after which, they meet again; for which reaſon it is called the *erratic* Year. And becauſe this return to the ſame day of the Julian Year, is performed in the ſpace of 1460 Julian Years, this circle is called the Sothic period.

This Year was applied by the Egyptians to civil uſes, till Anthony and Cleopatra were defeated; but the mathematicians and aſtronomers uſed it till the time of Ptolomy, who made uſe of it in his Almageſt; ſo that the knowledge of it is of great uſe in aſtronomy, for comparing the ancient obſervations with the modern.

The ancient Egyptians, we are told by Diodorus Siculus, (Plutarch, lib. 1, in the life of Numa, and Pliny, lib. 7, cap. 48) meaſured their Years by the courſe of the moon. At firſt they were only one month, then 3, then 4, like that of the Arcadians; and then 6, like that of the people of Acarnania. Thoſe authors add, that it is on this account that they reckon ſuch a vaſt number of Years from the beginning of the world; and that in the hiſtory of their kings, we meet with ſome who lived 1000, or 1200 Years. The ſame thing is maintained by Kircher; Oedip. Egypt. tom. 2, pa. 252. And a late author obſerves, that Varro has affirmed the ſame of all nations, that has been quoted of the Egyptians. By which means many account for the great ages of the more ancient patriarchs; expounding the gradual decreaſe in their ages, by the ſucceſſive increaſe of the number of months in their years.

Upon the Egyptians being ſubdued by the Romans, they received the Julian Year, though with ſome alteration; for they ſtill retained their ancient months, with the five additional days, and every 4th Year they intercalated another day, for the 6 hours, at the end of the Year, or between the 28th and 29th of Auguſt. Alſo, the beginning of their Year, or the firſt day of the month Thoth, anſwered to the 29th of Auguſt of the Julian Year, or to the 30th if it happened to be leap Year.

The Ancient Greek YEAR.—This was a lunar Year,

conſiſting of 12 months, which at firſt had each 30 days, then alternately 29 and 30 days, computed from the firſt appearance of the new moon; with the addition of an emboliſmic month of 30 days, every 3d, 5th, 8th, 11th, 14th, 16th, and 19th Year of a cycle of 19 Years; in order to keep the new and full moons to the ſame terms or ſeaſons of the Year.

Their Year commenced with that new moon which was neareſt to the ſummer ſolſtice And the order of the months, with the number of their days, were as follow : 1. Εκατομϐαιων, of 29 days; 2. Μηταγειτνιων 30; 3. Βοηδρομιων 29; 4. Μαιμακτηριων 30; 5. Πυανεψιων 29; 6. Ποσειδεων 30; 7. Γαμηλιων 29; 8. Ανθεςηριων 30; 9. Ελαφηϐολιων 29; 10. Μανυχιων 30; 11. Οαργηλιων 29; 12. Σκιροφοριων 30.— But many of the Greek nations had other names for their months.

The Ancient Jewiſh YEAR.—This is a lunar Year, uſually conſiſting of 11 months, containing alternately 30 and 29 days. And it was made to agree with the ſolar Year, by adding 11, and ſometimes 12 days, at the end of the Year, or by an emboliſmic month. The order and quantities of the months were as follow : 1. Niſan or Abib 30 days; 2. Jiar or Zius 29; 3. Siban or Sievan 30; 4. Thamuz or Tamuz 29; 5. Ab 30; 6. Elul 29; 7. Tiſri or Ethanim 30, 8. Marcheſvam or Bul 29; 9. Ciſleu 30; 10. Tebeth 29; 11. Sabat or Schebeth 30; 12. Adar 30 in the emboliſmic year, but 29 in the common year.—Note, in the defective Year, Ciſleu was only 29 days; and in the redundant Year, Marcheſvam was 30.

The Modern Jewiſh YEAR is likewiſe lunar, conſiſting of 12 months in common Years, but of 13 in emboliſmic Years; which, in a cycle of 19 Years, are the 3d, 6th, 8th, 11th, 14th, 17th, and 19th. Its beginning is fixed to the new moon next after the autumnal equinox. The names and order of the months, with the number of the days, are as follow : 1. Tiſri 30 days; 2. Marcheſvan 29; 3. Ciſleu 30; 4. Tebeth 29; 5. Schebeth 30; 6. Adar 29; 7. Veadar, in the emboliſmic year, 30; 8. Niſan 30; 9. Ilar 29; 10. Sivan 30; 11. Thamuz 29; 12. Ab 30; 13. Elul 29.

The Syrian YEAR, is a ſolar one, having its beginning fixed to the beginning of October in the Julian Year; from which it only differs in the names of the months, the quantities being the ſame; as follow : 1. Tiſhrin, anſwering to our October, and containing 31 days; 2. Latter Tiſhrin, containing, like November, 30 days; 3. Canun 31; 4. Latter Canun 31; 5. Shabat 28, or 29 in a leap-year; 6. Adar 31; 7. Niſan 30; 8. Aiyar 31; 9. Haziram 30; 10. Thamuz 31; 11. Ab 31; 12. Elul 30.

The Perſian YEAR, is a ſolar one, of 365 days, conſiſting of 12 months of 30 days each, with 5 intercalary days added at the end. The months are as follow : 1. Aſrudia meh; 2. Ardihaſcht meh; 3. Cardi meh; 4. Thir meh; 5. Merded meh; 6. Schabarir meh; 7. Mehar meh; 8. Aben meh; 9. Adar meh; 10. Di meh; 11. Behen meh; 12. Aſſirer meh. This Year is the ſame as the Egyptian Nabonaſſarean, and is called the *yezdegerdic Year*, to diſtinguiſh it from the fixed ſolar Year, called the Gelalean Year, which the Perſians began to uſe in the Year 1079, and which was
formed

formed by an intercalation, made six or seven times in four Years, and then once every 5th Year.

The Arabic, Mahometan, and Turkish Year, called also the Year of the *Hegira*, is a lunar Year, equal to 354^d 8^h 48^m, and consists of 12 months, containing alternately 30 and 29 days. Though sometimes it contains 13 months; the names &c being as follow: 1. Muharram of 30 days; 2. Saphar 29; 3. Rabia 30; 4. Latter Rabia 29; 5. Jomada 30; 6. Latter Jomada 29; 7. Rajab 30; 8. Shaaban 29; 9. Ramadan 30; 10. Shawal 29; 11. Dulkaadah 30; 12. Dulheggia 29, but in the embolismic year 30. An intercalary day is added every 2d, 5th, 7th, 10th, 13th, 15th, 18th, 21st,

24th, 26th, 29th, in a cycle of 29 Years. The months commence with the first appearance of the new moons after the conjunctions.

Ethiopic Year, is a solar Year perfectly agreeing with the Actiac, except in the names of the months, which are; 1. Mascaram; 2. Tykympt; 3. Hydar; 4. Tyshas; 5. Tyr; 6. Jacatil; 7. Magabit; 8. Mijazia; 9. Ginbat; 10. Syne; 11. Hamel; 12. Hahase. Intercalary days 5. It commences with the Egyptian Year, on the 29th of August of the Julian Year.

YESDEGERDIC Year. See *Persian* Year.

Z.

ZENITH, in Astronomy, the vertical point, or point in the heavens directly overhead. Or, the Zenith is a point in the surface of the sphere, from which a right line drawn through the place of any spectator, passes through the centre of the earth.

The Zenith of any place, is also the pole of the horizon, being 90 degrees distant from every point of it. And through the Zenith pass all the azimuths, or vertical circles.

The point diametrically opposite to the Zenith, is called the *nadir*, being the point in the sphere directly under our feet: and it is the Zenith to our antipodes, as our Zenith is their nadir.

ZENITH-*Distance*, is the distance of the sun or star from our Zenith; and is the complement of the altitude, or what it wants of 90 degrees.

ZENO, ELEATES, or of *Elea*, one of the greatest philosophers among the Ancients, flourished about 500 years before the Christian æra. He was the disciple of Parmenides, and even, according to some writers, his adopted son. Aristotle asserts that he was the inventor of logic: but his logic seems to have been calculated and employed to perplex all things, and not to clear up any thing. For Zeno employed it only to dispute against all comers, and to silence his opponents, whether they argued right or wrong. Among many other subtleties and embarrassing arguments, he proposed some with regard to motion, denying that ther was any such thing in nature; and Aristotle, in the 6th book of his physics, has preserved some of them, which are extremely subtile, especially the famous argument named Achilles; which was to prove this proposition, that the swiftest animal could never overtake the slowest, as a greyhound a tortoise, if the latter set out a little fore the former: for suppose the tortoise to be 100 yards before the dog, and that this runs 100 times as fast as the other; then while the dog runs the first 100 yards, the tortoise runs 1, and is therefore 1 yard

before the dog; again, while the dog runs over this yard, the tortoise will run the 100th part of a yard, and will be so much before the dog; and again, while the dog runs over this 100th part of a yard, the tortoise will have got the 100th part of that 100th part before him; and so on continually, says he, the dog will always be some small part behind the tortoise. But the fallacy will soon be detected, by considering where the tortoise will be when the dog has run over 200 yards; for as the former can have run only two yards in the same time, and therefore must then be 98 yards behind the dog, he consequently must have overtaken and passed the tortoise. It has been said that, to prove to him, or some disciple of his, that there is such a thing as motion, Diogenes the Cynic rose up and walked over the floor.—Zeno shewed great courage in suffering pain; for having joined with others to endeavour to restore liberty to his country, which groaned under the oppression of a tyrant, and the enterprize being discovered, he supported with extraordinary firmness the sharpest tortures. It is even said that he had the courage to bite off his tongue, and spit it in the tyrant's face, for fear of being forced, by the violence of his torments, to discover his accomplices. Some say that he was pounded to death in a mortar

ZENO, a celebrated Greek philosopher, was born at Citium, in the Isle of Cyprus, and was the founder of the Stoics; a sect which had its name from that of a portico at Athens, where this philosopher chose to hold his discourses. He was cast upon that coast by shipwreck; and he ever after regarded this as a great happiness, praising the winds for having so happily driven him into the port of Piræum.—Zeno was the disciple of Crates, and had a great number of followers. He made the sovereign good to consist in dying in conformity to nature, guided by the dictates of right reason. He acknowledged but one God; and admitted an inevitable destiny over all events. His

servant

servant taking advantage of this laft opinion, cried, while he was beating him for difhonefty, " I was deftined to fteal ;" to which Zeno replied, " Yes, and to be beaten too." This philofopher ufed to fay, " That if a wife man ought not to be in love, as fome pretended, none would be more miferable than beautiful and virtuous women, fince they would have none for their admirers but fools." He alfo faid, " That a part of knowledge confifts in being ignorant of fuch things as ought not to be known : that a friend is another felf : that a little matter gives perfection to a work, though perfection is not a little matter." He compared thofe who fpoke well and lived ill, to the money of Alexandria, which was beautiful, but compofed of bad metal.—It is faid that being hurt by a fall, he took that as a fign he was then to quit this life, and laid violent hands on himfelf, about 264 years before Chrift.

Cleanthes, Cryfippus, and the other fucceffors of Zeno maintained, that with virtue we might be happy in the midft even of difgrace and the moft dreadful torments. They admitted the exiftence of only one God, the foul of the world, which they confidered as his body, and both together forming a perfect being. It is remarked that, of all the fects of the ancient philofophers, this was one of thofe which produced the greateft men.

We ought not to confound the two Zenos above mentioned, with

ZENO, a celebrated Epicurean philofopher, born at Sidon, who had Cicero and Pomponius Atticus for his difciples, and who wrote a book againft the mathematics, which, as well as that of Poffidonius's refutation of it, is loft ; nor with feveral other Zenos mentioned in hiftory.

ZENSUS, or ZENZUS, in Arithmetic and Algebra, a name ufed by fome of the older authors, efpecially in Germany, for a fquare number, or the 2d power : being a corruption from the Italic *cenf*, of Pacioli, Tartalea, &c, or the Latin *cenfus*, which fignified the fame thing.

ZETETICE, or ZETETIC *Method*, in Mathematics, was the method made ufe of to inveftigate, or find out the folution of a problem ; and was much the fame thing as analytics, or the analytic method.

Vieta has an ingenious work of this kind in 5 books ; *Zeteticorum libri quinque*.

ZOCCO, ZOCCOLO, ZOCLE, or SOCLE, in Architecture, a fquare body, lefs in height than breadth, placed under the bafes of pedeftals, ftatues, vafes, &c. See SOCLE and PLINTH.

ZODIAC, in Aftronomy, an imaginary ring or broad circle, in the heavens, in form of a belt or girdle, within which the planets all make their excurfions. In the very middle of it runs the ecliptic, or path of the fun in his annual courfe ; and its breadth, comprehending the deviations or latitudes of the planets, is by fome authors accounted 16°, fome 18, and others 20 degrees.

The Zodiac, cutting the equator obliquely, makes with it the fame angle as the ecliptic, which is its middle line, which angle, continually varying, is now nearly equal to 23° 28' ; which is called the obliquity of the

Zodiac or ecliptic, and is alfo the fun's greateft declination.

The Zodiac is divided into 12 equal parts, of 30 degrees each, called the figns of the Zodiac, being fo named from the conftellations which anciently paffed them. But, the ftars having a motion from weft to eaft, thofe conftellations do not now correfpond to their proper figns ; from whence arifes what is called the *preceffion of the equinoxes*. And therefore when a ftar is faid to be in fuch a fign of the Zodiac, it is not to be underftood of that conftellation, but only of that dodecatemory or 12th part of it.

Caffini has alfo obferved a tract in the heavens, within whofe bounds moft of the comets, though not all of them, are obferved to keep, and which he therefore calls the *Zodiac of the comets*. This he makes as broad as the other Zodiac, and marks it with figns or conftellations, like that ; as Antinous, Pegafus, Andromeda, Taurus, Orion, the Leffer Dog, Hydra, the Centaur, Scorpion, and Sagittary.

ZODIACAL *Light*, a brightnefs fometimes obferved in the zodiac, refembling that of the galaxy or milky way. It appears at certain feafons, viz, towards the end of winter and in fpring, after funfet, or before his rifing, in autumn and beginning of winter, refembling the form of a pyramid, lying lengthways with its axis along the zodiac, its bafe being placed obliquely with refpect to the horizon. This phenomenon was firft defcribed and named by the elder Caffini, in 1683. It was afterwards obferved by Fatio, in 1684, 1685, and 1686; alfo by Kirch and Eimmart, in 1688, 1689, 1691, 1693, and 1694. See Mairan, Suite des Mem. de l'Acad. Royale des Sciences 1731, pa. 3.

The Zodiacal light, according to Mairan, is the folar atmofphere, a rare and fubtile fluid, either luminous by itfelf, or made fo by the rays of the fun furrounding its globe ; but in a greater quantity, and more extenfively, about his equator, than any other part.

Mairan fays, it may be proved from many obfervations, that the fun's atmofphere fometimes reaches as far as the earth's orbit, and there meeting with our atmofphere, produces the appearance of an Aurora borealis.

The length of the Zodiacal light varies fometimes in reality, and fometimes in appearance only, from various caufes.

Caffini often mentions the great refemblance between the Zodiacal light and the tails of comets. The fame obfervation has been made by Fatio : and Euler endeavoured to prove that they were owing to fimilar caufes. See Decouverte de la Lumiere Celefte que paroit dans le Zodiaque, art. 41. Lettre à M. Caffini, printed at Amfterdam in 1686. Euler, in Mem. de l'Acad. de Berlin, tom. 2.

This light feems to have no other motion than that of the fun itfelf : and its extent from the fun to its point, is feldom lefs than 50 or 60 degrees in length, and more than 20 degrees in breadth : but it has been known to extend to 100 or 103°, and from 8 to 9° broad.

It is now generally acknowledged, that the electric fluid is the caufe of the aurora borealis, afcribed by

Mairan

Mairan to the folar atmofphere, which produces the Zodiacal light, and which is thrown off chiefly and to the greateft diftance from the equatorial parts of the fun, by means of the rotation on his axis, and extending vifibly as far as the orbit of the earth, where it falls into the upper regions of our atmofphere, and is collected chiefly towards the polar parts of the earth, in confequence of the diurnal revolution, where it forms the aurora borealis. And hence it has been fuggefted, as a probable conjecture, that the fun may be the fountain of the electrical fluid, and that the Zodiacal light, and the tails of comets, as well as the aurora borealis, the lightning, and artificial electricity, are its various and not very diffimilar modifications.

ZONE; in Geography and Aftronomy, a divifion of the earth's furface, by means of parallel circles, chiefly with refpect to the degree of heat in the different parts of that furface.

The ancient aftronomers ufed the term Zone, to explain the different appearances of the fun and other heavenly bodies, with the length of the days and nights; and the geographers, as they ufed the climates, to mark the fituation of places; ufing the term climate when they were able to be more exact, and the term Zone when lefs fo.

The Zones were commonly accounted five in number; one a broad belt round the middle of the earth, having the equator in the very middle of it, and bounded, towards the north and fouth, by parallel circles paffing through the tropics of Cancer and Capricorn. This they called the *torrid Zone*, which they fuppofed not habitable, on account of its extreme heat. Though fometimes they divided this into two equal torrid Zones, by the equator, one to the north, and the other fouth; and then the whole number of Zones was accounted 6.

Next, from the tropics of Cancer and Capricorn, to the two polar circles, were two other fpaces called *temperate Zones*, as being moderately warm; and thefe they fuppofed to be the only habitable parts of the earth.

Laftly, the two fpaces beyond the temperate Zones, about either pole, bounded within the polar circles, and having the poles in the middle of them, are the two *frigid* or *frozen Zones*, and which they fuppofed not habitable, on account of the extreme cold there.

Hence, the breadth of the torrid Zone, is equal to twice the greateft declination of the fun, or obliquity of the ecliptic, equal to 46° 56', or twice 23° 28'. Each frigid Zone is alfo of the fame breadth, the diftance from the pole to the polar circle being equal to the fame obliquity 23° 28'. And the breadth of each temperate Zone is equal to 43° 4', the complement of twice the fame obliquity. See thefe Zones exhibited in plate 35, fig. 16.

The difference of Zones is attended with a great diverfity of phenomena. 1. In the torrid Zone, the fun paffes through the zenith of every place in it twice a year; making as it were two fummers in the year; and the inhabitants of this Zone are called *amphifcians*, becaufe they have their noon-day fhadows projected different ways in different times of the year, northward at one feafon, and fouthward at the other.

2. In the temperate and frigid Zones, the fun rifes and fets every natural day of 24 hours. Yet every where, but under the equator, the artificial days are of unequal lengths, and the inequality is the greater, as the place is farther from the equator. The inhabitants of the temperate Zones are called *heterofcians*, becaufe their noon-day fhadow is caft the fame way all the year round, viz, thofe in the north Zone toward the north pole, and thofe in the fouth Zone toward the fouth pole.

3. Within the frigid Zones, the inhabitants have their artificial days and nights extended out to a great length; the fun fometimes fkirting round a little above the horizon for many days together: and at another feafon never rifing above the horizon at all, but making continual night for a confiderable fpace of time. The inhabitants of thefe Zones are called *perifcians*, becaufe fometimes they have their fhadows going quite round them in the fpace of 24 hours.

ADDENDA

ADDENDA ET CORRIGENDA.

A.

ACCELERATED *Motion*, pa. 18, col. 1, line 17 from the bottom, *after* second inftant, *add*, or fmall part of time.—l. 6 from the bottom, *for* in every inftant, *read* at every moment.—l. 2 from bottom, *for* $16\frac{1}{12}$, *read* $32\frac{1}{8}$.—col. 2, l. 1 and 2, *for* $32\frac{1}{8}$, $48\frac{1}{4}$, $64\frac{1}{3}$, *read* $64\frac{1}{3}$, $96\frac{1}{2}$, $128\frac{2}{3}$.

ACCELERATING *Force*, pa. 21, col. 2, l. 27, *for* requires, *read* acquires.

Pa. 22, col. 2, l. 16 from the bottom, for $t = \dot{v}i$ read $\dot{s} = \dot{v}i$.—next line, for t and s, read $\dot{t}$ and $\dot{s}$.

ACHROMATIC, pa. 25, col. 2, l. 14, *for* fractions, *read* refractions.

Pa. 26, col. 1, l. 12, *for* Veritus, *read* Veritas.

After l. 9, *add*, Since this article was printed, I obferve, in the 3d volume of the Edinburgh Philofophical Tranfactions, an account of a curious fet of experiments, on the unequal refrangibility of light, with obfervations on Achromatic telefcopes, by Dr. Robert Blair. This ingenious gentleman fets out with obferving, " If the theory of the Achromatic telefcope is fo complete as it has been reprefented, may it not reafonably be demanded, whence it proceeds, that Hugenius and others could execute telefcopes with fingle object glaffes 8 inches and upwards in diameter, while a compound object glafs of half thefe dimenfions, is hardly to be met with ? or how it can arife from any defect in the execution, that reflectors can be made fo much fhorter than Achromatic refractors of equal apertures, when it is well known that the latter are much lefs affected by any imperfections in the execution of the lenfes compofing the object glafs, than reflectors are by equal defects in the figure of the great fpeculum ?—The general anfwer made by artifts to enquiries of this kind, is, that the fault lies in the imperfection of glafs, and particularly in that kind of glafs of which the concave lens of the compound object glafs is formed, called flint glafs.— It was in order to fatisfy myfelf concerning the reality of this difficulty, and to attempt to remove it, that I engaged in the following courfe of experiments."

Dr. Blair defcribes the apparatus and manner of making the experiments. He employed various prifms of different kinds of glafs ; alfo lenfes of glafs, and of

a great variety of fluid mediums, having different degrees of refraction. Having detailed the whole at confiderable length, for which a reference muft be made to the work itfelf, and it is very deferving of attentive perufal, he concludes with the following recapitulation of the contents and fcope of the whole difcourfe.

" The unequal refrangibility of light, as difcovered and fully explained by Sir Ifaac Newton, fo far ftands its ground uncontroverted, that when the refraction is made in the confine of any medium whatever, and a vacuum, the rays of different colours are unequally refracted, the red-making rays being the leaft refrangible, and the violet-making rays the moft refrangible.

" The difcovery of what has been called a different difperfive power in different refractive mediums, proves thofe theorems of Sir Ifaac Newton not to be univerfal, in which he concludes that the difference of refraction of the moft and leaft refrangible rays, is always in a given proportion to the refraction of the mean refrangible ray. There can be no doubt that this pofition is true with refpect to the mediums on which he made his experiments ; but there are many exceptions to it.

" For the experiments of Mr. Dollond prove, that the difference of refraction between the red and violet rays, in proportion to the refraction of the whole pencil, is greater in fome kinds of glafs than in water, and greater in flint-glafs than in crown-glafs.

" The firft fet of experiments above recited, prove, that the quality of difperfing the rays in a greater degree than crown-glafs, is not confined to a few mediums, but is poffeffed by a great variety of fluids, and by fome of thefe in a moft extraordinary degree. Solutions of metals, effential oils, and mineral acids, with the exception of the vitriolic, are moft remarkable in this refpect.

" Some confequences of the combinations of mediums of different difperfive powers, which have not been fufficiently attended to, are then explained. Although the greater refrangibility of the violet rays than of the red rays, when light paffes from any medium whatever into a vacuum, may be confidered as a law of nature, yet in the paffage of light from one medium into another, it depends entirely on the qualities of the mediums, which of thefe rays fhall be the moft refrangible, or whether there fhall be any difference in their refrangibility.

" The

" The application of the demonstrations of Hugenius to the correction of the aberration from the spherical figures of lenses, whether solid or fluid, is then taken notice of, as being the next step towards perfecting the theory of telescopes.

" Next it appears from trials made with object-glasses of very large apertures, in which both aberrations are corrected as far as the principles will admit, that the correction of colour which is obtained by the common combination of two mediums which differ in dispersive power, is not complete. The homogeneal green rays emerge most refracted, next to these the united blue and yellow, then the indigo and orange united, and lastly the united violet and red, which are least refracted.

" If this production of colour were constant, and the length of the secondary spectrum were the same in all combinations of mediums when the whole refraction of the pencil is equal, the perfect correction of the aberration from difference of refrangibility would be impossible, and would remain an insurmountable obstacle to the improvement of dioptrical instruments.

" The object of the next experiment is, therefore, to search, whether nature affords mediums which differ in the degree in which they disperse the rays composing the prismatic spectrum, and at the same time separate the several orders of rays in the same proportion. For if such could be found, the above-mentioned secondary spectrum would vanish, and the aberration from difference of refrangibility might be removed. The result of this investigation was unsuccessful with respect to its principal object. In every combination that was tried, the same kind of uncorrected colour was observed, and it was thence concluded, that there was no direct method of removing the aberration.

" But it appeared in the course of the experiments, that the breadth of the secondary spectrum was less in some combinations than in others, and thence an indirect way opened, leading to the correction sought after; namely by forming a compound concave lens of the materials which produce most colour, and combining it with a compound convex lens formed of the materials which produce least colour; and it was observed in what manner this might be effected by means of three mediums, though apparently four are required.

" In searching for mediums best adapted for the above purpose, a very singular and important quality was detected in the muriatic acid. In all the dispersive mediums hitherto examined, the green rays, which are the mean refrangible in crown-glass, were found among the less refrangible, and thence occasion the uncorrected colour which has been described. In the muriatic acid, on the contrary, these same rays make a part of the more refrangible; and in consequence of this, the order of the colours in the secondary spectrum, formed by a combination of crown glass with this fluid, is inverted, the homogeneal green being now the least refrangible, and the united red and violet the most refrangible.

" This remarkable quality found in the marine acid led to complete success in removing the great defect of optical instruments, that dissipation or aberration of the rays, arising from their unequal refrangibility, which has rendered it impossible hitherto to converge all of them to one point either by single or opposite refractions. A fluid in which the particles of marine acid and metalline particles hold a due proportion, at the same time that it separates the extreme rays of the spectrum much more than crown-glass, refracts all the orders of rays exactly in the same proportion as the glass does; and hence rays of all colours, made to diverge by the refraction of the glass, may either be rendered parallel by a subsequent refraction made in the confine of the glass and this fluid, or by weakening the refractive density of the fluid, the refraction which takes place in the confine of it and glass, may be rendered as regular as reflexion, while the errors arising from unavoidable imperfections of workmanship, are far less hurtful than in reflexion, and the quantity of light transmitted by equal apertures of the telescopes much greater.

" Such are the advantages which the theory presents. In reducing this theory to practice, difficulties must be expected in the first attempts. Many of these it was necessary to surmount before the experiments could be completed. For the delicacy of the observations is such as to require a considerable degree of perfection in the execution of the object-glasses, in order to admit of the phenomena being rendered more apparent by means of high magnifying powers. Great pains seem to have been taken by mathematicians to little purpose, in calculating the radii of the spheres requisite for Achromatic telescopes, from their not considering that the object-glass itself is a much nicer test of the optical properties of refracting mediums than the gross experiments made by prisms, and that the results of their demonstrations cannot exceed the accuracy of the data, however much they may fall short of it.

" I shall conclude this paper, which has now greatly exceeded its intended bounds, by enumerating the several cases of unequal refrangibility of light, that their varieties may at once be clearly apprehended.

" In the refraction which takes place in the confine of every known medium and a vacuum, rays of different colours are unequally refrangible, and the red-making rays are least refrangible, and the violet-making rays are most refrangible.

" This difference of refrangibility of the red and violet rays is not the same in all mediums. Those mediums in which the difference is greatest, and which, by consequence, separate or disperse the rays of different colours most, have been distinguished by the term dispersive, and those mediums which separate the rays least have been called indispersive. Dispersive mediums differ from indispersive, and still more from each other, in another very essential circumstance.

" It appears from the experiments which have been made on indispersive mediums, that the mean refrangible light is always the same, and of a green colour.

" Now, in by far the largest class of dispersive mediums, including flint glass, metallic solutions, essential oils, the green light is not the mean refrangible order, but forms one of the less refrangible orders of light, being found in the prismatic spectrum nearer to the deep red than the extreme violet.

" In another class of dispersive mediums, which includes the muriatic and nitrous acids, this same green light becomes one of the more refrangible orders, being now found nearer to the extreme violet than the deep red.

" These

" Thefe are the varieties in the refrangibility of light, when the refraction takes place in the confine of a vacuum ; and the phenomena will fcarce differ fenfibly in refractions made in the confine of denfe mediums and air.

" But when light paffes from one denfe medium into another, the cafes of unequal refrangibility are more complicated.

" In refractions made in the confine of mediums which differ only in ftrength, not in quality, as in the confine of water and crown-glafs, or in the confine of the different kinds of difperfive fluids more or lefs diluted, the difference of refrangibility will be the fame as above ftated in the confine of denfe mediums and air, only the whole refraction will be lefs.

" In the confine of an indifperfive medium, and a rarer medium belonging to either clafs of the difperfive, the red and violet rays may be rendered equally refrangible. If the difperfive power of the rare medium be then increafed, the violet rays will become the leaft refrangible, and the red rays the moft refrangible. If the mean refractive denfity of the two mediums be rendered equal, the red and violet rays will be refracted in oppofite directions, the one towards, the other from the perpendicular.

" Thus it happens to the red and violet rays, whichfoever clafs of difperfive mediums be employed. But the refrangibility of the intermediate orders of rays, and efpecially of the green rays, will be different when the clafs of difperfive mediums is changed.

" Thus, in the firft cafe, where the red and violet rays are rendered qually refrangible, the green rays will emerge moft refrangible if the firft clafs of difperfive mediums is ufed, and leaft refrangible if the fecond clafs is ufed. And in the other two cafes, where the violet becomes leaft refrangible, and the red moft refrangible, and where thefe two kinds of rays are refracted in oppofite directions, the green rays will join the red if the firft clafs of difperfive mediums be employed, and will arrange themfelves with the violet if the fecond clafs be made ufe of.

" Only one cafe more of unequal refrangibility remains to be ftated ; and that is, when light is refracted in the confine of mediums belonging to the two different claffes of difperfive fluids. In its tranfition, for example, from an effential oil, or a metallic folution, into the muriatic acid, the refractive denfity of thefe fluids may be fo adjufted, that the red and violet rays fhall fuffer no refraction in paffing from the one into the other, how oblique foever their incidence be. But the green rays will then fuffer a confiderable refraction, and this refraction will be from the perpendicular, when light paffes from the muriatic acid into the effential oil, and towards the perpendicular, when it paffes from the effential oil into the muriatic acid. The other orders of rays will fuffer fimilar refractions, which will be greateft in thofe adjoining the green, and will diminifh as they approach the deep red on the one hand, and the extreme violet on the other, where the refraction ceafes entirely.

" The manner of the production of thefe effects, by the attraction of the feveral mediums, may be thus explained. We fhall fuppofe the attractive forces, which

produce the refractions of the red, green and violet light, to be reprefented by the numbers, 8, 12, and 16, in glafs ; 6, 9, 14, in the metallic folution ; 6, 11, 14, in the muriatic acid ; and 6, 10, 14, in a mixture of thefe two fluids. The excefs of attraction of glafs for the red and violet light is equal to 2, whichfoever of the three fluids be employed. The refraction of thefe two orders of rays will therefore be the fame in all the three cafes. But the excefs of attraction for the green light is equal to 3, when the metallic folution is ufed, and therefore the green light will be more refracted than the red and violet, in this cafe. When the muriatic acid is ufed, the excefs of attraction of glafs for the green light is only 1, and therefore the green light will now be lefs refracted than the red and violet.

" We fhall next fuppofe the metallic folution and the acid to adjoin each other. The attractions of both thefe mediums, for the red light being 6, and for the violet light 14, thefe two orders of rays will fuffer no refraction in the confine of the two fluids, the difference of their attractions being equal to nothing.

" But the attractive force of the metallic folution for the green ray being only 9, and that of the muriatic acid for the fame ray being 11, the green light will be attracted towards the muriatic acid with the force 2 ; and therefore the difference between the refraction of the green light and the unrefracted red and violet light, which takes place in the confine of thefe fluids, will greatly exceed the difference of refraction of the green light, and equally refracted red and violet light, which is produced in the confine of glafs and either of the fluids.

" Laftly, in a mixture of the two kinds of fluids, the attraction for the red, green and violet rays, being 6, 10 and 14, and that of the glafs, 8, 12 and 16, the excefs of the attraction of the glafs for the green rays, is the fame which it is for the red and violet rays. Thefe three orders of rays will therefore fuffer an equal refraction, being each of them attracted towards the glafs with the force 2 ; and when this is the cafe, it appears, from the obfervations, that the indefinite variety of rays of intermediate colours and fhades of colours, which altogether compofe folar light, will alfo be regularly bent from their rectilinear courfe, conftituting what has been termed a planatic refraction."

In fhort, Dr. Blair fays, that he " ufes more tranfparent mediums than the common ones ; avoids or greatly diminifhes the reflections at the furfaces of the mediums ; applies fluid mediums more homogeneous than thick flint or crown glafs, which at the fame time difperfe the different coloured rays of light in the fame proportion, by which means an image is produced perfectly Achromatic, which is but imperfectly fo in Dollond's object glaffes made of flint and crown glafs combined.

ACOUSTICS, at the end, *add*, But this ftatute was repealed by the 15th of Geo. the 3d, cap. 32.

AEROSTATION, pa. 45, col. 2, l. 40, *for* 800, *read* 680.—l. 46, *for* 28¼ *read* 26.—l. 48, *for* balloon *read* parachute.—l. 51 and 52, *for* 28½ *read* 26, and *for* 13 *read* 12.—l. 55, *read* 2 feet 3 inches.

Pa. 46, col. 2, at the end of the article on *Aerostation*, *add*, See an ingenious and learned treatise on the mathematical and physical principles of Air-balloons, by the late Dr. Damen, professor of philosophy and mathematics in the University of Leyden, entitled, Physical and Mathematical Contemplations on Aerostatic Balloons, &c; in 8vo, at Utrecht, 1784.

Pa. 70, col. 1, l. 4. *dele* $-\sqrt{3-1} = 2.$—l. 5, at the end *add* $-\sqrt{3-1} = 2.$

Pa. 71, col. 1, l. 9, *for* $y^2 + 2y - 7$ *read* $y^2 + 2y - 7.$

AFFECTED *Equations*, *add* (from Francis Maseres, Esq.)—" This expression of *Affected Equations* seems to require some further explanation. It was introduced by the celebrated Vieta, the great father and restorer of Algebra. He has many expressions peculiar to himself, and which have not been adopted by subsequent Algebraists. Amongst these are the following ones. He calls a set of quantities in continual geometrical proportion, (such as the quantities 1, x, x^2, x^3, x^4, x^5, x^6 x^7, &c,) a set of *scalar* quantities, or *magnitudines scalares*; and, when there are several of these *scalar* quantities mentioned together, (as in the compound quantity $x^5 + ax^4 - b^2x^3$,) he calls the highest quantity, or that which is farthest in the scale of quantities 1, x, x^2, x^3, x^4, x^5, x^6, x^7, &c. (to wit, the quantity x^5 in the said compound quantity $x^5 + ax^4 - b^2x^3$,) *the power* of the fundamental quantity x, or of the second term in the said scale; and he calls the lower scalar quantities which are involved in the second and third terms of the said compound quantity $x^5 + ax^4 - b^2x^3$, to wit, the quantities x^4 and x^3, (or, in our present language, the inferiour powers of x,) scalar quantities of a *parodic* degree to x^5, or the power of the fundamental quantity x. This word *parodic* I take to be derived (though Vieta does not tell us so) from the Greek words πρὸς and ὁδὸς, which signify *near* and *a way* or *road*, because these inferior scalar quantities x^3 and x^4 lie *in the way* as you pass along in the scale of the aforesaid quantities 1, x, x^2, x^3, x^4, x^5, x^6 x^7, &c, from 1 to x^5, which he calls the power of x in the said compound quantity $x^5 + ax^4 - b^2x^3$. These inferiour scalar quantities x^3 and x^4 are therefore *parodic*, or *situated in the way to*, or are *leading to*, the higher scalar quantity x^5. He then proceeds to define *a pure power* and *an affected power*, and tells us that *a pure power* is a scalar quantity that is not affected with any *parodic*, or *inferiour* scalar quantity, and that *an affected power* is a scalar quantity that is connected by addition, or subtraction with one, or more, *inferiour*, or *parodic*, scalar quantities, combined with co-efficients that raise them to the same dimension as the power itself, or make them *homogeneous* to it, and consequently capable of being added to it, or subtracted from it. Thus x^5 alone is a *pure power* of x, namely, its fifth power; and $x^5 + ax^4 - b^2x^3$ is *an affected power* of x, namely, its fifth power *affected by*, or *connected with*, the two *parodic*, or *inferiour*, scalar quantities x^3 and x^4, which are multiplied into *bb* and *a*, in order to make

them *homogeneous* to, or *of the same dimension with*, x^5 itself, and capable of being added to it or subtracted from it. See Schooten's Edition of Vieta's works, published at Leyden in Holland in the year 1646, pages 3 and 4.

" This, then, being the meaning of the expression, *a pure power* and *an affected power*, the meaning of the corresponding expressions of *a pure equation* and *an affected equation* follows from it of course: *a pure equation* signifying an equation in which a pure power of an unknown quantity is declared to be equal to some known quantity; such as the equation $x^5 = 79$; and *an affected equation* signifying an equation in which a power of an unknown quantity affected by, or connected, either by addition or subtraction, with, some inferiour powers of the same unknown quantity, (multiplied into proper co-efficients in order to make them *homogeneous* to the said highest power of the said unknown quantity,) is declared to be equal to some known quantity; such as the equation $x^5 + ax^4 - b^2x^3 = 79$. This I take to be the original meaning of the expression *an affected equation*. But, as the language of *Vieta* has not been adopted by subsequent writers of Algebra, I should think it would be more convenient to call them by some other name. And, perhaps those of *binomial*, *trinomial*, *quadrinomial*, *quinquinomial*, and, in general, that of *multinomial* equations, would be as convenient as any. Thus, $xx + ax = rr$, and $x^3 + ax^2 = r^3$, and $x^3 + a^2x = r^3$, and $x^4 + a^3x = r^4$, and $x^4 + ax^3 = r^4$, might all be called *binomial* equations, because they would be equations in which a *binomial* quantity, or quantity consisting of two terms that involved the unknown quantity x, is declared to be equal to a known quantity; and, for a like reason, the equations $x^3 + ax^2 + b^2x = r^3$, and $x^4 - ax^3 + b^2x^2 = r^4$, and $x^4 - ax^3 + b^3x = r^4$, and $x^5 + ax^4 + b^2x^3 = r^5$, and $x^5 + ax^4 - b^2x^3 = r^5$, and $x^5 + b^2x^3 + c^4x = r^5$, might be called *trinomial* equations. And the like names might be given to equations of a greater number of terms. Dr. Hutton, I observe, in his excellent new Mathematical and Philosophical Dictionary, just now published, (Feb. 2. 1795,) calls them *compound* equations; which is likewise a very proper name for them, and less obscure than that of *affected* equations."

Pa. 76, col. 1, l. 25, *for* $\sqrt{3+1} - \sqrt{3-1}$, *read* $\sqrt{3+1} - \sqrt{3-1}$.

Pa. 94, col. 2, l. 34, *for* Spaniard, *read* Portuguese.

Pa. 95, col. 2, after l. 21, or the end of the paragraph relating to Dr. Barrow, add as follows:—Of these lectures, the 13th deserves the most special notice, being entirely employed upon—Equations, delivered in a very curious way. He there treats of the nature and number of their roots, and the limits of their magnitudes, from the description of lines accommodated to each, viz, treating the subject as a branch of the doctrine of maxima and minima, which, in the opinion of some persons, is the right way of considering them, and far preferable to the so much boasted invention of the generation of Equations from each other discovered by Harriot and Descartes.

Pa. 97, col. 2, after l. 3, *add*—Dr. Waring and the Rev. M. Vince, of Cambridge, have both given many improve-

improvements and difcoveries in feries and in other branches of analyfis. Thofe of Mr. Vince are chiefly contained in the latter volumes of the Philofophical Tranfactions; where alfo are feveral of Dr. Waring's; but the bulk of this gentleman's improvements are contained in his feparate publications, particularly the *Meditationes Algebraicæ*, publifhed in 1770; the *Proprietates Algebraicarum Curvarum*, 1772; and the *Meditationes Analyticæ*, 1776; an account of the chief contents of which, a friend has favoured me with, as follows.

Of Dr. Waring's Meditationes Algebraicæ.

The firft chapter treats of the transformation of algebraical equations into others, of which the roots have given algebraical relation to the roots of the given equations.

The general refolution of this problem requires the finding the aggregates of each of the values of algebraical functions of the roots of the given equation: for this purpofe the author begins with finding the fum of the mth power of each of the roots of the equation $x^n - px^{n-1} + qx^{n-2} - \&c = 0$ by a feries proceeding according to the dimenfions of p the fum of the roots: this feries (when continued in infinitum and converges) finds alfo the fum of any root of the above-mentioned quantities. From this feries is deduced the law of the reverfion of the feries $y = ax + bx^2 + cx^3 + \&c$, which finds x in terms of y; and alfo the law of a feries, which expreffes the greateft or leaft roots, and their powers or roots of a given algebraical equation, and which may be applied whether that root is poffible or impoffible, if the root be much greater or lefs than each of the remaining ones. All the powers and roots of this feries, when continued in infinitum, obferve the fame law.

On this fubject are further added fome elegant theorems; of which, one finds the fum of all quantities of this kind $\alpha^a\beta^b\gamma^c$, &c; where α, β, γ, &c, denote the roots of the given equation. This has been fince publifhed by the celebrated mathematician Mr. le Grange in the Academy of Sciences at Paris.

There is alfo added a method of confiderable utility in thefe matters; viz, the affuming equations whofe roots are known, and thence deducing the coefficients of the equations fought: and alfo from the terms of an inferior equation deducing the terms of a fuperior.

The fecond chapter principally treats of the limits and number of impoffible and affirmative and negative roots of algebraical equations.

Some new properties are added, of the limiting equations refulting from multiplying the fucceffive terms of the given equation into an arithmetical feries; and a method of finding limits between each of the roots of a given equation, fince publifhed in the Berlin Acts, and alfo fome new methods of finding equations whofe roots are limits between the roots of other equations. In theor. 4 and 5 are contained quantities which are always greater than certain others, when they are all poffible; from whence may be deduced Newton's and feveral other rules for finding the number of impoffible roots: thefe rules may be rendered fomewhat more general by multiplying the given equations into others, whofe roots are all poffible, and finding whether im-

poffible roots may be deduced by the rule in the refulting equation, which cannot from it be difcovered in the given one. A rule is given, deduced from each fucceffive four terms of the given equation, and confequently much more general than rules deduced from each fucceffive three terms. The former always difcovers the true number of impoffible roots contained in quadratic and cubic equations, the latter in quadratic only. There is alfo a rule given for finding the number of impoffible roots from an equation, of which the roots are the fquares, &c, of the roots of a given equation; and a fecond from an equation of which the roots are the fquares of the differences of the roots of a given equation; and a third rule for finding an equation, of which the root is $z = nx^{n-1} - \overline{n-1}\,px^{n-2} + \&c$; if $x^n - px^{n-1} + qx^{n-2} - \&c = 0$ be the given equation, &c, thefe latter refolutions always difcover the true number of impoffible roots contained in cubic, biquadratic and furfolid equations; and alfo whether or not any impoffible roots are contained in any given equation; and alfo from the laft term whether the number of impoffible roots contained be 2, 6, 10, &c, or 0, 4, 8, &c. The principle of a 4th rule is given by finding when two roots once, twice, thrice, &c, or four, &c, roots become equal. From a method given of finding the number of impoffible roots contained in an equation involving only one unknown quantity, is deduced a method of difcovering limits between which are contained any number of impoffible roots in an equation involving two or more unknown quantities. From the number of impoffible, affirmative and negative roots contained in a given equation, is delivered a method of finding the number of impoffible, &c roots contained in an equation of which the roots have a given algebraical relation to the roots of the given equation.

The principles are fubjoined of finding the number of affirmative and negative roots contained in an algebraical equation: but this neceffarily fuppofes a method of finding the number of its impoffible roots known. It is demonftrated, that if the equation $x^n - px^{n-1} + qx^{n-2} - \&c = 0$ be multiplied by $x - a$, then every change of figns in the given, will have one, or three, or five, &c in the refulting equation; and if it be multiplied by $x + a$, then every continuation from $+$ to $+$ or $-$ to $-$, will produce one, or three, or five, &c fuch continuations in the refulting, whence every equation $x^n - px^{n-1} + \&c = 0$ will contain at leaft fo many changes of figns in its fucceffive terms as there are affirmative roots, and fo many continued progreffes from $+$ to $+$ and $-$ to $-$, as there are negative. In a biquadratic $x^4 + px^3 + qx^2 + rx + s = 0$, of which two roots are impoffible, and s an affirmative quantity, then it is demonftrated that the two poffible ones will be both negative or both affirmative, according as $p^3 - 4pq + 8r$ is an affirmative or negative quantity, if the figns of the coefficients, p, q, r, s are neither all affirmative, nor alternately $-$ and $+$. The number of impoffible and affirmative and negative roots contained in the equation $x^n + Ax^m + B = 0$ is likewife given, &c. If $l\,x^m - px^{m-1} + qx^{m-2} - \&c = 0$ and $hx^n - ax^{n-1} + bx^{n-2} - \&c = v$, and further $hx^n - ax^{n-1} + bx^{n-2}\,\&c = 0$ and $lx^m - px^{m-1} + \&c = w$, then the content of all the values

of

of the quantity w will be to the content of all the values of the quantity $v :: \pm l^a : l^m$, from whence are deduced some properties of parabolic curves. *Ex. gr.* Let the equation expressing the relation between the abscifs x and ordinate y be $y = ax^n + bx^{n-1} + cx^{n-2} + \&c.$ Then will the content under the $(n - 1)$ greatest ordinates be to the square of the content of all the distances between any two points in which the abscifs cuts the curve $:: a^{n-1} : n^n - 2$. The quotient of the content of all the sines divided by the content of all the cosines to the points in which the abscifs cuts the curve, will be to the content of all the abovementioned greatest ordinates $:: n^n a : 1$. Similar propositions are deduced concerning the ordinates to the points of contrary flexure, &c

The third chapter is versant, concerning, 1st finding the roots of equations or irrational quantities, which have given relations to each other: this is performed by substitution or division and finding the common divisors of the quantities resulting; and 2d concerning more (n) equations containing a less number (m) of supposed unknown quantities, which consequently require $n-m$ equations, since named equations of condition; these are likewise deduced from the method of finding common divisors. 3dly, Concerning the resolution of equations; in this case is given, 1. The reduction or resolution of some recurring equations. 2. Some properties of the roots of the equation $x^n \pm 1 = 0$. 3. Resolution of a biquadratic $x^4 + px^3 + qx^2 + rx + s = 0$, by reducing it to an equation $z^4 + az^2 + b = 0$. 4. A resolution of the biquadratic $x^4 + 2px^3 = qx^2 + rx + s$ by adding $(p^2 + 2n)$ $x^2 + 2pnx + n^2$ to both sides of the equation, so as to complete the square; and the deducing that the values of n are $\dfrac{\alpha\beta + \gamma\delta}{2}, \dfrac{\alpha\gamma + \beta\delta}{2}, \dfrac{\alpha\delta + \beta\gamma}{2}$; the values of $\sqrt{(q + p^2 + 2n)}$ are $\dfrac{\alpha + \beta - \gamma - \delta}{2}$, $\dfrac{\alpha + \gamma - \beta - \delta}{2}$, &c, and the values of $\sqrt{(s + n^2)}$ are $\dfrac{\alpha\beta - \gamma\delta}{2}, \dfrac{\alpha\gamma - \beta\delta}{2}$, &c; if $\alpha, \beta, \gamma, \delta$, are the roots of the given equation. 5. A resolution of equations as general as any yet discovered, viz, the assuming $x = a\sqrt[n]{p} + b\sqrt[n]{p^2} + c\sqrt[n]{p^3} + \&c$; and exterminating the irrational quantities, viz, from assuming $x = a\sqrt[3]{p} + b\sqrt[3]{p^2}$ are deduced different resolutions of cubic; from $x = a\sqrt[4]{p} + b\sqrt[4]{p^2} + c\sqrt[4]{p^3}$ different resolutions of biquadric; from the equations $x = a\sqrt[n]{p} + b\sqrt[n]{p^2}$, $x = a\sqrt[m]{p} + b\sqrt[m]{p^{n-1}}$; $x = a\sqrt[n]{p} + b\sqrt[n]{p^3}$, $x = a\sqrt[n]{p} + \sqrt[n]{p^{n-2}}$, &c, are deduced De Moivre's equation, and several others of new formula not before delivered. 6. The resolution $x = \sqrt[n]{\alpha} + \sqrt[n]{\beta} + \sqrt[n]{\gamma} + \&c$, first given by Euler, shewn to be a very particular; but this is rendered here much more general by assuming a more general resolution. 7. The resolution and reduction of equations from exterminating irrational quantities. 8. Reduction of some equations, when they are deduced from others by reducing them to the

original equations. 9. The finding a quantity, which multiplied into a given irrational will produce a rational quantity, and thence deducing from a given equation involving irrational quantities the dimensions to which the equation freed from them will ascend. 10. Let P = a series either ascending or descending according to the dimensions of x, from thence is deduced the sum of a series consisting of its alternate terms, or terms at (n) distance from each other. 11. It is proved, that Cardan's resolution of a cubic, is a resolution of an equation of 9 dimensions or three different cubics: similar principles are applied to some other equations. 12. General principles are given for the deducing the function of the roots of the given, which constitute the coefficients or roots of the transformed equation. *E. g.* Let a cubic equation $x^3 + qx - r = 0$ and $z - \dfrac{q}{3z} = x$, thence is shewn the function of the roots of x, which constitute z, and further the cases of the cubic, which are resolvable by the transformed equation, whole root is z: the same principles are applied to biquadratics. 13. The correspondent impossible roots of a given irrational quantity are deduced; and also the different roots of a given resolution. 14. The biquadratic of the formula $x^2 - 2 (a + b\sqrt{-1})x - c - d\sqrt{-1} = 0$ is distinguished into two quadratic equations involving only possible quantities, and thence every algebraic equation is proved to consist of simple and quadratic divisors involving only possible quantities. 15. A method is delivered of transforming irrational quantities into others; but it is cautioned, that in reduction and transformation correspondent roots should be used, otherwise it is probable that we shall fall into errors, of which examples are given. 16. The convergency of a root found by the common method of approximations is given; and it is discovered that the convergency principally depends on the quantity assumed for the root being much more near to one root than to any other; and independent of it, not on how near it is to a root.

The fourth chapter is principally conversant concerning more algebraical equations and their reductions to one. 1. It gives the law of the resolution of any number of simple equations; and the reduction of n simple equations to $n-1$ by means of others. 2. The method of reducing more (n) equations into one so as to exterminate $n - 1$ unknown quantities by the method of common divisors, and further delivers the principles of investigating the roots or values of the unknown quantities, which result from this, or, which is much the same, from the common method of Erasmus Bartholinus, and which are not contained in the given equations. 3. If two algebraical equations of n and m dimensions of the unknown quantities x and y are reduced to one so as to exterminate one of the unknown quantities, the principles are given of finding the dimensions to which the other will ascend: if it ascends to $n \times m$ dimensions; then the sum of the roots depends on the terms of n and $n - 1$ dimensions in the one, and m and $m - 1$ in the other, and similarly of the products of every two; &c. From this principle are deduced several properties of algebraical curves.

The

The fame principles are applied to more equations involving more unknown quantities. 4. Some two equations of given formulæ are reduced to one fo as to exterminate one unknown quantity. 5. Two equations are likewife reduced to one fo as to exterminate unknown quantities by means of infinite feries. 6. A method of finding whether fome equations contain the fame roots of the unknown quantities as others. 7. From the correfpondent roots of the unknown quantities in given equations are found the conftitution of their co-efficients; and from thence the aggregates of the functions of the roots of two or more equations. 8. Some things are given concerning the transformations of more equations than one, of their impoffible roots, of their roots which have a given relation to each other. 9. Some reductions and refolutions of more equations involving more unknown quantities. 10. If two equations fimilarly involve two unknown quantities x and y; then the equation of which the root is x or y is demon-ftrated to have twice the dimenfions of the equation whofe root is any rational function of $x + y$ or $x^2 + y^2$ or any rational recurring function of x and y; and if for y be fubftituted $- y$; then in the equation whofe root is the refulting quantity the dimenfions will be the fame as in the equations whofe root is x or y, but its formula will be of half the number of dimenfions. The fame principles are applied to more equations fimilarly involving more unknown quantities. 11. If there are two equations involving two unknown quantities, one deduced from the other, by fome fubftitutions inveftigated from equations fimilarly involving two unknown quantities; then the equation whofe root is one of the unknown quantities will be recurring. 12. Let A and B be functions of x and y, a method is given of finding, whether A is a function of B. 13. Methods of approximations to the roots of equations when they are unequal, or two or more nearly equal, poffible or impoffible; and alfo fome remarks on the increments or decrements of the roots, in paffing from one equation to others of the fame number of dimenfions are given.

The fifth chapter treats of rational and integral values of the unknown quantities of given equations.

1. It finds the rational and integral fimple, quadratic, &c divifors (by a method different to Waeffaner's) of a given equation, which involves one or more unknown quantities. 2. If two equations involve two unknown quantities x and y; the fame irrationality, which is contained in x will likewife be contained in its correfpondent value of y, unlefs two or more values of the quantity (x or y) are equal, &c. 3. A method is given of finding integral correfpondent values of the unknown quantities of two or more equations involving as many unknown quantities. 4. A method is alfo delivered of deducing when a given equation can be refolved by means of fquare, cube, &c roots; and when by fimilar methods it can be reduced to equations of $\frac{1}{2}$, $\frac{1}{4}$, &c, its dimenfions. 5. A method is given of finding a quantity or number, in which are contained all the divifors of any given rational or integral quantities. 6. A method different from Schooten's, Newton's, and Euler's, of extracting the root of a binomial furd $a + \sqrt{b}$ is given, and the principle demonftrated on

which all the rules are founded given by Schooten, viz, the multiplying the binomial furd fo that the nth root of $A^2 - B$ can be extracted, where $A + \sqrt{B}$ is the refulting furd; and it is further proved that multiplying the given furd $a + \sqrt{b}$ into 2^n will render Newton's refolution as general as the others; and laftly the extraction of the (m^{th}) root of the quantity $A + B\sqrt[n]{p} + C\sqrt[n]{p^2} + \cdots + \sqrt[n]{p^{n-1}}$ is given. 7. The law of Dr. Wallis's approximations in terms of the fucceffive quotients, as alfo of continual fractions is deduced. 8. A method of deducing the integral values of each of the unknown quantities x, y, z, v, &c, contained in the equation $ax + by + cz + dv \pm$ &c $+ f = 0$ in terms of quantities, for which may be affumed any whole numbers. 9. Two or more equations are reduced to one, fo as to exterminate unknown quantities; and if the unknown quantities of the refulting equations be integral or fractional, then the unknown quantities of the given equations will alfo be integral or fractional. 10. Principles are delivered of deducing equations of which the unknown quantities admit of correfpondent and known integral or rational values. 11. Correfpondent integral or rational values of the unknown quantities in feveral equations are given, and from fome values of the abovementioned kind given, are deduced others. 12. A method of denoting any numbers either by fours, fives, fixes, &c, and their powers; and fimilar properties deduced as in decimal arithmetic. 13. It is demonftrated that the fum of the divifors of the number $1 . 2 . 3 \ldots x = N$ has to N a greater ratio than the fum of the divifors of any number L lefs than N has to L; and fome other fimilar properties. 14. In the Philofophical Tranfactions are given properties fimilar to Mr. Euler's of the fum of divifors of the natural numbers, and fome others. 15. Let $N = a^2 + rb^2$, where a, b, r, p and q are whole numbers, then $N2m + 1$ and $N2m + 2$ can be compounded by $(m + 1)$ different ways of the quantities $p^2 + rq^2$; the different ways were firft given in the Medit. 16. Every number confifts of 1, 2, 3 or 4 fquares, and of 1, 2, 3, 4, .. 9 cubes, and therefore if a number N is equal to 3 fquares or 8 cubes, the problem may not be poffible. 17. Let x and z be any whole numbers, and a and b numbers prime to each other, then $ax + bz$ can conftitute any number, which exceeds $a \times b - a - b$. 18. Let r the greateft common divifor of m and $n - 1$, where n is a prime number; the number of remainders from the divifion of the number $1^m, 2^m, 3^m,$ &c, in infinitum by n will be $\frac{n - 1}{r} + 1$: from which are deduced feveral propofitions. 19. Sir John Wilfon's property delivered and demonftrated, viz, $1 . 2 . 3 \ldots n - 1 + 1$ will be divifible by n, if n be a prime number. 20. The fum of the powers $1^r + 2^r + 3^r + \ldots x^r$ are found divifible by $x . \overline{x + 1}$, if r be a whole number; from whence is deduced an elegant property of all parabolas correfpondent to the property of Archimedes of the infcribed triangles in a conical parabola. 21. Some properties of exponential equations; feveral other new properties of algebraical quantities and equations are given in thefe Meditations. They were fent to the Royal Society in 1757, and fince publifhed in the years 1760, 62, and 69.

Properties

Properties of Algebraical Curves.

The equation expressing the relation between the abscifs and its correspondent ordinates of a curve is transformed into another which expresses the relation between different abscifsæ and their ordinates, from which is deduced, that there may be *n* and not more different diameters in a curve of *n* — 1 order, which cuts its ordinates in a given angle; and likewise that a diameter can have no more than *n* — 1 different inclinations of its ordinates, unless the diameter be a general one. 2. The formula of the equations to curves, all whose diameters are parallel, or cut each other in a given point, or which have a general diameter to which the lines any how inclined are ordinates. 3. It is proved that there cannot be more than $\frac{n}{m}$ different inclinations of parallel ordinates, which cut the curve in *n* — *m* points only, possible or impossible. 4. Something is added concerning diameters, which cut their ordinates on both sides into equal parts. 5. It is demonstrated that there are curves of any number of odd orders, that cut a right line in 2, 4, 6, &c, points only; and of any number of even orders that cut a right line in 3, 5, 7, &c points; and consequently that the order of the curve cannot be denounced from the number of points, in which it cuts a right line. 6. The principles are delivered of finding the asymptotes, parabolical legs, ovals, points, &c, of a curve, of which the equation marking the relation between the abscifs and its ordinates is given; and also given the number of asymptotes, parabolical legs of different kinds, ovals, points of different kinds, the least order of a curve, which receives them, is deduced. 7. An equation expressing the relation between an abscifs and its ordinates, is transformed into an equation expressing the relation between the distances from two or more points, the latter may be varied an infinite number of ways; and thence are deduced some properties. Many resolutions of this kind are only resolutions of a particular case contained in it; and consequently can never be deduced from any general reasoning; they are often deduced from some particular cases, which are known to answer several conditions of the problem. Transformations of a given curve into others by substitutions, and properties of the loci of some points are deduced, from which Mr. Cotes's property of algebraical curves, and others of a similar and somewhat different nature are derived. 8. Let a curve of *n* dimensions have *n* asymptotes, then the content of the *n* abscifsæ will be to the content of the *n* ordinates, in the same ratio in the curve and asymptotes, the sum of their *(n)* subnormals to ordinates perpendicular to their abscifsæ will be equal to the curve and the asymptotes; and they will have the same central and diametrial curves. 9. Some propositions are added concerning the construction of equations, and some equations are constructed from the principles of Slusius.—If two curves of *n* and *m* dimensions have a common asymptote; or the terms of the equations to the curves of the greatest dimensions have a common divisor, then the curves cannot intersect each other in *n* × *m* points, possible or impossible. If the two curves have a common general centre, and intersect each other in *n* × *m* points, then the sum of the

o

affirmative abscifsæ &c to those points will be equal to the sum of the negative; and the sum of the *n* subnormals to a curve which has a general centre will be proportional to the distance from that centre. 10. Something is added on the description of curves. 11 No curve which has an hyperbolical leg of the conical kind can in general be squared. 12. It is demonstrated that no oval figure, which does not intersect itself in a given point, can in general be expressed in finite algebraical terms. 13. Given an algebraical equation, and similarly equations expressing a relation between *x* and *y*, &c; and also a fluxional quantity which is an algebraical function *(z)* of *x* and *y* and their fluxions; a method is given of deducing an equation whose root is *z*; and thence some properties of curves. 14. Properties similar to the subsequent of conic sections, are extended to curves of superior orders, viz, if lines be drawn from given points in them in given angles to four lines inscribed in the conic section, then will the rectangle under two of those lines be to the rectangle under the other two in a given ratio. Several properties are added, which follow from the application of algebraical propositions invented in the Medit. Algebr. to curve lines.

The second chapter treats of curvoids and epicurvoids, or curves generated by the rotation of given curves on right lines or curves, and gives a method of rectifying and squaring them; and from the radii of curvature of the generating curves being given, it deduces the length and radius of curvature of the curve generated at the correspondent point; it also asserts that from them may be deduced the construction of the fluxional equations of the different orders.

The third chapter treats of algebraical solids. 1. It deduces the equation to every section of a solid generated by the rotation of a curve round its axis; and from thence the different sections generated by the rotation of conic sections round their axis. 2. The equation to solids contains the relation between the two abscifsæ and their ordinates, and the order of the solid may be distinguished according to the dimensions of the equation; or the solid may be defined by two equations expressing the relation between the three abovementioned quantities, and a fourth which may be the axis of the section: there is further given a method of deducing the equation to any section of these solids, and from it the equation to the curve projected on a plane by a given curve. 3. A method of deducing the projection of a curve or solid on each other. 4. If the equation be *x* — *a* = 0, (*x* being the distance from a given point) then it may denote the periphery of a circle if one plane, or the surface of a globe if it refers to a solid. 5. Let *x* and *y* denote the distances from two respective points, then an equation expressing the relation between *x* and *y* designs the periphery of a curve, if contained in the same plane, or the surface of a solid generated by the rotation of a curve round its axis, passing through the two given points, if a solid. 6. An equation expressing the relation between lines drawn from three or more points may denote an equation to a solid. 7. If *x*, *z* and *y* denote the two abscifses and correspondent ordinates to a solid, and the terms of *x* and *y*, or *x* and *z*, or *y* and *z*; or *x*, *z* and *y* be similarly involved; then may the solid be divided into two

or

or fix fimilar and equal parts; and if no unequal power of *x* or *y* or *z*; or *x* and *y*, &c; or *x*, *y* and *z* be contained in the equation, then the curve may further be divided in general into twice, four or fix times the preceding number of equal parts. 7. Curves of double curvature are defigned by two equations expreffing the relation between two abfciffæ and correfpondent ordinates, or between lines drawn from three or more points; fimilar properties may be deduced from thefe as from the equations to curves.

Chapter the 4th treats of the maxims and minims of polygons infcribed and circumfcribed about curves, and thence deduces certain quantities equal to each other, when maxims and minims are contained at every point of the curve: it further contains feveral properties of conic fections. 1. If any rectilinear figure circumfcribes an ellipfe, the content under the alternate fegments of the line made by the points in which the line touches the ellipfe will be equal. 2. If a right line cuts a conic fection, and the parts of the line without the conic fection on both fides are equal; and any rectilinear figure, which begins and ends at the bounds of the abovementioned line, be defcribed round the conic fection, then the contents under the alternate fegments of the circumfcribing lines as divided in the points of contact will be equal. 3. If two polygons be circumfcribed about an ellipfe, and the fides are cut by the points of contacts in the fame ratios in the one as in the other; then will the areas of the two polygons be equal. 4. If two lines cut a conic fection proportionally, i. e. they are divided by the conic fection in the fame ratio in the one as in the other, and if polygons be defcribed round the conic fection, terminated at the ends of thofe lines, of which the fides are divided by the points of contact in the fame ratio in the one as in the other, then will the area of the two polygons be equal, as likewife the curvilinear area. 5. If all the fides of two polygons infcribed in an ellipfe make the two angles at the fame point equal, and two polygons of this kind be infcribed in the curve, then will the fum of the fides of the one polygon be equal to the fum of the fides of the other. Several other fimilar properties are added, as alfo properties of folids generated by the rotation of a conic fection round its axis; to which I fhall mention the three or four following. 1. The diagonals of a parallelogram circumfcribing an ellipfe or hyperbola will be conjugate diameters. 2. The fections of a folid generated by the rotation of a conic fection round its axis, which pafs through its focus, will have that point for the focus of all the fections. 3. If 4 perpendiculars be drawn from any point in an hyperbola to its periphery; and two lines from the fame point to the afymptotes and the ordinates from the 4 points of the curve and the 2 of the afymptotes be drawn to the abfcifs; then will the fum of the refulting abfciffæ to the former be double to the fum of the abfciffæ to the latter. 4. If an arc of the periphery of a circle be divided into *n* equal parts, *a*, *2a*, *3a*, &c, and *p* = chord of the arc 180 − *na*, and *α* and *β* be the roots of the quadratic $x^2 - px + 1 = 0$ and radius 1: then will $\alpha^n + \beta^n$ = chord of the arc 180 − *na*, from whence may be deduced the divifors of the quantity $x^{2n} - Ax^n + 1$; and alfo the equation whofe roots are the diftances of a point in the circle from thofe points of equal divifion, and further may be deduced

the fum of all the values of any algebraical function of thofe lines.

Moft of the properties of circles given by Archimedes are extended to conic fections, and fome of the algebraical and geometrical properties of Pappus are rendered more general; and the principles invented applied to many other cafes. In the firft edition of this book publifhed in 1762 were nearly enumerated the lines of the fourth order on the fame principles as Newton's enumeration of lines of the third order; but this has fince been rejected by the author as not fufficiently diftinguifhing the curve, and as being of no great utility.

Meditationes Analyticæ.

The firft *chapter* treats of finding the fluxion of a fluent, when the quantity or fluent is confidered as generated by motion; or the parts from the whole when the whole or quantity is confidered as confifting of innumerable parts. It further gives the law of a feries, which expreffes the fluxion of an exponential of any order.

Chapter 2, is verfant about the fluents of fluxions. 1. It finds the general fluent of a fluxion P*ẋ*, when P is any algebraical function of *x* however irrational but not exponential; for which intent it inveftigates the common divifors of any two quantities contained under the different vincula; and thence the common divifors of the refulting divifors, and fo on; and likewife all the equal divifors contained in any of the abovementioned quantities; whence it fo reduces the quantity P, that no equal nor common divifors may be contained in any of the refulting quantities under the different vincula; and from the common method deduces the terms of a feries to the number, which the feries is fhewn to confift of, when it does not proceed in infinitum. 2. It demonftrates, that if the dimenfions of *x* in the denominator of P exceed its dimenfions in the numerator by 1, then the fluent cannot be expreffed in finite terms; and alfo if one factor of P be $(A \pm (A^2 + a)^{\frac{1}{2}})^\lambda$, where *a* is an invariable quantity, and in fome other cafes the fubftitution required muft be fomewhat different. 3. The fluents of fome fluential and exponential fluxions, or fluxions involving fluents and exponential quantities, are given. 4. A general method of difcovering whether the fluent of any fluxion of any order involving one, two or more variable quantities, and their fluxions, can be expreffed in terms of the variable quantities and their fluxions. 5. The correction of fluents of all orders, and thence the fluent contained between any values of the variable quantities and their fluxions, is given; in thefe corrections the fame roots of the irrational quantities are to be ufed in the correction as in the fluent. 6. From the transformation of equations and the principles before delivered, are deduced fluents equal to each other. 7. Some exponential quantities given which continually change from poffibility to impoffibility, and from impoffibility to poffibility. 8. Is a method of finding whether the fluent of any fluxion contained between any limits are finite or not. 9. The fum of the fluents of a fluxion which is an algebraical function of the letter *x* multiplied into *ẋ* can always be expreffed by finite terms, circular arcs and logarithms, the extraction of the roots of equations being granted.

10. Some

10. Some fluxions involving irrational quantities are reduced to others, in which no irrationality is contained. 11. The general principles of deducing whether the fluent of a given fluxion can generally be expressed by finite algebraical terms, their circular arcs and logarithms. 12 Some equal correspondent fluents are found by substitutions deduced from equations in which two variable quantities are similarly involved. 13. Some necessary corrections are given of finding the fluents of all the fluxions of the formula

$$x^{pn \pm \sigma n - 1} \dot{x} \times R^{m \pm \lambda} \times S^{o \pm \mu} \times T^{t} \times {}^{\nu} \times \&c,$$

(where σ, λ, μ, ν, &c denote any whole numbers,

and $R = e + f x^{n} + g x^{2n} + .. x^{\alpha n}$,

$S = h + k x^{n} + l x^{2n} + .. x^{\beta n}$,

$T = q + r x^{n} + s x^{2n} + .. x^{\gamma n}$, &c)

from $\alpha + \beta + \gamma +$ &c, independent fluents; but perhaps not from $\alpha + \beta + \gamma +$ &c fluents, which have different values of the quantities, σ, λ, μ, ν, &c. 14. The number of independent fluents of the formulæ $x^{\theta + \alpha n + \beta m} \times (a + b x^{n} + c x^{m})^{\lambda + \pi} \times \dot{x}$, where α, β and π denote whole affirmative numbers, &c ; and the number of independent fluents of the formulæ $X \int Y \dot{x}$, where X is a fluxion of which the fluent can be found, from which can be deduced all of the same formula, is immediately known from the number of independent fluents of the formula $Y \dot{x}$ and $XY \dot{x}$ which determine all of those formulæ. 15. Let $a + b x^{n} + c x^{2n} + ... k x^{\mu n} = p$, and from some fluents of the fluxions of the formulæ $p \times x^{\mu n - 1} \dot{x}$, where μ is a whole affirmative number, are determined the remaining ones of the same formula. 16. Something is added concerning finding the value of a fraction, when both the numerator and denominator vanish ; and lastly from the fluents of some fluxions being given, the method of deducing the fluents of others.

Chapter 3, principally treats of algebraical and fluxional equations. 1. It gives the method of transforming two or more fluxional equations into one so as to exterminate one or more variable quantities and their fluxions, and finds the order of the resulting equation. 2. It reduces some fluxional equations into more. 3. A method of reducing fluxional equations involving fluents so as to exterminate the fluents. 3. Some cases are given, in which the two variable quantities contained in a given equation are expressed in terms of a third. 4. Given an algebraical equation expressing the relation between x and y ; a method is given of finding the fluent of $y \dot{x}^{n}$ or other fluxions in finite terms of x and y, if they can be expressed by such ; or else by infinite series ; this was first taught in the Philosophical Transactions in the year 1764. 5. Something is added concerning the correction of fluxional equations. 6. A method of investigating, whether a given equation is the general fluent of a given fluxional equation. 7. The method of deducing, whether a given equation is a particular or general fluent of a given fluxional equation. In both by substituting for the fluxions their values deduced from the fluential equation their values &c in the

fluxional, the fluxional must result $= 0$; and in the general fluent there must be contained so many invariable quantities to be assumed at will independently as is the order of the fluent ; and in both all the variable quantities must necessarily be variable, and no function of them vanish out of the fluxional equation from the substitution ; for then all the conditions of the fluxional equation are answered by the fluential. 8. An investigation, when fluxional equations are integrable. 9. From some fluents are deduced others, *e. g.* if the area between any two ordinates to one abscissa can in general be found, then the area between any two ordinates of any other abscissa can be found &c. 10. From given fluxional equations and the fluents of some fluxions are deduced the fluents of many others. 11. The fluent of the first order of a fluxional equation of the nth order will have (n) different values and n different multipliers ; and the fluent of the second order $n \cdot \dfrac{n - 1}{2}$ different values, &c. 12. Let $\alpha = 0$, $\beta = 0$, $\gamma = 0$, &c, (n) general fluents of the fluxional equation, $\lambda = 0$, then will any function of the fluents α, β, γ, &c be a fluent of the same fluxional equation $\lambda = 0$. 13. From assuming equations, which contain only simple powers of the invariable quantities to be assumed at will, may easily be deduced fluxional equations, of which the general resolutions are known : 2. From assuming the values of any variable quantities and substituting then their fluxions for the variable quantities, &c. in any functions π, ϱ, &c of the variables assumed, let the quantities resulting be A, B, &c ; then generally will $\pi = A$, $\varrho = B$, &c. be fluxional equations, of which the particular fluentials are known. It may be observed in this place as before, that from no general reasoning can particular fluents be deduced. 14. In the resolution of fluxional equations it is observed, that from the logarithmic and exponential quantities contained in the fluxional, may be deduced by chapter 1 the exponentials &c contained in the fluential : 2, and in a similar manner from the irrational quantities and denominators contained in it, the correspondent irrational quantities and denominators contained in the fluential : 3, the greatest dimensions of y multiplied into x must be greater than those of y into y by unity ; when there are two of this kind &c, $\alpha y \dot{x} + \beta x \dot{y} = x^{m} y^{n} (\delta \dot{x} + i \dot{y})$ the resolution is given ; and so of more. 15. In the given equation, if the fluxion of the greatest order does not ascend to one dimension only ; then by extraction &c so reduce the equation, that it may ascend to one dimension only ; and thence find the fluent of any fluxion $P^{n} \dot{y} + Q^{n - 1} \dot{y} + $ &c, $+ R^{r \cdot n} \dot{z} + $ &c. 16. Let a fluxional equation be given involving x and y, in which x flows uniformly, a method is given of finding whether it admits of a multiplier, which is a function of $x \therefore$ and similarly of multipliers of other formulæ. 17. The method of deducing the multipliers of fluxional equations by infinite series. 18. Some fluxional equations are reduced by substitutions, which substitutions are commonly easily deducible from the fluxional equation given. 19. Somewhat concerning the reduction of some fluxional equations to homogeneous, and concerning homogeneous equations of different orders ; and of reducing an homogeneous fluxional equation of n order to a fluxional equation of $n - 1$ order : and also

also of reducing m fluxional equations of n order to one of $mn - 1$ orders, and so of all others to one degree less than the order generally occurring if they had not been homogeneous. 20. The substitution of an exponential for a variable quantity in equations which contain no exponential quantity; for sometimes n has been substituted for a quantity which flows uniformly, and then w supposed to flow uniformly, which leads to a false resolution. 21. A caution is given not to substitute homogeneous functions of no dimensions for variable quantities; and in the general resolution to observe, that there is contained an invariable quantity to be assumed at will, which is not contained in the fluxional equation. 22. Something more added concerning the fluents of $p^p\dot{y} + q^{-1}\dot{x}\dot{y} + r^{n-3}\dot{y}\dot{x}^2 + \&c, = 0$, where p, q, r, &c. are functions of x, and so of some other fluxional equations. 23. Fluxional equations are deduced, of which the variable quantities cannot be expressed in terms of each other, but both may be expressed in terms of a third. 24. Every fluxion or fluent which is a function x, y, z, and x, y &c. is expressed in terms of partial differences. 25. The resolution of some equations expressing the relation between partial differences &c is given. 26. Some observations on finding the fluents of fluxions, when the variable quantities become infinite.

The second book treats of increments and their integrals. 1. Some new laws of the increments are given. 2. The fluxion of the increment of P will be equal to the increment of the fluxion; where P is any function of x, if only the fluxion of the increment of x be equal to the increment of the fluxion. 3. Increments are reduced to others of given formulæ

$$e. g. \quad \alpha + \frac{\beta}{x} + \frac{\gamma}{x(x+\dot{x})} +, \&c,$$

and it is observed that if β be not $= 0$, then the integral cannot be found in finite terms of the variable quantity, &c. It may be observed, that Taylor, Monmort, &c, first found the integral of the two increments

$$\frac{1}{x \cdot x - \dot{x} \cdot x - 2\dot{x} \dots x - \overline{n-1}\dot{x}} \text{ and } \frac{1}{x \cdot x - \dot{x} \dots x - \overline{n-1}\dot{x}}$$

but did not proceed much further (correspondent to the finding the fluxion of the fluent x^n); the increments of fluents have been since deduced, &c. In this book are discovered propositions correspondent to most of the inventions in fluxions, e. g. a method of finding the integral of any increment expressed in algebraical or exponential terms of the variable quantity or quantities, and when the fluent cannot be expressed: it is observed that they cannot be expressed in finite terms of the variable x, &c, if the dimensions of x, &c, in the denominator exceed its dimensions in the numerator by 1; or if any factor in the denominator of the fraction reduced to its lower terms have not another contained likewise in the denominator, distant by a whole number, multiplied into the increment of x. —The increments of some integrals are deduced from the integrals of other increments; the integrals of some incremental equations from different methods; their general integrals, and particular corrections, &c, &c; but here it is to be observed, that the general problem of increments cannot be extended beyond the particular of fluxions,

but somewhat more may be added, when both are joined together. The third book is versant concerning infinite series. 1. It gives the ratio of the apparent and real convergency. 2. A method of finding limits between which the sum of the series consists; and also whether the sum of the series is finite or not, from the terms being given or equation between the terms. 3. The convergency of the whole series is judged from the ratio of convergency of the terms at an infinite distance. 4. The series from the fluent converges, if the series from the fluxion does, there are several propositions on infinite series deducible from the common algebra. 5. Let an equation $0 = a - bx + cx^2 - dx^3 + \&c$; and $\frac{b}{a}$ much greater than $\frac{c}{b}$, $\frac{c}{b}$ than $\frac{d}{c}$; &c. then will all the roots be possible, and $\frac{a}{b}$ an approximation to the least root, $\frac{b}{c}$ to the next, &c: if an equation $y^u + ay^{n-1} + \dots + fy^{n-m} + gy^{n-m-1} + = 0$, and if one root be much less than any m root, but much greater than the remaining; or if the equation be $x^n - px^{n-1} + qx^{n-2} \dots \pm gx^{n-m+1} \mp hx^{n-m} \pm i x^{n-m-1} \mp kx^{n-m-2} \pm \&c. = 0$, then will the approximation to the above root be $\frac{i}{h} - (\frac{k}{i} - \frac{gi^2}{h^3}) + \&c.$

6. Somewhat on the approximations when the approximation given is much more near to one, two, or more roots than to any other, and on the degree of convergency of the subsequent approximations deduced; and their ultimate approximations. 7. Given approximations to m roots of a given equation are deduced more near approximations to them. 8. The incremental equation given and applied to approximations. 9. From given approximations to two or more unknown quantities contained in two or more equations are deduced more near approximations to them, either when the approximations given are more near to one, or to two, or more roots of one or more of the unknown quantities than to any others, and so of infinite equations. 10. New series are given for the fluents of different fluxions. 1. Log. $\overline{x \pm e} = $ log. $x \pm \frac{e}{x} - \frac{e^2}{2x^2} \pm$ &c.; the number whose log. is $v \pm e$ (if N be log. of v) $= N \pm Ne \pm \frac{Ne^2}{2} \pm$ &c; the log. of $\frac{a+x+e}{a-x-e} = $ log. $\frac{a+x}{a-x} + \frac{e}{a^2-x^2} - $ &c. The sine of the arc $A \pm e$ is $S \pm Ce - \frac{1}{2} Se^2$, &c, and cosine of the same arc $= C \pm Se - \frac{1}{2.3} Ce^2 \pm$ &c. S and C being the sine and cosine of A, the fluent of the fluxion of an elliptical arc $\frac{\sqrt{(1 - cx^2)}\dot{x}}{\sqrt{(1 - x^3)}}$ which differs little from the arc of a circle when e is a very small quantity $= A - \frac{c}{2} \times \frac{1.A - x^P}{2} - \&c$, where $A = \int \frac{\dot{x}}{P}$, $B = \frac{1A - x^P}{2}$, $C =$

$C = \dfrac{3B - x^2 P}{3}$, &c. and $P = \sqrt{1-x^2}$, and $A =$ arc of a circle of which the sine is x.

A similar series may be applied from the arc of an hyperbola or ellipse, to find a correspondent arc of an hyperbola or ellipse not much different from the preceding. In this method the series proceeds according to the dimensions of some small quantities, and the first term of the series is generally a near value of the quantity sought. These series properly instituted will generally converge the swiftest. 11. Something new is added concerning the fluent of the fluxional equation $y = y\dot{z}^2$ viz $-y = E \times$ sin. arc: $(z) + F + $ cos. (ar. (z); E and F being any quantities to be assumed at will; and of correspondent equations to logarithms, and finding their values when z is increased, by e. 12. A series for the increase of the arc from a small increase of the tangent, sine, &c. 12. When the terms a and x of the binomial $a \pm x$ are equal, the cases are given in which the series $a^m \pm m a^{m-1} x + $ &c, $= \overline{a \pm x}{}^m$ or the series $a^m x \pm \dfrac{m}{2} a^{m-1} x + $ &c, &c. will ultimately converge. 13. If any algebraical quantity V a function of x be reduced into a series proceeding according to the dimensions of x, a general method of finding what are the limits between which it converges; or the series from $\int V x$, &c; and the method of interpolations so as to render them converging. 14. The convergency of different series are compared together.

$e. g.$ is given $\int \dfrac{\dot{x}}{1+x} = x - \frac{1}{2}x^2 + \frac{1}{3}x^3 -$, &c.
$= \dfrac{x}{1+x} + \dfrac{x^2}{2(1+x)^2} + \dfrac{x^3}{3(1+x)^3} + $ &c. or $\int \dfrac{\dot{x}}{1 \pm x} = \dfrac{x^1}{1 \pm x} \pm \dfrac{a^2}{2(1+x)^2} + $ &c: there is an erratum contained in this example, for $a -$ is sometimes printed instead of $a +$: this series is easily deduced from Bernouilli's method of deducing infinite series, and has been since printed in the Philosophical Transactions. 15. Given algebraical or fluxional equations, and a fluxional quantity, a method is given of finding a series, which expresses the fluent of the fluxional quantity, from which principles are deduced new series for the area of a segment of a circle, the periphery of the ellipse, hyperbola, &c. 16. It is shewn, that serieses proceeding according to the dimensions of a quantity x always diverge, when serieses for the same purpose proceeding according to the reciprocal of its dimensions converge; unless sometimes in the case when they both become the same. 17. As series proceeding in infinitum according to the dimensions of the quantity x were first invented or used for the finding the fluents of fluxions, it being reduced into terms, whose fluents were known: so in finding integrals of increments it may be necessary to reduce the quantity into an infinite series of terms, whose integrals are known, and which converges. Examples of formulæ of serieses of this kind are given. 18. Methods are given of finding the value of one unknown quantity contained in one or more equations involving more unknown quantities, and the law of their convergencies

and the interpolations necessary to render serieses for finding fluents converging, similar principles may be applied to incremental and fluxional equations. 19. It is observed, that in finding the value of any variable quantity in a series proceeding according to the dimensions of another, there will occur in a fluxional or incremental equation of (n) order in the series n invariable quantities to be assumed at will; and also the fluxional equations, &c. from whence they will arise.

20. The finding the integral of $\dfrac{z}{z}$, &c. 21. From the correspondent relation between the sums of two series resulting, which are functions of a variable quantity y, when the relation between x and z two values of y are given, is given a method of finding the coefficients of the series. 22. The rule generally called the reductio ad absurdum extended to more substitutions.

The fourth book treats of the summation of series, a method of correspondent values and several other problems. 1. Of finding the sum of a series expressed by a rational function of x into x^{nz}; where z denotes successively the numbers 1, 2, 3, &c, in infinitum. 2. Given an equation expressing the relation between the successive sums, the relation between the successive terms is known, and the *vice versa*, &c. 3. It is found from an equation expressing the relation between the successive sums, terms and z the distance from the first term of the series, whether the sum of the series is finite or not. 4. The difference between z^{-o} and $\overline{z+1}^{-o}$, where z denotes the distance from the first term of the series, will be $-o \times z^{-o-1}$, which is greater than the simple ratio let o be as small as possible, and consequently the sum of the series finite. 5. If a series $a + bx + cx^2 + x^3$, of which at an infinite distance the preceding coefficients have to the subsequent the ratio of $r : 1$, be multiplied into a function $= o$, when $x = \delta$, then if a be greater than r the series will diverge; if less converge. 6. From adding several terms of one or more series together may be formed a series, of which the sum from the sums of the preceding series is known. 6. Serieses are formed, of which the sums are known from varying the divisors, &c. 7. From given series are deduced others, of which the sums are known, and the sum of many series are deduced from finding the fluxions of fluents and fluents of fluxions. 8. From the relation between the different terms given is deduced the correspondent fluxional equation. 9. The finding the terms of any series, which can be deduced from given series; and thence deducing many series of which the sums can be found from the sum of the given series. 10. Series are given of which the sums can be found from finite terms, circular arcs, logarithms, elliptical and hyperbolical arcs. 11. From a general expression, when algebraical, fluxional, incremental, &c, for the sum of a series can be deduced a similar expression for the sum of every second, third, &c, terms. 12. An infinite series may be a particular resolution of infinite fluxional equations. 13. The terms of some series may be infinite and their sums known. 14. The general fluent of $y^n = y\dot{x}^n$ is given by a series of the same kind, and the same of some other fluxional equations. 15. A quantity is found which multiplied into a series

more

more swiftly converging gives a given series. 16. The first differences of the terms of some series are given; if the terms are in geometrical ratio to each other the abovementioned differences will also be in geometrical ratio to each other: whence it appears, that the series from this method of differences will converge least when the given series converges swiftest, &c, but not always the contrary. Several other propositions are added concerning the method of differences applied to series. 17. A parabolico-hyperbolical curve is drawn through any number of points, as also an algebraical solid —. 18. Something is given concerning the convergency &c. of series deduced from the differences of the numerators of a given series, of which the denominators constitute a geometrical progression. 19. A rule is given for rendering series converging, in which it is observed that the sum of so many terms should be found that z the distance from the first term of the series may exceed the greatest root of the equation resulting from the quantity which expresses the term made $= 0$. 20. An equation expressing the relation between the sums and terms is reduced to an infinite fluxional equation expressing the relation between the sum or term, its fluxions, and z the distance from the first term of the series. 21. From a method being known of finding the sum of a series, which involves one variable only, is given a method of finding the sum of series which involve more variable quantities: and from assuming sums of serieses of this kind are deduced their terms. 22. The sums of series are found consisting of irrational terms. 23. The principle of the convergency of the approximations found in drawing parabolical curves through given points. 24. Something new is given concerning the interpolations of quantities.

25. $\dfrac{e^{\alpha x} + e^{\beta x} + e^{\gamma x} + \&c}{n} = 1 + \dfrac{x^n}{1.2 .. n} + \dfrac{x^{2n}}{1.2!..2n} +$

&c. if α, β, γ, &c, are the roots of $x^n - 1 = 0$, &c. 26. Something is added concerning series from

$\int \dfrac{\dot{x}}{x} \int \dfrac{\dot{x}}{x} . \int \dfrac{\dot{x}}{x}$, &c, $\times \int \dfrac{\dot{x}}{1 + ax^n}$ 27. Nandens's Problems are somewhat extended. 28. Something is added on changing continual fractions into others. 29. A method of transforming series into continual factors.

30. A rule for finding the sine and cosine of $\dfrac{n}{m}$ the

arc; and transforming an algebraical equation into an equation expressed in terms of sines and cosines, and thence from an approximation to the sine is found one more near; the same might have been performed by tangents, cotangents, secants, cosecants, &c. 31. From some fluents given have been found others, and consequently by reducing the fluents to infinite series from some infinite series given

may others be deduced. 32. The fluent of $\dfrac{x^\alpha \dot{x}}{1 \pm x^n}$

is found by approximation, where α is an irrational quantity, which method of finding approximations to the indices may be applied to other cases. 33. The sum of the fractions are found when the denominators $= 0$, and consequently each particular in-

7

finite. 34. It is asserted, that the sum of certain fractions given become $= 0$, when the terms are expressed by a fraction of which the denominator is a rational function of the distance from the first term of the series. 35. $\int_x \alpha - \beta - 1 \,\dot{x} \int_x \beta - \gamma - 1 \,\dot{x}$

$\int_x \gamma - \delta - 1 \,\dot{x} \times$ P, where P $= Ax^n + Bx^{n+m} +$

$C x^{n+2m} + D x^{n+3m} +$, &c, will be to $\int_x \beta - \alpha - 1 \,\dot{x}$

$\int_x \alpha - \gamma - 1 \,\dot{x} \int_x \gamma - \delta - 1 \,\dot{x} \int$ &c. $\times$ P $:: x^\alpha : x^\beta$ if the fluents are contained between the same values of x. 36. Are given some series consisting of two, of which the one converges, when the other diverges, and consequently the sum of both diverges; &c. 37. From the law of a series being given, the law of the series which expresses the square, or some function of the given series, is found.

1. A method of differences, which deduces from the sums given any successive sums, e. g. Let S^1, S^2, S^3, S^4, be the logarithms of the ratios $r : r + p$, $r : r + 2p$, $r : r + 3p$, $r : r + 4p$, then will the logarithm of $r : r + 5p$ be $5 \times (S^4 - S^1) + 10 (S^2 - S^3)$ nearly: then rules are given in general, and likewise their errors from the true values.

2. A method of correspondent values is given, e. g. Let a, b, c, d, &c, be values of x; and S^a, S^b, S^c, S^d, &c, correspondent values of y; then may

$$ y = \frac{(x-b)\,(x-c)\,(x-d)\,\&c}{(a-b)\,(a-c)\,(a-d)\,\&c} \times S^a + $$

$$ \frac{(x-a)\,(x-c)\,(x-d)\,\&c}{(b-a)\,(b-c)\,(b-d)\,\&c} \times S^b + \&c. $$

3. If the formula of the series be $A + Bx + Cx^2 +$

&c $= y$; or $y = \dfrac{x}{a} \times \dfrac{(x-b)\,(x-c)\,\&c}{(a-b)\,(a-c)\,\&c} \times S^a +$

$\dfrac{x}{b} \times \dfrac{(x-a)\,(x-c)\,\&c}{(b-a)\,(b-c)\,\&c} \times S^b +$ &c; if the formula of the series be $Ax + Bx^2 +$ &c $= y$, which answers to Briggs's or Newton's method of interpolations; or the series will be

$$ \frac{x^h}{a^h} \times \frac{(x^k - b^k)\,(x^k - c^k)\,(x^k - d^k)\,\&c}{(a^k - b^k)\,(a^k - c^k)\,(a^k - d^k)\,\&c} \times S^a + $$

$$ \frac{x^h}{b^h} \times \frac{(x^k - a^k)\,(x^k - c^k)\,(x^k - d^k)\,\&c}{(b^k - a^k)\,(b^k - c^k)\,(b^k - d^k)\,\&c} \times S^b + $$

&c; if the formula of the series be $Ax^h + Bx^{h+k} + Cx^{h+2k} +$ &c, $= y$; a general formula, which includes the preceding.

5. The series is given for deducing others when the number of correspondent values given are either even or odd, and the values of x are equidistant from each other. 6. And also from correspondent values of x and y to a number of equidistant values of x is deduced the value of y to the next successive or any successive value of x. 7. Some arithmetical theorems are deduced from the preceding propositions. 8. Another method is given of resolving the preceding problem. 9. A method of correcting the solution from a solution given

given which finds (*n*) values of *y* to (*n*) given values of *x* true, and *m* falſe to (*m*) other values. 10. A ſimilar reſolution is added from correſpondent values of *x*, *y*, *z*, &c given; and more general reſolutions. 11. Given the reſolution of ſome caſes, a: d formula in which the general is contained, a method is given in ſome caſes of deducing it. 12. The principles of a method of deductions and reductions are added.

In a Pamphlet publiſhed at Cambridge, algebraical quantities are tranſlated into probable relations, and ſome theorems on probabilities thence deduced; to which are adjoined,

1. The theorem $\overline{a+b}\cdot\overline{a+b\pm l}\cdot\overline{a+b\pm 2l}$.

$$\overline{a+b\pm 2l}\ldots\overline{a+b\pm\overline{n-1}l}=a\cdot a\pm l\cdot a\pm 2l\ldots$$

$$\overline{a\pm\overline{n-1}l}+n\times a\cdot\overline{a\pm l}\cdot\overline{a\pm 2l}\ldots\overline{a\pm\overline{n-2}l}\times$$

$$b+n\cdot\frac{n-1}{2}\cdot a\cdot\overline{a\pm l}\cdot\overline{a\pm 2l}\ldots\overline{a\pm\overline{n-3}l}\times$$

$$b\cdot\overline{b\pm l}+n\cdot\frac{n-1}{2}\cdot\frac{n-2}{3}\times a\cdot\overline{a\pm l}\cdot\overline{a\pm 2l}\ldots$$

$$\overline{a\pm\overline{n-4}l}\times b\cdot\overline{b\pm l}\cdot\overline{b\pm 2l}+\text{&c};$$ this becomes the binomial theorem when $l=0$; and it will afford anſwers to ſimilar caſes when the whole number of chances are increaſed or diminiſhed conſtantly by *l*, as the binomial does when they remain the ſame, a ſimilar multinomial theorem is given. In the ſame pamphlet are further added ſome new propoſitions on chances, on the values of lives, ſurvivorſhips, &c. In theſe books are alſo contained the inventions of others on ſimilar ſubjects, which in the prefaces are aſcribed to their reſpective authors.

In the Philoſophical Tranſactions are given ſome properties of numbers, &c, of which ſome have been pub-liſhed in the books above mentioned; to which may be ſubjoined ſomething in mixed mathematics, viz, a paper on central forces, which extends not only to central forces, but alſo to forces applied in any other direction, as in the direction of the tangent, and conſe-quently includes reſiſtances, &c. It gives a rule for finding the forces tending to two or more given points when the curve deſcribed and velocity of the body in every point of it is given, *e. g.* Let the curve be an ellipſe, and the velocity the ſame at every point, and the two centres of force be the foci of the ellipſe; then will the forces tending to the two foci be equal, and vary as the ſquare of the ſine of the angle contained between the diſtance from the centre of force to the point in which the body is ſituated, and the tangent to the curve at that point.

The method of deducing the fluxional equations which expreſs the curve deſcribed by a body acted on by any forces tending to given points, or applied in any given directions; ſome other propoſitions are con-tained on ſimilar ſubjects. 2. A paper on the fluxions of the attractions of lines, ſurfaces, and ſolids, and from the different methods of deducing them are found different fluents equal to each other: a third paper gives a ſolution of Kepler's problem of cutting the area of a circle deſcribed round a point by approxima-tions, which alſo is applied to other caſes; this like-

9

wiſe contains ſome other problems. Many of theſe diſcoveries have ſince been publiſhed, ſome in the Lon-don, and other foreign tranſactions.

Let $e^l = N$, then will *l* denote the log. of N to the modulus *e*. If *e* the modulus $= 10$, then will the ſyſtem be the common or Briggs's ſyſtem of logarithms. Logarithms, and the ſums of ſome other ſerieſes, of the formulæ $ax^h + bx^h + k + \text{&c}$ may be deduced in a manner ſimilar to that which was uſed by the Ancients for finding the ſines of the arcs of circles.

To particulariſe the numerous propoſitions contained in theſe works, would exceed the limits of our deſign. Beſides thoſe already mentioned, others are interſperſed through the whole works.

ANEMOMETER, p. 111, col. 2, l. 1, *after* 12 ounces, *add* or $\frac{3}{4}$ of a pound. Owing to an overſight in the ſucceeding lines, of conſidering this 12 ounces as 12 pounds, in the calculations, ſeveral errors have been incurred, and the 3d column of the table of num-bers, in that page, or the column for the velocity, has the numbers only $\frac{1}{4}$ of what they ought to be, or they require to be all multiplied by 4, the ſquare-root of 16, the number of ounces in a pound. Hence, in line 6, *for* $\sqrt{12}$ r. $\sqrt{\frac{3}{4}}$; l. 7 and 8, *for* $22\frac{4}{5}$ r. $91\frac{1}{5}$; l. 8, *for* $15\frac{1}{2}$ r. 62. And the whole ſucceeding table cor-rected will be as follows:

Table of the correſponding Height of Water, Force on a Square Foot, and Velocity of Wind,

Height of Water.	Force of Wind.	Velocity of Wind per Hour
Inches.	Pounds.	Miles.
$0\frac{1}{4}$	1·3	18·0
$0\frac{1}{2}$	2·6	25·6
1	5·2	36·0
2	10·4	50·8
3	15·6	62·0
4	20·8	76·0
5	26·0	80·4
6	31·25	88·0
7	36·5	95·2
8	41·7	101·6
9	46·9	108·0
10	52·1	113·6
11	57·3	119·2
12	62·5	124·0

In one inſtance Dr. Lind found that the force of the wind was ſuch as to be equal $34\frac{9}{10}$ pounds, on a ſquare foot; and this by proportion, in the foregoing table, will be found to anſwer to a velocity of 93 miles per hour.

ARCH, p. 137, col. 1, l. 29, *for* ſuch caſes as they, *read* ſuch caſes they. Line 30, *after* hanches, *add*, See BRIDGE.

ARCHIMEDES, p. 139, col. 1, l. 52 and 53, *for* preface, a commentary, *read* preface. We find here alſo Eutocius's commentary. Pa. 59, *after* college, *add*, who had the ſole care of this edition.

ASSU-

ASSURANCE *on Lives.* Pa. 150, col. 2, in the 3d paragraph, for want of sufficient information concerning the London and Royal Exchange Assurance Offices, that paragraph gives an imperfect and, in some respect, erroneous account of them : it refers to their state 30 years ago, but the Companies have since that, altered their method of proceeding. Instead of that paragraph therefore, take the following account of their present constitution ; viz,

The London ASSURANCE, is a corporation established by a charter of king George the 1st, viz, in 1720; under power of which, Assurances are made from the risk of sea-voyages, and from the danger of fire to houses and goods ; the prices of which are regulated by the apparent risk to be assured. They also make Assurances on lives ; the prices of which are formed on an estimation of the probable duration of life at different ages, on the consideration of the apparent health of the persons to be assured, and of their avocations in life.

This corporation, and the Royal Exchange corporation, gave each the sum of 150,000 pounds to government, for an *exclusive right* of making Assurances as *corporate bodies.* They are known to possess a large and undeniable fund to answer losses. And the prudent management of these corporations has enabled them, of late years, to increase gradually their dividends to the proprietors of their stock. This *exclusive privilege* to make Assurances as corporate bodies, is of great advantage and convenience to the public ; and as they act under a common seal, the assured may have a speedy and easy mode of recovering losses, and cannot be subject to any calls or deductions whatever. When their charters were granted to them, it was enacted, that if a proprietor of the stock of one corporation should at the same time, directly or indirectly, be a proprietor of stock in the other corporation, the respective stock so held is to be forfeited, one moiety to the king, the other to the informer. This was evidently settled, to prevent their interest from becoming a joint one ; so that they should be made to act in competition to each other, for the greater benefit of the public.

The Royal Exchange ASSURANCE, is a corporation established by charter, as above, under the power of which, Assurances are made from the risk of sea voyages, and from the danger of fire to houses and goods ; the prices of which are regulated by the greater or less risk supposed to be assured. They also make Assurances on lives, the prices of which are formed on estimation of the probable duration of life at different ages, and under different circumstances. The present rates of Assurances on lives are as in the table below. And though a duty on these Assurances should take place on the plan lately proposed to the House of Commons, there is no great probability that these prices will be increased.

This corporation has also, like the former, been empowered to grant life annuities by an act of parliament, which requires that the prices of the annuities should be expressed in tables, hung up in some conspicuous place in their offices, for public inspection ; and no agreement for any price is valid, but such as shall be expressed in the tables last made and published by the corporation.

From the Office of the CORPORATION of the ROYAL EXCHANGE ASSURANCE, on the ROYAL EXCHANGE, LONDON.

RATES OF ASSURANCES ON LIVES.

SINGLE LIVES.

Age.	Premium per cent. for an assurance for one year.	Premium per cent. per annum, for an assurance for seven years.	Premium per ct. per ann. for an assurance for the whole continuance of life.
	£. s. d.	£. s. d.	£. s. d.
8 to 14	1 2 3	1 6 9	2 7 0
15	1 2 6	1 8 9	2 8 3
16	1 4 0	1 10 9	2 9 9
17	1 6 6	1 12 9	2 11 0
18	1 9 0	1 14 3	2 12 3
19	1 11 3	1 15 9	2 13 6
20	1 14 0	1 16 9	2 14 6
21	1 16 0	1 17 9	2 15 9
22	1 16 6	1 18 3	2 17 9
23	1 17 3	1 18 9	2 19 9
24	1 17 9	1 19 6	2 19 0
25	1 18 3	2 0 3	3 0 3
26	1 19 0	2 0 9	3 1 3
27	1 19 6	2 1 6	3 2 6
28	2 0 3	2 2 3	3 4 0
29	2 1 0	2 3 0	3 5 3
30	2 1 6	2 3 9	3 6 9
31	2 2 3	2 4 6	3 8 3
32	2 3 0	2 5 3	3 9 9
33	2 3 9	2 6 0	3 11 6
34	2 4 9	2 7 3	3 13 0
35	2 5 6	2 8 6	3 14 9
36	2 6 3	2 9 6	3 16 9
37	2 7 3	2 11 0	3 18 6
38	2 8 3	2 12 3	4 0 9
39	2 9 0	2 13 9	4 2 9
40	2 11 0	2 15 3	4 4 0
41	2 12 6	2 16 9	4 7 3
42	2 14 6	2 18 3	4 9 9
43	2 15 9	2 19 9	4 12 3
44	2 17 0	3 1 6	4 14 9
45	2 18 6	3 3 6	4 17 6
46	2 19 9	3 5 9	5 0 3
47	3 1 3	3 8 0	5 3 3
48	3 2 9	3 10 6	5 6 6
49	3 5 0	3 13 3	5 9 9
50	3 9 0	3 16 0	5 13 6
51	3 11 9	3 18 6	5 17 0
52	3 14 0	4 1 0	6 0 6
53	3 16 3	4 3 9	6 4 6
54	3 13 9	4 6 9	6 8 6
55	4 1 3	4 10 0	6 13 0
56	4 4 0	4 13 6	6 17 9
57	4 7 3	4 17 0	7 2 6
58	4 10 3	5 0 9	7 7 9
59	4 14 0	5 4 9	7 13 6
60	4 17 9	5 9 0	7 19 3
61	5 1 9	5 13 9	8 5 9
62	5 5 0	5 18 9	8 12 3
63	5 9 9	6 4 9	8 19 6
64	5 13 6	6 11 0	9 7 3
65	5 19 0	6 18 6	9 16 0
66	6 5 3	7 7 0	10 5 3
67	6 12 0	7 16 6	10 15 3

JOINT LIVES.

For the Assurance of a Gross Sum, payable when One of Two Joint Lives that shall be named shall drop.

Age of the life to be assured.	Age of the life against which the assurance is to be made.	Premium per cent. per ann.
		£. s. d.
10	10	1 15 9
	20	1 16 6
	30	1 15 6
	40	1 14 9
	50	1 13 9
	60	1 12 6
	70	1 11 3
	80	1 9 3
20	10	2 5 9
	20	2 6 3
	30	2 4 9
	40	2 3 6
	50	2 2 0
	60	2 0 3
	70	1 18 3
	80	1 15 3
30	10	2 16 9
	20	2 17 6
	30	2 15 9
	40	2 13 6
	50	2 11 3
	60	2 8 6
	70	2 5 9
	80	2 2 3
40	10	3 14 0
	20	3 14 9
	30	3 12 9
	40	3 10 0
	50	3 6 0
	60	3 1 9
	70	2 17 6
	80	2 12 3
50	10	5 1 3
	20	5 2 3
	30	5 0 3
	40	4 17 3
	50	4 12 3
	60	4 4 6
	70	3 17 0
	80	3 8 9
60	10	7 6 0
	20	7 7 9
	30	7 5 3
	40	7 2 6
	50	6 18 3
	60	6 8 0
	70	5 12 3
	80	4 17 6
67	10	10 1 3
	20	10 3 6
	30	10 1 0
	40	9 18 3
	50	9 14 6
	60	9 6 0
	70	8 3 6
	80	6 16 0

For the Assurance of a Gross Sum, payable when either of Two Joint Lives shall drop.

Age.	Age.	Premium per ct. per annum.
		£. s. d.
10	10	3 11 6
	15	3 16 6
	20	4 2 0
	25	4 6 0
	30	4 12 3
	35	4 19 6
	40	5 8 6
	45	5 19 0
	50	6 14 9
	55	7 13 6
	60	8 18 6
	67	11 12 9
15	15	4 1 3
	20	4 7 0
	25	4 11 6
	30	4 17 0
	35	5 4 0
	40	5 13 0
	45	6 4 3
	50	6 19 0
	55	7 17 9
	60	9 2 6
	67	11 16 9
20	20	4 12 6
	25	4 16 9
	30	5 2 3
	35	5 9 0
	40	5 18 3
	45	6 9 6
	50	7 4 3
	55	8 2 9
	60	9 7 9
	67	12 2 3
25	25	5 1 0
	30	5 6 3
	35	5 12 9
	40	6 1 9
	45	6 12 9
	50	7 7 3
	55	8 5 9
	60	9 10 6
	67	12 4 9
30	30	5 11 3
	35	5 17 9
	40	6 6 3
	45	6 17 0
	50	7 11 3
	55	8 9 3
	60	9 13 9
	67	12 7 9

Age.	Age.	Premium per cent. per ann.
		£. s. d.
35	35	6 3 9
	40	6 12 0
	45	7 2 3
	50	7 16 3
	55	8 14 0
	60	9 18 3
	67	12 11 6
40	40	6 19 9
	45	7 9 9
	50	8 3 6
	55	9 0 6
	60	10 4 3
	67	12 17 0
45	45	7 19 3
	50	8 12 3
	55	9 8 9
	60	10 12 0
	67	13 4 0
50	50	9 4 9
	55	10 0 3
	60	11 2 9
	67	13 13 6
55	55	10 15 3
	60	11 16 3
	67	14 5 6
60	60	12 16 0
	67	15 2 9
67	67	17 4 9

By whom the Assurance is made.	Name, age, and description of the life to be Assured.	Time for which the Assurance is made.	Conditions of Assurance made by Persons on their own Lives.	Sum assured.	Rate per cent per annum.
			The Assurance to be void if the person whose life is Assured shall depart beyond the limits of Europe, shall die upon the seas, or enter into or engage in any military or naval service whatever, without the previous consent of the company ; or shall come by death by suicide, duelling, or the hand of Justice ; or shall not be, at the time the Assurance is made, in good health.		

By whom the Assurance is made.	Name, age and description of the life to be Assured.	Time for which the Assurance is made.	Conditions of Assurance made by Persons on the Lives of others.	Sum assured.	Rate per cent. per annum.
			The Assurance to be void if the person whose life is Assured shall depart beyond the limits of Europe, shall die upon the seas, or enter into or engage in any military or naval service whatever, without the previous consent of the company ; or shall not be at the time the Assurance is made in good health.		

Place and date of birth.
If had the small-pox.
Whether in the army or navy.
The life Assured to appear at the office, or pay
  10s. *per cent.* on Assurances for one year.
  15s. *per cent.* for more than one year, and not exceeding seven years.
  20s. *per cent.* if for the whole continuance of life.
} In the first payment only.

Reference to be made to two persons of repute to ascertain his or her identity.
*** Attendance daily from ten to half past two o'clock and from five to seven, Saturday in the afternoon excepted.
☞ The lives of persons engaged in the army or navy may be Assured by special agreement.
N. B. THE CORPORATION ALSO GRANT ANNUITIES ON LIVES.

AUTOMATON. To the end of this article, in pa. 176, col. 2, may be added the following curious particulars, extracted from a letter of an ingenious gentleman since that article was published, viz, Thomas Collinson, Esq. nephew of the late ingenious Peter Collinson, Esq. F. R. S. "Turning over the leaves of your late valuable publication (says my worthy correspondent), part 1. of the Mathematical and Philosophical Dictionary, I observed under the article *Automaton*, the following:" 'But all these seem to be inferior to M. Kempell's chess-player, which may truly be considered as the greatest master-piece in mechanics that ever appeared in the world;' (upon which Mr. Collinson observes) "So it certainly would have been, had its scientific movements depended merely on mechanism. Being slightly acquainted with M. Kempell when he exhibited his chess-playing figure in London, I called on him about five years since at his house at Vienna; another gentleman and myself being then on a tour on the continent. The baron (for I think he is such) shewed me some working models which he had lately made—among them, an improvement on Arkwright's cotton-mill, and also one which he thought an improvement on Boulton and Watt's last steam-engine. I asked him after a piece of speaking mechanism, which he had shewn me when in London. It spoke as before, and I gave the same word as I gave when I first saw it, *Exploitation*, which it distinctly pronounced with the French accent. But I particularly noticed, that not a word passed about the chess-player; and of course I did not ask to see it.—In the progress of the tour I came to Dresden, where becoming acquainted with Mr. Eden, our envoy there, by means of a letter given me by his brother lord Auckland, who was ambassador when I was at Madrid, he obligingly accompanied me in seeing several things worthy of attention. And he introduced my companion and myself to a gentleman of rank and talents, named Joseph Freidrick Freyhere, who seems completely to have discovered the *Vitality* and soul of the chess-playing figure. This gentleman courteously presented me with the treatise he had published, dated at Dresden, Sept. 30, 1789, explaining its principles, accompanied with curious plates neatly coloured. This treatise is in the German language; and I hope soon to get a translation of it. A welltaught boy, very thin and small of his age (sufficiently so that he could be concealed in a drawer almost immediately under the chess-board), agitated the whole. Even after this abatement of its being strictly an automaton, much ingenuity remains to the contriver.—This discovery at Dresden accounts for the silence about it at Vienna; for I understand, by Mr. Eden, that Mr. Freyhere had sent a copy to baron Kempell: though he seems unwilling to acknowledge that Mr. F. has completely analysed the whole.

"I know that long and uninteresting letters are formidable things to men who know the value of time and science: but as this happens to be upon the subject, forgive me for adding one very admirable piece of mechanism to those you have touched upon. When at Geneva, I called upon Droz, son of the original Droz of la Chaux de Fonds (where I also was). He shewed me an oval gold snuff box, about (if I recollect right) 4 inches and a half long, by 3 inches broad, and about an inch and a half thick. It was double, having an horizontal partition; so that it may be considered as one box placed on another, with a lid of course to each box—One contained snuff—In the other, as soon as the lid was opened, there rose up a very small bird, of green enamelled gold, sitting on a gold stand. Immediately this minute curiosity wagged its tail, shook its wings, opened its bill of white enamelled gold, and poured forth, minute as it was (being only three quarters of an inch from the beak to the extremity of the tail) such a clear melodious song, as would have filled a room of 20 or 30 feet square with its harmony.—Droz agreed to meet me at Florence; and we visited the Abbé Fontana together. He afterwards joined me at Rome, and exhibited his bird to the pope and the cardinals in the Vatican palace, to the admiration, I may say to the astonishment of all who saw and heard it."

Another extract from a second letter upon the same subject, by Mr. Collinson, is as follows: "Permit me to speak of another Automaton of Droz's, which several years since he exhibited in England; and which, from my personal acquaintance, I had a commodious opportunity of particularly examining. It was a figure of a man, I think the size of life. It held in its hand a metal style; a card of Dutch vellum being laid under it. A spring was touched, which released the internal clockwork from its stop, when the figure immediately began to draw. Mr. Droz happening once to be sent for in a great hurry to wait upon some considerable personage at the west end of the town, left me in possession of the keys, which opened the recesses of all his machinery. He opened the drawing-master himself; wound it up; explained its leading parts; and taught me how to make it obey my requirings, as it had obeyed his own. Mr. Droz then went away. After the first card was finished, the figure rested. I put a second; and so on, to five separate cards, all different subjects: but five or six was the extent of its delineating powers. The first card contained, I may truly say, elegant portraits and likenesses of the king and queen, facing each other: and it was curious to observe with what precision the figure lifted up his pencil, in the transition of it from one point of the draft to another, without making the least flur whatever: for instance, in passing from the forehead to the eye, nose, and chin; or from the waving curls of the hair to the ear, &c. I have the cards now by me, &c, &c."

Pa. 177, col. 1, l. 2, *for* August *read* September.

B.

PAGE 195, col. 1, at the end of the article on Barometrical Measurements of Altitudes, *add*, See a learned paper in vol. 1. of the Transactions of the R. Soc. of Edinburgh, " On the Causes which affect the Accuracy of Barometrical Measurements; by John Playfair, A. M. F. R. S. Edin. and Professor of Mathematics in the University of Edinburgh." Also another by Dr. Damen, late Professor of Mathematics and Philosophy in the University of Leyden, intitled, " Dissertatio Physica & Mathematica de Montium Altitudine Barometro Metienda: Accedit Refractionis Astronomicæ Theoria; in 8vo, at the Hague, 1783.

Pa. 205, col. 1, after the life of Dan. Bernoulli, *add* the following life of James.

BERNOULLI (JAMES), another mathematical branch of the foregoing celebrated family. He was born at Basil in October 1759; being the son of John Bernoulli, and grandson of the first John Bernoulli, before mentioned, and the nephew of Daniel Bernoulli last noticed above. Our author's elder brother John, who still lives at Berlin, is also well known in the republic of science, particularly for his astronomical labours.

The gentleman to whom this article relates, was educated, as most of his relations had been, for the profession of law: but his genius led him very early into the study of mathematics; and at 20 years of age he read public lectures on experimental philosophy in the university of Basil, for his uncle Daniel Bernoulli, whom he hoped to have succeeded as professor. Being disappointed in this view, he resolved to leave his native place, and to seek his fortune elsewhere; hence he accepted the office of secretary to Count Breuner, the emperor's envoy to the republic of Venice; and in this city he remained till the year 1786, when, on the recommendation of his countryman, M. Fuss, he was invited to Petersburgh to succeed M. Lexell in the academy there, where he continued till his death, which happened the 3d of July 1789, at not quite 30 years of age, and when he had been married only two months, to the youngest daughter of John Albert Euler, the son of the so celebrated Leonard Euler.

Impossible. or Imaginary BINOMIAL. After this article, in pa. 208, the middle of col. 1, *add* what here follows.

In the foregoing article are given several rules for the roots of Binomials. Dr. Maskelyne, the Astronomer Royal, has also given a method of finding any power of an Impossible Binomial, by another like Binomial. This rule is given in his Introduction prefixed to Taylor's Tables of Logarithms, pa. 56; and is as follows.

The logarithms of a and b being given, it is required to find the power of the Impossible Binomial

$a \pm \sqrt{-b^2}$ whose index is $\frac{m}{n}$, that is, to find

$(a \pm \sqrt{-b^2})^{\frac{m}{n}}$ by another Impossible Binomial; and

thence the value of $(a+\sqrt{-b^2})^{\frac{m}{n}} + (a-\sqrt{-b^2})^{\frac{m}{n}}$, which is always possible, whether a or b be the greater of the two.

Solution. Put $\frac{b}{a} = $ tang. z. Then

$$(a \pm \sqrt{-b^2})^{\frac{m}{n}} = (a^2+b^2)^{\frac{m}{2n}} \times (\text{cof.}\frac{m}{n}z \pm \sqrt{-\text{fin.}^2\frac{m}{n}z}).$$

Hence $(a+\sqrt{-b^2})^{\frac{m}{n}} + (a-\sqrt{-b^2})^{\frac{m}{n}} = (a^2+b^2)^{\frac{m}{2n}} \times$

$2 \text{cof.}\frac{m}{n}z = (a \times \text{fec.}z)^{\frac{m}{n}} \times 2\text{cofin.}\frac{m}{n}z = (b \times \text{cofec}z.)^{\frac{m}{n}}$

$\times 2 \text{cofin.} \frac{m}{n}z$, where the first or second of these two last expressions is to be used, according as z is an extreme or mean arc; or rather, because $\frac{b}{a}$ is not only the tangent of z, but also of $z + 360°$, $z + 720°$, &c; therefore the factor in the answer will have several values, viz,

$2 \text{cof.} \frac{m}{n}z$; $2 \text{cof.} \frac{m}{n} (z + 360°)$; $2 \text{cof.} \frac{m}{n}(z + 720°)$; &c; the number of which, if m and n be whole numbers, and the fraction $\frac{m}{n}$ be in its least terms, will be equal to the denominator n; otherwise infinite.

By Logarithms. Put log $b +10 -$log. $a = $ log. tan. z. Then log $\left((a+\sqrt{-b^2})^{\frac{m}{n}} + (a-\sqrt{-b^2})^{\frac{m}{n}} \right) =$

$= \frac{m}{n} \times (l.a+10-l. \text{cof.} z) + l. z + l. \text{cof.}\frac{m}{n}z - 10$

$= \frac{m}{n} \times (l.b+10-l. \text{fin.} z) + l. z + l. \text{cof.}\frac{m}{n}z -10$;

where the first or second expression is to be used, according as z is an extreme or mean arc. Moreover by taking successively, l. cof. $\frac{m}{n}z$; l. cof. $\frac{m}{n}(z+360°)$;

l. cof. $\frac{m}{n}(z+720°)$; &c, there will arise several distinct answers to the question, agreeably to the remark above.

BINOMIAL *Theorem.* Francis Maseres, Esq. (Cursitor Baron of the Exchequer) has communicated the

the following obfervations on the Binomial theorem, and its demonftration ; viz, About the year 1666 the celebrated Sir Ifaac Newton difcovered that, if m were put for any whole number whatfoever, the coefficients of the term of the mth power of $1 + x$ would be

$$1, \quad \frac{m}{1}, \quad \frac{m}{1} \cdot \frac{m-1}{2}, \quad \frac{m}{1} \cdot \frac{m-1}{2} \cdot \frac{m-2}{3}, \quad \&c,$$

till we come to the term $\dfrac{m-(m-1)}{m}$, which will be the laft term. But how he difcovered this propofition, he has not told us, nor has he even attempted to give a demonftration of it. Dr. John Wallis, of Oxford, informs us (in his Algebra, chap. 85, pa. 319) that he had endeavoured to find this manner of generating thefe coefficients one from another, but without fuccefs ; and he was greatly delighted with the difcovery, when he found that Mr. Newton had made it. But he likewife has omitted to give a demonftration of it, as well as Sir Ifaac Newton ; and probably he did not know how to demonftrate it.

Sir Ifaac Newton, after he had difcovered this rule for generating the coefficients of the powers of $1 + x$ when the indexes of thofe powers were whole numbers, conjectured that it might poffibly be true likewife when they were fractions. He therefore refolved to try whether it was or not, by applying it to fuch indexes in a few eafy inftances, and particularly to the indexes $\frac{1}{2}$ and $\frac{1}{3}$, which, if the rule held good in the cafe of fractional indexes, would enable him to find feriefes equal to the values of $\overline{1 + x})^{\frac{1}{2}}$ and $\overline{1 + x})^{\frac{1}{3}}$, or the fquare-root and the cube-root of the Binomial quantity $1 + x$. And, when he had in this manner obtained a feries for $\overline{1 + x})^{\frac{1}{2}}$, which he fufpected to be equal to $\overline{1 + x})^{\frac{1}{2}}$, or the fquare root of $1 + x$, he multiplied the faid feries into itfelf, and found that the product was $1 + x$; and when he had obtained a feries for $\overline{1 + x})^{\frac{1}{3}}$ he multiplied the faid feries twice into itfelf, and found that the product was $1 + x$; and thence he concluded that the former feries was really equal to the fquare-root of $1 + x$, and that the latter feries was really equal to its cube-root. And from thefe and a few more fuch trials, in which he found the rule to anfwer, he concluded univerfally that the rule was always true, whether the index m ftood for a whole number or a fraction of any kind, as $\frac{1}{2}, \frac{1}{3}, \frac{2}{3}, \frac{3}{2}, \frac{5}{9}, \frac{9}{5},$ or, in general $\frac{p}{q}$.

After the difcovery of this rule by Sir Ifaac Newton, and the publication of it by Dr. Wallis, in his Algebra, chap. 85, in the year 1685, (which I believe was the firft time it was publifhed to the world at large, though it was inferted in Sir Ifaac Newton's firft letter to Mr. Oldenburgh, the fecretary to the Royal Society, dated June 13, 1676, and the faid letter was fhewn to Mr. Leibnitz, and probably to fome other of the learned mathematicians of that time it remained for fome years without a demonftration, either in the cafe of integral powers or of roots. At laft however it was demon-

ftrated in the cafe of integral powers by means of the properties of the figurate numbers, by that learned, fagacious, and accurate mathematician Mr. James Bernoulli, in the 3d chapter of the 2d part of his excellent treatife *De Arte Conjectandi*, or, *On the Art of forming reafonable Conjectures concerning Events that depend on Chance* ; which appears to me to be by much the beft written treatife on the doctrine of Chances that has yet been publifhed, though Mr. Demoivre's book on the fame fubject may have carried the doctrine fomething further. This treatife of Mr. James Bernoulli's was not publifhed till the year 1713, which was fome years after his death, which happened in Auguft 1705.; but there is reafon to think that it was compofed in the latter years of the preceding century, about the years 1696, 1697, 1698, 1699, and 1700, and even that fome parts of it, or fome of the propofitions inferted in it, had been found out by the author in the years 1689, 1690, 1691, and 1692. For the firft part of his very curious tract, intitled, *Pofitiones Arithmeticæ de Seriebus Infinitis* was publifhed at Bafil or Bafle in Switzerland (which was his native place, and in which he was at that time profeffor of mathematics) in the year 1689 ; and the fecond part of the faid *Pofitiones* (in the 19th Pofition of which thofe properties of the figurate numbers from which the Binomial theorem may be deduced, are fet down) was publifhed at the fame place in the year 1692. But the demonftrations of thofe properties of the figurate numbers, and of the Binomial theorem, which depends upon them, were never as I believe communicated to the public till the year 1713, when the author's pofthumous treatife *De Arte Conjectandi* made its appearance. Thefe demonftrations are founded on clear and fimple principles, and afford as much fatisfaction as can well be expected on the fubject. But the full difplay and explanation of thefe principles, and the deduction of the faid properties of the figurate numbers, and ultimately of the Binomial theorem, from them, is a matter of confiderable length. It will not therefore be amifs to give a fhorter proof of the truth of this important theorem, that fhall not require a previous knowledge of the properties of the figurate numbers, but yet fhall be equally conclufive with that which is derived from thofe properties. Now this may be done in the manner following.

Let us fuppofe that the coefficients of the terms of the firft fix powers of the Binomial quantity $1 + x$ have been found, upon trial, to be fuch as would be produced by the general expreffions

$$1, \quad \frac{m}{1}, \quad \frac{m}{1} \cdot \frac{m-1}{2}, \quad \frac{m}{1} \cdot \frac{m-1}{2} \cdot \frac{m-2}{3}, \quad \&c,$$

by fubftituting in them firft 1, then 2, then 3, then 4, then 5, and laftly 6, inftead of m. This may eafily be tried by raifing the faid firft fix powers of $1 + x$ by repeated multiplications by $1 + x$ in the common way, and afterwards finding the terms of the fame powers by means of the faid general expreffions above ; which will be found to produce the very fame terms as arofe from the multiplications. After thefe trials we of ll be fure that thofe general expreffions are the true values of the coefficients of the powers of $1 + x$ at leaft in the faid firft fix powers. And it will therefore only remain

remain to be proved that, since the rule is true in the said first six powers, it will also be true in the next following, or the 7th. power, and consequently in the 8th, 9th and 10th powers, and in all higher powers whatsoever.

Now, if the coefficients of the 1st, 2d, 3d, 4th, and other following terms of $\overline{1+x}^m$ be denoted by the letters a, b, c, d, &c, respectively, it is evident from the nature of multiplication, that the coefficients of the 1st, 2d, 3d, 4th, and other following terms of the next higher power of $1+x$, to wit, $\overline{1+x}^{m+1}$ will be equal to a, $a+b$, $b+c$, $c+d$, &c, respectively, or to the sums of every two contiguous coefficients of the terms of the preceding series which is $=\overline{1+x}^m$. This will appear from the operation of multiplication, which is as follows.

$$a + bx + cx^2 + dx^3 + ex^4 + \text{&c.}$$
$$1 + x$$
$$\overline{}$$
$$a + bx + cx^2 + dx^3 + ex^4 + \text{&c.}$$
$$ + ax + bx^2 + cx^3 + dx^4 + \text{&c.}$$

Therefore, if $\overline{1+x}^m$ is equal to the series
$$a + bx + cx^2 + dx^3 + ex^4 + \text{&c,}$$
then $\overline{1+x}^{m+1}$ will be equal to the series
$$a + \overline{a+b}.x + \overline{b+c}.x^2 + \overline{c+d}.x^3 + \text{&c.}$$

Now let n be $= m+1$. We shall then have to prove that, if the coefficients a, b, c, d, &c, be respectively equal to

$$1, \quad \frac{m}{1}, \quad \frac{m}{1}.\frac{m-1}{2}, \quad \frac{m}{1}.\frac{m-1}{2}.\frac{m-2}{3}, \quad \text{&c,}$$

the coefficients a, $a+b$, $b+c$, &c, will be respectively equal to

$$1, \quad \frac{n}{1}, \quad \frac{n}{1}.\frac{n-1}{2}, \quad \frac{n}{1}.\frac{n-1}{2}.\frac{n-2}{3}, \quad \text{&c.}$$

In order to prove this, there is nothing more to do than to collect together every two terms of the former of these two series, and then substitute into these sums, n instead of $m+1$, when there will immediately come out the terms of the latter series as above, viz,

$$\overline{1+x}^n = 1 + \frac{n}{1}x + \frac{n}{1}.\frac{n-1}{2}x^2 + \text{&c.} \qquad \text{Q. E. D.}$$

BINOMIAL *Theorem, Improvement of.* Mr. Bonnycastle, of the Royal Mil. Acad. has lately discovered the following ingenious improvement of this theorem, which is now published for the first time.

This celebrated theorem has been given under various forms, since the time of its first invention; but the following property of it is conceived to be new, and capable of an application of which the original series is not susceptible.

The Newtonian theorem, in one of its most commodious forms, is

$$\overline{1+p}^n = 1 + np + \frac{n.n-1}{2}p^2 + \frac{n.n-1.n-2}{2.3}p^3 + \frac{n.n-1.n-2.n-3}{2.3.4}p^4$$

&c; and the new theorem here alluded to, is

$$\overline{1+p}^n = 1 + sn + \frac{1}{2}s^2n^2 + \frac{1}{2.3}s^3n^3 + \frac{1}{2.3.4}s^4n^4 \text{ &c;}$$

where $s = p - \frac{1}{2}p^2 + \frac{1}{3}p^3 - \frac{1}{4}p^4 + \frac{1}{5}p^5$ &c.
Of which the investigation is as follows:

$$\overline{1+p}^n = 1 + np + \frac{n.n-1}{2}p^2 + \frac{n.n-1.n-2}{2.3}p^3 + \frac{n.n-1.n-2.n-3}{2.3.4}p^4 \text{ &c}$$

$$= 1 + np + (n^2-1)\frac{p^2}{2} + (n^3-3n^2+2n)\frac{p^3}{2.3} + (n^4-6n^3+11n^2-6n)\frac{p^4}{2.3.4}$$

$$+ (n^5-10n^4+35n^3-50n^2+24n)\frac{p^5}{2.3.4.5}$$

$$+ (n^6-15n^5+85n^4-225n^3+274n^2-120n)\frac{p^6}{2.3.4.5.6} \text{ &c.}$$

Then by connecting the several powers of p with all the like powers of n, the latter series will become

$$1 + (p - \frac{p^2}{2} + \frac{2p^3}{2.3} - \frac{6p^4}{2.3.4} + \frac{24p^5}{2.3.4.5} - \frac{120p^6}{2.3.4.5.6} \text{ &c})n$$

$$+ (\frac{p^2}{2} - \frac{3p^3}{2.3} + \frac{11p^4}{2.3.4} - \frac{50p^5}{2.3.4.5} + \frac{274p^6}{2.3.4.5.6} \text{ &c})n^2$$

$$+ (\frac{p^3}{2.3} - \frac{6p^4}{2.3.4} + \frac{35p^5}{2.3.4.5} - \frac{225p^6}{2.3.4.5.6} \text{ &c})n^3$$

$$+ (\frac{p^4}{2.3.4} - \frac{10p^5}{2.3.4.5} + \frac{85p^6}{2.3.4.5.6} \text{ &c})n^4$$

$$+ (\frac{p^5}{2.3.4.5} - \frac{15p^6}{2.3.4.5.6} \text{ &c})n^5$$

$$+ (\frac{p^6}{2.3.4.5.6} \text{ &c})n^6$$
$$\text{&c;}$$

which by abbreviation, &c, becomes

$$1 + (p - \frac{p^2}{2} + \frac{p^3}{3} - \frac{p^4}{4} + \frac{p^5}{5} - \frac{p^6}{6} \text{ &c})n$$

$$+ \frac{1}{2}(p^2 - \frac{3p^3}{3} + \frac{11p^4}{3.4} - \frac{50p^5}{3.4.5} + \frac{274p^6}{3.4.5.6} \text{ &c})n^2$$

$$+ \frac{1}{2.3}(p^3 - \frac{6p^4}{4} + \frac{35p^5}{4.5} - \frac{225p^6}{4.5.6} \text{ &c})n^3$$

$$+ \frac{1}{2.3.4}(p^4 - \frac{10p^5}{5} + \frac{85p^6}{5.6} \text{ &c})n^4$$

$$+ \frac{1}{2.3.4.5}(p^5 - \frac{15p^6}{6} \text{ &c})n^5$$

$$+ \frac{1}{2.3.4.5.6}(p^6 \text{ &c})n^6$$
$$\text{&c.}$$

In which last series, the literal parts of the coefficients of the 3d, 4th, 5th, &c terms, are the square, cube, biquadrate, &c, of the coefficient of the 2d term, as will appear either from the actual involution of

$$p - \frac{p^2}{2} + \frac{p^3}{3} - \frac{p^4}{4} \text{ &c, or by comparing its several}$$

powers with the multinomial theorem of Demoivre.
From hence it follows that,

$$\overline{1+p}^n$$

$$\overline{1+p}\,^n = 1 + (p - \frac{p^2}{2} + \frac{p^3}{3} \&c)n + \frac{1}{2}(p - \frac{p^2}{2} + \frac{p^3}{3} \&c)n^2$$

$$+ \frac{1}{2\cdot3}(p - \frac{p^2}{2} + \frac{p^3}{3}\&c)n^3 + \frac{1}{2\cdot3\cdot4}(p - \frac{p^2}{2} + \frac{p^3}{3}\&c)n^4$$

$$\&c.$$

And if $p - \frac{p^2}{2} + \frac{p^3}{3} - \frac{p^4}{4}$ &c be put $= s$, we shall have

$$\overline{1+p}\,^n = 1 + sn + \frac{1}{2}s^2n^2 + \frac{1}{2\cdot3}s^3n^3 + \frac{1}{2\cdot3\cdot4}s^4n^4 \&c,$$

as was to be shewn.

By a similar mode of deduction, it may also be proved that

$$\overline{1-p}\,^n = 1 - sn + \frac{1}{2}s^2n^2 - \frac{1}{2\cdot3}s^3n^3 + \frac{1}{2\cdot3\cdot4}s^4n^4\&c;$$

where in this case $s = p + \frac{p^2}{2} + \frac{p^3}{3} + \frac{p^4}{4}$ &c.

In each of which formulæ, the index n, may be considered either as a whole number, a fraction, a surd, a given or an unknown quantity, as the circumstance may require.

For the application of these theorems, see LOGARITHMS, and EXPONENTIAL *Equations*, following.

C.

CAN

CANAL, in general, denotes a long, round, hollow instrument, through which a fluid matter may be conveyed. In which sense, it amounts to the same as what is otherwise called a pipe, tube, channel, &c. Thus the Canal of an aqueduct, is the part through which the water passes; which, in the ancient works of this kind, is lined with a coat of mastic of a peculiar composition.

CANAL more particularly denotes a kind of artificial river, often furnished with locks and sluices, and sustained by banks or mounds. They are contrived for divers purposes; some for forming a communication between one place and another; as the Canals between Bruges and Ghent, or between Brussels and Antwerp: Others for the decoration of a garden, or house of pleasure; as the Canals of Versailles, Fontainbleau, St. James's Park, &c: And others are made for draining wet and marshy lands; which last however are more properly called water gangs, drains, ditches, &c.

It is needless to enumerate the many advantages arising from Canals and artificial navigations. Their utility is now so apparent, that most nations in Europe give the highest encouragement to undertakings of this kind wherever they are practicable. Nor did their advantages escape the observation of the Ancients. From the earliest accounts of society we read of attempts to cut through large isthmuses, to make communications by water, either between one sea and another, or between different nations, or distant parts of the same nation, where land-carriage was long and expensive.

Egypt is full of Canals, dug to receive and distribute the waters of the Nile, at the time of its inundation. They are dry the rest of the year, except the Canal of Joseph, and four or five others, which may be ranked as considerable rivers. There were also subterraneous Canals, or tunnels, dug by an ancient king of Egypt, by which those lakes, formed by the inundations of the Nile, were conveyed into the Mediterranean sea.

CAN

Herodotus relates, that the Cnidians, a people of Coria, in Asia Minor, designed to cut through the isthmus which joins that peninsula to the continent; but were superstitious enough to give up the undertaking, because it was interdicted by an oracle.

Several kings of Egypt attempted to join the Red-Sea to the Mediterranean; a project which Cleopatra was very fond of. This Canal was begun, according to Herodotus, by Necus son of Psammeticus, who desisted from the attempt on an answer from the oracle, after having lost 120 thousand men in the enterprise. It was resumed and completed by Darius son of Hystaspes, or, according to Diodorus and Strabo, by Ptolomy Philadelphus; who relate that Darius relinquished the work on a representation made to him by unskilful engineers, that the Red-Sea, being higher than the land of Egypt, would overflow and drown the whole country. It was wide enough for two galleys to pass abreast, and its length was four days sailing. Diodorus adds, that it was also called Ptolomy's river; that this prince built a city at its mouth on the Red-Sea, which he called Arsinoë, from the name of his favourite sister; and that the Canal might be either opened or shut, as occasion required. Diod. Sic. lib. 1; Strabo, Geog. lib. 17; Herod. lib. 2. Soliman the 2d, emperor of the Turks, employed 50 thousand men in this great work; which was completed under the caliphate of Omar, about the year 635; but was afterward allowed to fall into neglect and disrepair; so that it is now difficult to discover any traces of it. Hist. Acad. Scienc ann. 1703, pa. 110.

Both the Greeks and Romans intended to make Canal across the Isthmus of Corinth, which joins t Morea and Achaia, for a navigable passage bye Ionian sea into the Archipelago. Demetrius, Jus Cæsar, Caligula, and Nero, made several unsucceul efforts to open this passage. But as the Ancients re entirely ignorant of the use of water-locks, their ole attention

attention was employed in making level cuts, which is probably the chief reason why they so often failed in their attempts. Charlemagne formed a defign of joining the Rhine and the Danube, to make a communication between the Ocean and the Black-Sea, by a Canal from the river Almutz which difcharges itfelf into the Danube, to the Reditz, which falls into the Maine, which laft falls into the Rhine near Mayence or Mentz: for this purpofe he employed a prodigious number of workmen ; but he met with fo many obftacles from different quarters, that he was obliged to give up the attempt.

A new Canal for conveying the waters of the Nile from Ethiopia into the Red-Sea without paffing into Egypt, was projected by Albuquerque, viceroy of India for the Portuguefe, to render Egypt barren and unprofitable to the Turks.—M. Gaildereau attributes the frequency of the plague in Egypt, of late days, to the decay, or ftopping up of thefe Canals ; which happened upon the Turks becoming mafters of the country.

In China, there is fcarce a town or village without the advantage either of an arm of the fea, a navigable river, or a Canal, by which means navigation is rendered fo common, that there are almoft as many people on the water as the land. The great Canal of China, is one of the wonders of art, extending from north to fouth quite acrofs the empire, from Pekin to Canton, a diftance of 825 miles, and was made upwards of 800 years ago. Its breadth and depth are fufficient to carry barks of confiderable burden, which are managed by fails and mafts, as well as towed by hand. On this Canal it feems the emperor employs near ten thoufand fhips. It paffes through, or by, 41 large cities ; there are in it 75 vaft locks and fluices, to keep up the water, and pafs the fhips where the ground will not admit of fufficient depth of channel, befide feveral thoufand draw and other bridges. Indeed, F. Magaillane affures us, there are paffages from one end of China to the other, the fpace of 600 French leagues, either by Canals or rivers, except a fingle day's journey by land, neceffary to crofs a mountain.

The French at prefent have many fine Canals. That of Briere, otherwife called the Canal of Burgundy, was begun under Henry IV, and finifhed under the direction of cardinal Richelieu in the reign of Louis XIII. This Canal makes a communication between the Loire and the Seine, and fo to Paris. It extends 11 French great leagues from Briere to Montargis, and has 42 locks upon it.

The Canal of Orleans was begun in 1675, for eftablifhing a communication alfo between the Seine and the Loire. It is confiderably fhorter than that of Briere, and has only 20 fluices.

The Canal of Bourbon was but lately undertaken : its defign is to make a communication from the river Oife to Paris.

But the greateft and moft ufeful work of this kind, is the junction of the Ocean with the Mediterranean by the Canal of Languedoc, called alfo the Canal of the two feas. It was propofed in the reigns of Francis I and Henry IV, and was begun and finifhed under Louis XIV ; having been planned by Francis Riquet in the year 1666, and finifhed before his

death, which happened in 1680. It begins with a large refervoir 4000 paces in circumference, and 24 feet deep, which receives many fprings from the mountain Noire. The Canal is about 200 miles in length, extending from Narbonne to Tholoufe, being fupplied by a number of rivulets in the way, and furnifhed with 104 locks or fluices, of about 8 feet rife each. In fome places it is carried over bridges and aqueducts of vaft height, which give paffage underneath to other rivers ; and in fome places it is cut through folid rocks for a mile together.

The new Canal of the lake Ladoga, cut from Volhowa to the Neva, by which a communication is made between the Baltic, or rather Ocean, and the Cafpian-fea, was begun by the czar Peter the 1ft in 1719 : by means of which the Englifh and Dutch merchandize is eafily conveyed into Perfia, without being obliged to double the Cape of Good Hope.—There was a former Canal of communication between the Ladoga lake and the river Wolga, by which timber and other goods had been brought from Perfia to Peterfburg ; but the navigation of it was fo dangerous, that a new one was undertaken.

The Spaniards have feveral times had in view the digging a Canal through the Ifthmus of Darien, between North and South America, from Panama to Nombre de Dios, to make a ready communication between the Atlantic and the South Sea, and thus afford a ftraight paffage to China and the Eaft Indies.

In the Dutch, Auftrian, and French Netherlands, there is a great number of Canals : that from Bruges to Oftend carries veffels of 200 tons. But it would be an endlefs tafk to defcribe the numberlefs Canals in Holland, Germany, Ruffia, &c. We may therefore only take a view of thofe in our own country.

In England, that ancient Canal from the river Nyne, a little below Peterborough, to the river Witham, three miles below Lincoln ; called by the modern inhabitants Caerdike ; may be ranked among the monuments of the Roman grandeur, though it is now moft of it filled up. Morton will have it made under the emperor Domitian. Urns and medals have been difcovered on the banks of this Canal, which feem to confirm that opinion. Yet fome authors take it to be a Danifh work. It was 40 miles in length ; and, fo far as appears from the ruins, muft have been very broad and deep. Notwithftanding that early beginning, it is not long fince Canals have been revived in this country. They are now however become very numerous, particularly in the counties of York, Lincoln, and Chefhire. Moft of the counties between the mouth of the Thames and the Briftol channel are connected together either by natural or artificial navigations ; thofe upon the Thames and Ifis reaching within about 20 miles of thofe upon the Severn.

The Canal for fupplying London with water by means of the New River, was projected and begun by Mr. Edward Wright, author of the celebrated treatife on Navigation, about the year 1608 ; but finifhed by Mr. (afterwards Sir Hugh) Middleton, five years after. This Canal commences near Ware, in Hertfordfhire, and takes a courfe of 60 miles before it reaches the ciftern at Iflington, which fupplies the feveral water-pipes that convey it to the city and parts adjacent. In fome

places

places it is 30 feet deep, and in others it is conveyed over a valley between two hills, by means of a trough supported on wooden arches, and rising above 23 feet in height.

The Duke of Bridgwater's Canal, projected and executed under the direction of Mr. Brindley, was begun about the year 1759. It was first designed only for conveying coals to Manchester, from a mine in the duke's estate; but has since been applied to many other useful purposes of inland navigation. This Canal begins at a place called Worsley-mill, about 7 miles from Manchester, where a bason is made capable of holding all the boats, and a great body of water which serves as a reservoir or head to the navigation. The Canal runs through a hill by a subterraneous passage, large enough for admitting long flat-bottomed boats, which are towed by a rail on each hand, near three quarters of a mile, to the coal-works. There the passage divides into two channels, one of which goes off 300 yards to the right, and the other as many to the left; and both may be continued at pleasure. The passage is in some places cut through the solid rock, and in others arched over with brick; and air funnels, some of which are near 37 yards perpendicular, are cut, at certain distances, through the rock to the top of the hill. The arch at its entrance is about 6 feet wide, and about 5 feet high from the surface of the water; but widens within, so that in some places the boats may pass one another, and at the pits it is 10 feet wide. When the boats are loaded and brought out of the bason, five or six of them are linked together, and drawn along the Canal by a single horse, and thus reaching Manchester in a course of nine miles. It is broad enough for two barges to pass or go abreast; and on one side there is a good road for the passage of the people, and the horses or mules employed in the work. The Canal is raised over public roads by means of arches; and it passes over the navigable river Irwell near 50 feet above it; so that large vessels in full sail pass under the Canal, while the duke's barges are at the same time passing over them. This Canal joins that which passes from the river Mersey towards the Trent, taking in the whole a course of 34 miles.

The Lancaster Canal begins near Kendal, and terminates near Eccleston, comprehending the distance of $72\frac{1}{2}$ miles.

The Canal from Liverpool to Leeds is $108\frac{1}{2}$ miles: that from Leeds to Selby, $23\frac{1}{4}$ miles; from Chichester to Middlewich, $26\frac{3}{4}$ miles; from the Trent to the Mersey, 88 miles; from the Trent to the Severn, $46\frac{1}{2}$ miles. The Birmingham Canal joins this near Wolverhampton, and is $24\frac{1}{4}$ miles: the Droitwich Canal is $5\frac{1}{2}$ miles: the Coventry Canal, commencing near Lichfield, and joining that of the Trent, is $36\frac{1}{4}$ miles: the Oxford Canal breaks off from this, and is 82 miles: the Chesterfield Canal joins the Trent near Gainsborough, and is 44 miles.

A communication is now formed, by means of this inland navigation, between Kendal and London, by way of Oxford; between Liverpool and Hull, by the way of Leeds; and between the Bristol channel and the Humber, by the junction formed between the Trent and the Severn. Other schemes have been projected, which the present spirit of improvement will probably soon carry into execution, of opening a communication

3

between the German and Irish seas, so as to reduce a hazardous navigation of more than 800 miles by sea, into a little more than 150 miles by land, or inland navigation; and also of joining the Isis with the Severn.

In Scotland, a navigable Canal between the Forth and Clyde, which divides that country into two parts, was thought of more than a century since, for transports and small ships of war. It was again projected in the year 1722, and a survey made; but nothing more was done till 1761, when the then lord Napier, at his own expence, had a survey, plan, and estimate made on a small scale. In 1764, the trustees for fisheries, &c, in Scotland, procured another survey, plan, and estimate of a Canal 5 feet deep, which was to cost 79,000 pounds. In 1766, a subscription was obtained by a number of the most respectable merchants in Glasgow, for making a Canal 4 feet deep and 24 feet in breadth; but when the bill was nearly obtained in parliament, it was given up on account of the smallness of the scale, and a new subscription set on foot for a Canal 7 feet deep, estimated at 150,000 pounds. This obtained the sanction of parliament; and the work was begun in 1768, by Mr. Smeaton the engineer. The extreme length of the Canal from the Forth to the Clyde is 35 miles, beginning at the mouth of the Carron, and ending at Dalmure Burnfoot on the Clyde, 6 miles below Glasgow, rising and falling 160 feet by means of 39 locks, 20 on the east side of the summit, and 19 on the west, as the tide does not ebb so low in the Clyde as in the Forth by 9 feet; and it was deepened to upwards of 8 feet. This Canal was finished a few years since, after having experienced some interruptions and delays, for want of resources, and is esteemed the greatest work of the kind in this island. Vessels drawing 8 feet water, with 19 feet in the beam and 73 feet in length, pass with ease; and the whole enterprise displays the art of man in a high degree. To supply the Canal with water was of itself a very great work. There is one reservoir of 50 acres 24 feet deep, and another of 70 acres 22 feet deep, in which many rivers and springs terminate, which it is expected will afford sufficient supply of water at all times.

The Practice of Canal Digging and Inland Navigations.

The particular operations necessary for making artificial navigations, depend upon a number of circumstances. The situation of the ground; the vicinity or connection with rivers; the ease or difficulty with which a proper quantity of water can be obtained: these and many other circumstances necessarily produce great variety in the structure of artificial navigations, and augment or diminish the labour and expence of executing them. When the ground is naturally level, and unconnected with rivers, the execution is easy, and the navigation is not liable to be disturbed by floods: but when the ground rises and falls, and cannot be reduced to a level, artificial methods of raising and lowering vessels must be employed; which likewise vary according to circumstances.

Sometimes a kind of temporary sluices are employed, to raise boats over falls or shoals in rivers, by a very simple operation. Two pillars of mason-work, with grooves, are fixed, one on each bank of the river, at

some

Plate XXXVII.

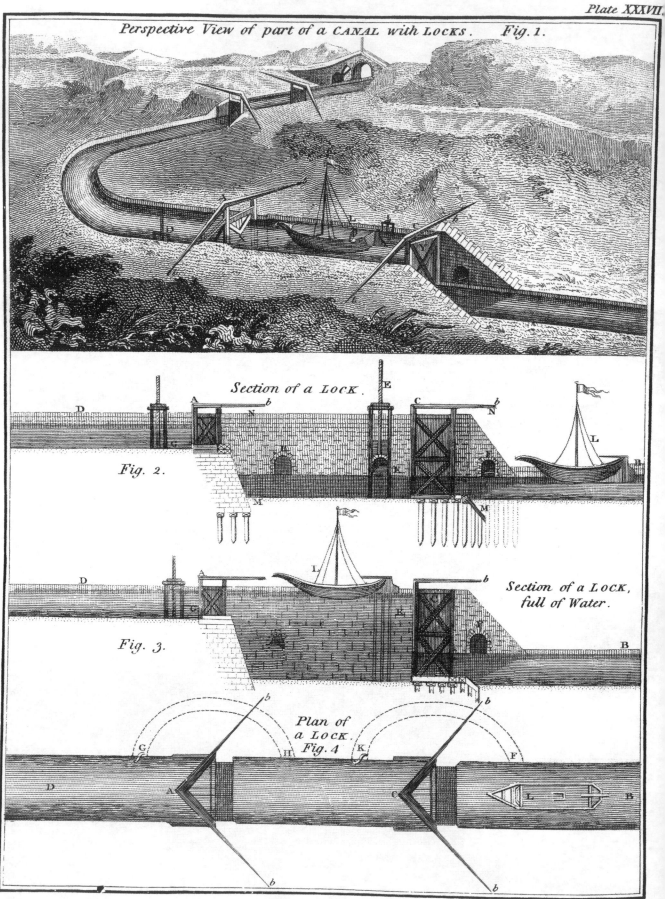

Perspective View of part of a CANAL with LOCKS. Fig. 1.

Section of a LOCK.

Fig. 2.

Section of a LOCK, full of Water.

Fig. 3.

Plan of a LOCK.
Fig. 4

fome diſtance below the ſhoal. The boat having paſſed theſe pillars, ſtrong planks are let down acroſs the river by pulleys into the grooves; by which means the water is dammed up to a proper height for allowing the boat to paſs up the river over the ſhoal.

The Dutch and Flemings at this day ſometimes, when obſtructed by caſcades, form an inclined plane or rolling-bridge upon dry land, along which their veſſels are drawn from the river below the caſcade, into the ri‑ ver above it. This it is ſaid was the only method em‑ ployed by the Ancients, and ſtill ſometimes uſed by the Chineſe. Theſe rolling-bridges conſiſt of a number of cylndrical rollers which turn eaſily on pivots. And a mill is commonly built near; ſo that the ſame machinery may ſerve the double purpoſe of working the mill and drawing up veſſels.

But in the preſent improved ſtate of inland naviga‑ tion, theſe falls and ſhoals are commonly ſurmounted by means of what are called locks or ſluices. A lock is a baſon placed lengthwiſe in a river or Canal, lined with walls of maſonry on each ſide, and terminated by two gates placed acroſs the Canal, where there is a caſcade or natural fall of the country; and ſo conſtruct‑ ed, that the baſon being filled with water by an upper ſluice to the level of the waters above, a veſſel may aſcend through the upper gate; or the water in the lock being reduced to the level of the water at the bot‑ tom of the caſcade, the veſſel may deſcend through the lower gate: for when the waters are brought to a level on either ſide, the gate on that ſide may be eaſily opened.

But as the lower gate is ſtrained in proportion to the depth of water it ſupports, when the perpendicular height of the water exceeds 12 or 13 feet, it becomes neceſſary to have more locks than one. Thus, if the fall be 16 feet, two locks are required, each of 8 feet fall; and if the fall be 25 feet, three locks are neceſſary, each having 8 feet 4 inches fall.—It is evident that the ſide‑ walls of locks ſhould be made very ſtrong: and where the natural foundation is bad, they ſhould be founded on piles and platforms of wood. They ſhould likewiſe ſlope outwards, in order to reſiſt the preſſure of the earth from behind.

To illuſtrate this by repreſentations: Plate 37, fig. 1, is a perſpective view of part of a Canal, with ſeveral locks &c; the veſſel L being within the lock AC.— Fig. 2 is an elevation or upright ſection along the Ca‑ nal; the veſſel L about to enter.—Fig. 3, a like ſec‑ tion of a lock full of water; the veſſel L being raiſed to a level with the water in the ſuperior Canal.—And fig. 4 is the plan or ground ſection of a lock: where L is a veſſel in the inferior Canal; C, the under gate; A, the upper gate; GH, a ſubterraneous-paſſage for let‑ ting water from the ſuperior Canal run into the lock; and KF, a ſubterraneous paſſage for water from the lock to the inferior Canal.

X and Y (fig. 1) are the two flood-gates, each of which conſiſts of two leaves, reſting upon one another, ſo as to form an obtuſe angle, the better to reſiſt the preſſure of the water. The firſt (X) prevents the wa‑ ter of the ſuperior Canal from falling into the lock; and the ſecond (Y) dams up and ſuſtains the water in the lock. Theſe flood-gates ought to be very ſtrong, and to turn freely upon their hinges. They ſhould alſo be

VOL. II.

made very tight and cloſe, that as little water as poſſible may be loſt. And, to make them open and ſhut with eaſe, each leaf is furniſhed with a long lever A*b*, A*b*; C*b*, C*b*.

By the ſubterraneous paſſage GH (fig. 2, 3, 4) which deſcends obliquely, by opening the ſluice G, the water is let down from the ſuperior Canal D into the lock, where it is ſtopped and retained by the gate C when ſhut, till the water in the lock comes to be on a level with the water in the ſuperior Canal D; as repreſented in fig. 3. When, on the other hand, the water contained by the lock is to be let out, the paſſage GH muſt be ſhut, by letting down the ſluice G; the gate A muſt alſo be ſhut, and the paſſage KF opened by raiſing the ſluice K. A free paſſage being thus given to the water, it deſcends through KF, into the inferior Canal, until the water in the lock be on a level with the water in the inferior Canal B; as repreſented in fig. 2.

Now ſuppoſe it be required to raiſe the veſſel L (fig. 2) from the inferior Canal B to the ſuperior one D. If the lock be full of water, the ſluice G muſt be ſhut, as alſo the gate A, and the ſluice K opened, ſo that the water in the lock may run out till it become to a le‑ vel with the water in the inferior Canal B. When the water in the lock comes to be on a level with the water at B, the leaves of the gate C are opened by the levers C*b*, which is eaſily performed, the water on each ſide of the gate being in equilibrio; the veſſel then ſails into the lock. After this, the gate C and the ſluice K are ſhut, and the ſluice G opened, in order to fill the lock, till the water in the lock, and conſequently the veſſel, be upon a level with the water in the ſuperior Canal D; as is repreſented in fig. 3. The gate A is then opened, and the veſſel paſſes into the Canal D.

Again let it be required to make a veſſel deſcend from the Canal D into the inferior Canal B. If the lock be empty, as in fig. 2, the gate C and ſluice K muſt be ſhut, and the upper ſluice G opened, ſo that the water in the lock may riſe to a level with the water in the upper Canal D. Then, opening the gate A, the veſſel will paſs through into the lock. This done, ſhut the gate A and the ſluice G; then open the ſluice K, till the water in the lock be on a level with the water in the inferior Canal; this done, the gate C is opened, and the veſſel paſſes along into the Canal B, as was re‑ quired.

CATENARY. Line 4, *for* ACB *read* BAC.— l. 6, *for* A and B *read* C and B. *After which add.* It is otherwiſe called the *Elaſtic Curve.*

CHALDRON. Line 4, *for* 2000 pounds, *read* 28 cwt. or 3136 pounds. *At the end add,* By act of parliament, a Newcaſtle Chaldron is to weigh 52½ cwt, or 3 waggons of 17½ cwt, or 6 carts of 8¾ cwt each, making 52½ cwt to the Chaldron. The ſtatute Lon‑ don Chaldron is to conſiſt of 36 buſhels heaped up, each buſhel to contain a Wincheſter buſhel and one quart, and to be 19½ inches diameter externally. Now it has been found by repeated trials, that 15 London Chaldrons are equal to 8 Newcaſtle Chaldrons, which, reckoning 52¼ cwt to the latter, gives 28 cwt to the former, or 3136 lbs to the London Chaldron.

This I find nearly confirmed by experiment. I

5 B   weighed

weighed one peck of coals, which amounted to 21¾ lb. Then 4 times this gives 87 lb for the weight of the bushel; and 36 times the bushel gives 3132 lb for the Chaldron; to which if the weight of the odd quart be added, or 3 lb nearly, it gives 3135 lb for the weight of the Chaldron, which is only one pound less than by statute.

Pa. 287, col. 2, l. 20, *for* YX — *a* — *x*, *read* YX = *a* — *x*.

CIRCLE *of Curvature.* To what is said of this article in the Dictionary, may be added what follows.

A circular arc is the only curve line that is equally curved in every point. In all other curve lines, such as the arc of an ellipse, or a parabola, or an hyperbola, or a cycloid, the curvature is different in different points, and the degree of curvature in any point is estimated by the curvature of a Circle which is said to have the same curvature as the proposed curve line in that point; by which is understood the Circle which, having the tangent of the proposed curve in the said point for its tangent, approaches so nearly to the proposed curve that no other Circle whatever can be drawn between it and that curve.

This Circle is also said to *osculate* the curve in the said point, and is therefore often called the *osculating Circle*, as well as the *Circle of equal curvature* with the curve in the said point. And the radius of this Circle is called the *radius of curvature* of the proposed curve in the said point; also its centre is called the *centre of curvature.*

Now there are some curve lines so very highly curved in some particular points, that every Circle, of how small a radius soever, having the tangent to the curve in one of those points for its tangent, will pass without the curve, or between the curve and its tangent. This, for example, is the case with the curve of a cycloid in the two points contiguous to its base, as also with the cissoid at its vertex. And in such points the curvature of these curves is said to be *infinite*, because it is greater than the curvature of any Circle, how small soever. Also the radius of the Circle of curvature in such points is nothing; the length of that radius being always inversely or reciprocally as the degree of curvature at any point.

The theory of these Circles of equal curvature with curves in particular points was first cultivated by Apollonius in his Conic Sections: and it has since been carried much farther by several great mathematicians of modern times; particularly by Mr. Huygens in his doctrine of Evolute Curves and Curves of Evolution, and by the great Sir Isaac Newton. See CURVATURE.

CLARKE (Dr. SAMUEL), a celebrated English divine, philosopher, and metaphysician, was the son of Edward Clarke, Esq. alderman of Norwich, and for several years one of its representatives in parliament; and was born there the 11th of October 1675. He was instructed in classical learning at the free-school of that town; and in 1691 removed thence to Caius college in Cambridge; where his uncommon abilities soon began to display themselves. Though the philosophy of Des Cartes was at that time the established philosophy of the university, yet Clarke easily mastered the new system of Newton; and in order to his first degree of arts, performed a public exercise in the schools upon a question taken from it. He greatly contributed to the establishment of the Newtonian philosophy by an excellent translation of Rohault's Physics, with notes, which he finished before he was 22 years of age: a book which had been for some time the system used in the university, and founded upon Cartesian principles. This was first published in the year 1697, and it soon after went through several other editions, all with improvements.

Mr. Whiston relates that, in that year, 1697, while he was chaplain to Dr. Moore bishop of Norwich, he met with young Clarke, then wholly unknown to him, at a coffee house in that city; where they entered into a conversation about the Cartesian philosophy, particularly Rohault's Physics, which Clarke's tutor, as he tells us, had put him upon translating. "The result of this conversation was, says Whiston, that I was greatly surprised that so young a man as Clarke then was, should know so much of those sublime discoveries, which were then almost a secret to all, but to a few particular mathematicians. Nor did I remember (continues he) above one or two at the most, whom I had then met with, that seemed to know so much of that philosophy as Mr. Clarke."

He afterwards turned his thoughts to divinity; and having taken holy orders, in 1698 he succeeded Mr. Whiston as chaplain to Dr. Moore bishop of Norwich, who was ever after his constant friend and patron. In 1699 he published two treatises: the one on Baptism, Confirmation, and Repentance; the other, Reflections on that part of a book called Amyntor, or a Defence of Milton's Life, which relates to the Writings of the Primitive Fathers, and the Canon of the New Testament. In 1701 he published A Paraphrase upon the Gospel of St. Matthew; which was followed in 1702 by the Paraphrases upon the Gospels of St. Mark and St. Luke, and soon after by a third volume upon St. John.

Mean while bishop Moore gave him the rectory of Drayton near Norwich, with a lectureship in that city. In 1704 he was appointed to preach Boyle's lecture; and the subject he chose was, The Being and Attributes of God. He succeeded so well in this, and gave so much satisfaction, that he was appointed to preach the same lecture the next year, when he chose for his subject, The Evidences of Natural and Revealed Religion. These sermons were first printed in two volumes, in 1705 and 1706; and contained some remarks on such objections as had been made by Hobbes and Spinoza, and other opposers of natural and revealed religion. In the 6th edition was added, A Discourse concerning the Connection of the Prophecies of the Old Testament, and the application of them to Christ.

About this time, Mr. Whiston informs us, he discovered that Mr. Clarke (having read much of the primitive writers) began to suspect that the Athanasian doctrine of the Trinity was not the doctrine of those early ages; and it was particularly remarked of him, that he never read the Athanasian Creed at his parish church.

In 1706 he published A Letter to Mr. Dodwell; answering all the arguments in his epistolary discourse against the immortality of the soul. Bishop Hoadley observes, obferves, that in this letter he anfwered Mr. Dodwell in fo excellent a manner, both with regard to the philofophical part, and to the opinions of fome of the primitive writers, upon whom thefe doctrines were fixed, that it gave univerfal fatisfaction. But this controverfy did not ftop here; for the celebrated Mr. Collins, coming in as a fecond to Dodwell, went much farther into the philofophy of the difpute, and indeed feemed to produce all that could be faid againft the immateriality of the foul, as well as the liberty of human actions. This enlarged the fcene of the difpute; into which our author entered, and wrote with fuch a fpirit of clearnefs and demonftration, as at once fhewed him greatly fuperior to his adverfaries in metaphyfical and phyfical knowledge; making every intelligent reader rejoice that fuch an incident had happened to provoke and extort from him fuch excellent reafoning and perfpicuity of expreffion.

In the midft of thefe labours, Mr. Clarke found time to fhew his regard to mathematical and philofophical ftudies, with his exact knowledge and fkill in them. And his natural affection and capacity for thefe ftudies were not a little improved by the friendfhip of Sir Ifaac Newton; at whofe requeft he tranflated his Optics into Latin in 1706. With this verfion Sir Ifaac was fo highly pleafed, that he prefented him with the fum of 500l. or 100l. to each of his five children.

The fame year alfo, bifhop Moore procured for him the rectory of St. Bennett's, Paul's Wharf, in London; and foon after carried him to court, and recommended him to the favour of queen Anne. She appointed him one of her chaplains in ordinary; and alfo prefented him to the rectory of St. James's, Weftminfter, when it became vacant in 1709. Upon this occafion he took the degree of D. D. when the public exercife which he performed for it at Cambridge was highly admired.

The fame year 1709, Dr. Clarke revifed and corrected Whifton's tranflation of the Apoftolical Conftitutions into Englifh, at his earneft requeft. In 1712 he publifhed a moft beautiful and pompous edition of Cæfar's Commentaries. And the fame year, his celebrated book called, The Scripture Doctrine of the Trinity. Whifton informs us, that fome time before the publication of this book, there was a meffage fent to the author by lord Godolphin, and others of queen Anne's minifters, importing, " That the affairs of the public were with difficulty then kept in the hands of thofe that were for liberty; that it was therefore an unfeafonable time for the publication of a book that would make a great noife and difturbance; and that therefore they defired him to forbear till a fitter opportunity fhould offer itfelf:" which meffage, fays he, the doctor paid no regard to, but went on according to the dictates of his own confcience with the publication of his book. The minifters however were very right in their conjectures; for the work made noife and difturbance cnough, and occafioned a great many books and pamphlets, written by himfelf and others. Nor were thefe the whole that his work occafioned: it rendered the author obnoxious to the ecclefiaftical power, and his book was complained of by the lower houfe of convention. The doctor drew up a preface, and afterwards gave in feve-

ral explanations, which feemed to fatisfy the upper houfe; at leaft the affair was not brought to any iffue, the members appearing defirous to prevent diffenfions and divifions.

In 1715 and 1716 he had a difpute with the celebrated Leibnitz, concerning the principles of natural philofophy and religion; and a collection of the papers which paffed between them, was publifhed in 1717. This work was addreffed to queen Caroline, then princefs of Wales, who was pleafed to have the controverfy pafs through her hands. It related chiefly to the fubjects of liberty and neceffity.

About the year 1718 he was prefented by the lord Lechmere, to the mafterfhip of Wigfton's hofpital in Leicefterfhire. In 1724 and 1725 he publifhed 18 fermons, preached on feveral occafions. In 1727, on the death of Sir Ifaac Newton, he had the offer of fucceeding him as Mafter of the Mint, a place worth from 12 to 15 hundred a year: but to this fecular preferment he could not reconcile himfelf; and therefore abfolutely refufed it.—In 1728 was publifhed, a Letter from Dr. Clarke to Mr. Benjamin Hoadley, occafioned by the Controverfy relating to the Proportion of Velocity and Force of Bodies in Motion; and printed in the Philofophical Tranfactions, num. 401.—In the beginning of 1729 he publifhed the firft 12 books of Homer's Iliad: a work which bifhop Hoadley calls an accurate performance; and his notes, a treafury of grammatical and critical knowledge. And the fame year came out, his Expofition of the Church Catechifm, and 10 volumes of Sermons: books fo well known and fo generally approved, that they need no recommendation. But the fame year, on Sunday the 11th of May, going to preach before the Judges at Serjeant's Inn, he was feized with a pain in his fide, which made it impoffible for him to perform his office. He was carried home and continued under his diforder till the 17th of the fame month, when he died, in the 54th year of his age, after long enjoying a vigorous ftate of health, having fcarce ever known ficknefs.

Three years after the doctor's death, appeared the other 12 books of the Iliad, publifhed in 4to by his fon, Mr. Samuel Clarke, who fays in the preface, that his father had finifhed the annotations to the firft three of thofe books, and as far as the 359th verfe of the 4th; and had revifed the text and verfion as far as verfe 510 of the fame book.

Dr. Clarke married Catherine, the only daughter of the Rev. Mr. Lockwood, rector of Little Miffingham in the county of Norfolk, by whom he had feven children, four of whom furvived him.

Queen Caroline took great pleafure in the doctor's converfation and friendfhip, feldom miffing a week in which fhe did not receive fome proof of the greatnefs of his genius, and the force of his underftanding.

As to the character of Dr. Clarke, he is reprefented as poffeffing one of the beft difpofitions in the world, remarkably humane and tender, free and eafy in his converfation, cheerful and even playful in his manner. Bifhop Hare fays of him, " He was a man who had all the good qualities that could meet together to recommend him. He was poffeffed of all the parts of learning that are valuable in a clergyman, in a degree that

few

few poffefs any fingle one. He has joined to a good fkill in the three learned languages, a great compafs of the beft philofophy and mathematics, as appears by his Latin works; and his Englifh ones are fuch a proof of his own piety, and of his knowledge in divinity, and have done fo much fervice to religion, as would make any other man, that was not under a fufpicion of herefy, fecure of the friendfhip of all good churchmen, efpecially the clergy. And to all this piety and learning was joined, a temper happy beyond expreffion; a fweet, eafy, modeft, obliging behaviour adorned all his actions; and neither paffion, vanity, infolence, or oftentation appeared either in what he faid or wrote. This is the learning, this the temper of the man, whofe ftudy of the Scriptures has betrayed him into a fufpicion of fome heretical opinions. Bifhop Hoadley too having remarked how great the doctor was in all branches of learning, adds, If in any one of thefe he had excelled only fo much as he did in all, he would have been juftly entitled to the character of a great man: but there is fomething fo very extraordinary, that the fame perfon fhould excel not only in thofe parts of knowledge which require the ftrongeft judgment, but in thofe which require the greateft memory too. So that, in a very high degree, divinity and mathematics, experimental philofophy and claffical learning, metaphyfics and critical fkill, were united in Dr. Clarke.—Much more may be feen, faid in his praife by bifhop Hoadley, Dr. Sykes, and Mr. Whifton, in their Memoirs of his life.

CLEF, or CLIFF, in Mufic, a mark at the beginning of the lines of a fong, which fhews the tone or key in which the piece is to begin. Or, it is a letter marked on any line, which explains and gives the name to all the reft.

Anciently, every line had a letter marked for a Clef; but now a letter on one line fuffices; fince by this all the reft are known; reckoning up and down, in the order of the letters.

It is called the Clef, or key, becaufe that by it are known the names of all the other lines and fpaces; and confequently the quantity of every degree, or interval. But becaufe every note in the octave is called a key, though in another fenfe, this letter marked is called peculiarly the *figned Clef*; becaufe, being written on any line, it not only figns and marks that one, but it alfo explains all the reft. By Clef, therefore, for diftinction fake, is meant that letter, figned on a line, which explains the reft; and by key, the principal note of a fong, in which the melody clofes.

There are three of thefe figned Clefs, *c, f, g*. The Clef of the higheft part in a fong, called *treble*, or *alt*, is *g*, fet on the fecond line counting upwards. The Clef of the bafs, or the loweft part, is *f* on the 4th line upwards. For all the other mean parts, the Clef is *c*, fometimes on one, fometimes on another line. Indeed, fome that are really mean parts, are fometimes fet with the *g* clef. It muft however be obferved, that the ordinary fignatures of Clefs bear little refemblance to thofe letters. Mr. Malcolm thinks it would be well if the letters themfelves were ufed. Kepler takes great pains to fhew, that the common fignatures are only corruptions of the letters they reprefent. The figures of thefe now are as follow:

Character of the treble Clef.

The mean Clef.

The bafs Clef.

The Clefs are always taken fifths to one another. So the Clef *f* being loweft, *c* is a fifth above it, and *g* a fifth above *c*.

When the place of the Clef is changed, which is not frequent in the mean Clef, it is with a defign to make the fyftem comprehend as many notes of the fong as poffible, and fo to have the fewer notes above or below it. So that, if there be many lines above the Clef, and few below it, this purpofe is anfwered by placing the Clef in the firft or fecond line: but if there be many notes below the Clef, it is placed lower in the fyftem. In effect, according to the relation of the other notes to the Clef note, the particular fyftem is taken differently in the fcale, the Clef line making one in all the variety.

But ftill, in whatever line of the particular fyftem any Clef is found, it muft be underftood to belong to the fame of the general fyftem, and to be the fame individual note or found in the fcale. By this conftant relation of Clefs, we learn how to compare the feveral particular fyftems of the feveral parts, and to know how they communicate in the fcale, that is, which lines are unifon, and which not: for it is not to be fuppofed, that each part has certain bounds, within which another muft never come. Some notes of the treble, for example, may be lower than fome of the mean parts, or even of the bafs. Therefore to put together into one fyftem all the parts of a compofition written feparately, the notes of each part muft be placed at the fame diftances above and below the proper Clef, as they ftand in the feparate fyftem: and becaufe all the notes that are confonant, or heard together, muft ftand directly over each other, that the notes belonging to each part may be diftinctly known, they may be made with fuch differences as fhall not confound, or alter their fignifications with refpect to time, but only fhew that they belong to this or that part. Thus we fhall fee how the parts change and pafs through one another; and which, in every note, is higheft, loweft, or unifon.

It muft here be obferved, that for the performance of any fingle piece, the Clef only ferves for explaining the intervals in the lines and fpaces: fo that it need not be regarded what part of any greater fyftem it is; but the firft note may be taken as high or low as we pleafe. For as the proper ufe of the fcale is not to limit the abfolute degree of tone; fo the proper ufe of the *figned* Clef is not to limit the pitch, at which the firft note of any part is to be taken; but to determine the tune of the reft, with refpect to the firft; and confidering all the parts together, to determine the relation of their feveral notes by the relations of their Clefs in the fcale: thus, their pitch of tune being determined in a certain note of one part, the other notes of that part are determined by the conftant relations of the
letters

letters of the scale, and the notes of the other parts by the relations of their Clefs.

In effect, for performing any single part, the Clef note may be taken in an octave, that is, at any note of the same name; provided we do not go too high, or too low, for finding the rest of the notes of a song. But in a concert of several parts, all the Clefs must be taken, not only in the relations, but also in the places of the system abovementioned; that every part may be comprehended in it.

The natural and artificial note expressed by the same letter, as *c* and *c*✗, are both set on the same line or space. When there is no character of flat or sharp, at the beginning with the Clef, all the notes are natural: and if in any particular place the artificial note be required, it is denoted by the sign of a flat or sharp, set on the line a space before that note.

If a sharp or flat be set at the beginning in any line or space with the Clef, all the notes on that line or space are artificial ones; that is, are to be taken a semitone higher or lower than they would be without such sign. And the same affects all their octaves above and below, though they be not marked so. In the course of the song, if the natural note be sometimes required, it is signified by the character ♮.

COMPASS. Pa. 314, col. 1, after l. 6 from the bottom, add, See also a new one in the Supplement to Cavallo's Treatise on Magnetism.

CONDORCET (JOHN-ANTHONY NICHOLAS *de* CARITAT, *Marquis of*), member of the Institute of Bologna, of the Academies of Turin, Berlin, Stockholm, Upsal, Philadelphia, Petersbourg, Padua, &c, and secretary of the Paris Academy of Sciences, was born at Ribemont in Picardie, the 17th of September 1743. His early attachment to the sciences, and progress in them, soon rendered him a conspicuous character in the commonwealth of letters. He was received as a member of the Academy of Sciences at 25 years of age, namely, in March 1769, as Adjunct-Mecanician; afterwards, he became Associate in 1770, Adjunct-Secretary in 1773, and sole Secretary soon after, which he enjoyed till his death, or till the dissolution of the Academy by the Convention.

Condorcet soon became an author, and that in the most sublime branches of science. He published his *Essais d'Analyse* in several parts; the first part in 1765 (at 22 years of age); the second, in 1767; and the third, in 1768. These works are chiefly on the Integral Calculus, or the finding of Fluents, and make one volume in 4to.

He published the Eloges of the Academicians or members of the Academy of Sciences, from the year 1666 till 1700, in several volumes. He wrote also similar Eloges of the Academicians who died during the time that he discharged the important office of Secretary to the Academy; as well as the very useful histories of the different branches of science commonly prefixed to the volumes of Memoirs, till the volume for the year 1783, when it is to be lamented that so useful a part of the plan of the Academy was discontinued.

His other memoirs contained in the volumes of the Academy, are the following.

1. Tract on the Integral Calculus; 1765.
2. On the problem of Three Bodies; 1767.
3. Observations on the Integral Calculus; 1767.
4. On the Nature of Infinite Series; on the Extent of the Solutions which they give; and on a new method of Approximation for Differential Equations of all Orders; 1769.
5. On Equations for Finite Differences; 1770.
6. On Equations for Partial Differences; 1770.
7. On Differential Equations; 1770.
8. Additions to the foregoing Tracts; 1770.
9. On the Determination of Arbitrary Functions which enter the Integrals of Equations to Partial Differences; 1771.
10. Reflexions on the Methods of Approximation hitherto known for Differential Equations; 1771.
11. Theorem concerning Quadratures; 1771.
12. Inquiry concerning the Integral Calculus; 1772.
13. On the Calculation of Probabilities, part 1 and 2; 1781.
14. Continuation of the same, part 3; 1782.
15. Ditto, part 4; 1783.
16. Ditto, part 5; 1784.

Condorcet had the character of being a very worthy honest man, and a respectable author, though perhaps not a first-rate one, and produced an excellent set of Eloges of the deceased Academicians, during the time of his secretaryship. A late French political writer has observed of him, that he laboured to succeed to the literary throne of d'Alembert, but that he cannot be ranked among illustrious authors; that his works have neither animation nor depth, and that his style is dull and dry; that some bold attacks on religion and declamations against despotism have chiefly given a degree of fame to his writings.

On the breaking out of the troubles in France, Condorcet took a decided part on the side of the people, and steadily maintained the cause he had espoused amid all the shocks and intrigues of contending parties; till, under the tyranny of Robespierre, he was driven from the convention, being one of those members proscribed on the 31st of May 1793, and he died about April 1794. The manner of his death is thus described by the public prints of that time. He was obliged to conceal himself with the greatest care for the purpose of avoiding the fate of Brissot and the other deputies who where executed. He did not, however, attempt to quit Paris, but concealed himself in the house of a female, who, though she knew him only by name, did not hesitate to risk her own life for the purpose of preserving that of Condorcet. In her house he remained till the month of April 1794, when it was rumoured that a domiciliary visit was to be made, which obliged him to leave Paris. Although he had neither passport nor civic card, he escaped through the Barrier, and arrived at the Plain of Mont rouge, where he expected to find an asylum in the country-house of an intimate friend. Unfortunately this friend had set out for Paris, where he was to remain for three days.—During all this period, Condorcet wandered about the fields and in the woods,

not

not daring to enter an inn on account of not having a civic card. Half dead with hunger, fatigue, and fear, and fcarcely able to walk on account of a wound in his foot, he paffed the night under a tree.

At length his friend returned, and received him with great cordiality ; but as it was deemed imprudent that he fhould enter the houfe in the day-time, he returned to the woods till night. In this fhort interval between morning and night his caution forfook him, and he refolved to go to an inn for the purpofe of procuring food. He went to an inn at Clamars, and ordered an omlette. His torn clothes, his dirty cap, his meagre and pale countenance, and the greedinefs with which he devoured the omlette, fixed the attention of the perfons in the inn, among whom was a member of the Revolutionary Committee of Clamars. This man conceiving him to be Condorcet, who had effected his efcape from the Bicetre, afked him whence he came, whither he was going, and whether he had a paffport ? The confufed manner in which he replied to thefe queftions, induced the member to order him to be conveyed before the Committee, who, after an examination, fent him to the diftrict of Boury la Reine. He was there interrogated again, and the unfatisfactory anfwers which he gave, determined the directors of the diftrict to fend him to prifon on the fucceeding day.—During

the night he was confined in a kind of dungeon. On the next morning, when his keeper entered with fome bread and water for him, he found him ftretched on the ground without any figns of life.

On infpecting the body, the immediate caufe of his death could not be difcovered, but it was conjectured that he had poifoned himfelf. Condorcet indeed always carried a dofe of poifon in his pocket, and he faid to the friend who was to have received him into his houfe, that he had been often tempted to make ufe of it, but that the idea of a wife and daughter, whom he loved tenderly, reftrained him. During the time that he was concealed at Paris, he wrote a hiftory of the Progrefs of the Human Mind, in two volumes.

CUBICS. The method of refolving all the cafes of Cubic equations by the tables of fines, tangents and fecants, are thus given by Dr. Mafkelyne, p. 57, Taylor's Logarithms.

" The following method is adapted to a Cubic equation, wanting the fecond term ; therefore, if the equation has the fecond term, it muft be firft taken away in the ufual manner. There are four forms of Cubic equations wanting the fecond term, whofe roots, according to known rules equivalent to Cardan's, are as follow :

$$\text{1ft. } x^3 + px - q = 0 \quad \cdots \quad x = \sqrt[3]{\frac{q}{2} + \sqrt{\frac{q^2}{4} + \frac{p^3}{27}}} - \sqrt[3]{-\frac{q}{2} + \sqrt{\frac{q^2}{4} + \frac{p^3}{27}}}$$

$$\text{2d. } x^3 + px + p = 0 \quad \cdots \quad x = \sqrt[3]{-\frac{q}{2} + \sqrt{\frac{q^2}{4} + \frac{p^3}{27}}} - \sqrt[3]{\frac{q}{2} + \sqrt{\frac{q^2}{4} + \frac{p^3}{27}}}$$

$$\text{3d. } x^3 - px - q = 0 \quad \cdots \quad x = \sqrt[3]{\frac{q}{2} + \sqrt{\frac{q^2}{4} - \frac{p^3}{27}}} + \sqrt[3]{\frac{q}{2} - \sqrt{\frac{q^2}{4} - \frac{p^3}{27}}}$$

$$\text{4th. } x^3 - px + q = 0 \quad \cdots \quad x = -\sqrt[3]{\frac{q}{2} - \sqrt{\frac{q^2}{4} - \frac{p^3}{27}}} - \sqrt[3]{\frac{q}{2} + \sqrt{\frac{q^2}{4} - \frac{p^3}{27}}}$$

The roots of the firft and fecond forms are negatives of each other ; and thofe of the third and fourth are alfo negatives of each other. The firft and fecond forms have only one root each. The third and fourth forms have alfo only one root each, when the quadratic furd $\sqrt{\frac{q^2}{4} - \frac{p^3}{27}}$ is poffible ; but have three roots each when that furd is impoffible.

The roots of all the four forms may, in all cafes, be eafily computed as follows :

Forms 1ft and 2d. Put $\left. \frac{q}{2} \times \frac{3}{p} \right|^{\frac{3}{2}} = $ tang. z ; and $\sqrt[3]{\tan. \overline{45° - \frac{1}{2}z}} = $ tan. u. Then $x = \pm \sqrt{\frac{4p}{3}} \times$ cot. $2u$; where the upper fign belongs to the firft form, and the lower fign to the fecond form.

Forms 3d and 4th. Put $\left. \frac{2}{q} \times \frac{p}{3} \right|^{\frac{3}{2}}$ if lefs than unity,

elfe its reciprocal $\left. \frac{q}{2} \times \frac{3}{4} \right|^{\frac{3}{2}} = $ cof. z. Then,

Cafe 1ft. $\left. \frac{2}{q} \times \frac{p}{3} \right|^{\frac{3}{2}} < $ unity. Put $\sqrt[3]{\tan. \overline{45° - \frac{1}{2}z}} = $ tan. u. Then $x = \pm \sqrt{\frac{4p}{3}} \times$ cofec. $2u$; where the upper fign belongs to the third form, and the lower fign to the fourth form.

Cafe 2d. $\left. \frac{2}{q} \times \frac{p}{3} \right|^{\frac{3}{2}} > $ unity. Then x has three values in each form, viz, $x = \pm \sqrt{\frac{4p}{3}} \times$ cof. $\frac{z}{3} = \mp \sqrt{\frac{4p}{3}} \times$ cof. $\overline{60° - \frac{z}{3}} = \mp \sqrt{\frac{4p}{3}} \times$ cof. $\overline{60° + \frac{z}{3}}$; where the upper figns belong to the third form, and the lower figns to the fourth form.

By

Plate XXXIII.

Plan & Elevation of a DOME, constructed without Centring.

Fig. 1.

Fig. 2.

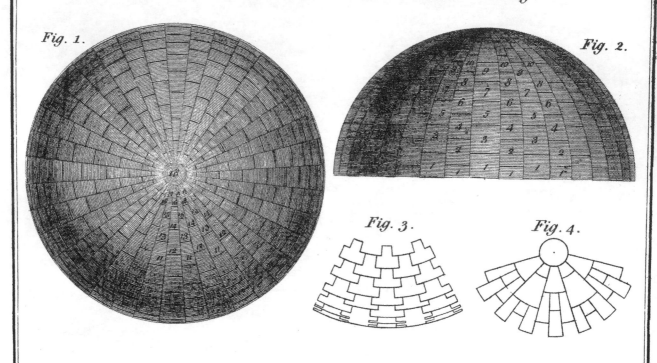

Fig. 3.

Fig. 4.

Jones's New Pocket MICROSCOPE

Fig. 5.

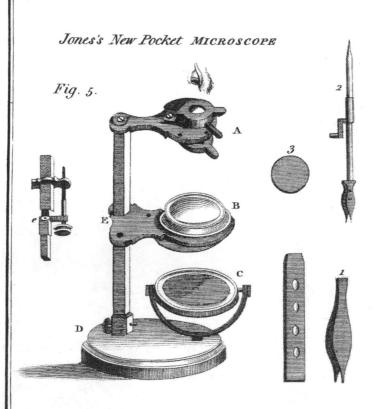

By Logarithms.

Forms 1*st and* 2*d.* Log. $\frac{q}{2}$ + 10 — $\frac{3}{2}$ × log. $\frac{p}{3}$ =

log. tan. z, and $\dfrac{\text{log. tan. } \overline{45° - \frac{1}{2}z} + 20}{3}$ = log. tan. u.

Then log. $x = \frac{1}{2}$ log. $\frac{4p}{3}$ + log. cot. $2u$ — 10; and x will be affirmative in the first form, and negative in the second form.

Forms 3*d and* 4*th.* $\frac{3}{2}$ × log. $\frac{p}{3}$ + 10 — log. $\frac{q}{2}$ being less than 10 (which is case first) or log. $\frac{q}{2}$ + 10 — $\frac{3}{2}$ × log. $\frac{p}{3}$ being less than 10 (which is case 2d) = log. cos. z.

Case 1*st.* $\dfrac{\text{Log. tan. } \overline{45° - \frac{1}{2}z} + 20}{3}$ = log. tan. u.

Then log. $x = \frac{1}{2}$ log. $\frac{4p}{3}$ + 10 — log. sin. $2u$; and x will be affirmative in the third form, and negative in the 4th form.

Case 2*d.* Here x has three values.

1st. Log. ± $x = \frac{1}{2}$ log. $\frac{4p}{3}$ + log. cos. $\frac{z}{3}$ — 10.

2d. Log. ∓ $x = \frac{1}{2}$ log. $\frac{4p}{3}$ + log. cos. $\overline{60° - \frac{z}{3}}$,

3d. Log. ∓ $x = \frac{1}{2}$ log. $\frac{4p}{3}$ + log. cos. $\overline{60° + \frac{z}{3}}$. 10;

where the upper signs belong to the third form, and the lower signs to the fourth form; that is, the first value of x in the third form is positive, and its second and third values negative; and the first value of x in the fourth form is negative, and its second and third values affirmative."

See also IRREDUCIBLE *Case.*

CURVE. Pa. 350, col. 2, l. 35, for $dx + x^2$ r. $\overline{dx + x^2}$.

D.

DIPPING *Needle.* Pa. 383, col. 2, after line 38 *add*, See a new Dipping-needle by Dr. Lorimer, in the Philos. Transf. 1775, also in the Supplement to Cavallo's Treatise on Magnetism.

DOME. In plate 33 is represented the plan and elevation of a Dome constructed without centring, by Mr. S. Bunce; viz, Fig. 1 the plan, and Fig. 2 the elevation. The first course consists of the stones marked 1, 1, 1, &c, of different sizes, the large ones exactly twice the height of the small ones, placed alternately, and forming intervals to receive the stones marked 2, 2, 2. The other courses are continued in the same manner, according to the order of the figures to the top.

It is evident, from the converging or wedgelike form of the intervals, that the stones they receive can only be inserted from the outside, and cannot fall through: therefore the whole Dome may be built without centring or temporary support. To break the upright joints, the stones may be cut of the form marked in Fig. 3; and those marked 16, 17, &c, near the keystones, may be enlarged as at Fig. 4.

Pa. 399, col. 2, line 10 from the bottom, *for* DYMANICS *read* DYNAMICS.

E.

EUTOCIUS, a respectable Greek mathematician, lived at Ascalon in Palestine about the year of Christ 550. He was one of the most considerable mathematicians that flourished about the decline of the sciences among the Greeks, and had for his preceptor Isidorus the principal architect of the church of St. Sophia at Constantinople. He is chiefly known however by his commentaries on the works of the two ancient authors, Archimedes and Apollonius. Those two commentaries are both excellent compositions, to which we owe many useful circumstances in the history of the mathematics.

His commentaries on Apollonius are published in Halley's edition of the works of that author; and those on Archimedes, first in the Basle edition, in Greek and Latin, in 1543, and since in some others, as the late Oxford edition. Of these commentaries, those rank the highest, which illustrate Archimedes's work on the Sphere and Cylinder; in one of which we have a recital of the various methods practised by the ancients in the solution of the Delian problem, or that of doubling the cube. The others are of less value; though it cannot but be regretted that Eutocius did not pursue his plan of commenting on all the works of Archimedes, with the same attention and diligence which he employed in his remarks on the sphere and cylinder.

Pa. 507, line 5 from the bottom, *for* $3 + 1$ *read* $3 + \frac{1}{7}$.

Pa. 551, line 22 from the bottom, *for* $\frac{7}{50}$ *read* $\frac{7}{30}$.

G.

GROIN, with Builders, is the angle made by the intersection of two arches. It is of two kinds, regular and irregular; viz, Regular when both the arches have the same diameter, but an Irregular Groin when one arch is a semicircle and the other a semiellipsis. Groins are chiefly used in forming arched roofs, where one hollow arched vault intersects with another; as in the roofs of most churches, and some cellars in large houses.

I.

IMPOSSIBLE *Binomial.* See BINOMIAL.

IRREDUCIBLE *Case*, in Algebra. Mr. Bonnycastle has communicated the following additional observations on this case, and, an improved solution by a table of sines. The

IRREDUCIBLE *Case, in Algebra,* is a cubic equation of the form $x^3 - ax = \pm b$, having $\frac{1}{27}a^3$ greater than $\frac{1}{4}b^2$, or $4a^3$ greater than $27b^2$; in which case, it is well known, that the solution cannot be generally obtained, either by Cardan's rule, or any other which has yet been devised.

One of the most convenient methods of determining the roots of equations of this kind, is by means of a Table of Natural Sines, &c, for which purpose the following formulæ will be found extremely commodious, the arc, in each case, being always less than a quadrant, and therefore attended with no ambiguity.

If

If the equation be $x^3 - ax = b$; let A be put $=$ arc whose cof. is $\frac{3b}{2a} \sqrt{\frac{3}{a}}$ to rad. 1, then the three roots, or values of x, will be as follows:

$$x = \quad 2\sqrt{\frac{a}{3}} \times \text{cofine } \frac{A}{3}$$

$$x = -2\sqrt{\frac{a}{3}} \times \text{fine } \frac{90° + A}{3}$$

$$x = -2\sqrt{\frac{a}{3}} + \text{fine } \frac{90° - A}{3}$$

And, if the equation be $x^3 - ax = -b$; let A be put $=$ arc whose fine is $\frac{3b}{2a}\sqrt{\frac{3}{a}}$ to rad. 1; then the three roots, or values of x, will be as follows.

$$x = \quad 2\sqrt{\frac{a}{3}} \times \text{fine } \frac{A}{3}$$

$$x = \quad 2\sqrt{\frac{a}{3}} \times \text{cof. } \frac{90° + A}{3}$$

$$x = -2\sqrt{\frac{a}{3}} \times \text{cof. } \frac{90° + A}{3}$$

Ex. 1. Let $x^3 - 3x = 1$, to find the 3 roots of the equation.

Here $\frac{3b}{2a}\sqrt{\frac{3}{a}} = \frac{3}{6}\sqrt{\frac{3}{3}} = \frac{1}{2} = 5 = \text{cof. } 60° = A.$

Hence
$$\begin{cases} x = \quad 2\text{cof. } \frac{60°}{3} = \quad 2\text{cof. } 20° = \quad 1\cdot8793852 \\ x = -2\text{fine } \frac{150°}{3} = -2\text{fine } 50° = -1\cdot5320888 \\ x = -2\text{fine } \frac{30°}{3} = -2\text{fine } 10° = -\cdot3472964 \end{cases}$$

Ex. 2d. Let $x^3 - 3x = -1$, to find the 3 roots of the equation.

Here $\frac{3b}{2a}\sqrt{\frac{3}{a}} = \frac{3}{6}\sqrt{\frac{3}{3}} = \frac{1}{2} = \cdot5 = \text{fine } 30° = A,$

Hence
$$\begin{cases} x = \quad 2\text{fine } \frac{30°}{3} = 2\text{fine } 10° = \quad \cdot3472964 \\ x = \quad 2\text{cof. } \frac{120°}{3} = 2\text{cof. } 40° = \quad 1\cdot5320888 \\ x = -2\text{cof. } \frac{60°}{3} = -2\text{cof. } 20° = -1\cdot8793852 \end{cases}$$

The investigation of this method is as follows:

It is shewn, by the writers on Trigonometry, that if c be the cosine of any arc to rad. 1, $4c^3 - 3c$ will be the cosine of 3 times that arc; and consequently c is the cosine of $\frac{1}{3}$ of the arc whose cosine is $4c^3 - 3c$, or any other equal quantity.

In order, therefore, to reduce the equation $x^3 - ax = b$ to this form, let $x = \frac{y}{z}$; then

$\frac{y^3}{z^3} - a \times \frac{y}{z} = b$, or $y^3 - az^2y = bz^3$, or $4y^3 - 4az^2y = 4bz^3$; whence if $4az^2$ be put $= 3$, we shall have

$z = \sqrt{\frac{3}{4a}} = \frac{1}{2}\sqrt{\frac{3}{a}}$, and consequently $4y^3 - 3y = \frac{3b}{2a}\sqrt{\frac{3}{a}}.$

From which last equation, it appears that $y = $ cof. $\frac{1}{3}$arc whose cof. is $\frac{3b}{2a}\sqrt{\frac{3}{a}}$; and therefore $x = \frac{y}{z} = \frac{y}{\frac{1}{2}\sqrt{\frac{3}{a}}}$

$= 2\sqrt{\frac{a}{3}} \times y = 2\sqrt{\frac{a}{3}} \times$ (cof. $\frac{1}{3}$arc whose cof. is $\frac{3b}{2a}\sqrt{\frac{3}{a}}$

or, if A be put $=$ arc whose cof. is $\frac{3b}{2a}\sqrt{\frac{3}{a}}$, x is

$= 2\sqrt{\frac{a}{3}} \times$ cof. $\frac{A}{3}.$

But the arc of which $\frac{3b}{2a}\sqrt{\frac{3}{a}}$ is the cofine, is either A, A $+ 360°$ or A $+ 720°$; whence $x = 2\sqrt{\frac{a}{3}} \times$ cof. $\frac{A}{3}$

or $2\sqrt{\frac{a}{3}} \times$ cof. $\frac{A + 360°}{3}$, or $2\sqrt{\frac{a}{3}} \times$ cof. $\frac{A + 720°}{3}$; the two latter of which being converted into fines, will give the same formulæ as in the rule.

In like manner, if s be the fine of any arc to rad. 1, $3s - 4s^3$ is well known to be the fine of 3 times that arc; and consequently s is the fine of $\frac{1}{3}$ of the arc whose fine is $3s - 4s^3$. Whence, to reduce the equation $x^3 - ax = -b$, to this form, let $x = \frac{y}{z}$, as before; then $\frac{y^3}{z^3} - a \times \frac{y}{z} = -b$, or $y^3 - az^2y = -bz^3$, or $az^2y - y^3 = bz^3$, or $4az^2y - 4y^3 = 4bz^3$; where, if $4az^2$ be put $= 3$, we shall have $z = \frac{1}{2}\sqrt{\frac{3}{a}}$, and consequently $3y - 4y^3 = \frac{3b}{2a}\sqrt{\frac{3}{a}}.$

From which last equation it appears that $y = $ fine $\frac{1}{3}$arc whose fine is $\frac{3b}{2a}\sqrt{\frac{3}{a}}$, and therefore $x = \frac{y}{z} = 2\sqrt{\frac{a}{3}} \times$ (fine $\frac{1}{3}$arc whose fine is $\frac{3b}{2a}\sqrt{\frac{3}{a}}$), which is the same as the rule, the other two roots being found as in the former calc.

L.

LOCK, for Canals, in Inland Navigations. See CANAL.

LOGARITHMS. Mr. Bonnycastle has communicated the following new method of making these useful numbers:

LOGARITHMS. The series now chiefly used in the computation of Logarithms were originally derived from the hyperbola, by means of which, and the logistic curve, the nature and properties of these numbers are clearly and elegantly explained.

The doctrine, however, being purely arithmetical, this mode of demonstrating it, by the intervention of certain curves, was considered, by Dr. Halley, as not conformable to the nature of the subject.

He

He has, accordingly, investigated the same series from the abstract principles of numbers; but his method, which is a kind of disguised fluxions, is, in many places, so extremely abstruse and obscure, that few have been able to comprehend his reasoning.

An easy and perspicuous demonstration, of this kind, was therefore still wanting; which may be obtained from the pure principles of Algebra, independently of the doctrine of Curves, as follows:

The Logarithm of any number, is the index of that power of some other number, which is equal to the given number.

Thus, if $r^x = a$, the logarithm of a is x, which may be either positive or negative, and r any number whatever, according to the different systems of Logarithms.

When $a = 1$, it is plain that x must be $= 0$, whatever be the value of r; and consequently the Logarithm of 1 is always 0 in every system.

If $x = 1$, it is also plain that a must be $= r$; and therefore r is always the number in every system, whose Logarithm in that system is 1.

To find the Logarithm of any number, in any system, it is only necessary, from the equation $r^x = a$, to find the value of x in terms of r and a.

This may be strictly effected, by means of a new property of the binomial theorem of Newton; which is given under its proper article in this Appendix. The general Logarithmic equation being $r^x = a$, let

$$a = 1 + p, \text{ and } \frac{1}{x} = z; \text{ then } r = a^{\frac{1}{x}} = \overline{1+p}^{\frac{1}{x}} = \overline{1+p}^{z} =$$

$$1 + \left(p - \frac{p^2}{2} + \frac{p^3}{3} - \frac{p^4}{4} \&c\right)z + \frac{1}{2}\left(p - \frac{p^2}{2} + \frac{p^3}{3} - \frac{p^4}{4} \&c\right)^2 z^2$$

$$+ \frac{1}{2\cdot3}\left(p - \frac{p^2}{2} + \frac{p^3}{3} - \frac{p^4}{4} \&c\right)^3 z^3 + \frac{1}{2\cdot3\cdot4}\left(p - \frac{p^2}{2} + \frac{p^3}{3} - \frac{p^4}{4}\&c\right)^4 z^4$$

$$+ \frac{1}{2\cdot3\cdot4\cdot5}\left(p - \frac{p^2}{2} + \frac{p^3}{3} - \frac{p^4}{4} \&c\right)^5 z^5, \&c.$$ See *Binomial* THEOREM, Appendix.

And if $p - \frac{p^2}{2} + \frac{p^3}{3} - \frac{p^4}{4} + \frac{p^5}{5} \&c$ be put $= s$, we shall have

$$1 + sz + \tfrac{1}{2}s^2 z^2 + \frac{1}{2\cdot3}s^3 z^3 + \frac{1}{2\cdot3\cdot4}s^4 z^4 + \frac{1}{2\cdot3\cdot4\cdot5}s^5 z^5 \&c = r,$$

or $sz + \tfrac{1}{2}s^2 z^2 + \frac{1}{2\cdot3}s^3 z^3 + \frac{1}{2\cdot3\cdot4}s^4 z^4 + \frac{1}{2\cdot5\cdot4\cdot5}s^5 z^5 \&c = r - 1,$

which let be put $= q$; then, by reverting the series z or $\frac{1}{x}$ will be found

$$= \frac{q - \tfrac{1}{2}q^2 + \tfrac{1}{3}q^3 - \tfrac{1}{4}q^4 + \tfrac{1}{5}q^5 \&c}{s} = \frac{q - \tfrac{1}{2}q^2 + \tfrac{1}{3}q^3 - \tfrac{1}{4}q^4 + \tfrac{1}{5}q^5 \&c}{p - \tfrac{1}{2}p^2 + \tfrac{1}{3}p^3 - \tfrac{1}{4}p^4 + \tfrac{1}{5}p^5 \&c}$$

and consequently $x = \dfrac{p - \tfrac{1}{2}p^2 + \tfrac{1}{3}p^3 - \tfrac{1}{4}p^4 + \tfrac{1}{5}p^5 \&c}{q - \tfrac{1}{2}q^2 + \tfrac{1}{3}q^3 - \tfrac{1}{4}q^4 + \tfrac{1}{5}q^5 \&c}.$

The Logarithm of a, or $1 + p$, is therefore

$$= \frac{p - \tfrac{1}{2}p^2 + \tfrac{1}{3}p^3 - \tfrac{1}{4}p^4 + \tfrac{1}{5}p^5 \&c}{q - \tfrac{1}{2}q^2 + \tfrac{1}{3}q^3 - \tfrac{1}{4}q^4 + \tfrac{1}{5}q^5 \&c};$$ or, since $p = a - 1$, and $q = r - 1$, the Logarithm of a is

$$= \frac{(a-1) - \tfrac{1}{2}(a-1)^2 + \tfrac{1}{3}(a-1)^3 - \tfrac{1}{4}(a-1)^4 + \tfrac{1}{5}(a-1)^5 \&c}{(r-1) - \tfrac{1}{2}(r-1)^2 + \tfrac{1}{3}(r-1)^3 - \tfrac{1}{4}(r-1)^4 + \tfrac{1}{5}(r-1)^5 \&c};$$

Which is a general expression for the Logarithm of any number, in any system of Logarithms, the radix r being taken of any value, greater or less than 1.

But as r in every system, is a constant quantity, being always the number whose Logarithm in the system to which it belongs is 1, the above expression may be simplified, either by assuming $r =$ to some particular number, and from thence finding the value of the series constituting the denominator; or by assuming this whole series $=$ to some particular number, and from thence finding the value which must be given to the radix r.

By the latter of these methods, the denominator may be made to vanish, by assuming the value of the series of which it consists $= 1$, in which case, the Logarithm

of $1 + p$ becomes $= p - \dfrac{p^2}{2} + \dfrac{p^3}{3} - \dfrac{p^4}{4} + \dfrac{p^5}{5}$ &c, or the Logarithm of

$$a = (a-1) - \tfrac{1}{2}(a-1)^2 + \tfrac{1}{3}(a-1)^3 - \tfrac{1}{4}(a-1)^4 + \tfrac{1}{5}(a-1)^5 \&c,$$

and r, by reversion of series is found $= 2.718\,818$ &c.

The system arising from this mode of determining the value of the radix r, is that which furnishes what have been usually called hyperbolic Logarithms; and appears to be the simplest form the general expression admits of.

If, on the contrary, the radix r be assumed $=$ to some particular number, as for instance 10, the value of the series $q - \tfrac{1}{2}q^2 + \tfrac{1}{3}q^3 - \tfrac{1}{4}q^4 + \tfrac{1}{5}q^5$ &c, or its equal $(r-1) - \tfrac{1}{2}(r-1)^2 + \tfrac{1}{3}(r-1)^3 - \tfrac{1}{4}(r-1)^4 + \tfrac{1}{5}(r-1)^5$&c will become $= 2.30258509$ &c, and the

$$\text{Log. of } 1 + p = \frac{1}{2.30258509} \times \left(p - \tfrac{1}{2}p^2 + \tfrac{1}{3}p^3 - \tfrac{1}{4}p^4 + \tfrac{1}{5}p^5 \&c\right)$$

or the Log. of a

$$= \frac{1}{2.30258509} \times (a-1) - \tfrac{1}{2}(a-1)^2 + \tfrac{1}{3}(a-1)^3 - \tfrac{1}{4}(a-1)^4 + \tfrac{1}{5}(a-1)^5$$

&c, which gives the system that furnishes Briggs's or the common Logarithms.

And, in like manner, by assuming any particular value for r, and thence determining the value of the series $q - \tfrac{1}{2}q^2 + \tfrac{1}{3}q^3 - \tfrac{1}{4}q^4 + \tfrac{1}{5}q^5$ &c, or its equal $(r-1) \tfrac{1}{2}(r-1)^2 + \tfrac{1}{3}(r-1)^3 - \tfrac{1}{4}(r-1)^4 + \tfrac{1}{5}(r-1)^5$ &c; or by assuming the same series of some particular value, and thence determining the value of r, any system of Logarithms may be derived.

The series $q - \tfrac{1}{2}q^2 + \tfrac{1}{3}q^3 - \tfrac{1}{4}q^4 + \tfrac{1}{5}q^5$ &c, or its equal $(r-1) \tfrac{1}{2}(r-1)^2 + \tfrac{1}{3}(r-1)^3 - \tfrac{1}{4}(r-1)^4 + \tfrac{1}{5}(r-1)^5$ &c, which forms the denominator of the above compound expression, exhibiting the Logarithms of numbers according to any system, is what was first called, by Cotes, the Modulus of the system, being always a constant quantity, depending only on the assumed value of r.

And, as the form of this series is exactly the same as that which constitutes the numerator, and which has been shewn to be the hyperbolic Logarithm of a, it follows that the Modulus of any system of Logarithms is equal to the hyperbolic Logarithm of the radix of that

system,

fyftem, or of the number whofe proper Logarithm in the fyftem to which it belongs is 1.

The form of the feries here obtained for the hyperbolic Logarithm of a, is the fame as that which was firft difcovered by Mercator; and if the feries of Wallis be required, it may be inveftigated in a fimilar manner as follows:

The general Logarithmic equation being $r^x = a$, as before, let $a = \frac{1}{1-p}$ and $z = \frac{1}{x}$; then $r = a^{\frac{1}{x}} =$

$$\overline{\frac{1}{1-p}}\Big|^{z} = \overline{\frac{1}{1-p}}\,z, \text{ and } \frac{1}{r} = \overline{1-p}\Big|^{z} = 1 - (p + \tfrac{p^2}{2} + \tfrac{p^3}{3} + \tfrac{p^4}{4} \&c)$$

$$+ \tfrac{1}{2}(p + \tfrac{p^2}{2} + \tfrac{p^3}{3} + \tfrac{p^4}{4} \&c)^2 z^2 - \tfrac{1}{2 \cdot 3}(p + \tfrac{p^2}{2} + \tfrac{p^3}{3} + \tfrac{p^4}{4} \&c)^3 z^3$$

$$+ \tfrac{1}{2 \cdot 3 \cdot 4}(p + \tfrac{p^2}{2} + \tfrac{p^3}{3} + \tfrac{p^4}{4} \&c)^4 z^4 - \tfrac{1}{2 \cdot 3 \cdot 4 \cdot 5}(p + \tfrac{p^2}{2} + \tfrac{p^3}{3} + \tfrac{p^4}{4} \&c)^5 z^5$$

&c.

And if $p + \tfrac{p^2}{2} + \tfrac{p^3}{3} + \tfrac{p^4}{4} + \tfrac{p^5}{5}$ &c be put $= s$, we fhall have $1 - sz + \tfrac{1}{2}s^2 z^2 - \tfrac{1}{2 \cdot 3}s^3 z^3 + \tfrac{1}{2 \cdot 3 \cdot 4}s^4 z^4$ &c $= \tfrac{1}{r}$,

or $sz - \tfrac{1}{2}s^2 z^2 + \tfrac{1}{2 \cdot 3}s^3 z^3 - \tfrac{1}{2 \cdot 3 \cdot 4}s^4 z^4$ &c $= 1 - \tfrac{1}{r}$, which let be put $= q$; then, by converfion of feries, z or $\frac{1}{x}$ will be found

$$= \frac{q + \tfrac{1}{2}q^2 + \tfrac{1}{3}q^3 + \tfrac{1}{4}q^4 + \tfrac{1}{5}q^5 \&c}{s} = \frac{q + \tfrac{1}{2}q^2 + \tfrac{1}{3}q^3 + \tfrac{1}{4}q^4 + \tfrac{1}{5}q^5}{p + \tfrac{1}{2}p^2 + \tfrac{1}{3}p^3 + \tfrac{1}{4}p^4 + \tfrac{1}{5}p^5}$$

and confequently $x = \frac{p + \tfrac{1}{2}p^2 + \tfrac{1}{3}p^3 + \tfrac{1}{4}p^4 + \tfrac{1}{5}p^5 \&c}{q + \tfrac{1}{2}q^2 + \tfrac{1}{3}q^3 + \tfrac{1}{4}q^4 + \tfrac{1}{5}q^5 \&c}$.

The Logarithm of a or $\frac{1}{1-p}$ is, therefore,

$$= \frac{p + \tfrac{1}{2}p^2 + \tfrac{1}{3}p^3 + \tfrac{1}{4}p^4 + \tfrac{1}{5}p^5 \&c}{q + \tfrac{1}{2}q^2 + \tfrac{1}{3}q^3 + \tfrac{1}{4}q^4 + \tfrac{1}{5}q^5 \&c};$$ or fince

$p = 1 - \tfrac{1}{a} = \tfrac{a-1}{a}$ and $q = 1 - \tfrac{1}{r} = \tfrac{r-1}{r}$, the Logarithm of a is $=$

$$\frac{\dfrac{a-1}{a} + \tfrac{1}{2}\left(\dfrac{a-1}{a}\right)^2 + \tfrac{1}{3}\left(\dfrac{a-1}{a}\right)^3 + \tfrac{1}{4}\left(\dfrac{a-1}{a}\right)^4 + \tfrac{1}{5}\left(\dfrac{a-1}{a}\right)^5 \&c}{\dfrac{r-1}{r} + \tfrac{1}{2}\left(\dfrac{r-1}{r}\right)^2 + \tfrac{1}{3}\left(\dfrac{r-1}{r}\right)^3 + \tfrac{1}{4}\left(\dfrac{r-1}{r}\right)^4 + \tfrac{1}{5}\left(\dfrac{r-1}{r}\right)^5 \&c}$$

Which is another general expreffion for the Logarithm of any number a, in any fyftem of Logarithms, that may be fimplified in the fame manner as the former, the denominator being ftill equal to the hyperbolic Logarithm of the radix r; or, which is the fame thing, to the Modulus of the fyftem.

For if the feries $q + \tfrac{1}{2}q^2 + \tfrac{1}{3}q^3 + \tfrac{1}{4}q^4 + \tfrac{1}{5}q^5$ &c, or its equal

$$\frac{r-1}{r} + \tfrac{1}{2}\left(\frac{r-1}{r}\right)^2 + \tfrac{1}{3}\left(\frac{r-1}{r}\right)^3 + \tfrac{1}{4}\left(\frac{r-1}{r}\right)^4 + \tfrac{1}{5}\left(\frac{r-1}{r}\right)^5 \&c,$$

be affumed $= 1$, the hyperbolic Logarithm of $\frac{1}{1-p}$

will be $= p + \tfrac{1}{2}p^2 + \tfrac{1}{3}p^3 + \tfrac{1}{4}p^4 + \tfrac{1}{5}p^5$ &c, or the hyperbolic Logarithm of a

$$= \frac{a-1}{a} + \tfrac{1}{2}\left(\frac{a-1}{a}\right)^2 + \tfrac{1}{3}\left(\frac{a-1}{a}\right)^3 + \tfrac{1}{4}\left(\frac{a-1}{a}\right)^4 + \tfrac{1}{5}\left(\frac{a-1}{a}\right)^5 \&c;$$

and r, by reverfion of feries will be found $= 2.7182818$, as before. And if, on the contrary, the radix r be affumed $= 10$, the value of the feries

$q + \tfrac{1}{2}q^2 + \tfrac{1}{3}q^3 + \tfrac{1}{4}q^4 + \tfrac{1}{5}q^5$ &c, or its equal

$$\frac{r-1}{r} + \tfrac{1}{2}\left(\frac{r-1}{r}\right)^2 + \tfrac{1}{3}\left(\frac{r-1}{r}\right)^3 + \tfrac{1}{4}\left(\frac{r-1}{r}\right) + \tfrac{1}{5}\left(\frac{r-1}{r}\right)^5 \&c,$$

will become $= 2.30258509$, as before; and the common Logarithm of

$$\frac{1}{1-p} = \frac{1}{2.30258509} \times (p + \tfrac{1}{2}p^2 + \tfrac{1}{3}p^3 + \tfrac{1}{4}p^4 + \tfrac{1}{5}p^5 \&c),$$

or the common Logarithm of $a = \dfrac{1}{2.30258509}$

$$\times \frac{a-1}{a} + \tfrac{1}{2}\left(\frac{a-1}{a}\right)^2 + \tfrac{1}{3}\left(\frac{a-1}{a}\right)^3 + \tfrac{1}{4}\left(\frac{a-1}{a}\right)^4 + \tfrac{1}{5}\left(\frac{a-1}{a}\right)^5 \&c.$$

Or the latter formula, for the Logarithm of $\frac{1}{1-p}$, or its equal a, may be more concifely derived from the firft, as follows:

The Logarithm of $1 + p$ has been fhewn to be $= \dfrac{p - \tfrac{1}{2}p^2 + \tfrac{1}{3}p^3 - \tfrac{1}{4}p^4 + \tfrac{1}{5}p^5 \&c}{q - \tfrac{1}{2}q^2 + \tfrac{1}{3}q^3 - \tfrac{1}{4}q^4 + \tfrac{1}{5}q^5 \&c}$, and if $-p$ be fubftituted in the place of $+p$, the logarithm of $1-p$ will become

$$= \frac{-p - \tfrac{1}{2}p^2 - \tfrac{1}{3}p^3 - \tfrac{1}{4}p^4 - \tfrac{1}{5}p^5 \&c}{q - \tfrac{1}{2}q^2 + \tfrac{1}{3}q^3 - \tfrac{1}{4}q^4 + \tfrac{1}{5}q^5 \&c},$$ whence the Logarithm of $\frac{1}{1-p} = $ Log. $1 -$ Log. $(1-p) = 0 -$

$$\left(\frac{p - \tfrac{1}{2}p^2 - \tfrac{1}{3}p^3 - \tfrac{1}{4}p^4 - \tfrac{1}{5}p^5 \&c}{q - \tfrac{1}{2}q^2 + \tfrac{1}{3}q^3 - \tfrac{1}{4}q^4 + \tfrac{1}{5}q^5 \&c}\right) = \frac{p + \tfrac{1}{2}p^2 + \tfrac{1}{3}p^3 + \tfrac{1}{4}p^4 + \tfrac{1}{5}p^5 \&c}{q - \tfrac{1}{2}q^2 + \tfrac{1}{3}q^3 - \tfrac{1}{4}q^4 + \tfrac{1}{5}q^5 \&c},$$

or Log. $a = \dfrac{\dfrac{a-1}{a} + \tfrac{1}{2}\left(\dfrac{a-1}{a}\right)^2 + \tfrac{1}{3}\left(\dfrac{a-1}{a}\right)^3 + \tfrac{1}{4}\left(\dfrac{a-1}{a}\right)^4 \&c}{(r-1) - \tfrac{1}{2}(r-1)^2 + \tfrac{1}{3}(r-1)^3 - \tfrac{1}{4}(r-1)^4 \&c};$

where the denominator is the fame as in the firft formula, q being here $= r-1$.

If the denominator, in either of thefe general formulæ, be put $= m$, the Logarithm of $1 + p$ will be denoted by $\frac{1}{m} \times (p - \tfrac{1}{2}p^2 + \tfrac{1}{3}p^3 - \tfrac{1}{4}p^4 + \tfrac{1}{5}p^5 \&c$, or the Logarithm of a by

$$\frac{1}{m} \times : (a-1) - \tfrac{1}{2}(a-1)^2 + \tfrac{1}{3}(a-1)^3 - \tfrac{1}{4}(a-1)^4 + \tfrac{1}{5}(a-1)^5 \&c.$$

And the Logarithm of $\frac{1}{1-p}$ will be denoted by

$$\frac{1}{m} \times (p + \tfrac{1}{2}p^2 + \tfrac{1}{3}p^3 + \tfrac{1}{4}p^4 + \tfrac{1}{5}p^5 \&c,$$

or the Logarithm of a by

$$\frac{1}{m} \times : \frac{a-1}{a} + \tfrac{1}{2}\left(\frac{a-1}{a}\right)^2 + \tfrac{1}{3}\left(\frac{a-1}{a}\right)^3 + \tfrac{1}{4}\left(\frac{a-1}{a}\right)^4 + \tfrac{1}{5}\left(\frac{a-1}{a}\right)^5 \&c.$$

And fince the fum of the Logarithms of any two numbers is equal to the Logarithm of their product, the Logarithm of $\frac{1+p}{1-p}$ will become

$$= \frac{2}{m}$$

$$= \frac{2}{m} \times (p + \tfrac{1}{3}p^3 + \tfrac{1}{5}p^5 + \tfrac{1}{7}p^7 \,\&c.),$$

or the Logarithm of a

$$= \frac{2}{m} \times : \frac{a-1}{a+1} + \tfrac{1}{3}\left(\frac{a-1}{a+1}\right)^3 + \tfrac{1}{5}\left(\frac{a-1}{a+1}\right)^5 + \tfrac{1}{7}\left(\frac{a-1}{a+1}\right)^7 \,\&c.$$

Which is a third general formula, that converges faster than either of the former.

The Logarithm of any number may, therefore, be exhibited universally, or according to any system of Logarithms, in the three following forms:

$$\text{Log.}(1+p) = \frac{1}{m} \times : p - \tfrac{1}{2}p^2 + \tfrac{1}{3}p^3 - \tfrac{1}{4}p^4 + \tfrac{1}{5}p^5 \,\&c.$$

$$\text{Log.}\frac{1}{1-p} = \frac{1}{m} \times : p + \tfrac{1}{2}p^2 + \tfrac{1}{3}p^3 + \tfrac{1}{4}p^4 + \tfrac{1}{5}p^5 \,\&c.$$

$$\text{Log.}\frac{1+p}{1-p} = \frac{2}{m} \times : p + \tfrac{1}{3}p^3 + \tfrac{1}{5}p^5 + \tfrac{1}{7}p^7 + \tfrac{1}{9}p^9 \,\&c.$$

Or

$$\text{Log.}a = \frac{1}{m} \times :(a-1) - \tfrac{1}{2}(a-1)^2 + \tfrac{1}{3}(a-1)^3 - \tfrac{1}{4}(a-1)^4 \,\&c.$$

$$\text{Log.}a = \frac{1}{m} \times : \frac{a-1}{a} + \tfrac{1}{2}\left(\frac{a-1}{a}\right)^2 + \tfrac{1}{3}\left(\frac{a-1}{a}\right)^3 + \tfrac{1}{4}\left(\frac{a-1}{a}\right)^4 \,\&c.$$

$$\text{Log.}a = \frac{2}{m} \times : \frac{a-1}{a+1} + \tfrac{1}{3}\left(\frac{a-1}{a+1}\right)^3 + \tfrac{1}{5}\left(\frac{a-1}{a+1}\right)^5 + \tfrac{1}{7}\left(\frac{a-1}{a+1}\right)^7 \,\&c.$$

And if $a+b$ be put $= s$, and $a \backsim b = d$, these general formulæ may be easily converted into the following:

$$\text{Log.}\frac{a}{b} = \frac{1}{m} \times : \frac{d}{b} - \frac{d^2}{2b^2} + \frac{d^3}{3b^3} - \frac{d^4}{4b^4} + \frac{d^5}{5b^5} \,\&c.$$

$$\text{Log.}\frac{a}{b} = \frac{1}{m} \times : \frac{d}{a} + \frac{d^2}{2a^2} + \frac{d^3}{3a^3} + \frac{d^4}{4a^4} + \frac{d^5}{5a^5} \,\&c.$$

$$\text{Log.}\frac{a}{b} = \frac{2}{m} \times : \frac{d}{s} + \frac{d^3}{3s^3} + \frac{d^5}{5s^5} + \frac{d^7}{7s^7} + \frac{d^9}{9s^9} \,\&c.$$

From which last expressions, if d or its equal $a \backsim b$ be put $= 1$, we shall have, by proper substitution, and the nature of Logarithms:

$$\text{Log.}a = \text{Log.}(a-1) + \frac{1}{m} \times : \frac{1}{a} + \frac{1}{2a^2} + \frac{1}{3a^3} + \frac{1}{4a^4} \,\&c.$$

$$\text{Log.}a = \text{Log.}(a-1) + \frac{1}{m} \times : \frac{1}{a-1} - \frac{1}{2(a-1)^2} + \frac{1}{3(a-1)^3} - \frac{1}{4(a-1)^4} \,\&c.$$

$$\text{Log.}a = \text{Log.}(a-2) + \frac{1}{m} \times : \frac{1}{a-1} + \frac{1}{3(a-1)^3} + \frac{1}{5(a-1)^5} + \frac{1}{7(a-1)^7} \,\&c.$$

And from the addition and subtraction of these series, several others may be derived; but in the actual computation of Logarithms they will be found to possess little or no advantage above those here given. The same general formula may be derived from the original Logarithmic equation $r^x = a$ in a different way, thus:

Let $r = 1+q$, then $r^x = \overline{1+q}\,|^x = 1 + (q - \tfrac{1}{2}q^2 + \tfrac{1}{3}q^3 - \tfrac{1}{4}q^4\,\&c.)x + \tfrac{1}{2}(q - \tfrac{1}{2}q^2 + \tfrac{1}{3}q^3 - \tfrac{1}{4}q^4\,\&c.)^2 x^2 + \frac{1}{2.3}(q - \tfrac{1}{2}q^2 + \tfrac{1}{3}q^3 - \tfrac{1}{4}q^4\,\&c.)^3 x^3 + \frac{1}{2.3.4}(q - \tfrac{1}{2}q^2 + \tfrac{1}{3}q^3 - \tfrac{1}{4}q^4\,\&c.)^4 x^4 \,\&c = a$; or if r be put $= \dfrac{1}{1-q}$, we shall have $\overline{1-q}^{\,x} = 1 - (q + \tfrac{1}{2}q^2 + \tfrac{1}{3}q^3 + \tfrac{1}{4}q^4\,\&c.)x + \tfrac{1}{2}(q + \tfrac{1}{2}q^2 + \tfrac{1}{3}q^3 + \tfrac{1}{4}q^4\,\&c.)^2 x^2 - \frac{1}{2.3}(q + \tfrac{1}{2}q^2 + \tfrac{1}{3}q^3 + \tfrac{1}{4}q^4\,\&c.)^3 x^3 + \frac{1}{2.3.4}(q + \tfrac{1}{2}q^2 + \tfrac{1}{3}q^3 + \tfrac{1}{4}q^4\,\&c.)^4 x^4 \,\&c = \dfrac{1}{a}$.

And by denoting $q - \tfrac{1}{2}q^2 + \tfrac{1}{3}q^3 - \tfrac{1}{4}q^4$ &c in the first case, or its equal $q + \tfrac{1}{2}q^2 + \tfrac{1}{3}q^3 + \tfrac{1}{4}q^4$, in the latter case, by m, these expressions will become

$$1 + mx + \tfrac{1}{2}m^2x^2 + \frac{1}{2.3}m^3x^3 + \frac{1}{2.3.4}m^4x^4 + \&c = a;$$

and $1 - mx + \tfrac{1}{2}m^2x^2 - \dfrac{1}{2.3}m^3x^3 + \dfrac{1}{2.3.4}m^4x^4 \,\&c = \dfrac{1}{a}$;

which are the two anti-Logarithmic series of Halley: from whence, by reversion of series, may be found the Logarithm of any number a, as before.

M.

MICROSCOPE. The following directions are given for using the New Universal Pocket Microscope, made and sold by W. and S. Jones, opticians, No. 135, Holborn, London. See fig. 4, pl. 33.

" This Microscope is adapted to the viewing of all sorts of objects, whether *transparent*, or *opake;* and for *insects, flowers, animalcules,* and the infinite variety of the *minutiæ* of Nature and Art, will be found the most complete and portable for the price, of any hitherto contrived.

Place the square pillar of the Microscope in the square socket at the foot D, and fasten it by the pin, as shewn in the figure. Place also in the foot, the reflecting mirror C. There are three lenses at the top shewn at A, which serve to magnify the objects. By using these lenses separately or combined, you make seven different powers. When transparent objects, such as are in the ivory sliders, number 4, are to be viewed, you place the sliders over the spring, at the underside of the stage B; then looking through the lens or magnifier, at A, at the same time reflect up the light, by moving the mirrour C below, and move gently upwards or downwards as may be necessary, the stage B, upon its square pillar, till you see the object illuminated and distinctly magnified; and in this manner for the other objects.

For animalcules, you unscrew the brass box that is fitted at the stage B, containing two glasses, and leave the undermost glass upon the stage, to receive the fluids. If you wish to view thereon any moving insect, &c, it may be confined by screwing on the cover: of the two glasses, the concave is best for fluids. Should the objects be opake, such as seeds, &c; they are to be placed upon the black and white ivory round piece, number 3, which is fitted also to the stage B. If the objects are of a dark colour, you place them contrastedly on the white side of the ivory. If they are of a white, or a light colour, upon the blackened side. Some objects

will

will be more conveniently viewed, by sticking them on the point of number 2 ; or between the nippers at the other end, which open by pressing the two little brass pins. This apparatus is also fitted to a small hole in the stage, made to receive the support of the wire.

The brass forceps, number 1, serve to take up any small object by, in order to place them on the stage for view. The instrument may be readily converted into an hand Microscope, to view objects against the common light ; and which, for some transparent ones, is better so. It is done by only taking out the pillar from its foot in D, turning it half round, and fixing it in again ; the foot then becomes a useful handle, and the reflector C is laid aside.

The whole apparatus packs into a fish-skin case, 4¼ inches long, 2¼ inches broad, and 1½ inches deep.

For persons more curious and nice in these sort of instruments, there is contrived a useful adjusting screw to the stage, represented at e. It is first moved up and down like the other, to the focus nearly, and made fast by the small screw. The utmost distinctness of the object is then obtained, by gently turning the long fine threaded screw, at the same time you are looking through the magnifiers A. In this case, there may be also added an extraordinary deep magnifier, and a concave silver speculum, with a magnifier to screw on at A, which will serve for viewing the very small, and opake objects, in the completest manner, and render the instrument as comprehensive in its uses and powers, as those formerly sold under the name of *Wilson's Microscope*."

MODULUS, and MODULAR *Ratio*. See p. 49 at the bottom.

N.

NUTATION, in Astronomy, a kind of libratory motion of the earth's axis ; by which its inclination to the plane of the ecliptic is continually varying, by a certain number of seconds, backwards and forwards. The whole extent of this change in the inclination of the earth's axis, or, which is the same thing, in the apparent declination of the stars, is about 19″, and the period of that change is little more than 9 years, or the space of time from its setting out from any point and returning to the same point again, about 18 years and 7 months, being the same as the period of the moon's motions, upon which it chiefly depends ; being indeed the joint effect of the inequalities of the action of the sun and moon upon the spheroidal figure of the earth, by which its axis is made to revolve with a conical motion, so that the extremity of it describes a small circle, or rather an ellipse, of 19·1 seconds diameter, and 14″·2 conjugate, each revolution being made in the space of 18 years 7 months, according to the revolution of the moon's nodes.

This is a natural consequence of the Newtonian system of universal attraction ; the first principle of which is, that all bodies mutually attract each other in the direct ratio of their masses, and in the inverse ratio of the squares of their distances. From this mutual attraction, combined with motion in a right line, Newton deduces the figure of the orbits of the planets, and particularly that of the earth. If this orbit were a circle, and if the earth's form were that of a perfect sphere, the attraction of the sun would have no other

effect than to keep the earth in its orbit, without causing any irregularity in the position of its axis. But neither is the earth's orbit a circle, nor its body a sphere ; for the earth is sensibly protuberant towards the equator, and its orbit is an ellipsis, which has the sun in its focus. Now when the position of the earth is such, that the plane of the equator passes through the centre of the sun, the attractive power of the sun acts only so as to draw the earth towards it, still parallel to itself, and without changing the position of its axis ; a circumstance which happens only at the time of the equinoxes. In proportion as the earth recedes from those points, the sun also goes out of the plane of the equator, and approaches that of the one or other of the tropics ; the semidiameter of the earth, then exposed to the sun, being unequal to what it was in the former case, the equator is more powerfully attracted than the rest of the globe, which causes some alteration in its position, and its inclination to the plane of the ecliptic : and as that part of the orbit, which is comprised between the autumnal and vernal equinox, is less than that which is comprised between the vernal and autumnal, it follows, that the irregularity caused by the sun, during his passage through the northern signs, is not entirely compensated by that which he causes during his passage through the southern signs ; and that the parallelism of the terrestrial axis, and its inclination to the ecliptic, is thence a little altered.

The like effect which the sun produces upon the earth, by his attraction, is also produced by the moon, which acts with greater force, in proportion as she is more distant from the equator. Now, at the time when her nodes agree with the equinoxial points, her greatest latitude is added to the greatest obliquity of the ecliptic. At this time therefore, the power which causes the irregularity in the position of the terrestrial axis, acts with the greatest force ; and the revolution of the nodes of the moon being performed in 18 years 7 months, hence it happens that in this time the nodes will twice agree with the equinoxial points ; and consequently, twice in that period, or once every 9 years, the earth's axis will be more influenced than at any other time.

That the moon has also a like motion, is shewn by Newton, in the first book of the Principia ; but he observes indeed that this motion must be very small, and scarcely sensible.

As to the history of the Nutation, it seems there have been hints and suspicions of the existence of such a circumstance, ever since Newton's discovery of the system of the universal and mutual attraction of matter ; some traces of which are found in his Principia, as above mentioned.

We find too, that Flamsteed had hoped, about the year 1690, by means of the stars near his zenith, to determine the quantity of the Nutation which ought to follow from the theory of Newton ; but he gave up that project, because, says he, if this effect exists, it must remain insensible till we have instruments much longer than 7 feet, and more solid and better fixed than mine. Hist. Cælest. vol. 3, pa. 113.

And Horrebow gives the following passage, extracted from the manuscripts of his master Roemer, who died in 1710, whose observations he published in 1753, under

der the title of *Bafis Aftronomiæ*. By this paragraph it appears that Roemer fufpected alfo a Nutation in the earth's axis, and had fome hopes to give the theory of it : it runs thus ; " Sed de altitudinibus non perinde certus reddebar, tam ob refractionum varietatem quam ob aliam nondum liquido perfpectam caufam ; fcilicet per hos duos annos, quemadmodum & alias, expertus fum effe quandam in declinationibus varietatem, quæ nec refractionibus nec parallaxibus tribui poteft, fine dubio ad vacillationem aliquam poli terreftris referendam, cujus me verifimilem dare poffe theoriam, obfervationibus munitam, fpero." Bafis Aftronomiæ, 1735, pa. 66.

These ideas of a Nutation would naturally prefent themfelves to thofe who might perceive certain changes in the declinations of the ftars ; and we have feen that the firft fufpicions of Bradley in 1727, were that there was fome Nutation of the earth's axis which caufed the ftar γ Draconis to appear at times more or lefs near the pole ; but farther obfervations obliged him to fearch another caufe for the annual variations (art. ABERRATION) : it was not till fome years after that he difcovered the fecond motion which we now treat of, properly called the Nutation. See the art. STAR, pa. 500 &c, where Bradley's difcovery of it is given at length ; to which may be farther added the following fummary.

For the better explaining the difcovery of the Nutation by Bradley, we muft recur to the time when he obferved the ftars in difcovering the aberration. He perceived in 1728, that the annual change of declination in the ftars near the equinoxial colure, was greater than what ought to refult from the annual preceffion of the equinoxes being fuppofed 50″, and calculated in the ufual way ; the ftar η Urfæ Majoris was in the month of September 1728, 20″ more fouth than the preceding year, which ought to have been only 18″ ; from whence it would follow that the preceffion of the equinoxes fhould be 55″½ inftead of 50″, without afcribing the difference between the 18 and 20″ to the inftrument, becaufe the ftars about the folftitial colure did not give a like difference. Philof. Tranf. vol. 35, pa. 659.

In general, the ftars fituated near the equinoctial colure had changed their declination about 2″ more than they ought by the mean preceffion of the equinoxes, the quantity of which is very well known, and the ftars near the folftitial colure the fame quantity lefs than they ought ; but, Bradley adds, whether thefe fmall variations arife from fome regular caufe, or are occafioned by fome change in the fector, I am not yet able to determine. Bradley therefore ardently continued his obfervations for determining the period and the law of thefe variations ; for which purpofe he refided almoft continually at Wanfted till 1732, when he was obliged to repair to Oxford to fucceed Dr. Halley ; he ftill continued to obferve with the fame exactnefs all the circumftances of the changes of declination in a great number of ftars. Each year he faw the periods of the aberration confirmed according to the rules he had lately difcovered ; but from year to year he found alfo other differences ; the ftars fituated between the vernal equinox and the winter folftice approached nearer to the north pole, while the oppofite ones receded farther from it : he began therefore to fufpect that the action of the moon upon the elevated equatorial parts of

the earth might caufe a variation or libration in the earth's axis : his fector having been left fixed at Wanfted, he often went there to make obfervations for many years, till the year 1747, when he was fully fatisfied of the caufe and effects, an account of which he then communicated to the world. Philof. Tranf. vol. 45, an. 1748.

" On account of the inclination of the moon's orbit to the ecliptic, fays Dr. Mafkelyne (Aftronomical Obfervations 1776, pa. 2), and the revolution of the nodes in antecedentia, which is performed in 18 years and 7 months, the part of the preceffion of the equinoxes, owing to her action, is not uniform : but fubject to an equation, whofe maximum is 18″ : and the obliquity of the ecliptic is alfo fubject to a periodical equation of 9″·55 ; being greater by 19·1″ when the moon's afcending node is in Aries, than when it is in Libra. Both thefe effects are reprefented together, by fuppofing the pole of the earth to defcribe the periphery of an ellipfis, in a retrograde manner, during each period of the moon's nodes, the greater axis, lying in the folftitial colure, being 19·1″, and the leffer axis, lying in the equinoctial colure, 14·2″ ; being to the greater, as the cofine of double the obliquity of the ecliptic to the cofine of the obliquity itfelf. This motion of the pole of the earth is called the Nutation of the earth's axis, and was difcovered by Dr. Bradley, by a feries of obfervations of feveral ftars made in the courfe of 20 years, from 1727 to 1747, being a continuation of thofe by which he had difcovered the aberration of light. But the exact law of the motion of the earth's axis has been fettled by the learned mathematicians d'Alembert, Euler, and Simpfon, from the principles of gravity. The equation hence arifing in the place of a fixed ftar, whether in longitude, right-afcenfion, or declination (for the latitudes are not affected by it) has been fometimes called Nutation, and fometimes Deviation." And again (fays the Doctor, pa. 8), the above " quantity 19·1″, of the greateft Nutation of the earth's axis in the folftitial colure, is what I found from a fcrupulous calculation of all Dr. Bradley's obfervations of γ Draconis, which he was pleafed to communicate to me for that purpofe. From a like examination of his obfervation of η Urfæ majoris, I found the leffer axis of the ellipfis of Nutation to be 14·1″, or only ₁₀th of a fecond lefs than what it fhould be from the obfervations of γ Draconis. But the refult from the obfervations of γ Draconis is moft to be depended upon."

Mr. Machin, fecretary of the Royal Society, to whom Bradley communicated his conjectures, foon perceived that it would be fufficient to explain, both the Nutation and the change of the preceffion, to fuppofe that the pole of the earth defcribed a fmall circle. He ftated the diameter of this circle at 18″, and he fuppofed that it was defcribed by the pole in the fpace of one revolution of the moon's nodes. But later calculations and theory, have fhewn that the pole defcribes a fmall ellipfis, whofe axes are 19·1″ and 14·2″, as above mentioned.

To fhew the agreement between the theory and obfervations, Bradley gives a great multitude of obfervations of a number of ftars, taken in different pofitions ; and out of more than 300 obfervations which he made, he found but 11 which were different from the mean by so much as 2″. And by the supposition of the elliptic rotation, the agreement of the theory with observation comes out still nearer.

By the observations of 1740 and 1741, the star η Ursæ majoris appeared to be 3″ farther from the pole than it ought to be according to the observations of other years. Bradley thought this difference arose from some particular cause; which however was chiefly the fault of the circular hypothesis. He suspected also that the situation of the apogee of the moon might have some influence on the Nutation. He invited therefore the mathematicians to calculate all these effects of attraction, which has been ably done by d'Alembert, Euler, Walmesley, Simpson and others; and the astronomers to continue to observe the positions of the smallest stars, as well as the largest, to discover the physical derangements which they may suffer, and which had been observed in some of them.

Several effects arise from the Nutation. The first of these, and that which is the most easily perceived, is the change in the obliquity of the ecliptic; the quantity of which ought to be varied from that cause by 18″ in about 9 years. Accordingly, the obliquity of the ecliptic was observed in 1764 to be 23° 28′ 15″, and in 1755 only 23° 28′ 5″: not only therefore had it not diminished by 8″, as it ought to have done according to the regular mean diminution of that obliquity; but it had even augmented by 10″; making together 18″, for the effect of the Nutation in the 9 years.

The Nutation changes equally the longitudes, the right ascensions, and the declinations of the stars, as before observed; it is the latitudes only which it does not affect, because the ecliptic is immoveable in the theory of the Nutation.

Dr. Bradley illustrates the foregoing theory of Nutation in the following manner. Let P represent the mean place of the pole of the equator, about which point, as a centre, suppose the true pole to move in the small circle ABCD, whose diameter is 18″. Let E be the pole of the ecliptic, and EP be equal to the mean distance between the poles of the equator and ecliptic; and suppose the true pole of the equator to be at A, when the moon's ascending node is in the beginning of Aries; and at B,

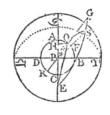

when the node gets back to Capricorn; and at C, when the same node is in Libra: at which time the north pole of the equator being nearer the north pole of the ecliptic, by the whole diameter of the little circle AC, equal to 18″; the obliquity of the ecliptic will then be so much less than it was, when the moon's ascending node was in Aries. The point P is supposed to move round E, with an equal retrograde motion, answerable to the mean precession arising from the joint actions of the sun and moon: while the true pole of the equator moves round P, in the circumference ABCD, with a retrograde motion likewise, in a period of the moon's nodes, or of 18 years and 7 months. By this means, when the moon's ascending node is in Aries, and the true pole of the equator, at A, is moving from A towards B; it will approach the stars that come to the

meridian with the sun about the vernal equinox, and recede from those that come with the sun near the autumnal equinox, faster than the mean pole P does. So that, while the moon's node goes back from Aries to Capricorn, the apparent precession will seem so much greater than the mean, as to cause the stars that lie in the equinoctial colure to have altered their declination 9″, in about 4 years and 8 months, more than the mean precession would do; and in the same time, the north pole of the equator will seem to have approached the stars that come to the meridian with the sun of our winter solstice about 9″, and to have receded as much from those that come with the sun at the summer solstice.

Thus the phenomena before recited are in general conformable to this hypothesis. But to be more particular; let S be the place of a star, PS the circle of declination passing through it, representing its distance from the mean pole, and ♈PS its mean right-ascension. Thus if O and R be the points where the circle of declination cuts the little circle ABCD, the true pole will be nearest that star at O, and farthest from it at R; the whole difference amounting to 18″, or to the diameter of the little circle. As the true pole of the equator is supposed to be at A, when the moon's ascending node is in Aries; and at B, when that node gets back to Capricorn; and the angular motion of the true pole about P, is likewise supposed equal to that of the moon's node about E, or the pole of the ecliptic; since in these cases the true pole of the equator is 90 degrees before the moon's ascending node, it must be so in all others.

When the true pole is at A, it will be at the same distance from the stars that lie in the equinoctial colure, as the mean pole P is; and as the true pole recedes back from A towards B, it will approach the stars which lie in that part of the colure represented by P♈, and recede from those that lie in P♎; not indeed with an equable motion, but in the ratio of the sine of the distance of the moon's node from the beginning of Aries. For if the node be supposed to have gone backwards from Aries 30°, or to the beginning of Pisces, the point which represents the place of the true pole will, in the mean time, have moved in the little circle through an arc, as AO, of 30° likewise; and would therefore in effect have approached the stars that lie in the equinoctial colure P♈, and have receded from those that lie in P♎ by 4½ seconds, which is the sine of 30° to the radius AP. For if a perpendicular fall from O upon AP, it may be conceived as part of a great circle, passing through the true pole and any star lying in the equinoctial colure. Now the same proportion that holds in these stars, will obtain likewise in all others; and from hence we may collect a general rule for finding how much nearer, or farther, any star is to, or from, the mean pole, in any given position of the moon's node.

For, If *from the right-ascension of the star, we subtract the distance of the moon's ascending node from Aries; then radius will be to the sine of the remainder, as 9″ is to the number of seconds that the star is nearer to, or farther from, the true, than the mean pole.*

This motion of the true pole, about the mean at P, will also produce a change in the right-ascension of the stars, ftars, and in the places of the equinoctial points, as well as in the obliquity of the ecliptic; and the quantity of the equations, in either of these cases, may be eafily computed for any given pofition of the moon's nodes.

Dr. Bradley then proceeds to find the exact quantity of the mean preceffion of the equinoctial points, by comparing his own obfervations made at Greenwich, with thofe of Tycho Brahe and others; the mean of all which he ftates at 1 degree in $71\frac{1}{2}$ years, or $50\frac{1}{3}''$ per year; in order to fhew the agreement of the foregoing hypothefis with the phenomena themfelves, of the alterations in the polar diftances of the ftars; the conclufions from which approach as near to a coincidence as could be expected on the foregoing circular hypothefis, the diameter of which is $18''$; inftead of the more accurate quantity $19\cdot1''$, as deduced by Dr. Mafkelyne, and the elliptic theory as determined by the mathematicians, in which the greater axis ($19\cdot1''$) is to the lefs axis ($14\cdot2''$), as the cofine of the greateft declination is to the cofine of double the fame.

To give an idea now of the Nutation of the ftars, in longitude, right-afcenfion, and declination; fuppofe the pole of the equator to be at any time in the point O, alfo S the place of any ftar, and OH perpendicular to AE: then, like as AE is the folftitial colure when the pole of the equator was at A, and the longitude of the ftar S equal to the angle AES; fo OE is the folftitial colure when that pole is at O, and the longitude is then only the angle OES; lefs than before by the angle AEO, which therefore is the Nutation in longitude: counting the longitudes from the folftitial inftead of the equinoctial colure, from which they differ equally by 90 degrees, and therefore have the fame difference AEO. Now the angle AEO will be as the line HO = fin. AO to radius PB = fin. AO × PB = fin. AO × $9''$; therefore as EO : HO :: radius 1 :

$$\frac{HO}{EO} = \frac{\text{fin. AO} \times 9''}{\text{fin. } 23^\circ 28'} = \frac{\text{fin. node} \times 9''}{\text{fin. } 23^\circ 28'},$$

fince AO is equal to longitude of the moon's node. This expreffion therefore gives the Nutation in longitude, fuppofing the maximum of Nutation, with Bradley, to be $18''$; and it is negative, or muft be fubtracted from the mean longitude of the ftars, when the moon's node is in the firft 6 figns of its longitude, but additive in the latter 6, to give the true apparent longitude.

This equation of the Nutation in longitude is the fame for all the ftars; but that for the declination and right afcenfion is various for the different ftars. In the foregoing figure, PS is the mean polar diftance, or mean codeclination, of the ftar S, when the true place of the pole is O; and SO the apparent codeclination; alfo, the angle SPE is the mean right-afcenfion, and SOE the apparent one, counted from the folftitial colure; confequently OPS or OPF the difference between the right-afcenfion of the ftar and that of the pole, which is equal to the longitude of the node increafed by 3 figns or 90 degrees; fuppofing OF to be a fmall arc perpendicular to the circle of declination PFS; then is SF = SO, and PF the Nutation in declination, or the quantity the declination of the ftar has increafed; but radius 1 : $9''$ :: cofin. OPF : PF = $9''$ × cof. OPF; fo that the equation of declination will be found by multiplying $9''$ by the fine of the ftar's right-afcenfion diminifhed by the longitude of the node; for that angle is the complement of the angle SPO. This Nutation in declination is to be added to the mean declination to give the apparent, when its argument does not exceed 6 figns; and to be fubtracted in the latter 6 figns. But the contrary for the ftars having fouth declination.

To calculate the Nutation in right-afcenfion, we muft find the difference between the angle SOE the apparent, and SPE the mean right-afcenfion, counted from the folftitial colure EO. Now the true rightafcenfion SOE is equal to the difference between the two variable angles GOE and GOS; the former of which arifes from the change of one of the variable circles EO, and depends only on the fituation of the node or of that of the pole O; the latter GOS depends on the angle GPS which is the difference between the right-afcenfion of the ftar and the place of the pole O. Now in the fpherical triangle GPE, which changes into GOE, the fide GE and the angle G remain conftant, and the other parts are variable; hence therefore the fmall variation PO of the fide next the conftant angle G, is to the fmall variation of the angle oppofite to the conftant fide GE, as the tangent of the fide PE oppofite to the conftant angle, is to the fine of the angle GPE oppofite to the conftant fide; that is, as

$$\text{tang. } 23^\circ 28' : \text{fin. OPE} :: 9'' : x = \frac{9'' \times \text{fin. OPE}}{\text{tang. } 23^\circ 28},$$

the difference between the angles GOE and GPE. This is the change which the Nutation PO produces in the angle GPE, being the firft part of the Nutation fought, and is common to all the ftars and planets. It is to be fubtracted from the mean right-afcenfion in the firft 6 figns of the longitude of the node, and added in the other fix.

In like manner is found the change which the Nutation produces in the other part of the right-afcenfion SPE, that is, in the angle SPG, which becomes SOG by the effect of the Nutation. This fmall variation will be calculated from the fame analogy, by means of the triangle SOG, in which the angle G is conftant, as well as the fide SG, whilft SP changes into SO. Hence therefore, tang. SP : fin. SPG :: $9''$: variation of SPG, that is, the cotangent of the declination is to the cofine of the diftance between the ftar and the node, as $9''$ are to the quantity the angle SPG varies in becoming the angle SOG, being the fecond part of the Nutation in right-afcenfion; and if there be taken for the argument, the right-afcenfion of the ftar minus the longitude of the node, the equation will be fubtractive in the firft and laft quadrant of the argument, and additive in the 2d and 3d, or from 3 to 9 figns. But the contrary for ftars having fouth declination.

This fecond part of the Nutation in right-afcenfion affects the return of the fun to the meridian, and therefore it muft be taken into the account in computing the equation of time. But the former part of the Nutation does not enter into that computation; becaufe it only changes the place of the equinox, without changing the point of the equator to which a ftar correfponds, and confequently without altering the duration of the returns to the meridian.

9

All

All these calculations of the Nutation, above explained, are upon Machin's hypothesis, that the pole describes a circle; however Bradley himself remarked that some of his observations differed too much from that theory, and that such observations were found to agree better with theory, by supposing that the pole, instead of the circle, describes an ellipse, having its less axis DB = 16″ in the equinoctial colure, and the greater axis AC = 18″, lying in the solstitial colure. But as even this correction was not sufficient to cause all the inequalities to disappear entirely, Dr. Bradley referred the determination of the point to theoretical and physical investigation. Accordingly several mathematicians undertook the task, and particularly d'Alembert, in his Recherches sur la précession des equinoxes, where he determines that the pole really describes an ellipse, and that narrower than the one assumed above by Bradley, the greater axis being to the less, as the

cosine of 23° 28′ to the cosine of double the same. And as Dr. Maskelyne found, from a more accurate reduction of Bradley's observations, that the maximum of the Nutation gives 19·1″ for the greater axis, therefore the above proportion gives 14·2″ for the less axis of it; and according to these data, the theory and observations are now found to agree very near together.

See La Lande's Astron. vol. 3, art. 2874 &c, where he makes the corrections for the ellipse. He observes however that by the circular hypothesis alone, the computations may be performed as accurately as the observations can be made; and he concludes with some corrections and rules for computing the Nutation in the elliptic theory.

The following set of general tables very readily give the effect of Nutation on the elliptical hypothesis; they were calculated by the late M. Lambert, and are taken from the Connoissance des Temps for the year 1788.

General Tables for Nutation in the Ellipse.

TABLE I.

Degrees	0·6 + −	1 7 + −	2·8 + −	
	″	″	″	
0	0·00	3·93	6·80	30
1	0·14	4·04	6·86	29
2	0·27	4·16	6·93	28
3	0·41	4·28	6·99	27
4	0·55	4·39	7·06	26
5	0·68	4·50	7·11	25
6	0·82	4·61	7·17	24
7	0·95	4·72	7·23	23
8	1·11	4·83	7·28	22
9	1·23	4·94	7·33	21
10	1·36	5·05	7·38	20
11	1·50	5·15	7·42	19
12	1·63	5·25	7·47	18
13	1·77	5·35	7·51	17
14	1·90	5·45	7·55	16
15	2·03	5·55	7·58	15
16	2·16	5·65	7·62	14
17	2·30	5·74	7·65	13
18	2·43	5·83	7·68	12
19	2·56	5·92	7·71	11
20	2·68	6·01	7·73	10
21	2·81	6·10	7·75	9
22	2·94	6·19	7·76	8
23	3·07	6·27	7·77	7
24	3·19	6·35	7·79	6
25	3·32	6·43	7·80	5
26	3·44	6·51	7·82	4
27	3·56	6·58	7·83	3
28	3·69	6·66	7·84	2
29	3·81	6·73	7·85	1
30	3·93	6·80	7·85	0
	+ −	+ −	+ −	Degrees
	5·11	4·10	3·9	

TABLE 2.

Degrees	0 6 + −	1·7· + −	2·8 + −	
	″	″	″	
0	0·00	0·58	1·00	30
1	0·02	0·59	1·01	29
2	0·04	0·61	1·02	28
3	0·06	0·63	1·02	27
4	0·08	0·64	1·03	26
5	0·10	0·66	1·04	25
6	0·12	0·68	1·05	24
7	0·14	0·69	1·06	23
8	0·16	0·71	1·07	22
9	0·18	0·72	1·07	21
10	0·20	0·74	1·08	20
11	0·22	0·75	1·09	19
12	0·24	0·77	1·09	18
13	0·26	0·78	1·10	17
14	0·28	0·80	1·11	16
15	0·30	0·81	1·11	15
16	0·32	0·83	1·12	14
17	0·34	0·84	1·12	13
18	0·35	0·85	1·13	12
19	0·37	0·87	1·13	11
20	0·39	0·88	1·13	10
21	0·41	0·89	1·14	9
22	0·43	0·91	1·14	8
23	0·45	0·92	1·14	7
24	0·47	0·93	1·14	6
25	0·49	0·94	1·15	5
26	0·50	0·95	1·15	4
27	0·52	0·96	1·15	3
28	0·54	0·97	1·15	2
29	0·56	0·99	1·15	1
30	0·58	1·00	1·15	0
	+ −	+ −	+ −	Degrees
	5·11	4·10	3·9	

TABLE 3.

Degrees	0·6 − +	1·7 − +	2·8 − +	
	″	″	″	
0	0·00	7·71	13·36	30
1	0·27	7·95	13·50	29
2	0·54	8·18	13·62	28
3	0·81	8·40	13·75	27
4	1·08	8·63	13·87	26
5	1·35	8·85	13·98	25
6	1·61	9·07	14·10	24
7	1·88	9·29	14·20	23
8	2·15	9·50	14·31	22
9	2·41	9·71	14·41	21
10	2·68	9·92	14·50	20
11	2·94	10·12	14·59	19
12	3·21	10·32	14·67	18
13	3·47	10·52	14·76	17
14	3·73	10·72	14·83	16
15	3·99	10·91	14·90	15
16	4·25	11·10	14·97	14
17	4·51	11·28	15·03	13
18	4·77	11·47	15·09	12
19	5·02	11·65	15·15	11
20	5·28	11·82	15·20	10
21	5·53	11·99	15·24	9
22	5·78	12·16	15·28	8
23	6·03	12·32	15 32	7
24	6·28	12·48	15·35	6
25	6·52	12·64	15·37	5
26	6·76	12·79	15·39	4
27	7·01	12·94	15·41	3
28	7·25	13·09	15·42	2
29	7·48	13·23	15·43	1
30	7·71	13·36	15·43	0
	− +	− +	− +	Degrees
	5·11	4·10	3·9	

The

The Ufe of the Tables.

The right-afcenfion of a ftar minus the moon's mean longitude, gives the argument of the firſt of thefe three tables. The fum of the fame two quantities gives the argument of the 2d table. Then the fum or the difference of the quantities found with thefe two arguments, will give the correction to be applied to the mean declination of the ftar, if it is north declination ; but if it is fouthern, the figns + or — are to be changed into — and +.

From each of thofe two arguments for the declination fubtracting 3 figns,' or 90°, gives the arguments for correcting the right-afcenfion ; the fum or difference of the quantities found, with thefe two arguments, in tables 1 and 2, is to be multiplied by the tangent of the ftar's declination, and to the product is to be added the quantity taken out of table 3, the argument of which is the mean longitude of the moon's afcending node : when the declination of the ftar is fouth, the tangent will be negative.

Example. To find the Nutation in right-afcenfion and declination for the ftar α Aquilæ, the 1ſt of July 1788.

Right-afcenfion of the ftar   9ˢ 25° 7ᶠ
Long. of the moon's node   8   15   40

Diff. being argument 1,   1   9   27   + 4·99
Sum, argument 2, - -   6   10   47   — 0·22

Correction of the declination - - -   + 4·77

The above two arguments being each diminifhed by 3 figns, give,

Argument 1 - - - - - - - 10   9   27   — 6·06
Argument 2 - - - - - - -   3   10   47   + 1·13

                      — 4·93
Declin. of ftar north, its tangent - - - 0·146

The product is - - - - - -   — 0·72
Long. of the ☾'s node, argum. 3 -   + 14·94

Correction of right-afcenfion - - -   + 14·22

In general, let ☊ denote the longitude of the moon's afcending node ; *r* the right-afcenfion of a ftar or planet ; *d* its declination ; the Nutation in declination and right-afcenfion will be expreffed by the two following formulæ ; viz, the Nutation in declination

$$= 7''·85 \times \text{fin.} (r - ☊) + 1''·15 \times \text{fin.} (r + ☊);$$

and the Nutation in right-afcenfion

$$= [7''·85 \times \text{fin.} (r - ☊ - 90°) + 1''·15 \times \text{fin.} (r + ☊ - 90°)] \times \text{tang.} d - 15''·43 \times \text{fin.} ☊.$$

For the mathematical inveſtigation of the effects of univerfal attraction, in producing the Nutation, &c, fee d'Alembert's Recherches fur la Preceſſion des Equinoxes ; Silvabelle's Treatife on the Preceſſion of the Equinoxes &c, in the Philof. Tranf. an. 1754, p. 385 ; Walmefley's treatife De Præceſſione Equinoctiorum et Axis Terræ Nutatione, in the Philof. Tranf. an. 1756,

pa. 700 ; Simpfon's Mifcellaneous Tracts, pa. 1 ; and other authors.

S

STEAM. The obfervations on the different degrees of temperature acquired by water in boiling, under different preffures of the atmofphere, and the formation of the vapour from water under the receiver of an air-pump, when, with the common temperatures, the preffure is diminifhed to a certain degree, have taught us that the expanfive force of vapour or Steam is different in the different temperatures, and that in general it increafes in a variable ratio as the temperature is raifed.

But there was wanting, on this important fubject, a feries of exact and direct experiments, by means of which, having given the degree of temperature in boiling water, we may know the expanfive force of the Steam rifing from it ; and vice verfa. There was wanting alfo an analytical theorem, expreffing the relation between the temperature of boiling water, and the preffure with which the force of its Steam is in equilibrium. Thefe circumftances then have lately been accomplifhed by M. Betancourt, an ingenious Spanifh philofopher, the particulars of which are defcribed in a memoir communicated to the French Academy of Sciences in 1790, and ordered to be printed in their collection of the Works of Strangers.

The apparatus which M. Betancourt makes ufe of, is a copper veffel or boiler, with its cover firmly fol dered on. The cover has three holes, which clofe up with fcrews : the firſt is to put the water in and out ; through the fecond paffes the ſtem of a thermometer, which has the whole of its fcale or graduations above the veffel, and its ball within, where it is immerfed either in the water or the Steam according to the different circumftances ; through the third hole paffes a tube making a communication between the cavity of the boiler and one branch of an inverted fyphon, which, containing mercury, acts as a barometer for meafuring the preffure of the elaftic vapour within the boiler. There is a fourth hole, in the fide of the veffel, into which is inferted a tube, with a turn-cock, making a communication with the receiver of an air-pump, for extracting the air from the boiler, and to prevent its return.

The apparatus being prepared in good order, and diftilled water introduced into the boiler by the firſt hole, and then ftopped, as well as the end of the inverted fyphon or barometer, M. Betancourt furrounded the boiler with ice, to lower the temperature of the water to the freezing point, and then extracting all the air from the boiler by means of the air-pump, the difference between the columns of mercury in the two branches of the barometer is the meafure of the fpring of the vapour arifing from the water in that temperature. Then, lighting the fire below the boiler, he raifed gradually the temperature of the water from 0 to 110 degrees of Reaumur's thermometer ; being the fame as from 32 to 212 degrees of Fahrenheit's ; and for each degree of elevation in the temperature, he obferved the height of the column of mercury which meafured the elaſticity or preffure of the vapour.

The refults of M. Betancourt's experiments are contained

tained in a table of four columns, which are but little different, according to the different quantities of water in the veffel. It is here obfervable, that the increase in the expanfive force of the vapour, is at firft very flow; but gradually increafing fafter and fafter, till at laft it becomes very rapid. Thus, the ftrength of the vapour, at 80 degrees, is only equal to 28 French inches of mercury; but at 110 degrees it is equal to no lefs than 98 inches, that is 3 times and a half more for the increafe of only 30 degrees of heat.

To exprefs analytically the relation between the degrees of temperature of the vapour, and its expanfive force, this author employs a method devifed by M. Prony. This method confifts in conceiving the heights of the columns of mercury, meafuring the expanfive force, to reprefent the ordinates of a curve, and the degrees of heat as the abfciffes of the fame; making the ordinates equal to the fum of feveral logarithmic ones, which contain two indeterminates, and determining thefe quantities fo that the curve may agree with a good number of obfervations taken throughout the whole extent of them. Then conftructing the curve which refults immediately from the experiments, and that given by the formula, thefe two curves are found to coincide almoft perfectly together; the fmall differences being doubtlefs owing to the little irregularities in the experiments and in dividing the fcale; fo that the phenomena may be confidered as truly reprefented by the formula.

M. Betancourt made alfo experiments with the vapour from fpirit of wine, fimilar to thofe made with water; conftructing the curve, and giving the formula proper to the fame. From which is derived this remarkable refult, that, for any one and the fame degree of heat, the ftrength of the vapour of fpirit of wine, is to that of water, always in the fame conftant ratio, viz, that of 7 to 3 very nearly; the ftrength of the former being always $2\frac{1}{3}$ times the ftrength of the latter, with the fame degree of heat in the liquid.

Of the Formula, or Equation to the Curve.

The equation to the curve of temperature and preffure, denoting the relation between the abfciffes and ordinates, or between the temperature of the vapour and its ftrength, is, for water,

$$y = b^{a+cx} - b^{a'+c'x} - b^{e+c''x} + b^{e+c'''x}.$$

Where x denotes the abfciffes of the curve, or the degrees of Reaumur's thermometer; and y the correfponding ordinates, or the heights of the column of mercury in Paris inches, reprefenting the ftrength or elafticity of the vapour anfwering to the number x of degrees of the thermometer. Then, by comparing this formula with a proper number of the experiments, the values of the conftant quantities come out as below:

$$b = 10\cdot$$
$$a = 0\cdot068831$$
$$c = 0\cdot019438$$
$$c' = 0\cdot013490$$

$$e = -4\cdot689760$$
$$c'' = 0\cdot058622$$
$$e' = -3\cdot937600$$
$$c''' = 0\cdot049220$$

Hence it is evident by infpection, that the terms of the equation are very eafy to calculate. For, b being the radix or root of the common fyftem of logarithms, and all the terms on the fecond fide of the equation being the powers of b, thefe terms are confequently the tabular natural numbers having the variable exponents for their logarithms. Now as x rifes only to the firft power, and is multiplied by a conftant number, and another conftant number being added to the product, gives the variable exponent, or logarithm; to which then is immediately found the correfponding natural number in the table of logarithms.

In the above formula, the two laft terms may be entirely omitted, as very fmall, as far as to the 90th degree of the thermometer; and even above that temperature thofe two terms make but a fmall part of the whole formula.

And for the fpirit of wine the formula is

$$y = b^{a+cx} + b^{a'+c'x} - b^{e+c''x} + b^{e'+c'''x} - A.$$

Where x and y, as before, denote the abfcifs and ordinate of the curve, or the temperature and expanfive force of the vapour from the fpirit of wine; alfo the values of the conftant quantities are as below:

$$b = 10\cdot$$
$$a = -0\cdot04853$$
$$c = 0\cdot02393$$
$$a' = -0\cdot63414$$
$$c' = -0\cdot096532$$
$$e = -2\cdot509542$$
$$c'' = 0\cdot046473$$
$$e' = -1\cdot790192$$
$$c''' = 0\cdot029448$$
$$A = 1\cdot12647$$

This formula is of the fame nature as the former, having alfo the like eafe and convenience of calculation; and perhaps more fo; as the fecond term $b^{a+c'x}$, having its exponent wholly negative, foon diminifhes to no value, fo as to be omitted from the 10th degree of temperature; alfo the difference between the laft two terms $-b^{e+c''x} + b^{e'+c'''x}$ may be omitted till the 70th degree, for the fame reafon. So that, to the 10th degree of temperature the theorem is only $y = b^{a+cx} + b^{a'+c'x} - A$; and from the 10th to the 70th degree it is barely $y = b^{a+cx} - A$; after which, for the laft 15 or 20 degrees, for great accuracy, the laft two terms may be taken in.

A compendium of the table of the experiments here follows, for the vapour of both water and fpirit of wine, the temperature by Reaumur's thermometer, and the barometer in French inches.

Table of the Temperature and Strength of the Vapour of Water and Spirit of Wine, by Reaumur's Thermometer, and French Inches.

Degr. of Reau. Ther.	Height of the Barometer for		Deg. of Reau. Ther.	Height of the Barometer for	
	Vapour of Water.	Vapour of Spirit of Wine.		Vapour of Water.	Vapour of Spirit of Wine.
1	0·0176	0·0043	56	7·6948	18 4420
2	0·0346	0·0208	57	8·1412	19·5081
3	0·0538	0·0478	58	8·6221	20·6286
4	0·0747	0·0837	59	9·1071	21·6071
5	0·1038	0·1279	60	9·6280	23·0544
6	0·1211	0·1794	61	10·1767	24·3451
7	0·1508	0·2377	62	10·7098	25·6107
8	0·1741	0·3024	63	11·3602	27·1444
9	0·2073	0·3733	64	11·9976	28·6483
10	0·2304	0·4502	65	12·6687	30·2262
11	0·2681	0·5130	66	13·3743	31·8795
12	0·3039	0·6058	67	14·1161	33·6114
13	0·3419	0·7040	68	14·8958	35·4258
14	0·3877	0·8077	69	15·7153	37 3232
15	0·4258	0·9172	70	16·577	39·3076
16	0·4778	1·0330	71	17·482	41·3807
17	0·5208	1·1553	72	18·433	43·5465
18	0·5730	1·2846	73	19·433	45·8042
19	0·6283	1·4212	74	20·485	48·1589
20	0·6872	1·5655	75	21·587	50·6096
21	0·7497	1·7180	76	22·746	53·1593
22	0·8159	1·8791	77	23·965	55·8095
23	0·8863	2·0494	78	25·260	58·3968
24	0·9610	2·2293	79	26·588	61·3057
25	1·0402	2·4194	80	28·006	64·3524
26	1·1239	2·6202	81	29·455	67·4095
27	1·2127	2·8325	82	30·980	70·4967
28	1·3068	3·0568	83	32·575	73·7647
29	1·4065	3·2937	84	34·251	77·0764
30	1·5019	3·5441	85	35·984	80·4708
31	1·6333	3·8087	86	37·800	83·9351
32	1·7413	4·0883	87	39·697	87·4625
33	1·8671	4·3837	88	41·642	91·1366
34	1·9980	4·6958	89	43·730	94·6580
35	2·1374	5·0256	90	45·870	98·2764
36	2·2846	5·3741	91	48·092	
37	2·4401	5·6423	92	50·408	
38	2·6045	6·1315	93	52·785	
39	2·7780	6·5426	94	55·253	
40	2·9711	6·9770	95	57·801	
41	3·1544	7·4360	96	60·423	
42	3·3583	7·9211	97	63·108	
43	3·5735	8·4336	98	65·877	
44	3·8005	8·9751	99	68·692	
45	4·0399	9·5476	100	71·552	
46	4·2922	10·1516	101	74·444	
47	4·5582	10·7906	102	77·359	
48	4·8386	11·4606	103	80·268	
49	5·1346	12·1800	104	83·259	
50	5·4453	12·9340	105	85·992	
51	5·7706	13·7300	106	88·735	
52	6·1194	14·5720	107	91·367	
53	6·4834	15·4610	108	93·815	
54	6·8667	16·4000	109	96·039	
55	7·2798	17·3930	110	98·356	

M. Betancourt deduces several useful and ingenious consequences and applications from this course of experiments. He shews, for instance, that the effect of Steam engines must, in general, be greater in winter than in summer; owing to the different degrees of temperature in the water of injection. And from the very superior strength of the vapour of spirit of wine, over that of water, he argues that, by trying other fluids, some may be found, not very expensive, whose vapour may be so much stronger than that of water, with the same degree of heat, that it may be substituted instead of water in the boilers of Steam-engines, to the great saving in the very heavy expence of fuel: nay, he even declares, that spirit of wine itself might thus be employed in a machine of a particular construction, which, with the same quantity of fuel, and without any increase of expence in other things, shall produce an effect greatly superior to what is obtained from the steam of water. He makes several other observations on the working and improvement of Steam-engines.

Another use of these experiments, deduced by M. Betancourt, is, to measure the height of mountains, by means of a thermometer, immersed in boiling water, which he thinks may be done with a precision equal, if not superior, to that of the barometer. As soon as I had obtained exact results of my experiments, says he, and was convinced that the degree of heat received by water depends absolutely on the pressure upon its surface, I endeavoured to compare my observations with such as have been made on mountains of different heights, to know what is the degree of heat which water can receive when the barometer stands at a determinate height; but from so few observations having been made of this kind, and the different ways employed in graduating instruments, it is difficult to draw any certain consequences from them.

The first observation which M. Betancourt compared with his experiments, is one mentioned in the Memoirs of the Academy of Sciences, anno 1740, page 92. It is there said, that M. Monnier having made water boil upon the mountain of Canigou, where the barometer stood at 20·18 inches, the thermometer immersed in this water stood at a point answering to 71 degrees of Reaumur: whereas in M. Betancourt's table of experiments, at an equal pressure upon the surface of the water, the thermometer stood at 73·7 degrees. This difference he thinks is owing partly to the want of precision in the observation, and partly to the different method of graduating the thermometer, and the neglect of purging the barometer tube of air.

M. Betancourt next compared his experiments with some observations made by M. De Luc on the tops of several mountains; in which, after reducing the scales of this gentleman to the same measures as his own, he finds a very near degree of coincidence indeed. The following table contains a specimen of these comparisons, the instances being taken at random from De Luc's treatise on the Modifications of the Atmosphere.

Degrees of Heat in Boiling Water upon the Tops of Mountains, observed by De Luc.				Heat of the Water in M. Betancourt's Experim.
Places of Observation.	Heat of the air.	Height of the Bar.	Heat of the Wa. by Th.	
Beaucaire -	14¼	28·248	80·37	80·29
Geneva - -	12½	27·056	79·33	79·33
Grange Town	16¼	24·510	77·11	77·42
Lans le Bourg		24·145	77·18	77·14
Grange le F.	15	24·089	76·76	77·09
Grenairon	10¼	20·427	73·25	73·89
Glaciere de B.	6½	19·677	72·56	73·24

Where it is remarkable, that the difference between the two is of no consequence in such matters.

Many other advantages might be deduced from the exact knowledge of the effect which the pressure of the atmosphere has upon the heat which water can receive: one of which, M. Betancourt observes, is of too great importance in physics not to be mentioned. As soon as the thermometer became known to philosophers, almost every one endeavoured to find out two fixed points to direct them in dividing the scale of the instrument; having found that those of the freezing and boiling of water were nearly constant in different places, they gave these the preference over all others: but having discovered that water is capable of receiving a greater or less quantity of heat, according to the pressure of the atmosphere upon its surface, they felt the necessity of fixing a certain constant value to that pressure, which it was almost generally agreed should be equal to a column of 28 French inches of mercury. This agreement however did not remove all the difficulties. For instance, if it were required to construct at Madrid a thermometer that might be comparable with another made at Paris, the thing would be found impossible by the means hitherto known, because the barometer never rises so high as 27 inches at Madrid; and it was not certainly known how much the scale of the thermometer ought to be increased to have the point of boiling water in a place where the barometer is at 28 inches. But by making use of the foregoing observations, the thing appears very easy, and it is to be hoped that by the general knowledge of them, thermometers may be brought to great perfection, the accurate use of which is of the greatest importance in physics.

Besides, without being confined to the height of the barometer in the open air, in a given place, we may regulate a thermometer according to any one assigned heat of water, by means of such an apparatus as M. Betancourt's. For, in order to graduate a thermometer, having a barometer ready divided; it is evident that by knowing, from the foregoing table of experiments, the degree of heat answering to any one expansive force, we can thence assign the degree of the thermometer corresponding to a certain height of the barometer. A determination admitting of great precision, especially in the higher temperatures, where the motion of the barometer is so considerable in respect to that of the thermometer.

F I N I S.

Printed in the United States
By Bookmasters